General Chemistry

Principles and Modern Applications

ABOUT THE COVER:
A small samarium-cobalt magnet is suspended above a superconducting disc of a barium-yttrium-copper-oxygen compound. The disc displays superconductivity at temperatures below 83 K. This magnetic levitation (Meissner effect) persists as long as the disc is maintained at the temperature of liquid nitrogen (77 K); see page 896. [Courtesy Edmund Scientific Co.]

General Chemistry

FIFTH EDITION

Principles and Modern Applications

RALPH H. PETRUCCI

California State University, San Bernardino

Macmillan Publishing Company
NEW YORK

Collier Macmillan Publishers
LONDON

Macmillan Publishing Company
866 Third Avenue, New York, New York 10022

Collier Macmillan Canada, Inc.

Library of Congress Cataloging in Publication Data

Petrucci, Ralph H.
    General chemistry: principles and modern applications / Ralph H.
    Petrucci.—5th ed.
      p.   cm.
    Includes index.
    ISBN 0-02-394791-8 (Hardcover Edition)
    ISBN 0-02-946560-5 (International Edition)
      1. Chemistry. I. Title.
    QD31.2.P48   1989
    540—dc19   88-25878
                                                                    CIP

Printing: 2 3 4 5 6 7 8   Year: 9 0 1 2 3 4 5 6 7 8

Preface

The majority of general chemistry students are not preparing to become professional chemists. Instead, they have career interests in biology, medicine, engineering, and environmental and agricultural sciences, to name but a few. Although some may take additional course work in chemistry (often organic chemistry), for many, general chemistry is the only broadly based college course in chemical principles that they will encounter—and their only opportunity to learn about the many practical applications of chemistry. This text is intended primarily for these "typical" students. Students of this text are most likely to have studied some chemistry previously, but those with no prior background in chemistry will find that the early chapters develop fundamental concepts from the most elementary of ideas. Students who plan to become professional chemists will find numerous opportunities to pursue their special interests.

The typical student, I think, needs help in identifying and applying principles and in visualizing their physical significance. The pedagogical features of this text are designed to provide this help, but at the same time I hope the text serves to sharpen student skills in problem solving and critical thinking. Throughout, I have attempted to strike the proper balances between principles and applications, qualitative and quantitative discussions, and rigor and simplification.

NEW FEATURES IN THIS EDITION

At first glance, the most obvious feature is the use of full-color art and photography. Not only does this use of color more accurately depict chemical compounds and their reactions, but in many instances it should help to convey ideas that are difficult to convey with words alone, for example, by enabling students to focus simultaneously on several important and coequal features of a diagram.

Highlighted Expressions. The most significant equations, concepts, and rules are labeled, numbered and screened with a yellow panel. These highlighted expressions are listed at the end of each chapter. Individual instructors are, of course, free to add to or delete from this list.

In-Text Summaries. To help students consolidate their knowledge, several chapters contain brief sections that summarize a set of ideas, rules, or problem-solving techniques before the next set of topics is introduced. Examples include summaries on thermochemical calculations (Chapter 7), key concepts of reaction kinetics (Chapter 15), and acid-base equilibrium calculations (Chapter 18).

Chemistry Evolving. Some of these brief essays trace the historical development of an idea or method. Others deal with current and projected developments in chemistry, such as supercritical fluids, superconductors, and advanced ceramics. All are intended to show that chemistry is an evolving discipline, and that concepts that students learn today may appear in a totally new context in the future.

v

Are You Wondering. In an attempt to clarify points that often puzzle students, occasional questions are posed and answered under this heading. These questions are cast in the form in which students often ask them. Many are designed to help students avoid common pitfalls.

Summarizing Examples. Each chapter concludes with a multipart example, usually of a practical nature. These examples link various important problem types introduced in the chapter with each other and often with problems types of earlier chapters. Emphasis is given to the stepwise manner in which many chemical problems can be solved.

ORGANIZATIONAL CHANGES

Major changes in organization in this edition include an early introduction (Chapter 5) of reactions in aqueous solutions, relocation of thermodynamics (Chapter 20) to follow rather than to precede the chapters on equilibrium in aqueous solutions, and a reordering of the descriptive chemistry chapters to place the chapter on representative metals (Chapter 22) ahead of the chapter on representative nonmetals (Chapter 23).

Other changes involve the addition of some topics and the reorganization or relocation of others. For example, the concept of the mole has been moved forward from Chapter 3 to Chapter 2. The first law of thermodynamics has been moved from the chapter on thermodynamics into the chapter on thermochemistry (Chapter 7). Electronegativity has been moved from the chapter on atomic properties and the periodic table to the one on basic concepts of chemical bonding (Chapter 10). Catalysis has been moved to the end of the chemical kinetics chapter (Chapter 15), following rather than preceding the section on reaction mechanisms. The quantitative treatment of the effect of temperature on the equilibrium constant has been moved from the equilibrium to the thermodynamics chapter (Chapter 20).

Descriptive chemistry receives expanded coverage and is more widely distributed than in previous editions. Some of the material in the new Chapter 5 is of a descriptive nature, Chapter 9 includes a section on the chemistry of the halogens, and the expanded Chapter 14 surveys much of the descriptive chemistry of the first 20 elements. Chapters 22, 23, and 24 are devoted to descriptive inorganic chemistry, and Chapter 25 to the chemistry of complex ions and coordination compounds. Chapters 27 and 28 deal with organic chemistry and biochemistry, respectively. In addition, descriptive chemistry and practical applications are introduced through the special features: *Focus on, Chemistry Evolving,* and *Summarizing Example.*

Descriptive topics can either be taken up as encountered or reorganized to suit the instructor's preference. For example, the individual sections of Chapter 14 can be regrouped with topics of other chapters. Thus, the light noble gases (Section 14-4) can be discussed together with the Focus feature on the atmosphere (Chapter 7), and hydrogen (Section 14-3) can be discussed in the section on sources and uses of energy (Section 7-11). Alternatively, Chapter 14 can be deferred and studied with Chapters 22 and 23. None of the material introduced in Chapter 14 is fundamental to the subject matter of Chapters 15-21.

FEATURES RETAINED AND REVISED FROM EARLIER EDITIONS

In-Text Examples. In each chapter most concepts—especially those that students will be expected to apply in homework assignments and examinations—are illustrated with worked-out examples. Often a line drawing or photograph accompanies an example to help students visualize what is "going on" in the problem. Many examples have been recast to emphasize a practical application, and each example has been given a descriptive title to identify the concept being illustrated. Each example concludes with a reference to one or several similar problems in the end-of-chapter exercises.

Focus on. Special sections on practical applications appear at the ends of certain chapters. Typical topics are key industrial chemicals (Chapter 1), aspects of industrial chemistry (Chapter 4), the greenhouse effect (Chapter 7), semiconductors (Chapter 11), the ozone layer (Chapter 14), hydrometallurgy (Chapter 22), and polymer chemistry (Chapters 9 and 27).

Summary/Key Terms/Glossary. Each chapter concludes with a comprehensive verbal summary of important concepts and factual information, followed by a list of key terms, which appear in **boldface** type in the text. Sections where key terms are introduced are noted, and the terms are also defined in the Glossary (Appendix F). Students can use the Key Terms lists and the Glossary to help them master the terminology of general chemistry.

End-of-Chapter Exercises. Each chapter has four categories of exercises: *Review Problems* require straightforward application of principles introduced in the chapter, each usually involving a single concept. *Exercises* are grouped by subject matter and are of a broader nature than the Review Problems. Those that either are more difficult or require an extension beyond the concepts of the particular chapter are designated by a star*. *Additional Exercises* are not grouped by type. *Self-Test Questions* present a group of multiple-choice items plus brief essay questions and/or problems typical of examination questions. Answers to all Review Problems and Self-Test Questions and to those Exercises designated by brown numbers or letters are provided in Appendix G.

Many of the end-of-chapter exercises have been rewritten to place them in a practical setting, and some are illustrated. Certain illustrations are themselves the subject of an exercise. In other cases the illustration is provided simply to help students to visualize the problem. Ideally, this practice will encourage students to get in the habit of routinely illustrating problem situations.

SUPPLEMENTS

The **Student Study Guide** is organized around a set of learning objectives for each chapter and features brief discussions of these objectives, drill problems, self quizzes, and sample tests.

The laboratory manual, **Experiments in General Chemistry,** contains 38 experiments that parallel the text, including a final group of six experiments on qualitative cation and anion analysis. There is an accompanying instructor's manual.

The **Solutions Manual** contains worked-out solutions to all the Review Problems, Exercises, and Self-Test Questions in the text.

The **Instructor's Manual** offers alternative organizational schemes for the general chemistry course, notes and comments on each chapter, and worked-out solutions to the Additional Exercises.

The **Test Generator** (available on IBM) can produce an almost infinite number of questions for worksheets, homework assignments, tests, and make-up examinations.

In addition, 74 full-color **Transparencies,** selected from the text, are available to adopters.

ACKNOWLEDGMENTS

The following colleagues from my campus helped by reviewing manuscript, passing along student comments, and being available for consultation on various matters that arose during the preparation of this edition: Arlo Harris, Dennis Pederson,

Kenneth Mantei, James Crum, and Lee Kalbus. As with the third and fourth editions, I owe a special debt of gratitude to Robert K. Wismer (Millersville University) who, in addition to his own efforts in producing many of the supplements to this text, commented extensively on the manuscript, read proof, and was my principal sounding board for many of the new features.

As in the fourth edition, I have been fortunate to have Carey Van Loon as my photographer. And, luckily for the users of this text, the more difficult the assignment, the more eager Carey was to tackle it. The same can be said for Jonathan Forman, who set up most of the photographs and checked many of the end-of-chapter exercises. Jon relished the opportunity to apply so much of what he had learned as a general chemistry student.

I am grateful to many people at Macmillan Publishing Company for their help and encouragement, starting with my editor Robert McConnin. Two people deserve special mention. Madalyn Stone, my development editor, read the entire manuscript in the spirit of an eager student and offered many helpful suggestions. She gets a grade of A in the hypothetical general chemistry course that she completed. For the third time, Elisabeth Belfer, my production supervisor, has managed to convert manuscript, cut pages, scribbled notes, rough sketches, and telephone conversations into a very attractive final product. Her organizational skills are truly amazing.

Throughout the five editions of this text, I have sought the forbearance of my wife, Ruth, and other family members for my preoccupation and neglect. I have received not only this but their full support and encouragement as well. For this I am most grateful of all.

REVIEWERS

I appreciate the special efforts of those who provided commentary on the fourth edition or reviewed manuscript chapters of this edition.

Stephen Carlson, *Lansing Community College*

J. Kenneth Cashion, *University of Wisconsin–Parkside*

Dean Cooke, *Western Michigan University*

James Paul Cowin, *University of California, Santa Barbara*

Walter K. Dean, *Lawrence Institute of Technology*

David O. Harris, *University of California, Santa Barbara*

Alton Hassell, *Baylor University*

J. E. House, Jr., *Illinois State University*

Larry M. Peck, *Texas A&M University*

Gordon Schlesinger, *University of California, San Diego*

R. Carl Stoufer, *University of Florida*

Billy L. Stump, *Virginia Commonwealth University*

REVIEWERS OF EARLIER EDITIONS

O. T. Beachley, *State University of New York, Buffalo*

Robert C. Brasted, *late of University of Minnesota*

Luther K. Brice, Jr., *Virginia Polytechnic Institute and State University*

Clark Bricker, *Emeritus, University of Kansas*

William Cass, *formerly Northeastern University*

Geoffrey Davies, *Northeastern University*

Norman V. Duffy, *Kent State University*

Jimmie G. Edwards, *University of Toledo*

Harry A. Eick, *Michigan State University*

Lawrence Epstein, *Emeritus, University of Pittsburgh*

Michael F. Farona, *University of Akron*

Kenneth R. Fountain, *Northeast Missouri State University*

Milton Fuller, *California State University, Hayward*

Michael F. Golde, *University of Pittsburgh*

Henry Heikkinen, *University of Maryland*

Joseph M. Kanamueller, *Western Michigan University*

Philip S. Lamprey, *University of Lowell*

William H. McMahan, *Mississippi State University*

Walter P. Menzies, Jr., *Patrick Air Force Base, Cocoa Beach, Florida*

Jules W. Moskowitz, *New York University*

R. Kent Murmann, *University of Missouri, Columbia*

Gardiner H. Myers, *University of Florida*

Galen L. Ober, *Clarion University*

Randall J. Remmel, *University of Alabama in Birmingham*

William R. Robinson, *Purdue University*

Donald E. Sands, *University of Kentucky*

W. J. Scaife, *Union College*

Curtis T. Sears, *Georgia State University*

Bassam Z. Shakhashiri, *University of Wisconsin–Madison*

Saul I. Shupack, *Villanova University*

Andrew Streitwieser, *University of California, Berkeley*

Billy L. Stump, *Virginia Commonwealth University*

W. P. Tannmeyer, *University of Missouri, Rolla*

Joseph Topich, *Virginia Commonwealth University*

Richard S. Treptow, *Chicago State University*

Carl Trindle, *University of Virginia*

Mary S. Vennos, *Essex Community College of Baltimore County, Maryland*

Kay O. Watkins, *Adams State College*

Milton J. Wieder, *Metropolitan State College*

Joseph Wolinsky, *Purdue University*

R. H. P.

Brief Contents

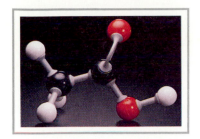

Detailed Contents

4 Chemical Reactions

90

5 Introduction to Reactions in Aqueous Solutions 123

6 Gases

157

16 Principles of Chemical Equilibrium   569

17 Acids and Bases   603

18 Additional Aspects of Acid–Base Equilibria   644

23 Chemistry of the Representative (Main-Group) Elements II: Nonmetals 831

24 The Transition Elements 870

25 Complex Ions and Coordination Compounds 906

26 Nuclear Chemistry 936

27 Organic Chemistry 967

28 Chemistry of the Living State 1008

Appendixes
A Mathematical Operations A1

To the Student

If you were to take a poll among former students of general chemistry, most would probably tell you that it was a demanding course. However, many would say that once you learn to think about chemistry in a certain way it is not that difficult—things do fall into place. So, how should you think about chemistry?

Primarily, you need to understand that chemical knowledge consists of a great deal of factual information that can be divided into a large number of specific facts, a much smaller number of general facts, and a still smaller number of fundamental principles. With a general fact we can correlate a large number of specific facts, and with fundamental principles we can explain general facts. In approaching new information, learn the general facts and how to apply them. Then, where possible, try to discover the fundamental principles underlying these general facts. Similarly, in situations that require solving numerical problems, concentrate on the general methods that can be used to handle large numbers of specific problem types.

To illustrate some of these distinctions, the observation that many copper compounds have a blue color is a *specific* fact. You will probably learn this fact early in your study of chemistry by observing copper compounds in the laboratory. Copper compounds belong to a broad category known as transition metal compounds. Most transition metal compounds are colored (but not necessarily blue). This is a *general* fact that you will find useful when studying transition metals. Finally, in Chapter 25 you will learn the fundamental *principle* underlying the origin of color in transition metal compounds.

This book contains a number of special features designed to help you master general facts and fundamental principles and apply them. Read about these features in the *Preface* and take full advantage of them as you proceed through the text.

WARNING: Many of the compounds described or pictured in this text are hazardous, as are many of the chemical reactions. The reader should not attempt any experiment pictured or implied in this text. Experiments should be performed only in authorized laboratory settings and under adequate supervision.

1 Matter—Its Properties and Measurement

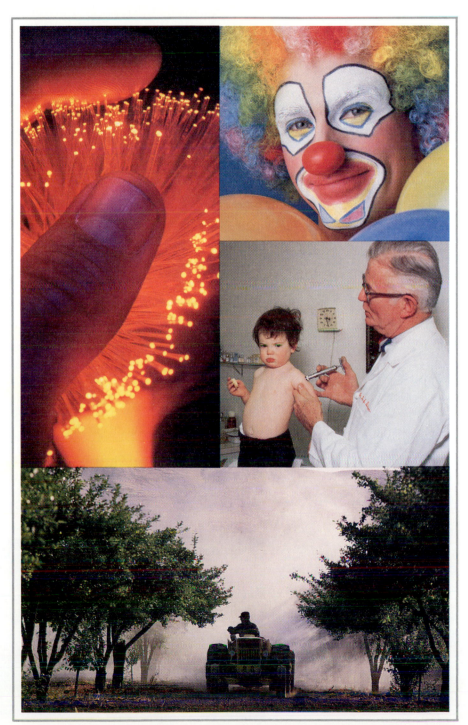

As suggested in this adaptation of a 1987 American Chemical Society poster, from paints and cosmetics, to medicinal products, to fertilizers and insecticides, to fiber optics and other high-technology materials—chemistry is everywhere. [Comstock, Inc./ clockwise from upper left: Michael Stuckey, Tom Grill, Russ Kinne, and George D. Lepp]

1

From the food label that reads "no chemicals added," to the clinic established to treat "chemical dependency," to the description of a performance as one in which the "chemistry was right," we are constantly reminded that chemistry is a part of our lives. Mostly, however, these everyday references to chemistry are misleading. For example, even if no substances are added to it during its preparation, a food consists entirely of chemicals. In fact, all material objects, including ourselves, are made up entirely of chemicals, and we should begin our study of chemistry with this thought clearly in mind.

By manipulating materials in their environment, people have always practiced chemistry. Until fairly recent times these manipulations have involved mostly minor modifications, such as extracting a metal from its ore. With modern chemical knowledge, though, we can literally decompose naturally occurring matter into its smallest components (atoms) and reassemble these components into materials that do not exist naturally. Thus, from petroleum we can produce motor fuels and a countless number of plastics, pharmaceuticals, and pesticides. We also need modern chemical knowledge to understand and control processes that are detrimental to the environment, such as the production of smog and the destruction of stratospheric ozone.

Early chemical knowledge consisted of the "how to" of chemistry. That is, early chemists learned by trial and error (experimentation) what chemical reactions could be used to produce new materials. Modern chemical knowledge answers the "why" as well as the "how to" of chemical change. It is based on principles and theory. Our study of chemistry will be broadly based, involving principles, theory, and applications.

1-1 Properties of Matter

Reduced to simplest terms, chemistry is a study of matter. In this study we emphasize the ways in which matter can be changed or transformed. **Matter** is any object or material that occupies space; the quantity of matter is measured by its **mass** (described further in Section 1-5). Thus, the gases of the atmosphere, because they occupy space and possess mass, are examples of matter, but sunlight (a form of electromagnetic radiation) is not. The characteristics that we can use to identify samples of matter and distinguish them from one another are called **properties.** Mass is an example of a property of matter. We can group properties into two broad categories: **physical** and **chemical.**

Physical Properties and Physical Change. A process in which an object changes its physical appearance but not its basic composition is called a **physical change.** While undergoing a physical change, an object of matter generally displays one or

FIGURE 1-1
Physical properties compared.

Copper (left) can be obtained as pellets, hammered into a thin foil, or drawn into a wire. Lump sulfur (right) crumbles into a fine powder when hammered.
[Carey B. Van Loon]

more **physical properties.** We can hammer a pellet of copper into a thin foil; copper is said to be *malleable*. We can draw copper into a fine wire; it is *ductile*. But whether in the form of a pellet, foil, or wire, the material remains the substance copper. One of the most useful physical properties of copper is that it is an excellent *electrical conductor*. Sulfur, unlike copper, has a yellow color and is a nonconductor of electricity. Also, if we strike a chunk of sulfur with a hammer it crumbles into a powder; sulfur is *brittle*. Thus, we can readily distinguish between copper and sulfur through their physical properties, as we see in Figure 1-1.

Metal and nonmetal are two fundamental terms used to describe the chemical elements. These terms are introduced in Section 3-4 and then described in more detail in Section 9-5 and elsewhere in the text.

Chemical Properties and Chemical Change. Paper burns, iron rusts, and wood rots. In each case the object changes not only in physical appearance but also in its basic composition. In a **chemical change** or **chemical reaction** a sample of matter is transformed into completely different materials. The types of chemical reactions that a material can undergo (e.g., whether it will burn or in some other way react with atmospheric oxygen) are determined by its **chemical properties.** Copper displays the chemical properties of a metal, whereas the chemical properties of sulfur are those of a typical nonmetal.

1-2 Classification of Matter

In this section we will discuss the scheme of classifying matter shown in Figure 1-2. As we will explore more fully in later chapters, matter is built up from inconceivably tiny units called **atoms.** Presently we know of 109 different types of atoms,

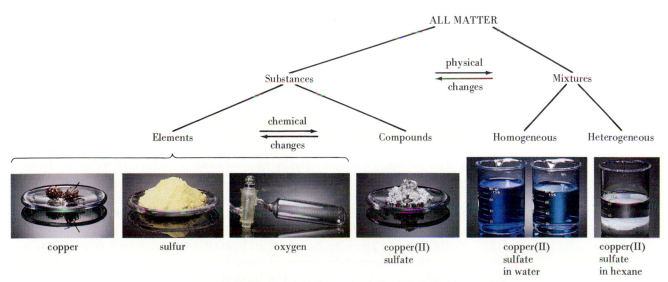

FIGURE 1-2
A classification scheme for matter.

Every sample of matter is either a substance or a mixture. If a substance, it is either an element or a compound; if a mixture, either homogeneous or heterogeneous. Transformations between elements and compounds involve chemical changes; conversions between substances and mixtures, physical changes.

The solid elements copper and sulfur and the gaseous element oxygen can undergo a chemical change (chemical reaction) to form the compound copper(II) sulfate. All samples of copper(II) sulfate contain the same proportions of copper, sulfur, and oxygen. Solid copper(II) sulfate dissolves in liquid water (physical change) to produce a homogeneous mixture or solution. The quantity of copper(II) sulfate in water can be varied from large (dark blue solution) to small (light blue solution). Copper(II) sulfate is insoluble in liquid hexane, and the mixture remains heterogeneous. [Carey B. Van Loon] and *all* matter is made up of just these 109! A substance comprised of just a *single* type of atom is called a **chemical element.** Elements cannot be decomposed into simpler substances, not by heating, crushing, exposure to acids, and so on. The 109 known elements range from such common substances as iron, copper, silver, and gold to uncommon ones such as lutetium and thulium. About 90 of the elements can be extracted from natural sources; the rest must be created through nuclear processes (discussed in Chapter 26). A complete listing of the elements and a special tabular arrangement known as the periodic table (discussed in Chapters 3 and 9) are presented on the inside front cover.

Atoms of different elements can combine with one another to form more complex substances called **chemical compounds.** The number of chemical compounds now known is in the millions. Compounds range in complexity from ordinary water, which consists of the elements hydrogen and oxygen, to the protein gamma globulin, consisting of the elements carbon, hydrogen, oxygen, and nitrogen. For many compounds we can isolate the smallest identifiable unit of the compound. We call it a **molecule.** A molecule of water consists of two hydrogen atoms joined to a single oxygen atom. By contrast, a molecule of gamma globulin is comprised of 19,996 atoms! Chemical compounds retain their identities during *physical* changes but can be separated into their component elements by appropriate *chemical* changes.

The composition and properties of an element or compound are uniform throughout a given sample *and from one sample to another.* Elements and compounds are called **substances.** (To a chemist, the term *substance* always means either an element or a compound.)

Samples of matter having compositions and properties that are uniform throughout a given sample *but that vary from one sample to another* are called **homogeneous mixtures** or **solutions.** A given sugar solution is uniformly "sweet" throughout the solution, but the "sweetness" of another sugar solution might be noticeably different.

A homogeneous mixture can be separated into its components by appropriate *physical* changes. Ordinary air is a solution of several gases, principally the *elements* nitrogen and oxygen. Seawater is a solution of the *compounds* water, sodium chloride (salt), and a host of others.

In some mixtures—sand and water, for example—the components separate into physically distinct regions. Mixtures in which the composition and physical properties vary from one part of the mixture to another are said to be **heterogeneous.** Samples of matter ranging from a glass of iced tea to a slab of concrete to the leaf of a plant are heterogeneous. Usually heterogeneous mixtures can be easily distinguished from homogeneous ones, but at times this may be more difficult.

Another classification scheme is based on the three **states of matter.** In a **solid,** atoms or molecules are in close contact, sometimes in a highly organized arrangement (called a *crystal*). A solid occupies a definite volume and has a definite shape. In a **liquid,** the atoms or molecules are separated by greater distances than in a solid. Movement of these atoms or molecules gives a liquid its most distinctive property—an ability to flow to cover the bottom and assume the shape of its container. In a **gas,** distances between atoms or molecules are greater still than in a liquid. Not only does a gas assume the shape of its container but it expands to *fill* the container. Depending on the conditions, a substance may exist exclusively in one state of matter or it may be distributed between two or three states. Thus, as the ice in a small pond begins to melt in the spring, water exists in two states: solid and liquid (actually, three states if we also consider water vapor in the air above the pond).

When viewed through a microscope, homogenized milk is seen to consist of globules of fat dispersed in a watery medium. Homogenized milk is a *heterogeneous* mixture. [© Bruce Iverson]

1-3 Separating Mixtures and Decomposing Compounds

Much of the work in chemical laboratories requires that mixtures be separated into their component substances. This is done through *physical* changes. The separation methods that we use in the general chemistry laboratory are mostly classical procedures of long standing.

The basis for establishing the name copper(II) sulfate is presented in Chapter 3. Systematic names for substances like hexane are introduced in Chapter 27.

Consider the heterogeneous mixture of copper(II) sulfate and hexane pictured in Figure 1-2. Suppose we pour the mixture through a funnel lined with porous filter paper. The liquid hexane passes through the pores of the paper, and we collect it in a beaker. The solid copper(II) sulfate is retained on the filter paper. (The hexane that wets the solid quickly evaporates, leaving a relatively pure, dry solid.) This separation of a solid from a liquid is called **filtration,** and is illustrated in Figure 1-3.

We cannot separate a homogeneous mixture (solution) of copper(II) sulfate in water by filtration; both components pass through the paper. However, as suggested by Figure 1-4, if we boil a copper(II) sulfate solution the vapor given off by the boiling solution contains only water. The vapor can be conducted away from the boiling solution and passed through a long tube, jacketed by a second tube through which cold water is circulated. The cold water removes heat from the vapor, permitting it to condense to liquid water, which can be collected. When all the water has boiled away, solid copper(II) sulfate remains in the flask. This process is called **simple distillation.** A more general procedure, called **fractional distillation,** permits a separation even when more than one component is present in the vapor; it is described in Chapter 13. In Chapter 13 we also describe a separation technique called *chromatography*. This method is based on the differing tendencies of two or more substances to be held up by a medium through which they move.

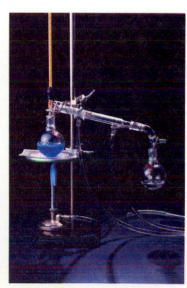

FIGURE 1-3
Separation of a heterogeneous mixture by filtration.

Solid copper(II) sulfate is retained on the filter paper as liquid hexane passes through.
[Carey B. Van Loon]

FIGURE 1-4
Separation of a homogeneous mixture by simple distillation.

Copper(II) sulfate remains in the flask on the left as water passes to the flask on the right, by first evaporating and then condensing back to liquid. [Carey B. Van Loon]

FIGURE 1-5

Obtaining an element from one of its compounds: a chemical change.

(a) When metallic zinc is added to a blue water solution of copper(II) sulfate, a chemical reaction occurs.

(b) The products of the reaction are a colorless water solution of zinc sulfate and a red-brown deposit of copper metal.

An important application of this type of reaction is in the extraction of metallic gold from natural sources. Gold is displaced from a solution of one of its compounds by zinc metal. [Carey B. Van Loon]

(a)   (b)

To break down or *decompose* a compound into its constituent elements requires *chemical* changes. As shown in Figure 1-5 we can partially decompose copper(II) sulfate to pure metallic copper by adding zinc to a water solution of copper(II) sulfate. The zinc displaces copper from solution, and the water solution becomes one of zinc sulfate. To recover the original sulfur and oxygen that were part of the copper(II) sulfate is a more difficult matter, however.

1-4   The Scientific Method

What makes science different from other intellectual activities is the way in which we acquire scientific knowledge and use this to *predict* future events. We can predict the time required for a rocket to reach the moon even more accurately than how long it will take us to drive from New York City to Washington, D.C. This is because the scientific basis of rocket propulsion is well understood, whereas science is very limited in predicting traffic conditions.

The ancient Greeks acquired knowledge by *deduction*. Starting from certain basic premises or assumptions they could figure out what conclusions must follow. This is the logical method that they used so successfully in the study of geometry. However, the deductive method is no better than the validity of the basic assumptions.

An example of deduction: if $a = b$ and if $b = c$, then $a = c$.

In the seventeenth century people began to appreciate the importance of experimentation in the discovery of scientific facts. Rather than to start with any assumptions, seventeenth-century scientists like Galileo, Francis Bacon, Descartes, Boyle, Hooke, and Newton made careful observations of phenomena (such as the motion of the planets) and then formulated natural laws to summarize them. We call this method of formulating a general statement or **natural law** from a series of observations *induction*.

Here is a problem with the method of induction: When have we made enough observations to justify a generalization (a natural law)? A city dweller has observed only a few dozen sheep in her lifetime and all have had a white coat. Is she justified in making the generalization "All sheep are white"?

To test a natural law a scientist designs a controlled situation, an *experiment*, to see if the conclusions deduced from the natural law agree with actual experience. If a natural law stands the test of repeated experimentation, our confidence in it grows. If the agreement between predicted and observed behavior is imperfect, the natural law must be modified or discarded. We judge the success of a natural law, then, by how effective it is in summarizing observations and predicting new phenomena. However, we can never accept a natural law as an *absolute* truth. There is always the possibility that someone may devise an experiment that would refute it.

We call a tentative explanation of a natural law a **hypothesis.** A hypothesis can be tested by experimentation, and if it survives this testing it is often referred to as a **theory.** We can also use the term *theory* in a broader sense: it is a conceptual framework or model (a way of looking at things) that can be used to explain and to make further predictions about natural phenomena. Sometimes, differing or conflicting theories are proposed to explain the same phenomenon. We usually choose the theory that is most successful in its predictions. Also, we generally prefer the theory that involves the smallest number of assumptions—the simplest theory. Over a period of time, as new evidence accumulates, most scientific theories undergo modification; some are discarded.

As an illustration, consider the development of John Dalton's atomic theory. Dalton's model was that all matter is composed of minute, indivisible particles called atoms. He based his theory on two natural laws of chemical combination. One was the *law of conservation of mass:* The total mass remains unchanged in a chemical reaction. The other was the *law of constant composition:* The proportions in which elements are combined in a compound are independent of the source of the compound. And with his theory Dalton was able to *predict* still another law of chemical combination (the law of multiple proportions). These laws and Dalton's theory are described in Chapter 2.

The sum of all the activities we have described—observations, experimentation, and the formulation of laws, hypotheses, and theories—is called the **scientific method.** Simply following the scientific method does not always guarantee success, however. Occasionally someone must break away from established patterns of thinking to discover the key to a scientific puzzle. These developments are called scientific breakthroughs (see, for example, the discussion of the quantum theory in Chapter 8). And always, we need to be alert to unexpected observations. A number of great discoveries (x rays, radioactivity, penicillin) were made by accident in the course of other investigations.

"Accidental" discoveries do not usually happen by accident. As noted by Louis Pasteur (1822–1895): "Chance favors the prepared mind."

1-5 The English and Metric Systems of Measurement

The scientific method is especially effective when we can assign numerical values to observations, that is, when observations are *quantitative*. To make quantitative observations we must compare objects to standards that have known values of a property. The most fundamental standards are those established for mass, length, and time.

The **English system** of measurement was once widely used in English-speaking countries, but now its use is limited mainly to the United States. This system has as its standard of mass, the standard **pound (lb),** and as its standard of length, the standard **yard (yd).** The system is sufficiently precise for modern manufacturing and commerce, but it is not particularly useful in scientific work.

Length. The **metric system** is based on a standard of length, called a **meter (m),** originally defined as 1/10,000,000 of the distance at sea level from the North Pole to the equator along a meridian passing through Paris. The metric system is a *decimal* system. The several multiple and submultiple units for expressing a measured property differ from one another by factors of *ten*. For example,

- **kilo** means *one thousand times* the base unit,
- **centi** means *one hundredth of* the base unit,
- **milli** means *one thousandth of* the base unit.

Thus, 1 *kilo*meter **(km)** is 1000 m (about 0.6 mile); 1 *centi*meter **(cm)** is 1/100 of 1 m (about 0.4 inch); and 1 *milli*meter **(mm)** is 1/1000 of 1 m (about the thickness

FIGURE 1-6
The metric and English systems compared.

The green ribbon is 1 centimeter (cm) wide and is wrapped around a stick that is 1 meter (m) long. The yellow ribbon is 1 inch (in.) wide and is wrapped around a stick 1 yard (yd) long. The meterstick is about 10% longer than the yardstick.

Of the two identical beakers shown, the left one contains 1 kilogram (kg) of candy and the right one, 1 pound (lb).

The two identical volumetric flasks hold 1 liter (L) when filled to the mark. The flask on the left and the carton behind it each contain 1 quart (qt) of milk. The flask on the right and the bottle behind it each contain 1 L of orange juice. (1 L = 1.057 qt.) [Carey B. Van Loon]

TABLE 1-1
English and Metric Equivalents

Metric		English
Mass		
1 kg	=	2.205 lb
453.6 g	=	**1 lb**
Length		
1 km	=	0.6214 mi
1 m	=	39.37 in.
2.54 cm[a]	=	**1 in.**
Volume		
3.785 L	=	**1 gal**
1 L	=	1.057 qt

[a] The English inch is now defined to be *exactly* 2.54 cm.

The symbol ∝ means "proportional to." It can always be replaced by an equality sign and a proportionality constant. The proportionality constant in equation (1.1), g, is called the acceleration due to gravity. Its significance is explored further in Appendix B.

of a dime). 1 *kilo*gram **(kg)** is 1000 grams **(g)** (about 2.2 pounds), and 1 *milli*gram **(mg)** is 1/1000 of 1 g (about the mass of a few inches of human hair).

Some comparisons between the metric and English systems are made in Figure 1-6. A few equivalent quantities in the two systems are listed in Table 1-1 and used in calculations in Sections 1-9.

Mass. Mass describes the quantity of matter in an object. The **kilogram (kg)** was originally defined as the mass of 1000 cubic centimeters (cm³) of water at 4 °C and standard atmospheric pressure. The current standard is a cylinder of platinum-iridium metal kept at the International Bureau of Weights and Measures in Sèvres, near Paris. The kilogram is a fairly large unit for most applications in chemistry, so the unit **gram (g)** is more commonly used.

Weight, which describes the force of gravity on an object, is directly proportional to mass, as represented through a simple mathematical equation.

$$W \propto m \qquad \text{and} \qquad W = gm \tag{1.1}$$

An object of matter has a fixed mass *(m)*, regardless of where or how the measurement is made. Its weight *(W)*, on the other hand, may vary because g varies slightly from one point on earth to another. (In outer space, where g becomes very small, we speak of an object as becoming "weightless," even though its mass is unchanged from that on earth.) The terms *weight* and *mass* are often used interchangeably, but only *mass* is a measure of the quantity of matter. (See Figure 1-8 for three different devices for measuring mass by the principle of weighing.)

Volume. Volume is an important property, but it is not as fundamental as mass, because the volume of an object varies with temperature and pressure. Its mass does not. Volume has the unit (length)³. The fundamental metric unit of volume is the **cubic meter (m³).** A more commonly used unit is the **cubic centimeter (cm³),** and still another is the **liter (L).** One liter is defined as a volume of 1000 cm³, which means that one **milliliter (1 mL)** is equal to one cubic centimeter **(1 cm³).** The liter is also equal to one **cubic decimeter (1 dm³).** These metric volume units are compared in Figure 1-7.

FIGURE 1-7
Some metric volume units compared.

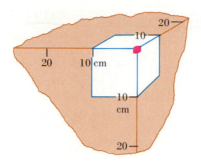

The largest volume, shown in part, is the metric standard of 1 cubic meter (m^3). A cube that has a length of 10 cm (1 dm) on edge (in blue) has a volume of 1000 cm^3 (1 dm^3) and is called 1 liter (1 L). The smallest cube is 1 cm on edge (red) and has a volume of 1 cm^3 = 1 mL. Note that 1 m^3 = 1000 L and 1 L = 1000 mL = 1000 cm^3.

CHEMISTRY EVOLVING

The Principle of Weighing

One of the most important laboratory measurements throughout the history of chemistry has been that of mass, through the method of weighing. The object being weighed experiences a force (gravity) directed toward the center of the earth that is proportional to the object's mass. This force is counterbalanced by an equal force whose magnitude is precisely measured with an instrument called a **balance.** The three types of balances shown in Figure 1-8 and described below parallel the development of chemistry from its earliest days to the present (and future).

(a) The Two-Pan Balance. A balanced condition is reached when the pointer on a horizontal beam rests at the center of a ruled scale. When this occurs the masses of the unknown and the standard weights are equal. Final adjustment of the balance point is made by moving a light metal wire called a rider along the beam. This type of balance was in use for hundreds of years, up to about the middle of the twentieth century.

(b) The Single-Pan Balance. This instrument differs from the two-pan balance in several ways: (1) Imagine that the balance in (a) is viewed "end on" so that only one pan is visible. (2) Now imagine that the pan out of view is replaced by a single large constant weight. (3) An appropriate set of weights is placed *above* the single pan to just balance the constant weight. (4) When an unknown is placed on the pan, the balance condition is upset. The balance condition can be restored, however, by *removing* from the set above the single pan appropriate weights whose combined mass is just equal to that of the unknown. The total mass removed is registered on the dials on the face of the balance. This type of balance, although first constructed in 1886, only came into common use in the 1950s. It is still in use, but it is quickly being replaced by electronic balances.

FIGURE 1-8
Three types of balances.

(a) A two-pan analytical balance.
(b) A single-pan balance.
(c) An electronic balance.
[Carey B. Van Loon]

(a)   (b)   (c)

(c) Electronic Balance. This type of balance uses a magnetic force to counterbalance the force of gravity on the object being weighed. The magnetic force is produced by an electromagnet. When an object is placed on the balance pan, the initial balance condition is upset and additional electric current must be passed through the electromagnet to restore the balance condition. The magnitude of the additional current is proportional to the mass of the object being weighed, and is translated into a mass reading that is displayed on the balance.   ■ ■ ■

1-6   SI Units

TABLE 1-2
Some Common SI Prefixes

Multiple	Prefix
10^9	giga (G)
10^6	mega (M)
10^3	kilo (k)
10^{-1}	deci (d)
10^{-2}	centi (c)
10^{-3}	milli (m)
10^{-6}	micro (μ)[a]
10^{-9}	nano (n)

[a] The Greek letter μ (pronounced "mew").

In time, SI units may be used exclusively throughout the world. However, in many activities in which American scientists, engineers, and technicians engage, older metric units or even English units take precedence over SI units.

There are two difficulties with the old metric standards: First is the difficulty in comparing objects with a standard when the standard is one of a kind (such as the standard kilogram in Sèvres). Second, the standards are subject to change. Thus, the metal bar that has served as the standard meter changes in length as the temperature changes.

These difficulties can be overcome by basing standards of measurement on *natural* universal constants, as is done in the International System of Units or **SI units.** In this system the unit of length corresponding to 1 meter is defined as a length equal to the distance traveled by light in a vacuum during a time interval of 1/299,792,458 of a second. The unit of time, the second, is defined as the duration of 9,192,631,770 periods of a particular radiation emitted by cesium-133 atoms. The unit of mass, the kilogram, has not been defined in terms of a natural constant, and so it remains the mass of a cylindrical bar of metal maintained at Sèvres.

Another aspect of the SI conventions is that, to facilitate communication among scientists, certain base units and derived units are preferred over others. Since the SI conventions have not yet been universally adopted, we use both familiar metric units and SI units in this text. Where they differ, we note this fact. Of the familiar units introduced to this point, the liter and milliliter are not SI units. Their SI counterparts are the cubic decimeter (dm^3) and the cubic centimeter (cm^3), respectively (recall Figure 1-7). A more complete description of SI units is presented in Appendix C, to which you may need to refer from time to time. Some common SI prefixes are listed in Table 1-2.

1-7   Uncertainties in Scientific Measurements

A quantity such as 299,792,458, which appears in the SI definition of the meter, is extremely precise: It is stated to *one part in 300 million*. On the other hand, a measurement made on a kitchen scale, e.g., the 205-g mass of a cup of sugar, is not so precisely known, perhaps to the nearest 10 g or to *one part in 20*.

All experimental measurements are subject to error. To some extent measuring instruments have built-in or inherent errors, called **systematic errors.** (Such would be the case if a kitchen scale consistently yielded a result that was 25 g too high, or a thermometer a reading that was 2° too low.) Other errors are introduced through limitations in the experimenter's skill or ability to read scientific instruments, leading to results that may be either too high or too low. These are called **random errors.**

Precision refers to the degree of reproducibility of a measured quantity, that is, the closeness of agreement among the values when the same quantity is measured several times. The precision of a series of repeated measurements is *high* (or good) if each of the individual measurements deviates by only a small amount from the average for the series. Conversely, if there is wide deviation among the measurements the precision is *poor* (or low). On the other hand, measurements are said to be **accurate** if they are close to the "correct" or most probable value. Measurements having high precision are not always accurate—a large systematic error could be present. Still, the likelihood of their being accurate is much greater than for measurements of low precision.

To illustrate some of these ideas, consider the mass of an object measured on two different balances. One is a relatively crude balance called a platform balance, and the other is a sophisticated analytical balance.

	Platform balance	Analytical balance
three measurements:	10.4; 10.2; 10.3 g	10.3107; 10.3108; 10.3106 g
their average:	10.3 g	10.3107 g
uncertainty:	±0.1 g	±0.0001 g
reported mass:	10.3 ± 0.1 g	10.3107 ± 0.0001 g
precision:	**low** or **poor**	**high** or **good**
	(about one part per 100)	(about one part per 100,000)

1-8 Significant Figures

In the preceding section we indicated the precision of some mass measurements by writing "10.3 ± 0.1 g" and "10.3107 ± 0.0001 g." You may see this type of notation in laboratory notebooks and scientific journals, but it is somewhat cumbersome to write and to use in numerical calculations.

As an alternative, assume that when a number is written down, all the digits preceding the last one are known with certainty and that there is an *uncertainty of about one unit in the last digit shown*. Thus, the number 10.3 is "between 10.2 and 10.4" and the number 10.3107 is "between 10.3106 and 10.3108." The number 10.3 consists of *three* significant figures, whereas 10.3107 consists of *six* significant figures. When we designate the number of significant figures in a measured quantity we indicate our confidence in the measurement. The greater the number of significant figures, the smaller the uncertainty (and the greater the precision) of the measurement.

These are the rules you should use in establishing how many significant figures are present in a number:

1. All nonzero digits are *significant* (e.g., 4.006, 12.012, 10.070).
2. Zeros placed between significant digits (called *interior* zeros) are *significant* (e.g., 4.006, 12.012, 10.07).
3. Zeros at the *end* of a number and to the *right* of the decimal point (*trailing* zeros following a decimal point) are *significant* (e.g., 10.070).
4. Zeros appearing at the *end* of a number and to the *left* of the assumed decimal point (*trailing* zeros preceding an assumed decimal point) *may or may not be significant*.
5. Zeros to the left of the first nonzero digit (*leading* zeros) are *not significant*. They simply locate the decimal point (e.g., 0.00002).

The two decimal numbers 0.00002 and .00002 are identical. The first zero to the *left* of the decimal point in 0.00002 simply improves the appearance of the written number; it is *not significant*.

To establish the number of significant figures in 4.006 and 12.012, we use Rules 1 and 2. There are *four* significant figures in 4.006 and *five* in 12.012.

To establish the number of significant figures in 10.070, we use Rules 1, 2, and 3. There are *five*.

According to Rules 1 and 5, there is only *one* significant figure in the number 0.00002. But according to Rules 1, 3, and 5, there are *two* significant figures in the number 0.000020.

To illustrate Rule 4, consider the numbers 750 and 20,000. We cannot be certain whether the number 750 is meant to indicate 750 ± 10 (in which case there are *two*

To summarize the treatment of zeros:

- interior zeros are significant,
- leading zeros are never significant,
- trailing zeros are significant in some cases and not in others.

significant figures) or 750 ± 1 (in which case there are *three*). One method to resolve this ambiguity is to write 750 if we want to represent *two* significant figures and 750. if we want to represent *three*. (That is, by using a decimal point where it is not otherwise required, we can signify that all digits to the left of the decimal point are significant.) This alternative is not sufficient for a number like 20,000, however. Here the number of significant figures might be anywhere from *one* to *five*. The best way of resolving this difficulty is to use exponential notation. (See Appendix A if you need to review exponential arithmetic.) The precision of the number is embodied in the coefficient and the power of ten simply locates the decimal point.

1 significant figure	2 significant figures	3 significant figures
2×10^4	2.0×10^4	2.00×10^4

Significant Figures in Numerical Calculations. There are ways to increase the precision of a result by repeated measurement and the appropriate handling of the data. However, once the best values of measured quantities have been established, *precision can be neither gained nor lost during arithmetic operations*. We can satisfy this requirement reasonably well by observing some simple rules involving significant figures:

Significant figure rule in multiplication/division.

> The result of multiplication and/or division may carry no more significant figures than the *least* precisely known quantity in the calculation.   (1.2)

In this chain multiplication the result should be rounded off to *three* significant figures.

$$14.79 \times 12.11 \times 5.05 = 904.48985 = 904 = 9.04 \times 10^2$$
(4 sig. fig.)   (4 sig. fig.)   (3 sig. fig.)   (3 sig. fig.)

Figure 1-9 should help you to understand why this result should be limited to three significant figures.

In adding and subtracting numbers the applicable rule is that

Significant figure rule in addition/subtraction.

> The result of addition and/or subtraction must be expressed with the same number of decimal places as the term carrying the *smallest* number of decimal places.   (1.3)

Consider the sum

$$
\begin{array}{r}
15.02 \\
9,986.0 \\
3.518 \\
\hline
10,004.538
\end{array}
$$

The sum has the same uncertainty, ± 0.1, as does the term carrying the smallest number of decimal places, 9,986.0. Note that this calculation is *not* limited by significant figures; in fact the sum has more significant figures (6) than do any of the terms in the addition. Nor is the calculation limited by the least precisely known term (9,986.0 is known with greater precision than either 15.02 or 3.518).

There are two situations when a number appearing in a calculation may be **exact.** This may occur

- By definition (e.g., 3 ft = 1 yd; 1 in. = 2.54 cm)

or as a result of

- Counting (e.g., *six* faces on a cube; *two* hydrogen atoms in a water molecule).

Significant figures for exact numbers.

> *Exact* numbers can be considered to have an unlimited number of significant figures.   (1.4)

$$14.79 \times 12.11 \times 5.04 \qquad 14.79 \times 12.11 \times 5.05 \qquad 14.79 \times 12.11 \times 5.06$$
$$= 902.69878 \qquad\qquad = 904.48985 = 904 \qquad\qquad = 906.28091$$

FIGURE 1-9
Handling significant figures in a multiplication.

The actual rule on multiplication and division is that the result should have about the same percent error as the least precisely known quantity. Usually the significant figure rule conforms to this requirement; sometimes it does not (see Exercise 67).

In forming the product $14.79 \times 12.11 \times 5.05$, the least precisely known quantity is 5.05, which is actually 5.05 ± 0.01. Shown on the above calculators are the products of 14.79 and 12.11 with 5.04, 5.05, and 5.06, respectively. In the three results only the first two digits, "90. . ," are identical. Variations begin in the third digit, and we are certainly not justified in carrying any digits beyond the third. Usually, instead of making a detailed analysis as is done here, we employ a simpler idea: *The result of a multiplication must contain no more significant figures than the least precisely known quantity.* [Carey B. Van Loon]

"Rounding Off" Numerical Results. To *three* significant figures, we should express 15.453 as 15.5 and 14,775 as 1.48×10^4. If our need is to drop just one digit, that is, to "round off" a number, the rule to follow is to increase the final digit by one unit if the digit dropped is greater than 5 and to leave the final digit unchanged if the digit dropped is less than 5. For example, to three significant figures, 15.56 rounds off to 15.6 and 15.54 rounds off to 15.5. If the digit dropped is 5, the usual rule is to increase the final remaining digit by one unit if necessary to make it *even;* otherwise, leave it unchanged. Thus, to three significant figures, 15.55 is rounded off to 15.6 and 15.45 is rounded off to 15.4.

Example 1-1

Applying the rules on significant figures in a calculation. Express the result of the following calculation with the correct number of significant figures.

$$(0.0300 \times 14.11) + (4.5 \times 10^3 \times 1.50 \times 10^{-4}) = ?$$

Solution. The product of the first two terms, 0.0300×14.11, is 0.4233. But this result should be rounded off to 0.423 because the least precisely known quantity, 0.0300, has only *three* significant figures. The product of the second two terms is 0.675, and again the rules on significant figures limit this result to *two* significant figures, that is, 0.68. The sum of these two quantities is,

$$0.423 + 0.68 = 1.103 = 1.10$$

If you do this calculation with an electronic calculator, you can store the intermediate results (0.4233 and 0.675) in the memory of the calculator. The final result that is displayed is 1.0983. To the correct number of decimal places this result is expressed as 1.10.

SIMILAR EXAMPLES: Exercises 14, 29, 62.

Are You Wondering:

Whether you should round off numerical results at each step of a calculation or only in the last step?

One of the nice features of electronic calculators is that you do not have to write down intermediate results. This means you have to round off a result just once, at the end of the calculation. On the other hand, because in many of the illustrative examples in this text we emphasize the stepwise manner of solving problems, we sometimes round off intermediate results. Usually, the two approaches will lead to identical results (as in Example 1-1), or at least to results that differ only in the last significant digit.

1-9   Problem Solving

Because of the importance of the cancellation of units, this problem-solving method is often called **unit analysis** or **dimensional analysis.**

The Conversion Factor Method (Dimensional Analysis). Where possible we will emphasize properties that can be expressed through numbers—quantitative measurements. But a number by itself is usually meaningless. A measured quantity must be accompanied by a **unit.** The unit indicates the standard against which the measured quantity is to be compared. Thus, a metal rod 9 m in length is nine times as long as the standard meter.

Many of the calculations of general chemistry simply require that we convert quantities from one set of units to another. We can do this by using conversion factors. Consider this well-known fact expressed as a simple mathematical equation.

1 yd = 36 in.

Divide each side of the equation by 1 yd.

$$\frac{1 \text{ yd}}{1 \text{ yd}} = \frac{36 \text{ in.}}{1 \text{ yd}}$$

The numerator and denominator on the left side are identical; they cancel.

$$1 = \frac{36 \text{ in.}}{1 \text{ yd}} \tag{1.5}$$

The numerator and denominator on the right side of equation (1.5) represent *same length*. It is for this reason that the ratio of the numerator (36 in.) and the denominator (1 yd) is equal to 1. *A conversion factor must always have the numerator and denominator representing equivalent quantities.*

Consider the question: How many inches are there in 6.00 yd? The measured quantity is 6.00 yd and multiplying this quantity by 1 does not change its value.

6.00 yd × 1 = 6.00 yd

Now replace the 1 by its equivalent—the conversion factor (1.5). Cancel the unit, yd, and carry out the required multiplication.

$$6.00 \text{ yd} \times \underbrace{\frac{36 \text{ in.}}{1 \text{ yd}}}_{\substack{\text{this factor} \\ \text{converts} \\ \text{yd to in.}}} = 216 \text{ in.}$$

Next consider the question: How many yards are there in 540. in.? We cannot use exactly the same factor (1.5) as previously; the result would be nonsensical.

$$540.\ \text{in.} \times \frac{36\ \text{in.}}{1\ \text{yd}} = 19{,}400\ \text{in.}^2/\text{yd}$$

Factor (1.5) must be rearranged to 1 yd/36 in.

$$540.\ \text{in.} \times \frac{1\ \text{yd}}{36\ \text{in.}} = 15.0\ \text{yd}$$

this factor
converts
in. to yd

This second illustration emphasizes two important points.

1. There are two ways of writing a conversion factor—in one form or its reciprocal (inverse). Since a conversion factor is equivalent to 1, its value is not changed by inversion, but
2. A conversion factor must be used in such a way as to produce the desired cancellation of units.

The basic setup for problem solving.

A convenient way to think about calculations involving conversion factors is that

information sought = information given × conversion factor(s)   (1.6)

Consider how expression (1.6) is used to answer this question: What is the length of 22 in., expressed in cm? The answer must consist of *two* parts—a number and a unit. The required unit is suggested in the statement of the problem—cm. The numerical part of the answer must be determined by calculation, and we refer to this as "number" or "no." The *information sought* is *no. cm*. The *information given*, that is, the quantity that is to be multiplied by a conversion factor, is also determined by a close reading of the problem; it is *22 in.* Thus, expression (1.6) takes the form

no. cm = 22 in. × conversion factor

Sometimes the statement of the problem includes the necessary conversion factor(s), and sometimes you are expected to know or to be able to derive the factor(s) you need. *The key to problem solving by the conversion factor method lies in knowing where to find and how to use conversion factors.*

In the present case the necessary relationship is 1 in. = 2.54 cm; the unit *in.* must cancel and the unit *cm* must remain.

$$\text{no. cm} = 22\ \text{in.} \times \frac{2.54\ \text{cm}}{1\ \text{in.}} = 56\ \text{cm}$$

this factor
converts
in. to cm

Example 1-2

Using conversion factors to convert a distance from metric to English units. **A 5.0-km run is planned through the center of a city. If there are 8.0 city blocks in one mile, how many blocks along the route must be closed to traffic to accommodate the run?**

Solution. **We start with expression (1.6). We are seeking information in the unit "blocks" and we are given the distance, 5.0 km.**

no. blocks = 5.0 km × conversion factors

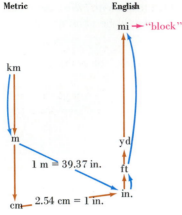

Metric   **English**

FIGURE 1-10
Alternatives in converting between metric and English systems—Example 1-2 visualized.

One possible route is outlined in blue. Another is in brown. The key to any problem of this type is to select a conversion factor that allows a crossover or bridge from one system of measurement to the other. The bridges shown here are between meters and inches and between centimeters and inches.

From this point we can solve the problem in a number of ways, depending on the conversion factors that we know, for example, km → m → cm → in. → ft → yd → mi → blocks. (This route is traced in brown in Figure 1-10.) An alternative route (blue in Figure 1-10) is km → m → in. → ft → mi → blocks.

$$\text{no. blocks} = 5.0 \; \text{km} \times \frac{1000 \; \text{m}}{1 \; \text{km}} \times \frac{39.37 \; \text{in.}}{1 \; \text{m}} \times \frac{1 \; \text{ft}}{12 \; \text{in.}}$$

$$(\text{km} \longrightarrow \text{m} \longrightarrow \text{in.} \longrightarrow \text{ft}$$

$$\times \frac{1 \; \text{mi}}{5280 \; \text{ft}} \times \frac{8.0 \; \text{blocks}}{1 \; \text{mi}}$$

$$\longrightarrow \text{mi} \longrightarrow \text{blocks})$$

$$= 25 \; \text{blocks}$$

SIMILAR EXAMPLES: Exercises 1, 2, 3, 32 through 35, 63.

In the next example you will discover three important new ideas.

- At times, units appear to a power (e.g., squared or cubed), and you may have to raise conversion factors to higher powers.
- Often you will be able to see more clearly how to solve a problem by drawing a sketch of the situation, as in Figure 1-11.
- Generally, there are alternate ways of solving a problem, and you can use any logical method.

Example 1-3

Writing conversion factors when units are squared (or cubed). How many square feet (ft^2) correspond to an area of 1.00 square meter (m^2)?

Solution. An area of 1.00 m^2 is represented in Figure 1-11. Think of it as a square with sides 1 m long. Figure 1-11 also shows the length, 1 ft, and an area of 1.00 ft^2. Can you see that there are somewhat more than 9 ft^2 in 1 m^2? This is what drawing a sketch can help you to see more clearly.

Expression (1.6) is written as follows:

$$\text{no. } ft^2 = 1.00 \; m^2 \times \underbrace{\left(\frac{39.37 \; \text{in.}}{1 \; m} \right)\left(\frac{39.37 \; \text{in.}}{1 \; m} \right)}_{\substack{\text{to convert} \\ m^2 \text{ to in.}^2}} \times \underbrace{\left(\frac{1 \; ft}{12 \; \text{in.}} \right)\left(\frac{1 \; ft}{12 \; \text{in.}} \right)}_{\substack{\text{to convert} \\ \text{in.}^2 \text{ to } ft^2}}$$

This is the same as writing

$$\text{no. } ft^2 = 1.00 \; m^2 \times \frac{(39.37)^2 \text{in.}^2}{1 \; m^2} \times \frac{1 \; ft^2}{(12)^2 \; \text{in.}^2} = 10.8 \; ft^2$$

Another way to look at the problem is to convert the length 1.00 m to feet

$$= 1.00 \; m \times \frac{39.37 \; \text{in.}}{1.00 \; m} \times \frac{1 \; ft}{12 \; \text{in.}} =$$

and square the result

$$\text{no. } ft^2 = 3.28 \; ft \times 3.28 \; ft = 10.8 \; ft^2$$

SIMILAR EXAMPLES: Exercises 4, 37, 38, 63.

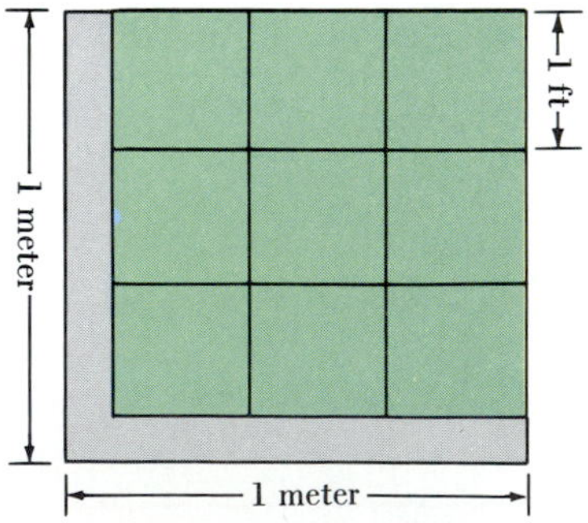

FIGURE 1-11
Comparison of one square foot and one square meter—Example 1-3 visualized.

1 m is slightly longer than 3 ft, and 1 m^2 is somewhat larger than 9 ft^2. The answer obtained in Example 1-3 is consistent with this observation.

Here are two more new ideas that you should find useful. They are illustrated in Example 1-4.

- In some calculations the answer must be expressed as a *ratio* of units. This may require you to do one series of conversions in the numerator, and another series in the denominator.
- At times the number of conversion factors in a setup may be rather large. In these cases you may follow the logic of the solution more easily by breaking down the problem into several steps.

Example 1-4

Making conversions in both the numerator and denominator of a setup. A baseball pitcher's fastest pitch is clocked by radar at 98 mi/h. What is this speed expressed in meters per second (m/s)?

Solution. What we need to do here is to convert from miles to meters in the *numerator* and hours to seconds in the *denominator*. We must be careful to set up the conversion factors so that each factor has a numerator and denominator that are equivalent, and that each factor produces the desired cancellation of units.

$$\text{no. } \frac{m}{s} = \frac{98 \text{ mi}}{1 \text{ h}} \times \frac{5280 \text{ ft}}{1 \text{ mi}} \times \frac{12 \text{ in.}}{1 \text{ ft}} \times \frac{1 \text{ m}}{39.37 \text{ in.}} \times \frac{1 \text{ h}}{60 \text{ min}} \times \frac{1 \text{ min}}{60 \text{ s}}$$

$$= 44 \frac{m}{s}$$

Here is an alternative way of looking at the problem: Speed is a ratio of distance traveled to the time required. Basically what we must do here is to (1) convert 98 miles to a distance in meters, (2) convert one hour to a time in seconds, and (3) express the speed as the ratio of the two new quantities.

Step 1

$$\text{no. m} = 98 \text{ mi} \times \frac{5280 \text{ ft}}{1 \text{ mi}} \times \frac{12 \text{ in}}{1 \text{ ft}} \times \frac{1 \text{ m}}{39.37 \text{ in.}} = 1.6 \times 10^5 \text{ m}$$

Step 2

$$\text{no. s} = 1 \text{ h} \times \frac{60 \text{ min}}{1 \text{ h}} \times \frac{60 \text{ s}}{1 \text{ min}} = 3600 \text{ s}$$

Step 3

$$\text{no. } \frac{m}{s} = \frac{1.6 \times 10^5 \text{ m}}{3600 \text{ s}} = 44 \frac{m}{s}$$

SIMILAR EXAMPLES: Exercises 36, 63, 66.

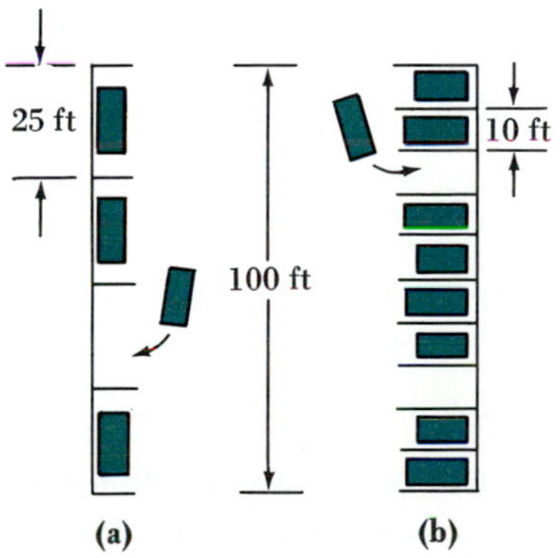

Equivalence and Equality. Figure 1-12 may give you additional insight into conversion factors. What is the relationship between an automobile and the curb space required to park it? In Figure 1-12a we need the conversion factor: 1 automobile ⇌ 25 ft, and we use an **equivalence sign** (⇌) rather than an equality sign (=). An automobile is *not identical* to 25 ft of curb space. *For the purposes of this calculation* we consider them to be **equivalent.** For every 25 ft of curb space we can park one automobile (regardless of its actual length). On the other hand, 25 ft and 7.6 m of curb space have an *identical* meaning, that is, 25 ft = 7.6 m. We

FIGURE 1-12
The concept of equivalence.

(a) For a parallel parking arrangement we can say that each automobile *is equivalent to* 25 ft. That is, 1 automobile ⇌ 25 ft.

no. automobiles

$$= 100 \text{ ft} \times \frac{1 \text{ automobile}}{25 \text{ ft}}$$

$$= 4 \text{ automobiles}$$

(b) In this parking arrangement (perpendicular) each automobile is equivalent to 10 ft of curb space. How many automobiles can be parked along the 100-ft section of curb?

will use the sign = when an equality exists and ⇌ for an equivalence, but conversion factors are set up and used in the same way in either case. For the perpendicular parking arrangement in Figure 1-12b there is a different equivalence between automobiles and curb space.

Percent as a Conversion Factor. Consider the literal meaning of *percent*. The word stem "cent" is derived from the Latin word *centum*, meaning 100. **Percent** refers to the number of parts of one constituent to 100 parts of the whole quantity. Thus, the statement that a seawater sample contains 3.5% sodium chloride, by mass, means that in a 100.0-g sample of the seawater there is present 3.5 g of sodium chloride. This statement can also be described through the equivalence

$$3.5 \text{ g sodium chloride} \mathrel{⇌} 100.0 \text{ g seawater} \tag{1.7}$$

and through the conversion factors

$$\frac{3.5 \text{ g sodium chloride}}{100.0 \text{ g seawater}} \quad \text{and} \quad \frac{100.0 \text{ g seawater}}{3.5 \text{ g sodium chloride}} \tag{1.8}$$

which are used in Examples 1-5 and 1-6.

Example 1-5

Using percent to calculate the quantity of a component in a mixture. 325 g of seawater containing 3.5% sodium chloride, by mass, is evaporated to dryness. What mass of sodium chloride will be present in the solid residue?

Solution. From expression (1.8) we need the conversion factor with "g sodium chloride" in the numerator and "g seawater" in the denominator.

$$\text{no. g sodium chloride} = 325 \text{ g seawater} \times \frac{3.5 \text{ g sodium chloride}}{100.0 \text{ g seawater}}$$

$$= 11 \text{ g sodium chloride}$$

SIMILAR EXAMPLES: Exercises 8, 52, 70.

Example 1-6

Calculating a mixture's quantity from the percent composition of its components. 87 g of sodium chloride is to be produced by evaporating to dryness a quantity of seawater containing 3.5% sodium chloride, by mass. How many *liters* of seawater must be taken for this purpose? (Assume that 1 cm³ of seawater has a mass of 1.03 g.)

Solution. Here we need the conversion factor from expression (1.8) that has "g seawater" in the numerator and "g sodium chloride" in the denominator. In addition, we must make the conversions: g seawater → cm³ seawater → L seawater.

$$\text{no. L seawater} = 87 \text{ g sodium chloride} \times \frac{100.0 \text{ g seawater}}{3.5 \text{ g sodium chloride}}$$

$$\times \frac{1 \text{ cm}^3 \text{ seawater}}{1.03 \text{ g seawater}} \times \frac{1 \text{ L seawater}}{1000 \text{ cm}^3 \text{ seawater}} = 2.4 \text{ L seawater}$$

SIMILAR EXAMPLES: Exercises 9, 53, 98.

> ## Are You Wondering:
>
> *In doing a problem with percentages, when to multiply and when to divide by percentage?*
>
> A common way of dealing with a percentage is to convert it to decimal form (3.5% becomes 0.035) and then to multiply or divide by this decimal, as appropriate. Students sometimes cannot decide which to do. A better method is the one used in Examples 1-5 and 1-6: Express percentage as a conversion factor that automatically leads to the correct cancellation of units. Also, you should note that
>
> - When you are seeking information about a component of a mixture, the quantity of the component must always be *less* than the quantity of the whole mixture.
> - When you are given information about one of the components and seek information about the entire mixture, the quantity of mixture must always be *greater* than the quantity of the component.

1-10 Density

Density is the ratio of the mass of an object or material to its volume.

The defining equation for density.

$$\text{density } (d) = \frac{\text{mass } (m)}{\text{volume } (V)} \tag{1.9}$$

A property whose magnitude depends on the quantity of material is an **extensive** property. Both mass and volume are extensive properties. Any property that is independent of the quantity of material is an **intensive** property. Density, the ratio of mass to volume, is an intensive property. Intensive properties are generally preferred for scientific work because they are independent of the quantity of matter being studied.

The mass of 1 liter of water at 4 °C is 1 kg. The density of water under these conditions is 1000 g/1000 cm^3 = 1.000 g/cm^3. Because volume varies with temperature while mass remains constant, *density is a function of temperature*. At 20 °C the density of water is 0.99823 g/cm^3. Table 1-3 lists the densities of a few liquids and solids at 20 °C.

Densities have been very precisely determined for many water solutions and these data are readily available in published tables. At times we can use the density of a solution as a means of determining its composition. At 20 °C, solutions of ethyl alcohol and water have densities ranging from 0.78934 to 0.99823 g/cm^3. An ethyl alcohol–water solution with a density of 0.898 g/cm^3, for example, consists of 57% ethyl alcohol and 43% water, by mass.

There is an old riddle that goes: "What weighs more, a ton of bricks or a ton of feathers?" The correct answer is that they weigh the same. If you gave this answer you have demonstrated a clear understanding of the meaning of mass—a measure of a quantity of matter. If you answered that the bricks weigh more than the feathers you have confused mass and density. Matter in a brick is more concentrated than in feathers, that is, confined to a smaller volume. Bricks are more dense than feathers.

TABLE 1-3
Some Densities at 20 °C

Liquids	
ethyl alcohol	0.789 g/cm^3
water	0.998
glycerol	1.261
carbon tetrachloride	1.595

Solids	
aluminum	2.70 g/cm^3
iron	7.87
lead	11.35
gold	19.32

FIGURE 1-13
Measuring the volume of a regularly shaped object—Example 1-7 visualized.

The volume of a parallelepiped is the product of its length, width, and height.

$V = l \times w \times h$

Example 1-7

Calculating the density of an object from its mass and volume. A block of wood with the dimensions 105 cm × 5.1 cm × 6.2 cm has a mass of 2.72 kg. What is the density of the wood, expressed in g/cm³?

Solution. As you can see from Figure 1-13 we can use a geometric formula to calculate the volume of a rectangular block.

$V = 105 \text{ cm} \times 5.1 \text{ cm} \times 6.2 \text{ cm} = 3300 \text{ cm}^3$

The mass of the block must be expressed in grams.

$m = 2.72 \text{ kg} \times \dfrac{1000 \text{ g}}{1 \text{ kg}} = 2720 \text{ g}$

The density of the wood is

$d = \dfrac{m}{V} = \dfrac{2720 \text{ g}}{3300 \text{ cm}^3} = 0.82 \text{ g/cm}^3$

SIMILAR EXAMPLES: Exercises 6, 16, 43, 44, 50.

Example 1-8

Measuring volume by the displacement of water. Several irregularly shaped pieces of zinc, weighing 30.0 g, are dropped into a graduated cylinder containing 18.0 cm³ of water. The water level rises to 22.2 cm³. What is the density of the zinc?

Solution. When added to the water, the pieces of zinc fall to the bottom of the cylinder and cause the liquid level to rise by an amount equal to their total volume, as illustrated in Figure 1-14.

volume of zinc = 22.2 cm³ − 18.0 cm³ = 4.2 cm³

$d = \dfrac{m}{V} = \dfrac{30.0 \text{ g}}{4.2 \text{ cm}^3} = 7.1 \text{ g/cm}^3$

SIMILAR EXAMPLES: Exercises 69, 76.

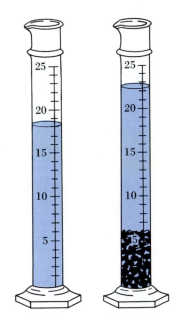

FIGURE 1-14
Measuring the volume of irregularly shaped objects—Example 1-8 visualized.

The irregularly shaped objects, which are insoluble in and more dense than water, sink to the bottom of the cylinder and displace a volume of water equal to their own. (If you are familiar with cooking techniques, you may recognize this as a common method for measuring the volumes of solid fats, such as butter.)

Example 1-9

Calculating the volume of a liquid from its mass and density. What is the volume, in liters, occupied by 50.0 kg of ethanol at 20 °C? The density of ethanol at 20 °C is 0.789 g/cm³.

Solution. The information given is *50.0 kg of ethanol* and what we are seeking is *number of liters of ethanol*. We need to convert kg → g → cm³ → L. We can think of density as a conversion factor between mass and volume: 1.00 cm³ of ethanol = 0.789 g of ethanol.

$\text{no. L ethanol} = 50.0 \text{ kg ethanol} \times \dfrac{1000 \text{ g ethanol}}{1 \text{ kg ethanol}} \times \dfrac{1.00 \text{ cm}^3 \text{ ethanol}}{0.789 \text{ g ethanol}}$

$\times \dfrac{1 \text{ L ethanol}}{1000 \text{ cm}^3 \text{ ethanol}}$

$= 63.4 \text{ L ethanol}$

Alternatively, we can solve the density equation for volume, $V = m/d$, and substitute the appropriate information.

$$V = \frac{50{,}000 \text{ g}}{0.789 \text{ g/cm}^3} = 63{,}400 \text{ cm}^3 = 63.4 \text{ L}$$

SIMILAR EXAMPLES: Exercises 7, 45, 47, 53.

1-11 Temperature

We consider more fundamental and intellectually satisfying descriptions of temperature in Chapters 6 and 7.

Temperature is rather difficult to define, even though we have an intuitive idea of what it is. If we say that temperature is the degree of "hotness" of an object, we are not being very precise, but we do convey a certain meaning. If we bring two objects of different temperature into contact, the warmer object becomes colder and the colder one becomes warmer. Eventually, both objects come to the same degree of "hotness"—the same temperature.

Temperature can be measured by its effect on some other measurable property. One common temperature-measuring device, a thermometer, is based on the length of a liquid column (e.g., mercury) in a thin capillary bore in a glass tube. As the temperature of the thermometer increases, the mercury expands and the length of the liquid column increases.

To set up a scale of temperatures we need to establish certain fixed points of temperature and a degree of temperature change. Two commonly used fixed points are the temperature at which ice melts (the ice point) and the temperature at which water boils (the steam point), both at standard atmospheric pressure.*

On the **Fahrenheit** temperature scale the ice point is 32 °F, the steam point is 212 °F, and the interval between is divided into 180 equal parts, called degrees Fahrenheit. On the **Celsius** (centigrade) scale the ice point is 0 °C, the steam point is 100 °C, and the interval is divided into 100 equal parts, called degrees Celsius. Figure 1-15 compares the Fahrenheit and Celsius temperature scales, including equations that can be used to convert between the two.

The SI convention calls for the use of an *absolute* temperature scale, one in which a value of zero is assigned to the lowest conceivable temperature. This absolute zero of temperature comes at −273.15 °C (see Section 6-3). The SI temperature scale is the **Kelvin** scale, which has its zero at the absolute zero of temperature. The degree interval on the Kelvin scale, called a **kelvin,** is identical to that of the Celsius scale. The relationship of the Kelvin to the Celsius scale is also outlined in Figure 1-15.

FIGURE 1-15

A comparison of temperature scales.

(a) Ice point. **(b)** Steam point. **(c)** Comparison of Fahrenheit (°F), Celsius (°C), and Kelvin (K) temperature scales. Shown below are equations for converting temperatures between scales.

$T \text{ (K)} = t \text{ (°C)} + 273.15$

$t \text{ (°C)} = \frac{5}{9}[t \text{ (°F)} - 32]$

$t \text{ (°F)} = \frac{9}{5}t \text{ (°C)} + 32$

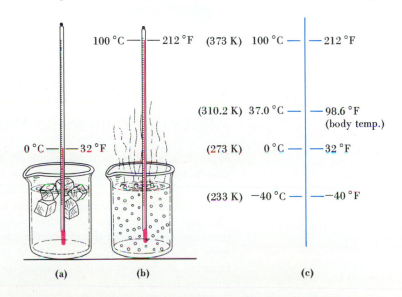

*We define standard atmospheric pressure in Section 6-2 and describe the effect of pressure on melting and boiling points in Chapter 12.

Temperature Conversions. Most laboratory measurements of temperature are made on the Celsius scale. Often these temperatures must be converted to the Kelvin scale (in describing the behavior of gases, for example). At other times, as in many engineering applications, temperatures must be converted between Celsius and Fahrenheit scales.

Celsius/Kelvin conversions are easy since the two scales differ only in their location of the zero degree mark, but we will defer a consideration of these to Chapter 6.

Celsius/Fahrenheit conversions are more difficult because both the zero degree marks and the sizes of the degree intervals differ in these two scales. Still, the equations shown in Figure 1-15 can be easily derived from the following considerations.

- 180 degrees of Fahrenheit temperature = 100 degrees of Celsius temperature

- number of degrees of F = number of degrees of C $\times \dfrac{180 \text{ degrees of F}}{100 \text{ degrees of C}}$
  above ice point   above ice point

- Fahrenheit temperature = number of degrees of F above ice point + 32

$$t(°F) = t(°C) \times \tfrac{180}{100} + 32$$

Equation for converting between Celsius and Fahrenheit temperatures.

$$t(°F) = \tfrac{9}{5} t(°C) + 32 \qquad\qquad (1.10)$$

Example 1-10

Converting between Fahrenheit and Celsius temperatures. A recipe calls for roasting a cut of meat in an oven at 350 °F. What is this temperature on the Celsius scale?

Solution. There are several ways to deal with a question of this type. One is to derive an equation to convert Fahrenheit to Celsius temperature (similar to what we did in deriving equation 1.10). Another is to use the equation for $t(°C)$ as a function of $t(°F)$ given in Figure 1-16. Still another is to derive this equation from equation (1.10), as follows.

$$t(°F) = \tfrac{9}{5} t(°C) + 32$$

$$t(°F) - 32 = \tfrac{9}{5} t(°C) + 32 - 32$$

$$t(°C) = \tfrac{5}{9}[t(°F) - 32]$$

$$t(°C) = \tfrac{5}{9}(350 - 32) = 177 \text{ °C}$$

SIMILAR EXAMPLES: Exercises 5, 40, 41, 42.

1-12   Algebraic Solutions to Problems

Most of the problems you will encounter in this text can be solved by the conversion-factor method introduced in Section 1-9. A few, like the temperature conversions of the preceding section, can be done either with conversion factors or by solving algebraic equations. Still others may *require* that you solve an algebraic equation. In these cases the simplest situation is one in which you are given the necessary equation and you must solve it by substituting numerical values of one or more variables, as in Example 1-11. At other times you may have to *rearrange* an equation and solve it for the unknown, as in Example 1-12. The most demanding situation is one in which you must *derive* the necessary equation for the particular problem and then solve it. Some applications of this type arise later in the text. You can review some basic algebra by studying Appendix A.

Example 1-11

Substituting data into an algebraic equation. A handbook gives the following equation for determining the density (d) of dry air (in g/cm^3) at standard atmospheric pressure as a function of Celsius temperature (t): $d = 0.001293/(1 + 0.00367t)$. What is the density of dry air at 25 °C?

Solution. Substitute $t = 25$ and solve for d.

$$d = \frac{0.001293}{1 + (25 \times 0.00367)} = \frac{0.001293}{1 + 0.092} = \frac{0.001293}{1.092} = 0.001184 \text{ g/cm}^3$$

SIMILAR EXAMPLES: Exercises 15, 55.

Example 1-12

Solving an algebraic equation for an unknown. Use the equation of Example 1-11 to determine the temperature at which the density of dry air at standard atmospheric pressure is 0.001200 g/cm^3.

Solution. Rearrange the density equation to solve for t.

$$d = \frac{0.001293}{(1 + 0.00367t)} \qquad 1 + 0.00367t = \frac{0.001293}{d}$$

$$0.00367t = (0.001293/d) - 1 \qquad t = \frac{(0.001293/d) - 1}{0.00367}$$

Substitute $d = 0.001200$.

$$t = \frac{(0.001293/0.001200) - 1}{0.00367} = \frac{1.078 - 1}{0.00367} = \frac{0.078}{0.00367} = 21 \text{ °C}$$

SIMILAR EXAMPLES: Exercises 56, 57.

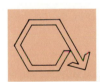

FOCUS ON
Commercially Important Chemicals

The production of key chemicals usually requires a large, complex, and costly industrial facility. One of the first steps before constructing such a facility is to build a pilot plant that can be used to test the process. [© Fred Ward/Black Star]

The known chemical substances number about five million. Of this large number, several tens of thousands are commercially important, but only a small number are produced in quantities measured in the millions of tons. These are called *commodity* chemicals. They are used to synthesize other chemicals and to produce materials important to modern society. In turn, these commodity chemicals are

TABLE 1-4
Some of the Top 50 Chemicals Produced in the United States (1986)

Rank	Billions of pounds	Chemical	Formula	Principal uses and end products
1	73.64	sulfuric acid	H_2SO_4	fertilizers (70%); metallurgy (10%); petroleum refining (5%); manufacture of chemicals (5%)
2	48.62	nitrogen	N_2	inert atmospheres for manufacture of chemicals (25%); electronics manufacturing (20%); petroleum recovery (20%); metals processing (5%); freezing foods (5%)
3	33.03	oxygen	O_2	steel-making and other metallurgical processes (55%); chemical manufacture (20%)
4	32.81	ethylene	C_2H_4	manufacture of plastics (70%), antifreeze (10%), fibers (5%), solvents (5%)
5	30.34	calcium oxide (lime)	CaO	metallurgy (40%); pollution control and wastewater treatment (15%); manufacture of chemicals (10%); municipal water treatment (10%)
6	28.01	ammonia	NH_3	fertilizers (80%); manufacture of plastics and fibers (10%) and explosives (5%)
7	22.01	sodium hydroxide	$NaOH$	manufacture of chemicals (50%), pulp and paper (20%), soaps and detergents (5%); oil refining (5%)
8	20.98	chlorine	Cl_2	manufacture of organic chemicals and plastics (65%), pulp and paper (15%)
9	18.41	phosphoric acid	H_3PO_4	fertilizers (95%)
10	17.34	propylene	C_3H_6	plastics (60%); fibers (15%); solvents (10%)
14	12.06	urea	$CO(NH_2)_2$	fertilizers (80%); animal feed (10%); manufacture of adhesives and plastics (5%)
18	8.50	carbon dioxide	CO_2	refrigeration (50%); carbonated beverages (20%); manufacture of chemicals (10%); oil recovery (5%)
19	8.42	vinyl chloride	CH_2CHCl	manufacture of plastics (~100%)
22	7.33	methyl alcohol	CH_3OH	manufacture of adhesives, fibers and plastics (50%), solvents (10%)
33	2.92	phenol	C_6H_5OH	manufacture of adhesives (60%), plastics (20%), fibers (10%)
34	2.59	carbon black	C	tires (65%); other rubber goods (25%); pigment for plastics and ink (10%)
43	1.94	acetone	$(CH_3)_2CO$	solvents (60%); manufacture of plastics (30%)
44	1.83	titanium dioxide	TiO_2	paints (50%); paper filler and coatings (25%); plastics and rubber filler (15%); ceramics (5%)

Sources: *Chemical and Engineering News:* July 13, 1987; June 24, 1987; June 8, 1987; Mar. 2, 1987; Jan. 5, 1987; Sept. 8, 1986; Apr. 28, 1986; Feb. 17, 1986; Nov. 4, 1985; Sept. 16, 1985; Mar. 4, 1985.

derived from a small number of basic raw materials, such as air, water, limestone, salt (sodium chloride), sulfur, phosphate rock, coal, petroleum, and natural gas. We refer to basic raw materials and commodity chemicals throughout this text. For example, we consider how commodity chemicals are produced from raw materials: nitrogen and oxygen from air (Chapter 6); sulfuric acid from sulfur, air, and water (Chapter 15); ammonia from air and natural gas (Chapter 16); etc. Moreover, we will discover that many of these chemicals nicely illustrate fundamental chemical principles.

Table 1-4 lists some commodity chemicals and their important uses. The entries in this table are referred to by name and by symbolic designations known as chemical formulas. We begin the task of relating names and formulas—referred to as chemical nomenclature—in the

next chapter and gradually expand our ability to do so throughout the text. All of the top ten chemicals, except ethylene, C_2H_4, and propylene, C_3H_6, are derived from inanimate mineral sources and are called *inorganic* chemicals. Ethylene, propylene, and most of the chemicals beyond the top ten are derived from a source (coal, petroleum, natural gas) that was living matter at one time. These carbon-and-hydrogen-based compounds are called *organic* chemicals. We comment further on these two broad categories of compounds in later chapters. In later chapters we also discover other important ways of categorizing compounds to help us understand their behavior. Some of the chemicals in Table 1-4 will be featured in those discussions.

The uses of commodity chemicals are generally dictated by what they contain, that is, their chemical composition.

Thus, ammonia and urea are widely used as fertilizers because of their high content of nitrogen—an essential plant nutrient. Another category of commercially important chemicals are those valued for what they can do, that is, for particular properties rather than for their chemical composition. These are called *specialty* chemicals. Although they are generally produced in smaller quantities than commodity chemicals, specialty chemicals play a crucial role in modern society, for example, as

- pesticides
- petroleum additives
- plastics additives
- food additives
- automotive chemicals
- photographic chemicals
- catalysts

Despite the fact that specialty chemicals may be mentioned less frequently than commodity chemicals in this book, you should not lose sight of their importance.

The chemical industry produces chemicals and products made from chemicals. Certain other sectors of industry deal with materials that are not usually thought of as chemicals but which are chemicals nevertheless, such as gasoline and liquefied petroleum gases (LPG) in the petroleum industry, and iron and steel, aluminum, magnesium, copper, and other metals in the metallurgical industry. In this text we deal with chemicals in a very broad sense.

Summary

Chemistry is a study of matter, and one of the first needs in this study is to classify matter into useful categories. One scheme introduced in this chapter considers matter as being either a substance—element or compound—or a mixture—homogeneous or heterogeneous. Chemists are interested in the distinctive properties of matter and how these may be changed by physical or chemical means. Compounds can be broken down into their constituent elements only by chemical changes, whereas mixtures can be separated into their components by physical changes. Another classification of matter is into the different possible states of matter—solids, liquids, and gases.

Like other branches of science, chemistry makes use of the scientific method. The scientific method consists of a series of activities that culminates in theories to explain and predict natural phenomena.

In chemistry, the need for precise measurement is essential, and a uniform system for expressing the results of measurements is also important. SI units provide this uniformity, but because they have not yet been universally adopted, other units of measurement are also used in this text. Still another need is to have a means of indicating the precision with which measurements are made. This can be done through the number of significant figures used to express measured quantities. Moreover, calculations based on measured quantities must be performed in such a way that the calculated result is no more or less precise than warranted by the measured quantities. This requirement is best met through a proper regard for the number of significant figures used to express the result of a calculation.

The principal problem-solving method used in this text is introduced in this chapter. It requires that relationships between quantities be expressed through conversion factors. The desired result of a calculation is obtained by multiplying a given quantity by one or more conversion factors. In this method cancellation of units is used as a guide to ensure that conversion factors are properly formulated. Because of these requirements, this problem-solving method is variously known as the conversion-factor method, unit analysis, or dimensional analysis. Although most general chemistry problems can be solved by the conversion-factor method, some problems require an algebraic solution. Algebraic solutions to problems are also briefly considered in the chapter.

Two important physical properties that are discussed and illustrated in this chapter are density and temperature. In problem-solving situations density can be thought of as a conversion factor between mass and volume. The important temperature calculations in this chapter involve conversions between Fahrenheit and Celsius temperature.

Summarizing Example

Methanol (methyl alcohol) is a potential future automotive fuel, either pure or mixed with gasoline. Pure methanol has an octane rating of 106, and a 85% methanol–15% gasoline mixture is about ten percent more efficient a motor fuel than straight gasoline.

1. The simple device pictured in Figure 1-16, a **pycnometer,** is used for precise density determinations. From the data presented **(a)** what is the volume capacity of the pycnometer? **(b)** what is the density of methanol at 20.0 °C?

Solution

(a) The mass of water required to fill the pycnometer at 20.0 °C is

$$35.552 \text{ g} - 25.601 \text{ g} = 9.951 \text{ g}.$$

The volume of water, and hence the volume capacity of the pycnometer, is

$$\text{no. cm}^3 = 9.951 \text{ g} \times \frac{1.0000 \text{ cm}^3}{0.99823 \text{ g}} = 9.969 \text{ cm}^3$$

(b) The mass of methanol required to fill the pycnometer at 20.0 °C is

$$33.490 \text{ g} - 25.601 \text{ g} = 7.889 \text{ g}$$

The density of methanol at 20.0 °C is

$$d = \frac{m}{V} = \frac{7.889 \text{ g}}{9.969 \text{ cm}^3} = 0.7914 \text{ g/cm}^3$$

(This example is similar to Examples 1-7 and 1-8.)

2. An automobile, modified to use methanol as a fuel, has a 15.6 gal fuel tank. What mass of methanol, expressed in kg, does this tank hold?

Solution. First, convert volume from gallons to liters and to cm³.

$$\text{no. cm}^3 = 15.6 \text{ gal} \times \frac{3.785 \text{ L}}{1 \text{ gal}} \times \frac{1000 \text{ cm}^3}{1 \text{ L}} = 5.90 \times 10^4 \text{ cm}^3$$

Then use density as a conversion factor between volume and mass.

$$\text{no. kg} = 5.90 \times 10^4 \text{ cm}^3 \times \frac{0.7914 \text{ g}}{1.000 \text{ cm}^3} \times \frac{1 \text{ kg}}{1000 \text{ g}} = 46.7 \text{ kg}$$

(This example is similar to Example 1-9.)

3. What mass of a 85.0% methanol–15.0% gasoline mixture can be produced per kilogram of methanol?

Solution. "Per kilogram" means for every 1.00 kg. We must find the mass of mixture containing 1.00 kg methanol. The required conversion from kg methanol $\longrightarrow$ kg mixture dictates the way in which percentage is used as a conversion factor.

$$\text{no. kg mixture} = 1.00 \text{ kg methanol} \times \frac{100.0 \text{ kg mixture}}{85.0 \text{ kg methanol}} = 1.18 \text{ kg mixture}$$

(This example is similar to Example 1-6.)

This municipal bus has been modified to burn methanol instead of diesel fuel. The exhaust from this bus is virtually smoke-free and contains much less carbon monoxide, nitrogen oxides, and benzene than does diesel exhaust. [Courtesy of Riverside Transit Agency, Riverside, CA]

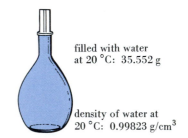

empty: 25.601 g

filled with water at 20 °C: 35.552 g

density of water at 20 °C: 0.99823 g/cm³

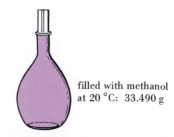

filled with methanol at 20 °C: 33.490 g

FIGURE 1-16
Determination of density with a pycnometer.

Key Terms

atom (1-2)
Celsius temperature (1-11)
chemical change (1-1)
chemical property (1-1)
chemical reaction (1-1)
compound (1-2)
conversion factor (1-9)
density (1-10)
dimensional analysis (1-9)
distillation (1-3)
element (1-2)
English system (1-5)
extensive property (1-10)
Fahrenheit temperature (1-11)

filtration (1-3)
gas (1-2)
heterogeneous mixture (1-2)
homogeneous mixture (solution) (1-2)
hypothesis (1-4)
intensive property (1-10)
Kelvin temperature (1-11)
liquid (1-2)
mass (1-5)
matter (1-1)
metric system (1-5)
mixture (1-2)
molecule (1-2)

natural law (1-4)
physical change (1-1)
physical property (1-1)
random error (1-7)
scientific method (1-4)
SI units (1-6)
significant figures (1-8)
solid (1-2)
substance (1-2)
systematic error (1-7)
theory (1-4)
unit analysis (1-9)
weight (1-5)

Highlighted Expressions

Significant figure rule in multiplication/division (1.2)
Significant figure rule in addition/subtraction (1.3)
Significant figures for exact numbers (1.4)
The basic setup for problem solving (1.6)

The defining equation for density (1.9)
Equation for converting between Celsius and Fahrenheit temperatures (1.10)

Review Problems

1. Perform the following conversions within the metric system.

(a) 2.14 kg = _____ g
(b) 6172 mm = _____ m
(c) 1316 mg = _____ kg
(d) 812 mL = _____ L
(e) 22.3 cm = _____ mm
(f) 0.0256 L = _____ mL
(g) 8.008 g = _____ mg
(h) 0.035 km = _____ m

2. Perform the following conversions within the English system.

(a) 32.08 ft = _____ in.
(b) 125.5 in. = _____ yd
(c) 2721 ft = _____ yd
(d) 2.65 mi = _____ ft
(e) 122 yd = _____ in.
(f) 2.5 h = _____ min

3. Perform the following conversions between the English and metric systems.

(a) 22.5 in. = _____ cm
(b) 126 ft = _____ m
(c) 2215 g = _____ lb
(d) 313 lb = _____ kg
(e) 26.5 mi = _____ m
(f) 825 yd = _____ km

4. Determine the number of

(a) square meters (m^2) in 1 square kilometer (km^2);
(b) square feet (ft^2) in 1 square mile (mi^2);
(c) square meters (m^2) in 1 square mile (mi^2);
(d) cubic centimeters (cm^3) in 1 cubic foot (ft^3);
(e) cubic nanometers (nm^3) in 1 cubic millimeter (mm^3).

5. Perform the following conversions between temperature scales.

(a) 42 °C = _____ °F
(b) 91 °F = _____ °C
(c) −31 °F = _____ °C
(d) −88 °C = _____ °F

6. A 2.50-L sample of pure glycerol has a mass of 3153 g. What is the density of glycerol?

7. Ethylene glycol, an antifreeze, has a density of 1.11 g/cm^3 at 20 °C.

(a) What is the mass, in grams, of 417 mL of this liquid?
(b) What is the mass, in kg, of 15.0 L of this liquid?
(c) What is the volume, in liters, occupied by 50.0 lb of this liquid?

8. A particular fertilizer is listed as containing 4.8% phosphorus, by mass. What mass of phosphorus, in grams, is contained in a 25.0-lb bag of this fertilizer?

9. Laboratory-grade ethyl alcohol is 95.5% ethyl alcohol, by mass. (The other component is water.) What mass of this material must be taken for an application requiring 125 g of ethyl alcohol?

10. Express the following numbers in exponential notation (see Appendix A). (a) 43,151; (b) 0.0500; (c) 0.200; (d) 63; (e) 0.0000087

11. Express the following numbers in common decimal form (see Appendix A.) (a) 3.12×10^4; (b) 6.17×10^{-1}; (c) 3.256×10^{-3}; (d) 612×10^{-2}; (e) 9.68×10^{-5}; (f) 371×10^{-3}; (g) 4.5×10^0; (h) 0.058×10^{-4}

12. How many significant figures are shown in each of the following numbers? If indeterminate, give the range of significant figures possible. (a) 625; (b) 320; (c) 0.033; (d) 820.03; (e) 0.04050; (f) 0.007; (g) 7820.0; (h) 47,000

13. Rewrite each of the following numbers to consist of *four* significant figures. (a) 7218.7; (b) 6.6×10^3; (c) 319.15; (d) 80×10^{-5}; (e) 918.74; (f) 186,000; (g) 35.6050; (h) 312,549

14. Perform the following calculations, expressing each number and the answer in exponential form and with the correct number of significant figures.

(a) $312 \times 601 =$

(b) $213 \times 357 \times 19.21 =$

(c) $0.092 \times 0.0300 =$

(d) $0.0150 \times 41.25 \times 0.0078 =$

(e) $\dfrac{4400 \times 13.7}{0.0090} =$

(f) $\dfrac{350 \times 0.10 \times 678,000}{0.0017 \times 6.3} =$

15. The following equation can be used to relate the density of liquid water to Celsius temperature in the range from 0 °C to about 20 °C.

$$d \ (\text{g/cm}^3) = \frac{0.99984 + (1.6945 \times 10^{-2}t) - (7.987 \times 10^{-6}t^2)}{1 + (1.6880 \times 10^{-2}t)}$$

To *four* significant figures determine the density of water at (a) 5 °C and (b) 10 °C.

16. The angle iron pictured is made of steel having a density of 7.78 g/cm³. What is the mass of this object? [*Hint:* What is its volume?]

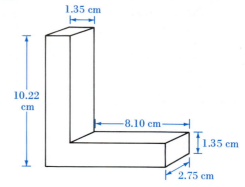

Problem 16

Exercises

Properties and classification of matter

17. State whether each property is physical or chemical.

(a) An iron nail is attracted to a magnet.

(b) Sugar caramelizes on being heated above its melting point.

(c) A bronze statue develops a green coating (patina) over time.

(d) A block of wood floats on water.

18. Indicate whether each sample of matter listed is a substance or a mixture, and if a mixture, whether homogeneous or heterogeneous.

(a) a cube of sugar   (b) chicken soup

(c) premium gasoline   (d) salad dressing

(e) tap water   (f) garlic salt

(g) hot chocolate   (h) ice

19. What type of change—physical or chemical—is necessary to bring about the following separations? [*Hint:* Refer to a listing of the elements.]

(a) nitrogen and oxygen gases from air

(b) chlorine gas from sodium chloride (salt)

(c) sulfur from sulfuric acid (battery acid)

(d) pure water from seawater

20. Suggest physical changes by which the following mixtures can be separated.

(a) iron filings and wood chips

(b) sugar and sand

(c) water and gasoline

(d) gold flakes and water

21. Indicate which are extensive and which are intensive quantities.

(a) the mass of air in a balloon

(b) the temperature of an ice cube

(c) the length of time to bring a beaker of water to the boiling point

(d) the color of light given off by a neon lamp.

Scientific method

22. Is it possible to predict how many experiments are required to verify a natural law? Explain.

23. What are the principal reasons why one theory might be adopted over a conflicting one?

24. An important premise of science is that there exists an underlying order to nature. Einstein described this belief in the words "God is subtle but He is not malicious." Explain more fully what he meant by this remark.

25. In an attempt to determine any possible relationship between the year in which a U.S. penny was minted and its current mass, students weighed an assortment of pennies and obtained the following data:

1968: 3.11, 3.08, 3.09 g   *1982:* 3.12, 3.08, 2.54, 2.53 g

1973: 3.14, 3.06, 3.07 g   *1983:* 2.51, 2.49, 2.47 g

1977: 3.13, 3.10, 3.06 g   *1985:* 2.54, 2.53, 2.53 g

1980: 3.12, 3.11 g

What valid conclusion(s) might they have drawn about the relationship between the current mass and the year of a penny?

Exponential arithmetic (see Appendix A)

26. Several measured or estimated quantities follow. Express each value in the stated unit but in exponential form.

(a) speed of light in vacuum: 186 thousand miles per second

(b) mass of air in the atmosphere: 5 to 6 quadrillion tons

(c) solar radiation received by the earth: 173 thousand trillion watts

(d) diameter of a typical aerosol smog particle: one millionth of a meter

(e) average diameter of a human cell: ten millionths of a meter

(f) estimated total recoverable natural gas resources in the U.S.: 900 to 1300 trillion cubic feet

27. Express the result of each of the following calculations in exponential form.

(a) $0.0056 + (33 \times 6.2 \times 10^{-4}) =$

(b) $\dfrac{(2.2 \times 10^3) + (4.7 \times 10^2)}{4.6 \times 10^{-2}} =$

(c) $\dfrac{3.15 \times 10^4 \times (2.6 \times 10^{-3})^2}{0.060 + (2.2 \times 10^{-2})} =$

Significant figures

28. Indicate whether each of the following is an exact number or a measured quantity subject to uncertainty.

(a) The number of oranges in one dozen

(b) The number of gallons of gasoline to fill an automobile gas tank

(c) The distance between the earth and the sun

(d) The number of days in the month of January

(e) The area of a city lot

29. Perform the following calculations, retaining the appropriate number of significant figures in each result.

(a) $622 \times 13.11 =$

(b) $45.6 \times 10^3 \times 1.25 \times 10^5 =$

(c) $2.11 \times 10^3 \times (16.2 \times 10^{-3})^2 =$

(d) $\dfrac{7.34 \times 10^2 \times 3.18 \times 10^{-4}}{(3.1 \times 10^{-3})^2}$

(e) $35.24 + 36.3 + 1.08 =$

(f) $(1.561 \times 10^3) - (1.80 \times 10^2) + (2.02 \times 10^4) =$

30. An object placed on a two-pan balance (see Figure 1-8a) requires use of these standard masses to achieve a balance condition: 20 g, 1 g, 500 mg, 200 mg, 100 mg, 50 mg, 20 mg, 5 mg, and 2 mg. Assuming that each mass is accurate to the nearest mg, what is the mass of the object, expressed with an appropriate number of significant figures?

31. A press release describing the 1986 nonstop, round-the-world trip by the aircraft *Voyager* included the following data:

flight distance: 25,012 mi

flight time: 9 days, 3 minutes, 44 seconds

fuel capacity: nearly 9000 lb

fuel remaining at end of flight: 14 gal

To the maximum number of significant figures permitted, calculate

(a) the average speed of the aircraft, in mi/h

★(b) the fuel consumption, in mi/lb fuel. (Assume a density of 0.70 g/cm³ for the fuel.)

Systems of measurement

32. The English unit, the hand (used in horsemanship), is 4 inches. What is the height, in cm, of a horse that stands 15 hands high?

33. A certain brand of coffee is offered for sale at $7.26 for a 3-lb can or $5.42 for a 1-kg can. Which is the better buy?

34. A sprinter runs the 100-yd dash in 9.3 s. At this same rate, how long would it take the sprinter to run 100 m?

35. The unit, the furlong, is used in horseracing. The units, chain and link, are used in surveying. There are 8 furlongs in 1 mi, 10 chains in 1 furlong, and 100 links in 1 chain. To three significant figures, what is the length of 1 link, in inches?

36. An English unit of mass used in pharmaceutical work is the grain (gr). 15 gr = 1.0 g. An aspirin tablet contains 5.0 gr of aspirin. A 155-lb arthritic person takes two aspirin tablets per day.

(a) What is the quantity of aspirin in two tablets, expressed in mg?

(b) What is the dosage rate of aspirin, expressed in mg aspirin per kg of body weight?

(c) At the given rate of consumption of aspirin tablets, how long would it take for the person to consume 1.0 lb of aspirin?

37. A block of ice measures 24 in. × 18 in. × 12 in.

(a) What is the volume of this block, in m³?

(b) What is the total surface area of the block, in cm²?

38. The English measure of land area, the acre, is represented in the accompanying illustration. Express the area of 1.0 acre in ft².

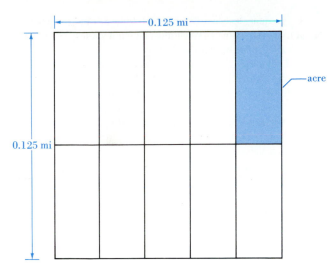

Exercise 38

★**39.** Table 1-1 lists that 1 in. = 2.54 cm and that 1 m = 39.37 in.

(a) Explain why only one of these statements can be *exact*.

(b) Given that 1 in. = 2.54 cm, *exactly*, how many inches are there in one meter, expressed to *six* significant figures?

Temperature scales

40. A table of climatic data lists the highest and lowest temperatures on record for San Bernardino, California, as 118 °F and 17 °F, respectively. What are these temperatures on the Celsius scale?

41. A class in home economics is given an assignment in candy making requiring that a sugar mixture be brought to a "soft ball" stage (234 to 240 °F). A student borrows a thermometer having a range from −10 to 110 °C from the chemistry laboratory to do this assignment. Will this thermometer serve the purpose? Explain.

42. The absolute zero of temperature is −273.15 °C. What is this temperature on the Fahrenheit scale?

Density

43. To determine the density of a liquid, a flask is weighed empty (108.6 g) and again when filled with 125 mL of a liquid (207.5 g). What is the density of the liquid?

44. A barrel contains 42.0 gal of petroleum having a mass of 298 lb. What is the density of this petroleum, in g/cm³?

45. To determine the volume of an irregularly shaped glass vessel, the vessel is weighed empty (121.3 g) and when filled with carbon tetrachloride (283.2 g). What is the volume capacity of the vessel, given that the density of carbon tetrachloride is 1.59 g/cm³?

46. Density, the ratio of mass to volume, can be expressed in various units. For a given substance, would the magnitude (numerical value) of its density be greatest if expressed in g/cm³, kg/m³, lb/ft³, or oz/gal? Explain. [*Hint:* 16 oz = 1 lb.]

47. The following densities are given at 20 °C: water, 0.998 g/cm³; iron, 7.86 g/cm³; aluminum 2.70 g/cm³. Arrange the following items in terms of *increasing* mass.

(a) a rectangular bar of iron, 71.8 cm × 1.8 cm × 1.2 cm.

(b) a sheet of aluminum foil, 11.35 m × 4.85 m × 0.002 cm.

(c) 2.716 L of water.

48. A square piece of aluminum foil, 8.0 in. on edge, is found to weigh 1.863 g. What must be the thickness of this foil? (The density of aluminum is 2.70 g/cm^3.)

49. Ethanol (ethyl alcohol) is a potential automotive fuel; its density at 20.0 °C is 0.78934 g/cm^3. Refer to the Summarizing Example. If the pycnometer pictured in Figure 1-16 were filled at 20.0 °C with ethanol instead of with methanol, what would be the mass observed? Could you easily distinguish between pure ethanol and pure methanol with a density determination of the precision achieved in Figure 1-16?

50. The regularly shaped wooden object pictured below weighs 28.6 g. Is it made of maple (density range = 0.62 to 0.75 g/cm^3) or white pine (density range = 0.35 to 0.50 g/cm^3)? [*Hint:* What is its volume?]

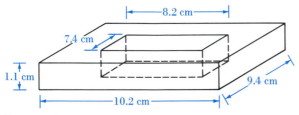

Exercise 50

Percent composition

51. In a class of 66 students the results of a particular examination were 8 A, 16 B, 30 C, 9 D, 3 F. What was the percent distribution of grades, that is, % A, % B, and so on?

52. A water solution that is 15% sucrose, by mass, has a density of 1.059 g/cm^3. What mass of sucrose, in grams, is contained in 3.05 L of this solution?

53. A solution containing 12.0% sodium hydroxide, by mass, has a density of 1.131 g/cm^3. What volume of this solution, in liters, must be used in an application requiring 1.00 kg of sodium hydroxide?

54. Referring to Exercise 49 and Figure 1-16, describe a method that might be used to relate density and percent composition of ethanol–methanol solutions of varying composition. How accurate would this method likely be? Explain.

Algebraic equations

55. A tabulation of data lists the following equation for the densities (*d*) of solutions of naphthalene in benzene (at 30 °C) as a function of the mass percent naphthalene (% N).

$$d \ (\text{g/cm}^3) = \frac{1}{1.153 - 0.00182 \ (\% \ N) + 1.08 \times 10^{-6} \ (\% \ N)^2}$$

What is the density of a solution (at 30 °C) with 1.15% naphthalene in benzene, by mass, i.e., with % N = 1.15?

56. For a benzene solution containing 6.38% para-dichlorobenzene, by mass, the density as a function of temperature (over the range 15 to 65 °C) is given by the equation

$$d \ (\text{g/cm}^3) = 1.5794 - 1.836 \times 10^{-3}(t - 15°)$$

At what temperature (*t*) will this solution have a density of 1.543 g/cm^3?

★57. It is desired to construct a cube with length *l* that will have the same volume as a sphere of radius *r*. What must be the ratio *l/r*?

★58. Use the equations in Figure 1-15 to find an *algebraic* solution to this question: At what single temperature are the numerical values on the Celsius and Fahrenheit scales equal? [*Hint:* A trial-and-error method is not an algebraic solution.]

★59. Refer to the equation given in Exercise 15 for the density of water as a function of Celsius temperature, and show that the density of water passes through a maximum somewhere in the temperature range for which the equation applies (i.e., from 0 °C to about 20 °C).

Additional Exercises

60. Human behavior cannot be studied quite as readily as the phenomena of natural science. Nevertheless, there are certain "laws" that are applicable to a variety of human activities. Explain what is meant by the "law of averages" and the "law of diminishing returns."

61. Use the concept of significant figures to criticize the way in which the following information was presented. "The estimated proved reserve of natural gas as of January 1, 1982, was 2,911,346 trillion cubic feet."

62. Express the result of each of the following calculations in exponential form and with the appropriate number of significant figures.

(a) $3.1 \times 10^5 \times 6.17 \times 10^{-3} \times 8.2 \times 10^{-4} \times 9.1 \times 10^{-3} =$

(b) $\dfrac{1711 \times (0.0033 \times 10^4) \times 1.25 \times 10^{-3}}{(6.15 \times 10^{-4})^3} =$

(c) $\dfrac{[(6.0 \times 10^3) + (4.2 \times 10^4)]^2}{(2.2 \times 10^3)^2 + 180,000} =$

63. Perform the following conversions between the English and metric systems of measurement.

(a) 65 mi/h = _____ km/h   (b) 33 ft/s = _____ km/h
(c) 235 in^3 = _____ cm^3
(d) 17.5 lb/in.^{2} = _____ kg/m^2

64. In scientific work densities are expressed in g/cm^3 and in engineering work, in lb/ft^3. The density of water at 20 °C is 0.998 g/cm^3. What is this density expressed in lb/ft^3?

65. In the English system land area is measured by the *acre*. In the metric system land area is measured by the *hectare*. There are 640 acres in 1 mi^2, and 1 hectare is defined as 1 hm^2. 1 hectometer (hm) = 100 m. How many acres correspond to 1 hectare?

66. An airplane flying at the speed of sound is said to be at Mach 1. Mach 1.5 is 1.5 times the speed of sound, and so on. If the speed of sound in air is given as 1130 ft/s, what is the speed of an airplane, in km/h, flying at Mach 1.27?

★67. According to the rules on significant figures, the product 99.9×1.008 should be expressed to three significant figures— 101. Yet, in this case it would be appropriate to express the result to *four* significant figures—1.007. Explain why this is so.

68. A European cheese-making recipe calls for 2.50 kg of whole milk. An American who wants to make the recipe has no scale and only volume-measuring containers marked in quarts, pints, cups, and fractions thereof. Assuming that the density of milk is 1.03 g/cm^3, what volume of milk is needed? (1 qt = 2 pt; 1 pt = 2 cups)

69. To determine the approximate mass of a small spherical shot of copper, the following experiment is performed. 100 pieces of the shot are counted out and added to 8.4 mL of water in a graduated cylinder; the total volume becomes 8.8 mL. The density of copper is 8.92 g/cm^3. Determine the approximate mass of a single piece of shot, assuming that all the pieces are of nearly the same dimensions.

70. An empty 3.00-L bottle weighs 1.70 kg. Filled with a certain wine it weighs 4.72 kg. The wine contains 11.0% ethyl alcohol, by mass. How many ounces of ethyl alcohol would be present in 400. mL of this wine? (1 lb = 16 oz.)

71. It is necessary to determine the density of a solution to *four* significant figures. The volume of solution can be measured to the nearest 0.1 mL.
   (a) What is the minimum volume of sample that can be used for the measurement?
   (b) Assuming the minimum volume determined in part (a), how accurately must the sample be weighed (i.e., to the nearest 0.1 g, 0.01 g, . . .) if the density of the solution is greater than 1.00 g/cm^3? If the density is less than 1.00 g/cm^3?

72. A solution used to chlorinate a home swimming pool contains 7% chlorine, by mass. An ideal chlorine level for the pool is one part per million (1 ppm). (Think of 1 ppm as being 1 g chlorine per million grams of water.) Assuming densities of 1.10 g/cm^3 for the chlorine solution and 1.00 g/cm^3 for the swimming pool water, what volume of the chlorine solution is required to produce a chlorine level of 1 ppm in a 18,000-gallon swimming pool?

★**73.** Use the equation of Exercise 15 to determine the temperature at which the density of water is 0.99916 g/cm^3. [*Hint:* Use the quadratic formula; see Appendix A.]

74. It is desired to relate Fahrenheit to Kelvin temperature.
   (a) Derive a *single* equation that can be used for this purpose.
   (b) What is the Kelvin temperature corresponding to 104 °F?

★**75.** A Fahrenheit and a Celsius thermometer are immersed in the same medium, whose temperature is to be measured. At what Celsius temperature will the numerical reading on the Fahrenheit thermometer be
   (a) twice that on the Celsius thermometer?
   (b) four times that on the Celsius thermometer?
   (c) one-sixth that on the Celsius thermometer?
   (d) 200° more than that on the Celsius thermometer?

★**76.** A pycnometer (recall Figure 1-16) weighs 25.60 g empty and 35.55 g when filled with water at 20 °C. The density of the water at 20 °C is 0.998 g/cm^3. When 10.20 g of lead is placed in the pycnometer and the pycnometer is again filled with water at 20 °C, the total mass is 44.83 g. What is the density of the lead?

★**77.** A standard kilogram mass is to be cut from a cylindrical bar of steel with a diameter of 1.50 in. The density of the steel is 7.70 g/cm^3. How long must the section be?

78. The volume of seawater on earth is estimated to be 330,000,000 mi^3. Assuming that seawater is 3.5% sodium chloride, by mass, and that the density of seawater is 1.03 g/cm^3, what is the approximate mass of sodium chloride, in tons, dissolved in the seawater on earth?

★**79.** The diameter of metal wire is often referred to by its American wire gauge number. A 16-gauge wire has a diameter of 0.05082 in. What length of wire, in meters, is there in a 1.00-lb spool of 16-gauge copper wire? The density of copper is 8.92 g/cm^3.

80. An important source of magnesium metal is seawater, from which it is extracted by the Dow process (described in Section 22-2). Magnesium occurs in seawater to the extent of 1.4 g magnesium per kilogram of seawater. The annual production of magnesium in the United States is about 10^5 tons. If all this magnesium were extracted from seawater, what volume of seawater, in m^3, would have to be processed?

81. The Antarctic, Greenland, and other ice caps contain approximately 7.2 million mi^3 of ice.
   (a) Given that the density of ice is 0.92 g/cm^3, together with other appropriate conversion factors, determine the mass of this ice, in tons.
   (b) When ice is melted, its volume decreases by about 10%. If all the polar ice were to melt completely, estimate the increase in sea level that would result from the additional liquid water entering the oceans. The oceans of the world cover about 1.4 × 10^8 mi^2.

★**82.** A typical rate of deposit of dust ("dustfall") from air that is not significantly polluted might be 10 tons per square mile per month. What is this dustfall, expressed in milligrams per square meter per hour?

★**83.** When water is used for irrigation purposes, its volume is often expressed in acre-feet. One acre-foot is a volume of water sufficient to cover 1 acre of land to a depth of 1 ft (640 acres = 1 mi^2). The principal lake in the California Water Project is Lake Oroville, whose water storage capacity is listed as 3.54 × 10^6 acre-feet. Express the volume of Lake Oroville in (a) ft^3; (b) m^3; (c) gal.

★**84.** The object pictured is made of brass stock 1.27 cm thick. Calculate its mass. (The density of brass = 8.75 g/cm^3).

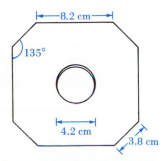

Exercise 84

Self-Test Questions

For questions 85 through 94 select the single item that best completes each statement.

85. Of the following masses, that which is expressed to the nearest milligram is (a) 32.7 g; (b) 32.71 g; (c) 32.707 g; (d) 32.7068 g.

86. The greatest length of the following group is (a) 4.0 m; (b) 140 in.; (c) 12 ft; (d) 0.001 km.

87. Of the following numbers, the one with three significant figures is (a) 22.03; (b) 0.0260; (c) 1.070; (d) 2000.

88. To the correct number of significant figures, the result of the calculation: $12.11 \times 0.0087 \times 302 =$

(a) 31.8; (b) 31.818; (c) 31.82; (d) 32.

89. The highest temperature of the following group is (a) 217 K; (b) 273 K; (c) 217 °F; (d) 105 °C.

90. Of the following substances, the greatest *density* is that of

(a) 1000 g water at 4 °C
(b) 100.0 cm^3 of chloroform, which weighs 148.9 g
(c) a 10.0 cm^3 piece of wood weighing 7.72 g
(d) an ethyl alcohol–water mixture of density 0.83 g/cm^3

91. The largest *volume* of the following group is that of

(a) 445 g of water at 4 °C
(b) 600 g of chloroform at 20 °C (density = 1.5 g/cm^3)
(c) 0.50 L of milk
(d) 155 cm^3 of steel (density = 7.70 g/cm^3)

92. A fertilizer contains 20% nitrogen, by mass. To provide a fruit tree with an equivalent of 1 lb of nitrogen, the quantity of *fertilizer* required is (a) 20 lb; (b) 0.20 lb; (c) 0.05 lb; (d) 5 lb.

93. An example of a *homogeneous* mixture or solution is (a) "7 Up"; (b) liquid oxygen; (c) distilled water; (d) chicken soup.

94. An example of a *chemical* change is (a) the melting of an ice cube; (b) the boiling of gasoline; (c) the frying of an egg; (d) all of these.

95. Describe briefly the distinction between the following pairs of terms:

(a) element and compound
(b) homogeneous and heterogeneous mixture
(c) mass and density.

96. Use exponential notation and the appropriate number of significant figures to express the result of the following calculation.

$$(19.541 + 1.03 - 3.6) \times 651 = ?$$

97. A 55.0-gal drum weighs 75.0 lb when empty. Filled with glycerol, the total mass is 653.4 lb. What is the density of glycerol, expressed in g/cm^3? (1 gal = 3.785 L; 1 lb = 454 g.)

98. A certain hydrochloric acid solution contains 26.0% hydrochloric acid (the rest is water) and has a density of 1.130 g/cm^3. How many liters of this solution must be taken to provide a sample containing 1.00 kg hydrochloric acid?

99. A certain paint is available at $5.00/quart or $14.00/gallon. The paint provides a coverage of 125 ft^2/quart. To paint a floor that measures 6.50 m × 4.80 m, is it cheaper to buy the necessary paint in quart cans or by the gallon?

2 Atoms and the Atomic Theory

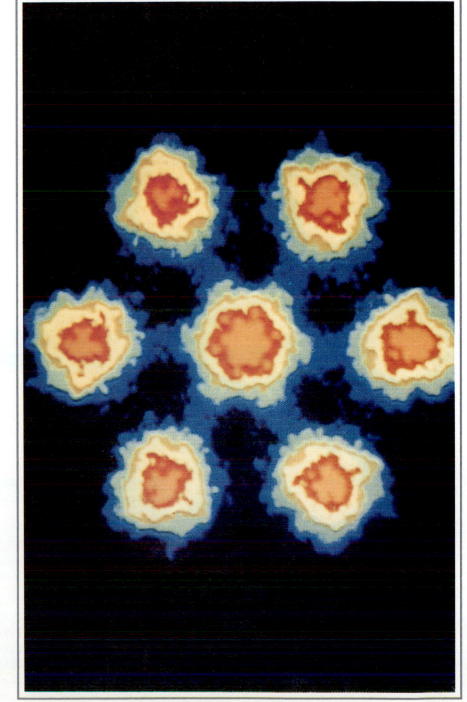

A uranyl acetate cluster on a thin film of carbon. The individual uranium atoms are the colored spots with the red-orange centers. Separation between the uranium atoms is 0.34 nm. The carbon film appears back in this color scheme. [Courtesy of M. Isaacson, Cornell University and M. Ohtsuki, University of Chicago]

Humans tend to think analytically—to break down a complex problem into simpler units. Very likely this inclination has been responsible for the growth of the idea that all matter can be broken down into ultimate units called atoms. The atomic nature of matter is one of the oldest scientific concepts. In fact, this idea is conveyed by the word atom itself [*atomos* (*a*, not + *tomos*, to cut): Greek, meaning indivisible—not capable of being cut].

In this chapter we begin with the first successful atomic theory—that of John Dalton. But we will quickly shift our attention from chemical to physical studies that established the existence of atoms. It was little more than 100 years ago that the first parts of atoms (electrons) were discovered. Today, with the help of sophisticated instruments, we can actually *see* images of individual atoms.

The existence of atoms, which for centuries could only be described through intuition or inferred from observations on bulk matter, is now a matter of fact.

2-1 Early Chemical Discoveries and the Atomic Theory

FIGURE 2-1

Two combustion reactions.

The apparent product of the combustion of the match, the ash, weighs *less* than the original match. The product of the combustion of the magnesium ribbon (provided all the "smoke" is collected) weighs *more* than the original ribbon. Actually, in *each* case the total mass of the materials involved in the combustion remains *unchanged*. To analyze these observations properly, we need to recognize that (1) oxygen gas is a reactant in both combustions and (2) water vapor and carbon dioxide are products of the combustion of the match that are ordinarily allowed to escape. [Carey B. Van Loon]

This and later chapters suggest that physicists have contributed more to our knowledge of the existence and structures of atoms than chemists have. Nevertheless, two centuries ago chemists were making discoveries that strongly suggested the atomic nature of matter.

Law of Conservation of Mass. One common phenomenon whose explanation was crucial to the development of the atomic theory is that of combustion or burning, illustrated in Figure 2-1. Antoine Lavoisier (1743–1794) established that oxygen is essential to combustion. That is, oxygen gas from air is consumed in the combustion and appears combined with other elements in the products of the combustion. By making careful measurements of mass (with an analytical balance similar to that pictured in Figure 1-8a) he formulated this fundamental generalization, which we now call the **law of conservation of mass:** *The mass of substances present after a chemical reaction is the same as the mass of substances entering into the reaction.* Stated in another way, this law says that matter can neither be created nor destroyed in a chemical reaction.

Law of Constant Composition. Joseph Proust (1754–1826) observed that samples of copper carbonate, whether he obtained them from natural sources or synthesized them in his laboratory, always had the same percent copper, by mass. His discovery became the basis of the generalization now known as the **law of constant composition** or the **law of definite proportions:** *All samples of a given compound have the same composition, that is, the same proportions by mass of the constituent elements.*

Dalton's Atomic Theory. From 1803 to 1808, John Dalton, an English schoolteacher, used the two fundamental laws of chemical combination just described as the basis of an atomic theory. His theory involved three assumptions.

1. Each chemical element is composed of minute, indestructible particles called atoms. Atoms can be neither created nor destroyed during a chemical change.
2. All atoms of an element are alike in mass (weight) and other properties, but the atoms of one element are different from those of all other elements.
3. In chemical compounds, atoms of different elements combine in simple numerical ratios: for example, one atom of A to one of B (AB), one atom of A to two of B (AB$_2$), ...

John Dalton (1766–1844)—developer of the atomic theory. Dalton was not a good experimenter (perhaps because of his color blindness), but he skillfully used the results of many other experimenters in formulating his atomic theory. [Burndy Library]

If the atoms of an element are indestructible (assumption 1), then the *very same* atoms must be present after a chemical reaction as were present before the reaction. The total mass of reactants and products must be the same. *Dalton's theory explains the law of conservation of mass.* If all atoms of an element are alike in mass (assumption 2), and if atoms unite in *fixed* numerical ratios (assumption 3), the percentage composition of a compound must have a unique value, regardless of the size of the sample analyzed or its origin. *Dalton's theory also explains the law of constant composition.*

The characteristic masses of the atoms of an element implied by assumption 2 above became known as **atomic weights,** and Dalton tried to establish a set of *relative* atomic weights. The difficulties he encountered are suggested by Figure 2-2. Despite these difficulties Dalton's theory did provide a basis for deducing (predicting) the **law of multiple proportions** (1805). *If two elements form more than a single compound, the masses of one element combined with a fixed mass of the second are in the ratio of small whole numbers.* Dalton's conception of this law is illustrated in Figure 2-3.

Elements		Compounds	
O	H	OH	HOH

at. wt.
H = 1

at. wt.
O = 7

at. wt.
O = 14

FIGURE 2-2
The "atomic weight problem."

Atoms of oxygen (O) and hydrogen (H) are shown combining to form water. If the combining ratio is 1:1, the formula is OH; if the combining ratio is two hydrogen atoms to one of oxygen, the formula is H_2O. Dalton *assumed* the formula OH and believed the mass of oxygen in water to be *seven* times the mass of hydrogen. He assigned oxygen an atomic weight of 7. If he had assumed the formula H_2O he would have assigned oxygen an atomic weight of 14 (since the ratio 14:2 is the same as 7:1). We now know that the mass ratio of oxygen to hydrogen is 8:1, that the formula of water is H_2O, and that the atomic weight of oxygen is 16.

FIGURE 2-3
The law of multiple proportions illustrated—with Dalton's symbols and atomic weights.

Per gram of H in olefiant gas there is 5 g of C, and per gram of H in carburetted hydrogen there is 2.5 g C, leading to the ratio of two small whole numbers (2:1).

$$\frac{5 \text{ g C/1 g H}}{2.5 \text{ g C/1 g H}} = \frac{2}{1}$$

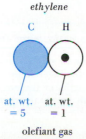

ethylene

C   H

at. wt.
= 5

at. wt.
= 1

olefiant gas

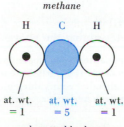

methane

H   C   H

at. wt.
= 1

at. wt.
= 5

at. wt.
= 1

carburetted hydrogen

FIGURE 2-4
Cathode rays and their deflection in a magnetic field.

Cathode rays are invisible. Only through their impact on a fluorescent material can they be detected. The beam of cathode rays originates at the cathode on the left and is deflected as it enters the field of the magnet situated slightly behind the anode. The rays are seen through the green fluorescence they produce as they strike the zinc sulfide-coated screen. The deflection corresponds to that expected of negatively charged particles. [Carey B. Van Loon]

2-2 Electrons and Other Discoveries in Atomic Physics

In his study of electrical discharge through evacuated tubes Michael Faraday (1791–1867) discovered a form of radiation called **cathode rays.** Cathode rays are emitted from the negative terminal or cathode and cross to the positive terminal or anode. Later workers found that cathode rays travel in straight lines, are invisible, and have properties that are independent of the material from which they originate (i.e., whether iron, platinum, etc.). Two additional properties are demonstrated in Figure 2-4: Cathode rays, upon striking glass or certain other materials, cause them to fluoresce (give off light). Cathode rays are deflected by electric and magnetic fields in the manner expected for *negatively charged* particles.

Charge-to-Mass Ratio of Electrons. Through experiments suggested by Figure 2-5, J. J. Thomson (1856–1940) established the ratio of electric charge (*e*) to mass

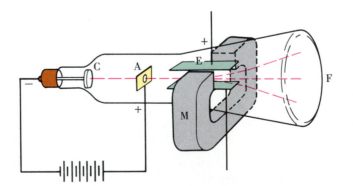

C, cathode; A, anode (perforated to allow the passage of a narrow beam of cathode rays); E, electrically charged condenser plates; M, magnet; F, fluorescent screen.

FIGURE 2-5
Determining the charge-to-mass ratio, e/m, for cathode rays.

The cathode ray beam can be made to strike the end screen, undeflected, if the forces on the particles exerted by the electric and magnetic fields are just counterbalanced. From values of the strengths of the electric and magnetic fields, together with other data, a value of e/m can be obtained. Precise measurements yield a value of

-1.7588×10^8 coulombs per gram

(Because cathode rays carry a negative charge, the sign of the charge-to-mass ratio is also negative.)

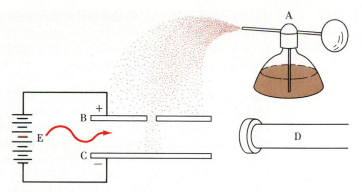

FIGURE 2-6
Millikan's oil drop experiment.

Oil droplets produced by the atomizer (A) enter the apparatus through a tiny hole in the top plate of an electrical condenser. The droplets are observed through a telescope equipped with a micrometer eyepiece (D). Ions (electrically charged atoms or molecules) are produced by ionizing radiation such as x rays (E). Some of the oil droplets acquire an electric charge by adsorbing ions (attaching ions to their surface).

The fall of a droplet between the condenser plates (B and C) is either speeded up or slowed down to an extent that depends on the sign and magnitude of the charge on the droplet. By analyzing data from a large number of droplets, Millikan concluded that the magnitude of the charge, q, on a droplet is always an integral multiple of the electronic charge, e. That is $q = ne$ (where $n = 1, 2, 3$, and so on).

(m) for cathode rays, that is, e/m. Also, Thomson concluded that cathode rays are negatively charged *fundamental* particles of matter found in *all* atoms. Cathode rays subsequently became known as **electrons,** a term first proposed by George Stoney in 1874.

Charge and Mass of the Electron. Robert Millikan (1868–1953) determined the electronic charge e through a series of ''oil drop'' experiments (1906–1914), illustrated in Figure 2-6. The currently accepted value of the electronic charge e (to five significant figures) is -1.6022×10^{-19} C. Combining this value with an accurate value of the charge-to-mass ratio for an electron yields the mass of an electron: 9.1094×10^{-28} g.

> The coulomb (C) is the SI unit of electrical charge.

Positive Ions (Canal Rays). Particles carrying *positive* charge were also discovered in modified cathode ray tubes. Unlike cathode rays, positive ray particles are *not* fundamental particles of matter. Their charge-to-mass ratios depend on the gases from which they are formed, and some positive rays carry a *multiple* of the fundamental unit of electric charge. Although the particle carrying a fundamental unit of positive charge, the **proton,** is found in positive rays produced from hydrogen, the proton was not isolated and characterized until 1919 (see Section 2-3). Nevertheless, the discovery of positive rays did lead to an atomic model (later shown to be incorrect) that can explain the formation of electrically charged atomic species or **ions** (see Figure 2-7). Positive rays are simply positively charged gaseous ions.

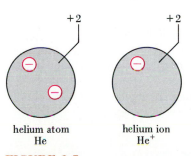

+2   +2

helium atom   helium ion
He   He⁺

FIGURE 2-7
The "plum pudding" model.

A hydrogen atom would consist of a "cloud" of positive charge (+1) and contain one electron (−1); helium, a +2 cloud and two electrons (−2); and so on. The loss of one electron by a helium atom results in the formation of an ion with a net charge of +1, that is, He⁺. (The loss of both electrons results in He²⁺.)

X Rays. Wilhelm Roentgen (1845–1923) was first to show (in 1895) that when cathode rays (electrons) strike materials within a cathode ray tube they produce a type of radiation that causes fluorescence *outside* the tube. Because of the unknown nature of this radiation, Roentgen coined the term **x ray,** a name that we still use.

Roentgen found that x rays are *not* deflected by electric and magnetic fields. X rays have a very high penetrating power through matter, and this accounts for

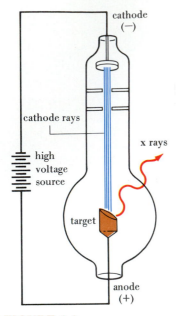

FIGURE 2-8
The production of x rays.

their widespread use in photographing the interior of objects. X rays are a form of electromagnetic radiation with wavelengths of approximately 1 angstrom unit ($1 \text{ Å} = 10^{-10}$ m).* An especially effective source of x rays is the impact of cathode rays on a dense metal anode called a *target*, illustrated in Figure 2-8.

Radioactivity. Antoine Becquerel (1852–1908) associated the emission of x rays with fluorescence and wondered if fluorescent materials would produce x rays (without the need for cathode rays). He designed some experiments to test this idea. On one occasion he noticed that a photographic plate, wrapped with black paper and covered with a uranium-containing fluorescent material, became strongly exposed, even though it had been kept in the dark where no fluorescence could occur. He found that radiation continuously emitted by the uranium had exposed the photographic plate. Becquerel had discovered radioactivity.

Ernest Rutherford (1871–1937) identified two types of radiation from radioactive materials, **alpha (α)** and **beta (β).** Alpha rays are *particles* carrying two fundamental units of positive charge and having the same mass as helium atoms. Alpha particles are identical to He^{2+} ions (see again Figure 2-7). Beta rays are *negatively* charged particles identical to electrons. A third form of radiation, **gamma (γ) rays,** is electromagnetic radiation of extremely high penetrating power.

By the early 1900s additional radioactive elements were discovered, principally through the work of Marie and Pierre Curie, and Rutherford and Frederick Soddy made another profound discovery: The chemical properties of a radioactive element *change* as it undergoes radioactive decay. This observation suggests that radioactivity involves fundamental changes at the *subatomic* level—that in radioactive decay one element is changed into another, a process known as **transmutation.** Additional aspects of radioactivity and nuclear chemistry are considered in Chapter 26.

2-3 The Nuclear Atom

In 1909, at Rutherford's suggestion, Hans Geiger and Ernest Marsden bombarded very thin foils of gold and other metals with α particles (see Figure 2-9). They detected the presence of α particles by the flashes of light (scintillations) that the α particles produced when they struck a zinc sulfide screen mounted on the end of a telescope. Geiger and Marsden observed that

- The majority of α particles penetrated the foil undeflected.
- Some α particles experienced slight deflections.
- A few (about one in every 20,000) suffered rather serious deflections as they penetrated the foil.
- A similar number did not pass through the foil at all but "bounced back" in the direction from which they had come.

"It is about as incredible as if you had fired a 15-in. shell at a piece of tissue paper and it came back and hit you."
                  Ernest Rutherford

FIGURE 2-9
The scattering of alpha particles by metal foil.

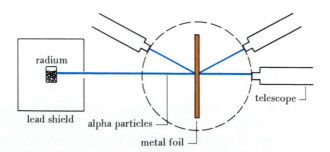

*Electromagnetic radiation is discussed in some detail in Chapter 8.

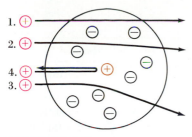

FIGURE 2-10
Rutherford's interpretation of the scattering of α particles by thin metal foils.

1. α particle passes through the atom undeflected. (This is the fate of the vast majority of α particles.)
2. α particle suffers a slight deflection by passing close to an electron.
3. α particle suffers severe deflection by passing close to atomic nucleus.
4. α particle bounces back as a result of approaching the atomic nucleus head-on.

Based on the "plum pudding" model, Rutherford reasoned that the positive charge of an atom was so diffuse that α particles should pass through this weak electric field largely undeflected. The severe deflections of some of the α particles astounded him. Rutherford interpreted the scattering of α particles in the manner suggested by Figure 2-10, and proposed this model of a **nuclear** atom.

1. Most of the mass and all of the positive charge of an atom are centered in a very small region called the **nucleus.** *The atom is mostly empty space.*
2. The magnitude of the positive charge is different for different atoms and is approximately one-half the atomic weight of the element.
3. There exist as many electrons outside the nucleus as there are units of positive charge on the nucleus. The atom as a whole is electrically neutral.

Protons and Neutrons. In studying the passage of α particles through air, Rutherford (1919) found that he could detect scintillations on a zinc sulfide screen much farther from a radium source than he expected α particles to be able to travel. He concluded that when an α particle strikes the nucleus of a nitrogen atom in the atmosphere a particle carrying a fundamental unit of positive charge is ejected from the nucleus—a **proton.** The scintillations that he detected were caused by these protons, not by α particles.

In 1932 James Chadwick showed that the properties of a very penetrating new type of radiation discovered in the early 1930s could best be explained by assuming a beam of *neutral* particles, called **neutrons,** that originate from the nuclei of atoms. So, it has been only for about the past 60 years that we have been able to represent an atom by the model suggested in Figure 2-11: The nucleus of an atom contains protons and neutrons. The nuclear charge is determined by the number of protons, and in an electrically neutral atom there must be the same number of electrons outside the nucleus as there are protons in the nucleus. The nucleus also contains a sufficient number of neutrons in its nucleus to account for the observed mass of the atom.

FIGURE 2-11
The nuclear atom—illustrated by the helium atom.

In this illustration electrons are pictured as being much closer to the nucleus than is actually the case. The actual situation is more nearly like representing an entire atom by a room, 5 m × 5 m × 5 m, and the nucleus by the period at the end of this sentence.

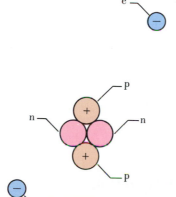

2-4   Properties of Fundamental Particles: A Summary

We now know of many fundamental particles of matter, but the behavior of an atom seems to depend just on its numbers of protons, neutrons, and electrons. Table 2-1 presents electric charges and masses of these three fundamental particles in two sets of units. One is the metric unit, and the other a special atomic unit.

An electron carries an atomic unit of negative electric charge. A proton carries an atomic unit of positive charge. The **atomic mass unit** (see page 42) is defined as

TABLE 2-1
Properties of Three Fundamental Particles

| | Electric charge | | Mass | |
	Metric (C)	Atomic	Metric (g)	Atomic (u)[a]
proton	$+1.602 \times 10^{-19}$	$+1$	1.673×10^{-24}	1.0073
neutron	0	0	1.675×10^{-24}	1.0087
electron	-1.602×10^{-19}	-1	9.109×10^{-28}	0.00055

[a]u is the SI unit for atomic mass unit (abbreviated amu).

exactly $\frac{1}{12}$ of the mass of the atom known as carbon-12 ("carbon twelve"). An atomic mass unit is referred to by the abbreviation, amu, and the unit, u. As we see from Table 2-1, the proton and neutron masses are just slightly greater than 1 u. By comparison the mass of an electron is seen to be extremely small.

The number of protons in the nucleus of an atom is called the **atomic number,** or the **proton number, Z.** The number of electrons in an electrically neutral atom is also equal to the atomic number Z. The mass of an atom is determined by the total number of protons and neutrons its nucleus. This total is called the **mass number, A.** The number of neutrons in an atom, the **neutron number,** is $A - Z$.

Example 2-1

Calculating the e/m ratio of an electron. From the data given in Table 2-1 calculate the *e/m* ratio for the electron and compare this with the value listed in Figure 2-5.

Solution. From Table 2-1 we extract these data about the electron.

charge: -1.602×10^{-19} C
mass: 9.109×10^{-28} g

charge-to-mass ratio: $e/m = \dfrac{-1.602 \times 10^{-19} \text{ C}}{9.109 \times 10^{-28} \text{ g}} = -1.759 \times 10^{8} \text{ C/g}$

To four significant figures, this ratio has the same value as that listed in Figure 2-5.

SIMILAR EXAMPLES: Exercises 30, 31.

2-5 Chemical Elements

All of the atoms of a given element have the same atomic number. At present the known elements are those with atomic numbers ranging from $Z = 1$ to $Z = 109$: Each element has a name and a distinctive symbol. For most elements the symbol is the abbreviated form of its English name, consisting of one or two letters. The first (but never the second) letter of the symbol is capitalized. For example,

carbon	oxygen	nitrogen	neon	sulfur	silicon
C	O	N	Ne	S	Si

Some elements known since ancient times have symbols based on their Latin names. For example,

iron *(ferrum)*	copper *(cuprum)*	lead *(plumbum)*
Fe	Cu	Pb

A few elements have symbols based on the Latin name of a compound, for example,

natrium (sodium carbonate)   *kalium* (potassium carbonate)

Na   K
sodium   potassium

The symbol for tungsten, W, is based on the German name, wolfram. A complete listing of the elements is on the inside front cover.

To represent the composition of any particular atom we need to specify the number of protons (p), neutrons (n), and electrons (e) in the atom. We can do this with the symbolism

Representing the composition of an atom.

$$\text{number p + number n} \longrightarrow {}^{A}_{Z}X \longleftarrow \text{symbol of element} \atop \text{number p} \longrightarrow \qquad (2.1)$$

This symbolism indicates that the atom is of the element X, that it has an atomic number Z, and that its mass number is A. For example, one of the atoms shown below, that of the element carbon, has six protons and six neutrons in its nucleus and six electrons outside the nucleus.

$${}^{12}_{6}C \qquad {}^{14}_{7}N \qquad {}^{16}_{8}O \qquad {}^{24}_{12}Mg \qquad {}^{56}_{26}Fe \qquad {}^{238}_{92}U \qquad \text{and so on.}$$

Contrary to Dalton's assumption, we now know that atoms of an element do not necessarily all have the same mass. In 1912, J. J. Thomson measured the charge-to-mass ratios of positive ions formed in neon gas (the gas responsible for the red light given off by electrical neon signs). He found that about 91% of the atoms had one mass and that the remaining atoms were about 10% heavier. All neon atoms have ten protons in their nuclei, and most have ten neutrons as well. Some neon atoms, however, have 11 neutrons and some have 12. We can represent these three different types of neon atoms as

Thomson identified the isotopes ${}^{20}_{10}Ne$ and ${}^{22}_{10}Ne$ but not ${}^{21}_{10}Ne$.

$${}^{20}_{10}Ne \qquad {}^{21}_{10}Ne \qquad {}^{22}_{10}Ne$$

Two or more atoms having the *same* atomic number *(Z)* but *different* mass numbers *(A)* are called **isotopes.** Of all Ne atoms on earth, 90.9% are ${}^{20}_{10}Ne$; the percentages of ${}^{21}_{10}Ne$ and ${}^{22}_{10}Ne$ are 0.3% and 8.8%, respectively. These percentages—90.9%, 0.3%, 8.8%—are called the **percent natural abundances** of the three neon isotopes. Sometimes the mass numbers of isotopes are incorporated into the names of elements, such as *neon-20* (read this as "neon twenty"). Some elements consist of just a single type of atom and therefore do not have isotopes.* Aluminum, for example, consists only of Al-27 atoms.

Percent natural abundances of isotopes are given on a *number* basis, not a mass basis. Thus, 909 out of every 1000 neon atoms are neon-20 atoms.

If an atom either loses or gains electrons, it acquires a net electric charge; it becomes an **ion.** The species ${}^{20}_{10}Ne^{+}$ and ${}^{22}_{10}Ne^{2+}$ are ions. The first one has ten protons, ten neutrons, and *nine* electrons; the second, ten protons, twelve neutrons, and *eight* electrons. To establish the numbers of protons, neutrons, and electrons in an ion, remember that an atom becomes a positive ion only by *losing* electrons or a negative ion only by *gaining* electrons. The number of protons never changes. The charge on an ion is equal to the number of protons *minus* the number of electrons. That is,

Representing the composition of an ion.

$$\text{number p + number n} \longrightarrow {}^{A}_{Z}X^{\pm ?} \longleftarrow \text{number p − number e} \atop \text{number p} \longrightarrow \qquad (2.2)$$

*A general term that is used to describe an atom with a particular atomic number and mass number is **nuclide.** If an element consists of two or more different nuclides, we can speak of these as being **isotopes.** If there is only a single nuclide of the element, it is improper to speak of this as being an isotope, although this is often done. After all, you do not speak of someone as being a twin unless he or she actually has a twin brother or sister.

Example 2-2

Relating the numbers of protons, neutrons, and electrons in atoms and ions to the symbolism $^A_Z X$. (a) Indicate the numbers of protons, neutrons, and electrons in $^{35}_{17}Cl$; and (b) write an appropriate symbol for the species consisting of 29 protons, 34 neutrons, and 27 electrons.

Solution. In using the symbolism $^A_Z X$ pay particular attention to whether the species is a neutral atom or an ion. If it is a neutral atom, the number p = number e = Z (the atomic number). If the species is an ion, determine whether the number of electrons is smaller (positive ion) or larger (negative ion) than the number of protons. Whether the species is an atom or ion, the number of neutrons is equal to $A - Z$.

(a) $^{35}_{17}Cl$: $Z = 17$, $A = 35$, a neutral atom.

number p = 17; number e = 17; number n = $A - Z$ = 35 − 17 = 18.

(b) The element with $Z = 29$ is copper (Cu). The mass number A = number p + number n = 29 + 34 = 63. Because the species has only 27 electrons, it must be a species with a net charge = number p − number e = 29 − 27 = +2. It should be represented as $^{63}_{29}Cu^{2+}$.

SIMILAR EXAMPLES: Exercises 6, 7, 9.

2-6 Atomic Weights

By international agreement a single atom of $^{12}_6C$ is *arbitrarily* assigned a mass of **12.00000 u.** Then, the mass of any other atom, relative to the value of 12.00000 assigned to carbon-12, can be established with a **mass spectrometer.** In this device we separate a beam of gaseous ions into components of differing mass by passing the beam through electric and magnetic fields. The separated components are focused on a measuring instrument, where their presence is detected and their

This definition also establishes that one atomic mass unit (1 u) is *exactly* $\frac{1}{12}$ the mass of a carbon-12 atom.

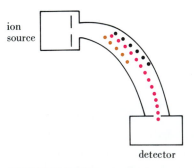

ion source

detector

FIGURE 2-12
A mass spectrometer.

Positive ions are produced in the ion source, for example, by bombardment of gaseous atoms with electrons. The ions are sent through an electric field where they are accelerated to some particular velocity. The beam of ions enters a curved chamber through a narrow opening or slit. A magnetic field is imposed perpendicular to the beam of ions (perpendicular to the page). The ions are deflected into a circular path by the magnetic field. For a given ion velocity, only ions with a particular mass-to-charge ratio will pass through the exit slit into the ion detector. Ions with a different mass-to-charge ratio will strike the chamber walls and be neutralized. By varying the velocity of the ion beam, ions having a large range of mass-to-charge ratios can be separated and detected.

FIGURE 2-13

Mass spectrum for mercury.

Separation of gaseous mercury into its seven naturally occurring isotopes is represented here. The signal from the ion detector (see Figure 2-12) has been converted to a scale of relative numbers of atoms. The percent natural abundances of the mercury isotopes are

$^{196}_{80}$Hg, 0.146% $\qquad$ $^{198}_{80}$Hg, 10.02%

$^{199}_{80}$Hg, 16.84% $\qquad$ $^{200}_{80}$Hg, 23.13%

$^{201}_{80}$Hg, 13.22% $\qquad$ $^{202}_{80}$Hg, 29.80%

$^{204}_{80}$Hg, 6.85%

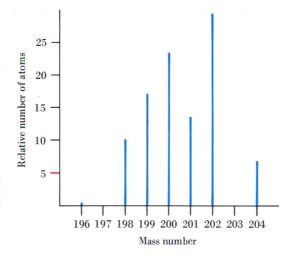

amounts recorded. The principle of mass spectrometry is illustrated in Figure 2-12, and a typical mass spectrum is pictured in Figure 2-13.

The relative masses of individual atoms, except for carbon-12, are never *exact* whole numbers, but they are very close in value to the corresponding mass numbers. This means, for example, that we should expect the relative mass of O-16 to be very nearly 16, and in Example 2-3 we see that it is.

Example 2-3

Establishing the relative masses of atoms. With mass spectral data we determine that the ratio of the mass of $^{16}_{8}$O to $^{12}_{6}$C is 1.3329. What is the mass of the $^{16}_{8}$O atom relative to a value of 12.00000 u assigned to the carbon-12 atom?

Solution. The ratio of the masses is $^{16}_{8}$O/$^{12}_{6}$C = 1.3329. The mass of the $^{16}_{8}$O atom is 1.3329 times the mass of $^{12}_{6}$C.

mass of $^{16}_{8}$O = 1.3329 × 12.00000 u = 15.9948 u

SIMILAR EXAMPLES: Exercises 10, 11.

The atomic weight of carbon is 12.011, but no atom of carbon has this relative mass; 12.011 is a weighted average.

In a table of atomic weights,* the value listed for carbon is 12.011, yet our atomic weight standard is 12.00000. How can this be? The atomic weight standard is based on a sample of carbon containing *only* atoms of carbon-12, whereas *naturally occurring* carbon contains some carbon-13 atoms as well. The existence of these two isotopes causes the *observed* atomic weight to be greater than 12. Tabulated atomic weights of the elements are *weighted averages* for *mixtures* of isotopes in their naturally occurring abundances.

The notion that the atomic weight of an element is a weighted average based on the naturally occurring mixture of isotopes can be expressed through the following general equation.

Expression relating atomic weight to isotopic masses and abundances.

$$
\begin{array}{l} \text{at. wt.} \\ \text{of an} \\ \text{element} \end{array} = \left(\begin{array}{l} \text{fractional} \\ \text{abundance of} \\ \text{isotope 1} \end{array} \times \begin{array}{l} \text{mass of} \\ \text{isotope} \\ 1 \end{array} \right) + \left(\begin{array}{l} \text{fractional} \\ \text{abundance of} \\ \text{isotope 2} \end{array} \times \begin{array}{l} \text{mass of} \\ \text{isotope} \\ 2 \end{array} \right) + \cdots \qquad (2.3)
$$

Equation (2.3) can be solved for any term if all the other values are known. In Example 2-4 we use it to determine the atomic weight of carbon, and in Example 2-5 we use it to establish the mass of one of the isotopes of the element bromine.

*From the discussion of mass and weight in Section 1-5, it is clear that the term **atomic mass** is more appropriate than atomic weight. However, atomic weight has become so ingrained in the vocabulary of chemists that its use is likely to continue for some time. Several terms that are used to describe the masses of atoms are compared in Table 3-1.

Are You Wondering:

Why some atomic weights (e.g., F = 18.9984032) are stated so much more precisely than others (e.g., Kr = 83.80)?

There is *one* naturally occurring type of fluorine atom: fluorine-19. Determining the atomic weight of fluorine means establishing the mass of this type of atom as precisely as possible. Krypton has *six* naturally occurring isotopes, and because the percentage distribution of these isotopes can vary slightly from one sample to another, the weighted average atomic weight of krypton cannot be stated with high precision.

Example 2-4

Calculating a weighted average atomic weight. The mass spectrum of carbon shows that 98.892% of carbon atoms are C-12 with a mass of 12.00000 u and 1.108% are C-13 with a mass of 13.00335 u. Calculate the atomic weight of naturally occurring carbon.

Solution. The *simple average* of the mass of C-12 and C-13 is (12.00000 + 13.00335)/2 = 12.50168. However, since a sample of carbon contains many more C-12 atoms than C-13, we must weigh the contribution of C-12 to the average atomic weight more heavily than that of C-13. We need to calculate a *weighted average*, and this is what equation (2.3) permits us to do. In the setup that follows we calculate the contribution of each isotope separately and then add these contributions together.

$$
\begin{array}{ccc}
\text{contribution} & \text{fraction of} & \text{mass of} \\
\text{to at. wt.} = \text{all C atoms} & \times\ \text{C-12} & = 0.98892 \times 12.00000 = 11.867 \\
\text{by C-12} & \text{that are C-12} & \text{atom}
\end{array}
$$

$$
\begin{array}{ccc}
\text{contribution} & \text{fraction of} & \text{mass of} \\
\text{to at. wt.} = \text{all C atoms} & \times\ \text{C-13} & = 0.01108 \times 13.00335 = 0.1441 \\
\text{by C-13} & \text{that are C-13} & \text{atom}
\end{array}
$$

$$
\begin{array}{l}
\text{atomic weight} \\
\text{of naturally} = (\text{contribution by C-12}) + (\text{contribution by C-13}) \\
\text{occurring carbon} \\
\qquad\qquad = \qquad 11.867 \qquad + \qquad 0.1441 \\
\qquad\qquad = 12.011
\end{array}
$$

SIMILAR EXAMPLES: Exercises 12, 44, 45.

Are You Wondering:

If you can assume that the most abundant isotopes of an element are those having mass numbers closest to the tabulated atomic weight?

This notion works for carbon (at. wt. 12.011): The most abundant isotope is carbon-12, and the second most abundant is carbon-13. For chlorine (at. wt. 35.453) the most abundant isotope (75.77%) is chlorine-35, but the second most abundant isotope (24.23%) is chlorine-37, not chlorine-36. In fact, chlorine-36 does not occur naturally at all. We consider factors that determine which atomic nuclei are stable and which are unstable (radioactive) in Chapter 26. For now, do not attempt to speculate on which isotopes occur naturally.

Example 2-5

Relating the masses and natural abundances of isotopes to the atomic weight of an element. Bromine has *two* naturally occurring isotopes. One of them, Br-79, has a mass of 78.9183 u and an abundance of 50.54%. What must be the mass and percent natural abundance of the other, Br-81?

Solution. It must always be true that *the percents natural abundance of all the isotopes of an element total 100.00%.* Thus the percent natural abundance of Br-81 is $100.00 - 50.54 = 49.46\%$. And, as required by equation (2.3),

$$\text{at. wt.} = \begin{pmatrix} \text{fraction of atoms} \\ \text{that are Br-79} \\ \times \text{ mass of Br-79} \end{pmatrix} + \begin{pmatrix} \text{fraction of atoms} \\ \text{that are Br-81} \\ \times \text{ mass of Br-81} \end{pmatrix}$$

Obtain the average atomic weight from the table on the inside front cover, and change percent abundances to fractional abundances. Solve for the unknown, which is mass of Br-81.

$$79.904 = (0.5054 \times 78.9183) + (0.4946 \times \text{mass of Br-81})$$
$$= 39.89 + (0.4946 \times \text{mass of Br-81})$$

$$\text{mass of Br-81} = \frac{79.904 - 39.89}{0.4946} = \frac{40.01}{0.4946} = 80.89$$

To four significant figures, the percent natural abundance and the mass of the isotope Br-81 are 49.46% and 80.89 u.

SIMILAR EXAMPLES: Exercises 41, 42, 43, 71.

CHEMISTRY EVOLVING

Jöns Berzelius (1779–1848)— a giant among chemists. Berzelius determined atomic weights, published an annual review of chemistry, and corresponded with most of the leading chemists of his day. [Culver Pictures, Inc.]

The Atomic Weight Scale

John Dalton assigned hydrogen an atomic weight of 1, and attempted to establish other atomic weights by comparison. He was not very successful, however, because he often assumed incorrect formulas for compounds. Jöns Berzelius (1779–1848) was a firm believer in Dalton's atomic theory and did countless chemical analyses to set up a table of relative atomic weights, most of which agree fairly well with present values. Berzelius took as his atomic weight standard oxygen, to which he assigned the value 16, and this value was used uniformly by scientists for many years.

With the discovery of the isotopes O-17 and O-18 in the late 1920s a dilemma arose—what to use now as the atomic weight standard. Physicists opted for the mass of the *pure* isotope O-16 and assigned it an atomic weight of exactly 16. Chemists stuck to the naturally occurring mixture of the oxygen isotopes and assigned that mixture a value of exactly 16. The two atomic weight scales differed by a factor of 1.000275.

This difference in atomic weight scales proved to be a continuing annoyance and in the late 1950s an international agreement was reached on a new atomic weight scale. Physicists required that the scale be based on a *pure* isotope having a mass number that is a multiple of four. Chemists required that their old atomic weights change by no more than 1 part in 10,000 in the changeover. Neither of the existing scales could meet these requirements and so agreement was reached on adopting C-12. With the adoption of the new atomic weight standard, the atomic weight of oxygen became 15.9994, a change of 0.6 part in 16,000 or 0.4 part in 10,000. ∎∎∎

2-7   The Avogadro Constant and the Concept of the Mole

Fundamental relationships among chemical quantities involve *numbers* of atoms, ions, or molecules. Yet, we cannot count atoms in the usual sense. We must resort to some other measurement, usually mass, that is related to numbers of atoms. And for this we need a relationship between the *measured* mass of an element and some *known* but *uncountable* number of atoms contained in that mass. Consider a practical example of mass substituting for a desired number of items: If you want to nail down new floorboards on the deck of your mountain cabin, you need a certain number of nails. However, you do not attempt to count out the number of nails you need—you buy them by the pound.

The atomic weight of an element can be established by comparing the mass of a large number of its atoms with an *equal* number of atoms of the atomic weight standard, carbon-12. The number that is used for this purpose is the number of atoms present in exactly 12.00000 g of C-12. This number, called the **Avogadro constant, N_A,** * has the value 6.02214×10^{23} mol^{-1}. The unit of the Avogadro constant (per mole) requires us to define an amount of substance called a **mole** (abbreviated **mol**).

Definition of a mole.

> A mole of substance is an amount of substance that contains the same number of elementary units as there are C-12 atoms in 12.00000 g C-12.   (2.4)

If a substance contains atoms of only a single isotope, we may write

1 mol *C-12* consists of 6.02214×10^{23} C-12 atoms and weighs 12.00000 g;
1 mol *O-16* consists of 6.02214×10^{23} O-16 atoms and weighs 15.9948 g;
and so on.

Most elements are composed of mixtures of two or more isotopes. The atoms to be "counted out" to yield one mole are not all of the same mass. They must be taken in the proportions in which they occur naturally. Thus, in 1 mol of carbon most of the atoms are carbon-12 but some are carbon-13. In 1 mol of oxygen most of the atoms are oxygen-16 but some are O-17 and some, O-18. As a result,

1 mol of *carbon* consists of 6.02214×10^{23} C atoms and weighs 12.011 g;
1 mol of *oxygen* consists of 6.02214×10^{23} O atoms and weighs 15.9994 g;
and so on.

Note that molar mass has the unit g/mol.

The mass of one mole of atoms, called the **molar mass, $\mathcal{M}$,** is easily obtained from a table of atomic weights, e.g., 6.941 g Li/mol Li. Figure 2-14 illustrates the meaning of 1 mol of atoms.

Thinking About One Mole. Avogadro's constant (6.02214×10^{23}) is an enormously large number and practically inconceivable in terms of ordinary experience. Suppose that the piles of spheres pictured in Figure 2-14 were garden peas instead of atoms. If the typical pea had a volume of about 0.1 cm^3, the required pile to constitute "one mole of peas" would cover the United States to a depth of about 6 km (4 mi). Or imagine that a mole of objects (perhaps grains of wheat) could be counted at the rate of 100 per minute. A given individual might be able to count out about 4 billion objects in a lifetime, but if all the people currently on earth were to spend their lives counting they could still not count out one mole. In fact, if all the

---

*The Avogadro constant is also called **Avogadro's number.** Although Avogadro recognized the significance of this numerical relationship (see page 166), he did not evaluate this number himself. Two methods of evaluating the Avogadro constant are presented in Chapter 3 (Exercises 88 and 89). Another method is discussed in Section 12-12.

FIGURE 2-14

An attempt to picture one mole of atoms.

If atoms could be piled up in the manner suggested here, it would take an enormously large pile to contain one mole of atoms.
(a) There is but a single naturally occurring nuclide of fluorine, so all the atoms are shown to be the same.
(b) In chlorine, 75.77% of the atoms are Cl-35 and the remainder are Cl-37.
(c) Magnesium has a principal isotope, Mg-24, and two minor ones, Mg-25 (10.00%) and Mg-26 (11.01%).
(d) Lead has four naturally occurring isotopes: Pb-204 (1.4%), Pb-206 (24.1%), Pb-207 (22.1%), and Pb-208 (52.4%).

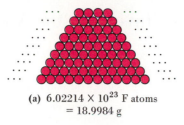

(a) 6.02214×10^{23} F atoms
= 18.9984 g

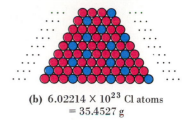

(b) 6.02214×10^{23} Cl atoms
= 35.4527 g

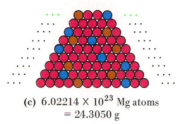

(c) 6.02214×10^{23} Mg atoms
= 24.3050 g

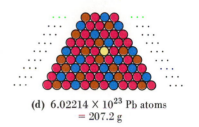

(d) 6.02214×10^{23} Pb atoms
= 207.2 g

people who have ever lived on earth were to have spent their lifetimes counting, the total would still be far less than one mole. (And one mole of grains of wheat is far more wheat than has been produced in human history.) Turning to a much more efficient counting method, a modern supercomputer is capable of counting about one billion per second. Even the supercomputer would take about 20 million years to count one mole!

The mole is clearly not a useful unit for measuring ordinary objects, but when this inconceivably large number is used in conjunction with inconceivably small objects such as atoms and molecules the result is quantities of materials that are easily within our grasp.

One mole of atoms. The watch glasses contain one mole of copper atoms (left) and one mole of sulfur atoms (right). The beaker contains one mole of liquid mercury, and the balloon contains one mole of helium gas. [Carey B. Van Loon]

2-8   Calculations Involving the Mole Concept

At various points in the text you will find that you must deal with the number of moles or the total number of atoms, ions, or molecules in a sample. This section considers the essential ideas that you will need in calculations involving these quantities.

Example 2-6

Determining the **total number** *of atoms if the number of moles of atoms is known.* How many Fe atoms are present in a sample containing 4.24×10^{-6} mol Fe?

Solution. The conversion factor we need is based on the fact that

1 mol Fe $\approx 6.022 \times 10^{23}$ Fe atoms.

$$\text{no. Fe atoms} = 4.24 \times 10^{-6} \text{ mol Fe} \times \frac{6.022 \times 10^{23} \text{ Fe atoms}}{1 \text{ mol Fe}}$$

$$= 2.55 \times 10^{18} \text{ Fe atoms}$$

SIMILAR EXAMPLES: Exercises 13, 47.

From this point on we will not routinely show the cancellation of units in calculations. However, you should always assure yourself that the proper cancellation will occur.

Example 2-7

Determining the number of **moles** *of atoms if the total number of atoms is known.* How many moles of Mg are present in a sample containing 1.00×10^{22} Mg atoms?

Solution. Again, the conversion factor we need is based on the Avogadro constant, but here the factor is *inverted* from the form used in Example 2-6.

$$\text{no. mol Mg} = 1.00 \times 10^{22} \text{ Mg atoms} \times \frac{1 \text{ mol Mg}}{6.022 \times 10^{23} \text{ Mg atoms}}$$

$$= 0.0166 \text{ mol Mg}$$

SIMILAR EXAMPLE: Exercise 14a.

The situations we described in Examples 2-6 and 2-7 were *hypothetical*. We have no way of directly counting atoms. We must always measure some other property. Molar mass is generally the key to these measurements, as illustrated in Examples 2-8 and 2-9.

Are You Wondering:

When to multiply and when to divide by the Avogadro constant, N_A?

Here is an idea to help ensure that you use N_A properly in a calculation: Any time you are required to find a total number of atoms, expect the answer to be a *very large* number; and, of course, it can *never* be smaller than 1. The number of *moles* of atoms, on the other hand, is generally a number of more modest size and will often be smaller than 1.

Example 2-8

Relating mass to the number of moles of an element. What is the mass of 1.00×10^{22} Mg atoms?

Molar mass is the conversion factor between the number of moles of an element and its mass in grams.

Solution. We begin as in Example 2-7. Then, the molar mass of Mg— 24.30 g Mg/1 mol Mg—allows us to convert from moles of Mg to a mass in grams.

$$\text{no. g Mg} = 1.00 \times 10^{22} \text{ Mg atoms} \times \frac{1 \text{ mol Mg}}{6.022 \times 10^{23} \text{ Mg atoms}} \times \frac{24.30 \text{ g Mg}}{1 \text{ mol Mg}}$$

$$= 0.404 \text{ g Mg}$$

Figure 2-15 shows how we could measure this quantity of magnesium.

SIMILAR EXAMPLES: Exercises 14, 49.

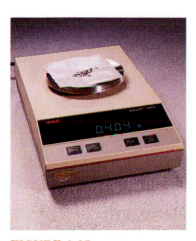

FIGURE 2-15
Measurement of 1.00×10^{22} Mg atoms (0.0166 mol Mg)—Example 2-8 illustrated.

The balance is set to zero (tared) when just the weighing paper is present. The sample of magnesium weighs 0.404 g. [Carey B. Van Loon]

Example 2-9

Combining several factors in a calculation—molar mass, the Avogadro constant, percent composition. Potassium-40 is one of the few naturally occurring radioactive isotopes of elements of low atomic number. Its percent natural abundance is 0.012%. How many K-40 atoms do you ingest by drinking one cup of whole milk containing 370. mg K?

Solution. Use a three-step approach. First, convert the mass of K to number of moles of K. For this use molar mass in the *inverse* manner to Example 2-8.

$$\text{no. mol K} = 370. \text{ mg K} \times \frac{1.00 \text{ g K}}{1000 \text{ mg K}} \times \frac{1 \text{ mol K}}{39.10 \text{ g K}} = 9.46 \times 10^{-3} \text{ mol K}$$

Then convert the number of moles of K to number of K atoms.

$$\text{no. K atoms} = 9.46 \times 10^{-3} \text{ mol K} \times \frac{6.022 \times 10^{23} \text{ K atoms}}{1 \text{ mol K}}$$

$$= 5.70 \times 10^{21} \text{ K atoms}$$

Finally, use the percent natural abundance of K-40 to formulate a factor to convert from number of K atoms to number of K-40 atoms.

$$\text{no. K-40 atoms} = 5.70 \times 10^{21} \text{ K atoms} \times \frac{0.012 \text{ K-40 atoms}}{100 \text{ K atoms}}$$

$$= 6.84 \times 10^{17} \text{ K-40 atoms}$$

SIMILAR EXAMPLES: Exercises 15, 75.

Summary

Dalton's atomic theory was based on the law of conservation of mass and the law of constant composition (definite proportions). It was successful in predicting the law of multiple proportions.

More fruitful in establishing the atomic nature of matter, however, were the efforts of nineteenth century physicists. Cathode ray research led to the discovery of the electron—a fundamental particle of all matter and the basic unit of negative electrical charge. The discovery of positive rays, x rays, and radioactivity were also a consequence of cathode ray research. Experiments with positive rays led to the characterization of isotopes and the development of modern mass spectrometry. Studies on the scattering of α particles by thin metal foils led to the concept of the nuclear atom. A more complete description of the atomic nucleus was made possible by the later discovery of the proton and the neutron.

By assigning an atomic mass of 12.00000 u to a C-12 atom, the masses of other atoms can be determined by mass spectrometry. From the masses of the different isotopes of an element and their percent natural abundances, the atomic weight of an element can be determined.

The Avogadro constant, $N_A = 6.02214 \times 10^{23}$ mol^{-1}, represents the number of C-12 atoms in 12.00000 g of C-12; or, more generally, it is the number of elementary units present in 1 mol of a substance. The mass of 1 mol of atoms of an element is called its molar mass. Molar mass and the Avogadro constant are used in a variety of calculations involving the mass, amount (in moles), or number of atoms in a sample of an element.

Summarizing Example

Steel is a metallic mixture having iron as its principal element. A key minor element in steel is carbon. The object pictured in Figure 2-16 is a stainless steel ball bearing. It has a radius of 6.35 mm and is made of steel with 0.25% carbon. The steel has a density of 7.75 g/cm^3.

1. Determine the volume of the ball bearing.

Solution. Use the formula for the volume of a sphere.

$$V = \tfrac{4}{3}\pi r^3 = \tfrac{4}{3}\pi \left(6.35 \text{ mm} \times \frac{1 \text{ cm}}{10 \text{ mm}}\right)^3 = \tfrac{4}{3}(3.14)(0.635)^3 \text{ cm}^3 = \mathbf{1.07 \text{ cm}^3}$$

(This example is similar to Example 1-7.)

2. What is the mass of carbon in the ball bearing?

Solution. Use density as the conversion factor between volume and mass, followed by percent composition.

$$\text{no. g carbon} = 1.07 \text{ cm}^3 \times \frac{7.75 \text{ g steel}}{1.00 \text{ cm}^3 \text{ steel}} \times \frac{0.25 \text{ g carbon}}{100.0 \text{ g steel}}$$

$$= \mathbf{0.021 \text{ g carbon}}$$

(This example is similar to Examples 1-5 and 1-9.)

3. What is the total number of carbon atoms in the ball bearing?

Solution. The mass of carbon from part **2** must be converted to mol C and then, through the Avogadro constant, to number of C atoms.

$$\text{no. C atoms} = 0.021 \text{ g C} \times \frac{1 \text{ mol C}}{12.0 \text{ g C}} \times \frac{6.02 \times 10^{23} \text{ C atoms}}{1 \text{ mol C}}$$

$$= \mathbf{1.1 \times 10^{21} \text{ C atoms}}$$

(This example is similar to Example 2-9.)

FIGURE 2-16
A stainless steel ball bearing.
[Carey B. Van Loon]

4. Given that the natural abundance of C-13 is 1.108%, how many C-13 atoms are present in the ball bearing?

Solution. Use the result of part **3** and percent natural abundance as a conversion factor.

$$\text{no. C-13 atoms} = 1.1 \times 10^{21} \text{ C atoms} \times \frac{1.11 \text{ C-13 atoms}}{100 \text{ C atoms}}$$

$$= 1.2 \times 10^{19} \text{ C-13 atoms}$$

(This example is similar to Example 2-9.)

5. The mass of the isotope C-13 is 13.00335 u. What is the mass of C-13 present in the ball bearing?

Solution. First convert the number of C-13 atoms found in part **4** to the number of moles of C-13. From the definition of the mole given in expression (2.4), the mass of one mole of C-13 is 13.00335 g. This provides us with a factor to convert from moles to grams.

$$\text{no. g C-13} = 1.2 \times 10^{19} \text{ C-13 atoms} \times \frac{1 \text{ mol C-13}}{6.02 \times 10^{23} \text{ atoms}} \times \frac{13.0 \text{ g C-13}}{1 \text{ mol C-13}}$$

$$= 2.6 \times 10^{-4} \text{ g C-13}$$

(This example is similar to Examples 2-7 and 2-8.)

Key Terms

alpha (α) particle (2-2)
atomic mass unit (u) (2-4)
atomic number, Z (2-4)
atomic weight (mass) (2-6)
Avogadro constant (2-7)
beta (β) particle (2-2)
cathode rays (2-2)
chemical symbols (2-5)
electron (2-2)

gamma (γ) ray (2-2)
ion (2-2)
isotope (2-5)
law of conservation of mass (2-1)
law of constant composition (2-1)
 (definite proportions)
law of multiple proportions (2-1)
mass number, A (2-4)
mass spectrometer (2-6)

molar mass (2-7)
mole (2-7)
neutron (2-3)
nuclide (2-5)
percent natural abundance (2-5)
positive (canal) ray (2-2)
proton (2-3)
radioactivity (2-2)
transmutation (2-2)

Highlighted Expressions

Representing the composition of an atom (2.1)
Representing the composition of an ion (2.2)

Expression relating atomic weight to isotopic masses and abundances (2.3)
Definition of a mole (2.4)

Review Problems

1. A 0.255-g sample of magnesium was allowed to react with oxygen, forming 0.423 g magnesium oxide. What mass of oxygen was consumed in the reaction?

2. A 7.12-g sample of zinc dust was mixed with 1.80 g of sulfur and the mixture was heated. All the sulfur was used up and 5.47 g of zinc sulfide was the only product. What mass of zinc remained unreacted?

3. Samples of pure carbon weighing 1.48, 2.06, and 3.17 g,

were burned in an excess of air. The masses of carbon dioxide obtained (the sole product in each case) were 5.42, 7.55, and 11.62 g, respectively.

(a) Do these data establish that carbon dioxide has a fixed composition?

(b) What is the composition of carbon dioxide, expressed in % C and % O, by mass?

4. Use modern atomic weights and the method outlined in

Figure 2-3 to show that the law of multiple proportions applies in each of the following cases: (a) SO_2 and SO_3; (b) H_2O and H_2O_2; (c) PCl_3 and PCl_5.

5. Use data from Table 2-1 to determine the *net* charge, in coulombs, associated with 3.16×10^{20} (a) hydrogen atoms; (b) sodium ions (Na^+); (c) sulfide ions (S^{2-}).

6. Complete the following table. What minimum amount of information is required to characterize completely an atom or ion?

Name	Symbol	Number protons	Number electrons	Number neutrons	Mass number
sodium	$^{23}_{11}Na$	11	11	12	23
silicon	—	—	—	14	—
—	—	37	—	—	85
—	^{40}K	—	—	—	—
—	—	—	33	42	—
—	$^{20}Ne^{2+}$	—	—	—	—
—	—	—	—	—	80
—	—	—	—	126	—

7. Arrange the following species in order of increasing (a) number of electrons; (b) number of neutrons; (c) mass.

$^{112}_{50}Sn, \quad ^{40}_{18}Ar, \quad ^{122}_{52}Te, \quad ^{59}_{29}Cu, \quad ^{120}_{48}Cd, \quad ^{58}_{27}Co, \quad ^{39}_{19}K$

8. For the nuclide $^{133}_{55}Cs$, express the percentage, *by number*, of the fundamental particles in the nucleus that are neutrons.

9. An isotope with mass number 63 has five more neutrons than protons. This is an isotope of what element?

10. The following data on atomic masses are given in a handbook. What is the ratio of each of these masses to that of $^{12}_{6}C$? (a) $^{27}_{13}Al$, 26.98153 u; (b) $^{40}_{20}Ca$, 39.96259 u; (c) $^{197}_{79}Au$, 196.9666 u.

11. The following ratios of masses were obtained with a mass spectrometer. (a) $^{19}_{9}F : ^{12}_{6}C = 1.5832$; (b) $^{35}_{17}Cl : ^{19}_{9}F = 1.8406$; (c) $^{81}_{35}Br : ^{35}_{17}Cl = 2.3140$. Determine the mass of an $^{81}_{35}Br$ atom in atomic mass units. [*Hint:* What is the mass of a C-12 atom?]

12. In naturally occurring uranium, 99.27% of the atoms are $^{238}_{92}U$ with mass 238.05 u; 0.72%, $^{235}_{92}U$ with mass 235.04 u; and 0.006%, $^{234}_{91}U$ with mass 234.04 u. Calculate the atomic weight of naturally occurring uranium.

13. What is the total number of atoms in each of the following samples? (a) 34.5 mol Mg; (b) 0.0123 mol He; (c) 6.1×10^{-12} mol Np.

14. Calculate the quantities indicated.
(a) The number of moles represented by 9.32×10^{25} Zn atoms
(b) The mass, in grams, of 3.27 mol Ar
(c) The mass, in mg, of a sample containing 3.07×10^{20} Ag atoms
(d) The number of atoms in 46.5 cm^3 of Fe (density of Fe = 7.86 g/cm^3)

15. How many Pb-204 atoms are present in a piece of lead shot weighing 1.57 g? The percent natural abundance of Pb-204 is 1.4%.

Exercises

Law of conservation of mass

16. When a strip of magnesium metal is burned in air (recall Figure 2-1), it produces a white powder that weighs more than the original metal. When a strip of magnesium is burned in a photoflash bulb, the bulb weighs the same before and after it is flashed. Explain the difference in these observations.

17. Within the limits of experimental error, show that the law of conservation of mass was obeyed in the following experiment: 10.00 g of calcium carbonate (found in limestone) was dissolved in 100.0 cm^3 of hydrochloric acid (density = 1.148 g/cm^3). The products were 120.40 g of solution (a mixture of hydrochloric acid and calcium chloride) and 2.22 L of carbon dioxide gas (density = 1.9769 g/L).

Law of constant composition

18. In one experiment 2.18 g of sodium was allowed to react with 16.12 g of chlorine. All the sodium was used up, and 5.54 g of sodium chloride (salt) was produced. In a second experiment 2.10 g of chlorine was allowed to react with 10.00 g of sodium. All the chlorine was used up, and 3.46 g of sodium chloride was produced. Show that these results are consistent with the law of constant composition.

19. The following data were obtained when magnesium (first value) was heated in air to produce magnesium oxide (second value): 0.62 g, 1.02 g; 0.48 g, 0.79 g; 0.36 g, 0.60 g. Show that these data are consistent with the law of constant composition.

20. Use the results of Exercise 19 to establish
(a) the mass of magnesium oxide that would be obtained from 0.26 g of magnesium;
(b) the mass of magnesium that must combine with oxygen to produce 0.56 g of magnesium oxide;
(c) the mass of oxygen that is required to convert 0.32 g of magnesium to magnesium oxide.

Dalton's atomic theory

21. A compound unknown in Dalton's time was hydrogen peroxide (used as a bleach). It consists of 94% oxygen and 6% hydrogen, by mass. Had Dalton known of this compound, what formula do you think he would have assigned to it, having assigned the formula OH to water?

22. Estimate the atomic weight of magnesium by using 16.0 for the atomic weight of oxygen, information presented in Exercise 19, and Dalton's idea that if there is but a single compound of two elements it should have the formula AB.

Law of multiple proportions

23. The formulas used by Dalton in establishing the law of multiple proportions were CH_2 for methane and CH for ethylene (see Figure 2-3). Show that the law is just as well established with modern formulas (methane CH_4 and ethylene C_2H_4) and modern atomic weights.

24. Dalton knew of three oxides of nitrogen. To one (nitrous gas) he assigned the formula NO. Using only Dalton's rule of adopting the *simplest formulas possible,* what formulas do you think he assigned to the other two? Show that the formulas for these three compounds are consistent with the law of multiple proportions.

25. There are two oxides of copper. One oxide has 20% oxygen, by mass. The second oxide has a *smaller* percent oxygen than the first. What is the probable percent oxygen in this second oxide?

Fundamental particles

26. Cite the evidence that most convincingly established that electrons are fundamental particles of all matter.

27. List several significant differences between cathode rays and positive (canal) rays, stressing the manner of production, electric charge, mass, and so on.

28. Why could not the same methods that had been used to characterize electrons be used to isolate and detect neutrons?

Fundamental charges and charge-to-mass ratios

29. These observations were made for a series of 10 oil drops in an experiment similar to Millikan's (see Figure 2-6). Drop 1 carried a charge of 1.28×10^{-18} C; drops 2 and 3 each carried $\frac{1}{2}$ the charge of drop 1; drop 4 carried $\frac{1}{4}$ the charge of drop 1; drop 5 had a charge four times that of drop 1; drops 6 and 7 had charges three times that of drop 1; drops 8 and 9 had charges twice that of drop 1; and drop 10 had the same charge as drop 1. Are these data consistent with the value of the electronic charge given in the text? Could Millikan have inferred the charge on the electron from this particular series of data? Explain.

30. It is now known that static electric charges are caused by the transfer of electrons.

(a) How many excess electrons are present on an object with a charge of -2.6×10^{-16} C?

(b) How many electrons are deficient from an object with a net charge of $+6.4 \times 10^{-14}$ C?

31. Use data from Table 2-1 to verify the following statements:

(a) The mass of electrons is about 1/2000 that of hydrogen atoms.

(b) The charge-to-mass ratio, e/m, for positive rays is considerably smaller than for the electron.

Atomic models

32. Use the atomic model of J. J. Thomson (see Figure 2-7) to draw pictures of the following gaseous atoms and ions. (a) Li; (b) C; (c) O^+; (d) F^-.

33. Represent each of the species given in Exercise 32 by the Rutherford model of the atom (see Figure 2-11). Describe the essential differences between the Thomson model of Exercise 32 and the Rutherford model.

Atomic number, mass number, and isotopes

34. Describe the significance of each term in the symbol $^A_Z X$.

35. For the atom $^{138}_{56}Ba$, with a mass of 137.9050 u, determine

(a) the numbers of protons, neutrons, and electrons in the atom;

(b) the ratio of the mass of this atom to that of an atom of $^{12}_6C$;

(c) the ratio of the mass of this atom to that of an atom of $^{16}_8O$ (refer to Example 2-3).

36. Explain why the symbols $^{35}_{17}Cl$ and ^{35}Cl actually convey the same information. Do the symbols $^{35}_{17}Cl$ and $_{17}Cl$ have the same meaning?

37. In each case, identify the element in question if [*Hint:* You will find it helpful to use algebra for some of these.]

(a) an atom has a *total* of 60 protons, neutrons, and electrons, with *equal* numbers of all three.

(b) the mass number of an atom is 234 and the atom has 60.0% more neutrons than protons.

(c) an ion with a 2+ charge has 10.0% more protons than electrons.

★(d) an ion with a mass number of 110 and a 2+ charge has 25.0% more neutrons than electrons.

Atomic mass units, atomic masses

38. What is the mass, in grams, corresponding to 1.000 u? [*Hint:* Refer to Table 2-1.]

39. What is the mass, in grams, of each of the following? [*Hint:* Use the result of Exercise 38.]

(a) 2.50×10^{18} atoms of bromine-79, with individual atoms having a mass of 78.9183 u.

(b) 2.50×10^{18} atoms of bromine-81, with individual atoms having a mass of 80.9163 u.

(c) 2.50×10^{18} atoms of a mixture of bromine-79 and bromine-81 in their naturally occurring abundances, 50.54% bromine-79 and 49.46% bromine-81.

Atomic weights

40. Which statement is probably true concerning the masses of *individual* copper atoms: that *all, some,* or *none* have a mass of 63.546? Explain.

41. There are *three* naturally occurring isotopes of magnesium. The masses and percent natural abundances of two of the isotopes are (24.98584 u, 10.13%) and 25.98259 u, 11.17%). Use the atomic weight listed for magnesium and determine for the *third* magnesium isotope (a) its percent natural abundance; (b) its mass in atomic mass units; (c) its mass number.

42. The two principal isotopes of lithium have masses of 6.01513 and 7.01601 u. The atomic weight of lithium is 6.941.

(a) Which of these two isotopes is the more abundant?

(b) What is the approximate ratio of atoms of the more abundant to the less abundant isotope, i.e., 2:1, 3:1, . . .?

★(c) Calculate the percent natural abundances of the two.

★**43.** The two naturally occurring isotopes of nitrogen have masses of 14.0031 and 15.0001 u, respectively. Use the atomic weight listed for nitrogen to determine the percentage of N-15 atoms in naturally occurring nitrogen.

Mass spectrometry

44. The masses of the naturally occurring mercury isotopes are Hg-196, 195.9658 u; Hg-198, 197.9668 u; Hg-199, 198.9683 u; Hg-200, 199.9683 u; Hg-201, 200.9703 u; Hg-202, 201.9706 u; and Hg-204, 203.9735 u. Use these data, together with data from Figure 2-13, to calculate the atomic weight of mercury.

45. Use the data suggested by the following mass spectrum to *estimate* the atomic weight of germanium. State two reasons why this result is only approximately correct.

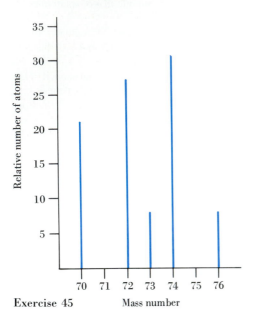

Exercise 45   Mass number

46. The three isotopes of hydrogen, H-1, H-2, and H-3, can all combine with chlorine to form simple diatomic molecules HCl. The percent natural abundances of the chlorine isotopes are Cl-35, 75.53%, and Cl-37, 24.47%. The percent natural abundances of H-2 and H-3 are 0.015% and <0.001%, respectively.

(a) How many HCl molecules of different mass are possible?

(b) What are the mass numbers of these different molecules (i.e., the sum of the mass numbers of the two atoms in the molecules)?

(c) Which is the most abundant of the possible HCl molecules? Which is the second most abundant?

(d) In the manner of Figure 2-13, sketch the mass spectrum you would expect to obtain for HCl molecules.

The Avogadro constant and the mole

47. How many Ag atoms are present in a piece of sterling silver jewelry weighing 38.7 g? Sterling silver contains 92.5% Ag, by mass.

48. Medical experts generally consider a lead level of 30 μg Pb per dL of blood to pose a significant health risk (1 dL = 0.1 L). Express this lead level (a) in the unit, mol Pb/L blood; (b) as the number of Pb atoms per cm^3 blood.

49. An alloy that melts at about the boiling point of water has Bi, Pb, and Sn atoms in the ratio 10:6:5, respectively. What is the mass of a sample of this alloy containing a total of one mole of atoms?

★**50.** How many Cu atoms are present in a 1.00-m length of 20-gauge copper wire? (A 20-gauge wire has a diameter of 0.03196 in.; density of Cu = 8.92 g/cm^3.)

★**51.** Monel metal is a corrosion resistant Ni–Cu alloy used in the electronics industry. The object pictured is made of a particular Monel metal having a density of 8.80 g/cm^3 and containing 0.022% Si, by mass. What is the total mass of Si-30 in this object? The percent natural abundance of Si-30 is 3.10%, and its isotopic mass is 29.97376 u. [Hint: Refer to the Summarizing Example.]

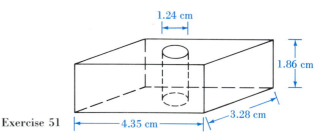

Exercise 51

Additional Exercises

52. When an iron object rusts its mass increases. When a match burns its mass decreases. Do these observations violate the law of conservation of mass? Explain.

53. When 3.06 g of hydrogen was allowed to react with an excess of oxygen, 27.35 g of water was obtained. In a second experiment a sample of water was decomposed by electrolysis, resulting in 1.45 g of hydrogen and 11.51 g of oxygen. Are these results consistent with the law of constant composition? Demonstrate why or why not.

54. In one experiment the burning of 0.312 g of sulfur produced 0.623 g of sulfur dioxide as the sole product. In a second experiment 0.842 g of sulfur dioxide was obtained. What mass of sulfur must have been burned in the second experiment?

55. 2.50 g of potassium perchlorate (used in fireworks and flares) is completely dissolved in 100.0 mL of water at 40 °C. When the solution is cooled to 20 °C, its volume is found to be 100.0 mL and some of the potassium perchlorate deposits from solution (crystallizes). Use the data in the sketch to

(a) Estimate, to two significant figures, the mass of potassium perchlorate that is deposited.

★(b) Why cannot this estimate be more precise?

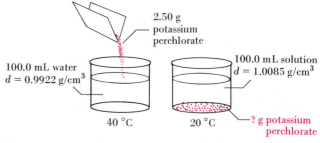

Exercise 55

56. Mercury and oxygen form two compounds. One contains 96.2% mercury, by mass, and the other, 92.6%. Show that these data conform to the law of multiple proportions. Speculate on the formulas of these two oxides.

57. Investigation of electrical discharge in gases led to the discovery of both negative and positive particles of matter. Explain why the negative particles *all* proved to be fundamental particles of matter but most of the positive particles did not.

58. Determine the approximate value of the charge-to-mass ratio, *e/m*, in coulombs/gram, for the ions $^{127}_{53}I^-$ and $^{32}_{16}S^{2-}$. Why are these values only approximate?

59. Arrange the following species in order of *increasing* absolute magnitude of charge-to-mass, *e/m*: proton, electron, neutron, α particle, the atom $^{40}_{18}Ar$, the ion $^{35}_{17}Cl^-$. Note whether any two of the species listed have identical *e/m* ratios. (The absolute magnitude refers to the value of the *e/m* ratio without regard to its sign.)

60. All of these radioactive isotopes have applications in medicine. Write their symbols in the form $^A_Z X$. **(a)** cobalt-60; **(b)** phosphorus-32; **(c)** iodine-131; **(d)** sulfur-35.

61. Given the following species: $^{24}Mg^{2+}$; ^{47}Cr; $^{59}Co^{2+}$; $^{35}Cl^-$; $^{124}Sn^{2+}$; ^{226}Th; ^{90}Sr. Which of these species

  (a) has equal numbers of neutrons and electrons?
  (b) has protons contributing more than 50% of the mass?
  (c) has a number of neutrons equal to the number of protons plus one-half the number of electrons?

62. Use 1×10^{-13} cm as the approximate diameter of the spherical nucleus of the H-1 atom, together with data from Table 2-1, to estimate the density of matter in a proton.

63. Osmium metal (used in catalysts) has about the highest density known: 22 g/cm³. What does a comparison of this value and the density of the proton estimated in Exercise 62 suggest about the amount of empty space in matter?

64. An isotope of silver has a mass that is 6.68374 times that of O-16. What is the mass (in u) of this isotope? What is the ratio of its mass to that of C-12? [*Hint:* Refer to Example 2-3.]

65. Prout (1815) advanced the hypothesis that all other atoms are built up of hydrogen atoms, suggesting that all elements should have integral atomic weights based on an atomic weight of 1 for hydrogen. This hypothesis appeared discredited by the discovery of atomic weights such as 24.3 for magnesium and 35.5 for chlorine. In terms of modern knowledge, explain why Prout's hypothesis is actually quite reasonable.

66. Determine the only possible 2+ ion for which *both* of the following are satisfied.

- The net ionic charge is *one-tenth* the nuclear charge.
- The number of neutrons is *four* more than the number of electrons.

★67. Determine the only possible isotope (X) for which the following conditions are *both* met.
  (1) The mass number of X is 2.50 times its atomic number.
  (2) The atomic number of X is equal to the mass number of the isotope Z of another element. The number of neutrons in Z is

    • equal to the number of electrons in the ion Sn^{2+}
    and *also*
    • 1.33 times the number of protons in Z.

68. The mass of a C-12 atom is taken to be exactly 12.00000 u. Are there likely to be any other atoms with an *exact* integral (whole number) mass, expressed in u? Explain.

69. The atomic weight of oxygen listed in this book is 15.9994. A textbook printed 30 years ago lists a value of 16.0000. How do you account for this discrepancy? Would you expect other atomic weights listed in the older text to be the same, generally higher, or generally lower than in this text? Explain.

70. Suppose we redefined the atomic weight scale by arbitrarily assigning to the naturally occurring *mixture* of chlorine isotopes an atomic weight of 35.00000.

  (a) What would be the atomic weights of helium, sodium, and iodine on this new atomic weight scale?

  ★(b) Why do these three elements have nearly integral (whole number) atomic weights based on C-12 but not based on naturally occurring chlorine?

★71. Silicon has one major isotope, Si-28 (27.97693 u, 92.21% natural abundance) and two minor ones, Si-29 (28.97649 u) and Si-30 (29.97376 u). What are the percent natural abundances of the two minor isotopes? Comment on the limitation of the precision of this calculation.

★72. From the densities of the lines in the mass spectrum of krypton gas, the following observations were made.
  (1) Somewhat more than 50% of the atoms were Kr-84.
  (2) The numbers of Kr-82 and Kr-83 atoms were essentially equal.
  (3) The number of Kr-86 atoms was 1.50 times greater than the number of Kr-82 atoms.
  (4) The number of Kr-80 atoms was 19.6% of the number of Kr-82 atoms.
  (5) The number of Kr-78 atoms was 3.0% of the number of Kr-82 atoms.
The isotopic masses are Kr-78, 77.9204 u; Kr-80, 79.9164 u; Kr-82, 81.9135 u; Kr-83, 82.9141 u; Kr-84, 83.9115 u; Kr-86, 85.9106 u. The atomic weight of Kr is 83.80. Use the data presented here to calculate the percent natural abundances of the six isotopes.

73. Refer to Figure 2-14 and determine the number of Mg-24 atoms in one mole of magnesium produced from naturally occurring sources.

74. Deuterium, H-2, is sometimes used to replace ordinary H-1 atoms in chemical studies. The percent natural abundance of H-2 is 0.015%. What mass of hydrogen gas would have to be processed to extract 1.00 g of pure H-2 atoms? The mass of the H-2 isotope is 2.0140 u.

75. During a severe air pollution episode the concentration of lead in air was observed to be 3.01 μg Pb/m³. How many Pb atoms would be present in a 0.500-L sample of this air (the approximate lung capacity of a human adult)?

76. A particular silver solder (used in the electronics industry to join together electrical components) is to be prepared containing the *atom* ratio: 5.00 Ag : 4.00 Cu : 1.00 Zn. What masses of the three metals must be melted together to prepare 1.00 kg of the solder?

★77. A low-melting Sn–Pb–Cd alloy called *eutectic alloy* is analyzed and the following results are obtained: The *mole* ratio of tin to lead is 2.73 : 1.00. The *mass* ratio of lead to cadmium is 1.78 : 1.00.

  (a) What is the percent composition of this alloy, by mass?

  (b) What is the mass of a sample of this alloy containing a *total* of 1.00 mol of atoms?

Self-Test Questions

For questions 78 through 89 select the single item that best completes each statement.

78. Of these assumptions or results of Dalton's atomic theory, the only one that remains essentially correct is
  (a) All atoms of an element are identical in mass.
  (b) Atoms are indivisible and indestructible.
  (c) Oxygen has an atomic weight of 7.
  (d) Atoms of elements combine in the ratios of small whole numbers to form compounds.

79. A 0.100-g sample of magnesium, when combined with oxygen, yields 0.167 g of magnesium oxide. A second magnesium sample weighing 0.150 g, when combined with oxygen, is expected to yield the following mass of magnesium oxide: (a) 0.167 g; (b) 0.25 g; (c) 0.267 g; (d) 2.5 g.

80. When 10.0 g zinc and 8.0 g sulfur are allowed to react, all the zinc is used up, 15.0 g zinc sulfide is formed, and some unreacted sulfur remains. The mass of *unreacted* sulfur is (a) 2.0 g; (b) 3.0 g; (c) 5.0 g; (d) impossible to determine from this information alone.

81. One oxide of rubidium has 0.187 g O per gram Rb. A possible O-to-Rb mass ratio for a second oxide of rubidium (atomic weights: $O = 16.0$; $Rb = 85.5$) is (a) $16:1$; (b) $16:85.5$; (c) $0.374:1$; (d) all of these.

82. Cathode rays (a) may be positively or negatively charged; (b) are a form of electromagnetic radiation; (c) have properties identical to β particles; (d) have masses that depend on the material that emits them.

83. The scattering of α particles by thin metal foils established that (a) the mass and positive charge of an atom are concentrated in a nucleus; (b) electrons are fundamental particles of all matter; (c) all electrons have the same charge; (d) atoms are electrically neutral.

84. The species that has the same number of electrons as $^{32}_{16}S$ is (a) $^{35}_{17}Cl^-$; (b) $^{34}_{16}S^+$; (c) $^{40}_{18}Ar^{2+}$; (d) $^{35}_{16}S^{2-}$.

85. All of the following masses are possible for an *individual* carbon atom except one. The impossible one is (a) 12.00000 u; (b) 12.01115 u; (c) 13.00335 u; (d) 14.00324 u.

86. There are *two* principal isotopes of indium (atomic weight = 114.82). One of these, $^{113}_{49}In$, has an atomic mass of 112.9043 u. The second isotope is most likely to be (a) $^{111}_{49}In$; (b) $^{112}_{49}In$; (c) $^{114}_{49}In$; (d) $^{115}_{49}In$.

87. The mass of the isotope $^{84}_{36}Kr$ is 83.9115 u. If the atomic weight scale were *redefined* so that $^{84}_{36}Kr = 84.00000$ u, *exactly,* the mass of $^{12}_{6}C$ would be (a) 11.9115 u; (b) 11.9874 u; (c) 12.0127 u; (d) 12.0885 u.

88. A 6.02×10^{23}-atom sample of magnesium from naturally occurring sources weighs (a) 24.3 g; (b) 24.0 g; (c) 6.02 g; (d) 1.00 g.

89. A 558.5 g sample of iron (Fe) (a) consists of 558.5 mol Fe; (b) contains 6.022×10^{23} atoms; (c) contains *10* times as many atoms as does 0.5200 g Cr; (d) contains *twice* as many atoms as does 60.06 g C.

90. In one experiment the reaction of 1.00 g of mercury and an excess of sulfur yields 1.16 g of a sulfide of mercury as the sole product. In a second experiment the same sulfide is produced in the reaction of 1.50 g of mercury and 1.00 g of sulfur.
  (a) How many grams of the sulfide of mercury are produced in the second experiment?
  (b) How many grams of sulfur remain *unreacted?*

91. Identify the isotope X if

• the nucleus contains one more neutron than protons
and
• the mass number is 9 times the charge on the ion X^{3+}.

92. What is the charge-to-mass ratio, e/m, in C/g, of a chloride ion, $^{37}_{17}Cl^-$? The mass of $^{37}_{17}Cl$ is 36.966 u. The electronic charge is -1.602×10^{-19} C, and the relationship between the units, u and g, is 1.673×10^{-24} g $= 1.0073$ u.

93. There are two principal isotopes of silver, $^{107}_{47}Ag$ and $^{109}_{47}Ag$. The atomic weight of silver is 107.87, with 51.82% of the atoms being $^{107}_{47}Ag$. The mass of an atom of $^{107}_{47}Ag$ is 106.9 u. What is the mass, in u, of an atom of $^{109}_{47}Ag$?.

94. Determine
  (a) the number of moles of Na in a 150.0-g sample;
  (b) the number of S atoms in 500.0 kg of sulfur;
  (c) the mass of a one trillion (1.0×10^{12}) atom sample of naturally occurring copper (Cu).

95. Gold occurs in a particular seawater to an extent of 0.15 mg Au/ton seawater. How many Au atoms are present in a glassful (300. g) of this seawater? (1 ton = 2000 lb; 1 lb = 454 g; atomic weight of Au = 197.)

3 Chemical Compounds

Brightly colored chemical compounds used in the manufacture of cosmetics. [Burt Glinn/Magnum Photos, Inc.]

Water, ammonia, carbon monoxide, carbon dioxide, and sulfuric acid—these substances are familiar to almost everyone. All of them are rather simple chemical compounds. Only slightly less familiar are substances like vitamin C, chlorophyll, and sucrose. These, too, are chemical compounds.

Sucrose, which is ordinary cane sugar, consists of 42.10% C, 6.48% H, and 51.42% O. How can we establish this fact by experiment? Once we have determined this percentage composition, how can we use it to deduce the chemical formula of sucrose? And what types of information can we derive from a chemical formula? The aspect of chemistry that deals with these fundamental questions is called *stoichiometry*, which means, literally, to measure the elements (Gr., *stoicheion*, element).

We begin a study of stoichiometry in this chapter. Another practical matter that we will start to consider in this chapter is how to relate the names of chemical compounds to their formulas (referred to as chemical nomenclature).

3-1 Formulas, Formula Weights, and Molecular Weights

Chemical compounds are represented by combinations of symbols called chemical formulas. A **chemical formula** indicates

1. The elements present in a compound.
2. The relative numbers of atoms of each element in the compound.

In the following formula the elements are denoted by their symbols and the relative numbers of atoms by *subscript* numerals (where no subscript is written, the number 1 is understood).

FIGURE 3-1
A molecule of CCl₄.

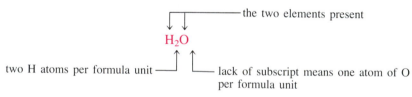

the two elements present

H_2O

two H atoms per formula unit — lack of subscript means one atom of O per formula unit

Here are three additional formulas.

CCl_4 $NaCl$ $MgCl_2$
carbon tetrachloride   sodium chloride   magnesium chloride

The group of five atoms represented by the formula CCl₄ and pictured in Figure 3-1 is called a molecule. A **molecule** is a group of bonded atoms that actually exists as a separate entity. With sodium chloride, NaCl, we encounter a different situation. As shown in Figure 3-2, each Na⁺ ion is surrounded by *six* Cl⁻ ions (and vice versa). We cannot say that any one of these six Cl⁻ ions belongs exclusively to a Na⁺ ion, so we *arbitrarily* select a combination of one Na⁺ ion and one Cl⁻ ion and call this a formula unit. A **formula unit** is the smallest collection of positive and negative ions from which we can derive the simplest formula (that is, the formula with the smallest subscripts). Because the formula unit NaCl is buried in a vast array of ions and cannot be obtained as a separate entity, we should not call it a molecule. The situation with MgCl₂ is similar to that of NaCl.

Formula Weight and Molecular Weight. Once we have identified a formula unit it is a simple matter to establish the formula weight of a compound. **Formula weight** is the mass of a formula unit relative to an assigned mass of 12.00000 for C-12. A more useful definition, though, is

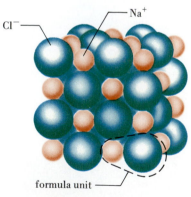

FIGURE 3-2
A formula unit of NaCl.

Sodium chloride consists of Na⁺ and Cl⁻ ions in a very large network called a crystal. The combination of one Na⁺ and one Cl⁻ ion, called a formula unit, is the smallest collection of ions from which we can deduce the formula NaCl.

Definition of formula weight.

Formula weight is the sum of the atomic weights of all the atoms in a formula unit. $\qquad$ (3.1)

Thus, for sodium chloride, NaCl, whose formula unit consists of one Na^+ ion and one Cl^- ion,

formula weight NaCl = at. wt. Na + at. wt. Cl

$$= \quad 22.99 \quad + \quad 35.45 \quad = \mathbf{58.44}$$

And for magnesium chloride, $MgCl_2$,

formula weight MgCl$_2$ = at. wt. Mg + (2 × at. wt. Cl)

$$= \quad 24.305 \quad + \quad (2 \times 35.453) \quad = \mathbf{95.21}$$

If a compound consists of discrete molecules, we can also speak of a molecular weight. **Molecular weight** is the mass of a molecule relative to an assigned mass of 12.00000 for C-12. Again, a more useful definition is

Definition of molecular weight.

Molecular weight is the sum of the atomic weights of all the atoms in a molecule. $\qquad$ (3.2)

To determine the molecular weight of carbon tetrachloride, for example, we note that one molecule of CCl_4 consists of one C atom and four Cl atoms, and

molecular weight CCl$_4$ = at. wt. C + (4 × at. wt. Cl)

$$= \quad 12.01 \quad + \quad (4 \times 35.45) \quad = \mathbf{153.8}$$

Although we can always refer to the formula weight of a compound, the term *molecular weight* is valid only if discrete molecules of the compound exist. If the formula unit and a molecule are identical (as in CCl_4), the formula weight and the molecular weight are identical.

Mole of a Compound. We can apply the concept of a mole to any species—atoms, ions, formula units, molecules. Thus, we can describe a mole of a compound as an amount of the compound containing 6.02214×10^{23} formula units or molecules. We can also extend the term *molar mass* to moles of formula units or molecules, so that

Recall the meaning of the equivalence sign introduced in Section 1-9.

1 mol $MgCl_2$ ≏ 95.21 g $MgCl_2$ ≏ 6.022×10^{23} $MgCl_2$ formula units

and

1 mol CCl_4 ≏ 153.8 g CCl_4 ≏ 6.022×10^{23} CCl_4 molecules

Example 3-1

Using molar mass and the Avogadro constant to determine the total number of ions in a sample of ionic compound. An analytical balance can detect a mass of 0.1 mg. What is the total number of ions present in this minimally detectable quantity of $MgCl_2$?

Solution. We use the molar mass to convert from mass to number of moles of $MgCl_2$. Then we follow with a conversion factor based on the Avogadro constant to convert from moles to number of formula units. Our final factor is based on the fact that there are *three* ions (*one* Mg^{2+} and *two* Cl^-) per formula unit (f.u.) of $MgCl_2$.

$$\text{no. ions} = 0.1 \text{ mg MgCl}_2 \times \frac{1 \text{ g MgCl}_2}{1000 \text{ mg MgCl}_2} \times \frac{1 \text{ mol MgCl}_2}{95 \text{ g MgCl}_2}$$

$$\times \frac{6.0 \times 10^{23} \text{ f.u. MgCl}_2}{1 \text{ mol MgCl}_2} \times \frac{3 \text{ ions}}{1 \text{ f.u. MgCl}_2}$$

$$= 2 \times 10^{18} \text{ ions}$$

SIMILAR EXAMPLES: Exercises 1, 4, 20, 21.

Example 3-2

Combining several factors in a calculation—molar mass, composition of a gaseous mixture, liquid density. The volatile liquid ethyl mercaptan, C_2H_6S, is one of the most odoriferous substances known. It is used in natural gas to make gas leaks detectable. We can detect as little as 9×10^{-4} μmol/m^3 air. What minimum volume of this liquid (through its vaporization) would we be able to detect in a chemistry lecture hall having a volume of 5000 m^3? (Density of $C_2H_6S = 0.84$ g/cm^3.)

Solution. In addition to the usual mole/mass conversion, we need to convert from *micro*moles to moles and from mass of C_2H_6S to volume.

$$\text{no. cm}^3 \text{ C}_2\text{H}_6\text{S} = 5000 \text{ m}^3 \text{ air} \times \frac{9 \times 10^{-4} \text{ } \mu\text{mol C}_2\text{H}_6\text{S}}{\text{m}^3 \text{ air}} \times \frac{1 \times 10^{-6} \text{ mol C}_2\text{H}_6\text{S}}{1 \text{ } \mu\text{mol C}_2\text{H}_6\text{S}}$$

$$\times \frac{62 \text{ g C}_2\text{H}_6\text{S}}{1 \text{ mol C}_2\text{H}_6\text{S}} \times \frac{1 \text{ cm}^3 \text{ C}_2\text{H}_6\text{S}}{0.84 \text{ g C}_2\text{H}_6\text{S}}$$

$$= 3 \times 10^{-4} \text{ cm}^3$$

The unit *micro*liter is useful for expressing very small volumes.

$$3 \times 10^{-4} \text{ cm}^3 = 3 \times 10^{-4} \text{ mL} = 3 \times 10^{-7} \text{ L} = 0.3 \times 10^{-6} \text{ L} = 0.3 \text{ } \mu\text{L}$$

SIMILAR EXAMPLES: Exercises 21, 22, 23, 68.

Given that a normal drop of liquid is about 0.05 mL, the volume of liquid calculated here is about 0.006 of a drop.

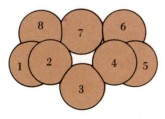

FIGURE 3-3

What is a "mole of sulfur"?

In solid sulfur, atoms are joined together into puckered rings with eight members. A molecule of sulfur is composed of eight atoms, S_8. In a sample of solid sulfur, there are eight times as many atoms as there are molecules. If by a "mole of sulfur" we mean 1 mol of sulfur atoms, we can measure out $\frac{1}{8}$ mol of the S_8 molecules and describe this as 1 mol S.

Mole of an Element—A Second Look. In Section 2-7 we defined a mole of an element in terms of a number of *atoms* (the Avogadro constant). This is the only definition possible for elements like iron, magnesium, sodium, and copper, in which enormous numbers of individual spherical atoms are clustered together, much like marbles in a can. But with some elements, atoms of the same kind are joined together to form molecules, and bulk samples of the elements are composed of collections of molecules (see Figure 3-3). For these elements we can refer to 1 mol of molecules and a molecular weight of the element. Some important examples that you should become familiar with are

$$H_2 \quad O_2 \quad N_2 \quad F_2 \quad Cl_2 \quad Br_2 \quad I_2 \quad P_4 \quad S_8$$

Thus, we can write molar masses such as 2.016 g H_2/mol H_2, 32.00 g O_2/mol O_2, and so on.

Atomic Weights, Molecular Weights, . . .: A Final Word. We have introduced a number of important terms related to the stoichiometry of elements and compounds, both in Chapter 2 and continuing through this section. Table 3-1 summarizes these terms. In studying this table, try to develop a clear understanding of the relationships among the terms, rather than simply memorizing definitions.

Are You Wondering:

When to use 1.008 g/mol for the molar mass of hydrogen and when to use 2.016 g/mol?

The statement, "a mole of hydrogen" is ambiguous; you should always say either a mole of hydrogen *atoms* or a mole of hydrogen *molecules*. Better still write **1 mol H** or **1 mol H₂.** If you do this is, then you will see that the molar masses should be expressed as **1.008 g H/mol H** and **2.016 g H₂/mol H₂.** This distinction is very much like the distinction between one dozen socks and one dozen pairs of socks (that is, the H atom is analogous to a single sock and the H_2 molecule to a pair of socks).

TABLE 3-1
A Summary of Terms Used in Stoichiometry

Term[a]	Definition or usage
atomic weight standard	An atomic weight of 12.00000 is arbitrarily assigned to $^{12}_6C$.
isotopic mass (nuclidic mass)	The mass, in atomic mass units, u, of a single atom, on a scale in which the mass of an atom of $^{12}_6C$ is arbitrarily defined as 12.00000 u (e.g., $^{35}_{17}Cl$ has an isotopic mass of 34.968852 u).
atomic weight (relative atomic weight)	A dimensionless (pure) number that expresses the mass of the naturally occurring mixture of isotopes of an element, relative to an arbitrarily assigned atomic weight of 12.00000 for carbon-12. These are the values listed in a table of atomic weights (e.g., atomic weight of Cl = 35.4527).
formula weight (relative formula weight)	A dimensionless (pure) number that expresses the mass of a formula unit of a compound, relative to the atomic weight standard, carbon-12 (e.g., formula weight of NaCl = 58.44). Tabulated atomic weights are used in computing formula weights.
molecular weight (relative molecular weight)	A dimensionless (pure) number that compares the mass of a molecule to the atomic weight standard, carbon-12 (e.g., molecular weight of CCl_4 = 153.82). Tabulated atomic weights are used in computing molecular weights. Use of this term should be limited to situations where discrete, identifiable molecules actually exist.
mole	An amount of substance containing the same number of elementary units (6.02214×10^{23}) as there are $^{12}_6C$ atoms in 12.00000 g $^{12}_6C$.
molar mass, $\mathcal{M}$ (molar weight, mole weight)	The mass of one mole of a substance, whether the substance is composed of individual atoms (e.g., 35.45 g Cl/mol Cl), formula units (e.g., 58.44 g NaCl/mol NaCl), or molecules (e.g., 153.82 g CCl_4/mol CCl_4). Often the terms atomic, formula, or molecular weight are used, even though molar mass is what is intended.

[a]Wherever the term "weight" occurs, the term "mass" is equally (or even more) appropriate, e.g., atomic weight = atomic mass.

3-2   Composition of Chemical Compounds

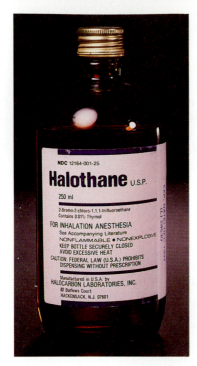

The volatile liquid known as halothane* has been used as a fire extinguisher and is also an inhalation anesthetic. Its formula is $C_2HBrClF_3$ and its molecular weight is 197.38. *Per mole* of halothane there are 2 moles of C atoms; one mole each of H, Br, and Cl atoms; and 3 moles of F atoms. We can also express this composition as

$$1 \text{ mol } C_2HBrClF_3 \backsimeq 2 \text{ mol C} \backsimeq 1 \text{ mol H} \backsimeq 1 \text{ mol Br} \backsimeq 1 \text{ mol Cl} \backsimeq 3 \text{ mol F}$$

$$(3.3)$$

All the atoms represented in a chemical formula are *combined* or *bonded* atoms. However, a formula often does not tell us which atoms in a compound are joined or bonded to which other atoms. That is, there is nothing in the formula $C_2HBrClF_3$ to suggest that all the other atoms are bonded directly to the C atoms. This kind of information is revealed through *structural formulas,* a topic explored fully in Chapters 10 and 11. Figure 3-4 represents the formula of halothane in several different ways, though for the present the simplest (empirical) formula will serve all our purposes.

We can use the equivalencies of expression (3.3) to establish the conversion factors that we need in Example 3-3.

Example 3-3

Deriving various relationships from a chemical formula.
(a) How many C atoms are present in 25.8 g halothane?
(b) How many grams of F are present *per gram* of C in halothane?

FIGURE 3-4
Several representations of the halothane molecule. [Carey B. Van Loon]

Empirical formula:	$C_2HBrClF_3$	Shows the different types of atoms and their relative numbers.
Condensed structural formula:	$CF_3CHBrCl$	Shows that the three F atoms are bonded to the same C atom and all other atoms are bonded to the second C atom.
Structural formula:	$F-\overset{\displaystyle F}{\underset{\displaystyle F}{C}}-\overset{\displaystyle H}{\underset{\displaystyle Cl}{C}}-Br$	Shows explicitly the bonding arrangements between atoms.
Molecular model: ("ball-and-stick")	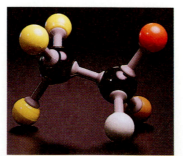	Shows the three-dimensional structure of the molecule, with the centers of the atoms represented by colored balls and the bonds between atoms by sticks.
Molecular model: ("space-filling")		The most accurate three-dimensional representation of the molecule.

*Halothane is a common or trivial name. Its systematic name is 2-bromo-2-chloro-1,1,1-trifluoroethane. This system for naming compounds like halothane is discussed in Chapter 27.

Calculations based on a chemical formula usually require that a formula weight or molecular weight be established. That of $C_2HBrClF_3$ is

$(2 \times 12.01) + 1.01 + 79.90 + 35.45 + (3 \times 19.00) = 197.4$

Solution

(a) As in similar questions from Section 2-8, we need to express the amount of substance in moles and also use the Avogadro constant. In addition we need a conversion factor based on expression (3.3), shown in blue.

$$\text{no. C atoms} = 25.8 \text{ g } C_2HBrClF_3 \times \frac{1 \text{ mol } C_2HBrClF_3}{197.4 \text{ g } C_2HBrClF_3}$$

$$\times \frac{2 \text{ mol C}}{1 \text{ mol } C_2HBrClF_3} \times \frac{6.022 \times 10^{23} \text{ C atoms}}{1 \text{ mol C}}$$

$$= 1.57 \times 10^{23} \text{ C atoms}$$

Anti Given.

(b) Perhaps the easiest approach is to start with the *mole* ratio, 3 mol F/2 mol C, and convert both numerator and denominator to mass, in grams.

$$\text{no. g F/g C} = \frac{3 \text{ mol F} \times \dfrac{19.00 \text{ g F}}{1 \text{ mol F}}}{2 \text{ mol C} \times \dfrac{12.01 \text{ g C}}{1 \text{ mol C}}} = \frac{57.00 \text{ g F}}{36.03 \text{ g C}} = 1.582 \text{ g F/g C}$$

SIMILAR EXAMPLES: Exercises 3, 29, 30, 64.

Calculating Percent Composition from a Chemical Formula. In a manner similar to Example 3-3(b), we can start with the mol ratio, 3 mol F/1 mol $C_2HBrClF_3$. Then, by using molar masses, we can convert this mole ratio to a mass ratio having the unit: g F/g $C_2HBrClF_3$. This mass ratio represents the number of g F per g $C_2HBrClF_3$. If we multiply this mass ratio by 100, we obtain the expression: number of g F per 100 g $C_2HBrClF_3$. But this is simply the definition of *percent;* our result is the percent fluorine in halothane, by mass (also called the mass percent fluorine). We can do this type of calculation for all the elements in halothane to obtain the *percent composition* of the compound.

Example 3-4 _____

Calculating the mass percent composition of a compound. What is the mass percent composition of halothane, $C_2HBrClF_3$?

Solution. Applying the reasoning outlined above we get

$$\% \text{ F, by mass} = \frac{3 \text{ mol F} \times \dfrac{19.00 \text{ g F}}{1 \text{ mol F}}}{1 \text{ mol } C_2HBrClF_3 \times \dfrac{197.4 \text{ g } C_2HBrClF_3}{1 \text{ mol } C_2HBrClF_3}} \times 100 = 28.88\%$$

Note that by expressing percentages to the nearest 0.01% all the values are given to four significant figures, *except % H*, which is given only to two. The relative error in % H is greater than in the others because of the very low atomic weight of hydrogen.

This procedure can be simplified somewhat, as in the following setup.

$$\% \text{ C, by mass} = \frac{(2 \times 12.01) \text{ g C}}{197.4 \text{ g } C_2HBrClF_3} \times 100 = 12.17\%$$

Applying this same procedure to Br, Cl, and H, we obtain:

$\% \text{ Br} = 40.48\%$ $\% \text{ Cl} = 17.96\%$ $\% \text{ H} = 0.51\%$

SIMILAR EXAMPLES: Exercises 5, 6, 31, 32, 65, 67.

The percentages of the elements in a compound should add up to 100.00, and we can use this fact in either of two ways.

1. Check the accuracy of the individual percentage computations by ensuring that the percentages do total 100.00. As applied to the results of Example 3-4: $28.88 + 12.17 + 40.48 + 17.96 + 0.51 = 100.00$
2. Determine the percentages of all the elements but one, and obtain that one by subtraction. Again, as applied to the results of Example 3-4: % H = 100.00 − % F − % C − % Br − % Cl = $100.00 − 28.88 − 12.17 − 40.48 − 17.96 = 0.51\%$

Establishing Formulas from the Experimentally Determined Percent Composition of a Compound. Percent composition establishes the relative proportions of the elements in a compound on a *mass* basis. A chemical formula requires that these proportions be in terms of *numbers* of atoms, i.e., on a *mole* basis. The principle of the method used in Example 3-5 is this: The relative numbers of atoms of each type are independent of whether we select a single formula unit, a mole, or any arbitrary mass of compound. Selecting a sample size of 100.0 g allows for the easiest conversion of percentages to masses of the elements in a compound.

The formula obtained by the method of Example 3-5 is the simplest possible formula—the **empirical formula.** From a separate experiment the true molecular weight of a compound can be measured (with methods introduced in Chapters 6 and 13). The true molecular weight is either equal to the empirical formula weight or some whole number multiple of it. As shown in Example 3-5, we get a **molecular formula** by multiplying all the subscripts in the empirical formula by the same factor that relates the empirical formula weight to the true molecular weight.

Example 3-5

Determining the empirical formula of a compound from its mass percent composition—a five-step approach. The compound methyl benzoate, used in the manufacture of perfumes, consists of 70.58% C, 5.93% H, and 23.49% O, by mass. Also, by experiment, the molecular weight is found to be 136. What are the empirical formula and the molecular formula of methyl benzoate?

Solution.
Step 1. Determine the mass of each element in a 100.0-g sample. In 100.0 g of compound (100 parts) there are 70.58 g C (70.58 parts), and so on. The masses are

70.58 g C;   5.93 g H;   23.49 g O

Step 2. Convert the mass of each element in the 100.0-g sample to number of moles.

$$\text{no. mol C} = 70.58 \text{ g C} \times \frac{1 \text{ mol C}}{12.011 \text{ g C}} = 5.876 \text{ mol C}$$

$$\text{no. mol H} = 5.93 \text{ g H} \times \frac{1 \text{ mol H}}{1.008 \text{ g H}} = 5.88 \text{ mol H}$$

$$\text{no. mol O} = 23.49 \text{ g O} \times \frac{1 \text{ mol O}}{15.994 \text{ g O}} = 1.468 \text{ mol O}$$

Step 3. Write a tentative formula based on the numbers of moles just determined.

$C_{5.88}H_{5.88}O_{1.47}$

Step 4. Attempt to convert the subscripts of Step 3 to small whole numbers. This requires dividing each of the subscripts by the smallest one (1.47).

$$C_{\frac{5.88}{1.47}}H_{\frac{5.88}{1.47}}O_{\frac{1.47}{1.47}} = C_{4.00}H_{4.00}O_{1.00}$$

Step 5. If the subscripts from Step 4 are not whole numbers but differ from whole numbers by only about 0.10 or less, round them off to whole numbers.

C_4H_4O (empirical formula)

The empirical formula weight of the compound is $[(4 \times 12.0) + (4 \times 1.01) + 16.0] = 68.0$. Since the experimentally determined molecular weight (136) is twice the empirical formula weight, the molecular formula is $C_8H_8O_2$.

SIMILAR EXAMPLES: Exercises 9, 35, 36, 75.

Example 3-6

Determining the empirical formula of a compound from its mass percent composition when a final adjustment of formula subscripts is required. Butylated hydroxyanisole (BHA) is used as a preservative in potato chips and other foods. Its composition is found to be 73.31% C, 8.94% H, and 17.75% O, by mass. What is the empirical formula of BHA?

Solution. The first five steps are the same as in Example 3-5, but here we find that we also need a sixth step.

Step 1. Determine the mass of each element in a 100.0-g sample.

73.31 g C;   8.94 g H;   17.75 g O

Step 2. Convert these masses to moles. Since these conversions are so similar to those of Example 3-5, only the results are shown here.

6.10 mol C;   8.87 mol H;   1.11 mol O

Step 3. Write a tentative formula based on the numbers of moles above.

$C_{6.10}H_{8.87}O_{1.11}$

Step 4. Attempt to convert the subscripts to small whole numbers by dividing each subscript by the smallest (1.11).

$$C_{\frac{6.10}{1.11}}H_{\frac{8.87}{1.11}}O_{\frac{1.11}{1.11}} = C_{5.50}H_{7.99}O_{1.00}$$

Step 5. Round off the subscripts to whole numbers, *where possible.* The deviation of the subscript of H (7.99) from a whole number (8) can be attributed to experimental error, but not the subscript of C (5.50).

$C_{5.50}H_8O$

Step 6. Multiply all subscripts by a small whole number chosen to make all subscripts integral. Here we must multiply by 2 (that is, $5.50 \times 2 = 11$).

$C_{5.50 \times 2}H_{8 \times 2}O_{1 \times 2} = C_{11}H_{16}O_2$

SIMILAR EXAMPLES: Exercises 7, 8, 37, 71, 76.

Combustion Analysis. Figure 3-5 illustrates an experimental method for establishing an empirical formula—combustion analysis. It is useful for compounds that are easily burned, such as those containing carbon and hydrogen or C and H in combi-

Are You Wondering:

When to round off subscripts (as in Step 5 of Example 3-5) and when to multiply through by a constant (as in Step 6 of Example 3-6)?

Because of experimental errors and/or minor computational discrepancies, subscripts in empirical formula calculations often deviate very slightly from integral numbers. A useful rule is that you can round off a subscript if it is within a few hundredths of a whole number. If the deviation from an integral value is more than about 0.1, then you need to adjust the subscripts to integral values by multiplying through by the appropriate constant. (In choosing this constant, you will find it helpful to recognize the decimal equivalents of some common fractions: $0.50 = \frac{1}{2}$; $0.333 \approx \frac{1}{3}$; $0.25 = \frac{1}{4}$; $0.20 = \frac{1}{5}$; and so on.)

FIGURE 3-5
Apparatus for combustion analysis.

Oxygen gas (A) is passed into the combustion tube containing the sample to be analyzed (B). This portion of the apparatus is enclosed within a high-temperature furnace (C). Products of the combustion are absorbed as they leave the furnace—water vapor by magnesium perchlorate (D) and carbon dioxide gas by sodium hydroxide (E) (to produce sodium carbonate).

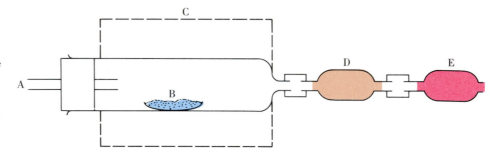

nation with O, N, and a few other elements (organic compounds, discussed in Chapter 27). A weighed sample of a compound is heated in a stream of oxygen gas. The water vapor and carbon dioxide gas produced in the combustion are absorbed by appropriate substances. The increases in mass of these absorbers correspond to the masses of water and carbon dioxide. The principle of combustion analysis is outlined in Figure 3-6.

Example 3-7

Determining an empirical formula from combustion analysis data. Vitamin C is essential for the prevention of scurvy (and large dosages may be effective in preventing colds). Combustion of a 0.2000-g sample of this carbon–hydrogen–oxygen compound yields 0.2998 g CO_2 and 0.0819 g H_2O. What is the empirical formula of vitamin C?

FIGURE 3-6
Principle of combustion analysis—Example 3-7 visualized.

$$C_xH_yO_z + O_2 \longrightarrow CO_2 + H_2O$$

The basis of combustion analysis is to trace what happens to all the atoms in the compound being analyzed (represented here as $C_xH_yO_z$). All the C atoms in the compound appear as C atoms in CO_2 (and there is no other source of C atoms for the CO_2). All the H atoms in the compound appear as H atoms in H_2O (and again, there is no other source of H atoms). Oxygen atoms in the CO_2 and H_2O come both from the compound being analyzed and from oxygen gas consumed in the combustion. The quantity of oxygen in the compound must be determined indirectly.

Solution. We can base our calculation directly on the 0.2000-g sample if we can establish the number of moles of C, H, and O present. And the key to doing this is to note that all the C in the vitamin C sample appears in 0.2998 g CO_2 and all the H, in 0.0819 g H_2O.

$$\text{no. mol C} = 0.2998 \text{ g } CO_2 \times \frac{1 \text{ mol } CO_2}{44.010 \text{ g } CO_2} \times \frac{1 \text{ mol C}}{1 \text{ mol } CO_2} = 0.006812 \text{ mol C}$$

$$\text{no. g C} = 0.006812 \text{ mol C} \times \frac{12.011 \text{ g C}}{1 \text{ mol C}} = 0.08182 \text{ g C}$$

$$\text{no. mol H} = 0.0819 \text{ g } H_2O \times \frac{1 \text{ mol } H_2O}{18.02 \text{ g } H_2O} \times \frac{2 \text{ mol H}}{1 \text{ mol } H_2O} = 0.00909 \text{ mol H}$$

$$\text{no. g H} = 0.00909 \text{ mol H} \times \frac{1.008 \text{ g H}}{1 \text{ mol H}} = 0.00916 \text{ g H}$$

We can obtain the mass of O in the vitamin C sample by difference.

no. g O = no. g cpd. − no. g C − no. g H

$$= 0.2000 \text{ g} - 0.08182 \text{ g} - 0.00916 \text{ g} = 0.1090 \text{ g O}$$

Now we can calculate the number of moles of O.

$$\text{no. mol O} = 0.1090 \text{ g O} \times \frac{1 \text{ mol O}}{15.999 \text{ g O}} = 0.006813 \text{ mol O}$$

As trial subscripts for the empirical formula we can write

$C_{0.006812}H_{0.00909}O_{0.006813}$

Next, we divide each subscript by the smallest—0.006812—to obtain

$CH_{1.33}O$

Finally, we multiply each subscript by 3 (since $3 \times 1.33 = 3.99 \approx 4.00$). **The empirical formula of vitamin C is $C_3H_4O_3$.**

SIMILAR EXAMPLES: Exercises 10, 50, 51, 52, 81, 85.

Example 3-8

Using combustion analysis data to establish the mass percent composition of a compound. Use the data of Example 3-7 to calculate the mass percent composition of vitamin C.

Solution. To obtain the % C and % H, use the results: 0.08182 g C and 0.00916 g H. Calculate the % O by difference.

$$\% \text{ C} = \frac{0.08182 \text{ g C}}{0.2000 \text{ g cpd.}} \times 100 = 40.91\%$$

$$\% \text{ H} = \frac{0.00916 \text{ g H}}{0.2000 \text{ g cpd.}} \times 100 = 4.58\%$$

$$\% \text{ O} = 100.00\% - 40.91\% - 4.58\% = 54.51\%$$

SIMILAR EXAMPLES: Exercises 10, 50, 81.

Precipitation Analysis. Not all samples can be burned easily. Another useful method of analysis is to cause one of the components of the sample to deposit from solution as an insoluble solid, a *precipitate*. This precipitate is then treated in such a way as to yield a *pure* solid of *known composition*. From the measured masses of this pure solid and of the original sample, we can determine the percentage of the component in the sample. Figure 3-7 outlines how we might determine the percent tin in a brass sample, and Figure 3-8 pictures part of this analysis. The calculations involved are illustrated through Example 3-9.

Example 3-9

Performing an analysis by measuring the mass of a precipitate. A 2.568-g sample of brass, when treated in the manner outlined in Figure 3-7, yields 0.1330 g SnO_2. What is the % Sn in the brass?

Solution. To determine the mass of tin in 0.1330 g SnO_2, proceed through these three steps.

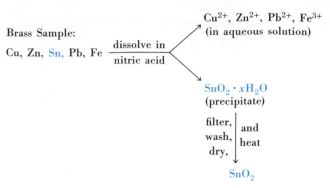

Brass Sample:

Cu, Zn, Sn, Pb, Fe $\xrightarrow[\text{nitric acid}]{\text{dissolve in}}$ Cu^{2+}, Zn^{2+}, Pb^{2+}, Fe^{3+} (in aqueous solution)

$SnO_2 \cdot xH_2O$ (precipitate)

filter, wash, dry, | and heat

SnO_2

FIGURE 3-7
Determination of tin in brass—Example 3-9 visualized.

Brass is an alloy (mixture) of copper and zinc with small amounts of tin, lead, and iron. When a weighed sample of brass is treated with nitric acid (a water solution of HNO_3), the copper, zinc, lead, and iron dissolve and appear in aqueous solution in their ionic forms. Tin is converted to an insoluble oxide with an unknown amount of water associated with it ($SnO_2 \cdot xH_2O$). This precipitate is filtered off from the solution, washed, dried, and then heated to drive off all the water. The result is *pure* SnO_2, which is weighed.

FIGURE 3-8
Isolating SnO_2 from a brass sample.

Except for tin, the several metals in a typical brass sample are readily soluble in concentrated nitric acid. The red-brown gaseous product is nitrogen dioxide (left). Tin is converted to a hydrated SnO_2, which remains as a white precipitate in the solution made blue by Cu^{2+} (right). The precipitate can be filtered off, purified, dried, and weighed, thereby providing a basis for determining % Sn in brass. [Carey B. Van Loon]

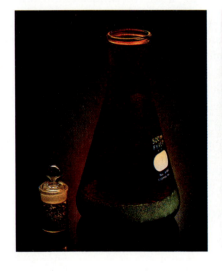
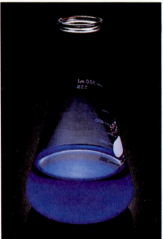

Step 1. Convert 0.1330 g SnO_2 to the number of moles of SnO_2.

no. mol SnO_2 = 0.1330 g $SnO_2 \times \dfrac{1 \text{ mol } SnO_2}{150.70 \text{ g } SnO_2}$ = 8.825×10^{-4} mol SnO_2

Step 2. Convert 8.825×10^{-4} mol SnO_2 to mol Sn.

no. mol Sn = 8.825×10^{-4} mol $SnO_2 \times \dfrac{1 \text{ mol Sn}}{1 \text{ mol } SnO_2}$ = 8.825×10^{-4} mol Sn

Step 3. Convert 8.825×10^{-4} mol Sn to g Sn.

no. g Sn = 8.825×10^{-4} mol Sn $\times \dfrac{118.71 \text{ g Sn}}{1 \text{ mol Sn}}$ = 0.1048 g Sn

    An alternate route from g SnO_2 to g Sn is based on the conversion factor 118.71 g Sn/150.70 g SnO_2. There is one mole of Sn (with mass, 118.71 g) for every mole of SnO_2 (with mass, 150.70 g).

no. g Sn = 0.1330 g $SnO_2 \times \dfrac{118.71 \text{ g Sn}}{150.70 \text{ g } SnO_2}$ = 0.1048 g Sn

The percent tin in the brass sample is

$$\% \text{ Sn} = \frac{0.1048 \text{ g Sn}}{2.568 \text{ g brass}} \times 100 = 4.08\%$$

SIMILAR EXAMPLES: Exercises 55, 56, 57.

Example 3-10

Determining an atomic weight by precipitation analysis. A 0.7718-g sample of the compound MCl yields 2.6091 g AgCl. What must be the atomic weight of the element M?

Solution. First we should determine the number of moles and the mass of Cl in the AgCl. This Cl all comes from the sample of MCl.

$$\text{no. mol Cl} = 2.6091 \text{ g AgCl} \times \frac{1 \text{ mol AgCl}}{143.321 \text{ g AgCl}} \times \frac{1 \text{ mol Cl}}{1 \text{ mol AgCl}}$$

$$= 0.018204 \text{ mol Cl}$$

$$\text{no. g Cl} = 0.018204 \text{ mol Cl} \times \frac{35.4527 \text{ g Cl}}{1 \text{ mol Cl}} = 0.6454 \text{ g Cl}$$

CHEMISTRY EVOLVING

Atomic Weight Determinations

Jean Stas (1813–1891) established a precise atomic weight of carbon by burning samples of pure carbon (diamond and graphite) and determining the masses of carbon dioxide produced. T. W. Richards (1868–1928) and his students determined precise atomic weights of about sixty elements. They used a variety of chemical methods, including the precipitation of various chlorides as AgCl, as illustrated in Example 3-10 and Figure 3-9.

Theodore W. Richards (1868–1928)—mentor to a generation of American chemists. Richards, who pioneered in introducing physical chemistry to the United States, launched the careers of many great American chemists. [Center for the History of Chemistry, University of Pennsylvania]

FIGURE 3-9
Precipitation of silver chloride.

The element M in Example 3-10 is lithium. The precipitate of silver chloride (AgCl) shown here is obtained by adding silver nitrate ($AgNO_3$) solution to a solution of lithium chloride (LiCl). [Carey B. Van Loon]

Richards pushed the precision of precipitation analysis to the point where he started to detect variations in the distributions of the isotopes of the elements. Richards won the Nobel Prize in 1914 for his atomic weight studies, the first American to win this award in chemistry. Interestingly, Richards' studies demonstrated the limitations of chemical methods of atomic weight determination. Atomic weight determinations are now done by physical methods (mass spectrometry). ■ ■ ■

The mass of M in the sample of MCl is

no. g M = 0.7718 g MCl − 0.6454 g Cl = 0.1264 g M

Since the formula of the compound is MCl,

no. mol M = no. mol Cl = 0.018204 mol

The molar mass is the ratio: no. g M/no. mol M.

$$molar\ mass = \frac{0.1264\ g\ M}{0.018204\ mol\ M} = 6.944\ g\ M/mol\ M$$

The atomic weight of M = 6.944.

SIMILAR EXAMPLES: Exercises 11, 60, 61, 62, 63, 95.

3-3   The Need to Name Chemical Compounds— Nomenclature

Throughout this chapter we have referred to compounds mostly by their formulas, but we also need to identify compounds by name. This allows us to look up properties in a handbook, locate a chemical on a storeroom shelf, or just discuss a laboratory experiment with a colleague. Later in the text we encounter situations in which different compounds have the same formula. In these situations it is essential to distinguish among compounds by name. No two substances can have the same name, yet at the same time there should be some similarities in the names of similar substances (see Figure 3-10). Otherwise, the task of naming millions of compounds would be next to impossible. Such would be the case if all compounds were referred to by a common or trivial name such as *water* (H_2O) or *ammonia* (NH_3).

What we need is a *systematic* method of assigning names—a system of **nomenclature.** Actually several such systems exist, and we introduce each at an appropriate point in the text. For the present we restrict ourselves to some of the simpler aspects of naming compounds. We begin by introducing two new concepts that are useful in nomenclature: metals/nonmetals and oxidation state.

FIGURE 3-10
Two oxides of lead.

These two compounds contain the same elements—lead and oxygen—but in different proportions. Their names and formulas must convey this fact.

lead(II) oxide = PbO (yellow)
lead(IV) oxide =
                 PbO_2 (red-brown)
[Carey B. Van Loon]

3-4   Metals and Nonmetals—An Introduction to the Periodic Table

Each chemical element has its own set of properties yet also bears similarities to certain other elements. The elements can be grouped into categories according to these similarities. This subject is explored fully in Chapter 9, through the tabular arrangement known as the periodic table of the elements. A periodic table is found on the inside front cover and is reproduced in Figure 3-11.

1 1A	2 2A	3 3B	4 4B	5 5B	6 6B	7 7B	8	9 8B	10	11 1B	12 2B	13 3A	14 4A	15 5A	16 6A	17 7A	18 8A
1 H																	2 He
3 Li	4 Be											5 B	6 C	7 N	8 O	9 F	10 Ne
11 Na	12 Mg											13 Al	14 Si	15 P	16 S	17 Cl	18 Ar
19 K	20 Ca	21 Sc	22 Ti	23 V	24 Cr	25 Mn	26 Fe	27 Co	28 Ni	29 Cu	30 Zn	31 Ga	32 Ge	33 As	34 Se	35 Br	36 Kr
37 Rb	38 Sr	39 Y	40 Zr	41 Nb	42 Mo	43 Tc	44 Ru	45 Rh	46 Pd	47 Ag	48 Cd	49 In	50 Sn	51 Sb	52 Te	53 I	54 Xe
55 Cs	56 Ba	57 La*	72 Hf	73 Ta	74 W	75 Re	76 Os	77 Ir	78 Pt	79 Au	80 Hg	81 Tl	82 Pb	83 Bi	84 Po	85 At	86 Rn
87 Fr	88 Ra	89 Ac†	104 Rf	105 Ha	106 Unh	107 Uns	108 Uno	109 Une									

*	58 Ce	59 Pr	60 Nd	61 Pm	62 Sm	63 Eu	64 Gd	65 Tb	66 Dy	67 Ho	68 Er	69 Tm	70 Yb	71 Lu
†	90 Th	91 Pa	92 U	93 Np	94 Pu	95 Am	96 Cm	97 Bk	98 Cf	99 Es	100 Fm	101 Md	102 No	103 Lr

FIGURE 3-11
Metals, nonmetals, and the periodic table.

Nonmetals have a blue background. Except for hydrogen, they are found in the right-hand portion of the table. The metallic elements have a yellow background; the two groups at the extreme left of the table are the most metallic of all. The elements with the green background, metalloids, have some of the properties of metals and some of the properties of nonmetals. The unique elements in group 8A—the noble gases (purple background)—the special case of hydrogen (a nonmetal grouped with the metals), and other details of the periodic table are discussed in Chapter 9. Refer to the periodic table on the inside front cover for the atomic weights of the elements.

Our current interest is in the two broad categories of **metal** and **nonmetal.** All metals (except mercury, a liquid) are solids at room temperature. Metals generally share the physical properties of being malleable (capable of being flattened into thin sheets) and ductile (capable of being drawn into fine wires). They are also good conductors of heat and electricity and have a lustrous or shiny appearance. Nonmetals generally have the "opposite" properties of metals, e.g., they are generally poor conductors of heat and electricity. Several of the nonmetals are gases at room temperature (e.g., N_2 and O_2); some are brittle solids (e.g., Si and S); one, bromine, is a liquid.

Metal atoms have a tendency to lose one or more electrons when they form compounds with nonmetal atoms. The nonmetal atoms in these combinations show a tendency to gain one or more electrons. Since the loss of electron(s) produces a *positive* ion **(cation)** and the gain of electron(s), a *negative* ion **(anion),** the combination of a metal and a nonmetal produces an **ionic compound.** Compounds formed between nonmetal atoms involve the *sharing* of electrons, not a loss and gain. These compounds are called **covalent compounds.** Metals and nonmetals are identified in Figure 3-11.

Another feature of the periodic table we need to note is that similar elements are arranged into vertical groups or families. For example, the elements in group **1A** (except H) are called the **alkali metals;** the elements in group **2A** are the **alkaline earth metals;** the nonmetals of group **7A** are the **halogens;** group **6A,** the **oxygen family (chalcogens);** and group **5A,** the **nitrogen family.**

3-5 Oxidation States

Oxidation states (oxidation numbers)* reflect, in a general way, how electrons are involved in compound formation. Consider the case of NaCl. In this compound a

*Because oxidation state refers to a number, the term *oxidation number* is often used synonymously. We will use the two terms interchangeably. Another term whose meaning is similar to oxidation state is *valence*.

Na atom, a metal, loses one electron to a Cl atom, a nonmetal. The compound consists of the ions Na^+ and Cl^-, as pictured in Figure 3-2. Na is in the oxidation state $+1$ and Cl^-, -1.

In $MgCl_2$, an Mg atom loses two electrons to become Mg^{2+}, and each Cl atom gains one electron to become Cl^-. The oxidation state of Mg is $+2$ and of Cl, -1. If we take the *total* of the oxidation states of all the atoms (ions) in a formula unit of $MgCl_2$, we obtain $+2 - 1 - 1 = 0$.

In the molecule H_2O, if we *arbitrarily* assign H the oxidation state $+1$, then for the total of the oxidation states of the atoms to be zero, the oxidation state of oxygen must be -2. In another covalent molecule, Cl_2, if the two Cl atoms are to have the *same* oxidation state and if their total is to be zero, each oxidation state must itself be 0.

From the examples just given you can see that some conventions or rules must be followed in assigning oxidation states. The six rules listed below are sufficient to deal with all cases in this text, with this important qualification:

Rules for assigning oxidation states.

*Whenever two rules appear to contradict one another (which they often will), follow the rule that appears **higher** in the list.*
1. The oxidation state (**O.S.**) of an atom in the pure (uncombined) element is **0.**
2. The total of the O.S. of all the atoms in a molecule or formula unit is **0.** For an *ion* this total is equal to *the charge on the ion*, both in magnitude and sign.
3. In their compounds the alkali metals (periodic table group 1A, i.e., Li, Na, K, Rb, Cs, Fr) have an O.S. of **+1** and the alkaline earth metals (group 2A), **+2.**
4. In its compounds, the O.S. of hydrogen is **+1;** that of fluorine, **−1.**
5. In its compounds, oxygen has an oxidation state of **−2.**
6. In their binary (two-element) compounds with metals, the elements of group 7A have an oxidation state of **−1;** those of group 6A, **−2;** and those of group 5A, **−3.**

(3.4)

Example 3-11

Assigning oxidation states. What is the oxidation state of the underlined atom in each of the following? (a) $\underline{P}_4$; (b) $\underline{Al}_2O_3$; (c) $Mn\underline{O}_4^-$; (d) $Na\underline{H}$; (e) $H_2\underline{O}_2$; (f) $K\underline{O}_2$; (g) $\underline{Fe}_3O_4$.

Solution

(a) P_4: This formula represents a molecule of elemental phosphorus. For an atom of a pure element the O.S. = 0 (rule 1). The oxidation state of P in P_4 is 0.

(b) Al_2O_3: This formula represents a formula unit of the compound Al_2O_3. The total of the oxidation numbers of all the atoms must be 0 (rule 2). The O.S. of oxygen is -2 (rule 5). The total for three O atoms is -6. The total for two Al atoms is $+6$. The oxidation state of Al is $+3$.

(c) MnO_4^-: This is the formula for the permanganate *ion.* The total of the oxidation numbers of all the atoms in this ion must be -1 (rule 2). The total O.S. of the four O atoms is -8. The oxidation state of Mn is $+7$.

(d) NaH: This formula represents a formula unit of the *ionic* compound sodium hydride. Rule 3 states that Na should have an O.S. of $+1$. Rule 4 indicates that the O.S. of H should also be $+1$. If both atoms had the O.S. of $+1$, the total for the formula unit would be $+2$. This would violate rule 2.

Are You Wondering:

What is the significance of the nonintegral (fractional) values of the oxidation states in parts (f) and (g) of Example 3-11?

Because it is a system of electron bookkeeping based on a set of arbitrary rules, there is a certain artificiality to the concept of oxidation state. In parts (f) and (g) certain atoms have a *nonintegral* (fractional) oxidation state. We can rationalize the O.S. of $-\frac{1}{2}$ for oxygen in KO_2 if a molecule of O_2 gains one electron from a K atom, forming K^+ and O_2^-. In Fe_3O_4, two of the Fe atoms have an O.S. of $+3$ (as in Fe_2O_3) the third one has an O.S. of $+2$ (as in FeO). This results in an average of $+2\frac{2}{3}$ per Fe atom. You should gain more insight into the significance of the oxidation state concept in later chapters.

Rules 2 and 3 take precedence over rule 4. The O.S. of Na is $+1$, the total for the formula unit is 0, and the oxidation state of H is -1.

(e) H_2O_2: This formula represents a molecule of hydrogen peroxide. Rule 4, stating that H has an O.S. of $+1$, takes precedence over rule 5 (which says that the O.S. of oxygen is -2). The sum of the oxidation numbers of the two H atoms is $+2$ and that of the two O atoms is -2. The oxidation state of O in H_2O_2 is -1.

(f) KO_2: This formula represents a formula unit of the *ionic* compound potassium superoxide. Rule 3 (requiring that the O.S. state of $K = +1$) takes precedence over rule 5 (requiring that the O.S. of oxygen is -2). The sum of the oxidation numbers of the two O atoms is -1. The oxidation state of oxygen in KO_2 is $-\frac{1}{2}$.

(g) Fe_3O_4: This is a formula unit of the mixed oxide $Fe_2O_3 \cdot FeO$. The total of the oxidation numbers of four O atoms is -8 (-2 for each atom). For three Fe atoms the total must be $+8$. The oxidation state per Fe atom in Fe_3O_4 is $+2\frac{2}{3}$.

SIMILAR EXAMPLES: Exercises 15, 40, 41, 42, 87.

3-6 Systematic Nomenclature of Inorganic Compounds

Compounds formed by carbon and hydrogen or carbon and hydrogen in combination with oxygen, nitrogen, and a few other elements are organic compounds. They are generally considered in a special branch of chemistry—organic chemistry (see Chapter 27). Compounds that do not fit this description are called **inorganic compounds.** At this time we consider only the naming of inorganic compounds. In later chapters we have much more to say about each of the several categories of inorganic compounds that follow.

Binary Ionic Compounds. **Binary compounds** are those formed between *two* elements. A binary *ionic* compound is formed between a metal and a nonmetal. To name a binary ionic compound

Rules for naming binary ionic compounds.

- write the *unmodified* name of the **metal,** followed by
- the name of the **nonmetal,** modified to end in **"ide."**

(3.5)

TABLE 3-2
Some Simple Ions

Name	Symbol
Positive ions (cations)	
lithium	Li^+
sodium	Na^+
potassium	K^+
rubidium	Rb^+
cesium	Cs^+
magnesium	Mg^{2+}
calcium	Ca^{2+}
strontium	Sr^{2+}
barium	Ba^{2+}
aluminum	Al^{3+}
chromium(II)	Cr^{2+}
chromium(III)	Cr^{3+}
iron(II)	Fe^{2+}
iron(III)	Fe^{3+}
cobalt(II)	Co^{2+}
cobalt(III)	Co^{3+}
copper(I)	Cu^+
copper(II)	Cu^{2+}
zinc	Zn^{2+}
silver	Ag^+
mercury(I)	Hg_2^{2+}
mercury(II)	Hg^{2+}
tin(II)	Sn^{2+}
lead(II)	Pb^{2+}
Negative ions (anions)	
hydride	H^-
nitride	N^{3-}
oxide	O^{2-}
sulfide	S^{2-}
fluoride	F^-
chloride	Cl^-
bromide	Br^-
iodide	I^-

$$MgI_2 \qquad = \qquad \text{magnesium iodide}$$

Ionic compounds, though comprised of positive and negative ions, must be *electrically neutral*. That is, the net or total charge of the ions in a formula unit must be *zero*. This means

one Na^+ to *one* Cl^- (in NaCl), *one* Mg^{2+} to *two* I^- (in MgI_2), and so on.

The metal iron forms *two* common ions, $\mathbf{Fe^{2+}}$ and $\mathbf{Fe^{3+}}$. To distinguish between them we call the first, *iron(II)* ion, and the second, *iron(III)* ion. The Roman numeral immediately following the name of the metal indicates its oxidation state or simply the charge on the ion. To distinguish between $FeCl_2$ and $FeCl_3$, we call the former *iron(II) chloride* and the latter, *iron(III) chloride*.

With the ionic charges presented in Table 3-2, you should be able to relate the names and formulas of a host of binary ionic compounds, as illustrated in Examples 3-12 and 3-13.

Example 3-12

Writing formulas when names of compounds are given. Write formulas for the compounds: barium oxide, calcium fluoride, and magnesium nitride.

Solution. In each case identify the cations and their charges: Ba^{2+}, Ca^{2+}, and Mg^{2+}; then the anions and their charges: O^{2-}, F^-, and N^{3-}. Combine the cations and anions in the relative numbers required to produce electrically *neutral* formula units.

barium oxide: *one* Ba^{2+} and *one* O^{2-} = $\mathbf{BaO}$

calcium fluoride: *one* Ca^{2+} and *two* F^- = $\mathbf{CaF_2}$

magnesium nitride: *three* Mg^{2+} and *two* N^{3-} = $\mathbf{Mg_3N_2}$

Note that in the first case an electrically neutral formula unit resulted from the combination of the charges +2 and −2; in the second case, +2 and 2 × (−1); and in the third case, 3 × (+2) and 2 × (−3).

SIMILAR EXAMPLES: Exercises 14, 16, 17, 45.

Example 3-13

Naming compounds when their formulas are given. Write acceptable names for the compounds: Na_2S, AlF_3, Cu_2O.

Solution. This task is generally easier than that of Example 3-12 because all you need to do is identify the names of the ions appearing in each compound. (Balancing of ionic charges to obtain neutral formula units has already been done.) However, you must recognize that the cation in Cu_2O is Cu^+, that copper forms *two* different ions, and that these facts must be reflected in the name you assign.

Na_2S: sodium sulfide

AlF_3:   aluminum fluoride

Cu_2O: copper(I) oxide

SIMILAR EXAMPLES: Exercises 13, 17, 43.

CHEMISTRY EVOLVING

Naming Chemical Compounds

Dalton used *pictorial* symbols for atoms of the various elements (see Figure 2-3, for example). Berzelius (1813) was first to propose using chemical symbols for the elements [H = hydrogen, O = oxygen, C = carbon, Cu = copper (cuprum), etc.] and combinations of symbols and subscript numerals for the formulas of compounds (e.g., H_2O = water).

With the discovery of ever-increasing numbers of compounds during the nineteenth century, systematic nomenclature became more difficult. One difficulty, for example, was with binary compounds containing the same two elements but in different proportions, such as the two oxides of copper, Cu_2O and CuO. In Cu_2O, the oxidation state of the copper is +1, in CuO it is +2. To distinguish between them, Cu_2O was assigned the name *cuprous* oxide and CuO, *cupric* oxide. That is, the **"ous"** ending was used for the *lower* oxidation state and **"ic"** for the *higher*. This system is applied to several ions of variable oxidation state in Table 3-3.

The old "ous/ic" system is quite inadequate, however. There is no systematic way to apply it, for example, to the oxides of vanadium: VO, V_2O_3, VO_2, and V_2O_5. Although the "ous/ic" system remains in use and you should be familiar with it, the currently preferred system is the one based on Roman numerals to represent oxidation states. It was proposed by Alfred Stock shortly after World War I and began to receive widespread acceptance after World War II. It is usually referred to as the *Stock system of nomenclature*. The Stock system works well for ionic compounds, but there are some difficulties in applying it to covalent compounds. ■ ■ ■

TABLE 3-3

Two Systems of Nomenclature Compared

Ion	Systematic (Stock) name	Old name
Cr^{2+}	chromium(II)	chromous
Cr^{3+}	chromium(III)	chromic
Co^{2+}	cobalt(II)	cobaltous
Co^{3+}	cobalt(III)	cobaltic
Cu^+	copper(I)	cuprous
Cu^{2+}	copper(II)	cupric
Fe^{2+}	iron(II)	ferrous
Fe^{3+}	iron(III)	ferric
Hg_2^{2+}	mercury(I)	mercurous
Hg^{2+}	mercury(II)	mercuric

Binary Covalent Compounds. Naming a binary covalent compound is similar to naming an ionic compound. For example,

HCl   hydrogen chloride

In both the formula and the name, we write first the element with the positive oxidation state, that is, HCl and *not* ClH.

Some pairs of nonmetals form more than a single binary covalent compound, and we need to distinguish among them. The Stock system can be used for this purpose, but it is generally not as useful as a system based on *prefixes* to designate relative numbers of atoms.

Prefixes used in naming compounds.

> mono = 1;   di = 2;   tri = 3;   tetra = 4;
> penta = 5;   hexa = 6

(3.6)

Are You Wondering:

Why we do not use the names sodium(I) chloride for NaCl and magnesium(II) chloride for $MgCl_2$?

Although each of these proposed names would clearly indicate the compound in question, we should follow the general rule of writing *as simple a name as possible*. The metals of periodic group 1A (including Na) and of periodic group 2A (including Mg) have *only one ionic form*, and Roman numerals to designate oxidation states would be superfluous. Later in the book you will acquire a better understanding of the factors that determine whether an element can exist in several oxidation states. For now, use the information in Table 3-2 as a guide.

TABLE 3-4
Naming Binary Covalent Compounds

Formula	Name with prefixes[a]	Stock system name
BCl_3	boron trichloride	boron(III) chloride
CCl_4	carbon tetrachloride	carbon(IV) chloride
CO	carbon monoxide	carbon(II) oxide
CO_2	carbon dioxide	carbon(IV) oxide
NO	nitrogen monoxide	nitrogen(II) oxide
NO_2	nitrogen dioxide	nitrogen(IV) oxide
N_2O	dinitrogen monoxide	nitrogen(I) oxide
N_2O_3	dinitrogen trioxide	nitrogen(III) oxide
N_2O_4	dinitrogen tetroxide	nitrogen(IV) oxide[b]
N_2O_5	dinitrogen pentoxide	nitrogen(V) oxide
SF_6	sulfur hexafluoride	sulfur(VI) fluoride

[a] Where the prefix ends in an "a" or "o" and the element name begins with an "a" or "o," the final vowel of the prefix is often *dropped for ease of pronunciation*. For example, carbon monoxide (*not* monooxide) and dinitrogen tetroxide (*not* tetraoxide). However, PI_3 is phosphorus triiodide (*not* triodide) and SI_4 is sulfur tetraiodide (*not* tetriodide).

[b] To distinguish it from NO_2, N_2O_4 can be called the dimer of nitrogen(IV) oxide. A dimer (N_2O_4) is formed from two simpler monomers (NO_2).

Thus, for the two principal oxides of sulfur we can write

SO_2 = sulfur dioxide = sulfur(IV) oxide

SO_3 = sulfur trioxide = sulfur(VI) oxide

But for the following boron–bromine compound we have difficulty with the Stock system. The Stock name would lead to the wrong formula.

B_2Br_4 = *di*boron *tetra*bromide = boron(II) bromide [BBr_2 (?)]

Several binary covalent compounds are named in Table 3-4. Although it is not incorrect to use prefixes in naming all binary covalent compounds, the prefix *mono-* is treated in a special way. It is not applied to the first-named element and is often omitted for the second element as well. Thus, NO might be called nitrogen monoxide or nitrogen oxide, but *not mono*nitrogen *mono*xide. Finally, several substances have common names that are so well established that their systematic names are almost never used. For example,

H_2O = water

NH_3 = ammonia

Binary Acids. Certain binary covalent compounds containing hydrogen are called **acids.** The characteristic they share is that under appropriate conditions, such as

Are You Wondering:

*Why $MgCl_2$ is not called magnesium di*chloride *and $FeCl_3$, iron tri*chloride?

These names give a clear indication of the compounds in question, but they are *not the simplest names that we can write*. Since the magnesium ion can only be Mg^{2+}, the simple name magnesium chloride leads to the correct formula. Similarly, the name iron(III) chloride adequately represents the formula $FeCl_3$. In general the prefix system (mono, di, tri, . . .) is used only for covalent compounds.

TABLE 3-5
Some Common Polyatomic Ions

Learn the information in this table as quickly as you can. You will find a great deal of use for it as we proceed through the text.

Name	Formula	Typical compound
Cation		
ammonium	NH_4^+	NH_4Cl
Anions		
acetate	$C_2H_3O_2^-$	$NaC_2H_3O_2$
carbonate	CO_3^{2-}	Na_2CO_3
hydrogen carbonate[a] (or bicarbonate)	HCO_3^-	$NaHCO_3$
hypochlorite	ClO^-	$NaClO$
chlorite	ClO_2^-	$NaClO_2$
chlorate	ClO_3^-	$NaClO_3$
perchlorate	ClO_4^-	$NaClO_4$
chromate	CrO_4^{2-}	Na_2CrO_4
cyanide	CN^-	$NaCN$
hydroxide	OH^-	$NaOH$
nitrite	NO_2^-	$NaNO_2$
nitrate	NO_3^-	$NaNO_3$
permanganate	MnO_4^-	$NaMnO_4$
phosphate	PO_4^{3-}	Na_3PO_4
hydrogen phosphate[a]	HPO_4^{2-}	Na_2HPO_4
dihydrogen phosphate[a]	$H_2PO_4^-$	NaH_2PO_4
sulfite	SO_3^{2-}	Na_2SO_3
hydrogen sulfite[a] (or bisulfite)	HSO_3^-	$NaHSO_3$
sulfate	SO_4^{2-}	Na_2SO_4
hydrogen sulfate[a] (or bisulfate)	HSO_4^-	$NaHSO_4$
thiosulfate	$S_2O_3^{2-}$	$Na_2S_2O_3$

[a]These anion names are sometimes written as a single word, i.e., hydrogencarbonate, hydrogen-phosphate, etc.

NH_3 belongs to a complementary category of substances called **bases**. These substances yield hydroxide ion (OH^-) in water solutions, either because they contain OH^-—as do NaOH and KOH, for example—or because they produce OH^- through a reaction with water, as does NH_3.

when dissolved in water, they produce hydrogen ions (H^+).* HCl, when dissolved in water, dissociates into hydrogen ions (H^+) and chloride ions (Cl^-). NH_3 in water is *not* an acid; it shows practically no tendency to produce H^+ under any conditions. The most important binary acids are listed below. They can be named as covalent compounds (e.g., hydrogen chloride), but more often they are given special names to emphasis their acidic character. These names use the prefix "hydro," followed by the nonmetal name modified to an "ic" ending.

HF = **hydro**fluor**ic** acid

HCl = **hydro**chlor**ic** acid

HBr = **hydro**brom**ic** acid

HI = **hydro**iod**ic** acid

H_2S = **hydro**sulfur**ic** acid

Polyatomic Ions. The ions listed in Table 3-2 (with the exception of Hg_2^{2+}) are monatomic ions; they consist of a single atom. Ions in which two or more atoms are bonded together, *poly*atomic ions, are also commonly encountered, especially among the nonmetals. A number of polyatomic ions and representative compounds containing these ions are listed in Table 3-5, from which you can see that

*The species produced in water solution is actually more complex than the simple ion H^+. In Chapter 17 we refer to this species as hydronium ion H_3O^+. For present purposes we will not pursue this distinction and will use the simpler notation H^+.

1. Polyatomic anions occur more frequently than polyatomic cations. A common polyatomic cation is the ammonium ion, NH_4^+.
2. Very few polyatomic anions carry the "ide" ending in their names. Of those listed only OH^- (hydroxide ion) and CN^- (cyanide ion) do. The common endings are **"ite"** and **"ate"**, and some names carry the prefixes **"hypo"** or **"per"**.
3. An element common to many polyatomic anions is *oxygen.* The oxygen is combined with another nonmetal. Such anions are called **oxoanions.**
4. Certain nonmetals (e.g., Cl, N, P, and S) form a series of oxoanions containing different numbers of oxygen atoms. Their names are related to the oxidation state of the nonmetal atom to which the O atoms are bonded, ranging from "hypo" (lowest) to "per" (highest) according to the scheme

Scheme for naming oxoanions.

$$\text{Increasing oxidation state} \longrightarrow$$
$$\text{hypo\_\_\_\_ite} \qquad \text{\_\_\_\_ite} \qquad \text{\_\_\_\_ate} \qquad \text{per\_\_\_\_ate} \qquad (3.7)$$
$$\text{Increasing number of oxygen atoms} \longrightarrow$$

5. All the common oxoanions of Cl carry a charge of -1; of S, -2.
6. Some series of oxoanions also contain varying numbers of H atoms and are named accordingly. For example, HPO_4^{2-} is the *hydrogen phosphate* ion and $H_2PO_4^-$ is the *dihydrogen phosphate* ion.
7. The prefix "thio" signifies that a sulfur atom has been substituted for an oxygen atom. (The sulfate ion has *one* S and *four* O atoms; thiosulfate ion has *two* S and *three* O atoms.)

Oxoacids. The majority of acids contain *three* different elements (**ternary** compounds)—hydrogen, *oxygen,* and another nonmetal. These are called **oxoacids.** Or think of oxoacids as resulting from a combination of hydrogen ions (H^+) and oxoanions. The scheme for naming oxoacids is similar to that outlined in expression (3.7), except that the ending "ous" is used instead of "ite" and "ic" instead of "ate." Several oxoacids are listed in Table 3-6. Also listed are the names and formulas of the compounds that result when the hydrogen of an oxoacid is replaced by a metal such as sodium. Although we will consider more precise definitions later, these compounds are called **salts.** Acids are covalent compounds and salts are ionic compounds.

TABLE 3-6
Nomenclature of Some Oxoacids and Their Salts

Oxidation state	Formula of acid[a]	Name of acid	Formula of salt	Name of salt
+1	HClO	**hypo**chlor**ous** acid	NaClO	sodium **hypo**chlor**ite**
+3	HClO$_2$	chlor**ous** acid	NaClO$_2$	sodium chlor**ite**
+5	HClO$_3$	chlor**ic** acid	NaClO$_3$	sodium chlor**ate**
+7	HClO$_4$	**per**chlor**ic** acid	NaClO$_4$	sodium **per**chlor**ate**
+3	HNO$_2$	nitr**ous** acid	NaNO$_2$	sodium nitr**ite**
+5	HNO$_3$	nitr**ic** acid	NaNO$_3$	sodium nitr**ate**
+4	H$_2$SO$_3$	sulfur**ous** acid	Na$_2$SO$_3$	sodium sulf**ite**
+6	H$_2$SO$_4$	sulfur**ic** acid	Na$_2$SO$_4$	sodium sulf**ate**

In general the "ic" and "ate" names are assigned to compounds in which the central nonmetal atom has an oxidation state equal to the periodic group number. Halogen compounds are exceptional in that the "ic" and "ate" names are assigned to compounds in which the halogen has an oxidation state of +5 (even though the group number is 7, that is, 7A).

[a] In all these acids H atoms are bonded to O atoms, not the central nonmetal atom. Often formulas are written to reflect this fact, such as HOCl instead of HClO and HOClO instead of HClO$_2$.

Example 3-14

Applying the various rules in naming compounds. Name the compounds
(a) $CuCl_2$; (b) ClO_2; (c) HIO_4; (d) $Ca(H_2PO_4)_2$.

Solution

(a) The oxidation state of Cu is $+2$. Since Cu can also exist in the oxidation state $+1$, we should use the Stock system to distinguish clearly between the two possible chlorides. $CuCl_2$ is copper(II) chloride.

(b) Both Cl and O are nonmetals. ClO_2 is a binary covalent compound called chlorine dioxide.

(c) The oxidation state of I is $+7$. By analogy to the chlorine-containing oxoacids in Table 3-6, we should name this compound periodic acid (pronounced "purr eye oh dic" acid).

(d) The polyatomic anion $H_2PO_4^-$ is dihydrogen phosphate ion. Two of these ions are present for every Ca^{2+} ion in the compound calcium dihydrogen phosphate.

SIMILAR EXAMPLES: Exercises 13, 14, 17, 43, 44.

Do not make the common mistake of calling ClO_2 *chlorite*. ClO_2^- is chlorite *ion*, and ClO_2 is the *compound* chlorine dioxide. There can be no compound consisting of just a single type of ion—there is no substance called chloride or chlorite or chlorate and so on.

Example 3-15

Applying the various rules in writing formulas. Write acceptable formulas for the following compounds: (a) tetranitrogen tetrasulfide; (b) iron(III) oxide; (c) ammonium chromate; (d) bromic acid; (e) calcium hypochlorite.

Solution

(a) This is a covalent molecule containing *four* N atoms and *four* S atoms. The formula is N_4S_4.

(b) In a formula unit there are *two* Fe^{3+} ions $[2 \times (+3) = +6]$ and *three* O^{2-} ions $[3 \times (-2) = -6]$. The formula is Fe_2O_3.

(c) Two ammonium ions (NH_4^+) must be present for every chromate ion (CrO_4^{2-}). Place parentheses around NH_4^+, followed by the subscript 2. The formula is $(NH_4)_2CrO_4$. (This formula is read as "N, H, 4, taken twice, C, R, O, 4.")

(d) The "ic" acid for the oxoacids of the halogens has the halogen in the oxidation state $+5$. Bromic acid is $HBrO_3$.

(e) Here there are *one* Ca^{2+} and *two* ClO^- in a formula unit, leading to the formula $Ca(ClO)_2$.

SIMILAR EXAMPLES: Exercises 16, 17, 45, 87.

Note the importance of the parentheses here. If we omitted them, we would have $CaClO_2$, an *incorrect* formula for both calcium hypochlorite and calcium chlorite.

3-7 Compounds of Greater Complexity

In closing this discussion of chemical compounds we consider two additional types of compounds. In particular, we want to be able to interpret chemical formulas that have a different appearance from what we are used to.

$CoCl_2 \cdot 6H_2O$, cobalt(II) chloride hexahydrate. [Carey B. Van Loon]

Hydrates. In substances called **hydrates** each formula unit has associated with it a certain number of water molecules. The formula shown below signifies *six* H_2O molecules per formula unit of $CoCl_2$.

In general, dots (·) are used to signify that a chemical formula is a composite of two or more simpler formulas.

$K_3[Fe(CN)_6]$, potassium ferricyanide or potassium hexacyanoferrate(III). [Carey B. Van Loon]

A complex ion is generally represented within bracket symbols ([ ]), as in $[Fe(CN)_6]^{3-}$ and $K_3[Fe(CN)_6]$.

$CoCl_2 \cdot 6H_2O$

Using the prefix notation of (3.6), we should call this compound cobalt(II) chloride *hexa*hydrate. Its formula weight is that of $CoCl_2$ *plus* that associated with six H_2O: $129.8 + (6 \times 18.02) = 237.9$. We can speak of the percent water in a hydrate. For $CoCl_2 \cdot 6H_2O$ this is

$$\% \ H_2O = \frac{(6 \times 18.02) \ g \ H_2O}{237.9 \ g \ CoCl_2 \cdot 6H_2O} \times 100 = 45.45\%$$

In Section 25-11 we learn of the different ways in which water of hydration may be incorporated in a solid compound.

Coordination Compounds. A coordination compound contains ions (called complex ions) that are themselves made up of simpler ions and/or molecules. In the complex anion $[Fe(CN)_6]^{3-}$ there are six CN^- ions bonded to one Fe^{3+} ion. When combined with cations, e.g., K^+, the result is the coordination compound $K_3[Fe(CN)_6]$. Complex ions and coordination compounds (including their nomenclature) are discussed in Chapter 25. For the present all that you need to understand is information of this type.

- The total number of atoms in a formula unit of $K_3[Fe(CN)_6]$ is 16, i.e., 3 K, 1 Fe, 6 C, and 6 N atoms.
- The formula weight of the compound is $(3 \times 39.10) + 55.85 + (6 \times 12.01) + (6 \times 14.01) = 329.3$.

FOCUS ON
Analytical Chemistry

Atomic absorption spectrophotometry—a modern analytical method. The sample to be analyzed is vaporized in the flame, and the light-absorbing ability of the vaporized atoms is measured. [Robert Frerck/Woodfin Camp & Associates]

To analyze something means to separate it into its constituent parts in order to learn more about these constituents. The branch of chemistry that deals with the analysis of chemical compounds and mixtures is called **analytical chemistry.** At times all that we seek is a knowledge of what substances are present in a sample, such as in the questions

- Does this brand of paint contain lead compounds?
- Is there nitrate ion in this sample of drinking water?
- What metals comprise this sample of stainless steel?

To answer these questions we need the methods of **qualitative analysis.** If our concern is with the *actual quantity* of one or more of the constituents of a sample, we need to do a **quantitative analysis.** Typical quantitative analyses might include determining

- the percentage composition of a physiologically active compound extracted from a tropical plant,
- the percentage of iron in an iron ore sample,
- the number of parts per million (ppm) of mercury in a fish.

There is almost no limit to the number of different methods employed by analytical chemists, but these methods can generally be categorized as chemical methods or as instrumental methods. **Chemical methods** are based on subjecting the sample of interest to chemical reactions. In some way the outcome of these reactions depends on the composition of the sample. Quite often it is possible simply to measure certain physical properties of the sample, usually with sophisticated instruments. This type of analysis, known as **instrumental analysis,** is especially important. It can yield accurate results on small samples, often when constituents are present only in trace amounts, and sometimes without having to destroy the sample being analyzed.

We can further subdivide chemical methods into two broad categories: volumetric and gravimetric analyses. In **volumetric analyses** chemical reactions are conducted in solutions, and the key measurements involve solution volumes. We introduce this subject in Section 5-3. In **gravimetric analyses** the key data are the masses of substances, measured on an analytical balance. We have already considered two examples of gravimetric analysis in this chapter—**combustion analysis** and **precipitation analysis.** In both of these cases we dealt with masses of substances present before and after chemical reactions.

An instrumental method we have already encountered is **mass spectrometry.** Although we emphasized its use in

atomic weight determinations (Section 2-6), it has a wide variety of analytical applications. Two other important instrumental methods are **spectroscopy** (analysis of samples as a result of their emission or absorption of radiant energy) and **chromatography** [separation of components of a mixture by their differing affinities for two different phases (states) of matter with which they come into contact]. Two instrumental methods discussed later in this book are based on the relationship between electricity and chemical change (**electrochemical methods**) and on phenomena associated with radioactivity (**radiochemical methods**). New instrumental methods continue to be developed, making possible the analysis of increasing numbers of substances and in ever smaller amounts. In some cases robots are used for the rapid handling of samples, and computers are used to control analytical instruments and to process analytical data. Analytical chemistry is a dynamic field!

Summary

Chemical formulas, once established, provide a wealth of information about chemical compounds, serving as the basis for determining

- formula weights,
- molecular weights,
- molar masses,
- percent compositions,
- various relationships among the elements present.

To establish chemical formulas requires *experimental* methods for determining percent compositions of compounds. For organic compounds this usually involves combustion analysis. A straightforward calculation can then be used to convert these percentages to empirical formulas. Other calculations based on chemical formulas, combined with data from precipitation analyses, can be used to establish atomic weights for certain elements.

Chemical formulas determined from experimental data are empirical formulas—the simplest formulas that can be written. In some cases the experimentally determined molecular weight is equal to the formula weight calculated from the empirical formula; the empirical formula is that of an actual molecule of the compound. In other cases the experimentally determined molecular weight is a multiple of the formula weight, and the molecular formula is the corresponding multiple of the empirical formula. Experimental methods for determining molecular weights are considered in later chapters (e.g., 6 and 13).

Chemical compounds are designated by names as well as by formulas. Three important ideas are introduced in this chapter for relating names and formulas (nomenclature): metals and nonmetals, the periodic classification of the elements, and oxidation states. These ideas are used in naming binary ionic and covalent compounds, polyatomic ions, and oxoacids and their salts. Also, formulas are written and interpreted for somewhat more complex chemical substances—hydrates and coordination compounds.

Summarizing Example

(a)   (b)   (c)   (d)

FIGURE 3-12

Decomposition of a hydrate of copper(II) sulfate.

(a) Original fully hydrated copper(II) sulfate.
(b) Product obtained by heating to 140 °C.
(c) Product obtained by heating to 400 °C.
(d) Product obtained by heating to 1000 °C.

[Carey B. Van Loon]

Copper sulfate is the most widely used copper compound, finding applications in electroplating, in pesticide mixtures, as a wood preservative, and as an algicide (for killing algae in lakes). When heated, the hydrated solid goes through the succession of changes suggested by Figure 3-12. Here are some typical student data.*

A 2.574-g sample of hydrated copper(II) sulfate, $CuSO_4 \cdot xH_2O$, was heated to 140 °C, cooled to room temperature, and reweighed. The resulting solid was reheated to 400 °C, cooled to room temperature and reweighed. Finally, this solid was heated to 1000 °C; the resulting black residue was cooled to room temperature and weighed for a final time. The measured masses were

original sample:	2.574 g
after heating to 140 °C:	1.833 g
after reheating to 400 °C:	1.647 g
after reheating to 1000 °C:	0.812 g

1. Assuming that all the water of hydration is driven off after heating to 400 °C, what must be the formula of the hydrate $CuSO_4 \cdot xH_2O$?

Solution. The 1.647-g mass corresponds to the *anhydrous* compound $CuSO_4$. The loss of mass $(2.574 - 1.647)$ g $= 0.927$ g is the mass of water of hydration in the original sample. Convert the masses of the anhydrous compound and of the water to number of moles.

$$\text{no. mol } CuSO_4 = 1.647 \text{ g } CuSO_4 \times \frac{1 \text{ mol } CuSO_4}{159.6 \text{ g } CuSO_4} = 0.01032 \text{ mol } CuSO_4$$

$$\text{no. mol } H_2O = 0.927 \text{ g } H_2O \times \frac{1 \text{ mol } H_2O}{18.02 \text{ g } H_2O} = 0.05144 \text{ mol } H_2O$$

Now, write a formula based on these molar amounts

$$(CuSO_4)_{0.01032} \cdot (H_2O)_{0.05144}$$

and reduce to the simplest formula.

$$(CuSO_4)_{\frac{0.01032}{0.01032}} \cdot (H_2O)_{\frac{0.05144}{0.01032}} = (CuSO_4) \cdot (H_2O)_{4.98} = CuSO_4 \cdot 5H_2O$$

(This example is similar to Example 3-7.)

2. What is the formula of the hydrate obtained after the original solid is heated to 140 °C?

*A. D. Harris and L. H. Kalbus, *J. Chem. Educ.*, **56**, 417 (1979).

Solution. The loss of water occurring at 140 °C is $(2.574 - 1.833) = 0.741$ g. The mass of water *remaining* at 140 °C is the *total* loss of water (0.927 g) minus the loss that occurs at 140 °C (0.741 g): $0.927 - 0.741 = 0.186$ g H_2O. Now express this amount of water on a mole basis.

$$\text{no. mol } H_2O = 0.186 \text{ g } H_2O \times \frac{1 \text{ mol } H_2O}{18.02 \text{ g } H_2O} = 0.01032 \text{ mol } H_2O$$

The number of moles of $CuSO_4$ was determined in part 1. It is 0.01032 mol. The $CuSO_4$ and the H_2O are in a 1:1 mol ratio. The formula of the hydrate obtained at 140 °C is $CuSO_4 \cdot H_2O$.

(This example is similar to Example 3-3.)

3. The black residue obtained at 1000 °C is an oxide of copper. What is its percentage composition and its empirical formula?

Solution. Recall that the 1.647-g anhydrous $CuSO_4$ sample corresponds to 0.01032 mol. The 0.01032 mol Cu present in this sample appears in the 0.812 g of copper oxide residue. The mass of Cu, its percentage, and the percent O in copper oxide are

$$\text{no. g Cu} = 0.01032 \text{ mol Cu} \times \frac{63.546 \text{ g Cu}}{1 \text{ mol Cu}} = 0.6558 \text{ g Cu}$$

$$\% \text{ Cu} = \frac{0.6558 \text{ g Cu}}{0.812 \text{ g cpd.}} \times 100 = 80.8\% \qquad \% \text{ O} = 100.0 - 80.8 = 19.2\%$$

We can readily determine the empirical formula from the percentage composition by the method established in this chapter. (That is, base the calculation on a 100.0-g sample.)

$$\text{no. mol Cu} = 80.8 \text{ g Cu} \times \frac{1 \text{ mol Cu}}{63.55 \text{ g Cu}} = 1.27 \text{ mol Cu}$$

$$\text{no. mol O} = 19.2 \text{ g O} \times \frac{1 \text{ mol O}}{16.00 \text{ g O}} = 1.20 \text{ mol O}$$

The empirical formula is $Cu_{1.27}O_{1.20} = Cu_{1.06}O = CuO$.

(This example is similar to Examples 3-5 and 3-8.)

Key Terms

acid (3-6)	formula unit (3-1)	nonmetal (3-4)
anion (3-4)	formula weight (3-1)	oxidation state (3-5)
binary acid (3-6)	hydrate (3-7)	oxoacid (3-6)
binary compound (3-6)	ionic compound (3-4)	oxoanion (3-6)
cation (3-4)	metal (3-4)	polyatomic ion (3-6)
chemical formula (3-1)	molecular formula (3-1)	precipitation analysis (3-2)
combustion analysis (3-2)	molecular weight (3-1)	salt (3-6)
coordination compound (3-7)	molecule (3-1)	stoichiometry (introduction)
covalent compound (3-4)	nomenclature (3-3)	ternary compound (3-6)
empirical formula (3-1)		

Highlighted Expressions

Definition of formula weight (3.1)	Rules for naming binary ionic compounds (3.5)
Definition of molecular weight (3.2)	Prefixes used in naming compounds (3.6)
Rules for assigning oxidation states (3.4)	Scheme for naming oxoanions (3.7)

Review Problems

1. Calculate the indicated quantities.
(a) The total *number* of molecules in 2.65 mol H_2O
(b) The *mass* of 3.65 mol SO_2
(c) The total *number* of F atoms in 1.25 kg $C_2HBrClF_3$

2. Determine the *mass* of each of the following: **(a)** 22.5 mol $CuSO_4 \cdot 5H_2O$; **(b)** a quantity of liquid ethanol consisting of 5.00×10^{25} molecules of C_2H_5OH; **(c)** a quantity of the essential amino acid lysine, $C_6H_{14}N_2O_2$, containing 1.50 mol N atoms.

3. The amino acid methionine is essential in human diets. It has the molecular formula $C_5H_{11}NO_2S$. Determine **(a)** its molecular weight; **(b)** the number of moles of H in 6.21 mol methionine; **(c)** the number of C atoms in 3.18 mol methionine; **(d)** the number of grams of O per gram of N in the compound.

4. If a sample of $MgBr_2$ is to contain 4.16×10^{25} Br^- ions,
(a) How many Mg^{2+} ions will the sample contain?
(b) How many formula units of $MgBr_2$ will be present?
(c) What will be the mass of the sample?

5. Determine the % O, by mass, in the mineral malachite, $Cu_2(OH)_2CO_3$.

6. Determine the percent, by mass, of the indicated element in each of the following.
(a) Pb in tetraethyllead, $Pb(C_2H_5)_4$, once extensively used as an antiknock additive in gasoline;
(b) Fe in Prussian blue, $Fe_4[Fe(CN)_6]_3$, a pigment used in paints and printing inks;
(c) S in sodium thiosulfate pentahydrate, $Na_2S_2O_3 \cdot 5H_2O$, used as a photographic fixing agent (to remove unexposed silver salts from photographic negatives);
(d) W in ammonium metatungstate, $(NH_4)_6H_2W_{12}O_{40}$, used in the preparation of high-purity tungsten metal.

7. An oxide of cobalt used in glazing pottery contains 71.06% Co and 28.94% O. What is the empirical formula of the oxide?

8. Rubbing alcohol, isopropyl alcohol, is a carbon–hydrogen–oxygen compound with 59.96% C and 13.42% H. What is its empirical formula?

9. A compound of C, H, and O, known as terephthalic acid, is used in the manufacture of Dacron. Its molecular weight is 166.1, and by combustion analysis it is found to have 57.83% C and 3.64% H. What is the *molecular* formula of terephthalic acid?

10. Ibuprofen is a carbon–hydrogen–oxygen compound used in painkillers. When a 1.235-g sample is burned completely, it yields 3.425 g CO_2 and 0.971 g H_2O.
(a) What is the percentage composition, by mass, of ibuprofen?
(b) What is the empirical formula of ibuprofen?

11. The iodide ion in a 1.552-g sample of the ionic compound MI is removed through precipitation. The precipitate is found to

contain 1.186 g I. What is the element M? [*Hint:* Follow the sequence of steps outlined in the figure to find the atomic weight of M.]

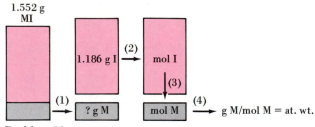

Problem 11

12. An oxide of chromium used in chrome plating has a formula weight of 100.0 and contains *four* atoms per formula unit. What must be the formula of this compound?

13. Name the following compounds: **(a)** LiI; **(b)** $CaCl_2$; **(c)** NaCN; **(d)** NH_4NO_3; **(e)** ICl_3; **(f)** N_2O_3; **(g)** PCl_5; **(h)** HClO; **(i)** $KBrO_3$.

14. Supply the missing information (name or formula) for each of the following ions.
(a) tin(II), _____
(b) _____, Co^{3+}
(c) _____, Mg^{2+}
(d) chromium(II), _____
(e) iodate, _____
(f) _____, ClO_2^-
(g) gold(III), _____
(h) _____, HSO_4^-
(i) hydrogen carbonate, _____
(j) hydroxide, _____

15. Indicate the oxidation state of the underlined element in each of the following: **(a)** $\underline{Al}$; **(b)** $Li_2\underline{S}$; **(c)** $\underline{N}O_2$; **(d)** $Br\underline{F}_5$; **(e)** $H\underline{N}O_2$; **(f)** $K_2\underline{Mn}O_4$; **(g)** $\underline{V}^{3+}$; **(h)** $\underline{Cr}_2O_7^{2-}$; **(i)** $H\underline{P}O_4^{2-}$.

16. Write correct formulas for the following compounds: **(a)** magnesium oxide; **(b)** strontium fluoride; **(c)** barium hydroxide; **(d)** cesium carbonate; **(e)** mercury(II) nitrate; **(f)** iron(III) sulfate; **(g)** magnesium perchlorate; **(h)** potassium hydrogen carbonate; **(i)** nitrogen trichloride; **(j)** bromine pentafluoride; **(k)** copper(I) sulfide; **(l)** potassium hydrogen phosphate; **(m)** ferrous chloride.

17. Supply the name or formula of each of the following acids.
(a) _____ = hydrobromic acid **(b)** $HClO_2$ = _____
(c) _____ = iodic acid **(d)** H_2SO_3 = _____
(e) _____ = phosphoric acid **(f)** H_2Se = _____
(g) _____ = chloric acid **(h)** HNO_2 = _____

18. What is the percent, by mass, of water in the hydrate $ZnSO_4 \cdot 7H_2O$?

Exercises

Terminology
19. Explain the distinction between the terms in each pair: **(a)** formula unit and molecule; **(b)** empirical formula and molec-

ular formula; **(c)** cation and anion; **(d)** binary and ternary compound; **(e)** oxoacid and oxoanion; **(f)** hypo _____ ous and per _____ ate.

The Avogadro constant and the mole

20. In 1.12 mol of the compound Li_2O
(a) What fraction of the total *number* of ions is O^{2-}?
(b) What fraction of the total *mass* is contributed by O^{2-}?
(c) What is the total *number* of Li^+ ions?

21. Refer to Example 3-2. How many molecules of C_2H_6S must be present in the 5000-m^3 lecture room to be detectable? [*Hint:* Start with the answer to Example 3-2.]

22. In rhombic sulfur, S atoms are joined into the molecules S_8 (see Figure 3-3). If the density of rhombic sulfur is 2.07 g/cm^3, determine for a crystal of volume 6.15 mm^3: **(a)** the number of moles of S_8 present and **(b)** the total number of S atoms.

23. A typical adult body contains about 6 liters of blood. The hemoglobin content of the blood is about 15.5 g/100 mL of blood. The approximate molecular weight of hemoglobin is 64,500, and there are four iron (Fe) atoms in a hemoglobin molecule. Approximately how many Fe atoms are present in the blood of a typical adult?

24. Imagine a supply of grains of sand, each grain having a roughly spherical shape with average radius of 0.25 mm. Sand (SiO_2) has a density of 2.3 g/cm^3.
(a) How many grains of sand must be taken to contain 1.00 mol SiO_2?
(b) What would be the mass of a quantity of sand described as "1.00 mol grains of sand?"
(c) Explain why "1.00 mol grains of sand" would not be a unique quantity, but "1.00 mol SiO_2" is?

25. Some substances that are only very slightly soluble in a liquid will spread over the surface of the liquid to produce a film that is *one molecule thick*. This film is called a *monomolecular layer* or *monolayer*. One practical use of this phenomenon is to cover ponds to reduce loss of water by evaporation. A molecule of stearic acid, $C_{18}H_{36}O_2$, has a cross-sectional area of $2.2 \times 10^{-15} cm^2$. What area of water surface, in m^2, would be covered by a monolayer made from 10.0 g stearic acid? [*Hint:* Refer to the sketch below.]

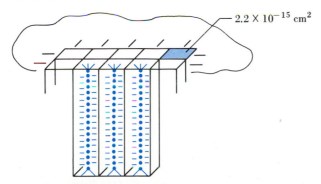

Exercise 25

26. If stearic acid has a density of 0.85 g/cm^3, what is the thickness, in nanometers, of the monolayer described in Exercise 25? This thickness corresponds to the length of a stearic acid molecule.

Chemical formulas

27. Indicate which of the following statements is(are) correct concerning glucose (blood sugar), $C_6H_{12}O_6$.
(a) The percentages, by mass, of C and O are the same as in CO.

(b) The ratio of C to H to O atoms is the same as in acetic acid, CH_3COOH.
(c) The proportions, by mass, of C and O are equal.
(d) The highest percentage, by mass, is that of H.

28. Each of the following formulas represents an actual substance. Which are molecular formulas? Can you tell whether the others are empirical or molecular formulas? Explain. **(a)** C_2H_6; **(b)** Cl_2O; **(c)** CH_4O; **(d)** N_2O_4.

29. For the compound $Ge[S(CH_2)_4CH_3]_4$, determine
(a) the total number of atoms in one formula unit
(b) the ratio, by *number*, of C atoms to H atoms
(c) the ratio, by *mass*, of Ge to S
(d) the number of g S in 1 mole of the compound
(e) the mass of compound required to contain 1.00 g Ge
(f) the number of C atoms in 33.1 g of the compound

30. Determine the percent, *by number of atoms*, of the indicated element in each of the following:
(a) N in tenormin, $C_{14}H_{22}N_2O_3$, used in the treatment of high blood pressure
(b) H in ammonium stearate, $C_{17}H_{35}COONH_4$, used in vanishing creams and other cosmetic products
(c) Cr in chrome alum, $CrK(SO_4)_2 \cdot 12H_2O$, used in tanning hides to produce leather
(d) N in glyceryl trinitrate (nitroglycerin), CH_2NO_3CH-$NO_3CH_2NO_3$, used in the treatment of heart ailments and the manufacture of high explosives

Percent composition of compounds

31. What is the percent, by mass, of boron in the mineral axinite, $HCa_3Al_2BSi_4O_{16}$?

32. All of the substances listed below are of value in fertilizers because they supply the element nitrogen. Which of these is the richest source of nitrogen on a percent by *mass* basis? urea, $CO(NH_2)_2$; ammonium nitrate, NH_4NO_3; or guanidine, $HNC(NH_2)_2$.

33. Three different brands of "liquid chlorine" for use in purifying water in home swimming pools all cost $1.50 per gallon and are water solutions of NaOCl. Brand A contains 10% OCl by mass; brand B, 7% available chlorine (Cl) by mass; and brand C, 14% NaOCl by mass. Which of the three brands is the best buy?

34. *Without performing detailed calculations* explain which of the following compounds has the greatest % S, by mass: SO_2, SO_3, $MgSO_4$, Li_2S.

Chemical formulas from percent composition

35. A compound of carbon and hydrogen consists of 93.71% C and 6.29% H, by mass. The molecular weight of the compound is found to be 128. What is its molecular formula?

36. Selenium, an element used in the manufacture of photoelectric cells and solar energy devices, forms two oxides. One has 28.8% O, by mass, and the other, 37.8% O. What are the formulas of these oxides? Propose acceptable names for them.

37. What is the empirical formula of
(a) the rodenticide (rat killer) warfarin, which consists of 74.01% C, 5.23% H, and 20.76% O;
(b) the chemical warfare agent, mustard gas, which consists of 30.20% C, 5.07% H, 44.58% Cl, and 20.16% S;
(c) benzo[*a*]pyrene, a cancer-causing agent found in cigarette smoke and smoke produced in the charcoal grilling of meat, which consists of 95.21% C and 4.79% H?

*38. A *binary* acid, H_xE, consists of 2.49% H. What is the formula of this acid? [*Hint:* What must the *nonmetal* E be? Note that there is but *one* atom of E per molecule, but there may be more than one atom of H.]

*39. Freons are compounds containing C, Cl, F, and possibly H. They are used as refrigerants and as blowing agents in forming foam plastics. They are derived from hydrocarbons by replacing some or all of the H atoms with F and Cl (e.g., $CHCl_2F$ from CH_4). What is the formula of a freon derived from CH_4 that has 31.43% F, by mass?

Oxidation states

40. Indicate the oxidation state of the underlined element in each of the following: (a) $\underline{C}H_4$; (b) $\underline{S}F_4$; (c) $Na_2\underline{O}_2$; (d) $\underline{C}_2H_3O_2^-$; (e) $\underline{Fe}O_4^{2-}$; (f) $\underline{S}_4O_6^{2-}$.

41. Arrange the following sulfur-containing anions in order of *increasing* oxidation state of S: SO_3^{2-}; $S_2O_3^{2-}$; $S_2O_8^{2-}$; HSO_4^-; HS^-; $S_4O_6^{2-}$.

42. Nitrogen forms five different compounds with oxygen. Write appropriate formulas for these compounds if the oxidation state of N in them is +1, +2, +3, +4, and +5, respectively.

Nomenclature

43. Name the following compounds: (a) BaS; (b) ZnO; (c) K_2CrO_4; (d) Cs_2SO_4; (e) Cr_2O_3; (f) $FeSO_4$; (g) $Mg(HCO_3)_2$; (h) $(NH_4)_2HPO_4$; (i) $Ca(HSO_3)_2$; (j) $Cu(OH)_2$; (k) HNO_3; (l) $KClO_4$; (m) KIO; (n) $LiCN$; (o) $HBrO_3$; (p) H_3PO_3.

44. Assign suitable names to the following compounds: (a) ICl; (b) ClF_3; (c) SF_4; (d) BrF_5; (e) CS_2; (f) $SiCl_4$; (g) N_4S_4; (h) SF_6.

45. Write correct formulas for these compounds: (a) aluminum sulfate; (b) potassium chromate; (c) silicon tetrafluoride; (d) lead(II) acetate; (e) iron(III) oxide; (f) tricarbon disulfide; (g) cobalt(II) nitrate; (h) strontium nitrite; (i) chlorine dioxide; (j) tin(IV) oxide; (k) calcium dihydrogen phosphate; (l) hydrobromic acid; (m) iodic acid; (n) aluminum phosphate; (o) phosphorous dichloride trifluoride; (p) tetrasulfur dinitride; (q) cupric sulfate; (r) chromous chloride.

Hydrates

46. *Without performing detailed calculations,* indicate which of the following hydrates has the greatest % H_2O, by mass.

(a) $CuSO_4 \cdot 5H_2O$ (b) $Cr_2(SO_4)_3 \cdot 18H_2O$
(c) $MgCl_2 \cdot 6H_2O$ (d) $LiC_2H_3O_2 \cdot 2H_2O$

47. Anhydrous $CuSO_4$ can be used to dry liquids in which it is insoluble. The $CuSO_4$ is converted to $CuSO_4 \cdot 5H_2O$, which can be filtered off from the liquid. What minimum mass of anhydrous $CuSO_4$ would be required to remove 8.5 mL H_2O that had inadvertently become mixed with a tankful of gasoline?

48. Refer to the Summarizing Example. Under certain conditions copper sulfate forms the *tri*hydrate $CuSO_4 \cdot 3H_2O$. If these conditions had been used in the Summarizing Example, what would have been the mass of trihydrate obtained?

49. A sample of $MgSO_4 \cdot xH_2O$ weighing 8.129 g is heated until all the water of hydration is driven off. The resulting anhydrous compound, $MgSO_4$, weighs 3.967 g. What is the formula of the hydrate?

Combustion analysis

50. An 0.1888-g sample of a hydrocarbon produces 0.6260 g CO_2 and 0.1602 g H_2O in combustion analysis. Its molecular weight is found to be 106. For this hydrocarbon, determine (a) its percent composition; (b) its empirical formula; (c) its molecular formula.

51. Para-cresol is used as a disinfectant and in the manufacture of herbicides and artificial food flavors. A 0.4039-g sample of this carbon–hydrogen–oxygen compound yields 1.1518 g CO_2 and 0.2694 g H_2O in combustion analysis. What is the empirical formula of para-cresol?

52. Dimethylhydrazine is a carbon–hydrogen–nitrogen compound with important uses in rocket fuels. When burned completely, a 0.312-g sample yields 0.458 g CO_2 and 0.374 g H_2O. From a separate 0.525-g sample, the nitrogen content is converted to 0.244 g N_2. What is the empirical formula of dimethylhydrazine?

53. Refer to Example 3-6. What mass of BHA should be burned to yield 0.500 g CO_2 in combustion analysis?

*54. In a hydrocarbon, C_xH_y, the carbon atoms are bonded to one another and the H atoms to the C atoms, never to other H atoms. Furthermore, a C atom always forms *four* bonds. For the bottled petroleum gas propane, C_3H_8, these ideas can be represented through the structural formula

$$
\begin{array}{ccccccc}
 & H & & H & & H & \\
 & | & & | & & | & \\
H- & C & - & C & - & C & -H \\
 & | & & | & & | & \\
 & H & & H & & H & \\
\end{array}
$$

(Recall, also, Figure 3-4.) Can there be a hydrocarbon which, on complete combustion, yields a greater mass of H_2O than of CO_2? Explain.

Precipitation analysis

55. A 0.5155-g sample of KI is dissolved in water, and all the iodide ion present is precipitated as AgI. How many grams of pure, dry AgI are obtained?

56. Refer to Example 3-9. A particular type of brass contains the elements Cu, Sn, Pb, and Zn. A sample weighing 1.713 g is treated in such a way as to convert the Sn to 0.245 g SnO_2, the Pb to 0.115 g $PbSO_4$, and the Zn to 0.246 g $Zn_2P_2O_7$. What is the percent, by mass, of each element in the brass?

57. A 110.520-g sample of mineral water is analyzed for its magnesium content. The Mg^{2+} in the sample is first precipitated as $MgNH_4PO_4$, and this precipitate is then converted to $Mg_2P_2O_7$, which is found to weigh 0.0549 g. Express the quantity of magnesium in the sample in parts per million (that is, in grams of Mg per million grams H_2O).

*58. A 1.013-g sample of $ZnSO_4 \cdot xH_2O$ is dissolved in water and the sulfate ion precipitated as $BaSO_4$. The mass of pure, dry $BaSO_4$ obtained is 0.8223 g. What is the formula of the zinc sulfate hydrate?

Atomic weight determinations

59. Two compounds of Cl and X are found to have molecular weights and % Cl, by mass, as follows: mol. wt. = 137, 77.5% Cl; mol. wt. = 208, 85.1% Cl. What is the element X? What is the formula of each compound?

60. A sample of the compound MSO_4 weighing 0.1131 g reacts with barium chloride and yields 0.2193 g $BaSO_4$. What must be the atomic weight of the metal M?

61. The chlorine present in an 0.5250-g sample of the compound MCl_2 is precipitated as 0.5070 g AgCl. What is the atomic weight of M? [*Hint:* Use the figure as a guide.]

62. The metal M forms the sulfate $M_2(SO_4)_3$. An 0.738-g sample of this sulfate is converted to 1.511 g $BaSO_4$. What is the atomic weight of M?

★**63.** An 0.622-g sample of a metal oxide with the formula M_2O_3 is converted to the sulfide, MS, yielding 0.685 g. What is the atomic weight of the metal M?

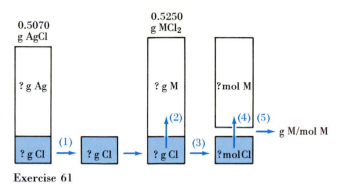

Exercise 61

<hr>

Additional Exercises

64. Refer to Example 3-3. In the substance halothane determine **(a)** the *total number* of atoms in one formula unit; **(b)** the *ratio* of F atoms to C atoms; **(c)** the ratio, *by mass*, of Br to F in the compound; **(d)** the mass of compound to contain 1.00 g F.

65. Determine the percent, *by mass*, of each of the elements in the antimalarial drug quinine, $C_{20}H_{24}N_2O_2$.

66. Table 1-4 lists the production of key chemicals on a *mass* basis. Reorder the first *ten* chemicals of that table on a *mole* basis. That is, of which chemical is the greatest number of moles produced? second greatest? etc.

67. Spodumene has the formula $Li_2O \cdot Al_2O_3 \cdot 4SiO_2$. Given that the percentage of Li-6 atoms in naturally occurring lithium compounds is 7.40%, how many Li-6 atoms are present in a 426-g sample of spodumene?

68. A public water supply was found to contain 1 part per billion (ppb) by mass of chloroform, $CHCl_3$. (Consider this to be essentially 1.00 g $CHCl_3$ per 10^9 g water.)

    (a) How many $CHCl_3$ molecules would be present in a glassful of this water (250 mL)?

    (b) If the $CHCl_3$ found in **(a)** could be isolated, would this quantity be detectable on an ordinary analytical balance that measures mass to about ±0.0001 g?

69. All of the following minerals are semiprecious or precious stones. Determine the percent, by mass, of the indicated element in each one. **(a)** Zr in zircon, $ZrSiO_4$; **(b)** Be in beryl (emerald), $Be_3Al_2Si_6O_{18}$; **(c)** Fe in almandine (garnet), $Fe_3Al_2Si_3O_{12}$; **(d)** S in lazurite (lapis lazuli), $Na_4SSi_3Al_3O_{12}$.

70. The most important natural sources of boron compounds [used in the manufacture of glass (Pyrex), cleaning compounds, fire retardants, and bleaches] are the minerals *kernite*, $Na_2B_4O_7 \cdot 4\,H_2O$, and *borax*, $Na_2B_4O_7 \cdot 10\,H_2O$. To obtain boron compounds containing 1.00 kg B, how much *additional* mass of mineral must be processed if borax rather than kernite is used as the source of boron?

71. The food-flavor enhancer monosodium glutamate (MSG) has the composition 13.6% Na, 35.5% C, 4.8% H, 8.3% N, 37.8% O, *by mass*. What is the empirical formula of MSG?

72. Chlorophyll (essential to the process of photosynthesis) contains Mg to the extent of 2.72% by mass. Assuming one Mg atom per chlorophyll molecule, what is the molecular weight of chlorophyll?

73. *Without actually determining percentage compositions,* arrange the following oxides of chromium in order of *increasing*

% Cr and explain your reasoning: CrO, Cr_2O_3, CrO_2, CrO_3.

74. Ammonium sulfate, $(NH_4)_2SO_4$, is a commonly used fertilizer. If a mature avocado tree requires 1.0 lb of actual nitrogen per year, what mass of ammonium sulfate would be required, per year?

75. A gaseous compound of boron and hydrogen consists of 78.5% B by mass. By an independent experiment the molecular weight of the compound is found to be 27.5. What is the *molecular* formula of the compound?

76. A certain hydrate is found to have the composition: 20.3% Cu, 8.95% Si, 36.3% F, and 34.5% H_2O, by mass. What is the empirical formula of this hydrate?

77. Refer to the Summarizing Example. Show from the data given that the high-temperature (1000 °C) residue could *not* have been CuS.

78. Anhydrous sodium sulfate, Na_2SO_4, absorbs water vapor and is converted to the *deca*hydrate, $Na_2SO_4 \cdot 10\,H_2O$. How much would the mass of 3.50 g of anhydrous Na_2SO_4 increase if converted completely to the decahydrate?

79. The element X forms the chloride XCl_4 containing 75.0% Cl, by mass. What is the atomic weight of X? What is the element X?

80. A 1.562-g sample of the hydrocarbon C_7H_{16} is burned in an excess of oxygen. What masses of CO_2 and H_2O should be obtained?

81. What is the percent composition of a carbon-hydrogen-oxygen compound if 1.3663 g of the compound yields 3.447 g CO_2 and 0.6049 g H_2O? What is the empirical formula of this compound?

82. A certain brand of lunch meat contains 0.10% sodium benzoate, $NaC_7H_5O_2$, by mass, as a preservative. If a person eats 2.52 oz of this meat, how many mg Na will that person consume?

83. Refer to the compound ethyl mercaptan described in Example 3-2. Assuming that the complete combustion of this compound produces CO_2, H_2O, and SO_2, what masses of each of these three products would be produced in the complete combustion of 1.50 mL of ethyl mercaptan?

★**84.** An oxoacid with the formula $H_xE_yO_z$ has a formula weight of 178.0, has 13 atoms in its formula unit, contains 34.80%, by mass, and 15.38%, by *number* of atoms, of the element E. What is the element E, and what is the formula of this oxoacid?

85. The insecticide dieldrin contains carbon, hydrogen, oxygen, and chlorine. Upon complete combustion a 1.510-g sample yields 2.094 g CO_2 and 0.286 g H_2O. The compound has a molecular weight of 381 and has half as many chlorine atoms as carbon atoms. What is the molecular formula of dieldrin?

★86. $MgCl_2$ often occurs as an impurity in table salt (NaCl), and is responsible for "caking" of the salt. A 0.5200-g sample of table salt is found to contain 61.10% Cl, by mass. What is the % $MgCl_2$ in the sample? Why is the precision of this calculation so poor?

87. Write a formula for
(a) a sulfate of iron with Fe in the oxidation state +3.
(b) an oxoacid of nitrogen with N in the oxidation state +3.
(c) an oxide of chlorine with Cl in the oxidation state +7.

88. To deposit exactly one mole of Ag from an aqueous solution containing Ag^+ requires 96,485 coulombs of electric charge to be passed through the solution. The electrodeposition requires that each Ag^+ ion that is converted to an Ag atom gain one electron. Use this information and data from Table 2-1 to obtain a value of the Avogadro constant, N_A.

★89. A very dilute solution of oleic acid, $C_{18}H_{34}O_2$, in liquid pentane is prepared by a series of dilutions, as shown. That is, 1.00 mL of oleic acid is added to 9.00 mL of pentane. Next, 1.00 mL of this solution (1) is added to another 9.00 mL of pentane, etc., until solution (4) is obtained. Then, 0.10 mL of solution (4) is spread in a *monolayer* on water. The area covered by the monolayer is found to be 85 cm^2. Assume that the cross-sectional area of an oleic acid molecule in the monolayer is 4.6×10^{-15} cm^2. The density of pure oleic acid is 0.895 g/cm^3. Use these data to obtain an approximate value of the Avogadro constant, N_A. [*Hint:* Refer also to Exercises 25 and 26.]

90. The compound potassium dichromate, $K_2Cr_2O_7$, is used in the chrome tanning of leather and in electroplating.
(a) What percent of all the atoms in $K_2Cr_2O_7$ are Cr atoms?
(b) The percent natural abundance of the isotope Cr-50 is 4.35%. What percent of all the atoms in $K_2Cr_2O_7$ are Cr-50 atoms?
★(c) Calculate the percent, *by mass,* of Cr-50 in $K_2Cr_2O_7$.

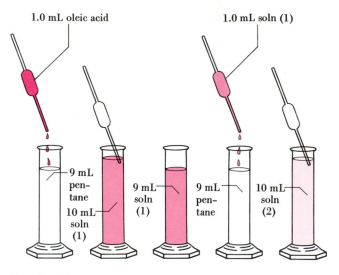

1.0 mL oleic acid   1.0 mL soln (1)

9 mL pentane
10 mL soln (1)

9 mL soln (1)

9 mL pentane

10 mL soln (2)

Exercise 89

★91. When 2.750 g of the oxide of lead Pb_3O_4 is strongly heated, decomposition occurs, producing 0.0640 g of oxygen gas and 2.686 g of a second oxide of lead. What is the empirical formula of this second oxide?

★92. A hydrocarbon mixture consists of 60.0% *by mass* of C_3H_8 and 40.0% of C_xH_y. When 10.0 g of this mixture is burned, it yields 29.0 g CO_2 and 18.8 g H_2O as the only products. What is the formula of the unknown hydrocarbon?

★93. A 0.732-g mixture of methane, CH_4, and ethane, C_2H_6, is burned, yielding 2.064 g CO_2. What is the percent composition of this mixture (a) by mass; (b) on a mole basis?

★94. A thoroughly dried 1.271-g sample of Na_2SO_4 is exposed to the atmosphere and found to gain 0.387 g in mass. What is the percent, by mass, of $Na_2SO_4 \cdot 10H_2O$ in the resulting mixture of anhydrous Na_2SO_4 and the decahydrate?

★95. The atomic weight of Bi is to be determined by converting the compound $Bi(C_6H_5)_3$ to Bi_2O_3. If 5.610 g $Bi(C_6H_5)_3$ yields 2.969 g Bi_2O_3, what is the atomic weight of Bi?

Self-Test Questions

For questions 96 through 105 select the single item that best completes each statement.

96. One *mole* of fluorine gas, F_2 (a) weighs 19.0 g; (b) contains 6.02×10^{23} F atoms; (c) contains 1.20×10^{24} F atoms; (d) weighs 6.02×10^{23} g.

97. Three of the following formulas might be either empirical or molecular formulas, but one of the four must be a molecular formula. That one is (a) N_2O; (b) N_2O_4; (c) NH_3; (d) Mg_3N_2

98. The compound $C_7H_7NO_2$ (a) contains 17 atoms per mole; (b) contains equal percentages of C and H atoms, *by mass;* (c) contains twice the percent, *by mass,* of O as of N; (d) contains twice the percent, *by mass,* of N as of H.

99. The greatest number of N atoms is found in (a) 50.0 g N_2O; (b) 17.0 g NH_3; (c) 150 cm^3 of liquid pyridine, C_5H_5N ($d = 0.983$ g/cm^3); (d) 1.0 mol N_2.

100. XF_3 is found to consist of 65% F, *by mass.* The atomic weight of X must be (a) 8; (b) 11; (c) 31; (d) 35.

101. The oxidation state of I in the ion $H_4IO_6^-$ is (a) −1; (b) +1; (c) +7; (d) +8.

102. The correct formula for calcium chlorite is (a) $Ca(ClO_2)_2$; (b) $CaClO_2$; (c) $Ca(ClO_3)_2$; (d) $Ca(ClO_4)_2$.

103. The hydrate with more than 50% H_2O, *by mass,* is
(a) $KAl(SO_4)_2 \cdot 12H_2O$;   (b) $MgCl_2 \cdot 6H_2O$;
(c) $Na_2SO_4 \cdot 7H_2O$;   (d) $LiNO_3 \cdot 3H_2O$

104. Of the following, the *greatest* mass of H_2O is produced in the complete combustion of 1.00 mol of (a) CH_4; (b) C_2H_5OH; (c) $C_{10}H_8$; (d) C_6H_5OH.

105. A formula unit of the coordination compound $[Cu(NH_3)_4]SO_4$ has *equal or nearly equal* masses of (a) S and O; (b) N and O; (c) H and N; (d) Cu and O.

106. The liquid $CHBr_3$ has a density of 2.89 g/cm^3. What volume of this liquid should be measured out to contain a total of 1.00 mol Br atoms?

107. Supply the missing name or formula.

(a) CaI_2 = _____

(b) _____ = iron(III) sulfate

(c) _____ = sulfur trioxide

(d) _____ = bromine pentafluoride

(e) NH_4CN = _____

(f) $Ca(ClO)_2$ = _____

(g) _____ = lithium hydrogen carbonate

108. An important copper-containing mineral is malachite, $CuCO_3 \cdot Cu(OH)_2$.

(a) What is the percent Cu, *by mass,* in malachite?

(b) When malachite is strongly heated, carbon dioxide and water are driven off, yielding copper(II) oxide. What mass of copper(II) oxide is produced *per kilogram* of malachite?

109. Hexachlorophene, used in making germicidal soaps, has the percent composition, by mass: 38.37% C, 1.49% H, 52.28% Cl, and 7.86% O. What is the empirical formula of hexachlorophene?

110. A hydrate of Na_2SO_3 contains almost exactly 50% H_2O, by mass. What is the formula of this hydrate?

4 Chemical Reactions

Sodium metal and chlorine gas exhibit striking visual evidence of their chemical reaction. [Carey B. Van Loon]

90

The burning of natural gas, the rusting of iron, and the production of acid rain are three chemical reactions with which you are probably familiar. Each is a process in which new chemical substances, called **products,** are produced from an original set of substances, called **reactants.** Chemical reactions are the central concern of chemistry.

We can rather conveniently divide our study of chemical reactions into two broad areas. In this chapter we will concentrate on some practical questions, such as

- How do we know when a chemical reaction has occurred?
- How can we describe a chemical reaction in the symbolic form known as a chemical equation?
- What kinds of calculations are possible with chemical equations?
- How are chemical reactions used to analyze and synthesize substances?

In later chapters of the text we will explore such fundamental matters as the conditions necessary for a chemical reaction to occur and the rate or speed of the reaction.

4-1   Chemical Reactions and the Chemical Equation

A chemical reaction is a process in which new substances, **products,** are produced from a set of original substances, **reactants.** Often the evidence of a chemical reaction can be as simple as

Some physical evidence of chemical reactions (illustrated in several figures in this chapter).

- a color change
- evolution of a gas
- formation of a precipitate
- evolution or absorption of heat

(4.1)

At times, however, chemical analysis, sometimes using sophisticated instruments, may be needed to prove that a reaction has occurred.

Just as we have symbolic representations for elements (chemical symbols) and compounds (chemical formulas), we have one for chemical reactions—the **chemical equation.** We write formulas of the reactants on the *left* side of the equation and formulas of the products, on the *right*. We join the two sides of the equation by an equal sign (=) or an arrow (→). The reactants are said to *yield* the products. Writing a chemical equation is a three-step procedure, although often we can just think about the first step without actually writing it down. This three-step procedure is illustrated below, for the reaction of colorless nitrogen monoxide and oxygen gases to form red-brown nitrogen dioxide gas, an important ingredient of photochemical smog. (Smog formation is discussed in Section 14-10.)

1. Write the *names* of the reactants and products to obtain a *word expression*.

   nitrogen monoxide + oxygen $\longrightarrow$ nitrogen dioxide

2. Substitute chemical *formulas* for names, resulting in a *formula expression*.

   $$NO + O_2 \longrightarrow NO_2$$

   3 O   2 O

3. *Balance* the formula expression to obtain a *chemical equation*.

   $$2\,NO + O_2 \longrightarrow 2\,NO_2$$

   2 N   2 O   2 O   2 N   4 O

The purpose of **balancing an equation*** is to establish that

Basis of equation balancing.

The total number of atoms of each type remains unchanged, since atoms can neither be created nor destroyed in a chemical reaction.   (4.2)

In step 2 above, *three* O atoms are represented on the left side (*one* in the molecule NO and *two* in the molecule O_2). On the right side there are only *two* O atoms (in the molecule NO_2). In the balancing step (3), we place a **coefficient** of 2, called a **stoichiometric coefficient,** in front of the formulas NO and NO_2. Where no coefficient is used, as with O_2, we understand this to be 1. We read the equation as

"two N, O plus O, 2 yield two N, O, 2"

which means that *two* molecules of NO and *one* molecule of O_2 are consumed and *two* molecules of NO_2 are produced. In the balanced equation there are *two* N atoms and *four* O atoms on each side. In balancing a chemical equation these are the points to keep in mind:

Additional ideas in equation balancing.

- The equation can be balanced *only* by adjusting the *coefficients* of formulas, as necessary.
- *Never* introduce extraneous formulas.
- *Never* change the subscripts of formulas.   (4.3)

These points are further illustrated through Figure 4-1.

The method of equation balancing introduced here and illustrated in Examples 4-1 and 4-2 is called **balancing by inspection.** (Balancing by inspection means to adjust coefficients by "trial and error" until a balanced condition is reached.)

Example 4-1 ———————————————————————————

Writing and balancing an equation: The combustion of a hydrocarbon. Propane gas, C_3H_8, is easily liquefied, stored and transported for use as a fuel. Write a balanced chemical equation to represent its complete combustion (see Figure 4-2).

FIGURE 4-1

Balancing the equation

$$2\ NO + O_2 \rightarrow 2\ NO_2$$

(a) *Incorrect:* There is no evidence of the presence of atomic oxygen (O) as a product. *Extraneous formulas cannot be introduced for the purpose of balancing an equation.*

(b) *Incorrect:* The product of the reaction is NO_2, *not* NO_3. *A formula cannot be altered for the purpose of balancing an equation.*

(c) *Correct:* An equation can be balanced only through the use of correct formulas and coefficients.

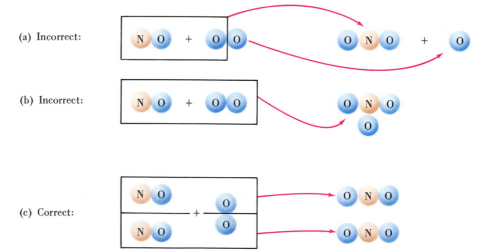

(a) Incorrect:

(b) Incorrect:

(c) Correct:

———

*An equation—whether mathematical or chemical—must have the left and right sides equal. Strictly speaking, a formula expression cannot be called an equation until it is balanced, and the term "chemical equation" signifies that this balance has been achieved. Nevertheless, to stress the importance of the balanced condition, the somewhat contradictory statement "unbalanced chemical equation" and the somewhat redundant statement "balanced chemical equation" are both used.

FIGURE 4-2
Combustion of propane.

Liquid propane vaporizes as it escapes through the nozzle, mixes with oxygen gas, and then burns. Physical evidence of this reaction is seen in the high-temperature flame. Heat is evolved in a combustion reaction. (Cylinders like the one pictured here are widely available in hardware stores and home improvement centers.) [Carey B. Van Loon]

Solution. We learned in Section 3-2 that when a hydrocarbon (carbon–hydrogen compound) is burned in an excess of oxygen the sole products are CO_2 and H_2O.

Word expression: propane + oxygen $\longrightarrow$ carbon dioxide + water

Formula expression: $C_3H_8 + O_2 \longrightarrow CO_2 + H_2O$

As each coefficient is established, it is kept fixed while the next one is being set, and so on, until a final balance results.

Balance C: $C_3H_8 + O_2 \longrightarrow 3\ CO_2 + H_2O$

Balance H: $C_3H_8 + O_2 \longrightarrow 3\ CO_2 + 4\ H_2O$

Balance O: $C_3H_8 + 5\ O_2 \longrightarrow 3\ CO_2 + 4\ H_2O$

Balanced equation: $C_3H_8 + 5\ O_2 \longrightarrow 3\ CO_2 + 4\ H_2O$ (4.4)

SIMILAR EXAMPLES: Exercises 2, 17, 19, 65.

Example 4-2 ────────────

Writing and balancing an equation: The combustion of a carbon–hydrogen–oxygen compound. Triethylene glycol, $C_6H_{14}O_4$, is used as a solvent and plasticizer for vinyl and polyurethane plastics. Write a balanced chemical equation for its complete combustion.

Solution. Carbon–hydrogen–oxygen compounds, like hydrocarbons, yield CO_2 and H_2O upon burning in oxygen gas.

Formula expression: $C_6H_{14}O_4 + O_2 \longrightarrow CO_2 + H_2O$

Balance C: $C_6H_{14}O_4 + O_2 \longrightarrow 6\ CO_2 + H_2O$

Balance H: $C_6H_{14}O_4 + O_2 \longrightarrow 6\ CO_2 + 7\ H_2O$ (4.5)

The right side of expression (4.5) has 19 O atoms. To obtain 19 on the left side, we start with 4 in $C_6H_{14}O_4$ and add 15 more. This requires a fractional coefficient of $\frac{15}{2}$ for O_2.

Balance O: $C_6H_{14}O_4 + \frac{15}{2}\ O_2 \longrightarrow 6\ CO_2 + 7\ H_2O$ (balanced)   (4.6)

Final Adjustment of Coefficients. Although fractional coefficients are acceptable in many circumstances, general practice is to remove them by multiplying all coefficients in a chemical equation by the same whole number—in this case "2."

$2\ C_6H_{14}O_4 + 15\ O_2 \longrightarrow 12\ CO_2 + 14\ H_2O$ (balanced)   (4.7)

SIMILAR EXAMPLES: Exercises 2, 17, 19, 65.

We can represent the state of matter or physical form in which reactants and products appear through the following symbols.

(g) = gas   **(l)** = liquid   **(s)** = solid   **(aq)** = aqueous (water) solution

Thus, for the reaction of hydrogen and oxygen *gases* to form *liquid* water

$2\ H_2(g) + O_2(g) \longrightarrow 2\ H_2O(l)$ (4.8)

Are You Wondering:

In what order to balance the elements in an equation?

In *balancing by inspection* you can balance the elements in any order that you choose. Often, however, you will find that a particular sequence will lead to the quickest result. Here are two important ideas.

- If an element occurs *in only one compound* on each side of the equation, you may start with that element. The elements C and H fit this description in Examples 4-1 and 4-2.
- When one of the reactants or products exists as the *free* element (as was the case with O_2 in Examples 4-1 and 4-2), it is generally best to balance this element *last*.

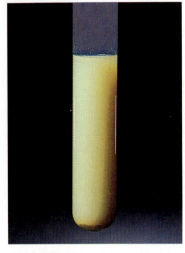

FIGURE 4-3
Precipitation of silver iodide.

When a water solution of $AgNO_3$ is added to one of NaI, insoluble yellow AgI precipitates from solution. [Carey B. Van Loon]

Net Ionic Equation. The reaction of water solutions of silver nitrate and sodium iodide yields a *precipitate* of silver iodide. This reaction is pictured in Figure 4-3 and represented through the equation

$$AgNO_3(aq) + NaI(aq) \longrightarrow AgI(s) + NaNO_3(aq) \tag{4.9}$$

Equation (4.9) is in the **molecular** form; formulas are written for electrically neutral formula units. The equation is more informative if we show that the substances involved in this reaction are *ionic* compounds, and that in their aqueous solutions they are dissociated into cations and anions. Equation (4.10) is in the **ionic** form.

$$Ag^+(aq) + \cancel{NO_3^-(aq)} + \cancel{Na^+(aq)} + I^-(aq) \longrightarrow AgI(s) + \cancel{Na^+(aq)} + \cancel{NO_3^-(aq)} \tag{4.10}$$

A still better alternative is to *eliminate any species that appear on both sides of an equation*. Species that appear on both sides of the equation are "spectators" of the reaction, not participants in the reaction. The **spectator** ions are cancelled out in equation (4.10). All that remains in equation (4.11) are the ions that actually participate in the reaction and the precipitate they form. It is called a **net ionic equation.**

Equation (4.11) simply states that when silver and iodide ions are present in the same solution they combine to form insoluble silver iodide.

$$Ag^+(aq) + I^-(aq) \longrightarrow AgI(s) \tag{4.11}$$

In the reaction pictured in Figure 4-4, copper metal reacts with a water solution of silver nitrate, yielding silver metal and a water solution of copper nitrate. In this reaction, nitrate ion is a spectator ion. The net ionic reaction can be represented as

$$Cu(s) + Ag^+(aq) \longrightarrow Cu^{2+}(aq) + Ag(s) \quad \text{(unbalanced)} \tag{4.12}$$

Although this ionic expression may appear balanced (equal numbers of both types of atoms on each side of the expression), *it is not*. An additional requirement is that

Basis of balancing net ionic equations.

==Electric charge is conserved in a chemical reaction. The same net electric charge must appear on both sides of a *net ionic equation*.== (4.13)

In expression (4.12) there is *one* unit of positive charge on the left and *two* on the right. This situation is corrected in equation (4.14).

$$Cu(s) + 2\,Ag^+(aq) \longrightarrow Cu^{2+}(aq) + 2\,Ag(s) \tag{4.14}$$

A verbal description of equation (4.14) is that copper metal *displaces* silver ion from solution; copper metal goes into solution as Cu^{2+} and Ag^+ comes out of solution as silver metal.

FIGURE 4-4
Displacement of Ag⁺(aq) by Cu(s).

A coil of copper wire is placed in an aqueous solution of silver nitrate (left), and immediately the displacement reaction (4.14) begins. The photograph at the right was taken after about 2 hours. The blue color of the solution indicates the presence of Cu^{2+}. [Carey B. Van Loon]

Writing net ionic equations is an important skill that we develop further throughout the text, including a more extensive treatment of the subject in the next chapter.

Example 4-3

Writing and balancing a net ionic equation. Many metal ions form insoluble sulfide precipitates; their presence in solution can be established by passing hydrogen sulfide gas into the solution. For example, Bi^{3+} ion forms a dark brown precipitate of bismuth sulfide, Bi_2S_3. The reaction is accompanied by an increase in the number of H^+ ions in solution. Write a balanced ionic equation for this reaction.

Solution. Electric charges are shown in the following expression (it is written in ionic form). The final equation must show a balance both in numbers of atoms *and in electric charges.* The stepwise balancing is suggested below.

$$Bi^{3+}(aq) + H_2S(aq) \longrightarrow Bi_2S_3(s) + H^+(aq)$$

Balance Bi: $\quad 2\,Bi^{3+}(aq) + H_2S(aq) \longrightarrow Bi_2S_3(s) + H^+(aq)$

Balance S: $\quad 2\,Bi^{3+}(aq) + 3\,H_2S(aq) \longrightarrow Bi_2S_3(s) + H^+(aq)$

Balance H: $\quad 2\,Bi^{3+}(aq) + 3\,H_2S(aq) \longrightarrow Bi_2S_3(s) + 6\,H^+(aq)$

Balanced equation: $\quad 2\,Bi^{3+}(aq) + 3\,H_2S(aq) \longrightarrow Bi_2S_3(s) + 6\,H^+(aq)$

Proof of balance of electric charge

$$\underbrace{2 \times (+3)}_{\text{left}} = \underbrace{6 \times (+1)}_{\text{right}}$$

charge on Bi^{3+} — charge on H^+

An aqueous solution is always electrically neutral. The reason that the solution here appears to carry a net charge of +6 is that spectator ions (e.g., 6 Cl⁻) are not included in the net ionic equation.

SIMILAR EXAMPLES: Exercises 3, 18, 22.

Reaction Conditions. Much of modern chemical research involves working out the conditions for carrying out a reaction. Just having the chemical equation is not enough. Reaction conditions are often written above or below the arrow. For example, the Greek capital letter delta, Δ, means that an elevated temperature is required; that is, the reaction mixture must be heated.

$$2 \ Ag_2O(s) \xrightarrow{\Delta} 4 \ Ag(s) + O_2(g) \tag{4.15}$$

An even more explicit statement of reaction conditions is shown below for the BASF (Badische Anilin- & Soda-Fabrik) process for the synthesis of methanol from CO and H_2. This reaction occurs at 350 °C, under a total pressure of 340 atm, and on the surface of a catalyst (a mixture of ZnO and Cr_2O_3).

$$CO(g) + 2 \ H_2(g) \xrightarrow[\substack{340 \ atm \\ ZnO, \ Cr_2O_3}]{350 \ °C} CH_3OH(g) \tag{4.16}$$

> Gas pressure is discussed in Chapter 6, and the role of a catalyst in speeding up a chemical reaction in Chapter 15.

4-2   Writing and Balancing Equations: A Summary

Most beginning students find that writing chemical equations is a challenging task. From time to time you will learn new approaches that may simplify this task for you. For the present, consider the subject in terms of these three possibilities:

1. **Balancing a Formula Expression.** If you are given the formulas of all the reactants and products (the formula expression), your task is to adjust the coefficients of these formulas to achieve a balance in the number of atoms of all types (and sometimes also of electric charge). The method introduced in this chapter is to balance by inspection (trial and error). Certain types of reactions (oxidation–reduction) are difficult to balance by inspection. We develop special methods for these types of reaction in the next chapter.

2. **Writing a Chemical Equation from a Word Expression.** At times a chemical reaction will be described through the *names* of the reactants and products. Here, you must (a) substitute *correct formulas* for names and (b) balance the formula expression. We considered the relationship of names and formulas of simple compounds in Chapter 3. Until we discuss their nomenclature, you can expect the formulas of more complex substances to be provided. You can practice these skills through Exercises 4, 19, 20, 21, 22, 66, 68, and 72.

3. **Completing a Chemical Equation.** This refers to the ability to *predict* whether substances will react, and, if so, what the products will be. This prediction is

TABLE 4-1
Some Types of Reactions.

Type	Fundamental change/Examples
Combustion	An element or compound combines with $O_2(g)$, producing simple oxygen-containing compounds, such as CO_2, H_2O, and SO_2. Reactions (4.4) and (4.7).
Combination (synthesis)	A more complex substance is formed from two or more simpler substances. Reactions (4.8) and (4.16).
Decomposition	A substance is broken down into simpler substances (e.g., into its elements). Reaction (4.15).
Displacement (single replacement)	One element replaces another in a compound. Reaction (4.14).
Metathesis (double replacement)	An exchange (usually of ions) occurs between two reactants. Reaction (4.9).

then followed by writing a formula expression and balancing an equation. To be able to predict chemical reactions requires the greatest knowledge of chemical principles, and it is these principles that we establish throughout the text. For the present you should be able to predict the outcome of the *combustion* (burning in oxygen) of hydrocarbons and carbon–hydrogen–oxygen compounds. [The products are $CO_2(g)$ and $H_2O(l)$.]

The terms listed in Table 4-1 are also useful in describing chemical reactions, although we introduce additional and still more useful terms in Chapter 5.

4-3 Quantitative Significance of the Chemical Equation

The coefficients in the chemical equation

$$2 H_2(g) + O_2(g) \longrightarrow 2 H_2O(l)$$

mean that

2 molecules H_2 + 1 molecule $O_2 \longrightarrow$ 2 molecules H_2O

or that

$2x$ molecules H_2 + x molecules $O_2 \longrightarrow 2x$ molecules H_2O

Suppose that we let $x = 6.02214 \times 10^{23}$ (the Avogadro constant). Then x molecules represents *1 mol*. Thus the chemical equation also means that

2 mol H_2 + 1 mol $O_2 \longrightarrow$ 2 mol H_2O

The chemical equation allows us to write the expressions

(1) 2 mol $H_2O \Leftrightarrow$ 2 mol H_2

(2) 2 mol $H_2O \Leftrightarrow$ 1 mol O_2

(3) 2 mol $H_2 \Leftrightarrow$ 1 mol O_2

Recall the meaning of the equivalence sign introduced in Section 1-9.

which have the following meanings.

1. *Two* moles of H_2O are *produced* for every *two* moles of H_2 *consumed*.
2. *Two* moles of H_2O are *produced* for every *one* mole of O_2 *consumed*.
3. *Two* moles of H_2 are *consumed* for every *one* mole of O_2 *consumed*.

These expressions (and hence the chemical equation from which they are derived) are the source of conversion factors, called **stoichiometric factors.** Stoichiometric factors relate the *molar* amounts of any two substances involved in a chemical reaction. In the three examples that follow the stoichiometric factors are shown in blue.

Example 4-4

Using a stoichiometric factor to determine the number of moles of a product from the number of moles of a reactant. How many moles of H_2O are produced by burning 2.72 mol H_2 in an excess of O_2?

Solution. The statement "an excess of O_2" means that there is more than enough O_2 available to permit the complete conversion of 2.72 mol H_2 to H_2O. The necessary conversion factor is derived from the expression, 2 mol $H_2O \Leftrightarrow$ 2 mol H_2, which in turn is obtained from the balanced equation: $2 H_2 + O_2 \longrightarrow 2 H_2O$.

$$\text{no. mol } H_2O = 2.72 \text{ mol } H_2 \times \frac{2 \text{ mol } H_2O}{2 \text{ mol } H_2} = 2.72 \text{ mol } H_2O$$

SIMILAR EXAMPLES: Exercises 5, 24.

Example 4-5

Calculating the mass of a reactant required to produce a certain mass of a product. What mass of H_2 must react with excess O_2 to produce 7.22 g H_2O?

Solution. Here (a) the unknown is a quantity of one of the *reactants* (H_2) instead of one of the products (H_2O), and (b) information is given and sought in the unit *gram* rather than mole. Even though the calculation can be performed in a single setup, we should think in terms of three steps.
Step 1. Convert the quantity of H_2O from grams to moles (molar mass).
Step 2. From the number of moles of H_2O in Step 1, calculate the number of moles of H_2 consumed (stoichiometric factor from the equation).
Step 3. Convert the quantity of H_2 in Step 2 from moles to grams (molar mass).

$$\text{no. g } H_2 = 7.22 \text{ g } H_2O \times \frac{1 \text{ mol } H_2O}{18.02 \text{ g } H_2O} \times \frac{2 \text{ mol } H_2}{2 \text{ mol } H_2O} \times \frac{2.016 \text{ g } H_2}{1 \text{ mol } H_2}$$

$$(\text{g } H_2O \longrightarrow \text{mol } H_2O \longrightarrow \text{mol } H_2 \longrightarrow \text{g } H_2)$$

$$= 0.808 \text{ g } H_2$$

SIMILAR EXAMPLES: Exercises 6, 7, 24.

Example 4-6

Relating the masses of two reactants to one another. What mass of $O_2(g)$ is consumed in the complete combustion of 6.86 g H_2?

Solution. Here we seek the relationship between two reactants (rather than a reactant and a product), but the required conversion factor (stoichiometric factor) is still drawn from the chemical equation.

Are You Wondering:

Whether you should ever multiply molar masses by stoichiometeric coefficients when doing a calculation based on a chemical equation (such as 2 × 2.016 g H_2/mol H_2)?

The answer is no! To avoid this and related errors that beginning students sometimes make, note that

- A molar mass is based solely on the *formula* of a substance, and is *independent* of the particular reaction in which the substance is involved. (Thus, 1 mol H_2 ⇌ 2.016 g H_2, *always*.)
- A chemical equation furnishes only *one* conversion factor in a stoichiometric calculation, relating the number of moles of one substance to another. Coefficients from the chemical equation appear in this one factor and *nowhere else* in the setup. (Once you have written the stoichiometric factor you have no more need for the chemical equation.)

FIGURE 4-5

The reaction

$$2\ Al(s) + 6\ HCl(aq) \rightarrow$$
$$2\ AlCl_3(aq) + 3\ H_2(g)$$

HCl(aq) is introduced to the flask on the left through a long funnel. The reaction of Al(s) and HCl(aq) to form $AlCl_3$(aq) occurs within the flask. The liberated H_2(g) is conducted to a gas-collection apparatus, where it displaces water. Hydrogen is only very slightly soluble in water.

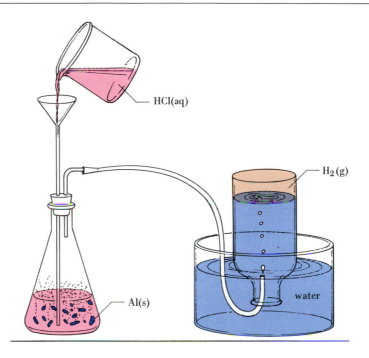

$$\text{no. g. O}_2 = 6.86\ \text{g H}_2 \times \frac{1\ \text{mol H}_2}{2.016\ \text{g H}_2} \times \frac{1\ \text{mol O}_2}{2\ \text{mol H}_2} \times \frac{32.00\ \text{g O}_2}{1\ \text{mol O}_2}$$

$$(\text{g H}_2 \longrightarrow \text{mol H}_2 \longrightarrow \text{mol O}_2 \longrightarrow \text{g O}_2)$$

$$= 54.4\ \text{g O}_2$$

SIMILAR EXAMPLES: Exercises 6, 7, 69, 70.

Now we shift our attention to the reaction pictured in Figure 4-5, a simple laboratory method of preparing small volumes of hydrogen gas.

$$2\ Al(s) + 6\ HCl(aq) \longrightarrow 2\ AlCl_3(aq) + 3\ H_2(g) \tag{4.17}$$

Examples 4-7, 4-8, and 4-9 are based on this reaction. Again, stoichiometric factors from equation (4.17) are shown in blue. Figure 4-6 is a general outline of calculations based on the chemical equation.

Example 4-7

Using additional conversion factors in a stoichiometric calculation: volume and density. A piece of pure Al(s) having a volume of 0.688 cm^3 reacts with an excess of HCl(aq). What is the mass of H_2 obtained? (The density of Al is 2.70 g/cm^3.)

Solution

Step 1. Use density to convert from volume to mass.

$$\text{no. g Al} = 0.688\ \text{cm}^3\ \text{Al} \times \frac{2.70\ \text{g Al}}{1\ \text{cm}^3\ \text{Al}} = 1.86\ \text{g Al}$$

Step 2. Express the quantity of Al from Step 1 in moles.

$$\text{no. mol Al} = 1.86\ \text{g Al} \times \frac{1\ \text{mol Al}}{26.98\ \text{g Al}} = 0.0689\ \text{mol Al}$$

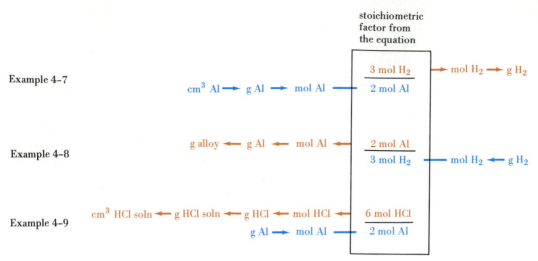

Example 4-7

Example 4-8

Example 4-9

FIGURE 4-6

Outlining a stoichiometric calculation.

Each of these examples starts with information about one substance (in blue) and seeks information about another (in brown). The chemical equation provides the *stoichiometric* factor to convert from one substance to another *on a mole basis. Only one factor in each calculation is derived from the chemical equation.* Other conversions use molar mass, density, or percent composition, as necessary.

Step 3. Use a stoichiometric factor from the chemical equation to determine the amount of H_2 that will be produced.

$$\text{no. mol } H_2 = 0.0689 \text{ mol Al} \times \frac{3 \text{ mol } H_2}{2 \text{ mol Al}} = 0.103 \text{ mol } H_2$$

Step 4. Convert the amount of H_2 from Step 3 to mass in grams.

$$\text{no. g } H_2 = 0.103 \text{ mol } H_2 \times \frac{2.016 \text{ g } H_2}{1 \text{ mol } H_2} = 0.208 \text{ g } H_2$$

SIMILAR EXAMPLES: Exercises 29, 44, 71.

Example 4-8

Using the percent composition of a mixture in a stoichiometric calculation. An alloy used in fabricating aircraft structures consists of 93.7% Al and 6.3% Cu. Assuming that *all* the Al and *none* of the Cu reacts with HCl(aq), what mass of the alloy is required to produce 4.84 g H_2?

Solution. The reactant whose quantity we are seeking is not pure. A conversion factor based on the percent composition of the alloy is required in the last step of a four-step solution.

Step 1. Describe the H_2 produced as an amount in moles.

$$\text{no. mol } H_2 = 4.84 \text{ g } H_2 \times \frac{1 \text{ mol } H_2}{2.016 \text{ g } H_2} = 2.40 \text{ mol } H_2$$

Step 2. Convert from mol H_2 to mol Al with a factor from equation (4.17).

$$\text{no. mol Al} = 2.40 \text{ mol } H_2 \times \frac{2 \text{ mol Al}}{3 \text{ mol } H_2} = 1.60 \text{ mol Al}$$

Step 3. Express the required quantity of Al as a mass, in grams.

$$\text{no. g Al} = 1.60 \text{ mol Al} \times \frac{26.98 \text{ g Al}}{1 \text{ mol Al}} = 43.2 \text{ g Al}$$

Step 4. If the Al were pure, the quantity required would be 43.2 g, but the sample is only 93.7% Al. The mass of alloy required is *greater* than 43.2 g.

$$\text{no. g alloy} = 43.2 \text{ g Al} \times \frac{100.0 \text{ g alloy}}{93.7 \text{ g Al}} = 46.1 \text{ g alloy}$$

SIMILAR EXAMPLES: Exercises 16, 25, 27.

Example 4-9

Using the density, percent composition, and volume of a solution in a stoichiometric calculation. A hydrochloric acid solution consists of 28.0% HCl, by mass, and has a density of 1.14 g/cm³. What is the volume of this solution required to dissolve 1.87 g Al in reaction (4.17)?

Solution. The several steps outlined in Figure 4-6 are performed below.
Step 1. Convert 1.87 g Al to no. mol Al. *Result:* 0.0693 mol Al.
Step 2. Determine the no. mol HCl required to dissolve the Al.

$$\text{no. mol HCl} = 0.0693 \text{ mol Al} \times \frac{6 \text{ mol HCl}}{2 \text{ mol Al}} = 0.208 \text{ mol HCl}$$

Step 3. Determine the mass of 0.208 mol HCl. *Result:* 7.58 g HCl.
Step 4. Calculate the mass of the acid solution containing 7.58 g HCl.

$$\text{no. g HCl soln} = 7.58 \text{ g HCl} \times \frac{100.0 \text{ g HCl soln}}{28.0 \text{ g HCl}} = 27.1 \text{ g HCl soln}$$

Step 5. Use density as a factor to convert from mass to volume of solution.

$$\text{no. cm}^3 \text{ HCl soln} = 27.1 \text{ g HCl soln} \times \frac{1 \text{ cm}^3 \text{ HCl soln}}{1.14 \text{ g HCl soln}}$$

$$= 23.8 \text{ cm}^3 \text{ HCl soln}$$

This example, as all others that we have considered in stepwise fashion, can also be solved through a single setup, if you are able to visualize each step within the setup.

$$\text{no. cm}^3 \text{ HCl soln} = 1.87 \text{ g Al} \times \frac{1 \text{ mol Al}}{26.98 \text{ g Al}} \times \frac{6 \text{ mol HCl}}{2 \text{ mol Al}} \times \frac{36.46 \text{ g HCl}}{1 \text{ mol HCl}}$$

$$\text{(g Al} \longrightarrow \text{mol Al} \longrightarrow \text{mol HCl} \longrightarrow \text{g HCl}$$

$$\times \frac{100.0 \text{ g HCl soln}}{28.0 \text{ g HCl}} \times \frac{1 \text{ cm}^3 \text{ HCl soln}}{1.14 \text{ g HCl soln}}$$

$$\longrightarrow \text{g HCl soln} \longrightarrow \text{cm}^3 \text{ HCl soln)}$$

$$= 23.8 \text{ cm}^3 \text{ HCl soln}$$

SIMILAR EXAMPLES: Exercises 30, 39.

Summary. If you stop to think about the illustrative examples we have done in this section, you will see that they all conform to the basic three-step approach summarized below. This is a useful scheme for you to keep in mind.

1. Convert information about the given substance to a mole basis. (The given information might be a mass in grams, a volume and density, etc.)
2. Use a stoichiometric factor from the balanced equation to convert from moles of the given substance to moles of the desired substance.
3. Convert from moles of the desired substance to whatever units are asked for (e.g., grams, mL of solution, etc.).

4-4 An Introduction to Solutions

Because so many chemical reactions are carried out in solution, we need to make some brief comments now about solutions. One component, called the **solvent,** determines whether the solution exists as a solid, liquid, or gas. The solvent is the component (usually the one present in greatest amount) in which the other component(s), called the **solute(s),** is (are) dissolved. NaCl(aq), for example, describes a solution in which water is the solvent and NaCl, the solute. Gasoline is a *nonaqueous* (no water) solution, though gasoline has so many components that no single one is easily labeled the solvent.

Because the quantity of solute present in a given quantity of solution can vary, it is usually necessary to specify the exact composition of a solution. In Example 4-9 the composition of an HCl solution was described in terms of solution density and percent composition. A description based on the mole is more useful.

Molarity. The composition or concentration of a solution expressed as the *number of moles of solute per liter of solution* is called **molarity (M).** *

Definition of molarity.

$$\text{molarity (M)} = \frac{\text{number of moles of solute}}{\text{number liters solution}} \quad (4.18)$$

If 0.455 mol of urea, $CO(NH_2)_2$, is dissolved in 1.000 L of water solution, the solution molarity is

Because the volume of a solution is rarely equal to the volume of the solvent used, molarity must be based on the volume of *solution*.

$$\frac{0.455 \text{ mol } CO(NH_2)_2}{1.000 \text{ L soln}} = 0.455 \text{ M } CO(NH_2)_2$$

This solution is also referred to as a "0.455 **molar**" solution.

Of course, molar quantities cannot be measured out directly; they must be related to other measurements, usually mass or volume, as we see in Examples 4-10 and 4-11.

Example 4-10

Calculating molarity concentration when the quantity of solute is given through its volume and density. A solution is prepared by dissolving 25.0 cm^3 of ethyl alcohol, C_2H_5OH ($d = 0.789$ g/cm^3), in a sufficient quantity of water to produce 250.0 mL of solution. What is the molarity of C_2H_5OH in this solution?

Solution. First we must calculate the number of moles of ethyl alcohol in a 25.0-cm^3 sample. For this we need conversion factors based on density and molar mass.

$$\text{no. mol } C_2H_5OH = 25.0 \text{ cm}^3 \text{ } C_2H_5OH \times \frac{0.789 \text{ g } C_2H_5OH}{1 \text{ cm}^3 \text{ } C_2H_5OH} \times \frac{1 \text{ mol } C_2H_5OH}{46.07 \text{ g } C_2H_5OH}$$

$$= 0.428 \text{ mol } C_2H_5OH$$

Now, we use the definition of molarity, expressed through equation (4.18). Note that 250.0 mL = 0.2500 L.

---

*In SI units the term liter (L) is discouraged—its equivalent, the cubic decimeter (dm^3), has been adopted. Thus, the unit of molarity concentration would be mol/dm^3. Also, use of the term "molar" is discouraged, since "molar" generally means "per mole" (as in g/mol for molar mass). Nevertheless, because they are so well established, we will continue to use the definitions introduced here.

$$\text{molarity} = \frac{0.428 \text{ mol } C_2H_5OH}{0.2500 \text{ L soln}} = 1.71 \text{ M } C_2H_5OH$$

SIMILAR EXAMPLES: Exercises 8, 32.

Example 4-11

Calculating the mass of solute in a solution of a given molarity. Potassium chromate, K_2CrO_4, is used in dyeing textiles, in manufacturing pigments for paints and inks, and as a reagent (reactant) in analytical chemistry. We want to prepare exactly 0.5000 L (500.0 mL) of an 0.250 M K_2CrO_4 solution in water. What mass of K_2CrO_4 should we use for this purpose? (The laboratory procedure for preparing this solution is suggested by Figure 4-7. Study this figure carefully.)

Solution

Method 1. Rearrange equation (4.18) to solve for the amount of solute.

no. mol K_2CrO_4 = molarity (mol/L) × volume (L)

$$= \frac{0.250 \text{ mol } K_2CrO_4}{1 \text{ L}} \times 0.5000 \text{ L} = 0.125 \text{ mol } K_2CrO_4$$

Then compute the mass of solute.

no. g K_2CrO_4 = 0.125 mol $K_2CrO_4 \times \dfrac{194.2 \text{ g } K_2CrO_4}{1 \text{ mol } K_2CrO_4} = 24.3 \text{ g } K_2CrO_4$

Method 2. Use solution molarity as a conversion factor between volume of solution and number of moles of solute. That is, for the solution here, 1 L soln ≏ 0.250 mol K_2CrO_4.

no. g K_2CrO_4 = 0.5000 L soln $\times \dfrac{0.250 \text{ mol } K_2CrO_4}{1 \text{ L soln}} \times \dfrac{194.2 \text{ g } K_2CrO_4}{1 \text{ mol } K_2CrO_4}$

(L soln ⟶ mol K_2CrO_4 ⟶ g K_2CrO_4)

= 24.3 g K_2CrO_4

SIMILAR EXAMPLES: Exercises 9, 33, 36.

FIGURE 4-7
Preparation of
0.250 M K_2CrO_4—Example
4-11 illustrated.

Because the volume of a solution is rarely equal to the volume of the solvent used, this solution cannot be prepared just by adding 24.3 g K_2CrO_4 to 500.0 mL of water. Instead, the 24.3 g solute is added to a small quantity of water in a container called a *volumetric flask*. When dissolving is complete, the flask is filled with water exactly to the 500.0 mL mark. [Carey B. Van Loon]

Solution Dilution. In many cases we need to use equation (4.18) two (or more) times, for example, if we mix two or more solutions and need to know the final concentration. The situation described in Example 4-12 is perhaps more common. Here we prepare a more dilute solution by adding water to a more concentrated solution. This procedure is often used in the laboratory, where solutions of fairly high concentrations are stored and other solutions are prepared by diluting them. The principle involved, illustrated through Figure 4-8, is that

Principle involved in dilution problems.

> All the solute in the initial, more concentrated solution appears in the final diluted solution. (4.19)

Statement (4.19) and the definition of molarity are all that you need in working out dilution problems. However, you may prefer a method based on a rearrangement of equation (4.18), that is,

no. mol solute = molarity (M) × volume (V, in liters)

FIGURE 4-8
Preparing a solution by dilution—Example 4-12 illustrated.

(a) A pipet is used to dispense 50.0 mL of 0.250 M K_2CrO_4 into a small quantity of water in a 250.0-mL volumetric flask.
(b) Following this, water is added to bring the solution to the mark on the flask. At this point the solution is 0.0500 M K_2CrO_4. [Carey B. Van Loon]

(a)   (b)

When a solution is diluted, the amount of solute *remains constant* between the initial *(i)* and final *(f)* solutions. Thus,

$$M_i \times V_i = \text{amount of solute (mol)} = M_f \times V_f$$

and

Equation for dilution problems.

$$M_i \times V_i = M_f \times V_f \tag{4.20}$$

Example 4-12

Preparing a solution by dilution. A particular analytical chemistry procedure requires us to use 0.0500 M K_2CrO_4. What volume of 0.250 *M* K_2CrO_4 must we dilute with water to prepare 250.0 mL of 0.0500 M K_2CrO_4? (Also see Figure 4-8.)

Solution. Consider the two solutions separately. First, calculate the amount of solute that must be present in the *final* solution.

$$\text{no. mol } K_2CrO_4 = 0.250 \text{ L soln} \times \frac{0.0500 \text{ mol } K_2CrO_4}{1 \text{ L soln}} = 0.0125 \text{ mol } K_2CrO_4$$

Since all the solute in the final, dilute solution must come from the initial, more concentrated solution, we must answer the question: What volume of 0.250 M K_2CrO_4 must be taken to contain 0.0125 mol K_2CrO_4?

$$\text{no. L soln} = 0.0125 \text{ mol } K_2CrO_4 \times \frac{1 \text{ L soln}}{0.250 \text{ mol } K_2CrO_4} = 0.0500 \text{ L soln}$$

Alternatively, we can use equation (4.20): $M_i \times V_i = M_f \times V_f$. Here we know the volume of solution that we wish to prepare ($V_f = 250.0$ mL) and the molarities of the final solution (0.0500 M) and the initial solution (0.250 M). We must solve for the initial volume, V_i.

$$V_i = V_f \times \frac{M_f}{M_i} = 250.0 \text{ mL} \times \frac{0.0500 \text{ M}}{0.250 \text{ M}} = 50.0 \text{ mL}$$

SIMILAR EXAMPLES: Exercises 10, 11, 34, 80, 82.

Solution Stoichiometry. Molarity concentrations are often required in calculations based on the chemical equation. Two possibilities are illustrated by Examples 4-13 and 4-14.

Example 4-13

Relating the mass of a product to the volume and molarity of a reactant solution. 25.00 mL of 0.1122 M $AgNO_3$ is added to an excess of $K_2CrO_4(aq)$. What mass of Ag_2CrO_4 will precipitate from the solution?

$$2 \text{ AgNO}_3(aq) + K_2CrO_4(aq) \longrightarrow Ag_2CrO_4(s) + 2 \text{ KNO}_3(aq)$$

Solution. The steps involved in this calculation are (1) determine the amount of $AgNO_3$ that reacts; (2) use a stoichiometric factor from the chemical equation to determine the number of moles of Ag_2CrO_4 produced; (3) convert the amount of $Ag_2CrO_4(s)$ to its mass in grams. These steps can be combined into a single setup, as follows:

$$\text{no. g Ag}_2CrO_4 = 25.00 \text{ mL} \times \frac{1 \text{ L}}{1000 \text{ mL}} \times \frac{0.1122 \text{ mol AgNO}_3}{1 \text{ L}}$$

$$\times \frac{1 \text{ mol Ag}_2CrO_4}{2 \text{ mol AgNO}_3} \times \frac{331.74 \text{ g Ag}_2CrO_4}{1 \text{ mol Ag}_2CrO_4}$$

$$= 0.4653 \text{ g Ag}_2CrO_4(s)$$

SIMILAR EXAMPLES: Exercises 12, 41, 43.

Example 4-14

Using a chemical reaction to determine the molarity of a solution. The electrolyte in a lead storage battery is dilute sulfuric acid, $H_2SO_4(aq)$. This acid must have a concentration between 4.8 M and 5.3 M if the battery is to be most effective. A 5.00-mL sample of a particular battery acid requires 46.40 mL of 0.875 M NaOH for its complete reaction (neutralization). Does the concentration of the battery acid fall within the range for the most effective operation of the battery?

$$H_2SO_4(aq) + 2 \text{ NaOH} \longrightarrow Na_2SO_4(aq) + 2 \text{ H}_2O(l)$$

Solution. Our essential task is to determine the molarity of the $H_2SO_4(aq)$. We again use a three-step approach: (1) Determine the number of mol NaOH in 46.40 mL of 0.875 M NaOH. (2) Determine the number of mol H_2SO_4 that react with this NaOH. (3) Calculate the molarity of the $H_2SO_4(aq)$.

$$\text{no. mol NaOH} = 46.40 \text{ mL} \times \frac{1 \text{ L}}{1000 \text{ mL}} \times \frac{0.875 \text{ mol NaOH}}{1 \text{ L}}$$

$$= 0.0406 \text{ mol NaOH}$$

$$\text{no. mol H}_2SO_4 = 0.0406 \text{ mol NaOH} \times \frac{1 \text{ mol H}_2SO_4}{2 \text{ mol NaOH}} = 0.0203 \text{ mol H}_2SO_4$$

Since the 0.0203 mol H_2SO_4 is derived from a 5.00-mL (0.00500-L) sample,

$$\text{molarity} = \frac{0.0203 \text{ mol } H_2SO_4}{0.00500 \text{ L soln}} = 4.06 \text{ M } H_2SO_4$$

This concentration falls below the range for the most effective battery operation.

SIMILAR EXAMPLES: Exercises 39, 42, 91.

The calculation involved in Example 4-14 is not difficult, but we are left with some interesting questions about what experimental method to use to carry out the reaction. How can we deliver the precise volume of NaOH(aq) (46.40 mL) into the reaction mixture? How do we determine when the reaction has just reached completion, so that we do not use an excess of NaOH(aq)? We explore this procedure, called *titration*, in more detail in the next chapter.

4-5   Determining the Limiting Reagent

When we carry out a reaction in such a way that the reactants are *simultaneously* consumed completely, we say that the reactants are in **stoichiometric proportions.** This condition is often required, especially in chemical analyses (e.g., the analysis of battery acid in Example 4-14). At other times, such as in a manufacturing process, the emphasis is on converting one of the reactants completely into products, by using an excess of all the other reactants. The reactant or reagent that is completely consumed—the **limiting reagent**—determines the quantity of products that form. Up to this point we have identified the reactant(s) in excess and, by implication, the limiting reagent. In some cases, however, the limiting reagent will not be explicitly indicated, and you must first do a calculation to identify it. The principle involved is suggested by the analogy in Figure 4-9.

> By an excess of a reactant we mean that more of the reactant is present than is consumed in the reaction.

Example 4-15

Determining the limiting reagent in a reaction. Phosphoryl chloride, $POCl_3$, is used in the manufacture of fire retardants, gasoline additives, and hydraulic fluids. One reaction by which it can be prepared is

$$6 \text{ } PCl_3(l) + 6 \text{ } Cl_2(g) + P_4O_{10}(s) \longrightarrow 10 \text{ } POCl_3(l)$$

If 1.00 kg each of PCl_3, Cl_2, and P_4O_{10} are allowed to react, how many kg of $POCl_3$ will be formed?

Solution. The method suggested by Figure 4-9 is to do three calculations. In each of the three calculations determine the amount of $POCl_3$ that would be produced from the available amount of a reactant, *assuming that the other two reactants are in excess.* Only one of the three results can be the correct answer, and this will always be the *smallest* value. The reactant that leads to the smallest amount of product is the limiting reagent.

Assuming the PCl_3 is the limiting reagent:

$$\text{no. mol } POCl_3 = 1000 \text{ g } PCl_3 \times \frac{1 \text{ mol } PCl_3}{137.3 \text{ g } PCl_3} \times \frac{10 \text{ mol } POCl_3}{6 \text{ mol } PCl_3}$$

$$= 12.1 \text{ mol } POCl_3$$

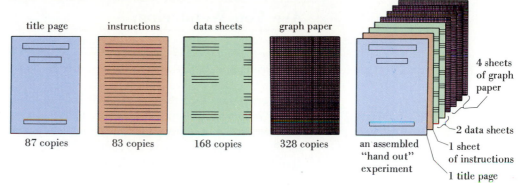

FIGURE 4-9

An analogy to determining the limiting reagent in a chemical reaction—assembling a hand out experiment.

Assembling the "handout" experiment is equivalent to a reaction whose equation is

title page + instructions + 2 data sheets + 4 graph paper → handout

From the number of copies of each type of sheet (analogous to moles of reactants) calculate how many complete handouts (analogous to moles of products) can be assembled. Do you get 82? Which is the "limiting reagent?"

Based on the title page we would say that no more than 87 complete copies of the handout are possible, but based on the instruction page we would say that no more than 83. Two data sheets are required per handout. There are enough data sheets for 168/2 = 84 handouts, but no more than 83 handouts are possible because of the limited number of instruction sheets. Finally, because four sheets of graph paper are required per handout, we conclude that only 328/4 = 82 handouts are possible. Graph paper is the "limiting reagent." The excess pages are the title page (5 copies), the instruction sheet (one copy), and the data page (4 copies).

Assuming that Cl_2 is the limiting reagent:

$$\text{no. mol POCl}_3 = 1000 \text{ g Cl}_2 \times \frac{1 \text{ mol Cl}_2}{70.91 \text{ g Cl}_2} \times \frac{10 \text{ mol POCl}_3}{6 \text{ mol Cl}_2}$$

$$= 23.5 \text{ mol POCl}_3$$

Assuming that P_4O_{10} is the limiting reagent:

$$\text{no. mol POCl}_3 = 1000 \text{ g P}_4\text{O}_{10} \times \frac{1 \text{ mol P}_4\text{O}_{10}}{283.9 \text{ g P}_4\text{O}_{10}} \times \frac{10 \text{ mol POCl}_3}{1 \text{ mol P}_4\text{O}_{10}}$$

$$= 35.2 \text{ mol POCl}_3$$

In this example PCl_3 is the limiting reagent, but in the commercial process it is kept in excess, and P_4O_{10} is the limiting reagent.

The smallest amount of $POCl_3$ is 12.1 mol. The limiting reagent must be PCl_3. The mass of $POCl_3$ corresponding to 12.1 mol is

$$\text{no. kg POCl}_3 = 12.1 \text{ mol POCl}_3 \times \frac{153.3 \text{ g POCl}_3}{1 \text{ mol POCl}_3} \times \frac{1 \text{ kg POCl}_3}{1000 \text{ g POCl}_3}$$

$$= 1.85 \text{ kg POCl}_3$$

Some people prefer to compare the amounts of reactants, *on a mole basis*, to identify the limiting reagent. This is followed by a single calculation based on the limiting reagent. For example, since PCl_3 and Cl_2 combine in the mole ratio 1:1 (the coefficients in the equation are 6 and 6), there is more than enough Cl_2 (14.1 mol) for the amount of PCl_3 available (7.28 mol). To combine with the 7.28 mol PCl_3 requires only about 1.2 mol P_4O_{10} (the coefficients in the equation are 6 and 1). The amount of P_4O_{10} available (3.52 mol) is greater than what is needed to react with the PCl_3. We conclude that Cl_2 and P_4O_{10} are in excess and that PCl_3 is the limiting reagent. The question then becomes one of determining the mass of $POCl_3$ that can be produced from 1000 g PCl_3 (7.28 mol) and an excess of the other reactants.

SIMILAR EXAMPLES: Exercises 13, 45, 46, 49, 85.

Why the mass of product in Example 4-15 isn't simply the sum of the masses of the starting materials—3.00 kg?

If the quantities of the three reactants in Example 4-15 had been in *stoichiometric proportions* (6 mol PCl_3 for every 6 mol Cl_2 and 1 mol P_4O_{10}), all reactants would have been consumed completely and indeed the quantity of the $POCl_3$ would have been the sum of the masses of the reactants (law of conservation of mass). In general, however, the reactants will not be in stoichiometric proportions. Some of the initial reactants will remain unreacted, that is, will be in *excess*. It is the mass of product *plus that of the unreacted reactants* that is equal to the initial mass.

4-6 Theoretical Yield, Actual Yield, and Percent Yield

When we calculate that a certain quantity of product should result from given quantities of reactants, this *calculated* quantity of product is called the **theoretical yield** of the reaction. The quantity of product that is *actually* produced is called the **actual yield.** The **percent yield** is defined as

Expression relating actual, theoretical, and percent yields.

$$\text{percent yield} = \frac{\text{actual yield}}{\text{theoretical yield}} \times 100 \qquad (4.21)$$

For many reactions the actual yield is almost exactly equal to the theoretical yield. Such reactions are said to be *quantitative;* they can be used in quantitative chemical analyses, for example. On the other hand, for some reactions the actual yield is *less than* the theoretical yield, and the percent yield is *less than* 100%. This is because the reaction may not go to completion (see Section 4-8), competing reactions may reduce the yield of product, or material may be lost in handling.

Example 4-16

Determining theoretical, actual, and percent yields. **Billions of pounds of urea, $CO(NH_2)_2$, are produced annually for use as a fertilizer. The principal reaction employed is**

$$2 NH_3 + CO_2 \longrightarrow CO(NH_2)_2 + H_2O$$

The typical reaction mixture contains NH_3 and CO_2 in a 3:1 mol ratio. If 47.7 g urea is obtained *per mol* CO_2 that reacts, what are the (a) theoretical yield, (b) actual yield, and (c) percent yield, in this reaction?

Solution

(a) **The stoichiometric proportions are 2 mol NH_3:1 mol CO_2. Since the mol ratio of NH_3 to CO_2 actually employed is 3:1, NH_3 is in excess and CO_2 is the limiting reactant. Base the calculation on 1.00 mol CO_2. The theoretical yield is**

$$\text{no. g } CO(NH_2)_2 = 1.00 \text{ mol } CO_2 \times \frac{1 \text{ mol } CO(NH_2)_2}{1 \text{ mol } CO_2} \times \frac{60.06 \text{ g } CO(NH_2)_2}{1 \text{ mol } CO(NH_2)_2}$$

$$= 60.1 \text{ g } CO(NH_2)_2$$

(b) The actual yield is 47.7 g $CO(NH_2)_2$.

(c) The percent yield is

$$\% \text{ yield} = \frac{47.7 \text{ g } CO(NH_2)_2}{60.1 \text{ g } CO(NH_2)_2} \times 100 = 79.4\%$$

SIMILAR EXAMPLES: Exercises 14, 50, 51, 87.

Example 4-17

Determining how the percent yield affects the quantities of reactants and products of a reaction. When heated with sulfuric or phosphoric acid, cyclohexanol, $C_6H_{12}O$, is converted to cyclohexene, C_6H_{10}.

$$C_6H_{12}O(l) \longrightarrow C_6H_{10}(l) + H_2O(l)$$

Additional procedures are required to purify the cyclohexene. The percent yield is 83%. To use in synthesizing other organic compounds, we need 25 g of pure cyclohexene. What mass of cyclohexanol that is 91% pure must we use to obtain this 25 g of pure cyclohexene?

Solution. First we have to answer this question: What should be the theoretical yield of C_6H_{10}, if the 25 g C_6H_{10} that we plan to obtain as an actual yield is only 83% of the theoretical yield? In the remaining steps we calculate the quantity of *pure* $C_6H_{12}O$ required to produce this theoretical yield of C_6H_{10}, and then the quantity of *impure* (91%) $C_6H_{12}O$ required.

Step 1. Rearrange equation (4.21) and calculate the theoretical yield.

> You may want to think of this first step as an algebra problem: If 83% of a number is equal to 25, what is the number? $0.83x = 25$; $x = 25/0.83 = 30$.

$$\text{theoretical yield} = \frac{\text{actual yield} \times 100}{\text{percent yield}} = \frac{25 \text{ g} \times 100}{83} = 30. \text{ g}$$

Step 2. Calculate the quantity of $C_6H_{12}O$ to produce 30. g C_6H_{10}.

$$\text{no. g } C_6H_{12}O = 30. \text{ g } C_6H_{10} \times \frac{1 \text{ mol } C_6H_{10}}{82.1 \text{ g } C_6H_{10}} \times \frac{1 \text{ mol } C_6H_{12}O}{1 \text{ mol } C_6H_{10}}$$

$$\times \frac{100.2 \text{ g } C_6H_{12}O}{1 \text{ mol } C_6H_{12}O}$$

$$= 37 \text{ g } C_6H_{12}O$$

> If we need 37 g $C_6H_{12}O$, the quantity of *impure* $C_6H_{12}O$ that we must use is *greater than* 37 g.

Step 3. Calculate the quantity of *impure* cyclohexanol required.

$$\text{no. g cyclohexanol (impure)} = 37 \text{ g } C_6H_{12}O \times \frac{100 \text{ g cyclohexanol (impure)}}{91 \text{ g } C_6H_{12}O}$$

$$= 41 \text{ g cyclohexanol (impure)}$$

SIMILAR EXAMPLES: Exercises 52, 73.

4-7 Simultaneous and Consecutive Reactions

Both in the chemical laboratory and in the industrial plant, two or more reactions may be required to obtain a desired product. In some cases the reactions may occur at the same time (**simultaneously**), and in others, they may occur in succession (**consecutively**). Example 4-18 considers the case of two metals found together in a mixture (alloy), each reacting independently to produce $H_2(g)$ when treated with HCl(aq). Example 4-19 considers a two-step process for purifying titanium dioxide.

Example 4-18

Calculating the quantity of a product formed by two reactions occurring simultaneously. Magnalium alloys are widely used in aircraft construction. Two typical alloys are A (70.0% Al, 30.0% Mg) and B (90.0% Al, 10.0% Mg). To determine whether a particular alloy is A or B, a 0.710-g sample is treated with excess HCl(aq) and 0.0734 g H_2 is collected. What is the alloy?

$$2\ Al(s) + 6\ HCl(aq) \longrightarrow 2\ AlCl_3(aq) + 3\ H_2(g)$$

$$Mg(s) + 2\ HCl(aq) \longrightarrow MgCl_2(aq) + H_2(g)$$

Solution. The simplest approach is to calculate the mass of H_2 that should be produced by 0.710 g of each alloy and compare these results with the measured mass of H_2—0.0734 g. Start with alloy A.

Step 1. Use percent composition to determine the mass of each metal in the alloy. *Result:* 0.497 g Al; 0.213 g Mg.

Step 2. Use molar masses to convert from mass to number of moles of each metal. *Result:* 0.0184 mol Al; 0.00877 mol Mg.

Step 3. Determine the no. mol H_2 produced by each metal.

$$\text{no. mol } H_2 = 0.0184 \text{ mol Al} \times \frac{3 \text{ mol } H_2}{2 \text{ mol Al}} = 0.0276 \text{ mol } H_2$$

$$\text{no. mol } H_2 = 0.00877 \text{ mol Mg} \times \frac{1 \text{ mol } H_2}{1 \text{ mol Mg}} = 0.00877 \text{ mol } H_2$$

Step 4. The total amount of H_2 is $0.0276 + 0.00877 = 0.0364$ mol H_2. Its mass is

$$\text{no. g } H_2 = 0.0364 \text{ mol } H_2 \times \frac{2.016 \text{ g } H_2}{1 \text{ mol } H_2} = 0.0734 \text{ g } H_2$$

The sample is alloy A. As a further check, repeat the calculation for alloy B. The value obtained will be 0.0776 g H_2.

SIMILAR EXAMPLES: Exercises 53, 54, 55.

Example 4-19

Calculating the quantity of a substance produced through a series of reactions occurring consecutively. Titanium dioxide, TiO_2, is the most widely used white pigment for paints, having mostly displaced lead-based pigments, which are environmental hazards. Before it can be used, however, naturally occurring TiO_2 must be freed of colored impurities. One process for doing this converts impure $TiO_2(s)$ to $TiCl_4(g)$, which is then converted back to pure $TiO_2(s)$. How many grams of carbon are required to produce 1.00 kg of pure $TiO_2(s)$ in this process?

$$2\ TiO_2(\text{impure}) + 3\ C(s) + 4\ Cl_2(g) \longrightarrow 2\ TiCl_4(g) + CO_2(g) + 2\ CO(g)$$

$$TiCl_4(g) + O_2(g) \longrightarrow TiO_2(s) + 2\ Cl_2(g)$$

Solution

Step 1. Determine the no. mol $TiCl_4$ that must be decomposed in the second reaction to produce 1000 g $TiO_2(s)$.

$$\text{no. mol } TiCl_4 = 1000 \text{ g } TiO_2 \times \frac{1 \text{ mol } TiO_2}{79.88 \text{ g } TiO_2} \times \frac{1 \text{ mol } TiCl_4}{1 \text{ mol } TiO_2}$$

$$= 12.5 \text{ mol } TiCl_4$$

Step 2. Determine the mass of carbon required to produce 12.5 mol $TiCl_4$ in the first reaction.

$$\text{no. g } C = 12.5 \text{ mol } TiCl_4 \times \frac{3 \text{ mol } C}{2 \text{ mol } TiCl_4} \times \frac{12.01 \text{ g } C}{1 \text{ mol } C} = 225 \text{ g } C$$

SIMILAR EXAMPLES: Exercises 57, 58, 59, 77.

4-8 An Introduction to Chemical Equilibrium

If we attempt to produce ammonia from N_2 and H_2

$$N_2(g) + 3 H_2(g) \longrightarrow 2 NH_3(g)$$

we discover that ammonia molecules also decompose back into N_2 and H_2

$$2 NH_3(g) \longrightarrow N_2(g) + 3 H_2(g).$$

Thus, *two* reactions occur simultaneously—a forward ($\rightarrow$) and a reverse ($\leftarrow$) reaction. The reaction is **reversible.** We can describe both of these reactions through a single equation by using a double arrow ($\rightleftharpoons$).

$$N_2(g) + 3 H_2(g) \rightleftharpoons 2 NH_3(g) \tag{4.22}$$

The existence of reversible reactions complicates the discussion of chemical reactions. For example, we are not easily able to answer this seemingly simple question: "How many moles of NH_3 are produced in the reaction of 0.10 mol N_2 with an excess of H_2?" Depending on the conditions used (e.g., the temperature and pressure), in most cases the result will not be the expected 0.20 mol NH_3.

When a point is reached in reaction (4.22) where each reacting species is consumed and reformed at a constant rate, the reaction is said to have reached a condition of **dynamic equilibrium.** *Dynamic* means that reaction among the molecules continues to occur, and *equilibrium* means that beyond this point there is no net change with time in the amounts of any of the reacting species. Since some of each species must be present at equilibrium, we cannot calculate the yield of a reversible reaction with stoichiometric factors from the chemical equation alone. The yield in the hypothetical question posed above would be *less than* 0.20 mol, since the N_2 is not completely consumed. Chapter 16 presents some new principles that, when combined with stoichiometric factors, will make possible an exact numerical answer.

Fortunately, in many cases the reverse reaction occurs to such a slight extent that we do not have to consider it. Reactions in which the reverse reaction is negligible are said to **go to completion,** and the various examples we have considered in this chapter are of this type. In Chapter 5 we consider more explicitly some of the situations under which reactions tend to go to completion.

FOCUS ON

Industrial Chemistry

Quite possibly hydrazine, N_2H_4, was used in the manufacture of the pesticide being applied here. Some 30 or 40 different pesticides require hydrazine for their synthesis. [Fred Ward/Black Star]

Large-scale processes for the manufacture of chemicals are based on the principles presented in this chapter and elsewhere in the text. Yet, factors beyond the usual textbook examples often need to be considered. We attempt here to convey the essence of industrial chemistry.

Hydrazine and Its Uses. Hydrazine, N_2H_4, is an oily, colorless liquid that freezes at 2.0 °C and boils at 113.5 °C. The nature of chemical bonding in hydrazine is discussed in Section 10-4, and its physical and chemical properties are explored further in Chapters 12 and 23.

The first important use of hydrazine was as a fuel in the German rocket-powered ME-163 fighter airplane in World War II, and it continues to find use in rocket engines. Currently its largest use (40%) is in the synthesis of some 30 to 40 different pesticides. Next in importance (33%) is the production of hydrazine-based chemicals used as blowing agents in the polymer industry. (A blowing agent decomposes into gases. The gases produce the holes or pores in products like sponge rubber and foamed plastics.) Hydrazine is also used (15%) in water treatment. When

added to boilers or hot-water heating systems, it scavenges (removes) dissolved oxygen, thereby reducing the rate of corrosion of metal parts exposed to the hot water.

A method of manufacturing hydrazine (the Raschig process) is outlined in Figure 4-10 in a form commonly used to represent an industrial process—a **flow diagram.**

Stepwise Reactions, Intermediates, and the Net Chemical Equation. An industrial process is usually carried out in stages or steps. The Raschig process involves three steps. Equations for these reactions are listed in Figure 4-11.

1. $Cl_2(g)$ and $NaOH(aq)$ react to produce sodium hypochlorite, $NaOCl(aq)$.
2. $NH_3(aq)$ reacts with $NaOCl(aq)$ to form chloramine, $NH_2Cl(aq)$.
3. Additional NH_3 reacts with $NH_2Cl(aq)$ to form hydrazine, $N_2H_4(aq)$.

$NaOCl(aq)$ and $NH_2Cl(aq)$ are **intermediates** in the process. Their presence is crucial to the overall process, but they are consumed immediately after their formation. Hydrazine is the **end product** of the process. The outcome of the overall process can be represented through a single equation, called the **net chemical equation,** which we obtain by adding together equations for the three consecutive reactions. In this addition (see Figure 4-11) intermediate species "cancel out."

By-products and Side Reactions. The net equation shows two products in addition to N_2H_4—NaCl and H_2O. Substances formed along with the desired end product are called **by-products** of a process. Sometimes the economic feasibility of a process depends on the value of the by-products. (For example, about 90% of all hydrochloric acid is produced as a by-product of other manufacturing processes.)

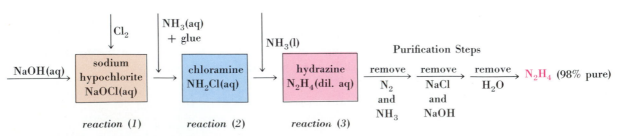

FIGURE 4-10
Flow diagram for the commercial production of hydrazine, N_2H_4.

The Net Chemical Reaction:

reaction (1) $2\,NaOH(aq) + Cl_2 \longrightarrow NaOCl(aq) + NaCl(aq) + H_2O$

reaction (2) $NaOCl(aq) + NH_3(aq) \longrightarrow NH_2Cl(aq) + NaOH(aq)$

reaction (3) $NH_2Cl(aq) + NH_3(l) + NaOH(aq) \xrightarrow{\Delta} N_2H_4(aq) + NaCl(aq) + H_2O$

(1) + (2) + (3) $2\,NaOH(aq) + Cl_2 + NaOCl(aq) + NH_3(aq) + NH_2Cl(aq) + NH_3(l) + NaOH(aq) \longrightarrow$
$NaOCl(aq) + NaCl(aq) + H_2O + NH_2Cl(aq) + NaOH(aq) + N_2H_4(aq) + NaCl(aq) + H_2O$

cancellations: $2\,NaOH(aq) + Cl_2 + \cancel{NaOCl(aq)} + NH_3(aq) + \cancel{NH_2Cl(aq)} + NH_3(l) + \cancel{NaOH(aq)} \longrightarrow$
$\cancel{NaOCl(aq)} + NaCl(aq) + H_2O + \cancel{NH_2Cl(aq)} + \cancel{NaOH(aq)} + N_2H_4(aq) + NaCl(aq) + H_2O$

Net equation: $2\,NaOH(aq) + Cl_2 + 2\,NH_3 \longrightarrow N_2H_4(aq) + 2\,NaCl(aq) + 2\,H_2O$

Side Reaction: $2\,NH_2Cl(aq) + N_2H_4(aq) \xrightarrow{Cu^{2+}} 2\,NH_4Cl(aq) + N_2(g)$

FIGURE 4-11
Reactions involved in the production of hydrazine, N_2H_4.

By-products may also form as a result of **side reactions** that compete with the main reaction. Some of the hydrazine formed in the third reaction combines with the intermediate chloramine to produce $NH_4Cl(aq)$ and $N_2(g)$. This side reaction, which reduces the yield of N_2H_4, is catalyzed (speeded up) by the presence of Cu^{2+} or other heavy metal ions. Removal of these ions helps to minimize the side reaction.

Purification of the Product. Rarely is the end product sufficiently pure for its intended uses—the product must be purified. In the Raschig process hydrazine is present in dilute aqueous solution, together with NH_3, NaCl, and traces of NaOH. The substances that can be obtained as solids, NaCl and NaOH, are crystallized from solution. NH_3 and H_2O are removed by distillation. Ultimately a product containing 98% or more N_2H_4 is obtainable. NH_3 recovered in the purification is reused in the main reactions, illustrating still another principle of industrial chemistry: *Materials are recycled whenever possible.*

Reaction Conditions. Even if the conditions of a chemical reaction are shown in a chemical equation, the reasons for these conditions are probably not. In the Raschig process the formation of chloramine (Step 2) proceeds rapidly, but the conversion of chloramine and ammonia to hydrazine (Step 3) does not. To speed up this third reaction, elevated temperatures (about 130 °C) are used.

At the same time that the main reaction is speeded up the side reaction must be minimized. This is accomplished in two ways.

1. A large excess of NH_3 is used, perhaps 20 to 30 mol NH_3 for every mol of N_2H_4 formed. This means that NH_2Cl is more likely to react with NH_3 than with N_2H_4.
2. A protein-based material—gelatin, albumin, glue—is added to the reaction mixture. Protein molecules bind

with metal ions in solution (e.g., Cu^{2+}) and reduce their ability to catalyze the side reaction.

In Step 3, NH_3 is introduced as the anhydrous (water-free) liquid rather than in aqueous solution, for two reasons.

1. Heat evolved when NH_3 dissolves in $NH_2Cl(aq)$ is sufficient to raise the reaction temperature to the desired 130 °C. (Energy savings are essential in industrial operations.)
2. The less water added to the reaction mixture, the less that has to be removed in the purification steps (again, resulting in energy savings).

Alternative Processes. In a recent modification of the Raschig process reactions are carried out in the organic solvent acetone. Acetone and hydrazine form a compound that is resistant to attack by NH_2Cl. The hydrazine–acetone compound is later decomposed into hydrazine and acetone, and the acetone is recycled. By eliminating the side reaction, the yield of hydrazine is close to 100%, compared to the 60 to 80% yield in the original Raschig process. Currently about 25% of the hydrazine-producing capacity in the United States is based on the original Raschig process and 75% on the modified process. Again two points are illustrated.

1. Industrial methods of producing chemicals undergo constant change.
2. The changeover from one process to another takes place over a period of time, as new plants are built or old ones converted.

Summary

Once the reactants and products of a chemical reaction are identified, the reaction can be represented through a chemical equation. The physical states or forms of the reactants and products and reaction conditions can also be indicated. A chemical equation must be balanced atomically, and, where necessary, for electric charge as well. The principal feature of calculations based on the chemical equation (stoichiometric calculations) is the use of conversion factors derived from the equation (stoichiometric factors). These calculations often require the use of molar masses, densities, and percent compositions as well.

Molarity concentration describes the composition of a solution in terms of the number of moles of solute per liter of solution. From this definition it is possible to perform calculations relating molarity, solution volume, and amount of solute. This may be done for individual solutions or for situations in which solutions are mixed, diluted by adding more solvent, or concentrated by removing solvent. For reactions occurring in solution, molarity

concentration also provides important conversion factors for stoichiometric calculations.

Additional features sometimes arise in stoichiometric calculations: It may be necessary to identify the single reactant, called the *limiting reagent,* that determines the amount of product formed (other reactants being in excess). The yield of a product in a reaction may be less than the calculated value. Two or more reactions may occur at the same time, or a series of reactions may occur in succession. All the computational methods of this chapter are based on the assumption that reactions "go to completion." Although this is often the case, many chemical reactions are *reversible,* requiring that special computational methods (considered in Chapter 16) be applied to the condition known as *equilibrium.* Finally, when fundamental principles are applied to actual industrial processes (industrial chemistry), other factors must be considered as well, such as reaction conditions, side reactions, and purification of product.

Summarizing Example

Sodium nitrite is used in the production of dyes for coloring fabrics, as a preservative in meat processing (to prevent botulism), as a bleach for fibers, and in photography. It is prepared by passing nitrogen monoxide and oxygen gases into an aqueous solution of sodium carbonate. Carbon dioxide gas is another product of the reaction.

1. Write a chemical equation for the reaction described above.

Solution. Use information from Tables 3-2, 3-4, and 3-5 to write formulas for the reactants and products.

formula expression: $Na_2CO_3(aq) + NO(g) + O_2(g) \longrightarrow NaNO_2(aq) + CO_2(g)$

In writing the chemical equation, start with an element that appears in only one compound on each side, Na; balance O atoms last (C atoms remain in balance throughout.)

Balance Na atoms:
$$Na_2CO_3(aq) + NO(g) + O_2(g) \longrightarrow 2\,NaNO_2(aq) + CO_2(g)$$

Balance N atoms:
$$Na_2CO_3(aq) + 2\,NO(g) + O_2(g) \longrightarrow 2\,NaNO_2(aq) + CO_2(g)$$

Balance O atoms:
$$Na_2CO_3(aq) + 2\,NO(g) + \tfrac{1}{2}\,O_2(g) \longrightarrow 2\,NaNO_2(aq) + CO_2(g)$$

Adjust coefficients:
$$2\,Na_2CO_3(aq) + 4\,NO(g) + O_2(g) \longrightarrow 4\,NaNO_2(aq) + 2\,CO_2(g)$$

(This example is similar to Examples 4-1 and 4-2.)

Sodium nitrite, $NaNO_2$, is used in low concentrations [less than 100 parts per million (ppm)] as a preservative in meat products, such as the frankfurters shown here. In higher concentrations sodium nitrite is a carcinogen (cancer-producing agent) in test animals. [Tom Tracy/FPG International]

2. What mass of $NaNO_2$ should result from the reaction of 748 g Na_2CO_3 with an excess of NO(g) and $O_2(g)$?

Solution. The series of conversions required here is

g Na_2CO_3 $\longrightarrow$ mol Na_2CO_3 $\longrightarrow$ mol $NaNO_2$ $\longrightarrow$ g $NaNO_2$

These conversions can be done on a step-by-step basis or in a single setup.

$$\text{no. g. } NaNO_2 = 748 \text{ g } Na_2CO_3 \times \frac{1 \text{ mol } Na_2CO_3}{106.0 \text{ g } Na_2CO_3} \times \frac{4 \text{ mol } NaNO_2}{2 \text{ mol } Na_2CO_3}$$

$$\text{(g } Na_2CO_3 \longrightarrow \text{mol } Na_2CO_3 \longrightarrow \text{mol } NaNO_2$$

$$\times \frac{69.00 \text{ g } NaNO_2}{1 \text{ mol } NaNO_2}$$

$$\longrightarrow \text{g } NaNO_2)$$

$$= 974 \text{ g } NaNO_2$$

(This example is similar to Examples 4-5 and 4-6.)

3. In another reaction the reactants are 225 mL of 1.50 M $Na_2CO_3(aq)$, 22.1 g NO, and a large excess of O_2. What mass of $NaNO_2$ should be produced if the reaction has a 95% yield?

Solution. This problem is best broken down into *three* smaller problems, each of a type considered in the chapter.

(a) Which is the limiting reagent? $O_2(g)$ is stated to be in excess. We must choose between Na_2CO_3 and NO.

Assuming Na_2CO_3 is the limiting reagent:

$$\text{no. mol } NaNO_2 = 0.225 \text{ L} \times \frac{1.50 \text{ mol } Na_2CO_3}{1 \text{ L}} \times \frac{4 \text{ mol } NaNO_2}{2 \text{ mol } Na_2CO_3}$$

$$= 0.675 \text{ mol } NaNO_2$$

Assuming NO is the limiting reagent:

$$\text{no. mol } NaNO_2 = 22.1 \text{ g NO} \times \frac{1 \text{ mol NO}}{30.01 \text{ g NO}} \times \frac{4 \text{ mol } NaNO_2}{4 \text{ mol NO}}$$

$$= 0.736 \text{ mol } NaNO_2$$

Since 0.675 mol $NaNO_2$ is the smaller of the two results, Na_2CO_3 is the limiting reagent.

(This example is similar to Examples 4-13 and 4-15.)

(b) What is the theoretical yield of $NaNO_2$? Simply convert the amount of $NaNO_2$ determined in **(a)** to a mass in grams.

$$\text{no. g } NaNO_2 = 0.675 \text{ mol } NaNO_2 \times \frac{69.00 \text{ g } NaNO_2}{1 \text{ mol } NaNO_2} = 46.6 \text{ g } NaNO_2$$

(This example is similar to Example 4-16.)

(c) What is the actual yield of $NaNO_2$?

$$\text{actual yield} = \frac{\% \text{ yield} \times \text{theoretical yield}}{100} = \frac{95 \times 46.6 \text{ g}}{100}$$

$$= 44.3 \text{ g } NaNO_2$$

(This example is similar to Example 4-17.)

Key Terms

actual yield (4-6)
balancing an equation (4-1)
by-product (Focus feature)
chemical equation (4-1)
chemical reaction (4-1)
dilution (4-4)
dynamic equilibrium (4-8)
intermediate (Focus feature)

limiting reagent (4-5)
molarity (M) (4-4)
net chemical reaction (Focus feature)
net ionic equation (4-1)
percent yield (4-6)
products (4-1)
reactants (4-1)
side reaction (Focus feature)

solute (4-4)
solvent (4-4)
spectator ions (4-1)
stoichiometric coefficients (4-1)
stoichiometric factor (4-3)
stoichiometric proportions (4-5)
theoretical yield (4-6)

Highlighted Expressions

Some physical evidence of chemical reactions (4.1)
Basis of equation balancing (4.2)
Additional ideas in equation balancing (4.3)
Basis of balancing net ionic equations (4.13)

Definition of molarity (4.18)
Principle involved in dilution problems (4.19)
Equation for dilution problems (4.20)
Expression relating actual, theoretical, and percent yield (4.21)

Review Problems

1. Write balanced equations for the following reactions.
(a) magnesium + oxygen → magnesium oxide
(b) sulfur + oxygen → sulfur dioxide
(c) methane (CH_4) + oxygen → carbon dioxide + water
(d) aqueous silver sulfate + aqueous barium iodide →
solid barium sulfate + solid silver iodide
2. Balance the following equations by inspection.
(a) $Na_2SO_4(s) + C(s) \rightarrow Na_2S(s) + CO_2(g)$
(b) $HCl(g) + O_2(g) \rightarrow H_2O(l) + Cl_2(g)$
(c) $PCl_3(l) + H_2O(l) \rightarrow H_3PO_3(aq) + HCl(aq)$
(d) $PbO(s) + NH_3(g) \rightarrow Pb(s) + N_2(g) + H_2O(l)$
(e) $Mg_3N_2(s) + H_2O(l) \rightarrow Mg(OH)_2(s) + NH_3(g)$
3. Balance the following equations written in *ionic* form.
(a) $Zn(s) + Ag^+(aq) \rightarrow Zn^{2+}(aq) + Ag(s)$
(b) $Mn^{2+}(aq) + H_2S(aq) \rightarrow MnS(s) + H^+(aq)$
(c) $Al(s) + H^+(aq) \rightarrow Al^{3+}(aq) + H_2(g)$
4. Write a balanced chemical equation to represent
(a) the complete combustion of C_5H_{12};
(b) the complete combustion of $C_2H_6O_2$;
(c) the reaction (neutralization) of hydroiodic acid with aqueous sodium hydroxide, producing sodium iodide and water;
(d) the precipitation of lead(II) iodide by the mixing of aqueous solutions of potassium iodide and lead(II) nitrate.
5. Iron metal reacts with chlorine gas as follows.

$$2 Fe(s) + 3 Cl_2(g) \rightarrow 2 FeCl_3(s)$$

How many moles of $FeCl_3$ are obtained when 6.12 mol Cl_2 reacts with excess Fe?
6. 0.426 mol PCl_3 is produced by the reaction

$$6 Cl_2 + P_4 \rightarrow 4 PCl_3$$

How many grams each of Cl_2 and P_4 are consumed?
7. The reaction of calcium hydride with water can be used to prepare small quantities of hydrogen gas, as in the filling of weather observation balloons.

$$CaH_2(s) + 2 H_2O(l) \rightarrow Ca(OH)_2(s) + 2 H_2(g)$$

(a) How many moles of $H_2(g)$ result from the reaction of 127 g CaH_2 with an excess of water?

(b) How many grams of H_2O are consumed in the reaction of 56.2 g CaH_2?
(c) What mass of CaH_2 must react with an excess of H_2O to produce 8.12×10^{24} molecules of $H_2(g)$?
8. What are the molarity concentrations of the solutes listed below when dissolved in water?
(a) 3.28 mol C_2H_5OH in 7.16 L of solution
(b) 8.67×10^{-3} mol CH_3OH in 50.00 mL of solution
(c) 22.3 g $(CH_3)_2CO$ in 125 mL of solution
(d) 14.5 mL of pure glycerol, $C_3H_8O_3$ ($d = 1.26$ g/cm³), in 425 mL of solution
9. How much
(a) KCl, in moles, is required to prepare 1.75×10^3 L of 0.115 M KCl(aq)?
(b) Na_2CO_3, in grams, is required to produce 535 mL of 0.412 M $Na_2CO_3(aq)$?
(c) NaOH, in mg, is present *per mL* of 0.183 M NaOH(aq)?
10. What volume of 0.750 M KOH must be diluted with water to prepare 250.0 mL of 0.448 M KOH?
11. Water is evaporated from 135 mL of 0.224 M $MgSO_4$ solution until the solution volume becomes 105 mL. What is the molarity of $MgSO_4$ in the solution that results?
12. Excess $NaHCO_3(s)$ is added to 325 mL of 0.305 M $Cu(NO_3)_2(aq)$. What mass of $CuCO_3(s)$ will precipitate from solution?

$$Cu(NO_3)_2(aq) + 2 NaHCO_3(s) \rightarrow$$
$$CuCO_3(s) + 2 NaNO_3(aq) + H_2O + CO_2(g)$$

13. A 0.496-mol sample of Cu is added to 128 mL of 6.0 M $HNO_3(aq)$. How many grams of NO are produced? [*Hint:* Which is the limiting reagent?]

$$3 Cu(s) + 8 HNO_3 \rightarrow 3 Cu(NO_3)_2(aq) + 4 H_2O + 2 NO(g)$$

14. In the following reaction, 100.0 g $C_6H_{12}O$ yielded 69.0 g C_6H_{10}.

$$C_6H_{12}O \rightarrow C_6H_{10} + H_2O$$

(a) What is the theoretical yield of the reaction?
(b) What is the percent yield?

(c) What mass of $C_6H_{12}O$ should have been used to produce 100.0 g C_6H_{10}, if the percent yield is that determined in part (b)?

15. Refer to Example 4-1 and equation (4.4). Suppose that the $CO_2(g)$ produced in the combustion of a sample of propane is absorbed in $Ba(OH)_2(aq)$, producing 0.506 g $BaCO_3(s)$. What must have been the mass of propane (C_3H_8) that was burned?

$$CO_2(g) + Ba(OH)_2(aq) \rightarrow BaCO_3(s)$$

16. Chalkboard chalk is made from calcium carbonate and calcium sulfate, with minor impurities such as SiO_2. Only the $CaCO_3$ is soluble in dilute $HCl(aq)$. What is the % $CaCO_3$ in a piece of chalk if a 2.58-g sample yields 0.818 g $CO_2(g)$?

$$CaCO_3(s) + 2\ HCl(aq) \rightarrow CaCl_2(aq) + H_2O + CO_2(g)$$

Exercises

Writing and balancing chemical equations

17. Balance the following equations by inspection.
(a) $P_2H_4(l) \rightarrow PH_3(g) + P_4(s)$
(b) $NO_2(g) + H_2O(l) \rightarrow HNO_3(aq) + NO(g)$
(c) $S_2Cl_2 + NH_3 \rightarrow N_4S_4 + NH_4Cl + S_8$
(d) $SO_2Cl_2 + HI \rightarrow H_2S + H_2O + HCl + I_2$

18. Balance the following equations written in *ionic* form.
(a) $Fe^{3+}(aq) + H_2S(aq) \rightarrow Fe_2S_3(s) + H^+(aq)$
(b) $Al^{3+} + NH_3 + H_2O \rightarrow Al(OH)_3(s) + NH_4^+(aq)$
(c) $S_2O_3^{2-} + H^+ \rightarrow H_2O + S(s) + SO_2(g)$
(d) $MnO_2(s) + H^+ + Cl^- \rightarrow Mn^{2+} + H_2O + Cl_2(g)$

19. Write balanced equations to represent the complete combustion of
(a) benzene, C_6H_6
(b) isopropyl alcohol, C_3H_7OH
(c) benzoic acid, C_6H_5COOH
(d) thiobenzoic acid, C_6H_5COSH [*Hint:* The sulfur is converted to sulfur dioxide.]

20. Following are some chemical reactions commonly carried out by students in a general chemistry laboratory. Write balanced equations for these reactions.
(a) Zinc metal dissolves in hydrochloric acid solution, producing zinc chloride solution and hydrogen gas.
(b) The solids, ammonium chloride and sodium hydroxide, are heated together. The reaction products are solid sodium chloride, liquid water, and ammonia gas.
(c) Solid potassium chlorate is heated, yielding solid potassium chloride and oxygen gas as products.
(d) Sodium metal reacts with liquid water, producing aqueous sodium hydroxide and hydrogen gas.
(e) Hydrochloric acid solution is added to solid sodium carbonate. The products are aqueous sodium chloride, water, and carbon dioxide gas.

21. Following are some reactions associated with common natural phenomena. Write balanced equations for these reactions.
(a) Nitrogen monoxide gas forms from nitrogen and oxygen gases (a reaction occurring during electrical storms).
(b) Nitrogen dioxide gas reacts with water to produce nitric acid and nitrogen monoxide gas (one reaction involved in the formation of acid fog and acid rain).
(c) Copper metal reacts with gaseous oxygen, carbon dioxide, and water to form green basic copper carbonate, $Cu_2(OH)_2CO_3$ (a reaction responsible for the formation of a green patina or coating on outdoor bronze statues).
(d) Limestone (calcium carbonate) dissolves in water containing dissolved carbon dioxide, producing calcium hydrogen carbonate (bicarbonate) (a reaction producing temporary hardness in drinking water).
★**(e)** Calcium dihydrogen phosphate reacts with sodium hydrogen carbonate (bicarbonate), producing (tri)calcium phos-

phate, (di)sodium hydrogen phosphate, carbon dioxide, and water (the principal reaction occurring in the action of ordinary baking powder in the baking of cakes, bread, and biscuits).

22. Write *net ionic equations* to represent the following common reactions.
(a) Magnesium ion reacts with hydroxide ion to precipitate magnesium hydroxide (the first step in the production of magnesium metal from seawater).
(b) Upon heating, hydrogen carbonate ion (bicarbonate ion) in aqueous solution decomposes to carbonate ion, carbon dioxide gas, and water.
(c) In aqueous solution, aluminum ion, ammonia, and water react to form ammonium ion and a precipitate of aluminum hydroxide.
(d) In aqueous solution, copper(II) ion and iodide ion react to form water insoluble iodine and copper(I) iodide.

23. A sulfide of iron, containing 36.5% S, by mass, is heated in $O_2(g)$, and the products are sulfur dioxide and an oxide of iron containing 27.6% O, by mass. Write a balanced chemical equation for this reaction. [*Hint:* The formulas of the sulfide and oxide will not necessarily be those that you would predict from Table 3-2. Use the method of Examples 3-5 and 3-6 to determine these formulas.)]

Stoichiometry of chemical reactions

24. A laboratory method of preparing $O_2(g)$ involves the decomposition of $KClO_3(s)$.

$$2\ KClO_3(s) \rightarrow 2\ KCl(s) + 3\ O_2(g)$$

A 12.8-g sample of $KClO_3(s)$ is decomposed. How many **(a)** mol $O_2(g)$; **(b)** molecules of $O_2(g)$; **(c)** g $KCl(s)$ are produced?

25. A 76.4-g sample of a $CaSO_3$–$CaSO_4$ mixture, containing 61.3% $CaSO_3$ by mass, is treated with excess $HCl(aq)$. Only the $CaSO_3$ reacts. What mass of SO_2 is produced?

$$CaSO_3(s) + 2\ HCl(aq) \rightarrow CaCl_2(aq) + H_2O(l) + SO_2(g)$$

26. Iron ore is impure Fe_2O_3. When Fe_2O_3 is heated with an excess of carbon (coke), iron metal is produced. From a sample of ore weighing 878 kg, 515 kg of pure iron is obtained. What is the % Fe_2O_3, by mass, in the ore sample? [*Hint:* How much Fe_2O_3 is required to produce the 515 kg iron?]

$$Fe_2O_3(s) + 3\ C(s) \xrightarrow{\Delta} 2\ Fe(l) + 3CO(g)$$

27. Silver oxide decomposes at temperatures in excess of 300 °C, yielding metallic silver and oxygen gas. A 2.95-g sample of *impure* silver oxide yields 0.183 g $O_2(g)$. Assuming that $Ag_2O(s)$ is the only source of $O_2(g)$, what is the percent, by mass, of Ag_2O in the sample? [*Hint:* First write an equation for the reaction that occurs.]

28. Refer to Example 4-6. What is the mass of $H_2O(l)$ that would be produced in the reaction described? Establish this quantity in *two* different ways.

29. A piece of aluminum foil measuring 4.84 in. × 3.05 in. × 0.0125 in. is dissolved in excess HCl(aq). What mass of $H_2(g)$ is produced? Use equation (4.17) and 2.70 g/cm^3 as the density of Al.

30. How many mL of a $KMnO_4$(aq) solution containing 12.6 g $KMnO_4$/L must be used to convert 9.13 g KI to I_2?

$$2\ KMnO_4 + 10\ KI + 8\ H_2SO_4 \rightarrow$$
$$6\ K_2SO_4 + 2\ MnSO_4 + 5\ I_2 + 8\ H_2O$$

31. Refer to Example 4-2 and equation (4.7). What would be the *total* mass of the products (i.e., CO_2 *and* H_2O) produced in the complete combustion of 0.875 g $C_6H_{14}O_4$ (triethylene glycol)?

Molarity

32. What are the molarities of the following solutes?
(a) urea, $CO(NH_2)_2$, if 87.8 g of the 98.7% pure solid is dissolved in 500.0 mL of aqueous solution
(b) diethyl ether, $(C_2H_5)_2O$, if 11.0 mg is dissolved in 3.00 gal of water solution (1 gal = 3.78 L)
★(c) NaCl, if a solution has 1.52 ppm of Na [Assume that NaCl is the only source of Na and that the solution density is 1.00 g/cm^3. Parts per million (ppm) can be taken to mean g Na per million g solution.]

33. How much
(a) methanol, CH_3OH (d = 0.792 g/cm^3), in cm^3, must be dissolved in water to produce 2.60 L of 0.724 M CH_3OH?
(b) ethanol, C_2H_5OH (d = 0.789 g/cm^3), in gal, must be dissolved in water to produce 55.0 gal of 1.65 M C_2H_5OH? (1 gal = 3.78 L.)
(c) $Ca(NO_3)_2$, in mg, must be present in 50.0 L of a solution with 2.35 ppm Ca? [*Hint:* See Exercise 32(c).]

34. Two sucrose solutions—125 mL of 1.50 M $C_{12}H_{22}O_{11}$ and 275 mL of 1.25 M $C_{12}H_{22}O_{11}$—are mixed. Assuming the solution volumes to be additive, what is the molarity of $C_{12}H_{22}O_{11}$ in the final solution?

35. What volume of concentrated hydrochloric acid solution (36.0% HCl, by mass; d = 1.18 g/mL) is required to produce 15.0 L of 0.225 M HCl?

36. Refer to Example 4-10. Which has the greater ethyl alcohol content, the solution described in the example or a sample of white wine (d = 0.95 g/cm^3) having an ethyl alcohol content of 11% C_2H_5OH, by mass?

37. Refer to Example 4-12 and Figure 4-8. Typical sizes of the volumetric flasks found in a general chemistry laboratory are 100.0, 250.0, 500.0, and 1000.0 mL, and typical sizes of volumetric pipets are 1.00, 5.00, 10.00, 25.00, and 50.00 mL. Given the 0.250 M K_2CrO_4 stock solution, what combination(s) of pipet and volumetric flask would you use to prepare a solution that is 0.0125 M K_2CrO_4?

38. Silver nitrate is a very common, yet very expensive laboratory chemical, particularly useful for carrying out quantitative analyses. To conduct a particular analysis, a student needs 100.0 mL of 0.0750 M $AgNO_3$(aq), but she has available only about 60 mL of 0.0500 M $AgNO_3$. Rather than to make up a fresh 100.0 mL of solution, she decides to (1) pipet exactly 50.00 mL of the solution that she does have into a 100.0 mL

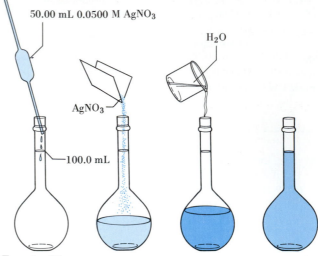

50.00 mL 0.0500 M $AgNO_3$

H_2O

$AgNO_3$

100.0 mL

Exercise 38

flask, (2) add an appropriate mass of $AgNO_3$, and (3) dilute the resulting solution to exactly 100.0 mL. Her procedure is pictured in the accompanying illustration. What mass of $AgNO_3$ must she weigh out?

Chemical reactions in solutions

39. For the reaction

$$Ca(OH)_2(s) + 2\ HCl(aq \rightarrow CaCl_2(aq) + 2\ H_2O$$

(a) How many grams of $Ca(OH)_2$ are required to react completely with (neutralize) 485 mL of 0.886 M HCl?
(b) What mass of $Ca(OH)_2$, in kilograms, is required to neutralize 465 L of an HCl solution that is 30.12% HCl, by mass, and has a density of 1.15 g/cm^3?

40. Refer to Example 4-9 and equation (4.17). For the conditions stated in Example 4-9, determine (a) the amount, in moles, of $AlCl_3$ and (b) the *molarity* of the $AlCl_3$(aq), if the solution volume is simply the 23.8 cm^3 calculated in the example.

41. Refer to the Summarizing Example. What mass of $O_2(g)$ is consumed if the reaction is found to produce 125 L of 2.05 M $NaNO_2$?

42. Refer to the Summarizing Example. If 175 g Na_2CO_3 in 1.15 L of aqueous solution is treated with an excess of NO(g) and $O_2(g)$, what should be the molarity of the $NaNO_2$(aq) solution that results (assuming that the reaction goes to completion)?

43. A method of adjusting the concentration of HCl(aq) is to allow the solution to react with a small quantity of Mg.

$$Mg(s) + 2\ HCl(aq) \rightarrow MgCl_2(aq) + H_2(g)$$

What mass of Mg must be added to 250.0 mL of 1.023 M HCl to reduce the concentration to exactly 1.000 M HCl?

44. A drop (0.05 mL) of 12.0 M HCl is spread over a sheet of thin aluminum foil (see figure). Assuming that all the acid dissolves through the foil, what will be the area, in cm^2, of the

Exercise 44
[Carey B. Van Loon]

hole produced? (Density of $Al = 2.70$ g/cm^3; thickness of the foil $= 0.10$ mm.)

$$2\ Al(s) + 6\ HCl(aq) \rightarrow 2\ AlCl_3(aq) + 3\ H_2(g)$$

Determining the limiting reagent

45. The following is a side reaction in the manufacture of rayon from wood pulp.

$$3\ CS_2 + 6\ NaOH \rightarrow 2\ Na_2CS_3 + Na_2CO_3 + 3\ H_2O$$

(a) How many moles each of Na_2CS_3, Na_2CO_3, and H_2O are produced by allowing 1.00 mol each of CS_2 and NaOH to react?

(b) How many grams of Na_2CS_3 are produced in the reaction of 88.0 cm^3 of liquid CS_2 ($d = 1.26$ g/cm^3) and 3.12 mol NaOH?

46. What mass of H_2 is produced by the reaction of 2.14 g Al with 75.0 mL of 2.90 M HCl? [*Hint:* Use equation (4.17).]

47. Refer to Example 4-15. How many kg of P_4O_{10} are left *unreacted* after the reaction producing $POCl_3$ has gone to completion?

48. Lithopone is a brilliant white pigment used in water-based interior paints. It is a mixture of $BaSO_4$ and ZnS produced by the reaction

$$BaS(aq) + ZnSO_4(aq) \rightarrow ZnS(s) + BaSO_4(s)$$
$$\text{lithopone}$$

What mass of lithopone is produced in the reaction of 275 mL of 0.350 M $ZnSO_4$ and 325 mL of 0.280 M BaS?

49. A mixture of 4.800 g H_2 and 36.40 g O_2 is allowed to react.

(a) Write a balanced equation for this reaction.

(b) What substances are present after the reaction, and in what quantities?

(c) Show that the total mass of substances present before and after the reaction is the same.

Theoretical, actual, and percent yields

50. A laboratory manual calls for 13.0 g C_4H_9OH, 21.6 g NaBr, and 33.8 g H_2SO_4 as reactants in this reaction.

$$C_4H_9OH + NaBr + H_2SO_4 \rightarrow C_4H_9Br + NaHSO_4 + H_2O$$

A student following these directions obtains 16.8 g C_4H_9Br. What are (a) the theoretical yield; (b) the actual yield; (c) the percent yield of this reaction?

51. Azobenzene, an intermediate in the manufacture of dyes, can be prepared from nitrobenzene by reaction with triethylene glycol in the presence of Zn and KOH. In one reaction 0.10 L of nitrobenzene ($d = 1.20$ g/cm^3) and 0.30 L of triethylene glycol ($d = 1.12$ g/cm^3) yielded 55 g azobenzene. What was the percent yield of this reaction?

$$2\ C_6H_5NO_2 + \ 4\ C_6H_{14}O_4 \ \xrightarrow[\text{KOH}]{\text{Zn}}$$

nitrobenzene   triethylene glycol

$$(C_6H_5N)_2 + 4\ C_6H_{12}O_4 + 4\ H_2O$$
$$\text{azobenzene}$$

52. How many grams of commercial acetic acid (97% $C_2H_4O_2$, by mass) must be allowed to react with an excess of PCl_3 to produce 50.0 g of acetyl chloride (C_2H_3OCl), if the reaction has a 70.% yield?

$$C_2H_4O_2 + PCl_3 \rightarrow C_2H_3OCl + H_3PO_3 \quad \text{(not balanced)}$$

Simultaneous reactions

53. How much HCl, in grams, is required to dissolve 387 g of a mixture containing 32.8% $MgCO_3$ and 67.2% $Mg(OH)_2$, by mass?

$$Mg(OH)_2 + 2\ HCl \rightarrow MgCl_2 + 2\ H_2O$$

$$MgCO_3 + 2\ HCl \ \rightarrow MgCl_2 + H_2O + CO_2(g)$$

54. A natural gas sample consists of 70.7% propane (C_3H_8) and 29.3% butane (C_4H_{10}), by mass. How many moles of $CO_2(g)$ would be produced by the complete combustion of 406 g of this gaseous mixture?

55. An organic liquid is either methyl alcohol (CH_3OH), ethyl alcohol (C_2H_5OH), or a mixture of the two. A 0.220-g sample of the liquid is burned in an excess of $O_2(g)$ and yields 0.352 g $CO_2(g)$. Is the liquid a pure alcohol or a mixture of the two?

*56. A 2.05-g sample of an iron–aluminum alloy (ferroaluminum) is dissolved in excess HCl(aq), producing 0.105 g $H_2(g)$. What is the percent composition, by mass, of the ferroaluminum?

$$Fe(s) + 2\ HCl(aq) \rightarrow FeCl_2(aq) + H_2(g)$$

$$2\ Al(s) + 6\ HCl(aq) \rightarrow 2\ AlCl_3(aq) + 3\ H_2(g)$$

Consecutive reactions

57. Dichlorodifluoromethane, a widely used refrigerant, can be prepared by the following reactions.

$$CH_4 + 4\ Cl_2 \rightarrow CCl_4 + 4\ HCl$$

$$CCl_4 + 2\ HF \rightarrow CCl_2F_2 + 2\ HCl$$

How many moles of Cl_2 must be consumed to produce 76.2 mol CCl_2F_2?

58. Malathion, a popular insecticide with farmers and home gardeners, partly because of its lower toxicity than comparable products, is synthesized in the following two-step process.

$$P_4S_{10} + 8\ CH_3OH \rightarrow 4\ (CH_3O)_2PSSH + 2\ H_2S$$
$$(CH_3O)_2PSSH + (CHCOOC_2H_5)_2 \rightarrow$$
$$(CH_3O)_2PSSCH(CH_2COOC_2H_5)COOC_2H_5$$
$$\text{malathion}$$

What is the minimum mass of P_4S_{10} consumed for every 1.00 kg of malathion produced? [*Hint:* Assume a 100% yield in each reaction.]

59. The following process has been used to obtain iodine from oil-field brines in California.

$$NaI + AgNO_3 \rightarrow AgI + NaNO_3$$

$$2\ AgI + Fe \rightarrow FeI_2 + 2\ Ag$$

$$2\ FeI_2 + 3\ Cl_2 \rightarrow 2\ FeCl_3 + 2\ I_2$$

How many grams of $AgNO_3$ are required in the first step for every kg I_2 produced in the third step?

60. NaBr, used to produce AgBr for use in photography, can itself be prepared as follows. How much Fe, in kg, is consumed to produce 2.50×10^3 kg NaBr?

$$Fe + Br_2 \rightarrow FeBr_2$$

$$FeBr_2 + Br_2 \rightarrow Fe_3Br_8 \qquad \text{(not balanced)}$$

$$Fe_3Br_8 + Na_2CO_3 \rightarrow NaBr + CO_2 + Fe_3O_4 \quad \text{(not balanced)}$$

Industrial chemistry

61. Described below are three important industrial processes for the production of chemicals. Each is carried out in successive steps. Write a series of equations for each process.

(a) Production of dinitrogen monoxide gas: Ammonia gas is passed into a nitric acid solution, producing ammonium nitrate solution. Ammonium nitrate crystallized from the solution is decomposed by heating, yielding gaseous dinitrogen monoxide and water.

(b) Recovery of sulfur from hydrogen sulfide gas (Claus process): Hydrogen sulfide gas is burned in air (oxygen), producing gaseous sulfur dioxide and water. The sulfur dioxide reacts with more hydrogen sulfide gas, forming gaseous water and liquid sulfur.

(c) Production of phosphoric acid: (Tri)calcium phosphate is heated with silicon dioxide and carbon, producing calcium silicate ($CaSiO_3$), phosphorus (P_4), and carbon monoxide. P_4 and oxygen react to form P_4O_{10}. P_4O_{10} reacts with water to form phosphoric acid.

62. Melamine, $C_3N_3(NH_2)_3$, is used in adhesives and in the manufacture of resins for use in paper and textiles. The production of melamine is a two-step reaction in which urea is the starting material and isocyanic acid, HNCO, an intermediate.

$$CO(NH_2)_2 \rightarrow HNCO + NH_3$$
$$6\ HNCO \rightarrow C_3N_3(NH_2)_3 + 3\ CO_2$$

(a) Write a single equation for the *net reaction* that occurs.

(b) What is the mass of melamine obtained from 100.0 kg $CO(NH_2)_2$ if the yield of the overall process is 84%?

63. In a particular plant using the Raschig process (see Figure 4-11), the mole ratio of NH_3 to NH_2Cl used in the third reaction is 30:1. What is the maximum quantity of NH_3 that is recoverable in the purification steps for every 1.00 kg Cl_2 that reacts?

64. Acrylonitrile, CH_2CHCN, is used in the production of synthetic fibers, plastics, and rubber goods. The Sohio process produces acrylonitrile from propylene, air, and ammonia. The unbalanced equation for the reaction is

$$CH_2CHCH_3 + NH_3 + O_2 \rightarrow CH_2CHCN + H_2O$$

The process yields 0.73 lb acrylonitrile per pound of propylene (CH_2CHCH_3).

(a) Balance the equation for the reaction.

(b) What is the percent yield of the process?

(c) At the percent yield calculated in (b), what is the minimum mass of NH_3 required to produce 1.00 ton (2000 lb) of acrylonitrile by the process?

Additional Exercises

65. Balance the following equations by inspection.

(a) $SiCl_4(l) + H_2O(l) \rightarrow SiO_2(s) + HCl(g)$

(b) $CaC_2(s) + H_2O(l) \rightarrow Ca(OH)_2(s) + C_2H_2(g)$

(c) $Na_2HPO_4(s) \rightarrow Na_4P_2O_7(s) + H_2O(l)$

(d) $NCl_3(g) + H_2O(l) \rightarrow NH_3(g) + HOCl(aq)$

(e) $Al_2O_3(s) + H^+(aq) \rightarrow Al^{3+}(aq) + H_2O(l)$

66. Following are some reactions commonly encountered in chemical industry. Write balanced equations for these reactions.

(a) Limestone rock (calcium carbonate) is heated and decomposes to calcium oxide and carbon dioxide gas.

(b) Zinc sulfide ore is heated in air (roasted) and is converted to zinc oxide and sulfur dioxide gas. (Note that oxygen gas in the air is also a reactant.)

(c) Propane gas (C_3H_8) reacts with gaseous water to produce a mixture of carbon monoxide and hydrogen gases (called *synthesis gas* and used in the chemical industry to produce a variety of other chemicals).

★ (d) Sulfur dioxide gas is passed into an aqueous solution containing sodium sulfide and sodium carbonate. The reaction products are carbon dioxide gas and a water solution of sodium thiosulfate.

67. $H_2(g)$ is passed over $Fe_2O_3(s)$ at 400 °C. Water vapor is formed, together with a black residue—a compound consisting of 72.3% Fe and 27.7% O. Write a balanced equation for this reaction.

68. The following commercial processes are carried out in a stepwise manner. First, write the equation for each step in the process, and then a *single* equation for the *net reaction* that occurs.

(a) The production of slaked lime (calcium hydroxide) from limestone (calcium carbonate). (1) Limestone is decomposed into lime (calcium oxide) and carbon dioxide gas, and (2) the lime reacts with water to form slaked lime.

(b) The production of methanol, CH_3OH. (1) Gaseous methane (CH_4) and water react to produce a mixture of carbon monoxide and hydrogen gases, and (2) the carbon monoxide and hydrogen gases, in the presence of a catalyst, are converted to methanol.

(c) The production of sulfuric acid from sulfur. (1) Sulfur and oxygen react to form sulfur dioxide; (2) sulfur dioxide and oxygen react to form sulfur trioxide; and (3) sulfur trioxide and water react to form sulfuric acid.

69. How many grams of Ag_2CO_3 must have been decomposed if 56.2 g Ag was obtained in this reaction?

$$Ag_2CO_3(s) \rightarrow Ag(s) + CO_2(g) + O_2(g)\quad \text{(not balanced)}$$

70. Given the reaction

$$3\ Fe(s) + 4\ H_2O(g) \rightarrow Fe_3O_4(s) + 4\ H_2(g)$$

(a) How many moles of $H_2(g)$ can be produced from 64.4 g Fe and an excess of $H_2O(g)$ [steam]?

(b) How many grams of H_2O would be consumed in the conversion of 76.3 g Fe to Fe_3O_4?

(c) If 9.02 mol $H_2(g)$ is produced, what mass of Fe_3O_4 must also be produced?

71. Which of the following metals yields the maximum amount of H_2 *per gram* of metal reacting with HCl(aq): Na, Mg, Al, or Zn? [*Hint:* Write equations similar to (4.17).]

72. A popular general chemistry laboratory experiment involves a series of reactions with copper as the original starting material and as the final product. The reactions involved are (1) copper is dissolved in nitric acid, producing copper(II) nitrate, water, and nitrogen dioxide gas; (2) copper(II) nitrate solution is treated with a solution of sodium hydrogen carbonate (bicarbonate), yielding solid copper(II) carbonate, aqueous sodium nitrate, water, and carbon dioxide gas; (3) copper(II) carbonate is dissolved in dilute sulfuric acid solution, forming copper(II) sulfate solution, water, and carbon dioxide gas; (4) zinc metal is added to the copper(II) sulfate solution, forming zinc sulfate solution and pure copper metal (recall Figure 1-5). Write balanced equations for these four reactions.

73. One of the objects of the experiment described in Exercise 72 is to see how much of the original copper metal can be recovered. If the yield in each of the steps outlined in Exercise 72 is 98%, how much copper will be recovered from an original 3.05-g sample? Explain your reasoning.

74. The mineral ilmenite, $FeTiO_3$, is an important source of TiO_2 for use as a white pigment (recall also, Example 4-19). In the first step of the conversion of this mineral to TiO_2, it is treated with sulfuric acid.

$$FeTiO_3 + 2 H_2SO_4 + 5 H_2O \rightarrow FeSO_4 \cdot 7H_2O + TiOSO_4$$

TiO_2 is obtained from $TiOSO_4$ in two further steps. How many kilograms of the hydrate $FeSO_4 \cdot 7H_2O$ are produced for every 1000. kg of ilmenite processed?

75. $FeSO_4 \cdot 7H_2O$ resulting from the processing of ilmenite (Exercise 74) is a waste material that cannot simply be dumped into the environment. Its further treatment involves dehydration, followed by decomposition to Fe_2O_3 (which can be used in iron and steel production).

$$FeSO_4 \cdot 7H_2O \xrightarrow{\Delta} FeSO_4 + 7 H_2O$$

$$4 FeSO_4 \xrightarrow{\Delta} 2 Fe_2O_3 + 4 SO_2(g) + O_2(g)$$

How many kilograms of Fe_2O_3 are obtained for every 1000. kg $FeSO_4 \cdot 7H_2O$ that is treated in this way?

76. $SO_2(g)$ released in the process described in Exercise 75 cannot be discharged into the atmosphere. Instead it is converted to sulfuric acid through the two reactions represented in simplified fashion below.

$$2 SO_2(g) + O_2(g) \rightarrow 2 SO_3(g)$$

$$SO_3(g) + H_2O(l) \rightarrow H_2SO_4(l)$$

How many kilogram of $H_2SO_4(l)$ can be produced for every 1000. kg $SO_2(g)$?

★77. Consider the overall process whereby $H_2SO_4(l)$ is produced in the processing of ilmenite ore in Exercises 74, 75, and 76. How many kilograms of $H_2SO_4(l)$ can be produced for every 1000. kg of ore that is processed?

78. Ammonia can be generated by heating together the solids NH_4Cl and $Ca(OH)_2$; $CaCl_2$ and H_2O are also formed. If a mixture containing 25.0 g each of NH_4Cl and $Ca(OH)_2$ is heated, how many grams of NH_3 will form? [*Hint:* Write a balanced equation for the reaction.]

79. A 25.00-mL sample of liquid benzene, C_6H_6 ($d = 0.879$ g/cm^3), is burned completely, yielding CO_2 and H_2O as the only products. The CO_2 formed is passed into a barium hydroxide solution. What mass of $BaCO_3$ is produced?

$$CO_2(g) + Ba(OH)_2(aq) \rightarrow BaCO_3(s) + H_2O$$

80. What volume of 0.668 M NH_4NO_3 solution must be diluted with water to produce 1.00 L of a solution with a concentration of 2.85 mg N per mL?

81. The urea [$CO(NH_2)_2$] solutions pictured are mixed, and the resulting solution evaporated to a final volume of 825 mL. What is the molarity of $CO(NH_2)_2$ in this final solution?

82. A seawater sample has a density of 1.03 g/cm^3 and

2.8% NaCl, by mass. A saturated solution of NaCl in water is 5.45 M NaCl. How much water would have to be evaporated from 1.00×10^6 L of the seawater before NaCl would precipitate? (A saturated solution contains the maximum amount of dissolved solute possible.)

83. Refer to Example 4-13. Suppose that the precipitated $Ag_2CrO_4(s)$ is filtered off and the $KNO_3(aq)$ evaporated to dryness. How many grams of KNO_3 would be obtained?

84. A small piece of zinc is dissolved in 50.00 mL of 1.050 M HCl.

$$Zn(s) + 2 HCl(aq) \rightarrow ZnCl_2(aq) + H_2(g)$$

At the end of the reaction, the concentration of the 50.00-mL sample is redetermined and found to be 0.776 M HCl. What must have been the mass of the piece of zinc that dissolved?

85. 99.8 mL of a solution containing 12.0% KI by mass and having a density of 1.093 g/mL is added to 96.7 mL of another solution that is 14.0% $Pb(NO_3)_2$ by mass and has a density of 1.134 g/mL. What mass of PbI_2 should form?

$$Pb(NO_3)_2(aq) + 2 KI(aq) \rightarrow PbI_2(s) + 2 KNO_3(aq)$$

86. Refer to Example 4-14. Suppose that the lead storage battery from which the 5.00-mL sample was taken contains a total of 1.2 L of battery acid. What approximate volume of concentrated H_2SO_4 (93.2% H_2SO_4, by mass; $d = 1.835$ g/cm^3) should be added to the battery to bring the acid concentration to an acceptable level, say 5.0 M H_2SO_4? (To simplify the calculation, assume that the volumes of the original battery acid and the added concentrated H_2SO_4 are additive.)

87. A chemical plant using the Raschig process obtains 0.299 kg of 98.0% N_2H_4 for every 1.00 kg Cl_2 that reacts with excess NaOH and NH_3. What are the (a) theoretical, (b) actual, and (c) percent yields of *pure* N_2H_4? Refer to the net reaction in Figure 4-11.

★88. An 0.155-g sample of an Al–Mg alloy is dissolved in an excess of HCl(aq), producing 0.0163 g H_2. What is the % Mg in the alloy? [*Hint:* Write equations similar to (4.17).]

★89. The manufacture of ethyl alcohol, C_2H_5OH, yields diethyl ether, $(C_2H_5)_2O$, as a by-product. The *complete combustion* of a 1.005-g sample of the product of this process yields 1.963 g CO_2. What must be the percent, by mass, of C_2H_5OH and of $(C_2H_5)_2O$ in this sample?

90. Hydrogen produced by the decomposition of water has considerable potential as a fuel. Key to its development is finding an appropriate series of chemical reactions that has as its net reaction: $2 H_2O \rightarrow 2 H_2 + O_2$. Demonstrate that this requirement is met by the Fe/Cl cycle (given on the next page), in which the maximum reaction temperature is 500 °C.

655 mL 0.852 M $CO(NH_2)_2$

432 mL 0.487 M $CO(NH_2)_2$

825 mL ? M $CO(NH_2)_2$

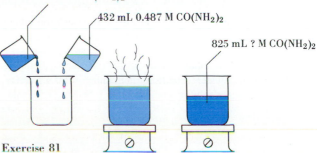

Exercise 81

$$FeCl_2 + H_2O \xrightarrow{500\ °C} Fe_3O_4 + HCl + H_2$$

$$2\ Fe_3O_4 + 12\ HCl + 3\ Cl_2 \xrightarrow{200\ °C} 6\ FeCl_3 + 6\ H_2O + O_2$$

$$FeCl_3 \xrightarrow{420\ °C} FeCl_2 + Cl_2$$

★**91.** $CaCO_3(s)$ reacts with HCl(aq) to form H_2O, $CaCl_2$(aq) and CO_2(g). If a 45.0-g sample of $CaCO_3$(s) is added to 1.25 L HCl(aq) that has a density of 1.13 g/cm^3 and contains 25.7% HCl, by mass, what will be the molarity of HCl in the solution after the reaction is completed? (Assume that the solution volume remains constant.)

★**92.** Under appropriate conditions, copper sulfate, potassium chromate, and water react to form a product containing Cu^{2+}, CrO_4^{2-}, and OH^- ions. Analysis of the compound yields 48.7% Cu^{2+}, 35.6% CrO_4^{2-}, and 15.7% OH^-.

(a) Determine the empirical formula of the compound.

(b) Write a plausible equation for the reaction.

Self-Test Questions

For questions 93 through 102 select the single item that best completes each statement.

93. The decomposition of potassium chlorate to produce potassium chloride and oxygen gas is expressed symbolically as

(a) $KClO_3(s) \rightarrow KCl(s) + O_2(g) + O(g)$

(b) $2\ KClO_3(s) \rightarrow 2\ KCl(s) + 3\ O_2(g)$

(c) $KClO_3(s) \rightarrow KClO(s) + O_2(g)$

(d) either (b) or (c), but not (a).

94. In the equation

$$?\ Fe^{2+} + O_2 + 4\ H^+ \rightarrow ?\ Fe^{3+} + 2\ H_2O$$

the missing coefficients (a) are each 4; (b) are each 2; (c) can have any values as long as they are the same; (d) must be determined by experiment.

95. For the reaction $2\ H_2S + SO_2 \longrightarrow 3\ S + 2\ H_2O$

(a) 3 mol S is produced per mole of H_2S;

(b) 1 mol SO_2 is consumed per mole of H_2S;

(c) 1 mol H_2O is produced per mole of H_2S;

(d) the number of moles of products is independent of how many moles of reactants are used.

96. *Per gram of reactant,* the maximum quantity of $O_2(g)$ is produced in the decomposition reaction

(a) $2\ NH_4NO_3 \xrightarrow{\Delta} 2\ N_2(g) + 4\ H_2O + O_2(g)$

(b) $2\ N_2O \xrightarrow{\Delta} 2\ N_2(g) + O_2(g)$

(c) $2\ Ag_2O(s) \xrightarrow{\Delta} 4\ Ag(s) + O_2(g)$

(d) $2\ Pb(NO_3)_2(s) \xrightarrow{\Delta} 2\ PbO(s) + 4\ NO_2(g) + O_2(g)$

97. 1.0 mol of calcium cyanamide ($CaCN_2$) and 1.0 mol of water are allowed to react.

$$CaCN_2 + 3\ H_2O \rightarrow CaCO_3 + 2\ NH_3$$

The number of moles of NH_3 produced is (a) 3.0; (b) 2.0; (c) 1.0; (d) less than 1.0.

98. If the reaction of 1.00 mol $NH_3(g)$ and 1.00 mol $O_2(g)$

$$4\ NH_3(g) + 5\ O_2(g) \rightarrow 4\ NO(g) + 6\ H_2O(l)$$

is carried to completion, (a) all the $O_2(g)$ is consumed; (b) 4.0 mol NO(g) is produced; (c) 1.5 mol $H_2O(l)$ is produced; (d) none of these.

99. To obtain a solution that is 1.00 M $NaNO_3$, prepare (a) 1.00 L of water solution containing 100 g $NaNO_3$; (b) 500 mL of water solution containing 85 g $NaNO_3$; (c) a solution containing 8.5 mg $NaNO_3$/mL; (d) 5.00 L of water solution containing 425 g $NaNO_3$.

100. To prepare a solution that is 0.50 M KCl starting with 100 mL of 0.40 M KCl (a) add 0.75 g KCl; (b) add 20 mL of water; (c) add 0.10 mol KCl; (d) evaporate 10 mL of water.

101. In the reaction of 2.0 mol CCl_4 with an excess of HF, 1.70 mol CCl_2F_2 is obtained.

$$CCl_4 + 2\ HF \rightarrow CCl_2F_2 + 2\ HCl$$

(a) The theoretical yield is 1.70 mol CCl_2F_2.

(b) The theoretical yield is 1.0 mol CCl_2F_2.

(c) The percent yield of the reaction is 85%.

(d) The theoretical yield of the reaction depends on how large an excess of HF is used.

102. Suppose that each of the following reactions has a 90% yield.

(1) $CH_4 + Cl_2 \rightarrow CH_3Cl + HCl$

(2) $CH_3Cl + Cl_2 \rightarrow CH_2Cl_2 + HCl$

Starting with 100 g CH_4 in reaction (1) and an excess of $Cl_2(g)$, the number of grams of CH_2Cl_2 formed in reaction (2) is (a) $100 \times 0.81 \times (85/16)$; (b) 100×0.90; (c) $100 \times 0.90 \times 0.90$; (d) $100 \times 0.90 \times 0.90 \times (16/50.5)(70.9/85)$.

103. Write a balanced chemical equation to represent each of the following reactions.

(a) The decomposition, by heating, of solid mercury(II) nitrate to produce pure liquid mercury, nitrogen dioxide gas, and oxygen gas.

(b) The reaction of aqueous sodium carbonate with aqueous hydrochloric acid (hydrogen chloride) to produce water, carbon dioxide gas, and aqueous sodium chloride.

(c) The complete combustion of malonic acid, a compound with 34.62% C, 3.88% H, and 61.50% O, by mass.

104. How many grams of Na must react with 125 mL H_2O to produce an NaOH solution that is 0.250 *M*? (Assume that the final solution volume is 125 mL.)

$$2\ Na(s) + 2\ H_2O(l) \rightarrow 2\ NaOH(aq) + H_2(g)$$

105. An essentially 100% yield is necessary for a chemical reaction to be used to *analyze* a compound, but is almost never expected for a reaction that is used to *synthesize* a compound. Explain why this is so.

5 Introduction to Reactions in Aqueous Solutions

Seawater is an aqueous (water) solution in which compounds of over one half the elements can be found, a few in abundance, such as NaCl and $MgCl_2$, but others mostly in trace quantities. [Alastair Black/FPG International]

Perhaps the most familiar experience of general chemistry students is in producing *precipitates* from solution. Often precipitation is used to synthesize a pure solid. $Mg(OH)_2(s)$ is precipitated from seawater as the first step in the manufacture of magnesium metal. Sometimes, the formation of a precipitate may serve as a *qualitative* test to determine the presence of a certain ion in solution. If the addition of HCl(aq) to a solution produces a white precipitate, Ag^+ may be present in the solution (very few other ions produce this result). Precipitation reactions are the first of three types considered in this chapter.

Magnesium hydroxide, $Mg(OH)_2$, although insoluble in water, dissolves in hydrochloric acid, HCl(aq). This dissolving is the second step in the production of magnesium from seawater. It is also the reaction whereby a suspension of $Mg(OH)_2(s)$ in water (called milk of magnesia) neutralizes excess stomach acid. In its reaction with HCl(aq), $Mg(OH)_2$ acts as a base. The reaction is an acid–base reaction. One of our main interests in studying this second type of reaction is to see how acid–base reactions are used to perform chemical analyses by a method called titration.

The third type of reaction presented in this chapter is oxidation–reduction. We describe these reactions in terms of the transfer of electrons and the resulting changes in oxidation states, characteristics from which we develop methods of balancing oxidation–reduction equations. We will also find that oxidation–reduction reactions are useful in chemical analyses, again by the method of titration.

5-1 The Nature of Aqueous Solutions

Although many important chemical reactions are carried out in each of the three states of matter—solid, liquid, and gas—reactions in *aqueous* (water) solutions are particularly common. This is because (1) water is able to dissolve a large variety of substances, and (2) in water solution many substances are dissociated into *ions,* and thus become available to participate in certain chemical reactions.

Figure 5-1 pictures a simple laboratory apparatus that tells us much about the nature of aqueous solutions. The three types of observations that are made are

FIGURE 5-1
Electrical conductivity of aqueous solutions.

For electric current to flow, electrical contact must be made between the two graphite rods immersed in solution. Electrical contact through the solution is possible only if the solution contains ions. The ability of the solution to conduct electricity is judged by the brilliance of the incandescent lamp.
(a) Nonelectrolyte.
(b) Strong electrolyte.
(c) Weak electrolyte.
[Carey B. Van Loon]

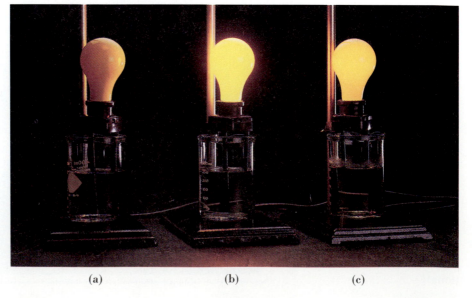

(a)   (b)   (c)

TABLE 5-1
Electrolytic Properties of Aqueous Solutions

Strong electrolytes		Weak electrolytes	Nonelectrolytes
Ionic compounds	**Covalent compounds**		
NaCl	HCl	$HCHO_2$ (formic acid)	H_2O (water)
$MgCl_2$	HBr	$HC_2H_3O_2$ (acetic acid)	C_2H_5OH (ethyl alcohol)
NaOH	HI	HOCl (hypochlorous acid)	$C_6H_{12}O_6$ (glucose)
KBr	HNO_3	HNO_2 (nitrous acid)	$C_{12}H_{22}O_{11}$ (sucrose)
$KClO_4$	H_2SO_4[a]	H_2SO_3 (sulfurous acid)	$CO(NH_2)_2$ (urea)
$Al_2(SO_4)_3$	$HClO_4$	NH_3 (ammonia)	$C_2H_6O_2$ (ethylene glycol)
$CuSO_4$		$C_6H_5NH_2$ (aniline)	$C_3H_8O_3$ (glycerol)
$LiNO_3$	plus		
	a *few*		
plus	others	plus	plus
		many	*many*
many		others	others
others			

[a]H_2SO_4(aq) is completely ionized into H^+ and HSO_4^- ions, but HSO_4^- undergoes only partial ionization to H^+ and SO_4^{2-}. This solution behavior of H_2SO_4(aq) is discussed in Section 17-7.

- *The lamp fails to light up.* There are essentially no ions present (or, if they are present, their concentration is extremely low). The substance in solution (the solute) is called a **nonelectrolyte.**
- *The lamp lights up brilliantly.* Even though the amount of solute may be small, the concentration of ions in solution is *high;* the electrical conductivity is *high.* The substance in solution (the solute) is called a **strong electrolyte.**
- *The lamp lights up only dimly.* Even though the amount of solute may be large, the concentration of ions in solution is *low;* the electrical conductivity is *low.* The substance in solution (the solute) is called a **weak electrolyte.**

Table 5-1 lists the electrolytic properties of a number of substances in aqueous solution and notes that

- Essentially all water-soluble ionic compounds and only a relatively few covalent compounds are *strong electrolytes.*
- Most covalent compounds are either *nonelectrolytes* or *weak electrolytes.*

Now consider how best to represent these three types of substances in chemical equations. If, for a soluble ionic compound we write,

$$NaCl(aq) \longrightarrow Na^+(aq) + Cl^-(aq)$$

what we mean is that in a water (aqueous) solution formula units of NaCl are dissociated into the separate ions. The best representation of NaCl(aq) is as $Na^+(aq) + Cl^-(aq)$.*

For a covalent compound that is a *strong electrolyte* we similarly can write,

$$HCl(aq) \longrightarrow H^+(aq) + Cl^-(aq)$$

which means that the best representation of HCl(aq) is as $H^+(aq) + Cl^-(aq)$.

For a compound that is a *weak electrolyte* the situation is best described as a *reversible* reaction, such as

$$HC_2H_3O_2(aq) \rightleftharpoons H^+(aq) + C_2H_3O_2^-(aq)$$

*To say that a strong electrolyte is completely dissociated into ions in aqueous solution is an oversimplification. Some of the positive and negative ions may remain combined in units called *ion pairs*. We discuss some of the consequences of this in Sections 13-9 and 19-3. In most instances, though, the assumption of complete dissociation will not seriously affect our results.

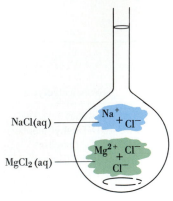

FIGURE 5-2
Determining molarity of Cl^- in mixture of $NaCl(aq)$ and $MgCl_2(aq)$—Example 5-1 visualized.

The key to calculating the chloride ion concentration in this solution is in noting that NaCl supplies one Cl^- per formula unit and $MgCl_2$ supplies two Cl^- per formula unit.

This equation indicates that only some of the molecules are ionized. The predominant species in solution is $HC_2H_3O_2$, and the solution is best represented as $HC_2H_3O_2(aq)$.

For a compound that is a *nonelectrolyte,* any ionization that occurs can generally be neglected and the molecular formula is used in chemical equations, such as $C_2H_5OH(aq)$.

One of the first uses to which we can put this new information is in calculating the concentrations of *ions* in solutions of *strong electrolytes,* as in Example 5-1. This calculation is illustrated in Figure 5-2.

Example 5-1

Calculating ion concentrations in a solution of a strong electrolyte. The main ions in seawater are Na^+, Mg^{2+}, and Cl^-. A reasonable approximation to seawater is a solution that is 0.438 M NaCl and 0.0512 M $MgCl_2$.
(a) What is the molarity of Cl^- in this solution?
(b) What is the concentration of Cl^-, expressed in mg/L?

Solution

(a) First, identify the two solutes as *strong electrolytes* and represent their dissociation. In this representation it is imperative that you recognize that there are *two* Cl^- ions per formula unit of magnesium chloride; this you should see from the formula, $MgCl_2$.

$$NaCl(aq) \longrightarrow Na^+(aq) + Cl^-(aq)$$

$$MgCl_2(aq) \longrightarrow Mg^{2+}(aq) + 2\ Cl^-(aq)$$

Now, determine the number of moles of Cl^- from each of the two sources. In the setups below the factors that convert from moles of compound to moles of an ion are shown in blue.

$$\text{no. mol } Cl^- \text{ in NaCl} = 1\ L \times \frac{0.438\ \text{mol NaCl}}{L} \times \frac{1\ \text{mol } Cl^-}{1\ \text{mol NaCl}}$$

$$= 0.438\ \text{mol } Cl^-$$

$$\text{no. mol } Cl^- \text{ in } MgCl_2 = 1\ L \times \frac{0.0512\ \text{mol } MgCl_2}{L} \times \frac{2\ \text{mol } Cl^-}{1\ \text{mol } MgCl_2}$$

$$= 0.102\ \text{mol } Cl^-$$

$$\text{total no. mol } Cl^- = 0.438 + 0.102 = 0.540\ \text{mol } Cl^-$$

$$\text{molarity of } Cl^- = \frac{0.540\ \text{mol } Cl^-}{1\ L} = 0.540\ M$$

Molarity concentrations are often expressed through bracket symbols, [ ]. The concentrations in this solution are [NaCl] = 0.438 M and [$MgCl_2$] = 0.0512 M. Also the solution can be described as having [Cl^-] = 0.540 M. This symbolism will be used extensively in later chapters (e.g., Chapters 15 and 16).

(b) Convert from mol/L → g/L → mg/L.

$$\text{no. mg } Cl^-/L = \frac{0.540\ \text{mol } Cl^-}{L} \times \frac{35.45\ \text{g } Cl^-}{1\ \text{mol } Cl^-} \times \frac{1000\ \text{mg } Cl^-}{1\ \text{g } Cl^-}$$

$$= 1.91 \times 10^4\ \text{mg } Cl^-/L$$

(The tabulated value of the chloride ion content of a typical seawater sample is about 19,000 mg Cl^-/L.)

SIMILAR EXAMPLES: Exercises 2, 3, 4, 5, 20, 22, 52, 53.

Simple stoichiometric calculations will not work for determining the concentrations of ions in a *weak electrolyte* solution, because the ionization of a weak acid is a *reversible* reaction (recall Section 4-8). We defer such calculations to Chapter 17.

5-2 Precipitation Reactions

Many reactions between ionic compounds in aqueous solution occur because certain combinations of cations and anions yield substances that are insoluble in water.* These are called **precipitation reactions.** Table 5-2 summarizes some important precipitation reactions encountered in quantitative chemical analysis and in industrial chemistry. Later in this chapter we examine the importance of precipitation reactions in *qualitative* analysis.

Predicting Precipitation Reactions. Suppose we are asked to *predict* what happens when the following aqueous solutions are mixed.

$$AgNO_3(aq) + KBr(aq) \longrightarrow ? \tag{5.1}$$

An appropriate beginning is to rewrite expression (5.1) to show the actual species present in solution, as described in the preceding section. Then consider *four* possibilities for combinations of cations and anions.

$$Ag^+(aq) + NO_3^-(aq) + K^+(aq) + Br^-(aq) \longrightarrow ? \tag{5.2}$$

1. no reaction
2. $AgBr(s) + K^+(aq) + NO_3^-(aq)$
3. $Ag^+(aq) + Br^-(aq) + KNO_3(s)$
4. $AgBr(s) + KNO_3(s)$

We know that the combinations $(Ag^+ + NO_3^-)$ and $(K^+ + Br^-)$ are soluble; these are the original reactants in expression (5.1). The first possibility listed in expression (5.2) is that all combinations of cations and anions produce water soluble compounds, that is, that there would be *no reaction*. Possibility (2) is that only $AgBr(s)$ precipitates; possibility (3), only $KNO_3(s)$ precipitates; and possibility (4), both $AgI(s)$ and $KNO_3(s)$ precipitate. The result that is actually observed, shown in Figure 5-3, is the *second* one. Expressed as an ionic equation this is

$$Ag^+(aq) + NO_3^-(aq) + K^+(aq) + Br^-(aq) \longrightarrow AgBr(s) + K^+(aq) + NO_3^-(aq)$$

and as a *net ionic equation*

$$Ag^+(aq) + Br^-(aq) \longrightarrow AgBr(s) \tag{5.3}$$

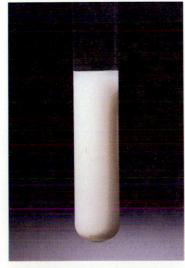

FIGURE 5-3
Precipitation of silver bromide.

Colorless $AgNO_3(aq)$ is added to colorless $KBr(aq)$; pale yellow $AgBr(s)$ precipitates. [Carey B. Van Loon]

*In principle all ionic compounds will dissolve in water to some extent, but some dissolve only to a very slight degree. For practical purposes, if the amount of compound that dissolves is less than about 0.01 mol per liter, we can consider the compound to be insoluble.

TABLE 5-2
Some Practical Examples of Precipitation Reactions

Reaction in aqueous solution	Applications
Analytical chemistry	
$Ag^+ + Cl^- \longrightarrow AgCl(s)$	Used to analyze for either Ag^+ or Cl^-. Typical sample for analysis: Waste water from photographic finishing plant.
$Ca^{2+} + C_2O_4^{2-} \longrightarrow CaC_2O_4(s)$	Used to analyze for Ca^{2+}. CaC_2O_4 is decomposed by heating, and the final product weighed is either $CaCO_3$ or CaO. Typical sample for analysis: Limestone.
$Mg^{2+} + NH_4^+ + PO_4^{3-} \longrightarrow$ $\qquad\qquad\qquad MgNH_4PO_4(s)$	Used to analyze for either Mg^{2+} or phosphorus. $MgNH_4PO_4$ is converted to $Mg_2P_2O_7$ by intense heating. Typical sample for analysis: Limestone (for Mg); municipal waste water (for P).
$Ba^{2+} + SO_4^{2-} \longrightarrow BaSO_4(s)$	Used to analyze for SO_4^{2-}. Other sulfur-containing species, e.g., S^{2-} or $SO_2(g)$, can be converted to SO_4^{2-} and then precipitated. Typical sample for analysis: $SO_2(g)$ in flue gases from a smelter or power plant.
Industrial chemistry	
$Mg^{2+} + 2\,OH^- \longrightarrow Mg(OH)_2(s)$	The first step in extracting magnesium metal from seawater.
$Ag^+ + NO_3^- + Cl^- \longrightarrow$ $\qquad\qquad AgCl(s) + NO_3^-$ $Ag^+ + NO_3^- + Br^- \longrightarrow$ $\qquad\qquad AgBr(s) + NO_3^-$	AgCl(s) and AgBr(s), precipitated from $AgNO_3(aq)$, are used in the manufacture of photographic film.
$Zn^{2+} + SO_4^{2-} + Ba^{2+} + S^{2-} \longrightarrow$ $\qquad\qquad ZnS(s) + BaSO_4(s)$	The mixture of precipitates, called **lithopone,** is a brilliant white pigment used in paints, paper, and white rubber goods.
$NH_4^+ + HCO_3^- + Na^+ + Cl^- \longrightarrow$ $\qquad\qquad NaHCO_3(s) + NH_4^+ + Cl^-$	$NaHCO_3$ is the least soluble of the possible ion combinations in cold water. Na_2CO_3 is obtained by heating $NaHCO_3$. Primary uses are in the food industry ($NaHCO_3$) and the manufacture of glass (Na_2CO_3).

To choose among the four possibilities on page 127 requires either that we do an experiment and analyze the results (as suggested through Figure 5-3) or, better still, that we *predict* the outcome, based on *solubility rules*. Table 5-3 lists some rules for common ionic compounds.

In later chapters we consider ideas to help us understand why, for example, NaOH is water soluble and $Mg(OH)_2$ is not. For now, if we simply accept the data in Table 5-3 at face value, we have a means of predicting large numbers of precipitation reactions. Figure 5-4 verifies several predictions made in Example 5-2.

Example 5-2 _____

Using the solubility rules to predict precipitation reactions. Predict whether a reaction will occur in each of the following cases. If so, write a *net ionic equation* for the reaction.

TABLE 5-3

Some General Rules for the Water Solubilities of Common Ionic Compounds

1. All common compounds of the alkali (1A) metals and the ammonium ion (NH_4^+) are **soluble.**
2. All nitrates (NO_3^-), chlorates (ClO_3^-), perchlorates (ClO_4^-), and acetates ($C_2H_3O_2^-$) are **soluble.** Silver acetate is only **moderately soluble.**
3. The chlorides (Cl^-), bromides (Br^-), and iodides (I^-) of most metals are **soluble.** The principal *exceptions* are those of Pb^{2+}, Ag^+, and Hg_2^{2+}.
4. All sulfates (SO_4^{2-}) are **soluble** *except* for those of Sr^{2+}, Ba^{2+}, Pb^{2+}, and Hg_2^{2+}. Sulfates of Ca^{2+} and Ag^+ are only **moderately soluble.**
5. All carbonates (CO_3^{2-}), chromates (CrO_4^{2-}), and phosphates (PO_4^{3-}) are **insoluble** *except* for those of the alkali metals (including ammonium).
6. The group IA metal hydroxides (OH^-) are **soluble.** The hydroxides of Ca^{2+}, Sr^{2+}, and Ba^{2+} are **moderately soluble.** The rest of the hydroxides are **insoluble.**
7. The sulfides of all metals are **insoluble** *except* for those of NH_4^+ and the 1A and 2A metals.

Generally speaking, if a saturated solution of an ionic solute is greater than about 0.10 M, the solute is said to be soluble; if a saturated solution of an ionic solute is less than about 0.01 M, the solute is said to be insoluble; and if the saturated solution is of an intermediate concentration (i.e., between about 0.01 M and 0.10 M), we say that the solute is moderately soluble.

Many compounds that are insoluble in *pure* water will dissolve under other conditions, however, as in the presence of acids.

(a) $NaOH(aq) + MgCl_2(aq) \rightarrow$?
(b) $BaS(aq) + CuSO_4(aq) \rightarrow$?
(c) $(NH_4)_2SO_4(aq) + ZnCl_2(aq) \rightarrow$?

Solution

(a) Since all common Na compounds are water soluble, Na^+ remains in solution. The combination of Mg^{2+} and OH^- produces *insoluble* $Mg(OH)_2$. With this information we can write

$$2\ \cancel{Na^+(aq)} + 2\ OH^-(aq) + Mg^{2+}(aq) + \cancel{2\ Cl^-(aq)} \longrightarrow$$
$$Mg(OH)_2(s) + \cancel{2\ Na^+(aq)} + \cancel{2\ Cl^-(aq)}$$

and with the cancellation of "spectator ions" we obtain

$$2\ OH^-(aq) + Mg^{2+}(aq) \longrightarrow Mg(OH)_2(s)$$

As you gain experience in making predictions, you can go directly to a net ionic equation (without showing the cancellation of "spectator" ions).

FIGURE 5-4

Verifying the predictions made in Example 5-2.

(a) Addition of NaOH(aq) to $MgCl_2$(aq) yields a white precipitate of magnesium hydroxide, $Mg(OH)_2$(s).
(b) When colorless BaS(aq) is added to blue $CuSO_4$(aq), a dark precipitate forms. The precipitate is a mixture of white $BaSO_4$(s) and black CuS(s). (A slight excess of $CuSO_4$ remains in solution.)
(c) No reaction occurs when $(NH_4)_2SO_4$(aq) is added to $ZnCl_2$(aq).
[Carey B. Van Loon]

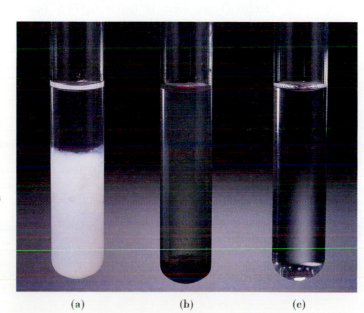

(a)   (b)   (c)

(b) The solubility rule for sulfides is that only those of periodic table groups 1A and 2A are soluble. CuS must be *insoluble*. Although most sulfates are water soluble, $BaSO_4$ is one of the few that are not. The reaction produces a mixture of precipitates.

$$Ba^{2+}(aq) + S^{2-}(aq) + Cu^{2+}(aq) + SO_4^{2-}(aq) \longrightarrow BaSO_4(s) + CuS(s)$$

(c) A careful review of Table 5-2 shows that all the possible ion combinations lead to water-soluble compounds.

$$2\ NH_4^+(aq) + SO_4^{2-}(aq) + Zn^{2+}(aq) + 2\ Cl^-(aq) \longrightarrow \textbf{no reaction}$$

SIMILAR EXAMPLES: Exercises 6, 24.

5-3   Acid–Base Reactions

Acids, bases, and reactions involving them are so important that we devote most of Chapters 17 and 18 to their study. Although not as general as we will ultimately need, the definitions of acids and bases first given in Section 3-6 are adequate for the present.

Acids. An **acid** is a substance that produces hydrogen ions (H^+)* in aqueous solution. Thus, when hydrogen chloride gas dissolves in water, complete ionization occurs (HCl is a strong electrolyte).

$$HCl(g) + water \longrightarrow HCl(aq) \longrightarrow H^+(aq) + Cl^-(aq) \tag{5.4}$$

A water solution of the oxoacid nitric acid, also a strong electrolyte, can be represented as

$$HNO_3(aq) \longrightarrow H^+(aq) + NO_3^-(aq) \tag{5.5}$$

Because they are completely ionized in water solutions, HCl and HNO_3 are called **strong acids.**

Because their ionizations are *reversible* and do not go to completion, the majority of acids are *weak electrolytes*.

$$HC_2H_3O_2(aq) \rightleftharpoons H^+(aq) + C_2H_3O_2^-(aq) \tag{5.6}$$
(acetic acid)

Partially ionized acids like acetic acid are called **weak acids.**

The case of sulfuric acid is somewhat different. From each molecule of $H_2SO_4(aq)$ *two* H^+ ions can be obtained, but these are produced in a stepwise fashion.

$$H_2SO_4(aq) \longrightarrow H^+(aq) + HSO_4^-(aq)$$

$$HSO_4^-(aq) \rightleftharpoons H^+(aq) + SO_4^{2-}(aq) \tag{5.7}$$

In its first ionization $H_2SO_4(aq)$ is a *strong acid,* but in the second ionization $HSO_4^-(aq)$ is a *weak acid.* (This aspect of the solution behavior of sulfuric acid is discussed further in Chapter 17.)

---

*The simple hydrogen ion, H^+, does not exist in aqueous solutions. Its actual form is as hydronium ion, H_3O^+, in which an H^+ ion is attached to an H_2O molecule. The hydronium ion is discussed in Chapter 17. For the present we follow a common practice among chemists by writing the simple formula H^+ instead of H_3O^+.

TABLE 5-4
The Common Strong Acids and Strong Bases

Acids	Bases
HCl	LiOH
HBr	NaOH
HI	KOH
HClO$_4$	RbOH
HNO$_3$	CsOH
H$_2$SO$_4$^{a}	Ca(OH)$_2$
	Sr(OH)$_2$
	Ba(OH)$_2$

^{a}H$_2$SO$_4$ ionizes in two distinct steps. It is a strong acid only in its first ionization step (see page 623).

Are You Wondering:

How you can tell if a substance is an acid or a base?

We generally identify any *ionizable* hydrogen atoms by the way we write the formula of an acid. For now we do this by writing ionizable H atoms first. Thus, HC$_2$H$_3$O$_2$ has *one* ionizable H atom (and *three* that are not), it *is* an acid. CH$_4$ has four H atoms, but they are *not* ionizable; CH$_4$ is *not* an acid. If the formula indicates a combination of OH$^-$ ions with metal ions (e.g., NaOH), expect the substance to be a base. To identify a weak base, you usually need a chemical equation for the ionization reaction, as in equation (5.9). Note that ethanol, C$_2$H$_5$OH, is *not* a base. The OH group is not present as OH$^-$ in either pure ethanol or in its aqueous solutions. Finally, make use of these facts: *There are only a few common strong acids and strong bases; these are listed in Table 5-4. The most common weak base is NH$_3$(aq).*

Bases. A definition that is suitable for the present is that a **base** is a substance capable of producing hydroxide ions (OH$^-$) in aqueous solution. Consider a soluble ionic hydroxide like NaOH. In the solid state this compound consists of Na$^+$ and OH$^-$ ions. When the solid dissolves in water the ions become dissociated from each other. NaOH, a *strong electrolyte,* is also a **strong base.**

$$NaOH(aq) \longrightarrow Na^+(aq) + OH^-(aq) \tag{5.8}$$

Certain substances produce OH$^-$ ions by *reacting* with water, not just by dissolving in it. Such substances are also bases, but if the reaction with water does not go to completion, as is usually the case, the substance is a **weak base.** In aqueous solution most NH$_3$ molecules remain nonionized—NH$_3$ is a weak base.

$$NH_3(aq) + H_2O \Longrightarrow NH_4^+(aq) + OH^-(aq) \tag{5.9}$$

Neutralization. Several properties of acids and bases are described in Figure 5-5. Perhaps the most significant property, though, is the ability of each to cancel or neutralize the properties of the other. In a **neutralization** reaction an acid and a base are converted to an aqueous solution of an ionic compound called a **salt.** Thus, in *ionic* form

$$\underset{\text{(acid)}}{H^+(aq) + Cl^-(aq)} + \underset{\text{(base)}}{Na^+(aq) + OH^-(aq)} \longrightarrow \underset{\text{(salt)}}{Na^+(aq) + Cl^-(aq)} + \underset{\text{(water)}}{H_2O(l)}$$

FIGURE 5-5
Properties of acids and bases.

Acids have a sour taste, dissolve certain metals and carbonate minerals, and produce color changes in acid–base indicators. Illustrated here is the acidic nature of lemon juice, shown by the *red* color of the indicator, *methyl red.*
Bases have a bitter taste and a slippery feel and also affect the color of acid–base indicators. Here we see the basic nature of soap, shown through the color change of methyl red from red to *yellow.* [Carey B. Van Loon]

By eliminating the "spectator" ions we discover the essential nature of a neutralization reaction:

$$H^+(aq) + OH^-(aq) \longrightarrow H_2O(l)$$

In a neutralization reaction H^+ ions from an acid and OH^- ions from a   (5.10)
base combine to form water.

The fundamental nature of a neutralization reaction.

At first, expression (5.10) seems not to apply to neutralizing the weak base $NH_3(aq)$. Hydroxide ion, OH^-, does not appear in the net ionic equation (5.11). However, if we think in terms of a stepwise process we do see the role played by H^+, OH^-, and H_2O. For the neutralization of $NH_3(aq)$ by $HCl(aq)$,

To stress the fact that $NH_3(aq)$ is a weak base, its formula is sometimes written as NH_4OH (ammonium hydroxide) and its ionization represented as

$$NH_4OH(aq) \rightleftharpoons$$
$$NH_4^+(aq) + OH^-(aq)$$

There is no evidence for the actual existence of NH_4OH, however.

$$NH_3(aq) + H_2O \longrightarrow NH_4^+(aq) + OH^-(aq)$$
$$H^+(aq) + Cl^-(aq) + OH^-(aq) \longrightarrow H_2O + Cl^-(aq)$$
$$\overline{H^+(aq) + NH_3(aq) \longrightarrow NH_4^+(aq)} \qquad (5.11)$$

Example 5-3

Writing equation(s) for a neutralization reaction. Strontium nitrate is used in highway flares and in the manufacture of fireworks. Strontium compounds impart a bright red color to flames. One method of preparing strontium nitrate is to neutralize strontium hydroxide with nitric acid. For this neutralization reaction, write (a) molecular, (b) ionic, and (c) net ionic equations.

Solution

(a) Substitute chemical formulas for the names of the substances involved in the reaction and then balance the equation.

$$HNO_3(aq) + Sr(OH)_2(aq) \longrightarrow Sr(NO_3)_2(aq) + H_2O(l) \quad \text{(unbalanced)}$$

$$2\,HNO_3(aq) + Sr(OH)_2(aq) \longrightarrow Sr(NO_3)_2(aq) + 2\,H_2O(l) \qquad (5.12)$$

(b) In equation (5.12) write chemical formulas in the ionic form for strong electrolytes and in molecular form for the nonelectrolyte water.

$$2\,H^+(aq) + 2\,NO_3^-(aq) + Sr^{2+}(aq) + 2\,OH^-(aq) \longrightarrow$$
$$Sr^{2+}(aq) + 2\,NO_3^-(aq) + 2\,H_2O(l)$$

(c) Remove the spectator ions (Sr^{2+} and NO_3^-) from the above equation.

$$2\,H^+(aq) + 2\,OH(aq) \longrightarrow 2\,H_2O(l)$$

or, more simply,

$$H^+(aq) + OH^-(aq) \longrightarrow H_2O(l) \qquad (5.13)$$

SIMILAR EXAMPLES: Exercises 7, 24, 27.

Example 5-4

Relating solution volumes and molarities in a neutralization reaction. How many mL of 0.0252 M $HNO_3(aq)$ are required to neutralize 25.00 mL (0.02500 L) of 0.0148 M $Sr(OH)_2(aq)$?

Solution. This question is similar to examples considered in Chapter 4 (e.g., Example 4-14). The main difference is that in Chapter 4 you were *given* chemical equations for reactions. Here you are learning how to *predict* these equations. As long as the equation is *balanced*, you can use either the molecular equation or the net ionic equation.

METHOD 1: Using the molecular equation. The calculation is set up in three steps below. It is a stoichiometric calculation. The first and third steps involve relating solution volume, molarity, and amount of solute. The middle step converts from moles of one reactant to the other. It is in this step that the stoichiometric factor from equation (5.12) is required (shown in blue).

$$\text{no. mol } Sr(OH)_2 = 0.02500 \text{ L} \times \frac{0.0148 \text{ mol } Sr(OH)_2}{\text{L}}$$

$$= 3.70 \times 10^{-4} \text{ mol } Sr(OH)_2$$

$$\text{no. mol } HNO_3 = 3.70 \times 10^{-4} \text{ mol } Sr(OH)_2 \times \frac{2 \text{ mol } HNO_3}{1 \text{ mol } Sr(OH)_2}$$

$$= 7.40 \times 10^{-4} \text{ mol } HNO_3$$

$$\text{no. mL } HNO_3 = 7.40 \times 10^{-4} \text{ mol } HNO_3 \times \frac{1 \text{ L } HNO_3}{0.0252 \text{ mol } HNO_3}$$

$$\times \frac{1000 \text{ mL } HNO_3}{1 \text{ L } HNO_3}$$

$$= 29.4 \text{ mL } HNO_3$$

METHOD 2: Using the net ionic equation. Here emphasis is on relating *ionic* concentrations to solution molarities, as in Example 5-1. The net ionic equation (5.13) simply states that one H^+ is required for every OH^- in the neutralization reaction. Again, this factor arises in the second step.

$$\text{no. mol } OH^- = 0.02500 \text{ L} \times \frac{0.0148 \text{ mol } Sr(OH)_2}{\text{L}} \times \frac{2 \text{ mol } OH^-}{1 \text{ mol } Sr(OH)_2}$$

$$= 7.40 \times 10^{-4} \text{ mol } OH^-$$

$$\text{no. mol } H^+ = 7.40 \times 10^{-4} \text{ mol } OH^- \times \frac{1 \text{ mol } H^+}{1 \text{ mol } OH^-} = 7.40 \times 10^{-4} \text{ mol } H^+$$

$$\text{no. mL } HNO_3 = 7.40 \times 10^{-4} \text{ mol } H^+ \times \frac{1 \text{ mol } HNO_3}{1 \text{ mol } H^+} \times \frac{1 \text{ L } HNO_3}{0.0252 \text{ mol } HNO_3}$$

$$\times \frac{1000 \text{ mL } HNO_3}{1 \text{ L } HNO_3} = 29.4 \text{ mL } HNO_3$$

SIMILAR EXAMPLES: Exercises 8, 9, 58, 85.

Titrations. In a neutralization reaction a point is reached where both acid and base are consumed and *neither* is in excess. This is called the **equivalence point** of the neutralization. To locate the equivalence point in a neutralization reaction requires that

- careful control be exercised over the addition of base to acid (or acid to base) and
- a means be found to signal the point at which the reaction mixture turns from acid to base (or base to acid).

Key to a successful titration is knowing how to select an appropriate indicator. We consider this matter in Section 18-3.

The first objective can be achieved by dispensing one solution into the other through a *buret*. The second objective is achieved by using an **acid–base indicator,** a substance that, if properly chosen, changes color when the solution changes from having a very slight excess of acid to having a very slight excess of base. This analytical technique, called **titration,** is illustrated in Figure 5-6.

FIGURE 5-6
The technique of titration—
Example 5-5 illustrated.

(a) A 5.00-mL sample of vinegar, a small quantity of water, and a few drops of phenolphthalein indicator are all added to a flask.
(b) 0.1000 M NaOH from a previously filled buret is slowly added to the flask.
(c) While the acid is in excess, the solution in the flask remains colorless. When the acid has just been neutralized, an additional drop of NaOH(aq) causes the solution to become slightly basic, and the phenolphthalein indicator turns to a light pink color. This is taken to be the equivalence point of the titration.
[Carey B. Van Loon]

(a)   (b)   (c)

Example 5-5

Using titration data to establish the concentrations of acids and bases. Vinegar is a dilute aqueous solution of acetic acid produced by the bacterial fermentation of apple cider, wine, or other carbohydrate material. The legal minimum acetic acid content of vinegar is 4%, by mass. A 5.00-mL sample of a particular vinegar was titrated with 38.08 mL (0.03808 L) of 0.1000 M NaOH(aq). Does this sample meet the legal limit? (Vinegar has a density of about 1.01 g/mL.)

Solution. Through ideas introduced earlier in this chapter, you should be able to write this net ionic equation for the neutralization reaction.

$$HC_2H_3O_2(aq) + OH^-(aq) \longrightarrow C_2H_3O_2^-(aq) + H_2O(l)$$

Proceed in a stepwise fashion similar to that used in Example 5-4.

$$\text{no. mol OH}^- = 0.03808 \text{ L} \times \frac{0.1000 \text{ mol NaOH}}{\text{L}} \times \frac{1 \text{ mol OH}^-}{1 \text{ mol NaOH}}$$

$$= 3.808 \times 10^{-3} \text{ mol OH}^-$$

$$\text{no. mol HC}_2\text{H}_3\text{O}_2 = 3.808 \times 10^{-3} \text{ mol OH}^- \times \frac{1 \text{ mol HC}_2\text{H}_3\text{O}_2}{1 \text{ mol OH}^-}$$

$$= 3.808 \times 10^{-3} \text{ mol HC}_2\text{H}_3\text{O}_2$$

$$\text{molarity of HC}_2\text{H}_3\text{O}_2 = \frac{3.808 \times 10^{-3} \text{ mol HC}_2\text{H}_3\text{O}_2}{5.00 \times 10^{-3} \text{ L}} = 0.762 \ M \text{ HC}_2\text{H}_3\text{O}_2$$

Now, use the definition of molarity, the molar mass of $HC_2H_3O_2$, and the density of the vinegar to determine the mass percent of $HC_2H_3O_2$.

$$\% \text{ HC}_2\text{H}_3\text{O}_2 = \frac{0.762 \text{ mol HC}_2\text{H}_3\text{O}_2}{1000 \text{ mL vinegar}} \times \frac{1.00 \text{ mL vinegar}}{1.01 \text{ g vinegar}}$$

$$\times \frac{60.05 \text{ g HC}_2\text{H}_3\text{O}_2}{1 \text{ mol HC}_2\text{H}_3\text{O}_2} \times 100 = 4.53\% \text{ HC}_2\text{H}_3\text{O}_2$$

The vinegar sample does exceed the legal minimum limit of 4% acetic acid, by mass.

SIMILAR EXAMPLES: Exercises 9, 29, 32, 33, 59.

5-4 Dissolution and Gas Evolution Reactions

As we saw in Example 5-4, $Sr(OH)_2$ is moderately soluble in water. $Mg(OH)_2$, on the other hand, is quite insoluble. We can readily dissolve $Mg(OH)_2(s)$ in an acidic solution, however. Here we can think in terms of (a) very slight dissolving of $Mg(OH)_2(s)$, (b) dissociation of $Mg(OH)_2(aq)$ into its ions, (c) neutralization of the OH^- by H^+ ion; followed by further dissolving of $Mg(OH)_2(s)$, and so on. By adding three equations together we obtain a simple *net ionic equation*.

$$Mg(OH)_2(s) \longrightarrow \cancel{Mg(OH)_2(aq)}$$
$$\cancel{Mg(OH)_2(aq)} \longrightarrow Mg^{2+}(aq) + \cancel{2\ OH^-(aq)}$$
$$\cancel{2\ OH^-(aq)} + 2\ H^+(aq) \longrightarrow 2\ H_2O(l)$$
$$\overline{Mg(OH)_2(s) + 2\ H^+(aq) \longrightarrow Mg^{2+}(aq) + 2\ H_2O(l)} \qquad (5.14)$$

Equation (5.14) indicates that the essential reactants are OH^- from $Mg(OH)_2(s)$ and H^+ from a strong acid. The essential product is $H_2O(l)$. It does not matter whether the strong acid is $HNO_3(aq)$ or $HCl(aq)$ or another strong acid. Because the final solution has $Mg^{2+}(aq)$ at a high concentration we can describe reaction (5.14) as the dissolving or dissolution of $Mg(OH)_2(s)$. In a *dissolution* reaction, a normally insoluble substance is made to dissolve as a result of a chemical reaction.

$Mg(OH)_2(s)$ can also be dissolved by *weak* acids, such as acetic acid. Here the necessary H^+ is produced by the partial ionization of the weak acid; this is followed by further ionization of the weak acid; and so on. In writing a net ionic equation in this case we use the *molecular* form rather than the ionic form for the weak acid.

$$Mg(OH)_2(s) + 2\ HC_2H_3O_2(aq) \longrightarrow$$
$$Mg^{2+}(aq) + 2\ C_2H_3O_2^-(aq) + 2\ H_2O(l) \quad (5.15)$$

Gas Evolution. Another solid that is insoluble in water but soluble in acids is calcium carbonate (the principal constituent of limestone). A simple way to think of the dissolution reaction is that

- H^+ and CO_3^{2-} combine to form unstable H_2CO_3, which decomposes to
- H_2O and CO_2, a gas that escapes from the reaction mixture.

If one of the products of a reaction (here, a gas) is removed from the reaction mixture, there is no opportunity for a reverse reaction to occur and the reaction goes to completion.

$$CaCO_3(s) + 2\ H^+(aq) \longrightarrow$$
$$Ca^{2+}(aq) + H_2CO_3(aq) \longrightarrow Ca^{2+}(aq) + H_2O + CO_2(g) \quad (5.16)$$

FIGURE 5-7

Action of baking powder—a gas evolution reaction.

When water is added to the dry baking powder, an acidic consitutent of the powder [$Ca(H_2PO_4)_2$] dissolves. The reaction of H^+ from this acid with bicarbonate ion (HCO_3^-) produces carbon dioxide gas.

$$HCO_3^-(aq) + H^+(aq) \rightarrow$$
$$H_2O + CO_2(g)$$

[Carey B. Van Loon]

TABLE 5-5
Some Gas Evolution Reactions

Ion	Gas-forming reaction
HSO_3^-	$HSO_3^- + H^+ \longrightarrow SO_2(g) + H_2O$
SO_3^{2-}	$SO_3^{2-} + 2\,H^+ \longrightarrow SO_2(g) + H_2O$
HCO_3^-	$HCO_3^- + H^+ \longrightarrow CO_2(g) + H_2O$
CO_3^{2-}	$CO_3^{2-} + 2\,H^+ \longrightarrow CO_2(g) + H_2O$
S^{2-}	$S^{2-} + 2\,H^+ \longrightarrow H_2S(g)$
NH_4^+	$NH_4^+ + OH^- \longrightarrow NH_3(g) + H_2O$

Ordinary baking powder is a solid mixture of sodium hydrogen carbonate (bicarbonate) and an acidic substance. When water is added, a reaction occurs in which $CO_2(g)$ is given off, providing the leavening action (see Figure 5-7 and Table 5-5).

Table 5-5 lists six examples of reactions in which a gas is produced, that is, *gas evolution reactions.* Five of the reactions involve an anion reacting with H^+ and one, a cation reacting with OH^-.

5-5   Oxidation–Reduction: Some Definitions

When an iron nail is exposed to the atmosphere (Figure 5-8), it rusts. A simplified equation for this reaction is

$$4\,Fe(s) + 3\,O_2(g) \longrightarrow 2\,Fe_2O_3(s) \tag{5.17}$$

Originally, the term "oxidation" was applied to a reaction in which a substance combines with oxygen (in this case, iron combines with oxygen).

Iron rust is an oxide of iron and so are most iron ores. In simplified fashion, the production of iron from iron ore is described as

$$Fe_2O_3(s) + 3\,CO(g) \longrightarrow 2\,Fe(l) + 3\,CO_2(g) \tag{5.18}$$

Reaction (5.18) involves the oxidation of $CO(g)$ to $CO_2(g)$. The O atoms required for this oxidation come from the Fe_2O_3, which is *reduced.* Originally, the term "reduction" was used to signify reactions in which O atoms are removed from a substance. Both oxidation and reduction occur in reactions (5.17) and (5.18), and they are called oxidation–reduction reactions. A definition of oxidation–reduction reactions based solely on the transfer of O atoms is too restrictive; it would limit us to reactions in which O atoms are involved. We need a broader definition.

One type of reaction that we should include in the category oxidation–reduction is the displacement of $Cu^{2+}(aq)$ by zinc metal, which we pictured in Figure 1-5. In addition, it is useful to separate the net reaction into two **half-reactions.** We call the half-reaction involving zinc, **oxidation,** and that involving copper, **reduction.**

Oxidation:	$Zn(s) \longrightarrow Zn^{2+}(aq) + 2\,e^-$	(5.19)
Reduction:	$Cu^{2+}(aq) + 2\,e^- \longrightarrow Cu(s)$	(5.20)
Net reaction:	$Zn(s) + Cu^{2+}(aq) \longrightarrow Zn^{2+}(aq) + Cu(s)$	(5.21)

In half-reaction (5.19) the oxidation state (O.S.) of zinc *increases* from 0 to +2 (corresponding to a *loss* of two electrons by each zinc atom). $Zn(s)$ is *oxidized.* In half-reaction (5.20) the oxidation state (O.S.) of copper *decreases* from +2 to 0 (corresponding to the *gain* of two electrons by each Cu^{2+} ion). $Cu^{2+}(aq)$ is *reduced.* To summarize,

Because it is easier to say, the term redox is often used instead of oxidation–reduction.

FIGURE 5-8
Rusting of iron nails.

The rusting of ordinary iron nails is an oxidation–reduction reaction. [Carey B. Van Loon]

FIGURE 5-9
Displacement of $H^+(aq)$ by iron metal—Example 5-6 illustrated.

(a) An iron nail is wrapped in a piece of copper gauze.
(b) The nail and gauze are placed in a hydrochloric acid solution. Hydrogen gas is evolved as the iron nail dissolves.
(c) The nail dissolves completely, producing $Fe^{2+}(aq)$, but the copper does not react.
[Carey B. Van Loon]

(a)   (b)   (c)

Important ideas concerning oxidation–reduction.

- Oxidation is a process in which the O.S. of some element *increases*, and in which electrons appear on the *right-hand* side of a half-equation.
- Reduction is a process in which the O.S. of some element *decreases*, and in which electrons appear on the *left-hand* side of a half-equation. (5.22)
- Oxidation and reduction half-reactions must *always* occur together, and the total number of electrons associated with the oxidation must be equal to the total number associated with the reduction.

Example 5-6

Expressing an oxidation–reduction reaction through half-equations and a net equation. Write net ionic equation(s) to represent the oxidation–reduction reaction(s) occurring in Figure 5-9.

Solution. The iron metal dissolves, displacing H^+ ions as hydrogen gas. The iron metal appears in solution as $Fe^{2+}(aq)$. The iron metal is oxidized and the $H^+(aq)$ is reduced.

Oxidation: $Fe(s) \longrightarrow Fe^{2+}(aq) + 2\ e^-$
Reduction: $2\ H^+(aq) + 2\ e^- \longrightarrow H_2(g)$

Net reaction: $Fe(s) + 2\ H^+(aq) \longrightarrow Fe^{2+}(aq) + H_2(g)$

The experimental evidence in Figure 5-9 is that copper does not react with hydrochloric acid. We can represent this as

$Cu(s) + H^+(aq) \longrightarrow$ no reaction

SIMILAR EXAMPLES: Exercises 7e, 7f, 39d.

TABLE 5-6
Behavior of Some Common Metals Toward Mineral Acid Solutions

Dissolve in mineral acids, producing $H_2(g)$	Do not dissolve in mineral acids
alkali metals (group 1A)[a]	Cu, Ag, Au
alkaline earth metals (group 2A)[a]	
Al, Zn, Fe, Sn, Pb	

A mineral acid (e.g., HCl, HBr, HI) is one in which the only possible reduction half-reaction is the reduction of H^+ to H_2. Some additional possibilities in metal/acid reactions are considered in Chapter 21.

[a] With the exception of Be and Mg, all group 1A and group 2A metals also react with cold water to produce $H_2(g)$. (The metal hydroxide is the other product.)

Example 5-6 suggests some fundamental questions about oxidation–reduction. For example,

- Why does Fe dissolve in HCl(aq), displacing $H_2(g)$, whereas Cu does not?
- Why does Fe form Fe^{2+} and not Fe^{3+} in this reaction (both ions exist)?

We deal extensively with the theory of oxidation–reduction reactions in Chapter 21, where we can answer such questions. In this chapter, our treatment of oxidation–reduction will be descriptive rather than theoretical. For now you should find the information in Table 5-6 helpful. It lists the common metals that do and do not displace $H_2(g)$ from acidic solution.

5-6   Balancing Oxidation–Reduction Equations

The Half-Reaction (Ion Electron) Method. Oxidation–reduction equations are often difficult to balance by inspection, but the way in which we obtained equation (5.21) suggests a fairly straightforward method.

Balancing redox equations by the half-reaction method.

- Write and balance separate half-equations for oxidation and reduction.
- Adjust coefficients in the two half-equations so that the same number of electrons appears in each half-equation.
- Add together the two half-equations (cancelling out electrons) to obtain the balanced net equation.

(5.23)

This procedure is applied in a stepwise fashion in Example 5-7.

Example 5-7

Using the half-reaction method to balance the equation for a redox reaction in acidic solution. A common laboratory method for determining the sulfite ion content of a sample (e.g., wastewater from a papermaking plant) is to titrate the $SO_3^{2-}(aq)$ with permanganate ion (usually as $KMnO_4$) in an acidic medium. Write a balanced equation for the reaction.

$$SO_3^{2-}(aq) + MnO_4^-(aq) + H^+ \longrightarrow SO_4^{2-}(aq) + Mn^{2+}(aq) + H_2O$$

(unbalanced)

Solution

Step 1. *Identify the species involved in oxidation state changes and write "skeleton" half-equations based on them.* The oxidation state of S increases from $+4$ in SO_3^{2-} to $+6$ in SO_4^{2-}. The oxidation state of Mn decreases from $+7$ in MnO_4^- to $+2$ in Mn^{2+}.

$$SO_3^{2-}(aq) \longrightarrow SO_4^{2-}(aq)$$

$$MnO_4^-(aq) \longrightarrow Mn^{2+}(aq)$$

Step 2. *Balance each half-equation "atomically," in this order.*

- atoms other than H and O
- O atoms by adding H_2O
- H atoms by adding H^+.

The "other" atoms (S and Mn) are already balanced in the skeleton half-equations. To balance O atoms we write

$$SO_3^{2-}(aq) + H_2O \longrightarrow SO_4^{2-}(aq)$$

$$MnO_4^-(aq) \longrightarrow Mn^{2+}(aq) + 4\ H_2O$$

and to balance H atoms

$$SO_3^{2-}(aq) + H_2O \longrightarrow SO_4^{2-}(aq) + 2\ H^+(aq)$$

$$MnO_4^-(aq) + 8\ H^+(aq) \longrightarrow Mn^{2+}(aq) + 4\ H_2O$$

Step 3. *Balance each half-equation "electrically."* Add the number of electrons necessary to get the same electric charge on both sides of the half-equation. By doing this you will see that the half-equation in which electrons appear on the *right* side is the *oxidation*; the other half-equation, with electrons on the *left*, is the *reduction*.

Oxidation: $SO_3^{2-}(aq) + H_2O \longrightarrow SO_4^{2-}(aq) + 2\ H^+(aq) + 2\ e^-$

(net charge on each side, -2)

Reduction: $MnO_4^-(aq) + 8\ H^+(aq) + 5\ e^- \longrightarrow Mn^{2+}(aq) + 4\ H_2O$

(net charge on each side, $+2$)

Step 4. *Obtain the net oxidation–reduction equation by combining the half-equations.* Multiply through the oxidation half-equation by *5* and through the reduction half-equation by *2*. This results in 10 e$^-$ on *each* side of the net equation. These terms cancel out. *Electrons must not appear in the final net equation.*

$$5 \, SO_3^{2-}(aq) + 5 \, H_2O \longrightarrow 5 \, SO_4^{2-}(aq) + 10H^+(aq) + \cancel{10 \, e^-}$$
$$2 \, MnO_4^-(aq) + 16 \, H^+(aq) + \cancel{10 \, e^-} \longrightarrow 2 \, Mn^{2+}(aq) + 8 \, H_2O$$
$$\overline{5 \, SO_3^{2-}(aq) + 2 \, MnO_4^-(aq) + 5 \, H_2O + 16 \, H^+(aq) \longrightarrow}$$
$$5 \, SO_4^{2-}(aq) + 2 \, Mn^{2+}(aq) + 8 \, H_2O + 10 \, H^+(aq)$$

Step 5 *Simplify.* The net equation should not contain the same species on both sides. Subtract *five* H_2O from each side of the equation in Step 4. This leaves *three* H_2O on the right. Also subtract *ten* H^+ from each side, leaving *six* on the left.

$$5 \, SO_3^{2-}(aq) + 2 \, MnO_4^-(aq) + 6 \, H^+(aq) \longrightarrow$$
$$5 \, SO_4^{2-}(aq) + 2 \, Mn^{2+}(aq) + 3 \, H_2O \quad (5.24)$$

Step 6. *Verify.* Check the final net equation to ensure that it is balanced both "atomically" and "electrically." For example, show that in equation (5.24) the net charge on each side of the equation is -6.

SIMILAR EXAMPLES: Exercises 11, 12, 35, 38, 63.

> To summarize: Identify skeleton half-equations. In the half-equations balance, in this order: (1) "other" atoms, (2) oxygen atoms, (3) hydrogen atoms, (4) electric charge. Adjust coefficients in half-equations to obtain equal numbers of electrons in oxidation and reduction half-equations. Add half-equations and simplify the net equation.

Although the method outlined in Example 5-7 should generally lead to a successfully balanced equation, there are some complications that you should be aware of. For example,

- The reaction might occur in *basic* solution.
- The same substance may undergo both oxidation and reduction; this is called a **disproportionation reaction.**
- In the final balanced equation, the added species H_2O and/or H^+ and/or OH^- may appear on a different side of the equation than in the unbalanced equation.

These three factors are all considered in Example 5-8. You should pay special attention to the matter of balancing a redox equation in basic solution. Since OH^- and H_2O *both* contain H and O atoms, it is hard to decide on which side of the half-equations to place them. One simple approach is to treat the reaction *as if* it occurs in acidic solution. Then, add to *each* side of the net equation a number of OH^- ions equal to the number of H^+ ions. Where H^+ and OH^- appear on the same side of the equation, replace them with H_2O molecules.

Example 5-8

Using the half-reaction method to balance the equation for a disproportionation reaction in basic solution. Phosphine, PH_3, a highly poisonous gas, is nevertheless useful as an intermediate in the synthesis of flame-retardant chemicals for use in cotton fabrics. One method of preparation involves the reaction of elemental phosphorus with $NaOH(aq)$, which can be represented as

$$P_4(s) + OH^-(aq) \longrightarrow H_2PO_2^-(aq) + PH_3(g) + H_2O$$

Balance this oxidation–reduction equation.

Solution.
Step 1. *Write the two skeleton half-equations and balance the P atoms.*

$$P_4(s) \longrightarrow 4 \, H_2PO_2^-(aq)$$

$$P_4(s) \longrightarrow 4 \, PH_3(g)$$

Step 2. *Balance each half-equation for H and O atoms. First add H_2O and then H^+, as required.*

$$P_4(s) + 8\ H_2O \longrightarrow 4\ H_2PO_2^-(aq) + 8\ H^+(aq)$$

$$P_4(s) + 12\ H^+(aq) \longrightarrow 4\ PH_3(g)$$

Step 3. *Balance each half-equation for electric charge by adding the appropriate numbers of electrons.*

$$P_4(s) + 8\ H_2O \longrightarrow 4\ H_2PO_2^-(aq) + 8\ H^+(aq) + 4\ e^-$$

$$P_4(s) + 12\ H^+(aq) + 12\ e^- \longrightarrow 4\ PH_3(g)$$

Step 4. *Combine the half-equations to obtain the net oxidation–reduction equation. (Multiply the oxidation half-equation by 3.)*

$$3\ P_4(s) + 24\ H_2O \longrightarrow 12\ H_2PO_2^-(aq) + 24\ H^+(aq) + \cancel{12\ e^-}$$
$$\underline{P_4(s) + 12\ H^+(aq) + \cancel{12\ e^-} \longrightarrow 4\ PH_3(g)}$$
$$4\ P_4(s) + 24\ H_2O \longrightarrow 12\ H_2PO_2^-(aq) + 12\ H^+(aq) + 4\ PH_3(g)$$

Step 5. *Add OH^- to each side of the equation to eliminate H^+.* By adding 12 OH^- to each side of the equation, the right side of the equation will have 12 H^+ and 12 OH^-, which can be combined into 12 H_2O.

$$4\ P_4(s) + 24\ H_2O + 12\ OH^- \longrightarrow 12\ H_2PO_2^-(aq) + 12\ H_2O + 4\ PH_3(g)$$

Step 6. *Simplify.* Subtract 12 H_2O from each side of the equation, and divide all coefficients by 4.

$$P_4(s) + 3\ H_2O + 3\ OH^-(aq) \longrightarrow 3\ H_2PO_2^-(aq) + PH_3(g)$$

SIMILAR EXAMPLES: Exercises 36, 62, 63.

Oxidation-State Change Method. Changes in oxidation states, a characteristic of all oxidation–reduction reactions, underlie a second general method for balancing oxidation–reduction equations. The method is illustrated in Example 5-9 for the reaction of $Fe^{2+}(aq)$ with $MnO_4^-(aq)$ in acidic solution (a reaction that can be used to establish the iron content of iron ores, for example).

Example 5-9

Using the oxidation-state change method to balance a redox equation. Balance the equation.

$$Fe^{2+}(aq) + MnO_4^-(aq) + H^+(aq) \longrightarrow Fe^{3+}(aq) + Mn^{2+}(aq) + H_2O$$

Solution

Step 1. Denote the oxidation states of all the elements involved in the reaction.

$$\overset{+2}{Fe^{2+}}(aq) + \overset{+7\ -2}{MnO_4^-}(aq) + \overset{+1}{H^+}(aq) \longrightarrow \overset{+3}{Fe^{3+}}(aq) + \overset{+2}{Mn^{2+}}(aq) + \overset{+1\ -2}{H_2O}$$

Step 2. Identify the elements whose oxidation states (O.S.) change and the number of units of change per atom.

$$\overset{+2}{Fe^{2+}}(aq) + \overset{+7}{MnO_4^-}(aq) + H^+(aq) \longrightarrow \overset{+3}{Fe^{3+}}(aq) + \overset{+2}{Mn^{2+}}(aq) + H_2O$$

increase of 1 in O.S. per Fe

decrease of 5 in O.S. per Mn

For a review of how oxidation states are assigned, see Section 3-5.

The changes in oxidation states noted here are the same as the numbers of electrons that would appear in the half-reaction method.

Step 3. Adjust coefficients so that the total increase in O.S. for oxidation is equal to the total decrease in O.S. for reduction.

$$\overset{+2}{5\ Fe^{2+}(aq)} + \overset{+7}{MnO_4^-(aq)} + H^+(aq) \longrightarrow \overset{+3}{5\ Fe^{3+}(aq)} + \overset{+2}{Mn^{2+}(aq)} + H_2O$$

total | increase of 5 in O.S.

total decrease of 5 in O.S.

Step 4. Adjust the remaining coefficients by inspection.

$$5\ Fe^{2+}(aq) + MnO_4^-(aq) + 8\ H^+(aq) \longrightarrow 5\ Fe^{3+}(aq) + Mn^{2+}(aq) + 4\ H_2O$$

SIMILAR EXAMPLES: Exercises 13, 37, 38, 63

The oxidation-state change method is a little more difficult to apply in Example 5-10 than in Example 5-9. In such cases you may find use for two additional points.

- In a disproportionation reaction, where the same substance is both oxidized and reduced, it may be helpful to write the formula for this substance *twice* on the left side of the equation.

- It may be easier at times to balance the net electrical charge on both sides of the equation *before* completing the balance of all the atoms.

Example 5-10

Using the oxidation-state change method to balance the equation for a disproportionation reaction in basic solution. $Cl_2(g)$ and OH^- [as $NaOH(aq)$] are both produced in the electrolysis of $NaCl(aq)$. If they are allowed to mix, some of the chlorine is converted to chlorate ion. This then becomes a method of producing $NaClO_3(aq)$. Sodium chlorate is used to make chlorine dioxide, an important bleach in the papermaking industry. The ionic equation for the formation of chlorate ion is

$$Cl_2(g) + OH^-(aq) \longrightarrow Cl^-(aq) + ClO_3^-(aq) + H_2O$$

Balance this equation.

Solution. In assessing changes in oxidation states we find that only Cl atoms undergo change. To simplify the equation balancing we write Cl_2 twice, once for oxidation and once for reduction.

$$\overset{0}{Cl_2(g)} + \overset{0}{Cl_2(g)} + OH^-(aq) \longrightarrow \overset{-1}{Cl^-(aq)} + \overset{+5}{ClO_3^-(aq)} + H_2O$$

decrease | of 1 in O.S. per Cl

increase of 5 in O.S. per Cl

To establish the total changes in O.S., we must work in multiples of 2 (since there are *two* Cl atoms per Cl_2), e.g., the oxidation of *one* Cl_2 must produce *two* ClO_3^-, with a total increase of 10 in O.S.

$$\overset{0}{5\ Cl_2(g)} + \overset{0}{Cl_2(g)} + OH^-(aq) \longrightarrow \overset{-1}{10\ Cl^-(aq)} + \overset{+5}{2\ ClO_3^-(aq)} + H_2O$$

total | decrease of 10 in O.S.

total increase of 10 in O.S.

Next, combine the two terms in $Cl_2(g)$ on the left and *balance the equation for electric charge*. Balancing for electric charge is easily done at this time. Add 12 OH^- to the left side, since the net charge on the right side (10 Cl^- and 2 ClO_3^-) is fixed at -12.

$$6 Cl_2(g) + 12 OH^-(aq) \longrightarrow 10 Cl^-(aq) + 2 ClO_3^-(aq) + H_2O$$

Then, adjust the remaining coefficient by inspection.

$$6 Cl_2(g) + 12 OH^-(aq) \longrightarrow 10 Cl^-(aq) + 2 ClO_3^-(aq) + 6 H_2O$$

Finally, *simplify*. (Divide all coefficients by 2.)

$$3 Cl_2(g) + 6 OH^-(aq) \longrightarrow 5 Cl^-(aq) + ClO_3^-(aq) + 3 H_2O$$

SIMILAR EXAMPLES: Exercises 13, 37, 63.

Additional Points to Consider. You now have two methods that you can use to balance oxidation–reduction equations. Here are some additional matters that you may need to consider when you attempt to balance these equations on your own (as in the Exercises).

- Some redox reactions involve more than one oxidation or reduction. In the half-equation

$$As_2S_3(s) + 8 H_2O \longrightarrow 2 H_3AsO_4 + 3 S(s) + 10 H^+ + 10 e^-$$

both As and S atoms are oxidized.
- Some redox reactions are not carried out in aqueous solution. The oxidation of ammonia to nitrogen monoxide (the first step in the commercial production of nitric acid) occurs in the *gaseous* state.

$$4 NH_3(g) + 5 O_2(g) \longrightarrow 4 NO(g) + 6 H_2O(g)$$

This equation can be balanced by the half-reaction method *as if* the reaction occurred in acidic solution. H^+ cancels out in the net equation. Alternatively, the oxidation-state change method also works well for gaseous reactions.
- Some equations, though of the redox type, are still most easily balanced by inspection. For example, $2 H_2 + O_2 \rightarrow 2 H_2O$.
- Some equations, though they may look plausible, cannot be balanced because they do not involve both oxidation and reduction. For example, *two* reductions and *no* oxidation are represented in the equation

$$\overline{MnO_4^-(aq)} + \overline{SO_4^{2-}(aq)} + \overline{H^+(aq)} \longrightarrow \overline{Mn^{2+}(aq)} + \overline{SO_3^{2-}(aq)} + \overline{H_2O}$$

The reaction cannot occur.

5-7 Oxidizing and Reducing Agents

In an oxidation–reduction reaction the substance that is oxidized makes it possible for some other substance to be *reduced*. The substance that is oxidized is called the **reducing agent.** Similarly, the substance that is reduced makes it possible for another substance to be *oxidized*. The substance that is reduced is the **oxidizing agent.** These statements are summarized diagrammatically in Figure 5-10.

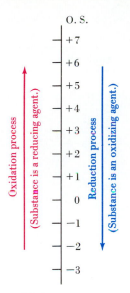

FIGURE 5-10

Identifying oxidizing and reducing agents.

A substance in which certain atoms have one of the higher oxidation states among the permitted values tends to be an *oxidizing* agent; in redox reactions the O.S. of these atoms is lowered. A substance in which certain atoms have one of the lower oxidation states among the permitted values tends to be a *reducing* agent; in redox reactions the O.S. of these atoms is raised. Certain substances in which atoms have an oxidation state at an intermediate point among the permitted values may act as oxidizing agents in some instances and reducing agents in others.

Are You Wondering:

How to keep straight in your mind the terms oxidation, reduction, oxidizing agent, and reducing agent?

Remember that in an oxidation process the oxidation state of an element *increases*. The species (atom, molecule, or ion) containing that element is *oxidized,* and the species that is oxidized is a *reducing* agent. Conversely, reduction requires that the oxidation state of an element *decrease*. The species containing that element is *reduced,* and the species is an *oxidizing* agent. (Also, reason by analogy: The oxidizing agent is *not* oxidized, just as the real estate agent does not move into the new home and the travel agent does not take the vacation trip.) Finally, notice that we refer to the oxidizing or reducing agent as being an entire *species* and not just the element undergoing the change in oxidation state. Thus, in Example 5-11b we refer to the oxidizing agent as the *permanganate ion,* MnO_4^- (even though the oxidation state of only the Mn changes in the redox reaction).

Example 5-11

Identifying oxidizing and reducing agents. Hydrogen peroxide, H_2O_2, is a versatile chemical. Its uses include bleaching wood pulp and fabrics, substituting for chlorine in water purification, and as an oxidizer in rocket fuels. One important reason for its versatility is that it can function as either an oxidizing or a reducing agent, depending on the particular reaction. Is hydrogen peroxide an oxidizing or a reducing agent in each of these reactions?

(a) $H_2O_2(aq) + 2 Fe^{2+}(aq) + 2 H^+(aq) \longrightarrow 2 H_2O + 2 Fe^{3+}$

(b) $5 H_2O_2(aq) + 2 MnO_4^-(aq) + 6 H^+(aq) \longrightarrow$
$$8 H_2O + 2 Mn^{2+}(aq) + 5 O_2(g)$$

(c) $H_2O_2(aq) + Cl_2(aq) + 2 OH^-(aq) \longrightarrow 2 H_2O + 2 Cl^- + O_2(g)$

Solution

(a) The oxidation state (O.S.) of oxygen in H_2O_2 is -1. In H_2O its O.S. is -2. Hydrogen peroxide is *reduced* and thereby acts as an **oxidizing agent**. (Fe^{2+} is the reducing agent.)

(b) Here the O.S. of oxygen atoms *increases* from -1 in H_2O_2 to 0 in O_2. Hydrogen peroxide is *oxidized* and thereby acts as a **reducing agent**. (MnO_4^- is the oxidizing agent.)

(c) Again the O.S. of the oxygen atoms *increases* from -1 to 0. Hydrogen peroxide is *oxidized,* thereby acting as a **reducing agent**. (Cl_2 is the oxidizing agent.)

SIMILAR EXAMPLES: Exercises 14, 40, 65.

Oxidizing and reducing agents play an important role throughout chemistry. Some typical examples are given in Table 5-7. In order to *predict* whether a particular combination of oxidizing and reducing agents results in an oxidation–reduction reaction, we will consider some additional principles in Chapter 21.

TABLE 5-7
Some Common Oxidizing and Reducing Agents

Substance	Typical reaction	Applications
Oxidizing agents		
$OCl^-(aq)$	$2 CN^- + 5 OCl^- + H_2O \longrightarrow$ $N_2 + 2 HCO_3^- + 5 Cl^-$	Hypochlorite ion is the active component of typical household bleaches. It is also used to oxidize cyanide ion in waste solutions from gold-mining operations.
$MnO_4^-(aq)$	$5 Fe^{2+} + MnO_4^- + 8 H^+ \longrightarrow$ $5 Fe^{3+} + Mn^{2+} + 4 H_2O$	Permanganate ion is a widely used oxidizing agent in the analytical chemistry laboratory, as in the quantitative analysis of iron.
$O_3(g)$	$C_6H_5OH + 14 O_3 \longrightarrow$ $6 CO_2 + 3 H_2O + 14 O_2$	Ozone, O_3, is used in water purification, as in the oxidation of phenol, C_6H_5OH.
Reducing agents		
$I^-(aq)$	$O_3 + 2 I^- + H_2O \longrightarrow$ $O_2 + I_2 + 2 OH^-$	Iodide ion is used to determine quantities of ozone (O_3) in air-pollution studies.
$S_2O_3^{2-}(aq)$	$S_2O_3^{2-} + 4 Cl_2 + 5 H_2O \longrightarrow$ $2 HSO_4^- + 8 H^+ + 8 Cl^-$	Thiosulfate ion is used as an "antichlor" to destroy residual chlorine from the bleaching of fibers.
$C(s)$	$ZnO(s) + C(s) \overset{\Delta}{\longrightarrow}$ $Zn(g) + CO(g)$	Carbon (as coke or coal) is the most important reducing agent for obtaining metals from their oxides.

5-8 Stoichiometry of Redox Reactions

Once we have balanced a redox equation, we can perform the same types of stoichiometric calculations as we discussed in Chapter 4. (In fact, many of the examples in that chapter were redox reactions.)

Oxidation–reduction reactions are often carried out by *titration*, very similar to what was outlined in Section 5-3. Example 5-12 applies a redox titration to a procedure known as standardization of a solution. The procedure is further illustrated in Figure 5-11.

Example 5-12

Standardizing a solution for use in redox titrations. Suppose for general use in an analytical laboratory we need a $KMnO_4(aq)$ solution of exactly known molarity, close to 0.02000 M. We cannot prepare this solution just by weighing out the required amount of $KMnO_4(s)$ and dissolving it in water. The solid is *not pure* and its actual purity (i.e., % $KMnO_4$) is *not known*. On the other hand, we can obtain iron wire in essentially pure form, and we can dissolve the wire in an acid, yielding $Fe^{2+}(aq)$. $Fe^{2+}(aq)$ is *oxidized* to $Fe^{3+}(aq)$ by $KMnO_4(aq)$ in an acidic solution. By determining the volume of $KMnO_4(aq)$ required to oxidize a known quantity of $Fe^{2+}(aq)$, we can calculate the exact molarity of the $KMnO_4(aq)$. This procedure is referred to as **standardizing a solution**.

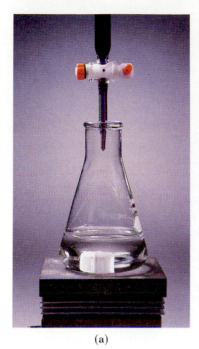
(a)

(b)

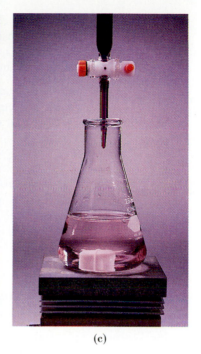
(c)

FIGURE 5-11
Standardizing a solution of an oxidizing agent through an oxidation–reduction titration.

(a) A solution containing a known amount of Fe^{2+} and a buret filled with the intensely colored $KMnO_4(aq)$ to be standardized.
(b) As the $KMnO_4(aq)$ is added to the solution of $Fe^{2+}(aq)$, the $MnO_4^-(aq)$ is immediately decolorized as a result of reaction (5.25).
(c) When all the Fe^{2+} has been oxidized to Fe^{3+}, additional $KMnO_4(aq)$ has nothing left to oxidize and the solution turns a distinctive pink color. [Even a fraction of a drop of the $KMnO_4(aq)$ beyond the equivalence point is sufficient to cause this pink coloration.]
[Carey B. Van Loon]

A piece of iron wire weighing 0.1568 g is converted to $Fe^{2+}(aq)$ and requires 26.24 mL of a $KMnO_4(aq)$ solution for its titration. What is the molarity of the $KMnO_4(aq)$?

$$5\ Fe^{2+}(aq) + MnO_4^-(aq) + 8\ H^+(aq) \longrightarrow$$
$$5\ Fe^{3+}(aq) + Mn^{2+}(aq) + 4\ H_2O \quad (5.25)$$

Solution. First, determine the no. mol Fe^{2+} used in the titration.

$$\text{no. mol } Fe^{2+} = 0.1568\ g\ Fe \times \frac{1\ mol\ Fe}{55.847\ g\ Fe} \times \frac{1\ mol\ Fe^{2+}}{1\ mol\ Fe}$$

$$= 2.808 \times 10^{-3}\ mol\ Fe^{2+}$$

Next, determine the no. mol $KMnO_4(aq)$ that must have been used.

$$\text{no. mol } KMnO_4 = 2.808 \times 10^{-3}\ mol\ Fe^{2+} \times \frac{1\ mol\ MnO_4^-}{5\ mol\ Fe^{2+}} \times \frac{1\ mol\ KMnO_4}{1\ mol\ MnO_4^-}$$

$$= 5.616 \times 10^{-4}\ mol\ KMnO_4$$

The volume of solution containing the 5.616×10^{-4} mol $KMnO_4$ is 26.24 mL = 0.02624 L, which means that

$$\text{molarity of } KMnO_4(aq) = \frac{5.616 \times 10^{-4}\ mol\ KMnO_4}{0.02624\ L} = 0.02140\ M\ KMnO_4(aq)$$

SIMILAR EXAMPLES: Exercises 15, 42, 43, 44, 71.

5-9 Conditions for a Reaction to Go to Completion: A Summary

In Section 4-8 we noted that, in general, the products of a reaction, once formed, can react to re-form the initial reactants. The reaction is *reversible* and reaches a condition of *equilibrium*. There are circumstances, however, in which the reverse reaction is insignificant and the reaction is said to *go to completion*. The essential requirement is that one or more of the products be effectively removed from the reaction mixture. This will occur, for example, if ions combine to form

Conditions for a reaction to go to completion.

- a precipitate from solution,
- a nonelectrolyte (e.g., H_2O),
- a gas that escapes from the reaction mixture.

(5.26)

Example 5-13

Predicting whether a reaction will go to completion. Each of the following reactions is encountered in an industrial process. Which would you expect essentially to go to completion? Explain.

(a) Synthesis of ammonia (used in the production of fertilizers, explosives, and other nitrogen-containing substances).

$$N_2(g) + 3\ H_2(g) \longrightarrow 2\ NH_3(g)$$

(b) Production of hydrogen fluoride (used in the production of chlorofluorocarbons for refrigeration systems).

$$CaF_2(s) + H_2SO_4(l) \longrightarrow CaSO_4(s) + 2\ HF(g)$$

(c) Production of zinc sulfate (used as a dietary supplement).

$$ZnO(s) + H_2SO_4(aq) \longrightarrow ZnSO_4(aq) + H_2O(l)$$

(d) Conversion of $Na_2CO_3(aq)$ to $NaOH(aq)$ (a reaction used in the pulp and paper industry).

$$Ca(OH)_2(s) + 2\ Na^+(aq) + CO_3^{2-}(aq) \longrightarrow$$
$$CaCO_3(s) + 2\ Na^+(aq) + 2\ OH^-(aq)$$

Solution

(a) Although the product of the reaction is a gas, so are both of the reactants. The reactants must be confined in order to react, and as long as the product is confined with them, the reverse reaction will occur. The reaction will *not* go to completion.

(b) Hydrogen fluoride is a gas, which can escape from the reaction mixture. The reaction will go to completion.

(c) The nonelectrolyte H_2O forms in a solution containing ions. The reaction goes to completion.

(d) Although the forward reaction produces a precipitate ($CaCO_3$), so does the reverse reaction [$Ca(OH)_2$]. The importance of the forward reaction compared to the reverse reaction depends on the relative solubilities of these two precipitates (a matter that we consider in Chapter 19). For the present, our best conclusion is that the reaction does *not* go to completion.

SIMILAR EXAMPLES: Exercises 34, 61.

In Chapter 21 we consider some additional criteria for predicting the conditions that lead to completeness in oxidation–reduction reactions.

5-10   Introduction to Qualitative Cation Analysis

As we learned in the Focus feature of Chapter 3, a *qualitative* analysis is used to determine what substances are present in a mixture, but *not* their quantities. If this

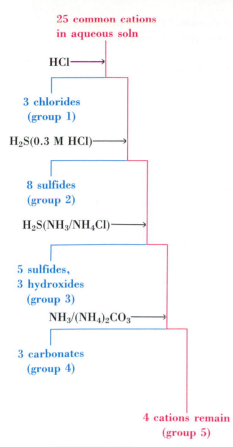

25 common cations
in aqueous soln

HCl⟶

**3 chlorides
(group 1)**

H_2S(0.3 M HCl)⟶

**8 sulfides
(group 2)**

H_2S(NH_3/NH_4Cl)⟶

**5 sulfides,
3 hydroxides
(group 3)**

NH_3/$(NH_4)_2CO_3$⟶

**3 carbonates
(group 4)**

4 cations remain
(group 5)

FIGURE 5-12
Outline of a qualitative analysis scheme for cations.

The original aqueous solution may contain any combination of about 25 common cations. The reagent added is in black. Some cations precipitate (blue), and others remain in solution (red). Following each group precipitation, a new reagent is added to precipitate the next group. Ultimately the ions are separated into five groups.

analysis is directed at discovering what cations are present in a mixture, it is called **qualitative cation analysis.** The types of reactions considered in this chapter—precipitation (and dissolution), acid–base, and oxidation–reduction—are fundamental to the procedures of qualitative cation analysis.

Figure 5-12 suggests that the basic approach is to separate the common cations into groups (usually five) by *precipitation*. For example, only the members of *group 1*, that is, Ag^+, Hg_2^{2+}, and Pb^{2+}, form insoluble chlorides. All other common metal chlorides are water soluble (recall Table 5-3). *Group 2* cations are precipitated as sulfides from acidic solution. Cations of *group 3* form insoluble sulfides in basic solution. The *group 4* cations are precipitated as carbonates. The cations of *group 5* remain in solution throughout the separation of the other four groups.

Following the separation of cations into the five major groups, further separation and testing must be done within each group. The end result is to establish for an unknown mixture the presence or absence of each ion in the scheme.

Example 5-14

Solving a qualitative analysis "unknown." You are given a *colorless* aqueous solution and told that it contains *none, one, two,* or *all three* of the following ions: Ag^+, Ba^{2+}, Cu^{2+}. You treat this unknown with $(NH_4)_2CO_3$(aq) and obtain a *white* precipitate. What conclusions can you draw from these observations? (Use data from Table 5-3, together with general observations that you can make in the laboratory.)

Solution. All three of the possible ions produce an insoluble carbonate (Table 5-3, item 5). The formation of a precipitate only allows you to conclude that at least one of the three ions is present. The general laboratory observation that you can make is of solutions and solids containing Ag^+, Ba^{2+}, and Cu^{2+} [such as $AgNO_3$, $Ba(NO_3)_2$, and $Cu(NO_3)_2$]. You will find that only solutions and solids with Cu^{2+} display a color (blue). The observation that neither the solution nor the precipitate is colored *suggests* that Cu^{2+} is *absent*. (It is still possible—although not likely—that Cu^{2+} is present in the solution. At very low concentrations its color cannot be detected.) Whether the white precipitate is Ag_2CO_3(s), $BaCO_3$(s), or a mixture of the two cannot be determined from the observations made.

SIMILAR EXAMPLES: Exercises 16, 48, 49, 87.

We could get an unambiguous answer to Example 5-14 by following the qualitative analysis procedure of Figure 5-12. If the sample were first treated with HCl(aq), formation of a precipitate would prove the presence of Ag^+ (the only one of the three ions that forms an insoluble chloride). Lack of a precipitate would establish the *absence* of Ag^+. The filtrate—the solution remaining after removal of any AgCl(s)—could now be treated with H_2S. Here, formation of a precipitate would prove the presence of Cu^{2+}, and lack of a precipitate, the absence of Cu^{2+}. (Cu^{2+} is in qualitative analysis group 2.) *After* removal of any Ag^+ or Cu^{2+} present in the unknown, a test could be performed for Ba^{2+} using the group 4 reagent. (Ba^{2+} is in qualitative analysis group 4.)

Precipitation reactions are used to separate cations between and within groups and for the final identification and confirmation of most of the cations. Other types of reactions also play an important role in the qualitative analysis scheme, as illustrated by the following examples.

Acid–Base Reactions. At several points in the procedure an acidic solution must be neutralized by adding NH_3(aq), or a basic solution by adding HCl(aq). For example, the solution to be analyzed for group 2 ions contains excess HCl (which was used in the group 1 precipitation). This solution must first be neutralized, so that its acidity can then be properly adjusted for the group 2 precipitation.

Dissolution Reactions. At one point in the procedure for group 3, Fe^{3+}, Ni^{2+}, and Co^{3+} are obtained together as a mixed hydroxide precipitate. In order to separate these three ions from one another, it is first necessary to get them back into solution. The hydroxide precipitates readily dissolve in HCl(aq). For example,

$$Fe(OH)_3(s) + 3\ H^+(aq) \longrightarrow Fe^{3+}(aq) + 3\ H_2O$$

Oxidation–Reduction Reactions. Oxidation and reduction enter into the qualitative analysis scheme in several ways. In the analysis of group 2 it is necessary at one point to dissolve several metal sulfides. This is accomplished with $HNO_3(aq)$, an *oxidizing* agent. Sulfide ion is oxidized to elemental sulfur and NO_3^- is reduced to NO(g). Because of this gas formation the reaction goes to completion. For example,

$$3\ PbS(s) + 8\ H^+(aq) + 2\ NO_3^-(aq) \longrightarrow$$
$$3\ Pb^{2+}(aq) + 3\ S(s) + 2\ NO(g) + 4\ H_2O$$

Another use of oxidation–reduction reactions is to obtain an ion in the proper oxidation state for further testing. In the group 2 analysis any Sn^{2+} present must be oxidized to Sn^{4+}, so that the sulfide precipitate obtained is SnS_2 and not SnS. Hydrogen peroxide, H_2O_2, is used as the oxidizing agent.

$$Sn^{2+}(aq) + H_2O_2(aq) + 2\ H^+(aq) \longrightarrow Sn^{4+}(aq) + 2\ H_2O$$

Summary

Substances in aqueous solution can be categorized as nonelectrolytes, weak electrolytes, or strong electrolytes, depending on the extent to which they produce ions. In solutions of strong electrolytes the solute is dissociated into ions, and the concentration of the solution can be expressed in terms of these ions.

Chemical reactions in solution often involve interactions between ions. A chemical equation that focuses solely on the species participating in a chemical reaction in solution is called a *net ionic equation*. One type of ionic reaction involves the combination of ions to yield water-insoluble solids—precipitates. To predict whether precipitates will form when certain ions are brought together requires some knowledge of which substances are water soluble and which are water insoluble. This information is most easily organized into the form of a small number of solubility rules (see Table 5-3).

Another type of ionic reaction occurs because of the combination of H^+ and OH^- ions to form H_2O (HOH). The source of H^+ is called an acid and the source of OH^-, a base. The reaction is an acid–base reaction, or more specifically a *neutralization* reaction. The acids and bases may be completely ionized, in which case they are called *strong*. Or, they may be only partially ionized. Then they are called *weak* acids and bases. In addition to H_2O, the other product of a neutralization reaction is called a *salt*. The most common method for conducting a neutralization reaction is by the technique known as *titration*. Titration data can be used in the types of stoichiometric calculations introduced in Chapter 4 to establish the molarities of acid and base solutions or to provide other information about the compositions of samples being analyzed.

The essential nature of an *oxidation–reduction (redox)* reaction is most readily seen by separating the net reaction into two half-reactions. In the oxidation half-reaction, certain atoms undergo an increase in oxidation state, and electrons appear on the right side of the half-equation. In the reduction half-reaction, the oxidation states of certain atoms decrease, and electrons appear on the left side of the half-equation. In a net oxidation–reduction equation, the same number of electrons must appear in each half-equation. This requirement is the basis of the half-reaction method of balancing oxidation–reduction equations. A second method is based on the changes in oxidation states that occur (oxidation state change method). To facilitate the discussion of oxidation–reduction reactions it is helpful to use the terms oxidizing and reducing agent. The oxidizing agent makes possible the oxidation process by itself becoming *reduced*. The reducing agent makes possible the reduction and is *oxidized*.

The types of reactions presented in this chapter find extensive application in *qualitative analysis*. In the qualitative analysis scheme for cations, the cations are first separated into groups based on differing solubilities of their compounds. Further separation and testing of ions is then done within each group.

Summarizing Example

Sodium dithionite, $Na_2S_2O_4$, is an important reducing agent used in the textile, pulp and paper, and ceramics industries. One interesting use is the reduction of chromate ion to insoluble chromium(III) hydroxide by dithionite ion, $S_2O_4^{2-}$, in basic solution. Sulfite ion is another product. The chromate ion may be present in wastewater from a chrome plating plant, for example.

White solid sodium dithionite, $Na_2S_2O_4$, is added to a yellow solution of potassium chromate, $K_2CrO_4(aq)$ (left). A product of the reaction is green chromium(III) hydroxide, $Cr(OH)_3(s)$ (right). [Carey B. Van Loon]

1. Write an ionic expression representing the reaction of chromate and dithionite ions describe above.

Solution. Our task is to relate names and formulas by the means described in Chapter 3. Chromate ion is CrO_4^{2-} and sulfite ion is SO_3^{2-}. The formula for chromium(III) hydroxide is $Cr(OH)_3$. Since the reaction occurs in basic solution, OH^- must also appear. (H_2O may also be required but this cannot be determined until the expression is balanced.)

$$CrO_4^{2-}(aq) + S_2O_4^{2-}(aq) + OH^- \longrightarrow Cr(OH)_3(s) + SO_3^{2-}(aq)$$

(This example is similar to Example 4-3.)

2. Use the half-reaction method to balance the expression written in Part **1.**

Solution. Write the dithionite–sulfite skeleton half-equation and balance the S atoms.

$$S_2O_4^{2-} \longrightarrow 2\ SO_3^{2-}$$

To balance O atoms, add 2 H_2O to the left,

$$S_2O_4^{2-} + 2\ H_2O \longrightarrow 2\ SO_3^{2-}$$

and to balance H atoms, add 4 H^+ to the right.

$$S_2O_4^{2-} + 2\ H_2O \longrightarrow 2\ SO_3^{2-} + 4\ H^+$$

To balance electric charge, add 2 e^- to the right. (This is an oxidation half-reaction.)

Oxidation: $S_2O_4^{2-} + 2\ H_2O \longrightarrow 2\ SO_3^{2-} + 4\ H^+ + 2\ e^-$

The second (reduction) half-equation is

$$CrO_4^{2-} \longrightarrow Cr(OH)_3(s)$$

To balance O atoms, add H_2O to the right,

$$CrO_4^{2-} \longrightarrow Cr(OH)_3(s) + H_2O$$

and to balance H atoms, add 5 H^+ to the left.

$$CrO_4^{2-} + 5\ H^+ \longrightarrow Cr(OH)_3(s) + H_2O$$

To balance electric charge, add 3 e^- to the left.

Reduction: $CrO_4^{2-} + 5\ H^+ + 3\ e^- \longrightarrow Cr(OH)_3(s) + H_2O$

Multiply the oxidation half-equation by 3 and the reduction half-equation by 2. Add the half-equations.

$$3\ S_2O_4^{2-} + 6\ H_2O \longrightarrow 6\ SO_3^{2-} + 12\ H^+ + 6\ e^-$$
$$2\ CrO_4^{2-} + 10\ H^+ + 6\ e^- \longrightarrow 2\ Cr(OH)_3(s) + 2\ H_2O$$

Net rxn: $3\ S_2O_4^{2-} + 2\ CrO_4^{2-} + 4\ H_2O \longrightarrow$
$$6\ SO_3^{2-} + 2\ Cr(OH)_3(s) + 2\ H^+$$

Add 2 OH^- to each side, and combine 2 H^+ and 2 OH^- into 2 H_2O on the right side.

$$3\ S_2O_4^{2-} + 2\ CrO_4^{2-} + 4\ H_2O + \textbf{2\ OH}^- \longrightarrow$$
$$6\ SO_3^{2-} + 2\ Cr(OH)_3(s) + \textbf{2\ H}_2\textbf{O}$$

Subtract 2 H_2O from each side to obtain the final balanced equation.

$$3\ S_2O_4^{2-} + 2\ CrO_4^{2-} + 2\ H_2O + 2\ OH^- \longrightarrow 6\ SO_3^{2-} + 2\ Cr(OH)_3(s)$$

(This example is similar to Example 5-8.)

3. What mass of $Na_2S_2O_4$ is consumed in reacting with 100.0 L of wastewater that is 0.0148 M in CrO_4^{2-}?

Solution. Determine the no. mol CrO_4^{2-} that reacts.

$$\text{no. mol } CrO_4^{2-} = 100.0\ L \times \frac{0.0148\ \text{mol } CrO_4^{2-}}{1\ L} = 1.48\ \text{mol } CrO_4^{2-}$$

Determine the no. mol $S_2O_4^{2-}$ that reacts.

$$\text{no. mol } S_2O_4^{2-} = 1.48\ \text{mol } CrO_4^{2-} \times \frac{3\ \text{mol } S_2O_4^{2-}}{2\ \text{mol } CrO_4^{2-}} = 2.22\ \text{mol } S_2O_4^{2-}$$

Determine the mass of $Na_2S_2O_4$.

$$\text{no. g } Na_2S_2O_4 = 2.22\ \text{mol } S_2O_4^{2-} \times \frac{1\ \text{mol } Na_2S_2O_4}{1\ \text{mol } S_2O_4^{2-}} \times \frac{174.1\ \text{g } Na_2S_2O_4}{1\ \text{mol } Na_2S_2O_4}$$

$$= 387\ \text{g } Na_2S_2O_4$$

(This example is similar to Example 5-12.)

4. Suggest an alternate means of removing chromate ion from wastewater by precipitation as an insoluble chromate.

Solution. Use the solubility rules in Table 5-3. With the exception of alkali metal chromates, most chromates are *insoluble*. A number of possibilities exist, for example

$$Zn^{2+}(aq) + CrO_4^{2-}(aq) \longrightarrow ZnCrO_4(s)$$

(This example is similar to Example 5-2.)

Key Terms

acid (5-3)
acid–base indicator (5-3)
base (5-3)
disproportionation reaction (5-6)
equivalence point (5-3)
half-reaction (5-5)
half-reaction method (5-6)
neutralization (5-3)
nonelectrolyte (5-1)

oxidation (5-5)
oxidation–reduction reaction (5-5)
oxidation-state change method (5-6)
oxidizing agent (5-7)
precipitation (5-2)
qualitative cation analysis (5-10)
reducing agent (5-7)
reduction (5-5)

standardization (5-8)
strong acid (5-3)
strong base (5-3)
strong electrolyte (5-1)
titration (5-3)
weak acid (5-3)
weak base (5-3)
weak electrolyte (5-1)

Highlighted Expressions

The fundamental nature of a neutralization reaction (5.10)
Important ideas concerning oxidation–reduction (5.22)

Balancing redox equations by the half-reaction method (5.23)
Conditions for a reaction to go to completion (5.26)

Review Problems

1. Identify each of the following substances using the terms: strong acid, weak acid, strong base, weak base, or salt. **(a)** Na_2SO_4; **(b)** KOH; **(c)** $CaCl_2$; **(d)** H_2SO_3; **(e)** HI; **(f)** HNO_2; **(g)** NH_3; **(h)** NH_4I; **(i)** $Ca(OH)_2$

2. Determine the molarity of the ion indicated in each of the following solutions. **(a)** K^+ in 0.215 M KNO_3; **(b)** NO_3^- in 0.041 M $Ca(NO_3)_2$; **(c)** Al^{3+} in 0.185 M $Al_2(SO_4)_3$; **(d)** Na^+ in 0.324 M Na_3PO_4.

3. A solution is prepared by dissolving 0.112 g of the hydrate, $Ba(OH)_2 \cdot 8H_2O$, in 225 mL of water solution. What is the molarity of OH^- in this solution?

4. A solution is 0.118 M in KCl and 0.186 M $MgCl_2$. What are the molarities of K^+, Mg^{2+}, and Cl^- in this solution?

5. How many mg MgI_2 must be added to 250.0 mL of 0.0998 M KI to produce a solution in which the molarity of I^- is 0.1000 M?

6. Complete each of the following as a *net ionic equation* by indicating whether a precipitate forms. If no reaction occurs, so state.

  (a) $Na^+ + Br^- + Pb^{2+} + 2 NO_3^- \rightarrow$
  (b) $Mg^{2+} + 2 Cl^- + Cu^{2+} + SO_4^{2-} \rightarrow$
  (c) $Fe^{3+} + 3 Cl^- + Na^+ + OH^- \rightarrow$
  (d) $Ca^{2+} + 2 NO_3^- + 2 K^+ + CO_3^{2-} \rightarrow$
  (e) $Ba^{2+} + S^{2-} + 2 Na^+ + CrO_4^{2-} \rightarrow$
  (f) $2 K^+ + S^{2-} + Mg^{2+} + 2 Cl^- \rightarrow$

7. Complete each of the following as a *net ionic equation*. If no reaction occurs, so state.

  (a) $Ba^{2+} + 2 OH^- + HC_2H_3O_2 \rightarrow$
  (b) $H^+ + Cl^- + HC_2H_3O_2 \rightarrow$
  (c) $Na^+ + HSO_4^- + K^+ + OH^- \rightarrow$
  (d) $K^+ + HCO_3^- + H^+ + I^- \rightarrow$
  (e) $Al(s) + H^+ \rightarrow$
  (f) $Ag(s) + H^+ \rightarrow$

8. What volume of 0.1572 M HCl is required exactly to neutralize 25.00 mL of 0.2185 M NaOH? [*Hint:* Write the net ionic equation for this reaction.]

9. The exact neutralization of a 10.00-mL sample of 0.06852 M H_2SO_4(aq) requires 22.19 mL of NaOH(aq). What must be the molarity of the NaOH(aq)?

$$H_2SO_4(aq) + 2 NaOH(aq) \rightarrow Na_2SO_4(aq) + 2 H_2O$$

10. What mass of MgO(s) can be dissolved in 275 mL of 3.50 M HNO_3(aq)?

$$MgO(s) + 2 H^+ \rightarrow Mg^{2+} + H_2O$$

11. Complete and balance the following half-equations, and indicate whether oxidation or reduction is involved.

  (a) $S_2O_8^{2-} \rightarrow SO_4^{2-}$
  (b) $HNO_3 \rightarrow N_2O(g)$ (acidic solution)
  (c) $Br^- \rightarrow BrO_3^-$ (acidic solution)
  (d) $NO_3^- \rightarrow NH_3$ (basic solution)

12. Balance the following equations by the half-reaction method.

  (a) $Cu(s) + H^+ + NO_3^- \rightarrow Cu^{2+} + NO(g) + H_2O$
  (b) $Zn(s) + H^+ + NO_3^- \rightarrow Zn^{2+} + NH_4^+ + H_2O$
  (c) $ClO_2 + OH^- \rightarrow ClO_3^- + Cl^- + H_2O$
  (d) $Fe_2S_3(s) + H_2O + O_2(g) \rightarrow Fe(OH)_3(s) + S(s)$

13. Balance the following by the oxidation-state change method.

  (a) $NO(g) + H_2(g) \rightarrow NH_3(g) + H_2O(g)$
  (b) $Cu(s) + H^+ + NO_3^- \rightarrow Cu^{2+} + H_2O + NO(g)$
  (c) $Zn(s) + H^+ + NO_3^- \rightarrow Zn^{2+} + H_2O + N_2O(g)$
  (d) $H_2O_2 + MnO_4^- + H^+ \rightarrow Mn^{2+} + H_2O + O_2(g)$

14. Identify the oxidizing and reducing agents in each of the reactions in Problems **(a)** 11, **(b)** 12, **(c)** 13.

15. Mn^{2+}(aq) can be determined by titration with MnO_4^-(aq).

$$Mn^{2+} + MnO_4^- + OH^- \rightarrow MnO_2(s) + H_2O \quad \text{(not balanced)}$$

A 25.00-mL sample of Mn^{2+}(aq) requires 34.77 mL of 0.05876 M $KMnO_4$(aq) for its titration. What is the molarity of the Mn^{2+}(aq)?

16. Addition of HCl(aq) to a solution containing several different cations produces a white precipitate. What conclusion can you draw from this single observation?

17. Assuming the solution volumes are additive, what is the molarity of Na^+ in the resulting 0.350-L solution pictured?

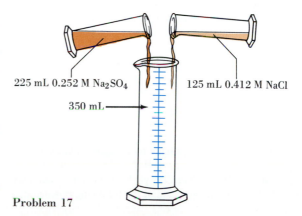

225 mL 0.252 M Na_2SO_4 125 mL 0.412 M NaCl

350 mL

Problem 17

Exercises

Strong electrolytes, weak electrolytes, and nonelectrolytes

18. None of the following substances is listed in Table 5-1; but using information from that table and elsewhere in the chapter, indicate whether you expect each of the following substances in water solution to be a nonelectrolyte, weak electrolyte, or strong electrolyte. Explain. **(a)** $HC_7H_5O_2$; **(b)** Cs_2SO_4; **(c)** $CaCl_2$; **(d)** $(C_2H_5)_2O$; **(e)** $H_2C_3H_2O_4$

19. NH_3(aq) conducts electric current only weakly. The same is true for $HC_2H_3O_2$(aq). When these solutions are mixed, however, the resulting solution conducts electric current very well. Propose a plausible explanation.

Ion molarities

20. Which of the following solutions has the highest concentration of Na^+? **(a)** 0.124 M Na_2SO_4; **(b)** a solution containing 1.50 g NaCl/100 mL; **(c)** a solution having 13.4 mg Na^+/mL.

21. Refer to Example 5-1. These data are given for several cations in seawater. Express them as molarity concentrations. **(a)** Ca^{2+} = 400. mg/L; **(b)** K^+ = 380. mg/L; **(c)** Zn^{2+} = 0.01 mg/L.

22. If the 0.350 L of solution described in Problem 17 were diluted to 0.500 L, what would be the final molarities of **(a)** Na^+, **(b)** Cl^-, and **(c)** SO_4^{2-}?

23. You have available a solution that is 0.0250 M $Ba(OH)_2$ and the following pieces of equipment: 1.00-, 5.00-, 10.00-, 25.00-, and 50.00-mL pipets and 100.0-, 250.0-, 500.0-, and 1000.0-mL volumetric flasks. Describe how you would use these materials to produce a solution that has a molarity of OH^- of 0.0100 M.

Writing ionic equations

24. Predict whether a reaction is likely to occur in each of the following cases. If so, write a net ionic equation.
- **(a)** $NaI(aq) + ZnSO_4(aq) \rightarrow$
- **(b)** $CuSO_4(aq) + Na_2CO_3(aq) \rightarrow$
- **(c)** $ZnO(s) + HCl(aq) \rightarrow$
- **(d)** $AgNO_3(aq) + CuCl_2(aq) \rightarrow$
- **(e)** $BaS(aq) + CuSO_4(aq) \rightarrow$
- **(f)** $Al(OH)_3(s) + HCl(aq) \rightarrow$
- **(g)** $NH_3(aq) + H_2SO_4(aq) \rightarrow$

25. In the chapter we described an acid as a substance capable of producing H^+ and a salt as the ionic compound formed in the neutralization of an acid by a base. Write ionic equations to show that sodium hydrogen sulfate has both the characteristics of an acid and of a salt (it is sometimes called an *acid salt*).

26. Sodium metal has varied uses, ranging from lamps for street and highway lighting to a heat transfer medium in breeder nuclear reactors. Write net ionic equations for this hypothetical series of reactions that start with sodium metal.
- **(a)** Sodium reacts vigorously with water (sometimes with explosive violence), producing sodium hydroxide and hydrogen gas.
- **(b)** To the solution produced in part (a) is added an excess of $FeCl_3(aq)$.
- **(c)** To the precipitate formed in part (b) is added an excess of $HCl(aq)$.

27. Every antacid contains one or more ingredients capable of reacting with excess stomach acid (HCl). The essential neutralization products are H_2O and/or $CO_2(g)$. Write net ionic equations to represent the neutralizing action of the following popular antacids.
- **(a)** Alka-Seltzer (sodium bicarbonate)
- **(b)** Tums (calcium carbonate)
- **(c)** Milk of Magnesia (magnesium hydroxide)
- **(d)** Maalox (magnesium hydroxide; aluminum hydroxide)
- **(e)** Rolaids [$AlNa(OH)_2CO_3$]

28. You are provided with the following: NaOH(aq), K_2SO_4(aq), $Mg(NO_3)_2$(aq), $BaCl_2$(aq), $AgNO_3$(aq), $Ca(NO_3)_2$(aq). Write ionic equations to show how you would use these reagents to obtain the following: **(a)** $CaSO_4$(s); **(b)** $Mg(OH)_2$(s); **(c)** KCl(aq); **(d)** $Ba(NO_3)_2$(aq).

Neutralization and acid–base titrations

29. Household ammonia, used as a window cleaner and for other cleaning purposes, is NH_3(aq). 31.08 mL of 0.9928 M HCl(aq) is required to neutralize the NH_3 present in a 5.00-mL sample; the net ionic equation for the neutralization is NH_3(aq) + H^+(aq) $\rightarrow$ NH_4^+(aq).
- **(a)** What is the molarity of NH_3 in the sample?
- **(b)** Assuming a density of 0.96 g/mL for the NH_3(aq), what is its mass percent NH_3?

30. For use in titrations it is necessary to prepare 20. L of HCl(aq) of a concentration known to *four* significant figures. A two-step procedure is used. First, a solution having a concentration of about 0.25 M HCl is prepared by dilution of concentrated HCl(aq). Then, a sample of the diluted HCl(aq) is titrated with an NaOH(aq) solution of known molarity. The molarity of the HCl(aq) is calculated from the titration data.
- **(a)** How many mL of concentrated HCl(aq) (d = 1.19 g/cm³; 38% HCl, by mass) must be diluted to 20. L with water to prepare 0.25 M HCl?
- **(b)** A 25.00-mL sample of the approximately 0.25 M HCl prepared in part (a) requires 29.87 mL of 0.2029 M NaOH for its titration. What is the molarity of the HCl(aq)?
- **(c)** Why is a titration necessary? That is, why was it not possible to prepare the final solution simply by an appropriate dilution of the concentrated HCl(aq)?

31. 25.00 mL of 0.144 M HNO_3 and 10.00 mL of 0.408 M KOH are mixed. Is the resulting solution acidic, basic, or exactly neutralized? [*Hint:* Which is the limiting reagent?]

32. A sample of battery acid is to be analyzed for its sulfuric acid content. A 1.00-mL sample weighs 1.239 g. This 1.00-mL sample is diluted to 250.0 mL and 10.00 mL of this diluted acid requires 32.44 mL of 0.00498 M $Ba(OH)_2$ for its titration. What is the mass percent of H_2SO_4 in the battery acid? [*Hint:* Assume that complete neutralization of the H_2SO_4 occurs, that is, that both H atoms are ionized and neutralized. What is the neutralization reaction?]

33. Refer to Example 5-5. Suppose the analysis of all vinegar samples uses 5.00 mL of the vinegar and 0.1000 M NaOH for the titration. What volume of the 0.1000 M NaOH would represent the legal 4.0%, by mass, acetic acid content of the vinegar? That is, calculate the volume of 0.1000 M NaOH such that if a titration requires more than this volume the legal limit is met (less than this volume and the limit is not met).

Completeness of reactions

34. Which of the following reactions would you expect to go to completion and which to reach a condition of equilibrium? Explain.
- **(a)** $FeS(s) + 2 H^+(aq) \rightarrow Fe^{2+}(aq) + H_2S(g)$
- **(b)** $2 NO(g) + O_2(g) \rightarrow 2 NO_2(g)$
- **(c)** $2 Na^+(aq) + SO_4^{2-}(aq) + Ba^{2+}(aq) + 2 Cl^-(aq) \rightarrow BaSO_4(s) + 2 Na^+(aq) + 2 Cl^-(aq)$
- **(d)** $Cl_2(aq) + H_2O(l) \rightarrow H^+(aq) + Cl^-(aq) + HOCl(aq)$

Oxidation–reduction equations

35. Balance, by the half-reaction method, the following equations for oxidation–reduction reactions occurring in *acidic* solution.

(a) $MnO_4^- + I^- + H^+ \rightarrow Mn^{2+} + I_2(s) + H_2O$

(b) $Cl_2 + I^- + H_2O \rightarrow IO_3^- + Cl^- + H^+$

(c) $BrO_3^- + N_2H_4 \rightarrow Br^- + N_2 + H_2O$

(d) $VO_4^{3-} + Fe^{2+} + H^+ \rightarrow VO^{2+} + Fe^{3+} + H_2O$

(e) $UO^{2+} + Cr_2O_7^{2-} + H^+ \rightarrow UO_2^{2+} + Cr^{3+} + H_2O$

36. Balance, by the half-reaction method, the following equations for oxidation–reduction reactions occurring in *basic* solution.

(a) $CN^- + MnO_4^- + OH^- \rightarrow MnO_2(s) + CNO^- + H_2O$

(b) $[Fe(CN)_6]^{3-} + N_2H_4 + OH^- \rightarrow$
$[Fe(CN)_6]^{4-} + N_2(g) + H_2O$

(c) $Fe(OH)_2(s) + O_2(g) + OH^- \rightarrow$
$Fe(OH)_3(s) + H_2O$

(d) $C_2H_5OH(aq) + MnO_4^- + OH^- \rightarrow$
$C_2H_3O_2^- + MnO_2(s) + H_2O$

37. Balance the following oxidation–reduction equations by the oxidation-state change method.

(a) $S_2O_3^{2-} + H_2O + Cl_2(g) \rightarrow SO_4^{2-} + Cl^- + H^+$

(b) $(NH_4)_2Cr_2O_7(s) \rightarrow Cr_2O_3(s) + N_2(g) + H_2O(g)$

(c) $P_4(s) + H^+ + NO_3^- + H_2O \rightarrow H_2PO_4^- + NO(g)$

(d) $MnO_4^- + H^+ + NO_2^- \rightarrow Mn^{2+} + NO_3^- + H_2O$

(e) $S_8(s) + OH^- \rightarrow S^{2-} + S_2O_3^{2-} + H_2O$

38. Balance the following oxidation–reduction equations either by the half-reaction or by the oxidation-state change method.

(a) $Pb(NO_3)_2(s) \rightarrow PbO(s) + NO_2(g) + O_2(g)$

(b) $S_2O_3^{2-} + MnO_4^- + H^+ \rightarrow SO_4^{2-} + Mn^{2+} + H_2O$

(c) $HS^-(aq) + HSO_3^-(aq) \rightarrow S_2O_3^{2-}(aq) + H_2O$

(d) $Fe^{3+} + (NH_3OH)^+ + H^+ \rightarrow Fe^{2+} + H_2O + N_2O(g)$

(e) $O_2^-(aq) + H_2O \rightarrow OH^-(aq) + O_2(g)$

39. Write a balanced oxidation–reduction equation for each of the following.

(a) The oxidation of $NH_3(g)$ to $NO(g)$ by $O_2(g)$. The $O_2(g)$ is reduced to $H_2O(g)$.

(b) The oxidation of nitrite ion to nitrate ion by permanganate ion, MnO_4^-, in acidic solution. (MnO_4^- ion is reduced to Mn^{2+}.)

(c) The reaction of H_2S, a reducing agent, with SO_2, an oxidizing agent. Both reactants are converted to elemental sulfur, and the other product is water.

(d) The reaction of sodium metal with hydroiodic acid.

Oxidizing and reducing agents

40. What are the oxidizing and reducing agents in the following redox reactions?

(a) $5 SO_3^{2-} + 2 MnO_4^- + 6 H^+ \rightarrow$
$5 SO_4^{2-} + 2 Mn^{2+} + 3 H_2O$

(b) $2 NO_2(g) + 7 H_2(g) \rightarrow 2 NH_3(g) + 4 H_2O(g)$

(c) $2 [Fe(CN)_6]^{4-} + H_2O_2 + 2 H^+ \rightarrow$
$2 [Fe(CN)_6]^{3-} + 2 H_2O$

41. Thiosulfate ion, $S_2O_3^{2-}$, is a reducing agent that can be oxidized to different products, depending on the strength of the oxidizing agent and other conditions. By adding H^+, H_2O, and/or OH^- as necessary, write redox equations to show the oxidation of $S_2O_3^{2-}$

(a) to $S_4O_6^{2-}$ by I_2; iodide ion is another product.

(b) to HSO_4^- by Cl_2; chloride ion is another product.

(c) to SO_4^{2-} by OCl^- in basic solution; chloride ion is another product.

Stoichiometry of oxidation-reduction reactions

42. Refer to Example 5-7. Assume that the only reducing agent present in a particular wastewater is SO_3^{2-}. If a 25.00-mL sample of this wastewater requires 34.08 mL of 0.01964 M $KMnO_4$ for its titration, what is the molarity of SO_3^{2-} in the wastewater?

43. An iron ore sample weighing 0.8765 g is dissolved in HCl(aq) and the iron is obtained as $Fe^{2+}(aq)$. This solution is then titrated with 29.43 mL of 0.04212 M $K_2Cr_2O_7(aq)$. What is the % Fe, by mass, in the ore sample?

$$6 Fe^{2+} + 14 H^+ + Cr_2O_7^{2-} \rightarrow 6 Fe^{3+} + 2 Cr^{3+} + 7 H_2O$$

44. Refer to Example 5-12. 25.8 mL of the 0.02140 M $KMnO_4$ solution is required for the titration of 5.00 mL of a saturated solution of sodium oxalate, $Na_2C_2O_4$. How many grams of $Na_2C_2O_4$ would be present in 1.00 L of this saturated solution? [*Hint:* First you must balance the equation for the titration reaction.]

$$C_2O_4^{2-} + MnO_4^- + H^+ \rightarrow Mn^{2+} + H_2O + CO_2(g)$$

45. Refer to the Summarizing Example. In the treatment of 1.00×10^2 L of a wastewater solution that is 0.015 M in CrO_4^{2-}

(a) How many grams of $Cr(OH)_3(s)$ would precipitate?

(b) How many grams of $Na_2S_2O_4$ would be consumed?

★46. Manganese is used in the steelmaking process to produce a tough steel for use in earth-moving machinery. It is derived from pyrolusite ore, an impure form of manganese dioxide, MnO_2. In addition MnO_2 is used in dry-cell batteries and as a decolorizer for glass. To analyze a pyrolusite ore for its MnO_2 content the following procedure is used. A 0.533-g sample is treated with 1.651 g of oxalic acid ($H_2C_2O_4 \cdot 2H_2O$) in an acidic medium. Following this reaction the *excess* oxalic acid is titrated with 0.1000 M $KMnO_4$, 30.06 mL being required. What is the % MnO_2, by mass, in this ore sample? The *unbalanced* equations are

$$H_2C_2O_4 + MnO_2 + H^+ \rightarrow Mn^{2+} + H_2O + CO_2$$

$$H_2C_2O_4 + MnO_4^- + H^+ \rightarrow Mn^{2+} + H_2O + CO_2$$

★47. Chile saltpeter is a natural source of $NaNO_3$ which also contains $NaIO_3$. The $NaIO_3$ can be used as a source of iodine. (Iodine is used in the manufacture of dyes, as an industrial catalyst, and as an antiseptic and germicide.) Iodine is produced from sodium iodate by the following two-step process.

$$IO_3^- + HSO_3^- \rightarrow I^- + H^+ + SO_4^{2-}$$

$$I^- + IO_3^- + H^+ \rightarrow I_2(s) + H_2O$$

(a) Balance the above redox equations.

(b) 1.00 L of the starting solution, which contains 5.80 g $NaIO_3/L$, is treated with the stoichiometric quantity of $NaHSO_3$ (no excess of either reactant). Then a further quantity of the starting solution is added to the reaction mixture to bring about the second reaction. How many grams of $NaHSO_3$ are required in step (1), and what additional volume of the starting solution must be added in step (2)? The process is pictured on the next page.

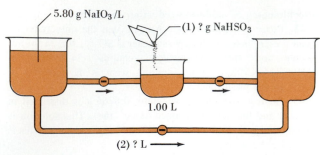

Exercise 47

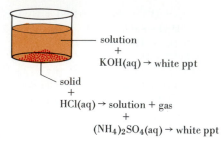

Exercise 49

Qualitative analysis

48. What reagent solution (including water) would you use to separate the cations in the following pairs, that is, with one appearing in solution and the other in a precipitate? [*Hint:* Refer to Table 5-3.]

   (a) $BaCl_2(s)$ and $NaCl(s)$
   (b) $MgCO_3(s)$ and $Na_2CO_3(s)$
   (c) $AgNO_3(s)$ and $KNO_3(s)$
   (d) $PbSO_4(s)$ and $Cu(NO_3)_2(s)$

49. A *white* solid unknown consists of *two* compounds, each containing a different cation. As suggested in the illustration, the unknown is partially soluble in water. The solution is treated with $KOH(aq)$ and yields a *white* precipitate. The part of the original solid unknown that is insoluble in water dissolves in $HCl(aq)$ with the evolution of a *gas*. This cation-containing $HCl(aq)$ is then treated with $(NH_4)_2SO_4(aq)$ and produces a *white* precipitate. State whether it is *possible* that any of the following cations were present in the original unknown? Explain your reasoning. **(a)** Mg^{2+}; **(b)** Cu^{2+}; **(c)** Ba^{2+}; **(d)** Na^+; **(e)** NH_4^+.

★50. Refer to Exercise 49. What might the compounds in the unknown mixture possibly be? (That is, what anions might be present?)

★51. Describe a qualitative test for NH_4^+ that should work no matter what other cations are present in a sample.

Additional Exercises

52. How many mL of 0.248 M $CaCl_2$ must be added to 335 mL of 0.186 M KCl to produce a solution with a concentration of 0.250 M Cl^-?

53. Assuming the volumes are additive, what is the molarity of NO_3^- in a solution obtained by mixing 315 mL of 0.211 M KNO_3, 625 mL of 0.570 M $Mg(NO_3)_2$, and 725 mL of H_2O?

54. Which of the following 0.010 M solutions would have the highest molarity of H^+? Explain. **(a)** $HC_2H_3O_2$; **(b)** HCl; **(c)** H_2SO_4; **(d)** NH_3

55. A handbook lists the solubility of barium hydroxide as 39 g $Ba(OH)_2 \cdot 8H_2O$ per liter of solution. Is it possible to make up a solution of $Ba(OH)_2$ having a concentration of OH^- equal to 0.1000 M?

56. Refer to Example 5-4. If the neutralized solution is evaporated to dryness, what mass of $Sr(NO_3)_2$ should be obtained?

57. What reagent solution might you use to separate the cations in the following pairs, that is, with one ion appearing in solution and the other in a precipitate? [*Hint:* Refer to Table 5-3, and consider water also to be a reagent.] **(a)** $MgCl_2(s)$ and NaCl(s); **(b)** $MgCO_3(s)$ and $NaHCO_3(s)$; **(c)** $AgNO_3(s)$ and $KCl(s)$; **(d)** $PbSO_4(s)$ and $CuCO_3(s)$.

58. A NaOH(aq) solution cannot be made up to an exact concentration simply by weighing out the required mass of NaOH. (The NaOH is not pure, and water vapor condenses on the solid as it is being weighed.) The solution must be standardized by titration. For this purpose a 25.00-mL sample of an NaOH(aq) solution requires 26.27 mL of 0.1107 M HCl. What is the molarity of the NaOH(aq)?

59. It is desired to determine the acetylsalicyclic acid content of a series of aspirin tablets by titration with NaOH(aq).

$$HC_9H_7O_4(aq) + OH^-(aq) \rightarrow C_9H_7O_4^-(aq) + H_2O(l)$$

Each of the tablets is expected to contain about 0.32 g of $HC_9H_7O_4$. What molarity NaOH should be used if titration volumes of about 23 mL are desired? (This procedure ensures good precision and allows the titration of two samples with the contents of a 50-mL buret.)

★60. When concentrated $CaCl_2(aq)$ is added to $Na_2HPO_4(aq)$, a white precipitate forms that is 38.7% Ca, by mass. Write a net ionic equation to represent the reaction that probably occurs. [*Hint:* What is the probable calcium-containing product?]

61. Indicate which of the following reactions you would expect to go to completion. [*Hint:* First write a net ionic equation for each reaction.]

   (a) The reaction of zinc metal with sulfuric acid, producing aqueous zinc sulfate and hydrogen gas.
   (b) The reaction of nitrogen and oxygen gases to form nitrogen monoxide gas.
   (c) The reaction of copper(II) carbonate with hydrochloric acid, forming copper(II) chloride, water and carbon dioxide gas.
   (d) The reaction of solid barium carbonate and aqueous sodium sulfate, producing solid barium sulfate and aqueous sodium carbonate.

62. Use the half-reaction method to balance the following equations for *disproportionation* reactions.

   (a) $Br_2(l) + OH^- \rightarrow Br^- + BrO_3^- + H_2O$
   (b) $S_2O_4^{2-} + H_2O \rightarrow S_2O_3^{2-} + HSO_3^-$
   (c) $MnO_4^{2-} + H_2O \rightarrow MnO_2(s) + MnO_4^- + OH^-$

63. Balance the following oxidation–reduction equations by an appropriate method. [*Hint:* Add H^+ (or OH^-) and/or H_2O as necessary.]

   (a) $Fe_2S_3(s) + H_2O + O_2(g) \rightarrow Fe(OH)_3(s) + S(s)$
   (b) $IBr + BrO_3^- + H^+ \rightarrow IO_3^- + Br^- + H_2O$
   (c) $As_2S_3 + OH^- + H_2O_2 \rightarrow AsO_4^{3-} + H_2O + SO_4^{2-}$

(d) $CrI_3(s) + H_2O_2 + OH^- \longrightarrow CrO_4^{2-} + IO_4^- + H_2O$

(e) $F_5SeOF + OH^- \rightarrow SeO_4^{2-} + F^- + O_2(g) + H_2O$

(f) $Ag(s) + CN^- + O_2(g) + OH^- \rightarrow$
$[Ag(CN)_2]^- + H_2O$

(g) $B_2Cl_4 + OH^- \rightarrow BO_2^- + Cl^- + H_2O + H_2(g)$

(h) $C_2H_5NO_3 + Sn + H^+ \rightarrow$
$NH_2OH + C_2H_5OH + Sn^{2+} + H_2O$

★ **(i)** $As_2S_3 + H^+ + NO_3^- + H_2O \rightarrow$
$H_3AsO_4 + S(s) + NO(g)$

64. In the procedure pictured, a piece of marble (assumed to be pure $CaCO_3$) dissolves completely in 2.00 L of 2.52 M HCl. After dissolution of the marble, a 10.00-mL sample of the remaining HCl(aq) is withdrawn, added to some water, and titrated with 24.87 mL of 0.9987 M NaOH.

(a) Write a net ionic equation for the dissolving of the marble.

(b) What must have been the mass of the piece of marble?

(c) Explain why the mass of the marble can only be stated with *one* significant figure, even though the titration data are given to *four* significant figures.

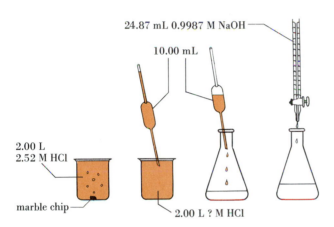

24.87 mL 0.9987 M NaOH

10.00 mL

2.00 L
2.52 M HCl

marble chip

2.00 L ? M HCl

Exercise 64

65. Refer to Example 5-11. In this example we saw that H_2O_2 can act as an oxidizing agent in some reactions and as a reducing agent in others. Write an equation for a reaction (disproportionation) in which H_2O_2 is both the oxidizing and the reducing agent. [*Hint:* What are the expected oxidation and reduction products?]

66. Refer to Example 5-8. An alternate method of producing phosphine, PH_3, from elemental phosphorus, P_4, involves heating the P_4 with H_2O. The other product is phosphoric acid, H_3PO_4. Write a balanced oxidation–reduction equation for this reaction.

67. Iron pyrites (e.g., FeS_2) are waste products in the mining of coal. When these pyrites are exposed to the environment over long periods of time oxidation of the sulfur to sulfuric acid oc-

curs. As water percolates through these wastes it leaches out the sulfuric acid and contributes to the environmental problem known as *acid mine drainage*. Balance the following redox equations describing the reactions that occur.

(a) $FeS_2(s) + O_2 + H_2O \rightarrow Fe^{3+} + SO_4^{2-} + H^+$

(b) $FeS_2(s) + Fe^{3+} + H_2O \rightarrow Fe^{2+} + SO_4^{2-} + H^+$

68. Following are some laboratory methods that are occasionally used for the preparation of small quantities of chemicals. Write balanced equations for each.

(a) Preparation of $H_2S(g)$: HCl(aq) is added to FeS(s)

(b) Preparation of $Cl_2(g)$: HCl(aq) is added to $MnO_2(s)$; $MnCl_2(aq)$ and H_2O are other products.

(c) Preparation of N_2: Br_2 and NH_3 react in aqueous solution; NH_4Br is another product.

(d) Preparation of chlorous acid: an aqueous suspension of solid barium chlorite is treated with dilute $H_2SO_4(aq)$. [*Hint:* What is the other probable product in addition to chlorous acid?]

69. The reaction of potassium dichromate and hydrochloric acid can be used as a laboratory method of preparing small quantities of $Cl_2(g)$. If a 58.9-g sample that is 97.3% $K_2Cr_2O_7$, by mass, is allowed to react with 345 mL of HCl(aq) with a density of 1.15 g/cm^3 and 30.1% HCl, by mass, how many g $Cl_2(g)$ are produced? [*Hint:* Balance the redox equation. Which reactant is in excess?]

$Cr_2O_7^{2-} + H^+ + Cl^- \rightarrow$
$\qquad\qquad Cr^{3+} + H_2O + Cl_2(g)$ (not balanced)

70. Refer to Example 5-10. If an excess of $Cl_2(g)$ is passed into 15.0 L of 2.50 M NaOH, how many kg of $NaClO_3$ could potentially be isolated from the solution? [*Hint:* Use the balanced equation from Example 5-10.]

71. Refer to Example 5-12. Suppose that the $KMnO_4(aq)$ described in this example were standardized by reaction with As_2O_3 instead of iron wire. If a 0.1256-g sample that is 99.97% As_2O_3, by mass, had been used in the titration, what volume of the $KMnO_4(aq)$ would have been required?

$As_2O_3 + MnO_4^- + H^+ + H_2O \rightarrow$
$\qquad\qquad H_3AsO_4 + Mn^{2+}$ (not balanced)

72. A new method for water treatment under development uses chlorine dioxide rather than chlorine itself. [Among the advantages of this treatment is a reduction in the quantity of halogenated organic compounds (potential carcinogens) that are formed.] One method for the production of ClO_2 involves passing $Cl_2(g)$ into a concentrated solution of sodium chlorite; NaCl(aq) is the other product. If the reaction has a 97% yield, how many moles of ClO_2 are produced per gallon of 2.0 M $NaClO_2(aq)$ treated in this way? [*Hint:* Write a net equation for the redox reaction. Also, 1 gal = 3.78 L.]

Self-Test Questions

For questions 73 through 82 select the single item that best completes each statement.

73. The *best* electrical conductor of the following aqueous solutions is (a) 0.10 M NaCl; (b) 0.10 M C_2H_5OH (ethyl alcohol); (c) 0.10 M $HC_2H_3O_2$ (acetic acid); (d) 0.10 M $C_3H_8O_3$ (glycerol).

74. The highest concentration of sulfate ion will be found in a solution that is (a) 0.10 M H_2SO_4; (b) 0.05 M Na_2SO_4; (c) 0.05 M $Al_2(SO_4)_3$; (d) 0.10 M $MgSO_4$.

75. The highest concentration of H^+ will be found in a solution that is (a) 0.10 M HCl; (b) 0.10 M H_2SO_4; (c) 0.10 M $HC_2H_3O_2$; (d) 0.10 M NH_3.

76. The number of moles of hydroxide ion, OH^-, in 0.300 L of 0.0050 M $Ba(OH)_2$ is (a) 0.0075; (b) 0.0015; (c) 0.0030; (d) 0.0050.

77. To complete the titration of 10.00 mL of 0.0500 M NaOH

$$2\ NaOH(aq) + H_2SO_4(aq) \rightarrow Na_2SO_4(aq) + 2\ H_2O$$

requires (a) 50.0 mL of 0.0100 M H_2SO_4; (b) 25.0 mL of 0.0100 M H_2SO_4; (c) 100.0 mL of 0.0100 M H_2SO_4; (d) 10.0 mL of 0.100 M H_2SO_4.

78. The only insoluble compound of the following is (a) BaS; (b) $ZnCl_2$; (c) $CuSO_4$; (d) $PbCrO_4$.

79. When treated with dilute HCl(aq), the compound that reacts to produce a gas is (a) ZnO; (b) NaBr; (c) $BaSO_3$; (d) Na_2SO_4.

80. To precipitate Ca^{2+} from an aqueous solution of $CaCl_2$ add (a) Na_2S; (b) NaI; (c) $NaNO_3$; (d) Na_2CO_3.

81. In the reaction $Cu(s) + 2\ H_2SO_4(aq) \rightarrow CuSO_4(aq) + 2\ H_2O + SO_2(g)$ (a) $H_2SO_4(aq)$ is oxidized; (b) Cu(s) is reduced; (c) $SO_2(g)$ is the reducing agent; (d) $H_2SO_4(aq)$ is reduced.

82. In the half-reaction in which NpO_2^+ is converted to Np^{4+}, the number of electrons appearing in the half-equation is (a) 1; (b) 2; (c) 3; (d) 4.

83. Balance the following oxidation–reduction equations.
(a) $S_2O_3^{2-} + H^+ + Cl_2(g) \rightarrow HSO_4^- + Cl^- + H_2O$
(b) $P(s) + H^+ + NO_3^- \rightarrow H_2PO_4^- + NO(g) + H_2O$

84. Write a balanced redox equation to represent the oxidation of PbO(s) to PbO_2(s) and the reduction of MnO_4^-(aq) to MnO_2(s), in basic solution.

85. What volume of 0.102 M $Ba(OH)_2$(aq) is required to titrate 50.00 mL of 0.0526 M HNO_3(aq)?

86. A $KMnO_4$(aq) solution is to be standardized by titration against As_2O_3(s). A 0.1156-g sample of As_2O_3 requires 27.08 mL of the $KMnO_4$(aq) for its titration. What is the molarity of the $KMnO_4$(aq)?

$$5\ As_2O_3 + 4\ MnO_4^- + 9\ H_2O + 12\ H^+ \rightarrow$$
$$10\ H_3AsO_4 + 4\ Mn^{2+}$$

87. Suppose you were given the four solids Na_2CrO_4, $BaCO_3$, MgO, and $ZnSO_4$ and the three solvents $H_2O(l)$, HCl(aq), and H_2SO_4(aq). Your task is to prepare four solutions, each containing one of the cations (i.e., one with Na^+, one with Ba^{2+}, etc.). Using water as your *first* choice, what solvent would you use to prepare each of the solutions? Explain.

6 Gases

These balloons being filled with helium gas can lift a 1300-kg payload of scientific instruments to an altitude of 35 km. [Dr. C. T. McElroy]

157

Stop to think of all that you already know, intuitively, about gases. You know (1) that you must not overinflate a bicycle tire with air and (2) that you should not dispose of an aerosol can in an incinerator. In either case the object might explode. Perhaps (3) you have traced the flow of carbon dioxide gas from a block of dry ice, watching it sink to the floor. And you probably know (4) that in warming a room with a space heater the air temperature is higher at the ceiling than at the floor. Finally, you know (5) that you should never search for a gas leak with an open flame.

After studying this chapter you should be able to relate these familiar observations to specific principles involving the properties of gases. The above observations are based, respectively, on

1. The relationship between the pressure and amount of a gas at a constant volume and temperature.
2. The relationship between the pressure and temperature of a fixed amount of gas at a constant volume.
3. The relationship of the density to the molecular weight of a gas.
4. The relationship of the density to the temperature of a gas.
5. The tendency of gases to diffuse into one another or to mix.

A description of diffusion is one of the results possible with the kinetic molecular theory of gases—a theory that we introduce here and use in several subsequent chapters of the text.

6-1   Properties of a Gas

Gases may be described in several ways. Gases expand to fill and assume the shapes of their containers. They diffuse into one another and mix in all proportions; that is, gaseous mixtures are *homogeneous* solutions. Although individual particles of a gas are invisible, some gases are colored, such as chlorine (greenish yellow), bromine (brownish red), and iodine (violet). Some gases are combustible, such as hydrogen and methane, and some are chemically inert, such as helium and neon.

In their gaseous states the halogen elements (group 7A) are distinctively colored: Cl_2 (greenish yellow); Br_2 (brownish red); I_2 (violet). Most other common gases, e.g., H_2, O_2, N_2, CO, CO_2, are colorless. [Carey B. Van Loon]

Four fundamental properties determine the physical behavior of a gas.

- amount of gas   • volume of gas   • temperature   • pressure

If we know any three of these properties, we can usually calculate a value of the remaining one, by using a mathematical equation called an **equation of state.** We see how this is done in Section 6-4. We have already discussed to some extent the properties of amount, volume, and temperature, but we need briefly to consider the notion of pressure.

6-2   The Concept of Pressure

We are all familiar with the fact that a balloon expands when it is inflated with air, but what keeps the balloon in its distended shape? A plausible hypothesis is that molecules of a gas are in constant motion, frequently colliding with one another and with the walls of their container. In their collisions the gas molecules exert a *force* on the container walls. This force keeps the balloon distended. However, it is not easy to measure the *total force* exerted by a gas. Instead of focusing on this total force, consider instead the **gas pressure. Pressure** is a *force per unit area,* that is, a force divided by the area over which the force is exerted.

$$P = \frac{F}{A}$$

(6.1)

Refer to Appendix B for a review of these fundamental physical quantities.

In SI units, force should be expressed in **newtons (N)** and area in **square meters** **(m^2)**; the corresponding pressure is in the unit called a **pascal (Pa).** Thus, a pascal is a pressure of **1 N/m^2.** A pascal is a rather small unit of pressure, and the unit **kilopascal (kPa)** is more commonly used. The pressure unit pascal honors Blaise Pascal (1623–1662), whose many contributions in physics and mathematics included ideas about pressures and the transmission of forces through liquids—the basis of modern hydraulics.

Example 6-1

Calculating a pressure from mass and geometric data. The steel cylinder pictured in Figure 6-1 has a diameter of 4.00 cm and a mass of 1.00 kg. What pressure, in N/m^2, does this cylinder exert on the surface beneath it?

Solution. Our main job is to calculate the area of contact between the cylinder and the surface beneath it. This area is that of a circle with a radius of 2.00 cm (half of the 4.00-cm diameter), but the area must be expressed in square meters.

$$\text{area} = \pi r^2 = \pi \left\{ 2.00 \text{ cm} \times \frac{1 \text{ m}}{100 \text{ cm}} \right\}^2 = 3.14 \times 4.00 \times 10^{-4} \text{ m}^2$$
$$= 1.26 \times 10^{-3} \text{ m}^2$$

Another important point to recognize is that the force exerted by the cylinder is equal not to its mass but to its *weight.* (Recall equation 1.1; weight is the force of gravity on an object, $W = gm$.) The acceleration due to gravity is $g = 9.8067$ m s^{-2}.

Force = weight = mass × g = 1.00 kg × 9.8067 m s^{-2} = 9.8067 N (newton)

Finally, pressure is force divided by area.

$$P = \frac{F}{A} = \frac{9.8067 \text{ N}}{1.26 \times 10^{-3} \text{ m}^2} = 7.78 \times 10^3 \text{ N/m}^2$$

SIMILAR EXAMPLES: Exercises 24, 73.

FIGURE 6-1
Determining a pressure—
Example 6-1 illustrated.

The pressure exerted by the metal cylinder on the surface supporting it is equal to a force (weight of the cylinder) divided by the area of contact between the cylinder and the surface. [Carey B. Van Loon]

Liquid Pressure. We usually measure the pressure of a gas *indirectly,* by comparing it to a liquid pressure. The concept of liquid pressure is illustrated in Figure 6-2 for a liquid with density d, contained in a cylinder with cross-sectional area A, filled to a height h. Equation (6.2) shows that

The pressure of a liquid depends only on the height of the liquid column and the density of the liquid.

To establish this fact, recall again that

- Weight (W) is a force.
- Weight (W) and mass (m) are proportional, with the proportionality constant being the acceleration due to gravity (g); $W = mg$.
- The mass of a liquid is equal to the product of its volume and density ($m = Vd$).

FIGURE 6-2
The concept of liquid pressure.

The pressure exerted by the liquid in the cylinder on the left is calculated in the text. Through equation (6.2), pressure is shown to depend only on the height of the liquid column *(h)* and the density of the liquid *(d)*. All of the interconnected vessels pictured here fill to the same height, and the liquid pressures are the same, despite the different shapes and volumes of the containers.

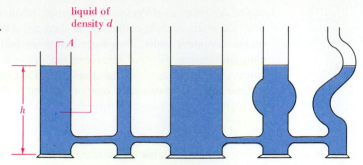

- The volume *(V)* of a cylinder is equal to the product of its height *(h)* and cross-sectional area *(A)*; $V = hA$.

$$P = \frac{F}{A} = \frac{W}{A} = \frac{mg}{A} = \frac{dVg}{A} = \frac{dAhg}{A} = ghd \qquad (6.2)$$

Measurement of Gas Pressure. In 1643 Evangelista Torricelli constructed the device pictured in Figure 6-3 to measure the pressure exerted by the atmosphere. This device is called a **mercury barometer.**

If a tube having *both ends open* is placed upright in a container of mercury (see Figure 6-3a), the mercury levels inside and outside the tube are the same. To create the situation in Figure 6-3b, first a long glass tube (say, about 1 m long) is *sealed at one end* and filled with Hg(l). Next, the open end is covered while the tube is inverted into a container of Hg(l). Then this end is reopened. The mercury level in the tube does not drop to that in the outside container. Instead, it falls to a certain height and remains there. Something must maintain the mercury at a greater height inside the tube than outside. Some early scientists tried to explain this phenomenon in terms of forces *within* the tube (e.g., an invisible thread between the mercury and the top of the tube). Torricelli understood that these forces exist *outside* the tube.

Torricelli developed the mercury barometer while working on an explanation of how a suction pump works.

In the open-end tube (Figure 6-3a) the atmosphere exerts the same pressure on the surface of the mercury both inside and outside the tube; the liquid levels are equal. With the closed-end tube (Figure 6-3b) there is no air inside the tube above the mercury (only a trace of mercury vapor). The atmosphere exerts a force on the surface of the mercury that is transmitted through the liquid, holding up mercury in the tube. The column of mercury in the tube exerts a downward pressure that depends on its height and density. For a particular height, the pressure at the bottom of the mercury column and that of the atmosphere are equal and the column is maintained.

FIGURE 6-3
Measurement of atmospheric pressure with a mercury barometer.

Arrows represent the pressure exerted by the atmosphere.
(a) The mercury levels are equal inside and outside the open-end tube.
(b) A column of mercury 760 mm high is maintained in the closed-end tube.

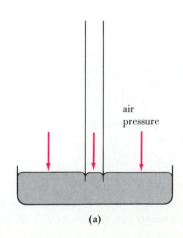

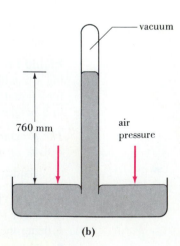

(a)

(b)

The height of mercury in a barometer is not constant but varies with atmospheric conditions and with altitude (decreasing about 3% for every 900-ft increase in altitude). The **standard atmosphere** is defined as the pressure exerted by a mercury column of exactly 760 mm height under conditions where the density of the mercury is 13.5951 g/cm³ and the acceleration due to gravity $g = 9.80665$ m s⁻². This statement relates two useful units of pressure, the standard atmosphere (**atm**) and the millimeter of mercury (**mmHg**).

Definition of a standard atmosphere.

$$1 \text{ atm} = 760 \text{ mmHg} \qquad (6.3)$$

To honor Torricelli, the pressure unit **torr** is also used. The torr is defined as exactly 1/760 of a standard atmosphere.

$$760 \text{ torr} = 1 \text{ atm}$$

Thus, the pressure units torr and mmHg can be used interchangeably.

Mercury is a relatively rare, expensive, and poisonous liquid. Why use it rather than water as the liquid in a barometer? As we see in Example 6-2, this is because of the extreme height of a water barometer.

Example 6-2

Determining the height, density, or pressure of a column of liquid. What is the height of a water column, in meters, that could be maintained by standard atmospheric pressure?

Solution. We can rephrase the question: What is the height of a column of water that exerts the same pressure as a column of mercury 76.0 cm (760 mm) high?

Whereas the pressure of the atmosphere can be measured with a mercury barometer less than 1 meter high, a water barometer would have to be as tall as a two-story building.

$$\text{pressure of Hg column} = gh_{Hg}d_{Hg} = g \times 76.0 \text{ cm} \times 13.6 \text{ g/cm}^3$$

$$\text{pressure of H}_2\text{O column} = gh_{H_2O}d_{H_2O} = g \times h_{H_2O} \times 1.00 \text{ g/cm}^3$$

$$\cancel{g} \times h_{H_2O} \times 1.00 \text{ g/cm}^3 = \cancel{g} \times 76.0 \text{ cm} \times 13.6 \text{ g/cm}^3$$

$$h_{H_2O} = 76.0 \text{ cm} \times \frac{13.6}{1.00} = 1.03 \times 10^3 \text{ cm} = 10.3 \text{ m}$$

SIMILAR EXAMPLE: Exercise 20.

Example 6-2 helps us to understand how an old-fashioned suction pump works. As shown in Figure 6-4, the pump action evacuates air from a cylindrical pipe. According to the ancient principle that "nature abhors a vacuum," Galileo Galilei (1564–1642) and others reasoned that water would be *pulled* into the evacuated space to fill it. However, according to this explanation there should be no limit to the height to which water can be raised by a suction pump. Torricelli explained that atmospheric pressure, acting on the surface of the water in the well, *pushes* a column of water up the evacuated pipe. Even if all the air in the pipe could be evacuated (which it cannot), the column of water could not be pushed higher than 10.3 m. The use of a straw for drinking liquids is based on the same principle as the suction pump.

In the laboratory, we usually measure gas pressures with a **manometer.** Figure 6-5 illustrates the principle of an open-end manometer. As long as the gas pressure being measured and the prevailing atmospheric (barometric) pressure are equal, the heights of the mercury columns in the two arms of the manometer are equal. A difference in height of the two arms means a difference between the gas pressure and barometric pressure.

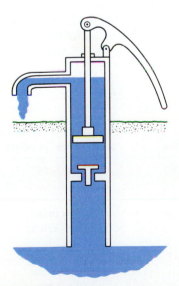

FIGURE 6-4
Pumping water by suction.

(a) Gas pressure equal to barometric pressure

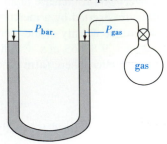

$$P_{gas} = P_{bar.}$$

(b) Gas pressure greater than barometric pressure

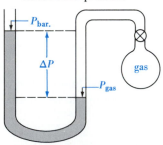

$$P_{gas} = P_{bar.} + \Delta P$$
$$(\Delta P > 0)$$

(c) Gas pressure less than barometric pressure

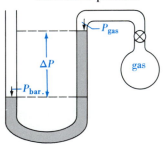

$$P_{gas} = P_{bar.} + \Delta P$$
$$(\Delta P < 0)$$

FIGURE 6-5

Measurement of gas pressure with an open-end manometer.

The possible relationships between a measured gas pressure and barometric pressure are pictured here.

Example 6-3

Using a mercury manometer to measure a gas pressure. What is the gas pressure, P_{gas}, if the conditions in Figure 6-5b are that barometric pressure is 748.2 mmHg and $\Delta P = 25.0$ mm?

Solution

$$P_{gas} = P_{bar.} + \Delta P = 748.2 \text{ mmHg} + 25.0 \text{ mmHg} = 773.2 \text{ mmHg}$$

SIMILAR EXAMPLES: Exercises 3, 71.

Example 6-4

Using a liquid other than mercury in a manometer. What is the gas pressure, P_{gas}, if the manometer in Figure 6-5c is filled with glycerol ($d = 1.26$ g/cm^3), $P_{bar.}$ is 762.4 mmHg, and ΔP is 8.2 mm?

Solution. First, we must convert ΔP, expressed as *8.2 mm of glycerol*, to an equivalent height of mercury. This is done in exactly the same way as in Example 6-2.

Density of glycerol

$$h_{Hg} = 8.2 \text{ mm} \times \frac{1.26}{13.6} = 0.76 \text{ mmHg}$$

Density of Hg

In the remaining calculation, $\Delta P = -0.76$ mmHg. (P_{gas} is less than $P_{bar.}$.)

$$P_{gas} = P_{bar.} + \Delta P = 762.4 \text{ mmHg} - 0.76 \text{ mmHg} = 761.6 \text{ mmHg}$$

SIMILAR EXAMPLES: Exercises 23, 72.

Units of Pressure: A Summary. Table 6-1 lists several different units used to express pressure. In this text we use primarily the first three units, shown in red. When we use the units mmHg and torr we substitute the height of a liquid column (mercury) for an actual pressure. Similarly, the fourth and fifth units are based on a mass (rather than a force) per unit area. The unit pound per square inch (psi) is used most commonly in engineering work. The next three units (in blue) are the ones preferred in the SI system. The last two units, bar and millibar, differ from the SI units by a factor of 100. The unit millibar is commonly used by meteorologists.

TABLE 6-1
Some Common Expressions of Pressure

Pressure unit	Abbreviation	Equivalent to standard atmospheric pressure
1. atmosphere	atm	1.000 atm
2. millimeter of mercury	mmHg	760 mmHg
3. torr	torr	760 torr
4. pound per square inch	lb/in.^{2}	14.696 lb/in.^{2}
5. kilogram per square centimeter	kg/cm^2	1.0333 kg/cm^2
6. newton per square meter	N/m^2	101,325 N/m^2
7. pascal	Pa	101,325 Pa
8. kilopascal	kPa	101.325 kPa
9. bar	bar	1.01325 bar
10. millibar	mb	1013.25 mb

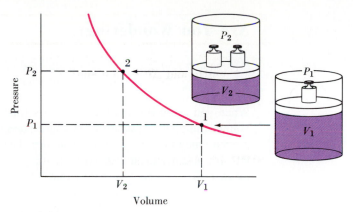

FIGURE 6-6

Relationship between gas volume and pressure—Boyle's law.

When the temperature and amount of gas are held constant, a doubling of the pressure causes the volume to decrease to one-half its original value. The situation here is like a hand-operated air pump with the needle plugged. The handle can be depressed slightly, and the air in the pump compressed to some extent. But it becomes increasingly difficult to reduce the gas volume further as more and more pressure (force per unit area) is required.

6-3 The Simple Gas Laws

Boyle's Law. In 1662 Robert Boyle discovered the first of several relationships among gas variables.

For a fixed amount of gas at a constant temperature, the gas volume is inversely proportional to the gas pressure.

Figure 6-6 should help to explain the meaning of this statement.

The gas in Figure 6-6 is contained in a cylinder closed off by a freely moving "weightless" piston. The pressure of the gas depends on the total weight placed on top of the piston. [This weight (a force), divided by the area of the piston, yields the gas pressure.] If the weight on the piston is doubled, the pressure *doubles* and the gas volume decreases to *one-half* of its original value; if the pressure of the gas is tripled, the volume decreases to *one-third*. On the other hand, if the pressure is reduced to *one-half*, the volume *doubles*, and so on.

The inverse relationship between pressure and volume is

Mathematical expressions of Boyle's law.

$$P \propto \frac{1}{V} \quad \text{or} \quad P = \frac{a}{V} \quad \text{or} \quad PV = a \quad \text{(where } a \text{ is a constant)} \qquad (6.4)$$

When we replace the proportionality sign in equation (6.4) by an equal sign, we see that the product of the pressure and volume of a fixed amount of gas at a constant temperature is a constant (a). The graph of the relationship $PV = a$, shown in Figure 6-6, is called an equilateral (or rectangular) hyperbola. By plotting P versus $1/V$, this hyperbola can be converted to a straight line passing through the origin and with a slope equal to a (see Exercise 76).

Example 6-5 ──────────────

Relating gas volume and pressure—Boyle's law. The volume of a large, irregularly shaped, closed tank is determined as follows. The tank is first evacuated, and then it is connected to a 50.0-L cylinder of compressed nitrogen gas. The gas pressure in the cylinder, originally at 21.5 atm, falls to 1.55 atm after it is

Initial condition

21.5 atm

50.0 L

Final condition

1.55 atm

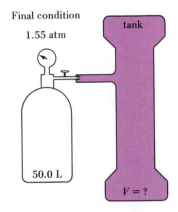

tank

50.0 L

$V = ?$

FIGURE 6-7
An application of Boyle's law—Example 6-5 visualized.

The final volume is the cylinder volume (50.0 L) plus that of the tank. The amount of gas remains constant when the cylinder is connected to the evacuated tank, but the pressure drops from 21.5 to 1.55 atm.

Are You Wondering:

If there is a simple way to tell whether the ratio of pressures (or volumes) is written correctly in a Boyle's law problem?

When you apply Boyle's law, you set a final gas volume equal to an initial volume multiplied by a ratio of pressures; or a final pressure equal to an initial pressure multiplied by a ratio of volumes. The ratio of pressures (or volumes) that you use must cause the volume (or pressure) to change in the *expected* manner. In Example 6-5 the pressure of a gas is *lowered;* the final volume must be *larger* than the initial volume. Of the two pressures given, only the ratio 21.5 atm/1.55 atm will produce a V_f that is *larger* than V_i.

$$V_f = 50.0 \text{ L} \times \frac{21.5 \text{ atm}}{1.55 \text{ atm}} = 694 \text{ L}$$

We can call this approach a "commonsense" check of the simple gas laws. We will use it on several other occasions in this chapter.

connected to the evacuated tank. What is the volume of the tank? (See Figure 6-7.)

Solution. Equation (6.4) is written for the initial condition i and the final condition, f.

$$P_i V_i = a = P_f V_f$$

Now solve for the final volume, V_f.

$$V_f = V_i \times \frac{P_i}{P_f} = 50.0 \text{ L} \times \frac{21.5 \text{ atm}}{1.55 \text{ atm}} = 694 \text{ L}$$

$$V_f = \frac{V_i \, P_i}{P_f}$$

Of this volume, 50.0 L is that of the cylinder. **The volume of the tank is 694 L − 50.0 L = 644 L.**

SIMILAR EXAMPLES: Exercises 4, 7, 25, 26, 77.

Charles's Law. The relationship between gas volume and temperature was discovered by the French physicist Jacques Charles in 1787 and, independently, by Joseph Gay-Lussac, who published it in 1802.

Figure 6-8 pictures a fixed amount of gas confined in a cylinder. The pressure is held constant while the temperature is varied. The volume of gas increases as the temperature is raised or decreases as the temperature is lowered. The relationship is linear (straight line). Three possibilities are indicated in the figure.

A common feature of the lines in Figure 6-8 is the point at which they intersect the temperature axis. Although they differ at every other temperature, the gas volumes for the three cases shown all appear to reach a value of zero at some temperature below −270 °C (actually at −273.15 °C). The temperature −273.15 °C corresponds to that at which the volume of a hypothetical gas would become zero.* This is the **absolute zero of temperature.**

If we shift the volume axis of Figure 6-8 to the left by 273.15 °C, as shown in

---

*All gases condense to liquids and solids before the temperature approaches absolute zero, and when we speak of the volume of a gas we mean the free volume among the gas molecules, not the volume of the molecules themselves. Thus, the *hypothetical* gas referred to here is one whose molecules are point masses and that does not condense to a liquid or solid.

FIGURE 6-8

Gas volume as a function of Celsius temperature.

Of many possible starting conditions, three are represented here: A, 10 cm³ of gas at 1 atm and 100 °C; B, 40 cm³ of gas at 1 atm and 200 °C; C, 100 cm³ of gas at 1 atm and 300 °C. When cooled from point C (300 °C) to about 70 °C, the gas volume decreases from 100 to 60 cm³. And, as pictured here, in the temperature interval from about 70 to −100 °C the gas volume decreases by half, i.e., from 60 to 30 cm³. The volume appears to become zero at about −270 °C.

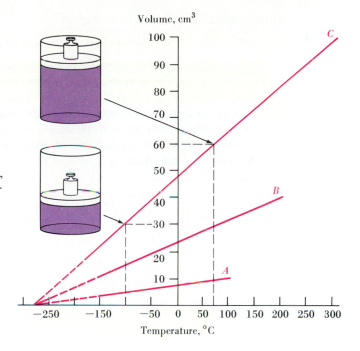

Figure 6-9, the straight lines now pass through the origin of the new axis. The origin corresponds to the hypothetical zero volume at the absolute zero of temperature. The further effect of shifting the volume axis in this way is that we must add 273.15 degrees to each temperature value. This leads to the following relationship between Celsius and **Kelvin** or **absolute** temperature.

Relationship between Kelvin and Celsius temperatures.

$$T \text{ (K)} = t \text{ (°C)} + 273.15 \tag{6.5}$$

Thus, Charles's law may be stated in this way.

The volume of a fixed amount of gas at constant pressure is directly proportional to the Kelvin (absolute) temperature.

FIGURE 6-9

Gas volume as a function of Kelvin temperature.

The vertical axis of Figure 6-8 (broken line) has been shifted 273.15° to the left. Note how the points A, B, and C, which were at 100, 200, and 300 °C in Figure 6-8, now appear at 373, 473, and 573 K, respectively.

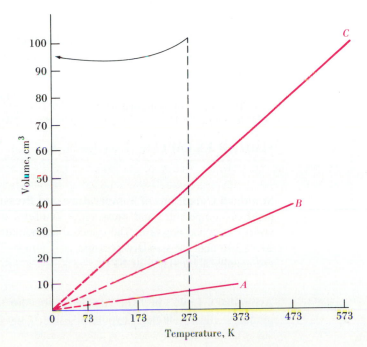

Mathematically, this may be written as

$$V \propto T \quad \text{or} \quad V = bT \quad \text{(where } b \text{ is a constant)} \tag{6.6}$$

From equation (6.6) we see that doubling the Kelvin (absolute) temperature of a gas causes its volume to double; reducing the Kelvin temperature to one-half its value (say from 300 K to 150 K) causes the volume to decrease to one-half; and so on. It is not difficult to see that the temperature used in equation (6.6) *must* be on an absolute (Kelvin) scale. After all, increasing the temperature of a gas from 1 °C to 2 °C or from 1 °F to 2 °F (in each case an apparent "doubling" of the temperature) would not cause its volume to double.

Example 6-6

Relating gas volume and temperature—Charles's law. A balloon is inflated with air in a warm living room (24 °C) to a volume of 2.50 L. Then it is taken outside on a very cold winter's day (−30 °C). Assuming that the quantity of air in the balloon and its pressure both remain constant, what will be the volume of the balloon when it is outdoors?

Solution. Since the final temperature is *lower* than the initial temperature, the final volume must also be lower than the initial volume, that is, less than 2.50 L. Write equation (6.6) twice, once for the initial condition (*i*) and once for the final condition (*f*).

$$V_i = b \times T_i \quad \text{or} \quad b = V_i/T_i$$

$$V_f = b \times T_f \quad \text{or} \quad b = V_f/T_f$$

Now equate the two expressions for the constant *b*.

$$\frac{V_i}{T_i} = b = \frac{V_f}{T_f}$$

Finally, solve the above equation for V_f, but in doing so you must remember to change temperatures to the Kelvin scale, i.e., 24 °C = 24 + 273 = 297 K and −30 °C = −30 + 273 = 243 K.

$$V_f = V_i \times \frac{T_f}{T_i} = 2.50 \text{ L} \times \frac{(273 - 30) \text{ K}}{(273 + 24) \text{ K}} = 2.50 \text{ L} \times \frac{243 \text{ K}}{297 \text{ K}} = 2.05 \text{ L}$$

As expected, the final volume is smaller than the original volume.

Alternatively, we can use a "commonsense" check to arrive at the same result. That is,

$$V_f = V_i \times \text{ratio of temperatures} = 2.50 \text{ L} \times \frac{243 \text{ K}}{297 \text{ K}} = 2.05 \text{ L}$$

SIMILAR EXAMPLES: Exercises 5, 6, 7, 78.

Standard Conditions of Temperature and Pressure. Because gas properties depend on temperature and pressure, it is useful at times to work at a particular temperature and pressure. The standard temperature for gases is defined as 0 °C = 273.15 K and the standard pressure as 1 atm = 760 mmHg. Standard conditions are usually abbreviated as **STP.**

Avogadro's Law. In 1811 Amadeo Avogadro stated a hypothesis (the "equal numbers–equal volumes" hypothesis) that is the basis of a third simple gas law.

FIGURE 6-10

Molar volume of a gas visualized.

The wooden cube has the same volume as a mole of gas at STP: 22.4 L. By contrast, the volume of air in the basketball is 7.5 L, in the soccer ball 6.0 L, and in the football 4.4 L. [See F. H. Jardine, "3 Basketballs = 1 Mole of Ideal Gas at STP," *J. Chem. Educ.*, 54, 112 (1977).] [Courtesy of Arlo Harris; photograph by Carey B. Van Loon]

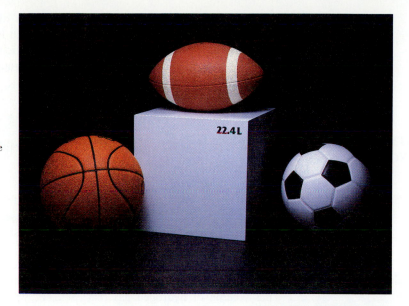

1. Equal volumes of *different* gases compared at the *same temperature and pressure* contain equal numbers of molecules.
2. Equal numbers of molecules of *different* gases compared at the *same temperature and pressure* occupy equal volumes.

The meaning of these statements, known as **Avogadro's law,** is that

At a fixed temperature and pressure, the volume of gas is directly proportional to the amount of gas (i.e., the number of molecules or number of moles of gas, n).

If the number of moles of gas is doubled, the volume doubles, and so on. (Of course, since the number of moles of a gas and its mass are also proportional, doubling the *mass* of gas will also double its volume.) A mathematical statement of this fact is

Mathematical expressions of Avogadro's law.

$$V \propto n \quad \text{and} \quad V = cn \quad \text{(where } c \text{ is a constant)} \tag{6.7}$$

At STP the number of molecules contained in 22.414 L of a gas is 6.02214×10^{23} or *1 mol*. Rounded off to three significant figures, we can express the **molar volume of a gas** through the relationship

Molar volume of a gas at STP.

$$1 \text{ mol gas} = 22.4 \text{ L gas (at STP)} \tag{6.8}$$

Figure 6-10 should help you to visualize 22.4 L of a gas.

It is important to keep in mind that Avogadro's law and statements derived from it apply *only to gases.* There is *no* similar relationship that applies to liquids or solids.

Example 6-7 ——————————

Using the molar volume of a gas at STP. What is the mass of 1.00 L of cyclopropane gas, C_3H_6 (used as an anesthetic), when measured at STP?

Solution. Expression (6.8) allows us to convert directly from volume at STP to number of moles of gas. The conversion from moles of gas to grams of gas requires the molar mass.

$$\text{no. g } C_3H_6 = 1.00 \text{ L} \times \frac{1 \text{ mol } C_3H_6}{22.414 \text{ L } C_3H_6} \times \frac{42.08 \text{ g } C_3H_6}{1 \text{ mol } C_3H_6} = 1.88 \text{ g } C_3H_6$$

SIMILAR EXAMPLES: Exercises 8, 28.

Relating the volume of a gas to the number of moles or mass of the gas is a straightforward matter if the gas is at STP, as in Example 6-7. If the gas is *not* at STP, this type of calculation becomes more difficult when based on the simple gas laws. However, with the ideal gas equation introduced in Section 6-4, all calculations of this type can be done rather easily.

CHEMISTRY EVOLVING

The Gas Laws and Development of the Atomic Theory

Avogadro's hypothesis could have been of great value in assigning atomic weights (recall Section 2-1), but for 50 years it was unused. Maybe his hypothesis was too bold for the time. In 1860 Stanislao Cannizzaro showed how Avogadro's hypothesis could be used. Here is how he reasoned.

Take the atomic weight of hydrogen to be exactly 1. Assume that hydrogen exists as diatomic molecules, H_2. The molecular weight of hydrogen is 2. Now, determine the volume of hydrogen gas that, at STP, weighs exactly 2 g. This volume is 22.4 L. Now we come to Avogadro's hypothesis. 22.4 L of some other gas at STP contains the same number of molecules as does 22.4 L H_2. The ratio of the mass of 22.4 L of this other gas to the mass of 22.4 L H_2 is the same as the ratio of their molecular weights.

Figure 6-11 illustrates this method for hydrogen and oxygen: 22.4 L of oxygen at STP weighs 32.00 g. The molecular weight of O_2 is 16 times as great as that of H_2. Assuming both gases exist as *diatomic* molecules, the atomic weight of O is 16.

The method can also be applied to gaseous *compounds*. Consider ammonia, which by chemical analysis has 82.5% N. At STP, 22.4 L of this gas weighs 17 g. The molecular weight of ammonia is 17, and the relative mass of N in the molecule is $0.825 \times 17 = 14$. Now consider nitrous oxide, which by chemical analysis has 63.7% N. At STP, 22.4 L of this gas weighs 44 g. The molecular weight of nitrous oxide is 44, and the relative mass of N in the molecule is $0.637 \times 44 = 28$. Cannizzaro's conclusion would be that the atomic weight of N is 14, and that there is one N atom per molecule of ammonia (NH_3) and two N atoms per molecule of nitrous oxide (N_2O). Can you see that we did not need to know the formulas of these compounds in order to determine the atomic weight of N? Cannizzaro resolved Dalton's "atomic weight dilemma." After 1860 the use of atomic weights became firmly established.

Amadeo Avogadro (1776–1856). Avogadro knew that his hypothesis required the existence of diatomic molecules such as H_2, O_2, and N_2. He was unable to persuade such people as Dalton and Berzelius, but Cannizzaro was more successful with the next generation of chemists. [Burndy Library]

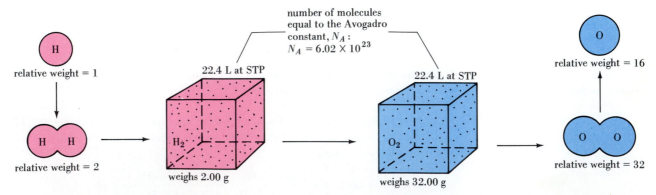

FIGURE 6-11
Cannizzaro's method illustrated.

Avogadro's hypothesis enters into the method with the statement that 22.4 L H_2 and 22.4 L O_2, both at STP, contain equal numbers of molecules.

6-4 The Ideal Gas Equation

The simple gas laws relating gas volume to pressure, temperature, and amount of gas, respectively, are stated again below.

Boyle's law: $\quad V \propto \dfrac{1}{P} \quad$ (n and T constant)

Charles's law: $\quad V \propto T \quad$ (n and P constant)

Avogadro's law: $\quad V \propto n \quad$ (P and T constant)

If all four variables are considered simultaneously, it seems reasonable that the volume of a gas should be *directly* proportional to the amount of gas and temperature and *inversely* proportional to pressure. That is,

The ideal gas equation.

$$V \propto \frac{nT}{P} \quad \text{and} \quad V = \frac{RnT}{P} \quad \text{or} \quad PV = nRT \tag{6.9}$$

Proving equation (6.9) is beyond the scope of this discussion,* but it has been amply demonstrated by experiment that any gas that obeys the three simple gas laws obeys equation (6.9) as well. Such a gas is said to be an **ideal gas** (or **perfect gas**), and equation (6.9) is known as the **ideal gas equation.** Real gases can only approach the behavior implied by the ideal gas equation, as we shall see in Section 6-10. Under suitable conditions, however, enough real gases do approach this behavior to make the equation very useful.

Before we can apply equation (6.9) to specific situations, we need a numerical value for *R,* called the **ideal gas constant.** One of the simplest ways to obtain this value is to substitute into equation (6.9) the molar volume of a gas at STP, 22.414 L.

The gas constant R.

$$R = \frac{PV}{nT} = \frac{1 \text{ atm} \times 22.414 \text{ L}}{1 \text{ mol} \times 273.15 \text{ K}} = 0.082057 \frac{\text{L atm}}{\text{mol K}} \tag{6.10}$$

A proportionality constant transforms a proportionality into an equality, but the value of the constant depends on the units used for the variables in the equation. The value of R given in (6.10) is the one that will mostly be used in this chapter (usually rounded off to 0.08206 or 0.0821). The units are "liter atmosphere per mol per K" and they can be written as in (6.10) or as **L atm mol^{-1} K^{-1}.** However, there are some applications where different units are required. R is expressed in several different units on the inside back cover.

In using the ideal gas equation, you should note that there are five terms in the equation—P, V, n, R, and T. The gas constant R is known, and if any three of the remaining four are also known, you can solve for the fourth. In the examples that follow, all five terms in the ideal gas equation are stated first (enclosed in braces). This should help you to identify the unknown. Finally, check to be certain that units cancel properly when you solve the ideal gas equation.

Example 6-8

Calculating a gas volume with the ideal gas equation. **What is the volume occupied by 13.7 g $Cl_2(g)$ at 45 °C and 745 mmHg?**

---

*Mathematically, the three proportionalities cannot simply be combined because each is stated for different conditions. For example, Boyle's law requires amount of gas and temperature to be held constant, but these are not the variables to be held constant in Charles's law. Several derivations of the ideal gas equation from the simple gas laws are possible. [See J. D. Herron, "Derivation of the Ideal Gas Law," *J. Chem. Educ.*, **56**, 315 (1979).]

Solution

$$P = 745 \text{ mmHg} \times \frac{1 \text{ atm}}{760 \text{ mmHg}} = \frac{745}{760} \text{ atm} = 0.980 \text{ atm}$$

$$V = ?$$

$$n = 13.7 \text{ g Cl}_2 \times \frac{1 \text{ mol Cl}_2}{70.91 \text{ g Cl}_2} = 0.193 \text{ mol Cl}_2$$

$$R = 0.08206 \text{ L atm mol}^{-1} \text{ K}^{-1}$$

$$T = 45 \text{ °C} + 273 = 318 \text{ K}$$

$$PV = nRT$$

Divide both sides by P.

To check the cancellation of units we have generally looked for the same unit in the numerator and the denominator of a setup, for example, atm/atm = 1. Here we need to note that a unit such as mol^{-1} is the same as 1/mol. Thus, $\text{mol} \times \text{mol}^{-1} = 1$. Also, $\text{K}^{-1} \times \text{K} = 1$.

$$\cancel{P}\frac{V}{\cancel{P}} = \frac{nRT}{P} \quad \text{and} \quad V = \frac{nRT}{P}$$

$$V = \frac{0.193 \text{ mol} \times 0.08206 \text{ L atm mol}^{-1} \text{ K}^{-1} \times 318 \text{ K}}{0.980 \text{ atm}}$$

$$= 5.14 \text{ L}$$

SIMILAR EXAMPLES: Exercises 9, 10, 31, 33, 81.

Example 6-9

Calculating an amount of gas with the ideal gas equation. Although we think of an evacuated system as one from which essentially all gas molecules have been removed, the number of remaining molecules is still considerable. In an ultrahigh vacuum system the pressure of the residual gas is reduced to 10^{-9} mmHg or less. How many molecules of N_2 remain in a system of 128-mL volume when the pressure is reduced to 5×10^{-10} mmHg at 25.0 °C?

Solution. The chief need here is to determine the number of moles of gas. For this we use the ideal gas equation. Then we can convert from number of moles to number of molecules with the Avogadro constant.

$$P = 5 \times 10^{-10} \text{ mmHg} \times \frac{1 \text{ atm}}{760 \text{ mmHg}} = 7 \times 10^{-13} \text{ atm}$$

$$V = 128 \text{ mL} \times \frac{1 \text{ L}}{1000 \text{ mL}} = 0.128 \text{ L}$$

$$n = ?$$

$$R = 0.0821 \text{ L atm mol}^{-1} \text{ K}^{-1}$$

$$T = 25 \text{ °C} + 273 = 298 \text{ K}$$

$$n = \frac{PV}{RT} = \frac{7 \times 10^{-13} \text{ atm} \times 0.128 \text{ L}}{0.0821 \text{ L atm mol}^{-1} \text{ K}^{-1} \times 298 \text{ K}} = 4 \times 10^{-15} \text{ mol}$$

$$\text{no. molecules N}_2 = 4 \times 10^{-15} \text{ mol N}_2 \times \frac{6.02 \times 10^{23} \text{ molecules N}_2}{1 \text{ mol N}_2}$$

$$= 2 \times 10^9 \text{ molecules N}_2$$

SIMILAR EXAMPLES: Exercises 32, 34.

The General Gas Law. Examples 6-8 and 6-9 involved a *single* set of conditions (P, V, n, and T), and we used the ideal gas equation *once*. Sometimes a gas is described under *two* different sets of conditions. Here we have to apply the ideal gas equation *twice*, to an *initial* condition and a *final* condition. That is,

Initial condition	Final condition
$P_iV_i = n_iRT_i$	$P_fV_f = n_fRT_f$
$\dfrac{P_iV_i}{n_iT_i} = R$	$\dfrac{P_fV_f}{n_fT_f} = R$

Since each of the above expressions is equal to the same constant, R, we can write

The general gas law.

$$\frac{P_iV_i}{n_iT_i} = \frac{P_fV_f}{n_fT_f} \tag{6.11}$$

Expression (6.11) is sometimes called the general gas law. It is used in Example 6-10 to establish the relationship between gas pressure and temperature, known as *Amonton's law*.

Example 6-10

Applying the general gas law. Pictured in Figure 6-12 is a 1.00-L flask of $O_2(g)$, first at STP and then at 100 °C. What is the pressure of the gas at 100 °C?

Solution. Identify the terms in the general gas law that *remain constant*. Cancel out these terms and solve the equation that remains. In this case the amount of $O_2(g)$ is constant ($n_i = n_f$) and the volume is constant ($V_i = V_f$).

$$\frac{P_i \cancel{V_i}}{\cancel{n_i}T_i} = \frac{P_f \cancel{V_f}}{\cancel{n_f}T_f} \qquad \text{and} \qquad \frac{P_i}{T_i} = \frac{P_f}{T_f} \qquad \text{and} \qquad P_f = P_i \times \frac{T_f}{T_i}$$

Since P_i is standard pressure = 1.00 atm,

$$P_f = 1.00 \text{ atm} \times \frac{(100 + 273) \text{ K}}{273 \text{ K}} = 1.00 \text{ atm} \times \frac{373 \text{ K}}{273 \text{ K}} = 1.37 \text{ atm}$$

Here is a "commonsense" way to check the result. Since the volume and amount of gas are held constant, an increase in temperature should produce an increase in pressure. The final pressure must be the initial pressure multiplied by a ratio of temperatures. The only ratio of temperatures that will produce this increase in pressure is 373 K/273 K. Thus, $P_f = 1.00$ atm $\times$ (373 K/273 K) = 1.37 atm.

SIMILAR EXAMPLES: Exercises 27, 29, 79.

FIGURE 6-12
Pressure of a fixed amount of gas in a fixed volume as a function of temperature—Example 6-10 visualized.

(a) 1.00 L $O_2(g)$ at STP.
(b) 1.00 L $O_2(g)$ at 100 °C.

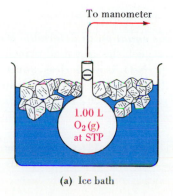

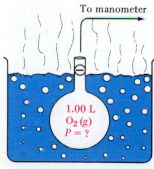

To manometer

1.00 L $O_2(g)$ at STP

(a) Ice bath

To manometer

1.00 L $O_2(g)$ $P = ?$

(b) Boiling water

Are You Wondering:

When to use the ideal gas equation, when to use the general gas law, and when to use the simple gas laws?

There are many instances in which you can use any one of the three, and your choice will probably depend on the method that seems clearest to for the particular problem. In general, however, you will want to use the *ideal gas equation* if

- you are dealing with a *single* set of conditions and three of the four gas variables (P, V, n, T) are given.

You may find the *general gas law* easiest to use if

- you need to deal with *two* sets of conditions (initial and final).

You can use the appropriate *simple gas law* if

- you have to deal only with the effect of one variable (e.g., P) on another (e.g., V).

6-5 Molecular Weight Determination

The ideal gas equation gives us a more direct approach to establishing molecular weights than does Cannizzaro's method (recall Figure 6-11). For this purpose it is helpful to alter the equation slightly. The number of moles of gas, usually expressed as n, is also equal to the mass of gas, m, divided by the molar mass, $\mathcal{M}$ (whose units are g/mol). That is, $n = m/\mathcal{M}$. The molecular weight (a dimensionless number) is numerically equal to the molar mass.

Ideal gas equation modified for molecular weight determination.

$$PV = \frac{mRT}{\mathcal{M}} \qquad (6.12)$$

To determine the molecular weight of a gas with equation (6.12) requires that we measure the volume (V) occupied by a known mass of gas (m) at a certain temperature (T) and pressure (P). The form of the ideal gas equation in (6.12) is not limited to determining molecular weights. You can use it in any application in which the quantity of gas is given or sought in grams rather than moles.

To establish that equation (6.12) is of the proper form, simply substitute the appropriate units for each term and check the cancellation of units.

$$(atm)(L) = (L)(atm)$$

This technique is known as *unit analysis* or *dimensional analysis*. It is especially useful in deriving and remembering equations.

Example 6-11

Using the ideal gas equation to determine a molecular weight. Nitric oxide, NO, is a highly toxic colorless gas used principally in the manufacture of nitric acid. Nitrous oxide, N_2O, is a colorless, sweet-tasting gas used as an anesthetic in dentistry. A 1.27-g sample of an oxide of nitrogen believed to be either NO or N_2O occupies a volume of 1.07 L at 25 °C and 737 mmHg pressure. Which oxide is it?

Solution. Use equation (6.12) to determine the molecular weight of the gas and compare this with the molecular weights of NO and N_2O.

$$
\begin{cases}
P = 737 \text{ mmHg} \times \dfrac{1 \text{ atm}}{760 \text{ mmHg}} = 0.970 \text{ atm} \\[2ex]
V = 1.07 \text{ L} \\[1ex]
m = 1.27 \text{ g} \\[1ex]
\mathcal{M} = ? \\[1ex]
R = 0.08206 \text{ L atm mol}^{-1} \text{ K}^{-1} \\[1ex]
T = 25 \,^{\circ}\text{C} + 273 = 298 \text{ K}
\end{cases}
$$

Rearrange equation (6.12) to

$$
\mathcal{M} = \frac{mRT}{PV}
$$

and substitute the known values.

$$
\mathcal{M} = \frac{1.27 \text{ g} \times 0.08206 \text{ L atm mol}^{-1} \text{ K}^{-1} \times 298 \text{ K}}{0.970 \text{ atm} \times 1.07 \text{ L}} = 29.9 \text{ g/mol}
$$

The molecular weight of the gas is 29.9. This agrees most closely with the molecular weight of NO (30.0; the mol. wt of $N_2O = 44.0$). The gas is NO.

SIMILAR EXAMPLES: Exercises 12, 37.

Example 6-12 _____

Determining a molecular weight with the ideal gas equation. Propylene is one of the most important chemicals produced (no. 10 among the top chemicals). It is used in the synthesis of other organic chemicals and in plastics production (polypropylene). A glass vessel weighs 40.1305 g when clean, dry, and evacuated; 138.2410 g when filled with water at 25.0 °C (density of water = 0.9970 g/cm^3); and 40.2959 g when filled with propylene gas at 740.4 mmHg and 24.0 °C. What is the molecular weight of propylene?

Solution. We must first determine the volume of the glass vessel (and hence the volume of the gas) and the mass of the gas.

mass of water to fill vessel = 138.2410 g − 40.1305 g = 98.1105 g

$$
\text{volume of water (volume of vessel)} = 98.1105 \text{ g H}_2\text{O} \times \frac{1 \text{ cm}^3 \text{ H}_2\text{O}}{0.9970 \text{ g H}_2\text{O}}
$$

$$
= 98.41 \text{ cm}^3 = 0.09841 \text{ L}
$$

mass of gas = 40.2959 g − 40.1305 g = 0.1654 g

temperature = 24.0 °C + 273.15 = 297.2 K

$$
\text{pressure} = 740.4 \text{ mmHg} \times \frac{1 \text{ atm}}{760.0 \text{ mmHg}} = 0.9742 \text{ atm}
$$

$$
\mathcal{M} = \frac{mRT}{PV} = \frac{0.1654 \text{ g} \times 0.08206 \text{ L atom mol}^{-1} \text{ K}^{-1} \times 297.2 \text{ K}}{0.9742 \text{ atm} \times 0.09841 \text{ L}}
$$

$$
= 42.08 \text{ g/mol}
$$

The molecular weight of propylene is 42.08.

SIMILAR EXAMPLES: Exercises 38, 39, 83.

The method of determining molecular weights outlined in Example 6-12 can be combined with an elemental analysis (e.g., combustion analysis) to yield the molecular formula of a gas. That is, if propylene is found to be 85.63% C and 14.37% H, by mass, what is its molecular formula? (See Exercise 36.)

6-6 Gas Densities

In Examples 6-11 and 6-12 we rearranged equation (6.12) to solve for molar mass, $\mathcal{M}$. A different rearrangement leads to

Ideal gas equation modified for gas density determination.

$$d = \frac{m}{V} = \frac{\mathcal{M}P}{RT} \qquad (6.13)$$

The term m/V is the mass of a gas divided by its volume—the **gas density (d).** Gas densities differ from those of solids and liquids in some important ways.

1. Gas densities are generally stated in g/L instead of g/cm^3.
2. Gas densities are strongly dependent on pressure and temperature, increasing as the gas pressure increases and decreasing as the temperature increases (see equation 6.13). Densities of liquids and solids do depend somewhat on temperature, but they are far less dependent on pressure.
3. The density of a gas is directly proportional to its molar mass. No simple relationship exists between the density and molar mass for liquids and solids.

The density of a gas at STP can be easily calculated by dividing its molar mass by the molar volume (22.414 L/mol). For $O_2(g)$ at STP, for example, the density is 32.0 g/22.4 L = 1.43 g/L. Under other conditions of temperature and pressure, equation (6.13) is required.

Example 6-13

Using the ideal gas equation to calculate a gas density. What is the density of oxygen gas (O_2) at 298 K and 0.987 atm?

Solution. We can get the terms for the right side of equation (6.13) easily. The density is simply the left side of the equation, m/V.

$$d = \frac{m}{V} = \frac{\mathcal{M}P}{RT} = \frac{32.0 \text{ g mol}^{-1} \times 0.987 \text{ atm}}{0.08206 \text{ L atm mol}^{-1} \text{ K}^{-1} \times 298 \text{ K}} = 1.29 \text{ g/L}$$

SIMILAR EXAMPLES: Exercises 13, 40, 42.

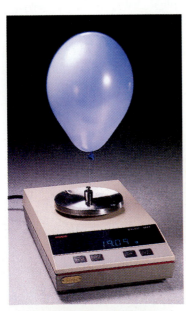

The helium-filled balloon exerts a lifting force on the 20.00-g weight, so that the balloon and weight together weigh only 19.09 g. [Carey B. Van Loon]

An important application of gas densities is in establishing conditions for lighter-than-air balloons. A gas-filled balloon will rise in the atmosphere only if the density of the gas is less than that of air. Since gas densities are directly proportional to molar masses, the lower the molecular weight of the gas the greater its lifting power. The lowest molecular weight gas is hydrogen, but hydrogen is flammable and forms explosive mixtures with air. The explosion of the dirigible *Hindenburg* in 1937 spelled the end of transoceanic travel by hydrogen-filled airships. Now, airships (such as the Goodyear blimps) use helium, which has a molar mass only twice that of hydrogen. Hydrogen continues to be used for weather and other observational balloons.

Another alternative for filling a balloon with a gas less dense than air is to fill the balloon with *hot* air. As indicated by equation (6.13), density is *inversely* proportional to temperature. The higher the temperature of the air, the lower its density.

However, because the density of the atmosphere decreases rapidly with altitude, there is a limit to the height to which balloons can rise. Some practical questions concerning the lifting of balloons are found in Exercises 42, 85, and 102.

6-7 Gases in Chemical Reactions

We now have a new tool to apply to calculations dealing with gaseous reactants and/or products of a chemical reaction—the ideal gas equation. Specifically, we can handle information about gases not only in grams and moles, but also in terms of gas volumes, temperatures, and pressures.

In many cases the best approach to stoichiometry problems involving gases is to use (a) stoichiometric factors to relate moles of a gas to moles of other reactants or products and (b) the ideal gas equation or the simple gas laws to relate the number of moles of gas to volume, temperature, and pressure. In Example 6-14, for example, we determine the number of moles of $N_2(g)$ from the stoichiometry of a reaction; then we use the ideal gas equation to determine a gas volume.

Example 6-14

Using the ideal gas equation in stoichiometric calculations based on the chemical equation. The decomposition of sodium azide, NaN_3, at high temperatures produces $N_2(g)$. Combined with the necessary devices to initiate the reaction and to trap the sodium metal produced, this reaction is used in "air-bag" safety systems in automobiles. How many liters of $N_2(g)$, measured at 735 mmHg and 26 °C, are produced when 125 g NaN_3 is decomposed?

$$2 \ NaN_3(s) \longrightarrow 2 \ Na(l) + 3 \ N_2(g)$$

An air-bag safety system being tested in a simulated car crash. [Insurance Institute for Highway Safety]

Solution

$$\text{no. mol } N_2 = 125 \text{ g NaN}_3 \times \frac{1 \text{ mol NaN}_3}{65.01 \text{ g NaN}_3} \times \frac{3 \text{ mol } N_2}{2 \text{ mol NaN}_3} = 2.88 \text{ mol } N_2$$

$$P = 735 \text{ mmHg} \times \frac{1 \text{ atm}}{760 \text{ mmHg}} = 0.967 \text{ atm}$$

$$V = ?$$

$$n = 2.88 \text{ mol}$$

$$R = 0.08206 \text{ L atm mol}^{-1} \text{ K}^{-1}$$

$$T = 26 \text{ °C} + 273 = 299 \text{ K}$$

$$V = \frac{nRT}{P} = \frac{2.88 \text{ mol} \times 0.08206 \text{ L atm mol}^{-1} \text{ K}^{-1} \times 299 \text{ K}}{0.967 \text{ atm}} = 73.1 \text{ L}$$

SIMILAR EXAMPLES: Exercises 14, 44, 45, 46, 86.

Law of Combining Volumes. Avogadro's hypothesis is useful for a second kind of calculation based on the balanced chemical equation. This relates to a situation in which either all the reactants and products are gases, or at least those involved in the particular calculation are gases. Consider this reaction.

$$2 \text{ NO(g)} + O_2(g) \longrightarrow 2 \text{ NO}_2(g)$$

$$2 \text{ mol NO(g)} + 1 \text{ mol } O_2(g) \longrightarrow 2 \text{ mol NO}_2(g)$$

Suppose the gases are compared at the same T and P. Under these conditions 1 mol of gas occupies a particular volume, call it V liters; 2 mol of gas, $2V$ liters; and so on.

$$2V \text{ L NO(g)} + V \text{ L } O_2(g) \longrightarrow 2V \text{ L NO}_2(g)$$

Now divide each coefficient by V:

$$2 \text{ L NO(g)} + 1 \text{ L } O_2(g) \longrightarrow 2 \text{ L NO}_2(g)$$

From this description of the chemical equation, we can write these statements.

$$2 \text{ L NO}_2(g) \backsimeq 2 \text{ L NO(g)}; \qquad 2 \text{ L NO(g)} \backsimeq 1 \text{ L } O_2(g); \qquad \text{and so on.}$$

What we have just done is to develop, in modern terms, Gay-Lussac's **law of combining volumes** (1808). This law is applied in Example 6-15.

Example 6-15

Illustrating the law of combining volumes. Zinc blende, ZnS, is the most important zinc ore. Roasting of ZnS is the first step in the production of Zn.

$$2 \text{ ZnS(s)} + 3 \text{ } O_2(g) \longrightarrow 2 \text{ ZnO(s)} + 2 \text{ SO}_2(g)$$

What volume of $SO_2(g)$ forms per liter of $O_2(g)$ consumed? Both gases are measured at 740 mmHg and 25 °C.

Solution. Since the reactant and product being compared are both gases, and both are measured at the same temperature and pressure, a ratio of combining volumes can be derived from the balanced equation and used as follows.

$$\text{no. L SO}_2(g) = 1.00 \text{ L } O_2(g) \times \frac{2 \text{ L SO}_2(g)}{3 \text{ L } O_2(g)} = 0.667 \text{ L SO}_2(g)$$

SIMILAR EXAMPLES: Exercises 15, 47a.

You should note carefully the following points concerning a calculation of the kind performed in Example 6-15.

1. It was not necessary to use the specific temperature (25 °C) and pressure (740 mmHg) at all. As long as a comparison is made at the *same* T and P, the relationship between volume and number of molecules (or moles) of a gas is the same for all gases.
2. If temperature and pressure are *not* identical for the gases being compared, the method of Example 6-15 will not work. Then it is best to convert information about the gases to a mole basis and to use mole rather than volume ratios.
3. If the relationship is between a *solid* (or *liquid*) and a gas, it is again necessary to use a mole ratio as a conversion factor (as we did in Example 6-14).

6-8 Mixtures of Gases

As a first approximation at least, all gases behave pretty much alike. The ideal gas equation is applicable to all gases under the appropriate conditions of temperature and pressure. As a result the ideal gas equation applies to a *mixture of gases* just as it does to a single gas (as long as the gases do not react with each other). In these applications it is only necessary to use for the value of n the *total* number of moles of molecules in the gaseous mixture ($n_{tot.}$).

Example 6-16

Applying the ideal gas equation to a mixture of gases. What is the pressure exerted by a mixture of 1.0 g H_2 and 5.0 g He when the mixture is confined to a volume of 5.0 L at 20 °C?

Solution

$$n_{tot.} = \left(1.0 \text{ g } H_2 \times \frac{1 \text{ mol } H_2}{2.02 \text{ g } H_2}\right) + \left(5.0 \text{ g He} \times \frac{1 \text{ mol He}}{4.00 \text{ g He}}\right)$$

$$= 0.50 \text{ mol } H_2 + 1.25 \text{ mol He} = 1.75 \text{ mol gas}$$

$$P = \frac{n_{tot.}RT}{V} \tag{6.14}$$

$$P = \frac{1.75 \text{ mol} \times 0.08206 \text{ L atm mol}^{-1} \text{ K}^{-1} \times 293 \text{ K}}{5.0 \text{ L}} = 8.4 \text{ atm}$$

SIMILAR EXAMPLES: Exercises 16, 51, 52, 84.

In addition to his contribution to the atomic theory, John Dalton also made an important contribution to the study of gaseous mixtures. He proposed that in a mixture of gases each gas expands to fill the container. Each gas exerts the same pressure, called its **partial pressure,** that it would if it were the only gas present in the container. *The sum of these partial pressures is equal to the total pressure of the mixture* (see Figure 6-13). For a mixture of gases, A, B, . . .

Dalton's law: Total gas pressure is the sum of partial pressures of individual gases.

$$P_{tot.} = P_A + P_B + \cdots \tag{6.15}$$

We can easily demonstrate that equations (6.14) and (6.15) are equivalent in the following way.

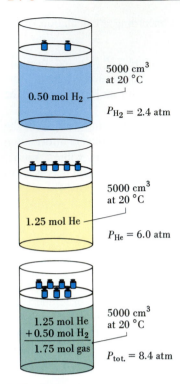

FIGURE 6-13
Dalton's law of partial pressures illustrated.

This figure indicates that the pressure of each gas is proportional to the number of moles of that gas. The total pressure of the mixture is the sum of the partial pressures of the individual gases.

Relating mole fraction, pressure fraction, and volume fraction of a gaseous mixture.

$$P_{tot.} = P_A + P_B + \cdots$$

$$= \frac{n_A RT}{V} + \frac{n_B RT}{V} + \cdots = \frac{RT}{V}(n_A + n_B + \cdots)$$

$$P_{tot.} = \frac{n_{tot.} RT}{V}$$

(where $n_{tot.} = n_A + n_B + \cdots$)

An alternative expression, known as *Amagat's law,* is useful in dealing with gaseous mixtures whose compositions are expressed in percent *by volume*. Here we begin with the expression

$$V_{tot.} = \frac{n_{tot.} RT}{P_{tot.}}$$

and again note that $n_{tot.} = n_A + n_B + \cdots$. This allows us to write

$$V_{tot.} = \frac{n_A RT}{P_{tot.}} + \frac{n_B RT}{P_{tot.}} + \cdots$$

$$= V_A + V_B + \cdots$$

The terms $V_A, V_B, \ldots$ are called partial volumes. The **partial volume** of a component in a gaseous mixture is the volume that would be occupied by that component if it existed alone at the total pressure of the mixture. *The total volume of a gaseous mixture is the sum of the partial volumes of its components.*

Still another useful expression for gaseous mixtures is obtained by taking the ratio of a partial pressure to a total pressure or a partial volume to a total volume.

$$\frac{P_A}{P_{tot.}} = \frac{n_A \dfrac{RT}{V_{tot.}}}{n_{tot.} \dfrac{RT}{V_{tot.}}} = \frac{n_A}{n_{tot.}} \quad \text{and} \quad \frac{V_A}{V_{tot.}} = \frac{n_A \dfrac{RT}{P_{tot.}}}{n_{tot.} \dfrac{RT}{P_{tot.}}} = \frac{n_A}{n_{tot.}}$$

which means that

$$\frac{n_A}{n_{tot.}} = \frac{P_A}{P_{tot.}} = \frac{V_A}{V_{tot.}} \tag{6.16}$$

The term $n_A/n_{tot.}$ is given a special name. It is the mole fraction of A in the gaseous mixture. The **mole fraction** of a component in a mixture is the number of moles of the component divided by the total number of moles in the mixture. The sum of the mole fractions of all the components in a mixture is 1.

As illustrated in Examples 6-17 (Method 2) and 6-18, expression (6.16) is an especially useful one for dealing with gaseous mixtures.

Example 6-17 ———————————

Calculating the partial pressures in a gaseous mixture—an application of Dalton's law of partial pressures. What are the partial pressures of H_2 and He in the gaseous mixture described in Example 6-16?

Solution

METHOD 1. From the number of moles of each gas and the conditions stated in Example 6-16, we may calculate the partial pressures directly.

$$P_{H_2} = \frac{n_{H_2} \cdot RT}{V} = \frac{0.50 \text{ mol} \times 0.0821 \text{ L atm mol}^{-1} \text{ K}^{-1} \times 293 \text{ K}}{5.0 \text{ L}} = 2.4 \text{ atm}$$

$$P_{He} = \frac{n_{He} \cdot RT}{V} = \frac{1.25 \text{ mol} \times 0.0821 \text{ L atm mol}^{-1} \text{ K}^{-1} \times 293 \text{ K}}{5.0 \text{ L}} = 6.0 \text{ atm}$$

As expected, these partial pressures, when added together, yield the total pressure calculated in Example 6-16—8.4 atm.

METHOD 2. Use expression (6.16), with mole fractions and total pressure obtained from Example 6-16.

$$P_{H_2} = \frac{n_{H_2}}{n_{tot.}} \times P_{tot.} = \frac{0.50}{1.75} \times 8.4 \text{ atm} = 2.4 \text{ atm}$$

$$P_{He} = \frac{n_{He}}{n_{tot.}} \times P_{tot.} = \frac{1.25}{1.75} \times 8.4 \text{ atm} = 6.0 \text{ atm}$$

SIMILAR EXAMPLE: Exercise 52.

Example 6-18

Relating partial pressures to the volume percent composition of a gaseous mixture. The major components of air, by volume, are nitrogen, 78.08%; oxygen, 20.95%; argon, 0.93%; and carbon dioxide, 0.03%. What are the partial pressures of these four gases in a sample of air at standard atmospheric pressure (1.000 atm)?

Solution. Volume percent means that in a total volume of 100.0 L of air, the partial volume of $N_2(g)$ is 78.08 L; $O_2(g)$, 20.95 L; and so on. We then substitute these values into equation (6.16).

$$P_{N_2} = \frac{V_{N_2}}{V_{tot.}} \times P_{tot.} = \frac{78.08 \text{ L}}{100.0 \text{ L}} \times 1.000 \text{ atm} = 0.7808 \text{ atm}$$

$$P_{O_2} = \frac{V_{O_2}}{V_{tot.}} \times P_{tot.} = \frac{20.95 \text{ L}}{100.0 \text{ L}} \times 1.000 \text{ atm} = 0.2095 \text{ atm}$$

$$P_{Ar} = \frac{V_{Ar}}{V_{tot.}} \times P_{tot.} = \frac{0.93 \text{ L}}{100.0 \text{ L}} \times 1.000 \text{ atm} = 0.0093 \text{ atm}$$

$$P_{CO_2} = \frac{V_{CO_2}}{V_{tot.}} \times P_{tot.} = \frac{0.03 \text{ L}}{100.0 \text{ L}} \times 1.000 \text{ atm} = 0.0003 \text{ atm}$$

SIMILAR EXAMPLES: Exercises 54, 90.

A gaseous mixture is sometimes described by its **apparent molar mass**—the mass of one mole of molecules of the gaseous mixture. The apparent molar mass can be determined by adding together the contributions of each component to the mass of one mole of the mixture.

Example 6-19

Determining the apparent molar mass of a gaseous mixture. From the data in Example 6-18 calculate the apparent molar mass of air.

Solution. The key to this calculation again lies in equation (6.16). The ratio of the number of moles of a gaseous component to the total number of moles of gas (i.e., $n_A/n_{tot.}$) is the same as the volume ratio (i.e., $V_A/V_{tot.}$). In "one mole of air," $n_{tot.} = 1.000$ and the numbers of moles of the individual gases are 0.7808 mol N_2, 0.2095 mol O_2, 0.0093 mol Ar, and 0.0003 mol CO_2. The apparent molar mass of air is

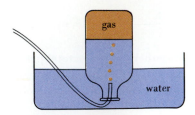

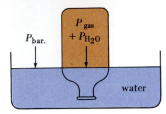

FIGURE 6-14
Collection of a gas over water.

To make the total pressure of the gaseous mixture in the bottle equal to barometric pressure, it is necessary to adjust the position of the bottle so that the water levels inside and outside the bottle are equal.

$$\left(0.7808 \text{ mol N}_2 \times \frac{28.013 \text{ g N}_2}{1 \text{ mol N}_2}\right) + \left(0.2095 \text{ mol O}_2 \times \frac{31.999 \text{ g O}_2}{1 \text{ mol O}_2}\right)$$

$$+ \left(0.0093 \text{ mol Ar} \times \frac{39.9 \text{ g Ar}}{1 \text{ mol Ar}}\right) + \left(0.0003 \text{ mol CO}_2 \times \frac{44 \text{ g CO}_2}{1 \text{ mol CO}_2}\right)$$

$$= 28.96 \text{ g/mol air}$$

SIMILAR EXAMPLES: Exercises 54, 90.

Collection of Gases over Water. A *pneumatic trough,* pictured in Figure 6-14, can be used to isolate gaseous products of chemical reactions. The method only works for gases that are insoluble in the liquid being displaced, but since so many important gases are insoluble in water (e.g., H_2, O_2, and N_2) the pneumatic trough has been extensively used in chemical laboratories.

The gas that is collected in a pneumatic trough is said to be "wet." It is a mixture of the desired gas and water vapor. The gas being collected expands to fill the container and exerts its partial pressure, P_{gas}. Water vapor, formed by the evaporation of liquid water, also fills the container and exerts a partial pressure, P_{H_2O}. The pressure of the water vapor depends only on the temperature of the water. Water vapor pressure data are readily available in tabulated form (see Table 12-2). We consider the concept of vapor pressure more fully in Chapter 12.

According to Dalton's law the *total* pressure of the "wet" gas is the sum of the two partial pressures. The total pressure can be made equal to the prevailing pressure of the atmosphere (barometric pressure).

$$P_{tot.} = P_{bar.} = P_{gas} + P_{H_2O} \qquad \text{or} \qquad P_{gas} = P_{bar.} - P_{H_2O} \qquad (6.17)$$

Example 6-20

Applying Dalton's law in a calculation based on the chemical equation—collecting a gas over a liquid (water). In the following reaction 81.2 cm³ of $O_2(g)$ is collected *over water* at 23 °C and barometric pressure 751 mmHg. What must have been the mass of $Ag_2O(s)$ decomposed? (Vapor pressure of water at 23 °C = 21.1 mmHg.)

$$2 \text{ Ag}_2\text{O(s)} \longrightarrow 4 \text{ Ag(s)} + \text{O}_2\text{(g)}$$

Solution. First we need to calculate the number of moles of $O_2(g)$; this we can do with the ideal gas equation. The key to this calculation is the fact that the gas collected is "wet," i.e., a *mixture* of $O_2(g)$ and water vapor.

$$P_{O_2} = P_{bar.} - P_{H_2O} = 751 \text{ mmHg} - 21.1 \text{ mmHg}$$

$$= 730 \text{ mmHg} \times \frac{1 \text{ atm}}{760 \text{ mmHg}} = 0.961 \text{ atm}$$

$$V = 81.2 \text{ cm}^3 = 0.0812 \text{ L}$$

$$n = \text{?}$$

$$R = 0.08206 \text{ L atm mol}^{-1} \text{ K}^{-1}$$

$$T = 23 \text{ °C} + 273 = 296 \text{ K}$$

$$n = \frac{PV}{RT} = \frac{0.961 \text{ atm} \times 0.0812 \text{ L}}{0.08206 \text{ L atm mol}^{-1} \text{ K}^{-1} \times 296 \text{ K}} = 0.00321 \text{ mol}$$

From the chemical equation we obtain a factor to convert from mol O_2 to mole Ag_2O. The molar mass of Ag_2O provides the final factor.

$$\text{no. g } Ag_2O = 0.00321 \text{ mol } O_2 \times \frac{2 \text{ mol } Ag_2O}{1 \text{ mol } O_2} \times \frac{231.7 \text{ g } Ag_2O}{1 \text{ mol } Ag_2O} = 1.49 \text{ g } Ag_2O$$

SIMILAR EXAMPLES: Exercises 17, 56, 91.

6-9 Kinetic Molecular Theory of Gases

The simple gas laws and the ideal gas equation are reasonably good for predicting gas behavior at usual temperatures and pressures. A theory that can be used to *explain* the gas laws was developed in the middle of the nineteenth century. It is called the **kinetic molecular theory of gases** and is based on the following model.

1. A gas is comprised of a very large number of extremely small particles (molecules or, in some cases, atoms) in *constant, random, straight-line* motion.
2. Molecules of a gas are separated by great distances. The gas is mostly empty space. (The molecules are treated as if they have mass but no volume, so-called "point masses.")
3. Molecules frequently collide with one another and with the walls of their container. However, these collisions occur very rapidly and most of the time molecules are not colliding.
4. There are assumed to be no forces between molecules (intermolecular forces) except very briefly during collisions. That is, each molecule acts independently of all the others, unaffected by their presence.
5. Individual molecules may gain or lose energy as a result of collisions; however, in a collection of molecules at constant temperature, *the total energy remains constant.*

Figure 6-15 should help you to visualize a gas in these terms.

Since pressure is force per unit area, central to the kinetic molecular theory is assessing the force of molecular collisions, and this depends on several factors.

One factor is the *frequency* of molecular collisions, that is, the number of collisions that occur in one second. The greater their frequency, the greater the total force of these collisions. Collision frequency increases with the number of molecules per unit volume and with molecular speeds.

A second factor is the amount of *translational kinetic energy* that molecules have. Translational kinetic energy is the energy associated with objects moving through space. Like speeding bullets, gas molecules have energy of motion. The translational kinetic energy of a molecule is represented as e_k and has the value, $e_k = \frac{1}{2}mu^2$, where m is the mass of the molecule and u is its speed (see also Appendix B). The faster molecules move, the greater their translational kinetic energies.

Still another factor is that molecules move in all directions and at different speeds, but this factor is too complicated for us to consider here. When all factors are properly considered, however, this basic equation of the kinetic molecular theory is obtained.

$$PV = \frac{n'm\overline{u^2}}{3} \tag{6.18}$$

In equation (6.18) n' represents the number of molecules of mass m in the volume V. $\overline{u^2}$ is the *average* of the *squares* of the molecular speeds. The product $\frac{1}{2} \times m \times \overline{u^2}$ is the *average* translational kinetic energy, $\overline{e}_k$, of a collection of gas molecules.

$$\overline{e}_k = \frac{1}{2}(m\overline{u^2}) \tag{6.19}$$

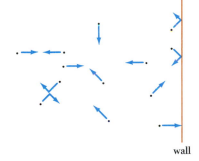

FIGURE 6-15
Visualizing molecular motion.

Molecules of a gas are in constant random motion and undergo collisions with each other and with the walls of their container.

wall

The bar over a quantity means that the quantity may have a range of values and that the *average* value is intended. Here, $\overline{u^2}$ refers to the average of the squares of molecular speeds.

By slightly rearranging equation (6.18) and combining it with equation (6.19), we can write

$$PV = \tfrac{2}{3}n'(\tfrac{1}{2}m\overline{u^2}) \qquad \text{and} \qquad PV = \tfrac{2}{3}n'\overline{e}_k \qquad (6.20)$$

The Meaning of Temperature. We can gain an important insight into the meaning of temperature by starting with equation (6.20) and considering

- 1 mol of gas (the number of molecules, $n' = N_A$, the Avogadro constant)
- that the gas is ideal ($PV = nRT$, and $PV = RT$ when $n = 1$).

$$PV = \tfrac{2}{3}N_A\overline{e}_k = RT$$

and, since R and N_A are constants,

$$\overline{e}_k = \frac{3}{2}\frac{R}{N_A}T = \text{constant} \times T \qquad (6.21)$$

> If the five postulates of the kinetic molecular theory hold, a gas is automatically an ideal gas.

Through equation (6.21) we have a new definition of temperature. *The Kelvin temperature of a gas is directly proportional to the average translational kinetic energy of its molecules.*[*]

Also, we have a new conception of what changes in temperature mean—changes in the intensity of molecular notion. When heat flows from one body to another, molecules in the hotter body (higher temperature) give up some of their kinetic energy through collisions with molecules in the colder body (lower temperature). The flow of heat continues until the average translational kinetic energies of the molecules become equal, i.e., the temperatures become equalized. Finally, we now have a way of looking at the absolute zero of temperature. *It is the temperature at which molecular motion ceases.* The absolute zero is unattainable, although current attempts have resulted in temperatures as low as 0.00001 K.

Using the Kinetic Theory of Gases to Derive Simple Gas Laws: Avogadro's Law. From equation (6.20), for two different gases, A and B,

$$P_A = \frac{2}{3}\frac{n'_A}{V_A}(\overline{e}_k)_A \qquad \text{and} \qquad P_B = \frac{2}{3}\frac{n'_B}{V_B}(\overline{e}_k)_B$$

If the two gases are compared at identical temperatures $(\overline{e}_k)_A = (\overline{e}_k)_B$, and if they have identical pressures, $P_A = P_B$. Under these conditions the number of molecules per unit volume must be the same for the two gases.

$$\frac{n'_A}{V_A} = \frac{n'_B}{V_B}$$

Thus, if equal volumes of gases are compared ($V_A = V_B$), the numbers of molecules of the two gases must be equal, $n'_A = n'_B$. If equal numbers of molecules are compared ($n'_A = n'_B$), the volumes must be equal, $V_A = V_B$.

In a similar fashion, we can derive the other simple gas laws from the kinetic molecular theory (see Exercise 60).

Distribution of Molecular Speeds. The statements we have made about an *average* kinetic energy imply that in a collection of molecules there is a *distribution* of energies, from very high to very low. There is a distribution of speeds as well, as shown in Figure 6-16.

Three different speeds are indicated on the curve of Figure 6-16. These are the **most probable** or **modal speed**, u_m, the **average speed**, $u_{av.} = \overline{u}$, and the **root-mean-square speed**, $u_{rms} = \sqrt{\overline{u^2}}$. More molecules have the most probable speed

---

[*]In gases and liquids temperature is a measure of the average translational kinetic energies of molecules. In solids, where molecules cannot move freely, temperature is a measure of the vibrational kinetic energy (see also Section 12-1).

FIGURE 6-16

Distribution of molecular speeds—hydrogen gas at 0 °C.

The ordinate values represent the percent of the molecules having a certain speed. The abscissas represent these speeds, based on an interval of 1 m/s. (For example, all molecules with speeds between 1499.5 and 1500.5 m/s are taken to have a speed of 1500 m/s.)

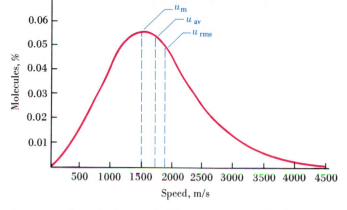

than any other single speed. The average speed refers to the simple average. The root-mean-square speed is the *square root* of the *average* of the *squares* of the speeds of all the molecules in a sample. The root-mean-square speed can be calculated by substituting for the value of e_k in equation (6.21).

$$\bar{e}_k = \frac{1}{2}m\overline{u^2} = \frac{3}{2}\frac{R}{N_A}T \qquad \text{and} \qquad \overline{u^2} = \frac{3RT}{mN_A}$$

Since the product mN_A represents the mass of 1 mol of molecules, we can replace it by the molar mass, $\mathcal{M}$.

Relating root-mean-square speed to temperature and molar mass.

$$u_{\text{rms}} = \sqrt{\overline{u^2}} = \sqrt{\frac{3RT}{\mathcal{M}}} \qquad\qquad (6.22)$$

Root-mean-square speed is the most important of these three speeds since the kinetic theory equations all involve velocity *squared*.

Expressions for u_m and $\bar{u}$ are harder to derive than for u_{rms} and we will forgo these. The analogy in Table 6-2 should help you to appreciate the distinction among most probable, average, and root-mean-square speed.

To calculate root-mean-square speed with equation (6.22) we have to express the gas constant as

TABLE 6-2
Analogy to the Distribution of Molecular Speeds

Consider ten automobiles on a highway traveling at these speeds:

	Speed, mi/h	(Speed)^2
	40	1,600
	42	1,764
	45	2,025
	48	2,304
most probable (modal)	50	2,500
speed = 50	50	2,500
	55	3,025
	57	3,249
	58	3,364
	60	3,600

sum of speeds = $\sum$speed = 505 $\qquad\qquad\qquad \sum(\text{speed})^2 = 25{,}931$

average speed = $\overline{\text{speed}} = \dfrac{\sum\text{speed}}{10} = 50.5$ $\qquad\qquad \overline{(\text{speed})^2} = \dfrac{\sum(\text{speed})^2}{10} = 2593.1$

root-mean-square speed

$= \sqrt{\overline{(\text{speed})^2}} = \sqrt{2593.1} = 50.9$

modal speed: 50 mi/h   average speed: 50.5 mi/h   rms speed: 50.9 mi/h

$$R = 8.314 \text{ J mol}^{-1} \text{ K}^{-1}$$

Furthermore, to produce the proper cancellation of units, we have to express the joule (J) in terms of mass, length, and time. Since kinetic energy is expressed as $e_k = \frac{1}{2}(mu^2)$, this means that the joule must have the units of (mass) × (velocity)^2 = (kg)(m/s)^2, leading to the value

$$R = 8.314 \text{ kg m}^2 \text{ s}^{-2} \text{ mol}^{-1} \text{ K}^{-1} \tag{6.23}$$

Example 6-21

Calculating a root-mean-square speed. Which is the greater speed, that of a bullet from a high-powered M-16 rifle (2180 mi/h) or the root-mean-square speed of H_2 molecules at 25 °C?

Solution. Determine u_{rms} of H_2 with equation (6.22). In doing so, note that R must have the units given in (6.23) and that the molar mass must be expressed in *kilograms* per mole.

$$u_{rms} = \sqrt{\frac{3 \times 8.314 \text{ kg m}^2 \text{ s}^{-2} \text{ mol}^{-1} \text{ K}^{-1} \times 298 \text{ K}}{2.016 \times 10^{-3} \text{ kg mol}^{-1}}}$$

$$= \sqrt{3.69 \times 10^6 \text{ m}^2/\text{s}^2} = 1.92 \times 10^3 \text{ m/s}$$

The remainder of the problem requires either converting 1.92×10^3 m/s to a speed in mi/h, or 2180 mi/h to m/s and comparing the two speeds. When this is done, we see that 1.92×10^3 m/s corresponds to 4.29×10^3 mi/h. The root-mean-square speed of H_2 molecules at 25 °C is greater than the speed of the high-powered rifle bullet.

SIMILAR EXAMPLES: Exercises 61, 62, 63, 93, 95.

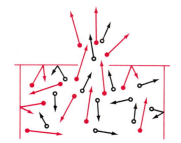

FIGURE 6-17
Effusion through an orifice.

Average speeds of the two different types of molecules are suggested by the lengths of the arrows. The faster molecules (shown in red) effuse more rapidly.

Effusion refers to the passage of gas molecules through a tiny opening in their container into an evacuated space. **Diffusion** refers to the mixing of gases with one another. Each gas spreads throughout the mixture until its partial pressure is the same everywhere.

Effusion and Diffusion. A molecular speed of 1500 m/s corresponds to about 1 mi/s or 3600 mi/h. However, when gases are allowed to mix or diffuse into one another, they do so much more slowly than suggested by their molecular speeds. This is because molecules undergo frequent collisions and change direction as a result of these collisions. Although molecules travel in straight lines from one collision to another, over long distances their path is very tortuous (zigzag). Still, the rate at which gases diffuse or mix does depend on the speeds of the molecules.

Figure 6-17 pictures a concept related to diffusion but simpler. Here molecules are allowed to escape from their container through a tiny orifice or pin hole. This escape through an orifice is called **effusion**. The rates of effusion of molecules are directly proportional to their speeds. That is, faster moving molecules effuse faster than slower molecules. For two different gases at the same temperature and pressure, we can first compare effusion rates with root-mean-square speeds and then substitute expression (6.22) for these speeds.

$$\frac{\text{rate of effusion of A}}{\text{rate of effusion of B}} = \frac{(u_{rms})_A}{(u_{rms})_B} = \frac{\sqrt{3RT/\mathcal{M}_A}}{\sqrt{3RT/\mathcal{M}_B}} = \sqrt{\frac{\mathcal{M}_B}{\mathcal{M}_A}} \tag{6.24}$$

The result shown in equation (6.24) is a kinetic-theory statement of a nineteenth-century law called **Graham's law of effusion.***

---

*Graham's law of effusion strictly applies only if certain conditions are met. For example, the gas pressure must be very low so that molecules escape individually and not as a jet of gas. The orifice must be very small so that molecules undergo no collisions as they pass through. A corresponding law dealing with diffusion is even more subject to qualification, because the diffusing molecules undergo collisions with each other and with the gas into which they are diffusing, and because some molecules move in the direction opposite to that of the net flow. Nevertheless, the idea that the rates of effusion and diffusion are inversely proportional to the square roots of molecular weights still leads to useful qualitative answers.

The rates of effusion of two different gases are inversely proportional to the square roots of their molecular weights.

At the same temperature, molecules of two different gases have the same average kinetic energy, $\frac{1}{2}mu^2$. Molecules with a larger mass (m) have a lower average speed (u_{rms}).

Stated more simply, low molecular weight gases effuse (or diffuse) faster than high molecular weight gases, but when comparisons are made they must be based on the *square roots* of molecular weights, not the molecular weights themselves.

Example 6-22

Comparing rates of effusion of gases. How does the rate of effusion of He atoms through a tiny hole compare with the rate of effusion of H_2 molecules at the same temperature and pressure?

Solution. In expression (6.24) we can take He as gas A and H_2 as gas B.

$$\frac{\text{rate of effusion of He}}{\text{rate of effusion of } H_2} = \sqrt{\frac{\mathcal{M}_{H_2}}{\mathcal{M}_{He}}} = \sqrt{\frac{2.016}{4.003}} = 0.710$$

rate of effusion of He = 0.710 × rate of effusion of H_2

SIMILAR EXAMPLE: Exercise 64.

When we use equation (6.24), we find that a lighter gas (e.g., H_2) effuses *faster* than a heavier gas (e.g., He). The gas that effuses *fastest* takes the *shortest* time to do so. Also the gas that effuses *fastest* travels *farthest* in a given period of time. We need to use variations of equation (6.24) to describe these alternative ways of looking at effusion. An effective method for doing this is to note that in every case a ratio of two terms (effusion rates, times, . . .) is equal to the *square root* of a ratio of molar masses. That is,

General expression to relate effusion properties to molar masses.

$$\text{ratio of} \begin{cases} (1) \text{ molecular speeds} \\ (2) \text{ rates of effusion} \\ (3) \text{ effusion times} \\ (4) \text{ amount of gas effused} \\ (\cdot) \ . \ . \ . \end{cases} = \sqrt{\frac{\text{ratio of two}}{\text{molar masses}}} \qquad (6.25)$$

In using equation (6.25) to answer questions like those posed in Examples 6-22, 6-23, and 6-24, you need to establish whether the ratio of molar masses should be $\mathcal{M}_A/\mathcal{M}_B$ or $\mathcal{M}_B/\mathcal{M}_A$. You can do this by reasoning whether the ratio of properties sought (effusion rate, effusion time, etc.) should be *greater or less than one*.

Example 6-23

Relating effusion times and molar masses. A sample of Kr(g) escapes through a tiny hole in 87.3 s, and an unknown gas requires 42.9 s under identical conditions. What is the molar mass of the unknown gas?

Solution. Since the unknown effuses faster than Kr, it must have a lower molar mass. The required ratio of molar masses should be *less than one*.

$$\frac{\text{effusion time for unknown}}{\text{effusion time for Kr}} = \frac{42.9 \text{ s}}{87.3 \text{ s}} = \sqrt{\frac{\mathcal{M}_{unk.}}{\mathcal{M}_{Kr}}} = 0.491$$

$$\frac{\mathcal{M}_{unk.}}{83.80} = (0.491)^2$$

$$\mathcal{M}_{unk.} = 20.2 \text{ g/mol}$$

SIMILAR EXAMPLES: Exercises 18, 65, 66.

Example 6-24 _____

Comparing distances traveled by molecules of different gases. If molecules effusing from a sample of N_2 travel a distance of 60.2 cm in a certain period of time, how far will molecules of O_2 travel under the same conditions?

Solution. O_2 molecules are slightly heavier than N_2 molecules and should travel a *shorter* distance.

$$\frac{\text{distance for } O_2}{\text{distance for } N_2} = \sqrt{\frac{\mathcal{M}_{N_2}}{\mathcal{M}_{O_2}}} = \sqrt{\frac{28.0}{32.0}} = 0.935$$

distance for O_2 = 0.935 × 60.2 cm = **56.3 cm**

SIMILAR EXAMPLE: Exercise 67.

Applications of Diffusion. That gases effuse through openings and that they diffuse into one another are commonly experienced phenomena. Natural gas and liquefied petroleum gas (LPG) are odorless, and for commercial use small quantities of a gaseous organic sulfur compound are added to them. The sulfur compound has an odor that can be detected in parts per billion (ppb) or less. When a leak occurs we rely on the diffusion of this odorous compound through air for detection of the leak (recall Example 3-2).

In the chemistry laboratory a white cloud of $NH_4Cl(s)$ is produced if bottles of concentrated $NH_3(aq)$ and $HCl(aq)$ in close proximity are opened at the same time. $NH_3(g)$ and $HCl(g)$ escape from the solutions, diffuse toward each other, and react to form $NH_4Cl(s)$ (see Exercise 67).

In the Manhattan Project during World War II one of the methods developed for separating the desired isotope, U-235, from the predominant species U-238 involved gaseous diffusion. The method is now also used to obtain U-235-enriched nuclear fuels for nuclear power plants. One of the few compounds of uranium that can be obtained as a gas at moderate temperatures is uranium hexafluoride, UF_6. When high-pressure $UF_6(g)$ is forced through a porous barrier, molecules containing the isotope U-235 pass through the barrier slightly faster than those containing U-238. As a result the $UF_6(g)$ contains a slightly higher ratio of U-235 to U-238 than it did previously; the gas has become "enriched" in U-235. By continuing this slight enrichment through several thousand stages, a product is finally obtained having a high proportion of U-235 (e.g., 90% or more of the uranium).

6-10 Nonideal (Real) Gases

On several occasions we have implied that real gases can be described by the ideal gas equation only under certain conditions. What are these conditions, and how serious are the departures from ideality for real gases? Figure 6-18 describes the situation. Here, PV/RT is plotted as a function of P, for 1 mol of gas at a fixed temperature (0 °C). If a gas is ideal, $PV/RT = 1$ for 1 mol of gas. The extent to which the measured value of PV/RT, called the *compressibility factor*, deviates from 1 is a measure of the nonideality of a gas. Figure 6-18 suggests that all gases behave ideally at sufficiently low pressures, say below 1 atm, but that deviations become significant at increased pressures. At very high pressures the compressibility factor is always greater than 1.

Here is how we might explain nonideal gas behavior: Boyle's law, for example, predicts that at very high pressures a gas volume becomes extremely small, approaching zero. This cannot be, however, because the molecules themselves occupy space and are practically incompressible. Also, we must allow for the fact that

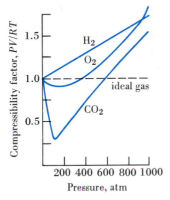

FIGURE 6-18

The behavior of real gases—compressibility factor for 1 mole of gas as a function of pressure at 0 °C.

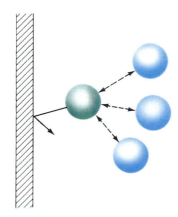

FIGURE 6-19

Intermolecular attractive forces.

Attractive forces of other molecules for the molecule shown in green cause that molecule to exert less force when it collides with the wall than if these attractions did not exist.

TABLE 6-3

Van der Waals Constants for Several Gases

Gas	a, L^2 atm mol^{-2}	b, L mol^{-1}
Ar	1.35	0.0322
Cl_2	6.49	0.0562
CO	1.49	0.0399
CO_2	3.59	0.0427
H_2	0.244	0.0266
He	0.034	0.0237
N_2	1.39	0.0391
O_2	1.36	0.0318
SO_2	6.71	0.0564

intermolecular forces do exist in gases. As suggested by Figure 6-19, if there is an attractive force among the molecules of a gas, the frequency and force of the collisions of gas molecules with the container walls is less than what we would expect for an ideal gas. Intermolecular forces of attraction account for compressibility factors that are less than 1; these forces become increasingly important at *low temperatures* where molecular motion is diminished in intensity. To summarize

- Gases tend to behave *nonideally* at *low temperatures and high pressures*.
- Gases tend to behave *ideally* at *high temperatures and low pressures*.

The van der Waals Equation. A number of equations of state have been proposed for real gases, equations that apply over a wider range of temperatures and pressures than does the ideal gas equation. Such equations must correct for the volume associated with the molecules themselves and for intermolecular forces of attraction. One equation that works well is the van der Waals equation.

$$\left(P + \frac{n^2a}{V^2}\right)(V - nb) = nRT \tag{6.26}$$

In equation (6.26) V represents the volume of n moles of gas. The term n^2a/V^2 is related to the intermolecular forces of attraction. It is added to the pressure because the measured pressure is lower than anticipated. The term b is related to the volume of the gas molecules and must be subtracted from the measured volume. Thus, $V - nb$ represents the *free* volume within the gas. The terms a and b have particular values for particular gases and vary somewhat with temperature and pressure (see Table 6-3). In Example 6-25 the pressure of a real gas is calculated with the van der Waals equation. Solving equation (6.26) for either n or V is more difficult (see Exercise 69).

Example 6-25

Using the van der Waals equation to calculate the pressure of a nonideal gas. Use the van der Waals equation to calculate the pressure exerted by 1.00 mol $Cl_2(g)$ when it is confined to a volume of 2.00 L at 273 K. Values of a and b are given in Table 6-3.

Solution. Substitute the following values into equation (6.26).

$$n = 1.00 \text{ mol}; \quad V = 2.00 \text{ L}; \quad T = 273 \text{ K}; \quad R = 0.08206 \text{ L atm mol}^{-1} \text{ K}^{-1};$$

$$n^2a = (1.00)^2 \text{ mol}^2 \times 6.49 \frac{L^2 \text{ atm}}{\text{mol}^2} = 6.49 \text{ } L^2 \text{ atm};$$

$$nb = 1.00 \text{ mol} \times 0.0562 \text{ L/mol} = 0.0562 \text{ L}$$

$$P = \left(\frac{nRT}{V - nb}\right) - \frac{n^2a}{V^2}$$

$$= \frac{1.00 \text{ mol} \times 0.08206 \text{ L atm mol}^{-1} \text{ K}^{-1} \times 273 \text{ K}}{(2.00 - 0.0562) \text{ L}} - \frac{6.49 \text{ } L^2 \text{ atm}}{(2.00)^2 \text{ } L^2}$$

$$= 11.5 \text{ atm} - 1.62 \text{ atm} = 9.9 \text{ atm}$$

The pressure of the $Cl_2(g)$ calculated with the ideal gas equation is 11.2 atm. If only the b term is used in the van der Waals equation, the calculated pressure is 11.5 atm. Including the a term reduces the calculated pressure by 1.62 atm. Under the conditions stated here, intermolecular forces are the main cause of the failure of $Cl_2(g)$ to behave ideally.

SIMILAR EXAMPLE: Exercise 68.

Are You Wondering:

When you can use the ideal gas equation and when you must use some other equation of state for a gas, such as the van der Waals equation?

With few exceptions, you need only the ideal gas equation in this text. This means that either the conditions given for the gas will be those at which the gas is essentially ideal or, if not, the results you calculate will be only approximately correct. To use the van der Waals equation, you must have additional data about the gas in question—van der Waals constants such as listed in Table 6-3. If you do not have such data, of course, the only gas equation that is *independent* of the particular gas being described is the ideal gas equation. Exercises at the end of the chapter that require you to use the van der Waals equation are clearly identified.

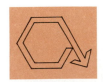

FOCUS ON
The Atmosphere

Clouds and lightning are familiar phenomena of the lower atmosphere (troposphere). [E. R. Degginger]

Air is a mixture of nitrogen and oxygen gases, with smaller quantities of argon, carbon dioxide, and a dozen or so other gases (see Table 6-4). Life on earth exists at the bottom of a "sea" of air called the **atmosphere,** and all objects on earth are subjected to the pressure (barometric pressure) exerted by this blanket of air.

Structure of the Atmosphere. If we take as the outer limit of the atmosphere the distance at which its composition becomes the same as that of interplanetary space (very, very low density atomic hydrogen gas, H) our atmosphere is about 10,000 km "thick." From another viewpoint the atmospheric blanket is very "thin": The total mass of the atmosphere is only about one millionth that of the earth itself, and of the mass of the atmosphere, about 90% is in the first 10 km above the surface of the earth.

TABLE 6-4
Composition of Dry Air (Near Sea Level)

Component	Volume percent
nitrogen (N_2)	78.084
oxygen (O_2)	20.946
argon (Ar)	0.934
carbon dioxide (CO_2)	0.033
neon (Ne)	0.001818
helium (He)	0.000524
methane (CH_4)	0.0002
krypton (Kr)	0.000114
hydrogen (H_2)	0.00005
nitrous oxide (N_2O)	0.00005
xenon (Xe)	0.000009
ozone (O_3)	
sulfur dioxide (SO_2)	
nitrogen dioxide (NO_2)	trace
ammonia (NH_3)	
carbon monoxide (CO)	
iodine (I_2)	

The first 80 km or so of the atmosphere is a region known as the homosphere, so called because the relative proportions of N_2, O_2, and other gases remain essentially constant (homogeneous) throughout this region. The portion of the atmosphere beyond the 80-km limit is called the heterosphere. It consists of four different layers of gases—molecular nitrogen (N_2), atomic oxygen (O), helium (He), and atomic hydrogen (H) (see Figure 6-20).

As suggested by Figure 6-20, the temperature of the atmosphere falls continuously for the first 10 km above the earth's surface. This 10-km layer of air is that most familiar to us—the **troposphere.** Temperatures in the

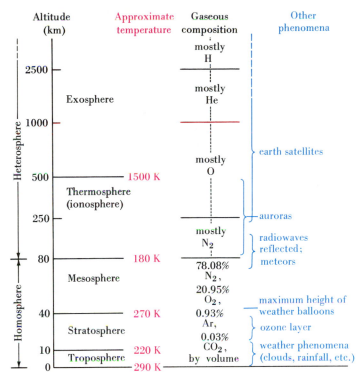

FIGURE 6-20

The atmosphere: structure, temperatures, composition, and other phenomena.

The values shown here are approximate. For example, the height of the troposphere varies from about 8 km at the poles to 16 km at the equator, and temperatures in the thermosphere vary greatly between day and night.

ter would give very low readings, even though in the midst of highly energetic, *but widely separated,* particles.

We can observe this unusual character of the temperature distribution in the upper atmosphere through the behavior of meteors. Meteors are extraterrestrial chunks of matter that are trapped in the earth's gravitational field, disintegrate, and give off light as they fall to earth (hence the name ''shooting stars''). Light emission from meteors is believed to be preceded by evaporation and ionization of atoms from the surface of the meteor. In turn, this evaporation of surface atoms results from collisions with molecules of air. Meteors do not start to give off light until they fall to within 110 km of the earth's surface. The majority of them are completely vaporized in the range from about 80 to 110 km. Thus, meteors pass through the higher temperatures of the thermosphere without vaporizing. Instead, they vaporize in a lower temperature region (about 220 K), but a region where gas densities are much higher.

The Atmosphere as a Source of Industrial Chemicals. Nitrogen and oxygen, the most abundant components in air, are also among the most widely produced and used industrial chemicals (No. 2 and No. 3, respectively, in the United States). Table 6-5 lists some important uses of nitrogen and oxygen. Particularly rapid growth is occurring in some of these uses, such as liquid nitrogen in the flash freezing of fast foods and oxygen gas in sewage and wastewater treatment. Nitrogen, oxygen, and argon (as well as neon, krypton, and xenon) are obtained principally by the fractional distillation of liquid air, a process that has

layer of air from about 10 to 40 km increase slowly from about 220 to 270 K. This is the region known as the **stratosphere.** (Supersonic aircraft fly in the lower regions of the stratosphere.) In the third atmospheric layer—the **mesosphere**—the temperature continues to rise to about 300 K and then falls to a minimum of about 180 K at 80 km. In the next layer, the **thermosphere** (or **ionosphere**), temperatures rise continuously to 1500 K. In this region absorption of ultraviolet radiation from the sun causes gas molecules to ionize and/or dissociate. Thus, at these altitudes the atmosphere consists of positive and negative ions, free electrons, neutral atoms, and molecules. The term *ionosphere* is suggestive of this ionization process, and the term *thermosphere* suggests high temperatures for this ionized gas.

We would consider a temperature of 1500 K, if achieved in bulk matter, to be quite high—enough to cause a bright red glow in iron. At high altitudes in the atmosphere high temperatures have a different significance. Molecular speeds are high, but to experience the effects of high temperature requires that heat be transferred through *frequent* molecular collisions. Because the gas density at these high altitudes is so low, molecular collisions occur only *infrequently.* An ordinary thermome-

TABLE 6-5
Uses of Nitrogen and Oxygen Gases

Nitrogen	
1986 production: 24.3 × 10⁶ tons	provide a blanketing (inert) atmosphere for the production of chemicals and electronic components; pressurized gas for enhanced oil recovery; metals treatment; refrigerant (e.g., fast freezing of foods).

Oxygen	
1986 production: 16.5 × 10⁶ tons	manufacture of iron and steel; manufacture and fabrication of other metals; chemicals manufacture and other oxidation processes; water treatment; oxidizer of rocket fuels; medicinal uses.

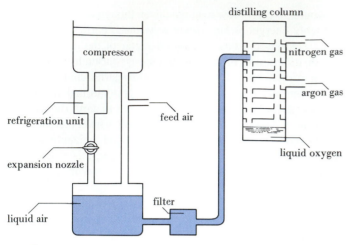

FIGURE 6-21

The separation of atmospheric gases—a simplified representation.

Clean air is fed into a compressor and then cooled by re-frigeration. The cold air expands through a nozzle, and as a result is cooled still further—sufficiently to cause it to liquefy. The liquid air is filtered to remove $CO_2(s)$ and hydrocarbons and then distilled. Liquid air enters the top of the column where nitrogen, the most volatile compo-nent, passes off as a gas. In the middle of the column gas-eous argon is removed, and liquid oxygen, the least vola-tile component, collects at the bottom. The normal boiling points of nitrogen, argon, and oxygen are −195.8, −185.7, and −183.0 °C, respectively.

been in use for nearly 100 years. The process, which in-volves only physical changes, not chemical reactions, is presented in a simplified form in Figure 6-21. Another important manufacturing process that uses nitrogen di-rectly from the atmosphere is the production of ammonia, from which a host of nitrogen-containing substances, chiefly fertilizers, are produced. (See Chapter 16 Focus feature.)

The Nitrogen Cycle. Nitrogen, like several other ele-ments—carbon, oxygen, phosphorus, potassium, and cal-cium, for example—is essential to life on earth. Unlike sunlight, which constantly streams to the earth, there is a limited pool of these nutrient elements present and they are continuously recycled. A description of the pathways that these nutrients follow between living (biotic) and non-living (abiotic) environments is called a nutrient cycle. One of the most familiar is the **nitrogen cycle,** pictured in Figure 6-22. As shown in the figure, (a) the reservoir of nitrogen for the cycle is the atmosphere, and (b) there are only two types of natural processes by which nitrogen from the atmosphere is converted to nitrogen compounds (called nitrogen fixation) and transported into the bio-sphere.

    The delicate balance of the nitrogen cycle can be easily upset by human activities. When land is extensively culti-vated, fixed nitrogen is removed at a greater rate than it

can be naturally replenished. This requires that nitrogen compounds be added to the soil as fertilizers. Also, the need for explosives, plastics, fibers, and industrial chemi-cals requires the artificial fixation of nitrogen.

Environmental Problems. The rate of extraction of ni-trogen and oxygen from the atmosphere for industrial pur-poses, though very large, is insignificant in comparison with the total quantities of these gases available in the atmosphere. However, the rate of artificial fixation of ni-trogen presently equals or exceeds the rate of natural fixa-tion. For example, it has been estimated that, worldwide, about 40% of all nitrogen supplied to agricultural land is through manufactured fertilizers. Currently, the total rate of fixation exceeds the rate of return of nitrogen to the atmosphere, which means that nitrogen compounds are

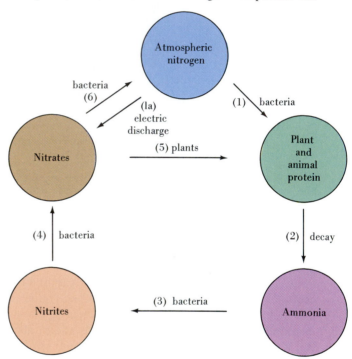

FIGURE 6-22

The nitrogen cycle in nature.

(1) Certain bacteria that reside as parasites in the root nodules of leguminous plants (beans, peas, clover, and al-falfa) can fix atmospheric nitrogen directly for conversion into plant proteins. Animals meet their requirements for nitrogen by feeding on these and other plants.
(2) The decay of plant and animal protein leads to the formation of ammonia.
(3) and **(4)** Successive bacterial actions convert ammonia into nitrates.
(5) The natural forms of fixed nitrogen that most plants require are the nitrates. This consumption again leads to plant and animal protein.
(6) Certain denitrifying bacteria are capable of decompos-ing nitrates into elemental nitrogen, which returns to the atmosphere.
(1a) As a result of electric discharges in rainstorms (light-ning), a series of chemical reactions results in the direct production of nitrates as nitric acid, HNO_3.

accumulating in soil, groundwater, surface water, and the oceans, with subsequent environmental effects. For example, nitrate ion (NO_3^-) occurs in significant concentrations in certain groundwater as a result of fertilizer runoff and drainage from cattle feedlots. When this water is drunk, nitrate-reducing bacteria in the gastrointestinal tract can convert NO_3^- to NO_2^-. Nitrite ion (NO_2^-) interferes with the oxygen transport ability of blood, which can lead to a condition known as *methemoglobinemia*. This disorder can be fatal in infants.

At various points elsewhere in the text we present topics in atmospheric chemistry and environmental problems involving the atmosphere. Some of these topics and the sections in which they are discussed are

- greenhouse effect (Chapter 7 Focus feature)
- industrial smog (14-13)
- photochemical smog (14-10)
- functions of ozone in the atmosphere (Chapter 14 Focus feature)
- chlorofluorocarbons and destruction of the ozone layer (Chapter 14 Focus feature)
- acid rain (Chapter 17 Introduction)

Summary

A gas is described through four variables—pressure, temperature, volume, and amount of gas. Gas pressure is most readily measured by comparing it to the pressure exerted by a liquid column, usually mercury. Atmospheric pressure is measured with a mercury barometer, and other gas pressures with a manometer. Pressure can be expressed through a variety of units (Table 6-1), though the ones chiefly used in this chapter are the atmosphere (atm) and the millimeter of mercury (mmHg). A standard atmosphere of pressure, 1 atm, is also equal to 760 mmHg and 760 torr.

Relationships between gas variables taken two at a time (with the remaining two held constant) are known as the simple gas laws. Most frequently encountered are Boyle's law, relating gas pressure and volume; Charles's law, relating gas volume and temperature; and Avogadro's law, relating volume and amount of gas. A number of important ideas originate with the simple gas laws. Among these are the concept of an absolute zero of temperature, a temperature scale (Kelvin) based on this absolute zero, a standard condition of temperature and pressure (STP), and the molar volume of a gas at STP—22.414 L/mol.

By combining Boyle's, Charles's, and Avogadro's laws, a more general statement of gas behavior is obtained, the ideal gas equation: $PV = nRT$. This equation can be solved for any one of the variables when values are known for the others. The ideal gas equation can also be applied to molecular weight and gas density determinations. Still other uses of the ideal gas equation are in describing (a) the gaseous reactants and/or products of a chemical reaction and (b) mixtures of gases. Partial pressures and partial volumes are also useful for dealing with gaseous mixtures. A particularly important statement is that of Dalton's law of partial pressures: The total pressure of a gaseous mixture is equal to the sum of the partial pressures of the components of the mixture. One common application of Dalton's law involves the collection of gases over water.

The kinetic molecular theory provides a theoretical basis for the various gas laws. With this theory a relationship is established between average molecular kinetic energy and Kelvin temperature. Another relationship involves the root-mean-square speed of molecules, temperature, and molar mass of a gas. Also, the effusion and diffusion of gases can be related to their molar masses through the kinetic molecular theory.

Real gases generally behave ideally only at high temperatures and low pressures. Nonideal behavior results chiefly from intermolecular attractions and the finite volume occupied by gas molecules. Alternative equations of state have been developed for real gases. The best known of these, perhaps, is the van der Waals equation.

Summarizing Example

Gasoline, used as a motor fuel, is a mixture of a large number of hydrocarbons. The principal ones are various octanes, which have the formula C_8H_{18}. When we burn gasoline in an automobile engine we try to get complete conversion of these hydrocarbons to CO_2 and H_2O. This requires that we use the correct mass ratio of air to gasoline (called the air/fuel ratio). If the mixture is fuel-rich there is insufficient air to complete the combustion of the gasoline and unburned hydrocarbons and carbon monoxide are produced. These are unwanted air pollutants. If the mixture is fuel-lean, there is an excess of air. This completes the combustion process, reducing the emissions of hydrocarbons and carbon

monoxide. However, under these conditions the formation of nitrogen oxides (by the combination of N_2 and O_2 in the air) is favored; these are also unwanted air pollutants. The ideal air/fuel ratio is one in which oxygen (from the air) and gasoline hydrocarbons are in *stoichiometric* proportions (meaning that neither is in excess). One of our objectives here will be to calculate this ideal ratio.

1. Write a chemical equation for the complete combustion of octane (C_8H_{18}).

Solution. The reactants are C_8H_{18} and O_2 and the products are CO_2 and H_2O.

$$2\ C_8H_{18}(l) + 25\ O_2(g) \longrightarrow 16\ CO_2(g) + 18\ H_2O(l)$$

(This example is similar to Examples 4-1 and 4-2.)

2. What is the volume of $CO_2(g)$, measured at 22 °C and 745 mmHg pressure, that is produced in the complete combustion of 100. g C_8H_{18}?

Solution. A two-step approach probably works best. Use the stoichiometric methods of Chapter 4 to determine the number of moles of $CO_2(g)$ produced. Follow this with an application of the ideal gas equation.

$$\text{no. mol } CO_2 = 100.\ \text{g } C_8H_{18} \times \frac{1\ \text{mol } C_8H_{18}}{114.2\ \text{g } C_8H_{18}} \times \frac{16\ \text{mol } CO_2}{2\ \text{mol } C_8H_{18}} = 7.01\ \text{mol } CO_2$$

$$V = \frac{nRT}{P} = \frac{7.01\ \text{mol} \times 0.08206\ \text{L atm mol}^{-1}\ \text{K}^{-1} \times (273 + 22)\ \text{K}}{(745/760)\ \text{atm}}$$

$$= 173\ \text{L } CO_2$$

(This example is similar to Example 6-14.)

3. What is the mass of $O_2(g)$ consumed in the complete combustion of 100. g C_8H_{18}?

Solution. This calculation is based solely on the chemical equation and does not require us to use any of the gas laws.

$$\text{no. g } O_2 = 100.\ \text{g } C_8H_{18} \times \frac{1\ \text{mol } C_8H_{18}}{114.2\ \text{g } C_8H_{18}} \times \frac{25\ \text{mol } O_2}{2\ \text{mol } C_8H_{18}} \times \frac{32.00\ \text{g } O_2}{1\ \text{mol } O_2}$$

$$= 350.\ \text{g } O_2$$

(This example is similar to Example 4-6.)

4. Using data from Examples 6-18 and 6-19, calculate the percent, by *mass*, of oxygen in air.

Solution. This question can be answered in two ways. One method is to consider the partial volumes and then the masses of the several gases present in a 100-L sample of air. These individual masses can be converted to percentages (see Exercise 55). A second method uses the data in Example 6-19. In one mole of air, with a mass of 28.96 g, there is present 0.2095 mol of O_2, weighing

$$0.2095\ \text{mol } O_2 \times \frac{31.999\ \text{g } O_2}{1\ \text{mol } O_2} = 6.704\ \text{g } O_2$$

The mass percent of O_2 is

$$\frac{6.704 \text{ g } O_2}{28.96 \text{ g air}} \times 100 = 23.15\% \text{ } O_2$$

5. Calculate the *mass* ratio of air to octane for a mixture in which the octane burns completely and there is no air in excess.

Solution. Base the calculation on the combustion of 100. g C_8H_{18}. The required mass of O_2 is 350. g (from Part 3). Use the percent composition of air from Part 4 to determine the mass of air containing 350. g O_2.

$$\text{no. g air} = 350. \text{ g } O_2 \times \frac{100. \text{ g air}}{23.15 \text{ g } O_2} = 1.51 \times 10^3 \text{ g air}$$

The air/octane ratio, by mass, is

$$\frac{1.51 \times 10^3 \text{ g air}}{100. \text{ g } C_8 H_{18}} = 15.1 \quad \text{(The ratio typically used in automobiles is about 14.5.)}$$

Key Terms

absolute zero of temperature (6-3)
atmosphere (6-2; Focus feature)
Avogadro's hypothesis (law) (6-3)
barometer (6-2)
Boyle's law (6-3)
Charles's law (6-3)
Dalton's law of partial pressures (6-8)
diffusion (6-9)
effusion (6-9)

equation of state (6-1, 6-4)
gas constant, *R* (6-4)
Graham's law (6-9)
ideal gas (6-4)
ideal gas equation (6-4)
kinetic molecular theory of gases (6-9)
manometer (6-2)
mole fraction (6-8)

nonideal gas (6-10)
nitrogen cycle (Focus feature)
partial pressure (6-8)
pressure (6-2)
stratosphere (Focus feature)
temperature (6-9)
troposphere (Focus feature)
van der Waals equation (6-10)

Highlighted Expressions

Definition of a standard atmosphere (6.3)
Mathematical expressions of Boyle's law (6.4)
Relationship between Kelvin and Celsius temperatures (6.5)
Mathematical expressions of Charles's law (6.6)
Mathematical expressions of Avogadro's law (6.7)
Molar volume of a gas at STP (6.8)
The ideal gas equation (6.9)
The gas constant, *R* (6.10)
The general gas law (6.11)

Ideal gas equation modified for molecular weight determination (6.12)
Ideal gas equation modified for gas density determination (6.13)
Dalton's law of partial pressures (6.15)
Relating mole fraction, pressure fraction and volume fraction of a gaseous mixture (6.16)
Relating root-mean-square speed to temperature and molar mass (6.22)
Expression to relate effusion (diffusion) properties to molar masses (6.25)

Review Problems

1. Convert the following pressures to an equivalent pressure in standard atmospheres: **(a)** 722 mmHg; **(b)** 66.4 cm Hg; **(c)** 1167 torr; **(d)** 47 psi.

2. Calculate the height of a mercury column required to produce a pressure **(a)** of 1.56 atm; **(b)** of 817 torr; **(c)** equal to that of a column of water 112 ft high.

3. What is P_{gas} for the manometer readings shown here (in mmHg) when barometric pressure is 737 mmHg?

4. A sample of $O_2(g)$ occupies a volume of 21.7 L at 742 mmHg. What is the new gas volume if, while the temperature and amount of gas are held constant, the pressure is **(a)** lowered to 356 mmHg; **(b)** increased to 2.18 atm?

5. A 166-cm³ sample of Ne(g) is initially at 742 mmHg and 30 °C. What will be the new volume if, while the pressure and

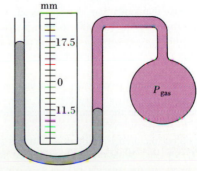

Problem 3

amount of gas are held constant, the temperature is **(a)** increased to 77 °C; **(b)** lowered to −10 °C?

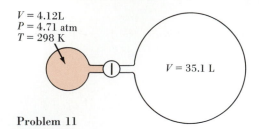

$V = 4.12$ L
$P = 4.71$ atm
$T = 298$ K

$V = 35.1$ L

Problem 11

6. It is desired to increase the volume of a fixed amount of gas from 77.6 to 124 cm^3 while holding the pressure constant. What must be the final temperature if the initial temperature is 21 °C?

7. Indicate how the final volume, V_f, is related to the initial volume, V_i, for a fixed amount of gas in each case.

(a) The pressure is decreased from 3 to 1 atm while the temperature is held at 25 °C.

(b) The temperature is lowered from 400 to 100 K while the pressure is held constant at 1 atm.

(c) The temperature is raised from 200 to 300 K while the pressure is increased from 2 to 3 atm.

8. What is the volume at STP of a 35.5-g sample of acetylene gas, C_2H_2 (used with oxygen to generate high-temperature flames for welding)?

9. What is the volume occupied by 68.5 g $CO_2(g)$ at 35 °C and 717 mmHg?

10. A 32.5-L cylinder contains 275 g $SO_2(g)$ at 24 °C. What is the pressure exerted by this gas?

11. The Ne(g) present in the bulb on the left is allowed to expand into the evacuated bulb pictured at the left. Assuming that the temperature remains constant at 25 °C during this expansion, what will be the final gas pressure? [*Hint:* What is the final volume?]

12. A 0.341-g sample of gas has a volume of 355 cm^3 at 98.7 °C and 743 mmHg. What is the molecular weight of this gas?

13. What is the density (in g/L) of $CO_2(g)$ at 26.8 °C and 764 mmHg?

14. A method of removing $CO_2(g)$ from a spacecraft is to allow the CO_2 to react with LiOH. What volume of $CO_2(g)$ at 23.6 °C and 735 mmHg can be removed per kg of LiOH?

$$2 \text{ LiOH(s)} + CO_2(g) \rightarrow Li_2CO_3(s) + H_2O(l)$$

15. What volume of $O_2(g)$ is consumed in the combustion of 45.8 L $C_3H_8(g)$ if both gases are measured at STP? [*Hint:* Write a balanced equation for the combustion reaction.]

16. What is the volume occupied by a mixture of 14.8 g Ne(g) and 37.6 g Ar(g) at 14.5 atm pressure and 31.2 °C?

17. A 58.8-cm^3 sample of "wet" $O_2(g)$ is collected over water at 22 °C and 738 mmHg barometric pressure. How much O_2, in moles, is present in the gas? Vapor pressure of H_2O at 22 °C = 19.8 mmHg.

18. A sample of $Cl_2(g)$ effuses through a tiny hole in 28.6 s. How long would it take for an equivalent sample of $N_2O(g)$ to effuse under the same conditions?

Exercises

Pressure and its measurement

19. Convert each of the following pressures to the equivalent pressure in standard atmospheres: (a) 1127 mmHg; (b) 6.78 kg/cm^2; (c) 231 kPa; (d) 912 mb; (e) 2.35×10^5 N/m^2.

20. Calculate the following quantities:

(a) The height of a column of liquid glycerol ($d = 1.26$ g/cm^3) required to exert the same pressure as 4.17 m of $CCl_4(l)$ ($d = 1.59$ g/cm^3).

(b) The height of liquid benzene ($d = 0.879$ g/cm^3) required to exert a pressure of 2.78×10^4 N/m^2.

(c) The density of a liquid if a 15.0-ft column is to exert a pressure of 12.5 lb/in.^{2}.

21. Explain why it is necessary to include the density of Hg(l) and the value of the acceleration due to gravity, g, in a precise definition of a standard atmosphere of pressure (page 161).

22. Synthetic diamonds can be produced from ordinary graphite by subjecting the graphite to high pressures and high temperatures. One process uses a temperature of 2500–3500 K and a pressure of 120. kbar. Convert this pressure into each of the units listed in Table 6-1.

23. A gas is collected over water when the barometric pressure is 753.5 mmHg, but, as shown in the drawing, the water levels inside and outside the container of gas differ by 3.8 cm. What is the total pressure of the gas inside the container?

24. The object pictured is made of a stainless steel having a density of 7.75 g/cm^3. What pressure, expressed in kg/cm^2, will this object exert on the surface below it, when the object is

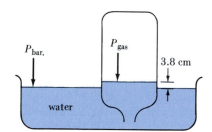

$P_{bar.}$

P_{gas}

3.8 cm

water

Exercise 23

placed on (a) face A; (b) face B; (c) face C? [*Hint:* What is the mass of the object? What is the area of contact between the object and supporting surface in each case? Consider that what you are calculating here is the pressure in addition to atmospheric pressure.]

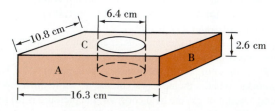

6.4 cm

10.8 cm

C

2.6 cm

B

A

16.3 cm

Exercise 24

The simple gas laws

25. A sample of $N_2(g)$ that occupies a volume of 533 cm^3 at 715 mmHg is compressed, at constant temperature, to 2.37 atm. What is the final gas volume?

26. A 15.5-L cylinder of Ar(g) is connected to an evacuated 2148-L tank. If the final pressure is 722 mmHg, what must have been the original gas pressure in the cylinder? [*Hint:* What is the final volume?]

27. A fixed amount of gas, maintained in a constant volume of 256 cm^3, exerts a pressure of 822 mmHg at 24.1 °C. At what temperature will the pressure of the gas become exactly 100. kPa?

28. A 27.6-cm^3 sample of $PH_3(g)$ (used in the manufacture of flame retardant chemicals) is obtained at STP.

(a) What is the mass of this gas, in milligrams?

(b) How many molecules of PH_3 are present?

29. A 12.5-g sample of gas is added to an evacuated, constant-volume container at 22 °C. The *pressure of the gas is to be held constant* as the temperature is raised. This requires that some gas be allowed to escape. What mass of gas must be released if the temperature is raised to 202 °C?

30. You purchase a bag of potato chips at an ocean beach to take on a picnic in the mountains. At the picnic you notice that the bag has become inflated, almost to the point of bursting. Use your knowledge of gas behavior to explain this phenomenon.

Ideal gas equation

31. A 35.6-L constant-volume cylinder containing 1.62 mol He is heated until the pressure reaches 3.17 atm. What is the temperature of the gas?

32. Kr(g) in an 18.5-L cylinder exerts a pressure of 7.19 atm at 33.5 °C. What is the mass of gas present?

33. A sample of gas has a volume of 4.22 L at 27.8 °C and 734 mmHg. What will be the volume of this gas at 24.6 °C and 755 mmHg?

34. A 34.0-L cylinder contains 212 g $O_2(g)$ at 21 °C. What mass of $O_2(g)$ must be released to reduce the pressure in the cylinder to 1.24 atm?

*__35.__ Use SI units and the ideal gas equation to

(a) Express the gas constant R in the units kPa dm^3 mol^{-1} K^{-1}.

(b) Use the value found in (a), together with information from Appendix B, to obtain R in the units J mol^{-1} K^{-1}.

(c) Calculate the pressure, in kPa, exerted by 1198 g CO(g) confined to a tank of 1.68 m^3 volume at 294 K.

Molecular weight determination

36. Refer to Example 6-12. By combustion analysis, the mass percent composition of propylene is found to be 85.63% C and 14.37% H. What is the *molecular* formula of propylene?

37. A gaseous hydrocarbon weighing 0.185 g occupies a volume of 110. cm^3 at 26 °C and 743 mmHg. What is the molecular weight of this compound? What conclusion can you draw about its molecular formula?

38. A 2.650-g sample of a gas occupies a volume of 428 cm^3 at 742.3 mmHg and 24.3 °C. Analysis of this compound shows it to be 15.5% C, 23.0% Cl, and 61.5% F. What is the molecular formula of this compound?

39. A glass vessel weighs 56.1035 g when evacuated; 264.2931 g when filled with Freon-113, a liquid with a density

of 1.576 g/cm^3; and 56.2445 g when filled with acetylene gas at 749.3 mmHg and 20.02 °C. What is the molecular weight of acetylene?

Gas densities

40. A particular application calls for $N_2(g)$ with a density of 1.45 g/L at 25 °C. What must be the pressure of the $N_2(g)$?

41. The density of phosphorus vapor at 310 °C and 775 mmHg is 2.64 g/L. What is the molecular formula of the phosphorus?

42. In order for a gas-filled balloon to rise in air, the density of the gas in the balloon must be less than that of air, and one way to reduce the density of a gas is to heat it.

(a) Use the result of Example 6-19 to determine the density of air at 25 °C and 1 atm pressure.

(b) Show, by calculation, that a balloon filled with nitrous oxide, N_2O, at 25 °C and 1 atm would not be expected to rise in air.

(c) To what minimum temperature would the N_2O have to be heated before a balloon filled with the gas at 1 atm would rise in air at 25 °C? (Neglect the mass of the balloon itself.)

Cannizzaro's method

43. The gases listed in the table all contain the element X. Use the method outlined on page 168 (Cannizzaro's method) to determine the atomic weight of X. What element do you think X is?

Compound	Molecular weight	X, %
nitryl fluoride	65.01	49.4
nitrosyl fluoride	49.01	32.7
thionyl fluoride	86.07	18.6
sulfuryl fluoride	102.07	31.4

Gases in chemical reactions

44. A particular coal sample contains 2.12% S, by mass. When the coal is burned, the sulfur is converted to $SO_2(g)$. What volume of $SO_2(g)$, measured at 25 °C and 738 mmHg, is produced by burning 3.5×10^6 lb of this coal?

45. 2.92-g sample of a KCl–KClO$_3$ mixture is decomposed by heating and produces 89.8 cm^3 $O_2(g)$, measured at 21.8 °C and 727 mmHg. What is the mass percent of KClO$_3$ in the mixture. [*Hint:* The KCl is unchanged.]

$$2 \text{ KClO}_3(s) \rightarrow 2 \text{ KCl}(s) + 3 \text{ O}_2(g)$$

46. Hydrogen peroxide, H_2O_2, is finding new uses as an oxygen source for the treatment of municipal water and industrial wastewater.

$$2 \text{ H}_2\text{O}_2(aq) \rightarrow 2 \text{ H}_2\text{O} + \text{O}_2(g)$$

Calculate the volume of $O_2(g)$ at 23 °C and 726 mmHg that could be liberated from 1.00 L of a water solution containing 30.% H_2O_2, by mass. The density of the aqueous solution of H_2O_2 is 1.11 g/cm^3.

47. The Haber process is the principal method for fixing nitrogen (converting N_2 to nitrogen compounds).

$$N_2(g) + 3 \text{ H}_2(g) \rightarrow 2 \text{ NH}_3(g)$$

Assuming complete conversion of the reactant gases to $NH_3(g)$, and that the gases behave ideally,

(a) How many liters of $NH_3(g)$ can be produced from 313 L of $H_2(g)$ if the gases are measured at 515 °C and 525 atm pressure?

(b) How many liters of $NH_3(g)$, *measured at 25 °C and 727 mmHg*, can be produced from 313 L $H_2(g)$ measured at 515 °C and 525 atm pressure?

48. 1.50-L $H_2S(g)$, measured at 23.0 °C and 735 mmHg, is mixed with 4.45 L $O_2(g)$, measured at 26.1 °C and 750. mmHg, and burned.

$$2 H_2S(g) + 3 O_2(g) \rightarrow 2 SO_2(g) + 2 H_2O(g)$$

(a) How much $SO_2(g)$, in moles, is produced?

(b) If the excess reactant and products of the reaction are collected at 748 mmHg and 120.0 °C, what volume will they occupy?

49. An explosive undergoes rapid decomposition when subjected to mechanical shock or rapid heating. Great quantities of gases and heat are given off. The explosion of nitroglycerin can be represented as

$$C_3H_5(NO_3)_3(l) \rightarrow$$
$$CO_2(g) + H_2O(g) + N_2(g) + O_2(g) \quad \text{(unbalanced)}$$

Estimate the total pressure, in atm, that would be produced by the explosion of 1.0 kg of nitroglycerin, if the products were confined to a volume of 1.0 L at 3300 °C. (The result can only be an estimate because additional reactions might occur and because the gases would not be ideal under these conditions.)

50. Refer to the Summarizing Example. Use C_8H_{18} for the "formula" of gasoline and 0.71 g/mL for its density. If an automobile obtains 31.2 mi/gal, how many liters of $CO_2(g)$, measured at 28 °C and 732 mmHg, are produced in a trip of 245 mi? (1 gal = 3.785 L.)

Mixtures of gases

51. A gas cylinder of 48.5 L volume contains $N_2(g)$ at a pressure of 31.8 atm and 23 °C. What mass of $Ne(g)$ must be added to this same cylinder to raise the total pressure to 75.0 atm?

52. A 1.76-L container of $H_2(g)$ at 758 mmHg and 24.8 °C is connected to a 2.76-L container of $He(g)$ at 715 mmHg and 24.8 °C. What is the *total* gas pressure after the gases have mixed, with the temperature remaining at 24.8 °C?

53. A mixture of 4.0 g $H_2(g)$ and an unknown quantity of $He(g)$ is maintained at STP. If 10.0 g $H_2(g)$ is added to the mixture while conditions are maintained at STP, the gas volume doubles. What mass of He is present?

54. Air that is exhaled (expired) differs from normal air. A typical analysis of expired air at 37 °C and 760. mmHg, expressed as percent *by volume*, is 74.2% N_2, 15.2% O_2, 3.8% CO_2, 5.9% H_2O, and 0.9% Ar.

(a) Refer to Example 6-19. What is the apparent molar mass of expired air?

(b) Would you expect the density of expired air to be greater or less than that of ordinary air at the same temperature and pressure? Explain.

(c) What is the ratio of the partial pressure of $CO_2(g)$ in expired air to that in ordinary air?

55. Refer to Example 6-18 and the Summarizing Example. What is the mass percent of N_2, O_2, Ar, and CO_2 in air. [*Hint:* Base your calculation on 1.00 L or 100. L of air.]

Collection of gases over liquids

56. A 1.76-g sample of aluminum reacts with excess HCl and the liberated H_2 is collected over water at 26 °C at a barometric pressure of 738 mmHg. What *total* volume of gas is collected? (Vapor pressure of water at 26 °C = 25.2 mmHg.)

$$2 Al(s) + 6 HCl(aq) \rightarrow 2 AlCl_3(aq) + 3 H_2(g)$$

57. A 243-cm^3 sample of $Ar(g)$ at 25 °C and at a barometric pressure of 751 mmHg is passed through water at 25 °C. What is the volume of gas when saturated with water vapor and again measured at 25 °C and 751 mmHg barometric pressure? (Vapor pressure of water at 25 °C = 23.8 mmHg.)

58. A sample of $O_2(g)$ is collected over water at 25 °C. The volume of gas is 1.28 L. In a subsequent experiment it is determined that the mass of O_2 present is 1.58 g. What must have been barometric pressure at the time the gas was collected? (Vapor pressure of water at 25 °C = 23.8 mmHg.)

59. A 1.072-g sample of $He(g)$ is found to occupy a volume of 8.446 L when collected over hexane at 25.0 °C and 738.6 mmHg barometric pressure. Use these data to determine the vapor pressure of hexane at 25 °C. [*Hint:* Refer to Figure 6-14.]

Kinetic molecular theory

60. A kinetic theory verification of Avogadro's law was provided in the text. Use equations from Section 6-9 to verify Boyle's and Charles's laws.

61. The root-mean-square speed, u_{rms}, of H_2 molecules at 273 K is 1.84×10^3 m/s.

(a) At what temperature is u_{rms} for H_2 equal to 3.68×10^3 m/s?

(b) What is u_{rms} for N_2 at 273 K?

62. Calculate u_{rms}, in m/s, for $Cl_2(g)$ molecules at 25 °C.

63. Refer to Example 6-21.

(a) What must be the molecular weight of a gas if the gas molecules are to have a root-mean-square speed at 25 °C equal to that of the M-16 rifle bullet?

(b) Noble gases (group 8A) are monatomic gases (existing as atoms, not molecules). Cite one noble gas whose u_{rms} at 25 °C is greater than the rifle bullet's, and one whose u_{rms} is smaller.

Effusion of gases

64. What are the ratios of the effusion rates for the following pairs of gases? **(a)** H_2 and O_2; **(b)** H_2 and D_2 (D = deuterium, i.e., ^2_1H); **(c)** $^{14}CO_2$ and $^{12}CO_2$.

65. If 0.00251 mol $N_2O(g)$ effuses through an orifice in a certain period of time, how much $NO_2(g)$ would effuse in the same time with the same conditions?

66. A sample of $N_2(g)$ effuses through a tiny hole in 38 s. What must be the molecular weight of a gas that requires 55 s to effuse under identical conditions?

67. A common laboratory demonstration of Graham's law of diffusion is pictured here. The diffusing species are $NH_3(g)$ and $HCl(g)$. Each is produced by vaporization from an aqueous solution. At the point where these gases meet, they react to form a white cloud of $NH_4Cl(s)$. If initially a few drops of $NH_3(aq)$ and a few drops of $HCl(aq)$ are placed one meter apart, at approximately what point between them will $NH_4Cl(s)$ first form?

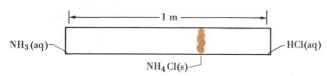

Exercise 67

$$V^3 - n\left(\frac{RT + bP}{P}\right)V^2 + \left(\frac{n^2a}{P}\right)V - \frac{n^3ab}{P} = 0$$

(b) What is the volume occupied by 132 g $CO_2(g)$ at a pressure of 10.0 atm and 280 K? Use data from Table 6-3 as necessary.

★70. The virial equation of state for $O_2(g)$ has the form

$$P\overline{V} = RT\left\{1 + \frac{B}{\overline{V}} + \frac{C}{\overline{V}^2}\right\}$$

where $\overline{V}$ is the molar volume, $B = -21.89$ cm^3/mol, and $C = 1230$ cm^6/mol^2.

(a) Use the equation to calculate the pressure exerted by 1 mol $O_2(g)$ confined to a volume of 500 cm^3 (i.e., having $\overline{V} = 500$ cm^3/mol) at 273 K.

(b) Is the result calculated in (a) consistent with that suggested for $O_2(g)$ by Figure 6-18? Explain.

Nonideal gases

68. Calculate the pressure exerted by 1.00 mol $CO_2(g)$ confined to a volume of 855 cm^3 at 30 °C. Use **(a)** the ideal gas equation and **(b)** the van der Waals equation. **(c)** Compare the results and explain.

★69. If the van der Waals equation is solved for volume, a cubic equation is obtained.

(a) Derive the following equation by rearranging equation (6.26).

Additional Exercises

71. The mercury level in the open arm of an open-end manometer is 283 mm above a reference point. In the arm connected to a container of gas, the level is 38 mm above the same reference point. If barometric pressure is 753.5 mmHg, what is the total pressure of the gas in the container? [*Hint:* Refer to the drawing in Review Problem 3.]

72. Refer to Examples 6-3 and 6-4. Describe circumstances under which you might measure a gas pressure with an open-end manometer filled with mercury and other circumstances where you might choose to use a liquid like glycerol instead.

73. Calculate the pressure, in kilopascals (kPa), exerted by the blade of a knife if one presses down with a force corresponding to 33 kg. The blade has a length of 10.3 cm and a thickness along its edge of 0.27 mm.

74. The pressure of He(g) in a cylinder of the compressed gas is given as 775 lb/in.^{2}. What is this pressure expressed in each of the units listed in Table 6-1? [*Hint:* Derive conversion factors based on the standard atmosphere of pressure.]

★75. Pictured below is the apparatus used by Boyle in establishing the relationship between pressure and volume of a fixed quantity of gas. (Boyle's data have been converted to modern units.) At the start of the experiment, the length of the air column (A) on the left was 30.5 cm and the heights of mercury in each arm of the tube were equal. When mercury was added to the right arm of the tube, a difference in mercury levels (B) was produced and the entrapped air on the left was compressed to a shorter length (smaller volume). For example, in the illustration A = 27.9 cm and B = 7.1 cm. Boyle's values of A and B are

A, cm	B, cm	A, cm	B, cm	A, cm	B, cm
30.5	0.0	20.3	38.3	12.7	105.6
27.9	7.1	17.8	53.8	10.2	147.6
25.4	15.7	15.2	75.4	7.6	224.6
22.9	25.7				

Barometric pressure at the time of the experiment was 739.8 mmHg. Assuming that the length of the air column (A) is proportional to the volume of air, show that these data do in fact conform reasonably well to Boyle's law.

★76. Figure 6-6 is the usual graphical representation of Boyle's law. An alternative, however, is to plot P against $1/V$. The resulting graph is a straight line passing through the origin. Use Boyle's data from Exercise 75 to draw such a straight-line graph. What factors would affect the *slope* of this straight line? Explain.

77. A sample of $N_2(g)$ occupies a volume of 58.0 cm^3 under the existing barometric pressure. Increasing the pressure by 125 mmHg reduces the volume to 49.6 cm^3. What is the prevailing barometric pressure?

78. Start with the conditions at points A, B, and C in Figure 6-8. Use Charles's law to calculate the volume of each gas at 0, −100, −200, −250, and −270 °C; and show that indeed the volume of each gas becomes zero at −273.15 °C.

★79. An automobile tire is inflated to a gauge pressure of 28 lb/in.^{2} at 60 °F. After the car has been driven for several hours at high speed, the tire is checked and found to have a gauge pressure of 32 lb/in.^{2}. What is the Fahrenheit temperature of the air in the tire? Assume that the volume of air in the tire remains constant. Note that gauge pressure means a pressure above atmospheric; that is, a gauge pressure of 28 lb/in.^{2} corresponds to an absolute pressure of 28 + 14.7 = 43 lb/in.^{2}.

80. A 10.0-g sample of a gas has a volume of 4.62 L at 35 °C and 762 mmHg. If to this *constant* 4.62-L volume is added 2.3 g of the same gas and the temperature raised to 51 °C, what is the new gas pressure?

81. A 14.7 L cylinder contains 46.7 g O_2 at 35 °C. What is the pressure of this gas?

82. A sample of $N_2(g)$ fills a 1.98-L container at 21.5 °C and 751 mmHg. How many grams of N_2 must be released from this container if the temperature is to be raised to 99.8 °C while the *pressure* and *volume* are held constant?

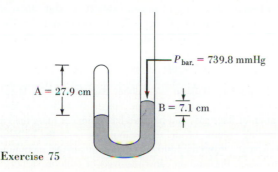

Exercise 75

83. A 0.312-g sample of a gaseous compound occupies 185 cm^3 at 25.0 °C and 745 mmHg. The compound consists of 85.6% C and 14.4% H, by mass. What is its molecular formula?

84. What mass of He(g) should be added to 1.87 L O$_2$(g) at 22 °C and 721 mmHg to increase the pressure to 2.77 atm? (The volume and temperature are held constant.)

⋆85. A balloon is inflated with 1.00 ft^3 of He(g) at STP and released. What is the gas pressure in the balloon when it has expanded to a volume of 75.0 L? Assume a temperature of −20 °C at this altitude.

86. A 2.37 g sample of NH$_4$NO$_3$(s) is introduced into an evacuated 1.97-L flask and then heated to 250 °C. What is the total gas pressure in the flask at 250 °C when the NH$_4$NO$_3$ has completely decomposed?

$$NH_4NO_3(s) \rightarrow N_2O(g) + 2\ H_2O(g)$$

87. In the reaction of CO$_2$(g) and sodium peroxide (Na$_2$O$_2$), sodium carbonate and oxygen gas are formed.

$$2\ Na_2O_2(s) + 2\ CO_2(g) \rightarrow 2\ Na_2CO_3(s) + O_2(g)$$

This reaction is used in submarines and space vehicles to eliminate expired CO$_2$(g) and to generate some of the O$_2$(g) required for breathing. Assume the following. Volume of gases exchanged in the lungs: 4.0 L/min; CO$_2$ content of expired air: 3.8% CO$_2$, by volume. If the CO$_2$(g) and O$_2$(g) in the above reaction are measured at the same temperature and pressure: (a) How many mL O$_2$(g) are produced per minute, if the source of CO$_2$ is expired air? (b) At what rate is the Na$_2$O$_2$(s) consumed, in g/h, assuming that the gases are at 25 °C and 735 mmHg pressure?

88. Refer to the Summarizing Example. Ethyl alcohol, C$_2$H$_5$OH, is widely used as a substitute for gasoline in motor fuels in Brazil.

(a) Write a balanced chemical equation for the complete combustion of C$_2$H$_5$OH.

(b) Assuming the same ideal conditions for the combustion of ethyl alcohol as described for the combustion of gasoline, what is the ideal mass ratio of air to alcohol?

89. What is the partial pressure of Cl$_2$(g) in a gaseous mixture at STP that consists of 50.0% N$_2$, 22.3% Ne, and 27.7% Cl$_2$, *by mass?*

90. Producer gas is a type of fuel gas made by passing air or steam through a bed of hot coal or coke. A typical producer gas has the following composition, in percent by volume: 8.0% CO$_2$, 23.2% CO, 17.7% H$_2$, 1.1% CH$_4$, and 50.0% N$_2$.

(a) What is the apparent molar mass of this gas?

(b) What is the density of this gas at 25 °C and 752 mmHg?

(c) What is the partial pressure of CO in this gaseous mixture at STP?

91. A *mixture* of H$_2$(g) and O$_2$(g) is prepared by electrolyzing 1.32 g water, and the mixture of gases is collected over water at 30 °C when the barometric pressure is 748 mmHg. The volume of "wet" gas obtained is 2.90 L. What must be the vapor pressure of water at 30 °C?

$$2\ H_2O(l) \xrightarrow{\text{electrolysis}} 2\ H_2(g) + O_2(g)$$

⋆92. An 0.168-L sample of O$_2$(g) is collected over water at 26 °C and a barometric pressure of 737 mmHg. In the gas that is collected, what is the *percent* water vapor, (a) by volume; (b) by number of molecules; (c) by mass? (Vapor pressure of water at 26 °C = 25.2 mmHg.)

93. At what temperature will u_{rms} for Ne(g) be the same as u_{rms} for He at 300 K?

94. Following the method in Table 6-2, determine $\bar{u}$ and u_{rms} for a group of six particles with the speeds: 9.8×10^3, 9.0×10^3, 8.3×10^3, 6.5×10^3, 3.7×10^3, and 1.8×10^3 m/s.

95. At an altitude of 300 km the principal gaseous species is *atomic* oxygen, O. Oxygen atoms at this altitude have a root-mean-square speed of 1.48×10^3 m/s. What is the approximate temperature associated with this molecular speed?

96. Refer to Example 6-25. Recalculate the pressure of Cl$_2$(g) using both the ideal gas equation and van der Waals equation for the data given, except change the temperature to (a) 100 °C, (b) 200 °C, and (c) 400 °C. From the results, confirm the statement that a gas tends to be more ideal at high temperatures than at low temperatures.

⋆97. Recall the composition of air (Example 6-18). What volume of air, measured at STP, is required to complete the combustion of 1.00×10^3 L of a natural gas (measured at 23 °C and 741 mmHg) having the composition, 77.3% CH$_4$, 11.2% C$_2$H$_6$, 5.8% C$_3$H$_8$, 2.3% C$_4$H$_{10}$ (and 3.4% noncombustible gases), by volume?

⋆98. Mixtures of the anesthetic gas cyclopropane, (CH$_2$)$_3$, and air with between 2.4 and 10.3%, (CH$_2$)$_3$, by volume, are explosive. A sealed 1500-ml cylinder of (CH$_2$)$_3$ (g) at 2.50 atm and 25 °C is placed in a fume hood of volume 72 ft^3 containing air at 755 mmHg and 25 °C. If the seal on the cylinder were to break and the (CH$_2$)$_3$ to mix with the air in the fume hood, would an explosive mixture result?

⋆99. A gaseous mixture of He and O$_2$ has a density of 0.518 g/L at 25 °C and 720 mmHg. What is the % He, by mass, in the mixture?

⋆100. The equation

$$\frac{m/V}{P} = \frac{d}{P} = \frac{\mathcal{M}}{RT}$$

suggests that the ratio of gas density (*d*) to gas pressure (*P*), at constant temperature, should be a constant. The following gas density data were obtained for *oxygen* gas at various pressures at 273.15 K: 1.000 atm, 1.428962 g/L; 0.750 atm, 1.071485 g/L; 0.500 atm, 0.714154 g/L; 0.250 atm, 0.356985 g/L.

(a) Calculate values of *d/P*, and with a graph or by other means determine the best value of the term *d/P* for oxygen at 273.15 K. (This is the value corresponding to oxygen as an ideal gas.)

(b) Use the value of *d/P* from part (a) to calculate a precise value of the atomic weight of oxygen and compare it with that listed in an atomic weight table.

⋆101. A particular limestone contains only CaCO$_3$ and MgCO$_3$. When a 0.4515-g sample of this limestone is decomposed by heating, 0.2398 g of mixed oxide (CaO and MgO) is obtained. What volume of CO$_2$(g) at 752.0 mmHg and 285.3 K will be produced by decomposing 50.0 lb of this limestone?

$$MCO_3(s) \rightarrow MO(s) + CO_2(g) \quad \text{(where M = Ca or Mg)}$$

⋆102. A sounding balloon is a rubber bag, filled with H$_2$(g) and carrying a set of instruments (the "payload"). Because this combination of bag, gas, and payload has a smaller mass than a corresponding volume of air, the balloon rises. As the balloon rises, it expands. From the following data estimate the maximum height to which the balloon can rise: mass of balloon, 1200 g;

payload, 1700 g; quantity of $H_2(g)$ in balloon, 120 ft^3 at STP; diameter of balloon at maximum height, 25 ft. Air pressure and temperature as a function of altitude are 0 km, 1.0×10^3 mb, 288 K; 5 km, 5.4×10^2 mb, 256 K; 10 km, 2.7×10^2 mb, 223 K; 20 km, 5.5×10^1 mb, 217 K; 30 km, 1.2×10^1 mb, 230 K; 40 km, 2.9×10^0 mb, 250 K; 50 km, 8.1×10^{-1} mb, 250 K; 60 km, 2.3×10^{-1} mb, 256 K.

*103. Atmospheric pressure as a function of altitude can be calculated with an equation known as the barometric formula.

$$P = P_0 \times 10^{-\mathcal{M}gh/2.303\,RT}$$

where P is the pressure, in atm, at an altitude of h meters. P_0 is the pressure at sea level (usually taken to be 1 atm), g is the acceleration due to gravity (9.81 m/s^2); and $\mathcal{M}$ is the molar mass of air, expressed in kg per mole. R is expressed as 8.314 J mol^{-1} K^{-1}, and T is in kelvins.

(a) Estimate barometric pressure at the top of Mt. Whitney in California. (Altitude: 14,494 ft; assume a temperature of 10 °C.)

(b) Use the barometric formula to show that barometric pressure decreases by one-thirtieth in value for every 900-ft increase in altitude.

Self-Test Questions

For questions 104 through 113 select the item that best completes each statement.

104. The greatest pressure of the following is that exerted by (a) a column of Hg(l) 75.0 cm high ($d = 13.6$ g/cm^3); (b) 10.0 g $H_2(g)$ at STP; (c) a column of air 10 mi high; (d) a column of $CCl_4(l)$ 60.0 cm high ($d = 1.59$ g/cm^3).

105. Of the following gases the one with the greatest density at STP is (a) Cl_2; (b) SO_3; (c) N_2O; (d) PF_3.

106. For a fixed amount of gas at a fixed pressure, changing the temperature from *100 °C* to *200 K* causes the gas volume (a) to decrease; (b) to double; (c) to increase, but not to twice its original value; (d) not to change.

107. The temperature and pressure at which Cl_2 is most likely to behave like an ideal gas are (a) 100 °C and 10.0 atm; (b) 0 °C and 0.50 atm; (c) 200 °C and 0.50 atm; (d) -100 °C and 10.0 atm.

108. A sample of $O_2(g)$ is collected over water at 23 °C at a barometric pressure of 751 mmHg (vapor pressure of water at 23 °C = 21 mmHg). The *partial* pressure of $O_2(g)$ in the sample collected is (a) 21 mmHg; (b) 751 mmHg; (c) 0.96 atm; (d) 1.02 atm.

109. A comparison is made at standard temperature and pressure (STP) of 0.50 mol $H_2(g)$ and 1.0 mol He(g). The two gases will (a) have equal average molecular kinetic energies; (b) have equal molecular speeds; (c) occupy equal volumes; (d) have equal effusion rates.

110. At 0 °C and 0.500 atm, 4.48 L $NH_3(g)$ (a) contains 0.20 mol NH_3; (b) weighs 1.70 g; (c) contains 6.02×10^{23} NH_3 molecules; (d) contains 0.40 mol NH_3.

111. In the reaction

$$2\ Al(s) + 6\ HCl(aq) \rightarrow 2\ AlCl_3(aq) + 3\ H_2(g)$$

(a) 67.2 L $H_2(g)$ at STP is produced for every mole Al that reacts; (b) 6 L HCl(aq) is consumed for every 3 L $H_2(g)$ produced; (c) 11.2 L $H_2(g)$ at STP is produced for every mol HCl consumed; (d) 33.6 L $H_2(g)$ is produced, *regardless of temperature and pressure*, for every mol Al that reacts.

112. A mixture of 5.0×10^{-5} mol $H_2(g)$ and 5.0×10^{-5} mol $SO_2(g)$ is introduced into a 10.0-L container at 25 °C. The container has a "pinhole" leak. After a period of time, the partial pressure of $H_2(g)$ in the *remaining mixture* (a) exceeds that of $SO_2(g)$; (b) is equal to that of $SO_2(g)$; (c) is less than that of $SO_2(g)$; (d) is the same as in the original mixture.

113. To establish a pressure of 2.00 atm in a 2.24-L cylinder containing 1.60 g $O_2(g)$ at 0 °C, (a) add 1.60 g O_2; (b) release 0.80 g O_2; (c) add 2.00 g He; (d) add 0.60 g He.

114. 0.10 mol He(g) is added to 2.24 L $H_2(g)$ at STP. This is followed by an increase in temperature to 100 °C while the pressure and amount of gas are held constant. What is the final gas volume?

115. How many grams of $O_2(g)$ must be present in a 2.10-L flask at 24 °C, if this $O_2(g)$ is to exert the same pressure as does 11.5 g N_2 in a 1.80-L flask at 18 °C?

116. Explain briefly why the height of the mercury column in a barometer is *independent* of the diameter of the barometer tube (i.e., whether the diameter is 1 mm, 1 cm, 10 cm, . . .).

117. Calculate the volume of $H_2(g)$ (expressed at 22 °C and 745 mmHg) required to react with 30.0 L CO(g) (measured at 0 °C and 760 mmHg) in the reaction

$$3\ CO(g) + 7\ H_2(g) \rightarrow C_3H_8(g) + 3\ H_2O(l)$$

118. A particular gaseous hydrocarbon that is 82.7% C and 17.3% H, by mass, has a density of 2.35 g/L at 25 °C and 752 mmHg. What is the molecular formula of this hydrocarbon?

7 Thermochemistry

A fuel fire test. Thermochemistry has important applications to combustion reactions. [Naval Research Laboratory photo by Dan Boyd]

200

Natural gas consists primarily of methane gas, CH_4. When methane burns completely in air, the products are carbon dioxide and water. Another "product" of this combustion is perhaps even more important—heat energy. We can use this heat energy to heat water in a water heater, to warm air in a space heater, or to cook food on a stove. Chemical reactions that give off heat energy are called *exothermic* reactions. But not all chemical reactions are exothermic. In some cases we must supply heat energy to a reaction mixture to make the reaction go; such a reaction is called *endothermic*. We need a way to predict whether a reaction is endothermic or exothermic.

First, however, we must explore some fundamental ideas about the energy forms called heat and work, including how to measure quantities of heat. Finally, and most important, we will learn how to use tabulated data to *calculate* heat quantities associated with chemical reactions. In fact, the great value of thermochemistry is in permitting us to make large numbers of calculations from a relatively small number of measurements.

7-1 Energy

You will find a more quantitative treatment of some of the topics in this section in Appendix B. For example, a derivation of equation (7.1) is given. You may find it helpful to read portions of Appendix B as you read this section, but the quantitative detail is not essential to the ideas presented in this chapter.

Like so many other fundamental terms in science *energy* is derived from Greek; it means "work within." **Energy** is a capacity to do work. Work is done when a force acts through a distance. Moving objects can do work when they are slowed down or stopped. Thus when one billiard ball strikes another and sets it in motion, work is done. The energy (or capacity to do work) of a moving object is called **kinetic energy.** Again, the word *kinetic* is of Greek origin, meaning "motion." The amount of kinetic energy (K.E.) of an object, as we saw in Section 6-9, is related to its mass (m) and velocity (u).

$$K.E. = \tfrac{1}{2}mu^2$$

(7.1)

The units in equation (7.1) are $kg\ m^2\ s^{-2}$ = joule (J). As shown in more detail in Appendix B, a force of 1 newton (N) has the units $kg\ m\ s^{-2}$. The product, one newton × one meter, a quantity of work, has the value $1\ kg\ m\ s^{-2} \times 1\ m = 1\ kg\ m^2\ s^{-2} = 1$ joule (J). Thus, the basic unit of energy (and work) is the **joule.**

Figure 7-1, which depicts a bouncing ball, suggests something about the nature of energy and work. First, to raise the ball to the starting position we have to apply a force through a distance. (That is, we have to apply a force to overcome the force of gravity.) The work we do is "stored" in the ball as energy. This "energy of position" is called **potential energy,** an energy associated with forces of attraction or repulsion between objects.

These hypothetical *elastic* collisions of the bouncing ball are exactly the type of collisions that are postulated in the kinetic molecular theory (Section 6-9).

When we release the ball it is pulled toward the center of the earth by the force of gravity—it falls. Potential energy is converted to kinetic energy during this fall. If the collision of the ball with an *immovable* surface were perfectly *elastic*, the ball would reverse direction and leave the surface with the same kinetic energy that it had just before the collision. Throughout its rise, the kinetic energy of the ball would decrease while its potential energy increased. At the top of its climb, all of the kinetic energy would have been converted to potential energy. When the ball

FIGURE 7-1
Potential energy (P.E.) and kinetic energy (K.E.).

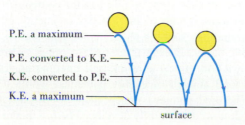

P.E. a maximum

P.E. converted to K.E.

K.E. converted to P.E.

K.E. a maximum

surface

The energy of the bouncing ball changes continuously from potential to kinetic energy, back to potential energy, etc. The sum P.E. + K.E. continuously decreases with each bounce as energy, in the form of heat, is transferred to the surface and to the air.

FIGURE 7-2

Distinguishing between heat and work.

The motion of molecules (or atoms) and their average kinetic energy are represented by arrows within the object.
(a) When the object gains heat energy, the intensity of the *random* or *chaotic* molecular motion (average kinetic energy) increases, indicated by the longer arrows. However, there is no net movement of the object itself.
(b) When work is done on the object, the intensity of molecular motion (average kinetic energy) is unchanged, but there is a net movement of the object.

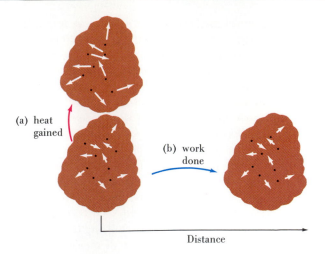

(a) heat gained

(b) work done

Distance

reached its original high position it would reverse direction and start toward its second bounce.

For the process just described, at every point in its path the *sum* of the potential and kinetic energy of the ball would have some *constant* value. And, of course, the ball would keep bouncing forever. However, we know that the ball will not bounce forever. In each bounce its maximum height is less than in the previous bounce, and eventually the ball comes to rest. The ball must lose energy to the surface (and to molecules of air) in every bounce. But, the energy transfer to the surface (and to the air) is not in the form of work. Instead, the energy transfer simply goes toward increasing the kinetic energies of the atoms in the surface (and molecules of air), slightly raising the temperature. An energy transfer that produces a change in temperature is called **heat.** Thus, in our discussion of energy we need to make a distinction between energy changes that result in the action of forces through distances—*work*—and energy changes associated with temperature changes—*heat*. Figure 7-2 may help you to understand this distinction.

7-2   Some Terminology

Before continuing with our discussion of energy, we must introduce some terms that are commonly used in the study of thermodynamics, and the branch of thermodynamics that we are investigating in this chapter—thermochemistry. In a narrow sense thermodynamics is the study of the relationships between heat ("thermo") and work ("dynamics"). In a broader sense it is concerned with all forms of energy and their interconversions, and with interconversions of matter as well. Thus, thermodynamics lies at the very core of modern chemistry. Thermochemistry is more limited in scope. Its main concern is the heat effects that accompany chemical reactions.

The portion of the universe that is the subject of a thermodynamic study is called a **system,** and the portions of the universe with which the system interacts are called the **surroundings** of the system. A thermodynamic system may be as simple as a beaker of water or as complex as the contents of a blast furnace or a polluted lake. **Interactions** refer to the transfer of energy or matter between a system and its surroundings, and it is these interactions that we generally focus on in a thermodynamic study. Energy transfers can occur as **heat (q)** or in several others forms, known collectively as **work (w).** Energy transfers occurring as heat or work affect the total amount of energy contained within a system, its **internal energy (E).**

Thermodynamics is independent of any particular theory of the structure of matter. It was, in fact, fully developed *before* modern atomic theory. Thus, the concept of the internal energy of a system can be handled without ever describing where this energy comes from. We now recognize, however, that internal energy represents the total energy (kinetic and potential) associated with the ultimate particles of matter in the system. This includes energy associated with the chemical bonds between atoms, energies of attractions between molecules, the kinetic energy of translational motion of molecules, and so on.

An important idea for you to understand is that the only energy *contained* within a system is internal energy. A system *does not* contain either work or heat! Work and heat are means by which a system can *exchange* energy with its surroundings. And in these exchanges a strict accounting must be kept of the internal energy within the system and heat and/or work flowing into or out of the system, but more about this later, in Section 7-5 on the first law of thermodynamics. First, we explore more fully some ideas about work and heat.

7-3 Work

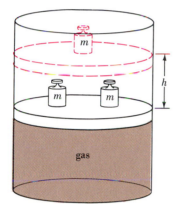

FIGURE 7-3
Pressure–volume work.

The most common type of work that we must consider is called **pressure–volume work.** It is work associated with the expansion or compression of gases. Our chief purpose here is to derive a simple mathematical equation for pressure–volume work. A hypothetical situation for the expansion of a gas is shown in Figure 7-3. A quantity of gas is confined in a cylinder by a freely moving piston. The gas is confined to its initial volume by two weights, each having a mass, m. When one of the weights is removed, the remaining weight is raised through the distance h. Now we consider two ways of expressing the quantity of work performed.

1. *Work as force times distance.* The force exerted by a weight of mass m is mg, where g is the acceleration due to gravity. This force, multiplied by the distance h, is the amount of work done.

$$w = mgh \qquad (7.2)$$

2. *Work as pressure times a volume change.* The pressure exerted by the remaining weight on the expanding gas is a force per unit area ($P = F/A$). The force of the expansion then is a pressure times the area, $F = PA$. The work performed is this force times the distance h: $w = PAh$. The area of the cylinder, A, times a height, h, is the volume of the portion of the cylinder that lies between the two positions of the piston. This is the *change in volume* that results from the expansion of the gas, ΔV. The amount of work done is

Equation for pressure–volume work (at constant pressure).

$$w = P \, \Delta V \qquad (7.3)$$

It is not difficult to show that equations (7.2) and (7.3) are equivalent. For the pressure, P, in (7.3), substitute a force (mg) divided by the area over which it is exerted (A); and for the change in volume, ΔV, substitute a cross-sectional area (A) times a height (h).

The Greek letter *delta*, Δ, is used to indicate a *change* in some quantity. By convention, it is always expressed as the *final* value minus the *initial* value, that is, $\Delta V = V_f - V_i$.

$$\text{work} = P \, \Delta V = \frac{mg}{A} \times Ah = mgh$$

To decide between equations (7.2) and (7.3) in a problem-solving situation, consider the form in which data are given, as illustrated by Examples 7-1 and 7-2.

Example 7-1

Calculating work as the product of a force and a distance. Refrigerator repair workers carry a plastic enclosure that can be slipped under the refrigerator and inflated with a foot pump. When the refrigerator has been lifted by a cushion of air, it can be easily moved. How much work is done in lifting a 345-lb refrigerator to a height of 5.0 mm?

Solution. To use equation (7.2) all we need are mass in kg and height in meters. It is not difficult to find these. Equation (7.3) is not useful in this situation because we have no information about gas pressures and volumes.

$$w = mgh = 345 \text{ lb} \times \frac{1.00 \text{ kg}}{2.20 \text{ lb}} \times \frac{9.81 \text{ m}}{\text{s}^2} \times 5.0 \text{ mm} \times \frac{1 \text{ m}}{1000 \text{ mm}}$$

$$= 7.7 \text{ kg m}^2 \text{ s}^{-2} = 7.7 \text{ J}$$

SIMILAR EXAMPLE: Exercise 6.

Example 7-2

Calculating work as the product of a pressure and a volume change. What is the quantity of work performed in the expansion of the gas in Figure 7-3 if the gas expands against a constant pressure of 3.55 kPa and the change in volume (ΔV) is 420. cm^3?

Solution. Here, since we are given a pressure and a volume change, equation (7.3) is the logical choice. The pressure unit, Pa, has the unit N/m^2. If this is multiplied by a volume change in m^3, the product has the unit N m = J. Thus, we need to convert kPa to Pa and cm^3 to m^3.

> If pressure is given in units such as atm, convert from atm to Pa (recall Table 6-1).

$$\text{work} = P\,\Delta V = 3.55 \text{ kPa} \times \frac{1000 \text{ Pa}}{1 \text{ kPa}} \times \frac{1 \text{ N/m}^2}{1 \text{ Pa}} \times 420. \text{ cm}^3 \times \frac{1 \text{ m}^3}{1 \times 10^6 \text{ cm}^3}$$

$$= 1.49 \text{ N m} = 1.49 \text{ J}$$

SIMILAR EXAMPLES: Exercises 19, 20.

7-4 Heat

> Recall from Section 6-9 that temperature is proportional to the average kinetic energies of molecules. A difference in temperature between two objects means that their molecules have different average kinetic energies.

Heat is energy that is transferred as a result of a temperature difference. Energy, as heat, passes from a warmer body (higher temperature) to a colder body (lower temperature). At the molecular level, molecules of the warmer body, through collisions, lose kinetic energy to those of the colder body. Energy is transferred—heat "flows"—until the average molecular kinetic energies of the two bodies become the same, until the temperatures become equal.* Heat, like work, describes energy in transit between an object and its surroundings.

It is reasonable for us to expect the quantity of heat energy, q, required to change the temperature of a substance to depend on

- how much the temperature is to be changed,
- the quantity of substance,
- the nature of the substance (type of atoms or molecules).

The quantity of heat required to change the temperature of one *gram* of water by one degree *Celsius* has historically been defined as one **calorie (cal).** The calorie is

---

*Although the ultimate result of heat flow is to equalize the temperatures of two bodies, the temperature of a body may remain constant for a time as heat enters or leaves it. This is what happens, for example, when a block of ice (0 °C) absorbs heat from the surrounding air. Its temperature does not change until all the ice has melted. Such processes, called phase transitions, are described in Chapter 12. A process occurring at a constant temperature is said to be *isothermal*.

a small unit of energy and the unit **kilocalorie (kcal)** has also been widely used. More fundamental, however is the unit the **joule (J)**, the SI energy unit. The relationship between the joule and calorie is

$$1 \text{ cal} = 4.184 \text{ J} \tag{7.4}$$

Although we use the joule almost exclusively in this book, you should be familiar with the calorie as well; it is widely encountered in older scientific literature and still used to some extent.

The quantity of heat required to change the temperature of a system by one degree is called the **heat capacity** of the system. If the system is a mole of substance, we can use the term **molar heat capacity;** if the system is one gram of substance, we can use the term **specific heat capacity** or more commonly **specific heat** (sp. ht.).*

The specific heat of water is somewhat temperature dependent, but over the range from 0 to 100 °C its value is approximately

The specific heat of water.

$$\frac{1.00 \text{ cal}}{\text{g } H_2O \text{ °C}} = 1.00 \text{ cal g}^{-1} \text{ °C}^{-1} = \frac{4.18 \text{ J}}{\text{g } H_2O \text{ °C}} = 4.18 \text{ J g}^{-1} \text{ °C}^{-1} \tag{7.5}$$

Example 7-3

Relating a quantity of heat to a mass and a temperature change. While running, a typical adult expends 5.0×10^5 J/mi. If this 5.0×10^5 J of energy could be transferred as heat to water, how many grams of water could be raised from room temperature (25 °C) to the boiling point (100 °C)?

In nutritional studies the unit called a Calorie (Cal, with capital C) is actually a kilocalorie. Thus, a "1 Calorie" soft drink actually has a food value of 1000 cal. The quantity of heat described here is 5.0×10^5 J = 120 kcal = 120 Calories. Thus, running "burns off" nutritional Calories at the rate of 120 Calories for every mile that is run.

Solution. To raise the temperature of one gram of water from 25 °C to the boiling point (100 °C) requires 75 times as much energy as to raise the temperature by just one degree, that is, 75×4.18 J, or

$$\text{no. J} = 1.00 \text{ g water} \times \frac{4.18 \text{ J}}{\text{g water °C}} \times (100 - 25) \text{ °C} = 3.1 \times 10^2 \text{ J}$$

With the entire 5.0×10^5 J, the mass of water that can be brought to the boiling point is

$$\text{no. g water} = 5.0 \times 10^5 \text{ J} \times \frac{1.00 \text{ g water}}{3.1 \times 10^2 \text{ J}} = 1.6 \times 10^3 \text{ g water}$$

SIMILAR EXAMPLES: Exercises 1, 22, 63.

The key calculation in Example 7-3 was based on the following equation.

Relationship between a quantity of heat and mass, specific heat, and temperature change.

$$\text{quantity of heat} = q$$
$$= \underbrace{\text{mass of substance} \times \text{sp. ht.}}_{\text{heat capacity}} \times \text{temperature change} \tag{7.6}$$

We should express the temperature change in equation (7.6) as

$$\Delta T = T_f - T_i$$

where T_f is the final temperature, T_i is the initial temperature, and ΔT is the temperature change.

---

*The original meaning of the term specific heat was that of a *ratio:* the quantity of heat required to change the temperature of a mass of substance divided by the quantity of heat required to produce the same temperature change in the same mass of water. This would make it a dimensionless quantity. The meaning given here is more commonly used, however.

FIGURE 7-4
Conceptualizing heat transfer.

The two quantities $q_{sys.}$ and $q_{surr.}$ are equal in magnitude and opposite in sign. That is, one is always positive and the other negative; so that $q_{sys.} + q_{surr.} = 0$. Heat effects are generally measured in the *surroundings*. If the temperature of the surroundings *increases*, $q_{surr.} > 0$ (positive) and $q_{sys.} < 0$ (negative). The process occurring in the system is **exothermic**. If the temperature of the surroundings *decreases*, $q_{surr.} < 0$ (negative) and $q_{sys.} > 0$ (positive). The process occurring in the system is **endothermic**.

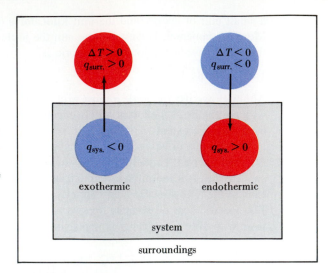

When the temperature of a system increases, T_f is greater than T_i (noted symbolically as $T_f > T_i$). ΔT is *positive* (i.e., $\Delta T > 0$), and so is q (i.e., $q > 0$). *A positive sign of q signifies that heat is absorbed or gained.* When the temperature of a system decreases, T_f is less than T_i (noted symbolically as $T_f < T_i$). ΔT is *negative* (i.e., $\Delta T < 0$), and so is q (i.e., $q < 0$). *A negative sign of q signifies that heat is evolved or lost.*

An additional idea that enters into heat energy calculations is the **law of conservation of energy.** In interactions between a system and its surroundings, the total energy remains *constant*. Thus, heat energy *lost* by a system is *gained* by its surroundings, and vice versa. Figure 7-4 illustrates this notion and gives additional detail on measuring quantities of heat. Note particularly that if a system *loses* heat the process is called **exothermic,** and if the system *gains* heat, the process is **endothermic.**

Figure 7-5 illustrates a simple laboratory method of determining the specific heat of a metal; the method is based on the law of conservation of energy. In the exchange of heat energy between the lead (q_{lead}) and the water (q_{water}), the total quantity of heat energy must be zero.

$$q_{lead} + q_{water} = 0 \tag{7.7}$$

or

$$q_{lead} = -q_{water} \tag{7.8}$$

That is, the two quantities of heat must be equal in magnitude and opposite in sign.

FIGURE 7-5
Determination of specific heat of lead—Example 7-4 illustrated.

(a) 150.0 g of lead at the temperature of boiling water (100.0 °C).
(b) 50.0 g of water in a thermally insulated beaker at 22.0 °C.
(c) Final lead–water mixture at a temperature of 28.8 °C.

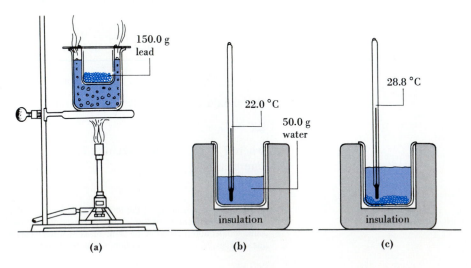

Are You Wondering:

What the specific heat of a substance tells you about the substance?

In Example 7-4 we find that only 0.13 J of heat is required to change the temperature of one gram of lead by one degree Celsius. Specific heats of other metals are of comparable magnitude. These low specific heats of metals mean that the temperatures of metals can be changed with relatively small quantities of heat. In general, metals can be heated quickly and cooled equally quickly.

Water, on the other hand, has a high specific heat (over 30 times as great as that of lead, for example). A much larger quantity of heat energy is required to change the temperature of a sample of water than of an equal mass of a metal. The moderating effects of large lakes on the temperatures of nearby communities can be understood in terms of the high specific heat of water. It takes much longer for a lake to heat up in the summer and to cool in the winter than does a land mass.

Example 7-4

Determining a specific heat from experimental data. Use data presented in Figure 7-5 to calculate the specific heat of lead.

If you know any four of the following five quantities, you can solve equation (7.6) for the remaining one: q, m, sp. ht., T_f, T_i.

Solution. First, use equation (7.6) to calculate q_{water}.

$$q_{water} = 50.0 \text{ g water} \times \frac{4.18 \text{ J}}{\text{g water } °C} \times (28.8 - 22.0) \text{ °C} = 1.4 \times 10^3 \text{ J}$$

From equation (7.8) we can write

$$q_{lead} = -q_{water} = -1.4 \times 10^3 \text{ J}$$

Now, from equation (7.6) again, we obtain

$$q_{lead} = 150.0 \text{ g lead} \times \text{sp. ht. lead} \times (28.8 - 100.0) \text{ °C}$$

$$= -1.4 \times 10^3 \text{ J}$$

$$\text{sp. ht. lead} = \frac{-1.4 \times 10^3 \text{ J}}{150.0 \text{ g lead} \times (28.8 - 100.0) \text{ °C}}$$

$$= \frac{-1.4 \times 10^3 \text{ J}}{150.0 \text{ g lead} \times (-71.2 \text{ °C})} = 0.13 \text{ J g}^{-1} \text{ °C}^{-1}$$

SIMILAR EXAMPLES: Exercises 4, 21, 23, 25, 65.

7-5   The First Law of Thermodynamics

The first law of thermodynamics is a restatement of the law of conservation of energy. In an isolated system (one that does not interact with its surroundings) the total energy remains constant. Or, if a system does exchange heat and/or work with its surroundings, this must occur in such a way that the total energy of the system and its surroundings remains constant. In terms of internal energy change (ΔE), heat (q), and work (w),

The first law of thermody-namics.

$$\Delta E = q + w \qquad (7.9)$$

In using equation (7.9) here are the important points to keep in mind.

Scientists vary in their use of thermodynamic notation. Sometimes the symbol U is used instead of E; a different sign convention may be followed for w; and a different form used for equation (7.9).

- Any energy *entering* the system carries a *positive* sign. Thus, if heat is *absorbed* by the system, $q > 0$. If work is done *on* the system (energy enters), this too carries a *positive* sign, $w > 0$.
- Any energy *leaving* the system carries a *negative* sign. Thus, if heat is *given off* by the system, $q < 0$. If work is done *by* the system (energy leaves), this too carries a *negative* sign, $w < 0$.
- In general there will be a net change in the total energy of a system as a result of energy entering or leaving the system as heat and/or work. This net change is the *change* in internal energy of the system, ΔE. If, on balance, more energy enters the system than leaves, ΔE is *positive;* if more energy leaves than enters, ΔE is *negative*.

Example 7-5

Relating ΔE, q, and w through the first law of thermodynamics. A gas, while expanding (recall Figure 7-3), *absorbs* 225 J of heat and *does* 243 J of work. What is ΔE for the gas?

Solution. The key to problem solving with equation (7.9) lies in assigning the *correct* signs to the quantities of heat and work. Since heat is absorbed by (enters) the system, q is *positive*, and since work done *by* the system represents energy *leaving* the system, w is *negative*. In substituting into equation (7.9), you may find it useful to represent the values of q and w, with their correct signs, within parentheses. Then, complete the algebra.

$$\Delta E = q + w = (+225\ \text{J}) + (-243\ \text{J}) = 225\ \text{J} - 243\ \text{J} = -18\ \text{J}$$

SIMILAR EXAMPLES: Exercises 7, 27, 28.

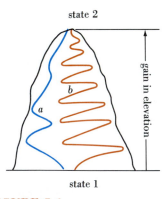

state 2

gain in elevation

state 1

FIGURE 7-6

An analogy to a thermodynamic function of state.

The gain in elevation in climbing from the base to the summit of the mountain is independent of the path chosen. This elevation gain is analogous to ΔE in a thermodynamic system.

Functions of State. We have described internal energy as the total energy associated with the ultimate particles of matter in a system. But internal energy is something that we *cannot measure*. Fortunately, internal energy depends only on the existing condition of a system and *not* on how this condition was achieved. By the condition of a system we mean what substances are present and in what amounts, what are the temperature and pressure? The condition of a system is more commonly called its **state,** and any property that depends only on the state of a system is called a **function of state** (or state function). The practical meaning of these statements is that as long as the state of a system is specified, its internal energy E (a state function) has a *unique* value, even though we cannot measure it. Another practical consequence of these definitions of state and state functions is that the *difference* in value of a state function between two states (e.g., ΔE) is also a *unique quantity*. Heat (q) and work (w) are *not* state functions. Their values depend on the path followed between two states of a system.

This entire matter of states and state functions should be more understandable in terms of Figure 7-6—a mountain-climbing analogy. The initial state or *state 1* is the base of the mountain and the final state or *state 2* is the summit. The elevation at any point on the mountain is analogous to internal energy E, and the *difference* in elevation between the base and summit of the mountain (known as the gain in elevation) is analogous to ΔE. Heat (q) and work (w) are analogous to the length of time required to climb the mountain. Path (a) is short but steep; path (b) is longer and more gradual. The length of time to climb the mountain depends on the path chosen, but the total elevation gain is *fixed*. Furthermore, the loss of elevation in climbing back down the mountain is analogous to $-\Delta E$. Thus, following a round trip to the top of the mountain (ΔE) and back down ($-\Delta E$), the total elevation gain is *zero*. To state this conclusion in thermodynamic terms, following a transition from *state 1* to *state 2* and back to *state 1* (the initial condition), all functions of state, including internal energy, must return to their initial values.

$$\text{state } 1 \xrightarrow{\Delta E} \text{state } 2 \xrightarrow{-\Delta E} \text{state } 1 \qquad (7.10)$$

Change in Internal Energy for a Chemical Reaction. We can think of the reactants of a chemical reaction as representing one state of a thermodynamic system, state 1, with an internal energy E_1. The products represent a different state, state 2, with an internal energy E_2.

$$\underset{\substack{\text{(state 1)}\\ E_1}}{\text{reactants}} \longrightarrow \underset{\substack{\text{(state 2)}\\ E_2}}{\text{products}}$$

Accompanying the reaction is the change in internal energy

$$\Delta E = E_2 - E_1$$

which according to the first law of thermodynamics can be represented as

$$\Delta E = q + w$$

To evaluate ΔE we need to measure q and w, and these quantities, of course, depend on how the reaction is carried out, as we shall see in the next two sections.

7-6 Heats of Reaction as Changes in Internal Energy, ΔE

When sucrose (ordinary cane sugar) is metabolized in the body, a complicated series of chemical reactions and energy conversions occurs. The net result of these reactions, though, is the same as in the complete combustion of sucrose—the production of $CO_2(g)$ and $H_2O(l)$.

$$C_{12}H_{22}O_{11}(s) + 12\ O_2(g) \longrightarrow 12\ CO_2(g) + 11\ H_2O(l) \qquad (7.11)$$

Antoine Lavoisier was the first to recognize that metabolism and combustion are closely related, that is, that the heat of reaction (7.11) is the same for both processes.

The caloric value of sucrose, when used as a food, is the same as the difference in internal energy between the products [12 mol $CO_2(g)$ and 11 mol $H_2O(l)$] and the reactants [1 mol $C_{12}H_{22}O_{11}(s)$ and 12 mol $O_2(g)$] in reaction (7.11). Now we demonstrate how we might carry out the combustion of sucrose so as to obtain a value of ΔE.

Bomb Calorimetry. A **calorimeter** is a laboratory device in which quantities of heat can be measured. The type of calorimeter shown in Figure 7-7 is ideally suited for measuring the heat evolved in the combustion reaction (7.11); it is called a **bomb calorimeter.**

In Figure 7-7 the *system* is just the *contents* of the bomb, that is, just the reactants (state 1) and the products (state 2). The steel bomb itself, the water in which the bomb is immersed, the thermometer, the stirrer, and so on, constitute the *surroundings*. The bomb confines the system to a *constant volume,* and as a consequence *there is no opportunity for the system to do work or to have work done on it.* As a result, $w = 0$, and $\Delta E = q + w = q + 0 = q_V$.

Heat lost or gained in a chemical reaction is called a **heat of reaction (q_{rxn});** and in the case of the bomb calorimeter, where the reaction occurs at *constant volume,* the heat of reaction can be designated as q_V and called the **heat of reaction at constant volume.** A heat of reaction at constant volume, then, is equal to ΔE.

Heat of reaction at constant volume.

$$\Delta E = q_V \qquad (7.12)$$

In the same manner as in expressions (7.7) and (7.8) we can describe the transfer of heat in the bomb calorimeter as

$$q_{rxn} + q_{calorim.} = 0$$

or

$$q_{rxn} = -q_{calorim.} \quad \text{(where } q_{rxn} = q_{sys.} \text{ and } q_{calorim.} = q_{surr.}) \qquad (7.13)$$

Heat evolved in the combustion reaction goes largely toward raising the temperature of the water surrounding the bomb. However, small quantities of heat are also required to raise the temperature of the bomb itself, of the stirrer, and of other parts of the calorimeter. Because the calorimeter assembly is not a pure substance, we cannot easily determine the heat absorbed by the calorimeter in the way that we did in Example 7-4 (where we had to account only for the heat absorbed by water). Fortunately, if the calorimeter is assembled in exactly the same way each time it is used—the same bomb, the same quantity of water, and so on—we can define a *heat capacity of the calorimeter*. This is the quantity of heat required to change the temperature of the calorimeter assembly by one degree Celsius. Multiplying this heat capacity by the observed temperature change leads directly to $q_{calorim.}$. (In an alternative procedure illustrated in Exercise 45, the heat capacity of the calorimeter assembly and of the water it contains are assessed separately.)

In Example 7-6 we complete our evaluation of the heat of reaction for the combustion of sucrose (called, simply, the **heat of combustion** of sucrose). In Example 7-7 we learn how the heat capacity of a bomb calorimeter can be established by experiment.

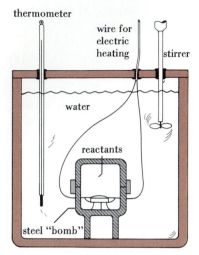

thermometer

wire for electric heating

stirrer

water

reactants

steel "bomb"

FIGURE 7-7
Experimental measurement of a heat of reaction.

The reaction is initiated by momentary electric heating of a length of iron wire covered with the reactants. The heat of reaction is determined by measuring the total quantity of heat absorbed by the surroundings. A separate experiment is required to establish the heat capacity of the calorimeter assembly. Every calorimeter assembly has its own distinctive heat capacity.

Example 7-6

Using bomb calorimetry data to determine a heat of reaction. The combustion of 1.010 g sucrose, $C_{12}H_{22}O_{11}$, in a bomb calorimeter causes the water temperature to rise from 24.92 to 28.33 °C. The heat capacity of the calorimeter assembly is 4.90 kJ/°C. **(a)** What is the heat of combustion of sucrose, expressed in kJ/mol $C_{12}H_{22}O_{11}$? **(b)** Verify the claim of sugar producers that one teaspoon of sugar (about 4.8 g) "contains only 19 Calories."

Solution

(a) First we can calculate $q_{calorim.}$.

$$q_{calorim.} = 4.90 \text{ kJ/°C} \times (28.33 - 24.92) \text{ °C} = 4.90 \times 3.41 = 16.7 \text{ kJ}$$

Now recall equation (7.13).

$$q_{rxn} = -q_{calorim.} = -16.7 \text{ kJ}$$

This is the heat of combustion of a 1.010-g sample.

Per gram $C_{12}H_{22}O_{11}$:

$$q_{rxn} = \frac{-16.7 \text{ kJ}}{1.010 \text{ g } C_{12}H_{22}O_{11}} = -16.5 \text{ kJ/g } C_{12}H_{22}O_{11}$$

Per mol $C_{12}H_{22}O_{11}$:

$$q_{rxn} = \frac{-16.5 \text{ kJ}}{\text{g } C_{12}H_{22}O_{11}} \times \frac{342.3 \text{ g } C_{12}H_{22}O_{11}}{1 \text{ mol } C_{12}H_{22}O_{11}}$$

$$= -5.65 \times 10^3 \text{ kJ/mol } C_{12}H_{22}O_{11}$$

The negative sign of the heat of reaction signifies that the reaction is *exothermic*.

(b) We need the heat of combustion per gram of sucrose determined in part (a), together with a factor to convert from kJ to kcal. (Since 1 cal = 4.184 J, 1 kcal = 4.184 kJ.)

$$\text{no. kcal} = \frac{-16.5 \text{ kJ}}{\text{g } C_{12}H_{22}O_{11}} \times 4.8 \text{ g } C_{12}H_{22}O_{11} \times \frac{1 \text{ kcal}}{4.184 \text{ kJ}} = -19 \text{ kcal}$$

Recall from the marginal note on page 205 that 1 food calorie (1 Calorie) is actually 1000 cal or 1 kcal. Therefore, 19 kcal = 19 Calories, and the producers' claim seems justified.

SIMILAR EXAMPLES: Exercises 10, 11, 42, 43, 71.

Example 7-7

Using a known heat of combustion to establish the heat capacity of a bomb calorimeter. The combustion of benzoic acid is often used to establish the heat capacity of a bomb calorimeter assembly. If the combustion of a 1.176-g sample of benzoic acid ($HC_7H_5O_2$) causes a temperature increase of 4.96 °C, what is the heat capacity of the bomb calorimeter assembly? (A handbook lists the heat of combustion of benzoic acid as −26.42 kJ/g.)

In precise studies the heat capacity of a calorimeter is established with an energy output from an electrical source.

Solution. First determine the total heat evolved by the benzoic acid sample.

$$q_{rxn} = 1.176 \text{ g } HC_7H_5O_2 \times \frac{-26.42 \text{ kJ}}{\text{g } HC_7H_5O_2} = -31.07 \text{ kJ}$$

Next, recall that $q_{calorim.} = -q_{rxn}$.

$$q_{calorim.} = -q_{rxn} = -(-31.07 \text{ kJ}) = +31.07 \text{ kJ}$$

Since $q_{calorim.} = \text{ht. capacity calorim.} \times \Delta T$

$$\text{ht. capacity calorim.} = \frac{31.07 \text{ kJ}}{4.96 \text{ °C}} = 6.26 \text{ kJ/°C}$$

SIMILAR EXAMPLES: Exercises 41, 44.

7-7 Heats of Reaction as Changes in Enthalpy, ΔH

For any combustion reaction carried out in a bomb calorimeter, that is, for a reaction at constant volume, the measured heat of reaction is $q_v = \Delta E$. Most chemical reactions, however, are not carried out in bomb calorimeters. The metabolism of sucrose occurs under the conditions present in the human body. The combustion of methane (natural gas) in a water heater occurs in an open flame. This question then arises: How does the heat of a reaction measured in a bomb calorimeter compare with the heat of reaction if the reaction is carried out in some other way? The usual "other" way is in beakers, flasks, and other containers open to the atmosphere and under the constant pressure of the atmosphere.

The heat of combustion of sucrose is the same, whether the reaction is carried out in a bomb calorimeter or in the open atmosphere. This is because, in either case, the only form of energy transfer between the reaction mixture and the surroundings is as heat. However, in many reactions carried out in the open atmosphere a small amount of pressure–volume work is done as the system expands or contracts. In these cases the measured heat of reaction is slightly different from ΔE. Because of this it is useful to define a new thermodynamic property that is closely related to internal energy but has this important advantage: Its change corresponds to the measured heat of reaction for a reaction carried out in the open atmosphere (more precisely, at constant pressure and with work limited to the pressure–volume type). The function that serves this purpose is called **enthalpy** and is denoted by the symbol H.

Relationship Between Enthalpy and Internal Energy. Consider a reaction that is carried out at *constant pressure*. The **heat of reaction at constant pressure** is designated as q_P. Consider that the system *does* a quantity of work that is of the pressure–volume type, and that this work is done at constant pressure. This quantity of work is $w = -P \, \Delta V$. (The negative sign signifies that this work is energy *leaving* the system, that is, work done *by* the system.) For these conditions the first law of thermodynamics can be written as

A term used to denote a process carried out at constant pressure is *isobaric*.

$$\Delta E = q + w = q_P - P \, \Delta V \quad \text{(at constant pressure)}$$

and then solved for q_P.

$$q_P = \Delta E + P \, \Delta V \quad \text{(at constant pressure)} \tag{7.14}$$

At this point let us formally introduce the enthalpy function, H, and define it as the sum of the internal energy and the pressure–volume product of a system.

$$H = E + PV$$

For a change occurring at *constant pressure,*

$$\Delta H = \Delta E + P \, \Delta V \tag{7.15}$$

Now, note that the right sides of equations (7.14) and (7.15) are identical. This makes their left sides also identical, that is

Heat of reaction at constant pressure.

$$\Delta H = q_P \tag{7.16}$$

Here, then, is the significance of the enthalpy function: *Its change in value for a reaction (ΔH) is equal to the heat of reaction at constant pressure (q_P) when work is limited to pressure–volume work.* A further comparison of internal energy and enthalpy is made in Table 7-1.

Representing ΔH in a Chemical Reaction. Since internal energy (E) and the pressure–volume product (PV) of a system are both functions of state, so too is their sum, the enthalpy (H) of a system. The situation for the combustion of sucrose is outlined below:

$$\underbrace{C_{12}H_{22}O_{11}(s) + 12 \, O_2(g)}_{\substack{\text{initial state,} \\ \text{having an enthalpy, } H_i. \\ H_i \text{ cannot be} \\ \text{measured but has} \\ \text{a unique value.}}} \longrightarrow \underbrace{12 \, CO_2(g) + 11 \, H_2O(l)}_{\substack{\text{final state,} \\ \text{having an enthalpy, } H_f. \\ H_f \text{ cannot be} \\ \text{measured but has} \\ \text{a unique value.}}}$$

For the reaction: $\quad \Delta H = H_f - H_i$

ΔH has a unique value, which is the measured heat of reaction when the reaction is carried out at constant pressure with work limited to pressure–volume work.

As we stated earlier in this section, for the combustion of sucrose $\Delta E \,(= q_V)$ and $\Delta H \,(= q_P)$ are identical. From the result of Example 7-6, $\Delta E = \Delta H = -5.65 \times 10^3$ kJ. The customary way of presenting this information in a chemical equation is

TABLE 7-1
Internal Energy (E) and Enthalpy (H) Compared

	E	$H = E + PV$
nature	fundamental	invented for convenience
most useful at:	constant volume (e.g., reaction in a bomb calorimeter)	constant pressure (e.g., reaction in an open beaker)
first-law statement under these conditions	$q_V = \Delta E$	$q_P = \Delta H$

$$C_{12}H_{22}O_{11}(s) + 12\ O_2(g) \longrightarrow 12\ CO_2(g) + 11\ H_2O(l)$$
$$\Delta H = -5.65 \times 10^3\ kJ \quad (7.17)$$

Example 7-8 illustrates how enthalpy changes may appear in stoichiometric factors for problem solving.

Example 7-8

Including quantities of heat in stoichiometric calculations based on the chemical equation. How much heat is evolved in the complete combustion of 1.00 kg of sucrose, $C_{12}H_{22}O_{11}$?

Solution. First, express the quantity of sucrose in moles.

$$\text{no. mol} = 1.00\ kg\ C_{12}H_{22}O_{11} \times \frac{1000\ g\ C_{12}H_{22}O_{11}}{1\ kg\ C_{12}H_{22}O_{11}} \times \frac{1\ mol\ C_{12}H_{22}O_{11}}{342.3\ g\ C_{12}H_{22}O_{11}}$$

$$= 2.92\ mol\ C_{12}H_{22}O_{11}$$

Now, formulate a conversion factor (shown in blue below) based on the information in equation (7.17), that is, that 1 mol $C_{12}H_{22}O_{11}$ yields 5.65×10^3 kJ of heat on combustion.

$$\text{no. kJ} = 2.92\ mol\ C_{12}H_{22}O_{11} \times \frac{-5.65 \times 10^3\ kJ}{1\ mol\ C_{12}H_{22}O_{11}} = -1.65 \times 10^4\ kJ$$

The negative sign denotes that heat is *given off* in the combustion.

SIMILAR EXAMPLES: Exercises 34, 35, 36, 70.

Heat of Reaction vs. Enthalpy Change of a Reaction. For reactions carried out at constant pressure and with work limited to pressure–volume work, ΔH and q_P are identical. Without this restriction they could be quite different. We restrict work to the pressure–volume type throughout this text (with one minor exception noted in Chapter 21). Consequently we use terms such as "heat of combustion" to refer to the enthalpy change that occurs during a combustion reaction. And, in referring to this enthalpy change, we may at times drop the word "change" and say "enthalpy of combustion." Our meaning, however, remains the enthalpy *change* for a combustion reaction.

Exothermic and Endothermic Reactions. The negative sign of ΔH in equation (7.17) means that the enthalpy of the products is lower than that of the reactants. This *decrease* in enthalpy appears as heat evolved to the surroundings. A reaction that *gives off* heat is an **exothermic** reaction. In the reaction

$$N_2(g) + O_2(g) \longrightarrow 2\ NO(g) \quad \Delta H = +180.50\ kJ \quad (7.18)$$

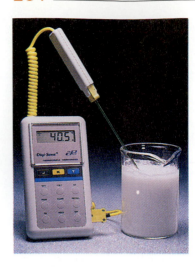

FIGURE 7-8

The slaking of lime—an exothermic reaction.

Slaked lime, $Ca(OH)_2$, is produced by adding water to quicklime, $CaO(s)$.

$$CaO(s) + H_2O \rightarrow Ca(OH)_2(s) \qquad \Delta H = -65.2 \text{ kJ}$$

This reaction is exothermic, as can be seen from the elevated temperature (40.5 °C) achieved by mixing the room-temperature reactants. This exothermic reaction accounts for the need to exercise care in the storage, handling, and slaking of quicklime. A mixture of slaked lime, sand, and water is extensively used as an exterior plaster or stucco in building construction. [Carey B. Van Loon]

the products have a *higher* enthalpy than the reactants; ΔH is positive. To produce this increase in enthalpy, heat is absorbed from the surroundings. A reaction that *absorbs* heat is an **endothermic** reaction. Figure 7-8 shows an exothermic reaction and Figure 7-9, an endothermic reaction. Figure 7-10 suggests a useful method of indicating exothermic and endothermic reactions diagrammatically.

Experimental Determination of ΔH. The simple calorimeter pictured in Figure 7-11 is much more commonly used in the general chemistry laboratory than is a bomb calorimeter. A chemical reaction is carried out in solution in a Styrofoam cup (generally in aqueous solution), and the temperature change is measured. Styrofoam is a good heat insulator, so that there is very little heat transfer between the cup and the surrounding air. Because the reaction mixture is maintained under atmospheric pressure, the quantity of heat energy measured is at constant pressure: $q_P = \Delta H$.

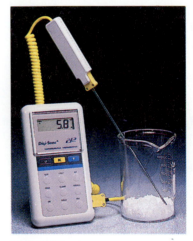

FIGURE 7-9

An endothermic reaction.

The solids $Ba(OH)_2 \cdot 8H_2O$ and NH_4Cl are mixed at room temperature. The following endothermic reaction occurs and the temperature falls far below room temperature (here, to 5.8 °C).

$$Ba(OH)_2 \cdot 8H_2O(s) + 2 NH_4Cl(s) \rightarrow BaCl_2 \cdot 2H_2O(s) + 2 NH_3(aq) + 8 H_2O$$
$$\Delta H = +63.6 \text{ kJ}$$

[Carey B. Van Loon]

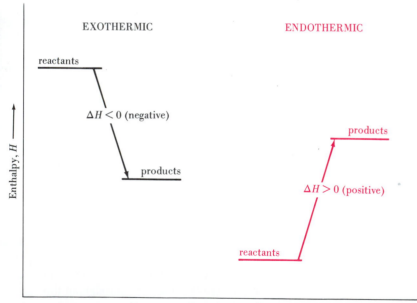

FIGURE 7-10

Exothermic and endothermic reactions represented through an enthalpy diagram.

Because an absolute value of H cannot be measured, we cannot assign numerical values on the enthalpy axis. What we can say is that enthalpy increases in the direction shown by the arrow (upward). For an exothermic reaction, such as the combustion of sucrose (reaction 7.17), the reactants have a higher enthalpy than the products and the reaction is accompanied by a decrease in enthalpy. The enthalpy change, ΔH, is negative, i.e., has a value less than zero. For an endothermic reaction, such as the synthesis of $NO(g)$ from its elements (reaction 7.18), the opposite situation prevails; i.e., ΔH is positive or greater than zero.

FIGURE 7-11

A calorimeter constructed from Styrofoam coffee cups.

The reaction mixture is in the inner cup. The outer cup provides additional thermal insulation from the surrounding air. The cup is closed off with a cork stopper through which a thermometer and a stirrer are immersed into the reaction mixture.

Calculations using data obtained from a "coffee-cup" calorimeter can generally be kept quite simple if certain assumptions are made, as in Example 7-9. If greater precision is required, additional factors can be considered, as illustrated in Exercise 40.

Example 7-9

Determining an enthalpy of reaction (ΔH) from calorimetric data. In the reaction of a strong acid with a strong base the essential reaction is a neutralization, the combination of $H^+(aq)$ and $OH^-(aq)$ to form water (recall equation 5.10).

$$H^+(aq) + OH^-(aq) \longrightarrow H_2O$$

100.0 ml of 1.00 M HCl(aq) and 100.0 mL of 1.00 M NaOH(aq), both initially at 21.1 °C, are added to a Styrofoam cup calorimeter and allowed to react. The temperature in the calorimeter is observed to rise to 27.9 °C. Determine ΔH for the neutralization reaction in which 1.00 mol H_2O is formed. Is the reaction endothermic or exothermic?

In addition to assuming that there is no exchange of heat between the contents of the calorimeter and the surrounding air, also make the following assumptions.

1. The "surroundings" of the reaction mixture is 200.0 mL of water. (This neglects the fact that 0.10 mol each of NaCl and H_2O are formed in the reaction and become mixed with the water in the calorimeter.)
2. The density of the solution in the calorimeter [NaCl(aq)] is the same as that of water: 1.00 g/mL.
3. The specific heat of the solution in the calorimeter [NaCl(aq)] is the same as that of water: 4.18 J g^{-1} °C^{-1}.

Solution. Call the two quantities of heat, $q_{neutr.}$ (heat of neutralization) and q_{water} (that is, $q_{surr.}$). To obtain q_{water} we proceed in exactly the same fashion as in Example 7-4, that is, by using equation (7.6).

$$q_{water} = 200.0 \text{ mL} \times \frac{1.00 \text{ g}}{\text{mL}} \times \frac{4.18 \text{ J}}{\text{g °C}} \times (27.9 - 21.1) \text{ °C} = 5.7 \times 10^3 \text{ J}$$

$$q_{neutr.} = -q_{water} = -5.7 \times 10^3 \text{ J} = -5.7 \text{ kJ}$$

But now it is essential that you understand the following point: In 100.0 mL of 1.00 M HCl the number of moles of H^+ is

$$\text{no. mol } H^+ = 0.1000 \text{ L} \times \frac{1.00 \text{ mol HCl}}{\text{L}} \times \frac{1 \text{ mol } H^+}{1 \text{ mol HCl}} = 0.100 \text{ mol } H^+$$

Similarly, in 100.0 mL of 1.00 M NaOH there is present 0.100 mol OH^-. Thus, the H^+ and the OH^- combine to form 0.100 mol H_2O. (The two are in *stoichiometric* proportions; neither is in excess.)

Per mole of H_2O produced:

$$\Delta H_{neutr.} = q_{neutr.} = \frac{-5.7 \text{ kJ}}{0.100 \text{ mol } H_2O} = -57 \text{ kJ/mol } H_2O$$

Since $q_{neutr.} = \Delta H_{neutr.}$ is a *negative* quantity, the neutralization reaction is *exothermic.*

SIMILAR EXAMPLES: Exercises 12, 38, 39, 74.

Are You Wondering:

Why the neutralization reaction in Example 7-9 is not considered to be an endothermic *reaction, since heat was* absorbed *by the water (its temperature increased)?*

The answer to this question lies in understanding clearly which is the system and which is the surroundings. The 200.0 mL of water in which the temperature increase was recorded is the *surroundings*. If the surroundings absorbed heat ($q_{surr.} > 0$), then the system (the reaction mixture) must have *lost* heat ($q_{sys.} < 0$). The terms endothermic and exothermic are always applied to what happens to the *system*, even though the measurements are made in the surroundings.

Another way to obtain ΔH values for certain reactions is to measure ΔE, say by bomb calorimetry, and then to calculate ΔH using equation (7.15). This is illustrated through Exercise 48. For reactions involving liquids and solids, the term $P \, \Delta V \approx 0$ and $\Delta H \approx \Delta E$. If gases are involved the difference between ΔH and ΔE is greater. In most cases, however, unless precise results are required, you can generally assume that the difference in value between ΔH and ΔE is negligibly small.

7-8 Relationships Involving ΔH

One of the most useful things we can do with the enthalpy concept is to calculate large numbers of heats of reaction from a relatively small number of measurements. The following statements about enthalpy changes (ΔH) are crucial in this regard.

1. ΔH is an Extensive Property. Enthalpy change is directly proportional to the amounts of substances involved in a process. If we take one-half the amounts of reactants and products in reaction (7.18)—that is, if we divide through the equation by 2—the enthalpy change is also halved.

$$\tfrac{1}{2} N_2(g) + \tfrac{1}{2} O_2(g) \longrightarrow NO(g) \qquad \Delta H = \tfrac{1}{2} \times 180.50 = 90.25 \text{ kJ}$$

As a consequence of this fact, when we state a ΔH value we should always write (or in some way imply) the chemical equation to which it applies.

2. ΔH Changes Sign When a Process is Reversed. Enthalpy is a function of state. As described in the mountain-climbing analogy in Figure 7-6, if we reverse the direction of a process the change in a function of state (such as ΔH) reverses sign ($-\Delta H$). Thus, if for the formation of *one mole* of NO(g) from its elements

$$\tfrac{1}{2} N_2(g) + \tfrac{1}{2} O_2(g) \longrightarrow NO(g) \qquad \Delta H = +90.25 \text{ kJ}$$

then for the decomposition of *one mole* of NO(g) into its elements

$$NO(g) \longrightarrow \tfrac{1}{2} N_2(g) + \tfrac{1}{2} O_2(g) \qquad \Delta H = -90.25 \text{ kJ}$$

Although we have previously avoided fractional coefficients in chemical equations, note that they are necessary here. We require that the coefficient of NO(g) be 1.

3. Hess's Law of Constant Heat Summation

If a process can be considered to occur in stages or steps (either actually or hypothetically), the enthalpy change for the overall process can be obtained by summing the enthalpy changes for the individual steps.

This statement again follows from the fact that enthalpy is a function of state. Let us turn again to the mountain-climbing analogy of Figure 7-6. Imagine that a trip from the base to the summit of the mountain is made in stages. We can determine the elevation gain (or loss) for each stage; the total elevation gain is the sum of the changes for each stage (e.g., +1000 m, −200 m, +400 m, etc.).

Suppose in the combination of $N_2(g)$ and $O_2(g)$ that, instead of stopping at $NO(g)$, we allow the reaction to proceed to $NO_2(g)$.

$$\tfrac{1}{2} N_2(g) + O_2(g) \longrightarrow NO_2(g) \qquad \Delta H = ? \qquad (7.19)$$

This reaction can be carried out in the two steps indicated below. When the two equations are summed, the net equation is (7.19). Hess's law states that the two enthalpy changes can also be summed, leading to Δ*H* for reaction (7.19).

$$
\begin{aligned}
\tfrac{1}{2} N_2(g) + \tfrac{1}{2} O_2(g) &\longrightarrow NO(g) & \Delta H &= +90.25 \text{ kJ} \\
NO(g) + \tfrac{1}{2} O_2(g) &\longrightarrow NO_2(g) & \Delta H &= -57.07 \text{ kJ} \\
\hline
\tfrac{1}{2} N_2(g) + O_2(g) &\longrightarrow NO_2(g) & \Delta H &= +90.25 - 57.07 \\
& & &= +33.18 \text{ kJ}
\end{aligned}
$$

Note that in this summation a species that would appear on both sides of the net equation (NO) cancels out (a feature previously illustrated in Figure 4-11). Figure 7-12 summarizes what we have just done through an enthalpy diagram.

As another example suppose we wish to determine the enthalpy change (Δ*H*) for the reaction

$$3 \text{ C(graphite)} + 4 H_2(g) \longrightarrow C_3H_8(g) \qquad \Delta H = ? \qquad (7.20)$$

How should we proceed? If we attempt to get graphite and hydrogen gas to react, some reaction will occur, but the reaction will not go to completion. Furthermore, the product will not be limited to propane (C_3H_8); several other hydrocarbons will form as well. The fact is Δ*H* *for this reaction cannot be measured directly*. Instead we must resort to an *indirect calculation* of the desired Δ*H* value from Δ*H* values that can be established by experiment. Heats of combustion are easily obtained from bomb calorimetry data, and extensive tabulations of these data can be found in handbooks. For example

$$
\begin{aligned}
\Delta H_{\text{combustion}}: \quad C_3H_8(g) &= -2220.1 \text{ kJ/mol } C_3H_8 \\
C(\text{graphite}) &= -393.5 \text{ kJ/mol C(graphite)} \\
H_2(g) &= -285.8 \text{ kJ/mol } H_2(g)
\end{aligned}
$$

FIGURE 7-12

Hess's law illustrated through an enthalpy diagram.

The conversion of $N_2(g)$ and $O_2(g)$ to $NO_2(g)$ is accompanied by the enthalpy change Δ*H* = +33.18 kJ/mol, whether the reaction is considered in a single step or in two steps [by way of NO(g)].

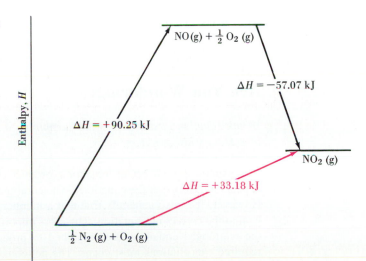

Example 7-10 ——————————————————————————

Applying Hess's law to determine a ΔH value. Use the heat of combustion data on page 217 to determine ΔH for reaction (7.20).

Solution. To determine an enthalpy change with Hess's law, we need to combine the appropriate chemical equations. A logical starting point is to write chemical equations for the given combustion reactions, based on *one mole* of the indicated reactant.

(a) $C_3H_8(g) + 5 O_2(g) \longrightarrow 3 CO_2(g) + 4 H_2O(l)$ $\Delta H = -2220.1$ kJ

(b) $C(graphite) + O_2(g) \longrightarrow CO_2(g)$ $\Delta H = -393.5$ kJ

(c) $H_2(g) + \frac{1}{2} O_2(g) \longrightarrow H_2O(l)$ $\Delta H = -285.8$ kJ

Since our objective in equation (7.20) is to *produce* $C_3H_8(g)$, we look for an equation in which $C_3H_8(g)$ is formed. This will be the *reverse* of equation (a).

$-$(a): $3 CO_2(g) + 4 H_2O(l) \longrightarrow C_3H_8(g) + 5 O_2(g)$ $\Delta H = -(-2220.1)$ kJ

$$= +2220.1 \text{ kJ}$$

Suppose that the $CO_2(g)$ required in equation $-$(a) is produced by the combustion of graphite and that the $H_2O(l)$ is produced by the combustion of $H_2(g)$. To get the proper number of moles of each we must multiply equation (b) by 3 and equation (c) by 4.

$3 \times$ (b): $3 C(graphite) + 3 O_2(g) \longrightarrow 3 CO_2(g)$ $\Delta H = 3 \times (-393.5 \text{ kJ})$

$$= -1180. \text{ kJ}$$

$4 \times$ (c): $4 H_2(g) + 2 O_2(g) \longrightarrow 4 H_2O(l)$ $\Delta H = 4 \times (-285.8 \text{ kJ})$

$$= -1143 \text{ kJ}$$

Now let us think about the net change we have described. 3 mol C(graphite) and 4 mol $H_2(g)$ have been consumed and 1 mol $C_3H_8(g)$ has been produced. This is exactly what is required in equation (7.20), and we can now combine the three modified equations.

$-$(a): $3 CO_2(g) + 4 H_2O(l) \longrightarrow C_3H_8(g) + 5 O_2(g)$ $\Delta H = +2220.1$ kJ
$3 \times$ (b): $3 C(graphite) + 3 O_2(g) \longrightarrow 3 CO_2(g)$ $\Delta H = -1180.$ kJ
$4 \times$ (c): $4 H_2(g) + 2 O_2(g) \longrightarrow 4 H_2O(l)$ $\Delta H = -1143$ kJ
——
     $3 C(graphite) + 4 H_2(g) \longrightarrow C_3H_8(g)$ $\Delta H = -103$ kJ

SIMILAR EXAMPLES: Exercises 15, 49, 50, 51, 78, 79.

Are You Wondering:

Which equation to start with when combining a series of chemical equations in a Hess's law problem?

It is usually a good bet to look for a product of the final net equation that appears in just one of the equations to be combined. Begin with that equation. However, it is also generally true that a number of different approaches will lead to the correct net equation. The most important requirement is that when the rearranged equations are combined to produce the net equation all the required cancellations must occur. The net equation must be *identical* to the one whose ΔH value is sought.

7-9 Standard Enthalpies of Formation

We have noted that *absolute values of E and H cannot be established*. Nevertheless, we have been successful in dealing with *changes* in these properties alone, e.g., ΔH.

Consider again the mountain-climbing analogy of Figure 7-6. We can determine the difference in elevation between the summit and the base of the mountain very precisely, but what is the *absolute* elevation of the mountain? Do we mean by this the vertical distance between the mountaintop and the center of the earth? Between the mountaintop and the deepest trench in the ocean? No, by common agreement we mean the vertical distance between the mountaintop and mean sea level. If we arbitrarily assign to mean sea level an elevation of 0, we can assign all other points on earth an elevation relative to this zero. The elevation of Mt. Everest is +8848 m; that of Badwater, Death Valley, California, is −86 m.

We can do this same thing with enthalpies, that is, assign a ''zero'' of enthalpy to certain substances and determine other enthalpies relative this *arbitrary* zero. As part of this procedure we need also to define a **standard state,** which for substances is simply *the pure substance at 1 atm pressure* at the specified temperature*. By convention, we assign a value of **zero** to the enthalpies of the elements in their most stable forms when in the standard state. Thus, the following elements, *in the forms indicated*, all have enthalpies of zero at 298 K.

$$Na(s) \qquad H_2(g) \qquad N_2(g) \qquad O_2(g) \qquad C(graphite) \qquad Br_2(l)$$

The situation with carbon is an interesting one. In addition to graphite, carbon can also exist in the form of diamond at 298 K and one atmosphere pressure. But of the two forms, graphite is the most stable under these conditions, and only graphite can be assigned an enthalpy of 0. Similarly, even though Br_2 can exist in the gaseous state at 298 K, the most stable form is the liquid. Only $Br_2(l)$ has an enthalpy of formation of zero at 298 K and 1 atm. [If we attempted to obtain $Br_2(g)$ at 298 K and 1 atm pressure, the gas would condense to the liquid. This is what we mean by the liquid being more stable than the gas.]

The **standard molar enthalpy of formation** (also called the **molar heat of formation**) is the difference in enthalpy between *one mole* of a compound in its standard state and its elements in their standard states. The standard molar enthalpy of formation of a compound is denoted as ΔH_f°. The superscript $^\circ$ signifies that all substances are in their standard states.

We can use standard enthalpies of formation to perform a variety of thermochemical calculations. A few typical data are presented in Table 7-2. A more extensive tabulation is provided in Appendix D.

Often in using enthalpy of formation data we need to write the chemical equation to which a ΔH_f° applies, as in Example 7-11.

TABLE 7-2
Some Standard Molar Enthalpies (Heats) of Formation at 298 K

Substance	$\Delta H_{f, 298}^\circ$, kJ/mol[a]
$CH_4(g)$	−74.81
$C_2H_2(g)$	226.7
$C_2H_4(g)$	52.26
$C_2H_6(g)$	−84.68
$C_3H_8(g)$	−103.8
$CO(g)$	−110.5
$CO_2(g)$	−393.5
$HCl(g)$	−92.31
$H_2O(l)$	−285.8
$NH_3(g)$	−46.11
$NO(g)$	90.25
$SO_2(g)$	−296.8

[a] Values are for reactions in which one mole of substance is formed. Data are given to four significant figures.

Example 7-11

Relating a standard molar enthalpy of formation to a chemical equation. The enthalpy of formation of ammonia at 298 K is $\Delta H_f^\circ = -46.11$ kJ/mol $NH_3(g)$. Write the chemical equation to which this value applies.

Solution. The equation must be written for the formation of *one mole* of *gaseous* NH_3 from the most stable forms of its elements at 298 K and 1 atm—

* The International Union of Pure and Applied Chemistry (IUPAC) recommends that the standard-state pressure be changed from 1 atm (101,325 Pa) to 1 bar (1×10^5 Pa). The immediate effects of this recommendation are minor, and we will continue to use 1 atm as the standard-state pressure in this text.

gaseous N_2 and H_2. Note that fractional coefficients are required in this equation.

$$\tfrac{1}{2} N_2(g) + \tfrac{3}{2} H_2(g) \longrightarrow NH_3(g) \qquad \Delta H_f^\circ = -46.11 \text{ kJ}$$

SIMILAR EXAMPLE: Exercise 13.

We can use enthalpies of formation to *calculate* enthalpy changes (heats) of reaction. If the reactants and products of a reaction are all in their standard states, we can call the enthalpy change for a reaction the **standard enthalpy of reaction** (or the **standard heat of reaction**). We use the symbols ΔH° and ΔH_{rxn}° to represent this quantity.

Let us apply Hess's law and other ideas from Section 7-8 to calculate the standard enthalpy (heat) of combustion of ethane, $C_2H_6(g)$, a component of natural gas.

$$C_2H_6(g) + \tfrac{7}{2} O_2(g) \longrightarrow 2 CO_2(g) + 3 H_2O(l) \qquad \Delta H_{rxn}^\circ = ? \qquad (7.21)$$

Three equations that can be added together to yield equation (7.21) are

(a) $\quad C_2H_6(g) \longrightarrow 2 C(graphite) + 3 H_2(g)$
$$\Delta H^\circ = -\Delta H_f^\circ[C_2H_6(g)]$$
(b) $2 C(graphite) + 2 O_2(g) \longrightarrow 2 CO_2(g) \qquad \Delta H^\circ = 2 \times \Delta H_f^\circ[CO_2(g)]$
(c) $\quad 3 H_2(g) + \tfrac{3}{2} O_2(g) \longrightarrow 3 H_2O(l) \qquad \Delta H^\circ = 3 \times \Delta H_f^\circ[H_2O(l)]$

$$\overline{C_2H_6(g) + \tfrac{7}{2} O_2(g) \longrightarrow 2 CO_2(g) + 3 H_2O(l)}$$
$$\Delta H_{rxn}^\circ = ? \qquad (7.21)$$

We should recognize that equation (a) is the *reverse* of the equation representing the formation of one mole of $C_2H_6(g)$ from its elements. ΔH for equation (a) is the *negative* of the enthalpy of formation of $C_2H_6(g)$. For equations (b) and (c) the ΔH values are two and three times the enthalpies of formation of $CO_2(g)$ and $H_2O(l)$, respectively. For reaction (7.21), then,

$$\Delta H_{rxn}^\circ = \{2 \times \Delta H_f^\circ[CO_2(g)] + 3 \times \Delta H_f^\circ[H_2O(l)]\} - \{\Delta H_f^\circ[C_2H_6(g)]\} \qquad (7.22)$$

Equation (7.22) is simply a specific application of a more general relationship that is expressed as

Relating an enthalpy of reaction to enthalpies of formation.

$$\Delta H_{rxn}^\circ = \left[\sum \nu_p \, \Delta H_f^\circ(\text{products})\right] - \left[\sum \nu_r \, \Delta H_f^\circ(\text{reactants})\right] \qquad (7.23)$$

The symbol Σ (Greek, sigma) means "the sum of." The terms that are added together are the products of the standard molar enthalpies of formation (ΔH_f°) and their stoichiometric coefficients, ν. One summation is required for the reaction products and another for the initial reactants. The enthalpy (heat) of reaction is the sum of terms for the products minus the sum of terms for the reactants.

Example 7-12

Calculating an enthalpy of reaction from tabulated values of ΔH_f°. Complete the calculation of ΔH_{rxn}° for reaction (7.21).

Solution. The relationship of ΔH_{rxn}° to enthalpies of formation is expressed through equation (7.22). All we need to do is substitute tabulated enthalpy of formation data (see Table 7-2) into this equation.

$$\Delta H_{rxn}^\circ = 2 \times \Delta H_f^\circ[CO_2(g)] + 3 \times \Delta H_f^\circ[H_2O(l)] - \Delta H_f^\circ[C_2H_6(g)]$$

$$= 2 \times (-393.5 \text{ kJ}) + 3 \times (-285.8 \text{ kJ}) - (-84.7 \text{ kJ})$$

$$\Delta H^\circ_{\mathrm{rxn}} = -787.0 \text{ kJ} - 857.4 \text{ kJ} + 84.7 \text{ kJ} = -1559.7 \text{ kJ}$$

SIMILAR EXAMPLES: Exercises 16, 54, 55.

Example 7-13

Calculating an enthalpy of formation from an enthalpy of reaction and tabulated values of ΔH°_f. The combustion of cyclopropane (an anesthetic) is represented as

$$(CH_2)_3(g) + \tfrac{9}{2} O_2(g) \longrightarrow 3 CO_2(g) + 3 H_2O(l) \qquad \Delta H^\circ_{\mathrm{rxn}} = -2091.4 \text{ kJ}$$

Use this value of $\Delta H^\circ_{\mathrm{rxn}}$ and other data from Table 7-2 to calculate the standard enthalpy (heat) of formation of cyclopropane.

Solution. We need to use equation (7.23), but we will solve for an unknown enthalpy (heat) of formation rather than for a heat of reaction. [Note that no ΔH°_f term appears for the $\tfrac{9}{2} O_2(g)$, since the enthalpy (heat) of formation of a free element in its standard state is zero.]

$$\Delta H^\circ_{\mathrm{rxn}} = 3 \Delta H^\circ_f[CO_2(g)] + 3 \Delta H^\circ_f[H_2O(l)] - \Delta H^\circ_f[(CH_2)_3(g)]$$

$$-2091.4 \text{ kJ} = 3 \times (-393.5 \text{ kJ}) + 3 \times (-285.8 \text{ kJ}) - \Delta H^\circ_f[(CH_2)_3(g)]$$

$$\Delta H^\circ_f[(CH_2)_3(g)] = -1180.5 \text{ kJ} - 857.4 \text{ kJ} + 2091.4 \text{ kJ} = +53.5 \text{ kJ}$$

SIMILAR EXAMPLES: Exercises 17, 56.

Since the enthalpy of formation data used in Examples 7-12 and 7-13 were for 298 K, the result of each calculation also applies only at 298 K. Generally, ΔH values are not significantly temperature dependent. This means that results obtained for 298 K are often applicable over a range of temperatures. (We discuss the temperature dependence of ΔH again in Chapter 20.)

7-10 Thermochemical Calculations: A Summary

You will generally find that thermochemical calculations are of two types:

1. Calculating heat effects from calorimetric data—heat capacities, specific heats, masses, temperature changes.
2. Manipulating given ΔH values to obtain other ΔH values.

In the *first* type of calculation think in terms of a system and its surroundings. Heat effects are generally measured in the *surroundings*, with $q_{\mathrm{surr.}}$ expressed as the product.

mass × specific heat × temperature change

or

heat capacity × temperature change

The heat effect in the *system* ($q_{\mathrm{sys.}}$) is $-q_{\mathrm{surr.}}$, and $q_{\mathrm{sys.}}$ is generally *not* broken down into the product: mass × specific heat × temperature change (except for situations like those presented in Examples 7-3 and 7-4). Usually, $q_{\mathrm{sys.}} = q_V = \Delta E$ (bomb calorimeter) or

$$q_{\mathrm{sys.}} = q_P = \Delta H.$$

In the *second* type of calculation expect that you may have to do one or more of the following:

- Write a chemical equation to correspond to a given ΔH value, such as a ΔH_f° or $\Delta H_{combustion}$.
- Reverse a chemical equation (and thereby change the sign of ΔH).
- Multiply the coefficients in a chemical equation by a constant factor (and thereby multiply ΔH by the same factor).
- Use Hess's law to combine several equations and their ΔH values into a net equation and its ΔH value.
- Use equation (7.23) to solve either for ΔH_{rxn}° or for an unknown ΔH_f°.

7-11 Sources and Uses of Energy

In the United States, daily per capita energy consumption exceeds by a hundred times or more the 10,000 to 12,000 kJ needed to sustain human life. With 6% of the world's population, the United States consumes about 30% of the world's energy production, for agriculture, industry, transportation, and material comforts. Total annual energy consumption in the United States currently amounts to about

$$7.8 \times 10^{16} \text{ kJ} = 2.9 \times 10^{16} \text{ kcal} = 7.4 \times 10^{16} \text{ Btu} = 74 \text{ quad.}$$

[One British thermal unit (Btu) is the quantity of heat required to change the temperature of one pound of water by one degree F. One quadrillion (10^{15}) Btu = 1 quad.]

The sources of this energy and the manner in which it is consumed are depicted in Figure 7-13.

Fossil Fuels. From Figure 7-13 we see that the bulk of our energy needs currently are being met by petroleum, natural gas, and coal—so-called fossil fuels. What these fuels have in common is that they are derived from plant and animal life of millions of years ago. The original source of energy that is locked into these fuels is *solar energy,* derived through the process of **photosynthesis.** In this process CO_2 and H_2O, in the presence of chlorophyll (a catalyst) and sunlight, are converted to **carbohydrates,** compounds having formulas $C_m(H_2O)_n$, where m and n are integers. For example, glucose is a carbohydrate in which $m = 6$ and $n = 6$, that is, $C_6(H_2O)_6 = C_6H_{12}O_6$. Its formation through photosynthesis is represented by the equation

$$6\ CO_2 + 6\ H_2O \xrightarrow[\text{sunlight}]{\text{chlorophyll}} C_6H_{12}O_6 + 6\ O_2 \qquad \Delta H = +2.8 \times 10^3 \text{ kJ} \quad (7.24)$$

Reaction (7.24) is *endothermic*. When the reaction is reversed, as in the combustion of glucose, heat is evolved; the combustion reaction is *exothermic*.

The complex carbohydrate cellulose (with molecular weights ranging up to 500,000) is the principal structural material of plants. When plant life decomposes in the absence of air and in the presence of bacteria, O and H atoms are removed and the approximate carbon content of the residue increases in the progression

peat $\longrightarrow$ lignite (32% C) $\longrightarrow$ subbituminous coal (40% C) $\longrightarrow$ bituminous coal (60% C) $\longrightarrow$ anthracite coal (80% C)

For the process to proceed all the way to anthracite coal may take about 300 million years. *Coal,* then, is a combustible organic rock consisting of carbon, hydrogen, and oxygen, together with small quantities of nitrogen, sulfur, and mineral matter (ash). (One proposed formula for a "molecule" of bituminous coal is $C_{153}H_{115}N_3O_{13}S_2$.)

Petroleum and natural gas have formed somewhat differently than coal. Plants

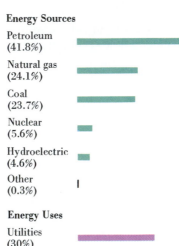

Energy Sources

Petroleum (41.8%)

Natural gas (24.1%)

Coal (23.7%)

Nuclear (5.6%)

Hydroelectric (4.6%)

Other (0.3%)

Energy Uses

Utilities (30%)

Transportation (26%)

Industry (24%)

Residential/ commercial (20%)

FIGURE 7-13
Sources and uses of energy in the United States.

Current energy consumption in the United States is about 7.8×10^{16} kJ/yr. The largest single consumer in the industrial sector is the chemical industry, followed closely by the primary metals industry.

TABLE 7-3

Approximate Heats of
Combustion of Some Fuels

Fuel	Heat of combustion, kJ/g
cellulose	−17.5
pine wood	−21.2
municipal waste	−12.7
methyl alcohol	−22.7
ethyl alcohol	−29.7
peat	−20.8
bituminous coal	−28.3
isooctane (a component of gasoline)	−36.5
natural gas	−49.5

and animals living in ancient seas were deposited on the ocean floor, decomposed by bacteria, and covered with sand and mud. Over a period of time, the sand and mud were converted to sandstone. The high pressures and temperatures resulting from this overlying sandstone rock transformed the original organic matter to petroleum and natural gas. The ages of these deposits range from about 250 to 500 million years.

A typical natural gas consists of about 85% methane (CH_4), 10% ethane (C_2H_6), 3% propane (C_3H_8), and small quantities of other combustible and noncombustible gases. A typical petroleum consists of several hundred different hydrocarbons ranging in complexity from C_1 molecules (that is, CH_4) to C_{40} or higher (e.g., $C_{40}H_{82}$). The composition of petroleum and its separation into different fractions is discussed in Section 27-11.

One method of comparing different fuels is in terms of their heats of combustion. In general, the higher the heat of combustion the better the fuel. Approximate heats of combustion for the fossil fuels are listed in Table 7-3. For comparison, the heats of combustion of pure cellulose, a typical soft wood, a municipal waste, and two alcohols are also given. These data show that *biomass* (living matter or materials derived from it) can be used as fuel, but that the fossil fuels contain a greater energy content per unit mass.

Problems Posed by Fossil Fuel Use. There are two fundamental problems associated with the use of fossil fuels. First is the fact that the formation of new fossil fuels, if it is occurring at all, cannot possibly match the rate at which existing resources are being depleted. We must consider the fossil fuels to be *nonrenewable* energy sources—once depleted they cannot be replaced. As shown in Figure 7-14, the remaining time during which fossil fuels will be available for human use is short (at least compared to the span of human history).

A second problem with fossil fuels is in their environmental impact. Sulfur impurities in fuels lead to oxides of sulfur. The high temperatures associated with combustion processes cause the reaction of N_2 and O_2 in air to form oxides of nitrogen. Some of the environmental issues associated with the oxides of nitrogen and sulfur are discussed in Sections 14-10 and 14-13. Another inevitable product of the combustion of fossil fuels is carbon dioxide. Its potential effect on the earth's climate is discussed in the Focus feature.

Synthetic Fuels from Coal. In the United States coal reserves have been estimated at between 5,000 and 21,000 quads. Corresponding reserves of petroleum and natural gas do not exceed 1000 quads each. Despite its relative abundance there has been

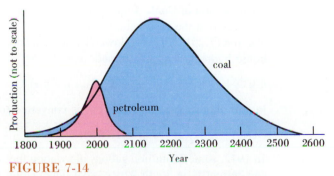

FIGURE 7-14

Estimated world production of fossil fuels.

The exact shape and time of exhaustion for each curve depend on the estimate used for the total quantity of recoverable fuel and on the rate of production. These curves are based on the assumption of no increases from current rates of consumption, probably an unrealistic assumption. Even a growth rate of just 1% per year would cause the rate of consumption to double in 70 years and would greatly shorten the times to exhaustion.

no significant increase in the use of coal in recent years. In addition to the environmental problems cited above, the expense and hazards involved in deep mining of coal are considerable. Strip mining, which is less expensive and less hazardous, is also more damaging to the environment. One promising possibility in using coal reserves is to convert coal to gaseous or liquid fuels, either in surface installations or while the coal is still underground.

Gasification of Coal. Before cheap natural gas became available in the 1940s, gas produced from coal (variously called producer gas, town gas, or city gas) was widely used in the United States. This gas is manufactured by passing steam and air through heated coal, leading to reactions such as

$$C(s) + H_2O(g) \longrightarrow CO(g) + H_2(g) \qquad \Delta H° = +118 \text{ kJ} \qquad (7.25)$$
$$CO(g) + H_2O(g) \longrightarrow CO_2(g) + H_2(g) \qquad \Delta H° = -54 \text{ kJ} \qquad (7.26)$$
$$2\ C(s) + O_2(g) \longrightarrow 2\ CO(g) \qquad \Delta H° = -221 \text{ kJ} \qquad (7.27)$$
$$C(s) + 2\ H_2(g) \longrightarrow CH_4(g) \qquad \Delta H° = -75 \text{ kJ} \qquad (7.28)$$

The principal gasification reaction (7.25) is highly endothermic and the heat requirements for this reaction are met by the carefully controlled partial burning of coal (reaction 7.27) and other exothermic reactions.

A typical producer gas consists of about 23% CO, 18% H_2, 8% CO_2, and 1% CH_4, by volume; but it also contains about 50% N_2, since air is used in the process. Because so much of the gas is noncombustible (that is, the N_2 and CO_2), producer gas has only about 10 to 15% of the heat value of natural gas.

Modern coal gasification processes

1. Use $O_2(g)$ instead of air [thereby reducing $N_2(g)$ in the product to mere traces].
2. Provide for the removal of noncombustible $CO_2(g)$ and of sulfur impurities. For example,

$$CaO(s) + CO_2(g) \longrightarrow CaCO_3(s) \qquad (7.29)$$

$$2\ H_2S(g) + SO_2(g) \longrightarrow 3\ S(s) + 2\ H_2O(g) \qquad (7.30)$$

3. Include a step (called *methanation*) to convert CO and H_2, in the presence of a catalyst, to CH_4.

$$CO(g) + 3\ H_2(g) \longrightarrow CH_4(g) + H_2O(g) \qquad (7.31)$$

With these modifications it is possible to obtain a **substitute natural gas (SNG),** a gaseous mixture with composition and heat value similar to that of natural gas.

Liquefaction of Coal. Liquid fuels can also be produced from coal. This generally involves first the gasification of coal to water gas—a mixture of CO and H_2—by reaction (7.25). This is followed by catalytic reactions (Fischer–Tropsch) in which liquid hydrocarbons are formed.

$$n\ CO + (2n + 1)\ H_2 \longrightarrow C_nH_{2n+2} + n\ H_2O \qquad (7.32)$$

In still another process water gas is converted to liquid methanol.

$$CO(g) + 2\ H_2(g) \longrightarrow CH_3OH(l) \qquad (7.33)$$

In 1942, some 32 million gallons of aviation fuel were made from coal in Germany, and currently, in South Africa, coal liquefaction is used to produce gasoline and a variety of other petroleum products and chemicals.

Methanol, Ethanol, and Hydrogen. Methanol, CH_3OH, can be obtained from coal by reaction (7.33). It can also be produced by thermal decomposition (pyrolysis) of wood, manure, sewage, or municipal waste. The heat of combustion of

This sugar farm–alcohol distillery is typical of several hundred such installations in Brazil. Juice is extracted from sugar cane in the mill in the foreground and pumped to the fermentation tanks and distillery in the center of the picture. Behind the distillery are three 5-million-liter storage tanks (and a 10-million-liter tank under construction). In the background and to the right is one of the many sugarcane fields that are part of this industrial complex. [Courtesy of Claudio Veiga de Brito, Usina Boa Vista S.A., Três Pontas, M.G., Brazil. Photograph by José Vieira de Mendonça Sobrinho.]

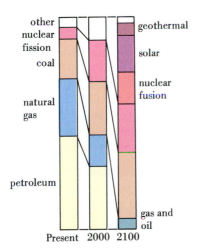

FIGURE 7-15
Relative importance of different energy sources—present and future (speculative).

The projections for A.D. 2000 and 2100 are highly speculative. The actual situation might be much different. Solar energy might become an important energy source by 2000, or nuclear fusion may not be developed as an energy source even by 2100. Or perhaps biological sources may become increasingly important. One fact that seems quite certain for the time period shown is that natural gas and petroleum (oil) will decrease in importance.

methanol is only about one-half that of a typical gasoline on a mass basis, but methanol has a high octane number—106. It has been tested and used as a fuel in internal combustion engines and found to burn more cleanly than gasoline. (For example, The State of California has operated a fleet of several hundred motor vehicles powered by a blend of 85% methanol–15% gasoline.) Methanol can also be used for space heating, electric power generation, fuel cells, and organic synthesis.

Ethanol, C_2H_5OH, is produced mostly from ethylene, C_2H_4, which in turn is derived from petroleum. Current interest centers on the production of ethanol by the fermentation of organic matter, a process known throughout recorded history. Ethanol production by fermentation is probably in its most advanced state in Brazil, where sugar cane and cassava (manioc) are the plant matter (biomass) principally employed. In its "Proalcool" program Brazil is attempting to shift from gasoline to alcohol as its primary motor fuel. By the early 1990s it is expected that 50% of Brazilian motor vehicles will operate on "hydrous" alcohol, a 95% ethanol–5% water mixture. Current production of alcohol in Brazil is 3 to 4 billion gallons per year, several times U.S. production. In the United States ethanol is used chiefly as a 90% gasoline–10% ethanol mixture called *gasohol*. Ethanol is also useful as an octane enhancer for gasoline.

Hydrogen is another fuel with great potential. Its most attractive features are that

- on a per gram basis, its heat of combustion is more than twice that of methane and about three times that of gasoline, and
- the product of its combustion is H_2O, not CO and CO_2 as with gasoline.

However, there are several serious problems in using hydrogen, such as

- it forms explosive mixtures with air;
- it is bulky to transport as a gas because of its very low density;
- it is difficult to transport as a liquid because of its very low boiling point (20 K);
- it dissolves in metals, causing metal parts to become brittle and fail.

Methods do exist to deal with these problems, however, leaving as the most serious drawback the fact that there is no cheap source of hydrogen. Currently, the bulk of hydrogen used commercially is made from petroleum and natural gas. Alternate methods of producing hydrogen, and the prospects of developing an economy based on hydrogen are discussed in Section 14-3.

Future Energy Sources. Although there is considerable uncertainty as to what the profile of future energy sources will be, there is no doubt that it will differ greatly from that which exists today. One possibility is suggested by Figure 7-15. (Nuclear fission and fusion are discussed in Chapter 26.)

FOCUS ON
The Greenhouse Effect

One result of a global warming through the "green-house effect" might be an increased production of icebergs from the polar ice caps. [John Eastcott/ Yva Momatiuk/Woodfin Camp & Associates]

We do not think of carbon dioxide as an air pollutant because it is essentially nontoxic. However, its ultimate effect on the environment is likely to be even greater than that of the oxides of nitrogen and sulfur. The problem is that a buildup of $CO_2(g)$ in the atmosphere is affecting the heat balance of the earth.

The earth's atmosphere is largely transparent to the visible light and ultraviolet radiation arriving from the sun. This radiation is absorbed at the earth's surface, warming it, but also some of this absorbed energy is reradiated as infrared radiation. Certain atmospheric gases, primarily CO_2 and water vapor, although transparent to visible and ultraviolet light, absorb some of this infrared radiation coming from the earth's surface. As a result a certain amount of energy is retained in the atmosphere, producing a warming effect on earth. This process is outlined in Figure 7-16. The warming effect described here is often compared to the retention of heat energy in a greenhouse, and is thus called the "greenhouse effect."* The natural greenhouse effect is crucial to maintaining the proper temperature to sustain life on earth. Without it, the earth would be permanently covered with ice.

From 1880 to 1980 the CO_2 content of the atmosphere increased from 275 to 339 ppm (parts per million). The current content is about 350 ppm. These increases are attributed to the burning of fossil fuels and the deforestation of tropical regions (trees, through photosynthesis help to reduce the CO_2 content of the atmosphere). The expected effect of this CO_2 buildup is to produce an increase in the earth's average temperature.

How large an increase in the average global temperature can we expect?

The effect can only be estimated, but best estimates are that a doubling of the CO_2 content in air from its preindustrial values could cause a global temperature increase of from 1.5 to 4.5 °C.

When can this increase be expected to occur?

The full temperature-enhancing effect of a buildup in CO_2 content of the atmosphere may be delayed for several decades, by mechanisms that are not clearly understood. At the present rate of CO_2 buildup, however, we seem committed to the effects produced by a doubling of atmospheric CO_2 as early as the middle of the next century.

How significant are global temperature increases of this magnitude (1.5 to 4.5 °C)?

A global temperature change of 3 °C is beyond the range previously experienced in human history, and the global temperature would be comparable to the climate on earth 13 million years ago. Among the potential climatic effects of a 1.5 to 4.5 °C global temperature rise would be

- local temperature changes (Washington, D.C. might experience 12 days a year of 100 °F temperatures instead of the current average of 1.);
- shifts in rainfall patterns (some areas, e.g., the U.S. Midwest, could become much drier; others, such as sub-Saharan Africa, might receive much more rainfall);

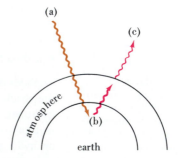

FIGURE 7-16
The "greenhouse effect."

(a) Sunlight received by earth.
(b) Infrared radiation reradiated by surface of earth.
(c) Infrared radiation emitted to space, following absorption of some of this radiation by $CO_2(g)$.

---

*Window glass, like CO_2, is also transparent to visible (and some ultraviolet) light and absorbs infrared. The functioning of a greenhouse, however, depends primarily on the glass surfaces preventing the bulk flow of warm air out of the greenhouse.

- shifts in ocean currents (displacement of the Gulf Stream could produce a cooling effect in Western Europe);
- rise in mean sea level (increased melting of polar ice caps could raise sea level by several meters, with profound effects on coastal cities).

Is the greenhouse effect actually likely to occur or is this mostly hypothetical?

Until recently much of the thinking about the greenhouse effect was speculative. Recent evidence does seem to support the theory's likelihood. For example, global mean temperatures (taken from the land and the oceans) show a warming trend beginning in 1890, with five of the nine warmest years since 1978. There has been a rise in the temperature of the Arctic permafrost of 2 to 4 °C over the past century. Sea level has risen about 10 cm over the last 50 years. Finally, analyses of tiny air bubbles trapped in the Antarctic ice cap provide evidence of a strong correlation between the atmospheric CO_2 content and temperature for the past 160,000 years—low temperatures during periods of low CO_2 levels and higher temperatures during periods of higher levels.

Is CO_2 the only atmospheric gas responsible for the greenhouse effect?

There are several gases that are even stronger infrared absorbers than CO_2 [e.g., methane (CH_4), ozone (O_3), nitrous oxide (N_2O), and chlorofluorocarbons (CFCs)]. About one-half of the anticipated greenhouse effect is due to CO_2 and the remainder to these other gases. Furthermore, the atmospheric concentrations of some of these other gases (especially CFCs) have been growing at a faster rate than that of CO_2.

What can be done to reverse or forestall the anticipated greenhouse effect?

An important starting point is a heightened awareness of the potential problem; this seems to be occurring. Strategies need to be developed to use energy more efficiently, and to shift from fossil fuels to other energy sources, such as nuclear energy (Sections 26-8 and 26-9) and the direct use of solar energy (see Focus feature, Chapter 11). Also, reductions must be achieved in the releases of other greenhouse gases (see discussion of CFCs and depletion of the ozone layer in the Focus feature of Chapter 14).

Summary

Thermochemistry is concerned primarily with heat effects accompanying chemical reactions. In dealing with this subject it is necessary to develop ideas about the nature of heat and its measurement and the distinction between heat and work. Heat effects are most commonly measured through temperature changes in a device called a calorimeter. The most common type of work in chemical reactions is that associated with the expansion and compression of gases—pressure–volume work.

The first law of thermodynamics relates the internal energy change (ΔE) in a system to exchanges of heat (q) and work (w) between a system and its surroundings, through the equation: $\Delta E = q + w$. A chemical reaction can be treated as a thermodynamic system. If a reaction is carried out at constant volume, with no opportunity for work to be performed, the heat of the reaction $q_V = \Delta E$. Reactions in which heat is *evolved* by the system to the surroundings are called *exothermic;* those in which heat is *absorbed* by the system are called *endothermic.*

Combustion reactions can be carried out in a bomb calorimeter, a device in which a heat of reaction is used to change the temperature of a quantity of water and other objects in the surroundings. Since these reactions occur at constant volume, the measured heats of reaction are $q_V = \Delta E$. Most often, however, chemical reactions are not conducted in bomb calorimeters. They are carried out in containers open to the atmosphere and under the constant pressure of the atmosphere. For these reactions it is advantageous to define a thermodynamic function of state called enthalpy (H). For a reaction conducted at constant pressure and with work limited to pressure–volume work, the heat of reaction is $q_P = \Delta H$. For reactions involving only solids and liquids, $\Delta H \approx \Delta E$; with gases, differences between ΔH and ΔE are larger, though still negligible in many cases. For a reaction at constant pressure ΔH and ΔE are related through the equation $\Delta H = \Delta E + P \, \Delta V$.

For certain processes ΔH values can be determined in a calorimeter of an especially simple design (see Figure 7-11). For many others it is possible to determine ΔH indirectly. Hess's law permits a process to be broken down into a series of steps and ΔH for the process to be obtained by summing ΔH values for the individual steps. The indirect determination of enthalpy changes is also facilitated by establishing an *arbitrary* zero of enthalpy for the elements in their most stable forms at 1 atm pressure. From this basis it is then possible to derive standard molar enthalpies (heats) of formation of compounds (ΔH_f°). These data can be compiled into extensive listings (see Appendix D) and used in computations.

One of the chief applications of thermochemistry is to the study of the combustion of fuels as energy sources. Currently, the principal fuels in use are the fossil fuels. Fossil fuels and other potential energy sources are also considered in this chapter.

Summarizing Example

The partial burning of coal in the presence of O_2 and H_2O produces a mixture of $CO(g)$ and $H_2(g)$ called **synthesis gas**. Synthesis gas can be used to produce a number of important organic compounds (e.g., methyl alcohol), burned as a fuel, or converted to a gaseous mixture, called **substitute natural gas (SNG)**, that is predominantly methane (CH_4).

A particular synthesis gas is found to consist of 55% $CO(g)$, 33% $H_2(g)$, and 11% noncombustible gases (mostly CO_2), *by volume*. The following calculations are based on a 1.00 L sample of the gaseous mixture, measured at STP.

1. How many moles of $CO(g)$ and $H_2(g)$ are present in the 1.00 L sample?

Solution. We need two expressions from Chapter 6. First is the molar volume of a gas at STP (expression 6.8).

$$\text{no. mol gas} = 1.00 \text{ L} \times \frac{1 \text{ mol gas}}{22.4 \text{ L}} = 0.0446 \text{ mol gas} \quad (n_{\text{tot.}})$$

Next, is the fact that the mol fraction of a gas in a mixture is the same as its volume fraction (expression 6.16). That is,

$$\frac{n_{CO}}{n_{\text{tot.}}} = \frac{V_{CO}}{V_{\text{tot.}}}$$

Thus, if there is 55 L CO for every 100 L of the gaseous mixture, there must also be 55 mol CO for every 100 mol of the mixture. As a result, the molar amounts of CO and H_2 must be

$$n_{CO} = 0.0446 \text{ mol gas} \times \frac{55 \text{ mol CO}}{100 \text{ mol gas}} = 0.0245 \text{ mol CO}$$

$$n_{H_2} = 0.0446 \text{ mol gas} \times \frac{33 \text{ mol H}_2}{100 \text{ mol gas}} = 0.0147 \text{ mol H}_2$$

(This example is similar to Examples 6-17 and 6-18.)

2. Use tabulated data, as necessary, to calculate the enthalpy (heat) of combustion of 0.0147 mol H_2.

Solution. If the equation for the chemical reaction on which an enthalpy calculation is based is not given, writing this equation is always a good starting point. For the combustion of $H_2(g)$,

$$H_2(g) + \tfrac{1}{2} O_2(g) \longrightarrow H_2O(l) \qquad \Delta H^\circ = \text{?}$$

Once you have written this equation, you should immediately recognize that it *also* represents the formation of $H_2O(l)$ from its elements. ΔH° for this reaction is the tabulated value of the enthalpy of formation of one mole of $H_2O(l)$: $\Delta H_f^\circ[H_2O(l)] = -285.8 \text{ kJ/mol } H_2O(l)$. That is, we can write

$$H_2(g) + \tfrac{1}{2} O_2(g) \longrightarrow H_2O(l) \qquad \Delta H^\circ = -285.8 \text{ kJ}$$

Now we can use ΔH° to write a conversion factor (in blue below).

$$\text{no. kJ} = 0.0147 \text{ mol H}_2 \times \frac{-285.8 \text{ kJ}}{1 \text{ mol H}_2} = -4.20 \text{ kJ}$$

(This example is similar to Examples 7-8 and 7-11.)

3. Use tabulated data, as necessary, to calculate the enthalpy (heat) of combustion of 0.0245 mol $CO(g)$.

This plant, located in the Mojave Desert of California, can convert 1000 tons of coal per day into a medium-Btu synthesis gas. Coal is stored in the two large silos in the background, and a coal-water slurry is partially burned in an oxygen atmosphere in the tower at the right of the picture. [Courtesy of Donald H. Watts, Program Manager, Cool Water Coal Gasification Program.]

Solution. Again we can start by writing the equation for the combustion of one mole of CO(g) to form one mole of CO_2(g).

$$CO(g) + \tfrac{1}{2} O_2(g) \longrightarrow CO_2(g) \qquad \Delta H° = ?$$

This time, however, the enthalpy change is not the enthalpy of formation of CO_2(g), since the CO_2(g) is not being formed from its elements. (O_2 is an element but not CO.) Here we apply equation (7.23).

$$\Delta H°_{rxn} = \Delta H°_f[CO_2(g)] - \Delta H°_f[CO(g)] = -393.5 \text{ kJ} - (-110.5 \text{ kJ})$$
$$= -283.0 \text{ kJ}$$

Now we can write

$$CO(g) + \tfrac{1}{2} O_2(g) \longrightarrow CO_2(g) \qquad \Delta H° = -283.0 \text{ kJ}$$

Finally,

$$\text{no. kJ} = 0.0245 \text{ mol CO} \times \frac{-283.0 \text{ kJ}}{1 \text{ mol CO}} = -6.93 \text{ kJ}$$

(This example is similar to Examples 7-8 and 7-12.)

4. What is the heat of combustion per liter (STP) of the synthesis gas?

Solution. In parts **2** and **3**, we calculated the heats of combustion of the two combustible gases in 1.00 L (STP) of the synthesis gas. We need simply to add them together.

heat of combustion of 1.00 L (STP) = -4.20 kJ $- 6.93$ kJ = -11.13 kJ
of synthesis gas

5. To what temperature can 1.00 kg of water at 25.0 °C be heated with the heat given off in the combustion of 1.00 L (STP) of the synthesis gas?

Solution. Here we need to apply equation (7.6) and solve for the temperature change, ΔT. In the setup below, $q_{water} = +11.13$ kJ.

$$q_{water} = \text{mass water} \times \text{sp. ht. water} \times \Delta T$$

$$11.13 \text{ kJ} = 11{,}130 \text{ J} = 1.00 \times 10^3 \text{ g} \times \frac{4.184 \text{ J}}{\text{g °C}} \times \Delta T$$

$$\Delta T = \frac{11{,}130 \text{ J}}{1.00 \times 10^3 \text{ g} \times 4.184 \text{ J g}^{-1} \text{ °C}^{-1}} = 2.66 \text{ °C}$$

Final temp. = initial temp. + ΔT = 25.0 °C + 2.66 °C = 27.7 °C

(This example is similar to Example 7-4.)

Key Terms

bomb calorimeter (7-6)
calorie (cal) (7-4)
calorimeter (7-6, 7-7)
endothermic (7-4)
enthalpy (7-7)
enthalpy change (7-7)
enthalpy (heat) of formation (7-9)
exothermic (7-4)

function of state (state function) (7-5)
heat (7-1, 7-4)
heat capacity (7-4)
heat of reaction (7-6, 7-7)
Hess's law (7-8)
internal energy (E) (7-2, 7-5)
joule (J) (7-1)

law of conservation of energy (7-4)
pressure–volume work (7-3)
specific heat (7-4)
standard state (7-9)
surroundings (7-2)
system (7-2)
work (7-1, 7-3)

Highlighted Expressions

Equation for pressure–volume work (at constant pressure) (7.3)
The specific heat of water (7.5)
Relationship between a quantity of heat, and mass, specific heat, and temperature change (7.6)

The first law of thermodynamics (7.9)
Heat of reaction at constant volume, q_V (7.12)
Heat of reaction at constant pressure, q_p (7.16)
Relating an enthalpy of reaction to enthalpies of formation (7.23)

Review Problems

1. Calculate the quantity of heat
(a) in *kcal*, required to raise the temperature of 12.0 L of water from 25.0 to 33.0 °C.
(b) in *kJ*, associated with a 25.0 °C *decrease* in temperature in a 15.0-kg iron bar (specific heat of iron = 0.473 J g^{-1} °C^{-1}).

2. Calculate the final temperature that results from each of the following processes.
(a) A 8.82-g sample of water at 20.3 °C *absorbs* 112 cal of heat.
(b) A 6.25-kg sample of solid sulfur at 45.3 °C *gives off* 165 kcal of heat. (Sp. ht. of S = 0.173 cal g^{-1} °C^{-1}.)
(c) A 125-L sample of toluene, C_7H_8(l), at 32.2 °C *gives off* 3.8×10^3 kJ of heat. [For C_7H_8(l), $d = 0.866$ g/cm^3; sp. ht. = 1.7 J g^{-1} °C^{-1}.]

3. From the information given, express each of the following properties.
(a) The *specific heat* of benzene (C_6H_6), given that 192 J of heat is required to raise the temperature of a 20.0-g sample from 25.2 to 30.8 °C.
(b) The *final temperature*, given that 3.50 kcal of heat is removed from a 1.50-kg sample of water initially at 25.2 °C.
(c) The *molar heat capacity* of the gasoline component octane, C_8H_{18}(l), if 1.10 kJ of heat is given off when a 62.5-g sample is cooled from 32.1 to 24.8 °C.

4. A 1.00-kg sample of magnesium at 40.0 °C is added to 1.00 L of water maintained at 20.0 °C in an insulated container. What will be the final temperature of the Mg–H_2O mixture? (Sp. ht. of Mg = 1.04 J g^{-1} °C^{-1}.)

5. What volume of 15.5-°C water must be added, together with the 1.05-kg piece of iron at 60.5 °C, so that the temperature of the water in the insulated container remains constant at 25.6 °C?

6. What is the quantity of work performed in the expansion of the gas in Figure 7-3 if each weight has a mass of 500. g and if the height $h = 16.8$ cm? [*Hint:* What mass is lifted by the expanding gas?)

7. What are the changes in internal energy of a system, ΔE, if the system
(a) Absorbs 58 J and does 58 J of work?
(b) Absorbs 125 J of heat and does 687 J of work?
(c) Loses 22 J of heat and has 111 J of work done on it?
(d) Absorbs no heat and does 117 J of work?

8. Upon complete combustion in a bomb calorimeter, the following quantities of heat are evolved by the substances indicated. Express each heat of combustion as kJ/mol of substance.
(a) 0.107 g of acetylene, C_2H_2(g), yields 5.36 kJ;
(b) 1.030 g of urea, CO(NH_2)$_2$, yields 2.610 kcal;
(c) 1.05 cm^3 of acetone, (CH_3)$_2$CO(l) ($d = 0.791$ g/cm^3), yields 26.0 kJ.

9. A sample that is burned in a bomb calorimeter gives off 20.9 kJ of heat. The temperature of the calorimeter assembly increases by 3.68 °C. Calculate the heat capacity of the calorimeter.

10. The following substances undergo complete combustion in a bomb calorimeter. The calorimeter assembly has a heat capacity of 5.011 kJ/°C. In each case, what is the *final temperature* if the initial water temperature is 24.98 °C? [The heats of combustion (ΔE) listed are per mole of substance burned.]
(a) 0.5060 g cyclohexanol, C_6H_{12}O(l); heat of combustion = −890.7 kcal/mol cyclohexanol.
(b) 0.853 g thymol, $C_{10}H_{14}$O(s); heat of combustion = -5.65×10^3 kJ/mol thymol.
(c) 1.25 mL of ethyl acetate, $C_4H_8O_2$(l) ($d = 0.901$ g/cm^3); heat of combustion = −2246 kJ/mol ethyl acetate.

11. A bomb calorimetry experiment is performed with isobutane [(CH_3)$_3$CH] as the combustible substance. The data obtained are

mass of isobutane burned:	1.036 g
heat capacity of calorimeter:	4.947 kJ/°C
initial calorimeter temperature:	24.88 °C
final calorimeter temperature:	35.17 °C

(a) What is the heat of combustion, *per gram* of isobutane?
(b) Assume that the heat of combustion can be represented as the enthalpy change, ΔH. (That is, assume that $\Delta H \approx \Delta E$.) What is ΔH, *in kJ*, for the combustion of one mole of (CH_3)$_3$CH?
(c) In the manner of equation (7.17), write the chemical equation for the complete combustion of isobutane and represent the value of ΔH in this equation.

12. A 1.50-g sample of NH_4NO_3 is added to 35.0 g H_2O in a Styrofoam cup and stirred until it dissolves. The temperature of

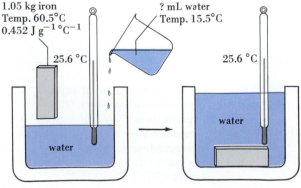

1.05 kg iron
Temp. 60.5°C
0.452 J g^{-1}°C^{-1}

? mL water
Temp. 15.5°C

25.6 °C

25.6 °C

water

water

Problem 5

the solution drops from 22.7 to 19.4 °C. **(a)** Is the process endothermic or exothermic? **(b)** What is the heat of solution of NH_4NO_3, expressed in kJ/mol NH_4NO_3?

13. Write the balanced chemical equations which have as their enthalpy changes:
  (a) $\Delta H_f^{\circ} = -209$ kJ/mol $COCl_2(g)$
  (b) $\Delta H_f^{\circ} = +50.63$ kJ/mol $N_2H_4(l)$
  (c) $\Delta H_{combustion}^{\circ} = -1.66 \times 10^3$ kJ/mol $C_3H_8O_3(l)$

14. Use Hess's law to determine ΔH for the reaction $CO(g) + \frac{1}{2} O_2(g) \rightarrow CO_2(g)$, given that
  (a) $C(graphite) + \frac{1}{2} O_2(g) \rightarrow CO(g)$ $\Delta H = -110.5$ kJ
  (b) $C(graphite) + O_2(g) \rightarrow CO_2(g)$ $\Delta H = -393.5$ kJ

15. Use Hess's law to determine ΔH for the reaction $C_3H_4(g) + 2 H_2(g) \rightarrow C_3H_8(g)$, given that
  (a) $H_2(g) + \frac{1}{2} O_2(g) \rightarrow H_2O(l)$ $\Delta H = -285.8$ kJ
  (b) $C_3H_4(g) + 4 O_2(g) \rightarrow 3 CO_2(g) + 2 H_2O(l)$
  $$\Delta H = -1941 \text{ kJ}$$
  (c) $C_3H_8(g) + 5 O_2(g) \rightarrow 3 CO_2(g) + 4 H_2O(l)$
  $$\Delta H = -2220. \text{ kJ}$$

16. Use enthalpies of formation from Appendix D in equation (7.23) to determine the heats of the following reactions:
  (a) $C_3H_8(g) + H_2(g) \rightarrow C_2H_6(g) + CH_4(g)$
  (b) $2 H_2S(g) + 3 O_2(g) \rightarrow 2 SO_2(g) + 2 H_2O(l)$
  (c) $Fe_2O_3(s) + 3 CO(g) \rightarrow 2 Fe(s) + 3 CO_2(g)$

17. Use the following information, together with data from Appendix D and equation (7.23), to calculate the enthalpy of formation, *per mole*, of $ZnS(s)$.

$$2 ZnS(s) + 3 O_2(g) \rightarrow 2 ZnO(s) + 2 SO_2(g) \quad \Delta H^{\circ} = -880. \text{ kJ}$$

18. The following list of heats of combustion of the substances were obtained from a handbook. Determine the enthalpy of formation (ΔH_f°) of each substance. [*Hint:* Write an equation for each combustion reaction; assume that the heat of combustion is ΔH° for the combustion of one mole of substance; and use equation (7.23) and data from Table 7.2.]
  (a) Dimethyl ether, $(CH_3)_2O(g)$, -1.454×10^3 kJ/mol
  (b) Benzaldehyde, $C_7H_6O(l)$, -3.520×10^3 kJ/mol
  (c) Glucose, $C_6H_{12}O_6(s)$, -2.816×10^3 kJ/mol

Exercises

Work

19. How much work, in joules, is performed when the indicated mass is removed from the piston in the cylinder of gas pictured? [*Hint:* What is the expanded volume of the gas? In what unit should pressure be expressed?]

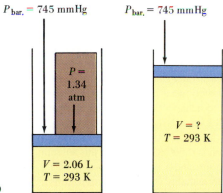

$P_{bar.} = 745$ mmHg $P_{bar.} = 745$ mmHg

$P = 1.34$ atm

$V = 2.06$ L
$T = 293$ K

$V = ?$
$T = 293$ K

Exercise 19

20. Calculate the quantity of work, in joules, associated with a 5.0-L expansion of a gas (ΔV) against a pressure of 735 mmHg in the units **(a)** liter atmospheres (L-atm); **(b)** joules (J); **(c)** calories (cal). [*Hint:* You can use the different units of the gas constant R (from the inside back cover) to establish conversion factors among energy units.]

Heat capacity (specific heat)

21. Refer to Example 7-4. The experiment is repeated with several different metals substituting for the lead. That is, the masses of metal and water and the initial temperatures of the metal and water are the same as in Figure 7-5. The final temperatures are given below. What is the specific heat of each metal, expressed as $J\ g^{-1}\ ^{\circ}C^{-1}$? **(a)** Zn, final temperature 39.0 °C; **(b)** Pt, 28.9 °C; **(c)** Al, 52.8 °C.

22. An electric range burner having a mass of 625 g is turned off after reaching a temperature of 592 °C. If the heat capacity of the burner is 0.4 J g^{-1} $^{\circ}C^{-1}$.
  (a) approximately how much heat will the burner exchange with the surroundings in cooling back to room temperature (20 °C)?
  (b) approximately what mass of water could be heated from room temperature to the boiling point if all this heat could be transferred to the water?

23. A piece of stainless steel (sp. ht. = 0.50 J g^{-1} $^{\circ}C^{-1}$) is transferred from an oven (152 °C) to 125 mL of water at 24.8 °C, into which it is immersed. The water temperature rises to 40.3 °C. What is the mass of the steel? How precise is this method of mass determination? Explain.

24. Magnesium metal has a specific heat of 1.04 J g^{-1} $^{\circ}C^{-1}$. A 70.0-g sample of this metal, at a temperature of 99.8 °C, is added to a beaker containing 50.0 g water at 30.0 °C. The final water temperature is found to be 47.2 °C. Is this result consistent with the law of conservation of energy? Explain.

25. Brass has a density of 8.40 g/cm^3 and a specific heat of 0.385 J g^{-1} $^{\circ}C^{-1}$. The brass cube in the figure is 2.25 cm on edge and at an initial temperature of 152 °C. What is the final temperature after the cube is immersed in 115 g water initially at 24.8 °C in an insulated container?

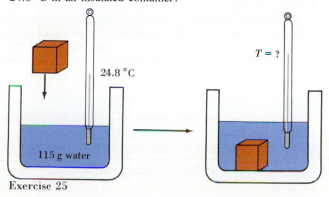

24.8 °C

$T = ?$

115 g water

Exercise 25

*26. Samples of two different metals, M_1 and M_2, are each maintained at the same initial temperature. If, in cooling to the same final temperature, the two samples are to lose the same quantity of heat, show that the masses of the samples must be *inversely* proportional to the specific heats of the metals.

First law of thermodynamics

27. *The internal energy of a fixed quantity of an ideal gas depends only on its temperature.* A sample of an ideal gas is allowed to expand at a *constant temperature* (isothermal expansion). (a) Does the gas do work? (b) Does the gas exchange heat with its surroundings? (c) What happens to the temperature of the gas? (d) What is ΔE for the gas?

28. In an *adiabatic* process a system is thermally insulated from its surroundings such that there is no exchange of heat ($q = 0$). For the adiabatic expansion of an ideal gas (a) Does the gas do work? (b) Does the internal energy of the gas increase, decrease, or remain constant? (c) What happens to the temperature of the gas? [*Hint:* Also refer to Exercise 27.]

29. State whether you think each of the following observations is in any way possible and indicate your reasons. [*Hint:* Also refer to the statements made in Exercises 27 and 28.]

(a) An ideal gas is expanded at *constant temperature* and is observed to do twice as much work as the heat it absorbs from its surroundings.

(b) A gas absorbs heat from the surroundings while being compressed.

*30. While on his honeymoon in Switzerland, James Joule (the discoverer of the first law of thermodynamics) did an experiment in which he measured the temperature of the water at the top of a waterfall and again at the bottom. What observation would you expect him to have made? Explain.

Functions of state

31. Explain how a thermodynamic function such as enthalpy (H) can be used even though its absolute value cannot be determined.

32. Both graphite and diamond are pure forms of carbon. The heat of combustion of graphite, yielding $CO_2(g)$ as the only product and with all reactants and products being at 25 °C and 1 atm pressure, is −393.51 kJ/mol. If diamond is substituted for graphite, would you expect the heat of combustion to be the same, larger, or smaller? Explain.

Heats of reaction

33. Refer to the Summarizing Example. What volume of the synthesis gas, measured at STP and burned in an open flame (constant-pressure process), is required to heat 30.0 gal of water from 22.3 to 61.6 °C?

34. The heat of reaction of quicklime (CaO) with water produces slaked lime [$Ca(OH)_2$], a substance widely used in the construction industry to make mortar, plaster, and stucco. The reaction is highly exothermic. How much heat, in kJ, is associated with the production of 1 ton (2000 lb) of slaked lime?

$$CaO(s) + H_2O(l) \rightarrow Ca(OH)_2(s) \qquad \Delta H = -65.2 \text{ kJ}$$

35. The combustion of hydrogen–oxygen mixtures is used to produce very high temperatures (ca. 2500 °C) needed for certain types of welding operations. Consider the reaction to be

$$H_2(g) + \tfrac{1}{2} O_2(g) \rightarrow H_2O(g) \qquad \Delta H = -241.8 \text{ kJ}$$

What is the quantity of heat evolved when a 100.-g mixture containing *equal* parts of H_2 and O_2, *by mass,* is burned?

36. The combustion of methane gas (the principal constituent of natural gas) is represented by the equation

$$CH_4(g) + 2 O_2(g) \rightarrow CO_2(g) + 2 H_2O(l) \qquad \Delta H = -890. \text{ kJ}$$

(a) What mass of methane must be burned to liberate 1.00×10^6 kJ of heat?

(b) What quantity of heat is liberated in the complete combustion of 1.03×10^3 L of $CH_4(g)$, measured at 21.8 °C and 748 mmHg?

(c) If the quantity of heat calculated in (b) could be transferred with 100% efficiency to water, what volume of water could be heated from 22.7 to 60.8 °C as a result?

37. Thermite mixtures are used for certain types of welding. The thermite reaction is highly exothermic.

$$Fe_2O_3(s) + 2 Al(s) \rightarrow Al_2O_3(s) + 2 Fe(l) \qquad \Delta H = -850 \text{ kJ}$$

1.00 mol Fe_2O_3 and 2.00 mol Al are mixed at room temperature (25 °C) and a reaction is initiated. The liberated heat is retained within the products, whose combined specific heat over a broad temperature range is about 0.8 J g^{-1} $°C^{-1}$. The melting point of iron is 1530 °C. Show that the quantity of heat liberated is sufficient to raise the temperature of the products to the melting point of iron.

38. A pellet of potassium hydroxide, KOH, weighing 0.150 g is added to 45.0 g of water in a Styrofoam coffee cup. The water temperature rises from 24.1 to 24.9 °C. [Assume that the specific heat of dilute KOH(aq) is the same as that of water.]

(a) What is the approximate heat of solution of KOH, expressed as kJ/mol KOH?

(b) How could the precision of this measurement be improved, *without* modifying the apparatus?

39. A lecture demonstration is being planned to illustrate an *endothermic* process. It is desired to lower the temperature of 1400 mL water in an insulated container from 25 to 10 °C. Approximately what mass of $NH_4Cl(s)$ should be dissolved in the water to achieve this result? The heat of solution of NH_4Cl is +14.8 kJ/mol NH_4Cl.

40. Refer to Example 7-9. The product of the neutralization reaction is 0.500 M NaCl. For this solution, assume a density of 1.02 g/mL and a specific heat of 4.02 J g^{-1} $°C^{-1}$ and recalculate the enthalpy (heat) of neutralization.

Bomb calorimetry

41. *o*-Phthalic acid ($C_8H_6O_4$) is sometimes used as a calorimetric standard. Its heat of combustion (ΔE) is -3.224×10^3 kJ/mol $C_8H_6O_4$. From the following data determine the heat capacity of a bomb calorimeter assembly (that is, of the bomb, water, stirrer, thermometer, wires . . .):

mass of *o*-phthalic acid burned: 1.078 g
initial calorimeter temperature: 24.96 °C
final calorimeter temperature: 30.76 °C

42. Refer to Example 7-6. Based on the heat of combustion (ΔE) of sucrose established in the example, what should be the temperature change (ΔT) produced by the combustion of 1.227 g $C_{12}H_{22}O_{11}$ in a bomb calorimeter assembly having a heat capacity of 3.87 kJ/°C?

43. The burning of 2.051 g of glucose, $C_6H_{12}O_6$, in a bomb calorimeter causes the temperature of the water to increase from

24.92 to 31.41 °C. The bomb calorimeter assembly has a heat capacity of 4.912 kJ/°C.

(a) What is the heat of combustion (ΔE) of the glucose expressed in kJ/mol $C_6H_{12}O_6$?

(b) Write a balanced equation for the combustion reaction assuming that $CO_2(g)$ and $H_2O(l)$ are the sole products of the reaction. Represent ΔH in this equation, in the manner of equation (7.17). [*Hint:* ΔH for this reaction is the same as ΔE—the heat of combustion per mole $C_6H_{12}O_6$.]

44. A 1.567-g sample of napththalene, $C_{10}H_8(s)$, is completely burned in a bomb calorimeter assembly and a temperature increase of 8.37 °C is noted. When a 1.227-g sample of thymol, $C_{10}H_{14}O(s)$, (a preservative and a mold and mildew preventative) is burned in the same calorimeter assembly, the temperature increase is 6.12 °C. If the heat of combustion (ΔE) of naphthalene is -5153.9 kJ/mol $C_{10}H_8$, what is the heat of combustion (ΔE) of thymol, expressed in kJ/mol $C_{10}H_{14}O$?

45. An alternative approach to calibrating a bomb calorimeter is to establish the heat capacity of the calorimeter, *exclusive* of the water it contains. In this way the calorimeter can be used with different quantities of water if so desired. However, in these cases the heat absorbed by the water and by the rest of the calorimeter must be calculated separately and then added together. Also, the quantity of water and its specific heat must be known.

A bomb calorimeter assembly containing 983.5 g of water is calibrated by the combustion of 1.354 g of anthracene; the temperature of the calorimeter rises from 24.87 to 35.63 °C. When 1.065 g of citric acid, $H_3C_6H_5O_7$ (used in confections and in soft drinks), is burned in the same assembly containing 968.6 g of water, the temperature increase is from 25.01 to 27.19 °C. The heat of combustion (ΔE) of anthracene, $C_{14}H_{10}(s)$, is -7163 kJ/mol $C_{14}H_{10}$. What is the heat of combustion (ΔE) of citric acid, $C_6H_8O_7$, expressed in *kJ/mol*?

Relating ΔH and ΔE

46. Only one of the following expressions can be used to describe the heat of a chemical reaction *regardless of how the reaction is carried out*. Which is the correct expression and why? **(a)** q_V; **(b)** q_P; **(c)** $\Delta E - w$; **(d)** ΔE; **(e)** ΔH

*★47.** Use the fact that $\Delta H = \Delta E + P\,\Delta V$ to determine whether ΔH is equal to, greater than, or less than ΔE for the following reactions. Recall that "greater than" means more positive (or less negative) and "less than" means less positive (or more negative). Assume that the only significant change in volume that occurs in a reaction is that associated with *gases*. From the ideal gas equation: $PV = nRT$, we can write the expression $P\,\Delta V = \Delta nRT$. Thus, you need to assess for each reaction whether Δn is zero, a positive quantity, or a negative quantity. Also, you must write a balanced equation for each reaction.

(a) The complete combustion of one mole of butanol, $C_4H_9OH(l)$.

(b) The complete combustion of one mole of glucose, $C_6H_{12}O_6(s)$.

(c) The decomposition of one mole of $NH_4NO_3(s)$ into liquid water and gaseous dinitrogen monoxide.

*★48.** An experimental determination of the heat of combustion (ΔE) of isopropyl alcohol (rubbing alcohol) in a bomb calorimeter yields a value of -33.41 kJ/g C_3H_7OH.

$$C_3H_7OH(l) + \tfrac{9}{2}\,O_2(g) \rightarrow 3\,CO_2(g) + 4\,H_2O(l)$$

Use ideas relating ΔE and ΔH outlined in Exercise 47 to determine for the combustion of *one mole* of $C_3H_7OH(l)$ **(a)** ΔE; **(b)** ΔH at 298 K.

Hess's law

49. Determine ΔH for the reaction

$$N_2H_4(l) + 2\,H_2O_2(l) \rightarrow N_2(g) + 4\,H_2O(l)$$

from these data:

$$N_2H_4(l) + O_2(g) \rightarrow N_2(g) + 2\,H_2O(l) \qquad \Delta H = -622.3 \text{ kJ}$$
$$H_2(g) + \tfrac{1}{2}\,O_2(g) \rightarrow H_2O(l) \qquad \Delta H = -285.8 \text{ kJ}$$
$$H_2(g) + O_2(g) \rightarrow H_2O_2(l) \qquad \Delta H = -187.8 \text{ kJ}$$

50. Use Hess's law and the following data

$$CH_4(g) + 2\,O_2(g) \rightarrow CO_2(g) + 2\,H_2O(g) \qquad \Delta H = -802 \text{ kJ}$$
$$CH_4(g) + CO_2(g) \rightarrow 2\,CO(g) + 2\,H_2(g) \qquad \Delta H = +206 \text{ kJ}$$
$$CH_4(g) + H_2O(g) \rightarrow CO(g) + 3\,H_2(g) \qquad \Delta H = +247 \text{ kJ}$$

to determine ΔH for the reaction

$$CH_4(g) + \tfrac{1}{2}\,O_2(g) \rightarrow CO(g) + 2\,H_2(g),$$

an important commercial source of hydrogen gas.

51. CCl_4, an important commercial solvent, is prepared by the reaction of $Cl_2(g)$ with a carbon compound. Determine ΔH for the reaction

$$CS_2(l) + 3\,Cl_2(g) \rightarrow CCl_4(l) + S_2Cl_2(l)$$

given these data:

$$CS_2(l) + 3\,O_2(g) \rightarrow CO_2(g) + 2\,SO_2(g) \qquad \Delta H = -1077 \text{ kJ}$$
$$2\,S(s) + Cl_2(g) \rightarrow S_2Cl_2(l) \qquad \Delta H = -60.2 \text{ kJ}$$
$$C(s) + 2\,Cl_2(g) \rightarrow CCl_4(l) \qquad \Delta H = -135.4 \text{ kJ}$$
$$S(s) + O_2(g) \rightarrow SO_2(g) \qquad \Delta H = -296.9 \text{ kJ}$$
$$C(s) + O_2(g) \rightarrow CO_2(g) \qquad \Delta H = -393.5 \text{ kJ}$$

52. The heats of combustion (ΔH) per mole of 1,3-butadiene [$C_4H_6(g)$], normal butane [$C_4H_{10}(g)$], and $H_2(g)$ are -2543.5, -2878.6, and -285.8 kJ, respectively. Use these data to calculate the heat of hydrogenation of 1,3-butadiene to normal butane.

$$C_4H_6(g) + 2\,H_2(g) \rightarrow C_4H_{10}(g) \qquad \Delta H = ?$$

[*Hint:* Write equations for the combustion reactions. In each combustion the products are $CO_2(g)$ and $H_2O(l)$.]

*★53.** The experiment that is pictured in part on the next page is used to confirm Hess's law. The following net reaction is carried out in two ways.

$$NH_3(\text{conc. aq}) + HCl(aq) \rightarrow NH_4Cl(aq)$$

First, the experiment is carried out directly by adding 8.00 mL of concentrated aqueous NH_3 to 100.0 mL of 1.00 M HCl in a calorimeter similar to that of Figure 7-11. (The NH_3 is in slight excess.) In a typical student experiment the initial temperature of the $HCl(aq)$ and $NH_3(\text{conc. aq})$ was 23.8 °C and the final temperature after the neutralization reaction, 35.8 °C.

In the second experiment, air is bubbled through 100.0 mL of concentrated aqueous ammonia, and ammonia vapor is swept into 100.0 mL of 1.00 M HCl. In this case the reactions are

$$NH_3(\text{conc. aq}) \rightarrow NH_3(g)$$

$$NH_3(g) + HCl(aq) \rightarrow NH_4Cl(aq)$$

In a typical student experiment, the temperature of the concentrated aqueous NH_3 in the gas bubbler bottle fell from 19.3 to 13.2 °C. At the same time the temperature of the $HCl(aq)$ rose from 23.8 to 42.9 °C as it was neutralized.

Show that the results of these two experiments are consistent with Hess's law. (Assume that all solutions have densities of 1.00 g/mL and specific heats of 4.18 J g^{-1} °C^{-1}.) Cite some probable sources of error in this experiment.

Exercise 53   [Carey B. Van Loon]

Enthalpies (heats) of formation

54. Use standard enthalpies of formation from Table 7-2 to determine the heat of the reaction

$$2 Cl_2(g) + 2 H_2O(l) \rightarrow 4 HCl(g) + O_2(g) \qquad \Delta H° = ?$$

55. What is the heat of combustion of $C_2H_5OH(l)$ if the reactants and products are maintained at 25 °C and 1 atm? [*Hint:* Use data from Appendix D.]

56. Given the heat of the following reaction, determine the enthalpy of formation of $CCl_4(g)$ at 25 °C and 1 atm. [*Hint:* Use data from Table 7-2.]

$$CH_4(g) + 4 Cl_2(g) \rightarrow CCl_4(g) + 4 HCl(g) \qquad \Delta H° = -397 \text{ kJ}$$

57. For the reaction

$$C_2H_4(g) + 3 O_2(g) \rightarrow 2 CO_2(g) + 2 H_2O(l)$$
$$\Delta H° = -1410.8 \text{ kJ}$$

if the H_2O were obtained as a gas rather than a liquid, **(a)** would the heat of reaction be greater (more negative) or smaller (less negative) than that indicated in the equation? **(b)** Explain your answer. **(c)** Now, calculate the value of ΔH in this case. [*Hint:* Refer to Appendix D.]

58. Use data from Appendix D, together with the fact that $\Delta H° = -3534$ kJ for the complete combustion of one mole of pentane, $C_5H_{12}(l)$, to calculate $\Delta H°$ for the synthesis of 1 mol $C_5H_{12}(l)$ from $CO(g)$ and $H_2(g)$.

$$5 CO(g) + 11 H_2(g) \rightarrow C_5H_{12}(l) + 5 H_2O(l) \qquad \Delta H° = ?$$

59. Refer to Example 7-7. If the bomb calorimeter assembly described in this example is used for the combustion of a 1.066-g sample of the bottled gas propane, C_3H_8, what temperature change will be produced? [*Hint:* You must determine a value of the heat of combustion of propane from tabulated data. Consider the difference between ΔH and ΔE to be insignificant.]

60. The decomposition of limestone, $CaCO_3(s)$, into quicklime, $CaO(s)$, and $CO_2(g)$ is accomplished at 900 °C in a gas-fired kiln. (Assume that heats of reaction under these conditions are the same as at 25 °C and 1 atm pressure.)

(a) Use data from Appendix D to determine how much heat is required to decompose 1.35×10^3 kg $CaCO_3(s)$.

(b) If the heat energy calculated in part (a) is supplied by the combustion of methane, $CH_4(g)$, what volume of the gas, measured at 24.6 °C and 756 mmHg, is required? [*Hint:* Again use data from Appendix D to determine the heat of combustion of $CH_4(g)$.]

61. Under the entry "H_2SO_4" a handbook lists several values for the enthalpy of formation, ΔH_f. For example, for pure $H_2SO_4(l)$, $\Delta H_f = -814$ kJ/mol H_2SO_4; for an aqueous solution that is 1.0 M H_2SO_4, $\Delta H_f = -888$ kJ/mol H_2SO_4; for 0.25M H_2SO_4, $\Delta H_f = -890$ kJ/mol H_2SO_4; for 0.020 M H_2SO_4, $\Delta H_f = -897$ kJ/mol H_2SO_4.

(a) Explain why these values are not all the same.

(b) When a concentrated aqueous solution of H_2SO_4 is diluted, does the solution temperature increase or decrease? Explain.

★(c) If 250 mL of 0.02 M H_2SO_4 is prepared by diluting pure $H_2SO_4(l)$ with water, estimate the change in temperature that occurs. [Assume that the $H_2SO_4(l)$ and the water used for its dilution are at the same temperature initially, and and that all liquids and solutions have a specific heat of 4.18 J g^{-1} °C^{-1}.]

★**62.** A 1.00-L sample (at STP) of a natural gas evolves, upon complete combustion at constant pressure, 43.6 kJ of heat. If the gas is a mixture of $CH_4(g)$ and $C_2H_6(g)$, what is its percent composition, *by volume?* [*Hint:* What are the heats of combustion of $CH_4(g)$ and $C_2H_6(g)$?]

Additional Exercises

63. Express the quantity of heat in the unit indicated, and designate whether this heat is gained or lost by the liquid.

(a) The quantity of heat, in *cal*, when 417 g H_2O undergoes a temperature *decrease* of 2.92 °C.

(b) The quantity of heat, in *kcal,* when 62.3 kg chloroform, $CHCl_3(l)$, has its temperature changed from 16.8 to 25.1 °C (sp. ht. $CHCl_3 = 0.232$ cal g^{-1} $°C^{-1}$).

(c) The quantity of heat, in *kJ,* when 22.5 L ethylene dichloride, $C_2H_4Cl_2(l)$, undergoes a temperature change from 44.3 to 24.4 °C. [For $C_2H_4Cl_2(l)$, $d = 1.253$ g/cm³; sp. ht. = 1.30 J g^{-1} $°C^{-1}$.]

*64. A 16-lb shot put is dropped from the top of a building 123 m high. What is the maximum temperature increase that could occur in the shot put? Assume a specific heat of 0.47 J g^{-1} $°C^{-1}$ for the shot put. Why would the actual measured temperature increase likely be less than the calculated value? [*Hint:* Use equations B.3 and B.4 from Appendix B.]

65. In 1818, Dulong and Petit observed that the molar heat capacities of elements in their solid states are approximately constant. They derived the expression:

atomic weight × specific heat (cal g^{-1} $°C^{-1}$) ≈ 6.4

The specific heats of silicon, phosphorus, and lead are 0.17, 0.19, and 0.031 cal g^{-1} $°C^{-1}$, respectively.

(a) Comment on the validity of the law of Dulong and Petit when applied to silicon, phosphorus, and lead.

(b) To raise the temperature of 75.0 g of a particular metal by 15 °C requires 107 cal of heat. What is the approximate atomic weight of the metal? What might the metal be?

(c) A sample of tin weighing 315 g and at a temperature of 65.0 °C is added to 100.0 cm³ of water at 25.0 °C. Estimate the final water temperature.

66. A British thermal unit (Btu) is defined as the quantity of heat required to change the temperature of 1 lb of water by 1 °F. Assuming the specific heat of water to be independent of temperature, how much heat is required to raise the temperature of the water in a 40-gal water heater from 71 to 151 °F? (a) in Btu; (b) in kcal; (c) in kJ. (1 L = 1.06 qt; 4 qt = 1 gal.)

67. A mixture of 235 g of copper and 265 g of water is heated in an open beaker from 24.7 to 87.6 °C. What is ΔH (in kJ) for the copper–water mixture? (Specific heats: water, 4.18 J g^{-1} $°C^{-1}$; copper, 0.393 J g^{-1} $°C^{-1}$?)

68. A 74.8-g sample of copper at 143.2 °C is added to an insulated vessel containing 165 cm³ of glycerol, $C_3H_8O_3(l)$ ($d = 1.26$ g/cm³), at 24.8 °C. The final temperature is 31.1 °C. The specific heat of copper is 0.393 J g^{-1} $°C^{-1}$. What is the *molar* heat capacity of glycerol?

*69. An old "trick" for cooling down a hot beverage quickly is to immerse a cold spoon into it. A silver spoon weighing 4.12 oz and at a temperature of 72 °F is placed in a cup of hot coffee (8 oz) at 162 °F. What will be the resulting liquid temperature? Assume that the specific heat of coffee is 1 cal g^{-1} $°C^{-1}$ and that of silver, 0.056 cal g^{-1} $°C^{-1}$. Comment on the effectiveness of this method of cooling a hot drink. Would a stainless steel spoon (sp. ht. = 0.15 cal g^{-1} $°C^{-1}$) be more or less effective than the silver spoon?

70. The combustion of normal octane in an excess of oxygen is represented by the equation

$C_8H_{18}(l) + \frac{25}{2} O_2(g) \rightarrow 8 CO_2(g) + 9 H_2O(l)$
$$\Delta H = -5.48 \times 10^3 \text{ kJ}$$

How much heat is liberated per gal $C_8H_{18}(l)$ burned at constant pressure? (density of octane = 0.703 g/cm³)

71. What increase in temperature would you expect to occur if a 0.242-g sample of naphthalene is burned in an excess of

$O_2(g)$ in a bomb calorimeter assembly having a heat capacity of 5.12 kJ/°C. The heat of combustion (ΔE) of naphthalene is -5.15×10^3 kJ/mol $C_{10}H_8$.

72. The method of Exercise 45 is used in some bomb calorimetry experiments. A 1.148-g sample of benzoic acid is burned in an excess of $O_2(g)$ in a bomb immersed in 1181 g of water. The temperature of the water rises from 24.96 to 30.25 °C. The heat of combustion (ΔE) of benzoic acid is -26.42 kJ/g. In a second experiment, a 0.895-g powdered coal sample is burned in the same calorimeter assembly. The temperature of 1162 g of water rises from 24.98 to 29.81 °C.

(a) What is the heat capacity of the bomb calorimeter assembly (exclusive of the water)?

(b) What is the heat of combustion of the coal, expressed in kJ/g?

(c) How many metric tons (1 metric ton = 1000 kg) of this coal could have to be burned to release 2.15×10^9 kJ of heat?

73. The heat of solution of KI(s) in water is +21.3 kJ/mol KI. If a quantity of KI is added to sufficient water at 23.5 °C in a Styrofoam cup to produce 150.0 cm³ of 2.50 M KI, what will be the final temperature? (Assume a density of 1.30 g/cm³ and a specific heat of 2.7 J g^{-1} $°C^{-1}$ for 2.50 M KI.)

74. Care must be taken in preparing solutions of solutes that liberate heat on dissolving. The heat of solution of NaOH is -42 kJ/mol NaOH. To what approximate temperature will a sample of water, originally at 21 °C, be raised in the preparation of 500 cm³ of 7.0 M NaOH? Assume that no effort is made to remove heat from the solution.

75. The heat of neutralization of HCl(aq) by NaOH(aq) is -55.90 kJ/mol H_2O produced. If 50.00 mL of 1.05 M NaOH is added to 25.00 mL of 1.86 M HCl, with both solutions originally at 24.72 °C, what will be the final solution temperature? Assume that no heat is lost to the surrounding air and that the aqueous solution produced in the neutralization reaction has a density of 1.02 g/cm³ and a specific heat of 3.98 J g^{-1} $°C^{-1}$.

*76. A particular natural gas consists, on a *molar* basis, of 83.0% CH_4, 11.2% C_2H_6, and 5.8% C_3H_8. The heats of combustion (ΔH) of these gases are $-890.$ kJ/mol CH_4, -1559 kJ/mol C_2H_6, and -2219 kJ/mol C_3H_8. A 432-L sample of this gas, measured at 23.8 °C and 756 mmHg, is burned at constant pressure in an excess of oxygen gas. How much heat is evolved in the combustion reaction?

77. Substitute natural gas (SNG) is a gaseous mixture containing $CH_4(g)$ that can be used as a fuel. One reaction for the production of SNG is

$4 CO(g) + 8 H_2(g) \rightarrow 3 CH_4(g) + CO_2(g) + 2 H_2O(l)$
$$\Delta H = ?$$

Use the following, as necessary, to determine ΔH for this SNG reaction.

$C(graphite) + \frac{1}{2} O_2(g) \rightarrow CO(g)$	$\Delta H = -110.52$ kJ
$CO(g) + \frac{1}{2} O_2(g) \rightarrow CO_2(g)$	$\Delta H = -282.97$ kJ
$H_2(g) + \frac{1}{2} O_2(g) \rightarrow H_2O(l)$	$\Delta H = -285.83$ kJ
$C(graphite) + 2 H_2(g) \rightarrow CH_4(g)$	$\Delta H = -74.81$ kJ

78. Methanol, a potential fuel source, can be prepared by heating CO and H_2 under pressure in the presence of a catalyst.

$CO(g) + 2 H_2(g) \rightarrow CH_3OH(l) \qquad \Delta H = ?$

Determine the heat of this reaction, using $\Delta H = -726.6$ kJ/mol $CH_3OH(l)$ for the heat of combustion of methanol, together with the following data.

$$C(\text{graphite}) + \tfrac{1}{2} O_2(g) \rightarrow CO(g) \qquad \Delta H = -110.5 \text{ kJ}$$

$$C(\text{graphite}) + O_2(g) \rightarrow CO_2(g) \qquad \Delta H = -393.5 \text{ kJ}$$

$$H_2(g) + \tfrac{1}{2} O_2(g) \rightarrow H_2O(l) \qquad \Delta H = -285.8 \text{ kJ}$$

79. For the reaction $C_2H_4(g) + Cl_2(g) \rightarrow C_2H_4Cl_2(l)$ determine ΔH, given that

$$2 Cl_2(g) + 2 H_2O(l) \rightarrow 4 HCl(g) + O_2(g)$$
$$\Delta H = +202.5 \text{ kJ}$$

$$2 HCl(g) + C_2H_4(g) + \tfrac{1}{2} O_2(g) \rightarrow C_2H_4Cl_2(l) + H_2O(l)$$
$$\Delta H = -319.6 \text{ kJ}$$

$$\tfrac{1}{2} H_2(g) + \tfrac{1}{2} Cl_2(g) \rightarrow HCl(g) \qquad \Delta H = -92.3 \text{ kJ}$$

$$H_2(g) + \tfrac{1}{2} O_2(g) \rightarrow H_2O(l) \qquad \Delta H = -285.8 \text{ kJ}$$

80. A net reaction for a coal gasification process is

$$2 C(s) + 2 H_2O(g) \rightarrow CH_4(g) + CO_2(g)$$

Show that this net equation can be obtained by an appropriate combination of equations (7.25), (7.26), and (7.31).

81. What is $\Delta H°$ for the net reaction if all reactants and products of the coal gasification process in Exercise 80 are measured at 25 °C and 1 atm?

82. A handbook lists two different values for the heat of combustion of hydrogen 34.18 kcal/g H_2 if $H_2O(l)$ is formed, and 29.15 kcal/g H_2 if $H_2O(g)$ (steam) is formed. Explain why these two values are different, and verify your conclusions with data from Appendix D.

83. Use data from Appendix D to calculate the enthalpy change ($\Delta H°$) for the following reaction at 25 °C.

$$Fe_2O_3(s) + 3 CO(g) \rightarrow 2 Fe(s) + 3 CO_2(g) \qquad \Delta H° = ?$$

84. Use data from Appendix D, Hess's law, or other appropriate means to complete the diagram below. That is, determine values of $\Delta H°$ represented by the red arrows and write chemical equations to which these $\Delta H°$ values apply. [*Hint:* You may find it helpful to refer to Figure 7-12.]

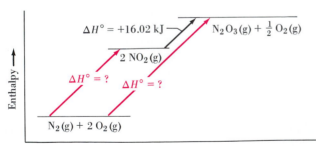

Exercise 84

***85.** Which of the following gases has the greater fuel value, on a per liter (STP) basis? That is, which has the greater heat of combustion? [*Hint:* The only combustible gases are CH_4, C_3H_8, CO and H_2. Also, use data from Appendix D.]

    (a) Coal gas: 49.7% H_2, 29.9% CH_4, 8.2% N_2, 6.9% CO, 3.1% C_3H_8, 1.7% CO_2, and 0.5% O_2, by volume.

    (b) Sewage gas: 66.0% CH_4, 30.3% CO_2, and 4.0% N_2, by volume.

86. A calorimeter that measures an exothermic heat of reaction by the quantity of ice that can be melted is called an *ice calorimeter*. Now consider that 0.100 L of methane gas, $CH_4(g)$, at 25.0 °C and 744 mmHg is burned completely at constant pressure in an excess of air. The heat liberated is captured and used to melt 10.7 g of ice at 0 °C. (The heat required to melt ice, called the heat of fusion, is 333.5 J/g.)

    (a) Write an equation for the combustion reaction.

    (b) What is ΔH for this reaction?

87. In 1850, Joule published the result that: "The quantity of heat capable of increasing the temperature of one pound of water by one degree Fahrenheit requires for its evolution the expenditure of a mechanical force represented by the fall of 772 pounds through the space of one foot." Show that this statement is equivalent to the fact that 4.18 J = 1.00 cal.

88. Some of the butane, $C_4H_{10}(g)$, in a 200.0-L cylinder at 26.0 °C is withdrawn and burned at constant pressure in an excess of air. As a result the pressure of the gas in the cylinder falls from 2.35 atm to 1.10 atm. The liberated heat is used to raise the temperature of 35.0 gal of water from 26.0 to 62.2 °C. Assuming that the combustion products are $CO_2(g)$ and $H_2O(l)$ exclusively, what is the efficiency of the water heater? (That is, what percent of the heat of combustion was absorbed by the water?)

89. One of the advantages of modern coal gasification processes over earlier ones is said to be that some of the heat required to bring about gasification (equation 7.25) is supplied by the methanation reaction (equation 7.31). Use data from Appendix D to show that this is indeed possible.

90. The metabolism of glucose, $C_6H_{12}O_6$, yields $CO_2(g)$ and $H_2O(l)$ as products. Heat released in the process is converted to useful work with about 70% efficiency. Calculate the mass of glucose metabolized by a 58.0-kg person in climbing a mountain with an elevation gain of 1450 m. Assume that the work performed in the climb is about four times that required simply to lift 58.0 kg by 1450 m. $\Delta H_f°$ of $C_6H_{12}O_6(s)$ is -1274 kJ/mol.

91. An alkane hydrocarbon has the formula C_nH_{2n+2}. Whatever the value of the subscript for carbon, n, the subscript for hydrogen is "twice n plus 2." The enthalpies of formation of the alkanes decrease (become more negative) as the numbers of C atoms increase. Starting with propane, C_3H_8, for each additional CH_2 group in the formula the enthalpy of formation, $\Delta H_f°$, changes by about -21 kJ/mol. Use this fact, and data from Appendix D, to estimate the heat of combustion of heptane, $C_7H_{16}(l)$.

Self-Test Questions

For questions 92 through 101 select the item that best completes each statement.

    92. 1.00 *kcal* of heat is (a) absorbed when 1.00 cm³ of water is heated from 14.5 to 15.5 °C; (b) absorbed when 1.00 L of water is heated from 20.0 to 30.0 °C; (c) given off when 100.0 cm³ of water is cooled from 20.0 to 10.0 °C; (d) equal to 1.0×10^6 cal.

    93. Each of four Styrofoam cups contains 75.0 g of water at

25 °C . To each cup is added 50.0 g of a different metal at 100 °C. The final temperature is *highest* in the cup to which the added metal is (a) Al (sp. ht. = 0.908 J g^{-1} °C^{-1}); (b) Ag (sp. ht. = 0.23 J g^{-1} °C^{-1}); (c) Fe (sp. ht. = 0.45 J g^{-1} °C^{-1}); (d) Cu (sp. ht. = 0.39 J g^{-1} °C^{-1})?

94. A 100-mL sample of water at 50.0 °C is added to 50.0 mL of water at 30.0 °C. The final temperature will be (a) 35.0 °C; (b) 37.0 °C; (c) 40.0 °C; (d) 43.4 °C.

95. The heat of solution of NaOH(s) is −41.6 kJ/mol NaOH. When NaOH is dissolved in water, the solution temperature (a) increases; (b) decreases; (c) remains constant; (d) either increases or decreases, depending on how much NaOH is dissolved.

96. A handbook lists the heat of combustion of CS_2(l) as −3.24 kcal/g CS_2. For the reaction,

$$CS_2(l) + 3\ O_2(g) \rightarrow CO_2(g) + 2\ SO_2(g)$$

ΔH is (a) −3.24 kcal; (b) −1.03 × 10^3 kJ; (c) −58.95 kJ; (d) −13.6 kJ.

97. The molar enthalpy of formation of CO_2(g) is equal to (a) 0; (b) the molar heat of combustion of C(graphite); (c) the sum of the molar enthalpies of formation of CO(g) and O_2(g); (d) the molar heat of combustion of CO(g).

98. The heat of combustion of propane is $\Delta H =$ −2220 kJ/mol C_3H_8(g). For the combustion of 1.00 L C_3H_8(g) measured at 25 °C and 1 atm pressure, $\Delta H =$ (a) −2220 kJ; (b) +(1.00/22.4) × 2220 kJ; (c) +(1.00/22.4) × (298/273) × 2220 kJ; (d) −(1.00/(0.0821 × 298)) × 2220 kJ.

99. Given the following enthalpies of reaction:

$$N_2(g) + 3\ H_2(g) \rightarrow 2\ NH_3(g) \qquad \Delta H_1$$

$$2\ NH_3(g) + \tfrac{5}{2}\ O_2(g) \rightarrow 2\ NO(g) + 3\ H_2O(l) \qquad \Delta H_2$$

$$2\ H_2(g) + O_2(g) \rightarrow 2\ H_2O(l) \qquad \Delta H_3$$

For the reaction $N_2(g) + O_2(g) \rightarrow 2\ NO(g) \qquad \Delta H_{net} = ?$
(a) $\Delta H_{net} = \Delta H_1 + \Delta H_2 + \Delta H_3$;
(b) $\Delta H_{net} = \Delta H_1 + \Delta H_2 - \Delta H_3$;
(c) $\Delta H_{net} = \Delta H_1 + \Delta H_2 - \tfrac{3}{2} \Delta H_3$;
(d) $\Delta H_{net} = -\tfrac{3}{2} \Delta H_1 + \Delta H_2 - \Delta H_3$;

100. The enthalpy of formation of CO_2(g) is −394 kJ/mol and that of H_2O(l) is −286 kJ/mol. The heat of combustion of C_5H_{12}(l) is −3534 kJ/mol, that is

$$C_5H_{12}(l) + 8\ O_2(g) \rightarrow 5\ CO_2(g) + 6\ H_2O(l) \quad \Delta H^o = -3534\ \text{kJ}$$

The enthalpy of formation of C_5H_{12}(l), in kJ/mol, is (a) +3534; (b) [−3534 − 5 × (−394) − 6 × (−286)]; (c) [5 × (−394) + 6 × (−286) − 3534]; (d) [5 × (−394) + 6 × (−286) + 3534].

101. $\Delta E = +100$ J for a system that *gives off* 100 J of heat and (a) does 200 J of work; (b) has 200 J of work done on it; (c) does no work; (d) has 100 J of work done on it.

102. Explain briefly the difference in meaning between
(a) specific heat and molar heat capacity of a substance;
(b) endothermic and exothermic reaction;
(c) enthalpy of formation and heat of combustion of the hydrocarbon C_4H_{10}(g).

103. A 1.35-kg piece of iron (sp. ht. = 0.45 J g^{-1} °C^{-1}) is dropped into 0.817 kg of water and the water temperature is observed to rise from 23.3 to 39.6 °C. What must have been the initial temperature of the iron?

104. The heat of combustion (ΔE) of phenol, C_6H_5OH(s), is determined in a bomb calorimeter and found to be −32.55 kJ/g. Write the chemical equation for the combustion of one mole of phenol, including a value of ΔH (which you can assume to be essentially equal to ΔE).

105. The enthalpy of formation of NH_3(g) is −46 kJ/mol NH_3. What is $\Delta H°$ for the following reaction?

$$2\ NH_3(g) \rightarrow N_2(g) + 3\ H_2(g) \qquad \Delta H° = ?$$

106. The complete combustion of propane, C_3H_8(g), is represented by the equation

$$C_3H_8(g) + 5\ O_2(g) \rightarrow 3\ CO_2(g) + 4\ H_2O(l) \quad \Delta H = -2220.\ \text{kJ}$$

(a) How much heat energy is evolved in the complete combustion of 1.100 g C_3H_8(g)?
(b) The quantity of heat calculated in part (a) is completely transferred to an iron–water mixture consisting of a 1.00-kg piece of iron submerged in 550. mL of water. The initial temperature of the iron–water mixture is 20.0 °C. What will be its final temperature? (Sp. ht. iron = 0.473 J g^{-1} °C^{-1}.)

107. The heats of combustion ($\Delta H°$) per mole of C(graphite) and CO(g) are −393.5 and −283.0 kJ/mol, respectively. In both cases CO_2(g) is the sole product. For the formation of the poisonous gas, phosgene,

$$CO(g) + Cl_2(g) \rightarrow COCl_2(g) \qquad \Delta H° = -108\ \text{kJ}$$

Use Hess's law to calculate the enthalpy of *formation* of $COCl_2$(g).

8 Electrons in Atoms

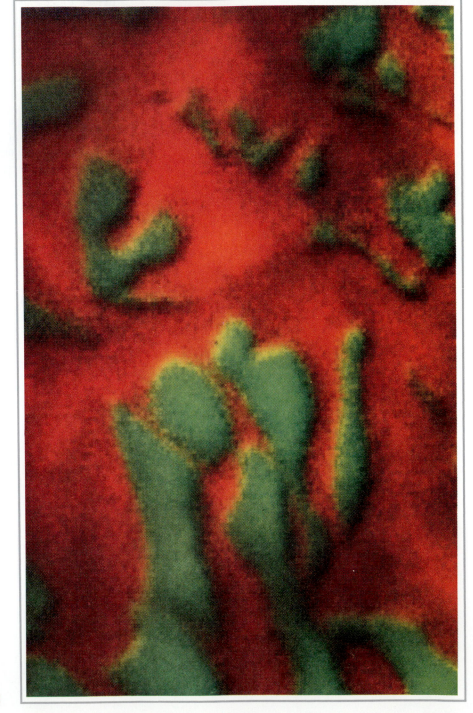

Probing a surface with electrons. This image was obtained from electrons emitted by surface atoms after they had been struck by a beam of electrons. In this image red = Al and green = Si. [Courtesy Perkin Elmer—Physical Electronics Division]

238

The elements argon ($Z = 18$) and potassium ($Z = 19$) have molar masses that are nearly the same—39.95 g Ar/mol Ar and 39.10 g K/mol K. We expect this to be the case because the mass of an atom is concentrated in its nucleus, and the nuclei of these two elements are so similar—18 protons and 22 neutrons for the principal isotope of Ar and 19 protons and 20 neutrons for the principal isotope of K. But molar mass (atomic weight) is about all that is similar between these two elements.

Potassium is a solid metallic element that reacts very vigorously with water and with nonmetallic elements such as O_2 and Cl_2. Argon, on the other hand, is an inert gaseous element that does not react with any of the other elements. What is there about these two elements that makes them so dissimilar in their physical and chemical properties? In this chapter we will find that the essential difference is in the number and arrangement of the electrons in the atoms of these two elements. Potassium and argon have distinctly different *electron configurations*.

The electronic structures of atoms that we consider in this chapter provide the basis of the periodic table of the elements (Chapter 9), underlie the different types of chemical bonding and molecular structures (Chapters 10 and 11), and, ultimately, account for the intermolecular forces leading to solids, liquids, and gases (Chapter 12).

8-1 Electricity and Magnetism

We have already discussed the following facts about electricity: If an object has *equal* numbers of electrons and protons, it is electrically neutral. If it has *unequal* numbers of electrons and protons, it carries a net electric charge. The charge is negative if there is an excess of electrons, positive if there is a deficiency of electrons. Oppositely charged objects exert an attractive force on one another; like-charged objects repel one another. The region surrounding electrically charged objects in which attractive and repulsive forces are experienced is called an **electric field.**

The two poles of a magnet are analogous to the two types of electric charge.

Attractive and repulsive forces are also the principal phenomena associated with magnetism. The places where these forces originate are called the poles of a magnet, one of which is called the **north pole** and the other, the **south pole.** Opposite poles of magnets exert attractive forces on one another; like poles repel. The region surrounding a magnetic pole in which attractive and repulsive forces are felt is called a **magnetic field.** One common method of picturing a magnetic field is in terms of imaginary "lines of force" that originate at the north pole and terminate at the south pole of a magnet, as shown in Figure 8-1.

FIGURE 8-1
A magnet and its magnetic field.

Shown here are the north and south poles and "lines of force" outlining the magnetic field of a bar magnet.

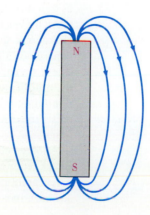

239

FIGURE 8-2

An electromagnetic phenomenon—the electromagnet.

Electric current from the battery passes through the coil of wire wrapped around the iron bar. The magnetic field induced by the electric current causes the bar to act as a magnet, attracting small iron objects. If the direction of the current is reversed, the poles of the magnet are also reversed. When the electric current is cut off, the magnetic field dissipates and the bar loses its magnetism. [Carey B. Van Loon]

Recall from our discussion in Chapter 7 that potential energy is due to the position of an object and kinetic energy is due to its motion.

Electricity and magnetism are closely related phenomena, referred to collectively as *electromagnetism*. One example of this close connection is that the passage of an electric current creates a magnetic field around the conductor. This is the principle of an *electromagnet,* illustrated in Figure 8-2. Another example is that the movement of an electrical conductor in a magnetic field creates a flow of electricity in the conductor. This is the principle on which an *electric power generator* is based. These phenomena demonstrate that potential energy is associated with electric and magnetic fields, just as it is associated, for example, with the earth's gravitational field.

8-2 Electromagnetic Radiation

When electrically charged particles move with respect to one another, alternating electric and magnetic fields are produced and transmitted through the space or medium surrounding the particles. The form in which these fields are transmitted is called a **wave.** Energy is associated with the electric and magnetic fields, and the wave is a means of transmitting energy through distances. This form of energy transmission is called **electromagnetic radiation.**

Electromagnetic wave motion is complex, but certain characteristics of these waves can be thought of in terms of vibrations in a string, pictured in Figure 8-3. Imagine that you tie one end of a string many meters long to a door knob and hold the other end of the string in your hand. Imagine further that you have colored one small segment (about 1 cm) of the string with red ink. Now consider what happens as you move your hand up and down. You set up a wave motion in the string. The wave moves from left to right along the string, but the red segment simply moves up and down, that is, perpendicular to the wave motion (see Figure 8-3). The wave consists of a number of regions in which the string is at a maximum or high point. These are called wave crests. The maximum height to which the wave rises is called the **amplitude** of the wave. Corresponding to each wave crest there is a region at a minimum or low point, called a wave trough. The distance between two successive crests (or troughs) is called the **wavelength,** usually designated by the Greek letter lambda, λ.

Another characteristic property of a wave is its **frequency,** designated by the Greek letter nu, ν. This is the number of wave crests or troughs that pass through a given point per second. That is, frequency can be expressed with the unit s^{-1} (per second), meaning the number of events or cycles per second. For the traveling wave in Figure 8-3, this is simply the number of times per second that the hand driving the wave goes through its up-and-down motion.

The product of the length of a wave (λ) and the number of cycles per second (that is, the frequency ν) shows how far the wave front has traveled down the string in one second. This is the **velocity** of the wave, c.

$$c = \nu\lambda \tag{8.1}$$

Now to return to electromagnetic waves. These waves are produced by the oscillations of charged particles, and the waves travel in all directions simultaneously—they are three-dimensional waves. Electromagnetic radiation actually involves an oscillating electric field and, perpendicular to this, an oscillating magnetic field. An interesting difference between electromagnetic radiation and other kinds of waves (such as sound waves or waves in the ocean) is that electromagnetic radiation requires no medium for its transmission. It can travel through a vacuum or empty space. Despite these complications, an electromagnetic wave is often represented in the simplified manner shown in Figure 8-4. This is adequate for our purposes.

Frequency, Wavelength, and Velocity. A variety of units are used to describe electromagnetic radiation. The SI unit for frequency, s^{-1}, is the **hertz (Hz).** Wavelength must have a unit of length, and logically this should be the meter (m). However, because so many kinds of electromagnetic radiation are of very short wavelength, some smaller units are necessary. The units listed below are among those commonly encountered. [The angstrom is named for the Swedish physicist, Anders Ångström (1814–1874), who used 1×10^{-10} m as a basic unit of wavelength in his studies of the solar spectrum; the angstrom is not an SI unit.]

1 centimeter (cm) = 1×10^{-2} m

1 nanometer (nm) = 1×10^{-9} m = 1×10^{-7} cm = 10 Å

1 angstrom (Å) = 1×10^{-10} m = 1×10^{-8} cm

A distinctive feature of electromagnetic radiation is that its velocity has a constant value of 2.997925×10^8 m s^{-1} in a vacuum (usually rounded off to 3.00×10^8 m s^{-1}). Since ordinary light is a form of electromagnetic radiation, this characteristic velocity is often called the **speed of light.** The frequencies and wavelengths of a number of different kinds of electromagnetic radiation are compared in Figure 8-5.

Relationship among frequency, wavelength, and velocity of electromagnetic radiation.

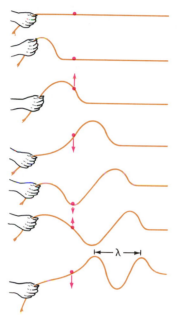

FIGURE 8-3

The simplest wave motion—traveling wave in a string.

The hypothetical string pictured here is infinitely long. As a result of the up-and-down hand motion (top to bottom), waves pass along the string from left to right. This one-directional moving wave is called a traveling wave.

FIGURE 8-4

An electromagnetic wave.

Although oversimplified, this representation does suggest that electromagnetic radiation results from the motion of electrically charged objects and shows one characteristic of the radiation—the wavelength λ. The wave in (a) has a longer wavelength than the one in (b).

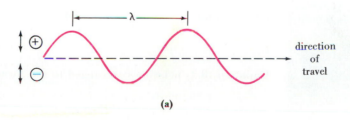

(a)

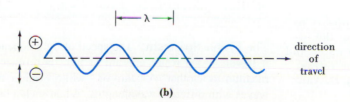

(b)

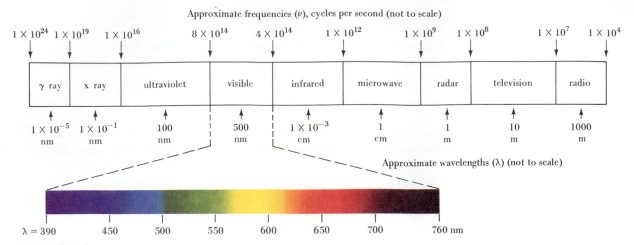

Approximate frequencies (ν), cycles per second (not to scale)

Approximate wavelengths (λ) (not to scale)

FIGURE 8-5
The electromagnetic spectrum.

Example 8-1

Calculating the wavelength of electromagnetic radiation from its frequency and velocity. An FM radio station broadcasts on a frequency of 91.5 megahertz (MHz). What is the wavelength of these radio waves, in meters?

Solution. The prefix *mega* denotes one million or 1×10^6. A frequency of 91.5 MHz $= 91.5 \times 10^6$ s^{-1}. Since radio waves are a form of electromagnetic radiation, $c = 2.998 \times 10^8$ m/s. Solve equation (8.1) for

$$\lambda = \frac{c}{\nu} = \frac{2.998 \times 10^8 \text{ m s}^{-1}}{91.5 \times 10^6 \text{ s}^{-1}} = 3.28 \text{ m}$$

SIMILAR EXAMPLES: Exercises 2, 4

Example 8-2

Calculating the frequency of electromagnetic radiation from its wavelength and velocity. Most of the light emitted by a sodium vapor lamp has a wavelength of 589 nm. What is the frequency of this radiation?

Solution. Note particularly that wavelength must be converted from nm to m before equation (8.1) can be used.

$$c = 2.998 \times 10^8 \text{ m/s}$$
$$\lambda = 589 \text{ nm} \times \frac{1 \times 10^{-9} \text{ m}}{1 \text{ nm}} = 5.89 \times 10^{-7} \text{ m}$$
$$\nu = ?$$

Equation (8.1) must be rearranged to the form $\nu = c/\lambda$.

$$\nu = \frac{c}{\lambda} = \frac{2.998 \times 10^8 \text{ m s}^{-1}}{5.89 \times 10^{-7} \text{ m}} = 5.09 \times 10^{14} \text{ s}^{-1}$$

SIMILAR EXAMPLES: Exercises 3, 4.

The thermometer appears to be bent or broken at the interface between the water and air as a result of the refraction of light. [Carey B. Van Loon]

The Visible Spectrum. The speed of light depends on the medium through which it travels. As a result a beam of light is refracted or bent as it passes from one medium to another. A beam of "white" light consists of a large number of light waves with different wavelengths. When such a beam is passed through a transpar-

FIGURE 8-6

The spectrum of "white" light.

Red light is refracted the least and violet light the most when "white" light is passed through a glass prism. The other colors of the visible spectrum are found between the red and violet.

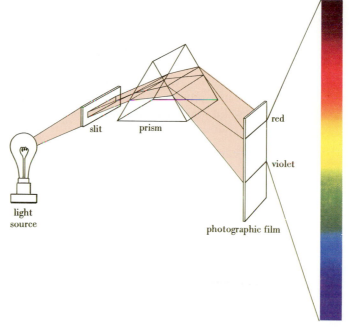

light source

slit   prism

red

violet

photographic film

ent medium, the different wavelength components of white light are refracted differently, causing the beam of light to be dispersed into a band or spectrum of colors. The shortest wavelength light that the human eye can detect corresponds to the color violet; the longest, red.

These points are illustrated in Figure 8-6. Here a beam of "white" light is passed through a narrow opening or slit and dispersed into its colored components by a glass prism. The colored components are then recorded on a photographic film. The device incorporating these features is called a **spectrograph.** Example 8-3 illustrates some important points about electromagnetic radiation, especially as they pertain to visible light.

Example 8-3

Understanding the relationship between the frequency and wavelength of electromagnetic radiation. Is the light emitted by a mercury vapor lamp principally of higher, lower, or equal frequency to the light emitted by a sodium vapor lamp (see Figure 8-7)?

FIGURE 8-7

Sodium and mercury vapor lamps.

A sodium vapor lamp (foreground) produces light of a distinctly yellow cast, whereas the light from a mercury lamp (background) is predominantly blue. [Carey B. Van Loon]

Solution. To answer this question you need to

1. Understand, from equation (8.1), that wavelength and frequency are inversely proportional. (The *longer* the wavelength, the *lower* the frequency; the *shorter* the wavelength, the *higher* the frequency.)
2. Have a sense, from Figure 8-5, of the ranking of the different colors in terms of their wavelengths. In the order of *decreasing* wavelength: red (R) > orange (O) > yellow (Y) > green (G) > blue (B) > indigo (I) > violet (V). A useful mnemonic (memory) device is to note that these letters spell out the hypothetical name *Roy G. Biv*.

Because of its *shorter* wavelength, the predominantly bluish light of the mercury vapor lamp is of higher frequency than the yellow light of the sodium vapor lamp.

SIMILAR EXAMPLE: Exercise 15.

8-3 Atomic Spectra

In Figure 8-6 the light source is "white" light. This could be sunlight or certain artificial light sources, such as the heated filament of an ordinary electric light bulb. Each wavelength component of the white light, after passing through the slit and prism, produces an image of the slit in the form of a line. There are so many of these lines that they all blend together into an unbroken band of color from red through orange, yellow, green, and blue to violet. The spectrum of white light is said to be *continuous*.

On the other hand, the light emitted by most heated gases produces only a limited set of images of the slit. These spectra, consisting of only a limited number of colored lines with dark spaces between them, are said to be *discontinuous*. Such spectra are called either *atomic* or *line* spectra. The production of a line spectrum is depicted in Figure 8-8.

The line spectrum obtained for each element differs from that of every other element. The line spectrum is like a fingerprint of the element. One of the first investigators to make extensive use of atomic spectra to identify chemical elements

FIGURE 8-8

Production of an atomic or line spectrum.

The light source depicted here is a helium lamp. The visible spectrum of helium consists of six lines bright enough to be seen with the unaided eye.

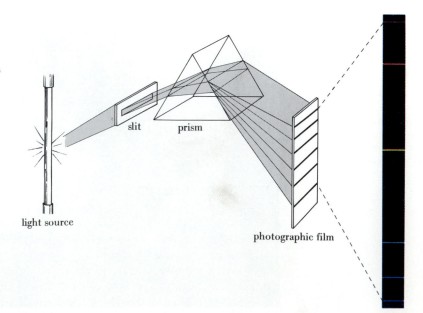

light source

slit prism

photographic film

410.1 nm   434.0 nm   486.1 nm   656.3 nm

FIGURE 8-9
The Balmer series for hydrogen—a line spectrum.

The four lines shown here are the only ones visible to the unaided eye. Additional closely spaced lines lie in the near ultraviolet but are invisible.

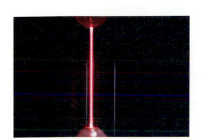

Light emission from a hydrogen lamp. [Carey B. Van Loon]

was the German chemist Robert Bunsen (1811–1899). During the latter half of the nineteenth century the spectra of most of the elements had been investigated, including even that of the element helium, which was found to exist on the sun before it was discovered on earth.

Among the most extensively studied atomic spectra in the nineteenth century was that of hydrogen. Light from a hydrogen lamp appears to the eye as a bright red color. The principal component of this light is indeed red light of wavelength 656.3 nm. However, three other lines appear in the visible spectrum of atomic hydrogen: a greenish blue line at 486.1 nm, a violet line at 434.0 nm, and another violet line at 410.1 nm. This spectrum for atomic hydrogen is displayed in Figure 8-9. In 1885, the Swiss physicist-schoolteacher Johann Balmer deduced a formula for the wavelengths of these spectral lines. He apparently did this by trial and error. Balmer's results were restated by Johannes Rydberg, who was simultaneously engaged in trying to develop general relationships for predicting spectral lines. Written in terms of the frequencies of the spectral lines, the Balmer equation is

$$\nu = Rc\left(\frac{1}{2^2} - \frac{1}{n^2}\right) = 3.2881 \times 10^{15} \text{ s}^{-1}\left(\frac{1}{2^2} - \frac{1}{n^2}\right) \tag{8.2}$$

R is a numerical constant, called the Rydberg constant, having a value of 10,967,800 m^{-1}; c is the velocity of light, 2.997925 × 10^8 m s^{-1}. To simplify calculations, the product $R \times c$ is also given in equation (8.2). To five significant figures, it is 3.2881 × 10^{15} s^{-1}. If $n = 3$ is substituted into Balmer's formula, the frequency of the red line is obtained; if $n = 4$ is substituted, the frequency of the greenish blue line is obtained; and so on.

By the early nineteenth century a wave theory of light (that is, that light is energy transmitted through wave motion) had been firmly established and was successful in explaining continuous spectra like the rainbow. But the existence of line spectra could not be explained by wave theory, not even when approached through the refinements introduced by James Maxwell in the 1860s (that light is energy transmitted through electric and magnetic fields). That such a simple equation as Balmer's could be used to correlate atomic spectral data suggested that there was some basic principle underlying all atomic spectra. Yet this principle was never discovered by the methods of nineteenth-century physics known as *classical physics*. The key to the small number of unsolved problems of classical physics lay in a great breakthrough of modern science—the quantum theory.

8-4   Quantum Theory

Max Planck (1858–1947) proposed the quantum theory in 1900 to explain a phenomenon known as blackbody radiation, which is the radiation given off by heated objects (such as a "red-hot" fireplace poker or "white hot" molten iron). Planck's revolutionary hypothesis was that, like matter, which is discontinuous and exists as discrete atoms, energy also is *discontinuous* and consists of large numbers of tiny *discrete* units called **quanta.** Whether a system loses or gains energy, it must do so in terms of these quanta. The energy associated with a quantum of electromagnetic radiation is proportional to the frequency, that is, the higher the frequency of the radiation, the greater the energy of the quantum. This relationship is expressed as

Planck's equation relating the energy of radiation to its frequency.

$$E = h\nu \qquad\qquad (8.3)$$

The proportionality constant, **h**, is called **Planck's constant** and has a value of 6.626×10^{-34} J s.

According to the laws of classical physics, the energy of a system should be able to assume any value and to change by any amount. By the principles of quantum theory, however, we find that the energy of a system can have only a unique set of values. This means that energy can change only by certain discrete amounts, called quantum jumps. A system may gain or lose one quantum of energy, or two quanta, or three, and so on; but it cannot gain or lose $\frac{1}{3}$, $\frac{1}{2}$, $1\frac{1}{3}$, or $2\frac{1}{2}$ quanta. Figure 8-10 suggests an analogy between classical and quantum theory that may prove helpful in understanding the essential difference between the two.

There have been instances in the history of science when a new hypothesis proved useful in explaining one phenomenon but was not generally applicable to any others. It was only in the discovery of other applications of the quantum hypothesis that it acquired status as a significant new theory of science. The first notable new success came in 1905 with Albert Einstein's (1879–1955) quantum explanation of the photoelectric effect.

The Photoelectric Effect. Figure 8-11 pictures an interesting observation made in 1887 by H. Hertz. A beam of electrons ("electric") is produced by shining a beam of light ("photo") on certain metal surfaces. This phenomenon is called the photoelectric effect. According to classical physics, the energy content of a light beam depends on the *brightness* of the light (as does the power of the sun to produce sunburn, for example). In turn, this suggests that in the photoelectric effect *both* the *number* of electrons ejected from the surface and the electron *energies* should depend on the *brightness* of the light beam. The number of ejected electrons does depend on the brightness of the light, but the electron energies do not! The electron energies depend on the frequency (color) of the light. Thus, the kinetic energies with which electrons leave the surface of a photoelectric material are greater if the light shining on the surface is a feeble blue light than a bright red one (see Figure 8-12). As with atomic spectra, here was another phenomenon that defied explanation by classical physics.

Einstein proposed that electromagnetic radiation has particlelike characteristics and that "particles" of light, called **photons,** possess a characteristic energy, given by Planck's equation, $E = h\nu$. We can think of the energy of a light wave as being concentrated into photons. In the photoelectric effect these photons transfer energy during collisions with electrons. In each collision a photon gives up its entire en-

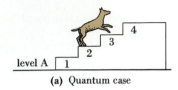

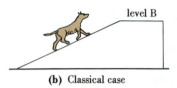

(a) Quantum case

(b) Classical case

FIGURE 8-10

An analogy between quantum mechanics and classical mechanics.

To climb from level A to level B the dog in (a) proceeds up steps 1, 2, 3, and 4. It may change its position (and hence its potential energy) only by these four discrete steps of fixed magnitude. The dog in (b) can change its position and potential energy from level A to level B in any number of steps of any size whatsoever. Case (a) corresponds to quantum changes in energy, case (b) to classical changes in energy.

The photoelectric effect has many practical applications, ranging from automatic door openers to light meters to light-sensitive elements in television cameras.

FIGURE 8-11

The photoelectric effect.

A light beam strikes the metal surface and knocks out electrons. The photoelectric metal, having lost electrons, acquires a positive charge. This positively charged metal draws electrons away from the electroscope, causing the metal-foil leaves of the electroscope to acquire a positive charge. Having like charges, the leaves repel one another. (See Appendix B for more on the electroscope.)

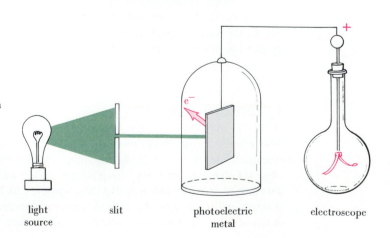

light source   slit   photoelectric metal   electroscope

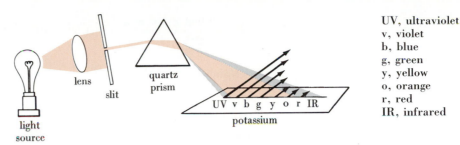

UV, ultraviolet
v, violet
b, blue
g, green
y, yellow
o, orange
r, red
IR, infrared

FIGURE 8-12
Dependence of the photoelectric effect on the frequency of light.

A beam of "white" light is dispersed into its wavelength components by the quartz prism. Light of the highest frequency (violet and ultraviolet) produces the most energetic photoelectrons (longest arrows). Light of lower frequencies (e.g., yellow) produces much less energetic photoelectrons (shorter arrows). Light with a frequency lower than that corresponding to a wavelength of 710 nm (red) produces no photoelectric effect at all. The lengths of the arrows represent the kinetic energies of the photoelectrons.

A light beam having this appearance

actually consists of "particles" called photons.

FIGURE 8-13
Photons of light visualized.

ergy—a quantum of energy—to an electron. The more energetic the photon, the more energy it transfers to an electron and the greater the kinetic energy of the ejected electron. Therefore, the kinetic energy of the electron should depend on the frequency of the light. The particlelike nature of light is suggested by Figure 8-13.

In summary, then, light consists of photons. Each photon possesses a quantity of energy called a quantum. The energy of a photon is proportional to the *frequency* of the light, so that photons of blue light (higher frequency) are more energetic than photons of red light (lower frequency). The *brightness* of light is related to the *number* of photons, regardless of the photon energy. The more photons in a beam of light, the brighter the light.

The product of Planck's constant *(h)* and frequency *(ν)* yields the energy of a single photon of electromagnetic radiation in the unit joules. Typically, this energy is only a tiny fraction of a joule. Often we deal with the much larger energy associated with a mole of photons (that is, with 6.02214×10^{23} photons).

Example 8-4

Using Planck's equation to calculate the energy of a photon. The lowest frequency light than can produce a photoelectric effect on potassium metal is 4.2×10^{14} s^{-1}. What is the energy of *one photon* of this light?

Solution.

This lowest or *threshold frequency* corresponds to $\lambda = 710$ nm. This is why in Figure 8-12 no electrons are emitted by infrared light.

$$E = h\nu = (6.626 \times 10^{-34} \text{ J s})(4.2 \times 10^{14} \text{ s}^{-1}) = 2.8 \times 10^{-19} \text{ J}$$

SIMILAR EXAMPLES: Exercises 6, 22.

Example 8-5

Planck's equation and the energy per mol of photons. What is the energy, in kJ/mol, associated with monochromatic radiation of wavelength 225 nm?

Solution. First we must use equation (8.1) to determine the frequency of this radiation. Then we apply the Planck equation (8.3) to determine the energy per photon, in kJ/photon. Finally, we use the Avogadro constant to convert from kJ/photon to kJ/mol photons.

$$\nu = \frac{c}{\lambda} = \frac{2.998 \times 10^8 \text{ m s}^{-1}}{225 \times 10^{-9} \text{ m}} = 1.33 \times 10^{15} \text{ s}^{-1}$$

$$\text{no. kJ/photon} = 6.626 \times 10^{-34} \frac{\text{J s}}{\text{photon}} \times 1.33 \times 10^{15} \text{ s}^{-1} \times \frac{1 \text{ kJ}}{1000 \text{ J}}$$

$$= 8.81 \times 10^{-22} \text{ kJ/photon}$$

$$\text{no. kJ/mol} = 8.81 \times 10^{-22} \frac{\text{kJ}}{\text{photon}} \times \frac{6.022 \times 10^{23} \text{ photons}}{1 \text{ mol}}$$

$$= 5.31 \times 10^2 \text{ kJ/mol}$$

SIMILAR EXAMPLES: Exercises 6, 22, 23, 68.

8-5 The Bohr Atom

FIGURE 8-14

An unsatisfactory atomic model.

According to this model, as energy is lost through light emission, the electron is drawn ever closer to the nucleus, eventually spiraling into it. This collapse of the atom would occur very quickly.

Niels Bohr (1885–1962). In addition to his work on the hydrogen atom, Bohr headed the Institute of Theoretical Physics in Copenhagen, which became a mecca for theoretical physicists in the 1920s and 1930s.

Even from classical physics it is clear that electrons in atoms must move, otherwise they would be attracted into the positively charged nucleus. Also, this movement of electrons around the nucleus can explain light emission, since light emission results from the relative motion of charged particles (recall Figure 8-4). But here is the dilemma for classical physics: In revolving around the nucleus of an atom an electron should experience a constant acceleration (change in velocity). As a result it should give off energy as light. Having lost energy, however, the electron should be drawn ever closer to the nucleus, soon spiraling into it. This unstable situation is pictured in Figure 8-14. We do not seem able to develop a model for a stable atom with the principles of classical physics.

In 1913, Niels Bohr (1885–1962) resolved this dilemma by introducing Planck's quantum of action (specifically, Planck's constant, h) into the description of atomic structure. Bohr's treatment of the hydrogen atom involved an interesting blend of classical and quantum theory. Bohr postulated that

1. There is only a *fixed set of allowable orbits* for an electron in a hydrogen atom. These orbits, also called stationary states, are circular paths about the nucleus. The motion of an electron in an orbit can be described by ordinary mechanics. However, even though classical theory would predict otherwise, *as long as an electron remains in a given orbit its energy remains constant and no light is emitted*.

2. An electron can pass only from one allowable orbit to another. In such transitions fixed, discrete quantities of energy (quanta) are involved, in accordance with Planck's equation, $E = h\nu$.

3. The allowable orbits are those in which certain properties of the electron have unique values. In particular, a property called the angular momentum must be an integral multiple of $h/2\pi$; that is, the angular momentum must be $nh/2\pi$, where n is an integer and h is Planck's constant. (Angular momentum is illustrated in Figure 8-15.)

The atomic model for hydrogen based on these ideas is pictured in Figure 8-16. The allowable states for the electron are numbered, $n = 1$, $n = 2$, $n = 3$, and so on. These *integral* numbers are called **quantum numbers.**

Several properties of the electron in a hydrogen atom can be calculated with the Bohr theory. For one, the radii of the allowable orbits can be expressed as $n^2 a_0$, where $a_0 = 0.53$ Å and n is an integer: 1, 2, 3, ... It is also possible to calculate the velocities associated with the electron in each of these orbits, and, most important, the energy. When the electron is free of the nucleus, its energy is taken to be *zero*. When the electron is attracted to the nucleus and confined to the orbit n, energy is emitted and the electron energy is lowered to the value

Electron energy in the Bohr hydrogen atom.

$$E_n = \frac{-B}{n^2} \tag{8.4}$$

B is a numerical constant with a value of 2.179×10^{-18} J.

One of the assumptions in Bohr's theory is that discrete quantities of energy are involved when an electron in a hydrogen atom passes from one allowable state to another. Let us use equation (8.4) to calculate the *difference in energy* between two states. As a case of special interest, we choose the states: $n = 3$ and $n = 2$.

$$\Delta E = E_3 - E_2 = \left(\frac{-B}{3^2}\right) - \left(\frac{-B}{2^2}\right) = \left(\frac{B}{2^2}\right) - \left(\frac{B}{3^2}\right) = B\left(\frac{1}{2^2} - \frac{1}{3^2}\right) \tag{8.5}$$

Equation (8.5) bears a striking resemblance to the Balmer equation (8.2)! Let us pursue this matter further. Normally the electron in a hydrogen atom is found in the orbit closest to the nucleus ($n = 1$). When the atom is excited the electron absorbs a quantum of energy and jumps to a higher numbered orbit. When the electron drops from a higher to a lower numbered orbit, a unique quantity of energy is emitted as a photon of light of a characteristic frequency.

Returning to the transition of an electron from $n = 3$ to $n = 2$, the difference in energy between these two orbits, ΔE, is the energy of the emitted photon. We can write two equations for ΔE.

$$\Delta E = h\nu \tag{8.3}$$

$$\Delta E = B\left(\frac{1}{2^2} - \frac{1}{3^2}\right) \tag{8.5}$$

and then set these equal to one another.

$$h\nu = B\left(\frac{1}{2^2} - \frac{1}{3^2}\right)$$

For the frequency of the photon of light, ν, we obtain

$$\nu = \frac{B}{h}\left(\frac{1}{2^2} - \frac{1}{3^2}\right) \tag{8.6}$$

The numerical value of the constant B/h can be determined readily.

$$\frac{B}{h} = \frac{2.179 \times 10^{-18} \text{ J}}{6.626 \times 10^{-34} \text{ J s}} = 3.289 \times 10^{15} \text{ s}^{-1}$$

Substituting this value of B/h back into equation (8.6), we obtain the Balmer equation!

$$\nu = 3.289 \times 10^{15} \text{ s}^{-1}\left(\frac{1}{2^2} - \frac{1}{3^2}\right) \tag{8.7}$$

FIGURE 8-15
Angular momentum.

The angular momentum of the electron orbiting around the nucleus is mvr. In a given orbit, m, the mass of the electron, and r, the radius of the orbit, are fixed. To say that the angular momentum must have a particular value (is quantized) is equivalent to saying that the orbiting electron can have only a certain velocity, v (represented by the arrow).

FIGURE 8-16
Bohr model of the hydrogen atom.

A portion of the hydrogen atom is pictured. The nucleus is at the center of the atom, and the electron is found in one of the discrete orbits $n = 1, 2, \ldots$. Electron transitions for excitation of the atom are shown as black arrows. Two transitions in which light is emitted are also indicated, in the approximate colors of the spectral lines they produce in the Balmer series.

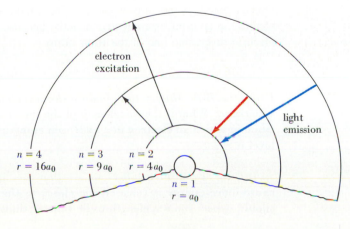

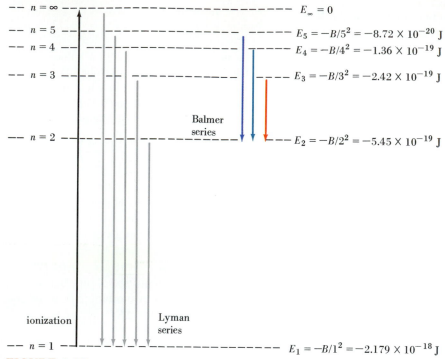

FIGURE 8-17

Energy-level diagram for the hydrogen atom.

Among the features brought out by this diagram are

- The zero of energy corresponds to the completely ionized atom ($n = \infty$).
- Ionization of the normal hydrogen atom requires moving the electron from the level $n = 1$ to $n = \infty$, a process requiring 2.179×10^{-18} J.
- Electron transitions from higher quantum levels to the level $n = 2$ produce lines in the Balmer series (note correspondence of colors to the spectrum in Figure 8-9). Transitions to the level $n = 1$ produce lines in the Lyman series.
- Energy differences between successive levels are smaller, the higher the values of n.

If we solve equation (8.7), we obtain the frequency of the red line in the Balmer series. Every transition from a higher energy state ($n = 3, 4, \ldots$) to the state $n = 2$ produces a line in the Balmer series.

In Figure 8-17 the energies associated with the different orbits or energy levels of the electron in a hydrogen atom are shown as a group of lines. This representation is called an **energy-level diagram.** When the electron is completely separated from the atom (ionized), it is in an orbit of infinitely high quantum number ($n = \infty$); there is no attraction between the electron and the nucleus (proton). This, as we have already noted, is the zero of energy. The energy of a bound electron, that is, an electron in an orbit, is a *negative* quantity because energy is given off when a free electron drops into one of the Bohr orbits.

Example 8-6

Using the Bohr theory to calculate the wavelengths of lines in the hydrogen spectrum. What is the wavelength of the spectral line associated with the transition in which an electron in a hydrogen atom drops from the level $n = 2$ to the level $n = 1$?

Solution. From Figure 8-17 we see that this line is in the Lyman series. Because more energy is released the closer to the nucleus an electron falls, we should expect the wavelength to be shorter than any of those observed for the

Balmer series (in which electrons fall to the level $n = 2$). First, use equation (8.4) to calculate the difference in energy between the levels $n = 2$ and $n = 1$.

$$E_1 = \frac{-B}{1^2} = -B \quad \text{and} \quad E_2 = \frac{-B}{2^2} = \frac{-B}{4}$$

$$\Delta E = E_2 - E_1 = \frac{-B}{4} - (-B) = \frac{3B}{4} = 0.7500\ B$$

$$= 0.7500 \times 2.179 \times 10^{-18}\ \text{J} = 1.634 \times 10^{-18}\ \text{J}$$

Now we can substitute this energy into Planck's equation (8.3) to obtain the frequency of the radiation.

$$\nu = \frac{E}{h} = \frac{1.634 \times 10^{-18}\ \text{J}}{6.626 \times 10^{-34}\ \text{J s}} = 2.466 \times 10^{15}\ \text{s}^{-1}$$

Finally, we use equation (8.1) to solve for the wavelength.

$$\lambda = \frac{c}{\nu} = \frac{2.9979 \times 10^8\ \text{m s}^{-1}}{2.466 \times 10^{15}\ \text{s}^{-1}} \times \frac{1\ \text{nm}}{1 \times 10^{-9}\ \text{m}} = 121.6\ \text{nm}$$

From Figure 8-5 we see that this line is in the ultraviolet region, as expected.

SIMILAR EXAMPLES: Exercises 7, 29, 64, 65.

The great success of the Bohr theory was in its ability to predict lines in the hydrogen atom spectrum. However, the theory was never very successful in describing atomic spectra other than those of hydrogen; nor, more importantly, could it account for the ability of atoms to form molecules through chemical bonds. A new quantum mechanical theory of atomic structure was required.

8-6 Wave–Particle Duality

Although most phenomena involving light could be adequately explained in terms of a wave theory of light, Albert Einstein (1905) had to propose a particlelike nature of light (photons of light) in order to explain the photoelectric effect. In this way the idea developed that light has a *dual* nature. In some instances the behavior of light is best understood in terms of the wave theory (e.g., dispersion of light into its spectrum by passage through a prism) and in others, in terms of particles (e.g., the photoelectric effect).

In 1924 Louis de Broglie, considering the nature of light and matter, offered a startling proposition: *Not only does light display particlelike characteristics, but small particles may at times display wavelike properties.* De Broglie's proposal was verified in 1927, through experiments that led to the discovery of the electron microscope. De Broglie described matter waves in mathematical terms. The de Broglie wavelength associated with a particle is related to the particle momentum, p, and Planck's constant, h. (Momentum is the product of mass, m, and velocity, v.)

De Broglie equation for the wavelength of matter waves.

$$\lambda = \frac{h}{p} = \frac{h}{mv} \tag{8.8}$$

In equation (8.8) wavelength is in meters, mass is in kilograms, and velocity is in meters per second. Planck's constant must also be expressed in units of mass, length, and time. This requires replacing the unit, joule, by the units $\text{kg m}^2\ \text{s}^{-2}$ (see Appendix B, equation B.8).

Example 8-7

Calculating the wavelength associated with a beam of particles. What is the wavelength associated with electrons traveling at one-tenth the speed of light?

Solution. The electron mass, expressed in kilograms, is 9.109×10^{-31} kg (recall Table 2-1). The electron velocity is $v = 0.100 \times c = 0.100 \times 3.00 \times 10^8$ m s^{-1} = 3.00×10^7 m s^{-1}. Planck's constant $h = 6.626 \times 10^{-34}$ J s = 6.626×10^{-34} kg m^2 s^{-2} s. Substituting these data into equation (8.8), we obtain

$$\lambda = \frac{6.626 \times 10^{-34} \text{ kg m}^2 \text{ s}^{-2} \text{ s}}{(9.109 \times 10^{-31} \text{ kg})(3.00 \times 10^7 \text{ m s}^{-1})} = 2.42 \times 10^{-11} \text{ m} = 0.0242 \text{ nm}$$

SIMILAR EXAMPLES: Exercises 43, 44, 73.

The wavelength calculated in Example 8-7, 0.0242 nm, is about one-half the radius of the first Bohr orbit of a hydrogen atom. (The radius of the first Bohr orbit = $n^2 a_0 = 1^2 \times 0.53$ Å = 0.53 Å = 0.053 nm.) We can think of this wavelength as being comparable to "atomic dimensions." It is only when wavelengths are comparable to atomic or nuclear dimensions that wave–particle duality is important. The concept has little meaning when applied to large (macroscopic) objects like baseballs and automobiles, because their wavelengths are too small to measure. For these macroscopic objects the laws of classical physics are perfectly adequate (see Exercise 45).

8-7   The Uncertainty Principle

With the laws of classical physics we can tell what types of physical behavior are possible and we can make predictions. For example we can *calculate* the exact point at which a rocket will land after it is fired. The more precisely we measure the variables that affect the rocket's trajectory (path), the more accurate our calculation (prediction) will be, and there is essentially no limit to the accuracy we can achieve. In classical physics nothing is left to chance—physical behavior can be predicted with certainty.

During the 1920s Niels Bohr and Werner Heisenberg thought about hypothetical experiments that would establish how precisely the behavior of subatomic particles could be determined. The two variables that must be measured are the position of the particle (x) and its momentum (p). [Recall that momentum is the product of mass (m) and velocity (v).] The conclusion they reached is that there must *always* be uncertainties in measurement such that the product of the uncertainty in position, Δx, and the uncertainty in momentum, Δp, is

The symbol ≥ stands for equal to or greater than.

$$\Delta x \, \Delta p \geq \frac{h}{4\pi} \tag{8.9}$$

The uncertainty principle is not easy for most people to accept. Einstein spent a good deal of time from the middle of the 1920s until his death in 1955 attempting, unsuccessfully, to disprove it. (See also, Exercises 40 and 41.)

What is significant about this expression, referred to as **Heisenberg's uncertainty principle,** is that we cannot measure position and momentum with great precision simultaneously. If we design an experiment to locate the *position* of a particle with great precision, we cannot measure its *momentum* precisely, and vice versa. In simpler terms, if we design an experiment to tell us precisely where a particle is, we cannot also know precisely where it has come from or where it is going. If we design an experiment that tells us precisely how a particle is moving, we cannot also know precisely where it is. In the subatomic world, things must always be "fuzzy." But why should this be so?

Suppose that we wish to learn something about an electron in a hydrogen atom by

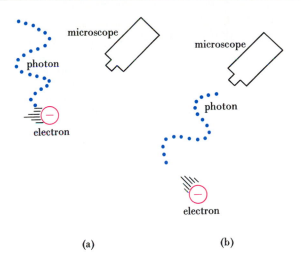

FIGURE 8-18

The uncertainty principle.

A free electron moves into focus of a hypothetical microscope (a). A photon of "light" strikes the electron and is reflected. In the collision the photon transfers some momentum to the electron. The reflected photon is seen through the microscope, but the electron has moved out of focus (b). Its exact position cannot be determined.

looking at it with a microscope. What kind of microscope should this be? The resolving power of a microscope is limited to objects that are about the same size as the wavelength of the light used. In an ordinary microscope with visible light the resolving power is about 1000 nm. With an electron microscope the resolving power is about 1 nm.

The first Bohr orbit in a hydrogen atom is 0.053 nm, and the diameter of the atom is about 0.1 nm (10^{-10} m). Electrons are much smaller than the atoms in which they are found; suppose we assume a diameter of about 10^{-14} m for an electron. Light of this wavelength would have a frequency of 3×10^{22} s^{-1} ($\nu = c/\lambda$) and an energy per photon of 2×10^{-11} J ($E = h\nu$). But from Figure 8-17 we can see that this energy is far, far in excess of what is required to ionize the electron in a hydrogen atom, that is, to strip it completely away from the atomic nucleus. In our attempt to see an electron in an atom, the measuring system (the light used) would greatly interfere with the measurement. We would knock the electron out of the atom just in trying to look at it! This point is illustrated further in Figure 8-18.

8-8 Wave Mechanics

There are two important implications of the ideas presented in the two preceding sections.

1. *We must discard the Bohr theory.* With the Bohr model of the hydrogen atom we can state precisely the radius of the allowable electron orbits and the velocity of the electron in these orbits. This is equivalent to a precise statement of the position and momentum of the electron. However, according to the Heisenberg uncertainty principle these two properties cannot both be known with high precision.

2. *Electrons in atoms have wavelike properties, that is, they are matter waves.*

The idea of treating electrons as matter waves was explored with great success by Irwin Schrödinger in 1927. But how are we to think of an electron in an atom according to this wave picture? Does the electron cease to exist as a particle? Is it literally smeared out into a wave? The ultimate answer to these questions, of course, is that we do not know and *cannot* know. This is the significance of the Heisenberg uncertainty principle. However, perhaps we can gain some useful insights by returning to a description with which we opened the chapter—wave motion in a string (Figure 8-3).

Standing Waves. The wave motion described through Figure 8-3 is called a *traveling* wave. In a traveling wave every portion of an infinitely long string goes through

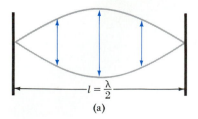

$l = \dfrac{\lambda}{2}$

(a)

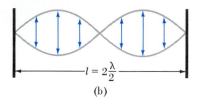

$l = 2\dfrac{\lambda}{2}$

(b)

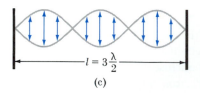

$l = 3\dfrac{\lambda}{2}$

(c)

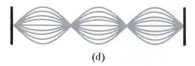

(d)

FIGURE 8-19

Standing waves in a string.

The blue arrows show the range of displacements at each point on the string for each standing wave. The patterns (a), (b), and (c) are for a given instant of time. The pattern in (d) represents an extended period of time. Note how the standing wave in (d) achieves its maximum displacement for constructive interference and zero displacement (straight line) for destructive interference.

an identical up-and-down motion and the wave transmits energy along the entire length of the string (essentially an infinite length).

Consider what happens in a short string with the far end *fixed,* for example, by being attached to an immovable wall (see Figure 8-19). When the wave motion reaches the wall it is reflected back toward the hand driving the wave. In this case, to describe the motion of the string as a whole we need to look simultaneously at what is happening to the forward wave and the reflected wave.

At a given instant of time, if everywhere the forward wave is at a crest the reflected wave is also at a crest, the two waves overlap or reinforce one another. The displacement of the string is at a maximum. (This situation is called *constructive* interference.) If there is a complete "mismatch" between the forward and reflected waves (that is, with the crests of one coinciding with the troughs of the other), the waves cancel one another and the string is not displaced at all. (This situation is called *destructive* interference.)

Over extended periods of time, each segment of the string vibrates between fixed limits (noted in Figure 8-19 by blue arrows), but these limits *differ* from point to point on the string. Some points on the string, called **nodes,** do not move at all. There is one node at each end of the standing wave, and there may be intermediate nodes as well. Since the nodes do not move, energy cannot be transmitted along the entire length of the string. Because of this the wave is called a *standing* wave.

There is an exact relationship between the length of the string and the permitted wavelengths of the standing waves. The length of the string *(l)* must be equal to a whole number *(n)* times one-half the wavelength ($\lambda/2$).

$l = n(\lambda/2)$ (where $n = 1, 2, 3, \ldots$
                    and the total number of nodes $= n + 1$).   (8.10)

The different standing waves pictured in Figure 8-19 are generated through the appropriate frequency of the up-and-down hand motion. The greater the frequency, the shorter the wavelength and the greater the number of nodes.

The one-dimensional standing wave in a string is the type of wave motion that we see in a plucked guitar string. The wave motion in a drum is more complex. It is a two-dimensional standing wave in the drumskin. Still more complex is a three-dimensional standing wave, since this wave motion extends throughout three-dimensional space. Yet, this is the type of wave motion we must use to describe a *matter wave,* such as an electron in an atom. Although we cannot give a complete description of a three-dimensional standing wave here, we can say that only certain wave patterns are allowable, as suggested through Figure 8-20.

When a guitar string is plucked, the frequency of the vibration that produces a standing wave as in Figure 8-19a is called the *fundamental.* That which produces the standing wave in Figure 8-19b is the *first overtone,* and so on.

Wave Functions. The allowable patterns for electron waves can be described through mathematical equations, but this task is well beyond the scope of our study. Let us just say that the acceptable solutions of these wave equations are called **wave functions,** denoted by the Greek letter psi, ψ. Wave functions contain three terms called **quantum numbers.** When specific values are assigned to these three quantum numbers, the result is called an **orbital.** An orbital is a mathematical function, but we can try to give it physical meaning. *An orbital represents a region in an atom where an electron might be found.* This orbital description of an electron permits us to think of the electron in one of two ways.

Assigning these quantum numbers in the wave function ψ is analogous to assigning values of n in equation (8.10) for standing waves in a string.

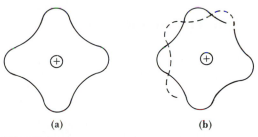

FIGURE 8-20
The electron as a matter wave.

These patterns are two-dimensional cross sections of a much more complicated three-dimensional wave. The wave pattern in (a), a *standing wave*, is an acceptable representation of a matter wave. It has an integral number of wavelengths (four) about the nucleus; successive waves reinforce one another. The pattern in (b) is unacceptable because the number of wavelengths is nonintegral (about 4.5), and successive waves show destructive interference. Each crest in one part of the wave is cancelled by a trough in another part of the wave.

- An electron is a cloud of negative electric charge, with the density of the charge varying from point to point.
- An electron is a particle, with the probability of finding the electron varying from point to point.

Whichever way we think of the electron, at every point in an atom the value of the *square* of the wave function, ψ^2, represents either the electron charge density or the probability of finding an electron at that point.

Perhaps the simplest orbital to describe is of the type known as $1s$. Figure 8-21a shows ψ^2 as a function of distance from the nucleus (r) along a line through the nucleus. It is a **probability distribution** plot, establishing that the highest electron probability or electron charge density in a $1s$ orbital exists at points in the vicinity of the nucleus. The pattern of dots in Figure 8-21b represents the distribution of electronic charge and electron probabilities in a plane with the nucleus at its center. Where the dots are closely spaced there is a higher charge density or a higher probability of finding an electron than where the spacing is greater. The pattern of dots in Figure 8-21b extends symmetrically in all directions and to all distances from the nucleus, with the spacing between dots constantly increasing. Because of this fact it is not possible to draw a picture that encompasses all such dots for all planes through the nucleus. Instead, we can only represent a region in which some portion (say 90%) of all the dots are found. This is a region in which there is a 90% probability of finding an electron or in which 90% of the charge density is contained. The spherical envelope in Figure 8-21c pictures a $1s$ orbital in this way.

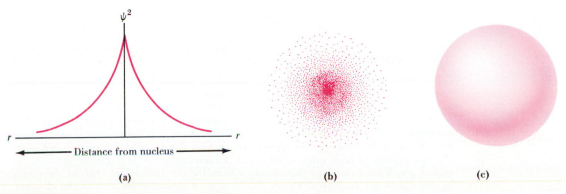

FIGURE 8-21
Three representations of the 1s orbital.

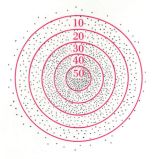

FIGURE 8-22

Dart board analogy to a 1s orbital.

Imagine that a single dart (analogous to the single electron in a hydrogen atom) is thrown at a dart board 1500 times. The board itself contains 90% of all the dart holes (1350 out of 1500). It is analogous to the 1s orbital, though of course the region described by a 1s orbital is a sphere, not a circular disc. Where is the dart most likely to be found?

The density of dart holes (number of holes per unit area) is greatest in the "50" region. But if we ask what is the most likely score for a dart throw, it is "30" (400 throws out of 1500), not "50." Even though the density of dart holes is less in the "30" ring, the total area of the "30" ring is much greater than that of the "50" ring.

Summary of scoring

200 darts score	"50"
300	"40"
400	"30"
250	"20"
200	"10"
150	off the board

1500 darts thrown

Are You Wondering:

What it means for the electron charge density or electron probability in a 1s orbital to be highest in the region of the nucleus? Does this mean that the electron is most likely to be at the nucleus itself?

The probability of finding an electron (or the electron charge density) at an isolated *point* in space is highest at the nucleus, that is, ψ^2 has its greatest value at this point. However, the only *meaningful* assessment of the electron probability is to add together the probabilities at all equivalent points. Thus, if we want to assess the probability of an electron in a 1s orbital being 0.1 nm from the nucleus, we must take the probability at *all* points 0.1 nm from the nucleus and add them together. All points 0.1 nm from the nucleus lie on a thin *spherical* shell with a radius of 0.1 nm. When we deal with probabilities in this way, the greatest probability of finding an electron in a 1s orbital is in a spherical shell with a radius of 0.053 nm (0.53 Å). This is the same as the first Bohr orbit.

Figure 8-22 presents an analogy that should help you understand this distinction between electron probability at a point and in a spherically symmetric region of equivalent points. The appearance of a well-used dart board is one in which the bullseye region is pretty well "punched out," yet in the hands of the average player the most probable dart hit will not be a bullseye.

Figure 8-23 represents the electron charge distribution in an orbital of the type 2s. A new feature seen in this orbital is a spherical shell in which the electron probability or electron charge density is *zero*. This is called a **spherical node.** It is equivalent to the nodes in a standing wave in a string.

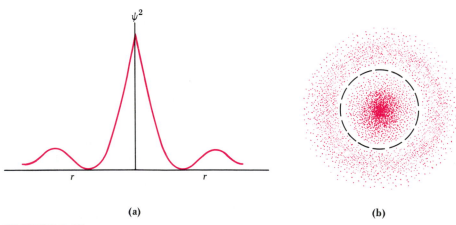

(a)   (b)

FIGURE 8-23

The 2s orbital.

(a) As with the 1s orbital, the electron charge density is highest in the vicinity of the nucleus. It drops to zero at a particular distance from the nucleus (a condition called a spherical node), rises to a second high value at a somewhat greater distance, and again gradually decreases, approaching zero at large distances from the nucleus.

(b) This diagram contains two regions with a high density of dots, corresponding to the two regions of high value of ψ^2 in (a). To encompass 90% of all the dots requires a spherical envelope that is larger than for the 1s orbital. Also, as discussed in the dartboard analogy of Figure 8-22, the most probable location of the electron is in a spherical shell corresponding to the second ring of dots.

8-9 Quantum Numbers and Electron Orbitals

To recapitulate from the preceding section, by introducing *three* quantum numbers into a wave function (ψ) we obtain a mathematical description of an orbital. Every different combination of the three quantum numbers corresponds to a different orbital, a relationship that we need to explore in some detail. First, though, let us simply consider the nature of these three quantum numbers and their relationship to one another.

The first of these three quantum numbers is the **principal quantum number, n.** This quantum number may have only a *positive, nonzero integral* value.

The principal quantum number, n.

$$n = 1, 2, 3, 4, \ldots \qquad (8.11)$$

The second quantum number is the **orbital (azimuthal) quantum number, l,** which may be zero or a positive integer. It cannot be negative and it cannot be larger than $n - 1$ (where n is the principal quantum number).

The orbital quantum number, l.

$$l = 0, 1, 2, 3, \ldots, n - 1 \qquad (8.12)$$

The third quantum number is called the **magnetic quantum number, m_l.** Its value may be positive or negative, may include zero, and may range from $-l$ to $+l$ (where l is the orbital quantum number.)

The magnetic quantum number, m_l.

$$m_l = -l, -l + 1, -l + 2, \ldots, 0, 1, 2, \ldots, +l \qquad (8.13)$$

Example 8-8

Determining l and m_l from n. What are the possible values of l and m_l for an electron with the principal quantum number, $n = 3$?

Solution. From expression (8.12) we see that the allowable values of l are 0, 1, and 2. The allowable values of m_l depend on the value of l (expression 8.13).

If $l = 0$, there can be but a single value of m_l: 0.

If $l = 1$, there are three allowable values of m_l: $-1, 0, +1$.

If $l = 2$, there are five allowable values of m_l: $-2, -1, 0, +1, +2$.

SIMILAR EXAMPLES: Exercises 8, 49, 50.

Example 8-9

Recognizing whether a set of quantum numbers is allowable. Can an electron have the quantum numbers $n = 2$, $l = 2$, and $m_l = 2$?

Solution. No; the l quantum number cannot be greater than $n - 1$. Thus, if $n = 2$, l can be only 0 or 1. And if l can be only 0 or 1, m_l cannot be 2, because m_l can never be greater than l (expression 8.13).

SIMILAR EXAMPLES: Exercises 10, 51.

All orbitals having the same value of the quantum number n are said to be in the same **principal electronic shell** or **principal level,** and all orbitals having the same l value are in the same **subshell** or **sublevel.**

The principal shells are numbered according to the value of n. The *first* principal electronic shell consists of orbitals with $n = 1$; the *second* principal shell, of orbitals

FIGURE 8-24
Three representations of a 2*p* orbital.

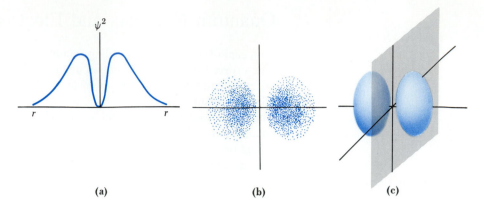

(a)   (b)   (c)

with $n = 2$; and so on. The value of the quantum number n relates to the energies of electrons and to their most probable distances from the nucleus.

The value of the l quantum number determines the *geometrical shape* of the electron cloud or electron probability distribution. All orbitals with the value $l = 0$ are *s* **orbitals.** If the s orbital is in the first principal electronic shell ($n = 1$), it is a 1s orbital; if it is in the second principal shell, it is a 2s orbital; and so on. The electron cloud or electron probability distribution for an s orbital is spherically symmetric, that is, it has the shape of a sphere with the nucleus at its center. Because when $l = 0$, m_l must also be 0, *there can be only one orbital of the* s *type for each principal shell.*

As we saw in Figure 8-21, in a 1s orbital there is some electron probability (no matter how slight) at all distances from the nucleus. In the 2s orbital, however, as noted in Figure 8-23, there is a *spherical node,* a spherical shell at some distance from the nucleus where the electron probability falls to zero. In a 3s orbital there are two concentric spherical nodes. In general for s orbitals the number of spherical nodes is $n - 1$ (where n is the principal quantum number).

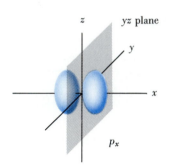

The orbital type corresponding to $l = 1$ is the *p* **orbital.** Because when $l = 1$, m_l can have one of three values (-1, 0, $+1$), p orbitals occur in sets of three. That is, there are three p orbitals in the p subshell. As shown in Figure 8-24a, for a 2p orbital the electron probability or electron charge density (ψ^2) is zero at the nucleus, rises to a maximum on either side of the nucleus, and then falls off with distance (r) along a line passing through the nucleus (that is, along the x, y, or z axis). The pattern of dots in Figure 8-24b represents the electron probability distribution in a plane passing through the nucleus (that is, the xy, xz, or yz plane). The electron clouds or electron probability distributions for p orbitals are *not* spherically symmetric. The greatest probability of finding an electron in a 2p orbital is within the dumbbell-shaped region of Figure 8-24c. This is the representation commonly used for a p orbital. Usually the three p orbitals are shown to be directed along the perpendicular axes through the nucleus and are represented by the symbols p_x, p_y, and p_z. The set of three p orbitals is shown in Figure 8-25.*

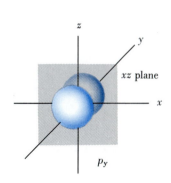

The two lobes of high electron probability or electron charge density of a p orbital are separated by a region in which the probability or charge density drops to zero. Because this region is a plane, it is called a **nodal plane.** The number of nodal planes for an orbital is equal to the l quantum number (and for a p orbital $l = 1$). The nodal planes for the three p orbitals are indicated in Figure 8-25.

There is a set of *five* orbitals having $l = 2$; these are the *d* **orbitals** and comprise the d subshell. The geometrical shapes of the d orbitals, which are more complex than those of s and p orbitals, are shown in Figure 8-26.* The d orbitals have *two* nodal planes; the ones for the d_{xy} orbital are indicated in Figure 8-26. (The nodal surfaces for the d_{z^2} orbital are actually cone-shaped.)

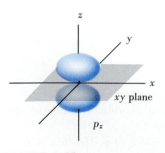

FIGURE 8-25
The three p orbitals.

*The orientation of a set of equivalent orbitals with respect to one another (e.g., the three p orbitals or the five d orbitals) is related to the values of the magnetic quantum number, m_l.

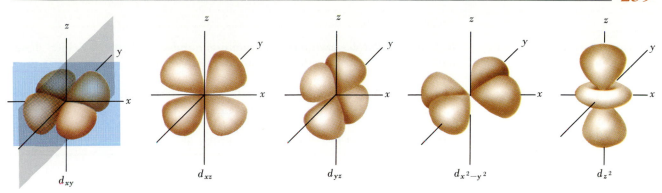

d_{xy} d_{xz} d_{yz} $d_{x^2-y^2}$ d_{z^2}

FIGURE 8-26
The five *d* orbitals.

The designations *xy*, *xz*, *yz*, and so on are related to the values of the quantum number m_l ($l = 2$ for all *d* orbitals), but this is a detail not pursued further in the text.

A fourth type of orbital is the **f orbital,** with $l = 3$. There are *seven* different *f* orbitals, corresponding to the seven possible values of m_l. We will not attempt to depict the *f* orbitals, although from the previous generalization about nodal planes we see that each *f* orbital has three nodal planes.

Some of the points discussed in the preceding paragraphs are illustrated through Table 8-1 and Example 8-10.

TABLE 8-1
Electronic Shells, Orbitals, and Quantum Numbers

Principal shell	1st	2nd			3rd							
$n =$	1	2	2	2	2	3	3	3	3	3	3	3	3	3
$l =$	0	0	1	1	1	0	1	1	1	2	2	2	2	2
$m_l =$	0	0	-1	0	$+1$	0	-1	0	$+1$	-2	-1	0	$+1$	$+2$
orbital designation	1s	2s	2p	2p	2p	3s	3p	3p	3p	3d	3d	3d	3d	3d
number of orbitals in subshell	1	1	3			1	3			5				
total number of orbitals $= n^2$	1	4				9								

When assigning the quantum number m_l, it is immaterial which value we assign first. We will follow the convention of beginning with the most negative of the permitted values and proceeding in the order: negative → zero → positive.

Are You Wondering:

How an electron gets from one lobe of a p *orbital to the other, given that there is zero probability of an electron being in the nodal plane?*

The difficulty here is in thinking of the electron as a discrete particle. In the wave-mechanical model the particle acts as a matter wave, and the wave is found in many places at the same time. The nodal plane is simply a region in which the wave function has a value of zero. This situation is analogous to the standing waves in a string (Figure 8-19). Think of the entire vibrating string as being equivalent to the electron. The particular points on the string that are not displaced, the nodal points, are equivalent to the nodal planes in the *p* orbitals.

Recall the relationship between the l quantum number and orbital designations: $l = 0$, s orbital; $l = 1$, p orbital; $l = 2$, d orbital; $l = 3$, f orbital.

Example 8-10

Relating orbital designations and quantum numbers. Write an orbital designation for an electron with the quantum numbers $n = 4$, $l = 2$, $m_l = 0$.

Solution. The type of orbital is determined by the l quantum number. If $l = 2$, the orbital is of the d type. Because $n = 4$, the designation is 4d

SIMILAR EXAMPLES: Exercises 9, 50, 52, 53.

8-10   Electron Spin—A Fourth Quantum Number

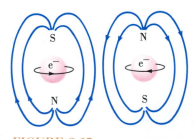

FIGURE 8-27
Electron spin visualized.

The spinning electron has an associated magnetic field. The poles and magnetic lines of force corresponding to each electron spin are shown. Electrons with opposing spins have opposing magnetic fields. Of course, this representation of electron spin only has meaning if we think in terms of the particlelike characteristics of an electron.

In 1925, George Uhlenbeck and Samuel Goudsmit proposed that some unexplained features of the hydrogen spectra could be understood by assigning electrons a fourth quantum number. The property of electrons associated with this fourth quantum number has become known as electron spin. An electron acts as if it spins on its axis, much as the earth spins on its axis as it revolves around the sun. There are two possibilities for electron spin, as suggested by Figure 8-27. The quantum number describing electron spin, m_s, may have a value of $+\frac{1}{2}$ (also denoted by an arrow ↑) or $-\frac{1}{2}$ (denoted by the arrow ↓). Unlike the quantum numbers n, l, and m_l, which have interrelated values, the value of m_s does not depend on the other three quantum numbers.

Throughout the text, we will discover important uses of the concept of electron spin, but what experimental evidence is there of electron spin, especially in light of the uncertainty principle? An experiment reported by Otto Stern and Walter Gerlach in 1922, though designed for a different purpose, seems to yield this proof. Stern and Gerlach vaporized silver metal in a furnace and passed a beam of silver atoms through an inhomogeneous (nonuniform) magnetic field. The beam split in two. A simplified explanation of this behavior is

1. An electron, by virtue of its spin, has associated with it a magnetic field.
2. A pair of electrons with opposing spins has no associated magnetic field.
3. In a silver atom, 23 electrons have a spin of one type and 24 of the opposite type. The direction in which this atom is deflected in a magnetic field depends only on the spin of the unmatched electron.
4. In a collection of a large number of silver atoms there is an equal chance that the unmatched electron will have a spin of $+\frac{1}{2}$ or $-\frac{1}{2}$. Thus, the beam of atoms is split into two beams.

8-11   Multielectron Atoms

The electron orbitals of Schrödinger's wave equation apply to the hydrogen atom or hydrogenlike atoms—species containing a single electron: H, He^+, Li^{2+}, ... For atoms containing many electrons a new idea enters—the mutual repulsions between electrons. Since exact electron positions are not known, these repulsions can only be approximated. As a result only approximate solutions can be obtained for the wave equation, but these solutions are in good agreement with experimental results. Solutions of the wave equation for multielectron atoms are obtained by considering the electrons one by one in an environment established by the nucleus and the other electrons. Any given electron experiences an attraction to a *net* nuclear charge that is less than the total positive charge on the nucleus. This is because other electrons partially neutralize the nuclear charge. The net nuclear charge that remains is called the **effective nuclear charge, Z_{eff}.** When wave equations are solved using effective nuclear charges, the electron orbitals are of the same

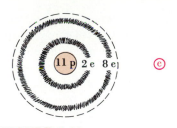

FIGURE 8-28
The shielding effect.

The atom pictured here is sodium. The charge on the nucleus is +11. If the nucleus and the two inner electronic shells (contained within the broken circle) acted as a single unit, the net charge on this unit would be +1. When the outermost electron is outside the broken circle, it would experience much the same force of attraction as if it were the only electron present in the atom and the nucleus consisted of a charge of +1. Shielding by the inner electrons is not perfect, however, and the effective nuclear charge, $Z_{eff.}$, is closer to +2 than +1.

types and shapes as the orbitals established for the hydrogen atom. They are called *hydrogenlike* orbitals.

Through equation (8.4) and Figure 8-17 we saw that the electron energies in the Bohr hydrogen atom are a function only of the Bohr orbit (that is, only a function of the distance of the electron from the nucleus). We get the same kind of result with wave mechanics, namely, in a hydrogen atom all orbitals having the same principal quantum number, n, have the same energy. Orbitals having the same energy are said to be **degenerate.** With multielectron atoms, however, orbital energies depend on the orbital type, as we will learn in the following discussion. Although the concept of orbitals suggests that we should work with wave functions, you may find the discussion easier to follow by thinking of the electron as a particle.

Nuclear Charge Effect. In general, the attractive force of the nucleus for a given electron increases as the nuclear charge increases, even taking into account that it is the effective nuclear charge, Z_{eff}, that we must consider. As a result, we find the energy of interaction between the electron and the nucleus—the orbital energy—to become lower (that is, more negative) with increasing atomic number.

Shielding Effect. Think about the attractive force of the atomic nucleus for the electron farthest away. The inner electrons reduce the effectiveness of the nucleus in attracting this electron. They *screen* or *shield* the outermost electron from the full effects of the nucleus by partially neutralizing and thereby reducing the nuclear charge, from Z to Z_{eff}. The shielding effect is illustrated in Figure 8-28.

The effectiveness of the shielding by inner electrons depends on the particular type of orbital in which the outermost electron is found. For example, an electron in an s orbital spends more time close to the nucleus than does an electron in a p orbital of the same electronic shell (compare the probability distributions for $2s$ and $2p$

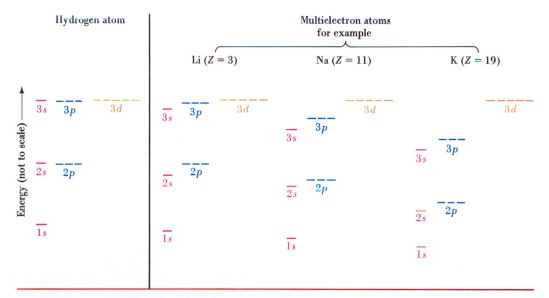

FIGURE 8-29
Orbital energy diagram for first three electronic shells.

Energy levels for the subshells of the first three principal quantum levels are shown for a hydrogen atom (left) and three typical multielectron atoms (right). Each multielectron atom has its own energy-level diagram. One of the distinctive features of the energy-level diagram for multielectron atoms is the splitting of each level into different energies for each subshell. (Note that for the hydrogen atom the orbital energies for $3s$, $3p$, and $3d$ are all alike, but in the multielectron atoms they become rather widely separated.) Another feature is a steady decrease in all the orbital energies with increasing atomic number. (Note that the energy level for an orbital like $2s$ drops steadily from left to right.)

orbitals in Figures 8-23 and 8-24). The electron in the *s* orbital is not as well screened by inner electrons as is the electron in the *p* orbital. The *s* electron experiences a higher Z_{eff} than does the *p* electron. It is held more tightly and is therefore at a lower energy than is the *p* electron. Just as a *s* orbital is lower in energy than a *p* orbital of the same shell, a *p* orbital is lower in energy than a *d* orbital. The effect of shielding, then, is to cause a splitting of the energy level of each principal shell into separate levels for each subshell. These levels are arranged in order of the combined value, $n + l$: The lower the $n + l$ value, the lower the energy level. Thus, in the third principal shell, for example, the *s* subshell ($n + l = 3 + 0 = 3$) is at a lower energy than the *p* subshell ($n + l = 3 + 1 = 4$). There is no further splitting of energies within a subshell, however. All three *p* orbitals of a principal shell have the same energy; all five *d* orbitals have the same energy; and so on.

The ideas presented here are illustrated through Figure 8-29, which represents orbital energies for the first three shells of the hydrogen atom and some typical multielectron atoms.

8-12   Electron Configurations

How are the electrons in a given atom distributed among the various orbitals? Do they all pile up into just a few orbitals or do they distribute themselves more widely? How does the expected distribution of electrons among available orbitals differ for atoms of different elements? To answer these fundamental questions we need to consider three important rules or principles governing electron configurations. The **electron configuration** of an atom is a designation of the distribution of its electrons among the different electronic shells and orbitals.

1. *Electrons occupy orbitals in such a way as to minimize the energy of the atom.* Figure 8-29 implies an order in which electrons occupy orbitals, first the 1*s*, then 2*s*, 2*p*, and so on. Actually, the energy of an atom is not minimized in most cases just by filling the principal electronic shells in succession. At higher quantum numbers, and for certain elements, an overlapping of sublevels occurs, for example, with 4*s* filling before 3*d* in K and Ca. As a result, we must establish the order of filling of orbitals by experiment. The order obtained is roughly the one listed below.

$$1s,\ 2s,\ 2p,\ 3s,\ 3p,\ 4s,\ 3d,\ 4p,\ 5s,\ 4d,\ 5p,\ 6s,\ 4f,\ 5d,\ 6p,\ 7s,\ 5f,\ 6d,\ 7p \quad (8.14)$$

Most students find the device pictured in Figure 8-30 a useful way to remember this order. [You should be able to reproduce the columns of orbitals and the diagonal arrows running from upper right to lower left, as the need arises.]

Another useful rule that we have already seen in Section 8-11 is that for any given pair of orbitals the one with the *lower* total of the *n* and *l* quantum numbers fills first. For example, the 3*s* orbital ($3 + 0 = 3$) fills before the 3*p* ($3 + 1 = 4$). If the $n + l$ totals are the same for two orbitals, the one with the *lower n* value generally fills first. Thus, the 3*d* orbital ($3 + 2 = 5$) fills before the 4*p* ($4 + 1 = 5$).

When we describe the relationship between electron configurations and the periodic table in Section 9-3, we consider still another method of establishing the order of filling of electron orbitals.

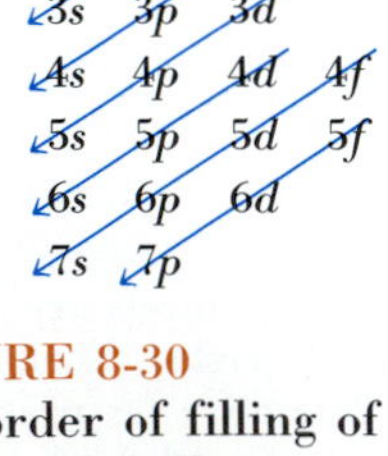

2. *No two electrons in an atom may have all four quantum numbers alike—the Pauli exclusion principle.* In 1926, Wolfgang Pauli noted that some lines that should have been present in emission spectra were not there. He was able to explain the *absence* of these lines by stating that no two electrons in an atom can have all four quantum numbers alike. The first three quantum numbers, n, l, and m_l determine a specific orbital. Two electrons may have these three quantum numbers alike; but if they do, they must have different values of m_s, the spin

1*s*
2*s*   2*p*
3*s*   3*p*   3*d*
4*s*   4*p*   4*d*   4*f*
5*s*   5*p*   5*d*   5*f*
6*s*   6*p*   6*d*
7*s*   7*p*

FIGURE 8-30
The order of filling of electronic subshells.

Follow the arrows from top to bottom, and the order obtained is the same as that in expression (8.14).

quantum number. Another way to state this result is that *only two electrons may exist in the same orbital and these electrons must have opposite spins.*

Because of this limit of two electrons per orbital, the capacity of a subshell for electrons can be obtained by *doubling* the number of orbitals in the subshell. Thus, the *s* subshell consists of *one* orbital having a capacity of *two* electrons; the *p* subshell consists of *three* orbitals having a total capacity of *six* electrons; and so on. The capacity of a principal shell is also *twice* the number of orbitals it contains (recall Table 8-1), leading to the expression

$$\text{maximum number of electrons in the electronic shell with principal quantum number } n = 2n^2$$

3. *The principle of maximum multiplicity—Hund's rule.* When orbitals of identical energy (those in the same subshell) are available, electrons occupy these singly rather than in pairs. As a result, an atom tends to have as many unpaired electrons as possible. This behavior can be rationalized by saying that electrons, because they all carry the same electric charge, seek out empty orbitals of similar energy in preference to pairing up with electrons in half-filled orbitals.

8-13 Electron Configurations of the Elements

To apply the principles of the preceding section it is handy to use shorthand designations. The electron configuration of an atom of carbon is represented in two different ways below.

spdf notation: C $1s^2 2s^2 2p^2$

orbital diagram: C   1s 2s 2p

In each of these designations we must assign a total of *six* electrons since the atomic number of carbon is 6. Two of these electrons are in the 1*s* subshell, two in the 2*s*, and two in the 2*p*. The *spdf* notation denotes the total number (the superscript) of electrons in each subshell. The orbital diagram breaks down each subshell into individual orbitals (drawn as boxes) and indicates the number of electrons for each orbital. This is done with arrows. An arrow pointing upward corresponds to one type of spin ($+\frac{1}{2}$) and an arrow pointing down to the other spin ($-\frac{1}{2}$). Electrons in the same orbital with opposing (opposite) spins are said to be *paired*. The electrons in the 1*s* and 2*s* orbitals of the carbon atom are paired. Electrons with the same type of spin are said to have parallel spins, as in the 2*p* orbitals of carbon.

The electron configuration described above for carbon is called the *ground-state* electron configuration. It is the most probable or the most energetically favored configuration. These are the types of configurations that we describe in this section, although from time to time later in the text we briefly mention some electron configurations that are not the most stable. We speak of atoms having such configurations as being in an "*excited*" state.

The two orbital diagrams for carbon shown below are not ground-state electron configurations. The first is incorrect because the two 2*p* electrons have been paired into a single 2*p* orbital in violation of Hund's rule. The second is incorrect because a configuration with singly occupied orbitals having opposing spins is not as energetically favorable as with parallel spins.

The number of unpaired electrons can also be denoted with *spdf* notation if the subshells are broken down into individual orbitals, such as

C $1s^2 2s^2 2p_x^1 2p_y^1$

(incorrect)   (incorrect)

1s 2s 2p   1s 2s 2p

C   C

The Aufbau Process. Consider the following *hypothetical* process—the building up of a more complex atom starting with the simplest atom, hydrogen. This hypothetical process is called the **Aufbau** process (meaning "building up" in German). In this process we proceed from an atom of one element to the next by adding a proton and the requisite number of neutrons to the nucleus and one electron to the appropriate orbital. We pay particular attention to this added electron, called the differentiating electron.

Hydrogen, Z = 1. The lowest energy state for the electron in a hydrogen atom is the $1s$ orbital. The electron configuration is $1s^1$.

Helium, Z = 2. In a helium atom a second electron goes into the $1s$ orbital. The two electrons must have opposing spins, $1s^2$.

Lithium, Z = 3. The differentiating electron cannot be accommodated in the $1s$ orbital (Pauli exclusion principle). It must be placed in the next lowest energy orbital, $2s$. The electron configuration is $1s^2 2s^1$.

Beryllium, Z = 4. The configuration is $1s^2 2s^2$.

Boron, Z = 5. Now the $2p$ subshell begins to fill. The electron configuration is $1s^2 2s^2 2p^1$.

Carbon, Z = 6. A second electron goes into the $2p$ subshell, but here the rule of maximum multiplicity (Hund's rule) applies. This second $2p$ electron goes into an empty orbital, and with a spin parallel to the first $2p$ electron.

Nitrogen, Z = 7, through Neon, Z = 10. In this series of four elements the filling of the $2p$ subshell is completed. Note how the number of unpaired electrons reaches a maximum (3) with nitrogen and then decreases to zero with neon.

Sodium, Z = 11, through Argon, Z = 18. The building up of the electron configurations of this series of eight elements closely parallels what we have already seen for the eight elements from Li through Ne, except that electrons go into $3s$ and $3p$ orbitals. To simplify the presentation of these electron configurations we will speak of the electrons in the first two shells as being **core electrons.** These core electrons are in the neon configuration ($1s^2 2s^2 2p^6$) and we will represent their contribution to the overall electron configuration by the symbol: [Ne]. Electrons that are added to the electronic shell of highest principal quantum number (the outermost shell) we will call **valence electrons.** Using these ideas we can then write

Na	*Mg*	*Al*	*Si*
[Ne]$3s^1$	[Ne]$3s^2$	[Ne]$3s^2 3p^1$	[Ne]$3s^2 3p^2$

P	*S*	*Cl*	*Ar*
[Ne]$3s^2 3p^3$	[Ne]$3s^2 3p^4$	[Ne]$3s^2 3p^5$	[Ne]$3s^2 3p^6$

Potassium, Z = 19, and Calcium, Z = 20. Here we encounter the first "irregularity" in the Aufbau process. Instead of continuing with the filling of the third principal shell (that is, of $3d$ orbitals) we find that after argon the differentiating electron goes into the $4s$ orbital. This is consistent, however, with the order of filling of orbitals outlined in expression (8.14) and Figure 8-30. Using the symbolism [Ar] to represent the *core* electrons in the configuration $1s^2 2s^2 2p^6 3s^2 3p^6$, for potassium and calcium we have

K $[Ar]4s^1$ and   Ca $[Ar]4s^2$

Scandium, Z = 21, through Zinc, Z = 30. This next series of elements is characterized by electrons filling the d orbitals of the third shell. The d subshell has a total capacity of 10 electrons; hence 10 elements are involved. We have to choose here whether to write the electron configuration of scandium as

(a) Sc $[Ar]3d^1 4s^2$ or   (b) Sc $[Ar]4s^2 3d^1$

Both methods are commonly used. Method (a) groups all the subshells of a principal shell and places last subshells of the highest principal quantum level. Method (b) lists orbitals in the apparent order in which they fill. In this text we use method (a).

The electron configurations of this series of ten elements are listed below in both the orbital diagram and the *spdf* notation.

An easy way to remember the electron configurations of Cr and Cu is that both have an [Ar] core and *one* $4s$ electron. The remaining electrons to bring the total up to the atomic number must be $3d$ electrons.

The d orbitals fill in a fairly regular fashion in this series but there are two exceptions: chromium (Cr) and copper (Cu). That is, the electron configuration of Cr is not the expected $[Ar]3d^4 4s^2$ and that of Cu is not $[Ar]3d^9 4s^2$. These irregularities can best be understood in terms of the closeness in energy between the $3d$ and $4s$ levels and the contribution of electron repulsions in determining the most stable electron configurations for multielectron atoms.

Gallium, Z = 31, to Krypton, Z = 36. In this series of six elements the $4p$ subshell is filled. The configuration for krypton is

Kr $1s^2 2s^2 2p^6 3s^2 3p^6 3d^{10} 4s^2 4p^6$

Rubidium, Z = 37, to Xenon, Z = 54. In this series of 18 elements the subshells fill in the order $5s$, $4d$, and $5p$, ending with the configuration of xenon.

Xe $1s^2 2s^2 2p^6 3s^2 3p^6 3d^{10} 4s^2 4p^6 4d^{10} 5s^2 5p^6$

FIGURE 8-31

Representing electron configurations—a summary.

The electron configuration of selenium is outlined here. Commonly used terms are printed in blue.

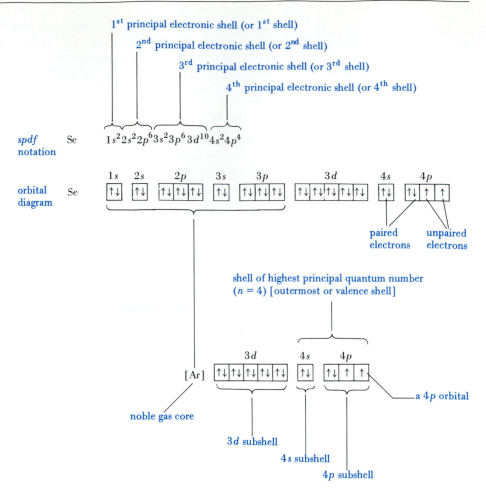

Cesium, $Z = 55$, to Radon, $Z = 86$. In this series of 32 elements, with a few exceptions, the subshells fill in the order $6s$, $4f$, $5d$, and $6p$. The configuration of radon is

Rn $1s^2 2s^2 2p^6 3s^2 3p^6 3d^{10} 4s^2 4p^6 4d^{10} 4f^{14} 5s^2 5p^6 5d^{10} 6s^2 6p^6$

Francium, $Z = 87$, to ?. Francium starts a series of elements in which the subshells that fill are $7s$, $5f$, $6d$, and presumably $7p$, though elements in which the $7p$ subshell is occupied are not yet known.

We have presented a brief outline of the assignment of electron configurations. We consider further aspects of this topic at appropriate points later in the text, such as the relationship between electron configurations and the periodic table in Section 9-3 and further details on the electron configurations of transition elements in Chapter 24. Ground-state electron configurations of the atoms of all the elements are presented in Appendix E. Figure 8-31 summarizes some of the more commonly used terms in describing electron configurations.

Some Limitations of the Aufbau Process. We have seen two examples, chromium and copper, where the electron configuration predicted by the Aufbau process is not the one found by experiment. There are several other examples of this type in the filling of d and f orbitals. In some cases the order of filling of orbitals is not in strict accordance with orbital energies. For example, the orbital energy of the $3d$ electron in Sc is lower than $4s$, even though in the Aufbau process we first fill the $4s$ subshell before adding electrons to the $3d$. Another complicating factor is that when considering the loss of electrons by an atom it is not always the last electrons added in the Aufbau process that are lost. The electron configuration of Sc^+ appears to be $[Ar]3d^14s^1$. The electron lost in forming Sc^+ from Sc is a $4s$ electron, not the $3d$ electron that was placed last in the Aufbau process. Despite the qualifications just cited, the Aufbau process does provide a good overview of the assignment of electron configurations.

Example 8-11

Using spdf notation for an electron configuration. Complete each of the following electron configurations.
(a) ? $1s^22s^22p^63s^23p^5$
(b) Nb (Z = 41) $[Kr]4d^?5s^1$
(c) As (Z = 33)   ?

Solution

(a) All electrons must be accounted for in an electron configuration. Add up the superscript numerals $(2 + 2 + 6 + 2 + 5)$ to obtain the atomic number (17). Look up the element with $Z = 17$ in the periodic table or a listing of the elements.

Cl (Z = 17) $1s^22s^22p^63s^23p^5$

(b) The electrons in this atom are those of the Kr configuration (36), the $4d$ electrons, and the *one* $5s$ electron. These must total to the atomic number, $Z = 41$. The number of $4d$ electrons is $41 - 36 - 1 = 4$.

Nb (Z = 41) $[Kr]4d^45s^1$

(c) The first 18 electrons are in the configuration $[Ar]$. The next two electrons go into the $4s$ subshell, and these are followed by ten electrons that go into the $3d$ subshell. At this point we have accounted for $18 + 2 + 10 = 30$ electrons. The remaining three electrons go into the $4p$ subshell, leading to the electron configuration

As (Z = 33) $[Ar]3d^{10}4s^24p^3$

SIMILAR EXAMPLES: Exercises 12, 14, 59.

Example 8-12

Using an orbital diagram for an electron configuration. The electron configuration of Nb was worked out in Example 8-11(b). Represent this electron configuration with an orbital diagram.

Solution. We need to represent all of the orbitals that appear in the *spdf* notation of Example 8-11(b) and fill these orbitals with the appropriate number of electrons. All of the orbitals, except the $4d$ and $5s$, contain pairs of electrons with opposing spins. The *four* $4d$ electrons must be distributed among the *five* $4d$ orbitals in accordance with Hund's rule. Four of the five $4d$ orbitals are half-filled and the fifth is empty.

SIMILAR EXAMPLES: Exercises 12, 14, 58, 59.

Summary

Ideas concerning the nature of electromagnetic radiation are useful starting points in the discussion of atomic structure. The dispersion of ''white'' light produces a continuous spectrum—a rainbow. However, most light originating from excited gaseous atoms produces a discontinuous or line spectrum—a series of colored lines. The simplest line spectrum is that of hydrogen, which can be described through an empirical equation (the Balmer equation).

Theoretical explanations of atomic spectra required a breakthrough in our thinking about energy. Planck proposed that energy exists as tiny discrete units called quanta. Einstein used Planck's quantum theory to explain the photoelectric effect, and Bohr applied classical and quantum mechanics to develop a model of the hydrogen atom. Bohr's model permits a calculation of the permissible energy levels for the electron in a hydrogen atom. Energies of photons emitted by excited hydrogen atoms correspond to differences in these energy levels. Frequencies of lines in the hydrogen spectrum can be predicted by combining Bohr's theory and the Planck equation.

To provide explanations of such phenomena as the formation of chemical bonds between atoms, a model of atomic structure must be based on a new form of quantum theory—wave mechanics. The essential ideas contributing to this newer quantum mechanics are de Broglie's concept of wave–particle duality (that is, the existence of matter waves) and Heisenberg's uncertainty principle. These ideas were used by Schrödinger to provide a new picture of the hydrogen atom.

The essential feature of the Schrödinger atom is to think of an electron as a cloud of negative electric charge having a certain geometrical shape. Also, the electron can be viewed as a particle whose probability of being found extends throughout space, though the probability is highest in certain three-dimensional regions. The variation of charge density or electron probability from point to point is described through expressions known as orbitals. The key parameters that distinguish among orbitals are a set of three quantum numbers, n, l, and m_l. The specific orbital types described in this chapter are known as s, p, and d. A fourth quantum number is known as the spin quantum number, m_s.

The wave mechanical description developed for the hydrogen atom can be modified to apply to other atoms (multielectron atoms). By using a set of rules, probable assignments of electrons to the orbitals of atoms can be made. These assignments, called electron configurations, are made for various elements as a conclusion to Chapter 8.

Summarizing Example

Microwave ovens are becoming increasingly popular in kitchens around the world. They provide a rapid, energy-efficient means of preparing foods. Microwave ovens are also useful in the chemical laboratory, particularly in drying samples for chemical analysis. A typical microwave oven uses microwave radiation with frequency $\nu = 2.45 \times 10^9 \text{ s}^{-1}$.

1. Calculate the wavelength of microwave radiation having a frequency of $2.45 \times 10^9 \text{ s}^{-1}$.

Solution. Microwaves are a form of electromagnetic radiation and thus travel at the speed of light, $2.998 \times 10^8 \text{ m s}^{-1}$. Use equation (8.1).

$$\lambda = \frac{c}{\nu} = \frac{2.998 \times 10^8 \text{ m s}^{-1}}{2.45 \times 10^9 \text{ s}^{-1}} = 0.122 \text{ m} = 12.2 \text{ cm}$$

(This example is similar to Example 8-1.)

$CuSO_4 \cdot 5H_2O$ (deep blue color) can be dehydrated to $CuSO_4 \cdot H_2O$ (pale blue) in a microwave oven. [Carey B. Van Loon]

2. What is the energy associated with one photon of this microwave radiation?

Solution. Planck's equation, $E = h\nu$ yields the answer directly.

$$E = h\nu = 6.626 \times 10^{-34} \text{ J s} \times 2.45 \times 10^9 \text{ s}^{-1} = 1.62 \times 10^{-24} \text{ J}$$

(This example is similar to Example 8-4.)

3. How many *moles* of this radiation would contain sufficient energy to raise the temperature of 225 g of water from 24.5 to 99.5 °C?

Solution. A mole of this radiation means one mole of photons. The energy content per mole of radiation is

$$\text{no. J/mol} = \frac{1.62 \times 10^{-24} \text{ J}}{\text{photon}} \times \frac{6.022 \times 10^{23} \text{ photons}}{1 \text{ mol photons}} = 0.976 \text{ J/mol}$$

To determine the quantity of heat energy required to raise the temperature of the water, we must return to Chapter 7 and use expression (7.6).

$$\text{no. J} = 225 \text{ g water} \times \frac{4.18 \text{ J}}{\text{g water } °C} \times (99.5 - 24.5) °C = 7.05 \times 10^4 \text{ J}$$

Now we can determine the amount of radiation required. In doing so, we use a conversion factor based on the fact that 1 mol photons $\approx$ 0.976 J.

$$\text{no. mol radiation} = 7.05 \times 10^4 \text{ J} \times \frac{1 \text{ mol photons}}{0.976 \text{ J}} = 7.22 \times 10^4 \text{ mol}$$

(This example is similar to Examples 7-3 and 8-5.)

4. Are there any electronic transitions in the hydrogen atom that could *conceivably* produce microwave radiation of the type described here?

Solution. We are being asked a *hypothetical* question. We do not have to do a detailed calculation, nor do we have to indicate just how we would go about getting this kind of radiation from a hydrogen atom, if it is theoretically possible to do so.

Look at Figure 8-17, the energy level diagram for hydrogen based on the Bohr atom. All of the energy levels shown are of the order 10^{-19} to 10^{-20} J, and this is orders of magnitude (10^4 to 10^5 times) greater than the energy per photon calculated in part 2. However, the energies of electron transitions (and hence the energies of photons of light emitted) correspond to *differences* between energy levels, and these are smaller quantities. Note also that

- for the higher numbered orbits the energy differences become progressively smaller, and
- for the orbit $n = \infty$ the energy is zero.

This means that some of the transitions among the high-numbered orbits of the hydrogen atom must correspond to microwave radiation. Microwave emission by electronic transitions in hydrogen atoms is theoretically possible. (See also Exercises 35 and 80.)

Key Terms

atomic (line) spectrum (8-3)
Aufbau process (8-13)
continuous spectrum (8-3)
effective nuclear charge, Z_{eff} (8-11)
electromagnetic radiation (8-2)
electron configuration (8-12)
frequency (8-2)
Hund's rule (8-12)
orbital (8-8, 8-9)

orbital diagram (8-13)
Pauli exclusion principle (8-13)
photoelectric effect (8-4)
photon (8-4)
principal shell (8-9)
quantum numbers (8-8, 8-9)
quantum theory (8-4)
shielding effect (8-11)

spdf notation (8-13)
standing wave (8-8)
subshell (8-9)
uncertainty principle (8-7)
valence electrons (8-13)
wave function (8-8)
wavelength (8-2)
wave mechanics (8-8)

Highlighted Expressions

Relationship among frequency, wavelength, and velocity of electromagnetic radiation (8.1)

Planck's equation relating the energy of radiation to its frequency (8.3)

Electron energy in the Bohr hydrogen atom (8.4)
De Broglie equation for the wavelength of matter waves (8.8)
The principal quantum number, n (8.11)

The orbital quantum number, l (8.12)
The magnetic quantum number, m_l (8.13)

Review Problems

1. Restate each of the following wavelengths in the unit indicated.
(a) 1875 Å = _____ nm
(b) 2.35 m = _____ cm
(c) 4.67 cm = _____ nm
(d) 376 nm = _____ m
(e) 1618 nm = _____ μm
(f) 4.57 m = _____ Å

2. What are the wavelengths, in meters, associated with radiation of the following frequencies? In what portion of the electromagnetic spectrum is each radiation found? (a) 3.4×10^{13} s^{-1}; (b) 7.1×10^{16} s^{-1}; (c) 4.82×10^5 s^{-1}.

3. What is the frequency associated with radiation of each of the following wavelengths? (a) 3.4×10^{-4} cm; (b) 6.78 μm; (c) 825 Å; (d) 246 cm.

4. In the hypothetical electromagnetic wave pictured, the distance between two representative points is noted. What is the frequency of this radiation?

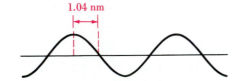

1.04 nm

Problem 4

5. Use the Balmer equation (8.2) to determine
(a) the frequency of the radiation corresponding to $n = 5$;
(b) the wavelength of the line in the Balmer series corresponding to $n = 7$;
(c) the value of n corresponding to the line in the Balmer series at 380 nm.

6. Use Planck's equation (8.3) to determine
(a) the energy, in J/photon, of radiation of frequency, $\nu = 6.75 \times 10^{15}$ s^{-1};
(b) the energy, in kJ/mol, of radiation of frequency, $\nu = 7.05 \times 10^{14}$ s^{-1};
(c) the frequency of radiation having an energy of 3.54×10^{-20} J/photon;
(d) the wavelength of radiation having an energy of 166 kJ/mol.

7. What is the wavelength of the spectral line resulting from the transition of an electron from $n = 5$ to $n = 3$ in a Bohr hydrogen atom?

8. Give a possible value for the missing quantum number(s) in each of the following sets: (a) $n = 3$, $l = 0$, $m_l = ?$; (b) $n = 3$, $l = ?$, $m_l = -1$; (c) $n = ?$, $l = 1$, $m_l = +1$; (d) $n = ?$, $l = 2$, $m_l = ?$.

9. Write appropriate values of the n, l, and m_l quantum numbers for each of the following orbital designations: (a) 4s; (b) 3p; (c) 5f; (d) 3d.

10. Indicate which of the following sets of quantum numbers is *not* allowed.
(a) $n = 3$, $l = 2$, $m_l = -1$
(b) $n = 2$, $l = 3$, $m_l = -1$
(c) $n = 4$, $l = 0$, $m_l = -1$
(d) $n = 5$, $l = 2$, $m_l = -1$
(e) $n = 3$, $l = 3$, $m_l = -3$
(f) $n = 5$, $l = 3$, $m_l = +2$

11. Arrange the following orbitals in the order in which electrons are assigned to them when writing electron configurations.

5s, 3p, 3d, 4p, 5f, 6p, 6s

12. Complete the following. Use part (a) as an example.
(a) Na ($Z = 11$) $1s^2 2s^2 2p^6 3s^1$
(b) _____ $1s^2 2s^2 2p^6 3s^2 3p^3$
(c) Zr ($Z = 40$) [Kr]$4d^{(?)} 5s^2$
(d) _____ [Kr]$4d^{(?)} 5s^2 5p^4$
(e) _____ [Ar]$3d^{(?)} 4s^{(?)} 4p^3$
(f) Bi ($Z = 83$) [Xe]$4f^{(?)} 5d^{(?)} 6s^{(?)} 6p^{(?)}$

13. Without consulting Appendix E, identify the atoms with the following electron configurations.
(a) $1s^2 2s^2 2p^1$
(b) [Ar]$3d^3 4s^2$

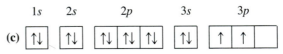

14. Without consulting Appendix E, use *spdf* notation to write the most probable electron configurations of the following atoms: (a) Al; (b) Rb; (c) Cd; (d) Sb; (e) Pb; (f) Xe

Exercises

Electromagnetic radiation

15. A certain radiation emitted by magnesium has a wavelength of 266.8 nm. Which of the following statements is (are) correct concerning this radiation?
(a) It has a higher frequency than radiation with wavelength 325 nm.
(b) It is visible to the eye.
(c) It has a greater speed in vacuum than does red light of wavelength, 7100 Å.
(d) Its wavelength is longer than that of x rays.

16. How long does it take light from the sun, 93 million miles away, to reach the earth?

17. In this chapter we learned that energy, as electromagnetic radiation, can be transmitted through vacuum or empty space. Can heat energy be similarly transferred through empty space? Explain.

Atomic spectra

18. Use equations (8.2) and (8.1) to calculate the wavelengths of the first four lines of the Balmer series of the hydrogen spectrum, starting with the *longest* wavelength component.

19. A line is detected in the hydrogen spectrum at 1880 nm. Is this line in the Balmer series? Explain.

20. What is the wavelength limit to which the Balmer series for hydrogen converges; that is, what is the *shortest* possible wavelength in the series?

21. The Lyman series of the hydrogen spectrum can be represented by the equation

$$v = 3.2881 \times 10^{15} \text{ s}^{-1} \left(\frac{1}{1^2} - \frac{1}{n^2} \right) \quad (\text{where } n = 2, 3, \ldots)$$

(a) Calculate the maximum and minimum wavelengths of lines in this series.

(b) In what portion of the electromagnetic spectrum will this series be found?

(c) What value of n corresponds to a spectral line at 95.0 nm?

★(d) Is there a line in the Lyman series at 108.5 nm? Explain.

Quantum theory

22. A certain radiation has a wavelength of 285 nm. What is the energy of (a) a single photon; (b) a mole of photons of this radiation?

23. What is the wavelength, in nm, of light that has an energy content of exactly 525 kJ/mol? In what portion of the electromagnetic spectrum is this light?

24. Figure 8-5 establishes regions of the electromagnetic spectrum in terms of frequency limits. What range of energies, in kJ/mol, corresponds to visible light?

★**25.** Infrared lamps are used in restaurants and cafeterias to keep food warm. The infrared radiation is strongly absorbed by water, raising its temperature and that of the foods in which it is incorporated. How many photons per second of infrared radiation are produced by an infrared lamp that consumes energy at the rate of 100. watts (100. J/s) and is 12% efficient in converting this energy to infrared radiation? Assume that the radiation has a wavelength of 1500. nm.

The photoelectric effect

26. The lowest frequency light that will produce the photoelectric effect on a material is called the *threshold frequency*.

(a) The threshold frequency for platinum is $1.3 \times 10^{15} \text{ s}^{-1}$. What is the energy of a quantum of this radiation?

(b) Will platinum display the photoelectric effect when exposed to ultraviolet light? infrared light? Explain.

27. In describing Einstein's quantum explanation of the photoelectric effect, Sir James Jeans made the following remark: "It not only prohibits killing two birds with one stone, but also the killing of one bird with two stones." Comment on the appropriateness of this analogy.

The Bohr atom

28. Use the description of the Bohr atom given in the text to determine (a) the radius, in nm, of the sixth Bohr orbit for hydrogen; (b) the energy of the electron when it is in this orbit (with the zero of energy being that of the free electron, that is, with orbit $n = \infty$).

29. What are (a) the frequency and (b) the wavelength of the light emitted when the electron in a hydrogen atom drops from the energy level $n = 6$ to $n = 4$? (c) In what portion of the electromagnetic spectrum is this light?

30. Which of the following electron transitions requires that the greatest quantity of energy be *absorbed* by a hydrogen atom? From (a) $n = 1$ to $n = 2$; (b) $n = 2$ to $n = 4$; (c) $n = 3$ to $n = 6$; (d) $n = \infty$ to $n = 1$. Explain.

31. What electron transition in a hydrogen atom, starting from the orbit $n = 7$, will produce infrared light of wavelength 2170 nm? [*Hint:* In what orbit must the electron end up?]

★**32.** The Bohr theory can be extended to one-electron species other than the hydrogen atom, e.g., He^+, Li^{2+}, and Be^{3+}. In these cases the energies are related to the quantum number, n, through the expression

$$E_n = \frac{-Z^2 B}{n^2}$$

where Z is the atomic number of the species and $B = 2.179 \times 10^{-18}$ J.

(a) What is the energy of the lowest level ($n = 1$) of a He^+ ion?

(b) What is the energy of the level $n = 3$ of a Li^{2+} ion?

★**33.** Both the Balmer series of the hydrogen spectrum and the energy-level diagram of Figure 8-17 feature lines that become very closely spaced in a certain region. Explain the relationship between these two phenomena.

34. Consider the following two possibilities for electron transitions in a hydrogen atom, pictured below: (1) The electron drops from the Bohr orbit $n = 3$ to the orbit $n = 2$, followed by the transition from $n = 2$ to $n = 1$. (2) The electron drops from the Bohr orbit $n = 3$ directly to the orbit $n = 1$.

(a) Show that the sum of the energies for the transitions $n = 3 \rightarrow n = 2$ and $n = 2 \rightarrow n = 1$ is equal to the energy of the transition $n = 3 \rightarrow n = 1$.

(b) Are either the wavelengths or the frequencies of the emitted photons additive in the same way as their energies? Explain.

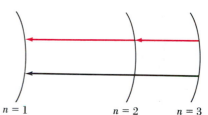

Exercise 34 $n = 1$ $n = 2$ $n = 3$

★**35.** Refer to the Summarizing Example. Assume that the microwave radiation described could be produced by an electronic transition from the Bohr orbit $(n + 1)$ to the Bohr orbit n of a hydrogen atom.

(a) Show that the value of n is 139.

(b) What would be the approximate size of this atom? [*Hint:* What are the radii of the various Bohr orbits?]

(c) How much energy would have to be absorbed to ionize the electron from the orbit $n = 139$—that is, to move the electron from this orbit to the orbit $n = \infty$?

[Hydrogen atoms meeting essentially these requirements have in fact been described. See R. F. Stebbins, "High Rydberg Atoms," *Science*, **193**, 537 (1976).]

36. Refer to the Summarizing Example (part 4). Is it possible through the appropriate electron transitions in a Bohr hydrogen atom to produce x rays? Explain. [*Hint:* Refer also to Figure 8-5.]

The uncertainty principle

37. Describe the ways in which the Bohr model of the hydrogen atom violates the Heisenberg uncertainty principle.

38. A proton is accelerated to one-tenth the velocity of light. Suppose that its velocity can be measured with a precision of ±1%. What must be the uncertainty in its position?

★39. Show that the uncertainty principle has little significance when applied to a large object like an automobile. [*Hint:* Assume that m is precisely known; assign a reasonable value to either Δx or Δv and estimate a value of the other.]

40. Although Einstein himself made some early contributions to quantum theory, he was never able to accept the Heisenberg uncertainty principle. He stated, "God does not play dice with the Universe." What do you suppose Einstein meant by this remark?

41. In reply to Einstein's remark in Exercise 40, Niels Bohr is supposed to have said, "Albert, stop telling God what to do." What do you suppose that Bohr meant by this remark?

Wave–particle duality

42. Which must possess a greater velocity to produce matter waves of the same wavelength (e.g., 1 nm), protons or electrons? Explain your reasoning.

43. What must be the velocity of a beam of electrons if they are to display a de Broglie wavelength of 1.00 nm?

44. Calculate the de Broglie wavelength associated with a 145-g baseball traveling at a speed of 95 mi/h. How does this wavelength compare with typical nuclear or atomic dimensions?

45. What is the wavelength associated with a 1000-kg automobile traveling at a speed of 25 m/s, that is, considering the automobile to be a "matter" wave? Comment on the feasibility of an experimental measurement of this wavelength.

Wave mechanics

46. A standing wave in a string 35 cm long has a *total* of six nodes (including those at the ends). What is the wavelength of this standing wave? [*Hint:* Make a sketch of the standing wave.]

47. Briefly describe the several differences between the orbits of the Bohr atom and the orbitals of the wave mechanical atom. Are there any similarities?

48. The greatest probability of finding the electron in a small volume element of the 1s orbital of the hydrogen atom is at the nucleus. Yet the most probable distance of the electron from the nucleus is 0.53 Å. How can you reconcile these two statements?

Quantum numbers and electron orbitals

49. Select the correct answer: An electron that has the quantum numbers $n = 3$ and $m_l = 2$ **(a)** must have the quantum number $m_s = +\frac{1}{2}$; **(b)** must have the quantum number $l = 1$; **(c)** may have the quantum number $l = 0, 1,$ or 2; **(d)** must have the quantum number $l = 2$.

50. With reference to Table 8-1, complete the entry for $n = 4$. The new subshell that arises is the f subshell. How many f orbitals are present in this subshell?

51. Which of the following sets of quantum numbers is (are) not allowable? Why not?
  (a) $n = 2, l = 1, m_l = 0$
  (b) $n = 2, l = 2, m_l = -1$
  (c) $n = 3, l = 0, m_l = 0$
  (d) $n = 3, l = 1, m_l = -1$
  (e) $n = 2, l = 0, m_l = -1$
  (f) $n = 2, l = 3, m_l = -2$

52. What type of electron orbital (i.e., $s, p, d,$ or f) is designated; **(a)** $n = 2, l = 1, m_l = -1$; **(b)** $n = 4, l = 2, m_l = 0$; **(c)** $n = 5, l = 0, m_l = 0$.

53. What are the n and l quantum number designations for the subshells $3s, 4p, 5f,$ and $6d$?

54. How many orbitals can there be of each of the following types? Explain. **(a)** $2s$; **(b)** $3f$; **(c)** $4p$; **(d)** $5d$.

55. Which of the following statements is (are) correct for an electron that has the quantum numbers $n = 4$ and $m_l = -2$? Explain your reasoning.
  (a) The electron is in the fourth principal shell.
  (b) The electron may be in a d orbital.
  (c) The electron may be in a p orbital.
  (d) The electron must have a spin quantum number, $m_s = +\frac{1}{2}$.

Electron configurations

56. Five electrons in an atom have the quantum numbers given below. Arrange these electrons in order of increasing energy. If any two have the same energy, so indicate.
  (a) $n = 5, l = 0, m_l = 0, m_s = +\frac{1}{2}$
  (b) $n = 3, l = 1, m_l = -1, m_s = -\frac{1}{2}$
  (c) $n = 3, l = 2, m_l = 0, m_s = +\frac{1}{2}$
  (d) $n = 3, l = 2, m_l = -2, m_s = -\frac{1}{2}$
  (e) $n = 3, l = 0, m_l = 0, m_s = -\frac{1}{2}$

57. State the basic idea(s) discussed in this chapter that is (are) violated by each of the following electron configurations, and replace each by the correct configuration. **(a)** B $1s^2 2s^3$ **(b)** Na $1s^2 2s^2 2p^6 2d^1$ **(c)** K [Ar]$3d^1$ **(d)** Ti [Ar]$4s^2 4p^2$ **(e)** Xe [Kr]$5s^2 5p^6 5d^{10}$ **(f)** Hg [Xe]$4f^{105} d^{10} 6s^2 6p^4$

58. Which of the following electron configurations is correct for phosphorus ($Z = 15$)? What is wrong with each of the others?

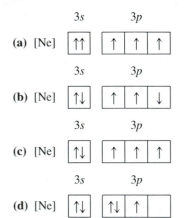

59. On the basis of rules for electron configurations, indicate the number of **(a)** unpaired electrons in an atom of Si; **(b)** $3d$ electrons in an atom of S; **(c)** $4p$ electrons in an atom of As; **(d)** $3s$ electrons in an atom of Sr; **(e)** $4f$ electrons in an atom of Au.

60. The electron configurations described in the text are all for normal atoms in their ground states. An atom may absorb a quantum of energy and promote one or more electrons to a higher energy level; it becomes an "excited" atom. The following configurations represent excited states. Indicate why this is so. **(a)** $1s^2 2s^1 2p^1$; **(b)** [Ne]$3s^2 3p^2 3d^2$; **(c)** [Ar]$3d^{10} 4s^1 4p^3$.

★61. What would be the electron configuration of the element Cs in each case?
  (a) If there were *three* possibilities for electron spin.
  (b) If the quantum number, l, could have the value, n, and if all the rules governing electron configurations were otherwise valid.

Additional Exercises

62. In what region of the electromagnetic spectrum would you expect to find radiation emitted by a hydrogen atom when the electron in the atom falls from the orbit $n = 6$ to $n = 3$?

63. A line is found in the Balmer series of hydrogen at 389.0 nm. To what value of n in equation (8.2) does this line correspond?

64. The spectrum of hydrogen consists of several different series of spectral lines. The Pfund series has as its *longest* wavelength component a line at 7400 nm. What are the electron transitions that produce this series? [*Hint:* Recall that the Balmer series involves transitions to the orbit $n = 2$ and lies in the visible region. Which is the longest wavelength component in the Balmer series?]

65. Calculate the increase in (a) distance from the nucleus and (b) energy when an electron is excited from the first to the fourth Bohr orbit. [*Hint:* What is the expression for the radii of the Bohr orbits?]

*66. Between which two orbits of the Bohr hydrogen atom must an electron fall to produce light having a wavelength of 1876 nm?

*67. Balmer seems to have deduced his formula for the visible spectrum of hydrogen just by manipulating numbers. A more common scientific procedure is to graph experimental data and then find a mathematical equation to describe the graph. Show that equation (8.2) describes a straight line. Indicate which variables must be plotted and determine numerical values of the slope and intercept of this line.

68. What is the energy, in kJ/mol, of light having a wavelength of 825 nm?

69. Determine (a) the energy of an H atom when its electron is in the orbit $n = 5$; (b) the total energy required to ionize 1 mol of *normal* H atoms.

70. Refer to the marginal note on page 254 and the discussion of standing waves. What would be the wavelengths corresponding to the fundamental frequency and the first and second overtones of a 24-in. guitar string? [*Hint:* Make sketches of these three standing waves.]

71. High-pressure sodium lamps are increasingly being used for street lighting. They are about twice as efficient as mercury vapor lamps (another type of street lamp) and several times more efficient than ordinary incandescent lamps. The two brightest lines (yellow) in the spectrum of sodium are at 589.00 nm and 589.59 nm. What is the *difference* in energy per photon of the radiation responsible for these two lines?

72. Ozone, O_3, a component of photochemical smog, absorbs ultraviolet radiation ($h\nu$) and dissociates into O_2 molecules and O atoms.

$$O_3 + h\nu \rightarrow O_2 + O$$

The O atoms are very reactive and participate in other atmospheric reactions. Assume that with radiation of 300.-nm wavelength each photon absorbed results in the dissociation of one O_3 molecule.

(a) What is the energy of this 300.-nm radiation, in J/photon?

(b) If a 1.00-L sample of air contains 0.25 parts per million (ppm) by volume of $O_3(g)$, how many photons of the 300.-nm ultraviolet light must be absorbed to dissociate all the O_3 molecules? Assume the sample is at STP.

(c) What is the total quantity of energy absorbed to produce the dissociation in part (b)?

73. Refer to Example 8-7. What would have to be the velocity of electrons if the wavelength of the radiation associated with them is to be made equal to the radius of the first Bohr orbit of the hydrogen atom?

*74. What must be the velocity of a beam of electrons if the wavelength of the radiation associated with this beam is to be equal to the *shortest* wavelength line in the Lyman series? [*Hint:* Refer to Figure 8-17.]

75. Use orbital diagrams to show the distribution of electrons among the orbitals in (a) the $4p$ subshell of Br; (b) the $3d$ subshell of Co^{2+}, given that the two electrons lost are both $4s$; (c) the $4d$ subshell of Sn.

76. The subshell that arises after f is called the g subshell (i.e., s, p, d, f, g).

(a) How many g orbitals are present in the g subshell?

(b) In what principal electronic shell would the g subshell first occur, and what is the total number of orbitals in this principal shell?

77. Complete the following assignments by writing an acceptable value for the missing quantum number. What type of orbital is described by each set?

(a) $n = ?, l = 2, m_l = 0, m_s = +\frac{1}{2}$

(b) $n = 2, l = ?, m_l = -1, m_s = -\frac{1}{2}$

(c) $n = 4, l = 2, m_l = 0, m_s = ?$

(d) $n = ?, l = 0, m_l = ?, m_s = ?$

78. Which of the following electron configurations is correct for molybdenum ($Z = 42$)? Comment on the errors in each of the others.

(a) $[Ar]3d^{10}3f^{14}$ (b) $[Kr]4d^55s^1$

(c) $[Kr]4d^55s^2$ (d) $[Ar]3d^{14}4s^24p^8$

(e) $[Ar]3d^{10}4s^24p^64d^6$

79. A light year is defined as the distance that electromagnetic radiation can travel in space in 1 year.

(a) What is this distance, in km?

(b) What is the distance, in km, to Alpha Centauri, the star closest to our solar system, if this distance is listed as 4.3 light years?

*80. An atom in which just one of the outer-shell electrons is excited to a very high quantum level (*n*) is called a "high Rydberg" atom. In some ways all these atoms resemble a Bohr hydrogen atom having its electron in a high-numbered orbit. Explain why you might expect this to be the case.

81. The *work function* of a photoelectric material is the energy that a photon of light must possess to just secure the release of an electron from the surface of a material. The corresponding frequency of the light is the threshold frequency. The higher the energy of the incident radiation, the more kinetic energy the ejected electrons have in moving away from the surface. The work function for the element mercury is equivalent to 435 kJ/mol of photons.

(a) What is the threshold frequency?

(b) What is the wavelength of light of this frequency?

(c) Can the photoelectric effect be obtained with mercury using visible light?

*(d) What is the maximum velocity of the electrons ejected when light of wavelength 215 nm strikes the mercury?

*82. Use the equation given in Exercise 32 to calculate the wavelength of the spectral line for the transition of an electron from the orbit $n = 3$ to $n = 2$ in a Li^{2+} ion.

*83. The angular momentum of an electron in the Bohr hydrogen atom is mvr, where m is the mass of the electron, v, its velocity, and r, the radius of the Bohr orbit. Bohr established that the angular momentum only can have values equal to $nh/2\pi$ (where n is an integer). Combine these ideas and other data given in the text to obtain for an electron in the third orbit ($n = 3$) of a hydrogen atom: **(a)** its velocity; **(b)** the number of revolutions about the nucleus it makes per second.

*84. Radio signals for *Voyager 1* spacecraft on its trip to Jupiter in the late 1970s were broadcast at a frequency of 8.4 giga-

hertz. On earth this radiation was received by a 64-m antenna capable of detecting signals as weak as 4×10^{-21} watt (1 watt = 1 J/s). Approximately how many photons per second did the antenna intercept from this signal?

*85. Certain metallic compounds, when heated in flames, impart characteristic colors to the flames: for example, sodium compounds, yellow; lithium, red; barium, green. "Flame tests" can be used to detect these elements.

(a) If the flame temperature is 800 °C, can collisions with other gaseous atoms or molecules possessing an average amount of kinetic energy supply the required energy to excite an atom to the point that visible light is emitted?

(b) If not, how do you account for the excitation energy?

Self-Test Questions

For questions 86 through 95 select the single item that best completes each statement.

86. The *shortest* wavelength radiation of the following is (a) 735 nm; (b) 6.3×10^{-5} cm; (c) 1.05 μm; (d) 3.5×10^{-6} m.

87. A particular electromagnetic radiation with wavelength 200 nm
(a) has a higher frequency than radiation with wavelength 400 nm;
(b) is in the visible region of the electromagnetic spectrum;
(c) has a greater speed in vacuum than does radiation of wavelength 400 nm;
(d) has a greater energy content per photon than does radiation with wavelength 100 nm.

88. The set of quantum numbers, $n = 2, l = 2, m_l = 0$ (a) describes an electron in a $2d$ orbital; (b) describes an electron in a $2p$ orbital; (c) describes one of five orbitals of a similar type; (d) is not allowed.

89. If the quantum number $n = 3$, (a) the quantum number m_l *must* be 0; (b) the quantum number, l, *cannot* be larger than $+2$; (c) the quantum number, m_s, *must* be $+\frac{1}{2}$; (d) there are *three* possible values of m_l.

90. The m_l quantum number for an electron in a $5d$ orbital (a) can have any value less than 5; (b) may be 0; (c) may be $+\frac{1}{2}$ or $-\frac{1}{2}$; (d) is 3.

91. The number of $2p$ electrons in an atom of Cl is (a) 0; (b) 2; (c) 5; (d) 6.

92. The number of unpaired electrons in an atom of scandium ($Z = 21$) is (a) 3; (b) 2; (c) 1; (d) 0.

93. Of the following electron transitions in the Bohr hydrogen atom, the one for which light of the *longest* wavelength is *emitted* is that from (a) $n = 4$ to $n = 3$; (b) $n = 1$ to $n = 2$; (c) $n = 2$ to $n = 3$; (d) $n = 2$ to $n = 1$.

94. An orbital for which the electron probability distribution is spherically symmetric about the nucleus is (a) $2p$; (b) $3s$; (c) $3p$; (d) $3d$.

95. If traveling at equal speeds, the *longest* wavelength of the following matter waves is that for a (a) electron; (b) proton; (c) neutron; (d) alpha particle (He^{2+}).

96. What is the energy content, in kJ/mol photons, of a red light with wavelength 755 nm?

97. The line at 434 nm in the Balmer series of the hydrogen spectrum corresponds to a transition of an electron from the nth to the second Bohr orbit. What is the value of n?

$$\nu = 3.2881 \times 10^{15} \ s^{-1}\left(\frac{1}{2^2} - \frac{1}{n^2}\right)$$

98. Write out the complete electron configuration of **(a)** tellurium ($Z = 52$), using *spdf* notation; **(b)** iodine ($Z = 53$), using an orbital diagram.

9

Atomic Properties and the Periodic Table

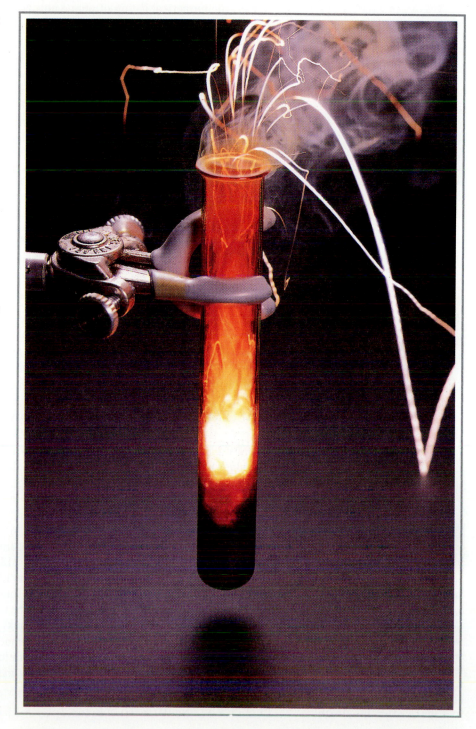

The reaction of aluminum, a metal, and bromine, a nonmetal. Some general properties of metals and nonmetals are explored in the context of the periodic table in this chapter. [Carey B. Van Loon]

Imagine the difficulty you would have in looking up a friend's number in a telephone book if the entries in the book were in a totally random fashion. You would have to inspect each entry until finally finding the one you were seeking. Now suppose that the telephone numbers were listed in numerical order, the smallest numbers first, followed by the larger numbers. Your job of finding the friend's number would be no easier than in the random phone book. It is only if the phone book is arranged alphabetically by the last name of the subscriber that you can readily find the phone number you are seeking.

The situation faced by nineteenth century chemists was similar. They had plenty of information about the behavior of elements and compounds, but they needed to find a way of organizing this information. Several false starts (analogous to the numerical listing of phone numbers) preceded the hugely successful invention of the periodic table in 1871. In this chapter we describe the periodic table and its theoretical link to electron configurations. Then we use the periodic table as a guide in establishing a set of properties associated with individual atoms—atomic radii, ionization energies, electron affinities. We will apply these properties many times in the chapters that follow. Also, in Chapters 14, 22, 23, and 24, we will use the periodic table as a basis for organizing factual information about the elements and their compounds.

9-1 Classifying the Elements: The Periodic Law and the Periodic Table

In 1869, independently and almost simultaneously, Dimitri Mendeleev and Lothar Meyer proposed the **periodic law.**

If the elements are listed in order of increasing atomic weight, certain sets of properties recur periodically in this listing.

An early example of a property conforming to the periodic law was that of **atomic volume**—the atomic weight of an element divided by its density. By recognizing that atomic weight and molar mass are numerically equal, we can reinterpret atomic volume. It is actually a **molar volume**—the volume occupied by one mole of atoms of an element.

$$\text{molar (atomic) volume (cm}^3\text{/mol)} = \text{molar mass (g/mol)} \times \frac{1}{d} \text{ (cm}^3\text{/g)} \qquad (9.1)$$

Figure 9-1 is a plot of atomic volumes as a function of atomic number. Notice how the values rise to a maximum periodically with the alkali (group 1A) metals Li, Na, K, Rb, and Cs. Several other properties, such as electrical conductivity, thermal conductivity, and hardness, show similar behavior when considered as a function of atomic number (see Exercise 17).

Mendeleev's Periodic Table. A tabular arrangement of the elements based on the periodic law is a **periodic table.** Mendeleev's 1871 table is reproduced in Table 9-1. In this table the elements are arranged in 12 horizontal rows and 8 vertical columns or groups. To bring similar elements into appropriate groups, Mendeleev left blank spaces for elements undiscovered at the time. He also made some assumptions about the correct values of atomic weights.

The elements within a particular group of Mendeleev's table have similar properties, and these properties change gradually from top to bottom in the group. As an example, the alkali metals (Group I) have low melting points that decrease in the order

Li (174 °C) > Na (97.8 °C) > K (63.7 °C) > Rb (38.9 °C) > Cs (28.5 °C)

Dimitri Mendeleev (1834–1907). Mendeleev's discovery came as a result of his attempt to systematize properties of the elements for presentation in a chemistry textbook. His highly influential book went through eight editions in his lifetime and five more after his death.

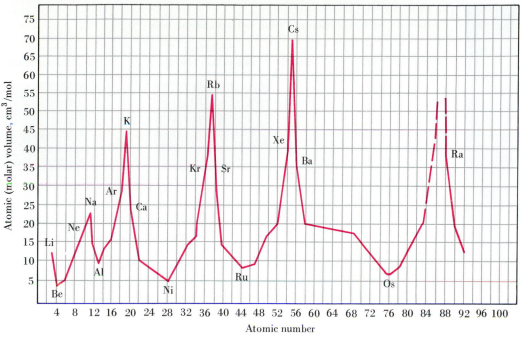

FIGURE 9-1

An illustration of the periodic law—atomic volume as a function of atomic number.

The alkali metals also have high atomic volumes (low densities) and are extremely reactive toward water, producing hydrogen gas. Both of these properties are illustrated in Figure 9-2.

At the top of Mendeleev's table are listed the formulas of the chlorides, hydrides, and oxides of the elements (R) in each group. Mendeleev was able to correlate these

TABLE 9-1

Mendeleev's Periodic Table of 1871

R	Group I	Group II	Group III	Group IV	Group V	Group VI	Group VII	Group VIII
O	R_2O	RO	R_2O_3	RO_2	R_2O_5	RO_3	R_2O_7	RO_4
W	RCl	RCl_2	RCl_3	RCl_4 RH_4	RH_3	RH_2	RH	
1	H = 1							
2	Li = 7	Be = 9.4	B = 11	C = 12	N = 14	O = 16	F = 19	
3	Na = 23	Mg = 24	Al = 27.3	Si = 28	P = 31	S = 32	Cl = 35.5	
4	K = 39	Ca = 40	___ = 44	Ti = 48	V = 51	Cr = 52	Mn = 55	Fe = 56, Co = 59, Ni = 59, Cu = 63
5	(Cu = 63)	Zn = 65	___ = 68	___ = 72	As = 75	Se = 78	Br = 80	
6	Rb = 85	Sr = 87	?Yt = 88	Zr = 90	Nb = 94	Mo = 96	___ = 100	Ru = 104, Rh = 104, Pd = 106, Ag = 108
7	(Ag = 108)	Cd = 112	In = 113	Sn = 118	Sb = 122	Te = 125	I = 127	
8	Cs = 133	Ba = 137	?Di = 138	?Ce = 140	. . .	. . .	. . .	
9	. . .	. . .	. . .	. . .	. . .	. . .	. . .	
10	. . .	. . .	?Er = 178	?La = 180	Ta = 182	W = 184	. . .	Os = 195, Ir = 197, Pt = 198, Au = 199
11	(Au = 199)	Hg = 200	Tl = 204	Pb = 207	Bi = 208	. . .	. . .	
12	. . .	. . .	. . .	Th = 231	. . .	U = 240	. . .	

FIGURE 9-2
Two properties of potassium illustrated.

From this photograph you can see that potassium has a low density (high atomic volume); it floats on water. Its reaction with water is so highly exothermic that the temperature is raised to the point that the liberated hydrogen gas ignites. [Carey B. Van Loon]

The term "eka" is derived from Sanskrit and means "first." That is, eka-silicon means, literally, first comes silicon (and then comes the unknown element).

Until 1962, no compounds of these elements were known and they were called the *inert gases*. Following the discovery of compounds of Xe and Kr, the group name was changed to *noble gases*.

TABLE 9-2
Properties of Germanium: Predicted and Observed

Property	Predicted: eka-silicon (1871)	Observed: germanium (1886)
atomic weight	72	72.6
density, g/cm^3	5.5	5.47
color	dirty gray	grayish white
density of oxide, g/cm^3	EsO$_2$: 4.7	GeO$_2$: 4.703
boiling point of chloride	EsCl$_4$: below 100 °C	GeCl$_4$: 86 °C
density of chloride, g/cm^3	EsCl$_4$: 1.9	GeCl$_4$: 1.887

formulas with the group numerals, e.g., sodium chloride has the formula NaCl; the arsenic–hydrogen compound arsine, AsH$_3$; and an oxide of molybdenum, MoO$_3$.

Correction of Atomic Weights. Mendeleev made adjustments in the previously accepted atomic weights of several elements so as to fit them properly into his table. One of these elements was indium, known to occur in zinc ores and at the time assumed to form the oxide InO (similar to that of zinc, ZnO). From the percent composition of the oxide (82.5% In) and its assumed formula, indium had been assigned an atomic weight of 76. This atomic weight would have placed indium, a metal, between arsenic and selenium, both nonmetals. Mendeleev proposed that indium formed the oxide In$_2$O$_3$. Using this formula he determined its atomic weight to be 113, and placed indium between cadmium and tin, both metals. Mendeleev also corrected the atomic weight of beryllium (from 13.5 to 9) and uranium (from 120 to 240).

Prediction of New Elements. The concept of the periodic table was readily accepted because Mendeleev was so successful in some (but not all) of his predictions. Mendeleev left blank spaces at atomic weights 44, 68, 72, and 100. The blank space at atomic weight 72 was for an element in the same group as silicon. Mendeleev called this element eka-silicon. Table 9-2 illustrates the remarkable agreement between the measured properties of germanium and its compounds and Mendeleev's predictions.

Discovery of the Noble (Inert) Gases. In 1785, Henry Cavendish, the discoverer of hydrogen gas, passed electrical discharges through air to form oxides of nitrogen (similar to what happens in lightning storms). He then dissolved these oxides in water to form nitric acid. Even though he added excess oxygen, Cavendish was unable to get all of the air to react. He suggested that air contained an unreactive gas constituting "not more than $\frac{1}{120}$ of the whole." John Rayleigh and William Ramsay isolated this gas one century later (1894) and named it argon ("the lazy one"). Ramsay (1895) also isolated helium, which had been observed in the sun's spectrum in 1868. Ramsay assigned argon and helium to a new group in the periodic table (in this text referred to as group 8A), since they did not resemble any of the existing elements. Once these first two noble gases were discovered, the periodic law predicted the existence of others. The remaining noble gases were soon isolated— neon, krypton, and xenon in 1898 and radon in 1900.

Atomic Number as the Basis for the Periodic Law. In the early periodic table it was necessary to place certain pairs of elements out of order. For example, argon (at. wt. 39.9) was placed ahead of potassium (at. wt. 39.1). If this were not done, potassium, an active metal, would appear among the inert gases and argon, an inert gas, among a group of active metals. At first, however, there was no theoretical basis for this reordering.

Moseley's brilliant career was cut short at 27 years of age—he was killed in action in Turkey in World War I.

In 1913, H. G. J. Moseley conducted experiments in which he used different elements as targets in x-ray tubes (recall Figure 2-8). Moseley found that the x-ray frequency varied with the target material, and he was able to correlate these frequencies through a mathematical equation by assigning to each element a unique integral (whole) number. These so-called atomic numbers were found to be identical to the nuclear charges described by Rutherford (page 39). The significance of Moseley's work can be measured by such successes as

- Prediction of three new elements with atomic numbers 43, 61, and 75. These elements were discovered in 1937, 1945, and 1925, respectively.
- Proof that in certain areas of the periodic table there could be no new elements, since all available atomic numbers had been assigned.
- Establishing, in a few days of experiments, atomic numbers for the lanthanide elements. It had taken 20 years to sort out these elements by chemical means and atomic weight measurements.

For our purposes, however, Moseley's most important accomplishment was in determining the true basis of the periodic law.

Similar properties recur periodically when elements are arranged by increasing atomic number.

Placing argon ($Z = 18$) before potassium ($Z = 19$) now makes perfect sense.

9-2 A Modern Periodic Table—The Long Form

Mendeleev's periodic table is called a "short" form; it consists of eight groups. Most modern periodic tables arrange the elements in 18 groups, referred to as a "long" form. Following are some features of the long form of the periodic table found on the inside front cover.

The vertical columns, which bring together elements with similar properties, are called **groups** or **families** (Figure 9-3). The horizontal rows of the table, which are arranged in order of increasing atomic number, are called **periods** (Figure 9-4). The first period of the table consists of just two elements, hydrogen and helium. This is followed by two periods of eight elements each, lithium through neon, and sodium through argon. The fourth and fifth periods comprise 18 elements each, ranging from potassium through krypton and from rubidium through xenon. The sixth period is a long one of 32 members. To fit this period into a table that is held to a maximum width of 18 members, we extract 14 members of the period and place them at the bottom of the table. This series of 14 elements follows lanthanum

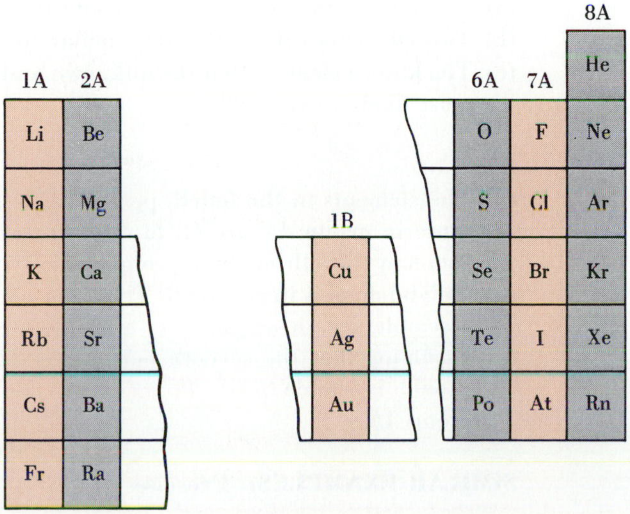

FIGURE 9-3

Groups of the periodic table.

Each group (family) is designated by an Arabic number followed by the letter A or B. The A groups are representative or main-group elements. The B groups are transition elements. Distinctive family names are

- 1A alkali metals
- 2A alkaline earth metals
- 1B coinage metals
- 6A oxygen family
- 7A halogens
- 8A noble gases

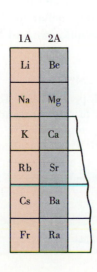

FIGURE 9-4
Periods of the periodic table.

The seven periods of the periodic table are outlined here. The lanthanide (Ce to Lu) and actinide (Th to Lr) elements belong to the sixth and seventh periods, respectively.

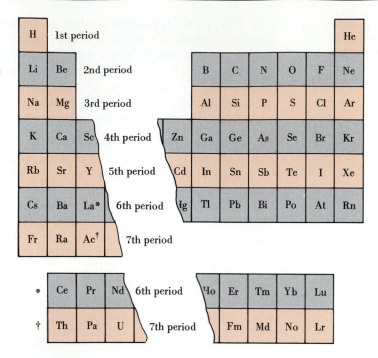

It has been proposed to use the terms *lanthanoid* (lanthanum-like) and *actinoid* (actinium-like) instead of lanthanide and actinide. The "ide" ending is generally reserved for the names of nonmetal ions (such as chloride ion). The "oid" proposal has not been widely adopted, however.

($Z = 57$) and is called the **lanthanide series.** The seventh and final period is incomplete (some members are yet to be discovered), but it is known to be a long one. A 14-member series is also extracted from the seventh period and placed at the bottom of the table. Because this series follows actinium ($Z = 89$), it is called the **actinide series.**

Different systems are currently being used for designating the vertical groups in the periodic table. Two of them are shown in the periodic table on the inside front cover and the relationship between them is described in Section 9-4. The system in this text is to use arabic numerals (1 through 8) and the letters A and B. Thus we speak of group 1A, 2A, . . . and 1B, 2B, ... Other important terms are **main-group elements** and **representative elements** for the "A" groups and **transition elements** for the "B" groups. In this classification scheme the lanthanide and actinide elements are also transition elements.

Example 9-1

Describing relationships based on the periodic table. Refer to the periodic table on the inside front cover and indicate
(a) An element that is in group 4A and the fourth period.
(b) Two elements with properties similar to molybdenum (Mo).
(c) The known element that the unknown and undiscovered element $Z = 114$ is most likely to resemble.

Solution

(a) The elements in the fourth period range from K ($Z = 19$) to Kr ($Z = 36$); those in group 4A are C, Si, Ge, Sn, and Pb. The only element that is common to both of these groupings is Ge ($Z = 32$).
(b) Molybdenum is in group 6B. The other members of this group that it should resemble are chromium (Cr) and tungsten (W).
(c) Assuming that the seventh period has 32 members, the element $Z = 114$ should resemble lead, Pb ($Z = 82$). Both Pb and element 114 are in group 4A.

SIMILAR EXAMPLES: Exercises 2, 19.

9-3 Electron Configurations and the Periodic Table

TABLE 9-3
Electron Configurations of Some Groups of Elements

Group	Element	Configuration
1A	Li	$[He]2s^1$
	Na	$[Ne]3s^1$
	K	$[Ar]4s^1$
	Rb	$[Kr]5s^1$
	Cs	$[Xe]6s^1$
	Fr	$[Rn]7s^1$
7A	F	$[He]2s^22p^5$
	Cl	$[Ne]3s^23p^5$
	Br	$[Ar]3d^{10}4s^24p^5$
	I	$[Kr]4d^{10}5s^25p^5$
	At	$[Xe]4f^{14}5d^{10}6s^26p^5$
8A	He	$1s^2$
	Ne	$[He]2s^22p^6$
	Ar	$[Ne]3s^23p^6$
	Kr	$[Ar]3d^{10}4s^24p^6$
	Xe	$[Kr]4d^{10}5s^25p^6$
	Rn	$[Xe]4f^{14}5d^{10}6s^26p^6$

Exceptions to the orderly filling of orbitals suggested by Figure 9-5 are found among a few of the d-block and some of the f-block elements.

To construct Table 9-3 we have extracted three typical groups of elements from the periodic table and then written their electron configurations. You should have no difficulty in identifying the similarity in electron configuration within each group. If we label the shell of highest principal quantum number (the outermost shell) as n,

- The group 1A atoms (alkali metals) have a *single* outer-shell (valence) electron in an s orbital, that is, ns^1.
- The group 7A atoms (halogens) have *seven* outer-shell (valence) electrons, in the configuration ns^2np^5.
- The group 8A atoms (noble gases), with the exception of helium which has only two electrons, have outermost shells with *eight* electrons, in the configuration ns^2np^6.

We will soon start to accumulate evidence that elements in the same group of the periodic table have similar properties because of the similarities in their electron configurations.

Although it is not correct in all details, Figure 9-5 relates the filling of orbitals (the Aufbau process of Section 8-13) to the periodic table. We can divide the table into four blocks of elements, depending on which orbitals are being filled in the Aufbau process.

- **s-block.** The s orbital of highest principal quantum number (n) is filled. The s-block includes groups 1A and 2A.
- **p-block.** The p orbitals of highest quantum number (n) are filled. The p-block includes groups 3A, 4A, 5A, 6A, 7A, and 8A.
- **d-block.** The d orbitals of the electronic shell $n - 1$ (the next-to-outermost) are filled. The d-block includes groups 3B, 4B, 5B, 6B, 7B, 8B, 1B, and 2B.
- **f-block.** The f orbitals of the electronic shell $n - 2$ are filled. The f-block elements are the lanthanides and the actinides. These elements are also called **inner-transition elements.**

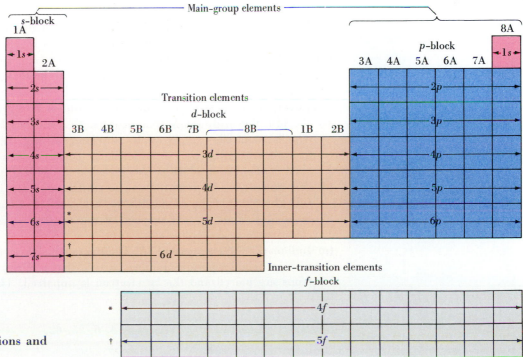

FIGURE 9-5
Electron configurations and the periodic table.

Other points to note are

- For A-group (*main-group* or *representative*) elements the group numeral (that is, *1*A, *2*A, . . .) is identical to the number of electrons in *s* and *p* orbitals of the *outermost* electronic shell (the shell with highest principal quantum number).
- For B-group (*transition*) elements the groups numerals correspond to the number of outershell electrons *only* for groups 1B and 2B. Most other B-group elements have the outer-shell electron configuration ns^2, and the group numeral (that is, *3*B, *4*B, *5*B, . . .) corresponds to the maximum number of electrons available for compound formation—the maximum oxidation state of the element in its compounds.
- The period number (*1*, *2*, . . .) is the same as the highest principal quantum number for the elements in the period, that is, the period number is the principal quantum number of the *outermost* electronic shell.

The properties of an element are determined largely by the electron configuration of the *outermost* or valence electronic shell. Adjacent members of a series of *representative* elements in the same period (such as, P, S, and Cl) have rather different properties because of differences in their outer-shell (valence) electron configurations. Within a transition series, differences in electron configurations are found mostly in *inner* shells. As a result, within a transition series there are similarities among adjacent members of the same *period* (e.g., Fe, Co, and Ni) *as well as* within the same vertical group.

Example 9-2

Relating electron configurations to the periodic table. Indicate how many (a) outer-shell electrons in an atom of bromine; (b) shells of electrons in an atom of strontium; (c) 5*p* electrons in an atom of tellurium; (d) 3*d* electrons in an atom of zirconium; (e) unpaired electrons in an atom of indium.

Solution. Determine the atomic number of each element and the location of the element in the periodic table. Then establish the significance of each location.

(a) Bromine ($Z = 35$) is a representative element in group 7A. There are *seven outer-shell electrons* in all the atoms in this group.

(b) The highest principal quantum number for atoms in the fifth period is $n = 5$. There are *five electronic shells* in the Sr atom.

(c) Tellurium ($Z = 52$) follows the *second* transition series of elements, in which the 4*d* subshell is filled. The next subshell to receive electrons after the 4*d* is the 5*p*. In the Aufbau process, Te is the fourth atom following Cd. The Te atom has *four 5p electrons*. (Alternatively, Te is in group 6A. All group 6A atoms have six outer-shell electrons, two *s* and four *p*. The outer-shell configuration of Te is $5s^2 5p^4$.)

(d) Zirconium ($Z = 40$) is in the *second* transition series, in which 4*d* orbitals fill. The filling of the 3*d* subshell occurs in the *first* transition series (from Sc to Zn). Thus, the 3*d* subshell of the Zr atom is filled. Zr has *ten 3d electrons*.

(e) Indium ($Z = 49$) is in group 3A. The In atom has three outer-shell electrons with principal quantum number, $n = 5$, that is, $5s^2 5p^1$. The two 5*s* electrons are paired and the 5*p* electron is unpaired. The In atom has *one unpaired electron*.

SIMILAR EXAMPLES: Exercises 3, 4, 6, 25, 62.

9-4   Group Designations in the Periodic Table

Starting about 1920 Niels Bohr began to promote the relationship between the periodic table and quantum theory. Many formats of the periodic table have been proposed since that time, with the main goal being to depict relationships based on electron configurations.

The format used in this text is sometimes called the "American" table: families designated with numerals 1 through 8 and the letters A for representative elements and B for transition elements. We considered its main features in Sections 9-2 and 9-3.

The International Union of Pure and Applied Chemistry (IUPAC) has recommended a new format of the periodic table. In this "IUPAC" table the groups are numbered consecutively from left to right—from 1 to 18. The letters A and B are not used. One advantage of the IUPAC system is that the groups headed by Fe, Co, and Ni have separate designations (8, 9, and 10, respectively) instead of all being lumped together as group 8B. Also, for the transition elements the group numbers correspond to the *total* of inner-shell d and outer-shell s electrons. (For example, cobalt, in group 9, has a total of nine electrons in the $3d$ and $4s$ orbitals—$3d^7 4s^2$.) This system does pose a problem for the p-block elements. Instead of representing the number of outer-shell electrons, here the group numerals are "10 + no. outer-shell electrons." Thus, oxygen, which has six outer-shell electrons ($2s^2 2p^4$) is in group 16. Adding "10" to the number of outer-shell electrons to obtain the group numbers has a certain significance for p-block elements of the fourth period and beyond. The group number represents the 10 d electrons of the next-to-outermost shell plus the s and p electrons in the outermost shell.

You are likely to encounter the two types of group designations described here (and perhaps others) in wall charts and in the chemical literature. To assist you in gaining some familiarity with alternate systems, particularly the IUPAC system, we will from time to time represent a group in two ways. For example, generally we call oxygen a group 6A element, but at times we also refer to it as a group 6A (16) element.

Roman numerals have also been used in group designations.

9-5   Metals and Nonmetals

In Section 3-4, to help us write chemical names and formulas, we divided the elements into the categories, **metal** and **nonmetal.** Figure 9-6 extends this categorization of the elements to include the noble gases and a group of elements known as metalloids. **Metalloids** have the appearance of metals but display nonmetallic properties as well. Having now considered the relationship between electron configurations and the location of elements in the periodic table, we might wonder if there is not also a connection between electron configurations and metallic/nonmetallic character. Generally speaking, there is.

We have placed the noble gases in group 8A of the periodic table. Helium has the outer-shell electron configuration $1s^2$, and the other noble gases, $ns^2 np^6$. These configurations are very stable and difficult to alter.

The electron configurations of the atoms of groups 1A and 2A—the most active metals—differ from those of a noble gas atom by only one or two electrons in the s orbital of a new electronic shell. This fact is brought out clearly when we write electron configurations in the form

Group 1A		**Group 2A**
K $[Ar]4s^1$	and	Ca $[Ar]4s^2$

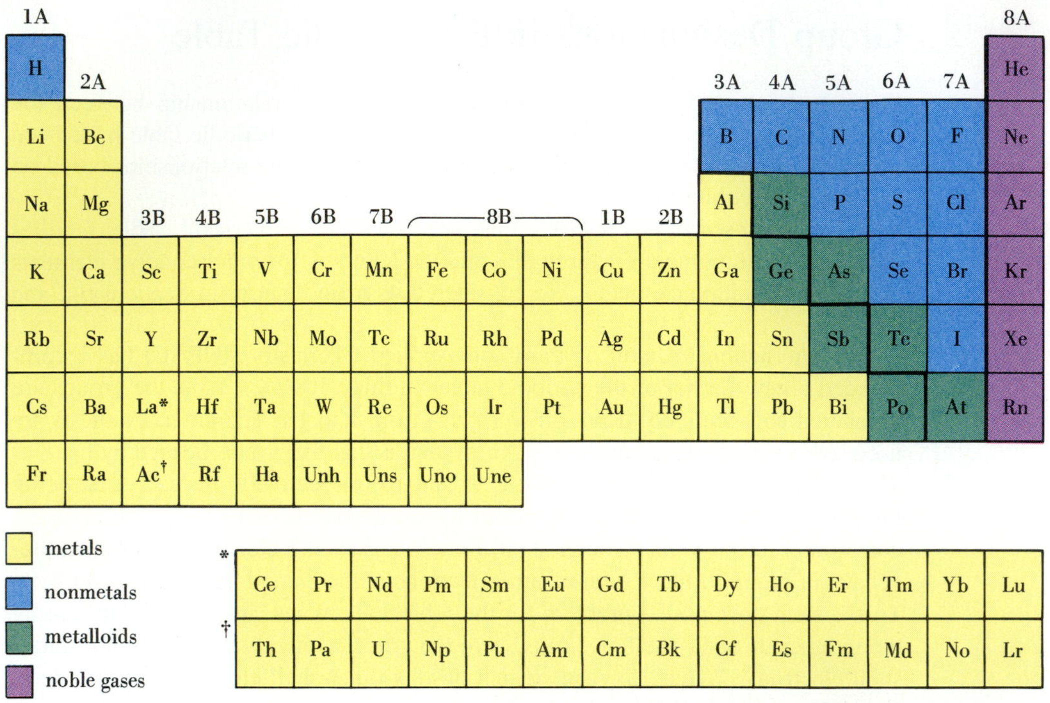

FIGURE 9-6

Metals, nonmetals, metalloids, and noble gases.

If a K atom is stripped of its outer-shell electron, it becomes the *positive ion* K^+, with the electron configuration [Ar]. A Ca atom acquires the [Ar] configuration following the removal of two electrons.

$$K \ ([Ar]4s^1) \longrightarrow K \ ([Ar]) + e^-$$

$$Ca \ ([Ar]4s^2) \longrightarrow Ca \ ([Ar]) + 2\ e^-$$

A characteristic of group 1A and 2A atoms—active metals—is the tendency to lose their outer-shell electrons to produce positive ions with the electron configurations of noble gas atoms.

The atoms of groups 7A (17) and 6A (16), the most active nonmetals, have one and two electrons fewer than the corresponding noble gas. Group 6A and 7A atoms can acquire the electron configurations of noble gas atoms by *gaining* the appropriate number of electrons. The Cl atom becomes the *negative ion* Cl^- by gaining *one* electron, and the S atom becomes S^{2-} by gaining *two* electrons.

$$Cl \ ([Ne]3s^23p^5) + e^- \longrightarrow Cl^- \ ([Ar])$$

$$S \ ([Ne]3s^23p^4) + 2\ e^- \longrightarrow S^{2-} \ ([Ar])$$

A characteristic of nonmetals is the tendency of their atoms to gain a small number of electrons (1, 2, occasionally 3) from metal atoms to acquire the electron configuration of a noble gas atom. In some situations nonmetal atoms share electrons with other nonmetal atoms rather than gaining them from metal atoms. This happens in the formation of covalent bonds, as discussed in Chapters 10 and 11.

Figure 9-6 shows that the transition elements are metals. A few transition metal atoms acquire the electron configuration of a noble gas by losing electrons (for example, the loss of three electrons by an Sc atom to form Sc^{3+}). Most transition metal atoms, however, *do not* acquire a noble gas configuration when they lose electrons. The electrons most readily lost by transition metal atoms are those of the *s* orbital of the outermost shell (the electron shell of highest principal quantum number); some *d* electrons may also be lost. Furthermore, some transition metals are able to form more than a single type of ion. Thus, an atom of iron may lose two *4s* electrons to form the ion Fe^{2+},

When writing electron configurations of transition metal ions, the outer-shell s electrons of the atoms are removed first and then d electrons, as necessary.

$$Fe\ ([Ar]3d^64s^2) \longrightarrow Fe^{2+}\ ([Ar]3d^6) + 2\ e^-$$

or it may lose the two $4s$ and one $3d$ electron to form the ion Fe^{3+}.

$$Fe\ ([Ar]3d^64s^2) \longrightarrow Fe^{3+}\ ([Ar]3d^5) + 3\ e^-$$

With information presented in this section you should be able to write electron configurations of the most commonly encountered monatomic cations and anions; these ions were listed in Table 3-2.

The stair-step diagonal line in the periodic table of Figure 9-6 separates the metals and nonmetals. Some of the elements in groups 4A and 5A are clear-cut nonmetals, some are predominantly metallic, and those adjacent to the line are metalloids, that is, they exhibit both metallic and nonmetallic properties. Metalloids are also found at the bottom of groups 6A and 7A.

In later sections of this chapter we consider some properties that can be used to assess the relative metallic and nonmetallic characters of the elements.

9-6 Additional Issues Concerning the Periodic Table

The Placement of Hydrogen. Although all the other elements have a definite place in the periodic table, there are some problems in deciding where to put hydrogen. Because it has the electron configuration $1s^1$, it would seem that hydrogen should be placed in group 1A, as we have done in this text. However, under normal conditions of temperature and pressure, hydrogen is a gas and does not at all resemble the alkali metals. Elsewhere we consider other properties in which hydrogen is markedly different from the alkali metals (such as ionization energy in Section 9-8). Sometimes hydrogen is placed in group 7A with the other elements whose outermost electronic shells contain just one electron less than a noble gas—the halogens. Still another alternative is to place hydrogen by itself at the top of the periodic table above carbon, an element it does resemble in some respects. The uniqueness of hydrogen stems from the fact that its atoms have only one electron, unlike any other atom.

There is also a small problem in the placement of He. It belongs logically with the other noble gases in group 8A; but, of course, it has only two electrons, not eight outer-shell electrons like the other noble gases.

Predicting Properties of the Heavy Elements. When we study certain elements in the sixth period (e.g., Au and Hg), we find that their properties differ from those of the preceding fifth-period element in ways that do not seem to conform with the periodic law. We will find a partial explanation of this observation in a phenomenon called the *lanthanide contraction,* which concerns the filling of the $4f$ subshell (see Section 24-1). Other evidence, however, suggests an even more fundamental basis for this behavior—Einstein's theory of relativity.

In the Schrödinger treatment of hydrogen and hydrogenlike atoms, the values of the mass and charge of the electron enter into establishing the various orbital energies. According to Einstein's theory of relativity, the mass of a particle *increases* if the particle travels at speeds approaching the speed of light. The value listed in Table 2-1 for the mass of an electron is called the **rest mass.** We are justified in neglecting relativistic effects (that is, effects based on the theory of relativity) and using the rest mass of an electron as long as it travels at moderate speeds. In atoms of high atomic number, because of the high nuclear charge, electrons are accelerated to high speeds. This is especially true for electrons that have rather high probabilities of being found near the nucleus—s electrons. As the electron mass increases, the electron is drawn closer to the nucleus, the orbital energy is lowered, and properties related to this orbital energy are affected. Since relativistic effects are found only in heavy elements, these effects may cause certain of the heavy elements to have properties that seem anomalous compared to lighter elements in the same

Naming the Elements

Until recent times the pattern for the discovery and naming of the elements was fairly simple. Usually one or a small group of scientists did the experiments leading to the discovery of a new element; other scientists then confirmed the discovery. Those making the discovery were entitled to name the new element. Some of these early names related to significant properties of the element, such as

hydrogen   (Greek, meaning ''water former'')
oxygen   (Greek, meaning ''acid former'')
chlorine   (Greek *chloros,* meaning ''greenish yellow'')

In some cases place names were used, such as the place where the discovery was made, as with

berkelium   (Berkeley, California)

Some names suggest strong patriotic feelings on the part of the discoverer, such as

gallium   (Latin name Gallia, for France)
germanium   (Latin name Germania, for Germany)

And some elements have been named after famous scientists, such as

curium, einsteinium, fermium, mendelevium, nobelium, and lawrencium.

Today, the discovery of new elements of high atomic number must be carried out by teams of researchers working with particle accelerators and producing radioactive atoms with very short half-lives (discussed in Chapter 26). Whether a new element actually is produced depends on a complex analysis of data based on only a few atoms having a fleeting existence. Inevitably controversies arise as to which group made the discovery, and, in fact, whether a discovery was made at all. To resolve these conflicts, the IUPAC has proposed that *provisional* names be used until the discovery is clearly confirmed. Then the discoverer(s) can propose a permanent name. The system for assigning these provisional names uses an ''ium'' ending and relates numbers and names as follows.

0	1	2	3	4	5	6	7	8	9
nil	un	bi	tri	quad	pent	hex	sept	oct	enn

By this scheme, for example, element 106 is ''unnilhexium'' and has the symbol Unh. However, some of the scientists who work in this field (''transactinide'' chemistry) do not care for this system of nomenclature and simply use atomic numbers in place of names.   ■ ■ ■

group. Some specific examples are presented in Chapter 22. Perhaps as more superheavy elements are synthesized and studied (see Section 26-7) we will discover even more significant departures from the periodic law.

9-7 Atomic Radius

We need to learn something about the sizes of atoms because a number of physical and chemical properties are related to atomic sizes. However, atomic size is rather hard to define. We have seen that the probability of finding an electron decreases with increasing distance from the nucleus, but nowhere does the probability equal zero. There is no precise outer boundary to an atom. Furthermore, we generally

Covalent radius:

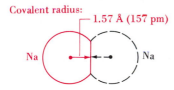

Ionic radius:

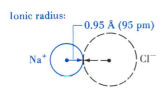

Metallic radius:

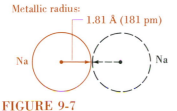

FIGURE 9-7
Covalent, ionic, and metallic radii compared.

The **covalent radius** is one-half the distance between the centers of two Na atoms in the gaseous molecule $Na_2(g)$. The **ionic radius** is based on the distance between centers of ions in an ionic compound, such as NaCl. Here, of course, the cation and anion are of different sizes. The **metallic radius** is taken as one-half the distance between the centers of adjacent atoms in solid metallic sodium.

Atomic size and position in a group of the periodic table.

observe atoms not in isolation but in contact with other atoms, either of the same or of a different kind. The size of an atom depends to a considerable extent on its environment. Despite these difficulties, let us think of the size of an atom as the distance from the nucleus to the outer-shell electron(s), and let us call this distance the *atomic radius*.

The unit that has long been used to describe atomic dimensions is the angstrom unit, Å. However, the angstrom is not a recognized SI unit. Appropriate SI units are either the nanometer (nm) or the picometer (pm).

$$1 \text{ Å} = 1 \times 10^{-10} \text{ m} = 1 \times 10^{-8} \text{ cm} = 0.10 \text{ nm} = 100 \text{ pm} \qquad (9.2)$$

Since the angstrom unit is still used by many scientists who study atomic and molecular dimensions, we use both the picometer and the angstrom in our discussion.

For *isolated* atoms the atomic radius is known as the **van der Waals radius.** For *bonded* atoms we customarily speak of a **covalent radius, ionic radius,** and, in the case of metals, a **metallic radius.** These three different radii for sodium are compared in Figure 9-7.

In Chapter 11 we picture a covalent bond as arising from the overlap of electron orbitals in the region between the centers of two atoms (see Figure 11-1). As a result of this overlap, the nuclei of bonded atoms approach each other more closely than do the nuclei of nonbonded atoms.

The covalent radius is one-half the distance between the nuclei of two identical atoms covalently bonded together.

Actually covalent radii vary with the character of the covalent bond between atoms. We learn about this in Chapter 10. Figure 9-8 compares covalent radii based on the bond type called single covalent. Throughout this text, unless statements are made to the contrary, we take the term atomic radius to mean *covalent* radius. Figure 9-8 suggests certain trends for atomic sizes—large for group 1A atoms, decreasing in size to those of the group 7A atoms, and so on. Let us briefly consider some factors that influence atomic sizes.

1. Variation of Atomic Sizes Within a Group of the Periodic Table. The distance of an electron from the nucleus of an atom depends primarily on its principal quantum number, n. The higher the principal quantum number of the outermost electronic shell, the larger we expect an atom to be. This idea certainly works well for the group members of lower atomic numbers, where the percent increase in size from one period to the next is large (as from Li to Na to K in group 1A). At higher atomic numbers the percent increase is much smaller (as from K to Rb to Cs in group 1A). In these elements of high atomic number, outer-shell electrons are held more tightly by the nucleus than would otherwise be expected. This is because inner-shell electrons in d and f subshells are less effective that s and p electrons in screening outer-shell electrons from the nucleus (recall Figure 8-28). Nevertheless, we can say that *in general*

The more electronic shells in an atom (the farther down a group of the periodic table) the larger the atom. $\qquad (9.3)$

2. Variation of Atomic Sizes Within a Period of the Periodic Table. Let us consider the hypothetical process of starting with sodium and building up, in succession, the other atoms in the third period. In this process the number of inner-shell or core electrons remains constant at 10, in the configuration $1s^2 2s^2 2p^6$.

First, *assume* that the core electrons are totally effective in shielding or screening the outer-shell electrons from the nucleus, and also *assume* that the outer-shell electrons do not screen one another. In this case, in sodium, with atomic number 11, the effective charge of the combination of the nucleus and the core electrons

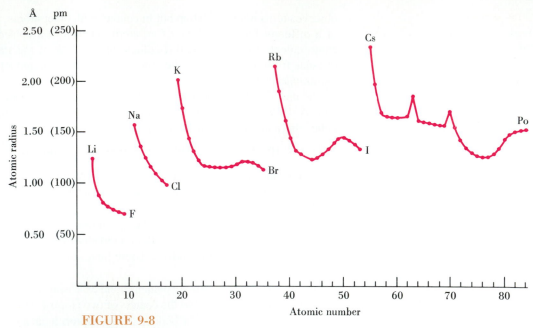

FIGURE 9-8
Covalent radii of atoms.

The values plotted here are for the bond type known as the single covalent bond. Radii are given both in angstroms and in picometers. Data for the noble gases are not shown in this graph because of the difficulty of measuring covalent radii for these atoms. (Only Kr and Xe compounds are known.) Indirect methods yield estimated values for the noble gas atoms that seem to fit the trends discussed in this section. Also, explanations have been given for the small peaks at $Z = 63$ and $Z = 70$, but these are beyond the scope of this text.

would be $+1$ (that is, $+11 - 10$). In magnesium, with atomic number 12, the effective charge would be $+2$ (that is, $+12 - 10$). In aluminum this effective charge would be $+3$, and so on. According to Coulomb's law (Appendix B-5), the force of attraction of the nucleus for outer-shell electrons increases with increased nuclear charge, and the atom becomes *smaller*. An Al atom is smaller than a Mg atom, which in turn is smaller than a Na atom.

Actually neither of the assumptions made above is correct. In sodium, the core electrons are not totally effective in shielding the outer-shell electron from the nucleus. The **effective nuclear charge, Z_{eff},** is more nearly $+2$ than $+1$. Also, outer-shell electrons do screen one another somewhat from the nucleus; they are about one-third effective in doing so. Thus, the effective nuclear charge that the two outer-shell electrons in magnesium experience is about $+2.7$; in aluminum, Z_{eff} is about 3.3. Z_{eff} increases from left to right in a period. However, whether we deal with Z_{eff} or more simply with the nuclear charge Z, the result is that

Atomic size and position in a period of the periodic table.

The atomic size decreases from left to right through a period of elements. (9.4)

3. Variation of Atomic Sizes Within a Transition Series. For the portions of the fourth and higher periods where transition elements are found the situation is a little different (see Figure 9-9). In a series of transition elements, additional electrons go into an *inner* electron shell, where they participate in shielding outer-shell electrons from the nucleus. At the same time the number of electrons in the *outer* shell tends to remain constant. Thus the outer-shell electrons experience a roughly comparable force of attraction to the nucleus throughout a transition series. Consider Fe, Co, and Ni. Fe has 26 protons in the nucleus and 24 inner-shell electrons. In Co ($Z = 27$) there are 25 inner-shell electrons, and in Ni ($Z = 28$) there are 26. In each case the two outer-shell electrons are under the influence of about the same nuclear charge (about $+2$). Thus, although there is an initial sharp decrease in atomic size

FIGURE 9-9
Variation of atomic sizes in the fourth period.

The transition elements are represented by the portion of the curve shown in blue.

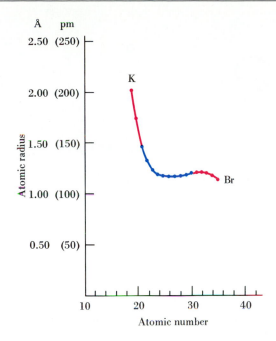

for the first two or three members of a transition series and some additional fluctuations within the series, atomic radii do not change very much within a transition series.

Example 9-3

Applying the generalizations relating atomic sizes to positions in the periodic table. With reference only to the periodic table, indicate which of the following atoms has the largest covalent radius: Sc, Ba, or Se.

Solution. Sc and Se are both in the fourth period and we should expect Sc to be larger than Se since it is much closer to the beginning of the period (to the left). Ba is in the sixth period and so has more electronic shells than either Sc or Se. Furthermore, it lies even closer to the left side of the table (group 2A) than does Sc (group 3B). We can state with confidence that the Ba atom is largest of the three.

SIMILAR EXAMPLES: Exercises 8, 31.

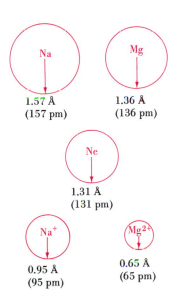

FIGURE 9-10
A comparison of atomic and ionic sizes.

The radii shown for Na and Mg are covalent radii, for Na^+ and Mg^{2+} ionic radii, and for Ne the van der Waals radius. The variations in size can be explained by the factors discussed in the text.

Ionic Radius. Electrons lost by a metal atom usually come from the outermost shell. Having one shell less, the metal atom (ion) becomes smaller. Also, because there is now an excess of nuclear charge over the number of electrons, the nucleus draws the electrons in closer. As a consequence

Sizes of cations.

Cations (positive ions) are **smaller** than the atoms from which they are formed.   (9.5)

Figure 9-10 compares five species: a Na atom, a Mg atom, a Na^+ ion, a Mg^{2+} ion, and a Ne atom. The Mg atom is smaller than the Na atom (expression 9.4), and the ions are smaller than the corresponding atoms (expression 9.5). Na^+, Mg^{2+}, and Ne are **isoelectronic**—they have equal numbers of electrons (10) in identical configurations, $1s^2 2s^2 2p^6$. Ne has a nuclear charge of $+10$. Na^+ is smaller than Ne because it has a nuclear charge of $+11$. Mg^{2+} is still smaller because its nuclear charge is $+12$.

When a nonmetal atom gains one or more electrons to form a negative ion

FIGURE 9-11
Covalent and anion radii compared.

The two Cl atoms in a Cl_2 molecule gain one electron each to form two Cl^- ions.

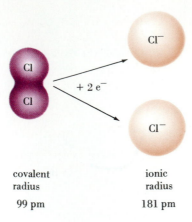

covalent radius   ionic radius

99 pm   181 pm

(anion), the nuclear charge remains constant and repulsions among the electrons increase. The electrons spread out more and the size of the atom increases, as suggested in Figure 9-11.

Sizes of anions.

> Anions (negative ions) are **larger** than the atoms from which they are formed.

(9.6)

In summary, for a series of isoelectronic ions *the greater the nuclear charge, the smaller the ion.* This generalization is illustrated in Figure 9-12, which lists the radii of a number of common cations and anions.

Example 9-4

Comparing the sizes of cations, anions, and neutral atoms. Arrange the following species in order of increasing size: Ar, K^+, Cl^-, S^{2-}, Ca^{2+}.

Solution. The key to this question lies in recognizing that the five species are *isoelectronic*, having the electron configuration of Ar, $1s^2 2s^2 2p^6 3s^2 3p^6$. The greater the nuclear charge, the more tightly these electrons are held and the smaller the size. This means that Ca^{2+} ($Z = 20$) is smaller than K^+ ($Z = 19$), which in turn is smaller than Ar ($Z = 18$). We can continue to think in terms of

Li^+ 60	Be^{2+} 31												B^{3+} 20	C –	N^{3-} 171	O^{2-} 140	F^- 136
Na^+ 95	Mg^{2+} 65												Al^{3+} 50	Si –	P^{3-} 212	S^{2-} 184	Cl^- 181
K^+ 133	Ca^{2+} 99	Sc^{3+} 81	Ti^{2+} 80	V^{2+} 88	Cr^{2+} 83	Mn^{2+} 80	Fe^{2+} 74	Co^{2+} 72	Ni^{2+} 69	Cu^{2+} 72	Zn^{2+} 74					Se^{2-} 198	Br^- 195
Rb^+ 148	Sr^{2+} 113															Te^{2-} 221	I^- 216
Cs^+ 169	Ba^{2+} 135																

FIGURE 9-12
Some representative ionic radii in picometers (pm).

Many of the elements form more than a single ion, and these different ions have different sizes. The data listed here are meant only to be representative.

this trend to conclude that Ar ($Z = 18$) is smaller than Cl^- ($Z = 17$), and in turn, Cl^- ($Z = 17$) is smaller than S^{2-} ($Z = 16$). Alternatively, for the anions we can also think in terms of the excess negative charge and the accompanying electron repulsions. The larger the excess negative charge the larger the size. This means that Cl^- is larger than Ar and S^{2-} is larger than Cl^-. Either way we are led to this order of *increasing* size

$$Ca^{2+} < K^+ < Ar < Cl^- < S^{2-}$$

SIMILAR EXAMPLES: Exercises 9, 33, 34.

9-8 Ionization Energy

There are several ways in which we can make atoms lose electrons. For example, we can bombard gaseous atoms with cathode rays. Whatever method we use, however, under normal conditions atoms do not simply release electrons. Atoms must *absorb* energy in order to ionize. The **ionization energy, I,** of an atom is the amount of energy that the *gaseous* atom must absorb so that its most loosely held electron may be stripped from the atom.

We can measure ionization energies in modified cathode ray tubes in which the atoms of interest are present as a gas at low pressure. Here are two typical values.

$$Mg(g) \longrightarrow Mg^+(g) + e^- \qquad I_1 = 738 \text{ kJ/mol} \qquad (9.7)$$

$$Mg^+(g) \longrightarrow Mg^{2+}(g) + e^- \qquad I_2 = 1451 \text{ kJ/mol} \qquad (9.8)$$

The symbol I_1 stands for the *first* ionization energy—the energy required to strip one electron from a *neutral* gaseous atom. I_2 stands for the *second* ionization energy—the energy to strip an electron from a gaseous ion with a charge of $+1$. Further ionization energies can be written as $I_3, I_4, \ldots$. Because ionization energies are small quantities, the values given in expressions (9.7) and (9.8) are based on *one mole* of atoms.* Invariably we find that each succeeding ionization energy is larger than the preceding ones. In the case of magnesium, for example, in the second ionization the ionized electron has to move away from an ion with a charge of $+2$ (Mg^{2+}). More energy must be invested than to get an ionized electron to move away from an ion with a charge of $+1$ (Mg^+). Thus, I_2 is greater than I_1.

To compare ionization energies of different elements we need to consider several factors. First, the farther the electron to be lost is from the nucleus, the more easily it can be extracted. This means that

Ionization energies and atomic sizes.

Ionization energies **decrease** as the sizes of atoms increase. (9.9)

By this measure, elements become *more metallic* (lose electrons more easily) as we trace a group of the periodic table from top to bottom. Similarly, the elements become *less metallic* as we trace a period from left to right. The relationship between ionization energy and atomic size is brought out for the group 1A metals in Table 9-4, and for all the elements in Figure 9-13. Collectively, atoms of group 1A are the largest of all the atoms, and their ionization energies are the lowest. Figure 9-13 also shows that the noble gases have the highest ionization energies. This is consistent with our previous statements about the stability of noble gas electron configurations.

TABLE 9-4
Ionization Energies of the Alkali Metal (Group 1A) Elements

	kJ/mol
Li	520.3
Na	495.9
K	418.9
Rb	403.0
Cs	375.7

*Ionization energies are also expressed in the unit **electron volt (eV)**. One electron volt is the energy acquired by an electron as it falls through an electrical potential difference of 1 volt. It is a very small energy unit, especially suited to describing processes involving individual atoms. 1 eV/atom = 96.49 kJ/mol.

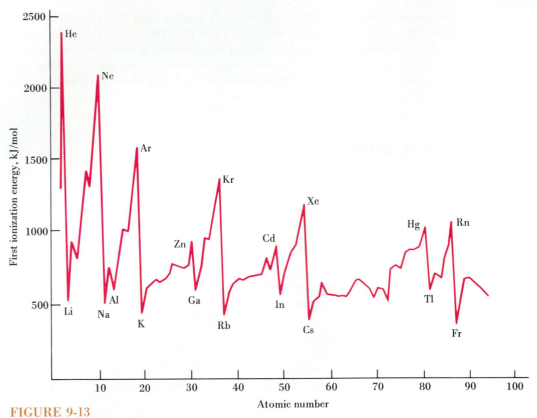

FIGURE 9-13

First ionization energies as a function of atomic number.

Table 9-5 lists ionization energies for the third-period elements. With minor exceptions the trend in moving across a period (follow the blue stripe) is that ionization energies increase from group 1A to group 8A. Table 9-5 also lists stepwise ionization energies (I_1, I_2, . . .). Note particularly the large differences that occur along the stairstep line. Consider magnesium as an example. We have already seen why I_2 is larger than I_1. To remove a *third* electron, as measured by I_3, requires breaking into the especially stable octet of electrons characteristic of a noble gas atom (ns^2np^6). I_3 is much larger than I_2, so much larger that Mg^{3+} cannot be produced in ordinary chemical processes. Similarly, we do not expect to find the ions Na^{2+} or Al^{4+}.

Are You Wondering:

Why I_1 for aluminum is smaller than I_1 for magnesium?

According to the general tendency for first ionization energies to *increase* progressively from left to right through a period, we would expect I_1 for Al to be *larger* than for Mg. The reason for the reversal lies in considering the electrons to be lost. In Mg this is an electron in a $3s$ orbital; in Al it is in a $3p$ orbital. The energy required to strip the electron from the lower energy $3s$ orbital (in Mg) is *greater* than from the $3p$ orbital (in Al). Recall the orbital energy diagrams in Figure 8-29. [I_1 for S is *slightly* lower than for P. In this case we can think of the electron pair in one of the $3p$ orbitals of an S atom as exerting a repulsive force on the unpaired electrons in the other $3p$ orbitals, leading to a slightly easier loss of the electron.]

TABLE 9-5
Ionization Energies of the Third-Row Elements (in kJ/mol)

	Na	Mg	Al	Si	P	S	Cl	Ar
I_1	495.8	737.7	577.6	786.5	1012	999.6	1251.1	1520.5
I_2	4562	1451	1817	1577	1903	2251	2297	2666
I_3		7733	2745	3232	2912	3361	3822	3931
I_4			11580	4356	4957	4564	5158	5771
I_5				16090	6274	7013	6542	7238
I_6					21270	8496	9362	8781
I_7						27110	11020	12000

Example 9-5

Relating ionization energies and atomic sizes. With reference only to the periodic table, arrange the following in order of *increasing* first ionization energy I_1: As, Sn, Br, Sr.

Solution. According to expression (9.9), if we can arrange these four atoms according to *decreasing* size we will automatically have arranged them according to *increasing* ionization energy. That is, the largest atom will have the smallest ionization energy, and the smallest atom, the largest ionization energy. The largest atoms are to the left and the bottom of the periodic table. Of the four atoms the one that best fits this category is *Sr*. The smallest atoms are to the right and toward the top of the periodic table. Although none of the four atoms is particularly close to the top of the table, *Br* is the best fit for this category. This fixes the two extremes: Sr with the lowest ionization energy and Br with the highest. In comparing As and Sn, we see that Sn is *below* and *to the left* of As in the periodic table. It should have a lower ionization energy than As. The order of *increasing* first ionization energies is **Sr < Sn < As < Br**.

SIMILAR EXAMPLES: Exercises 10, 38, 65.

9-9 Electron Affinity

Ionization energy deals with the *loss* of electrons. Electron affinity is a property that deals with the *gain* of electrons. **Electron affinity, EA,** is the amount of energy associated with the gain of an electron by a *gaseous* atom. For example,

$$F(g) + e^- \longrightarrow F^-(g) \qquad EA = -328 \text{ kJ/mol} \qquad (9.10)$$

Like ionization energy, electron affinities are small quantities on a per-atom basis and are usually expressed on a *per-mole* basis. According to thermochemical conventions a negative quantity of energy signifies that energy is *given off.** The gain of an electron by a F atom is an *exothermic* process.

Why should a neutral fluorine atom tend to gain an additional electron? When an additional electron approaches an atom from an "infinite" distance away, the electron "sees" a center of positive charge—the atomic nucleus—to which it is attracted. This attraction is offset to some extent by the repulsive effect of other electrons in the atom, but as long as the attractive force on the additional electron

---

*An alternative approach describes electron affinity as the energy associated with the process: $X^-(g) \rightarrow X(g) + e^-$. This approach leads to a reversal of the sign of the electron affinity. Also, electron affinities are sometimes expressed in eV/atom instead of kJ/mol: 1 eV/atom = 96.49 kJ/mol.

exceeds the repulsive force, the electron is gained and energy is given off. This is the case in equation (9.10). In terms of orbital diagrams

$$F \quad \boxed{\uparrow\downarrow}\ \boxed{\uparrow\downarrow}\ \boxed{\uparrow\downarrow\ |\uparrow\downarrow\ |\uparrow}\ \quad + e^- \longrightarrow$$

$$F^- \quad \boxed{\uparrow\downarrow}\ \boxed{\uparrow\downarrow}\ \boxed{\uparrow\downarrow\ |\uparrow\downarrow\ |\uparrow\downarrow}\ \quad EA = -328 \text{ kJ/mol}$$

In becoming F^-, fluorine acquires the electron configuration of a noble gas (Ne).

By the line of reasoning outlined above, we might expect that even metal atoms can form negative ions *in the gaseous state;* and this is so. The situation for gaseous Li atoms is

$$Li \quad \boxed{\uparrow\downarrow}\ \boxed{\uparrow}\ \quad + e^- \longrightarrow Li^- \quad \boxed{\uparrow\downarrow}\ \boxed{\uparrow\downarrow}\ \quad EA = -59.6 \text{ kJ/mol}$$

Of course, the tendency for a Li atom to form a negative ion is not nearly as great as for a F atom.

Are there some atoms for which the electron affinity is a *positive* quantity, that is, an *endothermic* process? This is the case for the noble gases, where the added electron must seek out the s orbital of the next electronic shell.

$$Ne \quad \boxed{\uparrow\downarrow}\ \boxed{\uparrow\downarrow}\ \boxed{\uparrow\downarrow\ |\uparrow\downarrow\ |\uparrow\downarrow}\ \quad + e^- \longrightarrow$$

$$Ne^- \quad \boxed{\uparrow\downarrow}\ \boxed{\uparrow\downarrow}\ \boxed{\uparrow\downarrow\ |\uparrow\downarrow\ |\uparrow\downarrow}\ \boxed{\uparrow}\ \quad EA = +29 \text{ kJ/mol}$$

The group 2A atoms, with their filled s subshells (ns^2), also have positive electron affinities.

Even better examples of *positive* electron affinities are those associated with gaining a *second* electron. Here the electron to be added is approaching not a neutral atom but a *negative ion*. A strong repulsion is felt and the energy of the system increases.

Some representative electron affinities are shown in Figure 9-14. From these data we see that, with some notable exceptions, the general trends are that

Electron affinities and positions in the periodic table.

> Electron affinities become smaller (*more negative*) from left to right across a period of the periodic table and become larger (*less negative*) (9.11) from top to bottom within a group.

We expect more negative electron affinities for smaller atoms (to the right of a period) because the electron to be gained can get closer to the nucleus. Similarly,

FIGURE 9-14
Some electron affinities.

Values are in kJ/mol for the process $X(g) + e^- \rightarrow X^-(g)$. Values listed as >0 are for unstable ions.

Li	Be	B	C	N	O	F	Ne
-59.6	>0	-26.7	-122	~0	-141	-328	>0
						Cl	
						-349	
						Br	
						-325	
						I	
						-295	

since the atoms at the bottom of a group are larger than at the top, we expect those at the bottom to have less negative electron affinities.

Most of the exceptions to the general trends have plausible explanations. For example, the fact that fluorine has a less negative electron affinity than chlorine seems to be due to the relatively greater effectiveness of 2p electrons in the small F atom to repel the additional electron entering the atom than do 3p electrons in the larger Cl atom.

9-10 Magnetic Properties

On several occasions we have noted how atoms may display magnetic properties (recall, for example, the Stern–Gerlach experiment in Section 8-10). One property, **diamagnetism,** is that displayed by species (atoms or ions) in which all electrons are *paired*. They are repelled by a magnetic field. Atoms or ions having one or more *unpaired* electrons are attracted into a magnetic field, a property called **paramagnetism.**

We can liken a spinning electron to a tiny electric current. As we learned in Section 8-1, an electric current induces a magnetic field around it (recall the electromagnet shown in Figure 8-1). In a species (atom or ion) in which all the electrons are paired, these individual magnetic effects cancel out. Such a species is slightly repelled by a magnetic field—it is *diamagnetic*. If a species has *unpaired* electrons the individual magnetic effects do not cancel out. The attractive force of a magnetic field for the unpaired electron(s) far outweighs the repulsion of paired electrons, and the species is attracted into a magnetic field—it is *paramagnetic*. The more unpaired electrons present the more strongly the species is attracted into the field.

A straightforward way to measure the magnetic properties of a substance is to weigh the substance "in" and "out" of a magnetic field, as illustrated in Figure 9-15. If the substance is diamagnetic, it weighs less in the magnetic field; if it is paramagnetic it weighs more.

The attraction of iron objects into a magnetic field is far stronger than can be accounted for by unpaired electrons alone. The special magnetic property possessed by iron and a few other metals and alloys, called *ferromagnetism,* is discussed in Section 24-7.

FIGURE 9-15
Paramagnetism illustrated.

(a) A sample is weighed in the absence of a magnetic field.
(b) When the field is turned on, the balanced condition is upset. The sample gains weight because it is attracted into the magnetic field.

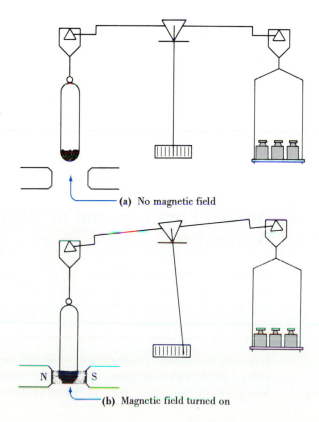

(a) No magnetic field

N   S

(b) Magnetic field turned on

In many cases we can nicely correlate measured magnetic properties with electron configurations. From its electron configuration we would predict that iron is paramagnetic (four unpaired electrons).

Fe [Ar] 3d   4s

When an iron atom loses *two* electrons it becomes the ion Fe^{2+}. Fe^{2+} is paramagnetic, and its paramagnetism corresponds to *four* unpaired electrons. This observation is consistent with the idea that the electrons lost are the two 4s electrons.

Fe^{2+} [Ar] 3d   4s

When an additional electron is lost to produce Fe^{3+}, we find that the ion has a paramagnetism corresponding to *five* unpaired electrons. This fact establishes that the additional electron lost is one of the pair of electrons in a 3d orbital.

Fe^{3+} [Ar] 3d   4s

Example 9-6

Determining the magnetic properties of an atom or ion. Which of the following species would you expect to be diamagnetic and which paramagnetic? **(a)** a Na atom; **(b)** a Mg atom; **(c)** a Cl^- ion; **(d)** a Ca^{2+} ion; **(e)** an Ag atom.

Solution

(a) Paramagnetic. The Na atom has a single 3s electron outside the Ne core. This electron is unpaired.

(b) Diamagnetic. The Mg atom has *two* 3s electrons outside the Ne core. They must be paired, as are all the other electrons.

(c) Diamagnetic. The Ar atom has all electrons paired ($1s^2 2s^2 2p^6 3s^2 3p^6$) and Cl^- is isoelectronic with Ar.

(d) Diamagnetic. Ca^{2+} is also isoelectronic with Ar.

(e) Paramagnetic. Even without considering the exact electron configuration of Ag, because the atom has 47 electrons—an odd number—at least one of the electrons must be unpaired.

SIMILAR EXAMPLES: Exercises 15, 48, 49, 67.

9-11   Atomic Properties and the Periodic Table: A Summary

In the several preceding sections we have investigated some atomic properties (e.g., atomic size, ionization energy) and described how they vary among the groups and periods of the periodic table. These variations are summarized in Figure 9-16.

Example 9-7

Relating the atomic properties of elements to their positions in the periodic table. Of the following pairs, which would you expect to have **(a)** the *lower* first ionization energy, Sr or Br? **(b)** the more *nonmetallic* behavior, Ga or P? **(c)**

FIGURE 9-16
Atomic properties and the periodic table—a summary.

- Atomic radius refers to covalent radius.
- Ionization energy refers to the first ionization energy.
- Direction of increasing electron affinity is the direction in which values become *less negative*.
- Metallic character refers generally to the ease of loss of electrons.
- Nonmetallic character refers to the ease of gain of electrons.

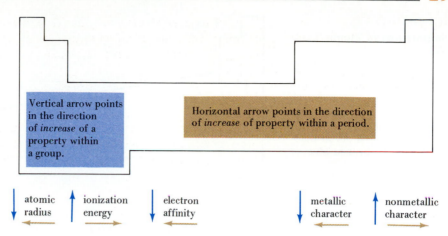

Vertical arrow points in the direction of *increase* of a property within a group.

Horizontal arrow points in the direction of *increase* of property within a period.

atomic radius   ionization energy   electron affinity   metallic character   nonmetallic character

the larger atomic radius, Mg or Sr? **(d)** the more negative electron affinity, Br or I?

Solution

(a) Figure 9-16 shows that ionization energy decreases toward the *bottom* of a group and the *left* end of a period, a position in which Sr is found relative to Br. Sr has a lower ionization energy than Br.

(b) The direction of increasing nonmetallic character is toward the *top* of a group and the *right* end of a period, a condition met more closely by P than by Ga. P has more nonmetallic character than Ga.

(c) Mg and Sr are both in group 2A. Atoms toward the bottom of a group (Sr) are larger than those toward the top (Mg). The atomic radius of Sr is larger than that of Mg.

(d) Electron affinities become more negative in the same directions that ionization energies increase. Within a group, more negative electron affinities are expected for the smaller atoms higher in the group. Br has a more negative electron affinity than I.

SIMILAR EXAMPLES: Exercises 13, 28, 66, 69.

Are You Wondering:

How to answer a question such as "Which has the larger atomic radius, Mg or I?"

The effect of being toward the *left* end of a period suggests that the Mg atom is larger than I. On the other hand, the fact that I is toward the *bottom* of a group and Mg toward the *top* suggests that the I atom is larger than Mg. These are two opposing effects and we do not know which one is the stronger. The issue here is that the outline in Figure 9-16 works for some qualitative comparisons of atomic properties and not others. (Figure 9-8 shows that in fact Mg has a somewhat larger atomic radius than I.)

The general pattern that you can expect when asked to make comparisons is that suggested by Figure 9-17. Comparisons are easiest when the elements are either within the same period or group or their relative positions are lower left to upper right of the periodic table. Qualitative comparisons in the direction upper left to lower right are very often not possible.

FIGURE 9-17
Comparisons of atomic properties.

In general, qualitative comparisons of atomic properties can only be made between pairs of atoms whose relative positions in the periodic table are as shown by the red arrows, not in the direction of the broken blue arrow.

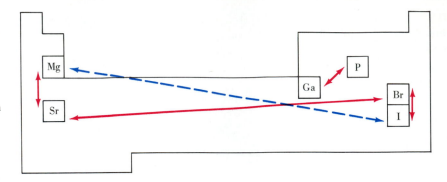

9-12 Applying The Periodic Table: The Group 7A (17) Elements

Chlorine is a yellow-green gas; bromine is a dark red liquid; and iodine is a grayish black solid. [Carey B. Van Loon]

We have seen how Mendeleev used his periodic table to make predictions about new elements. Now let us consider briefly how we can use the modern periodic table to help us describe the physical and chemical behavior of a group of elements: group 7A—the halogens. Table 9-6, which lists some properties of the group 7A elements, has a blank space, the value of the boiling point of bromine. Let us fill in this blank. To do so we use this *approximation*.

The value of a property of the elements tends to change smoothly from top to bottom in a group of the periodic table. (9.12)

Example 9-8

Using the periodic table to estimate physical properties. Estimate the boiling point of bromine.

Solution. The atomic number of bromine (35) is intermediate to those of chlorine (17) and iodine (53). Its atomic weight (79.90) is also about intermediate. [The average for Cl and I is $(35.45 + 126.90)/2 = 81.18$.] It is reasonable to expect that the boiling point of bromine, on the Kelvin scale, might also be intermediate to those of chlorine and iodine.

$$\text{b.p. } Br_2 \approx \frac{239 \text{ K} + 458 \text{ K}}{2} = 348 \text{ K}$$

(The actual boiling point is 332 K.)

SIMILAR EXAMPLES: Exercises 52, 53, 71.

TABLE 9-6
Some Properties of the Halogen (Group 7A) Elements

	Atomic number	Atomic weight	Electron configuration	Molecular form	Physical form at room temperature	Melting point, K	Boiling point, K
F	9	19.00	[He]$2s^2 2p^5$	F_2	yellow-green gas	53	85
Cl	17	35.45	[Ne]$3s^2 3p^5$	Cl_2	greenish yellow gas	172	239
Br	35	79.90	[Ar]$3d^{10} 4s^2 4p^5$	Br_2	dark red liquid	266	?
I	53	126.90	[Kr]$4d^{10} 5s^2 5p^5$	I_2	grayish black solid	387	458

TABLE 9-7
Boiling Points of the Hydrogen Halides

Hydrogen halide	Boiling point, K
HF	?
HCl	188
HBr	206
HI	238

Now consider the boiling points of the hydrogen compounds of the halogens (called the hydrogen halides). Values for HCl, HBr, and HI are listed in Table 9-7 and the value for HF is missing. If we use the method of the preceding example to estimate this missing boiling point, we conclude that it should be below the boiling points of all the other hydrogen halides, that is, below 188 K. Actually, the boiling point of HF is 293 K, *higher* than that of the other hydrogen halides. This "anomalous" behavior of HF is based on a phenomenon discussed in Section 12-7. The points to be learned from this example are that

- the generalization in (9.12) does not always apply; and
- where anomalous behavior seems to occur we can generally find an explanation, but it may require something more than the periodic law.

Writing Chemical Equations. The halogens are active nonmetals. They have high ionization energies (they lose electrons with difficulty) and very negative electron affinities (they gain electrons with relative ease). A typical reaction of nonmetals is the reaction with metals to form ionic compounds (also called salts). The name halogen means "salt former." To represent the reactions between a metal and a nonmetal we can write a

Specific equation: $2 \text{ Na(s)} + \text{Cl}_2(g) \longrightarrow 2 \text{ NaCl(s)}$ (9.13)

or a

General equations such as (9.14) are especially effective in summarizing periodic properties.

General equation: $2 \text{ M(s)} + \text{X}_2(g) \longrightarrow 2 \text{ MX(s)}$ (9.14)

[where M is an alkali (group 1A) metal and X_2, a halogen].

In the opening photograph in Chapter 4, we saw just how vigorous the reaction of sodium and chlorine is. We should expect similar reactions between the halogens and many other metals, as with the aluminum–bromine reaction in the opening photograph of this chapter.

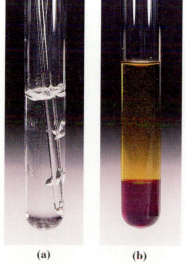

(a)   **(b)**

FIGURE 9-18
Displacement of I^- by Cl_2.

(a) $Cl_2(g)$ is bubbled through a colorless, dilute solution containing I^-.
(b) I_2 produced in the aqueous layer is extracted into $CCl_4(l)$, in which it is more soluble (purple layer). [Carey B. Van Loon]

TABLE 9-8
Selected Reactions of the Halogen Elements

Reaction with	Reaction equation	Remarks
alkali metals	$2 \text{ M} + \text{X}_2 \longrightarrow 2 \text{ MX}$	
alkaline earth metals	$\text{M} + \text{X}_2 \longrightarrow \text{MX}_2$	
other metals (e.g., Fe)	$2 \text{ Fe} + 3 \text{ X}_2 \longrightarrow 2 \text{ FeX}_3$	
hydrogen	$\text{H}_2 + \text{X}_2 \longrightarrow 2 \text{ HX}$	extremely rapid with F_2; rapid with Cl_2 when exposed to light; slow with Br_2 and I_2
sulfur	$4 \text{ X}_2 + \text{S}_8 \longrightarrow 4 \text{ S}_2\text{X}_2$	with Cl_2 and Br_2; F_2 forms SF_6
phosphorus	$6 \text{ X}_2 + \text{P}_4 \longrightarrow 4 \text{ PX}_3$	PX_5 also forms (except with I_2)
other halogens	$\text{I}_2 + \text{Cl}_2 \longrightarrow 2 \text{ ICl}$	other known interhalogen compounds include ClF, ClF$_3$, BrCl, BrF, BrF$_3$, BrF$_5$, IBr, ICl$_3$, IF$_5$, IF$_7$
water	(1) $2 \text{ X}_2 + 2 \text{ H}_2\text{O} \longrightarrow 4 \text{ H}^+ + 4 \text{ X}^- + \text{O}_2(g)$	reverse reaction with iodine
	(2) $\text{X}_2 + \text{H}_2\text{O} \longrightarrow \text{H}^+ + \text{X}^- + \text{XOH}$	with F_2, reaction (1) only

Return to Sections 5-5 and 5-7 for a review of these terms, if necessary.

Now we reintroduce a term that we have not used for a while—*oxidizing agent*. Since an oxidation process involves a loss of electrons, an oxidizing *agent* makes it possible for the oxidation to occur by *gaining* electrons. (The oxidizing agent is reduced.) We have seen that the halogens have a strong tendency to gain electrons, so it is reasonable to expect that they can act as oxidizing agents.

Of course, we expect Cl to have a greater tendency to gain an electron than I and therefore to be a better oxidizing agent. If we have a situation in which Cl atoms (actually Cl_2 molecules) and I^- ions are brought together, a competition occurs between Cl and I for the extra electron. Cl wins, and the following chemical reaction occurs.

$$Cl_2(aq) + 2\ I^-(aq) \longrightarrow I_2(aq) + 2\ Cl^-(aq) \tag{9.15}$$

We say that Cl_2 displaces I^- from solution (see Figure 9-18). Will Cl_2 displace Br^- from solution? Will I_2 displace Br^-? (See Exercise 55.)

You will find a selected set of general equations for reactions of the halogens in Table 9-8. We consider other aspects of the chemistry of the halogens in Chapters 14 and 23.

Summary

The *experimental* basis for the tabular arrangement of the elements known as the periodic table is the periodic law: Certain properties are found to recur periodically when the elements are listed according to increasing atomic number. Mendeleev produced the first successful periodic table, and was able to use it to correct atomic weights and predict the existence and properties of undiscovered elements.

The *theoretical* basis of the periodic table is that the properties of an element are determined by the electron configuration of its atoms. Elements in the same group of the periodic table have similar electron configurations. To bring out the relationship between electron configurations and properties of the elements, a long-form (18-column) periodic table and numbering/lettering systems are used to highlight (1) the order of filling of electron orbitals, (2) the numbers of electrons in outermost (valence) electronic shells, and (3) the distinction between *main-group* (representative) and *transition* elements. Another distinction that is embodied in the periodic table is between *metals* and *nonmetals*. Metal atoms tend to alter their electron configurations by losing a small number (1, 2, occasionally 3) of electrons. Nonmetal atoms tend to alter their electron configurations by gaining (or sometimes sharing) electrons. A small number of elements whose properties are intermediate between those of metals and nonmetals are called *metalloids*. Another small group, the *noble gases,* have stable electron configurations that are not easily altered.

To assess the extent of the metallic or nonmetallic character of an element, it is useful to measure certain properties of individual atoms. Among these atomic properties are atomic radius, ionization energy (the energy required to extract the most loosely held electron from a gaseous atom), and electron affinity (the energy associated with the gaining of an additional electron by a neutral gaseous atom). Large atomic radii and low ionization energies are associated with active metals. Small atomic radii, high ionization energies, and large negative electron affinities are associated with active nonmetals. The ways in which these properties vary with location in the periodic table are described in the chapter and serve as the basis for making comparisons among the elements. Other atomic properties useful in establishing the electron configurations of atoms and ions relate to their behavior when placed in a magnetic field. A substance in which all electrons are paired is *diamagnetic*—it is repelled by a magnetic field. A substance whose atoms or ions have one or more unpaired electrons is *paramagnetic*—it is attracted into a magnetic field.

Finally, as an example of how the periodic table can assist in describing the physical and chemical behavior of the elements, the group 7A (17) elements (halogens) are briefly considered in this chapter. In this discussion the practice of representing specific chemical reactions [e.g., $2\ Na(s) + Cl_2 \rightarrow 2\ NaCl(s)$] through more generalized equations [e.g., $2\ M(s) + X_2 \rightarrow 2\ MX(s)$] is introduced.

Summarizing Example

Francium (Z = 87) is an extremely rare radioactive element formed when actinium (Z = 89) undergoes alpha particle emission. Francium occurs in natural uranium minerals, but estimates are that there is no more than 15 g of francium in the top 1 km of the earth's crust. Few of francium's properties have been measured, but some can be inferred from its position in the periodic table.

1. What is the electron configuration of Fr, and in what group of the periodic table is francium found?

Solution. The element with $Z = 87$ is the thirty-second element after cesium ($Z = 55$). Since the sixth period is 32 members long and Cs starts the sixth period, francium should start the seventh period. Francium is in group 1A and has the electron configuration $[Rn]7s^1$.

(This example is similar to Example 9-2.)

2. Describe a few of the expected properties of francium, based on information in this chapter about other elements.

Solution. Fr is an alkali metal. It should have a relatively low density (large atomic volume) and low melting point. It should react vigorously with water, producing hydrogen gas (recall page 277), and form ionic compounds with the halogens (equation 9.14).

3. From data elsewhere in this chapter, estimate the **(a)** melting point, **(b)** density, and **(c)** covalent radius of francium.

Solution

(a) Recall the listing of melting points of the alkali metals on page 276. In going down the group, melting points decrease, but the *difference* in melting point between successive members of the group becomes progressively smaller (76 °C between Li and Na; 34 °C between Na and K; 25 °C between K and Rb; and 10 °C between Rb and Cs). Assuming a difference in melting point between Cs and Fr of about 5 °C, we would estimate a melting point of $28.5 - 5 \approx 24$ °C. (The measured value is 27 °C.)

(b) Recall Figure 9-1. The atomic volume at $Z = 87$ looks to be about 80 cm³/mol. Assuming a molar mass of 223 g/mol and rearranging equation (9.1), we can write

$$d = \frac{1 \text{ mol}}{80 \text{ cm}^3} \times \frac{223 \text{ g}}{1 \text{ mol}} \approx 2.8 \text{ g/cm}^3$$

(c) From expression (9.3) we expect the element at the bottom of a group to have the largest atoms. The covalent radius of Fr should be comparable to (probably slightly larger than) that of Cs. From Figure 9-8 we might predict about 250 pm.

(This example is similar to Examples 9-3, 9-7, and 9-8.)

Key Terms

actinides (9-2, 9-3)	ionization energy (9-8)	paramagnetism (9-10)
covalent radius (9-7)	isoelectronic (9-7)	*p*-block (9-3)
d-block (9-3)	lanthanides (9-2, 9-3)	period (9-2)
diamagnetism (9-10)	main-group elements (9-2)	periodic law (9-1)
electron affinity (9-9)	metal (9-5)	periodic table (9-1, 9-2)
family (9-2)	metallic radius (9-7)	representative elements (9-2)
f-block (9-3)	metalloid (9-5)	rest mass (9-6)
group (9-2)	noble gases (9-3)	*s*-block (9-3)
inner-transition elements (9-3)	nonmetal (9-5)	transition elements (9-2)
ionic radius (9-7)		

Highlighted Expressions

Atomic size and position in a group of the periodic table (9.3)
Atomic size and position in a period of the period table (9.4)
Sizes of cations (9.5)

Sizes of anions (9.6)
Ionization energies and atomic sizes (9.9)
Electron affinities and positions in the periodic table (9.11)

Review Problems

1. Without referring to *any* tables or listings in the text, mark the location of each of the following elements in the following blank periodic table. Use part (a) as an example. [*Hint:* Look for a clue for each element—its atomic number, the number of outer-shell electrons, the family to which it belongs.] **(a)** hydrogen; **(b)** sulfur; **(c)** aluminum; **(d)** iodine; **(e)** calcium; **(f)** argon; **(g)** potassium; **(h)** silicon.

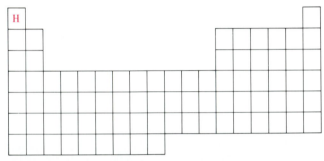

Exercise 1

2. Refer to the periodic table and identify
(a) an element that is both in group 3A and in the fifth period;
(b) an element similar to, and one unlike, sulfur;
(c) a highly reactive metal in the sixth period;
(d) the halogen element in the fifth period;
(e) an element with atomic number greater than 50 that is similar to the element with atomic number 18.

3. In what group of the periodic table are the elements with the following electron configurations found?
(a) $1s^2 2s^2 2p^6 3s^2 3p^5$ **(b)** $[Ar]3d^{10}4s^2 4p^2$
(c) $1s^2 2s^2 2p^6 3s^2$ **(d)** $[Ar]3d^{10}4s^1$
(e) $[Xe]4f^{14}5d^4 6s^2$

4. Use the periodic table as a guide to write electron configurations for **(a)** In; **(b)** Y; **(c)** Sb; **(d)** Au.

5. Write electron configurations of the following ions: **(a)** Rb^+; **(b)** Br^-; **(c)** O^{2-}; **(d)** Ba^{2+}; **(e)** Zn^{2+}; **(f)** Ag^+; **(g)** Bi^{3+}.

6. Use Figure 9-5 as a guide to indicate the number of **(a)** 4s electrons in a K atom; **(b)** 5p electrons in an I atom; **(c)** 3d electrons in an atom of Zn; **(d)** 2p electrons in an atom of S; **(e)** 4f electrons in an atom of Pb; **(f)** 3d electrons in an atom of Ni.

7. Match each of the lettered items in the column on the left with an appropriate numbered item in the column on the right. Some of the numbered items may be used more than once, and some not at all.

(a) Tl
(b) $Z = 70$
(c) Ni
(d) $[Ar]4s^2$
(e) a metalloid
(f) a nonmetal

1. an alkaline earth metal
2. element in the fifth period and group 5A
3. largest atomic radius of *all* the elements
4. an element in the fourth period and group 6A
5. $3d^8$
6. one p electron in the shell of highest n
7. a d-block element
8. an f-block element

8. For each of the following pairs, indicate the atom that has the *larger* size: **(a)** Br or As; **(b)** Sr or Mg; **(c)** Ca or Cs; **(d)** Ne or Xe; **(e)** C or O; **(f)** Hg or Cl.

9. Indicate the *smallest* and the *largest* species (atom or ion) in the following group: an Al atom, an F atom, an As atom, a Cs^+ ion, an I^- ion, an N atom.

10. Use principles established in this chapter to arrange the following atoms in order of *increasing* value of the first ionization energy: Sr, Cs, S, F, As.

11. How many joules of energy must be absorbed to convert to Li^+ all the atoms present in 1.00 mg of *gaseous* Li? The first ionization energy of Li is 520.3 kJ/mol.

12. Listed below are the electron configurations of five elements. Arrange these elements in order of *increasing* metallic character. **(a)** $[Ar]3d^{10}4s^2 4p^1$; **(b)** $[Ar]4s^2$; **(c)** $[Ar]3d^{10}4s^2 4p^6 5s^2$; **(d)** $[Ar]3d^{10}4s^2 4p^5$; **(e)** $[Ar]3d^{10}4s^2 4p^6 5s^1$

13. With reference only to the periodic table, indicate which of the atoms Bi, S, Ba, As, and Ca **(a)** is most metallic; **(b)** is most nonmetallic; **(c)** has the intermediate value when the five are arranged in order of increasing first ionization energy.

14. Arrange the following elements in order of *decreasing* metallic character: Sc; Fe; Rb; Br; O; Ca; F; Te.

15. Which of the following species would you expect to be diamagnetic and which paramagnetic: K^+; Cr^{3+}; Zn^{2+}; Cd; Co^{3+}; Sn^{2+}; Br?

Exercises

The periodic law

16. Use data from Figure 9-1 and equation (9.1) to estimate the density that will be exhibited by element 114, if and when it is discovered. Assume a mass number of 298. Why might this estimate be inaccurate?

17. The following melting points are in °C. Show that melting point is a periodic property of these elements: aluminum, 660; argon, −189; beryllium, 1278; boron, 2300; carbon, 3350; chlorine, −101; fluorine, −220; lithium, 179; magnesium, 651; neon, −249; nitrogen, −210; oxygen, −218; phosphorus, 590; silicon, 1410; sodium, 98; sulfur, 119.

The periodic table

18. There is every indication that the periodic table can be extended to elements of higher atomic numbers. What are the prospects of finding new elements within the existing periodic table at lower atomic numbers?

19. Assuming that the seventh period is 32 members long, what should be the atomic number of the noble gas following radon (Rn)? Of the alkali metal following francium (Fr)? What would you expect their approximate atomic weights to be?

20. Find the several pairs of elements that are ''out of order'' in the periodic table in terms of increasing atomic weight and explain why it is necessary to arrange them in inverse order of atomic weight.

21. Use Mendeleev's periodic table (Table 9-1) to predict formulas of the compounds listed below. Compare your results with the formulas you would have expected based on ideas introduced in Chapter 3 and explain any discrepancies.

  (a) The oxide of aluminum

  (b) The oxide of sulfur

  (c) The chloride of silicon

  (d) The chloride of phosphorus

  (e) The oxide of iron

22. Use the system of naming elements with atomic number greater than 100 outlined on page 286 to write names and three-letter symbols for the elements with atomic numbers 104 through 109.

Periodic table and electron configurations

23. Explain why the several periods in the periodic table do not all have the same number of members.

24. Sketch a periodic table that would permit inclusion of *all* the elements in the main body of the table. How many ''members'' wide would the table be? Explain.

25. Based on the relationship between electron configurations and the periodic table, give the number of **(a)** outer-shell electrons in an atom of Sb; **(b)** electrons in the fourth principal electronic shell of Pt; **(c)** elements whose atoms have six outer-shell electrons; **(d)** unpaired electrons in an atom of Te; **(e)** transition elements in the sixth period.

26. Refer to Example 9-1. There we concluded that the unknown and undiscovered element 114 would most closely resemble Pb.

  (a) Write the electron configuration of Pb.

  (b) Propose a plausible electron configuration for element 114.

27. Write probable electron configurations for the following ions: **(a)** Sr^{2+}; **(b)** Y^{3+}; **(c)** Se^{2-}; **(d)** Cu^{2+}; **(e)** Ni^{2+}; **(f)** Ga^{3+}; **(g)** Ti^{2+}.

28. Match each of the lettered items in the column on the left with an appropriate numbered item in the column on the right. All of the numbered items should be used at least once, and some must be used more than once.

  (a) $Z = 37$ 1. two unpaired *p* electrons

  (b) $Z = 9$ 2. diamagnetic

  (c) $Z = 16$ 3. more negative electron affinity

  (d) $Z = 30$ than elements on either side of it in

  (e) $Z = 82$ the same period

  (f) $Z = 12$ 4. lower first ionization energy than Ca

               but greater than Cs

29. Without referring to any tables or listings in the text, mark an appropriate location in the blank periodic table provided for each of the following: **(a)** the fifth period noble gas; **(b)** a sixth period element whose atoms have three unpaired *p* electrons; **(c)** a *d*-block element having one 4*s* electron; **(d)** a *p*-block element that is a metalloid; **(e)** a metal that forms the oxide M_2O_3.

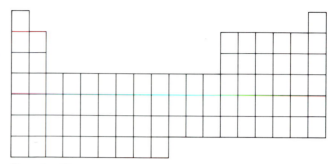

Exercise 29

Atomic sizes

30. Explain why the sizes of atoms do not simply increase uniformly with increasing atomic number.

31. Which is

  (a) the smallest atom in group 3A?

  (b) the smallest of the following atoms: Te, In, Sr, Po, Sb?

32. How would you expect the sizes of the hydrogen ion, H^+, and the hydride ion, H^-, to compare with that of the He atom? Explain.

33. The following species are *isoelectronic* with the noble gas krypton. Arrange them in order of *increasing* size and comment on the principles involved in doing so: Rb^+; Y^{3+}; Br^-; Kr; Sr^{2+}; Se^{2-}.

34. Arrange the following species in expected order of *increasing* size: Ge; Li^+; S; Br^-.

★**35.** With reference only to the periodic table, arrange the following atomic and ionic species in the expected order of *increasing* size: B; Br; Br^-; Cl; Li^+; P; Ag.

Ionization energies, electron affinities

36. Are there any atoms for which the second ionization energy (I_2) is smaller than the first (I_1)? Explain.

37. The ion Na^+ and the atom Ne are isoelectronic. The ease of loss of an electron by a gaseous Ne atom is measured by I_1 and has a value of 2081 kJ/mol. The ease of loss of an electron from a gaseous Na^+ ion is measured by I_2 for sodium and has a value of 4562 kJ/mol. Why are these values not the same?

38. Listed below are the locations of certain elements in groups and periods of the periodic table. Arrange these elements in the expected order of increasing first ionization energy.

  (a) Element in the fourth period and group 4A;

  (b) Element in the third period and group 6A;

  (c) Element in the sixth period and group 3A;

  (d) Element in the second period and group 8A;

  (e) Element in the fourth period and group 6A.

39. How much energy, in kJ, must be absorbed to ionize *completely* all the third-shell electrons in a mole of gaseous phosphorus atoms?

40. What is the maximum number of Cs^+ ions that can be produced per joule of energy absorbed by a sample of gaseous Cs atoms?

41. Use data from the energy level diagram in Figure 8-17 to determine the ionization energy of hydrogen, in kJ/mol.

42. The production of gaseous chloride ions from chlorine molecules can be considered to be a two-step process in which the first step is

$$Cl_2(g) \rightarrow 2\ Cl(g) \qquad \Delta H = +242.8\ kJ$$

Is the formation of $Cl^-(g)$ from $Cl_2(g)$ an endothermic or exothermic process?

43. Use ionization energies and electron affinities listed in the text to determine whether the following process is endothermic or exothermic.

$$Mg(g) + 2\ F(g) \rightarrow Mg^{2+}(g) + 2\ F^-(g)$$

44. From the data given in Figure 9-14 the formation of a gaseous *anion* Li^- appears energetically favorable. That is, energy is given off (59.6 kJ/mol) when gaseous Li atoms accept electrons. Comment on the likelihood of forming a stable compound containing the Li^- ion, such as Li^+Li^- or Na^+Li^-.

★45. With reference only to the periodic table, arrange the following ionization energies in order of *increasing* value. Explain the basis of any uncertainties in your arrangement. I_1 for F; I_2 for Ba; I_3 for Sc; I_2 for Na; I_3 for Mg.

46. In the electron configurations of the first ten elements, those of the noble gases He ($1s^2$) and Ne ($1s^2 2s^2 2p^6$) are especially stable. Another stable configuration is that in which both the $1s$ and $2s$ orbitals are filled ($1s^2 2s^2$). Still another is that in which the $2p$ orbitals are *half-filled*, that is, with unpaired electrons in each of the $2p$ orbitals: $1s^2 2s^2 2p^3$. Show how these facts help to explain the following order of the first ionization energies of the second-period elements: Li, 520 kJ/mol; Be, 899; B, 801; C, 1086; N, 1402; O, 1314; F, 1681; Ne, 2081. [*Hint:* That is, explain any departures from the expected constant increase of first ionization energy across the period.]

★47. Use the information from Exercise 46 to explain departures from the expected regular decrease in electron affinity in going across the second period of elements (see Figure 9-14). [*Hint:* That is, explain why certain electron affinities are 0 or >0; why the value for Li is more negative than for B; and why the value for C is nearly as negative as that for O.]

Magnetic properties

48. Unpaired electrons are found in only one of the following species. Indicate which is that one and explain why: F^-; Ca^{2+}; Fe^{2+}; S^{2-}.

49. Write electron configurations consistent with the following data on number of unpaired electrons: V^{3+}, two; Cu^{2+}, one; Cr^{3+}, three.

50. Must all atoms with an odd atomic number be paramagnetic? Must all atoms with an even atomic number be diamagnetic? Explain.

★51. The number of unpaired electrons in an atom does not show the same periodicity with atomic number as do such properties as atomic radius and ionization energy. Give a reason(s) for this difference. [*Hint:* You may find it useful to refer to the data in Appendix E.]

Predictions based on periodic relationships

52. In 1829, Dobereiner found that when certain similar elements are arranged in groups of three the atomic weight of the middle member of the group is roughly the average of the other two. Explain why Dobereiner's method works in the first two and fails in the other three cases for determining the atomic weight of **(a)** Na from those of Li and K; **(b)** Br from those of Cl and I; **(c)** Si from those of C and Ge; **(d)** Sb from those of As and Bi; **(e)** Ga from those of B and Tl.

53. Estimate the missing boiling point in the following series of compounds.

(a) CH_4, $-164\ °C$; SiH_4, $-112\ °C$; GeH_4, $-90\ °C$; SnH_4, ?.
(b) H_2O, ?; H_2S, $-61\ °C$; H_2Se, $-41\ °C$; H_2Te, $-2\ °C$.

Does your estimate of the boiling point of water, based on the data given here, agree with the known value? (An explanation is presented in Chapter 12.)

54. Gallium is currently a very important element (for fabricating gallium arsenide, a crucial material for the semiconductor industry). Gallium was unknown in Mendeleev's time. Mendeleev predicted properties of this element, which he called eka-aluminum. Predict the following for gallium: **(a)** its density; **(b)** the formula and percent composition of its oxide. [*Hint:* Use Figure 9-1, equation (9.1), and Table 9-1.]

55. Complete and balance the following equations. If no reaction occurs, so state. [*Hint:* Recall equation (9.15).]

(a) $Cl_2(aq) + Br^-(aq) \rightarrow$ **(b)** $I_2(aq) + F^-(aq) \rightarrow$
(c) $Br_2(aq) + I^-(aq) \rightarrow$

56. Complete and balance the following equations. [*Hint:* Use Table 9-8 as a guide.]

(a) $Sr + Cl_2 \rightarrow$ **(b)** $P_4 + I_2 \rightarrow$
(c) $Zn + Br_2 \rightarrow$ **(d)** $Br_2 + Cl_2 \rightarrow$

Additional Exercises

57. Use data from Figure 9-1 and equation (9.1) to estimate the density of silver, Ag.

58. The density of tellurium is $6.24\ g/cm^3$. Estimate its atomic weight using Figure 9-1 and equation (9.1).

59. Verify the statements on page 278 concerning Mendeleev's correction of the atomic weight of indium. That is, show that if indium oxide is assumed to have the formula InO the atomic weight of indium must be 76; and if In_2O_3, 113. (Recall that the oxide is 82.5% In, by mass.)

60. Studies conducted in 1880 showed that a chloride of uranium was 37.34% Cl, by mass, and had an approximate formula weight of 382. Uranium has a specific heat of 0.0276 cal g^{-1} $°C^{-1}$. Use these data, together with the law of Dulong and Petit given in Exercise 65 of Chapter 7, to calculate the atomic weight of uranium and compare it with the value assigned by Mendeleev.

61. Concerning the incomplete seventh period of the periodic table, what should be the atomic number of the element in the period **(a)** for which the filling of the $6d$ subshell is completed; **(b)** that should most closely resemble bismuth; **(c)** that you might expect to be a metalloid; **(d)** that should be a noble gas?

62. On the basis of the periodic table and rules for electron configurations, indicate the number of **(a)** $2p$ electrons in an atom of N; **(b)** $4s$ electrons in an atom of Rb; **(c)** $4d$ electrons in an atom of As; **(d)** $4f$ electrons in an atom of Au; **(e)** unpaired electrons in an atom of Pb; **(f)** elements in group 4A of the periodic table; **(g)** elements in the sixth period of the periodic table.

63. With reference to the periodic table, indicate **(a)** the most nonmetallic element; **(b)** the transition metal with lowest atomic number; **(c)** a metalloid whose atomic number places it exactly midway between two noble gas elements.

64. Refer to the periodic table and explain why
  (a) all nonmetals are representative elements, whereas among metals, some are representative and some are transition elements;
  (b) metalloids are found only among the representative elements and not among the transition elements.

65. With reference only to the periodic table, arrange the following atoms in terms of
  (a) increasing first ionization energies: O, Rb; Br; Ca; Sc; Se; F; Cs; He.
  (b) decreasing metallic character: I; O; Cs; K; Te; F; Mg; Al.

66. Listed below are two atomic properties of the element germanium. Refer to the periodic table and indicate probable values for each of the following elements, expressed as greater than, about equal to, or less than the value for Ge.

	Covalent radius	First ionization energy
(a) Ge	122 pm	762 kJ/mol
(b) Al	?	?
(c) In	?	?
(d) Se	?	?

67. Which of the following species has the greatest number of unpaired electrons: **(a)** Ge; **(b)** Cl; **(c)** Cr^{3+}; **(d)** Br^-? Explain.

68. Among the following ions, several pairs are *isoelectronic*. Identify these pairs. Fe^{2+}, Sc^{3+}, K^+, Br^-, Co^{2+}, Co^{3+}, Sr^{2+}, O^{2-}, Zn^{2+}, Al^{3+}.

69. For the following groups of elements select the one that has the property noted.
  (a) The largest atom: H; Ar; Ag; Ba; Te; Au.
  (b) The lowest first ionization energy: B; Sr; Al; Br; Mg; Pb.
  (c) The smallest (most negative) electron affinity: Na; I; Ba; Se; Cl; P.
  (d) The largest number of unpaired electrons: F; N; S^{2-}; Mg^{2+}; Sc^{3+}; Ti^{3+}.

70. Each of the following formulas represents an actual compound, but one formula is inconsistent with the others given (that is, not predictable from the others). Which one is inconsistent and what is the inconsistency? SiO_2; CO; H_2S; CCl_4; HCl; H_2O.

71. Refer to Example 9-8 and Table 9-6. In the molecule X_2, if the two X atoms are the same halogen (that is, both F, both Cl, and so on), the substance is a halogen element. If the two X atoms are different, (such as Cl and I) we describe an *interhalogen* compound. Devise an appropriate averaging method to predict the melting and boiling points of BrCl and ICl.

72. Use ideas presented in this chapter to indicate
  (a) three metals that you would expect to exhibit the photoelectric effect with visible light, and three that you would not.
  (b) the noble gas element that should have the highest density when in the liquid state;
  (c) the approximate first ionization energy of fermium (Z = 100);
  (d) the approximate density of solid radium (Z = 88).

⋆73. When sodium chloride is strongly heated in a flame, the flame takes on the yellow color associated with the emission spectrum of sodium atoms (recall Figure 8-7). The reaction that occurs in the *gaseous* state is

$$Na^+(g) + Cl^-(g) \rightarrow Na(g) + Cl(g).$$

Show that the reaction of Na^+ and Cl^- to produce Na(g) and Cl(g) is *exothermic*. [*Hint:* How are the energies involved here related to ionization energies and electron affinities?]

⋆74. Refer to Exercise 32 in Chapter 8 and calculate the *second* ionization energy (I_2) for the helium atom. Compare your result with the tabulated value of 5251 kJ/mol.

⋆75. Various values are given for the radius of a hydrogen atom, including 37, 53, 120, and 210 pm. Why do you suppose there is such variability in the values given?

⋆76. Use values of basic physical constants and other data from the appendices to establish the relationship that 1 eV/atom = 96.49 kJ/mol.

Self-Test Questions

For questions 77 through 86 select the single item that best completes each statement.

77. The element whose atoms have the electron configuration $[Kr]4d^{10}5s^25p^3$ (a) is in group 3A of the periodic table; (b) bears a similarity to the element Bi; (c) is similar to the element Te; (d) is a transition element.

78. An atom of As has (a) 5 electrons in the 4p subshell; (b) 10 electrons in the 4d subshell; (c) 6 electrons in the 3p subshell; (d) 3 electrons in the 4s subshell.

79. The fourth-period element with the largest atoms is (a) K; (b) Br; (c) Pb; (d) Kr.

80. The *largest* of the following species is (a) an Ar atom; (b) a K^+ ion; (c) a Ca^{2+} ion; (d) a Cl^- ion.

81. An example of a *d*-block element is (a) Ca; (b) Al; (c) Fe; (d) Cl.

82. The *highest* (first) ionization energy of the following elements is that of (a) Cs; (b) Cl; (c) I; (d) Li.

83. The most negative electron affinity of the following elements is that of (a) Br; (b) Sn; (c) Ba; (d) Li.

84. The *most* metallic of the following elements is (a) Mg; (b) Li; (c) K; (d) Ca.

85. The number of *unpaired* electrons in an atom of Br is (a) 0; (b) 1; (c) 2; (d) 5.

86. An ion that is *isoelectronic* with Se^{2-} is (a) S^{2-}; (b) I^-; (c) Xe; (d) Sr^{2+}.

87. For the atom $^{79}_{34}Se$ indicate the number of **(a)** protons in the nucleus; **(b)** neutrons in the nucleus; **(c)** electrons in the 3d subshell; **(d)** electrons in the 2s orbital; **(e)** 4p electrons; **(f)** electrons in the shell of highest principal quantum number.

88. Give the symbol of the element **(a)** in group 4A that has the smallest atoms; **(b)** in the fifth period that has the largest atoms; **(c)** in group 7A that has the lowest first ionization energy.

89. Refer to the periodic table and explain why
  (a) there are 2 elements in the first period, 8 in the third, 18 in the fifth; and 32 in the seventh.
  (b) Argon (Ar), with an atomic weight of 39.948, is placed ahead of potassium (K), with an atomic weight of 39.098.

10 Chemical Bonding I: Basic Concepts

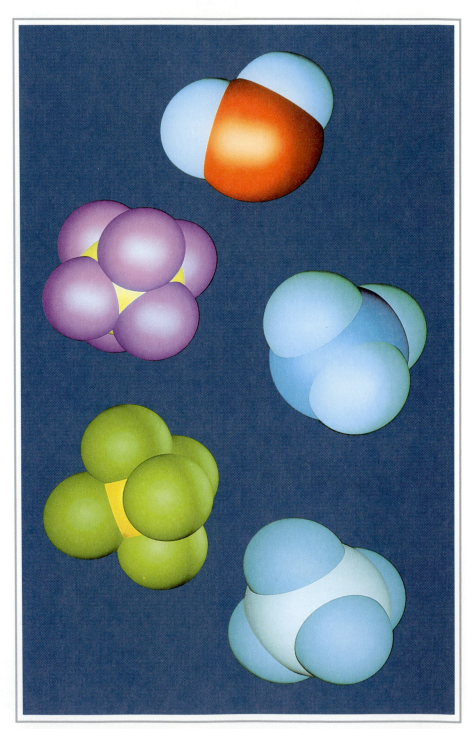

Molecular models of H_2O, NH_3, CH_4, PCl_5, and SF_6 (clockwise from top). These computer-generated images are based on quantum-mechanical descriptions of the molecules. In this chapter we use simpler approaches to bonding and structure that yield results very similar to those pictured here. [Courtesy Tripos Corporation, St. Louis, MO]

How can we prove that chlorine gas consists of Cl_2 molecules and not individual Cl atoms? We can do this rather simply by determining its molecular weight through measured properties of the gas (equation 6.12: $\mathcal{M} = mRT/PV$). Since the molecular weight is 70.9 (twice the atomic weight of Cl), we conclude that the gas exists as Cl_2. Similarly, we can deduce that the formula of sodium chloride is NaCl by determining its percentage composition (perhaps by precipitating the Cl as AgCl). In short, we can determine the composition of chemical compounds, write chemical formulas and equations, and perform stoichiometric and thermochemical calculations on chemical reactions without ever having to consider fundamental questions about the ultimate structure of matter.

Eventually, though, we run into questions that we cannot answer simply with experimental measurements on bulk samples of matter. Why is sodium chloride a solid at room temperature (m.p. = 801 °C), whereas hydrogen chloride is a gas (m.p. = −115 °C)? Why does sodium react so vigorously with chlorine gas but not at all with argon gas? We want to be able to answer questions of this type as well as to write chemical equations and do chemical calculations. It is for this reason that we need to develop some fundamental ideas about chemical bonding. In this chapter we consider the simpler aspects of chemical bonding, and in the next chapter, some concepts of a more advanced nature.

10-1 Lewis Theory—An Overview

The discovery of the noble (inert) gases around the turn of the twentieth century had an important spin-off. It stimulated thinking about **chemical bonds**—the forces that bind atoms together in elements and compounds. The noble (inert) gases were the only elements whose atoms seemed *unable* to form bonds to one another or to other atoms.* Explain why these elements are chemically inert, the thinking went, and one might then have answers as to the nature of the bonding between atoms of the other elements.

In the period 1916–1919, two Americans, G. N. Lewis and Irving Langmuir, and a German, Walther Kossel, made several important proposals about chemical bonding. Their basic idea was that something unique in the electron configurations of inert gas atoms prevents their combining with other atoms and that other atoms unite with one another so as to acquire electron configurations like the inert gas atoms. The theory that developed around this model became most closely associated with Lewis, and it is called the **Lewis theory.** In our application of the Lewis theory we will use the following ideas.

1. Electrons, especially those of the outermost (valence) electronic shell, play a fundamental role in chemical bonding.
2. In some cases chemical bonding results from the *transfer* of one or more electrons from one atom to another. This leads to the formation of positive and negative ions and a bond type called **ionic.**
3. In other cases chemical bonding involves the *sharing* of electrons between atoms. This leads to molecules having a bond type called **covalent.**
4. Electrons are transferred or shared to the extent that each atom acquires an especially stable electron configuration. Usually this configuration is that of a noble gas, that is, having *eight* outer-shell electrons, an arrangement called an **octet.**

The term *covalent* was introduced by Irving Langmuir.

*Chemists' belief in the uniqueness of noble gas electron configurations tended to discourage attempts to prepare noble gas compounds. Since 1962, a number of compounds of xenon (Xe) and krypton (Kr) have been synthesized, but this fact does not alter the usefulness of the ideas presented in this chapter. (See also, Section 23-6.)

FIGURE 10-1
Lewis symbols for the second period elements.

			Group				
1A (1)	2A (2)	3A (13)	4A (14)	5A (15)	6A (16)	7A (17)	8A (18)
Li·	Be·	·B·	·C·	·N·	:O·	:F·	:Ne:

Lewis Symbols. The Lewis symbol of an element consists of the common chemical symbol surrounded by a number of dots. The chemical symbol represents the kernel of the atom, consisting of the nucleus and *inner-shell* or *core* electrons. The dots represent the *outershell* or *valence* electrons. Thus, for silicon we can write the orbital diagram

$$\begin{array}{ccccc} 1s & 2s & 2p & 3s & 3p \end{array}$$

Si [↑↓] [↑↓] [↑↓][↑↓][↑↓] [↑↓] [↑][↑][ ]

A literal translation of this orbital diagram into a Lewis symbol produces

Si·
¨

where the pair of dots represents the pair of 3s electrons and the two lone dots represent the two unpaired 3p electrons. However, a representation that is more faithful to Lewis's original theory is

·Si·

The concept of electron spin had not been discovered at the time of Lewis's theory, and so there was no basis for thinking of electrons in the valence shell as existing in pairs. (In fact, in his original theory Lewis pictured the valence electrons as occupying the eight corners of a cubical atom.) The representation of Lewis symbols that we will find most useful is to place single dots on each side of the chemical symbol, up to a maximum of four, and then to pair up the dots until an octet is reached. This is the manner of writing Lewis symbols that is illustrated in Figure 10-1. This figure also illustrates again that for A-group atoms the number of outer-shell (valence) electrons equals the group number.

Example 10-1

Writing Lewis symbols of the elements. Refer to Figure 10-1 and write Lewis symbols for the following groups of elements: **(a)** N, P, As, Sb, Bi; **(b)** Ca, Ge, I.

Solution

(a) These are the elements of group 5A of the periodic table. Their atoms all have five valence electrons (ns^2np^3). The Lewis symbols feature five dots.

·N·   ·P·   ·As·   ·Sb·   ·Bi·

(b) Ca is in group 2A, Ge is in group 4A, and I is in group 7A.

Ca·   ·Ge·   :I:

SIMILAR EXAMPLE: Exercise 1.

Gilbert Newton Lewis (1875–1946). Lewis's contribution to the study of chemical bonding is evident throughout this text. Equally important, however, was his pioneering introduction of thermodynamics into chemistry. [E. F. Smith Memorial Collection, Center for the History of Chemistry, University of Pennsylvania]

Lewis Structures. Although Lewis's work dealt primarily with covalent bonding, we will use his ideas to depict both ionic and covalent bonds. A **Lewis structure** is a combination of Lewis symbols representing the transfer or sharing of electrons in a chemical bond.

Ionic bonding: Na× + ·Cl: ⟶ [Na]⁺[×Cl:]⁻   (10.1)

Lewis symbols   Lewis structure

$$\text{Covalent bonding:} \quad H_\times + \cdot \overset{\cdot\cdot}{\underset{\cdot\cdot}{Cl}}: \quad \longrightarrow \quad H \overset{\cdot\cdot}{\underset{\cdot\cdot}{\times}Cl}: \tag{10.2}$$

Lewis symbols   Lewis structure

In these two examples the electrons from one atom are denoted as × and from the other atom, as ·. However, it is impossible to distinguish among the electrons in bonded atoms. In all subsequent Lewis structures we will use only the dot symbol ·. Lewis structures are discussed more fully in several sections of this chapter.

10-2   Ionic Bonding—An Introduction

First, let us note some simple facts about ionic compound formation.

1. An ionic bond is formed when a *metal* atom transfers one or more electrons to a *nonmetal* atom. As a result of this transfer the metal atom becomes a positively charged ion (a cation) and the nonmetal, a negatively charged ion (anion).

2. The nonmetal atom gains a sufficient number of electrons to form an anion with the same electron configuration as a noble gas. In particular, the Lewis symbol for the anion will show an outer-shell *octet* of electrons. For some metals (e.g., those of group 1A and 2A) the loss of electrons produces a cation with a noble gas electron configuration; for others it does not. Several types of electron configurations of metal ions are listed in Table 10-1.

3. In the solid state each cation surrounds itself with anions, and each anion with cations. These very large numbers of ions are arranged in an orderly network called an **ionic crystal** (see Figure 10-2). Simple ion pairs or small clusters of ions are found only in the *gaseous* state, as in $(Na^+Cl^-)(g)$.

4. A formula unit of an ionic compound is the smallest collection of ions in an ionic crystal that would be electrically neutral (recall Figure 3-2). The formula unit is automatically obtained when the Lewis structure of an ionic compound is written.

Keep these ideas in mind and you should have no trouble in writing Lewis structures for ionic compounds. Also, note these two features of the Lewis structures in Example 10-2: (1) The Lewis symbols for the metal ions have no dots, since all the

FIGURE 10-2
Formation of an ionic crystal.

Each Na^+ ion (small sphere) is surrounded by six Cl^- ions (large spheres). In turn, each Cl^- ion is surrounded by six Na^+ ions. This structural arrangement is described further in Section 12-11.

TABLE 10-1
Electron Configurations for Some Metal Ions

"Octet"		"18"		"18 + 2"	"Various"	
Na^+	Mg^{2+}	Cu^+	Zn^{2+}	In^+	Cr^{2+}:	$[Ar]3d^4$
K^+	Ca^{2+}	Ag^+	Cd^{2+}	Tl^+	Cr^{3+}:	$[Ar]3d^3$
Rb^+	Sr^{2+}	Au^+	Hg^{2+}	Sn^{2+}	Mn^{2+}:	$[Ar]3d^5$
Cs^+	Ba^{2+}		Ga^{3+}	Pb^{2+}	Mn^{3+}:	$[Ar]3d^4$
Fr^+	Ra^{2+}		Sb^{3+}	Sb^{3+}	Fe^{2+}:	$[Ar]3d^6$
	Al^{3+}		Tl^{3+}	Bi^{3+}	Fe^{3+}:	$[Ar]3d^5$
	Sc^{3+}				Co^{2+}:	$[Ar]3d^7$
	Y^{3+}				Co^{3+}:	$[Ar]3d^6$
	La^{3+}				Ni^{2+}:	$[Ar]3d^8$
					Ni^{3+}:	$[Ar]3d^7$

The octet configuration is that of a noble gas—ns^2np^6. Li^+ and Be^{2+} also have a noble gas electron configuration but not an octet of outer-shell electrons. Theirs is the configuration of helium: $1s^2$.

In the configuration labeled "18" all outer-shell electrons of the atom are ionized. The ion produced has a new outer shell with 18 electrons in the configuration $ns^2np^6nd^{10}$. For a few post-transition elements all outer-shell electrons but the two s electrons (an "inert pair") are ionized, producing ions with the "18 + 2" configuration, $(n-1)s^2(n-1)p^6(n-1)d^{10}ns^2$.

In ion formation among the transition metals, the outershell s electron(s) and some number of d electrons are ionized, producing the configurations listed as "various."

valence electrons are lost by group 1A and 2A atoms, and (2) the Lewis structures of ionic compounds should always show the charges on the ions.

Example 10-2

Writing Lewis structures of ionic compounds. Write Lewis structures for the following ionic compounds: **(a)** BaO; **(b)** $MgCl_2$; **(c)** potassium sulfide.

Solution

(a) Write the Lewis symbol and determine the number of electrons to be lost or gained by each atom in acquiring a noble gas electron configuration. The barium atom must lose two electrons and the oxygen atom must gain two.

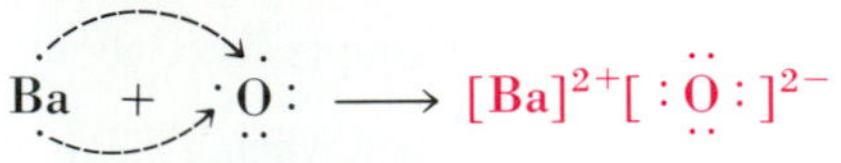

$$Ba \ + \ \ddot{O}: \ \longrightarrow \ [Ba]^{2+}[:\ddot{O}:]^{2-}$$

Lewis structure

(b) A Cl atom can accept but a single electron, whereas a Mg atom must lose two electrons. Two Cl atoms are required for each Mg atom.

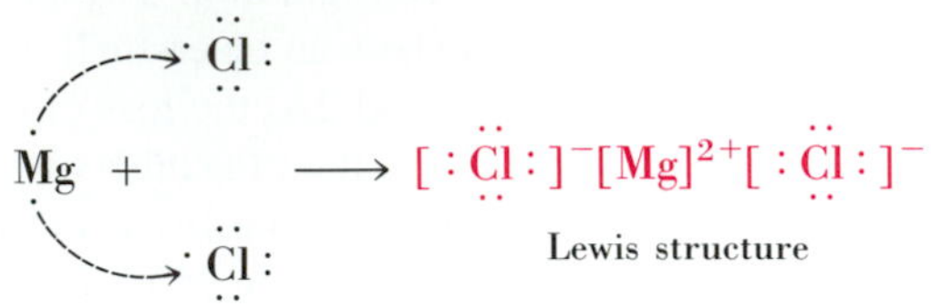

$$Mg \ + \ \begin{matrix} \cdot\ddot{Cl}: \\ \\ \cdot\ddot{Cl}: \end{matrix} \ \longrightarrow \ [:\ddot{Cl}:]^-[Mg]^{2+}[:\ddot{Cl}:]^-$$

Lewis structure

(c) We are not given the formula of potassium sulfide, but as noted earlier, the formula follows directly from the principles of writing Lewis structures.

$$\begin{matrix} K\cdot \\ \\ + \ \cdot\ddot{S}: \\ \\ K\cdot \end{matrix} \ \longrightarrow \ [K]^+[:\ddot{S}:]^{2-}[K]^+$$

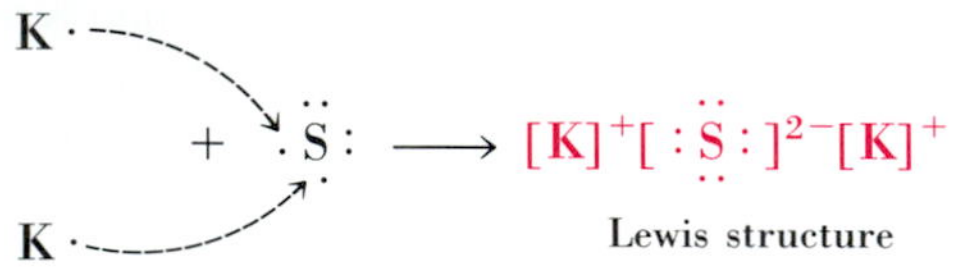

Lewis structure

SIMILAR EXAMPLES: Exercises 2, 20.

10-3 Energetics of Ionic Bond Formation

Figure 10-3 suggests how the ions Na^+ and Cl^- are formed from *isolated gaseous* atoms. First, the Na atom must *absorb* energy to release the outer-shell electron ($3s^1$). This energy requirement is $+496$ kJ per mole of Na, the first ionization energy, I_1, of sodium. Next, the Cl atom gains this electron and in the process *gives off* energy. The quantity of energy is -349 kJ per mole of Cl, the electron affinity (EA) of chlorine. Once formed, the ions, being oppositely charged, attract one another and move into contact. Energy is *given off* in this process. This quantity of energy is about -500 kJ. The *overall* process of forming the gaseous ion pair (Na^+Cl^-) from the gaseous atoms is *exothermic*, with an energy change of about -350 kJ per mole (Na^+Cl^-) [that is, $496 - 349 - 500 \approx -350$]. As a rough generalization let us think of a process accompanied by a loss of energy (evolution of heat) as being one that is likely to occur. Thus, we can account for the formation of gaseous ion pairs (Na^+Cl^-) from the gaseous atoms.

A more significant quantity of energy in describing ionic compound formation is *the energy given off when separated gaseous ions, positive and negative, come together to form 1 mol of the solid ionic compound*. This quantity of energy is called the **lattice energy.** It is possible to calculate the lattice energy of an ionic com-

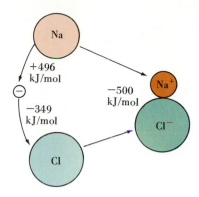

FIGURE 10-3
Energetics of ionic bond formation.

The processes shown here have already been described in part in Chapter 9. When a sodium atom loses its outer-shell electron to become a sodium ion, it decreases in size. Although the conversion of a *covalently bonded* chlorine atom to a chloride ion is accompanied by an increase in size, there is essentially no change in size when an *isolated nonbonded* chlorine atom gains an electron. The energy data refer to the loss of electrons by Na atoms (+496 kJ/mol), the gain of electrons by Cl atoms (−349 kJ/mol), and the interaction of the two ions (about −500 kJ/mol).

pound, though this calculation is rather difficult. The problem is that at the same time that oppositely charged ions are attracting each other, like charged ions are repelling one another. Rather than attempt a direct calculation of lattice energy, we take a different tack. Let us determine the lattice energy of NaCl *indirectly* by using Hess's law (recall Section 7-8).

The special application of Hess's law that we are about to undertake is called the Born–Fajans–Haber method, named after its originators Max Born, Kasimir Fajans, and Fritz Haber. In this application we will use the first ionization energy of sodium and the electron affinity of chlorine, as we did above. We will also need three other quantities:

- The *heat of formation* of 1 mol NaCl(s).
- The quantity of energy required to convert 1 mol of solid sodium to gaseous sodium, called the *heat of sublimation*.
- The quantity of energy required to dissociate ½ mol of $Cl_2(g)$ into 1 mol of gaseous Cl atoms, called the *heat of dissociation*.

The crux of the method is to consider two different ways to produce 1 mol NaCl(s): (a) Allow solid sodium and gaseous chlorine to react; the energy (enthalpy) change for this reaction is the heat (enthalpy) of formation of NaCl(s). (b) Proceed through the following five-step process, outlined in Figure 10-4. The sum of the five steps is the equation for the formation of NaCl(s) from its elements. All energy changes are shown as ΔH values and the lattice energy is denoted by the symbol U.

> Review the definition of enthalpy (heat) of formation (Section 7-9). Sublimation refers to the direct passage of a substance from the solid to the gaseous state.

1. Sublimation of solid Na:

   $$Na(s) \longrightarrow Na(g) \qquad \Delta H_1 = +108 \text{ kJ}$$

2. Ionization of gaseous Na:

   $$Na(g) \longrightarrow Na^+(g) + e^- \qquad \Delta H_2 = +496 \text{ kJ}$$

3. Dissociation of gaseous Cl_2:

   $$\tfrac{1}{2} Cl_2(g) \longrightarrow Cl(g) \qquad \Delta H_3 = +121 \text{ kJ}$$

4. Ion formation by gaseous Cl:

   $$Cl(g) + e^- \longrightarrow Cl^-(g) \qquad \Delta H_4 = -349 \text{ kJ}$$

5. Combination of gaseous ions:

   $$Na^+(g) + Cl^-(g) \longrightarrow NaCl(s) \qquad U = ?$$

   Formation of NaCl(s):

   $$Na(s) + \tfrac{1}{2} Cl_2(g) \longrightarrow NaCl(s) \qquad \Delta H_f^\circ = -411 \text{ kJ}$$

Since

$$\Delta H_f^\circ = \Delta H_1 + \Delta H_2 + \Delta H_3 + \Delta H_4 + U$$

$$U = \Delta H_f^\circ - \Delta H_1 - \Delta H_2 - \Delta H_3 - \Delta H_4 \qquad (10.3)$$

$$= (-411 - 108 - 496 - 121 + 349) \text{ kJ} = -787 \text{ kJ}$$

FIGURE 10-4

Enthalpy diagram for determining the lattice energy, U, of NaCl(s).

Two methods of preparing one mole of NaCl(s) from its elements are compared here. For the direct reaction the enthalpy change is $\Delta H_f^\circ = -411$ kJ (blue arrow). The overall enthalpy change for the five-step process is also equal to ΔH_f°. The enthalpy change for the fifth step in this process (red arrow) is what we are seeking.

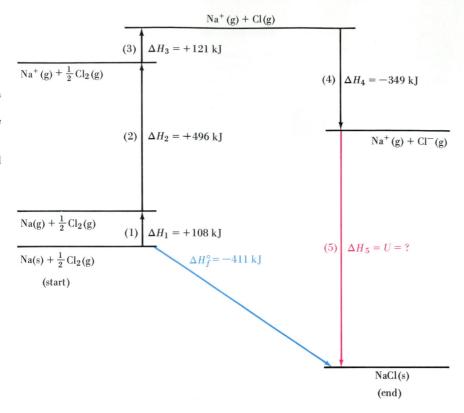

One application of the concept of lattice energy and the Born–Fajans–Haber method is to make predictions about the feasibility of synthesizing ionic compounds. In Example 10-3 we predict the enthalpy (heat) of formation of MgCl(s).

Example 10-3

Relating enthalpy of formation, lattice energy, and other energy quantities through the Born–Fajans–Haber method. Following are data corresponding to the five steps outlined above and in Figure 10-4, but applied to the formation of Mg^+, Cl^-, and MgCl(s). Use these data to calculate ΔH_f° per mol MgCl(s): (1) Heat of sublimation of 1 mol Mg: +150. kJ; (2) First ionization energy of 1 mol Mg(g): +738 kJ; (3) Heat of dissociation of $\frac{1}{2}$ mol Cl_2: +121 kJ; (4) Electron affinity of 1 mol Cl(g) = −349 kJ; (5) Lattice energy of 1 mol MgCl(s) = −676 kJ.

Solution. Following are the individual steps that must be combined to yield ΔH_f° (as in equation 10.3):

$$Mg(s) \longrightarrow Mg(g) \qquad\qquad \Delta H_1 = +150\ kJ$$
$$Mg(g) \longrightarrow Mg^+(g) + e^- \qquad \Delta H_2 = +738\ kJ$$
$$\tfrac{1}{2} Cl_2(g) \longrightarrow Cl(g) \qquad\qquad \Delta H_3 = +121\ kJ$$
$$Cl(g) + e^- \longrightarrow Cl^-(g) \qquad\quad \Delta H_4 = -349\ kJ$$
$$\underline{Mg^+(g) + Cl^-(g) \longrightarrow MgCl(s) \qquad U = -676\ kJ}$$
$$Mg(s) + \tfrac{1}{2} Cl_2(g) \longrightarrow MgCl(s) \qquad \Delta H_f^\circ = ?$$

$$\Delta H_f^\circ[MgCl(s)] = \Delta H_1 + \Delta H_2 + \Delta H_3 + \Delta H_4 + U$$

$$\Delta H_f^\circ[MgCl(s)] = (+150 + 738 + 121 - 349 - 676)kJ = -16\ kJ$$

SIMILAR EXAMPLES: Exercises 23, 24, 86, 87.

Example 10-3 suggests that we should be able to obtain MgCl(s) as a stable compound. Why have we been writing $MgCl_2$ as the formula of magnesium chloride all this time instead of MgCl? If we were to repeat the calculation of Example 10-3 for the formation of $MgCl_2(s)$, we would obtain an enthalpy of formation that is very much more negative than the value we obtained for MgCl (see Exercise 24). Even though the energy requirement to produce Mg^{2+} is larger than to produce Mg^+, the lattice energy U is very much greater for $MgCl_2(s)$ than for MgCl(s). This is because the attractive force exerted by the *doubly* positively charged Mg^{2+} ions on Cl^- ions is much greater than the force exerted by *singly* charged Mg^+ ions. As a result, the reaction between Mg and Cl atoms does not stop at MgCl but continues on to form the more stable $MgCl_2$. Is $NaCl_2$ a stable compound? Here the answer is no. The additional lattice energy that would be associated with $NaCl_2$ over NaCl is not nearly enough to compensate for the very high second ionization energy of sodium (see Exercise 87). To extract a second electron from a Na atom would require breaking into the very stable electron configuration of neon.

Perhaps the most important present application of the Born–Fajans–Haber method is in determining electron affinities from other experimental data. Equation (10.3) involves six terms. If any five of the six are known, the remaining one can be calculated. In particular, the equation can be solved for an unknown electron affinity (see Exercise 25).

10-4 Covalent Bonding—An Introduction

In the Lewis structures for NaCl and HCl presented in expressions (10.1) and (10.2), the Cl atom acquires the electron configuration of a noble gas (an outer-shell octet of electrons). A Cl atom shows a certain tendency to gain an electron, measured by its electron affinity (-349 kJ/mol), but from which atom, Na or H, can the electron most readily be extracted? Of course, neither an Na nor an H atom gives up an electron *freely*. However, the energy required to extract an electron from Na, the first ionization energy ($+496$ kJ/mol), is much smaller than for H ($+1312$ kJ/mol). If you recall that the lower its ionization energy the more metallic an element, you can see that *sodium is much more metallic than hydrogen*. In fact, because of its high ionization energy, we consider hydrogen to be a *nonmetal* under normal conditions. A hydrogen atom does *not* give up an electron to another nonmetal. Bonding between a hydrogen and a chlorine atom involves the *sharing* of electrons, leading to a **covalent bond.**

To emphasize the sharing of electrons and the octet rule we can think of the Lewis structure of HCl in the manner indicated below.

The broken circles represent the outermost electronic shells of the bonded atoms. The effective number of electrons in each outer shell is established by counting all the dots lying on or within each circle. For the H atom there are two dots, corresponding to the electron configuration of He. For the Cl atom the number of dots is eight, corresponding to the outer-shell configuration of Ar. Note that the two electrons between H and Cl (:) have been counted *twice*. These are the two electrons being shared by the H and Cl atoms. This shared pair of electrons constitutes the covalent bond. Some additional examples of simple Lewis structures are shown in Figure 10-5.

One early success of the Lewis theory was in explaining the existence of gaseous hydrogen and chlorine as *diatomic* molecules, H_2 and Cl_2. In each case a pair of

FIGURE 10-5
Some examples of covalent bonds.

Elements	H·	·Ö:	:C̈l·	·N̈·
Compounds	H:C̈l:	H:Ö: H	:C̈l:Ö: :C̈l:	H:N̈:H H
Names	hydrogen chloride	hydrogen oxide (water)	chlorine oxide	hydrogen nitride (ammonia)
Molecules	HCl	H_2O	Cl_2O	NH_3
Molecular weights	36.46	18.02	86.91	17.03

electrons is shared between the two atoms. The sharing of a *single* pair of electrons between bonded atoms produces a **single** covalent bond. To underscore the importance of electron *pairs* in the Lewis theory, we introduce the term **bond pair** to describe a pair of electrons that constitutes a covalent bond and **lone pair** to describe electron pairs not involved in bonding. Also it is customary to replace some electron pairs by dash signs (—), especially for bond pairs. These features are brought out in the Lewis structures below.

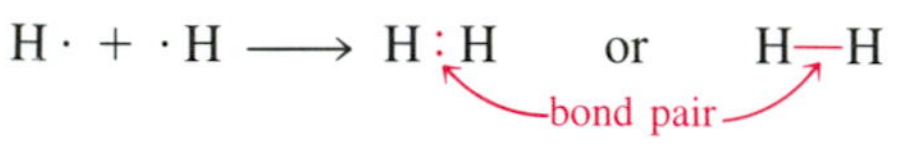

$$\text{H} \cdot \; + \; \cdot \text{H} \longrightarrow \text{H} \underset{\text{bond pair}}{:} \text{H} \quad \text{or} \quad \text{H—H} \tag{10.4}$$

$$:\ddot{\text{C}}\text{l} \cdot \; + \; \cdot \ddot{\text{C}}\text{l}: \longrightarrow :\ddot{\text{C}}\text{l} \underset{\text{bond pair}}{:} \ddot{\text{C}}\text{l}: \quad \text{or} \quad :\ddot{\text{C}}\text{l} \text{—} \ddot{\text{C}}\text{l}: \longleftarrow \text{lone pairs} \tag{10.5}$$

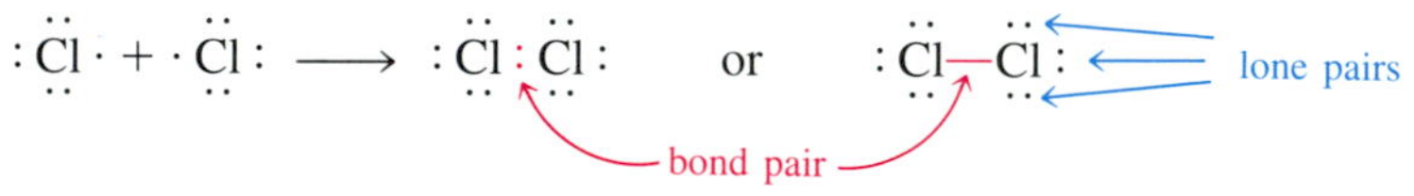

Multiple Covalent Bonds. If we try to extend what we just wrote for H_2 and Cl_2 to the molecule N_2, we have a problem.

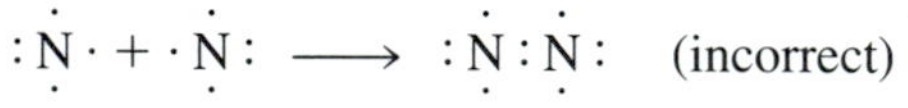

$$:\dot{\text{N}} \cdot \; + \; \cdot \dot{\text{N}}: \longrightarrow :\dot{\text{N}} : \dot{\text{N}}: \quad \text{(incorrect)} \tag{10.6}$$

This Lewis structure *violates the octet rule*. The N atoms appear to have only *six* outer-shell electrons. We can correct the situation by bringing the four unpaired electrons of structure (10.6) into the region between the N atoms and making them into additional bond pairs. In all, we now show the sharing of *three* pairs of electrons between the N atoms—*three* bond pairs. The bond between N atoms in N_2 is a **triple** covalent bond (≡).

$$:\text{N} : \text{N}: \longrightarrow \left(\text{N} \; \vdots \; \text{N} \right) \quad \text{or} \quad :\text{N} \equiv \text{N}: \tag{10.7}$$

Let us follow the same approach for O_2 and see what happens.

$$:\ddot{\text{O}} \cdot \; + \; \cdot \ddot{\text{O}}: \longrightarrow :\ddot{\text{O}} : \ddot{\text{O}}: \longrightarrow :\ddot{\text{O}} : : \ddot{\text{O}}: \quad \text{or} \quad :\ddot{\text{O}}\text{=}\ddot{\text{O}}: \quad \text{(incorrect)} \tag{10.8}$$

Why have we labeled the structure in (10.8) "incorrect"? For this reason: Although the structure satisfies all the requirements of the Lewis theory, it fails to account for the fact that oxygen is *paramagnetic*. The O_2 molecule has *unpaired* electrons. We propose a somewhat better way to represent the O_2 molecule in Section 10-9, and a still better way in Section 11-5. For now, just be certain that you understand this point.

The fact that you can write a plausible Lewis structure for a species does not prove that this is the true electronic structure.

A proposed Lewis structure must always be verified with experimental evidence.

We can write the Lewis structures required in Examples 10-4 and 10-5 with the following basic ideas:

Fundamental requirements of Lewis structures.

1. *All* the valence (outer-shell) electrons of the atoms in a Lewis structure must be shown in the structure.
2. *Usually*, each atom in a Lewis structure acquires an electron configuration with an outer-shell **octet.** Hydrogen, however, is limited to an outer-shell **duet** (two electrons).
3. *Usually*, all the electrons in a Lewis structure are paired.
4. *Sometimes* it is necessary to represent double or triple bonds in a Lewis structure.

(10.9)

Multiple covalent bonds are formed most readily by C, N, O, P, and S atoms.

Another requirement in writing a plausible Lewis structure is to start with the correct **skeleton structure.** This is an arrangement of the atoms in the order in which they are bonded together. For example, a water molecule has the skeleton structure H—O—H and *not* O—H—H. In a skeleton structure it is generally useful to distinguish between **central atom(s)** and **terminal atoms.** A central atom is bonded to two or more other atoms. A terminal atom is bonded to just one other atom. *Hydrogen atoms are always terminal atoms.**

Example 10-4

The preparation and additional uses of hydrazine were described on pages 112–13.

Writing a simple Lewis structure of a covalent molecule. Write a plausible skeleton structure, and then the Lewis structure for a molecule of hydrazine, N_2H_4, a rocket propellant.

Solution. Several skeleton structures are shown below, but only (c) is correct.

N—N—H—H—H—H   H—H—N—N—H—H
      (a)   (b)

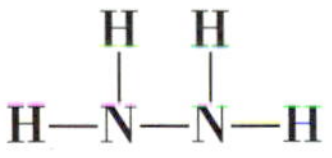

    H H
    | |
H—N—N—H   H—N—H—H—N—H
   (c)   (d)

H atoms can only be *terminal* atoms in a structure, never central atoms. This is because the H atom can accommodate only *two* electrons in its electronic shell, and in structures like (a), (b), and (d) some of the H atoms would have *four* electrons. The four electrons could not all be in the first electronic shell, and the H atoms would not have noble gas electron configurations. Another way to state all of this is that *a hydrogen atom can form only one single covalent bond.*

Now let us see how we can apply the basic ideas of expression (10.9) to the skeleton structure (c). Nitrogen is in group 5A of the periodic table. This means that N atoms have *five* valence electrons. Between them, the two N atoms contribute *ten* electrons to the Lewis structure. Each of the *four* H atoms contributes *one* electron. The total number of electrons in the Lewis structure must be *14* [(2 × 5) + (4 × 1)]. We can account for 10 of the 14 electrons through the five dash signs in the skeleton structure. The additional four electrons appear as two lone pairs, one pair on each N atom. The Lewis structure of hydrazine is

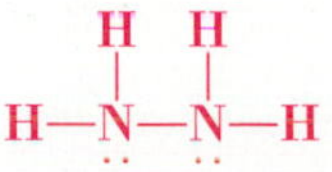

    H H
    | |
H—N—N—H
    ·· ··

To check the plausibility of this structure count the number of electrons around each atom. Each H atom forms one single covalent bond. This repre-

*We will encounter one important exception to this rule in molecules such as B_2H_6 in Section 23-5.

sents *two* electrons for each H atom, the required *duet*. Each N atom forms *three* single covalent bonds, accounting for *six* electrons. Add the contribution of the lone pair electrons on each N atom and this leads to a total of *eight* electrons for each N atom, the required *octet*.

SIMILAR EXAMPLES: Exercises 3, 26.

Rather than relying on the general ideas of expression (10.9), you may prefer a specific strategy for writing Lewis structures. The following strategy is used in Example 10-5.

Strategy for writing Lewis structures.

- Establish the total number of valence electrons in the structure.
- Write the skeleton structure, and join the atoms in this structure by *single* covalent bonds.
- For each single bond in the structure, subtract two electrons from the total number of valence electrons. (10.10)
- Arrange the remaining valence electrons around the terminal atoms, and then, to the extent possible, around the central atom(s).
- If the central atom(s) lacks an outer-shell octet, form multiple covalent bonds by shifting lone-pair electrons from terminal atoms.

Example 10-5

Representing multiple covalent bonds in a Lewis structure. A molecule of the poisonous gas hydrogen cyanide is represented by the skeleton structure HCN. Write the Lewis structure of HCN.

Solution. The number of valence electrons in the structure must be 10 (*one* from H, *four* from C, and *five* from N). Bonding in the skeleton structure accounts for four electrons.

H—C—N

Note that in strategy (10.10) it is not necessary to keep track of which valence electrons come from which atoms.

The remaining six electrons are placed around the terminal N atom.

$$\text{H—C—}\overset{..}{\underset{..}{\text{N}}}: \quad \text{(incorrect)} \tag{10.11}$$

In structure (10.11) the central C atom does not have an octet (only *four* electrons are shown around it). We can correct this situation by moving two lone pairs of electrons of the N atom into the region between the C and N atom, forming a *triple covalent bond*.

$$\text{H}:\text{C}:\text{N}: \longrightarrow \text{H}:\text{C}:::\text{N}: \quad \text{or} \quad \text{H}:\text{C}\equiv\text{N}: \tag{10.12}$$

SIMILAR EXAMPLES: Exercises 4, 27.

In Example 10-5 we were *given* the skeleton structure—HCN. Is there some way to establish that this is the skeleton structure (rather than HNC)? In Section 10-7 we will learn how to do this. There are some other aspects of writing Lewis structures that we must also explore, but we will find these easier to deal with if we first consider some other ideas about chemical bonding.

10-5 Bond Lengths and Bond Energies

If we think of electrons as the "glue" that binds atoms together in covalent bonds, it would seem reasonable that the more electrons present in the bond (the more

TABLE 10-2
Some Representative Bond Energies and Bond Lengths

Bond	Bond energy, kJ/mol	Bond length, pm	Bond	Bond energy, kJ/mol	Bond length, pm	Bond	Bond energy, kJ/mol	Bond length, pm
H—H	435	74	C—C	347	154	N—N	163	145
H—C	414	110	C=C	611	134	N=N	418	123
H—N	389	100	C≡C	837	120	N≡N	946	109
H—O	464	97	C—N	305	147	N—O	230	136
H—S	368	132	C=N	615	128	N=O	590	115
H—F	569	101	C≡N	891	116	O—O	142	145
H—Cl	431	136	C—O	360	143	O=O	498	121
H—Br	368	151	C=O	728	123	F—F	159	128
H—I	297	170	C—Cl	326	177	Cl—Cl	243	199
						Br—Br	192	228
						I—I	151	266

"glue") the more tightly the atoms are held together. That is, we would expect that two atoms joined by a double bond are held together more tightly than by a single bond, and that atoms joined by a triple bond would be held more tightly still. We can use the term **bond order** to indicate whether a covalent bond is *single* (bond order = 1), *double* (bond order = 2), or *triple* (bond order = 3).

There are two important properties of bonds that are related to the bond order. These are bond length and bond energy. **Bond length** is the distance between the centers of two atoms joined by a covalent bond. From the statements above, we would expect a certain bond length for a single bond between atoms, a shorter one for a double bond, and shorter still for a triple bond. You will see this relationship clearly in Table 10-2 if you compare the three different bond lengths for the nitrogen-to-nitrogen bond.

Perhaps you can also better understand now the meaning of a particular atomic radius that we introduced in Section 9-7—the single covalent radius. It is one-half the distance between the centers of identical atoms when they are joined by a *single* covalent bond. Thus, the single covalent radius of a chlorine atom, which from Figure 9-8 appears to be about 100 pm, is one-half the bond length given in Table 10-2, that is $\frac{1}{2} \times 199$ pm. Furthermore, as a *rough* generalization,

Relationship between bond length and covalent radii.

The length of the covalent bond between unlike atoms can be approximated as the sum of the covalent radii of the atoms.   (10.13)

Some of these ideas about bond length are applied in Example 10-6.

Example 10-6

Relating bond order and bond length. Estimate the indicated bond lengths:
(a) The nitrogen-to-nitrogen bond in N_2H_4.
(b) The bromine-to-chlorine bond in BrCl.

Solution
(a) Table 10-2 lists three nitrogen-to-nitrogen distances. The best approach is to draw a plausible Lewis structure for the molecule; see whether the nitrogen-to-nitrogen bond is single, double, or triple; and then select the appropriate value from Table 10-2. We have already drawn the Lewis structure for N_2H_4 (Example 10-4). The bond is a single bond and the expected bond length is 145 pm.

(b) No bromine-to-chlorine bond distance is given in Table 10-2. Here we are expected to calculate an approximate bond length using expression (10.13).

BrCl contains a Br—Cl single bond [just imagine substituting one Br atom for one Cl atom in structure (10.5)]. The length of the Br—Cl bond is one half the Cl—Cl bond length *plus* one-half the Br—Br bond length, which from Table 10-2 is $(\frac{1}{2} \times 199) + (\frac{1}{2} \times 228) = 214$ pm.

SIMILAR EXAMPLES: Exercises 11, 47, 48, 49.

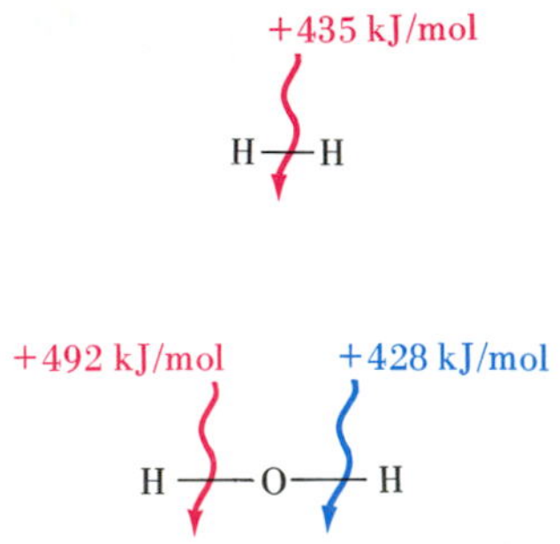

Bond Energy. Energy is *released* when isolated atoms join together to form a covalent bond, and energy must be *absorbed* to break apart covalently bonded atoms. Let us define **bond energy** as the quantity of energy required to break one mole of covalent bonds in a *gaseous* species. The SI units of bond energy are kilojoules per mole of bonds (kJ/mol). Bond energy is also called bond *dissociation* energy, for which the symbol, **D,** is sometimes used. Some representative data are presented in Table 10-2.

Bond energies are not difficult to conceptualize when considering *diatomic* molecules, since there is but one bond per molecule. In the manner introduced in Chapter 7, we can think of bond energy (or **bond enthalpy**) as an enthalpy change or a heat of reaction. For example,

Bond breakage: $H_2(g) \longrightarrow 2\ H(g)$ $\Delta H = +435$ kJ

Bond formation: $2\ H(g) \longrightarrow H_2(g)$ $\Delta H = -435$ kJ

With a polyatomic molecule such as H_2O, the situation is different (see Figure 10-6). The energy required to dissociate one mole of H atoms by breaking one O—H bond per molecule

$H—OH(g) \longrightarrow H(g) + OH(g)$ $\Delta H = +492$ kJ

is different from the energy required to dissociate a second mole of H atoms by breaking the remaining bonds in OH(g)

$O—H(g) \longrightarrow H(g) + O(g)$ $\Delta H = +428$ kJ

As you can readily see from Table 10-2 there is a relationship between bond order and bond energy. Double bonds between atoms have higher bond energies than do single bonds, and triple bonds have higher bond energies still. What is not readily apparent from the data in Table 10-2 is that bond energies depend on the environment of a given bond—the portion of the molecule close to the bond. Thus, the bond energy of the O—H bond is somewhat different in H—O—H than in H—O—O—H and in $H_3C—O—H$. It is customary to list *average* values, as is done in Table 10-2.

Experimental methods are available (such as those based on spectroscopy) for determining bond lengths and bond energies, but bond energies can also be determined indirectly through Hess's law (see Exercise 53).

Calculations Involving Bond Energies. For a reaction between *gaseous* species, we can visualize a process in which certain bonds in the reactants are broken and some new bonds are formed in the products. The sum of the enthalpy changes for bond breakage and bond formation yields the enthalpy change for the reaction. That is,

$$\Delta H_{rxn} = \sum \Delta H_{bonds\ broken} + \sum \Delta H_{bonds\ formed} \tag{10.14}$$

In using expression (10.14) do not forget that enthalpy changes for bond breakage are equal to bond energies, but enthalpy changes for bond formation are the *negatives* of bond energies.

FIGURE 10-6
Some bond energies compared.

The same quantity of energy (435 kJ/mol) is required to break all H—H bonds. In H_2O more energy is required to break the first O—H bond (492 kJ/mol) than to break the second (428 kJ/mol). The value of the O—H bond energy in Table 10-2 is an *average* value based on water and other compounds containing the O—H bond.

Bond energies and enthalpy of reaction.

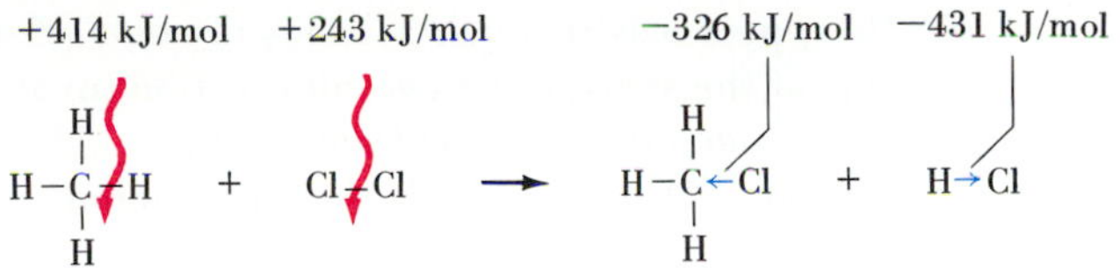

FIGURE 10-7
Bond breakage and formation
in a chemical reaction—
Example 10-7 illustrated.

$+414$ kJ/mol $+243$ kJ/mol -326 kJ/mol -431 kJ/mol

$$
\begin{array}{c}
\text{H} \\
| \\
\text{H}-\text{C}-\text{H} \\
| \\
\text{H}
\end{array}
\;+\; \text{Cl}-\text{Cl} \;\longrightarrow\;
\begin{array}{c}
\text{H} \\
| \\
\text{H}-\text{C}-\text{Cl} \\
| \\
\text{H}
\end{array}
\;+\; \text{H}-\text{Cl}
$$

Example 10-7

Calculating an enthalpy of reaction from bond energies. The reaction of methane (CH_4) and chlorine produces a mixture of products called chloromethanes. One of these is monochloromethane CH_3Cl (also called methyl chloride), used in the preparation of silicones (see Section 23-4). Calculate ΔH for the reaction

$$CH_4(g) + Cl_2(g) \longrightarrow CH_3Cl(g) + HCl(g)$$

Solution. To do this type of calculation you have to know which bonds are being broken and which new bonds are being formed. It helps to draw structural formulas (or Lewis structures), as in Figure 10-7. Bond energies used below are from Table 10-2.

ΔH for bond breakage:

```
1 mol C—H bonds  = +414 kJ
1 mol Cl—Cl bonds = +243 kJ
           sum:   +657 kJ
```

ΔH for bond formation:

Note again that for bond *formation* we must use the *negatives* of bond energies.

```
1 mol C—Cl bonds = −326 kJ
1 mol H—Cl bonds = −431 kJ
           sum:   −757 kJ
```

Heat of reaction:

$$\Delta H_{rxn} = \sum \Delta H_{\text{bond breakage}} + \sum \Delta H_{\text{bond formation}} = +657 - 757 = -100.\ \text{kJ}$$

SIMILAR EXAMPLES: Exercises 51, 52, 85.

As a rule, there is no advantage to calculating a heat of reaction using bond energies (as in Example 10-7) over using heat of formation data (recall Section 7-9). This is because heats of formation are known rather precisely, whereas bond energies are only average values. But there are times when heat-of-formation data are not known. Here bond energies can prove particularly useful.

Another important use of bond energies is in allowing us to predict whether a reaction will be *endothermic* or *exothermic*. In general,

Exothermic/endothermic reactions and bond energies.

if weak bonds $\longrightarrow$ strong bonds $\Delta H < 0$ *(exothermic)*
and   (10.15)
if strong bonds $\longrightarrow$ weak bonds $\Delta H > 0$ *(endothermic)*

Example 10-8 applies this idea to a reaction involving highly reactive, unstable gaseous species for which heats of formation are not normally tabulated.

Example 10-8

Using bond energies to predict whether a reaction involving gases is exothermic or endothermic. The formation of methyl chloride, described in Example

10-7, proceeds through a series of steps (discussed in detail in Section 27-2). One of the steps in the overall reaction involves the reaction of a gaseous chlorine *atom* with a molecule of methane. The products are an unstable methyl radical, CH_3, and $HCl(g)$. Is this reaction endothermic or exothermic?

$$CH_4(g) + Cl(g) \longrightarrow CH_3(g) + HCl(g)$$

Solution. In this reaction for every molecule of CH_4 that reacts *one* C—H bond is broken, requiring 414 kJ per mol of bonds. At the same time *one* H—Cl bond is formed, releasing 431 kJ per mol of bonds. Since more energy is released in forming new bonds than is absorbed in breaking old bonds, the reaction is *exothermic* (that is, $414 - 431 = -17$ kJ).

SIMILAR EXAMPLE: Exercise 12.

10-6 Partial Ionic Character of Covalent Bonds

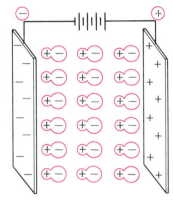

FIGURE 10-8

Behavior of polar molecules in the electric field of a condenser.

In the absence of an electric field the molecules are oriented randomly. However, when one plate is made negative and the other positive, polar molecules tend to orient themselves in the manner shown here.

FIGURE 10-9

Nonpolar and polar covalent bonds.

In H—H there is an even distribution of electronic charge between the atoms. In H—Cl electron charge density is displaced toward the Cl atom.

We have established two fundamental bond types, ionic and covalent, but classification schemes often involve intermediate possibilities. This is very much the case with bond type.

Polar Molecules and Dipole Moments. The device pictured in Figure 10-8 is called an electrical condenser (or capacitor). It consists of a pair of electrodes separated by a *nonconducting* medium, and it is able to store small quantities of electric charge—positive on one plate and negative on the other. Certain substances, when placed between the plates of an electrical condenser, increase the condenser's charge-storing ability. These are substances in whose molecules there is a slight separation of electric charge. Figure 10-8 suggests how such molecules line up in an electric field.

The separation of electric charge in a covalent bond involves a shift of electrons toward the atom with the more negative electron affinity. A bond in which such a charge separation is found is called a **polar covalent bond.** Note in Figure 10-9 how the electronic charge is evenly distributed between the two H atoms in H_2. Since both H atoms have the same electron affinity, there is no displacement of the electronic charge and the bond is **nonpolar.** In HCl, on the other hand, the Cl atom definitely has a more negative electron affinity than does the H atom. Electronic charge distribution is shifted toward the Cl atom. The H—Cl bond is *polar*. This distinction can also be shown through Lewis structures.

$$H : H \qquad {}^{\delta+}H : \overset{..}{\underset{..}{Cl}} : {}^{\delta-}$$

The magnitude of the charge displacement in a polar covalent bond is measured through a quantity called the **dipole moment μ.** The dipole moment is the product of the magnitude of the charges (δ) and the distance separating them *(d)*. (The symbol δ suggests a small magnitude of charge, less than the charge on an electron.)

$$\mu = \delta d \qquad\qquad (10.16)$$

If the product of the charge and distance of separation of the charge centers has a value of 3.34×10^{-30} coulomb · meter (C · m), the dipole moment, μ, has a value called 1 **debye (D).** Perhaps you are wondering how significant this matter of charge separation is for a molecule like HCl. Let us figure it out this way: The measured dipole moment of HCl is 1.03 D, and from Table 10-2 the H—Cl bond distance is 136 pm. This means that

magnitude of charge $\times 136 \times 10^{-12}$ m $= 1.03$ D $= 1.03 \times 3.34 \times 10^{-30}$ C · m

$$\text{magnitude of charge} = \frac{3.44 \times 10^{-30} \text{ C} \cdot \text{m}}{136 \times 10^{-12} \text{ m}} = 2.53 \times 10^{-20} \text{ C}$$

This charge is about 16% of the charge on an electron (1.60×10^{-19} C), suggesting that HCl is about 16% ionic in character.

Molecular Shapes and Dipole Moments. In the carbon monoxide molecule, CO, the O atom has a more negative electron affinity than does the C atom, and so there is a slight displacement of electrons toward the O atom. The CO molecule is slightly polar. In the representation below the cross-base arrow ($\leftrightarrow$) points to the atom that attracts electrons more strongly. (Not shown here is the fact that the carbon-to-oxygen bond is a multiple covalent bond.)

$$\text{C} \longrightarrow \text{O} \qquad \mu = 0.11 \text{ D}$$

Even though the difference in electron affinity between C and O produces a **bond dipole moment** in each carbon-to-oxygen bond in CO_2, *the CO_2 molecule is non-polar*. What this means is the effects of the two O atoms in attracting electrons from the C atom cancel each other out. We say that the **resultant dipole moment** is zero. And this happens because the O atoms lie along the same straight line through the central C atom.

The result described here for CO_2 is analogous to a tug-of-war contest between equally matched teams. Although there is a strong pull in each direction, the knot at the center of the rope does not move.

$$\text{O} \longleftarrow \text{C} \longrightarrow \text{O} \qquad \mu = 0$$

Because of the difference in electron affinities of H and O, there is a bond dipole moment of 1.51 D in the O—H bond. If H_2O were a linear molecule, the bond moments would be in opposite directions and there would be no resultant dipole moment. But the measured dipole moment of H_2O is 1.84 D. *The molecule is polar, and so it cannot be linear.* The two bond moments combine to yield a resultant dipole moment of 1.84 D for a particular bond angle, 104°.

$$\text{O} \longleftarrow \text{H} \qquad \mu = 1.84 \text{ D} \qquad \qquad (10.17)$$
$$104°$$
$$\text{H}$$

Measurements of dipole moments can assist us in describing molecular shapes, and we will return to a discussion of this topic in Section 10-10.

Electronegativity. Electronegativity describes the ability of an atom to compete for electrons with another atom to which it is bonded. Electronegativity is related to ionization energy (I) and electron affinity (EA), since these quantities reflect the ability of an atom to lose or gain electrons. The most widely used electronegativity scale is one devised by Linus Pauling (1901–   ), which is based on an evaluation of bond energies (see Exercise 60). Electronegativity (EN) values of the elements are given in Figure 10-10. As a *rough* rule, metals have electronegativity values below 2; metalloids, about 2; and nonmetals, above 2. From Figure 10-10 we also see that

Electronegativity and position in the periodic table.

==Electronegativity decreases from top to bottom in a group and increases from left to right in a period of the periodic table.== (10.18)

From electronegativity values Pauling has been able to assign a "percent ionic character" to a bond, with the result illustrated in Figure 10-11. If the electronegativity difference between atoms is very small, a bond is essentially covalent, and if this difference is large, a bond is essentially ionic. There is no particular electronegativity difference at which a bond type changes from covalent to ionic. From Figure 10-11 we see that at an electronegativity difference of 1.7 a bond is about 50% ionic

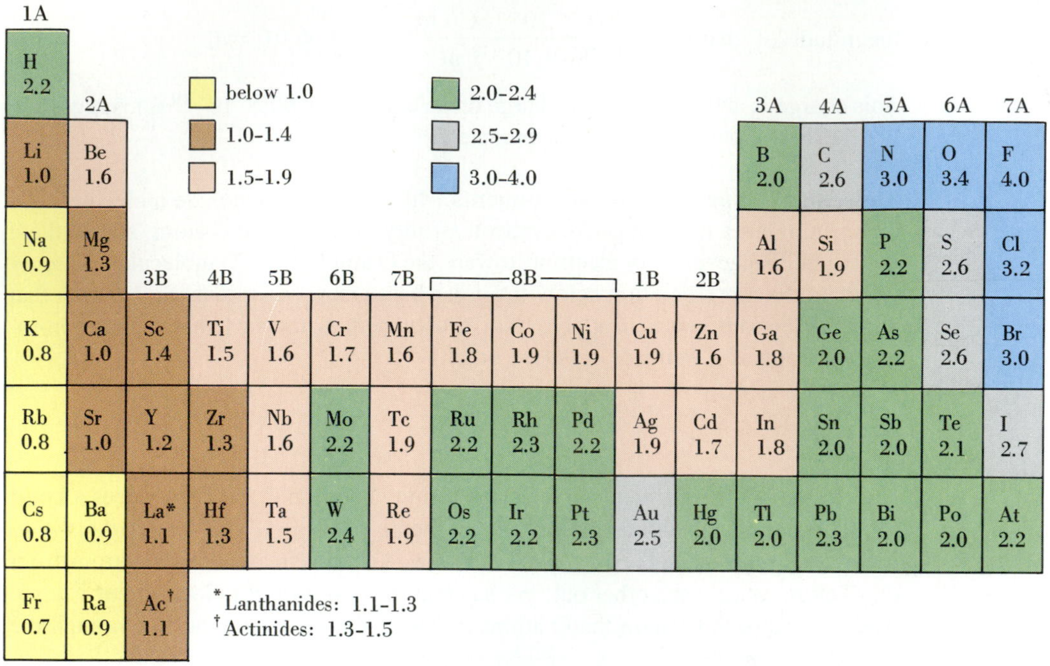

1A												3A	4A	5A	6A	7A
H 2.2	2A															
Li 1.0	Be 1.6											B 2.0	C 2.6	N 3.0	O 3.4	F 4.0
Na 0.9	Mg 1.3	3B	4B	5B	6B	7B		8B		1B	2B	Al 1.6	Si 1.9	P 2.2	S 2.6	Cl 3.2
K 0.8	Ca 1.0	Sc 1.4	Ti 1.5	V 1.6	Cr 1.7	Mn 1.6	Fe 1.8	Co 1.9	Ni 1.9	Cu 1.9	Zn 1.6	Ga 1.8	Ge 2.0	As 2.2	Se 2.6	Br 3.0
Rb 0.8	Sr 1.0	Y 1.2	Zr 1.3	Nb 1.6	Mo 2.2	Tc 1.9	Ru 2.2	Rh 2.3	Pd 2.2	Ag 1.9	Cd 1.7	In 1.8	Sn 2.0	Sb 2.0	Te 2.1	I 2.7
Cs 0.8	Ba 0.9	La* 1.1	Hf 1.3	Ta 1.5	W 2.4	Re 1.9	Os 2.2	Ir 2.2	Pt 2.3	Au 2.5	Hg 2.0	Tl 2.0	Pb 2.3	Bi 2.0	Po 2.0	At 2.2
Fr 0.7	Ra 0.9	Ac† 1.1														

Key:
below 1.0   2.0–2.4
1.0–1.4   2.5–2.9
1.5–1.9   3.0–4.0

*Lanthanides: 1.1–1.3
†Actinides: 1.3–1.5

FIGURE 10-10
Electronegativities of the elements.

As a general rule, electronegativities *increase* from left to right in a period of elements and *decrease* from top to bottom in a group. (There are several exceptions, however.)

and 50% covalent. Bonds in this general range of electronegativity difference are best referred to as covalent bonds with some ionic character (or ionic bonds with some covalent character).

Example 10-9

Assessing electronegativity differences and the polarity of bonds. Earlier in this section we made an assessment that the H—Cl bond is about 16% ionic. (a) What is the percent ionic character of the H—Cl bond according to Figures 10-10 and 10-11? (b) Which bond has the more polar character, H—Cl or H—O?

Solution

(a) Look up the electronegativities of H and Cl in Figure 10-9 and compute the electronegativity difference: $EN_H = 2.2$; $EN_{Cl} = 3.2$; EN difference = $3.2 - 2.2 = 1.0$. Now estimate the percent ionic character of the bond from Figure 10-11. An EN difference of 1.0 corresponds to about 20% ionic character.

(b) We would expect the covalent bond with the greater electronegativity difference to be the more polar bond. In (a) we determined the EN difference of

FIGURE 10-11
Percent ionic character of a chemical bond as a function of electronegativity difference.

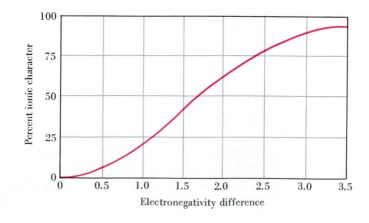

the H—Cl bond to be 1.0. For the H—O bond we obtain an EN difference = 3.4 − 2.2 = 1.2. The H—O bond is more polar than the H—Cl bond.

SIMILAR EXAMPLES: Exercises 58, 60.

10-7 Covalent Lewis Structures

One of the questions that we left unanswered in Section 10-4 was how to determine which should be the central atom in a Lewis structure. In Example 10-5 the basic rules for Lewis structures would have been met equally well by either (a) or (b). The concept of formal charge, which we introduce next, helps us to understand why the correct structure is (a).

$$H—C≡N\colon \qquad H—N≡C\colon \qquad\qquad (10.19)$$
$$\text{(a)} \qquad\qquad \text{(b)}$$

The Concept of Formal Charge. Suppose, in addition to establishing that atoms acquire outer-shell octets, we also count and assign electron dots in Lewis structures in this way (illustrated in Figure 10-12).

- Count all *lone pair* electrons as belonging entirely to the atom in which they are found.
- Count *bonding* electrons by dividing them *equally* between the bonded atoms.

After completing this assignment, determine whether any atoms in the Lewis structure have a **formal charge.**

Formal charge is the number of valence (outer-shell) electrons in an isolated atom minus the number of electrons assigned to that atom in a Lewis structure.

This definition of formal charge (FC) can also be expressed through a simple formula.

Equation for formal charge.

$$FC = \text{no. valence electrons} - \tfrac{1}{2}(\text{no. bonding electrons}) - \text{no. lone pair electrons} \qquad (10.20)$$

Where formal charges are found to exist they can be denoted as shown below for structure (b) of Figure 10-12.

$$\overset{\oplus}{H}—\overset{\ominus}{N}≡C\colon$$

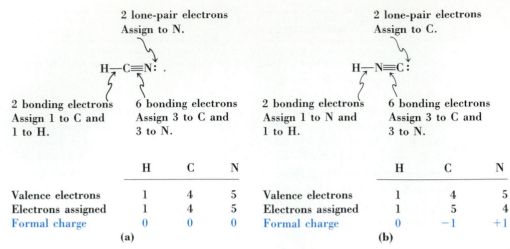

2 lone-pair electrons
Assign to N.

2 lone-pair electrons
Assign to C.

H—C≡N: .   H—N≡C :

2 bonding electrons   **6 bonding electrons**   **2 bonding electrons**   **6 bonding electrons**
Assign 1 to C and   Assign 3 to C and   Assign 1 to N and   Assign 3 to C and
1 to H.   3 to N.   1 to H.   3 to N.

	H	C	N			H	C	N
Valence electrons	1	4	5		Valence electrons	1	4	5
Electrons assigned	1	4	5		Electrons assigned	1	5	4
Formal charge	0	0	0		Formal charge	0	−1	+1

                    (a)   (b)

FIGURE 10-12
Selecting a plausible Lewis structure—the concept of formal charge illustrated.

Structure (a) is the more plausible because there are no formal charges on any of the atoms in the structure.

You should note, however, that individual atoms in a covalent molecule do not carry actual charges. To distinguish between formal and actual charges, we use small encircled numbers for formal charges.

If all atoms make equal contributions of electrons to covalent bonds, there will be no formal charges in a Lewis structure, and the basic idea in applying the formal charge concept is to keep formal charges to a minimum. We can achieve this goal by applying the following rules.

Rules for formal charge.

- Usually, the most plausible Lewis structure is one with no formal charges (that is, with formal charges of 0 on all atoms).
- Where formal charges are required, these should be as small as possible.
- Negative formal charges should appear on the most electronegative atoms.
- Formal charges on the atoms in a Lewis structure must total to *zero* for a neutral molecule (and to the net charge on a polyatomic ion).   (10.21)

In applying these rules you will find that

Selecting the central atom in a Lewis structure.

the central atom in a structure is generally the atom with the lowest electronegativity.   (10.22)

Example 10-10

Using formal charges in selecting a Lewis structure. Thionyl chloride, $SOCl_2$, is a toxic, irritating liquid used in the manufacture of pesticides. Write the most plausible Lewis structure for this molecule.

Solution. According to the way in which the formula is written, it appears that O is the central atom, with the other atoms bonded to it. Let us call this structure (a). However, according to statement (10.22) it would seem that S should be the central atom—it has the lowest electronegativity. Let us call this structure (b). Let us write both structure (a) and (b) and then use the rules in (10.21) to decide on the correct one. In both cases the first step will be the same. **Step 1.** Establish the number of electron dots that must appear in the structure. This is the total number of valence electrons.

From S   From O   From Cl

$$(1 \times 6) + (1 \times 6) + (2 \times 7) = 26$$

Step 2. Join each of the terminal atoms to the central atom by *single* covalent bonds. Then place the remaining electrons as lone pairs to establish octets, first around the terminal atoms and then around the central atom.

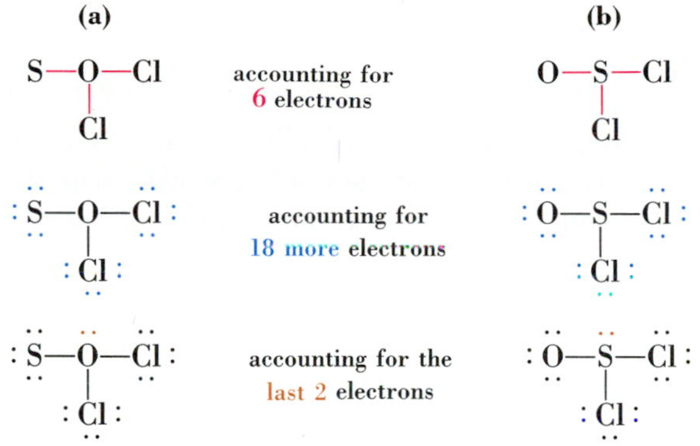

    (a)   **(b)**

S—O—Cl   accounting for   O—S—Cl
   | **6** electrons |
  Cl   Cl

:S—O—Cl: accounting for   :O—S—Cl:
   | **18 more** electrons |
  :Cl: :Cl:

:S—O—Cl: accounting for the   :O—S—Cl:
   | **last 2** electrons |
  :Cl: :Cl:

Step 3. Evaluate the formal charge on each atom. In both structures each Cl atom has a formal charge of $7 - \frac{1}{2}(2) - 6 = 0$. In structure (a) the O atom has a formal charge of $6 - \frac{1}{2}(6) - 2 = +1$, and the S atom has a formal charge of $6 - \frac{1}{2}(2) - 6 = -1$. In structure (b) the situation is reversed; the formal charge on $S = 6 - \frac{1}{2}(6) - 2 = +1$, and the formal charge on $O = 6 - \frac{1}{2}(2) - 6 = -1$.

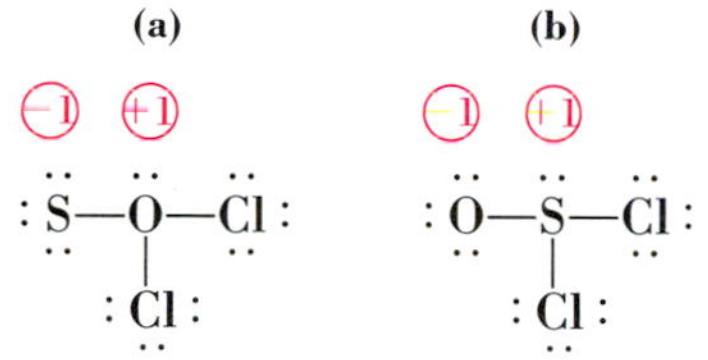

    (a)   **(b)**

⊖ ⊕   ⊖ ⊕

:S—O—Cl: :O—S—Cl:
   | |
  :Cl: :Cl:

Can you see that we might better represent thionyl chloride by writing the formula $OSCl_2$ rather than $SOCl_2$?

Step 4. Structure (a) has the negative formal charge on the S atom, whereas structure (b) has the negative formal charge on the O atom. Because O is a more electronegative element than S, we conclude that structure (b) is the correct one.

SIMILAR EXAMPLES: Exercises 6, 31, 32, 79.

Polyatomic Ions. Polyatomic ions consist of two or more atoms. The forces holding atoms together *within* such ions are covalent bonds. Consider, for example, the hydroxide ion, OH^-. Its Lewis structure must involve *eight* electrons: six from O, one from H, and one additional electron to account for the net charge of -1. (The "extra" electron is shown in color in structure 10.23, but of course there is no way of distinguishing among electrons.)

$$[:\!\ddot{O}\!:H]^- \qquad\qquad (10.23)$$

Thus, the *ionic* compound sodium hydroxide consists of simple Na^+ cations and the *polyatomic* anions, OH^-. The negative charge on OH^- arises from an electron gained from Na.

$$Na\cdot \ + \ \cdot\ddot{O}\!:H \longrightarrow [Na]^+[:\!\ddot{O}\!:H]^- \qquad\qquad (10.24)$$

Example 10-11

Writing a Lewis structure for a polyatomic ion. Write a plausible Lewis structure for the chlorite ion, ClO_2^-.

Solution

Step 1. Assess the number of valence electrons in the Lewis structure.

		To establish	
From Cl	From O	charge of −1	
$(1 \times 7) +$	$(2 \times 6) +$	1	$= 20$

Step 2. Write the plausible skeleton structure. For oxoanions (polyatomic anions containing oxygen) it is generally true that the O atoms are *terminal* atoms bonded to some other nonmetal as the central atom. The skeleton structure is

This is also the skeleton structure predicted by statement (10.22).

O—Cl—O

Step 3. Join the terminal atoms to the central atom with single covalent bonds and then use the remaining electrons to provide lone pair electrons, first around the terminal atoms and then around the central atom. Four electrons are used for bond pairs and the remaining 16 electrons provide an octet around each atom in the structure.

Step 4. Show the net charge on the ion and the formal charges on atoms.

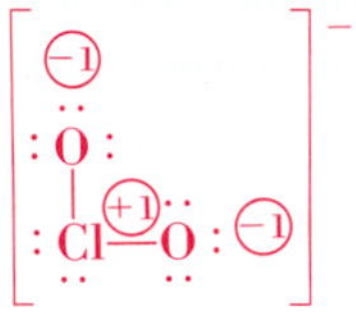

$$\left[\begin{array}{c} \overset{\ominus}{\underset{..}{\ddot{\mathrm{O}}}} \\ | \\ :\ddot{\mathrm{Cl}} \overset{\oplus}{-} \ddot{\mathrm{O}} : \ominus \end{array} \right]^{-} \qquad (10.25)$$

SIMILAR EXAMPLES: Exercises 7, 33, 34.

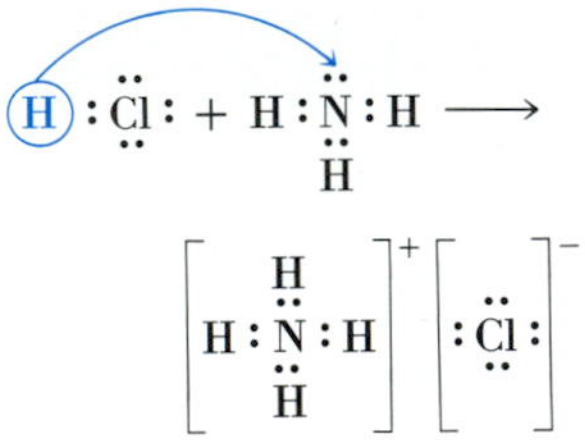

FIGURE 10-13
Formation of the ammonium ion, NH$_4^+$.

The formation of coordinate covalent bonds gives rise to formal charges in Lewis structures.

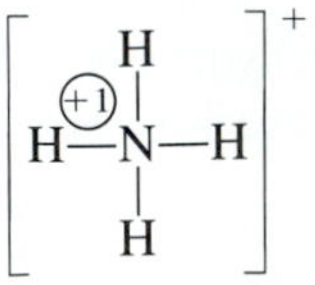

Figure 10-13 illustrates bonding in the polyatomic *cation* NH$_4^+$, the ammonium ion. It also shows the formation of the ionic compound ammonium chloride by the reaction of HCl and NH$_3$. In determining the number of valence electrons for the Lewis structure of NH$_4^+$, we find *five* from N, *one* each from *four* H atoms, *minus one* electron to account for the net charge +1. This gives a total of eight electrons for the Lewis structure. Another way to look at this situation is that in the formation of NH$_4^+$ (see again Figure 10-13) the fourth hydrogen atom that attaches itself to the N atom of the NH$_3$ molecule leaves its electron behind with the Cl atom (converting it to Cl$^-$). In the bond between the N atom and this fourth H atom the N atom supplies *both* electrons (its lone pair) for the covalent bond. A covalent bond in which one atom contributes both electrons is called a **coordinate covalent** bond. Once such a bond is formed, however, it is indistinguishable from a regular covalent bond. In other words, in the NH$_4^+$ ion all four N—H bonds are equivalent. Note also that the formal charge on N is +1 and that the sum of the formal charges for all the atoms is equal to the ionic charge, +1.

$$\left[\begin{array}{c} \mathrm{H} \\ \overset{\oplus}{|} \\ \mathrm{H} - \mathrm{N} - \mathrm{H} \\ | \\ \mathrm{H} \end{array} \right]^{+} \qquad (10.26)$$

10-8 Resonance

Oxygen gas commonly occurs as *diatomic* molecules O$_2$, but it can also exist as *triatomic* molecules O$_3$. This triatomic form of oxygen, called *ozone*, is found

The shapes of molecules and the concept of bond angle (117° for O_3) are discussed in Section 10-10.

naturally in the stratosphere and is also produced in the lower atmosphere as a constituent of smog (see Section 14-10). Let us try to draw a Lewis structure for the O_3 molecule from the following *experimental* data: O—O bond lengths, 128 pm; O—O—O bond angle, 117°. These data suggest the following structure.

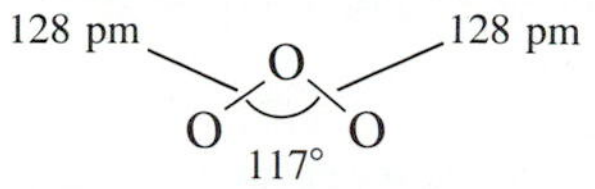

When we apply the usual rules for Lewis structures (deploy 18 valence electrons around the central and terminal atoms, give each atom an outer-shell octet, and so on) we come up with these two possibilities.

But there is something wrong with both of these structures. Each suggests that one oxygen-to-oxygen bond is single and the other is double. Yet, experimental evidence indicates that the two oxygen-to-oxygen bonds are the same. Furthermore, from Table 10-2 we see that the O—O bond length is 145 pm and the O=O bond length is 121 pm. The measured bond lengths in O_3 have an intermediate value (128 pm); the bonds are intermediate between a single and double bond.

We can resolve our difficulty by saying that the *true* Lewis structure for O_3 is *neither* of these structures. Instead the true structure is a composite or *hybrid* of these two structures, a fact that we can represent as

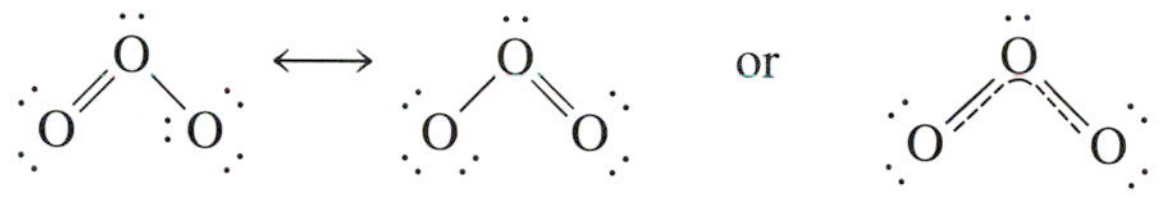

(10.27)

A mundane analogy is to consider a mule to be a resonance hybrid of a horse and a donkey. The mule has some of the characteristics of each, but it is a very distinctive animal. It is neither a horse part of the time and a donkey the rest, nor is it half-horse and half-donkey.

The phenomenon we have been describing is called **resonance.** This is a situation in which more than one plausible structure can be written for a species, and in which the true structure cannot be written at all. The true structure is called a **resonance hybrid** of plausible *contributing* structures. In (10.27) the two contributing structures on the left are joined by a double-headed arrow. The arrow does *not* mean that the molecule is in one structure part of the time and the other structure the rest of the time. It simply means that the true structure is a *hybrid* of the contributing structures shown. The structure on the right in (10.27) is an attempt to suggest the resonance hybrid with a single structure. Note that whichever representation we use we are suggesting that the oxygen-to-oxygen bonds are identical and that each has a bond order of 1.5 (halfway between a single and a double bond).

Example 10-12

Representing the Lewis structure of a resonance hybrid. **Write the Lewis structure of the nitrate ion, NO_3^-. All three nitrogen-to-oxygen bonds in the ion are identical.**

Solution. The structure we are seeking must contain 24 valence electrons (*five* from N, *six* each from *three* O atoms, and *one* extra electron to produce the net charge of −1). Trial 1 provides each terminal atom with an octet, but it leaves the central atom with an incomplete octet. This situation is remedied by moving a lone pair of electrons from one of the O atoms to form a double bond between it and the N atom, as in trial 2.

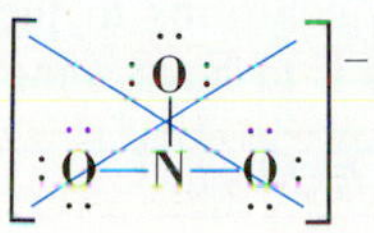

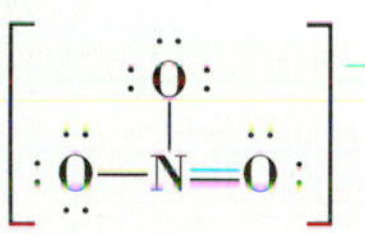

Are You Wondering:

Why the following structure or others like it cannot be written for the nitrate ion, together with or in place of the ones shown in (10.28)?

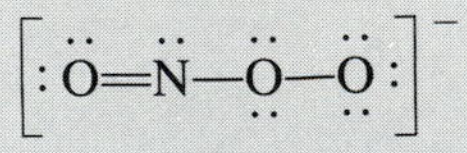

$$\left[\ :\ddot{O}=N-\ddot{O}-\ddot{O}:\ \right]^{-} \quad \textbf{(incorrect)}$$

1. This structure *cannot* be considered a contributing structure to the resonance hybrid of (10.28). *It has a different arrangement of its atoms.* Valid contributing structures must have the same atomic arrangement and differ *only* in the distribution of electrons.
2. Although the structure proposed here conforms to the formal charge rules even better than those of (10.28), it does not conform to the *experimental* evidence that all the O atoms are bonded directly to the N atom.
3. Although some molecules have several central atoms (that is, with atoms strung out in a chain), usually simple structures are compact. *Where possible,* write Lewis structures with a single central atom.

Trial 2 certainly produces a plausible structure, but it does not conform to the statement that all the nitrogen-to-oxygen bonds are *identical*. No single structure can be drawn to do this. Three contributing structures can be written, and the true structure is a resonance hybrid of these.

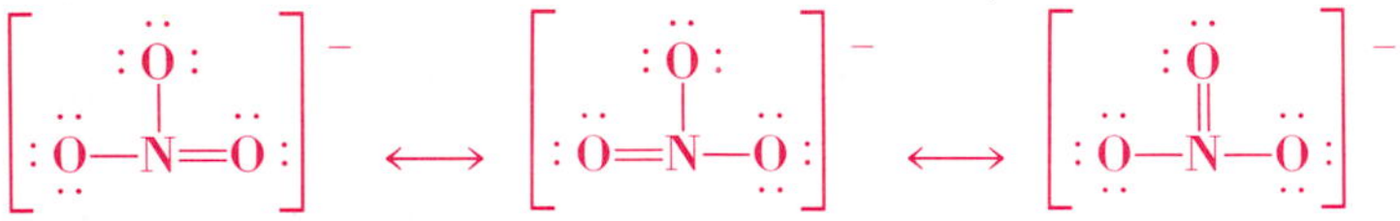

$$\left[\ :\ddot{O}-N=\ddot{O}:\ \right]^{-} \longleftrightarrow \left[\ :\ddot{O}=N-\ddot{O}:\ \right]^{-} \longleftrightarrow \left[\ :\ddot{O}-N-\ddot{O}:\ \right]^{-} \tag{10.28}$$

SIMILAR EXAMPLES: Exercises 36, 37, 38.

10-9   Exceptions to the Octet Rule

Odd-Electron Species. The molecule NO has 11 valence electrons, an *odd* number. If the number of valence electrons in a Lewis structure is *odd,* we can immediately say that

- There is at least one unpaired electron somewhere in the structure.
- At least one atom lacks a completed octet of electrons.

In the Lewis structure below there is an unpaired electron on the N atom, and the N atom has only *seven* outer-shell electrons.

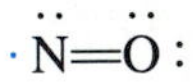

$$\cdot N=\ddot{O}: \tag{10.29}$$

The presence of unpaired electrons causes odd-electron species to be *paramagnetic*. NO is paramagnetic. Generally, molecules with an *even* number of electrons are expected to have all electrons paired and to be *diamagnetic*. Yet, O_2, with 12 valence electrons, is *paramagnetic. The O_2 molecule must have unpaired electrons.* This is why we labeled structure (10.8) "incorrect." It contained no unpaired electrons. A resonance hybrid of the following two contributing structures has unpaired electrons, and also conforms to the fact that the experimentally measured bond length in O_2 shows it to have some multiple bond character.

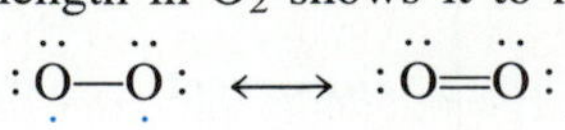

$$:\ddot{O}-\ddot{O}: \longleftrightarrow :\ddot{O}=\ddot{O}: \tag{10.30}$$

Incomplete Octets. Up to this point we have rejected any Lewis structure in which one or more atoms has not had an outer-shell octet of electrons. But, there are a few structures with incomplete octets that do appear to be correct. This is the case for boron trifluoride, BF_3. (For emphasis, the valence electrons of the B atom are shown in color.)

$$\begin{array}{c} \ddot{} \quad \ddot{} \\ :\!\ddot{F}\!:\!B\!:\!\ddot{F}\!: \\ :\!\ddot{F}\!: \\ \end{array} \tag{10.31}$$

Here is a fact that confirms the plausibility of structure (10.31): BF_3 readily reacts with NH_3 to form the compound $H_3N \cdot BF_3$. In this compound the N atom donates both electrons to the boron–nitrogen bond (a coordinate covalent bond) and the B atom completes its octet.

> To resolve the difficulty of an incomplete octet on the central atom we usually shift a lone pair of electrons from a terminal atom to a bond with the central atom. Here this would produce a B-to-F double bond and formal charges of -1 on B and $+1$ on F. This is a less acceptable structure than in (10.31).

$$\begin{array}{cc} \begin{array}{c} :\!\ddot{F}\!:\!H \\ :\!\ddot{F}\!:\!B\!:\!N\!:\!H \\ :\!\ddot{F}\!:\!H \end{array} & \text{or} & \begin{array}{c} :\!\ddot{F}\!:\ H \\ |\quad| \\ :\!\ddot{F}\!-\!B\!-\!N\!-\!H \\ |\quad| \\ :\!\ddot{F}\!:\ H \end{array} \end{array} \tag{10.32}$$

"Expanded" Octets. Phosphorus forms two chlorides, PCl_3 and PCl_5. The Lewis structure for PCl_3 meets the basic criterion of all atoms acquiring outer-shell octets of electrons. In PCl_5, because the five Cl atoms are bonded directly to the central P atom, *ten* electrons are found in the outer shell of the P atom. The octet has been "expanded" to 10 electrons. In the SF_6 molecule the octet is further expanded to 12 electrons. Octet expansion may be shown as follows, where the electrons to be counted around the central atom are shown in blue.

$$\tag{10.33}$$

octet	expanded octet	expanded octet

In Section 11-1 we will see that when an atom acquires an octet of electrons the *s* and *p* subshells of the outermost electronic shell are filled. Expansion to 10 or 12 electrons requires additional orbitals. The energy difference between the $2p$ and $3s$ orbitals is too great to provide this additional bonding possibility to the elements of the *second* period. Once the *d* subshell becomes available, however, expanded octets become possible. Thus the phenomenon of octet expansion is found in elements in the *third* and *higher* periods.

At times, by using expanded octets, we can write Lewis structures that correspond more closely to experimental evidence than do structures limited to octets. For example, *without* using an expanded octet we obtain a Lewis structure for SO_4^{2-} in which all bonds are single bonds and all atoms carry a formal charge.

$$\tag{10.34}$$

We can write a structure with fewer formal charges by using some sulfur-to-oxygen double bonds, and experimentally determined bond lengths indicate that there is

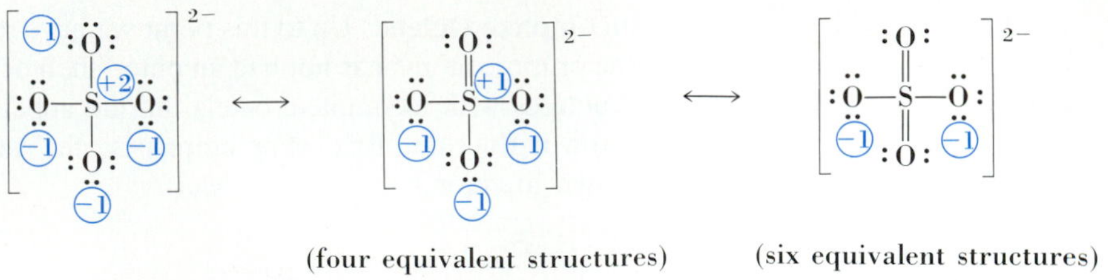

FIGURE 10-14
Representation of bonding in the sulfate ion.

The concepts of resonance and the expanded octet provide the best representation of the sulfate ion. The resonance hybrid of SO_4^{2-} involves contributions from all the structures shown here.

$$\left[\ \ \overset{\ominus}{:}\overset{..}{\underset{..}{O}}:\ \ \atop :\overset{..}{O}-\overset{\oplus 2}{S}-\overset{..}{O}: \atop \overset{\ominus}{:}\overset{..}{\underset{..}{O}}\overset{\ominus}{:} \right]^{2-} \longleftrightarrow \left[:\overset{..}{O}: \atop :\overset{..}{O}-\overset{\oplus 1}{\overset{\|}{S}}-\overset{..}{O}: \atop \overset{\ominus}{:}\overset{..}{\underset{..}{O}}\overset{\ominus}{:} \right]^{2-} \longleftrightarrow \left[:\overset{..}{O}: \atop :\overset{..}{O}-\overset{\|}{S}-\overset{..}{O}: \atop \overset{\ominus}{:}\overset{..}{\underset{..}{O}}\overset{\ominus}{:} \right]^{2-}$$

(four equivalent structures)　　　　(six equivalent structures)

some multiple bond character to these bonds. Moreover, since the sulfur-to-oxygen bonds in SO_4^{2-} are all the same, we find it necessary to invoke resonance in writing the Lewis structure, as shown in Figure 10-14.

Example 10-13

Using expanded octets in writing Lewis structures. Show that by using an expanded octet for the sulfur atom in thionyl chloride, $SOCl_2$ (Example 10-10), you can write a Lewis structure with *no* formal charges.

Solution. Follow the first two steps in the procedure for structure (b) in Example 10-10, to obtain this structure

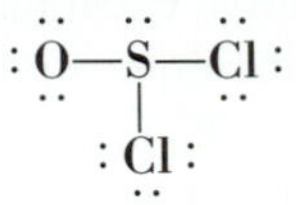

$$:\overset{..}{O}-\overset{..}{S}-\overset{..}{Cl}: \atop \ \ \ \ \ :\overset{..}{Cl}:$$

Then shift a pair of electrons from the terminal O atom to the S-to-O bond, making it a double bond. Note that the S atom now has *10* electrons in its outer shell.

Note that if the double bond were placed between S and Cl this would require formal charges: +1 on Cl and −1 on O. There are no formal charges in structure (10.35).

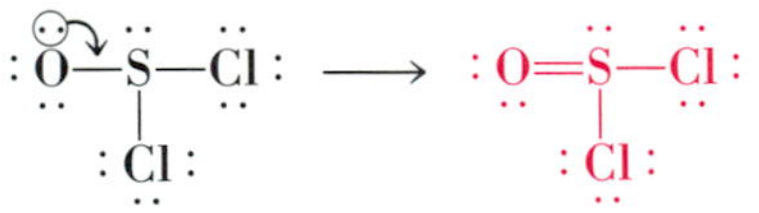

$$:\overset{\curvearrowright}{\overset{..}{O}}-\overset{..}{S}-\overset{..}{Cl}: \atop \ \ \ \ :\overset{..}{Cl}: \ \ \ \ \longrightarrow \ \ \ :O=\overset{..}{S}-\overset{..}{Cl}: \atop \ \ \ \ \ \ \ :\overset{..}{Cl}: \tag{10.35}$$

Determine the formal charges: For Cl, 7 valence electrons $-\frac{1}{2}(2$ bonding electrons$) - 6$ lone-pair electrons $= 7 - 1 - 6 = 0$. For S, 6 valence electrons $- \frac{1}{2}(8$ bonding electrons$) - 2$ lone-pair electrons $= 0$. For O, 6 valence electrons $- \frac{1}{2}(4$ bonding electrons$) - 4$ lone-pair electrons $= 0$.

SIMILAR EXAMPLES: Exercises 9, 44, 45.

10-10 Molecular Shapes

Since the distribution of electronic charge in an atom extends in all directions from the nucleus, an atom has a three-dimensional shape (we have pictured it as a sphere). Molecules are made up of atoms, so they too occupy three-dimensional space, but this three-dimensional space can come in different shapes. Let us agree that by molecular shape we mean the geometrical figure we get by joining the nuclei of bonded atoms by straight lines.

Figure 10-15 pictures a hypothetical *triatomic* (three-atom) molecule, represented through a "ball-and-stick" model. The balls represent the three atoms in the molecule and the straight lines (sticks) the bonds between atoms. In reality the atoms in the molecule are in close contact, but for clarity we show only the centers of the atoms. To have a complete description of the shape of a molecule, we need to know two quantities,

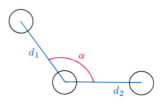

FIGURE 10-15

Geometrical shape of a molecule.

A hypothetical triatomic molecule is represented here. Although the atoms are actually in contact, for clarity only the centers of the atoms are shown (producing a so-called ball-and-stick model). To establish the shape of this molecule we must determine the distances between the centers of the bonded atoms (d_1 and d_2) and the angle between the adjacent bonds (α). Our primary concern is with the bond angle. For molecules with more than three atoms, additional bond distances and angles must be established, usually for a three-dimensional figure.

- **bond lengths,** the distances between the nuclei of bonded atoms. In Figure 10-15 the bond lengths are d_1 and d_2.
- **bond angles,** the angle between adjacent lines representing chemical bonds. The triatomic molecule in Figure 10-15 has just one bond angle, α.

Of these two quantities, we will concentrate on bond angles.

In a *diatomic* molecule there is only one bond and no bond angle. Since the geometrical figure determined by two points is a straight line, *all diatomic molecules are linear*. In a *triatomic* molecule there are two bonds and one bond angle. If the bond angle is 180° the three atoms lie on a straight line and the molecule is *linear*. For any other bond angle the triatomic molecule is said to be *angular* or *bent*. In Section 10-6 we found, through measurements of dipole moments, that the triatomic molecule CO_2 is *linear* and H_2O is *bent*. Some molecules with more than three atoms (*polyatomic* molecules) also have planar or even linear shapes. More commonly, however, the centers of the atoms in a polyatomic molecule define a three-dimensional geometrical figure.

The fact that the water molecule is bent makes it a *polar* molecule, and this in turn contributes to many unusual properties for water: the fact that it is a liquid (rather than a gas) at body temperature (see Section 12-7), its ability to dissolve so many different substances (see Section 13-2), and so on. Although water may provide the most striking example of the importance of molecular shape in determining the properties of a substances, we will see other examples later in the text.

Valence-Shell Electron-Pair Repulsion (VSEPR) Theory. Molecular shapes cannot be predicted from empirical formulas, and they cannot be predicted with Lewis structures. *They must be determined by experiment.* However, there is a way of thinking about molecules that allows us to predict their probable molecular shapes, and these predictions are usually borne out by experiment. This approach, called the valence-shell electron-pair repulsion theory (written VSEPR and pronounced ''vesper'') focuses on *pairs* of electrons in the outermost (*valence*) electronic shell. *Electron pairs, whether they are in chemical bonds (bond pairs) or unshared (lone pairs) repel one another. Electron pairs assume orientations in space to minimize repulsions.* This tendency to minimize repulsions between electron pairs causes chemical bonds from a central atom to assume certain orientations. This, in turn, results in particular geometrical shapes for molecules.

Consider a noble gas atom such as Ne, for example. What orientation will the four pairs of valence electrons ($2s^2 2p^6$) assume? As suggested by the ''balloon'' analogy in Figure 10-16, the electron pairs are farthest apart when they occupy the corners of a regular tetrahedron with the atomic nucleus at its center. Now consider the methane molecule, CH_4, in which the central C atom has acquired the Ne electron configuration by forming covalent bonds with four H atoms.

$$\begin{array}{c} \text{H} \\ \ddots \\ \text{H} \colon \text{C} \colon \text{H} \\ \ddots \\ \text{H} \end{array}$$

The method predicts that CH_4 should be a *tetrahedral* molecule, with a C atom at the center of the tetrahedron and H atoms at the corners. This structure agrees with that established by experiment.

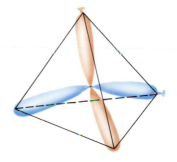

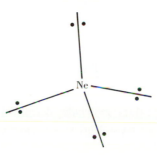

FIGURE 10-16

Balloon analogy to valence-shell electron-pair repulsion.

When two elongated balloons are twisted together at their centers, separation into four lobes occurs. To minimize mutual interferences, the lobes spread out into a tetrahedral shape. (A tetrahedron has four faces, each an equilateral triangle.) The lobes are analogous to valence-shell electron pairs. The distribution of the four pairs of valence-shell electrons of a neon atom is also shown.

FIGURE 10-17
Geometric shapes based on the tetrahedral distribution of four electron pairs—CH_4, NH_3, and H_2O.

The shape of the molecule is established by the blue lines. Lone-pair electrons are shown as brown dots along broken lines originating at the central atom. Space-filling models are shown at the top of the figure.
(a) All electron pairs about the central carbon atom are bond pairs. (The valence of the carbon atom is said to be *saturated*.) The blue lines that establish the shape of the CH_4 molecule are different from those representing carbon-to-hydrogen bonds (in black).
(b) The lone pair of electrons is directed toward the "missing" corner of the tetrahedron. The figure remaining is a trigonal pyramid. The nitrogen-to-hydrogen bonds form three of the edges of this pyramid.
(c) The H_2O molecule is an angular molecule outlined by the two oxygen-to-hydrogen bonds. [Photograph by Carey B. Van Loon]

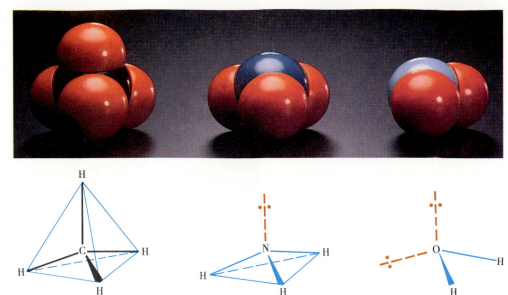

VSEPR notation:

AX_4 (a) AX_3E (b) AX_2E_2 (c)

In NH_3 and H_2O the central atom is also surrounded by four pairs of electrons,

$$H : \overset{..}{N} : H \quad \text{and} \quad H : \overset{..}{\underset{..}{O}} : H$$

but these molecules do not have a tetrahedral shape. The situation is this: VSEPR theory predicts the distribution of electron pairs. The geometrical shape of a molecule, however, is described in terms of the geometrical figure that results by *joining the appropriate atomic nuclei by straight lines.*

In the NH_3 molecule only *three* of the electron pairs are bond pairs; the fourth pair is a *lone pair*. When the nucleus of the N atom is joined to those of the H atoms, the geometrical figure obtained is a *pyramid* (called a trigonal pyramid), not a tetrahedron. The pyramid has H atoms at the base and an N atom at the apex. In the H_2O molecule two of the four electron pairs about the O atom are *bonding* pairs and two are *lone* pairs. The figure obtained by joining the nuclei of the two H atoms with that of the O atom is a *planar* figure. The molecule is *angular* or *bent*. The geometrical shapes of CH_4, NH_3, and H_2O are shown in Figure 10-17, together with "space-filling" molecular models. A special notation is used in Figure 10-17 that has this meaning: A represents the central atom, X represents a terminal atom bonded to the central atom, and E represents a lone pair of electrons. Thus, the symbolism AX_2E_2 signifies that *two* atoms (X) are bonded to the central atom (A). The central atom also has *two* lone pairs of electrons. H_2O is an example of a molecule of the AX_2E_2 type.

For a tetrahedral distribution of electron pairs we expect bond angles of 109.5°, known as the **tetrahedral bond angle.** In the CH_4 molecule the measured bond angles are, in fact, 109.5°. The bond angles in NH_3 and H_2O are slightly smaller: 107° for the H—N—H bond and 104.5° for the H—O—H bond. These less-than-tetrahedral bond angles can be explained by assuming that the charge cloud of the lone-pair electrons spreads out and forces the bonding electron pairs closer together, reducing the bond angles.

Possibilities for Electron-Pair Distributions. The three molecules discussed in some detail to this point—CH_4, NH_3, and H_2O—all have a molecular geometry

FIGURE 10-18
Several electron-pair distributions illustrated.

The electron-pair distributions pictured here are trigonal planar (orange), tetrahedral (green), trigonal bipyramidal (pink), and octahedral (yellow). The relationship between electron-pair distributions and molecular shapes is summarized in Table 10-3. [Courtesy of Arlo Harris; photograph by Carey B. Van Loon]

based on the distribution of four electron pairs, a *tetrahedral* distribution. In general, we will encounter situations in which central atoms may have 2, 3, 4, 5, or 6 electron pairs distributed about them. The geometrical orientations of electron pairs for these cases are

- **linear:**   2 electron pairs
- **trigonal planar:**   3 electron pairs
- **tetrahedral:**   4 electron pairs
- **trigonal bipyramidal:**   5 electron pairs
- **octahedral:**   6 electron pairs.

Some of these electron pair distributions are pictured through a continuation of the balloon analogy in Figure 10-18.

Now, just as we saw in comparing CH_4, NH_3, and H_2O, the geometrical shape of a molecule and the geometrical distribution of electron pairs around its central atom are the same *only* for cases where all the electron pairs are bond pairs. These are for the VSEPR notation AX_n (that is, AX_2, AX_3, . . .). In Table 10-3 the AX_n cases are illustrated through photographs of ''ball-and-stick'' models. If one or more of the electron pairs are lone pairs, the geometrical distribution of the electron pairs and the geometrical shape of the molecule are *different*, as in the other examples in Table 10-3.

To understand some of these other examples in Table 10-3 we need two additional ideas

- *The closer together two pairs of electrons are forced, the stronger the repulsion between them.* Thus the force of repulsion between two electron pairs is much stronger at an angle of 90° than at an angle of 120° or 180°.
- *Lone-pair electrons spread out more than do bond-pair electrons.* As a result the repulsion of one lone pair of electrons for another lone pair is greater than, say, between two bond pairs. In fact, we can set up the following order of repulsive forces, from strongest to weakest.

Comparative strengths of electron-pair repulsions.

lone pair–lone pair > lone pair–bond pair > bond pair–bond pair   (10.36)

Now consider an example from Table 10-3: SF_4 (having the notation AX_4E). There are two possible places to put the lone pair of electrons. In Table 10-3 this pair is placed into the central plane of the bipyramid, with the result that *two* of the lone pair–bond pair interactions are at 90°. If the lone-pair electrons were placed at the top of the bipyramid, above the central plane, there would be *three* lone pair–bond pair interactions at 90°. This is a less favorable arrangement.

TABLE 10-3
Molecular Geometry as a Function of Geometrical Distribution of Valence-Shell Electron Pairs and Number of Lone Pairs

Number of electron pairs	Geometrical distribution of electron pairs	Number of lone pairs	VSEPR notation	Molecular geometry	Ideal bond angles	Example[c]	
2	linear	0	AX_2	X—A—X (linear)	180°	$BeCl_2$	
3	trigonal planar	0	AX_3	(trigonal planar)	120°	BF_3	
	trigonal planar	1	AX_2E	(angular)	120°	SO_2[a]	(BF_3)
4	tetrahedral	0	AX_4	(tetrahedral)	109.5°	CH_4	
	tetrahedral	1	AX_3E	(trigonal pyramidal)	109.5°	NH_3	
	tetrahedral	2	AX_2E_2	(angular)	109.5°	OH_2	(CH_4)
5	trigonal bipyramidal	0	AX_5	(trigonal bipyramidal)	90°, 120°	PCl_5	
	trigonal bipyramidal	1	AX_4E[b]	(irregular tetrahedral)	90°, 120°	SF_4	(PCl_5)

TABLE 10-3 (Continued)

Number of electron pairs	Geometrical distribution of electron pairs	Number of lone pairs	VSEPR notation	Molecular geometry	Ideal bond angles	Example[c]
	trigonal bipyramidal	2	AX_3E_2	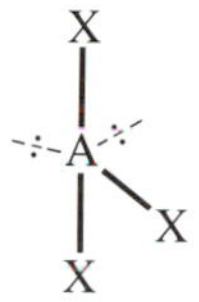 (T-shaped)	90°	ClF_3
	trigonal bipyramidal	3	AX_2E_3	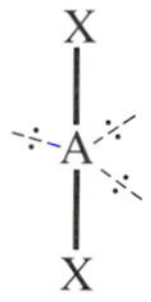 (linear)	180°	XeF_2
6	octahedral	0	AX_6	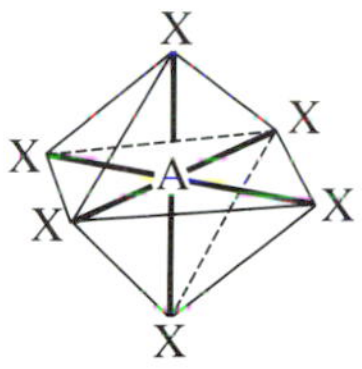 (octahedral)	90°	SF_6
	octahedral	1	AX_5E	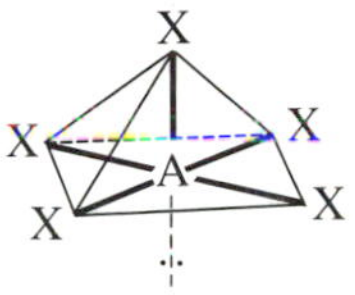 (square pyramidal)	90°	BrF_5
	octahedral	2	AX_4E_2	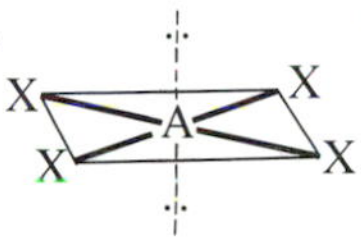 (square planar)	90°	XeF_4

(SF_6)

[a]For a discussion of the structure of SO_2, see page 337.
[b]For a discussion of the placement of the lone-pair electrons in this structure, see page 333.
[c][Photographs by Carey B. Van Loon]

Applying VSEPR Theory. In applying VSEPR theory let us use this four-step procedure for assessing molecular shape.

Procedure for applying VSEPR theory.

1. Draw a plausible Lewis structure of the species (molecule or polyatomic ion).*
2. Determine the number of electron pairs around the central atom and identify each pair as being either a *bond* pair or a *lone* pair.

(10.37)

*Strictly speaking, the VSEPR theory can be applied without ever writing Lewis structures. Refer to Exercise 97 to see how this is done. Our approach here will be to write Lewis structures, since we have found them so useful for other purposes as well.

3. Establish the geometrical orientation of the electron pairs around the central atom, that is, linear, trigonal planar, tetrahedral, trigonal bipyramidal, or octahedral.
4. Use the positions around the central atom that are occupied by other atoms (not lone-pair electrons) to specify the geometrical shape. That is, use information from Table 10-3.

(10.37)

Example 10-14

Using the VSEPR theory to predict a geometrical shape. Predict the shape of the chlorate ion, ClO_3^-.

Solution. Apply the four steps outlined in (10.37).
Step 1. Write the Lewis structure. The number of valence electrons is

		To establish	
From Cl	From O	charge of -1	
$(1 \times 7) +$	$(3 \times 6) +$	1	$= 26$

In the Lewis structure, the atom with the lower electronegativity, Cl, is the central atom. A plausible structure is

$$: \overset{..}{\underset{..}{O}} — \overset{..}{Cl} — \overset{..}{\underset{..}{O}} : \quad ^-$$
$$| $$
$$: \overset{}{\underset{..}{O}} :$$

Step 2. Count and identify the types of electrons around the central atom. There are four electron pairs around the central atom; 3 are *bond* pairs (red) and one is a *lone* pair (blue).
Step 3. Describe the orientation of the electron pairs. The orientation of *four* electron pairs is *tetrahedral*.
Step 4. Describe the geometrical shape of the ion. In VSEPR notation the ClO_3^- ion is AX_3E. **The ion has a *pyramidal* shape** (similar to that shown for NH_3 in Figure 10-17 and Table 10-3).

SIMILAR EXAMPLES: Exercises 10, 61, 82

Example 10-15

Using VSEPR theory to predict a geometrical shape. Predict the geometrical shape of the polyatomic anion ICl_4^-.

Solution. Again use the four-step approach.
Step 1. The Lewis structure of the anion has I as the central atom and a total of 36 electrons (seven from each of the halogen atoms and one extra electron for the -1 charge). To join four Cl atoms to the I atom and to provide octets for all the atoms requires 32 electrons. This means that four *additional* electrons must be placed around the central I atom. That is, the I atom has an expanded octet. Since these additional electrons are two lone pairs, it is immaterial exactly how we show them around the I atom in the Lewis structure.

$$\left[\begin{matrix} : \overset{..}{\underset{}{Cl}} : \ : \overset{..}{\underset{}{Cl}} : \\ : \overset{..}{\underset{..}{I}} : \\ : \overset{}{\underset{..}{Cl}} : \ : \overset{}{\underset{..}{Cl}} : \end{matrix} \right]^-$$

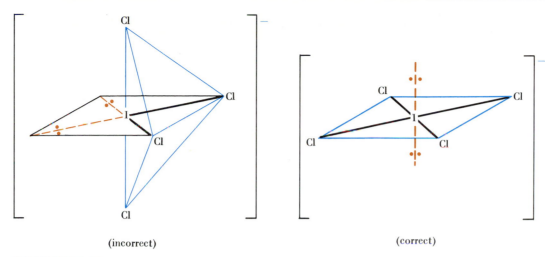

(incorrect)   (correct)

FIGURE 10-19
Two predictions of the structure of ICl_4^-—Example 10-15 illustrated.

The observed structure is the square planar structure shown on the right.

Step 2. The I atom is surrounded by six electron pairs; four are *bond* pairs and two are *lone* pairs.

Step 3. The orientation of six electron pairs around the central atom is *octahedral.*

Step 4. The ICl_4^- anion is of the type AX_4E_2, which according to Table 10-3 is *square planar.*

Figure 10-19 suggests that we might have considered two possibilities for the distribution of bond pairs and lone pairs in the anion ICl_4^-. The reason why the square planar structure is the correct one is that the lone pair–lone pair interaction is kept at 180°. In the incorrect structure this interaction is at 90°, leading to a strong repulsion.

SIMILAR EXAMPLES: Exercises 62, 67, 93.

Structures with Multiple Covalent Bonds. In a double or triple covalent bond, all the electrons in the bond are confined to the region between the bonded atoms. As a result, the electrons in a multiple bond, collectively, have the same effect on the shape of a molecule as does a single covalent bond. Let us see how to use this idea by considering the Lewis structure of sulfur dioxide, SO_2. In this structure S is the central atom, and the total number of valence electrons is 18 (3×6). One plausible structure is shown below

$$\text{:O} \overset{\cdot\cdot}{\underset{}{\diagup}} \overset{\text{S}}{\diagdown} \text{O:} \tag{10.38}$$

An equally plausible structure is

$$\overset{\cdot\cdot}{\underset{\text{:O}\diagdown\quad\diagup\text{O:}}{\text{S}}}$$

To apply the VSEPR theory it is necessary to consider only one contributing structure to the resonance hybrid.

Count the electrons shown in the double covalent bond as if this were a *single* bond. This bond and the S-to-O single bond account for *two* bond pairs. The third electron pair around the central S atom is a *lone* pair. Thus, the orientation of electron pairs around the central S atom corresponds to that of *three* electron pairs, *trigonal planar;* of the three electron pairs, two are bond pairs and one is a lone pair. This is the case of AX_2E (Table 10-3), and the molecule is *angular* or *bent,* with an expected bond angle of 120°. (The experimentally determined bond angle in SO_2 is 119°.)

Example 10-16

The industrial preparation of $POCl_3$ and some of its uses were described in Section 4-5.

Using VSEPR theory to predict the geometrical shape of a molecule with a multiple covalent bond. Predict the geometrical shape of a molecule of phosphoryl chloride, $POCl_3$.

Solution. In the Lewis structure we expect P as the central atom (lowest electronegativity) and a total of 32 valence electrons. The following is a plausible Lewis structure for $POCl_3$:

$$
\begin{array}{c}
\ominus \quad \overset{\displaystyle\;\;\;\;\ddot{C}l:}{\underset{\displaystyle\;\;\;\;\;}{\overset{\displaystyle\;\;\;\;\;\oplus|}{}}} \\
:\ddot{O}-\overset{|}{P}-\ddot{C}l: \\
\underset{\displaystyle:\ddot{C}l:}{|}
\end{array}
$$

Based on this Lewis structure, we conclude that the distribution of the four electron pairs around the P atom should be *tetrahedral*. And, because all four pairs are bond pairs, the structure of the molecule is also *tetrahedral*.

Note that the above Lewis structure has formal charges on the P and O atoms. We can write a better Lewis structure, one without formal charges, by allowing octet expansion of the central P atom. This requires shifting a lone pair of electrons from the O atom into the O—P bond.

$$
\begin{array}{c}
:\ddot{C}l: \\
| \\
:\ddot{O}=\overset{|}{P}-\ddot{C}l: \\
| \\
:\ddot{C}l:
\end{array}
$$

What molecular geometry should we predict based on this improved Lewis structure? By counting the O-to-P double bond *as if* it were a single bond with just one electron pair, we again count a total of four electron pairs around the P atom. The electron pair orientation and the molecular shape are both *tetrahedral*.

SIMILAR EXAMPLE: Exercise 63.

Molecules with More Than One Central Atom. VSEPR theory deals with the orientation of valence-shell electron pairs around a *central* atom. All the examples that we have discussed thus far have been based on a *single* central atom—C in CH_4, N in NH_3, O in H_2O, S in SO_2, and so on. But we can also apply the VSEPR theory to a molecule or polyatomic ion having two or more central atoms. In effect what we have to do is to work out the geometrical distribution of terminal atoms around *each* central atom and then combine the results into a single description of the molecular shape. We use this idea in Example 10-17.

Example 10-17 _____

Applying the VSEPR theory to a molecule with more than one central atom. Methyl isocyanate, H_3CNCO, a very toxic chemical, is used in the manufacture of insecticides, such as Sevin. In the H_3CNCO molecule, the three H atoms and the O atom are terminal atoms; the N and both C atoms are central atoms. Sketch this molecule.

Solution. To apply the VSEPR method we must begin with a plausible Lewis structure. The number of valence electrons in the structure is

From C   From N   From O   From H

$$(2 \times 4) + (1 \times 5) + (1 \times 6) + (3 \times 1) = 22$$

In drawing the skeleton structure and assigning valence electrons, we first obtain structure I. By shifting the indicated electrons, each atom acquires an octet (structure II).

Margin notes:

In applying the VSEPR theory we do not have to find the "best" Lewis structure. Any plausible Lewis structure will do.

The worst chemical accident in recorded history, killing over 2500 people, resulted from the release of methyl isocyanate (MIC) into the atmosphere at Bhopal, India, on December 3, 1984.

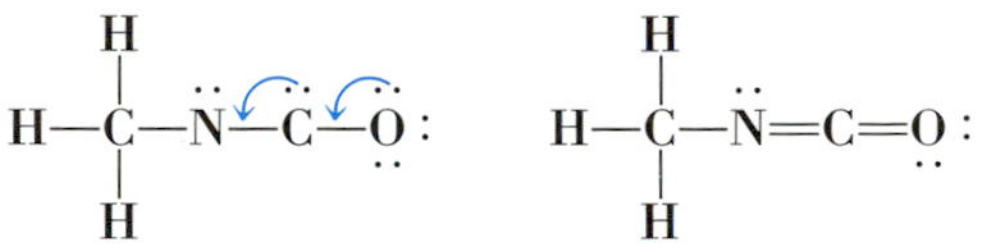

Structure I   **Structure II**

$$H-\overset{\overset{\displaystyle H}{|}}{\underset{\underset{\displaystyle H}{|}}{C}}-\ddot{N}\underset{}{\overset{\frown}{-}}\overset{\frown}{C}\underset{}{\overset{..}{-}}\ddot{O}: \qquad H-\overset{\overset{\displaystyle H}{|}}{\underset{\underset{\displaystyle H}{|}}{C}}-\ddot{N}=C=\ddot{O}:$$

The C atom to the left has four electron pairs distributed around it—all bond pairs. The geometrical shape of this end of the molecule is *tetrahedral*. The C atom to the right, forming two double bonds, is treated as if it had *two* pairs of electrons to be distributed. This distribution is *linear*. For the N atom we need to consider the distribution of *three* pairs of electrons—*trigonal planar*. The C—N—C bond angle should be about 120°.

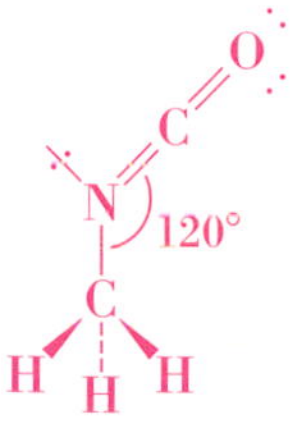

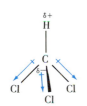

SIMILAR EXAMPLES: Exercises 69, 95.

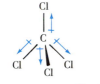

CCl$_4$: a nonpolar molecule
$\mu = 0$

CHCl$_3$: a polar molecule
$\mu = 1.92$ debye

FIGURE 10-20

Dipole moments and molecular shapes.

(a) The symmetrical distribution of the four carbon-to-chlorine bonds in CCl$_4$ causes a cancellation of all bond dipole moments; there is no resultant dipole moment in the molecule. CCl$_4$ is nonpolar.
(b) The carbon-to-hydrogen bond has a dipole moment of essentially zero because the electronegativities of C and H are quite similar. The three carbon-to-chlorine bond dipole moments cause the chlorine end of the molecule to develop a slight negative charge. The hydrogen end carries a slight positive charge. CHCl$_3$ is polar.

Molecular Shapes and Dipole Moments. According to our previous discussion of dipole moments and polar molecules (Section 10-6), we would expect the bond dipole moment of the C—Cl bond to be large. This is because the Cl atom is considerably more electronegative than the C atom. The experimentally measured bond dipole moment is large, 2.05 D. Since a molecule of carbon tetrachloride, CCl$_4$, has *four* C—Cl bonds, we might expect an even larger resultant dipole moment for the entire molecule. In fact, when we measure the resultant dipole moment for CCl$_4$ we find it to be *zero*. *CCl$_4$ is a nonpolar molecule.* The only way to account for this fact is if the Cl atoms are arranged around the C atoms in some symmetrical fashion in which the individual bond dipole moments cancel each other.

In CCl$_4$ there are four pairs of electrons, all bond pairs, distributed around the central C atom. The electron pair orientation and the shape of the molecule are both *tetrahedral*. From Figure 10-20 we see that a *symmetrical* tetrahedral orientation of the C—Cl bonds is in fact just what we need to account for the lack of a resultant dipole moment. Thus, we can use the VSEPR theory to predict whether molecules are *polar* (that is, have a resultant dipole moment) or *nonpolar* (a resultant dipole moment of 0). Our next prediction might be that the molecule will be polar if we replace one of the Cl atoms of CCl$_4$ with an atom having a much different electronegativity than Cl, say H. The molecule CHCl$_3$, in which there is now an imbalance in bond dipole moments, is indeed a polar molecule. The cases of CCl$_4$ and CHCl$_3$ are compared in Figure 10-20.

Example 10-18

Determining the relationship between geometrical shapes and the resultant dipole moments of molecules. Which of these molecules would you expect to be polar and which, nonpolar? Cl$_2$, ICl, BF$_3$, NO, SO$_2$, XeF$_4$.

Solution

Polar: ICl, NO, SO$_2$. ICl and NO are diatomic molecules with an electronegativity difference between the bonded atoms. SO$_2$ is a bent (nonlinear) molecule

The symmetrical structures in Table 10-3 are those with the VSEPR notations: AX_2, AX_3, AX_4, AX_5, AX_6, AX_2E_3, and AX_4E_2.

(structure 10.38) with an electronegativity difference between the S and O atoms.

Nonpolar: Cl_2, BF_3, XeF_4. Cl_2 is a diatomic molecule in which both atoms are the same; hence no electronegativity difference. For BF_3 and XeF_4 refer to Table 10-3. BF_3 is a symmetrical planar molecule (120° bond angles) and the individual bond dipole moments of the B—F bonds cancel each other. XeF_4 is a square planar molecule with the F atoms arranged symmetrically around the central Xe atom.

SIMILAR EXAMPLES: Exercises 16, 70, 71.

FOCUS ON
Polymers

In 1934, after four years of development work, Wallace Carrothers and his associates at E. I. du Pont de Nemours & Company succeeded in producing the first synthetic fiber—*nylon*. It is now possible for chemistry students to carry out a variation of this polymerization reaction in the general chemistry laboratory. [Courtesy of Arlo Harris; photograph by Carey B. Van Loon]

Let us use the Lewis symbols of C and H (Figure 10.1) to write a Lewis structure for ethylene, C_2H_4. Our first trial might be

$$\cdot \overset{\displaystyle H}{\underset{\displaystyle H}{C}} - \overset{\displaystyle H}{\underset{\displaystyle H}{C}} \cdot \qquad (10.39)$$

This is not an acceptable structure because it leaves each C atom with an incomplete octet and an unpaired electron. Let us try the usual remedy, that is, combine the two un-

paired electrons into an additional bond between the two C atoms. This is the accepted Lewis structure for ethylene.

$$\overset{\displaystyle H}{\underset{\displaystyle H}{C}} = \overset{\displaystyle H}{\underset{\displaystyle H}{C}} \qquad (10.40)$$

Something else we might have tried is to join several of the units represented by (10.39). Here is a grouping of three such units:

$$\cdot C - C - C - C - C - C \cdot \qquad (10.41)$$

In (10.41) the central unit (1) has a satisfactory Lewis structure, but we still have a problem with the end units (2 and 3). They have unpaired electrons and incomplete octets on the end C atoms. Let us add a couple more CH_2 units to (10.41). Now the structures of 1, 2, and 3 are satisfactory but the problem of unpaired electrons and incomplete octets has been shifted to the new end units, 4 and 5.

$$\cdot C - C - C - C - C - C - C - C - C - C \cdot \qquad (10.42)$$

What we are describing is the plausibility of joining together a large number of simple molecules called **monomers** (Gr. *monos*, single, and *meros*, parts) into a complex long-chain molecule called a **polymer** (Gr. *polys*, many, and *meros*, parts). We call the chemical reaction

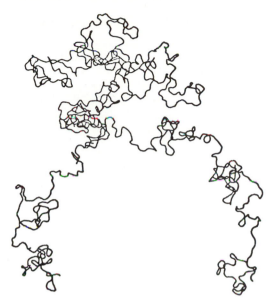

FIGURE 10-21

A hypothetical polyethylene molecule.

This is a randomly generated chain of 500 C_2H_4 monomer units. [From L. R. G. Treloar, *Introduction to Polymer Science*, Wykeham Publications (London) Ltd., 1970.]

that achieves this result a **polymerization reaction.** In the case we have been describing, the monomer is *ethylene* and the polymer is *polyethylene*, which is represented as

$$\left[\begin{array}{cc} H & H \\ | & | \\ -C-C- \\ | & | \\ H & H \end{array}\right]_n \qquad (10.43)$$

The subscript n signifies that the monomer unit within the brackets is repeated many times; n can have a range of values, say from several hundred to several thousand. A hypothetical molecule is pictured in Figure 10-21.

In Chapter 27 we consider how the double bond in C_2H_4 (structure 10.40) is "opened up" to initiate the polymerization reaction, how the polymer chains are propagated and terminated (to eliminate unpaired electrons), and the meaning of molecular weight when applied to a polymer.

Ubiquitous Polymers. The bulk of the polymer industry centers on *synthetic organic* polymers, like polyethylene. Ethylene is one of the principal starting materials for polymer production, and this accounts for its high ranking (fourth) among the key chemicals (recall Table 1-3). Another measure of the importance of polymers is that between 30 and 50% of all chemists and chemical engineers work with polymers.

In a general sense, any substance that can be thought of in terms of small units joined together into long chains or two- or three-dimensional networks is a polymer. This broad definition will allow us to consider certain *inorganic*

materials as polymers, e.g., diamond and graphite in Chapter 12 and liquid and plastic sulfur, red phosphorus, and SiO_2 in Chapter 14. This definition also helps us to recognize *natural* as well as synthetic polymers. Among the natural polymers that we will encounter (Chapter 28) are carbohydrates, proteins, and nucleic acids. One of the most familiar of the natural polymers is probably natural rubber.

Rubber. The milky white fluid called **latex** is collected as an exudate of the tree *Hevea brasiliensis*. When this fluid is coagulated with salt and acetic acid, the product is crude rubber. This rubber is refined and processed according to its final use. The name "rubber" is attributed to the chemist Joseph Priestley (1770), who found that it could be used to rub out pencil marks. Its chief use for several decades was in pencil erasers. The repeating unit of rubber is shown in brackets below.

$$\begin{array}{c} \underset{CH_3}{\diagdown}C=C\underset{CH_2}{\diagup}H \quad \left[\underset{CH_3}{\diagdown}C=C\underset{CH_2}{\diagup}H \right]_n \underset{CH_3}{\diagdown}C=C\underset{CH_2}{\diagup}H \\ \text{rubber} \end{array}$$

$$(10.44)$$

Early rubber products had many undesirable properties. The rubber tended to creep or flow, and the products were sticky in hot weather and stiff in cold weather. In 1839, Charles Goodyear, a Connecticut inventor, accidentally spilled a sulfur–rubber mixture on a hot stove; the resulting rubber was strong, more elastic, and more resistant to heat and cold. This treatment is now called **vulcanization** (after Vulcan, the Roman god of fire). The purpose of vulcanization is to form **crosslinks** between long polymer chains. For example, in structure (10.45) the link is through two sulfur atoms.

$$\begin{array}{c} CH_3 \\ | \\ \sim\!CH_2-C=CH-CH\sim \\ | \\ S \\ | \\ S \\ | \\ \sim\!CH_2-C=CH-CH\sim \\ | \\ CH_3 \end{array} \qquad (10.45)$$

Classification of Polymers. Some polymers have sites on their chains at which crosslinks between chains are formed when the polymer is heated for the first time. Such polymers can be heated to their softening points and molded into particular shapes. Because of crosslinking they retain these shapes on cooling. Reheating does not reverse the process; that is, instead of again becoming soft, the polymers degrade or decompose upon reheating.

TABLE 10-4
Some Synthetic Carbon-Chain Polymers

Name of polymer	Repeating unit	Some uses
Elastomers		
neoprene [polychloroprene]		wire and cable insulators, industrial hoses and belts, shoe soles and heels, gloves
silicone rubber		gaskets, electrical insulation, surgical membranes, medical devices for use in the body
Fibers		
nylon 66		hosiery, rope, tire cord, fish line, parachutes, artificial blood vessels
Acrilan, Orlon		fabrics, carpets, drapes, upholstery, electrical insulation
Plastics		
polyethylene		bags, bottles, tubing, packaging film, paper coating
polypropylene		laboratory and household ware, storage battery cases, artificial turf, surgical casts, toys
PVC, "vinyl" [poly(vinyl chloride)]		bottles, records, floor tile, food wrap, piping, hoses, linings for ponds and reservoirs
Teflon [poly(tetrafluoroethylene)]		bearings, insulation, gaskets, nonstick surfaces (ovenware, frying pans), heat-resistant industrial plastics

These are called **thermosetting** polymers. Other polymers, those unable to form crosslinks, also soften when heated, and they too can be molded into desired shapes that they retain on cooling. These polymers, however, are able to go through repeated cycles of heating, softening, reshaping, and cooling. These are called **thermoplastic** polymers. Another classification scheme is based on the properties of polymers described below. Examples are given in Table 10-4.

Elastomers. The chief characteristic of elastomers is their ability to be elongated under stress (stretched) and to regain their former shapes when the stress is relieved. In short, they are elastic. Rubber, whether natural or synthetic, is the best known example. Silicones are also elastomers (see Section 23-4). Current annual production of elastomers in the United States is about 3.5 million metric tons (1 metric ton = 1000 kg).

Fibers. Fibers are polymers oriented to provide optimum properties, such as high tensile strength, along one axis. These are threadlike polymers that can be woven into fabrics. Cotton, wool, and silk are natural fibers. Some synthetic fibers, such as nylon, Orlon, and Dacron, have additional desirable properties: increased tensile strength; lightness of weight; low moisture absorption; resistance to moths, mildew, rot, and fungus; and wrinkle resistance. Current annual United States production of fibers is about 10 million metric tons.

Plastics. Plastics, which are intermediate between elastomers and fibers, can be imparted with a variety of room-temperature properties. Thus polystyrene is stiff and brittle, whereas polypropylene is extremely tough, impact resistant, tear resistant, and flexible in thin sheets. Annually, about 15 million metric tons of plastics are produced in the United States.

Summary

The Lewis theory of chemical bonding focuses on the outer-shell (valence) electrons of the atoms joined together. The Lewis symbol of an atom represents these valence electrons through dots arranged around the chemical symbol. The Lewis structure of a compound is a combination of Lewis symbols of atoms, arranged in such a way as to satisfy a set of rules. The most important rules are that normally all the electrons in a Lewis structure are paired and each atom in the structure acquires eight electrons (an *octet*) in its outermost electronic shell.

Lewis structures can be used to obtain the formula unit of an ionic compound, but other aspects of ionic bonding are based on the clustering of large numbers of ions into a network called a *crystal*. Of special importance is the concept of *lattice energy,* the energy released when gaseous ions combine to produce one mole of a crystalline ionic compound. Lattice energies can be determined indirectly by combining other quantities, such as ionization energies and electron affinities.

For a covalent molecule, in addition to the basic principles of the Lewis theory, it is necessary to know the skeleton structure of the molecule, that is, which is the central atom and what atoms are bonded to it (terminal atoms). The concept of formal charge is useful in establishing the skeleton structure and in assessing the general plausibility of a Lewis structure.

At times, even though all the applicable rules are followed, it is impossible to write a single Lewis structure for a species. In these cases two or more plausible structures can be written and the true structure is said to be a *resonance hybrid* of the two or more contributing structures. Other times, the octet rule and/or the requirement that all electrons in a structure be paired may fail. This is the case for odd-electron species (e.g., NO), for electron-deficient species (e.g., BF_3), and for structures in which the central atom can accommodate an expansion of its valence shell to 10 or 12 electrons (e.g., PCl_5 and SF_6). Octet expansion cannot occur for second-period elements but is rather commonly encountered for nonmetals of the third and higher periods.

Molecular properties such as bond energy and bond length are not required in writing Lewis structures, but they are sometimes helpful in judging whether a structure is plausible. For example, they can be used to establish whether a covalent bond has multiple bond character. Also, bond energies can be used to calculate enthalpy changes (heats) of reactions involving gaseous species.

Although the discussion of chemical bonding is conveniently broken down into the categories of ionic and covalent, most bonds have both partial ionic and partial covalent character. One manifestation of the partial ionic character of certain covalent bonds is the degree of separation of electric charge that exists in the bond, referred to as the *bond dipole moment*. Another approach to the partial ionic character of covalent bonds is through the concept of *electronegativity*. Electronegativity measures the electron-attracting power of an atom when it is bonded to other atoms in a compound. In the chemical bond between two atoms, electrons are displaced toward the atom of higher electronegativity. Thus, the greater the electronegativity difference between two atoms, the more ionic is the bond between them. Electronegativities have been assigned to the elements and a relationship has been established between electronegativity difference and percent ionic character of a bond.

A powerful method for predicting the geometrical shapes of molecules is the valence-shell electron-pair repulsion (VSEPR) theory. This method requires determining the number of pairs of electrons surrounding the central atom in a structure. The shape of the molecule (or polyatomic ion) depends on the geometrical orientation of the valence-shell electron pairs and whether these pairs are bond-pairs or lone-pairs. A particularly important use of information about the shapes of molecules is in establishing whether a molecule is expected to have a resultant dipole moment. A resultant dipole moment is established by an appropriate combination of individual bond dipole moments. In a molecule of CCl_4, the individual C—Cl bond moments are large, but because of the symmetrical distribution of these bond moments in the tetrahedral CCl_4 molecule, the resultant dipole moment is *zero*. Molecules with a resultant dipole moment are said to be *polar*, and those with no resultant dipole moment are *nonpolar*.

Summarizing Example

Nitryl fluoride, FNO_2, is a reactive gas that has been used in rocket propellants and as a fluorinating agent (to introduce fluorine atoms into compounds).

1. Write a plausible Lewis structure for FNO_2, indicating any formal charges.

Solution. Both from the way in which the formula is written and from the fact that N has the lowest electronegativity of the atoms present, we expect N to be the central atom in this structure. Also, the number of valence electrons in the structure is 7 from F, 5 from N, and 6 each from the O atoms: $7 + 5 + (2 \times 6) = 24$.

$$:\!\overset{\ominus}{\underset{..}{O}}\!-\!N\!-\!\overset{..}{\underset{..}{F}}\!: \longrightarrow :O\!=\!N\!-\!\overset{..}{\underset{..}{F}}\!:$$
$$\qquad\quad\underset{..}{\overset{|}{:O:}} \qquad\qquad\qquad \underset{..}{\overset{|}{:O:}}$$

(This example is similar to Example 10-10.)

2. Show that the best representation of FNO_2 is through a resonance hybrid.

Solution. Assuming that the two N-to-O bonds are the same, we can write two contributing structures to a resonance hybrid.

$$:O\!=\!N\!-\!\overset{..}{\underset{..}{F}}\!: \longleftrightarrow :\!\overset{..}{\underset{..}{O}}\!-\!N\!-\!\overset{..}{\underset{..}{F}}\!:$$
$$\quad\;\underset{..}{\overset{|}{:O:}} \qquad\qquad\qquad\quad \underset{..}{\overset{||}{:O}}$$

(This example is similar to Example 10-12.)

3. Assume that the N-to-O bonds are essentially the same in FNO_2 and in NO_2. Also, the bond energy for the N—F bond in FNO_2 is 188 kJ/mol. Predict whether the following reaction is *endothermic* or *exothermic*.

$$NO_2(g) + \tfrac{1}{2} F_2(g) \longrightarrow FNO_2(g)$$

Solution. Based on the stated assumptions we need to consider only the breaking of $\tfrac{1}{2}$ mol F—F bonds and the formation of 1 mol of N—F bonds. The F—F bond energy is listed in Table 10-2. The energy requirement for breaking F—F bonds is $\tfrac{1}{2}(159) = +80$ kJ/mol. The energy change for the formation of the N—F bonds is -188 kJ/mol. Since considerably more energy is released in forming new bonds than in breaking old bonds, we expect the reaction to be exothermic, to the extent of -108 kJ/mol. (The tabulated value of the enthalpy of reaction is -113 kJ/mol.)

(This example is similar to Example 10-8.)

4. Predict the geometrical shape of the FNO_2 molecule.

Solution. Examine the Lewis structure in 1. There are *three* electron pairs to be distributed around the N atom. (Recall that a multiple covalent bond is counted *as if* it were a *single* bond.) The distribution of the electron pairs is *trigonal planar*. Since all the electron pairs are *bond pairs*, the geometrical structure of FNO_2 is also trigonal planar. The predicted bond angles are 120°.

(The experimentally determined F—N—O bond angle is 118°.)

(This example is similar to Examples 10-14 and 10-16.)

5. Is the FNO_2 molecule polar or nonpolar?

Solution. The trigonal planar structure is a symmetrical one. If all the bonds to the central N atom were of the same type, the individual bond dipole moments would cancel and the molecule would be nonpolar. However, we should expect a greater separation of charge in the N-to-F bond than in the N-to-O bonds, because F has a higher electronegativity than O. As a result there should be at least a small displacement of electron charge density toward the F atom and the molecule should be polar. (The measured resultant dipole moment of FNO_2 is $\mu = 0.47$ D.)

(This example is similar to Example 10-18.)

Key Terms

bond energy (10-5)
bond length (10-5)
bonding pair (10-4)
Born–Fajans–Haber method (10-3)
central atom (10-4)
coordinate covalent bond (10-7)
covalent bond (10-1, 10-4)
dipole moment (μ) (10-6)
double covalent bond (10-4)
electronegativity (10-6)
expanded octet (10-9)

formal charge (10-7)
incomplete octet (10-9)
ionic bond (10-1, 10-2)
lattice energy (10-3)
Lewis structure (10-1)
Lewis symbol (10-1)
lone pair (10-4)
monomer (Focus feature)
multiple covalent bond (10-4)
nonpolar molecule (10-6)
octet (10-1)

odd-electron species (10-9)
polar molecule (10-6)
polymer (Focus feature)
resonance (10-8)
single covalent bond (10-4)
skeleton structure (10-4)
terminal atom (10-4)
triple covalent bond (10-4)
valence-shell electron-pair repulsion
  (VSEPR) theory (10-10)

Highlighted Expressions

Fundamental requirements of Lewis structures (10.9)
Strategy for writing Lewis structures (10.10)
Relationship between bond length and covalent radii (10.13)
Bond energies and enthalpy of reaction (10.14)
Exothermic/endothermic reactions and bond energies (10.15)
Electronegativity and position in the periodic table (10.18)

Equation for formal charge (10.20)
Rules for formal charge (10.21)
Selecting the central atom in a Lewis structure (10.22)
Comparative strengths of electron-pair repulsions (10.36)
Procedure for applying the VSEPR theory (10.37)

Review Problems

1. Write Lewis symbols for the following atoms and ions: **(a)** H; **(b)** Kr; **(c)** Sn; **(d)** Ca^{2+}; **(e)** I^-; **(f)** Ga; **(g)** Sc^{3+}; **(h)** Rb; **(i)** Se^{2-}.

2. Write Lewis structures for the following ionic compounds: **(a)** KI; **(b)** CaS; **(c)** $BaBr_2$.

3. Write plausible structures for the following molecules, which contain only single covalent bonds: **(a)** I_2; **(b)** BrCl; **(c)** OF_2; **(d)** NI_3; **(e)** H_2Se.

4. The following molecules each contain a double covalent bond. Give a plausible structure for each. **(a)** CS_2 (that is, SCS); **(b)** H_2CO; **(c)** Cl_2CO.

5. Indicate what is wrong with each of the following structures.

(a) H—H—N̈—Ö—H   **(b)** :Ö—Cl—Ö:

(c) [·C=N̈:]⁻   **(d)** Ca—Ö:

6. Assign formal charges to the atoms in the species represented below. If there are no formal charges present for certain of these species, so indicate.

(a) :Ï—Ï: **(b)**

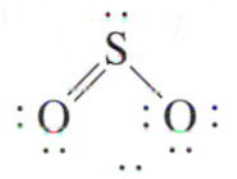

(c) 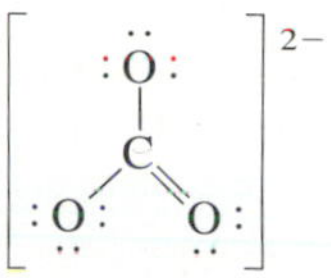   **(d)**

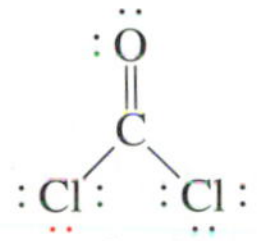

(e) [H—Ö—Ö:]⁻   **(f)**

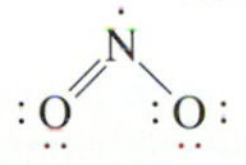

7. Each of the following ionic compounds consists of a combination of monatomic and polyatomic ions. Represent these compounds with Lewis structures: (a) $Mg(OH)_2$; (b) NH_4I; (c) $Ca(ClO_2)_2$. [*Hint:* Each of the polyatomic ions is described in Section 10-7.]

8. Which of the following species would you expect to be diamagnetic and which paramagnetic? (*Note:* Some of these species are not especially stable.) (a) OH^-; (b) OH; (c) NO_3; (d) SO_3; (e) SO_3^{2-}; (f) HO_2.

9. Draw plausible Lewis structures for the following species, using the notion of expanded octets where necessary: (a) BrF_5; (b) PF_3; (c) ICl_3; (d) SF_4.

10. Use the VSEPR theory to predict the geometrical shapes of the following species: (a) CO; (b) $SiCl_4$; (c) H_2Te; (d) ICl_3; (e) $SbCl_5$; (f) SO_2; (g) AlF_6^{3-}.

11. For each of the bonds shown in the following structure, indicate (a) the bond length and (b) the bond energy. (c) How much energy, in joules, would be required to break all the bonds in *one molecule* of $H_2ClCCHO$?

$$
\begin{array}{c c c}
& O & H \\
& \| & | \\
H-\!\!& C-\!\!& C-Cl \\
& & | \\
& & H
\end{array}
$$

12. Using data from Table 10-2, but without performing detailed calculations, indicate whether each of the following reactions is endothermic or exothermic.
 (a) $CH_4(g) + I(g) \rightarrow CH_3(g) + HI(g)$
 (b) $H_2(g) + I_2(g) \rightarrow 2\ HI(g)$

13. Without referring to tables or figures, except the periodic table, indicate which of the following atoms, Bi, S, Ba, As, or Mg, has the intermediate value when the five are arranged in order of increasing electronegativity.

14. Use your knowledge of electronegativites, but *without* referring to tables or figures in the text, arrange the following bonds in terms of *increasing* ionic character: C—H; F—H; Na—Cl; Br—H; K—F.

15. Two of the following species have the same shape. Which are these two? What is their shape? What are the shapes of the other two? NI_3, I_3^-, SO_3^{2-}, NO_3^-

16. Which of the following molecules would you expect to have a resultant dipole moment (μ)? Explain. (a) F_2; (b) NO_2; (c) BF_3; (d) HBr; (e) H_2CCl_2; (f) SiF_4; (g) OCS.

Exercises

Lewis theory

17. What are some of the essential differences in the ways in which Lewis structures are written for ionic and covalent bonds?

18. Cite several Lewis structures for which the following statement proves to be incorrect. "All atoms in a Lewis structure have an octet of electrons in their valence shells."

19. With reference to the Lewis theory and Lewis structures, what is meant by each of the following terms? (a) valence electrons; (b) octet; (c) lone-pair electrons; (d) multiple bond; (e) expanded octet; (f) resonance; (g) coordinate covalent bond; (h) odd-electron species.

Ionic bonding

20. Derive the correct formulas for the following ionic compound by writing Lewis structures: (a) lithium oxide; (b) sodium iodide; (c) barium fluoride; (d) scandium chloride.

21. In all simple binary ionic compounds the nonmetal atoms acquire the electron configurations of noble gas atoms. This is not always the case with the metal atoms. Explain why this is so.

22. Why is it inappropriate to speak of "molecules of NaCl" when describing *solid* sodium chloride? What would the term signify if one were describing *gaseous* sodium chloride?

Lattice energy

23. The enthalpy of formation of CsCl is -442.8 kJ/mol, and the enthalpy of sublimation of Cs(s) is 77.6 kJ/mol. Use these data, together with other data from the text, to calculate the lattice energy of CsCl.

24. Refer to Example 10-3. Together with data listed in the example, use the data given below to calculate ΔH_f° for 1 mol $MgCl_2(s)$. Explain why you would expect $MgCl_2$ to be much more stable a compound than MgCl(s). Second ionization energy of Mg, $I_2 = 1451$ kJ/mol; Lattice energy of $MgCl_2(s) =$

$-2500.$ kJ/mol $MgCl_2$. [*Hint:* Recall that I_2 is the energy requirement for the process $Mg^+ \rightarrow Mg^{2+} + e^-$.]

25. Hydrogen can form ionic compounds with some metals. In these compounds hydrogen exists as the hydride ion H^-. Determine the electron affinity of hydrogen [that is, ΔH for the reaction $H(g) + e^- \rightarrow H^-(g)$]. To do so, (a) use data from the Born–Fajans–Haber treatment of NaCl(s) on page 311; (b) use $+218$ kJ as the energy requirement to dissociate $\frac{1}{2}$ mol $H_2(g)$ to form 1 mol H(g); (c) use -812 kJ/mol NaH for the lattice energy of sodium hydride; and (d) use -57 kJ/mol NaH for the enthalpy (heat) of formation of sodium hydride.

Lewis structures

26. By means of Lewis structures represent bonding between the following pairs of elements. Your structures should clearly show whether the bonding is essentially ionic or covalent. Give the name, formula, and formula weight of each compound. (a) Cs and Br; (b) H and Se; (c) B and Cl; (d) Rb and S; (e) Mg and O; (f) F and O.

27. Write plausible Lewis structures for the following molecules: (a) H_2NOH; (b) N_2F_2; (c) $HONO$; (d) H_2NNO_2.

28. Indicate what is wrong with each of the following Lewis structures. Replace each with a more acceptable structure.
 (a) $[:\!\overset{..}{S}\!\!-\!\!C\!\!=\!\!\overset{..}{N}:]^-$ (b) $[:\!\overset{..}{\underset{..}{Cl}}\!:]^+[:\!\overset{..}{\underset{..}{O}}\!:]^{2-}[\overset{..}{\underset{..}{Cl}}\!:]^+$
 (c) $:\!\overset{..}{O}\!\!=\!\!N\!\!=\!\!\overset{..}{O}:$ (d) $:\!\overset{..}{Cl}\!\!-\!\!N\!\!=\!\!\overset{..}{Cl}:$
$$\qquad\qquad\qquad\qquad\qquad | \\ \qquad\qquad\qquad\qquad\qquad :\!\overset{..}{Cl}\!:$$

29. Suggest reasons why the following do not exist as stable molecules: (a) H_3; (b) HHe; (c) He_2; (d) H_3O.

30. Two molecules that have the same formulas but different structures are said to be *isomers*. (In isomers the same atoms are present but linked in different ways.) Draw plausible Lewis structures for *two* isomers of S_2F_2. [*Hint:* What is(are) the likely central atom(s)?]

Formal charge

31. Assign formal charges to the following species. If there are no formal charges present for certain of these species, so indicate.

(a) $\begin{bmatrix} :\ddot{O}: \\ | \\ :\ddot{O}-Cl-\ddot{O}: \end{bmatrix}^-$ (b) $:\ddot{O} \overset{\overset{S}{\|}}{} \ddot{O}:$

(c) $\begin{bmatrix} :\ddot{F}: \\ | \\ :\ddot{F}-B-\ddot{F}: \\ | \\ :\ddot{F}: \end{bmatrix}^-$ (d) $H-\ddot{O}-\overset{\overset{\textstyle :O:}{|}}{\underset{\underset{\textstyle H}{|}}{\overset{\|}{P}}}-\ddot{O}-H$

(e) $[:\ddot{N}=N=\ddot{N}:]^-$ (f) $[:\ddot{N}-N\equiv N:]^-$

(g) $[:\ddot{O}=N=\ddot{O}:]^+$

32. Use the concept of formal charge to select the more likely skeleton structure for each of the following molecules: **(a)** H₂NOH or H₂ONH; **(b)** SCS or CSS; **(c)** NOCl or ONCl; **(d)** SCN⁻ or CNS⁻ or CSN⁻.

Polyatomic ions

33. The following polyatomic anions involve covalent bonds between O atoms and the central nonmetal atom. Propose a plausible Lewis structure for each. **(a)** OCl⁻; **(b)** IO₄⁻; **(c)** BrO₃⁻.

34. Represent each of the following ionic compounds by an appropriate Lewis structure: **(a)** KIO₃; **(b)** Ca(OCl)₂; **(c)** NH₄ClO₄.

35. Propose Lewis structures for the following ionic species containing sulfur-to-sulfur bonds: **(a)** S₂²⁻; **(b)** S₃²⁻; **(c)** S₄²⁻; **(d)** S₅²⁻.

Resonance

36. In the manner used to establish the structures for O₃ shown in (10.27), demonstrate that there are *three* equivalent structures that can be written for SO₃.

37. In Example 10-12 the phenomenon of resonance was illustrated for the nitrate ion. Resonance is also involved in the nitrite ion, NO₂⁻. Represent this fact through appropriate Lewis structures.

38. Dinitrogen oxide (nitrous oxide), N₂O, is the common "laughing gas" used as an anesthetic in dental offices. Here are some data about the N₂O molecule: N—N bond length = 113 pm; N—O bond length = 119 pm. Use these data, data from Table 10-2, and other ideas established in this chapter to comment on the plausibility of each of the following Lewis structures. That is, are some of the structures unlikely? Is one structure more likely than all the rest? Is the molecule best represented by a resonance hybrid?

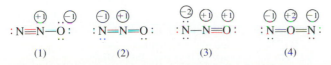

39. Nitric acid, HNO₃, can be represented as a resonance hybrid of the following structures. Which structure(s) seems most plausible? Explain.

(a) $H-\ddot{O}-\overset{\overset{\textstyle :O:}{}}{N}\overset{\diagdown}{\ddot{O}:}$

(b) $H-\ddot{O}-\overset{\overset{\textstyle :\ddot{O}:}{}}{N}\overset{\diagdown}{\ddot{O}:}$

(c) $H-\ddot{O}=N\overset{\overset{\textstyle :\ddot{O}:}{}}{\underset{\underset{\textstyle \ddot{O}:}{\diagdown}}{}}$

Odd-electron species

40. As with the case of NO described in the text, the molecule NO₂ is paramagnetic. Represent this molecule through a Lewis structure(s).

41. NO₂ (see Exercise 40) may *dimerize*. That is, two NO₂ molecules may join together to form N₂O₄. Write a plausible Lewis structure for N₂O₄. Do you think that N₂O₄ is diamagnetic or paramagnetic?

42. Write plausible Lewis structures for the following odd-electron species: **(a)** HO₂; **(b)** CH₃; **(c)** ClO₂; **(d)** NO₃.

Expanded octets

43. Phosphorus and sulfur atoms make use of expanded octets in many of their compounds. Nitrogen and oxygen never do. Would you expect As and Se to resemble more nearly P and S or N and O? Explain.

44. Draw plausible Lewis structures for the following species. (In some cases the best structures will involve expanded octets.) **(a)** SO₃²⁻; **(b)** HOClO₂; **(c)** HONO; **(d)** OP(OH)₃; **(e)** O₂SCl₂; **(f)** XeO₃.

45. Indicate the nature of the sulfur-to-nitrogen bond in F₃SN (i.e., single, double, or triple). [*Hint:* Use the notion of an expanded octet and ideas about formal charges.]

46. Exercise 36 refers to *three* equivalent structures for SO₃. To these, add *four* plausible structures based on an expanded octet for the sulfur atom. Of all seven structures (three from Exercise 36 and four here), which is the most plausible from the standpoint of formal charges? What experimental evidence would be helpful in assessing the relative importance of the various structures?

Bond lengths

47. Refer to the Summarizing Example. Use data from the chapter to estimate the length of the N—F bond in FNO₂.

48. A relationship between bond lengths and single covalent radii of atoms is given in expression (10.13). Use this relationship together with appropriate data from Table 10-2 to calculate the bond lengths: **(a)** H—Cl; **(b)** C—N; **(c)** C—Cl; **(d)** C—F; **(e)** N—I.

49. In which of the following molecules would you expect the nitrogen-to-nitrogen bond to be the *shortest:* **(a)** N₂H₄; **(b)** N₂; **(c)** N₂O₄; **(d)** N₂O? Explain.

50. Draw a sketch of the hydroxylamine molecule, H₂NOH, representing the shape of the molecule and, where possible, bond angles and bond lengths.

Bond energies

51. Use bond energies from Table 10-2 to estimate the enthalpy change (ΔH) for the following reaction. [*Hint:* Two bonds are broken and two are formed.]

$$C_2H_6(g) + Cl_2(g) \rightarrow C_2H_5Cl(g) + HCl(g) \qquad \Delta H = ?$$

52. Use bond energies to calculate the enthalpy (heat) of formation of $NH_3(g)$ and compare your result with the value listed in Appendix D.

$$N_2(g) + 3 H_2(g) \rightarrow 2 NH_3(g)$$

53. Equations (a) and (b) can be combined to yield the equation for the formation of $CH_4(g)$ from its elements.

(a)	$C(s) \rightarrow C(g)$	$\Delta H = 717$ kJ
(b)	$C(g) + 2 H_2(g) \rightarrow CH_4(g)$	$\Delta H = ?$

Net: $C(s) + 2 H_2(g) \rightarrow CH_4(g)$ $\Delta H_f^\circ = -75$ kJ

Use the above data and a bond energy of 435 kJ/mol for H_2 to estimate the C—H bond energy. Compare your result with the value listed in Table 10-2. [*Hint:* What is the value of ΔH for reaction (b)?]

54. The two most common oxides of carbon are carbon monoxide, CO, and carbon dioxide, CO_2.

   (a) Draw plausible Lewis structures for these molecules.

   (b) In which molecule would you expect the *greater* C—O bond length? In which would you expect the greater C—O bond energy? Explain.

55. If the polarity of the bond A—B is about the same as that of A—A and of B—B, the bond energy of A—B may be taken as the average of A—A and B—B. Assume that this situation applies to the diatomic molecule BrCl and estimate the Br—Cl bond energy, using data from Table 10-2.

★56. Use a value of 498 kJ/mol for the bond energy in $O_2(g)$ and other necessary data from the text to estimate the bond energy in NO(g).

Partial ionic character of covalent bonds

57. Estimate the percent ionic character of the HBr molecule, given that the dipole moment (μ) is 0.79 D.

58. The molecule H_2O_2 has a resultant dipole moment (μ) of 2.13 D. The bonding is H—O—O—H. Which of these bonds have bond dipole moments? Can the molecule be linear? Explain.

59. Plot the data of Figure 10-10 as a function of atomic number. Does the property of electronegativity conform to the periodic law? Do you think that it should?

★60. **(a)** Use the idea outlined in Exercise 55 to estimate the H—Cl bond energy, and compare your result to the experimentally determined H—Cl bond energy of 431 kJ/mol. **(b)** Determine the difference between the predicted and actual H—Cl bond energy found in part **(a)**. This difference is called the *ionic resonance energy*. It represents the extra stability of the H—Cl bond that results from the partial ionic character of the bond. **(c)** Pauling's electronegativity scale is related to ionic resonance energy (IRE) through the equation $(\Delta EN)^2 = IRE/96$, where EN is the electronegativity difference between two atoms that are bonded together and IRE is expressed in kJ/mol. Calculate the electronegativity difference between H and Cl. **(d)** From the electronegativity difference calculated in part **(c)** and Figure 10-11, estimate the percent ionic character of the H—Cl bond and compare your result with the values mentioned in the text (e.g., in Example 10-9).

Molecular shapes

61. Use VSEPR theory to predict the geometrical shapes of the following molecules and ions whose Lewis structures are presented in the text. N_2 (10.7); HCN (10.12); ClO_2^- (10.25); NH_4^+ (10.26); NO_3^- (10.28); PCl_3 (10.33); SO_4^{2-} (10.34); $SOCl_2$ (10.35).

62. One of the following ions has a *trigonal planar* shape. Which ion is it? Explain. SO_3^{2-}; PO_4^{3-}; CN^-; CO_3^{2-}.

63. Each of the following molecules contains one or more multiple covalent bonds. Draw plausible Lewis structures to represent this fact, and predict the shape of each molecule. **(a)** CO_2; **(b)** N_2O; **(c)** NSF; **(d)** $ClNO_2$.

64. Can you think of an example of a molecule in which the central atom has *one bonding pair* and *three lone pairs* of electrons? What must be the shape of this molecule?

65. The structure of BF_3 is shown in Table 10-3 to be planar. If a fluoride ion is attached to the B atom of BF_3 through a coordinate covalent bond, the ion BF_4^- results. What is the geometrical shape of this ion?

66. Three possible Lewis structures for SO_3 were described in Exercise 36, and another four (based on an expanded octet for S) in Exercise 46. Why is the geometrical shape predicted for SO_3 *independent* of whichever of these Lewis structures is used?

67. Use the VSEPR theory to predict the geometrical shape of **(a)** the molecule OSF_2; **(b)** the molecule O_2SF_2; **(c)** the molecule XeF_4; **(d)** the ion ClO_4^-; **(e)** the ion I_3^-.

68. In the gas phase the nitric acid molecule, HNO_3, is planar with the bond angles indicated below. What would you predict for these bond angles according to VSEPR theory, and how do you explain the discrepancies between the predicted and measured values?

69. Sketch the probable geometric shape of a molecule of hydrogen peroxide, H_2O_2. Is H_2O_2 planar?

Polar molecules

70. Predict the geometrical shapes of the following molecules, and then predict which molecules you would expect to have resultant dipole moments: **(a)** SO_2; **(b)** NH_3; **(c)** H_2S; **(d)** C_2H_4; **(e)** SF_6; **(f)** CH_2Cl_2.

71. Refer to the Summarizing Example. A compound related to nitryl fluoride is nitrosyl fluoride, FNO. For this molecule indicate **(a)** a plausible Lewis structure, and **(b)** the geometrical shape. **(c)** Explain why the measured resultant dipole moment for FNO, 1.81 D, is so much larger than the value for FNO_2 (0.47 D).

Polymers.

72. Describe the difference in meaning of the following pairs of terms: **(a)** monomer and polymer; **(b)** elastomer and fiber; **(c)** natural and synthetic rubber; **(d)** inorganic and organic polymer; **(e)** thermosetting and thermoplastic polymer.

73. For the polymer Teflon (see Table 10-4)

   (a) Draw the structure of a portion of this polymer chain consisting of four repeating units.

   (b) What is the percent of F in Teflon? Would you expect this to depend on the length of the polymer chains? Explain.

74. Formaldehyde, $\begin{matrix} H \\ \diagdown \\ C=O \\ \diagup \\ H \end{matrix}$ is the monomer of polyfor-

maldehyde, a polymer with carbon-to-oxygen bonds.
  (a) Draw the structure of a portion of this polymer chain consisting of ten repeating units.
  (b) What volume of $CO_2(g)$ measured at 25.0 °C and 751 mmHg would be produced by the complete combustion of 1.05 g of this polymer?

75. A 315-cm^3 sample of propylene gas (C_3H_6) at 20 °C and 748 mmHg is polymerized. If it were possible to produce poly-

mer molecules that all had the formula $\left(\begin{matrix} H & CH_3 \\ | & | \\ -C-C- \\ | & | \\ H & H \end{matrix} \right)_n$

where $n = 875$, how many polymer molecules would be formed?

Additional Exercises

76. Refer to Table 10-1 and write the complete electron configurations for the following ions: **(a)** Y^{3+}; **(b)** Cd^{2+}; **(c)** Sb^{3+}.

77. A compound is found to consist of 47.5% S and 52.5% Cl, by mass. Write a Lewis structure for this compound and comment on its deficiencies. Write a different structure with the same ratio of S to Cl that is more plausible.

78. A 1.65-g sample of a hydrocarbon, when completely burned in an excess of $O_2(g)$, yields 5.37 g CO_2 and 1.65 g H_2O. Draw a plausible Lewis structure for the hydrocarbon molecule. [*Hint:* There is more than one possible arrangement of the C and H atoms.]

79. What is the formal charge on the indicated atom in each of the following?
  (a) oxygen in OH^- (structure 10.23)
  (b) chlorine in ClO_2^- (structure 10.25)
  (c) boron in $H_3N \cdot BF_3$ (structure 10.32)
  (d) phosphorus in PCl_5 (structure 10.33)
  (e) iodine in ICl_4^- (Example 10-15)

80. What is the relationship, if any, between the concepts of oxidation state and formal charge? Explain.

81. What is the relationship between the shapes of **(a)** the ammonia molecule, NH_3 and the ammonium ion, NH_4^+; **(b)** sulfur trioxide, SO_3, and the sulfate ion, SO_4^{2-}?

82. One each of the following species is linear, angular, planar, tetrahedral, and octahedral. Indicate the correct structure for each: **(a)** H_2Te; **(b)** C_2Cl_4; **(c)** CO_2; **(d)** $SbCl_6^-$; **(e)** SO_4^{2-}.

83. Some of these statements regarding molecular shape are always true and some are not. Identify those that are not always true and explain why they are not.
  (a) Diatomic molecules have a linear shape.
  (b) Molecules in which four atoms are bonded to the same central atom have a tetrahedral shape.
  (c) Molecules with a planar shape consist of three atoms (triatomic).
  (d) Molecules with a nonmetal of the second period as the central atom cannot have an octahedral shape.

84. Draw Lewis structures for two different molecules having the formula C_3H_4. Is either of these molecules linear? Explain.

85. Use bond energies from Table 10-2 to estimate the enthalpy change (ΔH) for the following reaction. [*Hint:* What is the nature of the carbon-to-carbon bond in each compound?]

$$C_2H_4(g) + H_2(g) \rightarrow C_2H_6(g) \qquad \Delta H = ?$$

*86. The enthalpy (heat) of formation of $NaI(s)$ is -288 kJ/mol. Use this value, together with other data in the text, to calculate the lattice energy of $NaI(s)$. [*Hint:* An extra step is required in addition to those outlined in Figure 10-4. This is for the subli-

mation of iodine: $I_2(s) \rightarrow I_2(g)$. Determine the enthalpy of sublimation of iodine with data from Appendix D.]

*87. Show that the formation of $NaCl_2(s)$ is very unfavorable, that is, $\Delta H_f[NaCl_2(s)]$ is a large *positive* quantity. To do this, use data from the Born–Fajans–Haber treatment of $NaCl(s)$ on page 311; use $I_2 = +4562$ kJ/mol for the second ionization energy of sodium; and assume that the lattice energy for $NaCl_2$ would be about the same as that of $MgCl_2$ in Exercise 24, that is, $-2500.$ kJ/mol. [*Hint:* Recall the discussion immediately following Example 10-3.]

*88. The second electron affinity of oxygen cannot be measured directly,

$$O^-(g) + e^- \rightarrow O^{2-}(g) \qquad EA_2 = ?$$

The O^{2-} ion can exist in the solid state, however, where the high energy requirement for its formation is offset by the large lattice energies of ionic oxides.
  (a) Show that EA_2 can be calculated from the enthalpy of formation and lattice energy of $MgO(s)$, enthalpy of sublimation of $Mg(s)$, ionization energies of Mg, bond energy of O_2, and EA_1 for $O(g)$.
  (b) The enthalpy of sublimation of $Mg(s)$ is 150. kJ/mol, the bond energy of O_2 is 498 kJ/mol, and the lattice energy of MgO is -3925 kJ/mol. Combine these data with other values in the text to calculate a value of EA_2 for oxygen.

89. Indicate which of the following molecules you would expect to have a resultant dipole moment, and give reasons for your conclusions: **(a)** HCN; **(b)** SO_3; **(c)** CS_2; **(d)** OCS; **(e)** $SOCl_2$; **(f)** SiF_4; **(g)** POF_3; **(h)** XeF_2.

90. The oxide Cl_2O_7 has the structure shown below. The bonds in blue have a length of 1.46 Å, and those in tan, 1.72 Å. Use this information to write a plausible Lewis structure(s) for the Cl_2O_7 molecule.

91. A 0.212-g sample of a gaseous hydrocarbon occupies a volume of 127 cm^3 at 738 mmHg pressure and 24.7 °C. Show that there is only one possible structure for this hydrocarbon and draw its Lewis structure.

*92. Carbon suboxide has the formula C_3O_2. The carbon-to-carbon distances are found to be 130 pm, and the carbon-to-oxygen distances are 120 pm. Propose plausible Lewis structures to account for these bond distances, and predict the geometrical shape of the molecule.

93. Use VSEPR theory to predict the geometrical shapes of the following anions: **(a)** ClO_4^-; **(b)** $S_2O_3^{2-}$ (that is, SSO_3^{2-}); **(c)** SbF_4^-

94. In certain polar solvents PCl_5 undergoes an ionization reaction in which a Cl^- ion leaves one PCl_5 molecule and attaches itself to another. The products of the ionization are PCl_4^+ and PCl_6^-. Sketch the changes in geometrical shapes that occur in this ionization (that is, what are the shapes of PCl_5, PCl_4^+, and PCl_6^-)?

$$2\ PCl_5 \rightleftharpoons PCl_4^+ + PCl_6^-$$

95. Refer to Example 10-4. Is the N_2H_4 molecule linear in shape? planar? How might you best describe the geometrical shape of this molecule? What are the difficulties in describing its shape?

★96. Refer to Example 10-5. We want to estimate the enthalpy (heat) of formation of HCN using bond energies from Table 10-2. Think in terms of the combination of the following two equations.

(a) $C(s) \rightarrow C(g)$ $\Delta H = 717$ kJ
(b) $C(g) + \frac{1}{2} N_2(g) + \frac{1}{2} H_2(g) \rightarrow HCN(g)$ $\Delta H = \ ?$

Net: $C(s) + \frac{1}{2} N_2(g) + \frac{1}{2} H_2(g) \rightarrow HCN(g)$ $\Delta H_f^\circ = \ ?$

[Hint: Use bond energies from Table 10-2 to evaluate ΔH for reaction (b). Then use Hess's law to obtain ΔH_f° for the net reaction. Refer to structures (10.7) and (10.12) to decide which bond energies to use from Table 10-2.]

★97. By the following method you can predict the geometrical shapes of species with only one central atom, without first drawing Lewis structures.

1. Total no. electron pairs = (no. valence electrons ± electrons for ionic charge)/2
2. no. bond electron pairs = no. of atoms − 1
3. no. electron pairs around central atom
              = total no. electron pairs
               − 3[no. terminal atoms (excluding H)]
4. no. lone pairs = no. central electron pairs − no. bond pairs

Once you have evaluated items 2, 3, and 4, you can establish the VSEPR notation for the species, that is, AX_nE_m and determine the molecular shape from Table 10-3. Use this method to predict the geometrical shapes of (a) PCl_5; (b) NH_3; (c) ClF_3; (d) SO_2; (e) XeF_4; and check your results against those given in Table 10-3.

★98. Justify each of the steps of the procedure outlined in Exercise 97 and show why that procedure yields the same geometrical shapes as the VSEPR method based on Lewis structures. [Hint: Draw a hypothetical (or real) Lewis structure and relate steps 1 through 4 to this Lewis structure. How does the method of Exercise 97 deal with multiple bonds?]

99. The aldehyde propynal has the formula HCCCHO. Draw a sketch that represents bonding in the molecule and its shape. Include bond distances and bond angles.

100. The total bond energy associated with all the bonds in thiourethane

$$\overset{\displaystyle O}{\underset{\displaystyle \|}{H_2NCSCH_2CH_3}}$$

is 4780 kJ/mol. Use this value, together with data from Table 10-2, to estimate the energy of the C—S bond. How would you expect this energy to compare in value with the bond energies of the carbon-to-sulfur bonds in CS_2?

101. Estimate the enthalpies (heats) of formation of the following species at 25 °C and 1 atm: (a) $N_2H_4(g)$; (b) $OH(g)$. Use data from Table 10-2 as necessary.

102. Use a value of 498 kJ/mol for the O-to-O bond energy in the O_2 molecule, together with other data in Table 10-2, to estimate the heat of combustion of dimethyl ether, H_3COCH_3. [Hint: All bonds in the dimethyl ether molecule are single covalent.]

★103. The bond energy in the $O_2(g)$ molecule is 498 kJ/mol; and the heat of formation of $H_2O_2(g)$ is −136 kJ/mol. Use these values, together with other appropriate data from the text, to estimate the oxygen-to-oxygen single bond energy.

★104. The text states that the *bond* dipole moment of the O—H bond is 1.51 D; the H—O—H bond angle is 104°; and the resultant dipole moment of the H_2O molecule is 1.84 D. (See expression 10.17.)

(a) Show by an appropriate geometric calculation that the three statements made above for the H_2O molecule are mutually consistent.

(b) Use the same method as developed in part (a) to estimate the bond angle in H_2S, given that the H—S bond moment is 0.67 D and the resultant dipole moment of the H_2S molecule is 0.93 D.

Self-Test Questions

For questions 105 through 114 select the single item that best completes each statement.

105. The highest electronegativity of the following is for an atom of (a) S; (b) Cs; (c) Si; (d) Al.

106. Of the following, the bond with the greatest percent ionic character is (a) F—F; (b) Cl—F; (c) Al—F; (d) C—O.

107. Of the following species, the one containing a *triple* covalent bond is (a) NO_3^-; (b) CN^-; (c) CO_2; (d) $AlCl_3$.

108. The formal charge of the O atoms in the ion $[\ddot{:}\ddot{O}{=}N{=}\ddot{O}\ddot{:}\ ]^+$ is (a) −2; (b) −1; (c) 0; (d) +1.

109. In the ammonium ion, NH_4^+ (a) the four H atoms are situated at the corners of a square; (b) all bonds are ionic; (c) all

bonds are coordinate covalent; (d) the shape is that of a tetrahedron.

110. All of the following molecules are linear except one. That one is (a) SO_2; (b) CO_2; (c) HCN; (d) NO.

111. All the following molecules are polar except one. That one is (a) BCl_3; (b) CH_2Cl_2; (c) NH_3; (d) FNO.

112. The highest bond energy among the following diatomic molecules is found in (a) O_2; (b) N_2; (c) Cl_2; (d) I_2.

113. The greatest bond length among the following diatomic molecules is found in (a) O_2; (b) N_2; (c) Br_2; (d) BrCl.

114. Of the Lewis structures shown, one is plausible, but the other three are incorrect. The plausible one is

(a) cyanate ion, $[:\overset{..}{\underset{..}{O}}{-}C{=}\overset{..}{N}:]^-$

(b) carbide ion, $[C{\equiv}C:]^{2-}$

(c) hypochlorite ion, $[:\overset{..}{\underset{..}{Cl}}{-}\overset{..}{\underset{..}{O}}:]^-$

(d) nitrogen(II) oxide, $:\overset{..}{N}{=}\overset{..}{O}:$

115. A compound is found to have the following percent composition, by mass: 24.3% C, 71.6% Cl, and 4.1% H.
   (a) What is the *empirical* formula of this compound?
   (b) Draw a Lewis structure based on the empirical formula and comment on its inadequacies.
   (c) Propose a *molecular* formula for the compound that results in a more plausible Lewis structure.

116. Predict the shapes of the following sulfur-containing species: **(a)** SO_2; **(b)** SO_3; **(c)** SO_4^{2-}.

117. Given the bond energies: N-to-O bond in NO, 628 kJ/mol; H—H, 435 kJ/mol; N—H, 389 kJ/mol; O—H, 464 kJ/mol, calculate ΔH for the reaction

$$2\ NO(g) + 5\ H_2(g) \rightarrow 2\ NH_3(g) + 2\ H_2O(g)$$

118. The following statements are not made as carefully as they might be. Criticize each one.
   (a) Triatomic molecules have a planar shape.
   (b) Molecules in which there is an electronegativity difference between the bonded atoms are polar.
   (c) Lewis structures in which atoms carry formal charges are incorrect.

11 Chemical Bonding II: Additional Aspects

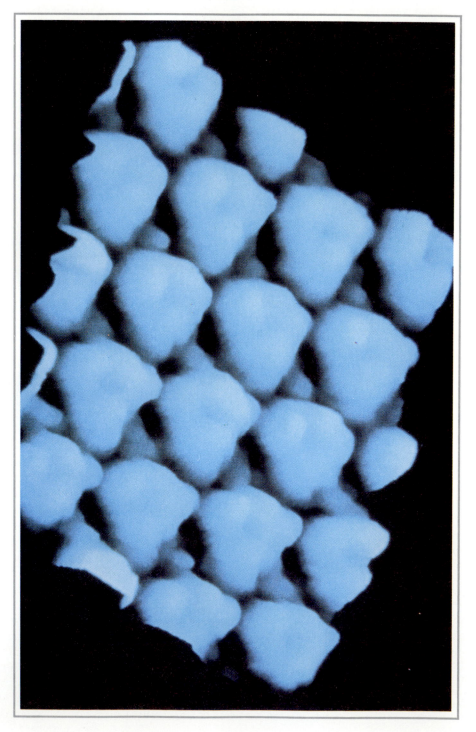

Various bonding models, from Friedrich Kekulé's proposal of 1865 to modern quantum theory, have suggested a ring-like structure for the benzene molecule, C_6H_6 (see page 373). In 1988, this photograph of individual C_6H_6 molecules on a surface, made with a scanning tunneling microscope, provided the first visual evidence of this structure. [Courtesy IBM Research]

In the preceding chapter the most important connection we made between atomic and molecular structure was through the number of valence electrons in atoms. Once we knew the electron configuration of the valence shell we were able to write Lewis structures and to predict geometrical shapes with the VSEPR theory.

However, there are additional aspects of atomic structure that we can apply to a description of chemical bonding, and this we will do in the current chapter. Our basic view in this chapter is that a covalent bond is a region between bonded atoms where the probability of finding electrons and the electron charge density are high.

One way to describe this region of high electron charge density is in terms of the interpenetration or overlap of electron orbitals of the bonded atoms. Because atomic orbitals have distinctive geometrical shapes, the regions in which orbitals overlap also have specific geometrical orientations, and this explains the distinctive shapes of molecules. This approach to chemical bonding, called the valence bond method, also gives a good description of multiple covalent bonds.

Another basic approach to covalent bonding is to consider that a molecule can be built up from the individual atomic nuclei and a set of allowable electron orbitals called molecular orbitals. Electrons are assigned to these orbitals in a similar fashion to the Aufbau process described for atoms in Chapter 8. This approach, called the molecular orbital method, yields good descriptions of bond order and magnetic properties.

11-1 An Introduction to Valence Bond Theory

In your mind's eye, picture two hydrogen atoms being brought close together. At some point their electron charge clouds begin to merge. We say that the 1s orbitals of the atoms *overlap*. As a result of this overlap, the electron charge density increases in the region between the atomic nuclei, and this increased density of negative charge can serve to hold the positively charged atomic nuclei together. Here is a new way to think about a covalent bond then: *The covalent bond is a region of high electron charge density (high electron probability) that results from the overlap of atomic orbitals between two atoms.* In general, the greater the amount of overlap between two orbitals, the stronger the bond. However, two atoms cannot be brought together too closely, for then the atomic nuclei would strongly repel one another, and the bond would become unstable. For each bond there is a condition of maximum atomic orbital overlap leading to a maximum bond strength (bond energy) at a particular internuclear distance (bond length).

This description of covalent bond formation is called the **valence bond approach** or the **valence bond method.** The overlap of two 1s orbitals in a hydrogen molecule is suggested in Figure 11-1. Note how, except in the region of overlap, the characteristic features of the 1s atomic orbitals first pictured in Figure 8-21c are retained in Figure 11-1. The valence bond theory is a *localized* electron model of bonding. Most of the electrons retain essentially the same orbital locations as in the separated atoms, and the bonding electrons are localized (fixed) in the region of atomic orbital overlap.

Figure 11-2 shows the overlap of atomic orbitals involved in the formation of hydrogen-to-sulfur bonds in hydrogen sulfide. Note especially that the maximum amount of overlap between the 1s orbital of a H atom and a 3p orbital of an S atom occurs along a line joining the centers of the H and S atoms. As a result of this requirement of a maximum degree of overlap between atomic orbitals, the molecule

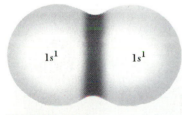

FIGURE 11-1

Bonding in H_2 represented by atomic orbital overlap.

Each atomic orbital contains one unpaired electron. As a result of the overlap of the two orbitals, the electrons become paired and a region of high electron charge density (high electron probability) results—a covalent bond.

353

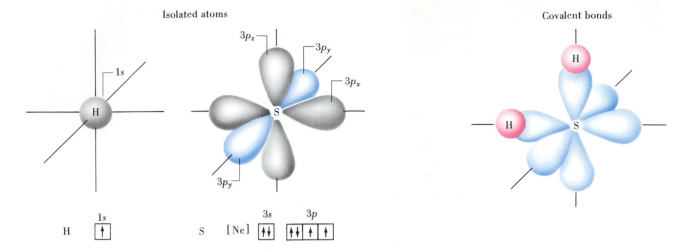

FIGURE 11-2
Covalent bonding in H_2S represented by atomic orbital overlap.

Orbitals having a single electron are grey; those with an electron pair are in color. For S, only $3p$ orbitals are shown. The $1s$ orbitals of two hydrogen atoms overlap with the $3p_x$ and $3p_z$ orbitals of the sulfur atom. Here and elsewhere in this chapter the shapes of p orbitals have been elongated to provide a better visualization of geometric structures.

acquires a distinctive geometrical shape. In summary, the basic ideas of the valence bond method are

1. Normally, orbital overlap occurs to the extent that all electrons become paired.
2. If we count valence electrons in the same way as in Lewis structures, each atom normally acquires a noble gas electron configuration (an outer-shell octet).
3. The shape of the molecule is determined by the geometric orientation of the overlapping atomic orbitals of the bonded atoms.

Basic ideas of the valence bond method.

The valence bond method predicts an H—S—H bond angle of 90°; the measured bond angle is 92°.

Are You Wondering:

Which is the correct method to use in predicting the geometrical shape of a molecule, the VSEPR method (Section 10-10) or the valence bond method?

In general, there is no "correct" method for describing molecular structures. The only "correct" information is the *experimental* evidence from which the structure is established. Once this experimental evidence is in hand you may find it easier to *rationalize* this evidence in terms of one method or the other. In the case of H_2S, the valence bond method, which predicts a bond angle of 90°, certainly does a better job than the VSEPR theory. For the Lewis structure, H—S̈—H, VSEPR theory predicts a tetrahedral angle, that is, 109.5°.

As a general observation, though, VSEPR theory gives reasonably good results in the majority of cases. You should continue to use it as your primary method.

The formation of PH₃ and some of its uses were described in Example 5-8.

Example 11-1

Using the valence bond method to describe the geometrical shape of a molecule. Describe the structure of the phosphine molecule, PH_3, by the valence bond method.

Solution. Here is a four-step approach to using the valence bond method:
Step 1. Draw orbital diagrams for the separate atoms that are to be united.

$$3s \qquad 3p \qquad\qquad 1s$$

P [Ne] [↑↓] [↑ | ↑ | ↑]   H [↑]

Step 2. Sketch the orbitals of the central atom (P) that are involved in the overlap (see Figure 11-3).
Step 3. Complete the structure by bringing together the bonded atoms and representing the orbital overlap.
Step 4. Describe the structure. PH_3 is a *trigonal pyramidal* molecule. The three H atoms lie in the same plane. The P atom is situated at the apex of the pyramid above the plane of the H atoms. The three H—P—H bond angles are predicted to be 90°.
(The experimentally measured H—P—H bond angles are 93–94°.)

SIMILAR EXAMPLE: Exercise 1.

bonding orbitals of P atom

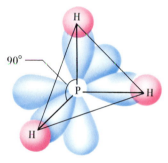

covalent bonds formed

FIGURE 11-3
Bonding and structure of the PH_3 molecule—Example 11-1 illustrated.

Orbitals with single electrons are grey; those with electron pairs are in color. Only bonding orbitals are shown. The 1s orbitals of three hydrogen atoms overlap with the three 3p orbitals of the phosphorus atom.

The Case of H₂O and NH₃. In the two examples considered thus far the central atom was a third period element—S from group 6A in H_2S and P from group 5A in PH_3. Let us consider the analogous molecules in which we use the *second* period elements, that is, H_2O and NH_3. The only difference here is that the central atoms employ $2p$ orbitals rather than $3p$. By the valence bond method, then, the H—O—H bond angle in H_2O and the H—N—H bond angles in NH_3 should both be 90°. The experimentally measured values are 104.5° for the H—O—H bond and 107° for the H—N—H bond. The valence bond method seems not to work as well in these cases.

However, consider these facts about the O and N atoms. They are small and have high electronegativities. In the O—H and N—H bonds electrons are displaced toward the central atom. This creates centers of positive charge at the H atoms. The H atoms, themselves small and in close proximity when bonded to O or N, repel one another. This forces the bond angle to enlarge to the observed values of 104.5° in water and 107° in ammonia. The corresponding effect in the H_2S and PH_3 molecules is practically negligible. The larger S and P atoms are not as electronegative as O and N, and the H atoms in the molecules do not come into close proximity.

11-2 Hybridization of Atomic Orbitals

The orbital diagrams we used in Section 11-1 were based on normal or *ground-state* electron configurations. If we use the valence bond method to describe the simplest carbon–hydrogen compound, our prediction, based on the ground-state electron configuration of carbon

$$1s \qquad 2s \qquad 2p$$

C [↑↓] [↑↓] [↑ | ↑ |]

is for a molecule with the formula CH_2 and a bond angle of 90°.

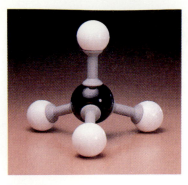

FIGURE 11-4
Ball-and-stick model of meth-ane, CH_4.

The H—C—H bond angles are 109.5°. [Carey B. Van Loon]

CH_2 does not exist as a stable molecule, however. The simplest stable hydrocarbon is methane, CH_4. To account for this fact we need an orbital diagram in which there are *four* unpaired electrons. We can get such a diagram by imagining that one of the 2s electrons in a normal C atom is *promoted* to the empty 2p orbital. Energy must be absorbed to bring this about, so the resulting electron configuration, which places unpaired electrons in the 2s and each of the 2p orbitals, is called an *excited state*. The situation can be represented as

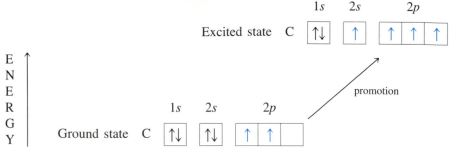

The geometrical shape that we would predict from the excited-state electron configuration would be a molecule with three mutually perpendicular C—H bonds (that is, bond angles of 90°) based on the 2p orbitals of the C atom and a fourth undirected bond based on the 2s orbital. The experimentally determined H—C—H bond angles, however, are found all to be the same and to be the tetrahedral angles of 109.5°. A ball-and-stick model of CH_4 is shown in Figure 11-4; this model is identical to the structure predicted by VSEPR theory (Table 10-3). A bonding scheme based on the excited-state electron configuration does a good job of predicting the formation of *four* C-to-H bonds, but it does not do a good job of predicting bond angles.

Our problem here is essentially this: We have been describing *bonded* atoms as if they possess the very same kinds of orbitals (that is, s, p, . . .) as *isolated unbonded* atoms. This assumption worked quite well for H_2S, PH_3, and (with some modification) H_2O and NH_3. But there is no reason for us to expect these unmodified simple atomic orbitals to work equally well in all cases.

Suppose we think of the 2s and three 2p orbitals of the excited-state electron configuration as being merged or blended together in some way so as to produce four new orbitals that are identical to each other. This process, which can be carried out as a mathematical calculation, is called **hybridization.** Figure 11-5 pictures the hybridization of an s and three p orbitals into the new set of four **hybrid orbitals** called **sp^3.** Figure 11-6 is an energy-level diagram of the pure atomic orbitals and sp^3 hybrid orbitals.

In a hybridization scheme *the number of hybrid orbitals is equal to the total number of atomic orbitals that are combined*. The symbolism used identifies the kinds and numbers of atomic orbitals combined. Thus, sp^3 signifies that *one s* and *three p* orbitals are combined. A useful orbital diagram to represent hybridization is

sp^3 hybridization: C   (11.2)

The use of sp^3 hybrid orbitals in bond formation in methane is pictured in Figure 11-7.

The term ''scheme'' (a systematic plan for attaining some objective) is appropriate for describing hybridization. The objective is an after-the-fact attempt to account for the *experimentally observed* geometrical shape of a molecule. Hybridization is not an actual physical phenomenon. We cannot observe electron charge distributions altering from those of simple orbitals to those of hybrid orbitals during bond formation. Moreover, for some covalent bonds no single hybridization scheme works well.

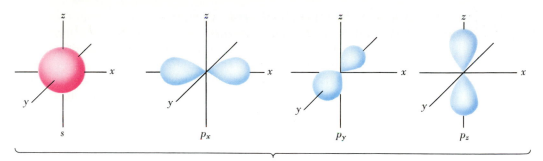

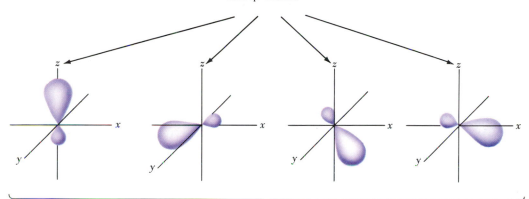

combine to generate
four sp^3 orbitals

FIGURE 11-5
The sp^3 hybridization
scheme.

which are represented
as the set

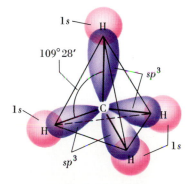

geometric structure

FIGURE 11-7
Bonding and structure of
CH_4.

The four carbon orbitals are sp^3
hybrid orbitals (violet). Those of
the hydrogen atoms (red) are $1s$.
The structure is tetrahedral,
with H—C—H bond angles of
about 109.5° (more exactly,
109°28′).

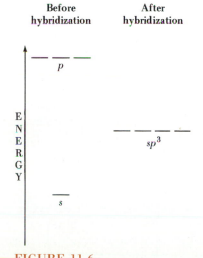

Before
hybridization

After
hybridization

E
N
E
R
G
Y

p

sp^3

s

FIGURE 11-6
Energy-level diagram for sp^3
hybridization.

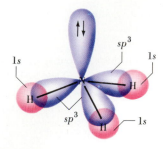

FIGURE 11-8
sp^3 hybrid orbitals and bonding in NH_3.

Although an sp^3 hybridization scheme yields a geometrical shape close to that observed experimentally, hybridization of atomic orbitals does not seem to occur in NH_3.

The Case of H_2O and NH_3 Reconsidered. sp^3 hybridization accounts for tetrahedral bond angles (109.5°), the same as predicted by VSEPR theory for the orientation of *four* electron pairs. If this hybridization scheme is used to describe bonding in H_2O and NH_3, we would predict angles of 109.5° for the H—O—H bond in water and the H—N—H bonds in NH_3. These angles are in closer agreement with the experimentally observed bond angles (104.5° in water; 107° in NH_3) than the 90° predictions in Section 11-1 based on the overlap of simple s and p orbitals. The case of NH_3, for example, can be described in terms of the following orbital diagram and the bonding scheme in Figure 11-8.

$$1s \qquad 2sp^3$$

$$N \quad \boxed{\uparrow\downarrow} \quad \boxed{\uparrow\downarrow}\,\boxed{\uparrow}\,\boxed{\uparrow}\,\boxed{\uparrow} \qquad\qquad (11.3)$$

Having said all this, we now need to observe that the weight of theoretical and experimental evidence still favors the description of bonding in H_2O and in NH_3 based on the overlap of simple orbitals, modified by H atom repulsions (page 355).* The energy required to promote fully a $2s$ electron to form sp^3 hybrid orbitals is simply too great in these cases. What this example can teach us is that, although some evidence may favor one bonding theory over another, the best description of bonding should conform to *all* the available evidence.

sp^2 Hybrid Orbitals. Let us turn our attention to the group 3A element boron. Since there are only *three* electrons and *four* orbitals in the valence shell of the B atom, however we decide to use these orbitals for bond formation, one orbital must remain empty. In most boron compounds the hybridization scheme combines the $2s$ and *two* $2p$ orbitals into *three* sp^2 hybrid orbitals, leaving one simple p orbital empty. A pictorial representation of this hybridization scheme is shown in Figure 11-9; an energy-level diagram of the pure and hybridized atomic orbitals is given in Figure 11-10; and orbital diagrams representing this hybridization scheme are shown below.

$$\begin{array}{ccc} 1s & 2s & 2p \\ B \quad \boxed{\uparrow\downarrow} & \boxed{\uparrow\downarrow} & \boxed{\uparrow}\,\boxed{\ }\,\boxed{\ } \end{array} \longrightarrow \begin{array}{ccc} 1s & 2sp^2 & 2p \\ B \quad \boxed{\uparrow\downarrow} & \boxed{\uparrow}\,\boxed{\uparrow}\,\boxed{\uparrow} & \boxed{\ } \end{array} \qquad (11.4)$$

$$\text{ground state} \qquad\qquad\qquad \text{hybridized}$$

FIGURE 11-9
The sp^2 hybridization scheme.

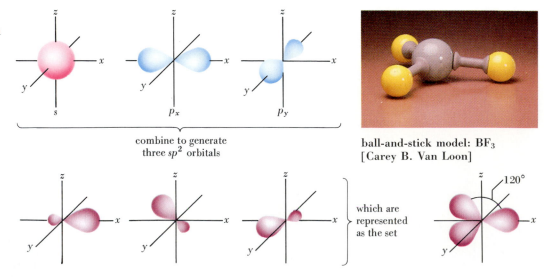

combine to generate three sp^2 orbitals

ball-and-stick model: BF_3
[Carey B. Van Loon]

which are represented as the set

120°

*See, M. Laing, "No Rabbit Ears on Water," *J. Chem. Educ.*, **64**, 124 (1987).

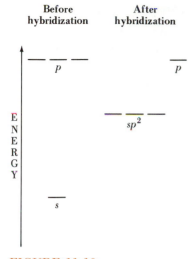

Before hybridization   **After hybridization**

ENERGY

_p_   _p_

sp^2

_s_

FIGURE 11-10
Energy-level diagram for sp^2 hybridization.

As pictured in Figure 11-9 the geometrical shape of a molecule with sp^2 hybrid bonding is trigonal planar with 120° bond angles, as in BF_3.

sp Hybrid Orbitals. If we turn next to a group 2A element like beryllium, we find that there are only *two* electrons and *four* orbitals in the valence shell. *Two* orbitals must be left vacant. The hybridization scheme that can be used to describe certain *gaseous* beryllium compounds involves sp hybridization. The 2s and one 2p orbital of Be are hybridized and the remaining two 2p orbitals are left unchanged. A pictorial representation of sp hybridization is presented in Figure 11-11, and an energy-level diagram showing the simple and hybridized orbitals is found in Figure 11-12. Orbital diagrams for this hybridization scheme are

$$1s \quad 2s \quad 2p \qquad\qquad 1s \quad 2sp \quad 2p$$

Be $\boxed{\uparrow\downarrow}$ $\boxed{\uparrow\downarrow}$ $\boxed{||}$ $\longrightarrow$ Be $\boxed{\uparrow\downarrow}$ $\boxed{\uparrow|\uparrow}$ $\boxed{|}$ (11.5)

ground state   hybridized

sp hybrid bonding leads to linear structures with 180° bond angles, as suggested by the molecular model of $BeCl_2(g)$ in Figure 11-11.

d Hybrid Orbitals. If extended to d orbitals, the concept of hybridization helps us to explain the existence of expanded octets. In the formation of the molecule PCl_5 we have to account for the formation of *five* covalent bonds, and this requires the availability of *five* half-filled orbitals in the central P atom. These five orbitals can be generated through sp^3d hybridization. The orbital diagrams below trace the hybridization scheme, from the ground-state electron configuration, through promotion of a 3s electron to 3d, to the formation of the hybrid orbitals.

$$3s \qquad 3p \qquad\qquad 3d$$

Ground state: P [Ne] $\boxed{\uparrow\downarrow}$ $\boxed{\uparrow|\uparrow|\uparrow}$ $\boxed{||||}$

$$3s \qquad 3p \qquad\qquad 3d$$

Promotion: P [Ne] $\boxed{\uparrow}$ $\boxed{\uparrow|\uparrow|\uparrow}$ $\boxed{\uparrow||||}$

$$sp^3d \qquad\qquad 3d$$

Hybridization: P [Ne] $\boxed{\uparrow|\uparrow|\uparrow|\uparrow|\uparrow}$ $\boxed{|||}$ (11.6)

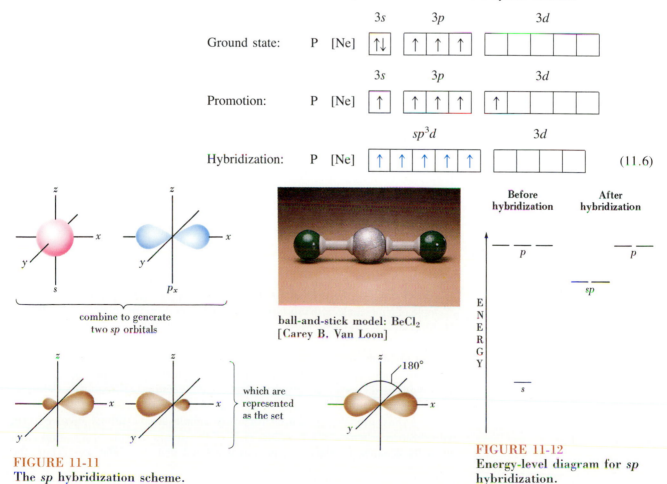

combine to generate two sp orbitals

ball-and-stick model: $BeCl_2$
[Carey B. Van Loon]

which are represented as the set

Before hybridization   **After hybridization**

ENERGY

_p_   _p_

sp

_s_

FIGURE 11-11
The sp hybridization scheme.

FIGURE 11-12
Energy-level diagram for sp hybridization.

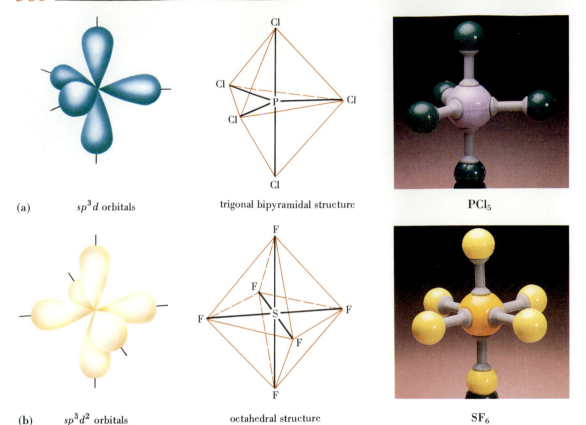

(a) sp^3d orbitals   trigonal bipyramidal structure   PCl₅

(b) sp^3d^2 orbitals   octahedral structure   SF₆

FIGURE 11-13
d **hybrid orbitals.**

(a) sp^3d **and (b)** sp^3d^2. [Photographs by Carey B. Van Loon]

The sp^3d orbitals and the *trigonal bipyramidal* structure associated with PCl_5 are shown in Figure 11-13a, as is a molecular model of PCl_5.

A similar type of hybridization can be used to describe bonding in the molecule SF_6, except that here *six* hybrid orbitals are required. These are obtained through the hybridization scheme sp^3d^2, for which the orbital diagrams are

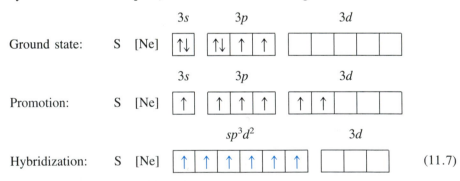

$$\text{(11.7)}$$

The sp^3d^2 orbitals, the *octahedral* structure associated with these orbitals, and an example of a molecule with this type of hybridization, SF_6, are all pictured in Figure 11-13b.

Hybrid Orbitals and the Valence-Shell Electron-Pair Repulsion Theory. The geometrical structures of molecules predicted by using hybrid orbitals generally agree with predictions made by the valence-shell electron-pair repulsion (VSEPR) theory. There is, of course, a connection between the two.

One hybrid orbital is produced for every simple atomic orbital involved in a hybridization scheme. In a molecule each of the hybrid orbitals of a central atom normally acquires an electron pair, either a bond pair or a lone pair. Thus, the

TABLE 11-1
Hybrid Orbitals and Their Geometric Orientation

Atomic orbitals	Hybrid orbitals	Orientation	Example	Predicted bond angle
$s + p$	sp	linear	$BeCl_2$	180°
$s + p + p$	sp^2	trigonal planar	BF_3	120°
$s + p + p + p$	sp^3	tetrahedral	CH_4	109.5°
$d + s + p + p$	$^a dsp^2$	square planar	$[Pt(NH_3)_4]^{2+}$	90°
$s + p + p + p + d$	$^b sp^3 d$	trigonal bipyramidal	PCl_5	120, 90°
$s + p + p + p + d + d$	$^b sp^3 d^2$	octahedral	SF_6	90°
$d + d + s + p + p + p$	$^a d^2 sp^3$	octahedral	$[Co(NH_3)_6]^{2+}$	90°

^{a}These hybrid orbitals involve d orbital(s) from the next-to-outermost shell, together with s and p orbitals of the outermost shell. They are encountered commonly in the structures of complex ions (see Chapter 25).
^{b}These hybrid orbitals involve s, p, and d orbitals, all from the outermost electronic shell. They are encountered in structures with an expanded octet that have nonmetals such as P, As, S, Cl, Br, and I as the central atom (see Chapter 23).

number of hybrid orbitals equals the number of valence-shell electron pairs. In most cases the orientation of hybrid orbitals is the same as that of electron pairs predicted by the VSEPR method, as outlined in Table 11-1. In one case, however, the valence bond method is superior to the VSEPR method. This occurs when four pairs of valence electrons appear in the hybridization scheme dsp^2 rather than sp^3. The resulting structure has a square planar geometry not predicted by the VSEPR method. Some complex ions with this geometrical structure are described in Chapter 25.

The following four-step approach generally works well to predict a plausible hybridization and bonding scheme for some unknown species. It is illustrated in Example 11-2.

Steps in predicting a hybridization scheme.

> 1. Write a plausible Lewis structure for the species of interest.
> 2. Use the VSEPR method to predict the probable geometrical shape of the species.
> 3. Select the hybridization scheme corresponding to the VSEPR prediction.
> 4. Sketch and describe the orbital overlap.

(11.8)

Example 11-2 _____

Proposing a hybridization scheme to describe covalent bonding. Describe hybridization schemes for the central atom and the orbital overlaps that occur in these two halogen compounds: **(a)** $SiCl_4$, used as a starting material in the production of silicone polymers; **(b)** XeF_2, one of the first noble gas compounds prepared.

Solution. In each case follow the four-step procedure outlined above.
(a) The Lewis structure of $SiCl_4$ is

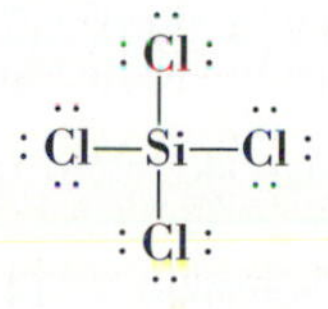

$$\begin{array}{c} :\ddot{Cl}: \\ | \\ :\ddot{Cl}-Si-\ddot{Cl}: \\ | \\ :\ddot{Cl}: \end{array}$$

FIGURE 11-14
Bonding scheme and structure for $SiCl_4$.

For simplicity, only the orbitals involved in the bonding scheme are shown.

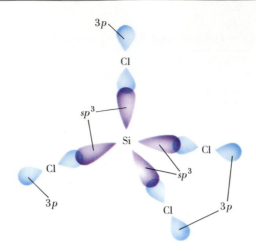

The geometrical shape predicted by the VSEPR method is tetrahedral, with Cl—Si—Cl bond angles of 109.5°. The hybridization scheme that produces a tetrahedral distribution of hybrid orbitals is sp^3. The four valence electrons of the Si atom, each unpaired, are placed in these sp^3 hybrid orbitals, and these orbitals overlap with the half-filled $3p$ orbitals of Cl atoms. The orbital diagrams for Si and Cl are

$$3sp^3$$

Si   [Ne] | ↑ | ↑ | ↑ | ↑ |

$$3s \qquad 3p$$

Cl   [Ne] | ↑↓ | | ↑↓ | ↑↓ | ↑ |

The bonding orbitals are sketched in Figure 11-14.

(b) A Xe atom has *eight* valence electrons and each F atom has *seven*. The total number of valence electrons in the Lewis structure is $8 + (2 \times 7) = 22$. Following the customary pattern for distributing valence electrons in a Lewis structure, we need to expand the outer shell of the Xe atom to *10* electrons.

$$:\ddot{F}-\overset{..}{\underset{..}{Xe}}-\ddot{F}:$$

The orientation of five electron pairs, two bond pairs, and three lone pairs is *trigonal bipyramidal*. The VSEPR prediction of the geometrical shape for this case (that is, AX_2E_3) is *linear* (see again, Table 10-3). The hybridization scheme for five hybrid orbitals in a trigonal bipyramidal distribution is sp^3d. The orbital diagrams for the central Xe atom and the terminal F atoms are

$$5sp^3d \qquad\qquad\qquad 5d$$

Xe   [Kr]$4d^{10}$ | ↑↓ | ↑↓ | ↑↓ | ↑ | ↑ | | | | |

$$1s \qquad 2s \qquad 2p$$

F | ↑↓ | | ↑↓ | | ↑↓ | ↑↓ | ↑ |

The bonding orbitals are sketched in Figure 11-15.

FIGURE 11-15
Bonding scheme and structure for XeF_2.

For simplicity, only the orbitals involved in the bonding scheme are shown.

SIMILAR EXAMPLES: Exercises 3, 14, 15, 21, 58.

11-3 Multiple Covalent Bonds

Ethylene, derived from petroleum, is the leading industrial
organic chemical. It is used
in making polyethylene and
other plastics (see page 340).

Let us see if what we have learned thus far about the valence bond method will help us to describe bonding in another simple hydrocarbon molecule, ethylene, C_2H_4. First, we can write a plausible Lewis structure.

$$
\begin{array}{cc}
H & H \\
| & | \\
H-C & = C-H
\end{array}
$$

This structure indicates a *double* covalent bond between the C atoms. Now, using the VSEPR method, we conclude that the orientation of electron pairs around each C atom is equivalent to that expected for *three* electron pairs—*trigonal planar*. Thus the ethylene molecule is a *planar* molecule with 120° H—C—H and H—C—C bond angles.

Recall that the sp^2 hybridization scheme described for boron (page 358) counted for 120° bond angles. Let us see if this hybridization scheme also works to describe bonding in C_2H_4. The relevant orbital diagrams for this scheme are

Ground state: C $1s$ [↑↓] $2s$ [↑↓] $2p$ [↑|↑|]

Promotion: C $1s$ [↑↓] $2s$ [↑] $2p$ [↑|↑|↑]

Hybridization: C $1s$ [↑↓] $2sp^2$ [↑|↑|↑] $2p$ [↑]   (11.9)

The orbital set suggested by orbital diagram (11.9) is pictured in Figure 11-16. The sp^2 orbitals are in purple and the p orbital in blue. One of the bonds between the carbon atoms results from the overlap of sp^2 hybrid orbitals from each atom. This overlap occurs along the line joining the nuclei of the two atoms. Orbitals that overlap in this "end-to-end" fashion produce a **sigma bond,** designated **σ bond.** Figure 11-16 shows that a second bond between the C atoms results from overlap of the unhybridized p orbitals. In this bond there is a region of high electron charge density (high electron probability) above and below the plane of the carbon and

FIGURE 11-16
sp^2 hybridization and bonding in C_2H_4.

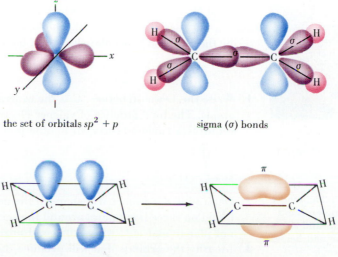

the set of orbitals $sp^2 + p$ sigma (σ) bonds

overlap of p orbitals leading to pi (π) bond

hydrogen atoms. The bond produced by this "side-by-side" overlap of p orbitals is called a **pi bond**, designated π **bond**.

Bonding in the ethylene molecule is illustrated through the ball-and-stick model of Figure 11-17, and there are three points that we should note particularly.

Always think of a double bond as consisting of one σ and one π bond.

1. The σ bond involves more extensive overlap than does the π bond. This means that a carbon-to-carbon double bond ($\sigma + \pi$) is stronger than a single bond (σ), but not twice as strong (from Table 10-2, C—C, 347 kJ/mol; C=C, 611 kJ/mol).

Now we can see the rationale in the VSEPR method of counting electron pairs in multiple covalent bonds as if there were only *one* bond pair. We want to count only the bond pairs that are involved in σ-bond formation, not in π bonds.

2. The shape of a molecule is determined only by the orbitals forming σ bonds (the σ-bond framework).

3. Rotation about the double bond is severely restricted. That is, the double bond is rigid. To twist one —CH$_2$ group out of the plane of the other (see Figure 11-17) would reduce the amount of overlap of the p orbitals, weakening the π bond.

To describe bonding in acetylene, C$_2$H$_2$, we could repeat the discussion of C$_2$H$_4$, but with these differences: The Lewis structure of C$_2$H$_2$ features a **triple covalent bond**. The molecule is *linear*. A hybridization scheme to account for linear bonds (180° bond angles) is *sp,* and *sp* hybridization results in *two* half-filled unhybridized p orbitals.

Ground state:

Promotion:

Hybridization: (11.10)

Always think of a triple bond as consisting of one σ and two π bonds.

In the triple bond in C$_2$H$_2$, one of the carbon-to-carbon bonds is a σ bond and *two* are π bonds, as suggested through Figure 11-18.

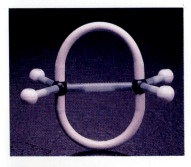

FIGURE 11-17

Ball-and-stick model of ethylene, C$_2$H$_4$.

The H—C—H and C—C—H bond angles are 120°. The model also distinguishes between the σ bond between C atoms and the π bond extending above and below the plane of the molecule.
[Carey B. Van Loon]

Example 11-3

Proposing hybridization schemes involving σ and π bonds. Formaldehyde gas, H$_2$CO, is used in the manufacture of plastics, and, in water solution, is familiar in the laboratory as the biological preservative called formalin. Describe the expected molecular shape and bonding scheme of the H$_2$CO molecule.

Solution

1. Write the Lewis structure. C is the central atom and H and O are terminal atoms. The total number of valence electrons is 12. Note that this structure requires a carbon-to-oxygen double bond.

2. Determine the geometrical shape of the molecule. The σ-bond framework is based on three bond pairs around the central C atom. This suggests a trigonal planar molecule with 120° bond angles.

3. Identify the orbitals that will produce the predicted shape. A trigonal planar structure is associated with sp^2 hybrid orbitals (recall Table 11-1).

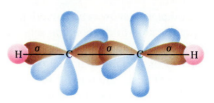

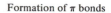

Formation of σ bonds   Formation of π bonds   Space-filling model

FIGURE 11-18
sp hybridization and bonding in C_2H_2.

4. Sketch the orbitals of the central atom that are involved in orbital overlap. The C atom (expression 11.9) uses two of its sp^2 hybrid orbitals to form σ bonds with two H atoms. The remaining sp^2 hybrid orbital is used to form a σ bond with oxygen. The simple p orbital of the C atom is used to form a π bond with O.

5. Sketch the bonding orbitals of the terminal atoms. The H atoms have only 1s orbitals available for bonding. The situation with oxygen is less clear. Either of the orbital diagrams shown below for O will work. Each provides an orbital for end-to-end overlap in the σ bond to carbon, and each has a half-filled p oribtal that can participate in the "sidewise" overlap leading to a π bond.

$$\text{O} \quad \boxed{\uparrow\downarrow}_{1s} \quad \boxed{\uparrow\downarrow}_{2s} \quad \boxed{\uparrow\downarrow}\,\boxed{\uparrow}\,\boxed{\uparrow}_{2p} \qquad (11.11)$$

or

$$\text{O} \quad \boxed{\uparrow\downarrow}_{1s} \quad \boxed{\uparrow\downarrow}\,\boxed{\uparrow\downarrow}\,\boxed{\uparrow}_{2sp^2} \quad \boxed{\uparrow}_{2p} \qquad (11.12)$$

The bonding and structure of the formaldehyde molecule are suggested by the three-dimensional sketch in Figure 11-19.

SIMILAR EXAMPLES: Exercises 4, 6, 7, 18, 20, 26, 55.

Are You Wondering:

When to use simple orbitals (such as expression 11.11) and when to use hybrid orbitals (such as expression 11.12) for terminal atoms in a structure?

As illustrated in Example 11-3, the primary results—geometrical shape of the molecule, bond angles—are affected only by the hybridization scheme chosen for a central atom(s), not for the terminal atoms. We usually do not have sufficient experimental information about the molecule (e.g., exact bond lengths and electron density distributions) to be certain which orbitals of the terminal atoms are best for a given situation.

On the one hand, a hybridization scheme for terminal atoms (such as expression 11.12) places lone-pair electrons farthest apart; this is good. On the other hand, we must consider the energy required to convert simple orbitals to hybrid orbitals (recall the discussion of H_2O and NH_3 on page 358). This energy requirement could be too high.

In conclusion, then, since our main concern is with the orbital scheme for central atoms, terminal atoms can generally be shown as employing either simple or hybrid atomic orbitals.

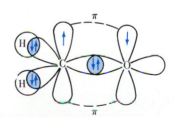

FIGURE 11-19
Bonding and structure of the H_2CO molecule—Example 11-3 illustrated.

For simplicity, only bonding orbitals of the valence shells are shown. Specifically, the 2s and $2p_y$ orbitals of the oxygen atom, each of which contains an electron pair, have been omitted.

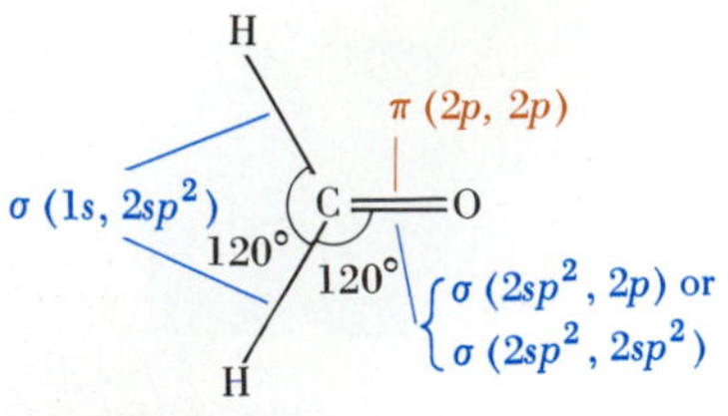

FIGURE 11-20
Bonding and structure in H_2CO—a schematic representation.

It is not always easy to draw a three-dimensional sketch to show orbital overlap as in Figure 11-19. A simpler, two-dimensional representation of the bonding scheme for formaldehyde is shown in Figure 11-20. In Figure 11-20 bonds between atoms are represented by straight lines; bonds are labeled σ or π; and the orbital overlap involved is indicated. (In Figure 11-20, the first orbital designation is for the atom on the left of the bond, and the second orbital designation is for the atom on the right.)

Although we have stressed drawing a Lewis structure as the first step in describing a bonding scheme, at times the starting point is a description of the species *obtained by experiment*. In such cases we must draw Lewis structures consistent with the experimental evidence and then proceed to plausible bonding schemes. This situation is explored through Example 11-4.

Example 11-4

Using experimental data to assist in selecting a hybridization and bonding scheme. **Dinitrogen oxide (nitrous oxide) is the common "laughing gas" used as an anesthetic in dental offices. A reference source on molecular structures lists the following data for N_2O:**

Bond lengths: N—N, 113 pm; N—O, 119 pm; bond angle: 180°

Describe a plausible hybridization and bonding scheme(s) for N_2O.

Structure I

Solution. Our usual first step is to draw a Lewis structure, but before doing so let us compare the experimental bond lengths with average values from Table 10-2. The relevant values from Table 10-2 are N=N, 123 pm; N≡N, 109 pm; N—O, 136 pm; and N=O, 115 pm. From these data we see that in N_2O the N-to-N bond is intermediate between a double and triple bond, and the N-to-O bond is intermediate between a single and a double bond. We can draw two plausible Lewis structures such that the hybrid of these two conforms well to the experimental data. Note that in each structure the distribution of 16 valence electrons requires the use of formal charges.

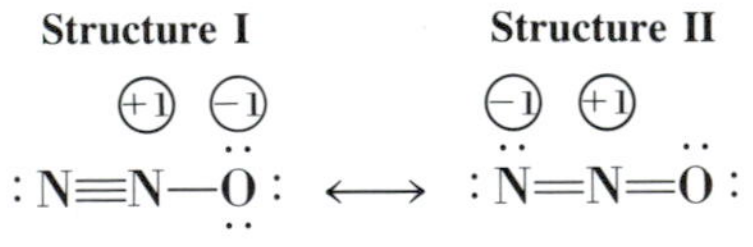

Structure II

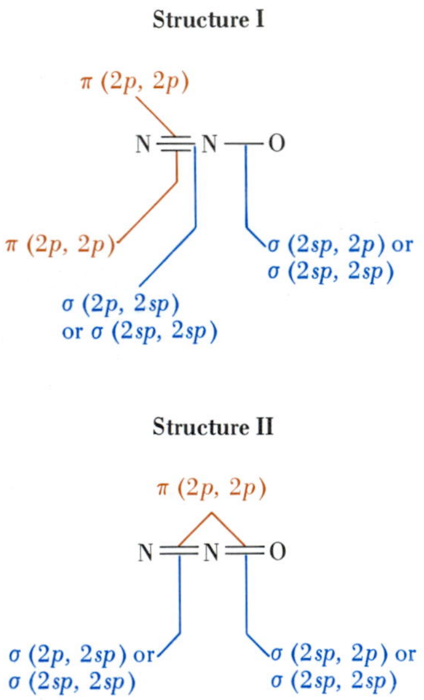

FIGURE 11-21
Bonding and structure in dinitrogen oxide, N_2O—Example 11-4 illustrated.

The true structure is a resonance hybrid of the two possibilities shown here.

Note that in both of these structures the hybridization scheme for the central N atom must produce a bond angle of 180°. The hybridization scheme is *sp*. In structure I there are one σ and two π bonds between the N atoms and a σ bond between the N and O atom. In structure II the central N atom is bonded to both other atoms with one σ and one π bond. The bonding schemes for these two structures are outlined in Figure 11-21, but the true structure, of course, is a hybrid of these two.

SIMILAR EXAMPLES: Exercises 11, 19, 28, 59, 61, 62.

11-4   An Introduction to Molecular Orbital Theory

Let us try to imagine producing a hydrogen molecule from two hydrogen atoms. Consider that initially the two atoms are "infinitely" far apart. This means far enough apart so that there is no energy of interaction between them—no attraction

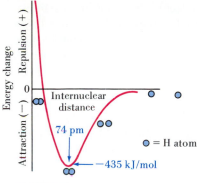

FIGURE 11-22

Energy of interaction of two hydrogen atoms as a function of internuclear distance.

Two H atoms form the molecule H_2 at a particular internuclear distance (74 pm).

and no repulsion. We will call this condition our zero of energy. Now imagine that the two atoms are allowed to approach each other and that any energy change observed is plotted as a function of the distance between the nuclei of the two atoms—the **internuclear distance.** As the internuclear distance decreases, the atoms attract each other and more and more energy is *released*. The energy of the system of two atoms is lowered from that of the widely separated atoms; the energy of the system becomes *negative*. The maximum energy release (435 kJ/mol), corresponding to the minimum in the curve of Figure 11-22, comes at an internuclear distance of 74 pm, the H—H bond length in H_2. To force the atoms closer together than 74 pm requires that energy now be *absorbed* (work must be done), and the curve rises very steeply from its minimum point.

The repulsion between H atoms in close proximity is easy to understand. It results from the mutual repulsion of the two positively charged nuclei. But what is the source of the attraction at intermediate distances?

Figure 11-23 shows two different arrangements of the protons and electrons in two H atoms when they are brought into close proximity. In the arrangement in which the electrons are located *away* from the internuclear region (blue), there is a strong repulsion between the protons. Energy is high, and the arrangement is unstable. The arrangement where the two electrons are located *between* the atomic nuclei (red) is of *lower* energy than in the separated atoms. Thus, even classical theory (Coulomb's law) predicts that two H atoms should combine to form an H_2 molecule. However, the energy of interaction predicted by classical theory is much less than the measured value. The calculated internuclear distance is also in error. We need to turn to wave mechanics.

Let us think of electrons in terms of charge densities or probabilities extending over an entire molecule, that is, in terms of **molecular orbitals.** One method of deriving molecular orbitals is by an appropriate mathematical combination of

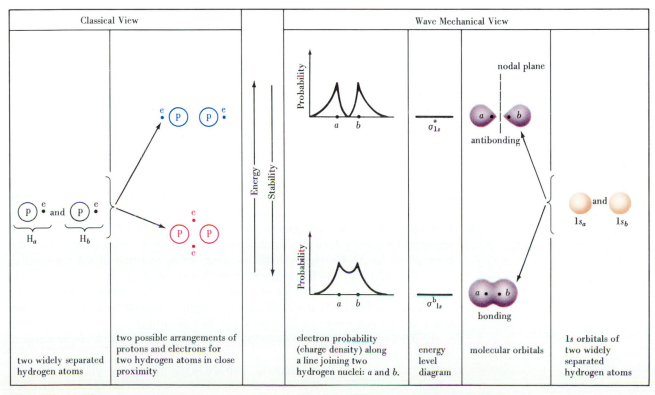

FIGURE 11-23

The interaction of two hydrogen atoms.

atomic orbitals of the atoms being united into a molecule. Wave mechanics allows two possibilities. One combination of two $1s$ orbitals represented in Figure 11-23 produces a **bonding molecular orbital,** designated as σ_{1s}^b. Another combination produces an **antibonding orbital,** σ_{1s}^*. Also depicted in Figure 11-23 is an energy-level diagram for these two molecular orbitals.

The bonding molecular orbital is at a lower energy than the separated atomic orbitals, and the antibonding orbital is at a higher energy.

In terms of electron probability or charge density, Figure 11-23 shows that the bonding molecular orbital has a high electron probability or charge density *between* the atomic nuclei. Electron charge density concentrated in the internuclear region reduces the repulsion between the positively charged nuclei. This permits bonding between the atoms—hence the term *bonding molecular orbital.* In the antibonding molecular orbital the electron probability or charge density in the internuclear region is much lower. In fact it falls to zero in a nodal plane midway between the nuclei. Electron charge density in an antibonding molecular orbital is concentrated *away* from the internuclear region, where it is ineffective in reducing internuclear repulsion—hence the term *antibonding.* The probability distributions for the bonding and antibonding molecular orbitals correspond roughly to the electron positions in the classical view in Figure 11-23. That is, the probability (wave mechanics) is high in the region where the electron is located (classical view).

Recall the discussion of a standing wave in a string and the significance of nodes in wave mechanics (Section 8-8).

Basic Ideas Concerning Molecular Orbitals. Here are some useful ideas concerning molecular orbitals and how electrons are assigned to them.

Basic ideas about molecular orbitals.

1. The number of molecular orbitals (MO) produced is equal to the number of atomic orbitals that are combined.
2. Of the two MOs produced when two atomic orbitals are combined, one is a *bonding* MO at a *lower* energy than the original atomic orbitals. The other is an *antibonding* MO at a *higher* energy.
3. Electrons normally seek the lowest energy MOs available to them in a molecule.
4. The maximum number of electrons that can be assigned to a given MO is *two* (Pauli exclusion principle).
5. Electrons enter MOs of identical energies *singly* before they pair up (Hund's rule).
6. Formation of a bond between atoms requires that the number of electrons in *bonding* MOs exceeds the number in *antibonding* MOs.

(11.13)

Diatomic Molecules of the First Period Elements. Let us see how we can use the rules in expression (11.13) to describe the possible molecular species formed by the first period elements, H and He. Figure 11-24 describes the one bonding and one antibonding molecular orbital formed by combining two $1s$ atomic orbitals in the following manner: The energies of the $1s$ atomic orbitals are represented by line segments on either side of the diagram. The energy of the bonding molecular orbital corresponds to the line segment at the lower level, and that of the antibonding molecular orbital by the line segment at the higher level.

H_2^+: This species has a single electron. It enters the σ_{1s}^b orbital and produces a bond between the two H atoms. We might call this a one-electron or "*half*" bond (since the normal *single* covalent bond has *two* electrons).

H_2: This molecule has two electrons, both in the σ_{1s}^b orbital. A *single* covalent bond is formed.

He_2^+: This ion has three electrons. Two are in the σ_{1s}^b orbital, and one in the σ_{1s}^* orbital. The net number of bonding electrons is $2 - 1 = 1$. This species exists as a stable ion, but with only a one-electron or "half" bond.

FIGURE 11-24
Molecular orbital diagrams for the diatomic molecules (or ions) formed from first period elements.

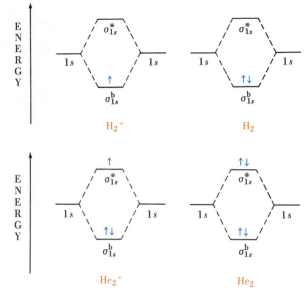

H_2: Two electrons are in the σ_{1s}^{b} orbital and two in the σ_{1s}^{*}. The net number of bonding electrons is $2 - 2 = 0$. No bond is produced. We should not expect to encounter He_2 as a stable species.

The discussion of bonding in He_2^{+} and He_2 suggests that electrons in antibonding orbitals cancel out the effectiveness of electrons in bonding orbitals. Thus, if a diatomic molecule is to have a stable covalent bond, it must have more electrons in bonding orbitals than it has in antibonding orbitals. For example, if the excess of bonding over antibonding electrons is *two*, this corresponds to a *single* covalent bond and a **bond order** of 1. *Bond order* is simply one-half the difference between the number of bonding and antibonding electrons.

Expression for calculating bond order.

$$\text{bond order} = \frac{\text{no. } e^- \text{ in bonding MOs} - \text{no. } e^- \text{ in antibonding MOs}}{2} \quad (11.14)$$

Example 11-5

Relating bond energy and bond order. The bond energy of H_2 is 435 kJ/mol. Estimate the bond energies of H_2^{+} and He_2^{+}.

Solution. The bond order in H_2 is one, that is, a single bond. In H_2^{+} and in He_2^{+} the bond order is $\frac{1}{2}$. We should expect the bonds in these two species to be only about one-half as strong as in H_2, that is, about 220 kJ/mol. (Actual values: H_2^{+}, 255 kJ/mol; He_2^{+}, 251 kJ/mol.)

SIMILAR EXAMPLES: Exercises 10, 31.

Example 11-6

Relating magnetic properties to the occupancy of molecular orbitals. Which of the four species in Figure 11-24 are *paramagnetic*?

Solution. Paramagnetism requires the presence of unpaired electrons. One unpaired electron is found in H_2^{+} and in He_2^{+}. These ions are paramagnetic.

SIMILAR EXAMPLE: Exercise 30.

FIGURE 11-25

Representation of the molecular orbitals formed by a combination of $2p$ atomic orbitals.

These diagrams are meant simply to suggest the nature of the electron charge distribution for the several molecular orbitals. They are not exact in all details. Nodal planes for the antibonding orbitals are represented by the broken lines.

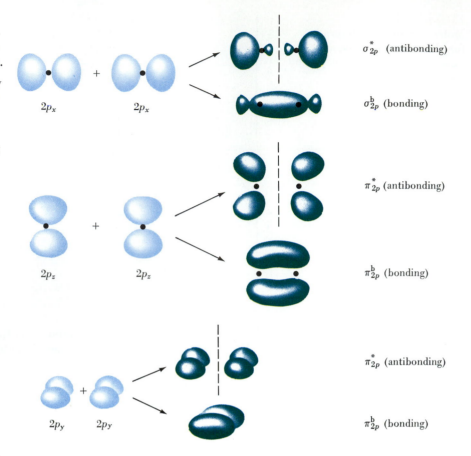

σ^*_{2p} (antibonding)

σ^b_{2p} (bonding)

$2p_x$ $+$ $2p_x$

π^*_{2p} (antibonding)

π^b_{2p} (bonding)

$2p_z$ $+$ $2p_z$

π^*_{2p} (antibonding)

π^b_{2p} (bonding)

$2p_y$ $+$ $2p_y$

11-5 Molecular Orbitals in Homonuclear Diatomic Molecules of the Second Period Elements

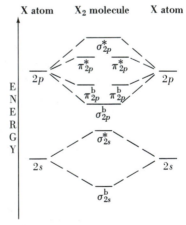

X atom X_2 molecule   X atom

σ^*_{2p}

π^*_{2p} π^*_{2p}

$2p$ $2p$

π^b_{2p} π^b_{2p}

σ^b_{2p}

E N E R G Y

σ^*_{2s}

$2s$ $2s$

σ^b_{2s}

FIGURE 11-26

Energy level diagram for a hypothetical second period diatomic molecule, X_2.

The diagram shown here is applicable to the heavier members of the second period, e.g., O_2 and F_2. For the lighter members the relative placement of the σ^b_{2p} and the π^b_{2p} orbitals is reversed (see Figure 11-27).

To describe the molecular orbitals of diatomic molecules of the second period elements (Li through Ne) we must first do two things: (1) describe the *eight* molecular orbitals formed by combining the atomic orbitals of the *second* principal electronic shell—$2s$, $2p_x$, $2p_y$, $2p_z$—and (2) construct an energy-level diagram for these molecular orbitals.

The molecular orbitals produced by combining $2s$ orbitals have the same characteristics as those derived from $1s$ orbitals, except that they are at a higher energy. In considering the possibilities for the formation of molecular orbitals by combining $2p$ atomic orbitals, we discover *two* possibilities, as illustrated in Figure 11-25. p orbitals that overlap along the same straight line (that is, end-to-end) combine to produce σ-type molecular orbitals: $\boldsymbol{\sigma^b_{2p}}$ and $\boldsymbol{\sigma^*_{2p}}$. Those that overlap in a parallel or sidewise fashion produce π-type molecular orbitals: $\boldsymbol{\pi^b_{2p}}$ and $\boldsymbol{\pi^*_{2p}}$. There are *two* molecular orbitals of each π type because there are *two* pairs of p orbitals ($2p_y$ and $2p_z$) that are arranged in a parallel fashion. In Figure 11-25 we again see that in bonding molecular orbitals there is a high electron charge density between the atomic nuclei, and that in antibonding molecular orbitals there are nodal planes between the atomic nuclei, where the electron charge density falls to zero.

The energy-level diagram for the molecular orbitals in a molecule X_2 (where X is a second period element) is related to the energy-level diagram for the corresponding atomic orbitals. That is, we start with the energy-level diagrams of the separated atoms, arranged in the fashion we first used in Figure 8-29. The energies of the molecular orbitals for the merged atoms (that is, the molecule) are arranged in the region between the diagrams for the separated atoms. Recall that the energy levels of the atomic orbitals in multielectron atoms *differ for different atoms* (an effect pictured in the right portion of Figure 8-29). Because of this the molecular orbital energy levels in a molecule of X_2 will also be different for different atoms, X. A hypothetical energy-level diagram for X_2 is pictured in Figure 11-26. The general features of Figure 11-26 are those that we might expect.

- The energies of bonding molecular orbitals are *lower* than those of antibonding molecular orbitals.
- The energies of molecular orbitals formed from $2s$ atomic orbitals are *lower* than those formed from $2p$ atomic orbitals. (This is the same relationship as between the $2s$ and $2p$ atomic orbitals.)
- The energies of σ-type bonding orbitals are lower than those of π-type orbitals. [End-to-end overlap of $2p$ orbitals (σ-type) is more extensive than sidewise overlap of $2p$ orbitals (π-type), resulting in a lower energy.]

In spectroscopy the first electronic shell ($n = 1$) is sometimes referred to as the K shell.

To describe the assignment of electrons to the molecular orbitals in X_2, we can proceed in this way: We start with the σ_{1s}^b and σ_{1s}^* orbitals filled. Then we add electrons, in order of increasing energy, to the available molecular orbitals of the second principal shell. Alternatively, we can think of the first shell (K shell) electrons as not involved in the bonding (that is, as *nonbonding* electrons). This allows us simply to consider an assignment of the valence-shell electrons of the two atoms. Figure 11-27 presents a generalized energy-level diagram for the actual and hypothetical homonuclear diatomic molecules of the second period elements and the assignment of electrons to their molecular orbitals.

A *homonuclear* diatomic molecule is one in which both atoms are of the same kind.

We can understand some of the previously unexplained features of the O_2 molecule in terms of its molecular orbital diagram. Each O atom brings *six* valence

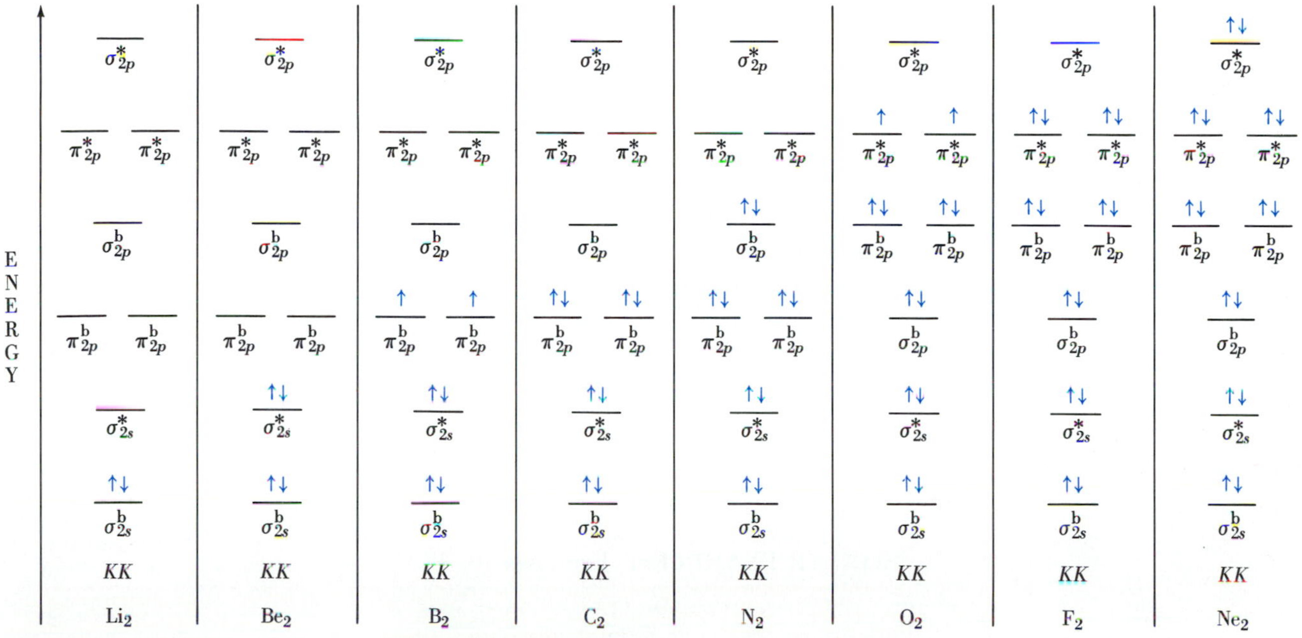

FIGURE 11-27
Molecular orbital diagrams for actual and hypothetical diatomic molecules of the second period elements.

Concerning the relative placement of the σ_{2p}^b and π_{2p}^b orbitals, more extensive overlap occurs when p atomic orbitals are combined end-to-end rather than side-by-side, suggesting that σ_{2p}^b lies at a lower energy than π_{2p}^b. This is the situation, confirmed by experiment, when the energy difference between $2s$ and $2p$ orbitals is large (as in O, F, and Ne). When this difference is smaller (as in Li through N), there is a mixing of $2s$ and $2p$ atomic orbitals in the formation of molecular orbitals that results in a reversal of the σ_{2p}^b and π_{2p}^b levels. This variation in the energy-level diagrams of the diatomic molecules of the second period elements has little effect on the matters discussed in this chapter.
The symbol KK means that electrons in the first electronic shells (KK) are not involved in the bonding.

electrons to the diatomic molecule O_2. There are *12* electrons to be assigned to molecular orbitals. From Figure 11-27 we see that

1. The O_2 molecule has *two unpaired electrons*, even though the total number of electrons is even. This explains the *paramagnetism* of oxygen.
2. The total number of valence electrons in bonding orbitals is *eight*, and the number of electrons in antibonding orbitals is *four*. The excess of bonding over antibonding electrons is $8 - 4 = 4$. At two electrons per bond (expression 11.14), this corresponds to a *double* covalent bond between oxygen atoms.

Example 11-7

Relating molecular properties to bond order. Which of the species indicated in Figure 11-27 would you expect **(a)** *not* to exist as a stable molecule, and **(b)** to have the largest bond energy?

Solution

(a) A stable diatomic molecule must have an excess of bonding electrons over antibonding electrons. All the species in Figure 11-27 do except Be_2 and Ne_2. These two have equal numbers of bonding and antibonding electrons and a bond order of *zero* (recall expression 11.14). Be_2 and Ne_2 do not exist as stable molecules.

(b) We expect the largest bond energy for N_2, the molecule with the highest bond order. From Figure 11-27 we see that N_2 has the greatest excess of bonding over antibonding electrons. The six excess bonding electrons form a triple covalent bond.

SIMILAR EXAMPLES: Exercises 10, 31, 32, 33.

Example 11-8

Writing a molecular orbital diagram. Represent bonding in $O_2{}^+$ with a molecular orbital diagram.

Solution. The O_2 molecule has 12 valence electrons. In the *ion* $O_2{}^+$ there are *11* valence electrons. These are assigned to the available molecular orbitals in accordance with the principles established on page 368. The representation below resembles the orbital diagrams that we first used for the electron configurations of atoms. The symbol KK means that the electrons in the first electronic shells (K shells) are not involved in the bonding.

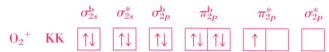

SIMILAR EXAMPLES: Exercises 10, 37.

11-6 Molecular Orbitals in Heteronuclear Diatomic Species

In a *heteronuclear* diatomic species the two atoms are *different*. The types of species covered by this discussion are molecules such as CO and NO or ions such as CN^-. There is an added difficulty in dealing with species of this type. In constructing an energy-level diagram of the molecular orbitals, the energies of the atomic

X atom   XY molecule   Y atom

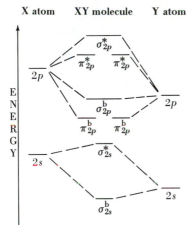

FIGURE 11-28
Molecular orbital energy-level diagram for a hypothetical heteronuclear diatomic molecule, XY.

orbitals of the separated atoms are different and the energy-level diagram becomes "distorted," as suggested in Figure 11-28. The greater the differences in atomic number and electronegativity of the atoms, the greater the "distortion."

In an extreme case like the HF molecule, for example, the $1s$ and $2s$ orbitals of the F atom are much lower in energy than the $1s$ orbital of the H atom. As a result the atomic orbitals that combine to produce molecular orbitals, one bonding and one antibonding, are the $1s$ of the H atom and a $2p$ of the F atom. Even at that, because a $2p$ orbital of the F atom is at a lower energy than the $1s$ of H, the bonding molecular orbital has a greater contribution from the $2p$ orbital of F than from the $1s$ of H. The electron pair in the bonding orbital spends more time near the F atom than the H atom and the bond is polar, just as we would have predicted in other ways in Chapter 10.

In this text we limit consideration of *heteronuclear* diatomic species to ones in which the two atoms are only one or two places removed in the second period of elements. In such cases we can assume that the orbital energies in the separated atoms are similar enough that the molecular orbital diagram for the diatomic species will closely resemble those in Figure 11-27.

Example 11-9

Writing molecular orbital diagrams for heteronuclear diatomic species. Write the molecular orbital diagram for the cyanide ion, CN^-, and determine the bond order in this ion. Compare your results with the predicted bond order from a Lewis structure.

Solution. The number of valence electrons to be assigned in the diagram is 10 (*four* from C, *five* from N, and *one* extra electron to convey the -1 charge on the ion). Use the molecular orbital diagram for either C_2 or N_2 from Figure 11-27 (they are the same), and introduce the 10 valence electrons.

$$\sigma_{2s}^b \quad \sigma_{2s}^* \quad \pi_{2p}^b \quad \sigma_{2p}^b \quad \pi_{2p}^* \quad \sigma_{2p}^*$$

CN^- KK $\boxed{\uparrow\downarrow}$ $\boxed{\uparrow\downarrow}$ $\boxed{\uparrow\downarrow}\boxed{\uparrow\downarrow}$ $\boxed{\uparrow\downarrow}$ $\boxed{}\boxed{}$ $\boxed{}$

The bond order (expression 11.14) is one-half the difference between the number of electrons in bonding orbitals and in antibonding orbitals. The number of electrons in the bonding orbitals, σ_{2s}^b, π_{2p}^b, and σ_{2p}^b, is $2 + 4 + 2 = 8$. The number of electrons in the antibonding orbital σ_{2s}^* is 2. The bond order is $(8 - 2)/2 = 3$. The CN^- ion has a *triple bond*. The Lewis structure of the cyanide ion is $[:C\equiv N:]^-$, which also indicates a triple covalent bond.

SIMILAR EXAMPLES: Exercises 37, 66.

11-7 Bonding in the Benzene Molecule

We can describe simple hydrocarbons like CH_4, C_2H_4, and C_2H_2 through Lewis structures, VSEPR theory, and the valence bond method. But there are some hydrocarbons that we cannot describe adequately by any of these approaches; to describe them we need the molecular orbital theory. These molecules are the so-called aromatic hydrocarbons, substances related to the compound benzene, C_6H_6. The term "aromatic" relates to the fragrant aromas associated with some (but by no means all) of these compounds.

The first good proposal of the structure of benzene was advanced by Friedrich Kekulé in 1865. Kekulé hypothesized that the benzene molecule consists of a flat,

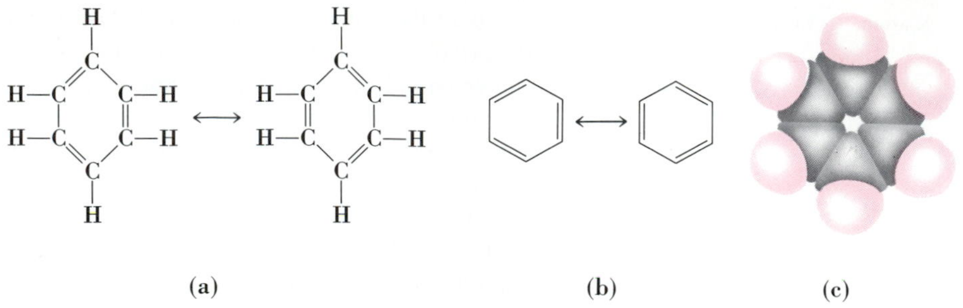

FIGURE 11-29

Resonance in the benzene molecule and the Kekulé structures.

(a) Lewis structures for C_6H_6, showing alternate carbon-to-carbon single and double covalent bonds.
(b) Two equivalent Kekulé structures for benzene. A carbon atom is at each corner of the hexagonal structure, and a hydrogen atom is bonded to each carbon. (The symbols for carbon and hydrogen are customarily *not* written in these structures.)
(c) A space-filling model.

(a)   (b)   (c)

hexagonal ring of six carbon atoms joined by alternating single and double covalent bonds. Each carbon atom is joined to two other carbon atoms and to one hydrogen atom. To explain the fact that the carbon-to-carbon bonds are all alike, Kekulé suggested that the single and double bonds in the ring of carbon atoms continually oscillated from one position to another. Other researchers suggested that there are two discrete structures possible for the C_6H_6 molecule and that benzene is a mixture of these. A more correct view is that the two possible Kekulé structures are actually contributing structures to a resonance hybrid. This view is suggested by Figure 11-29.

To account for a *planar* structure with 120° bond angles by the valence bond method, we need to use sp^2 hybrid orbitals. To account for double bonds we need to use p orbitals. End-to-end overlap involving the sp^2 orbitals produces the σ-bond framework. Sidewise overlap of $2p$ orbitals yields π bonds. However, to account for the fact that all the C-to-C bonds are identical we need to invoke the phenomenon of resonance. The valence bond interpretation of bonding in C_6H_6 is described in Figure 11-30.

Delocalized Molecular Orbitals. To find a better explanation of bonding in benzene we need to introduce the notion of delocalized molecular orbitals. In a **delocalized molecular orbital** high electron probability or electron charge density extends over three or more atoms instead of being limited (localized) to the internuclear region between two atoms.

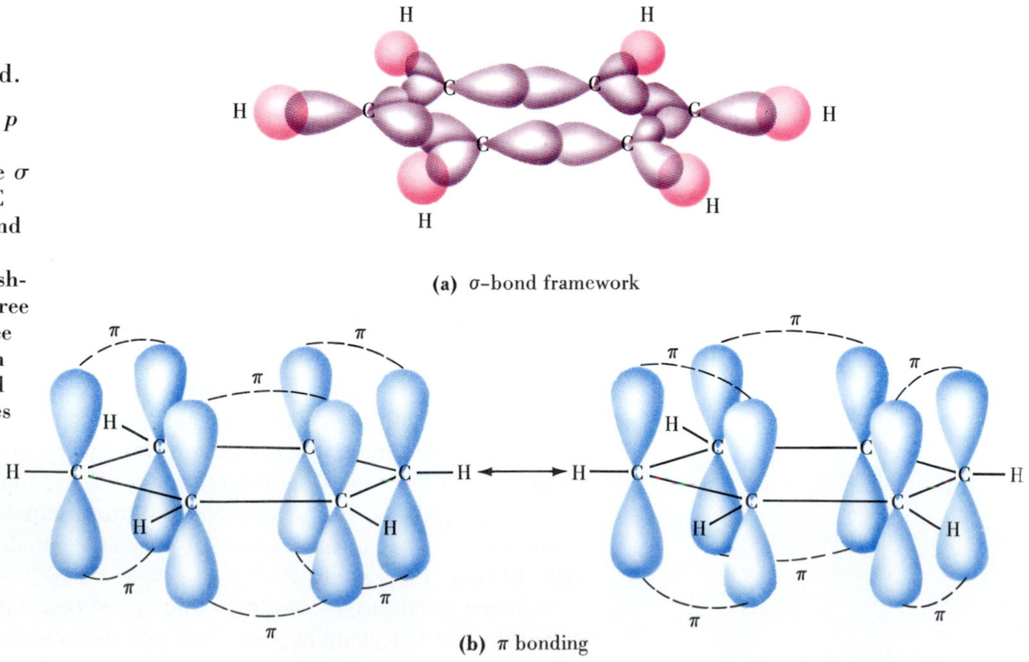

FIGURE 11-30

Bonding in benzene, C_6H_6, by the valence bond method.

(a) Carbon atoms use sp^2 and p orbitals (recall Figure 11-16). Each carbon atom forms three σ bonds, two with neighboring C atoms in the hexagonal ring and a third with an H atom.
(b) The overlap in sidewise fashion of $2p$ orbitals produces three π bonds. Thus, there are three double bonds ($\sigma + \pi$) between carbon atoms in the hexagonal ring. Two equivalent structures are possible.

(a) σ–bond framework

(b) π bonding

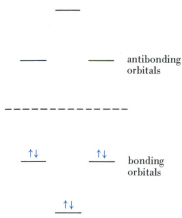

FIGURE 11-31
π molecular orbitals in C_6H_6.

The σ-bond framework of Figure 11-30 adequately describes the bonds formed within the plane of the hexagonal ring of C and H atoms in C_6H_6. Now let us consider combining the six $2p$ atomic orbitals of the C atoms into molecular orbitals. This process yields three bonding and three antibonding orbitals of the π type. The three bonding molecular orbitals fill with six electrons (one $2p$ electron from each C atom), and the three antibonding molecular orbitals remain empty. The bond order associated with the six electrons in bonding molecular orbitals is $(6 - 0)/2 = 3$. The three bonds are distributed among the six C atoms, amounting to $\frac{3}{6}$ or a half-bond between each pair of C atoms. Add to this the σ bonds in the σ-bond framework and we have a bond order of 1.5 for the C-to-C bonds. This is exactly what we also get by averaging the two Kekulé structures of Figure 11-29. This bonding scheme for benzene is outlined in Figure 11-31.

The combination of the three bonding π molecular orbitals in C_6H_6 describes the distribution of π electron charge in the molecule. We can think of this in terms of two donut-shaped regions, one above and one below the plane of the C and H atoms (see Figure 11-32). Since they are spread out among six C atoms, these molecular orbitals are *delocalized*. The concept of delocalized electrons is carried over into the symbolic representation of the benzene molecule shown in Figure 11-32b. The circle inscribed within the hexagon represents the multiple bond character displayed by all six carbon atoms.

FIGURE 11-32
Molecular orbital representation of bonding in benzene, C_6H_6.

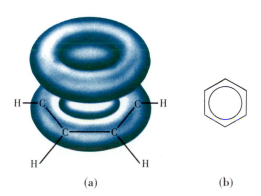

(a)   (b)

(a) A representation incorporating the σ-bond framework of Figure 11-30a and delocalized molecular orbitals (the doughnut-shaped regions) for the π bonds. (b) A symbolic representation suggesting the delocalized nature of the π bonds (circle inscribed in the hexagon) that is often used in place of the Kekulé structures of Figure 11-29b.

11-8 Other Structures with Delocalized Molecular Orbitals

We should look for the possibility of delocalized molecular orbitals whenever we encounter the phenomenon of resonance in structures with multiple covalent bonds, just as we did for benzene in Section 11-7. Consider, for example, the nitrate ion, NO_3^-, which we previously described in Example 10-12. In place of the resonance hybrid based on these contributing structures,

$$\left[\begin{array}{c} :O \\ \| \\ N \\ \diagup \diagdown \\ :O \quad O: \end{array} \right]^- \longleftrightarrow \left[\begin{array}{c} :O: \\ | \\ N \\ \diagup \diagdown \\ :O \quad O: \end{array} \right]^- \longleftrightarrow \left[\begin{array}{c} :O: \\ | \\ N \\ \diagup \diagdown \\ :O \quad O: \end{array} \right]^-$$

(11.15)

we can write a single structure with a σ-bond framework and delocalized electrons in π molecular orbitals, shown in Figure 11-33. As we did with benzene, suppose we thought about bonding in NO_3^- in the following way.

1. Assume sp^2 hybridization for each atom in the σ-bond framework.
2. Of the 24 valence electrons in NO_3^-, assign 18 to the sp^2 hybrid orbitals of the σ-bond framework. Six of these 18 electrons are bonding electrons (red in Figure 11-33). The other 12 (blue) are lone-pair electrons.
3. The four $2p$ orbitals not involved in the sp^2 hybridization scheme combine to form four molecular orbitals of the π type. Suppose we assume that two of the π orbitals are bonding orbitals and two are antibonding.
4. The remaining 6 of the 24 valence electrons are now assigned to these presumed π orbitals. Four electrons go into the bonding molecular orbitals and two into the antibonding.
5. The bond order associated with the π molecular orbitals is $(4 - 2)/2 = 1$. This π bond is distributed among *three* N—O bonds, amounting to $\frac{1}{3}$ of a π bond for each. The total bond order for each N—O bond is $1\frac{1}{3}$.

This description of NO_3^- is equivalent to averaging the three Lewis structures of (expression 11.15). Thus, with molecular orbital theory it is unnecessary to invoke the phenomenon of resonance.

The formation of delocalized π molecular orbitals can also be described symbolically, in a way that begins with the atomic orbital diagrams of the valence bond method. This method, illustrated through Example 11-10, also provides some additional insight into the notion of formal charge.

What we are actually doing here is describing the σ-bond framework by the valence bond method and π bonding by the molecular orbital method.

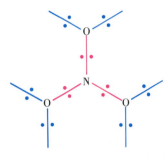

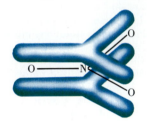

(a) σ-bond framework

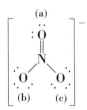

(b) Delocalized π molecular orbital

FIGURE 11-33
Structure of the nitrate anion, NO_3^-.

Example 11-10

Representing delocalized molecular orbitals with atomic orbital diagrams. Represent π molecular orbital formation, bonding, and orbital occupancy in the nitrate ion, NO_3^-.

Solution. First, we should recognize this fact: For this type of exercise we need not concern ourselves with resonance; we can use any one of the three structures in expression (11.15). Let us use the first structure shown, which is repeated below. In this representation we have labeled the O atoms (a), (b), and (c).

$$\left[\begin{array}{c} \overset{(a)}{\underset{}{\overset{..}{:}\text{O}}} \\ \| \\ \text{N} \\ \underset{(b)\quad(c)}{\overset{..}{:}\text{O}\diagup\quad\diagdown\overset{..}{\text{O}:}} \end{array} \right]^{-}$$

All the atoms in the structure use sp^2 orbitals in forming σ bonds and $2p$ orbitals in forming π bonds. In assigning valence shell electrons to the atoms, place one unpaired electron in a sp^2 for every σ bond that the atom forms and one unpaired electron in a p orbital for every π bond the atom forms. Then place lone-pair electrons in the orbitals that remain available. Because of the existence of formal charges in the structure, the number of valence electrons for each atom must be adjusted for the formal charge [e.g., O atom (b) has a formal charge of -1 and must show 7 electrons in its valence shell orbitals, not the customary 6].

Once you have assigned all the valence electrons to orbitals, show that all the unpaired electrons in sp^2 hybrid orbitals participate in σ-bond formation. Now, combine the p orbitals to form the appropriate number of π molecular orbitals (four). Assign the p electrons to these molecular orbitals in the usual fashion.

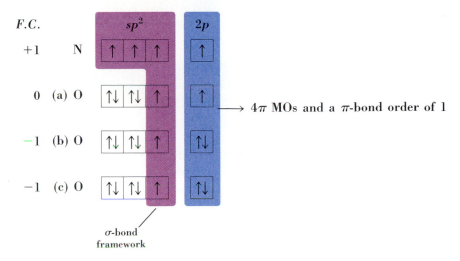

SIMILAR EXAMPLES: Exercises 41, 43, 63.

When you use the method illustrated in Example 11-10 at times you may find it necessary to postulate the presence of nonbonding molecular orbitals. A **nonbonding molecular orbital** has the same energy as the atomic orbitals from which it is formed and therefore neither adds to nor detracts from bond formation (that is, does not affect the bond order). In the SO_2 molecule the delocalized π bonding is based on three molecular orbitals, one bonding, one antibonding, and one *nonbonding* (see Exercise 43).*

One additional matter arises in describing π bonding in sulfate ion, SO_4^{2-}, another case involving resonance. We pictured several contributing Lewis structures to the resonance hybrid in Figure 10-14. Those structures with one or two S-to-O double bonds require octet expansion for the central S atom. However, we still predict a tetrahedral structure based on the VSEPR theory, and to account for this by the valence bond method we use sp^3 hybrid orbitals in the σ-bond framework. But what orbitals are involved in forming the π portions of the double bonds? These appear to involve a combination of $2p$ orbitals of the O atoms and $3d$ orbitals of the S atom. This type of bonding is called p_π–d_π.

11-9 Bonding in Metals

In nonmetal atoms the valence shells generally contain more electrons than they do vacant or partially filled orbitals. To illustrate, an F atom has *four* valence-shell orbitals ($2s, 2p_x, 2p_y, 2p_z$) and *seven* valence-shell electrons. By contrast, a Li atom has the same four valence-shell orbitals but only *one* valence-shell electron ($2s^1$). In solid lithium metal each Li atom is bonded, somehow, to eight nearest neighbors. There appear to be too few electrons to hold these atoms together. A bonding scheme for metals must also account for these distinctive properties that all metals share, more or less.

---

*Actually, bonding in NO_3^- also involves nonbonding molecular orbitals. Of the π molecular orbitals in Example 11-10, one is bonding, one is antibonding, and two are *nonbonding*. Of the six π electrons, two are bonding and four are *nonbonding*; the π-bond order is 1. In general, if you can establish the correct π-bond order by assuming only bonding and antibonding molecular orbitals, this will be satisfactory, even though the actual case may be more complicated.

FIGURE 11-34

A comparison of electron energy levels in insulators, metallic conductors, and semiconductors.

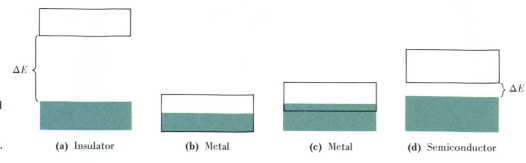

(a) Insulator   (b) Metal   (c) Metal   (d) Semiconductor

(a) In an insulator the valence band is filled with electrons, and a large energy gap, ΔE, separates the valence band and conduction band (outlined in black). Few electrons can make the transition when an electric field is applied, and the insulator does not conduct electricity.
(b) In some metals the valence band is only partially filled with electrons, and the valence band also serves as a conduction band (e.g., the half-filled 3s band in Na).
(c) In other metals the valence band is filled, but a conduction band overlaps it. In an electric field, electrons from the valence band can move through the conduction band (e.g., the empty 3p band of Mg overlaps with the filled 3s valence band).
(d) In a semiconductor the valence band is filled, and the conduction band is empty. The energy gap between the two, ΔE, is small enough, however, that some electrons make the transition simply by acquiring extra thermal energy.

- Ability to conduct electricity.
- Ability to conduct heat.
- Ease of deformation [that is, the ability to be flattened into sheets (malleability) and to be drawn into wires (ductility)].
- Lustrous appearance.

One oversimplified model that can account for some of these properties is the **electron-sea model.** The metal is pictured as a network of positive ions immersed in a "sea of electrons." In lithium the ions would be Li^+ and one electron per atom would be contributed to the sea. These free electrons account for the characteristic metallic properties. If the ends of a bar of metal are connected to a source of electric current, electrons from the external source enter the bar at one end. Free electrons pass through the metal and leave the other end at the same rate. In thermal conductivity no electrons enter or leave the metal, but those in the region being heated gain kinetic energy and transfer this to other electrons. According to the electron-sea model the ease of deformation of metals can be thought of in this way: If one layer of metal ions is forced across another, perhaps by hammering, the internal structure remains essentially unchanged as the sea of electrons rapidly adjusts to the new situation.

Band Theory.   The term *electron sea* is not very specific about the region occupied by free electrons in a metal. A more exact description is possible with molecular orbital theory. Recall the formation of molecular orbitals and the bonding between two Li atoms (Figure 11-27). Each Li atom contributes one 2s orbital to the production of two molecular orbitals—σ_{2s}^b and σ_{2s}^*. The electrons originally described as the $2s^1$ electrons of the Li atoms enter and half-fill these molecular orbitals; that is, they fill the σ_{2s}^b orbital and leave the σ_{2s}^* empty. If we extend this combination of Li atoms to a third Li atom, three molecular orbitals are formed, which contain three electrons, and again the set of molecular orbitals is half-filled. We can extend this process to an enormously large number of atoms, N—the total number of atoms in a crystal of Li. Here is the result we get: A set of N molecular orbitals is formed, with the difference in energy between the lowest and highest energy level being not much greater than between the σ_{2s}^b and the σ_{2s}^* orbitals of Figure 11-27. Because the number of individual molecular orbitals in this set is so large (N), the energy separation between each pair of successive levels is extremely small. This collection of very closely spaced molecular orbital energy levels is called a **band.**

In the band just described there are N electrons (a $2s^1$ electron from each Li atom) occupying, in pairs, $N/2$ molecular orbitals of lowest energy. These are the electrons responsible for bonding the Li atoms together. They are valence electrons, and the band in which they are found is called a **valence band.** However, because the energy difference between the occupied and unoccupied levels in the valence band is so small, electrons can be easily excited from the highest filled levels to the unfilled levels that lie immediately above them in energy. This excitation, which has the effect of producing mobile electrons, can be accomplished by heating the crystal or by applying a small electrical potential difference across the crystal. This is how the band theory explains the abilities of metals to conduct heat and electricity. The essential feature for electrical conductivity, then, is *an energy band that is only partly filled with electrons*. Such an energy band is called a **conduction band.** In lithium the $2s$ band is both a valence band and a conduction band.

Let us extend our discussion to N atoms of beryllium, which has the electron configuration $1s^22s^2$. We expect that the $2s$ valence band will be filled—N molecular orbitals and $2N$ electrons. But how can we reconcile this with the fact that Be is a good electrical conductor? At the same time that $2s$ orbitals are being combined into a $2s$ band, $2p$ orbitals combine to form an *empty* $2p$ band. The lowest levels of the $2p$ band are at a *lower* energy than the highest levels of the $2s$ band. The bands overlap. As a consequence, empty molecular orbitals are available to the valence electrons in beryllium.

In an **electrical insulator** like diamond or silica (SiO_2), not only is the valence band filled, but there is a large **energy gap** between the valence band and the conduction band. Very few electrons are able to make the transition between the two, and no electrons can be in the **forbidden zone** that separates the two bands. Still another possibility, found in silicon and germanium, for example, is that a filled valence band and an empty conduction band are separated by only a small energy gap. Electrons in the valence band may acquire enough energy, such as thermal energy, to jump to a level in the conduction band. The greater the thermal energy, the more electrons that can make the transition. A material of this type is called a **semiconductor.** Figure 11-34 offers a simple comparison of insulators, conductors, and semiconductors based on band theory. In the figure the valence band is shown in color and the conduction band is the open area with a black border.

FOCUS ON
Semiconductors

The GM Sunraycer uses an array of solar cells to power a rechargeable silver-zinc battery system. [GM and GM Hughes Electronics]

Let us consider an alternative description of semiconductor behavior to the one just given. In Figure 11-35 the normal condition is for all the valence electrons of the silicon atoms to participate in electron-pair covalent bond formation. The electrons are in the valence band—they are localized. Thermal energy is sufficient, however, to cause some electrons to be displaced from electron-pair bonds into the crystal as a whole, that is, to become *delocalized* and enter the conduction band. For every electron

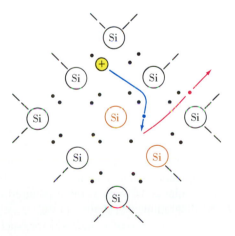

FIGURE 11-35
Formation of conduction electrons and positive holes in a semiconductor.

When an electron escapes from the bond between two Si atoms (shown in brown), it moves into the crystal at large. It becomes a conduction electron and leaves a vacancy at the original bond site. An electron from elsewhere in the crystal (blue line) enters this vacancy, creating a vacancy or positive hole at the bond site from which it came.

that enters the conduction band (a conduction electron) a vacancy or "hole" is left at the site of the broken bond (which means that a vacancy or hole is also created in the valence band). An electron may leave a bond elsewhere in the crystal to fill this vacancy, in turn creating a vacancy at the bond site from which it originated, and so on. A deficiency of an electron at a bond site is equivalent to a center of positive charge; thus these vacancies are called **positive holes.** When an electric field is applied, both conduction electrons and positive holes migrate.

We call semiconductors in which electrical conductivity involves the thermal promotion of valence electrons to the conduction band **intrinsic semiconductors.** Silicon and germanium are the two best known examples. More important, however, are **extrinsic semiconductors.** These are semiconductor materials to which have been added small and very carefully controlled quantities of impurity atoms—a process called **doping.** Suppose, for instance, that we introduce a trace of phosphorus (group 5A) into a crystal of silicon (group 4A). Because P atoms have *five* valence electrons compared to four for Si, one electron is promoted to the conduction band for every P atom we introduce. This is equivalent to creating one immobile positive charge (P^+) and one mobile conduction electron for every P atom present as a dopant.

Electrical conduction in this type of semiconductor results primarily from the movement of conduction electrons. We call this type of semiconductor **n-type.** The "n" refers to negative—the type of charge carried by an electron. If we dope with boron (group 3A), the semiconductor has a deficiency of one electron for every B atom. When an electron from elsewhere in the crystal moves into a vacancy at the site of a B atom, one immobile center of negative charge (B^-) and one positive hole are formed. In this case electrical conduction results primarily from the movement of positive holes, and we call the semiconductor **p-type.** The chemical makeup of n-type and p-type semiconductors is shown in Figure 11-36.

We can use appropriate combinations of n-type and p-type semiconductors, called **transistors,** to control electric currents. Transistors are the basic working components of modern solid-state electronic devices, whether they be found in television sets, electronic calculators, or high-speed computers. Another interesting use of semiconductors is in solar cells, such as those pictured in Figure 11-37 and diagrammed in Figure 11-38.

Figure 11-38 shows a p-type semiconductor, such as Si doped with B, in contact with an n-type semiconductor, such as Si doped with P. To a limited extent conduction electrons (from the n-type semiconductor) and positive holes (from the p-type semiconductor) can cross the boundary or junction between the two semiconductors. These migrations are very limited, however, because they tend to produce a separation of charge. For example, when positive holes leave the p-type semiconductor, a buildup

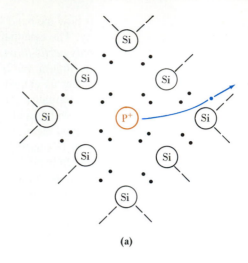

(a)

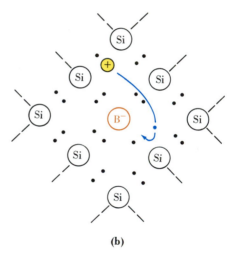

(b)

FIGURE 11-36

n-type and *p*-type semiconductors.

(a) The presence of the group 5A atom phosphorus in the crystal introduces an extra electron that is lost to the silicon crystal as a whole, that is, goes into the conduction band. This is an *n*-type semiconductor.
(b) The boron atom, having only three valence electrons, requires an additional electron from somewhere in the crystal in order to form bonds with each of its four neighboring Si atoms. As a result the boron atom becomes an immobile center of negative charge (B^-), and the electron vacancy created in the crystal becomes a positive hole. This is a *p*-type semiconductor.

of negative charge occurs because of the presence of the immobile B^- ions.

Now imagine that the p-type semiconductor is struck by a beam of light. Electrons in the valence band can absorb some of this light energy and be promoted to the conduction band (that is, the semiconductor now acts like an *intrinsic* semiconductor). Conduction electrons, unlike posi-

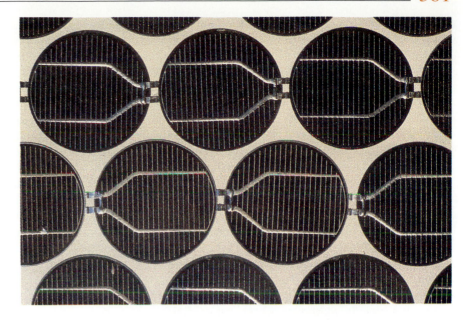

FIGURE 11-37
FIGURE 11-37
An array of photovoltaic cells.
[Courtesy Exxon Corporation]

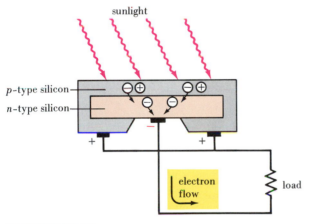

sunlight

p-type silicon
n-type silicon

electron flow   load

FIGURE 11-38
A photovoltaic (solar) cell using silicon-based semi-conductors.

tive holes, can easily cross the junction into the n-type semiconductor. A flow of electrons, an electric current, is set up. This type of device, which is called a **photovoltaic cell,** converts solar to electrical energy. Because the conduction electrons produced within the p-type semiconductor can so readily be neutralized by positive holes, the layer of p-type semiconductor must be kept very thin (about 10^{-4} cm) so that the electrons have a chance to cross the junction.

Summary

The valence bond method considers a covalent bond to result from the overlap of atomic orbitals of the bonded atoms. This produces a region of high electron charge density (electron probability) in the region of overlap. Some simple covalent molecules can be described adequately in terms of the overlap of simple s and p orbitals. In many cases, however, these simple orbitals must be *hybridized*. That is, they must be replaced by a set of hybrid orbitals whose properties depend on the number and types of simple atomic orbitals used to form them. The geometrical shape of a molecule is determined by the spatial distribution of the orbitals involved in bond formation. For the most part these geometrical shapes agree well with predictions made by the VSEPR theory.

In applying the valence bond method we recognize two types of orbital overlap. One type (σ) involves end-to-end overlap along the line joining the nuclei of the bonded atoms. The other type (π) requires a "sidewise" overlap of two p orbitals. Single covalent bonds are σ bonds; a double bond consists of one σ and one π bond; and a triple bond, one σ and two π bonds. The geometrical shape of a species is determined by its σ-bond framework, not its π bonds.

In molecular orbital theory, when atoms join to form a molecule new regions of high electron probability—molecular orbitals—are established for the molecule as a whole. *Bonding* molecular orbitals correspond to high electron probability or electron charge density in the internuclear region between atoms. *Antibonding* molecular orbitals concentrate electron probability or charge density in regions away from the internuclear region. *Nonbonding* molecular orbitals neither contribute to nor detract from bond formation. The numbers and kinds of molecular orbitals in a molecule are related to the corresponding atomic orbitals from which they arise. A procedure similar to the Aufbau process for the electron configurations of atoms can be used to describe the electronic structure of a molecule. A necessary part of this procedure requires a molecular-orbital energy-level diagram. These diagrams are established in the text for homonuclear (like atoms) diatomic molecules of the first and second period elements. The same diagrams can be used for heteronuclear (different atoms) diatomic molecules if the atoms do not differ greatly in atomic number and electronegativity.

Bond order follows directly from the assignment of electrons to molecular orbitals: It is one-half the difference between the numbers of electrons in bonding molecular orbitals and in antibonding orbitals.

A description of bonding in the benzene molecule, C_6H_6, requires the concept of delocalized molecular orbitals. These are regions of high electron probability that extend over *three or more* atoms in a molecule. Delocalized molecular orbitals also provide an alternative to the concept of resonance in dealing with species such as SO_2, SO_3, and NO_3^-. σ bonding in these species can be explained by the valence bond method and π bonding, by molecular orbital theory.

Finally, molecular orbital theory in the form called *band theory* can be applied to metals, semiconductors, and insulators, explaining, for example, their electrical conductivity properties.

Summarizing Example

Melamine, $C_3N_3(NH_2)_3$, is produced by the reaction of ammonia (NH_3) and urea [$CO(NH_2)_2$], the two most important manufactured nitrogen-containing chemicals. Melamine can undergo reactions with appropriate substances (e.g., formaldehyde, HCHO) to produce amino resins. Amino resins are used as adhesives, in the manufacture of protective coatings, and in textile finishing (such as in wrinkle-free, wash-and-wear fabrics).

Colorful, inexpensive dinnerware made from melamine plastic. [E. R. Degginger]

1. In the melamine molecule, $C_3N_3(NH_2)_3$, the three C atoms and three of the N atoms are joined together into a six-membered ring. The three NH_2 groups are attached to the three C atoms. Draw a plausible Lewis structure(s) for melamine.

Solution. The total number of valence electrons must be

From C   From N   From H

$(3 \times 4) + (6 \times 5) + (6 \times 1) = 48$

The skeleton structure in which all atoms are joined by single bonds accounts for 30 electrons (structure I). Suppose we next add lone-pair electrons to the N atoms outside the ring; this accounts for 6 more electrons (blue). Then let us complete the octets of the N atoms within the ring. This requires 12 more electrons (brown), and at that we have used up the total of 48 valence electrons. In structure II the three C atoms have incomplete octets.

We can correct the deficiency in structure II by shifting a lone pair of electrons from each N atom into one of its bonds to a C atom. The result is a six-membered ring with alternating single and double bonds. As shown below, however, there are two equally plausible structures that we can write. The true structure is a *resonance hybrid* of the two.

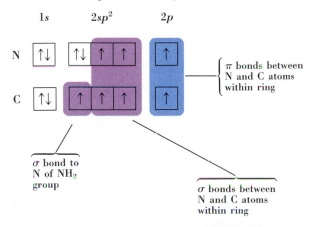

(This example is similar to Example 10-12.)

2. Describe bonding within the six-membered ring of melamine in terms of orbital overlap.

Solution. From VSEPR theory we would predict a planar ring with 120° bond angles, resulting from the trigonal planar distribution of three electron pairs. To achieve this geometry with the valence-bond method we must use sp^2 hybrid orbitals. The orbital diagrams for the N and C atoms and the orbital overlaps giving rise to the various σ and π bonds are indicated below. Note that to account for π-bond formation by the N atoms it is necessary to place their lone-pair electrons in one of the sp^2 hybrid orbitals (allowing for an unpaired electron to be placed in a $2p$ orbital).

(This example is similar to Example 11-10).

3. Describe bonding in the melamine ring through molecular orbital theory.

Solution. The situation here is very similar to that described for benzene on page 375. Each of the six ring atoms (three C atoms and three N atoms) has a singly occupied p orbital. These six p orbitals are combined to produce six *molecular* orbitals of the π type. Three of the six molecular orbitals are bonding orbitals and the other three, antibonding. All six electrons go into the bonding orbitals. The electron density distribution can be pictured as shown in Figure 11-31, and represented by a circle inscribed in a hexagon.

Key Terms

antibonding molecular orbital (11-4)	hybridization (11-2)	sp^3 hybrid orbital (11-2)
band theory (11-9)	molecular orbital theory (11-4)	sp^3d hybrid orbital (11-2)
bond order (11-4)	nonbonding molecular orbital (11-8)	sp^3d^2 hybrid orbital (11-2)
bonding molecular orbital (11-4)	pi (π) bond (11-3)	semiconductor (11-9; Focus feature)
delocalized molecular orbital (11-7)	sp hybrid orbital (11-2)	sigma (σ) bond (11-3)
hybrid orbital (11-2)	sp^2 hybrid orbital (11-2)	valence bond method (11-1)

Highlighted Expressions

Basic ideas of the valence bond method (11.1)
Steps in predicting a hybridization scheme (11.8)

Basic ideas about molecular orbitals (11.13)
Expression for calculating bond order (11.14)

Review Problems

1. In the manner of Example 11-1, describe the structure and bonding in (a) HCl; (b) ICl; (c) H_2Se; (d) NI_3.

2. In the manner of Figure 11-19, indicate the structures of the following simple molecules in terms of the overlap of simple atomic orbitals and hybrid orbitals: (a) H_2CCl_2; (b) $BeCl_2$; (c) BF_3.

3. Match each of the following species with one of these hybridization schemes: sp, sp^2, sp^3, sp^3d, sp^3d^2. (a) SF_6; (b) CS_2; (c) $SnCl_4$; (d) NO_3^-; (e) AsF_5. [*Hint:* You may find it useful to draw Lewis structures and to refer to the VSEPR theory.]

4. Formic acid, HCOOH, is an irritating substance released by ants when they sting (Latin, *formica,* ant). The structure of formic acid is indicated below. Propose a hybridization and bonding scheme consistent with this structure.

$$
\begin{array}{c}
\quad\quad\overset{\displaystyle O}{\underset{\displaystyle \|}{}} \, 124° \\
118° \; C \; 108° H \\
H \quad\quad O
\end{array}
$$

5. For the following molecules first write Lewis structures and then label each σ and π bond. (a) HCN; (b) C_2N_2; (c) $H_3CCHCHCCl_3$; (d) HONO.

6. Use the method of Figure 11-20 to represent bonding in the molecule, dimethyl ether, H_3COCH_3.

7. Propose a hybridization scheme to account for bonds formed by the carbon atom in each of the following molecules: (a) hydrogen cyanide, HCN; (b) chloroform, $CHCl_3$; (c) methyl alcohol, H_3COH;

$$(d) \text{ carbamic acid, } H_2N\overset{\displaystyle O}{\overset{\displaystyle \|}{C}}OH.$$

8. Indicate which of the following molecules are linear, which are planar, and which are neither. Propose appropriate hybridization schemes for the central atoms. (a) $HC\equiv N$; (b) $N\equiv C-C\equiv N$; (c) $F_3C-C\equiv N$; (d) $H_2C\equiv C\equiv O$.

9. Represent bonding in the carbon dioxide molecule, CO_2, by (a) a Lewis structure and (b) the valence bond method, identifying σ and π bonds, the necessary hybridization scheme, and orbital overlap.

10. For each of the following species: H_2^-; C_2^+; O_2^{2-}; F_2^+
(a) Write the molecular orbital diagram (in the manner of Example 11-8).
(b) Determine the bond order and state whether you expect the species to be stable or unstable.
(c) Determine if the species is diamagnetic or paramagnetic, and if paramagnetic, indicate the number of unpaired electrons.

11. Azobenzene, $(C_6H_5)_2N_2$, is an important starting material in the production of dyes (called azo dyes). Its structural formula is given below. [*Hint:* Complete the Lewis structure, and recall the symbolism for the benzene ring introduced in Figure 11-29.]

$$
\text{(benzene ring)} \quad \overset{\displaystyle \beta}{\underset{\displaystyle \alpha}{N}} = N^* \quad \text{(benzene ring)}^*
$$

(a) Describe the hybridization scheme for the atoms marked with an asterisk.
(b) Indicate the value of the bond angles labeled α and β.

12. Draw a Lewis structure(s) for the carbonate ion, CO_3^{2-}. Then propose a bonding scheme to describe the σ and π bonding in this ion. Describe π bonding through delocalized molecular orbitals.

Exercises

Valence bond theory

13. Indicate ways in which the valence bond method (atomic orbital overlap) is superior to Lewis structures in describing covalent bonds.

14. Predict the shape of the ammonium ion, NH_4^+, and describe a bonding scheme consistent with this.

15. For each of the following species identify the central atom and indicate the hybridization scheme for that atom: (a) CO_2; (b) $ClNO_2$; (c) ClO_3^-.

16. Identify the σ and π bonds in the carbon monoxide molecule, CO, and represent bonding through orbital overlap.

17. Acetic acid is a very common organic acid (5% by mass in vinegar). Sketch a three-dimensional structure of the molecule, indicating the type of orbital overlap that is consistent with this structural formula.

$$H_3C-\overset{\overset{\displaystyle O}{\displaystyle \|}}{C}-OH$$

18. Glycine is a common amino acid found in many proteins. It has the formula H_2NCH_2COOH. Indicate the orbitals involved in the bonding scheme. [*Hint:* The —COOH portion of the structure is as shown in Exercise 17.]

19. Propose a bonding scheme that is consistent with this structure for propynal. [*Hint:* You may find it useful to consult Table 10-2 to assess the multiple bond character in some of the bonds.]

$$\underset{\text{106 pm}}{H}\overset{}{\underset{}{-}}\underset{\text{120.4 pm}}{C}\overset{}{\underset{}{-}}\underset{\text{146 pm}}{C}\overset{123°}{\underset{120°}{-}}\overset{\displaystyle O}{\underset{\displaystyle H}{C}}\overset{}{\underset{\text{108 pm}}{-}}\text{121 pm}$$

20. Use the method of Figure 11-20 to represent bonding in each of the following molecules or ions: **(a)** CCl_4; **(b)** $ONCl$; **(c)** $HONO$; **(d)** H_3CCCH; **(e)** I_3^-; **(f)** C_3O_2; **(g)** $C_2O_4^{2-}$.

21. Refer to Example 11-2b. Another fluoride of xenon is XeF_4. Describe the geometrical structure of XeF_4 and the hybridization scheme for the Xe atom that can account for it.

22. Although they have similar formulas, the hybridization schemes for PF_5 and BrF_5 are different. Explain why this is so.

23. Based on the distinction between σ and π bonds and the data given, estimate the bond strength of a carbon-to-carbon triple bond. Compare your result with the value listed in Table 10-2. Bond energies: C—C, 347 kJ/mol; C=C, 611 kJ/mol.

24. The Lewis structure of N_2 indicates that the N-to-N bond is a triple covalent bond. Other evidence suggests that the σ bond in this molecule involves the overlap of *sp* hybrid orbitals.

  (a) Draw orbital diagrams for the two N atoms to describe bonding in N_2.
  (b) Could the hybridization of the N atoms be either sp^2 or sp^3? Could bonding in N_2 be described in terms of simple rather than hybrid orbitals? Explain.

25. The ion ClF_2^- is linear, but the ion ClF_2^+ is bent. Describe hybridization schemes for the central Cl atom to account for this difference in structure.

26. Shown below are ball-and-stick models. Describe hybridization and orbital overlap schemes to account for each of the

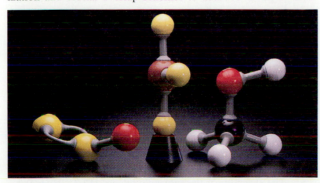

(a) S_2O (b) BrF_3 (c) CH_3OH

Exercise 26   [Carey B. Van Loon]

structures. [*Hint:* Begin by describing the geometrical shape of each molecule.] **(a)** S_2O; **(b)** BrF_3; **(c)** CH_3OH.

★ **27.** Refer to the Summarizing Example. Another method of producing melamine uses dicyandiamide, $NCNC(NH_2)_2$, as a starting material.

  (a) Show the C—N skeleton structure of the $NCNC(NH_2)_2$ molecule, labeling the various bond angles.
  (b) Label the σ and π bonds in the C—N skeleton and indicate the type of orbital overlaps producing these bonds.

28. Hydrogen azide (hydrazoic acid), HN_3, and salts derived from it (metal azides) tend to be unstable substances that find use in detonators for high explosives. [Sodium azide, NaN_3, is used in the ''air-bag'' safety systems in automobiles.] A reference source on molecular structures lists the following data for the hydrogen azide molecule (where the subscripts a, b, and c distinguish the three N atoms from one another).

Bond lengths: $N_a—N_b = 124$ pm; $N_b—N_c = 113$ pm
Bond angles: $H—N_a—N_b = 112.7°$; $N_a—N_b—N_c = 180°$

$$\underset{H}{\overset{}{\diagdown}}N_a{-}N_b{-}N_c$$

Show that the structure of HN_3 can be described as a resonance hybrid with two contributing structures. Describe a plausible hybridization and bonding scheme for each contributing structure.

29. Chloramphenicol, whose structural formula is shown below, is an important antibiotic. [*Hint:* Recall the symbolism for the benzene ring introduced in Figure 11-29.]

$$\overset{O\ \ \overset{\gamma}{\diagup}\ \ O}{\underset{N^*}{\diagdown}}$$

$$H—O—\overset{}{\underset{}{C}}—H\quad \overset{O}{\underset{}{\|}}\quad Cl$$
$$H—\overset{}{\underset{}{C}}—\underset{*}{N}{-}\overset{\alpha}{\underset{}{C^*}}{-}\overset{\beta}{\underset{}{C}}{-}Cl$$
$$\overset{}{\underset{H}{|}}\qquad \overset{}{\underset{H}{|}}$$
$$H—\overset{*}{\underset{}{C}}—O—H$$
$$\overset{}{\underset{H}{|}}$$

  (a) Describe the hybridization scheme for the atoms marked with an asterisk.
  (b) Indicate the value of the bond angles labeled α, β, and γ.

Molecular orbital method

30. With reference to the molecular orbital diagrams in Figure 11-27, which of the molecules are stable and which, unstable? Which of the stable molecules are diamagnetic and which, paramagnetic?

31. Would you expect either N_2^- or N_2^{2-} to be a stable ionic species in the gaseous state?

32. The molecular orbital diagram of O_2 is shown in Figure 11-27, and of O_2^+ in Example 11-8. Which species, O_2 or O_2^+, has the shorter bond length? Explain.

33. The paramagnetism of gaseous B_2 has been established. Explain how this observation confirms that the π_{2p}^b orbitals are at a lower energy than the σ_{2p}^b orbital for B_2.

34. Describe the bond order of diatomic carbon, C_2, in terms of Lewis theory and molecular orbital theory, and explain why the results are different.

35. In our discussion of bonding we have not encountered a bond order higher than triple. Use the energy-level diagrams of Figure 11-27 to show why this is to be expected.

36. Is it correct to say that when a diatomic molecule loses an electron the bond energy always decreases, that is, that the bond is always weakened? Explain.

37. Assuming that the energy-level diagrams of Figure 11-27 are applicable, write molecular orbital diagrams for the following heteronuclear diatomic species: (a) NO; (b) NO^+; (c) CO; (d) CN; (e) CN^-; (f) CN^+; (g) BN.

38. We have used the term isoelectronic to refer to atoms with identical electron configurations. In molecular orbital theory this term can be applied to molecular species as well. Which of the species in Exercise 37 are isoelectronic?

39. One of the characteristics of antibonding molecular orbitals is the presence of a nodal plane. Which of the *bonding* molecular orbitals considered in this chapter have nodal planes? Explain how a molecular orbital can have a nodal plane and still be a bonding molecular orbital.

Delocalized molecular orbitals

40. Explain how it is possible to get around the concept of resonance by using molecular orbital theory.

41. Represent chemical bonding in the molecule SO_3 (a) by writing a Lewis structure(s); (b) by using a combination of localized and delocalized orbitals.

42. In which of the following species would you expect to find delocalized molecular orbitals? Explain. (a) C_2H_4; (b) CO_3^{2-}; (c) NO_2; (d) H_2CO.

*★**43.** In a manner similar to that outlined in Section 11-8, propose a bonding scheme for SO_2 that is consistent with its Lewis structure(s). To do so requires the concept of a *nonbonding* molecule orbital, one in which the electron charge density neither adds to nor detracts from bond formation. Explain why this is necessary in describing the SO_2 molecule.

44. Refer to the Summarizing Example. Pyridine, C_5H_5N, is used in the synthesis of vitamins and drugs. The molecule can be thought of in terms of replacing a CH unit in benzene with an N atom. Draw orbital diagrams to show the orbitals of the C and N atoms that are involved in the σ and π bonding in pyridine. How

many bonding and antibonding π-type molecular orbitals are present? How many delocalized electrons are present?

Metallic bonding

45. Which of the following factors are especially important in determining whether an element has metallic properties? Explain. (a) atomic number; (b) atomic weight; (c) number of valence electrons; (d) number of vacant atomic orbitals; (e) total number of electronic shells in the atom.

46. Based on the ground-state electron configurations of the atoms, how would you expect the melting points and hardnesses of sodium, iron, and zinc to compare?

47. How many energy levels are present in the $3s$ conduction band of a single crystal of sodium weighing 26.8 mg? How many electrons are present in this band?

Semiconductors

48. Which of the following would you expect to be a *p*-type, which an *n*-type, and which an *intrinsic* semiconductor? (a) germanium; (b) germanium doped with Al; (c) silicon doped with arsenic; (d) silicon.

49. Based on the discussion of semiconductors in the text, explain why

(a) Even in an *n*-type semiconductor some of the electrical conductivity is due to positive holes.

(b) The rate at which the conductivity of a semiconductor increases with temperature is greater for an intrinsic semiconductor than for an extrinsic semiconductor.

50. The solar cell in Figure 11-38 is effective throughout the energy range of visible light.

(a) Estimate the maximum value of the energy gap, in kJ/mol, separating the valence and conduction bands in silicon.

(b) Why does the device operate over a broad range of wavelengths rather than at a single wavelength (as is so often the case when quantum effects are involved)?

*★**51.** A solar cell that is 15% efficient in converting solar to electrical energy is exposed to full sunlight, which produces an energy flow of 1.00 kW/m^2. If the cell has an area of 40. cm^2 that is exposed to sunlight

(a) What is the power output of the cell, in watts?

(b) If the power calculated in (a) is produced at 0.45 V, how much current does the cell deliver? [*Hint:* Refer to Appendix B.]

Additional Exercises

52. Explain how well each of the following methods describes the shape of the water molecule. (a) Lewis theory; (b) valence bond method using simple atomic orbitals; (c) valence bond method using hybridized atomic orbitals; (d) VSEPR theory.

53. Figure 11-22 represents the energy of interaction of two H atoms when the two electrons enter a bonding molecular orbital. Sketch a graph of energy vs. internuclear distance to represent the situation you would expect if the two electrons were to enter an antibonding orbital.

54. Show that both the valence bond method and molecular orbital theory provide an explanation for the existence of the covalent molecule Na_2 in the gaseous state.

55. The poisonous gas phosgene has the formula $COCl_2$. Propose an appropriate scheme of atomic orbital overlap for bonding in this molecule. What is its shape?

56. Lewis theory is satisfactory for explaining bonding in the ionic compound K_2O. However, it does not readily explain formation of the ionic compounds potassium superoxide, KO_2, and potassium peroxide, K_2O_2.

(a) Show that molecular orbital theory can provide this explanation.

(b) Write Lewis structures consistent with this explanation.

57. The structure of the molecule allene, H_2CCCH_2 is indicated in the figure. Propose hybridization schemes for the C atoms in this molecule.

Exercise 57

58. Describe a hybridization scheme for the central Cl atom in the molecule ClF_3 that is consistent with the geometrical shape pictured in Table 10-3. Which orbitals of the Cl atom are involved in overlap and which are occupied by lone-pair electrons?

59. All the atoms in the urea molecule $CO(NH_2)_2$ lie in the same plane and all the bond angles are 120°. Propose a hybridization and bonding scheme that will account for these observations.

60. Isocyanic acid, HNCO, is an intermediate in the manufacture of urethane plastics. Propose a plausible Lewis structure, hybridization scheme, and geometrical structure for this molecule.

61. The substance methyl nitrate, CH_3NO_3, is used as a rocket propellant. The skeleton structure of the molecule is H_3CONO_2. The N and three O atoms all lie in the same plane but the CH_3 group is not in the same plane as the NO_3 group. The bond angle C—O—N is 105° and the bond angle O—N—O is 125°. One N—O bond length is 136 pm and the other two are 126 pm.
   (a) Draw a sketch of the molecule, showing its geometrical shape.
   (b) Label all the bonds in the molecule as σ or π and indicate the probable orbital overlap involved.
   (c) Explain why all three N—O bond lengths are not the same.

62. Fluorine nitrate, $FONO_2$, is a strong oxidizing agent that is used as a rocket propellant. A reference source on molecular structures lists the following data for the FO_aNO_2 molecule (The subscript a indicates that this O atom is different from the other two.)
   Bond lengths: N—O = 129 pm; N—O_a = 139 pm; O_a—F = 142 pm
   Bond angles: O—N—O = 125°; F—O_a—N = 105°
   NO_aF plane is perpendicular to the O_2NO_a plane.
Use these data to write a Lewis structure(s), a bonding scheme showing hybridization and orbital overlap, and three-dimensional sketch of the molecule.

★**63.** Draw a Lewis structure(s) for the nitrite ion, NO_2^-. Then propose a bonding scheme to describe the σ and π bonding in this ion. Describe π bonding through delocalized molecular orbitals.

64. Toluene-2,4-diisocyanate, $C_9H_6N_2O_2$, is used in the manufacture of polyurethane foam. Its structural formula is represented below. [*Hint:* Recall the symbolism for the benzene ring introduced in Figure 11-29.]

   (a) Describe the hybridization scheme for the atoms marked with an asterisk. [*Hint:* Complete the portion of the Lewis structure associated with the NCO groups.)
   (b) Indicate the values of the bond angles labeled α and β.

★**65.** Think of the reaction below as involving the transfer of a fluoride ion from ClF_3 to AsF_5 to form the ions ClF_2^+ and AsF_6^-. As a result, the hybridization schemes of the central atoms must change. For the reactant molecules and product ions indicate **(a)** the geometrical structure and **(b)** the hybridization scheme for the central atom.

$$ClF_3 + AsF_5 \rightarrow (ClF_2^+)(AsF_6^-)$$

66. Use the molecular orbital theory to describe bonding in CO and CO^+ and indicate in which of these species you would expect the greater **(a)** bond energy; **(b)** bond length.

67. Suppose that the σ_{2p}^b orbitals were to fill before the π_{2p}^b orbitals in **(a)** B_2 and **(b)** C_2. How would the number of unpaired electrons and bond order in these molecules compare to that shown in Figure 11-27?

68. In terms of concepts introduced in this chapter, explain why you would hesitate to stir a boiling aqueous solution with a metal rod but not with a rod made of glass or wood.

★**69.** In the gaseous state, HNO_3 molecules have two nitrogen-to-oxygen bond distances of 121 pm and one of 140 pm. Draw a plausible Lewis structure(s) to represent this fact. Based on a plausible Lewis structure propose a bonding scheme in the manner of Figure 11-20.

★**70.** He_2 does not exist as a stable molecule, but there is evidence that such a molecule can be formed between electronically excited He atoms. Suggest a bonding scheme based on molecular orbitals to account for this.

★**71.** The molecule formamide, $HCONH_2$, has the following approximate bond angles: H—C—O, 123°; H—C—N, 113°; N—C—O, 124°; C—N—H, 119°; H—N—H, 119°. The C—N bond length is 138 pm. Two Lewis structures can be written for this molecule, with the true structure being a resonance hybrid of the two. Use the data given here (and refer to Table 10-2) to establish a bonding scheme for each structure.

★**72.** Furan, C_4H_4O, is a substance derivable from oat hulls, corn cobs, and other cellulosic waste. It is a starting material for the synthesis of other chemicals used as pharmaceuticals and herbicides. The furan molecule is planar and the C and O atoms are bonded into a five-membered pentagonal ring. The H atoms are attached to the C atoms. The chemical behavior of the molecule suggests that it is a resonance hybrid of several contributing structures.

These structures show that the double-bond character is associated with the entire ring in the form of a π electron cloud.
   (a) Draw orbital diagrams to show the orbitals that are involved in the σ and π bonding in furan. [*Hint:* You need use only one of the contributing structures, such as the one with no formal charges.]
   (b) How many π electrons are there in the furan molecule? Show that this number of π electrons is the same, regardless of the contributing structure used.

Self-Test Questions

For questions 73 through 81, select the single item that best completes each statement.

73. The bond angle in H_2Se is best described as (a) between 109° and 120°; (b) less than in H_2S; (c) less than in H_2S but not less than 90°; (d) less than 90°.

74. A molecule in which sp^2 hybrid orbitals are used by the central atom in forming covalent bonds is (a) PCl_5; (b) N_2; (c) SO_2; (d) He_2.

75. In carbon–hydrogen–oxygen compounds (a) all O-to-H bonds are π bonds; (b) all C-to-H bonds are σ bonds; (c) all C-to-C bonds consist of a σ and a π bond; (d) all C-to-C bonds are π bonds.

76. The hybridization scheme for the central atom includes a d orbital contribution in (a) I_3^-; (b) PCl_3; (c) NO_3^-; (d) H_2Se.

77. Of the following, the species with a bond order of 1 is (a) H_2^+; (b) Li_2; (c) He_2; (d) H_2^-.

78. Delocalized molecular orbitals are found in (a) H_2; (b) HS^-; (c) CH_4; (d) CO_3^{2-}.

79. The hybridization scheme for Xe in XeF_2 is (a) sp; (b) sp^3; (c) sp^3d; (d) sp^3d^2.

80. The best electrical conductor of the following materials is (a) $Li(s)$; (b) $Br_2(l)$; (c) $Ge(s)$; (d) $Si(s)$.

81. A substance in which a large energy gap separates the valence and conduction bands is a (a) metal; (b) metalloid; (c) semiconductor; (d) insulator.

82. What is the total number of **(a)** σ bonds and **(b)** π bonds in the molecule H_3CNCO? [*Hint:* Draw a plausible Lewis structure.]

83. Why does the hybridization scheme sp^3d *not* account for bonding in the molecule BrF_5? What hybridization scheme does work? Explain.

84. For the following diatomic ions: NO^+, N_2^+
(a) Use molecular orbital theory to predict the bond order.
(b) Which of these ions is paramagnetic? diamagnetic?
(c) Which of these ions has the greatest bond length? Explain.

85. Explain why the concept of delocalized molecular orbitals is essential to an understanding of bonding in the benzene molecule, C_6H_6.

12

Liquids, Solids, and Intermolecular Forces

Liquids form drops, even in samples as small as a few microliters, as pictured here. [Courtesy Beckman Instruments, Inc., a SmithKline Beckman Company]

There are several fundamental reasons to learn about physical properties of liquids and solids and about intermolecular forces. For example, this knowledge will help us to understand the basis of various techniques for separating mixtures. But there are other practical reasons too. Perhaps more so than in any chapter to this point, we can use the concepts studied here to explain a wide range of natural phenomena. Typical of the questions we address are

- Why does rubbing alcohol produce a cooling sensation on the skin?
- Why do boiled foods take longer to cook in the mountains than at sea level?
- Why does ordinary ice produce a liquid when it melts, whereas dry ice (solid CO_2) changes to a gas without melting?
- How can dry ice maintain a much lower temperature than ordinary ice?
- Why is diamond (a form of carbon) hard enough to scratch glass while graphite (another form of carbon) is soft enough to use in pencils?

Look for answers to these questions as you progress through the chapter. Also, see if you can find answers to some questions of your own.

12-1   Comparison of the States of Matter

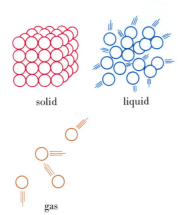

FIGURE 12-1

Comparison of the states of matter.

The structural particles of a solid are constrained to fixed points. They may vibrate or oscillate about these points but, ordinarily, may not move from them. In a liquid there is some free space among the structural particles, motions are more vigorous, and the structure is more random. In a gas there is a great deal of free space, motion is chaotic, and disorder is at a maximum.

Let us review some of the facts that we have already learned about the three states of matter in which a substance may exist. In a solid the constituent atoms, ions, or molecules are in close contact, and in many solids these structural units are found in a highly ordered network called a **crystal.** Crystals have geometric shapes with plane surfaces intersecting at definite angles. Not all solids are crystalline, however. If the structural units are disorganized, the solid is said to be **amorphous.** Whether a solid is crystalline or amorphous, it occupies a definite volume and maintains a definite shape. Solids are practically incompressible; they cannot easily be forced into smaller volumes by applying pressure to them. This is what we should expect if structural units are in close contact.

In liquids, the structural particles are also in close proximity, though we do not think of them as being in contact. Because of the free space among their structural particles, liquids are more compressible than solids. The attractive forces among molecules in a liquid are strong enough to maintain the liquid in a fixed volume but not strong enough to maintain it in a fixed shape. A liquid flows to cover the bottom and assume the shape of its container. Liquids are *fluid.*

Gases are also *fluid.* They expand to fill their containers and so have neither fixed shapes nor volumes. Also, because their structural units are ordinarily far apart, gases are highly compressible. (In a typical gas at STP the molecules occupy less than 1% of the total gas volume.)

To visualize the concepts we discuss in this chapter, try to form mental images of solids, liquids, and gases such as illustrated through Figure 12-1. Figure 12-1 brings out, in part, the fact that in each of these three states of matter the structural units move. In fluids (liquids and gases) this motion includes a translational component. Individual structural units may move throughout the bulk of the fluid. In solids the motion is mainly vibrational; structural units may move but only very short distances about fixed points. Because their structural units are confined to fixed points, solids have fixed shapes.

12-2   Two Properties of Liquids

Surface Tension. When describing the behavior of molecules in condensed matter (liquids or solids) we sometimes need to distinguish between molecules that are

390

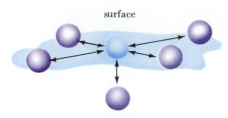

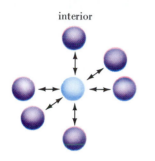

surface

interior

FIGURE 12-2

Intermolecular forces in a liquid.

Molecules in the interior experience attractive forces from neighboring molecules in all directions. Molecules at the surface are attracted only by other surface molecules and by molecules below the surface.

buried in the interior of a sample and those that are present at the surface. Figure 12-2 suggests why this may be necessary. Interior molecules experience forces of attraction exerted by surrounding molecules in all directions. Surface molecules experience essentially no forces from the direction above the surface. This imbalance between the forces at the surface and within the bulk of a liquid creates a measurable tension in the surface of the liquid, called **surface tension.** Think of the surface as if it were covered with a very thin, strong skin.

Or think of the matter in this way: Interior molecules, because they have a larger number of neighboring molecules to which they are attracted, are in a lower energy state than are surface molecules. In a liquid sample as many molecules crowd into the interior of the liquid as possible, leaving the liquid with the least possible surface area. A sphere is the geometrical shape that has the smallest ratio of surface area to volume, and this is why freely falling liquids form spherical drops, like raindrops in a shower.

Still another consequence of the difference between interior and surface molecules is that energy must be *expended* to *increase* the surface area of a liquid. Surface tension measures the quantity of energy required and has the units of energy per unit area, such as J/m^2. Surface tension is often represented by the Greek letter gamma (γ). Some typical values at 20 °C are

ethanol (CH_3CH_2OH): $\gamma = 2.28 \times 10^{-2}$ J/m^2

water (H_2O): $\gamma = 7.28 \times 10^{-2}$ J/m^2

mercury (Hg): $\gamma = 47.6 \times 10^{-2}$ J/m^2

As the intensity of molecular motion increases, the effectiveness of intermolecular forces of attraction is lessened. This means that surface tension *decreases* with *increased* temperature.

Phenomena Related to Surface Tension. Figure 12-3 dramatically illustrates the phenomenon of surface tension. As we well know, steel ($d \approx 7.8$ g/cm^3) is much more dense than water and a steel object should sink in water. Yet a clean steel needle can be floated on water. This is because the combined forces of buoyancy and surface tension of the water that support the needle are greater than the force of gravity on the needle. We observe additional examples of this phenomenon when we see water striders and other insects walk on water.

When a drop of liquid spreads into a film across a surface, we say that the liquid **wets** the surface. Energy (albeit a very small amount) must be expended for this to happen. Whether a drop of liquid will wet a surface or retain its spherical shape and stand on the surface depends on the comparative strengths of two types of intermolecular forces. **Cohesive forces** are the intermolecular forces between like molecules. **Adhesive forces** are intermolecular forces between *unlike* molecules, such as molecules of a liquid and molecules of the underlying surface. If cohesive forces are strong compared to adhesive forces, a liquid drop is maintained. If adhesive forces are sufficiently strong, the energy requirement for spreading the drop into a film is met; the liquid wets the surface.

Water wets many surfaces, such as glass and certain fabrics. This is essential if water is be used as a cleaning agent. If a glass surface is coated with a film of oil or grease, water is no longer able to wet the glass and water droplets stand on the glass rather than form a film. This difference in behavior is shown in Figure 12-4. When we clean glassware in the laboratory or in a dishwasher at home, we consider our job well done if water forms a uniform thin film on the glass. When we wax an automobile, we have done a good job if water uniformly beads up all along the surface.

Adding a detergent to water has two effects: the detergent solution dissolves grease, exposing the clean glass surface, and the detergent lowers the surface tension of the water. Lowering the surface tension means lowering the energy required

FIGURE 12-3

Surface tension illustrated.

The surface tension of the water supports the steel needle and keeps it from sinking. [Carey B. Van Loon]

to spread drops into a film. Other substances that reduce the surface tension of
water, allowing it to spread more easily, are known as **wetting agents.** They are
used in a variety of applications ranging from dishwashing to industrial processes.
In some applications we make use of the *inability* of water to wet and penetrate a
material. For instance, liquid water will not penetrate the microscopic holes of
"Gore-Tex" film. Sandwiched between layers of nylon fabric, "Gore-Tex" is used
in rainwear and boots. Its advantage over rubber is that the fabric "breathes."
Molecules of water vapor (from perspiration) escape through the microscopic holes
one by one, not as drops or a film.

Another observation we are all familiar with is illustrated in Figure 12-5. If the
liquid in the tube is water, for example, the interface between the water and the air
above it, called the **meniscus** (from Greek, meaning crescent), is curved upward
(concave). The water is drawn slightly up the glass walls by adhesive forces between the water and the glass. With liquid mercury the meniscus is curved downward (convex). The cohesive forces in mercury are strong and mercury does not wet
glass.

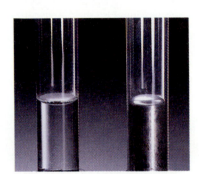

FIGURE 12-5
Meniscus formation.

If a liquid (here, water) wets
glass, the meniscus is concave. If
a liquid (here, mercury) does not
wet glass, the meniscus is convex. [Carey B. Van Loon]

The forces leading to meniscus formation are greatly magnified in tubes of very
small diameter, called *capillary* tubes. In the **capillary action** shown in Figure 12-6
the water inside the capillary tube is drawn to a noticeably greater height than in the
surrounding liquid. The action of a sponge in soaking up water depends on the rise
of water into capillary pores of a fibrous material such as cellulose. The penetration
of water into soils also depends in part on capillary action, making it possible, for
example, to water a potted plant simply by placing the pot in a shallow tray of
water.

Viscosity.   Another property that is related, at least in part, to intermolecular forces
is **viscosity**—a liquid's resistance to flow. The stronger the intermolecular forces of
attraction, the greater the viscosity. When a liquid flows, one portion of the liquid

FIGURE 12-6
Capillary action.

Water wets the inside walls of a
glass capillary, and a thin film of
water spreads up the tube. The
pressure inside the tube below
the meniscus falls slightly below
atmospheric pressure. Atmospheric pressure then pushes a
column of liquid up the tube to
eliminate the pressure difference.
The *greater* the surface tension
of the liquid and the *smaller* the
diameter of the capillary, the
higher the capillary rise.
[Carey B. Van Loon]

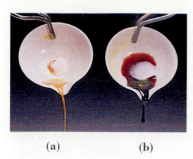

FIGURE 12-7
Viscosity of liquid sulfur.

(a) At its melting point of
119 °C, liquid sulfur is a mobile
liquid (low viscosity).
(b) At temperatures above
160 °C, S_8 molecules polymerize
into long-chain molecules and the
liquid becomes very viscous.
[Carey B. Van Loon]

moves with respect to neighboring portions and cohesive forces within the liquid create an "internal friction" which reduces the rate of flow. The effect is not noticeable in liquids of low viscosity like diethyl ether, ethyl alcohol, and water—they flow easily. Some liquids, however, like molasses, honey, and heavy motor oil flow much more sluggishly—we say that they are *viscous*. One method of measuring viscosity is to time the fall of a steel ball through a certain depth of liquid. The greater the viscosity of the liquid, the longer it takes for the ball to fall. Because intermolecular forces of attraction are offset by the greater average kinetic energy of the molecules at higher temperatures, viscosities of liquids generally *decrease* with temperature.

The viscosity of a liquid or liquid mixture is also affected by the sizes and shapes of the molecules present. At its melting point of 119 °C, sulfur produces a thin, mobile, straw-colored liquid comprised of the molecules S_8. At a *higher* temperature of about 160 °C the liquid darkens and becomes very viscous (see Figure 12-7). This results from the joining together of S_8 molecules into polymer chains with up to several thousand S atoms per chain. One way to assess the degree of polymerization (average length of chains) in a polymer is to measure the viscosities of polymer solutions.

12-3 Vaporization and Vapor Pressure

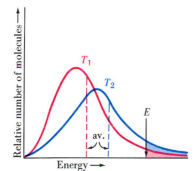

FIGURE 12-8
Distribution of molecular kinetic energies.

There is an average kinetic energy associated with a large group of molecules, and this average kinetic energy increases with temperature ($T_2 > T_1$). Individual molecules may have kinetic energies that range from very low to very high. Suppose that molecules must have a kinetic energy in excess of E in order to vaporize from a liquid. These plots show that many more molecules meet this energy requirement at the higher temperature (T_2) than at the lower one (T_1).

ΔH_{vap} values are somewhat temperature dependent (see Exercise 21).

As we learned in our study of the kinetic molecular theory (Section 6-9), the speeds and kinetic energies of the molecules in a liquid vary over a wide range (see Figure 12-8). Temperature is related to the *average kinetic energy*. Molecules with sufficiently high kinetic energies may overcome the attractive forces of neighboring molecules and escape from a liquid surface into the gaseous or vapor state. This phenomenon is called **vaporization** or **evaporation.** The tendency for a liquid to vaporize *increases* with

- *increased* temperature;
- *increased* surface area of liquid;
- *decreased* strength of intermolecular forces.

Enthalpy (Heat) of Vaporization. Since the molecules lost through evaporation are more energetic than average, the average kinetic energy of the molecules remaining in a liquid must fall. The temperature of the evaporating liquid *decreases*. This accounts for the cooling sensation if you spill a volatile liquid such as ether or ethyl alcohol on your skin.

Suppose we allow a liquid to vaporize but wish to keep the temperature of the liquid *constant*. What is required to do this? We must replace the excess kinetic energy carried away by the vaporizing molecules, and this we can do by adding heat energy to the liquid. The **enthalpy (heat) of vaporization** is the quantity of heat that must be absorbed if a certain quantity of liquid is vaporized at a constant temperature. That is, to convert a quantity of liquid to vapor,

$$\Delta H_{vaporization} = H_{vapor} - H_{liquid} \tag{12.1}$$

As is always the case, we cannot measure the absolute enthalpies H_{vapor} and H_{liquid}. We do know, however, that these enthalpies are functions of state; they have unique values. Their difference, that is, $\Delta H_{vaporization}$, or more simply, ΔH_{vap}, is also a unique and a *measurable* quantity. In this book, we express enthalpies of vaporization in terms of *one mole* of liquid vaporized, using the units kJ/mol. However, the units cal/g and kcal/mol are also frequently used. Table 12-1 lists some typical enthalpies (heats) of vaporization. All are positive quantities since vaporization is an *endothermic* process.

TABLE 12-1
Some Enthalpies (Heats) of
Vaporization at 298 K

Liquid	ΔH_{vap}, kJ/mol
diethyl ether, $(C_2H_5)_2O$	27.4
methyl alcohol, CH_3OH	38.0
ethyl alcohol, CH_3CH_2OH	42.6
water, H_2O	44.0

Example 12-1

Including ΔH_{vap} in calculations involving quantities of heat. How much heat energy is required to convert 175 g diethyl ether, $(C_2H_5)_2O$, from the liquid state at 20.0 °C to the gaseous (vapor) state at 30.0 °C? The specific heat capacity of liquid diethyl ether in the range 20.0 to 30.0 °C is 2.30 J g^{-1} °C^{-1}. Assume that the value of ΔH_{vap} given in Table 12-1 holds at 30.0 °C.

Solution. The heat required is the sum of two quantities: that needed to raise the temperature of the liquid from 20.0 to 30.0 °C (calculated as in Example 7-3) and that needed to vaporize the liquid at 30.0 °C.

Energy to raise the temperature of the liquid ether:

$$\text{no. kJ} = 175 \text{ g ether} \times \frac{2.30 \text{ J}}{\text{g ether °C}} \times \frac{1 \text{ kJ}}{1000 \text{ J}} \times (30.0 - 20.0) \text{ °C} = 4.02 \text{ kJ}$$

Energy to vaporize the ether:

$$\text{no. kJ} = 175 \text{ g } C_4H_{10}O \times \frac{1 \text{ mol } C_4H_{10}O}{74.12 \text{ g } C_4H_{10}O} \times \frac{27.4 \text{ kJ}}{1 \text{ mol } C_4H_{10}O} = 64.7 \text{ kJ}$$

Total heat energy required:

$$\text{no. kJ} = 4.02 + 64.7 = 68.7 \text{ kJ}$$

SIMILAR EXAMPLES: Exercises 1, 19, 20, 82.

When a vapor is converted back to a liquid, we call the process **condensation.** Heat is given off, and as expected

$$\Delta H_{condensation} = H_{liquid} - H_{vapor} = -\Delta H_{vaporization} \qquad (12.2)$$

Condensation, then, is an *exothermic* process. This explains why burns produced by steam (vaporized water) are much more severe than burns produced by hot water. Hot water burns by releasing heat as it cools. Steam releases a large quantity of heat when it condenses to liquid water, and this is followed by the further release of heat as the hot water cools.

FIGURE 12-9
Establishing liquid–vapor equilibrium at constant temperature.

o   molecules in vapor state
o→   molecules undergoing vaporization
o→   molecules undergoing condensation

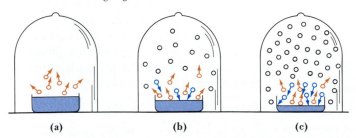

(a)   (b)   (c)

(a) A liquid is allowed to evaporate into a closed vapor volume. Initially only vaporization occurs.
(b) Condensation begins. However, because molecules are evaporating at a faster rate than they are condensing, the number of molecules in the vapor state continues to increase.
(c) The rate of condensation has become equal to the rate of vaporization. Dynamic equilibrium is established. The number of molecules present in the vapor state remains constant over time, as does the pressure exerted by the vapor.

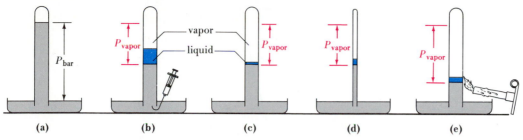

FIGURE 12-10
Measurement of vapor pressure.

(a) A mercury barometer.
(b) A small volume of liquid is introduced to the top of the mercury column. The pressure of the vapor in equilibrium with the liquid depresses the level of the mercury in the barometer tube.
(c) By comparing this with situation (b) we see that vapor pressure is independent of the volume of liquid used to establish liquid–vapor equilibrium.
(d) The volume of vapor here is quite small, but vapor pressure is independent of the volume of vapor involved in the liquid–vapor equilibrium.
(e) An increase in temperature causes an increase in vapor pressure.

A minor but nevertheless important source of hydrocarbons in the production of smog (see Section 14-10) is vaporized gasoline from oil refineries, automobile gas tanks and carburetors, filling-station operations, etc.

Vapor Pressure. When a liquid vaporizes into an unlimited vapor volume, evaporation continues until the liquid disappears. (No one doubts that water left in a beaker on a demonstration table will eventually evaporate away.) On the other hand, a different condition results if vaporization occurs into a *closed* vapor volume, as pictured in Figure 12-9.

When a vapor is in contact with a liquid, some molecules return from the vapor to the liquid. The extent of this condensation increases with the concentration of vapor molecules (the number of vapor molecules per unit volume) and with the area of contact between the liquid and its vapor. In a container with both liquid and vapor present, vaporization and condensation occur simultaneously. Although molecules continue to pass back and forth between liquid and vapor, if sufficient liquid is present, eventually a condition is reached in which no net *additional* vapor is formed. This condition is one of **dynamic equilibrium.** Dynamic equilibrium implies that two opposing processes are occurring simultaneously and at equal rates. As a result there is no net change with time once equilibrium has been established.

We give a special name to the pressure exerted by a vapor in dynamic equilibrium with its liquid; it is called the **vapor pressure.** Like so many other properties the values of vapor pressure vary widely. Liquids with high vapor pressures are said to be **volatile.** Those with very low vapor pressures are **nonvolatile.** Whether a liquid is volatile or nonvolatile at a given temperature is determined primarily by the strengths of intermolecular forces—the *weaker* these forces, the more volatile the liquid (the higher its vapor pressure). Diethyl ether and acetone are volatile liquids. Water at ordinary temperatures is a moderately volatile liquid; at 25 °C its vapor pressure is 23.8 mmHg. Gasoline is a mixture of hydrocarbons, most of which are somewhat less volatile than water.

As an excellent first approximation the vapor pressure of a liquid depends only on the particular liquid and its temperature. We can measure approximate values of vapor pressures with a simple mercury barometer by the method outlined in Figure 12-10. A graph of vapor pressure as a function of temperature is known as a **vapor pressure curve.** Vapor pressure curves always have the appearance of those in Figure 12-11; *vapor pressure increases with temperature.* The "steepness" of the vapor pressure curve is related to the heat of vaporization of the liquid; *the greater the heat of vaporization the sharper the rise.* Vapor pressures of water at different temperatures are also presented in Table 12-2.

Boiling and the Boiling Point. The temperature at which the vapor pressure of a liquid is equal to standard atmospheric pressure (1 atm = 760 mmHg) has special significance. It is called the **normal boiling point.*** Boiling occurs when a liquid is heated in a container *open to the atmosphere* and vaporization occurs *throughout* the liquid rather than simply at the surface. Pockets of vapor form within the bulk of the

---

*In a footnote on page 219 we noted that the IUPAC recommends that the standard-state pressure be changed from 1 atm (101,325 Pa) to 1 bar (1×10^5 Pa). There has been no corresponding recommendation that the pressure for normal boiling points be changed from 1 atm to 1 bar, although it is conceivable that such a recommendation could be made some time in the future. In this text we continue to use 1 atm as both the standard-state pressure and the normal-boiling-point pressure.

FIGURE 12-11
Vapor pressure curves of several liquids.

(a) Diethyl ether, $C_4H_{10}O$;
(b) benzene, C_6H_6; **(c)** water, H_2O; **(d)** toluene, C_7H_8; **(e)** aniline, C_6H_7N.

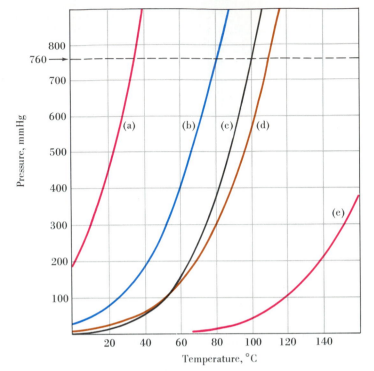

TABLE 12-2
Vapor Pressure of Water at Various Temperatures

Temperature, °C	Pressure, mmHg	Temperature, °C	Pressure, mmHg	Temperature, °C	Pressure mmHg	Temperature, °C	Pressure mmHg
0.0	4.6	26.0	25.2	70.0	233.7	96.0	657.6
10.0	9.2	27.0	26.7	80.0	355.1	97.0	682.1
20.0	17.5	28.0	28.3	90.0	525.8	98.0	707.3
21.0	18.7	29.0	30.0	91.0	546.0	99.0	733.2
22.0	19.8	30.0	31.8	92.0	567.0	100.0	760.0
23.0	21.1	40.0	55.3	93.0	588.6	110.0	1074.6
24.0	22.4	50.0	92.5	94.0	610.9	120.0	1489.1
25.0	23.8	60.0	149.4	95.0	633.9		

FIGURE 12-12
Boiling water in a paper cup.

An empty paper cup heated over a Bunsen burner flame quickly bursts into flame. If a paper cup is filled with water, it can be heated for an extended period of time as the water boils. This is made possible for three reasons: (1) Because of the large heat capacity of water, heat from the Bunsen burner goes primarily into heating the water, not the cup. (2) As the water begins to boil, large quantities of heat (ΔH_{vap}) are required to convert liquid water to its vapor. (3) The temperature of the cup and water do not rise above the boiling point as long as liquid water remains. [Carey B. Van Loon]

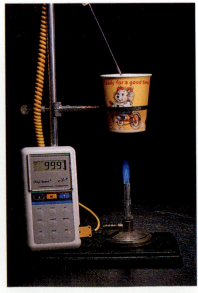

liquid, rise to the surface, and escape. During boiling, energy absorbed as heat is used only to convert molecules of liquid to vapor. The temperature remains constant until all the liquid has boiled away, a fact that is dramatically illustrated in Figure 12-12.

Figure 12-11 helps us to see that the boiling point of a liquid varies significantly with atmosphere pressure. Shift the line shown at $P = 760$ mmHg to higher or lower pressures and determine the temperature of the new points of intersection with the vapor pressure curves. Atmospheric pressures below one atm are commonly encountered at high altitudes. At an altitude of 1609 m (that of Denver, Colorado) atmospheric pressure is about 630 mmHg. The boiling point of water at this pressure is 95 °C (203 °F). To cook foods under these conditions of lower boiling temperatures, it is necessary to use longer cooking times. A ''three-minute'' boiled egg takes longer than 3 min to cook. We can counteract the effect of high altitudes by using a pressure cooker. In a pressure cooker the cooking water is maintained under higher than atmospheric pressure and its boiling temperature increases.

The Critical Point. In describing the phenomenon of boiling we made an important qualification: that this occurs ''in a container open to the atmosphere.'' If a liquid is heated in a *closed* container, boiling does not occur. Instead, the temperature and vapor pressure rise continuously. Pressures many times atmospheric pressure may be attained. If the appropriate quantity of liquid is sealed in a glass tube and heated, as in Figure 12-13, these observations are made.

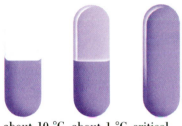

about 10 °C about 1 °C critical
below T_c below T_c temp. T_c

FIGURE 12-13
Attainment of the critical point.

The meniscus separating a liquid (bottom) from its vapor (top) disappears at the critical point. The liquid and vapor become indistinguishable.

- The density of the liquid decreases, that of the vapor increases, and eventually the two densities become equal.
- The surface tension of the liquid approaches zero; the meniscus between the liquid and vapor becomes less distinct and eventually disappears.

At the point where these conditions are reached, the **critical point,** the liquid and vapor become indistinguishable. The temperature at the critical point is the **critical temperature, T_c,** and the pressure, the **critical pressure, P_c.** The critical point is the highest temperature–pressure point on a vapor pressure curve and represents the highest temperature at which the liquid can exist. Several critical temperatures and pressures are listed in Table 12-3.

One significant consequence of the phenomenon of the critical point is that gaseous substances whose critical temperatures are *above* room temperature can be liquefied by applying sufficient pressure to the gas. To liquefy gaseous substances whose critical temperatures are *below* room temperature requires *both* the application of pressure *and* a lowering of the temperature (to a value below T_c). A distinction that is sometimes made when describing the gaseous state of a substance is to call it a *vapor* if the temperature is *below* the critical temperature, and a *gas* if the temperature is *above* the critical temperature. We can condense a vapor simply by applying pressure, but to liquefy a gas we must both apply pressure and lower the temperature.

TABLE 12-3
Some Critical Temperatures, T_c, and Critical Pressures, P_c

Substance	T_c, K	P_c, atm
H_2	33.3	12.8
N_2	126.2	33.5
O_2	154.8	50.1
CH_4	191.1	45.8
CO_2	304.2	72.9
HCl	324.6	82.1
NH_3	405.7	112.5
SO_2	431.0	77.7
H_2O	647.3	218.3

Experimental Determination of Vapor Pressure. The method of determining vapor pressure pictured in Figure 12-10 makes an effective lecture demonstration, but it is not widely used. For one thing, the method does not work for vapor pressures that are either very low or very high. Also the results are not very accurate. A somewhat more useful method (called the transpiration method) is suggested in Example 12-2. Here, we saturate an inert gas with the vapor under study and then use the ideal gas equation to calculate the vapor pressure.

Example 12-2 _____

Using the ideal gas equation to calculate vapor pressures. 113 L of helium gas at 1360. °C and prevailing barometric pressure is passed through molten silver at 1360. °C. The gas becomes saturated with silver vapor. As a result, the liquid silver loses 0.120 g in mass. What is the vapor pressure, in mmHg, of liquid silver at 1360. °C? Neglect any change in gas volume due to vaporization of the silver.

Solution.   Although the saturated gas is actually a mixture of He and Ag, we can deal with the Ag as if it were a single gas occupying a volume of 113 L. The necessary data follow.

$P = ?$

$V = 113$ L

$n = 0.120 \text{ g Ag} \times \dfrac{1 \text{ mol Ag}}{107.9 \text{ g Ag}} = 0.00111 \text{ mol Ag}$

$R = 0.08206 \text{ L atm mol}^{-1} \text{ K}^{-1}$

$T = 1360. \text{ °C} + 273 = 1633 \text{ K}$

$PV = nRT \qquad P = \dfrac{nRT}{V}$

$P = \dfrac{(0.00111 \text{ mol})(0.08206 \text{ L atm mol}^{-1} \text{ K}^{-1})(1633 \text{ K})}{113 \text{ L}} = 1.32 \times 10^{-3} \text{ atm}$

$= 1.32 \times 10^{-3} \text{ atm} \times \dfrac{760 \text{ mmHg}}{1 \text{ atm}} = 1.00 \text{ mmHg}$

SIMILAR EXAMPLES: Exercises 3, 25, 26.

An Application of Vapor Pressure Data—Predicting States of Matter. When liquid–vapor equilibrium is established in a closed container, the amount of vapor present depends on its volume and its pressure. The quantity of liquid remaining at equilibrium is immaterial; it can be quite large or very small, but there must be some. What if the total quantity of substance present in a container is less than enough to saturate the volume with vapor? In this case any liquid will evaporate completely and only vapor will exist in the container, at a pressure less than the equilibrium vapor pressure. To determine what the situation will be in any specific case, we need to use vapor pressure data and the ideal gas equation, as in Example 12-3.

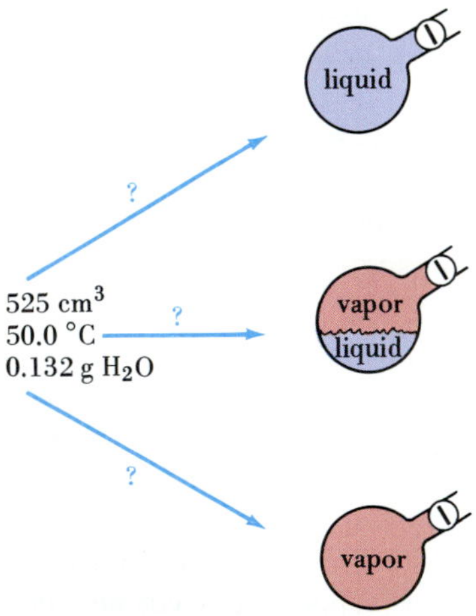

Example 12-3

Using vapor pressure data and the ideal gas equation to determine whether a sample will exist as a vapor only or as a liquid–vapor mixture. As a result of a chemical reaction 0.132 g H_2O is produced and maintained at a temperature of 50.0 °C in a closed container of 525 cm³ volume. The vapor pressure of water at 50.0 °C is 92.5 mmHg.

(a) Will the water in the container be present as liquid only, vapor only, or as a liquid and vapor in equilibrium?

(b) If both liquid and vapor are present, what is the mass of each?

Solution. The three possibilities are pictured in Figure 12-14.

(a) Let us first eliminate the possibility that water might be present as *liquid only* in this way: A 0.132-g sample of water can only occupy a volume of about 0.132 cm³ [since the density of $H_2O(l) \approx 1.0$ g/cm³]. Something has to occupy the remaining volume (525 cm³), and this can only be vapor. We can proceed in several ways at this point. One way is to *assume that*

FIGURE 12-14
Predicting states of matter—Example 12-3 illustrated.

For the conditions given on the left, which of the final conditions pictured on the right will result?

525 cm³
50.0 °C
0.132 g H_2O

liquid

vapor
liquid

vapor

both liquid and vapor are present at equilibrium, with the vapor exerting its equilibrium vapor pressure of 92.5 mmHg. Next we can use the ideal gas equation to calculate the mass of water vapor present in 525 cm^3 at 50.0 °C and 92.5 mmHg. If this calculated mass is *less than* the total mass of water available (0.132 g), then both liquid and water vapor are present. On the other hand, if the calculated mass of water vapor is greater than what is available, the water must be completely vaporized.

$$PV = \frac{m}{\mathcal{M}}RT \quad \text{and} \quad m = \frac{\mathcal{M}PV}{RT}$$

$$m = \frac{18.02 \text{ g mol}^{-1} \times (92.5/760) \text{ atm} \times 0.525 \text{ L}}{0.08206 \text{ L atm mol}^{-1} \text{ K}^{-1} \times 323 \text{ K}} = 0.0434 \text{ g}$$

Since the mass of water vapor is less than the total mass of water, the condition must be one of **liquid and vapor in equilibrium**.

(b) Now we can write that

mass $H_2O(g) = 0.0434$ g
mass of $H_2O(l) = 0.132 - 0.0434 = 0.089$ g

SIMILAR EXAMPLES: Exercises 7, 41, 42.

Collection of Gases over Water. We have already considered this additional situation where we need to use vapor pressures—the collection of gases over liquids. Review this subject by rereading portions of Section 6-8, in particular Example 6-20.

An Equation for Calculating Vapor Pressures. It is not easy to interpolate (estimate values) between points in a table or graph if the relationship is nonlinear. And the vapor pressure curve is decidedly not a straight line—it becomes ever steeper as the temperature is increased. At times we can convert a nonlinear expression into a linear one by using a different function of the variables.

Straight-line plot for vapor pressure data.

Logarithms of vapor pressures plotted against the reciprocals of Kelvin temperature (1/T) yield a straight line.   (12.3)

As described in Appendix A-4, a straight line can always be expressed through a simple mathematical equation. The equations for the lines in Figure 12-15 are of the form

$$\log P = -A\left(\frac{1}{T}\right) + B \tag{12.4}$$

equation of straight line: $y \quad = \quad m \times x \quad + b$

Whenever we find that an equation like (12.4) describes a phenomenon, we should suspect that there is a theoretical basis for the equation and that the fundamental expression involves natural logarithms (ln) rather than common logarithms (log). We will consider a derivation of equation (12.4) following a further discussion of thermodynamics in Chapter 20. For now let us start with the equation relating vapor pressure and temperature, known as the Clausius–Clapeyron equation and having the form

Natural (base-e) and common (base-10) logarithms are related through the factor $\log_e 10 = 2.303$. (See Appendix A-2).

$$\ln P = \frac{-\Delta H_{\text{vap}}}{RT} + B' \tag{12.5}$$

To calculate the vapor pressure (P) at some given temperature (T) with equation (12.5) requires that we have values for the molar enthalpy of vaporization (ΔH_{vap})

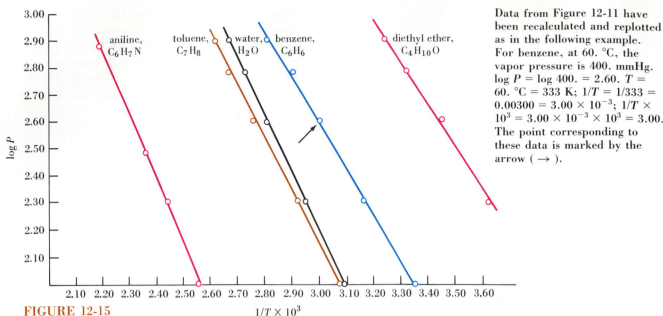

Data from Figure 12-11 have been recalculated and replotted as in the following example. For benzene, at 60. °C, the vapor pressure is 400. mmHg. log P = log 400. = 2.60. T = 60. °C = 333 K; $1/T$ = 1/333 = 0.00300 = 3.00 × 10^{-3}; $1/T$ × 10^3 = 3.00 × 10^{-3} × 10^3 = 3.00. The point corresponding to these data is marked by the arrow (→).

FIGURE 12-15
Vapor pressure data of Figure 12-11 replotted—log P versus $1/T$.

and the numerical constant (B'). We can sometimes find such data in handbooks, but a more usual approach is to eliminate the constant B'. Write equation (12.5) twice—once for vapor pressure P_1 at temperature T_1, and once for P_2 and T_2.

$$(1) \quad \ln P_1 = \frac{-\Delta H_{vap}}{RT_1} + B' \qquad (2) \quad \ln P_2 = \frac{-\Delta H_{vap}}{RT_2} + B'$$

Subtract equation (1) from equation (2).

$$\ln P_2 - \ln P_1 = \frac{-\Delta H_{vap}}{R} \left(\frac{1}{T_2} - \frac{1}{T_1} \right) + \cancel{B'} - \cancel{B'}$$

The Clausius–Clapeyron equation is usually written in the form

Clausius–Clapeyron equation relating vapor pressure, heat of vaporization, and temperature.

$$\ln \frac{P_2}{P_1} = \frac{\Delta H_{vap}}{R} \left(\frac{1}{T_1} - \frac{1}{T_2} \right) \tag{12.6}$$

In equation (12.6) temperatures must be expressed in kelvins, the same units of pressure must be used for P_1 and P_2, and the same units of energy for ΔH_{vap} and R. Generally this will require that R be expressed as 8.314 J mol^{-1} K^{-1}.

Are You Wondering:

Which values to choose for P_2 and P_1 and which to choose for T_2 and T_1 when using the Clausius–Clapeyron equation (12.6)?

The only essential requirement in this choice is that T_2 must be the temperature at which the liquid has a vapor pressure of P_2, and T_1 the temperature at which the vapor pressure is P_1. It does not matter whether P_2 and T_2 are chosen as the higher pressure and temperature or the lower ones. If P_2 and T_2 are chosen to be the higher values, the right side of equation (12.6) will be positive and the ratio P_2/P_1 on the left side will be greater than 1. If P_2 and T_2 are assigned the lower values, the right side of equation (12.6) will be negative and the ratio P_2/P_1 on the left side will be less than 1.

Example 12-4

Using the Clausius—Clapeyron equation to relate the heat of vaporization of a liquid and its vapor pressure at two different temperatures. A carefully measured volume of $O_2(g)$ is collected over water at 18.0 °C. In order to calculate the number of moles of $O_2(g)$, it is necessary to know the vapor pressure of water at 18.0 °C (recall Example 6-20). Calculate the vapor pressure of water at 18.0 °C using data from Tables 12-1 and 12-2.

Solution. Let P_1 be the unknown vapor pressure at temperature $T_1 = $ 18.0 °C = 291.2 K. From Table 12-2, choose for P_2 and T_2 known data at a temperature close to 18.0 °C. At $T_2 = 20.0$ °C = 293.2 K, $P_2 = 17.5$ mmHg. From Table 12-1 obtain a value of $\Delta H_{vap} = 44.0$ kJ/mol. Substitute these values into equation (12.6) to obtain

In Example 12-4 it is best to keep the temperature interval $T_2 - T_1$ small because ΔH_{vap} varies somewhat with temperature.

$$\ln \frac{17.5 \text{ mmHg}}{P_1} = \frac{44.0 \times 10^3 \text{ J mol}^{-1}}{8.314 \text{ J mol}^{-1} \text{ K}^{-1}} \left(\frac{1}{291.2} - \frac{1}{293.2} \right) \text{K}^{-1}$$
$$= 5.29 \times 10^3 (0.003434 - 0.003411) = 0.12$$

Next, determine the *antiln* of 0.12. (What is the number whose natural logarithm is 0.12?) The *antiln* is $e^{0.12} = 1.13$ (see Appendix A). Thus,

Are You Wondering:

When you should use natural logarithms and when you should use common logarithms in calculations?

In most cases you can use either logarithm function, and generally electronic calculators have both "log" and "ln" keys and their inverses (antilogs). Just remember that the natural logarithm of a number is 2.303 times its common logarithm. Thus,

$$\ln \frac{P_2}{P_1} = 2.303 \log \frac{P_2}{P_1} = \frac{\Delta H_{vap}}{R} \left(\frac{1}{T_1} - \frac{1}{T_2} \right)$$

and

$$\log \frac{P_2}{P_1} = \frac{\Delta H_{vap}}{2.303R} \left(\frac{1}{T_1} - \frac{1}{T_2} \right)$$

Since the ln function is the one that naturally arises in mathematical derivations, a good case can be made for emphasizing natural over common logarithms. On the other hand, a great deal of scientific data (such as vapor pressures) have been recorded in terms of common logarithms and common logarithms continue to be useful. And, in Chapters 17 and 18 we deal with one concept (pH) where common logarithms must be used instead of natural logarithms. In this text we use both logarithmic forms, more or less interchangeably. For example, with common logarithms the calculation of Example 12-4 becomes

$$\log \frac{17.5 \text{ mmHg}}{P_1} = \frac{44.0 \times 10^3 \text{ J mol}^{-1}}{2.303 \times 8.314 \text{ J mol}^{-1} \text{ K}^{-1}} \left(\frac{1}{291.2} - \frac{1}{293.2} \right) \text{K}^{-1}$$
$$= 0.054$$

$$\frac{17.5 \text{ mmHg}}{P_1} = \text{antilog } 0.054 = 1.13$$

$$P_1 = (17.5/1.13) \text{ mmHg} = 15.5 \text{ mmHg}$$

$$\frac{17.5 \ \text{mmHg}}{P_1} = 1.13$$

$$P_1 = 17.5 \ \text{mmHg}/1.13 = 15.5 \ \text{mmHg}$$

(The measured vapor pressure of water at 18.0 °C is 15.48 mmHg.)

SIMILAR EXAMPLES: Exercises 5, 31, 32.

12-4   Transitions Involving Solids

Melting, Melting Point, and Heat of Fusion. As the temperature of a crystalline solid is raised, the vibrations of the structural units become more vigorous. Eventually a temperature is reached at which the crystalline structure is destroyed by these vibrations. The solid is converted to a liquid and the process is called **melting.** The reverse process, the conversion of a liquid to a solid, is called **freezing.** The temperature at which a solid melts, the **melting point,** and the temperature at which the corresponding liquid freezes, the **freezing point,** are identical. At this temperature solid and liquid coexist in dynamic equilibrium.

If we add heat to a solid–liquid mixture at equilibrium, the solid gradually melts *while the temperature remains constant*. Only when all the solid has melted does the temperature begin to rise. Conversely, if we remove heat from a solid–liquid mixture at equilibrium the liquid freezes *at a constant temperature*. The quantity of heat required to melt a given amount of solid is called the **enthalpy (heat) of fusion.** Some typical enthalpies (heats) of fusion are listed in Table 12-4. Melting is an *endothermic* process. The reverse process—freezing—is *exothermic,* and the quantity of heat evolved is also equal to the heat of fusion. Perhaps the example of a melting (and freezing) point with which we are most familiar is that of water, 0 °C. This is the temperature at which liquid and solid water, when in contact with air and under standard atmospheric pressure, are in equilibrium.* The heat of fusion of water is 6.01 kJ/mol, which we can express as either of the following enthalpy changes.

When a large freshwater lake freezes, heat is given off and helps to moderate temperatures in areas immediately adjacent to the lake.

$$H_2O(s) \longrightarrow H_2O(l) \qquad \Delta H = +6.01 \ \text{kJ/mol}$$
$$H_2O(l) \longrightarrow H_2O(s) \qquad \Delta H = -6.01 \ \text{kJ/mol}$$

Here is an easy way to determine the freezing point of a liquid. Allow the liquid to cool and measure the liquid temperature as a function of time; temperature decreases with time. When freezing begins the temperature *remains constant* until all the liquid has frozen. Then the temperature is again free to fall as the solid cools. If we plot temperatures against time we get a graph known as a **cooling curve.** Figure 12-16 is a cooling curve for water. We can also run this process backward, that is, by starting with the solid and adding heat. Again the temperature remains constant while melting occurs. This temperature–time plot is called a **heating curve.**

Often an experimentally determined cooling curve does not look quite like the solid-line plot in Figure 12-16. The temperature may drop *below* the freezing point without any solid appearing. This is a condition known as **supercooling.** In order for a crystalline solid to start forming from a liquid at the freezing point, there must be present in the liquid some small particles (such as suspended dust particles) on which crystals can form. If a liquid contains a very limited number of nuclei on

TABLE 12-4

Some Enthalpies (Heats) of Fusion

Substance	Melting point, °C	ΔH_{fusion}, kJ/mol
mercury, Hg	−38.8	2.33
sodium, Na	97.8	2.64
methyl alcohol, CH$_3$OH	−97.9	3.18
ethyl alcohol, CH$_3$CH$_2$OH	−114.5	5.02
water, H$_2$O	0.0	6.01
diethyl ether, (C$_2$H$_5$)$_2$O	−116.3	7.27

*If air is excluded and solid and liquid water are in equilibrium with their own vapor (at a pressure of 4.58 mmHg), the equilibrium temperature is slightly different, +0.01 °C.

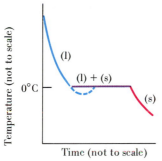

FIGURE 12-16
Cooling curve for water.

The broken line portion represents the condition of supercooling that occasionally occurs. (l) = liquid; (s) = solid.

which crystals can grow, it may supercool for a time before freezing. When a supercooled liquid does begin to freeze, however, the temperature rises again to the normal freezing point where freezing is completed. We can always recognize supercooling through a slight dip in a cooling curve just before the straight-line portion.

In the preceding section we did calculations in which liquids and vapor were the expected states of matter. We can now include the solid state in these calculations, as in Example 12-5.

Example 12-5

Using thermochemical data to determine whether a sample will exist as a solid, a liquid, or a solid–liquid mixture. A 25.0-g cube of ice at 0.0 °C is added to 100.0 g of liquid water at 22.0 °C in a thermally insulated container (e.g., a Styrofoam cup). The heat of fusion of ice is 6.01 kJ/mol and the specific heat of liquid water is 4.18 J g^{-1} °C^{-1}.

(a) What is the final condition reached—liquid only, solid only, or a mixture of solid and liquid water?

(b) What is the temperature in this final condition?

Solution

(a) As in Example 12-4 there is one condition that we can reject immediately. The final condition cannot be one of all solid. This would require the liquid water to cool down and then to freeze. In both of these processes heat is evolved, but the heat would have to be absorbed by something.

Let us call the heat associated with melting ice, q_{ice}, and that for cooling the *original* water, q_{water}.

$$q_{ice} = 25.0 \text{ g ice} \times \frac{1 \text{ mol ice}}{18.02 \text{ g ice}} \times \frac{6.01 \text{ kJ}}{1 \text{ mol ice}} = 8.34 \text{ kJ}$$

The maximum quantity of heat that the original liquid water could give off would be if it cooled from 22.0 to 0.0 °C.

$$q_{water} = 100.0 \text{ g water} \times \frac{4.18 \text{ J}}{\text{g water °C}} \times (0.0 - 22.0) \text{ °C} = -9196 \text{ J}$$

This quantity of heat (9.20 kJ) is greater than that required to melt all the ice. The final condition is one of "all liquid" at a temperature somewhat above 0 °C.

(b) Calculate the heat associated with (1) melting 25.0 g ice (as in part a), (2) warming the resulting 25.0 g of liquid water from 0 °C to a final (higher) temperature T, and (3) cooling the original 100.0 g liquid water from 22.0 °C to a final (lower) temperature T. The sum of these three quantities of heat must be zero. We have dropped the units from the expression that follows, but each of the three quantities of heat is expressed in joules.

$$\underbrace{8340}_{(1)} + \underbrace{25.0 \times 4.18 \times (T - 0)}_{(2)} + \underbrace{100.0 \times 4.18 \times (T - 22.0)}_{(3)} = 0$$

$$8340 + 104T + 418T - 9200 = 0$$

$$522T = 860$$

$$T = 1.6 \text{ °C}$$

An alternative approach is to note from part (a) that for a temperature of 0 °C the quantity of heat lost by the original liquid water (9200 J) exceeds that required to melt the ice (8340 J) by 860 J. Now determine the final temperature if 860 J were reabsorbed by 125.0 g liquid water at 0 °C. The final temperature is 1.6 °C.

$$125.0 \text{ g water} \times \frac{4.18 \text{ J}}{\text{g water °C}} \times (T - 0.0) \text{ °C} = 860 \text{ J}$$

$$T = 1.6 \text{ °C}$$

SIMILAR EXAMPLES: Exercises 43, 44.

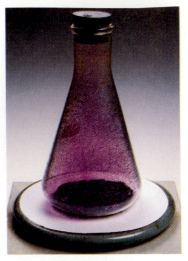

FIGURE 12-17
Sublimation of iodine.

Even at temperatures well below its melting point of 114 °C, solid iodine exhibits an appreciable sublimation pressure. Here, purple iodine vapor is produced at about 70 °C and redeposits as solid iodine on the colder walls of the flask. [Carey B. Van Loon]

Relationship of enthalpies (heats) of fusion, vaporization, and sublimation.

Sublimation. Solids also vaporize, though in general solids are not as volatile as liquids. The direct passage of molecules from the solid to the vapor state is called **sublimation.** The reverse process, the passage of molecules from the vapor to the solid state is called **deposition.** When sublimation and deposition occur at equal rates, a dynamic equilibrium is established between a solid and its vapor. The vapor exerts a characteristic pressure called the **sublimation pressure.** A plot of sublimation pressure as a function of temperature is called a **sublimation curve.** The **enthalpy (heat) of sublimation** is the quantity of heat needed to convert a given quantity of solid to vapor. It is related to enthalpies of fusion and vaporization in a simple way. If we combine the expressions for the melting of a solid and the vaporization of a liquid, the resulting expression represents sublimation. For example, for water at 0 °C

$$
\begin{array}{lll}
H_2O(s) \longrightarrow H_2O(l) & \Delta H_{fusion} = & 6.01 \text{ kJ} \\
H_2O(l) \longrightarrow H_2O(g) & \Delta H_{vap} = & 44.92 \text{ kJ} \\
\hline
H_2O(s) \longrightarrow H_2O(g) & \Delta H_{subl} = & 50.93 \text{ kJ}
\end{array}
$$

$$\Delta H_{subl} = \Delta H_{fusion} + \Delta H_{vap} \tag{12.7}$$

Two familiar solids with significant sublimation pressures are ice and dry ice (solid carbon dioxide). If you live in a cold climate you are aware of the fact that snow may disappear from the ground even though the temperature may fail to rise above 0 °C. Under these conditions the snow does not melt, it sublimes. The sublimation pressure of ice at 0 °C is 4.58 mmHg. The sublimation of iodine is pictured in Figure 12-17.

12-5 Phase Diagrams

At low temperatures and/or high pressures we expect a substance to exist in the solid state or as a solid phase. At high temperatures and/or low pressures we expect the gaseous state or phase. At intermediate values of temperature and pressure we expect a liquid phase. The collection of all the points in red in Figure 12-18 represents conditions under which a hypothetical substance exists as a solid phase. Similarly, the collection of blue points represents the area of existence of a liquid phase; and the brown points, a gaseous phase. The pressure–temperature plot that we are describing is called a **phase diagram.**

More interesting than the areas of a phase diagram representing single states or phases of matter—solid, liquid, or gas—are the curved lines where different areas adjoin, as illustrated in the phase diagram for iodine in Figure 12-19. The curve *OC* is the **vapor pressure curve** of liquid iodine, and *C* is the **critical point.** *OB* is the **sublimation curve** of solid iodine. The effect of pressure on the melting point of iodine is represented by the curve *OD*; it is called the **fusion curve.** The point, *O*, has a special significance. It gives the *unique* temperature and pressure at which solid, liquid, and vapor coexist in equilibrium. It is called a **triple point;** for iodine

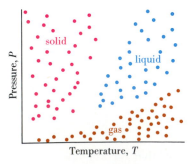

FIGURE 12-18
Pressures, temperatures, and states of matter.

FIGURE 12-19
Phase diagram for iodine.

FIGURE 12-19
Phase diagram for iodine.

Note that the melting point and triple point temperatures for iodine are essentially the same. Generally, large pressure increases are required to produce even small changes in solid–liquid equilibrium temperatures.

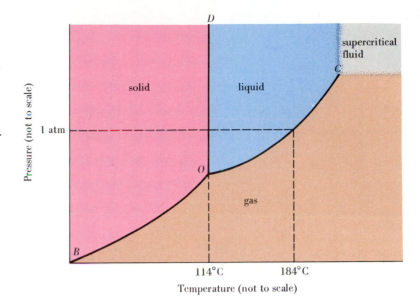

this is at 114 °C and 91 mmHg. The normal melting point (114 °C) and boiling point (184 °C) are the temperatures at which a line at $P = 1$ atm intersects the fusion and vapor pressure curves, respectively.

Let us look a little more closely at the region of the critical point, C. This is the point at which the liquid and gaseous phases of a substance become identical and indistinguishable. It is difficult to know what to call the phase that exists at temperatures and pressures above the critical point. For example, it has a high density (like a liquid), but a low viscosity (like a gas). One term that is finding increased usage is *supercritical fluid*. Although it is not customary to do so, we have marked off an area in gray in Figure 12-19 and called this the supercritical fluid region.

The phase behavior of carbon dioxide, shown in Figure 12-20, differs from that of iodine in one important respect—the pressure at the triple point (O) is greater than 1 atm. A line at $P = 1$ atm intersects the *sublimation curve,* not the vapor pressure curve. If solid CO_2 is heated in a container open to the atmosphere, it sublimes away at a *constant* temperature of −78.5 °C. *It does not melt* (and so is called "dry ice"). Because it is able to maintain a low temperature and does not produce a liquid by melting, dry ice is widely used in freezing and preserving foods.

FIGURE 12-20
Phase diagram for carbon dioxide.

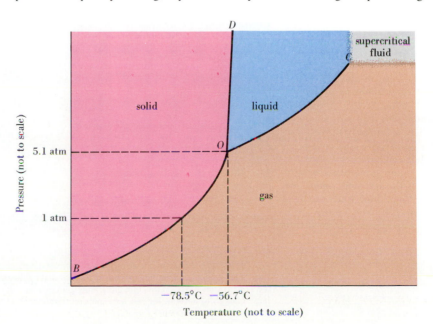

FIGURE 12-21
Phase diagram for water.

Point *O*, the triple point, is at +0.0098 °C and 4.58 mmHg. The critical point, *C*, is at 374.1 °C and 218.2 atm. The negative slope of the fusion curve *OD* is exaggerated in this diagram. An increase in pressure of about 125 atm is required to produce a decrease of 1 °C in the melting point of ice I. All of the high pressure forms of ice are more dense than liquid water. Ice I, ice III, and liquid water are at equilibrium (point *D*) at −22.0 °C and 2045 atm. Ice VII can be maintained at temperatures approximating the normal boiling point of water (100 °C), but only at pressures in excess of 25,000 atm. The significance of the broken straight lines is discussed in the text.

Bismuth and antimony are the only other common substances whose melting points decrease with increased pressure.

Liquid CO_2, which can be obtained at pressures above 5.1 atm, is most frequently encountered in CO_2 fire extinguishers. All three states of matter are involved in the action of these fire extinguishers. When the liquid CO_2 is released the bulk of it quickly vaporizes. The heat required for this vaporization is extracted from the remaining $CO_2(l)$, which has its temperature lowered to the point that it freezes and falls as a $CO_2(s)$ "snow." In turn the $CO_2(s)$ quickly sublimes to $CO_2(g)$.

The phase diagram for water (Figure 12-21) presents several new features. For one, the fusion curve *OD* has a *negative* slope; it slopes toward the pressure axis. The melting point of ice *decreases* with increased pressure, very unusual behavior for a solid.* Another feature is that of **polymorphism,** the existence of a solid substance in more than one crystalline form. Ordinary ice (ice I) exists under ordinary conditions of temperature and pressure. The other forms of ice exist only when maintained under very high pressures. Polymorphism is much more the rule than the exception among substances. Where polymorphism occurs a phase diagram has triple points in addition to the familiar solid–liquid–vapor triple point. For example, ice I, ice III, and $H_2O(l)$ are in equilibrium at −22.0 °C and 2045 atm.

FIGURE 12-22
Polymorphism of mercury(II) iodide.

The stable form of mercury(II) iodide at room temperature is a red solid (left). At 127 °C the red solid undergoes a phase change to a yellow solid (right). Mercury(II) iodide also has an appreciable sublimation pressure, with the vapor condensing to a mixture of the red and yellow solids on the colder walls of the flask. [Carey B. Van Loon]

*The decrease in melting point of ice with increased pressure is not easily observed in natural phenomenon. One example commonly given is that in ice skating the pressure of the skate blade melts the ice and the skater skims along on this thin lubricating film of water. It is doubtful that the skate pressure is really large enough to account for much of this (see Exercise 47). More likely the melting occurs primarily through ordinary frictional forces.

Are You Wondering:

What is the difference between a phase *and a* state *of matter?*

There are *three* states of matter—solid, liquid, and gas. A **phase** is a sample of matter with definite composition and uniform properties throughout the sample. In the equilibrium between ordinary ice and liquid water, we have a *two-phase* system. Many physical properties of the ice–water mixture change abruptly between the ice phase and the liquid water phase.

Because all gases are miscible, every gaseous mixture exists as a *single* phase. For substances (except helium) there is also only a *single* liquid phase; but substances commonly can exist in *several* different solid phases, as in the equilibrium involving ice II, ice III, and ice V. For *mixtures* of substances two or more phases are often possible, both in the solid and liquid states of matter.

Another well-known example of polymorphism, that of mercury(II) iodide, is illustrated in Figure 12-22.

Examples 12-6 and 12-7 illustrate how a phase diagram may be used to explain the physical behavior of a substance.

Example 12-6

Interpreting a phase diagram. A sample of ice is maintained at 1 atm and at a temperature represented by point *P* in Figure 12-21. Describe what happens when **(a)** the temperature is raised, at constant pressure, to point *R*, and **(b)** the pressure is raised, at constant temperature, to point *Q*. The conditions that exist at points *P*, *Q*, and *R* are shown in Figure 12-23.

Solution

(a) When the temperature reaches a point on the fusion curve *OD* (0 °C), ice I begins to melt. The temperature remains constant as ice I is converted to liquid. When melting is complete, the temperature increases again. No vapor appears in the cylinder until the temperature reaches 100 °C, the temperature at which the vapor pressure is 1 atm. When all the liquid has vaporized, the temperature is again free to rise, to a final value of *R*.

(b) Because of the very small compressibility of solids, essentially no change occurs until the pressure reaches a value corresponding to the point of

FIGURE 12-23
Predicting phase changes—
Example 12-6 illustrated.

These are the conditions that prevail at points *P*, *Q*, and *R* in Figure 12-21.

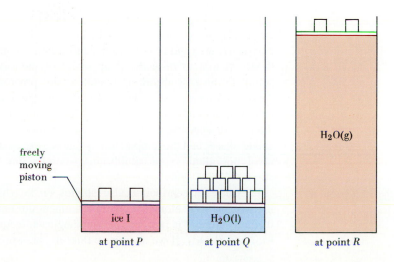

freely
moving
piston

ice I $H_2O(l)$ $H_2O(g)$

at point *P*   at point *Q*   at point *R*

intersection of the constant-temperature line, *PQ*, with the fusion curve, *OD*. At this pressure melting occurs. Melting is accompanied by a significant decrease in volume (10% or so) as ice I is converted to liquid water. Following melting, additional increases in pressure produce very little effect since liquids are not very compressible.

SIMILAR EXAMPLES: Exercises 8, 48, 49, 78.

Example 12-7

Using a phase diagram to sketch a heating curve. Sketch the heating curve that will be obtained if a sample of ice at point *P* in Figure 12-21 is heated until it becomes gaseous water at point *R*.

Solution. The phase changes that occur were described in Example 12-6(a). That is, ice → ice + liquid water → liquid water → liquid water + gaseous water → gaseous water. The key to constructing the heating curve is to recognize that the temperature is free to rise with time when there is only a *single* phase present (ice or liquid water or gaseous water), but that the temperature must remain constant with time when there are *two phases in equilibrium*. Thus, there is one straight-line portion of the heating curve at 0 °C and another at 100 °C.

Another feature that you should be able to incorporate in the heating curve is the length of time that the temperature remains constant in each straight-line portion. If we assume that the rate of heating is held constant throughout, it takes several times more heat to vaporize a quantity of water (Table 12-1; 44.0 kJ/mol) than to melt the same quantity of ice (Table 12-4; 6.0 kJ/mol). The straight-line portion of the heating curve for vaporization should be about seven times as long as for melting. A heating curve is sketched in Figure 12-24.

SIMILAR EXAMPLES: Exercises 48, 84, 85.

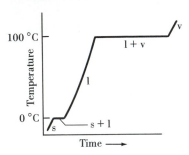

FIGURE 12-24
A heating curve for water—Example 12-7 illustrated.

The heating curve consists of three steeply rising portions, representing the heating of solid (s), liquid (l), and vapor (v), respectively. The straight-line portions represent two-phase equilibrium between a solid and liquid (s + l) and between a liquid and its vapor (l + v).

Le Châtelier's Principle: A Preview. Let us briefly consider an aid to describing phase behavior that we discuss in more detail in Chapter 16. It is called Le Châtelier's principle, and when applied to an equilibrium involving different phases of a substance states that

Le Châtelier's principle applied to phase equilibria.

> An attempt to change the temperature or pressure of a system in phase equilibrium stimulates changes within the system that preserve the condition of equilibrium for as long as possible.   (12.8)

Let us consider again the phase equilibria encountered in Example 12-6. In Figure 12-21, at the point where the constant-pressure line *PR* intersects the fusion curve *OD*, ice I and H₂O(l) are in equilibrium at constant temperature. At this point, if we attempt to raise the temperature of the system by adding more heat, we stimulate the heat-absorbing (endothermic) process—the further melting of ice I. As we add more heat, more ice I melts and the temperature remains constant. The equilibrium condition is maintained for as long as possible. Eventually, however, the ice I disappears; the two-phase equilibrium is destroyed; the system reverts to a single phase, and the temperature is free to rise. A similar situation occurs, this time involving liquid–vapor equilibrium, when the line *PR* intersects the vapor pressure curve *OC*.

Le Châtelier's principle also applies to the changes that occur as we slowly increase pressure along the constant-temperature line *PQ* in Figure 12-21. When the fusion curve *OD* is reached, melting begins and equilibrium is established between ice I and H₂O(l). If we attempt to raise the pressure at this point, we simply stimu-

CHEMISTRY EVOLVING

Supercritical Fluid Extraction

Although we do not ordinarily think of solids as being soluble in gases, a volatile solid will dissolve in a gas. The mole fraction solubility of a solid in a gas is simply the ratio of the vapor pressure (sublimation pressure) of the solid to the total gas pressure (recall equation 6.16). If a solid is in contact with a gas that is maintained above its critical pressure and temperature, the solid becomes much more soluble, however. This is due in part to the fact that the vapor pressure of a solid increases when it is subjected to a high pressure. But more important is the fact that the density of the supercritical fluid (SCF) is high, approaching that of a liquid. Molecules in supercritical fluids, being in much closer proximity than in ordinary gases, can exert strong attractive forces on the molecules of a solid. As a result, supercritical fluids display solvent properties similar to ordinary liquid solvents, and these fluids can be used to dissolve solids. In a like manner, supercritical fluids can also dissolve liquids, even nonvolatile ones.

The density of a substance in the critical point region is strongly dependent on temperature and pressure. To vary the pressure of a SCF, for example, means to vary its density and also its solvent properties. A given SCF, such as carbon dioxide, can be made to behave like many different solvents.

Until recently, the principal method of decaffeinating coffee has been to extract the caffeine with a solvent such as methylene chloride (CH_2Cl_2). But CH_2Cl_2 is a hazardous material in the workplace, and it is difficult to rid the product of all traces of the solvent. Now, methods have been developed that use supercritical fluid CO_2. In one process, green coffee beans are brought into contact with CO_2 at about 90 °C and 160 to 220 atm. The caffeine content of the coffee is reduced from its normal 1 to 3% to about 0.02%. When the temperature and pressure of the CO_2 are reduced, the caffeine precipitates and the CO_2 is recycled.

Extraction with supercritical fluid CO_2 has several advantages over conventional liquid extraction. Carbon dioxide is nontoxic, nonhazardous, unreactive toward food constituents, and inexpensive. Supercritical fluids have considerably lower viscosities than liquids, so they flow through materials more readily. And the energy requirements, including recycling of the solvent, are lower for SCF extraction than for ordinary liquid extraction.

Many future uses of SCF extraction are contemplated. These include

- extracting volatile components from coal;
- detoxifying wastewater;
- extracting flavors and fragrances from products such as lemons, black pepper, almonds, and nutmeg;
- extracting cooking oils from foods like potato chips, improving their nutritional value and shelf life;
- extracting cholesterol from butter and lard. ■ ■ ■

The crossing of a two-phase equilibrium curve in a phase diagram is called a **transition**. Specific names used for transitions are

    melting (s → l)
    freezing (l → s)
    vaporization (l → v)
    condensation (v → l)
    sublimation (s → v)
    deposition (v → s)
    transition (s_1 → s_2)

late the process in which the volume of the system decreases—the melting of ice I. Equilibrium is maintained until all the ice I has melted. Then the pressure is again free to rise.

The processes described in the two preceding paragraphs can be summarized and generalized by saying that in crossing a two-phase curve in a phase diagram

- from lower to higher temperatures along a *constant-pressure* line there is an *increase in enthalpy*, and
- from lower to higher pressure along a *constant-temperature* line there is a *decrease in volume* (increase in density).

These statements help us understand why a fusion curve generally has a positive slope (away from the pressure axis). Typical behavior is for a solid to be more dense than the corresponding liquid, and for increased pressure to favor the process, liquid → solid. These statements also suggest, for example, that for the transition ice II → ice III in Figure 12-21 $\Delta H_{transition} > 0$; that the densities of all forms of ice except ice I are greater than the density of $H_2O(l)$; etc.

12-6 Van der Waals Forces

The lightest of the group 8A elements, helium, forms no stable chemical bonds. Given the fact that He atoms do not bond to one another we might expect helium to remain a gas, right down to the absolute zero of temperature. In fact, helium remains gaseous to very low temperatures, but it does condense to a liquid at 4 K and freezes to a solid (at 25 atm pressure) at 1 K. These data suggest that intermolecular forces, albeit very weak, must exist among He atoms, and if the temperature is sufficiently low these forces overcome thermal agitation and cause helium to condense. What kind of force can this be?

Instantaneous and Induced Dipoles. In describing the electronic structures of atoms and molecules we always speak in terms of electron charge density or the *probability* of an electron being in a certain region at a given time. Included among the probabilities is one in which, purely by chance, at some particular instant electrons are concentrated in one region of the atom or molecule. This displacement of electrons causes a normally nonpolar species to become polar; an **instantaneous dipole** is formed. Following this, electrons in a neighboring atom or molecule may be displaced, also producing a dipole. This is a process of induction, and the newly formed dipole is called an **induced dipole.**

Taken together, these two events lead to an intermolecular force of attraction (see Figure 12-25). We can call this an instantaneous dipole-induced dipole attraction, but the names more commonly used are **dispersion force** or **London force.** (Fritz London offered a theoretical explanation of these forces in 1928.)

The ease with which electron charge density is distorted by an external electric field, that is, the ease with which a dipole can be induced in an atom or molecule, is called **polarizability.** Polarizability increases with increased numbers of electrons in a species, and the number of electrons is related in a general way to the molecular weight. Because dispersion forces become stronger as polarizability increases, we can say that for covalent substances, *in general*, melting points and boiling points increase with increasing molecular weight. As an example, radon, with a molecular (atomic) weight of 222, has a boiling point of 211 K, whereas helium (atomic weight, 4) has a boiling point of 4 K. As we shall see, however, because of additional factors there are many exceptions to this generalization. For example, the

The attraction of a balloon to a surface is a commonplace example of induction. The balloon is charged by rubbing, and the charged balloon induces an opposite charge on the surface. (See also Appendix B.) [Carey B. Van Loon]

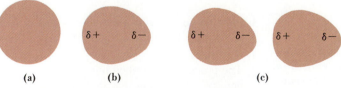

(a)   (b)   (c)

FIGURE 12-25

Instantaneous and induced dipoles.

(a) *Normal condition.* A nonpolar species has a symmetrical charge distribution.
(b) *Instantaneous condition.* A displacement of the electronic charge produces an instantaneous dipole, with charges of $\delta+$ and $\delta-$.
(c) *Induced dipole.* The instantaneous dipole on the left induces a charge separation in the species on the right. The result is a dipole–dipole interaction; the two dipoles are attracted to each other.

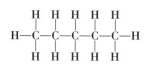

(a) Neopentane
b.p. = 9.5 °C, T_c = 160.6 °C

(b) Normal pentane
b.p. = 36.1 °C, T_c = 196.6 °C

FIGURE 12-26
Molecular shapes and polarizability.

The elongated normal pentane molecule is more easily polarized than the compact neopentane molecule. Intermolecular forces are stronger in normal pentane than in neopentane. As a result, normal pentane boils at a higher temperature than neopentane; its critical temperature is also higher than that of neopentane. [Photographs by Carey B. Van Loon]

strength of dispersion forces also depends on *molecular shape*. Electrons in elongated molecules are more easily displaced than are those in small, compact, symmetrical molecules. Thus two substances with identical numbers and kinds of atoms (isomers) but different molecular shapes may have different properties. This idea is illustrated in Figure 12-26.

Dipole–Dipole Interactions. In a *polar* substance molecules try to line up with the positive end of one dipole directed toward the negative ends of neighboring dipoles. An idealized situation is pictured in Figure 12-27. This additional partial ordering of molecules can cause a substance to persist as a solid or liquid at temperatures higher than otherwise expected. For example, compare normal pentane, C_5H_{12}, and normal butyl alcohol, C_4H_9OH, two substances with nearly equal molecular weights. Because of the significant difference in electronegativity between H and O, there is a large O—H bond dipole moment in C_4H_9OH that is not offset by other bond moments. Normal butyl alcohol is *polar*. The electronegativity difference between C and H is so small that the bond dipole moments in C_5H_{12} are essentially zero. Normal pentane is *nonpolar*. Physical properties of the two substances are compared below.

$\mu = 0$	C_5H_{12}	M.W. = 72	m.p. − 130 °C	b.p. 36.1 °C	
$\mu = 1.66$ D	C_4H_9OH	M.W. = 74	m.p. − 89.5 °C	b.p. 117.2 °C	

Example 12-8

Comparing physical properties of polar and nonpolar substances. Which would you expect to have the *higher* boiling point, the hydrocarbon fuel normal butane, C_4H_{10}, or the organic solvent acetone, $(CH_3)_2CO$?

Solution. A good starting point is to compare molecular weights to see if this is an important factor to consider. In this case, both substances have a molecular weight of 58. We must look for a factor other than molecular weight.

Next, determine whether either molecule is polar. The electronegativity different between C and H is so small that we generally expect hydrocarbons to be *nonpolar*. The acetone molecule contains an O atom, and we do expect a strong C-to-O bond dipole moment. At times it may also be necessary to sketch the structure of a molecule to see whether symmetrical features of a molecule cause bond dipole moments to cancel. This does not happen in the acetone molecule since there is only one bond with a large bond dipole moment. Acetone is a *polar* molecule. Given two substances with the same molecular weight, one polar and one nonpolar, we expect the *polar* substance to have the *higher* boiling point—acetone. (The measured boiling points are butane, −0.5 °C; acetone, 56.2 °C.)

FIGURE 12-27
Dipole–dipole interactions.

Dipoles tend to arrange themselves with the positive end of one dipole pointed toward the negative end of a neighboring dipole. Ordinarily, thermal motion upsets this orderly array, cancelling out most of the dipole–dipole attractions. Nevertheless, this tendency for dipoles to align themselves can affect physical properties, such as the melting points of solids.

SIMILAR EXAMPLES: Exercises 9, 53.

Summary of Van der Waals Forces. The intermolecular forces we have described in this section—dispersion (London) forces and forces between permanent dipoles— are commonly called **van der Waals forces.** They are the kinds of intermolecular forces that cause a gas to depart from ideal gas behavior and are accounted for in the van der Waals equation of state. To assess the relative contributions to van der Waals forces, you need to consider the following factors.

Assessing van der Waals (intermolecular) forces.

- *Dispersion (London) forces* are based on displacing all the electrons in a molecule. The strength of these forces increases with increased molecular weight and also depends on molecular shapes. Dispersion forces exist between all molecules.
- *Forces associated with permanent dipoles* involve displacements of electron pairs in bonds rather than in the molecule as a whole. They are found only for substances having resultant dipole moments. The existence of permanent dipoles *adds* to the effect of dispersion forces that are also present.
- *In comparing substances of roughly comparable molecular weights,* the presence of dipole forces can produce significant differences in such properties as melting point, boiling point, and enthalpy of vaporization.
- *In comparing substances of rather widely different molecular weights,* dispersion forces are likely to be more important than dipole forces in determining physical properties.

(12.9)

Let us see how the statements in (12.9) are supported by the data in Table 12-5. HCl and F_2 have comparable molecular weights, but because HCl is polar (and F_2 is not) it has a significantly larger ΔH_{vap} and a higher boiling point than does F_2. Within the series HCl, HBr, and HI, molecular weight increases sharply, and ΔH_{vap} and boiling points increase in the order HCl < HBr < HI. The higher polarities of HCl and HBr relative to HI are not sufficient to reverse the trends produced by the increasing molecular weights and strengths of dispersion forces—dispersion forces are the predominant intermolecular forces.

Example 12-9

Relating intermolecular forces and physical properties. Arrange the following substances in the order in which you would expect their boiling points to increase: CCl_4 (M.W. 153.82); Cl_2 (M.W. 70.91); ClNO (M.W. 65.46); N_2 (M.W. 28.01).

Solution. Three of these are nonpolar substances for which we should expect only dispersion (London) forces. The strengths of these forces, and hence the boiling points, should increase with increasing molecular weight, i.e., $N_2 < Cl_2 < CCl_4$. ClNO has a molecular weight that is comparable to that of Cl_2, but the ClNO molecule is polar (bond angle $\sim 120°$). This should result in stronger intermolecular forces and a higher boiling point for ClNO than for Cl_2. However, we would not expect the boiling point of ClNO to be higher than that of CCl_4 because of the considerable difference in molecular weight between the two. The expected order is

$$N_2 < Cl_2 < ClNO < CCl_4$$

(The observed boiling points are 77.4, 239.1, 266.8, and 349.8 K, respectively.)

SIMILAR EXAMPLES: Exercises 9, 10, 52, 53.

TABLE 12-5
Intermolecular Forces and Properties of Selected Substances

	M.W.	Dipole moment, D	Van der Waals forces		ΔH_{vap}, kJ/mol	Boiling point, K
			% dispersion	% dipole		
F_2	38.00	0	100	0	6.86	85.01
HCl	36.46	1.08	81.4	18.6	16.15	188.11
HBr	80.92	0.78	94.5	5.5	17.61	206.43
HI	127.91	0.38	99.5	0.5	19.77	237.80

12-7 Hydrogen Bonds

In Figure 12-28 we have plotted boiling points of a series of similar compounds as a function of molecular weight. The hydrogen compounds (hydrides) of the group 4A (14) elements display normal behavior, that is, the boiling point increases regularly with increasing molecular weight. But note the three striking exceptions in groups 5A, 6A, and 7A: ammonia (NH_3), water (H_2O), and hydrogen fluoride (HF). Their boiling points are higher than for any other hydride in their group, not lowest as we would expect. The cause of this exceptional behavior is the presence of a special kind of intermolecular force of attraction called **hydrogen bonding.** Figure 12-29 represents hydrogen bonding in hydrogen fluoride and illustrates three important ideas that we will find useful in describing hydrogen bonding, here and elsewhere in the text.

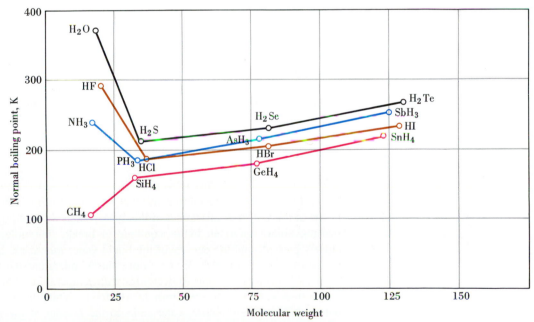

FIGURE 12-28
Comparison of boiling points of some hydrides of the elements of groups 4A, 5A, 6A, and 7A.

The values for NH_3, H_2O, and HF are unusually high compared to those of other members of their groups.

FIGURE 12-29

The hydrogen bond.

In the gaseous state, several polymeric forms of the HF molecule exist in which the individual molecules (monomers) are held together through hydrogen bonds. A pentagonal arrangement of five HF molecules is shown here.

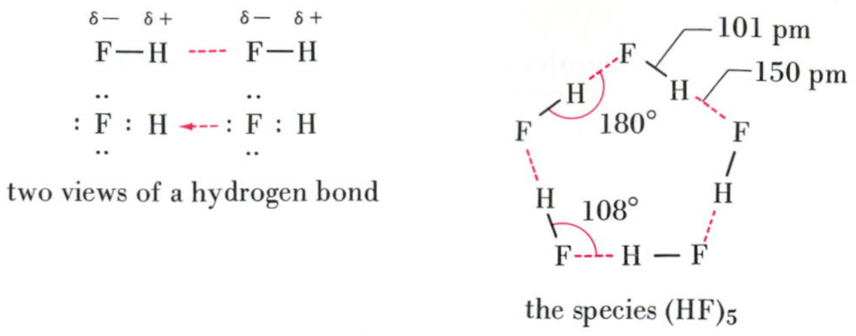

$$\delta- \quad \delta+ \qquad \delta- \quad \delta+$$
$$\text{F—H} \ \text{----} \ \text{F—H}$$

$$:\overset{..}{\text{F}}:\text{H} \ \longleftarrow \ :\overset{..}{\text{F}}:\text{H}$$

two views of a hydrogen bond

the species (HF)$_5$

1. The H atom in HF is a center of positive charge and the F atom is a center of negative charge. Dipoles align in the usual fashion, the positive end of one dipole directed toward the negative end of another. This alignment places an H atom between two F atoms. Because of the very small size of the H atom, the dipoles come into close proximity, producing a strong dipole–dipole attraction.
2. At the same time that an H atom is strongly bonded to one F atom it is weakly bonded to the F atom of a nearby HF molecule. This occurs through one of the lone pairs of electrons on the F atom. In hydrogen bonding an H atom acts as a bridge between two nonmetal atoms.
3. An essential aspect of hydrogen bonding is the tendency for hydrogen bonds to form throughout a cluster of molecules, producing so-called polymeric structures. The bond angle between two nonmetal atoms bridged by an H atom (that is, the bond angle, X—H $\cdots$ X) is usually about 180°. However, there are preferred orientations for hydrogen-bonded molecules, such as in the pentagonal structure (HF)$_5$.

By contrast typical bond energies range from about 150 to several hundred kJ/mol, and van der Waals interactions are of the order of 2 to 20 kJ/mol.

To summarize: The hydrogen bond is a rather strong intermolecular force, with energies of the order of 15 to 40 kJ/mol. Strong hydrogen bonding occurs when an H atom in one molecule is simultaneously attracted to a highly electronegative atom—F, O, or N—of a neighboring molecule. In some cases weak hydrogen bonding may also occur between an H atom of one molecule and the nonmetal atoms Cl or S of a neighboring molecule.

Ordinary water is certainly the most common substance in which hydrogen bonding occurs. Figure 12-30 shows how one water molecule is held to four neighbors in a tetrahedral arrangement by hydrogen bonds. This is the structural arrangement in crystalline water or ice. Hydrogen bonds hold the water molecules in a rigid but rather open structure. As ice melts, only a fraction of the hydrogen bonds are broken. Extensive hydrogen bonding persists in liquid water just above the melting point; liquid water retains an icelike structure. One indication of this is the relatively low heat of fusion of ice (6.01 kJ/mol); this is much less than we would expect if all the hydrogen bonds were to break during melting.

At the melting point, water molecules are packed more closely together in liquid water than in ice, and this is why the liquid is more dense than ice. It is highly unusual for a liquid to be more dense than the solid from which it is formed. Water is one of the very few substances of this type. When we heat liquid water above the melting point, hydrogen bonds continue to break, the molecules become more closely packed, and the density of the liquid water *increases*. Liquid water attains its maximum density at 3.98 °C. Above this temperature the water behaves in a ''normal'' fashion: its density decreases with temperature. This behavior explains why a freshwater lake freezes from the top down. When the water temperature falls below 4 °C the more dense water sinks to the bottom of the lake and the colder surface water freezes. The ice at the top of the lake then tends to insulate the water below the ice from further heat loss. Generally, a freshwater lake does not freeze solid in the winter time.

It is not an overstatement for us to say that hydrogen bonding makes life possible. Living organisms are maintained through a series of chemical reactions involving complex structures, such as DNA and proteins. Certain bonds in these structures

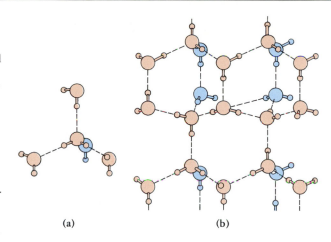

(a)　　　　　　(b)

FIGURE 12-30
Hydrogen bonding in water.

(a) Each water molecule is linked to four others through hydrogen bonds. The arrangement is tetrahedral. Each hydrogen atom is situated along a line joining two oxygen atoms, but somewhat closer to one oxygen atom (100 pm) than to the other (180 pm).

(b) The crystal structure of ice. Oxygen atoms are located in hexagonal rings arranged in layers. Molecules behind the plane of the page are light blue. Positions available to the hydrogen atoms lie between pairs of oxygen atoms, again closer to one than to the other. This characteristic pattern is revealed at the macroscopic level in the hexagonal shapes of snowflakes.

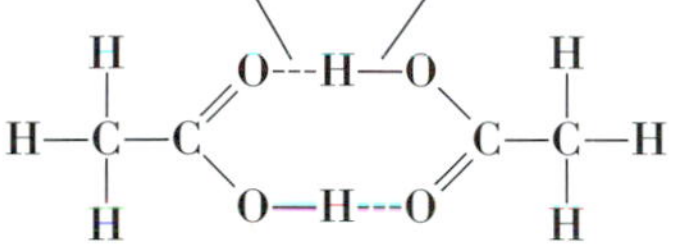

FIGURE 12-31
Dimerization of gaseous acetic acid.

Hydrogen bonding permits molecules to exist in stable pairs (dimers).

FIGURE 12-32
Hydrogen bonding and viscosity.

The unit of viscosity (η) used here is called a *centipoise* (cP). Values are at about 20 °C. By contrast, the viscosity of water at 20 °C is 1 cP.

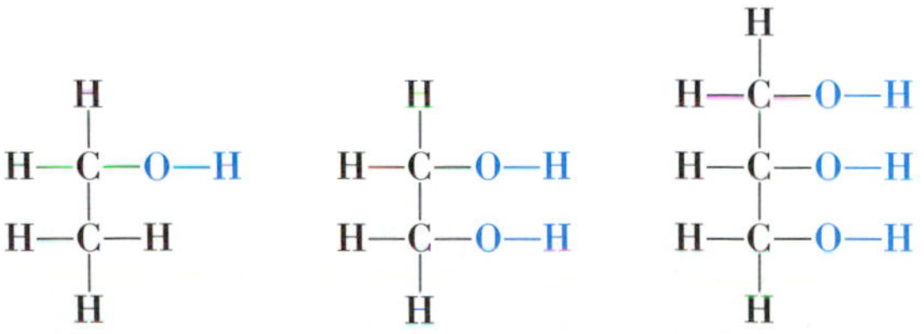

ethyl alcohol
$\eta = 1.2$ cP

ethylene glycol
$\eta = 17$ cP

glycerol
$\eta = 1070$ cP

must be easily able to break and reform. Only hydrogen bonds have just the right amount of energy to permit this. We look at the importance of hydrogen bonding in biological molecules in Chapter 28.

Liquids in which hydrogen bonding occurs exhibit stronger than usual intermolecular forces, and these liquids generally have high heats of vaporization. In acetic acid, H_3CCOOH, hydrogen bonding leads to the formation of dimers (double molecules) both in the liquid and *in the vapor state*. Not all the hydrogen bonds between molecules need to be broken to vaporize acetic acid and the heat of vaporization is abnormally *low*. Figure 12-31 shows a dimer of acetic acid.

Hydrogen bonding can also help us to understand certain trends in viscosity. Alcohols are carbon-based compounds containing the group O—H. The H atoms in alcohols can form hydrogen bonds with O atoms in neighboring molecules. The more O—H groups in a molecule, the more possibilities for hydrogen bonding, the more resistance to flow and the greater the viscosity of a liquid. Three alcohols are compared in Figure 12-32.

Throughout this section we have been describing hydrogen bonding between two molecules—*inter*molecular hydrogen bonding. Another possibility is for a hydrogen atom to bridge two nonmetal atoms within the *same* molecule. This is called *intra*molecular hydrogen bonding (see Exercises 58 and 96).

12-8　Network Covalent Solids

In most covalent substances the bonds between atoms within molecules are quite strong but the attractive forces between different molecules (intermolecular forces) are relatively weak. This is why many covalent substances of low molecular weight are gaseous at room temperature; others, usually of somewhat higher molecular weights, are liquids; still others are solids with moderately low melting points.

In some covalent substances, on the other hand, covalent bonds are not limited to bonds between atoms in small molecules. They extend throughout a crystalline

FIGURE 12-33
The diamond structure.

(a) A portion of the Lewis structure.
(b) Crystal structure. Each carbon atom is bonded to four others in a tetrahedral fashion. The segment of the entire crystal shown here is called a unit cell.

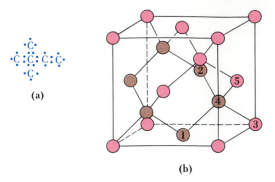

(a)

(b)

solid. In these cases *inter*molecular and *intra*molecular forces are indistinguishable and the entire crystal is held together by strong covalent bonds. Two of the best examples are the two forms in which pure carbon can occur—diamond and graphite.

The Diamond Structure. Figure 12-33 shows one way in which carbon atoms can bond one to another in a very extensive array or crystal. The two-dimensional Lewis structure in Figure 12-33a is useful only in pointing out that this bonding scheme involves ever-increasing numbers of carbon atoms leading to a giant molecule. It does not give us any insight into the three-dimensional structure of the molecule. On the other hand, the structure of the entire crystal can be inferred from the portion of the crystal shown in Figure 12-33b. Each atom is bonded to four others. Atoms 1, 2, and 3 lie in a plane with atom 4 above the plane. Atoms 1, 2, 3, and 5 define a tetrahedron with atom 4 inscribed at its center. The type of hybridization scheme that directs four bonds from a central atom to the corners of a tetrahedron, of course, is sp^3. When viewed from a particular direction, a nonplanar hexagonal arrangement of carbon atoms is also seen (shown in brown).

If silicon atoms are substituted for one-half the carbon atoms, the resulting structure is that of silicon carbide (carborundum). Both diamond and silicon carbide are extremely hard; in fact, diamond is the hardest substance known. To scratch or break diamond or silicon carbide crystals we have to break covalent bonds, and this accounts for their extensive use as abrasives. These two materials are also nonconductors of electricity and do not melt or sublime until very high temperatures are reached. SiC sublimes at 2700 °C, and diamond melts above 3500 °C.

Another silicon-containing network covalent solid we discuss in Section 14-9 is silicon dioxide, SiO_2 (silica).

The Graphite Structure. Carbon atoms can bond together in a different way to produce a solid with properties very much different from diamond. This bonding involves the orbital set $sp^2 + p$. The three sp^2 orbitals are directed in a plane at angles of 120°; the p orbital is perpendicular to the plane, directed above and below it. These are the same orbitals used by carbon atoms in benzene, C_6H_6 (described in Figure 11-30).

The crystal structure produced by this type of bonding is pictured in Figure 12-34; it is the **graphite** structure. Each carbon atom forms strong covalent bonds with three neighboring carbon atoms in the same plane, giving rise to layers of carbon atoms in a hexagonal arrangement. The p electrons of the carbon atoms are *delocalized* (recall the discussion in Section 11-7). Bonding within layers is strong, but between layers it is much weaker. We can see this through bond distances. The C—C bond distance within a layer is 142 pm (compared to 139 pm in benzene); between layers it is 335 pm.

Graphite has some distinctive properties because of its unique crystal structure. Because bonding between layers is weak, the layers can be forced to glide over one another rather easily. As a result graphite is a good lubricant, either in dry form or

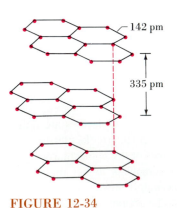

142 pm

335 pm

FIGURE 12-34
The graphite structure.

suspended in oil.* If we apply a mild pressure to a piece of graphite, layers of the graphite flake off, and this is exactly what we do when we use a graphite pencil. Because the p electrons are *delocalized* they will migrate through the planes of carbon atoms when an electric field is applied; graphite conducts electricity. An important use of graphite is as electrodes in batteries and in industrial electrolysis processes. Diamond is not an electrical conductor because all of its valence electrons are localized or permanently fixed into single covalent bonds.

12-9   The Ionic Bond as an Intermolecular Force

To describe the properties of an ionic solid often comes down to this question: How difficult is it to break up an ionic crystal and separate the ions it contains? We learned in Section 10-3 that the quantity that addresses this question is the *lattice energy* of the crystal. In Section 10-3 we learned how to calculate lattice energies by using thermochemical data and certain atomic properties. In a number of instances, though, we do not have to know exact values of lattice energies. We just need to be able to make qualitative comparisons of lattice energies for different ionic compounds. Although the lattice energy depends to some extent on the type of crystal structure, the following generalization is usually sufficient.

Predicting the strengths of interionic attractions.

The attractive force between a pair of oppositely charged ions increases with increased charge on the ions and decreased ionic sizes. (12.10)

The idea expressed through expression (12.10) is illustrated in Figure 12-35.

For most ionic compounds lattice energies are sufficiently high that ions do not readily detach themselves from the crystal and pass into the gaseous state. Ionic solids do not sublime at ordinary temperatures. All ionic solids can be melted by supplying enough thermal energy to destroy the crystalline lattice, and, in general, the higher the lattice energy the higher the melting point.

Two conditions are necessary for electrical conductivity: (1) charged particles must be present, and (2) the particles must be able to migrate in an electric field. In solid ionic compounds only the first requirement is met; these solids do not conduct electricity. However, whenever ions enter the fluid state, either by melting an ionic solid or by dissolving it in a suitable solvent, the liquid becomes a good electrical conductor. For the process of dissolving, the energy required to break up an ionic crystal comes as a result of the interaction of ions in the crystal with molecules of the solvent (for example, water molecules). The extent to which an ionic solid dissolves in a solvent is again determined, at least in part, by the lattice energy of

FIGURE 12-35
Interionic forces of attraction.

Because of the higher charges on the ions and the closer proximity of their centers, the interionic attractive force between Mg^{2+} and O^{2-} is about seven times as great as between Na^+ and Cl^-.

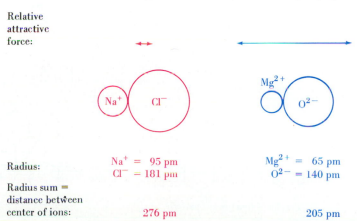

Relative attractive force:

Radius:	$Na^+ = 95$ pm $Cl^- = 181$ pm	$Mg^{2+} = 65$ pm $O^{2-} = 140$ pm
Radius sum = distance between center of ions:	276 pm	205 pm

*Apparently additional factors are involved in the lubricant properties of graphite, since these properties are greatly diminished when graphite is heated in vacuum to expel gases.

the ionic solid. The lower the lattice energy the greater the quantity of an ionic solid that can be dissolved in a given quantity of solvent (see Section 13-2).

Example 12-10

Predicting physical properties of ionic compounds. Of KI or CaO, which has the higher melting point, and which has the greater solubility in water?

Solution. Ca^{2+} and O^{2-} are more highly charged than K^+ and I^-. Also, Ca^{2+} is smaller than K^+ and O^{2-} is smaller than I^-. We certainly expect the lattice energy of CaO to be much larger than that of KI. As a result, CaO should have the higher melting point, but KI should have the greater solubility in water.

    [The observed melting points are 677 °C for KI and 2590 °C for CaO. The solubility of KI in water is several moles per liter. CaO reacts with water to form $Ca(OH)_2$, and $Ca(OH)_2$ is only slightly soluble in water.]

SIMILAR EXAMPLES Exercises 12, 63, 64.

12-10   Crystal Structures

Crystals—whether as ice, rock salt, quartz, or gemstones—have aroused our interest from earliest times. Yet, only in relatively recent times have we come to a fundamental understanding of the crystalline state. This understanding started with the invention of the optical microscope and was greatly expanded in the present century following the discovery of x rays. The key idea, now supported by countless experiments, is that the regularity we observe in crystals at the macroscopic level is due to an underlying regular pattern in the arrangement of atoms, ions, or molecules.

Crystal Lattices. You can probably think of a number of situations in which you have had to deal with repeating patterns in one or two dimensions. These might include sewing a decorative border on a piece of material, wallpapering a room, or creating a design with floor tiles. For example, the design shown in Figure 12-36a might be used as a fringe or border on a piece of material. No matter how long we want this border to be, we can generate it simply by displacing the repeating unit in *one* dimension, that is, along a line to the left and to the right.

FIGURE 12-36
One- and two-dimensional patterns.

(a) One-dimensional pattern
The repeating unit is in red.

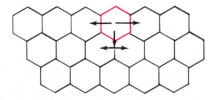

(b) Two-dimensional pattern

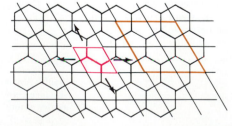

(c) Two-dimensional pattern against a grid of parallel lines
The repeating unit is in red. The hexagonal pattern becomes clear in a group of four parallelograms (brown).

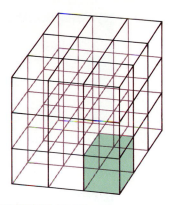

FIGURE 12-37
The cubic space lattice.

A typical parallelepiped formed by the intersection of mutually perpendicular parallel planes is shaded in green. It is a cube.

To choose a unit cell larger than the primitive cell is like choosing a group of several parallelograms in Figure 12-36c. Notice how the design of the hexagon is more clearly seen in a group of four parallelograms than it is in a single parallelogram.

FIGURE 12-38
Unit cells in the cubic crystal system.

In the top row only the centers of spheres (atoms) are shown at their respective positions in the unit cells. In the space-filling models in the bottom row, contacts between spheres (atoms) are shown. In the simple cubic cell, spheres come into contact along each edge. In the body-centered cubic cell, contact of the spheres is along the cube diagonal, and in the face-centered cubic cell, along the diagonal of a face. [Carey B. Van Loon]

In the two-dimensional array of Figure 12-36b—a design sometimes used in ceramic tiling of floors—we are accustomed to think of the hexagon as the repeating unit. But the hexagon fails an important test for a repeating unit used by scientists who study crystal structures (crystallographers). *The repeating unit must generate neighboring units and the entire array by single, straight-line displacements.* The hexagon outlined in red generates its neighbors in the same row by single displacements to the left and to the right. To generate its neighbors in the next row, however, requires *two* displacements, *down* and then to the *left* or to the *right*. The only two-dimensional figures that meet the crystallographer's test for a repeating unit are those having parallel sides—squares, rectangles, or general parallelograms. The design of Figure 12-36b is redrawn in Figure 12-36c against a grid of parallel lines, and a repeating unit is identified. Can you see that the repeating unit (red) will generate the entire array through single displacements in the directions of the arrows? Let us call the network of parallel lines in Figure 12-36c a two-dimensional *lattice,* and the points of intersection of the parallel lines, *lattice points.*

To describe the structures of crystals we have to consider *three* dimensions. We can form a basic lattice by constructing three sets of parallel *planes.* We have chosen a special case for Figure 12-37: The planes are equidistant and mutually perpendicular (intersect at 90° angles). This is called the **cubic lattice.** It can be used to describe some, but by no means all, crystals. For some crystals the appropriate lattice may involve planes that are not equidistant or that intersect at angles other than 90°. In all there are seven possibilities for crystal lattices, leading to seven crystallographic systems. We will emphasize only the cubic system.

Lattice planes intersect to produce three-dimensional figures having six faces arranged in three sets of parallel planes. These figures are called **parallelepipeds.** In Figure 12-37 these parallelepipeds are cubes. A parallelepiped that can be used to generate the entire lattice by simple translation is called a **unit cell.** Where possible it is convenient to arrange the three-dimensional space lattice so that structural units of the crystal (atoms, ions, or molecules) are situated at lattice points. If a unit cell has structural particles only at its corners, we call it a *primitive* unit cell. This is the simplest unit cell that we can consider. But sometimes we choose a unit cell that has more structural particles. In the **body-centered cubic (bcc)** structure, a structural particle of the crystal is found at the center of the cube as well as at each corner. In a **face-centered cubic (fcc)** structure there is a structural particle at the center of each face as well as at each corner. These unit cells are shown in Figure 12-38.

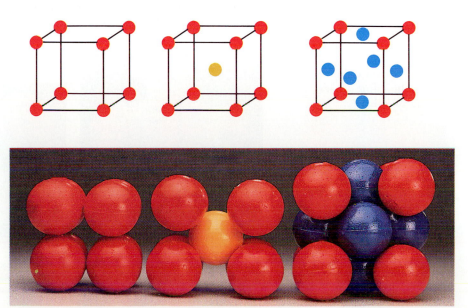

simple cubic   body-centered cubic   face-centered cubic

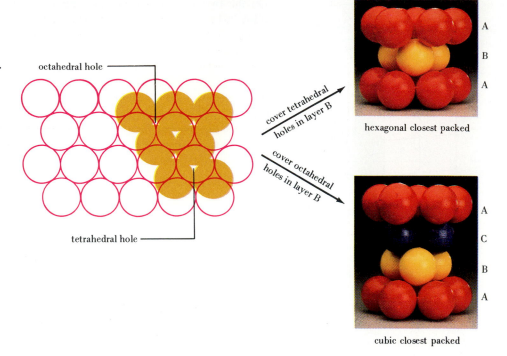

octahedral hole

tetrahedral hole

cover tetrahedral holes in layer B

cover octahedral holes in layer B

A
B
A

hexagonal closest packed

A
C
B
A

cubic closest packed

Closest Packed Structures. We can gain some insight into the structures of crystals by considering the ways in which identical spheres—marbles, cannonballs, atoms—can be stacked. Unlike the case of cubes, which can be stacked to fill all space, when spheres are stacked together there must always be some unfilled space. There are some arrangements of spheres, however, in which the spheres come into as close contact as possible and the holes or voids are kept to a minimum. These are known as closest packed structures. Two such structures are shown in Figure 12-39.

To analyze the structures in Figure 12-39, let us imagine one layer of spheres, layer A (red), in which each sphere is in contact with six others arranged in a hexagonal fashion around it. Among the spheres we see open spaces or **voids.** Once the first sphere is placed in the next layer, layer B (yellow), the entire pattern for that layer is fixed. Again, there are holes or voids in layer B, but the holes are of two different types. **Tetrahedral holes** fall directly over spheres in layer A and have this shape ▽. **Octahedral holes** fall directly over holes in layer A and have this

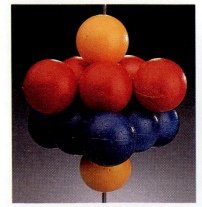

The 14 spheres on the left are extracted from a larger array of spheres in a cubic closest packed structure. The two middle layers each have six atoms; the top and bottom layers, one. Rotation of the 14 spheres reveals the fcc unit cell. [Carey B. Van Loon]

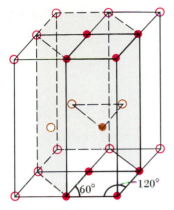

FIGURE 12-41
The hexagonal closest packed (hcp) crystal structure.

A unit cell is highlighted in heavy black and the atoms that are part of that cell are solid color. Note that the unit cell is not a cube. Three adjoining unit cells are also shown. The broken-line, shaded region, together with the highlighted unit cell, shows the layering (ABA) described in Figure 12-39.

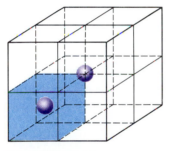

FIGURE 12-42
Apportioning atoms among bcc unit cells.

Eight unit cells are outlined. For clarity, only the centers of two atoms are pictured. Our attention is focused on the blue unit cell. The atom in the center of the cell belongs entirely to that cell. The corner atom is seen to be shared by all eight unit cells.

shape ✿. Now, there are two possibilities for the third layer. In one arrangement, called **hexagonal closest packed (hcp),** all the *tetrahedral* holes are covered; the third layer is identical to layer A; and the structure begins to repeat itself. In the other arrangement, called **cubic closest packed,** all the *octahedral* holes are covered. The spheres in layer C (blue) are out of line with those in layer A. Only when the *fourth* layer is added does the structure begin to repeat itself.

Study Figure 12-40 and you will see that the cubic closest packed structure has a face-centered cubic unit cell. The unit cell of the hexagonal closest packed structure is shown in Figure 12-41. In both the hcp and the fcc structures voids account for only 25.96% of the total volume. Another arrangement, in which the packing of spheres is not quite as close, has a body-centered cubic (bcc) unit cell. In this structure voids account for 31.98% of the total volume. (A method of calculating the percent voids in crystal structures is outlined in Exercise 71.) The best examples of crystal structures based on the closest packing of spheres are found among the metals. Some examples are listed in Table 12-6.

Crystal Coordination Number and Number of Atoms per Unit Cell. In close packed structures of atoms, each atom is in contact with several others. For example, can you see in the bcc unit cell in Figure 12-38 that the center atom is in contact with each corner atom? We call the number of atoms with which a given atom is in contact the **crystal coordination number.** For the bcc structure this is 8. For the fcc and hcp structures the crystal coordination number is 12. The easiest way to see this is from the layering of spheres described in Figure 12-39. Each sphere is in contact with *six* others in the same layer, *three* in the layer above, and *three* in the layer below.

Although we use nine atoms to draw the bcc unit cell, we are wrong if we conclude that the unit cell consists of nine atoms. As shown in Figure 12-42, only the center atom belongs *entirely* to the bcc unit cell. The corner atoms are shared among eight adjoining unit cells. Only one-eighth of each corner atom should be thought of as belonging to a given unit cell. Thus, the eight corner atoms collectively contribute the equivalent of *one* atom to the unit cell. The total number of atoms in a bcc unit cell, then, is *two* [that is, $1 + (8 \times \frac{1}{8})$]. For the hcp unit cell of Figure 12-41 we also get *two* atoms per unit cell if we use the correct counting procedure. The corner atoms account for $\frac{1}{8} \times 8 = 1$ atom and the central atom belongs entirely to the unit cell. In the fcc unit cell the corner atoms account for $\frac{1}{8} \times 8 = 1$ atom, and those in the center of the faces for $\frac{1}{2} \times 6 = 3$ atoms. The fcc unit cell contains *four* atoms.

X-Ray Diffraction. We can acquire knowledge of macroscopic objects by using visible light. To determine how atoms, ions, or molecules are arranged in a crystal, we need light of much shorter wavelength. When a beam of x rays encounters atoms, x rays interact with electrons in the atoms and the original beam is reradiated, diffracted, or scattered in all directions. The pattern of this scattered radiation is related to the distribution of electronic charge in the atoms. Unlike visible light, however, x rays cannot be perceived by our eyes and brains. The scattered x rays

TABLE 12-6
Some Features of Close-Packed Structures in Metals

	Crystal coordination number	No. atoms per unit cell	Examples
hexagonal closest packed (hcp)	12	2	Cd, Mg, Ti, Zn
face-centered cubic (fcc)	12	4	Al, Cu, Pb, Ag
body-centered cubic (bcc)	8	2	Fe, K, Na, W

FIGURE 12-43
Diffraction of x rays by a crystal.

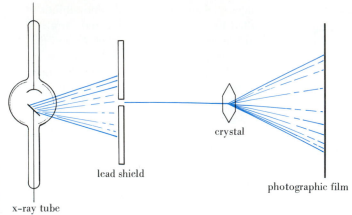

FIGURE 12-44
A portion of a Laue x-ray photograph of a crystal.

The crystal producing this pattern consists of Er^{3+} and $HEDTA^{3-}$ ions. (Ethylenediaminetetraacetic acid, EDTA, and ions derived from it are described in Chapter 25.) [Courtesy of James E. Benson, Ames Laboratory, Iowa State University.]

FIGURE 12-45
Determination of crystal structure by x-ray diffraction.

The two triangles outlined by the dotted lines are identical right triangles. The hypotenuse of each triangle is equal to the interatomic distance, d. The side opposite the angle θ thus has a length of $d \sin \theta$. Wave b travels farther than wave a by the distance $2d \sin \theta$.

must be made to produce a visible pattern, as on a photographic film. Then we have to infer the microscopic structure of the substance responsible for the x-ray scattering from the visible pattern. How successful we are in making inferences depends on the amount of the scattered radiation that we recover. The power of the x-ray diffraction method has been greatly increased by the use of high-speed computers. Computers can process vast amounts of x-ray data (in much the same way that the human eye processes visible light).

Figure 12-43 suggests a method of scattering x rays from a crystal, and Figure 12-44 shows a photographic record of the scattered x rays called a Laue pattern (after Max von Laue, who pioneered in the application of x rays to crystal-structure determination). Some aspects of scattering patterns can be explained by a geometric analysis proposed by W. H. Bragg and W. L. Bragg in 1912, illustrated in Figure 12-45.

Figure 12-45 pictures two rays in a monochromatic (single wavelength) x-ray beam, labeled a and b. Wave a is reflected by one plane of atoms or ions in a crystal and wave b from the next plane below. Wave b travels a greater distance than wave a. The additional distance is $2d \sin \theta$. The intensity of the scattered radiation will be greatest if waves a and b reinforce each other, that is, if their crests and troughs line up. And, to satisfy this requirement, the additional distance traveled by wave b must be an integral multiple of the wavelength of the x rays.

$$n\lambda = 2d \sin \theta \qquad (12.11)$$

So, by measuring the angle θ for which the intensity of the scattered x rays is at

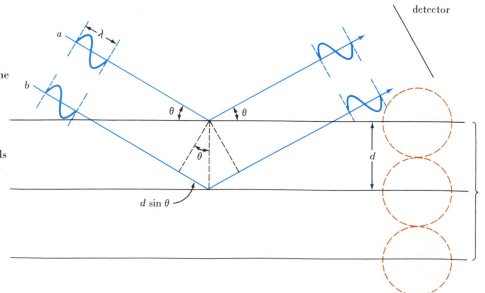

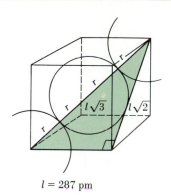

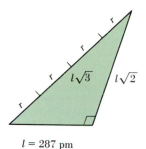

$l = 287$ pm

FIGURE 12-46
Determination of the atomic radius of iron—Example 12-11 illustrated.

The right triangle must conform to the Pythagorean formula: $a^2 + b^2 = c^2$. That is, $(l)^2 + (l\sqrt{2})^2 = (l\sqrt{3})^2$, or $l^2 + 2(l^2) = 3(l^2)$.

a maximum, and by also knowing the x-ray wavelength (λ), we can calculate the spacing (d) between atomic planes. With different orientations of the crystal we can determine atomic spacings and electron densities for different directions through the crystal, in short, the crystal structure.

Once we know the crystal structure we can determine certain other properties by calculation. In Example 12-11 we calculate a metallic radius, and in Example 12-12 we estimate the Avogadro constant. For both of these calculations we need to sketch, or in some way visualize, a unit cell of the crystal. In particular, we need to see which atoms are in direct contact.

Example 12-11

Using x-ray data to determine an atomic radius. At room temperature iron crystallizes in a bcc structure. By x-ray diffraction, the edge of the cubic cell corresponding to Figure 12-46 is found to be 287 pm. What is the radius of an iron atom?

Solution. Nine atoms are associated with a *bcc* unit cell. One atom is located at each of the eight corners of the cube and one at the center of the cube. The three atoms along a cube diagonal are in contact. The length of the cube diagonal (the distance from the farthest upper-right corner to the nearest lower-left corner) is four times the atomic radius. But also shown in Figure 12-46 is that the diagonal of a cube is equal to $\sqrt{3}$ times the length of an edge. The length l is what is given.

$$4r = l\sqrt{3} \qquad r = \frac{\sqrt{3} \times 287 \text{ pm}}{4} = \frac{1.73 \times 287 \text{ pm}}{4} = 124 \text{ pm } (1.24 \text{ Å})$$

SIMILAR EXAMPLE: Exercise 14.

Example 12-12

Determining the Avogadro constant. Use the fact that the density of iron is 7.86 g/cm^3 and that its molar mass is 55.85 g Fe/mol Fe, together with data from Example 12-11, to estimate the value of the Avogadro constant, N_A.

Solution. In Example 12-11, we are given the length of a unit cell of iron: $l = 287$ pm $= 287 \times 10^{-12}$ m $= 2.87 \times 10^{-8}$ cm. The volume of the unit cell is $V = l^3 = (2.87 \times 10^{-8})^3$ cm$^3 = 2.36 \times 10^{-23}$ cm^3. The mass of the unit cell is

$$m = V \times d = 2.36 \times 10^{-23} \text{ cm}^3 \times \frac{7.86 \text{ g}}{\text{cm}^3} = 1.85 \times 10^{-22} \text{ g}$$

If we divide the mass of a unit cell by the number of atoms that the cell contains, we obtain the mass of a single Fe atom. We have previously established that the number of atoms per unit cell for the *bcc* structure is *two* (see Table 12-6).

$$\text{mass of Fe atom} = \frac{1.85 \times 10^{-22} \text{ g}}{\text{unit cell}} \times \frac{1 \text{ unit cell}}{2 \text{ Fe atoms}}$$

$$= 9.25 \times 10^{-23} \text{ g Fe/Fe atom}$$

The molar mass of Fe, 55.85 g/mol, is the mass of Avogadro's number of atoms. The molar mass divided by the mass per atom is the Avogadro constant.

$$N_A = \frac{55.85 \text{ g Fe}}{1 \text{ mol Fe}} \times \frac{1 \text{ Fe atom}}{9.25 \times 10^{-23} \text{ g Fe}} = 6.04 \times 10^{23} \text{ Fe atoms/mol Fe}$$

SIMILAR EXAMPLES: Exercises 14, 69, 70, 72.

12-11   Ionic Crystal Structures

If we try to apply the packing-of-spheres model to an ionic crystal we run into two complications: (1) some of the ions are positively charged and some, negatively charged, and (2) the cations and anions are of different sizes. What we can expect, however, is that oppositely charged ions will come into close proximity; generally we think of them as being in contact. Like-charged ions, because of mutual repulsions, are not in direct contact. We can think of some ionic crystals in this way: There is a fairly closely packed arrangement of ions of one type with holes or voids filled by ions of the opposite charge. The relative sizes of cations and anions are important in establishing a particular packing arrangement. In defining a unit cell of an ionic crystal we must choose a unit cell that

Requirements for the unit cell of an ionic crystal.

- by simple translation in three dimensions generates the entire crystal;
- indicates the crystal coordination numbers of the ions;   (12.12)
- is consistent with the formula of the compound.

Unit cells of crystalline NaCl and CsCl are pictured in Figures 12-47 and 12-48.

To establish the crystal coordination number in an ionic crystal, count the number of nearest-neighboring ions of opposite charge to any given ion in the crystal. Note the Na^+ ion in the center of the unit cell of Figure 12-47. It is surrounded by *six* Cl^- ions. The crystal coordination numbers of both Na^+ and Cl^- are six. By contrast, the crystal coordination numbers of Cs^+ and Cl^- in Figure 12-48 are *eight*.

We must apportion the 27 ions in Figure 12-47 among the unit cell pictured and its neighboring unit cells in the following way: Each Cl^- ion in a corner position is shared by *eight* unit cells, and each Cl^- in the center of a face is shared by *two* unit cells. This leads to a total number of Cl^- ions in the unit cell of $(8 \times \frac{1}{8}) + (6 \times \frac{1}{2}) = 1 + 3 = 4$. There are 12 Na^+ ions along the edges of the unit cell, and each edge is shared by *four* unit cells. The Na^+ ion in the very center of the unit cell belongs entirely to that cell. Thus, the total number of Na^+ ions in a unit cell is $(12 \times \frac{1}{4}) + (1 \times 1) = 3 + 1 = 4$. The unit cell has the equivalent of 4 Na^+ and 4 Cl^- ions. The ratio of Na^+ to Cl^- is $4:4 = 1:1$, corresponding to the formula NaCl.

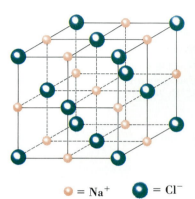

$\bigcirc$ = Na⁺ $\bullet$ = Cl⁻

FIGURE 12-47
The unit cell of sodium chloride.

For clarity, only the centers of the ions are shown. Oppositely charged ions are actually in contact.

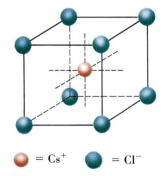

$\bullet$ = Cs⁺ $\bullet$ = Cl⁻

FIGURE 12-48
The cesium chloride unit cell.

The Cs^+ ion is in the center of a cube with Cl^- ions at the corners. In reality each Cl^- is in contact with the Cs^+ ion. An alternative unit cell has Cl^- at the center and Cs^+ at the corners.

Example 12-13

Relating ionic radii and the dimensions of a unit cell of an ionic crystal. The ionic radii of Na^+ and Cl^- are 95 and 181 pm, respectively. What is the length of the unit cell of NaCl?

Solution. Again the key to solving this problem lies in understanding geometrical relationships in the unit cell. Along each edge of the unit cell (Figure 12-47) two Cl^- ions are in contact with one Na^+. The edge length is equal to the radius of one Cl^-, plus the diameter of Na^+, plus the radius of another Cl^-, that is,

$$\text{edge length} = (r_{Cl^-}) + (r_{Na^+}) + (r_{Na^+}) + (r_{Cl^-})$$

$$= 2(r_{Na^+}) + 2(r_{Cl^-})$$

$$= [(2 \times 95) + (2 \times 181)] \text{ pm}$$

$$= 552 \text{ pm } (5.52 \text{ Å})$$

SIMILAR EXAMPLES: Exercises 75, 102.

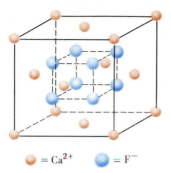

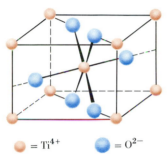

$= Ca^{2+}$ $= F^-$

Unit cell of CaF_2
the fluorite structure

$= Ti^{4+}$ $= O^{2-}$

Unit cell of TiO_2
the rutile structure

FIGURE 12-49
Some unit cells of greater complexity.

Example 12-14

Calculating the density of an ionic crystal from data about the unit cell. Use data about the unit cell of NaCl to estimate the density of NaCl.

Solution. The relevant data are the length of the unit cell that we established in Example 12-13 and the fact that the unit cell contains 4 formula units of NaCl (that is, 4 Na^+ ions and 4 Cl^- ions). From the length of the unit cell we can get the volume of the unit cell, in cm^3.

$$no.\ cm^3 = \left(552\ pm \times \frac{1\ m}{1 \times 10^{12}\ pm} \times \frac{100\ cm}{1\ m}\right)^3 = 1.68 \times 10^{-22}\ cm^3$$

Next, we can determine the mass of a unit cell; it is the mass of 4 formula units of NaCl.

$$no.\ g = 4\ f.u. \times \frac{1\ mol\ NaCl}{6.022 \times 10^{23}\ f.u.} \times \frac{58.44\ g\ NaCl}{1\ mol\ NaCl}$$

$$= 3.88 \times 10^{-22}\ g\ NaCl$$

Finally, density is the mass divided by the volume of the unit cell.

$$density = \frac{3.88 \times 10^{-22}\ g\ NaCl}{1.68 \times 10^{-22}\ cm^3}$$

$$= 2.31\ g\ NaCl/cm^3$$

SIMILAR EXAMPLES: Exercises 74, 75, 102, 103.

Ionic compounds of the type $M^{2+}X^{2-}$ (e.g., MgO, BaS, CaO) may form crystals of the NaCl type. For substances with the formulas MX_2 or M_2X, the crystal structures are more complex. Because the cations and anions occur in unequal numbers, the crystals have *two* coordination numbers, one for the cation and another for the anion. Two typical structures of a more complex type are shown in Figure 12-49.

In CaF_2 (the fluorite structure) there are twice as many fluoride ions as calcium ions. The crystal coordination number of Ca^{2+} is *eight*, that of F^- is *four*. In TiO_2 (the rutile structure) Ti^{4+} has a crystal coordination number of *six* and O^{2-}, *three*. In this structure two of the O^{2-} ions are within the interior of the cell, two are in the top face, and two in the bottom face of the cell. Ti^{4+} ions are at the corners and center of the cell.

12-12 Types of Solids: A Summary

We have learned that the structural particles of a crystalline solid can be atoms, ions, or molecules. In the preceding two sections we have emphasized crystal structures of metal atoms and of ions. To picture a crystal structure having molecules as its structural units, consider solid methane, CH_4. The crystal structure is *fcc*, which means that the unit cell has a CH_4 molecule at each corner and at the center of each face. The CH_4 molecules are in constant rotation, and the volume occupied by the rotating molecule approximates that of a sphere with a radius of 228 pm.

The intermolecular forces operating among the structural units of a crystal may be van der Waals forces, covalent bonds, interionic attractions, or metallic bonds. Table 12-7 lists the basic types of solid structures, the intermolecular forces within them, some of their characteristic properties, and typical examples of each.

TABLE 12-7
Types of Solids

Type of solid	Structural particles	Intermolecular forces	Typical properties	Examples
molecular	molecules (atoms of noble gases)	London and/or dipole–dipole and/or hydrogen bonds	soft; low melting points; nonconductors of heat and electricity; sublime easily in many cases	noble gas elements; CH_4; CO_2; P_4; S_8; I_2; H_2O
network covalent	atoms	covalent bonds	very hard; very high melting points; nonconductors of electricity	C(diamond); SiC; SiO_2
ionic	cations and anions	electrostatic attractions	hard; moderate to very high melting points; nonconductors of electricity (but good electrical conductors in the molten state)	NaCl; $NaNO_3$; MgO
metallic	cations plus delocalized electrons	metallic bonds	hardness varies from soft to very hard; melting point varies from low to very high; lustrous; ductile; malleable; very good conductors of heat and electricity	Na; Mg; Al; Fe; Zn; Cu; Ag; W

FOCUS ON
Liquid Crystals

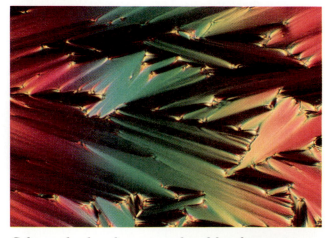

Color and color changes produced by changes in temperature or other variables are characteristic of liquid crystals. [Courtesy Dr. Mary Neubert, Liquid Crystal Institute, Kent State University]

In 1888 the Austrian botanist Reinitzer noticed that the substance cholesteryl benzoate melted sharply at 145.5 °C to produce a milky fluid. In turn, this fluid underwent a sharp transition to a clear liquid at 178.5 °C. Between 145.5 and 178.5 °C this substance is in a phase that has the fluid properties of a liquid, the optical properties of a crystalline solid, and some unique properties of its own. The term **liquid crystal** is now commonly used to describe

this state of matter that lies between that of a liquid and a solid. Figure 12-50 suggests how we can represent liquid crystals in a phase diagram.

Liquid crystals are observed most commonly in organic compounds comprised of cylindrically shaped (rodlike) molecules having molecular weights of 200 to 500 and having lengths that are four to eight times their diameters. (Potentially this represents about 0.5% of all organic compounds.) Figure 12-51 shows that such molecules, when they are in the liquid state, have the expected random or-

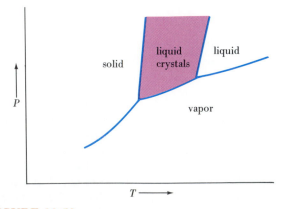

FIGURE 12-50
Representation of liquid crystals in a phase diagram.

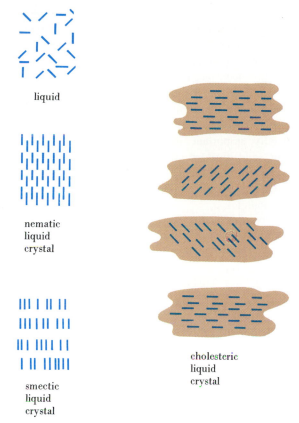

liquid

nematic
liquid
crystal

smectic
liquid
crystal

cholesteric
liquid
crystal

FIGURE 12-51
The liquid crystalline state.

rotation about the long axis. The **cholesteric** form is related to the nematic form, but molecules are stratified into layers, and the direction in which the parallel molecules are oriented shifts from one layer to the next.

As suggested by Figure 12-51, each layer in a cholesteric structure has a different orientation of molecules from the layers immediately above and below it. Over a span of several layers, however, each particular orientation is repeated. The distance between planes having molecules in the same orientation is a distinctive characteristic of a cholesteric liquid crystal. When a beam of light strikes a film of cholesteric liquid crystal, the properties of the reflected light depend on this characteristic distance. Since this distance is very temperature sensitive, the reflected light changes (e.g., in color) with changing temperature. This phenomenon is the basis of liquid-crystal temperature-sensing devices. Temperature changes as small as 0.01 °C can be detected with such devices with ordinary white light.

The orientation of molecules in a thin film of nematic liquid crystals is easily altered by pressure and by an electric field. The altered orientation affects the optical properties of the film, such as causing the film to become opaque. Suppose we arrange electrodes in certain patterns (say in the shape of numbers). When an electric field is imposed through these electrodes onto a thin film of liquid crystals, the patterns of the electrodes become visible. This is the principle used in liquid-crystal displays (LCD) in hand-held calculators and in digital watches.

Liquid crystals occur widely in living matter. Cell membranes and certain tissues have structures that can be described as liquid crystalline. Hardening of the arteries is caused by the deposition of liquid crystalline compounds of cholesterol. Liquid crystalline properties have also been identified in various polymers (such as Du Pont Kevlar fiber). As the range of materials found to have liquid crystalline properties is extended, so too will the range of possible applications.

dering. Liquid molecules are able to translate (move in three directions) and to rotate freely. In the liquid crystalline state molecules have both some mobility and some ordering among them.

In the form known as **nematic** (meaning thread-like) the rodlike molecules are arranged in a parallel fashion. They are free to move in three directions but they can rotate only on their long axes. (Imagine the possible ways in which you might move a particular pencil in a box of loosely packed pencils.) In the **smectic** form (meaning grease-like) the rodlike molecules are arranged in layers, with the long axes of the molecules perpendicular to the planes of the layers. The molecular motions possible here are translation in two directions (that is, within the layers) and

Summary

Molecules at the surface of a liquid are not subject to intermolecular forces of attraction to the same extent as are molecules in the interior. This results in an imbalance of forces at the surface of liquids and gives rise to surface tension and a variety of other phenomena related to it (drop formation, meniscus formation, capillary action).

Another result of the reduced effect of intermolecular forces of attraction at the surface of a liquid is the tendency of energetic molecules to leave the surface and pass into the vapor state.

When vaporization and condensation occur at the same rate in a closed container, dynamic equilibrium is established. The pressure exerted by the vapor in this liquid–vapor equilibrium is called the vapor pressure of the liquid. The magnitude of the vapor pressure depends very much on the particular liquid, but beyond this, vapor pressure is primarily a function only of temperature. Vapor pressure increases sharply with temperature, as shown through a pressure–temperature plot called the *vapor pressure curve*. A linear plot is obtained by plotting log P or ln P against $1/T$, and an equation of this straight line (the Clausius–Clapeyron equation) is useful for computational purposes. A point of special interest on the vapor pressure curve is the temperature at which the vapor pressure equals barometric pressure—the boiling point. As barometric pressure varies, so does the boiling point. Another point of interest on the vapor pressure curve is its high-temperature terminus—the critical point. At the critical point the liquid and gaseous states become indistinguishable. Other views of the critical point are that it is the highest temperature at which the liquid state of a substance can exist, or it is the highest temperature at which a gas can be condensed to a liquid by the application of pressure alone. (Above the critical point a gas must both be cooled and compressed to condense it to a liquid.)

The states or phases of matter that exist for various conditions of temperature and pressure can be represented through a *phase diagram*. The phase diagram represents liquid–vapor, solid–vapor (sublimation), and solid–liquid (fusion) equilibria through characteristic curves. The single point at which the three curves intersect is called the triple point of the substance. For substances that can exist in more than one solid form, the phase diagram has additional three-phase equilibrium points. Other points that can be located on a phase diagram are the critical point and melting and boiling points. Phase diagrams can be used to describe the changes that occur when a sample of matter is heated, cooled, or subjected to changes in pressure. A useful idea that assists in the interpretation of phase diagrams is Le Châtelier's principle.

The most common intermolecular forces of attraction are based on electrostatic attractions between instantane-ous and induced dipoles. The strengths of these forces generally increase with increased molecular weight. For polar substances dipole–dipole attractions may also be significant. A hydrogen atom that is covalently bonded to one highly electronegative nonmetal atom may be simultaneously attracted to another highly electronegative atom in the same or a neighboring molecule. This type of intermolecular force of attraction, called *hydrogen bonding*, is especially important in hydrogen-containing fluorine, oxygen, and nitrogen compounds. Hydrogen bonding is crucial to an understanding of several properties of solid and liquid water and also of many molecules found in living matter.

In network covalent solids, bonds extend throughout a crystalline structure. Network covalent solids have high melting points compared to other covalent substances. If all the electrons in a network covalent bond are localized into bonds (e.g., diamond), the solid is an electrical insulator. However, if some of the electrons are *delocalized* (as in graphite), the solid is an electrical conductor.

The strengths of intermolecular forces in ionic solids can be expressed through lattice energies, though for making qualitative predictions simply knowing ionic sizes and charges may be sufficient.

To describe the crystal structure of a solid, for example, a metal, it is useful to think in terms of a closely packed structure of spheres. The two closest packed arrangements are the hexagonal closest packed and the cubic closest packed. A third arrangement, the body-centered cubic structure, is not quite as closely packed. Interactions of x rays with crystals can provide the type of data necessary to determine the crystal structures of solids.

The crystal structures of ionic solids can also be described through a packing-of-spheres model, but the matter is complicated by the fact that the ions are not all of the same size or charge. An important concept in describing crystals of all types is the unit cell. If the unit cell of a crystal is known, a variety of calculations involving atomic radii, densities, and other properties is possible.

Summarizing Example

In Chapter 4 (Focus feature) we described the commercial production of hydrazine and some of its important uses. In Chapter 10 (page 315) we represented covalent bonding in the N_2H_4 molecule through a Lewis structure. Now let us consider some aspects of the physical behavior of hydrazine in relation to concepts introduced in this chapter. Several properties of hydrazine are listed in Table 12-8.

1. Assess the importance of hydrogen bonding as an intermolecular force of attraction in hydrazine.

Solution. First, let us see if the basic requirements for hydrogen bonding are met: hydrogen atoms bonded to small highly electronegative atoms and capable

TABLE 12-8
Physical Properties of Hydrazine, N_2H_4

Property	Value
Molecular weight	32.05
Freezing point	2.0 °C
Boiling point	113.5 °C
Critical temperature	380 °C
Critical pressure	145.4 atm
Heat of fusion	12.66 kJ/mol
Heat capacity of liquid	98.87 J mol^{-1} $°C^{-1}$
Density of liquid at 25 °C	1.0045 g/mL
Vapor pressure at 25 °C	14.4 mmHg

The upfiring thrusters of the Space Shuttle Columbia, shown here in an in-flight test, use a derivative of hydrazine (methylhydrazine, CH_3NHNH_2) as a fuel. [NASA]

of being simultaneously attracted to similar atoms in other molecules. Hydrazine fits this requirement well. The H atoms are bonded to N atoms, and hydrogen bonding can occur as represented through these two Lewis structures:

$$H-\underset{\underset{\cdot\cdot}{|}}{\overset{\overset{H}{|}}{N}}-\underset{\underset{\cdot\cdot}{|}}{\overset{\overset{H}{|}}{N}}-H---:\underset{\underset{|}{\underset{H}{}}}{\overset{\overset{H}{|}}{N}}-\underset{\underset{|}{\underset{H}{}}}{\overset{\overset{H}{|}}{N}}:$$

Not only are the basic requirements met, but the melting point (2.0 °C) and boiling point (113.5 °C) are *much* higher than we would expect for a substance with a molecular weight of only 32. Most substances with a molecular weight in this range, or even higher, are gases, such as CO, CO_2, NO, NO_2, H_2S, SO_2, and SO_3. Hydrogen bonding is an important intermolecular force of attraction in hydrazine.

(This example is similar to Example 12-9.)

2. Estimate the heat of vaporization of hydrazine.

Solution. The only *quantitative* relationship that we have seen in this chapter that includes a heat of vaporization is the Clausius–Clapeyron equation (12.6). If we have two vapor pressure values and their corresponding temperatures, we can solve the equation for ΔH_{vap}. Whenever we are given a normal boiling point we automatically have one value—$P = 760$ mmHg at the boiling point temperature (386.7 K). A second value is given at 25 °C—14.4 mmHg.

$$\ln \frac{P_2}{P_1} = \frac{\Delta H_{vap}}{R}\left(\frac{1}{T_1} - \frac{1}{T_2}\right) \quad \text{and} \quad \ln \frac{P_2}{P_1} = \frac{\Delta H_{vap}}{R} \times \left(\frac{T_2 - T_1}{T_2 T_1}\right)$$

$$\Delta H_{vap} = \frac{RT_2T_1}{T_2 - T_1} \times \ln(P_2/P_1)$$

$$\Delta H_{vap} = \frac{8.314 \text{ J mol}^{-1}\text{ K}^{-1} \times 386.7 \text{ K} \times 298.2 \text{ K}}{(386.7 - 298.2) \text{ K}} \times \ln(760/14.4)$$

$$\Delta H_{vap} = 1.08 \times 10^4 \text{ J mol}^{-1} \times 3.97 = 4.29 \times 10^4 \text{ J/mol} = 42.9 \text{ kJ/mol}$$

(This example is similar to Example 12-4.)

3. What is the heat of sublimation of hydrazine at its melting point?

Solution. From equation (12.7) we see that ΔH_{subl} is simply the sum of ΔH_{fusion} (from Table 12-8) and ΔH_{vap} (from part **2**).

$$\Delta H_{subl} = \Delta H_{vap} + \Delta H_{fusion} = 42.9 \text{ kJ/mol} + 12.66 \text{ kJ/mol} = 55.6 \text{ kJ/mol}$$

4. A sample of hydrazine is maintained in a closed container in an ice–water bath. What is the pressure of hydrazine vapor in this container?

Solution. An ice–water bath is at 0 °C. At this temperature hydrazine exists as a *solid* in equilibrium with its vapor. We need to calculate the *sublimation* pressure of hydrazine at 0 °C. Proceed in two steps: First use the Clausius–Clapeyron equation to calculate the vapor pressure of liquid hydrazine at its freezing point (essentially the triple point) [**Answer: 3.4 mmHg**]. At this point the solid and the liquid have the same vapor pressure, so this point is also on the sublimation curve. To calculate the sublimation pressure at 0 °C, use the Clausius–Clapeyron equation, but with ΔH_{subl} substituting for ΔH_{vap}. The necessary data are $\Delta H_{subl} = 55.6$ kJ/mol; $P_2 = 3.4$ mmHg (the vapor pressure of liquid [and solid] hydrazine at 2.0 °C); $T_2 = 2.0$ °C $= 275.2$ K; $P_1 = ?$ (the sublimation pressure at 0 °C); $T_1 = 273.2$ K. (See Exercise 39 for the completion of this example.)

5. Sketch a phase diagram for hydrazine, including relevant data from Table 12-8 and elsewhere in this Summarizing Example.

Solution. We should expect the diagram to be similar to that for iodine (Figure 12-19). The points that we can represent in the diagram are the triple point (essentially the melting point), the boiling point, and the critical point. The diagram is sketched in Figure 12-52.

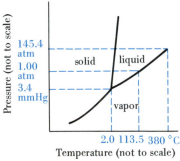

FIGURE 12-52
Phase diagram for hydrazine.

Key Terms

adhesive forces (12-2)
bcc (12-10)
body-centered cubic (12-10)
boiling (12-3)
capillary action (12-2)
cohesive forces (12-2)
condensation (12-3)
critical point (12-3)
crystal coordination number (12-10)
cubic closest packing (12-10)
deposition (12-4)
dispersion (London) forces (12-6)
face-centered cubic (12-10)

fcc (12-10)
freezing (12-4)
hcp (12-10)
hexagonal closest packing (12-10)
hydrogen bond (12-7)
induced dipole (12-6)
instantaneous dipole (12-6)
intermolecular force (12-6)
liquid crystals (Focus feature)
melting (12-4)
network covalent solid (12-8)
normal boiling point (12-3)
normal melting point (12-4)

phase diagram (12-5)
polarizability (12-6)
polymorphism (12-5)
sublimation (12-4)
surface tension (12-2)
triple point (12-5)
unit cell (**12-10**)
van der Waals forces (12-6)
vaporization (12-3)
vapor pressure (12-3)
viscosity (12-2)
x-ray diffraction (12-10)

Highlighted Expressions

Straight-line plot for vapor pressure data (12.3)
Clausius–Clapeyron equation relating vapor pressure, heat of vaporization, and temperature (12.6)
Relationship of enthalpies (heats) of fusion, vaporization, and sublimation (12.7)

Le Châtelier's principle applied to phase equilibria (12.8)
Assessing van der Waals (intermolecular) forces (12.9)
Predicting the strengths of interionic attractions (12.10)
Requirements for the unit cell of an ionic crystal (12.12)

Review Problems

1. At their normal boiling points the values of ΔH_{vap} of acetone, $(CH_3)_2CO$, chloroform, $CHCl_3$, and ethyl alcohol, C_2H_5OH, are 520.9, 247, and 854.8 J/g, respectively.

(a) What mass of acetone can be vaporized by the absorption of 7.15 kJ of heat?
(b) What is ΔH_{vap} of $CHCl_3$, expressed in kJ/mol?

(c) How much heat is *evolved* when 385 g $C_2H_5OH(g)$ condenses to $C_2H_5OH(l)$?

2. From Figure 12-11, estimate **(a)** the vapor pressure of C_6H_6 at 50 °C; **(b)** the normal boiling point of $C_4H_{10}O$.

3. Equilibrium is established between $Br_2(l)$ and $Br_2(g)$ at 25.0 °C. A 250.0-cm^3 sample of the vapor weighs 0.486 g. What is the vapor pressure of bromine at 25.0 °C, in mmHg?

4. Use data in Figure 12-15 to estimate **(a)** the normal boiling point of aniline; **(b)** the vapor pressure of toluene at 75 °C.

5. Cyclohexanol, used in the manufacture of nylon, has a vapor pressure of 10.0 mmHg at 56.0 °C and 100.0 mmHg at 103.7 °C. Calculate its **(a)** ΔH_{vap} and **(b)** normal boiling point.

6. At their normal melting points, ΔH_{fusion} of Mg, Pb, and Cu are 8.954, 4.774, and 13.05 kJ/mol, respectively.

(a) What mass will be melted if a sample of Mg(s) absorbs 1215 J of heat at its melting point?

(b) How much heat must be absorbed to melt a bar of lead that is 8.2 in. × 1.8 in. × 1.2 in.? (Assume a density of 11 g/cm^3 for Pb.)

(c) How much heat, in kJ, is *evolved* when a 3.74-kg sample of molten Cu freezes?

7. 0.180 g $H_2O(l)$ is sealed into an evacuated 2.50-L flask. What is the pressure of the vapor in the flask if the temperature is **(a)** 30.0 °C; **(b)** 50.0 °C; **(c)** 70.0 °C. [*Hint:* Use data from Table 12-2. Does any of the water remain as liquid or does it vaporize completely?]

8. In the following portion of the phase diagram for phosphorus

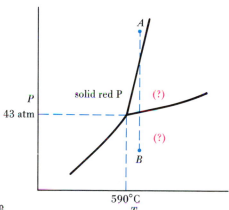

Problem 8

(a) Indicate the phases present in the regions labeled (?) and along each transition curve.

(b) A sample of solid red phosphorus cannot be melted by heating in a container open to the atmosphere. Explain why this is so.

(c) Trace the phase changes that occur when the pressure on a sample of phosphorus is reduced from point A to B, at constant temperature.

9. In each of the following pairs, which would you expect to have the higher boiling point? **(a)** normal C_7H_{16} or normal $C_{10}H_{22}$; **(b)** C_3H_8 or $H_3C-O-CH_3$; **(c)** H_3CCH_2-S-H or H_3CCH_2-O-H. (The term "normal" means that all C atoms are in a straight chain.)

10. One of the substances is out of order in the following list based on *increasing* boiling point. Identify it and put it in its proper place: N_2, O_3, F_2, Ar, Cl_2. Explain your reasoning.

11. Refer to Figure 12-26. A third isomer of pentane, called isopentane, has the structure shown below. Estimate the normal boiling point of liquid isopentane. Explain your basis for doing so.

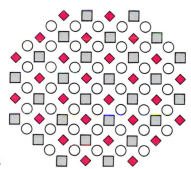

12. Arrange the following in the expected order of *increasing* melting point: CaO, MgF_2, CsI.

13. For the two-dimensional lattice pictured below **(a)** identify a unit cell. **(b)** How many of each of the following elements are in the unit cell: ◆, ■, and ◯? **(c)** Indicate some simpler units than the unit cell, but explain why they cannot function as a unit cell. [*Hint:* Adjacent unit shells must share common edges and corners. There can be no gaps between unit cells. Also, the left and right edges of a unit cell must be identical, as must be the top and bottom.]

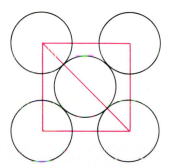

Problem 13

14. The fcc unit cell is a cube with atoms at each of the corners and in the center of each face, as shown in the following diagram. Copper has the fcc crystal structure.

Problem 14

Assuming an atomic (metallic) radius of 128 pm for a Cu atom,

(a) What is the length of the unit cell of Cu?

(b) What is the volume of the unit cell?

(c) How many atoms belong to the unit cell?

(d) What is the mass of a unit cell?

(e) Calculate the density of copper.

15. In the manner illustrated in the text for NaCl, show that the formula of CsCl is consistent with its unit cell pictured in Figure 12-48.

Exercises

Surface tension and related properties

16. Describe briefly the meaning of each term: **(a)** surface tension; **(b)** adhesive force; **(c)** capillary action; **(d)** wetting agent; **(e)** meniscus.

17. Silicone oils are often used in water repellents for treating tents, hiking boots, and similar items. Explain how they function.

****18.** When a wax candle is burned, the fuel consists of gaseous hydrocarbons that appear at the end of the candle wick. Describe the steps by which the solid wax is consumed.

Vaporization

19. When a sample of liquid benzene absorbs 1.00 kJ of heat at its normal boiling point of 80.1 °C, the volume of $C_6H_6(g)$ produced (at 80.1 °C and 1 atm) is 0.94 L. What is ΔH_{vap} per mole of C_6H_6?

20. For the combustion of $CH_4(g)$, $\Delta H = -890.$ kJ/mol CH_4 burned. What volume of the gas, measured at 23.4 °C and 756 mmHg, must be burned to provide for the vaporization of 2.73 L of liquid water at 100. °C? At 100. °C the density of $H_2O(l) = 0.958$ g/cm^3 and $\Delta H_{vap} = +40.6$ kJ/mol.

21. To vaporize 1.000 g of water at 20. °C requires 2447 J of heat. At 100. °C, 10.00 kJ of heat will convert 4.430 g $H_2O(l)$ to $H_2O(g)$. Does ΔH_{vap} of water increase or decrease with temperature? Does this observation conform to what you would expect in terms of intermolecular forces? Explain.

22. If a volatile liquid evaporates from an ordinary open container, the liquid temperature stays the same as that of the surroundings. If the same liquid evaporates from a thermally insulated container (a vacuum bottle or Dewar flask), its temperature falls below that of the surroundings. Explain this difference.

Vapor pressure and boiling point

23. Which of the following factors affect the vapor pressure of a liquid? Explain.
- **(a)** intermolecular forces in the liquid;
- **(b)** volume of liquid in the liquid–vapor equilibrium;
- **(c)** volume of vapor in the liquid–vapor equilibrium;
- **(d)** the size of the container in which the liquid–vapor equilibrium is established;
- **(e)** the temperature of the liquid.

24. Use data from Table 12-2 to estimate **(a)** the boiling point of water in Santa Fe, New Mexico, if the prevailing atmospheric pressure is 600 mmHg; **(b)** the prevailing atmospheric pressure at Lake Arrowhead, California, if the observed boiling point of water is 94 °C.

25. A 25.0-L volume of He(g) at 30.0 °C is passed through 6.220 g of liquid aniline ($C_6H_5NH_2$) at 30.0 °C. The liquid remaining after the experiment weighs 6.108 g. Assume that the He(g) becomes saturated with aniline vapor and that the total gas volume and temperature remain constant. What is the vapor pressure of aniline at 30.0 °C?

26. 7.53 L $N_2(g)$ at 742 mmHg and 45.0 °C is bubbled through $CCl_4(l)$ at 45.0 °C. Assuming the gas becomes saturated with $CCl_4(g)$, what is the volume of the resulting gaseous mixture if the total pressure remains at 742 mmHg and the temperature, 45.0 °C? The vapor pressure of CCl_4 at 45.0 °C is 261 mmHg.

27. A popular demonstration is performed by boiling a small quantity of water in a metal can, capping off the can, and allowing the can to cool. As it cools, the can suddenly collapses. Give an explanation of this crushing of the can.

28. Freon-12, dichlorodifluoromethane, CCl_2F_2, has been widely used in refrigeration and air-conditioning systems. When this highly volatile liquid is allowed to vaporize in a closed system, it cools its surroundings by absorbing heat (the heat of vaporization). The gas is converted back to a liquid by applying sufficient pressure to the vapor (in the compressor), heat (the heat of condensation) is given off, and the substance is then sent through the next evaporation–condensation cycle. The boiling point of Freon-12 is -29.8 °C and its critical point is at 111.5 °C and 39.6 atm. Other vapor pressure data are -12.2 °C, 2.0 atm; 16.1 °C, 5 atm; 42.4 °C, 10 atm; 74.0 °C, 20 atm.
- **(a)** Use the data presented here to plot the vapor pressure curve of Freon-12, in the manner of Figure 12-11.
- **(b)** Approximately what pressure would have to be developed in the compressor if it operates at 25 °C?

29. Explain what is wrong with the following approach used by a student to calculate the vapor pressure of a liquid at 50 °C from a measured vapor pressure of 80 mmHg at 20 °C.

$$\text{v.p. (at 50 °C)} = 80 \text{ mmHg} \times \frac{323 \text{ K}}{293 \text{ K}} = 88 \text{ mmHg}$$

****30.** Estimate the boiling point of water in Leadville, Colorado (elevation: 3170 m). Use the following two-step procedure:
- **(a)** Calculate normal atmospheric pressure in Leadville, if that at sea level is 1.00 atm. Use the barometric formula of Exercise 6-103. (Assume an air temperature of 20 °C.)
- **(b)** Use the Clausius–Clapeyron equation (12.6) to determine the temperature at which the vapor pressure of water has the value determined in (a). (Assume $\Delta H_{vap.} = 41$ kJ/mol.)

The Clausius–Clapeyron equation

31. By the method used to graph Figure 12-15, plot ln P vs. $1/T$ for liquid yellow phosphorus, and estimate **(a)** its normal boiling point and **(b)** its enthalpy of vaporization, ΔH_{vap}, in kJ/mol. Vapor pressure data: 76.6 °C, 1; 128.0 °C, 10; 166.7 °C, 40; 197.3 °C, 100; 251.0 °C, 400 mmHg.

32. The normal boiling point of acetone, an important laboratory and industrial solvent, is 56.2 °C and its ΔH_{vap} is 32.0 kJ/mol. At what temperature does acetone have a vapor pressure of 225 mmHg?

33. The normal boiling point of isooctane (a highly desired component in gasoline because of its excellent octane rating) is 99.2 °C and its ΔH_{vap} is 35.76 kJ/mol. Since isooctane and water have practically identical boiling points, can we assume that they probably have equal vapor pressures at room temperature? If not, which one would you expect to be more volatile at room temperature? Explain. [*Hint:* Use ΔH_{vap} for water from Table 12-1.]

Critical point

34. Which substances listed in Table 12-3 can exist as liquids at room temperature (about 20 °C)? Explain.

35. Can SO_2 be maintained as a liquid under a pressure of 100 atm at 0 °C? Can liquid methane be obtained under the same conditions? (Refer to Table 12-3.)

***36.** A supplier of cylinder gases cautions customers that to determine the quantity of gas remaining in a cylinder the customer should weigh the cylinder and compare this mass to the starting mass of the full cylinder. In particular, the customer is told not to try to establish the mass of gas available in the tank from the measured gas pressure. Explain the basis of this warning. Are there any cylinder gases for which a measurement of the gas pressure could be used as a measure of the remaining available gas?

Fusion/sublimation

37. How much heat is required to melt a cube of ice that measures 35.0 cm on an edge? The density of ice is 0.92 g/cm³ and the heat of fusion is 6.01 kJ/mol.

38. What is the total quantity of heat required to melt a 0.677-kg piece of lead, starting with the sample at 25.0 °C. Pb melts at 327.4 °C; its heat of fusion is 4.774 kJ/mol; its average specific heat from 25.0 to 327.4 °C is 0.134 J g⁻¹ °C⁻¹.

39. Refer to the Summarizing Example. Complete the calculation of the sublimation pressure of hydrazine at 0 °C (part **4**).

***40.** One handbook lists the sublimation pressure of *solid* benzene as a function of *Kelvin* temperature, T, as

$$\log_{10} P \text{ (mmHg)} = 9.846 - 2309/T$$

Another handbook lists the vapor pressure of *liquid* benzene as a function of *Celsius* temperature, t, as

$$\log_{10} P \text{ (mmHg)} = 6.90565 - 1211.033/(220.790 + t)$$

Use these equations to determine the normal melting point of benzene, and compare your result with the listed value of 5.5 °C. Assume that the normal melting point and triple point temperatures are the same.

States of matter and phase diagrams

41. 2.50 g $H_2O(l)$ is sealed in a 5.00-L flask at 120. °C.
   (a) Show that the sample exists completely as vapor.
   (b) Estimate the temperature to which the flask must be cooled before liquid water condenses.

42. A 20.0-L vessel contains 0.100 mol $H_2(g)$ and 0.050 mol $O_2(g)$. The mixture is ignited with a spark and the reaction

$$2 H_2(g) + O_2(g) \rightarrow 2 H_2O$$

goes to completion. The system is then cooled to 27 °C. What is the final pressure in the vessel? [*Hint:* Is the H_2O formed present as a gas, a liquid, or a mixture of the two?]

43. A block of ice that is 8.0 cm × 2.5 cm × 2.7 cm is taken from a freezer at −25.0 °C and added to 400.0 cm³ of $H_2O(l)$ at 32.0 °C. Assuming that the container is perfectly insulated from the surroundings, what will be the final temperature of the contents of the container? What state(s) of matter will be present? The specific heat of ice is 2.01 and that of $H_2O(l)$ is 4.18 J g⁻¹ °C⁻¹. The heat of fusion of ice is 6.01 kJ/mol. Use 0.917 and 0.998 g/cm³ for the densities of ice and $H_2O(l)$, respectively.

44. Refer to Example 12-5. What mass of steam (gaseous water), originally at 100 °C, should be passed into the mixture

described in part (b) to bring the temperature of the water to 25.0 °C? Assume a value of 4.18 J g⁻¹ °C⁻¹ for the specific heat of liquid water and 40.7 kJ/mol for the heat of vaporization of water at 100 °C.

45. Why is the triple point of water (ice I–liquid–vapor) a better fixed point for establishing a thermometric scale than either the melting point of ice or the boiling point of water?

46. Investigations in which quantities of heat are measured are carried out in calorimeters, and often these calorimeters are thermally insulated containers (such as Styrofoam coffee cups). Do you think that cooling curves (such as Figure 12-16) should also be determined in insulated containers? Explain.

***47.** Assume that a skater has a mass of 80 kg and that his skates make contact with 0.50 cm² of ice.
   (a) Calculate the pressure in atm exerted by the skates on the ice. [*Hint:* Review Example 6-1 and Table 6-1.]
   (b) If the melting point of ice decreases by 1.0 °C for every 125 atm of pressure, what would be the melting point of the ice under the skates?

48. The following data are given for mercury(II) iodide.

transition from red (α) to yellow (β) solid: 127 °C

vapor pressure at 127 °C: 0.2 mmHg

melting point of yellow (β) $HgI_2(s)$: 259 °C

vapor pressure at melting point: 92 mmHg

boiling point of liquid HgI_2: 354 °C

Assume that the triple point and the normal melting temperatures are the same. (Also refer to Figure 12-22.) In the following phase diagram
   (a) indicate the phase present in each area marked **?**.
   (b) indicate the two phases in equilibrium along each line marked **?**
   (c) on the temperature and pressure axes enter the values marked **?**
   (d) describe the phase changes that occur as a sample of HgI_2 is heated from the point represented by x to the point y (be as specific as you can about temperatures and pressures at which changes occur);
   ***(e)** indicate which solid form of HgI_2 has the higher density, α or β, and explain your conclusion.

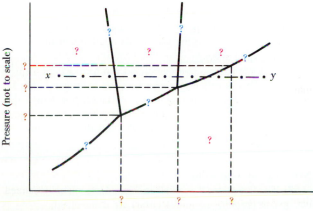

Exercise 48   Temperature (not to scale)

49. Describe what happens to the following samples in a device like that pictured in Figure 12-23. Be as specific as you can about the temperatures and pressures at which changes occur.

(a) A sample of H_2O is heated from -20 to $200\ °C$ at a constant pressure of 600 mmHg.

(b) The pressure on a sample of I_2 is increased from 90 mmHg to 100 atm at a constant temperature of 114.5 °C.

(c) A sample of CO_2 at 35 °C is cooled to $-100\ °C$ at a constant pressure of 50 atm. [*Hint:* Refer also to Table 12-3.]

50. Use the phase diagram in Figure 12-21 (and other information about water) to predict

(a) the state in which water exists at 600 K and 225 atm;

(b) whether the following transition at 3500 atm is endothermic or exothermic: ice V → ice II;

(c) whether the density of ice III at $-22.0\ °C$ and 2045 atm is greater or less than $1.00\ g/cm^3$.

Van der Waals forces

51. For each of the following substances describe the importance of dispersion (London) forces, dipole–dipole interactions, and hydrogen bonds: (a) HCl; (b) Br_2; (c) ICl; (d) HF; (e) CH_4.

52. A handbook lists the following normal boiling points for a series of normal (straight-chain) alkanes: propane, C_3H_8, $-42.1\ °C$; butane, C_4H_{10}, $-0.5\ °C$; pentane, C_5H_{12}, 36.1 °C; hexane, C_6H_{14}, 68.7 °C; heptane, C_7H_{16}, 98.4 °C; octane, C_8H_{18}, 125.6 °C. Estimate the normal boiling point of the normal alkane nonane, C_9H_{20}.

53. When another atom or group of atoms is substituted for one of the H atoms in benzene, C_6H_6, the boiling point changes. Explain the order of the following boiling points: C_6H_6, 80 °C; C_6H_5Cl, 132 °C; C_6H_5Br, 156 °C; C_6H_5OH, 182 °C.

Hydrogen bonding

54. Describe the conditions necessary for the formation of a hydrogen bond and how this bond differs from other intermolecular forces.

55. One of the following substances is a liquid at room temperature, whereas the others are gases. Which do you think is the liquid? Explain. CH_3OH; C_3H_8; N_2; CO.

56. Figure 12-30 shows how one H_2O molecule can be bonded to four others through hydrogen bonds. Based on information from Section 12-7, what would you expect the situation to be with H_2S?

57. If water were a normal liquid, what would you expect to find for its (a) boiling point; (b) freezing point; (c) temperature of maximum density of the liquid; (d) relative densities of the solid and liquid states?

58. In some cases a hydrogen bond can form *within* a single molecule. This is called an *intra*molecular hydrogen bond. Do you think this type of bonding is an important factor in (a) C_2H_6; (b) H_3CCH_2OH; (c) H_3CCOOH; (d) *ortho*-phthalic acid, $C_6H_4(COOH)_2$? (The structure of *ortho*-phthalic acid is given in Exercise 96.)

Network covalent solids

59. The text states that all electrons in diamond are localized, but that in graphite certain electrons are delocalized.

(a) Explain the meaning of *localized* and *delocalized*.

(b) Which electrons in graphite are delocalized?

60. Silicon carbide, SiC, crystallizes in a form similar to diamond, whereas the compound boron nitride, BN, crystallizes in a form similar to graphite.

(a) Sketch the SiC structure as in Figure 12-33.

(b) Propose a bonding scheme for BN.

★(c) BN can be obtained in a diamondlike form under high pressure, but SiC cannot be obtained in a graphitelike form. Why do you suppose that this is so?

61. Based on data presented in the text, would you expect diamond or graphite to have the greater density? Explain.

62. Diamond is often used as a cutting medium in glass cutters. What property of diamond makes this possible? Could graphite function as well?

Ionic properties and bonding

63. The melting points of NaF, NaCl, NaBr, and NaI are 988, 801, 755, and 651 °C, respectively. Are these data consistent with ideas developed in Section 12-9? Explain.

64. Which compound in each of the following pairs would you expect to be the more water soluble? (a) MgF_2 or BaF_2; (b) MgF_2 or $MgCl_2$.

★**65.** Use Coulomb's law to verify the conclusion concerning the relative strengths of the attractive forces in the ion pairs, Na^+Cl^- and $Mg^{2+}O^{2-}$, presented in Figure 12-35.

Crystal structures

66. Define what is meant by (a) closest packing of spheres; (b) tetrahedral holes; (c) octahedral holes.

67. Explain why there are *two* arrangements for the closest packing of spheres rather than a single one.

68. When cubes are stacked together they will fill all space; but as we saw in Section 12-10, the stacking of spheres always leaves open space or voids. Consider the corresponding situation in two dimensions: Squares can be arranged to cover an entire area, but circles cannot. For the arrangement of circles pictured below, what percentage of the area remains uncovered?

Exercise 68

69. Potassium has a body-centered cubic crystal structure. Using a metallic radius of 227.2 pm for the K atom, calculate the density of potassium.

70. In a manner similar to Example 12-12, derive a value of the Avogadro constant based on the facts that the crystal structure of Al is fcc, that the atomic radius of Al is 143.1 pm, and the density of Al is $2.6984\ g/cm^3$. [*Hint:* See also Problem 14.]

★**71.** Use the analyses of a bcc structure in Example 12-11 and the fcc structure in Problem 14 to determine the % voids in their respective packing-of-spheres arrangements: 31.98% for bcc and 25.95% for fcc. [*Hint:* What is the volume of the unit cell? How many atoms are present in the cell? What is their volume? Use the general case of the packing of spheres, without reference to a particular metal.]

72. Magnesium crystallizes in the *hcp* arrangement shown in Figure 12-41. The dimensions of the unit cell are height,

520. pm; length on an edge, 320. pm. Calculate the density of solid magnesium and compare with the measured value of 1.738 g/cm^3.

Ionic crystal structures

73. Show that the unit cells for CaF_2 and TiO_2 in Figure 12-49 are consistent with their formulas.

74. In a similar fashion to Example 12-14, calculate the density of CsCl. [*Hint:* What is the crystal structure of CsCl?]

75. The crystal structure of zinc sulfide (ZnS), is pictured. The length of the unit cell is 6.00×10^2 pm. For this structure, determine

(a) the crystal coordination numbers of Zn^{2+} and S^{2-};

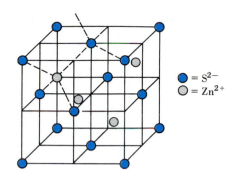

$\bullet = S^{2-}$
$\circ = Zn^{2+}$

Exercise 75

(b) the number of formula units in the unit cell;

(c) the density of ZnS.

Additional Exercises

76. A double boiler is used when a careful control of temperature is required in cooking. Water is boiled in an outside container to produce steam and the steam condenses on the outside walls of an inner container in which cooking occurs. (A related laboratory device is called a steam bath.)

(a) How is heat energy conveyed to the food to be cooked?

(b) What is the maximum temperature that can be reached in the inside container?

77. Is it likely that any of the following will occur naturally at or near the surface of the earth, anywhere on earth? Explain. (a) $CO_2(s)$; (b) $CH_4(l)$; (c) $SO_2(g)$; (d) $I_2(l)$; (e) $O_2(l)$. [*Hint:* Use the appropriate phase diagrams and data from Table 12-3.]

78. Trace the phase changes that occur as a sample of $H_2O(g)$, originally at 1.00 mmHg and -0.10 °C, is compressed at constant temperature until the pressure reaches 100. atm.

79. A television commercial claims that a product makes water "wetter." Is there any basis to this claim? Explain.

80. A dramatic lecture demonstration consists of continuously evacuating the water vapor from an open container of water in a bell jar. If the vacuum pump is sufficiently powerful, the water can be made to freeze. Describe the principles involved in this demonstration.

★**81.** The height, h, to which a liquid rises in a glass capillary tube depends on the density, d, and surface tension, γ, of the liquid and the radius of the capillary, r. The equation relating these quantities is $h = 2\gamma/dgr$ (where g is the acceleration due to gravity). As indicated in the following sketch, a capillary tube is placed in water and the water rises 4.3 cm. What must be the radius of the capillary bore? Assume that the surface tension of water is 7.28×10^{-2} J/m^2.

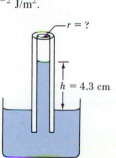

$r = ?$

$h = 4.3$ cm

Exercise 81

82. Using data from Appendix D, determine the quantity of

heat needed to vaporize 15.0 mL of liquid mercury at 25 °C. The density of Hg(l) = 13.6 g/mL. [*Hint:* What is ΔH for the process $Hg(l) \rightarrow Hg(g)$?]

★**83.** Refer to the footnote on page 395. What is the difference in the boiling point temperature of water at normal atmospheric pressure (1 atm) and at a pressure of 1 bar? [*Hint:* Use the Clausius–Clapeyron equation (12.6).]

84. Refer to the Summarizing Example and Table 12-8. A sample of hydrazine is maintained under a pressure of 0.500 atm in a device similar to that pictured in Figure 12-23.

(a) Describe the phase changes that would occur if the original sample is solid hydrazine at 0 °C and the sample is slowly heated to 100 °C. Be as specific as you can in your description.

★(b) What phase(s) would be present at a temperature of 75 °C? [*Hint:* At what temperature is the vapor pressure 0.500 atm?]

★**85.** Refer to Exercise 84. If 75.0 g of hydrazine is used, what minimum quantity of heat must be absorbed from the surroundings to produce the changes described in Exercise 84a. Why is the actual quantity of heat required greater than this? What data are lacking and how important are they for this calculation? [*Hint:* Use data from Table 12-8.]

86. You decide to cool a can of pop quickly in the freezer compartment of a refrigerator. When you take out the can, the pop is still liquid, but when you open the can, the pop immediately freezes. Explain why this happened.

87. Following are some values of ΔH_{vap} for several liquids at their normal boiling points: H_2, 0.92 kJ/mol; CH_4, 8.16 kJ/mol; C_6H_6, 31.0 kJ/mol; H_2O, 40.7 kJ/mol. Explain the differences among these values.

88. In their crystal structures, diamond, graphite, and ice all feature a hexagonal arrangement of atoms. How can this be reconciled with the facts that diamond is extremely hard and has a high melting point, graphite flakes easily and is a good electrical conductor, and ice has both a low density and low melting point?

★**89.** 100.0 L $H_2O(g)$ at 100. °C and 1 atm is passed into 1.00 kg $H_2O(l)$. The $H_2O(l)$ is maintained in a thermally insulated container (such as a vacuum bottle or Dewar flask) and is initially at 18.0 °C. What will be (a) the final mass of liquid; (b) the final temperature? ΔH_{vap} of water at 100. °C is 40.6 kJ/mol.

90. A 150.0-cm³ sample of $N_2(g)$ at 25.0 °C and 750 mmHg, is passed through $C_6H_6(l)$ until the gas becomes saturated with $C_6H_6(g)$. The new volume of the gas is 172 cm³ at a total pressure of 750. mmHg. What is the vapor pressure of C_6H_6 at 25.0 °C?

91. 525 cm³ Hg(l) at 20 °C is added to a large quantity of $N_2(l)$ kept at its boiling point in a thermally insulated container. What mass of $N_2(l)$ is vaporized as the Hg is brought to the temperature of the $N_2(l)$? For the specific heat of Hg(l) from 20 to −39 °C use 0.033 cal g⁻¹ °C⁻¹, and for Hg(s) from −39 to −196 °C, 0.030 cal g⁻¹ °C⁻¹. The density of Hg(l) is 13.6 g/cm³; the melting point of Hg is −39 °C, and its heat of fusion is 2.30 kJ/mol. The boiling point of $N_2(l)$ is −196 °C, and its $\Delta H_{vap} = 5.58$ kJ/mol.

⋆92. The vapor pressure of $NH_3(l)$ can be expressed as

$$\log_{10} P \text{ (mmHg)} = 9.95028 - 0.003863T - \frac{1473.17}{T}$$

What is the normal boiling point of $NH_3(l)$?

93. The following data are given for CCl_4: normal melting point, −23 °C; normal boiling point, 77 °C; density of liquid, 1.59 g/cm³; heat of fusion, 3.28 kJ/mol; vapor pressure at 25 °C, 110 mmHg.

   (a) How much heat must be absorbed to convert 10.0 g of solid CCl_4 to liquid at −23 °C?

  ⋆(b) How much heat is required to vaporize 20.0 L $CCl_4(l)$ at its normal boiling point? [*Hint:* What is ΔH_{vap}?]

   (c) What is the volume occupied by 1.00 mol of the saturated vapor of CCl_4 at 77 °C?

   (d) What phases—solid, liquid, and/or vapor—are present if 3.5 g CCl_4 is kept in an 8.21-L volume at 25 °C?

94. Place each of the following substances in the appropriate category in Table 12-7 and state your reasons for each placement. **(a)** Si; **(b)** CCl_4; **(c)** $CaCl_2$; **(d)** Ag; **(e)** HCl.

95. Arrange the following substances in the expected order of increasing melting point temperature: KI, Ne, K_2SO_4, C_3H_8, CH_3CH_2OH, MgO, $CH_2OHCHOHCH_2OH$. Explain your ranking.

96. *Ortho*-phthalic acid (used to synthesize the acid–base indicator phenolphthalein) and *para*-phthalic acid (also called terephthalic acid, and used in the manufacture of polyester fibers) are *isomers;* they have identical compositions but different structures. Because they have different structures they have different physical properties. Based on the structures shown, which of these two compounds would you expect to have the higher melting point? [*Hint:* See also Exercise 58.]

ortho-phthalic acid   *para*-phthalic acid

⋆97. Sketched here are two *hypothetical* phase diagrams for a substance, but neither of these diagrams is possible. Indicate what is wrong with each one.

98. Sketch a simple phase diagram for bismuth metal. Its triple point temperature is 544.5 K and its normal boiling point is 1832 K. (See the marginal note on page 406 for a description of the fusion curve for bismuth.)

99. Refer to Exercise 98. A sample of bismuth is heated under 1 atm pressure from 300 K to 2000 K.

   (a) Sketch the heating curve that would be obtained.

  ⋆(b) The heat of fusion of Bi is 10.9 kJ/mol and its heat of vaporization at the boiling point is 151.5 kJ/mol. The average molar heat capacity of the solid is 28 J mol⁻ K⁻¹, of the liquid, 31 J mol⁻¹ K⁻¹, and of the gas, 21 J mol⁻¹ K⁻¹. Assume a 1.00-mol sample of bismuth and a constant rate of heating of 1.00 kJ/min, and reconstruct the heating curve as much as possible to scale (that is, with the appropriate lengths of times and slopes of the segments of the curve).

⋆100. Sketch a phase diagram for tin by using the following information: There are three polymorphic forms of solid tin. The α form is stable below 13.2 °C, the β form from 13.2 to 161 °C, and the γ form from 161 to 232 °C, which is the melting point of tin. The densities of α, β, and γ tin are 5.8, 7.2, and 6.5 g/cm³, respectively. The vapor pressure of liquid tin is 1 mmHg at 1610 °C and its boiling point is 2750 °C. Provide as much detail as you can in your sketch. [*Hint:* How are the slopes of the solid–solid transition lines related to the densities of the solids?]

⋆101. Refer to Figures 12-45 and 12-47. Suppose that the two planes of ions pictured in Figure 12-45 correspond to the top and middle plane of ions in the NaCl unit cell in Figure 12-47. If the x rays used in the diffraction study have a wavelength of 153.9 pm, at what angle θ would the diffracted beam have its greatest intensity? (Use $n = 1$ in equation 12.11).

102. The crystal structure of magnesium oxide, MgO, is of the NaCl type (Figure 12-47). Use this fact, together with ionic radii from Figure 9-12, to establish **(a)** the crystal coordination numbers of Mg^{2+} and O^{2-}; **(b)** the number of formula units in the unit cell; **(c)** the length and volume of a unit cell; **(d)** the density of MgO.

103. Potassium chloride has the same crystal structure as NaCl. Careful measurement of the internuclear distance between K^+ and Cl^- ions gave a value of 314.54 pm. The density of KCl is 1.9893 g/cm³. Use these data to evaluate the Avogadro constant, N_A.

⋆104. A cylinder containing 151 lb Cl_2 has an inside diameter of 10. in. and a height of 45 in. The gas pressure is 100. psi (1 atm = 14.7 psi) at 20. °C. Cl_2 melts at −103 °C, boils at −35 °C, and has its critical point at 144 °C and 76 atm. In what state(s) of matter does the Cl_2 exist in the cylinder?

⋆105. Because solid *p*-dichlorobenzene, $C_6H_4Cl_2$, sublimes rather easily, it has been used as a moth repellant. From the data given below, estimate the sublimation pressure of $C_6H_4Cl_2(s)$ at 25 °C. For $C_6H_4Cl_2$: m.p. = 53.1 °C; vapor pressure of $C_6H_4Cl_2(l)$ at 54.8 °C is 10.0 mmHg; $\Delta H_{fusion} = 17.88$ kJ/mol;

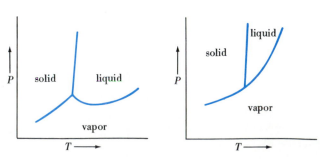

Exercise 97

$\Delta H_{vap} = 72.22$ kJ/mol. [*Hint:* Refer to the Summarizing Example.]

*106. In acetic acid vapor some molecules exist as monomers and some as dimers (see Figure 12-31). If the density of the vapor at 350. K and 1 atm is 3.23 g/L, what percent of the molecules must exist as dimers? Would you expect this percent to increase or decrease with temperature?

Self-Test Questions

For questions 107 through 116 select the single item that best completes each statement.

107. Of the following, the one with the highest normal boiling point is (a) $O_2(l)$; (b) $Ne(l)$; (c) $SO_3(l)$; (d) $Br_2(l)$.

108. The best electrical conductor of the following is (a) $CO_2(g)$; (b) $Si(s)$; (c) $NaCl(s)$; (d) $Br_2(l)$.

109. Of the compounds HF, CH_4, CH_3OH, and N_2H_4, hydrogen bonding as an important intermolecular force is expected in (a) none of these; (b) two of these; (c) all but one of these; (d) all of these.

110. Of the following properties, the magnitude of one must always increase with temperature. That property is (a) vapor pressure; (b) density; (c) ΔH_{vap}; (d) surface tension.

111. The intermolecular force of attraction between nonpolar molecules is called (a) hydrogen bonding; (b) dispersion forces; (c) interionic attractions; (d) adhesive forces.

112. Of these quantities, the one that we expect to be largest for a substance (expressed in kJ/mol) is (a) molar heat capacity of the liquid; (b) heat of fusion; (c) heat of vaporization; (d) heat of sublimation.

113. If the triple point pressure of a substance is greater than 1 atm, we expect (a) the solid to sublime without melting; (b) the boiling point temperature to be lower than the triple point temperature; (c) the melting point of the solid to come at a lower temperature than the triple point; (d) that the substance cannot exist as a liquid.

114. The form of carbon known as graphite (a) is harder than diamond; (b) contains a higher percentage of carbon than diamond; (c) is a better electrical conductor than diamond; (d) has equal carbon-to-carbon distances in *all* directions.

115. A metal that crystallizes in the body-centered cubic (bcc) structure has a crystal coordination number of (a) 6; (b) 8; (c) 12; (d) any value between 4 and 12.

116. A unit cell of an ionic crystal (a) is the same as the formula unit; (b) is any portion of the ionic crystal that has a cubic shape; (c) shares some its ions with other unit cells; (d) always contains the same number of cations and anions.

117. Argon, copper, sodium chloride, and carbon dioxide all crystallize in the fcc structure. How can this be when their physical properties are so different?

118. Summarize the characteristics of these four types of crystalline materials—ionic, molecular, network covalent, and metallic—in terms of intermolecular forces, structural particles of the crystal, and physical properties, such as melting point, boiling point, and electrical conductivity.

119. 10.0 g of steam at 100.0 °C and 100.0 g of ice at 0.0 °C are added to 100.0 g of liquid water at 20.0 °C in an insulated container. (For water: heat of fusion = 6.01 kJ/mol, heat of vaporization = 40.7 kJ/mol.) Which of these conditions will result? **(a)** The mixture will start to boil; **(b)** all of the liquid water will freeze; **(c)** all of the ice will melt; **(d)** a mixture of ice and liquid water will remain.

120. The triple point of CO_2 is at −56.7 °C and 5.1 atm. The critical point is at 31.1 °C and 72.9 atm. A 80.0-g piece of dry ice, $CO_2(s)$, is placed in an 8.00-L container and the container is sealed. If this container is held at 25 °C, what state(s) of matter must be present?

13 Solutions

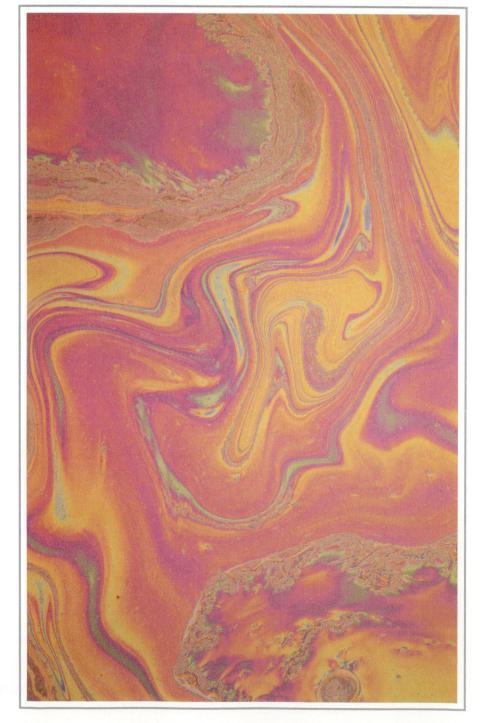

This oil slick reminds us that oil (hydrocarbons) and water do not mix. The conditions that do lead to solution formation are described in the chapter. [William Stanton]

438

No one can doubt the importance of solutions after thinking about all the ways we make use of solution properties. Most automobile engines are liquid cooled. The coolant is usually a solution of ethylene glycol (e.g., Prestone) in water. These solutions have boiling points that are higher and freezing points that are lower than for pure water. Thus, the coolant is less inclined than is water to boil away in hot weather and to freeze in cold weather.

In high concentrations fluoride ion is poisonous (for example, sodium fluoride is used as a rat poison). But, when present in drinking water in very low concentrations [about 1 part per million (ppm)], fluoride ion is effective in preventing tooth decay. However, at slightly higher concentrations (over 1.5 ppm) fluoride ion may cause teeth to become mottled. The fluoridation of municipal water supplies requires careful control of the fluoride ion concentration and is critically dependent on this characteristic of solutions: *Dissolved solutes are distributed uniformly throughout a solution, and the properties of the solution are fixed*.

Another application where careful control of a solution concentration is vital is in intravenous injections of fluids. Pure water cannot be used because it would enter blood cells, causing them to swell and burst. The fluids must be solutions having the correct value of a solution property known as osmotic pressure.

In this chapter we study, among other topics, factors that determine what mixtures of substances produce solutions, ways of expressing the concentrations of solutions, and ways of predicting solution properties such as vapor pressure, freezing point, boiling point, and osmotic pressure. We will also discover new methods of molecular weight determination based on the measurement of solution properties.

13-1 Types of Solutions: Some Terminology

TABLE 13-1
Some Common Solutions

Solution	Components
Gaseous solutions	
air	$N_2 + O_2 +$ several others
natural gas	$CH_4 + C_2H_6 +$ several others
Liquid solutions	
seawater	$H_2O + NaCl +$ many others
vinegar	$H_2O + HC_2H_3O_2$ (acetic acid)
soda pop	$H_2O + CO_2 +$ $C_{12}H_{22}O_{11}$ (sucrose)
Solid solutions	
yellow brass	$Cu + Zn$
palladium/ hydrogen	$Pd + H_2(g)$

First, let us repeat and extend some of the solution terminology we introduced in Chapters 1 and 4. A solution is a *homogeneous mixture*. It is a *mixture* because it contains two or more substances (and because their proportions can be varied). It is *homogeneous* because its composition and properties are uniform throughout the mixture. We generally use the term **solvent** for the component of a solution that is present in the greatest quantity or that determines the state of matter in which the solution exists. A **solute** is a solution component present in lesser quantity and is said to be *dissolved* in the solvent. For some solutions, such as mixtures of gases, the terms solvent and solute are not meaningful. A **concentrated** solution is one that has a relatively large quantity of dissolved solute(s), and a **dilute** solution has only a small quantity of solute(s). If a mixture of two or more substances separates into two physically distinct phases—such as solid sand (SiO_2) and liquid water in a sand–water mixture—the mixture is *heterogeneous;* it is not a solution.

If asked to name a solution most people would probably give an example like seawater (solid solutes in a liquid solvent) or vinegar (liquid solute in a liquid solvent). Some might mention wine (solid and liquid solutes in a liquid solvent) or soda pop (gaseous and solid solutes in a liquid solvent). Although solutions in the liquid state are the most common, solutions can exist in the gaseous and solid states as well. Table 13-1 lists a few common solutions and their components. The component present in greatest quantity is listed first.

In a solid solution the solvent is a *solid*. In some cases atoms or molecules of the solute substitute for those of the solvent in its crystalline lattice. These are called *substitutional* solid solutions; they require that the fundamental units of the solvent and solute be of very nearly the same size. Thus, copper (metallic radius, 128 pm) and nickel (metallic radius, 125 pm) form solid solutions in all proportions. In other

The familiar lead–tin alloy *solder* is a heterogeneous mixture. Although solder appears homogeneous to the eye, the separate solid phases of tin and lead can be distinguished under a microscope.

cases atoms of the solute may fill voids or holes (interstices) in the solvent lattice. These are called *interstitial* solid solutions and are most common if the solute atoms are very small, such as carbon or hydrogen dissolved in metals. In general, metallic mixtures are called **alloys.** Some alloys are solid solutions and some are heterogeneous mixtures. In still other cases metals may combine to form new intermetallic compounds.

13-2 Energetics of the Solution Process

On several occasions we have developed a better understanding of a process by analyzing its energy requirements. Such an approach will help us to understand which combinations of substances lead to solutions and which to heterogeneous mixtures. It will also enable us to make some predictions about physical properties of solutions. We will not meet with complete success, however. This is because there is an additional factor involved in solution formation that we will not be able to explore fully until later (Section 20-2). This factor (which we later introduce as the *entropy* factor) concerns the natural tendency for groups of atoms, ions, or molecules to become as randomized or disordered (mixed up) as possible. We have to consider this additional factor in one or two cases of solution formation.

Enthalpy (Heat) of Solution. In the solution process

$$\text{solute} + \text{solvent} \longrightarrow \text{solution} \tag{13.1}$$

the enthalpy change (heat) for the solution process is the difference between the enthalpy of the solution and the combined enthalpies of the pure solute and solvent.

$$\Delta H_{soln} = H_{solution} - (H_{solvent} + H_{solute})$$

As in previous cases we are not able to establish absolute enthalpies but we do know these to have unique values. As a consequence, the enthalpy (heat) of solution, ΔH_{soln}, also has a unique value. Moreover, we can usually measure ΔH_{soln} rather easily by experiment, as in the "coffee cup" calorimeter of Figure 7-11.

Now let us think about a stepwise approach for expressing ΔH_{soln} as a sum of other enthalpy changes. In a solution, *solvent* molecules must, on average, be far-

FIGURE 13-1
Enthalpy diagram for solution formation.

Depending on whether the broken arrow ends above, below, or on the line, the solution process is endothermic, exothermic, or has $\Delta H_{soln} = 0$, respectively.

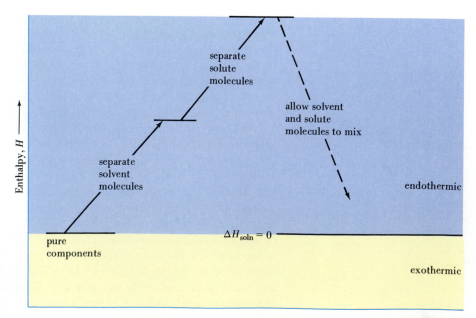

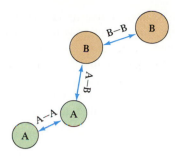

FIGURE 13-2

Representation of intermolecular forces in a solution.

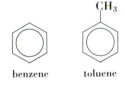

benzene   toluene

FIGURE 13-3

Two components of an ideal solution.

These two molecules are quite similar in structure. Think of the —CH₃ group in toluene as a small "bump" on the planar benzene ring (see also Figures 11-29 and 11-32).

Cl   CH₃
| |
Cl—C—H---O=C
| |
Cl   CH₃

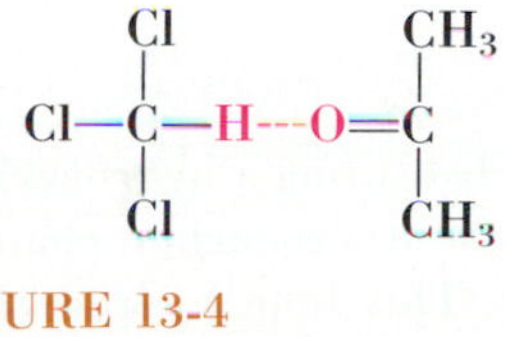

FIGURE 13-4

Intermolecular force between unlike molecules leading to nonideal solution.

Weak hydrogen bonding occurs between the H atom of a CHCl₃ (chloroform) molecule and the O atom of a (CH₃)₂CO (acetone) molecule.

ther apart than they are in the pure solvent. This is to make room for the solute molecules. Similarly, the *solute* molecules must be farther apart than they are in the pure solute. This is because the solute molecules are not packed as densely in the solution as in the pure solute. The first two steps to visualize in the solution process, then, involve increasing the distances between molecules, both in the solvent and in the solute. To do this means to overcome intermolecular forces of attraction. Work must be expended and the solvent and solute are raised to states of higher energy (enthalpy); $\Delta H > 0$ for both of these steps. Now, the separated molecules are allowed to mix randomly to form the solution. For this step we expect $\Delta H < 0$, because intermolecular forces of attraction between solute and solvent molecules bring the solution to a lower energy (enthalpy) state than that of the separated solute and solvent molecules. Depending on the relative values of the three enthalpy changes just described, ΔH for the overall process, that is, ΔH_{soln}, is either positive (endothermic) or negative (exothermic), or, in a few cases, zero. This three-step process is summarized below and also presented through Figure 13-1.

(a) pure solvent $\longrightarrow$ separated solvent molecules ΔH_a
(b) pure solute $\longrightarrow$ separated solute molecules ΔH_b
(c) separated solvent and solute molecules $\longrightarrow$ solution ΔH_c

net: pure solvent + pure solute $\longrightarrow$ solution

$$\Delta H_{\text{soln}} = \Delta H_a + \Delta H_b + \Delta H_c \tag{13.2}$$

Intermolecular Forces in Mixtures. We see from equation (13.2) that the magnitude (and sign) of ΔH_{soln} depends on the values of the three terms ΔH_a, ΔH_b, and ΔH_c. These in turn depend on the strengths of intermolecular forces of attraction of *three* kinds. Let us think in terms of two substances, **A** and **B**, one of which is the solvent and the other the solute; and let us represent intermolecular forces between like molecules as **A↔A** and **B↔B**, and between unlike molecules, **A↔B** (see Figure 13-2). Four possibilities are described below, with comparative forces denoted as: $\approx$ approximately equal to, $>$ greater than, $<$ smaller than, $\ll$ much smaller than.

1. A↔B ≈ A↔A ≈ B↔B. If all intermolecular forces of attraction are of about the same strength, a random intermingling of molecules occurs and a homogeneous mixture or solution results. We can generally predict properties of solutions of this type from a knowledge of the properties of the pure components. We call these **ideal** solutions. In particular, the volume of an ideal solution is the sum of the volumes of the solution components. There is no net change in volume in forming an ideal solution: $\Delta V_{\text{soln}} = 0$. There is also no net enthalpy change on forming an ideal solution from its components: $\Delta H_{\text{soln}} = 0$. This means that ΔH_c in equation (13.2) is equal in magnitude and opposite in sign to the sum of ΔH_a and ΔH_b. In general, mixtures of liquid hydrocarbons fit this description (see Figure 13-3).

2. A↔B > A↔A, B↔B. If intermolecular forces between unlike molecules *exceed* those between like molecules, a solution also forms. However, the properties of such solutions generally *cannot* be predicted from those of the pure components. These solutions are **nonideal.** The energy released in interactions between unlike molecules (ΔH_c) exceeds that required to separate the solvent and solute molecules ($\Delta H_a + \Delta H_b$). Energy is given off to the surroundings and the solution process is *exothermic* ($\Delta H_{\text{soln}} < 0$). Solutions of acetone and chloroform fit this type. As suggested by Figure 13-4, weak *hydrogen bonding* occurs between the unlike molecules, but the conditions for hydrogen bonding are not met in either of the pure liquids alone.*

*In most cases H atoms bonded to C atoms cannot participate in hydrogen bonding. In a molecule like CHCl₃, however, the three Cl atoms have a strong electron withdrawing effect on electrons in the C—H bond; the resultant dipole moment is $\mu = 1.92$ D. The H atom of CHCl₃ is thus able to form a hydrogen bond to the O atom of (CH₃)₂CO.

3. A↔B < A↔A, B↔B. If intermolecular forces between unlike molecules are *smaller than* between like molecules, complete mixing (dissolving) may occur, but the solutions are **nonideal.** The solution is in a higher energy state than the pure components; the process is *endothermic* (ΔH_c is less than the sum of ΔH_a and ΔH_b). This type of behavior is noted in mixtures of carbon disulfide (CS_2), a *nonpolar* liquid, and acetone ($\mu = 2.88$ D). In these mixtures the acetone molecules show some preference for other acetone molecules as neighbors, to which they are attracted by dipole–dipole interactions.

> Can you demonstrate through principles learned in Chapter 10 that carbon disulfide, CS_2, is nonpolar and that acetone, $(CH_3)_2CO$, is polar?

4. A↔B ≪ A↔A, B↔B. If intermolecular forces of attraction between unlike molecules are much smaller than those between like molecules, the components remain segregated into a **heterogeneous mixture;** dissolving does not occur to any significant extent. For example, in a mixture of water and octane (a constituent of gasoline) strong hydrogen bonds hold water molecules together in clusters. The nonpolar octane molecules are not able to exert a strong attractive force for the water molecules and the two liquids do not mix.

> An old adage with the same general meaning is "oil and water don't mix."

As a partial summary of these four cases, especially cases 1 and 4, we can state that "like dissolves like." That is, substances with similar molecular structures are likely to be soluble in one another; those with dissimilar structures are likely not to be. Of course, there are many cases in which the structures may be similar in part and dissimilar in part. Then it is a matter of trying to establish which part is most important. We explore this idea through Example 13-1.

Example 13-1 _____

Comparing intermolecular forces and predicting solution formation. Predict whether you would expect a solution to form in each of the following mixtures, and if so, whether the solution is likely to be ideal? **(a)** methyl alcohol (wood alcohol), CH_3OH, and water, (HOH); **(b)** the hydrocarbons hexane, $H_3C(CH_2)_4CH_3$, and octane, $H_3C(CH_2)_6CH_3$; **(c)** octyl alcohol (used in the manufacture of perfumes and flavoring agents), $H_3C(CH_2)_6CH_2OH$, and water (HOH).

Solution

> Although methyl alcohol and water are similar enough in structure that they can be mixed in any proportions to form solutions, they are not similar in all their properties. Water is an essential liquid to all life, and methyl alcohol is a poison when ingested.

(a) If we think of water as H—OH, methyl alcohol is similar to water. (Just substitute the group —CH_3 for one of the H atoms in water.) Both molecules meet the requirements of hydrogen bonding as an important intermolecular force. However, the strengths of the hydrogen bonds between like and between unlike molecules are likely to differ. In conclusion, we expect methyl alcohol and water to form *nonideal* solutions.

(b) The two hydrocarbons both consist of a carbon-chain skeleton with H atoms attached to the C atoms. In hexane the chain is six atoms long and in octane, eight. The molecules are nonpolar and intermolecular attractive forces (of the dispersion type) should be quite similar between like and between unlike molecules. We expect a solution to form and for it to be nearly *ideal.*

> Ideal or nearly ideal solutions are not too common. They require the solvent and solute to be quite similar in structure.

(c) At first sight, this case may seem similar to (a), substituting a hydrocarbon chain for an H atom in H—OH. However, here the hydrocarbon chain is *eight* members long, not just the group —CH_3. This long hydrocarbon chain is much more important than the terminal —OH group in establishing the physical properties of octyl alcohol. Viewed from this perspective, octyl alcohol and water are quite *dissimilar*. We do not expect a solution to form.

SIMILAR EXAMPLES: Exercises 8, 18, 80.

FIGURE 13-5
Dissolving of an ionic crystal
in water.

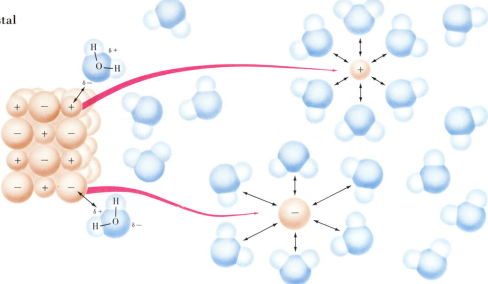

Formation of Ionic Solutions. Solutions of ionic solids in water are so important that we are going to devote considerable portions of several later chapters to them. Here, let us consider the energy requirements in forming such solutions. As pictured in Figure 13-5, water dipoles cluster around ions at the surface of a crystal. The negative ends of these dipoles are pointed toward the positive ions and the positive ends of dipoles, toward negative ions. If these ion–dipole forces of attraction are sufficiently strong to overcome the interionic forces of attraction in the crystal, dissolving will occur. Moreover, these ion–dipole forces persist in the solution. An ion surrounded by a cluster of water molecules is said to be *hydrated*. Energy is *released* when ions become hydrated. The greater the hydration energy in relation to the energy needed to separate ions from the ionic crystal, the more likely that the ionic solid will dissolve in water.

We can again use a hypothetical three-step process for the dissolving of an ionic solid. The energy requirement to dissociate a mole of the ionic solid into separate gaseous ions is just the negative of the lattice energy (that is, the process is *endothermic*). Energy is released in the next two steps—hydration of the gaseous cations and the gaseous anions. The enthalpy (heat) of solution is the sum of these three ΔH values, as illustrated below for NaCl(s).

NaCl(s) $\longrightarrow$ Na$^+$(g) + Cl$^-$(g)	ΔH_1($-$ lattice energy of NaCl)	
Na$^+$(g) + aq $\longrightarrow$ Na$^+$(aq)	ΔH_2 (hydration energy of Na$^+$)	
Cl$^-$(g) + aq $\longrightarrow$ Cl$^-$(aq)	ΔH_3 (hydration energy of Cl$^-$)	
NaCl(s) $\longrightarrow$ Na$^+$(aq) + Cl$^-$(aq)	$\Delta H_{soln} = \Delta H_1 + \Delta H_2 + \Delta H_3$ $\approx +5$ kJ/mol	

The dissolving of sodium chloride in water is *endothermic,* as it is also for the vast majority (about 95%) of soluble ionic compounds. How do we account for the fact that solutions form readily even though the energy state of so many aqueous ionic solutions is *higher* than for the pure solutes and solvents from which they are formed? Here is a case where the tendency for groups of atoms, ions, or molecules to achieve a more random or disordered arrangement plays an important role. The greater disorder resulting from the dissolving of NaCl(s) in H$_2$O(l) is sufficient to offset the +5 kJ/mol increase in energy in the solution process. In summary, if the three-step process outlined above is *exothermic* we expect dissolving of an ionic solid in water to occur. We also expect dissolving to occur if $\Delta H_{soln} > 0$ (endothermic), as long as it is not excessively large.

A similar line of argument is also required to account for endothermic solution formation in case 3 on page 442.

13-3 Solution Concentration

To describe a solution fully we need to state both what components are present and the solution *concentration*. We can express the concentration of a solution in several ways, but all of them must in some way describe the quantity of solute present in a given quantity of solvent (or solution). That is, we must establish

- The units used to measure the solute.
- Whether the second quantity measured is the solvent or the entire solution.
- The units used to measure this second quantity.

Percent by Mass, Percent by Volume, and Related Quantities. The statement, "5.00 g NaCl per 100.0 g of aqueous solution," means that 5.00 g NaCl is dissolved in 95.0 g H_2O, resulting in a solution mass of 100.0 g. We can also say that this solution is 5.00% NaCl, *by mass*. Mass percent concentration is widely used in industrial chemical applications. For example, the phosphoric acid produced by the action of sulfuric acid on phosphate rock is reported to be 28 to 32% H_3PO_4.

Because liquid volumes are so easily measured, some solutions are prepared on a percent basis *by volume*. Thus, a handbook lists a freezing point of -15.6 °C for a methyl alcohol–water antifreeze solution that is 25% CH_3OH, *by volume*. This is a solution prepared by dissolving 25 mL CH_3OH in enough water to prepare 100. mL of solution.

Another possibility is to measure the quantity of solute by mass and the quantity of solution by volume. For example, an aqueous solution with 0.9 g NaCl in 100. mL of solution is said to be 0.9% (*mass/vol*). This unit is extensively used in medicine and pharmacy.

Although they are of practical use, these several percent concentration units have little chemical significance. For example, they are not linked in any way to the concept of the mole. Let us consider some concentration units that are based on the mole.*

Molarity. In Section 4-4 we noted that

- The stoichiometry of chemical reactions is based on relative *numbers* of reacting atoms, ions, or molecules.
- Many chemical reactions are conducted in solution.

For these reasons we introduced at that time a solution concentration unit based on *number of solute particles*—the molarity concentration.

Definition of molarity concentration.

$$\text{molarity (M)} = \frac{\text{number of moles of solute}}{\text{number of liters of solution}} \qquad (13.3)$$

Molality. Suppose we prepare a solution at 20 °C by using a volumetric flask calibrated at 20 °C, but then suppose we use this solution at 25 °C. As the temperature increases from 20 to 25 °C, the amount of solute remains *constant* but the solution volume *increases* slightly. The number of moles of solute per liter—the molarity—*decreases* slightly.

At times we need to use a concentration unit that is *independent* of temperature. For this, *both* quantities, solute and solvent, should be measured by mass, since mass is *independent* of temperature. A useful unit is **molality,** expressed as *moles solute per kilogram solvent* (not of solution). A solution in which 1 mol of urea, $CO(NH_2)_2$, is dissolved in 1 kg of water is described as a 1 molal solution and designated by the symbol 1 *m* $CO(NH_2)_2$. Molal concentration is defined as

---

*A concentration scale based on the *chemical equivalent*, a quantity related to the mole, is presented in Sections 18-7 and 21-8. It is the *normality* concentration.

Definition of molality concentration.

$$\text{molality } (m) = \frac{\text{number of moles solute}}{\text{number of kilograms solvent}} \qquad (13.4)$$

An interesting use of molality concentration is in cases where the solvent, because it is a solid at room temperature, can be accurately measured only by its mass.

Mole Fraction. To relate certain physical properties of solutions (such as vapor pressure) to solution concentration we need a concentration unit in which all solution components are described on a mole basis. This we can do with the mole fraction. The mole fraction of component i, designated χ_i, is the fraction of all the molecules in a solution that are of type i. The mole fraction of component j is χ_j, and so on. The sum of the mole fractions of all the solution components is 1. The mole fraction of a solution component is defined by equation (13.5).

Definition of mole fraction concentration.

$$\chi_i = \frac{\text{moles of component } i}{\text{total moles of all solution components}} \qquad (13.5)$$

Mole percent is a concentration unit related to mole fraction. The mole percent of a solution component is the percent of all the molecules in a solution that are of a given type. Mole percents are mole fractions multiplied by 100.

Illustrative Examples. In Example 13-2 the concentration of a solution is expressed in several different concentration units. The calculation in Example 13-3 is perhaps more typical: A solution concentration is given in one scale (molarity) and you are asked to express it in a different one (mole fraction). In this type of calculation you should find it generally useful to identify

- the quantities needed for the final concentration unit;
- the quantities available through the original concentration unit;
- any additional factors that you may need (e.g., solution density);
- a useful quantity of solution on which to base the calculation.

Example 13-2

Expressing a solution concentration in various units. Ethanol (ethyl alcohol) is the common grain alcohol found in alcoholic beverages and also used as a fuel (as in the gasoline–alcohol solution called "gasohol"). Depending on how we plan to use a solution, we may need to express its concentration in different units. An ethanol–water solution is prepared by dissolving 10.00 cm³ of ethanol, C_2H_5OH ($d = 0.789$ g/cm³), in a sufficient volume of water to produce 100.0 cm³ of a solution having a density of 0.982 g/cm³ (see Figure 13-6). What is the concentration of this solution, expressed as **(a)** percent by volume; **(b)** percent by mass; **(c)** percent (mass/vol); **(d)** molarity; **(e)** molality; **(f)** mole fraction; **(g)** mole percent of ethanol?

Solution
(a) Percent ethanol, by volume:

$$\% \text{ ethanol, by volume} = \frac{10.00 \text{ cm}^3 \text{ ethanol}}{100.0 \text{ cm}^3 \text{ soln}} \times 100 = 10.00\%$$

(b) Percent ethanol, by mass:

$$\text{no. g ethanol} = 10.00 \text{ cm}^3 \text{ ethanol} \times \frac{0.789 \text{ g ethanol}}{1.00 \text{ cm}^3 \text{ ethanol}} = 7.89 \text{ g ethanol}$$

$$\text{no. g soln} = 100.0 \text{ cm}^3 \text{ soln} \times \frac{0.982 \text{ g soln}}{1.00 \text{ cm}^3 \text{ soln}} = 98.2 \text{ g soln}$$

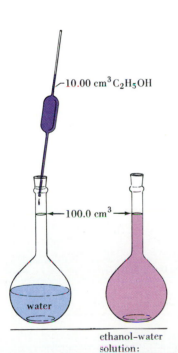

$-10.00 \text{ cm}^3 C_2H_5OH$

-100.0 cm^3-

water

ethanol–water
solution:
$d = 0.982$ g/cm³

FIGURE 13-6
Preparation of an ethanol–water solution–Example 13-2 illustrated.

$$\% \text{ ethanol, by mass} = \frac{7.89 \text{ g ethanol}}{98.2 \text{ g soln}} \times 100 = 8.03\%$$

(c) Mass/volume percent ethanol:

$$\% \text{ ethanol (mass/vol)} = \frac{7.89 \text{ g ethanol}}{100.0 \text{ cm}^3 \text{ soln}} \times 100 = 7.89\%$$

(d) Molarity of ethanol:

no. mol C_2H_5OH

$$= 10.00 \text{ cm}^3 \text{ ethanol} \times \frac{0.789 \text{ g ethanol}}{1.00 \text{ cm}^3 \text{ ethanol}} \times \frac{1 \text{ mol } C_2H_5OH}{46.07 \text{ g } C_2H_5OH}$$

$$= 0.171 \text{ mol } C_2H_5OH$$

$$\text{no. L soln} = 100.0 \text{ cm}^3 \text{ soln} \times \frac{1 \text{ L soln}}{1000 \text{ cm}^3 \text{ soln}} = 0.1000 \text{ L soln}$$

$$\text{molarity} = \frac{0.171 \text{ mol } C_2H_5OH}{0.1000 \text{ L soln}} = 1.71 \text{ M } C_2H_5OH$$

(e) Molality of ethanol:
The key to determining molal concentration often is in establishing the mass of solvent. Let us base our calculation on 100.0 cm^3 of solution.

mass soln = 98.2 g soln [see part (b)]

mass ethanol = 7.89 g ethanol [see part (b)]

no. mol ethanol = 0.171 mol C_2H_5OH [see part (d)]

mass H_2O = mass soln − mass ethanol = 98.2 g − 7.89 g = 90.3 g H_2O

$$\text{no. kg } H_2O = 90.3 \text{ g } H_2O \times \frac{1 \text{ kg } H_2O}{1000 \text{ g } H_2O} = 0.0903 \text{ kg } H_2O$$

$$\text{molality} = \frac{0.171 \text{ mol } C_2H_5OH}{0.0903 \text{ kg } H_2O} = 1.89 \text{ } m \text{ } C_2H_5OH$$

(f) Mole fraction of ethanol:
The no. of moles of ethanol in the solution has already been calculated [see part (d)]. Now we must calculate the no. of moles of water present, based on the mass of water determined in part (e).

$$\text{no. mol } H_2O = 90.3 \text{ g } H_2O \times \frac{1 \text{ mol } H_2O}{18.02 \text{ g } H_2O} = 5.01 \text{ mol } H_2O$$

$$\chi_{C_2H_5OH} = \frac{0.171 \text{ mol } C_2H_5OH}{0.171 \text{ mol } C_2H_5OH + 5.01 \text{ mol } H_2O} = \frac{0.171}{5.18} = 0.0330$$

(g) Mole percent ethanol:
$$\text{mole percent } C_2H_5OH = 100\chi_{C_2H_5OH} = 100 \times 0.0330 = 3.30\%$$

SIMILAR EXAMPLES: Exercises 2, 3, 21, 26, 28, 31, 33, 68.

Example 13-3

Converting molarity to mole fraction concentration. Laboratory ammonia is 14.8 M NH_3(aq) and has a density of 0.8980 g/cm^3. What is the mole fraction of NH_3 in this solution?

Solution. To express solution concentration in mole fractions we need to determine the numbers of moles of solute (NH_3) and of solvent (H_2O). If we base our calculation on 1.000 L (1000. cm^3) of solution, the number of moles of solute is simply 14.8 mol NH_3. To determine the number of moles of H_2O in the solution we can proceed as follows.

$$\text{mass of solution} = 1000.\ cm^3\ \text{soln} \times \frac{0.8980\ \text{g soln}}{cm^3\ \text{soln}} = 898.0\ \text{g soln}$$

$$\text{mass of } NH_3 = 14.8\ \text{mol } NH_3 \times \frac{17.03\ \text{g } NH_3}{1\ \text{mol } NH_3} = 252\ \text{g } NH_3$$

$$\text{mass of } H_2O = 898.0\ \text{g soln} - 252\ \text{g } NH_3 = 646\ \text{g } H_2O$$

$$\text{no. mol } H_2O = 646\ \text{g } H_2O \times \frac{1\ \text{mol } H_2O}{18.02\ \text{g } H_2O} = 35.8\ \text{mol } H_2O$$

$$\chi_{NH_3} = \frac{14.8\ \text{mol } NH_3}{14.8\ \text{mol } NH_3 + 35.8\ \text{mol } H_2O} = 0.292$$

SIMILAR EXAMPLES: Exercises 28, 71.

13-4 Solution Formation and Equilibrium

Figure 13-7 suggests what happens when a solute and solvent are mixed. At first only dissolving occurs, but after a time the reverse process becomes increasingly important. In this reverse process, called **precipitation,** the dissolved species (atoms, ions, or molecules) return to the undissolved state. When dissolving and precipitation occur at the same rate, *dynamic* equilibrium is established and the

FIGURE 13-7
Formation of a saturated solution.

The lengths of the arrows represent the rate of dissolving (→) and the rate of precipitation (←).
(a) When solute and solvent are first brought together only the process of dissolving occurs.
(b) After a time the rate of precipitation becomes significant.
(c) When dissolving and precipitation occur at the same rate, the solution is saturated. There is no further change in solution concentration with time.

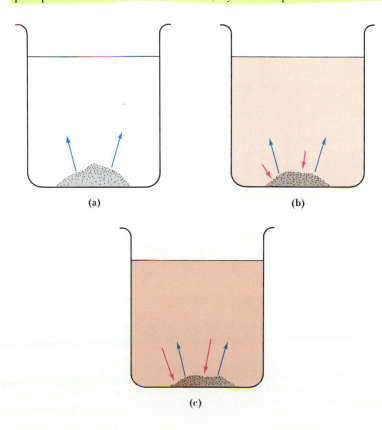

(a)   (b)

(c)

quantity of dissolved solute remains constant with time. We say that the solution is **saturated,** and we refer to the concentration of this saturated solution as the **solubility** of the solute in the given solvent. Thus, when a handbook lists the solubility of potassium chlorate ($KClO_3$) at 20 °C as 6.9 g per 100.0 g solution, this is a way of saying that 100.0 g of saturated solution at 20 °C contains 6.9 g $KClO_3$. Solubility varies with temperature, and a graph of solubility against temperature is called a **solubility curve.** Some typical solubility curves are shown in Figure 13-8.

If in the process pictured in Figure 13-7 we start with less solute than would be present in the saturated solution, the solute completely dissolves, no precipitation occurs, and the solution is said to be **unsaturated.** On the other hand, suppose that we prepare a saturated solution at one temperature and then change the temperature to a value where the solubility is lower (generally this means a lower temperature). Usually what we see is that excess solute precipitates from solution. But, occasionally, all the solute may remain in solution. Because in these cases the quantity of solute present in solution is greater than in a normal saturated solution, the solution is said to be **supersaturated.** If we add a few crystals of solute to serve as nuclei

In some solutions the solute and solvent are miscible (will mix and dissolve in one another) in all proportions. In these cases the solution never becomes saturated. Ethanol and water are miscible in all proportions, for example.

FIGURE 13-8
Water solubility of several salts as a function of temperature.

For each curve, as illustrated here for $KClO_4$ at 60 °C, regions above the curves (1) represent supersaturated solutions; points on the curves, S, saturated solutions; and regions below the curves (2), unsaturated solutions.

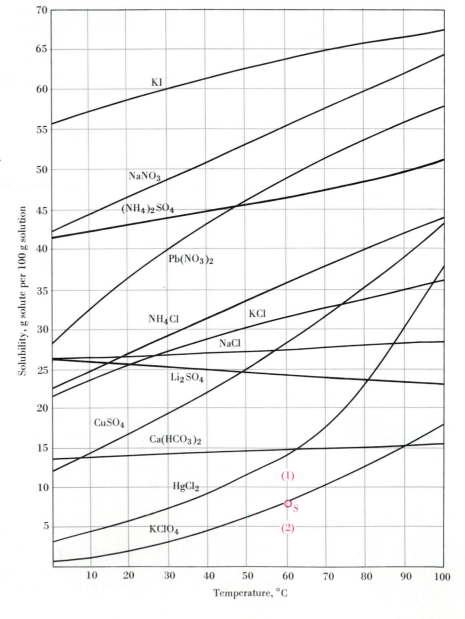

onto which precipitation can occur, the excess solute will usually precipitate from a supersaturated solution. Figure 13-8 shows how unsaturated and supersaturated solutions can be related to a solubility curve.

Solubility as a Function of Temperature. The solubilities of most ionic substances (about 95% of them) *increase* with temperature. The exceptions to this generalization are not easily predictable, although they tend to occur among compounds containing the anions SO_3^{2-}, SO_4^{2-}, SeO_4^{2-}, AsO_4^{3-}, and PO_4^{3-}.

In applying Le Châtelier's principle in Section 12-5, we learned that adding heat to a system at equilibrium (as in raising or attempting to raise the temperature) stimulates the heat-absorbing or endothermic reaction. This suggests that if $\Delta H_{soln} > 0$ (endothermic), raising the temperature will stimulate the dissolving process and *increase* the solubility of the solute. Conversely, if $\Delta H_{soln} < 0$ (exothermic), the solubility of the solute *decreases* with temperature. This is because, being endothermic, the precipitation process would be favored and not the dissolving process.

You should be careful, however, when you attempt to relate ΔH_{soln} and the effect of temperature on solubility. The value of ΔH_{soln} that you must use is that associated with dissolving a small quantity of solute in a solution that is already saturated or very nearly so. And this heat effect may be quite different from what you would observe when you add solute to pure solvent. For example, when NaOH is dissolved in water the process is highly *exothermic*. This observation might lead us to predict that the solubility of NaOH decreases with temperature. What we observe instead is that the solubility of NaOH *increases* with temperature. The explanation of this apparent discrepancy is that when a small quantity of NaOH is dissolved in a solution that is already nearly saturated in NaOH heat is *absorbed*, not evolved.*

We can put the fact that the solubilities of most compounds increase with temperature to practical use. Imagine that we have an *impure* compound that we are interested in purifying. Suppose we prepare a concentrated solution of this compound at an elevated temperature, and suppose that the impurities dissolve along with the compound. Now consider what happens when this concentrated solution is cooled. At some lower temperature the solution becomes saturated and the excess compound crystallizes from solution. As long as the solution remains unsaturated in the impurities and the impurities do not form solid solutions with the compound, the impurities remain behind in the solution and a crop of pure crystals of the compound can be harvested. This method of purifying a solid is called **recrystallization.**

Example 13-4

Applying the law of conservation of mass to the recrystallization of a solute from solution. A solution is prepared by dissolving 95 g NH_4Cl in 200.0 g of water at 60 °C. What mass of NH_4Cl will recrystallize from the solution when it is cooled to 20 °C?

Solution. According to the law of conservation of mass, the total mass remains constant at 295 g throughout. Also, after recrystallization occurs the total mass of NH_4Cl, partly in solution and partly undissolved, is the same as was originally present—95 g. From Figure 13-8, note that the concentration of the saturated solution at 20 °C is 27 g $NH_4Cl/100.0$ g soln.

Now, let us call the mass of NH_4Cl that recrystallizes x. This means that the mass of NH_4Cl that *remains* in solution at 20 °C is $95 - x$, and the total mass of solution is $295 - x$. The ratio of the mass of dissolved solute at 20 °C ($95 - x$) to the mass of solution ($295 - x$) must be the same as the solubility at 20 °C,

---

*The solid in equilibrium with saturated NaOH(aq) over a range of temperatures, including 25 °C, is actually NaOH·H_2O. It is this hydrate of NaOH whose solubility dependence on temperature is predicted by Le Châtelier's principle.

that is, 27 g $NH_4Cl/100.0$ g soln. This fact permits us to solve the following equation for x.

$$\frac{95 - x}{295 - x} = \frac{27}{100.0}$$

$$9.5 \times 10^3 - 100.0\,x = 8.0 \times 10^3 - 27\,x \qquad \text{and} \qquad 73\,x = 1.5 \times 10^3$$

$$x = 1.5 \times 10^3/73 = \text{21 g } NH_4Cl \text{ recrystallized}$$

SIMILAR EXAMPLES: Exercises 35, 36, 74.

13-5 Solutions Solubilities of Gases

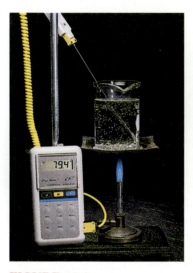

FIGURE 13-9
Effect of temperature on the solubility of gases.

Dissolved air is released as water is heated, even at temperatures below the boiling point. [Carey B. Van Loon]

Henry's law: effect of pressure on the solubilities of gases.

The dissolving of gases in liquids presents an interesting variation in the three-step solution process of Figure 13-1. This is in the step in which solute molecules are separated before being mixed with solvent molecules. In a gas, the molecules are already much farther apart than they will be in solution. We must think in terms of bringing the gas molecules *closer together* before they mix with the solvent. This is almost like saying that the gas must be condensed to a liquid before it dissolves in another liquid. The condensation of a gas is an *exothermic* process, and the ΔH value for this process is generally much greater than the energy requirement to separate solvent molecules to make room for the solute molecules.

As a consequence, we expect the formation of solutions of *gaseous* solutes generally to be *exothermic* processes, and we expect the solubilities of gases to *decrease* with increased temperature. We can observe this behavior when we see bubbles of dissolved air (a gaseous solute) escaping from heated water, even when the temperature is well below the normal boiling point (see Figure 13-9). This observation also helps us to understand why many fish cannot live in warm water; there is not enough dissolved air (oxygen) present in the water.

Henry's Law. Pressure generally affects the solubility of a gas in a liquid much more than does temperature, and, as noted in Figure 13-10, the effect is always the same: *The solubility of a gas increases as the gas pressure is increased.* In 1803 the English chemist William Henry (1797–1836) proposed the generalization that the concentration (C) of a dissolved gas is proportional to the gas pressure (P_{gas}) above the solution.

$$C = kP_{gas} \tag{13.6}$$

The value of k depends on the particular gas and on the units of C and P. For example, a handbook lists the solubility of $H_2S(g)$ as 437 cm^3 at STP per 100.0 g H_2O. We can convert the volume of H_2S dissolved to number of moles: 0.437 L $\times$ 1 mol/22.414 L = 0.0195 mol H_2S. The *molality* of the saturated solution is 0.0195 mol $H_2S/0.1000$ kg H_2O = 0.195 m. The Henry's law constant for H_2S dissolved in water is

$$k = C/P_{gas} = 0.195\ m/1\ \text{atm} = 0.195\ m\ \text{atm}^{-1} \tag{13.7}$$

We can rationalize Henry's law in this way: Equilibrium is reached between the gas above and the dissolved gas within a liquid when the rates of evaporation and condensation of the gas molecules become equal. The rate of evaporation depends on the number of molecules of dissolved gas per unit volume of solution, and the rate of condensation depends on the number of molecules per unit volume in the gas above the solution. To maintain equal rates of evaporation and condensation, as the number of molecules per unit volume increases in the gaseous state (through an

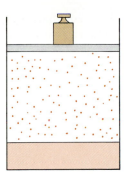

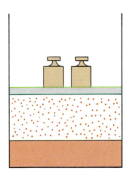

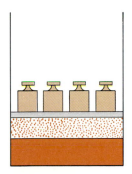

FIGURE 13-10

Effect of pressure on the solubility of a gas.

The concentration of dissolved gas is proportional to the pressure on the gas above the solution.

increase in the gas pressure), the number of molecules per unit volume must also increase in the solution (through an increase in concentration). When we apply Henry's law we assume that the gas does not react with the solvent. This assumption is valid in many cases.

We can see a practical application of Henry's law in soft drinks. The dissolved gas is carbon dioxide, and the higher the gas pressure above the soda pop, the more CO_2 that can be kept dissolved. When a can of pop is opened, the excess gas pressure is released and dissolved CO_2 escapes, usually rapidly enough to cause fizzing. In the production of sparkling wines, the gas dissolved under pressure is also CO_2, but the CO_2 is produced by a natural fermentation process within the bottle, rather than being introduced artificially as in soda pop.

Example 13-5

Using Henry's law to predict the solubility of a gas. Certain mineral waters contain dissolved hydrogen sulfide gas, which accounts for their faint "rotten egg" smell. A particular mineral water containing 0.5% by mass of H_2S is pumped from under ground and kept under an atmosphere in which the partial pressure of $H_2S(g)$ is 255 mmHg. Will the water dissolve more H_2S or lose some that is already dissolved?

Solution. First, let us convert 255 mmHg to a pressure in atmospheres.

$$P = 255 \text{ mmHg} \times \frac{1 \text{ atm}}{760 \text{ mmHg}} = 0.336 \text{ atm}$$

Next, let us use the expression, $C = kP$, where $P = 0.336$ atm and k is obtained from expression (13.7), that is, 0.195 m/atm.

$$C = \frac{0.195 \ m}{1 \text{ atm}} \times 0.336 \text{ atm} = 0.0655 \ m$$

The concentration of a solution *saturated* in H_2S and in equilibrium with gaseous H_2S at a partial pressure of 255 mmHg is 0.0655 m. If the mineral water has a molality *greater than* 0.0655 m, it will be supersaturated and *lose* H_2S to the gas above it. If its molality is *less than* 0.0655 m, it is unsaturated and will absorb additional H_2S. So the question now becomes "What is the molality of a solution that is 0.5% H_2S, by mass?" This is a question of establishing a relationship between two concentration scales.

A solution that is 0.5% H_2S, by mass, contains 0.5 g H_2S per 100.0 g of solution, or 0.5 g H_2S per 99.5 g (0.0995 kg) H_2O. Its molality is

$$\text{molality} = \frac{0.5 \text{ g } H_2S \times \dfrac{1 \text{ mol } H_2S}{34.08 \text{ g } H_2S}}{0.0995 \text{ kg } H_2O} = 0.1 \ m \ H_2S$$

Since the mineral water has a higher molality of H_2S than does a saturated solution under the given conditions, the mineral water will lose H_2S.

SIMILAR EXAMPLES: Exercises 11, 37, 38.

13-6 Vapor Pressures of Solutions

In this discussion we consider solutions that contain only two components (called *binary* solutions). Let us call the solvent A and the solute B. In the 1880s the French chemist F. M. Raoult found that a dissolved solute reduces the vapor pressure of the solvent. If we represent concentration of the solute through its mole fraction χ_B, the vapor pressure of the pure solvent as P_A°, and its vapor pressure above a solution as P_A, the *lowering* of the vapor pressure (ΔP) is

$$\Delta P = P_A^\circ - P_A = \chi_B P_A^\circ \tag{13.8}$$

Since in a solution of two components, $\chi_A + \chi_B = 1$, and $\chi_B = 1 - \chi_A$, we can rewrite equation (13.8) as

$$P_A^\circ - P_A = (1 - \chi_A)P_A^\circ \qquad \text{and} \qquad \cancel{P_A^\circ} - P_A = \cancel{P_A^\circ} - \chi_A P_A^\circ$$

Here, then, are two statements of Raoult's law—one mathematical and one verbal.

Raoult's law for vapor pressure lowering.

$$P_A = \chi_A P_A^\circ$$

The vapor pressure of the solvent above a solution (P_A) is equal to the product of the vapor pressure of the pure solvent (P_A°) and its mole fraction in solution (χ_A). $\tag{13.9}$

If the solute(s) in solution is (are) volatile, we can also write

$$P_B = \chi_B P_B^\circ$$

In an *ideal* solution all components—solvent and solute(s) alike—adhere to Raoult's law over the entire concentration range. In all *dilute* solutions in which there is no chemical interaction (reaction) among components, Raoult's law applies to the *solvent*. This is true whether the solution is ideal or not! However, Raoult's law does not apply to the solute(s) in a dilute nonideal solution. The reason for this difference is that in dilute solutions solvent molecules predominate and do not behave much differently than they would in the pure solvent. Solute molecules, on the other hand, are surrounded by an overwhelming number of solvent molecules. For them the environment is very much different than in the pure solute. Although solutes in dilute nonideal solutions do not obey Raoult's law, they do conform to Henry's law (equation 13.6).

We run into a problem if we try to explain Raoult's law from fundamental principles, though. At first, we might think that the presence of solute molecules introduces intermolecular forces of attraction for the solvent molecules, thereby reducing their ability to vaporize and lowering the vapor pressure of the solvent. But this cannot be because we defined an ideal solution as one in which the strengths of intermolecular forces of attraction between like and unlike molecules are essentially all the same, and this is just the type of solution to which Raoult's law applies. Another attractive explanation is that the presence of solute molecules at the surface of the solution reduces the availability of solvent molecules for vaporization, resulting in a lower liquid–vapor equilibrium pressure. This explanation should help you to remember and work with Raoult's law, but it still is inconsistent with certain other principles and observations that are beyond the scope of this text to explore.* Fundamentally, vapor pressure lowering is a thermodynamic phenomenon and can best be explained in thermodynamic terms. We briefly return to this question in Section 20-4.

Example 13-6 _____

Using Raoult's law to predict the vapor pressure of solvent and solute in an ideal solution. The vapor pressures of pure benzene (C_6H_6) and toluene (C_7H_8) at 25 °C are 95.1 and 28.4 mmHg, respectively. A solution is prepared in which the mole fractions of benzene and toluene are both 0.500. What are the vapor pressures of the benzene and toluene above this solution? What is the total vapor pressure?

---

*A discussion of these inconsistencies can be found in K. J. Mysels, *J. Chem. Educ., 32*, 179 (1955). For example, solutes of a certain type concentrate in the surface of a solution (surface-active solutes), but they are no more effective in lowering the vapor pressure of the solvent than are normal solutes.

Solution

$$P_{\text{benz.}} = \chi_{\text{benz.}} P^{\circ}_{\text{benz.}} = 0.500 \times 95.1 \text{ mmHg} = 47.6 \text{ mmHg}$$

$$P_{\text{tol.}} = \chi_{\text{tol.}} P^{\circ}_{\text{tol.}} = 0.500 \times 28.4 \text{ mmHg} = 14.2 \text{ mmHg}$$

$$P_{\text{total}} = P_{\text{benz.}} + P_{\text{tol.}} = 47.6 \text{ mmHg} + 14.2 \text{ mmHg} = 61.8 \text{ mmHg}$$

SIMILAR EXAMPLES: Exercises 12, 41.

Example 13-7

Calculating the composition of the vapor in equilibrium with a liquid solution. What is the composition of the vapor in equilibrium with the benzene–toluene solution of Example 13-6.

Solution. The ratio of each partial pressure to the total pressure is the mole fraction of that component in the vapor. (This is another application of equation 6.16.) The mole-fraction composition of the vapor is

$$\chi_{\text{benz.}} = \frac{P_{\text{benz.}}}{P_{\text{total}}} = \frac{47.6 \text{ mmHg}}{61.8 \text{ mmHg}} = 0.770$$

$$\chi_{\text{tol.}} = \frac{P_{\text{tol.}}}{P_{\text{total}}} = \frac{14.2 \text{ mmHg}}{61.8 \text{ mmHg}} = 0.230$$

SIMILAR EXAMPLES: Exercises 13, 43.

Liquid–Vapor Equilibrium: Ideal Solutions. The data calculated in Examples 13-6 and 13-7, together with similar data for other benzene–toluene solutions, are plotted in Figure 13-11. This figure consists of four lines—three straight and one curved—spanning the entire concentration range.

FIGURE 13-11
Liquid–vapor equilibrium for benzene–toluene mixtures at 25 °C.

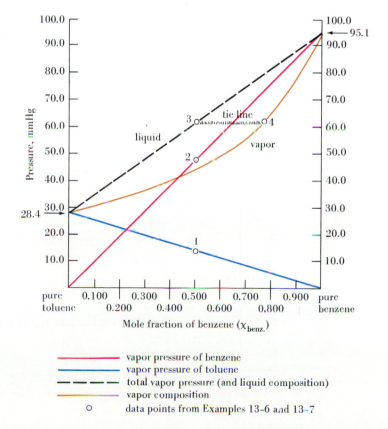

——————	vapor pressure of benzene
——————	vapor pressure of toluene
– – – – –	total vapor pressure (and liquid composition)
——————	vapor composition
○	data points from Examples 13-6 and 13-7

One straight line (red) originates at $P = 0$ and increases to $P = 95.1$ mmHg at $\chi_{benz.} = 1$. This straight line shows how the vapor pressure of benzene varies with the solution composition. It has the equation $P_{benz.} = \chi_{benz.}P°_{benz.}$, which means that benzene follows Raoult's law. Another straight line (blue) originates at $P = 28.4$ mmHg and falls to $P = 0$ when $\chi_{benz.} = 1$. This line shows how the vapor pressure of toluene varies with solution composition, and means that toluene also obeys Raoult's law. The broken straight line ranges from $P = 28.4$ mmHg for pure toluene to $P = 95.1$ mmHg for pure benzene. This line shows how the *total* vapor pressure varies with solution composition. Can you see that each pressure on this line is the sum of the pressures on the two straight lines that lie below it? For example, the pressure at point 3 is the sum of the pressures at points 1 and 2. Point 3 represents the total vapor pressure of a benzene–toluene solution in which $\chi_{benz.} = 0.500$.

As we calculated in Example 13-7, the vapor in equilibrium with a solution in which $\chi_{benz.} = 0.500$ is richer still in benzene; the vapor has $\chi_{benz.} = 0.770$ (point 4). The line joining points 3 and 4 is called a **tie line,** and we can think of it in this way: The tie line is plotted at a constant pressure equal to the total vapor pressure of the solution. One end of the tie line gives the composition of the liquid solution and the other end, the composition of the vapor. Imagine establishing a series of these tie lines throughout the composition range. The vapor ends of these ties lines can be joined by a smooth curve, the brown curve in Figure 13-11. From the relative placement of the liquid and vapor curves we see that for ideal solutions of two components, *the vapor phase is richer in the more volatile component.*

> Think of tie lines as joining or tying together the liquid and vapor composition curves in Figure 13-11.

Fractional Distillation. Let us look at liquid–vapor equilibrium in the benzene–toluene system in a somewhat different way. Suppose that for every solution composition we determine the boiling temperature—the temperature at which the *total* vapor pressure of the solution is 1 atm. We have plotted these temperatures as a function of solution composition in Figure 13-12. A graph such as Figure 13-12 is useful in explaining a procedure known as **fractional distillation,** a process for separating volatile liquids from one another.

Since benzene is more volatile than toluene, it is reasonable to expect the boiling points in Figure 13-12 to range from that of pure benzene to that of pure toluene. We should also expect the vapor above a boiling benzene–toluene solution to be richer in benzene (the more volatile component) than is the solution. In Figure 13-12 the lower curve (blue) represents liquid solutions of varying composition and their boiling points. The upper curve (brown) represents the composition of vapor in equilibrium with boiling solutions. For example, a benzene–toluene solution with

FIGURE 13-12
Liquid–vapor equilibrium for benzene–toluene mixtures: normal boiling temperature vs. composition.

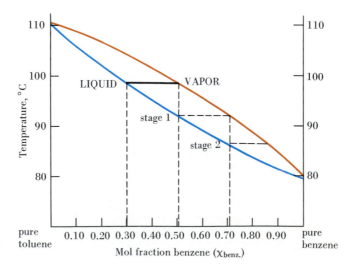

FIGURE 13-13
Fractional distillation of petroleum.

The various volatile components of petroleum are separated in fractional distillation columns such as these. [E. R. Degginger]

$\chi_{benz.} = 0.30$ boils at a temperature of 98.6 °C and is in equilibrium with a vapor in which $\chi_{benz.} = 0.51$. These two points are joined by the solid tie line.

Imagine that the vapor with $\chi_{benz.} = 0.51$ is removed from contact with the liquid solution and cooled to the point where the vapor completely condenses. The new liquid solution obtained from this vapor has $\chi_{benz.} = 0.51$ (labeled stage 1 in Figure 13-12). When this new liquid solution is boiled, the vapor in equilibrium with it has $\chi_{benz.} = 0.71$. If this vapor is removed and completely condensed, the resulting liquid will have $\chi_{benz.} = 0.71$ (stage 2). This boiling/condensation cycle can be repeated several times more.

Now let us return to the original solution with $\chi_{benz.} = 0.30$. Because the vapor is richer in benzene than is the boiling solution that produces it, as vapor is removed the remaining liquid solution becomes richer in the *less volatile* component, toluene. In the hypothetical boiling/condensation process that we have been describing, the ultimate vapor condensate is pure benzene and the liquid residue is pure toluene. The several boiling/condensation stages of the fractional distillation described through Figure 13-12 can be carried out continuously in a device called a **distillation column.**

We have previously described the separation of N_2, O_2, and Ar from liquid air by fractional distillation and pictured a distillation column schematically (see Figure 6-21). The most important industrial application of fractional distillation is in the manufacture of gasoline (see Figure 13-13).

Liquid–Vapor Equilibrium: Nonideal Solutions. For a *nonideal* binary solution, the curves representing the vapor pressures of the two components and the total vapor pressure of various solutions are *not straight lines*. As shown in Figure 13-14a, the vapor pressures for solutions of acetone and chloroform are *lower* than predicted for ideal solutions. Acetone–chloroform solutions exhibit *negative* deviations from Raoult's law (that is, from ideality). In solutions of acetone and carbon disulfide (Figure 13-14b) vapor pressures are higher than predicted for ideal solutions; acetone–carbon disulfide solutions display *positive* deviations from Raoult's law.

In our discussion of intermolecular forces and solution formation in Section 13-2, acetone–chloroform solutions were of the type (case 2) where intermolecular forces between unlike molecules exceed those between like molecules and $\Delta H_{soln} < 0$ (exothermic). It should seem reasonable that the components of such a solution would show a reduced tendency to vaporize, leading to *negative* deviations from Raoult's law. The situation for acetone–carbon disulfide solutions (case 3) is just

FIGURE 13-14
Vapor pressures of nonideal solutions.

Partial and total vapor pressures are plotted as a function of solution composition. Raoult's law applies in the green-shaded areas.

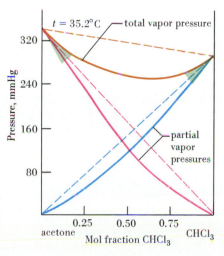

(a) Negative deviations from ideality.

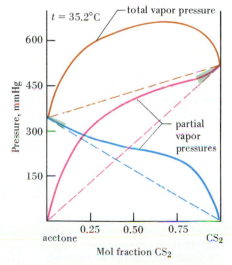

(b) Positive deviations from ideality.

the reverse: intermolecular forces of attraction between unlike molecules are weaker than between like molecules; $\Delta H_{soln} > 0$ (endothermic), and there is a greater tendency for vaporization than from an ideal solution, leading to *positive* deviations from Raoult's law.

Even though the departures from ideality for most solutions in Figure 13-14 are quite large, the vapor pressure curve of the *solvent* in *dilute* solutions (green shaded area) does fall nearly on the Raoult-law straight line. These graphs illustrate the statement on page 452 that the solvent does conform to Raoult's law, even in *dilute nonideal* solutions.

13-7 Freezing Point Depression and Boiling Point Elevation

An important class of solutions is one in which the solutes do not have an appreciable vapor pressure; that is, they are *nonvolatile*. For these solutions the solute still lowers the vapor pressure of the solvent, and the higher the solute concentration the greater the vapor pressure lowering. This effect is pictured in Figure 13-15. Here, the vapor pressure curve and the fusion curve for the solvent in a solution (red) are superimposed onto the phase diagram of the pure solvent. An important additional requirement in Figure 13-15 is that the solute be *insoluble* in the *solid* solvent. For many mixtures this requirement is easily met.

For a solvent containing a nonvolatile solute, the vapor pressure curve (red) intersects the sublimation curve at a lower temperature than is the case for the pure solvent. The fusion curve is also displaced to lower temperatures. Now let us recall how the melting point and the boiling point are established in a phase diagram. They are the temperatures at which a line at $P = 1$ atm intersects the fusion and vapor pressure curves, respectively. Four points of intersection are highlighted in Figure 13-15—the freezing points and the boiling points of the pure solvent and of the solvent in a solution. The freezing point of the solvent is *depressed* and its boiling point is *elevated*.

The amount by which the freezing point is lowered or the boiling point raised is proportional to the mole fraction of solute (just as is vapor pressure lowering). For *dilute* solutions the mole fraction of solute is proportional to molality concentration, and we can write

FIGURE 13-15

Vapor pressure lowering by a nonvolatile solute.

The phase diagram of the pure solvent is shown in blue, and for the solvent containing a nonvolatile solute, in red. The freezing point and boiling point of the pure solvent are f.p.$_0$ and b.p.$_0$. The corresponding points for the solution are f.p. and b.p. The freezing point depression, ΔT_f, and boiling point elevation, ΔT_b, are indicated. Because the solute is nonvolatile and assumed to be insoluble in the solid solvent, the sublimation curve of the solvent is unaffected by the presence of solute in the solution phase.

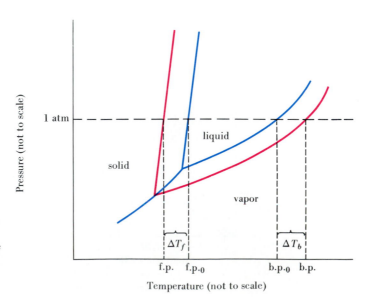

Equations for freezing point depression and boiling elevation.

$$\Delta T_f = K_f m \qquad (13.10)$$

$$\Delta T_b = K_b m \qquad (13.11)$$

ΔT_f and ΔT_b are the freezing point depression and boiling point elevation, respectively; m is the molality; K_f and K_b are proportionality constants. K_f is the **cryoscopic** or freezing point depression constant, and K_b is the **ebullioscopic** or boiling point elevation constant. The value of K_f is a function of the melting point, enthalpy of fusion, and molecular weight of the solvent; the value of K_b depends on the boiling point, enthalpy of vaporization, and molecular weight of the solvent. Another way to think about K_f and K_b is that they represent the freezing point depression and boiling point elevation for a 1 m solution. As a matter of fact, though, equations (13.10) and (13.11) often fail for solutions as concentrated as 1 m. Table 13-2 lists some typical values of K_f and K_b. Cooling curves for a pure solvent and a solution are compared in Figure 13-16.

Historically, the group of properties—vapor pressure lowering, freezing point depression, boiling point elevation, and osmotic pressure—has been used to establish molecular formulas. These properties, whose values depend only on the concentration of solute particles in solution and not on the identity of the solute, are called **colligative properties.**

As illustrated in Example 13-8, you can relate freezing point data and molecular formulas through a three-step procedure. To help you understand this procedure, these three steps are presented as the answers to three separate questions. In other cases, though, you should be prepared to work out your own stepwise procedure.

TABLE 13-2
Cryoscopic and Ebullioscopic Constants

Solvent	K_f	K_b
acetic acid	3.90	3.07
benzene	5.12	2.53
nitrobenzene	8.1	5.24
phenol	7.27	3.56
water	1.86	0.512

Values correspond to freezing point depressions and boiling point elevations, in degrees Celsius, due to 1 mol of solute particles dissolved in 1 kg of solvent. Units: °C kg solvent (mol solute)$^{-1}$.

Example 13-8

Establishing a molecular formula by using freezing point data. Nicotine, extracted from tobacco leaves, is a pale yellow liquid that is completely miscible with water at temperatures below 60 °C. **(a)** What is the *molality* of nicotine in an aqueous solution that begins to freeze at −0.450 °C? **(b)** If this solution is obtained by dissolving 1.921 g of nicotine in 48.92 g H_2O, what must be the molar mass of nicotine? **(c)** The analysis of nicotine shows it to consist of 74.03% C, 8.70% H, and 17.27% N, by mass. What is the molecular formula of nicotine?

Solution

(a) We can establish the molality of the nicotine by using equation (13.10) with the value of K_f for water listed in Table 13-2.

FIGURE 13-16
Cooling curves of a pure solvent and a solution compared.

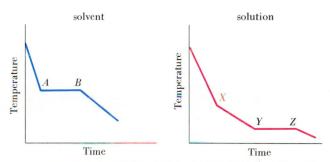

The cooling curve for the pure solvent has one horizontal break from A to B where complete freezing occurs. The cooling curve for the solution has a break at X, where the solvent begins to freeze from the solution (the freezing point). A second, horizontal break from Y to Z represents the freezing of both components from solution as a mixture of solids. The freezing points of solutions referred to in this section correspond to the point X. (The behavior described here assumes that the solute is insoluble in the solid solvent. Not represented is the fact that some supercooling is likely to occur at points A, X, and Y.)

$$m = \frac{\Delta T_f}{K_f} = \frac{0.450 \text{ °C}}{1.86 \text{ °C kg water (mol solute)}^{-1}} = 0.242 \frac{\text{mol solute}}{\text{kg water}}$$

(b) Here we can use the defining equation for molality concentration (equation 13.4), but with a known molality (0.242) and an unknown molar mass ($\mathcal{M}$) of solute. The number of moles of solute is just 1.921 g/$\mathcal{M}$.

$$m = \frac{1.921 \text{ g}/\mathcal{M}}{0.04892 \text{ kg water}} = 0.242 \frac{\text{mol}}{\text{kg water}}$$

$$\mathcal{M} = \frac{1.921 \text{ g}}{(0.04892 \times 0.242) \text{ mol}} = 162 \text{ g/mol}$$

(c) To establish the empirical formula of nicotine we need to use the method of Example 3-5. (This calculation is left as an exercise for you to do.) The result we obtain is C_5H_7N. The formula weight based on this empirical formula is 81. The experimentally determined molecular weight is exactly twice this value—162. The molecular formula is $C_{10}H_{14}N_2$.

SIMILAR EXAMPLES: Exercises 14, 46, 50, 53.

You need to understand some limitations to the method of molecular weight determination of Example 13-8. If we choose boiling point elevation as the method for determining the molality of a solution, we must hold the pressure constant. This is because the boiling point temperature depends on barometric pressure. Barometric pressure is not easy to hold constant, and as a consequence boiling point elevation is not often used to determine molecular weight. Because equation (13.10) is applicable only in dilute solutions (usually much less than 1 m), we have to determine freezing points quite precisely if we use water as the solvent, since its K_f is only 1.86. Notice that in Example 13-8 temperature was measured to ± 0.001 °C. This is much better precision than we can obtain with ordinary laboratory thermometers. We can achieve greater precision with the freezing point depression method by using a solvent with a larger value of K_f, such as cyclohexane ($K_f = 20$) or, better still, camphor ($K_f = 40$). If a solute has a high molecular weight, the number of moles present in a sample may be too small to affect the freezing point appreciably. In these cases measuring the osmotic pressure may be a better method.

There are many practical applications of the phenomenon of freezing point depression. Perhaps most familiar are methods used to lower the freezing point of water. An **antifreeze** (usually ethylene glycol), when added to the cooling system of an automobile, protects the coolant from freezing in cold weather. Salts such as NaCl or $CaCl_2$ can be used to prepare a freezing mixture for use in a home ice cream freezer, or to de-ice roads. In the de-icing of roads, as long as the outdoor temperature is *above* the lowest freezing point of a salt–water mixture, ice melts in the presence of that salt. If the outdoor temperature is below this temperature, no melting occurs.

Whatever solvent is used, it must have a high purity and a freezing point that is conveniently measured. The freezing point of camphor (177 °C) is rather high for usual laboratory operations, but camphor is still desirable as a solvent because of its large value of K_f.

The minimum temperature attainable with NaCl–water mixtures is -21 °C $= -6$ °F, and with $CaCl_2$–water mixtures, -55 °C $= -67$ °F.

13-8   Osmotic Pressure

Certain membranes appear to be continuous sheets or films, yet they contain a network of submicroscopic holes or pores. Small solvent molecules may pass through these pores, but the passage of dissolved solute molecules is severely restricted. Membranes that have this property are said to be **semipermeable.** These membranes may be of animal or vegetable origin, and they may occur naturally,

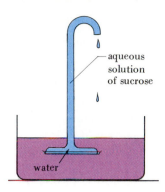

aqueous
solution
of sucrose

water

FIGURE 13-17
An illustration of osmosis.

Water molecules pass through the membrane, creating a pressure within the funnel. This pressure causes the liquid level to rise and the solution to overflow. As this process continues, the solution inside the funnel becomes more dilute and the pure water outside the funnel becomes a dilute sucrose solution. Liquid flow stops when the compositions of the solutions separated by the membrane become equal.

such as pig's bladder and parchment, or they may be synthetic materials, such as cellophane.

Figure 13-17 pictures an aqueous sucrose (sugar) solution in a long glass tube separated from pure water by a semipermeable membrane (permeable to water only). Water molecules can pass through the membrane from either direction, and they do. But because the concentration of water molecules is *greater* in the pure water than in the solution, there is a net flow *from* the pure solvent *into* the solution. This net flow, which is called **osmosis,** causes the solution to rise up the tube. The more concentrated the sucrose solution the higher the solution level rises. A 20% solution would rise to about 150 meters!

If we apply a pressure to the sucrose solution, this increases the tendency for water molecules to leave the solution. As a result the net flow of water into the sucrose solution is slowed down. If we apply a sufficiently high pressure to the sucrose solution we can stop the influx of water altogether. The necessary pressure to stop the osmotic flow is called the **osmotic pressure** of the solution. For the 20% sucrose solution this pressure is about 15 atm.

Osmotic pressure is a colligative property because its magnitude depends only on the *number* of solute particles per unit volume of solution. It does not depend on the identity of the solute. The expression written below, known as the van't Hoff equation, works quite well for calculating osmotic pressures of *dilute* solutions. The osmotic pressure is represented by the symbol π, R is the gas constant (0.08206 L atm mol^{-1} K^{-1}), and T is the Kelvin temperature. The term n represents the moles of *solute* and V is the volume (in liters) of *solution*. The ratio, n/V, then, is the *molarity* of the solution, M.

Osmosis is an apt term to describe this phenomenon. It is derived from a Greek word meaning "push."

Can you demonstrate that a column of sucrose solution 150 m high would itself exert 15 atm of pressure—enough to shut off osmosis through a semipermeable membrane? (See Exercise 56.)

Relationship between osmotic pressure and solution concentration.

$$\pi = \frac{n}{V} RT = MRT \tag{13.12}$$

Example 13-9

Calculating osmotic pressure. What is the osmotic pressure at 25 °C of an aqueous solution that is 0.0010 M $C_{12}H_{22}O_{11}$ (sucrose)?

Solution. We just need to substitute the data into equation (13.12).

$$\pi = \frac{0.0010 \text{ mol} \times 0.08206 \text{ L atm mol}^{-1} \text{ K}^{-1} \times 298 \text{ K}}{1.00 \text{ L}}$$

$$= 0.024 \text{ atm (18 mmHg)}$$

SIMILAR EXAMPLES: Exercises 58, 77.

The 0.0010 M sucrose solution in Example 13-9 has a molality of about 0.001 *m*. (In *dilute aqueous* solutions molarity and molality are essentially equal.) According to equation (13.10) we should expect a freezing point depression of about 0.00186 °C for this solution. Such a small temperature difference is extremely difficult to measure with any precision. On the other hand, the pressure difference of 18 mmHg that we calculated in Example 13-9 is easily measured. (It corresponds to a solution height of about 250 mm!) From this we see that measuring osmotic pressure is an especially useful method for determining molecular weights when we are dealing with (a) very dilute solutions and (b) solutes with high molecular weights.

Example 13-10 _____

Establishing a molar mass from a measurement of osmotic pressure. A solution is prepared by dissolving 1.08 g of human serum albumin, a protein obtained from blood plasma, in 50.0 cm³ of aqueous solution. The solution has an osmotic pressure of 5.85 mmHg at 298 K. What is the molar mass of the albumin?

Solution. First we need to express the osmotic pressure in atm.

$$\text{no. atm} = 5.85 \text{ mmHg} \times \frac{1 \text{ atm}}{760 \text{ mmHg}} = 7.70 \times 10^{-3} \text{ atm}$$

Now, we need to apply equation (13.12) in a slightly modified form [that is, with the number of moles of solute represented by the mass of solute (m) divided by the molar mass ($\mathcal{M}$).]

$$\pi = \frac{(m/\mathcal{M})RT}{V} \qquad \mathcal{M} = \frac{mRT}{\pi V}$$

$$\mathcal{M} = \frac{1.08 \text{ g} \times 0.08206 \text{ L atm mol}^{-1} \text{ K}^{-1} \times 298 \text{ K}}{7.70 \times 10^{-3} \text{ atm} \times 0.0500 \text{ L}} = 6.86 \times 10^4 \text{ g/mol}$$

SIMILAR EXAMPLES: Exercises 16, 58, 59.

Practical Applications of Osmosis. The most important examples of osmosis are those found in living organisms. For instance, consider red blood cells. If we place red blood cells in pure water, the cells expand and eventually burst as a result of water entering the cells through osmosis. The osmotic pressure associated with the fluid inside the cell is equivalent to that of an 0.9% (mass/vol) sodium chloride solution. Thus, if we place the cells in a sodium chloride solution (saline solution) of this concentration there will be no net flow of water through the cell walls and the cells will remain stable. A solution having the same osmotic pressure as body fluids is said to be **isotonic.** If we place cells in a solution with NaCl concentration greater than 0.9%, water flows out of the cells and the cells shrink. The solution is **hypertonic.** If the salt concentration is less than 0.9%, water flows into the cells; the solution is **hypotonic.** When fluids are intravenously injected into hospital patients, whether to combat dehydration or as a means of supplying nutrients, the fluids must be adjusted so that they are isotonic with blood. The osmotic pressure of the fluids must be the same as that of 0.9% (mass/vol) NaCl.

The walls (membranes) of red blood cells are approximately 10 nm thick and have pores (holes) about 0.8 nm in diameter. The diameters of water molecules are less than half this size. Water molecules easily pass through these pores. Potassium ions (K^+), which are found inside the cells, are also smaller than the pore diameters. But because the pore walls carry a positive electrical charge, K^+ ions are repelled. Thus, factors other than simple pore size may be involved in establishing what species can and cannot pass through the pores of a semipermeable membrane.

One recent application of osmosis is closely related to the very definition of osmotic pressure. If in the device shown in Figure 13-18 we apply a pressure to the right side (side B) that is *less than* the osmotic pressure of the salt water solution, the net flow of water molecules through the membrane will be from side A to side B. This is the process of *osmosis*. However, we have discussed the possibility of applying sufficient pressure to the liquid in side B to just offset the influx of water from side A; that is, to achieve a condition where there is no net flow of water. The

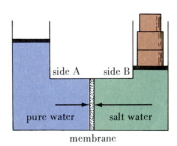

side A   side B

pure water   salt water

membrane

FIGURE 13-18
Desalinization of seawater by reverse osmosis.

The membrane pictured here is permeable to water but not to sodium or chloride ions. The normal flow of water through the membrane, in the absence of external pressures, is from side A (pure water) to side B (salt water). If a pressure is exerted on side B that exceeds the osmotic pressure of the salt water, a net flow of water can be created in the *reverse* direction, that is, from the salt water (side B) to the pure water (side A). The magnitudes of the flow of water molecules from each side are suggested by the lengths of the arrows.

required pressure is the *osmotic pressure*. If we apply a *higher* pressure than the osmotic pressure to side B, we can actually cause a net flow of water in the *reverse* direction, from the salt solution into the pure water. This is the condition known as **reverse osmosis.** One practical application of reverse osmosis is to render seawater fit to drink, either for emergency situations or as an actual source of municipal water. One reverse osmosis plant in Japan currently processes 5 million gallons of water per day. Another application of reverse osmosis is in the removal of dissolved materials from industrial or municipal wastewater, so that the wastewater can be discharged into the environment.

13-9 Solutions of Electrolytes

Svante Arrhenius (1859–1927). At the time Arrhenius was awarded the Nobel Prize in Chemistry (1903), his results were described in this way: "Chemists would not recognize them as chemistry; nor physicists as physics. They have in fact built a bridge between the two." The field of physical chemistry had its origins in Arrhenius's work. [Burndy Library, Norwalk, CT]

The van't Hoff factor for the colligative properties of electrolyte solutions.

Recall the three types of solutions that we introduced in Section 5-1: strong electrolytes, weak electrolytes, and nonelectrolytes. We related these three solution types to their abilities to conduct electricity, and in doing so we retraced some early work of Svante Arrhenius. For his doctoral dissertation (in 1883) Arrhenius undertook a careful investigation of the electrical conductivities of various aqueous solutions. Prevailing opinion at the time was that ions form only as a result of the passage of electric current. Arrhenius, however, reached the conclusion that ions may exist in a solute and become dissociated from one another simply by dissolving the solute in water. Electricity is not required to produce ions. He referred to the extent to which solute molecules became dissociated as the **degree of dissociation, α.**

For a nonelectrolyte, such as ethyl alcohol (C_2H_5OH), the electrical conductivity is extremely low; practically no ions exist in solution and $\alpha = 0$. For a weak electrolyte, such as acetic acid ($HC_2H_3O_2$), which exists in solution partly as ions and partly as undissociated molecules, α is a fractional number less than 1. For strong electrolytes, such as NaCl, dissociation of the solute into ions is essentially complete, and $\alpha = 1$. Furthermore, from the measured electrical conductivity of a solution Arrhenius was able to establish the number of ions produced per mole of solute. As we have already learned, for example, in NaCl, $MgCl_2$, and K_2SO_4 the numbers of moles of ions per mole of substance are *two, three,* and *three,* respectively.

Although Arrhenius developed his theory of electrolytic dissociation to explain the electrical conductivities of solutions, he was able to apply it more widely. One of his first successes was in explaining certain anomalous values of colligative properties described by Jacobus van't Hoff.

"Anomalous" Colligative Properties. Van't Hoff and others observed that certain solutes produce a greater effect on colligative properties than expected for a nonelectrolyte. To describe this effect van't Hoff defined the factor i as

$$i = \frac{\text{measured value for solute}}{\text{expected value for a nonelectrolyte}} \qquad (13.13)$$

For a large group of solutes, nonelectrolytes such as urea, glycerol, and sucrose, i has a value of 1. For another equally large group of weak and strong electrolytes, i has values greater than 1. In Example 13-11 we see how Arrhenius's theory of electrolytic dissociation can account for the observed values of the van't Hoff factor.

Example 13-11

Predicting colligative properties for nonelectrolytes, weak electrolytes, and strong electrolytes. The following freezing points are observed for 0.010 *m* aqueous solutions. Calculate the van't Hoff factor (i) for each solute and account for its value: urea, $-0.0185\,°C$; acetic acid, $-0.0193\,°C$; magnesium chloride, $-0.0514\,°C$.

Solution. First determine the freezing point depression expected for an 0.010 *m* aqueous solution of a nonelectrolyte. According to equation (13.10) it should be 0.0186 °C. Now, for the other solutes,

Urea [CO(NH$_2$)$_2$]: $i = \dfrac{0.0185\ °C}{0.0186\ °C} = 0.995 \approx 1.00$

Urea is a nonelectrolyte that remains undissociated in aqueous solution.

Acetic acid (HC$_2$H$_3$O$_2$): $i = \dfrac{0.0193\ °C}{0.0186\ °C} = 1.04$

Acetic acid is a weak electrolyte. About 4% of the molecules are dissociated, producing two ions (H$^+$ and C$_2$H$_3$O$_2$$^-$) per molecule. ($\alpha = 0.04$.)

Magnesium chloride (MgCl$_2$): $i = \dfrac{0.0514\ °C}{0.0186\ °C} = 2.76 \approx 3$

A van't Hoff factor of $i \approx 3$ suggests that MgCl$_2$ is dissociated in aqueous solution, producing three moles of ions per mole of compound. A reason why i is somewhat less than 3 is explained below.

SIMILAR EXAMPLES: Exercises 15, 17, 61, 62.

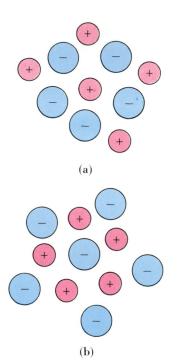

(a)

(b)

FIGURE 13-19
Interionic attractions in aqueous solution.

(a) A positive ion in aqueous solution is surrounded by a shell of negative ions.
(b) A negative ion attracts positive ions to its immediate surroundings.

Interionic Attractions. Arrhenius's theory stimulated great progress in physical chemistry for several decades. However, some observations made in the 40-year period from 1883 to 1923 suggested that his theory needed to be modified. The main problem was that the electrolytic conductances of concentrated solutions of strong electrolytes are less than is expected for complete dissociation of these solutes into ions. One suggestion was that in concentrated solutions strong electrolytes are not completely dissociated. Yet, x-ray diffraction studies indicate that salts exist completely as ions in the solid state. Should they not also exist as ions in solution?

The modern view of electrolyte solutions is based on a theory proposed by Peter Debye and Erich Hückel in 1923. This theory states that in aqueous solutions salts do exist in completely ionized form. The ions in solution, however, do not behave independently of one another. Instead, each positive ion is surrounded by a cluster of predominantly negative ions, and each negative ion by a cluster in which positive ions predominate. In short, each ion is enveloped by an ionic atmosphere with a net charge that is opposite in sign to that of the central ion, as pictured in Figure 13-19.

In an electric field the mobility of each ion is reduced because of the attraction or drag exerted by its neighbors in the ionic atmosphere. Similarly, the magnitudes of colligative properties are reduced, explaining, for example, why the value of i for MgCl$_2$ in Example 13-11 is 2.76 rather than the expected 3. What we can say then is that each type of ion in an aqueous solution has two "concentrations." One, called the *stoichiometric concentration,* is based on the amount of solute that is dissolved. Another is the "effective" concentration, called the *activity,* which takes into account interionic attractions. For certain purposes, such as stoichiometric calculations of the sort that we did in Chapters 4 and 5, we need the stoichiometric concentration. In other cases, such as to make precise predictions of colligative properties, activities should be used. We can relate the activity of a solution to its stoichiometric concentration through a factor called the **activity coefficient.** The Debye–Hückel theory gives us a theoretical basis for doing this.

Discrepancies between experimentally measured and predicted properties of solutions also arise because of ion-pair formation, which we explore more fully on page 687. In effect, for every ion pair (MgCl$^+$) formed by the association of one Mg^{2+} and one Cl$^-$ ion, we should count one "particle" in solution instead of two.

13-10 Colloidal Mixtures

At the beginning of this chapter we chose sand in water as an example of a heterogeneous mixture. Common experience tells us that sand should settle to the bottom of such a mixture, even if the quantity of sand (silica, SiO_2) is very small. SiO_2 is very insoluble in water. Yet, mixtures can be prepared in which large quantities of silica, up to 40% by mass, are dispersed in water and remain so for years! Such mixtures are clear, although with an opalescent or milky cast. Obviously, these dispersions do not involve ordinary grains of sand. Neither do they consist of dissolved ions or molecules. They are called **colloidal** mixtures. Let us see what makes them different from solutions.

The freezing points of colloidal mixtures of silica in water are only slightly below 0 °C. These mixtures must contain *small* numbers of particles compared to true solutions with comparable quantities of solutes. But if the numbers of particles in colloidal dispersions are relatively small, their physical dimensions and masses are huge compared to typical solute particles. Figure 13-20 compares hypothetical colloidal particles of different sizes and shapes with the more familiar particles of chemistry—atoms, ions, and molecules.

There are two basic methods for preparing colloidal mixtures. In one method, as in the formation of the colloidal silica that we have been discussing, large numbers of molecules aggregate through a process called *condensation*. The contrasting method that sometimes works is one of *dispersion*. Large particles are broken down, for example, by grinding, until the particles are sufficiently small to remain suspended.

The process of condensation, as it applies to the formation of colloidal mixtures of silica and water, is discussed in Section 14-9.

FIGURE 13-20

A comparison of colloidal, molecular, and atomic dimensions.

Some approximate sizes and shapes of hypothetical colloidal particles are represented here. For comparison, hypothetical particles with typical atomic and molecular dimensions (less than 1 nm) are also shown.

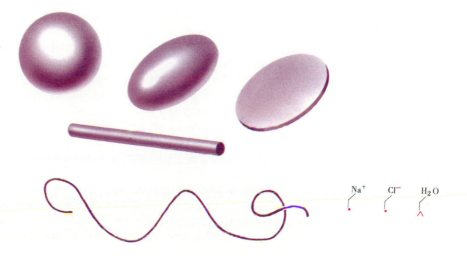

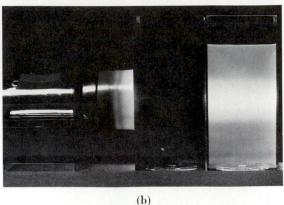

(a) The two containers have the same overall contents—30% SiO_2 in water, by mass. The one on the left is the expected heterogeneous mixture of sand in water. That on the right has SiO_2 in the colloidal state. (b) The Tyndall effect. The flashlight beam is not visible (left) as it passes through water (or a true molecular solution), but it is readily visible as it passes through colloidal silica (right).

(a)

(b)

FIGURE 13-21
Colloidal silica.

["Ludox" HS-30 Colloidal Silica: Courtesy of E. I. du Pont de Nemours & Company; photographs by Carey B. Van Loon.]

As so aptly put years ago by the American chemist Wilder Bancroft, "Colloid chemistry is the chemistry of bubbles, drops, grains, filaments, and films."

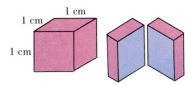

FIGURE 13-22
Surface area-to-volume ratio.

The cube on the left has a volume of 1 cm³ and a total surface area of 6 × (1 cm × 1 cm) = 6 cm². Its surface area-to-volume ratio is 6:1. When the cube is cut in half, two new faces are exposed, each with a surface area of 1 cm². The total volume of the two smaller cubes remains 1 cm³, but the surface area-to-volume ratio is now 8:1. Repeated subdivision of these cubes would eventually put them in the colloidal particle range. The total surface area would become very large, though the total volume would remain at 1 cm³. (See Exercise 67.)

In order for a material to be classified as colloidal, one or more of its dimensions (length, width, or thickness) must fall in the approximate range of 1 to 100 nm. If all the dimensions are smaller than 1 nm, the particles are in the molecular size range. If all the dimensions exceed 100 nm, the particles are of ordinary or macroscopic size (even if they are only visible under a microscope).

The colloidal particles in silica–water mixtures have a spherical shape. Some colloidal particles are rod-shaped, for example, tobacco mosaic virus. Some have disclike shapes, like the gamma globulin in human blood plasma. Thin films, such as an oil slick on water, are colloidal. And some colloids are randomly coiled filaments, such as cellulose fibers.

How can we tell if a mixture is a true solution or a colloid? One method that works well is suggested in Figure 13-21. When light is passed through a true solution, an observer viewing from a direction perpendicular to the light beam sees no light. But in a colloidal dispersion light is scattered in many directions and can easily be seen. This effect, first studied by John Tyndall in 1869, is known as the Tyndall effect. A common example is the scattering of light by dust particles in the light beam of a movie projector in a darkened room.

Stability of Colloids. An important characteristic of colloidal particles is that they have a high ratio of surface area to volume (see Figure 13-22). As we learned in Section 12-2, the atoms, ions, or molecules at the surface of a substance behave differently from those in the interior. This fact accounts for the existence of surface tension and related phenomena. For ordinary materials the proportion of atoms, ions, or molecules at the surface is very small compared to the interior, and the distinctive phenomena associated with surfaces are masked. In colloidal materials these surface phenomena are often quite pronounced.

One of the properties of surfaces is that of being able to attach species to themselves, a phenomenon called **adsorption.** During their formation some colloidal particles adsorb large numbers of ions from solution and become electrically charged. Silica particles in the colloidal silica mixtures mentioned previously adsorb hydroxide ions (OH^-) in preference to other ions, as suggested by Figure 13-23. As a result the silica particles all acquire a negative charge. Having like charges, the particles repel one another. These mutual repulsions overcome the force of gravity and keep the particles suspended.

Although the electrical charge factor can be very important in stabilizing a colloid, a high concentration of dissolved ions can bring about the *coagulation* or *precipitation* of a colloid. The ions responsible for this are those carrying the opposite charge to the colloidal particles themselves. The colloidal silica particles pictured in Figure 13-23 are precipitated by *cations*, and the higher the charge on the cation the more effective it is. The coagulation of colloidal iron oxide (Fe_2O_3) is pictured in Figure 13-24.

To stabilize a colloidal dispersion we sometimes use the method pictured in

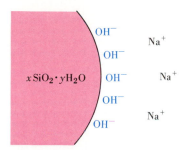

FIGURE 13-23
Surface of an SiO_2 particle in colloidal silica.

The points made in this simplified drawing are

- The SiO_2 particles are hydrated.
- OH^- ions are preferentially adsorbed on the surface.
- In the immediate vicinity of the particle, negative ions outnumber positive ions, and the particle has a net negative charge.

Not illustrated here are the facts that

- Some of the negative charge on the surface comes from the formation of silicate anions, e.g., SiO_3^{2-}.
- As a whole, the solution in which these particles are found is electrically neutral.

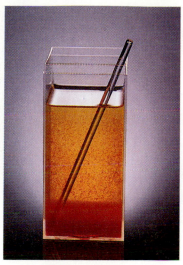

FIGURE 13-24
Coagulation of colloidal iron oxide.

The red mixture on the left is a colloidal dispersion of hydrous Fe_2O_3, obtained by adding a few drops of concentrated $FeCl_3(aq)$ to boiling water. The colloid is formed by a condensation method. When a few drops of $Al_2(SO_4)_3(aq)$ are added, the colloidal particles rapidly coagulate into a precipitate of $Fe_2O_3(s)$ (right). [Carey B. Van Loon]

Figure 13-25, called **dialysis.** Dialysis is similar to osmosis but is based on the ability of small solute particles, especially ions, to pass through a semipermeable membrane together with solvent molecules. The membrane is impermeable to large colloidal particles. We can achieve an even more effective removal of excess ions by placing the colloidal solution in an electric field. In this *electrodialysis*, ions are attracted to the electrode carrying the opposite charge from themselves. When a human kidney is functioning properly, blood, a colloidal solution, is dialyzed to remove excess electrolytes produced in metabolic processes and to restore proper

FIGURE 13-25
The principle of dialysis.

Water molecules, other solute molecules, and dissolved ions are free to pass through the pores of the membrane (e.g., cellophane) in either direction. The direction of net flow of these species depends on their relative concentrations on either side of the membrane. Colloidal particles cannot pass through the pores of the membrane.

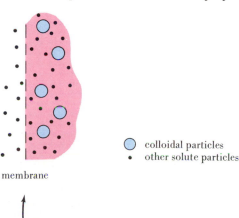
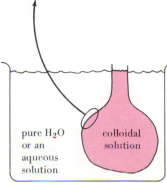

Chromatography

Chromatography is a technique for separating the components of a mixture present in one phase, called the **mobile phase,** as it moves in relation to another phase, called the **stationary phase.** As originally conceived by the Russian botanist Mikhail Tswett in 1903, the method is based on the ability of a column of finely powdered solid material, such as Al_2O_3, (the *stationary* phase) to *adsorb* substances from a solution (the *mobile* phase) that is allowed to trickle through it. In general, the attractive force of the solid surface differs for different species in solution. The substances that the solid adsorbs most strongly move through the stationary phase much more slowly than do those that are not so strongly adsorbed. This means that although the various components present in the mobile phase start out together when they first come in contact with the stationary phase, *they soon become separated.* Twsett used this method to separate plant pigments into colored components. His column developed bands of color, and he named this separation technique **chromatography,** which in Greek means "written in color."

A simple laboratory demonstration of the technique of chromatography is pictured in Figure 13-26. Here, the stationary phase is solid cellulose, in the form of ordinary filter paper. The mobile phase is water and the material undergoing chromatographic separation is common black ink.

Twsett's method remained largely unknown and unused until it was reintroduced by biochemists in the 1930s. This first form of chromatography is known as *adsorption* chromatography because its fundamental basis is the adsorption of substances on a solid surface. In 1942, Archer Martin and Richard Synge used as the stationary phase a liquid adsorbed on a solid, and as the mobile phase another liquid immiscible with the stationary phase. In this case, the components being separated distributed themselves between the two liquid phases. This type of chromatography is called *liquid–liquid partition* chromatography. In 1952, Martin and A. T. James substituted a gas for the liquid mobile phase and thus introduced *gas–liquid partition* chromatography. The effectiveness of liquid–liquid partition chromatography was greatly increased in the 1970s by the use of smaller particle sizes for the solid support and pumps to force the mobile phase through the stationary phase under high pressure. This method is now known as *high performance liquid chromatography* (HPLC). And, most recently, methods have been developed to use supercritical fluids as the mobile phases—*supercritical fluid chromatography* (SFC).

In the course of a few decades chromatographic methods, in their various forms, have evolved from a rediscovered technique to separate plant pigments into what is now unquestionably the chief separation technique in analytical chemistry. ∎∎∎

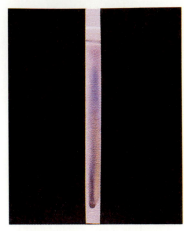

FIGURE 13-26
Separation of the components of black ink by paper chromatography.

As shown in this chromatogram, ordinary black ink is a mixture of colored components.
[Carey B. Van Loon]

Martin and Synge were awarded the Nobel Prize in 1952 for their work on chromatography.

electrolyte balance. In certain diseases the kidney loses its ability to purify blood. In these instances a dialysis machine, external to the body, can be used.

The electrostatic factor is particularly important in stabilizing a class of colloids referred to as lyophobic ("solvent fearing"). Other colloids, called lyophilic ("solvent loving"), owe their stabilities to an ability to swell in a solvent and remain suspended. If the suspending medium is water, the prefix "hydro" replaces "lyo" in these two terms (that is, hydrophobic and hydrophilic). A large number of hydrophilic colloidal materials are of biochemical interest and are considered again in Chapter 28.

Types of Colloids. Colloidal mixtures are classified according to the phases of matter involved. For example, a dispersion of solid particles in a liquid is called a **sol.** Other examples of colloidal mixtures are given in Table 13-3.

TABLE 13-3
Some Common Types of Colloids

Dispersed phase	Dispersion medium	Type	Examples
solid	liquid	sol	clay sols,[a] colloidal gold
liquid	liquid	emulsion	oil in water, milk, mayonnaise
gas	liquid	foam	soap and detergent suds, whipped cream, meringues
solid	gas	aerosol[b]	smoke, dust-laden air[c]
liquid	gas	aerosol[b]	fog, mist (as in aerosol products)
solid	solid	solid sol	ruby glass, certain natural and synthetic gems, blue rock salt, black diamond
liquid	solid	solid emulsion	opal, pearl
gas	solid	solid foam	pumice, lava, volcanic ash

[a] In water purification it is sometimes necessary to precipitate clay particles or other suspended colloidal materials. This is often done by treating the water with an appropriate electrolyte. Clay sols are also suspected of adsorbing organic substances, such as pesticides, and distributing them in the environment.

[b] Smogs are complex materials that are at least partly colloidal. The suspended particles are both solid (smoke) and liquid (fog): smoke + fog = smog. Other constituents of smog are molecular, such as sulfur dioxide, carbon monoxide, nitric oxide, and ozone.

[c] The bluish haze of tobacco smoke and the brilliant sunsets in desert regions are both attributable to the scattering of light by colloidal particles suspended in air.

FOCUS ON
Zone Refining

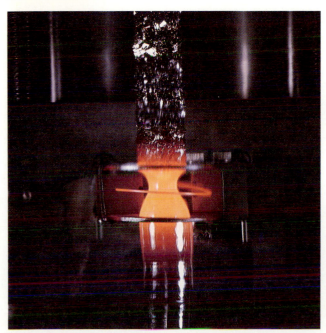

Zone Refining. As the heating coil moves up the rod of material, melting occurs. Impurities concentrate in the molten zone. The portion of the rod below the molten zone is purer than that in or above the zone. With each successive passage of the heating coil the rod becomes more pure. [Sol Mednick]

Recall from the Focus feature in Chapter 11 that the characteristics of a semiconductor are critically dependent on the quantity of dopant (boron, phosphorus, arsenic, . . .) used. Generally this is of the order of a few tenths of a part per million. The purities of the basic semiconductor materials to which these dopant "impurities" are added must be 100 to 1000 times greater. That is, the dopants are added to a material that initially has no more than about 10 parts per billion (ppb) of impurities. One method of preparing these ultrapure materials is **zone refining.**

Our discussion of the phenomenon of freezing point depression (Section 13-7) started with the assumption that a solute is *soluble* in a liquid solvent but *insoluble* in the solid solvent that freezes from solution. If this assumption were strictly valid, we would have, in principle, a method of preparing an ultrapure solid: Collect the solid that freezes from a solution, and the impurities will remain behind in the unfrozen solution. In actual practice this method is not so simple. Some impurities are entrapped in the frozen solvent, and this solid is also wet with the solution from which freezing occurred, further contaminating the solid. And, very likely some dissolving of the solute does occur in the solid solvent, even if this is slight. What we can say is that the solute distributes itself between the liquid and solid solvent, as suggested in Figure 13-27. Figure 13-27 bears a certain resemblance to the liquid–

FIGURE 13-27

The principle of zone refining illustrated.

The freezing point of pure A is T_A°. Addition of the impurity B lowers the freezing point of A; the greater the quantity of B, the greater the freezing point depression, as indicated by the red line. The compositions of the solids that separate from the various freezing solutions lie along the blue line. When a liquid solution at point l_1 is cooled to the temperature T_1, it begins to freeze, initially producing a solid with composition s_1. Suppose that a small quantity of this solid could be removed from contact with the liquid and then reheated to point l_2. As the liquid l_2 is allowed to cool to point T_2, it begins to freeze, initially producing a solid with composition s_2. From this solid the liquid l_3 would be produced by remelting; liquid l_3 initially produces solid s_3 on freezing, and so on. With each melting/freezing cycle the melting (and freezing) point is increased and the point representing the composition of the solid moves closer to pure A.

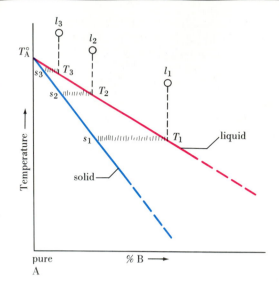

vapor equilibrium curves presented in Figures 13-11 and 13-12. Those liquid–vapor equilibrium curves were used to establish how a substance can be purified by repeated cycles of vaporization and condensation. Figure 13-27 describes how repeated melting/freezing cycles can be used to obtain a pure solid. The photograph featured at the beginning of this section suggests that successive passes of a molten zone through a rod of material sweeps impurities to the end of the rod. The end of the rod is cut off and the impurities are discarded. Wafers of semiconductor materials are obtained by cutting a rod of material purified by zone refining into thin slices.

Summary

By comparing the relative strengths of intermolecular forces of attraction between like and unlike molecules, we can often predict whether two substances will mix to form a solution. This prediction may also include the expected sign of ΔH_{soln} and ΔV_{soln}. If these quantities are zero, or nearly so, the solution is an *ideal* solution; otherwise it is *nonideal*. In predicting whether an ionic solid is soluble in water, the essential forces to compare are attractive forces among ions within a crystal *(lattice energy)* and attractive forces of water dipoles for ions *(ion–dipole* forces). That the criterion for solution formation involves more than the value of ΔH_{soln} is seen in the fact that for many solutions $\Delta H_{soln} > 0$, suggesting that solution formation is energetically unfavorable. The other criterion that we must consider is the tendency for groups of atoms, ions, or molecules to achieve a more random (mixed up) arrangement. (This is the entropy factory discussed in Section 20-2.)

To describe the composition of a solution we must state the relative proportions of solute(s) and of solvent. We can do this through a variety of concentration units. Some of these units are expressed on a percentage basis. More important from a chemical standpoint are concentration units based on molar quantities. *Molarity* is the number of moles of solute per liter of solution; *molality* is the number of moles of solute per kilogram of solvent; *mole fraction* is the number of moles of solute divided by the total number of moles of solution components.

In some cases a solvent and solute can be mixed in all proportions; the solution components are completely misciple. Usually, though, the solvent has a limited ability to dissolve solute. A solution containing this limiting quantity of solute is *saturated,* and the concentration of the saturated solution is referred to as the *solubility* of the solute. Solute solubilities are generally functions of temperature and can be represented through *solubility curves.* Le Châtelier's principle can be used to predict whether the solubility of a solute should increase or decrease with temperature. We can use the fact that the solubilities of most solutes increase with temperature as a basis for purifying them by *recrystallization:* A saturated solution is prepared at a high temperature and excess *pure* solute precipitates as the temperature is lowered. The solubilities of gases depend on pressure as well as temperature. The concentration of a gas in solution can be related to the pressure of the gas above the solution through *Henry's law.*

Solution properties that depend primarily on the number of solute particles in a solution but not on their identities are called *colligative properties.* One of these colligative properties is the tendency of each solution component to lower the vapor pressure of other components. In ideal solutions this vapor pressure lowering conforms to *Raoult's law*. Even in *nonideal* solutions, Raoult's law is applicable to the *solvent* if the solution is sufficiently dilute. We can use graphs of vapor pressures or boiling temperatures as a function of solution composition to describe the separation of the volatile components of a solution by *fractional distillation*.

An especially important class of solutions is that in which the solutes are *nonvolatile*. For these solutions freezing point depression and boiling point elevation can be used to determine the molecular weight of the solute. When we deal with solutes of high molecular weight (e.g., polymers) the preferred colligative property for molecular weight determination is osmotic pressure. *Osmosis* describes the net flow of solvent from a more dilute to a more concentrated solution when the solutions are separated by a semipermeable membrane. Osmosis and osmotic pressure are important phenomena observed in a number of biological systems.

The expressions relating colligative properties to solution concentration must be modified if complete or partial dissociation of the solute into ions occurs. This is done through the *van't Hoff factor*. In concentrated solutions of strong electrolytes interionic attractions cause the solute to behave as if its "effective" concentration in solution is less than its actual or stoichiometric concentration. This effective concentration is called the *activity*. Ordinary stoichiometric concentrations can be used in place of activity in sufficiently dilute solutions.

There are numerous instances, some of great practical importance, where a mixture does not fall neatly into the category of homogeneous (true solutions) or heterogeneous. These intermediate cases are called *colloidal mixtures*. In general, if a material contains particles having one or more dimensions in the range 1 to 100 nm, it is called *colloidal*. Many distinctive features of colloidal systems stem from their high surface area-to-volume ratios and their abilities to adsorb foreign molecules or ions to their surfaces.

Summarizing Example

Cinnamaldehyde is a yellowish oily liquid with a strong odor of cinnamon. It is the chief constituent of cinnamon oil, which is obtained from the twigs and leaves of cinnamon trees grown in tropical areas. Cinnamon oil is used in the manufacture of food flavorings, perfumes, and cosmetics. The structure of cinnamaldehyde is

Cinnamon trees in Sri Lanka. [A–Z Botanical Garden]

1. Which would you choose as a solvent to dissolve cinnamaldehyde—water or benzene? Explain.

Solution. The cinnamaldehyde molecule features a benzene ring and hydrocarbon side chain. We would not expect hydrogen bonding to occur in cinnamaldehyde (all H atoms are bonded to C atoms). These facts suggest that the molecule is much more similar to benzene than to water. We should expect cinnamaldehyde to dissolve in benzene but to be rather insoluble in water.

(This example is similar to Example 13-1.)

2. A solution is prepared by dissolving 1.00 g of cinnamaldehyde in 50.00 g of benzene (C_6H_6). If the vapor pressure of pure benzene at 25 °C is 95.1 mmHg, what is the partial pressure of benzene above this solution?

Solution. We need to use Raoult's law (13.9) which relates the partial pressure of a solution component to its mole fraction. The chief part of the problem is to express the solution concentration in mole fractions.

$$\text{no. mol } C_6H_6 = 50.00 \text{ g } C_6H_6 \times \frac{1 \text{ mol } C_6H_6}{78.114 \text{ g } C_6H_6} = 0.6401 \text{ mol } C_6H_6$$

$$\text{no. mol } C_9H_8O = 1.00 \text{ g } C_9H_8O \times \frac{1 \text{ mol } C_9H_8O}{132.2 \text{ g } C_9H_8O} = 0.00756 \text{ mol } C_9H_8O$$

$$\text{mol fraction } C_6H_6 = \frac{0.6401 \text{ mol } C_6H_6}{(0.6401 + 0.0076) \text{ mol total}} = 0.9883$$

The partial pressure of benzene above the solution is

$$P_{benz.} = \chi_{benz.}P^{\circ}_{benz.} = 0.9883 \times 95.1 \text{ mmHg} = \textbf{94.0 mmHg}$$

(This example is similar to Example 13-6.)

3. Does the calculation of the partial pressure of benzene in part **2** require that the solution be ideal? Explain.

Solution. In order for Raoult's law to apply to *all* solution components, the solution must be ideal. However, in a *dilute* solution Raoult's law will apply to the *solvent*, even if the solution is *nonideal*. Since in this case benzene is the solvent in a dilute solution, **Raoult's law should work reasonably well, whether the solution is ideal or not.**

4. The freezing point of benzene is 5.533 °C. What is the freezing point of the solution prepared by dissolving 1.00 g of cinnamaldehyde in 50.00 g of benzene?

Solution. Our main task is to determine the molality concentration of this solution, which we can do with equation (13.4). Note that we already know the number of moles of solute from part **2**.

$$\text{molality} = \frac{0.00756 \text{ mol } C_9H_8O}{0.05000 \text{ kg } C_6H_6} = 0.151 \text{ } m$$

Now, use the solution molality (0.151 *m*) and the freezing point depression constant of benzene from Table 13-2 ($K_f = 5.12$) to obtain the freezing point *depression*. That is, $\Delta T_f = K_f m$.

$$\Delta T_f = \frac{5.12 \text{ °C kg solvent}}{\text{mol solute}} \times \frac{0.151 \text{ mol solute}}{\text{kg solvent}} = 0.773 \text{ °C}$$

The freezing point of the solution is 0.773 °C *below* that of benzene, f.p. = 5.533 °C − 0.773 °C = **4.760 °C**

(This example is similar to Example 13-8.)

5. Cinnamaldehyde dissolves in water only to the extent of 1 part to about 700 parts of water, by mass. Estimate the osmotic pressure of a saturated solution of cinnamaldehyde in water at 25 °C.

Solution. The solubility is approximately 1 g of cinnamaldehyde in 700 g water, and 700 g of water occupies a volume of essentially 700 mL. The solution volume will be very close to that of the water—700 mL = 0.7 L. We have already determined the number of moles of cinnamaldehyde in 1.00 g (part **2**); it is 0.00756 mol C_9H_8O. The *molarity* of the saturated water solution is 0.00756 mol C_9H_8O/0.7 L = 0.01 M C_9H_8O. Now, we can calculate the osmotic pressure by using equation (13.12).

$$\pi = MRT = 0.01 \text{ mol L}^{-1} \times 0.0821 \text{ L atm mol}^{-1} \text{ K}^{-1} \times 298 \text{ K} = \textbf{0.2 atm}$$

(This example is similar to Example 13-9.)

Key Terms

adsorption (13-10)
alloy (13-1)
chromatography (13-10)
colligative properties (13-7)
colloidal mixture (13-10)
dialysis (13-10)
Henry's law (13-5)

ideal solution (13-2)
molality (*m*) (13-3)
mole fraction (13-3)
osmosis (13-8)
osmotic pressure (13-8)
Raoult's law (13 6)
reverse osmosis (13-8)

saturated solution (13-4)
semipermeable membrane (13-8)
solute(s) (13-1)
solvent (13-1)
supersaturated solution (13-4)
unsaturated solution (13-4)
zone refining (Focus feature)

Highlighted Expressions

Definition of molarity concentration (13.3)
Definition of molality concentration (13.4)
Definition of mole fraction concentration (13.5)
Henry's law: effect of pressure on the solubilities of gases (13.6)
Raoult's law for vapor pressure lowering (13.9)

Equations for freezing point depression and boiling point elevation (13.10; 13.11)
Relationship between osmotic pressure and solution concentration (13.12)
The van't Hoff factor for colligative properties of electrolyte solutions (13.13)

Review Problems

1. A saturated aqueous solution of KI at 20 °C contains 144 g KI/100 g H_2O. Express this composition in the more conventional percent by mass, that is, as g KI/100 g solution.

2. An aqueous solution with density 0.980 g/cm^3 at 20 °C is prepared by dissolving 11.3 mL CH_3OH ($d = 0.793$ g/cm^3) in enough water to produce 75.0 mL of solution. What is the % CH_3OH, expressed as **(a)** % by volume; **(b)** % by mass; **(c)** % (mass/vol)?

3. A solution that is 12.00% propylene glycol, $C_3H_8O_2$, in water, by mass, has a density of 1.007 g/cm^3. What is the *molarity* of $C_3H_8O_2$ in this solution?

4. A sodium chloride solution isotonic with blood is 0.9% NaCl (mass/vol).
   (a) What is the molarity of NaCl in this solution?
   (b) What is the total molarity of ions in this solution?
   (c) What is the approximate freezing point of this solution? (Assume that the solution has a density of 1.005 g/cm^3.)

5. You are asked to prepare 125.0 mL of 0.0108 M $AgNO_3$. What mass of a sample known to be 99.67% $AgNO_3$, by mass, would you need?

6. What is the molality of *para*-dichlorobenzene in a solution prepared by dissolving 3.73 g $C_6H_4Cl_2$ in 50.0 cm^3 of benzene ($d = 0.879$ g/cm^3)?

7. A solution is prepared by mixing the following hydrocarbons: 1.56 mol C_7H_{16}, 3.06 mol C_8H_{18}, and 2.17 mol C_9H_{20}. What are the **(a)** mole fraction and **(b)** mole percent of each component of the solution?

8. Two of the substances listed below are highly soluble in water; two are only slightly soluble in water; and two are insoluble in water. Indicate the situation you expect for each one.

(a) iodoform, CHI_3

(b) benzoic acid,

(c) formic acid, $H-\overset{O}{\overset{\|}{C}}-O-H$

(d) butyl alcohol, $CH_3CH_2CH_2CH_2OH$

(e) chlorobenzene, ⬡—Cl

(f) propylene glycol, $CH_3CHOHCH_2OH$

9. In light of the factors outlined on page 417, which of the following ionic fluorides would you expect to be most water soluble? Explain your reasoning. MgF_2; NaF; KF; CaF_2

10. A solution prepared by dissolving 0.80 mol NH_4Cl in 150.0 g H_2O is brought to a temperature of 25 °C. Use Figure 13-8 to determine whether this solution is unsaturated or whether excess solute will precipitate.

11. Under an $O_2(g)$ pressure of 1.00 atm, 28.31 cm^3 of $O_2(g)$ at 25 °C dissolves in 1.00 L H_2O at 25 °C. Assuming that the solution volume remains at 1.00 L,
   (a) What is the molarity of a saturated $O_2(aq)$ solution at 25 °C when $P_{O_2} = 1$ atm?
   (b) What is the aqueous solubility of O_2 at 25 °C under the pressure of the atmosphere, which is 20.95% O_2, by volume? [*Hint:* Recall equations (6.16) and (13.6).]

12. What are the partial and total vapor pressures of a solution obtained by mixing 25.5 g of benzene, C_6H_6, and 45.8 g of toluene, C_7H_8, at 25 °C? At 25 °C the vapor pressure of pure C_6H_6 is 95.1 mmHg and that of C_7H_8 is 28.4 mmHg.

13. Determine the composition of the vapor above the benzene–toluene solution described in Exercise 12.

14. Adding 1.10 g of an unknown compound reduces the freezing point of 75.22 g of benzene from 5.53 to 4.92 °C. What is the molecular weight of the compound?

15. The freezing point of an 0.010 *m* aqueous solution of a nonvolatile solute is −0.072 °C. What would you expect the normal boiling point of this same solution to be?

16. Polyvinyl chloride (PVC) is a plastic widely used in the manufacture of food wrap and phonograph records. An 0.61-g sample of PVC is dissolved in 250.0 cm³ of a suitable solvent at 25 °C. The resulting solution has an osmotic pressure of 0.79 mmHg. What is the molecular weight of the PVC?

17. Use your knowledge of strong, weak, and nonelectrolytes to arrange the following 0.0010 *m* aqueous solutions in order of *decreasing* freezing point: C_2H_5OH, NaCl, $MgBr_2$, $HC_2H_3O_2$, and $Al_2(SO_4)_3$.

Exercises

Homogeneous and heterogeneous mixtures

18. Although substances that dissolve in water do not generally dissolve in benzene, there are some substances that are moderately soluble in both solvents. One of the following four is such a substance. Which substance do you think this is and why?

(a) *para*-dichlorobenzene (a moth repellent), Cl—⟨⟩—Cl

(b) salicyl alcohol (a local anesthetic),

(c) diphenyl (a heat transfer agent),

(d) hydroxyacetic acid (used in textile dyeing),

19. Explain the observation that all metal nitrates are water soluble, whereas many metal sulfides are not water soluble. Among metal sulfides, which would you expect to be most soluble?

20. On page 443 we learned that ΔH_{soln} for NaCl(s) in water is about +5 kJ/mol. The formation of NaCl(aq) is an *endothermic* process. By contrast, when LiCl is dissolved in water, $\Delta H_{soln} = -35$ kJ/mol, an *exothermic* process. What factor can account for this difference in solution behavior? [*Hint:* Which step in the three-step process outlined on page 443 is responsible?]

Percent concentration

21. A sample of vinegar is 5.88% acetic acid ($HC_2H_3O_2$), by mass. What mass of $HC_2H_3O_2$ is contained in a 775-mL bottle of vinegar? Assume a density of 1.01 g/cm³.

22. The calculations in Example 13-2 show that the percent ethanol, by mass, in a particular aqueous solution is less than the percent by volume in the same solution. Explain why you would expect this to be also true for all aqueous solutions of ethanol. Would it be true of all ethanol solutions, regardless of the other component? Explain.

23. Is either concentration term, percent by mass or percent by volume, independent of temperature? Explain your answer.

Molarity concentration

24. Commercial concentrated sulfuric acid ($d = 1.831$ g/cm³) is 94.0% H_2SO_4, by mass. What is the molarity of H_2SO_4 in this solution?

25. How many mL of the ethanol–water solution described in Example 13-2 should be diluted with water to produce 1050 mL of 0.285 M C_2H_5OH?

26. A 10.00% by mass solution of ethanol, C_2H_5OH, in water has a density of 0.9831 g/cm³ at 15 °C and 0.9804 g/cm³ at 25 °C. What is the molarity of C_2H_5OH in this solution at each temperature?

Molality concentration

27. How many grams of iodine, I_2, must be dissolved in 225.0 mL of carbon disulfide, CS_2 ($d = 1.261$ g/cm³), to produce a 0.116 *m* solution?

28. An aqueous solution of nitric acid is 36.0% HNO_3, by mass, and has a density of 1.2205 g/cm³. What are the molarity and molality of HNO_3 in this solution?

29. How many grams of water should you add to 1.00 kg of 1.12 *m* CH_3OH(aq) to reduce the *molality* to 1.00 *m* CH_3OH?

★**30.** An aqueous solution has a concentration described as 109.2 g KOH/L solution. The solution density is 1.09 g/cm³. You are asked to use 100.0 cm³ of this solution to prepare 0.250 *m* KOH. What mass of which component, KOH or H_2O, would you add to the 100.0 cm³ solution?

Mole fraction, mole percent

31. Calculate the exact or approximate mole fraction of solute in the following aqueous solutions: **(a)** 16.8% C_2H_5OH, by mass; **(b)** 0.374 *m* $CO(NH_2)_2$ (urea); **(c)** 0.080 M $C_6H_{12}O_6$. Which calculation(s) is (are) exact, and which, approximate? Explain.

32. Refer to Example 13-2. What mass of C_2H_5OH must be added to 100.0 mL of the solution described in part (f) to increase the mole fraction of C_2H_5OH to 0.0500?

33. What volume of glycerol, $C_3H_8O_3$ (density = 1.26 g/cm³), must be added per kilogram of water to produce a solution with 6.20 mole percent $C_3H_8O_3$?

Solubility equilibrium

34. Refer to Figure 13-8 and estimate the temperature at which a saturated aqueous solution of $KClO_4$ is 1.00 *m*.

35. A solution prepared by dissolving 26.0 g $KClO_4$ in 500.0 g of water is brought to a temperature of 20 °C.
  (a) Refer to Figure 13-8 and determine whether the solution is unsaturated or supersaturated at 20 °C.
  (b) Approximately what mass of $KClO_4$ must be added to saturate the solution (if it is originally unsaturated) or what mass of $KClO_4$ can be crystallized from the solution (if it is originally supersaturated)?

36. A solid mixture consists of 95.0% NH_4Cl and 5.0% $(NH_4)_2SO_4$, by mass. A 50.0-g sample of this solid is added to 100.0 g of water at 90 °C. With reference to Figure 13-8
  (a) Will all of the solid dissolve at 90 °C?
  ★**(b)** If the resulting solution is cooled to 0 °C, approximately what mass of NH_4Cl can be crystallized?
  (c) Will $(NH_4)_2SO_4$ also crystallize at 0 °C?

Solubility of gases

37. Most natural gases consist of about 90% methane, CH_4. Assume that the solubility of natural gas at 20 °C and 1 atm gas pressure is about the same as that of CH_4, 0.02 g/kg water. If a sample of natural gas under a pressure of 20 atm is kept in contact with 1.00×10^3 kg of water, what mass of natural gas will dissolve?

38. At 25 °C and under 1 atm $N_2(g)$, the aqueous solubility of N_2 is 14.34 cm³ $N_2(g)$ (measured at 25 °C and 1 atm) per liter of solution. What is the molarity of N_2 in a water solution that is in equilibrium with N_2 in the atmosphere? (Air contains 78.08% N_2, by volume.)

39. Henry's law is often stated in this way: The mass of a gas dissolved by a given quantity of solvent at a fixed temperature is directly proportional to the pressure of the gas. Show how this statement is related to equation (13.6).

★40. Still another statement of Henry's law is this: A given quantity of liquid at a fixed temperature dissolves the same volume of gas at all pressures. What is the connection between this statement and the one given in Exercise 39? Under what conditions is this second statement not valid?

Raoult's law and liquid–vapor equilibrium

41. Calculate the vapor pressure at 25 °C of a solution containing 32.6 g of the *nonvolatile* solute urea, $CO(NH_2)_2$, in 685 g H_2O. The vapor pressure of water at 25 °C is 23.8 mmHg.

42. In the following illustration are two beakers containing a solution of sodium chloride, NaCl. *Solution 1* is saturated and has undissolved NaCl(s) present. *Solution 2* is unsaturated.

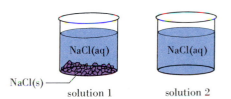

Exercise 42   solution 1   solution 2

(a) Above which solution is the vapor pressure of water, P_{H_2O}, greater? Explain.

(b) Above one of these solutions the vapor pressure of water, P_{H_2O}, remains *constant*, even as water evaporates from solution. Which solution is this? Explain.

(c) Which of the solutions has the *higher* boiling point? Explain.

43. Styrene is the monomer from which polystyrene plastics are made (such as used in toys, Styrofoam coffee cups, and kitchen wares). The styrene is manufactured from ethylbenzene by the extraction of hydrogen atoms. The primary product obtained contains 38% styrene (C_8H_8) and 62% ethylbenzene (C_8H_{10}), by mass. The mixture is separated by fractional distillation at 90 °C. Determine the mole fraction composition of the vapor in equilibrium with this 38%–62% mixture at 90 °C given the following vapor pressures of the two pure components.

ethylbenzene
v.p. = 182 mmHg at 90 °C

styrene
v.p. = 134 mmHg at 90 °C

★44. A small sample of the vapor phase described in Exercise 43 is condensed. The resulting liquid is again allowed to come to equilibrium with vapor at 90 °C. What is the mass percent of styrene (C_8H_8) in this new vapor phase? (*Hint:* Refer to Figure 13-12.)

★45. Calculate $\chi_{C_6H_6}$ in a benzene–toluene liquid solution that is in equilibrium at 25 °C with a vapor phase that contains 62.0 mol % C_6H_6. (Use data from Exercise 12.)

Freezing point depression and boiling point elevation

46. A compound is 42.4% C, 2.4% H, 16.6% N, and 37.8% O. The addition of 6.45 g of this compound to 50.0 mL of benzene, C_6H_6 (density = 0.879 g/cm³), lowers the freezing point from 5.53 to 1.37 °C. What is the molecular formula of this compound?

47. Adding 1.64 g of naphthalene, $C_{10}H_8$, to 80.00 g of cyclohexane, C_6H_{12}, reduces the freezing point of the cyclohexane from 6.5 to 3.3 °C.

(a) What is the value of K_f for cyclohexane?

(b) Which is the better solvent for molecular weight determinations by freezing point depression, benzene or cyclohexane? Explain.

48. What approximate proportions by volume of water (d = 1.00 g/cm³) and ethylene glycol (Prestone), $C_2H_6O_2$ (d = 1.12 g/cm³), must be mixed to protect an automobile cooling system to −10 °C? (Assume equation 13.10 applies.)

49. Suppose you have available 2.50 L of a solution with a density of 0.9767 g/mL that is 13.8%, by mass, ethanol (C_2H_5OH). From this solution you would like to make the *maximum* quantity of ethanol–water antifreeze solution that will offer protection to −2.0 °C. Would you add additional ethanol or water to the solution? What mass of liquid would you add?

50. Coniferin is a glycoside (a derivative of a sugar) found in conifers (e.g., fir trees). When a 1.205-g sample of coniferin is subjected to combustion analysis (recall Section 3-2) the products are 0.698 g H_2O and 2.479 g CO_2. Coniferin is not very soluble in cold water but it is appreciably soluble in hot water. A 2.216-g sample is dissolved in 48.68 g H_2O and the normal boiling point of this solution is found to be 100.068 °C. What is the molecular formula of coniferin?

51. Mixtures of ethylene glycol, $HOCH_2CH_2OH$ (Prestone), and water are the most commonly used coolants in automobile cooling systems. This coolant offers protection both against cooling system boilover in the summer and cooling system freeze-up in the winter. Explain how the same mixture can serve both functions.

52. Citrus growers know that it is not necessary to fire their smudge (smoke) pots even if the temperature is expected to drop several degrees Fahrenheit below the normal freezing point of water for several hours.

(a) Why does the citrus fruit not freeze at the normal freezing point of water?

(b) Why do you suppose that lemons freeze at a higher temperature than do oranges?

53. The boiling point of water at 756 mmHg is 99.85 °C. What percent sucrose ($C_{12}H_{22}O_{11}$), by mass, should be present to raise the boiling point to 100.00 °C at this pressure?

Osmotic pressure

54. The two solutions pictured here are separated by a semipermeable membrane that permits only the passage of water mol-

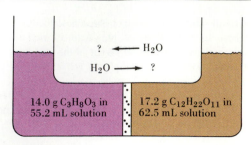

14.0 g C$_3$H$_8$O$_3$ in 55.2 mL solution

17.2 g C$_{12}$H$_{22}$O$_{11}$ in 62.5 mL solution

Exercise 54

ecules. In what direction will a net flow of water occur, that is, from left to right or right to left?

55. Refer to the Summarizing Example. Calculate the freezing point of a saturated solution of cinnamaldehyde in water. Which is the best method of determining the molecular weight of cinnamaldehyde, measuring the osmotic pressure or the freezing point depression of a saturated water solution? Explain. [*Hint:* Refer to Step 5.]

56. Verify that a 20.% sucrose solution [C$_{12}$H$_{22}$O$_{11}$(aq)] would rise to a height of about 150 m as a result of its osmotic pressure.

57. When the stems of cut flowers are held in a concentrated salt solution, the flowers wilt. When a fresh cucumber is placed in a similar solution, it shrivels up (becomes pickled). Explain the basis of these phenomena.

58. What mass of hemoglobin (M.W. 6.84 × 10^4) must be present per 100. cm^3 of a solution to exert an osmotic pressure of 6.15 mmHg at 25 °C?

59. An 0.50-g sample of polyisobutylene (a polymer used in synthetic rubber) in 100.0 cm^3 of benzene solution has an osmotic pressure at 25 °C that supports a 5.1-mm column of solution ($d = 0.88$ g/cm^3). What is the molecular weight of the polyisobutylene?

Strong electrolytes, weak electrolytes, and nonelectrolytes

60. Arrange the following aqueous solutions in order of increasing ability to conduct electric current. Comment on the rea-

sons for this arrangement. 0.01 M NaCl; 1.0 M C$_2$H$_5$OH; 0.01 M MgCl$_2$; 0.01 M HC$_2$H$_3$O$_2$.

61. Predict the approximate freezing points of 0.10 m solutions of the following solutes dissolved in water: (**a**) urea; (**b**) NH$_4$NO$_3$; (**c**) HCl; (**d**) CaCl$_2$; (**e**) MgSO$_4$; (**f**) ethanol; (**g**) HC$_2$H$_3$O$_2$ (acetic acid).

62. Calculate the van't Hoff factors of the following weak electrolyte solutions.

(**a**) 0.050 m HCHO$_2$, which freezes at −0.0986 °C
(**b**) 0.100 M HNO$_2$, which has a hydrogen ion (and nitrite ion) concentration of 6.91 × 10^{-3} M

63. NH$_3$(aq) conducts electric current only weakly. The same is true for acetic acid, HC$_2$H$_3$O$_2$(aq). When these solutions are mixed, however, the resulting solution conducts electric current very well. Propose an explanation.

Colloidal mixtures

64. Discuss some of the principal differences between colloidal mixtures and true solutions.

65. Describe what is meant by the terms (**a**) aerosol; (**b**) emulsion; (**c**) foam; (**d**) hydrophobic colloid; (**e**) dialysis.

66. The particles of a particular colloidal solution of arsenic trisulfide (As$_2$S$_3$) are negatively charged. Which 0.0005 M solution would be most effective in coagulating this colloidal solution: KCl, MgCl$_2$, AlCl$_3$, or Na$_3$PO$_4$? Explain.

*67. Refer to Figure 13-22. Suppose the original cube, 1.00 cm on edge, is cut in half in all three directions, so as to produce *eight* cubes, each 0.50 cm on edge. Then suppose these 0.50-cm cubes are each subdivided into *eight* identical cubes, 0.25 cm on edge, and so on.

(**a**) How many of these successive subdivisions are required before the cubes are reduced in size to colloidal dimensions? [*Hint:* Use the definition of colloidal dimensions given on page 464.]
(**b**) What is the total surface area-to-volume ratio of the cubes at this point where the colloidal dimensions have just been reached?

Additional Exercises

68. A solution has a density of 1.235 g/cm^3 and is 90.0% glycerol, C$_3$H$_8$O$_3$, and 10.0% H$_2$O, by mass. Determine (**a**) the molarity of C$_3$H$_8$O$_3$ (with H$_2$O as the solvent); (**b**) the molarity of H$_2$O (with C$_3$H$_8$O$_3$ as the solvent); (**c**) the molality of H$_2$O in C$_3$H$_8$O$_3$; (**d**) the mole fraction of C$_3$H$_8$O$_3$; (**e**) the mole percent H$_2$O.

69. A brine solution contains 2.52% NaCl, by mass. If a 75.0-mL sample is found to weigh 76.7 g, how many liters of this brine would be required to extract 1000. kg NaCl by evaporating this solution to dryness?

70. A typical root beer is described as containing 0.013% of a 75% H$_3$PO$_4$ solution, by mass. How many mg P are contained in a 12-oz can of this root beer (1 oz = 29.6 mL)? Assume a solution density of 1.00 g/mL.

71. Calculate the molality of the ethanol–water solution described in Exercise 26. Does the molality differ at the two temperatures (i.e., 15 and 25 °C)? Explain.

*72. Water and phenol are only partially miscible at temperatures below 66.8 °C. In a mixture prepared at 29.6 °C from 50.0 g of water and 50.0 g of phenol, 32.8 g of a phase consisting of 92.50% water and 7.50% phenol is obtained. This can be

considered a saturated solution of phenol in water. What is the percent by mass of water in the second phase—a saturated solution of water in phenol?

73. Benzoic acid, HC$_7$H$_5$O$_2$, is much more soluble in sodium hydroxide solution, NaOH(aq), than it is in pure water. Can you suggest a reason for this? The structural formula for benzoic acid is

74. One way to cause a solute to recrystallize from solution is to change the temperature of a saturated solution. Another way is to evaporate solvent from the solution. 335 mL of a saturated solution of KCl in water is prepared at 25.0 °C ($d = 1.16$ g/mL). If 55 g H$_2$O is evaporated from the solution, at the same time that the temperature is reduced from 25.0 to 0.0 °C, what mass of KCl(s) will recrystallize? (Refer to Figure 13-8.)

75. A benzene–toluene solution with $\chi_{benz.} = 0.300$ has a normal boiling point of 98.6 °C. The vapor pressure of pure toluene at 98.6 °C is 533 mmHg. What must be the vapor pressure of pure benzene at 98.6 °C? (Assume ideal solution behavior.)

76. You plan to do some molecular weight determinations by freezing point depression. You plan to use 50.0 g of benzene as the solvent and you want the freezing point depression, ΔT_f, to be between 2 and 3 °C. What mass of the unknown would you use if its anticipated molecular weight is **(a)** 85; **(b)** 125?

77. Use the concentration of an isotonic saline solution, 0.9% NaCl (mass/vol), to determine the osmotic pressure of blood at body temperature, 37.0 °C. [*Hint:* Recall that NaCl is completely dissociated in aqueous solutions.]

78. What pressure is required in the reverse osmosis depicted in Figure 13-18 if the salt water contains 3.0% NaCl, by mass? [*Hint:* Recall that NaCl is completely dissociated in aqueous solutions. Also, assume a temperature of 25 °C.]

79. An important test for the purity of an organic compound is to measure its melting point. Usually, if the compound is not pure it melts at a *lower* temperature than the pure compound.

  (a) Why is this the case, rather than the melting point being higher in some cases and lower in others?

★ **(b)** Are there any conditions where the melting point of the impure compound may be *higher* than that of the pure compound? Explain.

80. Vitamins are substances required in human diets to maintain normal health. Vitamins are classified in various ways; one is based on their solubilities. Some vitamins are soluble in water and some are soluble in fats. (Fats are substances whose molecules have long hydrocarbon chains.) Given below are the structural formulas of four common vitamins. Two are *water-soluble* and two are *fat-soluble*. Categorize the solubility of each of these vitamins and explain your reasoning.

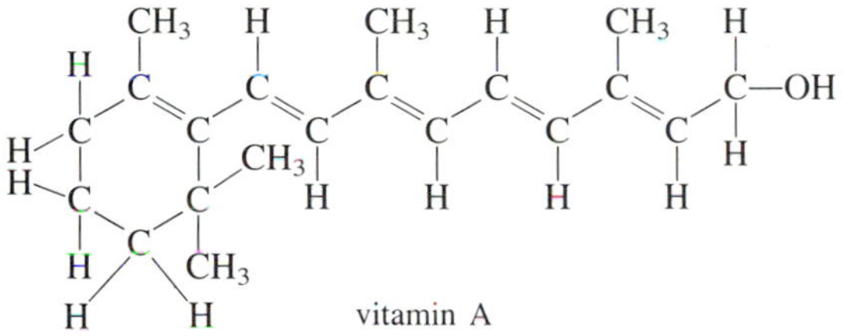

vitamin A

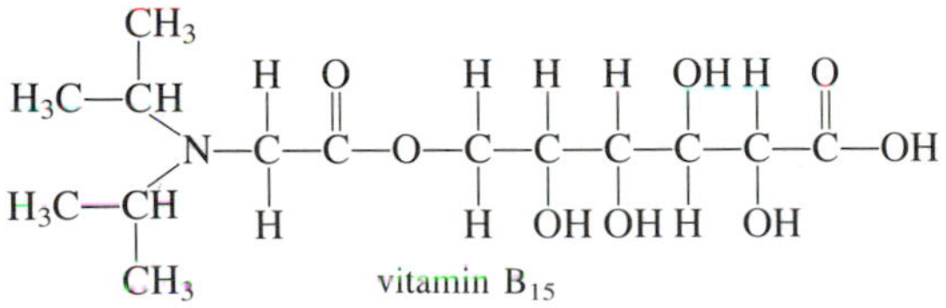

vitamin B$_{15}$

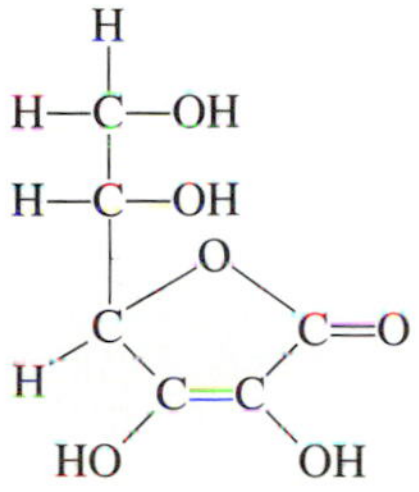

vitamin C

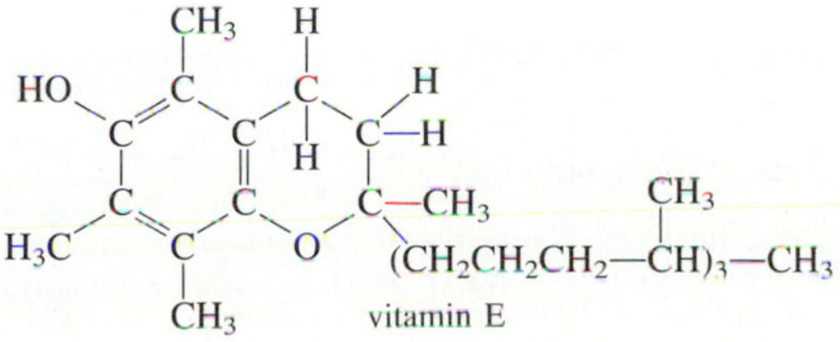

vitamin E

81. Instructions on a container of Prestone (ethylene glycol) give the following volumes of Prestone to be used in protecting a 12-qt cooling system against freeze-up at different temperatures (the remaining liquid is water): 10 °F, 3 qt; 0 °F, 4 qt; −15 °F, 5 qt; −34 °F, 6 qt. Since the freezing point of the coolant is successively lowered by using more Prestone, why not use still more than 6 qt of Prestone (and proportionally less water) to ensure the maximum protection against freezing? [*Hint:* What is the solvent in these solutions?]

82. The two compounds whose structures are depicted below are *isomers* (identical compositions but different structures). When derived from petroleum they always occur mixed together. *Meta*-xylene is used in aviation fuels and in the manufacture of dyes and insecticides. The principal use of *para*-xylene is in the manufacture of polyester resins and fibers (e.g., Dacron). Comment on the effectiveness of fractional distillation as a method of separating these two xylenes. What other method(s) might be used to separate them?

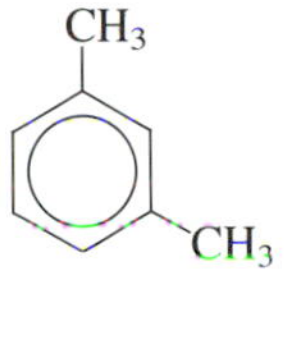

meta-xylene	*para*-xylene
f.p. −47.9 °C	f.p. 13.3 °C
b.p. 139.1 °C	b.p. 138.4 °C
$d = 0.864$ g/mL	$d = 0.861$ g/mL

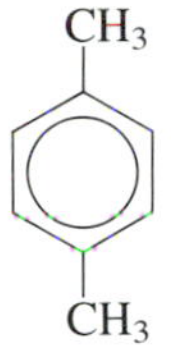

★ **83.** Stearic acid ($C_{18}H_{36}O_2$) is the most common fatty acid found in animal fat. Palmitic acid ($C_{16}H_{32}O_2$) is another common fatty acid. In fact, commercial grades of stearic acid usually contain palmitic acid as well. A 1.115-g sample of a commercial-grade stearic acid is dissolved in 50.00 mL of benzene ($d = 0.879$ g/mL) and the freezing point of the solution is found to be 5.072 °C. The freezing point of pure benzene is 5.533 °C. What must be the percent, by mass, of palmitic acid in the stearic acid sample?

84. Aluminum sulfate, $Al_2(SO_4)_3$, is widely used in water treatment plants. It helps to precipitate suspended particles in the water. Explain why this particular electrolyte is so effective for this purpose.

85. Suppose that 1.00 mg of gold is obtained in a colloidal dispersion in which the gold particles are assumed to be spherical, with a radius of 100. nm. (The density of gold is 19.3 g/cm^3.)

  (a) What is the total surface area of the particles?
  (b) What is the surface area of a single cube of gold weighing 1.00 mg?

86. In what volume of water must 1 mol of a nonelectrolyte be dissolved if the solution is to have an osmotic pressure of 1 atm at 273 K?

87. An aqueous solution that is 0.205 m in urea [$CO(NH_2)_2$] is observed to boil at 100.025 °C. Is the prevailing barometric pressure above or below 760. mmHg?

88. For the situation described in Exercise 87, what is the actual value of the prevailing barometric pressure? [*Hint:* You will need to use some data and equations from Chapter 12.]

89. Thiophene (f.p., −38.3 °C; b.p. 84.4 °C) is a sulfur-containing hydrocarbon that is sometimes used as a solvent in place of benzene because it has a wider liquid range. Combustion of a 2.348-g sample of thiophene produces 4.913 g CO_2, 1.005 g H_2O, and 1.788 g SO_2. When a 0.867-g sample of thiophene is dissolved in 44.56 g of benzene (C_6H_6), the freezing point is lowered by 1.183 °C. What is the molecular formula of thiophene?

90. Nitrobenzene, $C_6H_5NO_2$ (f.p. = 5.7 °C), and benzene, C_6H_6 (f.p. = 5.5 °C), are completely miscible in each other. With these two liquids it is possible to prepare *two* different solutions having a freezing point of 0.0 °C. What are the compositions of these two solutions, expressed as percent nitrobenzene, by mass? [*Hint:* Use K_f values from Table 13-2.]

***91.** Refer to Exercises 11 and 38. What is the composition of air dissolved in water, expressed as volume percents of N_2 and O_2? [*Hint:* Is the composition of dissolved air the same as that of atmospheric air? If not, which component of air is enriched in the dissolving process?]

***92.** Refer to the Summarizing Example. The normal boiling point of cinnamaldehyde is 246.0 °C, but at this temperature it begins to decompose. As a result, cinnamaldehyde cannot be easily purified by ordinary distillation. A method that can be used instead is *steam distillation*. A *heterogeneous* mixture of cinnamaldehyde and water is heated until the *sum* of the vapor pressures of the two liquids is equal to barometric pressure. At this point the temperature remains constant as the liquids vaporize. The mixed vapor condenses to produce *two immiscible* liquids; one liquid is essentially pure water and the other, pure cinnamaldehyde. The following vapor pressures of cinnamaldehyde are given: 1 mmHg at 76.1 °C; 5 mmHg at 105.8 °C; and 10 mmHg at 120.0 °C.

(a) What is the approximate temperature at which the steam distillation occurs? [*Hint:* Use data from Chapter 12 also.]

(b) The proportions of the two liquids condensed from the vapor is *independent* of the composition of the boiling mixture, as long as both liquids are present in the boiling mixture. Explain why this is so.

(c) Which of the two liquids, water or cinnamaldehyde, condenses in the greater quantity, by mass? Explain.

93. Solution A contains 0.515 g of urea, $CO(NH_2)_2$, dissolved in 85.0 g H_2O. Solution B contains 2.50 g of sucrose, $C_{12}H_{22}O_{11}$, dissolved in 92.5 g H_2O. Above which of these solutions is the water vapor pressure greater?

***94.** The two solutions described in Exercise 93 are placed in separated containers but in an enclosure where their vapors may mix freely (see the illustration). Water evaporates from the solution of higher vapor pressure and condenses into the solution of lower vapor pressure. This process continues until both solutions have the same water vapor pressure. What are the compositions of solutions A and B when this equilibrium is reached?

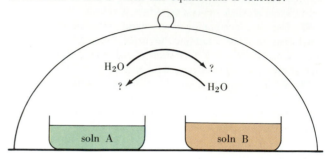

Exercise 94

***95.** At 20 °C liquid benzene has a density of 0.879 g/cm³; liquid toluene, 0.867 g/cm³. Assuming ideal solutions

(a) Calculate the densities of solutions containing 20, 40, 60, and 80 vol. % benzene.

(b) Plot a graph of density vs. volume % composition.

***96.** For the solution described in Exercise 95, let V = vol. % benzene and establish that

$$d = \frac{1}{100}[0.879\ V + 0.867(100 - V)]$$

***97.** The following data are given for the densities of ethanol-water solutions at 15 °C as a function of *volume* percent ethanol: 0%, 0.999 g/cm³; 20.0%, 0.977 g/cm³; 40.0%, 0.952 g/cm³; 60.0%, 0.914 g/cm³; 80.0%, 0.864 g/cm³; 100%, 0.794 g/cm³. Are ethanol-water mixtures ideal?

***98.** Demonstrate that for a *dilute aqueous* solution the molality is essentially equal to the molarity concentration.

***99.** Show that in a dilute solution the solute mole fraction is proportional to molality, and that in a dilute *aqueous* solution solute mole fraction is proportional to molarity.

***100.** A saturated solution is prepared at 70 °C containing 32.0 g $CuSO_4$ per 100.0 g soln. A 335-g sample of this solution is then cooled to 0 °C and $CuSO_4 \cdot 5H_2O$ crystallizes out. If the concentration of a saturated solution at 0 °C is 12.5 g $CuSO_4$/ 100.0 g soln, how many g $CuSO_4 \cdot 5H_2O$ would be obtained? [*Hint:* Note that the solution composition is stated in terms of $CuSO_4$ but that the solid that crystallizes is the hydrate, $CuSO_4 \cdot 5H_2O$.]

Self-Test Questions

For questions 101 through 110 select the single item that best completes each statement.

101. An *aqueous* solution is 0.01 M CH_3OH. The concentration of this solution is also very nearly (a) 0.01% CH_3OH (mass/vol); (b) 0.01 *m* CH_3OH; (c) $\chi_{CH_3OH} = 0.01$ (i.e., mole fraction $CH_3OH = 0.01$); (d) 0.99 M H_2O.

102. The most water soluble of the following is (a) $C_6H_6(l)$; (b) C(s); (c) $NH_2OH(s)$; (d) $C_{10}H_8(s)$.

103. The most likely of the following mixtures to be an ideal solution is (a) $NaCl–H_2O$; (b) $C_2H_5OH(l)–C_6H_6(l)$; (c) $C_7H_{16}(l)–H_2O(l)$; (d) $C_7H_{16}(l)–C_8H_{18}(l)$.

104. A compound that is moderately soluble both in water [$H_2O(l)$] and in benzene [$C_6H_6(l)$] is (a) phenol, C_6H_5OH; (b) naphthalene, $C_{10}H_8$; (c) hexane, C_6H_{14}; (d) sodium chloride, NaCl.

105. The *solubility* of a nonreactive gas in water increases with (a) an increase in temperature; (b) an increase in gas pressure; (c) increases both in temperature and gas pressure; (d) an increase in the volume of gas in equilibrium with the available water.

106. The *best* electrical conductor of the following aqueous solutions is (a) 0.10 M NaCl; (b) 0.10 M C_2H_5OH (ethanol);

(c) 0.10 M $HC_2H_3O_2$ (acetic acid); (d) 0.10 M $C_{12}H_{22}O_{11}$ (sucrose).

107. The aqueous solution with the *lowest* freezing point of the following group is (a) 0.01 *m* $MgSO_4$; (b) 0.01 *m* NaCl; (c) 0.01 *m* C_2H_5OH (ethanol); (d) 0.008 *m* MgI_2.

108. An ideal liquid solution has equal mole fractions of two *volatile* components A and B. In the vapor above the solution the mole fractions of A and B (a) are both 0.50; (b) are equal but not necessarily 0.50; (c) are not very likely to be equal; (d) are 1.00 and 0.00, respectively.

109. The best method for determining the molecular weight of a substance with high molecular weight generally involves measuring a (a) vapor density; (b) osmotic pressure; (c) freezing point depression; (d) boiling point elevation.

110. A method of removing excess solutes from a colloidal mixture is (a) distillation; (b) recrystallization; (c) dialysis; (d) gas chromatography.

111. A solution contains 0.865 g naphthalene, $C_{10}H_8$, in 38.7 g of benzene, C_6H_6.

    (a) What is the percent $C_{10}H_8$, by mass, in this solution?
    (b) What is the molality of $C_{10}H_8$ in this solution?
    (c) What is the freezing point of the solution? [The freezing point of pure C_6H_6 is 5.53 °C; K_f for C_6H_6 = 5.12 °C kg solvent (mol solute)$^{-1}$.]

112. Pure liquid HCl is a poor electrical conductor. So is pure liquid water. When these two liquids are mixed, however, the resulting solution conducts electric current very well. How do you explain this?

113. Solutions of isopropanol, $CH_3CHOHCH_3$, are prepared in which the concentration of isopropanol is **(a)** 0.10%, by mass; **(b)** 0.10 *m*; and **(c)** $\chi_{isopropanol}$ = 0.10. Which of these three solutions has the highest vapor pressure of water? the lowest freezing point?

114. Which aqueous solution from the column on the right has the property listed in the column of the left? Explain your choices. [*Hint:* The same solution may display more than one of the properties listed.]

Property	Solution
lowest electrical conductivity	0.10 *m* KCl
lowest boiling point	0.15 *m* $CO(NH_2)_2$ (urea)
highest vapor pressure of H_2O	0.10 *m* $HC_2H_3O_2$ (acetic acid)
lowest freezing point	0.05 *m* NaCl

14

An Introduction to Descriptive Chemistry: The First 20 Elements

The Voyager traveled nonstop around the world in 1986. Its construction featured organic polymers, graphite fibers, and advanced ceramics. These materials, as well as the fuel for the aircraft, are all comprised of elements among the first 20. [Mark Greenberg/Visions]

The day may come when hydrogen will be the most widely used fuel. Helium, the noble gas of lowest atomic weight, has physical properties that make it indispensable for achieving very low temperatures. Lithium could be a vital element in the development of practical fusion nuclear reactors. Nitrogen is the principal element in the atmosphere, and, together with such other elements as phosphorus and potassium, is an essential nutrient for plant growth. Carbon, in combination with H, O, N, and several other elements, such as sulfur, is the key element in living matter. Oxygen, in addition to its presence in water, which is surely one of the most important of all compounds, is the most abundant element in the earth's crust. And the element found most commonly with oxygen in mineral sources is silicon. Aluminum is the most abundant metal in the earth's crust and its uses have grown significantly in this century. There can be little doubt of the practical importance of the first 20 elements. Interestingly, these same 20 elements have also been important to us in establishing theoretical principles in earlier chapters.

An important bridge between theoretical and applied chemistry is descriptive chemistry, by which we mean a systematic study of the physical and chemical behavior of the elements and their compounds, based, where possible, on fundamental ideas. In this chapter, we apply principles of atomic and molecular structure, periodic relationships, thermochemistry, phase equilibria, and intermolecular forces to a few aspects of the descriptive chemistry of the first 20 elements. We also look at some typical reactions and uses of these elements and their compounds. When we return to a discussion of descriptive chemistry in several later chapters, we will apply additional principles developed in the text.

14-1 Occurrence and Abundances of the Elements

What is the most abundant element? This seemingly simple question does not have a simple answer. If we consider the entire universe, hydrogen accounts for about 90% of all the atoms and three-fourths of the mass; helium accounts for most of the rest. If we consider only the elements present on earth, iron is probably the most abundant element. However, most of this iron is believed to be in the core of the earth. For practical purposes we are limited to using the elements that are present in the atmosphere, oceans, and solid crust of the earth. The relative abundances of the elements in the earth's solid crust, which varies in depth from 5 to 40 km, are represented in Figure 14-1. The principal forms in which the eight most abundant elements appear are listed in Table 14-1.

The abundances of the elements on a *mole percent* basis, of course, are rather different from those of Figure 14-1 (see Exercise 16).

Not all the known elements are found in the earth's crust. Trace amounts of neptunium ($Z = 93$) and plutonium ($Z = 94$) are found in uranium minerals, but for all practical purposes elements with atomic numbers higher than 92, called *transuranium* elements, can only be produced by nuclear processes (described in Section 26-4). Moreover, most of the elements do not occur *free* in nature, that is, as the uncombined element. Only about one-fifth of them exist in the elemental form, and the remainder occur exclusively in chemical combination with other elements.

Do not assume that the ease and the cost of obtaining a pure element from its natural sources are necessarily related only to the relative abundance of the element. Aluminum is the most abundant of the metals, but it cannot be produced as cheaply as iron, primarily because concentrated deposits of iron-containing compounds, called iron **ores,** are more common than are ores of aluminum. Some elements whose percent abundances are quite low are nevertheless widely used because their ores are fairly common. This is the case with copper, whose abundance is only 0.005%. On the other hand, some elements have a fairly high abundance but no characteristic ores of their own. This is the case with rubidium, the sixteenth most abundant element.

479

FIGURE 14-1
Abundances of the elements in the earth's solid crust (mass percent).

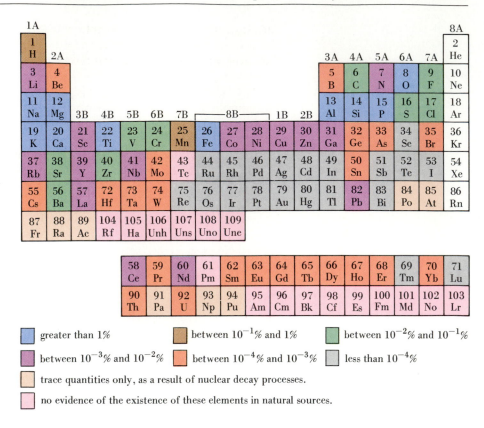

- ■ greater than 1%
- ■ between $10^{-1}\%$ and 1%
- ■ between $10^{-2}\%$ and $10^{-1}\%$
- ■ between $10^{-3}\%$ and $10^{-2}\%$
- ■ between $10^{-4}\%$ and $10^{-3}\%$
- □ less than $10^{-4}\%$
- ■ trace quantities only, as a result of nuclear decay processes.
- ■ no evidence of the existence of these elements in natural sources.

The availability of the elements found in seawater is based on their concentration and on whether we can find a chemically feasible and economically viable method of extracting them. The most abundant elements in seawater are listed in Table 14-2. Although we could get about one-half the elements from seawater, only magnesium and bromine are produced in commercially significant quantities.

We can obtain a few of the elements that occur *uncombined* in nature by physical separations. One example is the fractional distillation of liquid air for the production of oxygen, nitrogen, and argon. The majority of the elements, however, require

"Panning" for gold also involves a physical separation.

TABLE 14-1
Commonly Occurring Forms of the Eight Most Abundant Elements in the Earth's Solid Crust

Rank	Element	Mass percent	Some commonly occurring forms
1	oxygen	45.5	silica (SiO_2); silicates; metallic oxides [also in H_2O in oceans and $O_2(g)$ in the atmosphere]
2	silicon	27.2	silica (sand, quartz, agate, flint); silicates (feldspar, clay, mica)
3	aluminum	8.3	silicates (clay, feldspar, mica); oxide (bauxite)
4	iron	6.2	oxide (hematite, magnetite)
5	calcium	4.7	carbonate (limestone, marble, chalk); sulfate (gypsum); fluoride (fluorite); silicates (feldspar, zeolites)
6	magnesium	2.8	carbonate; chloride; sulfate (Epsom salts)
7	sodium	2.3	chloride (rock salt); silicates (feldspar, zeolites) (also as NaCl in seawater)
8	potassium	1.8	chloride; silicates (feldspar, mica)

TABLE 14-2
Principal Constituents of Seawater

Constituent	Concentration, mg/L
Cl^-	19,000
Na^+	10,500
SO_4^{2-}	2,650
Mg^{2+}	1,350
Ca^{2+}	400
K^+	380
HCO_3^-	140
Br^-	65
H_3BO_3	26
Sr^{2+}	8
F^-	1

chemical reactions to release them from their combined forms. Since the objective of these reactions is to obtain the uncombined element in the *zero* oxidation state, they must be oxidation–reduction reactions.

14-2 Trends Among the First 20 Elements

In this chapter we will see many examples of the fact that members of the same group of the periodic table have similar properties. However, we will also discover that the first member of a group differs somewhat from the heavier members below it. And, for certain of the lighter elements we will find "diagonal" similarities, that is, between Li (group 1A) and Mg (2A), Be (2A) and Al (3A), and B (3A) and Si (4A).

What we have also learned to expect is a change in properties as we move across a period of the periodic table, as suggested by Figure 14-2. Some properties increase (or decrease) regularly from left to right in a period, such as electronegativity. Some properties do not. Melting points, which are related to the strengths of intermolecular forces of attraction in a crystalline solid, rise to a maximum with C in the second period and Si in the third period. This is because C and Si form network covalent solids (recall Figure 12-33).

Example 14-1 _____

Using the periodic table to describe trends in properties. Describe the trends that you would expect among the third period elements, E, **(a)** in electrical conductivity of the solid element; **(b)** in percent ionic character of the bond E—Cl.

FIGURE 14-2
Two properties of the second and third period elements.

The trends outlined in these graphs are described in the text.

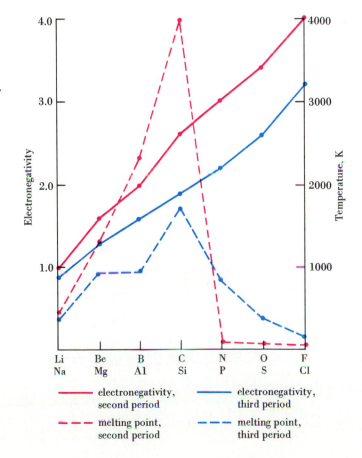

Solution

(a) Good electrical conductivity is something we associate with the metallic bond. We would expect the three metals of the third period—Na, Mg, and Al—to be good electrical conductors. However, we have not studied the metallic state in sufficient detail to say which of the three should be the best conductor. Silicon is a metalloid and also a semiconductor. This makes it not nearly as good an electrical conductor as the three metals preceding it. The remaining third period elements—P, S, Cl (and Ar)—are all nonmetals and, therefore, nonconductors.

(b) We have learned that the greater the electronegativity (EN) difference between bonded atoms the greater the percent ionic character of the bond (recall Section 10-6). We can get EN differences directly from Figure 14-2. They are the differences between the EN of Cl and those of the other third period elements. The greatest difference is between Cl and Na; the Na—Cl bond has the greatest percent ionic character. This is followed by the Mg—Cl bond, Al—Cl bond, and so on through the third period. Recall that an EN difference of about 2 or greater corresponds to a bond with more than 50% ionic character (see Figure 10-11), and you will see that only the Na—Cl and Mg—Cl bonds are essentially ionic. The other E—Cl bonds become progressively more covalent.

SIMILAR EXAMPLES: Exercises 10, 11, 18, 66.

14-3 Hydrogen

1A	
1 **H**	2A
3 **Li**	**4** **Be**
11 **Na**	**12** **Mg**

Since its discovery by Henry Cavendish in 1766, hydrogen has been crucial to the development of theories about the structure of matter. John Dalton based atomic weights on a value of 1 for hydrogen. Humphry Davy (1810) proposed that hydrogen is the key element in common acids. Niels Bohr chose the hydrogen atom for the first application of quantum mechanics to atomic structure. Erwin Schrödinger (1927) based his wave mechanics on the H atom, and the H_2 molecule is the usual starting point of theories of molecular structure. For all of its theoretical significance, though, hydrogen is also of great practical importance.

Occurrence and Preparation. *Hydrogen occurs in more compounds than any other element.* However, the free element can be easily produced only from a few of its compounds. As a first choice we probably would pick H_2O—the most abundant hydrogen-containing compound. To extract elemental hydrogen from water means reducing the oxidation state of H from $+1$ in H_2O to 0 in H_2. To do this we need an appropriate *reducing* agent. Three suitable reducing agents are carbon (as coal or coke), carbon monoxide, and hydrocarbons—particularly methane (natural gas). Reaction (14.1) is carried out at elevated temperatures. Reactions (14.2) and (14.3) require catalysts as well.

Water gas reactions: $$C(s) + H_2O(g) \longrightarrow CO(g) + H_2(g) \qquad (14.1)$$

$$CO(g) + H_2O(g) \longrightarrow CO_2(g) + H_2(g) \qquad (14.2)$$

Reforming of methane: $$CH_4(g) + H_2O(g) \longrightarrow CO(g) + 3\,H_2(g) \qquad (14.3)$$

The systematic naming of hydrocarbons is discussed in Section 27-1.

Another important source of $H_2(g)$ is as a by-product of petroleum refining operations, such as in the conversion of butane to 1-butene.

$$
\begin{array}{c}
\underset{\text{butane}}{
\begin{array}{c}
\;\;\;\;\text{H}\;\;\text{H}\;\;\text{H}\;\;\text{H}\\
\;\;\;\;|\;\;\;\;|\;\;\;\;|\;\;\;\;|\\
\text{H}-\text{C}-\text{C}-\text{C}-\text{C}-\text{H}\\
\;\;\;\;|\;\;\;\;|\;\;\;\;|\;\;\;\;|\\
\;\;\;\;\text{H}\;\;\text{H}\;\;\text{H}\;\;\text{H}
\end{array}}
\end{array}
\xrightarrow{\text{catalyst}}
\begin{array}{c}
\underset{\text{1-butene}}{
\begin{array}{c}
\;\;\;\;\text{H}\;\;\text{H}\\
\;\;\;\;|\;\;\;\;|\\
\text{H}-\text{C}-\text{C}-\text{C}=\text{C}-\text{H} + H_2(g)\\
\;\;\;\;|\;\;\;\;|\;\;\;\;|\;\;\;\;|\\
\;\;\;\;\text{H}\;\;\text{H}\;\;\text{H}\;\;\text{H}
\end{array}}
\end{array}
\qquad (14.4)
$$

FIGURE 14-3
The electrolysis of water.

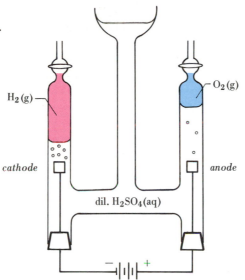

Electricity is passed through a dilute electrolyte solution, such as dilute $H_2SO_4(aq)$. H^+ ions are attracted to the negative electrode (cathode), where they gain electrons to form H atoms, which join to form H_2 molecules that escape from solution. Although SO_4^{2-} ions are attracted to the positive electrode (anode), they are not involved in the transfer of charge at the anode. Instead, a reaction occurs in which H_2O molecules decompose, replacing the H^+ ions lost at the cathode and releasing $O_2(g)$. The net reaction, then, is the production of the gases H_2 and O_2 at the expense of water molecules: $2 H_2O \rightarrow 2 H_2(g) + O_2(g)$. Electrode processes are discussed in Chapter 21. In this chapter our interest is just on the net outcome of electrolysis reactions, not on the details of what happens at the electrodes.

Electrolysis. The decomposition of water offers the most direct method of producing high-purity hydrogen (and oxygen as well). Some compounds can be decomposed into their elements simply by heating, but even at 2000 °C water is only about 1% decomposed. A technique long used by chemists when decomposition by heating is not possible is **electrolysis**—decomposition by electric current. The electrolysis of water is pictured in Figure 14-3, where a brief description of the process is also given. We study the principles of electrochemistry in Chapter 21 and consider details of electrolysis reactions at that time. For now, whenever electrolysis is required let us just note this by writing "electrolysis" above the arrow.

An example of a compound that readily decomposes on heating is silver oxide (recall equation 4.15).

Electrolysis of water.

$$2 H_2O(l) \xrightarrow{\text{electrolysis}} 2 H_2(g) + O_2(g) \qquad (14.5)$$

The principal drawback of the electrolysis of water is the demanding electrical energy requirement—about 0.1 kilowatt hour (kWh) per mol $H_2(g)$. As a consequence, electrolysis is a significant method of producing $H_2(g)$ only where hydroelectric power is cheap (such as in Canada and Norway).

A 100-watt light bulb requires 0.1 kWh of energy for every hour it is lit.

Laboratory Methods. Very often we can use methods in the chemical laboratory that are not feasible for commercial production. The electrolysis of water is a very useful laboratory method, and another method involves the reaction of certain metals with an aqueous solution of a mineral acid (recall the list in Table 5-6).

$$Zn(s) + 2 H^+(aq) \longrightarrow Zn^{2+}(aq) + H_2(g) \qquad (14.6)$$

The most active of the metals, those of group 1A and the heavier members of group 2A, are even able to displace $H_2(g)$ from pure water, where the concentration of H^+ is very low.

Reaction of active metals with water.

$$2 M(s) + 2 H_2O(l) \longrightarrow 2 M^+(aq) + 2 OH^-(aq) + H_2(g) \qquad (14.7)$$
$$\text{(where M is any group 1A metal)}$$

$$M(s) + 2 H_2O(l) \longrightarrow M^{2+}(aq) + 2 OH^-(aq) + H_2(g) \qquad (14.8)$$
$$\text{(where M is Ca, Sr, Ba, or Ra)}$$

Some alkali metals react with water with explosive violence (recall Figure 9-2), but the alkaline earth metals react more slowly, as shown in Figure 14-4.

Reactions of Hydrogen. Hydrogen forms binary compounds with most of the other elements. In part, this is because its electronegativity is intermediate in value between those of active metals and active nonmetals. Binary hydrides are usually

FIGURE 14-4
Reaction of calcium with
water.

Bubbles of $H_2(g)$ are clearly visible rising from the calcium
metal. Note that Ca, unlike several of the alkali metals, is more
dense than water. A few drops of
phenolphthalein (an acid–base
indicator) have been added to
the water. The pink color of the
phenolphthalein signals the
buildup of OH^- in the solution.
[Carey B. Van Loon]

grouped into the three broad categories: covalent, ionic, and metallic. In fact, however, many hydrides have properties that suggest a bonding type that is a mix of
these categories. Hydrides of the group 3A, 4A, 5A, 6A, and 7A elements are
essentially **covalent hydrides.** Some of these hydrides are the simple molecules that
you would predict from locations of the elements in the periodic table, such as HF,
H_2O, and NH_3. Others are more complex and will be considered elsewhere in the
text. Many hydrides form by the direct union of hydrogen and the second element,
such as

$$H_2(g) + Cl_2(g) \longrightarrow 2\ HCl(g) \tag{14.9}$$

$$3\ H_2(g) + N_2(g) \rightleftharpoons 2\ NH_3(g) \tag{14.10}$$

Ionic hydrides are formed between hydrogen and the most active metals, particularly those of groups 1A and 2A. In these compounds hydrogen exists as the
hydride *ion*, $\mathbf{H:^-}$.

$$2\ M(s) + H_2(g) \longrightarrow 2\ MH(s) \qquad M(s) + H_2(g) \longrightarrow MH_2(s) \tag{14.11}$$
$$\text{(M is any group 1A metal)} \qquad \text{(M is Ca, Sr, or Ba)}$$

Ionic hydrides react with water to liberate $H_2(g)$. CaH_2, a gray solid, has been used
as a ''portable'' source of $H_2(g)$ for filling weather observation balloons.

$$CaH_2(s) + 2\ H_2O \longrightarrow Ca^{2+}(aq) + 2\ OH^-(aq) + 2\ H_2(g) \tag{14.12}$$

Metallic hydrides are those commonly found among the transition elements.
A distinctive feature of these hydrides is that in many cases they are **nonstoichiometric**—the ratio of H atoms to metal atoms is variable, not fixed. This is because
H atoms can enter the voids or holes among the metal atoms in a crystalline lattice,
filling some but not others. We can think of some of these metal hydrides as alloys
or interstitial solid solutions of hydrogen in metals. For example, palladium can
dissolve up to 900 times its volume of hydrogen.

Uses of Hydrogen. Hydrogen was not listed among the top chemicals in Table 1-4
because only a small percentage of the hydrogen produced is ever sold to customers.
Most hydrogen is produced and used on the spot. Viewed in these terms, its most
important use (about 42%) is in the manufacture of NH_3 (reaction 14.10). Next in
importance (about 38%) is petroleum refining, where H_2 is produced in some operations and consumed in others, such as in the production of the high-octane gasoline
component, isooctane, from diisobutylene.

$$
\begin{array}{c}
\quad\quad CH_3 \\
\quad\quad | \\
CH_3\!-\!\overset{\displaystyle |}{\underset{\displaystyle |}{C}}\!-\!C\!=\!C\!-\!CH_3 + H_2(g) \\
\quad\quad CH_3 \ \ H \ \ CH_3 \\
\text{diisobutylene}
\end{array}
\xrightarrow{\text{catalyst}}
\begin{array}{c}
\quad\quad CH_3 \ H \ \ H \\
\quad\quad | \quad\ | \ \ \ | \\
CH_3\!-\!\overset{\displaystyle |}{\underset{\displaystyle |}{C}}\!-\!C\!-\!C\!-\!CH_3 \\
\quad\quad CH_3 \ H \ \ CH_3 \\
\quad\quad \text{isooctane}
\end{array}
\tag{14.13}
$$

In reactions similar to (14.13), called **hydrogenation** reactions, hydrogen atoms
can be added to double bonds in other molecules. This type of reaction, for example, will convert a liquid like oleic acid, $C_{17}H_{33}COOH$, to a solid—stearic acid
$C_{17}H_{35}COOH$. Similar reactions serve as the basis for converting oils such as corn
oil into solid or semisolid fats (e.g., Crisco).

Another important chemical manufacturing process that uses hydrogen is the
synthesis of methyl alcohol (methanol).

$$CO(g) + 2\ H_2(g) \longrightarrow CH_3OH(g) \tag{14.14}$$

Hydrogen gas is an excellent reducing agent and in some instances is used to
produce metals from their metal oxide ores. For example, at 850 °C,

$$WO_3(s) + 3\ H_2(g) \longrightarrow W(s) + 3\ H_2O(g) \tag{14.15}$$

Oleic acid has one carbon-
to-carbon double bond and
stearic acid has all single
bonds. Note that their formulas differ by just two H
atoms.

A Hydrogen Economy

A fundamental change in the twentieth century has been the shift from coal to petroleum as the chief fuel resource, just as coal supplanted wood a century or two before. But in Section 7-11 we learned that available deposits of coal, and especially of petroleum, are limited. What will be the principal fuel of future centuries? One candidate is hydrogen.

Hydrogen has several attractive features as a fuel. For example, an automobile engine is 25–50% more efficient when it burns hydrogen rather than gasoline, and the exhaust is low in pollutants. The range of supersonic aircraft could be increased if the aircraft used liquid hydrogen as a fuel; hypersonic aircraft would also become possible. The problems with hydrogen are in finding a cheap source and an effective means of storing it. If an essentially limitless, inexpensive energy source, such as fusion energy, can be developed (see Section 26-9), then hydrogen can be produced by electrolysis, probably of seawater. Another attractive alternative is to develop a thermochemical cycle of chemical reactions which has as its net reaction the decomposition of water. It is important, however, that no step in the cycle require a very high temperature, and that each step give a reasonably high yield. In the following cycle temperatures do not exceed 500 °C.

The thermally insulated storage system for liquid hydrogen in an automobile adapted to use hydrogen as a fuel. [BMW]

$$6\ FeCl_2 + 8\ H_2O \xrightarrow{500\ °C} 2\ Fe_3O_4 + 12\ HCl + 2\ H_2$$

$$2\ Fe_3O_4 + 12\ HCl + 3\ Cl_2 \xrightarrow{200\ °C} 6\ FeCl_3 + 6\ H_2O + O_2$$

$$\underline{6\ FeCl_3 \xrightarrow{450\ °C} 6\ FeCl_2 + 3\ Cl_2}$$

$$2\ H_2O \longrightarrow 2\ H_2 + O_2 \qquad\qquad (14.16)$$

Gaseous hydrogen, because of its bulk, is difficult to store. When liquefied, the hydrogen occupies a much smaller volume, but then it must be maintained at very low temperatures ($T_c = -240.\ °C$). And, in either case, hydrogen must be maintained out of contact with oxygen or air, with which it forms explosive mixtures. One of the most promising storage systems may be to dissolve $H_2(g)$ in a metal or metal alloy, for example an iron–titanium alloy. The gas could then be released by mild heating. In an automobile, for example, this storage system would replace the gasoline tank. The heat required to release hydrogen from the metal could come from the exhaust gases from the engine.

If the problems described here can be solved, not only can hydrogen be used to supplant gasoline as a fuel for transportation, but it could also replace natural gas for space heating. Moreover, because H_2 is a good reducing agent, it could replace carbon (as coal or coke) in many metallurgical processes. And, of course, it would be abundantly available for reaction with N_2 to produce NH_3 for fertilizer manufacture and other uses. Another workable option would be the widespread use of hydrogen/oxygen fuel cells as a source of electricity (see Section 21-5). The combination of all these potential uses of hydrogen could bring about a fundamental change in our way of life, giving rise to what is called a **hydrogen economy.** ■ ■ ■

Liquid hydrogen is used as a rocket fuel. The engines in the Space Shuttle rocket system, for example, involve the reaction of hydrogen and oxygen, both of which are stored as liquids. The potential of hydrogen as a fuel of the future is explored in "A Hydrogen Economy" above.

An interesting, specialized use of hydrogen is in producing very high temperatures for welding and metal cutting operations. When $H_2(g)$ is passed through an electric arc, it dissociates into atomic H. The recombination of H atoms into molecular hydrogen, H_2, is a highly exothermic reaction. When this is followed by the

TABLE 14-3
Some Uses of Hydrogen

syntheses of
  ammonia, NH_3
  hydrogen chloride, HCl
  methanol, CH_3OH
hydrogenation reactions in:
  petroleum refining
  converting oils to fats
reduction of metal oxides, such as
  iron, cobalt, nickel,
  copper, tungsten,
  molybdenum
metal cutting and welding with
  atomic and oxyhydrogen torches
rocket fuel, usually $H_2(l)$ in
  combination with $O_2(l)$
fuel cells for generating electricity,
  in combination with $O_2(g)$
achieving low temperatures
  [b.p. of $H_2(l) = -253\ °C$]

combustion of $H_2(g)$ in $O_2(g)$, the temperature exceeds that required to melt tungsten (3370 °C). The uses of hydrogen described here, together with a few others, are listed in Table 14-3.

Example 14-2

Using fundamental principles to explain the facts of chemistry. One of the ways you may be asked to demonstrate your knowledge of chemical principles is to explain some of the factual information of descriptive chemistry. Use chemical principles learned in earlier chapters to

(a) Predict the formula of the simple hydride of arsenic.
(b) Identify the oxidizing and reducing agents in reaction (14.15).
(c) Determine the quantity of heat evolved in the recombination of H atoms to form 1 mol $H_2(g)$.

Solution

(a) Arsenic is in group 5A of the periodic table and has the Lewis structure
$\cdot\ddot{As}\cdot$. The rules for acceptable Lewis structures are satisfied in the combination of three H atoms with a central As atom: $H—\ddot{As}—H$. **The formula of the simple covalent hydride is AsH_3.** $\overset{|}{H}$

(b) On the left side of equation (14.15) the W atom is in the oxidation state (O.S.) +6, and the H atoms have the O.S. 0. On the right, the W atom is in the O.S. 0, and the H is in the O.S. +1. **$H_2(g)$ is** *oxidized*, **and so it is the reducing agent. The WO_3 is** *reduced*, **and so it is the oxidizing agent.**

(c) The energy required to dissociate the H_2 molecule is the bond dissociation energy. From Table 10-2 we see that this is 435 kJ/mol. This same amount of energy is *released* when H atoms combine to form H_2 molecules, that is, **−435 kJ/mol H_2 formed.**

SIMILAR EXAMPLES: Exercises 10, 11, 65, 66.

14-4   Helium, Neon, and Argon

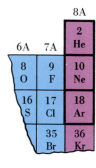

We surveyed the discovery of the noble gases in Section 9-1.

Group 8A of the periodic table is variously known as the rare gases, inert gases, or noble gases, although He and Ar are not rare and Kr, Xe, and Rn are not inert. Of the noble gases, He is of special interest because of its unique physical properties; Xe, because of what we can learn about molecular structure through its compounds (Section 23-6); Rn, because it is a radioactive element and its radioactivity poses significant environmental hazards (Section 26-2).

Occurrence. Air contains 0.000524% He, 0.001821% Ne, and 0.9340% Ar, by volume. Of the noble gases, Ar is the only one present in sufficient quantity to make its large-scale production from air practicable. There is an alternative source of helium, however. Certain natural gases in Texas, Oklahoma, and Kansas contain up to 2% He, by volume, and it is feasible to extract this helium from natural gas, even down to levels of about 0.3%. Underground helium accumulates as a result of alpha particle emission by radioactive elements in the earth's crust. We may ultimately have to extract He from the atmosphere when natural gas supplies are depleted.

Properties. The lighter noble gases are commercially important in part because they have no chemical properties! They are chemically inert. Also they are impor-

The fountain effect in liquid helium-II. A small quantity of heat is introduced into liquid He in the inner chamber. Liquid He enters from the outer chamber through a microporous material separating the two chambers and located below the heater. A jet of liquid is forced out the hole at the top of the inner chamber. [Dr. J. F. Allen, University of St. Andrews]

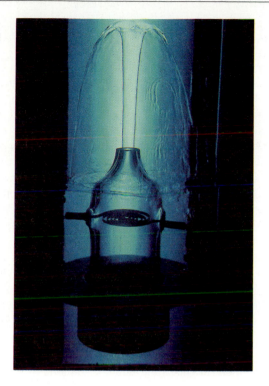

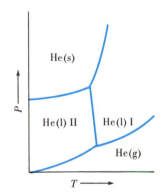

FIGURE 14-5
Phase diagram for helium-4.

The need for liquid helium will be greatly reduced if a new class of superconductors currently under investigation can be properly fabricated (see page 896).

tant because of their *physical* properties. Here are some of the unique properties of ^{4}He, the most common isotope of helium.

- A range of only one degree between the boiling point (4.2 K) and the critical point (5.3 K).
- The existence of *two* forms of liquid He: a normal liquid He-I, and a superfluid liquid He-II. The superfluid has essentially no resistance to flow (viscosity) and a very high thermal conductivity.
- The existence of a liquid phase (He-II) down to 0 K.
- No triple point involving solid, liquid, and gaseous He. Solid He can only be obtained by applying pressure (about 25 atm) to He(l).

We can represent this unusual behavior through a phase diagram, as in Figure 14-5, but it can only be explained in terms of quantum mechanics.

Uses. Because they are inert, both He and Ar can be used to blanket materials that need to be protected from nitrogen and oxygen in the air, such as in certain types of welding, in metallurgical processes, and in the preparation of ultrapure Si, Ge, and other semiconductor materials. Helium mixed with oxygen is used as a breathing mixture for deep-sea diving and in certain medical applications. An argon–nitrogen mixture is used to fill electric light bulbs to increase the efficiency and life of the bulb. Electric discharge through neon-filled glass or plastic tubes produces a distinctive red light ("neon light").

Large quantities of helium are used to maintain materials at low temperatures (cryogenics). Metals essentially lose their electrical resistivity at liquid He temperatures and become *superconductors*. Powerful magnets can be made by immersing the coils of electromagnets in liquid helium. Such magnets are used in particle accelerators and in fusion research (see Section 26-9). Helium is also used to fill airships (blimps).

14-5   Lithium, Sodium, and Potassium

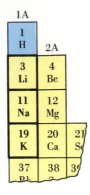

We have learned to think about metals in two ways: Their chemical behavior, such as the tendency to form ionic compounds, is associated with properties like large atomic radii, low ionization energies, and low electronegativities. Their physical properties, including hardness, malleability, ductility, and the ability to conduct heat and electricity, are attributable to the nature of metallic bonding.

The group 1A atoms have but a single valence electron (ns^1) that is rather easily lost. They display the chemical behavior of metals to the highest degree. On the other hand, because of their large atomic sizes and limited number of valence electrons, alkali metal atoms do not bond as strongly to one another in the solid state as do most metals. As a result they have low densitites, melting points, and hardnesses. Table 14-4 lists a few properties of the group 1A (alkali) metals.

The outer-shell electrons of the group 1A atoms can be promoted to higher energy levels relatively easily. This excitation can occur simply through collisions in a gas flame. When the excited atoms revert to their normal or ground-state electron configurations, specific amounts of energy are emitted as light. The flame acquires a characteristic color. The yellow color of a sodium flame involves primarily electrons dropping from the $3p$ to the $3s$ level, which can be described symbolically as

$$[Ne]3p^1 \longrightarrow [Ne]3s^1 \tag{14.17}$$

Figure 14-6 shows the lithium, sodium, and potassium flame colors.

A specimen of spodumene from Afghanistan. [Joel Arem]

Occurrence, Preparation, and Uses of the Metals. Because so many of their compounds are water soluble, some Li, Na, and K compounds can be extracted from natural brines, such as are found in Israel and in the western United States. These compounds are primarily the chlorides but also include the carbonates and the sulfates. The most important solid lithium-containing mineral from which lithium is extracted is spodumene, $LiAl(SiO_3)_2$. The principal compounds from which sodium and potassium metals are extracted are NaCl and KCl, whether derived from natural brines or mined as solids.

Alkali metal atoms lose electrons rather easily; the atoms are easily oxidized. By contrast, it is difficult to get alkali metal ions to gain electrons; the ions are *not* easily reduced through chemical reactions. The principal method of producing metallic lithium and sodium is by the electrolysis of a molten salt, usually the chloride.

TABLE 14-4
Some Properties of the Alkali (Group 1A) Metals

	Li	Na	K	Rb	Cs
atomic number	3	11	19	37	55
atomic weight	6.941	22.9898	39.0983	85.4678	132.905
atomic (covalent) radius, pm	123	157	202	216	235
electronegativity	1.0	0.9	0.8	0.8	0.8
first ionization energy, kJ/mol	520.3	495.9	418.9	403.0	375.7
melting point, °C	180.54	97.81	63.25	38.89	28.40
boiling point, °C	1342	882.9	760	686	669.3
density, g/cm³ at 20 °C	0.534	0.97	0.86	1.475	1.8785
hardness[a]	0.6	0.4	0.5	0.3	0.2
electrical conductivity[b]	19	36	24	13	8.4
flame color	carmine	yellow	violet	bluish red	blue

[a]Hardness measures the ability of substances to scratch, abrade, or indent one another. On the Mohs scale, hardnesses of 10 minerals range from that of talc (0) to diamond (10). Other values: wax (0 °C), 0.2; asphalt, 1–2; fingernail, 2.5; copper, 2.5–3; iron, 4–5; chromium, 9. Each substance is capable of scratching only others of hardness values less than its own.
[b]On a scale relative to silver, 100; copper, 94.5; gold, 70.8.

In an electrolysis reaction we force oxidation and reduction to occur in the manner that we choose. In reaction (14.18) we use an electric current to take electrons away from Cl^- ions and force them onto Na^+ ions in molten NaCl.

Preparation of Na: electrolysis of NaCl(l).

$$2 \ NaCl(l) \xrightarrow{\text{electrolysis}} 2 \ Na(l) + Cl_2(g) \tag{14.18}$$

Because they are so easily oxidized, the alkali metals are good reducing agents. We can, in fact, use one alkali metal to reduce ions of another alkali metal, as in this preparation of potassium.

$$KCl(l) + Na(l) \xrightarrow{850 \ °C} NaCl(l) + K(g) \tag{14.19}$$

Normally, reaction (14.19) is reversible and reaches a condition of equilibrium in which most of the KCl(l) remains unreacted. The reaction can be made to go to completion, however, by carefully selecting the reaction conditions, in this case a temperature of 850 °C. At this temperature Na is a liquid but K is a gas. As soon as it forms, K(g) escapes from the molten mixture and so it cannot participate in the reverse reaction. In fact, in modern processes a molten mixture of KCl(l) and Na(l) is continuously fed into a reaction vessel and K(g) is continuously removed and condensed.

When added in small quantities, lithium metal gives high-temperature strength to aluminum and ductility to magnesium. When alloyed with silver it is used for brazing (joining together) metals. It is currently finding important uses in the manufacture of light-weight electrical batteries (see Section 22-1), and a potential future use is in the production of tritium (^3H) for use in fusion reactors (see Section 26-9).

The most important use of sodium metal is as a reducing agent in producing refractory (high melting point) metals, such as titanium, zirconium, and hafnium.

$$TiCl_4 + 4 \ Na \longrightarrow Ti + 4 \ NaCl \tag{14.20}$$

Another important current use of sodium is as a heat transfer medium in liquid-metal-cooled fast breeder reactors (see Section 26-8). Liquid sodium is especially good for this purpose since the liquid has better thermal conductivity and a higher heat capacity (specific heat) than most metals, a low density and low viscosity (making it easy to pump), and a low vapor pressure even at high temperatures (550 °C). Sodium is used in sodium vapor lamps, which are very popular for outdoor lighting (recall Figure 8-6). However, since each lamp uses only a few mg Na, the total quantity consumed in this application is rather small.

Uses of potassium are limited to a few applications where sodium, the cheaper metal, cannot be used. One application is the combustion of K to form KO_2, which is used in life-support systems (see page 512).

Reactions of Li, Na, and K. The chemistry of the 1A metals reflects how easily the metal atoms are oxidized to the metal ions. As we learned in Section 9-12, the alkali metals (M) react directly with the halogens (X_2) to form ionic binary halides (MX). Oxides are formed by the reaction of $O_2(g)$ on the metals, but various products are formed, as described in Section 14-12. Ionic hydrides of the group 1A metals were discussed in Section 14-3. Of the alkali metals, only lithium forms a nitride by a direct reaction with $N_2(g)$.

$$6 \ Li(s) + N_2(g) \longrightarrow 2 \ Li_3N(s) \tag{14.21}$$

This nitride reacts with water to form $NH_3(g)$ and LiOH. Li, Na, and K are such active metals that they will displace $H_2(g)$ from acid solutions (reaction 14.6) and from water (reaction 14.7).

Important Compounds of Li, Na, and K. Among the most important lithium compounds are the carbonate, halides, hydroxide, and hydride. Li_2CO_3 is used to make some types of glass and ceramic ware. In high purity it is used in the treatment

Li

Na

K

FIGURE 14-6
Flame colors of lithium, sodium, and potassium. [Carey B. Van Loon]

Harvesting salt produced by the evaporation of seawater, at Cabo Frio, Brazil. [Nicholas deVore III/Bruce Coleman, Inc.]

of mental disorders. It is also a useful starting material in making other lithium compounds. For example

$$Li_2CO_3(aq) + Ca(OH)_2(aq) \longrightarrow CaCO_3(s) + 2\ LiOH(aq) \qquad (14.22)$$

Of the substances present in reaction (14.22), $CaCO_3$ is by far the least soluble. This favors the conversion of Li_2CO_3 to LiOH.

An important use of lithium hydroxide is to remove $CO_2(g)$ from expired air in confined quarters like submarines and space vehicles.

$$2\ LiOH(s) + CO_2(g) \longrightarrow Li_2CO_3(s) + H_2O \qquad (14.23)$$

LiOH is used because, per mole of CO_2 removed, a smaller mass of LiOH is required than of other, cheaper hydroxides (e.g., NaOH). Similarly, in its reaction with water, LiH produces a greater volume of $H_2(g)$ per unit mass of hydride than do other, cheaper hydrides.

Sodium chloride is easily the most important sodium compound. In fact, it is the most used of all minerals for the production of chemicals. It is not listed among the top chemicals because it is considered a raw material, not a manufactured chemical. Annual use of sodium chloride in the United States amounts to about 50 million tons. Salt is used in the dairy industry, in the treatment of hides, the preservation of meat and fish, the control of ice on roads, and the regeneration of water softeners. In the chemical industry, NaCl is a source of sodium metal, chlorine gas, sodium hydroxide, hydrochloric acid, sodium carbonate, sodium sulfate, and other sodium and chlorine compounds. Table 14-5 lists several of the top 50 chemicals whose commercial production uses NaCl as a raw material.

Sodium hydroxide is produced by the electrolysis of NaCl(aq). When the equation for this electrolysis is written in ionic form, we see that Na^+ goes through the electrolysis unchanged, Cl^- is oxidized to $Cl_2(g)$, and H_2O is reduced to $H_2(g)$ and OH^-.

TABLE 14-5
Some Chemicals Produced from NaCl

Chemical	U.S. production 1986	
	Billion lb	Ranking
NaOH	22.01	7
Cl_2	20.98	8
Na_2CO_3[a]	17.20	11
HCl[a]	5.97	23
Na_2SO_4	1.55	47

[a] Also produced from sources other than NaCl.

Preparation of NaOH(aq): electrolysis of NaCl(aq).

$$2\ Na^+ + 2\ Cl^- + 2\ H_2O \xrightarrow{\text{electrolysis}} 2\ Na^+ + 2\ OH^- + H_2 + Cl_2 \qquad (14.24)$$

NaOH is used in the manufacture of soap, rayon, and paper; in petroleum refining; in the textile and tanning industries; and in the production of various salts by the neutralization of acids.

Sodium sulfate is obtained partly from natural sources and partly through the following reaction, which was introduced by J. R. Glauber more than 300 years ago.

Preparation of a volatile acid from one of its salts.

$$H_2SO_4(\text{concd aq}) + 2\ NaCl(s) \xrightarrow{\Delta} Na_2SO_4(s) + 2\ HCl(g) \qquad (14.25)$$

The principle of reaction (14.25) is that of producing a *volatile* acid (HCl) by heating one of its salts (NaCl) with a *nonvolatile* acid (H_2SO_4). Several other acids can be produced by similar reactions. The major use of Na_2SO_4 is in the paper industry (about 70% of the annual U.S. consumption). In the kraft process for papermaking, undesirable lignin is removed from wood by digesting the wood in an alkaline solution of Na_2S. Na_2S is produced by reducing Na_2SO_4 with carbon.

$$Na_2SO_4 + 4\ C \longrightarrow Na_2S + 4\ CO \qquad (14.26)$$

About 100 lb of Na_2SO_4 is required for every ton of paper produced.

Sodium carbonate (soda ash) is used primarily (over 50%) in the manufacture of glass. The bulk of Na_2CO_3 produced in the United States comes from *natural sources*—dry lakes in California and immense deposits in western Wyoming. Elsewhere in the world the principal source is a process using NaCl, CO_2, and NH_3, introduced by the Belgian chemist Ernest Solvay in 1863 and described in Section 22-1. Other uses of sodium carbonate include the manufacture of chemicals (20%), pulp and paper (6%), soap and detergents (5%), and water treatment (3%).

Potassium compounds have some uses similar to those of sodium (e.g., K_2CO_3 in glass and ceramics), but their most important use by far (95%) is in fertilizers. Potassium is one of the three main nutrients required by plants (N and P are the other two). KCl is commonly used as fertilizer because this is the form in which most potassium is obtained from natural sources.

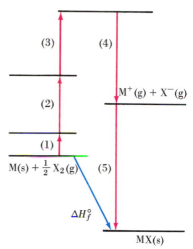

FIGURE 14-7

Generalized Born–Fajans–Haber method for alkali metal halides.

(1) $\qquad M(s) \rightarrow M(g)$
(2) $\qquad M(g) \rightarrow M^+(g) + e^-$
(3) $\qquad \frac{1}{2} X_2(g) \rightarrow X(g)$
(4) $\qquad X(g) + e^- \rightarrow X^-(g)$
(5) $M^+(g) + X^-(g) \rightarrow MX(s)$
See also Figure 10-4.

Properties of Alkali Metal Compounds. The lattice energies of alkali metal compounds are generally lower than those of many other ionic compounds. These relatively low lattice energies help explain why the vast majority of alkali metal compounds are water soluble and why the melting points of these compounds are moderately low (e.g., LiCl, 605 °C; NaCl, 801 °C; KCl, 770 °C). Figure 14-7 helps us to understand some of the physical properties of alkali metal compounds. If the metal M is lithium, the enthalpy of sublimation (1) and the ionization energy (2) are considerably larger than for the other alkali metals, because of the small size of the Li atom. But the lattice energy (5) also tends to be larger because of the small size of the Li^+ ion. The magnitude of the enthalpy of formation of a lithium compound relative to other alkali metal compounds depends, then, on the enthalpy changes of steps 1, 2, and 5, and most particularly on step 5.

In the formation of fluorides, for example, the lattice energy of LiF(s) is so large (because both the Li^+ and F^- ions are so small) that ΔH_f is more negative for LiF than for any other alkali metal fluoride. The high lattice energy of LiF makes it much less water soluble than the other alkali metal fluorides.

	LiF	NaF	KF	RbF	CsF
lattice energy, kJ/mol	-1043	-928	-826	-789	-758
solubility, g/100 g H_2O	0.27	4.22	92.3	130.6	366.5

The situation that we just described for LiF does not hold for all lithium compounds, however. Another factor that affects the solubilities of lithium compounds is the high energy of hydration of the Li^+ ion in aqueous solution (again because of the small size of the Li^+ ion). This effect, which causes the solubilities of some lithium compounds to be unusually high, is discussed further in Section 22-1.

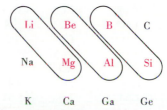

FIGURE 14-8

Diagonal relationships.

The encircled pairs of elements exhibit many similar properties.

Diagonal Relationships. As we have seen in this section, Li has several properties that set it apart from the rest of the alkali metals. In its ability to form a nitride, and in the very low solubility of its fluoride, carbonate, and phosphate, Li bears a resemblance to Mg. This similarity is called a **diagonal relationship**. Such relationships also exist between Be and Al and between B and Si (see Figure 14-8). The Li–

Mg similarity probably results from the roughly equal sizes of the Li and Mg atoms and of the Li^+ and Mg^{2+} ions.

Example 14-3

Writing equations for chemical reactions. An important skill in descriptive chemistry is being able to transform a verbal description of a reaction into an equation. Write equations to represent the following reactions, which are described in this and preceding sections of the chapter: **(a)** K(s) with F_2(g); **(b)** Li_3N with water; **(c)** LiH with H_2O.

Solution

(a) The direct action of a halogen (X_2) on an alkali metal (M) produces the ionic metal halide (MX).

$$2 \; K(s) + F_2(g) \longrightarrow 2 \; KF(s)$$

(b) The reaction of Li_3N with water produces LiOH and NH_3.

$$Li_3N(s) + 3 \; H_2O \longrightarrow 3 \; LiOH(s) + NH_3(g)$$

(c) The reaction of a metal hydride with water produces the metal hydroxide and hydrogen gas.

$$LiH(s) + H_2O \longrightarrow Li^+(aq) + OH^-(aq) + H_2(g)$$

SIMILAR EXAMPLES: Exercises 6, 12, 19, 26, 29, 31, 34, 42.

14-6   Beryllium, Magnesium, and Calcium

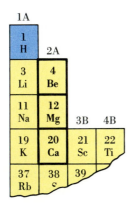

From a chemical standpoint, in their abilities to react with water and acids and to form ionic compounds, the heavier group 2A metals—Ca, Sr, Ba, and Ra—are nearly as active as the group 1A metals. In terms of certain physical properties (e.g., density, hardness, and melting point), the group 2A elements are more typically metallic than the 1A elements, as you can see by comparing data in Tables 14-4 and 14-6.

The Special Case of Beryllium. Be (and to some extent Mg) is different from the heavier members of group 2A. You can see from Table 14-6 that Be has a higher melting point and is much harder than the other group 2A metals. Its chemical properties also differ significantly. Some of its distinctive properties are

TABLE 14-6
Some Properties of the Alkaline Earth (Group 2A) Metals

	Be	Mg	Ca	Sr	Ba
atomic number	4	12	20	38	56
atomic weight	9.01218	24.305	40.08	87.62	137.33
atomic (covalent) radius, pm	89	136	174	192	198
electronegativity	1.6	1.3	1.0	1.0	0.9
first ionization energy, kJ/mol	899.5	737.8	589.8	549.5	502.9
melting point, °C	1278	648.8	839	769	725
boiling point, °C	2970[a]	1107	1484	1384	1640
density, g/cm³ at 20 °C	1.85	1.74	1.54	2.6	3.51
hardness[b]	ca. 5	2.0	1.5	1.8	ca. 2
electrical conductivity[b]	13	36	36	6.5	2.7
flame color	none	none	orange-red	scarlet	green

[a] Boiling point at 5 mmHg pressure.
[b] See footnotes to Table 14-4.

FIGURE 14-9
Covalent bonds in BeCl₂.

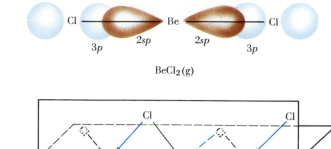

BeCl₂(g)

BeCl₂(s)

(a) In gaseous BeCl₂ discrete molecules exist with the bonding scheme shown in the figure. (b) In solid BeCl₂ two Cl atoms are bonded to a Be atom through normal covalent bonds. Two others are bonded by coordinate covalent bonds, using lone-pair electrons of the Cl atoms (bonds shown as arrows). The arrangement is essentially tetrahedral. Of course, once formed, these two types of bonds cannot be distinguished from one another. BeCl₂ units are linked into long chain-like polymeric molecules—(BeCl₂)ₙ.

- Be is quite unreactive toward air and water.
- BeO does not react with water. [For the other group 2A metal oxides, MO + $H_2O \longrightarrow M(OH)_2$.]
- Be and BeO dissolve in strongly basic solutions to form the ion BeO_2^{2-}.
- BeCl₂ and BeF₂ in the molten state are poor conductors of electricity; they are covalent substances.

We can best understand the chemical behavior of Be in terms of the small size and high ionization energy of the Be atom. Beryllium shows only a limited tendency to form the ion Be^{2+}; in fact, its ability to form covalent bonds is more pronounced. The fact that BeO reacts with strongly basic solutions shows that the oxide has *acidic* properties. This behavior, in turn, is associated with small ionic size and high charge (as discussed in Section 17-9). When it forms covalent bonds, beryllium appears to use hybrid orbitals. Figure 14-9 depicts bonding through *sp* hybrid orbitals in BeCl₂(g) and through sp^3 hybrid orbitals in BeCl₂(s).

Preparation and Uses of Be, Mg, and Ca. Reducing the compounds of Be, Mg, and Ca to the free metals is not quite as difficult as reducing the compounds of group 1A metals. Still, though, electrolysis is the preferred method of preparing Mg. For this, we generally choose the molten chloride, as in the Dow process for magnesium (described in detail in Section 22-2).

$$MgCl_2(l) \xrightarrow{\text{electrolysis}} Mg(l) + Cl_2(g) \qquad (14.27)$$

An important natural source of beryllium compounds is the mineral **beryl**, $3BeO \cdot Al_2O_3 \cdot 6SiO_2$. The metal itself is produced from BeF₂, with Mg as the reducing agent. Most calcium is now produced by the reduction of calcium oxide with aluminum.

Because it is able to withstand metal fatigue, an alloy of copper with about 2% Be is used in springs, clips, and electrical contacts. Other Be alloys are used for structural purposes where light weight is a primary requirement. Because the Be atom has so little stopping power for x rays or neutrons, beryllium is used to make "windows" for x ray tubes and for various components in nuclear reactors. A drawback to the use of beryllium is its extreme toxicity.

Magnesium has a lower density than any other structural metal and is valued for its light weight. Objects, such as aircraft parts, are manufactured from magnesium alloyed with aluminum and other metals. Magnesium is a good reducing agent and is used in a number of metallurgical processes (such as the production of beryllium mentioned above).

Calcium is used primarily as a reducing agent in the preparation, from their oxides or fluorides, of other, less common metals, such as Sc, W, Th, U, Pu, and

Many familiar gemstones, including aquamarine and emerald, are based on the mineral beryl. [Joel Arem]

most of the lanthanides. Calcium is also used in the manufacture of batteries, and in forming alloys with aluminum, silicon, and lead.

Reactions of Be, Mg, and Ca. Magnesium reacts with halogens (X_2) to form the halides, MgX_2; with oxygen to form the oxide, MgO; and with nitrogen to form the nitride, Mg_3N_2. Be and Ca give similar reactions.

Typical reactions of Mg.

$$Mg + X_2 \longrightarrow MgX_2 \tag{14.28}$$
$$\text{(where X is F, Cl, Br, I)}$$

$$2\,Mg + O_2 \longrightarrow 2\,MgO \tag{14.29}$$

$$3\,Mg + N_2 \longrightarrow Mg_3N_2 \tag{14.30}$$

Calcium reacts with cold water, as we saw in equation (14.8) and Figure 14-4. In the case of Mg, an impervious film of $Mg(OH)_2$ covers the surface and immediately stops the reaction. Mg does react with steam, however.

$$Mg(s) + H_2O(g) \longrightarrow MgO(s) + H_2(g) \tag{14.31}$$

Beryllium fails to react with either cold water or steam.

Important Compounds of Mg and Ca. Several magnesium compounds occur naturally, either in mineral form or in brine solutions. These include the carbonate, chloride, hydroxide, and sulfate. Other magnesium compounds can be prepared from these. A few important compounds of magnesium and some of their uses are listed in Table 14-7.

Limestone is a naturally occurring form of $CaCO_3$ containing some clay and other impurities. It is the most widely used type of rock. Its primary use is as a building stone (about 70%). Other uses include the manufacture of cement (15%), as a flux in metallurgical processes (5%), as a source of quicklime and slaked lime (5%), and as an ingredient in glass.

A metallurgical flux is a material that removes impurities during production of a metal through the formation of a freely flowing liquid called a slag.

Three steps are required to obtain pure $CaCO_3$ from limestone: (1) the thermal decomposition of limestone (called *calcination*), (2) the reaction of CaO with water (slaking), and (3) the conversion of $Ca(OH)_2$ to precipitated $CaCO_3$ (carbonation).

Important reactions of limestone and lime.

$$\text{Calcination:} \qquad CaCO_3(s) \xrightarrow{900\ ^\circ C} CaO(s) + CO_2(g) \tag{14.32}$$

$$\text{Slaking:} \qquad CaO(s) + H_2O(l) \longrightarrow Ca(OH)_2(s) \tag{14.33}$$

$$\text{Carbonation:} \quad Ca(OH)_2(aq) + CO_2(g) \longrightarrow CaCO_3(s) + H_2O(l) \tag{14.34}$$

Precipitated $CaCO_3$ is the chief mineral constituent (called a *filler*) in products as diverse as paper, paint, plastics, printing ink, rubber, putties, and adhesives. It is also used in dentrifices, food, cosmetics, and pharmaceuticals. For example, $CaCO_3$ is used in papermaking to impart brightness, opacity, smoothness, and good ink-absorbing qualities to paper.

Materials with very high melting points, like CaO (m.p. 2614 °C), emit light when heated to high temperatures. Before the advent of electric lighting, lime, heated to high temperatures with an oxyhydrogen flame, was used for theatrical lighting. This use gave rise to the expression "in the limelight."

You are probably familiar with the term "lime," but you may not be aware that it is used in reference to several different compounds. CaO is called **quicklime** and is produced by the calcination of limestone (reaction 14.32). $Ca(OH)_2$ is called **slaked lime** and is formed by the action of water on CaO (reaction 14.33). Reaction (14.32) is reversible, and at room temperature the *reverse* reaction occurs almost exclusively. In the calcination of $CaCO_3$ it is important that a high temperature be used and that the $CO_2(g)$ be continuously exhausted from the furnace (kiln) in which the reaction occurs. A noteworthy feature of reaction (14.33) that we have already encountered is the fact that the reaction is highly exothermic (recall Figure 7-8).

Quicklime, CaO, is used to make refractory materials, as a metallurgical flux, in the pulp and paper industry, for the removal of SO_2 from stack gases, and in treating wastewater effluents and sewage.

TABLE 14-7
Some Important Magnesium Compounds

Compound	Principal uses
$MgCO_3$	mfr. of refractory bricks; glass, inks, rubber reinforcing agent; dentrifices and cosmetics; antacid
$MgCl_2$	production of Mg metal; mfr. of textiles and paper; fireproofing agents; cements; refrigeration brine
MgO	refractories (furnace linings); ceramics; cements; SO_2 removal from stack gases
$MgSO_4$	fireproofing; textile mfr.; ceramics; fertilizers; cosmetics; dietary supplement

Slaked lime, $Ca(OH)_2$, is the cheapest commercial alkaline (basic) substance, and it is used in all applications where high water solubility is not essential. Among its uses are in the manufacture of other alkalis and bleaching powder, in the purification of sugar, in tanning hides, and in water softening. A mixture of slaked lime, sand, and water is the familiar mortar used in brick laying. In the initial setting of the mortar, bricks absorb excess water, which is then lost through evaporation. In the final setting, $CO_2(g)$ from the air reacts with $Ca(OH)_2$ and converts it back to $CaCO_3$.

$$Ca(OH)_2(s) + CO_2(g) \longrightarrow CaCO_3(s) + H_2O(g) \qquad (14.35)$$

Another important calcium-containing mineral is **gypsum**, $CaSO_4 \cdot 2H_2O$. In the United States, 50 million tons of gypsium are consumed annually. About one-half of this quantity is converted to the *hemihydrate* ($\frac{1}{2}$-hydrate), known as **plaster of Paris.**

$$CaSO_4 \cdot 2H_2O(s) \xrightarrow{\Delta} CaSO_4 \cdot \tfrac{1}{2}H_2O(s) + \tfrac{3}{2} H_2O(g) \qquad (14.36)$$
$$\text{gypsum} \qquad\qquad \text{plaster of Paris}$$

When it is mixed with water, plaster of Paris reverts to gypsum. Because it expands as it sets, a plaster of Paris–water mixture is useful in making castings where sharp details of an object must be retained. It is extensively used in jewelry making and in dental work. The most important application, though, is in producing gypsum wall board, which has all but supplanted other interior wall coverings in the construction industry.

14-7 Boron and Aluminum

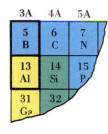

Boron and aluminum, both members of group 3A, have some similarities, but perhaps there are even more differences between the two. Also, from the diagonal relationship (Figure 14-8) we expect B to resemble Si and Al to resemble Be.

The tendency for B and Al atoms to lose their three outer-shell electrons to acquire noble gas electron configurations is much less pronounced than for the group 1A and 2A atoms. In fact, although we postulate its existence in some instances, the energy required to produce the ion B^{3+} is too high for it to exist. Even the ion Al^{3+} is not commonly encountered, except perhaps in $AlF_3(s)$. In aqueous solution Al^{3+} is found as the hydrated ion $[Al(H_2O)_6]^{3+}$ or in other complex forms.

Boric acid, $B(OH)_3$, is a weak *acid*. When it ionizes in aqueous solution $B(OH)_3$ produces H^+ rather than OH^-. Aluminum hydroxide, $Al(OH)_3$, is primarily basic, although it does act as an acid under certain conditions. The acid–base properties of $B(OH)_3$ and $Al(OH)_3$ are described in more detail in Section 17-9 and in Chapters 22 and 23.

Covalent Bonding. The covalent compounds of B and Al are characterized by a deficiency of electrons in the valence shells of the bonded B and Al atoms. Let us consider some of the consequences of this fact by describing the covalent halides.

In the molecule BX_3 (X = F, Cl, Br, I), the hybridization of the B atom is

	$1s$	sp^2	$2p$
B	↑↓	↑ ↑ ↑	

We can describe the B—X bonds as involving the overlap of sp^2 orbitals of the B atom with p orbitals of the X atoms to form σ bonds. These bonds are exceptionally strong for single bonds, however. For example, the B—F bond energy is 646 kJ/mol. The strength of these bonds is probably due in part to the small size of the B atom, but it seems also to involve some π bonding between the B atom and the X atoms. In fact, the π bonding can best be described through a π-type molecular orbital *delocalized* over all four atoms. Note that in this π bonding the B atom can contribute an empty p orbital but no electrons, since it is *electron deficient*.

The bonding scheme in the formula unit AlX_3 (X = Cl, Br, I) is different. The hybridization of the Al atom appears to be

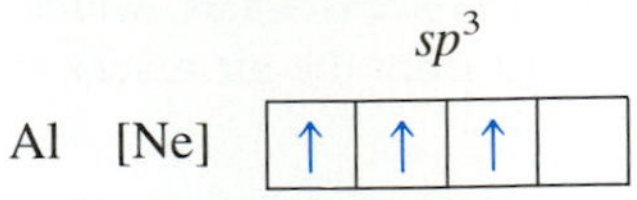

Al   [Ne] $\boxed{\uparrow | \uparrow | \uparrow | }$ *sp³*

When an Al atom bonds to three X atoms, the orbital overlap involves three sp^3 orbitals of the Al atom and p orbitals of the X atoms. The fourth sp^3 orbital of the Al atom remains vacant. Notice that in this bonding scheme the Al atom is electron deficient, having only *six* electrons in its valence shell instead of eight. To overcome this electron deficiency, two AlX_3 units join together into a *dimer*, Al_2X_6. An X atom of each AlX_3 unit makes a pair of p electrons available to the vacant sp^3 orbital of the Al atom of the other AlX_3 unit (see Figure 14-10).

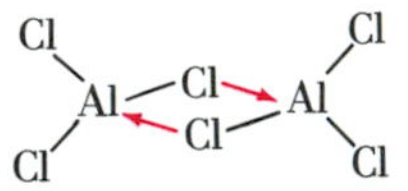

We have seen that the B atom in BX_3 and the Al atom in AlX_3 are both electron deficient. In BX_3 this is compensated for to some extent by π bonding, and in AlX_3 by formation of the dimers Al_2X_6. Another way that these molecules can overcome their electron deficiencies is to accept a pair of electrons and form an additional bond with some other species. The product is called an **adduct.** In Chapter 10 we saw how BF_3 and NH_3 form the adduct $F_3B:NH_3$ (structure 10.32) through the formation of a coordinate covalent bond between the B and N atoms. The halides of B and Al are used in organic chemistry where, through adduct formation, they can catalyze certain reactions.

> AlF₃ has considerable ionic character. Its melting point is 1040 °C, compared to 194 °C for AlCl₃ (at 5.2 atm), 97.5 °C for AlBr₃, and 191 °C for AlI₃.

FIGURE 14-10
Bonding in Al_2Cl_6.

Two Cl atoms bridge the AlCl₃ units, producing the dimer Al₂Cl₆. Electrons donated by these Cl atoms to Al atoms are denoted by arrows.

Example 14-4

Writing a Lewis structure. In the presence of diethyl ether, $(C_2H_5)_2O$, the dimers Al_2Cl_6 split and each $AlCl_3$ unit forms an adduct with an ether molecule. Represent this adduct through a Lewis structure.

Solution. First, draw Lewis structures of the two separate species.

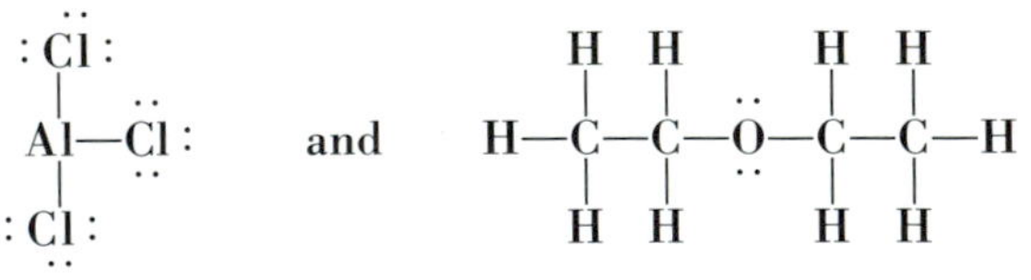

Now, note that an available pair of electrons to join these two species together can be found on the O atom of $(C_2H_5)_2O$. (The O atom has *two* lone pairs.)

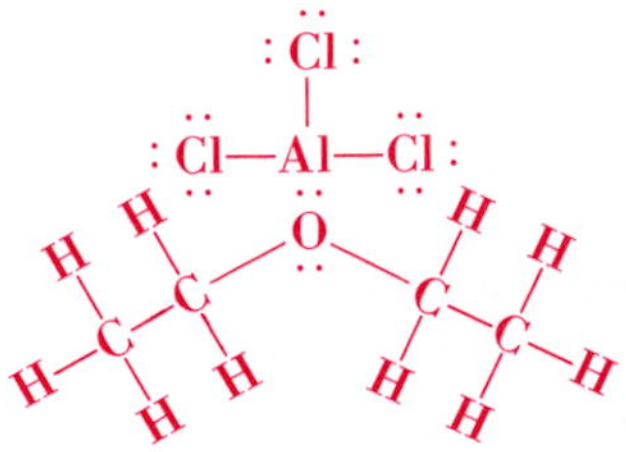

SIMILAR EXAMPLES: Exercises 32, 47, 68, 74.

Physical and Chemical Properties and Uses of Aluminum. Pure aluminum is a malleable, ductile, silvery-colored metal of low density. Its density is only about one-third that of steel. The metal is not very strong, but its strength increases considerably when it is alloyed with Cu, Mg, Mn, or Si. Aluminum–magnesium

Elevated levels of Al^{3+} have been detected in brain tissue of victims of Alzheimer's disease. It is not known, however, whether Al^{3+} is implicated in causing the disease or is a product of the disease.

alloys, because of their low densities, are extensively used in the aircraft industry. Aluminum alloys are also used in the construction and automotive industries.

Another important use of aluminum is as an electrical conductor. For any wire of a given diameter, Al has only about 60% of the conductivity of Cu, but because of its low density it is a better conductor than Cu on a mass basis. For this reason Al is increasingly used in electrical transmission lines. Perhaps the most familiar household use of aluminum is in pots, pans, and other cookware. Both the metal and its ion (Al^{3+}) are nonpoisonous.

Aluminum is a good reducing agent, which means that it is easily oxidized. For example, it dissolves in acids to produce $H_2(g)$.

$$2\ Al(s) + 6\ H^+(aq) \longrightarrow 2\ Al^{3+}(aq) + 3\ H_2(g) \tag{14.37}$$

If there is a stronger oxidizing agent than H^+ present in the solution, some species other than H^+ is reduced, such as in $H_2SO_4(aq)$, where SO_4^{2-} is reduced to $SO_2(g)$.

Aluminum is one of a small group of metals that dissolve in alkaline (basic) as well as acidic solution. This behavior depends on the acidic properties of $Al(OH)_3$ and is explained in Section 22-3.

$$2\ Al(s) + 2\ OH^-(aq) + 6\ H_2O \longrightarrow 2\ Al(OH)_4^-(aq) + 3\ H_2(g) \tag{14.38}$$

There are two consequences of reaction (14.38) that might be familiar to you. One is the fact that alkaline substances (such as some types of oven cleaners) should never be applied to aluminum surfaces; the Al dissolves. The other explains the action of certain drain cleaners that give off a gas to help unplug a stopped-up drain. The drain cleaner is a mixture of solid sodium hydroxide and powdered metallic Al. When the mixture is added to water, the resulting NaOH(aq) solubilizes fats and grease, and the Al reacts to give off gaseous hydrogen.

Powdered aluminum is easily oxidized in air in a highly exothermic reaction; this explains its use in rocket fuels and incendiary devices.

The thermite reaction. [E. R. Degginger]

$$2\ Al(s) + \tfrac{3}{2}\ O_2(g) \longrightarrow Al_2O_3(s) \qquad \Delta H = -1670\ kJ \tag{14.39}$$

We can use substances other than O_2 to supply O atoms for reaction (14.39). In the thermite reaction (also known as the Goldschmidt reaction), Al combines with oxygen from another metal oxide, reducing the metal to its free state.

The thermite reaction.

$$Fe_2O_3(s) + 2\ Al(s) \longrightarrow Al_2O_3(s) + 2\ Fe(l) \tag{14.40}$$

The thermite reaction can be used in on-site welding of large metal objects. It can also be used for the preparation of small quantities of metals (for example, chromium if Cr_2O_3 is substituted for Fe_2O_3).

An important reducing agent in organic chemistry is lithium aluminum hydride, $LiAlH_4$. The AlH_4^- ion can be thought of as an adduct of AlH_3 and H^-. The reduction of water by AlH_4^- produces $H_2(g)$.

$$AlH_4^- + 4\ H_2O \longrightarrow Al(OH)_4^-(aq) + 4\ H_2(g) \tag{14.41}$$

Acid–base reactions of $Al(OH)_3$, complex ion formation by Al^{3+}, the metallurgy of Al, and other aspects of the chemistry of aluminum are described later in the text.

14-8 Carbon

3A	4A	5A	6A
5 B	6 C	7 N	8 O
13 Al	14 Si	15 P	

Carbon is unique in the variety and complexity of compounds that it can form. We can attribute this to the pronounced ability of C atoms to bond to one another in long chains, in rings, and in combinations of rings and chains. Moreover, the ground-state electron configuration of carbon ($1s^2 2s^2 2p^2$) is easily hybridized to produce the orbital sets sp^3, or $sp^2 + p$, or $sp + p^2$. This means that carbon ring and chain structures can have multiple as well as single bonds between C atoms. Carbon, in

FIGURE 14-11
Phase diagram for carbon.

To include the vast range of pressures required (from 1 to 1,000,000 atm), pressure is plotted on a logarithmic scale. The arrow marks the point 1 atm, 25 °C. To represent carbon vapor at lower temperatures would require showing pressures approaching zero. (Recent experimental evidence suggests that the graphite region of the diagram may be considerably more complex than indicated here.)

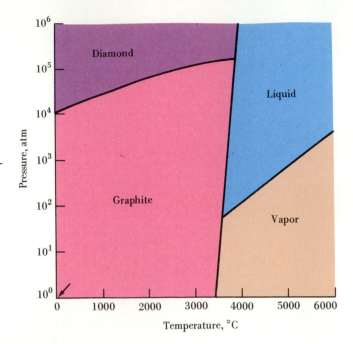

The ability of atoms of the same kind to join together into chains is called **catenation.** Carbon atoms possess this ability to a greater degree than any other atoms.

Industrial diamonds being pried from the iron matrix in which they were formed. [Fred Ward/Black Star]

combination with H, O, N, S, and a few other elements, produces such a variety of compounds that a specialized branch of chemistry called **organic chemistry** has developed around these compounds. We consider organic chemistry and the closely related biochemistry in Chapters 27 and 28. For the present we will examine a few aspects of what we might call the *inorganic* chemistry of carbon.

Graphite and Diamond. In Section 12-8 we took two different forms of carbon—diamond and graphite—as examples of network covalent bonding. We saw that the fundamental difference between these two solids is in the different bonding schemes for C atoms. In diamond, bonding is through the orbital set sp^3, whereas in graphite bonding involves the orbital set $sp^2 + p$. Two or more forms of an *element* that differ in the bonding or molecular structure of their fundamental units are called **allotropes.** Diamond and graphite are allotropes of carbon. Of these two allotropes, graphite is the more stable at ordinary temperatures and pressures (such as 298 K and 1 atm). We assign it an enthalpy of formation of zero. Diamond has a higher (positive) enthalpy of formation. The difference between the enthalpies of graphite and diamond is

$$C(\text{graphite}) \longrightarrow C(\text{diamond}) \qquad \Delta H = +1.88 \text{ kJ/mol} \qquad (14.42)$$

Another way to see that graphite is the more stable form of carbon is through the phase diagram in Figure 14-11. The point (1 atm, 298 K) falls in the graphite area. The area representing diamond is at much higher pressures. Thus, diamond should decompose to graphite at ordinary pressures. However, phase changes that require a rearrangement in bond type and crystal structure often occur extremely slowly. This is very much the case with the diamond → graphite transition. From the phase diagram we might predict that by applying a sufficiently high pressure to graphite we should be able to convert it to diamond. This is what is done in the production of artificial diamonds. The process is carried out at temperatures of 1000–2000 °C and pressures of up to 100,000 atm or more. Usually the graphite is mixed with a metal such as iron. The metal melts and the graphite is converted to diamond within the liquid metal.

Graphite has excellent lubricating properties, even when dry. We rely on this property in the graphite–clay mixture used in pencils. Because of its ability to conduct electric current, graphite is used for electrodes in batteries and industrial electrolysis. Its use in foundry molds, furnaces, and other high-temperature devices is based on graphite's ability to withstand high temperatures.

A newly developed use of graphite is in the manufacture of high-strength light-weight composites consisting of graphite fibers and other plastic materials. These composites are found in products ranging from tennis rackets to lightweight air-planes. The graphite fibers can be produced by heating a carbon-based fiber such as rayon to a high temperature. The other elements are driven off as gaseous products and a graphite fiber remains.

Aside from their value in jewelry, diamonds have two main industrial uses. They are used as industrial abrasives because of their extreme hardness (10 on the Moh hardness scale). Because of their hardness and also because of their high thermal conductivity (which means that they dissipate heat quickly), diamonds are used in drilling bits for cutting steel and other hard materials.

Other Forms of Carbon. Carbon can also be obtained in several forms known as **amorphous carbon.** (The term "amorphous" implies noncrystalline, but amorphous carbon is probably just microcrystalline graphite.) When coal is heated in the absence of air, volatile substances are driven off, leaving a high-carbon residue called **coke.** This same type of destructive distillation of wood produces **charcoal.** Incomplete combustion of natural gas (as in an improperly adjusted Bunsen burner in the laboratory) produces a smoky flame. This smoke can be deposited as a finely divided soot called **carbon black.**

A typical automobile tire contains several kilograms of carbon black.

Currently, coke is the principal metallurgical reducing agent. It is also used in the manufacture of fuel gases (recall reaction 7.25). The chief uses of carbon black are as a filler in rubber tires, as a pigment in printing inks, and as the transfer material in carbon paper, typewriter ribbons, and photocopying machines. Typical annual production of carbon black in the United States (over 1,000,000 tons) places it about 34th among industrial chemicals.

Some forms of carbon black are specially prepared by heating them to 800–1000 °C in the presence of steam to expel all volatile matter. These forms, called **activated carbons,** have highly porous internal structures (like sponges or honeycombs). They have a very high ratio of surface area to volume, and thus a great ability to *adsorb* substances from liquid solutions or from the gaseous state. Activated carbon is used in gas masks to adsorb poisonous gases from air, for water purification, as a decolorizer of sugar solutions, in air conditioning systems to control odors, and in industrial plants for the control and recovery of vapors.

Activated carbons may have surface areas as high as $1000 \ m^2$ per gram of carbon.

Oxides of Carbon. When carbon or carbon-containing compounds are burned in a *limited* quantity of air, the principal carbon-containing product is **carbon monoxide.** In an *excess* of air, **carbon dioxide** is formed.

Air pollution by CO comes chiefly from the incomplete combustion of fossil fuels, especially from automobile engines. CO is an inhalation poison because CO molecules bond to Fe atoms in hemoglobin in blood, displacing the O_2 molecules that the hemoglobin normally carries. This action of CO on hemoglobin is similar to the formation of a class of compounds known as metal carbonyls and discussed in Section 24-7. Carbon dioxide is not normally considered to be an air pollutant because it is nontoxic. However, its potential to alter the earth's climate through the greenhouse effect is disturbing (see Focus feature of Chapter 7).

Another minor and unstable oxide of carbon is carbon suboxide, C_3O_2.

There are three uses of carbon monoxide: One is in synthesizing other compounds. Often this is done through mixtures of CO and H_2 known as **synthesis gas.** One synthesis is that of methanol.

$$CO(g) + 2 \ H_2(g) \longrightarrow CH_3OH(l) \tag{14.43}$$

Another use of CO is as a reducing agent. In a blast furnace, for instance, coke is converted to CO and the CO reduces iron oxide to iron.

CO(g) as a reducing agent.

$$Fe_2O_3(s) + 3\ CO(g) \longrightarrow 2\ Fe(l) + 3\ CO_2(g) \tag{14.44}$$

A third use of CO is as a fuel, usually mixed with CH_4, H_2, and other combustible gases (recall Section 7-11).

The major use of carbon dioxide (about 50%) is as a refrigerant (usually in the form of dry ice) for freezing, preserving, and transporting food. Making carbonated beverages accounts for about 20% of CO_2 consumption. Other important uses are in fire extinguishing systems and in the recovery of oil in oil fields.

Carbonic Acid and Carbonates. $CO_2(g)$ dissolves in water and comes into equilibrium with carbonic acid, H_2CO_3. Carbonic acid is a weak acid that ionizes in two stages. Neutralization of the acid through its first ionization stage (for example, with NaOH) produces a salt variously called a hydrogen carbonate, an acid carbonate, or a bicarbonate. Neutralization through the second stage (that is, neutralization of the acid carbonate) yields a carbonate.

$$CO_2 + H_2O \rightleftharpoons H_2CO_3(aq) \tag{14.45}$$

$$H_2CO_3(aq) + Na^+ + OH^- \longrightarrow \underset{\text{sodium hydrogen carbonate}}{Na^+ + HCO_3^-} + H_2O \tag{14.46}$$

$$Na^+ + HCO_3^- + Na^+ + OH^- \longrightarrow \underset{\text{sodium carbonate}}{2\ Na^+ + CO_3^{2-}} + H_2O \tag{14.47}$$

We can reverse the above reactions by adding an acid to a carbonate. This is a simple way of preparing $CO_2(g)$ in the laboratory.

Action of an acid on a carbonate.

$$Na_2CO_3(aq) + 2\ H^+(aq) \longrightarrow 2\ Na^+(aq) + H_2O + CO_2(g) \tag{14.48}$$

Old fashioned baking powder consists of solid $NaHCO_3$ and one or more other solids that have acidic properties. When water is added to the mixture $CO_2(g)$ is evolved (recall Figure 5-7). Typically, the acidic solid is an acid phosphate. The net reaction is

$$H_2PO_4^-(aq) + HCO_3^-(aq) \longrightarrow H_2O + HPO_4^{2-} + CO_2(g) \tag{14.49}$$

The alkali metal carbonates are water soluble and occur in natural brines. Alkaline earth and most other metal carbonates are water insoluble and many of these are naturally occurring minerals. Calcite ($CaCO_3$) is one such mineral; dolomite ($CaCO_3 \cdot MgCO_3$) is another. We encounter other carbonate minerals later in the text. In later chapters we also consider other aspects of the chemistry of carbonic acid and carbonates. These range from their action in maintaining the proper acidity (pH) of blood to forming limestone caves, introducing hardness in water, and water softening.

14-9   Silicon

A silicon atom, like a carbon atom, can form four bonds simultaneously. It uses all of its valence electrons ($3s^2 3p^2$) in an sp^3 hybridization scheme, and crystallizes in an *fcc* structure similar to diamond (recall Figure 12-33). There is no allotrope of silicon equivalent to the carbon allotrope, graphite. Whereas diamond is an electrical insulator and graphite a moderately good electrical conductor, silicon is a *semiconductor*. In SiO_2 (silica) each Si atom is bonded to *four* O atoms and each O atom to *two* Si atoms. The structure is that of a network covalent solid, as suggested by Figure 14-12. Certain properties of silica resemble those of diamond. For example, quartz, a form of silica, has a Mohs hardness of 7 (compared to 10 for diamond), a high melting point (about 1700 °C), and is a nonconductor of electricity. Silica, of

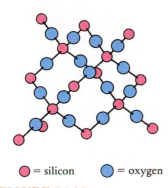

= silicon   = oxygen

FIGURE 14-12
Bonding in silica—SiO$_2$.

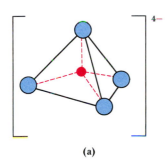

(a)

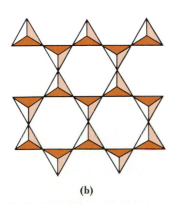

(b)

FIGURE 14-13
Silicate anions.

(a) The simple anion SiO$_4$$^{4-}$ consists of a central Si atom surrounded by four O atoms in a tetrahedral arrangement.
(b) A portion of the anion structure of mica. The cations required to balance the electric charge of the sheet anions are located above and below each sheet. Usually these are Al^{3+} and K$^+$. Talc and clay have similar structures.

Are You Wondering:

Why silica (SiO$_2$) does not exist as simple molecules as CO$_2$ does?

Because they are both in group 4A of the periodic table and have four valence electrons, we might expect carbon and silicon to form oxides with similar properties. In CO$_2$ the ability of C and O atoms to form π bonds through the sidewise overlap of their 2p orbitals is strong. The result is strong C-to-O double bonds and a very stable triatomic molecule, as suggested by the Lewis structure

$$: \ddot{O} = C = \ddot{O} :$$

Silicon, being in the third period, would have to use 3p orbitals to form double bonds with oxygen. The sidewise overlap of these orbitals with the 2p orbitals of oxygen is too limited for π bond formation. From an energy standpoint, a stronger bonding arrangement results if the Si atoms form *four* single bonds with O atoms (bond energy: 464 kJ/mol) rather than *two* double bonds (bond energy: 640 kJ/mol). Since each O atom must be simultaneously bonded to two Si atoms, the result is a network of —Si—O—Si— bonds.

Here, again, we see an example of how the second period member of a group (C) differs from the higher-period members (Si, Ge, Sn and Pb).

which there are more than a dozen polymorphic forms, is the basic raw material of the glass, ceramics, and refractory materials industries.

Silicate Minerals. Just as carbon is the key element of the living world, silicon is the key element of the inanimate or mineral world. The vast majority of silicates are *insoluble* in water; only the simple alkali metal silicates are soluble. A common feature of silicate minerals is the complexity of their silicate anions. Within these complex anions, however, we can generally find a basic structural unit: a tetrahedron with O atoms at its corners and a Si atom at its center. These SiO$_4$ tetrahedra may be

- separate units;
- joined into chains or rings in groups of 2, 3, 4, or 6;
- joined together into long single or double chains;
- arranged in sheets;
- linked into a three-dimensional network.

Two of these possibilities are illustrated in Figure 14-13, and several examples are listed in Table 14-8. To establish the charge on a silicate anion in this table, you will

TABLE 14-8
Some Representative Silicate Minerals

	Anion type	Mineral	Composition
(a) orthosilicate	SiO$_4$$^{4-}$	zircon, olivine	ZrSiO$_4$, Mg$_2$SiO$_4$
(b) polysilicate	Si$_2$O$_7$$^{6-}$	thortveitite	Sc$_2$Si$_2$O$_7$
	Si$_6$O$_{18}$$^{12-}$	beryl	Be$_3$Al$_2$Si$_6$O$_{18}$
(c) pyroxene	Si—O chains	spodumene	LiAl(SiO$_3$)$_2$
(d) amphibole	Si—O double chains	hornblende	(Ca,Na,K)$_{2-3}$(Mg,Fe,Al)$_5$ (Si,Al)$_8$Si$_6$O$_{22}$(OH)$_2$
(e) mica	Si—O sheets	muscovite	KAl$_2$(AlSi$_3$O$_{10}$)(OH)$_2$
(f) zeolite	Si—O three-D network	natrolite	Na$_2$Al$_2$Si$_3$O$_{10}$ · 2 H$_2$O

An aqueous solution of acetic acid containing a high concentration of an electrolyte (here, NH₄Cl) is mixed with sodium silicate solution. A stiff silica gel forms in a few seconds. [Carey B. Van Loon]

find it convenient to think of each Si atom as carrying a charge of +4, and each O, −2. Thus the charge on the anion SiO₄ is [+4 + (4 × −2)] = −4; and on the anion Si₂O₇, [(2 × +4) + (7 × −2)] = −6.

Colloidal Silica. Just as CO_2 dissolves in water and reacts with bases to produce carbonates, SiO_2 slowly dissolves in strong bases. It forms a variety of silicates, such as Na_2SiO_3 (sodium metasilicate) and Na_2SiO_4 (sodium orthosilicate). Aqueous sodium silicates are often referred to as "water glass."

When we acidify an aqueous carbonate solution, the product is H_2CO_3 (carbonic acid), which decomposes into H_2O and CO_2 (reaction 14.48). When we acidify sodium silicate solutions, we get silicic acids, which are also unstable; they decompose to silica. Depending on the acidity of the solution, we obtain the silica as a colloidal dispersion, a gelatinous precipitate, or a solidlike gel in which all of the liquid is entrapped. The collodial silica is produced as a result of water molecules being eliminated between neighboring molecules, producing a polymer of ever increasing size. The process is

$$SiO_4{}^{4-}(aq) + 4\ H^+(aq) \longrightarrow Si(OH)_4 \tag{14.50}$$

followed by

$$(xSiO_2 \cdot yH_2O) \tag{14.51}$$
colloidal silica

Glass. If we mix sodium and calcium carbonates with sand and fuse the mixture at about 1500 °C, we get a liquid mixture of sodium and calcium silicates. When this mixture is cooled it becomes more viscous and eventually ceases to flow; it becomes solid. This solid, even in sheets of considerable thickness, is transparent to visible light. This product is ordinary **glass.** By varying the proportions of the three basic ingredients, and by adding other substances, we can alter the properties of the glass. For example, by adding about 13% B_2O_3 we obtain a glass that is especially resistant to thermal shock. This is the borosilicate glass that you are probably familiar with under the brand name Pyrex.

Substances other than silicates can be obtained in a glasslike state, and collectively all of these materials are called glasses. A unique feature of glasses is that order exists over only short distances. Glasses lack the long-range order that we see in crystalline solids. X-ray diffraction patterns of glasses resemble those of liquids more than they do solids. Other ways in which glasses differ from both crystalline solids and liquids are suggested by Figure 14-14.

Generally a crystalline solid melts sharply at a fixed temperature (melting point) and the volume of a sample increases abruptly as it converts from solid to liquid. When the liquid is cooled it retraces these changes in the reverse order at the same fixed temperature (freezing point). This path is shown in red in Figure 14-14. At times a liquid can be cooled below its freezing point without solidifying. This is the condition of *supercooling* previously described (Section 12-4), and is noted in blue in Figure 14-14. The formation of a glass is represented by the paths in tan. Although these paths follow that of a supercooled liquid over a short temperature interval, a temperature called the glass transition temperature, T_g, is soon reached. At this temperature the volume–temperature plot for the glass departs significantly from that of the supercooled liquid. Moreover, this T_g value depends on the rate at which the liquid is cooled. When we retrace these paths for solid glass, we see that

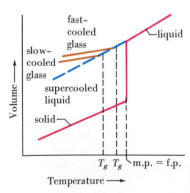

FIGURE 14-14
Glass behavior.

the solid glass does not have a definite melting point, but simply a range of temperatures over which it softens and then becomes liquid.

Portland Cement. Portland cement is a complex mixture of calcium silicates and aluminates that is formed by heating limestone to about 1500 °C with materials rich in silica (SiO_2) and alumina (Al_2O_3). The reactions of cement with water and its subsequent hardening are complicated. Certain reactions occur almost immediately; others require a period of months or years. Because only water is required for the setting of cement, it is referred to as *hydraulic* cement. It will even set when completely submerged in water, as in the construction of bridge piers. Pure cement does not have much strength, but when mixed with sand and gravel it sets to the hard mass we know as concrete.

Portland cement is produced in long coal-fired, firebrick-lined rotary kilns. A mixture of limestone, clay, and sand is heated to progressively higher temperatures as it slowly moves down the inclined kiln. First moisture, and then chemically bound water, are driven off. This is followed by calcination of $MgCO_3$ and $CaCO_3$ (reaction 14.32) and, finally, by the reaction of oxides to form silicates, aluminates, and other cement components.

CHEMISTRY
EVOLVING

A turbine rotor assembly made of a modern ceramics material. [Courtesy Garrett Turbine Engine Company]

A new class of superconducting ceramic materials was discovered in 1986. These ceramics are currently the subject of intensive research and development efforts (see page 896).

Ceramics

Traditionally, the field of ceramics has focussed on naturally occurring inorganic materials, chiefly silica (SiO_2), clay (rich in aluminum silicate), and feldspar (sodium, potassium, calcium, and aluminum silicates). The production of traditional ceramic products usually takes the form of (1) reducing the ceramic material to a powder; (2) molding the powder into the desired shape; (3) heating the molded object to a high temperature where the powder particles fuse (sinter) together; (4) finishing operations (polishing, applying a glaze, and so on).

Among the properties of traditional ceramic materials that govern their uses are hardness, rigidity, resistance to heat and chemical attack, and electrical insulating properties. The production of glass accounts for about one-half the output of the ceramics industry. Other products include bricks, tiles and other refractory materials; porcelain enamel coatings (as for stoves and refrigerators); dinnerware; and abrasives.

More recently, interest has grown in new or advanced ceramic materials made from a broader range of substances (some of very high purity) and having a broader range of properties. For example, while traditional oxide ceramics are usually electrically insulating, other ceramics have semiconducting or even electrically conducting properties. Some new ceramics materials have specially designed magnetic or optical properties. Also new are methods of fabricating ceramic materials, such as the *sol-gel* process. Here the starting point is a colloidal dispersion of particles ranging in size from 1 to 100 nm. The sol is poured into a mold and, by removing some of the liquid phase, the sol is converted to a gel. The gel is then processed into the final ceramic product. Some exceptionally lightweight ceramic materials can be produced in this way. (Recall the sol-gel process in colloidal silica presented on page 502.)

Uses of these advanced ceramics fall into two general categories: electrical, magnetic, or optical applications (as in the manufacture of integrated circuit components) and applications that takes advantage of the ceramic's mechanical and structural properties at high temperatures. These latter properties are currently being explored in developing ceramic components for gas turbines and automotive engines. Quite possibly the automobile engine of the twenty-first century will be a ceramic engine. There is some truth to the description of the present as the "new stone age." ∎∎∎

14-10   Nitrogen

4A	5A	6A
6 C	7 N	8 O
14 Si	15 P	16 S

Nitrogen has the electron configuration $1s^2 2s^2 2p^3$. In forming compounds with other atoms, the N atom is able to gain, or more likely share, three electrons to acquire a valence shell octet, $1s^2 2s^2 2p^6$. In other cases the N atom can use all five of its valence electrons in covalent bond formation. As a result, the oxidation state of N in its compounds can range from -3 to $+5$. The maximum oxidation state, $+5$, corresponds to its periodic table group number, 5A.

Because so many different oxidation states are available to N atoms in their compounds, the chemistry of nitrogen is quite varied. Interestingly, though, the molecule from which all nitrogen compounds are ultimately derived, N_2, is unusually stable. The reason for nitrogen's relative inertness is the great strength of the bond (a triple bond) between N atoms in N_2; 946.4 kJ of energy is required to rupture 1 mol of these bonds. That is,

$$N\!-\!N(g) \longrightarrow 2\,N(g) \qquad \Delta H = +946.4 \text{ kJ/mol} \tag{14.52}$$

Because this bond energy is so high, many nitrogen-containing compounds have positive enthalpies (heats) of formation. For example, the energy released in forming two moles of N-to-O bonds in NO is not as great as the energy required to break one mole of N-to-N bonds in N_2 and one mole of O-to-O bonds in O_2.

$$N\!-\!N(g) + O\!-\!O(g) \longrightarrow 2\,N\!-\!O(g) \qquad \Delta H = +181 \text{ kJ/mol} \tag{14.53}$$

If reaction (14.53) were highly exothermic instead of endothermic, $N_2(g)$ and $O_2(g)$ might not coexist so peacefully in the atmosphere.

Example 14-5

Relating bond energies and enthalpy change for a reaction. Use bond energies of 946 kJ/mol for N_2 and 498 kJ/mol for O_2, together with data from equation (14.53), to estimate the strength of the N-to-O bond (i.e., the bond energy) in NO.

Solution. In reaction (14.53) the energy required for bond breakage is

Bonds broken: N-to-N + O-to-O = 946 kJ/mol + 498 kJ/mol = 1444 kJ/mol

Two moles of N-to-O bonds are formed, and for this the energy *release* is $-2 \times$ (bond energy of NO). The sum of these two quantities is the measured enthalpy of reaction.

$$\Delta H = 1444 \text{ kJ/mol} - 2(\text{bond energy of NO}) = +181 \text{ kJ/mol}$$

bond energy of NO = 1263/2 = 632 kJ/mol

SIMILAR EXAMPLES: Exercises 22, 59.

Occurrence of Nitrogen and Nitrogen Compounds. The abundance of nitrogen in the earth's solid crust is only 0.002% by mass; very little nitrogen occurs in mineral deposits. The only important minerals are KNO_3 (niter or saltpeter) and $NaNO_3$ (sodaniter or Chile saltpeter) found in a few dry places on earth. At one time the only source of nitrogen compounds was from one of these minerals or from living matter or coal. All this changed about 100 years ago, first with the development of a process for liquefying air and fractionally distilling its components, and then with the development of a method (the Haber–Bosch process) of converting atmospheric nitrogen to ammonia. We discussed the extraction of nitrogen and oxygen from air and uses of these atmospheric gases in the Focus feature of Chapter 6. We also presented the nitrogen cycle—the succession of changes whereby atmospheric nitrogen is assimilated into living matter and then returned to the atmosphere.

Ammonia and Ammonium Compounds. In Chapter 16 we take a detailed look at the Haber–Bosch process. For the present we just note the net change—the conversion of N_2 and H_2 to NH_3.

Synthesis of ammonia.

$$N_2(g) + 3 H_2(g) \rightleftharpoons 2 NH_3(g) \qquad (14.54)$$

Ammonia is the starting point for the manufacture of all other nitrogen compounds, which accounts for its perennial ranking among the top half-dozen chemicals produced in the United States (about 15,000,000 tons per year). Ammonia also has some direct uses of its own. The most important is as a fertilizer. The most concentrated form in which nitrogen fertilizer can be applied to fields is as pure liquid NH_3, known as anhydrous ammonia. Another direct application of $NH_3(aq)$ is in a variety of household cleaning products.

A simple approach to producing other nitrogen compounds from NH_3 is to neutralize ammonia, a base, with an appropriate acid. The reaction forming **ammonium sulfate,** an important solid fertilizer, is

$$2 NH_3(aq) + H_2SO_4(aq) \longrightarrow (NH_4)_2SO_4(aq) \qquad (14.55)$$

Ammonium chloride is used in the manufacture of dry-cell batteries, in cleaning metal surfaces, and as a flux in soldering metals. It is made by the reaction of $NH_3(aq)$ and $HCl(aq)$. If we use $HNO_3(aq)$ to neutralize $NH_3(aq)$, we obtain **ammonium nitrate,** used both as a fertilizer and as a high explosive. The reaction of $NH_3(aq)$ and $H_3PO_4(aq)$ yields **ammonium phosphates** [such as $NH_4H_2PO_4$ and $(NH_4)_2HPO_4$]. These are good fertilizers because they supply two vital plant nutrients, both N and P. Ammonium phosphates are also used as fire retardants.

One of the most important chemical processes using ammonia as a starting material is the production of urea, $CO(NH_2)_2$. Carbon dioxide is the other reactant. (See Focus feature of Chapter 16.)

Nitrogen Oxides. Nitrogen forms a series of oxides in which the oxidation state of N can have every value ranging from +1 (N_2O) to +5 (N_2O_5). **Dinitrogen oxide** [nitrogen(I) oxide, nitrous oxide] has the structure

$$:N\equiv N-\overset{\cdot\cdot}{\underset{\cdot\cdot}{O}}: \quad \longleftrightarrow \quad :\overset{\cdot\cdot}{N}=N=\overset{\cdot\cdot}{O}:$$

We can prepare it in the laboratory by decomposing molten ammonium nitrate by gentle heating

$$NH_4NO_3(l) \xrightarrow{\Delta} N_2O(g) + 2 H_2O(g) \qquad (14.56)$$

You may have experienced one of its chief uses: as an anesthetic ("laughing gas") in dental offices.

The molecular orbital diagram for NO is the same as that shown for O_2^+ in Example 11-8.

Nitrogen oxide [nitrogen(II) oxide, nitric oxide] is an odd-electron molecule; it is *paramagnetic*. The best Lewis structure we can write for it is $:N=\overset{\cdot\cdot}{O}:$. With molecular orbital theory we can think of NO as being similar to N_2, but with an O atom substituting for one of the N atoms. The extra electron brought by the O atom to the molecular orbital diagram of N_2 in Figure 11-27 enters an antibonding π_{2p}^* orbital, suggesting a bond order of *2.5*. We can prepare NO(g) in the laboratory by the reaction of Cu(s) with cold dilute $HNO_3(aq)$.

$$3 Cu(s) + 8 H^+(aq) + 2 NO_3^-(aq) \longrightarrow 3 Cu^{2+}(aq) + 4 H_2O + 2 NO(g) \qquad (14.57)$$

Commercially, NO is produced by oxidizing NH_3 in the presence of a catalyst (the Ostwald process). This oxidation is the first step in converting NH_3 to other nitrogen compounds.

Ostwald process: oxidation of $NH_3(g)$.

$$4 NH_3(g) + 5 O_2(g) \xrightarrow{Pt} 4 NO(g) + 6 H_2O(g) \qquad (14.58)$$

Nitrogen dioxide [nitrogen(IV) oxide] is another *paramagnetic*, odd-electron molecule.

We can prepare $NO_2(g)$ by the reaction of $Cu(s)$ with warm concentrated $HNO_3(aq)$.

$$Cu(s) + 4\,H^+(aq) + 2\,NO_3^-(aq) \longrightarrow Cu^{2+}(aq) + 2\,H_2O + 2\,NO_2(g) \quad (14.59)$$

Often, however, when we see brown $NO_2(g)$ produced by a reaction involving $HNO_3(aq)$, the initial product may have been colorless $NO(g)$. NO is readily oxidized to NO_2 in air.

$$2\,NO(g) + O_2(g) \longrightarrow 2\,NO_2(g) \qquad (14.60)$$

At low temperatures NO_2 molecules tend to pair up to form *diamagnetic* N_2O_4, **dinitrogen tetroxide** [dimer of nitrogen(IV) oxide].

Recall that this reaction was illustrated in Figure 3-8.

$N_2O_4(l)$ is used to oxidize fuels (e.g., hydrazine, N_2H_4) in rocket systems such as the Titan.

There are several unusual features about the N_2O_4 structure. One of these is an N—N bond distance that is about 20% longer than the usual N—N single bond. We can think of the positive formal charges on the adjacent N atoms as exerting repulsive forces that lengthen the bond.

Nitric Acid and Nitrates. Reaction (14.58), followed by (14.60), followed by the reaction of $NO_2(g)$ with water describes the commercial preparation of nitric acid.

Preparation of nitric acid.

$$3\,NO_2(g) + H_2O(l) \longrightarrow 2\,HNO_3(aq) + NO(g) \qquad (14.61)$$

Pure nitric acid is a colorless liquid ($d = 1.50$ g/cm^3). It dissociates at temperatures slightly above its melting point (-42 °C) to N_2O_4, O_2, and water. Ordinary concentrated nitric acid is an aqueous solution with a density of 1.41 g/cm^3 and a concentration of about 15 M HNO_3. It generally has a yellow color due to the presence of dissolved oxides of nitrogen. We can make nitrate salts by neutralizing nitric acid with the appropriate bases [for example, KOH to produce KNO_3; $Ca(OH)_2$ to produce $Ca(NO_3)_2$]. In addition to its acidic properties, nitric acid (more specifically the NO_3^- ion in acidic solution) is a good oxidizing agent. With dilute $HNO_3(aq)$ and an active metal, the reduction product is N_2O (or even NH_4^+ in some cases).

$$4\,Zn(s) + 10\,H^+(aq) + 2\,NO_3^-(aq) \longrightarrow 4\,Zn^{2+}(aq) + 5\,H_2O + N_2O(g)$$
$$(14.62)$$

Air Pollution and Oxides of Nitrogen. About the turn of the twentieth century a new word entered the English language: **smog.** It was coined to describe a condition, common in London, in which a combination of smoke and fog obscured visibility and produced a number of minor and major health problems (including deaths). These conditions are commonly associated with industrial activities and this type of smog is now called **industrial smog.** Another type of air pollution, also referred to as smog, is common to southern California and to other large metropolitan areas such as Mexico City and Tokyo. This smog is formed by the action of sunlight on the products of combustion processes. Chemical reactions that are brought about by light are called *photochemical* reactions, and smog produced by photochemical reactions is called **photochemical smog.**

Normally, air temperature decreases regularly with increased altitude; the normal temperature–altitude profile is shown in Figure 14-15a. At times a different situa-

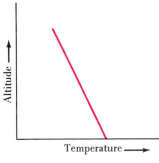

(a) Normal condition

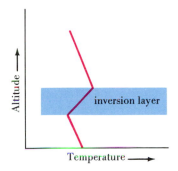

(b) Temperature inversion

FIGURE 14-15
Air temperature as a function of altitude.

(a) Under normal conditions, air temperature decreases with altitude.
(b) Under conditions of temperature inversion, a layer of warmer air is suspended aloft. At times the bottom of the inversion layer may be at ground level.

Smog over Mexico City. The inversion layer is visible in the center of the picture.
[Tony Morrison]

tion may prevail. For example, a mass of cool ocean air may move onshore and displace a layer of warmer air, holding it aloft, and producing a temperature–altitude profile like that shown in Figure 14-15b. This condition is known as a **temperature inversion.** When a temperature inversion exists, the volume of air into which pollutants can be dispersed is reduced. The inversion layer acts like a lid on the smoggy air below.

The reaction of $N_2(g)$ and $O_2(g)$ does not occur at low temperatures, but it becomes significant at high temperatures. Inevitably, in any high-temperature combustion process, as in an automobile engine, $N_2(g)$ and $O_2(g)$ in the air combine to form some $NO(g)$. Once $NO(g)$ is formed it is readily oxidized to $NO_2(g)$ (reaction 14.60). $NO_2(g)$ may then trigger the sequence of reactions listed in Table 14-9. The process begins when a molecule of NO_2 absorbs a quantum ($h\nu$) of near-ultraviolet light and dissociates to NO and O (reaction 1). This is followed by the reaction of O and O_2 to produce ozone, O_3 (reaction 2). The O_3 in turn reacts with NO to reform NO_2 and O_2 (reaction 3). If these were the only reactions that occurred, the ozone level under smog conditions would be limited. Reactions 4, 5, and 8 in Table 14-9 account for most of the O_3 formation. These reactions require the presence of hydrocarbons (RH), and unburned hydrocarbons are another important product of automobile engines.

If you ever experience photochemical smog, the features you will probably notice most are eye irritation and breathing difficulties. The eye irritation is caused by formaldehyde (HCHO) and acrolein ($H_2C{=}CHCHO$) produced in reaction 6 and peroxyacyl nitrates (PAN) produced in reaction 9. Ozone is largely responsible for the breathing difficulties. The incidence of all respiratory diseases (bronchitis, emphysema) is much higher than normal under smog conditions. Also, photochemical smog causes heavy crop losses (e.g., oranges) and the deterioration of rubber goods.

To control smog, automobiles are now provided with **catalytic converters.** CO and hydrocarbons are oxidized to CO_2 and H_2O in the presence of an *oxidation* catalyst (e.g., platinum or palladium metal). NO must be *reduced* to N_2, and this requires a *reduction* catalyst. A dual catalyst system is provided with both types of catalysts. Alternatively, the air–fuel ratio of the engine is set to produce some CO and unburned hydrocarbons. These then act as reducing agents to reduce NO to N_2. For example,

$$2\ CO(g) + 2\ NO(g) \longrightarrow 2\ CO_2(g) + N_2(g) \qquad (14.63)$$

Next, the exhaust gases are passed through an oxidation catalyst to oxidize the remaining CO and hydrocarbons to CO_2 and H_2O.

TABLE 14-9
Simplified Reaction Scheme for the Production of Photochemical Smog

nitrogen–oxygen reactions:	$NO_2 + h\nu \longrightarrow NO + O$	reaction 1
	$O + O_2 \longrightarrow O_3$	2
	$O_3 + NO \longrightarrow NO_2 + O_2$	3
hydrocarbon[a] reactions:	$RH + O \longrightarrow RO\cdot$	4
	$RO\cdot + O_2 \longrightarrow RO_3\cdot$	5
	$RO_3\cdot + RH \longrightarrow$ aldehydes + ketones	6
	(RCHO) (RCOR)	
	$RO_3\cdot + NO \longrightarrow RO_2\cdot + NO_2$	7
	$RO_3\cdot + O_2 \longrightarrow RO_2\cdot + O_3$	8
	$RO_3\cdot + NO_2 \longrightarrow$ peroxyacyl nitrates	9
	(PAN)	

[a]R = a hydrocarbon chain, e.g., C_8H_{17}, or ring, e.g., C_6H_5. Thus, RH = C_8H_{18}, C_6H_6, . . . ; RO$\cdot$, RO$_2\cdot$, RO$_3\cdot$ represent highly reactive, odd-electron species called radicals.

14-11   Phosphorus

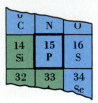

Phosphorus, the eleventh most abundant element in the earth's crust, occurs principally as deposits of phosphate rock (phosphorite) in several states of the U.S. (principally Florida), in Morocco, and in the U.S.S.R.. Phosphate rock was produced from the remains of marine animals deposited millions of years ago; it has approximately the composition of the mineral fluorapatite [$3Ca_3(PO_4)_2 \cdot CaF_2$].

The chief method of preparing elemental phosphorus is to heat phosphate rock, silica, and coke in an electric furnace. The net change that occurs is

Preparation of phosphorus.

$$2\ Ca_3(PO_4)_2(s) + 10\ C(s) + 6\ SiO_2(s) \xrightarrow{1500\ °C}$$
$$6\ CaSiO_3(l) + 10\ CO(g) + P_4(g) \qquad (14.64)$$

The $P_4(g)$ is condensed, collected, and stored under water as a solid.

Allotropy and Polymorphism of Phosphorus. The solid obtained from reaction (14.64) is a waxy, white, phosphorescent solid (a phosphorescent material glows in the dark). This solid, which can be cut with a knife and melts at 44.1 °C, is called **white phosphorus.** White phosphorus is a nonconductor of electricity, ignites spontaneously in air (hence the reason for storing it under water), and is insoluble in water but soluble in carbon disulfide (CS_2) and certain other nonpolar solvents. The form of white phosphorus found at room temperature, called *β-white* phosphorus, belongs to the cubic crystal system, but at −77 °C β-white converts to *α-white* phosphorus, which belongs to the hexagonal crystal system. The α-white and β-white are two *polymorphic* forms of white phosphorus.

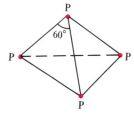

FIGURE 14-16
The P_4 molecule.

The basic structural units of white phosphorus are P_4 molecules. These molecules, pictured in Figure 14-16, are *tetrahedral;* a P atom is found at each corner of the tetrahedron. The P-to-P bonds in P_4 appear to involve the overlap of 3p orbitals almost exclusively. Such overlap normally produces 90° bond angles (recall Figure 11-3), but in P_4 the P—P—P bond angles are 60°. The bonds are said to be *strained,* and species with strained bonds are generally quite reactive, as is white P.

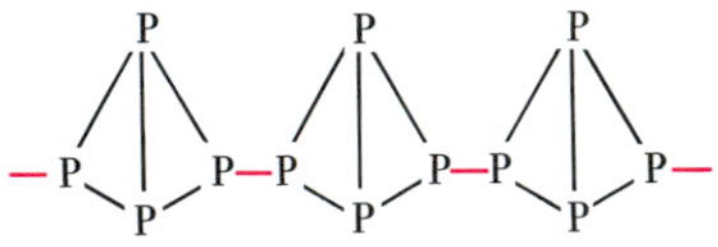

When white P is heated to about 300 °C, out of contact with air, it transforms to **red phosphorus.** What appears to happen is that one P—P bond per P_4 molecule breaks and the resulting fragments join together into long chains, as suggested by Figure 14-17. Red P is a more stable solid than white P and correspondingly less reactive. The triple point of red phosphorus is 590 °C and 43 atm. Thus, red phos-

FIGURE 14-17
The structure of red phosphorus.

Are You Wondering:

How allotropic and polymorphic forms differ.

Polymorphism applies *only to the solid state*. It refers to a situation in which the same substance (element or compound) may appear in different crystalline forms, but in which the bonding among atoms in the basic structural units is the same. Thus, in Figure 12-21 we considered several different polymorphic forms of water, all based on the structural units H_2O.

Allotropy applies *only to elements,* regardless of the state of matter. Allotropic forms differ in the bonding arrangements of atoms in the structural units. White, red, and black phosphorus are allotropes because they all have different P-to-P bonds. Moreover, we can also refer to *liquid* white P and *solid* red P as allotropes.

In common usage, however, a precise distinction between allotropy and polymorphism is often not made.

FIGURE 14-18
Molecular structures of P_4O_6
and P_4O_{10}.

🔴 phosphorus   🔵 oxygen

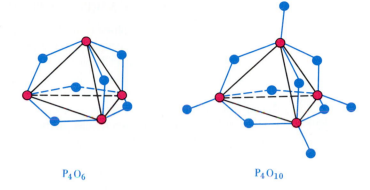

P_4O_6 P_4O_{10}

phorus sublimes without melting (at about 420 °C). Because they have a different atomic arrangement in their basic structural units, red and white phosphorus are allotropic forms of phosphorus.

The most stable form of phosphorus appears to be **black phosphorus,** which can be formed from white P under high pressures, or by heating white P in the presence of a catalyst (Hg) and "seed" crystals of black P. Black P has a layered crystalline structure, similar to graphite, but the layers are buckled. Black P is a semiconductor. Again, because the bonding between atoms differs in black P from white and red P, it too is an allotrope of phosphorus.

Oxides and Oxoacids of Phosphorus. The two most important oxides of phosphorus have P in the oxidation states +3 and +5, respectively. The simplest formulas we can write for these two oxidation states are P_2O_3 and P_2O_5. The corresponding names are phosphorus trioxide and phosphorus pentoxide. However the true formulas of the oxides are "double" those just written. The reason is that each oxide molecule is based on the P_4 tetrahedron and so must have *four* P atoms, not two. As shown in Figure 14-18, the oxide with P in the oxidation state +3 is P_4O_6. One O atom bridges each pair of P atoms in the P_4 tetrahedron, and this means *six* O atoms per P_4 tetrahedron. In the oxide with P in the oxidation state +5, P_4O_{10}, there is an O atom bonded to each corner P atom as well as the six bridging O atoms; this means a total of *ten* O atoms per P_4 tetrahedron.

The reaction of a limited quantity of $O_2(g)$ with P_4 produces P_4O_6; with an excess of O_2 we get P_4O_{10}. Both oxides react with water to form oxoacids.

Preparation of phosphorous and phosphoric acids.

$$P_4O_6 + 6\ H_2O \longrightarrow \underset{\text{phosphorous acid}}{4\ H_3PO_3}$$

$$P_4O_{10} + 6\ H_2O \longrightarrow \underset{\text{phosphoric acid}}{4\ H_3PO_4}$$

(14.65)

Phosphoric acid is used to treat metals to make them more corrosion resistant. In the food industry it is used to impart tartness to soft drinks. Calcium phosphates are used as dietary supplements, in baking powders, and as polishing agents in dentrifices. Sodium and potassium phosphates are used in dairy products, hams, cereals, and processed potatoes. In Section 23-3 we discuss some additional properties and uses of phosphoric acid and phosphates as fertilizers and detergents.

Other Compounds of Phosphorus. Phosphorus forms several hydrides, of which the best known is **phosphine, PH_3.** This compound is analogous to ammonia, for example, by acting as a base and forming phosphonium (PH_4^+) compounds. Unlike ammonia, PH_3 has a positive enthalpy of formation, is thermally unstable, and burns in air. It is extremely poisonous. The PH_3 molecule has a pyramidal shape with H—P—H bond angles of 93.6° (recall our prediction in Example 11-1). The commercial production of phosphine involves the reaction of P_4 and water in a basic solution.

We balanced this oxidation–reduction equation in Example 5-8.

Another important phosphorus-containing chemical is **POCl₃, phosphorus oxochloride (phosphoryl chloride).** We considered its production in Example 4-15.

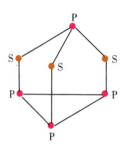

FIGURE 14-19
The molecular structure of P_4S_3.

$$P_4 + 3\ OH^- + 3\ H_2O \longrightarrow 3\ H_2PO_2^- + PH_3(g) \qquad (14.66)$$

Because of its toxicity, phosphine is used as a fumigant against rodents and insects. It is also used in the production of flame retardants.

Phosphorus trichloride, PCl₃, a colorless liquid (b.p. 76.1 °C), is the most important phosphorus halide. PCl_3 is produced by the direct action of $Cl_2(g)$ on elemental phosphorus. Although you will probably never see PCl_3, chemicals produced from it are found everywhere—soaps and detergents, plastics and synthetic rubber, nylon, motor oils, insecticides, herbicides. Like PH_3, PCl_3 has a pyramidal shape; its Cl—P—Cl bond angles are 100°.

All of the phosphorus sulfides are prepared in about the same way: the reaction of elemental sulfur and elemental phosphorus. Depending on the proportions of the two reactants and other conditions we obtain different sulfides. One sulfide, **P_4S_3 (phosphorus sesquisulfide),** has *three* S atoms bridging P atoms in the P_4 tetrahedron, unlike the *six* O atoms in P_4O_6 (see Figure 14-19). One of the most familiar uses of P_4S_3 comes in mixing it with an oxidizer like potassium chlorate ($KClO_3$) and forming it into the heads of "strike anywhere" matches. Another important sulfide is **P_4S_{10},** which has a structure analogous to P_4O_{10}. P_4S_{10} is commonly called phosphorus pentasulfide, but this is a misnomer (see Example 14-6). It is used in the production of insecticides and organic phosphates used as antiwear additives to motor oils.

Example 14-6

Systematic naming of binary covalent compounds. What are the *systematic* names of P_4S_3 and P_4S_{10}?

Solution. Apply the system of prefixes introduced as expression (3.6).

P_4S_3: *tetraphosphorus trisulfide*.

P_4S_{10}: *tetraphosphorus decasulfide*.

SIMILAR EXAMPLES: Exercises 2, 70.

14-12 Oxygen

Oxygen, in group 6A, has the electron configuration $1s^2 2s^2 2p^4$. Two of the $2p$ electrons are unpaired. If you remember our discussion of bonding in the O_2 molecule in Chapters 10 and 11, this is not easily explained. The bond order is two and the molecule is *paramagnetic,* and it is not possible to write a simple Lewis structure that incorporates both of these features. We do find a satisfactory description in the molecular orbital theory, however (recall Figure 11-27).

Oxygen is one of the most active nonmetals and one of the most important. It forms compounds with all the elements except the light noble gases. In general it forms ionic compounds with metals and covalent compounds with nonmetals. In a systematic study of the elements, we consider the formation and behavior of oxides as we consider each group of elements. Oxygen gas is itself an important commercial chemical, ranking third among the top chemicals (annual production, about 17,000,000 tons). We considered the extraction of O_2 from air and its commercial uses in the Focus feature of Chapter 6.

Ozone. Although the name oxygen conjures up the formula O_2, there are actually two *allotropes* of oxygen. Ordinary oxygen is **O_2.** The other allotrope is **ozone, O_3.** Ozone is a pale blue gas with a characteristic pungent odor. The two O—O bonds in

An ozone generator used for water purification. [Courtesy Los Angeles Department of Water and Power]

O_3 are equivalent and have a bond order of 1.5. We can represent ozone through Lewis structures, although the most satisfactory treatment of the molecule is with molecular orbital theory.

The quantity of O_3 in the atmosphere is quite limited at low altitudes, only about 0.04 part per million (ppm). As we saw in Section 14-10, however, its level increases (perhaps severalfold) in smog situations. The ozone content is appreciably greater at high altitudes, reaching about 10 ppm at about 25 to 30 km and existing in a belt called the *ozone layer* (see the Focus feature at the end of this chapter).

The reaction producing $O_3(g)$ directly from $O_2(g)$ is highly endothermic and does not occur under normal conditions in the lower atmosphere.

$$3 \; O_2(g) \longrightarrow 2 \; O_3(g) \qquad \Delta H° = +285 \; kJ \tag{14.67}$$

This reaction does occur in high-energy environments like electrical storms, and if you have ever smelled a pungent odor around heavy-duty electrical equipment it was probably ozone. The chief method of producing ozone in the laboratory is to pass a silent electric discharge through $O_2(g)$. Ozone is unstable and decomposes back to $O_2(g)$; for this reason it is always generated at the point where it is to be used.

Ozone is an excellent oxidizing agent. Its oxidizing ability is surpassed by only a few other substances (such as F_2 and OF_2). Its most important use is as a substitute for chlorine in purifying drinking water. Its advantages in this application are that it does not impart a taste to the water, nor does it form the potentially carcinogenic chlorination products that chlorine sometimes does. There are over 1200 installations, mostly in Europe, where ozone is used in municipal water treatment. An important oxidation–reduction reaction used to determine trace amounts of O_3 and other oxidants in polluted air involves oxidation of iodide ion.

$$2 \; I^-(aq) + O_3(g) + H_2O \longrightarrow 2 \; OH^-(aq) + I_2(aq) + O_2(g) \tag{14.68}$$

The solution is acidified and the $I_2(aq)$ is titrated with sodium thiosulfate solution using starch indicator (described in Section 23-1).

The severity of photochemical smog is measured through its oxidant level. Oxidants are the smog components capable of oxidizing I^- to I_2. In addition to O_3 the principal oxidants are NO_2 and organic peroxides.

Water and Hydrogen Peroxide. Certainly the most important compound of oxygen is water, H_2O. Scientists have probably measured more properties of water than of any other substance. Among the properties that we have already considered are its molecular structure, crystal structure, polymorphic forms, density, vapor pressure, heat of vaporization, specific heat, heat of fusion, melting point, triple point, boiling point, critical point, and solvent properties.

Although it is not as important as water, there is another hydride of oxygen—**hydrogen peroxide, H_2O_2.** This molecule features an O-to-O single bond and a nonlinear structure as seen in Figure 14-20.

Hydrogen peroxide enters into a wide variety of oxidation–reduction reactions. Because the O atoms in H_2O_2 are in the oxidation state -1, the H_2O_2 can either be reduced to H_2O (with O.S. of O = -2) or oxidized to O_2 (with O.S. of O = 0). This means that H_2O_2 may be an oxidizing agent in some instances and a reducing agent in others. For example,

$$H_2O_2 + 2 \; H^+ + 2 \; I^- \longrightarrow 2 \; H_2O + I_2 \tag{14.69}$$

$$5 \; H_2O_2 + 6 \; H^+ + 2 \; MnO_4^- \longrightarrow 2 \; Mn^{2+} + 8 \; H_2O + 5 \; O_2(g) \tag{14.70}$$

There are several ways in which we can prepare hydrogen peroxide. A simple laboratory method is to add barium peroxide to cold, dilute, aqueous sulfuric acid.

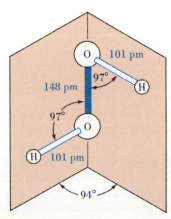

FIGURE 14-20
Geometric structure of H_2O_2.

$$BaO_2(s) + H_2SO_4(aq) \longrightarrow BaSO_4(s) + H_2O_2(aq) \tag{14.71}$$

Pure H_2O_2 is a pale blue liquid with a freezing point of -0.46 °C. The liquid is considerably more dense than water (1.47 g/cm³ at 0 °C). Pure H_2O_2 is unstable. Its decomposition is an exothermic reaction that is catalyzed by light and a variety of materials.

This reaction is described in detail in Chapter 15.

$$2 H_2O_2(l) \longrightarrow 2 H_2O(l) + O_2(g) \qquad \Delta H° = -197 \text{ kJ} \qquad (14.72)$$

Hydrogen peroxide has some important industrial uses. It is used as a bleaching agent for textiles and wood pulp, as a substitute for chlorine in water and sewage treatment, and in the manufacture of other chemicals. You will also find it in consumer products such as hair bleaches and antiseptics. In these consumer applications the hydrogen peroxide is usually in the form of a 3% solution in water. $H_2O_2(aq)$ is safe to use, especially at low concentrations, because its decomposition products are H_2O and $O_2(g)$.

Another classification scheme, that of acidic, basic, and amphoteric oxides, is considered in Section 17-9.

Ionic Oxides. From time to time we will consider ways of classifying oxides of the elements. Here, let us consider ionic oxides, that is, the compounds of oxygen with metals. **Normal oxides,** such as Li_2O and CaO, involve the simple oxide anion. The **peroxide** ion, as in Na_2O_2, features an O—O bond and an oxidation state of -1 for oxygen. The **superoxide** ion, such as in KO_2, also features and O—O bond. In addition, it has an unpaired electron, and an oxidation state of $-1/2$ for oxygen. Lewis structures of these ions are

Oxide ion	**Peroxide ion**	**Superoxide ion**
$\left[:\!\ddot{\underset{..}{O}}\!: \right]^{2-}$	$\left[:\!\ddot{\underset{..}{O}}\!:\!\ddot{\underset{..}{O}}\!: \right]^{2-}$	$\left[:\!\ddot{\underset{..}{O}}\!:\!\ddot{\underset{..}{O}}\!: \right]^{-}$

All three ions are unstable toward water and can only be found in the solid state. The decompositions of O_2^{2-} and O_2^- involve oxidation and reduction. Both of these ions, especially O_2^-, are good oxidizing agents.

Oxide: $\qquad\qquad O^{2-} + H_2O \longrightarrow 2 OH^-(aq)$

Peroxide: $\qquad 2 O_2^{2-} + 2 H_2O \longrightarrow O_2(g) + 4 OH^-(aq)$

Superoxide: $\quad 4 O_2^- + 2 H_2O \longrightarrow 3 O_2(g) + 4 OH^-(aq)$

In spacecraft, submarines, and emergency breathing apparatus it is necessary to have a ready source of oxygen gas. And for extended use there must also be a means of removing CO_2 from expired air. We noted in Section 14-5 the use of LiOH for removing CO_2. Oxygen-containing chemicals that can supply O_2 are chlorates and perchlorates (e.g., $KClO_3$ and $KClO_4$), peroxides, and superoxides. One of the most useful chemicals for these life-support systems is potassium superoxide, because it both *absorbs CO_2* and *produces $O_2(g)$.*

$$4 KO_2(s) + 2 CO_2(g) \longrightarrow 2 K_2CO_3(s) + 3 O_2(g) \qquad (14.73)$$

14-13 Sulfur

On the basis of electron configurations we expect sulfur and oxygen to be similar. Both elements form ionic compounds with active metals and both form similar covalent compounds, such as H_2S and H_2O, CS_2 and CO_2, SCl_2 and Cl_2O. However, there are important differences between oxygen and sulfur compounds. For example, H_2O has a very high boiling point (100 °C) for a compound of such low molecular weight (18), whereas the boiling point of H_2S (molecular weight, 34) is more normal (-61 °C). This is because hydrogen bonding is not a significant feature in sulfur compounds as it is in some oxygen compounds. Compared to the O atom, the S atom has a greater capacity to be bonded simultaneously to other atoms.

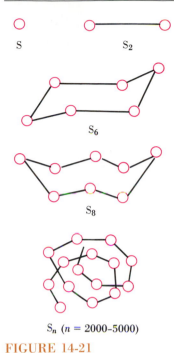

S   S₂

S₆

S₈

Sₙ (n = 2000–5000)

FIGURE 14-21
Different molecular forms of sulfur.

This is because the availability of $3d$ orbitals allows the S atom to have an expanded octet (as in SF_6).

Allotropy and Polymorphism of Sulfur. Many different molecular species are possible for elemental sulfur (see Figure 14-21), and this accounts for the existence of the large number of physical forms in which the element may appear. Allotropy is more complex for the element sulfur than for any other. A few of the familiar forms of sulfur are

- **rhombic sulfur (S_α),** which has sixteen S_8 rings in a unit cell and converts at 95.5 °C to
- **monoclinic sulfur (S_β).** Monoclinic sulfur is thought to have six S_8 rings in its unit cell. It melts at 119 °C, yielding
- **liquid sulfur (S_λ)** comprised of S_8 molecules. This is a yellow, transparent, mobile liquid. At 160 °C a remarkable transformation occurs. The S_8 rings open up and join together into long spiral-chain molecules, resulting in
- **liquid sulfur (S_μ)** which is dark in color and very thick and viscous. The chain length and viscosity of the liquid reach a maximum at about 180 °C. At higher temperatures the chains break up and the viscosity decreases again. At 445 °C, this liquid boils, producing
- **sulfur vapor,** which consists of molecules ranging from S_2 to S_{10}, but predominately S_8. At higher temperatures S_2 predominates.
- **Plastic sulfur** forms if liquid S_μ is poured into cold water. It consists of chain-like molecules and has rubber-like properties when first formed. On standing, however, it becomes brittle and eventually converts to rhombic sulfur.

Some of these forms of sulfur are pictured in Figure 14-22. We can summarize the allotropy and polymorphism of sulfur as a function of temperature in this way.

$$S_\alpha \xrightarrow{95.5\ °C} S_\beta \xrightarrow{119} S_\lambda \xrightarrow{160} S_\mu \xrightarrow{445}$$
$$S_8(g) \longrightarrow S_6 \longrightarrow S_4 \xrightarrow{1000} S_2 \xrightarrow{2000} S$$

Because of the sluggishness of some of the transformations involved, especially those in the solid state, we at times observe some additional phenomena. For example, if rhombic sulfur is heated rapidly it fails to convert to monoclinic sulfur and melts at 113 °C.

FIGURE 14-22
Several modifications of sulfur.

(a) Rhombic sulfur. [E. R. Degginger]
(b) Monoclinic sulfur. [E. R. Degginger]
(c) Liquid S_μ is poured into water to produce plastic sulfur. [Carey B. Van Loon]

(a)

(b)

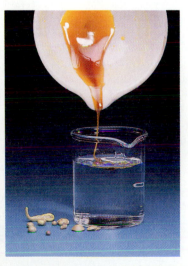

(c)

FIGURE 14-23
The Frasch process for mining sulfur.

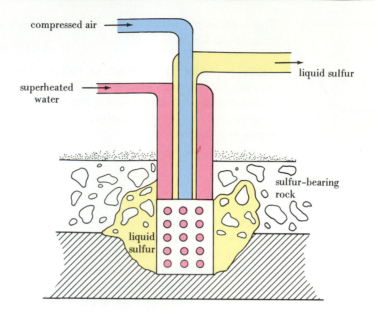

A reserve of solid sulfur. The sulfur is melted down when needed. Most sulfur is shipped and handled in liquid form. [Courtesy Freeport McMoRan, Inc.]

Production and Uses of Sulfur. Sulfur occurs abundantly in the earth's crust—as elemental sulfur, as mineral sulfides and sulfates, as $H_2S(g)$ in natural gas, and as organic sulfur compounds in oil and coal. Extensive deposits of elemental sulfur are found in Texas and Louisiana, some of them in offshore sites. This sulfur is mined in an unusual way, known as the Frasch process (Figure 14-23). A mixture of superheated water and steam (at about 160 °C and 16 atm) is forced down the outermost of three concentric pipes into an underground bed of sulfur-containing rock. The sulfur melts and forms a liquid pool. Compressed air (at 20–25 atm) is pumped down the innermost pipe and forces liquid sulfur up the remaining pipe. The liquid sulfur is either stored and shipped as a liquid (by tank car, ship, or pipeline) or allowed to solidify to a solid.

Although the Frasch process was once the principal source of elemental sulfur, this is no longer the case. This change has been brought about by the need to control sulfur emissions from industrial operations. In one process to remove H_2S from a gaseous mixture, the gas stream is split into two parts. One part (consisting of about one-third of the stream) is burned to convert H_2S to SO_2. The streams are rejoined in a catalytic converter at 200 to 300 °C, where this reaction occurs.

$$2 \, H_2S(g) + SO_2(g) \longrightarrow 3 \, S(g) + 2 \, H_2O(g) \tag{14.74}$$

About 90% of all the sulfur produced is burned to form $SO_2(g)$, and in turn, most $SO_2(g)$ is converted to sulfuric acid, H_2SO_4 (see Focus feature of Chapter 15). Elemental sulfur does have a few uses of its own, however. One of these is in vulcanizing rubber (discussed on page 341). Sulfur is also used in agriculture as a fertilizer and a pesticide (for example, in dusting grape vines).

Oxides and Oxoacids of Sulfur. More than a dozen oxides of sulfur have been reported, but only **sulfur dioxide, SO_2,** and **sulfur trioxide, SO_3,** are commonly encountered. Their structures and that of S_2O are outlined in Figure 14-24. The main commercial methods of producing $SO_2(g)$ are the direct combustion of sulfur and the "roasting" of metal sulfides.

Preparation of $SO_2(g)$.

$$S(s) + O_2(g) \longrightarrow SO_2(g) \tag{14.75}$$

$$2 \, ZnS(s) + 3 \, O_2(g) \longrightarrow 2 \, ZnO(s) + 2 \, SO_2(g) \tag{14.76}$$

Sulfur trioxide is produced by oxidizing $SO_2(g)$, but a catalyst is needed to speed the attainment of equilibrium in the reaction.

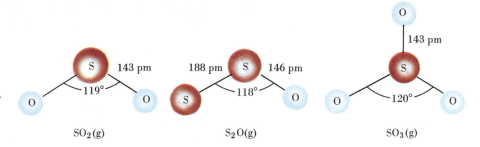

FIGURE 14-24
Structures of some sulfur oxides.

The SO_2 molecule owes its shape to bonding based on sp^2 hybrid orbitals about the S atom. The molecule S_2O has a similar structure but with an S atom substituted for one O atom. In the gaseous state SO_3 molecules are planar with 120° bond angles.

Preparation of SO_3(g).

$$2 SO_2(g) + O_2(g) \rightleftharpoons 2 SO_3(g) \qquad (14.77)$$

When SO_2(g) reacts with water, it produces H_2SO_3(aq), but this acid, **sulfurous acid,** has never been isolated in pure form. By contrast, **sulfuric acid, H_2SO_4,** which is produced by the reaction of SO_3(g) with water, can be obtained as a pure liquid ($d = 1.86$ g/cm^3; m.p. = 10 °C).

Sulfurous acid and its salts (sulfites) are good reducing agents,

$$Cl_2(g) + SO_3{}^{2-}(aq) + H_2O \longrightarrow 2 Cl^-(aq) + SO_4{}^{2-}(aq) + 2 H^+(aq) \qquad (14.78)$$

but they can also act as oxidizing agents, as in this reaction with H_2S.

$$2 H_2S(g) + 2 H^+(aq) + SO_3{}^{2-}(aq) \longrightarrow 3 H_2O + 3 S(s) \qquad (14.79)$$

Dilute sulfuric acid, H_2SO_4(aq), enters into all the common reactions of a strong mineral acid, such as neutralizing bases, dissolving metals to produce H_2(g), and dissolving carbonates to liberate CO_2(g).

Concentrated sulfuric acid has some distinctive properties. It has a very strong affinity for water, strong enough that it will even remove H and O atoms (in the proportion H_2O) from some compounds. In the reaction of concentrated sulfuric acid with a carbohydrate like sucrose, all the H and O atoms are removed and a residue of pure carbon is left.

All acids produce a stinging sensation, but concentrated H_2SO_4 must be handled with extreme care because it produces severe burns as a result of reactions like reaction (14.80).

$$C_{12}H_{22}O_{11}(s) \xrightarrow{H_2SO_4 \text{ (concd)}} 12 C(s) + 11 H_2O \qquad (14.80)$$

The concentrated acid is a moderately good oxidizing agent and is able, for example, to dissolve copper.

$$Cu(s) + 2 H_2SO_4 \longrightarrow Cu^{2+}(aq) + SO_4{}^{2-}(aq) + 2 H_2O + SO_2(g) \qquad (14.81)$$

In keeping with its perennial number 1 ranking among manufactured chemicals, sulfuric acid has many uses (discussed in the Focus feature of Chapter 15).

Concentrated H_2SO_4 is added to sucrose (cane sugar) at the left. The carbon produced in the reaction is seen at the right. [Carey B. Van Loon]

Oxides of Sulfur and Air Pollution. **Industrial smog** consists primarily of particles (ash and smoke), SO_2(g), and H_2SO_4 mist. Among the industrial operations that generate significant quantities of SO_2(g) are oil refineries, smelters, coke plants, and sulfuric acid plants. The main contributors to atmospheric releases of SO_2(g), however, are power plants burning coal or high-sulfur fuel oils. SO_2 can undergo oxidation to SO_3, especially when the reaction is catalyzed on the surfaces of airborne particles or through reaction with NO_2. In turn, SO_3 can react with water to produce H_2SO_4 mist, a component of **acid rain.** Also, the reaction of H_2SO_4 with airborne NH_3 produces particles of $(NH_4)_2SO_4$. The exact physiological effects of low concentrations of SO_2 and H_2SO_4 are not well understood, but these substances are respiratory irritants. Levels above 0.10 ppm are considered unhealthful.

The control of industrial smog hinges on removing sulfur from fuels and controlling SO_2(g) emissions where they occur. Dozens of processes have been proposed for removing SO_2 from smokestack gases. None has yet proved superior to all others, and none, in fact, is totally effective in removing SO_2 under all circumstances. One simple method involves the reactions

$$SO_2(g) + MgO(s) \underset{750\ °C}{\overset{150\ °C}{\rightleftharpoons}} MgSO_3(s) \tag{14.82}$$

The $SO_2(g)$ released when $MgSO_3(s)$ is heated can be used in the manufacture of H_2SO_4. Removing SO_2 from exhaust gases adds considerable expense to the burning of fossil fuels. It may be that processes designed to remove sulfur from fuels before they are burned will prove to be more feasible.

14-14   Fluorine and Chlorine

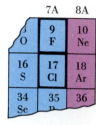

These two elements, having the valence-shell electron configurations ns^2np^5, belong to periodic group 7A, the halogens. Their atoms have high ionization energies—electrons are lost with difficulty—and large (negative) electron affinities—electrons are readily gained. Fluorine and chlorine are very active nonmetals; in fact, fluorine is the most active of all nonmetals.

Occurrence, Preparation, and Uses of Fluorine and Chlorine. Fluorine and chlorine do not occur free in nature. Because of their strong tendency to gain electrons, they are found principally as F^- (for example, in the mineral fluorite, CaF_2) and as Cl^- (as in NaCl). To obtain the free halogen element (X_2) requires oxidation of the halide ion (X^-).

Although the existence of fluorine had been known since early in the nineteenth century, no one was able to devise a chemical reaction to extract the free element from its compounds. H. Moissan was successful in preparing $F_2(g)$ in 1886, but he used an *electrolysis* reaction, not a straight chemical reaction. Moissan's method, which is still the only important commercial method, involves the electrolysis of HF dissolved in molten KHF_2.

> Notice that Mendeleev (1871) included fluorine in his periodic table and assigned it an atomic weight of 19 (see Table 9-1).

> The hydrogen difluoride ion in KHF_2 features a strong hydrogen bond, with an H^+ midway between two F^- ions, $[F—H—F]^-$.

$$2\ H^+ + 2\ F^- \xrightarrow{\text{electrolysis}} H_2(g) + F_2(g) \tag{14.83}$$

In 1986 a straight chemical method of producing fluorine was announced (see Chapter 23, Exercise 65).

Chlorine is not so difficult to produce by chemical reactions as is fluorine; an electron can be extracted from Cl^- more easily than from F^-. C. W. Scheele (1774) used the reaction of HCl(aq) and $MnO_2(s)$ to produce $Cl_2(g)$. This reaction is occasionally used in the laboratory, but the only significant commercial methods of preparing $Cl_2(g)$ involve electrolysis. The electrolysis may involve molten NaCl or $MgCl_2$ (recall reactions 14.18 and 14.27). More commonly, however, an aqueous solution of NaCl is used. [Na^+(aq) does not take part in the electrolysis of NaCl(aq).]

> Preparation of $Cl_2(g)$: electrolysis of NaCl(aq).

$$2\ Cl^-(aq) + 2\ H_2O \xrightarrow{\text{electrolysis}} 2\ OH^-(aq) + H_2(g) + Cl_2(g) \tag{14.84}$$

Reaction (14.84), which is the basis of the *chlor-alkali* industry, is perhaps the most important commercial electrolysis process. It is the source of three top chemicals—NaOH, H_2, and Cl_2. This process is explored more fully in the Focus feature of Chapter 21.

Elemental fluorine has three commercial uses.

1. The production of $UF_6(g)$ by the reaction of $UF_4(s)$ and $F_2(g)$. The UF_6 is subjected to gaseous diffusion to separate the U-235 and U-238 isotopes in the preparation of nuclear fuels (see Section 6-9).

2. The production of $SF_6(g)$, which forms when sulfur is burned in fluorine gas. $SF_6(g)$ is a nonflammable, nontoxic, unreactive gas that is useful as an insulating gas in high-voltage electrical equipment.

3. The production of **interhalogen compounds,** e.g., ClF_3, BrF_3. These can substitute for elemental fluorine in many reactions.

Elemental chlorine is the eighth ranking chemical, with about 10,000,000 tons produced annually. It also has three main commercial uses.

1. The production of chlorinated organic compounds (about 70%).
2. As a bleach in the paper and textile industries and for the treatment of swimming pools, municipal water, and sewage (about 20%).
3. The production of chlorine-containing inorganic chemicals (about 10%).

Reactions of Fluorine and Chlorine. Fluorine reacts with all the other elements except He, Ne, and Ar. The reactions of F_2 with other elements tend to be highly exothermic because the F—F bond energy is rather low (159 kJ/mol), whereas the energies of the bond between F and other atoms are quite large. Chlorine is less reactive than fluorine, but it is still able to form stable compounds with most of the other elements, except the noble gases. We have already considered a few typical reactions of the halogen elements in relation to the periodic table (see Table 9-8).

An important and interesting compound of fluorine is hydrogen fluoride, HF, the chemical most often used in preparing other fluorine compounds. We can make it by the reaction of a nonvolatile acid (H_2SO_4) on the salt (CaF_2) of a volatile acid (HF).

$$CaF_2(s) + H_2SO_4(\text{concd aq}) \longrightarrow CaSO_4(s) + 2\ HF(g) \tag{14.85}$$

One well-known property of HF is its ability to etch (and ultimately to dissolve) glass.

$$SiO_2(s) + 4\ HF(aq) \longrightarrow 2\ H_2O + SiF_4(g) \tag{14.86}$$

Because of this reaction, HF must be stored in special containers, such as wax or Teflon.

Uses of Fluorine and Chlorine Compounds. We encounter fluorine and chlorine compounds in many materials. Table 14-10 lists several important inorganic compounds of fluorine and their uses. A number of the important inorganic compounds of chlorine are discussed in some detail in Section 23-1. Equally or more important are *organic* compounds containing chlorine. Several of these are produced by the reaction of chlorine with hydrocarbons. The products are called **chlorinated hydrocarbons.** For example, in the direct reaction of $Cl_2(g)$ with $CH_4(g)$ chlorine atoms substitute for hydrogen atoms and products ranging from CH_3Cl to CCl_4 are formed. HCl is also a product of the reaction. Similarly, the reaction of ethane, C_2H_6, with $Cl_2(g)$ yields products with all degrees of substitution of Cl for H. Many of these chlorinated hydrocarbons are good solvents for nonpolar solutes, but they have other uses as well, as suggested in Table 14-11.

Fluorinated hydrocarbons are also of commercial importance, with one of the most important being **tetrafluoroethylene.** This substance is a monomer that can be polymerized to form **polytetrafluoroethylene,** a plastic material known by a number of brand names, including Teflon.

The small F—F bond energy is probably the result of the small size of the F atoms. Repulsions between the nuclei and between lone-pair electrons on the F atoms are greater than in the other halogens.

Etching glass with HF.

TABLE 14-10
Some Important Inorganic Compounds of Fluorine

Compound	Uses
Na_3AlF_6	mfr. of aluminum
BF_3	catalyst
CaF_2	optical components, mfr. of HF, metallurgical flux
ClF_3	fluorinating agent, reprocessing nuclear fuels
HF	mfr. of F_2, AlF_3, Na_3AlF_6, and fluorocarbons
LiF	ceramics mfr., welding and soldering
NaF	fluoridating water, dental prophylaxis, insecticide
SnF_2	mfr. of toothpaste
UF_6	mfr. of uranium fuel for nuclear reactors

Teflon

TABLE 14-11
Some Chlorinated Hydrocarbons

Compound		Uses[a]
methylene chloride	H | Cl—C—Cl | H	aerosols, paint removing, urethane foams, metal degreasing, electronics, pharmaceutical mfr.
1,1,1-trichloroethane	Cl H | | Cl—C—C—H | | Cl H	metal degreasing, cold cleaning, aerosols, adhesives, electronics, coatings and inks
trichloroethylene	Cl H | | Cl—C=C—Cl	metal degreasing, polyvinyl chloride mfr., adhesives
tetrachloroethylene	Cl Cl | | Cl—C=C—Cl	drycleaning, chlorofluorocarbon mfr., vapor degreasing

[a] In decreasing order of importance.

Teflon plastics are highly valued for applications that require resistance to high temperature or to chemical attack. Their uses range from nonstick frying pans and other kitchen utensils to electrical insulation to tubing, pipes, gaskets, and other industrial chemical equipment.

Another important class of organic compounds has both F and Cl atoms incorporated into hydrocarbon molecules. These are the so-called **chlorofluorocarbons (Freons).** The most widely used are $CFCl_3$, CF_2Cl_2, $CHClF_2$, and $C_2Cl_3F_3$. Because these substances are volatile liquids or easily condensable gases, they are extensively used as refrigerants (such as in air-conditioning systems in automobiles). They are also used as blowing agents in making Styrofoam products. The role that chlorofluorocarbons appear to play in the destruction of stratospheric ozone is discussed in the Focus feature.

Environmental Concerns over Fluorine and Chlorine Compounds. We have come to learn in recent years of some serious problems posed by organic chlorine and fluorine compounds. For one thing, many chlorinated hydrocarbons are believed to be carcinogenic (cancer-producing). One of the first compounds identified as an environmental hazard was the popular insecticide DDT ($C_{14}H_9Cl_5$), which proved so effective in the worldwide fight against malaria after World War II. Polychlorinated biphenyls (PCBs) are especially useful in electrical transformers and capacitors because they are stable at high temperatures. PCBs have also been used in printing inks, carbonless carbon paper, and as plasticizers in plastics. Some PCBs are still in use (principally in sealed electrical devices), but their production has been discontinued in the United States. The structure of biphenyl and a typical chlorinated biphenyl are shown below.

Teflon equipment used in well water sampling. To avoid contaminating the samples with trace impurities, it is imperative that objects coming into contact with the water be made of chemically inert materials, like Teflon. [Modern Industrial Plastics, Inc.]

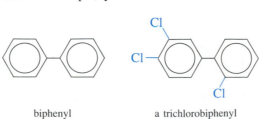

biphenyl   a trichlorobiphenyl

FOCUS ON
The Ozone Layer

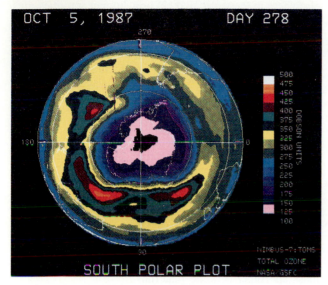

A map of stratospheric ozone levels over Antarctica on October 5, 1987. The levels vary widely, with the lowest levels being over the South Pole itself. [NASA–Goddard Space Flight Center/Total Ozone Mapping Project]

At altitudes between 25 and 35 km, in the stratosphere, the concentration of ozone, O_3, is about 10 parts per million. This belt of the atmosphere is called the **ozone layer.** (At sea level the ozone level is only 0.04 ppm.) Stratospheric ozone plays a vital role in permitting life to exist on earth, because of the ability of O_3 molecules to absorb ultraviolet radiation in the range 240 to 320 nm.

Ozone's role is twofold. First, ultraviolet radiation is damaging to humans, causing skin cancer and eye damage, and to other biological organisms as well. Second, in absorbing ultraviolet radiation O_3 molecules undergo dissociation with the evolution of heat, thereby helping to maintain a heat balance in the atmosphere.

The principal chemical reactions that produce $O_3(g)$ in the upper atmosphere are

$$O_2 + h\nu \longrightarrow O + O \tag{14.87}$$

$$O_2 + O + M \longrightarrow O_3 + M \tag{14.88}$$

Equation (14.87) describes the dissociation of O_2 following the absorption of ultraviolet radiation ($h\nu$). Atomic and molecular oxygen then react to form ozone. M, the "third body" in reaction (14.88), is needed to carry off excess collision energy; otherwise the O_3 produced would be so energetic as to decompose immediately. This "third body" might be $N_2(g)$.

Equation (14.89) demonstrates the first of the two atmospheric roles of ozone: that of absorbing ultraviolet ra-

diation. The second role, that of releasing heat into the atmosphere, is illustrated by equation (14.90).

$$O_3 + h\nu \longrightarrow O_2 + O \tag{14.89}$$

$$O_3 + O \longrightarrow 2\,O_2 \qquad \Delta H = -389.8 \text{ kJ} \tag{14.90}$$

O_3 is also consumed by other processes involving naturally occurring gases, such as oxides of nitrogen.

Thus, natural events account for the continuous formation and destruction of stratospheric ozone, maintaining a "steady state" concentration of about 10 ppm. In recent years it has been observed that certain gases of anthropogenic origin (that is, produced by human activities) appear to be reacting with stratospheric ozone, causing its destruction and threatening the integrity of the ozone layer. For example, Table 14-12 shows how oxides of nitrogen, produced through combustion processes in air, can contribute to ozone destruction.

Most worrisome at this time are gaseous chlorofluorocarbons. These molecules have a long lifetime in the atmosphere and eventually appear in the stratosphere in low concentrations. When these chlorofluorocarbons react with atomic oxygen (O), reactions occur in which various reactive products such as Cl and ClO are produced. These products, in turn, react with ozone, as suggested in Table 14-12.

Currently, the most convincing evidence that excessive destruction of stratospheric ozone is indeed occurring comes from studies in Antarctica. There it is found that with the coming of spring (October) a large depletion of

TABLE 14-12
Ozone-Destroying Reactions[a]

Based on oxides of nitrogen

$$NO + O_3 \longrightarrow NO_2 + O_2$$
$$O_3 + h\nu \longrightarrow O_2 + O$$
$$NO_2 + O \longrightarrow NO + O_2$$

net reaction: $2\,O_3 + h\nu \longrightarrow 3\,O_2$

Based on reactive chlorine species

$$Cl + O_3 \longrightarrow ClO + O_2$$
$$O_3 + h\nu \longrightarrow O + O_2$$
$$ClO + O \longrightarrow Cl + O_2$$

net reaction: $2\,O_3 + h\nu \longrightarrow 3\,O_2$

[a]These are simplified reaction schemes. The actual processes are much more complex.

O_3 occurs over a period of several weeks, before returning to more normal levels. The chemical reactions postulated to account for this are much more complex than the simplified scheme outlined in Table 14-12. In addition, meteorological conditions play an important role. It is now believed that ice crystals in stratospheric clouds over the South Pole provide a surface on which certain chemical reactions occur that normally would not. For example, hydrogen chloride (HCl) and chlorine nitrate ($ClONO_2$) are both known to be present in trace amounts in the stratosphere, but they do not react directly with ozone. On the surface of ice crystals, however, they react with each other, as in the following reaction in which chlorine is produced.

$$HCl + ClONO_2 \longrightarrow HNO_3 + Cl_2 \qquad (14.91)$$

Cl_2 dissociates into Cl atoms by absorbing solar radiation, and once Cl atoms are formed ozone-destroying reactions take place.

Important questions that remain to be answered are (1) whether appreciable ozone destruction is occurring elsewhere in the world and (2) how human activities must change in order to reverse or at least slow down ozone destruction. The most significant control measure taken to date is an international agreement among the industrialized nations to reduce the production and use of chlorofluorocarbons by up to 50% over the next decade.

Summary

The first 20 elements are among the most abundant in the earth's crust. They and their compounds have important uses. $H_2(g)$ is obtainable from a variety of sources and is used in the synthesis of a number of other materials. Its possible future uses as a fuel and a metallurgical reducing agent hinge on finding cheaper means of producing it. The lighter noble gases are noted for their inertness. In addition liquid He has a number of special uses based on its unique properties.

Li, Na, K, and the other group 1A elements are very active metals. Their atoms lose electrons easily to form the ions M^+. This oxidation occurs when the metals react with the halogens, oxygen, hydrogen, acids, or water. Most alkali metal compounds are highly water soluble, except for some Li compounds. In this and certain other respects Li resembles Mg. Such resemblances are sometimes called diagonal relationships.

Be, Mg, Ca, and the other group 2A elements are also metals, but here again the lightest member of the group, Be, differs in properties from the heavier members and resembles Al. Mg and Ca engage in most of the same reactions as the 1A metals do. Many of their compounds, however, are insoluble in water. A number of important calcium compounds are derived from the mineral limestone ($CaCO_3$).

The chemistry of boron is largely that of a nonmetal. Although Al exhibits some similarities to boron (e.g., in forming a covalent chloride), aluminum's most important uses are based on its metallic properties.

The inorganic chemistry of carbon deals primarily with the physical forms of carbon, the oxides of carbon, carbonic acid, and various carbonates. The inorganic chemistry of silicon focuses on SiO_2 and the variety of silicates that are the backbone of the mineral world. Important silicon-containing products are cement and concrete and glass and other ceramic materials.

A practical aspect of the chemistry of nitrogen involves converting the stable N_2 molecule into nitrogen-containing compounds. The oxides of nitrogen are important in these conversions, for example, in the production of nitric acid. Also, oxides of nitrogen are implicated in the formation of photochemical smog. Much of the chemistry of phosphorus centers on materials that can be made from phosphoric acid (H_3PO_4). Certain chlorides, oxochlorides, and sulfides of phosphorus are also commercially important.

The chemistry of oxygen is based primarily on O_2 and water, but ozone, O_3, and hydrogen peroxide, H_2O_2, are also important, particularly in oxidation–reduction reactions. Sulfur is of interest because of its variety of molecular and physical forms and because of its use in the manufacture of sulfuric acid, H_2SO_4, the most widely produced commercial chemical. Oxides of sulfur are common air pollutants produced by power plants and other industrial operations.

Fluorine and chlorine are the first two members of the halogen family (group 7A), the most active nonmetals. Their chemistry includes forming ionic halides with metals, covalent halides with nonmetals, and compounds among themselves (interhalogen compounds). Elemental chlorine and fluorine both have important uses, as do several of their compounds. Compounds of carbon and chlorine are some of the most important industrial solvents, and compounds of carbon, chlorine, and fluorine (called chlorofluorocarbons) have important uses as refrigerants and in the plastics industry. Environmental concerns over chlorocarbons and chlorofluorocarbons are discussed in the chapter and in its Focus feature.

Summarizing Example

Magnesium oxide is the most extensively used magnesium compound, and its principal use, because of its high melting point, is as a refractory material—one able to withstand very high temperatures. It is used to line furnaces in steel-making and in electrical insulators and ceramic materials.

A technician inspects refractory bricks made of magnesium oxide. [Courtesy Allied Chemical]

1. Why should magnesium oxide have such a high melting point?

Solution. Magnesium oxide is an ionic compound comprised of the ions Mg^{2+} and O^{2-}. These ions have small radii and high charges (recall Figure 12-35). Interionic attractive forces are especially strong and account for the very high melting point (2800 °C) of MgO.

(This example is similar to Example 12-10.)

2. One method of preparing magnesium oxide is to burn pure magnesium in air. In one laboratory experiment 0.200 g Mg yielded 0.315 g of product. *Assuming that there was no loss of material,* could this product have been pure MgO?

Solution. Write the chemical equation for the combustion reaction and calculate the theoretical yield of MgO.

$$2\ Mg(s) + O_2(g) \longrightarrow 2\ MgO(s)$$

$$\text{no. g MgO} = 0.200\ \text{g Mg} \times \frac{1\ \text{mol Mg}}{24.30\ \text{g Mg}} \times \frac{2\ \text{mol MgO}}{2\ \text{mol Mg}} \times \frac{40.30\ \text{g MgO}}{1\ \text{mol MgO}}$$

$$= 0.332\ \text{g MgO}$$

We should have obtained 0.332 g MgO but obtained only 0.315 g of product. The product cannot be pure MgO. It must contain a substance having a higher mass percent of Mg than does MgO (so that all of the original 0.200 g Mg can be combined into a smaller total mass than is possible in pure MgO).

(This example is similar to Example 4-6.)

3. What is the likely impurity in the product described in part 2.

Solution. Since the magnesium was *pure*, the second compound must be derived from a reaction of Mg with a component of air other than oxygen. The expected reaction would be that with nitrogen (reaction 14.30).

$$3\ Mg(s) + N_2(g) \longrightarrow Mg_3N_2(s)$$

To demonstrate that this is the likely reaction, show that the mass percent Mg in Mg_3N_2 is greater than in MgO (see Exercise 30).

(This example is similar to Example 3-4.)

4. Is the burning of magnesium metal an economically feasible method of preparing MgO?

Solution. Magnesium only occurs in compounds—never as the free metal. To prepare the pure metal from one of its compounds is an expensive, energy-intensive process (such as electrolysis of molten $MgCl_2$, reaction 14.27). It is far cheaper to convert other magnesium compounds directly to MgO than to convert them to Mg metal and then to MgO.

5. What alternate reaction can be used to produce MgO from a natural source?

Solution. Like calcium, magnesium forms various carbonate minerals (e.g., $MgCO_3$, magnesite). Calcination of magnesium carbonate produces MgO, in a reaction similar to (14.32).

$$MgCO_3(s) \longrightarrow MgO(s) + CO_2(g)$$

Key Terms

adduct (14-7)
allotropy (14-8)
calcination (14-6)
diagonal relationship (14-5)

electrolysis (14-3)
industrial smog (14-13)
interhalogen compounds (14-14)
nonstoichiometric compounds (14-3)

peroxide ion (14-12)
photochemical smog (14-10)
superoxide ion (14-12)

Highlighted Expressions

Electrolysis of water (14.5)
Reaction of active metals with water (14.7, 14.8)
Preparation of Na: electrolysis of NaCl(l) (14.18)
Preparation of NaOH(aq): electrolysis of NaCl(aq) (14.24)
Preparation of a volatile acid from a salt (14.25)
Typical reactions of Mg (14.28, 14.29, 14.30)
Important reactions of limestone and lime (14.32, 14.33, 14.34)
The thermite reaction (14.40)
CO(g) as a reducing agent (14.44)
Action of an acid on a carbonate (14.48)

Synthesis of ammonia (14.54)
Ostwald process: oxidation of $NH_3(g)$ (14.58)
Preparation of nitric acid (14.61)
Preparation of phosphorus (14.64)
Preparation of phosphorous and phosphoric acids (14.65)
Preparation of $SO_2(g)$ (14.75, 14.76)
Preparation of $SO_3(g)$ (14.77)
Preparation of $Cl_2(g)$: electrolysis of NaCl(aq) (14.84)
Etching glass with HF (14.86)

Review Problems

1. Briefly describe the meaning of each of the following terms, citing specific examples wherever possible: **(a)** allotropy; **(b)** an ore; **(c)** diagonal relationship; **(d)** PCB; **(e)** thermite reaction; **(f)** noble gas; **(g)** peroxide.

2. Supply a name or formula for each of the following: **(a)** SF_4; **(b)** magnesium nitride; **(c)** potassium superoxide; **(d)** $Ca(OH)_2$; **(e)** H_2O_2; **(f)** potassium perchlorate; **(g)** uranium hexafluoride; **(h)** hydrogen difluoride ion; **(i)** $LiAlH_4$; **(j)** O_3; **(k)** ammonium dihydrogen phosphate; **(l)** $NaHSO_3$; **(m)** N_2O_4; **(n)** bromine trifluoride.

3. Describe the following industrial materials by their chemical names. **(a)** coke; **(b)** limestone; **(c)** quicklime; **(d)** silica; **(e)** slaked lime; **(f)** gypsum; **(g)** synthesis gas; **(h)** water glass.

4. Complete the following equations for the reactions of substances with water.

    (a) $Ca(s) + H_2O \rightarrow$ **(b)** $LiH(s) + H_2O \rightarrow$

    (c) $Mg(s) + H_2O \xrightarrow{\Delta}$ **(d)** $CaO(s) + H_2O \rightarrow$

    (e) $C(s) + H_2O \xrightarrow{\Delta}$ **(f)** $Na_2O_2(s) + H_2O \rightarrow$

5. Complete the following equations for the reactions of substances with acids.

    (a) $Mg(s) + HCl(aq) \rightarrow$
    (b) $NH_3(g) + HNO_3(aq) \rightarrow$
    (c) $CaCO_3(s) + HCl(aq) \rightarrow$

    (d) $NaF(s) + H_2SO_4(conc. aq) \xrightarrow{\Delta}$
    (e) $Ag(s) + HNO_3(dil. aq) \rightarrow$
    (f) $NaHCO_3(s) + HC_2H_3O_2(aq) \rightarrow$

6. Write equations to show how H_2O_2 **(a)** oxidizes NO_2^- to NO_3^- in acidic solution; **(b)** oxidizes $SO_2(g)$ to $SO_4^{2-}(aq)$ in basic solution; **(c)** reduces MnO_4^- to Mn^{2+} in acidic solution; **(d)** reduces $Cl_2(g)$ to Cl^- in basic solution. [*Hint:* In acidic solutions you may write H^+ and/or H_2O on either side of the equation, as required; for a basic solution, add OH^- and/or H_2O.]

7. Write the formula of a compound mentioned in this chapter that has **(a)** N in the oxidation state (O.S.) +4; **(b)** O in the O.S. -1; **(c)** C in the O.S. +4; **(d)** S in the O.S. -2; **(e)** P in the O.S. -3; **(f)** N in the O.S. -3.

8. Identify the oxidizing and reducing agents in the following reactions found in this chapter: (14.3); (14.7); (14.19); (14.21); (14.31); (14.59); (14.63); (14.76).

9. Use data from Appendix D to establish whether each of the following reactions is endothermic or exothermic at 298 K and 1 atm: (14.2); (14.3); (14.10); (14.43); (14.58); (14.60); (14.63).

10. State which of the first 20 elements
(a) are gases at STP
(b) are semiconductors
(c) exhibit allotropy
(d) has the highest melting point
(e) form compounds with oxygen
(f) has the lowest critical temperature
(g) is the best oxidizing agent

11. For the following groupings of substances, select
(a) the most metallic of K, Be, and Ca;

(b) the least water soluble of Li_2CO_3, Na_2CO_3, and $CaCO_3$;

(c) the best oxidizing agent from H_2O_2, O_2, and O_3;

(d) the most volatile of the liquids $H_2O(l)$, $H_2S(l)$, $NH_3(l)$;

(e) the hardest of the solids $Na(s)$, $SiO_2(s)$, $C(graphite)$;

(f) the best electrical conductor from $LiF(l)$, $OF_2(l)$, $CF_4(l)$, and $BeF_2(l)$;

(g) the highest-melting among $LiCl(s)$, $NaCl(s)$, $LiF(s)$, $NaF(s)$.

12. Write an equation to represent the reaction of gypsum, $CaSO_4 \cdot 2H_2O$, with ammonium carbonate to produce ammonium sulfate (a fertilizer), calcium carbonate and water.

13. A reference work describing the production of phosphorus by reaction (14.64) states that for every 8.00 tons of phosphate rock used 1.00 ton of elemental phosphorus is obtained.

The phosphorus content of the phosphate rock is given as 31.0% P_2O_5. What is the percent yield of phosphorus in this reaction? [*Hint:* Is it necessary to use equation (14.64)?]

14. Imagine that the sulfur present in seawater as SO_4^{2-} (see Table 14-2) could be recovered as elemental sulfur. If this sulfur were then converted to H_2SO_4, how many km^3 of seawater would have to be processed to yield the average U.S. annual consumption of about 38 million tons of H_2SO_4?

15. In 1968, before automotive emission pollution controls were introduced, 75 billion gal of gasoline were used in the U.S. as a motor fuel. Assuming an emission of oxides of nitrogen of 5 g per vehicle mile and an average mileage of 15 mi/gal of gasoline, what mass of nitrogen oxides, in tons, was released to the atmosphere?

Exercises

Occurrence and abundance of the elements

16. Rearrange the eight most abundant elements listed in Table 14-1 on a mole basis. That is, of which of the elements is the greatest number of atoms to be found in the solid crust of the earth, the second greatest, and so on? Is (are) there any element(s) not listed in Table 14-1 that might be among the eight most abundant on a mole basis? [*Hint:* Refer also to Figure 14-1.]

17. The practicality of extracting an element from seawater strongly depends on the concentration of that element in seawater. The practicality of extracting the same element from the solid crust of the earth is much less dependent on the percent abundance of the element in the earth's crust. Why are these two situations different?

Group and period trends

18. Use data from the text to plot, in a manner similar to Figure 14-2, the following properties of the third period elements: (a) single covalent radius; (b) first ionization energy; (c) number of unpaired electrons in the neutral atoms. Explain the significance of each graph.

Hydrogen

19. Write chemical equations to represent the following reactions that are described in part in Section 14-3.

(a) The displacement of $H_2(g)$ from $HCl(aq)$ by aluminum metal.

(b) The reforming of ethane gas (C_2H_6) with steam.

(c) The complete hydrogenation of methylacetylene, $H_3C-C\equiv CH$. [*Hint:* What is the maximum number of H atoms that can be present in a three-carbon hydrocarbon?]

(d) The reduction of $MnO_2(s)$ to $Mn(s)$ with $H_2(g)$.

(e) The reaction of cesium metal with liquid water.

20. The text refers to the high temperatures achieved with an oxygen–hydrogen torch as being due in large part to the quantity of energy released when H atoms combine into H_2 molecules (page 485). Use appropriate data from Chapter 10 and Appendix D to compare the energy released when 1 mol H_2 is formed from H atoms with that released in the combustion of 1 mol H_2. Which quantity is larger?

21. What mass of $CaH_2(s)$ is required to generate sufficient $H_2(g)$ to fill a 215-L weather observation balloon at 726 mmHg and 20.8 °C?

22. The hydrogenation of acetylene to ethane is represented by the equation $HC\equiv CH(g) + 2 H_2(g) \rightarrow H_3C-CH_3(g)$.

(a) Use bond energies from Table 10-2 to calculate the enthalpy change for this reaction.

(b) Calculate ΔH for the reaction using data from Appendix D and compare your result with that obtained in part (a).

Helium, neon, argon

23. A 55-L cylinder contains Ar at 137 atm and 23 °C. What minimum volume of air at STP must have been liquefied and distilled to produce this Ar? Air contains 0.934% Ar, by volume.

24. A typical natural gas from which He can be extracted contains 0.3% He, by volume. Air contains 5 ppm He, by volume. How much more abundant is He in the natural gas than in air?

25. Suppose that a breathing mixture is prepared in which He is substituted for N_2, that is, a gas is prepared that is 79% He and 21% O_2, by volume. What is

(a) the density of this mixture, in g/L, at STP?

(b) the apparent molar mass of this mixture (recall Example 6-19)?

★ (c) the temperature at which this gas mixture would have the same density as does O_2 at 0 °C? (Assume that the gases are compared at 1 atm pressure.)

Lithium, sodium, potassium

26. Write chemical equations to represent the following reactions that are described in Section 14-5.

(a) The reaction of lithium metal with chlorine gas.

(b) The reaction of potassium metal with water.

(c) The reaction of lithium metal with $HCl(aq)$.

(d) The combustion of potassium metal to form potassium superoxide.

27. A pure white solid is either LiCl or KCl. Describe a simple test(s) to determine which it is.

28. Comment on the feasibility of using a reaction similar to (14.19) to produce (a) lithium metal from LiCl and (b) cesium metal from CsCl with Na(l) as the reducing agent. [*Hint:* Consider both physical properties, such as boiling point, and atomic properties, such as ionization energies.]

Beryllium, magnesium, calcium

29. Write chemical equations to represent the following reactions that are described in part in Section 14-6.

(a) The reduction of BeF_2 to Be metal with Na as a reducing agent.

(b) The reaction of calcium metal with chlorine gas.

(c) The reduction of uranium(IV) oxide to uranium metal with calcium as a reducing agent.

(d) The calcination of dolomite, which is a mixed calcium magnesium carbonate ($MgCO_3 \cdot CaCO_3$).

(e) The complete neutralization of sulfuric acid using slaked lime.

30. Refer to the Summarizing Example.

(a) Describe a chemical reaction that you could use to establish if the product contains Mg_3N_2.

★(b) If the product in part 2 is a mixture of MgO and Mg_3N_2, what must be the mass percent Mg_3N_2 in the mixture?

Boron, aluminum

31. Write chemical equations to represent the following reactions that are described in part in Section 14-7.

(a) The ionization of aluminum hydroxide as a base.

(b) The oxidation of Al to Al^{3+}(aq) by an aqueous solution of sulfuric acid (H_2SO_4). The reduction product is SO_2(g).

(c) The production of Cr metal from Cr_2O_3(s) by the thermite reaction, using Al as the reducing agent.

32. Write Lewis structures for the following species, both of which involve coordinate covalent bonding.

(a) fluoborate ion, BF_4^-, used in metal cleaning and in electroplating baths;

(b) boron trifluoride monoethylamine, used in curing epoxy resins (monoethylamine is $C_2H_5NH_2$).

33. Concerning the thermite reaction

(a) Use data from Appendix D to calculate ΔH at 298 K for $2\ Al(s) + Fe_2O_3(s) \rightarrow 2\ Fe(s) + Al_2O_3(s)$.

(b) Write an equation for the reaction when MnO_2(s) is substituted for Fe_2O_3(s), and calculate ΔH for this reaction.

(c) Show that if MgO were substituted for Fe_2O_3 the reaction would be *endothermic*. (Al does not reduce MgO to Mg.)

Carbon

34. Write chemical equations to represent the following reactions that are described in part in Section 14-8.

(a) The action of vinegar (aqueous acetic acid) on baking soda (sodium hydrogen carbonate).

(b) Carbon monoxide as a reducing agent in the reduction of zinc oxide to zinc metal.

(c) The production of ethylene glycol (CH_2OHCH_2OH), used as an antifreeze, from synthesis gas.

35. Comment on the accuracy of a jeweler's advertising that "diamonds last forever." In what sense is the statement true, and in what ways is it false?

36. Methane and sulfur vapor react to form carbon disulfide and hydrogen sulfide. Carbon disulfide reacts with Cl_2(g) to form carbon tetrachloride and S_2Cl_2. Further reaction of carbon disulfide and S_2Cl_2 produces additional carbon tetrachloride and sulfur. Write a series of equations for the reactions described here.

37. The combustion of 5 billion metric tons of fossil fuels (the approximate annual rate, worldwide) raises the atmospheric content of CO_2 by 0.7 ppm. Starting with the current level of 350 ppm, and assuming a fossil fuel reserve of 10,000 billion metric tons, what is the limit to which the CO_2 content of air could be raised by burning all this fuel?

Silicon

38. With reference to Table 14-8, show that the formula given for mica is consistent with the usual oxidation states and ionic charges of its constituents.

39. Write chemical equations for the following reactions. [*Hint:* The essential substances have been named. You may have to supply an additional reactant and/or product to complete each equation.]

(a) The reduction of silica to elemental silicon by aluminum metal.

(b) The preparation of potassium metasilicate by the high-temperature fusion of silica and potassium carbonate.

(c) The combustion of trisilane (Si_3H_8), yielding silica.

40. Describe and explain the similarities and differences between the reaction of a silicate with an acid and that of a carbonate with an acid.

Nitrogen

41. Write balanced equations to represent

(a) the complete neutralization of H_2SO_4(aq) with NH_3(aq);

(b) the dissolving of Ag in conc. HNO_3(aq);

(c) the reaction of hot conc. HNO_3(aq) with carbon;

(d) the role of NO_2(g) in the formation of O_3(g) in photochemical smog.

42. Step 1a in the nitrogen cycle in Figure 6-22 results in the production of nitric acid during lightning storms. Write a series of equations to represent this nitric acid formation.

★**43.** Mg can reduce NO_3^- to NH_3(g) in basic solution. (The following equation is *not* balanced.)

$$NO_3^- + Mg(s) + H_2O \rightarrow Mg(OH)_2(s) + OH^-(aq) + NH_3(g) \tag{14.92}$$

The NH_3 can be neutralized with an excess of HCl(aq), and the unreacted HCl titrated with NaOH. In this way a quantitative determination of NO_3^- is achieved. A 25.00-mL sample of solution was treated according to equation (14.92). The NH_3(g) was passed into 50.00 mL of 0.1500 M HCl. The excess HCl required 32.10 mL of 0.1000 M NaOH for its titration. What was the molarity of NO_3^- in the original sample?

Phosphorus

44. Rewrite equation (14.64) as a two-step process in which P_4O_{10} is formed in the first step and then is reduced to P_4 by C(s) in the second step. [*Hint:* The net reaction is the same as in equation (14.64).]

45. A certain phosphate rock is 58.0% $Ca_3(PO_4)_2$. What mass of this rock is required to produce 115 kg of phosphorus, assuming no loss of material in reaction (14.64).

★**46.** The following process can be used to obtain solid red P. Liquid white P is boiled at 287 °C and phosphorus vapor condensed to solid red P at 350 °C. Explain how it is possible for the vapor to deposit as a solid at a temperature *higher* than the boiling point of the liquid.

Oxygen

47. Use Lewis structures or other information from this chapter to explain the facts that

(a) H_2S is a gas at room temperature whereas H_2O is a liquid;

(b) O_3 is diamagnetic;

(c) The O-to-O bond lengths in O_2, O_3, and H_2O_2 are 121, 128, and 148 pm, respectively.

48. Recall that in a disproportionation reaction the same substance is both oxidized and reduced. Write a plausible equation for the disproportionation of H_2O_2.

49. $O_3(g)$ is a powerful oxidizing agent. Write equations to represent oxidation of **(a)** I^- to I_2 in acidic solution; **(b)** sulfur in the presence of moisture to sulfuric acid; **(c)** $[Fe(CN)_6]^{4-}$ to $[Fe(CN)_6]^{3-}$ in basic solution. In each case $O_3(g)$ is reduced to $O_2(g)$. H^+ and/or H_2O, and OH^- and/or H_2O can be included as necessary.

50. Chlorine and certain chlorine compounds are often used as bleaching agents because of their oxidizing power. After bleaching, quite often an antichlor is used. An antichlor is an agent that enters into an oxidation–reduction reaction with the excess chlorine or chlorine compound. Hydrogen peroxide is often used as an antichlor. On the other hand, hydrogen peroxide is also used as a bleach. Explain how hydrogen peroxide can work in both ways.

51. A statement is made on page 511 that the H_2O_2 molecule is not linear. Describe the geometrical shape that you would predict for this molecule using VSEPR theory and compare your prediction with the molecular model shown in Figure 14-20.

★52. It has been estimated that if all the O_3 in the atmosphere were brought to sea level at STP, the gas would form a layer 0.3 cm thick. Estimate the number of O_3 molecules in the earth's atmosphere. (Assume that the radius of the earth is 4000 mi.)

Sulfur

53. Sulfur can occur naturally as sulfates, but not as sulfites. Explain why this is so.

54. 25.0 L of a natural gas, measured at 25 °C and 740. mmHg, is bubbled through $Pb^{2+}(aq)$; a precipitate weighing 0.535 g is obtained. What is the % H_2S, by volume, in the natural gas? $(Pb^{2+} + H_2S \rightarrow PbS(s) + 2\ H^+)$

55. The Frasch process for mining sulfur uses superheated water at 160 °C.
 (a) What is superheated water?
 ★(b) Under what minimum pressure must the water be maintained at 160 °C? [*Hint:* Use data from Chapter 12.]

56. A portion of the phase diagram of sulfur follows. With respect to this diagram and the discussion on page 513,

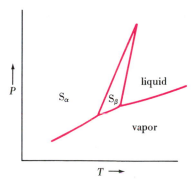

Exercise 56

 (a) Label the points corresponding to the following transitions and indicate the temperatures at which they occur: $S_\alpha \rightarrow S_\beta$; $S_\beta \rightarrow S_\lambda(liquid)$; $S_\mu(liquid) \rightarrow S_8(g, 1\ atm)$
 ★(b) How would you represent the melting of *rhombic* sulfur (S_α) that is sometimes observed at 113 °C? [*Hint:* What would the phase diagram look like if the monoclinic sulfur (S_β) did not form?]
 ★(c) How do you account for the fact that if rhombic sulfur (S_α) melts at 113 °C the liquid formed itself freezes at 119 °C?

Fluorine, chlorine

57. Write balanced equations for the following reactions.
 (a) $Ca(s) + Cl_2(g) \rightarrow$ **(b)** $P_4(s) + Cl_2(g) \rightarrow$
 (c) $Cl_2(aq) + Br^-(aq) \rightarrow$ **(d)** $Al(s) + F_2(g) \rightarrow$

58. Although fluorite, CaF_2, is its chief mineral source, fluorine can also be obtained as a by-product of the production of phosphate fertilizers. These fertilizers are derived from phosphate rock $[3\ Ca_3(PO_4)_2 \cdot CaF_2]$. What is the maximum mass of fluorine that could be extracted as by-product from 1000. kg of phosphate rock?

★59. One reaction of a chlorofluorocarbon implicated in the destruction of stratospheric ozone is $CFCl_3 + h\nu \rightarrow CFCl_2 + Cl$.
 (a) What is the energy of the photons ($h\nu$) required to bring about this reaction, expressed in kJ/mol?
 (b) What is the frequency and wavelength of the light necessary to produce the reaction? In what portion of the electromagnetic spectrum will this light be found?

Additional Exercises

60. Write equations to show how each of the following substances can be used in the preparation of $H_2(g)$: **(a)** H_2O; **(b)** $HI(aq)$; **(c)** $Mg(s)$; **(d)** $CO(g)$; **(e)** $NaOH(aq)$. Use other reactants as necessary—water, acids or bases, metals, etc.

61. Write equations to represent the reactions of the following with H_2O: **(a)** $K(s)$; **(b)** $KH(s)$; **(c)** $KO_2(s)$.

62. Write equations for the reactions you would expect to occur when
 (a) $MgCO_3(s)$ is heated to a high temperature;
 (b) $CaCl_2(l)$ is electrolyzed;
 (c) $Be(s)$ is dissolved in concentrated $NaOH(aq)$;
 (d) $Ca(s)$ is added to cold dilute $HCl(aq)$.

63. Write equations for the reactions that you would expect to occur when $Al(s)$ is allowed to react with **(a)** $HI(aq)$; **(b)** $H_2SO_4(aq)$; **(c)** $KOH(aq)$; **(d)** Cr_2O_3 at high temperatures; **(e)** $O_2(g)$ at high temperatures; **(f)** $Cl_2(g)$.

64. Write equations for the reactions that you would expect to occur when
 (a) C_8H_{18} is burned in an excess of air;
 (b) $CO(g)$ is heated with $PbO(s)$;
 (c) $CO_2(g)$ is bubbled into $KOH(aq)$;
 (d) $MgCO_3(s)$ is added to $HCl(aq)$.

65. Explain why the compound SF_6 exists, but OF_6 does not; PCl_5, but not NCl_5.

66. With reference to Figure 14-2, give a reasonable explanation of the following observations.
 (a) The melting point of C is higher than that of Si.
 (b) The group 1A, 2A, and 3A metals of the third period have *lower* melting points than do those of the second period. On the other hand, the elements of the third period groups 5A, 6A, and 7A have *higher* melting points than do the second period elements.

67. If the abundances of the elements are listed in terms of the entire crust of the earth (including oceans and atmosphere) instead of just the solid crust, some of the relative rankings change. Name *three* elements whose relative abundances you would expect to increase significantly over those listed in Figure 14-1 based on this expanded definition of the earth's crust.

68. For carbon suboxide, C_3O_2 (see marginal note on page 499),

(a) Write a plausible Lewis structure. (The molecule has a three-carbon chain skeleton, with O atoms as terminal atoms.)

(b) Describe the hybridization of the orbitals of the C atoms.

(c) Write a plausible chemical equation for its combustion. C_3O_2 burns in air with a blue *smoky* flame to produce $CO_2(g)$. [*Hint:* What is the smoke?]

69. The process referred to on page 514 for the removal of H_2S from a gaseous mixture is the Claus process. It consists of two steps, of which the *second* step is given by equation (14.74).

(a) What is the first step in this process?

(b) What is the overall or net reaction, that is, the sum of the two steps?

70. Write a systematic name or formula for each of the following binary covalent compounds. **(a)** N_2F_4; **(b)** dinitrogen trioxide; **(c)** P_4S_4; **(d)** disulfur decafluoride; **(e)** B_2Cl_4.

71. Suppose that 90% of $SO_2(g)$ emissions of an electric power plant are converted to H_2SO_4. What volume of sulfuric acid (98% H_2SO_4, by mass; $d = 1.84$ g/cm^3) could be produced from the 2.2×10^6 tons of coal (3.5% S, by mass) used by the plant annually?

72. Suppose that, unlike in Figure 14-3, no attempt is made to separate the $H_2(g)$ and $O_2(g)$ produced by the electrolysis of water (reaction 14.5).

(a) What volume of an H_2/O_2 mixture at 23 °C and 748 mmHg could be obtained as a result of electrolyzing 12.5 g of water?

(b) Recalculate the volume in part (a) by taking into account the fact that the H_2/O_2 mixture is saturated with $H_2O(g)$. Assume a total pressure of 748 mmHg and a water vapor pressure of 20.5 mmHg for the dilute electrolyte solution.

(c) Express the composition of the gas in part (b) in mole fractions of $H_2(g)$, $O_2(g)$, and $H_2O(g)$.

73. For every volume of O_2 that a person inhales he or she expires 0.82 volumes of $CO_2(g)$. Oxygen-generating systems for enclosed spaces (e.g., spacecraft) should have the capacity to consume 0.82 L $CO_2(g)$ for every liter of $O_2(g)$ produced. Show that the KO_2 system described through reaction (14.73) meets this requirement.

74. Use Lewis structures or other information from this chapter to explain the fact that

(a) S_2Cl_2 has a structure similar to H_2O_2.

(b) SO_2 possesses a dipole moment but SO_3 does not.

★(c) there exists an unstable, paramagnetic, purple solid, S_2.

75. Show that the structure of the O_3 molecule given in the text is consistent with the prediction made with VSEPR theory.

★**76.** Use the lattice energy of LiF given in Section 14-5, together with an enthalpy of sublimation of Li(s) = 160.7 kJ/mol Li and other data from the text, to calculate ΔH_f for LiF(s).

★**77.** The composition of a phosphate mineral can be expressed as % P, % P_2O_5, or % BPL [bone phosphate of lime, $Ca_3(PO_4)_2$].

(a) Show that % P = $0.436 \times$ (% P_2O_5) and % BPL = $2.185 \times$ (% P_2O_5).

(b) What is the significance of a % BPL greater than 100?

(c) What is the % BPL of a typical phosphate rock?

★**78.** Propose a plausible bonding scheme for O_3 involving hybrid orbitals.

★**79.** Assuming a similar packing of spherical atoms in the crystalline alkali metals, account for the fact that Na has a higher density than both Li and K. [*Hint:* Use data from Table 14-4.]

Self-Test Questions

For questions 80 through 89 select the item that best completes each statement.

80. All the following metals react with cold water *except* (a) K; (b) Ca; (c) Al; (d) Li.

81. All the following molten compounds are good electrical conductors *except* (a) $BeCl_2$; (b) KF; (c) CsI; (d) NaCl.

82. Under the appropriate conditions, each of the following can act as an oxidizing agent *except* (a) H_2O_2; (b) Cl^-; (c) F_2; (d) SO_2.

83. The mass percent H is greatest in (a) natural gas (CH_4); (b) seawater; (c) the atmosphere; (d) the universe as a whole.

84. Of the following liquids, the *highest* boiling point is expected for (a) Cl_2; (b) CS_2; (c) CCl_4; (d) H_2S.

85. All of the following substances react with water. The pair that yield the same gaseous products are (a) Ca and CaH_2; (b) Na and Na_2O_2; (c) K and KO_2; (d) Li_3N and LiH.

86. The most abundantly occurring metal in the earth's crust is (a) Ca; (b) Al; (c) Fe; (d) Na.

87. The *cheapest* raw material from which an alkaline (basic) medium can be prepared is (a) coke; (b) limestone; (c) phosphate rock; (d) gypsum.

88. Of the following, the one that is *unimportant* in the production of fertilizers is (a) phosphate rock; (b) HNO_3; (c) NH_3; (d) Na_2CO_3.

89. An important component of photochemical smog is (a) O_3; (b) CO; (c) SO_2; (d) SO_3.

90. Supply a name or formula for each of the following: **(a)** lithium hydride; **(b)** $NaHSO_4$; **(c)** ClO^-; **(d)** magnesium perchlorate; **(e)** sodium nitrite; **(f)** N_2O_5; **(g)** BaO_2.

91. Write balanced equations to represent **(a)** the decomposition of $MgCO_3$ by heating; **(b)** the action of HCl(aq) on $NaHCO_3$(s); **(c)** the formation of H_3PO_4 from P_4; **(d)** the reduction in acidic solution of Cl_2 to Cl^- by H_2O_2; **(e)** the oxidation of NH_3 to NO by O_2; **(f)** the reaction of Cu with dilute HNO_3(aq).

92. A chemistry magazine listed the 1986 production of $O_2(g)$ as 3.99×10^{11} ft^3 at STP. What is this quantity of $O_2(g)$ expressed in kg?

93. Explain why the air pollution control measures required for automobiles are not the same as those required for fossil-fuel electric power plants.

15 Chemical Kinetics

The decomposition of hydrogen peroxide, H_2O_2, to H_2O and O_2 is a highly exothermic reaction that is strongly catalyzed by platinum metal. The kinetics of this decomposition and the nature of catalysis are both explored in this chapter. [Carey B. Van Loon]

527

Think, for a moment, about all the principles we can apply to the reaction of hydrogen and oxygen to form water. We know how to describe bonding and structure in the molecules H_2, O_2, and H_2O, and we have a good idea why H_2 and O_2 are gases at room temperature and pressure, whereas H_2O is a liquid. We can write the balanced equation and also express the quantity of heat that is evolved in this reaction.

$$2\ H_2(g) + O_2(g) \longrightarrow 2\ H_2O(l) \qquad \Delta H = -572\ kJ$$

Using principles of stoichiometry (Chapter 4) and the gas laws (Chapter 6), we can perform a variety of calculations concerning masses and/or volumes of the reactants and products of this reaction. But there is a question we cannot yet answer: "How long will it take to convert two moles of $H_2(g)$ and one mole of $O_2(g)$ into two moles of $H_2O(l)$ at room temperature and pressure?" The answer is essentially "forever." Yet, if we introduce a spark or flame, the mixture explodes. To explain these facts about the *rate* at which a chemical reaction occurs we must turn to the subject of *chemical kinetics*.

Now think about other situations where a knowledge of the rates of chemical reactions is important. When we use a high explosive we want a chemical reaction that releases gaseous products and a great deal of energy in an extremely short period of time. In burning a rocket fuel we also seek a rapid release of gaseous products and energy to obtain maximum thrust for the rocket, but we do not want an explosive reaction. When we store milk in a refrigerator our objective is the opposite—to slow down the chemical reactions by which milk is spoiled. These cases illustrate our need to be able to measure, control, and where possible, *predict* the rates of chemical reactions. We can predict rates of reaction through mathematical equations known as *rate laws*. And to understand rate laws we will find it helpful to construct step-by-step descriptions, called *reaction mechanisms*, of how initial reactants are converted to final products. Chemical kinetics, then, is the study of the rates of reactions, mathematical rate laws, and reaction mechanisms.

15-1 The Rate of a Chemical Reaction

Since our principal concern will be with the rates of chemical reactions, we begin with an explanation of what this term means. Rate (or speed, or velocity) suggests that something happens in a unit of time. For a car traveling at 60 mph, what happens is that a distance of 60 miles is traveled, and the unit of time in which this occurs is 1 hour. What happens in chemical reactions is that amounts of reactants and products change. These changes are most commonly expressed through changes in *molarity* concentrations. Thus, by the rate of the hypothetical reaction

$$A + 3\ B \longrightarrow 2\ C + 2\ D \tag{15.1}$$

we might mean the rate of decrease of the molarity of A. Since molarity has the units mol L^{-1} (mol per liter), the rate of reaction has the units mol $L^{-1}\ s^{-1}$ (mol per liter per second), for example.

We can also describe the rate of reaction (15.1) in terms of the disappearance of B or the formation of C or D, but now we have a problem: These rates are not the same as the rate of disappearance of A. From the coefficients in equation (15.1) we see that *three* moles of B are consumed for every mole of A. That is, B disappears three times as fast as A, or

rate of disappearance of B = 3 × rate of disappearance of A

Similarly, we can write that

rate of formation of C = rate of formation of D
$$= 2 \times \text{rate of disappearance of A}$$

528

Here is still another problem: Rates of disappearance are *negative* (concentrations decrease with time) and rates of formation are *positive* (concentrations increase with time). All these matters are nicely taken care of if we define a *single, positive* **rate of reaction,** which for reaction (15.1) is

rate of reaction (reaction rate) $= -$(rate of disappearance of A)

$$= -\tfrac{1}{3}\text{(rate of disappearance of B)}$$
$$= \tfrac{1}{2}\text{(rate of formation of C)} \qquad (15.2)$$
$$= \tfrac{1}{2}\text{(rate of formation of D)}$$

Thus, we can relate the rate of reaction (reaction rate) either to the rate of disappearance of a reactant or the rate of formation of a product, so long as we divide by its stoichiometric coefficient and use the correct sign ($-$ for disappearance and $+$ for formation).

Example 15-1

Expressing the rate of a reaction. Suppose that at some particular point in the hypothetical reaction (15.1) we find that the concentration of A is 1.0000 M, i.e., [A] = 1.0000 M. Suppose that exactly 1.00 minute later [A] = 0.9982 M. **(a)** What is the rate of the reaction, expressed in mol L^{-1} min^{-1}? **(b)** What is the rate of formation of C? **(c)** What is the rate of reaction expressed in mol L^{-1} s^{-1}?

Recall from page 126 that the symbol [ ] means "the molarity of"; thus [A] is the molarity of A.

Solution

As in other instances describing a change, the delta symbol (Δ) means the final value minus the initial value.

(a) The rate of disappearance of A is given by the *change* in molarity, $\Delta[A]$, divided by the time interval, Δt, over which this change occurs. The change in molarity is $\Delta[A] = 0.9982$ M $- 1.0000$ M $= -0.0018$ M. The time interval is $\Delta t = 1.00$ min.

$$\text{rate of disappearance of A} = \frac{\Delta[A]}{\Delta t} = \frac{-0.0018 \text{ M}}{1.00 \text{ min}}$$

$$= -1.8 \times 10^{-3} \text{ M min}^{-1}$$

$$= -1.8 \times 10^{-3} \text{ mol L}^{-1} \text{ min}^{-1}$$

$$\text{rate of reaction} = -(\text{rate of disappearance of A})$$

$$= -(-1.8 \times 10^{-3} \text{ mol L}^{-1} \text{ min}^{-1})$$

$$= 1.8 \times 10^{-3} \text{ mol L}^{-1} \text{ min}^{-1}$$

(b) From expression (15.2) we see that the rate of reaction $= \tfrac{1}{2}$ rate of formation of C, or

rate of formation of C $= 2 \times$ rate of reaction

$$= 2 \times 1.8 \times 10^{-3} \text{ mol L}^{-1} \text{ min}^{-1}$$

$$= 3.6 \times 10^{-3} \text{ mol L}^{-1} \text{ min}^{-1}$$

(c) The decrease in concentration used in the rate calculation in (a) occurs over a 1-min time interval. A much smaller change in concentration will occur over a 1-s interval. Specifically, we must convert min^{-1} to s^{-1}. (Note that $min^{-1} \times min = 1$.)

$$\text{rate of reaction} = 1.8 \times 10^{-3} \text{ mol L}^{-1} \text{ min}^{-1} \times \frac{1 \text{ min}}{60 \text{ s}}$$

$$= 3.0 \times 10^{-5} \text{ mol L}^{-1} \text{ s}^{-1}$$

SIMILAR EXAMPLES: Exercises 1, 2, 17.

The combustion of flour dust has been responsible for many flour mill explosions. The combustion reaction occurs so rapidly because of the large surface area of the dust. [Lee Dodds]

Factors Affecting the Rate of a Reaction. Most of the discussion in this chapter deals with the several factors that affect the rate or speed of a chemical reaction. Here are the principal factors.

- *Concentrations of reactants.* Reaction rates generally increase with the concentrations of the reactants.
- *Temperature.* Reaction rates generally increase very rapidly as the temperature increases.
- *Surface area.* Some reactions occur on a surface rather than within a homogeneous solution. For surface reactions the rate of reaction increases with surface area.
- *Catalysis.* Some substances, called *catalysts,* enter into a reaction in such a way that they are regenerated unchanged, but they have the effect of speeding up the reaction.

15-2 Measuring Reaction Rates: Concentrations as a Function of Time

Hydrogen peroxide in a 3% aqueous solution is a common household product available at any pharmacy. It is a useful mild antiseptic because it produces oxygen gas when it decomposes. Let us choose this decomposition reaction to illustrate further a number of ideas about reaction rates.

$$H_2O_2(aq) \longrightarrow H_2O + \tfrac{1}{2} O_2(g) \tag{15.3}$$

"Following a Chemical Reaction." In the course of reaction (15.3) $O_2(g)$ escapes from the reaction mixture and the reaction goes to completion.* We can follow the progress of the reaction ("follow the reaction") by focusing our attention either on the formation of $O_2(g)$ or the disappearance of H_2O_2. That is, we can do either of the following.

- Measure the volume of $O_2(g)$ evolved as a function of time and relate this volume to the decrease in concentration of H_2O_2. (See Figure 15-1 and Exercise 19.)

_____

* Although reaction (15.3) goes to completion, it does so very slowly. Generally a catalyst is used to speed up the reaction. The function of a catalyst in this reaction is described in Section 15-11.

Permanganate ion, MnO_4^-, is the essential reactant in $KMnO_4(aq)$. Equation (15.4) is written in the net ionic form.

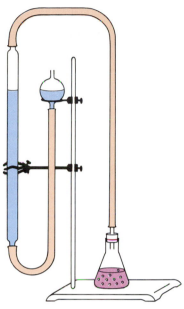

FIGURE 15-1
Experimental setup for determining the rate of decomposition of H_2O_2.

In this simple laboratory setup oxygen gas given off from the reaction mixture is trapped, and its volume is measured in the gas buret. The rate of H_2O_2 decomposition can be related to the rate of production of $O_2(g)$ (see Exercise 19).

TABLE 15-1
Decomposition of H_2O_2

Time, s	$[H_2O_2]$, M
0	2.32
300	1.86
600	1.49
1200	0.98
1800	0.62
3000	0.25

• Remove small samples of the reaction mixture from time to time and analyze these samples for their H_2O_2 content, for example, by titration with $KMnO_4(aq)$.

$$2\ MnO_4^- + 5\ H_2O_2 + 6\ H^+ \longrightarrow 2\ Mn^{2+} + 8\ H_2O + 5\ O_2(g) \qquad (15.4)$$

Typical data for the concentration of H_2O_2 as a function of time are listed in Table 15-1 and graphed in Figure 15-2.

In many instances we can "follow a chemical reaction" by measuring a physical property whose value is directly proportional to the concentration of a reactant or product. This we can do without disturbing the reaction mixture.

Example 15-2

"Following a chemical reaction": A method of obtaining concentration data as a function of time. Exactly 300 s after the start of reaction (15.3) a 5.00-mL portion of the reaction mixture is removed and immediately titrated with 0.1000 M $KMnO_4$ in acidic solution (reaction 15.4). 37.1 mL of 0.1000 M $KMnO_4$ is required. What is $[H_2O_2]$ at the 300-s point in the reaction?

Solution. $[H_2O_2]$ in the 5.00-mL portion is the same as $[H_2O_2]$ in the larger sample that is undergoing decomposition. For this 5.00-mL portion

$$\text{no. mol } H_2O_2 = 0.0371\ L \times \frac{0.1000\ \text{mol } MnO_4^-}{L} \times \frac{5\ \text{mol } H_2O_2}{2\ \text{mol } MnO_4^-}$$

$$= 0.00928\ \text{mol } H_2O_2$$

$$[H_2O_2] = \frac{0.00928\ \text{mol } H_2O_2}{0.00500\ L} = 1.86\ M\ H_2O_2$$

SIMILAR EXAMPLES: Exercises 34a, b, 75, 76.

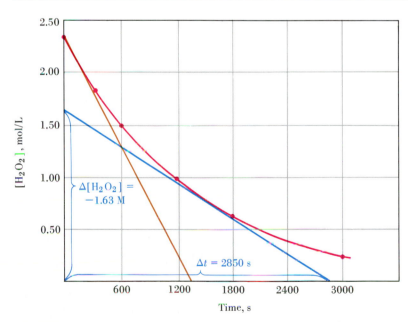

FIGURE 15-2
Graphical representation of the rate of the reaction

$$H_2O_2 \rightarrow H_2O + \tfrac{1}{2} O_2$$

This is the usual form in which kinetic data are plotted: concentration of a reactant (or product) as a function of time. Reaction rates are then determined from the slopes of the tangent lines (see Examples 15-3 and 15-4).

Are You Wondering:

Why the rate of reaction in the remaining reaction mixture is not affected when a portion is removed for analysis (as in Example 15-2)?

The rate of a reaction depends on the *concentrations* of reactants, the amounts present *per unit volume*. Removal of a sample of the reaction mixture for analysis reduces the total amount of reactant(s) in the mixture, but it reduces the volume of the reaction mixture as well. The amount of reactant(s) per unit volume is unchanged. As a consequence the rate of a reaction is unaffected by the removal of a portion of the mixture.

Rate of Reaction: A Variable Quantity. From Figure 15-2 we see that the concentration of H_2O_2 decreases with time at a rate that is initially rapid but then slows down. The reaction *decelerates*. For example, in the first 600 s $[H_2O_2]$ decreases from about 2.30 to 1.50 M, a decrease of 0.80 mol H_2O_2/L. In the next 600 s the decrease is only about 0.50 mol H_2O_2/L. Ordinarily the rate of a reaction changes during the course of a reaction, and we need to specify further just how the rate of reaction should be expressed.

Rate of Reaction Expressed as $-\Delta[H_2O_2]/\Delta t$. Let us extract some data from Figure 15-2 and list them in Table 15-2 (in red). Column III of Table 15-2 lists the molarities of H_2O_2 at the times shown in Column I. Column II states the arbitrary time interval between data points—400 s. Column IV reports the concentration changes that occur for each 400-s interval. The figures in Column V represent the reaction rates; their values decrease continuously with time.

Starting with $[H_2O_2] = 2.32$ M at $t = 0$, with each succeeding time interval the concentration becomes smaller; $\Delta[H_2O_2]$ is a *negative* quantity. But, as was the case in expression (15.2), the rate of reaction is the *negative* of the rate of disappearance of H_2O_2.

$$\text{rate of reaction} = \frac{-\Delta[H_2O_2]}{\Delta t} \tag{15.5}$$

TABLE 15-2
Decomposition of H_2O_2—Derived Rate Data

I	II	III	IV	V
Time, s	Δt, s	$[H_2O_2]$, mol/L	$\Delta[H_2O_2]$, mol/L	Reaction rate = $-\Delta[H_2O_2]/\Delta t$, mol L^{-1} s^{-1}
0		2.32		
	400		-0.60	15.0×10^{-4}
400		1.72		
	400		-0.42	10.5×10^{-4}
800		1.30		
	400		-0.32	8.0×10^{-4}
1200		0.98		
	400		-0.25	6.2×10^{-4}
1600		0.73		
	400		-0.19	4.8×10^{-4}
2000		0.54		
	400		-0.15	3.8×10^{-4}
2400		0.39		
	400		-0.11	2.8×10^{-4}
2800		0.28		

In calculus notation, the ratio $-\Delta[H_2O_2]/\Delta t$ in the limit where $\Delta t \rightarrow 0$ can be replaced by the derivative $-d[H_2O_2]/dt$. That is,

$$\lim_{\Delta t \rightarrow 0} \frac{-\Delta[H_2O_2]}{\Delta t} = \frac{-d[H_2O_2]}{dt}$$

Rate of Reaction Expressed as the Slope of a Tangent Line. When expressed as $-\Delta[H_2O_2]/\Delta t$, the reaction rate is simply an *average* value during the time interval Δt. For example, the rate averages 15.0×10^{-4} mol L^{-1} s^{-1} in the interval from 0 to 400 s (see first entry in Column V of Table 15-2). We might think of this as the reaction rate at the middle of the interval—200 s. We could also describe the reaction rate at 200 s by choosing concentration data at 100 s and 300 s and using $\Delta t = 200$ s. The result would differ slightly from 15.0×10^{-4} mol L^{-1} s^{-1}. To obtain a unique value of the reaction rate at 200 s we should use a time interval approaching zero, that is, $\Delta t \rightarrow 0$. Under these circumstances the rate of reaction becomes equal to the *negative of the slope of the tangent line* to the graph of $[H_2O_2]$ as a function of time. This rate of reaction is called the **instantaneous rate of reaction.** Two tangent lines (and hence two instantaneous rates of reaction) are shown in Figure 15-2—one at $t = 0$ and one at $t = 1500$ s.

Example 15-3

Obtaining a rate of reaction from a graph of concentration versus time. From Figure 15-2, determine the rate of decomposition of H_2O_2 at 1500 s.

Solution. Draw a tangent line (blue) to the graph at $t = 1500$ s and determine the slope of this line.

Even though we can try to read points from the graph to three significant figures, because of uncertainties in constructing the tangent line we are probably not justified in carrying more than two figures in the slope.

$$\text{rate of reaction} = -(\text{slope of tangent line}) = \frac{-\Delta[H_2O_2]}{\Delta t} = \frac{1.63 \text{ mol/L}}{2850 \text{ s}}$$

$$= 5.7 \times 10^{-4} \text{ mol } L^{-1} \text{ s}^{-1}$$

SIMILAR EXAMPLES: Exercises 3, 15, 34d, e.

Initial Rate of Reaction. To use the method of Example 15-3 we have to collect data over an extended time. Sometimes all we need to know is the rate of reaction immediately after the reactants are brought together—the **initial rate of reaction.** We can obtain the initial reaction rate by dividing the change in reactant concentration (Δ[reactant]) by a *brief* time interval (Δt) immediately after the reaction is started. Alternatively, the initial rate can be related to the slope of the tangent line at $t = 0$ in a graph such as Figure 15-2. The key to these approaches is to choose as a Δt the time interval over which the tangent line and the "concentration vs. time" curve practically coincide. In Figure 15-2 this is about 200 s. Put in another way, initial rate calculations work over a time span in which no more than a few percent of the available reactants are consumed.

Example 15-4

Determining and using an initial rate of reaction. For the decomposition of H_2O_2 represented by the data in Table 15-1, (a) determine the initial rate of reaction, and (b) $[H_2O_2]$ at $t = 100$ s.

Solution

Note that the curve and the brown tangent line of Figure 15-2 start to deviate significantly at $t = 300$ s.

(a) Since the shortest time interval in Table 15-1 is 300 s, the method of dividing an initial change in concentration by a *short* time interval probably will not work well. An alternate method is to determine the slope of the tangent line at $t = 0$. This we can do from Figure 15-2 by using the intersections of the brown tangent line with the axes: $t = 0$, $[H_2O_2] = 2.32$ M; $t = 1330$ s, $[H_2O_2] = 0$.

$$\text{initial rate} = -(\text{slope of tangent line}) = \frac{-(0 - 2.32) \text{ mol/L}}{1330 \text{ s}}$$

$$= 1.7 \times 10^{-3} \text{ mol } L^{-1} \text{ s}^{-1}$$

(b) Assume that the rate determined in (a) remains essentially constant for at least 100 s. Since

$$\text{rate of reaction} = \frac{-\Delta[H_2O_2]}{\Delta t} \quad \text{then}$$

$$\Delta[H_2O_2] = -(\text{rate of reaction}) \times \Delta t$$

$$\Delta[H_2O_2] = -1.7 \times 10^{-3} \text{ mol L}^{-1} \text{ s}^{-1} \times 100 \text{ s} = -1.7 \times 10^{-1} \text{ mol L}^{-1}$$

$$[H_2O_2]_{t=100 \text{ s}} = [H_2O_2]_{t=0} + \Delta[H_2O_2]$$

$$= 2.32 \text{ M} - 0.17 \text{ M} = 2.15 \text{ M}$$

SIMILAR EXAMPLES: Exercises 14, 18, 34d, 61, 62.

15-3 Effect of Concentrations on Rates of Reactions: The Rate Law

In many cases we can express reaction rates through an experimentally determined mathematical equation. This equation is called a **rate law** or **rate equation.** Consider the hypothetical reaction

$$a\,A + b\,B + \cdots \longrightarrow g\,G + h\,H + \cdots \tag{15.6}$$

where $a, b, \ldots$, stand for coefficients in the balanced equation. The rate of this reaction can be represented as*

The rate law for a chemical reaction.

$$\text{rate of reaction} = k[A]^m[B]^n \cdots \tag{15.7}$$

In expression (15.7) [A], [B], . . . represent molarities. The exponents, m, n, . . . , are generally small whole numbers, although in some cases they may be fractional and/or negative. And here is an important point: There is *no* relationship between the exponents, m, n, . . . , and the corresponding coefficients in the balanced equation, $a, b, \ldots$. If in some cases they happen to be identical (that is, with $m = a$, $n = b$, etc.), this is just a matter of chance. You cannot expect it to happen.

The exponents in the rate equation determine the **order of the reaction.** If $m = 1$, we say that the reaction is *first order in A.* If $n = 2$, the reaction is *second order in B,* and so on. The *total* of all the exponents, $m + n + \ldots$, is called the **overall order** of the reaction. The term k is called the **rate constant.** It is a proportionality constant that is characteristic of the particular reaction. Its value also depends on the temperature and the presence of catalysts. *The faster a reaction proceeds, the larger the value of* k. The units of k depend on the order of the reaction (that is, on the values of the exponents n, m, . . .).

To establish a rate law, then, we need to determine the order of a reaction and a value of the rate constant. Once we have done this, we can

- calculate rates of reaction corresponding to known concentrations of the reactants.
- derive an equation, called the **integrated rate equation,** with which we can express reactant concentrations as a function of time.

——————

*To apply equation (15.7) we must assume that reaction (15.6) goes to completion, that is, that the reverse reaction is insignificant. If the reaction is reversible, the rate equation is considerably more complex than (15.7). Even for reversible reactions, however, equation (15.7) will apply to describing the *initial rate* of reaction. This is because in the early stages of the reaction there are not enough products to produce a reverse reaction. In this text we limit ourselves to situations where the general rate law expression (15.7) does apply.

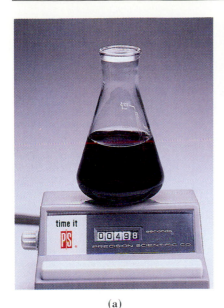

(a)

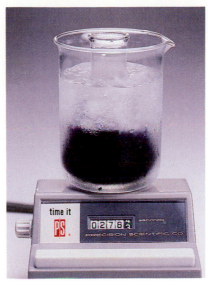

(b)

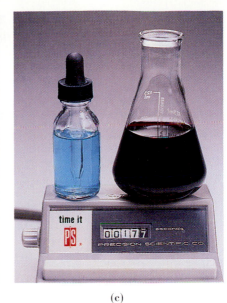

(c)

FIGURE 15-3
Kinetics of the peroxodisulfate–iodide ion reaction.

The reactants, $S_2O_8^{2-}$ and I^-, are mixed, in the presence of fixed amounts of sodium thiosulfate and starch, in a solution that is originally colorless. I_3^-, a product of the main reaction (15.8), reacts with thiosulfate ion ($S_2O_3^{2-}$). When the fixed amount of thiosulfate ion is consumed, I_3^- reacts with starch to produce a blue color. The rate of the reaction is related to the time required for the blue color to appear. (The *faster* the reaction, the more quickly the solution turns a blue color.)
(a) Here, at room temperature, the color appears in 49.8 s.
(b) When the same reaction is carried out in ice water it occurs much more slowly—276.3 s.
(c) A few drops of Cu^{2+}(aq) catalyst speeds up the reaction significantly—17.7 s. [Carey B. Van Loon]

Method of Initial Rates. In this method of establishing the order of a reaction the initial rates of reaction are compared for different initial concentrations of reactants. To illustrate this method we describe the reaction between peroxodisulfate ($S_2O_8^{2-}$) and iodide (I^-) ions in aqueous solution.

$$S_2O_8^{2-} + 3\ I^- \longrightarrow 2\ SO_4^{2-} + I_3^- \tag{15.8}$$

The rate law for reaction (15.8) has the form

$$\text{rate of reaction} = k[S_2O_8^{2-}]^m[I^-]^n \tag{15.9}$$

Figure 15-3 describes how the reaction is carried out and Table 15-3 lists some typical experimental results. The method of initial rates is applied in Example 15-5.

Example 15-5

Establishing the order of a reaction by the method of initial rates. Use data from Table 15-3 to establish (a) the order of reaction (15.8) with respect to $S_2O_8^{2-}$ and I^- and (b) the overall order of the reaction.

Solution

(a) What we need to do is determine the values of m and n in equation (15.9). In comparing Expt. 2 with Expt. 1, note that the concentration of $S_2O_8^{2-}$ is doubled while that of I^- is held constant at 0.060 M. Note also that $R_2 = 2R_1$. In the final step: $(0.076)^m = (2 \times 0.038)^m = 2^m(0.038)^m$.

TABLE 15-3
Experimental Data for the Reaction $S_2O_8^{2-} + 3\ I^- \longrightarrow 2\ SO_4^{2-} + I_3^-$

Expt.	Initial concentrations, M		Initial rate of reaction,[a] mol L^{-1} s^{-1}
	$[S_2O_8^{2-}]$	$[I^-]$	
1	$[S_2O_8^{2-}]_1 = 0.038$	$[I^-]_1 = 0.060$	$R_1 = 1.4 \times 10^{-5}$
2	$[S_2O_8^{2-}]_2 = 0.076$	$[I^-]_2 = 0.060$	$R_2 = 2.8 \times 10^{-5}$
3	$[S_2O_8^{2-}]_3 = 0.076$	$[I^-]_3 = 0.030$	$R_3 = 1.4 \times 10^{-5}$

[a]For example, this might be $-$(rate of disappearance of $S_2O_8^{2-}$) or rate of formation of I_3^-.

$$\frac{R_2}{R_1} = \underbrace{\frac{2.8 \times 10^{-5}}{1.4 \times 10^{-5}}}_{2} = \frac{k[S_2O_8{}^{2-}]_2^m[I^-]_2^n}{k[S_2O_8{}^{2-}]_1^m[I^-]_1^n} = \frac{k(0.076)^m(0.060)^n}{k(0.038)^m(0.060)^n} = \underbrace{\frac{2^m(0.038)^m}{(0.038)^m}}_{2^m}$$

In order that $2^m = 2$, $m = 1$. The reaction is first order in $S_2O_8{}^{2-}$.

In comparing Expt. 2 to Expt. 3, $[S_2O_8{}^{2-}]$ is held constant at 0.076 M, $[I^-]$ is doubled, and $R_2 = 2R_3$.

$$\frac{R_2}{R_3} = \underbrace{\frac{2.8 \times 10^{-5}}{1.4 \times 10^{-5}}}_{2} = \frac{k[S_2O_8{}^{2-}]_2^m[I^-]_2^n}{k[S_2O_8{}^{2-}]_3^m[I^-]_3^n} = \frac{k(0.076)^m(0.060)^n}{k(0.076)^m(0.030)^n} = \underbrace{\frac{2^n(0.030)^n}{(0.030)^n}}_{2^n}$$

If $2^n = 2$, then $n = 1$. The reaction is first order in I^-.

(b) The overall order of the reaction is $m + n = 1 + 1 = 2$—second order.

SIMILAR EXAMPLES: Exercises 5, 26, 27, 28, 66, 67.

The basic idea here is to double one initial concentration and hold the other(s) constant. Then look to see what happens to the initial rate of reaction.

In Example 15-5 we established that if doubling the initial concentration of a reactant causes the initial rate to double, the reaction is *first order* in that reactant. Similarly, we could show that if doubling an initial concentration causes a *fourfold* increase in the initial reaction rate, the reaction is *second order* in that reactant. For *third order*, the initial rate would increase *eightfold* with a doubling of the reactant concentration.

Now that we know the exponents in the peroxodisulfate–iodide rate equation (15.9), we can determine the value of the rate constant k. What we need to do is to establish simultaneously $[S_2O_8{}^{2-}]$, $[I^-]$, and the rate of reaction.

Example 15-6

Using rate law data to determine a rate constant k. Use the results of Example 15-5 and data from Table 15-3 to establish the value of k in the rate equation (15.9).

Solution. We can use the data for any one of the three experiments of Table 15-3, together with the values $m = n = 1$. Equation (15.9) is solved for k.

$$k = \frac{\text{rate of reaction}}{[S_2O_8{}^{2-}]^m[I^-]^n} = \frac{R_1}{[S_2O_8{}^{2-}][I^-]} = \frac{1.4 \times 10^{-5} \text{ mol L}^{-1}\text{ s}^{-1}}{0.038 \text{ mol L}^{-1} \times 0.060 \text{ mol L}^{-1}}$$

$$= 6.1 \times 10^{-3} \text{ L mol}^{-1}\text{ s}^{-1}$$

SIMILAR EXAMPLE: Exercise 26b.

Example 15-7

Using a rate equation to calculate an instantaneous rate of reaction. What is the rate of reaction (15.8) at a point where $[S_2O_8{}^{2-}] = 0.050$ M and $[I^-] = 0.025$ M?

From the results of Examples 15-5 and 15-6 we can write

$$\text{rate of reaction} = k[S_2O_8{}^{2-}][I^-]$$
$$= 6.1 \times 10^{-3} \text{ L mol}^{-1}\text{ s}^{-1} \times 0.050 \text{ mol/L} \times 0.025 \text{ mol/L}$$
$$= 7.6 \times 10^{-6} \text{ mol L}^{-1}\text{ s}^{-1}$$

SIMILAR EXAMPLE: Exercise 26c.

Are You Wondering:

How to distinguish clearly between the rate of a reaction and the rate constant of a reaction?

Many students have difficulty with this distinction. Remember that the *rate of a reaction* describes a change in some property of the reaction mixture with time, usually the molarity of one of the reactants, e.g., $-\Delta[A]/\Delta t$. The rate of a reaction usually *changes* with reactant concentrations. The *rate constant* of a reaction (k) relates the rate of a reaction to reactant concentrations. Once the value of k has been established for a reaction at a given temperature, *its value is fixed*. The units of the rate of a reaction are the same regardless of the reaction (e.g., $\text{mol L}^{-1}\text{ s}^{-1}$), whereas the units of k differ for reactions of different orders.

Example 15-7 illustrates the main purpose served by rate laws. They allow us to *calculate* rates of reaction. To summarize how we got to this point in this section, we started with

- experimental data (as in Figure 15-3 and Table 15-3)

which were

- summarized compactly in a rate law (as in Examples 15-5 and 15-6)

which was

- used to *predict* an instantaneous rate of reaction for given reactant concentrations (as in Example 15-7).

15-4 Zero-Order Reactions

We have seen that generally the concentrations of reactants strongly affect the rate of a chemical reaction, but there are a few reactions whose rates are *independent* of concentration. This is the situation if some variable other than concentration controls the rate of reaction, such as light intensity in a photochemical reaction or the availability of an enzyme in an enzyme-catalyzed reaction. In such cases the reaction proceeds at a *constant* rate.

Rate law for a zero-order reaction.

rate of reaction = k = constant   (15.10)

Equation (15.10) is a special case of the rate equation (15.7) in which each of the exponents m, n, . . . , equals *zero*. The reaction is **zero order.** Here are the important features of a zero-order reaction.

1. A plot of concentration of reactant vs. time is a *straight line* with a *negative* slope (see Figure 15-4).

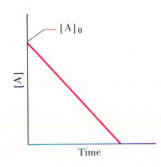

FIGURE 15-4
A straight-line plot for the zero-order reaction

A → products

[A] decreases from a maximum value of $[A]_0$ at time $t = 0$ to $[A] = 0$ at a time $t = [A]_0/k$. The rate constant $k = -$slope.

2. The rate of the reaction, which remains *constant* throughout the reaction, is the *negative* of the slope of the straight-line graph described in 1.

3. The units of k are the same as the units of the rate of a reaction: $mol\ L^{-1}$ $(time)^{-1}$, for example, $mol\ L^{-1}\ s^{-1}$.

15-5 First-Order Reactions

The decomposition of H_2O_2 (reaction 15.3) is a first-order reaction, meaning that $[H_2O_2]$ appears in the rate equation to the first power.

$$\text{rate of reaction} = k[H_2O_2] \qquad (15.11)$$

From the measured rate of reaction at a particular $[H_2O_2]$, we can calculate the rate constant k. Or, if we know the value of k we can calculate the rate of reaction for some particular $[H_2O_2]$. These calculations are similar to those in Examples 15-6 and 15-7.

But, is there not a problem here? How do we know that the decomposition of H_2O_2 is a first-order reaction? We have studied one method for establishing the order of a reaction—the *method of initial rates* (recall Example 15-5). We next consider some equally useful methods.

An Integrated Rate Equation for a First-Order Reaction. The rate equation (15.11) expresses a *rate of reaction* as a function of *concentration* of a reactant. From the rate equation we can derive another equation that expresses *concentration of reactant* as a function of *time*. We can do this using the integration concept from calculus, and the equation we obtain is called an **integrated rate equation.** Let us do this for a hypothetical first-order reaction

$$A \longrightarrow products$$

for which the rate law is

$$\text{rate of reaction} = -(\text{rate of disappearance of A}) = k[A] \qquad (15.12)$$

The result of this derivation is*

Integrated rate equation for a first-order reaction: natural log form.

$$\ln \frac{[A]_t}{[A]_0} = -kt \qquad \text{or} \qquad \ln [A]_t = -kt + \ln [A]_0 \qquad (15.13)$$

Expressed in common logarithms, equation (15.13) becomes

Integrated rate equation for a first-order reaction: common log form.

$$\log \frac{[A]_t}{[A]_0} = \frac{-kt}{2.303} \qquad \text{or} \qquad \log [A]_t = \frac{-kt}{2.303} + \log [A]_0 \qquad (15.14)$$

In equations (15.13) and (15.14), $[A]_t$ represents the concentration of A at some time t; $[A]_0$ represents the concentration of A at $t = 0$; k is the rate constant for the reaction. Only the time t and the corresponding concentration $[A]_t$ are variables in these equations. Both equations are seen to be equations of a straight line. For example, if $\ln [A]_t$ is plotted as a function of t,

$$\underbrace{\ln [A]_t}_{} = -kt + \underbrace{\ln [A]_0}_{} \qquad (15.15)$$

equation of straight line: $\qquad y \quad = \quad mx + \quad b$

---

*The steps in this derivation are

(1) Replace the rate of disappearance of A in equation (15.12) by $d[A]/dt$.

(2) Rearrange (15.12) to the form, $d[A]/[A] = -kdt$.

(3) Integrate between the limits $[A]_0$ at time $t = 0$ and $[A]_t$ at time t.

$$\int_{[A]_0}^{[A]_t} \frac{d[A]}{[A]} = -k \int_0^t dt, \qquad \text{yielding } \ln \frac{[A]_t}{[A]_0} = -kt$$

FIGURE 15-5
Test for a first-order reaction: decomposition of H_2O_2.

The data plotted here are from Table 15-1, and the slope of the line is used in the text.

t, s	$[H_2O_2]$, M	ln $[H_2O_2]$
0	2.32	0.842
300	1.86	0.621
600	1.49	0.399
1200	0.98	−0.020
1800	0.62	−0.48
3000	0.25	−1.39

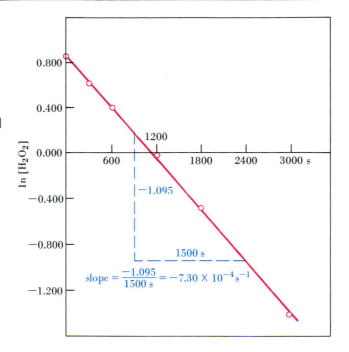

$$slope = \frac{-1.095}{1500 \text{ s}} = -7.30 \times 10^{-4} \text{ s}^{-1}$$

A test for a first-order reaction, then, is to plot the logarithm of a reactant concentration vs. time and determine if the graph is linear. Data from Table 15-1 for the first-order decomposition of H_2O_2 are plotted in Figure 15-5. The value of the rate constant k can be derived from the slope of the line.

$$k = -m = -\text{slope} = -(-7.30 \times 10^{-4} \text{ s}^{-1}) = 7.30 \times 10^{-4} \text{ s}^{-1}$$

Once we have established that a reaction is first order, we can use either equation (15.13) or (15.14) to calculate one of the four quantities appearing in the equations—$[A]_0$, $[A]_t$, k, and t—from values of the other three. In Example 15-8 we use both equations (15.13) and (15.14) to show that they always yield identical results. After that you may use whichever equation you prefer, which may depend on how particular functions are handled on your electronic calculator.

Example 15-8

Using the integrated rate equations for a first-order reaction to calculate *$[A]_t$*. $H_2O_2(aq)$, initially at a concentration of 2.32 M, is allowed to decompose. What will be $[H_2O_2]$ at $t = 1200$ s? Use $k = 7.30 \times 10^{-4}$ s^{-1} for this first-order decomposition.

Solution. For substitution into equations (15.13) and (15.14) we have

$$[H_2O_2]_t = ? \qquad [H_2O_2]_0 = 2.32 \text{ M}$$

$$k = 7.30 \times 10^{-4} \text{ s}^{-1} \qquad t = 1200 \text{ s}$$

USE OF EQUATION (15.13)

$$\ln [H_2O_2]_t = -7.30 \times 10^{-4} \text{ s}^{-1} \times 1200 \text{ s} + \ln 2.32$$

$$= \qquad -0.876 \qquad + 0.842 = -0.034$$

$$[H_2O_2]_t = \text{antiln}(-0.034) = e^{-0.034} = 0.97 \text{ M}$$

(To find the number whose natural logarithm is −0.034, raise e to the −0.034 power.)

USE OF EQUATION (15.14)

$$\log [H_2O_2]_t = \frac{-7.30 \times 10^{-4} \text{ s}^{-1} \times 1200 \text{ s}}{2.303} + \log 2.32$$

$$= \quad -0.380 \quad + \quad 0.365 \quad = -0.015$$

$$[H_2O_2]_t = \text{antilog}(-0.015) = 10^{-0.015} = 0.97 \text{ M}$$

The value listed in Table 15-1 for $[H_2O_2]$ at 1200 s is 0.98 M. Note how well the values calculated here agree with the experimental value.

(To find the number whose common logarithm is -0.015, raise 10 to the -0.015 power.)

SIMILAR EXAMPLES: Exercises 8, 30, 64.

The data used in the equations of first-order kinetics need not always be in terms of molarities. Sometimes masses of reactants, or just the fraction of reactant consumed, are sufficient, as in Example 15-9.

Example 15-9

Relating time and the fraction (or percent) of reactant consumed in a first-order reaction. Use a value of $k = 7.30 \times 10^{-4} \text{ s}^{-1}$ for the first-order decomposition of $H_2O_2(aq)$ to determine **(a)** the % H_2O_2 that has decomposed in the first 500 s after the reaction begins and **(b)** the time required for one-half of the sample to decompose.

Solution

(a) Regardless of the value of $[H_2O_2]_0$, the ratio, $[H_2O_2]_t/[H_2O_2]_0$, represents the fractional part of the initial quantity of A that remains *unreacted* at time t. Our problem essentially is to determine this ratio at $t = 500$ s. The following form of equation (15.13) is most useful for this purpose.

$$\ln \frac{[H_2O_2]_t}{[H_2O_2]_0} = -kt = -7.30 \times 10^{-4} \text{ s}^{-1} \times 500 \text{ s} = -0.365$$

$$\frac{[H_2O_2]_t}{[H_2O_2]_0} = \text{antiln}(-0.365) = 0.694 \quad \text{and} \quad [H_2O_2]_t = 0.694[H_2O_2]_0$$

The fractional part of the original H_2O_2 that remains *undecomposed* is 0.694 (69.4%). The fractional part that must have *decomposed* is $1.000 - 0.694 = 0.306$. The **% H_2O_2 decomposed is 30.6%.**

(b) What we are seeking here is the time at which $[H_2O_2]_t = \frac{1}{2}[H_2O_2]_0$. Note that in the expression below $\ln \frac{1}{2} = \ln 1 - \ln 2 = 0 - \ln 2$.

$$\ln \frac{[H_2O_2]_t}{[H_2O_2]_0} = \ln \frac{\frac{1}{2}[H_2O_2]_0}{[H_2O_2]_0} = \ln \frac{1}{2} = -\ln 2 = -0.693 = -kt$$

$$t = \frac{0.693}{k} = \frac{0.693}{7.30 \times 10^{-4} \text{ s}^{-1}}$$

$$= 9.49 \times 10^2 \text{ s} = 949 \text{ s}$$

SIMILAR EXAMPLES: Exercises 7, 31, 34g, h, 64.

Half-life of a Reaction. The time calculated in Example 15-9(b) is known as the **half-life** of the reaction. This is the time required for the concentration (or quantity) of a reactant to decrease to one-half of a previous value. Whatever the reactant and regardless of its initial concentration, *if a reaction is first order, the half-life,* $t_{1/2}$, *is constant and depends only on the value of* k.

From Example 15-9(b) we see that the relationship between $t_{1/2}$ and k is

Relationship between half-life and rate constant for a first-order reaction.

$$t_{1/2} = \frac{0.693}{k} \qquad (15.16)$$

In Example 15-9(b), the time required for the quantity of H_2O_2 to be reduced to $\frac{1}{4}$ of its original value would be (949 + 949) s; to $\frac{1}{8}$ of its original value, (949 + 949 + 949) s, and so on. This test of constancy of half-life can be applied to a simple plot of concentration against time. Try this with Figure 15-2; that is, starting with $[H_2O_2] = 2.32$ M at $t = 0$, at what time is $[H_2O_2] \approx 1.16$ M? $[H_2O_2] \approx 0.58$ M? $[H_2O_2] \approx 0.29$ M? Here are two useful features about the half-life of a first-order reaction:

The half-life is constant *only* for first-order reactions.

- Proving that the half-life of a reaction is a constant means proving that the reaction is *first order*.
- If the half-life of a reaction is known, the rate constant is easily obtainable ($k = 0.693/t_{1/2}$).

Reactions Involving Gases. In reactions involving gases reaction rates are often measured in terms of gas pressure. For the hypothetical reaction, A(g) → products, the initial partial pressure of A, $(P_A)_0$, and the partial pressure of A at some time t, $(P_A)_t$, are related through the expression

Integrated rate equation for a first-order reaction based on gas pressures.

$$\ln \frac{(P_A)_t}{(P_A)_0} = -kt \qquad or \qquad \log \frac{(P_A)_t}{(P_A)_0} = \frac{-kt}{2.303} \qquad (15.17)$$

Di-t-butyl peroxide (DTBP) is used as a catalyst in polymerization reactions. In the gaseous state DTBP decomposes into acetone and ethane by a first-order reaction.

$$\underset{\text{DTBP}}{C_8H_{18}O_2(g)} \longrightarrow 2 \underset{\text{acetone}}{C_3H_6O(g)} + \underset{\text{ethane}}{C_2H_6(g)} \qquad (15.18)$$

FIGURE 15-6
Decomposition of di-t-butyl peroxide (DTBP).

The rate of a gas-phase reaction can be followed by measuring the partial pressure of a gaseous species. The decomposition of di-t-butyl peroxide (DTBP) is described through equation (15.18). Three successive half-life intervals of 80 min each are indicated.

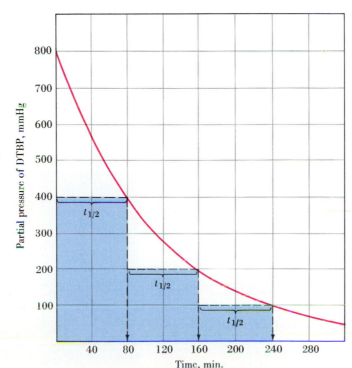

The partial pressure of DTBP as a function of time is plotted in Figure 15-6, and the half-life of the reaction is indicated on the graph.

Example 15-10

Determining the half-life, rate constant, and relationship of total pressure to partial pressure of reactant in a first-order reaction involving gases. Reaction (15.18) is started with pure DTBP at 147 °C and 800. mmHg pressure in a flask of constant volume. **(a)** What is the half-life of this reaction? **(b)** At what time will the partial pressure of DTBP be 100. mmHg? **(c)** What is the value of the rate constant k? **(d)** What will be the *total* gas pressure when the partial pressure of DTBP is 700. mmHg?

Solution

(a) From Figure 15-6 we see that the partial pressure of DTBP falls from 800 mmHg to 400 mmHg in 80 min., to 200 mmHg in 160 min., and to 100 mmHg in 240 min. Thus at the end of every 80-min interval the partial pressure of DTBP falls to one-half of its value at the start of the interval. The half-life is *constant* (as it should be for a first-order reaction); $t_{1/2} = 80.$ min.

(b) A DTBP partial pressure of 100. mmHg is $\frac{1}{8}$ of the starting pressure of 800. mmHg; that is, $P_{DTBP} = (\frac{1}{2})^3 \times 800. = 100.$ mmHg. The reaction must go through *three* half-lives: $t = 3 \times t_{1/2} = 3 \times 80. = 2.4 \times 10^2$ min.

(c) For a first-order reaction, $t_{1/2} = 0.693/k$, or

$$k = 0.693/t_{1/2} = 0.693/80.\text{ min} = 8.7 \times 10^{-3}\text{ min}^{-1}$$

(d) According to equation (15.18) three moles of gaseous products are produced for every mole of DTBP consumed. The amount of DTBP decomposed corresponds to a drop in its partial pressure of 100. mmHg (that is, from 800. mmHg to 700. mmHg). Products are formed in sufficient amount to produce a partial pressure of $3 \times 100.$ mmHg $= 300.$ mmHg.

$$\text{total pressure} = P_{DTBP} + P_{prod.} = 700. + 300. = 1000.\text{ mmHg}$$

SIMILAR EXAMPLES: Exercises 33, 77, 86.

Examples of First-Order Reactions and the Significance of k. One of the best examples of a first-order process is radioactive decay. For example, the isotope P-32, which is used in biochemical studies, has a half-life of 14.3 days. Whatever

TABLE 15-4
Some Typical First-Order Processes

Process	Half-life, $t_{1/2}$	Rate constant k, s^{-1}
radioactive decay of $^{238}_{92}U$	4.51×10^9 years	4.87×10^{-18}
radioactive decay of $^{14}_{6}C$	5.73×10^3 years	3.83×10^{-12}
radioactive decay of $^{32}_{15}P$	14.3 days	5.61×10^{-7}
radioactive decay of $^{26}_{11}Na$	1.0 s	6.9×10^{-1}
$(CH_2)_2O(g) \xrightarrow{415\,°C} CH_4(g) + CO(g)$ ethylene oxide	56.3 min	2.05×10^{-4}
$2\,N_2O_5 \xrightarrow[45\,°C]{\text{in } CCl_4} 2\,N_2O_4 + O_2(g)$	18.6 min	6.21×10^{-4}
$C_{12}H_{22}O_{11}(aq) + H_2O \xrightarrow{15\,°C} C_6H_{12}O_6(aq) + C_6H_{12}O_6(aq)$ sucrose $\qquad\qquad$ glucose $\qquad$ fructose	8.4 h	2.3×10^{-5}
$HC_2H_3O_2(aq) \longrightarrow H^+(aq) + C_2H_3O_2^-(aq)$	8.9×10^{-7} s	7.8×10^5

number of P-32 atoms are at hand at the moment, there will be half this number present in 14.3 days, one quarter of this number in $14.3 + 14.3 = 28.6$ days, etc. The rate constant for the decay is $k = 0.693/t_{1/2}$, and equations (15.13) and (15.14) apply, with numbers of atoms substituting for concentrations, that is, N_t for $[A]_t$ and N_0 for $[A]_0$. A few radioactive decay processes and several other examples of first-order kinetics are summarized in Table 15-4.

Table 15-4 brings out the extreme range of values possible for the rate constants in first-order processes. The *larger* the value of k or the *shorter* the half-life, the *faster* the process occurs.

Radioactive decay is a nuclear process, not a chemical reaction. Nuclear processes are discussed in Chapter 26.

15-6   Second-Order Reactions

In Section 15-3 we found that the peroxodisulfate–iodide ion reaction (15.8) is first order in each reactant, and second order overall. This reaction is of the type A + 3 B → products, and its rate law is of the form

$$\text{rate of reaction} = k[A][B] \tag{15.19}$$

Some reactions of the type A → products are also second order, and their rate laws are of the form

$$\text{rate of reaction} = k[A]^2 \tag{15.20}$$

These facts remind us that balanced chemical equations themselves tell us nothing about the order of a reaction. The order of a reaction can be established *only by experiment*. What we need to consider here is how to treat kinetic data to see if a reaction is second order. In general the task is more difficult than for first-order reactions, so we limit ourselves to the simplest case: *a reaction involving a single reactant and following the rate law (15.20)*.

Again, by using a calculus procedure (which we will not pursue) the rate law (15.20) can be converted into an integrated rate equation, and the equation can be written in the form of a straight-line graph.

$$\underbrace{\frac{1}{[A]_t}}_{y} = \underbrace{k}_{} \times \underbrace{t}_{} + \underbrace{\frac{1}{[A]_0}}_{} \tag{15.21}$$

equation of straight line: $y \;\; = m \times x + \;\; b$

If we plot $1/[A]_t$ as a function of time we obtain a straight line with a slope of k (see Figure 15-7). Since the *reciprocal* of a concentration ($1/[A]$) must have the units L mol^{-1}, each of the three terms in equation (15.21) must have the units L mol^{-1}. If the product kt has the units L mol^{-1}, then k must have the units L mol^{-1} (time)$^{-1}$, such as *L mol^{-1} s^{-1}* or *L mol^{-1} min^{-1}*.

To derive the half-life of the second-order reaction   A → products,   we need to substitute $t = t_{1/2}$ and $[A]_t = \frac{1}{2}[A]_0$ in equation (15.21).

$$\frac{1}{[A]_0/2} = kt_{1/2} + \frac{1}{[A]_0} \qquad \frac{2}{[A]_0} - \frac{1}{[A]_0} = kt_{1/2} \qquad \frac{1}{[A]_0} = kt_{1/2}$$

and

$$t_{1/2} = \frac{1}{k[A]_0} \tag{15.22}$$

From equation (15.22) we see that the half-life depends on both the rate constant and the initial concentration $[A]_0$. The half-life is *not constant;* its value depends on the concentration of reactant at the start of each half-life interval.

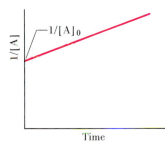

FIGURE 15-7
A straight-line plot for the second-order reaction

A → products

The reciprocal of the concentration, 1/[A], has its lowest value at the start of the reaction. As the reaction proceeds, [A] decreases and 1/[A] increases, in a straight-line fashion. The slope of the line is the rate constant k.

Integrated rate equation for a second-order reaction of the type
A → products

Relationship between half-life and rate constant for a second-order reaction of the type
A → products

TABLE 15-5

Kinetic Data for Example 15-11

Time min	[A], M	ln [A]	1/[A]
0	1.00	0.00	1.00
5	0.63	−0.46	1.6
10	0.46	−0.78	2.2
15	0.36	−1.02	2.8
25	0.25	−1.39	4.0

Example 15-11

Using a graphical method to determine the order of a reaction. The data listed in Table 15-5 were obtained for the decomposition reaction: $A \rightarrow$ products. (a) Establish the order of the reaction. (b) What is the rate constant, k. (c) What is the half-life, $t_{1/2}$, if $[A]_0 = 1.00$ M?

Solution

(a) Plot the following three graphs.

   1. [A] vs. time. (If a straight line, reaction is zero order.)
   2. ln [A] vs. time. (If a straight line, reaction is first order.)
   3. 1/[A] vs. time. (If a straight line, reaction is second order.)

These graphs are shown in Figure 15-8; the reaction is second order.

(b) The slope of graph 3 in Figure 15-8 is

$$k = \frac{(4.00 - 1.00)\ \text{L/mol}}{25\ \text{min}} = 0.12\ \text{L mol}^{-1}\ \text{min}^{-1}$$

(c) According to equation (15.22)

$$t_{1/2} = \frac{1}{k[A]_0} = \frac{1}{0.12\ \text{L mol}^{-1}\ \text{min}^{-1} \times 1.00\ \text{mol/L}} = 8.3\ \text{min}$$

SIMILAR EXAMPLES: Exercises 12, 24a, 35, 38, 39.

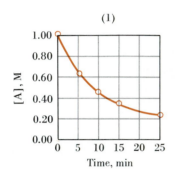

(1)

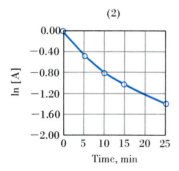

(2)

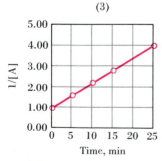

(3)

FIGURE 15-8

Testing for the order of a reaction—Example 15-11 illustrated.

The straight-line plot is that shown in (3). The reaction is second order.

Pseudo-First-Order Reactions. Sometimes higher-order reactions can be made to behave like reactions of a lower order. Then their rate laws become easier to work with. Consider, for example, the hydrolysis of ethyl acetate, a second-order reaction

$$CH_3COOC_2H_5 + HOH \longrightarrow CH_3COOH + C_2H_5OH$$

  ethyl acetate   acetic acid   ethanol

for which the rate law is

rate of reaction $= k[CH_3COOC_2H_5][H_2O]$.

Suppose we follow the hydrolysis of 1 L of aqueous 0.01 M ethyl acetate to completion. $[CH_3COOC_2H_5]$ decreases from 0.01 M to essentially zero. This means that 0.01 mol $CH_3COOC_2H_5$ is consumed, and along with it, 0.01 mol H_2O.

Are You Wondering:

If you can determine the order of a reaction just from the single graph of [A] vs. time?

For reactions of the type $A \rightarrow$ products and orders limited to zero, first, and second, the answer is "yes." If the [A] vs. t graph is a straight line, the reaction is *zero order*. If the graph is not a straight line, the reaction must be either first or second order. If you apply the half-life test as in Figure 15-6 and find that $t_{1/2}$ is a constant, the reaction is *first order*. If the [A] vs. t graph is not a straight line and if $t_{1/2}$ is not constant, the reaction is *second order*. (In fact, for a second-order reaction you will find that $t_{1/2}$ *doubles* for every successive half-life period.)

But consider what happens to the molarity of the H_2O in this reaction. Initially the solution contains about 1000 g H_2O, which is 55.5 mol H_2O (that is, $1000/18 = 55.5$). After the reaction is completed there is still 55.5 mol H_2O ($55.5 - 0.01 \approx 55.5$). Since the amount of water remains essentially constant throughout the reaction, so does its molarity—55.5 M. The rate of reaction does not appear to depend on $[H_2O]$. So, the reaction appears to be zero order in H_2O, first order in $CH_3COOC_2H_5$, and *pseudo-first* order overall. The reaction can be treated by the methods of first-order reaction kinetics.

15-7 Reaction Kinetics: A Summary

The preceding sections presented several ways of handling data and solving kinetics problems. Often a problem can be solved by several different methods, but in general you may find the most direct approaches to be the following.

1. If a question deals with the *rate of a reaction,* and if you have the necessary information (that is, values of $k, m, n, \cdots$), use the *rate law:* **rate of reaction = $k[A]^m[B]^n \cdots$** .

2. If a question deals with a *rate of reaction,* and you do *not* have values of $k, m, n, \cdots$, you may establish the rate of reaction from

   • the slope of an appropriate tangent line to the graph of [A] vs t. This is the *instantaneous* rate of reaction;

TABLE 15-6
Reaction Kinetics: A Summary
For the hypothetical reaction A $\longrightarrow$ products

Order	Rate law	Integrated rate equation	Straight line	$k =$	Units of k	Half-life
0	rate $= k$	$[A]_t = -kt + [A]_0$		$-$slope	mol L^{-1} s^{-1}	$[A]_0/2k$
1	rate $= k[A]$	$\ln [A]_t = -kt + \ln[A]_0$		$-$slope	s^{-1}	$0.693/k$
2	rate $= k[A]^2$	$\dfrac{1}{[A]_t} = kt + \dfrac{1}{[A]_0}$		slope	L mol^{-1} s^{-1}	$\dfrac{1}{k[A]_0}$

- the expression: *rate of reaction* = $-\Delta[A]/\Delta t$, if the time interval Δt is *short*. When applied to the initial stages of a reaction, this rate of reaction is the *initial* rate of reaction.

3. If you are asked to determine the *order* of a reaction, depending on the data given, use

- the *method of initial rates;*
- a graph of rate data to yield a *straight line* (as indicated in Table 15-6);
- a test for constancy of the half-life (good only for first-order).

4. If a question asks you to relate reactant *concentrations* and *times*, whether for a zero-, first-, or second-order reaction, use the appropriate *integrated rate equation*, first determining k (see also, Table 15-6).

15-8   Theories of Chemical Kinetics

The statement that *chemical reactions result from collisions between molecules* may seem so obvious as not to require mention. Yet, a theory explaining chemical reactions in terms of molecular collisions did not become firmly established until the early decades of this century. The kinetic molecular theory had to be developed first (recall Section 6-9).

Collision Theory. Although it is beyond the scope of this text to do so, the kinetic-molecular theory can be used to derive the number of collisions between molecules per unit time. This number is called the **collision frequency.** For example, consider 0.01 mol of a gas in a 1-L vessel at 25 °C, that is, a gas at an initial concentration of 0.01 M. Assuming a molecular weight of 100 for the gas, the collision frequency is of the order of 10^{30} collisions per second. If each of these collisions led to chemical reaction, the rate of reaction would be of the order of 10^6 mol L^{-1} s^{-1}.

$$\frac{10^{30} \text{ collisions } L^{-1} \text{ s}^{-1}}{6 \times 10^{23} \text{ collisions/mol}} \approx 10^6 \text{ mol } L^{-1} \text{ s}^{-1} \tag{15.23}$$

The rate of reaction suggested by (15.23) is extremely fast. There are some reactions that do proceed this rapidly, such as certain reactions involving ions in aqueous solution. However, if the hypothetical gas reacted at the rate of (15.23) it would be completely consumed in 10^{-8} s, and a more usual gas-phase reaction rate is of the order of 10^{-4} mol L^{-1} s^{-1}. This must mean that, typically, only a fraction of the collisions among gaseous molecules lead to chemical reaction, and this observation can be accounted for by two factors.

- Only the more energetic molecules in a mixture undergo reaction as a result of collisions.

- The probability of a particular collision resulting in a reaction depends on the orientation of the colliding molecules.

The extra energy that molecules require in order to react is commonly referred to as the **activation energy.** With the kinetic-molecular theory it is possible to establish what fraction of all the molecules in a collection possess energies in excess of any particular value. Figure 15-9 shows a distribution of molecular energies and notes the fraction of all the molecules possessing energies in excess of some hypothetical activation energy. So, now consider that the rate of a reaction depends on the product of the collision frequency *and* the fraction of activated molecules. Because the fraction of activated molecules is generally so small, the rate of reaction is usually much smaller than the collision frequency. Moreover, the *larger* the activa-

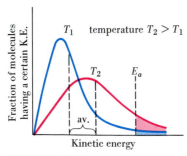

FIGURE 15-9

Distribution of molecular energies.

At the higher temperature, T_2, the distribution is broadened, the average kinetic energy increases, and many more molecules possess energies greater than the activation energy, E_a.

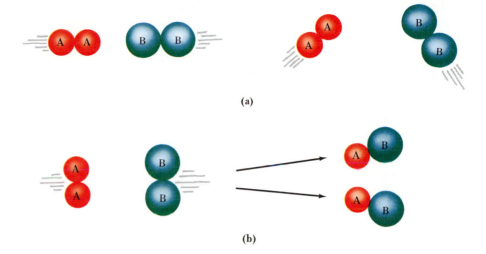

(a)

(b)

tion energy, the *smaller* the fraction of activated molecules and the more slowly a reaction proceeds.

To visualize the reaction

$$A_2(g) + B_2(g) \longrightarrow 2\ AB(g)$$

in terms of collision theory, assume that during a collision between molecules of A_2 and B_2 the bonds A—A and B—B break and the bonds A—B form. The result is the conversion of the reactants A_2 and B_2 to the product AB. But as pictured in Figure 15-10, this assumption does not hold for every collision. A particular orientation of the molecules may be required if the collision is to be effective in producing chemical reaction. Figure 15-10 suggests that the number of unfavorable collisions often exceeds the number of favorable ones. This means that the probability of a particular collision being favorable to reaction is generally small (that is, the probability is less than 1).

If we represent collision frequency as Z, the fraction of activated molecules as f, and the probability factor (also called the *steric* factor) as p, the rate of a chemical reaction has the form

$$\text{rate of reaction} = pfZ \tag{15.24}$$

Collision frequency is proportional to the concentrations of the molecular species involved in the collisions (say that these species are A and B). That is $Z \propto [A][B]$ and

$$\text{rate of reaction} \propto pf[A][B] = k[A][B] \tag{15.25}$$

The collision theory seems to lead to a general rate equation for chemical reactions, but there are some serious shortcomings to this result. Equation (15.25) describes a reaction that is second order overall, yet we know that other reaction orders are possible. At times the probability (steric) factor is smaller than can be explained just in terms of collision orientations; in a few cases it even appears to be much greater than 1. In some reactions, although molecules acquire activation energy through molecular collisions, the actual reaction (that is, the rearrangement of chemical bonds) does not occur at the time of the collision. The theoretical basis of chemical kinetics must involve more than just the concepts associated with simple collision theory.

Transition State Theory. An important extension of collision theory made by Henry Eyring (1901–1981) and others focuses on the details of a collision. Of particular concern is a *hypothetical* species with properties somewhere between the reactants and the products, called an **activated complex.** This transitory species

FIGURE 15-11
Reaction profile for the reaction

$H_2(g) + I_2(g) \rightarrow 2\,HI(g)$

The course of the reaction, from reactants through hypothetical activated complex to products, is represented in this reaction profile. The enthalpy change (ΔH) and the activation energies (E_a) of the forward and reverse reactions are shown.

Because their activation energies are so large, neither the reaction of H_2 and I_2 nor the decomposition of HI is likely to occur exactly in the manner suggested here. A more likely mechanism is discussed on page 553.

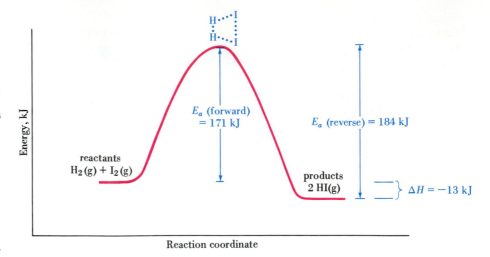

either dissociates back into the original reactants (in which case there is no reaction) or into product molecules. We might represent an activated complex in this way.

$$\begin{array}{ccccc} A & B & A\cdots B & & A\!-\!B \\ | & + \; | & \vdots \quad \vdots & \longrightarrow & \\ A & B & A\cdots B & & A\!-\!B \\ \text{reactants} & & \text{activated} & & \text{products} \\ & & \text{complex} & & \end{array} \qquad (15.26)$$

The energy that colliding molecules must invest into producing this high-energy activated complex is the activation energy.

Another way of looking at activation energy is presented in Figure 15-11, which is a **reaction profile** for the reaction of $H_2(g)$ and $I_2(g)$ to form HI(g). In Figure 15-11 energies of the species involved are plotted on the vertical axis and a quantity called the reaction coordinate, on the horizontal axis. The reaction coordinate can be thought of as representing the extent of the reaction. The reaction starts with reactants on the left, passes through an activated complex, and ends with products on the right. The difference in energies between the reactants and products is ΔH for the reaction. The formation of HI(g) is a slightly exothermic reaction. The difference in energy between the activated complex and the reactants, 171 kJ, is the activation energy. Thus, a large energy barrier separates the reactants from the products. Only especially energetic molecules can pass over this barrier (by forming an activated complex that dissociates into product molecules).

We can use Figure 15-11 to describe the reverse process too, the decomposition of HI(g) into $H_2(g)$ and $I_2(g)$. The activation energy for the reverse reaction is 184 kJ. Figure 15-11 also illustrates these two relationships: that the enthalpy change of a reaction is equal to the difference in activation energies of the forward and reverse reactions

Relationship between the enthalpy change and the activation energies of forward and reverse reactions.

$$\Delta H = E_a(\text{forward}) - E_a(\text{reverse}) \qquad (15.27)$$

for HI formation: $\Delta H = 171\text{ kJ} - 184\text{ kJ} = -13\text{ kJ}$

and that for an *endothermic* reaction the activation energy must be greater than the endothermic heat of reaction.

An Analogy to Activation Energy. You may find it helpful to think of activation energy in terms of this analogy: Imagine that a hike is being planned between two mountain valleys separated by a ridge through which there is a pass (Figure 15-12). The elevations of the valleys correspond to the energies of the reactants and products, and the elevation of the pass above the starting point to the activation energy.

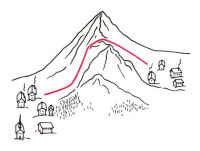

FIGURE 15-12
An analogy to activation energy.

The height of the pass is analogous to the activation energy of a reaction. The path of the hike (reaction profile) is traced in red.

Whether the hike is easy or difficult depends not so much on the difference in elevation of the two valleys as it does on the elevation of the pass above the starting point. If the climb from the starting point to the pass is very long and steep, not many people may care to take the hike, no matter how much "downhill" there is on the other side.

Attempts at purely theoretical predictions of reaction rate constants have not been very successful. The principal value of reaction rate theories is to facilitate the discussion of *experimentally observed* reaction rate data. In the next section we will see how the concept of activation energy enters into a discussion of the effect of temperature on reaction rates.

15-9 Effect of Temperature on Reaction Rates

As a practical matter we know that chemical reactions tend to go faster at higher temperatures. Thus, we raise the temperature to speed up the biochemical reactions involved in cooking. On the other hand, we slow down some reactions by lowering the temperature, such as in refrigerating milk to prevent it from souring. And now we have a useful explanation for the effect of temperature on reaction rates.

Increasing the temperature increases the fraction of the molecules that have energies in excess of the activation energy (recall Figure 15-9).

This factor is so important that it can lead to a severalfold increase in reaction rates for temperature increases of only 10 or 20 °C.

Data on the effect of temperature on the rate of decomposition of N_2O_5 are presented in Table 15-7. Even without plotting these data we see that the variation of the rate constant k with temperature is not linear. This situation should remind us of that encountered with vapor pressure. In that case we converted a steeply rising curve (Figure 12-11) to a straight line (Figure 12-15) by plotting the logarithm of vapor pressure versus the reciprocal of the Kelvin temperature. Let us try a similar plot here: $\ln k$ vs. $1/T$. The necessary data are given in the blue panels of Table 15-7 and plotted in Figure 15-13. The graph is indeed linear! It has the form

Expressed in common logarithms, this equation is

$$\log k = \frac{-E_a}{2.303R} \times \frac{1}{T} + B'$$

$$\ln k = \frac{-E_a}{R} \times \frac{1}{T} + B \qquad (15.28)$$

equation of straight line: $\quad y \;=\; m \;\times\; x \;+\; b$

TABLE 15-7
Rate Constant, k, at Several Temperatures for the
Reaction $N_2O_5(\text{in } CCl_4) \longrightarrow N_2O_4(\text{in } CCl_4) + \frac{1}{2} O_2(g)$

t, °C	T, K	$1/T$, K^{-1}	k, s^{-1}	$\ln k$
0	273	3.66×10^{-3}	7.87×10^{-7}	-14.055
25	298	3.36×10^{-3}	3.46×10^{-5}	-10.272
35	308	3.25×10^{-3}	1.35×10^{-4}	-8.910
45	318	3.14×10^{-3}	4.98×10^{-4}	-7.605
55	328	3.05×10^{-3}	1.50×10^{-3}	-6.502
65	338	2.96×10^{-3}	4.87×10^{-3}	-5.325

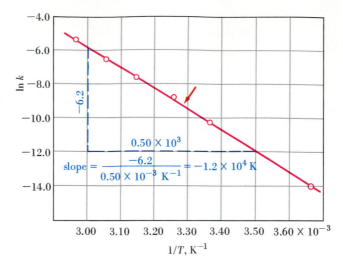

This graph can be used to establish the activation energy E_a for the reaction (see equation 15-28).

$$\text{slope of line} = \frac{-E_a}{R} = -1.2 \times 10^4 \text{ K}$$

$$E_a = 8.314 \text{ J mol}^{-1}\text{ K}^{-1} \times 1.2 \times 10^4 \text{ K} = 1.0 \times 10^5 \text{ J/mol} = 1.0 \times 10^2 \text{ kJ/mol}$$

(A more precise plot of the data of Table 15-7 yields a value of $E_a = 106$ kJ/mol.) The arrow is referred to in Example 15-12.

We can use equation (15.28) in various ways. One is to establish the activation energy of a reaction graphically, as shown in Figure 15-13. A variation of the equation we frequently encounter is one in which the constant B is eliminated by writing the equation for two sets of temperatures and k values; this treatment is identical to the one we used on page 400. The result is

The Arrhenius equation: effect of temperature on the rate constant.

$$\ln \frac{k_2}{k_1} = \frac{E_a}{R}\left(\frac{1}{T_1} - \frac{1}{T_2}\right) \quad \text{or} \quad \log \frac{k_2}{k_1} = \frac{E_a}{2.303R}\left(\frac{1}{T_1} - \frac{1}{T_2}\right) \qquad (15.29)$$

In equations (15.29), T_2 and T_1 are two Kelvin temperatures; k_2 and k_1 are the rate constants at these temperatures; E_a is the activation energy in J/mol; R is the gas constant expressed as 8.314 J mol^{-1} K^{-1}. Svante Arrhenius (1859–1927) was the first to develop this relationship between reaction rate constants and temperatures and the equations in (15.29) are referred to as the **Arrhenius equation.**

Example 15-12

Applying the Arrhenius equation. Use data from Table 15-7 and Figure 15-13 to determine the temperature at which the half-life for the first-order decomposition of N_2O_5 is 2.00 h.

Solution. First we need to find the rate constant k corresponding to a 2.00 h half-life. For a *first-order reaction*

$$k = \frac{0.693}{t_{1/2}} = \frac{0.693}{2.00 \text{ h}} = \frac{0.693}{7200 \text{ s}} = 9.62 \times 10^{-5} \text{ s}^{-1}$$

We can now proceed in one of two ways.

GRAPHICAL METHOD

We are seeking the temperature at which $k = 9.62 \times 10^{-5}$ and $\ln k = \ln 9.62 \times 10^{-5} = -9.249$. This point can be located directly on the straight-

Are You Wondering:

If there is a theoretical basis for the Arrhenius equation?

Let us follow through a bit with the kinetic theory treatment of the collision theory. Equation (15.24), which relates the rate of a reaction to collision frequency, Z, the probability (steric) factor, p, and the fraction of activated molecules, f, can be converted to an expression that yields the rate constant.

$$k = Zpe^{-E_a/RT} = Ae^{-E_a/RT} \qquad (15.30)$$

In equation (15.30) the product Zp is replaced by the symbol A, called the *frequency factor*. If we take the natural logarithms of both sides of equation (15.30) we get

$$\ln k = \frac{-E_a}{R} \times \frac{1}{T} + \ln A$$
$$\quad y \;\; = \;\; m \;\; \times \; x \; + \;\; b \qquad \text{(equation of a straight line)}$$

Thus we can find a basis for the straight-line relationship between $\ln k$ and $1/T$ in the collision theory of chemical reactions.

line graph of Figure 15-13 (marked by the arrow). Corresponding to $\ln k = -9.249$, $1/T = 3.28 \times 10^{-3}$ K^{-1}.

$T = 1/(3.28 \times 10^{-3})$ K $= 305$ K $= 32\ °C$

WITH EQUATION (15.29)

Take T_2 to be the temperature at which $k = k_2 = 9.62 \times 10^{-5}$ s^{-1}. T_1 is some other temperature at which a value of k is known. Suppose we take $T_1 = 298$ K and $k_1 = 3.46 \times 10^{-5}$ s^{-1}. The activation energy is 106 kJ/mol $= 1.06 \times 10^5$ J/mol (the more precise value given in Figure 15-13). Now we can solve equation (15.29) for T_2. (For simplicity we have omitted units below, but the temperature is obtained in kelvins.)

$$\ln \frac{k_2}{k_1} = \ln \frac{9.62 \times 10^{-5}}{3.46 \times 10^{-5}} = \ln 2.78 = 1.022$$

$$1.022 = \frac{E_a}{R}\left(\frac{1}{T_1} - \frac{1}{T_2}\right) = \frac{1.06 \times 10^5}{8.314}\left(\frac{1}{298} - \frac{1}{T_2}\right)$$

$$= 1.27 \times 10^4 \left(0.00336 - \frac{1}{T_2}\right)$$

$1.27 \times 10^4/T_2 = 42.7 - 1.022 = 41.7 \qquad T_2 = (1.27 \times 10^4)/41.7 = 305$ K

SIMILAR EXAMPLES: Exercises 9, 44, 46, 47, 80, 81.

15-10 Reaction Mechanisms

The reaction between gaseous iodine monochloride and hydrogen produces iodine and hydrogen chloride as gaseous products.

$$H_2(g) + 2\ ICl(g) \longrightarrow I_2(g) + 2\ HCl(g) \qquad (15.31)$$

The rate law for this reaction is found to be

$$\text{rate of reaction} = k[\text{H}_2][\text{ICl}] \tag{15.32}$$

Again remember that there is no necessary correspondence between the exponents in a rate law and the coefficients in the balanced chemical equation. (The exponent of the rate-law term [ICl] is not 2, but 1.) Suppose, however, that reaction (15.31) takes place in a series of steps, and that the steps are so chosen that the rate law for each step *does* have its exponents equal to the coefficients in the chemical equation for that step. Thus, let us *postulate* the following two steps, which when added together yield the *net* reaction.

(1) Slow: $\text{H}_2 + \text{ICl} \longrightarrow \text{HI} + \text{HCl}$
(2) Fast: $\text{HI} + \text{ICl} \longrightarrow \text{I}_2 + \text{HCl}$
Net: $\text{H}_2 + 2\,\text{ICl} \longrightarrow \text{I}_2 + 2\,\text{HCl}$

The rate law expressions we can write are

$$\text{rate (1)} = k_1[\text{H}_2][\text{ICl}] \qquad \text{and} \qquad \text{rate (2)} = k_2[\text{HI}][\text{ICl}]$$

Now, let us further postulate that step (1) occurs *slowly* but step (2) occurs *rapidly*. This suggests that HI produced in the first step will be consumed just as fast as it can be formed, and that the rate of the overall reaction will be governed just by the rate at which HI is formed in the first step. Thus we can explain why the observed rate law for the net reaction is: rate of reaction = $k[\text{H}_2][\text{ICl}]$.

The two-step sequence that we have just proposed is called a **reaction mechanism.** It is a sequence of events that

- is consistent with the stoichiometry of the overall (net) reaction;
- accounts for the *experimentally observed* rate law.

The individual steps in the mechanism are called **elementary processes.** Before continuing with a discussion of elementary processes and reaction mechanisms, let us briefly consider the following analogies.

Analogies to a Reaction Mechanism. A person in Los Angeles meets an old friend from New York City and asks, ''When did you leave New York?'' The friend answers, ''This morning.'' The friend must have made the principal part of the journey by airplane. If the friend had answered, ''Two weeks ago,'' the questioner could not be sure of the friend's mode of travel. Further questioning (experimentation) would have been required.

At times, the fact that a reaction occurs very rapidly will suggest a plausible mechanism, as in the preceding analogy. More often, however, we gain insight into a reaction mechanism by identifying the *slow* or *rate-determining* step, as in this analogy: A business traveler lands at an airport, rents a car, and then takes one of two possible routes to visit a client. The trip normally takes 20 minutes by either route, but it takes the traveler 50 minutes. Upon learning of this, the client remarks, ''You must have come on Route 330 and been delayed by the one-way construction traffic.''

Elementary Processes. Each molecular event that significantly alters a molecule's energy or geometry is called an **elementary process.** The combination of elementary processes in a reaction mechanism yields the equation for the net reaction. By writing rate equations for the elementary processes and combining them in the appropriate way, we can establish a rate equation for the net reaction. A crucial test of the *plausibility* of a proposed reaction mechanism is that it yield the same rate equation as that determined experimentally. We must emphasize the word ''plausible'' because there is no way to *prove* a reaction mechanism. It is often possible to propose several mechanisms that are consistent with the observed rate law.

Now let us consider some important characteristics of elementary processes, beginning with the one we used in working out a mechanism for reaction (15.31).

The border station at San Ysidro/Tijuana is a ''bottle-neck to travelers. Note the tie-up of traffic on the Mexican side of the border (top) and the relatively few cars on the United States side (bottom). [Aerial Fotobank, Inc., San Diego, CA]

1. The exponents of the concentration terms in the rate equation for an *elementary* process are the same as the coefficients in the balanced equation for the process.
2. Elementary processes in which a single molecule dissociates—**unimolecular**—or two molecules collide—**bimolecular**—are more probable than a process requiring the simultaneous collision of three molecules—**termolecular.**
3. Elementary processes are reversible, and some may reach a condition of equilibrium, in which the rates of the forward and reverse processes are equal.
4. Certain species are produced in one elementary process and consumed in another. Such species are called **intermediates.** In a proposed reaction mechanism, intermediates must not appear in either the net chemical equation or the overall rate equation.
5. One elementary process may occur much more slowly than all the others, and in some cases may determine the rate at which the overall reaction proceeds. Such a process is called the **rate-determining step.**

The mechanism for reaction (15.31) involved a slow step followed by a fast step.

A Mechanism with a Fast, Reversible First Step, Followed by a Slow Step. Gaseous hydrogen and iodine combine to produce gaseous hydrogen iodide in a second-order reaction.

$$H_2(g) + I_2(g) \longrightarrow 2 HI(g) \qquad \text{rate of reaction} = k[H_2][I_2]$$

A mechanism consistent with these findings is

Fast: $\qquad\qquad\qquad I_2(g) \underset{k_2}{\overset{k_1}{\rightleftharpoons}} 2 I(g)$ $\qquad\qquad$ (15.33)

Slow: $2 I(g) + H_2(g) \overset{k_3}{\longrightarrow} 2 HI(g)$ $\qquad\qquad$ (15.34)

Net: $\qquad I_2(g) + H_2(g) \longrightarrow 2 HI(g)$ $\qquad\qquad$ (15.35)

The slow or *rate-determining* step in this mechanism is pictured in Figure 15-14. Its rate equation is

$$\text{rate of reaction} = k_3[I]^2[H_2] \qquad\qquad (15.36)$$

However, since atomic I is an intermediate in the reaction, we must eliminate it from the rate equation. We can do this by assuming that the fast, reversible step (15.33) reaches a condition in which I_2 is consumed and produced at equal rates and from which atomic I is withdrawn only very slowly through the rate-determining step. Then, since

$$-(\text{rate of disappearance of } I_2) = \text{rate of formation of } I_2$$

$$k_1[I_2] = k_2[I]^2$$

and

$$[I]^2 = \frac{k_1}{k_2} [I_2]$$

Now, if we substitute this value of $[I]^2$ into the rate equation (15.36), we obtain for the net reaction

$$\text{rate of reaction} = \frac{k_1 k_3}{k_2} [H_2][I_2] = k[H_2][I_2] \qquad (\text{where } k = k_1 k_3/k_2) \qquad (15.37)$$

Equation (15.37) agrees with the observed rate law. Whether the mechanism proposed here is the actual reaction path we cannot say. All we can say is that it is *plausible*.

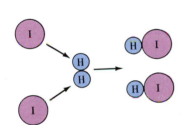

FIGURE 15-14
Rate-determining step in the reaction

$$H_2(g) + I_2(g) \to 2 HI(g)$$

The rate-determining step for this reaction appears to involve the simultaneous collision of two I atoms with an H_2 molecule.

In this text we emphasize testing the plausibility of a mechanism. Proposing a mechanism for a reaction can be somewhat more difficult.

Example 15-13

Testing the plausibility of a reaction mechanism. The reaction $2 NO + Cl_2 \to 2 NOCl$ has the rate law: rate of reaction = $k[NO]^2[Cl_2]$. Show that the following mechanism is consistent with the observed rate law.

Fast: $NO + Cl_2 \underset{k_2}{\overset{k_1}{\rightleftharpoons}} NOCl_2$

Slow: $NO + NOCl_2 \xrightarrow{k_3} 2 NOCl$

Solution. Assume that the second step, the slow step, is rate determining.

rate of reaction = $k_3[NO][NOCl_2]$

Also assume that in the fast reversible reaction (the first step),

rate of formation of $NOCl_2$ = $-$(rate of disappearance of $NOCl_2$)

$$k_1[NO][Cl_2] = k_2[NOCl_2]$$

Determine the concentration of $NOCl_2$

$$[NOCl_2] = \frac{k_1[NO][Cl_2]}{k_2}$$

and substitute this concentration into the rate-determining step.

$$\text{rate of reaction} = k_3[NO]\frac{k_1[NO][Cl_2]}{k_2} = \frac{k_1 k_3[NO]^2[Cl_2]}{k_2} = k[NO]^2[Cl_2]$$

SIMILAR EXAMPLES: Exercises 10, 11, 56, 57, 58.

15-11 Catalysis

We have seen that one way to make a reaction go faster is to raise the temperature. This increases the fraction of molecules with energies in excess of the activation energy. Another way to increase this fraction, *without raising the temperature*, is to find a route or mechanism for the reaction that has a *lower* activation energy. The function of a catalyst in a chemical reaction is to provide this alternate pathway. A catalyst enters into a chemical reaction in such a way that it undergoes *no permanent change*. As a result its formula does not appear in the net chemical equation (although its presence is generally indicated by an appropriation notation above the arrow sign). The success of a commercial process for producing a chemical substance often hinges on finding appropriate catalysts. The ranges of temperatures and pressures that can be used in industrial processes are not possible with biochemical reactions. The presence of appropriate catalysts (enzymes) is absolutely crucial to the existence of living matter.

To continue the analogy introduced on page 548, a catalyst is like a guide who can ease the climb for a party of hikers by showing them an easier (less steep) route to their objective.

Homogeneous Catalysis. Figure 15-15 represents the decomposition of formic acid (HCOOH). In the uncatalyzed reaction an H atom must be transferred from one part of the formic acid molecule to another before the C—O bond can break. The energy requirement for this atom transfer is high, resulting in a high activation energy and a slow reaction.

The acid-catalyzed decomposition of formic acid can be represented as

$$\overset{\displaystyle O}{\underset{\displaystyle \|}{H-C-O-H}} \xrightarrow{H^+} H_2O + CO \qquad (15.38)$$

In this catalyzed reaction a hydrogen ion from solution attaches itself to the O atom that is singly bonded to the C atom, forming $(HCOOH_2)^+$. The C—O bond ruptures, and an H atom attached to a *carbon* atom in the intermediate species $(HCO)^+$ is released to the solution as H^+. This reaction pathway does not require an atom

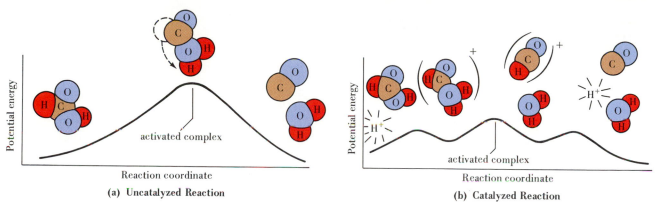

(a) Uncatalyzed Reaction   (b) Catalyzed Reaction

FIGURE 15-15

An example of homogeneous catalysis.

The potential energy of the activated complex—the activation energy—is lowered in the presence of a catalyst. [From George C. Pimentel, ed., *Chemistry: An Experimental Science.* Freeman, San Francisco, 1963. Reproduced by permission of the Chemical Education Material Study.]

Recall the description of adsorption given on page 464.

transfer within the formic acid molecule. It has a lower activation energy than does the uncatalyzed reaction and proceeds at a faster rate. Because the reactants, the products, and the catalyst of this reaction all exist within the same solution or *homogeneous* mixture, this type of catalysis is called **homogeneous catalysis.**

Heterogeneous Catalysis. Many reactions can be catalyzed by allowing them to occur on an appropriate surface. In these reactions the reactants and products may both be found in the same homogeneous mixture—gas or liquid. Essential intermediates in the reaction, however, are found in a separate phase—on a solid surface. This type of catalysis is called **heterogeneous catalysis.** The precise mechanism of heterogeneous catalysis is imperfectly understood, but it appears that the availability of *d* electrons and *d* orbitals in surface atoms may play an important role. Catalytic activity is associated with a large number of transition elements and their compounds.

The key requirement in heterogeneous catalysis is that reactants be *adsorbed* from a gaseous or solution phase onto the surface of the catalyst. Not all surface atoms are equally effective for catalysis; those that are, are called **active sites.** Basically, then, heterogeneous catalysis involves

1. *Adsorption* of reactants.
2. Diffusion of reactants along the surface.
3. Reaction at an active site to form adsorbed product.
4. *Desorption* of the product.

These steps are suggested through Figure 15-16 for the reaction of carbon monoxide and nitrogen oxide in the presence of a rhodium catalyst to produce carbon dioxide

FIGURE 15-16

Heterogeneous catalysis in the reaction

$$2\ CO + 2\ NO \xrightarrow{\ Rh\ } 2\ CO_2 + N_2$$

(a) Two molecules each of CO and NO are adsorbed to the rhodium surface.
(b) The adsorbed NO molecules dissociate into adsorbed N and O atoms.
(c) Adsorbed CO molecules and O atoms combine to form CO_2 molecules, which become desorbed and pass into the gaseous state. Two N atoms combine and are desorbed as an N_2 molecule.

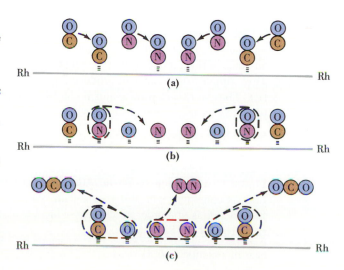

TABLE 15-8
Some Applications of Catalysis

Process	Product(s)	Typical catalyst(s)	Text reference
polymerization of propylene	polypropylene	titanium(IV) halides	Table 10-4
methane reforming with steam	$H_2(g)$ and $CO(g)$	Ni on K_2O/Al_2O_3	reaction (14.3)
methanol synthesis	$CH_3OH(g)$	ZnO/Cr_2O_3	reaction (4.16)
oxidation of $SO_2(g)$	$SO_3(g)$	V_2O_5 plus K_2SO_4 on SiO_2	reaction (15.41)
ammonia synthesis	$NH_3(g)$	Fe with Al_2O_3, MgO, CaO, and K_2O	reaction (16.36)
ammonia oxidation	NO(g)	90% Pt–10% Rh wire gauze	reaction (16.37)
alkylation of benzene	$C_6H_5C_2H_5$	H_3PO_4 on SiO_2	Figure 27-6

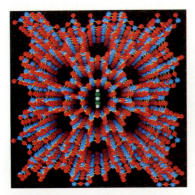

Model of a zeolite catalyst used in the synthesis of gasoline from methanol (CH_3OH). Removal of O atoms and conversion to hydrocarbons occurs in the channels of the zeolite. Zeolite structures were described in Table 14-8. [Chemical Design, Ltd.]

and nitrogen. This is an important reaction that occurs in automotive catalytic converters.

Heterogeneous catalysts must be carefully prepared and maintained, for they are easily "poisoned." Trace amounts of certain impurities may bind to active sites and destroy their catalytic activity. This is the reason why automobiles equipped with catalytic converters cannot be fueled with leaded gasoline.

Table 15-8 lists a few important industrial catalysts, some for processes that we have already encountered and others for processes we will study later.

Enzymes as Catalysts. Some catalysts, for example platinum metal, can catalyze a wide variety of reactions. Unlike platinum, the catalysts associated with chemical reactions in a living organism must be very specific. These catalysts, called **enzymes,** are high-molecular-weight proteins. Many enzymes catalyze one particular reaction and no others. For example, in alcoholic fermentation the six-carbon compound glucose is broken down into two molecules of ethanol and two of carbon dioxide.

$$C_6H_{12}O_6 \longrightarrow 2\ C_2H_5OH + 2\ CO_2$$

This process requires 12 enzymatic steps. In the last of these acetaldehyde is reduced to ethanol through the action of the enzyme alcohol dehydrogenase. (The necessary hydrogen atoms are furnished by other species in the reaction.)

$$\underset{\overset{|}{H}}{\overset{\overset{H}{|}}{H-C}}-\overset{\overset{O}{||}}{C}-H \xrightarrow[\text{dehydrogenase}]{\text{alcohol}} \underset{\overset{|}{H}}{\overset{\overset{H}{|}}{H-C}}-\underset{\overset{|}{H}}{\overset{\overset{H}{|}}{C}}-O-H$$

The simplest mechanism of enzyme action, known as the Michaelis–Menten mechanism, involves a reactant species, called the **substrate** (S), attaching itself to an active site on the enzyme (E). The result is an enzyme–substrate complex (ES). This complex dissociates to produce a product species (P) and the original enzyme (E). Thus, a two-step mechanism can be written, with each step being reversible (denoted by the double arrow $\rightleftharpoons$).

$$S + E \rightleftharpoons ES$$

$$ES \rightleftharpoons E + P$$

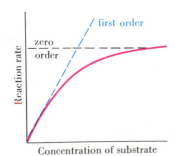

FIGURE 15-17
The Michaelis–Menton mechanism of enzyme action.

Figure 15-17, which shows how the reaction rate varies with substrate concentration, is consistent with this mechanism. Along the ascending portion of the curve the reaction is first order in S, because the rate at which the complex ES is formed is proportional to [S].

rate of reaction = k[S]

At high concentrations of the substrate the reaction is zero order. The enzyme is saturated, and adding more substrate cannot accelerate the reaction.

$$\text{rate of reaction} = k'[\text{S}]^0 = k'$$

The Catalyzed Decomposition of Hydrogen Peroxide. We close this chapter by taking another look at the reaction with which we opened the chapter, the decomposition of H_2O_2. We have previously noted (footnote on page 530) that this reaction is slow, and generally must be catalyzed. HBr is a good catalyst, and its catalytic activity can be represented by this two-step mechanism

Slow: $H_2O_2 + H^+ + Br^- \longrightarrow HOBr + H_2O$

Fast: $\underline{ H_2O_2 + HOBr \longrightarrow H_2O + H^+ + Br^- + O_2(g)}$

Net: $ 2\,H_2O_2 \longrightarrow 2\,H_2O + O_2(g)$

As required for a catalyzed reaction, the formula of the catalyst does not appear in the net equation. The intermediate species HOBr is consumed in the second step just as rapidly as it is formed in the first step. The rate of disappearance of H_2O_2, then, is determined by the rate of the slow first step.

$$\text{reaction rate} = -(\text{rate disappearance of } H_2O_2) = k[H_2O_2][H^+][Br^-] \qquad (15.39)$$

The opening photograph of this chapter shows the Pt-catalyzed decomposition of H_2O_2, an example of heterogeneous catalysis.

Because HBr is constantly regenerated, the concentrations of H^+ and Br^- are unchanged (constant) throughout a given reaction. If we replace the product of the constant terms $k[H^+][Br^-]$ by a new constant k', we can rewrite the rate law as

$$\text{rate of reaction} = k'[H_2O_2] \qquad (15.40)$$

Equation (15.40) is the rate law we used in Section 15-5. However, the rate equation (15.39) does indicate that the rate of decomposition will be affected by the concentration of HBr introduced initially. For each different initial concentration of HBr the rate constant in equation (15.40) has a different value.

FOCUS ON
Sulfuric Acid

Tank car loading sulfuric acid. Sulfuric acid is the leading manufactured chemical in the United States. [Courtesy Stauffer Chemical Company]

Sulfuric acid has a long history; it has been produced for at least 500 and possibly 1000 years. The sulfuric acid industry can be dated from about the middle of the eighteenth century, following the invention of a process in which a mixture of S and KNO_3 was burned and the gaseous products dissolved in water in lead-lined chambers, yielding aqueous sulfuric acid. Later it was discovered that oxides of nitrogen formed by heating KNO_3 acted as intermediates in the process, and that they could be recovered and reused. We now consider intermediates of this type to be catalysts. The manufacture of H_2SO_4 by the "lead chamber" process is one of the earliest examples of *homogeneous catalysis*.

The key step in the manufacture of sulfuric acid is the conversion of $SO_2(g)$ to $SO_3(g)$.

$$SO_2(g) + \tfrac{1}{2} O_2(g) \longrightarrow SO_3(g) \qquad (15.41)$$

Reaction (15.41) occurs very slowly unless it is catalyzed. A simplified description of catalysis by the oxides of nitrogen is

$$NO(g) + \tfrac{1}{2} O_2(g) \longrightarrow NO_2(g) \qquad (15.42)$$
$$NO_2(g) + SO_2(g) \longrightarrow SO_3(g) + NO(g) \qquad (15.43)$$

H_2SO_4 produced by the lead chamber process is impure and diluted with water to the point that it cannot be made more than 80% H_2SO_4, by mass. The method is now obsolete. Interestingly, though, the use of oxides of nitrogen in sulfuric acid production is being revived as an environmental control measure, for example, in removing SO_2 from gaseous emissions from ore smelters.

A method of converting $SO_2(g)$ to $SO_3(g)$ in the presence of platinum was patented in England in 1831. Here *heterogeneous catalysis* is involved. Adsorption of $SO_2(g)$ and $O_2(g)$ on the Pt is followed by reaction at active sites and desorption of the $SO_3(g)$ (recall Figure 15-16). The net reaction is (15.41). The advantages of this alternate route to sulfuric acid production, called the **contact process,** were not fully appreciated until later in the century when the synthetic organic chemicals industry began to require large quantities of a material variously called *oleum* and *fuming sulfuric acid*. Oleum is a solution of SO_3 in pure sulfuric acid. In a sense it is "greater than 100% H_2SO_4." In the contact process pure $SO_3(g)$ is passed into 98 or 99% H_2SO_4 to form oleum, which is then diluted with water to the desired acid strength. If we use the formula $H_2S_2O_7$ (pyrosulfuric acid) for a particular oleum, the reactions are

$$SO_3(g) + H_2SO_4(l) \longrightarrow H_2S_2O_7(l) \qquad (15.44)$$

$$H_2S_2O_7(l) + H_2O \longrightarrow 2\, H_2SO_4(l) \qquad (15.45)$$

$$H_2SO_4(l) + water \longrightarrow H_2SO_4(aq) \qquad (15.46)$$

Platinum metal catalyst, in a finely divided form called platinum black, is easily "poisoned" (for example, by as little as 1×10^{-8} g of arsenic compounds per liter of gas coming in contact with the catalyst). Because of this difficulty other catalysts have been developed. The principal one is vanadium pentoxide (V_2O_5) mixed with alkali metal sulfates. Other modifications in the contact process have been introduced to reduce the emissions of noxious gases (principally SO_2).

The chief source of $SO_2(g)$ for the contact process comes from the burning of sulfur. In the past this sulfur has been obtained mainly from the Frasch mining process (Figure 14-23). Now, with the need to remove sulfur compounds from natural gas and the smokestack gases in power plants and smelters, much of the sulfur used in the manufacture of sulfuric acid is recovered sulfur (recall reaction 14.82). It is likely that by the end of the century most of the sulfur used will be recovered sulfur.

Because the potential uses of sulfuric acid are so varied, it was once said that the quantity of sulfuric acid produced is a good indicator of the degree of industrialization of a nation and of its general economic conditions. This statement is no longer applicable in the United States, however, since the bulk of H_2SO_4 production (70%) is used in the manufacture of fertilizers (see Section 23-3). Other uses include metals processing (5%), oil refining (5%), ores processing (5%), manufacture of the white pigment titanium dioxide (2%), and various minor uses, such as in automobile storage batteries.

Summary

The rate of a chemical reaction is related to how fast the concentration of a reactant decreases or that of a product increases with time. An *initial* rate of reaction is determined by dividing the change in concentration of a reactant at the start of a reaction by the short time interval over which this change occurs. Beyond this initial stage, the *instantaneous* rate of reaction is given by the slope of a tangent line to a concentration vs. time graph. One goal of a kinetic study is to be able to express the reaction rate through a rate law having the form

$$reaction\ rate = k[A]^m[B]^n \cdots$$

The order of a reaction is related to the exponents in the rate law. Reactions that proceed at a constant rate, *independent* of the concentrations of the reactants, are zero-order reactions. A first-order reaction most often has a single concentration term appearing in the rate law and that term is raised to the first power. The most common forms of a second-order rate law have either a single concentration term raised to the second power or two concentration terms each raised to the first power. One method of establishing the order of a reaction requires measuring the *initial* reaction rate in a series of experiments. A second method requires plotting appropriate functions of reactant concentration against time so as to obtain a straight-line graph (see Table 15-6). A third method involves substituting experimentally determined concentrations and times into appropriate mathematical equations (see Table 15-6).

An important characteristic associated with reaction rates is the half-life of a reaction. For a first-order reaction the half-life is *independent* of the concentration of the reactant. For other reaction orders the half-life is concentration dependent (see Table 15-6).

The theoretical basis of chemical kinetics includes these

essential ideas: Chemical reactions occur as a result of collisions between molecules. Only collisions in which molecules possess sufficient energy and a proper geometrical orientation are effective in yielding a product. The course of a chemical reaction can be depicted through an energy diagram, called a reaction profile, in which the energies of reactants, products, and activated complex(es) are represented. Such a profile permits a visualization of the enthalpy change and activation energy of a reaction.

Raising the temperature of a reaction mixture increases the fraction of the molecules possessing energies in excess of the activation energy—the reaction speeds up. The increase in rate of reaction with temperature can be expressed through a mathematical equation. Another method of speeding up a reaction is to employ a catalyst. Some catalyst molecules participate as intermediates that are regenerated in a homogeneous reaction. In another type of catalysis—heterogeneous catalysis—the catalyst provides a surface on which the desired reaction proceeds at a faster rate. Biochemical reactions are promoted by catalysts called enzymes.

It is sometimes possible to propose a mechanism for a reaction. This is done by postulating a series of elementary processes, writing rate equations for them, and combining these elementary rate equations into a rate law for the net reaction.

Summarizing Example

Peroxyacetyl nitrate (PAN) is an air pollutant produced in photochemical smog by the reaction of hydrocarbons, oxides of nitrogen, and sunlight (recall Section 14-10). It produces eye irritations and causes plant damage. PAN is unstable and dissociates into peroxyacetyl radicals and $NO_2(g)$. Its presence in polluted air, then, is a kind of reservoir for NO_2 storage.

Smog damage to grape leaves. Ozone, sulfur dioxide, and PAN are common air pollutants responsible for extensive plant injury. [Ted Spiegel/Black Star]

$$
\underset{\text{PAN}}{CH_3\overset{\overset{\displaystyle O}{\|}}{C}OONO_2} \longrightarrow \underset{\substack{\text{peroxyacetyl} \\ \text{radical}}}{CH_3\overset{\overset{\displaystyle O}{\|}}{C}OO} + NO_2 \tag{15.47}
$$

1. A sample of polluted air is analyzed for its PAN content, which is reported as molecules PAN per liter of air. From the following data at 25 °C, determine the order of the PAN decomposition reaction (15.47).

Time:	0.0	10.0	20.0	30.0	40.0	50.0	60.0	min
Molecules of PAN/L air:	5.0	4.0	3.2	2.6	2.1	1.7	1.3	$\times 10^{14}$

Solution. We have considered several methods of determining the order of a reaction (see, for example, Table 15-6), and you might try any of these. The simplest approach, though, is to note that the number of PAN molecules drops by just about one-half in the first 30 min (from 5.0×10^{14} to 2.6×10^{14}). In the next 30 min the number drops by exactly one-half (from 2.6×10^{14} to 1.3×10^{14}). Between 10.0 and 40.0 min and between 20.0 and 50.0 min the number of PAN molecules/L also drops by essentially one-half. The half-life has a *constant* value of 30. min. **The reaction must be *first order*.**

(This example is similar to Example 15-11.)

2. What is the value of the rate constant k for reaction (15.47) at 25 °C?

Solution. Since the reaction is first order, we can use the expression

$k = 0.693/t_{1/2} = 0.693/30.$ min $= 0.023$ min^{-1}

(This example is similar to Example 15-10c.)

3. What is the activation energy (E_a) for the decomposition of PAN, if the half-life at 0 °C is found to be 35 hours?

Solution. From the half-life at 0 °C we can get the rate constant

$$k = \frac{0.693}{t_{1/2}} = \frac{0.693}{(35 \times 60) \text{ min}} = 3.3 \times 10^{-4} \text{ min}^{-1}$$

Now we have values of k at two different temperatures (0 and 25 °C), and we can use the Arrhenius equation (15.29) to solve for E_a.

$$\ln \frac{k_2}{k_1} = \frac{E_a}{R} \left(\frac{1}{T_1} - \frac{1}{T_2} \right)$$

$$\ln \frac{0.023}{3.3 \times 10^{-4}} = 4.24 = \frac{E_a}{8.314} \left(\frac{1}{273} - \frac{1}{298} \right) = \frac{E_a}{8.314} (0.00366 - 0.00336)$$

$$E_a = \frac{8.314 \times 4.24}{0.00030} = 1.2 \times 10^5 \text{ J} \ (1.2 \times 10^2 \text{ kJ})$$

(This example is similar to Example 15-12.)

4. An air sample contains 5.0×10^{14} molecules of PAN per liter of air. *At what temperature* will the rate of decomposition of PAN molecules in this sample be 1.0×10^{12} PAN molecules per liter of air per min?

Solution. First, note that the term "5.0×10^{14} PAN molecules/L" represents the concentration of PAN, and that the term "1×10^{12} PAN molecules L^{-1} min^{-1}" is a *rate of reaction*. The rate law allows us to relate the rate of reaction and reactant concentration: rate of reaction = k[PAN]. Let us solve for the value of k consistent with the information given.

$$k = \frac{\text{reaction rate}}{[\text{PAN}]} = \frac{1.0 \times 10^{12} \text{ PAN molecules } L^{-1} \text{ min}^{-1}}{5.0 \times 10^{14} \text{ molecules } L^{-1}} = 0.0020 \text{ min}^{-1}$$

Now we need to use the Arrhenius equation again. This time we have a value of E_a (from part 3), and values of k at other temperatures (0 and 25 °C). We are seeking the unknown temperature (T) at which the rate constant is 0.0020 min^{-1}. Completion of this calculation is the subject of Exercise 45. [*Hint:* Can you see that the temperature must be between 0 and 25 °C?]

Key Terms

activated complex (15-8)	half-life (15-5)	rate of a chemical reaction (15-1)
activation energy (15-8)	initial rate of reaction (15-2)	reaction mechanism (15-10)
active sites (15-11)	instantaneous rate of reaction (15-2)	reaction profile (15-8)
Arrhenius equation (15-9)	integrated rate equation (15-3)	second-order reaction (15-6)
bimolecular process (15-10)	method of initial rates (15-3)	substrate (15-11)
catalysis (15-11)	order of reaction (15-3)	termolecular process (15-10)
collision frequency (15-8)	pseudo-first-order reaction (15-6)	transition state theory (15-8)
collision theory (15-8)	rate constant, k (15-3)	unimolecular process (15-10)
elementary process (15-10)	rate-determining step (15-10)	zero-order reaction (15-4)
first-order reaction (15-5)	rate law (rate equation) (15-3)	

Highlighted Expressions

The rate law for a chemical reaction (15.7)
Rate law for a zero-order reaction (15.10)
Integrated rate equation for a first-order reaction: natural log form (15.13)

Integrated rate equation for a first-order reaction: common log form (15.14)
Relationship between half-life and rate constant for a first-order reaction (15.16)

Integrated rate equation for a first-order reaction based on gas pressures (15.17)

Integrated rate equation for a second-order reaction of the type A → products (15.21)

Relationship between half-life and rate constant for a second-order reaction of the type A → products (15.22)

Relationship between the enthalpy change and the activation energies of forward and reverse reactions (15.27)

The Arrhenius equation: effect of temperature on the rate constant (15.29)

Review Problems

1. In the reaction A → products, the initial concentration of A is 0.1108 M, and 44 s later, 0.1076 M. What is the initial rate of this reaction, expressed in **(a)** mol L^{-1} s^{-1}; **(b)** mol L^{-1} min^{-1}?

2. In the reaction 2 A + B → C + 3 D, $\Delta[A]/\Delta t$ is found to be -4.6×10^{-4} mol L^{-1} s^{-1}. **(a)** What is the rate of the reaction? **(b)** What is the rate of formation of D?

3. From Figure 15-2 estimate the rate of decomposition of H_2O_2 at **(a)** $t = 1000$ s; **(b)** the point in the reaction where $[H_2O_2] = 1.25$ M.

4. The decomposition of acetaldehyde is found to be second order. Write the rate law for this decomposition:

$CH_3CHO \rightarrow CH_4 + CO$.

5. The initial rate of the reaction 2 A + 2 B → C + D is determined for several different initial conditions. The results obtained are tabulated below. **(a)** What is the order of the reaction with respect to A and to B? **(b)** What is the overall reaction order?

Expt.	[A], M	[B], M	Initial rate, mol L^{-1} s^{-1}
1	0.210	0.115	6.30×10^{-4}
2	0.210	0.230	1.25×10^{-3}
3	0.420	0.115	2.51×10^{-3}
4	0.420	0.230	5.13×10^{-3}

6. A *zero-order* reaction is 50% complete in 30.0 min. How long after the start of the reaction will it be 80% complete?

7. Substance A decomposes by a *first-order* reaction. Starting initially with [A] = 2.00 M, after 159 min [A] = 0.250 M. For this reaction what is **(a)** $t_{1/2}$; **(b)** k?

8. In the *first-order* reaction A → products, [A] = 0.620 M initially and 0.520 M after 15.0 min.
   (a) What is the value of the *rate constant, k*?
   (b) What is the *half-life* of this reaction?
   (c) At what time will [A] = 0.155 M?
   (d) What will be [A] after 3.0 h?

9. The rate constant for the reaction

$H_2(g) + I_2(g) \rightarrow 2 HI(g)$

has been determined at the following temperatures: 556 K, $k = 1.2 \times 10^{-4}$ L mol^{-1} s^{-1}; 666 K, $k = 3.8 \times 10^{-2}$ L mol^{-1} s^{-1}.
   (a) Calculate the activation energy for the reaction.
   (b) At what temperature will the rate constant have the value $k = 1.0 \times 10^{-3}$ L mol^{-1} s^{-1}?

10. For the reaction A + 2 B → C + D, the rate law is rate = $k[A][B]$.

(a) Show that the following mechanism is consistent both with the stoichiometry of the net reaction and the rate law.

   A + B → I   (slow)

   I + B → C + D   (fast)

(b) Show that the following mechanism is consistent with the stoichiometry but *not* the rate law of the net reaction.

   $2 B \overset{k_1}{\underset{k_2}{\rightleftharpoons}} B_2$ (fast)

   $A + B_2 \overset{k_3}{\longrightarrow} C + D$ (slow)

(c) Show that the following mechanism is consistent with *neither* the stoichiometry *nor* the rate law of the net reaction.

   $2 A \overset{k_1}{\underset{k_2}{\rightleftharpoons}} A_2$ (fast)

   $A_2 + 2 B \overset{k_3}{\longrightarrow} C + D$ (slow)

11. The reaction A + 2 B → C + 2 D is found to be first order in A and in B. A proposed mechanism for the reaction involves the following first step.

A + B → I + D   (slow)

   (a) Write a plausible second step in a two-step mechanism.
   (b) Is the second step slow or fast? Explain.

Three different sets of data of [A] vs. time are given in the table below for the reaction A → products. [*Hint:* There are several ways to arrive at answers for each of the following five questions.]

Problems 12–16

I		II		III	
Time, s	[A], M	Time, s	[A], M	Time, s	[A], M
0	1.00	0	1.00	0	1.00
25	0.78	25	0.75	25	0.80
50	0.61	50	0.50	50	0.67
75	0.47	75	0.25	75	0.57
100	0.37	100	0.00	100	0.50
150	0.22			150	0.40
200	0.14			200	0.33
250	0.08			250	0.29

12. Which of these sets of data corresponds to a **(a)** zero-order, **(b)** first-order, **(c)** second-order reaction?

13. What is the approximate half-life of the first-order reaction?

14. What is the approximate initial rate of the second-order reaction?

15. What is the approximate rate at $t = 75$ s for the **(a)** zero-order, **(b)** first-order, **(c)** second-order reaction?

16. What is the approximate concentration of A remaining after 110 s in the **(a)** zero-order, **(b)** first-order, **(c)** second-order reaction?

Exercises

Rates of reactions

17. In the reaction A → products the initial concentration of A is 0.1503 M. After 1.00 min, $[A] = 0.1455$ M, and after 2.00 min, $[A] = 0.1409$ M.

  (a) Calculate the rate of the reaction during the first minute and during the second minute.

  (b) Why are these two rates not equal?

18. If the reaction A + 2 B → 2 C has a rate of 1.76×10^{-5} mol L^{-1} s^{-1}, at the time when $[A] = 0.4000$ M,

  (a) What is the rate of formation of C?

  (b) What will be $[A]$ 1.00 min later?

  (c) Assuming that the rate remains at 1.76×10^{-5} mol L^{-1} s^{-1}, how long would it take for $[A]$ to change from 0.4000 M to 0.3900 M?

19. Refer to Example 15-3. Consider a rate of decomposition of $H_2O_2(aq)$ of 5.7×10^{-4} mol L^{-1} s^{-1}. What is the rate of production of $O_2(g)$ from 1.00 L of the $H_2O_2(aq)$ at this point, expressed as **(a)** mol O_2 s^{-1}; **(b)** mol O_2 min^{-1}; **(c)** cm^3 O_2 at STP per minute?

20. For the decomposition reaction A → B + C, what is the meaning of each of the following terms: **(a)** $[A]_0$; **(b)** $[A]_t$; **(c)** $\Delta[A]$; **(d)** Δt; **(e)** $-\Delta[A]/\Delta t$; **(f)** $\Delta[B]/\Delta t$; **(g)** $t_{1/2}$?

21. Explain the difference in meanings of the terms *rate of a reaction* and *rate constant of a reaction*. Is there any type of reaction for which the two are the same? Explain.

22. Refer to Experiment 2 of Table 15-3. Exactly 1 min after the reaction is started, what are **(a)** $[S_2O_8^{2-}]$ and **(b)** $[I^-]$ in the mixture?

23. In the hypothetical reaction A(g) → 2 B(g) + C(g), the *total* pressure increases while the *partial* pressure of A(g) decreases. If the initial pressure of A(g) in a vessel of constant volume is 1000. mmHg,

  (a) What is the total pressure when the reaction has gone to completion?

  (b) What is the total gas pressure when the partial pressure of A(g) has fallen to 800. mmHg?

24. For the reaction A → products $[A]_t$ as a function of time is plotted in the accompanying graph. Use data from this graph to determine

  (a) the *order* of the reaction;

  (b) the *rate constant k;*

  (c) the *rate of the reaction* at $t = 3.5$ min, using the results of parts (a) and (b);

  (d) the *rate of the reaction* at $t = 5.0$ min, from the slope of the tangent line;

  (e) the *initial rate* of the reaction.

Method of initial rates

25. What would you expect for the initial rate of reaction for a fourth experiment in Table 15-3 in which $[S_2O_8^{2-}] = 0.019$ M and $[I^-] = 0.015$ M?

26. The rate of the reaction

$$2\ HgCl_2 + C_2O_4^{2-} \rightarrow 2\ Cl^- + 2\ CO_2(g) + Hg_2Cl_2(s)$$

is followed by measuring the number of moles of Hg_2Cl_2 that precipitate per liter per minute.

Expt.	$[HgCl_2]$, M	$[C_2O_4^{2-}]$, M	Initial rate, mol L^{-1} min^{-1}
1	0.105	0.15	1.8×10^{-5}
2	0.105	0.30	7.1×10^{-5}
3	0.052	0.30	3.5×10^{-5}
4	0.052	0.15	8.9×10^{-6}

  (a) Determine the order of the reaction with respect to $HgCl_2$, with respect to $C_2O_4^{2-}$, and overall.

  (b) What is the value of the rate constant k?

  (c) What would be the initial rate of reaction if $[HgCl_2] = 0.020$ M and $[C_2O_4^{2-}] = 0.22$ M?

27. The following data are obtained for the initial rates of reaction in the reaction A + 2 B + C → 2 D + E.

Expt.	[A], M	[B], M	[C], M	Initial rate
1	1.40	1.40	1.00	R_1
2	0.70	1.40	1.00	$R_2 = \frac{1}{2}R_1$
3	0.70	0.70	1.00	$R_3 = \frac{1}{4}R_2$
4	1.40	1.40	0.50	$R_4 = 16R_3$
5	0.70	0.70	0.50	$R_5 = ?$

  (a) What is the order of the reaction with respect to A, to B, and to C?

  (b) What is the value of R_5 in terms of R_1?

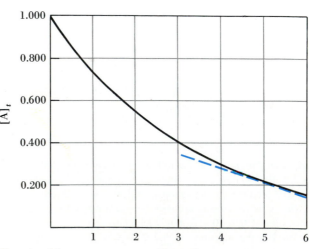

Exercise 24

Time, min

28. Listed below are *initial rates*, expressed as the rate of decrease of partial pressure of a reactant, at 826 °C.

$$2 NO(g) + 2 H_2(g) \rightarrow N_2(g) + 2 H_2O(g)$$

With initial $P_{H_2} = 400$ mmHg		With initial $P_{NO} = 400$ mmHg	
Initial P_{NO}, mmHg	Rate, mmHg/s	Initial P_{H_2}, mmHg	Rate, mmHg/s
359	0.750	289	0.800
300	0.515	205	0.550
152	0.125	147	0.395

(a) What is the order of the reaction with respect to NO, with respect to H_2, and overall?

(b) Write the rate equation for this reaction.

First-order reactions

29. Some of the following statements are true and some are not, regarding the *first-order* reaction $2 A \rightarrow B + C$. Indicate which are true and which are false. Explain your reasoning.

(a) The rate of the reaction decreases as more and more of B and C are formed.

(b) The time required for one half of substance A to react is directly proportional to the quantity of A.

(c) A plot of [A] vs. time yields a straight line.

(d) The rate of formation of C is one-half the rate of disappearance of A.

30. Acetoacetic acid (CH_3COCH_2COOH), a reagent used in organic synthesis, decomposes in acidic solution producing acetone and $CO_2(g)$.

$$CH_3COCH_2COOH \rightarrow CH_3COCH_3 + CO_2$$

This first-order decomposition has a half-life of 144 min. Starting with a concentration of 0.135 M, how long will it take for the acetoacetic acid to be 65% decomposed?

31. In the first-order reaction $A \rightarrow$ products, it is found that 99% of the original amount of reactant A decomposes in 137 min. What is the half-life, $t_{1/2}$, of this decomposition reaction?

32. The following first-order reaction is conducted in $CCl_4(l)$ at 45 °C.

$$N_2O_5 \rightarrow N_2O_4 + \tfrac{1}{2} O_2(g)$$

The rate constant is $k = 6.2 \times 10^{-4}$ s^{-1}. An 80.0-g sample of N_2O_5 is dissolved in $CCl_4(l)$ and allowed to decompose at 45 °C.

(a) How long will it take for the quantity of N_2O_5 to be reduced to 2.5 g?

(b) What volume of O_2 (at STP) is produced up to this point?

33. The decomposition of dimethyl ether at 504 °C is

$$(CH_3)_2O(g) \rightarrow CH_4(g) + H_2(g) + CO(g)$$

The following data are partial pressures of dimethyl ether (DME) as a function of time: $t = 0$ s, $P_{DME} = 312$ mmHg; 390 s, 264 mmHg; 777 s, 224 mmHg; 1195 s, 187 mmHg; 3155 s, 78.5 mmHg.

(a) Show that the reaction is first order.

(b) What is the value of the rate constant k?

(c) What is the total gas pressure at 390 s?

(d) What is the total gas pressure when the reaction has gone to completion?

(e) What is the total gas pressure at $t = 1000.$ s?

34. Benzenediazonium chloride decomposes by a first-order reaction in water, yielding $N_2(g)$ as a product.

$$C_6H_5N_2Cl \rightarrow C_6H_5Cl + N_2(g)$$

The reaction can be followed by measuring the volume of $N_2(g)$ evolved as a function of time. The following data were obtained for the decomposition of an 0.071 M solution at 50 °C. ($t = \infty$ corresponds to the completed reaction.) To convert from volume of $N_2(g)$ to [$C_6H_5N_2Cl$], note that after 3 min 10.8 cm^3 of a total 58.3 cm^3 $N_2(g)$ is evolved, corresponding to this fraction of the total reaction: $10.8/58.3 = 0.185$. The fraction of $C_6H_5N_2Cl$ remaining is $1.000 - 0.185 = 0.815$. [$C_6H_5N_2Cl$] at 3 min = 0.815×0.071 M = 0.058 M.

Time, min	$N_2(g)$, cm^3	Time, min	$N_2(g)$, cm^3
0	0	18	41.3
3	10.8	21	44.3
6	19.3	24	46.5
9	26.3	27	48.4
12	32.4	30	50.4
15	37.3	∞	58.3

(a) Determine [$C_6H_5N_2Cl$] remaining after 21 min.

(b) Construct a table similar to Table 15-2, with a time interval of $\Delta t = 3$ min. That is, determine [$C_6H_5N_2Cl$] at 3, 6, 9, . . . , min; Δ[$C_6H_5N_2Cl$] over every 3-min interval; and Δ[$C_6H_5N_2Cl$]/Δt for each 3-min interval.

(c) Plot a graph similar to Figure 15-2 showing both the formation of $N_2(g)$ and the disappearance of $C_6H_5N_2Cl$ as a function of time.

(d) What is the initial rate of the reaction?

(e) From the graph of part (c), estimate the rate of the reaction through the slope of the tangent to the curve at $t = 21$ min. Compare with the reported value of 1.1×10^{-3} mol $C_6H_5N_2Cl$ L^{-1} min^{-1}.

(f) Write the rate law for the first-order decomposition of $C_6H_5N_2Cl$ and estimate a value of k based on the rate determined in parts (d) and (e).

(g) Determine $t_{1/2}$ for this reaction by calculation (equation 15.16) and by estimation from the graph of the rate data.

(h) At what time should the decomposition of the sample be three-fourths completed?

(i) Plot ln [$C_6H_5N_2Cl$] vs. time (as in Figure 15-5) and show that the reaction is indeed first order.

(j) Determine k from the slope of the graph of part (i).

Second-order reactions

35. The following data were obtained for the dimerization of 1,3-butadiene at 600 K.

$$2 C_4H_6(g) \rightarrow C_8H_{12}(g)$$

$t = 0$ min, [C_4H_6] = 0.0169 M; 12.18 min, 0.0144 M; 24.55 min, 0.0124 M; 42.50 min, 0.0103 M; 68.05 min, 0.0085 M.

(a) Show that this reaction is second order.

(b) At what time would [C_4H_6] = 0.0050 M?

36. The decomposition of acetaldehyde is a second-order reaction with $k = 0.558$ L mol^{-1} s^{-1} at 800. K. Starting with a sample of acetaldehyde that is 0.100 M, how long would it take for 90.0% of the gas to decompose?

$$CH_3CHO(g) \rightarrow CH_4(g) + CO(g)$$

Establishing the order of a reaction

37. For the reaction A → products, the following data are obtained

Experiment 1	Experiment 2
[A] = 1.512 M $t = 0$ min	[A] = 3.024 M $t = 0$ min
[A] = 1.490 M $t = 1.0$ min	[A] = 2.935 M $t = 1.0$ min
[A] = 1.469 M $t = 2.0$ min	[A] = 2.852 M $t = 2.0$ min

(a) Determine the *initial rate* of reaction ($-\Delta[A]/\Delta t$) in each of the two experiments.

(b) Determine the order of the reaction with respect to A.

38. For the reaction A → 2 B + C, the following data are obtained for [A] as a function of time: $t = 0$ min, [A] = 0.80 M; 8 min, 0.60 M; 24 min, 0.35 M; 40 min, 0.20 M.

(a) By suitable means establish the order of the reaction.

(b) What is the value of the rate constant k?

(c) Calculate the rate of formation of B at $t = 30$ min.

39. In three different experiments the following results were obtained for the reaction A → products: $[A]_0 = 1.00$ M, $t_{1/2} = 50$ min; $[A]_0 = 2.00$ M, $t_{1/2} = 25$ min; $[A]_0 = 0.50$ M, $t_{1/2} = 100$ min. Write the rate equation for this reaction and indicate the value of k.

★**40.** Ammonia decomposes on the surface of a hot tungsten wire. Following are the half-lives that were obtained at 1100 °C for different initial concentrations of NH$_3$: $[NH_3]_0 = 0.0031$ M, $t_{1/2} = 7.6$ min; 0.0015 M, 3.7 min; 0.00068 M, 1.7 min. For this decomposition reaction what is (a) the order of the reaction; (b) the rate constant k?

Collision theory; activation energy

41. Explain why

(a) A reaction rate cannot be calculated from collision frequency alone.

(b) The rate of a chemical reaction may increase so dramatically with temperature while the collision frequency increases much more slowly.

(c) The addition of a catalyst to a reaction mixture can have such a pronounced effect on the rate of a reaction, even if the temperature is held constant.

42. For the reversible reaction A + B ⇌ C + D, the enthalpy change of the forward reaction is +21 kJ/mol. The activation energy of the forward reaction is 84 kJ/mol.

(a) What is the activation energy of the reverse reaction?

(b) In the manner of Figure 15-11 sketch the reaction profile of this reaction.

43. By an appropriate sketch indicate why there is some relationship between the enthalpy change and the activation energy for an *endothermic* reaction but not for an exothermic reaction.

Effect of temperature on reaction rate

44. Experiment 3 of Table 15-3 is repeated at several temperatures and the following rate constants are established: 3 °C,

$k = 1.4 \times 10^{-3}$ L mol^{-1} s^{-1}; 13 °C, $k = 2.9 \times 10^{-3}$; 24 °C, 6.2×10^{-3}; 33 °C, 1.20×10^{-2}.

(a) Construct a graph of ln k vs. $1/T$.

(b) What is the activation energy E_a of the reaction?

(c) Calculate a value of the rate constant k at 40 °C.

★(d) What would be the *initial rate* of reaction for Expt. 3 at 50 °C?

45. Refer to the Summarizing Example. Complete the determination referred to in part **4**.

46. The reaction 2 NO$_2$(g) → 2 NO(g) + O$_2$(g) has a rate constant $k = 1.0 \times 10^{-10}$ at 300 K and an activation energy of 111 kJ/mol. At what temperature will this reaction have a rate constant $k = 1.0 \times 10^{-5}$? (Assume there is no change in the reaction mechanism.)

47. At room temperature (20 °C) milk turns sour in about 64 hours. In a refrigerator at 3 °C milk can be stored *three times as long* before it sours. [*Hint:* For the souring reaction the rate constant k is *inversely* proportional to the souring time; i.e., the shorter the souring time, the greater the value of k. At each temperature assume that $k = c$/souring time (where c is a constant).]

(a) Estimate the activation energy of the reaction that causes the souring of milk.

(b) How long should it take milk to sour at 40 °C?

48. A commonly stated rule of thumb is that reaction rates double for a temperature increase of about 10 °C. (This rule is very often wrong.)

(a) What must be the approximate activation energy for this statement to be true for reactions at about room temperature?

(b) Would you expect this rule of thumb to apply to the formation of HI from its elements at room temperature? Explain. [*Hint:* Refer to Figure 15-11.]

49. Concerning the rule of thumb stated in Exercise 48, estimate how much faster cooking will occur in a pressure cooker in which the vapor pressure of water is 2.00 atm than in water under normal boiling conditions. [*Hint:* Refer to Table 12-2.]

Catalysis

50. The following statements are sometimes encountered with reference to catalysis, but they are not stated as carefully as they might be. What slight modifications should be made in each of them?

(a) A catalyst is a substance that speeds up a chemical reaction but does not take part in the reaction.

(b) The function of a catalyst is to lower the activation energy for a chemical reaction.

51. What is the principal difference between the catalytic activity of platinum metal and of an enzyme?

52. In a particular enzyme reaction the following data are obtained on the rate of disappearance of substrate S: $t = 0$ min, [S] = 1.00 M; 20 min. 0.90 M; 60 min, 0.70 M; 100 min, 0.50 M; 160 min, 0.20 M. What is the order of this reaction with respect to S in the concentration range studied?

Reaction mechanisms

53. We have used the terms "order of a reaction" and "molecularity of an elementary process" (i.e., unimolecular, bimolecular, termolecular). What is the relationship, if any, between these two terms?

54. The rate-determining step in a reaction mechanism is sometimes referred to as a "bottleneck." Comment on the appropriateness of this analogy.

55. The collision theory states that chemical reactions occur as a result of collisions between molecules. A unimolecular elementary process in a reaction mechanism involves dissociation of a *single* molecule. How can these two ideas be compatible? Explain.

56. For this reversible elementary process, obtain an expression for the concentration of N_2O_2 at the time when forward and reverse reactions proceed at the same rate.

$$2 \text{ NO} \underset{k_2}{\overset{k_1}{\rightleftharpoons}} \text{N}_2\text{O}_2$$

57. For the reaction $2 \text{ NO(g)} + \text{O}_2\text{(g)} \rightarrow 2 \text{ NO}_2\text{(g)}$, the rate law, expressed as the rate of formation of NO_2, is found to be rate = $k[\text{NO}]^2[\text{O}_2]$.
(a) Is the rate law consistent with this one-step mechanism:

$$2 \text{ NO} + \text{O}_2 \rightarrow 2 \text{ NO}_2?$$

(b) Why is this one-step mechanism not very plausible?
58. Show that the rate law in Exercise 57 is consistent with the following mechanism.

$$2 \text{ NO} \underset{k_2}{\overset{k_1}{\rightleftharpoons}} \text{N}_2\text{O}_2 \quad \text{(fast)}$$

$$\text{N}_2\text{O}_2 + \text{O}_2 \xrightarrow{k_3} 2 \text{ NO}_2 \quad \text{(slow)}$$

[*Hint*: Use the result of Exercise 56.]

★**59.** Propose a two-step mechanism for the reaction $2 \text{ O}_3\text{(g)} \rightarrow 3 \text{ O}_2\text{(g)}$ that is consistent with the observed rate law.

$$\text{rate} = k \frac{[\text{O}_3]^2}{[\text{O}_2]}$$

Additional Exercises

60. In the reaction $A \rightarrow$ products, [A] is found to be 0.550 M at $t = 60.2$ s and 0.540 M at $t = 80.3$ s. What is the rate of the reaction during this time interval?

61. In the reaction $A \rightarrow$ products, 4.50 min after the reaction is started [A] = 0.800 M. The rate of reaction at this point is found to be rate = $-\Delta[\text{A}]/\Delta t = 1.0 \times 10^{-2}$ mol L^{-1} min^{-1}. Assuming that this rate remains constant for a short period of time,
(a) What is [A] 5.00 min after the reaction is started?
(b) At what time after the reaction is started will [A] = 0.775 M?

62. The initial rate of decomposition of H_2O_2 (reaction 15.3) in a particular experiment is 1.7×10^{-3} mol L^{-1} s^{-1}. Assume that this rate holds for about 2 min. Starting with 175 mL of 1.55 M H_2O_2(aq) at $t = 0$,
(a) What is [H_2O_2] after 10 s?
(b) At what time after the start of the experiment would you expect [H_2O_2] = 1.50 M?
★**(c)** What volume of O_2(g), measured at STP, is released from solution in the first minute of the reaction? [*Hint*: Determine the decrease in [H_2O_2]. Then determine the actual decrease in amount of H_2O_2 (in moles), to which the amount of O_2 can be related.]

63. For the reaction $A \rightarrow$ products, what are the units of the rate constant k if the reaction is **(a)** zero order, **(b)** first order, **(c)** second order in A?

64. The first-order reaction $A \rightarrow$ products has $t_{1/2} = 120$ s.
(a) What percent of a sample of A remains *unreacted* 600 s after a reaction has been started?
(b) What is the reaction rate when [A] = 0.50 M? [*Hint*: What is the value of k?]

65. What is the initial rate of disappearance of I$^-$ in reaction (15.8) if the initial concentrations are [$S_2O_8^{2-}$] = 0.15 M and [I$^-$] = 0.010 M? Use data from Examples 15-6 and 15-7.
66. For the reaction $2 \text{ A} + \text{B} \rightarrow \text{C} + \text{D}$ the following initial rates of reaction were found. What is the rate law for this reaction?

Expt.	[A], M	[B], M	Initial rate, mol L^{-1} min.$^{-1}$
1	0.50	1.50	4.2×10^{-3}
2	1.50	1.50	1.3×10^{-2}
3	3.00	3.00	5.2×10^{-2}

★**67.** Hydroxide ion is involved in the mechanism of the following reaction but is not consumed in the net reaction

$$\text{OCl}^- + \text{I}^- \xrightarrow{\text{OH}^-} \text{OI}^- + \text{Cl}^-$$

(a) From the data given, determine the order of the reaction with respect to OCl$^-$, I$^-$, and OH$^-$.
(b) What is the overall reaction order?
(c) Write the rate equation and determine the value of the rate constant k.

[OCl$^-$], M	[I$^-$], M	[OH$^-$], M	Rate of formation of OI$^-$, mol L^{-1} s^{-1}
0.0040	0.0020	1.00	4.8×10^{-4}
0.0020	0.0040	1.00	5.0×10^{-4}
0.0020	0.0020	1.00	2.4×10^{-4}
0.0020	0.0020	0.50	4.6×10^{-4}
0.0020	0.0020	0.25	9.4×10^{-4}

68. The following rate data are obtained for the reaction $A \rightarrow$ products: $t = 0$ min, [A] = 1.000 M; 1.00 min, 0.909 M; 2.00 min, 0.833 M; 3.00 min, 0.769 M; 4.00 min, 0.714 M; 5.00 min, 0.667 M; 6.00 min, 0.625 M; 7.00 min, 0.588 M; 8.00 min, 0.555 M; 9.00 min, 0.526 M; 10.00 min, 0.500 M.
(a) The rate of reaction at $t = 5.00$ min can be approximated by the expression

$$\text{rate} = \frac{-\Delta[\text{A}]}{\Delta t} = \frac{-(0.500 - 1.000) \text{ M}}{(10.00 - 0) \text{ min}}$$

$$= 0.0500 \text{ mol L}^{-1} \text{ min}^{-1}$$

Show that this rate is only approximately correct by recalculating its value using different time intervals and corresponding concentration changes. That is, use several time intervals having $t = 5.00$ min as their midpoint, such as $\Delta t = (9.00 - 1.00)$, $(8.00 - 2.00)$, $(7.00 - 3.00)$, and $(6.00 - 4.00)$ min. Find a value of the rate of reaction at $t = 5.00$ min that is more exact than 0.0500 mol L^{-1} min^{-1}.

(b) Through an appropriate graph, establish the order of this reaction and evaluate the rate constant k.

(c) Use the rate law established in part (b) to calculate the *exact* rate of the reaction at $t = 5.00$ min. Compare this result with that obtained in part (a).

69. The following data are obtained in two separate experiments in the reaction A → products.

Experiment 1		Experiment 2	
[A], M	Time, s	[A], M	Time, s
0.800	0	0.400	0
0.775	40	0.390	64
0.750	83	0.380	132
0.725	129	0.370	203
0.700	179	0.360	278
0.650	288	0.350	357
0.600	417	0.300	833
0.500	750	0.250	1500
0.400	1250	0.200	2500
0.300	2083	0.150	4167
0.200	3750	0.100	7500
0.100	8750		

(a) Calculate the initial rate of reaction in Expt. 2.

(b) Establish the order of this reaction.

70. The reaction A → products is *first order* in A. Initially, [A] = 0.800 M; and after 54 min, [A] = 0.100 M. At what time will [A] = 0.025 M?

71. The following data were obtained for the decomposition of N_2O_5 in CCl_4(l) at 45 °C: $t = 0$ s, $[N_2O_5] = 1.46$ M; 423 s, 1.09 M; 753 s, 0.89 M; 1116 s, 0.72 M; 1582 s, 0.54 M; 1986 s, 0.43 M; 2343 s, 0.35 M. Determine a value of k for the reaction $N_2O_5 \rightarrow N_2O_4 + \frac{1}{2} O_2$(g).

72. The half-life of the radioactive isotope P-32 is 14.3 days. How long would it take for a sample of P-32 to lose 99% of its radioactivity?

73. The half-life for the first-order decomposition of nitramide NH_2NO_2(aq) → N_2O(g) + H_2O(l) is 123 min at 15 °C. If 165 mL of a 0.105 M NH_2NO_2 solution is allowed to decompose, how long must the reaction proceed to produce 50.0 cm^3 of "wet" N_2O(g) measured at 15 °C and a barometric pressure of 756 mmHg? (The vapor pressure of water at 15 °C is 12.8 mmHg.)

74. In the reaction A → products the time required for [A] to decrease to one-half of its initial value *doubles* if the initial concentration is *doubled*. What is the order of this reaction?

75. Example 15-2 illustrates how $[H_2O_2]$ in reaction (15.3) can be followed by titration with MnO_4^-(aq). (a) As the reaction proceeds, does the volume of MnO_4^-(aq) required for each successive titration increase or decrease? (b) Determine the volume of 0.1000 M MnO_4^-(aq) that must have been required for each

of the remaining titrations in Table 15-1, i.e., at 600 s, 1200 s, etc.

76. Use the volumes of MnO_4^-(aq) calculated in Exercise 75 and show that the number of mL MnO_4^-(aq) required for each titration plotted as a function of time yields the same type of curve as Figure 15-2. Show that from the tangent to the curve at $t = 1500$ s you can obtain the same value for the rate of reaction as that determined in Example 15-3.

77. Refer to Example 15-10. Concerning the decomposition of di-t-butyl peroxide (DTBP),

(a) Determine the time at which the *partial* pressure of DTBP = 700. mmHg.

★(b) Determine the times at which the *total* gas pressure is 2000. mmHg and 2100. mmHg.

78. For the reaction A → products, the following data were obtained. $t = 0$ s, [A] = 0.715 M; 22 s, 0.605 M; 74 s, 0.345 M; 132 s, 0.055 M. (a) What is the order of this reaction? (b) What is the half-life of the reaction?

79. If even a tiny spark is introduced into a mixture of H_2(g) and O_2(g), a highly exothermic explosive reaction occurs. Without the spark the mixture remains unreacted indefinitely.

(a) Explain this difference in behavior.

(b) Why is the nature of the reaction independent of the size of the spark?

80. The first-order reaction A → products has the following half-lives at different temperatures: $T = 25$ °C, $t_{1/2} = 46.2$ min; 77 °C, 5.8 min; 102 °C, 2.6 min.

(a) Calculate the activation energy of this reaction.

(b) At what temperature would the half-life be 10.0 min?

81. The rate-determining step in the hydrogen–iodine reaction is thought to be $2 I$(g) + H_2(g) → $2 HI$(g). The rate constant for this step has been determined at 520 and 710 K, yielding values of 3.96×10^{-5} and 1.61×10^{-4} L^2 mol^{-2} s^{-1}, respectively. What is the activation energy, E_a, for this elementary process?

★**82.** With reference to the data in Table 15-7 and Figure 15-13, to what temperature must a solution with $[N_2O_5] = 0.15$ M be heated to have an *initial rate* of decomposition equal to that of a 1.25 M solution at 0 °C?

83. A statement is made in the text that in enzyme-catalyzed reactions the reaction is first order at low substrate concentrations and becomes zero order at high concentrations. Certain gas-phase reactions on a heterogeneous catalyst are found to be first order at low gas pressures and zero order at high pressures. What is the connection between these two situations?

84. The decomposition of acetaldehyde, CH_3CHO, can be catalyzed by I_2(g).

Uncatalyzed reaction: $E_a = 190$ kJ/mol

CH_3CHO(g) → CH_4(g) + CO(g)

Catalyzed reaction: $E_a = 136$ kJ/mol

CH_3CHO(g) + I_2(g) → CH_3I(g) + HI(g) + CO(g)

CH_3I(g) + HI(g) → CH_4(g) + I_2(g)

The following enthalpies of formation are also given: $\Delta H_f^\circ[CH_3CHO$(g)$] = -166$ kJ/mol; $\Delta H_f^\circ[CH_4$(g)$] = -74.9$ kJ/mol; $\Delta H_f^\circ[CO$(g)$] = -110.5$ kJ/mol. Sketch the reaction profiles for the catalyzed and uncatalyzed reactions, representing ΔH and E_a for each.

85. From the result of Example 15-4a, determine the initial rate of evolution of $O_2(g)$, in cm^3 (STP)/s, from 0.100 L of 2.32 M $H_2O_2(aq)$.

*86. The decomposition of ethylene oxide at 690 K is followed by measuring the *total* gas pressure as a function of time. The data obtained are $t = 10$ min, $P_{tot.} = 139.14$ mmHg; 20 min, 151.67 mmHg; 40 min, 172.65 mmHg; 60 min, 189.15 mmHg; 100 min, 212.34 mmHg; 200 min, 238.66 mmHg; ∞, 249.88 mmHg. For the reaction

$$(CH_2)_2O(g) \rightarrow CH_4(g) + CO(g)$$

(a) What must be the initial total pressure (i.e., the pressure of pure ethylene oxide at $t = 0$)?

(b) What is the order of the reaction?

*87. The following data were obtained for the reaction $2 A + B \rightarrow$ products. Use appropriate methods introduced in this chapter to establish the order of the this reaction with respect to A and to B.

Experiment 1, [B] = 1.00 M		Experiment 2, [B] = 0.50 M	
Time, min	[A], M	Time, min	[A], M
0	1.000×10^{-3}	0	1.000×10^{-3}
1	0.951×10^{-3}	1	0.975×10^{-3}
5	0.779×10^{-3}	5	0.883×10^{-3}
10	0.607×10^{-3}	10	0.779×10^{-3}
20	0.368×10^{-3}	20	0.607×10^{-3}

*88. These data were obtained for the decomposition of acetaldehyde at 518 °C, $CH_3CHO(g) \rightarrow CH_4(g) + CO(g)$: $t = 0$ s, $P_{tot.} = 363$ mmHg; 42 s, 397 mmHg; 73 s, 417 mmHg; 105 s,

437 mmHg; 190 s, 477 mmHg; 310 s, 517 mmHg; 480 s, 557 mmHg; 665 s, 587 mmHg.

(a) By an appropriate graphical method, establish the order of the reaction.

(b) Obtain a value of the rate constant k.

*89. The peroxodisulfate–iodide reaction (15.8) can be followed by the series of reactions

$$S_2O_8^{2-}(aq) + 3 I^-(aq) \rightarrow 2 SO_4^{2-}(aq) + I_3^-(aq)$$

$$2 S_2O_3^{2-}(aq) + I_3^-(aq) \rightarrow S_4O_6^{2-}(aq) + 3 I^-(aq)$$

$$I_3^-(aq) + starch(aq) \rightarrow blue complex$$

Solutions of $S_2O_8^{2-}$ and I^- are mixed in the presence of a fixed amount of $S_2O_3^{2-}$ and starch indicator. The I_3^- produced in the main reaction reacts very rapidly with $S_2O_3^{2-}$ until all the $S_2O_3^{2-}$ is consumed. Then the I_3^- combines with the starch to produce a deep blue color. The rate of reaction can be related to the appearance of this blue color. In an experiment, 25.0 mL of 0.20 M $(NH_4)_2S_2O_8$, 25.0 mL of 0.20 M KI, 10.0 mL of 0.010 M $Na_2S_2O_3$, and 5.0 mL of starch solution are mixed. How long after mixing will the blue color appear? [*Hint:* Use the rate constant from Example 15-6.]

*90. Show that the following mechanism is consistent with the rate law established for the iodide–hypochlorite reaction in Exercise 67.

$$OCl^- + H_2O \underset{k_2}{\overset{k_1}{\rightleftharpoons}} HOCl + OH^- \quad (fast)$$

$$I^- + HOCl \overset{k_3}{\rightarrow} HOI + Cl^- \quad (slow)$$

$$HOI + OH^- \underset{k_5}{\overset{k_4}{\rightleftharpoons}} H_2O + OI^- \quad (fast)$$

Self-Test Questions

For questions 91 through 100 select the single item that best completes each statement.

91. For the reaction $A \rightarrow$ products, a plot of [A] vs. time is found to be a straight line. The order of this reaction is (a) zero; (b) first; (c) second; (d) impossible to determine from this graph.

92. The reaction $2 A + B \rightarrow C + 2 D$ is *first* order in A and *first* order in B. For this reaction

(a) rate of reaction = $k[A]^2[B]$;

(b) rate of reaction = $k[A]^2$;

(c) rate of disappearance of A = rate of disappearance of B;

(d) rate of formation of C = −(rate of disappearance of B).

93. A *first-order* reaction $A \rightarrow$ products has a half-life of 100. s. Whatever the quantity of A involved in a particular reaction,

(a) the reaction goes to completion in 200. s;

(b) the quantity of A remaining after 200. s is half of what remains after 100. s;

(c) the same quantity of A is consumed for every 100. s of the reaction;

(d) 100. s elapses before the reaction begins.

94. The rate equation for the reaction $2 A + B \rightarrow C$ is found to be rate = $k[A][B]$.

(a) The unit of k must be s^{-1}.

(b) $t_{1/2}$ is a constant.

(c) The value of k is *independent* of the initial concentrations of A and B.

(d) The rate of formation of C is twice the rate of disappearance of A.

95. A *first-order* reaction $A \rightarrow$ products has a half-life of 13.9 min. The rate at which this reaction proceeds when [A] = 0.40 M is (a) 0.020 mol L^{-1} min^{-1}; (b) 5.0×10^{-2} mol L^{-1} min^{-1}; (c) 8.0 mol L^{-1} min^{-1}; (d) 0.125 mol L^{-1} min^{-1}.

96. The decomposition of substance A is *second order*: $A \rightarrow$ products. The initial rate of decomposition when $[A]_0 = 0.50$ M is

(a) the same as the initial rate for any other value of $[A]_0$;

(b) half as great as when $[A]_0 = 1.00$ M;

(c) five times as great as when $[A]_0 = 0.10$ M;

(d) four times as great as when $[A]_0 = 0.25$ M.

97. The rate of a chemical reaction generally increases rapidly, even for small temperature increases, because of a rapid increase with temperature in the (a) collision frequency; (b) fraction of molecules with energies in excess of the activation energy; (c) activation energy; (d) average kinetic energy of molecules.

98. The reaction $A + B \rightarrow C + D$ has $\Delta H = +25$ kJ and an activation energy, $E_a = $ (a) -25 kJ; (b) less than $+25$ kJ; (c) more than $+25$ kJ; (d) either less than $+25$ kJ or more than $+25$ kJ, which can only be determined by experiment.

99. A catalyst speeds up a chemical reaction by increasing the (a) average kinetic energy of molecules; (b) frequency of molecular collisions; (c) activation energy of the reaction; (d) proportion of molecules with energies in excess of the activation energy.

100. For the net reaction $A + B \rightarrow 2\,C$, which proceeds by a *single-step bimolecular mechanism,* the following equation is applicable.

(a) $t_{1/2} = 0.693/k$;
(b) rate of reaction = $k[A][B]$;
(c) rate of appearance of C = rate of disappearance of A;
(d) $\ln [A] = -kt + \ln [A]_0$.

101. Briefly describe the meaning of each of the following terms.

(a) initial rate of a reaction;
(b) instantaneous rate of reaction;
(c) zero-order reaction;
(d) half-life of a reaction.

102. A kinetic study of the reaction $A \rightarrow$ products yields the data $t = 0$ s, $[A] = 2.00$ M; 500 s, 1.00 M; 1500 s, 0.50 M; 3500 s, 0.25 M. In the simplest way possible determine whether this reaction is zero-, first-, or second-order, and indicate your method of reasoning.

103. For the reaction $A \rightarrow$ products the following data are obtained.

Experiment 1	Experiment 2
$[A] = 1.204$ M $t = 0$ min	$[A] = 2.408$ M $t = 0$ min
$[A] = 1.180$ M $t = 1.0$ min	$[A] = $? $t = 1.0$ min
$[A] = 0.602$ M $t = 35$ min	$[A] = $? $t = 30$ min

(a) Determine the *initial* rate of reaction in the *first* experiment (i.e., $-\Delta[A]/\Delta t$).
(b) If the reaction is *second* order in A, what will be $[A]$ at $t = 1.0$ min in the *second* experiment?
(c) If the reaction is *first* order in A, what will be $[A]$ at $t = 30$ min in the *second* experiment?

104. In the *first-order* decomposition of substance A the following concentrations are found to exist at the indicated times following the start of the reaction: $t = 0$ s, $[A] = 0.88$ M; 50 s, 0.62 M; 100 s, 0.44 M; 150 s, 0.31 M. Calculate the *instantaneous* rate of reaction at 100 s.

105. The reaction $A \rightarrow$ products is *first-order* in A.

(a) If 1.60 g A is allowed to decompose for 20. min, the mass of A remaining undecomposed is found to be 0.40 g. What is the half-life, $t_{1/2}$, of this reaction?
(b) Starting with 1.60 g A, what is the mass of A remaining undecomposed after 33.2 min?

106. For the net reaction $CO + NO_2 \rightarrow CO_2 + NO$ the rate law is rate = $k[NO_2]^2$. In a proposed *two-step* mechanism, the second, fast step is $NO_3 + CO \rightarrow NO_2 + CO_2$. What must be the *first* step of this mechanism?

16 Principles of Chemical Equilibrium

During electrical storms $N_2(g)$ and $O_2(g)$ combine to produce small quantities of $NO(g)$ in a reversible chemical reaction. The condition of equilibrium in such reactions is the subject of this chapter. [Kennan Ward]

We have noted that $H_2(g)$ and $O_2(g)$ do not react to form water at room temperature and pressures, but they do react, and with explosive violence, at high temperatures (see introduction to Chapter 15). $N_2(g)$ and $O_2(g)$ are also unreactive at room temperature and normal pressures, and they too react at high temperatures, to form nitrogen monoxide (NO). But there is an important difference between the two reactions: Once formed, H_2O shows very little tendency to dissociate back into its elements; the reaction of $H_2(g)$ and $O_2(g)$ *goes to completion*. NO, on the other hand, is strongly dissociated back into its elements, even at high temperatures. The formation of NO(g) from its elements is a *reversible* reaction. Instead of going to completion the reaction attains a state of *equilibrium*, in which only a relatively small number of NO molecules is present (about 1 to 2% in a sample of air at 2000 K). In this chapter we introduce a concept to account for these differences. We say that the *equilibrium constant K* for the synthesis of H_2O from its elements is *extremely large*, and that for the synthesis of NO the equilibrium constant is *very small*.

These facts have an enormous practical significance. The reaction of $N_2(g)$ and $O_2(g)$ to form NO occurs in all high-temperature processes in air, such as in electrical storms, or in the combustion of coal in power plants and gasoline in automobiles. If the reaction went to completion instead of reaching a condition of equilibrium, much larger quantities of oxides of nitrogen would enter the atmosphere through electrical storms and the natural nitrogen cycle would be seriously affected. In addition, the quantities of oxides of nitrogen emitted during combustion would be much greater than we now observe. As a consequence, the environmental problems of smog and acid rain would be much more severe.

16-1 The Condition of Dynamic Equilibrium

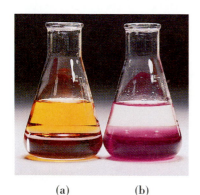

(a) **(b)**

FIGURE 16-1
Dynamic equilibrium in a physical process.

(a) A saturated solution of I_2 in water (top layer) is brought in contact with liquid carbon tetrachloride (bottom layer).
(b) I_2 molecules distribute themselves between the aqueous solution and the $CCl_4(l)$, in which they are much more soluble. When equilibrium is reached, the concentration of I_2 in the CCl_4 (violet layer) is about 85 times greater than in the water. [Carey B. Van Loon]

Our emphasis in this chapter is on the condition of equilibrium in reversible chemical reactions. But before we proceed to chemical reactions let us consider a couple of simpler, *physical* phenomena involving **dynamic equilibrium**—a condition in which two opposing processes occur at equal rates and in which there is no net change with time.

1. When a liquid vaporizes into a closed container, there comes a time when molecules return to the liquid state at the same rate at which they leave it. Vapor *condenses* at the same rate at which liquid *vaporizes*. Even though molecules continue to pass back and forth between liquid and vapor (a *dynamic* process), a condition is reached in which the pressure exerted by the vapor remains constant with time. *The vapor pressure of a liquid is a property associated with an equilibrium condition.*

2. When a solute dissolves in a solvent, a point is reached where the rate at which additional solute particles dissolve is just matched by the rate at which dissolved solute precipitates. Even though solute particles continue to pass back and forth between the solution and the undissolved solute, the concentration of dissolved solute remains constant with time. *The solubility of a solute is a property associated with an equilibrium condition.*

3. When an aqueous solution of I_2 is shaken with carbon tetrachloride, $CCl_4(l)$, I_2 molecules pass back and forth between the water and the $CCl_4(l)$ (see Figure 16-1). Solute molecules *distribute* themselves between the two immiscible liquids. When the solute molecules pass between the two liquids at equal rates, the concentration of I_2 in $CCl_4(l)$ is about 85 times as great as in water, and both concentrations remain constant with time. This ratio of concentrations of the solute in the two immiscible solvents is called the *distribution coefficient. The distribution coefficient of a solute between two immiscible solvents is a property associated with an equilibrium condition.*

┌───┐
Are You Wondering:
└───┘

How we know that an equilibrium is dynamic, *that is, that forward and reverse reactions continue even after equilibrium is reached?*

Consider the formation of a saturated aqueous solution of silver iodide.

$$AgI(s) \rightleftharpoons AgI(satd\ aq)$$

Before equilibrium is reached we can easily show that silver and iodide ions are passing from the solid into the solution phase. $[Ag^+]$ and $[I^-]$ continue to increase in the solution. But once equilibrium is achieved, there is no further change in $[Ag^+]$ or $[I^-]$ with time.

Now suppose we add to an *equilibrium* mixture of $AgI(s)$ and its saturated solution some $AgI(s)$ containing a trace of radioactive I-131. If the equilibrium condition were *static,* that is, if both the forward and reverse processes stopped when equilibrium was reached, the radioactivity would be confined to the solid solute. What we find to be the case, though, is that radioactivity shows up in the saturated solution as well as in the solid solute. The only way that the radioactive I-131 atoms can get into the saturated solution is if dissolving of the solid solute continues even after equilibrium is reached (and precipitation from the saturated solution continues as well). The equilibrium condition is *dynamic*.

The properties described in these three situations—vapor pressure, solubility, distribution coefficient—are all examples of a general quantity known as an *equilibrium constant*.

16-2   The Equilibrium Constant Expression

On two previous occasions (equations 4.16 and 7.33) we have mentioned the synthesis of methanol (methyl alcohol) from a carbon monoxide–hydrogen mixture called synthesis gas. This is an important industrial reaction today and could become even more important in the future. Methanol is a good motor fuel with a high octane rating, and its combustion produces much less air pollution than does gasoline. This synthesis is a *reversible* reaction, which means that at the same time $CH_3OH(g)$ is being formed

$$CO(g) + 2\ H_2(g) \longrightarrow CH_3OH(g) \tag{16.1}$$

it decomposes in the reverse reaction

$$CH_3OH(g) \longrightarrow CO(g) + 2\ H_2(g) \tag{16.2}$$

When we mix $CO(g)$ and $H_2(g)$, initially the forward reaction (16.1) occurs rapidly and the reverse reaction (16.2) is negligible. As CH_3OH starts to form, the reverse reaction begins to occur. As the amount of CH_3OH increases, the forward reaction slows down (because of the decreasing concentrations of CO and H_2) and the reverse reaction speeds up. Eventually the forward and reverse reactions proceed at equal rates and the reaction mixture reaches a condition of dynamic equilibrium, which we can represent as

$$CO(g) + 2\ H_2(g) \rightleftharpoons CH_3OH(g) \tag{16.3}$$

TABLE 16-1

Three Approaches to Equilibrium in the Reaction
$CO(g) + 2 H_2(g) \rightleftharpoons CH_3OH(g)$
(Reaction carried out in a 10.0-L vessel at 500 K)

	CO(g)	H₂(g)	CH₃OH(g)
Experiment 1			
initial amounts, mol	1.000	1.000	0.000
equil. amounts, mol	0.910	0.820	0.090
equil. concns, mol/L	0.0910	0.0820	0.0090
Experiment 2			
initial amounts, mol	0.000	0.000	1.000
equil. amounts, mol	0.753	1.506	0.247
equil. concns, mol/L	0.0753	0.151	0.0247
Experiment 3			
initial amounts, mol	1.000	1.000	1.000
equil. amounts, mol	1.380	1.760	0.620
equil. concns, mol/L	0.138	0.176	0.0620

One consequence of the equilibrium condition is that the amounts of the reactants and products remain constant with time. These equilibrium amounts, however, depend on the particular quantities of reactants and/or products that were present initially. To illustrate this point, we have listed the data for three different experiments in Table 16-1.

Expt. 1:   1.000 mol each of CO(g) and H₂(g) are allowed to react in a 10.0-L vessel at 500 K. There is no CH₃OH(g) initially.

Expt. 2:   1.000 mol of CH₃OH(g) is allowed to react in a 10.0-L vessel at 500 K. There is no CO(g) nor H₂(g) initially.

Expt. 3:   1.000 mol each of CO(g) and H₂(g) *and* of CH₃OH(g) are present initially in a 10.0-L vessel at 500 K.

The data from Table 16-1 are also plotted in Figure 16-2, and from these graphs

- We see that in no case is any reacting species completely consumed.
- There appears to be nothing in common among the equilibrium amounts of reactants and products in these three cases.

Although you probably cannot see this just by looking at the data, the following ratio of *equilibrium concentrations* is the same for each of the three experiments:

Expt. 1: $K_c = \dfrac{[CH_3OH]}{[CO][H_2]^2} = \dfrac{0.0090}{(0.0910)(0.0820)^2} = 15$

Expt. 2: $K_c = \dfrac{[CH_3OH]}{[CO][H_2]^2} = \dfrac{0.0247}{(0.0753)(0.151)^2} = 14.4$ (16.4)

Expt. 3: $K_c = \dfrac{[CH_3OH]}{[CO][H_2]^2} = \dfrac{0.0620}{(0.138)(0.176)^2} = 14.5$

The ratio of molarity concentrations in expression (16.4) is called an **equilibrium constant expression,** and the numerical constant K_c is the **equilibrium constant** of the reaction. The primary value of the equilibrium constant expression is that it allows us to calculate the equilibrium concentrations of reactants and products. We will look at this application in detail in Section 16-7, but as a preview, consider Example 16-1.

FIGURE 16-2
Three approaches to equilibrium in the reaction

$$CO(g) + 2 H_2(g) \rightleftharpoons CH_3OH(g)$$

t_e = time for equilibrium to be reached.

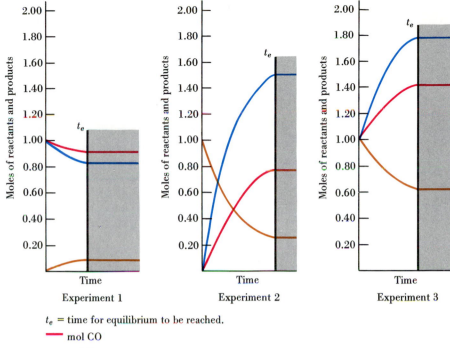

t_e = time for equilibrium to be reached.

— mol CO
— mol H_2
— mol CH_3OH

Example 16-1

Relating equilibrium concentrations of reactants and products through an equilibrium constant expression. Equilibrium is established in reaction (16.3). These equilibrium concentrations are measured: [CO] = 1.03 M and [CH_3OH] = 1.56 M. What is the equilibrium concentration of H_2?

Solution. Substitute the known concentrations into the equilibrium constant expression (16.4). Then solve for [H_2].

$$K_c = \frac{[CH_3OH]}{[CO][H_2]^2} = \frac{1.56}{1.03 \times [H_2]^2} = 14.5$$

$$[H_2]^2 = \frac{1.56}{1.03 \times 14.5} = 0.104 \qquad [H_2] = \sqrt{0.104} = 0.322 \text{ M}$$

SIMILAR EXAMPLES: Exercises 5, 7, 25, 26.

A General Expression for K_c. The equilibrium constant expression we have written for the methanol synthesis reaction (16.4) is just a specific example of a more general case. For the hypothetical, generalized reaction

$$a A + b B + \cdots \rightleftharpoons g G + h H + \cdots \qquad (16.5)$$

The equilibrium constant expression has the form

General expression for an equilibrium constant, K_c.

$$K_c = \frac{[G]^g[H]^h \cdots}{[A]^a[B]^b \cdots} \qquad (16.6)$$

The numerator of an equilibrium constant expression is the product of the concentrations of the species on the right side of the equation ([G], [H], . . .), with each concentration term raised to a power given by the coefficient in the balanced equation (g, h, . . .). The denominator is the product of the concentrations of the

Are You Wondering:

What units to use for an equilibrium constant, K_c?

From the general expression (16.6) it looks like the units of K_c should be some power of molarity, depending on the exponents of the concentration terms in the numerator and denominator. These units can be rather complex, such as mol L^{-1}, mol^2 L^{-2}, mol^3 L^{-3}, L mol^{-1}, L^2 mol^{-2}, and so on. These units are frequently used, but in this text we will *not* write units for K_c values. In Section 20-6 we introduce a quantity known as the thermodynamic equilibrium constant; this is a *dimensionless* number, that is, a quantity with *no units*. In anticipation of switching to this form of the equilibrium constant in a later chapter, we will not use units for K_c in this chapter. For the present, just remember that all the quantities used in K_c expressions must be molarity concentrations.

species on the left *side of the equation ([A], [B], . . .), with each term raised to a power given by the coefficient in the balanced equation* (a, b, . . .).

The numerical value of an equilibrium constant, K_c, depends on the particular reaction and on the temperature.

16-3   Important Relationships Involving Equilibrium Constants

This section presents *seven* ideas that you will find useful, and sometimes necessary, in working with the concept of chemical equilibrium. First we will state these ideas and then explain and illustrate them.

Relationship of K_c to the Balanced Chemical Equation

Relationship of K_c to the balanced chemical equation.

1. Be sure to match the expression used for K_c to the corresponding balanced equation.
2. If you *reverse* an equation, *invert* the value of K_c.
3. If you *multiply* each of the coefficients in a balanced equation by the same factor (2, 3, . . .), *raise* the equilibrium constant to the corresponding power (2, 3, . . .).   (16.7)
4. If you *divide* each of the coefficients in a balanced equation by the same factor (2, 3, . . .), take the corresponding root of the equilibrium constant (i.e., square root, cube root, . . .).

For the synthesis of $CH_3OH(g)$ from $CO(g)$ and $H_2(g)$, if we choose to write the balanced equation in the form

$$CO(g) + 2\,H_2(g) \rightleftharpoons CH_3OH(g) \qquad K_c = 14.5 \qquad (16.3)$$

then, according to the general expression (16.6), we must write the equilibrium constant expression as

$$K_c = \frac{[CH_3OH]}{[CO][H_2]^2} = 14.5 \qquad (16.4)$$

If, on the other hand, we choose to write the equation in terms of the decomposition of $CH_3OH(g)$,

$$CH_3OH(g) \rightleftharpoons CO(g) + 2\,H_2(g) \qquad K_c' = ?$$

then, according to expression (16.6), we should write

$$K'_c = \frac{[CO][H_2]^2}{[CH_3OH]} = \frac{1}{[CH_3OH]/[CO][H_2]^2} = \frac{1}{K_c} = \frac{1}{14.5} = 0.0690 \qquad (16.8)$$

Suppose for a certain application we need to write an equation based on *two* moles of $CH_3OH(g)$.

$$2\,CO(g) + 4\,H_2(g) \rightleftharpoons 2\,CH_3OH(g) \qquad K''_c = ?$$

For the equilibrium constant expression we should write

$$K''_c = \frac{[CH_3OH]^2}{[CO]^2[H_2]^4} = \left(\frac{[CH_3OH]}{[CO][H_2]^2}\right)^2 = (K_c)^2 = (14.5)^2 = 210. \qquad (16.9)$$

Additional applications of these ideas are illustrated in Example 16-2.

Example 16-2

Relating K_c to the balanced chemical equation. One of the most commercially important equilibrium reactions is the synthesis of NH_3 from its elements (see Focus feature). The following value is given for $K_c(a)$ at 298 K.
(a) $N_2(g) + 3\,H_2(g) \rightleftharpoons 2\,NH_3(g) \qquad K_c(a) = 3.6 \times 10^8$
What is the value of $K_c(b)$ at 298 K for the reaction
(b) $NH_3(g) \rightleftharpoons \frac{1}{2}\,N_2(g) + \frac{3}{2}\,H_2(g) \qquad K_c(b) = ?$

Solution. First, we must reverse equation (a) to get $NH_3(g)$ on the left side of the equation; the equilibrium constant becomes $1/K_c(a)$.

$$2\,NH_3(g) \rightleftharpoons N_2(g) + 3\,H_2(g)$$
$$K_c(a)' = 1/K_c(a) = 1/3.6 \times 10^8 = 2.8 \times 10^{-9}$$

Then, we need to *divide* all coefficients by 2 [so as to base the equation on 1 mol $NH_3(g)$]. This requires us to take the *square root* of $K_c(a)'$.

$$NH_3(g) \rightleftharpoons \frac{1}{2}\,N_2(g) + \frac{3}{2}\,H_2(g)$$
$$K_c(b) = \sqrt{K_c(a)'} = \sqrt{2.8 \times 10^{-9}} = 5.3 \times 10^{-5}$$

SIMILAR EXAMPLES: Exercises 2, 19.

Combining Equilibrium Constant Expressions

Procedure for combining equilibrium constants into a value for a net reaction.

5. When you combine (that is, add) individual equations *multiply* their equilibrium constants to obtain the equilibrium constant for the net reaction. (16.10)

Nitrogen and oxygen can combine to form either $NO(g)$ or $N_2O(g)$.

$$N_2(g) + O_2(g) \rightleftharpoons 2\,NO(g) \qquad K_c = 4.1 \times 10^{-31} \qquad (16.11)$$

$$N_2(g) + \tfrac{1}{2}\,O_2(g) \rightleftharpoons N_2O(g) \qquad K_c = 2.4 \times 10^{-18} \qquad (16.12)$$

Also, $N_2O(g)$ can react with oxygen to produce $NO(g)$.

$$N_2O(g) + \tfrac{1}{2}\,O_2(g) \rightleftharpoons 2\,NO(g) \qquad K_c = ? \qquad (16.13)$$

As we learned in Section 7-8, we can obtain equation (16.13) by an appropriate combination of equations (16.11) and (16.12). To do this, we need to *reverse* equation (16.12), and when we do, we must also take the *reciprocal* of its K_c value.

(a) $N_2(g) + O_2(g) \rightleftharpoons 2\,NO(g) \qquad K_c(a) = 4.1 \times 10^{-31}$
(b) $\qquad N_2O(g) \rightleftharpoons N_2(g) + \tfrac{1}{2}\,O_2(g) \qquad K_c(b) = 1/2.4 \times 10^{-18} = 4.2 \times 10^{17}$

net: $N_2O(g) + \tfrac{1}{2}\,O_2(g) \rightleftharpoons 2\,NO(g) \qquad K_c(net) = ? \qquad (16.13)$

According to the general expression (16.6), the equilibrium constant for reaction (16.13) is

$$K_c(\text{net}) = \frac{[NO]^2}{[N_2O][O_2]^{1/2}} = \underbrace{\frac{[NO]^2}{[N_2][O_2]}}_{K_c(a)} \times \underbrace{\frac{[N_2][O_2]^{1/2}}{[N_2O]}}_{K_c(b)} = K_c(a) \times K_c(b)$$

$$= 4.1 \times 10^{-31} \times 4.2 \times 10^{17} = 1.7 \times 10^{-13}$$

Equilibria Involving Gases: The Equilibrium Constant K_p

Expressing an equilibrium constant through partial pressures of gases: K_p.

> **6.** When describing equilibria involving *gases* you can write an equilibrium constant expression based on the *partial pressures* of the gaseous reactants and products, and called the partial pressure equilibrium constant, K_p.   (16.14)

To derive this alternative form of an equilibrium constant, consider the formation of $SO_3(g)$, the key step in the manufacture of sulfuric acid (recall Focus feature of Chapter 15).

$$2\,SO_2(g) + O_2(g) \rightleftharpoons 2\,SO_3(g) \tag{16.15}$$

$$K_c = \frac{[SO_3]^2}{[SO_2]^2[O_2]} = 2.8 \times 10^2 \text{ (at 1000 K)}$$

Now, use the ideal gas law, $PV = nRT$, to write

$$[SO_3] = \frac{n_{SO_3}}{V} = \frac{P_{SO_3}}{RT} \qquad [SO_2] = \frac{n_{SO_2}}{V} = \frac{P_{SO_2}}{RT} \qquad [O_2] = \frac{n_{O_2}}{V} = \frac{P_{O_2}}{RT}$$

and substitute the terms in blue for concentration terms in K_c.

$$K_c = \frac{(P_{SO_3}/RT)^2}{(P_{SO_2}/RT)^2(P_{O_2}/RT)} = \frac{(P_{SO_3})^2}{(P_{SO_2})^2(P_{O_2})} \times RT \tag{16.16}$$

The ratio of partial pressures shown in brown in (16.16) is the equilibrium constant, K_p. The relationship between K_p and K_c for reaction (16.15) is

$$K_c = K_pRT \qquad \text{and} \qquad K_p = \frac{K_c}{RT} = K_c(RT)^{-1} \tag{16.17}$$

If we carried out a similar derivation for the general reaction

$$a\,A(g) + b\,B(g) + \cdots \rightleftharpoons g\,G(g) + h\,H(g) + \cdots$$

our result would be

Relationship between K_p and K_c.

$$K_p = K_c(RT)^{\Delta n_{\text{gas}}} \tag{16.18}$$

where Δn_{gas} is the difference in the stoichiometric coefficients of *gaseous* products and reactants; that is, $\Delta n_{\text{gas}} = (g + h + \cdots) - (a + b + \cdots)$. In reaction (16.15), $\Delta n_{\text{gas}} = 2 - (2 + 1) = -1$, just as we found in equation (16.17).

Although we will not use units for numerical values of equilibrium constants, we do need to use the correct units for terms *within* equilibrium constant expressions. These should be *molarity* for K_c expressions, and pressures in *atm* for K_p expressions. These choices then require that we use a value of $R = 0.08206$ L atm mol^{-1} K^{-1} in equation (16.18).

Example 16-3

Relating K_c and K_p. Complete the calculation of K_p for reaction (16.15) from the data given.

Solution

$$K_p = K_c(RT)^{-1} = 2.8 \times 10^2(0.08206 \times 1000)^{-1} = \frac{2.8 \times 10^2}{0.08206 \times 1000} = 3.4$$

SIMILAR EXAMPLES: Exercises 4, 17, 20.

Example 16-4

Determining conditions under which $K_p = K_c$. What is the value of K_p for this reaction at 445 °C?

$$H_2(g) + I_2(g) \rightleftharpoons 2\,HI(g) \qquad K_c = 50.2 \text{ at } 445 °C$$

Solution. For this reaction $\Delta n_{gas} = 0$. In any case where $\Delta n_{gas} = 0$, the expression $K_p = K_c(RT)^{\Delta n_{gas}}$ becomes $K_p = K_c$ (since any number raised to the power 0 has a value of 1).

$$K_p = K_c = 50.2$$

SIMILAR EXAMPLES: Exercises 4, 20.

Equilibria Involving Pure Liquids and Solids (Heterogeneous Reactions)

Treatment of solids and liquids in equilibrium constant expressions.

7. Do not include concentration terms for *pure solids* and *pure liquids* in equilibrium constant expressions. (16.19)

An equilibrium constant expression contains terms only for substances whose concentrations can vary during the course of a reaction. Since their compositions *cannot* vary, even while participating in a chemical reaction, *pure* solids and *pure* liquids are not represented in equilibrium constant expressions.

All the examples considered to this point in the chapter have been reactions in a single phase, specifically the gaseous phase. They are *homogeneous* reactions, and their equilibrium constant expressions have terms for each reactant and product. The water–gas reaction, used to make combustible gases from coal, has reacting species in both gaseous and *solid* phases; it is a *heterogeneous* reaction.

$$C(s) + H_2O(g) \rightleftharpoons CO(g) + H_2(g) \qquad (16.20)$$

The equilibrium constant expression for reaction (16.20) contains terms only for the species in the homogeneous gas phase—H_2O, CO, and H_2.

$$K_c = \frac{[CO][H_2]}{[H_2O]}$$

The decomposition of calcium carbonate (limestone) is another example of a heterogeneous reaction; the equilibrium constant expression contains just a single term.

$$CaCO_3(s) \rightleftharpoons CaO(s) + CO_2(g) \qquad K_c = [CO_2] \qquad (16.21)$$

We can write K_p for reaction (16.21) by using equation (16.18), with $\Delta n_{gas} = 1$.

$$K_p = P_{CO_2} \qquad \text{and} \qquad K_p = K_c(RT) \qquad (16.22)$$

According to equation (16.22), the equilibrium pressure of $CO_2(g)$ in contact with $CaCO_3(s)$ and CaO(s) is a *constant*, equal to K_p, and its value is *independent* of the quantities of CaO_3 and CaO (as long as *both* solids are present).

In Section 16-1 we suggested that the equilibrium constant principle applies to *physical* equilibria (no chemical reactions involved). Consider the liquid–vapor equilibrium for water.

$$H_2O(l) \rightleftharpoons H_2O(g)$$

$$K_c = [H_2O(g)] \quad \text{and} \quad K_p = P_{H_2O} \quad \text{and} \quad K_p = K_c(RT) \quad (16.23)$$

Thus, equilibrium vapor pressures are simply values of K_p, and, as we have seen before, these values do *not* depend on the quantity of liquid in the equilibrium.

Example 16-5

Writing equilibrium constant expressions for reactions involving pure solids and/or liquids. Equilibrium is established in the following reaction at 60 °C, and the gas partial pressures are found to be $P_{HI} = 3.65 \times 10^{-3}$ atm and $P_{H_2S} = 9.96 \times 10^{-1}$ atm. What is the value of K_p for the reaction?

$$H_2S(g) + I_2(s) \rightleftharpoons 2\,HI(g) + S(s) \qquad K_p = ?$$

Solution. Recall that terms for pure solids do *not* appear in an equilibrium constant expression. The value of K_p is

$$K_p = \frac{(P_{HI})^2}{P_{H_2S}} = \frac{(3.65 \times 10^{-3})^2}{9.96 \times 10^{-1}} = 1.34 \times 10^{-5}$$

SIMILAR EXAMPLES: Exercises 1, 16.

The Thermodynamic Equilibrium Constant, K: A Preview. In Chapter 20 we will discuss equilibrium from the standpoint of thermodynamics, and this is the most fundamental description that we can give. In that discussion we will introduce a form of the equilibrium constant called the *thermodynamic equilibrium constant, K*. The terms used in this expression are *dimensionless* quantities known as *activities* (recall Section 13-9). Usually, molarity concentrations are substituted for solute activities in homogeneous solutions, pressures in atm are substituted for activities of gases, and pure solids and pure liquids are assigned activities of 1.

With this preliminary information about the thermodynamic equilibrium constant, you should find it understandable why

- Only a numerical value—no unit—is used to express an equilibrium constant;
- We have introduced K_p as well as K_c (though in Chapter 20 both will be replaced by K);
- There are no terms for pure solids and liquids in K_c or K_p (their activities are 1).

16-4 Significance of the Magnitude of an Equilibrium Constant

In principle every chemical reaction has an equilibrium constant, but we need not use them in every case. Table 16-2 lists equilibrium constants for several reactions previously encountered in the text. The first of these reactions is the synthesis of H_2O from its elements—a reaction we described in the introduction to this chapter. We have always assumed that reaction (4.8) *goes to completion*, that is, that it proceeds only in the forward direction. If a reaction goes to completion, at least one of the reactants is used up and a term in the *denominator* of the equilibrium constant expression approaches *zero*, making the value of the equilibrium constant very large.

The significance of a very large numerical value of K_c or K_p.

A very large numerical value of K_c or K_p signifies that the forward reaction, as written, *goes to completion* or very nearly so.　　(16.24)

TABLE 16-2
Some Equilibrium Reactions

Reaction	Equilibrium Constant, K_p	Text reference
$2 H_2(g) + O_2(g) \rightleftharpoons 2 H_2O(l)$	1.4×10^{83} at 298 K	reaction (4.8)
$N_2(g) + O_2(g) \rightleftharpoons 2 NO(g)$	5.3×10^{-31} at 298 K	Section 7-8
	1.3×10^{-4} at 1800 K	
$2 NO(g) + O_2(g) \rightleftharpoons 2 NO_2(g)$	1.6×10^{12} at 298 K	reaction (14.60)
$2 SO_2(g) + O_2(g) \rightleftharpoons 2 SO_3(g)$	3.4 at 1000 K	reaction (16.15)
$C(s) + H_2O(g) \rightleftharpoons CO(g) + H_2(g)$	1.6×10^{-21} at 298 K	reaction (14.1)
	10 at 1100 K	

Because the value of K_p for reaction (4.8) is 1.4×10^{83}, we are entirely justified in saying that the reaction goes to completion at 298 K.

The other reaction that we described in the introduction to this chapter is the synthesis of NO(g) from N_2(g) and O_2(g). From Table 16-2 we see that its value of K_p (or K_c) is very small at 298 K (only 5.3×10^{-31}). To account for a very small numerical value of an equilibrium constant, the *numerator* must be very small (approaching zero).

The significance of a very small numerical value of K_c or K_p.

A very small numerical value of K_c or K_p signifies that the forward reaction, as written, *does not occur to any significant extent.*   (16.25)

At 1800 K the value of K_p for the synthesis of NO(g) from its elements is 1.3×10^{-4}, which is much larger than K_p at 298 K; and this is the reason why some NO(g) is produced in high-temperature combustion processes, such as in automobile engines. When NO(g) comes into contact with O_2(g) at 298 K it is converted to NO_2(g) in a reaction (14.60) that goes nearly to completion ($K_p = 1.6 \times 10^{12}$).

The conversion of SO_2(g) and O_2(g) to SO_3(g) at 1000 K has an equilibrium constant that is neither very large nor very small. Both the forward and reverse reactions are important and we expect significant amounts of both reactants and products to be present at equilibrium. A similar situation exists for the water gas reaction (14.1) at 1100 K. However, at 298 K we would not expect the forward reaction to occur to any significant extent ($K_p = 1.6 \times 10^{-21}$). In conclusion then

Conditions under which an equilibrium constant value is significant.

A reaction is most likely to reach a state of equilibrium in which both reactants and products are present (and in which the equilibrium constant expression must be used) if the numerical value of K_c or K_p is *neither very large nor very small.*   (16.26)

The ideas presented in this section should help you, at times, to decide whether an equilibrium calculation is called for (see Section 16-7).

16-5   The Reaction Quotient, Q: Predicting the Direction of a Reaction

Recall the synthesis of CH_3OH from CO and H_2 discussed in Section 16-2.

$$CO(g) + 2 H_2(g) \rightleftharpoons CH_3OH(g) \qquad K_c = 14.5$$

We described three situations through Table 16-1 and Figure 16-2. When we started initially with just the reactants CO and H_2 (Expt. 1), a net change had to occur in which some CH_3OH was formed. Only in this way could an equilibrium condition be reached in which all reacting species would be present. We say that a net reaction occurs in the *forward* direction (*to the right*).

In another situation (Expt. 2) we started with the product, CH_3OH. Here a net change had to occur in which some CH_3OH decomposed back to CO and H_2. We say that a net reaction occurs in the *reverse* direction (*to the left*). In the third situation we started with all the reacting species present—CO, H_2, CH_3OH. Here, we need some other means of assessing the direction of net change required to establish equilibrium. Being able to predict the direction in which a net change will occur to establish equilibrium is important for two reasons.

• At times a detailed equilibrium calculation is unnecessary, and a qualitative prediction about the changes that occur in establishing equilibrium is enough.
• For certain equilibrium calculations it is necessary to determine the direction in which a net change must occur to establish equilibrium (see Example 16-14 in Section 16-7).

The Reaction Quotient. At each point in a reaction we can formulate a ratio of concentration terms having the same form as the equilibrium constant expression. This ratio is called the **reaction quotient**, usually designated by the symbol Q. For the hypothetical generalized reaction (16.5), the reaction quotient is

Formulation of the reaction quotient, Q.

$$Q = \frac{[G]^g[H]^h \cdots}{[A]^a[B]^b \cdots} \tag{16.27}$$

If a reaction is at equilibrium, $Q = K_c$. Otherwise, depending on the relationship of Q to K_c, a net reaction proceeds either in the forward direction (*to the right*) or in the reverse direction (*to the left*). We can illustrate the relationship between Q and K_c by referring again to the three situations described above.

A ratio of concentrations formed in the manner of the reaction quotient Q is also called the mass-action expression.

In Expt. 1 (Table 16-1) the *initial* concentrations of CO and H_2 are 1.00 mol/10.0 L. Initially there is no CH_3OH. In this case the reaction quotient Q_1 is

$$Q_1 = \frac{[CH_3OH]_{init.}}{[CO]_{init.}[H_2]_{init.}^2} = \frac{0}{(1.00/10.0)(1.00/10.0)^2} = 0 \tag{16.28}$$

We have already established that a net change should occur in the forward direction (*to the right*). As a net change occurs in the forward direction, the numerator in expression (16.28) increases, the denominator decreases, the value of Q *increases* and eventually $Q = K_c$.

Criterion for a net change in the forward direction.

A net reaction proceeds from left to right (the forward reaction) if $Q < K_c$. $\tag{16.29}$

In Expt. 2 (Table 16-1) the *initial* concentration of CH_3OH is 1.00 mol/10.0 L. Initially there is no CO nor H_2. The value of Q_2 is

$$Q_2 = \frac{[CH_3OH]_{init.}}{[CO]_{init.}[H_2]_{init.}^2} = \frac{(1.00/10.0)}{0 \times 0} = \infty \tag{16.30}$$

We have already established that a net change should occur in the reverse direction (*to the left*). As a net change occurs in the reverse direction, the numerator in expression (16.30) decreases, the denominator increases, the value of Q *decreases* and eventually $Q = K_c$.

Criterion for a net change in the reverse direction.

A net reaction proceeds from the right to the left (the reverse direction) if $Q > K_c$. $\tag{16.31}$

Now let us turn to a case where these new criteria for the direction of net change are really needed. In Expt. 3 (Table 16-1) the *initial* concentrations of all three species are 1.00/10.0. The value of Q_3 is

$$Q_3 = \frac{[CH_3OH]_{init.}}{[CO]_{init.}[H_2]_{init.}^2} = \frac{(1.00/10.0)}{(1.00/10.0)(1.00/10.0)^2} = 100.$$

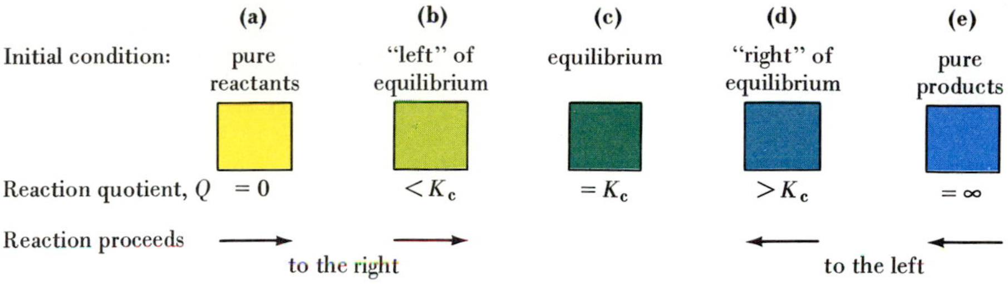

FIGURE 16-3
Predicting the direction of change in a reversible reaction. Five possibilities for the relationship of the initial conditions to the equilibrium conditions.

From Table 16-1 and Figure 16-2, Expt. 1 corresponds to initial condition (a), Expt. 2 to condition (e), and Expt. 3 to (d). The situation in Example 16-6 also corresponds to condition (d).

	(a)	(b)	(c)	(d)	(e)
Initial condition:	pure reactants	"left" of equilibrium	equilibrium	"right" of equilibrium	pure products
Reaction quotient, Q	$= 0$	$< K_c$	$= K_c$	$> K_c$	$= \infty$

Reaction proceeds → to the right → ← to the left ←

Since $Q > K_c$ (100. compared to 14.5), a net reaction must occur in the *reverse direction* in Expt. 3. (Note how you can verify this conclusion from Figure 16-2. The amounts of CO and H_2 at equilibrium are greater than they were initially and the amount of CH_3OH is less.)

The criteria for predicting the *direction* of a net chemical change in a reversible reaction are illustrated in Figure 16-3 and applied in Example 16-6.

Example 16-6

Predicting the direction in which a net change occurs to establish equilibrium. To increase the yield of $H_2(g)$ produced by the water-gas reaction (16.20), an additional reaction called the "water-gas shift reaction" is usually employed. For the water-gas shift reaction

$$CO(g) + H_2O(g) \rightleftharpoons CO_2(g) + H_2(g)$$

$K_c = 1.00$ at about 1100 K. The following amounts of substances are brought together and allowed to react at 1100 K: 1.00 mol CO, 1.00 mol H_2O, 2.00 mol CO_2, and 2.00 mol H_2. Compared to their initial amounts, which of the substances will be present in greater amount and which, in lesser amount, when equilibrium is established?

Solution. Basically, we need to determine the direction in which a net change occurs. This means evaluating Q. To substitute concentrations into the expression for Q, we can assume an arbitrary volume V. Its value is immaterial since the volume cancels out.

Volume terms will cancel from a reaction quotient or an equilibrium constant expression only if the total of the exponents of the concentrations terms in the numerator equals that in the denominator.

$$Q = \frac{[CO_2][H_2]}{[CO][H_2O]} = \frac{(2.00/V)(2.00/V)}{(1.00/V)(1.00/V)} = 4.00$$

Because $Q > K_c$ (that is, $4.00 > 1.00$), a net reaction occurs to the *left*. When equilibrium is established, the amounts of CO and H_2O will be *greater* than initially and the amounts of CO_2 and H_2 will be *less*.

SIMILAR EXAMPLES: Exercises 28, 32, 61.

16-6 Altering Equilibrium Conditions: Le Châtelier's Principle

There are times when all we need to do is make some qualitative statements about a reversible reaction—whether a net reaction should occur in the forward or reverse direction, whether the amount of a substance will increase or decrease when equilibrium is reached, and so on. Also, we may encounter situations where we do not have enough data to make a quantitative calculation. In these cases we will find useful a statement attributed to the French chemist Henri Le Châtelier (1884). Le Châtelier's principle is difficult to state unambiguously, but its essence is that

Le Châtelier's principle.

When a system at equilibrium is subjected to a change in temperature, pressure, or concentration of a reacting species, the system reacts in a way that *partially* offsets the change while reaching a new state of equilibrium.

(16.32)

The way in which an equilibrium mixture responds to changes in these variables involves, in some cases, a shift of the equilibrium condition "to the right" (meaning favoring a net forward reaction) and in others, a shift "to the left" (favoring a net reverse reaction). Generally, it is not difficult to predict the outcome of changing a system variable.

Effect of Changing the Amounts of Reacting Species. Let us return to the reaction

$$2\ SO_2(g) + O_2(g) \rightleftharpoons 2\ SO_3(g) \qquad K_c = 2.8 \times 10^2 \text{ at 1000 K} \qquad (16.15)$$

Suppose we start with a certain combination of amounts of SO_2, O_2, and SO_3 that is at equilibrium, as in Figure 16-4a. Now let us introduce a disturbance to the equilibrium mixture by forcing in an additional 1.00 mol SO_3 (Figure 16-4b). How will the amounts of the reacting species change to establish a new condition of equilibrium? One approach is to evaluate the reaction quotient immediately after adding the SO_3.

Original equilibrium

$$Q = \frac{[SO_3]^2}{[SO_2]^2[O_2]} = K_c$$

Following disturbance

$$Q = \frac{[SO_3]^2}{[SO_2]^2[O_2]} > K_c$$

Adding any quantity of SO_3 to a constant-volume equilibrium mixture makes the value of Q larger than that of K_c. A net reaction must occur in the direction that reduces $[SO_3]$, that is, to the left, the *reverse* direction. The amount of each species in the new equilibrium condition can be calculated by methods introduced in the next section. These amounts are pictured in Figure 16-4c.

An alternative *qualitative* approach is based on Le Châtelier's principle. If the system is to partially offset an action that increases the concentration of one of the reacting species, it must do so by favoring the reaction in which that species is consumed. Here, this means the *reverse* reaction—conversion of *some* of the added SO_3 to SO_2 and O_2. In the new equilibrium there will be greater amounts of all the substances than in the original equilibrium, although, as explained in the caption to Figure 16-4, the additional amount of SO_3 will be less than the 1.00 mol that was added.

Example 16-7

Applying Le Châtelier's Principle: Effect of adding more of one of the reactants to an equilibrium mixture. Predict the effect of adding more $H_2(g)$ to a constant-volume equilibrium mixture of N_2, H_2, and NH_3.

$$N_2(g) + 3\ H_2(g) \rightleftharpoons 2\ NH_3(g)$$

Solution. The action of increasing $[H_2]$ stimulates a shift of the equilibrium condition "to the right." However, only a portion of the added H_2 is consumed in this reaction. When a new equilibrium is established, there will be more H_2 than present originally. The amount of NH_3 will also be greater, but the amount of N_2 will be *smaller*. Some of the original N_2 must be consumed in converting some of the added H_2 to NH_3.

SIMILAR EXAMPLES: Exercises 15, 49, 50, 70.

FIGURE 16-4
FIGURE 16-4
Changing equilibrium conditions by increasing the amount of one of the reactants

$$2 SO_2(g) + O_2(g) \rightleftharpoons 2 SO_3(g)$$

$$K_c = 2.8 \times 10^2 \text{ at } 1000 \text{ K}$$

(a) The original equilibrium condition.
(b) Disturbance caused by the addition of 1.00 mol SO_3.
(c) The new equilibrium condition. (See Exercise 35 for calculation.)

Note that the amount of SO_3 in the new equilibrium of (c)—1.46 mol—is greater than in the original equilibrium of (a)—0.68 mol—but not as great as immediately after the addition of 1.00 mol SO_3 in (b)—1.68 mol.

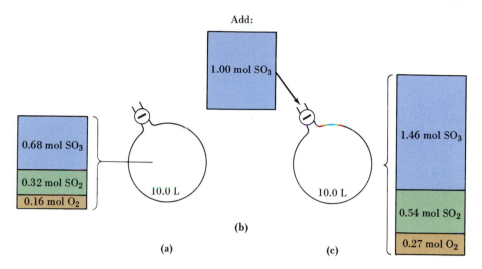

Add:
1.00 mol SO_3

0.68 mol SO_3
0.32 mol SO_2
0.16 mol O_2
10.0 L

10.0 L

1.46 mol SO_3
0.54 mol SO_2
0.27 mol O_2

(a)

(b)

(c)

Example 16-8

Applying Le Châtelier's Principle: Effect on an equilibrium mixture of adding or removing a reactant. What is the effect on equilibrium in the calcination (decomposition) of limestone, $CaCO_3(s) \rightleftharpoons CaO(s) + CO_2(g)$, produced by (a) adding a small quantity of $CaCO_3(s)$ and (b) removing some $CO_2(g)$.

Solution

(a) Adding or removing substances will affect an equilibrium condition *only if this produces changes in concentration terms appearing in an equilibrium constant expression*. An additional small quantity of $CaCO_3$ does not affect the reaction volume, and there is no concentration term for $CaCO_3(s)$ in the equilibrium constant expression. Adding more $CaCO_3(s)$ will not affect the equilibrium condition.

(b) Removing $CO_2(g)$ has the effect of reducing its concentration (or partial pressure). More $CaCO_3(s)$ will decompose to replace the $CO_2(g)$ that has been removed in an attempt to restore equilibrium. If $CO_2(g)$ is removed *continuously*, equilibrium will never be established and the reaction will go to completion; this is the way that the calcination of limestone is carried out commercially (recall reaction 14.32).

SIMILAR EXAMPLES: Exercises 15, 49, 50, 70.

Effect of Change in Pressure. Let us turn again to the reaction of $SO_2(g)$ and $O_2(g)$ to form $SO_3(g)$.

$$2 SO_2(g) + O_2(g) \rightleftharpoons 2 SO_3(g) \qquad K_c = 2.8 \times 10^2 \text{ at } 1000 \text{ K} \qquad (16.15)$$

The equilibrium mixture in Figure 16-5a has its volume reduced to one-tenth of its original value by increasing the external pressure. Again an adjustment of equilibrium amounts must occur in accord with the expression for K_c.

$$K_c = \frac{[SO_3]^2}{[SO_2]^2[O_2]} = \frac{(n_{SO_3}/V)^2}{(n_{SO_2}/V)^2(n_{O_2}/V)} = \frac{(n_{SO_3})^2}{(n_{SO_2})^2(n_{O_2})} \times V = 2.8 \times 10^2 \qquad (16.33)$$

From equation (16.33) it follows that, if V is reduced by a factor of 10, the ratio

$$\frac{(n_{SO_3})^2}{(n_{SO_2})^2(n_{O_2})}$$

FIGURE 16-5
Effect of pressure change on equilibrium condition in the reaction

$$2 SO_2(g) + O_2(g) \rightleftharpoons 2 SO_3(g)$$

An increase in external pressure forces a shift in equilibrium conditions "to the right." (See Exercise 35 for calculation.)

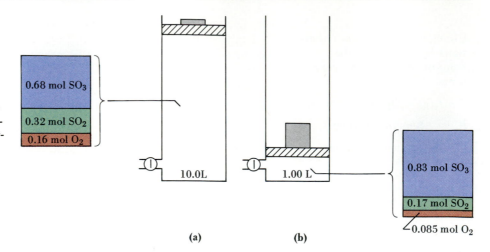

(a)　　　　(b)

must increase by a factor of 10. The equilibrium amount of SO_3 must increase and the amounts of SO_2 and O_2 must decrease. The new equilibrium amounts are given in Figure 16-5b.

An equilibrium system responds to an increase in external pressure by shrinking into a smaller volume. In reaction (16.15) 3 mol of *gases* on the left produce 2 mol of *gases* on the right. The product of the reaction, SO_3, occupies a smaller volume than the reactants from which it is formed. Thus an increase in pressure results in the production of additional SO_3.

Effect of pressure on an equilibrium condition.

When the pressure on an equilibrium mixture involving *gases* is increased, a net reaction occurs in the direction in which the number of moles of gases becomes smaller. If the pressure is decreased, a net reaction occurs in the direction producing a larger number of moles of gases. (16.34)

The effect of pressure on an equilibrium condition is not limited to reactions involving gases, but because gases are so much more compressible than liquids or solids the effect is much more pronounced with gases. Statement (16.34) stresses gases because we generally do not have a way of assessing the comparative volumes of solids and liquids through a chemical equation alone.

Effect of an Inert Gas. Since an inert gas (e.g., helium) does not participate in the reaction we should not expect it to affect the equilibrium condition. Actually, however, whether an inert gas does affect an equilibrium condition depends on how the inert gas is introduced. If the gas is introduced, *while the volume is held constant,* the *total* gas pressure will increase, but the concentrations (or partial pressures) of all the reactants will remain *constant. When added to a system at constant volume, an inert gas has no effect on an equilibrium condition.* On the other hand, if the inert gas is added to an equilibrium mixture, *while the pressure is held constant,* the reaction volume will *increase,* and the equilibrium condition will shift to the side of the reaction involving the greater number of moles of gas. In other words, *when added to a system at constant pressure, an inert gas has the same effect on the equilibrium condition as does increasing the system volume.* In summary, the presence of an inert gas affects an equilibrium condition only if this produces changes in concentrations (or partial pressures) of the reactants or products.

Effect of Temperature. When we change the temperature of an equilibrium mixture we can think of this as adding heat (raising the temperature) or removing heat (lowering the temperature). According to Le Châtelier's principle adding heat favors the heat-absorbing (*endothermic*) reaction and removing heat favors the heat-evolving (*exothermic*) reaction. In the first case the system attempts to accommodate to the heat energy entering the system by absorbing some of it, and in the second, to replace heat energy lost by the system by giving off heat.

Effect of temperature on an equilibrium condition.

Raising the temperature of an equilibrium mixture causes the equilibrium condition to shift in the direction of the **endothermic** reaction. Lowering the temperature causes a shift in the direction of the **exothermic** reaction. Or, the numerical value of K_c or K_p *increases* with temperature for an *endothermic* reaction and *decreases* with temperature for an *exothermic* reaction.

(16.35)

The effect of temperature on equilibrium constants is such an important matter than we reconsider this topic in Chapter 20. At that time we will learn how to *calculate* equilibrium constants as a function of temperature. For the present, however, we limit ourselves to the qualitative predictions that are possible with expression (16.35).

Example 16-9

Applying Le Châtelier's principle: effect of temperature on equilibrium. Is the conversion of $SO_2(g)$ to $SO_3(g)$ favored at high or low temperatures?

$$2 SO_2(g) + O_2(g) \rightleftharpoons 2 SO_3(g) \qquad \Delta H° = -180 \text{ kJ}$$

Solution. Raising the temperature favors the endothermic reaction, the reverse reaction above. To favor the forward (exothermic) reaction requires that we lower the temperature. Therefore, an equilibrium mixture would have a higher concentration of SO_3 at lower temperatures; **the conversion of SO_2 to SO_3 is favored at *low* temperatures.**

SIMILAR EXAMPLES: Exercises 14, 52, 53, 69.

Effect of a Catalyst on Equilibrium. When a catalyst is added to a reaction mixture it speeds up *both* the forward and reverse reactions. Equilibrium is achieved more rapidly, but the equilibrium amounts are *unchanged* by the catalyst. Consider again the reaction

$$2 SO_2(g) + O_2(g) \rightleftharpoons 2 SO_3(g) \qquad K_c = 2.8 \times 10^2 \text{ at } 1000 \text{ K} \qquad (16.15)$$

For a given set of reaction conditions the equilibrium amounts of SO_2, O_2, and SO_3 have fixed values. This is true whether the reaction is carried out by a slow homogeneous reaction, catalyzed in the gas phase, or conducted as a heterogeneous reaction on the surface of a catalyst (as described in the Focus feature of Chapter 15). Or, stated in another way, the presence of a catalyst does not change the numerical value of the equilibrium constant.

We can now make two statements about a catalyst.

- The function of a catalyst is to change the mechanism of a reaction to one having a lower activation energy.
- A catalyst has no effect on the condition of equilibrium in a reversible reaction.

In Chapter 20 we discover that the equilibrium condition is essentially a thermodynamic condition.

Taken together these two statements must mean that an equilibrium condition is *independent* of the reaction mechanism. Thus, even though we have been describing equilibrium in terms of opposing reactions occurring at equal rates, we do not have to concern ourselves with the kinetics of chemical reactions to work with the equilibrium concept.

16-7   Equilibrium Calculations: Some Illustrative Examples

We have learned in this chapter that, although many chemical reactions go to completion, many others do not; they are reversible reactions that reach a condition of equilibrium. We have also established that the changes that occur in reversible reactions cannot be handled with the principles of stoichiometry alone; the equilibrium constant expression is needed. These facts suggest that calculations involving equilibrium constants are of great importance. We encounter such calculations throughout the next several chapters, but to conclude this chapter we consider a few examples that apply the general equilibrium principles described earlier.

Each of the following examples includes a brief section labeled ''comments'' and printed on a tan color panel. The complete set of these ''comments'' constitutes the basic methodology of equilibrium calculations. You may want to refer back to these comments from time to time in later chapters.

Example 16-10

Determining a value of K_c when the equilibrium quantities of substances are known. Dinitrogen tetroxide, N_2O_4, is an important component of rocket fuels. For example, N_2O_4 mixed with hydrazine is the fuel used in the Titan rocket. $N_2O_4(l)$ boils at 21.1 °C; this means that N_2O_4 exists as a gas at 25 °C. N_2O_4, a colorless gas, dissociates into NO_2, a red-brown gas. Mixtures of the two gases vary in color depending on their relative proportions in the equilibrium mixture. Equilibrium is established in the reaction $N_2O_4(g) \rightleftharpoons 2\ NO_2(g)$ at 25 °C. The quantities of the two gases present in a 3.00-L vessel are 7.64 g N_2O_4 and 1.56 g NO_2. What is the value of K_c for this reaction?

Solution

Equilibrium amounts, mol	Equilibrium concentrations, mol/L
$7.64 \text{ g N}_2\text{O}_4 \times \dfrac{1 \text{ mol N}_2\text{O}_4}{92.01 \text{ g N}_2\text{O}_4} = 0.0830 \text{ mol N}_2\text{O}_4$	$[\text{N}_2\text{O}_4] = \dfrac{0.0830 \text{ mol N}_2\text{O}_4}{3.00 \text{ L}} = 0.0277 \text{ M}$
$1.56 \text{ g NO}_2 \times \dfrac{1 \text{ mol NO}_2}{46.01 \text{ g NO}_2} = 0.0339 \text{ mol NO}_2$	$[\text{NO}_2] = \dfrac{0.0339 \text{ mol NO}_2}{3.00 \text{ L}} = 0.0113 \text{ M}$

$$K_c = \frac{[\text{NO}_2]^2}{[\text{N}_2\text{O}_4]} = \frac{(0.0113)^2}{0.0277} = 4.61 \times 10^{-3}$$

SIMILAR EXAMPLES: Exercises 6, 8, 10.

Comments

1. In using an equilibrium constant, K_c, be sure that you substitute equilibrium *concentrations* in *mol/L, not* equilibrium amounts in moles or grams. You will find it helpful to organize all the equilibrium data and carefully label each item. A tabular format is often a good one.

Example 16-11

Determining a value of K_c when data about both the initial and equilibrium amounts of substances are given. Relating K_p and K_c. The equilibrium involving $SO_2(g)$, $O_2(g)$, and $SO_3(g)$, as we have noted before, is important in sulfuric acid production. Here we wish to determine (a) K_c and (b) K_p for the *decomposition* of $SO_3(g)$ at 900 K.

$$2 SO_3(g) \rightleftharpoons 2 SO_2(g) + O_2(g)$$

The known data are that when a 0.0200-mol sample of SO_3 is introduced into an evacuated 1.52-L vessel at 900 K, 0.0142 mol SO_3 is found to be present at equilibrium.

Solution. In the table of data below the *key* item is the change in amount of SO_3: $(0.0142 - 0.0200)$ mol $SO_3 = -0.0058$ mol SO_3. (The negative sign signifies that this amount of SO_3 is *consumed* to establish equilibrium.) In the row labeled "change" we must relate the changes in amounts of SO_2 and O_2 to the change in amount of SO_3. For this we use the stoichiometric coefficients from the balanced equation: 2, 2, and 1. That is, two moles of SO_2 and one mole of O_2 are produced for every two moles of SO_3 that dissociate.

the reaction:	$2 SO_3(g)$	$\rightleftharpoons$	$2 SO_2(g)$	$+$	$O_2(g)$
initial amounts:	0.0200 mol		0.00 mol		0.00 mol
change:	-0.0058 mol		$+0.0058$ mol		$+0.0029$ mol
equil. amounts:	0.0142 mol		0.0058 mol		0.0029 mol
equil. concns:	$[SO_3] = 0.0142$ mol/1.52 L $= 9.34 \times 10^{-3}$ M		$[SO_2] = 0.0058$ mol/1.52 L $= 3.8 \times 10^{-3}$ M		$[O_2] = 0.0029$ mol/1.52 L $= 1.9 \times 10^{-3}$ M

(a) $K_c = \dfrac{[SO_2]^2[O_2]}{[SO_3]^2} = \dfrac{(3.8 \times 10^{-3})^2(1.9 \times 10^{-3})}{(9.34 \times 10^{-3})^2} = 3.1 \times 10^{-4}$

(b) $K_p = K_c(RT)^{\Delta n_{gas}} = 3.1 \times 10^{-4}(0.0821 \times 900)^{(2+1)-2}$

$= 3.1 \times 10^{-4}(0.0821 \times 900)^1 = 2.3 \times 10^{-2}$

SIMILAR EXAMPLES: Exercises 6, 10, 21, 22, 60.

Comments

2. The chemical equation for a reversible reaction serves *both* to establish the form of the equilibrium constant expression *and* to provide the conversion factors (stoichiometric factors) to relate the equilibrium amounts or concentrations to the initial conditions.

3. Whether working with K_c or K_p, or the relationship between them, you must always base these expressions on the chemical equation that is given, not on what you may have used in other situations.

K_p and K_c were related for the sulfur dioxide–oxygen–sulfur trioxide reaction in Example 16-3, but the result here is different because in Example 16-3 the forward reaction represented the *synthesis* of SO_3 from SO_2 and O_2 and here it represents the *decomposition* of SO_3.

Example 16-12

Determining equilibrium partial and total pressures when a value of K_p is known. Ammonium compounds dissociate when heated. Ammonium hydrogen sulfide, $NH_4HS(s)$ (used as a photographic developer and in the textile industry), dissociates appreciably even at room temperature.

$$NH_4HS(s) \rightleftharpoons NH_3(g) + H_2S(g) \qquad K_p(atm) = 0.108 \text{ at } 25 \text{ °C}$$

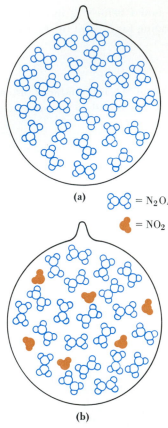

$= N_2O_4$

$= NO_2$

FIGURE 16-6
Equilibrium in the reaction

$N_2O_4(g) \rightleftharpoons 2\ NO_2(g)$

at 25 °C—Example 16-13 illustrated.

Each "molecule" illustrated represents 0.001 mol.
(a) Initially, pure N_2O_4 is introduced into an evacuated glass bulb and the bulb is sealed. The illustration shows 24 "molecules" corresponding to 0.024 mol N_2O_4.
(b) At equilibrium, some molecules of N_2O_4 have dissociated to NO_2 (shown in brown). The illustration contains 21 N_2O_4 and 6 NO_2 "molecules," corresponding to 0.021 mol N_2O_4 and 0.006 mol NO_2.

For a discussion of quadratic equations see Appendix A.

A sample of $NH_4HS(s)$ is introduced into an evacuated flask and allowed to establish equilibrium at 25 °C. What is the total gas pressure at equilibrium?

Solution. K_p for this reaction is just the product of the equilibrium partial pressures of $NH_3(g)$ and $H_2S(g)$, each stated in atm. (There is no term for the solid NH_4HS). Moreover, since these gases are produced in equimolar amounts, $P_{NH_3} = P_{H_2S}$; and $P_{tot.} = P_{NH_3} + P_{H_2S} = 2 \times P_{NH_3}$. Let us first find P_{NH_3} and then $P_{tot.}$.

$$K_p = (P_{NH_3})(P_{H_2S}) = (P_{NH_3})(P_{NH_3}) = (P_{NH_3})^2 = 0.108$$

$$P_{NH_3} = \sqrt{0.108} = 0.329\ \text{atm}$$

$$\mathbf{P_{tot.}} = 2 \times P_{NH_3} = 2 \times 0.329\ \text{atm} = \mathbf{0.658\ atm}$$

SIMILAR EXAMPLES: Exercises 42, 43, 63.

Comments

4. When using K_p expressions look for relationships among partial pressures of the reactants. If you need to relate the total pressure to reactant partial pressures, you should be able to do this with equations presented in Chapter 6 (e.g., 6.14, 6.15, and 6.16).

Example 16-13

Calculating equilibrium concentrations from initial concentrations and a value of K_c. Expressing the results of a calculation as a percent dissociation. This is perhaps most typical of the equilibrium calculations you will encounter in this text. Concerning the dissociation of $N_2O_4(g)$ (recall Example 16-10), an 0.0240-mol sample of $N_2O_4(g)$ is allowed to dissociate and come to equilibrium with $NO_2(g)$ in an 0.372-L flask at 25 °C. What is the percent dissociation of the N_2O_4?

$$N_2O_4(g) \rightleftharpoons 2\ NO_2(g) \qquad K_c = 4.61 \times 10^{-3}\ \text{at 25 °C}$$

Solution. By percent dissociation we mean the percent of the N_2O_4 molecules present initially that are converted to NO_2 (see Figure 16-6). This requires us to determine the number of moles of N_2O_4 that dissociate to establish equilibrium. For the first time we need to introduce an algebraic unknown, x. Suppose we let x = the number of moles of N_2O_4 that dissociate. We enter this value into the row labeled "change" in the table below. The amount of NO_2 produced is $2x$.

the reaction:	$N_2O_4(g)$	$\rightleftharpoons$	$2\ NO_2(g)$
initial amounts:	0.0240 mol		0.00 mol
change:	$-x$ mol		$+2x$ mol
equil. amounts:	$(0.0240 - x)$ mol		$2x$ mol
equil. concns:	$[N_2O_4] = (0.0240 - x\ \text{mol})/0.372\ \text{L}$		$[NO_2] = 2x\ \text{mol}/0.372\ \text{L}$

$$K_c = \frac{[NO_2]^2}{[N_2O_4]} = \frac{\left(\dfrac{2x}{0.372}\right)^2}{\dfrac{0.0240 - x}{0.372}} = \frac{4x^2}{0.372(0.0240 - x)} = 4.61 \times 10^{-3}$$

$$4x^2 = 4.12 \times 10^{-5} - (1.71 \times 10^{-3})x$$

The symbol $\pm$ in this equation signifies that there are two possible roots. You can decide between the two by considering their physical meaning. In this problem x must be a *positive* quantity, smaller than 0.0240.

$$x^2 + (4.28 \times 10^{-4})x - 1.03 \times 10^{-5} = 0$$

$$x = \frac{-4.28 \times 10^{-4} \pm \sqrt{(4.28 \times 10^{-4})^2 + 4 \times 1.03 \times 10^{-5}}}{2}$$

$$x = \frac{-4.28 \times 10^{-4} \pm \sqrt{(1.83 \times 10^{-7}) + (4.12 \times 10^{-5})}}{2}$$

$$x = \frac{-4.28 \times 10^{-4} \pm \sqrt{4.14 \times 10^{-5}}}{2}$$

$$x = \frac{-4.28 \times 10^{-4} \pm 6.43 \times 10^{-3}}{2}$$

$$= \frac{-4.28 \times 10^{-4} + 6.43 \times 10^{-3}}{2} = \frac{6.00 \times 10^{-3}}{2}$$

$$x = 3.00 \times 10^{-3} \text{ mol } N_2O_4 \quad \text{(dissociated)}$$

The percent dissociation of the N_2O_4 is given by the expression

$$\% \text{ dissoc. of } N_2O_4 = \frac{3.00 \times 10^{-3} \text{ mol } N_2O_4 \text{ dissoc.}}{0.0240 \text{ mol } N_2O_4 \text{ initially}} \times 100$$

$$= 12.5\%$$

SIMILAR EXAMPLES: Exercises 44, 45, 46.

Comments

5. When you need to introduce an algebraic unknown, x, into an equilibrium calculation,

- Introduce x into the tabular setup in the row labeled "change."
- Decide which change to label as x (the amount of a reactant consumed or of a product formed).
- Use stoichiometric factors to relate the other changes to x (that is, $2x$, $3x$, . . .).
- Remember that substances *consumed* are represented through a *negative* sign (e.g., $-x$, $-2x$, . . .). Substances *produced* are represented through a *positive* sign (e.g., $+x$, $+2x$, . . .).
- Consider that equilibrium amounts = initial amounts + "change." (If you have assigned the correct signs to the changes, then equilibrium amounts will also be correct.)

Example 16-14

Here is a straightforward way of assessing when you need to evaluate Q: If *all* of the reacting species are present initially, use Q to determine the direction of net change. If one or more reactants or products is missing initially, a net reaction must occur to produce some of that substance(s).

Using the reaction quotient, Q, in an equilibrium calculation. A solution is prepared having these initial concentrations: $[Fe^{3+}] = [Hg_2^{2+}] = 0.5000$ M; $[Fe^{2+}] = [Hg^{2+}] = 0.03000$ M. The following reaction occurs among these ions at 25 °C.

$$2 \text{ Fe}^{3+}(aq) + Hg_2^{2+}(aq) \rightleftharpoons 2 \text{ Fe}^{2+}(aq) + 2 \text{ Hg}^{2+}(aq) \quad K_c = 9.14 \times 10^{-6}$$

What will be the ionic concentrations when equilibrium is established?

Solution. Since all reactants and products are present initially, we do not know whether a net reaction will occur "to the left" or "to the right." This is where the reaction quotient Q can help us.

$$Q = \frac{[Fe^{2+}]^2[Hg^{2+}]^2}{[Fe^{3+}]^2[Hg_2^{2+}]} = \frac{(0.03000)^2(0.03000)^2}{(0.5000)^2(0.5000)} = \frac{8.10 \times 10^{-7}}{1.25 \times 10^{-1}}$$
$$= 6.48 \times 10^{-6}$$

Since Q (6.48×10^{-6}) is smaller than K_c (9.14×10^{-6}), a net reaction must proceed *to the right* (recall criterion 16.29). Let us define x as the changes in the numbers of moles per liter of Fe^{2+}, Fe^{3+}, and Hg^{2+}. Since only half as much Hg_2^{2+} is consumed as Fe^{3+}, the change in its concentration is $x/2$. By knowing that a net reaction occurs *to the right*, the changes for the species on the left side of the equation are *negative* and on the right side, *positive*. The relevant data are tabulated below.

the reaction:	2 Fe^{3+}(aq)	+	Hg_2^{2+}(aq)	⇌	2 Fe^{2+}(aq)	+	2 Hg^{2+}(aq)
initial concns:	0.5000 M		0.5000 M		0.03000 M		0.0300 M
change:	$-x$ M		$-x/2$ M		$+x$ M		$+x$ M
equil. concns:	$(0.5000 - x)$ M		$(0.5000 - x/2)$ M		$(0.03000 + x)$ M		$(0.03000 + x)$ M

$$K_c = \frac{(0.03000 + x)^2(0.03000 + x)^2}{(0.5000 - x)^2(0.5000 - x/2)} = 9.14 \times 10^{-6}$$

Solving this equation can be simplified greatly if the following assumption is made, and *if the assumption proves valid*. If x is much smaller than 0.5000, then $(0.5000 - x) \simeq 0.5000$, and $(0.5000 - x/2) \simeq 0.5000$. This assumption leads to the expression

$$K_c = \frac{(0.03000 + x)^2(0.03000 + x)^2}{(0.5000)^2(0.5000)} = \frac{(0.03000 + x)^4}{0.1250} = 9.14 \times 10^{-6}$$

$$(0.03000 + x)^4 = (0.1250)(9.14 \times 10^{-6}) = 1.14 \times 10^{-6}$$

Take the *fourth* root of each side of this equation (i.e., take the square root twice).

$$(0.03000 + x)^2 = 10.7 \times 10^{-4}$$
$$(0.03000 + x) = 3.27 \times 10^{-2}$$
$$x = 3.27 \times 10^{-2} - 0.03000 = 2.7 \times 10^{-3}$$

The simplifying assumption appears to be valid: 2.7×10^{-3} is considerably smaller than 0.5000.

Equilibrium concentrations

$[Fe^{2+}] = 0.03000 + x = 0.03000 + 2.7 \times 10^{-3} = 3.27 \times 10^{-2}$ M

$[Hg^{2+}] = 0.03000 + x = 0.03000 + 2.7 \times 10^{-3} = 3.27 \times 10^{-2}$ M

$[Fe^{3+}] = 0.5000 - x = 0.5000 - 2.7 \times 10^{-3} = 4.973 \times 10^{-1}$ M

$[Hg_2^{2+}] = 0.5000 - x/2 = 0.5000 - 1.4 \times 10^{-3} = 4.986 \times 10^{-1}$ M

Checking the results

If we have done the calculation correctly, we should obtain the known value of K_c when we substitute the *calculated* equilibrium concentrations into the reaction quotient, Q.

$$Q = \frac{[Fe^{2+}]^2[Hg^{2+}]^2}{[Fe^{3+}]^2[Hg_2^{2+}]} = \frac{(3.27 \times 10^{-2})^2(3.27 \times 10^{-2})^2}{(0.4973)^2 \times 0.4986} = 9.27 \times 10^{-6}$$

This value of Q (9.27×10^{-6}) is in reasonably good agreement with the value of K_c (9.14×10^{-6}). Our calculation appears to be correct. (The reason the agreement between Q and K_c is not still better is because of the simplifying assumption we made in our algebraic solution.)

SIMILAR EXAMPLES: Exercises 12, 30, 38, 39, 62, 75.

Comments

6. It is sometimes useful to compare the reaction quotient, Q, to the equilibrium constant, K_c or K_p, to determine in which direction a net reaction will occur.

7. In some equilibrium calculations—often those in aqueous solutions—you can work with molarity concentrations directly, without having to work with moles of reactants and solution volumes.

8. Often, you may be able to simplify algebraic calculations by recognizing relationships among the terms in the equilibrium constant expression. Usually these involve x being much smaller than some other numerical value to which it is added or from which it is subtracted.

9. Where possible, devise ways of checking your calculation. Often this can take the form of substituting *calculated* equilibrium concentrations into the reaction quotient, Q, to see if it equals K_c.

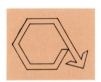

FOCUS ON
The Synthesis of Ammonia and Other Nitrogen Compounds

A nitrogen-containing fertilizer being applied to a bean field. Liquid ammonia and chemicals derived from ammonia—ammonium nitrate, ammonium sulfate, and urea—are all used as fertilizers. [Roy Morsch/The Stock Market]

Elemental nitrogen occurs in the atmosphere to the extent of 78% by volume, but because of nitrogen's relative inertness, nitrogen-containing compounds do not occur extensively in nature. Industrial processes of synthesizing nitrogen compounds, referred to as artificial fixation of nitrogen, are of utmost importance. Chief among these processes is the reaction of N_2 and H_2 to form ammonia. The ammonia can be converted to oxides of nitrogen, nitric acid, and nitrate salts. Urea, another important nitrogen-containing chemical, is formed by the reaction of NH_3 and CO_2. The production scheme for these several chemicals is outlined in Figure 16-7.

Ammonia. Fritz Haber (1908) worked out the theoretical basis of the ammonia synthesis reaction and tested it in the laboratory. Haber's laboratory apparatus was capable of producing about 1 kg NH_3 per day. The development work required to convert Haber's process into a commercial operation was one of the most challenging chemical engineering problems of its time, requiring that methods be devised to conduct gaseous reactions at high temperatures and pressures in the presence of an appropriate catalyst. This work was done under the leadership of Carl Bosch at Badische Anilin & Soda-Fabrik (BASF). By 1913 a plant capable of producing 30,000 kg NH_3 per day was in operation. Modern ammonia plants have about 50 times this capacity.

Here are some relevant data on the ammonia synthesis reaction

$$N_2(g) + 3\ H_2(g) \rightleftharpoons 2\ NH_3(g) \qquad (16.36)$$

$$\Delta H^\circ = -92.22 \text{ kJ} \qquad K_p = 6.2 \times 10^5 \text{ at 298 K}$$

For every 4 mol of gases that react (1 mol N_2 and 3 mol H_2), 2 mol NH_3 is produced. Increasing the pressure forces the reaction mixture into a smaller volume and favors the reaction producing the smaller number of moles of gas—the production of NH_3. The forward reaction is *exothermic*. The exothermic reaction is favored if the tem-

FIGURE 16-7

Production scheme for some important nitrogen-containing chemicals.

Those shown in brown have important uses as fertilizers.

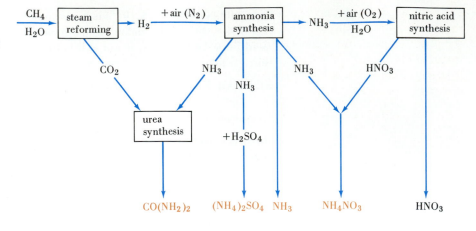

perature is *lowered*. Thus, the optimum conditions for the production of NH_3 appear to be *high pressure* and *low temperatures*.

These "optimum" conditions do not take into account the rate of reaction, however. Although the equilibrium production of NH_3 is favored at low temperatures, the rate of its formation is so slow as to make the method unfeasible. One way to speed up the reaction is to raise the temperature (even though the equilibrium concentration of NH_3 is decreased when this is done). Another way is to use a catalyst. The usual operating conditions of the Haber–Bosch process are about 550 °C, pressures ranging from 150 to 350 atm, and a catalyst—usually iron in the presence of Al_2O_3, MgO, CaO, and K_2O. The dramatic difference between the theoretical optimum conditions and the actual operating conditions is suggested through Figure 16-8.

Another way to increase the rate of production of NH_3 is to remove NH_3 continuously as it is formed. This is done by liquefying the NH_3 and recycling the N_2 and H_2, which are not easily liquefied. The disturbance caused by the continuous removal of NH_3 displaces the equilibrium toward the production of more NH_3. In fact, the mixture need not be allowed to come to equilibrium at all. In this way practically 100% conversion of N_2 and H_2 to NH_3 is possible. A schematic representation of a portion of the ammonia synthesis reaction is given in Figure 16-9.

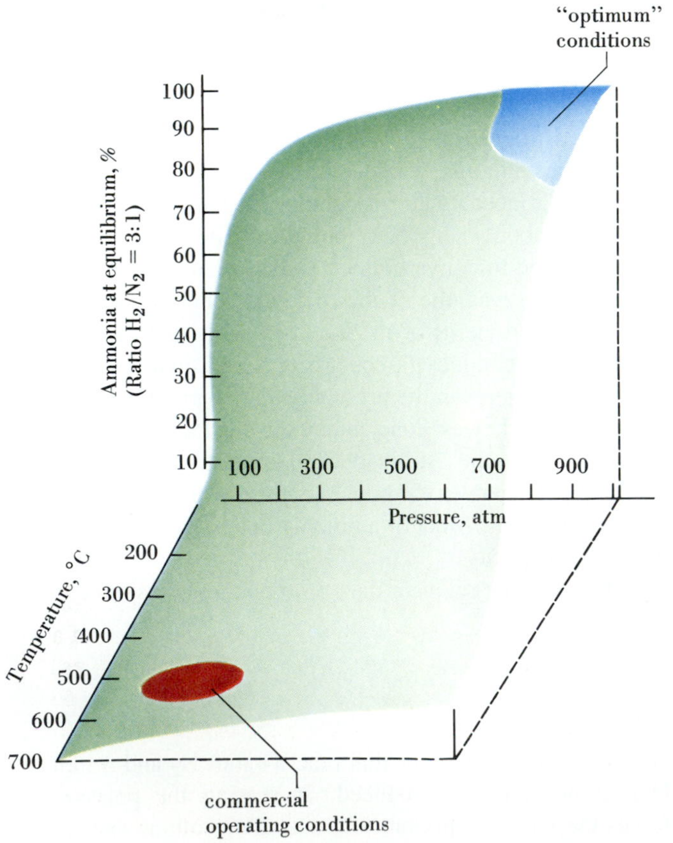

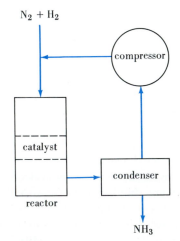

FIGURE 16-9

Ammonia synthesis reaction.

The gaseous N_2–H_2 mixture is introduced into a reactor at high temperature and pressure in the presence of a catalyst. The gaseous N_2–H_2–NH_3 mixture leaves the reactor and is cooled as it passes through a condenser. Liquefied NH_3 is removed, and the remaining N_2–H_2 mixture is compressed and returned to the reactor.

FIGURE 16-8

Equilibrium conversion of $N_2(g)$ and $H_2(g)$ to $NH_3(g)$ as a function of temperature and pressure.

A critical part of the ammonia synthesis reaction is the source of $H_2(g)$. In the original Haber–Bosch process, $H_2(g)$ was produced from coke and steam by the water gas reactions (14.1 and 14.2). The common method now used is steam reforming of natural gas (methane) or other hydrocarbons (reaction 14.3).

$$CH_4(g) + H_2O(g) \rightleftharpoons CO(g) + 3 H_2(g) \qquad (14.3)$$

followed by

$$CO(g) + H_2O(g) \rightleftharpoons CO_2(g) + H_2(g) \qquad (14.2)$$

Because natural gas is so necessary to the production of ammonia, the cost of ammonia-based fertilizers is closely linked to the cost of natural gas and other hydrocarbon fuels.

In addition to its use in producing nitric acid and urea, NH_3 is used in the preparation of various nitrogen-containing monomers used in the manufacture of nylon, acrylic polymers, and polyurethane foams. Ammonia is also used in the manufacture of pharmaceuticals, various organic and inorganic chemicals, detergents, and cleansers. Another important use is its direct application to fields as a fertilizer (under the name "anhydrous ammonia"). It is no wonder that ammonia ranks among the top half-dozen chemicals in annual production in the United States.

Nitric Acid. The key to the production of nitric acid is the oxidation of $NH_3(g)$ to $NO(g)$. The conditions developed by Wilhelm Ostwald early in the present century—brief contact (0.01 s) of $NH_3(g)$ and an excess of $O_2(g)$ with a Pt catalyst at about 1000 °C—are essentially those in use in the Ostwald process today.

$$4 NH_3(g) + 5 O_2(g) \rightleftharpoons 4 NO(g) + 6 H_2O(g) \quad (16.37)$$

The $NO(g)$ is cooled to near room temperature, where its conversion to $NO_2(g)$ is favorable (recall Table 16-2).

$$2 NO(g) + O_2(g) \rightleftharpoons 2 NO_2(g) \qquad (16.38)$$

The $NO_2(g)$ is dissolved in water, and the $NO(g)$ is recycled.

$$3 NO_2 + H_2O \longrightarrow 2 HNO_3(aq) + NO(g) \qquad (16.39)$$

Nitric acid is used to prepare nitrate salts (chiefly NH_4NO_3), the explosives nitroglycerine and nitrocellulose, and various nitrogen-containing organic compounds. Currently, nitric acid is about thirteenth among the top chemicals produced in the United States.

Urea. The reaction of $NH_3(g)$ and $CO_2(g)$ at high temperatures (190 °C) yields an intermediate called ammonium carbamate ($NH_4CO_2NH_2$), which dehydrates (loses H_2O) to form urea. The urea and H_2O form a concentrated liquid solution. The net equilibrium reaction, for which no catalyst is required, is

$$2 NH_3(g) + CO_2(g) \rightleftharpoons \underset{\text{urea}}{CO(NH_2)_2} + H_2O \qquad (16.40)$$

As expected, the forward reaction is favored by high pressures; the usual operating condition is about 200 atm. A small excess of NH_3 beyond the 2:1 molar proportions of equation (16.40) is also used.

Urea contains 46% nitrogen, by mass. It is an excellent fertilizer, used as a pure solid, as the solid mixed with ammonium salts, or a very concentrated aqueous solution mixed with NH_4NO_3 and/or NH_3. Urea is also used as a feed supplement for cattle and in the production of polymers and pesticides. Currently urea ranks about fourteenth among the top chemicals produced in the United States.

Summary

When two competing reactions—a forward and a reverse reaction—occur simultaneously, a reaction does not go to completion. The *reversible* reaction proceeds to the point where the rates of the forward and reverse reactions become equal. In this dynamic equilibrium no further net change occurs and the amounts of the reacting species remain constant with time. The chemical equation describes the proportions in which reactants participate in a reversible reaction, but a quantitative description of the equilibrium condition is provided through the equilibrium constant expression.

The *form* of the equilibrium constant expression, K_c, is established through the balanced chemical equation. The *numerical value* of the constant is determined by experiment. If a reaction involves gases, an equilibrium constant expression, K_p, can be based on partial pressures of the gases. K_p and K_c are related to each other through a simple mathematical equation. Only substances whose concentrations or partial pressures can change during the course of a reaction are represented in an equilibrium constant expression. This means that terms for pure solids and liquids do not appear. The equilibrium constant expression is inverted when a chemical equation is written in the reverse direction. If a chemical equation is multiplied by a factor, the equilibrium constant is raised to the same power. If two or more equations are added, the equilibrium constant for the resulting net reaction is the product of the constants for the individual reactions.

A ratio of *initial* reactant concentrations can be formed in the same manner as the equilibrium constant expression. By comparing the numerical value of this ratio, called the reaction quotient, Q, to the corresponding value of K_c or K_p, one can determine the *direction* in which a net reaction proceeds. An algebraic equation can be written and solved to determine the *extent* of reaction, that is, to relate the final equilibrium to the initial conditions. A variety of possibilities exists for equilibrium calculations, but the basic principles and algebraic techniques involved are few in number, and these are illustrated in the chapter.

Qualitative predictions of the effects of different variables on an equilibrium condition can be made with Le Châtelier's principle, which states that an equilibrium condition is modified or "shifts" whenever an equilibrium mixture is disturbed by adding or removing reagents or by changing the pressure or temperature. The equilibrium condition shifts in the direction that consumes some of a reagent if more of that reagent is added to an equilibrium mixture; it shifts in the direction of the smaller number of moles of gas when pressure is increased; it shifts in the direction of the *endothermic* reaction if the temperature is raised and the *exothermic* reaction if the temperature is lowered. Addition of an inert gas to a *constant volume* equilibrium mixture has no effect on the equilibrium condition. Although the presence of a catalyst speeds the attainment of equilibrium, because the catalyst speeds up the forward and reverse reactions equally, it has no effect on the amounts of reactants and products present at equilibrium.

Summarizing Example

In the manufacture of ammonia the chief source of hydrogen gas is this reaction for the reforming of methane at elevated temperatures (described in the Focus feature).

$$CH_4(g) + 2\,H_2O(g) \rightleftharpoons CO_2(g) + 4\,H_2(g) \tag{16.41}$$

The following data are available
(a) $CO(g) + H_2O(g) \rightleftharpoons CO_2(g) + H_2(g)$
$$\Delta H = -40.\ \text{kJ} \quad K_c = 1.4 \text{ at } 1000\text{ K}$$
(b) $CO(g) + 3\,H_2(g) \rightleftharpoons H_2O(g) + CH_4(g)$
$$\Delta H = -230.\ \text{kJ} \quad K_c = 190. \text{ at } 1000\text{ K}$$

1. Determine the value of K_c for reaction (16.41) at 1000 K.

Solution. To combine the given equations we must use (a) as written and reverse (b). The values of ΔH must be handled as described in Section 7-8, and the values of K_c as described in Section 16-3.

(a) $CO(g) + H_2O(g) \rightleftharpoons CO_2(g) + H_2(g)$
$$\Delta H = -40.\ \text{kJ} \quad K_c = 1.4$$
$-$(b) $CH_4(g) + H_2O(g) \rightleftharpoons CO(g) + 3\,H_2(g)$
$$\Delta H = +230.\ \text{kJ} \quad K_c = 1/190.$$

net: $CH_4(g) + 2\,H_2O(g) \rightleftharpoons CO_2(g) + 4\,H_2(g)$
$$\Delta H = +190.\ \text{kJ} \quad K_c = 1.4/190.$$
$$= 7.4 \times 10^{-3} \tag{16.42}$$

(This example is similar to Example 7-10 and Example 16-2.)

2. Determine the value of K_p for reaction (16.42) at 1000 K.

Solution. Here we need to use the expression $K_p = K_c(RT)^{\Delta n_\text{gas}}$, with a value of $\Delta n_\text{gas} = 2$ (5 moles of gases on the right and 3 moles on the left).

$$K_p = K_c(RT)^{\Delta n_\text{gas}} = 7.4 \times 10^{-3}\,(0.08206 \times 1000)^2 = 50.$$

(This example is similar to Example 16-3.)

Hydrogen-producing facility at an ammonia synthesis plant. [Photo: Bayer AG]

3. 1.00 mol each of $CH_4(g)$ and $H_2O(g)$ are allowed to come to equilibrium in a 10.0-L vessel at 1000 K. Calculate the number of moles of $H_2(g)$ present at equilibrium.

Solution. In the tabulation below, x represents the change in the number of moles of CH_4 and of CO_2. In accordance with the stoichiometry of the reaction, changes in the numbers of moles of H_2O and H_2 are $2x$ and $4x$, respectively. Also, increases in amounts of substances are denoted by positive signs ($+x$, $+4x$) and decreases in amounts by negative signs ($-x$, $-2x$).

the reaction:	$CH_4(g)$	$+$	$2 H_2O(g)$	$\rightleftharpoons$	$CO_2(g)$	$+$	$4 H_2(g)$
initial amts:	1.00 mol		1.00 mol		0.0 mol		0.0 mol
changes:	$-x$ mol		$-2x$ mol		$+x$ mol		$+4x$ mol
equil. amts:	$(1.00 - x)$ mol		$(1.00 - 2x)$ mol		x mol		$4x$ mol
equil. concns, M:	$(1.00 - x)/10.0$		$(1.00 - 2x)/10.0$		$x/10.0$		$4x/10.0$

$$K_c = \frac{[CO_2][H_2]^4}{[CH_4][H_2O]^2} = \frac{\left(\dfrac{x}{10.0}\right)\left(\dfrac{4x}{10.0}\right)^4}{\left(\dfrac{1.00 - x}{10.0}\right)\left(\dfrac{1.00 - 2x}{10.0}\right)^2} = \frac{x\,(4x)^4}{100(1.00 - x)(1.00 - 2x)^2}$$

$$= 7.4 \times 10^{-3}$$

$$256x^5 = 0.74\,[(1.00 - x)(1.00 - 2x)^2]$$

$$256x^5 - 0.74\,[(1.00 - x)(1.00 - 2x)^2] = 0$$

The above equation looks impossibly difficult to solve but is not. The trick in finding a simple solution is in using the **method of successive approximations** illustrated in Appendix A. This equation is solved on pages A6–A7; the result is $x = 0.23$. The number of moles of $H_2(g)$ at equilibrium is $4x = 0.92$.

(This example is similar to Example 16-13.)

4. Describe the *qualitative* effect that each of the following actions would have on the amount of $H_2(g)$ in the 10.0-L equilibrium mixture of part 3: **(a)** add a catalyst; **(b)** add 24.0 g CH_4; **(c)** add 10.0 g CO_2; **(d)** remove 6.00 g H_2O; **(e)** raise the temperature to 1200 K; **(f)** transfer the mixture to a 15.0-L vessel; **(g)** add 10.0 g He(g) at constant volume.

Solution. To describe these effects qualitatively means simply to indicate whether the amount of $H_2(g)$ would increase, decrease, or remain unchanged when equilibrium is reestablished. We can do this most simply by referring to equation (16.42) and applying Le Châtelier's principle.

The amount of $H_2(g)$ will

(a) Add a catalyst   not change
(b) Add 24.0 g CH_4 increase
(c) Add 10.0 g CO_2 decrease
(d) Remove 6.00 g H_2O decrease
(e) Raise the temperature to 1200 K   increase
(f) Transfer the mixture to a 15.0-L vessel   increase
(g) Add 10.0 g He(g) at constant volume   not change

(This example is similar to Examples 16-7, 16-8, and 16-9.)

Key Terms

equilibrium (16-1)
equilibrium constant (16-2)
equilibrium constant expression (16-2)

K_c (16-2)
K_p (16-3)
Le Châtelier's principle (16-6)

mass-action expression (16-5)
reaction quotient, Q (16-5)

Highlighted Expressions

General expression for an equilibrium constant, K_c (16.6)
Relationship of K_c to the balanced chemical equation (16.7)
Procedure for combining equilibrium constants into a value for a net reaction (16.10)
Expressing an equilibrium constant through partial pressures of gases: K_p (16.14)
Relationship between K_p and K_c (16.18)
Treatment of solids and liquids in equilibrium constant expressions (16.19)
The significance of a very large numerical value of K_c or K_p (16.24)

The significance of a very small numerical value of K_c and K_p (16.25)
Conditions under which an equilibrium constant value is significant (16.26)
Formulation of the reaction quotient, Q (16.27)
Criterion for a net change in the forward direction (16.29)
Criterion for a net change in the reverse direction (16.31)
Le Châtelier's principle (16.32)
Effect of pressure on equilibrium condition (16.34)
Effect of temperature on equilibrium condition (16.35)

Review Problems

1. Write an equilibrium constant expression, K_c, for each of the following reactions.

(a) $CO(g) + Cl_2(g) \rightleftharpoons COCl_2(g)$
(b) $2 NO(g) + O_2(g) \rightleftharpoons 2 NO_2(g)$
(c) $CO_2(s) \rightleftharpoons CO_2(g)$
(d) $CS_2(g) + 4 H_2(g) \rightleftharpoons CH_4(g) + 2 H_2S(g)$
(e) $2 NaHCO_3(s) \rightleftharpoons Na_2CO_3(s) + CO_2(g) + H_2O(g)$

2. From the values of K_c given

$$CO(g) + H_2O(g) \rightleftharpoons CO_2(g) + H_2(g) \qquad K_c = 23.2 \text{ at } 600 \text{ K}$$

$$SO_2(g) + \tfrac{1}{2} O_2(g) \rightleftharpoons SO_3(g) \qquad K_c = 56 \text{ at } 900 \text{ K}$$

$$2 H_2S(g) \rightleftharpoons 2 H_2(g) + S_2(g) \qquad K_c = 2.3 \times 10^{-4} \text{ at } 1405 \text{ K}$$

$$2 NO_2(g) \rightleftharpoons 2 NO(g) + O_2(g) \qquad K_c = 1.8 \times 10^{-6} \text{ at } 457 \text{ K}$$

determine values of K_c for the following reactions.

(a) $CO_2(g) + H_2(g) \rightleftharpoons CO(g) + H_2O(g)$
(b) $2 SO_2(g) + O_2(g) \rightleftharpoons 2 SO_3(g)$
(c) $H_2S(g) \rightleftharpoons H_2(g) + \tfrac{1}{2} S_2(g)$
(d) $NO(g) + \tfrac{1}{2} O_2(g) \rightleftharpoons NO_2(g)$

3. Determine K_c for the reaction

$$\tfrac{1}{2} N_2(g) + \tfrac{1}{2} O_2(g) + \tfrac{1}{2} Br_2(g) \rightleftharpoons NOBr(g)$$

from the following information (at 298 K).

$$2 NO(g) \rightleftharpoons N_2(g) + O_2(g) \qquad K_c = 2.4 \times 10^{30}$$

$$NO(g) + \tfrac{1}{2} Br_2(g) \rightleftharpoons NOBr(g) \qquad K_c = 1.4$$

4. Determine numerical values of K_p corresponding to the values of K_c listed for the four reactions in Exercise 2.

5. Equilibrium is established in the reversible reaction

$$A + B \rightleftharpoons 2 C$$

Following are the equilibrium concentrations: $[A] = 0.47$ M, $[B] = 0.55$ M, $[C] = 0.36$ M. What is the value of K_c for this reaction?

6. The reaction $2 A(g) + B(g) \rightleftharpoons C(g)$ was allowed to come to equilibrium. The *initial* amounts of the reactants present in a 1.80-L vessel were 1.18 mol A and 0.78 mol B. The amount of A, at equilibrium, was found to 0.92 mol. What is the value of K_c for this reaction?

7. An equilibrium mixture of SO_2, SO_3, and O_2 gases is maintained in a 2.52-L flask at a temperature at which $K_c = 76.0$ for the reaction

$$2 SO_2(g) + O_2(g) \rightleftharpoons 2 SO_3(g)$$

(a) If the numbers of moles of SO_2 and SO_3 in the flask are equal, how much O_2 is present?
(b) If the number of moles of SO_3 in the flask is twice the number of moles of SO_2, how much O_2 is present?

8. An equilibrium mixture at 1000 K contains 0.276 mol H_2, 0.276 mol CO_2, 0.224 mol CO, and 0.224 mol H_2O.

$$CO_2(g) + H_2(g) \rightleftharpoons CO(g) + H_2O(g)$$

(a) Show that for this reaction K_c is independent of the reaction volume, V.
(b) Determine the value of K_c and of K_p.

9. Both common chlorides of phosphorus, PCl_3 and PCl_5, are important starting materials for the commercial production of other phosphorus compounds. The two chlorides can exist in equilibrium through the reaction $PCl_3(g) + Cl_2(g) \rightleftharpoons PCl_5(g)$. At 250 °C, an equilibrium mixture in a 2.50-L flask contains 0.105 g PCl_5, 0.220 g PCl_3, and 2.12 g Cl_2. What is the value of (a) K_c and (b) K_p for this reaction?

10. When 1.00 mol $I_2(g)$ is introduced into an evacuated 1.00-L flask at 1200 °C, it is 5% dissociated into iodine atoms. For the reaction $I_2(g) \rightleftharpoons 2 I(g)$, what is the value of (a) K_c; (b) K_p?

11. 0.100 mol H_2 and 0.100 mol I_2 are sealed in a 1.50-L vessel and the mixture is allowed to come to equilibrium at

445 °C. What are the equilibrium concentrations of H_2, I_2, and HI?

$$H_2(g) + I_2(g) \rightleftharpoons 2\ HI(g) \qquad K_c = 50.2 \text{ at } 445 \text{ °C}$$

12. 0.100 mol H_2 and 0.100 mol HI are sealed in a 1.50-L vessel and the mixture is allowed to come to equilibrium at 445 °C. How many moles of I_2 will be present when equilibrium is established?

$$H_2(g) + I_2(g) \rightleftharpoons 2\ HI(g) \qquad K_c = 50.2 \text{ at } 445 \text{ °C}$$

13. What effect would increasing the external pressure on the reaction have on the equilibrium condition in each of the following reactions?
   (a) $C(s) + H_2O(g) \rightleftharpoons CO(g) + H_2(g)$
   (b) $CO(g) + H_2O(g) \rightleftharpoons CO_2(g) + H_2(g)$
   (c) $4\ HCl(g) + O_2(g) \rightleftharpoons 2\ H_2O(g) + 2\ Cl_2(g)$
14. For which of the following reactions would you expect the % dissociation to increase with increasing temperature? Explain.

(a) $NO(g) \rightleftharpoons \frac{1}{2} N_2(g) + \frac{1}{2} O_2(g)$	$\Delta H° = -90.2$ kJ
(b) $SO_3(g) \rightleftharpoons SO_2(g) + \frac{1}{2} O_2(g)$	$\Delta H° = +98.9$ kJ
(c) $N_2H_4(g) \rightleftharpoons N_2(g) + 2\ H_2(g)$	$\Delta H° = -95.4$ kJ
(d) $COCl_2(g) \rightleftharpoons CO(g) + Cl_2(g)$	$\Delta H° = +108.3$ kJ

15. The Deacon process is used to make chlorine gas from hydrogen chloride, especially in situations where a large amount of by-product HCl is available from other chemical processes.

$$4\ HCl(g) + O_2(g) \rightleftharpoons 2\ H_2O(g) + 2\ Cl_2(g) \qquad \Delta H° = -114 \text{ kJ}$$

A mixture of HCl, O_2, H_2O, and Cl_2 is brought to equilibrium at 400 °C. What will be the effect on the equilibrium amount of $Cl_2(g)$ if
   (a) Additional $O_2(g)$ is added to the mixture at constant volume?
   (b) HCl(g) is removed from the reaction mixture at constant volume?
   (c) The mixture is transferred to a vessel of twice the volume as the original equilibrium mixture?
   (d) A catalyst is added to the reaction mixture?
   (e) The temperature is raised to 500 °C?

Exercises

Writing equilibrium constant expressions

16. Based on the following descriptions of reversible reactions, write a balanced equation and then the K_c expression for each.
   (a) Oxygen gas oxidizes gaseous ammonia to gaseous nitrogen and water vapor.
   (b) Hydrogen gas reduces gaseous nitrogen dioxide to gaseous ammonia and water vapor.
   (c) Liquid acetone [$(CH_3)_2CO$] is in equilibrium with its vapor.
   (d) Chlorine gas reacts with liquid carbon disulfide to produce the liquids CCl_4 and S_2Cl_2.
   (e) Nitrogen gas reacts with the solids, sodium carbonate and carbon, to produce solid sodium cyanide and carbon monoxide gas.
17. Determine values of K_c from the K_p values given.
   (a) $SO_2Cl_2(g) \rightleftharpoons SO_2(g) + Cl_2(g)$
   $$K_p = 2.9 \times 10^{-2} \text{ at } 303 \text{ K}$$
   (b) $2\ NO(g) + O_2(g) \rightleftharpoons 2\ NO_2(g)$
   $$K_p = 1.48 \times 10^4 \text{ at } 184 \text{ °C}$$
   (c) $Sb_2S_3(s) + 3\ H_2(g) \rightleftharpoons 2\ Sb(s) + 3\ H_2S(g)$
   $$K_p = 0.429 \text{ at } 713 \text{ K}$$
18. The vapor pressure of water at 25 °C is 23.8 mmHg. Write K_p for the vaporization of water, in the unit, atm. What is the value of K_c for the vaporization process?
19. Given the equilibrium constant values

$$N_2(g) + \frac{1}{2} O_2(g) \rightleftharpoons N_2O(g) \qquad K_c = 3.4 \times 10^{-18}$$

$$N_2O_4(g) \rightleftharpoons 2\ NO_2(g) \qquad K_c = 4.6 \times 10^{-3}$$

$$\frac{1}{2} N_2(g) + O_2(g) \rightleftharpoons NO_2(g) \qquad K_c = 4.1 \times 10^{-9}$$

Determine a value of K_c for the reaction

$$2\ N_2O(g) + 3\ O_2(g) \rightleftharpoons 2\ N_2O_4(g)$$

20. Use the following equilibrium data at 1200 K to estimate a value of K_p at 1200 K for the reaction

$$2\ H_2(g) + O_2(g) \rightleftharpoons 2\ H_2O(g)$$

[*Hint:* Note that some of the values are K_p and some, K_c.]
   (a) C(graphite) + $CO_2(g) \rightleftharpoons 2\ CO(g) \qquad K_c = 0.64$
   (b) $CO_2(g) + H_2(g) \rightleftharpoons CO(g) + H_2O(g) \qquad K_p = 1.4$
   (c) C(graphite) + $\frac{1}{2} O_2(g) \rightleftharpoons CO(g) \qquad K_c = 1 \times 10^8$

Experimental determination of equilibrium constants

21. 1.00 g PCl_5 is introduced into a 250.-mL flask and the flask is heated to 250 °C, where the dissociation of PCl_5 is allowed to reach equilibrium: $PCl_5(g) \rightleftharpoons PCl_3(g) + Cl_2(g)$. The quantity of $Cl_2(g)$ present at equilibrium is found to be 0.25 g. What is the value of K_c for the dissociation reaction at 250 °C?
22. An experiment used to establish principles of chemical equilibrium was the homogeneous reaction of ethanol (C_2H_5OH) and acetic acid (CH_3COOH) to produce ethyl acetate and water.

$$C_2H_5OH + CH_3CO_2H \rightleftharpoons CH_3CO_2C_2H_5 + H_2O \qquad (16.43)$$

The reaction can be followed by analyzing the equilibrium mixture for its acetic acid content.

$$2\ CH_3CO_2H(aq) + Ba(OH)_2(aq) \rightarrow Ba(CH_3CO_2)_2(aq) + 2\ H_2O$$

An experiment is performed in which 1.000 mol of acetic acid and 0.500 mol of ethanol are mixed and allowed to come to equilibrium. A sample representing exactly one-hundredth of the total equilibrium mixture requires 28.85 mL of 0.1000 M $Ba(OH)_2$ for its titration. Show that the equilibrium constant, K_c, for reaction (16.43) is 4.0. [*Hint:* It is not necessary to know the volume of the reaction mixture.]
★23. The decomposition of HI(g) is represented by the equation $2\ HI(g) \rightleftharpoons H_2(g) + I_2(g)$. HI(g) is introduced into five identical 400-cm³ glass bulbs, and the five bulbs are maintained at 623 K. Each bulb is opened after a period of time and analyzed for I_2 by titration with 0.0150 M $Na_2S_2O_3(aq)$.

$$I_2(aq) + 2\ Na_2S_2O_3(aq) \rightarrow Na_2S_4O_6(aq) + 2\ NaI(aq)$$

What is the value of K_c at 623 K?

Bulb number	Original mass of HI(g), g	Bulb opened after; h	Volume of 0.0150 M $Na_2S_2O_3$ required for titration, mL
1	0.300	2	20.96
2	0.320	4	27.90
3	0.315	12	32.31
4	0.406	20	41.50
5	0.280	40	28.68

Equilibrium relationships

24. Explain briefly the relationship between
(a) the rates of chemical reactions and the condition of chemical equilibrium.
(b) the reaction quotient, Q, and the equilibrium constant expression, K_c.
(c) the equilibrium constants K_c and K_p.

25. Equilibrium is established at a temperature at which $K_c = 375$ for the reaction $2 NO(g) + O_2(g) \rightleftharpoons 2 NO_2(g)$. The equilibrium amount of $O_2(g)$ found in a 0.755-L vessel is 0.0148 mol. What is the ratio of [NO] to [NO_2] in this equilibrium mixture?

26. For the dissociation of $I_2(g)$ at about 1200 °C,

$$I_2(g) \rightleftharpoons 2 I(g)$$

$K_c = 1.1 \times 10^{-2}$. What volume vessel is required if it is desired that 1.00 mol I_2 and 0.50 mol I be present at equilibrium?

27. The Ostwald process for the oxidation of ammonia on a Pt catalyst is a key step in the manufacture of nitric acid (see Focus feature). A variety of oxidation products is possible—N_2, N_2O, NO, and NO_2—depending on the conditions used. One description of the process lists this reaction.

$$NH_3(g) + \tfrac{5}{4} O_2(g) \rightleftharpoons NO(g) + \tfrac{3}{2} H_2O(g)$$
$$K_p(atm) = 2.11 \times 10^{19} \text{ at 700 K}$$

Another source describes the decomposition of NO_2 at 700 K as

$$NO_2(g) \rightleftharpoons NO(g) + \tfrac{1}{2} O_2(g) \qquad K_p(atm) = 0.524$$

(a) Write a chemical equation for the oxidation of $NH_3(g)$ to $NO_2(g)$.
(b) Determine K_p for the chemical equation you have written.

Direction and extent of chemical change

28. Can a mixture of 3 mol O_2, 2 mol SO_2, and 6 mol SO_3 be maintained indefinitely in a 8.50-L flask at a temperature at which $K_c = 100$ in this reaction?

$$2 SO_2(g) + O_2(g) \rightleftharpoons 2 SO_3(g)$$

If not, in what direction will a net reaction occur?

29. Starting with 1.00 mol each of CO(g) and $COCl_2$(g) in a 1.75-L reaction vessel at 668 K, what is the number of moles of Cl_2(g) produced at equilibrium?

$$CO(g) + Cl_2(g) \rightleftharpoons COCl_2(g) \qquad K_c = 1.2 \times 10^3 \text{ at 668 K.}$$

30. Refer to Example 16-6. What will be the amounts of CO(g), H_2O(g), CO_2(g), and H_2(g) when equilibrium is established?

31. 3.00 mol $SbCl_3$ and 1.00 mol Cl_2 are introduced into an

evacuated 5.00-L vessel, and equilibrium is established at 248 °C. How many moles of $SbCl_5$, $SbCl_3$, and Cl_2 are present at equilibrium?

$$SbCl_5(g) \rightleftharpoons SbCl_3(g) + Cl_2(g) \qquad K_c = 2.5 \times 10^{-2} \text{ at 248 °C}$$

32. For the following reaction, $K_c = 2.00$ at 1000. °C.

$$2 COF_2(g) \rightleftharpoons CO_2(g) + CF_4(g)$$

If a 5.00-L mixture contains 0.105 mol COF_2, 0.220 mol CO_2, and 0.055 mol CF_4 at 1000 °C,
(a) Will the mixture be at equilibrium?
(b) If the gases are not at equilibrium, in what direction will a net reaction occur?
(c) What is the amount of each gas present when equilibrium is established?

33. A mixture of 1.00 g *each* of CO, H_2O, and H_2 is sealed in a 1.41-L vessel and brought to equilibrium at 600 K. What mass of CO_2 will be present in the equilibrium mixture?

$$CO(g) + H_2O(g) \rightleftharpoons CO_2(g) + H_2(g) \qquad K_c = 23.2$$

34. A mixture of 0.250 mol *each* of H_2(g) and I_2(g) is allowed to establish equilibrium with HI(g) in a 4.10-L reaction flask at 445 °C. What will be the mole % HI in the equilibrium mixture?

$$H_2(g) + I_2(g) \rightleftharpoons 2 HI(g) \qquad K_c = 50.2$$

★ **35.** Derive, by calculation, the equilibrium amounts of SO_2, O_2, and SO_3 listed in (a) Figure 16-4(c); (b) Figure 16-5(b).

36. The N_2O_4–NO_2 equilibrium mixture in the flask on the left is allowed to expand into the evacuated flask on the right. What is the composition of the gaseous mixture when equilibrium is reestablished in the system consisting of the two flasks?

$$N_2O_4(g) \rightleftharpoons 2 NO_2(g) \qquad K_c = 4.61 \times 10^{-3} \text{ at 25 °C}$$

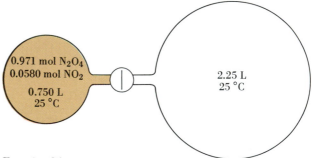

0.971 mol N_2O_4
0.0580 mol NO_2

0.750 L
25 °C

2.25 L
25 °C

Exercise 36

37. One of the key reactions in the gasification of coal is the methanation reaction, in which methane is produced from synthesis gas—a mixture of CO and H_2.

$$CO(g) + 3 H_2(g) \rightleftharpoons CH_4(g) + H_2O(g)$$
$$\Delta H° = -230 \text{ kJ} \quad K_c = 190. \text{ at 1000 K}$$

(a) Is the equilibrium conversion of synthesis gas to methane favored at higher or lower temperatures? Higher or lower pressures?
★ (b) Assume you have 4.00 mol of synthesis gas with a 3:1 mol ratio of H_2(g) to CO(g) in a 15.0-L reaction vessel. What will be the mole fraction of CH_4(g) when equilibrium is established in this mixture at 1000 K?

38. An aqueous solution is made 1.00 M in $AgNO_3$ and 1.00 M in $Fe(NO_3)_2$ and allowed to come to equilibrium. What are the values of $[Ag^+]$, $[Fe^{2+}]$, and $[Fe^{3+}]$ when equilibrium is established?

$$Ag^+(aq) + Fe^{2+}(aq) \rightleftharpoons Fe^{3+}(aq) + Ag(s) \qquad K_c = 2.98$$

★ 39. Solid iron metal is added to a solution having the concentrations $[Cr^{3+}] = 0.250$ M, $[Cr^{2+}] = 0.0500$ M, and $[Fe^{2+}] = 0.00100$ M. What are the concentrations of these ions when equilibrium is established?

$$2 Cr^{3+}(aq) + Fe(s) \rightleftharpoons 2 Cr^{2+}(aq) + Fe^{2+}(aq) \qquad K_c = 10.34$$

Partial pressure equilibrium constant, K_p

40. A mixture of 1.00 mol *each* of SO_2 and Cl_2 is introduced into an evacuated 1.75-L flask and the following equilibrium is established at 303 K.

$$SO_2Cl_2(g) \rightleftharpoons SO_2(g) + Cl_2(g) \qquad K_p = 2.9 \times 10^{-2}$$

For this equilibrium, calculate **(a)** the partial pressure of SO_2Cl_2; **(b)** the total gas pressure.

41. Starting with $SO_3(g)$ at 1.00 atm pressure, what will be the *total* pressure when equilibrium is reached in the following reaction at 700 K?

$$2 SO_3(g) \rightleftharpoons 2 SO_2(g) + O_2(g) \qquad K_p = 1.6 \times 10^{-5}$$

42. A sample of air with an original mole ratio of N_2 to O_2 of 79:21 is heated to 2500 K. When equilibrium is established, the mole percent of NO is found to be 1.8%. Calculate K_p for the reaction

$$N_2(g) + O_2(g) \rightleftharpoons 2 NO(g) \qquad K_p \text{ at } 2500 \text{ K} = ?$$

[*Hint:* The result is independent of both volume and total pressure. The presence of other gases can be neglected.]

43. In the manufacture of sodium carbonate by the Solvay process, $NaHCO_3(s)$ is decomposed by heating.

$$2 NaHCO_3(s) \rightleftharpoons Na_2CO_3(s) + CO_2(g) + H_2O(g)$$
$$K_p = 0.23 \text{ at } 100 \text{ °C}$$

(a) If a sample of $NaHCO_3(s)$ is brought to a temperature of 100 °C in a closed container, what will be the total gas pressure (in atm) at equilibrium?

(b) A mixture of 1.00 mol each of $NaHCO_3(s)$ and $Na_2CO_3(s)$ is introduced into a 2.50-L flask in which $P_{CO_2} = 2.10$ atm and $P_{H_2O} = 715$ mmHg. When equilibrium is established (at 100 °C), will the partial pressures of $CO_2(g)$ and $H_2O(g)$ be greater or less than their initial partial pressures? Explain.

★ (c) Starting with the initial conditions of part (b), what will be the partial pressures of $CO_2(g)$ and $H_2O(g)$ when equilibrium is established at 100 °C?

Dissociation reactions

44. Refer to Example 16-13. If the reaction mixture is transferred to a 10.0-L vessel, will the % dissociation increase, decrease, or remain the same? Explain.

45. Calculate the % dissociation referred to in Exercise 44.

46. What is the % dissociation of HI(g) into its gaseous elements at 340 °C?

$$H_2(g) + I_2(g) \rightleftharpoons 2 HI(g) \qquad K_p = 6.9 \times 10^1$$

★ 47. A sample of pure $PCl_5(g)$ is introduced into an evacuated flask and allowed to dissociate.

$$PCl_5(g) \rightleftharpoons PCl_3(g) + Cl_2(g)$$

If the fraction of the PCl_5 molecules that dissociate is denoted by α, and if the total gas pressure at equilibrium is P, show that

$$K_p = \frac{\alpha^2 P}{1 - \alpha^2}$$

[*Hint:* Start with 1.00 mol PCl_5, and use equation (6.16) in this derivation.]

48. With reference to the equation established in Exercise 47, if $K_p = 1.78$ at 250 °C,

(a) What is the % dissociation of PCl_5 at 250 °C and 1 atm total pressure?

(b) Under what total pressure must the gaseous mixture be maintained to limit dissociation of PCl_5 to 10.0%?

Le Châtelier's principle

49. Continuous removal of one of the products of a chemical reaction has the effect of causing the reaction to go to completion. Explain this fact in terms of Le Châtelier's principle.

50. The *endothermic* reaction $A(g) + B(g) \rightleftharpoons 2 C(g)$ proceeds to an equilibrium condition at 200 °C. Which of the following statements is true? Explain. [*Hint:* There may be more than one correct statement.]

(a) If the mixture is transferred to a reaction vessel of twice the volume, the amounts of reactants and products will remain unchanged.

(b) Addition of an appropriate catalyst will result in the formation of a greater amount of C(g).

(c) Lowering the reaction temperature to 100 °C will result in the formation of a greater amount of C(g).

(d) Addition of an inert gas at constant volume will have little or no effect on the equilibrium.

51. Show that the % dissociation in reaction (1) depends on the volume of the reaction vessel and in reaction (2) it does not. Explain this difference from the standpoint of Le Châtelier's principle.

(1) $SO_2Cl_2(g) \rightleftharpoons SO_2(g) + Cl_2(g)$

(2) $CS_2(g) \rightleftharpoons C(s) + S_2(g)$

52. Explain why all dissociation reactions of the type $A_2(g) \rightleftharpoons 2 A(g)$ proceed to a greater extent at higher temperatures [e.g., $I_2(g) \rightleftharpoons 2 I(g)$].

53. The reaction $N_2(g) + O_2(g) \rightleftharpoons 2 NO(g)$, $\Delta H° = +181$ kJ/mol, occurs whenever a substance is burned in air. This reaction occurs in internal combustion engines, leading to the formation of oxides of nitrogen that are involved in the production of photochemical smog. High-compression engines, characteristic of large automobiles, operate at high temperatures.

(a) What effect do these high temperatures have on the equilibrium production of NO(g)?

(b) What effect does high temperature have on the rate of this reaction?

54. The freezing of $H_2O(l)$ at 0 °C can be represented as $H_2O(l, d = 1.00 \text{ g/cm}^3) \rightleftharpoons H_2O(s, d = 0.92 \text{ g/cm}^3)$. Explain why the application of pressure to ice at 0 °C causes the ice to melt. Is this behavior to be expected of solids in general?

★ 55. The reaction $A(s) \rightleftharpoons B(s) + 2 C(g) + \frac{1}{2} D(g)$ is found to have $\Delta H° = 0$.

(a) Will K_p increase, decrease, or remain constant with temperature? Explain.

(b) If a constant-volume mixture originally at equilibrium at 298 K is heated to 400 K, will the amount of D(g) present increase, decrease, or remain constant? Explain.

Kinetics and equilibrium

56. In both the ammonia synthesis reaction

$$N_2(g) + 3 H_2(g) \rightleftharpoons 2 NH_3(g)$$

and the conversion of $SO_2(g)$ to $SO_3(g)$

$$2 SO_2(g) + O_2(g) \rightleftharpoons 2 SO_3(g)$$

the mole fraction at equilibrium of the desired product (NH_3, SO_3) is greater at lower temperatures. Yet, in the commercial production of these substances relatively high temperatures are used. Explain why this is so.

57. Suppose that the conversion of hydrogen and iodine to hydrogen iodide proceeded by the simple *one-step* mechanism suggested below.

$$H_2(g) + I_2(g) \underset{k_2}{\overset{k_1}{\rightleftharpoons}} 2 HI(g)$$

Show how the requirement that the rates of the forward and reverse reactions are equal at equilibrium could be used to derive the equilibrium constant expression for hydrogen iodide formation.

Additional Exercises

58. Write an equilibrium constant expression, K_c, for the formation of *1 mol* of each of the following *gaseous* compounds from its *gaseous* elements. **(a)** NO; **(b)** HCl; **(c)** NH_3; **(d)** ClF_3; **(e)** NOCl.

59. Use data from Exercise 3 to determine K_p at 298 K for the reaction

$$\tfrac{1}{2} N_2(g) + \tfrac{1}{2} O_2(g) + \tfrac{1}{2} Br_2(g) \rightleftharpoons NOBr \qquad K_p = ?$$

60. A mixture of 1.00 g H_2 and 1.06 g H_2S is introduced into a 0.500-L flask and the mixture is allowed to come to equilibrium at 1670 K.

$$2 H_2(g) + S_2(g) \rightleftharpoons 2 H_2S(g)$$

The equilibrium amount of $S_2(g)$ is found to be 8.00×10^{-6} mol. Determine the value of K_p.

61. If 0.390 mol SO_2, 0.156 mol O_2, and 0.657 mol SO_3 are introduced simultaneously into a 1.90-L vessel at 1000 K,

(a) is this mixture at equilibrium?

(b) if not, in what direction must a net reaction occur to establish equilibrium?

$$2 SO_2(g) + O_2(g) \rightleftharpoons 2 SO_3(g) \qquad K_c = 2.8 \times 10^2$$

62. Calculate the actual amounts of SO_2, O_2, and SO_3 at equilibrium in Exercise 61.

63. A sample of $NH_4HS(s)$ is introduced into a 1.60-L flask containing 0.170 g NH_3. What will be the total gas pressure when equilibrium is established?

$$NH_4HS(s) \rightleftharpoons NH_3(g) + H_2S(g) \qquad K_p(atm) = 0.108 \text{ at } 25 \text{ °C}$$

64. Refer to Example 16-5. A 1.05-g sample of solid iodine in placed in a 515-mL flask, air is evacuated from the flask, and then $H_2S(g)$ is introduced at a pressure of 758.2 mmHg. The sealed flask is held at a temperature of 60 °C until equilibrium is reached. What will be the *total pressure* in the flask at this time? [*Hint:* How does the quantity of $I_2(s)$ affect the equilibrium condition?]

$$H_2S(g) + I_2(s) \rightleftharpoons 2 HI(g) + S(s)$$
$$K_p(atm) = 1.34 \times 10^{-5} \text{ at } 60 \text{ °C}$$

65. The high-temperature dissociation of salicylic acid is represented by the equation

$$C_7H_6O_3(g) \rightleftharpoons C_6H_6O(g) + CO_2(g)$$

As a result of an experiment carried out at 200. °C, an initial sample of 0.300 g $C_7H_6O_3$ in a 50.0-cm³ vessel yielded an equilibrium mixture in which the partial pressure of $CO_2(g)$ was found to be 1.50 atm. What are **(a)** K_c and **(b)** K_p for this reaction at 200. °C?

66. Formamide is used as an intermediate and solvent in the manufacture of pharmaceuticals, dyes, and agricultural chemicals. At elevated temperatures it decomposes to $NH_3(g)$ and CO(g).

$$HCONH_2(g) \rightleftharpoons NH_3(g) + CO(g) \qquad K_c = 4.84 \text{ at } 400. \text{ K}$$

If 0.100 mol $HCONH_2(g)$ is allowed to dissociate in a 1.50-L flask at 400. K, what will be the *total* pressure at equilibrium?

67. With reference to the reaction described in Exercise 22,

$$C_2H_5OH + CH_3CO_2H \rightleftharpoons CH_3CO_2C_2H_5 + H_2O \qquad K_c = 4.0$$

15.5 g C_2H_5OH, 25.0 g CH_3CO_2H, 45.5 g $CH_3CO_2C_2H_5$, and 52.0 g H_2O are mixed and allowed to react.

(a) In what direction will a net reaction occur?

(b) What will be the equilibrium quantities of each of the reacting species?

68. A solution is prepared having $[Fe^{3+}] = 0.4000$ M and $[Hg_2^{2+}] = 0.2500$ M. What are the values of $[Fe^{3+}]$, $[Fe^{2+}]$, $[Hg_2^{2+}]$, and $[Hg^{2+}]$ when equilibrium is established?

$$2 Fe^{3+}(aq) + Hg_2^{2+}(aq) \rightleftharpoons 2 Fe^{2+}(aq) + 2 Hg^{2+}(aq)$$
$$K_c = 9.14 \times 10^{-6}$$

69. Use data from Appendix D to determine if the forward reaction is favored by high or low temperatures.

(a) $PCl_3(g) + Cl_2(g) \rightleftharpoons PCl_5(g)$

(b) $SO_2(g) + 2 H_2S(g) \rightleftharpoons 2 H_2O(g) + 3 S(rhombic)$

(c) $2 N_2(g) + 3 O_2(g) + 4 HCl(g) \rightleftharpoons$
$$4 NOCl(g) + 2 H_2O(g)$$

70. The steam-iron process is used to generate hydrogen gas, mostly in connection with the hydrogenation of oils (Section 14-3).

$$3 Fe(s) + 4 H_2O(g) \rightleftharpoons Fe_3O_4(s) + 4 H_2(g) \qquad \Delta H° = -150. \text{ kJ}$$

How would the following factors affect the amount of H_2 present in an equilibrium mixture? Explain.

(a) Raise the temperature of the mixture.

(b) Introduce more $H_2O(g)$.

(c) Double the volume of the container holding the mixture.

(d) Add an appropriate catalyst.

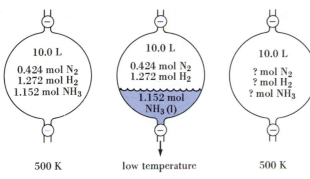

500 K   low temperature   500 K

Exercise 76

71. Use Le Châtelier's principle to make qualitative predictions about

(a) the effect on the amount of $Cl_2(g)$ at equilibrium if the volume of the reaction vessel in Exercise 29 is increased from 1.75 L to 2.50 L;

(b) the effect on the equilibrium amounts of $SbCl_5$, $SbCl_3$, and Cl_2 if a catalyst is used in the reaction in Exercise 31;

(c) the effect on the % dissociation of $HI(g)$ in Exercise 46 if an inert gas is added until the total pressure exerted by the gaseous mixture is increased from 1.0 atm to 10.0 atm.

72. What is the % dissociation of $H_2S(g)$ if 1.00 mol H_2S is introduced into an evacuated 1.10-L vessel at 1000 K?

$$2 H_2S(g) \rightleftharpoons 2 H_2(g) + S_2(g) \qquad K_c = 1.0 \times 10^{-6}$$

73. Equilibrium is established in a 2.50-L flask at 250 °C involving the reaction

$$PCl_5(g) \rightleftharpoons PCl_3(g) + Cl_2(g) \qquad K_c = 3.8 \times 10^{-2}$$

How many moles of PCl_5, PCl_3, and Cl_2 are present at equilibrium, if

(a) 1.50 mol each of PCl_5 and PCl_3 are initially introduced into the flask?

(b) 0.500 mol PCl_5 alone is introduced into the flask?

74. Refer to Example 16-13. The percent dissociation of $N_2O_4(g)$ depends on the total gas pressure. What must be the total pressure of an equilibrium mixture if the $N_2O_4(g)$ is to be 10.0% dissociated at 298 K?

$$N_2O_4(g) \rightleftharpoons 2 NO_2(g) \qquad K_p(atm) = 0.113 \text{ at } 298 K$$

[*Hint:* Work directly with K_p instead of K_c. How should the total pressure here compare with that in Example 16-13? What is the total pressure in Example 16-13?]

75. Refer to Example 16-14. Suppose that 0.100 L of the *equilibrium* mixture is diluted to 0.250 L with water. What will be the new concentrations when equilibrium is reestablished? [*Hint:* In what direction must a net reaction occur?]

★76. The method of extracting liquid ammonia from equilibrium mixtures in the ammonia synthesis reaction (recall Figure 16-9) is suggested in the illustration at the left. A particular *equilibrium* mixture is obtained at 500 K. The mixture is quickly chilled to a temperature at which the NH_3 liquefies, and the $NH_3(l)$ is removed. Then the mixture is returned to 500 K and equilibrium is reestablished. How many moles of $NH_3(g)$ will be present in the new equilibrium mixture?

$$N_2(g) + 3 H_2(g) \rightleftharpoons 2 NH_3(g) \qquad K_c = 152 \text{ at } 500 K$$

★77. Nitrogen dioxide obtained as a cylinder gas is always a mixture of $NO_2(g)$ and $N_2O_4(g)$. A 5.00-g sample obtained from such a cylinder is sealed in a 0.500-L flask at 298 K. What is the mole fraction of NO_2 in this mixture?

$$N_2O_4(g) \rightleftharpoons 2 NO_2(g) \qquad K_c = 4.61 \times 10^{-3}$$

★78. What is the apparent molar mass of the gaseous mixture that results when $COCl_2(g)$ is allowed to dissociate at 395 °C and a total pressure of 3.00 atm?

$$COCl_2(g) \rightleftharpoons CO(g) + Cl_2(g) \qquad K_p = 4.44 \times 10^{-2} \text{ at } 395 °C$$

79. At 2000 K the reaction

$$2 CH_4(g) \rightleftharpoons C_2H_2(g) + 3 H_2(g) \quad \text{has } K_c = 0.154.$$

If a reaction mixture at equilibrium at 2000 K contains 0.10 mol each of $CH_4(g)$ and $H_2(g)$ in a 1.00-L vessel, (a) what is the mole fraction of $C_2H_2(g)$ present? (b) Is the conversion of $CH_4(g)$ to $C_2H_2(g)$ favored by high or low pressures? Explain.

★80. Show that in terms of mole fractions of gases and *total* gas pressure, the equilibrium constant expression for

$$N_2(g) + 3 H_2(g) \rightleftharpoons 2 NH_3(g) \quad \text{is}$$

$$K_p = \frac{(\chi_{NH_3})^2}{(\chi_{N_2})(\chi_{H_2})^3} \times \frac{1}{(P_{tot.})^2}$$

★81. For the synthesis of ammonia at 500 K,

$$N_2(g) + 3 H_2(g) \rightleftharpoons 2 NH_3(g) \qquad K_p = 9.06 \times 10^{-2}$$

If N_2 and H_2 are allowed to react in the mole ratio 1:3 and the total pressure is maintained at 1.00 atm, what is the mol % NH_3 at equilibrium? [*Hint:* Use the result of the preceding exercise and a method of successive approximations.]

★82. A mixture of $H_2S(g)$ and $CH_4(g)$ in the mole ratio 2:1 was allowed to come to equilibrium at 700 °C and a total gas pressure of 1 atm. The *equilibrium* mixture was analyzed for the amount of H_2S present; 9.54×10^{-3} mol H_2S was found. The CS_2 present at equilibrium was converted successively to H_2SO_4 and then to $BaSO_4$; 1.42×10^{-3} mol $BaSO_4$ was obtained. Use these data to determine K_p at 700 °C for the reaction

$$2 H_2S(g) + CH_4(g) \rightleftharpoons CS_2(g) + 4 H_2(g) \qquad K_p \text{ at } 700 °C = ?$$

Self-Test Questions

For questions 83 through 90 select the single item that best completes each statement.

83. In the reaction $H_2(g) + I_2(g) \rightleftharpoons 2 HI(g)$, a mixture that initially contains 2 mol H_2 and 1 mol I_2 produces, at equilibrium, (a) 1 mol HI; (b) 2 mol HI; (c) more than 2 but less than 4 mol HI; (d) less than 2 mol HI.

84. In the reaction $2 SO_2(g) + O_2(g) \rightleftharpoons 2 SO_3(g)$ equilibrium is established at a temperature at which $K_c = 100$. If the number of moles of SO_3 in the equilibrium mixture is equal to the number of moles of SO_2,

(a) the number of moles of O_2 is also equal to the number of moles of SO_2;

(b) the number of moles of O_2 is half the number of moles of SO_2;

(c) $[O_2] = 0.01$ M;

(d) $[O_2]$ may have any of several different values.

85. The volume of the reaction vessel containing an equilibrium mixture in the reaction $SO_2Cl_2(g) \rightleftharpoons SO_2(g) + Cl_2(g)$ is increased. When equilibrium is reestablished,

(a) the amount of $Cl_2(g)$ will have increased;

(b) the amount of $SO_2(g)$ will have decreased;

(c) the amount of $Cl_2(g)$ will have remained unchanged;

(d) the amount of $SO_2Cl_2(g)$ will have increased.

86. Equilibrium is established in the reaction $A + B \rightleftharpoons C + D$; $K_c = 10.0$. At this point (a) $[C][D] = [A][B]$; (b) $[C] = [A]$ and $[B] = [D]$; (c) $[A][B] = 0.10 \times [C][D]$; (d) $[A] = [B] = [C] = [D] = 10.0$ M.

87. Equilibrium in a mixture of $CO(g)$, $H_2O(g)$, $CO_2(g)$, and $H_2(g)$ is established in a 1.00-L container at 1000 K. The following data are given.

$$CO(g) + H_2O(g) \rightleftharpoons CO_2(g) + H_2(g)$$
$$\Delta H° = -42 \text{ kJ} \quad K_c = 0.66$$

The equilibrium amount of $H_2(g)$ can be increased by (a) adding a catalyst; (b) increasing the temperature; (c) transferring the mixture to a 10.0-L container; (d) none of the methods described in (a), (b), or (c).

88. A mixture of 1.00 mol *each* of $CO(g)$, $H_2O(g)$, and $CO_2(g)$ is introduced into a 10.0-L flask at a temperature at which $K_c = 10.0$ for the reaction

$$CO(g) + H_2O(g) \rightleftharpoons CO_2(g) + H_2(g) \quad K_c = 10.0$$

When the reaction reaches a state of equilibrium,

(a) the amount of H_2 will be 1.00 mol;

(b) the amount of $CO_2(g)$ will be greater than 1.00 mol, and the amounts of $CO(g)$, $H_2O(g)$, and $H_2(g)$ will each be less than 1.00 mol;

(c) the amounts of all reactants and products will be greater than 1.00 mol;

(d) the amounts of $CO_2(g)$ and $H_2(g)$ will each be greater than 1.00 mol, and the amounts of $CO(g)$ and $H_2O(g)$ will each be less than 1.00 mol.

89. For the reaction $2 NO_2(g) \rightleftharpoons 2 NO(g) + O_2(g)$, $K_c = 1.8 \times 10^{-6}$ at 184 °C. At 184 °C, the value of K_c for the reaction $NO(g) + \frac{1}{2} O_2(g) \rightleftharpoons NO_2(g)$ is (a) 0.9×10^6; (b) 7.5×10^2; (c) 5.6×10^5; (d) 2.8×10^5.

90. For the dissociation reaction

$$2 H_2S(g) \rightleftharpoons 2 H_2(g) + S_2(g) \quad K_p = 1.2 \times 10^{-2} \text{ at } 1065 \text{ °C}$$

For this same reaction, when they are compared at 298 K, (a) K_c is less than K_p; (b) K_c is greater than K_p; (c) $K_c = K_p$; (d) whether K_c is less than, equal to, or greater than K_p depends on the total gas pressure.

91. Describe how the balanced chemical equation for a reversible reaction is used in equilibrium calculations. Explain why the balanced equation *alone* cannot be used for determining the composition of an equilibrium mixture.

92. A 0.0010-mol sample of $S_2(g)$ is allowed to dissociate in an 0.500-L flask at 1000 K. When equilibrium is reached, 1.0×10^{-11} mol $S(g)$ is present. What is K_c for the reaction $S_2(g) \rightleftharpoons 2 S(g)$ at 1000 K?

93. Into a 1.00-L vessel at 1000 K are introduced 0.100 mol each of $NO(g)$ and $Br_2(g)$ and 0.0100 mol $NOBr(g)$.

$$2 NO(g) + Br_2(g) \rightleftharpoons 2 NOBr(g) \quad K_c = 1.32 \times 10^{-2} \text{ at } 1000 \text{ K}$$

(a) In what direction must a net reaction occur?

★ **(b)** What is the partial pressure of $NOBr(g)$ in the vessel when equilibrium is established?

17

Acids and Bases

A lake in Ontario, Canada being tested for its acidity. [Ted Spiegel/Black Star]

In recent years the substances we call acids and bases have attracted a great deal of public attention because of the environmental problem of acid deposition (more familiarly called *acid rain*). Oxides of sulfur (SO_x) and of nitrogen (NO_x) react with atmospheric moisture to form acids that are damaging lakes and forests. Some soils and lake beds contain materials with *basic* properties (such as limestone) that can neutralize this acidic material from non-natural sources. Others do not. Rainfall in the northeastern United States regularly has an acidity described as "pH 4.2," which is considerably more acidic than the more normal pH 5.6 for rainwater. Some measurements of rainfall have yielded values as low as pH 2.1, and there have been reports of pH values as low as 1.8 for fog in the San Francisco bay area.

Several topics we explore in this chapter provide the background for a better understanding of the chemistry underlying the acid deposition problem and other practical examples of acid–base chemistry. For example, we consider three different theories of acid–base behavior; the nature of the reaction between element oxides and water that produces acids in some cases and bases in others; several factors that determine the strengths of acids and bases; what is measured by the pH scale, and how to calculate the pH values of different aqueous solutions; and the underlying principle of neutralization reactions. Acid deposition is a complex societal problem that does not respect national boundaries. Although this problem cannot be solved by chemists alone, a clear understanding of its chemical aspects is essential to its long-range solution.

17-1 Acids and Bases: A Brief Introduction and Overview

You can often tell that certain substances have been known for a long time by the names used to describe them. The term *acid* is derived from the Latin word *acidus* (sour), which is related to *acer* (sharp) and *acetum* (vinegar). Vinegar is simply an aqueous solution of acetic acid. The term *alkali* (base) is derived from the Arabic *al-qali*, which describes the ashes of certain plants associated with salt marshes and desert regions. Until relatively recent times, the principal source of bases or alkalis has been wood ashes. Also known for a very long time is that acids and bases can neutralize each other's distinctive properties and produce a type of substance known as a *salt*.

The properties that are most commonly associated with acids are their sour taste, the prickling sensation they produce on the skin, their ability to dissolve most metals, and their ability to dissolve limestone and other carbonate minerals. Bases have a bitter taste and a slippery feel; basic properties are commonly found in soap and other household cleaners. Both acids and bases have the ability to affect the color of certain naturally occurring plant constituents. For example, litmus, obtained from lichens, is red in acidic solutions and blue in basic solutions.

Theoretical attempts to explain acid–base behavior form an important chapter in the history of chemistry. Antoine Lavoisier (1777) proposed that all acids contain a common element—oxygen. In 1810, Humphry Davy proposed that *hydrogen*, not oxygen, is the common element in acids. He did this by showing that muriatic acid (hydrochloric acid) contains only H and Cl and no O.

Arrhenius Theory. The description of acids and bases introduced in Chapter 5 and used from time to time up to this point was developed by Svante Arrhenius (1884) in connection with his theory of electrolytic dissociation (Section 13-9). A strong electrolyte is completely dissociated into ions in a water solution; a weak electrolyte is only partially ionized. An acid is a substance that dissociates to produce hydrogen ions (H^+), for example,

Litmus is commonly used in the general chemistry laboratory as an acid–base indicator.

The name oxygen, proposed by Lavoisier, is derived from Greek and means "acid former."

604

$$HCl(aq) \longrightarrow H^+(aq) + Cl^-(aq) \tag{17.1}$$

A base dissociates to produce hydroxide ions (OH^-).

$$NaOH(aq) \longrightarrow Na^+(aq) + OH^-(aq) \tag{17.2}$$

And, as we noted in Section 5-3, the reaction of an acid and a base (neutralization) can be represented through an *ionic equation*

$$\underset{\text{an acid}}{H^+(aq) + Cl^-(aq)} + \underset{\text{a base}}{Na^+(aq) + OH^-(aq)} \longrightarrow \underset{\text{a salt}}{Na^+(aq) + Cl^-(aq)} + \underset{\text{water}}{H_2O}$$

$$\tag{17.3}$$

or through a *net ionic equation*.

$$\underset{\text{an acid}}{H^+(aq)} + \underset{\text{a base}}{OH^-(aq)} \longrightarrow \underset{\text{water}}{H_2O} \tag{17.4}$$

The Arrhenius theory, as represented through equation (17.4), brings out this essential point.

A neutralization reaction involves the combination of hydrogen and hydroxide ions to form water.

Arrhenius's description of neutralization explains why the enthalpies of neutralization of strong acids and strong bases are essentially constant at -55.9 kJ per mol H_2O: In every case $\Delta H_{\text{neutr.}}$ is simply that for reaction (17.4). Arrhenius's theory also helps to explain the catalytic activity of acids in certain reactions. The most effective catalysts are those with the best electrical conductivity—the strongest acids. The stronger the acid, the more complete its ionization in aqueous solution, the higher the concentration of ions, and the better the electrical conductivity. As we saw in the acid-catalyzed reaction of formic acid in Section 15-11, the actual catalyst is H^+.

Limitations of Arrhenius's Theory. Despite its early successes and continued usefulness, the Arrhenius theory does have serious limitations. One of the most glaring is in its treatment of the weak base ammonia, NH_3. According to Arrhenius a substance must *contain* OH^- in order to be a base. Where is the OH^- in NH_3? To get around this difficulty, the idea was proposed that in aqueous solution NH_3 forms the compound NH_4OH (ammonium hydroxide), which then dissociates as a weak base into NH_4^+ and OH^-.

$$NH_3 + H_2O \longrightarrow NH_4OH(aq)$$

$$NH_4OH(aq) \rightleftharpoons NH_4^+(aq) + OH^-(aq)$$

A holdover of the Arrhenius theory. Although there is no evidence for the existence of molecules of NH_4OH, solutions of $NH_3(aq)$ are commonly labeled this way. [Carey B. Van Loon]

The problem with this formulation is that NH_4OH does not exist! We should always be troubled by a hypothesis or theory that postulates the existence of hypothetical substances. As noted in the next section, the essential failure of the Arrhenius theory is in not recognizing the key role of the *solvent* in the ionization of the solute.

17-2 Brønsted–Lowry Theory of Acids and Bases

J. N. Brønsted in Denmark and T. M. Lowry in Great Britain independently proposed a new definition of acids and bases in 1923. In their theory an acid is a **proton* donor** and a base is a **proton acceptor.** To describe the basic behavior of ammonia, which we had difficulty doing with Arrhenius's theory, we can write

---

*In acid–base theory a proton refers to a particular H atom that has lost an electron, i.e., H^+. Since the ion H^+ is just the nucleus of the H atom —a proton—we speak of the transfer of a proton. You must never think of proton transfer as involving the extraction of a proton from the nucleus of an atom.

$$NH_3 + H_2O \longrightarrow NH_4^+ + OH^- \tag{17.5}$$
$$\text{base}\text{acid}$$

In reaction (17.5) H_2O acts as an *acid*. It donates a proton (H^+) which is gained by NH_3, a *base*. As a result of this transfer the ions NH_4^+ and OH^- are formed—the same ions produced by the *hypothetical* NH_4OH of the Arrhenius theory.

Now let us deal with this fact about $NH_3(aq)$: It is a *weak* base. We can do this by writing the *reverse* of (17.5); when we do we recognize NH_4^+ as an *acid* and OH^- as a *base*.

$$NH_4^+ + OH^- \longrightarrow NH_3 + H_2O \tag{17.6}$$
$$\text{acid}\text{base}$$

The conventional approach for a reversible reaction, as we learned in Chapter 16, is to use the double arrow notation. When we do, we then see that we should put our "acid" and "base" labels under all four species in the equation.

$$NH_3 \; + \; H_2O \; \rightleftharpoons \; NH_4^+ \; + \; OH^- \tag{17.7}$$
$$\text{base(2)}\quad\text{acid(1)}\qquad\text{acid(2)}\quad\text{base(1)}$$

In the margin: **In the designations (1) and (2), it does not matter which conjugate pair we call (1) and which we call (2).**

In this notation we treat NH_3/NH_4^+ as combination (2) and H_2O/OH^- as combination (1). Each combination is called a **conjugate pair.** NH_3 acts as a base by accepting a proton, and NH_4^+ is the **conjugate acid** of NH_3. Similarly, in reaction (17.7) H_2O is an acid and OH^- is its **conjugate base.** Figure 17-1 may help you to visualize the proton transfer involved in the forward and reverse reactions in (17.7).

For the ionization of acetic acid we can write

$$HC_2H_3O_2 + H_2O \rightleftharpoons H_3O^+ + C_2H_3O_2^- \tag{17.8}$$
$$\text{acid(1)}\qquad\text{base(2)}\qquad\text{acid(2)}\qquad\text{base(1)}$$

Here, acetate ion, $C_2H_3O_2^-$, is the conjugate base of the acid $HC_2H_3O_2$. And this time H_2O acts as a base. Its conjugate acid is called **hydronium ion, H_3O^+.**

We can represent the ionization of HCl in much the same way as we did for acetic acid, but with this important difference: The reverse reaction shows practically no tendency to occur.

$$HCl + H_2O \rightleftharpoons H_3O^+ + Cl^- \tag{17.9}$$

We can rationalize this behavior by saying that because HCl is a *very strong acid*, Cl^- is a *very weak base*. That is, if the HCl molecule has a strong tendency to donate a proton, the Cl^- ion has practically no tendency to accept a proton. The forward reaction goes essentially to completion.

To complete our discussion of the Brønsted–Lowry theory, we have to consider the relative strengths of acids and bases and the factors that affect acid and base strength. This we will do in Section 17-9, but for the present the fundamental ideas we need are

In the margin: **Features of the Brønsted–Lowry theory.**

- An **acid** donates a proton to another substance called a
- **base,** which accepts the proton.
- In general, acid–base reactions are **reversible,** meaning that   (17.10)
- every acid has a **conjugate base** and
- every base has a **conjugate acid.**

These additional points about the Brønsted–Lowry theory are illustrated in Example 17-1.

- Any species that is an acid by the Arrhenius theory remains an acid in the Brønsted–Lowry theory. The same is true of bases.
- Certain species, because they do not contain the hydroxide group, would not be classified as bases by the Arrhenius theory. Yet they are so classified by the Brønsted–Lowry theory, e.g., OCl^- and $H_2PO_4^-$.

FIGURE 17-1
Brønsted–Lowry acid–base reaction.

This figure represents the proton transfer in reaction (17.7). The solid arrows represent the forward reaction and the broken arrows, the reverse reaction. NH_4^+ is a stronger acid than H_2O and OH^- is a stronger base than NH_3. The reverse reaction proceeds to a greater extent than the forward reaction.

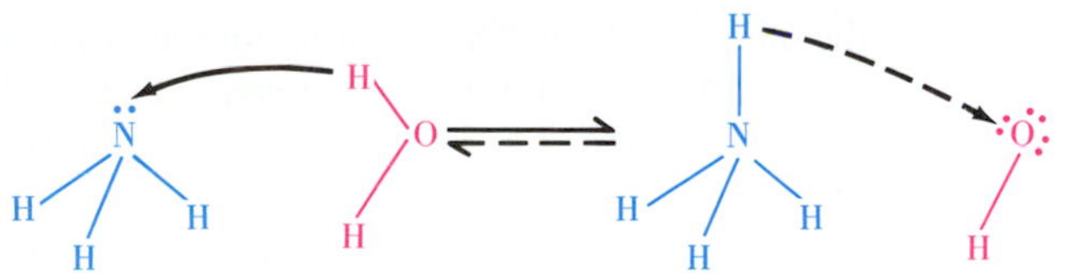

$$NH_3 \ + \ H_2O \ \rightleftharpoons \ NH_4^+ \ + \ OH^-$$
$$\text{base(2)} \qquad \text{acid(1)} \qquad \text{acid(2)} \qquad \text{base(1)}$$

- The Brønsted–Lowry theory accounts for a substance that can act either as an acid or a base (*amphiprotic*). The Arrhenius theory does not as easily account for this behavior.

Example 17-1

Writing equations for Brønsted–Lowry acid–base reactions: Identifying acids and bases and their conjugates. For each of the following identify the acids and bases involved in both the forward and reverse reactions.
(a) $HClO_2 + H_2O \rightleftharpoons H_3O^+ + ClO_2^-$
(b) $OCl^- + H_2O \rightleftharpoons HOCl + OH^-$
(c) $NH_3 + H_2PO_4^- \rightleftharpoons NH_4^+ + HPO_4^{2-}$
(d) $HCl + H_2PO_4^- \rightleftharpoons H_3PO_4 + Cl^-$

Solution. Consider $HClO_2$ in reaction (a). It is converted to ClO_2^- by losing a proton (H^+). $HClO_2$ must be an acid, and ClO_2^- is its conjugate base. H_2O accepts the proton lost by $HClO_2$. Then H_2O must be a base and H_3O^+, its conjugate acid. In reaction (b), H_2O acts as an acid and OH^- is its conjugate base. Reactions (c) and (d) illustrate another species that can either donate or accept a proton—an **amphiprotic** species. In this case the amphiprotic species is $H_2PO_4^-$.

(a) $HClO_2 \ + \ H_2O \ \rightleftharpoons \ H_3O^+ \ + \ ClO_2^-$
   $\quad$ acid(1) $\qquad$ base(2) $\qquad$ acid(2) $\qquad$ base(1)

(b) $H_2O \ + \ OCl^- \ \rightleftharpoons \ HOCl \ + \ OH^-$
   $\quad$ acid(1) $\qquad$ base(2) $\qquad$ acid(2) $\quad$ base(1)

(c) $H_2PO_4^- \ + \ NH_3 \ \rightleftharpoons \ NH_4^+ \ + \ HPO_4^{2-}$
   $\quad$ acid(1) $\qquad$ base(2) $\qquad$ acid(2) $\qquad$ base(1)

(d) $HCl \ + \ H_2PO_4^- \ \rightleftharpoons \ H_3PO_4 \ + \ Cl^-$
   $\quad$ acid(1) $\qquad$ base(2) $\qquad$ acid(2) $\quad$ base(1)

SIMILAR EXAMPLES: Exercises 1, 17, 18, 19.

17-3 Self-ionization of Water

Even in pure water there is a very low concentration of ions, detectable by precise measurements of electrical conductivity. These ions can arise only by the ionization of the water molecules themselves (self-ionization). According to the Arrhenius theory we would write

$$H_2O \rightleftharpoons H^+ + OH^- \tag{17.11}$$

In the Brønsted–Lowry theory the process involves one water molecule donating a proton to another water molecule. That is, one water molecule acts as an acid and the other as a base. The resulting ions are hydronium ion, H_3O^+ (a *conjugate acid*), and hydroxide ion, OH^- (a *conjugate base*). The reaction is reversible, and in the reverse reaction H_3O^+ donates a proton to OH^-. In fact, the reverse reaction is far more significant than the forward reaction. *Equilibrium is displaced far to the left.* We can say that acid (2) and base (1) are *much* stronger than are acid (1) and base (2).

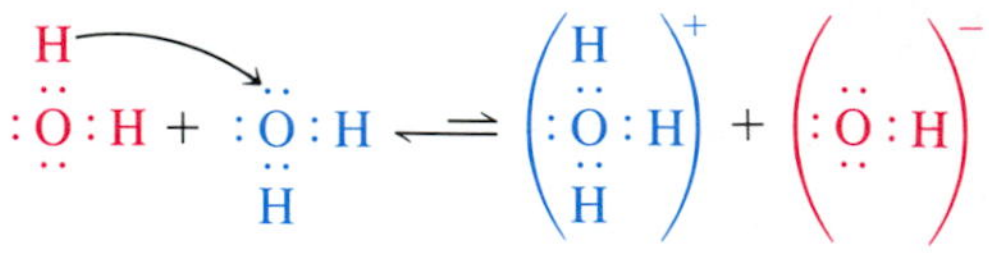

$$
H{-}\!\!\!\longrightarrow \\
\overset{..}{\underset{..}{:}}\!O\!:\!H + \overset{..}{\underset{..}{:}}\!O\!:\!H \rightleftharpoons \left(\overset{H}{\underset{H}{\overset{..}{:}O\overset{..}{:}H}}\right)^+ + \left(\overset{..}{\underset{..}{:}}\!O\!:\!H\right)^- \qquad (17.12)
$$

acid(1)   base(2)   acid(2)   base(1)

We can describe equilibrium in the self-ionization of water through an equilibrium constant expression which includes concentration terms for H_3O^+ and OH^-, but not for the liquid water itself.

$$K = [H_3O^+][OH^-]$$

From equation (17.12) we can see that the concentrations of H_3O^+ and OH^- are equal in pure water, and there are several experimental methods of determining these concentrations. All lead to this result.

At 25 °C in pure water: $[H_3O^+] = [OH^-] = 1.0 \times 10^{-7}$ M

The equilibrium constant for the self-ionization of water is called the **ion product of water,** and is represented as K_w. At 25 °C

The ion product of water, K_w.

$$K_w = [H_3O^+][OH^-] = 1.0 \times 10^{-14} \qquad (17.13)$$

Like other equilibrium constants, the ion product of water varies with temperature. At 60 °C, $K_w = 9.6 \times 10^{-14}$; at 100 °C, $K_w = 5.5 \times 10^{-13}$. The significance of expression (17.13) is that it applies to *all* aqueous solutions, not just to pure water, as we see in the sections that follow.

Calculations of $[H^+]$ in water yield results as low as 10^{-130} M. As described by N. V. Sidgwick (1950), this corresponds to *one* free proton in 10^{70} universes filled with a 1 M acid solution.

The Nature of the Proton in Aqueous Solution. The Arrhenius theory postulates H^+ ions in aqueous solutions (for example, through reactions like 17.11), but there is no evidence for their existence. Recall that the H^+ ion is just a lone proton (the nucleus of a hydrogen atom). Because of their very small size and high positive charge density, we should expect H^+ ions to seek out centers of negative charge with which to form bonds. This is what happens when an H^+ ion attaches itself to a lone pair of electrons in a water molecule through a *coordinate covalent* bond (as in reaction 17.12).

What is the situation with H_3O^+, whose existence is postulated by the Brønsted–Lowry theory? Does it exist? For many years this seemed an unanswerable question, and H_3O^+ was thought to be just one of a series of *hydrated* protons $[H(H_2O)_n]^+$. That is, if $n = 1$, the species is $[H(H_2O)]^+ = H_3O^+$—the hydronium ion. Recently, however, experimental evidence has been found for the existence of H_3O^+ in the gaseous state, in solids, and in aqueous solution. For example, what was once thought to be a monohydrate of perchloric acid, $HClO_4 \cdot H_2O$, has been shown through x-ray studies to be $H_3O^+ClO_4^-$. This salt, which we might call hydronium perchlorate, is structurally quite similar to ammonium perchlorate,

FIGURE 17-2
The hydronium ion in aqueous solution.

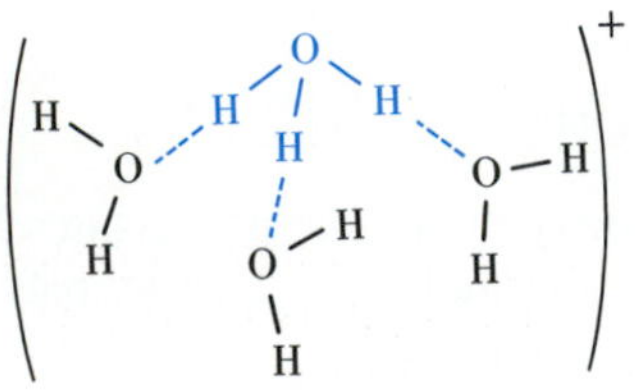

Represented here is a probable species that exists in aqueous solution. The central H_3O^+ is hydrogen-bonded to three H_2O molecules.

$NH_4^+ClO_4^-$. The current view of the hydronium ion in aqueous solution is something like that pictured in Figure 17-2—a central hydronium ion hydrogen-bonded to three H_2O molecules. Throughout this chapter we refer to the hydronium ion in solution as H_3O^+ or $H_3O^+(aq)$.

17-4 Strong Acids and Strong Bases

When an acid is added to water, as in aqueous solutions of hydrogen chloride, in addition to the self-ionization of water

$$H_2O + H_2O \rightleftharpoons H_3O^+ + OH^- \qquad (17.14)$$
$$\text{acid} \quad \text{base} \qquad \text{acid} \quad \text{base}$$

the acid also ionizes.

$$HCl + H_2O \longrightarrow H_3O^+ + Cl^- \qquad (17.15)$$
$$\text{acid} \quad \text{base} \qquad \text{acid} \quad \text{base}$$

The self-ionization of water (the forward reaction in 17.14) occurs only to a slight extent. By contrast, the ionization of HCl, a strong acid, goes essentially to completion (reaction 17.15).* As a result we can conclude that

$[H_3O^+]$ in aqueous solutions of strong acids.

In calculating $[H_3O^+]$ in an aqueous solution of a strong acid, the strong acid is the only significant source of H_3O^+, *unless the solution is* (17.16) *extremely dilute* (e.g., less than 10^{-6} M).

Example 17-2

Calculating ion concentrations in an aqueous solution of a strong acid. Calculate $[H_3O^+]$, $[Cl^-]$, and $[OH^-]$ in 100.0-mL of 0.015 M HCl(aq).

Solution. First, you need to understand that molarities are *independent* of solution volume. That is, $[H_3O^+]$ in 0.015 M HCl(aq) is the same whether we are describing 1.00 L, 10.0 L, or 100.0 mL of solution. The volume of solution does not enter into this calculation (but see Example 17-4 for a calculation where it does).

We can assume that HCl is completely ionized and is the sole source of H_3O^+ in solution (equation 17.15). Therefore,

$[H_3O^+] = 0.015$ M

Furthermore, since one Cl^- is produced for every H_3O^+,

$[Cl^-] = 0.015$ M

To calculate $[OH^-]$ these are the facts that we must use

- all the OH^- is derived from the self-ionization of water (17.14);
- $[OH^-]$ and $[H_3O^+]$ must have values consistent with K_w for water.

$$K_w = [H_3O^+][OH^-] = 1.0 \times 10^{-14}$$

$$(0.015)[OH^-] = 1.0 \times 10^{-14}$$

$$[OH^-] = \frac{1.0 \times 10^{-14}}{1.5 \times 10^{-2}} = 0.67 \times 10^{-12} = 6.7 \times 10^{-13} \text{ M}$$

SIMILAR EXAMPLES: Exercises 2, 21, 59.

*In very concentrated HCl(aq), ionization of HCl does not go to completion. (We can smell HCl in the vapor above such solutions.) In all situations described in this text, however, the assumption of complete ionization of a strong acid will be valid.

Corresponding to the statement made in (17.16), for solutions of strong bases in water we can say that

In calculating $[OH^-]$ in an aqueous solution of a strong base, the strong base is the only significant source of OH^-, *unless the solution is extremely dilute* (e.g., less than 10^{-6} M). (17.17)

Table 17-1 should remind you that the common strong acids and strong bases are really quite small in number.

TABLE 17-1
The Common Strong Acids and Strong Bases

Acids	Bases
HCl	LiOH
HBr	NaOH
HI	KOH
$HClO_4$	RbOH
HNO_3	CsOH
H_2SO_4[a]	$Ca(OH)_2$
	$Sr(OH)_2$
	$Ba(OH)_2$

[a] H_2SO_4 ionizes in two distinct steps. It is a strong acid only in its first ionization step (see page 623).

Example 17-3

Calculating ion concentrations in an aqueous solution of a strong base. Calcium hydroxide (slaked lime) is the cheapest base available and is generally used in all industrial operations where high concentrations of OH^- are not required, for $Ca(OH)_2(s)$ is soluble in water only to the extent of 0.165 g $Ca(OH)_2$/100. mL solution at 20 °C. What are $[Ca^{2+}]$, $[OH^-]$, and $[H_3O^+]$ in saturated $Ca(OH)_2(aq)$ at 20 °C?

Solution. Our first need is to express the solubility of $Ca(OH)_2$ on a molarity basis. Then we can relate the molarities of Ca^{2+} and OH^- to the molarity of $Ca(OH)_2$.

$$[Ca(OH)_2] = \frac{0.165 \text{ g } Ca(OH)_2 \times \dfrac{1 \text{ mol } Ca(OH)_2}{74.09 \text{ g } Ca(OH)_2}}{0.100 \text{ L}} = 0.0223 \text{ M}$$

$$[Ca^{2+}] = \frac{0.0223 \text{ mol } Ca(OH)_2}{L} \times \frac{1 \text{ mol } Ca^{2+}}{1 \text{ mol } Ca(OH)_2} = 0.0223 \text{ M } Ca^{2+}$$

$$[OH^-] = \frac{0.0223 \text{ mol } Ca(OH)_2}{L} \times \frac{2 \text{ mol } OH^-}{1 \text{ mol } Ca(OH)_2} = 0.0446 \text{ M } OH^-$$

To calculate $[H_3O^+]$ we need to recognize that the H_3O^+ comes from the self-ionization of water, and that the ion product of water must be satisfied.

$$K_w = [H_3O^+][OH^-] = [H_3O^+](0.0446) = 1.0 \times 10^{-14}$$

$$[H_3O^+] = 2.2 \times 10^{-13} \text{ M}$$

SIMILAR EXAMPLES: Exercises 2, 25, 59.

A Different Way of Looking at Neutralization Reactions. In Chapter 5 we simply assumed that a neutralization reaction goes to completion and applied basic principles of stoichiometry in numerical calculations. Now we are in a better position to see why neutralization reactions go to completion.

Suppose we mix 100.0 mL of the solution in Example 17-2 (having $[H_3O^+] = 0.015$ M) with 50.0 mL of the solution in Example 17-3 (having $[OH^-] = 0.0446$ M). Each solution dilutes the other in this mixing. Since the final solution volume = 100.0 mL + 50.0 mL = 150.0 mL = 0.1500 L, the concentrations of H_3O^+ and OH^- in this mixture are

$$[H_3O^+] = \frac{\text{no. mol } H_3O^+}{\text{no. L soln.}} = \frac{0.1000 \text{ L} \times 0.015 \text{ mol } H_3O^+/L}{0.1500 \text{ L}}$$

$$= \frac{0.0015 \text{ mol } H_3O^+}{0.1500 \text{ L}} = 0.010 \text{ M} \qquad (17.18)$$

$$[OH^-] = \frac{\text{no. mol } OH^-}{\text{no. L soln.}} = \frac{0.0500 \text{ L} \times 0.0446 \text{ mol } OH^-/L}{0.1500 \text{ L}}$$

$$= \frac{0.00223 \text{ mol OH}^-}{0.1500 \text{ L}} = 0.0149 \text{ M} \qquad (17.19)$$

The product of $[H_3O^+]$ and $[OH^-]$ in the solution would be $(0.010) \times (0.0149) = 1.49 \times 10^{-4}$. However, we have learned that in *all* aqueous solutions at 25 °C the product of $[H_3O^+]$ and $[OH^-]$ must be $K_w = 1.0 \times 10^{-14}$. *A solution cannot be simultaneously 0.010 M in H_3O^+ and 0.0149 M in OH^-.* A chemical reaction *must* occur in which H_3O^+ and OH^- combine to form water until the point is reached where their ion product becomes 1.0×10^{-14}. This means that the reaction goes essentially to completion.

$$H_3O^+ + OH^- \longrightarrow 2 H_2O(l) \qquad (17.20)$$

Two additional aspects of the neutralization described here are presented in Example 17-4. Some practical matters, such as selecting an appropriate indicator for an acid–base titration, are discussed in Chapter 18.

Example 17-4

Determining the outcome of the reaction of a strong acid and a strong base: neutralization. When 100.0 mL of the HCl(aq) of Example 17-2 and 50.0 mL of the $Ca(OH)_2$(aq) of Example 17-3 are mixed, **(a)** is the final solution acidic or basic? **(b)** What is $[H_3O^+]$?

Solution

(a) From equation (17.20) we see that H_3O^+ and OH^- combine in a 1:1 mol ratio. We need to compare the no. mol H_3O^+ present in 100.0 mL of 0.015 M HCl with the no. mol OH^- in 50.0 mL of saturated $Ca(OH)_2$(aq). These quantities were calculated above and are compared below.

no. mol $H_3O^+ = 0.0015$ and no. mol $OH^- = 0.00223$

H_3O^+ is the limiting reagent, OH^- is in excess, and **the final solution is basic**.

(b) We need to find the amount of OH^- that remains in excess.

$$\text{no. mol OH}^- \text{ consumed} = 0.0015 \text{ mol } H_3O^+ \times \frac{1 \text{ mol OH}^-}{1 \text{ mol } H_3O^+}$$

$$= 0.0015 \text{ mol OH}^-$$

no. mol OH^- in excess $= 0.00223 - 0.0015 = 7 \times 10^{-4}$ mol OH^-

This excess OH^- is found in the total 150.0 mL of solution, so

$$[OH^-] = \frac{7 \times 10^{-4} \text{ mol OH}^-}{0.1500 \text{ L soln.}} = 5 \times 10^{-3} \text{ M}$$

Now that we know $[OH^-]$, we can use K_w to solve for $[H_3O^+]$.

$$[H_3O^+][OH^-] = [H_3O^+](5 \times 10^{-3}) = 1.0 \times 10^{-14}$$

$$[H_3O^+] = 2 \times 10^{-12} \text{ M}$$

SIMILAR EXAMPLES: Exercises 6, 26, 67.

17-5 pH and pOH

In Chapter 1 we adopted the system of exponential notation as a convenient means of expressing very large and very small numbers. When expressing $[H_3O^+]$ and $[OH^-]$ in aqueous solutions, we will generally be working with numbers smaller than 1, sometimes much smaller than 1. We can, of course, express such numbers in exponential form, like $[H_3O^+] = 2.5 \times 10^{-3}$. But there is a still easier way. Here is a convention proposed by the Danish biochemist Søren Sørensen in 1909 that has really caught on. Define the **pH** of a solution as the *negative of the logarithm of* $[H_3O^+]$. (Sørensen meant for pH to represent the "potential of hydrogen ion.")

$$pH = -\log [H_3O^+]^* \qquad (17.21)$$

Thus, in a solution that is

0.0010 M HCl, $[H_3O^+] = 1.0 \times 10^{-3}$ M and

$$pH = -(\log 1.0 \times 10^{-3}) = 3.00$$

0.0025 M HCl, $[H_3O^+] = 2.5 \times 10^{-3}$ M and

$$pH = -(\log 2.5 \times 10^{-3}) = 2.60$$

pH = 2.60 is expressed to *two* **significant figures. The number "2" in effect "locates the decimal point," just as does the exponent "−3" in** $[H_3O^+] = 2.5 \times 10^{-3}$ **(see also Appendix A).**

To determine the $[H_3O^+]$ corresponding to a given pH value, we have to carry out the inverse calculations. Thus, in a solution with

pH = 4.00, $\log [H_3O^+] = -4.00$ and

$$[H_3O^+] = \text{antilog}(-4.00) = 10^{-4.00} = 1.0 \times 10^{-4}$$

pH = 4.50, $\log [H_3O^+] = -4.50$ and

$$[H_3O^+] = \text{antilog}(-4.50) = 10^{-4.50} = 3.2 \times 10^{-5}$$

Corresponding to expression (17.21) we can define pOH as

$$pOH = -\log [OH^-] \qquad (17.22)$$

And we can derive still another very useful expression by taking the *negative logarithm* of the K_w expression (written for 25 °C).

$$K_w = [H_3O^+][OH^-] = 1.0 \times 10^{-14}$$

$$-\log K_w = -(\log [H_3O^+][OH^-]) = -\log(1.0 \times 10^{-14})$$

$$-(\log [H_3O^+] + \log [OH^-]) = -(-14.00)$$

$$-\log [H_3O^+] - \log [OH^-] = 14.00$$

$$pH + pOH = 14.00 \qquad (17.23)$$

To summarize the important points that you should know about the pH and pOH concepts

Important ideas concerning pH and pOH.

- $pH = -\log [H_3O^+]$
- $pOH = -\log [OH^-]$
- $pH + pOH = 14.00$
- As a solution becomes *more acidic*, its $[H_3O^+]$ *increases* and its pH *decreases*.
- As a solution becomes *more basic*, its $[OH^-]$ *increases*, its pOH *decreases*, and its pH *increases*.
- A solution with pH = 7.00 is pH neutral; with pH < 7.00, a solution is *acidic*; with pH > 7.00, a solution is *basic*.

(17.24)

The pH values of a number of familiar substances are depicted through Figure 17-3. These values, together with numerous examples that we encounter in this chapter and the next, should help you become more familiar with the pH concept. Later we consider two experimental methods for measuring pH, with acid–base indicators (Section 18-3) and through electrical measurements (Section 21-4).

---

*In thermodynamic treatments, concentrations are replaced by activities, and pH is defined as the negative logarithm of the activity of H_3O^+ ($a_{H_3O^+}$).

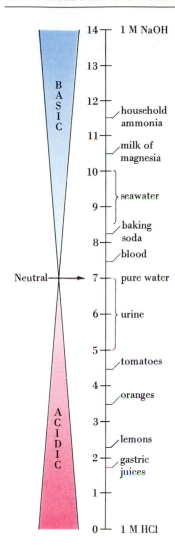

FIGURE 17-3
The pH scale and pH values of some common materials.

It is important to note that a change of pH of one unit represents a tenfold change in $[H_3O^+]$. For example, orange juice is about 10 times more acidic than tomato juice.

Finally, you should realize that the pH concept provides no new theoretical ideas for acid–base reactions. It is used simply because it is so convenient.

Example 17-5

Relating $[H_3O^+]$, $[OH^-]$, pH, and pOH. In a general chemistry laboratory experiment, students measured the pH values of the following: (1) A rainwater sample. (2) A solution prepared by dissolving one pellet of NaOH weighing 235 mg in 500.0 mL of distilled water solution. **(a)** If the measured pH of the rainwater is 4.35, what is $[H_3O^+]$ in the water? **(b)** What pH would you have expected in experiment (2)?

Solution

(a) By definition, $pH = -\log[H_3O^+]$, or

$$\log[H_3O^+] = -pH = -4.35$$

$$[H_3O^+] = \text{antilog } -4.35 = 10^{-4.35} = 4.5 \times 10^{-5} \text{ M}$$

(b) First, calculate $[OH^-]$, which in a solution of a strong base (NaOH) ordinarily is determined exclusively by the amount of the base. (That is, the self-ionization of water is an insignificant source of OH^-.)

$$\text{no. mol } OH^- = 235 \text{ mg NaOH} \times \frac{1.00 \text{ g NaOH}}{1000 \text{ mg NaOH}} \times \frac{1 \text{ mol NaOH}}{40.00 \text{ g NaOH}}$$

$$= 5.88 \times 10^{-3} \text{ mol } OH^-$$

$$\text{molarity of } OH^- = \frac{5.88 \times 10^{-3} \text{ mol } OH^-}{0.5000 \text{ L}} = 1.18 \times 10^{-2} \text{ M}$$

Now, use the definition of pOH.

$$pOH = -\log[OH^-] = -\log(1.18 \times 10^{-2}) = 1.93$$

Finally, since $pH + pOH = 14.00$

$$pH = 14.00 - pOH = 14.00 - 1.93 = 12.07$$

SIMILAR EXAMPLES: Exercises 4, 5, 20, 22, 60, 66.

A common error is to assume that you have calculated the pH when in fact you have found the pOH. You know that in this case the pH cannot be 1.93. The solution must be *basic*; it must have a pH > 7. Use qualitative facts to avoid simple errors.

17-6 Weak Acids and Weak Bases

Figure 17-4 shows two ways to observe the results of acid ionizations in aqueous solution. In Figure 17-4a, from the color of the indicator we judge the pH of 0.10 M HCl to be *less than 1.2*. The pH meter registers a value of 1.0 for the pH, and this is what we would expect for a *strong* acid solution with $[H_3O^+] = 0.10$ M. In Figure 17-4b, the color of the indicator suggests that in 0.10 M $HC_2H_3O_2$ (acetic acid) the pH is 2.8 *or greater*. The pH meter registers a value of 2.8, corresponding to $[H_3O^+] = 1.6 \times 10^{-3}$ M.

(a)   **(b)**

FIGURE 17-4

Strong and weak acids compared.

Thymol blue indicator, which is present in both solutions, imparts a color that depends on the pH of the solution.

pH < 1.2   1.2 < pH < 2.8
  red   orange

pH > 2.8
  yellow

The operation of the electrical meter, called a pH meter, is discussed in Section 21-4.
(a) 0.10 M HCl has pH ≈ 1.
(b) The pH of 0.10 M $HC_2H_3O_2$ is almost 3. [Carey B. Van Loon]

What we see from Figure 17-4, then, is that two acids can have identical molarity concentrations but *different* concentrations of H_3O^+ and pH values. The molarity concentration simply tells us what we put into the solution, but $[H_3O^+]$ depends on what *happens* in the solution. In both solutions some self-ionization of water occurs, but this is negligible. Ionization of HCl, a strong acid, goes to completion; ionization of $HC_2H_3O_2$, a weak acid, is a reversible reaction that reaches a condition of equilibrium.

Negligible: $\quad \underset{\text{acid}}{H_2O} + \underset{\text{base}}{H_2O} \rightleftharpoons \underset{\text{acid}}{H_3O^+} + \underset{\text{base}}{OH^-}$ (17.14)

Goes to completion: $\quad \underset{\text{acid}}{HCl} + \underset{\text{base}}{H_2O} \longrightarrow \underset{\text{acid}}{H_3O^+} + \underset{\text{base}}{Cl^-}$ (17.15)

Reversible reaction: $\quad \underset{\text{acid}}{HC_2H_3O_2} + \underset{\text{base}}{H_2O} \rightleftharpoons \underset{\text{acid}}{H_3O^+} + \underset{\text{base}}{C_2H_3O_2^-}$ (17.25)

The equilibrium constant expression for reaction (17.25) is

$$K_a = \frac{[H_3O^+][C_2H_3O_2^-]}{[HC_2H_3O_2]} = 1.74 \times 10^{-5}$$ (17.26)

K_a is called the **acid ionization constant** of acetic acid. Its value, 1.74×10^{-5} must be determined *by experiment*. Ionization constants for a few weak acids and weak bases are listed in Table 17-2. The symbol K_a represents the ionization constant of a weak acid, and K_b that of a weak base. **pK** is a commonly used shorthand designation for an equilibrium constant: $pK = -\log K$. Thus, for acetic acid,

$$pK_a = -\log K_a = -\log(1.74 \times 10^{-5}) = -(-4.76) = 4.76$$

As with other equilibrium constants, the *larger* the value of K_a (or K_b) the farther the forward reaction proceeds before equilibrium is reached, that is, the more extensive the ionization and the higher the concentrations of the ions produced.

TABLE 17-2
Ionization Constants for Some Weak Acids and Weak Bases in Water at 25 °C

	Ionization equilibrium	Ionization constant K	pK[a]
Acid		$K_a =$	$pK_a =$
acetic	$HC_2H_3O_2 + H_2O \rightleftharpoons H_3O^+ + C_2H_3O_2^-$	1.74×10^{-5}	4.76
benzoic	$HC_7H_5O_2 + H_2O \rightleftharpoons H_3O^+ + C_7H_5O_2^-$	6.3×10^{-5}	4.20
chlorous	$HClO_2 + H_2O \rightleftharpoons H_3O^+ + ClO_2^-$	1.2×10^{-2}	1.92
formic	$HCHO_2 + H_2O \rightleftharpoons H_3O^+ + CHO_2^-$	1.8×10^{-4}	3.74
hydrocyanic	$HCN + H_2O \rightleftharpoons H_3O^+ + CN^-$	4.0×10^{-10}	9.40
hydrofluoric	$HF + H_2O \rightleftharpoons H_3O^+ + F^-$	6.7×10^{-4}	3.17
hypochlorous	$HOCl + H_2O \rightleftharpoons H_3O^+ + OCl^-$	2.95×10^{-8}	7.53
monochloroacetic	$HC_2H_2ClO_2 + H_2O \rightleftharpoons H_3O^+ + C_2H_2ClO_2^-$	1.35×10^{-3}	2.87
nitrous	$HNO_2 + H_2O \rightleftharpoons H_3O^+ + NO_2^-$	5.13×10^{-4}	3.29
phenol	$HOC_6H_5 + H_2O \rightleftharpoons H_3O^+ + C_6H_5O^-$	1.6×10^{-10}	9.80
Base		$K_b =$	$pK_b =$
ammonia	$NH_3 + H_2O \rightleftharpoons NH_4^+ + OH^-$	1.74×10^{-5}	4.76
aniline	$C_6H_5NH_2 + H_2O \rightleftharpoons C_6H_5NH_3^+ + OH^-$	4.30×10^{-10}	9.37
ethylamine	$C_2H_5NH_2 + H_2O \rightleftharpoons C_2H_5NH_3^+ + OH^-$	4.4×10^{-4}	3.36
hydroxylamine	$HONH_2 + H_2O \rightleftharpoons HONH_3^+ + OH^-$	9.1×10^{-9}	8.04
methylamine	$CH_3NH_2 + H_2O \rightleftharpoons CH_3NH_3^+ + OH^-$	4.2×10^{-4}	3.38
pyridine	$C_5H_5N + H_2O \rightleftharpoons C_5H_5NH^+ + OH^-$	2.0×10^{-9}	8.70

[a]Although some of these pK values could be expressed with an additional significant figure, the circumstances of a calculation often do not warrant this.

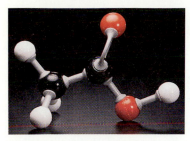

Acetic acid, CH_3COOH.
[Carey B. Van Loon]

How to Identify a Weak Acid. As we know, the key element in an acid is hydrogen, but just because a substance contains H atoms does not make it an acid. One or more H atoms must be ionizable. Let us explore this idea further with acetic acid, whose formula is written in five different ways below.

Empirical formula: CH_2O

Molecular formula: $C_2H_4O_2$

"Acid" formula: $HC_2H_3O_2$

Condensed structural formula: CH_3COOH

Structural formula:

$$\begin{array}{c} \quad\quad H \quad\; O \\ \quad\quad | \quad\;\; \| \\ H-C-C-O-H \\ \quad\quad | \\ \quad\quad H \end{array}$$

The *empirical* formula is based simply on an elemental analysis; it is what we would establish by combustion analysis, for example. There is no clue in this formula as to what makes acetic acid an acid. We would obtain the same empirical formula for the sugar glucose! The *molecular* formula shows that the molecular weight of acetic acid is 60 (twice the value based on the empirical formula), but again there is no indication of the origin of the acidity. In the *"acid" formula*, $HC_2H_3O_2$, we do demonstrate that one of the four H atoms is different in some way from the other three. This is the H atom that ionizes when acetic acid is added to water. $HC_2H_3O_2$ is the formula that we will use for the most part in this chapter. In the *condensed structural formula*, we can be more specific about what is different about the one H atom—it is bonded to an O atom whereas the other three are bonded to a C atom. The complete *structural formula* is even more specific in showing the bonding arrangements. We will work with structural formulas when discussing factors affecting the strengths of acids (Section 17-9).

If the ionizable hydrogen atom(s) is(are) written first in a formula, it is an easy matter to recognize an acid. To distinguish between a weak acid and a strong acid, note that the half-dozen strong acids listed in Table 17-1 are the ones you are most likely to encounter. Unless informed to the contrary, you can generally assume that any other acid you have to deal with is a weak acid. And this means that to describe the composition of the acid solution you will have to use an ionization constant expression, K_a.

Lavoisier was almost right about all acids containing oxygen. Most acids do, with ionizable H atoms bonded to O atoms.

Weak Bases and How to Identify Them. At first glance, weak bases will seem more difficult to identify than weak acids—there is no distinctive element written first in the formula. On the other hand, if you study the bases in Table 17-2 you will see that all but one of them (pyridine) can be viewed as an ammonia molecule in which some other group ($-C_6H_5$, $-C_2H_5$, $-OH$, $-CH_3$) has been substituted for one of the H atoms. The substitution of a methyl group, $-CH_3$, for an H atom is suggested below.

$$\begin{array}{cc} \quad\quad\;\; & \quad\; H \\ \quad\quad\;\; & \quad\; | \\ H-N-H \quad & H-C-N-H \\ \quad | \quad\quad & \quad\;\; |\;\;\; | \\ \quad H \quad\quad & \quad\;\; H\; H \end{array}$$

ammonia   methylamine

The ionization of methylamine as a Brønsted–Lowry base is

Just as CH_3NH_2 is analogous to NH_3, $CH_3NH_3^+$ is analogous to NH_4^+.

$$H_3C-\overset{\cdots}{N}-H + H-\overset{\cdots}{O}: \rightleftharpoons \left[H_3C-\overset{H}{\underset{H}{N}}-H \right]^+ + [H-\overset{\cdots}{\underset{\cdots}{O}}:]^-$$

base   acid   acid   base and the ionization constant expression is

$$K_b = \frac{[CH_3NH_3^+][OH^-]}{[CH_3NH_2]} = 4.2 \times 10^{-4} \tag{17.27}$$

Of course, not all weak bases contain N, but so many of them do that the similarity to NH_3 outlined here is certainly well worth remembering.

Illustrative Examples. For many students, calculations involving equilibria in solution are among the most difficult in their study of general chemistry. Where difficulties arise it is generally a matter of not being able to sort out what is relevant to a given problem-solving situation and what is not. The number of different types of calculations seems very large (although in fact it is quite limited). The key to solving solution equilibrium problems is in being able to picture in your mind's eye what is going on. That is,

- which are the principal species present in solution?
- what are the chemical reactions that produce them?
- are there some reactions (e.g., the self-ionization of water) that can be neglected?
- what is a reasonable answer to the problem? For example, should the final solution be acidic (pH < 7), basic (pH > 7)?

In short, think through a problem *qualitatively* first. At times you may not even have to do a calculation. Another point that cannot be overemphasized is to organize the relevant data for the problem in a clear, logical manner. This action alone can get you on the right track. Look for other helpful hints as you proceed through this and the following two chapters, but for now, in the matter of weak acid and weak base ionizations

Ion concentrations in solutions of weak acids and weak bases.

- In a solution of a **weak acid** in water, unless the solution is unusually dilute or the value of K_a is very small (approaching that of K_w), assume that all the H_3O^+ in solution is produced by the *weak acid*. All the OH^- is produced by the self-ionization of *water*.
- In a solution of a **weak base** in water, again unless the solution is unusually dilute or the value of K_b is very small, assume that all the OH^- is produced by the *weak base*. All the H_3O^+ is produced by the self-ionization of *water*. (17.28)

Example 17-6

Calculating the pH (or pOH) of a weak acid (or weak base) solution from the ionization constant. Show by calculation that the pH of 0.100 M $HC_2H_3O_2$ should be about the value shown on the pH meter in Figure 17-4, that is, pH ≈ 2.8.

Solution. According to the ideas expressed in (17.28), we will consider that all the H_3O^+ comes from the $HC_2H_3O_2$. Also, as suggested by the format below, we will treat the situation as if first all the $HC_2H_3O_2$ dissolves in molecular form and then the molecules ionize until equilibrium is reached.

$$HC_2H_3O_2 + H_2O \rightleftharpoons H_3O^+ + C_2H_3O_2^-$$

placed in solution:	0.100 M	—	—
changes:	$-x$ M	$+x$ M	$+x$ M
at equilibrium:	$(0.100 - x)$ M	x M	x M

$$K_a = \frac{[H_3O^+][C_2H_3O_2^-]}{[HC_2H_3O_2]} = \frac{x \cdot x}{(0.100 - x)} = 1.74 \times 10^{-5}$$

To solve this equation, let us assume that x is very small compared to 0.100; that is, that $(0.100 - x) \approx 0.100$.

$$x^2 = 0.100 \times 1.74 \times 10^{-5} = 1.74 \times 10^{-6}$$

$$x = [H_3O^+] = 1.32 \times 10^{-3}$$

Now, *we must check our assumption:* $(0.100 - 0.00132) = 0.099 \approx 0.100$. Our assumption is good to about 1 part per 100 (1%) and is valid for a calculation involving two or three significant figures.

$$pH = -\log [H_3O^+] = -\log(1.32 \times 10^{-3}) = -(-2.88) = 2.88$$

SIMILAR EXAMPLES: Exercises 8, 9, 29, 31, 32, 63, 69.

Example 17-7

Determining a value of K_b (or K_a) from a measurement of the pH of a solution of a weak base (or weak acid). The drug cocaine has some legitimate therapeutic uses but it is far better known for its abuse. It is a naturally occurring alkaloid found in the leaves of the coca bush. Alkaloids are noted for their bitter taste, an indication of their basic properties. Cocaine, $C_{17}H_{21}O_4N$, is soluble in water to the extent of 1.7 g/L solution, and a saturated solution has a pH = 10.08. What is the value of K_b for cocaine?

$$C_{17}H_{21}O_4N + H_2O \rightleftharpoons C_{17}H_{21}O_4NH^+ + OH^- \qquad K_b = ?$$

Solution. Since the ionization constant expression requires molarity concentrations, a good beginning is to express the concentration of the saturated solution in molarity.

$$\frac{\text{no. mol } C_{17}H_{21}O_4N}{L} = \frac{1.7 \text{ g } C_{17}H_{21}O_4N \times \dfrac{1 \text{ mol } C_{17}H_{21}O_4N}{303 \text{ g } C_{17}H_{21}O_4N}}{1.00 \text{ L}}$$

$$= 5.6 \times 10^{-3} \text{ M}$$

Now organize the data as in Example 17-6.

	$C_{17}H_{21}O_4N$	$+ H_2O \rightleftharpoons$	$C_{17}H_{21}O_4NH^+$	$+ OH^-$
placed in solution:	5.6×10^{-3} M		—	—
changes:	$-x$ M		$+x$ M	$+x$ M
at equilibrium:	$(5.6 \times 10^{-3} - x)$ M		x M	x M

Here, x is not unknown; it is $[OH^-]$ in solution. We can determine $[OH^-]$ from the pOH of the solution.

$$pOH = 14.00 - pH = 14.00 - 10.08 = 3.92$$

$$\log [OH^-] = -pOH = -3.92 \quad \text{and}$$

$$[OH^-] = \text{antilog}(-3.92); \quad [OH^-] = x = 1.2 \times 10^{-4} \text{ M}$$

Now we can solve the following expression for K_b, substituting in the known value for x.

$$K_b = \frac{[C_{17}H_{21}O_4NH^+][OH^-]}{[C_{17}H_{21}O_4N]} = \frac{x \cdot x}{(5.6 \times 10^{-3} - x)}$$

$$= \frac{(1.2 \times 10^{-4})(1.2 \times 10^{-4})}{(5.6 \times 10^{-3} - 1.2 \times 10^{-4})} = 2.6 \times 10^{-6}$$

SIMILAR EXAMPLES: Exercises 7, 27, 70.

Example 17-8

What to do when a simplifying assumption fails. Nitrous acid, HNO_2, cannot be isolated as a pure substance but it can be readily formed in aqueous solution by allowing a nitrite salt (e.g., $NaNO_2$) to react with a strong acid. Nitrous acid and its salts are used in the manufacture of azo dyes and pharmaceuticals. What is the nitrite ion concentration $[NO_2^-]$ in a solution described as 0.00250 M $HNO_2(aq)$?

Solution. We can begin this problem by describing the ionization equilibrium, just as we did for $HC_2H_3O_2$ in Example 17-6 and $C_{17}H_{21}O_4N$ in Example 17-7.

$$HNO_2 + H_2O \rightleftharpoons H_3O^+ + NO_2^- \quad K_a = 5.13 \times 10^{-4}$$

placed in solution:	0.00250 M	—	—
changes:	$-x$ M	$+x$ M	$+x$ M
at equilibrium:	$(0.00250 - x)$ M	x M	x M

$$K_a = \frac{[H_3O^+][NO_2^-]}{[HNO_2]} = \frac{x \cdot x}{0.00250 - x} = 5.13 \times 10^{-4} \qquad (17.29)$$

Let us make the usual assumption, that is, x is very much less than 0.00250 and $(0.00250 - x) \approx 0.00250$.

$$\frac{x^2}{0.00250} = 5.13 \times 10^{-4} \quad x^2 = 1.28 \times 10^{-6} \quad [NO_2^-] = x = 1.13 \times 10^{-3} \text{ M}$$

This value of x is about 45% as large as 0.00250.

$$\frac{0.00113}{0.00250} \times 100 = 45.2\%$$

The value of x is much too large to neglect. If $x = 0.00113$, then $0.00250 - 0.00113 = 0.00137$, and we would certainly agree that $0.00137 \neq 0.00250$.

Since our assumption failed, we must return to equation (17.29) and seek an exact solution. This means solving a *quadratic* equation.

$$\frac{x^2}{0.00250 - x} = 5.13 \times 10^{-4}$$

$$x^2 + 5.13 \times 10^{-4} x - 1.28 \times 10^{-6} = 0$$

$$x = \frac{-5.13 \times 10^{-4} \pm \sqrt{(5.13 \times 10^{-4})^2 + 4 \times 1.28 \times 10^{-6}}}{2}$$

$$= \frac{-5.13 \times 10^{-4} \pm 2.32 \times 10^{-3}}{2}$$

$$x = [NO_2^-] = 9.04 \times 10^{-4} \text{ M}$$

SIMILAR EXAMPLES: Exercises 31, 63.

Degree of Ionization (Percent Ionization). Consider the ionization of a hypothetical weak acid, HA.

$$HA + H_2O \rightleftharpoons H_3O^+ + A^-$$

The following ratio describes the fraction of the acid molecules that ionize; it is called the **degree of ionization.**

$$\text{degree of ionization} = \frac{\text{molarity of } H_3O^+ \text{ (or } A^-)}{\text{original molarity of acid}}$$

Thus, if in an original 1.00 M HA ionization produces $[H_3O^+] = [A^-] = 0.05$ M, then the degree of ionization = 0.05 M/1.00 M = 0.05. If the fraction of molecules

Are You Wondering:

If there is a way to know before doing a calculation whether a simplifying assumption will work (as in Example 17-6) or not work (as in Example 17-8)?

The simplifying assumption referred to here amounts to saying: treat the weak acid or weak base as if all of it remains essentially nonionized. Whether this assumption proves valid depends on two factors.

- the value of K_a (or K_b)
- the molarity concentration of the weak acid (or weak base).

In general, the assumption is likely to be valid if the molarity of the acid exceeds the value of the ionization constant by a factor greater than 1000, that is, if

$$\frac{\text{molarity of acid (or base)}}{K_a \text{ (or } K_b)} > 1000$$

In Example 17-8, the value of K_a was 5.13×10^{-4} and the molarity of $HNO_2(aq)$ was only 0.00250 M. The ratio $[HNO_2]/K_a = 0.00250/(5.13 \times 10^{-4}) = 4.9$. We might have predicted that the simplifying assumption would fail.

In the final analysis, the best approach is always to test the validity of any assumption that you make. If it is good to within a few percent (say, less than 5%), then assume that the assumption is valid. In Example 17-6 the simplifying assumption was good to about 1%, but in Example 17-8 it was off by 45%.

Simplifying assumptions are also more likely to be valid if only two significant figures are carried in a calculation, and this is often the case.

ionizing is expressed as a percent, we can refer to this as the **percent ionization.** A weak acid that has a degree of ionization of 0.05 has a percent ionization of 5.0%.

As we show by a simple calculation in Example 17-9,

Degree of ionization of weak acids and weak bases.

The degree of ionization (and percent ionization) of a weak acid or a weak base *increases* as the solution becomes *more dilute.* (17.30)

Example 17-9

Determining percent ionization as a function of weak acid concentration.

(a) What is the percent ionization of acetic acid in 1.0 M, 0.10 M, and 0.010 M $HC_2H_3O_2$?

(b) Write a simple mathematical equation that could be used to predict the percent ionizations established in (a).

Solution

(a) We use our "standard" format to describe 1.0 M $HC_2H_3O_2$.

$$HC_2H_3O_2 + H_2O \rightleftharpoons H_3O^+ + C_2H_3O_2^-$$

placed in solution:	1.0 M		
changes:	$-x$ M	$+x$ M	$+x$ M
at equilibrium:	$(1.0 - x)$ M	x M	x M

We need to calculate $x = [H_3O^+] = [C_2H_3O_2^-]$. In doing so, let us make the usual assumption that $(1.0 - x) \approx 1.0$.

$$K_a = \frac{[H_3O^+][C_2H_3O_2^-]}{[HC_2H_3O_2]} = \frac{x \cdot x}{(1.0 - x)} = \frac{x^2}{1.0} = 1.74 \times 10^{-5} \quad (17.31)$$

$$x = [H_3O^+] = [C_2H_3O_2^-] = \sqrt{1.74 \times 10^{-5}} = 4.2 \times 10^{-3} \text{ M}$$

The percent ionization of the 1.0 M $HC_2H_3O_2$ is

$$\% \text{ ionization} = \frac{[H_3O^+]}{[HC_2H_3O_2]} \times 100 = \frac{4.2 \times 10^{-3} \text{ M}}{1.0 \text{ M}} \times 100 = 0.42\%$$

The calculations for 0.10 M $HC_2H_3O_2$ and 0.010 M $HC_2H_3O_2$ are very similar and yield the results

0.10 M $HC_2H_3O_2$ is 1.3% ionized   0.010 M $HC_2H_3O_2$ is 4.2% ionized

(b) Refer to equation (17.31). If we replace the molarity of 1.0 by the general molarity **M**, and make the assumption that $(\mathbf{M} - x) \approx \mathbf{M}$, we get the general expression

$$K_a = \frac{x^2}{\mathbf{M}}; \qquad x^2 = \mathbf{M}K_a; \qquad x = [H_3O^+] = \sqrt{\mathbf{M}K_a}$$

$$\% \text{ ionization} = \frac{x}{\mathbf{M}} \times 100 = \frac{\sqrt{\mathbf{M}K_a}}{\mathbf{M}} \times 100 = \sqrt{\frac{K_a}{\mathbf{M}}} \times 100$$

This equation, of course, will work only if the assumption $(\mathbf{M} - x) \approx \mathbf{M}$ is valid.

If you substitute the values 1.0, 0.10, and 0.010 for **M** in this expression, you will obtain the same results as in part (a).

SIMILAR EXAMPLES: Exercises 10, 35, 36.

Other Aspects of Weak Acid/Weak Base Behavior. From an experimental measurement that yields the degree of ionization of a weak acid or weak base, we can work backward to evaluate K_a or K_b. This method is illustrated in Exercise 75. Another matter presented through the Exercises concerns *mixtures of weak acids* (Exercises 37–39).

17-7   Polyprotic Acids

All of the acids listed in Table 17-2 are weak *monoprotic* acids. They produce only one proton (H^+) per acid molecule, even if at times there is more than one H atom in the molecule. Thus, in acetic acid only the H atom attached to the O atom is ionizable.

$$
\underset{\substack{|\\ \text{H}}}{\overset{\substack{\text{H}\quad\text{O}\\ |\quad\;\parallel}}{\text{H}-\text{C}-\text{C}-\text{O}-\text{H}}} + \text{H}_2\text{O} \rightleftharpoons \text{H}_3\text{O}^+ + \left[\underset{\substack{|\\ \text{H}}}{\overset{\substack{\text{H}\quad\text{O}\\ |\quad\;\parallel}}{\text{H}-\text{C}-\text{C}-\text{O}}}\right]^-
$$

There are a number of acids, however, that contain *more than one* ionizable H atom per molecule. These are called **polyprotic acids.** Ionization constants for several polyprotic acids are listed in Table 17-3.

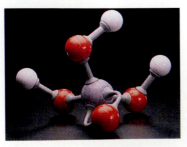

Phosphoric acid, H_3PO_4.
[Carey B. Van Loon]

Phosphoric Acid. Phosphoric acid ranks second only to sulfuric acid among the important commercial acids. Annual production in the United States typically runs about 9 to 10 million tons. Its principal use is in the manufacture of phosphate fertilizers, but large quantities are also consumed in the manufacture of detergents. Various sodium, potassium, and calcium phosphates are used in the food industry. Phosphoric acid, H_3PO_4, has *three* ionizable H atoms, and these H atoms are lost in three distinct steps. H_3PO_4 is a **triprotic acid.** In the first ionization step an H_3PO_4 molecule ionizes to produce H_3O^+ and $H_2PO_4^-$. In the second step, $H_2PO_4^-$ ionizes to produce H_3O^+ and HPO_4^{2-}. In the final step, HPO_4^{2-} forms H_3O^+ and PO_4^{3-}. For each step we can write an ionization constant expression having a distinctive value of K_a. These steps are

TABLE 17-3
Ionization Constants of Some Common Polyprotic Acids

Acid	Ionization equilibria	Ionization constants, K	pK
carbonic[a]	$H_2CO_3 + H_2O \rightleftharpoons H_3O^+ + HCO_3^-$ $HCO_3^- + H_2O \rightleftharpoons H_3O^+ + CO_3^{2-}$	$K_{a_1} = 4.2 \times 10^{-7}$ $K_{a_2} = 5.6 \times 10^{-11}$	$pK_{a_1} = 6.38$ $pK_{a_2} = 10.25$
hydrosulfuric[b]	$H_2S + H_2O \rightleftharpoons H_3O^+ + HS^-$ $HS^- + H_2O \rightleftharpoons H_3O^+ + S^{2-}$	$K_{a_1} = 1.1 \times 10^{-7}$ $K_{a_2} = 1.0 \times 10^{-14}$	$pK_{a_1} = 6.96$ $pK_{a_2} = 14.00$
phosphoric	$H_3PO_4 + H_2O \rightleftharpoons H_3O^+ + H_2PO_4^-$ $H_2PO_4^- + H_2O \rightleftharpoons H_3O^+ + HPO_4^{2-}$ $HPO_4^{2-} + H_2O \rightleftharpoons H_3O^+ + PO_4^{3-}$	$K_{a_1} = 7.11 \times 10^{-3}$ $K_{a_2} = 6.34 \times 10^{-8}$ $K_{a_3} = 4.22 \times 10^{-13}$	$pK_{a_1} = 2.15$ $pK_{a_2} = 7.20$ $pK_{a_3} = 12.37$
phosphorous	$H_3PO_3 + H_2O \rightleftharpoons H_3O^+ + H_2PO_3^-$ $H_2PO_3^- + H_2O \rightleftharpoons H_3O^+ + HPO_3^{2-}$	$K_{a_1} = 5.0 \times 10^{-2}$ $K_{a_2} = 2.5 \times 10^{-7}$	$pK_{a_1} = 1.30$ $pK_{a_2} = 6.60$
sulfurous[c]	$H_2SO_3 + H_2O \rightleftharpoons H_3O^+ + HSO_3^-$ $HSO_3^- + H_2O \rightleftharpoons H_3O^+ + SO_3^{2-}$	$K_{a_1} = 1.3 \times 10^{-2}$ $K_{a_2} = 6.3 \times 10^{-8}$	$pK_{a_1} = 1.89$ $pK_{a_2} = 7.20$
sulfuric[d]	$H_2SO_4 + H_2O \longrightarrow H_3O^+ + HSO_4^-$ $HSO_4^- + H_2O \rightleftharpoons H_3O^+ + SO_4^{2-}$	$K_{a_1} = $ very large $K_{a_2} = 1.29 \times 10^{-2}$	$pK_{a_1} < 0$ $pK_{a_2} = 1.89$

[a]H_2CO_3 cannot be isolated. It is in equilibrium with H_2O and dissolved CO_2. The value given for K_{a_1} is actually for the reaction

$CO_2(aq) + 2 H_2O \rightleftharpoons H_3O^+ + HCO_3^-$

[b]The value of K_{a_2} for H_2S has always been subject to doubt. Latest evidence seems to suggest that the best value may be as low as 10^{-19}. [see R. J. Myers, "The New Low Value for the Second Ionization Constant for H_2S," *J. Chem. Educ.*, **63**, 687 (1986).]
[c]H_2SO_3 is a hypothetical, nonisolable species produced in the reaction $SO_2(aq) + H_2O \rightleftharpoons H_2SO_3(aq)$.
[d]H_2SO_4 is completely ionized in the first step.

$$(1) \quad H_3PO_4 + H_2O \rightleftharpoons H_3O^+ + H_2PO_4^- \qquad K_{a_1} = \frac{[H_3O^+][H_2PO_4^-]}{[H_3PO_4]}$$
$$= 7.11 \times 10^{-3}$$

$$(2) \quad H_2PO_4^- + H_2O \rightleftharpoons H_3O^+ + HPO_4^{2-} \qquad K_{a_2} = \frac{[H_3O^+][HPO_4^{2-}]}{[H_2PO_4^-]}$$
$$= 6.34 \times 10^{-8}$$

$$(3) \quad HPO_4^{2-} + H_2O \rightleftharpoons H_3O^+ + PO_4^{3-} \qquad K_{a_3} = \frac{[H_3O^+][PO_4^{3-}]}{[HPO_4^{2-}]}$$
$$= 4.22 \times 10^{-13}$$

There is a ready explanation for the fact that the second ionization constant is smaller than the first ($K_{a_2} < K_{a_1}$), and that the third is smaller than the second ($K_{a_3} < K_{a_2}$). When ionization occurs in step (1), a proton (H^+) is being separated from an ion with a charge of -1 ($H_2PO_4^-$). In step (2) the proton is being separated from an ion with a charge of -2 (HPO_4^{2-}), and we should expect this to be more difficult. This accounts for the smaller ionization constant in the second step than in the first. Ionization is more difficult still in the third step.

Let us list several generalizations about polyprotic acids (as applied to phosphoric acid) and then we will illustrate some of them through appropriate calculations.

Generalizations concerning polyprotic acids.

1. All species involved in the ionization equilibria (that is, H_3PO_4, H_3O^+, $H_2PO_4^-$, HPO_4^{2-}, PO_4^{3-}) appear together in the same solution phase.
2. There is only *one* concentration for each species in the solution, regardless of its source(s), and each concentration value must be consistent with all ionization constant expressions in which that value appears. (For example, there is only one $[H_3O^+]$ in $H_3PO_4(aq)$; this same value of $[H_3O^+]$ is used in K_{a_1}, K_{a_2}, and K_{a_3}.)   (17.32)
3. The relative magnitudes of the ionization constants are always $K_{a_1} > K_{a_2} > K_{a_3}$.
4. Assume that ionization in the first step is far more important than in

Example 17-10

Calculating the concentrations of ions present in a polyprotic acid. For a 3.0 M H_3PO_4 solution, calculate (a) $[H_3O^+]$; (b) $[HPO_4^{2-}]$; (c) $[PO_4^{3-}]$.

Solution

(a) Since K_{a_1} is so much larger than K_{a_2}, we can assume that all the H_3O^+ is produce in the *first* ionization step. This is equivalent to thinking of H_3PO_4 as if it were a monoprotic acid, ionizing only through the first step.

$$H_3PO_4 + H_2O \rightleftharpoons H_3O^+ + H_2PO_4^-$$

placed in solution:	3.0 M	—	—
changes:	$-x$ M	$+x$ M	$+x$ M
after first ionization:	$(3.0 - x)$ M	x M	x M

In the assumption $(3.0 - x) \approx 3.0$, $x = 0.15$, which is 5.0% of 3.0. This is about the maximum error we can tolerate for an acceptable assumption.

Following the usual assumption that x is much smaller than 3.0 and that $(3.0 - x) \approx 3.0$, we obtain

$$K_{a_1} = \frac{[H_3O^+][H_2PO_4^-]}{[H_3PO_4]} = \frac{x \cdot x}{(3.0 - x)} = \frac{x^2}{3.0} = 7.11 \times 10^{-3}$$

$$x^2 = 0.0213 \qquad x = [H_3O^+] = 0.15 \text{ M}$$

(b) Even though for the purposes of determining $[H_3O^+]$ we made the assumption that the second ionization is insignificant, here we do have to consider the second ionization, no matter how slight. Otherwise, we would have no source of the ion HPO_4^{2-}. We can represent the second ionization as shown below. *Note especially how the results of the first ionization enter in.* We start with a solution in which $[H_2PO_4^-] = [H_3O^+] = 0.15$ M.

$$H_2PO_4^- + H_2O \rightleftharpoons H_3O^+ + HPO_4^{2-}$$

from first ionization:	0.15 M	0.15 M	—
changes:	$-y$ M	$+y$ M	$+y$ M
after second ionization:	$(0.15 - y)$ M	$(0.15 + y)$ M	y M

Finally, we can make the usual assumption: If y is much smaller than 0.15, $(0.15 + y) \approx (0.15 - y) \approx 0.15$. Now substitute into the K_a expression.

$$K_{a_2} = \frac{[H_3O^+][HPO_4^{2-}]}{[H_2PO_4^-]} = \frac{(0.15 + y)(y)}{(0.15 - y)} = 6.34 \times 10^{-8}$$

Note how we could have written this result directly just by using Statement 5 of (17.32).

$$y = [HPO_4^{2-}] = 6.34 \times 10^{-8} \text{ M} = K_{a_2}$$

(c) Let us try a simple approach here. Since PO_4^{3-} can only be formed in the third ionization step, let us start by writing the ionization constant expression for this step. After we do we will see that we have already calculated concentrations of all the species other than $[PO_4^{3-}]$. We can then simply solve the K_{a_3} expression for $[PO_4^{3-}]$.

$$K_{a_3} = \frac{[H_3O^+][PO_4^{3-}]}{[HPO_4^{2-}]} = \frac{0.15 \times [PO_4^{3-}]}{6.34 \times 10^{-8}} = 4.22 \times 10^{-13}$$

$$[PO_4^{3-}] = \frac{4.22 \times 10^{-13} \times 6.34 \times 10^{-8}}{0.15} = 1.8 \times 10^{-19} \text{ M}$$

SIMILAR EXAMPLES: Exercises 11, 40, 41, 42, 43.

Are You Wondering:

How good is the assumption that all the H_3O^+ in a polyprotic acid solution comes just from the first ionization step?

If the solution in question has a polyprotic acid as the only solute, this assumption will generally work well, regardless of the magnitudes of K_{a_1} and K_{a_2}. Think of the matter in this way: Suppose we started with a 1 M solution of the acid and that the percent ionization were the same in the first and second ionization steps, say 1%. In the first step the $[H_3O^+]$ produced would be 0.01×1 M $= 0.01$ M, but in the second step it would be only 0.01×0.01 M $= 0.0001$ M. Thus, even in this case, the second ionization would produce only 1% of the H_3O^+ in solution. In most situations the second ionization is even less important than this.

When we discuss the titration of polyprotic acids in Chapter 18, we will discover situations where the relative magnitudes of K_{a_1} and K_{a_2} are of greater significance.

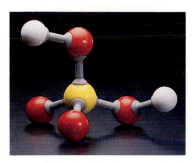

Sulfuric acid, H_2SO_4.
[Carey B. Van Loon]

A Somewhat Different Case—H_2SO_4. Sulfuric acid differs from other polyprotic acids in this important respect. It is a *strong* acid in its first ionization step and a weak acid only in the *second* step. Ionization is complete in the first step, and this means that in most $H_2SO_4(aq)$ solutions we can assume that $[H_2SO_4] \approx 0$. Thus, if a solution is 0.50 M H_2SO_4, we can treat it as if it were 0.50 M H_3O^+ and 0.50 M HSO_4^- initially and then determine how much ionization of HSO_4^- occurs to produce additional H_3O^+ and SO_4^{2-}.

Example 17-11

Calculating ion concentrations in sulfuric acid solutions: Strong acid ionization followed by weak acid ionization. Calculate $[H_3O^+]$, $[HSO_4^-]$, and $[SO_4^{2-}]$ in 0.50 M H_2SO_4.

Solution. Let us proceed in a similar fashion to what we did in Example 17-10 for the first two steps of the phosphoric acid ionization.

$$H_2SO_4 + H_2O \longrightarrow H_3O^+ + HSO_4^-$$

placed in solution:	0.50 M	—	—
changes:	−0.50 M	+0.50 M	+0.50 M
after first ionization:	≈ 0	0.50 M	0.50 M

$$HSO_4^- + H_2O \rightleftharpoons H_3O^+ + SO_4^{2-}$$

from first ionization:	0.50 M	0.50 M	—
changes:	−x M	+x M	+x M
after second ionization:	(0.50 − x) M	(0.50 + x) M	x M

We need to deal only with the ionization constant expression for K_{a_2}. If we assume that x is much smaller than 0.50, then $(0.50 + x) \approx (0.50 - x) \approx 0.50$ and

$$K_{a_2} = \frac{[H_3O^+][SO_4^{2-}]}{[HSO_4^-]} = \frac{(0.50 + x) \times x}{(0.50 - x)} = \frac{\cancel{0.50} \times x}{\cancel{0.50}} = 1.29 \times 10^{-2}$$

Our results, then, are

$$[H_3O^+] = 0.50 + x = 0.51 \text{ M}; \qquad [HSO_4^-] = 0.50 - x = 0.49 \text{ M}$$

$$[SO_4^{2-}] = x = K_{a_2} = 0.0129 \text{ M}$$

SIMILAR EXAMPLES: Exercises 44, 71(c).

The assumption made in Example 17-11, leading to the conclusion that $[SO_4^{2-}] = 0.0129$ M, is valid only for solutions whose concentrations are *greater* than about 0.25 M H_2SO_4. If a solution is *less than about 0.25 M H_2SO_4*, you will have to solve a quadratic equation to obtain exact values of $[H_3O^+]$, $[HSO_4^-]$, and $[SO_4^{2-}]$. On the other hand, if solutions are *sufficiently dilute*, say below 0.001 M H_2SO_4, you can assume that *both* ionization reactions go to completion. In these cases $[H_3O^+]$ is twice the molarity of the H_2SO_4.

17-8 Ions as Acids and Bases

In our discussion to this point we have emphasized the behavior of *neutral molecules* as acids (e.g., HCl, $HC_2H_3O_2$, H_3PO_4) or as bases (e.g., NH_3, CH_3NH_2). We have also seen, however, that ions can act as acids or bases. For instance, in the second and subsequent ionization steps of a polyprotic acid an anion acts as an acid.

$$H_2PO_4^- + H_2O \rightleftharpoons H_3O^+ + HPO_4^{2-} \qquad K = K_{a_2} = 6.34 \times 10^{-8} \qquad (17.33)$$

Let us think about how each of the following can be described as an acid–base reaction.

$$NH_4^+ + H_2O \rightleftharpoons NH_3 + H_3O^+ \qquad\qquad\qquad (17.34)$$

$$C_2H_3O_2^- + H_2O \rightleftharpoons HC_2H_3O_2 + OH^- \qquad\qquad (17.35)$$

Also, as we saw in equation (17.7), NH_4^+ is the conjugate acid of NH_3.

Equation (17.34) suggests that NH_4^+ is an *acid,* able to donate a proton to water, a *base*. Equilibrium in this reaction is described through the *acid ionization constant* of the ammonium ion, NH_4^+.

$$K_a = \frac{[NH_3][H_3O^+]}{[NH_4^+]} = 5.7 \times 10^{-10}$$

Equation (17.35) shows $C_2H_3O_2^-$ acting as a *base* by accepting a proton from water, an *acid*. Here, equilibrium is described through the *base ionization constant* of the acetate ion, $C_2H_3O_2^-$.

$$K_b = \frac{[HC_2H_3O_2][OH^-]}{[C_2H_3O_2^-]} = 5.7 \times 10^{-10}$$

Table 17-4 presents ionization constant data for several ions.

Hydrolysis. Acid–base reactions like (17.34) and (17.35) will occur in any solution in which NH_4^+ and $C_2H_3O_2^-$ ions, respectively, are present. We have a special interest, however, in situations where these ions *alone* are responsible for establishing the pH of a solution.

TABLE 17-4
Acid and Base Ionization Constants for Several Ions at 25 °C

	Ionization equilibrium	Ionization constant, K	pK
Acid		$K_a =$	p$K_a =$
ammonium ion	$NH_4^+ + H_2O \rightleftharpoons H_3O^+ + NH_3$	5.7×10^{-10}	9.24
anilinium ion	$C_6H_5NH_3^+ + H_2O \rightleftharpoons H_3O^+ + C_6H_5NH_2$	2.3×10^{-5}	4.64
methylammonium ion	$CH_3NH_3^+ + H_2O \rightleftharpoons H_3O^+ + CH_3NH_2$	2.4×10^{-11}	10.62
Base		$K_b =$	p$K_b =$
acetate ion	$C_2H_3O_2^- + H_2O \rightleftharpoons OH^- + HC_2H_3O_2$	5.7×10^{-10}	9.24
chlorite ion	$ClO_2^- + H_2O \rightleftharpoons OH^- + HClO_2$	8.3×10^{-13}	12.08
cyanide ion	$CN^- + H_2O \rightleftharpoons OH^- + HCN$	2.5×10^{-5}	4.60

Ions as acids and bases.

Each of these 1 M solutions contains bromothymol blue indicator, which has the following colors.

pH < 7	pH = 7	pH > 7
yellow	green	blue

(a) $NH_4Cl(aq)$ is acidic.
(b) $NaCl(aq)$ is pH neutral.
(c) $NaC_2H_3O_2(aq)$ is basic.
[Carey B. Van Loon]

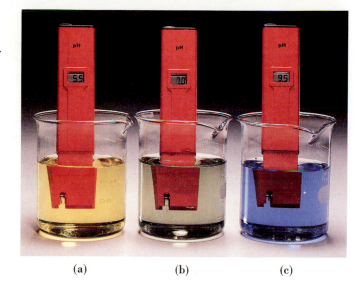

(a)   (b)   (c)

We have learned that in pure water at 25 °C, $[H_3O^+] = [OH^-] = 1.0 \times 10^{-7}$ M and pH = 7.00. *Pure water is pH neutral.* When a salt such as NaCl is added to water, complete dissociation into Na^+ and Cl^- ions occurs, but these ions do not influence the self-ionization of water. The pH of the solution remains at 7.00.

$$Na^+ + Cl^- + H_2O \longrightarrow \text{no reaction}$$

As shown in Figure 17-5, when NH_4Cl is added to water the pH falls below 7. This means that $[H_3O^+]$ in the solution increases and $[OH^-]$ decreases. A reaction producing H_3O^+ must occur. Cl^- cannot act as an acid—it has no proton to donate. However, we have already seen that a reaction occurs between NH_4^+ and water.

$$Cl^- + H_2O \longrightarrow \text{no reaction}$$
$$NH_4^+ + H_2O \rightleftharpoons NH_3 + H_3O^+$$

(17.34)

Reaction (17.34) is fundamentally no different from other acid–base reactions. However, a reaction between an ion and water, especially if water is the only other acid or base present, is often called a **hydrolysis** reaction. We say that *ammonium ion hydrolyzes* (and chloride ion does not).

When sodium acetate is added to water, the pH rises above 7 (see Figure 17-5). This means that $[OH^-]$ in the solution increases and $[H_3O^+]$ decreases. Sodium ion has neither acidic nor basic properties, but *acetate ion hydrolyzes.*

$$Na^+ + H_2O \longrightarrow \text{no reaction}$$
$$C_2H_3O_2^- + H_2O \rightleftharpoons HC_2H_3O_2 + OH^-$$

(17.35)

From this discussion we can make some *qualitative* statements about hydrolysis in aqueous solution. Note that these statements all reflect the general fact that *hydrolysis will occur only if a chemical reaction can produce a weak acid or weak base.*

Generalizations concerning hydrolysis reactions.

- Salts of strong acids and strong bases (e.g., NaCl) *do not hydrolyze.* The solution *pH = 7.*
- Salts of *weak* acids and strong bases (e.g., $NaC_2H_3O_2$) *hydrolyze: pH > 7.* (The *anion* acts as a *base.*)
- Salts of strong acids and *weak* bases (e.g., NH_4Cl) *hydrolyze: pH < 7.* (The *cation* acts as an *acid.*)
- Salts of *weak* acids and *weak* bases (e.g., $NH_4C_2H_3O_2$) *hydrolyze.* (The cations act as acids and the anions as bases, but whether the solution is acidic or basic depends on the relative values of K_a and K_b for the ions.)

(17.36)

Example 17-12

Making qualitative predictions about hydrolysis reactions. Predict whether you would expect each of the following solutions to be acidic, basic, or pH neutral: **(a)** NaCN(aq); **(b)** KCl(aq); **(c)** NH_4CN(aq).

Solution

(a) The ions in solution are Na^+, which does not hydrolyze, and CN^-, which does. CN^- is the conjugate base of HCN and should form a basic solution. $CN^- + H_2O \rightleftharpoons HCN + OH^-$.

(b) Neither K^+ nor Cl^- undergoes hydrolysis. KCl(aq) is neutral—pH = 7.

(c) *Both* NH_4^+ and CN^- hydrolyze in aqueous solution, one to produce H_3O^+ and the other, OH^-. With values from Table 17-4 we can write

$$NH_4^+ + H_2O \rightleftharpoons NH_3 + H_3O^+ \qquad K_a = 5.7 \times 10^{-10}$$

$$CN^- + H_2O \rightleftharpoons HCN + OH^- \qquad K_b = 2.5 \times 10^{-5}$$

Because K_b is larger than K_a, we should expect CN^- to hydrolyze to a greater extent than NH_4^+. This means that $[OH^-] > [H_3O^+]$ and the solution will be basic.

SIMILAR EXAMPLES: Exercises 12, 45, 73.

To predict that NH_4CN(aq) is basic (pH > 7) is easy. To calculate the *actual* pH of NH_4CN(aq) is much more difficult. Several algebraic equations must be solved simultaneously (see Exercise 88).

In Example 17-12 we predicted *qualitatively* that a solution of NaCN should be basic (pH > 7). To establish the exact pH of this solution we must do a *quantitative* calculation. For this we need to use K_b for CN^-, as illustrated in Example 17-13.

Example 17-13

Calculating the pH of a solution in which hydrolysis occurs. Sodium cyanide is an extremely poisonous substance, but it is useful nevertheless in such applications as gold and silver metallurgy and in electroplating baths. [It is the ability of CN^- to form complex ions with metal ions that is important in these applications (see Section 25-11).] Aqueous solutions of cyanides are especially hazardous if they become acidified because of the release of toxic hydrogen cyanide, HCN. Normally, as we saw in Example 17-12, NaCN(aq) solutions are basic. What is the pH of 0.50 M NaCN(aq)?

Solution. The concentrations of the several species involved in the hydrolysis reaction are summarized below, where $x = [OH^-]$.

	CN^-	$+ H_2O \rightleftharpoons$	HCN	$+ OH^-$
placed in solution:	0.50 M		—	—
changes:	$-x$ M		$+x$ M	$+x$ M
equilibrium concentrations:	$(0.50 - x)$ M		x M	x M

$$K_b = \frac{[OH^-][HCN]}{[CN^-]} = 2.5 \times 10^{-5}$$

$$\frac{x \cdot x}{0.50 - x} = \frac{x^2}{0.50 - x} = 2.5 \times 10^{-5}$$

Next a familiar assumption can be made. If x is much smaller than 0.50, then $0.50 - x \simeq 0.50$.

$$x^2 = (0.50)(2.5 \times 10^{-5}) = 1.2 \times 10^{-5} = 12 \times 10^{-6}$$

$$x = [OH^-] = (12 \times 10^{-6})^{1/2} = 3.5 \times 10^{-3}$$

$$[H_3O^+] = \frac{K_w}{[OH^-]} = \frac{1.0 \times 10^{-14}}{3.5 \times 10^{-3}} = 2.9 \times 10^{-12}$$

$$pH = -\log[H_3O^+] = -\log(2.9 \times 10^{-12}) = 11.54$$

SIMILAR EXAMPLES: Exercises 13, 46, 47, 48.

Relating Ionization Constants of Weak Acids (or Bases) and Their Conjugates.
Tabulated data are sometimes listed as K_a values for weak acids and K_b values for
weak bases (as in Table 17-2). Sometimes all that are listed are values of K_a for
weak acids and conjugate acids of weak bases. By using the method of combining
equilibrium constants expressions described in Section 16-3, you should always be
able to come up with the value you need. Let us see how this is done.

Written below are equations for the ionization of (a) acetic acid (HAc) as a weak
acid and (b) acetate ion (Ac^-) as a base. The sum of these two equations represents
the self-ionization of water. When equations are added, we learned in Section 16-3,
the equilibrium constant of the net reaction is the *product* of the equilibrium constants of the reactions that are combined.

(a) $HAc + H_2O \rightleftharpoons H_3O^+ + Ac^-$ $K_a(HAc)$
(b) $Ac^- + H_2O \rightleftharpoons OH^- + HAc$ $K_b(Ac^-)$

Net: $2\,H_2O \rightleftharpoons H_3O^+ + OH^-$ $K_w = K_a(HAc) \times K_b(Ac^-)$ (17.37)

A general statement of the result of equation (17.37) is that

$$K_a(\text{acid}) \times K_b(\text{its conjugate base}) = K_w \qquad (17.38)$$

Stated below are three ideas (two of which are just a restatement of equation 17.38).
These three ideas should enable you to deal with any situation in which you need
ionization constant data.

Ionization constants for weak acids, weak bases, and their conjugates.

- If for the equilibrium reaction of interest both the chemical equation
  and a K value are given, simply use these in writing the ionization
  constant expression.
- If you have K_a data and you need K_b for a conjugate base, $K_b = K_w/K_a$.   (17.39)
- If you have K_b data and you need K_a for a conjugate acid, $K_a = K_w/K_b$.

Example 17-14

Relating the ionization constant of a weak acid and its conjugate base.
Sodium nitrite, $NaNO_2$, and sodium benzoate, $NaC_7H_5O_2$, are both used as
food preservatives. If solutions of these two salts are compared, each at the
same concentration, which will have the *higher* pH?

Solution. Each of these substances is the salt of a *weak acid* and a strong base
(NaOH). The anions should ionize as *bases*, making their solutions somewhat
basic.

$$NO_2^- + H_2O \rightleftharpoons HNO_2 + OH^- \qquad K_b(NO_2^-) = ?$$

$$C_7H_5O_2^- + H_2O \rightleftharpoons HC_7H_5O_2 + OH^- \qquad K_b(C_7H_5O_2^-) = ?$$

Our essential task is to determine the K_b values, neither of which is listed in
a table in the chapter. We do have listings of K_a for the weak acids, however
(Table 17-2). We can use equation (17.38) to write

$$K_b(NO_2^-) = \frac{K_w}{K_a(HNO_2)} = \frac{1.0 \times 10^{-14}}{5.13 \times 10^{-4}} = 1.9 \times 10^{-11}$$

$$K_b(C_7H_5O_2^-) = \frac{K_w}{K_a(HC_7H_5O_2)} = \frac{1.0 \times 10^{-14}}{6.3 \times 10^{-5}} = 1.6 \times 10^{-10}$$

FIGURE 17-6

Acidic properties of hydrated metal ions.

The yellow color of bromothymol blue indicator in $Al_2(SO_4)_3(aq)$ indicates that the solution is somewhat acidic (see caption to Figure 17-5). A more precise value of the pH is indicated by the pH meter. [Carey B. Van Loon]

Comparing these two ionization constants, we see that $K_b(C_7H_5O_2^-)$ is about 10 times larger than $K_b(NO_2^-)$. We should expect the benzoate ion to hydrolyze to a somewhat greater extent than the nitrite ion, yielding a solution with a higher $[OH^-]$.

The sodium benzoate solution should have a higher pH than the sodium nitrite solution.

SIMILAR EXAMPLES: Exercises 14, 47, 51.

Hydrated Metal Ions as Acids. With the exception of NH_4^+, the cations in Table 17-4 are not familiar (they are derived from organic molecules). Among simple metal ions, we have learned that Na^+ does not hydrolyze. Neither do other familiar cations of the group 1A and 2A metals. Many other metal ions (especially of the transition metals) do hydrolyze, however. These cations ionize as *acids*—proton donors. They do this through the ionization of water molecules that are attached to the central metal ion. Thus, it is *hydrated* metal ions that hydrolyze, as in the case of $Al^{3+}(aq)$.

$$[Al(H_2O)_6]^{3+} + H_2O \rightleftharpoons [Al(H_2O)_5OH]^{2+} + H_3O^+ \qquad K_a = 1.1 \times 10^{-5}$$

One of the six H_2O molecules attached to Al^{3+} in the complex ion $[Al(H_2O)_6]^{3+}$ loses a proton to a free H_2O molecule, converting it to H_3O^+. The H_2O molecule that has ionized is converted to OH^-, which remains attached to the Al^{3+} ion and reduces the charge of the complex ion from 3+ to 2+. The extent of ionization of $[Al(H_2O)_6]^{3+}$, as measured by its K_a value and as pictured in Figure 17-6, is essentially the same as that of acetic acid ($K_a = 1.74 \times 10^{-5}$). Hydrated metal ions are important acids, as we will see when we discuss this matter more fully in Chapters 22–25.

17-9   Molecular Structure and Acid–Base Behavior

Here are some very fundamental questions that we still need to answer.

- Why is HCl a strong acid, whereas HF is a weak acid?
- Why is monochloroacetic acid ($HC_2H_2ClO_2$) a stronger acid than acetic acid ($HC_2H_3O_2$)?
- Why is phenol (C_6H_5OH) a much stronger acid than ethanol (C_2H_5OH)?

In the discussion that follows, when we say that one acid is stronger than another, we mean that it has a larger value of K_a.

Strengths of Brønsted–Lowry Acids and Bases. If we study the following reactions we find that the first goes to completion and the other proceeds hardly at all in the forward direction.

$$\begin{array}{ccccccc}
HI & + & OH^- & \rightleftharpoons & H_2O & + & I^- \\
\text{acid(1)} & & \text{base(2)} & & \text{acid(2)} & & \text{base(1)} \\
\text{very strong} & & \text{strong} & & \text{weak} & & \text{very weak}
\end{array} \qquad (17.40)$$

$$\begin{array}{ccccccc}
HCO_3^- & + & Br^- & \rightleftharpoons & HBr & + & CO_3^{2-} \\
\text{acid(1)} & & \text{base(2)} & & \text{acid(2)} & & \text{base(1)} \\
\text{weak} & & \text{very weak} & & \text{very strong} & & \text{strong}
\end{array} \qquad (17.41)$$

Each reaction conforms to the principle that

Predicting the direction of an acid–base reaction.

A Brønsted–Lowry acid–base reaction is always favored in the direction from the *stronger* to the *weaker* acid/base combination.   (17.42)

TABLE 17-5
Relative Strengths of Some Common Brønsted–Lowry Acids and Bases

	Acid		Conjugate base	
	perchloric acid	$HClO_4$	perchlorate ion	ClO_4^-
	hydroiodic acid	HI	iodide ion	I^-
	hydrobromic acid	HBr	bromide ion	Br^-
	hydrochloric acid	HCl	chloride ion	Cl^-
	sulfuric acid	H_2SO_4	hydrogen sulfate ion	HSO_4^-
	nitric acid	HNO_3	nitrate ion	NO_3^-
	hydronium ion[a]	H_3O^+	water[a]	H_2O
	hydrogen sulfate ion	HSO_4^-	sulfate ion	SO_4^{2-}
	nitrous acid	HNO_2	nitrite ion	NO_2^-
	acetic acid	$HC_2H_3O_2$	acetate ion	$C_2H_3O_2^-$
	carbonic acid	H_2CO_3	bicarbonate ion	HCO_3^-
	ammonium ion	NH_4^+	ammonia	NH_3
	bicarbonate ion	HCO_3^-	carbonate ion	CO_3^{2-}
	water	H_2O	hydroxide ion	OH^-
	methanol	CH_3OH	methoxide ion	CH_3O^-
	ammonia	NH_3	amide ion	NH_2^-

increasing acid strength (increasing K_a) →

increasing base strength (increasing K_b) →

[a] The hydronium ion/water combination refers to the ease with which a proton is passed from one water molecule to another; that is, $H_3O^+ + H_2O \rightleftharpoons H_3O^+ + H_2O$.

Thus, if HI has a very strong tendency to lose a proton and OH^- a strong tendency to gain one, we should expect proton transfer to be from HI to OH^-, the forward reaction in (17.40). To be able to apply expression (17.42), of course, we must have some idea of the relative strengths of acids and bases, such as listed in Table 17-5. One fact brought out by this listing is that *the stronger an acid, the weaker its conjugate base.*

The relative placement of some of the entries in Table 17-5 follows directly from what we have learned elsewhere in the chapter. Acetic acid ($K_a = 1.74 \times 10^{-5}$), for example, is a stronger acid than is carbonic acid ($K_{a_1} = 4.5 \times 10^{-7}$). Also, as we would expect, carbonic acid ($K_{a_1} = 4.5 \times 10^{-7}$) is a stronger acid than its anion, HCO_3^- ($K_{a_2} = 4.7 \times 10^{-11}$). But why do we rank $HClO_4$ ahead of HCl? In water solution both of these acids are so strong that they are completely ionized. Water is said to have a **leveling effect** on these two acids. Water is a strong enough base to accept protons from either acid, blurring whatever differences may exist between them.

To distinguish between the strengths of $HClO_4$ and HCl we need to use a **differentiating solvent.** This is a solvent that is such a weak base that it will accept protons from the stronger of the two acids more readily than from the weaker one. Diethyl ether ($C_2H_5OC_2H_5$) is such a solvent. In diethyl ether, $HClO_4$ is still completely ionized, but HCl is only partially ionized. Although the solvent plays an important role in establishing the relative strengths of acids and bases, we will still be concerned mostly with solutions in which water is the solvent—aqueous solutions.

Here is a simple analogy to explain the leveling effect and differentiating solvents. Suppose two people are pointed out to you; one is very wealthy and the other is of more modest means. You are asked to figure out who is the wealthy one, based on what the two people order when dining out. You might not be able to distinguish between them based on what they order in a drive-in fast-food restaurant, but you would have less difficulty in an exclusive, elegant restaurant.

Bond Strength and Acid Strength. Since the strength of an acid is related to the ease with which a proton is lost, we might expect some relationship between bond strength and acid strength. However, it is an oversimplification to try to relate the acidities of binary acids HX to the mere breaking of the bond H—X. For one thing, bond energies are based on the dissociation of gaseous species, and here we are dealing with species in solution. Nevertheless, it does seem reasonable that the stronger the H—X bond, the *weaker* the acid. This generalization does work for the series of acids HF, HCl, HBr, and HI, for which the bond energies *increase* in the order

$$HI < HBr < HCl < HF$$
$$297 < 368 < 431 < 569 \ kJ/mol$$

whereas acid strengths *decrease* in the order

$$HI > HBr > HCl > HF$$
$$K_a = \underbrace{10^9 > 10^8 > 1.3 \times 10^6}_{strong} > \underset{weak}{6.7 \times 10^{-4}}$$

> This generalization also works for other binary acids (see Exercise 82).

A different way to look at the matter, leading to the same conclusion, is that as the anion (X^-) *decreases* in size, its affinity for a proton *increases*; it becomes a *stronger base*. Since the anion is the conjugate base of the acid, as the conjugate base gets stronger the acid strength *decreases*.

Acidic, Basic, and Amphoteric Oxides. Most of the elements form compounds with oxygen, and one classification scheme for oxides uses the terms, acidic, basic, and amphoteric. To see how the scheme works, consider the hypothetical oxide, E_xO_y, formed by the element E. Consider also that this oxide reacts with water to form a hydroxo compound containing one or more bonds, E—O—H, and other bonds (---) to O atoms, E.

> The fact that HF is a weak acid, whereas the other hydrogen halides are very strong, has always seemed an anomaly. One proposal is that in HF(aq) ion pairs are held together by strong hydrogen bonds. This keeps the concentration of free H_3O^+ from being as large as otherwise expected.
>
> $$HF + H_2O \rightarrow$$
> $$(^-F \cdots H_3O^+) \rightleftharpoons$$
> $$H_3O^+ + F^-$$

$$E_xO_y + H_2O \longrightarrow \ \ E—O—H$$

Any factor that draws electrons toward the atom E strengthens the E—O bond, weakens the O—H bond, and causes the hydroxo compound to ionize as an acid. The ability of the compound to donate protons is increased in the presence of a strong base such as OH^-.

$$E—O(H + OH^-) \longrightarrow (E—O)^- + H_2O$$

This type of behavior is favored by *small size and high charge* (as indicated by high oxidation state or high formal charge) on the atom E, and is expected if the element E is nonmetallic. In these situations the oxide E_xO_y is called an **acid anhydride** (from the Greek, "without water"), and the hydroxo compound is an **oxoacid.**

A large size and small charge on the atom E favors rupture of the E—O bond. The hydroxo compound ionizes to produce OH^- and the atom E becomes a cationic species. This tendency to ionize as a base is increased when the hydroxo compound is placed in an acidic solution. Behavior of this type is associated with metallic character in the element E. In these instances the oxide E_xO_y is referred to as a **base anhydride.**

$$E(O—H + H_3O^+) \longrightarrow (E)^+ + 2 H_2O$$

In some cases the hydroxo compound may act either as an acid or a base. This behavior is known as **amphoterism** and is associated with elements having both metallic and nonmetallic properties.

TABLE 17-6
Classification of Some Oxides

Acidic		Basic	Amphoteric	
Representative Elements				
Cl_2O	P_4O_{10}	Na_2O	BeO	GeO
SO_2	CO_2	K_2O	Al_2O_3	Sb_2O_3
SO_3	SiO_2	MgO	SnO	
N_2O_5	B_2O_3	CaO	PbO	
Transition Elements				
CrO_3		Sc_2O_3	ZnO	
MoO_3		TiO_2	Cr_2O_3	
WO_3		ZrO_2		
Mn_2O_7				

A more quantitative description of the effect of size and charge on the acid–base character of a hydroxo compound is based on charge density—the ratio of the charge of the cation to its radius. The higher the charge density, the more acidic the compound. Conversely, the lower the charge density the more basic the compound. For the following hydroxo compounds the charge densities and order of increasing acidity are

$$Mg(OH)_2 < Be(OH)_2 \simeq Al(OH)_3 < B(OH)_3$$

	basic	amphoteric		acidic
ionic radius, pm:	Mg^{2+}, 65	Be^{2+}, 31	Al^{3+}, 50	B^{3+}, 20
charge density:	2/65 =	2/31 =	3/50 =	3/20 =
	0.031 <	0.065 ≃	0.060 <	0.15

This classification scheme for element oxides is summarized in Table 17-6.

Strengths of Oxoacids. An oxoacid has at least one E—O—H bond and other bonds that are either E—O—H or E—O (or occasionally, E—H). The formation and acidic behavior of an oxoacid are pictured in Figure 17-7.

Factors that promote the withdrawal of electrons from the O—H bond toward the nonmetal atom, E, favor breakage of the O—H bond and increased acid strength. These factors include a high electronegativity of the central atom, *E*, and an increased number of oxygen atoms (not OH groups) bonded directly to the *E* atom. The fact that HNO_3 is a strong acid whereas H_3PO_4 is weak is consistent with these statements:

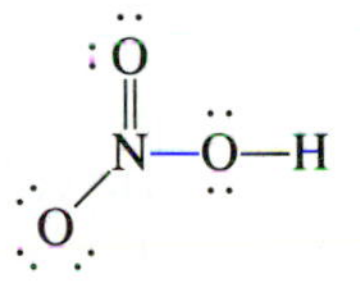

strong, $K_a \approx 30$ 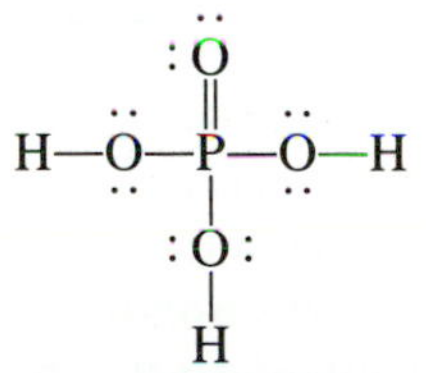   weak, $K_a = 7.11 \times 10^{-3}$

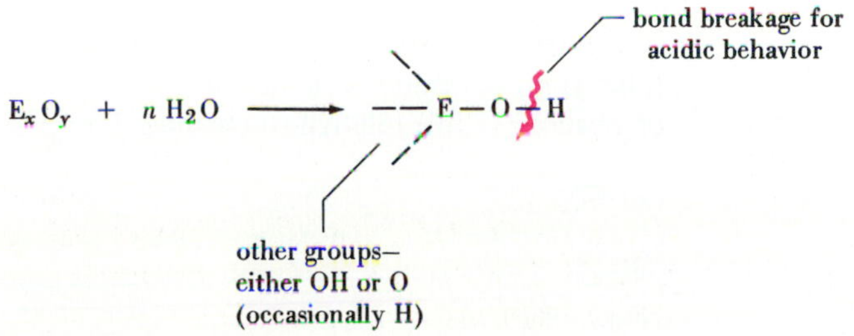

FIGURE 17-7
Formation and acidic behavior of an oxoacid.

$$E_xO_y + n\,H_2O \longrightarrow \;-\!\!-E-O-H$$

bond breakage for acidic behavior

other groups— either OH or O (occasionally H)

Strengths of Organic Acids. Ethanol and acetic acid both have an O—H group bonded to a carbon atom, but acetic acid is a much stronger acid than is ethanol.

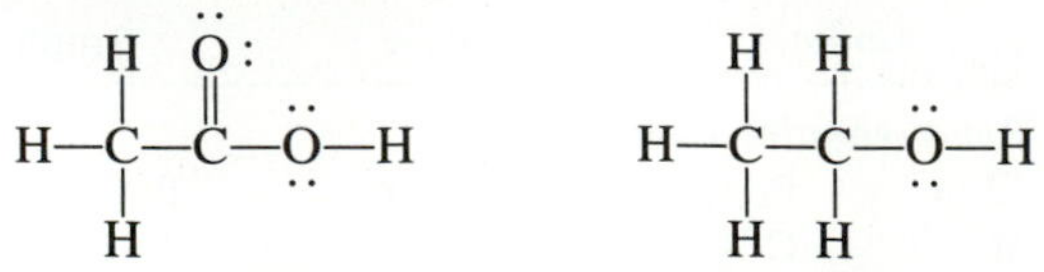

$$
\begin{array}{cc}
\text{acetic acid, p}K_a = 4.76 & \text{ethanol, p}K_a = 15.9
\end{array}
\tag{17.43}
$$

We might rationalize that the carbonyl oxygen atom in acetic acid (>C=O), being highly electronegative, withdraws electrons from the O—H bond. The result is that the proton can be lost more readily. Another reason for acetic acid being the stronger acid is based on the anions that are formed.

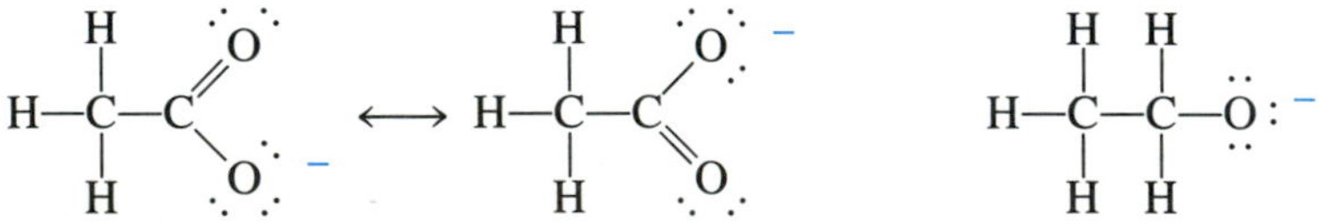

$$
\text{acetate ion} \qquad\qquad \text{ethoxide ion}
\tag{17.44}
$$

In acetate anion resonance occurs. Two plausible structures can be written in which the double bond is shifted from one O atom to the other. The net effect is that each carbon-to-oxygen bond is a "$\frac{3}{2}$" bond and each O atom carries "$\frac{1}{2}$" negative charge. In short, the excess unit of negative charge is spread out. This reduces the ability of either O atom to attract a proton and makes acetate ion a moderately weak Brønsted–Lowry base. Ethoxide ion, on the other hand, has no resonance possibilities. The unit of negative charge is centered on a single O atom. The ion is a much stronger base than acetate ion. If a conjugate base is strong, then the corresponding acid is weak (recall Table 17-5).

The substitution of groups can also have an effect on the strength of an organic acid. Replacement of one H atom by a Cl atom in acetic acid produces monochloroacetic acid.

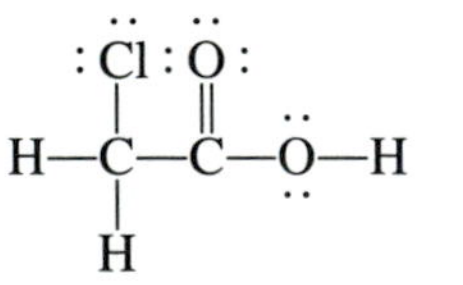

$$
\tag{17.45}
$$

monochloroacetic acid, p$K_a = 2.87$

The highly electronegative Cl atom helps to draw electrons away from the O—H bond. The bond is weakened, the proton is lost more readily, and the acid is a stronger acid than acetic acid. Viewed from the standpoint of the anion, the electronegative Cl atom causes an additional spreading out of the unit of negative charge beyond that described in (17.44). Monochloroacetate ion has less attraction for a proton than does acetate ion; it is a weaker base.

The electron-withdrawing power of certain groups in molecules is called the **inductive effect.** The inductive effect is strongest when the substituent group is adjacent to the carboxyl group ($-\text{C} \overset{\text{O}}{\underset{\text{O—H}}{\diagup}}$). The effect falls off rapidly as the substituent is moved father away on a hydrocarbon chain. The benzene ring system (phenyl group) exhibits an inductive effect and accounts for the greater acid strength of phenol (17.46) relative to ethanol.

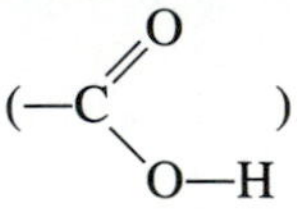

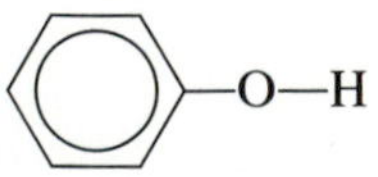

$$
\tag{17.46}
$$

phenol, p$K_a = 10.0$

Electronegative substituents on the phenyl group may also exhibit an inductive effect, causing the acid to become somewhat stronger.

In Example 17-15 we must apply several factors affecting bond strength. To summarize some of these factors,

Summary of factors affecting acid strength.

1. Acid strengths of binary acids HX (or H_nX) *increase* as the H—X bond strengths *decrease*.
2. The strengths of oxoacids *increase* as the
   - electronegativity of the central atom *increases*.
   - number of O atoms (not OH groups) bonded to the central atom *increases*.
3. Acids are made stronger by the presence of electron-withdrawing groups in the vicinity of the O—H bond to be broken (*inductive effect*).

(17.47)

Example 17-15

Factors that affect the strengths of acids. Explain which member of each of the following pairs is the stronger acid.

(a) O=N—O—H or O—Cl—O—H
     (I)   (II)

(b) CH_2FC (O, OH) or CH_2ClC (O, OH)
     (I)   (II)

(c) $ClCH_2CH_2C$ (O, OH) or CH_3CHClC (O, OH)
     (I)   (II)

(d) ⬡—OH or ⬡—OH with Br
     (I)   (II)

Solution

(a) There are *two* factors suggesting that chloric acid (II) is a stronger acid than nitrous acid (I). Chlorine has a slightly higher electronegativity than nitrogen, and there are *two* O atoms bonded directly to the central Cl atom in chloric acid and only *one* O atom bonded directly to the N atom in nitrous acid.

(b) Because F is more electronegative than Cl, it should exert a stronger pull on the electrons in the O—H bond than does Cl. Monofluoroacetic acid (I) is a stronger acid than monochloroacetic acid (II).

(c) The location from which the Cl atom exerts the strongest inductive effect is directly adjacent to the carboxyl group. Compound II (2-chloropropanoic acid) is more acidic than compound I (3-chloropropanoic acid).

(d) The electronegative Br atom draws electrons away from the O—H group. o-Bromophenol (II) is a stronger acid than phenol (I).

SIMILAR EXAMPLES: Exercises 15, 54, 80, 81.

17-10   Lewis Acids and Bases

To apply the ideas presented in the previous section we had to think about molecular structures of acids and bases. Another acid–base theory closely related to bonding and structure was proposed by G. N. Lewis in 1923. In the Lewis theory an *acid* is an atom, ion, or molecule that is an *electron pair acceptor* and a *base* is an atom, ion, or molecule that is an *electron pair donor*. A reaction between a Lewis acid and a Lewis base results in the formation of a covalent bond between them. In general, to identify Lewis *acids* we should look for a species with *available orbitals to accommodate additional electrons,* and for Lewis *bases,* species having *lone-pair electrons available for sharing.*

An important feature of Lewis acid–base reactions is that they do not have to involve H^+ and/or OH^-. Another feature is that the reactions are not limited to liquid solutions; Lewis acid–base reactions can occur among solids and in the gaseous state as well. HCl, a Brønsted–Lowry acid, is not a Lewis acid; it itself is not an electron pair acceptor. On the other hand, we can think of HCl as being a *source* of the Lewis acid, H^+. OH^- is a Lewis base because of the presence of lone pair electrons. So, too, is NH_3 a Lewis base. Lewis theory acid–base reactions involving some of these species are

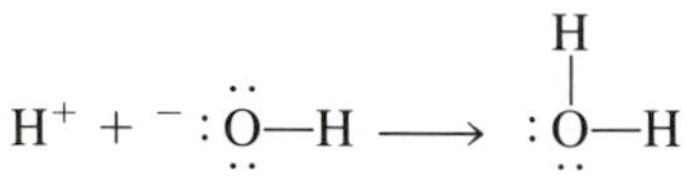

$$H^+ + {}^-:\!\ddot{O}\!-\!H \longrightarrow :\!\ddot{O}\!-\!H \quad \begin{matrix} H \\ | \\ \end{matrix} \tag{17.48}$$

$$H^+ + \begin{matrix} H-\ddot{N}-H \\ | \\ H \end{matrix} \longrightarrow \left[\begin{matrix} H \\ | \\ H-N-H \\ | \\ H \end{matrix} \right]^+ \tag{17.49}$$

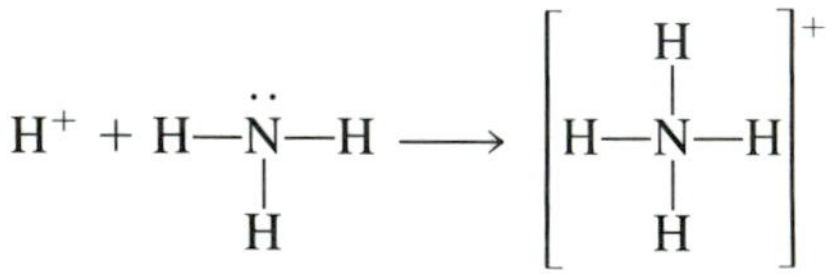

We should expect an "electron deficient" species to be a strong Lewis *acid,* and to react with a Lewis *base* to form an electron-pair bond (a coordinate covalent bond). The reaction between boron trifluoride and ammonia that we first described in Section 10-9 is a good example.

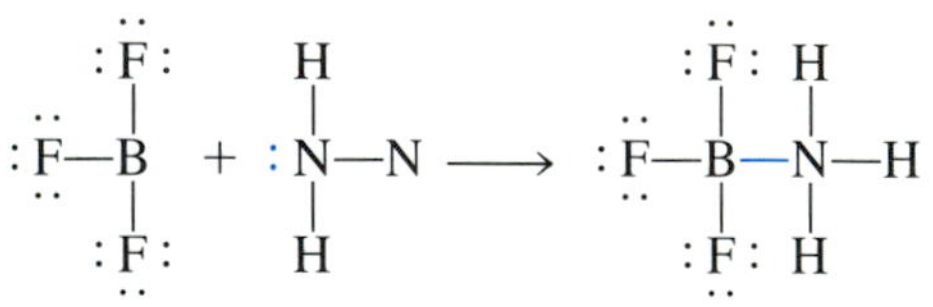

$$\begin{matrix} :\ddot{F}: & H \\ | & | \\ :\ddot{F}-B & + :N-N \\ | & | \\ :\ddot{F}: & H \end{matrix} \longrightarrow \begin{matrix} :\ddot{F}: & H \\ | & | \\ :\ddot{F}-B-N-H \\ | & | \\ :\ddot{F}: & H \end{matrix} \tag{17.50}$$

It is possible to use the Lewis theory of acids and bases in describing complex ion formation. This matter will be discussed later in the text (Section 19-7; Chapter 25). For now, consider the complex ion consisting of a central Zn^{2+} ion to which are bonded four NH_3 molecules, that is $[Zn(NH_3)_4]^{2+}$. Shown below are (a) the ground-state electron configuration of Zn^{2+}, (b) an sp^3 hybridization scheme for Zn^{2+}, and (c) a bonding scheme showing the donation of electron pairs by NH_3 molecules (base) and their acceptance by Zn^{2+} (acid).

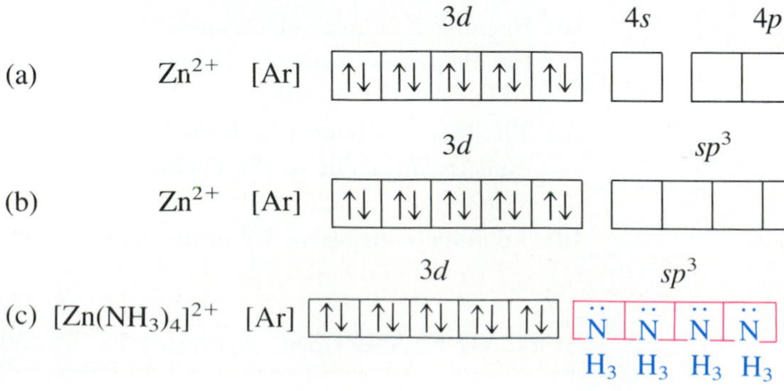

(a) Zn^{2+} [Ar]   (3d) ↑↓ ↑↓ ↑↓ ↑↓ ↑↓   (4s) □   (4p) □ □ □

(b) Zn^{2+} [Ar]   (3d) ↑↓ ↑↓ ↑↓ ↑↓ ↑↓   (sp^3) □ □ □ □

(c) $[Zn(NH_3)_4]^{2+}$ [Ar]   (3d) ↑↓ ↑↓ ↑↓ ↑↓ ↑↓   (sp^3) :N :N :N :N H_3 H_3 H_3 H_3 (17.51)

Example 17-16

Identifying Lewis acids and bases. Each of the following is an acid–base reaction in the Lewis sense. Which species is the acid and which is the base?

(a) $BF_3 + F^- \longrightarrow BF_4^-$

(b) $Ag^+(aq) + 2\,CN^-(aq) \longrightarrow [Ag(CN)_2]^-(aq)$

(c) $Ca^{2+}\; :\!\ddot{O}\!:^{2-} +$ [CO₂ structure] $\longrightarrow Ca^{2+}$ [CO₃²⁻ structure]

Solution

(a) Note in equation (17.50) that BF_3 is an electron-deficient molecule, with a vacant orbital on the B atom. The fluoride ion has an outer-shell octet of electrons. BF_3 is the electron pair acceptor—the acid. F^- is the electron pair donor—the base.

(b) Consider the Lewis structure of the cyanide ion, $:C\equiv N:^-$. Here we see lone-pair electrons on both the C and the N atoms. The electron configuration of Ag^+ is $[Kr]4d^{10}$. This configuration is similar to that outlined for Zn^{2+} in (17.51). An s and a p orbital in the outermost (fifth) shell can be hybridized and the *empty sp* hybrid orbitals can accept electron pairs from the CN^- ions. Ag^+ is a Lewis *acid* and CN^- is a Lewis *base*.

(c) The acid–base reaction is actually between O^{2-} and $CO_2(g)$ to form CO_3^{2-}. The oxide ion, O^{2-}, can only function as a Lewis *base*. It donates a pair of electrons, forming an additional O-to-C bond. The CO_2 is a Lewis *acid*.

SIMILAR EXAMPLES: Exercises 16, 57, 58, 79.

Summary

Where appropriate, one way to categorize substances is as *acids* and *bases*. The Arrhenius acid–base theory describes acids as substances that produce H^+ ions and bases as substances that produce OH^- ions when they dissociate. A more recent theory, the Brønsted–Lowry theory, describes acid–base behavior in terms of proton transfer. An acid is a *proton donor* and a base is a *proton acceptor*. The ion produced when an acid loses a proton is itself a base in the reverse reaction; it is called a *conjugate base* of the acid. Conversely, the product formed when a base accepts a proton is itself an acid, the *conjugate acid* of the base. The tendencies of substances to lose or gain protons, that is, their relative acid and base strengths, determine the direction in which an acid–base reaction occurs: from the *stronger* to the *weaker* acid/base combination. The Lewis acid–base theory views an acid–base reaction in terms of covalent bond formation between an *electron pair acceptor* (acid) and an *electron pair donor* (base). It is sometimes applied in situations where an acid contains no ionizable H atoms or where there is no solvent.

In the self-ionization of water a proton is transferred from one water molecule to another, producing the ions H_3O^+ and OH^-. The equilibrium constant for this self-ionization is the ion product of water, $K_w = [H_3O^+][OH^-]$. Constancy of the ion product permits a calculation of $[OH^-]$ from a known value of $[H_3O^+]$, and vice versa. Some useful shorthand designations are

$$pH = -\log [H_3O^+]$$

$$pOH = -\log [OH^-]$$

$$pH + pOH = 14.00 \text{ (at 25 °C)}$$

Strong acids ionize completely in aqueous solutions, producing H_3O^+. In aqueous solution, a strong base ionizes completely, producing OH^-. With weak acids and weak bases ionization of the molecular acid or base does not go to completion. Equilibrium exists between the nonionized acid or base and its ions, and is described through the ionization constant, K_a or K_b. Some weak acids produce more than one proton per molecule, and they do so in a stepwise fashion. That is, polyprotic acid molecules first lose one proton, acid anions then lose a second proton, and so on. Distinct ionization constants $K_{a_1}, K_{a_2}, \ldots$ apply to each step. Calculations involving ionization equilibria are in many ways similar to those introduced for gas-phase equilibria in Chapter 16, although some additional considerations are necessary for polyprotic acids.

In reactions between ions and water—sometimes called *hydrolysis* reactions—the ions react as weak acids or weak bases. An ion undergoing hydrolysis is itself the *conjugate base* or *conjugate acid* of another Brønsted–Lowry acid or base. The ionization constant needed to describe a hydrolysis reaction, if not directly available, can be obtained with the expression: K_a(acid) $\times$ K_b(conjugate base) = K_w. Pure water and solutions of ions that do *not* hydrolyze have a pH = 7.00. Hydrolysis of the salt of a weak acid and a strong base (hydrolysis of an *anion*) produces a solution with pH > 7. Hydrolysis of the salt of a strong acid and a weak base (hydrolysis of a *cation*) produces a solution with pH < 7. In the first case, water acts as an acid, and in the second, as a base.

Molecular structure is the key factor in determining whether a substance has acidic, basic, or amphoteric properties. In addition, molecular structure affects whether an acid or base is strong or weak. Differences in molecular structures among similar compounds often can be used to predict relative values of K_a or K_b. In assessing acid strength, for example, one must consider the strength of the bond between an H atom and another atom (usually O). Different factors affect the strength of this bond.

Summarizing Example

Oxalic acid, $H_2C_2O_4$, is present in many plants, usually as the potassium or calcium salt. It is poisonous to humans because it forms calcium oxalate which can cause kidney failure. Oxalic acid is a weak *diprotic* acid that is moderately soluble in water.

$$H_2C_2O_4 + H_2O \rightleftharpoons H_3O^+ + HC_2O_4^- \qquad K_{a_1} = ? \qquad (17.52)$$

$$HC_2O_4^- + H_2O \rightleftharpoons H_3O^+ + C_2O_4^{2-} \qquad K_{a_2} = ? \qquad (17.53)$$

A saturated aqueous solution of $H_2C_2O_4$ at 25 °C is found to have pH = 0.67 and $[C_2O_4^{2-}] = 5.4 \times 10^{-5}$ M.

1. A 5.00-mL sample of this saturated $H_2C_2O_4$(aq) is added to 100. mL of water and then titrated. 28.77 mL of 0.3650 M NaOH is required in the following titration

$$H_2C_2O_4(aq) + 2 OH^-(aq) \longrightarrow 2 H_2O(l) + C_2O_4^{2-}(aq) \qquad (17.54)$$

What is the molarity of saturated $H_2C_2O_4$(aq) at 25 °C?

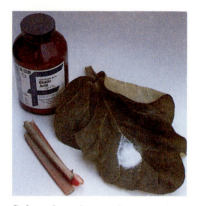

Salts of oxalic acid are concentrated in the leaves of rhubarb plants, making the leaves poisonous. The stalks, on the other hand, are edible. [Robert K. Wismer]

Solution. An important point for you to understand about acid–base titrations is that, in the titration of a *strong* acid by a strong base, all the H_3O^+ to be neutralized is present in the solution initially. In the titration of a *weak* acid by a strong base, the H_3O^+ to be neutralized is *not* just what is present in the solution initially. That is, think of the neutralization reaction in this way: as the H_3O^+ present initially is converted to H_2O, more acid ionizes, until eventually all of it is ionized and neutralized. As indicated in equation (17.54), with oxalic acid this means that ionization occurs through the *second* step, yielding a final solution of $Na_2C_2O_4$.

no. mol OH$^-$ consumed

$$= 28.77 \text{ mL} \times \frac{1 \text{ L}}{1000 \text{ mL}} \times \frac{0.3650 \text{ mol NaOH}}{1 \text{ L}} \times \frac{1 \text{ mol OH}^-}{1 \text{ mol NaOH}}$$

$$= 0.01050 \text{ mol OH}^-$$

no. mol neutr. $= 0.01050 \text{ mol OH}^- \times \dfrac{1 \text{ mol } H_2C_2O_4}{2 \text{ mol OH}^-} = 0.005250 \text{ mol } H_2C_2O_4$

molarity $= \dfrac{0.005250 \text{ mol } H_2C_2O_4}{0.00500 \text{ L}} =$ **1.05 M $H_2C_2O_4$**

(This example is similar to Examples 5-5 and 17-4.)

2. Determine the value of K_{a_1} for oxalic acid.

Solution. Here we should (a) write the equation for the *first* ionization step of oxalic acid, (b) describe the ionization equilibrium in a "standard" format, (c) substitute known data into this format, and (d) solve for a numerical value of K_{a_1}. Let us work with the saturated aqueous solution, for which we now know the total solution molarity (from part 1) and the pH. Note that for pH = 0.67; $\log [H_3O^+] = -0.67$; and $[H_3O^+] = 0.21$ M.

$$H_2C_2O_4 + H_2O \rightleftharpoons H_3O^+ + HC_2O_4^-$$

placed in solution:	1.05 M	—	—
changes:	−0.21 M	+0.21 M	+0.21 M
after first ionization:	0.84 M	0.21 M	0.21 M

$$K_{a_1} = \frac{[H_3O^+][HC_2O_4^-]}{[H_2C_2O_4]} = \frac{(0.21)(0.21)}{0.84} = 5.2 \times 10^{-2}$$

(This example is similar to Example 17-7.)

3. Determine the value of K_{a_2} for oxalic acid.

Solution. The key idea is that if K_{a_1} is considerably larger than K_{a_2}, then the concentration of the anion produced in the *second* ionization step is simply equal to K_{a_2} [recall statements (17.32)]. Thus, we conclude that $K_{a_2} = 5.4 \times 10^{-5}$, the experimentally determined concentration of oxalate ion in the saturated aqueous solution.

4. What is the concentration of $C_2O_4^{2-}$ in the neutralized solution produced in the titration in part 1?

Solution. At the point where complete neutralization occurs we have present a solution of sodium oxalate, with neither unreacted oxalic acid nor excess NaOH present. The solution volume is the 5.00 mL of acid + 100. mL of water + 28.77 mL of NaOH(aq) = 134 mL. [Note that the 100. mL water used to dilute the 5.00-mL acid sample has no effect on the volume of NaOH(aq) required for the titration, but it does affect the concentrations of the species present in the solution.] The amount of $H_2C_2O_4$ that is neutralized is 0.005250 mol, and one $C_2O_4^{2-}$ ion is produced for every $H_2C_2O_4$ molecule that is neutralized in reaction (17.54).

$$\text{no. mol } C_2O_4^{2-} = 0.005250 \text{ mol } H_2C_2O_4 \times \frac{1 \text{ mol } C_2O_4^{2-}}{1 \text{ mol } H_2C_2O_4}$$

$$= 0.005250 \text{ mol } C_2O_4^{2-}$$

$$[C_2O_4^{2-}] = \frac{0.005250 \text{ mol } C_2O_4^{2-}}{0.134 \text{ L}} = 0.0392 \text{ M}$$

5. Is the sodium oxalate solution described in part 4 acidic, basic, or pH neutral? What is the pH of the solution?

Solution. The oxalate ion in solution hydrolyzes.

$$C_2O_4^{2-} + H_2O \longrightarrow HC_2O_4^- + OH^-$$

Since hydrolysis produces OH^-, the solution is basic. (Or, simply note that sodium oxalate is the salt of a weak acid and strong base.) Calculating the pH of 0.0392 M $Na_2C_2O_4$ is the subject of Exercise 46.

(This example is similar to Examples 17-12 and 17-13.)

Key Terms

acid anhydride (17-9)
acid ionization constant, K_a (17-6)
amphoterism (17-9)
Arrhenius acid–base theory (17-1)
base anhydride (17-9)
base ionization constant, K_b (17-6)
Brønsted–Lowry theory (17-2)
conjugate acid (17-2)

conjugate base (17-2)
degree of ionization (17-6)
differentiating solvent (17-9)
hydrolysis (17-8)
hydronium ion, H_3O^+ (17-2, 17-3)
inductive effect (17-9)
ion product of water, K_w (17-3)

Lewis acid–base theory (17-10)
oxoacid (17-9)
pH (17-5)
pK (17-6)
pOH (17-5)
polyprotic acid (17-7)
self-ionization (17-3)

Highlighted Expressions

Features of the Brønsted–Lowry theory (17.10)
The ion product of water, K_w (17.13)
$[H_3O^+]$ in aqueous solutions of strong acids (17.16)
$[OH^-]$ in aqueous solutions of strong bases (17.17)
Important ideas concerning pH and pOH (17.24)
Ion concentrations in solutions of weak acids and weak bases (17.28)

Degree of ionization of weak acids and weak bases (17.30)
Generalizations concerning polyprotic acids (17.32)
Generalizations concerning hydrolysis reactions (17.36)
Ionization constants for weak acids, weak bases, and their conjugates (17.39)
Predicting the direction of an acid–base reaction (17.42)
Summary of factors affecting acid strength (17.47)

Review Problems

1. For each of the following identify the acids and bases involved in both the forward and reverse directions.
  (a) $HOBr + H_2O \rightleftharpoons H_3O^+ + OBr^-$
  (b) $HSO_4^- + H_2O \rightleftharpoons H_3O^+ + SO_4^{2-}$
  (c) $HS^- + H_2O \rightleftharpoons H_2S + OH^-$
  (d) $C_6H_5NH_3^+ + OH^- \rightleftharpoons C_6H_5NH_2 + H_2O$
2. Calculate $[H_3O^+]$ and $[OH^-]$ for each solution.
  (a) 0.0060 M HCl   **(b)** 0.082 M NaOH
  (c) 0.0015 M $Sr(OH)_2$ **(d)** 4.2×10^{-3} M HNO_3
3. What is the pH of each of the following solutions?
(a) 3.0×10^{-3} M HCl; **(b)** 0.000780 M HBr; **(c)** 4.75×10^{-3} M NaOH; **(d)** 4.6×10^{-4} M NaOH.
4. What is $[H_3O^+]$ in a solution with **(a)** pH = 6.0; **(b)** pH = 8.15; **(c)** pH = 0.65; **(d)** pOH = 4.40; **(e)** pOH = 10.67?
5. What is the pH of a water solution containing 2.16 g $Ba(OH)_2 \cdot 8H_2O$ in 815 mL of solution?
6. What is $[H_3O^+]$ in a solution obtained by mixing 24.10 mL of 0.150 M HNO_3 and 11.20 mL of 0.412 M KOH? [*Hint:* Is the solution acidic, basic, or neutral?]
7. In an aqueous solution prepared by dissolving 0.355 mol butyric acid ($HC_4H_7O_2$) in 715 mL of solution, it is found that $[H_3O^+] = [C_4H_7O_2^-] = 2.73 \times 10^{-3}$ M. What is the value of K_a for the ionization of butyric acid?

$$HC_4H_7O_2 + H_2O \rightleftharpoons H_3O^+ + C_4H_7O_2^- \qquad K_a = ?$$

8. Ionization constants of three acids are listed below.

$$HC_2H_3O_2 + H_2O \rightleftharpoons H_3O^+ + C_2H_3O_2^- \qquad K_a = 1.74 \times 10^{-5}$$
acetic acid

$$HC_8H_7O_2 \;\; + H_2O \rightleftharpoons H_3O^+ + C_8H_7O_2^- \qquad K_a = 4.9 \times 10^{-5}$$
phenylacetic acid

$$HC_6H_4ClO \;\; + H_2O \rightleftharpoons H_3O^+ + C_6H_4ClO^- \qquad K_a = 3.2 \times 10^{-9}$$
o-chlorophenol

  (a) What is $[C_2H_3O_2^-]$ in 0.624 M $HC_2H_3O_2$?
  (b) What is the pH of 0.105 M $HC_8H_7O_2$?
  (c) What molar concentration of o-chlorophenol is necessary to produce a solution with pH = 4.86? [*Hint:* What are the values of $[H_3O^+]$ and $[C_6H_4ClO^-]$ in this solution?]
9. Calculate $[(CH_3)_3NH^+]$ in an aqueous solution that is 1.02 M $(CH_3)_3N$ (trimethylamine).

$$(CH_3)_3N + H_2O \rightleftharpoons (CH_3)_3NH^+ + OH^- \qquad K_b = 6.2 \times 10^{-5}$$

10. What is the **(a)** degree of ionization and **(b)** percent ionization of propionic acid in a solution that is 0.45 M $HC_3H_5O_2$?

$$HC_3H_5O_2 + H_2O \rightleftharpoons H_3O^+ + C_3H_5O_2^- \qquad K_a = 1.34 \times 10^{-5}$$

11. For an 0.025 M solution of the weak diprotic acid H_2CO_3, calculate **(a)** $[H_3O^+]$, **(b)** $[HCO_3^-]$, and **(c)** $[CO_3^{2-}]$. (Use data from Table 17-3.)
12. Predict whether each of the following aqueous solutions is acidic, basic, or neutral: **(a)** KCl; **(b)** NH_4NO_3; **(c)** $NaNO_2$; **(d)** NaI; **(e)** $Ca(OCl)_2$.
13. Calculate the pH of an aqueous solution that is 1.15 M NH_4Cl.

$$NH_4^+ + H_2O \rightleftharpoons H_3O^+ + NH_3 \qquad K_a = 5.75 \times 10^{-10}$$

14. From data in Table 17-2 determine a value of **(a)** K_a for $C_5H_5NH^+$; **(b)** K_b for CHO_2^-; **(c)** K_b for $C_6H_5O^-$.
15. Which is the more acidic of each of the following pairs of acids? **(a)** HBr *or* HI; **(b)** HOClO *or* HOI; **(c)** H_3CCH_2COOH *or* Cl_3CCH_2COOH; **(d)** $I_3CCH_2CH_2COOH$ *or* $H_3CCH_2CF_2COOH$.
16. Indicate whether each of the following is a Lewis acid or base: **(a)** OH^-; **(b)** $B(OH)_3$; **(c)** $AlCl_3$; **(d)** CH_3NH_2. [*Hint:* You may find it helpful to draw Lewis structures.]

Exercises

Brønsted–Lowry theory of acids and bases

17. According to the Brønsted–Lowry theory, which of the following would you expect to be acidic and which basic? (a) HNO_2; (b) OCl^-; (c) NH_2^-; (d) NH_4^+; (e) $CH_3NH_3^+$.

18. The following acids are all *monoprotic* (yielding a single proton). Write the formula of the conjugate base of each acid; (a) HIO_4; (b) $HC_3H_5O_2$; (c) C_6H_5COOH; (d) $C_6H_5NH_3^+$.

19. Substances that can either lose or gain protons are said to be *amphiprotic*. Which of the following are amphiprotic? (a) OH^-; (b) NH_4^+; (c) H_2O; (d) HS^-; (e) NO_3^-; (f) HCO_3^-; (g) HNO_3.

Strong acids, strong bases, and pH

20. What is the pH of a solution obtained by mixing 365 mL of 3.25×10^{-3} M HI and 415 mL of 3.46×10^{-2} M HCl?

21. What is $[H_3O^+]$ in a solution obtained by dissolving 187 cm^3 HCl(g), measured at 22 °C and 742 mmHg, in 4.17 L of water solution?

22. What volume of concentrated HCl(aq) that is 36.0% HCl, by mass, and has a density of 1.18 g/cm^3, is required to produce 8.25 L of a solution with a pH = 1.77?

★ 23. Can a solution of pH 8 be prepared by dissolving HCl in water? If it is possible to do so, indicate how. If it is not possible, indicate why not.

Neutralization reactions

24. 25.00 mL of a HNO_3(aq) solution with a pH of 2.52 is mixed with 25.00 mL of a KOH(aq) solution with a pH of 12.05. What is the pH of the final solution? [*Hint:* Is the final solution acidic, basic, or pH neutral?]

25. A 10.0-mL sample of *saturated* $Ca(OH)_2$(aq) is diluted to 250.0 mL in a volumetric flask. The solubility of $Ca(OH)_2$(aq) is 0.17 g/100 mL soln. What is the pH of the solution in the flask?

26. A 10.0-mL sample of the $Ca(OH)_2$(aq) solution in the volumetric flask of Exercise 25 is transferred to a beaker and some water is added. The resulting solution requires 25.1 mL of an HCl solution for its titration. What is the *molarity* of this HCl solution? [*Hint:* Determine the amount of OH^- in the solution being titrated from data in Exercise 25. Also, write the net chemical equation for the neutralization.]

Weak acids, weak bases, and pH (Use data from Table 17-2 as necessary.)

27. Normal caproic acid, $HC_6H_{11}O_2$, found in small amounts in coconut and palm oils, is used in making artificial flavors. A saturated aqueous solution of the acid contains 11 g/L and has pH = 2.94. Calculate K_a for the acid.

$$HC_6H_{11}O_2 + H_2O \rightleftharpoons H_3O^+ + C_6H_{11}O_2^- \qquad K_a = ?$$

28. The compound *o*-nitrophenol, $HC_6H_4NO_3$, is slightly soluble in water and ionizes as a weak acid. A saturated solution of *o*-nitrophenol has pH = 4.53. What is the solubility of *o*-nitrophenol in water, expressed in g/L?

$$HC_6H_4NO_3 + H_2O \rightleftharpoons H_3O^+ + C_6H_4NO_3^- \qquad K_a = 5.9 \times 10^{-8}$$

29. The active ingredient in aspirin is acetylsalicyclic acid.

$$HC_9H_7O_4 + H_2O \rightleftharpoons H_3O^+ + C_9H_7O_4^- \qquad K_a = 2.75 \times 10^{-9}$$

What is the pH of the solution obtained by dissolving two aspirin tablets in 250 mL of water? Assume that each tablet contains 0.32 g of acetylsalicylic acid.

30. A particular vinegar is found to contain 5.7% acetic acid ($HC_2H_3O_2$), by mass. What mass of this vinegar should be diluted with water to produce 0.750 L of a solution with pH = 4.48?

31. The organic base piperidine is found in small amounts in black pepper. What is the pH of 315 mL of a water solution containing 114 mg of piperidine?

$$C_5H_{11}N + H_2O \rightleftharpoons C_5H_{11}NH^+ + OH^- \qquad K_b = 1.6 \times 10^{-3}$$

32. Phenylacetic acid, used in the manufacture of synthetic perfumes, is slightly soluble in water. 25.00 mL of a saturated solution of phenylacetic acid requires 17.70 mL of 0.1850 M NaOH for its neutralization. What is the pH of this saturated solution?

$$C_6H_5CH_2COOH(aq) + H_2O \rightleftharpoons$$
$$H_3O^+(aq) + C_6H_5CH_2COO^-(aq) \qquad K_a = 5.56 \times 10^{-5}$$

33. A handbook states that the solubility of methylamine, CH_3NH_2(g), in water at 1 atm pressure and 25 °C is 959 volumes of CH_3NH_2(g) per volume of water.

   (a) Estimate the maximum pH that can be attained by dissolving methylamine in water.

   (b) What molarity NaOH(aq) would be required to yield the same pH?

Degree of ionization

34. The % ionization found for acetic acid in Example 17-9 was 0.42% in 1.0 M $HC_2H_3O_2$, 1.3% in 0.10 M $HC_2H_3O_2$, and 4.2% in 0.010 M $HC_2H_3O_2$. Should we expect to find that the percent ionization is 13% in 0.0010 M $HC_2H_3O_2$ and 42% in 0.00010 M $HC_2H_3O_2$? Explain. [*Hint:* What are the limitations of the equation derived in Example 17-9(b)?]

35. What is the percent ionization of trichloroacetic acid in an 0.050 M $HC_2Cl_3O_2$ solution?

$$HC_2Cl_3O_2 + H_2O \rightleftharpoons H_3O^+ + C_2Cl_3O_2^- \qquad K_a = 2.0 \times 10^{-1}$$

36. A handbook lists the following formula that can be used to calculate the percent ionization of a weak acid.

$$\% \text{ ionized} = \frac{100}{1 + 10^{(pK - pH)}}$$

★ (a) Derive this equation. What assumptions must you make in this derivation?

   (b) Use the equation to determine the % ionization of a formic acid solution having a pH of 2.50.

★ (c) A solution in which propionic acid is 0.94% ionized has a pH of 2.85. What is K_a for propionic acid?

$$HC_3H_5O_2 + H_2O \rightleftharpoons H_3O^+ + C_3H_5O_2^- \qquad K_a = ?$$

Mixtures of acids and mixtures of bases (Use data from Table 17-2 as necessary.)

If two or more acids are mixed in water, the *strongest* acid (the one with the largest value of K_a) is most significant in determining $[H_3O^+]$ in the solution. If K_a for the strongest acid exceeds the K_a values of the other acids by a factor of 10^3 or more, you can generally treat the solution as if the strongest acid is the *only* source of H_3O^+. But, of course, the other acids will be the only source of their conjugate bases. Recall also, that a few especially strong acids are completely ionized in water solution (see Table 17-1). The situation with weak bases is essentially the same; the *strongest* base is most significant in determining $[OH^-]$. If K_a (or K_b) for the strongest acid (or base) does not exceed all the other K_a (or K_b) values by a factor of 10^3 or more, then you may have to solve two or more algebraic equations simultaneously.

37. What is the pH of a solution that is
(a) 0.0105 M HCl and 1.02 M $HC_2H_3O_2$ (acetic acid)?
(b) 0.0030 M KOH and 0.0018 M $Ca(OH)_2$?
(c) 0.55 M $HC_2H_3O_2$ and 0.16 M HOC_6H_5 (phenol)?
(d) 0.68 M H_2SO_4 and 1.5 M $HCHO_2$ (formic acid)?

38. A saturated aqueous solution of the weak base aniline contains 36.0 g $C_6H_5NH_2$ per liter.

$$C_6H_5NH_2 + H_2O \rightleftharpoons C_6H_5NH_3^+ + OH^- \qquad K_b = 4.30 \times 10^{-10}$$

(a) What is the pH of this saturated solution?
(b) If a sufficient quantity of NaOH is added to a sample of this saturated solution to raise the pH to 12.25, what will be the concentration of *anilinium* ion, $[C_6H_5NH_3^+]$, in solution?

★ **39.** Acetic acid is becoming increasingly important as an intermediate in the production of organic chemicals from synthesis gas (a CO and H_2 mixture). In some processes formic acid, $HCHO_2$, is formed as well. (Formic acid is the agent that causes irritation from ant stings; Latin, *formica*, ant.) What is the pH of a solution that is 0.315 M $HC_2H_3O_2$ and 0.250 M $HCHO_2$?

Polyprotic acids (Use data from Table 17-3 as necessary)

40. For the following solutions of $H_2S(aq)$, determine $[H_3O^+]$, $[HS^-]$, and $[S^{2-}]$: (a) 0.075 M H_2S; (b) 0.0050 M H_2S; (c) 1.0×10^{-5} M H_2S.

41. The antimalarial drug quinine, $C_{20}H_{24}O_2N_2$, a *diprotic base*, has a water solubility of 1.00 g per 1900. mL of solution. $pK_{b_1} = 6.0$ and $pK_{b_2} = 9.8$.
(a) Write equations for the ionization equilibria corresponding to pK_{b_1} and pK_{b_2}.
(b) What is the pH of saturated aqueous quinine?

42. Cola drinks have a phosphoric acid content that is described as "from 0.057 to 0.084% of 75% phosphoric acid, by mass." Estimate the pH range of cola drinks corresponding to this range of H_3PO_4 content.

43. Citric acid, $H_3C_6H_5O_7$, is the main acidic component of citrus juices. Its ionization as a *triprotic* acid is represented below, where $H_3Cit = H_3C_6H_5O_7$.

$$H_3Cit + H_2O \rightleftharpoons H_3O^+ + H_2Cit^- \qquad pK_a = 3.13$$

$$H_2Cit^- + H_2O \rightleftharpoons H_3O^+ + HCit^{2-} \qquad pK_a = 4.76$$

$$HCit^{2-} + H_2O \rightleftharpoons H_3O^+ + Cit^{3-} \qquad pK_a = 6.40$$

Determine the molarity of a citric acid solution that would have the same pH as lemon juice (see Figure 17-3).

44. Verify the statement made on page 624 following Example 17-11 concerning ion concentrations in $H_2SO_4(aq)$ solutions of *moderate, low,* and *very low* molarities. Specifically, calculate $[H_3O^+]$, $[HSO_4^-]$, and $[SO_4^{2-}]$ in (a) 0.75 M H_2SO_4; (b) 0.075 M H_2SO_4; (c) 7.5×10^{-4} M H_2SO_4.

Ions as acids and bases (hydrolysis)

45. Complete the following equations in those instances in which a reaction (hydrolysis) will occur. If no reaction occurs, so state.
(a) $NH_4^+(aq) + NO_3^-(aq) + H_2O \rightarrow$
(b) $Na^+(aq) + NO_2^-(aq) + H_2O \rightarrow$
(c) $K^+(aq) + C_7H_5O_2^-(aq) + H_2O \rightarrow$
(d) $K^+(aq) + Cl^-(aq) + Na^+(aq) + I^-(aq) + H_2O \rightarrow$
(e) $C_6H_5NH_3^+(aq) + Cl^-(aq) + H_2O \rightarrow$

46. Refer to the Summarizing Example. Complete the calculation of the pH of 0.0392 M $Na_2C_2O_4$ in part 5.

47. Sorbic acid, $HC_6H_7O_2$, is widely used in the food industry as a preservative. For example, in the form of its potassium salt, it is added to cheese to inhibit the formation of mold.

$$HC_6H_7O_2 + H_2O \rightleftharpoons H_3O^+ + C_6H_7O_2^- \qquad K_a = 1.73 \times 10^{-5}$$

What is the pH of an 0.55 molar solution of potassium sorbate?

48. It is desired to produce an aqueous solution of pH = 8.75 by dissolving one of the following salts in water. Which salt would you use and at what molarity? (a) NH_4Cl; (b) $KHSO_4$; (c) KNO_2; (d) $NaNO_3$.

49. Pyridine, C_5H_5N, forms a salt, pyridinium hydrochloride, as a result of a reaction with HCl. Write an ionic equation to represent the hydrolysis of the pyridium ion, and calculate the pH of 0.0482 M $C_5H_5NH^+Cl^-$.

50. For each of the following ions write two equations, one showing its ionization as an acid and the other as a base. Then use data from Table 17-3 to predict whether each ion makes the solution acidic or basic. (a) HSO_3^-; (b) HS^-; (c) HPO_4^{2-}.

51. In some tables of ionization constants you will find no K_b values. Everything is listed as K_a. Reconstruct the portion of Table 17-2 describing weak bases by listing the K_a values of the conjugate acids of the bases (that is, K_a for the cations formed by the ionization of the bases). Arrange your table in terms of increasing value of K_a.

Molecular structure and acid strength

52. Based on the Brønsted–Lowry theory, explain why
(a) Acetic acid is a strong acid in $NH_3(l)$ but a weak acid in $H_2O(l)$.
(b) Ammonia is a strong base in $HC_2H_3O_2(l)$ but a weak base in $H_2O(l)$.

53. Predict the direction favored in each of the following acid–base reactions. That is, does the reaction tend to go more in the forward or in the reverse direction?
(a) $NH_4^+ + OH^- \rightleftharpoons H_2O + NH_3$
(b) $HSO_4^- + NO_3^- \rightleftharpoons HNO_3 + SO_4^{2-}$
(c) $CH_3OH + C_2H_3O_2^- \rightleftharpoons HC_2H_3O_2 + CH_3O^-$
(d) $HC_2H_3O_2 + CO_3^{2-} \rightleftharpoons HCO_3^- + C_2H_3O_2^-$
(e) $HNO_2 + ClO_4^- \rightleftharpoons HClO_4 + NO_2^-$
(f) $H_2CO_3 + CO_3^{2-} \rightleftharpoons HCO_3^- + HCO_3^-$

54. Arrange the following in the order of increasing acid strength.

(a) HI (b) HCl

(c)
$$H-\overset{\overset{\displaystyle H}{|}}{\underset{\underset{\displaystyle H}{|}}{C}}-\overset{\overset{\displaystyle O}{\|}}{C}-O-H$$
(d)
$$Cl-\overset{\overset{\displaystyle H}{|}}{\underset{\underset{\displaystyle H}{|}}{C}}-\overset{\overset{\displaystyle O}{\|}}{C}-O-H$$

(e)
$$F-\overset{\overset{\displaystyle F}{|}}{\underset{\underset{\displaystyle F}{|}}{C}}-\overset{\overset{\displaystyle O}{\|}}{C}-O-H$$
(f)
$$I-\overset{\overset{\displaystyle H}{|}}{\underset{\underset{\displaystyle H}{|}}{C}}-\overset{\overset{\displaystyle O}{\|}}{C}-O-H$$

$\star$55. Given below are pK_a values for four *diprotic* acids. The acid molecules differ in the length of the hydrocarbon chain separating the two ionizable H atoms (on the —COOH groups). Explain the trend in the *differences* between pK_{a_1} and pK_{a_2} among these acids.

oxalic acid, HOOCCOOH: $pK_{a_1} = 1.25$; $pK_{a_2} = 3.81$

succinic acid, HOOC(CH$_2$)$_2$COOH: $pK_{a_1} = 4.21$; $pK_{a_2} = 5.64$

adipic acid, HOOC(CH$_2$)$_6$COOH: $pK_{a_1} = 4.41$; $pK_{a_2} = 5.41$

suberic acid, HOOC(CH$_2$)$_8$COOH: $pK_{a_1} = 4.51$; $pK_{a_2} = 5.40$

$\star$56. Phosphorous acid has the formula H_3PO_3, but it is listed in Table 17-3 as a *diprotic* acid. Propose a Lewis structure for

H_3PO_3 that is consistent with this fact. [*Hint:* To what atom are ionizable H atoms usually bonded?]

Lewis theory of acids and bases

57. The three reactions below are acid–base reactions according to the Lewis theory. Draw Lewis structures and identify the Lewis acid and Lewis base in each reaction.

(a) $B(OH)_3 + OH^- \longrightarrow [B(OH)_4]^-$
(b) $N_2H_4(aq) + H_2O \rightarrow N_2H_5^+(aq) + OH^-(aq)$
(c) $(C_2H_5)_2O + BF_3 \rightarrow (C_2H_5)_2OBF_3$

58. Show that in each of the following practical cases a Lewis acid–base reaction is involved. Identify the acid and the base. [*Hint:* Draw Lewis electronic structures as necessary.]

(a) $SO_2(g)$ can be removed from the exhaust gases of a power plant by allowing it to combine with lime, $CaO(s)$. The result is the ionic compound $CaSO_3(s)$.

(b) $CO_2(g)$ can be removed from confined quarters (such as spacecraft) by allowing it to combine with an alkali metal hydroxide, e.g.,

$$CO_2(g) + LiOH(s) \rightarrow LiHCO_3(s).$$

Additional Exercises

59. Calculate $[H_3O^+]$ and pH in (a) 0.0052 M HI(aq); (b) 0.0014 M Ca(OH)$_2$(aq); (c) saturated Ba(OH)$_2$(aq) [containing 39 g Ba(OH)$_2 \cdot 8H_2O$ per liter].

60. What volume of 0.218 M NaOH must be diluted to 1.00 L to produce a solution with pH = 11.22?

61. In the manner used in equation (17.12), represent the self-ionization of the following liquid solvents: (a) NH$_3$; (b) HF; (c) CH$_3$OH; (d) HC$_2$H$_3$O$_2$; (e) H$_2$SO$_4$.

62. Draw Lewis structures corresponding to the cations and anions formed in each of the self-ionization reactions of the preceding exercise.

63. What are the pH values of the following solutions of benzoic acid? $pK_a = 4.20$. (a) 0.40 M HC$_7$H$_5$O$_2$; (b) 1.0×10^{-3} M HC$_7$H$_5$O$_2$; (c) 1.4×10^{-5} M HC$_7$H$_5$O$_2$.

64. The pH of a saturated solution of phenol at room temperature is 4.90. What is the solubility of phenol in grams per liter of saturated solution?

$$HC_6H_5O + H_2O \rightleftharpoons H_3O^+ + C_6H_5O^- \qquad K_a = 1.6 \times 10^{-10}$$

65. With information from this chapter, but without performing calculations, arrange the following 0.01 M solutions in order of *increasing* pH: NH$_3$; KOH; H$_2$SO$_4$; HC$_2$H$_3$O$_2$; Ba(OH)$_2$; HCl.

66. A saturated aqueous solution of Mg(OH)$_2$ has a pH of 10.53. What is the solubility of Mg(OH)$_2$, expressed in mg/L?

67. 55.6 L of HCl(g), measured at 748 mmHg and 25.0 °C, is dissolved in 1515 mL water. What volume of NH$_3$(g), measured at 757 mmHg and 22.0 °C, must be absorbed by the same solution to exactly neutralize the HCl?

$$NH_3(aq) + H^+(aq) + Cl^-(aq) \rightarrow NH_4^+(aq) + Cl^-(aq)$$

$\star$68. The solubility of CO$_2$(g) in H$_2$O at 25 °C and under a CO$_2$(g) pressure of 1 atm is 1.45 g CO$_2$/L. CO$_2$ comprises 0.033% of air, by volume. Use this information, together with

data from Table 17-3, to show that rainwater saturated with CO$_2$ has a pH $\approx$ 5.6 (referred to in the chapter introduction as the "normal" pH for rainwater). [*Hint:* Recall Henry's law. What is the partial pressure of CO$_2$(g) in air?]

69. One handbook lists a value of 9.5 for the pK of quinoline, a weak base used as a preservative for anatomical specimens and as an antimalarial. Another handbook lists the solubility of quinoline in water at 25 °C as 0.6 g/100 mL. Use this information and that below to calculate the pH of a saturated solution of quinoline in water.

$$C_9H_7N(aq) + H_2O \rightleftharpoons C_9H_7NH^+(aq) + OH^-(aq) \qquad pK_b = 9.5$$

70. Fluoroacetic acid occurs in Gifblaar, one of the most poisonous of all plants. Its sodium salt is used as a rodenticide (e.g., rat killer). A 0.500 M solution of the acid is found to have a pH = 1.46. Calculate K_a for fluoroacetic acid.

$$CH_2FCOOH(aq) + H_2O \rightleftharpoons H_3O^+(aq) + CH_2FCOO^-(aq)$$
$$K_a = ?$$

71. What molarity of each of the following acids is required to produce a solution with a pH = 3.15? (Use data from Tables 17-1 and 17-2, as necessary.) (a) HCl; (b) HC$_2$H$_3$O$_2$; $\star$(c) H$_2$SO$_4$

72. The pH scale is a logarithmic scale. Each change of one unit represents a change in $[H_3O^+]$ of 10. Another familiar logarithmic scale is the Richter scale (ranging from 1 to 8 or more) used to establish the energy (and destructive force) of an earthquake.

(a) How much greater is $[H_3O^+]$ in an "acid rain" sample with pH = 4.20 than in a more "normal" rainwater with pH = 5.60?

(b) How much more destructive force is associated with an earthquake of Richter magnitude 6.5 than one of magnitude 5.0?

73. Indicate which of the following 0.10 M solutions has the highest pH and which, the lowest pH. Explain the basis of your reasoning. **(a)** $NH_4Cl(aq)$; **(b)** $NH_3(aq)$; **(c)** $NaC_2H_3O_2(aq)$; **(d)** $KCl(aq)$

74. You are given a laboratory assignment in which you are asked to prepare a 100.0-mL sample of a water solution having a pH of 5.5 by dissolving the appropriate amount of a solute in water having a pH of 7.0. Which one of these solutes would you use, and in what quantity? Explain your choice, since there is more than one possibility.

(a) 15 M $NH_3(aq)$; **(b)** 12 M $HCl(aq)$; **(c)** $NH_4Cl(s)$; **(d)** glacial (pure) acetic acid, $HC_2H_3O_2$.

★**75.** From the observation that 0.0500 M vinylacetic acid has a freezing point of −0.096 °C, determine K_a for this acid.

$$HC_4H_5O_2 + H_2O \rightleftharpoons H_3O^+ + C_4H_5O_2^- \qquad K_a = ?$$

76. Adipic acid is among the top 50 manufactured chemicals in the U.S. (nearly 1 million tons annually). Its chief use is in the manufacture of nylon (see Focus feature, Chapter 27). It is a *diprotic* acid.

$$HOOC(CH_2)_4COOH + H_2O \rightleftharpoons H_3O^+ + HOOC(CH_2)_4COO^-$$
$$K_{a_1} = 3.9 \times 10^{-5}$$

$$HOOC(CH_2)_4COO^- + H_2O \rightleftharpoons H_3O^+ + {}^-OOC(CH_2)_4COO^-$$
$$K_{a_2} = 3.9 \times 10^{-6}$$

A saturated solution of adipic acid is about 0.10 M $HOOC(CH_2)_4COOH$. Calculate the concentrations of all the species present in this solution, that is, $[H_3O^+]$, $[OH^-]$, $[H_2C_6H_8O_4]$, $[HC_6H_8O_4^-]$, and $[C_6H_8O_4^{2-}]$. [*Hint:* Demonstrate that the usual simplifying assumptions work in this case too.]

77. Label the Brønsted–Lowry acids and bases in the following reactions and indicate whether the forward or reverse reaction is favored. Explain your reasoning. [*Hint:* Use data from tables in the text, as necessary.]

(a) $H_3PO_4(aq) + OH^-(aq) \rightleftharpoons H_2O + H_2PO_4^-(aq)$
(b) $NH_4^+(aq) + Cl^-(aq) + H_2O \rightleftharpoons$
$$H_3O^+(aq) + Cl^-(aq) + NH_3(aq)$$
(c) $HCO_3^-(aq) + HNO_2(aq) \rightleftharpoons NO_2^-(aq) + H_2O + CO_2(g)$
(d) $NH_4^+(aq) + C_2H_3O_2^-(aq) \rightleftharpoons NH_3(aq) + HC_2H_3O_2(aq)$

78. The Brønsted–Lowry theory can be applied to acid–base reactions in *nonaqueous* solvents. Indicate whether each of the following would be an acid, a base, or either in pure liquid acetic acid, $HC_2H_3O_2$, as a solvent: **(a)** $C_2H_3O_2^-$; **(b)** H_2O; **(c)** $HC_2H_3O_2$; **(d)** $HClO_4$. [*Hint:* Think of analogous situations in water.]

79. Each of the following is a Lewis acid–base reaction. Which reactant is the acid and which is the base? Explain.

(a) $SO_3 + H_2O \rightarrow H_2SO_4$
(b) $Al(OH)_3(s) + OH^-(aq) \rightarrow [Al(OH)_4]^-$

80. Which compound is the more *acidic* in each of the following pairs? Explain your reasoning. **(a)** $ClCH_2CH_2OH$ *or* CH_3CH_2OH; **(b)** CH_3NH_2 *or* $ClCH_2NH_2$; **(c)** $ClCH_2CH_2COOH$ *or* CH_2FCOOH.

81. Which is the stronger base, $CH_3CH_2CH_2NH_2$ (propylamine), or ⬡—NH_2 (aniline)?

82. The table gives some data for binary acids of the group 6A elements, H_2X. Explain the trends in the K_a values. That is, why is $K_{a_2} < K_{a_1}$ for H_2S? Why is K_{a_1} for $H_2Te > K_{a_1}$ for H_2S? Does the acid–base behavior of H_2O conform to the trends shown here?

	H_2S	H_2Se	H_2Te
K_{a_1}	1.1×10^{-7}	1.3×10^{-4}	2.3×10^{-3}
K_{a_2}	1.0×10^{-14}	$\approx 10^{-11}$	1.6×10^{-11}

★**83.** It is possible to write simple equations to relate pH, pK, and molarities (M) of various solutions. Three such equations are
Weak acid: $pH = \frac{1}{2}pK_a - \frac{1}{2}\log M$
Weak base: $pH = 14.00 - \frac{1}{2}pK_b + \frac{1}{2}\log M$
Salt of weak acid (pK_a) and strong base:

$$pH = 14.00 - \frac{1}{2}pK_w + \frac{1}{2}pK_a + \frac{1}{2}\log M$$

(a) Derive these three equations, and point out what assumptions are involved in the derivations.
(b) Use these equations to determine the pH of 0.10 M $HC_2H_3O_2(aq)$, 0.10 M $NH_3(aq)$, and 0.10 M $NaC_2H_3O_2$. Verify that the equations give correct results by determining these pH values in the "usual" way.

★**84.** Often this generalization applies to oxoacids with the formula $EO_m(OH)_n$ (E is the central atom): If $m = 0$, $K_a \approx 10^{-7}$; $m = 1$, $K_a \approx 10^{-2}$; $m = 2$, K_a is large; $m = 3$, K_a is very large.
(a) Show that this generalization works well for the oxoacids of chlorine: $HOCl$, $pK_a = 7.52$; $HOClO$, $pK_a = 1.92$; $HOClO_2$, $pK_a = -3$; $HOClO_3$, $pK_a = -8$.
(b) Estimate the value of K_{a_1} for H_3AsO_4.
(c) Write a Lewis structure for hypophosphorous acid, H_3PO_2, for which $pK_a = 1.1$.

★**85.** Show that when $[H_3O^+]$ of a solution is reduced to one-half of its original value, the pH value increases by 0.30 unit, *regardless of the initial pH*. Can it also be said that when any solution is diluted to one-half of its original concentration its pH value increases by 0.30 unit? Explain.

★**86.** What total concentration of $HC_2H_3O_2$ must be placed in aqueous solution if the solution is to have the same freezing point as 0.150 M $HC_2H_2ClO_2$ (monochloroacetic acid)?

★**87.** What is the pH obtained by dissolving 1.55 g CH_3NH_2 and 12.5 g NH_3 in 375 mL of a water solution?

★**88.** Calculate the pH of 1.0 M $NH_4CN(aq)$. [*Hint:* Identify the six species (excluding H_2O) whose concentrations in this solution are "unknown," and find six equations relating these unknowns. Three equations are ionization constant expressions—K_w, K_a for HCN, and K_b for NH_3. Another two conditions are that $[NH_3] + [NH_4^+] = 1.0$ M = $[HCN] + [CN^-]$. A simplifying assumption is that $[NH_4^+] = [CN^-]$.]

Self-Test Questions

For questions 89 through 98 select the single item that best completes each statement.

89. The number of moles of OH^- in 0.300 L of 0.0050 M $Ba(OH)_2$ is (a) 0.0075; (b) 0.0015; (c) 0.0030; (d) 0.0050.

90. A solution has a pH of 5.00. In this solution, $[OH^-]$ must be (a) 1.0×10^{-9} M; (b) 1.0×10^{-7} M; (c) greater than 10^{-5} M; (d) 1.0×10^{-5} M.

91. $[H_3O^+]$ in 0.10 M $HC_3H_5O_2$ (propionic acid) must be
(a) equal to $[H_3O^+]$ in 0.10 M HNO_2 (nitrous acid);
(b) less than $[H_3O^+]$ in 0.10 M HI (hydroiodic acid);
(c) greater than $[H_3O^+]$ in 0.10 M HBr (hydrobromic acid);
(d) equal to 0.10 mol H_3O^+/L.

92. An aqueous solution is 0.10 M in the weak base methylamine, CH_3NH_2. In this solution (a) $[H_3O^+] = 0.10$ M; (b) $[OH^-] = 0.10$ M; (c) pH < 7; (d) pH < 13.

93. Of the following solutions, the one that is most alkaline (basic) is (a) NaCl(aq); (b) NH_4Cl(aq); (c) $KC_2H_3O_2$(aq); (d) KNO_3(aq).

94. Of the following solutions the one that is the most acidic is (a) $NaHSO_4$(aq); (b) NaCl(aq); (c) $NaC_2H_3O_2$(aq); (d) Na_2S(aq).

95. One of the following ions is *amphiprotic*, that is, able to lose or gain a proton in aqueous solution. The amphiprotic ion is (a) HCO_3^-; (b) CO_3^{2-}; (c) Cl^-; (d) NH_4^+.

96. The reaction of acetic acid, $HC_2H_3O_2$, with a base will proceed furthest toward completion (to the right) when that base is (a) H_2O; (b) NH_3; (c) Cl^-; (d) $HClO_4$.

97. For a 0.10 M solution of H_2SO_3, a *diprotic* acid, $K_{a_1} = 1.3 \times 10^{-2}$ and $K_{a_2} = 6.3 \times 10^{-8}$. In this solution

(a) $[H_3O^+] = 0.10$ M; (b) $[H_3O^+] = 0.013$ M; (c) $[HSO_3^-] = 0.013$ M; (d) $[SO_3^{2-}] = 6.3 \times 10^{-8}$ M.

98. In 0.10 M H_2SO_4, $[H_3O^+] =$ (a) 0.050 M; (b) 0.10 M; (c) 0.11 M; (d) 0.20 M.

99. Arrange the following 0.01 M aqueous solutions in order of *decreasing* $[H_3O^+]$: NH_3(aq); HNO_3(aq); $NaNO_2$(aq); H_2SO_4(aq); NaOH(aq); NaCl(aq); $Ba(OH)_2$(aq); NH_4ClO_4(aq); $HC_2H_3O_2$(aq).

100. What mass of benzoic acid, $HC_7H_5O_2$, must be dissolved in 250.0 mL of water to produce a pH = 2.60?

$$HC_7H_5O_2 + H_2O \rightleftharpoons H_3O^+ + C_7H_5O_2^- \qquad K_a = 6.3 \times 10^{-5}$$

101. Explain why
(a) $HClO_4$ is a stronger acid than HNO_3.
(b) trichloroacetic acid, $HC_2Cl_3O_2$, is a stronger acid than acetic acid, $HC_2H_3O_2$.
(c) *o*-chloroaniline is a weaker base than aniline.

o-chloroaniline   aniline

(d) $[H_3O^+]$ in a strong acid solution *doubles* as the total acid concentration doubles, whereas in a weak acid solution $[H_3O^+]$ increases only by a factor of $\sqrt{2}$.
(e) $[PO_4^{3-}]$ in 0.30 M H_3PO_4 is *not* simply $\frac{1}{3}[H_3O^+]$ but much, much less than $\frac{1}{3}[H_3O^+]$.

18

Additional Aspects of Acid–Base Equilibria

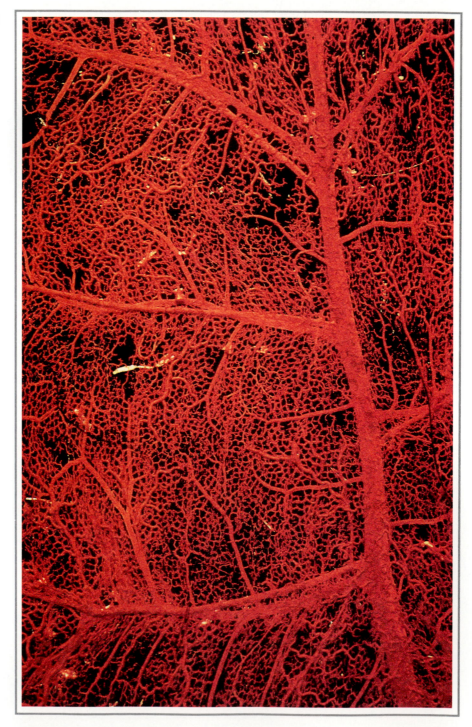

The blood carried by these vessels is maintained at a pH of 7.4. Blood is a highly buffered system (see page 656). [Biophoto Associates/Photo Researchers, Inc.]

An essential part of a modern physical examination is the chemical analysis of blood samples; many people know, for example, their blood cholesterol level. One characteristic of blood that we rarely give thought to is its pH, but the pH of blood is crucial to normal health and to life itself. The pH of blood must be held remarkably constant, varying only by a few hundredths of a pH unit—from pH 7.36 to 7.40. Blood pH is kept constant through a complex interaction of substances, some with acidic and some with basic properties. Blood is said to be a *buffered* mixture. Among the most important components responsible for maintaining the pH of blood are bicarbonate ion, HCO_3^-, and carbon dioxide (often represented in aqueous solution as carbonic acid, H_2CO_3).

In this chapter, before taking a further look at the pH of blood, we begin with a fundamental aspect of solution equilibria known as the *common ion effect*. We demonstrate how the common ion effect comes into play in buffered solutions and describe the qualities that make a buffered solution resist changes in its pH. We also explain that the common ion effect accounts for the color changes associated with acid–base indicators, and show how to use indicators to measure the pH of solutions and to follow the progress of acid–base neutralization reactions carried out by titration.

18-1 The Common Ion Effect

Let us begin by introducing a phenomenon that is key to much of what we will do in this chapter. In *pure water*, $[OH^-] = [H_3O^+] = 1.0 \times 10^{-7}$ M. In Example 17-2 we found that in 0.015 M HCl, $[H_3O^+] = [Cl^-] = 0.015$ M and $[OH^-] = 6.7 \times 10^{-13}$ M; that is

$$HCl + H_2O \longrightarrow H_3O^+ + Cl^- \qquad (18.1)$$
$$ \underset{0.015\,M}{} \quad \underset{0.015\,M}{}$$

$$H_2O + H_2O \rightleftharpoons H_3O^+ + OH^- \qquad [OH^-] = \frac{K_w}{0.015} = 6.7 \times 10^{-13}\text{ M} \quad (18.2)$$
$$ \underset{0.015\,M}{}$$

H_3O^+ is produced *both* by the complete ionization of HCl and by the self-ionization of water. We say that it is a **common ion** to these two reactions. The high concentration of H_3O^+ produced by the ionization of the *strong* acid HCl (18.1) displaces reaction (18.2) *to the left* (Le Châtelier's principle). This in turn causes the hydroxide ion concentration to be reduced from its value in pure water. The effect we have described, *the repression of the ionization of a weak electrolyte by the presence of a common ion from a strong electrolyte*, is called the **common ion effect.**

If we add the strong base NaOH to water, OH^- is the common ion in the water self-ionization equilibrium and the hydronium ion concentration, $[H_3O^+]$, is greatly reduced. Actually, even weak acids and weak bases produce much more H_3O^+ and OH^- than does water and will repress the self-ionization of water. In summary,

$$H_2O + H_2O \rightleftharpoons H_3O^+ + OH^-$$

In the presence of an acid or base, equilibrium condition shifts.

Solutions of Weak Acids and Strong Acids. Just as it represses the self-ionization of water, a strong acid represses the ionization of a weak acid, through the common ion H_3O^+. For example,

$$HC_2H_3O_2 + H_2O \rightleftharpoons H_3O^+ + C_2H_3O_2^- \qquad K_a = 1.74 \times 10^{-5} \qquad (18.3)$$

In the presence of a strong acid, equilibrium shifts.

645

FIGURE 18-1

A mixture of a strong acid and a weak acid.

The solution here is 0.100 M HCl–0.100 M $HC_2H_3O_2$. The red color of the thymol blue indicator shows that this solution has about the same pH as does 0.100 M HCl. Essentially all the H_3O^+ is produced by the strong acid alone. Compare this figure with Figure 17-4, where the separate acids 0.100 M HCl and 0.100 M $HC_2H_3O_2$ are shown. [Carey B. Van Loon]

The magnitude of this common ion effect is illustrated through Figure 18-1 and Example 18-1.

Example 18-1 _____

Demonstrating the common ion effect: A solution of a strong acid and a weak acid. (a) Determine $[H_3O^+]$ and $[C_2H_3O_2^-]$ in 0.100 M $HC_2H_3O_2$. (b) Then determine these same quantities in a solution that is 0.100 M in *both* $HC_2H_3O_2$ and HCl.

Solution

(a) This calculation is the same as in Example 17-6. In the setup below we make the usual assumption that x is very small and that $(0.100 - x) \approx 0.100$.

$$HC_2H_3O_2 + H_2O \rightleftharpoons H_3O^+ + C_2H_3O_2^-$$

placed in solution:	0.100 M	—	—
changes:	$-x$ M	$+x$ M	$+x$ M
at equilibrium:	$(0.100 - x)$ M	x M	x M

$$K_a = \frac{[H_3O^+][C_2H_3O_2^-]}{[HC_2H_3O_2]} = \frac{x^2}{0.100 - x} \approx \frac{x^2}{0.100} = 1.74 \times 10^{-5}$$

$$x^2 = 1.74 \times 10^{-6} \qquad x = [H_3O^+] = 1.32 \times 10^{-3} \text{ M}$$

$$x = [C_2H_3O_2^-] = 1.32 \times 10^{-3} \text{ M}$$

In this setup we use y instead of x to emphasize that the contribution to the $[H_3O^+]$ by $HC_2H_3O_2$ is different (less) in the presence of HCl than when $HC_2H_3O_2$ is present alone.

(b) Our setup must now include information about the common ion, H_3O^+.

$$HC_2H_3O_2 + H_2O \rightleftharpoons H_3O^+ + C_2H_3O_2^-$$

placed in solution:			
weak acid	0.100 M	—	—
strong acid	—	0.100 M	—
ioniz'n of weak acid:	$-y$ M	$+y$ M	$+y$ M
at equilibrium:	$(0.100 - y)$ M	$(0.100 + y)$ M	y M

Because the strong acid represses the ionization of the weak acid, we should expect the H_3O^+ concentration produced by the weak acid (y) to be very small. Thus, $(0.100 - y) \approx (0.100 + y) \approx 0.100$.

Note how $[C_2H_3O_2^-]$ in 0.100 M $HC_2H_3O_2$ decreases about 100-fold in the presence of 0.100 M HCl.

$$K_a = \frac{[H_3O^+][C_2H_3O_2^-]}{[HC_2H_3O_2]} = \frac{(0.100 + y)(y)}{(0.100 - y)} \approx \frac{0.100(y)}{0.100} = 1.74 \times 10^{-5}$$

$$y = [C_2H_3O_2^-] = 1.74 \times 10^{-5} \text{ M} \qquad 0.100 + y = [H_3O^+] = 0.100 \text{ M}$$

SIMILAR EXAMPLES: Exercises 1, 2, 18.

To summarize the effect of the strong acid HCl on the ionization of the weak acid $HC_2H_3O_2$ calculated in Example 18-1,

Concerning mixtures of strong and weak acids.

> • Assume that all the H_3O^+ in a mixture of a strong acid and a weak acid comes from the strong acid. (This assumption is not valid if the strong acid is very dilute and/or if K_a of the weak acid is large.) (18.4)

(a)   (b)

FIGURE 18-2
A mixture of a weak acid and its salt.

Bromophenol blue indicator, which is present in both solutions, imparts a color that depends on the pH of the solution.

pH < 3.0   3.0 < pH < 4.6

yellow   green

pH > 4.6

blue-violet

(a) 0.100 M $HC_2H_3O_2$ has a pH of about 3, but (b) a solution that is 0.100 M $HC_2H_3O_2$– 0.100 M $NaC_2H_3O_2$ has a pH of about 5. [Carey B. Van Loon]

- In the presence of a *strong* acid, the concentration of the weak acid anion (A^-) is very much less than it is when only the weak acid (HA) is present. $\qquad$ (18.4)

In a similar fashion, OH^- from a strong base represses the ionization of a weak base.

$$NH_3 + H_2O \;\overset{\longleftarrow}{\rightleftharpoons}\; NH_4^+ + OH^- \qquad K_b = 1.74 \times 10^{-5} \qquad (18.5)$$

In the presence of a strong base, equilibrium shifts.

Solutions of Weak Acids and Their Salts. The salt of a weak acid is a strong electrolyte—it is completely dissociated into ions in aqueous solution. One of the ions produced by the dissociation of the salt, the *anion,* is a common ion in the ionization equilibrium of the weak acid. The presence of this common ion represses the ionization of the weak acid.* For example, we can represent the effect of acetate salts on the acetic acid equilibrium as

$$NaC_2H_3O_2(aq) \longrightarrow Na^+ + C_2H_3O_2^-$$

$$HC_2H_3O_2 + H_2O \;\overset{\longleftarrow}{\rightleftharpoons}\; H_3O^+ + C_2H_3O_2^- \qquad (18.6)$$

In the presence of acetate ion, equilibrium shifts.

The effect of sodium acetate on the ionization of acetic acid is pictured in Figure 18-2 and illustrated through Example 18-2. In solving "common ion" problems like Example 18-2, assume that ionization of the weak acid (or base) does not begin until both the weak acid (or base) and its salt have been placed in solution. Then consider that ionization occurs until an equilibrium condition results.

Example 18-2

Demonstrating the common ion effect: A solution of a weak acid and a salt of the weak acid. Calculate $[H_3O^+]$ and $[C_2H_3O_2^-]$ in a solution that is 0.100 M both in $HC_2H_3O_2$ and in $NaC_2H_3O_2$.

Solution. The setup below is very similar to that in Example 18-1(b), except that here $NaC_2H_3O_2$ is the source of the common ion.

$$HC_2H_3O_2 \; + \; H_2O \; \rightleftharpoons \; H_3O^+ \; + \; C_2H_3O_2^-$$

placed in solution:			
weak acid	0.100 M	—	—
salt	—	—	0.100 M
ioniz'n of weak acid:	$-x$ M	$+x$ M	$+x$ M
at equilibrium:	$(0.100 - x)$ M	x M	$(0.100 + x)$ M

Because the salt represses the ionization of $HC_2H_3O_2$, we should expect $[H_3O^+] = x$ to be very small; and $(0.100 - x) \approx (0.100 + x) \approx 0.100$.

$[H_3O^+]$ in 0.100 M $HC_2H_3O_2$ decreases about 100-fold in the presence of 0.100 M $NaC_2H_3O_2$.

$$K_a = \frac{[H_3O^+][C_2H_3O_2^-]}{[HC_2H_3O_2]} = \frac{(x)(0.100 + x)}{(0.100 - x)} \approx \frac{(x)0.100}{0.100} = 1.74 \times 10^{-5}$$

$$x = [H_3O^+] = 1.74 \times 10^{-5} \text{ M} \qquad 0.100 + x = [C_2H_3O_2^-] = 0.100 \text{ M}$$

SIMILAR EXAMPLES: Exercises 17, 18, 53.

---

*Another approach is to consider the effect of a weak acid on the ionization (hydrolysis) of an anion. For this we can work with an expression such as $CN^- + H_2O \rightleftharpoons HCN + OH^-$. This approach is especially appropriate if K_b for the anion is much larger than K_a for the acid, but we will not consider such cases in this text.

FIGURE 18-3
A mixture of a weak base and its salt.

Thymolphthalein indicator is colorless below a pH of about 10 and blue above pH 10. **(a)** The pH of 0.10 M NH_3 is above 10 (actually about 11.1), and **(b)** that of the NH_3–NH_4Cl mixture is below 10 (actually about 9.2). The ionization of NH_3 is repressed in the presence of NH_4^+ and the pH is lowered. [Carey B. Van Loon]

(a)   (b)

Again, to summarize the effect of the *salt* $NaC_2H_3O_2$ on the ionization of the *weak acid* $HC_2H_3O_2$ calculated in Example 18-2,

Concerning mixtures of weak acids and their salts.

- Assume that all the *anion* in a mixture of a weak acid and its salt comes from the salt. (This assumption is not valid if the salt concentration is very low and/or if K_a of the weak acid is large.)
- In the presence of a *salt* of the weak acid, the concentration of hydronium ion, $[H_3O^+]$, is very much less than it is when only the weak acid is present.

(18.7)

The effect of ammonium salts on the ionization of the weak base ammonia is similar to the weak acid/salt situation just described: the *cation*, NH_4^+, is the common ion and $[OH^-]$ is reduced as reaction (18.8) shifts *to the left*.

$$NH_4Cl(aq) \longrightarrow NH_4^+ + Cl^-$$

$$NH_3 + H_2O \rightleftharpoons NH_4^+ + OH^-$$

(18.8)

⬅ In the presence of ammonium ion, equilibrium shifts.

Figure 18-3 illustrates ionization equilibrium in a weak base/salt mixture.

18-2   Buffer Solutions

If even a very small quantity of either an acid or a base is added to pure water, the pH changes dramatically. Pure water, as illustrated in Figure 18-4, has no resistance to a change in pH—it has no buffer capacity.

An acetic acid–sodium acetate solution, such as that described in Example 18-2, *does* have a capacity to resist a change in pH. It is a **buffer solution** (also called a **buffered solution**). *A buffer (buffered) solution is one whose pH changes only very slightly upon the addition of small amounts of either an acid or a base.* In order for a solution to act as a buffer it must have two components, one of which is able to neutralize acids, and the other, bases. As we demonstrate below, *common buffer solutions are mixtures containing a*

- *weak acid and its conjugate base (one of its salts)* or a
- *weak base and its conjugate acid (one of its salts).*

Suppose, for example, that we take a solution that has equal concentrations of $HC_2H_3O_2$ and $NaC_2H_3O_2$ (as in Example 18-2). In such a solution $[H_3O^+] = 1.74 \times 10^{-5}$ and pH = 4.76. That is,

FIGURE 18-4

Effect of small quantities of acids and bases on the pH of water.

Pure water has a pH = 7. Addition of 0.001 mol HCl (1.00 mL of 1 M HCl) to 1.00 L of water produces $[H_3O^+]$ = 0.001 M and pH = 3. Addition of 0.001 mol OH^- (40 mg NaOH) to 1.00 L of water produces $[OH^-]$ = 0.001 M and pH = 11. Bromothymol blue indicator is green at pH 7, yellow at pH < 7, and blue at pH > 7. [Photographs by Carey B. Van Loon]

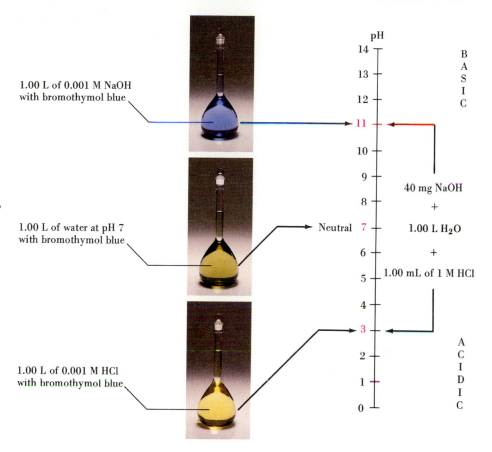

1.00 L of 0.001 M NaOH with bromothymol blue

1.00 L of water at pH 7 with bromothymol blue — Neutral

1.00 L of 0.001 M HCl with bromothymol blue

pH

B A S I C

40 mg NaOH
+
1.00 L H₂O
+
1.00 mL of 1 M HCl

A C I D I C

$$K_a = \frac{[H_3O^+][C_2H_3O_2^-]}{[HC_2H_3O_2]} = 1.74 \times 10^{-5} \tag{18.9}$$

Now consider adding a small quantity of either an acid (H_3O^+) or a base (OH^-) to this solution. The following reactions will occur in the buffer solution.

$$C_2H_3O_2^- + \underset{\text{(small added amt.)}}{H_3O^+} \longrightarrow HC_2H_3O_2 + H_2O \tag{18.10}$$

$$HC_2H_3O_2 + \underset{\text{(small added amt.)}}{OH^-} \longrightarrow C_2H_3O_2^- + H_2O \tag{18.11}$$

In the first instance (reaction 18.10) a small amount of the conjugate base ($C_2H_3O_2^-$) is converted to the acid ($HC_2H_3O_2$). In the second instance (reaction 18.11) a small amount of the acid is converted to its conjugate base. In both cases, all that occurs is a very slight change in the ratio $[C_2H_3O_2^-]/[HC_2H_3O_2]$ in equation (18.9). This change is so slight that the pH remains essentially constant at 4.76. For example, suppose that the initial concentrations of the weak acid and its salt were both 1.00 M and that we added enough acid or base to react with 1% of the appropriate buffer component. The changes in pH would be

- *initial buffer:* pH = 4.76
- *add acid* (convert 1% of salt to acid in reaction 18.10):

$$[H_3O^+] = 1.74 \times 10^{-5} \times [HC_2H_3O_2]/[C_2H_3O_2^-] = 1.74 \times 10^{-5} \times (1.01/0.99)$$

$$= 1.78 \times 10^{-5} \qquad pH = 4.75$$

- *add base* (convert 1% of acid to salt in reaction 18.11):

$$[H_3O^+] = 1.74 \times 10^{-5} \times [HC_2H_3O_2]/[C_2H_3O_2^-] = 1.74 \times 10^{-5} \times (0.99/1.01)$$

$$= 1.71 \times 10^{-5} \qquad pH = 4.77$$

The acetic acid–sodium acetate buffer is only good for maintaining a constant pH at about pH = 5. To obtain buffer solutions in other pH ranges other components must be used. If the buffer is to function in *basic* solutions, the necessary components are a weak base and its conjugate acid (salt), as illustrated in Example 18-3.

Example 18-3

Predicting whether a solution is a buffer **solution.** At times in the qualitative analysis procedure, to achieve complete separation of certain cations (recall Section 5-10), an NH_3–NH_4Cl solution is required. Show that such a solution is a buffer solution.

Solution. Whenever you have to decide whether a particular solution has buffer properties, the key question is, can you identify a component in the solution capable of neutralizing small added amounts of acid and a component that will neutralize added base? In the present case the components are NH_3 and NH_4^+, respectively.

$$NH_3 + \underset{\text{(small added amt.)}}{H_3O^+} \longrightarrow NH_4^+ + H_2O$$

$$NH_4^+ + \underset{\text{(small added amt.)}}{OH^-} \longrightarrow NH_3 + H_2O$$

In *all* aqueous solutions containing NH_3 and NH_4^+ we know that

$$NH_3 + H_2O \rightleftharpoons NH_4^+ + OH^-$$

and

$$K_b = \frac{[NH_4^+][OH^-]}{[NH_3]} = 1.74 \times 10^{-5}$$

If approximately equal concentrations of NH_4^+ and NH_3 are maintained in a solution, $[OH^-] \approx 1 \times 10^{-5}$ M; pOH ≈ 5; and pH ≈ 9. Thus, ammonia–ammonium chloride solutions are *alkaline* (basic) buffer solutions.

SIMILAR EXAMPLES: Exercises 3, 21, 22a.

Calculating the pH of Buffer Solutions. To calculate the pH of a buffer solution you need simply to substitute the equilibrium *concentrations* of a weak acid (or weak base) and its salt into the ionization equilibrium expression. Sometimes, however, you may need to do a preliminary calculation before you can use the K_a or K_b expression. In Example 18-4 this preliminary calculation involves diluting solutions.

Example 18-4

Calculating the pH of a buffer solution from information about the solution components. When a laboratory worker needs to prepare a buffer solution, he or she usually turns to a handbook in which explicit instructions are given for this purpose. For example, one such handbook states that to prepare a particular buffer solution you should mix 63.0 mL of 0.200 M $HC_2H_3O_2$ and 37.0 mL of 0.200 M $NaC_2H_3O_2$. What is the pH of this buffer solution?

Solution. It is easiest to deal with this problem in two distinct phases. First, calculate the stoichiometric concentrations of the weak acid and its salt in the 100.0 mL of buffer solution. Then consider the ionization equilibrium that occurs in this mixed solution.

Dilution calculation:

$$\text{molarity of } HC_2H_3O_2 = \frac{0.0630 \text{ L} \times 0.200 \text{ mol } HC_2H_3O_2/L}{0.1000 \text{ L}}$$

$$= 0.126 \text{ M } HC_2H_3O_2$$

$$\text{molarity of } C_2H_3O_2{}^- = \frac{0.0370 \text{ L} \times 0.200 \text{ mol } C_2H_3O_2{}^-/L}{0.1000 \text{ L}}$$

$$= 0.0740 \text{ M } C_2H_3O_2{}^-$$

Equilibrium calculation:

$$HC_2H_3O_2 + H_2O \rightleftharpoons H_3O^+ + C_2H_3O_2{}^-$$

placed in solution:			
weak acid	0.126 M	—	—
salt	—	—	0.0740 M
ioniz'n of weak acid:	$-x$ M	$+x$ M	$+x$ M
at equilibrium:	$(0.126 - x)$ M	$+x$ M	$(0.0740 + x)$ M

In our customary fashion, let us assume that x is very small and that $(0.126 - x) \approx 0.126$ and $(0.0740 + x) \approx 0.0740$.

$$K_a = \frac{[H_3O^+][C_2H_3O_2{}^-]}{[HC_2H_3O_2]} = \frac{(x)(0.0740)}{0.126} = 1.74 \times 10^{-5}$$

$$x = [H_3O^+] = (0.126/0.0740) \times 1.74 \times 10^{-5} = 2.96 \times 10^{-5}$$

$$pH = -\log [H_3O^+] = -\log(2.96 \times 10^{-5}) = 4.53$$

SIMILAR EXAMPLES: Exercises 4, 22c, 55a, 62a.

A Useful Equation for Buffer Solutions: The Henderson–Hasselbalch Equation. When dealing with buffer solutions, even though the complete problem may require several calculations of different types, there is usually one part of the problem that requires relating the pH (or pOH) of a buffer solution to the concentrations of the buffer components. This part of the overall problem can generally be easily solved with a simple equation, known as the Henderson–Hasselbalch equation. We will derive the equation for a hypothetical weak acid, HA (e.g., $HC_2H_3O_2$), and its salt, NaA (e.g., $NaC_2H_3O_2$). We start with the familiar

$$HA + H_2O \rightleftharpoons H_3O^+ + A^-$$

$$K_a = \frac{[H_3O^+][A^-]}{[HA]}$$

and rearrange the right side of the equation to obtain

$$K_a = [H_3O^+] \times \frac{[A^-]}{[HA]} \tag{18.12}$$

Now let us take the *negative logarithm* of each side of equation (18.12)

$$-\log K_a = -\log [H_3O^+] - \log \frac{[A^-]}{[HA]}$$

Recall that $pH = -\log [H_3O^+]$ and $pK_a = -\log K_a$.

$$pK_a = pH - \log \frac{[A^-]}{[HA]}$$

Finally, rearrange the equation to

pH of a buffer solution of a weak acid and its salt.

$$pH = pK_a + \log \frac{[A^-]}{[HA]} \quad or \quad pH = pK_a + \log \frac{[\text{conjugate base}]}{[\text{acid}]} \qquad (18.13)$$

A very similar derivation for a buffer solution of a weak base B and its salt, e.g., BH^+Cl^-, starts with the expressions

$$B + H_2O \rightleftharpoons BH^+ + OH^-$$

$$K_b = \frac{[BH^+][OH^-]}{[B]}$$

and results in the final equation

pH of a buffer solution of a weak base and its salt.

$$pOH = pK_b + \log \frac{[BH^+]}{[B]} \quad or \quad pOH = pK_b + \log \frac{[\text{conjugate acid}]}{[\text{base}]} \qquad (18.14)$$

What makes equations (18.13) and (18.14) especially useful is that ordinarily we can substitute stoichiometric concentrations (the concentrations of substances placed in solution) for equilibrium concentrations. This amounts to assuming that *the weak acid (or weak base) is essentially nonionized in the presence of its salt.* This assumption is valid as long as neither buffer component is too dilute and K_a (or K_b) is not too large.

Applied to the buffer solution of Example 18-4, equation (18.13) yields

$$pH = 4.76 + \log \frac{0.074}{0.126}$$

$$= 4.76 - 0.23 = 4.53$$

Preparing Buffer Solutions. Suppose we need a buffer solution with pH = 4.76. How shall we go about preparing this? If we recognize that $pK_a = 4.76$ for acetic acid, then according to equation (18.13) all we need to do is make up a solution having equal concentrations of $HC_2H_3O_2$ and $C_2H_3O_2^-$.

$$pH = pK_a + \log \frac{[C_2H_3O_2^-]}{[HC_2H_3O_2]} = 4.76 + \log 1 = 4.76$$

Suppose, however, that we need a buffer solution with pH 5.10. Now what do we do? We have two alternatives: (1) Use some ratio $[C_2H_3O_2^-]/[HC_2H_3O_2]$ having just the right value to make the solution pH = 5.10. (2) Find a weak acid that has a $pK_a = 5.10$ and prepare a solution that has equimolar concentrations of the acid and its salt.

Although the second alternative is simple in concept, generally it is not useful. What is the likelihood that we will find some *commonly available* weak acid that is *water-soluble* and has $pK_a = 5.10$? Not good. The first alternative is a more practical one, as illustrated in Example 18-5.

Example 18-5

Preparing a buffer solution of known pH: Using the Henderson–Hasselbalch equation. What mass of $NaC_2H_3O_2$ must be dissolved in 0.300 L of 0.250 M $HC_2H_3O_2$ to produce a solution with pH = 5.10? (Assume that the solution volume remains constant at 0.300 L.)

Solution. The conditions for using equation (18.13) appear to be met here. The value of K_a is not particularly large, and the solution is not very dilute. Solve equation (18.13) for $[C_2H_3O_2^-]$.

$$pH = pK_a + \log \frac{[C_2H_3O_2^-]}{[HC_2H_3O_2]}, \quad \text{where pH} = 5.10 \text{ and } pK_a = 4.76$$

$$5.10 = 4.76 + \log \frac{[C_2H_3O_2^-]}{0.250} \quad \text{and} \quad \log \frac{[C_2H_3O_2^-]}{0.250} = 5.10 - 4.76$$

$$= 0.34$$

$$\frac{[\text{C}_2\text{H}_3\text{O}_2{}^-]}{0.250} = \text{antilogarithm } 0.34 = 2.2$$

$$[\text{C}_2\text{H}_3\text{O}_2{}^-] = 0.250 \times 2.2 = 0.55 \text{ M}$$

$$\text{no. g NaC}_2\text{H}_3\text{O}_2 = 0.300 \text{ L} \times \frac{0.55 \text{ mol NaC}_2\text{H}_3\text{O}_2}{\text{L}} \times \frac{82.0 \text{ g NaC}_2\text{H}_3\text{O}_2}{1 \text{ mol NaC}_2\text{H}_3\text{O}_2}$$

$$= 14 \text{ g NaC}_2\text{H}_3\text{O}_2$$

SIMILAR EXAMPLES: Exercises 5, 6, 23, 24, 56.

Another alternative for preparing the buffer solution of pH = 5.10 involves reaction between a *strong acid* (HCl) and the *salt of the weak acid* ($\text{NaC}_2\text{H}_3\text{O}_2$). In reaction (18.15), acetate ion is a base and accepts protons from H_3O^+ to form molecules of the weak acid $\text{HC}_2\text{H}_3\text{O}_2$. Equilibrium is displaced far to the right.

$$\text{H}_3\text{O}^+ + \text{Cl}^- + \text{Na}^+ + \text{C}_2\text{H}_3\text{O}_2{}^- \longrightarrow \text{HC}_2\text{H}_3\text{O}_2 + \text{Na}^+ + \text{Cl}^- + \text{H}_2\text{O} \quad (18.15)$$

If HCl is in excess in reaction (18.15) we obtain a mixture of a strong and a weak acid. If $\text{NaC}_2\text{H}_3\text{O}_2$ is in excess, the result is a mixture of a *weak acid and its salt—a buffer solution*. This method is outlined in Figure 18-5.

FIGURE 18-5
Formation and action of a buffer solution.

Thymol blue indicator:

pH < 1.2   1.2 < pH < 2.8
  red   orange

pH > 2.8
  yellow

(a) 0.240 mol $\text{NaC}_2\text{H}_3\text{O}_2$ is added to 0.300 L of 0.250 M HCl (pH = 0.6).
(b) As a result of reaction (18.15), all the HCl is converted to 0.0750 mol $\text{HC}_2\text{H}_3\text{O}_2$ and 0.165 mol $\text{NaC}_2\text{H}_3\text{O}_2$ is left in excess. The solution is a buffer solution with a pH of 5.10. Pictured with the buffer solution are 0.0060 mol H_3O^+ (5.00 mL of 1.2 M HCl) and 0.0060 mol OH^- (240 mg NaOH).
(c) Adding 0.0060 mol H_3O^+ to the original buffer causes little change in pH.
(d) Adding 0.0060 mol OH^- to the original buffer also causes little change in pH. [Photographs by Carey B. Van Loon]

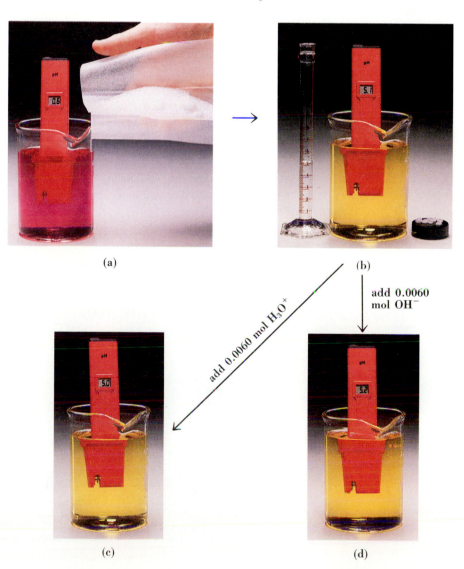

(a)

(b)

add 0.0060 mol H_3O^+

add 0.0060 mol OH^-

(c)

(d)

Calculating pH Changes in Buffer Solutions. We should make explicit here an idea that we have used several times before.

A calculation based on an equilibrium constant expression often must be preceded by a stoichiometric calculation.

To calculate the change in pH produced by adding a small amount of acid or base to a buffer solution, we must first use *stoichiometric* principles to establish how much of one buffer component is consumed and how much of the other is produced. Then we can use the new concentrations of weak acid (or weak base) and its salt to calculate the pH of the buffer solution. This method is applied in Example 18-6, and the results are shown in Figure 18-5.

Example 18-6

Calculating pH changes in a buffer solution. What is the effect of adding **(a)** 0.0060 mol HCl and **(b)** 0.0060 mol NaOH on the pH of 0.300 L of a buffer solution that is 0.250 M $HC_2H_3O_2$ and 0.550 M $NaC_2H_3O_2$? (This is the solution described in Example 18-5.)

Solution

(a) *Stoichiometric calculation:* To neutralize the 0.0060 mol H_3O^+, 0.0060 mol $C_2H_3O_2^-$ must be converted to 0.0060 mol $HC_2H_3O_2$.

	$C_2H_3O_2^-$	+	H_3O^+	$\longrightarrow$	$HC_2H_3O_2$	+ H_2O
original buffer:	0.300 L × 0.550 M				0.300 L × 0.250 M	
	0.165 mol				0.0750 mol	
add:			0.0060 mol			
changes:	−0.0060 mol		−0.0060 mol		+0.0060 mol	
final buffer:						
moles	0.159 mol		(?)		0.0810 mol	
concns	0.159 mol/0.300 L		(?)		0.0810 mol/0.300 L	
	0.530 M				0.270 M	

Equilibrium calculation: The conditions required by the Henderson–Hasselbalch equation are all met in this buffer solution.

$$pH = pK_a + \log \frac{[C_2H_3O_2^-]}{[HC_2H_3O_2]} = 4.76 + \log \frac{0.530}{0.270} = 4.76 + 0.29 = 5.05$$

(b) *Stoichiometric calculation:* To neutralize the added OH^-, 0.0060 mol $HC_2H_3O_2$ is converted to 0.0060 mol $C_2H_3O_2^-$.

	$HC_2H_3O_2$	+	OH^-	$\longrightarrow$	$C_2H_3O_2^-$	+ H_2O
original buffer:	0.300 L × 0.250 M				0.300 L × 0.550 M	
	0.0750 mol				0.165 mol	
add:			0.0060 mol			
changes:	−0.0060 mol		−0.0060 mol		+0.0060 mol	
final buffer:						
moles	0.0690 mol		(?)		0.171 mol	
concns	0.0690 mol/0.300 L		(?)		0.171 mol/0.300 L	
	0.230 M				0.570 M	

Equilibrium calculation:

$$pH = pK_a + \log \frac{[C_2H_3O_2^-]}{[HC_2H_3O_2]} = 4.76 + \log \frac{0.570}{0.230} = 4.76 + 0.39 = 5.15$$

SIMILAR EXAMPLES: Exercises 27, 28, 29, 57, 62.

Are You Wondering:

When to use molarities and when to use number of moles of components in buffer solution calculations?

Remember that when you use a K_a or K_b expression or the Henderson–Hasselbalch equation the terms substituted into these equations must be solution concentrations, that is, *molarities*. Before substituting data into these equations, however, you may need to do a stoichiometry calculation, and for this you may be able to use either "moles" or "molarities," depending on how you can best visualize the problem.

In the stoichiometry calculations in Example 18-6 we worked with numbers of moles of solution components, but we could have used molarities as well. Consider Example 18-6(a): When 0.0060 mol H_3O^+ is dissolved in 0.300 L solution this produces an immediate $[H_3O^+] = 0.0060$ mol/0.300 L = 2.0×10^{-2} M. Thus, we can set up the stoichiometry calculation as

$$C_2H_3O_2^- + H_3O^+ \longrightarrow HC_2H_3O_2 + H_2O$$

	$C_2H_3O_2^-$	H_3O^+	$HC_2H_3O_2$
original buffer:	0.550 M		0.250 M
add:		0.020 M	
changes:	−0.020 M	−0.020 M	+0.020 M
final buffer:	0.530 M	(?)	0.270 M

$[C_2H_3O_2^-]$ and $[HC_2H_3O_2]$ are exactly the same as we obtained in Example 18-6(a), and so too will be the pH of the buffer solution.

There is one case where you will generally find it best to work with moles instead of molarities in the stoichiometry portion of a calculation. This occurs when the solution volume changes, as in adding one solution to another.

Diluting Buffer Solutions. Not only does a buffer solution resist changes in pH upon the addition of small amounts of acid or base, but a buffer solution also *resists pH changes on dilution*. This additional property can be very useful at times, such as when a buffer solution becomes diluted as a result of a process occurring within it.

Consider the 0.250 M $HC_2H_3O_2$–0.550 M $NaC_2H_3O_2$ buffer solution with pH = 5.10 (Example 18-5). Suppose we dilute this solution to 10 times its original volume. The concentrations become 0.0250 M $HC_2H_3O_2$ and 0.0550 M $NaC_2H_3O_2$. If we compare these two cases we see

Original buffer solution

$$pH = pK_a + \log \frac{[C_2H_3O_2^-]}{[HC_2H_3O_2]}$$

$$pH = 4.76 + \log \frac{0.550}{0.250}$$

$$pH = 4.76 + \log 2.20$$

$$= 5.10$$

Diluted buffer solution

$$pH = pK_a + \log \frac{[C_2H_3O_2^-]}{[HC_2H_3O_2]}$$

$$pH = 4.76 + \log \frac{0.0550}{0.0250}$$

$$pH = 4.76 + \log 2.20$$

$$= 5.10$$

From this example we can see that, in dilution, whatever is done to the concentration of one buffer component, the concentration of the other buffer component changes by the same factor and the pH is unchanged. There is a limit to which this relationship can be carried, however. If the buffer solution is diluted too much the assumption in the Henderson–Hasselbalch equation that most of the weak acid (or base) remains nonionized fails (see Exercise 28).

Buffer Capacity and Buffer Range. In Example 18-6 if we were to add more than 0.0750 mol OH^-, the $HC_2H_3O_2$ would be completely converted to $C_2H_3O_2^-$. The result would be a mixture of $NaC_2H_3O_2(aq)$ and $NaOH(aq)$. The solution would become strongly basic, and the buffering action would be destroyed. The buffer solution in Example 18-6 is somewhat more able to neutralize added acid since there is 0.165 mol $C_2H_3O_2^-$ available to react with it. However, if more than this amount of H_3O^+ is added, the buffering action is again destroyed and the solution becomes strongly acidic.

    Buffer capacity refers to the amount of acid or base that can be added to a buffer solution before its pH changes appreciably. In general, the maximum capacity to resist pH changes exists when the concentrations of weak acid (or weak base) and its salt are kept *large* and *approximately equal to one another*. A buffer has its maximum capacity at pH = pK_a (or pOH = pK_b). Whenever the ratio of salt to weak electrolyte concentration is either less than about 0.10 or greater than about 10, the buffer begins to lose its effectiveness. Since log 0.10 = -1 and log 10 = $+1$, this means that the effective buffer range is about one pH unit on either side of the value of pK. For acetic acid–sodium acetate buffers the effective range is from about pH 3.8 to 5.8; for ammonia–ammonium chloride, about pH 8.2 to 10.2.

The Importance of pH Maintenance in Blood. Maintenance of the proper pH in blood and in intracellular fluids is absolutely crucial to the processes that occur in living organisms. This is primarily because the functioning of enzymes—the catalysts for these processes—is sharply pH-dependent. The normal pH value of blood plasma is 7.4. Severe illness or death can result from sustained variations of a few tenths of a pH unit.

    Among the factors that can lead to a condition of acidosis, in which there is a decrease in the pH of blood, are heart failure, kidney failure, diabetes mellitus, persistent diarrhea, or a long-term high-protein diet. A temporary condition of acidosis may result from prolonged, intensive exercise.

    Alkalosis, characterized by an increase in the pH of blood, may occur as a result of severe vomiting, hyperventilation (overbreathing, sometimes caused by anxiety or hysteria), or exposure to high altitudes (altitude sickness). In studies performed on blood samples collected by climbers who reached the summit of Mount Everest (8848 m = 29,028 ft) without supplemental oxygen, the pH of arterial blood was found to be between 7.7 and 7.8.* Extreme hyperventilation is required to compensate for the very low partial pressures of O_2 (about 43 mmHg) at this altitude.

Blood as a Buffered Solution. Here is a good measure of the buffer capacity of human blood. The addition of 0.01 mol HCl to one liter of blood lowers the pH only from 7.4 to 7.2. The same amount of HCl added to a saline (NaCl) solution isotonic with blood lowers the pH from 7.0 to 2.0. The saline solution has no buffer capacity.

    Several factors are involved in the control of the pH of blood. A particularly important one is the ratio of dissolved HCO_3^- (bicarbonate ion) to H_2CO_3 (carbonic acid). $CO_2(g)$ is moderately soluble in water, and in aqueous solution reacts only to a limited extent to produce H_2CO_3. Nevertheless, in using K_{a_1} we treat the dissolved CO_2 as if it were completely converted to H_2CO_3. Moreover, although H_2CO_3 is a weak diprotic acid, in the carbonic acid–bicarbonate ion buffer system we deal only with the first ionization step: H_2CO_3 is the weak acid and HCO_3^- is the conjugate base (salt).

$$CO_2(aq) + H_2O \rightleftharpoons H_2CO_3(aq)$$

$$H_2CO_3 + H_2O \rightleftharpoons H_3O^+ + HCO_3^- \qquad K_{a_1} = 4.2 \times 10^{-7}$$

*J. B. West, "Human Physiology at Extreme Altitudes on Mount Everest," *Science,* **223,** 784 (1984).

$$HCO_3^- + H_2O \rightleftharpoons H_3O^+ + CO_3^{2-} \qquad K_{a_2} = 5.6 \times 10^{-11}$$

Carbon dioxide enters the blood from tissues as the by-product of metabolic reactions. In the lungs $CO_2(g)$ is exchanged for $O_2(g)$, which is transported throughout the body by the blood.

With an equation similar to equation (18.13), a value of $pK_{a_1} = -\log(4.2 \times 10^{-7}) = 6.4$, and a normal blood pH of 7.4, we can write

$$pH = pK_{a_1} + \log \frac{[HCO_3^-]}{[H_2CO_3]} = 6.4 + \log \left(\frac{10}{1}\right) = 6.4 + 1.0 = 7.4 \qquad (18.16)$$

The large ratio of $[HCO_3^-]$ to $[H_2CO_3]$ (10:1) seems to place this buffer outside the range of its maximum buffer capacity. The situation is rather complex, but some of the factors involved are

1. The need to neutralize excess acid (lactic acid produced by exercise) is generally greater than the need to neutralize excess base. The high proportion of HCO_3^- helps in this regard.
2. If additional H_2CO_3 is needed to neutralize excess alkalinity, $CO_2(g)$ in the lungs can be reabsorbed to build up the H_2CO_3 content of the blood.
3. Other components, such as the phosphate buffer, $H_2PO_4^-/HPO_4^{2-}$, and some plasma proteins, contribute to maintaining the pH of blood at 7.4.

Additional Applications of Buffer Solutions. Buffer solutions have other important applications, albeit less dramatic than maintaining the pH of blood. Certain chemical reactions consume, produce, or are catalyzed by H_3O^+ (recall Figure 15-15). To study the kinetics of these reactions, or simply to control their reaction rates, requires that the pH be controlled. This control can be achieved by conducting reactions in buffered solutions. Enzyme reactions are particularly sensitive to pH changes, and protein studies must often be performed in buffered media because the magnitude and kind of electrical charge carried by the protein molecules depend on the pH (see Section 28-4). The solubilities of many substances (e.g., hydroxides, carbonates, and sulfides) depend on $[H_3O^+]$. We consider the importance of buffer solutions in controlling the pH of solubility/precipitation processes in the next chapter.

18-3 Acid–Base Indicators

From time to time in this and the preceding chapter we have referred to measuring the pH of a solution with an acid–base (pH) indicator. Let us now briefly describe how an acid–base indicator works.

An **acid–base indicator** is a *weak acid* for which the nonionized acid (HIn) has one color [color(1)], and the conjugate base, the anion (In$^-$), has a different color [color(2)]. When a small amount of the indicator is added to a solution, depending on whether the ionization equilibrium of the indicator is displaced toward the acid or anion form, the solution acquires either color(1) or color(2). The direction of displacement of equilibrium in reaction (18.17) depends on the concentration of H_3O^+, and therefore, on the pH.

According to Le Châtelier's principle, increasing $[H_3O^+]$ displaces reaction (18.17) *to the left* [color(1)]. Increasing $[OH^-]$ (thereby decreasing $[H_3O^+]$) displaces the reaction *to the right* [color(2)].

$$\underset{\text{color(1)}}{HIn} + H_2O \rightleftharpoons H_3O^+ + \underset{\text{color(2)}}{In^-} \qquad (18.17)$$

We can write an equation relating pH, pK_a, and concentrations of indicator molecules and anions in exactly the same way we formulated buffer equations [that is, the Henderson–Hasselbalch equation (18.13)].

Relating pH and pK_a of an acid–base indicator.

$$pH = pK_a + \log \frac{[In^-]}{[HIn]} \qquad (18.18)$$

In general, if 90% or more of an indicator is in the form HIn, the solution will take on the acid color, that is, color(1). If 90% or more is in the form of the indicator anion, a base, the solution takes on the base color, that is, color(2). If the concentrations of HIn and In$^-$ are about equal, the indicator is in the process of changing from its acid to its base form and has a color intermediate to color(1) and color(2). The complete change from color(1) to color(2) takes place over a range of about *2 pH units,* with pH = pK_a at about the middle of the range. The colors and pH ranges of several acid–base indicators are shown in Figure 18-6. To summarize,

Acid color	Intermediate color	Base color
$[In^-]/[HIn] < 0.10$	$[In^-]/[HIn] \approx 1$	$[In^-]/[HIn] > 10$
$pH < pK_a + \log 0.10$	$pH \approx pK_a + \log 1$	$pH > pK_a + \log 10$
$pH < pK_a - 1$	$pH \approx pK_a$	$pH > pK_a + 1$

Example: bromothymol blue, pK_a = 7.1

pH < 6.1 (yellow)	pH ≈ 7.1 (green)	pH > 8.1 (blue)

Bromothymol blue (pK_a = 7.1) was chosen for Figure 18-4 because we needed an indicator whose color change comes at the neutral point—pH = 7.0. Most indicators do *not* change color in neutral solutions, however. In Figure 18-2 we chose bromophenol blue (pK_a = 3.85) because we needed a color change between pH 3 and 5, and in Figure 18-3 we chose thymolphthalein (pK_a = 10.0), whose color change occurs in a basic solution.

Thymol blue, used in Figures 18-1 and 18-5, is a weak *diprotic* acid and undergoes color changes in *two* pH ranges. One range is from pH 1.2 to 2.8, where the color changes from red to yellow. The other range is from pH 8.0 to 9.6, where the color changes from yellow to blue. Ionization equilibria for thymol blue can be represented as

$$H_2In + H_2O \rightleftharpoons H_3O^+ + HIn^-$$
color(1)   color(2)

$$HIn^- + H_2O \rightleftharpoons H_3O^+ + In^{2-}$$
color(2)   color(3)

An acid–base indicator is usually prepared as a solution in water, ethanol, or some other solvent. In acid–base titrations a few drops of the indicator solution are added to the solution being titrated. In another form, porous paper is impregnated with an indicator solution and dried. When this paper is moistened with the solution being tested, it acquires a color determined by the pH of the solution. This paper is usually called pH paper.

Applications. Acid–base indicators find their greatest application where a precise determination of pH is not necessary. For example, they are used in soil-testing kits to establish the pH of soils. Soils are usually acidic in regions of high rainfall and heavy vegetation and alkaline in more arid regions, but the pH can vary considerably with local conditions. If a soil is found to be too acidic for a certain crop, its pH can be raised by adding lime (CaO). To reduce the pH of a soil, gypsum (CaSO$_4 \cdot$ 2H$_2$O) or organic matter may be added. In swimming pools, a particular pH is required for the most effective action of chlorinating agents, to prevent the growth of algae, to avoid corrosion of the pool plumbing, etc. The preferred pH is about 7.4, and phenol red is a common indicator used in testing swimming pool

FIGURE 18-6

pH and color changes for some common acid–base indicators.

The indicators pictured and the pH values at which they change color are **(a)** methyl violet (pH 0 to 2); **(b)** phenol red (pH 6 to 8); **(c)** thymol blue (pH 8 to 10). [Photographs by Carey B. Van Loon]

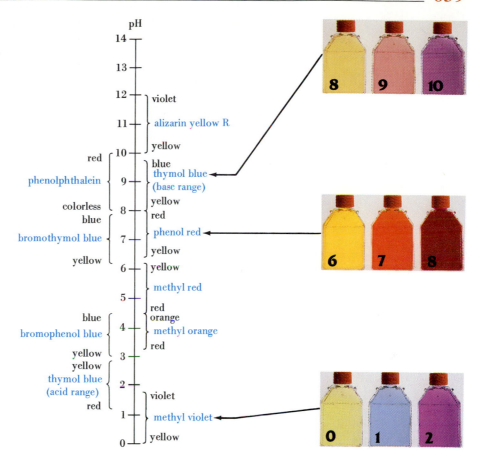

water. If chlorination is carried out with $Cl_2(g)$, the pool water becomes acidic because of the reaction of Cl_2 with H_2O (see Table 9-8). In this case basic substances, such as lime or sodium carbonate, are used to raise the pH. If sodium hypochlorite, $NaOCl(aq)$, is used as the chlorinating agent, the pool water becomes basic and the pH is adjusted by adding an acid (e.g., HCl or H_2SO_4).

18-4 Neutralization Reactions and Titration Curves

Neutralization means the reaction of an acid and a base to form a salt and water, and we have seen that the fundamental reaction that occurs during neutralization is

$$H_3O^+ + OH^- \longrightarrow 2\,H_2O$$

An important condition in a neutralization reaction is one in which *both* acid and base are consumed, that is, in which *neither* is in excess. This condition is called the **equivalence point** of the neutralization. To locate the equivalence point in a neutralization we must exercise very careful control over the addition of base to acid (or acid to base). This, as we first learned in Section 5-3, we can do through a procedure known as **titration.** We can now solve the problem of how to locate the equivalence point during a titration.

In a titration, one of the solutions to be neutralized, say the acid, is placed in a flask or beaker, together with a few drops of an acid–base indicator. The other solution, the base, is contained in a buret and is added to the acid, first rapidly and then dropwise, up to the equivalence point (recall Figure 5-6). We locate the equivalence point through the color change of the acid–base indicator. The point in a

*In a titration, the solution added from the buret is known as the **titrant.***

titration where the indicator changes color is called the **end point** of the indicator. What we need to do is to match the indicator end point with the equivalence point of the neutralization. We can do this if we choose an indicator whose color change occurs over a pH range that includes the pH of the equivalence point.

Our emphasis in the discussion that follows will be on calculating the pH at various points in a titration and representing this variation in pH graphically. We will call such a graph a **titration curve.** And one of the points of special interest on a titration curve is the equivalence point.

The Millimole. In a typical titration the volume of solution delivered from a buret is less than 50 mL, and the molarity of the solution used for the titration is generally less than 1 M. This means that the typical amount of OH^- (or H_3O^+) delivered from the buret during a titration is generally only a few thousandths of a mole, for example 5.00×10^{-3} mol. In performing the calculations necessary to construct a titration curve, we will find it easier to express 5.00×10^{-3} mol as *5.00 mmol*. The symbol **mmol** stands for a **millimole,** that is, one thousandth of a mole or 10^{-3} mol.

When we express amounts of a solute in millimoles we can also redefine molarity concentration in this way.

$$\text{molarity} = \frac{\text{no. mol solute}}{\text{no. L soln}} = \frac{\text{no. mol solute}/1000}{\text{no. L soln }/1000}$$

$$= \frac{\text{no. mmol solute}}{\text{no. mL soln}}$$

(18.19)

Molarity concentration and millimoles of solute.

Thus, just as the product, molarity $\times$ volume (L), yields the number of moles of a solute, the product, molarity $\times$ volume *(mL)*, yields the number of *millimoles* of solute.

Titration of a Strong Acid by a Strong Base. Suppose we place 25.00 mL 0.1000 M HCl (a *strong* acid) in a small flask or beaker and then add to it from a buret 0.1000 M NaOH (a *strong* base). We can readily calculate the pH of the accumulated solution at different points in this titration, and we can plot these pH values as a function of the volume of NaOH added. It is this graph that we call a titration curve. As we shall see, from this curve we can establish the pH at the equivalence point and this will allow us to select an appropriate indicator for the titration. Some typical calculations are outlined in Example 18-7.

Example 18-7 ───────────────────────

Calculating a titration curve: Strong acid titrated by a strong base. What is the pH at each of the following points in the titration of 25.00 mL 0.1000 M HCl by 0.1000 M NaOH?
(a) Before the addition of any NaOH (*initial pH*).
(b) After the addition of 24.00 mL of 0.1000 M NaOH (*before the equivalence point*).
(c) After the addition of 25.00 mL of 0.1000 M NaOH (*at the equivalence point*).
(d) After the addition of 26.00 mL of 0.1000 M NaOH (*beyond the equivalence point*).

Solution
(a) Before any NaOH is added we are dealing with 0.1000 M HCl. This solution has $[H_3O^+] = 0.1000$ M and pH $= 1.00$.
(b) The total number of millimoles of H_3O^+ to be titrated is

$$\text{no. mmol } H_3O^+ = 25.00 \text{ mL} \times \frac{0.1000 \text{ mmol}}{\text{mL}} = 2.500 \text{ mmol } H_3O^+$$

The number of mmol of OH^- present in 24.00 mL 0.1000 M NaOH is

$$\text{no. mmol } OH^- = 24.00 \text{ mL} \times \frac{0.1000 \text{ mmol}}{\text{mL}} = 2.400 \text{ mmol } OH^-$$

We can represent the neutralization reaction in the following format.

$$H_3O^+ \quad + \quad OH^- \quad \longrightarrow \quad 2 H_2O$$

initially present:	2.500 mmol	—
add:		2.400 mmol
changes:	−2.400 mmol	−2.400 mmol
after reaction:	0.100 mmol	~0

The 0.100 mmol of H_3O^+ is present in a solution volume of 49.00 mL (25.00 mL original acid + 24.00 mL added base).

$$[H_3O^+] = \frac{0.100 \text{ mmol } H_3O^+}{49.00 \text{ mL}} = 2.04 \times 10^{-3} \text{ M}$$

$$pH = -\log [H_3O^+] = -\log(2.04 \times 10^{-3}) = 2.69$$

(c) The equivalence point is the point at which the HCl is completely neutralized, with no excess of NaOH present. What we have at the equivalence point is simply NaCl(aq). And, as we learned in Section 17-8, since neither Na^+ nor Cl^- hydrolyzes in water, pH = 7.00.

(d) To determine the pH of the solution beyond the equivalence point we can return to the format we used in (b), except that now OH^- is present in excess. The no. mmol OH^- added is 26.00 mL × 0.1000 mmol/mL = 2.600 mmol.

$$H_3O^+ \quad + \quad OH^- \quad \longrightarrow \quad 2 H_2O$$

initially present:	2.500 mmol	—
add:		2.600 mmol
changes:	−2.500 mmol	−2.500 mmol
after reaction:	~0	0.100 mmol

The 0.100 mmol of excess NaOH is present in a solution volume of 51.00 mL (25.00 mL original acid + 26.00 mL added base). The concentration of OH^- in this solution is

$$[OH^-] = \frac{0.100 \text{ mmol}}{51.00 \text{ mL}} = 1.96 \times 10^{-3} \text{ M}$$

$$pOH = -\log(1.96 \times 10^{-3}) = 2.71 \qquad pH = 14.00 - 2.71 = 11.29$$

SIMILAR EXAMPLES: Exercises 10a, 11, 38a, 41.

Titration data and the titration curve for HCl titrated by NaOH are presented in Figure 18-7. These, then, are the principal features of the titration curve of a *strong acid by a strong base*.

Features of the titration curve: strong acid–strong base.

- The pH has a low value at the beginning of the titration.
- The pH changes slowly until just before the equivalence point.
- At the equivalence point the pH rises very sharply, perhaps by 6 units for an addition of only 0.10 mL (2 drops) of base.
- Beyond the equivalence point the pH again rises only slowly.
- Any indicator whose color changes in the range from about pH = 4 to pH = 10 is suitable for this titration.

(18.20)

FIGURE 18-7

Titration curve for a strong acid by a strong base—25.00 mL of 0.1000 M HCl by 0.1000 M NaOH.

Titration Data

mL NaOH(aq)	pH
0.00	1.00
10.00	1.37
20.00	1.95
22.00	2.19
24.00	2.69
25.00	7.00
26.00	11.29
28.00	11.75
30.00	11.96
40.00	12.36
50.00	12.52

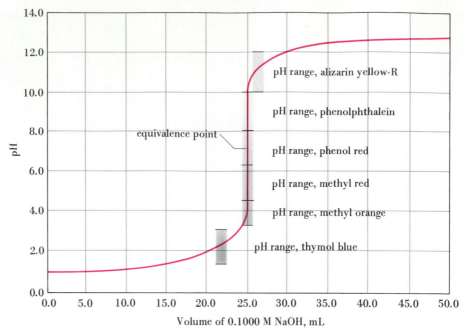

The indicators whose color-change ranges fall along the steep portion of the titration curve are all suitable for this titration. Thymol blue changes color too soon (that is, before 25.00 mL of base has been added); alizarin yellow-R, too late.

Titration of a Weak Acid by a Strong Base. There are several important differences between the titration of a weak acid and that of a strong acid. For one, initially most of the weak acid is present as nonionized molecules, HA, rather than as H_3O^+ and A^-. Another difference is in the pH of the solution at the equivalence point. But there is one important feature that is *unchanged* when comparing the titration of a weak acid with that of a strong acid. *For equal volumes of acid solutions of the same molarity, the volume of base required to titrate to the equivalence point is independent of whether the acid is strong or weak.* This is because in either case the amount of acid to be neutralized is the same, whether the acid is present initially as H_3O^+ (strong acid) or as nonionized molecules (weak acid). We can think of the neutralization of a weak acid as involving the direct transfer of protons from HA molecules to OH^-. Thus, we can represent the neutralization of $HC_2H_3O_2$ by NaOH through the net equation

$$HC_2H_3O_2 + OH^- \longrightarrow H_2O + C_2H_3O_2^-$$

In Example 18-8 and Figure 18-8 we consider the titration of 25.00 mL of 0.1000 M $HC_2H_3O_2$ by 0.1000 M NaOH.

Example 18-8

Calculating a titration curve: Weak acid titrated by a strong base. What is the pH at each of the following points in the titration of 25.00 mL 0.1000 M $HC_2H_3O_2$ by 0.1000 M NaOH?

(a) Before the addition of any NaOH (*initial pH*).

(b) After the addition of 10.00 mL of 0.1000 M NaOH (*before the equivalence point*).

(c) After the addition of 12.50 mL of 0.1000 M NaOH (*half-neutralization point*).

(d) After the addition of 25.00 mL of 0.1000 M NaOH (*at the equivalence point*).

(e) After the addition of 26.00 mL of 0.100 M NaOH (*beyond the equivalence point*).

FIGURE 18-8
Titration curve for a weak acid by a strong base— 25.00 mL of 0.1000 M $HC_2H_3O_2$ by 0.1000 M NaOH.

Titration Data

mL NaOH(aq)	pH
0.00	2.88
5.00	4.16
10.00	4.58
12.50	4.76
15.00	4.94
20.00	5.36
24.00	6.14
25.00	8.72
26.00	11.29
30.00	11.96
40.00	12.36
50.00	12.52

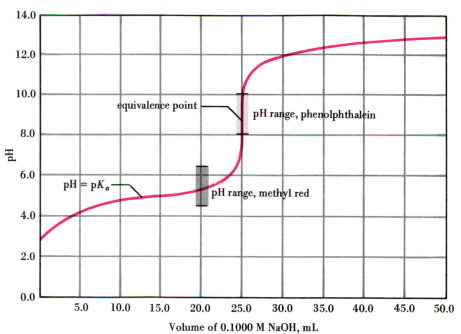

Phenolphthalein is a suitable indicator for this titration, but methyl red is not. When exactly one-half of the acid is neutralized. $[HC_2H_3O_2] = [C_2H_3O_2^-]$ and $pH = pK_a = 4.76$.

Solution

(a) The initial pH is obtained by the calculation in Example 18-1(a).

$$pH = -\log(1.32 \times 10^{-3}) = 2.88$$

(b) The total number of mmol $HC_2H_3O_2$ to be neutralized is

$$\text{no. mmol } HC_2H_3O_2 = 25.00 \text{ mL} \times \frac{0.1000 \text{ mmol } HC_2H_3O_2}{\text{mL}}$$

$$= 2.500 \text{ mmol } HC_2H_3O_2$$

At this point in the titration the amount of OH^- added is

$$\text{no. mmol } OH^- = 10.00 \text{ mL} \times \frac{0.1000 \text{ mmol } OH^-}{\text{mL}} = 1.000 \text{ mmol } OH^-$$

The total solution volume = 25.00 mL acid + 10.00 mL base = 35.00 mL. Information from these preliminary *stoichiometric* calculations is entered at appropriate points into the set up below.

	$HC_2H_3O_2$	$+$	OH^-	$\longrightarrow$	$C_2H_3O_2^-$	$+ H_2O$
initially present:	2.500 mmol		—		—	
add:			1.000 mmol			
changes:	−1.000 mmol		−1.000 mmol		+1.000 mmol	
after reaction:						
mmol	1.500 mmol				1.000 mmol	
concns	1.500 mmol/35.00 mL		~0		1.000 mmol/35.00 mL	
	0.0429 M				0.0286 M	

The concentrations of the weak acid and its salt can be substituted into the "buffer" equation (18.13).

$$pH = pK_a + \log \frac{[C_2H_3O_2^-]}{[HC_2H_3O_2]} = 4.76 + \log \frac{0.0286}{0.0429} = 4.76 - 0.18 = 4.58$$

(c) When we have added 12.50 mL of 0.1000 M NaOH we have added $12.50 \times 0.1000 = 1.250$ mmol OH^-. As we see from the setup below this is enough base to neutralize exactly *one-half* of the acid. In the resulting solution $[HC_2H_3O_2] = [C_2H_3O_2^-]$ and $pH = pK_a$.

	$HC_2H_3O_2$	$+$	OH^-	$\longrightarrow$	$C_2H_3O_2^-$	$+$	H_2O
initially present:	2.500 mmol		—		—		
add:			1.250 mmol				
changes:	−1.250 mmol		−1.250 mmol		+1.250 mmol		
after reaction:							
mmol	1.250 mmol				1.250 mmol		
concns	1.250 mmol/37.50 mL		~0		1.250 mmol/37.50 mL		
	0.0333 M				0.0333 M		

$$pH = pK_a + \log \frac{[C_2H_3O_2^-]}{[HC_2H_3O_2]} = 4.76 + \log \frac{0.0333}{0.0333} = 4.76 + \log 1 = 4.76$$

(d) At the equivalence point neutralization is complete and 2.500 mmol $NaC_2H_3O_2$ has been produced in 50.00 mL of solution, leading to a molarity of 0.05000 M $NaC_2H_3O_2$. The question becomes, "What is the pH of 0.05000 M $NaC_2H_3O_2$?" To answer this question we need to recognize that $C_2H_3O_2^-$ hydrolyzes (and Na^+ does not). We can use the method of Example 17-13 for the hydrolysis calculation.

	$C_2H_3O_2^-$	$+$	H_2O	$\rightleftharpoons$	$HC_2H_3O_2$	$+$	OH^-
placed in solution:	0.05000 M				—		—
changes	−x M				+x M		+x M
equilibrium concns:	(0.05000 − x) M				x M		x M

In the equilibrium calculation below, assume that $(0.05000 - x) \approx 0.05000$. Also, $K_b(C_2H_3O_2^-) = K_w/K_a(HC_2H_3O_2) = 1.0 \times 10^{-14}/1.74 \times 10^{-5} = 5.7 \times 10^{-10}$.

$$K_b = \frac{[HC_2H_3O_2][OH^-]}{[C_2H_3O_2^-]} = \frac{x \cdot x}{0.05000} = 5.7 \times 10^{-10} \qquad x^2 = 2.8 \times 10^{-11}$$

$$x = [OH^-] = 5.3 \times 10^{-6} \text{ M} \qquad pOH = -\log(5.3 \times 10^{-6}) = 5.28$$

$$pH = 14.00 - pOH = 14.00 - 5.28 = 8.72$$

(e) We can set this calculation up in the same format as in (b). Here the amount of OH^- added is 26.00 mL × 0.1000 mmol/mL = 2.600 mmol. The volume of solution is 25.00 mL acid + 26.00 mL base = 51.00 mL.

	$HC_2H_3O_2$	$+$	OH^-	$\longrightarrow$	$C_2H_3O_2^-$	$+$	H_2O
initially present:	2.500 mmol		—		—		
add:			2.600 mmol				
changes:	−2.500 mmol		−2.500 mmol		+2.500 mmol		
after reaction:	~0		0.100 mmol		2.500 mmol		

The solution beyond the equivalence point is that of a *strong base*, and its pOH and pH are easily calculated.

$$[OH^-] = 0.100 \text{ mmol } OH^-/51.00 \text{ mL} = 1.96 \times 10^{-3} \text{ M}$$

$$pOH = -\log(1.96 \times 10^{-3}) = 2.71 \qquad pH = 14.00 - 2.71 = 11.29$$

SIMILAR EXAMPLES: Exercises 10b, c, 12, 38b, 40.

Here, then, are the principal features of the titration curve for a weak acid titrated by a strong base.

Features of the titration curve: weak acid–strong base.

1. The initial pH is higher than in the titration of a strong acid. (The weak acid is only partially ionized).   (18.21)
2. There is a rather sharp increase in pH at the start of the titration.

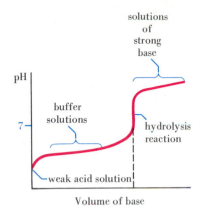

solutions
of
strong
base

pH

buffer
solutions

7—

hydrolysis
reaction

weak acid solution

Volume of base

FIGURE 18-9

Constructing the titration curve for a weak acid by a strong base.

The necessary calculations (also illustrated in Example 18-8) can be divided into *four* groups.

- pH of a pure weak acid *(initial pH).*
- pH of a buffer solution of a weak acid and its salt *(over a broad range before the equivalence point).*
- pH of a salt solution undergoing hydrolysis *(equivalence point).*
- pH of a solution of a strong base *(over a broad range beyond the equivalence point).*

3. Over a long section of the curve preceding the equivalence point the pH changes quite gradually. (Solutions for this portion of the curve are buffer solutions.)

4. At the point of half-neutralization $pH = pK_a$. (At the half neutralization point, $[HA] = [A^-]$.)

5. The pH at the equivalence point is greater than 7. (A salt of a weak acid and a strong base hydrolyzes.)

6. Beyond the equivalence point the titration curve is identical to that of a strong acid by a strong base. (In this portion of the titration the pH is established by the concentration of unreacted OH^-.)

7. The steep portion of the titration curve at the equivalence point is over a relatively short pH range (e.g., from about pH 7 to 10).

8. The selection of indicators available for the titration is more limited than in a strong acid–strong base titration.

(18.21)

Calculations like these can also be used to construct the titration curve for a weak base titrated by a strong acid (see Exercise 40).

As illustrated in Example 18-8 and suggested by Figure 18-9, the necessary calculations for a weak acid–strong base titration curve are of four distinct types, depending on the portion of the titration curve being described.

Titration of a Weak Polyprotic Acid. We can find the most striking evidence that a polyprotic acid ionizes in distinctive steps through its titration curve. In the neutralization of phosphoric acid by sodium hydroxide, essentially all the H_3PO_4 molecules are first converted to the salt, NaH_2PO_4. Then NaH_2PO_4 is converted to Na_2HPO_4. Finally Na_2HPO_4 is converted to Na_3PO_4.

This stepwise neutralization of a polyprotic acid will be observed only if successive ionization constants (K_{a_1}, K_{a_2}, . . .) differ by a factor of 10^3 or more. Otherwise, neutralization of the second H atom begins before neutralization of the first is completed, and so on.

$$H_3PO_4 + OH^- \longrightarrow H_2PO_4^- + H_2O \qquad (18.22)$$

$$is\ followed\ by \quad H_2PO_4^- + OH^- \longrightarrow HPO_4^{2-} + H_2O \qquad (18.23)$$

$$which\ is\ followed\ by \quad HPO_4^{2-} + OH^- \longrightarrow PO_4^{3-} + H_2O \qquad (18.24)$$

Corresponding to these three distinctive stages are three equivalence points. For every mole of H_3PO_4, 1 mol NaOH is required to reach the first equivalence point. At this first equivalence point, the solution is essentially one of NaH_2PO_4. This is an *acidic* solution because the acid ionization constant of $H_2PO_4^-$, K_{a_2},

$$H_2PO_4^- + H_2O \rightleftharpoons H_3O^+ + HPO_4^{2-} \qquad K_{a_2} = 6.34 \times 10^{-8} \qquad (18.25)$$

is larger than its base ionization (hydrolysis) constant, K_b,

$$H_2PO_4^- + H_2O \rightleftharpoons H_3PO_4 + OH^- \qquad K_b = 1.41 \times 10^{-12}$$

An additional mole of NaOH is required to titrate the acid to its second equivalence point. At this second equivalence point, the solution is *basic* because the base ionization (hydrolysis) constant, K_b,

$$HPO_4^{2-} + H_2O \rightleftharpoons H_2PO_4^- + OH^- \qquad K_b = 1.58 \times 10^{-7} \qquad (18.26)$$

is larger than the acid ionization constant, K_{a_3},

FIGURE 18-10
Titration of a weak poly-protic acid—10.00 mL of 0.1000 M H_3PO_4 by 0.1000 M NaOH.

Exactly 10.00 mL of 0.1000 M NaOH is required to reach the first equivalence point. During this portion of the titration only reaction (18.22) occurs. The additional volume of 0.1000 M NaOH required to reach the second equivalence point is also 10.00 mL. During this portion of the titration only reaction (18.23) occurs.

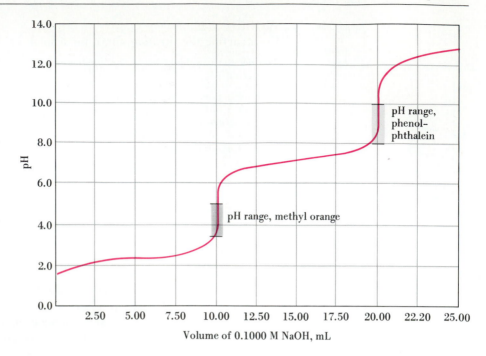

$$HPO_4^{2-} + H_2O \rightleftharpoons H_3O^+ + PO_4^{3-} \qquad K_{a_3} = 4.22 \times 10^{-13}$$

A titration curve for the titration of 10.00 mL of 0.1000 M H_3PO_4 by 0.1000 M NaOH is pictured in Figure 18-10. Notice that the two equivalence points come at equal intervals on the volume axis (at 10.00 mL and 20.00 mL).

18-5   Solutions of Salts of Polyprotic Acids

In discussing the neutralization of phosphoric acid by a strong base, we made qualitative predictions of the pH at each of the three equivalence points. Of these three equivalence points, the easiest pH value to calculate is that for $Na_3PO_4(aq)$.
    PO_4^{3-} cannot ionize further as an acid, but it can ionize as a base.

$$PO_4^{3-} + H_2O \rightleftharpoons HPO_4^{2-} + OH^- \qquad K_b = \frac{K_w}{K_{a_3}} = \frac{1.0 \times 10^{-14}}{4.2 \times 10^{-13}} = 2.4 \times 10^{-2}$$

Are You Wondering:

Why we have not shown a third equivalence point in Figure 18-10?

We cannot reach the third equivalence point in the titration of 0.1000 M H_3PO_4 with 0.1000 M NaOH, even though there are three ionizable H atoms in H_3PO_4. The pH of the strongly hydrolyzed Na_3PO_4 solution at the third equivalence point—approaching pH 13—is higher than the pH we can reach by adding 0.1000 M NaOH to water. $Na_3PO_4(aq)$ is nearly as basic as the NaOH(aq) used in the titration.

Example 18-9

Determining the pH of a solution when the anion (A^{n-}) of a polyprotic acid can act only as a base. Trisodium phosphate (TSP), Na_3PO_4, is a common ingredient of preparations used to clean painted walls before they are re-painted. The cleaner is an alkaline material that removes a surface layer of oxidized paint and solubilizes grease. What is the pH of 1.0 M $Na_3PO_4(aq)$?

Solution. In the usual fashion we can write

$$PO_4^{3-} + H_2O \rightleftharpoons HPO_4^{2-} + OH^- \qquad K_b = 2.4 \times 10^{-2}$$

placed in solution:	1.0 M	—	—
changes:	$-x$ M	$+x$ M	$+x$ M
at equilibrium:	$(1.0 - x)$ M	x M	x M

$$K_b = \frac{[HPO_4^{2-}][OH^-]}{[PO_4^{3-}]} = \frac{x \cdot x}{1.0 - x} = 2.4 \times 10^{-2}$$

Because K_b is relatively large, the usual assumption that x is much smaller than 1.0 is not likely to work well here. Solution by the quadratic formula leads to $x = [OH^-] = 0.14$ M.

$$pOH = -\log [OH^-] = -\log 0.14 = +0.85$$

$$pH = 14.00 - 0.85 = 13.15$$

SIMILAR EXAMPLES: Exercises 47, 48.

The result of Example 18-9 helps us to understand why we were unable to show the third equivalence point in Figure 18-10. We cannot use 0.100 M NaOH to titrate a solution to a pH of 13 or higher.

$NaH_2PO_4(aq)$ and $Na_2HPO_4(aq)$ are the solutions present at the first and second equivalence points, respectively, in the titration of H_3PO_4. It is more difficult to calculate the *exact* pH values of these solutions than of $Na_3PO_4(aq)$. This is because with both $H_2PO_4^-$ and HPO_4^{2-} we have to consider two equilibria *simultaneously*—ionization as an acid and ionization as a base (hydrolysis). These calculations are left as an exercise for interested students (Exercise 79), but the results are of suffi-cient importance for us to state them without proof. For solutions that are reasona-bly concentrated (say 0.10 M or greater), the pH values are *independent* of the solution concentration and are given by the following expressions. (pK_a values are from Table 17-3.)

Note how the pH values listed here correspond to the first and second equivalence points in Figure 18-10.

For NaH_2PO_4: $\quad pH = \frac{1}{2}(pK_{a_1} + pK_{a_2}) = \frac{1}{2}(2.15 + 7.20) = 4.68 \qquad (18.27)$

For Na_2HPO_4: $\quad pH = \frac{1}{2}(pK_{a_2} + pK_{a_3}) = \frac{1}{2}(7.20 + 12.37) = 9.78 \qquad (18.28)$

We can write expressions like (18.27) and (18.28) for salts of other polyprotic acids, as illustrated in Example 18-10.

Example 18-10

Determining the pH of a solution when the anion (HA^-) of a polyprotic acid can act either as an acid or a base. Sodium bicarbonate, $NaHCO_3$, produces mildly basic solutions, and this makes it useful as an antacid. More basic solu-tions, such as Na_2CO_3, NaOH, $Ca(OH)_2$, and NH_3, are very damaging to tis-sues. Show that $NaHCO_3(aq)$ solutions are only mildly basic.

Solution. The pH of a $NaHCO_3(aq)$ solution is governed by two reactions, the ionization of HCO_3^- as an acid, which is reflected through K_{a_2} of H_2CO_3, and its ionization as a base, which is reflected through $K_b = K_w/K_{a_1}$. It is for such cases that we can use a relationship of the type in (18.27) and (18.28). (pK_a values are from Table 17-3.)

$$pH = \frac{1}{2}[pK_{a_1}(H_2CO_3) + pK_{a_2}(H_2CO_3)] = \frac{1}{2}(6.38 + 10.25) = 8.32$$

SIMILAR EXAMPLES: Exercises 13, 67, 79.

18-6   Acid–Base Equilibrium Calculations: A Summary

In this and the preceding chapter we have considered a wide variety of equilibrium calculations. Here are some points to think about when you are faced with deciding what method to use to solve a particular problem.

1. What are the species potentially present in solution, and how significant are their concentrations likely to be?

Consider all the species that are potentially present in a solution containing phosphoric acid and/or a phosphate salt: H_3PO_4, $H_2PO_4^-$, HPO_4^{2-}, PO_4^{3-}, OH^-, H_3O^+, and possibly other cations. However, if the solution is described as $H_3PO_4(aq)$, the only species present in significant concentrations are those associated with the first ionization step: H_3PO_4, H_3O^+, and $H_2PO_4^-$. On the other hand, if the solution is described as $Na_3PO_4(aq)$ the significant species are Na^+ and the ions associated with the hydrolysis reaction: PO_4^{3-}, HPO_4^{2-}, and OH^-. In a solution of HCl and $HC_2H_3O_2$, the only significant *ionic* species are H_3O^+ and Cl^-. HCl is a completely ionized strong acid, and, in the presence of a strong acid, the weak acid $HC_2H_3O_2$ would be only very slightly ionized (common ion effect). In a mixture of two *weak* acids, such as $HC_2H_3O_2$ and $HC_3H_5O_2$, each acid ionizes and all of these species would be significant: $[HC_2H_3O_2]$, $[HC_3H_5O_2]$, $[C_2H_3O_2^-]$, $[C_3H_5O_2^-]$, and $[H_3O^+]$.

2. Are there reactions possible among any of the solution components, and what is their stoichiometry?

If you are asked to calculate $[OH^-]$ in a solution made 0.10 M NaOH and 0.20 M NH_4Cl, before answering that $[OH^-] = 0.10$ M stop to consider whether a solution can be simultaneously 0.10 M in OH^- and 0.20 M in NH_4^+. It cannot, because any solution containing both NH_4^+ and OH^- must also contain NH_3. What happens is that OH^- and NH_4^+ react in a 1:1 mole ratio to produce NH_3.

$$NH_4^+ + OH^- \longrightarrow NH_3 + H_2O$$

The question then becomes one of describing a *buffer* solution that is 0.10 M NH_3– 0.10 M NH_4^+.

3. What are the equilibrium expressions that must be obeyed? Which are the most significant?

One equilibrium expression that must be obeyed in all aqueous solutions is that $K_w = [H_3O^+][OH^-] = 1.0 \times 10^{-14}$. However, there are many calculations in which this expression will not be significant compared to others. One situation where it *will* be significant is if you are asked to calculate $[OH^-]$ in an *acidic* solution or $[H_3O^+]$ in a *basic* solution. After all, an acid does not produce OH^- and a base does not produce H_3O^+. Another situation where the ion product of water is likely to be important is in a calculation in which the pH is near 7.

If a question concerning the *triprotic* acid H_3PO_4 asks simply that you calculate $[H_3O^+]$ in $H_3PO_4(aq)$, only the *first* ionization step is important, since this is the step in which essentially all the H_3O^+ is produced. On the other hand, if you need to establish the pH of a solution containing the ions $H_2PO_4^-$ and HPO_4^{2-}, the important expression will be K_{a_2} of H_3PO_4. The *second* ionization step is the one in which both of these ions appear.

4. Can the situation be *readily* described as of a certain type?

If you recognize the given situation as corresponding to one of the important types discussed in these chapters, then you can use the method introduced for that type (sometimes with and sometimes without simplifying assumptions). If you cannot *readily* recognize the type of problem, then carefully consider the questions listed above. Some of the problem types that we have encountered in these two chapters are

- ionization of a strong acid (Example 17-2);
- ionization of a strong base (Example 17-3);
- relating $[H_3O^+]$, $[OH^-]$, pH, and pOH (Example 17-5);
- ionization of a weak monoprotic acid (Examples 17-6, 17-8);
- ionization of a weak monoprotic base (Example 17-7);
- ionization of a weak polyprotic acid (Example 17-10);
- hydrolysis of the salt of a weak monoprotic acid or base (Example 17-13);
- ionization of a weak acid (or base) in the presence of a strong acid or base— common ion effect, (Example 18-1);
- ionization of a weak acid (or base) in the presence of its salt—common ion effect, buffer solution (Examples 18-2, 18-4, 18-5);
- pH changes in a buffer solution (Example 18-6);
- titration involving strong acids and strong bases (Examples 17-4, 18-7);
- titration of weak acid (or base) by strong base (or acid) (Example 18-8);
- pH of solutions of salts of polyprotic acids (Examples 18-9; 18-10).

18-7 Equivalent Weight and Normality

Before the atomic theory was introduced, with its concept of atomic weights and, later, the mole, chemists described the combining ratios of chemical reactants through equivalent weights. Equivalent weight and the related concentration unit called normality (N) still find widespread practical use.

Equivalent Weight in Acid–Base Reactions. For an acid–base reaction, *an equivalent weight (equivalent) is a quantity of substance that will produce or react with one mole of H^+.* Since 1 mol HCl *produces* 1 mol H^+ when it ionizes, *1 mol HCl = 1 equiv. HCl.* Now consider NaOH. Since 1 mol NaOH produces 1 mol OH^-, which *reacts with* 1 mol H^+, *1 mol NaOH = 1 equiv. NaOH.* One mol H_2SO_4 produces 2 mol H^+ when it reacts with NaOH; *1 mol H_2SO_4 = 2 equiv. H_2SO_4.* Since 1 mol H_2SO_4 has a mass of 98.08 g, 1 equiv. H_2SO_4, which is $\frac{1}{2}$ mol, has a mass of 49.04 g. The equivalent weight of H_2SO_4 can be expressed as 49.04 g H_2SO_4/equiv. H_2SO_4.

The situation with phosphoric acid, H_3PO_4, is more complicated. If H_3PO_4 participates in reaction (18.29) its equivalent weight is equal to its molar mass. In reaction (18.30) the equivalent weight of H_3PO_4 is *one-half* the molar mass, and in reaction (18.31), *one-third* the molar mass. The essential principle of equivalency represented through these equations is that *a chemical reaction involves equal numbers of equivalents of all reactants.*

The titration of H_3PO_4 occurs in three distinct stages: H_3PO_4 has *three* equivalent weights. The titration of H_2SO_4 cannot be separated into two distinct stages: H_2SO_4 has a *single* equivalent weight (M.W./2) based on its complete neutralization.

$$H_3PO_4(aq) + NaOH(aq) \longrightarrow NaH_2PO_4(aq) + H_2O \qquad (18.29)$$
1 equiv.   1 equiv.

$$H_3PO_4(aq) + 2\,NaOH(aq) \longrightarrow Na_2HPO_4(aq) + 2\,H_2O \qquad (18.30)$$
2 equiv.   2 equiv.

$$H_3PO_4(aq) + 3\ NaOH(aq) \longrightarrow Na_3PO_4(aq) + 3\ H_2O \qquad (18.31)$$
3 equiv.   3 equiv.

Normality. Normality concentration is similar to molarity, but the quantity of solute is expressed in equivalents rather than moles.

$$\text{normality (N)} = \frac{\text{number of equiv. solute}}{\text{number liters soln}} \qquad (18.32)$$

To prepare a *1 normal* solution of HCl we would dissolve 36.5 g HCl (1 mole = 1 equiv.) in 1 L of water solution. This solution would also be *1 molar* in HCl.

$$1.00\ \text{M HCl} = \frac{1\ \text{mol HCl}}{1.00\ \text{L soln}} = \frac{36.5\ \text{g HCl}}{1.00\ \text{L soln}} = \frac{1\ \text{equiv. HCl}}{1.00\ \text{L soln}} = 1.00\ \text{N HCl}$$

Normality and molarity are *not* equal for $Ba(OH)_2(aq)$, however.

$$0.010\ \text{M Ba(OH)}_2 = \frac{0.010\ \text{mol Ba(OH)}_2}{1.00\ \text{L soln}} = \frac{0.020\ \text{equiv. Ba(OH)}_2}{1.00\ \text{L soln}}$$

$$= 0.020\ \text{N Ba(OH)}_2$$

The relationship between normality and molarity concentration is

$$\text{normality (N)} = n \times \text{molarity (M)} \qquad (18.33)$$

where n represents the number of moles of H^+ per mole of compound that a solute is capable of releasing (acid) or reacting with (base).

Example 18-11 _____

Determining equivalent weight and normality concentration. **What mass of H_2SO_4 is required to produce 250.0 mL of 0.107 N H_2SO_4?**

Solution. In a similar fashion to what we would do if working with solution molarity, we can write that

$$\text{no. equiv. } H_2SO_4 = 0.2500\ \text{L} \times \frac{0.107\ \text{equiv. } H_2SO_4}{\text{L}} = 0.0268\ \text{equiv. } H_2SO_4$$

As we have previously seen, 1 mole H_2SO_4 = 2 equiv. H_2SO_4. This means that the equivalent weight of H_2SO_4 is $\frac{1}{2}$ the molar mass; that is, 49.04 g H_2SO_4/equiv. H_2SO_4. The required mass of H_2SO_4 is

$$\text{no. g } H_2SO_4 = 0.0268\ \text{equiv. } H_2SO_4 \times \frac{49.04\ \text{g } H_2SO_4}{1\ \text{equiv. } H_2SO_4} = 1.31\ \text{g } H_2SO_4$$

SIMILAR EXAMPLES: 14, 15, 49, 50.

Normality Concentration and Solution Stoichiometry. We can combine two ideas to describe the stoichiometry of reactions in solution through the concept of chemical equivalents.

- The number of equivalents of solute in a solution = *normality* of solution × solution volume (in liters).
- In a chemical reaction the reactants combine in a 1:1 ratio of equivalents.

These ideas are applied in Example 18-12.

Example 18-12 _____

Describing an acid–base titration in terms of equivalents and normality concentration. The $H_2SO_4(aq)$ described in Example 18-11 is used to titrate a

25.00-mL sample of a KOH(aq) solution; 23.06 mL of the 0.107 N H_2SO_4 solution is required. What are the *normality* and the *molarity* of the KOH(aq)?

Solution. First calculate the number of equivalents of H_2SO_4 used in the titration.

$$\text{no. equiv. } H_2SO_4 = 0.02306 \text{ L} \times \frac{0.107 \text{ equiv. } H_2SO_4}{L} = 2.47 \times 10^{-3} \text{ equiv.}$$

$$\text{no. equiv. KOH} = \text{no. equiv. } H_2SO_4 = 2.47 \times 10^{-3}$$

$$\text{normality of KOH} = \frac{2.47 \times 10^{-3} \text{ equiv. KOH}}{0.02500 \text{ L}} = 0.0988 \text{ N KOH}$$

Since 1 mol KOH produces 1 mol OH^- (which is equivalent to 1 mol H^+), the equivalent weight and molar mass of KOH are identical. This means that molarity and normality concentration are also identical for KOH. Molarity of KOH = normality of KOH = 0.0988 M KOH = 0.0988 N KOH.

SIMILAR EXAMPLES: 16, 51, 52.

Did you notice that in Example 18-12 *we wrote no chemical equations?* This represents the principal advantage of the concept of chemical equivalents—all reactions occur with reactants combining in a 1:1 ratio. It is because they were so often unable to write chemical formulas and equations that early chemists were attracted to this system. This advantage of the concept is offset by an important disadvantage: *The equivalent weight of a substance and normality concentration of its solutions depend on the particular reaction in which the substance participates* (recall equations 18.29 through 18.31). Equivalent weights often do *not* have unique values, as do formula (molecular) weights. Furthermore, the concept of chemical equivalents does not offer any new problem-solving capabilities. We could just as easily have described the titration of Example 18-12 by using moles and molarity concentration. Although equivalent weight and normality concentration will continue to be used for some time, their use is likely to decline in importance. This is especially true now that the mole has been established as the base unit for amount of substance in the SI system.

Summary

A strong acid, because of the H_3O^+ produced by its complete ionization, reduces the extent to which a weak acid ionizes when the strong and weak acid are together in the same solution. The H_3O^+ produced by the strong acid is called a *common ion* in the weak acid equilibrium. A salt of a weak acid also represses the ionization of the weak acid; in this case the acid anion, A^-, is the common ion. The weak acid/salt combination is especially important since a solution with these two components resists changes in pH. It is a *buffer* (*buffered*) solution. The combination weak base/salt is also a buffer solution.

Typical buffer solutions have one component to react with small amounts of acid and another to react with small amounts of bases. The pH at which the buffer solution functions is determined by the pK value of the weak acid or weak base and the *ratio* of the molarity of the salt to that of the weak acid or base. That is, pH = pK_a + log([A^-]/[HA]) and pOH = pK_b + log([BH^+]/[B]). The molarities of the salt and the weak acid or weak base also establish the *buffer capacity*—the amount of added acid or base that the buffer is capable of neutralizing. Buffer action plays a crucial role in the functioning of blood and other body fluids.

In a titration, solutions of an acid and a base are allowed to react to the point of exact neutralization. At this point, called the *equivalence point*, there is neither excess acid nor base but simply the salt produced by their neutralization. Whether this salt solution is acidic, basic, or pH neutral is determined by whether ions of the salt can ionize (hydrolyze) as acids or bases. A *titration curve* is a graph of pH of the solution being titrated as a function of the volume of titrating agent (titrant) added. Generally, a titration curve shows a sharp change in solution pH at the equivalence point. In the titration of a weak acid (or base) by a strong base (or acid) another point of interest is the *half-neutralization* point, where pH = pK_a.

An acid–base indicator is a weak acid that has one color when present mostly as the nonionized acid, HIn, and a different color when present mostly as the anion, In^-. The particular color assumed by the indicator depends on the pH of the solution, being governed by the expression: $pH = pK_a + \log([In^-]/[HIn])$. In selecting an acid–base indicator for a titration one chooses an indicator whose pK value is close to the pH at the equivalence point of the titration.

In the titration of a polyprotic acid, one steep portion of the titration curve is generally observed for each ionizable H atom in the acid. That is, there is a separate equivalence point for each of the ionizable H atoms (if the K_a values of successive ionization steps differ by a factor of 10^3 or

more). If one of the ionization steps for the polyprotic acid has an unusually small value of K_a, it may not be possible to carry the neutralization through this step in a titration. Thus, in the titration of the *tri*protic acid, H_3PO_4, only the first two equivalence points can be reached in an acid–base titration (Figure 18-10). To establish the pH values of the salt solution at the equivalence points in the titration of a polyprotic acid, the further ionization of anions, both as acids and as bases, must be considered.

The stoichiometry of acid–base reactions can be described through the concepts of chemical equivalents and normality concentration. However, these concepts provide no particular advantages or insights over the use of the mole and molarity concentration.

Summarizing Example

p-Hydroxybenzoic acid, HOC_6H_4COOH, is a weak *diprotic* acid. It is used as a food preservative (up to 0.1% by mass is approved by the FDA). Other compounds derived from the acid are used as preservatives in cosmetics and pharmaceuticals. Its structure is shown in the illustration.

1. Write equations for the ionization of *p*-hydroxybenzoic acid, indicating which H atom is ionized in each step.

Solution. Since the acid is *diprotic* there are *two* ionizable H atoms, that is, HOC_6H_4COOH. In Section 17-9, where we considered factors affecting acid strength, we learned that the H atom on a —COOH (carboxyl) group is more easily ionized than the H atom of an —OH group (expressions 17.43 and 17.44). This suggests that the ionization reactions are

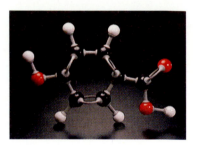

p-Hydroxybenzoic acid, HOC_6H_4COOH. [Photograph by Carey B. Van Loon]

$$HOC_6H_4COOH + H_2O \rightleftharpoons H_3O^+ + HOC_6H_4COO^- \qquad K_{a_1}$$

$$HOC_6H_4COO^- + H_2O \rightleftharpoons H_3O^+ + {}^-OC_6H_4COO^- \qquad K_{a_2}$$

(This example is similar to Example 17-15.)

2. 25.00 mL of a dilute aqueous solution of *p*-hydroxybenzoic acid is titrated with NaOH(aq); 16.24 mL of 0.0200 M NaOH is required to reach the first equivalence point. What is the molarity of the *p*-hydroxybenzoic acid solution?

Solution. This is a straightforward titration of a weak acid by a strong base. The neutralization reaction to the first equivalence point is

$$HOC_6H_4COOH + OH^- \longrightarrow HOC_6H_4COO^- + H_2O$$

$$\text{no. mmol } HOC_6H_4COOH = 16.24 \text{ mL} \times \frac{0.0200 \text{ mmol } OH^-}{mL}$$

$$\times \frac{1 \text{ mmol } HOC_6H_4COOH}{1 \text{ mmol } OH^-}$$

$$= 0.325 \text{ mmol } HOC_6H_4COOH$$

$$\text{molarity of } HOC_6H_4COOH = \frac{0.325 \text{ mmol } HOC_6H_4COOH}{25.00 \text{ mL}} = 0.0130 \text{ M}$$

(This example is similar to Examples 5-5 and 17-4.)

3. The pH was determined at several points during the titration referred to in part **2.** Two of the values obtained were *pH = 4.57* when *8.12 mL* of 0.0200 M NaOH had been added, and *pH = 7.02* after *16.24 mL* had been added (the equivalence point). Use these data to determine K_{a_1} and K_{a_2} for *p*-hydroxybenzoic acid.

Solution. The key to this question is in recognizing the special significance of the equivalence point (16.24 mL) and the *half-neutralization* point (8.12 mL). At the half-neutralization point, since $[HOC_6H_4COO^-] = [HOC_6H_4COOH]$, then $pH = pK_{a_1}$. Recall that this conclusion is based on the condition

$$pH = pK_{a_1} + \log \frac{[HOC_6H_4COO^-]}{[HOC_6H_4COOH]} = pK_{a_1} + \log 1 = pK_{a_1}$$

$$pK_{a_1} = 4.57; \qquad \log K_{a_1} = -4.57; \qquad K_{a_1} = \text{antilog } -4.57 = 2.7 \times 10^{-5}$$

To establish pK_{a_2}, note that the anion present at the first equivalence point, $HOC_6H_4COO^-$, can ionize further as an acid or it can ionize (hydrolyze) as a base. For a solution containing such anions we have established the expression: $pH = \frac{1}{2}(pK_{a_1} + pK_{a_2})$. The pH at the first equivalence point is 7.02, and the value of pK_{a_1} from above is 4.57.

$$pK_{a_2} = (2 \times pH) - pK_{a_1} = (2 \times 7.02) - 4.57 = 9.47$$
$$K_{a_2} = \text{antilog } -9.47 = 3.4 \times 10^{-10}$$

(This example is similar to Example 18-10.)

4. What is the pH at the point in the titration where 10.00 mL 0.0200 M has been added to the 25.00 mL of *p*-hydroxybenzoic acid solution?

Solution. This point should lie within the range in which the solution being titrated acts as a buffer solution, and we can calculate its pH with the "buffer equation" [Henderson–Hasselbalch equation, (18.13)]. First, however, we have to work out the details of the stoichiometry of the neutralization to this point. The no. mmol of weak acid present initially is 25.00 mL × 0.0130 mmol HOC_6H_4COOH/mL = 0.325 mmol. The amount of OH^- added through this point in the titration, in millimoles, is 10.00 mL × 0.0200 mmol OH^-/mL = 0.200 mmol OH^-.

	HOC_6H_4COOH	+	OH^-	$\longrightarrow$	$HOC_6H_4COO^-$	+ H_2O
initially present:	0.325 mmol		—		—	
add:			0.200 mmol			
changes:	−0.200 mmol		−0.200 mmol		+0.200 mmol	
after reaction:						
mmol	0.125 mmol		~0		0.200 mmol	
concns	0.125 mmol/35.00 mL				0.200 mmol/35.00 mL	
	3.57×10^{-3} M				5.71×10^{-3} M	

$$pH = pK_{a_1} + \log \frac{[HOC_6H_4COO^-]}{[HOC_6H_4COOH]} = 4.57 + \log \frac{5.71 \times 10^{-3}}{3.57 \times 10^{-3}}$$

$$= 4.57 + 0.20 = 4.77$$

(This example is similar to Example 18-8.)

Key Terms

acid–base indicator (18-3)

buffer capacity (18-2)

buffer range (18-2)

buffer solution (18-2)

common ion effect (18-1)

end point of a titration (18-4)

equivalence point (18-4)

equivalent weight (18-7)

millimole (18-4)

normality (N) (18-7)

titration curve (18-4)

Highlighted Expressions

Concerning mixtures of strong and weak acids (18.4)

Concerning mixtures of weak acids and their salts (18.7)

pH of a buffer solution of a weak acid and its salt (18.13)

pH of a buffer solution of a weak base and its salt (18.14)

Relating pH and pK_a of an acid–base indicator (18.18)

Molarity concentration and millimoles of solute (18.19)

Features of the titration curve: strong acid–strong base (18.20)

Features of the titration curve: weak acid–strong base (18.21)

Review Problems

1. For a solution that is 0.218 M $HC_3H_5O_2$ ($K_a = 1.3 \times 10^{-5}$) and 0.0852 M HI, calculate **(a)** $[H_3O^+]$; **(b)** $[OH^-]$; **(c)** $[C_3H_5O_2^-]$; **(d)** $[I^-]$.

2. For a solution that is 0.106 M NH_3 ($K_b = 1.74 \times 10^{-5}$) and 0.0742 M NH_4Cl, calculate **(a)** $[OH^-]$; **(b)** $[NH_4^+]$; **(c)** $[Cl^-]$; **(d)** $[H_3O^+]$.

3. Write equations to show how each of the following buffer solutions reacts with a small added amount of acid or base. **(a)** $HCHO_2/KCHO_2$; **(b)** $C_6H_5NH_2/C_6H_5NH_3^+Cl^-$; **(c)** KH_2PO_4/Na_2HPO_4.

4. Calculate the pH of a buffer that is **(a)** 0.0814 M $HC_7H_5O_2$ ($K_a = 6.3 \times 10^{-5}$) and 0.148 M $NaC_7H_5O_2$; **(b)** 0.246 M NH_3 ($K_b = 1.74 \times 10^{-5}$) and 0.0954 M NH_4Cl.

5. What concentration of formate ion, $[CHO_2^-]$ should be present in 0.472 M $HCHO_2$ to produce a buffer solution with pH = 4.12?

$$HCHO_2 + H_2O \rightleftharpoons H_3O^+ + CHO_2^- \qquad K_a = 1.8 \times 10^{-4}$$

6. What concentration of ammonia, $[NH_3]$, should be present in a solution that has $[NH_4^+] = 0.812$ M to produce a buffer solution with pH = 9.15? $[K_b(NH_3) = 1.74 \times 10^{-5}]$

7. Lactic acid, $HC_3H_5O_3$, is found in sour milk. A solution that contains 1.00 g $NaC_3H_5O_3$ in 100.0 mL of 0.0500 M $HC_3H_5O_3$ has pH = 4.11. What is K_a of lactic acid?

8. A handbook lists the following data.

		Color change
bromophenol blue	$K_a = 1.41 \times 10^{-4}$	yellow → blue
		(acid) (anion)
bromocresol green	$K_a = 2.09 \times 10^{-5}$	yellow → blue
bromothymol blue	$K_a = 7.9 \times 10^{-8}$	yellow → blue
2,4-dinitrophenol	$K_a = 1.26 \times 10^{-4}$	colorless → yellow
chlorophenol red	$K_a = 1.0 \times 10^{-6}$	yellow → red
thymolphthalein	$K_a = 1.0 \times 10^{-10}$	colorless → blue

(a) Which of these indicators change color in acidic solution which in basic solution, and which near the neutral point?

(b) What is the approximate pH of a solution if bromocresol green indicator assumes a green color; if chlorophenol red assumes an orange color?

9. With reference to the indicators listed in Exercise 8, what would be the color of each combination?

(a) 2,4-dinitrophenol when placed in 0.100 M HCl(aq)

(b) chlorophenol red in 1.00 M NaCl

(c) thymolphthalein in 1.00 M NH_3

(d) bromothymol blue in 1.00 M NH_4NO_3

(e) bromocresol green in seawater (recall Figure 17-3).

10. Sketch the titration curves (pH versus volume of titrant) that you would expect to obtain in the following titrations.

(a) NaOH(aq) is titrated with HNO_3(aq);

(b) NH_3(aq) is titrated with HCl(aq);

(c) $HC_2H_3O_2$(aq) is titrated with KOH(aq);

(d) NaH_2PO_4 is titrated with KOH(aq).

11. Calculate the pH at the points in the titration of 20.00 mL of 0.350 M KOH at which **(a)** 15.00 mL and **(b)** 20.00 mL of 0.425 M HCl have been added.

12. Calculate the pH at the points in the titration of 25.00 mL of 0.108 M HNO_2 at which **(a)** 10.00 mL and **(b)** 20.00 mL of 0.162 M NaOH have been added. For HNO_2, $K_a = 5.1 \times 10^{-4}$.

$$HNO_2 + OH^- \rightarrow H_2O + NO_2^-$$

13. Arrange the following 0.10 M solutions in order of *decreasing* acidity: $NaC_2H_3O_2$, $NaHSO_3$, $NaHSO_4$. Explain your reasoning. [*Hint:* Refer to Tables 17-2 and 17-3.]

14. Calculate the equivalent weights of the following substances for use in acid–base reactions: **(a)** $HClO_4$; **(b)** $Mg(OH)_2$; **(c)** $HC_3H_5O_2$ (propionic acid).

15. What are the normality concentrations corresponding to the following molarities? Indicate any cases where more than one normality seems possible. **(a)** 0.18 M KOH; **(b)** 0.086 M $H_2C_2O_4$; **(c)** 1.8×10^{-3} M $Ca(OH)_2$.

16. What volume of 0.1216 N KOH is required for the *complete* neutralization of **(a)** 25.00 mL of 0.1824 N H_2SO_4; **(b)** 20.00 mL of 0.06480 M H_2SO_4?

Exercises

The common ion effect (Use data from Table 17-2, as necessary.)

17. Describe the effect on the pH of the solution produced by adding **(a)** $NaNO_2$ to $HNO_2(aq)$; **(b)** $NaNO_3$ to $HNO_3(aq)$. Why are the effects not the same? Explain.

18. Calculate the concentration of the ionic species indicated.
(a) $[H_3O^+]$ in a solution that is 0.080 M HCl and 0.080 M HOCl;
(b) $[NO_2^-]$ in a solution that is 0.100 M $NaNO_2$ and 0.0100 M HNO_2;
(c) $[Cl^-]$ in a solution that is 0.185 M HCl and 0.0626 M $HC_2H_3O_2$;
(d) $[C_2H_3O_2^-]$ in a solution that is 0.100 M HCl and 0.350 M $HC_2H_3O_2$
(e) $[OH^-]$ in a solution that is 0.200 M $(NH_4)_2SO_4$ and 0.500 M NH_3.

19. What is the pH of a solution obtained by adding 1.15 mg of aniline hydrochloride ($C_6H_5NH_3^+Cl^-$) to 3.18 L of 0.105 M aniline ($C_6H_5NH_2$)?

$$C_6H_5NH_2 + H_2O \rightleftharpoons C_6H_5NH_3^+ + OH^- \quad K_b = 4.30 \times 10^{-10}$$

★20. You are given 250.0 mL of 0.100 M $HC_3H_5O_2$ (propionic acid, $K_a = 1.35 \times 10^{-5}$). You wish to adjust its pH by adding an appropriate solution. What volume would you add of **(a)** 1.00 M HCl to lower the pH to 1.00; **(b)** 1.00 M $NaC_3H_5O_3$ to raise the pH to 4.00; **(c)** water to raise the pH by 0.15 unit?

Buffer solutions (Use data from Tables 17-2 and 17-3, as necessary.)

21. Indicate which of the following aqueous solutions are buffer solutions. Explain your reasoning. [*Hint:* You must also consider whether any reactions occur between the solution components listed.]
(a) 0.100 M NaCl
(b) 0.100 M NaCl–0.100 M NH_4Cl
(c) 0.100 M CH_3NH_2–0.150 M $CH_3NH_3^+Cl^-$
(d) 0.100 M HCl–0.050 M $NaNO_2$
(e) 0.100 M HCl–0.200 M $NaC_2H_3O_2$
(f) 0.100 M $HC_2H_3O_2$–0.125 M $NaC_3H_5O_2$

22. In the text the $H_2PO_4^-/HPO_4^{2-}$ combination is mentioned as playing a role in maintaining the pH of blood.
(a) Write equations to show how a solution containing these ions functions as a buffer.
(b) Verify that this buffer is most effective at pH 7.2.
(c) Calculate the pH of a buffer solution in which $[H_2PO_4^-] = 0.050$ M and $[HPO_4^{2-}] = 0.150$ M. [*Hint:* Recall that phosphoric acid ionizes in three distinct steps and focus on the second step.]

23. What mass of $(NH_4)_2SO_4$ must be added to 320.0 mL of 0.105 M NH_3 to yield a solution with pH 9.35?

24. You are given the task of preparing a buffer solution with pH 3.50 and have available the following solutions, all 0.100 M: $HCHO_2$, $HC_2H_3O_2$, H_3PO_4, $NaCHO_2$, $NaC_2H_3O_2$, and NaH_2PO_4. Describe how you would prepare this buffer solution. [*Hint:* What volumes of which solutions would you use?]

★25. Suppose that the task in the preceding exercise were modified to require exactly 1.00 L of the buffer solution with pH 3.50 and that the solutions available were 0.100 M $NaCHO_2$, 0.100 M $NaC_2H_3O_2$, 0.100 M NaH_2PO_4, and 1.00 M HCl. Describe how you would prepare this buffer solution.

26. Compare the following buffers with respect to their **(a)** pH and **(b)** buffer capacities: 0.010 M $HC_2H_3O_2$–0.010 M $NaC_2H_3O_2$ and 0.100 M $HC_2H_3O_2$–0.50 M $KC_2H_3O_2$.

27. Refer to Examples 18-5 and 18-6. You are asked to reduce the pH of the 0.300-L of buffer solution in Example 18-5 from 5.10 to 5.00.
(a) Which of the following substances could you use to do this: 0.100 M NaCl; 0.150 M HCl; 0.100 M $NaC_2H_3O_2$; 0.125 M NaOH; 0.050 M $HC_2H_3O_2$? Explain your reasoning.
(b) Determine the volume (in mL) of the appropriate solution(s) for lowering the pH of the buffer solution from 5.10 to 5.00.

28. A handbook lists various procedures for preparing buffer solutions. To obtain a pH = 9.00, the handbook says to mix 36.00 mL of 0.200 M NH_3 with 64.00 mL of 0.200 M NH_4Cl.
(a) Show by calculation that the pH of this solution is indeed 9.00.
(b) Would you expect the pH of this solution to remain at pH = 9.00 if the 100.00 mL of buffer solution were diluted to 1.00 L? 1000. L? Explain.
(c) What will be the pH of the original 100.00 mL of buffer solution if 0.10 mL of 1.00 M HCl is added to it?
(d) What is the maximum volume of 1.00 M HCl that can be added to 100.00 mL of the original buffer solution so that the pH does not drop below 8.90?

29. An acetic acid–sodium acetate buffer can be prepared by allowing an *excess* of $NaC_2H_3O_2$ to react with HCl. The strong acid HCl is converted to the weak acid $HC_2H_3O_2$.

$$\underset{\text{(from } NaC_2H_3O_2)}{C_2H_3O_2^-} + \underset{\text{(from HCl)}}{H_3O^+} \rightarrow HC_2H_3O_2 + H_2O$$

(a) If 10.0 g $NaC_2H_3O_2$ is added to 0.300 L of 0.200 M HCl, what is the pH of the resulting solution?
(b) If 1.00 g $Ba(OH)_2$ is added to the solution in part (a), what happens to the pH?
(c) What is the capacity of the buffer solution in part (a) toward $Ba(OH)_2$?
(d) What is the pH of the solution in part (a) following the addition of 5.20 g $Ba(OH)_2$?

★30. Even though the carbonic acid–bicarbonate buffer system is crucial to the maintenance of the pH of blood, it has no practical use as a buffer in laboratory work. Can you think of a reason(s) for this? [*Hint:* Refer to the data in Exercise 68 of Chapter 17.]

Acid–base indicators (Use data from Tables 17-2 and 17-3, as necessary.)

31. In the use of acid–base indicators
(a) Why is it generally sufficient to use a *single* indicator in an acid–base titration but often necessary to use several indicators to establish the approximate pH of a solution?
(b) Why must the quantity of indicator used in a titration be kept as small as possible?

32. Phenol red indicator changes from yellow to red in the pH range from 6.6 to 8.0. *Without making detailed calculations,* state what color the indicator will assume in each of the following solutions. **(a)** 0.10 M KOH; **(b)** 0.10 M $HC_2H_3O_2$; **(c)** 0.10 M NH_4NO_3; **(d)** 0.10 M HBr; **(e)** 0.10 M NaCN.

33. The indicator methyl red has a $pK_a = 4.95$. It changes color from red to yellow over the pH range 4.4 to 6.2.

(a) If the indicator is placed in a buffer solution that is 0.10 M $HC_2H_3O_2$–0.10 M $NaC_2H_3O_2$, what percent of the indicator will be in the acid form and what percent in the anion form?

(b) Which form of the indicator do you think has the "stronger" (more visible) color, the acid form (red) or the anion form (yellow)? Explain your reasoning.

Neutralization reactions

34. Excess $Ca(OH)_2(s)$ is shaken with water to produce a saturated solution. A 50.00-mL sample of the clear, saturated solution is withdrawn and titrated. 10.7 mL of 0.1032 M HCl is required for this titration. What is the solubility of $Ca(OH)_2$, expressed as g $Ca(OH)_2$ per L soln.?

35. Two solutions are mixed: (A) 100.0 mL of an HCl(aq) with pH 2.50 and (B) 100.0 mL of an NaOH(aq) with pH 11.00. What is the pH of the solution that results from mixing A and B?

★36. With reference to Exercise 35, if solution A were 100.0 mL of a weak acid solution with pH 2.50 and solution B, 100.0 mL of a weak base solution with pH 11.00,

(a) Would the final pH on mixing the two solutions be greater or less than that calculated in Exercise 35? Explain.

(b) What additional information would be required to calculate the actual pH of the mixed solution?

Titration curves

37. In the text several differences were pointed out between the titration curves for a strong acid by a strong base and a weak acid by a strong base (compare Figures 18-7 and 18-8). One point that is identical in the two curves, however, is the volume of 0.1000 M NaOH required to reach the equivalence point. Explain why this should be so.

38. Sketch the following titration curves. Indicate the initial pH and the pH corresponding to the equivalence point. Indicate the volume of titrant required to reach the equivalence point, and select a suitable indicator from Figure 18-6.

(a) 25.0 mL of 0.100 M KOH by 0.200 M HI

(b) 10.0 mL of 1.00 M NH_3 by 0.250 M HCl

39. Determine, by calculation, the pH at the point where the original acid is 90.0% neutralized in the titration of

(a) 25.00 mL of 0.1000 M HCl by 0.1000 M NaOH (see Figure 18-7).

(b) 25.00 mL of 0.1000 M $HC_2H_3O_2$ by 0.1000 M NaOH (see Figure 18-8).

40. Several points are indicated on the following sketch of the

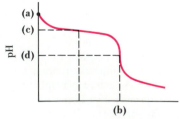

Exercise 40   Volume of 0.350 M HI

titration curve for 20.00 mL of 0.250 M NH_3(aq) titrated with 0.350 M HI. Indicate the pH at point **(a)**; the volume of 0.350 M HI at point **(b)**; the pH at point **(c)**; the pH at point **(d)**.

★41. The method used in the text for establishing points on a titration curve was to calculate the pH corresponding to the points at which different volumes of the titrating agent (titrant) had been added to the solution being titrated. An alternate procedure is to calculate the volume of titrant solution required to reach certain pH values. Use this procedure to determine the volumes of 0.1000 M NaOH required to reach the following pH values in the titration of 20.00 mL of 0.1500 M HCl. Then plot the titration curve. **(a)** pH = 2.00; **(b)** 3.50; **(c)** 5.00; **(d)** 10.50; **(e)** 12.00.

★42. Use the method outlined in Exercise 41 to determine the volume of titrant required to reach the following points on the titration curves.

(a) 25.00 mL of 0.250 M NaOH titrated with 0.300 M HCl, at the points where pH = 13.00; 12.00; 10.00; 4.00; 3.00.

(b) 50.00 mL of 0.0100 M benzoic acid, $HC_7H_5O_2$ ($K_a = 6.3 \times 10^{-5}$), titrated with 0.0500 M KOH, at the points where pH = 4.50; 5.50; 11.50.

★43. The following titration curve was obtained when 10.00 mL of a solution containing *both* HCl and H_3PO_4 was titrated with 0.216 M NaOH. What is the molarity of **(a)** HCl and **(b)** H_3PO_4 in this solution? [*Hint:* In what portion of the titration curve is the HCl neutralized? What neutralization occurs between the two steep portions of the titration curve?]

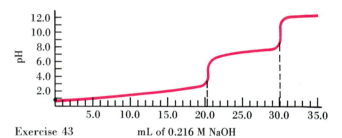

Exercise 43   mL of 0.216 M NaOH

44. Sketch a series of titration curves for the following three hypothetical weak acids when titrated with 0.1000 M NaOH. Select suitable indicators for the titrations. [*Hint:* Select a few key points at which to estimate the pH of the solution; for example, the initial pH, the condition pH = pK_a, and the pH at the equivalence point.]

(a) 10.00 mL of 0.1000 M HX, $K_a = 1.0 \times 10^{-3}$

(b) 10.00 mL of 0.1000 M HY, $K_a = 1.0 \times 10^{-5}$

(c) 10.00 mL of 0.1000 M HZ, $K_a = 1.0 \times 10^{-7}$

★45. Thymol blue in its acid range is not a suitable indicator for the titration of HCl by NaOH. Suppose that a student uses thymol blue by mistake in the titration of Figure 18-7, and suppose that the indicator end point is taken to be pH = 2.0.

(a) Would there be a sharp color change [that is, produced by a single drop of NaOH(aq)]?

(b) What percent of the HCl remains unneutralized at this point?

pH of salts of polyprotic acids

46. Is a solution that is 0.10 M Na_2S(aq) likely to be acidic, basic, or pH neutral? Explain.

47. Calculate the pH of **(a)** 1.0 M Na_2CO_3(aq) and **(b)** 0.010 M Na_2CO_3(aq).

48. Trisodium phosphate is made commercially by neutralizing phosphoric acid with sodium carbonate to obtain Na_2HPO_4. The Na_2HPO_4 is further neutralized to Na_3PO_4 with NaOH.

(a) Write net ionic equations for the reactions described here.

★(b) Na_2CO_3 is a much cheaper base than is NaOH. Why do you suppose that NaOH must be used together with Na_2CO_3 to produce Na_3PO_4? [*Hint:* Compare the pH values of equimolar solutions of OH^-, CO_3^{2-}, and PO_4^{3-}.]

Equivalent weight and normality concentration

49. In aqueous solutions carbonate ion, CO_3^{2-}, can react with *two* H^+ ions yielding one molecule of H_2O and one of CO_2. What mass of $Na_2CO_3 \cdot 10H_2O$, in grams, is required to produce 2.00 L of 0.175 N $Na_2CO_3(aq)$?

50. It is desired to prepare a standard solution of $Ba(OH)_2$ for use in acid–base titrations. What is the approximate maximum *normality* solution that can be prepared if the solubility of barium hydroxide is 3.89 g $Ba(OH)_2$ per 100 g of solution at 20 °C?

51. A 25.00-mL sample of $H_3PO_4(aq)$ requires 31.15 mL of 0.242 N KOH for its titration in reaction (18.30). What is the normality of this $H_3PO_4(aq)$ if it is always to be used (a) in reaction (18.29); (b) in reaction (18.30); (c) in reaction (18.31)?

52. A sample of battery acid is to be analyzed for its sulfuric acid content. A 1.00-mL sample weighs 1.239 g. This sample is diluted to 250.0 mL and 10.00 mL of the diluted acid requires 32.40 mL of 0.0100 N $Ba(OH)_2$ for its titration. What is the % H_2SO_4, by mass, in the battery acid?

Additional Exercises

53. In Example 17-9 we calculated the percent ionization of $HC_2H_3O_2$ in (a) 1.0 M; (b) 0.10 M, and (c) 0.010 M $HC_2H_3O_2$ solutions. Recalculate these percent ionizations if each solution is also made 0.10 M $NaC_2H_3O_2$.. Explain why the results obtained here differ from those of Example 17-9.

54. Although we say that a buffered solution has a constant pH and $[H_3O^+]$, small changes in pH do translate into significant changes in $[H_3O^+]$. What is the *percent* increase in $[H_3O^+]$ when the pH of blood drops from 7.4 to 7.3?

55. A buffer solution is prepared by dissolving 1.50 g each of benzoic acid, $HC_7H_5O_2$, and sodium benzoate, $NaC_7H_5O_2$, in 150.0 mL of water solution.

$$HC_7H_5O_2 + H_2O \rightleftharpoons H_3O^+ + C_7H_5O_2^- \qquad K_a = 6.3 \times 10^{-5}$$

(a) What is the pH of this buffer solution?

(b) Which buffer component, and in what quantity, must be added to the solution to change its pH to 4.00?

56. A solution is 0.298 M $HCHO_2$ (formic acid). What mass of sodium formate ($NaCHO_2$) must be added to 250.0 mL of this solution to produce a buffer solution of pH = 3.68? [$K_a(HCHO_2) = 1.8 \times 10^{-4}$.]

57. If to 100.0 mL of the buffer solution prepared in Exercise 56 is added 0.55 mL of 15 M NH_3, what will be the pH of the resulting solution?

58. In what approximate pH range would you expect each of the following buffer solutions to be most effective? (a) HNO_2/$NaNO_2$; (b) NH_3/$(NH_4)_2SO_4$; (c) CH_3NH_2/$CH_3NH_3^+Cl^-$.

★**59.** You are asked to bring the pH of 0.500 L of 0.500 M $NH_4Cl(aq)$ to a value of 7.00. How many drops (1 drop = 0.05 mL) of which of the following solutions would you use: 10.0 M HCl; 10.0 M NH_3?

60. What aqueous concentrations of the following substances are required to obtain solutions with the pH values indicated? (a) $Ba(OH)_2$ for pH 12.22; (b) aniline, $C_6H_5NH_2$, for pH 8.91; (c) $HC_2H_3O_2$ in 0.294 M $NaC_2H_3O_2$ for pH 4.48; (d) NH_4Cl for pH 5.05.

61. Use appropriate equilibrium constants to determine whether a solution can be simultaneously

(a) 0.10 M NH_3 and 0.10 M NH_4Cl, with pH = 6.07;

(b) 0.10 M $NaC_2H_3O_2$ and 0.058 M HI;

(c) 0.10 M KNO_2 and 0.25 M KNO_3.

62. A buffer solution is prepared by dissolving 1.51 g NH_3 and 3.85 g $(NH_4)_2SO_4$ in 0.500 L of water solution.

(a) What is the pH of this solution?

(b) If 1.00 g NaOH is added to this solution what is the pH?

(c) How many mL of 12 M HCl must be added to 0.500 L of the original buffer solution to change its pH to 9.00?

63. Refer to the Summarizing Example. Use the data presented or developed in the example to determine the pH at these additional points of the titration of *para*-hydroxybenzoic acid by NaOH(aq).

(a) The initial pH.

(b) Following the addition of 28.36 mL of 0.0200 M NaOH.

(c) At the *second* equivalence point.

(d) Following the addition of 40.00 mL of 0.0200 M NaOH.

64. Refer to the Summarizing Example. Use pH data from the example, together with the values established in Exercise 63, to sketch the complete titration curve for the titration of para-hydroxybenzoic acid by NaOH(aq).

65. Calculate the volume of 0.1000 M NaOH that must be added to just reach the following.

(a) a pH of 3.00 in the titration of 25.00 mL of 0.1000 M HCl (Figure 18-7).

(b) a pH of 5.25 in the titration of 25.00 mL of 0.1000 M $HC_2H_3O_2$ (Figure 18-8).

(c) a pH of 7.50 in the titration of 10.00 mL of 0.1000 M H_3PO_4 (Figure 18-10).

66. A 10.00-mL solution that is 0.0400 M H_3PO_4 and 0.0150 M NaH_2PO_4 is titrated with 0.0200 M NaOH. Sketch the titration curve obtained. [*Hint:* How many equivalence points are there? Are they equally spaced along the volume axis as in Figure 18-10?]

67. To measure the pH of a solution with a pH meter it is necessary to standardize the meter (compare it) against a buffer solution of known pH. One solution used for this purpose is 0.0500 M potassium hydrogen phthalate, $KHC_8H_4O_4$. What is the approximate pH of this solution?

$$H_2C_8H_4O_4 + H_2O \rightleftharpoons H_3O^+ + HC_8H_4O_4^- \qquad K_a = 1.14 \times 10^{-3}$$

$$HC_8H_4O_4^- + H_2O \rightleftharpoons H_3O^+ + C_8H_4O_4^{2-} \qquad K_a = 3.70 \times 10^{-6}$$

★**68.** A buffer solution can be prepared by starting with a weak acid, HA, and converting some of the weak acid to its salt, NaA, by titration with a strong base. The *fraction* of the original acid that is converted to the salt is designated as *f*.

(a) Derive an equation similar to equation (18.13) but expressed in terms of f rather than concentrations.

(b) What is the pH at the point in the titration of phenol, HC_6H_5O, where $f = 0.27$? (pK_a of phenol = 9.80)

69. Sodium hydrogen sulfate, $NaHSO_4$, is an acidic salt with a number of uses, such as for metal pickling (removal of surface deposits). (In the household it is used as a toilet bowl cleaner.) $NaHSO_4$ is made by the reaction of H_2SO_4 with NaCl. To determine the % NaCl impurity in a sodium hydrogen sulfate, a 1.028-g sample is titrated with NaOH(aq); 38.56 mL of 0.2150 M NaOH is required.

(a) Write the net equation for the neutralization reaction.

(b) What is the % NaCl in the sample titrated? [*Hint:* What mass of $NaHSO_4$ is present? What happens to NaCl during the titration?]

(c) Select a suitable indicator(s) from Figure 18-6.

70. You are asked to prepare a KH_2PO_4–Na_2HPO_4 solution that has the same pH as human blood, 7.40.

(a) What should be the *ratio* of concentrations $[HPO_4{}^{2-}]/[H_2PO_4{}^-]$ in this solution?

★(b) Suppose you had to prepare 1.00 L of the solution described in (a), and suppose that this solution had to be isotonic with blood (having the same osmotic pressure as blood). What mass of KH_2PO_4 and of $Na_2HPO_4 \cdot 12H_2O$ would you use? [*Hint:* Refer to the definition of isotonic on page 460; recall that a solution of NaCl having 9.0 g NaCl/L soln. is also isotonic with blood, and do not forget that the NaCl is completely ionized in water solution.]

★71. Piperazine is a diprotic weak base that finds use as a corrosion inhibitor, an insecticide, and an anthelmintic (expelling intestinal worms). Its ionization is described by the equations:

$pK_{b_1} = 4.22$

$pK_{b_2} = 8.67$

The piperazine used commercially is a hexahydrate, $C_4H_{10}N_2 \cdot 6H_2O$. A 1.00-g sample of this hexahydrate is dissolved in 100. mL of water and titrated with 0.500 M HCl. Sketch a titration curve for this titration, indicating **(a)** the initial pH; **(b)** the pH when the *first* step of the neutralization is *half* completed; **(c)** the volume of HCl(aq) required to reach the *first* equivalence point; **(d)** the pH at the *first* equivalence point; **(e)** the pH when the *second* step of the neutralization is *half* completed; **(f)** the volume of HCl(aq) required to reach the *second* equivalence point; **(g)** the pH at the *second* equivalence point.

★72. Since an acid–base indicator is a weak acid, it can be titrated with a strong base just like other weak acids. Suppose you titrate 25.00 mL of a 0.0100 M solution of the indicator p-nitrophenol, $HC_6H_4NO_3$, with 0.0200 M NaOH. The pK_a of p-nitrophenol is 7.15, and it changes from colorless to yellow over the pH range 5.6 to 7.6.

(a) Sketch the titration curve for this titration.

(b) Show the pH range over which p-nitrophenol changes color.

(c) Explain why p-nitrophenol cannot serve as its own indicator in this titration.

(d) Choose an indicator for this titration from those given in Figure 18-6 and Problem 8. [*Hint:* Consider carefully the colors involved.]

★73. The neutralization of NaOH by HCl is represented in equation (1) below, and the neutralization of NH_3 by HCl, in equation (2).

(1) $OH^- + H_3O^+ \rightleftharpoons 2\,H_2O \qquad K = ?$
(2) $NH_3 + H_3O^+ \rightleftharpoons NH_4{}^+ + H_2O \qquad K = ?$

(a) Determine the equilibrium constant K for each reaction.

(b) Explain why each neutralization reaction can be considered to go to completion.

74. The single equilibrium equation written below can be applied to different phenomena described in this or the preceding chapter.

$$HC_2H_3O_2 + H_2O \rightleftharpoons H_3O^+ + C_2H_3O_2{}^- \qquad K_a = 1.74 \times 10^{-5}$$

Indicate the phenomenon—ionization of pure acid, common ion effect, buffer formation, hydrolysis—for each of the following combinations of concentrations.

(a) $[H_3O^+]$ and $[HC_2H_3O_2]$ are high; $[C_2H_3O_2{}^-]$ is very low.

(b) $[C_2H_3O_2{}^-]$ is high; $[HC_2H_3O_2]$ and $[H_3O^+]$ are very low.

(c) $[HC_2H_3O_2]$ is high; $[H_3O^+]$ and $[C_2H_3O_2{}^-]$ are low.

(d) $[HC_2H_3O_2]$ and $[C_2H_3O_2{}^-]$ are high; $[H_3O^+]$ is low.

★75. The titration of a weak acid by a weak base is not a particularly satisfactory procedure because the pH does not increase sharply at the equivalence point. Demonstrate this fact by sketching a titration curve for the neutralization of 10.00 mL of 0.100 M $HC_2H_3O_2$ by 0.100 M NH_3.

★76. At times a salt of a weak base can be titrated by a strong base. Use appropriate data from the text to sketch a titration curve for the titration of 20.00 mL of 0.0500 M $C_6H_5NH_3{}^+Cl^-$ by 0.1000 M NaOH.

★77. Carbonic acid is a weak diprotic acid (H_2CO_3), having $K_{a_1} = 4.2 \times 10^{-7}$ and $K_{a_2} = 5.6 \times 10^{-11}$. The equivalence points for the titration of this acid come at approximately pH 4 and pH 9. Suitable indicators for use in titrating carbonic acid or carbonate solutions are methyl orange and phenolphthalein.

(a) Sketch the titration curve that would be obtained in titrating a sample of $NaHCO_3$(aq) with 1.00 M HCl.

(b) Sketch the titration curve for Na_2CO_3(aq) by 1.00 M HCl.

(c) What volume of 0.100 M HCl is required for the neutralization of 1.00 g $NaHCO_3$(s)?

(d) What volume of 0.100 M HCl is required for the complete neutralization of 1.00 g Na_2CO_3(s)?

(e) A sample of NaOH contains a small amount of Na_2CO_3. For titration to the phenolphthalein end point, 0.1000 g of this sample requires 23.98 mL 0.1000 M HCl. An additional 0.78 mL is required to the methyl orange end point. What is the % Na_2CO_3, by mass, in the sample?

★**78.** Thymol blue indicator has *two* pH ranges. It changes color from red to yellow in the pH range 1.2 to 2.8, and from yellow to blue in the pH range 8.0 to 9.6. What is the color of the indicator in the following situations?

   (a) The indicator is placed in 350. mL of 0.205 M HCl.
   (b) To the solution in part (a) is added 250. mL of 0.500 M $NaNO_2$.
   (c) To the solution in part (b) is added 150. mL of 0.100 M NaOH.
   (d) To the solution in part (c) is added 5.00 g $Ba(OH)_2$.

★**79.** Consider a solution of $NaH_2PO_4(aq)$ of molarity, M, and derive equation (18.27) by showing that the pH is independent of M.

★**80.** A solution is prepared that is 0.150 M $HC_2H_3O_2$ and 0.250 M $NaCHO_2$.

   (a) Show that this is a buffer solution.
   (b) Calculate the pH of this buffer solution.
   (c) What is the final pH if to 1.00 L of this buffer solution is added 1.00 L of 0.100 M HCl?

[*Hint:* Write equilibrium constant expressions for the two acids, $HC_2H_3O_2$ and $HCHO_2$. In parts (b) and (c), the total acetate concentration = $[HC_2H_3O_2] + [C_2H_3O_2^-]$ and total formate concentration = $[HCHO_2] + [CHO_2^-]$. Because each solution must be electrically neutral, in (b), $[Na^+] + [H_3O^+] = [C_2H_3O_2^-] + [CHO_2^-] + [OH^-]$; and in (c); $[Na^+] + [H_3O^+] = [C_2H_3O_2^-] + [CHO_2^-] + [OH^-] + [Cl^-]$. Because the buffer solution is acidic, a useful simplifying assumption is that $[OH^-] \simeq 0$.]

Self-Test Questions

For questions 81 through 90 select the single item that best completes each statement.

81. To *repress* the ionization of formic acid, $HCHO_2(aq)$, add to the solution (a) NaCl; (b) NaOH; (c) $NaCHO_2$; (d) $NaNO_3$.

82. To *increase* the ionization of formic acid, $HCHO_2(aq)$, add to the solution (a) NaCl; (b) $NaCHO_2$; (c) H_2SO_4; (d) $NaHCO_3$.

83. To *raise* the pH of 1.00 L of 0.50 M HCl(aq) *significantly*, add (a) 0.50 mol $HC_2H_3O_2$; (b) 1.00 mol NaCl; (c) 0.60 mol $NaC_2H_3O_2$; (d) 0.10 mol NaOH.

84. To convert $NH_4^+(aq)$ to $NH_3(aq)$, (a) add H_3O^+; (b) raise the pH; (c) add $KNO_3(aq)$; (d) add NaCl.

85. The effect of adding 0.001 mol KOH to 1.00 L of a solution that is 0.10 M NH_3–0.10 M NH_4Cl is to (a) raise the pH very slightly; (b) lower the pH very slightly; (c) raise the pH by several units; (d) lower the pH by several units.

86. The most acidic of the following 0.10 M salt solutions is that of (a) Na_2S; (b) $NaHSO_4$; (c) $NaHCO_3$; (d) Na_2HPO_4.

87. If an indicator is to be used in an acid–base titration having an equivalence point in the pH range 8 to 10, the indicator must (a) be a weak base; (b) have $K_a = 1 \times 10^{-9}$, (c) ionize in two steps; (d) be added to the solution only after the solution has become alkaline.

88. When a solution of a weak monoprotic acid has been *half-neutralized* by a strong base, (a) the pH = $\frac{1}{2}$ pK_a; (b) pH = $\frac{1}{2}$ of the pH value at the equivalence point; (c) pH = twice the initial pH value; (d) pH = pK_a.

89. In the titration of a *weak base* by a *strong acid*, the high-est pH on the titration curve is (a) the initial pH; (b) the pH at the half-neutralization point; (c) the pH at the equivalence point; (d) the pH beyond the equivalence point.

90. The ionization constants of citric acid, H_3Cit, are pK_{a_1} = 3.13; pK_{a_2} = 4.76; pK_{a_3} = 6.40. We expect a water solution of sodium dihydrogen citrate, NaH_2Cit, (a) to have a pH = pK_{a_1}; (b) to have a pH = pK_{a_2}; (c) to be acidic; (d) to be basic.

91. Explain the difference in meaning between an indicator *end point* and the *equivalence point* of a titration.

92. Indicate whether you would expect the equivalence point of each of the following titrations to be below, above, or at pH 7. Explain your reasoning. **(a)** $NaHCO_3(aq)$ is titrated with NaOH(aq); **(b)** HCl(aq) is titrated with $NH_3(aq)$; **(c)** KOH(aq) is titrated with HI(aq).

93. A $HCHO_2$–$NaCHO_2$ buffer solution is to be prepared; $K_a(HCHO_2$, formic acid) = 1.8×10^{-4}.

   (a) What mass of $NaCHO_2$ must be dissolved in 0.500 L of 0.650 M $HCHO_2$ to produce a pH of 3.90?
   (b) If to the 0.500 L of buffer solution produced in part (a) is added one small pellet of NaOH (about 0.20 g), what will the new pH value be?

94. 25.0 mL of 0.01000 M $HC_7H_5O_2$ (the monoprotic acid, benzoic acid, $K_a = 6.3 \times 10^{-5}$) is titrated by 0.01000 M $Ba(OH)_2$. Calculate the pH **(a)** of the initial acid solution; **(b)** after the addition of 6.25 mL of 0.0100 M $Ba(OH)_2$; **(c)** at the equivalence point; **(d)** after the addition of a total of 15.0 mL of 0.0100 M $Ba(OH)_2$.

19

Solubility and Complex Ion Equilibria

Precipitates of silver chromate (red-brown), basic copper chromate (brown), basic copper carbonate (blue), and lead iodide (yellow) formed in silica gels. The formation and physical form of a precipitate depend on a number of factors, including the concentrations of the reactant solutions. [Carey B. Van Loon]

The solubility rules (Table 5-3) suggest that calcium carbonate should precipitate from a solution containing Ca^{2+} and CO_3^{2-}. The net ionic equation for the reaction is certainly easy enough to write

$$Ca^{2+}(aq) + CO_3^{2-}(aq) \longrightarrow CaCO_3(s)$$

Yet there are conditions under which $CaCO_3$ does *not* precipitate. Whether precipitation occurs depends mainly on the concentrations of the Ca^{2+} and CO_3^{2-} ions. In turn, the CO_3^{2-} ion concentration depends on the *pH* of the solution. To develop a better understanding of the solubility of $CaCO_3$ and the conditions under which it precipitates, we need to consider *equilibrium* relationships between Ca^{2+} and CO_3^{2-}, and between CO_3^{2-}, H_3O^+, and HCO_3^-. This suggests that we will have to combine ideas about acid–base equilibria from Chapters 17 and 18 with new ideas to be introduced in this chapter. The behavior of $CaCO_3$ (limestone) toward dissolving and precipitation underlies several natural phenomena, ranging from the formation of limestone caverns (page 800), to the neutralization of acid rain in limestone soils, to the removal of boiler scale from an automatic coffee maker with a vinegar treatment (reaction 19.16).

Silver chloride is a familiar precipitate in the general chemistry laboratory, and its formation is reasonably independent of pH. However, we find that silver chloride does *not* precipitate from a solution having a moderate to high concentration of $NH_3(aq)$. This is because silver ion forms a stable complex ion with ammonia and remains in solution. Complex ion formation and equilibria involving complex ions are additional topics that we discuss in this chapter.

19-1 The Solubility Product Constant, K_{sp}

Gypsum, $CaSO_4 \cdot 2H_2O$, is an important calcium mineral found throughout the world. It is slightly soluble in water, and groundwater that comes into contact with gypsum often contains some dissolved calcium sulfate. This water is objectionable for certain applications, such as evaporative cooling systems in power plants, because solid deposits can form (much like boiler scale in a tea kettle). Equilibrium between $Ca^{2+}(aq)$ and $SO_4^{2-}(aq)$ and undissolved $CaSO_4$ can be represented as

$$CaSO_4(s) \rightleftharpoons Ca^{2+}(aq) + SO_4^{2-}(aq) \qquad (19.1)$$

To write an equilibrium constant expression for reaction (19.1), recall that we do not include a term for $CaSO_4(s)$.

$$K_c = [Ca^{2+}][SO_4^{2-}] \qquad (19.2)$$

Rather than K_c, the symbol generally used for the equilibrium constant that represents the equilibrium between an undissolved solute and its ions in a saturated solution is K_{sp}, called the **solubility product constant.**

$$K_{sp} = [Ca^{2+}][SO_4^{2-}] = 9.1 \times 10^{-6} \quad \text{(at 25 °C)} \qquad (19.3)$$

Some typical solubility product constants are listed in Table 19-1. Like other equilibrium constants, K_{sp} values depend on temperature. As illustrated in Example 19-1,

Formulating a K_{sp} expression.

A solubility product constant expression is the product of the concentrations of the ions appearing in the chemical equation for a solubility equilibrium, *with each term raised to the power given by the coefficent in the chemical equation.* $\qquad (19.4)$

681

TABLE 19-1
Solubility Product Constants at 25 °C

Solute	Solubility equilibrium	K_{sp}
aluminum hydroxide	$Al(OH)_3(s) \rightleftharpoons Al^{3+}(aq) + 3\ OH^-(aq)$	1.3×10^{-33}
barium carbonate	$BaCO_3(s) \rightleftharpoons Ba^{2+}(aq) + CO_3^{2-}(aq)$	5.1×10^{-9}
barium hydroxide	$Ba(OH)_2(s) \rightleftharpoons Ba^{2+}(aq) + 2\ OH^-(aq)$	5×10^{-3}
barium sulfate	$BaSO_4(s) \rightleftharpoons Ba^{2+}(aq) + SO_4^{2-}(aq)$	1.1×10^{-10}
bismuth(III) sulfide	$Bi_2S_3(s) \rightleftharpoons 2\ Bi^{3+}(aq) + 3\ S^{2-}(aq)$	1×10^{-97}
cadmium sulfide	$CdS(s) \rightleftharpoons Cd^{2+}(aq) + S^{2-}(aq)$	8×10^{-27}
calcium carbonate	$CaCO_3(s) \rightleftharpoons Ca^{2+}(aq) + CO_3^{2-}(aq)$	2.8×10^{-9}
calcium fluoride	$CaF_2(s) \rightleftharpoons Ca^{2+}(aq) + 2\ F^-(aq)$	5.3×10^{-9}
calcium hydroxide	$Ca(OH)_2(s) \rightleftharpoons Ca^{2+}(aq) + 2\ OH^-(aq)$	5.5×10^{-6}
calcium sulfate	$CaSO_4(s) \rightleftharpoons Ca^{2+}(aq) + SO_4^{2-}(aq)$	9.1×10^{-6}
chromium(III) hydroxide	$Cr(OH)_3(s) \rightleftharpoons Cr^{3+}(aq) + 3\ OH^-(aq)$	6.3×10^{-31}
cobalt(II) sulfide	$CoS(s) \rightleftharpoons Co^{2+}(aq) + S^{2-}(aq)$	4×10^{-21}
copper(II) sulfide	$CuS(s) \rightleftharpoons Cu^{2+}(aq) + S^{2-}(aq)$	6×10^{-36}
iron(II) sulfide	$FeS(s) \rightleftharpoons Fe^{2+}(aq) + S^{2-}(aq)$	6×10^{-18}
iron(III) hydroxide	$Fe(OH)_3(s) \rightleftharpoons Fe^{3+}(aq) + 3\ OH^-(aq)$	4×10^{-38}
lead(II) chloride	$PbCl_2(s) \rightleftharpoons Pb^{2+}(aq) + 2\ Cl^-(aq)$	1.6×10^{-5}
lead(II) chromate	$PbCrO_4(s) \rightleftharpoons Pb^{2+}(aq) + CrO_4^{2-}(aq)$	2.8×10^{-13}
lead(II) iodide	$PbI_2(s) \rightleftharpoons Pb^{2+}(aq) + 2\ I^-(aq)$	7.1×10^{-9}
lead(II) sulfate	$PbSO_4(s) \rightleftharpoons Pb^{2+}(aq) + SO_4^{2-}(aq)$	1.6×10^{-8}
lead(II) sulfide	$PbS(s) \rightleftharpoons Pb^{2+}(aq) + S^{2-}(aq)$	8×10^{-28}
lithium phosphate	$Li_3PO_4(s) \rightleftharpoons 3\ Li^+(aq) + PO_4^{3-}(aq)$	3.2×10^{-9}
magnesium carbonate	$MgCO_3(s) \rightleftharpoons Mg^{2+}(aq) + CO_3^{2-}(aq)$	3.5×10^{-8}
magnesium fluoride	$MgF_2(s) \rightleftharpoons Mg^{2+}(aq) + 2\ F^-(aq)$	3.7×10^{-8}
magnesium hydroxide	$Mg(OH)_2(s) \rightleftharpoons Mg^{2+}(aq) + 2\ OH^-(aq)$	1.8×10^{-11}
magnesium phosphate	$Mg_3(PO_4)_2(s) \rightleftharpoons 3\ Mg^{2+}(aq) + 2\ PO_4^{3-}(aq)$	1×10^{-25}
manganese(II) sulfide	$MnS(s) \rightleftharpoons Mn^{2+}(aq) + S^{2-}(aq)$	2×10^{-13}
mercury(I) chloride	$Hg_2Cl_2(s) \rightleftharpoons Hg_2^{2+}(aq) + 2\ Cl^-(aq)$	1.3×10^{-18}
mercury(II) sulfide	$HgS(s) \rightleftharpoons Hg^{2+}(aq) + S^{2-}(aq)$	2×10^{-52}
nickel(II) sulfide	$NiS(s) \rightleftharpoons Ni^{2+}(aq) + S^{2-}(aq)$	3×10^{-19}
silver bromide	$AgBr(s) \rightleftharpoons Ag^+(aq) + Br^-(aq)$	5.0×10^{-13}
silver carbonate	$Ag_2CO_3(s) \rightleftharpoons 2\ Ag^+(aq) + CO_3^{2-}(aq)$	8.1×10^{-12}
silver chloride	$AgCl(s) \rightleftharpoons Ag^+(aq) + Cl^-(aq)$	1.8×10^{-10}
silver chromate	$Ag_2CrO_4(s) \rightleftharpoons 2\ Ag^+(aq) + CrO_4^{2-}(aq)$	2.4×10^{-12}
silver iodide	$AgI(s) \rightleftharpoons Ag^+(aq) + I^-(aq)$	8.5×10^{-17}
silver sulfate	$Ag_2SO_4(s) \rightleftharpoons 2\ Ag^+(aq) + SO_4^{2-}(aq)$	1.4×10^{-5}
silver sulfide	$Ag_2S(s) \rightleftharpoons 2\ Ag^+(aq) + S^{2-}(aq)$	6×10^{-50}
strontium carbonate	$SrCO_3(s) \rightleftharpoons Sr^{2+}(aq) + CO_3^{2-}(aq)$	1.1×10^{-10}
strontium sulfate	$SrSO_4(s) \rightleftharpoons Sr^{2+}(aq) + SO_4^{2-}(aq)$	3.2×10^{-7}
tin(II) sulfide	$SnS(s) \rightleftharpoons Sn^{2+}(aq) + S^{2-}(aq)$	1×10^{-25}
zinc sulfide	$ZnS(s) \rightleftharpoons Zn^{2+}(aq) + S^{2-}(aq)$	1×10^{-21}

Example 19-1

Writing solubility product constant expressions for slightly soluble solutes.
Write the solubility product constant expression associated with a saturated
solution of

(a) calcium fluoride, CaF_2 (one of the products formed when a fluoride treat-
    ment is applied to teeth);
(b) bismuth sulfide, Bi_2S_3 [formed in small quantities when bismuthyl salicy-
    late ("Pepto-Bismol") passes through the intestinal tract].

Solution. The key to writing correct solubility product constant expressions is
to start with the solubility equilibrium equation for one mole of solute. Then,
simply write the equilibrium constant expression for that equation.

(a) $CaF_2(s) \rightleftharpoons Ca^{2+}(aq) + 2 F^-(aq)$ $K_{sp} = [Ca^{2+}][F^-]^2$

(b) $Bi_2S_3(s) \rightleftharpoons 2 Bi^{3+}(aq) + 3 S^{2-}(aq)$ $K_{sp} = [Bi^{3+}]^2[S^{2-}]^3$

SIMILAR EXAMPLES: Exercises 1, 2.

19-2   Relationship Between Solubility and K_{sp}

Because of the term *solubility,* we might expect some relationship between the *solubility* product constant (K_{sp}) of a solute and its **molar solubility** (its molarity concentration in a saturated aqueous solution). As shown in Examples 19-2 and 19-3, there is a relationship between them, since each can be calculated from the other. There are implicit assumptions involved in these calculations, though, such as the assumption that dissolved solute exists only as simple free cations and anions. Unlike the assumptions we have made in other cases, however, these assumptions are very often *not* valid. In Section 19-3 we will say more about the shortcomings of the K_{sp} concept, but for now let us consider the simplified calculations that are possible.

Example 19-2

Calculating K_{sp} of a slightly soluble solute from its experimentally measured solubility. A 100.0-mL sample is removed from a water solution saturated with $CaSO_4$ at 25 °C. The water is completely evaporated from the sample and a deposit of 0.24 g $CaSO_4$ is obtained. What is K_{sp} for $CaSO_4$ at 25 °C?

$CaSO_4(s) \rightleftharpoons Ca^{2+}(aq) + SO_4{}^{2-}(aq)$ $K_{sp} = ?$

Solution. The first step is to express the solubility in molarity.

$$\text{mol } CaSO_4/\text{L satd. soln.} = \frac{0.24 \text{ g } CaSO_4}{0.100 \text{ L soln.}} \times \frac{1 \text{ mol } CaSO_4}{136 \text{ g } CaSO_4}$$

$$= 0.018 \text{ M } CaSO_4$$

The key factors in the setup below (shown in blue) indicate that one mole of Ca^{2+} and one mole of $SO_4{}^{2-}$ appear in solution for each mole of $CaSO_4$ that dissolves.

$$[Ca^{2+}] = \frac{0.018 \text{ mol } CaSO_4}{L} \times \frac{1 \text{ mol } Ca^{2+}}{1 \text{ mol } CaSO_4} = 0.018 \text{ M } Ca^{2+}$$

$$[SO_4{}^{2-}] = \frac{0.018 \text{ mol } CaSO_4}{L} \times \frac{1 \text{ mol } SO_4{}^{2-}}{1 \text{ mol } CaSO_4} = 0.018 \text{ M } SO_4{}^{2-}$$

$K_{sp} = [Ca^{2+}][SO_4{}^{2-}] = (0.018)(0.018) = 3.2 \times 10^{-4}$

SIMILAR EXAMPLES: Exercises 16, 22, 23, 72.

Example 19-3

Calculating the solubility of a slightly soluble solute from its K_{sp} value. Lead iodide is a dense, golden yellow, "insoluble" solid used in bronzing and in ornamental work requiring a golden color (e.g., mosaic gold). Calculate the solubility of lead iodide in water at 25 °C, given that

$PbI_2(s) \rightleftharpoons Pb^{2+}(aq) + 2 I^-(aq)$ $K_{sp} = 7.1 \times 10^{-9}$

Solution. The solubility equilibrium equation shows that *one* mole of Pb^{2+} and *two* moles of I^- appear in solution for every mole of $PbI_2(s)$ that dissolves. If we let S represent the number of moles of PbI_2 that have dissolved per liter of saturated solution, then in this solution

$$[Pb^{2+}] = S \quad \text{and} \quad [I^-] = 2S$$

These concentrations must also satisfy the K_{sp} expression.

$$K_{sp} = [Pb^{2+}][I^-]^2 = (S)(2S)^2 = 7.1 \times 10^{-9}$$

$$4S^3 = 7.1 \times 10^{-9} \quad S^3 = 1.8 \times 10^{-9} \quad S = 1.2 \times 10^{-3}$$

S = solubility = 1.2×10^{-3} M PbI_2

SIMILAR EXAMPLES: Exercises 3, 17, 18, 19, 70.

Comparing Solubilities. In working with solubility equilibria, often you will be asked simply to compare the solubilities of different solutes, such as deciding which of a group of solutes is the most soluble. To do this, just comparing the magnitudes of K_{sp} values may not be sufficient. The basic idea is that

- If solutes are of the *same* type, the relationship of solubility to K_{sp} will be the same for each solute; their relative solubilities will be the same as their relative K_{sp} values. Solutes with the largest K_{sp} values should have the greatest solubilities. By the "same" type of solute we mean, for example, all solutes with the formula MX (such as AgCl, $CaSO_4$, and $AlPO_4$).
- If solutes are of *different types, molar solubilities must be obtained from K_{sp} values and then compared, as in Example 19-4.*

Example 19-4

Comparing solubilities for different solutes. You wish to prepare a saturated solution of a silver compound and you want it to have the highest concentration of silver ion possible. Which of the following compounds would you use? **(a)** AgCl, $K_{sp} = 1.8 \times 10^{-10}$; **(b)** Ag_2CrO_4, $K_{sp} = 2.4 \times 10^{-12}$; **(c)** AgBr(s), $K_{sp} = 5.0 \times 10^{-13}$.

Solution. Since AgCl and AgBr are solutes of the same type, the solubility of AgBr ($K_{sp} = 5.0 \times 10^{-13}$) should be less than that of AgCl ($K_{sp} = 1.8 \times 10^{-10}$). We can eliminate AgBr. However, AgCl and Ag_2CrO_4 are solutes of different types.

Suppose we let the solubility of AgCl = x. Then $[Ag^+] = [Cl^-] = x$, and $K_{sp} = [Ag^+][Cl^-] = x^2 = 1.8 \times 10^{-10}$. $x = [Ag^+] = \sqrt{1.8 \times 10^{-10}} \approx 1 \times 10^{-5}$ M.

If we let the solubility of $Ag_2CrO_4 = y$, then $[CrO_4^{2-}] = y$ and $[Ag^+] = 2y$. $K_{sp} = [Ag^+]^2[CrO_4^{2-}] = (2y)^2(y) = 4y^3 = 1.1 \times 10^{-12}$. $y^3 = 2.8 \times 10^{-13}$. We really do not have to carry the calculation beyond this point to see that $y > 1 \times 10^{-5}$ M and $[Ag^+] = 2y$. The silver ion concentration in saturated Ag_2CrO_4 should be highest of the three.

SIMILAR EXAMPLE: Exercise 17.

The Common Ion Effect. In Examples 19-2 and 19-3, the ions in saturated solutions came from a *single* source, the pure solid solute. Just as it represses the ionization of weak acids and weak bases, a *common ion reduces the solubility of a slightly soluble solute.* Suppose that to the saturated solution of PbI_2 in Example 19-3 we add some I^-—a *common ion*—from a source such as KI(aq).

(a)   (b)

Le Châtelier's principle reminds us that an equilibrium mixture responds to a forced increase in the concentration of one of its reactants by shifting in the direction in which that reactant is consumed. In the lead iodide solubility equilibrium,

$$PbI_2(s) \rightleftharpoons Pb^{2+}(aq) + 2\,I^-(aq)$$

if some of the common ion, I^-, is added the reverse reaction is favored,

leading to a new equilibrium in which

| some PbI_2 precipitates | $[Pb^{2+}]$ is less than in the original equilibrium | $[I^-]$ is greater than in the original equilibrium. |

The common ion effect in solubility equilibrium.

*The solubility of a slightly soluble ionic compound is lowered in the presence of a second solute that furnishes a **common ion**.*   (19.5)

The common ion effect is illustrated in Figure 19-1.

Example 19-5

Calculating the solubility of a slightly soluble solute in the presence of a common ion. What is the solubility of PbI_2 in 0.10 M KI(aq)?

Solution. Think of the situation here as being one in which we produce a saturated solution of PbI_2, but instead of using pure water as the solvent we use 0.10 M KI(aq). As in Example 19-3, we can let S represent the number of moles of PbI_2 per liter of saturated solution. If S moles of PbI_2 dissolve, then S moles of Pb^{2+} and $2S$ moles of I^- will appear in solution. However, the solution has an *additional* 0.10 mole of I^- per liter derived from the 0.10 M KI(aq). We can summarize this information in a familiar format.

$$PbI_2 \rightleftharpoons Pb^{2+}(aq) + 2\,I^-(aq)$$

from PbI_2:	S mol/L	$2S$ mol/L
from 0.10 M KI:		0.10 mol/L
equilibrium concentrations:	S mol/L	$(0.10 + 2S)$ mol/L

The usual K_{sp} relationship must be satisfied, that is,

$$K_{sp} = [Pb^{2+}][I^-]^2 = (S)(0.10 + 2S)^2 = 7.1 \times 10^{-9}$$

To simplify the solution to this equation, let us assume that S is much smaller than 0.10 M, so that $(0.10 + 2S) \approx 0.10$.

$$S(0.10)^2 = 7.1 \times 10^{-9}$$

$$S = 7.1 \times 10^{-7}\ M$$

Are You Wondering:

Why, in writing the K_{sp} expression for PbI$_2$ in Example 19-5, do we not double the I$^-$ concentration derived from the common ion, that is, why do we not write $K_{sp} = S(0.20 + 2S)^2$ instead of $K_{sp} = S(0.10 + 2S)^2$?

This is a common student error that you do not want to make. The expression $K_{sp} = [Pb^{2+}][I^-]^2$ tells us to (a) write the concentrations of Pb^{2+} and I^- in the saturated solution, (b) square $[I^-]$, and (c) take the product. In *any* solution of PbI$_2$, $[I^-]$ *derived from PbI$_2$* must be *twice* the molarity of the PbI$_2$, i.e., $2S$. But $[I^-]$ that comes from KI is just equal to the molarity of the KI, i.e., 0.10 M. We add these two concentrations together to get the total $[I^-]$ in the saturated solution—$(0.10 + 2S)$. Then, in the K_{sp} expression we replace $[I^-]^2$ with its equivalent: $(0.10 + 2S)^2$. In short, there simply is no relationship between the stoichiometry of the PbI$_2$ dissociation (which requires a factor of 2) and the KI dissociation (which does not).

Since we defined S as the number of moles of PbI$_2$ that dissolve per liter of the 0.10 M KI(aq), S is the quantity we are seeking.

$$S = \text{solubility} = 7.1 \times 10^{-7} \text{ M PbI}_2$$

SIMILAR EXAMPLES: Exercises 5, 26, 29, 32.

The solubility of PbI$_2$ in the presence of 0.10 M I$^-$ calculated in Example 19-5 is about 2000 times less than its value in pure water (Example 19-3). The effect of Pb^{2+} in reducing the solubility of PbI$_2$ is not as striking as that of I$^-$ (see Exercise 27), but it is significant nevertheless.

19-3 Limitations of the K_{sp} Concept

As we stated at the start of the previous section, calculations based on K_{sp} values often involve assumptions that are not valid. We need to examine some of the shortcomings of the K_{sp} concept and point out how we will deal with them.

Limitation of K_{sp} to Slightly Soluble Solutes. We have repeatedly used the term "slightly soluble" in describing the solutes used as examples in this chapter. Cannot similar expressions be written for moderately or highly soluble ionic compounds, such as NaCl, KNO$_3$, and NaOH? For these solutes we can write *thermodynamic* equilibrium constant expressions, but what we cannot do is substitute ionic concentrations for ionic activities. Saturated solutions of moderately or highly soluble ionic compounds are much too concentrated to permit the assumption that activities and molarity concentrations are equal. Without this assumption, much of the value of the solubility product concept is lost. Thus, you can expect that whenever a K_{sp} value is given it will be for a "slightly soluble" (essentially insoluble) solute.

The thermodynamic equilibrium constant, which is based on activities, was introduced on page 578, and is described more fully in Section 20-6.

The Diverse ("Uncommon") Ion Effect—The Salt Effect. We know what to expect when common ions are involved in a solubility equilibrium; they reduce the solubility of the slightly soluble solute. Next, we might ask, do ions *different* from those involved in the equilibrium ("uncommon" ions) have any effect on the solubilities of sparingly soluble ionic compounds? They do, but their effect is not as

Comparison of the common ion effect and the salt effect.

The presence of the common ion CrO_4^{2-}, derived from $K_2CrO_4(aq)$, reduces the solubility of Ag_2CrO_4 by a factor of about 35 over the concentration range shown (from 0 to 0.10 M added salt). Over this same range, the solubility of Ag_2CrO_4 is increased by the presence of the "uncommon" or diverse ions from KNO_3, but only by about 25% or so.

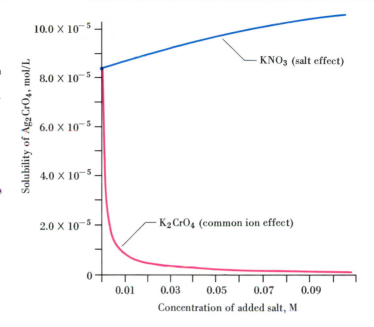

striking as the common ion effect. Moreover, the presence of "uncommon" ions tends to *increase* rather than decrease solubility. As the total ionic concentration of a solution increases, interionic attractions become more important (recall Section 13-9). Activities (effective concentrations) become smaller than the stoichiometric or measured concentrations. For the ions involved in the solution process this means that higher concentrations must appear in solution before equilibrium is established— *the solubility increases.* Figure 19-2 compares the common ion effect and the effect of "uncommon" ions.

The "uncommon" or diverse ion effect is more commonly called the **salt effect.** Because of the salt effect, the numerical value of K_{sp} based on molarity concentrations will vary depending on the ionic atmosphere. Tabulated values of K_{sp} usually are thermodynamic solubility product constants, that is, based on activities. When these K_{sp} values are used with molarities instead of activities, they often given results that are not accurate.

Other Factors Affecting the Solubilities of Sparingly Soluble Ionic Compounds. As we noted in performing calculations in Section 19-2, we *assumed* that all of the dissolved solute appears in solution as separated cations and anions, and this assumption is often *not* valid. For example, in a saturated solution of magnesium fluoride, in addition to individual Mg^{2+} and F^- ions there exist combinations consisting of one Mg^{2+} and one F^-; these are the **ion pairs, MgF^+.** To the extent that ion-pair formation occurs, the *free* ion concentrations are reduced. This means that the amount of solute that must dissolve to maintain the required *free* ion concentrations to satisfy the K_{sp} expression increases. *Solubility increases when ion-pair formation occurs in solution.*

Other factors that can *significantly* affect solute solubilities are acid–base or complex ion equilibria that may occur *simultaneously* with solubility equilibria. We consider these specific possibilities in some detail later in the chapter.

Adjusting to the Limitations of K_{sp}. Let us assess the importance of the various effects discussed in this section—molarities substituting for activities, salt effect, ion-pair formation. All of these apply to $CaSO_4$. Recall that we *calculated* K_{sp} for $CaSO_4$ in Example 19-2, based on its measured solubility. Our result was $K_{sp} = 3.2 \times 10^{-4}$. This value is about 35 times *larger* than the value listed in Table 19-1: $K_{sp} = 9.1 \times 10^{-6}$. Clearly, our result was correct only in its general order of magnitude (that is, by a factor of 10 to 100). If we had been required to obtain a more precise result, we would have had to take several additional factors into account, but calculations involving these factors are beyond the scope of this text. Moreover, "order-of-magnitude" results are usually sufficient for our purposes. They give us a general idea of how soluble slightly soluble solutes are; they allow us to compare relative solubilities of related compounds; and they are generally satisfactory for predicting the conditions under which a precipitate will form (Section 19-4).

19-4   Precipitation Reactions

Silver iodide (a light sensitive material) is used in photographic film and in cloud seeding to produce rain. Its solubility equilibrium can be represented as

$$AgI(s) \rightleftharpoons Ag^+(aq) + I^-(aq)$$

$$K_{sp} = [Ag^+][I^-] = 8.5 \times 10^{-17}$$

Suppose we mix solutions of $AgNO_3(aq)$ and $KI(aq)$, so as to obtain a mixed solution that has $[Ag^+] = 0.010$ M and $[I^-] = 0.015$ M. Is this solution unsaturated, saturated, or supersaturated? Recall the reaction quotient, Q, introduced in Chapter 16. It has the same form as an equilibrium constant expression but uses actual (initial) concentrations rather than equilibrium concentrations. Here,

$$Q = [Ag^+]_{init.} \times [I^-]_{init.} = (0.010)(0.015) = 1.5 \times 10^{-4} > K_{sp}$$

The fact that $Q > K_{sp}$ indicates that the concentrations of Ag^+ and I^- are higher than they would be in a saturated solution. The solution is *supersaturated,* and as is generally the case with supersaturated solutions, excess AgI should precipitate until the solution is saturated. If we had found $Q < K_{sp}$, the solution would have been *unsaturated;* no precipitate would form from such a solution.

When applied to solubility equilibria, Q is generally called the **ion product,** since its form is that of a product of ion concentrations raised to appropriate powers. A criterion for determining whether ions in a solution will combine to form a precipitate is to compare the ion product with K_{sp}. This leads to the criterion

Precipitation *should occur* if $Q > K_{sp}$.

Precipitation *cannot occur* if $Q < K_{sp}$.   (19.6)

A solution is just saturated if $Q = K_{sp}$.

This criterion is illustrated in Figure 19-3 and Example 19-6. This example emphasizes that *any possible dilutions that may occur must be considered before the criterion for precipitation is applied.*

Example 19-6

Applying the criterion for precipitation of a slightly soluble solute. Three drops of 0.20 M KI are added to 100.0 mL of 0.010 M $Pb(NO_3)_2$. Will a precipitate of lead iodide form? (Assume 1 drop = 0.05 mL.)

$$PbI_2(s) \rightleftharpoons Pb^{2+}(aq) + 2\,I^-(aq) \qquad K_{sp} = 7.1 \times 10^{-9}$$

Solution. We need to compare the product $[Pb^{2+}][I^-]^2$, based on the initial concentrations, with K_{sp} for PbI_2. For $[Pb^{2+}]$ we can simply use 0.010 M. For $[I^-]$, however, we must consider the great reduction in concentration that occurs when the three drops of 0.20 M KI are diluted to 100.0 mL.

(a)

(b)

FIGURE 19-3
Applying the criterion for precipitation from solution—Example 19-6 illustrated.

(a) When three drops of 0.20 M KI are added to 100.0 mL of 0.010 M Pb(NO₃)₂, at first a precipitate forms because K_{sp} is exceeded in the immediate vicinity of the drops. (b) When the KI becomes uniformly mixed in the Pb(NO₃)₂(aq), K_{sp} is no longer exceeded and the precipitate disappears. The criterion for precipitation must be applied *after* dilution has occurred. [Carey B. Van Loon]

Dilution calculation:

$$\text{no. mol I}^- = 3 \text{ drops} \times \frac{0.05 \text{ mL}}{1 \text{ drop}} \times \frac{1 \text{ L}}{1000 \text{ mL}} \times \frac{0.20 \text{ mol KI}}{\text{L}} \times \frac{1 \text{ mol I}^-}{1 \text{ mol KI}}$$

$$= 3 \times 10^{-5} \text{ mol I}^-$$

$$[\text{I}^-] = \frac{3 \times 10^{-5} \text{ mol I}^-}{0.1000 \text{ L}} = 3 \times 10^{-4} \text{ mol I}^-/\text{L}$$

Applying precipitation criterion:

$$Q = [\text{Pb}^{2+}][\text{I}^-]^2 = (0.010)(3 \times 10^{-4})^2 = 9 \times 10^{-10}$$

Since Q (9×10^{-10}) is *smaller than* K_{sp} (7.1×10^{-9}), we conclude that **PbI₂(s)** should *not* precipitate.

SIMILAR EXAMPLES: Exercises 6, 33, 34, 36.

Completeness of Precipitation. In applying the criterion for precipitation, some important questions are left unanswered, such as how much precipitate forms, and how much solute remains in solution? A useful rule of thumb in most applications is that precipitation is complete if 99.9% or more of the solute precipitates and less than one part per thousand (0.1%) remains in solution. Example 19-7 considers a case where precipitation is complete and Example 19-8, a case in which it is not. There is another difference between these two examples. In Example 19-7 the concentration of one of the ions (OH⁻) is held constant throughout a precipitation; in Example 19-8, the concentrations of both of the ions involved in the precipitation decrease. Example 19-7 is easier than 19-8, and you will find it useful to be able to recognize if an ion concentration remains essentially constant in a precipitation.

Example 19-7

Assessing the completeness of a precipitation reaction, when one of the ion concentrations is held constant. The first step in the extraction of magnesium metal from seawater involves precipitating Mg^{2+} as $\text{Mg(OH)}_2(s)$. The magnesium ion concentration in seawater is about 0.059 M. If a seawater sample is treated so that its [OH⁻] is maintained at 2.0×10^{-3} M, **(a)** what will be [Mg²⁺] remaining in solution after precipitation has occurred? **(b)** Can we say that precipitation of $\text{Mg(OH)}_2(s)$ is complete under these conditions?

The precipitation of Mg(OH)₂(s) from seawater is carried out in large vats. This is the first step in the Dow process for extracting magnesium from seawater. [Courtesy The Dow Chemical Company]

Solution

(a) There is no question that precipitation will occur since the product, $[Mg^{2+}][OH^-]^2 = (0.059)(2.0 \times 10^{-3})^2 = 2.4 \times 10^{-7}$, exceeds K_{sp}. We need to determine the $[Mg^{2+}]$ remaining in a solution from which some solid $Mg(OH)_2$ has precipitated. For the saturated solution at equilibrium, we substitute $[OH^-]$ (maintained constant at 2.0×10^{-3} M) into the solubility product constant expression and solve for $[Mg^{2+}]$.

$$K_{sp} = [Mg^{2+}][OH^-]^2 = [Mg^{2+}] \times (2.0 \times 10^{-3})^2 = 1.8 \times 10^{-11}$$

$$[Mg^{2+}] = \frac{1.8 \times 10^{-11}}{4.0 \times 10^{-6}} = 4.5 \times 10^{-6} \text{ M}$$

(b) $[Mg^{2+}]$ in the seawater is reduced from 0.059 M to 4.5×10^{-6} M as a result of the precipitation reaction. Since 4.5×10^{-6} M is less than 0.1% of 0.059 M, we can conclude that precipitation is complete.

SIMILAR EXAMPLES: Exercises 39a, 40.

Example 19-8

Assessing the completeness of a precipitation reaction, when both ion concentrations change during the precipitation. A 50.0-mL sample of 0.0152 M $Na_2SO_4(aq)$ is added to 50.0 mL of 0.0125 M $Ca(NO_3)_2(aq)$. (a) Should precipitation of $CaSO_4(s)$ occur? (b) Will precipitation of Ca^{2+} be complete?

$$CaSO_4(s) \rightleftharpoons Ca^{2+}(aq) + SO_4^{2-}(aq) \qquad K_{sp} = 9.1 \times 10^{-6}$$

Solution

(a) The concentrations of ions present *after* mixing (i.e., in the final 100.0 mL solution) are

$$[Ca^{2+}] = 0.0500 \text{ L} \times \frac{0.0125 \text{ mol } Ca^{2+}}{L} \times \frac{1}{0.1000 \text{ L}} = 6.25 \times 10^{-3} \text{ M}$$

$$[SO_4^{2-}] = 0.0500 \text{ L} \times \frac{0.0152 \text{ mol } SO_4^{2-}}{L} \times \frac{1}{0.1000 \text{ L}} = 7.60 \times 10^{-3} \text{ M}$$

$$Q = [Ca^{2+}][SO_4^{2-}] = (6.25 \times 10^{-3})(7.60 \times 10^{-3})$$
$$= 4.75 \times 10^{-5} > K_{sp} = 9.1 \times 10^{-6}$$

Precipitation of $CaSO_4(s)$ should occur.

With the value of K_{sp} for $CaSO_4$ from Example 19-2 (3.2×10^{-4}), our conclusion would be no precipitation. Because of the factors discussed in Section 19-3, at times predictions based on K_{sp} values may not be confirmed by experiment, especially if Q is not greatly different from K_{sp}.

(b) The course of the precipitation reaction is outlined below. In this outline the molarity of Ca^{2+} in equilibrium with $CaSO_4(s)$ is set equal to x, and we must solve for x.

	$CaSO_4(s) \rightleftharpoons$	$Ca^{2+}(aq)$ +	$SO_4^{2-}(aq)$
initial concentrations from (a):		0.00625 M	0.00760 M
changes:		$-(0.00625 - x)$	$-(0.00625 - x)$
at equilibrium:		x	$0.00760 - (0.00625 - x)$
			$0.00760 - 0.00625 + x$
			$0.00135 + x$

What we have indicated above is that if $[Ca^{2+}]$ falls from 0.00625 M to x, then the precipitation of $CaSO_4(s)$ must have consumed a concentration of Ca^{2+} equal to $0.00625 - x$. One SO_4^{2-} is removed from solution for every Ca^{2+} ion. The decrease in $[SO_4^{2-}]$ must also be $0.00625 - x$. The $[SO_4^{2-}]$ *remaining* in solution at equilibrium is

$$0.00760 - (0.00625 - x) = 0.00135 + x$$

Are You Wondering:

If there is an easy way to assess the completeness of a precipitation, that is, to know in advance whether to use the method of Example 19-7 or 19-8?

The key factors are the numerical value of K_{sp} and the concentration of the common ion in the saturated solution. Completeness of precipitation is favored by a

- *small value of K_{sp}*
- *high concentration of common ion.*

We cannot say exactly how small K_{sp} nor how large the common ion concentration must be. However, in Example 19-8 K_{sp} has one of the *largest* values in Table 19-1 and $[SO_4^{2-}]$ (the common ion) is low. These conditions certainly *do not* favor completeness of precipitation. Note also that when the concentration of common ion is *large* you may be able to treat it as a *constant*. This allows you to use the simpler method of Example 19-7.

We can now substitute the equilibrium concentrations into the K_{sp} expression and solve for x.

$$K_{sp} = [Ca^{2+}][SO_4^{2-}] = (x)(0.00135 + x) = 9.1 \times 10^{-6}$$

The result is a quadratic equation; $x^2 + 0.00135x - 9.1 \times 10^{-6} = 0$. Solution of this equation by the quadratic formula leads to the result

The usual simplifying assumption—that x is very small compared to a number to which it is added or from which it is subtracted—does not work here. The number 0.00135 is itself quite small. The calculated value of x turns out to be larger, not smaller, than 0.00135.

$$x = \frac{-0.00135 \pm \sqrt{(0.00135)^2 + 4 \times 9.1 \times 10^{-6}}}{2}$$

$$= \frac{-0.00135 \pm 6.2 \times 10^{-3}}{2}$$

$$[Ca^{2+}] = x = \frac{4.8 \times 10^{-3}}{2} = 2.4 \times 10^{-3} \text{ M}$$

The percentage of calcium ion left in solution can be expressed as

$$\% \text{ Ca}^{2+} \text{ remaining} = \frac{2.4 \times 10^{-3} \text{ M Ca}^{2+}}{6.25 \times 10^{-3} \text{ M Ca}^{2+}} \times 100 = 38\%$$

Precipitation of Ca^{2+} is incomplete.

SIMILAR EXAMPLES: Exercises 38, 41, 43.

19-5 Precipitation Reactions in Quantitative Analysis

In Chapter 3 we described how the precipitation of a solid can be used to determine the exact composition of a sample of matter (see, for example, Example 3-9). What does it take to conduct a successful precipitation analysis? We assessed one key factor through Examples 19-7 and 19-8: *completeness of precipitation*. Another solubility equilibrium principle is used when purifying a precipitate by washing. Often this is done, not with pure water, but with a water solution containing the common ion. This reduces the solubility of the precipitate. Figure 19-4 applies these ideas to the gravimetric determination (determination by weighing) of calcium.

FIGURE 19-4
Outline of a gravimetric analysis for calcium.

Ca^{2+}(aq) [from sample being analyzed]

add excess
(NH$_4$)$_2$C$_2$O$_4$(aq)

CaC$_2$O$_4 \cdot$H$_2$O(s) [impure]

wash with dilute (NH$_4$)$_2$C$_2$O$_4$(aq);
filter and dry

CaC$_2$O$_4 \cdot$H$_2$O(s) [pure]

heat to 500°C

CaCO$_3$(s) [pure]

A weighed sample is dissolved [usually in HCl(aq)] to obtain Ca^{2+}(aq). The solution is treated with excess (NH$_4$)$_2$C$_2$O$_4$(aq), and the hydrate, CaC$_2$O$_4 \cdot$H$_2$O, precipitates. The precipitate is purified by washing with (NH$_4$)$_2$C$_2$O$_4$(aq). When heated strongly, CaC$_2$O$_4 \cdot$H$_2$O(s) is converted to CaCO$_3$(s), in which form the calcium is finally weighed.

In the washing process it is important to use a solution that provides a common ion to reduce the solubility of CaC$_2$O$_4 \cdot$H$_2$O(s) but one that does not leave a residue that would contaminate the final CaCO$_3$(s). Upon heating, (NH$_4$)$_2$C$_2$O$_4$ undergoes decomposition to gaseous products—NH$_3$, H$_2$O, CO, and CO$_2$.

A reagent is a substance or solution that participates in a chemical reaction.

Fractional Precipitation. In this method two or more ions, each capable of being precipitated by the same reagent, are *separated* by the use of that reagent: *One ion is precipitated while the other(s) remains in solution.* The primary condition for a successful fractional precipitation is that there be a significant difference in the solubilities of the substances being separated. (Usually this means a significant difference in their K_{sp} values.)

Example 19-9 considers the separation of CrO$_4^{2-}$(aq) and Br$^-$(aq) through the use of Ag$^+$(aq). The data needed to describe this separation are the solubility equilibrium equations and K_{sp} values of Ag$_2$CrO$_4$ and AgBr.

$$Ag_2CrO_4(s) \rightleftharpoons 2\ Ag^+(aq) + CrO_4^{2-}(aq) \qquad K_{sp} = 2.4 \times 10^{-12}$$

$$AgBr(s) \rightleftharpoons Ag^+(aq) + Br^-(aq) \qquad K_{sp} = 5.0 \times 10^{-13}$$

Example 19-9

Separating ions by fractional precipitation. To a solution that has [CrO$_4^{2-}$] = 0.010 M and [Br$^-$] = 0.010 M is slowly added 0.10 M AgNO$_3$(aq) (see Figure 19-5)

(a) Show that AgBr(s) precipitates before Ag$_2$CrO$_4$(s).

(b) At the point where Ag$_2$CrO$_4$(s) begins to precipitate, what is [Br$^-$] remaining in solution?

(c) Is separation of Br$^-$(aq) and CrO$_4^{2-}$(aq) by fractional precipitation feasible?

FIGURE 19-5

Fractional precipitation—
Example 19-9 illustrated.

(a) 0.10 M AgNO$_3$(aq) is slowly added to a solution that is 0.010 M in Br$^-$ and 0.010 M in CrO$_4^{2-}$.
(b) At this point the red-brown Ag$_2$CrO$_4$(s) is just about to precipitate. Essentially all the Br$^-$ has precipitated as pale yellow AgBr(s), leaving [Br$^-$] = 3.3 × 10^{-8} M in the solution. The Br$^-$ and CrO$_4^{2-}$ have been separated.

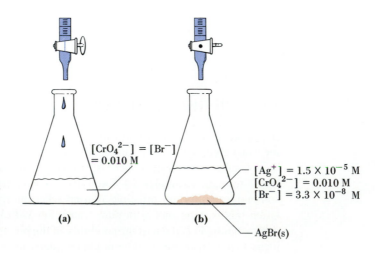

[CrO$_4^{2-}$] = [Br$^-$]
= 0.010 M

(a)

[Ag$^+$] = 1.5 × 10^{-5} M
[CrO$_4^{2-}$] = 0.010 M
[Br$^-$] = 3.3 × 10^{-8} M

(b)

AgBr(s)

Solution

(a) As a drop of the $AgNO_3(aq)$ enters the solution, $[Ag^+]$ builds up from a value of zero to a point where one of the ion products Q just exceeds the corresponding K_{sp}. Then precipitation begins. The required values of $[Ag^+]$ for precipitation are

AgBr ppt: $\quad Q = [Ag^+][Br^-] = [Ag^+](0.010) = 5.0 \times 10^{-13} = K_{sp}$
$\qquad\qquad [Ag^+] = 5.0 \times 10^{-11}\ M$

Ag_2CrO_4 ppt: $\quad Q = [Ag^+]^2[CrO_4^{2-}] = [Ag^+]^2(0.010) = 2.4 \times 10^{-12} = K_{sp}$
$\qquad\qquad [Ag^+]^2 = 2.4 \times 10^{-10} \qquad [Ag^+] = 1.5 \times 10^{-5}\ M$

Since $[Ag^+]$ required to start the precipitation of $AgBr(s)$ is much less than that for $Ag_2CrO_4(s)$, AgBr(s) precipitates first. As long as AgBr(s) is forming, the free silver ion concentration is not able to reach the value required for the precipitation of $Ag_2CrO_4(s)$.

(b) As more and more AgBr(s) precipitates, $[Br^-]$ gradually decreases, and this permits $[Ag^+]$ to increase. When $[Ag^+]$ reaches $1.5 \times 10^{-5}\ M$ precipitation of $Ag_2CrO_4(s)$ begins. Next, we need to answer the question: What is $[Br^-]$ at the point where $[Ag^+] = 1.5 \times 10^{-5}\ M$? For this we use K_{sp} for AgBr and solve for $[Br^-]$.

$$K_{sp} = [Ag^+][Br^-] = (1.5 \times 10^{-5})[Br^-] = 5.0 \times 10^{-13}$$

$$[Br^-] = 5.0 \times 10^{-13}/1.5 \times 10^{-5} = 3.3 \times 10^{-8}\ M$$

(c) Before $Ag_2CrO_4(s)$ begins to precipitate, $[Br^-]$ will have been reduced from $1.0 \times 10^{-2}\ M$ to $3.3 \times 10^{-8}\ M$. Essentially all of the Br^- will have precipitated from solution as AgBr(s) while the CrO_4^{2-} remains in solution. Fractional precipitation is feasible for separating mixtures of Br^- and CrO_4^{2-}.

SIMILAR EXAMPLES: Exercises 7, 44, 45, 46.

How might the titration illustrated in Figure 19-5 and Example 19-9 be carried out? That is, how can we stop the titration just as $Ag_2CrO_4(s)$ starts to precipitate? One possibility, suggested in Figure 19-5, is to look for a color change in the precipitate from pale yellow (AgBr) to red-brown (Ag_2CrO_4). A more effective method is to follow $[Ag^+]$ during the titration. $[Ag^+]$ increases very rapidly between the point where AgBr has finished precipitating and Ag_2CrO_4 is about to begin (see Exercise 85). We discuss an electrometric method of determining very low ion concentrations in solution in Chapter 21.

19-6 pH and Solubility

$Mg(OH)_2(s)$, when suspended in water, is known as "milk of magnesia"; it is a popular antacid. Its action as an antacid involves OH^- ions, derived from the solubility equilibrium, reacting with H_3O^+ (excess stomach acid) to form H_2O.

$$Mg(OH)_2(s) \rightleftharpoons Mg^{2+}(aq) + 2\ OH^-(aq) \qquad K_{sp} = 1.8 \times 10^{-11} \qquad (19.7)$$

$$OH^-(aq) + H_3O^+(aq) \longrightarrow 2\ H_2O \qquad\qquad\qquad\qquad (19.8)$$

Le Châtelier's principle suggests that reaction (19.7) is displaced to the right—$Mg(OH)_2$ dissolves to replace OH^- ions drawn off by the neutralization reaction

(19.8). In fairly acidic solutions, reactions (19.7) and (19.8) both go to completion; $Mg(OH)_2$ is highly soluble. The *net* reaction is

$$Mg(OH)_2(s) + 2 H_3O^+ \longrightarrow Mg^{2+}(aq) + 4 H_2O \qquad (19.9)$$

In summary, *slightly soluble salts having basic anions* [such as $Mg(OH)_2$, $ZnCO_3$, MgF_2, $Ca_2C_2O_4$] *become more soluble in acidic solutions.*

Net Ionic Equations. Although we have stressed quantitative problem solving to this point in the chapter, you should be aware of all the information about solution equilibria that can be packed into chemical equations, especially when they are written as net ionic equations. Equation (19.9) is a net ionic equation showing that $Mg(OH)_2(s)$ is soluble in acidic solutions. Example 19-10 illustrates a case in which *three* different types of solution equilibria are involved: solubility equilibrium, acid–base equilibrium, and gas formation. When you write a net ionic equation as in Example 19-10,

> - Use *ionic* formulas (that is, show separated ions) for *strong electrolytes* in aqueous solution—strong acids, strong bases, and salts.
> - Use *molecular* formulas for *nonelectrolytes* (e.g., H_2O) *weak electrolytes* (e.g., $HC_2H_3O_2$), *gases* (e.g., CO_2), and *undissolved solids* (e.g., $CaCO_3$).
> - If equilibrium is displaced far to the right, use a *single* arrow to indicate that the reaction "goes to completion."
> - If significant amounts of both reactants and products are likely to be present at equilibrium, use a *double* arrow. (19.10)

We first discussed net ionic equations in Chapters 4 and 5.

Important ideas regarding net ionic equations.

Example 19-10

Writing a net ionic equation to describe solution equilibria. Write a net ionic equation to represent the dissolving of calcium carbonate, $CaCO_3(s)$, in acetic acid, $HC_2H_3O_2(aq)$.

Solution. Carbonate ions are produced by the reaction

$$CaCO_3(s) \rightleftharpoons Ca^{2+}(aq) + CO_3^{2-}(aq) \qquad (19.11)$$

Hydronium ion is furnished by the ionization of acetic acid.

$$HC_2H_3O_2(aq) + H_2O \rightleftharpoons H_3O^+(aq) + C_2H_3O_2^-(aq) \qquad (19.12)$$

Carbonate ion, CO_3^{2-}, is a stronger base than acetate ion, $C_2H_3O_2^-$ (recall Table 17-5) and accepts protons from H_3O^+.

$$CO_3^{2-}(aq) + H_3O^+(aq) \longrightarrow HCO_3^-(aq) + H_2O \qquad (19.13)$$

Reaction (19.13) promotes both the ionization of $HC_2H_3O_2$ (19.12) and the dissolving of $CaCO_3$ (19.11). Further reaction occurs between HCO_3^- and H_3O^+.

$$HCO_3^-(aq) + H_3O^+(aq) \longrightarrow H_2CO_3(aq) + H_2O \qquad (19.14)$$

Finally, H_2CO_3 decomposes.

$$H_2CO_3(aq) \longrightarrow H_2O + CO_2(g) \qquad (19.15)$$

As you gain experience in writing net ionic equations, you may be able to write an equation like (19.16) directly, without breaking it down into as many steps as we did here.

The net reaction that occurs when calcium carbonate dissolves in acetic acid, obtained by combining equations (19.11) through (19.15), is

$$CaCO_3(s) + 2 HC_2H_3O_2(aq) \longrightarrow$$
$$Ca^{2+}(aq) + 2 C_2H_3O_2^-(aq) + H_2O + CO_2(g) \qquad (19.16)$$

SIMILAR EXAMPLES: Exercises 8, 66, 81.

Recall the discussion of buffer solutions in Section 18-2.

Illustrative Examples. There are essentially three types of calculations involved in situations where both a solubility equilibrium and an acid–base equilibrium are involved. Examples 19-11 through 19-13 consider these three types. You will probably find Example 19-11 to be easiest and 19-13, a little more difficult. In Example 19-12, where we are interested in preventing precipitation of $Mg(OH)_2$, note how the solution containing NH_3 and an ammonium salt is a *buffer* solution. This example illustrates how we can use a buffer solution to *control* a precipitation reaction.

Example 19-11

Determining whether a precipitate will form in a solution in which there is also an acid–base equilibrium. Should $Mg(OH)_2$ precipitate from a solution that is 0.010 M $MgCl_2$ if the solution is also made 0.10 M in NH_3?

Solution. Whenever we encounter a question that asks whether a precipitate will form, we should expect to compare Q and K_{sp}. The key here is in understanding that the hydroxide ion concentration is established by the ammonia ionization reaction.

Step 1. Consider the ionization equilibrium in $NH_3(aq)$.

$$NH_3 + H_2O \rightleftharpoons NH_4^+ + OH^- \qquad K_b = 1.74 \times 10^{-5}$$

If we let $x = [NH_4^+] = [OH^-]$ and $[NH_3] = (0.10 - x) \simeq 0.10$, we obtain

$$K_b = \frac{[NH_4^+][OH^-]}{[NH_3]} = \frac{x \cdot x}{0.10} = 1.74 \times 10^{-5}$$

$$x^2 = 1.7 \times 10^{-6} \qquad x = [OH^-] = 1.3 \times 10^{-3} \text{ M}$$

Step 2. Now we can rephrase the question as: Should $Mg(OH)_2(s)$ precipitate from a solution in which $[Mg^{2+}] = 1.0 \times 10^{-2}$ M and $[OH^-] = 1.3 \times 10^{-3}$ M? We must compare the ion product, Q, with K_{sp}.

$$Q = [Mg^{2+}][OH^-]^2 = (1.0 \times 10^{-2})(1.3 \times 10^{-3})^2$$

$$= 1.7 \times 10^{-8} > K_{sp} = 1.8 \times 10^{-11}$$

Precipitation should occur.

SIMILAR EXAMPLES: Exercises 9, 48, 80.

Example 19-12

Controlling the concentration of a solute species, either to cause precipitation or to prevent it. What concentration of NH_4^+ must be maintained to prevent the precipitation of $Mg(OH)_2$ from a solution that is 0.010 M $MgCl_2$ and 0.10 M NH_3?

Solution. The maximum value that the ion product, Q, can achieve before precipitation occurs is $Q = K_{sp} = 1.8 \times 10^{-11}$. This allows us to determine the maximum concentration of OH^- that can be tolerated.

$$[Mg^{2+}][OH^-]^2 = (1.0 \times 10^{-2})[OH^-]^2 = 1.8 \times 10^{-11}$$

$$[OH^-]^2 = 1.8 \times 10^{-9} \qquad [OH^-] = 4.2 \times 10^{-5} \text{ M}$$

Next we determine what $[NH_4^+]$ must be present in 0.10 M NH_3 to maintain $[OH^-] = 4.2 \times 10^{-5}$ M.

$$K_b = \frac{[NH_4^+][OH^-]}{[NH_3]} = \frac{[NH_4^+](4.2 \times 10^{-5})}{0.10} = 1.74 \times 10^{-5}$$

$[NH_4^+] = 0.041 \text{ M}$

SIMILAR EXAMPLES: Exercises 10, 75.

Example 19-13

Determining the solubility of a solute when an acid–base reaction occurs simultaneously with the solubility equilibrium. Calculate the solubility of $CaF_2(s)$ in a buffer solution with pH = 3.00.

Solution. CaF_2 dissolves to the extent that its K_{sp} expression is obeyed. At the same time the ionization equilibrium of HF also must be satisfied. The most direct approach in situations where two or more equilibria must be considered simultaneously is to look for a way of combining these into a single equilibrium constant expression that can be solved. In combining the expressions below we make use of ideas first introduced in Section 16-3.

$$CaF_2(s) \rightleftharpoons Ca^{2+}(aq) + 2\,\cancel{F^-(aq)} \qquad\qquad K_{sp} = 5.3 \times 10^{-9}$$
$$2\,H_3O^+(aq) + 2\,\cancel{F^-(aq)} \rightleftharpoons 2\,HF(aq) + 2\,H_2O \qquad K = 1/(K_a)^2$$
$$= 1/(6.7 \times 10^{-4})^2$$

$$\overline{CaF_2(s) + 2\,H_3O^+(aq) \rightleftharpoons Ca^{2+}(aq) + 2\,HF(aq) + 2\,H_2O}$$
$$K = K_{sp}/(K_a)^2 = 1.2 \times 10^{-2}$$
$$(19.17)$$

We can now deal with expression (19.17) in a familiar fashion. That is, if we let the solubility of $CaF_2 = x$ mol/L and note that $[H_3O^+]$ remains constant at 1.0×10^{-3} (corresponding to pH = 3.00), we obtain

$$CaF_2 + 2\,H_3O^+(aq) \rightleftharpoons Ca^{2+}(aq) + 2\,HF(aq) + 2\,H_2O$$

initial concentrations:	1.0×10^{-3} M	—	—
changes:	—	$+x$ M	$+2x$ M
at equilibrium:	1.0×10^{-3} M	x M	$2x$ M

$$K = \frac{[Ca^{2+}][HF]^2}{[H_3O^+]^2} = \frac{x \cdot (2x)^2}{(1.0 \times 10^{-3})^2} = 1.2 \times 10^{-2}$$

$$4x^3 = 1.2 \times 10^{-8} \qquad x^3 = 3.0 \times 10^{-9} \qquad x = 1.4 \times 10^{-3}$$

The molar solubility of CaF_2 in a buffer solution of pH 3.00 is 1.4×10^{-3} M.

SIMILAR EXAMPLES: Exercises 49, 50, 51, 53, 78.

19-7   Complex Ions and Coordination Compounds— An Introduction

In the previous section we saw that some slightly soluble ionic compounds become much more soluble in the presence of H_3O^+—a Brønsted–Lowry acid. There are other compounds that become more soluble in the presence of *Lewis bases*. These are compounds in which a metal ion can combine with a Lewis base to form *complex ions*. In this section we study aspects of complex ion formation that affect equilibrium processes in aqueous solutions.

    Cobalt forms a simple ionic chloride, $CoCl_3$, in which three electrons are transferred from a Co atom to Cl atoms. In the presence of $NH_3(aq)$, however, cobalt(III) chloride forms a series of coordination compounds with the formulas

$$CoCl_3 \cdot 6NH_3 \qquad CoCl_3 \cdot 5NH_3 \qquad CoCl_3 \cdot 4NH_3 \qquad\qquad (19.18)$$
$$\textbf{(a)} \qquad\qquad\qquad \textbf{(b)} \qquad\qquad\qquad \textbf{(c)}$$

The compounds in (19.18) differ in appearance; two are shown here. Compound (a) is pictured on the left, and compound (b) on the right. [Courtesy of Arlo Harris; photography by Carey B. Van Loon]

Here is a way we can tell that these three compounds are different: if we add an excess of $AgNO_3(aq)$ to an aqueous solution of each one, compound (a) yields *3 mol* AgCl(s) per mole of compound; compound (b) yields only *2 mol* AgCl(s); and compound (c), only *1 mol.*

In 1891, the Swiss chemist Alfred Werner proposed that certain metal atoms (primarily those of the transition metals) have *two types of valence.* One, the *primary* valence, is based on the number of electrons the atom loses in forming the metal ion. A *secondary* or auxiliary valence is responsible for bonding other groups, called **ligands,** to the central metal ion. Werner's theory describes the compounds listed in (19.18) in this way.

$$[Co(NH_3)_6]Cl_3 \qquad [Co(NH_3)_5Cl]Cl_2 \qquad [Co(NH_3)_4Cl_2]Cl \qquad (19.19)$$
$$\text{(a)} \qquad\qquad \text{(b)} \qquad\qquad\quad \text{(c)}$$

With the formulas in (19.19) we show that six ligands (in red) are bonded directly to the central Co^{3+} ion. We call the combination of a central metal ion and its ligands a **complex ion.** We call a neutral compound containing complex ions a **coordination compound.** The region surrounding the central metal ion and containing the ligands is known as the **coordination sphere.** The **coordination number** of the central metal ion is the number of positions in the coordination sphere at which ligands can attach themselves. Some of these terms are further illustrated in the margin.

In compound (a) all six ligands are NH_3 molecules. The three Cl^- ions are free anions (brown), and so one mole of compound (a) yields *three* mol AgCl. In compound (b) five NH_3 molecules and one Cl^- ion are the ligands and *two* Cl^- ions are free anions. Compound (b) is able to produce only *two* mol AgCl per mole of compound. Compound (c) has only *one* free Cl^- anion and produces only *one* mol AgCl per mole of compound.

We explore bonding, structures, properties, and uses of complex ions in Chapter 25.

complex ion   anions

$$[Co(NH_3)_6]^{3+} \quad 3\ Cl^-$$

ligands   coordination
number

a coordination
compound

19-8 Equilibria Involving Complex Ions

If we add moderately concentrated $NH_3(aq)$ to solid AgCl, the solid dissolves.

$$AgCl(s) + 2\ NH_3(aq) \longrightarrow [Ag(NH_3)_2]^+(aq) + Cl^-(aq) \qquad (19.20)$$

Ag^+ from AgCl combines with NH_3 to form the complex ion $[Ag(NH_3)_2]^+$, and the coordination compound $[Ag(NH_3)_2]Cl$ is soluble. If helps to think of reaction (19.20) as involving two equilibria simultaneously.

$$AgCl(s) \rightleftharpoons Ag^+(aq) + Cl^-(aq) \qquad (19.21)$$

$$Ag^+(aq) + 2\ NH_3(aq) \rightleftharpoons [Ag(NH_3)_2]^+(aq) \qquad (19.22)$$

Equilibrium in reaction (19.22) is shifted far to the right—$[Ag(NH_3)_2]^+$ is a stable complex ion. The equilibrium concentration of free $Ag^+(aq)$ in reaction (19.22) is kept so low that the ion product $[Ag^+][Cl^-]$ fails to exceed K_{sp}, and AgCl remains in solution. The situation is different with AgBr(s) and AgI(s), which have smaller

K_{sp} values than AgCl. AgBr(s) is only slightly soluble in NH_3(aq) and AgI(s) is essentially insoluble.

Example 19-14

Predicting reactions involving complex ions. In one point in the procedure for the qualitative analysis of cations nitric acid is added to a solution of $[Ag(NH_3)_2]Cl$ in NH_3(aq). Predict what happens.

Solution. Nitric acid neutralizes free ammonia in the solution.

$$HNO_3(aq) + NH_3(aq) \longrightarrow NH_4^+(aq) + NO_3^-(aq)$$

To replace free NH_3 lost in the neutralization, equilibrium in reaction (19.22) shifts to the *left*. This causes an increase in $[Ag^+]$. When $[Ag^+]$ increases to the point that the ion product $[Ag^+][Cl^-]$ exceeds K_{sp}, AgCl(s) precipitates.

SIMILAR EXAMPLES: Exercises 54, 55.

Formation Constants of Complex Ions. To deal with complex ion equilibria *quantitatively* we have to use a property known as the **formation constant, K_f.** This is an equilibrium constant describing the formation of a complex ion from a central metal ion and its ligands. For reaction (19.22)

$$Ag^+(aq) + 2\ NH_3(aq) \rightleftharpoons [Ag(NH_3)_2]^+(aq) \qquad (19.22)$$

the equilibrium constant expression is

> The values listed in Table 19-2 are called *overall* formation constants. In Section 25-8 we will describe the formation of complex ions in a *stepwise* fashion and will introduce formation constants for each step.

$$K_f = \frac{[[Ag(NH_3)_2]^+]}{[Ag^+][NH_3]^2} = 1.6 \times 10^7 \qquad (19.23)$$

Table 19-2 lists some representative formation constants, K_f.

Sometimes complex ion equilibria are written for the *dissociation* of a complex ion, that is, the *reverse* of the formation reaction. In this case, the equilibrium constant is called the *dissociation* constant, K_D, or *instability* constant, K_i. Thus, for the reaction

TABLE 19-2
Formation Constants for Some Complex Ions

Complex ion	Equilibrium reaction	K
$[AlF_6]^{3-}$	$Al^{3+} + 6\ F^- \rightleftharpoons [AlF_6]^{3-}$	6.7×10^{19}
$[Cd(CN)_4]^{2-}$	$Cd^{2+} + 4\ CN^- \rightleftharpoons [Cd(CN)_4]^{2-}$	7.1×10^{18}
$[Co(NH_3)_6]^{3+}$	$Co^{3+} + 6\ NH_3 \rightleftharpoons [Co(NH_3)_6]^{3+}$	4.5×10^{33}
$[Cu(CN)_3]^{2-}$	$Cu^+ + 3\ CN^- \rightleftharpoons [Cu(CN)_3]^{2-}$	2×10^{27}
$[Cu(NH_3)_4]^{2+}$	$Cu^{2+} + 4\ NH_3 \rightleftharpoons [Cu(NH_3)_4]^{2+}$	1.1×10^{13}
$[Fe(CN)_6]^{4-}$	$Fe^{2+} + 6\ CN^- \rightleftharpoons [Fe(CN)_6]^{4-}$	1×10^{37}
$[Fe(CN)_6]^{3-}$	$Fe^{3+} + 6\ CN^- \rightleftharpoons [Fe(CN)_6]^{3-}$	1×10^{42}
$[PbCl_3]^-$	$Pb^{2+} + 3\ Cl^- \rightleftharpoons [PbCl_3]^-$	2.4×10^1
$[HgCl_4]^{2-}$	$Hg^{2+} + 4\ Cl^- \rightleftharpoons [HgCl_4]^{2-}$	1.2×10^{15}
$[HgI_4]^{2-}$	$Hg^{2+} + 4\ I^- \rightleftharpoons [HgI_4]^{2-}$	1.9×10^{30}
$[Ni(CN)_4]^{2-}$	$Ni^{2+} + 4\ CN^- \rightleftharpoons [Ni(CN)_4]^{2-}$	1×10^{22}
$[Ag(NH_3)_2]^+$	$Ag^+ + 2\ NH_3 \rightleftharpoons [Ag(NH_3)_2]^+$	1.6×10^7
$[Ag(CN)_2]^-$	$Ag^+ + 2\ CN^- \rightleftharpoons [Ag(CN)_2]^-$	5.6×10^{18}
$[Ag(S_2O_3)_2]^{3-}$	$Ag^+ + 2\ S_2O_3^{2-} \rightleftharpoons [Ag(S_2O_3)_2]^{3-}$	1.7×10^{13}
$[Zn(NH_3)_4]^{2+}$	$Zn^{2+} + 4\ NH_3 \rightleftharpoons [Zn(NH_3)_4]^{2+}$	4.1×10^8
$[Zn(CN)_4]^{2-}$	$Zn^{2+} + 4\ CN^- \rightleftharpoons [Zn(CN)_4]^{2-}$	1×10^{18}
$[Zn(OH)_4]^{2-}$	$Zn^{2+} + 4\ OH^- \rightleftharpoons [Zn(OH)_4]^{2-}$	4.6×10^{17}

$$[Ag(NH_3)_2]^+ \rightleftharpoons Ag^+ + 2\,NH_3$$
$$K_D = K_i = 1/K_f = 1/1.6 \times 10^7 = 6.2 \times 10^{-8}$$

Illustrative Examples. As with the illustrative examples of the preceding section, essentially there are three types of calculations that combine solubility and complex ion formation equilibria. And, also as in the preceding section, the first two calculations are easiest. The third is a little more difficult because we have to work with two equilibrium expressions simultaneously.

Example 19-15

Determining whether a precipitate will form in a solution containing complex ions. A 0.10-mol sample of $AgNO_3$ is dissolved in 1.00 L of 1.00 M NH_3. If 0.010 mol NaCl is added to this solution, will $AgCl(s)$ precipitate?

Solution. The total silver concentration in the solution is 0.10 mol/L, found partly as Ag^+ but mostly as $[Ag(NH_3)_2]^+$. In the setup shown we assume that practically all the silver is tied up in the complex ion (complexed). If 0.10 mol/L Ag^+ is complexed, then twice this concentration of NH_3 (0.20 mol/L) must also be complexed.

$$Ag^+ \quad + 2\,NH_3 \rightleftharpoons [Ag(NH_3)_2]^+$$

	Ag^+	NH_3	$[Ag(NH_3)_2]^+$
dissolve:	0.10 M	1.00 M	
at equilibrium (assume $x \ll 0.10$):	$x = [Ag^+]$	0.80 M	0.10 M

$$\frac{[[Ag(NH_3)_2]^+]}{[Ag^+][NH_3]^2} = \frac{0.10}{[Ag^+](0.80)^2} = 1.6 \times 10^7$$

$$[Ag^+] = 9.8 \times 10^{-9}\ M$$

We must compare $[Ag^+][Cl^-]$ with $K_{sp}(AgCl) = 1.8 \times 10^{-10}$.

$$[Ag^+] = 9.8 \times 10^{-9}\ M \qquad [Cl^-] = 1.0 \times 10^{-2}\ M$$

$$(9.8 \times 10^{-9})(1.0 \times 10^{-2}) < 1.8 \times 10^{-10}$$

AgCl will not precipitate.

SIMILAR EXAMPLES: Exercises 11, 56, 76.

Example 19-16

Controlling the concentration of a ligand to cause precipitation or to prevent it. What is the *minimum* concentration of NH_3 required to *prevent* $AgCl(s)$ from precipitating from 1.00 L of a solution containing 0.10 mol $AgNO_3$ and 0.010 mol NaCl?

Solution. The $[Cl^-]$ that must be maintained in solution is 1.0×10^{-2} M. If no precipitation is to occur, $[Ag^+][Cl^-] \leq K_{sp}$.

$$[Ag^+](1.0 \times 10^{-2}) \leq K_{sp} = 1.8 \times 10^{-10} \qquad [Ag^+] \leq 1.8 \times 10^{-8}\ M$$

The maximum concentration of *free, uncomplexed* Ag^+ permitted in solution is 1.8×10^{-8} M. This means that essentially all the Ag^+ (0.10 mol/L) must be tied up (complexed) in the complex ion, $[Ag(NH_3)_2]^+$. We need to solve the following expression for $[NH_3]$.

$$K_f = \frac{[[Ag(NH_3)_2]^+]}{[Ag^+][NH_3]^2} = \frac{1.0 \times 10^{-1}}{1.8 \times 10^{-8}[NH_3]^2} = 1.6 \times 10^7$$

$$[NH_3]^2 = 0.35 \qquad [NH_3] = 0.59\ M$$

The concentration calculated above is that of *free*, *uncomplexed* NH_3. Considering as well the 0.20 mol NH_3/L complexed in $[Ag(NH_3)_2]^+$, the total concentration of $NH_3(aq)$ required is

$$[NH_3]_{tot.} = 0.59 + 0.20 = 0.79 \text{ M}$$

SIMILAR EXAMPLES: Exercises 12, 57.

Example 19-17

Determining the solubility of a solute when complex ion formation occurs. What is the molar solubility of $AgCl(s)$ in 0.100 M $NH_3(aq)$?

Solution. As in Example 19-13, the most direct approach is to write a single equilibrium constant expression by combining the separate equilibrium reactions involved. The equilibrium constant for reaction (19.20) is the product of K_{sp} for AgCl and K_f for $[Ag(NH_3)_2]^+$.

$AgCl(s) \rightleftharpoons Ag^+(aq) + Cl^-(aq)$	$K_{sp} = 1.8 \times 10^{-10}$
$Ag^+(aq) + 2 NH_3(aq) \rightleftharpoons [Ag(NH_3)_2]^+(aq)$	$K_f = 1.6 \times 10^7$
$AgCl(s) + 2 NH_3(aq) \rightleftharpoons [Ag(NH_3)_2]^+(aq) + Cl^-(aq)$	$K = K_{sp} \times K_f$
	$= 2.9 \times 10^{-3}$

Actually, the molar solubility x is the *total* silver concentration in solution, that is, $x = [Ag^+] + [[Ag(NH_3)_2]^+]$. By setting $[[Ag(NH_3)_2]^+] = x$ we are assuming that $[Ag^+]$ is so small as to be negligible. This assumption is usually valid for complex ions with large values of K_f. (See Exercise 88 for a situation where this assumption is not valid.)

According to equation (19.20), if x mol $AgCl(s)$ dissolves per liter of solution (the molar solubility), the expected concentrations of $[Ag(NH_3)_2]^+$ and Cl^- are also equal to x.

	$AgCl(s) +$	$2 NH_3(aq)$	$\rightleftharpoons$	$[Ag(NH_3)_2]^+(aq) +$	$Cl^-(aq)$
initial concentrations:		0.100 M			
changes:		$-2x$ M		$+x$ M	$+x$ M
at equilibrium:		$(0.100 - 2x)$ M		x M	x M

$$K = \frac{[[Ag(NH_3)_2]^+][Cl^-]}{[NH_3]^2} = \frac{x \cdot x}{(0.100 - 2x)^2} = \left\{ \frac{x}{0.100 - 2x} \right\}^2 = 2.9 \times 10^{-3}$$

We can solve this equation without making any approximations and also without using the quadratic formula. All we need to do is to take the square root of both sides.

$$\frac{x}{0.100 - 2x} = \sqrt{2.9 \times 10^{-3}} = 5.4 \times 10^{-2} \qquad x = 5.4 \times 10^{-3} - 0.11x$$

$$1.11x = 5.4 \times 10^{-3} \qquad x = 4.9 \times 10^{-3}$$

The solubility of $AgCl(s)$ in 0.100 M $NH_3(aq)$ is 4.9×10^{-3} mol AgCl/L.

SIMILAR EXAMPLES: Exercises 58, 86, 87.

19-9 Qualitative Analysis Revisited

In Section 5-10 we described a general scheme for establishing the presence or absence of some 25 common cations that might occur together in solution. We called this the **qualitative analysis scheme.** At that time we learned that the principal types of reactions involved in the analysis were precipitations, acid–base and oxidation–reduction reactions, and complex ion formation. Let us explore the subject a bit further now that we know more about these reaction types.

FIGURE 19-6
Cation group 1 (chloride
group) of the qualitative
analysis scheme.

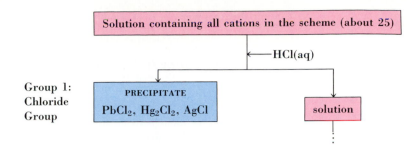

Cation Group 1: The Chloride Group. Figure 19-6 reproduces a portion of the flow diagram first presented as Figure 5-12. This figure shows that when HCl(aq) is added to a solution containing the 25 possible cations, if a precipitate forms one or more of these ions must be present: Pb^{2+}, Hg_2^{2+}, Ag^+. To establish the presence or absence of each of these three ions, we filter off the chloride precipitate and subject it to further testing.

Of the three chloride precipitates, $PbCl_2(s)$ is the most soluble. By washing the precipitate with hot water, we can dissolve a sufficient quantity of $PbCl_2$ to test for Pb^{2+} in the solution. In this test we precipitate a lead compound that is less soluble than $PbCl_2$, such as *yellow* lead chromate.

$$Pb^{2+}(aq) + CrO_4^{2-}(aq) \longrightarrow PbCrO_4(s) \qquad (19.24)$$

The portion of the chloride group precipitate that is insoluble in hot water is then treated with $NH_3(aq)$. Two things happen. One is that any AgCl(s) present dissolves, forming the complex ion $[Ag(NH_3)_2^+]$.

$$AgCl(s) + 2\ NH_3(aq) \longrightarrow [Ag(NH_3)_2]^+(aq) + Cl^-(aq) \qquad (19.20)$$

At the same time, any $Hg_2Cl_2(s)$ undergoes an *oxidation–reduction* reaction. One of the products of the reaction is finely divided, *black* mercury.

$$Hg_2Cl_2(s) + 2\ NH_3(aq) \longrightarrow \underline{Hg(l) + HgNH_2Cl(s)} + NH_4^+(aq) + Cl^-(aq)$$
$$\text{dark gray} \qquad\qquad\qquad (19.25)$$

The appearance of a dark gray mixture of black mercury and white $HgNH_2Cl$ [mercury(II) amidochloride] is the qualitative analysis test for mercury(I).

When the solution from reaction (19.20) is acidified with $HNO_3(aq)$, if there is any silver present it reprecipitates as AgCl(s). This is the reaction we predicted in Example 19-14.

Example 19-18

Solving a qualitative analysis group 1 "unknown." A qualitative analysis unknown gives a white precipitate with HCl(aq). The white precipitate is partially soluble in hot water, yielding a solution that gives a yellow precipitate with $K_2CrO_4(s)$. When the undissolved portion of white precipitate is treated with $NH_3(aq)$, there is no change in color of the precipitate. Which chloride group cation(s) is(are) present, which is(are) absent, and about which is there some doubt?

Solution. The fact that there is a chloride group precipitate indicates that at least one of the chloride group cations is present. The test with K_2CrO_4 indicates that Pb^{2+} is present. The failure to detect any color change in the chloride group precipitate when it is treated with $NH_3(aq)$ (reaction 19-25) indicates that Hg_2^{2+} is absent. We can only determine if some AgCl(s) dissolved in the $NH_3(aq)$ by acidifying the $NH_3(aq)$ to see if a white solid (AgCl) reprecipitates.

Since this test was not performed, we are uncertain whether Ag^+ was present in the unknown or not.

SIMILAR EXAMPLES: Exercises 63, 64.

Equilibria Involving Hydrogen Sulfide—H_2S. As shown in Figure 19-7, aqueous hydrogen sulfide (hydrosulfuric acid) is the key reagent in the analysis of qualitative analysis cation groups 2 and 3. Hydrogen sulfide gas has a familiar "rotten egg" odor that is especially noticeable around sulfur hot springs and in volcanic areas. In aqueous solution H_2S is a *weak diprotic acid.*

$$H_2S(aq) + H_2O \rightleftharpoons H_3O^+(aq) + HS^- \qquad K_{a_1} = 1.1 \times 10^{-7} \qquad (19.26)$$

$$HS^-(aq) + H_2O \rightleftharpoons H_3O^+(aq) + S^{2-} \qquad K_{a_2}* = 1.0 \times 10^{-14} \qquad (19.27)$$

> H_2S is an extremely toxic substance. It can cause headaches and nausea at concentrations of 10 ppm in air, and it can produce paralysis or death at 100 ppm. Fortunately, the gas is detectable at levels of less than 1 ppm, although exposure to the gas also deadens the sense of smell.

Based on what we learned about the ionization of weak polyprotic acids in Section 17-7, we can state that in a solution containing H_2S as the only source of H_3O^+ *the sulfide ion concentration is independent of the molarity of the acid;* it is $[S^{2-}] = K_{a_2} = 1.0 \times 10^{-14}$. With this value of $[S^{2-}]$ we can predict which metal ions should precipitate as sulfides.

Example 19-19

Predicting the formation of sulfide precipitates in $H_2S(aq)$ solutions. Which of the following ions will precipitate from a solution that is kept saturated in H_2S and 0.010 M in each ion: Cd^{2+}; Zn^{2+}; Mn^{2+}?

Solution. In saturated $H_2S(aq)$, as in any other aqueous solution of H_2S, $[S^{2-}] = K_{a_2} = 1.0 \times 10^{-14}$ M. We need to compare ion products with the values of K_{sp} (listed in Table 19-1). Since each ion concentration is 1.0×10^{-2} M, each ion product, $[M^{2+}][S^{2-}] = (1.0 \times 10^{-2})(1.0 \times 10^{-14}) = 1.0 \times 10^{-16}$.

$Q_{CdS} = [Cd^{2+}][S^{2-}] = 1.0 \times 10^{-16} > K_{sp} \ (8 \times 10^{-27})$ CdS precipitates.

$Q_{ZnS} = [Zn^{2+}][S^{2-}] = 1.0 \times 10^{-16} > K_{sp} \ (1 \times 10^{-21})$ ZnS precipitates.

$Q_{MnS} = [Mn^{2+}][S^{2-}] = 1.0 \times 10^{-16} < K_{sp} \ (2 \times 10^{-13})$ MnS does not ppt.

SIMILAR EXAMPLES: Exercises 13a, 59a.

The Common Ion Effect in H_2S Solutions. According to Example 19-19 both CdS and ZnS should precipitate from $H_2S(aq)$. Yet, in Figure 19-7 we see that CdS is found in cation group 2 and ZnS in cation group 3. To achieve this separation of the two ions, we need to add HCl to the $H_2S(aq)$ from which precipitation occurs. In a mixture of a strong acid and a weak acid, the *common ion* H_3O^+ is furnished by the strong acid, and the ionization of the weak acid is repressed. This reduces $[S^{2-}]$ in solution. By controlling $[H_3O^+]$ we can reduce $[S^{2-}]$ to the point where CdS still precipitates, but ZnS does not.

In the HCl–$H_2S(aq)$ solution that we are describing, we can combine the ionization constants of H_2S in this familiar way

$$
\begin{array}{ll}
H_2S + H_2O \rightleftharpoons H_3O^+ + \cancel{HS^-} & K_{a_1} = 1.1 \times 10^{-7} \\
\cancel{HS^-} + H_2O \rightleftharpoons H_3O^+ + S^{2-} & K_{a_2} = 1.0 \times 10^{-14} \\
\hline
H_2S + 2\,H_2O \rightleftharpoons 2\,H_3O^+ + S^{2-} & K = K_{a_1} \times K_{a_2}
\end{array}
\qquad (19.28)
$$

and

---

*As pointed out in the footnote to Table 17-3, there has been considerable doubt concerning the value of K_{a_2} for H_2S. It may be several powers of 10 smaller than the value given here.

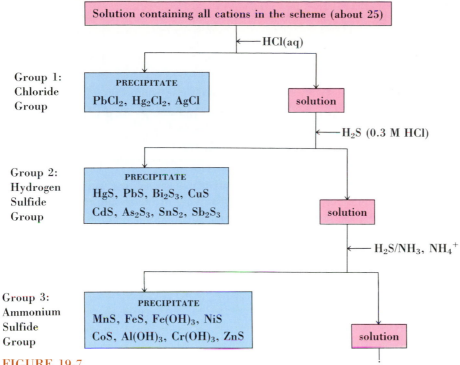

Group 1:
Chloride
Group

Group 2:
Hydrogen
Sulfide
Group

Group 3:
Ammonium
Sulfide
Group

FIGURE 19-7
Cation groups 1, 2, and 3 of
the qualitative analysis
scheme.

$$K_{a_1} \times K_{a_2} = \frac{[H_3O^+]^2[S^{2-}]}{[H_2S]} = (1.1 \times 10^{-7})(1.0 \times 10^{-14}) = 1.1 \times 10^{-21} \quad (19.29)$$

Equation (19.28) states that for every molecule of H_2S that manages to undergo *complete* ionization, two H_3O^+ ions and one S^{2-} ion are formed. Remember, though, that in H_2S solutions only a *very small fraction* of the H_2S molecules actually ionize through the second step. Most of the ionization stops at the first step. Therefore, the concentrations of H_3O^+ and HS^- are both much larger than is the concentration of S^{2-}. Equation (19.29) indicates that if any *two* of the three terms— $[H_2S]$, $[H_3O^+]$, $[S^{2-}]$—are known, the remaining one can be calculated.

Selective Precipitation of Metal Sulfides. ZnS is the *least* soluble of the cation group 3 sulfides and CdS is the *most* soluble of the cation group 2 sulfides. The conditions we must use to precipitate group 2 sulfides and leave group 3 cations unprecipitated, then, are those that will separate Cd^{2+} and Zn^{2+}.

Example 19-20 _____

Separating metal ions by selective precipitation of metal sulfides through control of $[H_3O^+]$ in $H_2S(aq)$. What $[H_3O^+]$ must be maintained in a saturated H_2S solution (0.10 M H_2S) to precipitate CdS, but not ZnS, if Cd^{2+} and Zn^{2+} are both present initially at a concentration of 0.10 M?

$$CdS(s) \rightleftharpoons Cd^{2+}(aq) + S^{2-}(aq) \qquad K_{sp} = 8 \times 10^{-27}$$

$$ZnS(s) \rightleftharpoons Zn^{2+}(aq) + S^{2-}(aq) \qquad K_{sp} = 1 \times 10^{-21}$$

Solution. If ZnS is *not* to precipitate,

$$[Zn^{2+}][S^{2-}] < K_{sp} = 1 \times 10^{-21}$$

$$(0.10)[S^{2-}] < 1 \times 10^{-21}$$

$$[S^{2-}] < 1 \times 10^{-20} \text{ M}$$

The maximum value of $[S^{2-}]$ that we can tolerate before ZnS precipitates is 1×10^{-20} M. The $[H_3O^+]$ required to maintain this $[S^{2-}]$ is

$$K_{a_1} \times K_{a_2} = \frac{[H_3O^+]^2[S^{2-}]}{[H_2S]} = \frac{[H_3O^+]^2(1 \times 10^{-20})}{0.10} = 1.1 \times 10^{-21}$$

$$[H_3O^+]^2 = \frac{0.10 \times 1.1 \times 10^{-21}}{1 \times 10^{-20}} = 1.1 \times 10^{-2} \quad [H_3O^+] = 1 \times 10^{-1} = 0.1 \text{ M}$$

In the margin:

In actual practice the cation group 2 precipitating reagent is kept at about pH = 0.5 ($[H_3O^+]$ = 0.3 M).

We can see that CdS will precipitate under these conditions.

$$[Cd^{2+}][S^{2-}] = (0.10)(1.0 \times 10^{-20}) = 1.0 \times 10^{-21} > K_{sp} = 8 \times 10^{-27}$$

SIMILAR EXAMPLES: Exercises 14, 59b, 77.

Dissolving Metal Sulfides. In the qualitative analysis scheme we both precipitate and redissolve sulfides. To discuss the dissolving of metal sulfides, let us again combine two equilibrium reactions. Consider dissolving CdS(s) in HCl(aq).

$$CdS(s) \rightleftharpoons Cd^{2+}(aq) + S^{2-}(aq) \qquad K_{sp} = 8 \times 10^{-27} \qquad (19.30)$$

$$\underset{\text{(from HCl)}}{2\ H_3O^+} + \underset{\text{(from CdS)}}{S^{2-}} \rightleftharpoons H_2S(aq) + 2\ H_2O \qquad \begin{aligned} K &= 1/(K_{a_1} \times K_{a_2}) \\ &= 9.1 \times 10^{20} \end{aligned} \qquad (19.31)$$

$$CdS(s) + 2\ H_3O^+ \rightleftharpoons Cd^{2+}(aq) + H_2S(aq) + 2\ H_2O$$

$$\begin{aligned} K &= K_{sp}/(K_{a_1} \times K_{a_2}) \\ &= 7 \times 10^{-6} \end{aligned} \qquad (19.32)$$

Sulfide ion from CdS in reaction (19.30) combines with H_3O^+ from HCl to produce H_2S in reaction (19.31). The equilibrium constant for reaction (19.31) is the *reciprocal* of the combined equilibrium constants in reaction (19.28). The net equation for the dissolving process indicates that CdS is converted to H_2S(aq) and Cd^{2+} in solution—the CdS dissolves. Because the value of K for reaction (19.32) is small, we should not expect very much CdS to dissolve. In fact, we know that the reverse of reaction (19.32) occurs in 0.3 M HCl; Cd^{2+} precipitates with the group 2 cations in the qualitative analysis scheme. Nevertheless, high concentrations of H_3O^+ will promote some dissolving of even the metal sulfides of cation group 2, as illustrated in Example 19-21.

Example 19-21

Determining the solubility of metal sulfides in acidic solutions. What is the molar solubility of CdS in 1.0 M HCl(aq)?

Solution. The relevant data are listed below equation (19.32), where the molar solubility = $[Cd^{2+}] = x$

$$CdS(s) + \quad 2\ H_3O^+ \quad \rightleftharpoons \quad Cd^{2+} + H_2S(aq) + 2\ H_2O \qquad (19.32)$$

initial concentrations:	1.0 M		
changes:	$-2x$ M	$+x$ M	$+x$ M
at equilibrium:	$(1.0 - 2x)$ M	x M	x M

$$K = \frac{[Cd^{2+}][H_2S]}{[H_3O^+]^2} = \frac{x \cdot x}{(1.0 - 2x)^2} = 7 \times 10^{-6}$$

Extract the square root of each side of the equation.

$$\frac{x}{1.0 - 2x} = 3 \times 10^{-3}$$

$$x + 6 \times 10^{-3}x = 3 \times 10^{-3}$$

$$x = [Cd^{2+}] = 3 \times 10^{-3} \text{ M}$$

SIMILAR EXAMPLES: Exercises 61, 62.

We will return to a discussion of some of the separations and tests used in cation group 2 (the hydrogen sulfide group) and 3 (the ammonium sulfide group) in Chapters 23 and 24.

Summary

Equilibrium between a slightly soluble ionic compound and its ions in solution is expressed through the *solubility product constant*, K_{sp}. In using K_{sp} expressions two situations are commonly encountered: (1) The ions in a saturated solution may be derived *solely* from the slightly soluble solute, or (2) an additional salts(s) may be present that contributes ions either *common to* or *different from* those of the slightly soluble solute. The solubility of the solute is *greatly reduced* in the presence of common ions derived from a dissolved salt. A salt whose ions are different from those of the slightly soluble solute causes a small *increase* in the solubility of the solute. The formation of ion pairs in solution may also cause the solubility of a solute to be higher than predicted from the K_{sp} value.

A comparison of the *ion product* (reaction quotient) Q with K_{sp} provides a criterion for precipitation: If $Q > K_{sp}$, precipitation should occur; if $Q < K_{sp}$, the solution remains unsaturated. Another matter of interest concerns the completeness of a precipitation reaction. If a precipitation is carried out in the presence of a high concentration of a common ion, it generally goes to completion. Combinations of factors, such as lack of sufficient common ion and a moderately high value of K_{sp}, can result in incomplete precipitation. At times, ions in solution can be separated by *fractional precipitation*. One type of ion is removed by precipitation while the others remain in solution.

By using the common ion effect to alter equilibria of weak acids and bases, solution concentrations of such ions as OH^-, CO_3^{2-}, and S^{2-} can be controlled. Such control allows for the selective precipitation or dissolving of certain ionic compounds whose solubilities are pH dependent.

The formation of a complex ion from a central metal ion and ligands is an equilibrium process with an equilibrium constant called the *formation constant*, K_f. In general, if the formation constant of a complex ion is large, the concentration of *free* metal ion in equilibrium with the complex ion is very small. Complex ion formation can render certain insoluble materials quite soluble in appropriate aqueous solutions, such as $AgCl(s)$ in $NH_3(aq)$.

Precipitation, acid–base, oxidation–reduction, and complex ion formation reactions all find extensive application in *qualitative analysis*. The basis of separating cation group 1 (Pb^{2+}, Hg_2^{2+}, and Ag^+) from the other groups is the insolubility of the group 1 chlorides. Further separations and tests within cation group 1 are based on the greater water solubility of $PbCl_2(s)$ than $Hg_2Cl_2(s)$ and $AgCl(s)$; the formation of the stable complex ion, $[Ag(NH_3)_2]^+$; and an oxidation–reduction reaction between $NH_3(aq)$ and $Hg_2Cl_2(s)$ that leads to a gray-black precipitate (a mixture of Hg and $HgNH_2Cl$). Separation of cation groups 2 and 3 is accomplished with $H_2S(aq)$ in which $[S^{2-}]$ is carefully controlled through control of the pH.

Summarizing Example

Lime is pumped into lake in Sweden to neutralize acidity. [Ted Spiegel/Black Star]

Lime (quicklime), CaO, is the cheapest basic substance available. It is obtained by heating limestone ($CaCO_3$). When treated with water, CaO produces $Ca(OH)_2$ (slaked lime). Important environmental uses of lime are in sewage treatment, neutralizing acid mine drainage, and removing oxides of sulfur from stack gases (see Section 14-13). Until World War II, the principal method of producing sodium hydroxide, NaOH, involved conversion of $Ca(OH)_2$ to $CaCO_3$ using Na_2CO_3.

$$Ca(OH)_2(s) + 2\,\cancel{Na^+(aq)} + CO_3^{2-}(aq) \rightleftharpoons$$
$$CaCO_3(s) + 2\,\cancel{Na^+(aq)} + 2\,OH^-(aq) \quad (19.33)$$

(The current method of producing NaOH is discussed in Chapter 21.)

1. What is the maximum pH that can be established in $Ca(OH)_2(aq)$?

$$Ca(OH)_2(s) \rightleftharpoons Ca^{2+}(aq) + 2\,OH^-(aq) \qquad K_{sp} = 5.5 \times 10^{-6}$$

Solution. The maximum pH will be found in the solution with the highest $[OH^-]$. This will be a saturated solution of $Ca(OH)_2$. We must calculate $[OH^-]$ in this saturated solution. Let us set $[Ca^{2+}] = S$ and $[OH^-] = 2S$.

$$K_{sp} = [Ca^{2+}][OH^-]^2 = (S)(2S)^2 = 4S^3 = 5.5 \times 10^{-6}$$

$$S^3 = 1.4 \times 10^{-6} \qquad S = 0.011\ M \qquad [OH^-] = 2S = 0.022\ M$$

$$pOH = -\log[OH^-] = -\log 0.022 = 1.66$$

$$pH = 14.00 - pOH = 14.00 - 1.66 = 12.34$$

The maximum pH attainable in an aqueous solution of $Ca(OH)_2$, is 12.34.

(This example is similar to Example 19-3.)

2. Show that $Ca(OH)_2(s)$ is more soluble in $NH_4Cl(aq)$ than it is in pure water.

Solution. Note that NH_4Cl is the salt of a weak base and a strong acid. This salt should hydrolyze to produce an acidic solution (that is, NH_4^+ can ionize as an acid). $Ca(OH)_2$, a base, should be more soluble in an acidic solution.

An alternative approach is to write a net equation for the dissolving reaction.

$$Ca(OH)_2(s) \rightleftharpoons Ca^{2+}(aq) + 2\,OH^-(aq) \qquad K_{sp} = 5.5 \times 10^{-6}$$
$$2\,NH_4^+(aq) + 2\,OH^-(aq) \rightleftharpoons 2\,NH_3(aq) + 2\,H_2O$$
$$K = 1/(K_b)^2 = 1/(1.74 \times 10^{-5})^2$$
_____
$$Ca(OH)_2(s) + 2\,NH_4^+(aq) \rightleftharpoons Ca^{2+}(aq) + 2\,NH_3(aq) + 2\,H_2O$$
$$K = 1.8 \times 10^4$$

With such a large value of K, we can conclude that equilibrium in the dissolving reaction is displaced far to the right. $[Ca^{2+}]$ should be much larger in this solution than in pure water saturated with $Ca(OH)_2$.

(This example is similar to Examples 19-10 and 19-13.)

3. Suppose that excess $Ca(OH)_2(s)$ is stirred into a 1.0 M Na_2CO_3 solution. What will be the pH of the NaOH solution produced as a result of reaction (19.33)? [This calculation illustrates how slightly soluble $Ca(OH)_2$ is converted to the much more soluble NaOH.]

Solution. To do an equilibrium calculation on reaction (19.33), we first need an equilibrium constant value. When we are dealing with a single equilibrium situation, we generally can find the equilibrium constant easily. It is a K_{sp}, K_a, K_b, ... But in reaction (19.33) there are two equilibria occurring simultaneously. Some $Ca(OH)_2$ dissolves and some $CaCO_3$ precipitates. In such cases the simplest approach is to try to write a net equation.

$$Ca(OH)_2(s) \rightleftharpoons Ca^{2+}(aq) + 2\,OH^-(aq) \qquad K_{sp} = 5.5 \times 10^{-6}$$
$$Ca^{2+}(aq) + CO_3^{2-}(aq) \rightleftharpoons CaCO_3(s) \qquad K = 1/K_{sp} = 1/2.8 \times 10^{-9}$$
_____
$$Ca(OH)_2(s) + CO_3^{2-}(aq) \rightleftharpoons CaCO_3(s) + 2\,OH^-(aq) \qquad K = 2.0 \times 10^3$$

Now we can proceed in a familiar fashion.

$$Ca(OH)_2(s) + CO_3^{2-}(aq) \rightleftharpoons CaCO_3(s) + 2\ OH^-$$

initial concentrations:	1.0 M		—
changes:	$-x$ M		$+2x$ M
at equilibrium:	$(1.0 - x)$ M		$2x$ M

$$K = \frac{[OH^-]^2}{[CO_3^{2-}]} = \frac{(2x)^2}{(1.0 - x)} = 2.0 \times 10^3$$

At this point our usual **approach** would be to assume that x is very small compared to 1.0, but that assumption definitely will *not* work here. Notice that K is quite a *large* quantity (2.0×10^3). Equilibrium will be displaced far *to the right*; x will be *very nearly equal to 1.0*. We can obtain an exact solution with the quadratic formula, however.

$$4x^2 + (2.0 \times 10^3)x - 2.0 \times 10^3 = 0$$

$$x = \frac{-2.0 \times 10^3 \pm \sqrt{4.0 \times 10^6 + 3.2 \times 10^4}}{2 \times 4} = 0.998 \approx 1$$

The hydroxide ion concentration $[OH^-] = 2x \approx 2$ M

$$pOH = -\log 2 = -0.3 \quad \text{and} \quad pH = 14.00 - pOH = 14.3$$

Reaction (19.33) goes essentially to completion, and this is why it was such a useful method of producing NaOH(aq).

Key Terms

complex ion (19-7)
coordination compound (19-7)
coordination number (19-7)
coordination sphere (19-7)

formation constant, K_f (19-8)
fractional precipitation (19-5)
ion-pair formation (19-3)
ligands (19-7)

molar solubility (19-2)
quantitative analysis (19-5)
salt effect (19-3)
solubility product constant, K_{sp} (19-1)

Highlighted Expressions

Formulating a K_{sp} expression (19.4)
The common ion effect in solubility equilibrium (19.5)

Criterion for precipitation from solution (19.6)
Important ideas regarding net ionic equations (19.10)

Review Problems

(Use data from tables in Chapters 17, 18, and 19, as necessary.)

1. Write K_{sp} expressions for the following equilibria. For example, $AgCl(s) \rightleftharpoons Ag^+(aq) + Cl^-(aq)$ $K_{sp} = [Ag^+][Cl^-]$
  (a) $Ag_2SO_4(s) \rightleftharpoons 2\ Ag^+(aq) + SO_4^{2-}(aq)$
  (b) $Ra(IO_3)_2(s) \rightleftharpoons Ra^{2+}(aq) + 2\ IO_3^-(aq)$
  (c) $Ni_3(PO_4)_2(s) \rightleftharpoons 3\ Ni^{2+}(aq) + 2\ PO_4^{3-}(aq)$
  (d) $Hg_2C_2O_4(s) \rightleftharpoons Hg_2^{2+}(aq) + C_2O_4^{2-}(aq)$
  (e) $PuO_2CO_3(s) \rightleftharpoons PuO_2^{2+}(aq) + CO_3^{2-}(aq)$

2. Write solubility equilibrium equations that are described by the following K_{sp} expressions. For example, $K_{sp} = [Ag^+][Cl^-]$ represents $AgCl(s) \rightleftharpoons Ag^+(aq) + Cl^-(aq)$.
  (a) $K_{sp} = [Fe^{3+}][OH^-]^3$ **(b)** $K_{sp} = [BiO^+][OH^-]$
  (c) $K_{sp} = [Hg_2^{2+}][I^-]^2$ **(d)** $K_{sp} = [Pb^{2+}]^3[AsO_4^{3-}]^2$
  (e) $K_{sp} = [Cu^{2+}]^2[Fe(CN)_6^{4-}]$
  (f) $K_{sp} = [Mg^{2+}][NH_4^+][PO_4^{3-}]$

3. Calculate the water solubility, in mol/L, of each of the fol-

lowing: **(a)** $BaCrO_4(s)$, $K_{sp} = 1.2 \times 10^{-10}$; **(b)** $PbBr_2$, $K_{sp} = 4.0 \times 10^{-5}$; **(c)** CeF_3, $K_{sp} = 8 \times 10^{-16}$; **(d)** $Mg_3(AsO_4)_2$, $K_{sp} = 2.1 \times 10^{-20}$.

4. The following aqueous solubility data, expressed in mol solute/L, are derived from a handbook. What are the values of K_{sp} for these solutes? **(a)** $CsMnO_4$, 3.8×10^{-3} M; **(b)** $Pb(ClO_2)_2$, 2.8×10^{-3} M; **(c)** Li_3PO_4, 2.9×10^{-3} M.

5. Calculate the molar solubility of $Mg(OH)_2(s)$ in **(a)** pure water; **(b)** 0.015 M $MgCl_2(aq)$; **(c)** 0.217 M KOH(aq).

6. Predict whether a precipitate is expected to form in a solution with the ion concentrations listed.
  (a) $[Mg^{2+}] = 0.015$ M, $[CO_3^{2-}] = 0.0072$ M
  (b) $[Ag^+] = 0.0038$ M, $[SO_4^{2-}] = 0.0105$ M
  (c) $[Cr^{3+}] = 0.041$ M, $[H_3O^+] = 0.0016$ M
    (*Hint:* What is $[OH^-]$?)

7. KI(aq) is slowly added to a solution that is 0.10 M in both Pb^{2+} and Ag^+.

(a) Which precipitate should form first, PbI_2 or AgI?

(b) What $[I^-]$ is required for the *second* cation to begin to precipitate?

(c) What concentration of the first cation to precipitate remains in solution at the point where the second cation begins to precipitate?

(d) Can these two cations be effectively separated by fractional precipitation?

8. Complete and balance the following equations. If no reaction occurs, so state:

(a) $Ag^+(aq) + NO_3^-(aq) + Na^+(aq) + Br^-(aq) \rightarrow$

(b) $Cu^{2+}(aq) + NO_3^-(aq) + H_3O^+(aq) + Cl^-(aq) \rightarrow$

(c) $Fe^{2+}(aq) + H_2S(aq, \text{ in } 0.3 \text{ M HCl}) \rightarrow$

(d) $Cu(OH)_2(s) + NH_3(aq) \rightarrow$

(e) $Fe^{3+}(aq) + NH_4^+ + OH^- \rightarrow$

(f) $Ag_2SO_4(s) + NH_3(aq) \rightarrow$

(g) $CaSO_3(s) + H_3O^+(aq) \rightarrow$

9. A buffer solution is prepared which is 0.50 M $HC_2H_3O_2$ and 0.25 M $NaC_2H_3O_2$. Which of the following ions can be maintained at a concentration of 0.10 M or greater without precipitating as the hydroxide from this solution? (a) Ca^{2+}; (b) Al^{3+}; (c) Cr^{3+}.

10. To 0.350 L of 0.100 M NH_3 is added 0.150 L of 0.100 M $MgCl_2$.

(a) Show that $Mg(OH)_2$ should precipitate.

(b) What mass of $(NH_4)_2SO_4$ should be added to cause the $Mg(OH)_2$ to redissolve. [*Hint:* What mass of $(NH_4)_2SO_4$ would have prevented precipitation in the first place?]

11. Which of the following ion concentrations can be maintained in the same solution *without* a precipitate forming? [*Hint:* What is the concentration of the *free* cation in each case?]

(a) $[[Ag(CN)_2]^-] = 0.012$ M, $[CN^-] = 1.05$ M, and $[I^-] = 2.0$ M

(b) $[[Ag(S_2O_3)_2]^{3-}] = 0.012$ M, $[S_2O_3^{2-}] = 1.05$ M, and $[I^-] = 2.0$ M

(c) $[[Cu(NH_3)_4]^{2+}] = 0.055$ M, $[NH_3] = 1.8$ M, and $[S^{2-}] = 3.2 \times 10^{-5}$ M

12. What concentration of *free* CN^- must be maintained in a solution that is 1.8 M $AgNO_3$ and 0.327 M NaCl to prevent $AgCl(s)$ from precipitating? [*Hint:* The silver ion must be kept in the form of $[Ag(CN)_2]^-$.]

13. A solution is 0.10 M in Cd^{2+}, in Cu^{2+}, and in Fe^{2+}. Which sulfides will precipitate if the solution is also made to be (a) 0.010 M $H_2S(aq)$; (b) 0.01 M $H_2S(aq)$–0.010 M $HCl(aq)$?

14. Should FeS precipitate from a solution that is saturated in H_2S (0.10 M) and 0.0022 M in Fe^{3+} at pH = 3.55?

15. A solution is saturated in H_2S (0.10 M H_2S) and 0.015 M $FeSO_4(aq)$. What is the minimum pH that can be maintained and still have FeS(s) precipitate from this solution?

Exercises

(Use data from tables in Chapters 17, 18, and 19, as necessary.)

K_{sp} and solubility

16. A handbook lists the water solubility of cadmium hydroxide, $Cd(OH)_2$, as 0.00026 g/100 mL. What is the value of K_{sp} for $Cd(OH)_2$?

17. Which of the following saturated aqueous solutions would you expect to have the highest concentration of Mg^{2+} ion? Explain. (a) $MgCO_3$; (b) MgF_2; (c) $Mg_3(PO_4)_2$.

18. Fluoridated drinking water contains about 1 part per million (ppm) of F^-. Is CaF_2 sufficiently soluble in water to be used as the source of fluoride ion for the fluoridation of drinking water? Explain. [*Hint:* Think of 1 ppm as signifying 1 g F^- per 10^6 g solution.]

19. In the qualitative analysis scheme Bi^{3+} is detected by the appearance of a white precipitate of bismuthyl hydroxide, BiOOH(s).

$$BiOOH(s) \rightleftharpoons BiO^+(aq) + OH^-(aq) \qquad K_{sp} = 4 \times 10^{-10}$$

Calculate the pH of a saturated aqueous solution of BiOOH.

$\star$ **20.** One of the substances sometimes responsible for the "hardness" of water is $CaSO_4$. A particular water sample has 131 ppm of $CaSO_4$ (131 g $CaSO_4$ per 10^6 g of water). If this water is boiled in a tea kettle, approximately what fraction of the water must be evaporated before $CaSO_4(s)$ begins to deposit? Assume that the solubility of $CaSO_4$ does not change with temperature in the range 0 to 100 °C.

21. A 25.00-mL sample of a clear *saturated* solution of PbI_2 requires 13.3 mL of a certain $AgNO_3(aq)$ for its titration. What is the molarity of this $AgNO_3(aq)$?

$$I^-(aq) + Ag^+(aq) \rightarrow AgI(s)$$
[from satd $PbI_2(aq)$] [from $AgNO_3(aq)$]

22. Saturated $CaC_2O_4(aq)$ is prepared and 250.0 mL of the solution is withdrawn and titrated with 6.3 mL of 0.00102 M $KMnO_4(aq)$. What is the value of K_{sp} for CaC_2O_4 obtained from these data? The titration reaction is

$$5 \, C_2O_4^{2-}(aq) + 2 \, MnO_4^-(aq) + 16 \, H^+(aq) \rightarrow$$
$$2 \, Mn^{2+}(aq) + 8 \, H_2O + 10 \, CO_2(g)$$

$\star$ **23.** To precipitate as $Ag_2S(s)$ all the Ag^+ present in 338 mL of a saturated solution of $AgBrO_3$ requires 30.4 cm^3 of $H_2S(g)$ measured at 23 °C and 748 mmHg. What is K_{sp} for $AgBrO_3$?

$$2 \, Ag^+(aq) + H_2S(g) \rightarrow Ag_2S(s) + 2 \, H^+(aq)$$

The common ion effect

24. The salts KI and KNO_3 are found to have different effects on the solubility of AgI in water. Describe the effect of each and explain why the effects are different.

25. A 0.200 M Na_2SO_4 solution is saturated with Ag_2SO_4 and is found to have $[Ag^+] = 9.2 \times 10^{-3}$ M. What is the value of K_{sp} for Ag_2SO_4 obtained from these data?

26. Refer to Example 19-5.

(a) What $[Pb^{2+}]$ should be maintained in a $Pb(NO_3)_2(aq)$ solution to produce a solubility of 1.0×10^{-4} mol PbI_2/L?

(b) What $[I^-]$ should be maintained in a KI(aq) solution to produce a solubility of 1.0×10^{-5} mol PbI_2/L?

(c) Can the solubility of PbI_2 be lowered to 1.0×10^{-7} mol PbI_2/L by using Pb^{2+} as the common ion? Explain.

27. Demonstrate, by calculation, the truth of the statement made in the text (page 686) that, "The effect of Pb^{2+} in reducing the solubility of PbI_2 is not as striking as that of I^-, but it is significant nevertheless." For example, what is the solubility of PbI_2 in 0.10 M $Pb(NO_3)_2$?

28. Plot a graph similar to Figure 19-2 showing how the solubility of lead iodate varies with concentration of KIO_3(aq), ranging from pure water to 0.10 M KIO_3(aq). For $Pb(IO_3)_2$, $K_{sp} = 3.2 \times 10^{-13}$.

29. A particular water sample is saturated in CaF_2 and has a total Ca^{2+} content of 115 ppm (that is, 115 g Ca^{2+} per 10^6 g of water sample). What is the F^- ion content of the water in ppm?

30. Refer to Example 19-2. If the 100.0 mL of solution saturated with $CaSO_4$ were 0.0010 M Na_2SO_4 instead of pure water, what mass of $CaSO_4$ would be present in the solution? [*Hint:* Does the usual simplifying assumption hold?]

31. Assume that to be visible to the unaided eye a precipitate must weigh more than 1 mg. If you add 1.0 mL of 1.0 M NaCl(aq) to 100.0 mL of a clear saturated aqueous AgCl solution, will you be able to see AgCl(s) precipitated as a result of the common ion effect? Explain.

32. Calculate $[Mg^{2+}]$ in the final equilibrium solution in each of the situations described and pictured here.

(a) 500.0 mL of saturated solution is in contact with $Mg(OH)_2$(s).

(b) 500.0 mL of water is added to the 500.0 mL of solution in (a). Undissolved $Mg(OH)_2$(s) remains.

(c) 100.0 mL of the clear saturated solution in (b) is added to 500.0 mL of water.

(d) 25.00 mL of the clear saturated solution in (b) is added to 250.0 ml of 0.065 M $MgCl_2$(aq).

(e) 50.00 mL of the clear saturated solution in (b) is removed and added to 150.0 mL of 0.150 M KOH(aq).

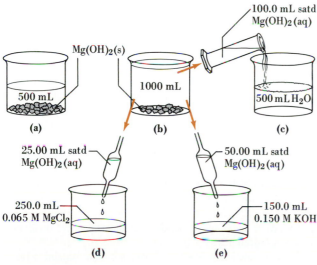

Exercise 32

Criterion for precipitation from solution

33. Should precipitation of MgF_2 occur if a 17.5-mg sample of $MgCl_2 \cdot 6H_2O$ is added to 325 mL of 0.045 M KF?

34. Should precipitation of $PbCl_2$(s) occur when 155 mL of 0.016 M KCl(aq) is added to 245 mL of 0.175 M $Pb(NO_3)_2$(aq)?

35. What must be the pH in a solution that is 0.17 M in Fe^{3+} to just cause the precipitation of $Fe(OH)_3$(s)?

36. Should precipitation occur in the following cases?

(a) 1.0 mg NaCl is added to 1.0 L of 0.10 M $AgNO_3$(aq).

(b) One drop (0.05 mL) of 0.20 M KBr is added to 200 mL of a saturated solution of AgCl.

(c) One drop (0.05 mL) of 0.0150 M NaOH(aq) is added to 5.0 L of a solution with 2.0 mg Mg^{2+} per liter.

$\star$**37.** 100.0 mL of a clear saturated solution of Ag_2SO_4 is added to 250.0 mL of a clear saturated solution of $PbCrO_4$. Will any precipitate form? [*Hint:* Refer to Table 19-1. What are the possible precipitates?]

Completeness of precipitation

38. When 200.0 mL of 0.350 M K_2CrO_4(aq) is added to 200.0 mL of 0.100 M $AgNO_3$(aq),

(a) Should a precipitate form?

(b) What $[Ag^+]$ is left unprecipitated?

39. A qualitative method of analyzing for lead involves precipitating lead chloride from aqueous solution.

$$PbCl_2(s) \rightleftharpoons Pb^{2+}(aq) + 2\ Cl^-(aq) \qquad K_{sp} = 1.6 \times 10^{-5}$$

(a) If a constant concentration of chloride ion, $[Cl^-] = 0.100$ M, is maintained in a solution in which the initial $[Pb^{2+}] = 0.050$ M, what percent of the original $[Pb^{2+}]$ will remain *unprecipitated?*

$\star$ (b) What $[Cl^-]$ should be maintained in the solution of part (a) to ensure that only 1.0% of the Pb^{2+} remains unprecipitated?

40. Refer to Example 19-7. For the precipitation of $Mg(OH)_2$ described,

(a) Can a constant $[OH^-]$ greater than 2.0×10^{-3} M be supplied by a saturated solution of $Ca(OH)_2$? ($K_{sp}[Ca(OH)_2] = 5.5 \times 10^{-6}$.)

(b) What is the *minimum* $[OH^-]$ at which precipitation is still complete (99.9% precipitated)?

41. What percent of the Ba^{2+} in solution is precipitated as $BaCO_3$(s) if *equal volumes* of 0.0020 M Na_2CO_3(aq) and 0.0010 M $BaCl_2$(aq) are mixed?

$\star$**42.** What is $[Pb^{2+}]$ remaining in solution if 225 mL of 0.15 M KCl(aq) is added to 135 mL of 0.12 M $Pb(NO_3)_2$(aq)? [*Hint:* Look for a simplifying assumption, but not the usual one.]

43. Refer to Example 19-8. You have available 1.0 M solutions of NaF, NaOH, Na_2SO_4, and Na_2CO_3. You may add 10.0 mL of one of these solutions to the solution described in part (b) of Example 19-8, in an attempt to complete as much as possible the precipitation of the remaining Ca^{2+}. Which of these solutions should you use? Explain.

Fractional precipitation

44. Assume that the seawater sample described in Example 19-7 contains approximately 440 g Ca^{2+} per metric ton of seawater (1 metric ton = 1000 kg; density of seawater ≈ 1.03 g/cm³).

(a) Should $Ca(OH)_2$(s) precipitate from seawater under the conditions stated, that is, with $[OH^-]$ maintained at 2.0×10^{-3} M?

(b) Is the separation of Ca^{2+} and Mg^{2+} from seawater by fractional precipitation feasible?

45. To a solution that is 0.250 M in NaCl(aq) and 0.0022 M in KBr(aq) is slowly added $AgNO_3$(aq).

(a) Which should precipitate first, AgCl(s) or AgBr(s)?

(b) Can the Cl^- and Br^- be separated effectively by this fractional precipitation?

46. Which of the following reagents would work best to separate Ba^{2+} and Ca^{2+} from a solution in which both are present at a concentration of 0.05 M? Explain. **(a)** 0.10 M NaCl(aq); **(b)** 0.50 M Na_2SO_4(aq); **(c)** 0.001 M NaOH(aq); **(d)** 0.50 M Na_2CO_3(aq).

47. Refer to Example 19-9 and the discussion following it. Could I^-(aq) and Cl^-(aq) be separated by fractional precipitation with $AgNO_3$(aq)? If so, could the fact that AgI(s) is yellow whereas AgCl(s) is white be used to determine the point at which to stop a titration of the type pictured in Figure 19-5? Explain.

Solubility and pH

48. Should the following precipitates form under the given conditions?

(a) PbI_2(s), from a solution that is 1.05×10^{-3} M HI, 1.05×10^{-3} M NaI, and 1.1×10^{-3} M $Pb(NO_3)_2$.

(b) $Mg(OH)_2$(s), from 2.50 L of 0.0150 M $Mg(NO_3)_2$ to which is added 1 drop (0.05 mL) of 1.00 M NH_3.

(c) $Al(OH)_3$(s), from a solution that is 1.0×10^{-2} M in Al^{3+}, 0.01 M $HC_2H_3O_2$, and 0.01 M $NaC_2H_3O_2$.

49. The solubility of $Mg(OH)_2$ in a particular buffer solution is found to be 0.95 g/L. What is the pH of the buffer solution?

★50. Calculate the molar solubility of $Mg(OH)_2$ in 1.00 M NH_4Cl(aq). [*Hint:* Consider the reaction to be

$$Mg(OH)_2(s) + 2 NH_4^+(aq) \rightleftharpoons$$
$$Mg^{2+}(aq) + 2 NH_3(aq) + 2 H_2O.]$$

51. Reaction (19.16) shows that acetic acid dissolves "boiler scale" in the form of $CaCO_3$(s). Show that acetic acid will *not* dissolve a "boiler scale" composed of $CaSO_4$. [*Hint:* Combine two chemical equations into the net equation: $CaSO_4$(s) + H_3O^+(aq) $\rightleftharpoons$ Ca^{2+}(aq) + HSO_4^-(aq) + H_2O.]

★52. A handbook lists the solubility of $CaHPO_4$ as 0.32 g $CaHPO_4 \cdot 2H_2O$/L and lists a K_{sp} value of

$$CaHPO_4(s) \rightleftharpoons Ca^{2+}(aq) + HPO_4^{2-}(aq) \qquad K_{sp} = 1 \times 10^{-7}$$

(a) Are these data consistent? (That is, are the molar solubilities the same when derived in two different ways?)

(b) How do you account for the "discrepancy"? [*Hint:* Recall the nature of phosphate species in solution.]

★53. The chief compound in marble is $CaCO_3$. Marble has been widely used for statues and ornamental work on buildings. Marble is readily attacked by acids. Determine the solubility of marble (that is, the concentration of Ca^{2+} in a saturated solution) in **(a)** normal rainwater of pH = 5.6; **(b)** "acid" rainwater of pH 4.20. Assume that the net reaction is

$$CaCO_3(s) + H_3O^+(aq) \rightleftharpoons Ca^{2+}(aq) + HCO_3^-(aq) + H_2O$$

Complex ion equilibria

54. $PbCl_2$(s) is found to be somewhat soluble in HCl(aq) but not in HNO_3(aq). Explain this difference in behavior.

55. Write equations to represent the following observations: $Zn(OH)_2$(s) is readily soluble in dilute HCl(aq), $HC_2H_3O_2$(aq), NH_3(aq), and NaOH(aq).

56. A solution is prepared that is 0.10 M in *free* NH_3, 0.10 M

NH_4Cl, and 0.15 M in $[Cu(NH_3)_4]^{2+}$. Should $Cu(OH)_2$(s) precipitate from this solution? K_{sp} for $Cu(OH)_2$ is 1.6×10^{-19}.

57. Refer to Example 19-15. What mass of KI could be dissolved in 1.00 L of the original $AgNO_3–NH_3$ solution before AgI(s) would precipitate?

58. What mass of AgBr can be dissolved in 1.00 L of each of the following solutions? **(a)** 1.50 M NH_3; **(b)** 0.10 M NaCN; **(c)** 0.50 M $Na_2S_2O_3$.

Precipitation and solubilities of metal sulfides

59. A buffer solution is 0.25 M $HC_2H_3O_2$–0.15 M $NaC_2H_3O_2$, saturated in H_2S (0.10 M) and has $[Mn^{2+}]$ = 0.015 M.

(a) Show that MnS will not precipitate from this solution.

(b) Which buffer component would you increase in concentration, and to what minimum value, to ensure that precipitation of MnS would begin? (*Note:* The concentration of the other buffer component is held constant.)

★60. In the qualitative analysis of a mixture of cations, Pb^{2+} is first precipitated as $PbCl_2$. Later, PbS is precipitated when the saturated $PbCl_2$(aq) is saturated with H_2S (0.10 M) and its pH adjusted to about 0.5. Show that in fact the precipitation of PbS should occur under these conditions.

61. What is the solubility of FeS, in g/L, in a buffer solution that is 0.500 M $HC_2H_3O_2$–0.250 M $NaC_2H_3O_2$? [*Hint:* Assume that $[H_3O^+]$ remains constant in the dissolving reaction.]

★62. Calculate the molar solubility of CoS(s) in 0.100 M $HC_2H_3O_2$. [*Hint:* Combine the appropriate equilibrium constant expressions.]

Qualitative anlaysis

63. Addition of HCl(aq) to a solution containing several different cations produces a white precipitate. The filtrate is removed and treated with H_2S(aq) in 0.3 M HCl. No precipitate forms. Which of the following conclusions is(are) valid? Explain. **(a)** Ag^+ and/or Hg_2^{2+} probably present; **(b)** Mg^{2+} probably not present; **(c)** Pb^{2+} probably not present; **(d)** Fe^{2+} probably not present. [*Hint:* Study Figure 19-7 carefully.]

64. Suppose you did a cation group 1 analysis and treated the chloride precipitate with NH_3(aq), without first treating it with hot water. What might you observe and what valid conclusions could you reach about cations present, cations absent, and cations in doubt?

65. In the cation group 1 analysis, $PbCl_2$(s) is separated from the other chlorides by dissolving it in hot water. Given the K_{sp} values for $PbCl_2$ listed below, assess the water solubility of $PbCl_2$(s) at low and high temperatures. K_{sp} for $PbCl_2$ = 1.6×10^{-5} at 25 °C; 3.3×10^{-3} at 80 °C.

66. Write net ionic equations for the following qualitative analysis procedures.

(a) Precipitation of $PbCl_2$(s) from a solution containing Pb^{2+}.

(b) Dissolving of $Zn(OH)_2$(s) in NaOH(aq).

(c) Dissolving of $Fe(OH)_3$(s) in HCl(aq).

(d) Precipitation of CuS(s) from an acidic solution of Cu^{2+} and H_2S.

(e) Precipitation of antimony sulfide from a solution containing H_2S and the complex ion $SbCl_4^-$.

Additional Exercises

(Use data from tables in Chapters 17, 18 and 19, as necessary.)

67. A handbook lists the solubilities 1.0×10^{-5} mol $BaSO_4$/L and 7.1×10^{-5} mol $BaCO_3$/L. When 0.50 M Na_2CO_3(aq) is added to a saturated solution of $BaSO_4$, a precipitate of $BaCO_3$(s) forms. How do you account for this fact, given that $BaCO_3$ has a larger K_{sp} than does $BaSO_4$?

68. A solution is saturated with $PbSO_4$ at 50 °C ($K_{sp} = 2.3 \times 10^{-8}$). How many mg $PbSO_4$ will precipitate from 815 mL of this solution if it is cooled to 25 °C? ($K_{sp} = 1.6 \times 10^{-8}$.)

69. A 25.00-mL sample of a saturated solution of $SrCrO_4$ requires 25.5 mL of 0.0138 M $Fe(NO_3)_2$(aq) for its titration in the reaction

$$CrO_4^{2-}(aq) + 3\ Fe^{2+}(aq) + 8\ H^+(aq) \rightarrow$$
$$Cr^{3+}(aq) + 3\ Fe^{3+}(aq) + 4\ H_2O$$

Calculate K_{sp} for $SrCrO_4$.

70. Suppose a certain groundwater comes into contact with the mineral gypsum ($CaSO_4 \cdot 2H_2O$). What is the maximum SO_4^{2-} content that you would expect to find in this water, expressed as parts per million (ppm) by mass?

71. Ba^{2+}(aq) is poisonous when ingested. The lethal dosage in mice is about 12 mg Ba^{2+} per kg of body mass. Despite this fact $BaSO_4$ is widely used in medicine to obtain x ray photographs of the gastrointestinal tract.

(a) Explain why $BaSO_4$(s) is safe to take internally, even though Ba^{2+}(aq) can be poisonous.

(b) How many mg Ba^{2+} would be present in 1.00 L water that is saturated with $BaSO_4$?

★(c) Would the solubility of $BaSO_4$ be affected by coming in contact with stomach acid? If so, how significant would this be? Assume that stomach acid is 0.1 M HCl. [*Hint:* Write an equation to represent any dissolving that might occur.]

72. An experiment to measure the solubility product constant of $CaSO_4$ [D. Masterman, *J. Chem. Educ.* **64**, 409 (1987)] involves passing a saturated solution of $CaSO_4$(aq) through an ion exchange column, pictured here (and described in Section 22-2). Ca^{2+} ion is retained on the column and *two* H_3O^+ ions appear in the solution leaving the column for every Ca^{2+} ion retained. A 25.00-mL sample of saturated $CaSO_4$(aq) is passed through the column, followed by a small volume of water. The solution col-

lected is then diluted to 100.0 mL in a volumetric flask. Finally, 10.00-mL of the solution from the volumetric flask is titrated with 8.25 mL of 0.0105 M NaOH. Use these data to obtain a value of K_{sp} for $CaSO_4$.

73. Show that in qualitative analysis group 1, if you obtain 1.00 mL of saturated $PbCl_2$(aq) at 25 °C, there should be sufficient Pb^{2+} present to produce a precipitate of $PbCrO_4$(s). Assume that you use 1 drop (0.05 mL) of 1.0 M K_2CrO_4 for the test.

74. Predict whether precipitation will occur under the following conditions.

(a) Will Ag_2CrO_4 precipitate from a solution that is 0.0050 M in Ag^+ and 0.0100 M in CrO_4^{2-}?

(b) Will $PbCl_2$(s) precipitate if 1.00 g NaCl is added to 250.0 mL of 0.0200 M $Pb(NO_3)_2$(aq)?

(c) Will AgI(s) precipitate from a solution that is 2.0 M in *free* $S_2O_3^{2-}$, 0.10 M in $[Ag(S_2O_3)_2]^{3-}$, and 0.040 M in I^-?

75. For the equilibrium

$$Al(OH)_3(s) \rightleftharpoons Al^{3+}(aq) + 3\ OH^-(aq),\ K_{sp} = 1.3 \times 10^{-33}$$

(a) What is the *minimum pH* at which $Al(OH)_3$(s) will precipitate from a solution that is 0.050 M in Al^{3+}?

★(b) A solution has $[Al^{3+}] = 0.050$ M and $[HC_2H_3O_2] = 1.00$ M. What is the maximum quantity of $NaC_2H_3O_2$ that can be added to 200.0 mL of this solution before precipitation of $Al(OH)_3$(s) begins? [*Hint:* The acetic acid–sodium acetate mixture is a buffer solution.]

76. Calculate

(a) K_f for $[Cu(CN)_4]^{3-}$, if it is found that in a solution that is 0.0500 M in $[Cu(CN)_4]^{3-}$ and 0.80 M in *free* CN^-, the concentration of *free* Cu^+ is 6.1×10^{-32} M.

$$Cu^+(aq) + 4\ CN^-(aq) \rightleftharpoons [Cu(CN)_4]^{3-}(aq) \qquad K_f = ?$$

(b) $[Cu^{2+}]$ in a 0.10 M $CuSO_4$(aq) solution that is also 6.0 M in *free* NH_3.

$$Cu^{2+}(aq) + 4\ NH_3(aq) \rightleftharpoons [Cu(NH_3)_4]^{2+}(aq)$$
$$K_f = 1.1 \times 10^{13}$$

77. What range of pH values must be maintained to be able to separate Fe^{2+} and Mn^{2+} by selective precipitation of FeS(s) from a solution that is 0.050 M Fe^{2+}, 0.050 M Mn^{2+}, and also saturated in H_2S (0.10 M H_2S)? [*Hint:* Obtain solubility product values for FeS and MnS from Table 19-1. Also, use equation 19.29.]

78. Refer to the Summarizing Example. What is the solubility of $Ca(OH)_2$(s) in a buffer solution with pH = 11.50? [*Hint:* Determine a value of K for this reaction.]

$$Ca(OH)_2(s) + 2\ H_3O^+(aq) \rightleftharpoons Ca^{2+}(aq) + 4\ H_2O$$

Calculate $[Ca^{2+}]$.]

★**79.** Refer to the Summarizing Example. When applied to the qualitative analysis scheme for *anions*, a reaction such as (19.33) is called a *carbonate transposition*. Anions from a slightly soluble compound (such as a hydroxide or a sulfate) are replaced by CO_3^{2-}. As a result these anions (OH^-, SO_4^{2-}, . . .) are obtained in a sufficient concentration in water solution so that identification tests can be done on them. Suppose we use 3 M

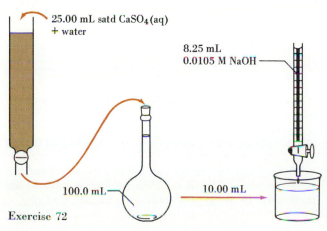

25.00 mL satd $CaSO_4$(aq) + water

8.25 mL 0.0105 M NaOH

100.0 mL

10.00 mL

Exercise 72

Na_2CO_3, and suppose that an anion concentration of 0.050 M is sufficient for detection of the anion. Predict whether carbonate transposition will be effective in the following conversions.

(a) $BaSO_4(s) + CO_3^{2-}(aq) \rightleftharpoons BaCO_3(s) + SO_4^{2-}(aq)$
(b) $Mg(OH)_2(s) + CO_3^{2-}(aq) \rightleftharpoons MgCO_3(s) + 2\ OH^-(aq)$
(c) $2\ AgCl(s) + CO_3^{2-}(aq) \rightleftharpoons Ag_2CO_3(s) + 2\ Cl^-(aq)$

80. The electrolysis of $MgCl_2(aq)$ can be represented as

$$Mg^{2+}(aq) + 2\ Cl^-(aq) + 2\ H_2O \rightarrow$$
$$Mg^{2+}(aq) + 2\ OH^-(aq) + H_2(g) + Cl_2(g)$$

The electrolysis of a 315-mL sample of 0.220 M $MgCl_2$ is continued until 1.04 L $H_2(g)$ at 23 °C and 748 mmHg has been collected. Will $Mg(OH)_2(s)$ precipitate when electrolysis is carried to this point? [*Hint:* Notice that $[Mg^{2+}]$ remains constant throughout the electrolysis while $[OH^-]$ *increases*.]

81. Predict whether a reaction is likely to occur in each of the following cases. If so write a net ionic equation. [*Hint:* Convert each of the molecular equations to a net ionic equation, using the ideas stated on page 694.]

(a) $NaI(aq) + ZnSO_4(aq) \rightarrow$
(b) $CuSO_4(aq) + Na_2CO_3(aq) \rightarrow$
(c) $AgNO_3(aq) + CuCl_2(aq) \rightarrow$
(d) $BaS(aq) + CuSO_4(aq) \rightarrow$
(e) $Al(OH)_3(s) + HCl(aq) \rightarrow$
(f) $CaC_2O_4(s) + HCl(aq) \rightarrow$
(g) $CdS(s) + HC_2H_3O_2(aq) \rightarrow$

82. The following pertain to the precipitation or dissolving of metal sulfides. Predict whether a reaction proceeds to a significant extent in the forward direction in each case.

(a) $Cu^{2+}(aq) + H_2S(satd\ aq) \rightarrow$
(b) $Mg^{2+}(aq) + H_2S(satd\ aq) \xrightarrow{0.3\ M\ HCl}$
(c) $PbS(s) + HCl(0.3\ M) \rightarrow$
(d) $Ag^+(aq) + H_2S(sat\ aq) \xrightarrow{0.3\ M\ HCl}$

83. Determine if 1.50 g $H_2C_2O_4$ (oxalic acid) can be dissolved in 0.200 L of 0.150 M $CaCl_2$ without the formation of $CaC_2O_4(s)$ (for which $K_{sp} = 1.3 \times 10^{-9}$). For $H_2C_2O_4$, $K_{a_1} = 5.4 \times 10^{-2}$ and $K_{a_2} = 5.4 \times 10^{-5}$.

* **84.** Write net ionic equations to represent the following observations.

(a) When concentrated $CaCl_2(aq)$ is added to $Na_2HPO_4(aq)$, a white precipitate is formed that is 38.7% Ca, by mass.
(b) When a piece of dry ice $[CO_2(s)]$ is placed into a clear dilute solution of "lime water" $[Ca(OH)_2(aq)]$, bubbles of gas are evolved. At first a white precipitate forms, but then it redissolves.

* **85.** For the titration that is the subject of Example 19-9 and illustrated in Figure 19-5, verify the assertion on page 693 that $[Ag^+]$ increases very rapidly between the point where AgBr has finished precipitating and Ag_2CrO_4 is about to begin.

* **86.** The solubility of AgCN(s) in 0.200 M $NH_3(aq)$ is 8.8×10^{-6} mol/L. What is the value of K_{sp} for AgCN?

* **87.** The solubility of $CdCO_3(s)$ in 1.00 M KI(aq) is 1.2×10^{-3} mol/L. Given that K_{sp} for $CdCO_3$ is 5.2×10^{-12}, what is K_f for CdI_4^{2-}?

* **88.** Use K_{sp} for $PbCl_2$ and K_f for $[PbCl_3]^-$ to determine the molar solubility of $PbCl_2$ in 0.10 M HCl(aq). [*Hint:* What is the total concentration of lead species in solution?]

* **89.** In the Mohr titration, Cl^- is determined by titration with $AgNO_3(aq)$. The solution also contains a small volume of $K_2CrO_4(aq)$ as an indicator. After sufficient $AgNO_3(aq)$ has been added to precipitate all of the Cl^-, a red precipitate of Ag_2CrO_4 forms. Explain the basis of this titration. That is, why does the red color not appear while AgCl(s) is still precipitating? Why does the red precipitate appear *immediately* after the AgCl(s) has finished precipitating?

* **90.** A mixture of $PbSO_4(s)$ and $PbS_2O_3(s)$ is shaken with pure water until a saturated solution is formed. Both solids remain in excess. What is $[Pb^{2+}]$ in the saturated solution? [*Hint:* Both of the following equilibrium expressions are required, and a third equation as well.]

$$PbSO_4(s) \rightleftharpoons Pb^{2+}(aq) + SO_4^{2-}(aq) \qquad K_{sp} = 1.6 \times 10^{-8}$$

$$PbS_2O_3(s) \rightleftharpoons Pb^{2+}(aq) + S_2O_3^{2-}(aq) \qquad K_{sp} = 4.0 \times 10^{-7}$$

* **91.** Use the method of the preceding exercise to determine $[Pb^{2+}]$ in a saturated solution in contact with a mixture of $PbCl_2(s)$ and $PbBr_2(s)$.

$$PbCl_2(s) \rightleftharpoons Pb^{2+}(aq) + 2\ Cl^-(aq) \qquad K_{sp} = 1.6 \times 10^{-5}$$

$$PbBr_2(s) \rightleftharpoons Pb^{2+}(aq) + 2\ Br^-(aq) \qquad K_{sp} = 4.0 \times 10^{-5}$$

* **92.** 2.50 g $Ag_2SO_4(s)$ is added to a beaker containing 0.150 L of 0.025 M $BaCl_2$.

(a) Write an equation for any reaction that occurs.
(b) Describe the final contents of the beaker, that is, the masses of any precipitates present and the concentrations of the ions in solution.

* **93.** Calculate the molar solubility of MnS in water (a) based only on the solubility product expression, K_{sp}, for MnS and (b) taking into account the hydrolysis of S^{2-} to HS^-. (c) Explain the difference in the results obtained.

Self-Test Questions

For questions 94 through 103 select the single item that best completes each statement.

94. Pure water is saturated with slightly soluble PbI_2. The most accurate statement that we can make about $[Pb^{2+}]$ in this solution is that (a) $[Pb^{2+}] = [I^-]$; (b) $[Pb^{2+}] = K_{sp}$ of PbI_2; (c) $[Pb^{2+}] = \sqrt{K_{sp}}$ of PbI_2; (d) $[Pb^{2+}] = 0.5\ [I^-]$.

95. The addition of 1.058 g Na_2SO_4 to 375.0 mL of saturated $BaSO_4(aq)$ has the following effect on the saturated solution: (a) reduces $[Ba^{2+}]$; (b) reduces $[SO_4^{2-}]$; (c) increases the solubility of $BaSO_4$; (d) no effect.

96. The slightly soluble solute Ag_2CrO_4 is expected to be *most* soluble in (a) pure water; (b) 0.10 M K_2CrO_4; (c) 0.25 M KNO_3; (d) 0.40 M $AgNO_3$.

97. Cu^{2+} and Ag^+ are both present in the same solution. To precipitate one of the ions and leave the other in solution, add (a) $H_2S(aq)$; (b) HCl(aq); (c) $HNO_3(aq)$; (d) $NH_4NO_3(aq)$.

98. The addition of $K_2CO_3(aq)$ to the following solutions is expected to produce a precipitate in every case but one. That one is (a) $BaCl_2(aq)$; (b) $CaBr_2(aq)$; (c) $Na_2SO_4(aq)$; (d) $Pb(NO_3)_2(aq)$.

99. A *large* excess of $MgF_2(s)$ is maintained in contact with 1.00 L of pure water to produce a saturated solution. When an additional 1.00 L of pure water is added to the solid–liquid mixture and equilibrium reestablished, compared to its value in the original saturated solution, $[Mg^{2+}]$ will be (a) the same; (b) twice as large; (c) half as large; (d) some unknown multiple of the original $[Mg^{2+}]$.

100. To increase the molar solubility of $CaCO_3(s)$ in a saturated aqueous solution, add (a) more water; (b) Na_2CO_3; (c) NaOH; (d) $NaHSO_4$.

101. $Cu(OH)_2$ is highly soluble in all of the following except one. The exception is (a) H_2O; (b) $NH_3(aq)$; (c) $HCl(aq)$; (d) $HNO_3(aq)$.

102. To increase *significantly* the concentration of *free* Zn^{2+} ion in a solution of the complex ion $[Zn(NH_3)_4]^{2+}$,

$$Zn^{2+}(aq) + 4\ NH_3(aq) \rightleftharpoons [Zn(NH_3)_4]^{2+}(aq) \qquad K_f = 4.1 \times 10^8$$

add to the solution some (a) H_2O; (b) $HCl(aq)$; (c) $NH_3(aq)$; (d) $NH_4Cl(aq)$.

103. The best way to ensure complete precipitation from *saturated* $H_2S(aq)$ of a metal ion M^{2+}, as its sulfide, $MS(s)$, is to (a) add an acid; (b) increase $[H_2S]$ in the solution; (c) raise the pH; (d) heat the solution.

104. A solution is saturated with Ag_2SO_4 ($K_{sp} = 1.4 \times 10^{-5}$).

(a) What is $[Ag^+]$ in this saturated solution?

(b) What $[SO_4^{2-}]$ must be present in the solution (through added Na_2SO_4) to reduce $[Ag^+]$ in the saturated solution to 2.0×10^{-3} M?

105. Which of the following solids are likely to be more soluble in acidic solution, and which in basic solution? Which are likely to have a solubility that is essentially independent of pH? Explain. (a) $H_2C_2O_4$; (b) $MgCO_3$; (c) CdS; (d) KCl; (e) $NaNO_3$; (f) $Ca(OH)_2$.

106. A solution is 0.010 M in CrO_4^{2-} and 0.010 M in SO_4^{2-}. To this solution is slowly added 0.10 M $Pb(NO_3)_2(aq)$. (K_{sp} for $PbCrO_4 = 2.8 \times 10^{-13}$; K_{sp} for $PbSO_4 = 1.6 \times 10^{-8}$.)

(a) What should be the first anion to precipitate from the solution?

(b) What is $[Pb^{2+}]$ at the point where the second anion begins to precipitate?

(c) Are the two anions effectively separated by this fractional precipitation? Explain.

20

Spontaneous Change: Entropy and Free Energy

As suggested by this common rural scene, evidence of the natural tendency toward increased disorder can be found everywhere. In thermodynamics, this tendency serves as the basis of a criterion for spontaneous change. [John Schultz/PAR/NYC]

In the introduction to Chapter 16 we described the reaction of nitrogen and oxygen gases. We noted that the reaction does not occur at all to speak of at room temperature, but it becomes increasingly significant at high temperatures.

(1) $N_2(g) + O_2(g) \longrightarrow 2\ NO(g)$

Another reaction involving oxides of nitrogen that we considered in Chapter 16 is the conversion of $NO(g)$ to $NO_2(g)$.

(2) $2\ NO(g) + O_2(g) \longrightarrow 2\ NO_2(g)$

This reaction, unlike the first, is favored at *low* temperatures. The combination of the two reactions—the first occurring at high temperatures and the second at lower temperatures—accounts for the production of $NO_2(g)$ in high-temperature combustion processes. And, as we learned in Section 14-10, NO_2 is an important starting material in the formation of photochemical smog.

What is there about these two reactions that causes the first to be favored at high temperatures and the second at low temperatures? This is one of the fundamental questions about chemical reactions that we hope to answer in this chapter. Specifically, we will develop criteria with which to predict what changes will or will not occur. Also, we will derive a *quantitative* expression for the variation of equilibrium constants with temperature. This will allow us to use tabulated data (as in Appendix D) to calculate the compositions of equilibrium mixtures at a variety of temperatures.

Together with ideas presented in Chapter 7, this chapter will demonstrate the great power of thermodynamics to contribute to our understanding of chemical phenomena.

20-1 Spontaneity: The Meaning of Spontaneous Change

Most of us have played with spring-wound toys, whether a toy automobile, a top, or a music box. In every case, once the wound-up toy is released, it keeps running until the energy stored in the spring has been released, then the toy stops. The toy never rewinds itself, human intervention is necessary (winding by hand). The running down of a wound-up spring is an example of a *spontaneous* process. The rewinding of the spring is a *nonspontaneous* process. Now let us explore the scientific meaning of these two terms.

A **spontaneous** or natural process is a process that occurs in a system left to itself; once started, no action from outside the system (external action) is necessary to make the process continue. On the other hand, a **nonspontaneous** process will not occur *unless* some external action is continuously applied. Consider the rusting of an iron pipe exposed to the atmosphere. Although the process may occur quite slowly, it does so *continuously* and always in the *same* direction. As a result, the amount of iron decreases and the amount of rust increases, until a final state of *equilibrium* is reached where essentially all the iron has been converted to iron(III) oxide. We say that the reaction

$2\ Fe(s) + 3\ O_2(g) \longrightarrow Fe_2O_3(s)$

is *spontaneous*. Now consider the reverse situation: converting iron(III) oxide into an iron pipe. We cannot say that the process is *impossible*, but it is certainly *nonspontaneous*. In fact, this nonspontaneous reverse process is essentially what we use to manufacture iron objects from iron ore.

Spontaneous . . . "arising from internal forces or causes; independent of external agencies; self-acting."—*The Random House Dictionary of the English Language, 2nd ed., 1987.*

715

Example 20-1 _____

Identifying spontaneous and nonspontaneous processes. Use your general knowledge to indicate whether each of the following processes is spontaneous or nonspontaneous.

(a) The action of NaOH(aq) on HCl(aq).
(b) The electrolysis of liquid water.
(c) The melting of an ice cube.

Solution

(a) This is a neutralization reaction. The net equation is

$$H_3O^+ + OH^- \rightarrow 2\ H_2O.$$

There is little tendency for the reverse reaction (self-ionization of water) to occur. We can say that the neutralization reaction is spontaneous.

(b) We have previously learned that we need to use electric current (an external action) to decompose liquid water into its elements.

$$2\ H_2O(l) \xrightarrow{\text{electrolysis}} 2\ H_2(g) + O_2(g) \qquad (14.5)$$

The electrolysis of liquid water is a *nonspontaneous* process.

(c) Here is an interesting case. From experience we know that the melting of an ice cube is *spontaneous* if the temperature is *above* the melting point (0 °C), but it is *nonspontaneous* if the temperature is *below* the melting point. Certain processes are spontaneous at some temperatures and nonspontaneous at other temperatures, as we shall see in Section 20-3.

SIMILAR EXAMPLE: Exercise 17.

_____

From our discussion of spontaneity to this point we can conclude that

- *If a process is found to be spontaneous, the reverse process is nonspontaneous.*
- *Both spontaneous and nonspontaneous processes are possible, but only spontaneous processes will occur naturally. Nonspontaneous processes require the system to be acted upon in some way.*

However, we need something that will allow us to *predict* whether the forward or the reverse direction is the direction of spontaneous change. We cannot simply base predictions on "common knowledge" as we did in Example 20-1. We need *a criterion for spontaneous change.* To begin, let us look to mechanical systems for a clue. A ball rolls downhill and water flows to a lower level. Figure 20-1 illustrates a common feature of these processes: *potential energy decreases.*

FIGURE 20-1

Direction of spontaneous change in a mechanical system.

Whether we consider the ball on the left or the one on the right, the direction of spontaneous change is downhill. The ball reaches a position of equilibrium when it comes to rest at the bottom, the point of lowest potential energy.

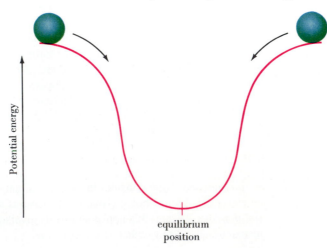

equilibrium
position

FIGURE 20-2

Search for a criterion for spontaneous chemical change.

We are looking for a thermodynamic property, here called the "chemical potential," that has a minimum value somewhere between the pure reactants and the pure products. At this point the reaction is at equilibrium, with the reaction quotient Q equal to the equilibrium constant K. For a condition on either side of the equilibrium point, spontaneous reaction occurs in the direction of the equilibrium point.

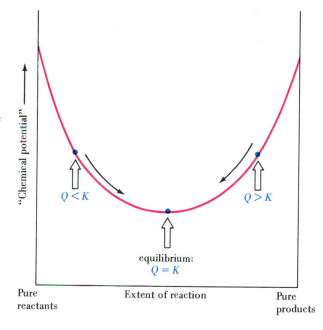

Devising a thermodynamic function having special characteristics is a common practice in thermodynamics. Recall how we devised the function called enthalpy in Section 7-7 so that we could describe reactions at constant pressure.

For chemical systems the property analogous to the potential energy of a mechanical system is the internal energy (E) or the closely related property of enthalpy (H). In the 1870s, the chemists P. Berthelot and J. Thomsen proposed that the direction of spontaneous change is that in which the enthalpy of a system decreases. An enthalpy decrease means that heat is given off by the system to the surroundings. Bertholet and Thomsen concluded that exothermic reactions should be spontaneous. In fact, exothermic processes generally are spontaneous, but so are some *endothermic* ones! Here are three examples of *spontaneous endothermic* processes.

• the melting of ice at room temperature;
• the evaporation of liquid ether from an open beaker;
• the dissolving of ammonium nitrate in water.

In abandoning enthalpy change (ΔH) as our criterion for spontaneous change, let us agree that it would be useful to find some other thermodynamic function having the properties suggested by Figure 20-2.

20-2 Spontaneity and Disorder: The Concept of Entropy

To continue our search for a criterion for spontaneous change we turn again to a mechanical analogy. Suppose you had 10 identical balls, each colored half red and half white, and released them at the top of a ramp. Which of the three arrangements of the balls pictured in Figure 20-3 would you expect to see? You would *not* expect arrangement (a) because this is not a condition in which the potential energy has been minimized. Arrangements (b) and (c) both have this minimum in potential energy, but still you would predict that arrangement (c) is the expected final arrangement. The scrambled or *disordered* arrangement in (c) is far more likely to occur than the very orderly arrangement in (b). This example suggests that we need to consider *two* factors when assessing spontaneous change—does the system gain or lose energy, and what happens to the degree of order or disorder in the system?

Figure 20-4 depicts a situation in which one ideal gas, labeled A, fills a glass bulb at a pressure of 1.00 atm. A second ideal gas, B, fills a second bulb, which is

FIGURE 20-3
**The importance of random-
ness or disorder.**

If 10 red-and-white balls are released at the top of the ramp,
the arrangement in (c) is what
we expect to find. The arrangements in (a) and (b) are extremely unlikely.

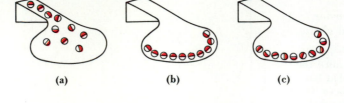

(a)   (b)   (c)

identical to the first. Again the pressure is 1.00 atm. The two bulbs are joined by a
stopcock valve. This initial condition is pictured in Figure 20-4a. Assume that the
two gases do not react and imagine what happens when the valve between the two
bulbs is opened. We expect the condition pictured in Figure 20-4b: each gas expands into the bulb containing the other. The mixing continues until the partial
pressure of each gas becomes 0.50 atm in each bulb.

One of the characteristics of an ideal gas is that the internal energy (E) and
enthalpy (H) depend only on temperature and not on the gas pressure or volume.
Because there are no intermolecular forces, when ideal gases mix at constant temperature, $\Delta E = \Delta H = 0$. Thus, energy change is *not* the driving force behind the
spontaneous mixing of gases in Figure 20-4. The driving force is simply the tendency of the molecules of the two gases to achieve the maximum state of mixing or
disorder possible. We call the thermodynamic property related to the degree of
disorder in a system **entropy,** and we designate it by the symbol S. That is,

**Relationship between entropy
and disorder.**

the greater the degree of randomness or disorder in a system, the greater   (20.1)
the entropy of the system.

To represent the mixing of gases symbolically, we can write

$$A(g) + B(g) \longrightarrow \text{mixture of } A(g) \text{ and } B(g)$$

$$\Delta S = S_{\text{final state}} - S_{\text{initial state}}$$

$$\Delta S = S_{\text{mixt. of gases}} - [S_{A(g)} + S_{B(g)}]$$

$$\Delta S > 0$$

In the mixing of gases, disorder and entropy *increase* and ΔS is a *positive* quantity,
that is, $\Delta S > 0$. Now, what can we say about entropy changes in the three endothermic processes listed at the end of Section 20-1?

First, in the melting of ice a crystalline solid (recall Figure 12-30) is replaced by
a less structured liquid. *Disorder and entropy increase in the process of melting.* In
the second case, molecules in the gaseous state, because of the large free volume in
which they can move, have a much higher entropy than in the liquid state. *The
process of evaporation is accompanied by an increase in entropy.* Finally, in the
dissolving of an ionic solid such as ammonium nitrate in water, a crystalline solid
and a pure liquid are replaced by a mixture of ions and water molecules in the liquid
(solution) state (recall Figure 13-5). There is a tendency for some ordering of water
molecules around the ions in solution (referred to as hydration of the ions). However, this ordering tendency is not as great as the disorder produced by destroying
the original crystalline solid. *Disorder and entropy increase in the dissolving*

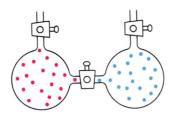

(a) Before mixing

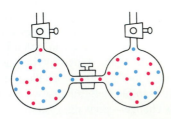

(b) After mixing

• gas A   • gas B

FIGURE 20-4
The mixing of ideal gases.

The total volume of the system and the total gas pressure are the same in each case.
The net change that occurs is this.
(a) Before mixing, each gas is confined to one half of the total volume (a single
bulb) at a pressure of 1.00 atm.
(b) After mixing, each gas has expanded into the total volume (both bulbs) yielding
a partial pressure of 0.50 atm.

FIGURE 20-5
Entropy and disorder—three processes in which entropy increases.

Each of the processes pictured here results in greater disorder and an increased entropy.

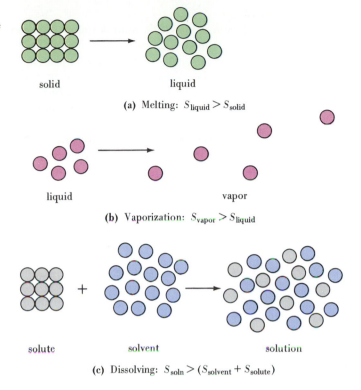

(a) Melting: $S_{liquid} > S_{solid}$

(b) Vaporization: $S_{vapor} > S_{liquid}$

(c) Dissolving: $S_{soln} > (S_{solvent} + S_{solute})$

solid   liquid

liquid   vapor

solute   solvent   solution

process. The increased disorder accompanying these processes is pictured in Figure 20-5. In each of these three cases, the increase in disorder (entropy) is more than sufficient to overcome the energy increase and the process is *spontaneous*. In general,

> entropy *increases* are expected to accompany processes in which
> - pure liquids or liquid solutions are formed from solids;
> - gases are formed, either from solids or liquids;
> - the number of molecules of gases increases in the course of a chemical reaction;
> - the temperature of a substance is increased. (Increased temperature means increased molecular motion, whether it be vibrational motion of atoms or ions in a solid or translational motion of molecules in a liquid or gas.)

(20.2)

Conditions leading to entropy increases.

Example 20-2 _____

Making qualitative predictions of entropy changes in chemical reactions. Predict whether each of the following processes involves an increase or decrease in entropy, or if the matter is uncertain.

(a) The decomposition of ammonium nitrate (a fertilizer and a high explosive).

$$2\ NH_4NO_3(s) \longrightarrow 2\ N_2(g) + 4\ H_2O(g) + O_2(g)$$

(b) The conversion of SO_2 to SO_3 (key step in the contact method for sulfuric acid manufacture).

$$2\ SO_2(g) + O_2(g) \longrightarrow 2\ SO_3(g)$$

(c) The extraction of salt from seawater.

$$NaCl(aq) \longrightarrow NaCl(s)$$

(d) The "water gas shift" reaction (involved in the gasification of coal).

$$CO(g) + H_2O(g) \longrightarrow CO_2(g) + H_2(g)$$

Solution

(a) Here a solid yields a large quantity of gas. *Entropy increases.*

(b) Three moles of gaseous reactants produce 2 moles of gaseous products. The *loss* of one mole of gas suggests that the S and O atoms are in a state of greater organization in SO_3 than in SO_2 and O_2. *Entropy decreases.*

(c) The Na^+ and Cl^- ions achieve a high degree of order when they leave the solution and arrange themselves into the crystalline state. *Entropy decreases.*

(d) The entropies of the four gases are likely to be different because of differences in their molecular structures. There should be an entropy change. However, because the number of moles of gases is the same on both sides of the equation, the entropy change is likely to be small. Furthermore, we cannot say whether entropy increases or decreases based on expression (20.2) alone.

SIMILAR EXAMPLES: Exercises 1, 18, 19, 20.

In many cases we are able to apply the entropy concept in a qualitative way, as in Example 20-2, but we also need to say something about how entropy changes are measured and expressed. Entropy, like enthalpy, must be defined in such a way as to be a function of state. It must have a *unique* value for each state or condition of a system, so that the entropy change between two states, ΔS, will also have a *unique* value. A further requirement is that entropy must be based on *measurable* quantities. Two measurable quantities that affect the amount of disorder in a system are a quantity of heat and temperature. For example, when a solid at its melting point absorbs a quantity of heat, it is converted to the more disordered liquid state (entropy increases). And, the ability of a given quantity of heat to produce disorder is much greater if the heat is absorbed by a highly ordered system (low temperature) than by a system that is already highly disordered (high temperature). Thus, an entropy change should be directly proportional to the quantity of heat absorbed (q) and inversely proportional to the temperature (T), as indicated by this equation.

> **You may find it helpful to review the discussion of functions of state in Section 7-5.**

> **Defining equation for entropy change.**

$$\Delta S = \frac{q_{rev.}}{T} \tag{20.3}$$

> **A reversible process is one that can be made to reverse its direction by making just an infinitesimal change in some system property (see Section 20-8).**

Equation (20.3) is deceptively simple in appearance. The difficulty is that ΔS for a process must be *independent* of path (since S is a function of state), but the value of q depends on the path chosen. Thus equation (20.3) only holds for a carefully defined path. The path must be of a type called *reversible*, for which $q = q_{rev}$. Fortunately, we do not need to pursue this matter further in this text. Since q has the unit J and $1/T$ has the unit K^{-1}, the unit of entropy change in equation (20.3) is J/K or $J\ K^{-1}$. Also, ΔS is an extensive property; its value depends on the quantities of substances involved in a process.

20-3 Criteria for Spontaneous Change: The Second Law of Thermodynamics

If you think about it for a moment, we still have not found a *single* criterion that alone will tell us whether a process is spontaneous. For example, suppose we use as a criterion for spontaneous change the tendency of entropy to increase. Then, *based on the entropy change of the water alone*, how do we explain the spontaneous freezing of water at $-10\ °C$? Since crystalline ice is a more ordered state than liquid water, the freezing of ice is a process for which entropy *decreases*. The answer to this puzzle lies in the fact that we must always simultaneously consider *three* entropy changes: the entropy change of the system itself, the entropy change of the surroundings, and the total of the two, called the entropy change of the "universe." The relationship among the three is

$$\Delta S_{\text{total}} = \Delta S_{\text{universe}} = \Delta S_{\text{system}} + \Delta S_{\text{surroundings}} \tag{20.4}$$

Although it is beyond the scope of this discussion to derive expression (20.5), this expression provides the basic criterion for spontaneous change. It is one of the many forms for stating **the second law of thermodynamics.**

The second law of thermodynamics.

$$\Delta S_{\text{univ.}} = \Delta S_{\text{tot.}} = \Delta S_{\text{syst.}} + \Delta S_{\text{surr.}} > 0$$

All spontaneous or natural processes produce an increase in the entropy of the universe. $\tag{20.5}$

According to expression (20.5), if a process produces *positive* entropy changes in *both* the system *and* its surroundings the process is surely *spontaneous*. And if *both* these entropy changes are *negative,* the process is just as surely *nonspontaneous*. The freezing of water produces a *negative* entropy change in the *system,* but the entropy change in the *surroundings* is *positive*. As long as the temperature is below 0 °C, the entropy of the surroundings increases more than the entropy of the system decreases, the *total* entropy change is *positive,* and the freezing of liquid water is indeed *spontaneous*.

Free Energy and Free Energy Change. We could use expression (20.5) as our basic criterion for spontaneity (spontaneous change), but we would find it very difficult to apply. The difficulty is that to evaluate a total entropy change ($\Delta S_{\text{univ.}}$) we always have to evaluate ΔS for the surroundings. At best, this is a tedious process, and in many cases it is not even possible because we cannot figure out all the interactions between a system and its surroundings. Surely it would be preferable to have a criterion that we could apply *just to the system itself,* without having to worry about the surroundings. Then we would immediately be able to recognize the freezing of supercooled water at −10 °C, which we described above, as a spontaneous process.

To develop this new criterion, we will explore a hypothetical process conducted at constant temperature and pressure and with work limited to pressure–volume work. This process is accompanied by a heat effect, q_p, which, as we saw in Section 7-7, is equal to ΔH for the system ($\Delta H_{\text{syst.}}$). The heat effect experienced by the surroundings is the *negative* of that for the system. That is $q_{\text{surr.}} = -q_p = -\Delta H_{\text{syst.}}$. Furthermore, if we make the hypothetical surroundings large enough, the path by which heat enters or leaves the surroundings can be made *reversible*. That is, the quantity of heat would produce only an infinitesimal change in the temperature of the surroundings. In this case, according to equation (20.3), the entropy change in the *surroundings* will be $\Delta S_{\text{surr.}} = -\Delta H_{\text{syst.}}/T$.* Now we substitute this value of $\Delta S_{\text{surr.}}$ into equation (20.4).

$$\Delta S_{\text{univ.}} = \Delta S_{\text{syst.}} - \frac{\Delta H_{\text{syst.}}}{T}$$

Multiply by T to obtain

$$T\Delta S_{\text{univ.}} = T\Delta S_{\text{syst.}} - \Delta H_{\text{syst.}} = -(\Delta H_{\text{syst.}} - T\Delta S_{\text{syst.}})$$

and then multiply both sides by −1,

$$-T\Delta S_{\text{univ.}} = \Delta H_{\text{syst.}} - T\Delta S_{\text{syst.}} \tag{20.6}$$

*Are you wondering whether we can also substitute $\Delta H_{\text{syst.}}/T$ for $\Delta S_{\text{syst.}}$? The answer lies in equation (20.3). If the process under discussion is spontaneous, it is a natural process, and all natural processes are *irreversible*. We cannot substitute q_p for an irreversible process into equation (20.3).

This is the significance of equation (20.6). The equation has on its right side terms involving *only the system*. On the left side appears the term $\Delta S_{\text{univ.}}$, which embodies our criterion for a spontaneous process: $\Delta S_{\text{univ.}} > 0$.

Equation (20.6) is generally cast in a somewhat different form, which requires that we introduce a new thermodynamic function called the **Gibbs free energy, G.** The Gibbs free energy for a system is defined as

$$G = H - TS$$

and, for a change at constant T,

Gibbs free energy change related to enthalpy and entropy changes.

$$\Delta G = \Delta H - T\Delta S \qquad (20.7)$$

Note that in equation (20.7) *all* the terms apply to measurements *on the system*. All reference to the surroundings has been eliminated. Also, when we compare equations (20.6) and (20.7), we see that

$$\Delta G_{\text{syst.}} = -T\Delta S_{\text{univ.}}$$

Now, by noting that when $\Delta S_{\text{univ.}}$ is *positive* $\Delta G_{\text{syst.}}$ is *negative*, we have our final criterion for spontaneous change, based on properties just of the system itself.

ΔG as a criterion for spontaneous change.

For a process occurring at constant T and P, if

$\Delta G < 0$ (*negative*) the process is *spontaneous*.

$\Delta G > 0$ (*positive*) the process is *nonspontaneous*.   (20.8)

$\Delta G = 0$ (*zero*) the process is *at equilibrium*.

Sometimes the term "$T\Delta S$" is referred to as "organizational energy," since ΔS is related to the creation of disorder (or order) in a process.

We can see that free energy is indeed an energy term by evaluating the units in equation (20.7). ΔH has the unit joules (J), and the product $T\Delta S$ has the units $K \times J/K = J$. ΔG is the difference in two quantities having the units of energy.

Applying the Free Energy Criterion for Spontaneous Change. Later, we look at some quantitative applications of equation (20.7), but for now let us just use equa-

---

Are You Wondering:

Just what the term "free energy" signifies?

We have just established that a free energy change, ΔG, has the units of energy; it is an energy quantity. If we exclude pressure–volume work, any other work that can be performed as a result of a chemical reaction can be obtained in a useful form (for example, as electrical work). *ΔG is the maximum amount of useful work that a spontaneous reaction can perform at constant temperature and pressure:* $\Delta G = w_{max(useful)}$.

If ΔG is *positive*, work must be done on a system to make the reaction occur (as is done in electrolysis, for example). Thus, ΔG represents the *maximum* amount of work that can be obtained from a spontaneous reaction and the *minimum* amount of work required to produce a nonspontaneous reaction.

We learned in Chapter 7 that work is *not* a function of state; it depends on the path taken. To realize the maximum amount of work from a spontaneous reaction, the reaction must be carried out by a path called *reversible*. Such a path can only be hypothetical because a truly reversible process would take an infinite time to complete. But by conducting a spontaneous process by a nearly reversible path a quantity of work can be *freely* obtained that would otherwise have been lost as heat. We say more about the connection between free energy change and useful work in Section 20-8 and in Chapter 21, where we discuss electrochemistry.

J. Willard Gibbs (1839–1903)—the United States' great "unknown" scientist. Gibbs, a Yale University professor of mathematical physics, lived most of his career in obscurity, partly because of the abstractness of his work and partly because his important publications were in little-read journals. Yet, today, Gibbs's ideas serve as the basis of most of chemical thermodynamics. [Yale University Library]

tion (20.7), together with the criterion for spontaneous change (20.8), to make some *qualitative* predictions.

Can you see that if ΔH is *negative* and ΔS is *positive* the expression

$$\Delta G = \Delta H - T\Delta S$$

will be *negative* at all temperatures? A process that both gives off heat and becomes more disordered will be spontaneous at all temperatures. This corresponds to the situation we noted previously in which $\Delta S_{\text{syst.}}$ and $\Delta S_{\text{surr.}}$ are both positive, making $\Delta S_{\text{univ.}}$ also positive.

Unquestionably, if a process is accompanied by an *increase* in enthalpy (heat is absorbed) and a *decrease* in entropy (increased order), ΔG will be *positive* at all temperatures and the process will be *nonspontaneous*. This corresponds to a situation in which both $\Delta S_{\text{syst.}}$ and $\Delta S_{\text{surr.}}$ are negative, making $\Delta S_{\text{univ.}}$ also negative.

The questionable cases are those in which the entropy and enthalpy factors work in opposition, that is, with ΔH and ΔS both negative or *both positive*. Here, whether a reaction is spontaneous or nonspontaneous, that is, whether ΔG is positive or negative, will depend on temperature. In general, the ΔH factor dominates at *low* temperature and the $T\Delta S$ term at *high* temperature.

Altogether there are *four* possibilities for ΔG based on the signs of ΔH and ΔS. These possibilities are outlined in Table 20-1.

Example 20-3

Using enthalpy and entropy changes to predict the direction of spontaneous change. Under what conditions would you expect the following reactions to occur spontaneously?
(a) $2\,NH_4NO_3(s) \rightarrow 2\,N_2(g) + 4\,H_2O(g) + O_2(g)$ $\Delta H = -236$ kJ
(b) $I_2(g) \rightarrow 2\,I(g)$

Solution

(a) This is an exothermic reaction with $\Delta H = -236$ kJ. In Example 20-2 we concluded that this same reaction has $\Delta S > 0$, because of the large quantities of gases produced in the reaction. With $\Delta H < 0$ and $\Delta S > 0$ we conclude that this reaction should be spontaneous at all temperatures (Case 1 of Table 20-1).

(b) Here, we should again expect an increase in entropy, since two moles of gaseous product are produced from one mole of gaseous reactant. But what of the sign of ΔH? Think of this reaction as simply involving the breaking of covalent bonds in $I_2(g)$, with no new bonds formed in the products. Since energy must be absorbed to break chemical bonds, $\Delta H > 0$. With $\Delta H > 0$ and $\Delta S > 0$, this is an example of Case 3 in Table 20-1. At low temperatures, ΔH (positive) exceeds the $-T\Delta S$ term (negative) and the re-

TABLE 20-1
Criterion for Spontaneous Change: $\Delta G = \Delta H - T\Delta S$

Case	ΔH	ΔS	ΔG	Result[a]	Example
1	−	+	−	spontaneous at all temp.	$2\,N_2O(g) \longrightarrow 2\,N_2(g) + O_2(g)$
2	−	−	− +	spontaneous at low temp. nonspontaneous at high temp.	$H_2O(l) \longrightarrow H_2O(s)$
3	+	+	+ −	nonspontaneous at low temp. spontaneous at high temp.	$2\,NH_3(g) \longrightarrow N_2(g) + 3\,H_2(g)$
4	+	−	+	nonspontaneous at all temp.	$3\,O_2(g) \longrightarrow 2\,O_3(g)$

[a] If either ΔH or ΔS changes sign in the temperature range under consideration, then one of the other cases applies.

action is nonspontaneous. At a sufficiently high temperature the $T\Delta S$ term becomes larger than ΔH, ΔG becomes negative and the reaction becomes spontaneous.

SIMILAR EXAMPLES: Exercises 2, 27, 28, 29, 31.

_____

Example 20-3(b) helps us to understand this interesting fact: There is an upper limit for the stabilities of chemical compounds. No matter how positive the value of ΔH for dissociation of a compound into its atoms, the magnitude of the term $T\Delta S$ will eventually exceed that of ΔH as the temperature is increased. Known temperatures range from near absolute zero to the interior temperatures of the stars (about 3×10^7 K). Molecules exist only at limited temperatures (up to about 1×10^4 K or about 0.03% of this complete temperature range).

20-4   Evaluating Entropy and Entropy Changes

In Chapter 7 and elsewhere, we worked only with *changes* in the thermodynamic functions of internal energy and enthalpy. We had to, since we cannot establish their absolute values. Often we need only to deal with *changes* in entropy, too. An example of this might be in determining an entropy change during a phase transition.

Phase Transitions. Starting with equation (20.7) and applying the criterion for equilibrium ($\Delta G = 0$), we can state that

$$\left. \begin{array}{l} \Delta G = \Delta H - T\Delta S = 0 \\[2mm] \Delta S = \dfrac{\Delta H}{T} \end{array} \right\} \text{ at equilibrium}$$

For transitions between phases at constant temperature—melting, freezing, vaporization, condensation, and so on—we can also use descriptive subscripts, such as "tr." (transition), "fus." (fusion), "vap." (vaporization), ... If the transitions involve standard-state conditions (1 atm pressure), we also use the ° sign. Thus, for the normal melting point of ice we have

$$H_2O(s, 1 \text{ atm}) \rightleftharpoons H_2O(l, 1 \text{ atm}) \qquad \Delta H° = 6.02 \text{ kJ at } 273.15 \text{ K}$$

and

$$\Delta S°_{\text{fus.}} = \frac{\Delta H°_{\text{fus.}}}{T} = \frac{6.02 \text{ kJ/mol}}{273.15 \text{ K}} = 2.20 \times 10^{-2} \text{ kJ mol}^{-1} \text{ K}^{-1}$$
$$= 22.0 \text{ J mol}^{-1} \text{ K}^{-1}$$

Entropy changes are *extensive* properties; they depend on the quantities of substances involved. For this reason we wrote $\Delta S°_{\text{fus.}}$ on a "per mole" basis.

Example 20-4 _____

Determining the entropy change for a phase transition. What is the standard molar entropy of vaporization of water at 100.00 °C, given that the standard molar enthalpy of vaporization is 40.7 kJ/mol?

Solution. The primary challenge here is to translate the verbal description of the process into a chemical equation and then to base the calculation of $\Delta S°_{\text{vap.}}$ on this equation.

$$H_2O(l, 1 \text{ atm}) \rightleftharpoons H_2O(g, 1 \text{ atm}) \quad \Delta H°_{vap.} = 40.7 \text{ kJ/mol } H_2O \quad \Delta S°_{vap.} = ?$$

$$\Delta S°_{vap.} = \frac{\Delta H°_{vap.}}{T_{b.p.}} = \frac{40.7 \text{ kJ/mol}}{373.15 \text{ K}} = 0.109 \text{ kJ mol}^{-1} \text{ K}^{-1}$$

$$= 109 \text{ J mol}^{-1} \text{ K}^{-1}$$

SIMILAR EXAMPLES: Exercises 6, 25, 26.

A useful generalization, known as Trouton's rule, that works for many liquids at their normal boiling points is that the standard molar *entropy of vaporization* has a value of about 88 J mol^{-1} K^{-1}.

Trouton's rule for the standard molar entropy of vaporization.

$$\Delta S°_{vap.} = \frac{\Delta H°_{vap.}}{T_{b.p.}} \approx 88 \text{ J mol}^{-1} \text{ K}^{-1} \quad (20.9)$$

If the degree of disorder produced in transferring 1 mol of molecules from liquid to vapor at 1 atm pressure is roughly comparable for different liquids, then we should expect similar values of $\Delta S°_{vap.}$. Instances where Trouton's rule fails are equally understandable. In water and ethanol, for example, hydrogen bonding among molecules produces a greater degree of order than would otherwise be expected in the liquid state. The degree of disorder produced in the vaporization process is greater than normal, and $\Delta S°_{vap.} > 88$ J mol^{-1} K^{-1}.* There is no regularity in entropies of fusion, other than that they are smaller than entropies of vaporization and generally smaller for metals than for most other substances.

The Third Law of Thermodynamics. Unlike the cases of internal energy and enthalpy, it *is* possible to establish actual numerical values of entropy. The method requires us to start with a condition in which a substance has the greatest degree of order; this is taken as the zero of entropy. Then we can evaluate the entropy changes as the substance is brought to another condition of temperature and pressure—the one of interest to us. Add together these entropy changes and we have a numerical value of the entropy. The principle that enables us to do what we have just outlined is the **third law of thermodynamics,** which we state as follows.

The third law of thermodynamics.

The entropy of a pure perfect crystal at 0 K is zero. $\quad (20.10)$

There are small entropy changes at the temperatures where a solid substance changes from one polymorphic form to another. Another sharp increase in entropy comes at the melting point of the solid, and a still larger increase at the boiling point of the liquid. We can calculate entropies of transition with equation (20.8). Over temperature ranges where there are no phase transitions, entropy changes are based on specific heats. Figure 20-6 illustrates the variation of entropy with temperature.

Standard molar entropies of several substances at 25 °C are tabulated in Appendix D. To use these values to *calculate* the entropy change of a reaction we use an equation similar in form to the one developed for $\Delta H°$ (equation 7.23).

$$\Delta S° = \left[\sum \nu_p S°(\text{products})\right] - \left[\sum \nu_r S°(\text{reactants})\right] \quad (20.11)$$

The symbol Σ means "the sum of," and the terms that are added are the products of the standard molar entropies and their stoichiometric coefficients, ν. Example 20-5 illustrates the use of this equation.

*Describing vaporization in terms of disorder helps to explain Raoult's law. Suppose the molar entropy of vaporization of a solvent is the same, whether vaporization occurs from the pure solvent or from an ideal solution (say, one containing a nonvolatile solute). Since the solution state is more disordered than the pure solvent, the equilibrium vapor produced from the solution also has to be more disordered than that produced by the pure solvent. Therefore, the vapor molecules above a solution are more widely separated or at a lower pressure than those above the pure solvent: $P_A = \chi_A P°_A$.

FIGURE 20-6
Molar entropy as a function of temperature.

The standard molar entropy of a hypothetical substance is represented here. By the third law of thermodynamics, an entropy of zero is expected at 0 K. However, experimental measurements cannot be carried to this temperature—extrapolation techniques are required (broken-line portion of graph). A transition at T_{tr} is noted from the solid(I) to solid(II) form of the substance. Melting occurs at the normal melting point, $T_{m.p.}$, and vaporization at the normal boiling point, $T_{b.p.}$.

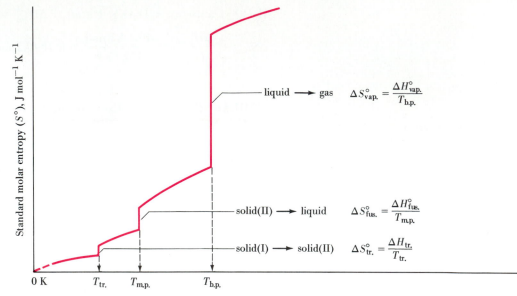

Temperature, K

Example 20-5

Calculating entropy changes from standard molar entropies. Use data from Appendix D to calculate the standard molar entropy for the conversion of nitrogen monoxide to nitrogen dioxide (an important reaction in the formation of photochemical smog).

$$2\ NO(g) + O_2(g) \longrightarrow 2\ NO_2(g) \qquad \Delta S^\circ_{298\ K} = ?$$

Are You Wondering:

What factors determine the magnitude of a molar entropy, for example, why is the molar entropy of $NO_2(g)$ greater than that of $NO(g)$?

In expression (20.2) we learned to expect gases generally to have higher entropies than liquids and liquids to have higher entropies than solids. Our reasoning there was based on the increased degree of molecular disorder in progressing from solid to liquid to gas. In comparing the molar entropies of $NO_2(g)$ [240.0 J K^{-1}] and $NO(g)$ [210.6 J K^{-1}], *both gases,* we have to look for other factors.

The molar entropy of a substance increases as a result of the substance absorbing heat (recall that $\Delta S = q_{rev.}/T$). Some of this heat goes simply into raising the average kinetic energies of molecules. But there are other ways for heat energy to be distributed among molecules. One possibility, pictured in Figure 20-7, is that the vibrational energies of molecules can be increased. In the *diatomic* molecule $NO(g)$ there is only one type of vibration possible, but in the *triatomic* molecule $NO_2(g)$ there are *three* types of vibration. Since there are more possible ways of distributing energy among NO_2 molecules than among NO molecules, $NO_2(g)$ has a higher molar entropy than does $NO(g)$ at the same temperature. We should add the following to the ideas stated in (20.2).

In general, the more complex their molecules (e.g., the greater the number of atoms present), the greater the molar entropies of substances.

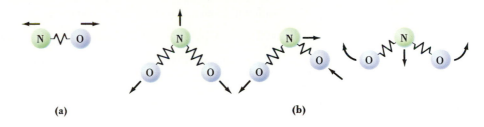

(a)   (b)

Solution. Equation (20.11) takes the form shown below.

$$\Delta S^\circ = 2S^\circ_{NO_2(g)} - 2S^\circ_{NO(g)} - S^\circ_{O_2(g)}$$

$$\Delta S^\circ = (2 \times 240.0) - (2 \times 210.6) - 205.1 = -146.3 \text{ J K}^{-1}$$

As a useful check on this calculation we can apply the *qualitative* reasoning based on expression (20.2) and used in Example 20-2. Since three moles of gaseous reactants produce only two moles of gaseous products, we should expect the entropy to *decrease*; ΔS° should be *negative*.

SIMILAR EXAMPLES: Exercises 10, 14a, 34a, 67.

20-5 Standard Free Energy Change, ΔG°

Because free energy is related to enthalpy ($G = H - TS$), we cannot establish absolute values of G, any more than we can for H. Also, we have a special need to work with free energy *changes* when the reactants and products are in their standard states—the standard free energy change, ΔG°. The standard state conventions that we will use are

Standard state conventions.

For a solid:	the pure substance at 1 atm external pressure
For a liquid:	the pure substance at 1 atm external pressure
For a gas:	an ideal gas at 1 atm partial pressure
For a solute:	an ideal solution at 1 M concentration

(20.12)

For a reaction in which a compound in its standard state is formed from its elements in their standard states, the free energy change is the **standard free energy of formation, ΔG°f.** And, as was the case with enthalpies of formation, we *arbitrarily* assign values of *zero* to the free energies of formation of the elements when in their most stable forms at 1 atm pressure. Standard free energies of formation are generally tabulated per mole of compound (see Appendix D).

Some additional relationships that we need to use when working with the free energy function are similar to those presented for the enthalpy function in Section 7-8.

Characteristics of ΔG.

1. ΔG is an extensive property
2. ΔG changes sign when a process is reversed
3. ΔG for a net or overall process can be obtained by summing the ΔG values for the individual steps
4. For a chemical reaction,

$$\Delta G^\circ = \left[\sum \nu_p \Delta G^\circ_f(\text{products})\right] - \left[\sum \nu_r \Delta G^\circ_f(\text{reactants})\right]$$

(20.13)

To illustrate the first three ideas in (20.13), consider the formation of copper(I) oxide from its elements. From a graph of ΔG° as a function of temperature, the following value was obtained at about 400 °C.

$$4 \text{ Cu(s)} + \text{O}_2(\text{g}) \longrightarrow 2 \text{ Cu}_2\text{O(s)} \qquad \Delta G° = -250 \text{ kJ}$$

To base the reaction on 1 mol $\text{Cu}_2\text{O(s)}$, we divide all stoichiometric coefficients *and* the free energy change by two.

$$2 \text{ Cu(s)} + \tfrac{1}{2} \text{O}_2(\text{g}) \longrightarrow \text{Cu}_2\text{O(s)} \qquad \Delta G° = -125 \text{ kJ}$$

Also, we can describe the *reverse* reaction, the decomposition of Cu_2O.

$$\text{Cu}_2\text{O(s)} \longrightarrow 2 \text{ Cu(s)} + \tfrac{1}{2} \text{O}_2(\text{g}) \qquad \Delta G° = +125 \text{ kJ}$$

Can we obtain Cu(s) from $\text{Cu}_2\text{O(s)}$ simply by heating it to 400 °C? From our criterion for spontaneous change, we see that $\text{Cu}_2\text{O(s)}$ will *not* decompose into Cu(s) and $\text{O}_2(\text{g})$ in their standard states [i.e., a partial pressure of $\text{O}_2(\text{g})$ of 1 atm] at 400 °C. Suppose that we combine this decomposition reaction with another that has a *negative* $\Delta G°$ at 400 °C: the partial oxidation of carbon to carbon monoxide.

$$
\begin{array}{ll}
\text{Cu}_2\text{O(s)} \longrightarrow 2 \text{ Cu(s)} + \tfrac{1}{2} \text{O}_2(\text{g}) & \Delta G° = +125 \text{ kJ} \\
\underline{\text{C(s)} + \tfrac{1}{2} \text{O}_2(\text{g}) \longrightarrow \text{CO(g)}} & \underline{\Delta G° = -175 \text{ kJ}} \\
\text{Cu}_2\text{O(s)} + \text{C(s)} \longrightarrow 2 \text{ Cu(s)} + \text{CO(g)} & \Delta G° = -50 \text{ kJ}
\end{array}
$$

Now we have a net reaction that produces the desired product Cu(s) whereas the single decomposition reaction would not. This combination of reactions, one with a *positive* $\Delta G°$ and one with a *negative* $\Delta G°$, yielding a spontaneous net reaction is referred to as a **coupled reaction.** Such reactions are important in producing metals from metal ores in metallurgy, especially for those using some form of carbon as a reducing agent (see Section 22-4). Coupled reactions also account for the production of complex molecules in living organisms. The formation of highly ordered structures (for which there is a large entropy decrease) is made possible through coupling with simpler reactions having negative free energy changes.

The fourth idea listed in expression (20.13) is one that we will use frequently. It is illustrated in Example 20-6.

Example 20-6

Calculating $\Delta G°$ for a reaction. Determine $\Delta G°$ for the following reaction discussed in the chapter introduction using **(a)** the basic definition of ΔG (equation 20.7) and **(b)** the expression listed in (20.13).

$$2 \text{ NO(g)} + \text{O}_2(\text{g}) \longrightarrow 2 \text{ NO}_2(\text{g}) \quad \text{(at 298.15 K)} \qquad \begin{aligned} \Delta H° &= -114.1 \text{ kJ} \\ \Delta S° &= -146.3 \text{ J K}^{-1} \end{aligned}$$

Solution

(a) If values of $\Delta H°$ and ΔS are given, as they are here, the most direct method of calculating $\Delta G°$ is to use the expression $\Delta G° = \Delta H° - T\Delta S°$, but when you do, be sure that you convert all the data to a common energy unit (e.g., kJ).

$$\Delta G° = -114.1 \text{ kJ} - (298.15 \text{ K} \times -0.1463 \text{ kJ K}^{-1})$$

$$= -114.1 \text{ kJ} + 43.62 \text{ kJ}$$

$$= -70.5 \text{ kJ}$$

(b) This method is most direct if $\Delta G_f°$ values are available. In the setup below, of course, the free energy of formation of the element $\text{O}_2(\text{g})$ is zero. The values for NO(g) and $\text{NO}_2(\text{g})$ are taken from Appendix D.

$$\Delta G° = 2\Delta G_f°[\text{NO}_2(\text{g})] - 2\Delta G_f°[\text{NO(g)}] - \Delta G_f°[\text{O}_2(\text{g})]$$

$$\Delta G° = 2(51.30 \text{ kJ}) - 2(86.57 \text{ kJ}) - 0$$

$$= -70.54 \text{ kJ}$$

The agreement between the two methods is excellent.

SIMILAR EXAMPLES: Exercises 13a, 14c, 15a, 33, 34c, 66.

Two Familiar Reactions Revisited. Written below are equations for the two reactions featured in the chapter introduction, this time including thermodynamic properties at 25 °C. [We calculated the properties for reaction (2) in Examples 20-5 and 20-6.]

(1) $N_2(g) + O_2(g) \longrightarrow 2 NO(g)$
$$\Delta H° = +180.5 \text{ kJ} \quad \Delta S° = +24.6 \text{ J K}^{-1} \quad \Delta G° = +173.2 \text{ kJ}$$

(2) $2 NO(g) + O_2(g) \longrightarrow 2 NO_2(g)$
$$\Delta H° = -114.1 \text{ kJ} \quad \Delta S° = -146.3 \text{ J K}^{-1} \quad \Delta G° = -70.5 \text{ kJ}$$

In terms of the criteria listed in Table 20-1, we see that reaction (1) corresponds to case 3. The $T\Delta S$ term dominates at *high* temperatures and the forward reaction is *entropy driven*. Reaction (2), on the other hand, corresponds to case 2. The ΔH term dominates at *low* temperatures and the forward reaction is *enthalpy driven*. Here, then, is a thermodynamic explanation of why reaction (1) is favored at high temperatures and reaction (2) at low temperatures.

20-6 Free Energy Change and Equilibrium

We have established that for a spontaneous process $\Delta G < 0$ and for a nonspontaneous process, $\Delta G > 0$. If $\Delta G = 0$, the reaction is neither spontaneous nor nonspontaneous; the forward and reverse reactions show an equal tendency to occur, and the reaction is at *equilibrium*. At this point, even an infinitesimal change in one of the system variables (e.g., temperature or pressure) will cause a net change to occur. But, as long as a system at equilibrium is left undisturbed, there is no net change with time.

An interesting comparison of spontaneous, nonspontaneous, and equilibrium processes is possible with Figure 20-8. The blue lines represent the temperature variation of ΔH and of the product $T\Delta S$. The ΔH line may have a positive slope in some cases and negative in others, but the slope is generally quite gradual. This means that ΔH is not especially temperature-dependent. The slope of the $T\Delta S$ line can also be positive or negative, but we expect it to be rather steep. This is because of the factor "T" in the $T\Delta S$ product. The vertical distance from the ΔH line to the $T\Delta S$ line (shaded in brown) represents

$$\Delta G = \Delta H - T\Delta S.$$

Now, let us consider the hypothetical process outlined in Figure 20-8. Starting at the left side of the figure, ΔG is positive and large; the process is *nonspontaneous*. The magnitude of ΔG decreases with increasing temperature. At the right side of the figure, $T\Delta S$ exceeds ΔH and ΔG is negative. At the temperature where the two lines intersect, $\Delta G = 0$, and the system is at equilibrium.

For the vaporization of water, *with both liquid and vapor in their standard states*, the intersection of the two lines in Figure 20-8 comes at $T = 373.15$ K.

$$H_2O(l, 1 \text{ atm}) \rightleftharpoons H_2O(g, 1 \text{ atm}) \qquad \Delta G° = 0 \text{ at } 373.15 \text{ K}$$

FIGURE 20-8
Free energy change as a function of temperature.

The value of ΔG is determined by the distance between the two lines. If the $T\Delta S$ value is greater than ΔH, the reaction is spontaneous. If ΔH is greater than $T\Delta S$, the reaction is nonspontaneous. The reaction is at equilibrium at the temperature at which the lines intersect.

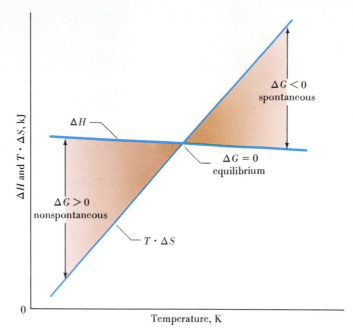

At 25 °C, the $\Delta H°$ line lies above the $T\Delta S°$ line in Figure 20-8, this means that $\Delta G° > 0$. In fact,

$$\text{H}_2\text{O}(1, 1 \text{ atm}) \longrightarrow \text{H}_2\text{O}(g, 1 \text{ atm}) \qquad \Delta G° = +8.6 \text{ kJ at } 298.15 \text{ K} \qquad (20.14)$$

What is the significance of the positive sign in $\Delta G° = +8.6$ kJ for the vaporization of water at 298.15 K (25 °C)? It cannot mean that the vaporization of water is nonspontaneous at 25 °C, for experience tells us that water will spontaneously evaporate at room temperature. What it does mean is that liquid water will not spontaneously produce $\text{H}_2\text{O}(g)$ at *1 atm pressure* at 25 °C. The equilibrium vapor pressure of water at 25 °C is 23.76 mmHg = 0.03126 atm, which we can represent as

$$\text{H}_2\text{O}(l, 0.03126 \text{ atm}) \rightleftharpoons \text{H}_2\text{O}(g, 0.03126 \text{ atm})$$

$$\Delta G = 0 \text{ at } 298.15 \text{ K} \qquad (20.15)$$

Figure 20-9 suggests the difference between the two situations described by equations (20.14) and (20.15).

Relationship of $\Delta G°$ to the Equilibrium Constant, K. If you think about the situation we just described for the vaporization of water, there does not seem to be much value in describing equilibrium in a process through its $\Delta G°$ value. There is only one temperature at which the reactants in their standard states are in equilibrium with products in their standard states, that is, only one temperature for which $\Delta G° = 0$. However, we want to be able to describe equilibrium for a variety of conditions, typically *nonstandard* conditions. For this we need to work with the quantity ΔG, not $\Delta G°$.

But here is an interesting prospect. We can relate ΔG for a reaction under *any* set of conditions to its value for standard conditions, that is, to $\Delta G°$. The key term in relating the two is the *reaction quotient, Q*, formulated for the actual, nonstandard conditions. Unfortunately, a derivation of this relationship is beyond the scope of this text, but we can make use of the result, which is

Relationship of ΔG, $\Delta G°$, and $\ln Q$.

$$\Delta G = \Delta G° + RT \ln Q \qquad (\text{or} \quad \Delta G = \Delta G° + 2.303RT \log Q) \qquad (20.16)$$

In Chapter 16 we learned that if a system is at equilibrium, $Q = K_c$ (or $Q = K_p$). And in this chapter we have seen that at equilibrium $\Delta G = 0$. These two facts allow us to write that, *at equilibrium*,

$$\Delta G = \Delta G° + RT \ln K = 0$$

FIGURE 20-9
Liquid–vapor equilibrium and the direction of spontaneous change.

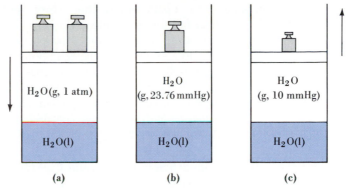

(a) At 25 °C and 1 atm, the direction of spontaneous change is the condensation of $H_2O(g)$. For the vaporization process $H_2O(l, 1 \text{ atm}) \rightarrow H_2O(g, 1 \text{ atm})$,
$\Delta G = \Delta G° = +8.6 \text{ kJ/mol}$.
(b) At 25 °C and 23.76 mmHg, the equilibrium $H_2O(l) \rightleftharpoons H_2O(g)$ is established;
$\Delta G = 0$.
(c) At 25 °C and 10 mmHg, the vaporization of water is spontaneous; for
$H_2O (1, 10 \text{ mmHg}) \rightarrow H_2O (g, 10 \text{ mmHg})$, $\Delta G < 0$.
In each case the arrow suggests the direction of spontaneous change.

which means that

Relationship of $\Delta G°$ to the equilibrium constant, K.

$$\Delta G° = -RT \ln K \qquad (\text{or} \quad \Delta G° = -2.303RT \log K) \qquad (20.17)$$

This is the significance of equation (20.17). We can derive values of $\Delta G°$ from tabulated thermodynamic data (Appendix D). If we have a value of $\Delta G°$ at a given temperature, we can *calculate* an equilibrium constant K. A tabulation of thermodynamic data, then, amounts to a useful tabulation of equilibrium constants.

Criterion for Spontaneous Change: Our Search Concluded. If you recall how we began this chapter, we were seeking a thermodynamic function that would reach

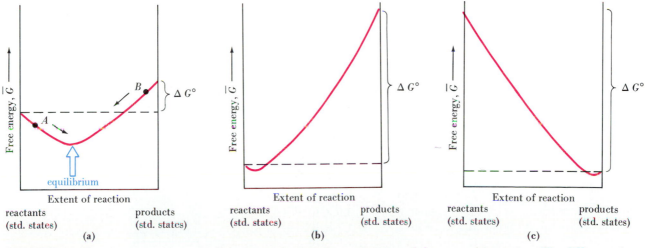

FIGURE 20-10
Free energy change, equilibrium, and the direction of spontaneous change.

(a) Free energy is plotted as a function of the extent of reaction. The difference between the standard molar free energies of reactants and products is the standard molar free energy change, $\Delta G°$. The equilibrium point lies somewhere between pure reactants and pure products. The free energy of the equilibrium mixture is lower than those of the pure reactants, of the pure products, and of any other mixture of the two. Mixtures A and B will each undergo spontaneous change in the direction of the equilibrium mixture.
(b) If $\Delta G°$ is large and positive, the minimum in free energy lies very close to the pure reactant side, and very little reaction occurs before equilibrium is reached.
(c) If $\Delta G°$ is large and negative, the minimum in free energy lies very close to the product side, and the reaction goes essentially to completion.

a minimum at a point in a reaction where equilibrium is reached. We plotted this hypothetical function in Figure 20-2. Now look at Figure 20-10, where we have plotted free energy as a function of the extent of a reaction. The free energy of the system falls to a minimum at the equilibrium point. The principal point made by Figure 20-10 is that whether a reaction goes to completion, occurs hardly at all, or reaches an equilibrium condition with significant amounts of both reactants and products present depends on the sign and magnitude of $\Delta G°$. Equation (20.17), of course, suggests the same thing, and these conclusions are presented somewhat differently in Table 20-2.

The Thermodynamic Equilibrium Constant. We want to do some calculations involving equations (20.16) and (20.17), but to do so we need to write equilibrium constants in a particular way. One reason for this is that when we take the logarithm of a reaction quotient (Q) or an equilibrium constant (K), as we must do in equations (20.16) and (20.17), it must be the logarithm of a *dimensionless* number (a number lacking units). We can achieve this result in the following way. For the general reaction

$$a \text{ A} + b \text{ B} + \cdots \rightleftharpoons g \text{ G} + h \text{ H} + \cdots \qquad (20.18)$$

the *thermodynamic equilibrium constant, K,* is

$$K = \frac{(a_G)^g (a_H)^h \cdots}{(a_A)^a (a_B)^b \cdots} \qquad (20.19)$$

Recall that we introduced activities in Section 13-9 as "effective concentrations" that give better agreement between the compositions and physical properties of solutions.

In equation (20.19) the *a* symbols represent the *activities* of the reactants and products. Each *a* symbol is actually a ratio of the equilibrium activity of a substance to its activity in its standard state. Standard states have been defined to have unit activity—the denominator in each of these ratios is 1. Thus, as long as the units used to express equilibrium activities are the same as those used to define standard states, activity units in the *a* ratios cancel. This leaves each *a* symbol *unitless* but numerically equal to the equilibrium activity. To establish activities, let us use the following conventions.

Conventions for assigning activities.

For pure solids and liquids: The activity is taken as unity, $a = 1$.
For gases: Assume ideal gas behavior. Then the activity of a gas is equal to its pressure in atm.
For components in solution: Assume ideal solution behavior (e.g., no interionic attractions). Then, if the standard state is taken to be 1 M, the activity is equal to the molarity concentration.

(20.20)

TABLE 20-2
Significance of the Magnitude of $\Delta G°$ (*at* 298.15 *K*)

Value of $\Delta G°$	Value of K		Significance
+200 kJ	9.1×10^{-36}		No reaction
+100	3.0×10^{-18}		
+50	1.7×10^{-9}	$K < 1$	
+10	1.8×10^{-2}		Equilibrium
+1.0	6.7×10^{-1}		
			calculation
0	1.0	$K = 1$	
			is
−1.0	1.5		
−10	5.6×10^{1}		necessary
−50	5.8×10^{8}	$K > 1$	
−100	3.3×10^{17}		Reaction goes
−200	1.1×10^{35}		to completion

In some instances thermodynamic equilibrium constants, K, will prove identical to the K_c and K_p values described in Chapter 16, in other instances not. This matter is illustrated through Example 20-7.

Example 20-7

Writing thermodynamic equilibrium constant expressions. For the following reversible reactions, write thermodynamic equilibrium constant expressions, make appropriate substitutions for activities using the conventions in (20.20), and equate K to K_c or K_p where this is appropriate.

(a) The water gas reaction.

$$C(s) + H_2O(g) \rightleftharpoons CO(g) + H_2(g)$$

(b) The formation of a saturated aqueous solution of lead iodide, a very slightly soluble solute.

$$PbI_2(s) \rightleftharpoons Pb^{2+}(aq) + 2\,I^-(aq)$$

(c) The oxidation of sulfide ion by oxygen gas (used in removing sulfides from waste water, as in pulp and paper mills).

$$O_2(g) + 2\,S^{2-}(aq) + 2\,H_2O(l) \rightleftharpoons 4\,OH^-(aq) + 2\,S(s)$$

Solution. In each case, once the appropriate substitutions have been made for activities, if all terms are molarity concentrations, the thermodynamic equilibrium constant is the same as K_c. If all terms are partial pressures, $K = K_p$. However, if both molarity concentrations *and* partial pressures appear in the expression, the equilibrium constant expression can only be designated as K.

(a) $K = \dfrac{(a_{CO(g)})(a_{H_2(g)})}{(a_{C(s)})(a_{H_2O(g)})} = \dfrac{(P_{CO})(P_{H_2})}{(1)(P_{H_2O})} = \dfrac{(P_{CO})(P_{H_2})}{(P_{H_2O})} = K_p$

(b) $K = \dfrac{(a_{Pb^{2+}})(a_{I^-})^2}{a_{PbI_2(s)}} = \dfrac{[Pb^{2+}][I^-]^2}{1} = [Pb^{2+}][I^-]^2 = K_c$

(c) $K = \dfrac{(a_{S(s)})^2(a_{OH^-})^4}{(a_{O_2(g)})(a_{S^{2-}})^2(a_{H_2O})^2} = \dfrac{(1)^2[OH^-]^4}{(P_{O_2})[S^{2-}]^2(1)^2} = \dfrac{[OH^-]^4}{(P_{O_2})[S^{2-}]^2}$

SIMILAR EXAMPLES: Exercises 8, 44.

Illustrative Examples. We have now acquired all the tools with which to perform some of the most practical calculations of chemical thermodynamics—*determining equilibrium constants and establishing the direction of net chemical change in reversible reactions.*

Example 20-8

Calculating the equilibrium constant of a reaction from the standard free energy change: Applying the equation $\Delta G° = -RT \ln K$. Nitrosyl chloride, NOCl, is one of the oxidizing agents present in *aqua regia* (a mixture of nitric and hydrochloric acids that can be used to dissolve gold). Another method of preparing nitrosyl chloride involves the reversible reaction of NO(g) and $Cl_2(g)$. What is the value of the equilibrium constant, K_p, for this reaction at 25.00 °C?

$$2\,NO(g) + Cl_2(g) \rightleftharpoons 2\,NOCl(g) \quad \text{at } 25.00\ °C, \qquad K_p = ?$$

Solution. The key to solving this problem is to find a value of $\Delta G°$, and then to use the expression, $\Delta G° = -RT \ln K$. To find a value of $\Delta G°$, we generally have two options: (1) Use the expression, $\Delta G° = \Delta H° - T\Delta S°$, or (2) obtain $\Delta G°$

from standard free energies of formation. Since ΔG_f° values for all three substances are listed in Appendix D, this second method is the most direct.

$$\Delta G^\circ = 2\Delta G_f^\circ[\text{NOCl(g)}] - 2\Delta G_f^\circ[\text{NO(g)}] - \Delta G_f^\circ[\text{Cl}_2\text{(g)}]$$

$$= 2(+66.07 \text{ kJ mol}^{-1}) - 2(+86.57 \text{ kJ mol}^{-1}) - 0$$

$$= -41.0 \text{ kJ mol}^{-1}$$

Now, solve for ln K and K.

$$\Delta G^\circ = -RT \ln K = -41.0 \text{ kJ mol}^{-1} = -41.0 \times 10^3 \text{ J mol}^{-1}$$

$$\ln K = \frac{-\Delta G^\circ}{RT} = \frac{-(-41.0 \times 10^3 \text{ J mol}^{-1})}{8.314 \text{ J mol}^{-1} \text{ K}^{-1} \times 298.15 \text{ K}} = 16.5$$

antiln $16.5 = e^{16.5} =$
 1.5×10^7

$K = $ antiln $16.5 = 1.5 \times 10^7$

The value of K obtained here is the *thermodynamic equilibrium constant*. But according to the conventions we have established, we can write

$$K = \frac{(a_{\text{NOCl(g)}})^2}{(a_{\text{NO(g)}})^2(a_{\text{Cl}_2\text{(g)}})} = \frac{(P_{\text{NOCl}})^2}{(P_{\text{NO}})^2(P_{\text{Cl}_2})} = K_p = 1.5 \times 10^7$$

SIMILAR EXAMPLES: Exercises 7, 9, 13, 47, 66.

Example 20-9

Determining the direction of spontaneous change: Applying the equation $\Delta G = \Delta G^\circ + RT \ln Q$. Concerning the reversible reaction involving NO, Cl₂,

---

Are You Wondering:

What is the purpose of the term "mol^{-1}" in the units of ΔG *and* ΔG° *in Examples 20-8 and 20-9? For example, why do we write* $-41.0 \text{ kJ mol}^{-1}$ *instead of just* -41.0 kJ?

Consider Example 20-8 where we used the equation $\Delta G^\circ = -RT \ln K$. K is a thermodynamic equilibrium constant and has no units, neither does ln K. The product "RT" on the right side of the equation has the units $(\text{J mol}^{-1} \text{ K}^{-1}) \times \text{K} = \text{J} \times \text{mol}^{-1}$. The term on the left side of the equation, ΔG°, must have the same units as the right side of the equation—J mol^{-1}. But what, you may wonder, does the "per mole" mean? It means "per mole of reaction." And one mole of reaction is simply the reaction based on the particular stoichiometric coefficients used in the balanced chemical equation. Thus, we can write

$$2 \text{ NO(g)} + \text{Cl}_2\text{(g)} \longrightarrow 2 \text{ NOCl(g)} \qquad \Delta G^\circ = -41.0 \text{ kJ mol}^{-1}$$

or we can write

$$\text{NO(g)} + \tfrac{1}{2} \text{Cl}_2\text{(g)} \longrightarrow \text{NOCl(g)} \qquad \Delta G^\circ = -20.5 \text{ kJ mol}^{-1}$$

In short, whenever we use the term "mol^{-1}" we need to specify the equation to which it applies. And, as long as we have the chemical equation, we should have no difficulty in determining any thermodynamic properties based on the equation (see Exercise 49).

    Quite often the "mol^{-1}" portion of a unit is simply dropped. This is what we have done in most cases up until now, and we will continue to drop it unless we need it for a proper cancellation of units.

and NOCl (see Example 20-8), in what direction does the value of ΔG predict that a net reaction will occur if these three substances are mixed together at 25.00 °C at the partial pressures shown below:

$$2 \text{ NO(g, } 1 \times 10^{-5} \text{ atm)} + \text{Cl}_2(\text{g, } 1 \times 10^{-2} \text{ atm)} \rightleftharpoons 2 \text{ NOCl(g, } 1 \times 10^{-2} \text{ atm)}$$

Solution. Since we determined $\Delta G°$ for this reaction in Example 20-8, we can use it, together with the partial pressure data, in the equation $\Delta G = \Delta G° + RT \ln Q$. We need to solve this equation for ΔG. If $\Delta G < 0$, the forward reaction will proceed spontaneously until equilibrium is reached. If $\Delta G > 0$, the forward reaction is nonspontaneous, and the reverse reaction will proceed spontaneously until equilibrium is reached.

$$\Delta G = -41.0 \times 10^3 \text{ J mol}^{-1} + RT \ln \frac{(1 \times 10^{-2})^2}{(1 \times 10^{-5})^2(1 \times 10^{-2})}$$

$$= -41.0 \times 10^3 \text{ J mol}^{-1} + 8.314 \text{ J mol}^{-1} \text{ K}^{-1} \times 298.15 \text{ K} \times \ln 1 \times 10^8$$

$$= -4.10 \times 10^4 \text{ J mol}^{-1} + 4.57 \times 10^4 \text{ J mol}^{-1}$$

$$= +0.47 \times 10^4 \text{ J mol}^{-1}$$

Since $\Delta G > 0$, a net change occurs in the *reverse* direction—from right to left.

An alternative, and perhaps simpler, method of answering this question would be just to compare the reaction quotient, $Q = 1 \times 10^8$, with the equilibrium constant calculated in Example 20-8, $K = 1.5 \times 10^7$. Since $Q > K$, a net reaction should occur in the *reverse* direction.

SIMILAR EXAMPLES: Exercises 12, 45, 46, 69, 70.

When you are confronted with a question that requires you to use thermodynamic properties, it is a good idea to think *qualitatively* about the problem before diving into calculations. Quite often you can come up with an acceptable solution with a minimum of calculations. In many ways, questions that can be answered qualitatively may test your understanding of fundamental principles more deeply than those that can be solved simply by "plugging" data into an equation. This point is illustrated in Example 20-10.

Example 20-10

Finding qualitative answers to thermodynamics questions. Does the decomposition of $CaCO_3(s)$ occur to any significant extent at 25 °C?

$$CaCO_3(s) \longrightarrow CaO(s) + CO_2(g)$$

Solution. The greatest difficulty in answering a question of this sort is usually in trying to establish the fundamental principles that apply. For a reaction to occur to a "significant extent" means for a spontaneous reaction to occur in the forward reaction until an equilibrium condition is reached. In the equilibrium condition there should be significant amounts of products present. The thermodynamic criterion that we have developed for the direction of spontaneous change is the free energy change, ΔG. Since no data are given in the question, all the data we have available are those listed in Appendix D, and the data there are for *standard states*. We should be able to establish $\Delta G°$ for this reaction rather easily.

$$\Delta G° = \Delta G_f°[\text{CaO(s)}] + \Delta G_f°[\text{CO}_2(\text{g})] - \Delta G_f°[\text{CaCO}_3(\text{s})]$$

$$= -604.0 - 394.4 - (-1128)$$

$$= +130. \text{ kJ}$$

From the *large, positive* value of $\Delta G°$ we can conclude directly that the decomposition does *not* occur to any significant extent; there is essentially no reaction (recall Table 20-2). Of course, if you are uncomfortable reaching this qualitative conclusion at this point, you can go on to calculate K through the expression $\Delta G° = -RT \ln K$. For this reaction $K = K_p = P_{CO_2}$, and this will prove to be an extremely low partial pressure.

Finally, you may recognize that limestone is a natural form of $CaCO_3$. You might then be inclined to say that the reaction cannot occur to any significant extent because you know that limestone does not spontaneously decompose at ordinary temperatures. You cannot use this fact by itself to answer the question, but having this knowledge can keep you from giving a *wrong* answer. (For example, if you mistakenly reversed the signs of the $\Delta G_f°$ values, you would obtain $\Delta G° = -130.$ kJ and falsely conclude that the reaction goes essentially to completion.) Never forget to use "common sense" as one of the tools for solving thermodynamics questions.

SIMILAR EXAMPLES: Exercises 11, 35, 51, 76a.

20-7 ΔG and K as Functions of Temperature

In Section 16-6 we learned how to make *qualitative* predictions of the effect of temperature on an equilibrium condition. We are now in a position to describe a *quantitative* relationship between the equilibrium constant and temperature. In the method illustrated in Example 20-11, we have to make certain assumptions about the temperature-dependence of values of thermodynamic properties. From Figure 20-8 we can see that these assumptions are reasonable: ΔH *changes only slightly with temperature*. Although $T\Delta S$ changes rapidly with temperature, this is because of the "T" term in the $T\Delta S$ product. ΔS *itself changes only slightly with temperature*. ΔG is the vertical distance between the ΔH and $T\Delta S$ lines in Figure 20-8, and we can see that ΔG *is strongly dependent on temperature*.

Example 20-11

Determining the relationship between equilibrium constant and temperature using equations for free energy change. In Example 20-8 we described the formation of NOCl from NO and Cl_2. Data for this reaction at 25 °C are

$2\ NO(g) + Cl_2(g) \longrightarrow 2\ NOCl(g)$
$\Delta G° = -41.0\ \text{kJ mol}^{-1} \quad \Delta H° = -77.1\ \text{kJ mol}^{-1} \quad \Delta S° = -121.0\ \text{J mol}^{-1}\ \text{K}^{-1}$

At what temperature will the value of the equilibrium constant for this reaction be $K = K_p = 1.0 \times 10^3$?

Solution. Since we have a *known* equilibrium constant and an *unknown* temperature, it should seem reasonable that we will need an equation in which both of these terms appear. The only equation that we have for this purpose is $\Delta G° = -RT \ln K$. If we knew the value of $\Delta G°$ at the unknown temperature, we could simply solve this equation for T. We know the value of $\Delta G°$ at 25 °C, but we *cannot* assume that this value will hold at other temperatures. What we can assume, though, is that the values of $\Delta H°$ and $\Delta S°$ will not change much with temperature; their values will be roughly the same as at 25 °C. Thus, we can obtain a value of $\Delta G°$ at the unknown T through the equation $\Delta G° = \Delta H° - T\Delta S°$. So, we have two equations in the unknown T that we can set equal to each other.

$$\Delta G^\circ = \Delta H^\circ - T\Delta S^\circ = -RT \ln K$$

$$T\Delta S^\circ - RT \ln K = \Delta H^\circ$$

$$T(\Delta S^\circ - R \ln K) = \Delta H^\circ$$

$$T = \frac{\Delta H^\circ}{\Delta S^\circ - R \ln K}$$

$$= \frac{-77.1 \times 10^3 \text{ J mol}^{-1}}{-121.0 \text{ J mol}^{-1} \text{ K}^{-1} - (8.314 \text{ J mol}^{-1} \text{ K}^{-1} \times \ln 1 \times 10^3)}$$

$$= \frac{-77.1 \times 10^3 \text{ J mol}^{-1}}{-121.0 \text{ J mol}^{-1} \text{ K}^{-1} - (8.314 \times 6.908) \text{ J mol}^{-1} \text{ K}^{-1}}$$

$$= \frac{-7.71 \times 10^4 \text{ J mol}^{-1}}{-178.4 \text{ J mol}^{-1} \text{ K}^{-1}} = 432 \text{ K} \approx 4.3 \times 10^2 \text{ K}$$

Note the following points about the result we just calculated.

- Although we carried a minimum of three significant figures throughout the calculation, we rounded the final result to just *two* significant figures. The assumption we made about the constancy of ΔH° and ΔS° is probably no more valid than this.
- In Example 20-8 we calculated that $K_p = 1.5 \times 10^7$ for this reaction at 25 °C. Since the reaction is *exothermic*, we should expect the value of the equilibrium constant to decrease with temperature. The temperature at which $K_p = 1.0 \times 10^3$ should be greater than 25 °C.

SIMILAR EXAMPLES: Exercises 15, 53, 54, 77.

An alternative to the method outlined in Example 20-11 is to relate the equilibrium constant and temperature directly, without specific reference to free energy change. Our starting point and basic assumptions are the same as in Example 20-11:

$$\Delta G^\circ \approx \Delta H^\circ - T \times \Delta S^\circ$$

Regardless of the temperature in question, assume that these quantities have the same values as at some other temperature at which they are known or can be determined.

In the derivation below, the substitution $\Delta G^\circ = -RT \ln K$ is made at the appropriate point.

$$\Delta G^\circ = \Delta H^\circ - T\Delta S^\circ$$

$$\frac{\Delta G^\circ}{T} = \frac{\Delta H^\circ}{T} - \Delta S^\circ$$

$$\frac{-RT \ln K}{T} = \frac{\Delta H^\circ}{T} - \Delta S^\circ$$

$$\ln K = \frac{-\Delta H^\circ}{RT} + \frac{\Delta S^\circ}{R}$$

Note that by assuming ΔH° and ΔS° are constant, the above equation is the equation of a straight line.

$$\ln K = \frac{-\Delta H^\circ}{R} \times \frac{1}{T} + \frac{\Delta S^\circ}{R} \qquad (20.21)$$

equation of straight line: $y \;=\; m \;\times\; x \;+\; b$

Table 20-3 lists equilibrium constants for the reaction of $SO_2(g)$ and $O_2(g)$ to form $SO_3(g)$ as a function of temperature. The data in the blue panels are plotted in Figure 20-11 and yield a straight line with a slope equal to $-\Delta H^\circ/R$.

TABLE 20-3

Equilibrium Constants, K_p, for the Reaction
$2\ SO_2(g) + O_2(g) \rightleftharpoons 2\ SO_3(g)$
at Several Temperatures

T, K	$1/T$, K^{-1}	K_p	$\ln K_p$
800	12.5×10^{-4}	9.1×10^2	6.81
850	11.8×10^{-4}	1.7×10^2	5.14
900	11.1×10^{-4}	4.2×10^1	3.74
950	10.5×10^{-4}	1.0×10^1	2.30
1000	10.0×10^{-4}	3.2×10^0	1.16
1050	9.52×10^{-4}	1.0×10^0	0.00
1100	9.09×10^{-4}	3.9×10^{-1}	−0.94
1170	8.55×10^{-4}	1.2×10^{-1}	−2.12

At this point we can follow a procedure we have used twice before (see, for example, page 550). We write equation (20.21) twice, for two different temperatures and with the corresponding equilibrium constants. Then, we subtract one equation from the other and obtain the result shown below.

The van't Hoff equation relating equilibrium constant and temperature.

$$\ln \frac{K_2}{K_1} = \frac{\Delta H°}{R}\left(\frac{1}{T_1} - \frac{1}{T_2}\right) \quad \text{or} \quad \log \frac{K_2}{K_1} = \frac{\Delta H°}{2.303R}\left(\frac{1}{T_1} - \frac{1}{T_2}\right) \quad (20.22)$$

In equation (20.22), T_2 and T_1 are two Kelvin temperatures; K_2 and K_1 are the equilibrium constants at these temperatures; $\Delta H°$ is the enthalpy of reaction, expressed in J mol^{-1} K^{-1}; R is the gas constant expressed as 8.314 J mol^{-1} K^{-1}. Jacobus van't Hoff (1852–1911) derived equation (20.22) and this equation is often referred to as the **van't Hoff equation.**

Example 20-12

Relating equilibrium constants and temperature through the van't Hoff equation. Use data from Table 20-3 and Figure 20-11 to estimate the temperature at which $K_p = 1.0 \times 10^6$ for the reaction

$2\ SO_2(g) + O_2(g) \rightleftharpoons 2\ SO_3(g)$

FIGURE 20-11

Temperature dependence of the equilibrium constant K_p for the reaction

$2\ SO_2(g) + O_2(g) \rightleftharpoons 2\ SO_3(g)$

This graph can be used to establish the heat of reaction, $\Delta H°$ (see equation 20.21).

$$\text{slope} = \frac{-\Delta H°}{R} = 2.2 \times 10^4\ \text{K}$$

$$\Delta H° = -8.314\ \text{J mol}^{-1}\ \text{K}^{-1}$$
$$\times\ 2.2 \times 10^4\ \text{K}$$

$$= -1.8 \times 10^5\ \text{J/mol}$$

$$= -1.8 \times 10^2\ \text{kJ/mol}$$

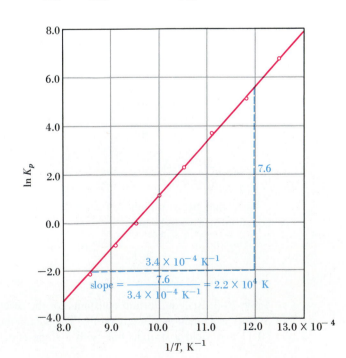

Solution. Select one known temperature and equilibrium constant from Table 20-3 and the enthalpy change of the reaction, $\Delta H°$, from Figure 20-11. To ensure that you substitute data correctly into equation (20.22), take the time to organize the necessary data.

$$T_2 = 800 \text{ K} \qquad K_2 = 9.1 \times 10^2$$

$$T_1 = ? \qquad K_1 = 1.0 \times 10^6 \qquad \Delta H° = -1.8 \times 10^5 \text{ J mol}^{-1}$$

$$\ln \frac{K_2}{K_1} = \frac{\Delta H°}{R} \left(\frac{1}{T_1} - \frac{1}{T_2} \right)$$

$$\ln \frac{9.1 \times 10^2}{1.0 \times 10^6} = \ln 9.1 \times 10^{-4} = -7.00$$

For simplicity we have dropped the units in this setup, but you should be able to show that units cancel properly.

$$-7.00 = \frac{-1.8 \times 10^5}{8.314} \left(\frac{1}{T_1} - \frac{1}{800} \right)$$

$$\frac{-7.00 \times 8.314}{-1.8 \times 10^5} + \frac{1}{800} = \frac{1}{T_1}$$

$$\frac{1}{T_1} = (3.2 \times 10^{-4}) + (1.25 \times 10^{-3}) = 1.57 \times 10^{-3}$$

$$T_1 = 1/(1.57 \times 10^{-3}) = 6.4 \times 10^2 \text{ K}$$

SIMILAR EXAMPLES: Exercises 16, 57, 58, 59.

20-8 The Scope of Thermodynamics

Sadi Carnot, a French military engineer, deduced the underlying principle of the second law of thermodynamics in a study of the efficiencies of heat engines (1824). William Thomson (Lord Kelvin) recognized the significance of Carnot's work and saw in it the basis of the second law of thermodynamics and an absolute temperature scale. Ever since that time scientists and nonscientists alike have speculated on ways in which the second law may apply to broader issues.

One of the larger questions has been whether the existence of life itself is consistent with the second law. Consider our own existence. Our growth and development as human beings require the assembly of highly organized structures from much simpler substances (e.g., O_2, N_2, CO_2, and H_2O). If we think of a living organism as a system, growth and development are accompanied by large *decreases* in entropy within the system. We have noted (page 728) that thermodynamics does explain how nonspontaneous reactions ($\Delta G > 0$) can occur if they are *coupled* with other reactions that are spontaneous ($\Delta G < 0$). Furthermore, we know that to stay alive we need to ingest foods for the processes of metabolism. In these processes complex structures (fats, carbohydrates, and proteins) are broken down into simpler end products (e.g., CO_2 and H_2O). These processes produce an *increase* in entropy. Of course, the original production of foods from simple molecules is a process in which entropy *decreases*. Then again, all life on earth is dependent on radiation from the sun, so a total assessment of the entropy changes associated with life on earth also must take into account processes for the production and radiation of energy from the sun to the earth. Can you see the difficulties in applying the second law of thermodynamics to large systems having complex interactions with very extensive surroundings?

The laws of thermodynamics are generalizations based on *experimental* evidence; they are not absolute truths. They are really applicable only to situations in which

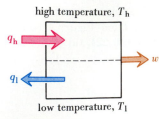

high temperature, T_h

low temperature, T_l

FIGURE 20-12
Schematic representation of a heat engine.

The efficiency of the engine is determined by the quantity of heat q_h that is converted to work w. The smaller the quantity of heat released to the surroundings at the lower temperature q_1, the more efficient the engine. The second law of thermodynamics places a theoretical maximum efficiency on every heat engine; it is always less than 100% (see equation 20.23).

experiments can be performed to test them. Although it is interesting to speculate on the application of thermodynamics to such matters as how an economy works or other societal activities, we cannot come to definite conclusions because we cannot perform controlled experiments. One area of practical interest where we do have solid experimental verification deals with the operation of heat engines—the stimulation for the development of thermodynamics in the first place.

Heat Engines. The basic principle of a heat engine is that heat (q_h) is absorbed by the working substance of the engine (for example, water) at a high temperature (T_h). This heat is partly converted to work (w), and the remainder (q_l) is released to the surroundings at a lower temperature (T_l). The process is pictured in Figure 20-12. An alternative statement of the *second law of thermodynamics* based on the operation of heat engines is that

No heat engine can operate without rejecting waste heat to lower temperature surroundings. That is, a heat engine cannot operate at a constant temperature.

The efficiency of a heat engine is governed by the ratio w/q_h. If all the heat absorbed could be converted to work, the engine would be 100% efficient. The second law of thermodynamics places an absolute limit on the efficiency of a heat engine, *and it is never 100%*. The expression obtained is

$$\text{efficiency} = \frac{w}{q_h} = \frac{T_h - T_l}{T_h} \tag{20.23}$$

where temperature is in kelvins.

According to equation (20.23), if a steam engine operates between 100 °C (the boiler temperature) and 25 °C (the condenser temperature), it can have an efficiency of only 0.20, that is, 20%; 80% of the heat supplied to the boiler is given off to the surroundings.

$$\text{efficiency} = \frac{373 - 298}{373} = \frac{75}{373} = 0.20$$

Conventional electric power plants are based on the burning of a fossil fuel, which provides the heat energy to convert water to steam. The steam powers a turbine that drives an electric generator. The process is not highly efficient. The efficiency can be improved, however, by using a higher working temperature for the turbine (T_h). This can be accomplished by operating the system under high pressure, at temperatures well in excess of 100 °C, using superheated steam. Still, though, most electric power plants operate at thermal efficiencies of less than 40%. Thus, more than half the heat required to produce electric power is waste heat. The implications of this fact are twofold: wasted heat means wasted fuel, and as the waste heat enters the environment it leads to thermal pollution.

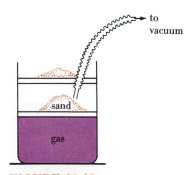

to vacuum

sand

gas

FIGURE 20-13
Pressure–volume work conducted reversibly.

A reversible process produces the maximum amount of work, but takes an infinitely long time to conduct.

Other Limitations to Energy Conversions. As described on page 722, the *maximum useful work* obtainable from a spontaneous reaction is equal to ΔG for the reaction. But there is a limitation here as well: To achieve the maximum useful work, the process must be carried out *reversibly*. A reversible process is one that can be made to reverse its direction through just an *infinitesimal* change in a system variable. In Figure 20-13 the process being carried out is the expansion of a gas. The external pressure on the gas is determined by the quantity of sand on the piston. With the aid of the vacuum connection, this sand can be removed one grain at a time. However, if at any instant we return two grains of sand to the piston instead of removing one, the piston will reverse direction—the gas will be compressed. In a *reversible* process a system is always within a minute or infinitesimal step of equilibrium.

Even though the maximum useful work obtainable is for a reversible process, such a process would take an infinitely long time to complete. For this reason, real processes must be conducted *irreversibly,* and some of the available (free) energy is dissipated as heat instead of appearing as work. The typical automobile battery delivers about 72% of the maximum possible useful work as electricity (see also Section 21-3).

FOCUS ON
Thermal Pollution

Electric power plants require large volumes of cooling water and are often located near large bodies of water, as is this nuclear power plant on the California coast. [George D. Lepp]

The origin of thermal pollution in a power plant is suggested by Figure 20-14. The best known effects of thermal pollution are those produced on fish and other aquatic life. The differential growth rate of algae with changing temperatures can cause one type of algae, which is ideal food for fish, to be displaced by another, which is a poor food or even toxic to fish. Also, because their metabolic rate goes up with temperature, fish need more food and oxygen as the temperature of their water rises, but the solubility of air in water *decreases* with increasing temperature. Temperature changes affect other physiological processes too. Some fish are killed by the thermal shock of even relatively small temperature changes.

Even if all other pollution sources could be controlled, release of heat energy to the surroundings by heat engines would still be unavoidable. Means must be found to minimize the amount of waste heat or to put the waste heat to useful purposes. The first objective can be met through more efficient heat engines and through the direct conversion of other energy forms to electricity (such as with fuel cells and solar energy devices). The second objective can be met by using waste heat from power plants to supply hot water for industrial, commercial, or residential purposes, or using this water for agricultural purposes. None of these methods to deal with thermal pollution is well

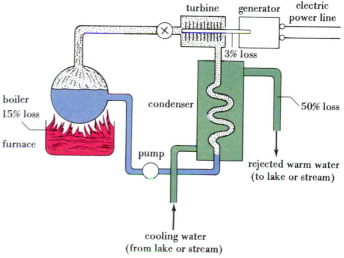

FIGURE 20-14
Efficiency of a fossil-fuel electric power plant and the origin of thermal pollution.

Water is heated in a boiler and converted to superheated steam. The steam is allowed to expand in the turbine where it does work by turning the rotor blades. The turbine shaft drives the electric generator, which produces electric power. Conversion of mechanical to electrical energy in the turbine-generator combination is nearly 100% efficient.

If the steam were to be rejected to the surroundings, all the makeup water in the boiler would have to be cold water. This would require more fuel for heating than if the steam is returned to the boiler. However, because of its greatly expanded volume after performing work in the turbine, the steam must be condensed back to liquid before being returned to the boiler. Condensation of the steam occurs in the condenser. The water that is used for cooling carries away up to 50% of the heat energy that was released in burning the fuel. It is this waste heat that produces thermal pollution.

established currently, and the problem is likely to continue to grow before solutions are found.

Summary

Although most exothermic reactions are spontaneous, many *endothermic* reactions are spontaneous too. A common feature shared by systems in which spontaneous endothermic processes occur is that they undergo an increase in disorder or randomness. The thermodynamic function related to the degree of disorder or randomness in a system is *entropy, S.* Entropy change is defined as a quantity of heat exchanged *reversibly* with the surroundings, $q_{rev.}$, divided by the Kelvin temperature, T. That is, $\Delta S = q_{rev.}/T$. For a phase transition at equilibrium, $q_{rev.}$ can be replaced by the enthalpy change for the transition, $\Delta H_{tr.}$, and $\Delta S_{tr.} = \Delta H_{tr.}/T_{tr.}$. A generalization known as Trouton's rule is applicable to the vaporization of many liquids: $\Delta S_{vap.}^\circ = \Delta H_{vap.}^\circ/T_{bp.} \approx 88$ J mol^{-1} K^{-1}. Not only can entropy changes be measured, but numerical values can be assigned to entropies of individual substances. This is made possible by the *third law of thermodynamics,* which takes as the zero of entropy that of a pure perfect crystal at 0 K.

In a spontaneous change, the *total* entropy change of the system *and* its surroundings must increase. This is one statement of the *second law of thermodynamics.* If entropy change alone is used as a criterion for spontaneous change, it is necessary to assess the entropy change both of the system *and* the surroundings. Such an assessment is inconvenient and often difficult to make.

A thermodynamic property that leads to a criterion for spontaneous change based *just on the system itself* is the *Gibbs free energy, G.* A change in free energy at constant temperature is related to enthalpy and entropy changes through the expression: $\Delta G = \Delta H - T\Delta S$. The criterion for spontaneous change at constant temperature and pressure is that there be a *decrease in free energy,* that is, $\Delta G < 0$. The free energy criterion for equilibrium is that $\Delta G = 0$.

The standard free energy change, ΔG°, is based on the conversion of reactants in their standard states to products in their standard states. Tabulated free energy data are usually standard molar free energies of formation, ΔG_f°. Two useful relationships involving free energy changes are

$$\Delta G = \Delta G^\circ + RT \ln Q \qquad \text{and} \qquad \Delta G^\circ = -RT \ln K$$

For example, if a value of ΔG° can be obtained from tabulated data, the corresponding equilibrium constant, K, can be calculated. The constant, K, is called a *thermodynamic equilibrium constant.* It is based on the *activities* of reactants and products, but these activities can be related to molarity concentrations and gas partial pressures through a few simple conventions.

Another derivation that starts with the relationship between standard free energy change and the equilibrium constant is of an equation (called the van't Hoff equation) relating the equilibrium constant and temperature (equation 20.22). With this equation it is possible to use tabulated data at 25 °C to determine equilibrium constants not just at 25 °C but at other temperatures as well.

Summarizing Example

In Chapter 16 we used as our first example of an equilibrium reaction the synthesis of methanol from CO and H$_2$ at 500. K (equation 16.3). This synthesis is of great importance because methanol can be used directly as a motor fuel, converted to gasoline for fuel use, or converted to other organic compounds. The synthesis reaction is

$$CO(g) + 2 H_2(g) \longrightarrow CH_3OH(g)$$

1. Would you expect this reaction to take place to any significant extent at 25 °C? Explain.

Solution. As we saw in Example 20-10, our first thought should be whether any significant amount of CH$_3$OH(g) will be present at equilibrium. For this, we need a value of the equilibrium constant K, or at least, of the standard free energy change, ΔG°.

$$\Delta G^\circ = \Delta G_f^\circ[CH_3OH(g)] - \Delta G_f^\circ[CO(g)]$$

$$= -162.0 \text{ kJ mol}^{-1} - (-137.2 \text{ kJ mol}^{-1}) = -24.8 \text{ kJ mol}^{-1}$$

The *negative* value of ΔG° signifies that the conversion of CO(g) and H$_2$(g) in their standard states to CH$_3$OH(g) in its standard state is indeed a spontaneous reaction. We should expect the forward reaction to proceed to a considerable

A large modern plant for the manufacture of methanol. CO and H$_2$ required in the synthesis are derived from natural gas (CH$_4$). [Courtesy Alberta Gas Chemicals, Ltd.]

extent before equilibrium is reached. Even though the production of CH_3OH is *thermodynamically* favorable at 25 °C, the reaction occurs very slowly. The reaction is *kinetically* controlled, and this is why the reaction is carried out at higher temperatures (e.g., 500 K).

(This example is similar to Examples 20-6 and 20-10.)

2. What is the value of K for the methanol synthesis reaction at 25.00 °C?

Solution. Here we simply need to use equation (20.17).

$$\Delta G° = -RT \ln K \quad \text{and} \quad \ln K = \frac{-\Delta G°}{RT}$$

$$\ln K = \frac{24.8 \times 10^3 \text{ J mol}^{-1}}{8.314 \text{ J mol}^{-1} \text{ K}^{-1} \times 298.15 \text{ K}} = 10.0$$

$K = $ antiln $10.0 = 2.2 \times 10^4$

(This example is similar to Example 20-8.)

3. What is the value of the thermodynamic equilibrium constant, K, at 500. K, given that $K_c = 14.5$?

Solution. We need to recognize that the thermodynamic equilibrium constant for this reaction is the same as K_p, that is,

$$K = \frac{a_{CH_3OH(g)}}{(a_{CO(g)})(a_{H_2(g)})^2} = \frac{P_{CH_3OH(g)}}{(P_{CO(g)})(P_{H_2(g)})^2} = K_p$$

The question then is one of finding K_p at 500. K, and for this we need equation (16.18), with the change in number of moles of gas, $\Delta n_{gas} = -2$.

$$K = K_p = K_c(RT)^{\Delta n_{gas}} = 14.5 \times (0.08206 \times 500.)^{-2} = 8.61 \times 10^{-3}$$

(This example is similar to Examples 20-7 and 16-3.)

4. Use the values of K at 298.15 K and 500. K to calculate the standard enthalpy change in this reaction, $\Delta H°$.

Solution. The quantities in question here are related through the van't Hoff equation (20.22). The unknown is $\Delta H°$.

$$K_2 = 8.61 \times 10^{-3} \quad T_2 = 500. \text{ K}$$
$$\Delta H° = ?$$
$$K_1 = 2.2 \times 10^4 \quad T_1 = 298.15 \text{ K}$$

$$\ln \frac{8.61 \times 10^{-3}}{2.2 \times 10^4} = \frac{\Delta H°}{8.314 \text{ J mol}^{-1} \text{ K}^{-1}} \left(\frac{1}{298.15 \text{ K}} - \frac{1}{500. \text{ K}} \right) = -14.8$$

$$-14.8 = \frac{\Delta H°}{8.314 \text{ J mol}^{-1} \text{ K}^{-1}} (3.35 \times 10^{-3} - 2.00 \times 10^{-3}) \text{ K}^{-1}$$

$$\Delta H° = \frac{-14.8 \times 8.314}{1.35 \times 10^{-3}} \text{ J mol}^{-1} = -9.11 \times 10^4 \text{ J mol}^{-1} = -91.1 \text{ kJ mol}^{-1}$$

(This example is similar to Example 20-12.)

Key Terms

coupled reactions (20-5)
entropy S (20-2)
entropy change, ΔS (20-3)
entropy change of the universe (20-3)
free energy, G (20-3)
free energy change, ΔG (20-3)

heat engine (20-8)
reversible process (20-8)
second law of thermodynamics (20-3)
spontaneous process (20-1)
standard free energy change, $\Delta G°$ (20-5)

standard free energy of formation, $\Delta G_f°$ (20-5)
thermodynamic equilibrium constant, K (20-6)
third law of thermodynamics (20-4)
Trouton's rule (20-4)

Highlighted Expressions

Relationship between entropy and disorder (20.1)
Conditions leading to entropy increases (20.2)
Defining equation for entropy change (20.3)
The second law of thermodynamics (20.5)
Gibbs free energy change related to enthalpy and entropy changes (20.7)
ΔG as a criterion for spontaneous change (20.8)
Trouton's rule for the standard molar entropy of vaporization (20-9)

The third law of thermodynamics (20.10)
Standard state conventions (20.12)
Characteristics of ΔG (20.13)
Relationship of ΔG, $\Delta G°$, and ln Q (20.16)
Relationship of $\Delta G°$ to the equilibrium constant, K (20.17)
Conventions for assigning activities (20.20)
The van't Hoff equation relating equilibrium constant and temperature (20.22)

Review Problems

1. Indicate whether you would expect the entropy of the system to increase or decrease in each of the following reactions. If you cannot be certain simply by inspecting the equation, explain why.

  (a) $CCl_4(l) \rightarrow CCl_4(g)$
  (b) $CuSO_4 \cdot 3H_2O(s) + 2\ H_2O(g) \rightarrow CuSO_4 \cdot 5H_2O(s)$
  (c) $SO_3(g) + H_2(g) \rightarrow SO_2(g) + H_2O(g)$
  (d) $H_2S(g) + O_2(g) \rightarrow H_2O(g) + SO_2(g)$ (not balanced)

2. From the data given, indicate which of the four cases in Table 20-1 applies for each of the following reactions

  (a) $H_2(g) \rightarrow 2\ H(g)$ $\Delta H° = +435.9$ kJ
  (b) $2\ SO_2(g) + O_2(g) \rightarrow 2\ SO_3(g)$ $\Delta H° = -197.8$ kJ
  (c) $N_2H_4(g) \rightarrow N_2(g) + 2\ H_2(g)$ $\Delta H° = -95.4$ kJ
  (d) $N_2(g) + 3\ Cl_2(g) \rightarrow 2\ NCl_3(g)$ $\Delta H° = +230$ kJ

3. A particular handbook lists values of $\Delta G_f°$ and $\Delta H_f°$ but not molar entropies. From the data given below, determine $\Delta S°$ for the reaction $NH_3(g) + HCl(g) \rightarrow NH_4Cl(s)$. All data are at 298.15 K.

	$\Delta H_f°$, kJ/mol	$\Delta G_f°$, kJ/mol
$NH_3(g)$	−46.1	−16.5
$HCl(g)$	−92.3	−95.3
$NH_4Cl(s)$	−314.4	−203.0

4. Use data from Appendix D to determine values of $\Delta G°$ for the following reactions at 25 °C.

  (a) $C_2H_2(g) + 2\ H_2(g) \rightarrow C_2H_6(g)$
  (b) $2\ SO_3(g) \rightarrow 2\ SO_2(g) + O_2(g)$
  (c) $Fe_3O_4(s) + 4\ H_2(g) \rightarrow 3\ Fe(s) + 4\ H_2O(g)$
  (d) $MgO(s) + 2\ HCl(g) \rightarrow MgCl_2(s) + H_2O(l)$

5. If a graph similar to Figure 20-8 were drawn for the process $I_2(s, 1\ atm) \rightarrow I_2(l, 1\ atm)$

  (a) At what temperature would the two lines intersect? [Hint: Refer also to Figure 12-19.]
  (b) What is the value of ΔG at this temperature? Explain.

6. From the following data determine $\Delta S_{tr}°$ (in J mol^{-1} K^{-1}) for each transition.

  (a) The boiling of HCl(l) at −85.05 °C; $\Delta H_{vap.}° = 3.86$ kcal/mol.
  (b) The melting of Na(s) at 97.82 °C; $\Delta H_{fus.}° = 27.05$ cal/g.
  (c) The transition of rhombic to monoclinic sulfur at 95.5 °C; $\Delta H_{tr.}° = 96$ cal/mol.

7. For the reaction $Cl_2(g) \rightleftharpoons 2\ Cl(g)$, $K_p = 2.45 \times 10^{-7}$ at 1000. K. What is $\Delta G°$ for this reaction at 1000. K?

8. Write thermodynamic equilibrium constant expressions for the following reactions. Do any of these expressions correspond to K_c or K_p?

  (a) $2\ NO(g) + O_2(g) \rightleftharpoons 2\ NO_2(g)$
  (b) $MgSO_3(s) \rightleftharpoons MgO(s) + SO_2(g)$
  (c) $HC_2H_3O_2(aq) \rightleftharpoons H^+(aq) + C_2H_3O_2^-(aq)$
  (d) $2\ NaHCO_3(s) \rightleftharpoons Na_2CO_3(s) + H_2O(g) + CO_2(g)$
  (e) $MnO_2(s) + 4\ H^+(aq) + 2\ Cl^-(aq) \rightleftharpoons$
               $Mn^{2+}(aq) + 2\ H_2O(l) + Cl_2(g)$

9. Use data from Appendix D to determine K_p at 298 K for the reaction

$$N_2O(g) + \tfrac{1}{2}\ O_2(g) \rightleftharpoons 2\ NO(g)$$

10. In Example 20-2 we were unable to conclude, by inspection, whether $\Delta S°$ for the reaction

$CO(g) + H_2O(g) \rightarrow CO_2(g) + H_2(g)$

should be positive or negative. Use data from Appendix D to obtain $\Delta S°$ at 298 K.

11. Using data from Appendix D but without performing detailed calculations, indicate whether any of the following reactions is expected to occur to some extent at 298.15 K.
  (a) Conversion of ordinary oxygen to ozone: $3 O_2(g) \rightarrow 2 O_3(g)$
  (b) Dissociation of N_2O_4 to NO_2: $N_2O_4(g) \rightarrow 2 NO_2(g)$
  (c) Formation of BrCl: $Br_2(l) + Cl_2(g) \rightarrow 2 BrCl(g)$

12. For the reaction $2 SO_2(g) + O_2(g) \rightleftharpoons 2 SO_3(g)$, $K_c = 2.8 \times 10^2$ at 1000. K.
  (a) What is the value of K_p for this reaction at 1000. K?
  (b) What is the value of $\Delta G°$ at 1000. K?
  (c) In what direction will a net reaction occur if 0.40 mol SO_2, 0.18 mol O_2, and 0.72 mol SO_3 are mixed in a 2.50-L vessel at 1000. K?

13. Use data from Appendix D to establish for the reaction

$2 NO(g) + O_2(g) \rightarrow 2 NO_2(g)$

  (a) $\Delta G°$ at 298 K for the reaction as written.
  (b) The value of K_p at 298 K.

14. The following are thermodynamic data at 298 K.

	$\Delta H_f°$, kJ/mol	$S°$, J mol^{-1} K^{-1}
$NaHCO_3(s)$	−947.7	102
$Na_2CO_3(s)$	−1131	136
$H_2O(l)$	−285.8	69.9
$CO_2(g)$	−393.5	213.6

Use these data to establish at 298 K for the reaction

$2 NaHCO_3(s) \rightarrow Na_2CO_3(s) + H_2O(l) + CO_2(g)$

  (a) $\Delta S°$; (b) $\Delta H°$; (c) $\Delta G°$; (d) K.

15. A possible reaction for converting methanol to ethanol is

$CO(g) + 2 H_2(g) + CH_3OH(g) \rightarrow CH_3CH_2OH(g) + H_2O(g)$

  (a) Use data from Appendix D to calculate $\Delta H°$, $\Delta S°$, and $\Delta G°$ for this reaction at 25 °C.
  (b) Is this reaction thermodynamically favored at high or low temperatures? High or low pressures?
  (c) Estimate a value of K_p for the reaction at 750. K.

16. Estimate the value of K_p at 100. °C for the reaction $2 SO_2(g) + O_2(g) \rightleftharpoons 2 SO_3(g)$. Use data from Table 20-3 and Figure 20-11.

Exercises

Spontaneous change, entropy, and disorder

17. Using common knowledge alone, which of the following is *not* a spontaneous change? Explain. (a) The formation of sour cream from fresh cream; (b) obtaining gold nuggets by "panning"; (c) forming a green patina (surface coating) on an outdoor bronze statue.

18. Based on the relationship of entropy to the degree of order in a system, indicate whether each of the following changes represents an increase or decrease in entropy in a system: (a) the freezing of ethanol; (b) the sublimation of dry ice; (c) the burning of a rocket fuel.

19. Arrange the entropy changes of the following processes in the expected order of increasing ΔS, and explain your reasoning.
  (a) $H_2O(l, 1 \text{ atm}) \rightarrow H_2O(g, 1 \text{ atm})$
  (b) S_8 (rhombic, 1 atm) $\rightarrow S_8$ (monoclinic, 1 atm)
  (c) $CO_2(s, 1 \text{ atm}) \rightarrow CO_2(g, 0.5 \text{ atm})$
  (d) $H_2O(l, 1 \text{ atm}) \rightarrow H_2O(g, 0.5 \text{ atm})$

20. For each of the following reactions, indicate whether ΔS for the reaction is positive or negative. If it is not possible to determine the sign of ΔS from the information given, indicate why.
  (a) $Na_2SO_4(s) + 4 C(s) \rightarrow Na_2S(s) + 4 CO(g)$
  (b) $2 Hg(l) + O_2(g) \rightarrow 2 HgO(s)$
  (c) $2 H_2O(g) \rightarrow 2 H_2(g) + O_2(g)$
  (d) $(NH_4)_2Cr_2O_7(s) \rightarrow Cr_2O_3(s) + N_2(g) + 4 H_2O(g)$
  (e) $Fe_2O_3(s) + 3 H_2(g) \rightarrow 2 Fe(s) + 3 H_2O(g)$
  (f) $Ni(s) + 4 CO(g) \rightarrow Ni(CO)_4(l)$

21. Use ideas from this chapter to explain the meaning of a famous remark attributed to Rudolf Clausius (1865). "Die Energie der Welt ist konstant; die Entropie der Welt strebt einem Maximum zu. [The energy of the world is constant; the entropy of the world increases toward a maximum.]"

22. Comment on the difficulties in combating environmental pollution, based on the entropy changes associated with the formation of pollutants and with their removal from the environment.

Phase transitions

23. Refer to Example 20-4. In this example we dealt with $\Delta H°_{vap.}$ and $\Delta S°_{vap.}$ for water at 100.00 °C.
  (a) Determine corresponding values at 25.00 °C. [*Hint:* Use data from Appendix D.]
 ★(b) Explain the differences in values of $\Delta H°_{vap.}$ and $\Delta S°_{vap.}$ between these two temperatures.

24. Which of the following substances would you expect to obey Trouton's rule most closely? (a) HF; (b) $C_6H_5CH_3$ (toluene); (c) CH_3OH (methanol). Explain your reasoning.

25. Estimate the normal boiling point of bromine, Br_2, in the following two-step procedure, and compare your result with the measured value of 58.8 °C.
  (a) Determine $\Delta H°_{vap.}$ for Br_2 with data from Appendix D.
  (b) Estimate the normal boiling point, assuming that $\Delta H°_{vap.}$ remains constant and that Trouton's rule is obeyed.

★**26.** The boiling point of $SO_2(l)$ is −10.01 °C. Use this fact, together with data from Appendix D and this chapter, to estimate a value of $\Delta H_f°[SO_2(l)]$ at 298 K.

Free energy and spontaneous change

27. For the following reactions, indicate whether the forward reaction tends to be spontaneous at low temperatures, high temperatures, all temperatures, or whether it tends to be nonspontaneous at all temperatures. If the information given is not sufficient to allow a prediction, state why this is so. [*Hint:* What do you predict for the sign of the entropy change in each reaction?]

(a) $PCl_3(g) + Cl_2(g) \rightleftharpoons PCl_5(g)$ $\Delta H° = -87.9$ kJ

(b) $CO_2(g) + H_2(g) \rightleftharpoons CO(g) + H_2O(g)$ $\Delta H° = +41.2$ kJ

(c) $NH_4CO_2NH_2(s) \rightleftharpoons 2\ NH_3(g) + CO_2(g)$

$$\Delta H° = +159.2 \text{ kJ}$$

(d) $H_2O(g) + \frac{1}{2} O_2(g) \rightleftharpoons H_2O_2(g)$ $\Delta H° = +105.5$ kJ

(e) $C_6H_6(l) + \frac{15}{2} O_2(g) \rightleftharpoons 6\ CO_2(g) + 3\ H_2O(g)$

$$\Delta H° = -3135 \text{ kJ}$$

28. For the process pictured in Figure 20-4, what are the values (positive, negative, or zero) for ΔH, ΔS, ΔG? Explain your reasoning.

29. What values of ΔH, ΔS, and ΔG would you expect for the formation of an ideal solution of liquid components (that is, are the values positive, negative, or zero)?

30. Explain why

(a) Some exothermic reactions do not occur spontaneously.

(b) Some reactions in which the entropy of the system increases also do not occur spontaneously.

31. Explain why you would expect a reaction of the type $AB(g) \rightarrow A(g) + B(g)$, always to be more spontaneous at *high* rather than low temperatures.

32. At what temperature will the following reaction have $\Delta G = -777.8$ kJ, if $\Delta H = -843.7$ kJ and $\Delta S = -165$ J/K?

$$2\ Pb(s) + 3\ O_2(g) \rightarrow 2\ PbO(s) + 2\ SO_2(g)$$

Standard free energy change

33. For the reaction $2\ PCl_3(g) + O_2(g) \rightarrow 2\ POCl_3(l)$, $\Delta H° = -555$ kJ at 298 K. The molar entropies at 298 K are $PCl_3(g)$, 312 J/K; $O_2(g)$, 205 J/K; and $POCl_3(l)$, 222 J/K. What is the value of $\Delta G°$ for this reaction at 298 K?

34. Hydrazine, N_2H_4, as we have noted previously, is used in great quantities as a rocket fuel, in the manufacture of pesticides, and in the manufacture of foam plastics. Since hydrazine is closely related to ammonia, perhaps it can be made by the following simple reaction.

$$2\ NH_3(g) \rightarrow N_2H_4(g) + H_2(g)$$

Assess the feasibility of this reaction by determining

(a) $\Delta S°$ for the reaction at 25 °C. (The molar entropy of $N_2H_4(g)$ is 238.4 J/K.)

(b) $\Delta H°$ for the reaction at 25 °C. [*Hint:* Use bond energies from Table 10-2.]

(c) $\Delta G°$

(d) Whether the reaction is favored at high or low temperatures.

35. The following standard free energy changes are given for 25 °C.

$N_2(g) + 3\ H_2(g) \rightarrow 2\ NH_3(g)$ $\Delta G° = -33.0$ kJ

$4\ NH_3(g) + 5\ O_2(g) \rightarrow 4\ NO(g) + 6\ H_2O(l)$ $\Delta G° = -1011.$ kJ

$N_2(g) + O_2(g) \rightarrow 2\ NO(g)$ $\Delta G° = +173.1$ kJ

$N_2(g) + 2\ O_2(g) \rightarrow 2\ NO_2(g)$ $\Delta G° = +102.6$ kJ

$2\ N_2(g) + O_2(g) \rightarrow 2\ N_2O(g)$ $\Delta G° = +208.4$ kJ

Combine the above equations, as necessary, to obtain $\Delta G°$ values for the following reactions.

(a) $N_2O(g) + \frac{3}{2} O_2(g) \rightarrow 2\ NO_2(g)$ $\Delta G° = ?$

(b) $2\ H_2(g) + O_2(g) \rightarrow 2\ H_2O(l)$ $\Delta G° = ?$

(c) $2\ NH_3(g) + 2\ O_2(g) \rightarrow N_2O(g) + 3\ H_2O(l)$ $\Delta G° = ?$

Which of the reactions (a), (b), and (c) would tend to go to completion at 25 °C, and which would reach an equilibrium condition with significant amounts of all reactants and products present?

36. The molar entropies of $F_2(g)$ and $F(g)$ at 298 K are 203.3 and 158.7 J K^{-1}, respectively. Use these data, together with that given below, to estimate the bond energy of the F_2 molecule. Compare your result with that listed in Table 10-2.

$$F_2(g) \rightarrow 2\ F(g) \qquad \Delta G° = +123.8 \text{ kJ}$$

37. $\Delta G°$ for the combustion of *n*-octane (a component of gasoline) is given below. What would be the standard free energy change if the water were produced as a gas at 1 atm pressure instead of as a liquid? [*Hint:* Use data from Appendix D.]

$$C_8H_{18}(l) + \tfrac{25}{2} O_2(g) \rightarrow 8\ CO_2(g) + 9\ H_2O(l)$$
$$\Delta G° = -5.30 \times 10^3 \text{ kJ at 298 K}$$

38. Following are some standard free energies of formation, $\Delta G_f°$, per mole of metal oxides at 1000. K; NiO, -115 kJ; MnO, -280 kJ; TiO_2, -630 kJ. The standard free energy of formation of CO at 1000 K is -250 kJ per mol CO. Use the method of coupled reactions outlined on page 728 to determine which of these metal oxides can be reduced to the metal by a spontaneous reaction at 1000. K.

39. Titanium metal is used extensively in the aerospace industry because the metal imparts strength to structures but does not unduly add to their masses. The metal is produced by the reduction of $TiCl_4(l)$, which in turn is produced from the mineral rutile (TiO_2).

(a) Can the following reaction for the production of $TiCl_4(l)$ be carried out at 25 °C?

$$TiO_2(s) + 2\ Cl_2(g) \rightarrow TiCl_4(l) + O_2(g)$$

	$\Delta H_f°$, kJ/mol	$S°$, J mol^{-1} K^{-1}
$TiO_2(s)$	-944.7	50.3
$TiCl_4(l)$	-804.2	252.3
$Cl_2(g)$	0.0	223.0
$O_2(g)$	0.0	205.1

(b) Show that the conversion of $TiO_2(s)$ to $TiCl_4(l)$ is spontaneous at 25 °C if the reaction in (a) is coupled with the reaction

$$2\ CO(g) + O_2(g) \rightarrow 2\ CO_2(g)$$

[*Hint:* Use data from Appendix D.]

Free energy change and equilibrium

40. At how many different temperatures and in what temperature range can this equilibrium be established? $H_2O(l, 0.50$ atm$) \rightleftharpoons H_2O(g, 0.50$ atm$)$

41. Refer to Figures 12-19 and 20-8. Does solid or liquid iodine have the lower free energy at 110 °C and 1 atm pressure?

42. Refer to Figures 12-20 and 20-8. Which has the lowest free energy at 1 atm and -60 °C, solid, liquid, or gaseous carbon dioxide? Explain.

The thermodynamic equilibrium constant

43. Why must the thermodynamic equilibrium constant K, be used in the expression $\Delta G° = -RT \ln K$, rather than simply K_c?

Under what circumstances can K_c and/or K_p be used in place of K?

44. A method of preparing $H_2(g)$ involves passing steam over hot iron.

$$3 \, Fe(s) + 4 \, H_2O(g) \rightleftharpoons Fe_3O_4(s) + 4 \, H_2(g)$$

(a) Write an expression for the thermodynamic equilibrium constant for this reaction.

(b) Explain why the partial pressure of $H_2(g)$ is independent of the amounts of $Fe(s)$ and $Fe_3O_4(s)$ present.

(c) Can we conclude that the production of $H_2(g)$ from $H_2O(g)$ could be accomplished regardless of what proportions of $Fe(s)$ and $Fe_3O_4(s)$ are used? Explain.

Relationships involving ΔG, $\Delta G°$, Q, and K

45. Refer to the situations in Figure 20-9 and calculate ΔG at 298 K for the process

$$H_2O(l, 10 \, mmHg) \rightarrow H_2O(g, 10 \, mmHg)$$

46. The following data are given for the vaporization of toluene at 298 K.

$$C_7H_8(l) \rightarrow C_7H_8(g) \qquad \Delta S° = 99.7 \, J \, mol^{-1} \, K^{-1}$$
$$\Delta H_f°[C_7H_8(l)] = 12.0 \, kJ \, mol^{-1}$$
$$\Delta H_f°[C_7H_8(g)] = 50.0 \, kJ \, mol^{-1}$$

(a) Show that $C_7H_8(l)$ does *not* spontaneously vaporize to $C_7H_8(g)$ at 0.50 atm pressure.

(b) Calculate the equilibrium vapor pressure of C_7H_8 at 298 K.

47. For the following equilibrium reactions discussed in Chapter 16, calculate the value of $\Delta G°$ at the indicated temperature. [*Hint:* How is each equilibrium constant related to a *thermodynamic* equilibrium constant?]

(a) $H_2(g) + I_2(g) \rightleftharpoons 2 \, HI(g) \qquad K_c = 50.2$ at 445 °C

(b) $N_2O(g) + \frac{1}{2} O_2(g) \rightleftharpoons 2 \, NO(g)$
$$K_c = 1.7 \times 10^{-13} \text{ at } 25 °C$$

(c) $N_2O_4(g) \rightleftharpoons 2 \, NO_2(g) \qquad K_c = 4.61 \times 10^{-3}$ at 25 °C

(d) $2 \, SO_2(g) + O_2(g) \rightleftharpoons 2 \, SO_3(g)$
$$K_p = 9.1 \times 10^2 \text{ at } 800. °C$$

(e) $2 \, Fe^{3+}(aq) + Hg_2^{2+}(aq) \rightleftharpoons 2 \, Fe^{2+}(aq) + 2 \, Hg^{2+}(aq)$
$$K_c = 9.14 \times 10^{-6} \text{ at } 25 °C$$

48. The equilibrium constant at 298 K for the reaction $CO(g) + Cl_2(g) \rightleftharpoons COCl_2(g)$ is $K_p = 1.6 \times 10^{12}$. What is the standard free energy of formation of $COCl_2(g)$ at 298 K, that is, $\Delta G_f°[COCl_2(g)]$? [*Hint:* You will need to use some data from Appendix D.]

49. On page 734 two different equations were written for the formation of NOCl from NO and Cl_2.

$$2 \, NO(g) + Cl_2(g) \rightarrow 2 \, NOCl(g) \qquad \Delta G° = -41.0 \, kJ/mol$$

$$NO(g) + \frac{1}{2} Cl_2(g) \rightarrow NOCl(g) \qquad \Delta G° = -20.5 \, kJ/mol$$

(a) Explain why these two equations have different values for $\Delta G°$.

(b) Will the values of K determined for these two equations be the same or different? Explain.

(c) Will the equilibrium partial pressures calculated for this reaction depend on which equation is used? Explain.

★**50.** Dinitrogen pentoxide, N_2O_5, is a solid with a high vapor pressure. Its vapor pressure at 7.5 °C is 100 mmHg, and the solid sublimes at a pressure of 1.00 atm at 32.4 °C. What is the

standard free energy change for the process $N_2O_5(s) \rightarrow N_2O_5(g)$ at 25.0 °C?

51. To establish the law of conservation of mass, Lavoisier carefully studied the decomposition of mercury(II) oxide. At 25 °C,

$$HgO(s) \rightarrow Hg(l) + \frac{1}{2} O_2(g) \qquad \begin{aligned} \Delta H° &= +90.83 \, kJ \\ \Delta G° &= +58.56 \, kJ \end{aligned}$$

(a) Show that the partial pressure of $O_2(g)$ in equilibrium with $HgO(s)$ and $Hg(l)$ at 25 °C is extremely low.

(b) What conditions do you suppose Lavoisier used to obtain significant quantities of oxygen?

★**52.** 1.00 mol $BrCl(g)$ is introduced into a 10.0-L vessel at 298 K and the following equilibrium is established:

$$BrCl(g) \rightleftharpoons \frac{1}{2} Br_2(g) + \frac{1}{2} Cl_2(g)$$

Calculate the amounts of each of the three gases present when equilibrium is established. [*Hint:* You must establish your own value of K_p or K_c using data from Appendix D.]

$\Delta G°$ as a function of temperature

53. Use data from Appendix D to determine (a) $\Delta H°$, $\Delta S°$, and $\Delta G°$ at 298 K and (b) K_p at 1025 K for the water gas shift reaction, used commercially to produce $H_2(g)$.

$$CO(g) + H_2O(g) \rightleftharpoons CO_2(g) + H_2(g)$$

[*Hint:* Assume that $\Delta H°$ and $\Delta S°$ are essentially unchanged over this temperature interval.]

54. Refer to Example 20-12. There we determined the approximate temperature at which $K_p = 1.0 \times 10^6$ for the reaction $2 \, SO_2(g) + O_2(g) \rightleftharpoons 2 \, SO_3(g)$ by using the van't Hoff equation. Obtain another estimate of this temperature with data from Appendix D and equation (20.7). Compare your result with that obtained in Example 20-12.

★**55.** Oxides of nitrogen are produced in high-temperature combustion processes. The essential reaction is

$$N_2(g) + O_2(g) \rightarrow 2 \, NO(g)$$

At what approximate temperature will an *equimolar* mixture of $N_2(g)$ and $O_2(g)$ be 1.0% converted to $NO(g)$? [*Hint:* Use data from Appendix D.]

★**56.** Refer to Example 20-10. For the dissociation of $CaCO_3(s)$ at 25 °C,

$$CaCO_3(s) \rightarrow CaO(s) + CO_2(g) \qquad \Delta G° = +129.6 \, kJ/mol$$

A sample of pure $CaCO_3(s)$ is placed in a flask and connected to an ultrahigh vacuum system capable of reducing the pressure to 10^{-9} mmHg (see illustration).

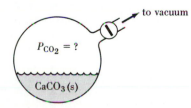

Exercise 56

(a) Is $CO_2(g)$ detectable in the vacuum system at 25 °C?

(b) What additional information do you need in order to determine P_{CO_2} as a function of temperature?

(c) With necessary data from Appendix D, determine the minimum temperature to which the $CaCO_3(s)$ would have to be heated for $CO_2(g)$ to become detectable in the vacuum system.

(d) At what temperature is $P_{CO_2} = 1$ atm?

$\Delta G°$ as a function of temperature: the van't Hoff equation

57. For the reaction $N_2O_4(g) \rightleftharpoons 2\ NO_2(g)$, $\Delta H° = +57.2$ kJ/mol and $K_p = 0.113$ at 298 K.

(a) What is the value of K_p at 273 K?

(b) At what temperature will $K_p = 1.00$?

58. Use data from Table 16-2 and Appendix D to estimate a value of K_p at 100. °C for the reaction

$$2\ NO(g) + O_2(g) \rightleftharpoons 2\ NO_2(g).$$

59. Sodium carbonate is an important chemical used in the production of glass. It can be made by the Solvay process (see Section 22-1). The data listed below are given for the temperature variation of K_p (partial pressures in atm) for this reaction, which is used in the process.

$$2\ NaHCO_3(s) \rightleftharpoons Na_2CO_3(s) + CO_2(g) + H_2O(g)$$

(a) Plot a graph similar to Figure 20-11 and determine $\Delta H°$ for the reaction.

(b) Calculate the temperature at which the total gas pressure above a mixture of $NaHCO_3(s)$ and $Na_2CO_3(s)$ is 2.00 atm.

t, °C	K_p
30.	1.66×10^{-5}
50.	3.90×10^{-4}
70.	6.27×10^{-3}
100.	2.31×10^{-1}

60. Refer to the Summarizing Example. In part 4 of the example we determined $\Delta H°$ for the reaction

$$CO(g) + 2\ H_2(g) \rightarrow CH_3OH(g)$$

by using the van't Hoff equation (20.22). Determine $\Delta H°$ for this reaction with data from Appendix D and compare the two re-

sults. How good is the assumption that $\Delta H°$ is essentially independent of temperature in this case?

Heat engines

61. Why is it not possible to develop a heat engine that is 100% efficient in converting heat to work?

62. What is the maximum efficiency of a heat engine that absorbs heat at 305 °C and discharges waste heat at 21 °C?

63. If a steam electric power plant discharges condensate at 41 °C and is found to operate at 36% efficiency,

(a) What is the minimum temperature of the steam used in the plant?

(b) Why is the actual steam temperature probably higher than that calculated in part (a)?

64. Since the beginning of the industrial age, people have been seeking ways of delivering useful energy at ever lower costs. The scheme pictured below is proposed for propelling a ship on the ocean. Water is drawn from one side of the ship into a heat engine where some heat is extracted from the water and converted to work to propel the ship. The slightly cooled water is then discharged over the other side of the ship. (a) Does the process proposed here violate the first law of thermodynamics? Explain. (b) Do you think this scheme will work? Explain your reasoning.

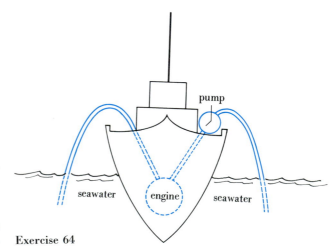

Exercise 64

Additional Exercises

65. Which of the following pairs of substances would you expect to have the greater entropy? Explain your reasoning.

(a) at 50. °C: 1 mol $H_2O(l, 1$ atm$)$ *or* 1 mol $H_2O(g, 1$ atm$)$

(b) at 5 °C: 50.0 g Fe(s, 1 atm) *or* 0.80 mol Fe(s, 1 atm)

(c) 1 mol $Br_2(l, 1$ atm, 8 °C$)$ *or* 1 mol $Br_2(s, 1$ atm, 2 °C$)$

(d) 0.225 mol $SO_2(g, 0.107$ atm, 325 °C$)$ *or* 0.100 mol $O_2(g, 10.00$ atm, 25 °C$)$

66. Use data from Appendix D to determine values at 298 K of $\Delta G°$ and K for the following reactions. (*Note:* The equations are not balanced.)

(a) $NO(g) + O_2(g) \rightleftharpoons NO_2(g)$

(b) $HCl(g) + O_2(g) \rightleftharpoons H_2O(g) + Cl_2(g)$

(c) $Fe_2O_3(s) + H_2(g) \rightleftharpoons Fe_3O_4(s) + H_2O(g)$

67. Unlike the situation with enthalpy and free energy, with

entropy we deal with an absolute molar entropy. We could, however, refer to an entropy of formation of a compound. By analogy to $\Delta H_f°$ and $\Delta G_f°$, how would you define an entropy of formation? Which do you think has the largest entropy of formation (a) $CH_4(g)$; (b) $C_2H_5OH(l)$; (c) $CS_2(l)$? First make a qualitative prediction and then test your prediction with data from Appendix D. [*Hint:* What are the reactions by which these compounds are formed from their elements?]

68. Use data from Appendix D to estimate (a) the normal boiling point of mercury and (b) the vapor pressure of mercury at 25 °C.

69. An *equilibrium* mixture at 1000. K contains 0.276 mol H_2, 0.276 mol CO_2, 0.224 mol CO, and 0.224 mol H_2O.

$$CO_2(g) + H_2(g) \rightleftharpoons CO(g) + H_2O(g)$$

(a) What is the value of K_p at 1000. K?

(b) Calculate $\Delta G°$ at 1000. K.

(c) In what direction will a spontaneous net reaction occur if one brings together at 1000. K: 0.0500 mol CO_2, 0.070 mol H_2, 0.0400 mol CO, and 0.0850 mol H_2O?

70. Indicate for each of the following whether a net reaction occurs to the left or to the right at 298 K.

(a) $SO_2(g, 0.010 \text{ atm}) + \frac{1}{2} O_2(g, 0.0010 \text{ atm}) \rightleftharpoons$
$$SO_3(g, 2.0 \text{ atm})$$

(b) $C_2H_2(g, 1.2 \text{ atm}) + 2 H_2(g, 0.020 \text{ atm}) \rightleftharpoons$
$$C_2H_6(g, 3.0 \text{ atm})$$

(c) $Br_2(l, 1 \text{ atm}) \rightleftharpoons Br_2(g, 0.10 \text{ atm})$

71. Use data from Appendix D and other information from this chapter to estimate the temperature at which the dissociation of $I_2(g)$ becomes appreciable [e.g., with the $I_2(g)$ being 50% dissociated into I(g) at 1 atm total pressure].

72. At 298 K the following standard enthalpies of formation are given for *cyclopentane*, $\Delta H_f°[C_5H_{10}(l)] = -105.9$ kJ; $\Delta H_f°[C_5H_{10}(g)] = -77.2$ kJ.

(a) Estimate the normal boiling point of cyclopentane.

(b) Estimate $\Delta G°$ for the vaporization of cyclopentane at 298 K.

(c) Comment on the significance of the sign of $\Delta G°$ at 298 K.

73. In Figure 12-22 we pictured the transition from red mercury(II) iodide (the stable form at room temperature) to yellow mercury(II) iodide at 127 °C: $HgI_2(\text{red}) \rightarrow HgI_2(\text{yellow})$. A handbook lists the following data for the two solid forms of HgI_2 at 298 K.

	$\Delta H_f°$, kJ/mol	$\Delta G_f°$, kJ/mol	$S°$, J mol^{-1} K^{-1}
HgI_2 (red)	-105.4	-101.7	180.
HgI_2 (yellow)	-102.9	(?)	(?)

Estimate values for the two missing entries. To do this, assume that $\Delta H°$ and $\Delta S°$ for this transition have the same values at 127 °C as they do at 25 °C.

74. Use the data you estimated in the preceding exercise to sketch ΔH and $T\Delta S$ as a function of T. That is, sketch a graph similar to Figure 20-8.

75. Here are the free energies and enthalpies of formation of three different metal oxides at 25 °C.

	$\Delta H_f°$, kJ/mol	$\Delta G_f°$, kJ/mol
PbO	-217.3	-187.9
Ag_2O	-31.1	-11.2
ZnO	-348.3	-318.4

(a) Which of these oxides can be most readily decomposed to the free metal and $O_2(g)$?

★(b) For the oxide that is most easily decomposed, to what temperature must it be heated to produce $O_2(g)$ at 1.00 atm pressure?

76. At several points in the text we have mentioned that synthesis gas (a mixture of CO and H_2) is an important starting material for the manufacture of substitute natural gas and a variety of organic compounds. Currently, investigations are underway testing the feasibility of using CO_2 as the source of carbon

atoms rather than CO. One possible reaction involves the conversion of CO_2 to acetylene, C_2H_2, from which many other compounds can be easily synthesized.

$$2 CO_2(g) + 5 H_2(g) \rightarrow C_2H_2(g) + 4 H_2O(g)$$

(a) Will this reaction proceed to any significant extent at 25 °C?

(b) Is the production of $C_2H_2(g)$ favored by raising or lowering the temperature?

(c) What is the value of K_p for this reaction at 1000. K?

(d) If CO(g) and $H_2(g)$, each initially at a partial pressure of 1 atm, react at 1000. K, what is the partial pressure of $C_2H_2(g)$ at equilibrium?

77. The decomposition of nitrosyl chloride can be represented by the equation

$$NOCl(g) \rightleftharpoons \frac{1}{2} N_2(g) + \frac{1}{2} O_2(g) + \frac{1}{2} Cl_2(g)$$

Using data from Appendix D and assuming that $\Delta H°$ and $\Delta S°$ are essentially unchanged in the temperature interval from 25 to 100. °C, estimate the value of K_p at 100. °C.

★**78.** 0.100 mol of $PCl_5(g)$ is introduced into a 1.50-L flask and the flask is held at a temperature of 227 °C until equilibrium is established. What is the total pressure of the gases in the flask at this point?

$$PCl_5(g) \rightleftharpoons PCl_3(g) + Cl_2(g)$$

[*Hint*: Use data from Appendix D and appropriate relationships from this chapter to obtain a value of K_p at 227 °C.]

79. The following equilibrium constants have been determined for the reaction $H_2(g) + I_2(g) \rightleftharpoons 2 HI(g)$: $K_p = 50.0$ at 448 °C and 66.9 at 350. °C. Use these data to estimate $\Delta H°$ for the reaction and compare your result with that given in the reaction profile of Figure 15-11.

★**80.** The following data are given for the loss of water by the trihydrate $CuSO_4 \cdot 3H_2O$ at 298 K.

$$CuSO_4 \cdot 3H_2O(s) \rightleftharpoons CuSO_4 \cdot H_2O(s) + 2 H_2O(g)$$
$$\Delta H° = +113 \text{ kJ} \quad K_p = 5.43 \times 10^{-5}$$

At what temperature will the partial pressure of water vapor above the two hydrates be 75 mmHg?

81. Many statements have been made about the first and second laws of thermodynamics that draw upon everyday life. One such statement, in gambler's parlance, is "You can't win and you can't even break even." Explain the basis of this statement.

★**82.** The neutralization of a strong acid and a strong base can be expressed as

$$H^+(aq) + OH^-(aq) \rightarrow H_2O(l)$$

By convention, for $H^+(aq)$ at a concentration of 1 M: $\Delta H_f° = 0$; $\Delta G_f° = 0$, and $S° = 0$. Compared to these values for $H^+(aq)$, the values for $OH^-(aq)$ at 1 M are: $\Delta H_f° = -229.99$ kJ/mol, $\Delta G_f° = -157.29$ kJ/mol, and $S° = 10.75$ J mol^{-1} K^{-1}. Data for $H_2O(l)$ are listed in Appendix D.

(a) By calculation, determine the enthalpy of neutralization, $\Delta H°_{neutr.}$, and compare with the value of -55.9 kJ/mol given in Chapter 17.

(b) Determine $\Delta H°_{neutr.}$ by an alternate method based on the temperature variation of K_w (data on page 608), that is, using the van't Hoff equation (20.22). Compare your result with that of part (a).

(c) Determine $\Delta G^\circ_{neutr.}$ by relating it to K_w and also by using equation (20.7).

*83. A plausible reaction for the production of ethylene glycol (used as an antifreeze) is

$$2\ CO(g) + 3\ H_2(g) \rightarrow CH_2OHCH_2OH(l).$$

These thermodynamic properties of $CH_2OHCH_2OH(l)$ at 25 °C are given: $\Delta H^\circ_f = -387.1$ kJ/mol $CH_2OHCH_2OH(l)$, and $\Delta G^\circ_f = -298.2$ kJ/mol $CH_2OHCH_2OH(l)$. Use these data, together with values from Appendix D, to obtain a value of the standard molar entropy of $CH_2OHCH_2OH(l)$, S°, at 25 °C.

*84. The normal boiling point of cyclohexane, C_6H_{12}, is 80.7 °C. Estimate the temperature at which the vapor pressure of cyclohexane is 100. mmHg.

*85. From the data given in Exercise 59, estimate a value of ΔS° at 298 K for the reaction

$$2\ NaHCO_3(s) \rightarrow Na_2CO_3(s) + H_2O(g) + CO_2(g).$$

*86. The decomposition of the poisonous gas phosgene is represented by the equation $COCl_2(g) \rightleftharpoons CO(g) + Cl_2(g)$. Values of K_p for this reaction are $K_p = 6.7 \times 10^{-9}$ at 100. °C and $K_p = 4.44 \times 10^{-2}$ at 395 °C. At what temperature is $COCl_2$ 15% dissociated when the total gas pressure is maintained at 1.00 atm?

*87. Assume that the steam [$H_2O(g)$] in the power plant referred to in Exercise 63 is in equilibrium with liquid water. What is the steam pressure corresponding to the minimum steam temperature calculated in Exercise 63(a)?

Self-Test Questions

For questions 88 through 95 select the single item that best completes each statement.

88. For a process to occur spontaneously,
(a) the entropy of the system must increase.
(b) the entropy of the surroundings must increase.
(c) both the entropy of the system and of the surroundings must increase.
(d) the entropy of the universe must increase.

89. The free energy change of a reaction is a measure of
(a) the quantity of heat given off to the surroundings.
(b) the direction in which a net reaction occurs.
(c) the increased molecular disorder that occurs in the system.
(d) how rapidly the reaction occurs.

90. If a reaction has $\Delta H < 0$ and $\Delta S < 0$, the reaction proceeds furthest in the forward direction at (a) low temperatures; (b) high temperatures; (c) all temperatures; (d) no temperature.

91. If it is necessary to employ electric current (electrolysis) to carry out a chemical reaction, then for that reaction (a) $\Delta H > 0$; (b) $\Delta G = \Delta H$; (c) $\Delta G > 0$; (d) $\Delta S > 0$.

92. For the reaction $Br_2(g) \rightarrow 2\ Br(g)$, we should expect that at all temperatures (a) $\Delta H < 0$; (b) $\Delta S > 0$; (c) $\Delta G < 0$; (d) $\Delta S < 0$.

93. If $\Delta G^\circ = 0$ for a reaction, then (a) $\Delta H^\circ = 0$; (b) $\Delta S^\circ = 0$; (c) $K = 0$; (d) $K = 1$.

94. For the reaction $2\ NO(g) + O_2(g) \rightarrow 2\ NO_2(g)$ all but one of the following equations is correct. The *incorrect* equation is (a) $K = K_p$; (b) $K = K_c$; (c) $K_p = $ antiln$(-\Delta G^\circ/RT)$; (d) $\Delta G = \Delta G^\circ + RT \ln Q$.

95. In a heat engine
(a) More heat is evolved at a low temperature than is absorbed at a high temperature.

(b) The amount of work performed is the ratio of the quantity of heat absorbed at the high temperature to the quantity of heat ejected at the low temperature.
(c) The efficiency of the engine is given by this ratio of Kelvin temperatures: low $T/($high $T -$ low $T)$.
(d) The quantity of heat absorbed at the high temperature is the sum of the quantity of heat ejected at the low temperature and the work performed.

96. Explain briefly why
(a) The change in entropy in a *system* is not always a suitable criterion for spontaneous change.
(b) ΔG° is so important in dealing with the question of spontaneous change, even though the conditions employed in a reaction are usually *nonstandard*.

97. Listed below are data at 298 K for the reaction $NH_4NO_3(s) \rightarrow N_2O(g) + 2\ H_2O(l)$.

	ΔH°_f, kJ/mol	S°, J mol^{-1} K^{-1}
$NH_4NO_3(s)$	-365.6	151.1
$N_2O(g)$	$+82.05$	219.7
$H_2O(l)$	-285.8	69.92

(a) Is the forward reaction endothermic or exothermic?
(b) What is the value of ΔG° at 298 K?
(c) What is the value of the equilibrium constant, K?
(d) Does the forward reaction tend to occur spontaneously at temperatures above 25 °C, below 25 °C, both or neither?

21 Electrochemistry

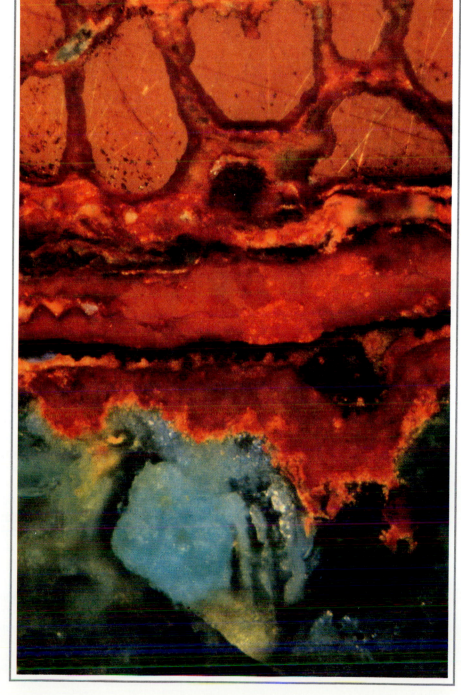

Corrosion of bronze. In this photomicrograph of an ancient bronze from Thailand, the interface between the metal and its green corrosion layer is clearly seen. As we learn in this chapter, corrosion is an electrochemical phenomenon. [William Marin/Brookhaven National Laboratory]

When automobiles were being developed in the late nineteenth century, scientists and engineers understood well that the ability of the internal combustion engine to convert heat to mechanical work is inherently limited by the second law of thermodynamics (recall page 740).

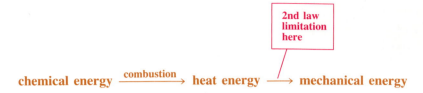

chemical energy $\xrightarrow{\text{combustion}}$ heat energy $\longrightarrow$ mechanical energy

They were equally aware that this particular limitation can be avoided if chemical energy is converted directly to electricity.

chemical energy $\longrightarrow$ electricity $\longrightarrow$ mechanical energy

From the standpoint of energy use, an electric-powered auto should be much more fuel efficient than a conventional automobile.

Unfortunately, early devices for converting chemical energy to electricity could not be made to perform at their theoretical efficiencies. This fact, combined with the availability of high-quality gasoline at a low cost, resulted in electric-powered cars being quickly displaced by the internal combustion automobile.

Now, with concern about long-term energy supplies and environmental pollution, there is a renewed interest in electric-powered automobiles. Moreover, the inherent high efficiency of electrochemical energy conversion devices may soon be realized.

In this chapter we will see how chemical reactions can be used to produce electricity and how electricity can be used to produce chemical reactions. The practical applications of electrochemistry are countless, ranging from batteries and fuel cells as electric power sources, to the manufacture of key chemicals, to the refining of metals, to methods of controlling corrosion. Equally important, however, are the theoretical applications. Because electricity involves a flow of electrons, studying the relationship between chemistry and electricity will give us more insight into reactions in which electrons are transferred—*oxidation–reduction reactions.*

21-1 Electrode Potentials and Their Measurement

Figure 21-1 shows the behavior of copper metal toward $AgNO_3(aq)$ and toward $Zn(NO_3)_2(aq)$. An oxidation–reduction reaction occurs between $Cu(s)$ and $Ag^+(aq)$, but not between $Cu(s)$ and $Zn^{2+}(aq)$. How can we explain this difference in behavior? What we need is a criterion of the tendency for metal ions to gain electrons and become reduced to the free metal. This criterion should show that Ag^+ ions are more easily reduced than are Zn^{2+} ions. The criterion we need is based on a new property called the *electrode potential*. In this section we explore the idea of electrode potentials and how they are measured.

When used in electochemical studies, a strip of metal, M, as shown in Figure 21-2, is called an **electrode.** The metal strip is immersed in a solution containing the metal ions, M^{n+}. The combination of the electrode and solution is called a **half-cell.*** Three kinds of interactions are possible between metal atoms on the electrode and metal ions in solution.

*Electrochemical nomenclature is not always consistent. Sometimes a half-cell assembly is referred to simply as an electrode.

752

(a)   (b)

FIGURE 21-1

Differences in behavior of copper toward $Ag^+(aq)$ and $Zn^{2+}(aq)$.

(a) Copper metal displaces Ag^+ from $AgNO_3(aq)$, producing silver metal.

$Cu(s) + 2\ Ag^+(aq) \rightarrow$
    colorless
        $Cu^{2+}(aq) + 2\ Ag(s)$
          blue

(b) Copper metal *does not* displace Zn^{2+} from $Zn(NO_3)_2(aq)$.

$Cu(s) + Zn^{2+}(aq) \rightarrow$ no reaction
        colorless

[Carey B. Van Loon]

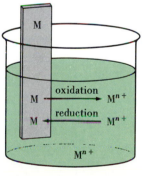

FIGURE 21-2
An electrochemical half-cell.

The half-cell consists of a metal electrode, M, immersed in an aqueous solution of its ions, M^{n+}. (The anions required to maintain electrical neutrality in the solution are not shown.) The situation illustrated here is limited to metals that do not react with water.

- A metal ion M^{n+} may collide with the electrode and undergo no change.
- A metal ion M^{n+} may collide with the electrode, *gain n electrons*, and be converted to a metal atom M. *The ion is reduced.*
- A metal atom M on the electrode may *lose n electrons* and enter the solution as the ion M^{n+}. *The metal atom is oxidized.*

An equilibrium is quickly established between the metal and the solution, which we can represent as

$$M(s) \underset{\text{reduction}}{\overset{\text{oxidation}}{\rightleftharpoons}} M^{n+}(aq) + n\ e^- \qquad (21.1)$$

If the tendency toward *oxidation* is strong, we should expect a very slight negative charge to accumulate on the electrode (from the electrons left behind). And, the greater the tendency for the metal to become oxidized, the greater should be this negative charge. In turn, the solution should have a very slight buildup in concentration of M^{n+} ions and should carry a slight positive charge. If the tendency toward *reduction* is especially strong, we should expect the reverse situation—a very slight positive charge on the electrode and negative charge in the solution. Unfortunately, we cannot make a direct measurement of the electric charges we are describing here.

But here is what we can do. If we connect two different electrodes, electrons will flow *from* the electrode of *higher* negative electric charge density *to* the electrode with a *lower* negative electric charge density. A property closely related to the density of negative electric charge is called the **electrode potential.** Electric current (electrons) flows from a high electrode potential to a lower electrode potential. Even the slightest *difference* in electrode potential is enough to set up an electric current. This is similar to the flow of water pictured in Figure 21-3.

To measure a *difference* in potential between two electrodes, we need to connect *two* half-cells. The electrical connection must be done in a special way. *Both* the metal electrodes *and* the solutions have to be connected, so that a continuous circuit is formed through which charged particles can flow. The electrodes can simply be connected by a metal wire to permit the flow of electrons. The flow of electric current between the solutions must be in the form of a migration of ions. This cannot occur through a wire. In some electrochemical cells contact between solutions is through a porous plug or membrane that separates the two solutions. In other cases this contact is established through a third solution, usually in a U-tube, that "bridges" the two half-cells. This connection is called a **salt bridge.** When these electrical connections have been properly made the combination of two half-cells is called an **electrochemical cell.**

Consider an electrochemical cell consisting of one half-cell in which a copper electrode is immersed in a solution of $Cu^{2+}(aq)$ and another half-cell in which a silver electrode is immersed in a solution of $Ag^+(aq)$. The solutions are joined by a salt bridge and the electrodes by a metal wire. This electrochemical cell is pictured in Figure 21-4.

FIGURE 21-3
Water flow—an analogy to electric current.

The heights of the water columns are analogous to electric charge densities or *electric potentials.* The difference in the two heights corresponds to an electric *potential difference.* Water flows from the higher to the lower level until the levels are equalized. Then, the flow of water stops.

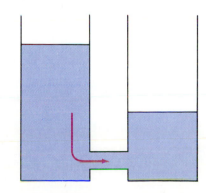

FIGURE 21-4

FIGURE 21-4

Measurement of the electromotive force of an electrochemical cell.

An electrochemical cell consists
of two half-cells with electrodes
joined by a wire and solutions
by a salt bridge. (The ends of
the salt bridge are plugged with
a porous material that allows
ions to migrate but prevents the
bulk flow of liquid.) The potentiometer measures the difference
in electric potential between the
two electrodes.

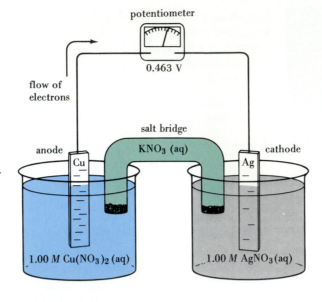

The migration of ions in an
electrochemical cell is always
anions toward the anode;
cations toward the cathode.

These are the changes that occur within the cell: Cu atoms lose electrons at the
Cu(s) electrode and enter the solution as Cu^{2+}(aq) ions. The electrons lost by the Cu
atoms pass through the wire and the electrical measuring circuit to the Ag(s) electrode. Here Ag^+ ions from solution gain electrons and deposit as silver metal.
Without a salt bridge, the solution in the copper half-cell would acquire excess
Cu^{2+} and a net *positive* charge. In the silver half-cell, there would be a deficiency of
Ag^+ ions, an excess of anions, and a *negative* charge buildup in the solution.
Because of the high energy requirement to create a separation of electric charges,
electric current would not continue to flow. However, anions (NO_3^-) from the salt
bridge migrate into the copper half-cell to neutralize the positive charge of the
excess Cu^{2+} ions. Cations (K^+) from the salt bridge migrate into the silver half-cell
to neutralize the negative charge of the excess NO_3^- ions. In short, the salt bridge
allows electric current to flow through the solutions. The net reaction that occurs as
electric current flows through the electrochemical cell is

Recall from the fundamental
definitions in Section 5-5 how
oxidation and reduction can
be represented through half-
equations.

Oxidation: $Cu(s) \longrightarrow Cu^{2+}(aq) + 2e^-$
Reduction: $2\{Ag^+(aq) + e^- \longrightarrow Ag(s)\}$
_____
Net: $Cu(s) + 2\ Ag^+(aq) \longrightarrow Cu^{2+}(aq) + 2\ Ag(s)$ (21.2)

The reading on the meter in the electrical circuit (0.463 V) is significant. It is the
potential difference between the two half-cells. Since this potential difference is
the "driving force" for electrons, it is often called the **electromotive force (emf)** of
the cell or the **cell potential.** The unit for measuring electric potential is the **volt,** so
the cell potential is also called the **cell voltage.**

Now let us return to the question posed in connection with Figure 21-1. That is,
why does copper *not* displace Zn^{2+} from solution? If we set up an electrochemical
cell consisting of a Zn(s)/Zn^{2+}(aq) half-cell and a Cu^{2+}(aq)/Cu(s) half-cell, we find
that electrons flow *from the Zn to the Cu*. The reaction that occurs spontaneously in
the electrochemical cell pictured in Figure 21-5 is

Here is a way to relate the
volt to other units: Passage of
one coulomb of electric
charge through a potential
difference of one volt pro-
duces one joule of work.

1 joule (J) = 1 volt (V)
                  × 1 coulomb (C)

Oxidation: $Zn(s) \longrightarrow Zn^{2+}(aq) + 2\ e^-$
Reduction: $Cu^{2+}(aq) + 2\ e^- \longrightarrow Cu(s)$
_____
Net: $Zn(s) + Cu^{2+}(aq) \longrightarrow Zn^{2+}(aq) + Cu(s)$ (21.3)

Since (21.3) is a reaction that occurs spontaneously, the displacement of Zn^{2+}(aq)
by Cu(s)—the *reverse* of (21.3)—does *not* occur spontaneously. This is the observation we made in Figure 21-1.

FIGURE 21-5
The reaction

$$Zn(s) + Cu^{2+}(aq) \rightarrow$$
$$Zn^{2+}(aq) + Cu(s)$$

occurring in an electrochemical cell.

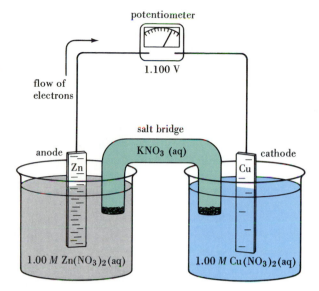

Precise Measurement of Cell Potentials. To get a precise value of a cell voltage we need to make the measurement in such a way as to draw practically no current through the cell. One way to do this is to oppose the current coming from the voltaic cell with that produced by another cell whose voltage can be set to a precisely known value. When the two opposing cells have voltages of the same magnitude, essentially no electric current flows. A device that makes this type of comparison is called a **potentiometer.** Another, simpler, alternative for measuring cell potentials is to use special electronic voltmeters that draw very little current.

Cell Diagrams and Terminology. You should generally find it helpful when describing electrochemical cells to draw sketches, as in Figures 21-4 and 21-5. (Your sketches, of course, can be more crudely drawn.) Another useful representation that chemists use is called a **cell diagram.** Listed below are the parts of the electrochemical cell that should appear in the cell diagram and the conventions you should use in showing them. The list is followed by the cell diagram for Figure 21-5.

Writing cell diagrams.

- The **anode.** This is the electrode at which *oxidation* occurs. Place the anode at the *left* side of the diagram.
- The **cathode.** This is the electrode at which *reduction* occurs. Place the cathode at the *right* side of the diagram.
- Boundaries (interfaces) between electrodes and solutions. Use a *single* vertical line ($|$).   (21.4)
- Boundaries (interfaces) between solutions. If this is a **salt bridge,** represent it with a *double* vertical line ($\|$). If this is a porous plug, use a series of vertical dots ($\vdots$).

Luigi Galvani (1737–1798) discovered current electricity, although he incorrectly considered it to be of animal origin. Alessandro Volta (1745–1827) gave a correct explanation of current electricity and made the first cells of the type we have been describing. His cells (called voltaic piles) consisted of alternating discs of copper, zinc, and blotting paper soaked in NaCl(aq): Cu, Zn, paper, Cu, Zn, paper, ...

anode $\longrightarrow$ Zn(s)$|$Zn^{2+}(aq) $\quad\|\quad$ Cu^{2+}(aq)$|$Cu(s) $\longleftarrow$ cathode

$\qquad$ half-cell $\qquad$ salt $\qquad$ half-cell
$\qquad$ (oxidation) $\quad$ bridge $\quad$ (reduction)

The electrochemical cells we have considered to this point are of a type that *produce* electricity as a result of a chemical reaction. They are called **voltaic (galvanic) cells.** In another type of cell called an **electrolytic cell** (Section 21-7) we use electricity to produce *nonspontaneous* chemical changes.

Example 21-1 ――――――――――――――――――――

Representing an oxidation–reduction reaction through a cell diagram. Aluminum metal displaces zinc(II) ion from aqueous solution.

(a) Write oxidation and reduction half-equations and a net equation for this oxidation–reduction reaction.

(b) Draw a cell diagram for an electrochemical (voltaic) cell in which this reaction would occur.

Solution

(a) The term "displaces" means that the aluminum goes into solution as ions (Al^{3+}) and Zn^{2+} comes out of solution as solid zinc. The Al is oxidized and Zn^{2+} is reduced. Note that in writing the net equation we must adjust coefficients so that equal numbers of electrons are involved in oxidation and in reduction. (This is the half-reaction method of balancing redox equations that we studied in Section 5-6.)

$$\text{Oxidation:} \quad 2\,\{Al(s) \longrightarrow Al^{3+}(aq) + 3\,e^-\}$$
$$\text{Reduction:} \quad 3\,\{Zn^{2+}(aq) + 2\,e^- \longrightarrow Zn(s)\}$$
$$\text{Net:} \quad 2\,Al(s) + 3\,Zn^{2+}(aq) \longrightarrow 2\,Al^{3+}(aq) + 3\,Zn(s)$$

Combinations such as $Al|Al^{3+}$ and $Zn^{2+}|Zn$ are often called couples.

(b) We need to write the combination $Al(s)|Al^{3+}(aq)$ for the anode (oxidation) half-cell and the combination $Zn^{2+}|Zn(s)$ for the cathode (reduction) half-cell. The two half-cells can be joined by a salt bridge.

$$Al(s)|Al^{3+}(aq)\|Zn^{2+}(aq)|Zn(s)$$

SIMILAR EXAMPLES: Exercises 1, 3, 25, 26.

21-2   Standard Electrode Potentials

Consider the advantages of making a measurement for a particular half-cell (or couple) just once and then using that value in every electrochemical cell in which that couple appears. This would allow us to *calculate* cell voltages in many cases. We can do this by *arbitrarily* choosing a particular couple to which we assign a potential of *zero*. We can then compare all other couples to this *reference* electrode.

Note that the standard hydrogen "electrode" is actually a half-cell: it consists of both an electrode and a solution.

The reference electrode for electrode potential measurements is the **standard hydrogen electrode (S.H.E.)**, pictured in Figure 21-6. The standard hydrogen electrode involves H_3O^+ ions in solution at unit activity ($a = 1$), but for simplicity we will write H^+ for H_3O^+ and we will assume that unit activity is essentially *1 M*. H_2 molecules in the gaseous state are at 1 atm pressure. The oxidized (H^+) and reduced (H_2) forms of hydrogen are in contact on an inert platinum metal surface and impart a characteristic potential to the surface. The temperature is exactly 25 °C. All of these conditions can be represented through an equation

$$2\,H^+(a = 1) + 2\,e^- \xrightleftharpoons{\text{on Pt}} H_2(g,\ 1\ atm) \qquad E° = 0.0000 \text{ volt (V)} \qquad (21.5)$$

or through a half-cell couple

$$H^+(a = 1)|H_2(g,\ 1\ atm),\ Pt \qquad E° = 0.0000 \text{ V} \qquad (21.6)$$

By international agreement, a **standard electrode potential, $E°$,** is based on the tendency for a **reduction** process to occur at the electrode. For other standard electrodes we can write expressions of this sort.

$$Cu^{2+}(1\ M) + 2\,e^- \rightleftharpoons Cu(s) \qquad E° = ? \qquad (21.7)$$

$$Cl_2(g,\ 1\ atm) + 2\,e^- \rightleftharpoons 2\,Cl^-(1\ M) \qquad E° = ? \qquad (21.8)$$

In all cases the ionic species are present in aqueous solution at unit activity (approximately 1 M); gases are at 1 atm pressure. Where no solid substance is indicated, the potential is established on an inert electrode such as platinum.

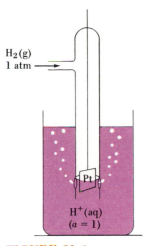

FIGURE 21-6
The standard hydrogen electrode (S.H.E.).

H₂(g) 1 atm

Pt

H⁺(aq) (a = 1)

To determine values of $E°$ for the electrodes in (21.7) and (21.8), let us compare each of these electrodes with a standard hydrogen electrode (S.H.E.). In the voltaic cell diagrammed below we measure a potential difference of 0.337 V, with electrons flowing from the H_2 to the Cu electrode. Since this is the potential of the cell formed from two *standard* electrodes, it is referred to as the **standard cell potential, $E°_{cell}$**.

$$Pt, H_2(g, \ 1 \ M)|H^+(1 \ M)\|Cu^{2+}(1 \ M)|Cu(s) \qquad E°_{cell} = 0.337 \ V \qquad (21.9)$$

The reaction that occurs in the voltaic cell of (21.9) is

Oxidation: $H_2(g, \ 1 \ atm) \longrightarrow 2 \ H^+(1 \ M) + 2 \ e^-$
Reduction: $Cu^{2+}(1 \ M) + 2 \ e^- \longrightarrow Cu(s)$

Net: $\quad H_2(g, \ 1 \ atm) + Cu^{2+}(1 \ M) \longrightarrow 2 \ H^+(1 \ M) + Cu(s)$
$$E°_{cell} = 0.337 \ V \quad (21.10)$$

The cell reaction (21.10) indicates that $Cu^{2+}(1 \ M)$ is more easily reduced than is $H^+(1 \ M)$. The standard electrode potential representing the reduction of $Cu^{2+}(aq)$ to $Cu(s)$ is $+0.337$ V.

$$Cu^{2+}(1 \ M) + 2 \ e^- \longrightarrow Cu(s) \qquad E° = +0.337 \ V \qquad (21.11)$$

When a standard hydrogen electrode is combined with a standard zinc electrode, electrons are found to flow in the opposite direction, that is, *from* the zinc *to* the hydrogen electrode. The S.H.E. acts as the *cathode* (reduction) and the zinc electrode as the *anode* (oxidation). The measured value of $E°_{cell}$ is 0.763 V.

$$Zn(s)|Zn(1 \ M)\|H^+(1 \ M)|H_2(g, \ 1 \ atm), \ Pt \qquad E°_{cell} = 0.763 \ V \qquad (21.12)$$

The reaction that occurs in the voltaic cell (21.12) is

Oxidation: $Zn(s) \longrightarrow Zn^{2+}(1 \ M) + 2 \ e^-$
Reduction: $2 \ H^+(1 \ M) + 2 \ e^- \longrightarrow H_2(g, \ 1 \ atm)$

Net: $\quad Zn(s) + 2 \ H^+(1 \ M) \longrightarrow Zn^{2+}(1 \ M) + H_2(g, \ 1 \ atm)$
$$E°_{cell} = 0.763 \ V \quad (21.13)$$

Because oxidation, not reduction, occurs at the zinc electrode, the reduction of $Zn^{2+}(1 \ M)$ must occur with *greater difficulty* than that of $H^+(1 \ M)$. $E°_{cell}$ of equation (21.13) describes the tendency for zinc to become *oxidized*. If we consider that *the reduction tendency is the opposite of the oxidation tendency,* then

$$Zn^{2+}(1 \ M) + 2 \ e^- \longrightarrow Zn(s) \qquad E° = -0.763 \ V \qquad (21.14)$$

In summary

Electrode potential conventions.

- The potential of the standard hydrogen electrode is set at 0.
- Any electrode at which a *reduction* half-reaction shows a *greater* tendency to occur than does

  $$2 \ H^+(1 \ M) + 2 \ e^- \longrightarrow H_2(g, \ 1 \ atm)$$

  has a **positive** value for its electrode potential, $E°$.
- Any electrode at which a *reduction* half-reaction shows a *lesser* tendency to occur than does   (21.15)

  $$2 \ H^+(1 \ M) + 2 \ e^- \longrightarrow H_2(g, \ 1 \ atm)$$

  has a **negative** value for its electrode potential, $E°$.
- If the tendency for a reduction process is $E°$, the *oxidation* tendency is simply the negative of this value, that is $-E°$.
- Tabulations such as Table 21-1 are generally written for *reduction* half-reactions.

TABLE 21-1
Some Selected Standard Electrode Potentials

Reduction half-reaction	$E°$, V
Acidic solution	
$F_2(g) + 2 e^- \longrightarrow 2 F^-(aq)$	+2.87
$O_3(g) + 2 H^+(aq) + 2 e^- \longrightarrow O_2(g) + H_2O$	+2.07
$S_2O_8^{2-}(aq) + 2 e^- \longrightarrow 2 SO_4^{2-}(aq)$	+2.01
$H_2O_2(aq) + 2 H^+(aq) + 2 e^- \longrightarrow 2 H_2O$	+1.77
$MnO_4^-(aq) + 8 H^+(aq) + 5 e^- \longrightarrow Mn^{2+}(aq) + 4 H_2O$	+1.51
$PbO_2(s) + 4 H^+(aq) + 2 e^- \longrightarrow Pb^{2+}(aq) + 2 H_2O$	+1.455
$Cl_2(g) + 2 e^- \longrightarrow 2 Cl^-(aq)$	+1.360
$Cr_2O_7^{2-}(aq) + 14 H^+(aq) + 6 e^- \longrightarrow 2 Cr^{3+}(aq) + 7 H_2O$	+1.33
$O_2(g) + 4 H^+(aq) + 4 e^- \longrightarrow 2 H_2O$	+1.229
$MnO_2(s) + 4 H^+(aq) + 2 e^- \longrightarrow Mn^{2+}(aq) + 2 H_2O$	+1.224
$2 IO_3^-(aq) + 12 H^+(aq) + 10 e^- \longrightarrow I_2(s) + 6 H_2O$	+1.195
$Br_2(l) + 2 e^- \longrightarrow 2 Br^-(aq)$	+1.065
$NO_3^-(aq) + 4 H^+(aq) + 3 e^- \longrightarrow NO(g) + 2 H_2O$	+0.96
$Ag^+(aq) + e^- \longrightarrow Ag(s)$	+0.800
$Fe^{3+}(aq) + e^- \longrightarrow Fe^{2+}(aq)$	+0.771
$O_2(g) + 2 H^+(aq) + 2 e^- \longrightarrow H_2O_2(aq)$	+0.682
$I_2(s) + 2 e^- \longrightarrow 2 I^-(aq)$	+0.535
$Cu^+(aq) + e^- \longrightarrow Cu(s)$	+0.52
$H_2SO_3(aq) + 4 H^+(aq) + 4 e^- \longrightarrow S(s) + 3 H_2O$	+0.45
$Cu^{2+}(aq) + 2 e^- \longrightarrow Cu(s)$	+0.337
$SO_4^{2-}(aq) + 4 H^+(aq) + 2 e^- \longrightarrow 2 H_2O + SO_2(g)$	+0.17
$Sn^{4+}(aq) + 2 e^- \longrightarrow Sn^{2+}(aq)$	+0.154
$S(s) + 2 H^+(aq) + 2 e^- \longrightarrow H_2S(g)$	+0.141
$2 H^+(aq) + 2 e^- \longrightarrow H_2(g)$	0.0000
$Pb^{2+}(aq) + 2 e^- \longrightarrow Pb(s)$	−0.126
$Sn^{2+}(aq) + 2 e^- \longrightarrow Sn(s)$	−0.136
$Cr^{3+}(aq) + e^- \longrightarrow Cr^{2+}(aq)$	−0.407
$Fe^{2+}(aq) + 2 e^- \longrightarrow Fe(s)$	−0.440
$Zn^{2+}(aq) + 2 e^- \longrightarrow Zn(s)$	−0.763
$Al^{3+}(aq) + 3 e^- \longrightarrow Al(s)$	−1.66
$Mg^{2+}(aq) + 2 e^- \longrightarrow Mg(s)$	−2.375
$Na^+(aq) + e^- \longrightarrow Na(s)$	−2.714
$Ca^{2+}(aq) + 2 e^- \longrightarrow Ca(s)$	−2.76
$K^+(aq) + e^- \longrightarrow K(s)$	−2.925
$Li^+(aq) + e^- \longrightarrow Li(s)$	−3.045
Basic solution	
$O_3(g) + H_2O + 2 e^- \longrightarrow O_2(g) + 2 OH^-$	+1.24
$OCl^-(aq) + H_2O + 2 e^- \longrightarrow Cl^- + 2 OH^-$	+0.89
$O_2(g) + 2 H_2O + 4 e^- \longrightarrow 4 OH^-(aq)$	+0.401
$CrO_4^{2-}(aq) + 4 H_2O + 3 e^- \longrightarrow Cr(OH)_3(s) + 5 OH^-$	−0.13
$S(s) + 2 e^- \longrightarrow S^{2-}(aq)$	−0.48
$2 H_2O + 2 e^- \longrightarrow H_2(g) + 2 OH^-(aq)$	−0.828
$SO_4^{2-}(aq) + H_2O + 2 e^- \longrightarrow SO_3^{2-}(aq) + 2 OH^-(aq)$	−0.93

In the voltaic cell (21.12), H^+ is reduced at the S.H.E. and the measured cell potential is 0.763 V. Since Cu^{2+} is reduced even more readily than is H^+, we should expect the cell potential difference to be greater if the S.H.E. is replaced by the Cu^{2+}/Cu couple. In fact, we should predict $E°_{cell}$ to be $0.337 - (-0.763) =$ 1.100 V, the value *measured* in Figure 21-5. The simplest method of combining electrode potentials to *predict* $E°_{cell}$ values is outlined below and illustrated in Examples 21-2 and 21-3.

Relating electrode potentials and $E_{cell}°$.

> 1. Write the proposed *reduction* half-equation and a standard potential to describe it. This will be a value taken *directly* from Table 21-1: $E°$.
> 2. Write the proposed *oxidation* half-equation and a standard potential to describe it. This will be the *negative* of a value found in Table 21-1: $-E°$.
> 3. Combine the half-equations into a net oxidation–reduction equation. *Add* the half-cell potentials to obtain $E_{cell}°$.

(21.16)

$E°$ values are not changed when half-equations are multiplied by coefficients.

In completing Step 3, note that an electrode potential is an *intensive* property of an electrode system. $E°$ values are *unaffected* by multiplying half-equations by constant coefficients.

Example 21-2

Combining $E°$ values into $E_{cell}°$ for a reaction. A new battery system currently under study for possible use in electric vehicles is the zinc–chlorine battery. The net reaction producing electricity in this voltaic cell is

$$Zn(s) + Cl_2(g) \longrightarrow ZnCl_2(aq)$$

What is $E_{cell}°$ of this voltaic cell?

Solution. From the net reaction we see that $Zn(s)$ is oxidized and $Cl_2(g)$ is reduced. We can write equations for each half-reaction and assign them the appropriate potentials from Table 21-1. $E_{cell}°$ is the sum of these potentials.

Oxidation:	$Zn(s) \longrightarrow Zn^{2+}(aq) + 2\ e^-$	$-E° = -(-0.763\text{ V})$
		$= +0.763\text{ V}$
Reduction:	$Cl_2(g) + 2\ e^- \longrightarrow 2\ Cl^-(aq)$	$E° = +1.360\text{ V}$
Net:	$Zn(s) + Cl_2(g) \longrightarrow ZnCl_2(aq)$	$E_{cell}° = +2.123\text{ V}$

SIMILAR EXAMPLE: Exercise 3.

Example 21-3

Determining an unknown $E°$ from an $E_{cell}°$ measurement. Cadmium is usually found in small quantities wherever zinc is found. Unlike zinc, which is an essential element in trace amounts, cadmium is an environmental poison. To measure cadmium ion concentrations by potentiometric methods (Section 21-4) we need to know the standard electrode potential for the Cd^{2+}/Cd couple. The following voltaic cell is constructed and its voltage is measured.

$$Cd(s)|Cd^{2+}(1\text{ M})||Cu^{2+}(1\text{ M})|Cu(s) \qquad E_{cell}° = 0.740\text{ V}$$

What is the standard potential for the Cd^{2+}/Cd electrode?

Solution. We know one half-cell potential and $E°_{cell}$ for the net reaction. We can solve for the unknown electrode potential, which we denote as $E°$. Note that we enter this unknown potential as its *negative* ($-E°$) in the setup below. This is because we are applying it to an oxidation half-reaction.

Oxidation:	$Cd(s) \longrightarrow Cd^{2+}(1\text{ M}) + 2\ e^-$	$-E°$
Reduction:	$Cu^{2+}(1\text{ M}) + 2\ e^- \longrightarrow Cu(s)$	$E° = +0.337\text{ V}$
Net:	$Cd(s) + Cu^{2+}(1\text{ M}) \longrightarrow Cd^{2+}(1\text{ M}) + Cu(s)$	

$$E_{cell}° = -E° + 0.337\text{ V}$$
$$= 0.740\text{ V}$$

$$0.740\text{ V} = -E° + 0.337\text{ V}$$

$$E° = 0.337\text{ V} - 0.740\text{ V} = -0.403\text{ V}$$

SIMILAR EXAMPLES: Exercises 2, 19, 20.

21-3 Electrical Work, $\Delta G°$, and Spontaneous Change

On page 722 we gave physical meaning to a free energy change. We said that ΔG represents the maximum amount of *useful* work that a constant-pressure process can perform. (Useful work is all work other than pressure–volume work.)

$$\Delta G = w_{\text{max(useful)}} \tag{21.17}$$

Now recall that work done *by* a system is *negative,* and that $\Delta G < 0$ is our principal criterion of *spontaneous* change. Thus, spontaneous changes are capable of doing useful work. Conversely, work must be performed to produce *nonspontaneous* changes, since $\Delta G > 0$.

When a reaction occurs in a voltaic cell it does work—electrical work. Think of this as the work in moving electric charges. The quantity of electrical work done is the product of the number of moles of electrons, the electric charge per mole of electrons (called the **Faraday constant, $\mathscr{F}$),** and the emf (voltage) of the cell. In expression (21.18), the value of the Faraday constant is 96,500 coulombs per mol electrons (96,500 C/mol e⁻) and E_{cell} is in volts (V). The negative sign signifies work done *by* the cell.

> A more precise value of $\mathscr{F}$ is 9.6485×10^4 C/mol e⁻.

$$w_{\text{elec.}} = -n\mathscr{F}E_{\text{cell}} \tag{21.18}$$

> The idea of a reversible process was discussed on page 740.

The electrical work can be done in such a way (reversibly) as to represent the maximum amount of useful work. This allows us to equate expressions (21.17) and (21.18) to obtain

$$\Delta G = -n\mathscr{F}E_{\text{cell}} \tag{21.19}$$

If the reactants and products are in their standard states,

> Relationship of $E°_{\text{cell}}$ to $\Delta G°$.

$$\Delta G° = -n\mathscr{F}E°_{\text{cell}} \tag{21.20}$$

Example 21-4

Determining the free energy change of a reaction from electrode potentials. Use electrode potential data to determine $\Delta G°$ for the reaction

$$Zn(s) + Cl_2(g, 1 \text{ atm}) \longrightarrow ZnCl_2(aq, 1 \text{ M})$$

Solution. This is the net reaction occurring in the voltaic cell described in Example 21-2. In this type of problem, usually the best approach is to separate the net equation into two half-equations. This allows us to determine the value of $E°_{\text{cell}}$ and the number of moles of electrons (n) involved in the cell reaction. From Example 21-2 we see that $E°_{\text{cell}} = +2.123$ V and $n = 2$ mol e⁻. Now we can use equation (21.20). (Recall that $1 \text{ V} \cdot \text{C} = 1 \text{ J}$.)

$$\Delta G° = -n\mathscr{F}E°_{\text{cell}} = -2 \text{ mol e}^- \times \frac{96,485 \text{ C}}{\text{mol e}^-} \times 2.123 \text{ V}$$

$$= -4.097 \times 10^5 \text{ J} = -409.7 \text{ kJ}$$

SIMILAR EXAMPLES: Exercises 7, 27.

Spontaneity in Oxidation–Reduction Reactions. The criterion for spontaneous change in oxidation–reduction reactions is the same as that for all reactions: $\Delta G < 0$. However, according to equation (21.19), if $\Delta G < 0$ then $E_{\text{cell}} > 0$. (That is, E_{cell} must be *positive* if ΔG is negative.) If reactants and products are in their standard states, we must work with $E°_{\text{cell}}$. To predict the direction of spontaneous change in an oxidation–reduction reaction, we can add two steps to the three listed on page 759.

E_{cell} as a criterion for spontaneity.

Example 21-5

Applying the criterion for spontaneity in a redox reaction, when a balanced equation is given. Will aluminum metal displace Cu^{2+} ion from aqueous solution? That is, does a spontaneous reaction occur in the forward direction?

$$2\ Al(s) + 3\ Cu^{2+}(1\ M) \longrightarrow 3\ Cu(s) + 2\ Al^{3+}(1\ M)$$

Solution. Since the balanced net equation is given, all you need do is separate this equation into two half-equations, and then recombine them.

Oxidation: $2\{Al(s) \longrightarrow Al^{3+}(1\ M) + 3\ e^-\}$ $-E° = -(-1.66)$
$$= +1.66\ V \quad (21.22)$$

Reduction: $3\{Cu^{2+}(1\ M) + 2\ e^- \longrightarrow Cu(s)\}$ $E° = +0.337\ V \quad (21.23)$

Net: $2\ Al(s) + 3\ Cu^{2+}(1\ M) \longrightarrow 3\ Cu(s) + 2\ Al^{3+}(1\ M)$
$$E°_{cell} = +2.00\ V \quad (21.24)$$

Since $E°_{cell}$ is *positive*, the direction of spontaneous change is the forward direction.

SIMILAR EXAMPLE: Exercise 5.

Reaction of Al(s) and Cu^{2+}(aq). Notice the holes in the foil where dissolving of Al has occurred. Notice also the red-brown deposit (Cu metal) at the bottom of the beaker. The gas bubbles are $H_2(g)$, formed by the reduction of H^+(aq), which also occurs to a limited extent. [Carey B. Van Loon]

Example 21-6

Applying the criterion for spontaneity in a redox reaction, when no equation is given. Peroxodisulfate salts (e.g., $Na_2S_2O_8$) are strong oxidizing agents used as bleaching agents for fats, oils, and fabrics. The usual starting material for the preparation of peroxodisulfates is sodium sulfate. If it works, one of the cheapest ways to produce peroxodisulfates would be to pass oxygen gas through a solution of sulfate ion. Will oxygen gas oxidize sulfate ion to peroxodisulfate ion ($S_2O_8^{2-}$) in acidic solution, with the $O_2(g)$ being reduced to water?

Solution. What you must do in cases like this is identify the oxidation and reduction half-equations and their half-cell potentials. Then, after adjusting the coefficients of the half-equations (if necessary), combine the half-equations into a net equation and the electrode potentials into a value of $E°_{cell}$. Can you see from the way the question is asked that the oxidation half-reaction involves SO_4^{2-} and $S_2O_8^{2-}$? Look for this half-equation in Table 21-1. Then look for the half-equation in which $O_2(g)$ is reduced to water.

Oxidation: $2\ \{2\ SO_4^{2-} \rightarrow S_2O_8^{2-} + 2\ e^-\}$ $-E° = -(+2.01\ V)$
$$= -2.01\ V \quad (21.25)$$

Reduction: $O_2(g) + 4\ H^+ + 4\ e^- \rightarrow 2\ H_2O$ $E° = +1.229\ V \quad (21.26)$

Net: $4\ SO_4^{2-} + O_2(g) + 4\ H^+ \rightarrow 2\ S_2O_8^{2-} + 2\ H_2O$ $E°_{cell} = -0.78\ V$
$$(21.27)$$

This result also indicates that the *reverse* reaction—the oxidation of H_2O by $S_2O_8^{2-}$—is spontaneous.

The large negative value of $E°_{cell}$ indicates that $O_2(g)$ will not oxidize SO_4^{2-} to $S_2O_8^{2-}$ to any significant extent.

SIMILAR EXAMPLES: Exercises 6, 21, 24, 61.

The Behavior of Metals Toward Acids. In our first encounter with oxidation–reduction reactions in Chapter 5 we noted that most metals dissolve in a mineral acid like HCl, but that a few do not (recall Table 5-6). We did not attempt to explain this difference in behavior at that time but now we can. When a metal, M, dissolves in an acid like HCl, the metal is oxidized to the metal ion, e.g., M^{2+}. The reduction involves H^+ being reduced to $H_2(g)$. We can express these ideas as

Oxidation:	$M(s) \longrightarrow M^{2+}(aq) + 2\ e^-$	$-E°$
Reduction:	$2\ H^+(aq) + 2\ e^- \longrightarrow H_2(g)$	$E° = 0.0000\ V$
Net:	$M(s) + 2\ H^+(aq) \longrightarrow M^{2+}(aq) + H_2(g)$	$E°_{cell} = -E°$

According to our new criterion for spontaneous change, any metal having a *negative* electrode potential should dissolve in an acid like HCl. This will make $E°_{cell} > 0$ (positive). Thus, all the metals listed *below* hydrogen in Table 21-1 should dissolve in acids.

In acids like HCl the oxidizing agent is H_3O^+ (or H^+). Certain metals that will not dissolve in HCl (for example, Cu) will dissolve in an acid if there is present an *anion* that is a better oxidizing agent than is H_3O^+. Nitrate ion in an acidic solution is a good oxidizing agent, and copper metal readily dissolves in nitric acid (see Exercise 24).

Relationship Between $E°_{cell}$ and K. We related $\Delta G°$ and $E°_{cell}$ through equation (21.20). In Chapter 20 we related $\Delta G°$ and K (through equation 20.17). The three quantities are thus related in this way.

> To be able to use the factor "0.0592" in equations (21.29) and (21.32), we must use common logarithms (log) and not natural logarithms (ln).

$$\Delta G° = -2.303RT \log K = -n\mathscr{F}E°_{cell}$$

and

$$E°_{cell} = \frac{2.303RT}{n\mathscr{F}} \log K \tag{21.28}$$

In equation (21.28) R has a value of 8.314 J mol^{-1} K^{-1}, and n represents the number of moles of electrons involved in the reaction. If we specify a temperature of 25.00 °C = 298.15 K (the temperature at which electrode potentials are generally determined and tabulated), the value of $E°_{cell}$ (in volts) is given by

> Relationship of $E°_{cell}$ to K.

$$E°_{cell} = \frac{0.0592}{n} \log K \tag{21.29}$$

Are You Wondering:

How to achieve the proper cancellation of units in converting from equation (21.28) to (21.29)?

$E°_{cell}$ must have the unit *volt* (V = J/C), and K and log K must both be dimensionless. For this to be the case the term "0.0592/n" must have the unit *volt*. Recall that the "0.0592/n" represents $2.303RT/n\mathscr{F}$. The key to obtaining the proper cancellation of units is in expressing the units of n as *mol e$^-$/mol*. This unit is moles of electrons *per mole of reaction*. (Recall the description of a "mole of reaction" on page 734. It is the reaction as represented by its balanced equation.)

$$\frac{2.303RT}{n\mathscr{F}} = \frac{2.303 \times 8.314\ \text{J mol}^{-1}\ \text{K}^{-1} \times 298.15\ \text{K}}{n(\text{mol e}^-/\text{mol}) \times 96{,}485\ \text{C/mol e}^-} = \frac{0.0592}{n}\ \frac{\text{J}}{\text{C}}$$

$$= \frac{0.0592}{n}\ \text{V}$$

Example 21-7

Relating K to E_{cell}° for a redox reaction. What is the value of the equilibrium constant, K, for the reaction between copper(II) and tin(II) ions in aqueous solution at 25 °C?

$$Cu^{2+}(aq) + Sn^{2+}(aq) \longrightarrow Sn^{4+}(aq) + Cu(s) \qquad K = ?$$

Solution. First, determine E_{cell}° for the reaction.

Oxidation:	$Sn^{2+}(aq) \longrightarrow Sn^{4+}(aq) + 2\ e^-$	$-E^{\circ} = -(+0.154)$
		$= -0.154$ V
Reduction:	$Cu^{2+}(aq) + 2\ e^- \longrightarrow Cu(s)$	$E^{\circ} = +0.337$ V
Net:	$Cu^{2+}(aq) + Sn^{2+}(aq) \longrightarrow Sn^{4+}(aq) + Cu(s)$	$E_{cell}^{\circ} = +0.183$ V

The number of moles of electrons per mole of cell reaction is 2.

$$E_{cell}^{\circ} = \frac{0.0592}{2} \log K = 0.183 \qquad \log K = \frac{2 \times 0.183}{0.0592} = 6.18$$

$$K = \text{antilog } 6.18 = 1.5 \times 10^6$$

SIMILAR EXAMPLES: Exercises 8, 28.

21-4 E_{cell} as a Function of Concentrations

We can establish the *standard* emf of the voltaic cell of Figure 21-5 by combining *standard* electrode potentials.

$$Zn(s) + Cu^{2+}(1\text{ M}) \longrightarrow Zn^{2+}(1\text{ M}) + Cu(s) \qquad E_{cell}^{\circ} = +1.100\text{ V}$$

But what if we use *nonstandard* conditions in a voltaic cell, such as

$$Zn(s) + Cu^{2+}(2\text{ M}) \longrightarrow Zn^{2+}(0.10\text{ M}) + Cu(s) \qquad E_{cell} = ?$$

From Le Châtelier's principle we might predict that *increasing* the concentration of a reactant (Cu^{2+}) while simultaneously *decreasing* the concentration of a product (Zn^{2+}) should favor the forward reaction. The reaction should become even more spontaneous and $E_{cell} > 1.100$ V. This is indeed what happens, and E_{cell} is found to vary linearly with $\log [Zn^{2+}]/[Cu^{2+}]$, as illustrated in Figure 21-7 and equation (21.30).

$$E_{cell} = 1.10 - 0.03 \log \frac{[Zn^{2+}]}{[Cu^{2+}]} \qquad (21.30)$$

Relationships of this type were first studied by Walter Nernst (1864–1941). Equation (21.30) is a specific example of a more general equation now called the **Nernst equation.** This equation can be established by experiment, but it also can be derived from thermodynamics. For the general reaction

$$a\,A + b\,B + \cdots \longrightarrow g\,G + h\,H + \cdots \qquad (21.31)$$

we can write

$$\Delta G = \Delta G^{\circ} + 2.303RT \log Q \qquad (20.16)$$

For ΔG and ΔG° we can substitute $-n\mathscr{F}E_{cell}$ and $-n\mathscr{F}E_{cell}^{\circ}$, respectively.

$$-n\mathscr{F}E_{cell} = -n\mathscr{F}E_{cell}^{\circ} + 2.303RT \log Q$$

and, after dividing through by $-n\mathscr{F}$, we obtain

$$E_{cell} = E_{cell}^{\circ} - \frac{2.303RT}{n\mathscr{F}} \log Q$$

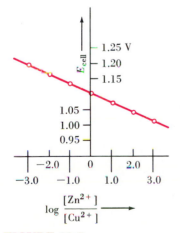

FIGURE 21-7
Variation of E_{cell} with ion concentrations for the cell reaction

$$Zn(s) + Cu^{2+}(aq) \longrightarrow Zn^{2+}(aq) + Cu(s)$$

Q is the reaction quotient, and as we have seen, the value of the term $2.303RT/n\mathscr{F}$ at 298.15 K is $0.0592/n$. Again n is the number of moles of electrons transferred in the reaction. In complete form, the Nernst equation is

The Nernst equation: E_{cell} as a function of concentrations.

$$E_{cell} = E_{cell}^{\circ} - \frac{0.0592}{n} \log \frac{(a_G)^g (a_H)^h \cdots}{(a_A)^a (a_B)^b \cdots} \tag{21.32}$$

In equation (21.32) we can make the usual substitutions for activities: $a = 1$ for pure solids and liquids; a = partial pressures (atm) for gases; and a = molarity concentrations for solution components.

Example 21-8

Applying the Nernst equation to a voltaic cell at nonstandard conditions. In the voltaic cell pictured in Figure 21-8 the solution in the anode half-cell is at nonstandard conditions. That is, the solution concentrations are not 1 M. What is the value of E_{cell}?

$$Pt|Fe^{2+}(0.10\ M),\ Fe^{3+}(0.20\ M)\|Ag^+(1.0\ M)|Ag(s) \qquad E_{cell} = ?$$

Solution. First use data from Table 21-1 to determine E_{cell}°. Then apply the Nernst equation.
Determining E_{cell}°:

Oxidation: $Fe^{2+}(aq) \longrightarrow Fe^{3+}(aq) + e^-$ $-E^{\circ} = -(+0.771)$
                                                                                                                      $= -0.771$ V

Reduction: $\underline{Ag^+(aq) + e^- \longrightarrow Ag(s)}$ $E^{\circ} = +0.800$ V

Net: $Fe^{2+}(aq) + Ag^+(aq) \longrightarrow Fe^{3+}(aq) + Ag(s)$ $E_{cell}^{\circ} = +0.029$ V
                                                                                                                     (21.33)

Nernst equation:

$$E_{cell} = E_{cell}^{\circ} - \frac{0.0592}{n} \log \frac{[Fe^{3+}]}{[Fe^{2+}][Ag^+]}$$

Substitute: $E_{cell}^{\circ} = +0.029$ V; $n = 1$;
$[Fe^{2+}] = 0.10$ M; $[Fe^{3+}] = 0.20$ M; $[Ag^+] = 1.0$ M.

$$E_{cell} = 0.029 - \frac{0.0592}{1} \log \frac{0.20}{0.10 \times 1.0}$$

Walther Nernst (1864–1941). Nernst was only 25 years old when he formulated his equation relating concentrations and cell voltages. He is also credited with proposing the solubility product concept in the same year. In 1906 he announced his Heat Theorem, which we now know as the third law of thermodynamics. [Courtesy German Information Center]

FIGURE 21-8
A voltaic cell with nonstandard conditions—Example 21-8 illustrated.

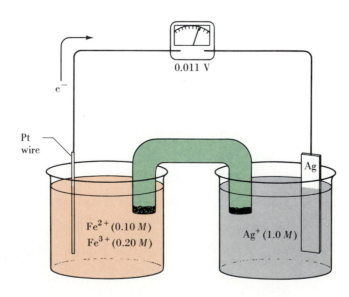

$$= 0.029 - (0.0592 \log 2.0)$$

$$E_{cell} = 0.029 - (0.0592 \times 0.301)$$

$$= 0.029 - 0.018 = 0.011 \text{ V}$$

SIMILAR EXAMPLES: Exercises 9, 10, 32, 66.

Example 21-9

Using the Nernst equation to determine conditions for spontaneous change.
What is the *minimum* $[Ag^+]$ at which the cell of Figure 21-8 is able to function
as a voltaic cell with Fe^{2+}/Fe^{3+} as the anode? That is, what is the *minimum*
$[Ag^+]$ possible before the electron flow reverses direction? (Continue to use
$[Fe^{2+}] = 0.10$ M and $[Fe^{3+}] = 0.20$ M.)

Solution. In Example 21-8 we established the value of E°_{cell} for this reaction.

$$Fe^{2+}(aq) + Ag^+(aq) \longrightarrow Fe^{3+}(aq) + Ag(s) \qquad E^\circ_{cell} = +0.029 \text{ V} \qquad (21.33)$$

To assess the effect of
changes in ion concentrations
on the value of E_{cell}, use

• **Le Châtelier's principle for**
qualitative predictions and
• **the Nernst equation for**
quantitative calculations.

Since the value of E°_{cell} is not very large, we should be able to find ways to
reduce E_{cell} to a value of 0. According to Le Châtelier's principle we can do this
by increasing $[Fe^{3+}]$ or by decreasing $[Fe^{2+}]$ and/or $[Ag^+]$. In this problem we
are asked to decrease $[Ag^+]$. When $E_{cell} = 0$ neither the forward nor the reverse reaction is favored and the reaction is at equilibrium. (If we reduce $[Ag^+]$
too much, the *reverse* reaction will be favored and the current in the voltaic cell
will reverse its direction.)

$$E_{cell} = 0 = 0.029 - \frac{0.0592}{1} \log \frac{0.20}{0.10 \times [Ag^+]}$$

$$= 0.029 - 0.0592(\log 2.0 - \log [Ag^+])$$

$$0.029 = 0.0592(0.301 - \log [Ag^+]) \qquad 0.0592 \log [Ag^+] = 0.018 - 0.029$$

$$\log [Ag^+] = \frac{-0.011}{0.0592} = -0.19 \qquad [Ag^+] = 0.65 \text{ M}$$

SIMILAR EXAMPLES: Exercises 38, 39.

Measurement of pH. The voltaic cell in Figure 21-9 consists of two hydrogen
electrodes. One is a *standard hydrogen electrode* (S.H.E.) and the other is a hydrogen electrode immersed in a solution of unknown $[H^+]$. The cell diagram is

$$\text{Pt, } H_2(g, 1 \text{ atm})|H^+(x \text{ M})||H^+(1 \text{ M})|H_2(g, 1 \text{ atm}), \text{ Pt} \qquad (21.34)$$

FIGURE 21-9
A concentration cell, consisting of two hydrogen elec-
trodes, for measuring pH.

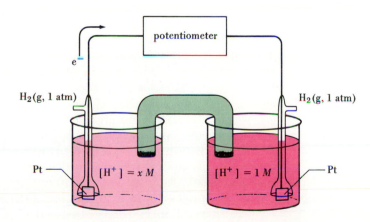

$H_2(g, 1 \text{ atm})$ potentiometer $H_2(g, 1 \text{ atm})$

e^-

Pt — $[H^+] = x M$ $[H^+] = 1 M$ — Pt

The reaction occurring in this cell is

Oxidation: $H_2(g, 1\text{ atm}) \longrightarrow 2 H^+(x, M) + 2e^-$
Reduction: $\underline{2 H^+(1\text{ M}) + 2e^- \longrightarrow H_2(g, 1\text{ atm})}$
Net: $2 H^+(1\text{ M}) \longrightarrow 2 H^+(x\text{ M})$ (21.35)

Any voltaic cell in which the net cell reaction involves only a change in the concentration of some species (here $[H^+]$) can be called a concentration cell. A **concentration cell** consists of two half-cells with *identical electrodes* but differing ion concentrations. Because the electrodes are identical, the half-cell potentials will be numerically equal and opposite in sign. This makes $E^\circ_{cell} = 0$. However, because the ion concentrations do differ, there is a potential difference between the two half-cells. The Nernst equation for reaction (21.35) takes the form

$$E_{cell} = E^\circ_{cell} - \frac{0.0592}{2} \log \frac{(x)^2}{(1)^2}$$

which simplifies to

$$E_{cell} = 0 - \frac{0.0592}{2} \times 2 \log \frac{x}{1} = -0.0592 \log x$$

Since x is $[H^+]$ in the unknown solution, and $-\log x = -\log [H^+] = pH$, our final result is

$$E_{cell} = 0.0592\ pH \qquad\qquad (21.36)$$

If an unknown solution has a pH of 3.50, for example, the measured cell voltage in Figure 21-9 will be $E_{cell} = 0.0592 \times 3.50 = 0.207$ V.

The Glass Electrode. Constructing and using a hydrogen electrode is difficult. The Pt metal surface must be specially prepared and maintained, gas pressures must be controlled, and the electrode cannot be used in the presence of strong oxidizing or reducing agents. A better approach to pH measurement than that of Figure 21-9 replaces the S.H.E. with some other reference electrode of precisely known E° value. The second hydrogen electrode is replaced by a **glass electrode.** The basic feature of this electrode is a thin glass membrane of a carefully regulated chemical composition. When the electrode is dipped into a solution, depending on the type and concentration of ions present, a potential is established on the outer surface of the membrane. This potential is registered through a reference electrode immersed in a solution inside the membrane. The glass electrode pictured in Figure 21-10, immersed in an unknown solution, serves as a half-cell. When combined with another reference half-cell, the assembly functions as a voltaic cell.

The most commonly used glass electrodes are those whose potentials are determined by $[H^+]$ in solution. These are the glass electrodes used in common laboratory pH meters. Other glass electrodes have been developed that can function with Na^+, K^+, or other cations. These so-called **ion-selective electrodes** are especially valuable because ordinary half-cells cannot use very active metals as electrodes. These metals react with water to liberate $H_2(g)$.

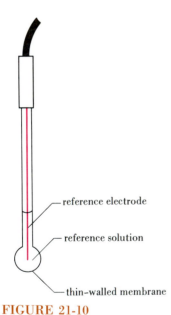

reference electrode

reference solution

thin-walled membrane

FIGURE 21-10
A glass electrode for pH measurements.

Potentiometric Titrations. In Chapter 18 we *calculated* the variation of pH during an acid–base titration. We displayed the results graphically, as a *titration curve*. We saw that pH changes very sharply with titrant volume at the equivalence point. From the titration curve we learned how to select an appropriate acid–base indicator for the titration. But what if we need to titrate an acid or base in a solution that is itself colored or very cloudy? We would not be able to see an indicator color change in such a mixture.

The voltaic cell in Figure 21-9 gives us a different approach to titration. According to equation (21.36) the measured value of E_{cell} is directly proportional to the pH.

Suppose we place an unknown acid in the anode half-cell (on the left) in Figure 21-9 and then titrate it with a strong base. We can measure E_{cell} during the titration and plot E_{cell} (or pH) against the volume of titrant. The graph will be a titration curve, and from the curve we can establish the volume of strong base required to reach the equivalence point.

Since the instrument used to measure a cell voltage traditionally has been a potentiometer, a titration based on the measurement of E_{cell} is called a **potentiometric titration.** A potentiometric titration can be used in any situation where (a) the concentration of one of the reactants changes extremely rapidly at the equivalence point and (b) an electrode can be found whose potential depends on the concentration of that reactant. The potentiometric titration of Ag^+ with I^- (forming a precipitate of AgI) is described in Exercise 83. Modern developments in potentiometric titrations include instrumentation by which titrations can be performed by machine and titration curves plotted and analyzed automatically.

Measurement of K_{sp}. Many of the solubility product constants that are listed in tables of data have been obtained by measuring cell voltages. Consider this concentration cell.

$$Ag(s)|Ag^+(\text{satd AgI(aq)})\|Ag^+(0.100)\ M|Ag \qquad E_{cell} = +0.417\ V \qquad (21.37)$$

In the anode half-cell, a silver electrode is placed in a saturated aqueous solution of silver iodide. In the cathode half-cell a second silver electrode is placed in a solution with $[Ag^+] = 0.100$ M. The two half-cells are connected by a salt bridge or porous plug, and the cell voltage is measured. It is 0.417 V. The cell reaction occurring in this *concentration* cell is

Oxidation:	$Ag(s) \longrightarrow Ag^+(\text{satd AgI(aq)}) + e^-$	$-E° = -0.800\ V$
Reduction:	$Ag^+(0.100\ M) + e^- \longrightarrow Ag(s)$	$E° = +0.800\ V$
Net:	$Ag^+(0.100\ M) \longrightarrow Ag^+(\text{satd, AgI(aq)})$	$E°_{cell} = 0.000\ V$
	and	$E_{cell} = +0.417\ V$

$$(21.38)$$

We complete the calculation of K_{sp} of silver iodide in Example 21-10.

Example 21-10

Using the Nernst equation to determine K_{sp} of a slightly soluble solute. With the data listed in expression (21.38), calculate K_{sp} for AgI.

$$AgI(s) \rightleftharpoons Ag^+(aq) + I^-(aq) \qquad K_{sp} = ?$$

Solution. Let us represent the unknown silver ion concentration in saturated AgI(aq) simply as $[Ag^+]$. Then we can apply the Nernst equation to the cell reaction (21.38). In doing this we should note that $E°_{cell} = 0$. This is a *concentration* cell and has the same type of electrode in each half-cell (a silver electrode).

$$E_{cell} = E°_{cell} - \frac{0.0592}{1} \log \frac{[Ag^+]}{0.100}$$

$$0.417 = 0 - 0.0592\ (\log [Ag^+] - \log 0.100)$$

$$\log [Ag^+] - \log 0.100 = \frac{-0.417}{0.0592}$$

$$\log [Ag^+] - (-1.00) = -7.04$$

$$\log [Ag^+] = -8.04$$

$$[Ag^+] = \text{antilog}\ (-8.04) = 10^{-8.04} = 9.1 \times 10^{-9}\ M$$

Note that in saturated $AgI(aq)$ $[Ag^+] = [I^-]$ and

$$K_{sp} = [Ag^+][I^-] = (9.1 \times 10^{-9})(9.1 \times 10^{-9}) = 8.3 \times 10^{-17}$$

SIMILAR EXAMPLES: Exercises 37, 39.

21-5 Batteries: Producing Electricity Through Chemical Reactions

To underscore the importance of batteries in the modern world, the annual production of batteries in western nations has been estimated to be as high as 10 batteries per person per year.

The terms *battery* and *cell* tend to be used interchangeably, but sometimes the following distinction is made: A cell is the simple voltaic cell consisting of two electrodes and the appropriate electrolyte(s). Batteries are two or more cells joined in series fashion, + to −, to increase the voltage obtained.

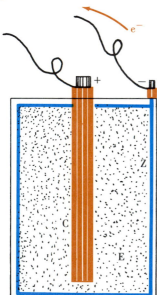

FIGURE 21-11

The Leclanché (dry) cell.

C: carbon rod serving as cathode—reduction occurs.
Z: zinc container serving as anode—oxidation occurs.
E: electrolyte—a moist paste of MnO_2, $ZnCl_2$, NH_4Cl, and carbon black.

A reaction for which $\Delta G < 0$ is capable of doing useful work. The chief method we use to extract this useful work is as electricity, from a voltaic cell. We can think of the reactants in a voltaic cell as containing stored chemical energy that can be released as electricity. In common usage we call a device that stores chemical energy for later release as electricity a **battery.** In this section we consider three types of batteries.

- **Primary** batteries (or primary cells). The cell reaction is not reversible. When the reactants have for the most part been converted to products, no more electricity is produced and the battery is "dead."
- **Secondary** batteries (or secondary cells). The cell reaction can be reversed by passing electricity through the battery ("charging"). This means that a battery may be used through several hundred or more cycles of discharging followed by charging.
- **Flow** batteries and **fuel cells.** Materials (reactants, products, electrolytes) pass through the battery, which is simply an electrochemical converter that converts chemical to electrical energy.

Leclanché (Dry) Cell. Perhaps the most familiar of all batteries is the flashlight battery outlined in Figure 21-11. In this battery oxidation occurs at a zinc anode and reduction at an inert carbon (graphite) cathode. The electrolyte is a moist paste of MnO_2, $ZnCl_2$, NH_4Cl, and carbon black. The maximum cell voltage is 1.55 V. The battery is called a "dry" cell because there is no free liquid present. The anode (oxidation) half-reaction is simple.

Oxidation: $Zn(s) \longrightarrow Zn^{2+}(aq) + 2\ e^-$ (21.39)

The reduction is more complex. Essentially, it involves the reduction of MnO_2 to a series of compounds having Mn in a +3 oxidation state, for example, Mn_2O_3.

Reduction: $2\ MnO_2(s) + H_2O + 2\ e^- \longrightarrow Mn_2O_3(s) + 2\ OH^-(aq)$ (21.40)

An acid–base reaction occurs between OH^- and NH_4^+ derived from NH_4Cl.

$$NH_4^+(aq) + OH^-(aq) \longrightarrow NH_3(g) + H_2O(l) \tag{21.41}$$

Since it would disrupt the current, a buildup of $NH_3(g)$ cannot be permitted to occur around the cathode. This is prevented by a reaction between Zn^{2+} and $NH_3(g)$ to form the complex ion $[Zn(NH_3)_2]^{2+}$, which crystallizes as the chloride salt.

$$Zn^{2+}(aq) + 2\ NH_3(g) + 2\ Cl^-(aq) \longrightarrow [Zn(NH_3)_2]Cl_2(s) \tag{21.42}$$

The Leclanché cell is a *primary* cell. The several reactions involved cannot be reversed by passing electricity back through the cell. The Leclanché cell is cheap to make, but it has two significant disadvantages. When current is drawn rapidly from the cell, products build up on the electrodes [e.g., $NH_3(g)$] and this causes the voltage to drop. Also, since the electrolyte medium is *acidic* zinc metal slowly dissolves.

A superior form of the Leclanché cell is the **alkaline** battery, which uses NaOH

FIGURE 21-12
A lead–acid (storage) cell.

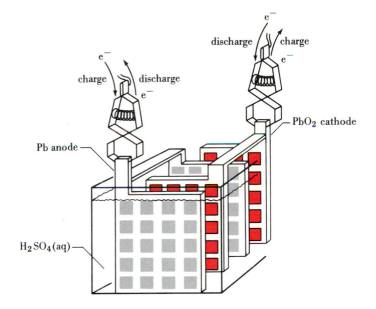

or KOH in place of NH_4Cl as the electrolyte. In this battery the reduction half-reaction is essentially that of (21.40), but the oxidation involves the formation of $Zn(OH)_2(s)$

Oxidation: $Zn(s) + 2\ OH^-(aq) \longrightarrow Zn(OH)_2(s) + 2\ e^-$ (21.43)

The advantages of the alkaline battery are that zinc does not dissolve as readily in a *basic* medium, and the battery does a better job of maintaining its voltage as current is drawn through it.

Lead–Acid (Storage) Battery. The most common *secondary* battery is the typical automobile storage battery. The electrodes are plates of a lead–antimony alloy. The anodes are impregnated with spongy lead metal and the cathodes with red-brown lead dioxide. The electrolyte is dilute sulfuric acid. When the cell pictured in Figure 21-12 is allowed to discharge, the following reactions occur. [Think of both of these half-reactions as producing Pb^{2+}, which combines with SO_4^{2-} from solution to deposit $PbSO_4(s)$.]

Oxidation: $Pb(s) + SO_4^{2-}(aq) \longrightarrow PbSO_4(s) + 2\ e^-$
Reduction: $PbO_2(s) + 4\ H^+(aq) + SO_4^{2-} + 2\ e^- \longrightarrow PbSO_4(s) + 2\ H_2O$
Net: $Pb(s) + PbO_2(s) + 4\ H^+ + 2\ SO_4^{2-} \longrightarrow 2\ PbSO_4(s) + 2H_2O$
$E_{cell} = +2.05\ V$ (21.44)

When the plates become partially coated with $PbSO_4(s)$ and the electrolyte has been diluted with water produced in the cell reaction, the cell is in a discharged condition. To recharge it, we can force electrons to flow in the opposite direction by connecting the battery to an external electric energy source. The reverse of reaction (21.44) occurs when the battery is recharged.

To prevent short circuiting, alternating anode and cathode plates are separated by sheets of insulating material. A group of anodes is connected electrically, as is a group of cathodes. This "parallel" connection increases the electrode area in contact with the electrolyte solution and increases the current-delivering capacity of the cell. Cells are then joined in a "series" fashion, + to −, to produce a battery. In a 6-V battery there are *three* cells; in a 12-V battery there are *six* cells.

The Silver–Zinc Cell—A Button Battery. The cell diagram of a silver–zinc cell is

$Zn, ZnO(s)|KOH(sat'd)|Ag_2O(s), Ag$

FIGURE 21-13
A silver–zinc button (miniature) cell.

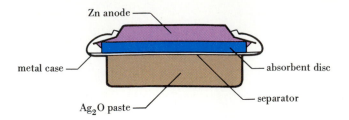

The half-reactions on discharging are

Anode: $Zn(s) + 2\ OH^-(aq) \longrightarrow ZnO(s) + H_2O + 2\ e^-$
Cathode: $Ag_2O(s) + H_2O + 2\ e^- \longrightarrow 2\ Ag(s) + 2\ OH^-(aq)$

Net: $Zn(s) + Ag_2O(s) \longrightarrow ZnO(s) + 2\ Ag(s)$ (21.45)

Because no solution species is involved in the net reaction, the quantity of electrolyte can be kept very small. The electrodes can be maintained very close together
and the cell kept nearly dry. The storage capacity of a silver–zinc cell is about six
times as great as a lead–acid cell of the same size. These characteristics make
batteries like the silver–zinc cell useful in "button" batteries. These miniature
batteries have many applications, such as in watches, electronic calculators, hearing
aids, and cameras (see Figure 21-13).

Fuel Cells. A voltaic cell converts chemical to electric energy with a high efficiency, perhaps as high as 90%. By contrast, the typical efficiency of the combus-
tion-steam-electric generator power plant is only about 30–40%. For 150 years or
more people have been intrigued by the possibility of converting the chemical energy of fuels directly to electricity. This is the function of a fuel cell. The essential
process in a fuel cell is *fuel + oxygen* $\longrightarrow$ *oxidation products*. Applied to natural
gas (methane), the half-reactions and cell reaction would be

Oxidation: $CH_4(g) + 2\ H_2O \longrightarrow CO_2(g) + 8\ H^+ + 8\ e^-$
Reduction: $2\ \{O_2(g) + 4\ H^+ + 4\ e^- \longrightarrow 2\ H_2O\}$

Net: $CH_4(g) + 2\ O_2(g) \longrightarrow CO_2(g) + 2\ H_2O(l)$
                            $\Delta H^\circ = -890\ kJ \quad \Delta G^\circ = -818\ kJ$ (21.46)

The potential usefulness of a fuel cell reaction is often stated in terms of the *efficiency value*, $\epsilon = \Delta G^\circ/\Delta H^\circ$. For the methane fuel cell this would be $\epsilon =$
$-818\ kJ/-890\ kJ = 0.92$.

The methane fuel cell has not been developed commercially because of the difficulty of finding electrodes at which the half-reactions can be carried out at a high
current density while retaining the theoretical voltage of the cell.

One of the simplest and most successful fuel cells, shown schematically in Figure
21-14, involves the reaction of $H_2(g)$ and $O_2(g)$ to form water. In an alkaline medium (e.g., 25% KOH) these reactions occur.

Oxidation: $2\ H_2(g) + 4\ OH^-(aq) \longrightarrow 4\ H_2O + 4\ e^-$
Reduction: $O_2(g) + 2\ H_2O + 4\ e^- \longrightarrow 4\ OH^-(aq)$

Net: $2\ H_2(g) + O_2(g) \longrightarrow 2\ H_2O(l)$ (21.47)

We should refer to a fuel cell as an *energy converter* rather than a battery. As long
as fuel and $O_2(g)$ are available, the fuel cell will produce electricity. It does not have
the limited capacity of a primary battery, but neither does it have the storage capacity of a secondary battery.

Fuel cells based on reaction (21.47) have had their most notable successes as
energy sources in space vehicles. (Water produced in this reaction is also a valuable
product of the fuel cell.) Other potential uses include the conversion of coal and
petroleum to CO and H_2 and oxidizing these gases in a fuel cell. Some small
commercial electrical power installations are already in use. Interestingly, though,

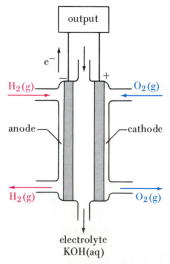

FIGURE 21-14
Schematic representation of a
hydrogen–oxygen fuel cell.

Are You Wondering:

Why the efficiency value, ϵ, of the methane fuel cell (21.46) is not 1.00?

We have implied that the only limitation imposed by the second law of thermodynamics is in converting *heat* energy to work. Since in reaction (21.46) the decrease in chemical energy ($\Delta H°$) is -890 kJ, why cannot the maximum useful work also be -890 kJ?

Recall the relationship between $\Delta G°$ and $\Delta H°$.

$$\Delta G° = \Delta H° - T\Delta S°$$

In the methane combustion reaction the $T\Delta S°$ term is *negative* (-72 kJ). This is because $\Delta S°$ of the reaction is *negative* (three moles of gaseous reactants produce one mole of gaseous product and two moles of liquid). Some of the chemical energy that might have appeared as electric energy is required to produce the increased order implied by the decrease in entropy. The maximum useful work is $\Delta H°$, *less* the $T\Delta S$ term, namely, $\Delta G°$ (-818 kJ).

the hydrogen–oxygen fuel cell has been known since 1842. The cell was never exploited in the last century because of the difficulty of developing electrodes for carrying out the energy conversion efficiently.

21-6 Corrosion: Unwanted Voltaic Cells

We have seen how useful oxidation–reduction reactions can be when carried out in voltaic cells (batteries). They are important sources of electricity. Similar reactions underlie corrosion processes, and here they are unwanted. First, let us consider the electrochemical basis of corrosion, and then we will see how electrochemical principles can also be applied to controlling corrosion.

Figure 21-15 demonstrates the basic processes in the corrosion of an iron nail. The nail is embedded in a gel of agar in water. Incorporated in the gel are the acid–base indicator phenolphthalein and the substance $K_3[Fe(CN)_6]$ (potassium ferricyanide). Here is what we see within hours of starting the experiment (Figure

FIGURE 21-15

Demonstration of corrosion and methods of corrosion protection.

Corrosion (oxidation) of the nail is marked by the blue precipitate. Corrosion occurs at strained regions: the head and tip (a) and a bend (b) in the nail. Contact with zinc (c) protects the nail from corrosion, but contact with copper (d) does not. [Carey B. Van Loon]

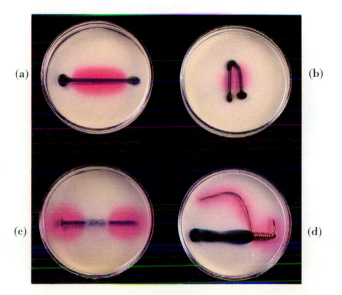

FIGURE 21-16

Protection of iron against electrolytic corrosion.

In the anodic reaction (oxidation), the metal that is more easily oxidized loses electrons to produce metal ions. In case (a) this is zinc; in case (b), iron. In the cathodic reaction (reduction), oxygen gas, which is dissolved in a thin film of adsorbed water, is reduced to hydroxide ion. Rusting of iron does not occur in (a), but it does in (b).

$$Fe^{2+} + 2\ OH^- \rightarrow Fe(OH)_2(s)$$

$$4\ Fe(OH)_2(s) + O_2 + 2\ H_2O \rightarrow 4\ Fe(OH)_3(s)$$

$$2\ Fe(OH)_3(s) \rightarrow Fe_2O_3 \cdot H_2O + 2\ H_2O$$
rust

A dented automobile fender begins to rust, partly because the painted finish of the metal is removed, but also because the strained metal is more active (more easily oxidized) than unstrained metal. Mechanical energy stored in the dented fender as a result of the collision tends to reappear as electrical energy.

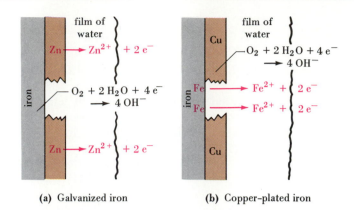

(a) Galvanized iron   (b) Copper-plated iron

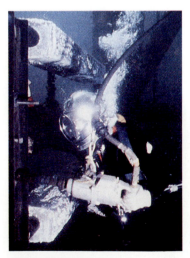

Diver installing two sacrificial aluminum anodes on an offshore oil platform. The anodes are the large metal blocks at the top and bottom of the photograph. [Courtesy Delta Corrosion Offshore, Belle Chasse, Louisiana]

21-15a): At the head and tip of the nail a deep blue precipitate forms. Along the body of the nail the agar gel acquires a pink color. The blue precipitate, known as Turnbull's blue, establishes the presence of iron(II). The pink color, of course, is that of phenolphthalein in *basic* solution. From these observations we can write two simple half-equations.

Oxidation: $2\ Fe(s) \longrightarrow 2\ Fe^{2+}(aq) + 4\ e^-$
Reduction: $O_2 + 2\ H_2O + 4\ e^- \longrightarrow 4\ OH^-(aq)$

In the corroding nail, oxidation occurs at the head and tip of the nail. Electrons given up in the oxidation pass along the body of the nail where they are used to reduce dissolved O_2. The reduction product, OH^-, is detected by the phenolphthalein. With a bent nail (Figure 21-15b), oxidation occurs at *three* points, the head and tip of the nail and the bend. The nail is preferentially oxidized at these points because the strained metal is more active (more anodic) than the unstrained metal.

Some metals, such as aluminum, form corrosion products that adhere tightly to the underlying metal and protect it from further corrosion. Iron oxide (rust), on the other hand, flakes off an object, constantly exposing fresh surface. It is this difference in corrosion behavior that explains why cans made of iron deteriorate rapidly in the environment, whereas aluminum cans have an almost unlimited lifetime. The simplest method of protecting a metal from corrosion is to cover it with paint or some other impervious protective coating. An iron surface is protected in this way only as long as the coating does not chip or peel off.

Another method of protecting an iron surface is to plate it with a thin layer of a second metal. Iron can be plated with copper by electroplating or with tin by dipping the iron into the molten metal. In either case the underlying metal is protected only as long as the coating remains intact. If the coating is cracked, as when a "tin" can is dented, the underlying iron is exposed and corrodes. Iron, being more active than copper and tin, undergoes oxidation; the reduction half-reaction occurs on the plating (see Figures 21-15d and 21-16b).

When iron is coated with zinc (galvanized iron) the situation is different. Zinc is more active than iron. If a break occurs in the zinc plating, the iron is still protected. Zinc is oxidized instead of the iron, and corrosion products protect the zinc from further corrosion (see Figures 21-15c and 21-16a).

Still another method can be used to protect large iron and steel objects—ships, storage tanks, pipelines, plumbing systems. This involves connecting to the object, either directly or through a wire, a chunk of magnesium, aluminum, zinc, or some other active metal. Oxidation occurs at the active metal and it slowly dissolves. The iron surface acquires electrons from the oxidation of the active metal; the iron acts as a cathode and supports a *reduction* half-reaction. As long as some of the active metal remains, the iron is protected. This type of protection is called **cathodic protection,** and the active metal is called, appropriately, a **sacrificial anode.** About 12 million pounds of magnesium is used annually in the United States in sacrificial anodes.

21-7 Electrolysis: Nonspontaneous Chemical Change

Let us return to the electrochemical cell consisting of the half-cells Zn/Zn^{2+} and Cu^{2+}/Cu. This cell was pictured in Figure 21-5. When the cell functions *spontaneously*, electrons flow from the zinc to the copper and the net chemical change in the *voltaic* cell is

Voltaic cell: $Zn(s) + Cu^{2+}(aq) \longrightarrow Zn^{2+}(aq) + Cu(s)$ $E^{\circ}_{cell} = +1.100 \text{ V}$

Now suppose we take the same cell and connect it to an electric energy source of voltage greater than 1.100 V. This could be an electric generator or a lead–acid storage battery. And suppose we make this connection in such a way that electrons are forced onto the zinc electrode and removed from the copper electrode. As shown below, the net reaction in this case is the *reverse* of the voltaic cell reaction and E°_{cell} is *negative*.

Oxidation: $Cu(s) \longrightarrow Cu^{2+}(aq) + 2\ e^{-}$ $-E^{\circ} = -0.337 \text{ V}$
Reduction: $Zn^{2+}(aq) + 2\ e^{-} \longrightarrow Zn(s)$ $E^{\circ} = -0.763 \text{ V}$
Electrolysis: $Cu(s) + Zn^{2+}(aq) \longrightarrow Cu^{2+}(aq) + Zn(s)$ $E^{\circ}_{cell} = -1.100 \text{ V}$

When the cell reaction in a voltaic cell is reversed by reversing the direction of the electron flow, the *voltaic* cell is changed into an *electrolysis* cell, as shown in Figure 21-17.

Predicting Electrolysis Reactions. We have seen that in Figure 21-17 electrolysis occurs, with zinc as the cathode and copper as the anode, if we apply a voltage exceeding 1.100 V. We can make similar calculations for other electrolyses. How-

---

Are You Wondering:

How to apply the terms cathode and anode to an electrolysis cell?

Regardless whether a cell is a voltaic or an electrolysis cell,

- The *anode* is the electrode at which *oxidation* occurs.
- The *cathode* is the electrode at which *reduction* occurs.

(An easy way to remember this convention is that in the anode/oxidation pair each word begins with a *vowel*, and in the cathode/reduction pair each word begins with a *consonant*.) Also, in either type cell anions migrate to the anode and cations to the cathode.

Something that changes when we switch from a voltaic to an electrolysis cell is the signs of the electrodes (terminals). If we think of the *negative* terminal as being the electrode at which there is a greater concentration or density of electrons, this will be the anode in a voltaic cell and the cathode in an electrolysis cell. In summary,

	Voltaic cell
Anode	oxidation, negative (−) terminal
Cathode	reduction, positive (+) terminal

	Electrolysis cell
Anode	oxidation, positive (+) terminal
Cathode	reduction, negative (−) terminal

FIGURE 21-17

The reaction

$$Cu(s) + Zn^{2+}(aq) \rightarrow Cu^{2+}(aq) + Zn(s)$$

occurring in an electrolytic cell.

The reaction occurring here is the *reverse* of that occurring in the voltaic cell of Figure 21-5. The direction of electron flow is reversed. The zinc electrode becomes the cathode and the copper electrode, the anode.

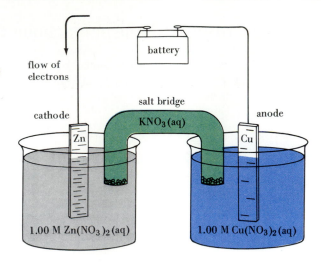

ever, what actually happens may not always correspond to these calculations.

In many cases the voltage necessary to bring about a particular electrode reaction may exceed what we calculate theoretically. Interactions called **polarization** may occur between the electrode surface and the species involved in an electrode reaction. This may require that an overpotential be applied in order for the electrode reaction to occur. An **overpotential** is a voltage in excess of that calculated to produce electrolysis. Overpotentials are particularly common when gases are involved. For example, the overpotential for the discharge of $H_2(g)$ at a mercury cathode is approximately 1.5 V, whereas on a platinum cathode it is practically zero.

A second complicating factor is that if the material being electrolyzed contains several species capable of undergoing oxidation and reduction, competing electrode reactions may occur. In the electrolysis of *molten* sodium chloride only one oxidation and one reduction are possible.

Oxidation: $2\ Cl^- \longrightarrow Cl_2(g) + 2\ e^-$
Reduction: $2\ Na^+ + 2\ e^- \longrightarrow 2\ Na(l)$

In the electrolysis of *aqueous* sodium chloride *two* oxidation and *two* reduction half-reactions must be considered.

Oxidation: $2\ Cl^- \longrightarrow Cl_2(g) + 2\ e^-$ $-E° = -1.36$ V   (21.48)
           $2\ H_2O \longrightarrow O_2(g) + 4\ H^+ + 4\ e^-$ $-E° = -1.23$ V   (21.49)

Reduction: $2\ Na^+ + 2\ e^- \longrightarrow 2\ Na(s)$ $E° = -2.71$ V   (21.50)
           $2\ H_2O + 2\ e^- \longrightarrow H_2(g) + 2\ OH^-$ $E° = -0.83$ V   (21.51)

The electrode potentials for half-reactions (21.48) and (21.49) are similar in magnitude. Exact values depend on $[Cl^-]$ in the one case and $[H^+]$ in the other. If the NaCl solution is concentrated, the oxidation half-reaction (21.48) is favored; if it is quite dilute, (21.49) is favored.

As far as the reduction half-reaction is concerned, the reduction of water occurs much more readily than that of Na^+. Generally, only the half-reaction (21.51) occurs. The principal exception is if liquid mercury is used as a cathode. In this case, because of the high overpotential of hydrogen on mercury and the solubility of sodium metal in liquid mercury, half-reaction (21.50) actually is observed.

Example 21-11

Predicting electrode half-reactions and net electrolysis reactions. Refer to Figure 21-18. Predict the electrode reactions and the net electrolysis reaction when the anode is made of (a) copper and (b) platinum.

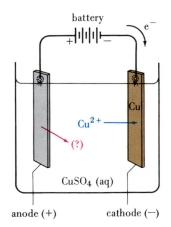

FIGURE 21-18

Predicting electrode reactions in electrolysis—Example 21-11 illustrated.

Electrons are forced onto the copper cathode by the external source (battery). Cu^{2+} ions are attracted to the cathode and are reduced to Cu(s). The oxidation half-reaction at the anode depends on the metal used for the anode.

Solution. In both cases the reduction of $Cu^{2+}(aq)$ to $Cu(s)$ seems quite feasible as a reduction process.

Reduction: $Cu^{2+}(aq) + 2\ e^- \longrightarrow Cu(s)$ $E° = +0.337$ V

(a) At the anode $Cu(s)$ can be oxidized to $Cu^{2+}(aq)$, as represented by

Oxidation: $Cu(s) \longrightarrow Cu^{2+}(aq) + 2\ e^-$ $-E° = -0.337$ V

If we add the oxidation and reduction half-equations we see that $Cu^{2+}(aq)$ cancels out. The net electrolysis reaction is simply

$$Cu(s)[anode] \longrightarrow Cu(s)[cathode] \qquad E°_{cell} = 0 \qquad (21.52)$$

Only a very slight voltage is required for this electrolysis. For every Cu atom that enters the solution at the anode, a Cu^{2+} ion deposits as a Cu atom at the cathode. All that happens here is that copper is transferred from the anode to the cathode (as Cu^{2+} through the solution). *The solution concentration remains unchanged.*

(b) Platinum metal is inert. It is not at all easily oxidized. The oxidation of SO_4^{2-} to $S_2O_8^{2-}$ also is not feasible here (see Table 21-1; $-E° = -2.01$ V). The oxidation that occurs most readily is that of water, as shown in equation (21.49).

Oxidation: $2\ H_2O \longrightarrow O_2(g) + 4\ H^+(aq) + 4\ e^-$ $-E° = -1.23$ V

The net electrolysis reaction and its $E°_{cell}$ are

$$2\ Cu^{2+}(aq) + 2\ H_2O \longrightarrow 2\ Cu(s) + 4\ H^+(aq) + O_2(g) \qquad E°_{cell} = -0.89 \text{ V}$$
$$(21.53)$$

SIMILAR EXAMPLES: Exercises 12, 49, 50, 51.

In Example 21-11b, where no other oxidation half-reaction seemed feasible, we turned to the oxidation of water. In other cases the only reduction process possible may be the reduction of water. And in still other instances (for example, the electrolysis of aqueous H_2SO_4 or Na_2SO_4), oxidation *and* reduction of water may be the only feasible half-reactions. In these cases the net electrolysis reaction is simply the decomposition of water into its elements: $2\ H_2O \rightarrow 2\ H_2(g) + O_2(g)$.

Faraday's Laws of Electrolysis. Michael Faraday (1791–1867) established the quantitative basis of electrolysis reactions and expressed his findings through these two statements.

Faraday's laws of electrolysis.

- The amount of chemical change produced is proportional to the quantity of electric charge passing through an electrolysis cell.
- A given quantity of electricity produces the same number of equivalents of any substance in electrolysis.

(21.54)

An *equivalent* of substance is associated with *1 mole of electrons* in a half-reaction (see also Section 21-8). Suppose we rewrite the electrolysis equation (21.53) in terms of half-equations based on the transfer of *1 mole of electrons* between the anode and cathode.

Anode (oxid.): $\frac{1}{2} H_2O \longrightarrow \frac{1}{4} O_2(g) + H^+(aq) + e^-$
Cathode (red.): $\frac{1}{2} Cu^{2+}(aq) + e^- \longrightarrow \frac{1}{2} Cu(s)$

From these half-equations we would define one (electrochemical) equivalent as equal to $\frac{1}{2}$ mol H_2O, $\frac{1}{4}$ mol O_2, 1 mol H^+, $\frac{1}{2}$ mol Cu^{2+}, and $\frac{1}{2}$ mol $Cu(s)$. Thus the passage of 1 mol electrons through the electrolysis cell of Figure 21-18 is signaled by the deposition of $\frac{1}{2}$ mol Cu (31.77 g) at the cathode.

One ampere (A) of electric current represents the passage of 1 coulomb of charge per second (C/s). The product of current and time yields the total quantity of charge transferred.

Quantity of electric charge related to current and time.

$$\text{current (C/s)} \times \text{time (s)} = \text{charge (C)} \tag{21.55}$$

With the Faraday constant, 96,485 coulombs/mol e^-, we can convert from a quantity of charge to a number of moles of electrons. Then we can do calculations regarding electrolysis processes.

Example 21-12

Using Faraday's laws of electrolysis to determine the amount of chemical change from current and time. The electrolysis reaction described in Example 21-11(b) is often used in a quantitative analysis laboratory to determine the copper content of a sample. The sample is dissolved to produce $Cu^{2+}(aq)$, and the solution is electrolyzed. At the cathode the reduction half-reaction is $Cu^{2+}(aq) + 2 e^- \rightarrow Cu(s)$. What mass of copper will be deposited in 1.00 hour by a current of 1.50 A?

Solution. From the electrode reaction we see that 1 mol Cu $\backsim$ 2 mol e^-. The calculation can be done through three simple steps.

$$\text{no. C} = 1.00 \text{ h} \times \frac{60 \text{ min}}{1 \text{ h}} \times \frac{60 \text{ s}}{1 \text{ min}} \times \frac{1.50 \text{ C}}{1 \text{ s}} = 5.40 \times 10^3 \text{ C}$$

$$\text{no. mol } e^- = 5.40 \times 10^3 \text{ C} \times \frac{1 \text{ mol } e^-}{9.648 \times 10^4 \text{ C}} = 5.60 \times 10^{-2} \text{ mol } e^-$$

$$\text{no. g Cu} = 5.60 \times 10^{-2} \text{ mol } e^- \times \frac{1 \text{ mol Cu}}{2 \text{ mol } e^-} \times \frac{63.55 \text{ g Cu}}{1 \text{ mol Cu}} = 1.78 \text{ g Cu}$$

SIMILAR EXAMPLES: Exercises 13, 52, 71.

Commercial Electrolysis Processes. We are not overstating the case to say that modern industry (and hence modern society) could not function without electrolysis reactions. A number of important substances produced almost exclusively by elec-

The refining of copper by electrolysis. [Courtesy of Kennecott/BP]

TABLE 21-2
**Some Substances Produced
by Electrolysis**

aluminum
magnesium
sodium
chlorine
fluorine
sodium hydroxide
potassium dichromate
potassium permanganate
sodium peroxodisulfate

trolysis are listed in Table 21-2. We will consider several of these electrolysis reactions in the chapters that follow, but let us preview the subject here.

Electrorefining of metals involves the deposition of pure metal at a cathode, from a solution containing the metal ion. The most important example is probably the electrorefining of copper. Copper metal produced by metallurgical smelting is of sufficient purity for some uses, such as plumbing. For other uses, however, the copper must be purified to a level of better than 99.9%. This is particularly so if copper is to be used as an electrical conductor. Trace quantities of impurities lower copper's electrical conductivity.

Electrolysis reaction (21.52) is the one used commercially to obtain high purity copper. A large chunk of impure copper is taken as the *anode* and a thin sheet of pure copper metal as the cathode. During the electrolysis, copper is transported continuously through the solution (as Cu^{2+}), from the anode to the cathode. The pure copper cathode increases in size as the impure chunk of copper is consumed. Gold and silver are commonly found as impurities in copper. These metals are much less active than copper (they are much more difficult to oxidize), and deposit at the bottom of the electrolysis tank as a sludge called *anode mud*. The value of the Ag and Au in the anode mud is generally sufficient to pay for the cost of the electrolysis.

Electrosynthesis is a method of producing substances through *nonspontaneous* reactions carried out by electrolysis. It is becoming an increasingly important method of synthesizing certain organic compounds. Usually, reaction conditions can be very precisely controlled. Manganese dioxide, MnO_2, can be obtained naturally as the mineral pyrolusite. The quality of this natural material, with regard to properties like crystal size and lattice imperfections, is not sufficient for some modern applications, as in the construction of alkaline batteries. A synthetic MnO_2 is superior to the naturally occurring compound, and the principal method of synthesis is by electrolysis.

The electrosynthesis of MnO_2 is carried out in a solution of $MnSO_4$ in $H_2SO_4(aq)$. MnO_2 is produced at an inert *anode* (graphite, for example) by the *oxidation* of Mn^{2+}. The anode half-reaction is

$$\text{Anode (oxid.):} \quad Mn^{2+} + 2\,H_2O \longrightarrow MnO_2(s) + 4\,H^+ + 2\,e^- \tag{21.56}$$

The cathode reaction is the reduction of $H^+(aq)$ to $H_2(g)$

$$\text{Cathode (red.):} \quad 2\,H^+(aq) + 2\,e^- \longrightarrow H_2(g) \tag{21.57}$$

and the net electrolysis reaction is

$$Mn^{2+}(aq) + 2\,H_2O \longrightarrow MnO_2(s) + 2\,H^+(aq) + H_2(g) \tag{21.58}$$

In the Focus feature of this chapter we consider the most important of all commercial electrolysis processes—the chlor-alkali process for the manufacture of $NaOH$, H_2, and Cl_2.

Example 21-13

Using Faraday's laws of electrolysis to determine the time required for an electrolysis process. A current of 25.5 A is used in the synthesis of $MnO_2(s)$ by reaction (21.58), in an electrolysis that uses 85% of the current passing through the cell. How long would it take to produce 1.000 kg $MnO_2(s)$?

Solution. It is easiest to think in terms of several questions: How many mol $MnO_2(s)$ are in 1.000 kg $MnO_2(s)$? How many mol e^- are transferred in the electrolysis? How does the efficiency of the current use enter in? Each question can be answered in a stepwise fashion.

$$\text{no. mol } MnO_2 = 1000.\text{ g } MnO_2(s) \times \frac{1 \text{ mol } MnO_2(s)}{86.94 \text{ g } MnO_2(s)} = 11.5 \text{ mol } MnO_2(s)$$

From half-equation (21.56) we see that 2 mol e$^-$ must be transferred for every mol $MnO_2(s)$ produced.

$$\text{no. mol e}^- \text{ for electrolysis} = 11.5 \text{ mol } MnO_2 \times \frac{2 \text{ mol e}^-}{1 \text{ mol } MnO_2} = 23.0 \text{ mol e}^-$$

We can introduce the efficiency information at this point. Only 85% of the electrons transferred participate in the desired electrolysis. Others are involved in side reactions.

$$\text{total no. mol e}^- = 23.0 \text{ mol e}^- \text{ (electrolysis)} \times \frac{100 \text{ mol e}^-}{85 \text{ mol e}^- \text{ (electrolysis)}}$$

$$= 27 \text{ mol e}^-$$

With a current of 25.5 A there is a passage of 25.5 C per second. This fact, together with the Faraday constant, allows us to complete the calculation.

$$\text{no. s} = 27 \text{ mol e}^- \times \frac{96,500 \text{ C}}{1 \text{ mol e}^-} \times \frac{1 \text{ s}}{25.5 \text{ C}} = 1.0 \times 10^5 \text{ s (28 hr)}$$

SIMILAR EXAMPLES: Exercises 52, 55.

21-8 Equivalent Weight and Normality Revisited

We have used balanced oxidation–reduction equations in stoichiometric calculations just as we have any other balanced equation. However, some people prefer to treat oxidation–reduction reactions from the standpoint of equivalent weight and normality rather than the mole and molarity concentration. Let us briefly explore how to do this.

For oxidation–reduction reactions *an equivalent is the amount of substance associated with 1 mol of electrons in a half-reaction.* Consider the equation

$$5 \text{ Fe}^{2+} + MnO_4^- + 8 \text{ H}^+ \longrightarrow 5 \text{ Fe}^{3+} + Mn^{2+} + 4 \text{ H}_2O \qquad (21.59)$$

Expressed as balanced half-equations, it becomes

Oxidation: $5 \text{ Fe}^{2+} \longrightarrow 5 \text{ Fe}^{3+} + 5 \text{ e}^-$
Reduction: $MnO_4^- + 8 \text{ H}^+ + 5 \text{ e}^- \longrightarrow Mn^{2+} + 4 \text{ H}_2O$

We conclude that 5 mol Fe^{2+} is 5 equiv Fe^{2+}, or that 1 mol Fe^{2+} is 1 equiv Fe^{2+}. The situation with MnO_4^- is that 1 mol MnO_4^- is 5 equiv MnO_4^-, or $\frac{1}{5}$ mol MnO_4^- is 1 equiv MnO_4^-. Stated in another way:

$$1 \text{ mol Fe}^{2+} \backsimeq \tfrac{1}{5} \text{ mol } MnO_4^- \qquad \text{and} \qquad 1 \text{ equiv Fe}^{2+} \backsimeq 1 \text{ equiv } MnO_4^-$$

Example 21-14

The end point of a permanganate titration is signaled by the first lasting pink color in solution.

Determining the normality concentration of a solution of an oxidizing agent from titration data. A piece of pure iron weighing 0.1568 g is dissolved in acidic solution and titrated with 26.24 mL of a $KMnO_4(aq)$ solution according to equation (21.59). What is the *normality* of the $KMnO_4(aq)$?

Solution. As in every case involving equivalent weights and normality concentrations, the reactants combine in a 1:1 ratio, *by equivalents.* The number of equivalents of MnO_4^- in 26.24 mL of the $KMnO_4(aq)$ is the same as the number of equivalents of Fe in the 0.1568-g iron sample.

$$\text{no. equiv Fe}^{2+} = \text{no. equiv Fe} = \text{no. mol Fe} = 0.1568 \text{ g Fe} \times \frac{1 \text{ mol Fe}}{55.847 \text{ g Fe}}$$

$$= 2.808 \times 10^{-3} \text{ equiv Fe}^{2+}$$

$$\text{no. equiv KMnO}_4 = \text{no. equiv MnO}_4^- = \text{no. equiv Fe}^{2+} = 2.808 \times 10^{-3}$$

$$\text{normality KMnO}_4 = \frac{2.808 \times 10^{-3} \text{ equiv KMnO}_4}{0.02624 \text{ L}} = 0.1070 \text{ N KMnO}_4$$

SIMILAR EXAMPLES: Exercises 14, 57.

Example 21-15

Performing a chemical analysis with the equivalent weight and normality concepts. 25.8 mL of the 0.1070 N KMnO$_4$(aq) described in Example 21-14 is used to titrate 5.00 mL of a saturated solution of sodium oxalate, Na$_2$C$_2$O$_4$. What is the solubility of Na$_2$C$_2$O$_4$, in g/L?

$$5 \text{ C}_2\text{O}_4^{2-} + 2 \text{ MnO}_4^- + 16 \text{ H}^+ \longrightarrow 2 \text{ Mn}^{2+} + 8 \text{ H}_2\text{O} + 10 \text{ CO}_2(\text{g}) \quad (21.60)$$

Solution. The number of equivalents of MnO$_4^-$ used in the titration is

$$\text{no. equiv MnO}_4^- = 0.0258 \text{ L} \times \frac{0.1070 \text{ equiv MnO}_4^-}{\text{L}}$$

$$= 2.76 \times 10^{-3} \text{ equiv MnO}_4^-$$

Using the basic idea that 1 equiv Na$_2$C$_2$O$_4$ $\backsimeq$ 1 equiv C$_2$O$_4^{2-}$ $\backsimeq$ 1 equiv MnO$_4^-$, we can express the solubility of Na$_2$C$_2$O$_4$ in equiv/L, that is, in *normality* concentration.

$$\frac{2.76 \times 10^{-3} \text{ equiv C}_2\text{O}_4^{2-}}{0.00500 \text{ L}} = 5.52 \times 10^{-1} \text{ N Na}_2\text{C}_2\text{O}_4$$

The final step is to convert from equiv Na$_2$C$_2$O$_4$ to g Na$_2$C$_2$O$_4$. For this we turn to equation (21.60). The reduction of 2 mol of MnO$_4^-$ to Mn^{2+} involves 10 mol of electrons (recall the reduction half-equation in 21.59). Ten moles of electrons must also be associated with the oxidation of 5 mol Na$_2$C$_2$O$_4$ to CO$_2$(g). The amount of Na$_2$C$_2$O$_4$ associated with *1 mol of electrons* is 0.500 mol Na$_2$C$_2$O$_4$: The equivalent weight of Na$_2$C$_2$O$_4$ is one half its molar mass, or 0.500 × 134.0 = 67.00 g Na$_2$C$_2$O$_4$/equiv Na$_2$C$_2$O$_4$.

$$\text{solubility} = \frac{5.52 \times 10^{-1} \text{ equiv Na}_2\text{C}_2\text{O}_4}{\text{L}} \times \frac{67.00 \text{ g Na}_2\text{C}_2\text{O}_4}{1 \text{ equiv Na}_2\text{C}_2\text{O}_4}$$

$$= 37.0 \text{ g Na}_2\text{C}_2\text{O}_4/\text{L}$$

SIMILAR EXAMPLES: Exercises 15, 58, 59.

If the KMnO$_4$(aq) of Examples 21-14 and 21-15 is used in a reaction in which MnO$_4^-$ is reduced to MnO$_2$(s) rather than to Mn^{2+}, its normality concentration would not be 0.1070 N. This is because the reduction of MnO$_4^-$ to MnO$_2$ involves *3 mol e$^-$* per mol MnO$_4^-$, whereas reduction to Mn^{2+} involves 5 mol e$^-$ per mol MnO$_4^-$. *Molarity is independent* of the reaction in which a solution is used. At times *normality* may not be. One of the drawbacks of equivalent weight and normality is that you must have prior knowledge of the specific reaction in which a solution is to be used.

FOCUS ON
The Chlor-Alkali Process

A bank of diaphragm cells for the electrolytic production of $Cl_2(g)$, $H_2(g)$, and $NaOH(aq)$ from $NaCl(aq)$. [Courtesy BASF AG, Ludwigshafen, FRG]

In Section 21-7 we considered these to be the most likely half-reactions in the electrolysis of a *concentrated* aqueous solution containing chloride ion.

Oxidation:

$$2\ Cl^-(aq) \longrightarrow Cl_2(g) + 2\ e^-$$
$$-E° = -1.36\ V \quad (21.48)$$

Reduction:

$$2\ H_2O + 2\ e^- \longrightarrow 2\ OH^-(aq) + H_2(g)$$
$$E° = -0.83\ V \quad (21.51)$$

Net:

$$2\ Cl^-(aq) + 2\ H_2O \longrightarrow 2\ OH^-(aq) + H_2(g) + Cl_2(g)$$
$$E°_{cell} = -2.19\ V \quad (21.61)$$

If the source of chloride ion is NaCl, the products of reaction (21.61) are $H_2(g)$, $Cl_2(g)$, and the alkali NaOH. A process yielding these products is called a **chlor-alkali process.** We can gauge the importance of this process by the fact that 0.5% of all the electric power generated annually in the United States is used by the chlor-alkali industry.

Diaphragm Cell. One type of chlor-alkali electrolysis cell, called a diaphragm cell, is pictured in Figure 21-19. The cell consists of an anode compartment, where $Cl_2(g)$ is produced, and a cathode compartment, where the production of $H_2(g)$ and $NaOH(aq)$ occurs. The Cl_2 and H_2 must be kept separated because they form explosive mixtures. Contact between $Cl_2(g)$ and $NaOH(aq)$ must also be

prevented, and this is the purpose of the diaphragm. (The reaction of Cl_2 with NaOH to form NaOCl and $NaClO_3$ is described in Section 23-1.) The NaCl(aq) in the anode compartment is kept at a slightly higher level than in the cathode compartment. This creates a gradual flow of NaCl(aq) between the compartments and reduces the back flow of NaOH(aq) into the anode compartment. The product obtained, however, is a mixture of about 10 to 12% NaOH(aq) and 14 to 16% NaCl(aq). The solution from the cathode compartment is concentrated and purified by evaporating water and crystallizing NaCl(s). Typically, the final product of the process is 50% NaOH(aq) with up to 1% NaCl. The $Cl_2(g)$ may contain up to 1.5% $O_2(g)$ because of the alternate oxidation process described by equation (21.49).

From the $E°_{cell}$ value in (21.61) we see that the electrolysis requires a source of direct current with voltage exceed-

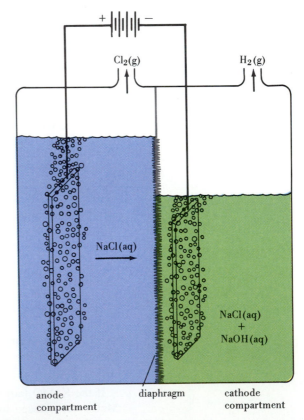

FIGURE 21-19
A diaphragm chlor-alkali cell.

The anode may be made of graphite or, in more modern technology, of specially treated titanium metal. The diaphragm and cathode are generally fabricated as a composite unit consisting of asbestos or an asbestos–polymer mixture deposited on a steel wire mesh or perforated steel cathode.

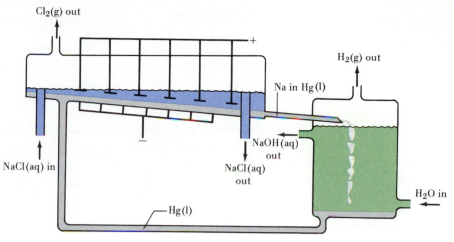

FIGURE 21-20
The mercury-cell chlor-alkali process.

The cathode is a layer of Hg(l) that flows along the bottom of the tank. Anodes, at which $Cl_2(g)$ forms, are immersed in NaCl(aq) just above the Hg(l). Sodium formed at the cathode dissolves in the Hg(l), and the sodium amalgam is decomposed with water, producing NaOH(aq) and $H_2(g)$. The regenerated Hg(l) is recycled.

ing 2.19 V. Actually, because of the internal resistance of the electrolysis cell and overpotentials at the electrodes (recall page 774), a somewhat higher voltage is required, about 3.5 V. If a current of 1.00 A were to pass through a diaphragm cell continuously for 24 hr, the quantity of Cl_2 produced would be about

$$\text{no. g } Cl_2 = 24 \text{ h} \times \frac{60 \text{ min}}{1 \text{ h}} \times \frac{60 \text{ s}}{1 \text{ min}} \times \frac{1.00 \text{ C}}{\text{s}}$$

$$\times \frac{1 \text{ mole e}^-}{96,500 \text{ C}} \times \frac{1 \text{ mol } Cl_2}{2 \text{ mole e}^-} \times \frac{70.9 \text{ g } Cl_2}{1 \text{ mol } Cl_2}$$

$$= 32 \text{ g } Cl_2$$

This is a minuscule rate of production for a commercial process. In order for the cell to produce about 1 metric ton (1000 kg) of Cl_2 per day, a current of approximately 31,000 A is required.

Mercury Cell. The products of the diaphragm cell process are suitable for most purposes, but not if high-purity NaOH(aq) is required (as in rayon manufacture). An electrolytic process that can produce this higher purity NaOH(aq) uses the mercury cell pictured in Figure 21-20. It is based on the fact that a mercury cathode has a high overpotential for the reduction of H_2O to OH^- and $H_2(g)$. The reduction that occurs instead is that of $Na^+(aq)$ to Na, which dissolves in Hg(l) to form an amalgam with about 0.5% Na, by mass.

Oxidation:

$$2 Cl^-(aq) \longrightarrow Cl_2(g) + 2 e^- \qquad -E° = -1.36 \text{ V}$$

Reduction:

$$2 Na^+(aq) + 2 e^- \longrightarrow 2 Na(\text{in Hg}) \qquad E° = -1.77 \text{ V}$$

Net:

$$2 Na^+(aq) + 2 Cl^-(aq) \longrightarrow 2 Na(\text{in Hg}) + Cl_2(g)$$
$$E°_{\text{cell}} = -3.13 \text{ V}$$
$$(21.62)$$

When the Na amalgam is removed from the cell and treated with water, NaOH(aq) is formed,

$$2 Na(\text{in Hg}) + 2 H_2O \longrightarrow$$
$$2 Na^+(aq) + 2 OH^-(aq) + H_2(g) + Hg(l) \quad (21.63)$$

and the liquid mercury is recycled back to the electrolysis cell. Notice that in the mercury cell chlor-alkali process NaOH(aq) and $Cl_2(g)$ never come in contact.

The mercury cell has many advantages over the diaphragm cell, particularly in being able to produce concentrated high-purity NaOH(aq) without extensive follow-up procedures. There are disadvantages to the mercury cell, however. It requires a higher voltage (about 4.5 V) than the diaphragm cell and consumes more electric energy, about 3100 kWh/ton Cl_2 in a mercury cell compared to 2700 in a diaphragm cell. Another serious drawback is the need to control mercury effluents to the environment. Prior to the establishment of environmental regulations, mercury losses were about 200 g Hg per metric ton Cl_2 produced. Mercury losses are now limited to 0.28 g Hg per metric ton of Cl_2 in existing plants and half this amount in new plants. About 25% of the chlor-alkali production in the United States is by the mercury-cell process, but this percentage is not likely to increase because of the difficulty in controlling mercury losses.

Membrane Cells. The ideal chlor-alkali process is one that is energy efficient and that does not use mercury. A type of cell that offers these advantages is one in which the porous diaphragm of Figure 21-19 is replaced by a cation exchange membrane. This is a separator between the two half-cells that permits hydrated cations (Na^+ and H_3O^+) to pass between the anode and cathode compartments but severely restricts the back flow of Cl^- and OH^- ions. Certain of these membranes are made of fluorocarbon polymers. Membrane cells do not have the production capacity of mercury cells, and it will be some time before mercury cells are phased out, although this has already happened in Japan.

Summary

To determine the direction of spontaneous change in an oxidation–reduction reaction it is necessary to establish the tendencies for oxidation and reduction processes to occur. This is most readily done by measuring the electric potential established at an electrode when it is in contact with its ions in solution. This combination of an electrode and solution is called a *half-cell*. An *electrochemical cell* consists of two half-cells in which the electrodes are joined by a wire and solutions through a *salt bridge* or a *porous plug*. The chemical reaction that occurs in an electrochemical cell involves oxidation at one electrode (called the *anode*) and reduction at the other electrode (the *cathode*). The difference in electric potential between the two electrodes is the *voltage* of the cell. An electrochemical cell that produces electricity from a spontaneous oxidation–reduction reaction is called a *voltaic* cell.

The standard hydrogen electrode (S.H.E.) is arbitrarily assigned a potential of *zero*. When some electrode and its ions serve as one half cell and the S.H.E. as the other, the voltage of the cell gives the standard potential of the electrode. Electrode potentials are written in terms of the tendency of *reduction* to occur. For any half-cell reaction having a *positive* electrode potential, the tendency for reduction to occur is greater than for

$$2 \, H^+(1 \, M) + 2 \, e^- \longrightarrow H_2(g, \, 1 \, atm)$$

If a standard electrode potential is *negative*, the reduction tendency is less than that for $H^+(aq)$ to $H_2(g)$. If a half-cell is written as an *oxidation*, the sign of the electrode potential is changed. The voltage of a voltaic cell, E_{cell}, is the sum of $-E$ for the oxidation half-reaction and E for the reduction half-reaction. If $E_{cell} > 0$, the cell reaction is *spontaneous;* if $E_{cell} < 0$, the reaction is *nonspontaneous.*

Cell voltages predicted from tabulated values of *standard* electrode potentials are $E°_{cell}$ values. Important relationships exist between $E°_{cell}$ and $\Delta G°$ and between $E°_{cell}$ and K.

$$\Delta G° = -n\mathscr{F}E°_{cell} \quad \text{and}$$
$$E°_{cell} = (0.0592/n) \log K \quad \text{(at 298.15 K)}$$

The Nernst equation relates E_{cell} for *nonstandard* conditions, $E°_{cell}$, and activities of reactants and products.

$$E_{cell} = E°_{cell} - (0.0592/n) \log Q$$

where Q is the reaction quotient, n = number of mol electrons transferred in the cell reaction, and the temperature = 298.15 K. Among the many applications of cell potential measurements and the Nernst equation are potentiometric titrations and the determination of pH and K_{sp} values.

Electric batteries are voltaic cells. In a *primary* battery (such as a flashlight battery) a spontaneous oxidation–reduction reaction produces electricity until the reaction has gone essentially to completion. In a *secondary* battery (e.g., a lead–acid storage battery) the cell reaction can be reversed through an external source of electricity. In a *fuel cell* the reactants are passed continuously through the cell. A reaction equivalent to a combustion reaction occurs, but with energy released as electricity instead of heat.

Corrosion and methods of corrosion protection can also be described in terms of voltaic cells because corrosion reactions are oxidation–reduction reactions, with electrons passing from an anode to a cathode region. One effective method of protection from corrosion is to bring the metal to be protected in contact with a more active metal, which corrodes preferentially. This type of action is called cathodic protection because the protected metal acts as a cathode, where reduction occurs.

In an electrolysis cell an external source of electricity is used to force electrons to flow in a direction opposite to that in which they would flow spontaneously. As a result, the reaction occurring in the electrolysis cell is a *nonspontaneous* reaction. The amount of chemical change produced in an electrolysis cell is directly proportional to the quantity of electric charge passing through the cell, as stated through Faraday's laws of electrolysis. Many important industrial operations rely on electrolytic processes.

Summarizing Example

A battery system that may be used to power automobiles in the future is the aluminum–air battery. This is a *flow battery* in which oxidation occurs at an aluminum anode and reduction at a carbon–air cathode. The electrolyte circulated through the battery is $NaOH(aq)$; the ultimate reaction product is $Al(OH)_3(s)$, which is removed from the battery as it is formed. In operation the battery can be kept charged by feeding Al anode slugs and water into the battery; oxygen is drawn from the air (see Figure 21-21). The battery can power an automobile several hundred miles between charges. The $Al(OH)_3(s)$ removed from the battery can be converted back to aluminum in an aluminum manufacturing facility.

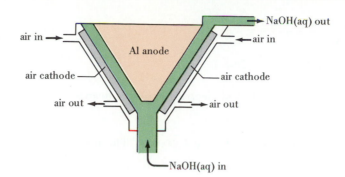

FIGURE 21-21
Simplified representation of aluminum–air battery.

1. In actual practice Al^{3+} produced at the anode does not precipitate as $Al(OH)_3(s)$ but is obtained as the complex ion $[Al(OH)_4]^-$ in the presence of $NaOH(aq)$. [$Al(OH)_3$ is precipitated from the circulating $NaOH(aq)$ electrolyte *outside* the battery.] Write plausible equations for the oxidation and reduction half-reactions and the net reaction that occur in the battery.

Solution. From the description of the anode (oxidation) half-reaction, we should expect it to involve Al and OH^- as *reactants* and $[Al(OH)_4]^-$ as a *product*. The reduction half-reaction must involve $O_2(g)$ as a *reactant* and H_2O and/or OH^- as needed to complete and balance the equation (since the solution is basic). The following half-equations meet these requirements.

Oxidation: $\quad 4\,\{Al(s) + 4\,OH^-(aq) \longrightarrow [Al(OH)_4]^- + 3\,e^-\}$
Reduction: $\quad 3\,\{O_2(g) + 2\,H_2O + 4\,e^- \longrightarrow 4\,OH^-(aq)\}$

Net: $\quad 4\,Al(s) + 3\,O_2(g) + 6\,H_2O + 4\,OH^- \longrightarrow 4\,[Al(OH)_4]^-$ (21.64)

(This example is similar to Example 21-1.)

2. The theoretical voltage of the aluminum–air cell is $+2.73$ V. Use this information and data from Table 21-1 to obtain $E°$ for the reduction

$$[Al(OH)_4]^- + 3\,e^- \longrightarrow Al(s) + 4\,OH^- \qquad E° = ?$$

Solution. We can simply return to the setup in part 1 and substitute the information we have available. We must solve for $E°$.

Oxidation: $\quad -E°$
Reduction: $\quad E° = +0.401$ V (from Table 21-1)
Net: $\quad E°_{cell} = -E° + 0.401$ V $= +2.73$ V
$E° = 0.401$ V $-\ 2.73$ V $= -2.33$ V

(This example is similar to Example 21-3.)

3. Given that $E°_{cell}$ for reaction (21.64) is $+2.73$ V, that $\Delta G_f°[OH^-] = -157$ kJ/mol, and that $\Delta G_f°[H_2O(l)] = -237.2$ kJ/mol, determine the free energy of formation, $\Delta G_f°$, of the aluminate ion, $[Al(OH)_4^-]$.

Solution. First we must make use of the relationship between $\Delta G°$ and $E°_{cell}$ for a reaction. [Note that reaction (21.64) is written for *12* moles of electrons.]

$$\Delta G° = -nFE°_{cell} = -12 \text{ mol } e^- \times 96{,}500 \text{ C/mol } e^- \times 2.73 \text{ V}$$
$$= -3.16 \times 10^6 \text{ V} \cdot \text{C} = -3.16 \times 10^6 \text{ J} = -3.16 \times 10^3 \text{ kJ}$$

Now we can use the familiar setup (from Chapters 7 and 20).

$$\Delta G° = 4\Delta G_f°[Al(OH)_4]^- - 6\Delta G_f°[H_2O(l)] - 4\Delta G_f°[OH^-]$$

$$-3.16 \times 10^3 \text{ kJ} = 4\Delta G_f°[Al(OH)_4]^- - 6(-237 \text{ kJ}) - 4(-157 \text{ kJ})$$

A large model airplane powered by an aluminum–air battery. [Courtesy Alupower, Inc.]

$$\Delta G_f^\circ[\text{Al(OH)}_4]^- = \frac{-3.16 \times 10^3 - 1422 - 628}{4} = -1.30 \times 10^3 \text{ kJ/mol}$$

(This example is similar to Examples 21-4 and 20-6.)

4. What mass of aluminum is consumed if 10.0 A of electric current is drawn from the battery for 4.00 h?

Solution. Think of this as you would an electrolysis calculation. The total amount of charge transferred is

$$10.0\, \frac{\text{C}}{\text{s}} \times 4.0\text{ h} \times \frac{60\text{ min}}{1\text{ h}} \times \frac{60\text{ s}}{1\text{ min}} = 1.44 \times 10^5 \text{ C}$$

This corresponds to

$$1.44 \times 10^5 \text{ C} \times \frac{1\text{ mol e}^-}{96{,}485\text{ C}} = 1.49\text{ mol e}^-$$

The quantity of Al that must be oxidized to produce this number of electrons is

$$\text{no. g Al} = 1.49\text{ mol e}^- \times \frac{1\text{ mol Al}}{3\text{ mol e}^-} \times \frac{26.98\text{ g Al}}{1\text{ mol Al}} = 13.4\text{ g Al}$$

(This example is similar to Example 21-12.)

Key Terms

anode (21-1)
battery (21-5)
cathode (21-1)
cathodic protection (21-6)
cell potential, E_{cell} (21-1, 21-2)
chlor-alkali process (Focus feature)
concentration cell (21-4)
corrosion (21-6)
electrochemical cell (21-1)
electrochemical equivalent (21-8)

electrode (21-1)
electrode potential (21-1)
electrolytic cell (21-1, 21-7)
electromotive force, emf (21-1)
electrorefining (21-7)
electrosynthesis (21-7)
Faraday constant, $\mathscr{F}$ (21-3, 21-7)
Faraday's laws of electrolysis (21-7)
flow battery (21-5)
fuel cell (21-5)

half-cell (21-1)
Nernst equation (21-4)
potentiometer (21-1)
primary battery (21-5)
salt bridge (21-1)
secondary battery (21-5)
standard cell potential, E_{cell}° (21-2)
standard electrode potential, E° (21-2)
standard hydrogen electrode, S.H.E. (21-2)
voltaic cell (21-1)

Highlighted Expressions

Writing cell diagrams (21.4)
Electrode potential conventions (21.15)
Relating electrode potentials and E_{cell}° (21.16)
Relationship of E_{cell}° to ΔG° (21.20)
E_{cell} as a criterion for spontaneity (21.21)

Relationship of E_{cell}° to K (21.29)
Nernst equation: E_{cell} as a function of concentrations (21.32)
Faraday's laws of electrolysis (21.54)
Quantity of electric charge related to current and time (21.55)

Review Problems

1. Write equations for the oxidation and reduction half-reactions and the net reaction describing
  (a) the displacement of Cu^{2+}(aq) by Fe(s);
  (b) the oxidation of Br^- to Br_2(aq) by Cl_2(aq);
  (c) the reduction of Fe^{3+}(aq) to Fe^{2+}(aq) by Al(s);
  (d) the oxidation of Cl^-(aq) to ClO_3^-(aq) by MnO_4^- in acidic solution (MnO_4^- is reduced to Mn^{2+});
  (e) the oxidation of S^{2-}(aq) to SO_4^{2-}(aq) by O_2(g) in basic solution.

2. The diagram shows a voltaic cell consisting of a standard nickel electrode and a standard electrode of a second metal, M. For the half-reaction

$$Ni^{2+}(1\text{ M}) + 2\text{ e}^- \rightarrow Ni(s) \qquad E^\circ = -0.246\text{ V}$$

From these measured values of E_{cell}°, in each case determine E° for the half-reaction

$$M^{2+}(1\text{ M}) + 2\text{ e}^- \rightarrow M(s) \qquad E^\circ = ?$$

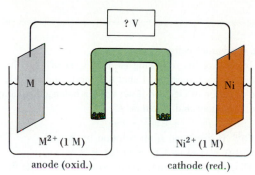

? V

M

Ni

M^{2+} (1 M)

Ni^{2+} (1 M)

anode (oxid.)   cathode (red.)

Problem 2

(a) Cr, $E^{\circ}_{cell} = +0.664$ V   (b) Co, $E^{\circ}_{cell} = +0.31$ V
(c) Cd, $E^{\circ}_{cell} = +0.157$ V   (d) Sn, $E^{\circ}_{cell} = -0.110$ V

3. Write the cell reactions for the electrochemical cells diagrammed below, and use data from Table 21-1 to calculate E°_{cell} for each reaction.

(a) $Zn(s)|Zn^{2+}(aq)||Sn^{2+}(aq)|Sn(s)$
(b) $Pt(s)|Fe^{2+}(aq), Fe^{3+}(aq)||Sn^{4+}(aq), Sn^{2+}(aq)|Pt(s)$
(c) $Cu(s)|Cu^{2+}(aq)||Cl^{-}(aq)|Cl_2(g), Pt(s)$

4. From the data listed below, together with data from Table 21-1, determine the quantity indicated.

(a) E°_{cell} for the cell

$$Pt(s), Cl_2(g)|Cl^{-}(aq)||Pb^{2+}(aq), H^{+}(aq)|PbO_2(s)$$

(b) E° for the couple, Sc^{3+}/Sc, given that

$$Mg(s)|Mg^{2+}(aq)||Sc^{3+}(aq)|Sc(s) \qquad E^{\circ}_{cell} = +0.35 \text{ V}$$

(c) E° for the couple, Cu^{2+}/Cu^{+}, given that

$$Pt(s)|Cu^{+}(aq), Cu^{2+}(aq)||Ag^{+}(aq)|Ag(s)$$
$$E^{\circ}_{cell} = +0.647 \text{ V}$$

5. Assume that all reactants and products are in their standard states and use data from Table 21-1 to predict whether a spontaneous reaction will occur in the forward direction in each case.

(a) $Sn(s) + Zn^{2+} \rightarrow Sn^{2+} + Zn(s)$
(b) $2 Fe^{3+} + 2 I^{-} \rightarrow 2 Fe^{2+} + I_2(s)$
(c) $4 NO_3^{-} + 4 H^{+} \rightarrow 3 O_2(g) + 4 NO(g) + 2 H_2O$
(d) $O_2(g) + 2 Cl^{-} \rightarrow 2 OCl^{-}$ (basic soln.)

6. Use data from Table 21-1 to predict whether to any significant extent

(a) Mg(s) will displace Sn^{2+} from aqueous solution;
(b) tin metal will dissolve in 1 M HCl;
(c) SO_4^{2-} will oxidize Fe^{2+} to Fe^{3+} in acidic solution;
(d) SO_4^{2-} will oxidize Fe^{2+} to Fe^{3+} in basic solution;
(e) I_2 will displace $Cl^{-}(aq)$ to produce $Cl_2(g)$.

7. Determine the value of ΔG° for the following reactions carried out in voltaic cells.

(a) $2 Al(s) + 3 Cu^{2+}(aq) \rightarrow 2 Al^{3+}(aq) + 3 Cu(s)$
(b) $O_2(g) + 4 I^{-}(aq) + 4 H^{+}(aq) \rightarrow 2 H_2O + 2 I_2(s)$
(c) $Cr_2O_7^{2-}(aq) + 14 H^{+} + 6 Ag(s) \rightarrow$
$$2 Cr^{3+}(aq) + 6 Ag^{+}(aq) + 7 H_2O$$

8. Write the equilibrium constant expression for each of the following reactions and determine the numerical value of K at 25 °C. Use data from Table 21-1 as necessary.

(a) $Fe^{3+}(aq) + Ag(s) \rightarrow Fe^{2+}(aq) + Ag^{+}(aq)$
(b) $MnO_2(s) + 4 H^{+}(aq) + 2 Cl^{-}(aq) \rightarrow$
$$Mn^{2+}(aq) + 2 H_2O + Cl_2(g)$$
(c) $2 OCl^{-}(aq) \rightarrow 2 Cl^{-}(aq) + O_2(g)$ (basic solution)

9. For each of the following combinations of electrodes and solutions, indicate

• the net cell reaction;
• the direction in which electrons flow spontaneously (from A to B or from B to A);
• the magnitude of the voltage read on the voltmeter, V.

Elec- trode A	Solution A	Elec- trode B	Solution B
(a) Cu	1.0 M Cu^{2+}	Fe	1.0 M Fe^{2+}
(b) Pt	1.0 M Sn^{2+}/1.0 M Sn^{4+}	Ag	1.0 M Ag^{+}
(c) Zn	0.10 M Zn^{2+}	Fe	1.0 × 10^{-3} M Fe^{2+}

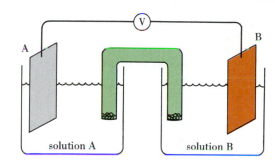

V

A

B

solution A   solution B

Problem 9

10. A voltaic cell represented by the following cell diagram has $E_{cell} = 1.000$ V. What must be $[Ag^{+}]$ in the cell?

$$Zn(s)|Zn^{2+}(1.00 \text{ M})||Ag^{+}(x \text{ M})|Ag(s)$$

11. For the cell pictured in Figure 21-9, what is E_{cell} if the unknown solution in the half-cell on the left (a) has a pH = 5.12; (b) is 0.00185 M HCl (aq); (c) is 0.357 M $HC_2H_3O_2$?

12. Use data from Table 21-1 to predict the probable products when Pt electrodes are used in the electrolysis of (a) $CuCl_2(aq)$; (b) HCl(aq); (c) $Na_2SO_4(aq)$; (d) $BaCl_2(l)$; (e) KI(aq); (f) KOH(aq).

13. What mass of metal would be deposited by the passage of 1.56 A of current for 2.25 h in the electrolysis of a solution containing (a) Zn^{2+}; (b) Al^{3+}; (c) Ag^{+}; (d) Ni^{2+}?

14. Calculate the equivalent weights of the following substances when they are used in the reaction indicated.

(a) Al, reaction (21.24)
(b) PbO_2, reaction (21.44)
(c) O_2, reaction (21.46)
(d) Mn^{2+}, reaction (21.58)

15. What is the % Fe, by mass, in an iron ore if the Fe^{2+} derived from a 0.8312-g ore sample requires 25.13 mL of 0.2821 N $KMnO_4(aq)$ for its titration by reaction (21.59)?

Exercises

(Use data from Table 21-1, as necessary.)

Definitions and terminology

16. Indicate the essential *difference* in meanings of the following pairs of terms: (a) half-reaction and net reaction; (b) voltaic (galvanic) cell and electrolytic cell; (c) anode and cathode; (d) E_{cell} and E_{cell}°.

17. Explain the role of each of the following in electrochemical studies: (a) a standard hydrogen electrode; (b) a salt bridge; (c) a potentiometer; (d) a platinum electrode; (e) a glass electrode.

Standard electrode potentials

18. From the observations listed, estimate the approximate value of the standard electrode potential for the half-reaction

$$M^{2+}(aq) + 2\,e^- \rightarrow M(s)$$

(a) The metal M dissolves in $HNO_3(aq)$ but not in $HCl(aq)$; it displaces $Ag^+(aq)$ but not $Cu^{2+}(aq)$.

(b) The metal M dissolves in $HCl(aq)$ producing $H_2(g)$, but displaces neither $Zn^{2+}(aq)$ nor $Fe^{2+}(aq)$.

19. You are given the task of estimating the standard electrode potential of indium.

$$In^{3+}(aq) + 3\,e^- \rightarrow In(s) \qquad E^{\circ} = ?$$

You do not have any electrical equipment but you do have all of the metals listed in Table 21-1 and water solutions of their ions. Describe the experiments you would perform and indicate the accuracy you would expect in your results.

*20. Two electrochemical cells are assembled in which these reactions occur.

$$V^{2+} + VO^{2+} + 2\,H^+ \rightarrow 2\,V^{3+} + H_2O \qquad E_{cell}^{\circ} = 0.616\ \text{V}$$

$$V^{3+} + Ag^+ + H_2O \rightarrow VO^{2+} + 2\,H^+ + Ag(s)$$
$$E_{cell}^{\circ} = 0.439\ \text{V}$$

Use these data and other values from Table 21-1 to calculate the standard electrode potential for the half-reaction

$$V^{3+} + e^- \rightarrow V^{2+}$$

Predicting oxidation–reduction reactions

21. According to standard electrode potentials, Na metal seemingly should displace Mg^{2+} from aqueous solution. Yet if this reaction is attempted it is found not to occur. Explain. [*Hint:* What reaction does occur?]

22. Zinc will react with 1 M HCl to displace $H_2(g)$, but copper will not. As pictured, if a piece of copper is joined to one of zinc and the pair of metals immersed in HCl(aq), some bubbles of $H_2(g)$ appear at the copper metal.

(a) Does this mean that copper reacts with HCl(aq)?

(b) What is the reaction that occurs?

23. What observations would you expect to make if copper and silver metal are joined in the manner described in Exercise 22 and the metals are immersed in (a) HCl(aq); (b) HNO_3(aq)?

24. Predict whether the following metals will dissolve in the acid indicated. If a reaction does occur write the net ionic equation for the reaction. Assume that reactants and products are in their standard states. (a) Ag in HNO_3(aq); (b) Zn in HI(aq); (c) Au in HNO_3 (for the couple Au^{3+}/Au, $E^{\circ} = +1.50$ V).

Voltaic (galvanic) cells

25. In each of the following examples, sketch a voltaic cell that uses the given reaction. Label the anode and cathode; indicate the direction of electron flow; write a balanced equation for each cell reaction; and calculate E_{cell}°.

(a) $Cu(s) + Fe^{3+} \rightarrow Cu^{2+} + Fe^{2+}$

(b) Pb^{2+} is displaced from solution by Al(s);

(c) $Cl_2(g) + H_2O \rightarrow Cl^- + O_2(g) + H^+$

26. Diagram the cell and, where possible, calculate the value of E_{cell}° for a voltaic cell in which

(a) $Cl_2(g)$ is reduced to Cl^- and Fe(s) is oxidized to Fe^{2+};

(b) Pb^{2+} is displaced from solution by Al(s);

(c) The net cell reaction is $2\,Cu^+ \rightarrow Cu^{2+} + Cu(s)$.

ΔG°, E_{cell}°, and K

27. Use data from Table 21-1 and appropriate equations from the text to determine ΔG° for the following reactions. [*Hint:* The equations are not balanced.]

(a) $Al(s) + Zn^{2+} \rightarrow Al^{3+} + Zn(s)$

(b) $Pb^{2+} + MnO_4^- + H^+ \rightarrow PbO_2(s) + Mn^{2+} + H_2O$

(c) $H^+ + Cl^- + MnO_2(s) \rightarrow Mn^{2+} + H_2O + Cl_2(g)$

28. Determine the value of K at 298 K for reaction (a) in Exercise 27. Does the displacement of Zn^{2+} by Al(s) go to completion? [*Hint:* What is $[Zn^{2+}]$ remaining at equilibrium if Al metal is added to a solution with $[Zn^{2+}] = 1.0$ M?]

29. The following data are available for $Ag_2O(s)$: $\Delta H_f^{\circ} = -31.05$ kJ/mol; $\Delta G_f^{\circ} = -11.21$ kJ/mol; and $S^{\circ} = 121$ J mol^{-1} K^{-1}. Use these data, together with values from Appendix D, to calculate a theoretical voltage of the silver–zinc button cell described on page 770.

30. The hydrazine fuel cell is based on the reaction

$$N_2H_4(aq) + O_2(g) \rightarrow N_2(g) + 2\,H_2O(l)$$

The theoretical voltage of this fuel cell is listed as 1.559 V. Use this information, together with data from Appendix D, to obtain a value of $\Delta G_f^{\circ}[N_2H_4(aq)]$. [*Hint:* You may need to break down the net reaction into half-reactions to determine the number of moles of electrons involved in the reaction.]

*31. The standard electrode potential corresponding to the reduction $Cr^{3+} + e^- \rightarrow Cr^{2+}$ is $E^{\circ} = -0.407$ V. If excess Fe(s) is added to a solution in which $[Cr^{3+}] = 1.00$ M, what will be $[Fe^{2+}]$ when equilibrium is established at 298 K?

$$Fe(s) + 2\,Cr^{3+} \rightleftharpoons Fe^{2+} + 2\,Cr^{2+}$$

Exercise 22

[Carey B. Van Loon]

Concentration dependence of E_{cell}—the Nernst equation

32. Use the Nernst equation and data from Table 21-1 to calculate E_{cell} for each of the following electrochemical cells. [*Hint:* Some of these have positive and some, negative E_{cell} values.]

(a) $Fe(s)|Fe^{2+}(0.35\ M)||Sn^{2+}(0.070\ M)|Sn(s)$

(b) $Pt(s),\ Cl_2(g,\ 0.25\ atm)|Cl^-(1.2\ M)||Ag^+(0.46\ M)|Ag(s)$

(c) $Mg(s)|Mg^{2+}(0.016\ M)||OH^-(0.65\ M)|H_2(g,\ 0.75\ atm)$, $Pt(s)$

33. Write an equation to represent the oxidation of Mn^{2+} to $MnO_2(s)$ by $O_2(g)$ in acidic solution. Will this reaction occur spontaneously as written if all other reactants and products are in their standard states and (a) $[H^+] = 6.6\ M$; (b) $[H^+] = 1.0\ M$; (c) $[H^+] = 0.060\ M$; (d) pH = 9.22?

34. Show that the oxidation of Cl^- to $Cl_2(g)$ by $Cr_2O_7^{2-}$ in acidic solution will not occur spontaneously with reactants and products in their standard states. Nevertheless, this method can be used to produce $Cl_2(g)$ in the laboratory. Explain why this is so. [*Hint:* What experimental conditions would you use?]

35. The following oxidizing agents are all listed in Table 21-1:

$Cl_2(g),\ O_2(g),\ MnO_4^-(aq),\ H_2O_2(aq),\ F_2(g)$.

(a) For which of them is the oxidizing power dependent on pH?

(b) For which is the oxidizing power independent of pH?

(c) Where the oxidizing power is pH dependent, which are better oxidizing agents in acidic and which in basic solution?

36. If $[Zn^{2+}]$ is maintained at 1.0 M

(a) What is the minimum $[Cu^{2+}]$ for which reaction (21.3) is still spontaneous in the forward direction?

(b) Does the displacement of $Cu^{2+}(aq)$ by $Zn(s)$ go to completion?

37. A voltaic cell is constructed as follows.

$Ag(s)|Ag^+(satd\ Ag_2SO_4(aq))||Ag^+(0.125\ M)|Ag(s)$

What is the value of E_{cell}? [*Hint:* Refer to Table 19-1.]

38. The following voltaic cell is constructed.

$Sn(s)|Sn^{2+}(0.150\ M)||Pb^{2+}(0.550\ M)|Pb(s)$

(a) What is E_{cell} initially?

(b) If the cell is allowed to operate spontaneously, will E_{cell} increase, decrease, or remain constant with time? Explain.

(c) What will be E_{cell} when $[Pb^{2+}]$ has fallen to 0.500 M?

(d) What will be $[Sn^{2+}]$ at the point where $E_{cell} = 0.020\ V$?

(e) What are the concentrations of each of the reacting species when $E_{cell} = 0$?

★39. It is desired to construct the following voltaic cell to have $E_{cell} = 0.0860\ V$. What $[Cl^-]$ must be present in the cathode half cell to achieve this result? [*Hint:* What are the K_{sp} values of AgCl and AgI?]

$Ag(s)|Ag^+(satd\ AgI(aq))||Ag^+(satd\ AgCl,\ x\ M\ Cl^-)|Ag(s)$

Batteries and fuel cells

40. For each of the following potential battery systems, describe the electrode reactions and the net cell reaction you would expect. Also determine the theoretical voltage of the battery. (a) $Zn-Br_2$; (b) $Li-F_2$; (c) $Mg-I_2$; (d) Fe–air.

41. Refer to the discussion of the Leclanché cell (page 768).

(a) Combine the several equations written for the operation of the Leclanché cell into a single net equation.

(b) Given that the theoretical voltage of the Leclanché cell is 1.55 V, determine E for the reduction half-reaction.

★42. Refer to the Summarizing Example. The concept behind the aluminum–air cell can be extended to other metals as well, for example zinc–air or iron–air. One of the attractive features of the aluminum–air battery is the relatively large quantity of charge that can be transferred per unit mass of Al consumed. Assume that the cell reactions for the zinc–air and iron–air batteries are similar to that for aluminum–air, and show that, of the three, the aluminum–air battery produces the greatest amount of charge per unit mass of metal.

43. What is the efficiency value, $\epsilon = \Delta G°/\Delta H°$, for the reaction occurring in the hydrogen–oxygen fuel cell (21.47)? What is the theoretical cell voltage, $E_{cell}°$?

44. One type of "breathalyzer" (alcohol meter) is based on the principle of the fuel cell. Ethanol vapor in a 1 cm³ sample of breath is passed through a cell in which the ethanol (C_2H_5OH) is oxidized to CO_2, and oxygen gas from air is reduced to H_2O. The quantity of ethanol in the sample is related to the amount of electric current produced by the fuel cell as a function of time. Assume that the cell reaction is

$C_2H_5OH(g) + 3\ O_2(g) \rightarrow 2\ CO_2(g) + 3\ H_2O(l)$

(a) Write the half-equation for the reduction of $O_2(g)$ to $H_2O(l)$, assuming an acidic solution. [*Hint:* Refer to Table 21-1.]

(b) Propose a plausible half-equation for the oxidation of $C_2H_5OH(g)$ to $CO_2(g)$.

(c) Determine $\Delta G°$ for the cell reaction by using free energy of formation data from Appendix D.

(d) Determine $E_{cell}°$ for the cell reaction.

(e) Estimate a value of $E°$ for the reduction of $CO_2(g)$ to $C_2H_5OH(g)$.

45. One type of fuel cell uses a mixture of molten carbonates as an electrolyte. In this mixture some decomposition of carbonate ion occurs.

$CO_3^{2-} \rightarrow CO_2 + O^{2-}$

Write the electrode reactions corresponding to the net cell reaction $CH_4(g) + 2\ O_2(g) \rightarrow CO_2(g) + 2\ H_2O(g)$. [*Hint:* Use O^{2-} in writing electrode reactions.]

Electrochemical mechanism of corrosion

46. Comment on the appropriateness of the statement, "A corroding metal is like a voltaic cell."

47. Natural gas transmission pipes are sometimes protected against corrosion by maintaining a small potential difference between the pipe and an inert electrode buried in the ground. Describe how the method works. [*Hint:* Is the inert electrode maintained positively or negatively charged with respect to the pipe?]

48. In the original construction of the Statue of Liberty, a framework of iron ribs was covered with thin sheets of copper less than 2.5 mm thick. The copper "skin" and iron framework were separated by a layer of asbestos. The asbestos wore away with time, and the iron ribs corroded to the extent that some of them lost more than half their mass in the 100 years before the statue was restored. The copper skin lost only about 4% of its thickness in these 100 years. Use electrochemical principles to explain these observations.

Electrolysis reactions

49. Indicate which of the following reactions can only be brought about through electrolysis, and indicate the *minimum* voltage required, assuming that all reactants and products are in their standard states.

(**a**) $2 H_2O \rightarrow 2 H_2(g) + O_2(g)$ [in 1 M $H^+(aq)$]

(**b**) $Zn(s) + Fe^{2+}(aq) \rightarrow Zn^{2+}(aq) + Fe(s)$

(**c**) $2 Br^-(aq) + O_2(g) + 2 H^+(aq) \rightarrow Br_2(l) + H_2O_2(aq)$

(**d**) $2 H_2O \rightarrow 2 H_2(g) + O_2(g)$ (in basic solution)

(**e**) $O_2(g) + 2 S^{2-}(aq) + 2 H_2O \rightarrow 2 S(s) + 4 OH^-(aq)$

50. The commercial production of magnesium involves the electrolysis of molten $MgCl_2$. Why cannot the simpler electrolysis of $MgCl_2(aq)$ be used instead?

51. If a lead storage battery is charged at too high a voltage, gases are produced at each electrode. (It is possible to recharge a lead storage battery only because of the high overpotential for gas formation on the electrodes.)

(**a**) What are these gases?

(**b**) Write a cell reaction to describe their formation.

Faraday's laws of electrolysis

52. Calculate the quantity indicated for each of the following electrolyses:

(**a**) The mass of Zn deposited at the cathode in 815 s when 1.77 A of electric current is passed through a solution of $Zn^{2+}(aq)$.

(**b**) The time required to produce 1.96 g I_2 at the anode if a current of 3.14 A is passed through KI(aq).

(**c**) $[Cu^{2+}]$ remaining in 335 mL of a solution that was originally 0.215 M $CuSO_4$, after the passage of 2.17 A for 235 s and the deposition of Cu at the cathode.

(**d**) The time required to reduce $[Ag^+]$ in 255 mL of $AgNO_3(aq)$ from 0.185 to 0.175 M by electrolyzing the solution with Pt electrodes and a current of 1.92 A.

53. Electrolysis is carried out in the cell to the right for 2.00 hours. The platinum *cathode*, which has a mass of 25.0782 g, weighs 25.2794 g after the electrolysis. The platinum *anode* weighs the same before and after the electrolysis.

(**a**) Write plausible equations for the half-reactions occurring at the two electrodes.

(**b**) What must have been the magnitude of the current used in the electrolysis (assuming a constant current throughout)?

(**c**) A gas is collected at the *anode*. What is this gas, and what volume should it occupy if (when dry) it is measured at 23 °C and 755 mmHg pressure?

54. Concentrated $Na_2SO_4(aq)$ is electrolyzed with a current of 1.75 A for 1.32 h. Assuming that the anode reaction is ex-

lusively (21.49), what volume of $O_2(g)$ is produced at 25 °C and 747 mmHg barometric pressure if (**a**) the gas is dry; (**b**) the gas is saturated with water vapor?

55. In a silver coulometer, Ag^+ is reduced to Ag(s) at a Pt cathode. If 0.918 g Ag is deposited in 787 s by a certain quantity of electricity, (**a**) how much electric charge (in C) must have passed, and (**b**) what was the magnitude (in A) of the electric current?

★**56.** A test for completeness of electrodeposition of Cu from a solution of $Cu^{2+}(aq)$ is to add $NH_3(aq)$. A blue color signifies the formation of the complex ion $[Cu(NH_3)_4]^{2+}$ ($K_f = 1.1 \times 10^{13}$). 250.0 mL of 0.1000 M $CuSO_4(aq)$ is electrolyzed with a 3.512 A current for 1368 s. A sufficient quantity of $NH_3(aq)$ is added to complex any remaining Cu^{2+} and to maintain a free $[NH_3] = 0.10$ M. If $[Cu(NH_3)_4]^{2+}$ is detectable at concentrations as low as 1×10^{-5} M, would you expect the blue color to form in this case? [*Hint:* Use the more precise value of $\mathscr{F}$ given in the marginal note on page 760.)

Stoichiometry of oxidation–reduction reactions

57. 40.10 mL $K_2Cr_2O_7$ is required for the titration of an 0.2050-g sample of $FeSO_4$ that is 99.72% pure. What are (**a**) the molarity and (**b**) the normality of the $K_2Cr_2O_7(aq)$?

$$6 Fe^{2+} + Cr_2O_7^{2-} + 14 H^+ \rightarrow 6 Fe^{3+} + 2 Cr^{3+} + 7 H_2O$$

★**58.** 25.00 mL of the $K_2Cr_2O_7(aq)$ described in Exercise 57 is added to 25.00 mL of an acidic solution that is 0.0101 M $H_2O_2(aq)$. What is $[Cr_2O_7^{2-}]$ following chemical reaction? [*Hint:* What is the reaction?]

59. The $KMnO_4$ solution described in Example 21-14 is used in the following titration. What volume of the solution is required to react with 0.0417 g $Na_2S_2O_3$?

$$S_2O_3^{2-} + MnO_4^- + H^+ \rightarrow MnO_2(s) + SO_4^{2-} + H_2O$$
(not balanced)

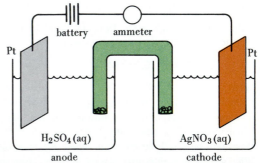

Exercise 53

Additional Exercises

60. Dichromate ion ($Cr_2O_7^{2-}$) in acidic solution is a very good oxidizing agent. Which of the following oxidations can be accomplished with dichromate ion in acidic solution? Explain.

(**a**) $Sn^{2+}(aq)$ to $Sn^{4+}(aq)$

(**b**) $I_2(s)$ to $IO_3^-(aq)$

(**c**) $Mn^{2+}(aq)$ to $MnO_4^-(aq)$

61. Predict whether, to any significant extent,

(**a**) Zn(s) will displace Al^{3+} from solution;

(**b**) MnO_4^- will oxidize Cl^- to $Cl_2(g)$ in acidic solution;

(**c**) $O_2(g)$ will oxidize S^{2-} to S in basic solution;

(**d**) Ag(s) will dissolve in 1 M HCl;

(**e**) the reaction $Mn^{2+} + O_2(g) + 2 H_2O \rightarrow MnO_2(s) + H_2O_2 + 2 H^+$ will occur as written.

62. E°_{cell} is measured for the reaction

$$3 U^{4+} + 2 NO_3^- + 2 H_2O \rightarrow 3 UO_2^{2+} + 2 NO(g) + 4 H^+$$
$E^\circ_{cell} = 0.63$ V

What is E° for the reduction of UO_2^{2+} to U^{4+} in acidic solution?

63. A $KMnO_4(aq)$ solution is to be standardized by titration against $As_2O_3(s)$. A 0.1097-g sample of As_2O_3 requires 26.10 mL of the $KMnO_4(aq)$ for its titration. What are the molarity and normality of the $KMnO_4(aq)$?

$$As_2O_3 + MnO_4^- + H^+ \rightarrow H_3AsO_4 + Mn^{2+} + H_2O$$
$$\text{(not balanced)}$$

64. The concentration of $Mn^{2+}(aq)$ can be determined by titration with $MnO_4^-(aq)$.

$$Mn^{2+} + MnO_4^- + OH^- \rightarrow MnO_2(s) + H_2O \quad \text{(not balanced)}$$

A 15.00-mL sample of $Mn^{2+}(aq)$ requires 20.06 mL of the $KMnO_4(aq)$ of Exercise 63 for its titration. What are the molarity and normality of the $Mn^{2+}(aq)$?

65. In a similar fashion to Exercise 36, would you say that the displacement of Pb^{2+} from a 1.0 M $Pb(NO_3)_2$ solution can be carried to completion by tin metal?

66. A voltaic cell is constructed based on the following reaction

$$Fe^{2+}(0.0050 \text{ M}) + Ag^+(2.0 \text{ M}) \rightarrow Fe^{3+}(0.0050 \text{ M}) + Ag(s)$$

(a) Use data from Table 21-1 to establish $E°_{cell}$ for this reaction.

(b) Use the Nernst equation to calculate E_{cell} for the stated concentrations.

★(c) The reaction proceeds in the forward direction until equilibrium is reached (at which time E_{cell} becomes zero). Calculate the concentration of Fe^{2+} remaining in solution when equilibrium is reached.

67. It is sometimes possible to separate two metal ions through electrolysis. One ion is reduced to the free metal at the cathode and the other remains in solution. Comment on the effectiveness of this method in the following cases. (That is, how complete would be the separation of the two ions?) (a) Cu^{2+} and K^+; (b) Cu^{2+} and Ag^+; (c) Pb^{2+} and Sn^{2+}.

★**68.** Show that for some fuel cells the efficiency value, $\epsilon = \Delta G°/\Delta H°$, can have a value greater than 1.00. [*Hint:* What are the conditions under which $\Delta G°$ for a reaction is greater (more negative) than $\Delta H°$? Can you identify one such reaction with data in Appendix D?]

69. The iron–chromium redox battery makes use of the reaction

$$Cr^{2+}(aq) + Fe^{3+}(aq) \rightarrow Cr^{3+}(aq) + Fe^{2+}(aq)$$

occurring at a chromium anode and an iron cathode.

(a) Write a cell diagram for this battery.

(b) Calculate the theoretical voltage of the battery.

70. When an iron pipe is partly submerged in water, as in a wooden boat dock supported on iron pipes, dissolving of the iron occurs more readily farther below the water line than it does at the water line. Explain this observation. [*Hint:* Where do the anode and cathode reactions occur? Compare this case with those pictured in Figure 21-15.]

71. Suppose that a fully charged battery contains 1.50 L of 5.00 M H_2SO_4. What will be the concentration of H_2SO_4 in the battery after 2.50 amp of current is drawn from the battery for 6.0 hours? Does the result depend on the mass of lead in the electrodes? Explain. [*Hint:* Refer to equation 21.44.]

★**72.** The amount of chemical change produced in an electrolysis depends only on the amount of electric charge (coulombs) transferred. The energy consumption, on the other hand, depends on the *product* of the charge and the voltage [volt × coulomb = V · C = J (joules)]. Determine the theoretical energy consumption *per ton of NaOH* produced in a diaphragm chloralkali cell that operates at 3.45 V. Express this energy in (a) kJ; (b) kilowatt-hours, kWh [*Hint:* To relate kWh and J, note that a watt (W) is 1 J s^{-1}.]

73. A galvanic cell is constructed of two hydrogen electrodes, one immersed in a solution with H^+ at 1.0 M and the other in 0.85 M KOH.

(a) Determine E_{cell} for the reaction that occurs.

(b) Compare your result with $E°$ for the reduction of H_2O to $H_2(g)$ in basic solution and explain.

74. If the 0.85 M KOH solution of Exercise 73 is replaced by 0.85 M NH_3

(a) Will E_{cell} be higher or lower than in 0.85 M KOH?

(b) What will be the value of E_{cell}?

★**75.** 1.00 L of a buffer solution is prepared that is 1.00 M NaH_2PO_4 and 1.00 M Na_2HPO_4. The solution is divided in half between the two compartments of an electrolysis cell. The electrodes used are Pt. Assume that the only reaction is the electrolysis of water. If electrolysis is carried out for 212 min, with a constant current of 1.25 A, what will be the pH in each cell compartment at the end of the electrolysis?

★**76.** A solution is prepared by saturating 100.0 mL of 1.00 M $NH_3(aq)$ with AgBr. A silver electrode is immersed in this solution, which is joined by a salt bridge to a S.H.E. What will be the measured E_{cell}? Is the S.H.E. the anode or cathode? [*Hint:* Review Example 19-17.]

★**77.** The electrolysis of $Na_2SO_4(aq)$ is conducted in two separate half-cells joined by a salt bridge, as suggested by the cell diagram.

$$Pt|Na_2SO_4(aq)||Na_2SO_4(aq)|Pt$$

(a) In one experiment the solution in the anode half-cell is found to become more acidic and that in the cathode compartment, more basic during the electrolysis. When the electrolysis is discontinued and the two solutions mixed, the resulting solution has pH = 7. Write half-equations and the net electrolysis equation.

(b) In a second experiment, a 10.00-mL sample of $H_2SO_4(aq)$ of unknown concentration and phenolphthalein indicator are added to the $Na_2SO_4(aq)$ in the cathode compartment. Electrolysis is carried out with a current of 21.5 mA (milliamperes) for 683 s, at which point the solution in the cathode compartment acquires a lasting pink color. What is the molarity of the unknown $H_2SO_4(aq)$?

★**78.** An Ni anode and a Fe cathode are placed in a solution with $[Ni^{2+}] = 1.0$ M and then connected to a battery.

(a) Write the equation for the electrolysis reaction.

(b) The iron has a surface area of 165 cm^2. How long must electrolysis be continued with a current of 1.50 A to build a 0.050-mm-thick deposit of nickel on the iron? (Density of nickel = 8.90 g/cm^3.)

★**79.** 100.0-mL solutions with ion concentrations of 1.000 M were placed in each of the half-cell compartments of the cell pictured in Figure 21-5. The cell was operated as an *electrolysis* cell, with copper as the anode, and zinc as the cathode. A current of 0.500 A was used. Assume that the only electrode reactions occurring were those involving Cu/Cu^{2+} and Zn/Zn^{2+}. Electrolysis was stopped after 10.00 h and the cell was allowed to function as a *voltaic* cell. What was E_{cell}?

★80. Show that if reaction (21.46) is carried out to produce $H_2O(g)$ rather than $H_2O(l)$, the methane fuel cell would have a higher efficiency value, $\epsilon = \Delta G°/\Delta H°$, than the 0.92 calculated for (21.46). On the other hand, producing $H_2O(l)$ instead of $H_2O(g)$ in the fuel cell would still produce more energy per mol CH_4 consumed. Explain why.

★81. A common reference electrode consists of a silver wire coated with AgCl(s) and immersed in 1 M KCl.

$$AgCl(s) + e^- \rightarrow Ag(s) + Cl^-(1\ M) \qquad E° = +0.2223\ V$$

  (a) What is $E°_{cell}$ when this electrode is a *cathode* in combination with a zinc electrode as an *anode?*

  (b) Cite several reasons why this electrode is easier to use than a standard hydrogen electrode.

  (c) By comparing the potential of the silver-silver chloride electrode with that of the standard silver electrode, determine K_{sp} for AgCl.

★82. Recovery as a by-product in the metallurgy of lead is an important source of Ag. The percentage of Ag in lead was determined as follows. A 1.050-g sample was dissolved in nitric acid to produce $Pb^{2+}(aq)$ and $Ag^+(aq)$. The solution was diluted to 355 mL with water, an Ag electrode was immersed in the solution, and the potential difference between this electrode and a S.H.E. was found to be 0.503 V. What was the % Ag, by mass, in the lead metal?

★83. The course of a precipitation titration can be followed by measuring the emf of a cell. The resulting curve of E_{cell} vs. volume of titrant shows a sharp change in E_{cell} at the equivalence point. The following cell is set up.

$$Hg(l)|Hg_2Cl_2(s)|Cl^-(0.10\ M)\|Ag^+(x\ M)|Ag(s)$$

The potential of the anode half-reaction $(-E°)$ remains constant at -0.2802 V.

  (a) Write an equation for the cell reaction.

  (b) Calculate E_{cell} at various points in the titration of 50.0 mL of 0.0100 M AgNO₃ carried out in the cathode half-cell compartment by adding 0.01000 M KI.

  (c) Sketch a titration curve for the titration.

Self-Test Questions

For questions 84 through 93 select the single item that best completes each statement.

84. The conversion of NpO_2^+ to Np^{4+} (a) is an oxidation half-reaction; (b) can only occur in an electrochemical cell; (c) is a reduction half-reaction; (d) is an oxidation–reduction reaction.

85. All of these metals should be soluble in HCl(aq), *except* (a) Mg; (b) Ag; (c) Zn; (d) Fe.

86. In an electrochemical cell that functions as a *voltaic* cell, (a) electrons move from the cathode to the anode; (b) electrons move through a salt bridge; (c) electrons can move either from the cathode to the anode or from the anode to the cathode, depending on the electrodes used; (d) reduction occurs at the cathode.

87. For the half-reaction

$$Hg^{2+}(aq) + 2\ e^- \rightarrow Hg(l),$$

$E° = 0.851$ V, which means that (a) Hg is more readily oxidized than H_2; (b) Hg^{2+} is more readily reduced than H^+; (c) Hg(l) will dissolve in 1 M HCl; (d) Hg(l) will displace Zn^{2+} from aqueous solution.

88. The reaction

$$Cu^{2+}(aq) + 2\ Cl^-(aq) \rightarrow Cu(s) + Cl_2(g)$$

has $E°_{cell} = -1.02$ V. This reaction (a) can be made to produce electricity in a voltaic cell; (b) can be made to occur in an electrolysis cell; (c) occurs whenever Cu^{2+} and Cl^- are brought together in aqueous solution; (d) can occur in acidic solution but not in basic solution.

89. The value of $E°_{cell}$ for the oxidation–reduction reaction $Zn(s) + Pb^{2+}(1.0\ M) \rightarrow Zn^{2+}(1.0\ M) + Pb(s)$ is +0.66 V. For the reaction $Zn(s) + Pb^{2+}(0.10\ M) \rightarrow Zn^{2+}(0.10\ M) + Pb(s)$, $E_{cell} =$ (a) +0.72 V; (b) +0.69 V; (c) +0.66 V; (d) +0.63 V.

90. For the reaction

$$Co(s) + Ni^{2+} \rightarrow Co^{2+} + Ni(s)$$

$E°_{cell} = +0.03$ V. If cobalt metal is added to a water solution in which $[Ni^{2+}] = 1.0$ M, (a) the reaction will not proceed in the forward direction at all; (b) the displacement of Ni^{2+} from solution by Co will go to completion; (c) the displacement of Ni^{2+} from solution by Co will proceed to a considerable extent, but the reaction will reach equilibrium before Ni^{2+} is completely displaced; (d) only the reverse reaction will occur.

91. The gas evolved at the *anode* when $K_2SO_4(aq)$ is electrolyzed between Pt electrodes is most likely (a) O_2; (b) H_2; (c) SO_2; (d) SO_3.

92. A quantity of electrical charge that brings about the deposition of 4.5 g Al from Al^{3+} at a cathode will also produce the following volume (STP) of $H_2(g)$ from H^+ at a cathode: (a) 44.8 L; (b) 22.4 L; (c) 11.2 L; (d) 5.6 L.

93. If a chemical reaction is carried out in a fuel cell, the maximum amount of useful work that can be obtained from the cell is (a) ΔG; (b) ΔH; (c) $\Delta G/\Delta H$; (d) $T\Delta S$.

94. Silver does not dissolve in HCl(aq), but it does dissolve in $HNO_3(aq)$, producing $Ag^+(aq)$.

  (a) Explain the difference in the behavior of silver toward these two acids.

  (b) Write an equation for the reaction of Ag with $HNO_3(aq)$.

95. Diagram a voltaic (galvanic) cell in which the following reaction occurs. Label the anode and cathode; use a table of standard electrode potentials to determine $E°_{cell}$; and balance the equation for the cell reaction.

$$Zn(s) + H^+ + NO_3^- \rightarrow Zn^{2+} + H_2O + NO(g)$$

96. The voltaic (galvanic) cell indicated below registers an $E_{cell} = +0.108$ V. What is the pH of the unknown solution?

$$Pt,\ H_2(g,\ 1\ atm)|H^+(x\ M)\|H^+(0.10\ M)|H_2(g,\ 1\ atm),\ Pt$$

97. $E°_{cell} = -0.0050$ V for the reaction

$$2\ Cu^+ + Sn^{4+} \rightarrow 2\ Cu^{2+} + Sn^{2+}$$

  (a) Can a solution be prepared that is 0.500 M in each of the four ions at 298 K?

  (b) If not, in what direction must a net reaction occur?

22

Chemistry of the Representative (Main-Group) Elements I: Metals

Several representative metals and their compounds are used in fireworks. The light emitted by these fireworks is produced by burning magnesium metal. [E. R. Degginger]

791

The representative metals range from the most metallic elements—the heavier members of group 1A—to the rather noble mercury, Hg. Here, as in our introduction to descriptive chemistry (Chapter 14), we look for similarities in properties, especially as they relate to locations in the periodic table. However, we also encounter wide variations in certain properties, such as reactivity toward water, acids, and bases. Some metal salts have high water solubility, whereas other metal salts are very insoluble. In ionic form, some of the representative metals (e.g., Na, K, and Mg) are essential to humans; other metal ions (e.g., those of Pb, Cd, and Hg) are highly toxic. Some of the representative metal compounds are easily reduced to the free metals with a reducing agent such as carbon; others require electrolysis of a molten compound.

To sort out this rich variety of observations, to the extent possible, we will use fundamental principles—periodic relationships, bonding theory, enthalpy and free-energy changes, electrode potentials, equilibrium constants. As you develop an understanding of why metals behave as they do, you should also gain some insights into why certain commercial methods have developed for producing metals from their compounds and an appreciation of why metals and their compounds are used in some of the ways that they are.

22-1 Group 1A—The Alkali Metals

We discussed several properties of the alkali metals in Section 14-5 and listed those properties in Table 14-4. Table 22-1 lists some additional properties. The main conclusions we can draw from these two tables is that the alkali metals are a group of active (the most active) metals. For example, they have low ionization energies and large negative electrode potentials, and they displace hydrogen from water. Another conclusion is that, in general, properties vary uniformly within the group in accordance with the periodic law. There are some irregularities, though, and these are found mostly with the *first* member, Li. Here are some of the ways in which lithium and its compounds differ from the other alkali metals.

- Low solubility of lithium carbonate, fluoride, hydroxide, and phosphate.
- Ability to form a nitride (Li_3N).
- Formation of a normal oxide (Li_2O) rather than a peroxide or superoxide.
- On heating, decomposition of lithium carbonate and hydroxide to the oxide.

In some of these properties, lithium and its compounds resemble magnesium and its compounds. This is an example of the *diagonal relationship* described on page 491. Another point made repeatedly in Chapter 14 applies here as well. *The first member of a group differs from heavier members of the group in some significant ways.* Among the alkali metals the primary reason for this difference is the high charge density of the Li^+ ion compared to the other alkali metal ions. The charge density is the ratio of the ionic charge to the ionic radius (see Table 22-1).

Because of its high charge density, Li^+ has a large *negative* enthalpy of hydration, $\Delta H^\circ_{hydr.}$. **Hydration energy** is the enthalpy change that accompanies the dissolving of 1 mol of gaseous ions in water. As pictured in Figure 22-1, a Li^+ ion in water is surrounded by water molecules. Energy is released because of the attraction of the negative ends of H_2O dipoles to the positive Li^+ ion. Other alkali metal ions are also hydrated in aqueous solutions, but Li^+ is more strongly hydrated because of its especially high positive charge density.

Electrode Potentials. What is the significance of the large negative electrode potentials of the alkali metals in Table 22-1? They mean that the metal ions M^+ can be reduced to the free metals M only *with great difficulty*. On the other hand, the reverse process—oxidation of the metals—occurs *with great ease*.

The earlier discussion of the alkali metals in Section 14-5 dealt with

- periodic relationships;
- occurrence, preparation, and uses of the metals;
- reactions of the metals;
- preparation and uses of important compounds.

792

TABLE 22-1
Properties of the Alkali Metals

	Li	Na	K	Rb	Cs
metallic radius, pm	155	190	235	248	267
ionic radius, pm	60	95	133	148	169
ionic charge density	+1.67	+1.05	+0.75	+0.68	+0.59
[ionic charge/(ionic radius, Å)]					
sublimation energy, kJ/mol	155	109	90	86	79
[$M(s) \rightarrow M(g)$; $\Delta H^\circ_{subl.}$]					
first ionization energy, kJ/mol	520	496	419	403	376
[$M(g) \rightarrow M^+(g) + e^-$; $\Delta H^\circ_{ioniz.}$]					
hydration energy, kJ/mol	−506	−397	−318	−289	−259
[$M^+(g) \rightarrow M^+(aq)$; $\Delta H^\circ_{hydr.}$]					
the sum $\{\Delta H^\circ_{subl.} + \Delta H^\circ_{ioniz.} + \Delta H^\circ_{hydr.}\}^a$	169	208	191	200	196
electrode potential E°, V	−3.045	−2.714	−2.925	−2.925	−2.923
[$M^+(aq) + e^- \rightarrow M(s)$]					

Several physical properties are listed in Table 14-4.
[a]The significance of this sum of terms is discussed below.

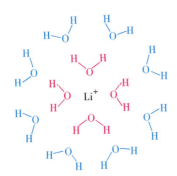

FIGURE 22-1
Hydration of a Li$^+$ ion.

The Li$^+$ ion has a small number of H_2O molecules held to it by electrostatic forces in a primary hydration sphere (in red). These molecules, in turn, hold other H_2O molecules, but more weakly, in a secondary hydration sphere (blue). In all, the Li$^+$ ion may have about 25 H_2O molecules surrounding it in a three-dimensional envelope (not the planar arrangement suggested here).

$$M(s) \longrightarrow M^+(aq) + e^- - E^\circ = +3.045, +2.714, +2.925, +2.925, +2.923 \text{ V}$$
$$\text{for Li, Na, K, Rb, and Cs, respectively.}$$

If we use the magnitude of the first ionization energy (Table 22-1) as an indicator of ease of oxidation, Li is the most difficult of the alkali metals to oxidize. On the other hand, if we use $-E^\circ$ as our criterion, Li is the *easiest* to oxidize. How can we explain this discrepancy?

First, the discrepancy is only apparent. These two criteria do not measure the same thing. Ionization energy measures the tendency for a *gaseous* atom to ionize. $-E^\circ$ measures the tendency for a metal atom to go from the *solid* state to the ion in *aqueous solution*.

Another way to look at the matter is that the process $M(s) \rightarrow M^+(aq) + e^-$ is actually the sum of *three* hypothetical steps, and the ionization energy describes just one of these steps. We learned in Section 21-3 that electrode potentials and E_{cell} values are related to free energy changes, ΔG. Also, we know that $\Delta G = \Delta H - T\Delta S$. If we assume that the entropy changes are small for the processes involved here, then $\Delta G \approx \Delta H$. We can assess the matter through ΔH values.

Oxidation of an alkali metal, M, to produce the ions $M^+(aq)$ involves the hypothetical steps

1. Subliming the solid metal to the gaseous state.
2. Ionizing the gaseous metal atoms to gaseous metal ions.
3. Hydrating the ions by dissolving $M^+(g)$ in water to produce $M^+(aq)$.

These three steps are pictured in Figure 22-2. The sum of their ΔH values is the ΔH for the oxidation half-reaction.

$$\Delta H^\circ_{ox.} = \Delta H^\circ_1 + \Delta H^\circ_2 + \Delta H^\circ_3 = \Delta H^\circ_{subl.} + \Delta H^\circ_{ioniz.} + \Delta H^\circ_{hydr.} \quad (22.1)$$

The sums of these three ΔH values for the alkali metals are listed in Table 22-1. We see that they are positive quantities, but that the value is smallest for Li (169 kJ). This is primarily because of the large negative ΔH for the hydration of Li$^+$. Thus, it appears that Li(s) should be the easiest to oxidize of the alkali metals.*

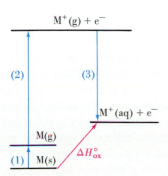

FIGURE 22-2
Representation of ΔH°_{ox} of an alkali metal,

$$M(s) \rightarrow M^+(aq) + e^-$$

(1) Sublimation of metal atoms.
(2) Ionization of gaseous metal atoms. (3) Hydration of gaseous metal ions.

Preparation of the Alkali Metals. Because they are so easily oxidized, the alkali metals themselves are excellent reducing agents. As a result one alkali metal can be

*To complete the discussion of why Li is the most easily oxidized of the alkali metals in aqueous solution, we should also consider the reduction process (the reduction of H^+ to H_2). This is offered as an exercise to interested students (Exercise 68).

used to produce another, as we outlined in Section 14-5 for the production of potassium.

$$KCl(l) + Na(l) \xrightarrow{850\ °C} NaCl(l) + K(g) \tag{14.19}$$

The problem in using reaction (14.19), however, is how do we get sodium metal? The most feasible method of preparing sodium metal is by the electrolysis of one of its molten salts.

The first commercial electrolytic method (the Castner process) involved the electrolysis of molten NaOH. The most widely used method now, however, is the electrolysis of molten NaCl (the Downs process).

Example 22-1

Predicting electrolysis reactions. The electrolysis of $NaCl(l)$ produces $Na(l)$ and $Cl_2(g)$ (see equation 14.18). What products would you expect from the electrolysis of molten NaOH?

Solution. The only ions present in molten NaOH are Na^+ and OH^-. The *reduction* of Na^+ yields Na as a product. In the oxidation process, one of the elements in the OH^- ion must *increase* in its oxidation state (O.S.). In OH^- the O.S. of O is -2 and that of H is $+1$. The O.S. of H cannot be greater than $+1$, but the O.S. of O can increase to 0 (in O_2).

Oxidation: $4\ OH^- \longrightarrow O_2(g) + 2\ H_2O(g) + 4\ e^-$
Reduction: $4\ \{Na^+ + e^- \longrightarrow Na(l)\}$

Net: $4\ Na^+ + 4\ OH^- \xrightarrow[\text{electrolysis}]{\text{molten}} 4\ Na(l) + 2\ H_2O(g) + O_2(g) \tag{22.2}$

SIMILAR EXAMPLES: Exercises 12, 18.

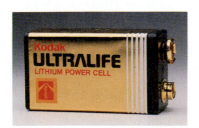

A modern lithium battery.
[Carey B. Van Loon]

TABLE 22-2
Some Li Battery Systems

Li/FeS	1.3 V
Li/CuO	1.5
Li/SO₂	2.8
Li/SOCl₂	3.6
Li/SO₂Cl₂	3.8

Alkali Metals in Modern Technology. Some of the most important uses of the alkali metals are based on special properties of these elements. Both Li and Na are destined to have important roles in the nuclear industry. Liquid sodium is an excellent heat transfer medium that can be used as a coolant in a nuclear power plant (Section 26-8). Lithium also has promise as a heat transfer medium and as a source of tritium for nuclear fusion reactors (Section 26-9).

Another interesting use of Li and Na is in modern batteries referred to as "advanced" batteries. Because of the ease with which it is oxidized, lithium makes a good anode in a battery system. What also makes lithium valuable in batteries is its low equivalent weight. Only 6.94 g Li (1 mol) needs to be consumed to produce 1 mol electrons. Lithium batteries are particularly useful where the installed battery must have a high reliability and lifetime, such as in computer memory systems and in cardiac pacemakers. Table 22-2 lists several lithium battery systems that are either presently available or under development. In each case Li is oxidized to Li^+ at the anode and the other battery component undergoes a reduction half-reaction. Generally, water cannot be part of the battery system since lithium metal reacts if it is in direct contact with water. Various solvents (mostly organic compounds) and electrolytes are used within the battery; some lithium batteries (e.g., Li/FeS) use molten salts as electrolytes and they must be maintained at high temperatures.

Preparation of Alkali Metal Compounds. Figure 22-3 introduces a new format for summarizing reaction chemistry. It deals with sodium compounds, but similar diagrams can be written for other elements. In this diagram we note a compound of

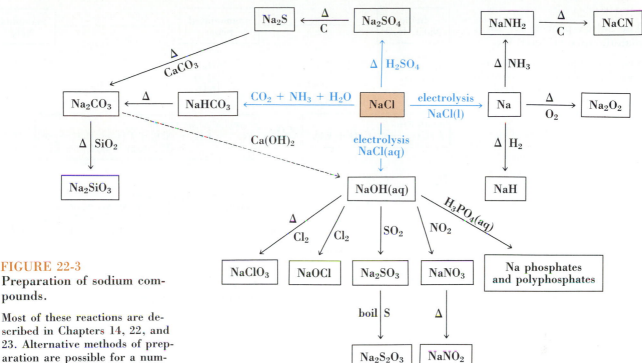

FIGURE 22-3
Preparation of sodium compounds.

Most of these reactions are described in Chapters 14, 22, and 23. Alternative methods of preparation are possible for a number of these compounds. The conversion of Na_2CO_3 to NaOH (broken line arrow) is no longer of commercial importance.

central importance (NaCl in Figure 22-3) and then show how several other compounds can be derived from it. Some of these conversions are single-step reactions, such as the reaction of NaCl to form Na_2SO_4; others may involve two or more consecutive reactions, as does the preparation of sodium silicate, Na_2SiO_3. The principal reactants required for the conversions are written with the arrow ($\rightarrow$), and the need to heat a reaction mixture may be noted with a Δ symbol. The reactions may also produce other products that are not noted in the diagram.

Example 22-2

Writing chemical equations from a summary diagram of reaction chemistry. Propose a method of synthesizing sodium carbonate from sodium chloride.

Solution. First turn to Figure 22-3 and see if you can find a route from NaCl to Na_2CO_3. One such route involves the conversions $NaCl \rightarrow Na_2SO_4 \rightarrow Na_2S \rightarrow Na_2CO_3$. Other necessary reactants (H_2SO_4, C, and $CaCO_3$) are also noted. In the set of equations written below you must, of course, propose other plausible reaction products as needed (such as HCl, CO, and CaS).

$$NaCl(s) + H_2SO_4 \text{ (conc. aq)} \xrightarrow{\Delta} NaHSO_4(s) + HCl(g)$$

$$NaCl(s) + NaHSO_4(s) \xrightarrow{\Delta} Na_2SO_4(s) + HCl(g)$$

Next, the Na_2SO_4 is reduced to Na_2S with carbon.

$$Na_2SO_4(s) + 4 C(s) \xrightarrow{\Delta} Na_2S(s) + 4 CO(g)$$

The final step is a reaction between Na_2S and $CaCO_3$.

We saw a reaction similar to the first two in equation (14.25), and the third reaction is reaction (14.26).

$$Na_2S(s) + CaCO_3(s) \xrightarrow{\Delta} CaS(s) + Na_2CO_3(s) \qquad (22.3)$$

SIMILAR EXAMPLES: Exercises 9, 11, 17.

FIGURE 22-4

The Solvay process for the manufacture of NaHCO$_3$.

The main reaction sequence is traced by solid arrows. Recycling reactions are shown by broken arrows.

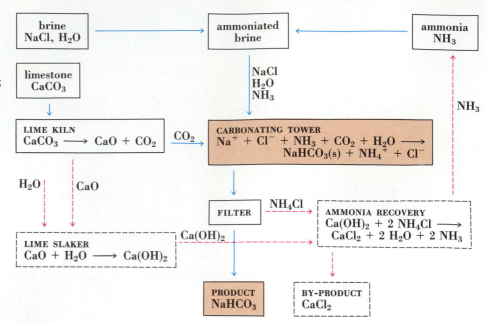

The Solvay Process. Until the advent of the chlor-alkali processes in the present century, the key conversion in the production of sodium compounds (such as NaOH) was that of NaCl to Na$_2$CO$_3$. The best method was that devised by the Belgian engineer Ernest Solvay in 1865, outlined in Figure 22-4.

The key step in the Solvay process involves the reaction of NH$_3$(g) and CO$_2$(g)

CHEMISTRY EVOLVING

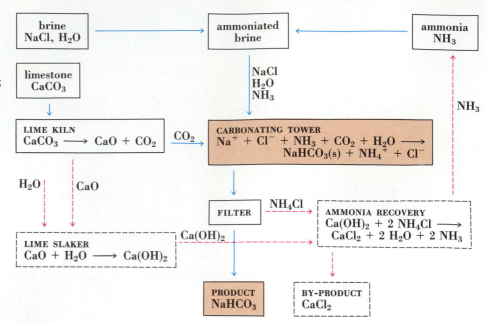

FIGURE 22-5

United States production of sodium carbonate.

During the past two decades the synthetic sodium carbonate industry (Solvay process) has disappeared in the United States. Mining of natural sodium carbonate has grown to about 8 million tons per year (about 25% of which is exported). The bulk of Na$_2$CO$_3$ produced is used in the manufacture of glass.

Manufacture of Sodium Carbonate

One of the chief uses of sodium carbonate has been in the conversion of Ca(OH)$_2$(s) to concentrated aqueous solutions of NaOH. This was the topic of the Summarizing Example in Chapter 19.

$$Ca(OH)_2(s) + 2\ Na^+(aq) + CO_3^{2-}(aq) \rightleftharpoons$$
$$CaCO_3(s) + 2\ Na^+(aq) + 2\ OH^-(aq) \quad (22.4)$$

The demand for Na$_2$CO$_3$ began to skyrocket in the eighteenth century, primarily for use in reaction (22.4). The NaOH(aq), in turn, was used in the manufacture of soap.

In 1775, in order to gain independence from the importation of natural Na$_2$CO$_3$, the French government offered a prize to anyone who could devise a process to prepare Na$_2$CO$_3$ from NaCl. Nicolas Leblanc responded with the process described through Example 22-2. By the turn of the twentieth century, however, the Leblanc process was almost completely displaced by the more economical Solvay process.

The great success of the Solvay process is in the efficient use of raw materials through recycling. Only one ultimate by-product results from the process—CaCl$_2$. The demand for CaCl$_2$ is quite limited, however, and only a small percentage of the several million tons produced annually in the United States could ever be consumed. Most of the CaCl$_2$ was dumped, generally into streams, but environmental restrictions no longer permit this. Partly for this reason, Na$_2$CO$_3$ is no longer being manufactured in the United States by the Solvay process (although the process continues to be used elsewhere in the world). In the United States it is more economical to extract Na$_2$CO$_3$ from two natural sources—certain salt lakes and brine deposits, primarily in California, and huge underground deposits in Wyoming of the mineral *trona* (Na$_2$CO$_3 \cdot$ NaHCO$_3 \cdot$ 2H$_2$O). Figure 22-5 shows how rapidly the method of producing sodium carbonate has changed because of economic and environmental factors. ∎∎∎

with *cold* concentrated NaCl(aq). Of the possible ionic compounds that could conceivably precipitate from such a mixture (NaCl, NH_4Cl, $NaHCO_3$, and NH_4HCO_3), the least soluble is $NaHCO_3$.

The Solvay process for $NaHCO_3$.

$$Na^+ + Cl^- + NH_3 + CO_2 + H_2O \longrightarrow NaHCO_3(s) + NH_4^+ + Cl^- \quad (22.5)$$

Sodium carbonate is produced by heating sodium bicarbonate.

$$2\, NaHCO_3(s) \xrightarrow{\Delta} Na_2CO_3(s) + H_2O(g) + CO_2(g) \quad (22.6)$$

One thing you should notice about the Solvay process in Figure 22-4 is the way in which fairly simple precipitation and acid–base reactions are used. Another important factor is the efficient way in which materials produced in one step are recycled or used in a subsequent step. Thus, when limestone ($CaCO_3$) is heated to produce the principal reactant, CO_2, the other reaction product, CaO, is also used. It is converted to $Ca(OH)_2$, which is then used to convert NH_4Cl (another reaction by-product) to $NH_3(g)$. The $NH_3(g)$ is recycled into the production of ammoniated brine.

22-2 Group 2A—The Alkaline Earth Metals

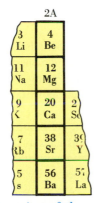

Table 22-3 lists some of the same properties for group 2A that we saw for group 1A in Table 22-1. The data from these two tables allow us to make certain comparisons between the two groups of metals.

Atomic (metallic) radii are smaller for the group 2A elements than are the corresponding radii for group 1A elements, for the reasons we learned in Chapter 9. The radii of the group 2A metal ions (M^{2+}) are also smaller than the radii of the group 1A ions (M^+). This is primarily because the 2A ions carry a charge of +2 instead of +1. Because of their high charges and small radii, the 2A metal ions have high positive charge densities (for example, +3.08 for Mg^{2+} compared to +1.05 for Na^+). This suggests that the ions M^{2+} should have high hydration energies, which indeed they do.

As seen in Table 22-3, the energy required to strip a group 2A metal atom of its two valence electrons—the sum of the first and second ionization energies—is very large. Since the energy required to remove just one electron is much less, you may

The earlier discussion of the alkaline earth metals in Section 14-6 considered

- occurrence, preparation, and uses of the metals;
- reactions of the metals;
- preparation and uses of important compounds;
- special properties of beryllium.

TABLE 22-3
Properties of the Alkaline Earth Metals

	Be	Mg	Ca	Sr	Ba
metallic radius, pm	111	160	197	215	217
ionic radius, pm	31	65	99	113	135
ionic charge density [ionic charge/(ionic radius, Å)]	+6.45	+3.08	+2.02	+1.77	+1.48
1st ionization energy, kJ/mol [$M(g) \rightarrow M^+(g) + e^-$]	+900	+738	+590	+550	+503
2nd ionization energy, kJ/mol [$M^+(g) \rightarrow M^{2+}(g) + e^-$]	+1757	+1451	+1145	+1064	+965
1st + 2nd ionization energy, kJ/mol [$M(g) \rightarrow M^{2+}(g) + 2\, e^-$]	+2657	+2189	+1735	+1614	+1468
hydration energy, kJ/mol [$M^{2+}(g) + aq \rightarrow M^{2+}(aq)$]	−2385	−1940	−1600	−1460	−1320
electrode potential $E°$, V [$M^{2+}(aq) + 2\, e^- \rightarrow M(s)$]	−1.70	−2.375	−2.76	−2.89	−2.90

Additional properties are listed in Table 14-6.

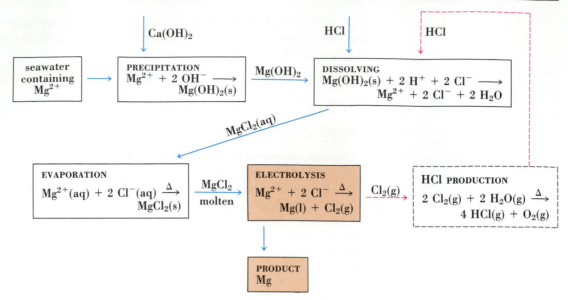

FIGURE 22-6
The Dow process for the production of Mg.

The main reaction sequence is traced by solid arrows. Recycling of $Cl_2(g)$ is shown by broken arrows.

question why alkaline earth metals (M) do not form compounds like MCl instead of MCl_2. The answer to this question lies in considering *all* the energy terms involved in the production of MCl and MCl_2 through the Born–Fajans–Haber method. We find that the formation of MCl_2 is energetically much more favorable (see page 313).

Because of their large negative electrode potentials, the group 2A ions are difficult to reduce to the free metal, though not quite so difficult as the group 1A ions. Although the alkaline earth metals can be produced by chemical reduction, electrolysis is the preferred route for magnesium, as described next.

Production of Magnesium—The Dow Process. Mg^{2+} is the only common ion in seawater and underground brine solutions that forms an insoluble hydroxide. The first step in the Dow process is the precipitation of $Mg(OH)_2(s)$. The source of OH^- is slaked lime $[Ca(OH)_2]$, which is formed by the reaction of quicklime (CaO) with water. Quicklime, in turn, is produced from limestone. In Example 19-7 we demonstrated by calculation that the precipitation of Mg^{2+} should go to completion.

These reactions, starting with limestone, are discussed on page 494.

The precipitated $Mg(OH)_2(s)$ is washed, filtered, and dissolved in HCl(aq). The resulting concentrated $MgCl_2(aq)$ is evaporated nearly to dryness. The $MgCl_2$ is then melted and electrolyzed, yielding pure Mg metal and $Cl_2(g)$. The $Cl_2(g)$ is recycled by conversion to HCl. The Dow process is outlined in Figure 22-6, and the electrolysis of $MgCl_2(l)$ is pictured in Figure 22-7.

Industrial operations that consume the greatest quantities of energy are those involving electrolysis and/or heating materials to high temperatures. The Dow process requires that both be done. The production of 1 kg Mg requires about 300 MJ of energy (300 MJ = 300×10^6 J). The process could be modified to reduce this requirement to perhaps 200 MJ. By contrast, the energy requirement to melt and recast recycled Mg is about 7 MJ/kg Mg. So, we see that recycling, where possible, can be very cost effective in the production and use of materials.

In addition to its many applications in lightweight alloys, Mg is used in batteries, in fireworks and flash photography, in the formation of important reagents for organic chemical reactions, and as a reducing agent in metallurgy.

FIGURE 22-7

The electrolysis of molten MgCl$_2$.

The actual electrolyte is a mixture of molten NaCl, CaCl$_2$, and MgCl$_2$. This mixture has a lower melting point and higher electrical conductivity than does MgCl$_2$ alone, but only Mg^{2+} is reduced under the conditions employed in the electrolysis.

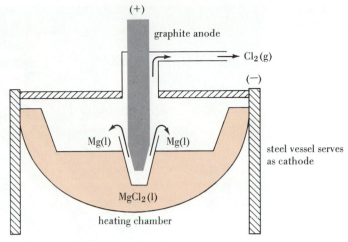

Oxidation: $2 Cl^- \rightarrow Cl_2(g) + 2 e^-$

Reduction: $Mg^{2+} + 2 e^- \rightarrow Mg(l)$

Alkaline Earth Metal Compounds. We have noted that alkaline earth metal ions (M^{2+}) have high positive charge densities. In combination with certain anions this leads to high lattice energies and compounds that are either insoluble or only moderately soluble in water (e.g., carbonates, fluorides, hydroxides, and oxides). Where lattice energy is not so great, the large hydration energies of the cations may result in high water solubility and salts that crystallize from solution as hydrates. Typical hydrates are MX$_2 \cdot$ 6H$_2$O (where M = Mg, Ca, or Sr, and X = Cl or Br).

The greater the water solubility of a solute, the more concentrated its saturated solution, and the greater the lowering of the vapor pressure of water above the solution. Sometimes the partial pressure of water vapor in the air is greater than it would be above saturated solutions of certain solutes. In these cases, water vapor from the air condenses on the solid solute and the solid dissolves in the liquid water, forming a solution. The salt is said to **deliquesce.** The MX$_2 \cdot$ 6H$_2$O hydrates mentioned above are all deliquescent. In fact, CaCl$_2 \cdot$ 6H$_2$O (the by-product of the Solvay process) has been used to control dust on dirt roads. The hydrate deliquesces, and the solution of CaCl$_2$ that is formed keeps the road wet.

Saturated solutions of deliquescent solutes can be used to maintain a constant humidity. Consider a saturated solution of MgCl$_2 \cdot$ 6H$_2$O at 25 °C. The partial pressure of water vapor above this solution is 7.84 mmHg. Above pure water at 25 °C the water vapor pressure is 23.76 mmHg. The **relative humidity** of air is the ratio of the actual water vapor pressure to the water vapor pressure in air that is saturated with water vapor, expressed on a percent basis. The relative humidity of air above a saturated solution of MgCl$_2 \cdot$ 6H$_2$O at 25 °C is

$$\text{relative humidity} = \frac{7.84 \text{ mmHg}}{23.76 \text{ mmHg}} \times 100 = 33.0\%$$

An equilibrium mixture of solid MgCl$_2 \cdot$ 6H$_2$O(s), its saturated solution, and water vapor at 25 °C, maintains a constant relative humidity of 33.0%. Saturated solutions of other solutes can be used to establish other constant relative humidities. (In addition, certain solid hydrates must be maintained at constant humidities so that they neither lose water of hydration nor gain additional water of hydration.)

A property of the alkaline earths not shared by the alkali metals (except Li) is the instability of their carbonates at high temperature.

Decomposition of group 2A metal carbonates.

$$MCO_3(s) \xrightarrow{\Delta} MO(s) + CO_2(g) \qquad (22.7)$$

We can use this instability as the basis of a method of preparing the metal oxides. And we can easily form the hydroxide from the oxide.

The decomposition (calcination) of limestone is carried out in long rotary kilns, whether for the production of lime (reaction 22.7) or the manufacture of Portland cement (page 503). [Courtesy National Lime Association]

Reaction of group 2A metal oxides with water.

$$MO(s) + H_2O \longrightarrow M(OH)_2(s) \tag{22.8}$$

Because calcium carbonate (limestone) is so abundant, $Ca(OH)_2$ is the cheapest base available to the chemical industry.

Calcium Ion in Natural Waters. Carbonates, especially $CaCO_3$, are involved in some chemical processes that are responsible for a number of natural phenomena. One process begins when rainwater dissolves atmospheric $CO_2(g)$. This makes the water slightly acidic, through the formation and ionization of carbonic acid.

$$CO_2(g) + H_2O \rightleftharpoons H_2CO_3 \tag{22.9}$$

$$H_2CO_3 + H_2O \rightleftharpoons H_3O^+ + HCO_3^- \qquad K_{a_1} = 4.2 \times 10^{-7} \tag{22.10}$$

$$HCO_3^- + H_2O \rightleftharpoons H_3O^+ + CO_3^{2-} \qquad K_{a_2} = 5.6 \times 10^{-11} \tag{22.11}$$

Rainwater that is saturated with $CO_2(aq)$ has a pH of about 5.6 as a result of these reactions. As rain falls through air that is polluted with oxides of sulfur and nitrogen and possibly some organic acids, its pH falls still lower. In the northeast and midwest regions of the United States rainfall typically averages in pH from 3.0 to 5.0.

As rainwater charged with CO_2 and possibly other acids seeps through limestone beds, insoluble $CaCO_3$ is converted to soluble $Ca(HCO_3)_2$.

Reaction involved in the formation of limestone caverns.

$$CaCO_3(s) + H_2O + CO_2 \rightleftharpoons Ca(HCO_3)_2(aq) \tag{22.12}$$

Over time this dissolving action can produce a large cavity in the limestone bed—a limestone cave. Reaction (22.12) is reversible, however, and the evaporation of a solution of $Ca(HCO_3)_2$ causes a loss of both water and CO_2 and conversion of $Ca(HCO_3)_2$ back to $CaCO_3(s)$. This process occurs very slowly, but over a period of many years, as $Ca(HCO_3)_2(aq)$ drips from the ceiling of a cave, $CaCO_3(s)$ remains as icicle-like deposits called **stalactites.** Some of the dripping $Ca(HCO_3)_2(aq)$ may hit the floor of the cave before decomposition occurs and limestone ($CaCO_3$) deposits build up from the floor in formations called **stalagmites.** Eventually, some stalactites and stalagmites grow together into limestone columns (see Figure 22-8).

Example 22-3 ────────────────────────────

Describing simultaneous equilibria in qualitative terms. Without performing detailed calculations, show how the data of equations (22.10) and (22.11), to-

FIGURE 22-8
Stalactites, stalagmites, and columns in a limestone cavern. [E. R. Degginger]

gether with K_{sp} for $CaCO_3$ (2.8×10^{-9}), can be used to establish that reaction (22.12) describes the dissolving action of rainwater on limestone ($CaCO_3$).

Solution. We need to use one idea from Section 17-7 and another from Section 19-2.

1. In $H_2CO_3(aq)$, a *diprotic* acid, since $K_{a_1} \gg K_{a_2}$, $[H_3O^+] = [HCO_3^-]$ and $[CO_3^{2-}] = K_{a_2} = 5.6 \times 10^{-11}$ M (see again, pages 621–22).
2. When pure water is saturated with $CaCO_3$, $[Ca^{2+}] = [CO_3^{2-}]$; $K_{sp} = [Ca^{2+}][CO_3^{2-}] = [CO_3^{2-}]^2 = 2.8 \times 10^{-9}$. In this solution $[CO_3^{2-}] = \sqrt{2.8 \times 10^{-9}} = 5.3 \times 10^{-5}$ M.

$[CO_3^{2-}]$ in saturated $CaCO_3(aq)$, 5.3×10^{-5} M, is much higher than $[CO_3^{2-}]$ normally produced by the ionization of H_2CO_3, 5.6×10^{-11} M. Carbonate ion derived from $CaCO_3$ acts as a common ion and displaces equilibrium (22.11) *to the left*, converting CO_3^{2-} to HCO_3^- and consuming H_3O^+. Removal of H_3O^+ in this way stimulates equilibrium (22.10) to shift *to the right* to produce more H_3O^+. In turn, equilibrium (22.9) also shifts *to the right* to replace the H_2CO_3 that has ionized. The net effect is that CO_2, H_2O, and CO_3^{2-} (from $CaCO_3$) are consumed and HCO_3^- [as $Ca(HCO_3)_2$] is produced.

SIMILAR EXAMPLES: Exercises 16, 66, 67.

Rainwater, because of its dissolving action on atmospheric gases, soil, and rocks, is not chemically pure. It may pick up anywhere from a few parts to perhaps 1000 parts per million (ppm) of dissolved substances. If the water contains ions capable of yielding significant quantities of a precipitate, we say that the water is **hard**. Water that owes its hardness primarily to bicarbonate ion, HCO_3^-, and associated cations is said to be **temporary hard water.**

The decomposition of $HCO_3^-(aq)$, the reverse of reaction (22.12), is greatly accelerated when water is heated. To describe what happens, in stepwise fashion, we should also think of the reverse of reactions (22.9) and (22.10). That is, H_2CO_3 breaks down into H_2O and $CO_2(g)$, which escapes. Some bicarbonate ion, HCO_3^-, combines with H_3O^+ to replace the H_2CO_3. And, to replenish the H_3O^+ consumed in this way, some HCO_3^- dissociates further to H_3O^+ and CO_3^{2-}. The overall effect of all these equilibrium shifts is a net reaction in which HCO_3^- decomposes to CO_3^{2-}, H_2O, and CO_2.

$$H_2CO_3 \xrightarrow{\Delta} H_2O + CO_2(g)$$
$$HCO_3^- + H_3O^+ \longrightarrow H_2O + H_2CO_3$$
$$HCO_3^- + H_2O \longrightarrow H_3O^+ + CO_3^{2-}$$

Effect of heating on temporary hard water.

$$2\ HCO_3^- \xrightarrow{\Delta} CO_3^{2-} + H_2O + CO_2(g) \tag{22.13}$$

Reaction (22.13) outlines the most undesirable effect of temporary hard water. When the water is heated, regenerated CO_3^{2-} reacts with multivalent cations in the water to form a mixed precipitate of $CaCO_3$, $MgCO_3$, and $FeCO_3$ called **boiler scale.** The formation of boiler scale can be a very serious problem in industries that use steam-driven electric power plants. Formation of boiler scale can cause a boiler to overheat and presents an explosion hazard.

Water Softening. In addition to temporary hardness (also called bicarbonate hardness), we may encounter **permanent hard water.** This is water containing significant concentrations of anions other than HCO_3^-, such as SO_4^{2-}, together with associated cations. "Water softening" refers to the removal of natural mineral impurities, and there are several ways we can do this. We can soften temporary hard

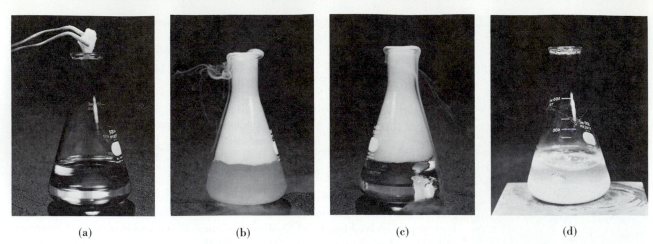

Reactions of $Ca^{2+}(aq)$ and CO_2. (a) CO_2 (dry ice) is added to dilute $Ca(OH)_2(aq)$. (b) $CaCO_3(s)$ precipitates. (c) The precipitate redissolves in the presence of excess $CO_2(aq)$ to form $Ca(HCO_3)_2(aq)$. (d) $Ca(HCO_3)_2$ decomposes on heating and reprecipitates as $CaCO_3(s)$. [Carey B. Van Loon]

water just by boiling it, but when we do boiler scale forms. Permanent hard water cannot be softened by boiling.

A more satisfactory way than boiling for softening temporary hard water is to treat it with a base and filter off the precipitated metal carbonate.

Chemical treatment of temporary hard water.

$$HCO_3^- + OH^- \longrightarrow H_2O + CO_3^{2-} \tag{22.14}$$

$$CO_3^{2-} + M^{2+} \longrightarrow MCO_3(s) \tag{22.15}$$

The source of OH^- for reaction (22.14) may be slaked lime [$Ca(OH)_2$] or *washing soda*, Na_2CO_3. (Can you write an equation to show how Na_2CO_3 produces OH^- by hydrolysis?) We can also soften permanent hard water with Na_2CO_3. Cations such as Ca^{2+} and Mg^{2+} precipitate as carbonates and Na_2SO_4 remains in solution.

Ion Exchange. One of the best ways to soften water is through **ion exchange,** that is, to exchange the undesirable ions in hard water for ions that are not objectionable. The ion exchange medium may be a natural sodium aluminosilicate called a **zeolite** or a synthetic resinous material. Ion exchange materials consist of macromolecular (polymer) particles that can ionize to produce two types of ions: *fixed* ions that remain attached to the polymer surface and free or mobile *counterions*. The counterions are the ones that exchange places with the undesirable ions when a sample of hard water is passed over the resin.

In the resin pictured in Figure 22-9, the fixed ions R are negatively charged and the counterions are positively charged. At the start the counterions in the resin bed are Na^+. When hard water is passed through the bed, the more highly charged Ca^{2+}, Mg^{2+}, and Fe^{2+} displace Na^+ as counterions. To regenerate the resin, we can pass concentrated $NaCl(aq)$ through the bed. When present in high concentration, Na^+ is able to displace multivalent cations from the bed and restore the ion exchange resin to its original condition. The ion exchange material has an indefinite lifetime.

The only material consumed in water softening by the process just described is the NaCl used to regenerate the ion exchange medium. Water softening by ion exchange does have the disadvantage that the treated water has a high concentration of Na^+ and would not be appropriate for drinking by a person on a low-sodium diet.

Ion exchange processes can be illustrated through simple chemical equations. For a zeolite, we use can the symbol Z, and for a synthetic resin, the symbol R.

Ion exchange reactions.

$$Na_2Z + M^{2+} \longrightarrow MZ + 2 Na^+ \tag{22.16}$$

$$Na_2R + M^{2+} \longrightarrow MR + 2 Na^+ \tag{22.17}$$

We can use H^+ instead of Na^+ as the counterions in an ion exchange resin. We can do this by flushing the resin with concentrated HCl(aq), for example. When hard water is passed through this H^+-charged resin, the exchange reaction is

$$H_2R + M^{2+} \longrightarrow MR + 2 H^+ \tag{22.18}$$

The ion-exchange reaction (22.18) presents us with two interesting possibilities.

1. We can titrate the H^+ displaced from the ion exchange resin with a standard base. This gives us a *quantitative* measurement of the hardness of the water, as illustrated in Example 22-4.
2. We can replace all the cations in a water sample with H^+ by the exchange reaction (22.18). Then we can pass the water through a second ion exchange resin that exchanges OH^- for all the *anions* present. The H^+ and OH^- combine to form H_2O. Because it is essentially free of all ions, we can call this water **deionized water.** You should be familiar with deionized water from your laboratory experiences. It, rather than water from municipal water lines, is used because ions present in ordinary water may interfere with chemical reactions (e.g., by forming precipitates or catalyzing reactions).

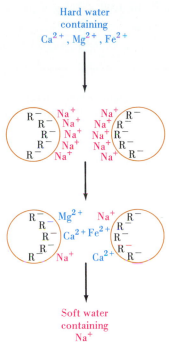

Hard water
containing
$Ca^{2+}, Mg^{2+}, Fe^{2+}$

Soft water
containing
Na^+

FIGURE 22-9
Ion exchange process.

The resin pictured here is a cation exchange resin: multivalent cations ($Ca^{2+}, Mg^{2+}, Fe^{2+}$) are exchanged for univalent cations (Na^+). This resin can be represented as RNa. Other ion exchange resins, designated as ROH, exchange OH^- for other anions; these are anion exchange resins.

Example 22-4

Determining the hardness of water by ion exchange. A 100.0-mL sample of hard water is passed through a column of the ion exchange resin H_2R. The water coming off the column requires 15.17 mL of 0.0265 M NaOH for its titration. What is the hardness of the water, expressed as ppm Ca^{2+}?

Solution

Step 1. Determine the no. mmol H^+ present in the 100.0-mL water sample. The titration reaction is simply $H^+ + OH^- \rightarrow H_2O$. The no. mmol H^+ is equal to the no. mmol OH^- used in the titration.

$$\text{no. mmol } H^+ = 15.17 \text{ mL NaOH soln} \times \frac{0.0265 \text{ mmol } OH^-}{\text{mL NaOH soln.}} \times \frac{1 \text{ mmol } H^+}{1 \text{ mmol } OH^-}$$

$$= 0.402 \text{ mmol } H^+$$

Step 2. Determine the no. mmol Ca^{2+} originally in the 100.0-mL water sample. According to equation (22.18), each Ca^{2+} ion in the hard water releases two H^+ ions from the ion exchange resin.

$$\text{no. mmol } Ca^{2+} = 0.402 \text{ mmol } H^+ \times \frac{1 \text{ mmol } Ca^{2+}}{2 \text{ mmol } H^+} = 0.201 \text{ mmol } Ca^{2+}$$

Step 3. Determine the mass of Ca^{2+} in the 100.0-mL water sample.

Many U.S. government agencies report water analyses in terms of individual ions (such as ppm Ca^{2+}). Another widely used method is to report hardness in terms of a $CaCO_3$ equivalent. Thus, 80.6 ppm Ca^{2+} is equivalent to $(80.6 \times 100.1/40.08) =$ 201 ppm $CaCO_3$.

$$\text{no. g } Ca^{2+} = 0.201 \text{ mmol } Ca^{2+} \times \frac{1 \text{ mol } Ca^{2+}}{1000 \text{ mmol } Ca^{2+}} \times \frac{40.08 \text{ g } Ca^{2+}}{1 \text{ mol } Ca^{2+}}$$

$$= 8.06 \times 10^{-3} \text{ g } Ca^{2+}$$

Step 4. Express the result of Step 3 as parts per million (ppm). The sample used in Steps 1, 2, and 3 was 100.0 mL = 100 g (density of water = 1.00 g/mL). That is,

$$\frac{8.06 \times 10^{-3} \text{ g } Ca^{2+}}{100 \text{ g water}} = \frac{8.06 \times 10^{-5} \text{ g } Ca^{2+}}{\text{g water}}$$

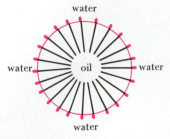

FIGURE 22-10
Representation of the cleaning action of soap.

Each soap molecule has both a long nonpolar portion and a polar carboxyl group (in red). The interface between an oil droplet and the aqueous medium in which it is suspended is lined with soap molecules having their nonpolar ends in the oil and polar ends in the water. The oil droplet is solubilized.

(a)   (b)

A sodium soap is added to (a) soft (distilled) and (b) hard water. A calcium soap precipitates from the hard water. [Carey B. Van Loon]

The quantity of Ca^{2+} present is 8.06×10^{-5} g Ca^{2+} *per g of water*. In 1 million (1.00×10^6) g of water the mass of Ca^{2+} is

$$\text{no. g } Ca^{2+} = 1.00 \times 10^6 \text{ g water} \times \frac{8.06 \times 10^{-5} \text{ g } Ca^{2+}}{\text{g water}} = 80.6 \text{ g } Ca^{2+}$$

But this is the answer we are seeking: 80.6 g Ca^{2+} in 1.00×10^6 g water is **80.6 ppm Ca^{2+}**.

SIMILAR EXAMPLES: Exercises 6, 26, 61.

Soaps and Detergents. Soaps and detergents are cleaning agents whose function is to solubilize or emulsify oils with water. Figure 22-10 suggests how they are able to do this.

One of the easiest ways to identify a sample of hard water is to watch the effect it has on soap. Suppose we represent a typical soap as the ionic compound $RCOO^-Na^+$. As suggested by the equation below, alkali metal soaps are water soluble, but the soaps of multivalent cations are not.

$$2 \text{ Na}^+ + 2 \text{ RCOO}^- + Ca^{2+} \longrightarrow Ca(RCOO)_2(s) + 2 \text{ Na}^+ \qquad (22.19)$$
$$\text{soap} \qquad\qquad\qquad\qquad \text{``bathtub ring''}$$

(a) Fatty acids and soaps

$$R-\overset{\overset{\displaystyle O}{\|}}{C}-O-H + Na^+ + OH^- \longrightarrow R-\overset{\overset{\displaystyle O}{\|}}{C}-O^- Na^+ + H_2O$$

fatty acid   soap

where $R = H_3C(CH_2)_nCH_2-$ and $n = 1$ to 19

e.g., $H_3C(CH_2)_{15}CH_2-\overset{\overset{\displaystyle O}{\|}}{C}-O-H$ and $H_3C(CH_2)_{15}CH_2-\overset{\overset{\displaystyle O}{\|}}{C}-O^- Na^+$

stearic acid   sodium stearate (a soap)

(b) An alkylbenzenesulfonate (ABS) detergent

$$H_3C-\overset{\overset{\displaystyle CH_3}{|}}{CH}-CH_2-\overset{\overset{\displaystyle CH_3}{|}}{CH}-CH_2-\overset{\overset{\displaystyle CH_3}{|}}{CH}-CH_2-\overset{\overset{\displaystyle CH_3}{|}}{CH}-\underset{}{\bigcirc}-\overset{\overset{\displaystyle O}{\|}}{\underset{\underset{\displaystyle O}{\|}}{S}}-O^- Na^+$$

(c) A linear alkanesulfonate (LAS) detergent

$$H_3C(CH_2)_nCH_2-\overset{\overset{\displaystyle O}{\|}}{\underset{\underset{\displaystyle O}{\|}}{S}}-O^- Na^+ \qquad \text{where } n = 12 \text{ to } 16$$

FIGURE 22-11
Structural features of fatty acids, soaps, and detergents.

(a) Fatty acids are based on hydrocarbon chains with from 4 to 22 C atoms. The carbon atom at one end of the chain is in a carboxyl group, $-\overset{\overset{\displaystyle O}{\|}}{C}-O-H$. Water-soluble soaps are the sodium or potassium salts of fatty acids. Most other salts of fatty acids are insoluble in water.
(b) An alkylbenzenesulfonate (ABS) detergent is a derivative of sulfuric acid, $HO-SO_2-OH$. A benzene ring with its attached hydrocarbon chain substitutes for one of the $-OH$ groups, and Na replaces the remaining H atom. Because of branching in the hydrocarbon chain, these molecules resist the action of microorganisms—they are not biodegradable.
(c) A linear alkanesulfonate (LAS) detergent is also a derivative of sulfuric acid, but the hydrocarbon portion has no chain branching. These molecules are biodegradable.

Soaps are effective (though expensive) water softeners, but unfortunately the object that is washed becomes coated with the precipitated heavy metal soaps. One solution to this soap/hard water problem is to use softened water. Another solution is to use a synthetic detergent. Synthetic detergents provide the same kind of emulsifying action as soap, but without forming precipitates with Ca^{2+} or other multivalent cations. [Although detergents do not form precipitates in hard water, they do not function well unless a builder is present to sequester multivalent cations (see page 854).]

Some fundamental information about the molecular structures of ordinary soaps and two different types of detergents is presented in Figure 22-11.

22-3 Group 3A Metals: Al, Ga, In, Tl

Boron, the first element in group 3A, is a nonmetal. Aluminum, in its appearance and physical properties and in much of its chemical behavior, is metallic. So are the remaining members of group 3A—gallium, indium, and thallium. Some properties of the group 3A metals are listed in Table 22-4.

The ionization energies of the group 3A metals are comparable to one another, but the hydration energy of Al^{3+} is largest among the group 3A cations. This accounts for the fact that Al^{3+} has the most negative electrode potential and is the most active of the group 3A metals.

Here is an interesting feature that we can see in Table 22-4: Ga, In, and Tl—but not Al—are able to form unipositive ions, M^+. That is, a Ga atom can lose its $4p$ electron and retain its $4s$ electrons to form the ion Ga^+, with the electron configuration $[Ar]3d^{10}4s^2$. The heavier members of group 3A show an even greater tendency to form a +1 ion. In fact, with thallium the +1 oxidation state is more stable in aqueous solution than is the +3 oxidation state. We can see this through the large *positive* potential for the reduction of Tl^{3+} to Tl^+.

$$Tl^{3+}(aq) + 2\ e^- \longrightarrow Tl^+(aq) \qquad E° = +1.25\ V \qquad (22.20)$$

The outer-shell pair of electrons (ns^2) retained in the formation of the +1 ions is sometimes called an **inert pair.** We will also observe this retention of the ns^2 pair of electrons during ion formation in other post-transition elements (e.g., Sn and Pb).

The chemistry of gallium is similar to that of aluminum. Thallium bears some similarity to lead, for instance, in its high density (11.85 g/cm^3), softness, and the

TABLE 22-4
Some Properties of the Group 3A Metals

	Al	Ga	In	Tl
melting point, °C	660.4	29.8	156.6	303.5
boiling point, °C	2467	2403	2080	1457
electron configuration	[Ne]$3s^23p^1$	[Ar]$3d^{10}4s^24p^1$	[Kr]$4d^{10}5s^25p^1$	[Xe]$4f^{14}5d^{10}6s^26p^1$
ionization energy, kJ/mol				
first [M(g) → M$^+$(g) + e$^-$]	577.6	578.8	558.3	589.4
second [M$^+$(g) → M^{2+}(g) + e$^-$]	1764	1979	1821	1971
third [M^{2+}(g) → M^{3+}(g) + e$^-$]	2745	2963	2705	2878
metallic radius, pm	143	141	166	171
ionic radius, pm				
M$^+$	—	113	132	140
M^{3+}	50	62	81	95
electrode potential, $E°$, V				
[M^{3+}(aq) + 3 e$^-$ → M(s)]	−1.66	−0.56	−0.34	+0.72
[M$^+$(aq) + e$^-$ → M(s)]	—	—	−0.25	−0.34

Gallium Arsenide

Of paramount importance in some semiconductor devices is the speed with which electrons can travel through them. For example, high-speed computers may require a current to be switched on or off in less than a billionth of a second. There are essentially two ways to speed up semiconductor devices—make them smaller or use a material through which electrons travel faster.

As we learned in Chapter 11 (Focus feature) the element silicon is the foundation of the semiconductor industry. Semiconductor devices based on silicon are relatively easy to make and inexpensive. However, these devices have now been pushed nearly to their limit in speed. Gallium arsenide (GaAs), which is also a semiconductor, allows the passage of electrons to occur about 5 to 10 times faster than does silicon. This increased electron speed, together with some desirable optical properties, have made GaAs an important new material for use in supercomputers, radar devices, and communications satellites.

Gallium bars and arsenic nuggets being fused into gallium arsenide. Techniques for producing semiconductors are much more difficult, however, requiring the deposition of thin films of GaAs. [Hank Morgan/Photo Researchers, Inc.]

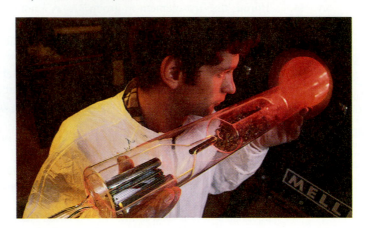

The method used to fabricate GaAs semiconductors is very much different from most chemical manufacturing processes. The reactants are trimethylgallium, $Ga(CH_3)_3$, and highly toxic arsine gas, AsH_3. The reaction is carried out at a high temperature, and the GaAs is deposited as a thin film.

In the early 1980s there was practically no market for GaAs semiconductor materials. By the mid-1990s, the worldwide market is projected to exceed 10 billion dollars. A manufacturing facility for GaAs in a space station is also a future possibility. ■ ■ ■

poisonous nature of its compounds. Thallium(I) compounds resemble alkali metal compounds in some respects—TlOH is a strong base and the Tl^+ ion (radius 140 pm) can crystallize together with the K^+ ion (radius 133 pm) in chlorates, perchlorates, sulfates, and various phosphates. In other respects Tl^+ resembles Ag^+, as in forming a light-sensitive, insoluble chloride, TlCl.

Gallium remains a liquid over one of the longest temperature ranges of any substance (over 2000 °C), and because of this property it finds some use in high-temperature thermometry. Ga, In, and Tl and their compounds have a number of commercial uses—alloys, transistors, photoconductors and specialty glasses.

Aluminum Ion in Solution. We must consider two factors to assess the water solubilities of aluminum compounds: the small size and high charge of the Al^{3+} ion and its very high hydration energy (-4613 kJ/mol). When Al^{3+} is combined with small highly charged anions, the high lattice energies of the resulting solids render them insoluble in water. Such is the case with Al_2O_3. Even AlF_3, a combination of Al^{3+} with the univalent F^- anion, has only a low water solubility (about 0.07 M).

$AlCl_3$, $AlBr_3$, and AlI_3 have considerable covalent character (see again, Section 14-7). These compounds are quite water soluble.

A number of aluminum salts, like those of the group 2A metals, crystallize from solution as hydrates. Some of these hydrates are highly soluble and therefore deliquescent, such as $AlX_3 \cdot 6H_2O$ (where $X = Cl$, Br, I, or ClO_3) and $Al(NO_3)_3 \cdot 9H_2O$. Moreover, certain aspects of the aqueous chemistry of aluminum compounds derive from the nature of the hydrated aluminum ion—$[Al(H_2O)_6]^{3+}$. One distinctive feature is that aqueous solutions of aluminum salts are *acidic*, which comes about in this way: Because of the attractive power of the small, highly charged Al^{3+} ion for electrons, an O—H bond in a ligand H_2O molecule breaks. A proton is released to an H_2O molecule outside the coordination sphere. The original H_2O ligand is converted to OH^-, and the complex ion to $[Al(H_2O)_5OH]^{2+}$.

The acidic nature of $[Al(H_2O)_6]^{3+}$(aq) is illustrated on page 628.

$$[Al(H_2O)_6]^{3+} + H_2O \rightleftharpoons H_3O^+ + [Al(H_2O)_5OH]^{2+} \qquad (22.21)$$

Reaction (22.21) is reversible. In strongly acidic solutions the ionization of $[Al(H_2O)_6]^{3+}$ is repressed, but in alkaline solution its ionization is favored. If a solution is made sufficiently alkaline, reaction (22.21) can be carried to completion. This is followed by further ionization of other H_2O ligand molecules.

$$[Al(H_2O)_5OH]^{2+} + H_2O \rightleftharpoons H_3O^+ + [Al(H_2O)_4(OH)_2]^+ \qquad (22.22)$$

$$[Al(H_2O)_4(OH)_2]^+ + H_2O \rightleftharpoons H_3O^+ + Al(H_2O)_3(OH)_3 \qquad (22.23)$$

The product of reaction (22.23) is **hydrated aluminum hydroxide,** $Al(OH)_3 \cdot 3H_2O$. Actually, what we have demonstrated through the preceding three equations is the precipitation of Al^{3+} as its hydroxide. We can describe this precipitation as

$$[Al(H_2O)_6]^{3+}(aq) + 3\ OH^-(aq) \longrightarrow Al(OH)_3(s) + 6\ H_2O \qquad (22.24)$$

or, still more simply, as

$$Al^{3+}(aq) + 3\ OH^-(aq) \longrightarrow Al(OH)_3(s) \qquad (22.25)$$

In a strongly basic solution, such as in NaOH(aq), reaction (22.23) can be carried a step further.

$$Al(H_2O)_3(OH)_3(s) + OH^- \rightleftharpoons [Al(H_2O)_2(OH)_4]^-(aq) + H_2O \qquad (22.26)$$

$[Al(H_2O)_2(OH)_4]^-$ is a complex *anion* called the **aluminate ion.** In this complex anion, Al^{3+} is the central metal ion and the ligands are *two* H_2O molecules and *four* OH^- ions. Sometimes, for simplicity, the two H_2O molecules are dropped and the formula is written as $[Al(OH)_4]^-$. Sometimes, an additional two H_2O molecules are removed (that is, four H and two O atoms), leading to the formula AlO_2^-. The fact that $Al(OH)_3(s)$ is soluble in strongly basic solutions can also be represented as

$$Al(OH)_3(s) + OH^-(aq) \longrightarrow [Al(OH)_4]^-(aq) \qquad (22.27)$$

In summary, Al_2O_3 and $Al(OH)_3$ are *amphoteric*. The principal species we expect to encounter in the aqueous chemistry of Al^{3+} are

Aluminum ion in aqueous solution.

$$[Al(H_2O)_6]^{3+} \text{ (or } Al^{3+}) \underset{H_3O^+}{\overset{OH^-}{\rightleftharpoons}} Al(OH)_3(s) \underset{H_3O^+}{\overset{OH^-}{\rightleftharpoons}} [Al(OH)_4]^- \qquad (22.28)$$

Equation (22.28) can also help us to understand a statement made in Chapter 14 that aluminum is one of the few metals that is soluble in alkaline as well as in acidic solutions. With its strong tendency to be oxidized to Al^{3+}, we should expect Al(s) to displace $H_2(g)$ from water, just as do the 1A and the heavier 2A metals.

$$2\ Al(s) + 6\ H_2O \longrightarrow 2\ Al(OH)_3(s) + 3\ H_2(g) \qquad (22.29)$$

This reaction probably does occur on a clean aluminum surface, but the surface film of $Al(OH)_3$ (or hydrated Al_2O_3) ordinarily protects the underlying metal. However, in acidic solution or in the presence of a strong base this surface film dissolves and the metal undergoes further attack.

$$2 \text{ Al(OH)}_3(s) + 2 \text{ OH}^-(aq) \longrightarrow 2[\text{Al(OH)}_4]^-(aq) \qquad (22.30)$$

The net reaction when Al is brought into contact with a strongly basic solution is the sum of reactions (22.29) and (22.30), and this proves to be identical to the reaction that we first described in Chapter 14.

Aluminum cookware should never be used to contain strongly basic substances. Reaction (14.38) will occur.

$$2 \text{ Al}(s) + 2 \text{ OH}^-(aq) + 6 \text{ H}_2\text{O} \longrightarrow 2[\text{Al(OH)}_4]^-(aq) + 3 \text{ H}_2(g) \qquad (14.38)$$

Aluminum Compounds. Most aluminum compounds are derived from the **oxide,** Al_2O_3, and various **hydrated oxides,** e.g., $\text{Al}_2\text{O}_3 \cdot \text{H}_2\text{O}$ and $\text{Al}_2\text{O}_3 \cdot 3\text{H}_2\text{O}$. The mineral **corundum,** which is used as an abrasive, is Al_2O_3. The ore **bauxite,** used in the manufacture of aluminum metal, is a mixture of the mono- and trihydrates of Al_2O_3. Many familiar gemstones have Al_2O_3 as their principal constituent. The stones acquire their distinctive colors and brilliance from small quantities of other oxides as impurities (see Table 22-5). Artificial gemstones can be made by fusing corundum with the appropriate metal oxides. Other important uses of Al_2O_3 (alumina) are as refractory linings for high-temperature furnaces and as catalyst supports in industrial chemical processes.

Aluminum sulfate is the most important aluminum compound used commercially. It is prepared by the action of hot, concentrated $\text{H}_2\text{SO}_4(aq)$ on Al_2O_3.

$$\text{Al}_2\text{O}_3(s) + 3 \text{ H}_2\text{SO}_4(\text{concd aq}) \longrightarrow \text{Al}_2(\text{SO}_4)_3(aq) + 3 \text{ H}_2\text{O} \qquad (22.31)$$

The product that crystallizes from solution is $\text{Al}_2(\text{SO}_4)_3 \cdot 18\text{H}_2\text{O}$. About 1 million tons of $\text{Al}_2(\text{SO}_4)_3$ are produced annually, with about one-half of this quantity used in water treatment. In this application $\text{Al(OH)}_3(s)$ is precipitated by controlling the

TABLE 22-5
Some Gemstones Based on Al_2O_3

Gem	Impurity
white sapphire	none
blue sapphire	Fe, Ti
green sapphire	Co
yellow sapphire	Ni, Mg
star sapphire	Ti
ruby	Cr

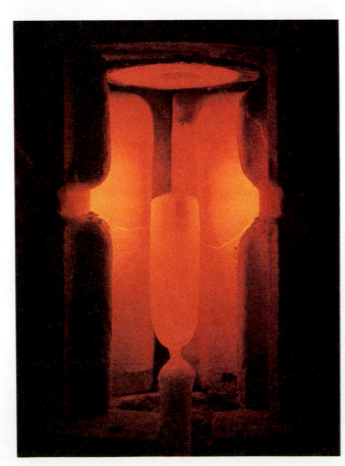

Manufacture of artificial gemstones. A mixed powder is sprayed from above, melts in the hottest regions of the furnace, and deposits as a liquid layer that then solidifies. The solidified material is gradually withdrawn from the furnace as layer after layer is added from above. Gemstones made in this way have many industrial applications, such as rubies used in lasers. [Courtesy Hrand Djevahirdjian, S.A.]

pH. As it settles, the finely divided $Al(OH)_3(s)$ removes suspended solids from the water. Another important use of $Al_2(SO_4)_3$ is in the sizing of paper. Sizing refers to making paper more resistant to penetration by water by incorporating into the paper materials such as waxes, glues, or synthetic resins. $Al(OH)_3(s)$ precipitated from an aluminum sulfate solution helps to deposit the sizing agent in the paper. Other uses of $Al_2(SO_4)_3$ include the waterproofing, fireproofing, and dyeing of fabrics.

An aqueous solution containing equimolar amounts of $Al_2(SO_4)_3$ and K_2SO_4 crystallizes as potassium aluminum sulfate, $KAl(SO_4)_2 \cdot 12H_2O$. This salt, which is variously known as potash alum, common alum, or ordinary alum, belongs to a large general class of substances called alums. **Alums** have the formula $M(I)M(III)(SO_4)_2 \cdot 12H_2O$ [where M(I) is almost any unipositive cation, except Li^+, and M(III) is a tripositive cation—Al^{3+}, Ti^{3+}, V^{3+}, Cr^{3+}, Mn^{3+}, Fe^{3+}, Co^{3+}, Ga^{3+}, In^{3+}, Re^{3+}, and Ir^{3+}]. Alums contain the ions $[M(H_2O)_6]^+$, $[M(H_2O)_6]^{3+}$, and SO_4^{2-} in the ratio 1:1:2. The most common alums have $M(I) = K^+$ or NH_4^+ and $M(III) = Al^{3+}$. Li^+ does not form alums because the ion is too small to meet the structural requirements of the crystal.

Alums are used for some of the same purposes as the simpler salts from which they are derived. One important use of potash alum is as a mordant in dyeing. The fabric to be dyed is dipped in a solution of the alum and heated with steam. Hydrolysis of $[Al(H_2O)_6]^{3+}$ deposits $Al(OH)_3(s)$ into the fibers of the material and the dye is adsorbed on the $Al(OH)_3$.

A number of aluminum compounds feature covalent bonds. These compounds and some of their uses were noted in Section 14-7.

Production of Aluminum. When the aluminum cap was placed atop the Washington Monument in 1884, aluminum was still a semiprecious metal. It cost several dollars a pound to produce and was used mostly in jewelry and artwork. But just two years later, all this changed. In 1886 Charles Martin Hall in the United States and Paul Heroult in France discovered an economically feasible method of producing aluminum by electrolysis.

The manufacture of Al involves several interesting principles that we have already studied. First, the chief ore, bauxite, contains Fe_2O_3 and TiO_2 as impurities, and these must be removed. The principle used is that Al_2O_3 is *amphoteric* and dissolves in NaOH(aq). The other oxides are *basic* oxides and therefore insoluble in NaOH(aq).

$$Al_2O_3(s) + 2\ OH^-(aq) + 3\ H_2O \longrightarrow 2\ [Al(OH)_4]^-(aq) \qquad (22.32)$$

When the solution containing $[Al(OH)_4]^-$ is diluted with water or slightly acidified, $Al(OH)_3(s)$ precipitates.

Charles Martin Hall was a student at Oberlin College at the time he invented an electrolytic process for the production of aluminum.

FIGURE 22-12
The Bayer process for purifying bauxite.

When an excess of $OH^-(aq)$ is added to a solution containing $Al^{3+}(aq)$ and $Fe^{3+}(aq)$, the Fe^{3+} precipitates as $Fe(OH)_3(s)$ and the $Al(OH)_3(s)$ first formed redissolves to produce $[Al(OH)_4]^-(aq)$ (left). The $Fe(OH)_3(s)$ is filtered off, and the $[Al(OH)_4]^-(aq)$ is made slightly acidic through the action of CO_2, here added as dry ice (center). The precipitated $Al(OH)_3(s)$ collects at the bottom of a clear, colorless solution (right). [Carey B. Van Loon]

$$[Al(OH)_4]^-(aq) + H_3O^+(aq) \longrightarrow Al(OH)_3(s) + 2\ H_2O \tag{22.33}$$

Pure Al_2O_3 is obtained by heating the $Al(OH)_3$.

$$2\ Al(OH)_3(s) \xrightarrow{\Delta} Al_2O_3(s) + 3\ H_2O(g) \tag{22.34}$$

This process for purifying bauxite, known as the Bayer process, is illustrated in part through Figure 22-12.

Al_2O_3 has a very high melting point (2020 °C) and produces a liquid that is a poor electrical conductor. Its electrolysis is not feasible. By contrast, Hall and Heroult found that up to 15% Al_2O_3, by mass, can be dissolved in the molten mineral **cryolite,** Na_3AlF_6, at about 1000 °C, and the liquid is a good electrical conductor. Their process, then, involves the electrolysis of Al_2O_3 in molten cryolite. The electrolysis cell pictured in Figure 22-13 is operated at about 950 °C. Aluminum of 99.6 to 99.8% purity is obtained. The electrode reactions are not known with certainty, but the net electrolysis reaction is

Oxid: $\quad 3\{C(s) + 2\ O^{2-} \longrightarrow CO_2(g) + 4\ e^-\}$
Red: $\quad 4\{Al^{3+} + 3\ e^- \longrightarrow Al(l)\}$

Electrolytic production of Al.   Net: $\quad 3\ C(s) + 4\ Al^{3+} + 6\ O^{2-} \longrightarrow 4\ Al(l) + 3\ CO_2(g) \tag{22.35}$

In Section 21-6 we referred to the Al_2O_3 coating that protects aluminum metal from corrosion. Its maximum resistance to corrosion is exhibited between pH 4.5 and 8.5. Much of the aluminum used commercially is treated in such a way as to build up its oxide coating. In one method, called **anodizing,** an aluminum object is made the anode and a graphite rod, the cathode, in an electrolyte bath of $H_2SO_4(aq)$. The anode half-reaction is

$$2\ Al(s) + 3\ H_2O \longrightarrow Al_2O_3(s) + 6\ H^+ + 6\ e^-$$

Al_2O_3 coatings of varying porosities and thicknesses can be obtained. Also, the oxide can be made to absorb coloring matter or other additives.

The energy consumed to produce aluminum by electrolysis is very high, about 15,000 kWh per ton Al [compared, for example, to 3000 kWh per ton of Cl_2 by the electrolysis of NaCl(aq)]. This means that aluminum production facilities are generally found near low-cost power sources, typically hydroelectric power. The energy required to recycle Al is only about 5% of that to produce the metal from bauxite, and currently about 20% of the Al produced in the United States is by the recycling of scrap aluminum.

FIGURE 22-13
Electrolysis cell for aluminum production.

The cathode is a carbon lining in a steel tank. The anodes are also made of carbon. Liquid aluminum is more dense than the electrolyte medium and collects at the bottom of the tank. A crust of frozen electrolyte forms at the top of the cell.

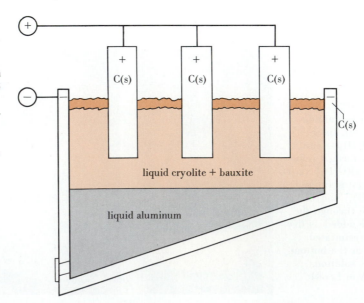

Are You Wondering:

Why so much energy is consumed in the electrolytic production of aluminum?

Any process that must be carried out at high temperature requires large amounts of energy for heating. In the electrolytic production of Al, the electrolysis bath must be kept at about 1000 °C, and this is done through electrical heating. But two other factors are also involved in the large energy consumption. One is that in order to produce one mole of Al, *three* moles of electrons must be transferred: $Al^{3+} + 3 e^- \rightarrow Al(l)$. Additionally, the molar mass of Al is relatively low, 27 g Al/mol Al. Thus, electric current equivalent to the passage of one mole of electrons produces only 9 g Al. By contrast, one mole of electrons produces 12 g Mg, 20 g Ca, or 35.5 g Cl_2. On the other hand, the same factors that make Al production a high energy consumer make Al a high energy producer when it is used in a battery (recall the aluminum–air battery described on page 782).

22-4 Extractive Metallurgy

One of the common features of the more active metals—groups 1A, 2A, and Al—is that the favored method of producing most of them is electrolysis. To produce them by chemical reactions requires expensive reducing agents, such as Na to prepare K (equation 14.19). Less active metals, on the other hand, can generally be prepared economically by chemical methods, often with carbon (coke or coal) as a reducing agent.

We use the term metallurgy for the general study of metals and **extractive metallurgy** for the winning of metals from their ores. There is no single method of extractive metallurgy, but there are a few basic operations that are generally used. Let us consider these general methods, using the extractive metallurgy of zinc as a specific example.

Concentration. In mining operations the desired mineral from which a metal is to be extracted often constitutes only a few percent (or even just a fraction of a percent) of the material mined. It is necessary to separate the desired ore from waste rock before proceeding with other metallurgical operations. One useful method, *flotation,* is described in Figure 22-14.

Roasting. An ore is roasted (heated to a high temperature) to convert a metal compound to its oxide, which can then be reduced. The commercially important ores of zinc are the sulfide (sphalerite) and the carbonate (smithsonite). When strongly heated, sulfides liberate $SO_2(g)$; carbonates liberate $CO_2(g)$. In modern smelting operations $SO_2(g)$ is converted to sulfuric acid rather than vented to the atmosphere.

Metallurgical roasting reactions.

$$2 \, ZnS(s) + 3 \, O_2(g) \xrightarrow{\Delta} 2 \, ZnO(s) + 2 \, SO_2(g) \qquad (22.36)$$

$$ZnCO_3(s) \xrightarrow{\Delta} ZnO(s) + CO_2(g) \qquad (22.37)$$

Reduction. If possible, carbon, in the form of coke or powdered coal, is used as the reducing agent. Several reactions occur simultaneously in which both C(s) and CO(g) act as reducing agents. The reduction of ZnO is carried out at about 1100 °C,

FIGURE 22-14
Concentration of an ore by flotation.

Powdered ore is suspended in water in a large vat, together with suitable additives, and the mixture is agitated with air. Particles of ore become attached to air bubbles, rise to the top of the vat, and are collected in the overflow froth. Particles of the undesired waste rock (gangue) fall to the bottom.

The success of this method depends on the use of proper additives—a material that will produce a stable foam (frother) and a substance (collector) that coats the particles of ore but does not "wet" the particles to be rejected. Pine oil is widely used as a frother and sodium ethyl xanthate as a collector.

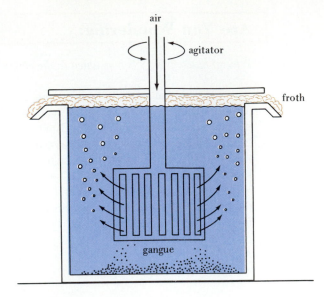

Metallurgical reduction reactions.

a temperature above the boiling point of zinc. The zinc is obtained as a vapor and condensed to the liquid.

$$ZnO(s) + C(s) \xrightarrow{\Delta} Zn(g) + CO(g) \tag{22.38}$$

$$ZnO(g) + CO(g) \xrightarrow{\Delta} Zn(g) + CO_2(g) \tag{22.39}$$

$$C(s) + CO_2(g) \rightleftharpoons 2\ CO(g) \tag{22.40}$$

Refining. Metal produced by chemical reduction is usually not pure enough for its intended uses. Impurities must be removed: the metal must be refined. The impurities in zinc are mostly Cd and Pb. They can be removed by the fractional distillation of liquid zinc.

Most of the zinc produced worldwide, however, is refined electrolytically, usually in a process that combines reduction and refining. ZnO from the roasting step is dissolved in $H_2SO_4(aq)$.

$$ZnO(s) + 2\ H^+(aq) + SO_4^{2-}(aq) \longrightarrow Zn^{2+}(aq) + SO_4^{2-}(aq) + H_2O \tag{22.41}$$

Powdered Zn is added to the solution to displace less active metals, such as Cd. Then the solution is electrolyzed. The electrode reactions are

Cathode: $Zn^{2+}(aq) + 2\ e^- \longrightarrow Zn(s)$
Anode: $H_2O \longrightarrow \frac{1}{2} O_2(g) + 2\ H^+(aq) + 2\ e^-$
Unchanged: $SO_4^{2-}(aq) \longrightarrow SO_4^{2-}(aq)$

Net: $Zn^{2+} + SO_4^{2-} + H_2O \longrightarrow Zn(s) + 2\ H^+ + SO_4^{2-} + \frac{1}{2} O_2(g)$
$$\tag{22.42}$$

Note that in the net electrolysis reaction Zn^{2+} is reduced to pure metallic zinc and sulfuric acid is regenerated. The acid is reused in reaction (22.41).

Example 22-5 ————————————————————————

Writing chemical equations for metallurgical processes. Write chemical equations to represent **(a)** roasting of galena, PbS; **(b)** reduction of $Cu_2O(s)$, using charcoal as a reducing agent; **(c)** deposition of pure silver from an aqueous solution of Ag^+.

Solution

(a) We expect this process to be essentially the same as reaction (22.36).

$$2 \; PbS(s) + 3 \; O_2(g) \xrightarrow{\Delta} 2 \; PbO(s) + 2 \; SO_2(g)$$

(b) The simplest possible equation is

$$Cu_2O(s) + C(s) \xrightarrow{\Delta} 2 \; Cu(l) + CO(g)$$

(c) This process involves a reduction half-reaction. The accompanying oxidation half-reaction is not specified. Neither is it specified whether this is an electrolysis process or whether silver is displaced by a more active metal. In either case the half-reaction is

$$Ag^+(aq) + e^- \longrightarrow Ag(s)$$

SIMILAR EXAMPLES: Exercises 8, 53.

Thermodynamics of Extractive Metallurgy. It is interesting to think of the reduction of ZnO with C as involving a competition between Zn and C for O atoms. The equations written below suggest that C has a greater tendency to become oxidized at 1100 °C than does Zn. $\Delta G°$ is more negative for reaction (a) than for (b).

(a) $2 \; C(s) + O_2(g) \longrightarrow 2 \; CO(g)$ $\Delta G° \approx -460$ kJ

(b) $2 \; Zn(g) + O_2(g) \longrightarrow 2 \; ZnO(s)$ $\Delta G° \approx -360$ kJ

If we reverse (b) and add the two equations, we obtain the net equation for the metallurgical reduction (22.38) and its $\Delta G°$ value.

(a)	$2 \; C(s) + O_2(g) \longrightarrow 2 \; CO(g)$	$\Delta G° \approx -460$ kJ
$-$(b)	$2 \; ZnO(s) \longrightarrow 2 \; Zn(g) + O_2(g)$	$\Delta G° \approx +360$ kJ
	$2 \; ZnO(s) + 2 \; C(s) \longrightarrow 2 \; Zn(g) + 2 \; CO(g)$	$\Delta G° \approx -100$ kJ

and

$$ZnO(s) + C(s) \longrightarrow Zn(g) + CO(g) \qquad \Delta G° \approx -50 \text{ kJ} \qquad (22.38)$$

Because of the large *negative* value of $\Delta G°$, we expect the forward reaction to be favored in reaction (22.38).

FIGURE 22-15
$\Delta G°$ as a function of temperature for some reactions of extractive metallurgy.

The points noted by arrows are the melting and boiling points of the metals zinc and magnesium.

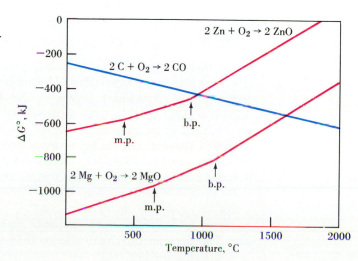

Why must we carry out the reduction of ZnO with C at 1100 °C? From an energy standpoint it would certainly be more economical to carry it out at a lower temperature. The graphical representation of Figure 22-15 is a useful tool to help answer such questions. The lines (2 C + O_2 → 2 CO) and (2 Zn + O_2 → 2 ZnO) cross at about 950 °C. At about this temperature Zn and C have equal affinities for O atoms: $\Delta G°$ for reaction (22.38) is 0. Above this temperature $\Delta G°$ is *negative*, and below this temperature $\Delta G°$ is *positive*. To make reaction (22.38) go essentially to completion we need a temperature somewhat in excess of 950 °C.

One of the reasons why carbon is such a good reducing agent is that its *descending* line in Figure 22-15 eventually intersects the *ascending* lines for most metals. Of course, for some metals the intersection comes at too high a temperature to make reduction of its oxide with carbon a feasible process. In some other cases, even though reduction with carbon is energetically feasible, we may decide against using carbon because it forms metal carbides that may be objectionable in the metal product.

Example 22-6

Predicting the free energy change of a metallurgical reduction with carbon.
Would you expect the reaction   MgO(s) + C(s) → Mg(g) + CO(g)   to occur to any significant extent at 1500 °C?

Solution.   We can extract these approximate data from Figure 22-15.

2 C(s) + O_2(g) ⟶ 2 CO(g)	$\Delta G° \approx -530$ kJ
2 MgO(s) ⟶ 2 Mg(g) + O_2(g)	$\Delta G° \approx +610$ kJ
2 MgO(s) + 2 C(s) ⟶ 2 Mg(g) + 2 CO(g)	$\Delta G° \approx +80$ kJ

The positive value of $\Delta G°$ indicates that the reduction of MgO(s) with C(s) does not occur to a significant extent at 1500 °C. (The value of K is much smaller than 1.) At temperatures above about 1600 °C, however, $\Delta G°$ becomes negative.

SIMILAR EXAMPLES: Exercises 10, 34, 35.

Are You Wondering:

Why $\Delta G°$ decreases *with temperature for the reaction*   2 C(s) + O_2(g) → 2 CO(g),   *whereas it* increases *for reactions like*   2 Zn + O_2(g) → 2 ZnO(s)?

Recall from Chapter 20 that $\Delta G° = \Delta H° - T\Delta S°$, and that in this expression $\Delta H°$ does not change appreciably with temperature. The change in $\Delta G°$ with temperature is determined largely by the $T\Delta S°$ term. For the reaction 2 Zn(s) + O_2(g) → 2 ZnO(s),   we expect the entropy change $\Delta S°$ to be *negative* because of the disappearance of one mole of gas. If $\Delta S°$ is negative then $-T\Delta S°$ is positive, and the higher the temperature the more positive it becomes. In turn this causes $\Delta G°$ to become larger (less negative). The line in Figure 22-15 has a positive slope; it is an *ascending* line.

By contrast, for the reaction   2 C(s) + O_2(g) → 2 CO(g)   there is an increase in number of moles of gas. $\Delta S°$ is *positive*. The term, $-T\Delta S°$, is *negative*. $\Delta G°$ decreases (becomes more negative) with increasing temperature. The line representing this reaction in Figure 22-15 has a negative slope; it is a *descending* line.

22-5 Group 4A Metals: Tin and Lead

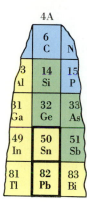

Both tin and lead have been known since ancient times and were thought to be related. At one time the Romans called tin *plumbum album*, and lead *plumbum nigrum*.

Carbon, the first member of group 4A (14), is a nonmetal. The next two members, silicon and germanium, are metalloids. Among the more interesting properties of Si and Ge is semiconductor behavior, which we have discussed previously. Tin and lead are mostly metallic in their behavior.

The data in Table 22-6 suggest that tin and lead are rather similar to each other. Both are soft, malleable, and melt at low temperatures. The ionization energies of the two metals are about the same, as are their standard electrode potentials. This means that their tendencies to be oxidized to the +2 oxidation state are comparable; both are slightly more metallic than hydrogen.

The fact that both tin and lead can exist in two oxidation states, +2 and +4, is a result of the inert pair effect mentioned on page 805. In the +2 oxidation state the inert pair, ns^2, is not involved in bond formation, and in the +4 oxidation state it is. Tin displays a stronger tendency to exist in the +4 oxidation state than does lead, a fact that we can infer in this way: The reduction of Sn(IV) to Sn(II) has a value of $E°$ that is only slightly positive (+0.15 V). For the reduction of Pb(IV) to Pb(II) the value of $E°$ is large (1.5 V). This means that Pb(IV) is readily reduced to Pb(II) whereas Sn(IV) is not so easily reduced to Sn(II).

Another difference between tin and lead is that tin exists in two allotropic forms (α and the β), whereas lead has but a single solid form. The α (gray) or "nonmetallic" form of tin is stable below 13 °C, and the β (white) or "metallic" form of tin is stable above 13 °C. Ordinarily, when a sample of β (white) tin is cooled, it must be kept below 13 °C for a long time before the transition to α (gray) tin occurs. Once it does begin, however, the transformation takes place rather rapidly and with dramatic results. The tin expands and crumbles to a powder. This is because α (gray) tin is less dense than β (white) tin. This transformation has led to the disintegration of organ pipes, buttons, medals, and other objects made of tin, and has been called *tin disease, tin pest,* or *tin plague.*

Metallurgy and Uses of Tin and Lead. The chief tin ore is **cassiterite, SnO_2.** The ore is concentrated by flotation and then roasted. The purpose of roasting is to oxidize metallic impurities and to remove sulfur and arsenic as their volatile oxides. Next, the oxide is reduced with carbon (coal).

$$SnO_2(s) + 2\ C(s) \xrightarrow{\Delta} Sn(l) + 2\ CO(g) \qquad (22.43)$$

The tin is purified by remelting. The easily melted tin is poured off from unmelted impurities. Impurities that are soluble in liquid tin are oxidized and removed by skimming off the oxide film that forms.

TABLE 22-6
Some Properties of Tin and Lead

	Tin	Lead
density, g/cm³	5.77 (α)	11.34
	7.29 (β)	
melting point, °C	232	327
electron configuration	$[Kr]4d^{10}5s^25p^2$	$[Xe]4f^{14}5d^{10}6s^26p^2$
principal oxidation states	+2, +4	+2, +4
ionization energy, kJ/mol		
first	709	716
second	1412	1450
electrode potential $E°$, V		
$[M^{2+}(aq) + 2\ e^- \rightarrow M(s)]$	−0.136	−0.126
$[M^{4+}(aq) + 2\ e^- \rightarrow M^{2+}(aq)]$	+0.15	+1.5

Lead is found chiefly as **galena, PbS.** The ore is concentrated by flotation. Then the ore is roasted. The main reaction is

$$2\ PbS(s) + 3\ O_2(g) \xrightarrow{\Delta} 2\ PbO(s) + 2\ SO_2(g) \qquad (22.44)$$

Reduction is carried out with coke and involves the reactions

$$PbO(s) + C(s) \xrightarrow{\Delta} Pb(l) + CO(g) \qquad (22.45)$$

$$PbO(s) + CO(g) \xrightarrow{\Delta} Pb(l) + CO_2(g) \qquad (22.46)$$

Nearly half of the tin metal produced is used in tin plate, especially in plating iron for use in cans for storing foods. The next most important use (about 25%) is in the manufacture of **solders**—low-melting-point alloys used to join wires or pieces of metal. Other important alloys of tin are *bronze* (90% Cu, 10% Sn) and *pewter* (85% Sn, 7% Cu, 6% Bi, 2% Sb).

Over half of the lead produced is used in lead–acid (storage) batteries. Other uses include the manufacture of solder and other alloys, ammunition, and radiation shields (to protect against x rays and γ rays). Uses that are of declining importance are in the manufacture of tetraethyllead as an antiknock additive for gasoline and insoluble lead compounds as paint pigments.

Reactions of Tin and Lead. Although tin and lead are similar in some of their physical and chemical properties, there are also differences in their chemical behaviors. These differences stem from (1) the stability of the +4 oxidation state of tin and (2) the fact that so many lead compounds are insoluble.

Tin dissolves slowly in dilute HCl(aq) and more rapidly in the concentrated acid. In concentrated $HNO_3(aq)$, tin is oxidized to insoluble $SnO_2(s)$.

$$Sn(s) + 4\ HNO_3(concd\ aq) \longrightarrow SnO_2(s) + 2\ H_2O + 4\ NO_2(g) \qquad (22.47)$$

Reaction (22.47) is the first step in the analysis of brass and bronze samples by the method described on page 68. Tin also dissolves in concentrated NaOH(aq), through a reaction similar to that described for aluminum (reaction 14.38). In this reaction tin is oxidized to the +4 oxidation state in the complex ion $[Sn(OH)_6]^{2-}$; $H_2(g)$ is the other product.

Lead reacts with dilute HCl(aq) and $H_2SO_4(aq)$, but the insoluble reaction products, $PbCl_2(s)$ and $PbSO_4(s)$, protect the metal from further attack after a brief initial reaction. However, because of the formation of the complex ion $[PbCl_3]^-$, $PbCl_2(s)$ dissolves in concentrated HCl(aq). In turn, this means that lead will dissolve.

$$Pb(s) + 3\ HCl(conc.\ aq) \longrightarrow [PbCl_3]^-(aq) + H^+(aq) + H_2(g) \qquad (22.48)$$

Lead dissolves in nitric acid to form Pb^{2+} and various oxides of nitrogen.

When tin is heated in air, the product is $SnO_2(s)$, whereas the air oxidation of lead produces PbO(s). Metallic tin reacts with $Cl_2(g)$ to form $SnCl_4$ (a reaction used to recover tin from scrap tin plate). Lead reacts with $Cl_2(g)$ to form $PbCl_2$. Both metals yield sulfides when heated with sulfur. With tin, the SnS formed can react further to produce SnS_2. Lead forms only PbS.

Compounds of Tin and Lead. Tin forms two primary oxides, SnO and SnO_2. SnO can be converted to SnO_2 by heating in air. When heated in the *absence* of air, SnO disproportionates to Sn and SnO_2. Lead forms a number of oxides and the chemistry of these oxides is not completely understood. The best known oxides of lead are yellow **PbO (litharge),** red-brown **lead dioxide, PbO₂,** and a mixed-valency oxide known as **red lead, Pb₃O₄.** Pb_3O_4 can be obtained from PbO by air oxidation at 400 °C, but the reverse reaction is favored at higher temperatures.

$$6\ PbO(s) + O_2(g) \rightleftharpoons 2\ Pb_3O_4(s) \qquad (22.49)$$

Tin and lead, together with bismuth and cadmium, are common constituents of low-melting-point alloys. The alloy pictured here has a melting point below the boiling point of water. [Carey B. Van Loon]

Pb_3O_4, which we can also represent as $2PbO \cdot PbO_2$, has lead in two oxidation states (mixed-valency). When Pb_3O_4 is treated with $HNO_3(aq)$, the Pb(II) forms $Pb^{2+}(aq)$ and the Pb(IV) precipitates as $PbO_2(s)$.

$$Pb_3O_4(s) + 4 HNO_3(aq) \longrightarrow 2 Pb(NO_3)_2(aq) + PbO_2(s) + 2 H_2O \qquad (22.50)$$

Simple hydroxides of tin and lead cannot be prepared from aqueous solutions: we need to discuss acid–base properties in terms of the oxides. SnO and PbO form salts with acids, and in doing so display *basic* properties. But they also have *acidic* character. They dissolve in NaOH(aq), forming $[Sn(OH)_3]^-$ and $[Pb(OH)_3]^-$.

$$SnO(s) + OH^-(aq) + H_2O \longrightarrow [Sn(OH)_3]^-(aq) \qquad (22.51)$$

The higher oxides, SnO_2 and PbO_2, are also amphoteric. They display *acidic* properties by dissolving in NaOH(aq).

$$SnO_2(s) + 2 OH^-(aq) + 2 H_2O \longrightarrow [Sn(OH)_6]^{2-} \qquad (22.52)$$

Complex anions such as $[Sn(OH)_3]^-$ and $[Sn(OH)_6]^{2-}$ are often named like oxoanions. If the oxidation state of the central atom has the oxidation state +2, the *ite* ending is used; if +4, *ate*. There are several *ite* and *ate* species. Some exist in solution and some only in solids. The best characterized are

stannite ion, $[Sn(OH)_3]^-$ (also written as $HSnO_2^-$ and SnO_2^{2-})

stannate ion, $[Sn(OH)_6]^{2-}$ (also written as SnO_3^{2-} and SnO_4^{4-})

plumbite ion, $[Pb(OH)_3]^-$ (also written as $HPbO_2^-$ and PbO_2^{2-})

plumbate ion, $[Pb(OH)_6]^{2-}$ (also written as PbO_3^{2-} and PbO_4^{4-})

Figure 22-16 shows the structures of the stannite and stannate ions. Stannite ion is a good reducing agent because it is easily oxidized to stannate ion.

$$[Sn(OH)_3]^-(aq) + 3 OH^-(aq) \longrightarrow [Sn(OH)_6]^{2-}(aq) + 2 e^-$$
$$-E° = +0.96 \text{ V} \qquad (22.53)$$

Lead oxides are used in the manufacture of lead–acid (storage) batteries, glass, ceramic glazes, cements (PbO), metal-protecting paints (Pb_3O_4), matches (PbO_2), and explosives (PbO_2). Other lead compounds are generally made from the oxides. One of the few soluble lead compounds is lead nitrate, $Pb(NO_3)_2$. It is formed in the reaction of Pb_3O_4 with nitric acid (reaction 22.50). Addition of a chromate salt to $Pb(NO_3)_2(aq)$ produces lead chromate, $PbCrO_4$, a pigment known as **chrome yellow.** Another lead-based pigment used in ceramic glazes and once extensively used in the manufacture of paint is **basic lead carbonate,** $2PbCO_3 \cdot Pb(OH)_2$, also known as **white lead.**

The chlorides of tin—$SnCl_2$ and $SnCl_4$—both have important industrial uses. In addition, $SnCl_2$ is a good reducing agent and is used in the laboratory to reduce Fe(III) to Fe(II), Hg(II) to Hg(I), and Cu(II) to Cu(I). Tin(II) fluoride, SnF_2 (*stannous fluoride*), has an important use as an anticavity additive to toothpaste. SnO_2 is used as a jewelry abrasive, and SnS_2 (tin bronze) is used as a pigment and for imitation gilding.

Lead Poisoning. Beginning with the ancient Romans and continuing to fairly recent times, lead has been used in plumbing systems, including those designed to transport water. Exposure to lead has also occurred through cooking and eating utensils and pottery glazes. In colonial times lead poisoning was clearly diagnosed as the cause of "dry bellyache" suffered by North Carolinians who consumed rum made in New England. The distilling equipment used in the manufacture of the rum was found to have components made from lead. Figure 22-17 records the dramatic increase in exposure to lead that has occurred in modern times.

Mild forms of lead poisoning produce nervousness and mental depression.

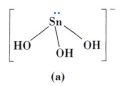

FIGURE 22-16
Structures of the (a) stannite and (b) stannate ions.

FIGURE 22-17

Deposition of lead in Greenland ice cap.

Fallen snow is converted to ice, in which the deposited lead is trapped. In 800 B.C. the level was less than 0.001 μg Pb/kg ice. Deposition of lead in the ice appears to be closely linked to human activities, with the lead being transported through the atmosphere. [Data from M. Murozumi, T. J. Chow, and C. C. Patterson. *Geochim. Cosmochim. Acta*, **33**, 1247 (1969).]

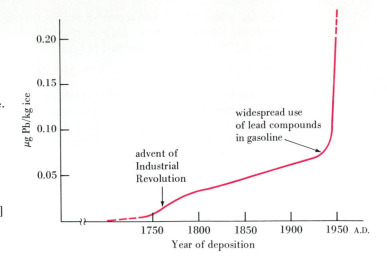

FIGURE 22-18

Lead in gasoline (—) and in blood (—).

The level of lead in the blood of a representative population has declined with the decline in use of lead in gasoline. Medical experts generally consider a lead level of 30 μg Pb/dL blood to pose a significant health risk, but recent evidence suggests measurable effects at lower levels. Changes in brain-wave patterns of children have been detected at levels of 10–15 μg Pb/dL blood.) [Data from Environmental Protection Agency Office of Policy Analysis, 1984.]

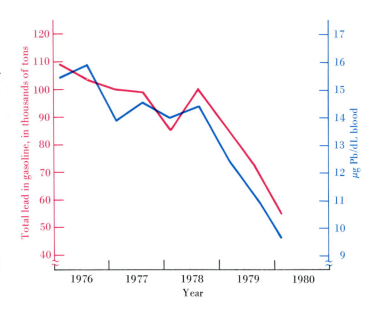

More severe cases can lead to permanent nerve, brain, and kidney damage. Lead interferes with the biochemical reactions that produce the iron-containing heme group in hemoglobin. The phasing out of leaded gasoline has resulted in a dramatic drop in average lead blood levels, as shown in Figure 22-18. The principal sources of lead contamination now seem to be lead-based painted surfaces in old buildings and soldered joints in plumbing systems.

22-6   Group 2B: Zinc, Cadmium, and Mercury

1B	2B	
29 Cu	30 Zn	31 Ga
47 Ag	48 Cd	49 In
79 Au	80 Hg	81 Tl

Until now we have discussed only elements that belong to A groups in the periodic table. These are the main-group or representative elements. In Chapter 24 we will consider transition elements, all of which belong to B subgroups. Subgroup 2B (or group 12) does not fit the description of transition elements that we will give in Chapter 24. These elements—Zn, Cd, Hg—are in fact a group of representative metals.

We might expect the group 2B elements to be similar to the transition elements that they immediately follow, and they are in such respects as an ability to form a

TABLE 22-7
Some Properties of the Group 2B Metals

	Zn	Cd	Hg
density, g/cm^3	7.14	8.64	13.59 (ℓ)
melting point, °C	419.6	320.9	−38.87
boiling point, °C	907	765	357
electron configuration	[Ar]$3d^{10}4s^2$	[Kr]$4d^{10}5s^2$	[Xe]$4f^{14}5d^{10}6s^2$
atomic radius, pm	133	149	150
ionization energy, kJ/mol			
first	906	867	1006
second	1703	1631	1809
principal oxidation state(s)	+2	+2	+1, +2
electrode potential $E°$, V			
[M^{2+}(aq) + 2 e$^-$ → M]	−0.763	−0.403	+0.851
[M_2^{2+}(aq) + 2 e$^-$ → 2 M]	—	—	+0.796

variety of complex ions. Also, we might expect some similarity to the group 2A metals, since the 2A and 2B elements are found in the same group in Mendeleev's periodic table (page 277). And these similarities do exist, mostly for Zn and Cd. For example, both Zn and Cd exhibit a single principal oxidation state (+2), their ions are colorless in solution, their solid compounds are mostly without strong color, and Zn^{2+} and Cd^{2+} can replace (are isomorphous with) Mg^{2+} in some of its compounds, such as, $Zn(Mg)SO_4 \cdot 7H_2O$.

As seen from Table 22-7, the group 2B metals have low melting and boiling points. This can probably be attributed to the relatively weak metallic bonding associated with the "18 + 2" electron configuration. Mercury is the only metal that exists as a liquid at room temperature and below (although liquid gallium can easily be supercooled to room temperature). Mercury differs from Zn and Cd in a number of ways in addition to its physical appearance.

- Mercury has little tendency to combine with oxygen; the oxide, HgO, is thermally unstable.
- Very few mercury compounds are water soluble and most are not hydrated.
- Many mercury compounds are covalent. The stability of the Hg—C bond makes possible a large number of organic mercury compounds. Mercury halides, except HgF_2, are only slightly ionized in aqueous solution.
- Mercury(I) forms a diatomic ion with a metal–metal covalent bond, Hg_2^{2+}.
- Mercury will not displace $H_2(g)$ from H^+(aq).

Some of these differences found in mercury can probably be attributed to the fact that $4f$ electrons are not as effective in shielding outer-shell electrons as are electrons in other inner subshells. This leads to a higher effective nuclear charge and a smaller size than we would otherwise expect for the Hg atom. As a result, ionization energies for Hg are somewhat higher than for Zn and Cd. In addition, hydration energies for Hg_2^{2+} and Hg^{2+} are also not as large as for Zn^{2+} and Cd^{2+}. The net result of these factors is that the electrode potentials for Hg_2^{2+} and Hg^{2+} are positive, whereas those of Zn^{2+} and Cd^{2+} are negative. Other explanations take into account the increased importance of the inert pair effect with the bottom member of a group. However, perhaps the most significant factor of all in explaining differences between the fifth and sixth period members of a group is relativistic effects on orbital energies in heavy atoms (see again Section 9-6).

Metallurgy and Uses of Zn, Cd, and Hg. We used zinc as a specific example of extractive metallurgy in Section 22-4. Cadmium generally occurs together with zinc in its ores, so that cadmium is obtained as a by-product in the production of zinc. Cadmium, because of its lower boiling point, can be separated from zinc by frac-

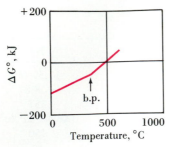

FIGURE 22-19
ΔG° as a function of temperature for the reaction

$$2 \, Hg + O_2 \rightarrow 2 \, HgO \quad (22.57)$$

At about the temperature where $\Delta G^\circ = 0$, spontaneous decomposition of HgO occurs. [The arrow marks the boiling point of Hg(l).]

tional distillation. If electrolytic reduction and refining of zinc is the chosen method, just prior to reaction (22.42) the solution containing Zn^{2+} is treated with powdered zinc, which goes into solution as Zn^{2+} and displaces Cd^{2+}. The Cd(s) is filtered off, dissolved in acidic solution, and electrolyzed to yield pure cadmium.

Mercury is one of the simplest metals to extract from its ores. When cinnabar (HgS) is heated, it decomposes directly to Hg(g), which condenses to Hg(l).

$$HgS(s) + O_2 \xrightarrow{\Delta} Hg(g) + SO_2(g) \qquad (22.54)$$

Alternative processes that eliminate the emission of $SO_2(g)$ involve roasting HgS with Fe or with CaO.

$$HgS(s) + Fe(s) \xrightarrow{\Delta} FeS(s) + Hg(g) \qquad (22.55)$$

$$4 \, HgS(s) + 4 \, CaO(s) \xrightarrow{\Delta} 3 \, CaS(s) + CaSO_4(s) + 4 \, Hg(g) \qquad (22.56)$$

The reason the roasting of HgS does not yield HgO is that HgO is unstable at high temperatures, decomposing to Hg(g) and $O_2(g)$, as suggested by Figure 22-19.

Mercury is purified by treatment with dilute $HNO_3(aq)$, which oxidizes most of the impurities. Insoluble products float to the surface of the liquid and are removed. Final purification is by distillation. Mercury can be easily obtained with a purity exceeding that of most metals (e.g., 99.9998% Hg or better).

About one-third of the metallic zinc produced is used in coating iron to give it corrosion protection (Section 21-6); the product is called *galvanized iron.* Large quantities of Zn are consumed in the manufacture of alloys. For example, about 20% of the production of Zn is used in *brass,* a copper alloy having 20 to 45% Zn and small quantities of Sn, Pb, and Fe. Brass is a good electrical conductor and it is corrosion resistant. Zinc is also used in the manufacture of dry cells, in printing (lithography), in the construction industry (roofing materials), and as sacrificial anodes in corrosion protection (Section 21-6).

Cadmium is used instead of zinc to protect iron in special applications. It is used in bearing alloys, in low-melting solders, in aluminum solders, and as an additive to impart strength to copper. Another important use, based on its neutron absorbing capacity, is in control rods and shielding for nuclear reactors (see Section 26-8).

The principal uses of mercury take advantage of its metallic and liquid properties and its high density. It is used in thermometers, barometers, gas pressure regulators, and electrical relays and switches and for electrodes, such as in the chlor-alkali process (Chapter 21). Mercury vapor is used in fluorescent tubes and street lamps. Mercury forms alloys, called **amalgams,** with most metals, and some of these amalgams are of commercial importance. Dental amalgam consists of 70% Hg and 30% Cu. Mercury compounds are used as pharmaceuticals, germicides, and fungicides.

Hydrated zinc ion ionizes in the same manner as described for $[Al(H_2O)_6]^{3+}$ on page 807. This makes aqueous solutions of Zn^{2+} acidic. If we add small quantities of OH^- to $Zn^{2+}(aq)$, $Zn(OH)_2(s)$ precipitates. With an excess of OH^- we produce $[Zn(OH)_4]^{2-}$. In summary,

Iron is one of the few metals that does not form an amalgam. Iron containers are used to store and ship mercury.

Zinc ion in aqueous solution.

$$[Zn(H_2O)_6]^{3+}(aq) \underset{H_3O^+}{\overset{OH^-}{\rightleftharpoons}} Zn(OH)_2(s) \underset{H_3O^+}{\overset{OH^-}{\rightleftharpoons}} [Zn(OH)_4]^{2-}(aq) \qquad (22.58)$$

Not only is $Zn(OH)_2(s)$ amphoteric, but so too is ZnO(s). In acidic solution it dissolves to produce $Zn^{2+}(aq)$ and in basic solution, $[Zn(OH)_4]^{2-}$. Furthermore, Zn, like aluminum, dissolves in NaOH(aq).

$$Zn(s) + 2 \, OH^-(aq) + 2 \, H_2O \longrightarrow [Zn(OH)_4]^{2-}(aq) + H_2(g) \qquad (22.59)$$

Some Important Compounds of the Group 2B Metals

Compound	Uses
ZnO	reinforcing agent in rubber; pigment; cosmetics; dietary supplement; photoconductors in copying machines.
ZnS	phosphors in x-ray and television screens; pigment; luminous paints.
$ZnSO_4$	rayon manufacture; animal feeds; wood preservative.
CdO	electroplating; batteries; catalyst; nematocide.
CdS	solar cells; photoconductor in xerography; phosphors; pigment.
$CdSO_4$	electroplating; standard voltaic cells (Weston cell).
HgO	polishing compounds; dry cells; antifouling paints; fungicide; pigment.
$HgCl_2$	manufacture of Hg compounds; disinfectant; fungicide; insecticide; wood preservative.
Hg_2Cl_2	electrodes; pharmaceuticals; fungicide.

$Cd(OH)_2$ displays basic properties by dissolving in acids. Its acidic character is much weaker than that of $Zn(OH)_2$, and it dissolves only in very concentrated $NaOH(aq)$. $Zn(OH)_2$ and $Cd(OH)_2$ are both soluble in $NH_3(aq)$ because of complex ion formation. For example,

$$Cd(OH)_2(s) + 4\ NH_3(aq) \longrightarrow [Cd(NH_3)_4]^{2+}(aq) + 2\ OH^-(aq) \qquad (22.60)$$

An interesting aspect of the chemistry of mercury centers on the disproportionation reaction

$$Hg_2^{2+}(aq) \longrightarrow Hg^{2+}(aq) + Hg(l) \qquad E^\circ_{cell} = -0.15\ V \qquad (22.61)$$

Although this reaction does not occur spontaneously under standard-state conditions, many situations arise in which the concentration of Hg^{2+} is brought to a very low value. When this happens, in accordance with Le Châtelier's principle, we can expect a displacement of equilibrium to the right, perhaps enough to drive the reaction essentially to completion. This is the case, for example, when $Hg_2^{2+}(aq)$ is treated with $S^{2-}(aq)$. $HgS(s)$ is so insoluble that reaction (22.61) does in fact occur. That is,

$$Hg_2^{2+}(aq) + S^{2-}(aq) \longrightarrow HgS(s) + Hg(l) \qquad (22.62)$$

Table 22-8 lists a few important compounds of the group 2B metals and some of their uses.

Mercury and Cadmium Poisoning. Accumulations of mercury in the body affect the nervous system and cause brain damage. "Hatter's disease" (which afflicted the Mad Hatter in *Alice's Adventures in Wonderland*) was a form of chronic mercury poisoning. Mercury compounds were used to convert fur to felt for making hats. One proposed mechanism of mercury poisoning involves interference with the functioning of sulfur-containing enzymes; Hg has a high affinity for sulfur. Organic mercury compounds are generally more poisonous than inorganic ones. An insidious aspect of mercury poisoning is that certain microorganisms have the ability to convert mercury to methylmercury (CH_3Hg^+) compounds, which then concentrate in the food chains of fish and other aquatic life.

In the free state, mercury is most poisonous as a vapor. Levels of mercury that exceed 10 μg Hg/m^3 air are considered unsafe. Even though we think of mercury as having a low vapor pressure, the concentration of Hg in its saturated vapor far exceeds this limit. Mercury vapor levels exceeding safe limits are sometimes found where mercury is used—chlor-alkali plants, thermometer factories, smelters, and dental laboratories.

Although zinc is an essential element in trace amounts, cadmium, which so

Disproportionation of mercury(I) ion, Hg_2^{2+}.

Hazardous materials on the painter's palette. A variety of popular yellow pigments are based on CdS, and red pigments, on CdSe and HgS; these are toxic materials. For ordinary use, as in house paints, these pigments have been largely replaced by iron oxides, which are environmentally safe. [Barbara Schultz/PAR/NYC]

resembles zinc, is a poison. One effect of cadmium poisoning is an extremely painful skeletal disorder known as "itai-itai kyo" (Japanese for "ouch-ouch" disease). This disorder was discovered in an area of Japan where effluents from a zinc mine became mixed with irrigation water used in rice fields. Cadmium poisoning was discovered in people who ate the rice. Cadmium poisoning can also cause liver damage, kidney failure, and pulmonary disease. Cd also appears to be a contributing factor in high blood pressure. The mechanism of Cd poisoning may involve substitution in certain enzymes of Cd, a poison, for Zn, an essential element. Concern over cadmium poisoning has increased with an awareness that some Cd is almost always found in zinc and zinc compounds, materials that have many commercial uses.

FOCUS ON

Hydrometallurgy

Large spherical autoclaves used in the leaching of ilmenite ore ($FeTiO_3$) with HCl at 120 °C and 350 kPa (3.5 atm). This is a step in the production of pure TiO_2, which is used as a white pigment in paint, paper, plastics, and rubber. [Courtesy United McGill Corporation]

The metallurgical procedure based on the roasting of an ore, followed by its reduction (usually with carbon), is known as **pyrometallurgy.** As we saw through Figure 22-15, a key consideration in pyrometallurgy is the temperature at which the reduction reaction becomes spontaneous. As the term "pyro" suggests, often this temperature is quite high. Some of the characteristics of pyrometallurgical processes are

- large quantities of waste materials generated in concentrating low-grade ores;
- high energy consumption to maintain high temperatures necessary for roasting and reduction of ores;
- gaseous emissions, such as $SO_2(g)$ in roasting, that must be controlled.

An alternative metallurgical procedure that is attractive for processing certain ores is **hydrometallurgy.** In *hydro*metallurgy the materials handled are water and aqueous solutions at moderate temperatures rather than dry

materials at high temperatures. Hydrometallurgy generally involves *three* steps.

1. *Leaching:* Metal ions are extracted from the ore by a liquid. Leaching agents include water, acids, bases, and salt solutions. Oxidation may also be required in the leaching process.
2. *Purification and/or concentration:* Impurities are separated from the desired metallic ions and/or the solution produced by leaching is made more concentrated. Methods used include evaporation of water, ion exchange, solvent extraction (distribution of a substance between two solvents), or adsorption on the surface of activated charcoal.
3. *Precipitation.* The desired metal is precipitated as an ionic solid, or metal ions are reduced to metal atoms. Reduction can also be carried out electrolytically.

One advantage of hydrometallurgy is the elimination of the traditional steps in concentrating an ore (such as flotation) before extracting a metal. In fact, at times, leaching of a metal ore may be carried out in place, that is, without even mining the ore (also called solution mining). Another advantage includes the ability to process low-grade ores and complex ores that would be difficult or uneconomic to process by pyrometallurgy. In still other cases, hydrometallurgy may reduce emissions of pollutants (particularly air pollutants). However, hydrometallurgy does produce liquid and solid products that present disposal problems.

Let us consider the hydrometallurgy of zinc, a process that has been developed to a commercial stage. Leaching of zinc sulfide ore is accomplished with a sulfuric acid solution at 150 °C and 700 kPa (7 atm) oxygen pressure. The net reaction is

$$ZnS(s) + H_2SO_4(aq) + \tfrac{1}{2} O_2(g) \longrightarrow$$
$$ZnSO_4(aq) + S(s) + H_2O \quad (22.63)$$

In this process there is no $SO_2(g)$ emission. Also mercury impurities in the ZnS ore are retained in the leaching solution rather than being emitted with $SO_2(g)$ as in the traditional roasting process.

Following the leaching reaction (22.63), $ZnSO_4(aq)$ is electrolyzed to produce pure $Zn(s)$ and to regenerate $H_2SO_4(aq)$ for use in the leaching process. The electrolysis reaction is the one we considered in Section 22-4.

$$Zn^{2+} + SO_4^{2-} + H_2O \xrightarrow{\text{electrolysis}} Zn(s) + 2 H^+ + SO_4^{2-} + \tfrac{1}{2} O_2(g) \quad (22.42)$$

In describing the metallurgy of several metals in Chapter 24 we will have occasion to note additional examples of both *pyro*metallurgy and *hydro*metallurgy.

Summary

The alkali (group 1A) metals are the most active of the metals, as indicated by their low ionization energies and large negative electrode potentials. Li is rather different from the other 1A metals, and these differences can be explained for the most part in terms of the high positive charge density on Li^+. The alkali metals are prepared mostly by the electrolysis of their molten salts. Electrolysis of $NaCl(aq)$ produces $NaOH(aq)$, from which other sodium compounds can then be prepared. $NaOH$ can also be prepared from Na_2CO_3. In turn, Na_2CO_3 can be produced from $NaCl$, NH_3, and CO_2 by the Solvay process. In the United States, however, the Solvay process has been phased out because of environmental and economic problems, and Na_2CO_3 is obtained from mineral deposits.

The alkaline earth (group 2A) metals are also very active; some are prepared by the electrolysis of a molten salt and some by chemical reduction. In the case of magnesium, the salt is $MgCl_2$, derived either from seawater or from natural brine solutions. The comparative strengths of lattice energies and cation hydration energies are important in establishing the solubilities of group 2A metal salts. A large number of these salts are insoluble in water, but others are quite soluble and crystallize from solution as deliquescent hydrates. Among the most important of the insoluble salts are the carbonates, especially $CaCO_3$. Reversible reactions involving CO_3^{2-}, HCO_3^-, $CO_2(g)$, and H_2O account for the formation of both limestone caves and temporary hard water. Water may also have noncarbonate or permanent hardness, through the presence of such ions as SO_4^{2-}. One of the objections to hard water is its action on soaps. Water can be softened either through chemical reactions or by ion exchange. Alternatively, synthetic detergents can be used even in hard water.

The principal metal of group 3A is aluminum, whose large-scale use has been made possible through an effective method of production. The amphoteric nature of Al_2O_3 is used as the basis for separating Al_2O_3 from its impurities (mostly Fe_2O_3). The electrolysis is carried out in molten Na_3AlF_6 with Al_2O_3 as a solute. Much of the solution chemistry of aluminum centers on the hydrated aluminum ion, $[Al(H_2O)_6]^{3+}$, the amphoteric hydroxide, $Al(OH)_3$, and the aluminate ion, $[Al(OH)_4]^-$, which forms in basic solution. For example, this solution behavior accounts for the fact that Al metal dissolves in $NaOH(aq)$.

None of the other group 3A metals has been of commercial importance until recent times. With the discovery of some highly desirable semiconductor properties of gallium arsenide (GaAs), the element gallium may become of great significance.

The traditional methods of extractive metallurgy involve concentrating an ore (often by the process of flotation), roasting the ore to convert it to an oxide, and reducing the oxide to the metal, with carbon as the preferred reducing agent. In a final step the metal is purified, sometimes electrolytically. A useful tool in establishing the required reduction temperature is a graph of free energy change for oxide formation as a function of temperature. In some instances this method of *pyrometallurgy* is being replaced by a system of metallurgy in which metal ions are leached from an ore, the metal ion solution purified, and pure metal then deposited from solution. This newer method, which relies on water solutions as chemical reaction media, is called *hydrometallurgy*.

Tin and lead in group 4A have some similarities, but they also exhibit some differences. The chief difference is that tin acquires the oxidation state +4 rather easily, whereas with lead the +2 oxidation state is favored. This means that lead(IV) compounds are easily reduced and are, therefore, powerful oxidizing agents.

In subgroup 2B the heaviest member, Hg, differs somewhat from the other two—Zn and Cd. These differences can be explained from several standpoints. The solution behavior of Zn^{2+} is similar to that of Al^{3+}, for instance, $Zn(OH)_2$ is amphoteric and Zn is soluble in $NaOH(aq)$.

Summarizing Example

Lime is used to soften water in this water treatment plant in Altoona, PA. [Courtesy National Lime Association]

The most widely used method of water softening is treatment with lime. One of the interesting features of this method is that even though Ca^{2+} is *added* to the water through the lime treatment, the overall $[Ca^{2+}]$ of the water is *reduced*. The essential role played by the lime is to increase the alkalinity of the water so that HCO_3^- is converted to CO_3^{2-}, which is then precipitated as $CaCO_3(s)$. Lime is so often chosen because it is the cheapest alkaline substance available.

1. A particular water sample is found to contain 88.2 ppm of SO_4^{2-} and 149 ppm HCO_3^-, with Ca^{2+} as the only cation. How many ppm of Ca^{2+} does this water contain?

Solution. When a problem deals with concentrations expressed in ppm, a useful starting point is to consider a 1-million-g sample. Then the masses of solution components are simply equal to their ppm values: 88.2 g SO_4^{2-} and 149 g HCO_3^-. Since the relationship between the amount of Ca^{2+} and the amounts of SO_4^{2-} and HCO_3^- must be on a *mole* basis, we should first express the quantities of all ions in moles.

$$\text{no. mol } SO_4^{2-} = 88.2 \text{ g } SO_4^{2-} \times \frac{1 \text{ mol } SO_4^{2-}}{96.06 \text{ g } SO_4^{2-}} = 0.918 \text{ mol } SO_4^{2-}$$

$$\text{no. mol } HCO_3^- = 149 \text{ g } HCO_3^- \times \frac{1 \text{ mol } HCO_3^-}{61.02 \text{ g } HCO_3^-} = 2.44 \text{ mol } HCO_3^-$$

$$\text{no. mol } Ca^{2+} = \left(0.918 \text{ mol } SO_4^{2-} \times \frac{1 \text{ mol } Ca^{2+}}{1 \text{ mol } SO_4^{2-}}\right)$$
$$+ \left(2.44 \text{ mol } HCO_3^- \times \frac{1 \text{ mol } Ca^{2+}}{2 \text{ mol } HCO_3^-}\right)$$
$$= 2.14 \text{ mol } Ca^{2+}$$

$$\text{no. ppm } Ca^{2+} = \frac{2.14 \text{ mol } Ca^{2+}}{1.00 \times 10^6 \text{ g water}} \times \frac{40.08 \text{ g } Ca^{2+}}{1 \text{ mol } Ca^{2+}} = 85.8 \text{ ppm } Ca^{2+}$$

2. Write equations to show how the addition of lime (CaO) brings about the precipitation of $CaCO_3$ from temporary hard water.

Solution. $CaO(s)$ reacts with water to produce $Ca(OH)_2$. The $Ca(OH)_2$ produces OH^- ions that convert HCO_3^- to CO_3^{2-}. Ca^{2+} and CO_3^{2-} combine to form the insoluble $CaCO_3(s)$.

$$CaO(s) + H_2O \longrightarrow Ca^{2+}(aq) + 2 \text{ } OH^-(aq)$$

$$OH^-(aq) + HCO_3^-(aq) \longrightarrow H_2O + CO_3^{2-}(aq)$$

$$Ca^{2+}(aq) + CO_3^{2-}(aq) \longrightarrow CaCO_3(s)$$

3. How many moles of CaO are consumed in removing HCO_3^- from 1.00×10^6 g of the water described above?

Solution. The mass of HCO_3^- to be converted to CO_3^{2-} is 149 g, and as shown in part **1**, this is 2.44 mol HCO_3^-. The question becomes that of determining the amount of CaO that produces enough OH^- to consume 2.44 mol HCO_3^- in the reactions in part **2**.

$$\text{no. mol CaO} = 2.44 \text{ mol } HCO_3^- \times \frac{1 \text{ mol } OH^-}{1 \text{ mol } HCO_3^-} \times \frac{1 \text{ mol CaO}}{2 \text{ mol } OH^-}$$

$$= 1.22 \text{ mol CaO}$$

4. What is the concentration of Ca^{2+} remaining in the water after the treatment described in part **3**?

Solution. First, calculate the total amount of Ca^{2+} in moles. This consists of the 2.14 mol Ca^{2+} originally present in the water (part **1**) *plus* the Ca^{2+} associated with 1.22 mol CaO (part **3**).

$$\text{no. mol } Ca^{2+} = 2.14 \text{ mol } Ca^{2+} + \left(1.22 \text{ mol CaO} \times \frac{1 \text{ mol } Ca^{2+}}{1 \text{ mol CaO}} \right)$$

$$= 3.36 \text{ mol } Ca^{2+}$$

Next, determine the number of moles of Ca^{2+} that were removed by precipitation as $CaCO_3(s)$ in part **3**. That is, how many mol Ca^{2+} are equivalent to 2.44 mol CO_3^{2-}?

$$\begin{aligned}\text{no. mol } Ca^{2+} \\ \text{precipitated}\end{aligned} = 2.44 \text{ mol } HCO_3^- \times \frac{1 \text{ mol } CO_3^{2-}}{1 \text{ mol } HCO_3^-} \times \frac{1 \text{ mol } Ca^{2+}}{1 \text{ mol } CO_3^{2-}}$$

$$= 2.44 \text{ mol } Ca^{2+}$$

Then,

$$\text{no. mol } Ca^{2+} \text{ remaining} = 3.36 \text{ mol } Ca^{2+} - 2.44 \text{ mol } Ca^{2+} = 0.92 \text{ mol } Ca^{2+}$$

This amount of Ca^{2+} is found in 1.00×10^6 g of water. Convert this amount to a mass in grams, which is also the Ca^{2+} concentration in ppm.

$$\text{no. ppm } Ca^{2+} = \frac{0.92 \text{ mol } Ca^{2+}}{1.00 \times 10^6 \text{ g water}} \times \frac{40.08 \text{ g } Ca^{2+}}{1 \text{ mol } Ca^{2+}} = 37 \text{ ppm } Ca^{2+}$$

Note how this partial softening of the water reduces the concentration of Ca^{2+} from about 86 ppm (part **1**) to 37 ppm (part **4**).

Key Terms

alums (22-3)
amalgams (22-6)
anodizing (22-3)
concentration (22-4)
deionized water (22-2)
deliquescence (22-2)
detergents (22-2)
extractive metallurgy (22-4)
galvanizing (22-6)

hard water (22-2)
hydration energy (22-1)
hydrometallurgy (Focus feature)
inert pair effect (22-3)
ion exchange (22-2)
leaching (Focus feature)
permanent hard water (22-2)
pyrometallurgy (Focus feature)
reduction (22-4)

refining (22-4)
relative humidity (22-2)
roasting (22-4)
soap (22-2)
solders (22-5)
stalactite (22-2)
stalagmite (22-2)
temporary hard water (22-2)

Highlighted Expressions

The Solvay process for $NaHCO_3$ (22.5)
Decomposition of group 2A metal carbonates (22.7)
Reaction of group 2A metal oxides with water (22.8)
Reaction involved in the formation of limestone caverns (22.12)
Effect of heating on temporary hard water (22.13)
Chemical treatment of temporary hard water (22.14, 22.15)
Ion exchange reactions (22.16, 22.17)

Aluminum ion in aqueous solution (22.28)
Electrolytic production of Al (22.35)
Metallurgical roasting reactions (22.36, 22.37)
Metallurgical reduction reactions (22.38, 22.39)
Zinc ion in aqueous solution (22.58)
Disproportionation of mercury(I) ion, Hg_2^{2+} (22.61)

Review Problems

1. Provide an acceptable name or formula for each of the following.
  (a) PbO_2 **(b)** Hg_2Br_2
  (c) $CaCl_2 \cdot 6H_2O$ **(d)** zincate ion
  (e) $NaAl(OH)_4$ **(f)** $K_2Sn(OH)_6$
  (g) magnesium hydrogen carbonate
  (h) aluminum ammonium alum

2. Complete and balance the following; write the simplest equation possible. If no reaction occurs, so state.

  (a) $MgCO_3(s) \xrightarrow{\Delta}$
  (b) $CaO(s) + HCl(aq) \rightarrow$
  (c) $Al(s) + KOH(aq) + H_2O \rightarrow$
  (d) $Pb(s) + HNO_3(aq) \rightarrow$
  (e) $Hg(l) + HCl(\text{dil. aq}) \rightarrow$

  (f) $ZnO(s) + CO(g) \xrightarrow{\Delta}$

3. Assuming the availability of water, common reagents (acids, bases, salts), and simple laboratory equipment, give a practical method that could be used to prepare **(a)** $MgCl_2 \cdot 6H_2O$ from $MgCO_3(s)$; **(b)** $NaAl(OH)_4$ from $Na(s)$ and $Al(s)$; **(c)** $ZnS(s)$ from $ZnO(s)$; **(d)** $HgS(s)$ from $Hg(l)$.

4. Write the simplest chemical equation to represent the reaction of **(a)** $K_2CO_3(aq)$ and $Ba(OH)_2(aq)$; **(b)** $Mg(HCO_3)_2(aq)$ upon heating; **(c)** tin(II) oxide when heated with carbon; **(d)** $HNO_3(\text{concd aq})$ and metallic tin; **(e)** $H_2SO_4(aq)$ and $CdO(s)$; **(f)** $PbO_2(s)$ and $HI(aq)$.

5. Write *plausible* balanced equations to represent these conversions outlined in Figure 22-3. **(a)** Na_2SiO_3 from Na_2CO_3; **(b)** Na_2SO_4 from $NaCl$; **(c)** K_2SO_3 from KOH; **(d)** K_2HPO_4 from KOH.

6. A sample of water whose hardness is expressed as 185 ppm Ca^{2+} is passed through an ion exchange column and the Ca^{2+} is replaced by Na^+. What is $[Na^+]$ in the water that has been so treated? [*Hint:* Base your calculation on a 1000-L sample, which weighs 1.00×10^6 g.]

7. Predict whether each of the following is expected to produce an aqueous solution that is acidic, basic, or neutral: **(a)** $NaHSO_4$; **(b)** $KAl(SO_4)_2 \cdot 12H_2O$; **(c)** $KAl(OH)_4$; **(d)** $ZnSO_4$; **(e)** NH_4Cl.

8. Write chemical equations to represent the most probable outcome in each of the following. If no reaction is likely to occur, so state.

  (a) $CdCO_3(s) \xrightarrow{\Delta}$
  (b) $MgO(s) \xrightarrow{\Delta}$
  (c) $SnO_2(s) + CO(g) \xrightarrow{\Delta}$
  (d) $Cd^{2+}(aq) + SO_4^{2-}(aq) \xrightarrow{\text{electrolysis}}$
  (e) $HgO(s) \xrightarrow{\Delta}$
  (f) $MgO(s) + Zn(s) \xrightarrow{\Delta}$

9. A chemical dictionary gives the following descriptions of the production of some compounds. Write plausible chemical equations based on these descriptions.

  (a) Lithium bromide: reaction of hydrobromic acid with lithium carbonate.
  (b) Lithium carbonate: reaction of lithium oxide with ammonium carbonate solution.
  (c) Magnesium sulfite: action of sulfurous acid on magnesium hydroxide.
  (d) Mercury(I) chloride: sublimation from a heated mixture of mercury and mercury(II) chloride.
  ★**(e)** Lead(IV) oxide: action of an alkaline solution of calcium hypochlorite on lead(II) oxide.

10. Use data from Figures 22-15 and 22-19 to answer the following.

  (a) Will $C(s)$ reduce $HgO(s)$ to Hg at 100 °C?
  (b) Will $Hg(l)$ reduce $ZnO(s)$ to $Zn(s)$ at 200 °C?
  (c) Will $Mg(s)$ reduce $ZnO(s)$ to $Zn(s)$ at 200 °C?

Exercises

Alkali (group 1A) metals

11. Write equations for the following reactions of lithium compounds, referred to on page 792. **(a)** formation of Li_3N; **(b)** formation of the oxide (Li_2O); **(c)** decomposition, on heating, of lithium carbonate to lithium oxide; **(d)** decomposition, on heating, of lithium hydroxide to lithium oxide.

12. Refer to Example 22-1. The melting point of $NaCl(s)$ is

801 °C, and this is much higher than that of $NaOH$ (322 °C). More energy is consumed to produce and maintain molten $NaCl$ than $NaOH$. Yet the preferred commercial process for the production of sodium is electrolysis of $NaCl(l)$. Can you think of a reason(s) for this?

13. A particular lithium battery used in a cardiac pacemaker has a voltage of 3.0 V. The capacity of this battery is listed as

0.5 A h (ampere hour). Assume that to regulate the heartbeat requires 5 μW of power. [*Hint:* 1 μW = 1 microwatt = 1 × 10^{-6} watt, and 1 W = 1 J/s.]

(a) How long will the battery last once it is implanted?

(b) What minimum mass of lithium must be present in the battery for the lifetime calculated in (a)?

14. An analysis of a Solvay plant shows that for every 1.00 ton of NaCl consumed 1.03 tons of $NaHCO_3$ is obtained. Only 1.5 lb NH_3 is consumed in the overall process.

(a) What is the % efficiency of this process for converting NaCl to $NaHCO_3$?

(b) Why is so little NH_3 required?

15. What is the pH of the resulting solution if 0.445 L NaCl(aq) is electrolyzed for 137 s with a current of 1.08 A? Does the result depend on the concentration of NaCl present? Explain.

16. We wish to consider the feasibility of manufacturing NaOH(aq) by the reaction

$$Ca(OH)_2(s) + Na_2SO_4(aq) \rightleftharpoons CaSO_4(s) + 2\,NaOH(aq)$$

(a) Write a net ionic equation for the reaction.

(b) Will the reaction go essentially to completion? [*Hint:* What is K for the reaction written in (a)? Use data from Chapter 19, as necessary.]

(c) What will be $[SO_4^{2-}]$ and $[OH^-]$, *at equilibrium*, if a slurry of $Ca(OH)_2(s)$ is mixed with 1.00 M $Na_2SO_4(aq)$?

Alkaline earth (group 2A) metals

17. In the manner used to construct Figure 22-3 complete the diagram outlined. Specifically, indicate the reactants (and conditions) required to produce the indicated chemicals from $Ca(OH)_2$.

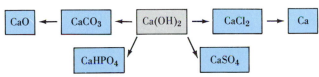

Exercise 17

18. Predict the most probable products when the following liquids are electrolyzed with platinum electrodes. [*Hint:* What are the most probable oxidation and reduction half-reactions?] (a) $CaCl_2(l)$; (b) $MgSO_4(aq)$; (c) $BaCl_2$ in HCl(aq); (d) $SrI_2(aq)$.

19. From these two statements estimate the quantity of Mg that has been produced since the metal was discovered: All the Mg produced in the world could have been made from as little as 4 km^3 of seawater. The quantity of Mg^{2+} in seawater is 1272 g Mg^{2+}/ton seawater. (The density of seawater is 1.03 g/cm^3.)

★**20.** The electrolysis of 0.250 L of 0.220 M $MgCl_2$ is conducted until 104 cm^3 of gas (a mixture of H_2 and water vapor) is collected at 23 °C and 748 mmHg. Will $Mg(OH)_2(s)$ precipitate if electrolysis is carried to this point? (Use 21 mmHg as the vapor pressure of the solution.)

21. The dissolving of $MgCO_3(s)$ in $NH_4^+(aq)$ can be represented as

$$MgCO_3(s) + NH_4^+(aq) \rightleftharpoons$$
$$Mg^{2+}(aq) + HCO_3^-(aq) + NH_3(aq)$$

(a) Arrange the following three solubilities in the expected order, from most to least soluble: $MgCO_3(s)$ in 1.00 M

$NH_4Cl(aq)$; $MgCO_3(s)$ in the solution 1.00 M NH_3–1.00 M NH_4Cl; $MgCO_3(s)$ in the solution 0.100 M NH_3–1.00 M NH_4Cl.

★(b) Calculate each of the solubilities referred to in part (a).

Hard water

22. Write chemical equations for the reactions represented by the photographs on page 802.

23. A particular hard water contains 120.0 ppm HCO_3^-. What mass of CaO is required to soften 1.00 × 10^6 gal of this water. [*Hint:* Recall equations (22.14) and (22.15).]

24. Suppose that all the cations associated with HCO_3^- in Exercise 23 are Ca^{2+}.

(a) What is the total mass of $CaCO_3$ that would be precipitated in softening 1.00 × 10^6 gal of this water?

(b) Show that in the $CaCO_3$, half the Ca^{2+} is derived from the CaO used in the water softening and half from the water itself.

25. A particular water sample has a hardness of 117 ppm SO_4^{2-} (as $CaSO_4$).

(a) Show how this water can be softened with Na_2CO_3.

(b) What mass of Na_2CO_3 is required to soften 162 L of this water?

26. Describe, using equations similar to (22.17), how a sample of hard water containing Fe^{3+}, Ca^{2+}, Mg^{2+}, HCO_3^-, and SO_4^{2-} can be deionized by passing it first through a cation exchange resin and then an anion exchange resin.

Aluminum

27. In some foam-type fire extinguishers the reactants are $Al_2(SO_4)_3(aq)$ and $NaHCO_3(aq)$. When the extinguisher is activated, these reactants are allowed to mix, producing $Al(OH)_3(s)$ and $CO_2(g)$. The $Al(OH)_3$–CO_2 foam extinguishes the fire. Write a net ionic equation to represent this reaction.

28. The maximum resistance to corrosion of Al metal is between pH 4.5 and 8.5. Explain how this observation is consistent with other facts about the behavior of Al presented in the text.

29. Describe a series of *simple* chemical reactions with which you could determine whether a particular metal sample is Aluminum 2S (99.2% Al) or magnalium (70% Al, 30% Mg). You are permitted to destroy the metal sample in the testing.

30. In the purification of bauxite ore as a preliminary step in the production of aluminum, $[Al(OH)_4]^-(aq)$ can be converted to $Al(OH)_3(s)$ by passing $CO_2(g)$ through it. Write an equation for the reaction that occurs.

31. Concerning the compound $NaAl(SO_4)_2 \cdot 12H_2O$

(a) Why is "sodium alum" not a sufficient name for the compound?

(b) What is a more appropriate name for the compound?

(c) Is an aqueous solution of this compound acidic, basic, or neutral? Explain.

32. Use information from the chapter to explain why neither the compound $Al(HCO_3)_3$ nor the compound $Al_2(CO_3)_3$ exists.

★**33.** Use the relationship $\Delta G° = -n\mathscr{F}E°_{cell}$ to estimate the minimum voltage required to electrolyze Al_2O_3 in the Hall–Heroult process (22.35). For this purpose use $\Delta G_f°[Al_2O_3(l)] = -1520$ kJ/mol and $\Delta G_f°[CO_2(g)] = -394$ kJ/mol. Show that the oxidation of the graphite anode to $CO_2(g)$ permits the electrolysis to occur at a lower voltage than if the electrolysis reaction were $Al_2O_3(l) \rightarrow 2\,Al(l) + \frac{3}{2}\,O_2(g)$.

Extractive metallurgy

34. The following approximate data are given for the reaction

$$2 H_2(g) + O_2(g) \rightarrow 2 H_2O(g)$$

$\Delta G° = -465$ kJ at 100 °C; -420 kJ at 500 °C; -360 kJ at 1000 °C; -240 kJ at 2000 °C.

(a) Sketch $\Delta G°$ as a function of temperature on the same axes as used in Figure 22-15.

(b) At what approximate temperature does the water gas reaction $C(s) + H_2O(g) \rightarrow CO(g) + H_2(g)$ become spontaneous?

35. Calcium metal will reduce MgO(s) to Mg at all temperatures from 0 to 2000 °C. Use this fact to sketch a plausible graph of $\Delta G°$ versus temperature for the reaction

$$2 Ca + O_2 \rightarrow 2 CaO$$

For calcium, m.p. = 839 °C; b.p. = 1484 °C. [*Hint:* Relate your sketch to Figure 22-15.]

*36. According to Figure 22-15, $\Delta G°$ decreases with temperature for the reaction $2 C(s) + O_2(g) \rightarrow 2 CO(g)$. Describe how you would expect $\Delta G°$ to vary with temperature for the following reactions.

(a) $C(s) + O_2(g) \rightarrow CO_2(g)$

(b) $2 CO(g) + O_2(g) \rightarrow 2 CO_2(g)$

*37. Explain why (a) breaks in the straight lines of Figure 22-15 occur at the melting points and boiling points of the metals; (b) the slopes of the lines become more positive at these breaks; (c) the break at the boiling point is sharper than at the melting point.

Tin and lead

38. Write plausible equations for the following reactions that are described in the chapter.

(a) The dissolving of lead in nitric acid.

(b) The dissolving of PbO in NaOH(aq).

(c) The conversion of SnO to SnO_2 by heating in air.

(d) The disproportionation of SnO to Sn and SnO_2.

39. The names dilead(II)lead(IV) oxide and plumbous orthoplumbate have both been used to describe red lead, Pb_3O_4. Show that these names are consistent with remarks made in the text about this oxide of lead.

40. Write chemical equations to represent the following reactions.

(a) The reaction of Sn with hot concentrated HNO_3(aq).

(b) The oxidation of HCl(aq) to Cl_2(g) by PbO_2(s).

(c) The reduction of Fe^{3+} to Fe^{2+} by Sn^{2+} in aqueous solution.

(d) The production of basic lead carbonate (white lead), $2PbCO_3 \cdot Pb(OH)_2$, by the action of H_2O and CO_2 on PbO.

41. With the reagents (acids, bases, salts) and equipment commonly available in a chemical laboratory indicate how you would prepare (a) $PbCrO_4$(s) from Pb_3O_4(s); (b) $PbCl_2$(s) from PbS(s).

*42. At 20 °C a saturated aqueous solution of $Pb(NO_3)_2$ maintains a relative humidity of 97%. What must be the concentration of this saturated solution, expressed as g $Pb(NO_3)_2$/100. g H_2O? [*Hint:* What is the relationship between the concentration of a solution and its vapor pressure? Vapor pressure data for water are given in Table 12-2.]

43. Scrap tin plate can be recycled by dissolving the tin in NaOH(aq), followed by electrolysis. Write plausible equations for the reactions involved.

*44. Use information from this chapter and elsewhere in the text to explain why neither the compound $PbBr_4$ nor the compound PbI_4 exists.

*45. To prevent the air oxidation of aqueous solutions of Sn^{2+} to Sn^{4+}, sometimes metallic tin is kept in contact with the Sn^{2+}(aq). Suggest how this helps to prevent the oxidation.

Group 2B metals

46. Write chemical equations to represent (a) the dissolving of Cd in H_2SO_4 (conc. aq); (b) the dissolving of Hg in dilute HNO_3(aq); (c) the dissolving of ZnO(s) in $HC_2H_3O_2$(aq); (d) the action of excess NaOH(aq) on $ZnSO_4$(aq).

47. Use data from Table 22-7 and equation (22.61) to determine $E°$ for the reduction $2 Hg^{2+}(aq) + 2 e^- \rightarrow Hg_2^{2+}(aq)$.

48. The following equilibrium data are given

$HgO(s) \rightarrow Hg(l) + \frac{1}{2} O_2(g)$ $\quad \Delta G° = 4.25$ kJ at 800 K
$\quad\quad\quad\quad\quad\quad\quad\quad\quad\quad\quad\quad\quad = -2.83$ kJ at 900 K

$HgO(s) \rightarrow Hg(g) + \frac{1}{2} O_2(g)$ $\quad \Delta G° = -12.19$ kJ at 800 K

(a) What is the value of $\Delta H°$ for the reaction

$$HgO(s) \rightarrow Hg(l) + \frac{1}{2} O_2(g) \quad \Delta H° = ?$$

[*Hint:* Can you establish two values of K_p?]

(b) What is the value of K_p at 700 K for the reaction

$$HgO(s) \rightarrow Hg(l) + \frac{1}{2} O_2(g) \quad K_p = ?$$

(c) What is $\Delta G°$ for the vaporization of Hg(l) at 800 K?

$$Hg(l) \rightarrow Hg(g) \quad \Delta G° = ?$$

49. The text notes that in small quantities Zn is an essential metal (though it is toxic in higher concentrations). Tin is considered among the toxic metals. Can you think of reasons why tin plate is used rather than galvanized iron in cans used for food storage?

50. The vapor pressure of Hg as a function of temperature is

$$\log p \text{ (mmHg)} = \frac{-0.05223a}{T} + b$$

where $a = 61,960$; $b = 8.118$; T = Kelvin temperature. Show that the concentration of Hg(g) in equilibrium with Hg(l) greatly exceeds the maximum permissible level of 0.05 mg Hg/m³ air. (Assume a temperature of 25 °C.)

Additional Exercises

51. Write chemical equations for the following reactions.

(a) The decomposition of $Mg(HCO_3)_2$ on heating.

(b) The recovery of NH_3 in the Solvay process.

(c) Production of an alkaline solution for the precipitation of $Mg(OH)_2$(s) from seawater.

(d) The softening of temporary hard water with NH_3(aq).

(e) The separation of Fe_2O_3 impurity from bauxite ore.

(f) The formation of lead(II) sulfate during the high-temperature roasting of lead(II) sulfide.

(g) The successive conversion of KCl to K_2SO_4 to K_2S to K_2CO_3.

52. Write chemical equations to represent the dissolving of **(a)** Zn in HCl(aq); **(b)** Pb in HCl(conc. aq); **(c)** cadmium oxide in dilute HNO_3(aq); **(d)** cadmium hydroxide in NH_3(aq); **(e)** Al in KOH(aq); **(f)** sodium aluminum alum in NH_3(aq).

53. Oersted (1825) produced aluminum chloride by passing Cl_2(g) over a heated mixture of aluminum oxide and carbon. Wöhler (1827) prepared Al by heating aluminum chloride with potassium. Write plausible equations for these reactions.

54. Exercise 30 describes how $Al(OH)_3$(s) can be obtained by treating $[Al(OH)_4]^-$(aq) with CO_2(aq). Could HCl(aq) be used instead of CO_2(g)? Explain.

55. A description for preparing potassium aluminum alum calls for dissolving aluminum foil in KOH(aq). The solution obtained is treated with H_2SO_4(aq) and the alum is crystallized from the resulting solution. Write plausible equations for the reactions just described.

56. Some baking powders contain the solids $NaHCO_3$ and $NaAl(SO_4)_2$. When water is added to this mixture of compounds, CO_2(g) and $Al(OH)_3$(s) are two of the products. Write a plausible equation(s) for this reaction.

★57. In a chemical storeroom it is not uncommon to find that a chemical that should exist as a crystalline solid at room temperature is found to be a mixture of a solid and a liquid. Explain what might have happened. Should the chemical be discarded or can it still be used for certain purposes? Explain.

58. Lead dioxide, PbO_2, is a good oxidizing agent whose main use is in the lead–acid battery. Use appropriate data from Chapter 21 to determine whether PbO_2(s) in a solution with $[H_3O^+] = 1$ M is a sufficiently good oxidizing agent to carry the following oxidations essentially to completion.

(a) Fe^{2+}(1 M) to Fe^{3+}

(b) SO_4^{2-}(1 M) to $S_2O_8^{2-}$

★(c) Mn^{2+}(1×10^{-4} M) to MnO_4^-

[*Hint:* Assume that the reaction has gone to completion if the concentration of the species being oxidized decreases to one-thousandth of its initial value.]

59. Sn^{2+} is a good reducing agent. Use data from Chapter 21 to determine whether Sn^{2+} is a sufficiently good reducing agent to reduce **(a)** I_2 to I^-; **(b)** Fe^{2+} to Fe(s); **(c)** Cu^{2+} to Cu(s); **(d)** Fe^{3+}(aq) to Fe^{2+}(aq).

60. In this chapter several examples were cited of the first member of a group differing from the second. Also in this chapter differences were cited between the fifth period and the sixth period members of a group. What are the fundamental reasons for these differences?

61. A sample of water has a hardness expressed as 77.5 ppm Ca^{2+}. This sample is passed through an ion exchange column and the Ca^{2+} is replaced by H^+. What is the pH of the water after it has been so treated? [*Hint:* Base your calculation on 1000 L (1.00×10^6 g) of water.]

62. What mass of "bathtub ring" will form if 25.5 L of water having 88 ppm Ca^{2+} is treated with an excess of the soap potassium stearate, $CH_3(CH_2)_{16}COO^-K^+$?

63. In the Dow process (Figure 22-6), the starting material is Mg^{2+} and the final product is Mg. This process seems to violate the principle of conservation of electric charge. Explain why this is not so.

64. Refer to the Summarizing Example.

(a) Show that the Ca^{2+} remaining in the water after the partial softening can be removed by adding Na_2CO_3. [*Hint:* What reaction occurs?]

(b) What mass of Na_2CO_3 is required to precipitate this remaining Ca^{2+} after the partial softening of 1.00×10^6 g of the water?

65. Concerning the reactions represented by the photographs on page 802, and assuming saturated CO_2(aq) is 0.034 M.

(a) If the $Ca(OH)_2$(aq) were replaced by $CaCl_2$(aq), would a precipitate form in the presence of CO_2(aq)? Explain.

★(b) The $CaCO_3$(s) precipitate redissolves if the $Ca(OH)_2$(aq) used is about 0.005 M, but fails to redissolve if the $Ca(OH)_2$(aq) is saturated. By calculation, demonstrate that this difference in behavior is to be expected.

66. For each of the following indicate if equilibrium is displaced either far to the left or far to the right. Detailed calculations are not required. Use data from Chapter 19, as necessary.

(a) $Ba(OH)_2$(s) + SO_4^{2-}(aq) $\rightleftharpoons$ $BaSO_4$(s) + 2 OH^-(aq)

(b) $Mg(OH)_2$(s) + CO_3^{2-}(aq) $\rightleftharpoons$ $MgCO_3$(s) + 2 OH^-(aq)

(c) $Ca(OH)_2$(s) + 2 F^-(aq) $\rightleftharpoons$ CaF_2(s) + 2 OH^-(aq)

67. Without performing detailed calculations, indicate why you would expect each of the following reactions to occur to a significant extent as written. Use equilibrium constants from earlier chapters, as necessary.

(a) $SrCO_3$(s) + 2 $HC_2H_3O_2$(aq) $\rightarrow$
$$Sr(C_2H_3O_2)_2(aq) + H_2O + CO_2(g)$$

(b) $Ba(OH)_2$(s) + 2 NH_4^+(aq) $\rightarrow$
$$Ba^{2+}(aq) + 2 NH_3(aq) + 2 H_2O$$

(c) ZnS(s) + 2 H^+(conc. aq) $\rightarrow$ Zn^{2+}(aq) + H_2S(g)

★68. To complete the explanation on page 793 of why Li is the most easily oxidized of the alkali metals, we should also consider the reduction half-reaction H^+(aq) + e^- $\rightarrow$ $\frac{1}{2}$ H_2(g). Consider that this reduction can also be described in three steps: (1) the reverse of the hydration of H^+(g), for which $\Delta H° = 1079$ kJ; (2) H^+(g) + e^- $\rightarrow$ H(g); (3) the recombination of H atoms, that is, 2 H(g) $\rightarrow$ H_2(g). Add the $\Delta H°$ values of these three steps to obtain $\Delta H°_{red.}$. Then add $\Delta H°_{red.}$ to $\Delta H°_{ox.}$ to obtain $\Delta H°_{rxn}$ for the overall reaction between an alkali metal and 1 M H^+(aq). Show that this $\Delta H°_{rxn}$ is *most negative* for Li among the alkali metals. [*Hint:* Look for data yielding ΔH for steps 2 and 3 elsewhere in the text.]

★69. Use Figure 22-15 to estimate for the reaction

$$ZnO(s) + C(s) \rightleftharpoons Zn(l) + CO(g)$$

(a) a value of K and **(b)** the equilibrium pressure of CO(g) at the boiling point of zinc.

★70. An Al production cell of the type pictured in Figure 22-13 operates at a current of 1.00×10^5 A and a voltage of 4.5 V. The cell is 38% efficient in converting electric energy to chemical change. (The rest of the electric energy is dissipated as heat energy in the cell.)

(a) What mass of Al can be produced by this cell in 8.00 h?

(b) If the electric energy required to power this cell is produced by burning coal (85% C; heat of combustion of C = 32.8 kJ/g) in a power plant with 35% efficiency, what mass of coal must be burned to produce the mass of Al determined in part (a)?

Self-Test Questions

For questions 71 through 80 select the single item that best completes each statement.

71. All but one of these metals will dissolve both in acidic and basic solution. The exception is (a) Na; (b) Zn; (c) Al; (d) Hg.

72. Of the following, the ion with the greatest hydration energy is (a) Na^+; (b) Al^{3+}; (c) Ba^{2+}; (d) Cs^+.

73. Production of the metal by reduction with carbon at moderate temperatures is not feasible with (a) CdO; (b) Al_2O_3; (c) PbO; (d) HgO.

74. All but one of the following can be used to soften temporary hard water. That one is (a) NH_3; (b) Na_2CO_3; (c) NH_4Cl; (d) NaOH.

75. Of the following sulfides the one that produces the free metal directly on roasting is (a) HgS; (b) PbS; (c) Na_2S; (d) SnS_2.

76. Of the following oxides, all are soluble in NaOH(aq) except (a) ZnO; (b) Al_2O_3; (c) Fe_2O_3; (d) SnO_2.

77. The most difficult of the following ions to reduce to the free metal in aqueous solution is (a) Zn^{2+}; (b) Cd^{2+}; (c) Pb^{2+}; (d) Hg^{2+}.

78. Of the following solutions the one that is acidic is (a) $ZnSO_4(aq)$; (b) $NaAl(OH)_4(aq)$; (c) $NaHCO_3(aq)$; (d) $KNO_3(aq)$.

79. The best oxidizing agent of the following oxides is (a) SnO_2; (b) PbO_2; (c) HgO; (d) MgO.

80. All of the following find appreciable use in low melting alloys except (a) Sn; (b) Pb; (c) Cd; (d) Mg.

81. Write chemical equations to represent
(a) the reaction of tin with concentrated $HNO_3(aq)$;
(b) the reaction of $MgCO_3(s)$ with $CO_2(aq)$;
(c) the dissolving of zinc in concentrated NaOH(aq);
(d) the roasting of cadmium sulfide followed by reduction with carbon.

82. Why is **(a)** the melting point of MgO (2800 °C) so much higher than that of BaO (1920 °C)? **(b)** the water solubility of $MgCl_2$ much greater than that of MgF_2?

83. Indicate how you would establish through a simple laboratory test(s) whether a particular metal sample is a Pb–Sn or a Pb–Cd alloy.

84. A sample of water has its hardness due only to $CaSO_4$. When this water is passed through an *anion* exchange resin, SO_4^{2-} ions are replaced by OH^-. A 25.00-mL sample of the water so treated requires 21.58 mL of 1.00×10^{-3} M H_2SO_4 for its titration. What is the hardness of the water, expressed in ppm of $CaSO_4$? Assume the density of the water is 1.00 g/mL.

23 Chemistry of the Representative (Main-Group) Elements II: Nonmetals

Bromine is a typical nonmetal. Here it is being used in the synthesis of flame retardants for use in plastics. [Courtesy Ethyl Corporation]

We have been referring to distinctions between metals and nonmetals throughout this text. In Chapter 3 we used this distinction primarily to assist us in naming compounds. In Chapter 9 we found a strong relationship between atomic properties of the elements and their metallic–nonmetallic characters. A principal objective of that chapter was to show how this relationship underlies the periodic table of the elements.

In Chapter 14 we surveyed aspects of the chemistry of several of the lighter nonmetallic elements—B, C, N, O, F, Si, P, S, and Cl. There we emphasized bonding, structure, and the reactions of these elements and their important compounds. Also, we looked at sources and uses of the elements and their compounds. In returning to the study of nonmetals, we consider a number of additional topics made possible by some of the newer concepts we have learned, such as criteria for spontaneous change (based both on ΔG and E_{cell}), acid–base behavior, and solubility and complex ion equilibria.

In studying the representative metals in Chapter 22 we began with the most active of metals (group 1A) and moved through the periodic table (from left to right) to progressively less active metals. Our approach in this chapter is to begin with the most active nonmetals (group 7A) and move through the periodic table (from right to left) toward the less active nonmetals. At the end of the chapter we discuss the chemistry of a group of elements that until recent times were believed to have no chemical behavior at all—the noble gases (group 8A).

23-1 Group 7A—The Halogen Elements

In Section 9-12, we chose the halogens as a typical group of elements with which to illustrate trends in physical and chemical properties in the periodic table. For example, we showed that the melting and boiling points of the halogen elements increase progressively from the lightest member of the group—fluorine—to the heaviest, iodine. Chemical reactivity toward other elements and compounds, on the other hand, progresses in the *opposite* order, with fluorine being the most reactive and iodine, least reactive.

A great deal of the reaction chemistry of the halogens can be explained in terms of the criteria for oxidation–reduction reactions that we introduced in Chapter 21—*standard electrode potentials.* Among the properties of the halogens listed in Table 23-1 are potentials for the half-reaction

$$X_2 + 2\,e^- \longrightarrow 2\,X^-(aq)$$

F_2 has the highest standard electrode potential of any substance ($E° = +2.87$ V). Of all the elements, it shows the greatest tendency to gain electrons; it is the most easily reduced. This accounts for the fact that fluorine occurs naturally only in combination with other elements, as the fluoride ion, F^-. The cases of chlorine and bromine are similar. On the other hand, we can find naturally occurring compounds in which iodine is in a *positive* oxidation state (such as the iodate ion, IO_3^-). This is not surprising. The tendency for I_2 to be reduced to I^- is not particularly great ($E° = +0.535$ V).

Preparation of the Halogens. If the reduction of $F_2(g)$ to $F^-(aq)$ has the greatest tendency to occur of all reduction processes, we can see that the reverse process, $2\,F^- \rightarrow F_2(g)$, must have the least tendency to occur among oxidation processes. As a result, the oxidation required for the preparation of $F_2(g)$ is carried out by electrolysis. As we learned in Chapter 14, the electrolyte used is molten KHF_2.

$$2\,H^+ + 2\,F^- \xrightarrow{\text{electrolysis}} H_2(g) + F_2(g) \tag{14.83}$$

(A chemical synthesis of F_2 is described in Exercise 65.)

According to the periodic relationship we should expect astatine (At; $Z = 85$) also to be a halogen element. Because it is a radioactive element with only short-lived isotopes, only about 0.05 μg has ever been prepared. Indications are, however, that its chemical behavior is similar to that of iodine (see Exercise 20).

832

TABLE 23-1
Group 7A Elements—The Halogens

	Fluorine (F)	Chlorine (Cl)	Bromine (Br)	Iodine (I)
physical form at room temperature	pale yellow gas	yellow-green gas	dark red liquid	violet-black solid
melting point, °C	-220	-101	-7.2	114
boiling point, °C	-188	-35	58.8	184
electron configuration	$[He]2s^22p^5$	$[Ne]3s^23p^5$	$[Ar]3d^{10}4s^24p^5$	$[Kr]4d^{10}5s^25p^5$
covalent radius, pm	64	99.5	114	133
ionic radius, X^-, pm	136	181	195	216
first ionization energy, kJ/mol	1681	1251	1140	1008
electron affinity, kJ/mol	-328	-349	-325	-295
electronegativity	4.0	3.2	3.0	2.7
standard electrode potential, V $(X_2 + 2\,e^- \longrightarrow 2\,X^-)$	$+2.87$	$+1.360$	$+1.065$	$+0.535$

The electrolysis of NaCl(aq) is the subject of the Focus feature of Chapter 21. The electrolyses of NaCl(l) and $MgCl_2(l)$ are discussed in Sections 14-5, 14-6, and 22-2.

The commercial production of **chlorine** is by electrolysis, either of NaCl(aq) or of a molten salt such as NaCl(l) or $MgCl_2(l)$. The oxidation of Cl^- to $Cl_2(g)$ by chemical reactions is also a possibility, as long as the reduction half-reaction has $E > 1.360$ V. One laboratory method uses MnO_4^- in acidic solution as the oxidizing agent. For this reaction $E^\circ_{cell} = +0.15$ V.

$$2\,MnO_4^-(aq) + 10\,Cl^-(aq) + 16\,H^+(aq) \longrightarrow$$
$$2\,Mn^{2+}(aq) + 8\,H_2O + 5\,Cl_2(g) \quad (23.1)$$

Another long-used laboratory method involves heating $MnO_2(s)$ and concentrated HCl(aq).

$$MnO_2(s) + 4\,H^+(aq) + 2\,Cl^-(aq) \longrightarrow Mn^{2+}(aq) + 2\,H_2O + Cl_2(g) \quad (23.2)$$

However, for reaction (23.2) $E^\circ_{cell} = -0.13$ V. Why does this method work even though $E^\circ_{cell} < 0$? It works because heating the solution drives off $Cl_2(g)$ and because the HCl(aq) is concentrated. According to Le Châtelier's principle, both of these conditions should favor the forward reaction.

$F_2(g)$ cannot be used in halogen displacement reactions since it oxidizes H_2O to $O_2(g)$.

In still another method, a more active (more electronegative) halogen can be used to displace a less active halogen from a solution of its halide ions. The most important commercial use of this reaction is in the preparation of **bromine.** Bromine can be extracted from seawater, where it occurs to an extent of about 70 parts per million (ppm) as Br^-. First the seawater is adjusted to pH 3.5 and treated with $Cl_2(g)$, which oxidizes Br^- to Br_2, in this displacement reaction.

A halogen displacement reaction.

$$Cl_2(g) + 2\,Br^-(aq) \longrightarrow Br_2(l) + 2\,Cl^-(aq) \quad (23.3)$$

The liberated Br_2 is swept from the seawater with a current of air. There are various ways to concentrate the Br_2 obtained. One method involves passing the bromine-laden air into an alkaline solution [usually $Na_2CO_3(aq)$] where a disproportionation reaction occurs (that is, a simultaneous oxidation and reduction of Br_2).

$$3\,Br_2 + 6\,OH^- \rightleftharpoons 5\,Br^- + BrO_3^- + 3\,H_2O \quad (23.4)$$

Reaction (23.4) is reversed by acidifying the solution.

$$5\,Br^- + BrO_3^- + 6\,H^+ \longrightarrow 3\,Br_2(l) + 3\,H_2O \quad (23.5)$$

An important inorganic compound of bromine is AgBr, the primary light-sensitive agent used in photographic film. Lithium and calcium bromides are used as dessicants (drying agents) and alkali and alkaline earth metal bromides are used as sedatives. Organic bromine compounds are used as dyes, pharmaceuticals, fumigants, and pesticides. Bromine compounds are also used as fire extinguishers and fire retardants.

Salt formations in the Dead Sea. The high concentrations of salts in the Dead Sea make it a good source of bromine and certain other chemicals. [Courtesy Dead Sea Bromine Group, Israel]

Iodine is obtainable in small quantities from dried seaweed, since certain marine plants absorb and concentrate I^- selectively in the presence of Cl^- and Br^-. Low concentrations of $I^-(aq)$ are also found in some natural brines (salt solutions) associated with oil fields. From such sources oxidation of I^- by a variety of oxidizing agents is possible. A more abundant natural source of iodine is $NaIO_3$, found in large deposits in Chile. To convert IO_3^- to I_2 requires the use of a *reducing* agent, e.g., sodium hydrogen sulfite (bisulfite). A two-step procedure is used in which $IO_3^-(aq)$ is first reduced to $I^-(aq)$.

$$IO_3^- + 3\ HSO_3^- \longrightarrow I^- + 3\ SO_4^{2-} + 3\ H^+ \tag{23.6}$$

The reaction of $I^-(aq)$ with additional $IO_3^-(aq)$ in acidic solution produces I_2 (similar to the $Br^- - BrO_3^-$ reaction of 23.5).

$$5\ I^- + IO_3^- + 6\ H^+ \longrightarrow 3\ I_2 + 3\ H_2O \tag{23.7}$$

Iodine is of less commercial importance than chlorine and bromine, although it and its compounds do have applications as catalysts, in medicine, and in the preparation of photographic emulsions (as AgI). Iodine and its compounds also have important uses in the analytical chemistry laboratory (see page 840).

Electrode (Reduction) Potential Diagrams. From a listing of reduction half-reactions arranged simply by decreasing magnitudes of $E°$ (as in Table 21-1) we cannot readily visualize redox relationships among related species. Figure 23-1 is a representation of standard electrode potentials that allows us to summarize the oxidation–reduction chemistry of chlorine compounds in a compact format. In this diagram several chlorine-containing species are joined, two at a time, by lines. Written above each line is the standard potential for the reduction of the species at the left to the one at the right. For example,

the symbolism $ClO_4^- \xrightarrow{\ 1.19\ V\ } ClO_3^-$

signifies that

$$ClO_4^- + 2\ H^+ + 2\ e^- \longrightarrow ClO_3^- + H_2O \qquad E° = 1.19\ V$$

and

the symbolism $ClO_3^- \overset{\ \ }{\underset{1.47\ V}{\rule{0pt}{0pt}}} Cl_2$

signifies that

$$2\ ClO_3^- + 12\ H^+ + 10\ e^- \longrightarrow Cl_2(g) + 6\ H_2O \qquad E° = 1.47\ V$$

CHEMISTRY EVOLVING

Ethylene Dibromide (EDB)

As recently as the mid-1970s, ethylene dibromide, $C_2H_4Br_2$, was the most widely used bromine compound. At that time about 40% of the bromine produced was used to make this one compound. Its most important use was as an additive to gasoline. Its function was to react with lead from the antiknock additive tetraethyllead to produce volatile lead(II) bromide, which was carried off with the automotive exhaust. Environmental restrictions on the use of lead additives in gasoline have greatly reduced the demand for ethylene dibromide. Another use of ethylene dibromide (EDB) that has been banned in the United States is as a fumigant for grain products and fruit. EDB is a known carcinogen in rats. With gallium arsenide (page 806), we saw how a chemical can greatly increase in importance in a matter of a few years. In the case of EDB, we see how circumstances can cause a chemical to decline in importance. ■ ■ ■

FIGURE 23-1

Standard electrode potential diagram for chlorine.

The numbers in red are the oxidation states of the chlorine atom. The potentials written in blue above the line segments are for the *reduction* of the species on the left to the one on the right. All reactants and products are at unit activity. The data for acidic solution are at $[H^+]$ = 1 M; in basic solution $[OH^-]$ = 1 M. Because of the basic properties of ClO_2^- and OCl^-, the weak acids $HClO_2$ and $HOCl$ are formed in acidic solution.

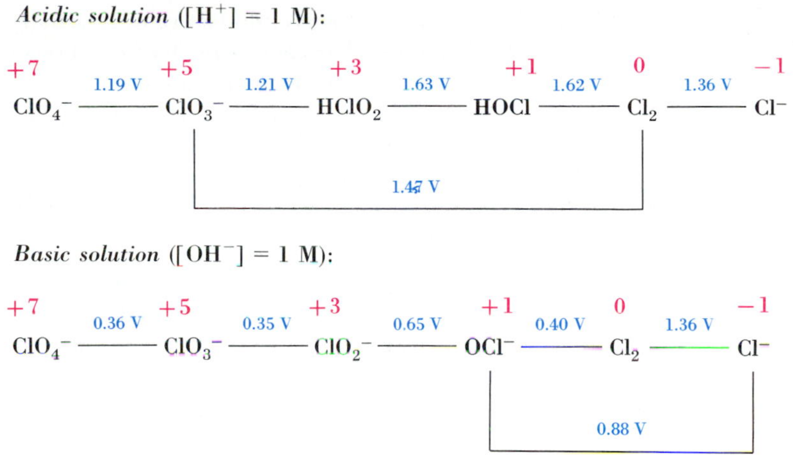

Acidic solution ($[H^+]$ = 1 M):

$$\overset{+7}{ClO_4^-} \underset{1.19\ V}{\rule{1.5cm}{0.4pt}} \overset{+5}{ClO_3^-} \underset{1.21\ V}{\rule{1.5cm}{0.4pt}} \overset{+3}{HClO_2} \underset{1.63\ V}{\rule{1.5cm}{0.4pt}} \overset{+1}{HOCl} \underset{1.62\ V}{\rule{1.5cm}{0.4pt}} \overset{0}{Cl_2} \underset{1.36\ V}{\rule{1.5cm}{0.4pt}} \overset{-1}{Cl^-}$$

1.47 V

Basic solution ($[OH^-]$ = 1 M):

$$\overset{+7}{ClO_4^-} \underset{0.36\ V}{\rule{1.5cm}{0.4pt}} \overset{+5}{ClO_3^-} \underset{0.35\ V}{\rule{1.5cm}{0.4pt}} \overset{+3}{ClO_2^-} \underset{0.65\ V}{\rule{1.5cm}{0.4pt}} \overset{+1}{OCl^-} \underset{0.40\ V}{\rule{1.5cm}{0.4pt}} \overset{0}{Cl_2} \underset{1.36\ V}{\rule{1.5cm}{0.4pt}} \overset{-1}{Cl^-}$$

0.88 V

Example 23-1

Using an electrode potential diagram to predict the direction of spontaneous change. Use data from the electrode potential diagram for chlorine (Figure 23-1) to predict whether the following reaction will occur to any significant extent in the forward direction.

$$Cl_2(g) + H_2O \longrightarrow Cl^-(aq) + ClO_3^-(aq) + H^+(aq)$$

Solution. We need to take one $E°$ value from the line segment joining Cl_2 with Cl^-. This is a reduction half-reaction for which $E° = +1.36$ V. The other $E°$ value is taken from the line segment joining Cl_2 and ClO_3^-. This is an *oxidation* half-reaction for which $-E° = -1.47$ V. We then combine the two half-reactions in the familiar manner.

Oxid:	$Cl_2 + 6\ H_2O \longrightarrow 2\ ClO_3^- + 12\ H^+ + 10\ e^-$	$-E° = -1.47$ V
Red:	$5\{Cl_2 + 2\ e^- \longrightarrow 2\ Cl^-\}$	$E° = +1.36$ V
Net:	$3\ Cl_2(g) + 3\ H_2O \longrightarrow 5\ Cl^- + ClO_3^- + 6\ H^+$	$E°_{cell} = -0.11$ V

The negative value of $E°_{cell}$ indicates that the reaction does *not* occur to a significant extent in the forward direction.

SIMILAR EXAMPLES: Exercises 5, 6, 18b, 26.

Look again at Figure 23-1. From the electrode potential diagram in *basic* solution we find the values that link ClO_3^- to ClO_2^- and ClO_2^- to OCl^-. What should we do if we need the potential for the reduction of ClO_3^- to OCl^-, since these two species are not directly linked by a line segment? Example 23-2 shows how we can establish the electrode potential for an unknown *half-reaction* from known values of related half-reactions. When half-reactions of the *same* kind are combined (either both oxidation or both reduction),

- electrons *do not* cancel;
- half-cell potentials *are not* additive; but
- free energy changes *are* additive.

The key to Example 23-2, then, is to use the relationship between free energy change and electrode potential, $\Delta G° = -n\mathscr{F}E°$.

Example 23-2

Using an electrode potential diagram to determine $E°$ for a half-reaction. Determine the standard electrode potential for the reduction of ClO_3^- to OCl^- in basic solution.

Solution. We must find two reduction half-reactions whose sum is the desired half-reaction. For each of the half-reactions being combined, we obtain the free energy change through the equation $\Delta G° = -n\mathscr{F}E°$.

$$ClO_3^- + H_2O + 2\ e^- \longrightarrow ClO_2^- + 2\ OH^- \qquad \Delta G° = -2\mathscr{F}(0.35)$$
$$ClO_2^- + H_2O + 2\ e^- \longrightarrow OCl^- + 2\ OH^- \qquad \Delta G° = -2\mathscr{F}(0.65)$$

$$ClO_3^- + 2\ H_2O + 4\ e^- \longrightarrow OCl^- + 4\ OH^- \qquad \Delta G° = -2\mathscr{F}(0.35 + 0.65)$$
$$= -n\mathscr{F}E°$$

We now have the desired reduction half-reaction and its $\Delta G°$ value. To obtain its $E°$ value we simply write that $E° = -\Delta G°/n\mathscr{F}$. In using this expression we must be careful to substitute $n = 4$ (not $n = 2$ as in the two half-equations that are combined).

$$E° = \frac{-\Delta G°}{n\mathscr{F}} = \frac{2\mathscr{F}(0.35 + 0.65)}{4\mathscr{F}} = 0.50\ V$$

SIMILAR EXAMPLES: Exercises 7, 18a.

Hydrogen Halides. We have encountered the hydrogen halides from time to time throughout the text. In aqueous solution they are typical mineral acids whose acid strengths decrease in the direction of increasing bond energy, that is

$$\underbrace{HI > HBr > HCl}_{\substack{\text{strong} \\ \text{acids}}} > \underset{\substack{\text{weak} \\ \text{acid}}}{HF}$$

Strong hydrogen bonding in HF seems to account for some of its rather unusual properties compared to the other hydrogen halides. For example, HF is a weak acid, whereas the other hydrogen halides are strong acids, and HF has a higher boiling point than the other hydrogen halides, even though its molecular weight is lowest (recall Table 9-7).

The hydrogen halides can be prepared by direct combination of the elements

Synthesis of hydrogen halides from their elements.

$$H_2(g) + X_2(g) \longrightarrow 2\ HX(g) \tag{23.8}$$

but in reactions that differ in their speeds and degrees of completion. The reaction of $H_2(g)$ and $F_2(g)$ is very fast, occurring with explosive violence under some conditions. With $H_2(g)$ and $Cl_2(g)$ the reaction also proceeds rapidly (explosively) in the presence of light. With Br_2 and I_2 the reaction occurs more slowly.

From the data in Table 23-2 we see that the free energies of formation of HF, HCl, and HBr are large and negative, suggesting that for them reaction (23.8) goes to completion. For HI(g), on the other hand, $\Delta G_f°$ is small and positive. This suggests that even at room temperature HI should dissociate to some extent into its elements (see Exercise 80). However, because of the high *activation energy* the dissociation occurs only very slowly in the absence of a catalyst. As a result, HI(g) is quite stable at room temperature. A chemist would say that the room-temperature decomposition of HI(g) is *kinetically* controlled (rather than thermodynamically controlled).

TABLE 23-2
Free Energy of Formation of Hydrogen Halides at 298 K

	$\Delta G_f°$, kJ/mol
HF(g)	−273.2
HCl(g)	−95.30
HBr(g)	−53.43
HI(g)	+1.72

Example 23-3

Determining K_p for a dissociation reaction from free energies of formation. What is the value of K_p for the dissociation of HI(g) into its elements at 298 K?

Solution. Dissociation of HI(g) into its elements is the *reverse* of the formation reaction, and $\Delta G°$ for the dissociation is the *negative* of $\Delta G_f°$ for HI(g) listed in Table 23-2.

$$HI(g) \longrightarrow \tfrac{1}{2} H_2(g) + \tfrac{1}{2} I_2(s) \qquad \Delta G° = -1.72 \text{ kJ}$$

Now we can use the relationship $\Delta G° = -RT \ln K$

$$\ln K_p = \frac{-\Delta G°}{RT} = \frac{-(-1.72 \times 10^3 \text{ J mol}^{-1})}{8.314 \text{ J mol}^{-1} \text{ K}^{-1} \times 298 \text{ K}} = 0.694$$

$K_p = \text{antiln } 0.694 = 2.00$

SIMILAR EXAMPLE: Exercise 21.

Another commonly used method of preparing hydrogen halides is to heat a halide salt with a *nonvolatile* acid like concentrated H_2SO_4. This is the method of preparing $HF(g)$ that we first considered in Chapter 14.

$$CaF_2(s) + H_2SO_4(\text{concd aq}) \longrightarrow CaSO_4(s) + 2 HF(g) \qquad (14.85)$$

This method also works for preparing $HCl(g)$ but not for $HBr(g)$ and $HI(g)$. This is because concentrated H_2SO_4 is a sufficiently good oxidizing agent to oxidize Br^- to Br_2 and I^- to I_2. The hydrogen halide first formed is oxidized, as in the reaction

$$2 HBr + H_2SO_4(\text{concd aq}) \longrightarrow 2 H_2O + Br_2(g) + SO_2(g) \qquad (23.9)$$

To get around this difficulty, we can use a *nonoxidizing* nonvolatile acid, such as phosphoric acid.

$$NaBr(s) + H_3PO_4(\text{concd aq}) \longrightarrow NaH_2PO_4(s) + HBr(g) \qquad (23.10)$$

Oxoacids and Oxoanions of the Halogens. The oxygen-containing acids and their salts are important halogen compounds. Table 23-3 lists these oxoacids. Chlorine forms a complete set, but for bromine and iodine the acids $HBrO_2$ and HIO_2 have not been observed. In Section 3-6 we learned to write names and formulas of oxoacids and oxoanions.

Hypochlorous acid is formed in the reaction of Cl_2 with water.

Reaction of $Cl_2(g)$ with water.

$$Cl_2(g) + H_2O \rightleftharpoons HOCl(aq) + H^+(aq) + Cl^-(aq) \qquad (23.11)$$

A solution of "chlorine water," commonly used as an oxidizing agent in the laboratory, is thus a mixture of $Cl_2(aq)$, $HOCl(aq)$, and $HCl(aq)$. The concentration of $HOCl(aq)$ can be increased by adding a reactant to remove Cl^-, displacing equilibrium to the right in reaction (23.11). Although we can obtain $HOCl$ in aqueous solution, we cannot isolate it in the pure state.

If $Cl_2(g)$ is dissolved in a basic solution, equilibrium in reaction (23.11) is also displaced far to the right. $HCl(aq)$ and $HOCl(aq)$ are neutralized and an aqueous solution of a **hypochlorite** salt is formed.

$$Cl_2(g) + 2 OH^-(aq) \longrightarrow OCl^-(aq) + Cl^-(aq) + H_2O \qquad (23.12)$$

TABLE 23-3
Oxoacids of the Halogens

Oxidation state of the halogen	Chlorine	Bromine	Iodine
+1	HOCl	HOBr	HOI
+3	$HClO_2$	—	—
+5	$HClO_3$	$HBrO_3$	HIO_3
+7	$HClO_4$	$HBrO_4$	HIO_4; H_5IO_6

In all these acids H atoms are bonded to O atoms, not to the central halogen atom. More accurate representations would be $HOClO$ (instead of $HClO_2$), $HOClO_2$ (instead of $HClO_3$), and so on.

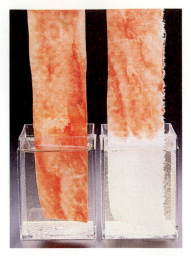

Both strips of cloth are heavily stained with tomato sauce. Pure water (left) has little ability to remove the stain. NaOCl(aq) (right) rapidly bleaches (oxidizes) the colored components of the sauce to colorless products. [Carey B. Van Loon]

Reaction (23.12) is carried out commercially in a chlor-alkali cell in which $Cl_2(g)$ from the anode is allowed to mix with NaOH(aq) from the cathode (recall Figure 21-19).

Common household bleaches and swimming pool "chlorine" are aqueous alkaline solutions of NaOCl. The product of the reaction of Cl_2 with $Ca(OH)_2(s)$ is a complex mixture of salts containing calcium hypochlorite, $Ca(OCl)_2$, and known as *bleaching powder*. Unlike HOCl, hypochlorite salts can be obtained in the pure state. Hypochlorous acid and its salts are strong oxidizing agents, capable of oxidizing I^- to I_2, Fe^{2+} to Fe^{3+}, Mn^{2+} to MnO_4^-, and Pb^{2+} to PbO_2.

Chlorite salts are generally produced by reducing chlorine dioxide with peroxide ion in aqueous solution.

$$2\ ClO_2(g) + O_2^{2-}(aq) \longrightarrow 2\ ClO_2^-(aq) + O_2(g) \tag{23.13}$$

Sodium chlorite is used as a bleaching agent for textiles. $ClO_2(g)$ for reaction (23.13) can be obtained by reducing $NaClO_3$ with $SO_2(g)$ in sulfuric acid.

Chlorate salts are formed when hot alkaline solutions are treated with $Cl_2(g)$.

Preparation of chlorate salts.

$$3\ Cl_2(g) + 6\ OH^-(aq) \longrightarrow 5\ Cl^-(aq) + ClO_3^-(aq) + 3\ H_2O \tag{23.14}$$

The chloride and chlorate salts are separated by fractional crystallization. Chlorates are good oxidizing agents. Also, solid chlorates decompose to produce oxygen gas, which makes them useful in matches and fireworks. A simple laboratory method of producing $O_2(g)$ involves heating $KClO_3(s)$ in the presence of $MnO_2(s)$ as a catalyst.

Decomposition of chlorate salts.

$$2\ KClO_3(s) \xrightarrow{\Delta} 2\ KCl(s) + 3\ O_2(g) \tag{23.15}$$

Perchlorate salts are mainly prepared by electrolyzing chlorate solutions. Oxidation of ClO_3^- occurs at a Pt anode through the half-reaction

$$ClO_3^-(aq) + H_2O \longrightarrow ClO_4^-(aq) + 2\ H^+(aq) + 2\ e^- \tag{23.16}$$

An interesting laboratory use of perchlorate salts is in solution studies where complex ion formation is to be avoided. ClO_4^- has the least tendency of any anion to act as a ligand in complex ion formation.

We can predict the geometric structures of the oxoanions of chlorine rather easily from their Lewis structures and the VSEPR theory. Figure 23-2 shows the results we obtain.

FIGURE 23-2
Structures of the chlorine oxoanions.

Lone-pair electrons are shown only for the central Cl atom.

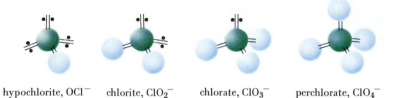

hypochlorite, OCl^- (linear) chlorite, ClO_2^- (angular) chlorate, ClO_3^- (trigonal pyramidal) perchlorate, ClO_4^- (tetrahedral)

FIGURE 23-3
Structures of some interhalogen molecules.

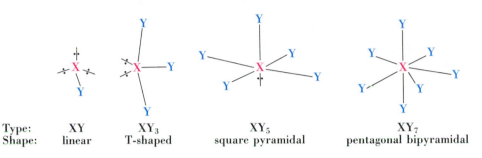

Type:	XY	XY$_3$	XY$_5$	XY$_7$
Shape:	linear	T-shaped	square pyramidal	pentagonal bipyramidal

Interhalogen Compounds. Two different halogen elements can combine to produce interhalogen compounds. Many compounds of this type are known. These compounds and their physical states at 25 °C and 1 atm are listed in Table 23-4. The molecular structures of the interhalogen compounds feature the larger, less electronegative halogen as the central atom and the smaller halogen atoms as the terminal atoms. Molecular shapes of the interhalogens agree quite well with predictions based on the VSEPR theory. Some representative structures are shown in Figure 23-3, including one type (IF$_7$) that we have not considered previously. This is a structure with *seven* electron pairs distributed about the central atom in the form of a pentagonal bipyramid (sp^3d^3 hybridization).

All the interhalogen compounds, with the exception of IF$_5$, are very reactive. ClF$_3$ and BrF$_3$, for example, react with explosive violence with water, organic materials, and some inorganic materials. These two trifluorides are used to fluorinate compounds, as in the preparation of UF$_6$ for use in the separation of uranium isotopes by gaseous diffusion (recall page 186). ICl is used as an iodination reagent in organic chemistry.

Polyhalide Ions. Iodine is only slightly soluble in pure water (0.3 g/L) but rather highly soluble in KI(aq). This increased solubility results from the formation of the triiodide ion.

Formation of triiodide ion.

$$I_2(s) + I^-(aq) \rightleftharpoons I_3^-(aq) \qquad K_c = 710 \text{ (at 298 K)} \qquad (23.17)$$
colorless   yellow-brown

TABLE 23-4
Some Interhalogen Compounds

Type			
XY	**XY$_3$**	**XY$_5$**	**XY$_7$**
ClF(g)	ClF$_3$(g)	ClF$_5$	
BrF(g)	BrF$_3$(l)	BrF$_5$(l)	
BrCl(g)			
ICl(s)	ICl$_3$(s)	IF$_5$(l)	IF$_7$(g)
IBr(s)			

Some interhalogen compounds, particularly of the type XY, are unstable (e.g., at 298 K ΔG_f° for BrCl(g) is only -0.96 kJ/mol).

Potassium triiodide, KI_3(aq).
[Carey B. Van Loon]

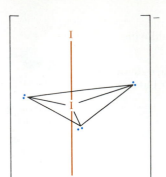

FIGURE 23-4
Structure of the I_3^- ion.

The bond-pair electrons are shown as brown lines. The lone-pair electrons of the central I atom are represented by bold dots.

Triiodide ion is one of a group of species, called **polyhalide ions,** that are produced by the reaction of a halide ion with a halogen or interhalogen molecule. In this reaction the halide ion acts as a *Lewis base* (electron pair donor) and the molecule as a *Lewis acid* (electron pair acceptor). The structure of the I_3^- ion is shown in Figure 23-4. The familiar iodine solutions used as antiseptics often contain triiodide ion, and triiodide ion solutions are extensively used in analytical chemistry.

Analytical Chemistry of the Halogens. We can easily detect Cl^-, Br^-, and I^- in aqueous solutions with $AgNO_3$(aq). The silver halides are all insoluble. Moreover, because of the differences in their K_{sp} values, first AgI, then AgBr, and finally AgCl precipitate as $AgNO_3$(aq) is slowly added to a solution containing Cl^-, Br^-, and I^- (recall Section 19-5).

Another way to detect Br^- and I^- is by their reaction with Cl_2(aq). The Cl_2 is reduced to Cl^-, and the Br^- or I^- is oxidized to Br_2 or I_2. The liberated Br_2 or I_2 is extracted from aqueous solution with a solvent such as CS_2 (see Figure 23-5). In CS_2 iodine has a violet color and Br_2 has a red color.

To detect the oxoanions of Cl, Br, and I, oxidation–reduction reactions are used to reduce the oxoanion to the free halogen or halide ion. Then the tests outlined above can be applied.

Iodine and iodide ion are among the most versatile reagents used in redox reactions in analytical chemistry. Thus, a standard solution of I_2 might be used to titrate a solution of a reducing agent such as sulfite ion.

$$SO_3^{2-} + I_2 + H_2O \longrightarrow SO_4^{2-} + 2 I^- + 2 H^+ \tag{23.18}$$

Iodide ion can be used in the titration of strong oxidizing agents.

$$2 MnO_4^- + 10 I^- + 16 H^+ \longrightarrow 2 Mn^{2+} + 5 I_2 + 8 H_2O \tag{23.19}$$

In some cases we can add an *excess* of iodide ion to a strong oxidizing agent and titrate the liberated I_2 with sodium thiosulfate solution.

$$I_2 + 2 S_2O_3^{2-} \longrightarrow 2 I^- + S_4O_6^{2-} \tag{23.20}$$

To mark the end point of the titration, we can add starch solution. In the presence of starch, I_2 forms a deep blue complex. When all the I_2 has been consumed in reaction (23.20), the blue color of the starch–iodine complex disappears.

Br₂, of course, is not a strong enough oxidizing agent to oxidize Cl^-(aq) to Cl_2(g), nor can I_2 oxidize either Cl^-(aq) or Br^-(aq).

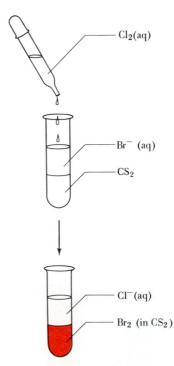

Cl₂(aq)

Br⁻ (aq)

CS₂

Cl⁻(aq)

Br₂ (in CS₂)

FIGURE 23-5
A test for bromide ion.

23-2 Group 6A Elements—The Oxygen Family

5A	6A	7A
7 N	8 O	9 F
15 P	16 S	1 C
33 As	34 Se	3 B
51 Sb	52 Te	5 I
33 Bi	84 Po	85 At

Polonium, Po, is a member of the oxygen family—group 6A. However, because all of its isotopes are radioactive, interest in polonium centers on its nuclear properties (see Chapter 26).

Some representative properties of the group 6A elements are listed in Table 23-5. Oxygen and sulfur are clearly nonmetallic in their behavior, but some metallic properties begin to appear toward the bottom of the group. In fact, you may recall that in Figure 9-6 we labeled the bottom two members, Te and Po, as metalloids.

One of the points about oxygen that we stressed in Chapter 14 is that as the first member of a group it differs in some important ways from other group members. These differences stem primarily from the (1) small size, (2) high electronegativity, and (3) unavailability of d orbitals for bonding. The data in Table 23-6 summarize some similarities and differences between oxygen and sulfur.

Oxygen. Oxygen is so central to a study of chemistry that inevitably we find ourselves considering physical and chemical properties of this element and its compounds as we attempt to gain an understanding of chemical principles. Some of the first examples in our discussion of stoichiometry dealt with combustion reactions, in which substances react with $O_2(g)$ to yield products such as $CO_2(g)$, $H_2O(l)$, and $SO_2(g)$. Combustion reactions also figured prominently in the study of thermochemistry. Many of the molecules and ions used as examples in the chapters on chemical bonding were oxygen-containing species. Water was a primary subject in the discussion of liquids, solids, and intermolecular forces, as well as in the study of acid–base and other solution equilibria. The extraction of $O_2(g)$ from air and the commercial uses of $O_2(g)$ were discussed in the Focus feature of Chapter 6. Ozone and hydrogen peroxide were considered in Chapter 14, and the acidic, basic, and amphoteric properties of element oxides in Chapter 17. To illustrate the frequency with which aspects of oxygen chemistry are encountered in other contexts, consider the following methods of preparing $O_2(g)$. Each appeared in the context cited.

Heating oxides of metals of low reactivity:

$$2\ HgO(s) \xrightarrow{\Delta} 2\ Hg(l) + O_2(g) \qquad \text{(Figure 22-19)}$$

$$2\ Ag_2O(s) \xrightarrow{\Delta} 4\ Ag(s) + O_2(g) \qquad \text{(Example 6-20)}$$

Heating certain oxygen-containing salts:

$$2\ KClO_3 \xrightarrow{\Delta} 2\ KCl(s) + 3\ O_2(g) \qquad \text{(reaction 23.15)}$$

Decomposition of water by a thermochemical cycle:

$$2\ H_2O(g) \xrightarrow{\Delta} 2\ H_2(g) + O_2(g) \qquad \text{(net reaction 14.16)}$$

TABLE 23-5
Group 6A Elements

	Oxygen (O)	Sulfur (S)	Selenium (Se)	Tellurium (Te)
physical form at room temperature	colorless gas	yellow solid	gray solid	silvery white solid
melting point, °C	−218	119	217	452
boiling point, °C	−183	445	684	1390
electron configuration	$[He]2s^22p^4$	$[Ne]3s^23p^4$	$[Ar]3d^{10}4s^24p^4$	$[Kr]4d^{10}5s^25p^4$
covalent radius, pm	66	104	117	137
ionic radius, X^{2-}, pm	140	184	198	221
first ionization energy, kJ/mol	1314	1000	941	869
electron affinity, kJ/mol	−141	−200	−195	−183
electronegativity	3.4	2.6	2.5	2.1
standard electrode potential, V ($X + 2\ e^- \longrightarrow X^{2-}$)	—	−0.48	−0.92	−0.84

TABLE 23-6
Some Comparisons of Oxygen and Sulfur

Oxygen	Sulfur

$O_2(g)$ at 298 K and 1 atm
Two allotropes, $O_2(g)$ and $O_3(g)$
 	
Possible oxidation states: $-2, -1, 0$
$O_2(g)$ and $O_3(g)$ are very good oxidizing agents
Forms, with metals, oxides that are mostly ionic in character
O^{2-} completely decomposes in water, producing OH^-
 	
O not often the central atom in a structure, and can never have more than 4 atoms bonded to it. More commonly has two (as in H_2O) or three (as in H_3O^+)
Can form only two- and three-atom chains, as in H_2O_2 and O_3. Compounds with O—O bonds decompose readily
Forms the oxide CO_2, which reacts with $NaOH(aq)$ to produce $Na_2CO_3(aq)$
Forms the hydride H_2O, which
  is a liquid at 298 K and 1 atm
  is extensively hydrogen-bonded
  has a large dipole moment
  is an excellent ionizing solvent
  forms hydrates and aqua complexes
  is oxidized with difficulty
	$S_8(s)$ at 298 K and 1 atm
	Two solid polymorphic forms and many different molecular species in liquid and gaseous states
	Possible oxidation states: all values from -2 to $+6^a$
	$S(s)$ is a poor oxidizing agent
	Forms ionic sulfides with the most active metals but many metal sulfides have partial covalent character
	S^{2-} strongly hydrolyzes in water to HS^- (and OH^-) but some free S^{2-} remains
	S the central atom in many structures. Can easily accommodate up to 6 atoms around itself (e.g., SO_3, SO_4^{2-}, SF_6)
	Can form molecules with up to six S atoms per chain in compounds such as H_2S_n, Na_2S_n, $H_2S_nO_6$
	Forms the sulfide CS_2, which reacts with $NaOH(aq)$ to produce $Na_2CS_3(aq)$
	Forms the hydride H_2S, which
	  is a (poisonous) gas at 298 K and 1 atm
	  is not hydrogen-bonded
	  has a small dipole moment
	  is a poor solvent
	  forms no complexes
	  is easily oxidized

^{a}For an assessment of the oxidation state of S in the ion $S_2O_8^{2-}$, see page 847.

Decomposition of water by electrolysis:

$$2\,H_2O(l) \xrightarrow{\text{electrolysis}} 2\,H_2(g) + O_2(g) \qquad \text{(reaction 14.5)}$$

$$2\,H_2O \longrightarrow 4\,H^+(aq) + O_2(g) + 4\,e^- \qquad \text{(half-reaction 21.49)}$$

Reaction of an ionic peroxide with water:

$$2\,O_2^{2-} + 2\,H_2O \longrightarrow 4\,OH^-(aq) + O_2(g) \qquad \text{(page 512)}$$

Reaction of a superoxide with water:

$$4\,O_2^- + 2\,H_2O \longrightarrow 4\,OH^-(aq) + 3\,O_2(g) \qquad \text{(page 512)}$$

Catalytic decomposition of hydrogen peroxide:

$$2\,H_2O_2(aq) \longrightarrow 2\,H_2O + O_2(g) \qquad \text{(reactions 14.72, 15.3)}$$

Oxidation of hydrogen peroxide:

$$5\,H_2O_2 + 2\,MnO_4^- + 6\,H^+ \longrightarrow 2\,Mn^{2+} + 8\,H_2O + 5\,O_2(g)$$
$$\text{(reaction 14.70)}$$

Oxidation–Reduction Reactions of O_3, O_2, and H_2O_2. The reactions of these substances are almost all of a redox type and account for their principal uses in water and sewage treatment, as bleaching agents, and in metallurgical processes. We considered several typical reactions in Chapter 14, but now we are able to look at such reactions from the standpoint of electrode potentials. An electrode potential diagram such as Figure 23-6 helps us to understand why $O_3(g)$ can act only as an oxidizing agent (being reduced to O_2). On the other hand, we can see that in some

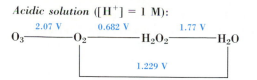
Acidic solution ([H^+] = 1 M):

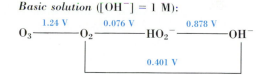
Basic solution ([OH^-] = 1 M):

FIGURE 23-6
Electrode potential diagram for oxygen.

The ion HO_2^- results from the acid ionization of H_2O_2 when it is placed in a strongly alkaline solution.

instances H_2O_2 may act as an oxidizing agent (being reduced to H_2O), and in other cases it acts as a reducing agent (being oxidized to O_2).

Electrode Potential Diagram for Sulfur. Some of the sulfur-containing species listed in the electrode potential diagram of Figure 23-7 are probably unfamiliar to you (e.g., $S_2O_3^{2-}$, $S_2O_6^{2-}$, and $S_4O_6^{2-}$). We will discuss these species in the following paragraphs, and Figure 23-7 will help us to establish relationships among them. For example, we will learn how some of these new species can be prepared by redox reactions.

Example 23-4 _____

Using electrode potentials to explain chemical phenomena. **Zinc dissolves in HCl(aq) to liberate $H_2(g)$, but from H_2SO_4(concd aq), $SO_2(g)$ is evolved. Explain this difference in the behavior of zinc toward the two acids.**

Solution. When zinc dissolves in an acid, the *oxidation* half-reaction is $Zn(s) \rightarrow Zn^{2+}(aq) + 2 e^-$, for which $-E° = +0.763$ V. If H^+ is the oxidizing agent, the *reduction* half-reaction is $2 H^+(aq) + 2 e^- \rightarrow H_2(g)$ for which $E° = 0.00$ V. Thus, when Zn dissolves in HCl(aq) we have the spontaneous redox reaction

$$Zn(s) + 2 H^+(aq) \longrightarrow Zn^{2+}(aq) + H_2(g) \qquad E°_{cell} = +0.763 \text{ V}$$

In concentrated sulfuric acid, sulfate ion is a better oxidizing agent than is H^+; $E°$ is larger than that of H^+ (see Figure 23-7).

$$SO_4^{2-}(aq) + 4 H^+(aq) + 2 e^- \longrightarrow 2 H_2O + SO_2(g) \qquad E° = +0.17 \text{ V}$$

For the net spontaneous reaction

$$Zn(s) + 4 H^+(aq) + SO_4^{2-}(aq) \longrightarrow Zn^{2+}(aq) + 2 H_2O + SO_2(g)$$

$E°_{cell} = +0.93$ V, and this reaction should predominate.

SIMILAR EXAMPLES: Exercises 11, 26.

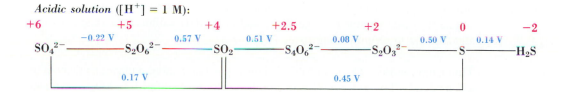
Acidic solution ([H^+] = 1 M):

Basic solution ([OH^-] = 1 M):

FIGURE 23-7
Electrode potential diagram for sulfur.

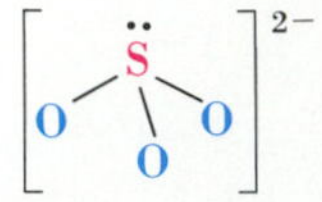

trigonal pyramidal

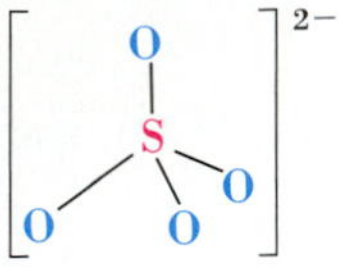

tetrahedral

FIGURE 23-8
Structures of the sulfite
(SO_3^{2-}) and sulfate (SO_4^{2-})
ions.

Sulfites and Sulfates. Both H_2SO_3 and H_2SO_4 are *diprotic* acids. They ionize in two steps and produce two types of salts, one in each ionization step. The term **acid salt** is sometimes used for salts like $NaHSO_3$ and $NaHSO_4$, because their anions undergo a further *acid* ionization. As we have learned before, and as suggested below, H_2SO_3 is a weak acid in both ionization steps, whereas H_2SO_4 is strong in the first step and weak in the second. If a solution of H_2SO_4 is sufficiently dilute however (less than about 0.001 M), we can treat the acid as if both ionization steps go to completion.

$$H_2SO_3(aq) + H_2O \rightleftharpoons H_3O^+ + HSO_3^- \qquad K_{a_1} = 1.3 \times 10^{-2}$$
hydrogen sulfite
(bisulfite) ion

$$HSO_3^- + H_2O \rightleftharpoons H_3O^+ + SO_3^{2-} \qquad K_{a_2} = 6.3 \times 10^{-8}$$
sulfite ion

$$H_2SO_4 + H_2O \rightleftharpoons H_3O^+ + HSO_4^- \qquad K_{a_1} = \text{large}$$
hydrogen sulfate
(bisulfate) ion

$$HSO_4^- + H_2O \rightleftharpoons H_3O^+ + SO_4^{2-} \qquad K_{a_2} = 1.29 \times 10^{-2}$$
sulfate ion

We have used both SO_3^{2-} and SO_4^{2-} as examples of ions whose structures can be predicted by the VSEPR theory. These structures are shown in Figure 23-8. What these structures do not show is that both SO_3^{2-} and SO_4^{2-} have some multiple bond character in their sulfur-to-oxygen bonds. This is made possible through an expanded octet in the central S atom.

Except for the ammonium and alkali metal compounds, sulfites are insoluble in water. Most sulfates, on the other hand, are water soluble. The main exceptions are $SrSO_4$, $BaSO_4$, $PbSO_4$, and Hg_2SO_4. Two sulfates that are only moderately soluble are $CaSO_4$ and Ag_2SO_4. It is a relatively simple matter to distinguish between SO_3^{2-}(aq) and SO_4^{2-}(aq). If Ba^{2+}(aq) is added to a solution containing either of these ions, a white precipitate forms. However, $BaSO_3$(s), as do all sulfites, decomposes in acidic solution with the evolution of SO_2(g).

A test for sulfite ion.

$$BaSO_3(s) + 2 H_3O^+(aq) \longrightarrow Ba^{2+}(aq) + 3 H_2O + SO_2(g) \qquad (23.21)$$

$BaSO_4$ is unaffected by acids.

The chief use of sulfites is in the pulp and paper industry. When wood is digested in an aqueous sulfite solution, chemical reactions occur that break down the cellulosic constituents of wood and free it of certain undesired materials such as lignin. The resulting wood pulp is the starting material for the manufacture of paper and rayon. Other uses of sulfite ion depend on its properties as a reducing agent, such as in photography and as a scavenger for O_2(aq) in treating boiler water.

Sulfates have many important uses. We have mentioned, for example, calcium sulfate as gypsum and plaster of Paris in the building industry (page 495); aluminum sulfate in water treatment (page 808); and copper(II) sulfate as a fungicide and algicide and in electroplating (page 82).

Thio Compounds. The term "thio" signifies that an S atom replaces an O atom in a compound. Thus, if we replace one of the O atoms of sulfate ion, SO_4^{2-}, with an S atom, the result is **thiosulfate ion, $S_2O_3^{2-}$.** The formal oxidation state of S in $S_2O_3^{2-}$ is +2, but as we can see from Figure 23-9, the two S atoms are not equivalent: The central S atom is in the oxidation state +6 and the terminal S atom, −2. Also shown in Figure 23-9 are anions derived from the series of acids called *thionic* acids, $H_2S_nO_6$ (where n can be 2, 3, 4, 5, or 6).

Thiosulfates can be prepared by boiling elemental sulfur in an alkaline solution of sodium sulfite. The sulfur is oxidized and the sulfite ion is reduced, both to thiosulfate ion.

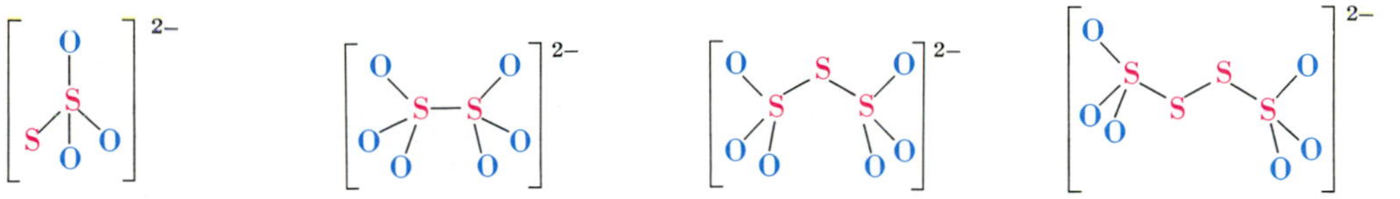

$S_2O_3^{2-}$ thiosulfate ion $S_2O_6^{2-}$ dithionate ion $S_3O_6^{2-}$ trithionate ion $S_4O_6^{2-}$ tetrathionate ion

FIGURE 23-9
Structures of some thio ions.

The electrode potential diagram of Figure 23-7 does not list an $E°$ value for the $SO_3^{2-}/S_2O_3^{2-}$ couple in basic solution. To obtain this value, use the method of Example 23-2 (see Exercise 38).

Oxid: $2\,S + 6\,OH^- \longrightarrow S_2O_3^{2-} + 3\,H_2O + 4\,e^-$ $-E° = +0.74\ V$
Red: $2\,SO_3^{2-} + 3\,H_2O + 4\,e^- \longrightarrow S_2O_3^{2-} + 6\,OH^-$ $E° = -0.57\ V$

Net: $SO_3^{2-} + S \longrightarrow S_2O_3^{2-}$ $E°_{cell} = +0.17\ V$

$$(23.22)$$

Thiosulfate solutions are common analytical reagents, often used in conjunction with iodine. For example, in the analysis of Cu^{2+} an excess of iodide ion is added,

$$2\,Cu^{2+} + 4\,I^- \longrightarrow 2\,CuI(s) + I_2 \qquad (23.23)$$

and the liberated iodine is titrated with a standard solution of $Na_2S_2O_3$ (see Exercise 37).

Thiosulfate–iodine titration.

$$I_2 + 2\,S_2O_3^{2-} \longrightarrow 2\,I^- + S_4O_6^{2-} \qquad (23.24)$$

Starch can be used as an indicator in this titration (see again, page 840).

Sulfides and Qualitative Analysis. All metals form sulfides, and most sulfides are insoluble in water. However, most sulfides can be dissolved in acidic solutions, and a few are soluble in a strongly basic solution. We can use the differing solubilities of the metal sulfides as a basis of separating metal ions into groups. As we have already seen, this is the basis of the qualitative analysis scheme for cations (Figure 19-7).

To predict whether a particular sulfide will precipitate from a solution we must use the ionization constants of H_2S and the solubility product constant of the metal sulfide, as we learned in Section 19-9. Furthermore, we have seen that the outcome depends very much on the pH. For the simple metal ion M^{2+}, which does precipitate from an aqueous solution saturated in H_2S, we can write

Precipitation of a metal sulfide in acidic solution.

$$H_2S(aq) + M^{2+}(aq) \longrightarrow MS(s) + 2\,H^+(aq) \qquad (23.25)$$

We can represent the dissolving of a metal sulfide through the reverse of equation (23.25), that is,

Dissolving of a metal sulfide in acidic solution.

$$MS(s) + 2\,H^+(aq) \longrightarrow M^{2+}(aq) + H_2S(g) \qquad (23.26)$$

Some metal sulfides have such low values of K_{sp} that they do not dissolve in ordinary mineral acids (like HCl). However, they can be partially dissolved in a strong oxidizing acid. The sulfide ion is oxidized to insoluble elemental sulfur, and the metal ion appears in solution.

Action of $HNO_3(aq)$ on a metal sulfide.

$$3\,CuS(s) + 8\,H^+ + 2\,NO_3^- \longrightarrow$$
$$3\,Cu^{2+} + 4\,H_2O + 2\,NO(g) + 3\,S(s) \qquad (23.27)$$

Mercury(II) sulfide, which is about the least soluble of all metal sulfides, will dissolve in a mixture of nitric and hydrochloric acids called **aqua regia.** Here dissolving occurs through oxidation of S^{2-} to $S(s)$ and formation of the complex ion $[HgCl_4]^{2-}$.

$$HgS(s) + 4\,H^+ + 4\,Cl^- + 2\,NO_3^- \longrightarrow$$
$$[HgCl_4]^{2-} + 2\,H_2O + 2\,NO_2(g) + S(s) \qquad (23.28)$$

A few metal sulfides have an acidic nature and will dissolve in a basic solution having a high concentration of S^{2-} or HS^-, just as acidic oxides dissolve in solutions with high $[OH^-]$. One such sulfide is SnS_2, which forms the thio anion SnS_3^{2-}. Some of these points are further illustrated in Table 23-7.

TABLE 23-7
Solubilities of Some Metal Sulfides

Soluble in H_2O	Soluble in 0.3 M HCl (K_{sp})	Soluble in 3 M HNO_3 (K_{sp})	Soluble in aqua regia (K_{sp})	Soluble in KOH(aq)
K_2S	MnS (2×10^{-13})	CdS (8×10^{-27})	HgS (2×10^{-52})	SnS_2
Na_2S	FeS (6×10^{-18})	PbS (8×10^{-28})		As_2S_3
CaS	CoS (4×10^{-21})	CuS (6×10^{-36})		Sb_2S_3
	ZnS (1×10^{-21})			

We can obtain H_2S for the precipitation of metal sulfides in several ways. One method involves the action of HCl(aq) on FeS(s).

$$FeS(s) + 2 HCl(aq) \longrightarrow FeCl_2(aq) + H_2S(g) \tag{23.29}$$

The generally preferred laboratory method is the hydrolysis of thioacetamide. This hydrolysis is quite slow at room temperature, but it proceeds at an appreciable rate at higher temperatures.

$$\underset{\text{thioacetamide}}{CH_3\overset{\overset{\displaystyle S}{\|}}{C}NH_2} + H_2O \longrightarrow \underset{\text{acetamide}}{CH_3\overset{\overset{\displaystyle O}{\|}}{C}NH_2} + H_2S(g) \tag{23.30}$$

Other Sulfur Compounds. The variety and complexity of sulfur compounds is impressive, largely as a result of the possibility of octet expansion for a central S atom. Consider, for example, compounds of sulfur and fluorine. There are two different compounds with the formula S_2F_2, as well as the compounds SF_4, SF_6, and S_2F_{10}. Substitution of other atoms for some of the F atoms accounts for the existence of compounds such as SF_5Cl, SF_5Br, and $(SF_5)_2O$.

Compounds with the characteristic group $OS{\large\diagup}$ can be thought of as derived from sulfurous acid $[OS(OH)_2]$ by substituting other atoms for the —OH groups. These are known as **thionyl** compounds, e.g., $OSCl_2$ = thionyl chloride. Compounds with the group $O_2S{\large\diagup}$, in turn, are derived from sulfuric acid $[O_2S(OH)_2]$. These are known as **sulfuryl** compounds, e.g., O_2SCl_2 = sulfuryl chloride. Thionyl and sulfuryl chlorides have a number of commercial uses in synthesizing organic compounds and as catalysts. Another interesting use of these compounds is as solvents and reactants in advanced batteries of the lithium type (recall Table 22-2).

Normally, the electrolysis of dilute $H_2SO_4(aq)$ produces $H_2(g)$ and $O_2(g)$ as products. That is, the net reaction is the electrolysis of water. However, the electrolysis can be carried out in a way that creates a high overvoltage for $O_2(g)$ formation, and in this way the anode half-reaction is

$$2 HSO_4^- \longrightarrow H_2S_2O_8 + 2 e^- \tag{23.31}$$

(H_2 is produced in the accompanying reduction half-reaction.)

$H_2S_2O_8$ is called **peroxodisulfuric acid** and has the structure

$$HO-\overset{\overset{\displaystyle O}{\|}}{\underset{\underset{\displaystyle O}{\|}}{S}}-O-O-\overset{\overset{\displaystyle O}{\|}}{\underset{\underset{\displaystyle O}{\|}}{S}}-OH$$

The presence of the —O—O— bond is characteristic of **peroxo** compounds. Salts of $H_2S_2O_8$ contain the ion $S_2O_8^{2-}$ and are called **peroxodisulfates** (often referred to more simply as persulfates). Peroxodisulfate ion is one of the most powerful oxidizing agents.

$$S_2O_8^{2-} + 2 e^- \longrightarrow 2 SO_4^{2-} \qquad E° = +2.01 \text{ V} \tag{23.32}$$

The formal oxidation state of S in $S_2O_8^{2-}$ is $+7$, although a more accurate representation of the situation is to consider that the six terminal O atoms are in the oxidation state -2 and the two atoms in the —O—O— linkage, -1. This leaves the oxidation state of S as $+6$ and indicates that the reduction involves breaking the —O—O— bond and converting these two O atoms from the oxidation state -1 to -2.

Selenium, Tellurium, and Polonium. In many ways these elements exemplify the fundamental idea that elements within the same group of the periodic table have similar properties, but also that properties change progressively in moving down a group of the periodic table. For example, oxygen and sulfur in group 6A are electrical insulators. Selenium and tellurium are semiconductors, and polonium is an electrical conductor. Polonium also has other metallic properties, such as an ability to form cations.

Se and Te are found as selenides and tellurides in sulfide ores (e.g., as Cu_2Se and Cu_2Te). The principal source of these elements is the anode mud obtained in the electrolytic refining of copper (see again page 777). Polonium is a very rare radioactive element; it occurs only to the extent of about 0.1 mg Po per ton of uranium ore.

Selenium and tellurium form unpleasant smelling gaseous hydrides, H_2Se and H_2Te, that are similar to but less stable than, H_2S. For example, the heats of formation of H_2Se and H_2Te are positive, whereas that of H_2S is negative. H_2Se and H_2Te ionize in aqueous solution somewhat more readily that does H_2S. They produce HSe^-, Se^{2-}, HTe^-, and Te^{2-} ions, and they precipitate metal ions as selenides and tellurides. Se and Te also form acids in which the nonmetal is in a positive oxidation state—H_2SeO_3, H_2TeO_3, H_2SeO_4, and H_6TeO_6. Selenic acid, H_2SeO_4, resembles H_2SO_4 very closely; the ionization constants of the two acids are almost identical.

Selenium plays an important role in modern technology, largely through its property of *photoconductivity:* The electrical conductivity of selenium increases in the presence of light. This property is used, for example, in photocells in cameras. In modern photocopying machines, the light-sensitive element is a thin film of Se deposited on aluminum. A distribution of electrostatic potential on the light-sensitive element is related to the dark and light areas of the image being copied. A dry black powder (toner) is distributed in accordance with the electrostatic image on the light-sensitive element, and then this image is transferred to a sheet of paper. Next, the dry powder is fused to the paper. In a final step, the electrostatic charge on the light-sensitive element is neutralized to prepare it for the next cycle.

The semiconductor form of selenium (gray selenium) is also used in the manufacture of rectifiers (devices used to convert alternating to direct electric current). Both Se and Te find some use in the preparation of alloys, and their compounds are used as additives to control the color of glass.

Polonium was the first new radioactive element isolated from uranium ore by Marie Curie and her husband, Pierre, in 1898. She named it after her native Poland.

23-3 Group 5A Elements—The Nitrogen Family

4A	5A	6A
6 C	7 N	8 O
14 Si	15 P	16 S
32 Ge	33 As	34 Se
50 Sn	51 Sb	52 Te
82 Pb	83 Bi	84 Po

Perhaps the most interesting feature of the group 5A elements is that, even more so than the group 6A elements just considered, properties of both nonmetals and metals are displayed within the group. The electron configurations of the elements provide only a limited clue to their metallic–nonmetallic behavior. Their outer-shell electron configurations are ns^2np^3. There are a number of ways in which this electron configuration may be altered when group 5A atoms form compounds. Several possibilities are worth noting.

One possibility is the gain or, more likely, sharing of three electrons in the outer shell to produce the noble gas configuration ns^2np^6 and an oxidation state of -3. This is especially likely for the smaller atoms N and P. For the larger atoms—As, Sb, and Bi—the p^3 set of electrons may be lost. This leads to an electron configuration involving a next-to-outermost shell of 18 and an outer shell of 2. This so-called

TABLE 23-8
Selected Properties of Group 5A Elements

Element	Covalent radius, pm	Electro-negativity	First ionization energy, kJ/mol	Common physical form(s)	Density of solid, g/cm^3	Comparative electrical conductivity[a]
N	70	3.0	1402	gas	1.03 ($-252°C$)	—
P	110	2.2	1012	wax-like white solid	1.82	—
				red solid	2.36	10^{-15}
As	121	2.2	947	yellow solid	2.03	—
				gray solid with metallic luster	5.73	4.6
Sb	141	2.0	834	yellow solid	5.3	—
				silvery white metallic solid	6.69	4.2
Bi	152	2.0	703	pinkish white metallic solid	9.75	1.4

[a]These values are relative to an assigned value of 100 for silver.

"18 + 2" configuration is adopted by a number of ionic species derived from metals. In some cases, all five outer-shell electrons may be involved in compound formation, leading to the oxidation state of +5.

Assessment of Metallic–Nonmetallic Character in Group 5A. To aid in this discussion a number of properties have been listed in Table 23-8. For group 5A the usual decrease of ionization energy with increasing atomic number is noted. This establishes the order of metallic character within the group. Nitrogen is least metallic and bismuth is most metallic. Of course, all these ionization energies are high compared to the 1A and 2A elements. The electronegativities indicate a high degree of nonmetallic character for nitrogen and less so for the remaining members of the group. None of the elements in group 5A is highly metallic, however.

Three of the elements—phosphorus, arsenic, and antimony—exhibit allotropy. For phosporus the stable form at room temperature is red phosphorus. Its physical properties are those of a nonmetal. For example, its triple point is at 590 °C and 43 atm pressure; red phosphorus sublimes without melting. For arsenic and antimony the stable allotropic forms are the "metallic" ones. They have high densities, fair thermal conductivities, and limited abilities to conduct electricity. Bismuth has no nonmetallic allotropic forms.

The increase in metallic behavior from top to bottom in group 5A is reflected by the data listed in Table 23-9. Covalent bond formation with hydrogen is expected for nonmetallic elements. The negative value of the free energy of formation of $NH_3(g)$ suggests it is a stable molecule that forms spontaneously from its elements in their standard states. The values for the other hydrides are positive and increase in magnitude with increasing atomic number, suggesting decreasing stabilities. In

TABLE 23-9
Free Energies of Formation of Group 5A Hydrides

Hydride	ΔG_f° (298 K), kJ/mol
NH_3	-16.5
PH_3	$+25.5$
AsH_3	$+68.9$
SbH_3	$+147.7$
BiH_3	$+230(?)$

TABLE 23-10
Acid–Base Behavior of Some Group 5A Oxides

Oxide	Nature of oxide	Principal product(s) obtained when dissolved in	
		Acidic soln	Basic soln
N_2O_3	acidic	HNO_2	NO_2^-
P_4O_6	acidic	H_3PO_3	$H_2PO_3^-$, HPO_3^{2-}
As_4O_6	acidic	$H_3AsO_3(HAsO_2)$	AsO_3^{3-} (AsO_2^-)
Sb_2O_3	amphoteric	SbO^+	SbO_2^-
Bi_2O_3	basic	BiO^+ and Bi^{3+}	insoluble

Acidic solution ($[H^+] = 1$ M):

$$NO_3^- \xrightarrow{+0.81\text{ V}} NO_2 \xrightarrow{+1.07\text{ V}} HNO_2 \xrightarrow{+0.99\text{ V}} NO \xrightarrow{+1.59\text{ V}} N_2O \xrightarrow{+1.77\text{ V}} N_2 \xrightarrow{-1.87\text{ V}} NH_3OH^+ \xrightarrow{+1.46\text{ V}} N_2H_5^+ \xrightarrow{+1.24\text{ V}} NH_4^+$$

Basic solution ($[OH^-] = 1$ M):

$$NO_3^- \xrightarrow{-0.85\text{ V}} NO_2 \xrightarrow{+0.88\text{ V}} NO_2^- \xrightarrow{-0.46\text{ V}} NO \xrightarrow{+0.76\text{ V}} N_2O \xrightarrow{+0.94\text{ V}} N_2 \xrightarrow{-3.04\text{ V}} NH_2OH \xrightarrow{+0.74\text{ V}} N_2H_4 \xrightarrow{+0.10\text{ V}} NH_3$$

FIGURE 23-10
Electrode potential diagram for nitrogen.

fact, BiH_3 is so unstable that its properties have not been measured with any accuracy.

The group 5A elements form a number of different oxides. Table 23-10 describes the solubilities of the oxides X_2O_3 or X_4O_6 in acidic and basic solutions. The oxides of nitrogen, phosphorus, and arsenic are acidic; antimony(III) oxide is amphoteric; and the oxide of bismuth has only basic properties. The acid–base behavior of the oxides establishes the gradation of properties within group 5A, from nonmetallic to metallic, as well as any other criterion.

Electrode Potential Diagram for Nitrogen. Figure 23-10 lists some important nitrogen-containing species through an electrode potential diagram. The variability possible in the oxidation state of N suggests the existence of a large number of nitrogen compounds.

Hydrides of Nitrogen. There are three nitrogen hydrides, NH_3, N_2H_4, and HN_3. A fourth compound, NH_2OH, though containing oxygen, is closely related to NH_3.

Ammonia, NH_3, is produced by the Haber–Bosch process (described in Chapter 16) and is the ultimate source from which other nitrogen compounds are synthesized. NH_3 is the commonest weak base.

$$NH_3(aq) + H_2O \rightleftharpoons NH_4^+(aq) + OH^-(aq) \qquad K_b = 1.74 \times 10^{-5} \qquad (23.33)$$

Neutralization of NH_3 by acids is the source of several ammonium compounds. The properties of the NH_4^+ ion are similar to those of the alkali metal ions, which means that all common ammonium salts are water soluble. Because N exists in its lowest possible oxidation state (-3) in NH_3, in oxidation–reduction reactions NH_3 is always a reducing agent (it is oxidized).

When in the presence of a very strong base, such as H^- or O^{2-}, NH_3 may act as an *acid,* that is, a proton donor.

$$NaH + NH_3(l) \longrightarrow NaNH_2 + H_2(g) \qquad (23.34)$$

The NH_2^- ion is the **amide ion.** Sodium amide (sodamide) is used in a variety of reactions designed to synthesize organic molecules.

If an H atom in NH_3 is replaced by the group $-NH_2$, the resulting molecule is H_2N-NH_2 or N_2H_4, **hydrazine.** Replacement of an H atom in NH_3 by $-OH$ produces NH_2OH, **hydroxylamine.** Both these compounds are weak bases. Because of its two N atoms, N_2H_4 can accept two protons in a stepwise fashion. However, the ion $N_2H_6^{2+}$ can be obtained in appreciable concentrations only in strongly acidic solutions.

$$NH_2OH(aq) + H_2O \rightleftharpoons NH_3OH^+ + OH^- \qquad K_b = 9.1 \times 10^{-9}$$

$$N_2H_4(aq) + H_2O \rightleftharpoons N_2H_5^+ + OH^- \qquad K_{b_1} = 8.5 \times 10^{-7}$$

$$N_2H_5^+(aq) + H_2O \rightleftharpoons N_2H_6^{2+} + OH^- \qquad K_{b_2} = 8.9 \times 10^{-16}$$

Hydrazine and hydroxylamine form salts analogous to ammonium salts, such as $[NH_3OH]Cl$, $N_2H_5NO_3$, and $N_2H_6SO_4$. These salts all hydrolyze in water to yield acidic solutions.

Hydrazine and some of its derivatives burn in air with the evolution of much heat; they are used as rocket fuels.

$$N_2H_4(l) + O_2(g) \longrightarrow N_2(g) + 2\ H_2O(l) \qquad \Delta H° = -622.2\ \text{kJ} \qquad (23.35)$$

Reaction (23.35) is used as the basis of a fuel cell (see Exercise 45), and is also the reaction that occurs when hydrazine is used to remove dissolved O_2 from boiler water. Hydrazine is especially valuable for boiler water treatment because no salts (ionic compounds) are produced. These would be objectionable in the boiler water. The industrial preparation of hydrazine was presented in detail in the Focus feature of Chapter 4.

Both hydrazine and hydroxylamine can act either as oxidizing or reducing agents (usually the latter), depending on the pH and the substances with which they react (see Exercises 43 and 44).

The oxidation of hydrazine in acidic solution by nitrite ion produces **hydrogen azide,** HN_3.

$$N_2H_5{}^+ + NO_2{}^- \longrightarrow HN_3 + 2\ H_2O \qquad (23.36)$$

The structure of the HN_3 molecule is

Pure HN_3 is a colorless liquid that boils at 37 °C. It is very unstable and will detonate when subjected to shock. In aqueous solution HN_3 is a weak acid, called **hydrazoic acid;** its salts are called **azides.** Azides resemble chlorides in some properties (for instance, AgN_3 is insoluble in water), but they are extremely unstable. Some azides (e.g., lead azide) are used to make detonators. The release of $N_2(g)$ by the decomposition of sodium azide, NaN_3, is the basis of the air-bag safety system currently used in some automobiles (recall Example 6-14).

Oxides and Oxoacids of Nitrogen. An interesting feature of the oxides of nitrogen listed in Table 23-11 is that their enthalpies (heats) of formation are mostly *positive*. This is because the energy released in the formation of nitrogen-to-oxygen bonds in these molecules is not enough to compensate for the large energy investment required to break the very strong $N\equiv N$ bonds in $N_2(g)$.

Example 23-5

Using bond energies to estimate an enthalpy of formation. The bond energies of N_2, O_2, and NO are 946, 498, and 632 kJ/mol, respectively. Use these data to estimate the enthalpy (heat) of formation of NO(g).

TABLE 23-11
Oxides of Nitrogen

Oxidation state	Oxide	At 298 K and 1 atm		
		Physical state	$\Delta H_f°$, kJ/mol	$\Delta G_f°$, kJ/mol
+1	dinitrogen monoxide, N_2O (nitrous oxide)	colorless gas	+82.05	+104.2
+2	nitrogen monoxide, NO (nitric oxide)	colorless gas	+90.25	+86.57
+3	dinitrogen trioxide, N_2O_3	—	—	—
+4	nitrogen dioxide, NO_2	brown gas	+33.18	+51.30
+4	dinitrogen tetroxide, N_2O_4	colorless gas	+9.16	+97.82
+5	dinitrogen pentoxide, N_2O_5	colorless solid	-43.1	+113.8

Solution. The formation of one mole of NO(g) can be represented as

$$\tfrac{1}{2} N_2(g) + \tfrac{1}{2} O_2(g) \longrightarrow NO(g) \qquad \Delta H^\circ_f = ?$$

In this reaction $\tfrac{1}{2}$ mol of bonds in N_2 and $\tfrac{1}{2}$ mol of bonds in O_2 must be broken. One mole of bonds in NO is formed.

$$\Delta H^\circ_f = \Delta H^\circ_{\text{bond breakage}} + \Delta H^\circ_{\text{bond formation}}$$

$$= [\tfrac{1}{2}(946) + \tfrac{1}{2}(498)] \text{ kJ/mol} - 632 \text{ kJ/mol} = +90.$$

SIMILAR EXAMPLES: Exercises 12, 28.

Free energies of formation of the oxides are also positive quantities. Conversely, the free energies of their decompositions are *negative,* suggesting that some of these oxides should decompose rather easily, such as

$$2 N_2O(g) \longrightarrow 2 N_2(g) + O_2(g) \qquad \Delta G^\circ = -208 \text{ kJ/mol (at 298 K)} \qquad (23.37)$$

Actually, N_2O is quite stable at room temperature. This is because the activation energy for the decomposition reaction is very high—about 250 kJ/mol. At elevated temperatures (about 600 °C) the rate of decomposition becomes appreciable. Reaction (23.37) accounts for the ability of $N_2O(g)$ to support combustion. Once a high enough temperature has been reached, the material undergoing combustion combines with the $O_2(g)$ produced in reaction (23.37).

Combustion reactions in $N_2O(g)$.

$$H_2(g) + N_2O(g) \xrightarrow{\Delta} H_2O(l) + N_2(g) \qquad (23.38)$$

$$Cu(s) + N_2O(g) \xrightarrow{\Delta} CuO(s) + N_2(g) \qquad (23.39)$$

Some common methods of preparing the oxides of nitrogen are outlined in Table 23-12. All nitrates decompose on heating, but only NH_4NO_3 produces $N_2O(g)$. Nitrates of active metals (e.g., $NaNO_3$) yield the corresponding nitrite and $O_2(g)$. Thermal decomposition of the nitrates of less active metals [e.g., $Pb(NO_3)_2$] yields the metal oxide, $NO_2(g)$, and $O_2(g)$.

N_2O_3 is the acid anhydride of nitrous acid

$$N_2O_3 + H_2O \longrightarrow 2 HNO_2 \qquad (23.40)$$

and N_2O_5, of nitric acid.

$$N_2O_5 + H_2O \longrightarrow 2 HNO_3 \qquad (23.41)$$

NO_2 produces both HNO_3 and $NO(g)$ when it reacts with water.

Preparation of $HNO_3(aq)$ from $NO_2(g)$.

$$3 NO_2(g) + H_2O \longrightarrow 2 H^+ + 2 NO_3^- + NO(g) \qquad (23.42)$$

$HNO_3(aq)$ is both a strong acid and a strong oxidizing agent capable of yielding a variety of reduction products depending on the reducing agent chosen and the con-

TABLE 23-12
Preparation of Oxides of Nitrogen

Oxide	A method of preparation
N_2O	$NH_4NO_3(s) \xrightarrow{\Delta} N_2O(g) + 2 H_2O(l)$
NO	$3 Cu(s) + 8 H^+ + 2 NO_3^- \longrightarrow 3 Cu^{2+} + 2 NO(g) + 4 H_2O$
N_2O_3	$NO(g) + NO_2(g) \rightleftharpoons N_2O_3(g) \qquad K_p = 0.48 \text{ (at 298 K)}$
NO_2	$2 Pb(NO_3)_2(s) \xrightarrow{\Delta} 2 PbO(s) + 4 NO_2(g) + O_2(g)$
	$2 NO(g) + O_2(g) \rightleftharpoons 2 NO_2(g) \qquad K_p = 1.6 \times 10^{12} \text{ (at 298 K)}$
N_2O_4	$2 NO_2(g) \rightleftharpoons N_2O_4(g) \qquad K_p = 8.84 \text{ (at 298 K)}$
N_2O_5	$4 HNO_3(l) + P_4O_{10}(s) \longrightarrow 4 HPO_3 + 2 N_2O_5(s)$

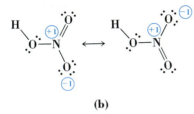

FIGURE 23-11

Structures of two oxoacids of nitrogen.

The N and O atoms are in the same plane and the H atom is out of the plane.
(a) Nitrous acid, HONO.
(b) Nitric acid, HONO₂.

Pyrophosphoric acid is also called diphosphoric acid.

TABLE 23-13
Nitric Acid as an Oxidizing Agent

Nitric acid concentration	Reducing agent	Principal reduction product[a]
15 M	Cu; Ag	$NO_2(g)$
8 M	Cu; Ag	$NO(g)$
dilute[b]	Zn; Fe	$N_2O(g)$

[a]The oxidation products are Cu^{2+}, Ag^+, Zn^{2+}, and Fe^{3+}, respectively.
[b]If $[H_3O^+]$ is increased by adding $H_2SO_4(aq)$, NH_3OH^+ or NH_4^+ might be obtained. In *basic* solution, reduction of NO_3^- by Zn produces $NH_3(g)$.

centration of the acid. Some of the many possibilities are listed in Table 23-13. $HNO_2(aq)$ is a weak acid that can function either as an oxidizing or reducing agent. Thus, it is reduced to $NO(g)$ by I^- and oxidized to NO_3^- by MnO_4^-.

Figure 23-11 presents structures of nitrous and nitric acids. From these structures it is clear that although we write the formulas HNO_2 and HNO_3, more appropriate representations would be $HONO$ and $HONO_2$.

Oxoacids of Phosphorus. In Section 14-11 we discussed the two principal oxides of phosphorus, P_4O_6 and P_4O_{10}, and noted that they are the acid anhydrides of H_3PO_3 and H_3PO_4, respectively. These oxoacids are listed in Table 23-14, but we need to explore more fully a complication suggested in this table—the existence of ortho, meta, and pyro acids.

Orthophosphoric acid forms when P_4O_{10} reacts with an excess of H_2O.

$$P_4O_{10}(s) + 6\ H_2O \longrightarrow 4\ H_3PO_4(aq) \qquad (23.43)$$

When orthophosphoric acid is heated to temperatures in excess of 215 °C, **pyrophosphoric acid** is formed.

$$2\ H_3PO_4 \xrightarrow{\Delta} H_4P_2O_7 + H_2O \qquad (23.44)$$

When either the ortho or pyro acid is heated to temperatures in excess of about 300 °C, a glassy product is formed. This is a polymerized form of **metaphosphoric acid**: $(HPO_3)_n$, where $n = 2, 3, 4, 6$. The relationship of these three types of acids to one another is suggested by Figure 23-12, and is applied to some acids of arsenic in Example 23-6.

TABLE 23-14
Oxoacids of Phosphorus

Oxidation state	Name	Formula	Structure
+3	orthophosphorous acid	H_3PO_3	HO—P—OH with =O above and H below
+5	orthophosphoric acid	H_3PO_4	HO—P—OH with =O above and OH below
+5	pyrophosphoric acid	$H_4P_2O_7$	HO—P—O—P—OH with =O above each P and OH below each P
+5	metaphosphoric acid	$(HPO_3)_n$	(see expression 23.46)

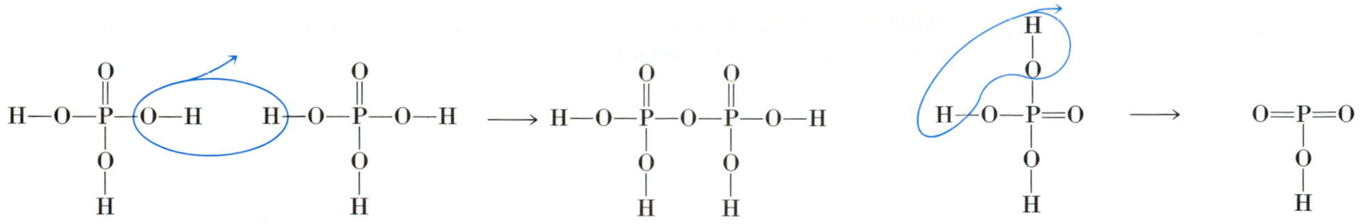

orthophosphoric acid H_3PO_4 — orthophosphoric acid H_3PO_4 — pyrophosphoric acid $H_4P_2O_7$ — orthophosphoric acid H_3PO_4 — metaphosphoric acid HPO_3

FIGURE 23-12
Some oxoacids of phosphorus(V).

Example 23-6

Relating names and formulas of ortho, meta, and pyro acids and their salts. By analogy to corresponding phosphorus compounds, supply names and/or formulas of the following arsenic compounds: **(a)** H_3AsO_4; **(b)** metaarsenious acid; **(c)** magnesium pyroarsenate.

Solution

(a) H_3AsO_4 is analogous to H_3PO_4. It is an ortho acid with As in the +5 oxidation state. This acid is called orthoarsenic acid.

(b) Since this is an "ous" acid it should be analogous to a *phosphorous* acid, that is, with As in the +3 oxidation state. Orthoarsenious would be H_3AsO_3. If we split out one H_2O molecule from this formula we obtain the formula of metaarsenious acid, $HAsO_2$.

(c) By analogy to pyrophosphoric acid, we should expect pyroarsenic acid to be $H_4As_2O_7$. Replace the H atoms with Mg to obtain a salt consisting of Mg^{2+} and $As_2O_7{}^{4-}$ ions: $Mg_2As_2O_7$.

SIMILAR EXAMPLE: Exercise 13.

Polyphosphoric Acids and Polyphosphates. As we noted in Figure 23-12, we can form pyrophosphoric acid (diphosphoric acid) by eliminating one molecule of H_2O from two molecules of H_3PO_4; it has 34.8% P, by mass. If a H_2O molecule is split out from between a molecule of H_3PO_4 and one of $H_4P_2O_7$, the resulting molecule is $H_5P_3O_{10}$, with 36.0% P. This process of eliminating H_2O can be continued until the % P reaches about 39%. The species obtained in this process are called **polyphosphoric acids.** Of the various phosphoric acids, only H_3PO_4 and $H_4P_2O_7$ can be obtained as pure substances (e.g., as crystalline solids). Commercial syrupy phosphoric acid is a mixture of H_3PO_4, $H_4P_2O_7$, and the polyphosphoric acids illustrated below.

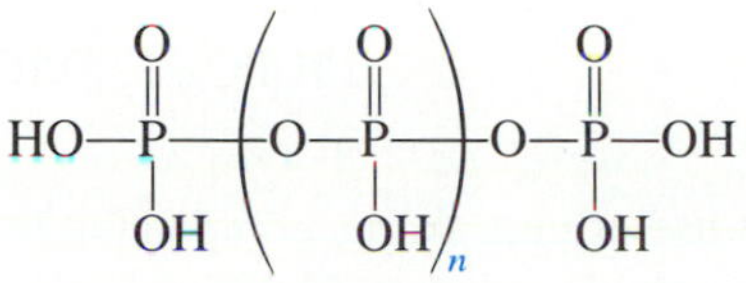

or $H_2PO_3(HPO_3)_nPO_4H_2$ (23.45)

repeating unit = HPO_3; if
$n = 1$, $H_5P_3O_{10}$ (tripolyphosphoric acid);
$n = 2$, $H_6P_4O_{13}$ (tetrapolyphosphoric acid); etc.

Salts of these acids are called **polyphosphates.** The most common of these polyphosphates is probably $Na_5P_3O_{10}$, called sodium tripolyphosphate.

Suppose we remove an additional H_2O molecule from a molecule of a polyphosphoric acid, using an H atom from one end of the chain and an OH group from the other end. The result will be a ringlike structure with the formula $(HPO_3)_n$. Acids with this formula and structure are called **metaphosphoric acids.** They are named according to the number of P atoms they contain. The structure shown below is called *tri*metaphosphoric acid.

To emphasize their ring structures, the IUPAC recommends that metaphosphoric acids be given names such as *cyclo*-triphosphoric acid and *cyclo*-tetraphosphoric acid.

$$
\begin{array}{c}
\text{metaphosphoric acid} = (HPO_3)_n \\
n = 3, \text{ trimetaphosphoric acid;} \\
n = 4, \text{ tetrametaphosphoric acid; etc.}
\end{array}
\qquad (23.46)
$$

FIGURE 23-13
Sequestering action of polyphosphates on metal ions.

Salts of these acids are called **metaphosphates,** for example, $Na_3P_3O_9$ is sodium trimetaphosphate.

The commercial production of sodium tripolyphosphate involves heating a mixture of powdered Na_2HPO_4 and NaH_2PO_4 in a 2:1 mole ratio.

$$2\,Na_2HPO_4 + NaH_2PO_4 \xrightarrow{\Delta} Na_5P_3O_{10} + 2\,H_2O \qquad (23.47)$$

Sodium metaphosphate is obtained as a glassy solid when NaH_2PO_4 is heated to temperatures above 620 °C and the melt rapidly cooled.

$$n\,NaH_2PO_4 \xrightarrow{\Delta} (NaPO_3)_n + n\,H_2O \qquad (23.48)$$

Polyphosphoric acids containing 36 to 37% P are used to produce high-strength fertilizers and as catalysts in petroleum refining. Sodium tripolyphosphate is used in cement manufacturing and in oil well drilling. Because a solution of tripolyphosphate ions is strongly basic through hydrolysis, sodium tripolyphosphate is used in a number of cleaning agents. Its most extensive use, however, has been as a builder in detergents. Detergents lose their effectiveness when used in water containing dipositive metal ions such as Ca^{2+}. The function of a builder is to complex or *sequester* these ions, reducing the concentration of the free ions to the point where they do not interfere with the detergent action. The sequestering action of $P_3O_{10}^{5-}$ ions is illustrated in Figure 23-13. Metaphosphates exhibit a similar action on metal ions, and they are used in water softening.

Phosphate rock being mined with a water cannon and drag line. [Courtesy International Minerals and Chemicals]

The method of Section 14-11 is called the thermal process. The method outlined here is the wet process. Over 90% of H_3PO_4 is produced by the wet process, both in the United States and worldwide.

Phosphate Rock. The principal source of phosphorus compounds is phosphate rock, a complex material containing fluorapatite $[3Ca_3(PO_4)_2 \cdot CaF_2]$. $Ca_3(PO_4)_2$ can be extracted from fluorapatite and used in the preparation of P_4 by the method outlined in Section 14-11. If P_4 (usually as a liquid) is burned in air, the product is P_4O_{10}, which is the acid anhydride of H_3PO_4. An impure form of phosphoric acid can be prepared by the direct action of H_2SO_4 on phosphate rock.

$$[3Ca_3(PO_4)_2 \cdot CaF_2] + 10\,H_2SO_4 + 20\,H_2O \rightarrow$$
$$\qquad\qquad 6\,H_3PO_4 + 10\,[CaSO_4 \cdot 2H_2O] + 2\,HF \qquad (23.49)$$

fluorapatite

gypsum

The principal use of phosphorus compounds is in fertilizers. A mixture of $CaSO_4$ and the more soluble calcium dihydrogen phosphate, called **normal superphosphate,** has a phosphorus content of 7 to 9% P, by mass. It is produced by treating phosphate rock with H_2SO_4 in a reaction similar to (23.49) but employing different proportions of the rock and acid.

$$[3Ca_3(PO_4)_2 \cdot CaF_2] + 7\,H_2SO_4 + 3\,H_2O \longrightarrow$$
$$\qquad\qquad 3\,[Ca(H_2PO_4)_2 \cdot H_2O] + 7\,CaSO_4 + 2\,HF \qquad (23.50)$$

normal superphosphate

If phosphate rock is treated with H_3PO_4 (derived from reaction 23.49) instead of H_2SO_4, the product is known as **triple superphosphate.** This process eliminates

$CaSO_4$, and the product has a much higher phosphorus content than normal superphosphate, about 20 to 21% P.

$$[3Ca_3(PO_4)_2 \cdot CaF_2] + 14\ H_3PO_4 + 10\ H_2O \longrightarrow$$

$$\underbrace{10\ [Ca(H_2PO_4)_2 \cdot H_2O]}_{\text{triple superphosphate}} + 2\ HF \qquad (23.51)$$

Environmental Problems of Phosphorus. HF is a byproduct of superphosphate production. In some instances this HF is recovered, but in the past it was mostly released into waterways. Now, because of strict environmental controls, it has become necessary to neutralize the HF, usually with lime. Large settling ponds are required for this. Another product of phosphate fertilizer manufacture is a great quantity of waste rock, perhaps two-thirds of the rock processed.

Changes that occur in freshwater bodies as a result of enrichment by nutrients are referred to by the term **eutrophication.** This is a natural process that occurs over geologic time periods, but it is greatly accelerated by human activities. Large quantities of nutrients such as nitrates and phosphates encourage algae to grow excessively in bodies of water. The algae, in turn, deplete the oxygen content. This kills fish and promotes the growth of anerobic bacteria, among other undesirable side effects.

Natural sources of nutrients include animal wastes, decomposition of dead organic matter, and natural nitrogen fixation. Human sources include industrial wastes, municipal sewage plant effluents, and fertilizer runoff, but one of the main contributors has been builders from detergents (e.g., $Na_5P_3O_{10}$). Actions limiting the quantities of phosphates discharged to the environment seem to be effective in combating eutrophication. However, the case against phosphorus is not conclusive. It may be that in some cases the eutrophication process is governed by some other nutrient or the availability of CO_2. (Photosynthetic activity by algae is dependent on the quantity of CO_2 produced by microorganisms that metabolize organic compounds.)

One means of reducing phosphate discharges into the environment is through their removal in sewage treatment plants. In the processing of sewage, polyphosphates are decomposed to orthophosphates, and organic phosphorus compounds are degraded to orthophosphates by bacterial action. The orthophosphates may then be precipitated, either as iron(III) phosphates, aluminum(III) phosphates, or as $Ca_3(PO_4)_2$ or $Ca_5OH(PO_4)_3$. The precipitating agents are generally aluminum(III) sulfate, iron(III) chloride, or calcium hydroxide. In a fully equipped modern sewage treatment plant, up to 98% of the phosphates in sewage can be removed.

Eutrophication of a freshwater pond. [John Schultz/ PAR/NYC]

Arsenic, Antimony, and Bismuth. At the beginning of this section we explored the trend from nonmetallic to metallic properties within group 5A of the periodic table. We found the first two members, N and P, to be typical nonmetals. In the remaining three members we saw increasing evidence of metallic character. In fact, one of the main uses of these elements is in the manufacture of alloys. For example, the addition of As and Sb to lead produces an alloy that has desirable properties for use in electrodes in lead–acid batteries. Arsenic and antimony are also used in semiconductor materials. (Recall the discussion of GaAs on page 806.) Because of their low melting points, antimony and bismuth find some use in low-melting alloys for use as solders or in fire-protection systems.

The most important compound of arsenic is the oxide, As_4O_6, known as "white arsenic." It can be made by burning arsenic in air, and is the acid anhydride of orthoarsenious acid.

$$As_4O_6(s) + 6\ H_2O \longrightarrow 4\ H_3AsO_3(aq) \qquad (23.52)$$

When orthoarseni*ous* acid reacts with NaOH(aq), the product is an orthoarsen*ite*.

$$H_3AsO_3(aq) + 3\ OH^-(aq) \longrightarrow AsO_3^{3-} + 3\ H_2O \qquad (23.53)$$

Both inorganic and organic compounds derived from As_4O_6 have found extensive use as insecticides.

As, Sb, and Bi form insoluble sulfides when solutions of their ions are treated with $H_2S(aq)$. These sulfides are also insoluble in mildly acidic solutions, which means that they precipitate in qualitative analysis cation group 2 (recall Figure 19-7). In addition, because of their acidic properties As_2S_3 and Sb_2S_3 dissolve in concentrated $KOH(aq)$ (see Table 23-7).

A sensitive test for arsenic that has been popularized through mystery novels is the **Marsh test.** The suspected arsenic-containing material is treated with zinc in an acidic solution. First, zinc metal displaces H^+, producing $H_2(g)$. Next, the freshly prepared (nascent) $H_2(g)$ reduces the arsenic compound to arsine, AsH_3. Finally, the AsH_3 is decomposed by heating it in a glass tube; the As deposits as a thin film (a mirror). Even trace quantities of arsenic can be detected by this method.

Example 23-7

Writing chemical equations from a description of chemical reactions. Write the chemical equations for the three reactions referred to in the description of the Marsh test for arsenic.

Solution. First, we have the reaction of Zn and $H^+(aq)$.

$$Zn(s) + 2\,H^+(aq) \longrightarrow Zn^{2+}(aq) + H_2(g)$$

Let us choose as a typical arsenic-containing compound, As_4O_6. The half-reaction for reducing As_4O_6 to $AsH_3(g)$ in acidic solution is

Red: $As_4O_6 + 24\,H^+ + 24\,e^- \longrightarrow 4\,AsH_3(g) + 6\,H_2O$

Since $H_2(g)$ is the reducing agent, the oxidation half-reaction must involve oxidizing $H_2(g)$ to the $+1$ oxidation state. In an acidic solution this would be to H^+.

Oxid: $H_2(g) \longrightarrow 2\,H^+(aq) + 2\,e^-$

For the complete redox equation, we must multiply the oxidation half-equation by 12 and add it to the reduction half-equation.

Redox rxn: $As_4O_6 + 12\,H_2(g) \longrightarrow 4\,AsH_3(g) + 6\,H_2O$

For the decomposition of $AsH_3(g)$ we can simply write

$$2\,AsH_3(g) \longrightarrow 2\,As(s) + 3\,H_2(g)$$

SIMILAR EXAMPLES: Exercises 8, 25, 31, 40, 64.

23-4 Group 4A Nonmetals: Carbon and Silicon

The differences between carbon and silicon, as outlined in Table 23-15, are perhaps the most striking to be found between the second and third period elements of a group in the periodic table. As suggested by the approximate bond energies, strong C—C and C—H bonds account for the central role of carbon-atom chains and rings in establishing the chemical behavior of carbon. A study of these chains and rings and their attached atoms is the focus of organic chemistry (Chapter 27) and biochemistry (Chapter 28). The relatively weaker Si—Si and Si—H bonds imply a less important "organic chemistry" of silicon, and the strength of the Si—O bond accounts for the predominance of the silicates among silicon compounds.

In this section we consider a few carbon compounds that, in addition to the oxides and carbonates discussed in Section 14-8, constitute the inorganic chemistry of

TABLE 23-15
Some Comparisons of Carbon and Silicon

Carbon	Silicon
Two principal allotropes, graphite and diamond	One stable, diamond-type crystalline modification
Forms two stable *gaseous* oxides, CO and CO_2, and everal less stable ones, such as C_3O_2	Forms only one *solid* oxide (SiO_2) that is stable at room temperature; a second oxide (SiO) is stable only in the temperature range 1180–2480 °C
Insoluble in alkaline medium	Dissolves in alkaline medium and forms $H_2(g)$ and $SiO_4^{4-}(aq)$
Principal oxoanion is CO_3^{2-}, which has a planar shape	Principal oxoanion is SiO_4^{4-}, which has a tetrahedral shape
Strong tendency for catenation, with straight and branched chains and rings containing up to hundreds of C atoms[a]	Less tendency for catenation, with silicon atom chains limited to about six Si atoms[a]
Readily forms multiple bonds through use of the orbital sets $sp^2 + p$ and $sp + p^2$	Multiple bond formation much less common than with carbon and limited to $p_\pi - d_\pi$ type (recall Section 11-8)
Approximate single bond energies, kJ/mol	Approximate single bond energies, kJ/mol
C—C, 347	Si—Si, 226
C—H, 414	Si—H, 318
C—O, 360	Si—O, 368
C—Cl, 326	Si—Cl, 250–335

[a]Catenation refers to the joining together of like atoms into chains.

carbon. To complement the inorganic chemistry of silicon, which was presented in Section 14-9, we briefly consider some organosilicon compounds.

Carbon is strictly nonmetallic in its behavior. Silicon is also essentially nonmetallic, though we have noted its semiconductor properties. We have also explored the semiconductor behavior of the metalloid germanium (Chapter 11 Focus feature). Tin and lead, although having certain properties that resemble the lighter elements in group 4A, are essentially metallic elements; we discussed them in Chapter 22.

Some Inorganic Carbon Compounds. Carbon combines with metals to form carbides. With the transition metals these are interstitial carbides—carbon atoms take up positions in the holes or voids in the crystalline structure of the metal. With active metals the carbides are ionic. **Calcium carbide** is formed by the high-temperature reaction of lime and coke.

$$CaO(s) + 3\ C(s) \xrightarrow{2000\ °C} CaC_2(s) + CO(g) \tag{23.54}$$

Calcium carbide is an important product because its reaction with water produces acetylene, a gas useful in the synthesis of organic compounds.

$$CaC_2(s) + 2\ H_2O \longrightarrow Ca(OH)_2(s) + C_2H_2(g) \tag{23.55}$$
$$\text{acetylene}$$

Two other important inorganic compounds of carbon are **carbon disulfide**, CS_2, and **carbon tetrachloride**, CCl_4. The modern method of preparing CS_2 involves the reaction of methane and sulfur vapor in the presence of a catalyst.

$$CH_4(g) + 4\ S(g) \xrightarrow{\Delta} CS_2(l) + 2\ H_2S(g) \tag{23.56}$$

CS_2 is a highly flammable, volatile liquid useful as a solvent for sulfur, phosphorus, bromine, iodine, fats, and oils. Its uses as a solvent are decreasing, however, because of its poisonous nature. Other important uses are in the manufacture of rayon and cellophane. CCl_4 can be prepared by the direct chlorination of methane.

$$CH_4(g) + 4\ Cl_2(g) \longrightarrow CCl_4(l) + 4\ HCl(g) \tag{23.57}$$

Although CCl_4 has found extensive use as a solvent, drycleaning agent, and fire extinguisher, these uses are now declining. CCl_4 causes liver and kidney damage and is a known carcinogen. We discussed compounds of carbon with H, Cl, and F—so-called chlorofluorocarbons—in Chapter 14, where we emphasized some serious environmental problems associated with their use.

Pseudohalogens. Certain groupings of atoms, several containing C atoms, have some of the characteristics of a halogen atom. These are called pseudohalogens and include

—CN (cyanide)　　　—OCN (cyanate)　　　—SCN (thiocyanate)　　　—N$_3$ (azide)

The **cyanide** ion, CN^-, is similar to the halide ions, X^-, in that it forms an insoluble silver salt, AgCN, and the acid, HCN, though this acid (hydrocyanic acid) is very weak. CN^- differs from the halide ions in being extremely poisonous. Despite its extreme toxicity, HCN, a liquid that boils at about room temperature, has industrial uses in the manufacture of plastics.

As a further similarity to the halogens, we find that two cyanide groups can combine to form $(CN)_2$, analogous to Cl_2. This compound is called **cyanogen,** and its formula is often represented as C_2N_2. To show one more similarity to a halogen, the reaction of $(CN)_2$ in basic solution

$$(CN)_2(g) + 2\,OH^-(aq) \longrightarrow CN^-(aq) + OCN^-(aq) + H_2O \tag{23.58}$$

is analogous to the reaction of $Cl_2(g)$ in basic solution (reaction 23.12). Cyanogen is used in organic synthesis, as a fumigant, and as a rocket propellant.

Organosilicon Compounds. Single bonds of the type Si—Si are not strong enough to allow for very long silicon-atom chains. Nevertheless, a series of **silanes** can be prepared, up to a limit of six Si atoms per chain.

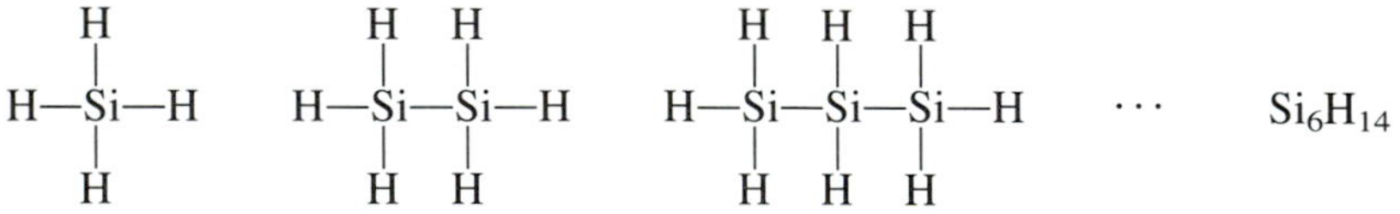

monosilane	disilane	trisilane	hexasilane

Monosilane is produced by the reaction of lithium aluminum hydride and silicon tetrachloride.

$$LiAlH_4 + SiCl_4 \longrightarrow LiCl + AlCl_3 + SiH_4(g) \tag{23.59}$$

The silanes are thermally unstable. With moderate heating the higher silanes decompose to the lower silanes, and above 500 °C, to the elements. Like the hydrocarbons, the silanes are combustible. In fact, they ignite spontaneously in air.

Combustion of a silane.

$$SiH_4 + 2\,O_2(g) \longrightarrow SiO_2(s) + 2\,H_2O \tag{23.60}$$

We can substitute other atoms for H atoms in the silanes rather easily. If we substitute Cl for H atoms in SiH_4, we obtain successively SiH_3Cl, SiH_2Cl_2, $SiHCl_3$, and $SiCl_4$.

The reaction of $(CH_3)_2SiCl_2$ with water produces an interesting compound, **dimethylsilanol,** $(CH_3)_2Si(OH)_2$. Dimethylsilanol undergoes a polymerization reaction in which H_2O molecules are eliminated from among large numbers of silanol molecules. The result of this polymerization is a material consisting of long-chain molecules: **silicones.**

$$(CH_3)_2SiCl_2 + 2\,H_2O \longrightarrow (CH_3)_2Si(OH)_2 + 2\,HCl$$

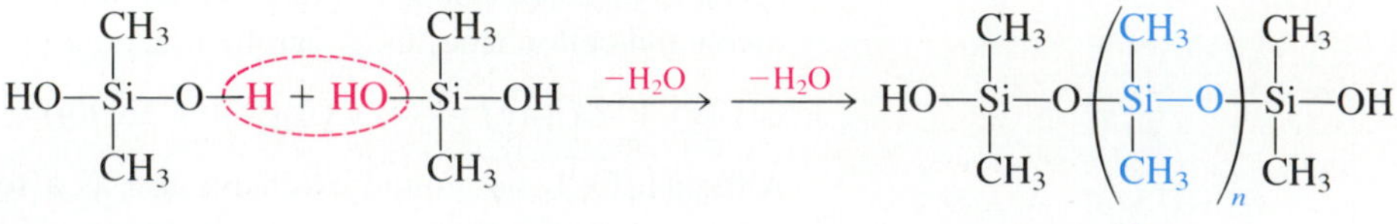

a silicone

Depending on the length of the chains and the degree of crosslinking between chains, silicones are obtained either as oils or as rubberlike materials. Silicone oils are not volatile; they may be heated without decomposition; they can be cooled to low temperatures without becoming viscous or solidifying. (Hydrocarbon oils become very viscous at low temperatures.) Silicone rubbers retain their elasticity at low temperatures; they are chemically resistant and thermally stable.

23-5 Group 3A: Boron

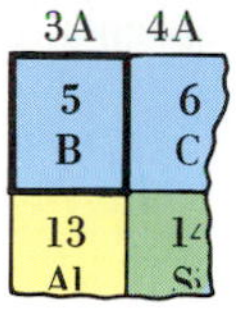

The only group 3A element that is almost exclusively nonmetallic in its physical and chemical properties is boron. We considered the remaining members of group 3A—Al, Ga, In, Tl—in our discussion of the representative metals in Chapter 22. A factor that governs much of the chemistry of boron is the electron deficiency associated with many of its compounds. This makes them strong Lewis acids. This electron deficiency leads to bonding of a type that we have not previously encountered. It is the bonding that occurs in boron hydrides.

Boron Hydrides. The molecule BH_3 (borane) may exist as a reaction intermediate, but it cannot be isolated as a stable compound. The simplest boron hydride that can be isolated is **diborane, B_2H_6.** To resolve the structure of B_2H_6 required fundamental contributions to bonding theory, in particular, to molecular orbital theory. The problem is this: In the B_2H_6 molecule there are 14 valence shell atomic orbitals (four each from the 2 B atoms and one each from the 6 H atoms). But there are only *12* valence electrons (three each from the 2 B atoms and one each from the 6 H atoms).

The currently accepted structure of diborane is pictured in Figure 23-14. Here are the key features of the structure.

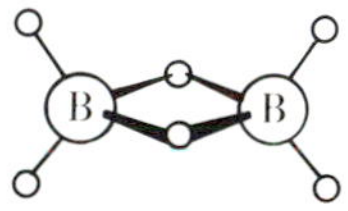

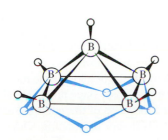

FIGURE 23-14
Structure of diborane, B_2H_6.

- The two B atoms and four of the H atoms lie in the same plane (the plane of the page in Figure 23-14).
- The orbitals used by the B atoms to bond these particular 4 H atoms are sp^2. (The H—B—H bond angles are 121.8°.) Eight electrons are involved in these four bonds.
- Each of the two remaining H atoms is *simultaneously* bonded to *two* B atoms. These bonds are variously called "three-center," "bridge," or "banana" bonds.
- *Six* atomic orbitals (an sp^2 and a p orbital from each B atom and an s orbital from each H atom) are combined into *six* molecular orbitals in these two "bridge" bonds.
- Of these six molecular orbitals, *two* are bonding orbitals, and these are the orbitals into which the remaining four electrons are placed.

If we extend the concept of "bridge" bonds to bonds of the type B—B—B, we can also describe the structure of higher boranes, such as B_5H_9 shown in Figure 23-15.

FIGURE 23-15
Structure of pentaborane, B_5H_9.

The boron atoms are joined through multicenter B—B—B bonds. Five of the H atoms are bonded directly to B atoms. The other four H atoms bridge pairs of B atoms.

Other Boron Compounds. Boron compounds are widely distributed in the earth's crust, but concentrated ores are found in only a few locations—in Italy, U.S.S.R., Tibet, Turkey, and the desert regions of California. Typical of these ores is the hydrated borate **borax, $Na_2B_4O_7 \cdot 10H_2O$.** Figure 23-16 suggests how borax can be converted to a wide variety of boron compounds.

Although elemental boron can be prepared by the reduction of B_2O_3 with an active metal, the product is rather impure. Higher purity boron can be obtained by the reduction of boron halides (usually BBr_3) with $H_2(g)$. This higher purity boron can be brought to the ultrapure levels required for semiconductor applications by zone refining (recall Figure 13-27).

One of the key compounds used in the synthesis of other boron compounds is **boric acid, $B(OH)_3$.** As discussed in Section 17-9, the high ratio of ionic charge to

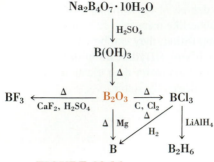

FIGURE 23-16

Preparation of some boron compounds.

$Na_2B_4O_7 \cdot 10H_2O$ (borax) is converted to $B(OH)_3$ by reaction with H_2SO_4. When heated strongly, $B(OH)_3$ is converted to B_2O_3. A variety of boron-containing compounds and boron itself can be prepared from B_2O_3 (see also Example 23-8).

ionic radius for B^{3+} causes $B(OH)_3$ to ionize as a weak acid. The ionization reaction involves formation of the ion $B(OH)_4^-$.

$$B(OH)_3(aq) + 2 H_2O \rightleftharpoons H_3O^+ + B(OH)_4^- \qquad K_a = 5.6 \times 10^{-10} \qquad (23.61)$$

In solutions more concentrated than about 0.1 M, $B(OH)_3$ and $B(OH)_4^-$ combine to form a series of polyborate ions (similar to the formation of polyphosphate ions described in Section 23-3).

As expected of the salts of a weak acid, borate salts produce basic solutions by hydrolysis, and this accounts for their use in cleaning agents. However, of the several million tons of borates produced annually, worldwide, only a small proportion is used in this way. Boron compounds are also used in products as varied as adhesives, cement, disinfectants, fertilizers, fire retardants, glass, herbicides, metallurgical fluxes, and textile bleaches and dyes.

Example 23-8

Writing chemical equations from a summary diagram of reaction chemistry. Based on Figure 23-16, write chemical equations for the successive conversions of borax to **(a)** boric acid, **(b)** B_2O_3, and **(c)** impure boron metal.

Solution. Figure 23-16 lists the key substances involved in each reaction. Our task is to identify other plausible reactants and/or products.

(a) Conversion of a salt to the corresponding acid is an acid–base reaction and does not involve changes in oxidation states. Balancing should be possible by inspection. The additional substances required in the equation are Na_2SO_4 and H_2O.

$$Na_2B_4O_7 \cdot 10H_2O + H_2SO_4 \longrightarrow 4 B(OH)_3 + Na_2SO_4 + 5 H_2O$$

(b) Conversion of a hydroxo compound to an oxide requires that H_2O be driven off.

$$2 B(OH)_3 \xrightarrow{\Delta} B_2O_3 + 3 H_2O$$

(c) Since the boron need not be of high purity, direct reduction of B_2O_3 with Mg is possible.

$$3 Mg + B_2O_3 \xrightarrow{\Delta} 2 B + 3 MgO$$

SIMILAR EXAMPLES: Exercises 51, 55, 64.

23-6 Group 8A—The Noble Gases

	8A
	2 He
9 F	10 Ne
17 Cl	18 Ar
35 Br	36 Kr
53 I	54 Xe
85 At	86 Rn

To conclude this two-chapter survey of the representative elements, we need to say more about the unusual elements of group 8A. We discussed sources, physical properties, and uses of the lighter noble gases in Section 14-4. For a long time these were the only aspects of the noble gases that could be considered, since no chemical compounds of these elements were known. In fact, the inertness of the noble gases helped to provide a theoretical framework for the Lewis theory of bonding: Other atoms tend to form bonds so that they may acquire noble gas electron configurations (ns^2np^6). Moreover, the unquestioning attitude of most chemists toward the stability of noble gas electron configurations tended to retard rather than to stimulate the search for noble gas compounds.

In 1962, N. Bartlett and D. H. Lohmann discovered that O_2 and PtF_6 would join in a 1:1 mole ratio to form the compound O_2PtF_6. Properties of this compound (for instance, its paramagnetism) suggested the bonding to be ionic: $(O_2)^+(PtF_6)^-$. The quantity of energy required to extract an electron from O_2 is 1177 kJ/mol. Bartlett noted that the first ionization energy of Xe (1170 kJ/mol) is almost identical to that

New York City skyline represented through neon lights. [Roy Morsch/The Stock Market]

of O_2 and reasoned that the compound $XePtF_6$ might also exist. He was able to prepare a yellow crystalline solid with a composition corresponding to this formula.* Soon thereafter a considerable number of noble gas compounds were synthesized by chemists around the world. For example, XeF_4 was prepared by the reaction of Xe and F_2 in the mole ratio $1:5$ in a nickel vessel at 400 °C and at about 6 atm. By varying the Xe/F_2 ratio and other reaction conditions, XeF_2 and XeF_6 can also be obtained. All are colorless, crystalline solids that sublime easily.

A fluoride of krypton, KrF_2, has been prepared, and compounds of radon also exist, though experiments with Rn are difficult to perform because of the intense radioactivity of all its isotopes. No compounds of He, Ne, or Ar are known. In general, the conditions necessary for the formation of noble gas compounds are

- a readily ionizable (therefore, heavy) noble gas atom and
- a highly electronegative group (e.g., F or O) to bond to the noble gas atom.

Compounds have been synthesized with Xe in the oxidation states of

	+2	+4	+6	+8
examples:	XeF_2	$XeF_4, XeOF_2$	XeF_6, XeO_3	XeO_4, H_4XeO_6

Because it is difficult to oxidize Xe to these positive oxidation states, we should expect Xe compounds to be very strong oxidizing agents, to be easily reduced. For example, in aqueous acidic solution

$$XeF_2(aq) + 2 H^+ + 2 e^- \longrightarrow Xe + 2 HF \qquad E° = +2.64 \text{ V} \qquad (23.62)$$

The hydrolysis of XeF_2 in basic solution is an oxidation–reduction reaction in which OH^- is oxidized to $O_2(g)$.

$$2 XeF_2 + 4 OH^- \longrightarrow 2 Xe + 4 F^- + 2 H_2O + O_2(g) \qquad E°_{cell} = +2.05 \text{ V} \qquad (23.63)$$

The fluorides XeF_2, XeF_4, and XeF_6 are stable if kept from contact with water. XeO_3 is an explosive white solid and XeO_4 is an explosive colorless gas.

Bonding in Noble Gas Compounds. The valence bond theory permits bonding schemes for Xe compounds that are consistent with their observed molecular shapes. These require sp^3d hybridization for XeF_2 and sp^3d^2 for XeF_4.

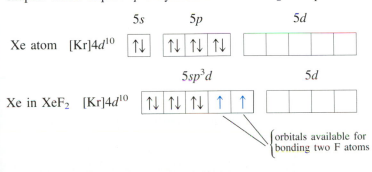

distribution of sp^3d orbitals: *trigonal bipyramidal*
geometric structure of XeF_2: *linear*

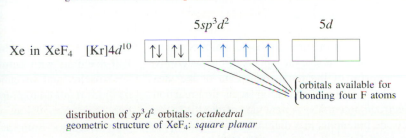

distribution of sp^3d^2 orbitals: *octahedral*
geometric structure of XeF_4: *square planar*

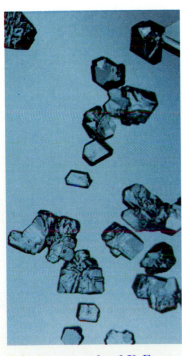

Colorless crystals of XeF_4 at $100\times$ magnification. [Argonne National Laboratory]

*It has since been established that this solid has the formula $Xe(PtF_6)_n$, where n is between 1 and 2.

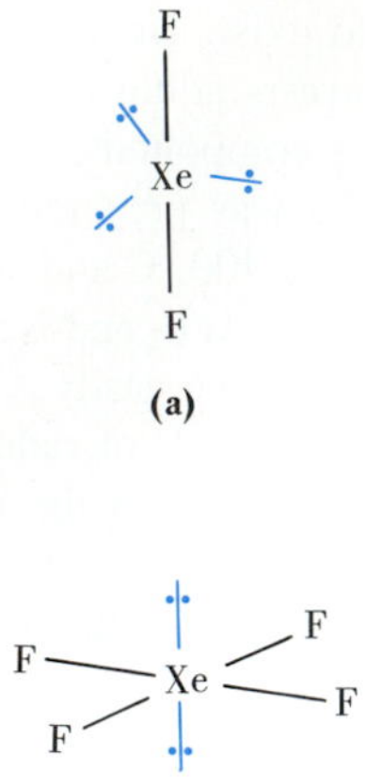

FIGURE 23-17
Geometric structures of
(a) XeF_2 and (b) XeF_4.

Lone-pair electrons are shown
only for the central Xe atoms.

Based on the high promotion energy for an electron from a $5p$ to a $5d$ orbital and on the observed bond strengths and bond distances in these fluorides, there is some doubt, however, whether d orbitals are involved in bond formation. A molecular orbital description of XeF_2 seems to provide better agreement with observed properties by proposing that a single electron pair bonds both F atoms to Xe (similar to the B—H—B bonds in B_2H_6 described in the preceding section).

What is clear is that VSEPR theory does do a satisfactory job of predicting molecular shapes for most Xe compounds and this is the approach that we will follow. The molecular shapes of XeF_2 and XeF_4 predicted by VSEPR theory are shown in Figure 23-17.

Example 23-9

Using VSEPR theory to predict geometric shapes of noble gas compounds.
Predict the geometric shapes of (a) XeF_2 and (b) XeF_4.

Solution

(a) The number of valence electrons in the XeF_2 molecule is 22 (eight from Xe and seven each from the two F atoms). Let us first draw a Lewis structure with these 22 valence electrons.

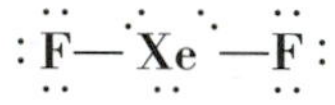

$$:\ddot{F}—\ddot{X}e\,—\ddot{F}:$$

This Lewis structure places 10 electrons (five electron pairs) in the valence shell of the central Xe atom. The distribution of these five electron pairs will be trigonal bipyramidal. Of the five electron pairs, *two* are bond pairs and three are lone pairs. The VSEPR designation for this distribution is AX_2E_3, and from Table 10-3 we see that this structure will be linear.

(b) In XeF_4 the number of valence electrons is 36 $[8 + (7 \times 4)]$. Here, the Lewis structure we can draw is

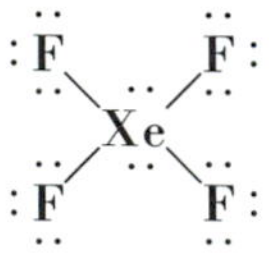

$$\begin{array}{cc} :\ddot{F} & \ddot{F}: \\ & \diagdown \quad \diagup \\ & Xe \\ & \diagup \quad \diagdown \\ :\ddot{F} & \ddot{F}: \end{array}$$

This Lewis structure places 12 electrons (six electron pairs) in the valence shell of the Xe atom. Four pair of electrons are bond pairs and two are lone pairs. Their distribution is octahedral, and the VSEPR designation is AX_4E_2. To minimize repulsions between lone pairs, they should be placed at opposite corners of the octahedron. This leaves the four bond pairs at the corners of a square having the Xe atom at its center (see Table 10-3). The structure is square planar.

SIMILAR EXAMPLES: Exercises 14, 56.

Summary

Principles developed in the first 21 chapters of this text are applied in this chapter to several aspects of the descriptive inorganic chemistry of the nonmetals. One of the ideas stressed is the discussion of oxidation–reduction reactions through electrode potentials. A new idea offered is to represent electrode potentials in a diagrammatic fashion, and electrode potential diagrams are given for Cl, O, S, and N. A new type of calculation based on these diagrams is to establish $E°$ for one reduction half-reaction by combining half-reactions with known values of $E°$.

Oxoacids and oxoanions are studied in terms of the methods required to prepare them, their acid–base properties, their strengths as oxidizing or reducing agents, and their structures. Some additional possibilities for oxoacids of sulfur are the substitution of S for O atoms (thiosulfates) and the presence of —O—O— bonds (peroxosulfates). With phosphorus, an additional possibility is the elimination of H_2O from simpler acid molecules to produce pyrophosphoric, polyphosphoric, and metaphosphoric acids and their salts.

The solubility characteristics of a number of compounds are noted in this chapter. Particular attention is given to the effect of acids, bases, and oxidizing agents on the solubilities of metal sulfides and the relationship of these solubilities to the qualitative analysis scheme for cations. Also noted at several points is how Le Châtelier's principle is used to control equilibrium processes.

The differences between the first and higher members of a group of the periodic table is encountered again in this chapter, as it was in Chapter 14. Examples cited are the failure of fluorine to form oxoacids and the numerous differences between O and S and between C and Si. Also, by considering several criteria, the progression of properties from nonmetallic to metallic within group 5A is established. Bond energies and thermodynamic stabilities are applied to a study of nitrogen oxides, and additional insights into chemical bonding are provided through a study of the boron hydrides, interhalogen compounds, and noble gas compounds.

Throughout the chapter practical uses of the nonmetals and their compounds are described. Quite often it is found that a particular use for a substance is a consequence of some special property that it possesses.

Summarizing Example

The most important thiosulfate compound is sodium thiosulfate pentahydrate, $Na_2S_2O_3 \cdot 5H_2O$. We have referred to its use as a reagent in analytical chemistry (page 845). In photographic work, where it is used as a "fixing agent," sodium thiosulfate is usually called "hypo." Its function is to dissolve unreacted silver halides in the photographic film through complex ion formation (see Section 25-11). Another important use is as an antichlor, that is, to react with excess Cl_2 in bleaching solutions. In this use $S_2O_3{}^{2-}$ acts as a reducing agent.

Decomposition of thiosulfate ion. When an aqueous solution of $Na_2S_2O_3$ is acidified, reaction (23.64) occurs. The sulfur is in the colloidal state when first formed (right). [Carey B. Van Loon]

1. For use in analytical chemistry, sodium thiosulfate solutions must be prepared carefully. In particular, the solution must be kept from becoming acidic. In acidic solutions thiosulfate ion *disproportionates* into $SO_2(g)$ and $S(s)$. Write a chemical equation for this disproportionation reaction.

Solution. A disproportionation reaction is a redox reaction. Our usual approach is to write half-equations for the oxidation and reduction half-reactions and combine these into the net redox equation. In the present case, the equation is quite simple and we can write it directly.

$$S_2O_3{}^{2-}(aq) + 2\,H^+(aq) \longrightarrow H_2O + SO_2(g) + S(s) \qquad (23.64)$$

(This example is similar to Examples 23-7 and 23-8.)

2. The electrode potential diagram of Figure 23-7 does not include a value for the reduction of $SO_2(g)$ to $S_2O_3{}^{2-}(aq)$ in acidic solution. Establish this value.

Solution. Figure 23-7 does have $E°$ values for the successive reductions of SO_2 to $S_4O_6{}^{2-}$ and $S_4O_6{}^{2-}$ to $S_2O_3{}^{2-}$. We can use the method of Example 23-2 to establish the desired $E°$. To do this, we must work through free energy changes: $\Delta G° = -n\mathscr{F}E°$.

$$
\begin{array}{ll}
4\,SO_2(g) + 4\,H^+ + 6\,e^- \longrightarrow S_4O_6{}^{2-}(aq) + 2\,H_2O & \Delta G° = -6\mathscr{F}(0.51)\ V \\
S_4O_6{}^{2-}(aq) + 2\,e^- \longrightarrow 2\,S_2O_3{}^{2-}(aq) & \Delta G° = -2\mathscr{F}(0.08)\ V \\
\hline
4\,SO_2(g) + 4\,H^+(aq) + 8\,e^- \longrightarrow 2\,S_2O_3{}^{2-}(aq) + 2\,H_2O &
\end{array}
$$

$\Delta G°$ for the net reaction can be expressed in two ways: $-6\mathscr{F}(0.51) - 2\mathscr{F}(0.08)$ and $-8\mathscr{F}E°$. Thus,

$$\Delta G° = -\mathscr{F}[(6 \times 0.51) + (2 \times 0.08)] = -8\mathscr{F}E°$$

$$E° = \frac{-\mathscr{F}(3.22)}{-8\mathscr{F}} = 0.40\ V$$

(This example is similar to Example 23-2.)

3. Determine $E°_{cell}$ for the disproportionation reaction in part **1**.

Solution. Here, we need to separate the net redox equation into two half-equations and write an $E°$ value for each of the half-equations. We will find one of these $E°$ values in Figure 23-7. The other one we can obtain from part **2** (that is, $-E° = -0.40\ V$).

Oxid:	$S_2O_3^{2-} + H_2O \longrightarrow 2\ SO_2 + 2\ H^+ + 4\ e^-$	$-E° = -0.40$ V
Red:	$S_2O_3^{2-} + 6\ H^+ + 4\ e^- \longrightarrow 2\ S + 3\ H_2O$	$E° = +0.50$ V
Net:	$S_2O_3^{2-} + 2\ H^+ \longrightarrow H_2O + SO_2 + S$	$E°_{cell} = +0.10$ V

(This example is similar to Examples 21-2 and 23-1.)

4. Thiosulfate ion, $S_2O_3^{2-}$, does not undergo decomposition (disproportionation) in *basic* solution. Explain why this should be so.

Solution. The positive value of $E°_{cell}$ determined in part **3** is for standard-state conditions, that is, with $[S_2O_3^{2-}] = [H^+] = 1$ M and a partial pressure of SO_2 of 1 atm. According to Le Châtelier's principle, the decomposition of $S_2O_3^{2-}$ (the forward reaction) is favored by high concentrations of H^+ (strongly acidic solutions). Also, the escape of $SO_2(g)$ favors the decomposition reaction. In *basic* solution, where $[H^+]$ is very low and where SO_2 is retained in solution as SO_3^{2-}, conditions should favor the *reverse* reaction. Thiosulfate ion should be stable in basic solutions.

Key Terms

acid salt (23-2)
electrode potential diagram (23-1)
eutrophication (23-3)
interhalogen compound (23-1)
meta acid (23-3)
metaphosphate (23-3)
metaphosphoric acid (23-3)

organosilicon compound (23-4)
ortho acid (23-3)
peroxo compound (23-2)
polyhalide ion (23-1)
polyphosphate (23-3)
polyphosphoric acid (23-3)
pseudohalogen (23-4)

pyro acid (23-3)
silicone (23-4)
sulfuryl compound (23-2)
superphosphate (23-3)
thio compound (23-2)
thionyl compound (23-2)
three-center (bridge, banana) bond (23-5)

Highlighted Expressions

A halogen displacement reaction (23.3)
Synthesis of hydrogen halides from their elements (23.8)
Reaction of $Cl_2(g)$ with water (23.11)
Preparation of chlorate salts (23.14)
Decomposition of chlorate salts (23.15)
Formation of triiodide ion (23.17)
A test for sulfite ion (23.21)

Thiosulfate–iodine titration (23.24)
Precipitation of a metal sulfide in acidic solution (23.25)
Dissolving of a metal sulfide in acidic solution (23.26)
Action of $HNO_3(aq)$ on a metal sulfide (23.27)
Combustion reactions in $N_2O(g)$ (23.38, 23.39)
Preparation of $HNO_3(aq)$ from $NO_2(g)$ (23.42)
Combustion of a silane (23.60)

Review Problems

1. Provide an acceptable name or formula for each of the following.

(a) $KBrO_3$
(b) ClF_3
(c) sodium hypoiodite
(d) trisilane
(e) KOCN
(f) sodium dithionate
(g) silver azide
(h) NaH_2PO_4

2. Complete and balance equations for the following reactions. If no reaction occurs, so state.

(a) $CaCl_2(s) + H_2SO_4(\text{conc. aq}) \xrightarrow{\Delta}$
(b) $I_2(s) + Cl^-(aq) \rightarrow$
(c) $PbS(s) + HCl(aq) \rightarrow$
(d) $NH_3(aq) + HClO_4(aq) \rightarrow$
(e) $NO(g) + O_2(g) \rightarrow$
(f) $Cu(NO_3)_2(s) \xrightarrow{\Delta}$

3. Give a practical method that could be used in the laboratory to prepare (a) $O_2(g)$; (b) $HCl(aq)$; (c) $N_2O(g)$; (d) $BaSO_3(s)$.

4. Write a chemical equation to represent the reaction of (a) $Cl_2(g)$ with cold $NaOH(aq)$; (b) $NaI(s)$ with hot H_2SO_4 (conc. aq); (c) $Cl_2(g)$ with $Br^-(aq)$; (d) $CdS(s)$ with $HNO_3(aq)$; (e) $Fe(s)$ with very hot $N_2O(g)$; (f) $Pb(s)$ with 8 M $HNO_3(aq)$.

5. Write half-equations for the following reduction half-reactions.

Acidic solution		**Basic solution**	
H_5IO_6 $\xrightarrow{+1.60\text{ V}}$ IO_3^-		OCl^- $\xrightarrow{+0.88\text{ V}}$ Cl^-	
H_3PO_2 $\xrightarrow{-0.51\text{ V}}$ P_4		B_2H_6 $\xrightarrow{+0.78\text{ V}}$ BH_4^-	
Sb_2O_5 $\xrightarrow{+0.58\text{ V}}$ SbO^+		$HXeO_4^-$ $\xrightarrow{+1.24\text{ V}}$ Xe	

6. Use data from the electrode potential diagrams indicated to predict whether the stated reaction is likely to occur to a significant extent in the forward direction.

(a) $H_2O_2 + HClO_2 \rightarrow H^+ + ClO_3^- + H_2O$

(Figures 23-1, 23-6)

(b) $2\ Cl^- + 2\ NO_3^- + 4\ H^+ \rightarrow$
$\qquad 2\ H_2O + 2\ NO_2(g) + Cl_2(g)$ (Figures 23-1, 23-10)

(c) $Cl_2(g) + S^{2-} \rightarrow 2\ Cl^- + S(s)$ in basic solution

(Figures 23-1, 23-7)

(d) $3\ OCl^- + H_2O \rightarrow ClO_2^- + Cl_2(g) + 2\ OH^-$

(Figure 23-1)

7. Use data from Figure 23-10 to obtain the standard electrode potential for the reduction of (a) NO_3^- to HNO_2 in acidic solution and (b) N_2 to N_2H_4 in basic solution.

8. If Br^- and I^- occur together in an aqueous solution, I^- can be oxidized to IO_3^- with an excess of $Cl_2(aq)$. Simultaneously, Br^- is oxidized to Br_2, which is extracted with $CS_2(l)$. Write chemical equations for the reactions that occur.

9. You are given a white solid to identify in the qualitative analysis laboratory and told that it is one of the following: $NaCl(s)$, $AgCl(s)$, $BaSO_4(s)$, $CaSO_3(s)$, $MgCO_3(s)$, or $Pb(NO_3)_2(s)$. Tell which solids are eliminated as possibilities at the conclusion of each of the following tests. (a) The solid is insoluble in water. (b) The solid dissolves in $HCl(aq)$ with the evolution of a gas. (c) The gas produced in (b) decolorizes an acidic solution containing MnO_4^-. What must the given solid be?

10. Use data from this chapter and Table 21-1 to predict which of the following are strong enough oxidizing agents in acidic solutions to oxidize H_2O_2 to $O_2(g)$. Explain the basis of your predictions. (a) $Cl_2(g)$; (b) $H_2SO_4(aq)$; (c) $Cr_2O_7^{2-}(aq)$; (d) $MnO_2(s)$; (e) $I_2(s)$

11. Use electrode potential data from this chapter or Table 21-1 to predict which of the following outcomes is the more likely. Explain your reasoning in each case.

(a) When $Cl_2(g)$ is added to an aqueous solution containing I^-, which is more likely to be produced, $O_2(g)$ or I_2?

(b) When added to an acidic solution containing NH_4^+, is H_2O_2 more likely to be oxidized to $O_2(g)$ or reduced to H_2O?

(c) When $Ag(s)$ is added to a solution that is 6 M in both H_2SO_4 and HNO_3, is the gaseous product most likely to be NO, SO_2, or H_2?

12. Given the bond energies at 298 K: O_2, 499; N_2, 946; F_2, 159; Cl_2, 243; ClF, 251; OF (in OF_2), 213; OCl (in OCl_2), 205; and NF (in NF_3), 280 kJ/mol, respectively, calculate ΔH_f° at 298 K for 1 mol of (a) $ClF(g)$; (b) $OF_2(g)$; (c) $OCl_2(g)$; (d) $NF_3(g)$.

13. Use the scheme of Figure 23-12 to supply plausible names or formulas for the following (a) calcium orthophosphate; (b) potassium pyrophosphate; (c) $NaSbO_2$; (d) $NaBiO_3$; (e) sodium orthobismuthate.

14. Use VSEPR theory to predict the probable geometric structures of the molecules (a) XeO_3; (b) XeO_4; (c) $OXeF_4$.

<hr>

Exercises

The Halogens

15. Make a general statement about which of the elements Cl_2, Br_2, and I_2 displaces other halogens from a solution of halide ions. That is, will the reaction $Br_2 + 2\ I^- \rightarrow 2\ Br^- + I_2$ occur? Will the reaction $Br_2 + 2\ Cl^- \rightarrow 2\ Br^- + Cl_2$ occur? And so on.

16. Fluorine was not mentioned in the halogen displacement series in Exercise 15.

(a) In principle would you expect F_2 to be able to displace Cl^-, Br^-, and I^- from solution (producing Cl_2, Br_2, and I_2)?

(b) What difficulty would be encountered in attempting these displacements with F_2?

(c) Is there any reagent that can displace $F^-(aq)$, producing $F_2(g)$? Explain.

17. Refer to Example 23-1. In this example we concluded that the disproportionation of $Cl_2(g)$ to $Cl^-(aq)$ and $ClO_3^-(aq)$ does not occur spontaneously in acidic solution. Yet in equation (23.14) we see that disproportionation of $Cl_2(g)$ is a method of producing chlorate salts. These observations appear to be inconsistent. Are they? Explain.

18. Refer to Examples 23-1 and 23-2. Determine (a) the standard electrode potential for the reduction of $HOCl$ to Cl^-; (b) whether the reaction $2\ HOCl \rightarrow HClO_2 + H^+ + Cl^-$ will go essentially to completion as written.

19. Freshly prepared solutions containing iodide ion are colorless, but over time they usually develop a yellow color. Can you describe chemical reaction(s) to account for this observation?

20. The following properties of astatine have been measured or estimated: (a) covalent radius; (b) ionic radius (At^-); (c) first ionization energy; (d) electron affinity; (e) electronegativity; (f) standard reduction potential ($At_2 + 2\ e^- \rightarrow 2\ At^-$). Based on periodic relationships and data in Table 23-1, what values would you expect for these properties?

21. Use data from Table 23-2 to determine for the dissociation of $HCl(g)$ into its elements at 298 K (a) K_p and (b) the % dissociation.

22. You have available: H_2O, $CaO(s)$, $NaCl(s)$, $NaBr(s)$, $KOH(aq)$, H_2SO_4(concd aq), and H_3PO_4(concd aq). How would you use these, together with common laboratory equipment, to prepare (a) $CaCl_2$; (b) KBr; (c) $KBrO_3$?

23. Use VSEPR theory to predict the geometric structures of (a) BrF_3; (b) IF_5; (c) ICl_2^-; (d) Cl_3IF^-.

★**24.** The following data are given.

$$IO_3^- + 3\ H_2SO_3(aq) \rightarrow I^- + 3\ SO_4^{2-} + 6\ H^+$$
$$E_{cell}^\circ = 0.92\ V$$

$$HOI(aq) + H^+ + I^- \rightarrow I_2(s) + H_2O \qquad E_{cell}^\circ = 0.91\ V$$

Use these data together with values from Table 21-1 to complete the standard electrode potential diagram shown.

$$IO_3^- \xrightarrow{(?)} HOI \xrightarrow{(?)} I_2(s) \xrightarrow{0.54\ V} I^-$$
$$\underset{(?)}{\lfloor\qquad\qquad\qquad\rfloor}$$

Oxygen

25. Each of the following compounds produces $O_2(g)$ when strongly heated. Write a plausible equation for the reactions that occur. (a) $HgO(s)$; (b) $KClO_4(s)$; (c) $Hg(NO_3)_2(s)$; (d) $H_2O_2(l)$.

26. Which of the following reactions using $O_3(g)$, $O_2(g)$, or $H_2O_2(aq)$ as an oxidizing agent are likely to go to completion? [*Hint:* Refer to appropriate electrode potential data.]

  (a) $H_2O_2(aq) + 2\ I^- + 2\ H^+ \rightarrow I_2 + 2\ H_2O$
  (b) $O_2(g) + 2\ H_2O + 4\ Cl^- \rightarrow 2\ Cl_2(g) + 4\ OH^-$
  (c) $O_3(g) + Pb^{2+} + H_2O \rightarrow PbO_2(s) + 2\ H^+ + O_2(g)$
  (d) $HO_2^-(aq) + 2\ Br^- + H_2O \rightarrow 3\ OH^- + Br_2(l)$

27. Hydrogen peroxide is a somewhat stronger acid than is water. For the ionization

$$H_2O_2(aq) + H_2O \rightarrow H_3O^+(aq) + HO_2^-(aq),$$

$pK_a = 11.75$. Calculate the expected pH of a 1 M $H_2O_2(aq)$ solution.

28. For the conversion of $O_2(g)$ to $O_3(g)$, which can be accomplished in an electric discharge, $3\ O_2(g) \rightarrow 2\ O_3(g)$, $\Delta H^\circ = +285$ kJ/mol. The bond energy in O_2, which is essentially the $O{=}O$ double bond energy, is 499 kJ/mol. The $O{-}O$ single bond energy is 142 kJ/mol.

  (a) Calculate the average $O{-}O$ bond energy in $O_3(g)$.
  (b) Estimate the average bond energy in $O_3(g)$ from the structure on page 511 and compare this result with that of part (a).

★**29.** Refer to Figure 11-27 and arrange the following species in the expected order of increasing (a) bond distance and (b) bond strength (energy). State the basis of your prediction. O_2, O_2^+, O_2^-, O_2^{2-}.

Sulfur

30. Give an appropriate name to each of the following: (a) ZnS; (b) $KHSO_3$; (c) $K_2S_4O_6$; (d) OSF_2.

31. Through a chemical equation give a specific example that illustrates

  (a) the dissolving of a metal sulfide in HCl(aq);
  (b) the action of a *nonoxidizing* acid on a metal sulfite;
  (c) the oxidation of $SO_2(aq)$ to $SO_4^{2-}(aq)$ by $MnO_2(s)$ in acidic solution;
  (d) a plausible reaction of $O_2(g)$ with a basic solution containing S^{2-}.

32. Available in a chemical laboratory are elemental sulfur, chlorine gas, metallic sodium, and water. Show how you would use these substances (and air) to produce aqueous solutions containing (a) Na_2SO_3; (b) Na_2SO_4; (c) $Na_2S_2O_3$. [*Hint:* You will have to use information from other chapters as well as this one, e.g., Chapter 14.]

33. We have learned that the thiosulfate and sulfate ions are closely related (see Figures 23-8 and 23-9). Yet the ions are easily distinguished from one another. Describe a chemical test that you could use to determine whether a crystalline white solid is Na_2SO_4 or $Na_2S_2O_3$. Explain the basis of this test, preferably by writing a chemical equation.

34. We have said that salts like $NaHSO_4$ are called *acid* salts because their anions undergo further ionization. What should be the pH of a solution in which 12.5 g $NaHSO_4$ is dissolved in 250.0 mL of water solution? [*Hint:* Use data from Chapter 17, as necessary.]

35. The text states that HgS is the least soluble of all metal sulfides; yet K_{sp} for Bi_2S_3 is 1×10^{-96} compared to 1.6×10^{-52} for HgS. Explain this apparent discrepancy.

36. What mass of Na_2SO_3 must have been present in a sample that required 26.50 mL of 0.0510 M $KMnO_4$ for its oxidation to Na_2SO_4 in an acidic solution? MnO_4^- is reduced to Mn^{2+}.

37. A 1.100-g sample of copper ore is dissolved and the $Cu^{2+}(aq)$ is treated with excess KI. The liberated iodine requires 12.12 mL of 0.1000 M $Na_2S_2O_3$ for its titration. What is the % copper, by mass, in the ore? [*Hint:* Recall equations 23.23 and 23.24.]

38. Derive the value of E° for the reduction of SO_3^{2-} to $S_2O_3^{2-}$ in basic solution that was used in establishing E_{cell}° for reaction (23.22).

★**39.** Refer to the Summarizing Example. In part **4** we reasoned qualitatively that $S_2O_3^{2-}$ should not disproportionate (decompose) in basic solution. Show that this should indeed be the case by applying the Nernst equation to the net reaction written in part **3**. As reasonable conditions, assume that $[S_2O_3^{2-}] = 1$ M, that the pH = 9, and that the partial pressure of $SO_2(g)$ is 1×10^{-6} atm.

Nitrogen

40. Write chemical equations to represent the following.

  (a) Equilibrium between nitrogen dioxide and dinitrogen tetroxide in the gaseous state
  (b) The decomposition of $NaNO_3(s)$ by heating
  (c) The neutralization of $NH_3(aq)$ by $H_2SO_4(aq)$
  (d) The dissolving of silver metal in 8 M $HNO_3(aq)$
  (e) The complete combustion of the rocket fuel, dimethyl hydrazine, $(CH_3)_2NNH_2$

41. What is the pH of an aqueous solution that is (a) 0.032 M in NH_2OH; (b) 0.018 M in $[NH_3OH]Cl$?

42. Draw plausible Lewis structures for (a) N_2O_3 (which has an $N{-}N$ bond) and (b) N_2O_5 (which has an $N{-}O{-}N$ bond).

43. In acidic solution hydroxylamine reduces Fe^{3+} to Fe^{2+}. The hydroxylamine, which in acidic solution is present as NH_3OH^+, is oxidized to $N_2O(g)$

  (a) Write half-equations and a net equation for this redox reaction.
  (b) Use the method of Example 23-2 and data from Figure 23-10 to determine E° for the N_2O/NH_3OH^+ couple.
  (c) Determine E_{cell}° for the net redox reaction.

44. In basic solution NH_2OH oxidizes $Fe(OH)_2(s)$ to $Fe(OH)_3(s)$ and the NH_2OH is reduced to NH_3.

  (a) Write half-equations and a net equation for this redox reaction.
  (b) Use the method of Example 23-2 and data from Figure 23-10 to determine E° for the reduction half-reaction.

(c) If $E°_{cell}$ for the reaction is $+0.98$ V, what must be $E°$ for the $Fe(OH)_3(s)/Fe(OH)_2(s)$ couple?

★45. Calculate the theoretical voltage of the hydrazine–oxygen fuel cell described through equation (23.35), given that the absolute entropy of $N_2H_4(l)$ is 121.2 J mol^{-1} K^{-1}. [*Hint:* You must use data from equation (23.35) and Appendix D, calculate $\Delta G°$, and establish the number of moles of electrons transferred in the reaction.]

Phosphorus

46. Write chemical equations to show why
(a) A solution of Na_3PO_4 is strongly basic.
(b) The first equivalence point in the titration of H_3PO_4 is on the acid side of pH 7.

47. Supply an appropriate name for each of the following:
(a) HPO_4^{2-}; **(b)** $Ca_2P_2O_7$; **(c)** $H_6P_4O_{13}$; **(d)** $(NaPO_3)_4$.

48. You have available concentrated aqueous solutions of H_3PO_4 and NaOH. With these solutions as starting materials, indicate how you would prepare the solids **(a)** $Na_5P_3O_{10}$ and **(b)** $(NaPO_3)_n$.

49. Various glassy metaphosphates have been called sodium metaphosphate. Explain why this name is less specific than most chemical names in describing the composition of a substance.

50. Write a series of equations to show how triple superphosphate could be produced without the use of H_2SO_4 anywhere in the process.

Carbon, silicon

51. Write chemical equations for the reactions you would expect to occur when
(a) KCN(aq) is added to $AgNO_3$(aq);
(b) Al_4C_3 reacts with water to produce CH_4(g);
(c) Si_3H_8 is burned in an excess of air.

52. Describe what is meant by the terms *silane* and *silanol*. What is their role in the preparation of silicones?

53. In a manner similar to that outlined on page 858,
(a) write equations to represent the reaction of $(CH_3)_3SiCl$ with water, followed by the elimination of H_2O from the resulting silanol molecules.
(b) Does a silicone polymer form?
(c) What would be the corresponding product obtained from H_3CSiCl_3?

Boron

54. The molecule tetraborane has the formula B_4H_{10}.
(a) Show that this is an electron deficient molecule.
(b) How many bridge bonds must occur in the molecule?
(c) Show that the carbon analog, butane, C_4H_{10}, is not electron deficient.

55. Write chemical equations to represent
(a) the preparation of boron from BBr_3;
(b) the formation of BF_3 from B_2O_3;
(c) the reaction at high temperatures of boron with N_2O(g).

Noble gas compounds

56. Predict plausible molecular shapes for **(a)** O_2XeF_2; **(b)** O_3XeF_2; **(c)** O_2XeF_4; **(d)** XeF_5^+.

★57. Use VSEPR theory to predict a plausible structure for XeF_6, and comment on the difficulty in applying the valence bond method of page 861 to a description of this structure.

★58. The bond energies of Cl_2 and F_2 are 243 and 155 kJ/mol, respectively. Use these data to explain why XeF_2 is a much more stable compound than $XeCl_2$. [*Hint:* Recall that Xe exists as a monatomic gas.]

★59. Write plausible half-equations and a balanced oxidation–reduction equation for the disproportionation of XeF_4 to Xe and XeO_3 in aqueous acidic solution. Xe and XeO_3 are produced in a $2:1$ mol ratio and O_2(g) is also produced.

Additional Exercises

60. Supply a name for each of the following. **(a)** $H_2S_2O_8$; **(b)** K_2HPO_4; **(c)** $Sr(ClO_4)_2$; **(d)** HIO_3; **(e)** BaO_2; **(f)** $Mg_2P_2O_7$; **(g)** $Hg(SCN)_2$; **(h)** MgTe; **(i)** $Ba(N_3)_2$; **(j)** $K_2S_3O_6$; **(k)** As_2S_3; **(l)** CSe_2; **(m)** $Pb_3(AsO_4)_2$; **(n)** $Ag_2S_2O_3$.

61. Supply a formula for each of the following. **(a)** magnesium nitrite; **(b)** bromine pentafluoride; **(c)** calcium hydrogen sulfide; **(d)** potassium cyanate; **(e)** phosphorus dichloride trifluoride; **(f)** hydrazinium chloride; **(g)** lithium dithionate; **(h)** calcium telluride; **(i)** iron(II) orthophosphate; **(j)** lead(II) metaarsenite; **(k)** calcium telluride; **(l)** copper(I) azide.

62. Complete and balance each of the following equations. If no reaction occurs, so state.
(a) $KI(s) + H_3PO_4(conc. aq) \rightarrow$
(b) $NaClO_3(s) \xrightarrow{\Delta}$
(c) $K_2O_2(s) + H_2O \rightarrow$
(d) $CuS(s) + HCl(aq) \rightarrow$
(e) $I_2(s) + KI(aq) \rightarrow$
(f) $SO_3^{2-}(aq) + MnO_4^-(aq) + H^+(aq) \rightarrow$
(g) $Br_2(l) + Cl^-(aq) \rightarrow$

63. Write equations to show how H_2O_2 **(a)** oxidizes NO_2^- to NO_3^- in acidic solution; **(b)** oxidizes SO_2(g) to SO_4^{2-} in basic solution; **(c)** reduces MnO_4^- to Mn^{2+} in acidic solution; **(d)** reduces Cl_2(g) to Cl^- in basic solution.

64. From the description given, write plausible equations for the preparation of the halogen compound indicated.
(a) $TiCl_4$, by the high-temperature reaction of TiO_2(s) with C(s) and Cl_2(g).
(b) ClO_2, by the disproportionation of $HClO_3$, with $HClO_4$ and H_2O as the other products.
(c) O_2SF_2, by the reaction of sulfur hexafluoride and sulfur trioxide.

65. In 1986, F_2 was first prepared by a *chemical* method (that is, not involving electrolysis). The reactions used were that of hexafluoromanganate(IV) ion, MnF_6^{2-}, with antimony pentafluoride to produce manganese(IV) fluoride and SbF_6^-, followed by the disproportionation of manganese(IV) fluoride to manganese(III) fluoride and fluorine gas. Write chemical equations for these two reactions.

66. The following properties of polonium have been measured or estimated: **(a)** covalent radius; **(b)** ionic radius; **(c)** first ionization energy; **(d)** electronegativity. Based on periodic relationships and data in Table 23-5, what values would you expect for these properties?

67. Although relatively rare, all of the following compounds exist. Based on what you know about related compounds (e.g., from the periodic table), propose a plausible name or formula for each compound.

    (a) silver astatide   **(b)** CSe_2
    (c) magnesium polonide   **(d)** H_2TeO_3
    (e) potassium thioselenate **(f)** $KAtO_4$

68. Polonium is the only element known to crystallize in the simple cubic form. In this structure, the interatomic distance between a Po atom and each of its six nearest neighbors is 335 pm. Use this description of the crystal structure to estimate the density of polonium.

69. Nitramide and hyponitrous acid both have the formula $H_2N_2O_2$. Hyponitrous acid is a weak diprotic acid; nitramide contains the amide group ($-NH_2$). Based on this information write plausible Lewis structures for these two substances.

70. *Peroxonitrous acid* is an unstable intermediate formed in the oxidation of HNO_2 by H_2O_2. It has the same formula as *nitric acid*, HNO_3. Show how you would expect these two acids to differ in structure.

71. The structure of $N(SiH_3)_3$ involves a planar arrangement of N and Si atoms, whereas that of the related compound $N(CH_3)_3$ has a pyramidal arrangement of N and C atoms. Propose bonding schemes for these molecules that are consistent with this observation.

★72. We learned in Chapter 15 that the decomposition of hydrogen peroxide can be catalyzed in various ways.

$$2 H_2O_2(aq) \xrightarrow{\text{catalyst}} 2 H_2O + O_2(g)$$

One mechanism explains the action of certain catalysts through two redox reactions: **(a)** The catalyst is reduced by the H_2O_2 (which itself is oxidized to O_2). **(b)** The catalyst is oxidized back to its initial form by H_2O_2 (which itself is reduced to H_2O). Based on this mechanism, and using appropriate electrode potential data, show that Fe^{3+} should catalyze the decomposition of H_2O_2. [*Hint:* Fe^{3+} is reduced to Fe^{2+} in reaction (a).]

★73. Refer to Exercise 72. What are the minimum and maximum values of $E°$ for the redox reactions involving the catalyst in the decomposition of H_2O_2? [*Hint:* The essential requirement is that *both* reactions (a) and (b) described in Exercise 72 must be spontaneous.]

★74. The text states that in the extraction of bromine from seawater (reaction 23.3), the seawater is first brought to a pH of 3.5

and then treated with $Cl_2(g)$. In actual practice, the pH of the seawater is adjusted with H_2SO_4 and the mass of chlorine used is 15% in excess of the theoretical. Assuming a seawater sample with an initial pH of 7.0, a density of 1.03 g/cm^3, and a bromine content of 70 ppm by mass, what mass of H_2SO_4 and of Cl_2 would be used in the extraction of bromine from 1000 L of seawater? [*Hint:* How would you describe the ionization of H_2SO_4 in a solution with pH = 3.5? That is, does it go to completion?]

★75. The unstable species HO_2 has O in the oxidation state $-\frac{1}{2}$. For the half-reaction

$$O_2 + H^+ + e^- \rightarrow HO_2$$

$E° = -0.131$ V. Reconstruct the electrode potential diagram for oxygen in acidic solution (Figure 23-6) to include the species HO_2 and fill in any missing electrode potential values.

★76. Use the data developed in Exercise 75 to show that HO_2 undergoes spontaneous disproportionation in acidic solution.

★77. Verify the value of $E°_{cell}$ (+2.05 V) given for reaction (23.63). [*Hint:* Use data from equation (23.62), Table 21-1, Table 17-2, and K_w.]

★78. The total solubility of $Cl_2(g)$ in water is 6.4 g/L at 25 °C. At this temperature the hydrolysis reaction

$$Cl_2(aq) + H_2O \rightleftharpoons HOCl(aq) + H^+(aq) + Cl^-(aq)$$

has a value of $K_c = 4.4 \times 10^{-4}$. For a saturated aqueous solution of Cl_2 in water, calculate $[Cl_2]$, $[HOCl]$, $[H^+]$, and $[Cl^-]$.

★79. The concentration of a saturated solution of I_2 in water is 1.33×10^{-3} M. Also

$$I_2(aq) \rightleftharpoons I_2(CCl_4) \qquad K = \frac{[I_2]_{CCl_4}}{[I_2]_{aq}} = 85.5$$

A 10.0-mL sample of saturated $I_2(aq)$ is shaken with 10.0 mL CCl_4. After equilibrium is established, the two liquid layers are separated.

    (a) What mass of I_2, in mg, remains in the water layer?
    (b) If the 10.0-mL water layer in (a) is extracted with a second 10.0-mL portion of CCl_4, what will be the number of mg I_2 remaining in the water?
    (c) If the 10.0-mL sample of saturated $I_2(aq)$ had originally been extracted with 20.0 mL CCl_4, would the quantity of I_2 remaining in the aqueous solution have been less than, equal to, or greater than in part (b)? Explain.

★80. Use data from Example 23-3 to calculate the % dissociation of HI(g) into its elements at 298 K.

★81. Estimate the % dissociation of $Cl_2(g)$ into Cl(g) at 1 atm total pressure and 1000 K. Use data from Appendix D and equations established elsewhere in the text, as necessary.

Self-Test Questions

For questions 82 through 91 select the single item that best completes each statement.

82. To displace Br_2 from an aqueous solution containing Br^-, add (a) $I_2(aq)$; (b) $Cl_2(aq)$; (c) $Cl^-(aq)$; (d) $I_3^-(aq)$.

83. All of the following compounds yield $O_2(g)$ when heated strongly (e.g., to about 1000 K) except (a) $KClO_3$; (b) HgO; (c) N_2O; (d) $CaCO_3$.

84. The expected gaseous product when Cu is dissolved in concentrated $HNO_3(aq)$ is (a) NO_2; (b) H_2; (c) N_2; (d) NH_3.

85. All of the following are bases except (a) N_2H_4; (b) NH_2OH; (c) NH_3; (d) HN_3.

86. To dissolve mercuric sulfide, HgS ($K_{sp} = 1.6 \times 10^{-52}$) use (a) $HNO_3(aq)$; (b) HCl(aq); (c) a mixture of $HNO_3(aq)$ and HCl(aq); (d) NaOH(aq).

87. The best reducing agent of the following is (a) H_2S; (b) $Cl^-(aq)$; (c) $SO_4^{2-}(aq)$; (d) O_3.

88. All of the following have a tetrahedral shape except (a) SO_4^{2-}; (b) XeF_4; (c) ClO_4^-; (d) XeO_4.

89. The term "thio" is used in the names of all of the following compounds except (a) $Na_2S_2O_3$; (b) $NaCS_3$; (c) $NaSCN$; (d) Na_2SO_3.

90. An example of an "ortho" acid is (a) $HAsO_2$; (b) H_3AsO_4; (c) $(HPO_3)_n$; (d) $H_4As_2O_7$.

91. The term that best describes I_3^- is (a) polyhalide ion; (b) interhalogen compound; (c) pseudohalogen; (d) oxoanion.

92. Write chemical equations to represent
(a) the thermal decomposition of $Pb(NO_3)_2(s)$;
(b) the reaction of $Cl_2(g)$ with cold $NaOH(aq)$;
(c) the neutralization of $H_3PO_4(aq)$ to the second equivalence point with $KOH(aq)$;
(d) the action of hot concentrated $H_2SO_4(aq)$ on $KBr(s)$;
(e) the formation of pyrophosphoric acid from orthophosphoric acid.

93. Use principles from this chapter and elsewhere in the text to explain why
(a) Ag will dissolve in HNO_3(concd aq) but not in HCl(concd aq).
(b) I_2 is much more soluble in $KI(aq)$ than it is in pure water.
(c) H_2S, which has nearly twice the molecular weight of H_2O, is a gas at room temperature whereas H_2O is a liquid.

94. The abundance of F^- in seawater is 1 g F^- per ton of seawater. Suppose that a commercially feasible method could be found to extract fluorine from seawater.
(a) What mass of F_2 could be obtained from 1 km^3 of seawater ($d = 1.03$ g/cm^3)?
(b) Do you think the process would resemble that for extracting bromine from seawater? Explain.

95. A portion of the standard electrode (reduction) potential diagram of selenium is given below. What is the $E°$ value for the reduction of H_2SeO_3 to H_2Se?

$$SeO_4^{2-} \xrightarrow{1.15\ V} H_2SeO_3 \xrightarrow{0.74\ V} Se \xrightarrow{-0.35\ V} H_2Se$$
$$\underset{(?)}{\underline{\hspace{6cm}}}$$

24 The Transition Elements

Production of steel in an electric arc furnace. The principal element in steel is Fe, and important minor elements include V, Cr, Mn, Co, Ni, Mo, and W. All are transition elements. [Courtesy Union Carbide Corporation]

Elements in the main transition series—the *d*-block elements—have atoms or ions with partially filled *d* orbitals. In the inner-transition series—the *f*-block elements—atoms or ions have partially filled *f* orbitals. All the elements in the middle section of the periodic table fit one or another of these descriptions. More than half the elements belong either to a transition or inner-transition series.

The chemistry of these elements has both theoretical and practical significance. In their ability to form complex ions, to catalyze chemical reactions, and to display certain types of magnetism, the transition elements offer insights into the fundamental nature of these phenomena. At the same time, these elements provide our main structural metal (Fe); important alloying metals in the manufacture of steel (V, Cr, Mn, Co, Ni, Mo, W); our best electrical conductors (Ag, Cu); the primary metallic constituents of paint pigments (Ti, Fe, Cr); the element essential to photography (Ag); and specialized materials for modern applications like color television screens (lanthanide oxides).

We begin with a survey of the general properties of transition elements. This is followed by discussions—some brief and some more lengthy— of the individual members of the first transition series. We continue our discussion of metallurgy, begun in Chapter 22, by considering the production of iron and steel. We also look at the metallurgy of copper, silver, and gold, together with other aspects of their chemical behavior. We close the chapter with a discussion of the qualitative analysis of some of the cations of the first transition series.

An important characteristic of the transition elements, referred to in this chapter and explored more fully in the next, is the ability to form complex ions.

24-1 General Properties

Figure 24-1 shows the *d*-block (tan) and *f*-block (gray) elements. These are the types of elements that are discussed in this chapter. The transition elements are often referred to simply as the *d*-block and *f*-block elements. A more exact definition, however, is that a transition element must have partially filled *d* or *f* orbitals in *either* its free (uncombined) atoms *or* one or more of its ions. In applying this more exact definition, we see that copper, in group 1B, does not have *d* orbital vacancies in the free atom nor in Cu^+. It does have *d*-orbital vacancies in the Cu^{2+} ion, however. Copper and the other members of group 1B are transition elements. Zinc, on the other hand, does not have *d*-orbital vacancies in either the free atom or in Zn^{2+}. The group 2B elements—Zn, Cd, and Hg—are representative metals and were discussed in Chapter 22.

The properties listed in Table 24-1 for elements in the first transition series ($Z = 21$ to $Z = 29$) are clearly those of a group of metallic elements. High melting points, good electrical conductivity, and moderate to extreme hardness result from the ready availability of electrons and orbitals for metallic bonding.

Atomic Radii. In a transition series the chief difference in atomic structure between successive elements involves one unit of positive charge on the nucleus and one electron in an orbital of an *inner* electronic shell. This is not a major difference and does not cause much of a change in atomic size, especially in the middle of a series. As a result, we see some similarities among *horizontal* groupings of elements. For example, the three elements, Fe, Co, and Ni—*the iron triad*—are usually discussed as a single group (see Section 24-7).

Lanthanide Contraction. When an element in the first transition series is compared with those of the second and third series within the same group, important

Zn, Cd, and Hg, because they are *d*-block elements, and because in some ways they resemble transition elements, are often considered together with the transition elements.

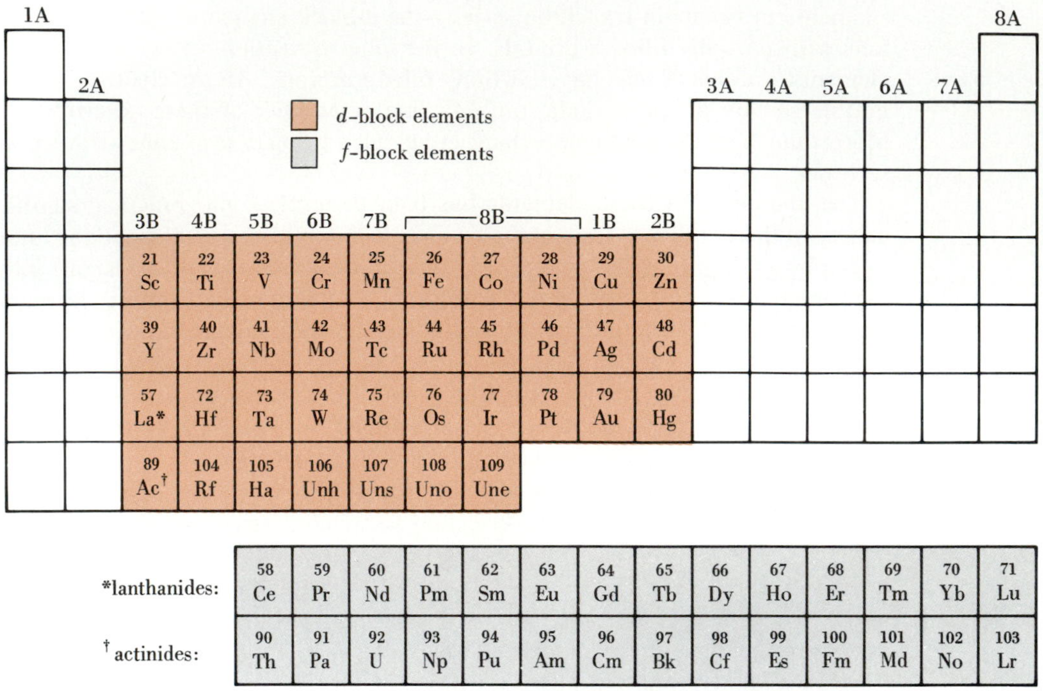

FIGURE 24-1
Locating the transition
elements.

differences appear. Table 24-2 lists representative data for the members of group 6B—Cr, Mo, and W. The most notable feature of the table is that the atomic (metallic) radii of Mo and W are the same. Along with the usual filling of s, p, and d subshells in the interval of elements separating Mo and W, the $4f$ subshell is also filled. Electrons in an f subshell are not very effective in screening outer-shell electrons from the nucleus. As a result, these outer-shell electrons are held more tightly than we would otherwise expect. Atomic size does not increase as expected. In fact, in the series of elements in which the $4f$ sublevel is filled, atomic sizes

TABLE 24-1
Selected Properties of Elements of the First Transition Series

	Sc	Ti	V	Cr	Mn	Fe	Co	Ni	Cu
atomic number	21	22	23	24	25	26	27	28	29
electron config.[a]	$3d^14s^2$	$3d^24s^2$	$3d^34s^2$	$3d^54s^1$	$3d^54s^2$	$3d^64s^2$	$3d^74s^2$	$3d^84s^2$	$3d^{10}4s^1$
metallic radius, pm	161	145	132	127	124	124	125	125	128
ioniz. energy, kJ/mol									
first	631	658	650	653	718	759	758	737	746
second	1235	1310	1414	1592	1509	1561	1646	1753	1958
third	2389	2653	2828	2987	3249	2957	3232	3394	3554
electrode potential,[b] V	−2.08	−1.63	−1.18	−0.91	−1.19	−0.44	−0.28	−0.23	+0.34
common oxidation states	3	2, 3, 4	2, 3, 4, 5	2, 3, 6	2, 3, 4, 7	2, 3	2, 3	2	1, 2
m.p., °C	1397	1672	1710	1900	1244	1530	1495	1455	1083
density, g/cm³	2.99	4.49	5.96	7.20	7.20	7.86	8.90	8.91	8.92
hardness[c]	—	—	—	9.0	5.0	4.5	—	—	—
electrical conductivity[d]	—	2	3	10	2	17	24	24	97

[a]Each atom has an argon inner core configuration.
[b]For the reduction process $M^{2+}(aq) + 2 e^- \rightarrow M(s)$ [except for scandium, where the ion is $Sc^{3+}(aq)$].
[c]Hardness values are on the Mohs scale (see Table 14-4).
[d]Compared to an arbitrarily assigned value of 100 for silver.

TABLE 24-2
Some Properties of Group 6B—Cr, Mo, W

Transition series	Element	Atomic number	Electron configuration	Atomic radius, pm	Standard electrode potential,[a] V	Oxidation states[b]
first	Cr	24	$[Ar]3d^54s^1$	127	-0.744	2, 3, 6
second	Mo	42	$[Kr]4d^55s^1$	139	-0.20	2, 3, 4, 5, 6
third	W	74	$[Xe]4f^{14}5d^46s^2$	139	-0.11	2, 3, 4, 5, 6

[a]For the reduction process $M^{3+} + 3 e^- \rightarrow M(s)$.
[b]The most common oxidation state(s) is shown in red.

The terms "lanthanide" and "lanthanoid" are used interchangeably.

actually decrease. This phenomenon occurs in the lanthanide series ($Z = 58$ to 71), so the phenomenon is called the **lanthanide contraction.** One of the ways in which Cr differs from Mo and W is that Cr can exist in aqueous solution as the simple hydrated ions $Cr^{2+}(aq)$ and $Cr^{3+}(aq)$, whereas the ionic forms of Mo and W are polyatomic (oxoanions). Also, the oxidation state $+3$ is very common for Cr, whereas higher oxidation states ($+5$, $+6$) are favored with Mo and W.

Electron Configurations and Oxidation States. All the elements of the first transition series have an electron configuration with the following characteristics.

- an inner core of electrons in the argon configuration;
- two electrons in the $4s$ orbital for seven members and one $4s$ electron for the remaining two (Cr and Cu);
- a number of electrons in $3d$ orbitals ranging from one in Sc to ten in Cu.

Sc has the electron configuration $[Ar]3d^14s^2$. In combination with nonmetals, the loss of three electrons by Sc atoms is energetically favorable and Sc^{3+} ions are formed. Ti atoms, with the electron configuration $[Ar]3d^24s^2$, can use four electrons in compound formation and display the oxidation state $+4$. It is also possible for Ti atoms to use a smaller number of electrons, such as through the loss of the $4s^2$ electrons to form the ion Ti^{2+}. With Ti, then, we note two features: (1) a variety of possible oxidation states and (2) a maximum oxidation state corresponding to the group number 4B. These two features continue with V, Cr, and Mn, for which the maximum oxidation states are $+5$, $+6$, and $+7$, respectively. A shift in behavior is noted in group 8B, however. Although Fe, Co, and Ni can all exist in more than one oxidation state, they do not display the variety found in the earlier members of the first transition series. Neither do they exhibit a maximum oxidation state corresponding to their group number. As we traverse the first transition series, the nuclear charge, number of d electrons, and energy requirement for the successive ionization of d electrons all increase. The use of a large number of d electrons becomes energetically unfavorable, and only the lower oxidation states are commonly encountered. Because of its single $4s$ electron in the configuration $[Ar]3d^{10}4s^1$, copper can exhibit an oxidation state of $+1$ as well as $+2$. These ideas on the variability of oxidation states of the transition elements are illustrated through Figure 24-2.

Although the transition elements display a variety in their oxidation states, the possible oxidation states differ in the ease with which they can be attained and in their stabilities. These stabilities depend on a number of factors—other atoms to which the transition metal is bonded, whether the compound is in crystalline form or in solution, the pH of an aqueous solution, etc. For example, $TiCl_2$ is a well-characterized compound, but the ion Ti^{2+} is unstable in aqueous solution; $Co^{3+}(aq)$ is unstable, but can be stabilized in complex ions with the appropriate ligands. Generalizations are difficult, beyond noting that stable higher oxidation states are gener-

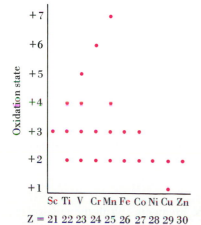

FIGURE 24-2
Oxidation states of the elements of the first transition series (plus Zn).

The common oxidation states are represented here. Additional, less common oxidation states are also found for some of these elements.

ally found for oxoanions, fluorides, and oxofluorides. (Stabilization of higher oxidation states by F is probably due to the combined effects of the weak F—F bond, the high electronegativity of fluorine, and the small size of F^-, which lead to high lattice energies.) In Table 24-1 and elsewhere, when the term "common" is used to describe an oxidation state, this can be taken to mean an oxidation state found in a number of compounds and/or in aqueous solution.

Ionization Energies and Electrode Potentials. Ionization energies are fairly constant across the first transition series of elements. Values of the first ionization energies are about the same as for the group 2A metals. Standard electrode potentials increase in value gradually across the transition series. However, with the exception of the oxidation of Cu to Cu^{2+}, all these elements are more readily oxidized than hydrogen. This means that they displace $H^+(aq)$, forming $H_2(g)$.

Ionic and Covalent Compounds. We tend to think of metals as forming ionic compounds with nonmetals. This is certainly the case with 1A metal compounds and most 2A metal compounds. On the other hand, we have seen that some metal compounds have significant covalent character, e.g., $BeCl_2$ and $AlCl_3$ (Al_2Cl_6). The transition metal compounds display both ionic and covalent character. In general, compounds with the transition metal in lower oxidation states are essentially ionic, and those in higher oxidation states (such as $TiCl_4$) have covalent character.

Color and Magnetism. As explained more fully in Section 25-7, electronic transitions that occur within partially filled d subshells impart color to solid transition metal compounds and their solutions. Absence of these transitions, in turn, accounts for the fact that so many representative metal compounds are without color. Because individual orbitals in d subshells are half-filled before electron pairing begins, many transition metal compounds are paramagnetic. A special form of magnetism is displayed by Fe, Co, Ni, and some of their alloys (see Section 24-7).

Catalytic Activity. Another consequence of d orbital availability in transition metals and transition metal compounds is their catalytic activity. An unusual ability to adsorb gaseous species makes the transition metals (e.g., Ni and Pt) good heterogeneous catalysts. The possibility of multiple oxidation states accounts for the catalytic effect of some transition metal ions on certain oxidation–reduction reactions. Many homogeneous chemical reactions seem to involve the exchange of ligands in complex ions, and complex ion formation is one of the distinctive features of the transition metals.

In Table 15-8 we noted the catalytic activity of Ni, Fe, Pt, Rh, Cr_2O_3, V_2O_5, and the titanium halides. Further examples in this chapter include $TiCl_4$, V_2O_5, and MnO_2. Catalysis is a key phenomenon in about 90% of all chemical manufacturing processes, and the transition elements are often the key elements in the catalysts used.

Comparison of Transition and Representative Elements. With the representative elements the s and p orbitals of the outermost electronic shell are the most important in establishing the nature of chemical bonding. Participation by d orbitals is essentially nonexistent for the second period elements and the group 1A and 2A metals and of limited importance in the heavier nonmetals. With the transition elements d orbitals are of primary importance in chemical bond formation and s and p orbitals are of less consequence. Most of the observed behavioral differences between the transition and representative elements—multiple versus single oxidation states, complex ion formation, color, magnetic properties, and catalytic activity—can be traced to this issue of which orbitals are most involved in bond formation.

24-2 Scandium

3B	4B	5B	6B	7B	⌐──8B──⌐			1B	2B
21 Sc	22 Ti	23 V	24 Cr	25 Mn	26 Fe	27 Co	28 Ni	29 Cu	30 Zn

Scandium is a rare element. It has been estimated to constitute only from 5 to 30 ppm of the earth's crust. Its principal mineral form is *thortveitite*, $ScSi_2O_7$. However, most scandium produced is extracted from uranium ores, where it occurs to the extent of only about 0.01% Sc, by mass. The commercial uses of scandium are very limited, and its production is measured in gram or kilogram quantities, not tonnages. One recent use has been as a component in high-intensity lamps. The pure metal is usually prepared by the electrolysis of a fused mixture of $SrCl_3$ with other chlorides.

Because of its noble gas electron configuration, the Sc^{3+} ion lacks some of the characteristic properties of transition metal ions. For instance, the ion is colorless and diamagnetic, as are most of its salts. In its chemical behavior, Sc^{3+} most closely resembles Al^{3+}, as in the

- hydrolysis of $[Sc(H_2O)_6]^{3+}(aq)$ to yield acidic solutions;
- formation of an amphoteric gelatinous hydroxide, $Sc(OH)_3$;
- formation in alkaline solution of such ions as $[Sc(H_2O)_2(OH)_4]^-$;
- formation of a fluoro compound, Na_3ScF_6 (similar to cryolite, Na_3AlF_6), in which Sc_2O_3 can be dissolved and electrolyzed to produce metallic Sc.

24-3 Titanium

3B	4B	5B	6B	7B	⌐──8B──⌐			1B	2B
21 Sc	22 Ti	23 V	24 Cr	25 Mn	26 Fe	27 Co	28 Ni	29 Cu	30 Zn

Titanium is the ninth most abundant element, comprising 0.6% of the earth's crust. Three physical properties underlie most current uses of the metal:

- low density,
- high structural strength,
- corrosion resistance.

The first two properties account for its extensive use in the aircraft industry, and the third for its applications in the chemical industry, where it is used in pipes, component parts of pumps, and reaction vessels.

Compounds of Titanium. Titanium tetrachloride, $TiCl_4$, is one of the most important compounds of titanium. It is the starting material for preparing other Ti compounds; it plays a central role in the metallurgy of titanium; it is used in formulating catalysts for the production of polyethylene and other plastics. The usual method of preparing $TiCl_4$ involves reaction of the naturally occurring ore **rutile** (TiO_2) with carbon and $Cl_2(g)$.

$$TiO_2(s) + 2\ C(s) + 2\ Cl_2(g) \xrightarrow{\Delta} TiCl_4(g) + 2\ CO(g) \tag{24.1}$$

$TiCl_4$ is a colorless liquid (m.p. -24 °C; b.p. 136 °C). Molecules of $TiCl_4$ have a tetrahedral shape with Cl—Ti—Cl bond angles of 109.5°. The hydrolysis of $TiCl_4$,

$$TiCl_4(l) + 2\ H_2O \longrightarrow TiO_2(s) + 4\ HCl(aq) \tag{24.2}$$

when carried out in moist air, is the basis of certain types of smoke grenades.

In the +4 oxidation state all the valence-shell electrons of Ti atoms are employed in bond formation. In this oxidation state Ti bears a strong resemblance to the group 4A elements. The physical properties and molecular shape of $TiCl_4$ are similar to those of CCl_4 and $SiCl_4$. $SiCl_4$ also fumes in moist air in a reaction like (24.2).

To produce pure TiO_2, **titanium dioxide,** a gaseous mixture of $TiCl_4$ and O_2 is passed through a silica tube at about 700 °C.

$$TiCl_4(g) + O_2(g) \xrightarrow{\Delta} TiO_2(s) + 2\ Cl_2(g) \tag{24.3}$$

TiO_2 is amphoteric, though it is actually not very soluble in either acids or alkalis. It dissolves slowly in hot concentrated $H_2SO_4(aq)$, from which a sulfate can be crystallized. With molten alkalis it forms *titanates,* such as K_2TiO_3.

Because of its whiteness, opacity, inertness, nontoxicity, and relative cheapness, TiO_2 is now the most widely used white pigment for paints. In this use TiO_2 has largely displaced toxic basic lead carbonate (see page 817). TiO_2 is also used as a paper whitener and in glass, ceramics, floor coverings, and cosmetics.

Metallurgy of Titanium. Extensive production of Ti is a recent development, spurred at first by the needs of the military and then by the aircraft industry. Ti is a good alternative to Al and steel in aircraft because Al loses its strength at high temperature and steel is too dense.

The first step in the production of Ti is the conversion of rutile ore to $TiCl_4$ (reaction 24.1). The purified $TiCl_4$ is next reduced to Ti by using a good reducing agent. The **Kroll process** uses Mg.

The Kroll process for producing titanium.

$$TiCl_4(g) + 2\ Mg(l) \xrightarrow[\text{He}]{850\ °C} Ti(s) + 2\ MgCl_2(l) \qquad (24.4)$$

The reaction is carried out in a steel vessel. The $MgCl_2(l)$ is removed and electrolyzed to produce Mg and Cl_2, both of which are recycled (reactions 24.1 and 24.4). The Ti is obtained as a sintered (fused) mass called titanium sponge. The sponge must be subjected to further treatment and alloying with other metals before it can be used. One of the most difficult problems in the commercial development of Ti was that of devising new metallurgical techniques for fabricating the metal. These techniques required working with powdered solid metals rather than a molten metal. This field of metallurgy has become known as **powder metallurgy.**

In 1947, the United States production of Ti was only 2 tons. Today, it exceeds 50,000 tons. Now there is concern that continued widespread use of the metal will deplete the world's known reserves of rutile (estimated to be equivalent to 10 to 70 million tons of Ti). Fortunately, titanium and its compounds can be produced from **ilmenite** ($FeTiO_3$), although not as cheaply as from rutile. With titanium and its oxide we see another example of how materials practically unknown in one decade may become major production items in the next.

Example 24-1

Writing a chemical equation from a description of a chemical reaction. The conversion of ilmenite ore, $FeTiO_3$, to $TiCl_4$ is carried out in a similar manner

Vacuum-distilled metallic titanium sponge produced by the Kroll process. [Courtesy Teledyne Wah Chang, Albany, OR]

to reaction (24.1). That is, the ore is heated with carbon and chlorine. Iron(III) chloride and carbon monoxide are also produced. Write a balanced equation for this reaction.

Solution. As in several other cases in the preceding two chapters, this type of problem requires you to identify all the reactants and products, substitute correct formulas for names, and balance the equation. The reactants are $FeTiO_3$, C, and Cl_2. The products are $TiCl_4$, $FeCl_3$, and CO.

$$FeTiO_3 + C + Cl_2 \xrightarrow{\Delta} TiCl_4 + FeCl_3 + CO \quad \text{(unbalanced)}$$

$$FeTiO_3 + 3\ C + \tfrac{7}{2}\ Cl_2 \xrightarrow{\Delta} TiCl_4 + FeCl_3 + 3\ CO \quad \text{(balanced)}$$

or

$$2\ FeTiO_3 + 6\ C + 7\ Cl_2 \xrightarrow{\Delta} 2\ TiCl_4 + 2\ FeCl_3 + 6\ CO \quad \text{(balanced)}$$

SIMILAR EXAMPLES: Exercises 6, 37, 52.

24-4 Vanadium

3B	4B	5B	6B	7B		8B		1B	2B
21	22	23	24	25	26	27	28	29	30
Sc	Ti	V	Cr	Mn	Fe	Co	Ni	Cu	Zn

Vanadium is a fairly abundant element (0.02% of the earth's crust) and is found in several dozen ores. Its principal ores are rather complex, such as *vanadinite*, $3Pb_3(VO_4)_2 \cdot PbCl_2$. The metallurgy of vanadium is not simple, but vanadium of high purity (99.99%) is obtainable. For most of its applications, though, V is prepared as an iron–vanadium alloy, **ferrovanadium,** containing from 35 to 95% V. Ferrovanadium is produced by reducing V_2O_5 with silicon in the presence of iron. SiO_2 combines with CaO to form a liquid slag of calcium silicate.

$$2\ V_2O_5 + 5\ Si\ (+\ Fe) \xrightarrow{\Delta} 4\ V\ (+\ Fe) + 5\ SiO_2 \qquad (24.5)$$
$$\text{ferrovanadium}$$

$$SiO_2(s) + CaO(s) \xrightarrow{\Delta} CaSiO_3(l) \qquad (24.6)$$

About 80% of the vanadium produced is used in the manufacture of steel. Vanadium-containing steels are used in applications requiring strength and toughness, such as in springs and high-speed machine tools.

Vanadium Oxides. As outlined in Table 24-3, in each of its oxidation states vanadium forms an oxide. These oxides behave in accordance with the factors we discussed in Section 17-9: If the metal is in a low oxidation state (having a low charge density) the oxide acts as a base. With a higher oxidation state (and charge density) on the central atom, acidic properties become important. In the case of vanadium, the oxides having V in the oxidation states +2 and +3 are basic; in the +4 and +5 oxidation states, the oxides are amphoteric.

The most important oxide is V_2O_5, and its most important use is as a catalyst, such as in the conversion of $SO_2(g)$ to $SO_3(g)$ in the contact method for the manufacture of sulfuric acid (recall Focus feature of Chapter 15). The activity of V_2O_5 as an oxidation catalyst may be linked to the reversible loss of oxygen that occurs from 700 to 1100 °C.

Oxidation States. Figure 24-3 summarizes oxidation–reduction relationships among some of the species listed in Table 24-3. In general, compounds with vana-

TABLE 24-3
Characteristics of Oxides and Some Ions of Vanadium

O.S.	Oxide	Behavior	Ion[a]	Name of ion	Color of ion
+2	VO	basic	V^{2+}	vanadium(II) (vanadous)	violet
+3	V_2O_3	basic	V^{3+}	vanadium(III) (vanadic)	green
+4	VO_2	amphoteric	VO^{2+}	oxovanadium(IV) (vanadyl)	blue
			[b]	hypovanadate (vanadite)	brown
+5	V_2O_5	amphoteric	VO_2^+	dioxovanadium(V)[c]	yellow
			VO_4^{3-}	orthovanadate[d]	colorless

[a]Some of these ions are hydrated in aqueous solution, specifically, $[V(H_2O)_6]^{2+}$, $[V(H_2O)_6]^{3+}$, $[VO(H_2O)_4]^{2+}$, and $[VO_2(H_2O)_4]^+$.

[b]There is no simple anionic species of vanadium(IV). One formulation of this ion is $V_4O_9^{2-}$.

[c]This ion is obtained only in strongly acidic solutions (pH <1.5).

[d]Orthovanadates can be obtained only in highly alkaline solutions (pH >13). At lower pH values the anionic species are more complex, e.g., pyrovanadates ($V_2O_7^{4-}$) from pH 10 to 13 and metavanadates, $(VO_3^-)_n$, from pH 7 to 10.

FIGURE 24-3
Electrode potential diagram for vanadium.

$$+5 \xrightarrow{+1.00 \text{ V}} +4 \xrightarrow{+0.361 \text{ V}} +3 \xrightarrow{-0.255 \text{ V}} +2 \xrightarrow{-1.18 \text{ V}} 0$$

$$VO_2^+(aq) \longrightarrow VO^{2+}(aq) \longrightarrow V^{3+}(aq) \longrightarrow V^{2+}(aq) \longrightarrow V(s)$$

(yellow)　　　　(blue)　　　　(green)　　　　(violet)

dium in its highest oxidation state (+5) are good oxidizing agents; in its lowest oxidation state, vanadium (as V^{2+}) is a good reducing agent.

Example 24-2

Using electrode potential data to predict an oxidation–reduction reaction.
We wish to oxidize $VO^{2+}(aq)$ to $VO_2^+(aq)$ in an acidic solution. Can $MnO_4^-(aq)$ be used as the oxidizing agent? If so, write a balanced equation for the redox reaction.

Solution. We should start by writing two half-equations, one for the oxidation of VO^{2+} to VO_2^+ and the other for the reduction of MnO_4^- to Mn^{2+}. Both half-reactions occur in acidic solution. We can find one $E°$ value in Figure 24-3 and the other in Table 21-1. We can then combine the $E°$ values into $E°_{cell}$.

Oxid: $5 \{VO^{2+} + H_2O \longrightarrow VO_2^+ + 2 H^+ + e^-)$　　　$-E° = -1.00$ V

Red: $MnO_4^- + 8 H^+ + 5 e^- \longrightarrow Mn^{2+} + 4 H_2O$　　　$E° = +1.51$ V

Net: $5 VO^{2+} + MnO_4^- + H_2O \longrightarrow 5 VO_2^+ + Mn^{2+} + 2 H^+$

$$E°_{cell} = +0.51 \text{ V}$$

Since $E°_{cell}$ is positive, we predict that MnO_4^- should oxidize VO^{2+} to VO_2^+ in acidic solution.

SIMILAR EXAMPLES: Exercises 18, 21, 33, 57.

24-5 Chromium

3B	4B	5B	6B	7B	8B			1B	2B
21 Sc	22 Ti	23 V	24 Cr	25 Mn	26 Fe	27 Co	28 Ni	29 Cu	30 Zn

Although it is found only to the extent of 122 ppm in the earth's crust, chromium is one of the most important industrial metals. The principal ore is **chromite,** $Fe(CrO_2)_2$, from which an alloy of Fe and Cr called **ferrochrome** is obtained by the reduction

Reduction of chromite ore.

$$Fe(CrO_2)_2 + 4\,C \xrightarrow{\Delta} \underbrace{Fe + 2\,Cr}_{ferrochrome} + 4\,CO(g) \qquad (24.7)$$

Ferrochrome may be added directly to iron, together with other metals, to produce steel. Chromium metal is very hard and maintains a bright surface through the protective action of an invisible oxide coating. It is extensively used in plating other metals.

Chrome Plating. Chrome plating of steel is accomplished from an aqueous solution containing CrO_3 and H_2SO_4 in a mass ratio of about $100:1$. The plating obtained is thin and porous and tends to develop cracks. Usual practice is first to plate the steel with copper or nickel, which is the true protective coating. Then chromium is plated over this for decorative purposes. The technical art of chrome plating is well understood, but the mechanism of the electrodeposition has not been established. The efficiency of chrome plating is limited by the fact that reduction of Cr(VI) to Cr(0) produces only $\frac{1}{6}$ mol Cr per faraday: Large quantities of electric energy are required for chrome plating relative to other types of metal plating.

Oxides of Chromium. The oxides of chromium (like those of vanadium) illustrate the general principles of acid–base behavior of element oxides discussed in Section 17-9. Their behavior is summarized in Table 24-4. The amphoterism of $Cr(OH)_3$ is pictured in Figure 24-4 and can be represented by a sequence of reactions similar to that used for Al in expression (22.28), that is, starting with the ionization of hydrated $Cr^{3+}(aq)$.

Acid ionization of $[Cr(H_2O)_6]^{3+}$.

$$[Cr(H_2O)_6]^{3+}(aq) + H_2O \rightleftharpoons [Cr(H_2O)_5OH]^{2+}(aq) + H_3O^+(aq) \qquad (24.8)$$

TABLE 24-4
Characteristics of Oxides and Some ions of Chromium

O.S.	Oxide[a]	Hydroxo compound	Behavior	Ion	Name of ion	Color of ion
+2	CrO	$Cr(OH)_2$	basic	Cr^{2+b}	chromium(II) (chromous)	light blue
+3	Cr_2O_3	$Cr(OH)_3{}^c$	amphoteric	Cr^{3+d}	chromium(III) (chromic)	violet (green)
				$Cr(OH)_4{}^{-e}$	chromite[f]	green
+6	CrO_3	$H_2CrO_4{}^g$ $H_2Cr_2O_7{}^h$	acidic	$CrO_4{}^{2-}$	chromate	yellow
				$Cr_2O_7{}^{2-i}$	dichromate	orange

[a]The oxide CrO_2 is also well known. It is a ferromagnetic electrical conductor used in high-quality recording tape.
[b]The hydrated ion is $[Cr(H_2O)_6]^{2+}$.
[c]The hydroxide is probably a hydrated oxide, $Cr_2O_3 \cdot xH_2O$.
[d]The hydrated ion $[Cr(H_2O)_6]^{3+}$ is violet. Substitution of other ligands for H_2O molecules causes a change in color (see Section 25-7).
[e]This ion is actually $[Cr(H_2O)_2(OH)_4]^-$ and is also sometimes represented in dehydrated form as $CrO_2{}^-$.
[f]A more systematic name, diaquatetrahydroxochromate(III), is based on complex ion nomenclature (see Section 25-3).
[g]The hydroxo compound is $CrO_2(OH)_2$.
[h]The hydroxo compound is $Cr_2O_5(OH)_2$.
[i]See page 882 for the relationship between $CrO_4{}^{2-}$ and $Cr_2O_7{}^{2-}$.

FIGURE 24-4
Amphoterism of $Cr(OH)_3(s)$.

Freshly precipitated $Cr(OH)_3(s)$ (center) is amphoteric. It dissolves in acidic solution [here, $HNO_3(aq)$] to produce violet $[Cr(H_2O)_6]^{3+}(aq)$ (left). $Cr(OH)_3(s)$ also dissolves in $NaOH(aq)$, forming green $[Cr(OH)_4]^-(aq)$ (right). [Carey B. Van Loon]

Other Compounds of Chromium. Pure chromium dissolves in dilute $HCl(aq)$ or $H_2SO_4(aq)$ to produce Cr^{2+} ion. Nitric acid and other oxidizing agents alter the surface of the metal (perhaps by formation of an oxide coating) and render the metal resistant to further attack—the metal becomes *passive*. The best source of chromium compounds, then, is not the pure metal but alkali metal chromates, which can be obtained from chromite ore by reactions such as

$$4\ Fe(CrO_2)_2 + 8\ Na_2CO_3 + 7\ O_2(g) \xrightarrow{\Delta} 2\ Fe_2O_3 + 8\ Na_2CrO_4 + 8\ CO_2(g)$$
$$(24.9)$$

A pure dichromate can be crystallized from an acidified solution of the chromate.

$$2\ Na_2CrO_4(aq) + H_2SO_4(aq) \longrightarrow Na_2Cr_2O_7(aq) + Na_2SO_4(aq) + H_2O \quad (24.10)$$

Reduction of a dichromate with carbon yields Cr_2O_3.

$$K_2Cr_2O_7 + 2\ C \xrightarrow{\Delta} Cr_2O_3 + K_2CO_3 + CO(g) \tag{24.11}$$

Ammonium dichromate yields Cr_2O_3 simply upon being heated (see Figure 24-5).

$$(NH_4)_2Cr_2O_7(s) \xrightarrow{\Delta} Cr_2O_3(s) + N_2(g) + 4\ H_2O(g) \tag{24.12}$$

In applications where pure Cr is required, $Cr_2O_3(s)$ can be reduced by aluminum in the thermite reaction.

Thermite reaction for the preparation of chromium.

$$Cr_2O_3(s) + 2\ Al(s) \longrightarrow Al_2O_3(s) + 2\ Cr(l) \tag{24.13}$$

Other Cr(III) compounds can be obtained by dissolving $Cr_2O_3(s)$ in acids or bases, or by the reduction of $Cr_2O_7{}^{2-}$, as in the preparation of **chrome alum.**

$$K_2Cr_2O_7(aq) + H_2SO_4(aq) + 3\ SO_2(g) \xrightarrow{\Delta} \underbrace{K_2SO_4(aq) + Cr_2(SO_4)_3(aq)}_{\substack{\text{crystallizes as} \\ KCr(SO_4)_2 \cdot 12H_2O}} + H_2O$$
$$(24.14)$$

Chrome alum is used in tanning hides, dyeing textiles, and in fixing baths in photography.

Chromium(II) compounds can be prepared by the reduction of Cr(III) compounds, with zinc in acidic solution or electrolytically at a lead cathode. The most distinctive feature of chromium(II) compounds is their reducing power.

$$Cr^{3+}(aq) + e^- \longrightarrow Cr^{2+}(aq) \qquad E° = -0.41\ V \tag{24.15}$$

FIGURE 24-5
Decomposition of $(NH_4)_2Cr_2O_7$.

Ammonium dichromate (left) contains both an oxidizing agent, $Cr_2O_7^{2-}$, and a reducing agent, NH_4^+. The products of the reaction between these two ions are $Cr_2O_3(s)$, $N_2(g)$, and $H_2O(g)$. Considerable heat and light are also evolved (center). The product is pure $Cr_2O_3(s)$ (right). [Carey B. Van Loon]

The reverse of (23.15)—oxidation of $Cr^{2+}(aq)$—has $-E° = +0.41$ V. An interesting use of Cr(II) compounds is in purging gases to remove traces of $O_2(g)$, through the reaction

$$4\ Cr^{2+}(aq) + O_2(g) + 4\ H^+(aq) \longrightarrow 4\ Cr^{3+}(aq) + 2\ H_2O \qquad E°_{cell} = +1.64\ V \tag{24.16}$$

Figure 24-6 shows the relationship between $Cr^{2+}(aq)$ and $Cr^{3+}(aq)$.

The Chromate–Dichromate Equilibrium. The red oxide CrO_3 dissolves in water to produce a strongly acidic solution. Although chromic acid H_2CrO_4 might be postulated as a product of the reaction, such a compound has never been isolated in the pure state. The observed reaction is

$$2\ CrO_3(s) + H_2O \longrightarrow 2\ H^+ + Cr_2O_7^{2-}$$

It is possible to crystallize a dichromate salt, such as $Na_2Cr_2O_7$ or $K_2Cr_2O_7$, from a water solution of CrO_3. If the solution is made basic, the color turns from orange to yellow. From basic solutions only chromate salts can be crystallized, for example, Na_2CrO_4 or K_2CrO_4. CrO_3 can be obtained by the action of concentrated sulfuric acid on a chromate or dichromate. This red solid is a powerful oxidizing agent and, in conjunction with concentrated sulfuric acid, is commonly used as a cleaning solution for laboratory glasswater. Its principal mode of action is to oxidize grease.

FIGURE 24-6
Relationship between Cr^{2+} and Cr^{3+}.

The solution on the left, containing blue $Cr^{2+}(aq)$, is prepared by dissolving chromium in HCl(aq). Within minutes, the $Cr^{2+}(aq)$ is oxidized to green $Cr^{3+}(aq)$ by atmospheric oxygen (right). [Carey B. Van Loon]

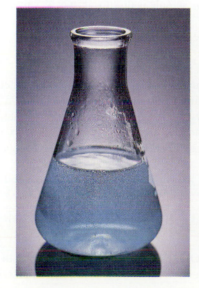

In the oxidation state +6, chromium exists as $Cr_2O_7^{2-}$ in acidic solution (left) and as CrO_4^{2-} in basic solution (right). [Carey B. Van Loon]

Whether a solution contains Cr(VI) as dichromate or chromate ion is a function of pH, since the equilibrium between these ions depends on the concentration of H^+.

Chromate ion–dichromate ion equilibrium.

$$2\ CrO_4^{2-} + 2\ H^+ \rightleftharpoons Cr_2O_7^{2-} + H_2O \tag{24.17}$$

$$K_c = \frac{[Cr_2O_7^{2-}]}{[CrO_4^{2-}]^2[H^+]^2} = 3.2 \times 10^{14} \tag{24.18}$$

$$\frac{[Cr_2O_7^{2-}]}{[CrO_4^{2-}]^2} = 3.2 \times 10^{14}[H^+]^2 \tag{24.19}$$

We can use equation (24.19) to calculate the relative amounts of the two ions as a function of $[H^+]$ (see Exercise 27), but also we can make qualitative predictions with equation (24.17). For this we use Le Châtelier's principle. In acidic solutions the forward reaction of (24.17) is favored and the predominant species in solution is $Cr_2O_7^{2-}$. In basic solution, H^+ ions are removed from the equilibrium in (24.17); the reverse reaction is favored; and the principal species is CrO_4^{2-}. We must exercise careful control over the pH of a solution in which $Cr_2O_7^{2-}$ is to be used as an oxidizing agent or CrO_4^{2-} as a precipitating agent for metal chromates.

$K_2Cr_2O_7$ is one of the most extensively used oxidizing agents in the analytical laboratory. For example, it can be used to determine the quantity of iron present in a sample. The sample is dissolved in acid, any iron that appears as Fe^{3+} is reduced back to Fe^{2+} [using Zn(s) or Sn^{2+}(aq) as a reducing agent], and the following titration is performed.

Titration of Fe^{2+}(aq) with $Cr_2O_7^{2-}$(aq).

$$6\ Fe^{2+} + Cr_2O_7^{2-} + 14\ H^+ \longrightarrow 6\ Fe^{3+} + 2\ Cr^{3+} + 7\ H_2O \tag{24.20}$$

Dichromates are also used as oxidizing agents in chemical industry. In the chrome tanning process, hides are immersed in $Na_2Cr_2O_7$(aq), which is then reduced by SO_2(g) to soluble basic chromic sulfate, $Cr(OH)SO_4$. Collagen, a protein in hides, reacts to form an insoluble complex chromium compound. The hides become tough, pliable, and resistant to biological decay. They are converted to **leather.**

Chromate ion in basic solution is not a good oxidizing agent; it is not readily reduced.

$$CrO_4^{2-}(aq) + 4\ H_2O + 3\ e^- \longrightarrow Cr(OH)_3(s) + 5\ OH^-(aq) \qquad E° = -0.13\ V \tag{24.21}$$

In fact, $Cr(OH)_3$(s) is rather easily oxidized to CrO_4^{2-}(aq); H_2O_2 is a suitable oxidizing agent. With CrO_4^{2-} precipitation reactions are more important than oxi-

FIGURE 24-7
Structures of CrO_4^{2-} and
$Cr_2O_7^{2-}$.

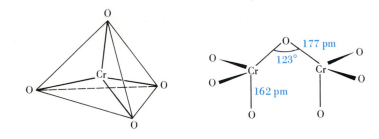

dation–reduction reactions. Insoluble $PbCrO_4$ and $ZnCrO_4$ both find use as yellow pigments.

The structures of the CrO_4^{2-} and $Cr_2O_7^{2-}$ ions are suggested by Figure 24-7. The Cr atom is at the center of a tetrahedron with O atoms at the corners. In $Cr_2O_7^{2-}$ two tetrahedra share an O atom. The Cr—O distance in the Cr—O—Cr link is somewhat greater than the other Cr—O distances.

24-6 Manganese

3B	4B	5B	6B	7B	8B			1B	2B
21	22	23	24	25	26	27	28	29	30
Sc	Ti	V	Cr	Mn	Fe	Co	Ni	Cu	Zn

Like V and Cr, the most important uses of Mn are in steel production, and for this the iron–manganese alloy, **ferromanganese,** is generally used. Ferromanganese is produced by reducing a mixture of iron and manganese oxides with carbon. The principal manganese ore is **pyrolusite,** MnO_2.

$$MnO_2 + Fe_2O_3 + 5\ C \xrightarrow{\Delta} \underbrace{Mn + 2\ Fe}_{ferromanganese} + 5\ CO(g) \qquad (24.22)$$

In steel production, Mn participates in the purification of iron by reacting with sulfur and oxygen and removing them through slag formation. An added function is to increase the hardness of steel. Steel containing high proportions of Mn is extremely tough and wear-resistant and is used in such applications as railroad rails, bulldozers, and road scrapers.

Oxidation States. The electron configuration of Mn is $[Ar]3d^5 4s^2$. By employing first the two $4s$ electrons and then, consecutively, up to all five of its unpaired $3d$ electrons, manganese exhibits all oxidation states from +2 to +7. The important reactions of manganese compounds then are oxidation–reduction reactions, which can be summarized through the electrode potential diagram in Figure 24-8.

FIGURE 24-8
Electrode potential diagram
for manganese.

Acidic solution ([H^+] = 1 M):

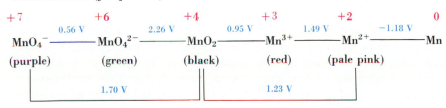

Basic solution ([OH^-] = 1 M):

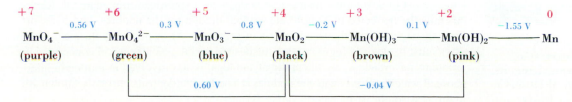

A number of important conclusions can be drawn from Figure 24-8. For example,

Recall that disproportionation refers to a redox reaction in which the same substance undergoes both oxidation and reduction.

1. $Mn^{3+}(aq)$ is unstable. Its disproportionation is spontaneous.

$$2\ Mn^{3+} + 2\ H_2O \longrightarrow Mn^{2+} + MnO_2(s) + 4\ H^+ \qquad E^\circ_{cell} = +0.54\ V$$
(24.23)

2. Manganate ion, MnO_4^{2-}, is unstable in acidic solution. The disproportionation reaction (24.24) is spontaneous.

$$3\ MnO_4^{2-} + 4\ H^+ \longrightarrow MnO_2(s) + 2\ MnO_4^- + 2\ H_2O \qquad E^\circ_{cell} = +1.70\ V$$
(24.24)

3. But MnO_4^{2-} can be obtained in alkaline solution.

$$3\ MnO_4^{2-} + 2\ H_2O \longrightarrow MnO_2(s) + 2\ MnO_4^- + 4\ OH^- \qquad E^\circ_{cell} = +0.04\ V$$
(24.25)

That is, if $[OH^-]$ is kept sufficiently high, the reverse of reaction (24.25) becomes important, and significant concentrations of MnO_4^{2-} can be maintained in solution.

Manganese Compounds. The principal source of manganese compounds is MnO_2. When MnO_2 is heated in the presence of an alkali and an oxidizing agent, a manganate salt is produced.

$$3\ MnO_2 + 6\ KOH + KClO_3 \longrightarrow 3\ K_2MnO_4 + KCl + 3\ H_2O(g) \qquad (24.26)$$

K_2MnO_4 is extracted from the fused mass with water, and can then be oxidized to $KMnO_4$ (with Cl_2 as an oxidizing agent, for instance). Alternatively, if $MnO_4^{2-}(aq)$ is acidified, MnO_4^- is produced by reaction (24.24).

Potassium permanganate, $KMnO_4$, is an important oxidizing agent. For chemical analyses it is generally used in acidic solutions, where it is reduced to $Mn^{2+}(aq)$. In the analysis of iron by MnO_4^-, a sample is prepared in the same manner as described for reaction (24.20) and titrated with $MnO_4^-(aq)$.

Titration of $Fe^{2+}(aq)$ with $MnO_4^-(aq)$

$$5\ Fe^{2+} + MnO_4^- + 8\ H^+ \longrightarrow 5\ Fe^{3+} + Mn^{2+} + 4\ H_2O \qquad (24.27)$$

$Mn^{2+}(aq)$ has a very pale pink color that is barely discernible. $MnO_4^-(aq)$ is intensely colored (purple). At the end point of titration (24.27) the solution being titrated acquires a lasting deep pink color with just a single drop of excess $MnO_4^-(aq)$. $MnO_4^-(aq)$ is less satisfactory for titrations in alkaline solutions because the insoluble reduction product $MnO_2(s)$ obscures the end point. Other titrations performed with MnO_4^- include the determination of nitrites, H_2O_2, and calcium (after precipitation as the oxalate). In organic chemistry $MnO_4^-(aq)$ is used to oxidize alcohols and unsaturated hydrocarbons (see Chapter 27). Manganese dioxide, MnO_2, is used in dry cells (recall Figure 21-11), in glass and ceramic glazes, and as a catalyst.

Manganese nodules on the ocean floor. [Courtesy of Lawrence Sullivan, Lamont-Doherty Geological Observatory of Columbia University.]

Manganese Nodules. These are rocklike objects found on the ocean floor. They are composed of layers of oxides of Mn and Fe, with small quantities of other metals such as Co, Cu, and Ni. The nodules are roughly spherical in shape with diameters ranging from a few millimeters to about 15 cm. They are believed to grow at a rate of a few millimeters per million years. It has been proposed that marine organisms may play a role in their formation. Estimates of the total quantity of these nodules are very great, perhaps billions of tons. However, numerous challenges exist to developing manganese nodules as a significant raw material. Methods must be perfected to explore the seabed, dredge for the nodules, and transport them through several thousand meters of seawater. Also, new metallurgical processes must be devised for extracting the desired metals. Nontechnical concerns, but problems nevertheless, are the political and legal issues involved in mining in international waters. (Who owns the nodules?) The largest deposits currently known are in an area southeast of the Hawaiian Islands.

24-7 Iron, Cobalt, Nickel: The Iron Triad

3B	4B	5B	6B	7B		8B		1B	2B
21 Sc	22 Ti	23 V	24 Cr	25 Mn	26 Fe	27 Co	28 Ni	29 Cu	30 Zn

Iron, with annual worldwide production approaching one billion tons, is, of course, the most important metal in modern civilization. It is found widely distributed in the earth's crust in an abundance of 4.7%. Cobalt is among the rarer metals. It constitutes only 20 ppm of the earth's crust, but it occurs in sufficiently concentrated deposits (ores) so that its annual production runs into the millions of pounds. Cobalt is used primarily in alloys with other metals. Nickel ranks twenty-fourth in abundance among the elements in the earth's crust. Its ores are mainly the sulfides, oxides, silicates, and arsenides. Particularly large deposits are found in Canada. Of the 300 million pounds or so of nickel consumed annually in the United States, about 80% is used in the production of alloys. Another 15% is used for electroplating, and the remainder for miscellaneous purposes.

As expected, there are some similarities between Fe, Co, and Ni and other elements in the first transition series. Various combinations of these metals form alloys with one another, and all of the simple metal ions have orbitals available for complex ion formation. But there are differences between the Fe–Co–Ni group and the elements preceding them in the first transition series. For example, Fe, Co, and Ni do not form stable oxoanions like VO_3^-, CrO_4^{2-}, and MnO_4^-. Also, they do not exhibit the same wide range of oxidation states. The strongest similarities for elements of group 8B in the periodic table occur horizontally among the three elements in each period of the group. For this reason Fe, Co, and Ni are generally considered a group of three—a triad.

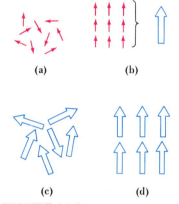

FIGURE 24-9

The phenomenon of ferro-magnetism.

(a) In ordinary paramagnetism the magnetic moments of the atoms or ions are randomly distributed.
(b) In a ferromagnetic material the magnetic moments are aligned into domains.
(c) In an unmagnetized piece of the material the domains are randomly oriented.
(d) In a magnetic field the domains are oriented in the direction of the field and the material becomes magnetized.

Ferromagnetism. One property unique to Fe, Co, and Ni among the common chemical elements is that of ferromagnetism. Although Fe, Co, and Ni atoms all have unpaired electrons, the property of ferromagnetism cannot be accounted for by the paramagnetism of these atoms alone. In the solid state the metal atoms are thought to be grouped together into small regions containing rather large numbers of atoms. These regions are called **domains.** Instead of the individual magnetic moments of the atoms within a domain being randomly oriented, they are all directed in the same way. In an unmagnetized piece of iron the domains are oriented in several directions and their magnetic effects cancel. When the metal is placed in a magnetic field, however, the domains are lined up and a strong resultant magnetic effect is produced. This alignment of domains may actually involve the growth of domains with favorable orientations at the expense of those with unfavorable orientations. The ordering of domains persists when the object is removed from the magnetic field, and thus permanent magnetism results (see Figure 24-9).

The key to ferromagnetism involves two basic factors: that the species involved have unpaired electrons (a property possessed by many species) and that interatomic distances be of just the right magnitude to make possible the ordering of atoms into domains. If atoms are too large, interactions among them are too weak to produce this ordering. With small atoms the tendency is for atoms to pair and their magnetic moments to cancel. This critical factor of atomic size is just met in Fe, Co, and Ni. However, it is possible to prepare alloys of metals other than these three in which this condition is met (e.g., Al–Cu–Mn, Ag–Al–Mn, and Bi–Mn). Also, certain rare earth elements, for example, gadolinium (Gd) and dysprosium (Dy), are ferromagnetic at low temperatures.

Metal Carbonyls. The transition metals, with few exceptions, form compounds with carbon monoxide. These compounds are called metal carbonyls. In the simple metal carbonyls listed in Table 24-5,

1. Each CO molecule contributes an electron pair to an empty orbital of the metal atom.

TABLE 24-5
Three Metal Carbonyls

	Number of e^-		
	From metal	From CO	Total
$Cr(CO)_6$	24	12	36
$Fe(CO)_5$	26	10	36
$Ni(CO)_4$	28	8	36

2. All electrons are paired (most metal carbonyls are diamagnetic).

3. The metal atom acquires the electron configuration of the noble gas Kr.

The structures of the simple carbonyls in Figure 24-10 are those that we would predict from VSEPR theory (based on a number of electron pairs equal to the number of CO molecules).

In the species $Mn(CO)_5$ the number of electrons associated with the Mn atom is 35 (25 of its own and two each from the five CO molecules). This would make it an odd-electron (paramagnetic) species. When two $Mn(CO)_5$ units join by forming an Mn—Mn bond between them, each Mn atom acquires the equivalent of 36 electrons (a noble gas electron configuration) and all electrons are paired. The *binuclear* carbonyl $Mn_2(CO)_{10}$ is shown in Figure 24-11.

Metal carbonyls are produced in several ways. Nickel metal combines with $CO(g)$ at ordinary temperatures and pressures in a reversible reaction.

$$Ni(s) + 4\ CO(g) \rightleftharpoons Ni(CO)_4(l) \tag{24.28}$$

With iron it is necessary to use higher temperatures (200 °C) and CO pressures (100 atm).

$$Fe(s) + 5\ CO(g) \rightleftharpoons Fe(CO)_5(g) \tag{24.29}$$

In other cases the carbonyl is obtained by reducing a metal compound in the presence of $CO(g)$.

Carbon monoxide poisoning results from an action similar to carbonyl formation. CO molecules coordinate with Fe atoms in hemoglobin, displacing the O_2 molecules normally carried by the hemoglobin. The metal carbonyls themselves are also very poisonous.

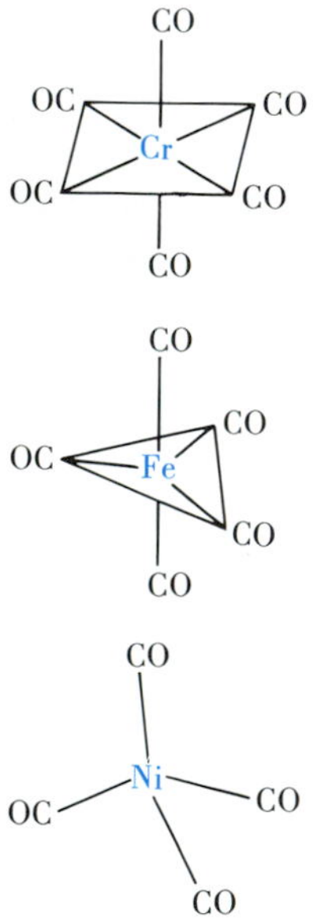

Oxidation States. Variability of oxidation states is encountered within the iron triad elements, even if not to the same degree as with certain other transition elements like vanadium and manganese. The oxidation state +2 is commonly encountered with all three elements.

$$Fe^{2+}\quad [Ar]3d^6 \qquad Co^{2+}\quad [Ar]3d^7 \qquad Ni^{2+}\quad [Ar]3d^8$$

For cobalt and nickel the oxidation state +2 is the most stable, but for iron the most stable is the oxidation state +3.

3d

Fe^{2+} [Ar]   ↑↓ | ↑ | ↑ | ↑ | ↑

3d

Fe^{3+} [Ar]   ↑ | ↑ | ↑ | ↑ | ↑

An electron configuration in which the *d* subshell is half-filled, with all electrons unpaired, has a special stability. This fact suggests that Fe(II) can be oxidized to

In the metallurgy of nickel (Mond process), $CO(g)$ is passed over a mixture of metal oxides. Nickel is carried off as $Ni(CO)_4(g)$ while the other metal oxides are reduced to the metals.

FIGURE 24-10
Structures of some simple carbonyls.

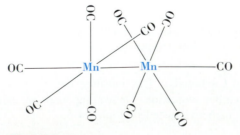

FIGURE 24-11
Structure of a binuclear carbonyl, $Mn_2(CO)_{10}$.

Fe(III) without great difficulty; for example, in the presence of oxygen at 1 atm pressure and with $[H^+] = 1$ M,

$$4\ Fe^{2+} + O_2(g) + 4\ H^+ \longrightarrow 4\ Fe^{3+} + 2\ H_2O \qquad E^{\circ}_{cell} = +0.44\ V \qquad (24.30)$$

But even at lower partial pressures of oxygen, as in the atmosphere, and in less acidic media, reaction (24.30) may still be spontaneous (see Exercise 60).

For Co(II) and Ni(II) the loss of an additional electron does not lead to an electron configuration with half-filled d orbitals.

$$3d$$

$$Co^{3+} \quad [Ar] \quad \boxed{\uparrow\downarrow} \; \boxed{\uparrow} \; \boxed{\uparrow} \; \boxed{\uparrow} \; \boxed{\uparrow}$$

$$3d$$

$$Ni^{3+} \quad [Ar] \quad \boxed{\uparrow\downarrow} \; \boxed{\uparrow\downarrow} \; \boxed{\uparrow} \; \boxed{\uparrow} \; \boxed{\uparrow}$$

As a result the conversion of Co(II) to Co(III) and Ni(II) to Ni(III) is accomplished not nearly so easily as in the case of iron. Consider, for example, this standard electrode potential.

$$Co^{3+}(aq) + e^- \longrightarrow Co^{2+}(aq) \qquad E^{\circ} = 1.82\ V \qquad (24.31)$$

The *reduction* of Co^{3+} to Co^{2+} occurs readily; Co(III) compounds tend to be very good oxidizing agents. For reasons explained in Section 25-11, the $+3$ oxidation state of cobalt can be stabilized, however, if the Co^{3+} is part of a complex ion. The oxidation state $+3$ is also difficult to achieve with nickel and is not commonly encountered. Nickel compounds in higher oxidation states do find use, however, as electrode materials in the nickel–cadmium cell.

Because of the stability attributed to the electron configuration of Fe^{3+}, iron does not commonly occur in higher oxidation states. The oxidation state $+6$ is attainable, however, under very strong oxidizing conditions.

$$2\ Fe(OH)_3(s) + 3\ OCl^- + 4\ OH^- \rightleftharpoons 2\ FeO_4{}^{2-} + 3\ Cl^- + 5\ H_2O \qquad (24.32)$$

The ion $FeO_4{}^{2-}$ is called **ferrate ion,** and a few ferrate salts have been prepared. For example, barium ferrate, $BaFeO_4$, is a purple colored solid. As might be expected the ferrates are unstable and are powerful oxidizing agents.

Some Reactions of the Iron Triad Elements. The reactions of the iron triad elements are many and varied. The metals are all more active than hydrogen and liberate $H_2(g)$ from an acidic solution.

$$Ni(s) + 2\ H^+ + 2\ Cl^- \longrightarrow Ni^{2+} + 2\ Cl^- + H_2(g) \qquad (24.33)$$

Hydrated, colored ions are characteristic of the iron triad elements. Co^{2+} and Ni^{2+} are red and green, respectively. In aqueous solution hydrated Fe^{2+} is pale green, and fully hydrated Fe^{3+} is pale purple. Generally, solutions of Fe^{3+} are yellow to brown in color, but this color is probably due to the presence of species formed in the hydrolysis of $Fe^{3+}(aq)$. As we have seen before with $Al^{3+}(aq)$, $Zn^{2+}(aq)$, and $Cr^{3+}(aq)$, this hydrolysis proceeds in a stepwise fashion, beginning with

Acid ionization of $[Fe(H_2O)_6]^{3+}(aq)$.

$$[Fe(H_2O)_6]^{3+} + H_2O \rightleftharpoons [Fe(H_2O)_5OH]^{2+} + H_3O^+$$
$$K_a = 8.9 \times 10^{-4} \qquad (24.34)$$

However, with Fe^{3+} the reaction does not proceed beyond $Fe(OH)_3(s)$, even in the presence of excess $OH^-(aq)$—$Fe(OH)_3$ is not amphoteric.

Salts of the iron triad elements usually crystallize from solution as hydrates. The hydrate of Co(II) chloride has an interesting application. When exposed to atmospheric moisture, depending on the partial pressure of $H_2O(g)$, the hydrate assumes different colors. In dry air, water of hydration is lost and the solid acquires a blue

TABLE 24-6
Some Qualitative Tests for Iron(II) and Iron(III)

Reagent	Iron(II)	Iron(III)
NaOH(aq)	green precipitate	red-brown precipitate
$K_4[Fe(CN)_6]$	white precipitate, turning blue rapidly	Prussian blue precipitate
$K_3[Fe(CN)_6]$	Turnbull's blue precipitate	red-brown color (no precipitate)
KSCN	no color	deep red color

color; as the humidity increases, the solid undergoes a gradual color change to pink. This reaction has been used as an inexpensive, if somewhat crude, method of determining moisture levels.

$$CoCl_2 \cdot 6H_2O(s) \rightleftharpoons CoCl_2(s) + 6\ H_2O \qquad (24.35)$$
$$\text{(pink)} \qquad\qquad \text{(blue)}$$

The system for writing names and formulas of complex ions such as these is taken up in Chapter 25.

The **hexacyanoferrates** are compounds containing the complex ions $[Fe(CN)_6]^{4-}$ and $[Fe(CN)_6]^{3-}$. Their systematic names are hexacyanoferrate(II) and hexacyanoferrate(III), respectively, but they are also commonly called **ferrocyanide** and **ferricyanide.** Iron(III) in aqueous solution yields a dark blue precipitate, **Prussian blue,** when treated with potassium hexacyanoferrate(II). The exact structure of Prussian blue is not known, but it is probably quite complex. It has the empirical formula $Fe_7C_{18}N_{18} \cdot 10H_2O$. A simplified representation of this reaction is

$$4\ Fe^{3+} + 3[Fe(CN)_6]^{4-} \longrightarrow Fe_4[Fe(CN)_6]_3 \qquad (24.36)$$

Iron(II) compounds yield a blue precipitate, **Turnbull's blue,** when treated with potassium hexacyanoferrate(III). The reaction appears to proceed in two stages. The first is an oxidation–reduction reaction in which iron(II) is oxidized to iron(III) and hexacyanoferrate(III) is reduced to hexacyanoferrate(II).

$$Fe^{2+} + [Fe(CN)_6]^{3-} \longrightarrow Fe^{3+} + [Fe(CN)_6]^{4-} \qquad (24.37)$$

This is followed by reaction (24.36). Turnbull's blue is a lighter shade than Prussian blue and this may be caused by the admixture of some white $K_2\{Fe[Fe(CN)_6]\}$.

The qualitative tests for iron just described, together with two others, are summarized in Table 24-6. The dark red coloration resulting from the reaction of thiocyanate ion, SCN^-, with Fe^{3+} is the basis of an extremely sensitive test for iron (see Section 24-11). The composition of the product appears to be $[Fe(H_2O)_5SCN]^{2+}$.

A very distinctive reaction of Ni^{2+} that can be used for both its qualitative detection and its quantitative determination is the formation of a neutral complex with dimethylglyoxime, which precipitates from solution as a brilliant scarlet precipitate. In addition to the coordination bonds between N atoms and the Ni^{2+} ion, this complex features hydrogen bonds.

$$(24.38)$$

24-8 Metallurgy of Iron and Steel

A type of steel called wootz steel was first produced in India about 3000 years ago. This is the same steel that became famous in ancient times as Damascus steel, renowned for its suppleness, its ability to hold a cutting edge, and its use in making swords. Many technological advances have been made since ancient times. These include introduction of the blast furnace in about 1300 A.D., the Bessemer converter in 1856, the open hearth furnace in the 1860s, and most recently, the basic oxygen furnace. However, a true understanding of the iron- and steel-making process has developed only within the past few decades. This understanding is based on concepts of thermodynamics, equilibrium, and kinetics.

Pig Iron. The reactions that occur in a blast furnace are complex. A highly simplified representation involves reduction of iron ore to impure iron and the removal of certain impurities as a slag.

Blast furnace reactions: Reduction of Fe_2O_3. Slag formation.

$$Fe_2O_3(s) + 3\ CO(g) \xrightarrow{\Delta} 2\ Fe(l) + 3\ CO_2(g) \tag{24.39}$$

$$CaO(s) + SiO_2(s) \xrightarrow{\Delta} CaSiO_3(l) \tag{24.40}$$

A more complete description of the blast furnace reactions is given in Table 24-7. Approximate temperatures are given for these reactions so that you may key them to regions of the blast furnace pictured in Figure 24-12a.

The blast furnace charge consists of iron ore, coke, a slag-forming flux, and perhaps some scrap iron. The exact proportions depend on the composition of the iron ore and its impurities. The formulas of some typical ores are listed in Table 24-8. The purpose of the flux is to maintain the proper ratio of acidic oxides (SiO_2, Al_2O_3, and P_2O_5) to basic oxides (CaO, MgO, and MnO) to obtain an easily liquefied silicate, aluminate, or phosphate slag. Since in most ores the acidic oxides

TABLE 24-7
Some Blast Furnace Reactions

Formation of reducing agents, principally $CO(g)$ and $H_2(g)$:

$2\ C + O_2 \longrightarrow 2\ CO\ (1700\ °C)$

$C + CO_2 \longrightarrow 2\ CO\ (>1000\ °C)$

$C + H_2O \longrightarrow CO + H_2\ (>600\ °C)$

Reduction of iron oxide:

$3\ CO + Fe_2O_3 \longrightarrow 2\ Fe + 3\ CO_2\ (900\ °C)$

$3\ H_2 + Fe_2O_3 \longrightarrow 2\ Fe + 3\ H_2O\ (900\ °C)$

Slag formation to remove impurities from ore:

$CaCO_3 \longrightarrow CaO + CO_2\ (800–900\ °C)$

$CaO + SiO_2 \longrightarrow CaSiO_3(l)\ (1200\ °C)$

$3\ CaO + P_2O_5 \longrightarrow Ca_3(PO_4)_2(l)\ (1200\ °C)$

Impurity formation in the iron:

$MnO + C \longrightarrow Mn + CO\ (1400\ °C)$

$SiO_2 + 2\ C \longrightarrow Si + 2\ CO\ (1400\ °C)$

$P_2O_5 + 5\ C \longrightarrow 2\ P + 5\ CO\ (1400\ °C)$

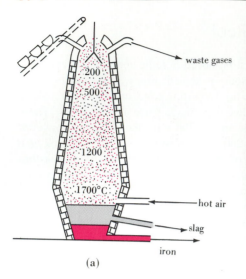

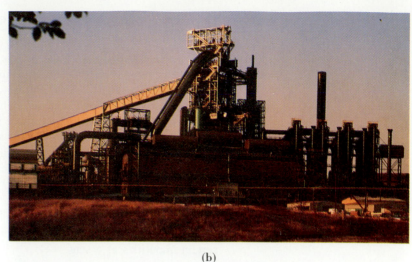

(a)   (b)

FIGURE 24-12
Typical blast furnace.

(a) Iron ore, coke, and limestone are added at the top of the furnace, and hot air is introduced through the bottom. Maximum temperatures are attained near the bottom of the furnace where molten iron and slag are drained off.
(b) The blast furnace shown here is capable of producing nearly 1×10^7 kg of pig iron per day. It stands about 90 m high, is computer-controlled, and is equipped with the latest environmental control devices. [Courtesy Bethlehem Steel.]

predominate, the flux generally employed is limestone, $CaCO_3$, or dolomite, $CaCO_3 \cdot MgCO_3$.

The iron obtained from a blast furnace is called **pig iron**. It contains about 95% Fe, 3 to 4% C, and varying quantities of other impurities. **Cast iron** can be obtained by pouring pig iron directly into molds of the desired shape. Cast iron is very hard and brittle and can be used only where it will not be subjected to mechanical or thermal shock. For example, it is used in engine blocks, brake drums, and transmission housings in automobiles.

Steel. The fundamental changes that must be made to convert pig iron to steel are

1. Reduce the carbon content from 3–4% in pig iron to 0–1.5% in steel.
2. Remove, through slag formation, Si, Mn, and P (each present in pig iron to the extent of 1% or so), together with other minor impurities.
3. Add alloying elements (such as Cr, Ni, Mn, V, Mo, and W) to give the steel its desired end properties.

The most important method of steel making today is the **basic oxygen process.** The process is carried out in a steel vessel with a refractory lining (usually made of dolomite, $CaCO_3 \cdot MgCO_3$). Oxygen gas at about 10 atm pressure and a stream of powdered limestone are fed through a water-cooled lance and discharged above the molten pig iron (see Figure 24-13). The reactions that occur accomplish the first two objectives listed above (see Table 24-9). A typical reaction time is 22 minutes. The reaction vessel is tilted to remove the liquid slag floating on top of the iron, and then the desired alloying elements are added.

Basic oxygen furnace being charged with molten iron from a blast furnace. [Courtesy Bethlehem Steel]

TABLE 24-8
Some Typical Iron Ores

Type of ore	Formula
oxide	
magnetite	Fe_3O_4
hematite	Fe_2O_3
ilmenite	$FeTiO_3$
limonite	$HFeO_2$
carbonate	
siderite	$FeCO_3$
sulfide	
pyrite	FeS_2
pyrrhotite	FeS

TABLE 24-9
Some Reactions Occurring in
Steel-Making Processes

$$2\,C + O_2 \longrightarrow 2\,CO$$
$$2\,FeO + Si \longrightarrow 2\,Fe + SiO_2$$
$$FeO + Mn \longrightarrow Fe + MnO$$
$$FeO + SiO_2 \longrightarrow FeO \cdot SiO_2$$
$$\text{slag}$$
$$MnO + SiO_2 \longrightarrow MnO \cdot SiO_2$$
$$\text{slag}$$
$$4\,P + 5\,O_2 \longrightarrow 2\,P_2O_5$$
$$3\,CaO + P_2O_5 \longrightarrow Ca_3(PO_4)_2$$
$$\text{slag}$$

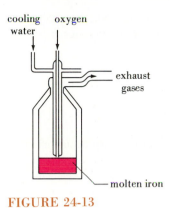

FIGURE 24-13
A basic oxygen furnace.

Steel making has been undergoing rapid technological change. In 1962 only about 4% of the steel produced in the United States was by the basic oxygen process. Today, it is the principal steel-making method. The ultimate goal may be to make iron and steel directly from iron ore in a single-step (perhaps continuous) process. Iron ore can be reduced to iron at temperatures below the melting point of any of the materials used in the process. In the direct reduction of iron (DRI) there is no need for blast furnaces. DRI processes use as reducing agents $CO(g)$ and $H_2(g)$ obtained in the reaction of steam with natural gas (recall reaction 14.3). The economic viability of DRI processes depends on an abundant supply of natural gas. Currently only a small percentage of the world's iron production is by direct reduction, but this is a fast-growing component of the iron and steel industry, particularly in the Middle East and South America.

24-9 Copper, Silver, Gold: The Coinage Metals

Throughout the ages, Cu, Ag, and Au have been the preferred metals for use in coins because these metals are so durable and resistant to corrosion. The data in Table 24-10 help us to understand why this is so. The metal ions are easy to reduce to the free metals, which also means that the metals are difficult to oxidize. It is this resistance to oxidation that imparts "nobility" to the group 1B metals.

In Mendeleev's periodic table the alkali metals (1A) and the coinage metals (1B) appear together as group I, but the only similarity between the two subgroups is that of having a single s electron in the valence shells of their atoms. However, as you can see through Tables 22-1 and 24-10, the first ionization energies for the 1B metals are much larger than for the 1A metals.

The similarity of the group 1B elements to the other transition elements that precede them in the periodic table is in being able to use d electrons in chemical bonding. Thus, the 1B elements can exist in different oxidation states, exhibit paramagnetism and color in some of their compounds, and form complex ions. They also possess to a high degree some distinctive physical properties of the metallic state—malleability, ductility, and excellent electrical and thermal conductivity.

In comparing the 1B metals among themselves, we find trends similar to those noted for 2B metals in Section 22-6, and for essentially the same reasons. The lightest metal (Cu) is the most active of the three; the heaviest metal (Au) is the least active. Bonding in most Au(III) compounds is covalent (just as it is in many mercury compounds).

TABLE 24-10
Some Properties of Cu, Ag, and Au

	Cu	Ag	Au
electron configuration	$[Ar]3d^{10}4s^1$	$[Kr]4d^{10}5s^1$	$[Xe]4f^{14}5d^{10}6s^1$
metallic radius, pm	128	144	144
first ioniz. energy, kJ/mol	745	731	890
electrode potential, V			
$M^+(aq) + e^- \rightarrow M(s)$	+0.522	+0.800	+1.68
$M^{2+}(aq) + 2 e^- \rightarrow M(s)$	+0.337	+1.39	—
$M^{3+}(aq) + 3 e^- \rightarrow M(s)$	—	—	+1.42
oxidation states[a]	+1, +2	+1, +2	+1, +3

[a]The most common oxidation state is shown in red.

Oxidation States. We can learn a lot about the stabilities of coinage metal compounds by using electrode potential data. For the different oxidation states of copper we can write

The value of $E°$ for half-reaction (24.41) is obtained by a method similar to that first introduced in Example 23-2 (see Exercise 38).

$$Cu^+(aq) \longrightarrow Cu^{2+}(aq) + e^- \qquad -E° = -0.152 \text{ V} \qquad (24.41)$$
$$Cu^+(aq) + e^- \longrightarrow Cu(s) \qquad E° = +0.522 \text{ V} \qquad (24.42)$$
$$\overline{2\ Cu^+(aq) \longrightarrow Cu^{2+}(aq) + Cu(s) \qquad E°_{cell} = +0.370 \text{ V} \qquad (24.43)}$$

According to equation (24.43) the disproportionation of $Cu^+(aq)$ occurs spontaneously under standard-state conditions. This does not mean, however, that we cannot obtain aqueous solutions of Cu(I) compounds. It just means that we may have to establish special conditions to do so. To see what these conditions might be, let us derive a value of the equilibrium constant for reaction (24.43). To do this, remember that we must use the relationship

$$E°_{cell} = \frac{0.0592}{n} \log K \qquad (21.29)$$

By substituting $n = 1$ and $E°_{cell} = +0.370$ into equation (21.29), we obtain

$$\log K = \frac{1 \times 0.370}{0.0592} = 6.25 \qquad \text{and} \qquad K = 1.8 \times 10^6$$

Thus, for the disproportionation reaction (24.43)

$$K = \frac{[Cu^{2+}]}{[Cu^+]^2} = 1.8 \times 10^6 \qquad \text{and} \qquad [Cu^{2+}] = 1.8 \times 10^6 \times [Cu^+]^2 \qquad (24.44)$$

Here is how we can interpret equation (24.44): If we attempt to maintain any appreciable concentration of Cu^+ in aqueous solution, the concentration of Cu^{2+} would have to be an impossibly large number. This is because $[Cu^{2+}]$ would have to be about 2 million times the square of $[Cu^+]$. Disproportionation of $Cu^+(aq)$ would go to completion.

Remember, the square of a number *less* than 1 is *smaller* than the number, such as $(1 \times 10^{-3})^2 = 1 \times 10^{-6}$.

On the other hand, if we can keep the concentration of Cu^+ very low (as in a very slightly soluble solute or a stable complex ion), the concentration of Cu^{2+} would be *exceedingly* small. Thus the copper(I) species would be stable.

Example 24-3 ———————————————————

Establishing equilibrium conditions through a value of K. Can a solution be prepared that is **(a)** 0.020 M in Cu^+; **(b)** 1.0×10^{-10} M in Cu^+?

Solution

(a) In order to maintain $[Cu^+] = 0.020$ M, we see from equation (24.44) that the solution would simultaneously need to have $[Cu^{2+}] = 1.8 \times 10^6 \times (0.020)^2 = 7.2 \times 10^2$ M. This is an impossibly high concentration of Cu^{2+}. No solution can be prepared that is 0.020 mol/L in a copper(I) salt.

(b) In a solution with $[Cu^+] = 1.0 \times 10^{-10}$ M, the $[Cu^{2+}]$ that simultaneously needs to be present is $[Cu^{2+}] = 1.8 \times 10^6 \times (1.0 \times 10^{-10})^2 = 1.8 \times 10^{-14}$ M. In this instance $[Cu^{2+}]$ is about 6000 times *smaller* than $[Cu^+]$. In other words, disproportionation of Cu^+ occurs hardly at all. **A solution with $[Cu^+] = 1.0 \times 10^{-10}$ M can be prepared.** This is essentially the situation one finds in a saturated solution of CuCN, which has $K_{sp} = 3.2 \times 10^{-20}$.

SIMILAR EXAMPLES: Exercises 11, 42, 43.

Do any of the other group 1B metal ions disproportionate? For silver the situation analogous to reaction (24.43) is

$$2\,Ag^+(aq) \longrightarrow Ag^{2+}(aq) + Ag(s) \qquad E^\circ_{cell} = -1.18\ V \qquad (24.45)$$

The large negative E°_{cell} value shows us that the disproportionation of Ag^+ is a very *unfavorable* reaction, accounting for the fact that silver is found mostly in Ag(I) compounds. Ag(II) compounds are formed under very strong oxidizing conditions, and these compounds are themselves very good oxidizing agents because of their tendency to be reduced to Ag(I).

The situation with gold is that

$$3\,Au^+(aq) \longrightarrow Au^{3+}(aq) + 2\,Au(s) \qquad E^\circ_{cell} = +0.39\ V \qquad (24.46)$$

$Au^+(aq)$ does disproportionate. Although the +3 oxidation state is most common for gold, this is not in the form of Au^{3+} ion. Instead, the Au is bonded to other atoms, producing a complex species in which Au has the oxidation state +3, such as $[AuCl_4]^-$.

Compounds of Cu, Ag, and Au. The 1B metals do not dissolve in HCl(aq), but Cu and Ag both dissolve in concentrated $H_2SO_4(aq)$ and $HNO_3(aq)$. The metals are oxidized to Cu^{2+} and Ag^+, respectively, and the reduction products are $SO_2(g)$ in $H_2SO_4(aq)$ and either NO(g) or $NO_2(g)$ in $HNO_3(aq)$ (recall Table 23-13). Au does not dissolve in either acid, but it does dissolve in aqua regia (1 part HNO_3 and 3 parts HCl). $HNO_3(aq)$ oxidizes the metal and Cl^- from the HCl(aq) promotes the formation of the stable complex ion $[AuCl_4]^-$.

Action of aqua regia on gold.

$$Au(s) + 4\,H^+ + NO_3^- + 4\,Cl^- \longrightarrow [AuCl_4]^- + 2\,H_2O + NO(g) \qquad (24.47)$$

The coinage metals are generally resistant to air oxidation, although in moist air copper corrodes to produce green basic copper carbonate. This is the green color associated with copper roofing and gutters and bronze statues. Fortunately this corrosion product forms a tough adherent coating that protects the underlying metal. The corrosion reaction is complex but may be summarized as

$$2\,Cu + H_2O + CO_2 + O_2 \longrightarrow \underset{\text{basic copper carbonate}}{Cu_2(OH)_2CO_3} \qquad (24.48)$$

If we add OH^- to an aqueous solution of Cu^{2+}, we obtain a precipitate of $Cu(OH)_2(s)$. This precipitate is soluble in $NH_3(aq)$ owing to the formation of the deep blue complex ion $[Cu(NH_3)_4]^{2+}$.

$$Cu^{2+}(aq) + 2\,OH^-(aq) \longrightarrow Cu(OH)_2(s) \qquad (24.49)$$

Formation of the complex ion $[Cu(NH_3)_4]^{2+}$.

$$Cu(OH)_2(s) + 4\,NH_3(aq) \longrightarrow [Cu(NH_3)_4]^{2+}(aq) + 2\,OH^-(aq) \qquad (24.50)$$

If we add OH^- to $Ag^+(aq)$ we obtain a precipitate of silver(I) oxide.

$$2\,Ag^+(aq) + 2\,OH^-(aq) \longrightarrow Ag_2O(s) + H_2O \qquad (24.51)$$

Left: Precipitation of $Cu(OH)_2(s)$ (reaction 24.49). Right: $[Cu(NH_3)_4]^{2+}(aq)$ formed as a result of reaction (24.50). [Carey B. Van Loon]

Trace amounts of Cu are essential to life, but larger quantities are toxic, especially to bacteria, algae, and fungi. Among the many copper compounds used as pesticides are the basic acetate, carbonate, chloride, hydroxide, and sulfate. Commercially, the most important copper compound is $CuSO_4 \cdot 5H_2O$. In addition to its agricultural applications, $CuSO_4$ is used in batteries and electroplating, in preparing other copper salts, and in the petroleum, rubber, and steel industries.

Silver nitrate is the principal silver compound of commerce and is also an important laboratory reagent for the precipitation of anions (most anions form insoluble silver salts). These precipitation reactions can be used for the quantitative determination of anions, either gravimetrically (by weighing precipitates) or volumetrically (by titration). $AgNO_3$ is the source from which most other Ag compounds are derived. Ag compounds are used in electroplating, in the manufacture of batteries,

in medicinal chemistry, as catalysts, and in cloud seeding (AgI). Their most important use by far (mostly as silver halides) is in photography (see Section 25-11).

Gold compounds are used in electroplating, photography, medicinal chemistry,
and the manufacture of special glasses and ceramics.

Metallurgy and Uses of Copper. The extraction of copper from its ores (generally
as sulfides) is rather complicated. The chief reason for this complexity is that copper
ores usually contain iron sulfides. If we simply follow the general scheme of extractive metallurgy outlined in Section 22-4, we obtain copper contaminated with iron.
To avoid this requires a five-step procedure: (1) concentration, (2) roasting;
(3) smelting; (4) converting; (5) refining.

Concentration is generally done by the flotation method, as shown in Figure
24-14.

Roasting serves to convert iron sulfides to iron oxide. The copper remains as the
sulfide if the temperature is kept below 800 °C.

$$2 \, \text{FeS(s)} + 3 \, \text{O}_2\text{(g)} \xrightarrow{\Delta} 2 \, \text{FeO(s)} + 2 \, \text{SO}_2\text{(g)} \qquad (24.52)$$

Smelting of the roasted ore is done in a furnace at 1400 °C. In this furnace the
charge melts and separates into two layers. The bottom layer is copper matte,
consisting chiefly of the molten sulfides of copper and iron. The top layer is a
silicate slag formed by the reaction of oxides of Fe, Ca, and Al with SiO_2 (which
either is present in the ore or is added).

$$\text{FeO(s)} + \text{SiO}_2\text{(s)} \xrightarrow{\Delta} \text{FeSiO}_3\text{(l)} \qquad (24.53)$$

Converting occurs in another furnace where air is blown through the molten
copper matte. First, the remaining iron sulfide is converted to the oxide (reaction
24.52), followed by slag formation (reaction 24.53). The slag is poured off and air
is again blown through the furnace. Now the following reactions occur, yielding a
product that is about 98 to 99% Cu.

$$2 \, \text{Cu}_2\text{S} + 3 \, \text{O}_2\text{(g)} \xrightarrow{\Delta} 2 \, \text{Cu}_2\text{O} + 2 \, \text{SO}_2\text{(g)} \qquad (24.54)$$

$$2 \, \text{Cu}_2\text{O} + \text{Cu}_2\text{S} \xrightarrow{\Delta} 6 \, \text{Cu(l)} + \text{SO}_2\text{(g)} \qquad (24.55)$$

The product of reaction (24.55) is called **blister copper** because of the presence of
frozen bubbles of SO_2(g). It can be used where high purity is not required (as in
plumbing).

Copper was once widely available in the free state. Now free copper is being mined in the United States only in Michigan.

Refining to obtain high purity copper is done electrolytically, by the method outlined in Section 21-7. High purity is essential in electrical applications.

Currently most copper-bearing ores being mined have only about 0.5% Cu. Both to process low-grade ores more easily and to eliminate SO_2 emissions, there has been some shift from the *pyro*metallurgical production of copper, described above, to *hydro*metallurgical processes (recall Focus feature in Chapter 22). For example, copper ores can be leached with $H_2SO_4(aq)$ and the resulting $CuSO_4(aq)$ electrolyzed to produce copper.

About one-half of the copper used in the United States is for electrical applications. Another 20% is used in plumbing and other aspects of the construction industry. The third important use of copper is in the manufacture of alloys, chiefly brass and bronze.

Metallurgy and Uses of Silver and Gold. Silver and gold both can be found free in nature, but easily accessible known deposits have all but disappeared. A typical gold ore may contain only about 10 g Au per ton. In the **cyanidation** process $O_2(g)$ oxidizes the free metal to Au^+, which complexes with CN^-.

Cyanidation process for the extraction of gold.

$$4\ Au(s) + 8\ CN^- + O_2(g) + 2\ H_2O \longrightarrow 4[Au(CN)_2]^- + 4\ OH^- \qquad (24.56)$$

The pure metal is then displaced from solution by an active metal.

Displacement of Au(I) from solution.

$$2\ [Au(CN)_2]^-(aq) + Zn(s) \longrightarrow 2\ Au(s) + [Zn(CN)_4]^{2-}(aq) \qquad (24.57)$$

The extraction of silver generally requires dissolving an ore in $CN^-(aq)$, followed by displacement of the silver.

$$Ag_2S(s) + 4\ CN^-(aq) \longrightarrow 2[Ag(CN)_2]^-(aq) + S^{2-}(aq) \qquad (24.58)$$

Ag_2S is highly insoluble, and to repress the reverse of reaction (24.58) air is blown through the mixture to oxidize S^{2-} to SO_4^{2-}.

Ag and Au are also obtained during the refining of other metals. They occur, for example, in the anode sludge produced in the electrolytic refining of Cu (Section 21-7).

Metallic silver and gold have uses in addition to being a source of their compounds. They both have served monetary purposes since ancient times. Millions of kg Au are currently held in the monetary reserves of nations throughout the world, and although Ag is much less used in the production of coins than in the past, a reserve of millions of kg Ag exists in old U.S. coins alone. In the past the most significant use of Ag was in silverware and jewelry. These uses are now declining relative to the use of Ag (the best electrical conductor among metals) in electrical contacts, printed circuits, and batteries. Silver is also widely used in alloys, such as silver solder and dental amalgams. About three-fourths of the world's consumption of gold is for jewelry. Other important uses are in dentistry, coins and medals, and specialized industrial applications. Gold is important in the electronics industry because of its excellent electrical and thermal conductivity, its resistance to oxidation, and the ease with which it can be electroplated onto other metals.

24-10 The Lanthanides

The elements from cerium (Z = 58) through lutetium (Z = 71) are *inner-transition elements*—their electron configurations feature the filling of $4f$ orbitals. These elements, together with lanthanum (Z = 57), which closely resembles them, are variously called the **lanthanide, lanthanoid,** or **rare earth** elements. "Rare earth" is a misnomer, though, because these elements are not rare. La, Ce, and Nd are more abundant than lead; Tm is about as abundant as iodine. The lanthanides occur

Superconductivity illustrated. The small magnet induces an electric current in the superconducting pellets, which are surrounded by liquid nitrogen. Associated with this electric current is another magnetic field that opposes the field of the magnet, causing the magnet to be repelled. The magnet remains suspended above the superconductor pellets for as long as the superconducting current is present, and the current persists as long as the temperature is maintained at the boiling point of liquid nitrogen (77 K). [Photo by Jim Bailey, courtesy University of Arkansas]

Superconductors

The vibrations of atoms in a metallic lattice scatter electrons and thus interfere with their flow as an electric current. Lattice vibrations increase with increased temperature, so we should expect a metal to become a *better* electrical conductor as its temperature is *lowered*. Shortly after its discovery at the close of the 19th century, helium was found to be the ideal medium in which to carry out low-temperature research. This is because helium remains liquid at temperatures approaching 0 K (recall Figure 14-5). And one of the properties first studied at liquid helium temperatures was electrical conductivity. The electrical resistance of metals was indeed found to be much lower at very low temperatures than at room temperature.

In fact, it was believed that the electrical resistance of a metal should continuously decrease as its temperature is lowered, so that for a pure perfect crystal at 0 K the electrical resistance should drop essentially to zero. In 1911, Heike Onnes found, unexpectedly, that some metals abruptly lose their electrical resistance at temperatures that are still well above 0 K. We now use the term **superconductivity** to describe this behavior. Once an electric current is established in a superconducting metal maintained below a certain critical temperature, the current flows indefinitely, with no loss of energy. Over the years, numerous theories have been proposed to explain superconductivity in metals, with a successful one finally appearing in the 1950s. The explanation of superconductivity in metals is based on quantum theory.

The larger the electric current carried through the coils of an electromagnet, the stronger the magnet (recall Figure 8-2). The ideal electric current to use is one that encounters no resistance, and this is what happens if the metal in the coils is kept in a superconducting condition. Thus, the chief use of superconducting metals has been in the construction of powerful electromagnets for use in medical imaging and in particle accelerators (Section 26-4) and nuclear fusion reactors (Section 26-9).

The drawback to using metallic superconductors is that they will only function at very low temperatures in liquid helium. The highest critical temperature among this group is that of a niobium–germanium alloy, 23 K. It is not hard to imagine many additional applications of superconductivity, if the phenomenon can be produced and maintained at higher temperatures. One suggestion has been of trains propelled by magnetic levitation.

In 1986, the first of a new class of materials was discovered that display superconducting properties at higher temperatures than traditional superconducting metals. Several of these new superconductors have been characterized; three are listed in Table 24-11. For example, the superconductor $YBa_2Cu_3O_7$, called a "1-2-3" superconductor, can be prepared by heating a mixture of yttrium oxide, barium carbonate, and copper oxide to 700 °C. At the time of this writing, the highest temperature at which superconductivity has definitely been observed is 125 K.

One feature shared by these new superconductors is that they are *ceramic* materials, not metals. Another feature is that they appear to involve copper and oxygen atoms that occur in sheet-like structures in the material. The great advantage of

TABLE 24-11
Some Ceramic Superconductors

Critical temperature for superconductivity, K	Elements present	Representative formula
40	La, Sr, Cu, O	$La_{2-x}Sr_xCuO_4$[a]
90	Y, Ba, Cu, O	$YBa_2Cu_3O_7$
105	Th, Ca, Ba, Cu, O	$Th_2CaBa_2Cu_2O_8$

[a] $x < 0.3$.

ceramic superconductors, of course, is that they can function at much higher temperatures than metallic semiconductors. The ideal would be to find a superconductor that functions at room temperature (about 300 K), but even those that are superconducting above 77 K are of great potential use. They will function in liquid nitrogen (b.p. 77 K), a much cheaper medium than liquid helium. Disadvantages of these new materials are that they are brittle and not easily fabricated into wires with a high current-carrying capacity. As yet, no practical devices have been made from these new superconductor materials, but research with ceramic superconductors is currently one of the most active fields in materials science.   ■ ■ ■

primarily as oxides, and mineral deposits containing them are found in various locations. Large deposits near the California–Nevada border are being developed to provide rare earth oxides used in phosphors in color television sets.

Some Properties of the Lanthanides. Because their differences in electron configuration are mainly in $4f$ orbitals, and because $4f$ electrons play a minor role in chemical bonding, strong similarities are found among the lanthanides. For example, for the reduction process

$$M^{3+}(aq) + 3 e^- \longrightarrow M(s) \tag{24.59}$$

$E°$ values do not show much variation; all fall between -2.52 V (La) and -2.25 V (Lu). The differences in properties that do exist among the lanthanides mostly arise from the lanthanide contraction discussed in Section 24-1. This contraction is best illustrated in the radii of the ions M^{3+}. These decrease regularly by about 1 to 2 pm for each unit increase in atomic number, from a radius of 106 pm for La^{3+} to 85 pm for Lu^{3+}.

Separation of the Lanthanides. The lanthanide elements are extremely difficult to extract from their natural sources and to separate from one another. All the methods for doing so are based on this principle: Species that are *strongly dissimilar* can often be completely separated in a one-step process, such as separating $Ag^+(aq)$ and $Cu^{2+}(aq)$ by adding $Cl^-(aq)$—AgCl is insoluble. Species that are very *similar* can at best be *fractionated* in a one-step process. That is, the ratio of the concentration of one species to that of another can be altered slightly. To achieve a complete separation may require repetition of the same basic step hundreds, or even thousands, of times.

Procedures that have been used to separate the lanthanides include fractional crystallization, fractional precipitation, solvent extraction, and ion exchange. The details of these procedures are beyond the scope of this text, but we should note that a number of advances in chemical technology came about through development of these procedures.

Reactions of the Lanthanides. The lanthanides (Ln) are active metals that liberate $H_2(g)$ from hot water and from dilute acids by undergoing oxidation to $Ln^{3+}(aq)$. The lanthanides combine with $O_2(g)$, sulfur, the halogens, $N_2(g)$, $H_2(g)$, and carbon, in much the same way as expected for metals about as active as the alkaline earths. Preparation of the pure metals can be achieved by electrolytic reduction of Ln^{3+} in a molten salt.

The most common oxidation state for the lanthanides is $+3$. About half the lanthanides can also be obtained in the oxidation state $+2$, and about half, $+4$. The special stability associated with an electron configuration involving half-filled f orbitals (f^7) may account for some of the observed oxidation states, but the reason for the predominance of the $+3$ oxidation state is less clear. Most of the lanthanide ions are paramagnetic and colored in aqueous solution.

The chlorides, bromides, iodides, nitrates, and perchlorates of the lanthanides are all water soluble; the sulfates are of variable solubility. The oxides and hydroxides are basic.

24-11 Qualitative Analysis of Some Transition Metal Ions

Here are some aspects of the qualitative analysis scheme that we have already explored

- A general description of the qualitative analysis scheme and the five cation groups (Section 5-10).
- Qualitative analysis cation group 1 (Section 19-9).
- Precipitation and dissolving of metal sulfides (Section 19-9).
- Sulfides and qualitative analysis (Section 23-2).

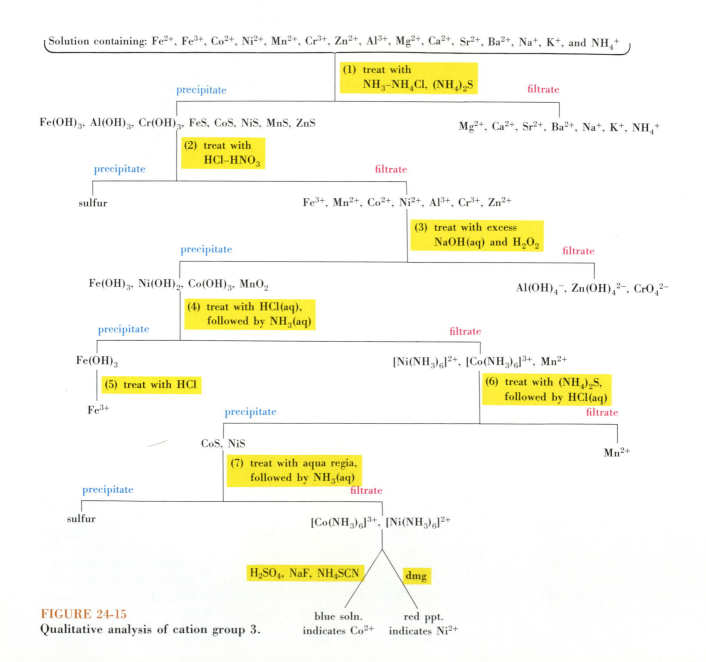

FIGURE 24-15
Qualitative analysis of cation group 3.

Qualitative tests for Co^{2+} and Fe^{3+}. In the presence of SCN^-, Co^{2+} forms a blue thiocyanato complex ion, $[Co(SCN)_4]^{2-}$ (top). Fe^{3+}, even if present only in trace amounts, produces the strongly colored, blood-red complex ion $[Fe(H_2O)_5SCN]^{2+}$ (center).

However, if a solution containing both Co^{2+} and Fe^{3+} is treated with an excess of F^-, the Fe^{3+} is converted to the extremely stable, pale yellow $[FeF_6]^{3-}$, and the Co^{2+} can be detected through its thiocyanato complex ion, which now appears as a blue-green color (bottom). [Carey B. Van Loon]

In this return to the subject of qualitative analysis we stress

- How cations are separated and identified following precipitation of a cation group (cation group 3).
- How the procedures of qualitative analysis can be used to illustrate chemical principles and some descriptive chemistry of the metals.

In Chapter 25 we cite some examples of complex ion formation in qualitative analysis.

Figure 24-15 outlines one possible scheme for cation group 3. The seven numbered steps in the outline are described below.

1. A solution containing the cations of groups 3, 4, and 5 (see Figure 5-12) is treated with $(NH_4)_2S$ in an NH_3–NH_4Cl buffer solution. Under these conditions Fe^{3+}, Cr^{3+}, and Al^{3+} precipitate as hydroxides and Fe^{2+}, Co^{2+}, Ni^{2+}, Mn^{2+}, and Zn^{2+} as sulfides. The solution contains cations of groups 4 and 5.
Note: Since the sulfides of the ammonium sulfide group are more soluble than most sulfides, we must use a basic solution to maintain a high $[S^{2-}]$ (or $[HS^-]$) for their precipitation. The pH of the solution must not be too high, however, or Mg^{2+} will precipitate as $Mg(OH)_2$. We described the use of a NH_3–NH_4Cl buffer to regulate the precipitation of $Mg(OH)_2$ in Example 19-12.

2. The precipitate from step 1 is treated with aqua regia (HNO_3–HCl), and any precipitated sulfur is filtered off.
Note: All components of the precipitate in step 1, except NiS and CoS, are readily soluble in HCl(aq). We must use aqua regia to dissolve NiS and CoS (recall equation 23.28).

3. The solution from step 2 is treated with an excess of NaOH. The hydroxides of Al, Zn, and Cr form and then dissolve; they are *amphoteric*. The hydroxides of Fe, Co, Ni, and Mn are *basic* and precipitate. The mixture of solution and precipitate is next treated with H_2O_2, which oxidizes $[Cr(OH)_4]^-$ to CrO_4^{2-} in the solution, and Co(II) to Co(III) and Mn(II) to Mn(IV) in the precipitate. The solution is subjected to further separations and tests for Al, Zn, and Cr. A yellow color in the solution indicates the presence of chromium (as CrO_4^{2-}). This can be confirmed by precipitating $BaCrO_4(s)$ from a slightly acidic solution.
Note: Because CrO_4^{2-} in alkaline solution is not a good oxidizing agent (reaction 24.21), we can easily oxidize $[Cr(OH)_4]^-$ to CrO_4^{2-}.

4. The precipitate from step 3 can be analyzed in a numbers of ways. In one method we dissolve the precipitate in HCl(aq) and then add $NH_3(aq)$. This causes $Fe(OH)_3$ to reprecipitate and Co^{2+} and Ni^{2+} to be converted to $[Co(NH_3)_6]^{3+}$ and $[Ni(NH_3)_6]^{2+}$. Manganese also appears in solution as Mn^{2+}.
Note: $MnO_2(s)$ is reduced to $Mn^{2+}(aq)$ by HCl(aq). In an alkaline solution Co(II) is oxidized to Co(III) by $O_2(g)$, and the presence of NH_3 stabilizes the Co(III) through the formation of the complex ion $[Co(NH_3)_6]^{3+}$.

5. The $Fe(OH)_3(s)$ is filtered off and dissolved in HCl(aq). The presence of Fe^{3+} is confirmed with the tests outlined in Table 24-6.

6. The filtrate containing $[Co(NH_3)_6]^{3+}$, $[Ni(NH_3)_6]^{2+}$, and Mn^{2+} is treated with $(NH_4)_2S$ to reprecipitate the sulfides. Because MnS is considerably more soluble than CoS and NiS, it dissolves in HCl(aq), leaving a residue of CoS and NiS. Tests for Mn^{2+} are performed on the filtrate.

7. The sulfide precipitate is dissolved in aqua regia, followed by treatment with $NH_3(aq)$. Thus, the species $[Co(NH_3)_6]^{3+}$ and $[Ni(NH_3)_6]^{2+}$ are obtained once again. A portion of the solution containing these ions is treated with dimethylglyoxime (dmg). Formation of a scarlet precipitate indicates the presence of nickel (see structure 24.38). A test for cobalt is performed on a second portion of the solution which is treated with H_2SO_4, NaF, and NH_4SCN. The presence of cobalt is disclosed by the formation of the blue complex ion $[Co(SCN)_4]^{2-}$.
Note: Acidification with H_2SO_4 converts Co(III) back to Co(II). The presence of NaF is required to complex Fe^{3+} as $[FeF_6]^{3-}$, so that any traces of Fe^{3+} present in the solution will not react with SCN^- and interfere with the test for cobalt.

Summary

More than half the elements are transition elements. The transition elements are all metals (most are more active than hydrogen). They tend to be found in several different oxidation states in their compounds, and they readily form complex ions (discussed further in Chapter 25).

Within a group of *d* block elements, the members of the *second* and *third* transition series resemble each other more than they do the group member from the first transition series. This is a consequence of the phenomenon known as the *lanthanide contraction*. Another type of resemblance is that which occurs among certain adjacent members of the same transition series. One such group considered in this chapter is the iron triad—Fe, Co, Ni. A particular property that these elements share is ferromagnetism.

The possibility of a variety of oxidation states for the transition metals means that oxidation and reduction are involved in many of their reactions. Some of the more important oxidizing agents used in the laboratory are transition metal compounds, such as dichromates and perman-

ganates. Dichromate ion participates in acid–base and hydration–dehydration equilibria with chromate ion. Chromate ion is a good precipitating agent. Most of the oxides and hydroxides of the transition metals are basic. This is typical of fairly active metals. However, certain oxides and hydroxides are amphoteric, and several examples are presented in the chapter.

The transition elements are among the most widely used metals. Their desirable properties range from the ready availability, low cost, and structural strength of iron to the excellent electrical conductivity of copper. Several of the transition metals—V, Cr, Mn, Co, Ni, Mo, W—are added to iron in steel making. The metallurgies of several transition metals are discussed in this chapter.

Several of the first transition series metals fall within the same group of cations in the qualitative analysis scheme. Many of the acid–base, oxidation–reduction, and precipitation reactions of this chapter are essential to the experimental procedures for separating and identifying these cations.

Summarizing Example

In aqueous solution, vanadium can exist in several different oxidation states, distinguished by their colors. One method of preparing these different species is to start with vanadium in the +4 oxidation state and to produce other species by appropriate oxidation–reduction reactions. In the following example the starting material is $VOSO_4$. We have already considered (Example 24-2) how the +5 oxidation state can be obtained using $MnO_4^-(aq)$ in acidic solution as the oxidizing agent.

1. Select a *reducing* agent that could be used to produce a solution of $V^{2+}(aq)$ starting with VO^{2+} in acidic solution. Write a balanced equation for this redox reaction.

Solution. Let us consider that the reduction occurs in two steps. First VO^{2+} is reduced to V^{3+}. From data in Figure 24-3 we can write

$$VO^{2+}(aq) + 2\ H^+(aq) + e^- \longrightarrow V^{3+}(aq) + H_2O \qquad E° = +0.361\ V$$

Then the $V^{3+}(aq)$ is reduced to $V^{2+}(aq)$,

$$V^{3+}(aq) + e^- \longrightarrow V^{2+}(aq) \qquad E° = -0.255\ V$$

In order to carry out the second step we must use an oxidation half-reaction for which $-E° > +0.255\ V$. Then the net redox reaction will have $E° > 0$. Whatever oxidizing agent will serve this purpose will also be effective in reducing $VO^{2+}(aq)$ to $V^{3+}(aq)$.

From Table 21-1 we might select reducing agents such as Fe, Zn, Al, or some other active metal. Let us choose zinc. The first step will be

Some vanadium species in solution. The yellow solution has vanadium in the +5 oxidation state, as VO_2^+. In the blue solution the oxidation state is +4, in VO^{2+}. The green solution contains V^{3+}, and the violet (amber) solution contains V^{2+}. [Courtesy Teledyne Wah Chang, Albany, OR]

Oxid:	$Zn(s) \longrightarrow Zn^{2+}(aq) + 2\ e^-$	$-E° = -(-0.763\ V) = +0.763\ V$
Red:	$2\{VO^{2+}(aq) + 2\ H^+(aq) + e^- \longrightarrow V^{3+}(aq) + H_2O\}$	$E° = +0.361\ V$
Net:	$Zn + 2\ VO^{2+} + 4\ H^+ \longrightarrow Zn^{2+} + 2\ V^{3+} + 2\ H_2O$	$E°_{cell} = 1.124\ V$

The second step will be

Oxid: $Zn(s) \longrightarrow Zn^{2+}(aq) + 2\ e^-$ $\qquad -E° = -(-0.763\ V) = +0.763\ V$
Red: $2\{V^{3+}(aq) + e^- \longrightarrow V^{2+}(aq)\}$ $\qquad E° = -0.255\ V$

Net: $Zn(s) + 2\ V^{3+}(aq) \longrightarrow Zn^{2+}(aq) + 2\ V^{2+}(aq)$ $\qquad E°_{cell} = +0.508\ V$

Each step of this two-step process is spontaneous, as is the overall reaction, which is obtained by adding together the two net equations.

$$2\ Zn(s) + 2\ VO^{2+}(aq) + 4\ H^+(aq) \longrightarrow 2\ Zn^{2+}(aq) + 2\ V^{2+}(aq) + 2\ H_2O$$

or

$$Zn(s) + VO^{2+}(aq) + 2\ H^+(aq) \longrightarrow Zn^{2+}(aq) + V^{2+}(aq) + H_2O$$

2. Select a *reducing* agent that could be used to produce a solution of $V^{3+}(aq)$ starting with VO^{2+} in acidic solution. Write a balanced equation for this redox reaction.

Solution. The important difference between this and part 1 is that here we want the reduction process to stop when V^{3+} has been produced. That is, we should use a reducing agent that is good enough to reduce VO^{2+} to V^{3+} but not good enough to reduce V^{3+} to V^{2+}. This reducing agent must have a value of $-E°$ that does not exceed $+0.255\ V$. If the first step is to occur, however, $-E°$ will have to exceed $-0.361\ V$. Thus, for the oxidation half-reaction: $-0.361\ V < -E° < +0.255\ V$. Since Table 21-1 lists reduction half-reactions, we must look for a value such that $-0.255\ V < E° < +0.361\ V$. $Sn^{2+}(aq)$ is a suitable reducing agent, as we see from the half-equations and the net equation.

Oxid: $Sn^{2+}(aq) \longrightarrow Sn^{4+}(aq) + 2\ e^-$ $\qquad -E° = -(+0.154\ V) = -0.154\ V$
Red: $2\{VO^{2+}(aq) + 2\ H^+ + e^- \longrightarrow V^{3+} + H_2O\}$ $\qquad E° = +0.361\ V$

Net: $Sn^{2+} + 2\ VO^{2+} + 4\ H^+ \longrightarrow Sn^{4+} + 2\ V^{3+} + 2\ H_2O$ $\qquad E°_{cell} = +0.207\ V$

Key Terms

ferromagnetism (24-7)
iron triad (24-7)
lanthanide contraction (24-1)

lanthanide elements (24-10)
metal carbonyls (24-7)
pig iron (24-8)

steel (24-8)
superconductivity (24-10)

Highlighted Expressions

The Kroll process for producing titanium (24.4)
Reduction of chromite ore (24.7)
Acid ionization of $[Cr(H_2O)_6]^{3+}$ (24.8)
Thermite reaction for the preparation of chromium (24.13)
Chromate ion–dichromate ion equilibrium (24.17)
Titration of $Fe^{2+}(aq)$ with $Cr_2O_7{}^{2-}(aq)$ (24.20)
Titration of $Fe^{2+}(aq)$ with $MnO_4{}^-(aq)$ (24.27)

Acid ionization of $[Fe(H_2O)_6]^{3+}$ (24.34)
Blast furnace reaction: reduction of Fe_2O_3 (24.39)
Blast furnace reaction: slag formation (24.40)
Action of aqua regia on gold (24.47)
Formation of the complex ion $[Cu(NH_3)_4]^{2+}$ (24.50)
Cyanidation process for the extraction of gold (24.56)
Displacement of Au(I) from solution (24.57)

Review Problems

1. Provide an acceptable name or formula for each of the following.
  (a) barium dichromate (b) $Sc(OH)_3$
  (c) chromium trioxide (d) MnO
  (e) iron(II) silicate (f) $Fe(CO)_5$

2. Describe the chemical composition of the materials described as (a) ferromanganese; (b) cast iron; (c) chromite ore; (d) chrome alum; (e) aqua regia; (f) Prussian blue.

3. Complete and balance the following equations. If no reaction occurs, so state.

(a) $TiCl_4(g) + Na(l) \xrightarrow{\Delta}$

(b) $Cr_2O_3(s) + Al(s) \xrightarrow{\Delta}$

(c) $Ag(s) + HCl(aq) \rightarrow$

(d) $K_2Cr_2O_7(aq) + KOH(aq) \rightarrow$

(e) $MnO_2(s) + C(s) \xrightarrow{\Delta}$

(f) $Fe(OH)_3(s) + NaOH(aq) \rightarrow$

4. Balance the following oxidation–reduction equations.

(a) $Fe_2S_3(s) + H_2O + O_2(g) \rightarrow Fe(OH)_3(s) + S(s)$

(b) $Mn^{2+} + S_2O_8^{2-} + H_2O \rightarrow MnO_4^- + SO_4^{2-} + H^+$

(c) $Ag(s) + CN^- + O_2(g) + H_2O \rightarrow [Ag(CN)_2]^- + OH^-$

5. Suggest chemical reactions that might be used to obtain the following compounds from Na_2CrO_4: (a) $Na_2Cr_2O_7$; (b) Cr_2O_3; (c) $CrCl_3$; (d) $NaCr(OH)_4$.

6. Through a chemical equation, give an example to represent the dissolving of

(a) a transition metal in an oxidizing acid;

(b) a transition metal oxide in $NaOH(aq)$;

(c) a transition metal sulfide in $HCl(aq)$;

(d) an iron triad metal hydroxide in $NH_3(aq)$;

(e) an inner transition metal in $HCl(aq)$.

7. Through orbital diagrams, write electron configurations for the following transition element atoms and ions: (a) Ti; (b) V^{3+}; (c) Cr^{2+}; (d) Mn^{2+}; (e) Fe^{3+}.

8. Arrange the following species according to the number of unpaired electrons they contain, starting with the one that has the greatest number: (a) Fe; (b) Sc^{3+}; (c) Ti^{2+}; (d) Mn^{4+}; (e) Cr; (f) Cu^{2+}.

9. A qualitative analysis test for Mn^{2+} referred to in step 6 on page 899 involves its oxidation to MnO_4^- in acidic solution by sodium bismuthate, $NaBiO_3$. The bismuthate ion is reduced to Bi^{3+}. Write a balanced equation for this test.

10. What are the relative concentrations of $Cr_2O_7^{2-}$ and CrO_4^{2-} in a solution with a pH of (a) 5.0; (b) 9.12?

11. Equilibrium is established in reaction (24.43) at 25 °C with all three reactants—Cu, Cu^+, and Cu^{2+}—present.

(a) What must be the molarity concentrations of Cu^+ and Cu^{2+} if they are found to be equal to each other?

(b) What percent of the copper ion is present as Cu^+ if the total copper ion concentration at equilibrium is 1.00×10^{-4} M?

12. What single reagent solution (including water) could be used to effect the separation of the following pairs of solids? (a) NaOH and $Fe(OH)_3$; (b) $Ni(OH)_2$ and $Fe(OH)_3$; (c) Cr_2O_3 and $Fe(OH)_3$; (d) MnS and CoS.

Exercises

Properties of the transition elements

13. Describe how the transition elements compare with representative metals (e.g., group 2A) with respect to oxidation states, formation of complexes, colors of compounds, and magnetic properties.

14. Why do the atomic radii vary so much more for two representative elements that differ by one unit in atomic number than they do for two transitions elements that differ by one unit?

15. With but minor irregularities, the melting points of the first series of transition metals rise from that of Sc to Cr and then fall to that of Cu (also note that the melting point of Zn = 420 °C). Give a plausible reason for this observation, based on atomic structure.

16. With but minor exceptions the standard electrode potentials of the first transition series metals increase regularly, from that of Sc (-2.08 V) to that of Cu ($+0.34$ V). In the lanthanide series the potentials increase by less than 0.3 V, from that of La (-2.52 V) to that of Lu (-2.25 V).

(a) Why is the variation of $E°$ values so much smaller for the lanthanides than for the first transition series?

(b) Why do you suppose that the lanthanides are more active metals than those of the first transition series?

Oxidation–reduction

17. Write plausible half-equations to represent (a) $VO^{2+}(aq)$ as an oxidizing agent in acidic solution; (b) $Cr^{2+}(aq)$ as a reducing agent; (c) the oxidation of $Fe(OH)_3(s)$ to FeO_4^{2-} in basic solution; (d) the reduction of $[Ag(CN)_2]^-$ to silver metal.

18. Use electrode potential data from this chapter and from Table 21-1 to predict whether each of the following reactions will occur to any significant extent as written.

(a) $2\ VO_2^+ + 6\ Br^- + 8\ H^+ \rightarrow 2\ V^{2+} + 3\ Br_2 + 4\ H_2O$

(b) $VO_2^+ + Fe^{2+} + 2\ H^+ \rightarrow VO^{2+} + Fe^{3+} + H_2O$

(c) $MnO_2(s) + H_2O_2 + 2\ H^+ \rightarrow Mn^{2+} + 2\ H_2O + O_2(g)$

19. When Zn is added to an acidified solution of $K_2Cr_2O_7$, the color of the solution changes from orange to green, then to blue, and, over a period of time, back to green. Write equations for this series of reactions.

20. The electrode potential diagram of Figure 24-8 does not include a value of $E°$ for the reduction of MnO_4^- to Mn^{2+} in acidic solution. Use other data in the figure to establish this value, and compare your result with the value listed in Table 21-1.

21. You are given these three reducing agents: Zn, Sn^{2+}, and I^-. For each of the reducing agents, determine whether it is capable of reducing (a) $Cr_2O_7^{2-}$ to Cr^{3+}; (b) Cr^{3+} to Cr^{2+}; (c) $SO_4^{2-}(aq)$ to $SO_2(g)$

22. Both $Cr_2O_7^{2-}(aq)$ and $MnO_4^-(aq)$ can be used to titrate $Fe^{2+}(aq)$ (reactions 24.20 and 24.27). Suppose you have available 0.1000 M solutions of each. For a given sample of $Fe^{2+}(aq)$,

(a) with which solution, $Cr_2O_7^{2-}(aq)$ or $MnO_4^-(aq)$, would the greater volume of titrant (titrating solution) be required? Explain.

(b) If a given titration requires 24.50 mL of 0.1000 M $Cr_2O_7^{2-}(aq)$, how many mL of 0.1000 M $MnO_4^-(aq)$ would have been required if it had been used instead?

Chromium and chromium compounds

23. In qualitative analysis, a confirmatory test for chromium involves the formation of a blue transient peroxo species, CrO_5, which is better represented as $CrO(O_2)_2$. It is formed when H_2O_2 is added to an acidic solution containing $Cr_2O_7^{2-}$. What is the oxidation state of Cr in $CrO(O_2)_2$?

24. Use data from the text and the method of Example 23-2 to supply electrode potentials for the following diagram in acidic solution.

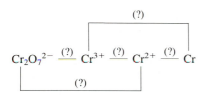

25. Explain why it is reasonable to expect the principal chemistry of dichromate ion to involve oxidation–reduction reactions and that of chromate ion, precipitation reactions.

26. If $CO_2(g)$ under pressure is passed into $Na_2CrO_4(aq)$, $Na_2Cr_2O_7(aq)$ is formed. What is the function of the $CO_2(g)$? Write a plausible equation for the net reaction.

27. If a solution is prepared by dissolving 0.100 mol Na_2CrO_4 in 1.00 L of a buffer solution at pH 7.00, what will be $[CrO_4^{2-}]$ and $[Cr_2O_7^{2-}]$ in the solution?

28. How long would an electric current of 4.2 A have to pass through a chrome-plating bath (see page 879) to produce a deposit 0.0010 mm thick on an object with a surface area of 22.7 cm²? (The density of Cr is 7.14 g/cm³.)

29. In some qualitative analysis procedures Ba^{2+} is separated from Sr^{2+} and Ca^{2+} by precipitating $BaCrO_4(s)$ while the Sr^{2+} and Ca^{2+} remain in solution. Show that this separation will work if a solution is 0.10 M in Ba^{2+}, 0.10 M in Sr^{2+}, 0.10 M in Ca^{2+}, 1.0 M in $HC_2H_3O_2$, 1.0 M in $NH_4C_2H_3O_2$, and 0.0010 M in $Cr_2O_7^{2-}$. [Use data from this and previous chapters, as necessary; $K_{sp}(BaCrO_4) = 1.2 \times 10^{-10}$, $K_{sp}(SrCrO_4) = 2.2 \times 10^{-5}$.]

★30. Equation (24.17), which represents the chromate–dichromate equilibrium, is actually the sum of two equilibrium expressions. The first is an acid–base reaction, $H^+ + CrO_4^{2-} \rightleftharpoons HCrO_4^-$. The second reaction involves elimination of a water molecule from between two $HCrO_4^-$ ions (a dehydration reaction), $2\,HCrO_4^- \rightleftharpoons Cr_2O_7^{2-} + H_2O$. If the ionization constant, K_a, for $HCrO_4^-$ is 3.2×10^{-7}, what is the value of K for the dehydration reaction? [*Hint:* You must also use the value of K for reaction 24.17.]

The iron triad

31. Explain why the iron triad elements resemble each other so strongly.

32. How does the property of ferromagnetism differ from ordinary paramagnetism? Why is this property so limited in its occurrence among the elements?

33. Which of the iron triad ions, M^{2+}, would you expect to be oxidized to M^{3+} by $O_2(g)$ in acidic solution?

$$4\,M^{2+} + O_2(g) + 4\,H^+ \rightarrow 4\,M^{3+} + 2\,H_2O$$

Carbonyls

34. Use methods outlined in the chapter to predict formulas and structures of the carbonyls of **(a)** molybdenum; **(b)** osmium; **(c)** rhenium.

35. Iron and nickel carbonyls are liquids at room temperature, but that of cobalt is a solid. Why should this be so?

36. The compound $Na[V(CO)_6]$ has been reported. Discuss the probable nature of chemical bonding in this compound.

The coinage metals

37. Described below are three reactions encountered in the hydrometallurgy of the coinage metals. Write plausible chemical equations for these reactions.
   (a) Copper is precipitated from a solution of copper(II) sulfate by treatment with $H_2(g)$.
   (b) Gold is precipitated from a solution of Au^+ by adding iron(II) sulfate.
   (c) Copper(II) chloride solution is reduced to copper(I) chloride when treated with $SO_2(g)$ in acidic solution.

38. Use data from Table 24-10, as necessary, to calculate $E°$ for the reduction half-reaction $Cu^{2+}(aq) + e^- \rightarrow Cu^+(aq)$.

39. In the metallurgical extraction of Ag and Au often an alloy of the two metals is obtained. They can be separated either with concentrated HNO_3 or boiling concentrated H_2SO_4, in a process called *parting*. Write chemical equations to show how these separations work.

40. In acidic solution silver(II) oxide first dissolves to produce $Ag^{2+}(aq)$. This is followed by the oxidation of H_2O to $O_2(g)$ and the reduction of Ag^{2+} to Ag^+.
   (a) Write equations for the dissolution and oxidation–reduction reactions.
   ★(b) Show that the oxidation-reduction reaction is indeed spontaneous. [*Hint:* You will have to use the method of Example 23-2 for part of the solution.]

41. Determine the values of K for reactions (24.45) and (24.46).

★42. If one attempts to make CuI_2 by the reaction of $Cu^{2+}(aq)$ and $I^-(aq)$, $CuI(s)$ and I_2 are obtained instead. Without performing detailed calculations, show why this reaction should occur as written.

$$2\,Cu^{2+} + 4\,I^- \rightarrow 2\,CuI(s) + I_2$$

[*Hint:* Use electrode potentials and $K_{sp}(CuI) = 1.1 \times 10^{-12}$.]

43. Without performing detailed calculations, show that significant decomposition of AuCl occurs if one attempts to make a saturated aqueous solution. Use equation (24.46) and $K_{sp}(AuCl) = 2.0 \times 10^{-13}$.

Qualitative analysis

44. With respect to the qualitative analysis scheme for the ammonium sulfide group outlined in Section 24-11, write equations to represent **(a)** dissolving FeS in HCl(aq); **(b)** dissolving CoS in aqua regia; **(c)** the action of excess NaOH(aq) and H_2O_2 on $Cr^{3+}(aq)$; **(d)** the action of $NH_3(aq)$ on a mixture of $Fe(OH)_3(s)$ and $Ni(OH)_2(s)$; **(e)** a qualitative test for $Fe^{2+}(aq)$; **(f)** the action of HCl(aq) on $MnO_2(s)$.

45. Describe what might happen in the qualitative analysis of the ammonium sulfide group if one did the following, in error.
   (a) failed to include NH_4Cl in the reagent used to treat the original solution in the flowchart of Figure 24-15.
   (b) failed to add H_2O_2 to the reagent used in step 3.
   (c) used NaOH(aq) in step 4 rather than $NH_3(aq)$.
   (d) used aqua regia in step 6 instead of HCl(aq).
   (e) neglected to use NaF in the test for cobalt in step 7.

46. A particular stainless steel sample is known to contain some combination of Ni, Cr, and Mn alloyed with Fe. Outline a simplified qualitative analysis scheme that would enable you to determine which of these metals are present.

47. A solution may contain any of the following ions: Fe^{3+}, Ni^{2+}, Cr^{3+}, Zn^{2+}, Mn^{2+}. The solution is treated as outlined below, with the results indicated. Which of the ions is(are) probably present? Which of the ions is(are) probably absent? Are there some ions about which you have insufficient information to reach a decision? Explain.

(1) The original unknown solution is treated with $(NH_4)_2S(aq)$ in a buffered basic solution. A *dark* precipitate is obtained.
(2) With the exception of a small residue of sulfur, the precipitate from (1) is completely soluble in aqua regia (HNO_3–HCl).
(3) The solution from (2) is treated with NaOH(aq) and $H_2O_2(aq)$. A *dark* precipitate separates from the solution. The solution is *colorless*.
(4) The precipitate from (3) dissolves in HCl to produce a *colored* solution.
(5) The solution from (4) is treated with $NH_3(aq)$. A *dark* precipitate forms.
(6) The precipitate from (5) is soluble in HCl(aq) and the solution develops an intense *red* color when treated with $SCN^-(aq)$.

48. An unknown solution contains only group 3 ions as its possible cations. The solution is first treated with an NH_3–NH_4Cl buffer solution and no precipitate is observed. A light-colored precipitate does form when $(NH_4)_2S$ is added. This precipitate is completely soluble in HCl(aq). Indicate for each of the group 3 cations whether it is present or absent or whether its presence is uncertain.

★49. A metal sample, when dissolved and subjected to qualitative analysis, yields a cation group 3 precipitate that is partially soluble in NaOH(aq). The remainder dissolves in aqua regia. The solution from the aqua regia reaction again yields a precipitate when treated with $H_2O_2(aq)$ and NaOH(aq). This precipitate is completely soluble in $NH_3(aq)$. From these observations indicate what conclusions are possible concerning cation group 3 metals that may have been present in the original sample.

Quantitative analysis

50. Nickel can be determined by the precipitation of nickel dimethylglyoximate.
(a) What is the formula of this compound (see structure 24.38)?
(b) A 1.502-g sample of steel yields 0.259 g of nickel dimethylglyoximate. What is the percent Ni in the steel?

51. A 0.589-g sample of pyrolusite ore (impure MnO_2) is treated with 1.651 g of oxalic acid ($H_2C_2O_4 \cdot 2H_2O$) in an acidic medium. Following this reaction the excess oxalic acid is titrated with 0.1000 M $KMnO_4$, 30.06 mL being required. What is the % MnO_2 in the ore?

$$H_2C_2O_4 + MnO_2 + 2\,H^+ \rightarrow Mn^{2+} + 2\,H_2O + 2\,CO_2$$

$$5\,H_2C_2O_4 + 2\,MnO_4^- + 6\,H^+ \rightarrow$$
$$2\,Mn^{2+} + 8\,H_2O + 10\,CO_2$$

Additional Exercises

52. Write chemical equations for the following reactions of scandium and its compounds that are described in the text.
(a) the dissolving of $Sc(OH)_3(s)$ in HCl(aq);
(b) the dissolving of $Sc_2O_3(s)$ in NaOH(aq);
(c) the electrolysis of Sc_2O_3 dissolved in Na_3ScF_6.

53. Describe a simple chemical test to distinguish between $Fe^{2+}(aq)$ and $Fe^{3+}(aq)$.

54. Suggest a series of reactions, using common chemicals, by which each of the following syntheses can be performed: (a) $Fe(OH)_3(s)$ from FeS(s); (b) $BaCrO_4(s)$ from $BaCO_3(s)$ and $K_2Cr_2O_7(aq)$; (c) $CrCl_3(aq)$ from $(NH_4)_2Cr_2O_7(s)$; (d) $MnCO_3(s)$ from $MnO_2(s)$.

55. When a soluble lead compound is added to a solution containing primarily *orange* dichromate ion, *yellow* lead chromate precipitates. Describe the equilibria involved.

56. When *yellow* $BaCrO_4$ is dissolved in HCl(aq), a *green* solution is obtained. Write a chemical equation to account for the color change.

57. In the text three conclusions were reached about certain ions of manganese by using the electrode potential diagrams of Figure 24-8. Show that the following statements are true as well.
(a) MnO_4^- in acidic solution oxidizes Mn^{2+} to $MnO_2(s)$; in basic solution MnO_4^- oxidizes $MnO_2(s)$ to MnO_4^{2-}.
(b) MnO_4^- will slowly liberate $O_2(g)$ from water, either in acidic or basic solution.

58. Although Au is soluble in aqua regia (3 parts HCl + 1 part HNO_3), Ag is not. What is the likely reason(s) for this difference?

59. The only important compounds of Ag(II) are AgF_2 and AgO. Explain why you would expect these two compounds to be stable, but not others such as $AgCl_2$, $AgBr_2$, AgS, etc.

60. Reaction (24.30) occurs spontaneously when all reactants and products are in their standard states.
(a) What is the value of K for reaction (24.30)?
(b) If a solution that is 0.10 M Fe^{2+} has a pH = 5.0 and is in contact with $O_2(g)$ in air at a partial pressure of 0.20 atm, will the oxidation of Fe^{2+} to Fe^{3+} occur to a significant extent?

61. A solution is believed to contain one or more of the following ions: Cr^{3+}, Zn^{2+}, Fe^{3+}, Ni^{2+}. When the solution is treated with NaOH(aq) and H_2O_2, a precipitate is formed. The solution separated from this precipitate is colorless. The precipitate is dissolved in HCl(aq) and the resulting solution is treated with $NH_3(aq)$. No precipitation occurs. Based solely on these observations, what conclusions can you draw about the ions in the solution? That is, are any proved to be present? to be absent? Are further tests needed?

62. Write a plausible equation for the disproportionation reaction brought about by passing $CO_2(g)$ into $MnO_4^{2-}(aq)$.

★63. Show that the corrosion reaction in which Cu is converted to its basic carbonate (reaction 24.48) can be thought of in terms of a combination of oxidation–reduction, acid–base, and precipitation reactions.

★64. In an atmosphere with industrial smog Cu corrodes to a basic sulfate, $Cu_2(OH)_2SO_4$. Propose a series of chemical reactions to describe this corrosion.

★65. Calculate $[Au^+]$, $[Au^{3+}]$, and $[Cl^-]$ in a saturated solution obtained by dissolving AuCl in water. [*Hint:* Neglect complex ion formation and use data from Exercise 43.]

*66. Suppose a solution is 0.50 M in OH^- and 0.10 M in MnO_4^- in the presence of $MnO_2(s)$. Show that a significant concentration of MnO_4^{2-} exists in the solution.

*67. Use data from Figure 24-8, together with equations from elsewhere in the text, to estimate K_{sp} for $Mn(OH)_2$.

*68. A steel sample is to be analyzed for Cr and Mn, simultaneously. By suitable treatment the Cr is oxidized to $Cr_2O_7^{2-}$ and the Mn to MnO_4^-. A 10.000-g sample of steel is used to produce

250.0 mL of a solution containing $Cr_2O_7^{2-}$ and MnO_4^-. A 10.00-mL portion (aliquot) of this solution is added to a $BaCl_2$ solution, and by proper adjustment of the acidity, the chromium is completely precipitated as $BaCrO_4$; 0.0549 g is obtained. A second 10.00-mL aliquot of this solution requires exactly 15.95 mL of 0.0750 M standard Fe^{2+} solution for its titration (in acid solution). Calculate the % Mn and % Cr in the steel sample. [*Hint:* Both MnO_4^- and $Cr_2O_7^{2-}$ are reduced by Fe^{2+}.]

Self-Test Questions

For questions 69 through 78 select the single item that best completes each statement.

69. A property generally expected of the transition elements is (a) low melting points; (b) high ionization energies; (c) variable oxidation states; (d) positive standard electrode (reduction) potentials.

70. Only one of the following ions is diamagnetic. That ion is (a) Cr^{2+}; (b) Fe^{3+}; (c) Cu^{2+} (d) Sc^{3+}.

71. Of the following elements the one that is *not* expected to display an oxidation state of +6 in any of its compounds is (a) Ti; (b) Cr; (c) Mn; (d) Fe.

72. One might expect $Cl_2(g)$ to be produced if an HCl(aq) solution is heated strongly in the presence of (a) CuO; (b) MnO_2; (c) Cr^{3+}; (d) Fe^{2+}.

73. Of the following ions the one most likely to disproportionate in aqueous solution is (a) Fe^{3+}; (b) Ag^+; (c) K^+; (d) Cu^+.

74. The best oxidizing agent among the following ions is (a) Na^+; (b) Al^{3+}; (c) Ag^{2+}; (d) Cu^{2+}.

75. To increase the water solubility of $BaCrO_4(s)$, add to the water (a) NH_4Cl; (b) $Ba(OH)_2$; (c) Na_2CrO_4; (d) $BaCl_2$.

76. Of the following materials, the highest % C is found in (a) electrolytic copper; (b) cast iron; (c) steel; (d) stainless steel.

77. In its compounds the element that most often has an oxidation state of +3 is (a) Au; (b) Ag; (c) Ni; (d) Cu.

78. To separate Fe^{3+} and Ni^{2+}, obtaining one in a precipitate with the other remaining in solution, add to the solution (a) NaOH(aq); (b) $H_2S(g)$; (c) $NH_3(aq)$; (d) HCl(aq).

79. Why is +3 the most stable oxidation state for Fe, whereas it is +2 for Co and Ni?

80. What products are obtained when $Mg^{2+}(aq)$ and $Cr^{3+}(aq)$ are each treated with a limited amount of NaOH(aq)? With an excess of NaOH(aq)? Why are the results different in these two cases?

81. Which of these reagents—H_2O, NaOH(aq), HCl(aq)— can be used to dissolve the following compounds? **(a)** $CuSO_4$; **(b)** $Ni(OH)_2$; **(c)** $AgNO_3$; **(d)** FeS; **(e)** $Cr(OH)_3$?

25 Complex Ions and Coordination Compounds

The green solid, $CrCl_3 \cdot 6H_2O$, produces a green solution. After one to two days the solution color becomes violet. If the violet solution is evaporated to dryness, the green solid is recovered. This and other phenomena involving complex ions are described in this chapter. [Carey B. Van Loon].

906

In the opening photograph, the *green* compound in the watch glass is chromium(III) chloride hexahydrate, $CrCl_3 \cdot 6H_2O$. If we dissolve this green solid in water we obtain a *green* solution. Within a few hours the solution color becomes *blue-green,* and after a day or two the solution color is *violet.* If this violet solution is evaporated to dryness, we recover a *green* solid identical to the one we started with.

In the preceding chapter we discussed several situations involving a succession of color changes, and we attributed these to changes in oxidation state (recall the Summarizing Example). The color changes here, however, are not caused by oxidation–reduction reactions. The oxidation state of chromium remains at +3 throughout.

To explain these observations, we need a greater understanding of coordination compounds and complex ions. First, we need to develop some ideas about the geometrical structures of complex ions and the existence of coordination compounds with identical compositions but different structures and properties (called isomers). Then, we can consider the nature of bonding and the origin of color in complex ions.

At the end of the chapter we describe some of the many places where complex ions and coordination compounds are encountered in chemistry and biology.

25-1 Werner's Theory of Coordination Compounds: An Overview

In Section 19-7 we learned that the series of coordination compounds $CoCl_3 \cdot 6NH_3$, $CoCl_3 \cdot 5NH_3$, and $CoCl_3 \cdot 4NH_3$ are better represented as

$$[Co(NH_3)_6]Cl_3 \qquad [Co(NH_3)_5Cl]Cl_2 \qquad [Co(NH_3)_4Cl_2]Cl \qquad (25.1)$$
$$\text{(a)} \qquad\qquad\qquad \text{(b)} \qquad\qquad\qquad \text{(c)}$$

Alfred Werner (1866–1919). Werner's success in explaining coordination compounds came in large part through his application of new ideas: the theory of electrolytic dissociation and principles of structural chemistry. [German Information Center]

The Swiss chemist Alfred Werner explained these compounds in terms of a primary valence of three and a secondary valence of six. The primary valence is shown in the initial formation of the Co^{3+} ion. The secondary valence is displayed in the attachment of a fixed number of ligands directly to this central ion (six in the case of Co^{3+}). The ionization of a coordination compound can be represented as

$$[Co(NH_3)_6]Cl_3(aq) \longrightarrow [Co(NH_3)_6]^{3+}(aq) + 3\ Cl^-(aq)$$

and this ionization scheme helps to explain why an aqueous solution of compound (a) is a good electrical conductor and yields *three* moles of AgCl(s) per mole of compound when treated with $AgNO_3(aq)$. Compound (b), on the other hand, yields only *two* moles of AgCl(s), and compound (c), *one* mole of AgCl(s).

Werner's theory also accounts for a coordination compound such as

$$[Co(NH_3)_3Cl_3] \qquad (25.2)$$

which is a *nonelectrolyte* and yields no precipitate with $AgNO_3(aq)$, and a compound such as

$$Na[Co(NH_3)_2Cl_4] \qquad (25.3)$$

in which the complex ion has a net *negative* charge, that is, $[Co(NH_3)_2Cl_4]^-$.

The species $[Co(NH_3)_6]^{3+}$, $[Co(NH_3)_5Cl]^{2+}$, and $[Co(NH_3)_4Cl_2]^+$ are all **complex ions,** and since they carry a positive charge, they are *cations*. $[Co(NH_3)_2Cl_4]^-$ is also a complex ion, but it is an *anion*. $[Co(NH_3)_3Cl_3]$ is a neutral molecule. To describe any species involving coordination of ligands to a metal ion, whether the species is a cation, anion, or neutral molecule, we can use the term **complex.**

907

FIGURE 25-1

Structures of some complex ions.

Attachment of the NH_3 molecules occurs through the lone-pair electrons on the N atoms.

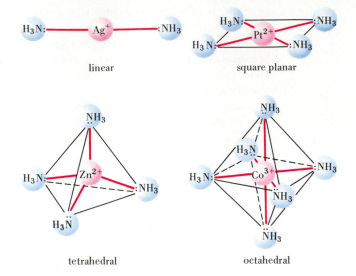

linear

square planar

tetrahedral

octahedral

An important postulate in Werner's theory is that the secondary valencies of a central metal ion are directed to specific positions in the region surrounding the ion (called the coordination sphere). This explains the fact that complex ions possess distinctive geometrical shapes. The four most commonly observed structures are those shown in Figure 25-1.

The number of secondary valencies displayed by a central ion is the coordination number of the ion. Coordination numbers ranging from 2 to 12 have been observed in complexes, although the number 6 is by far the most common, followed by 4. Coordination number 2 is mostly limited to complexes of Cu(I), Ag(I), and Au(I). Coordination numbers greater than 6 are not often found in members of the first transition series but are more common in those of the second and third series. Stable complexes with coordination numbers 3 and 5 are rarely encountered. The coordination number(s) exhibited by an ion depends on a number of factors, such as the ratio of the radius of the central metal ion to that of the attached ligands.

Coordination numbers of some common ions are listed in Table 25-1. One practical use of the coordination number is to assist in writing and interpreting formulas of complexes.

TABLE 25-1

Some Common Coordination Numbers of Metal Ions

Cu^+	2, 4	Zn^{2+}	4
Ag^+	2	Al^{3+}	4, 6
Au^+	2, 4	Sc^{3+}	6
Ca^{2+}	6	Cr^{3+}	6
Fe^{2+}	6	Fe^{3+}	6
Co^{2+}	4, 6	Co^{3+}	6
Ni^{2+}	4, 6	Au^{3+}	4
Cu^{2+}	4, 6		

Example 25-1

Relating the formula of a complex to the coordination number and oxidation state of the central metal ion. What are the coordination number and oxidation state of Al in the complex ion $[Al(H_2O)_4(OH)_2]^+$?

Solution. The complex ion has as ligands *four* H_2O molecules and *two* OH^- ions. **The coordination number is *six*.** Of these six ligand groups, *two* carry a charge of -1 each (the OH^- ions) and four are *neutral* (the H_2O molecules). The total contribution of the OH^- ions to the net charge on the complex ion is -2. Since the net charge on the complex ion is $+1$, the charge of the central aluminum ion, and hence **its oxidation state, must be $+3$.** Diagrammatically, we can write

charge = x

charge of -1 on OH^-; total negative charge = -2

$$[Al(H_2O)_4(OH)_2]^+$$

net charge on complex ion

coordination number = 6

$x - 2 = +1$
$x = +3$

SIMILAR EXAMPLES: Exercises 1, 2.

In the structure of $[Co(NH_3)_6]^{3+}$ depicted in Figure 25-1, all NH_3 ligands are equidistant from the Co^{3+} ion and the structure is octahedral. Figure 25-2 represents the structure of the complex ion $[Cu(NH_3)_4(H_2O)_2]^{2+}$. Here the two H_2O molecules in the so-called axial positions are more distant from the Cu^{2+} ion than are the four NH_3 molecules in the square planar (equatorial) arrangement. The ion has a distorted octahedral structure (called a tetragonal structure). If the axial ligands in such a structure are sufficiently far from the central ion, the octahedral structure with coordination number 6 gives way to the square planar structure with coordination number 4. The ion in Figure 25-2 is generally written as $[Cu(NH_3)_4]^{2+}$ and described as having square planar geometry.

25-2 Ligands

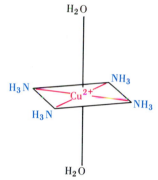

FIGURE 25-2
Structure of
$[Cu(NH_3)_4(H_2O)_2]^{2+}$.

A ligand is a species that is capable of donating an electron pair(s) to a central ion. It is a **Lewis base.** In accepting electron pairs, the central ion acts as a **Lewis acid.** A ligand that has only one pair of electrons that it can donate is called a **unidentate** ligand. Some examples of unidentate ligands are monoatomic anions like the halide ions, polyatomic anions like nitrite ion, simple molecules such as ammonia (ammine), and more complex molecules like methylamine, CH_3NH_2.

ligand name:	chloro	nitro	ammine	methylamine

Table 25-2 lists some common unidentate ligands.

Chelates. Some ligands are capable of donating more than a single electron pair, from different atoms in the ligand and to different sites in the geometric structure of a complex. These are called **multidentate** ligands. The molecule **ethylenediamine (en)** can donate *two* electron pairs, one from each N atom. It is a **bidentate** ligand.

$$H—\overset{\uparrow}{\underset{H}{N}}—CH_2CH_2—\overset{\uparrow}{\underset{H}{N}}—H$$

TABLE 25-2
Some Common Unidentate Ligands

Formula	Name as ligand	Formula	Name as ligand	Formula	Name as ligand
Neutral molecules		Anions		Anions	
H_2O	aqua[a]	F^-	fluoro	SO_4^{2-}	sulfato
NH_3	ammine	Cl^-	chloro	$S_2O_3^{2-}$	thiosulfato
CO	carbonyl	Br^-	bromo	NO_2^-	nitro[b]
NO	nitrosyl	I^-	iodo	ONO^-	nitrito[b]
CH_3NH_2	methylamine	O^{2-}	oxo	SCN^-	thiocyanato[c]
C_5H_5N	pyridine	OH^-	hydroxo	NCS^-	isothiocyanato[c]
		CN^-	cyano		

[a] Until recently, the term aquo was used.
[b] If the nitrite ion is attached through the N atom (—NO_2), the name *nitro* is used; if attached through an O atom (—ONO), *nitrito*.
[c] If the thiocyanate ion is attached through the S atom (—SCN), the name *thiocyanato* is used; if attachment is through the N atom (—NCS), *isothiocyanato*.

TABLE 25-3
Some Common Multidentate Ligands (Chelating Agents)

Abbreviation	Name	Formula
en	ethylenediamine	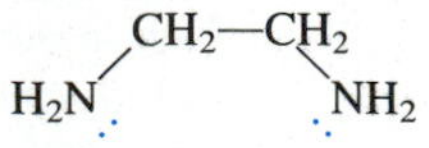 H_2N—CH_2—CH_2—NH_2
ox	oxalato	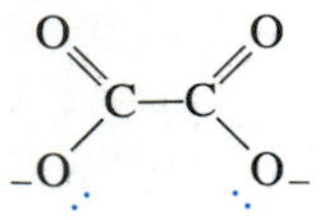 oxalate ion structure
EDTA	ethylenediaminetetraacetato	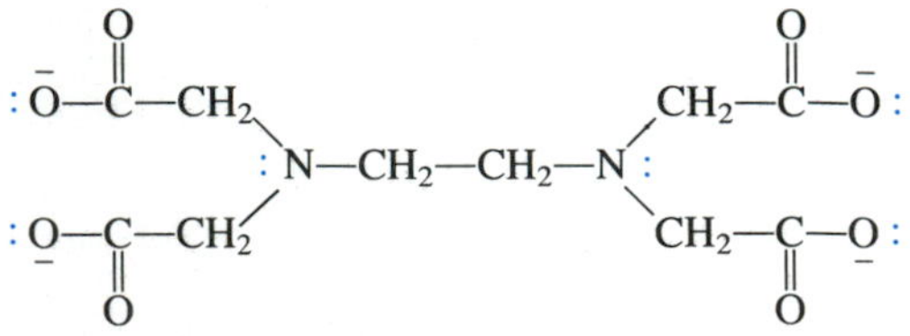 EDTA structure

FIGURE 25-3

Two representations of the chelate $[Pt(en)_2]^{2+}$.

The ligands attach at adjacent corners along an edge of the square. They do *not* bridge the square by attaching to opposite corners.

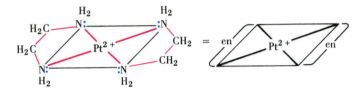

Three common multidentate ligands are shown in Table 25-3.

Figure 25-3 represents the attachment of two ethylenediamine (en) ligands to a Pt^{2+} ion. Here is how we can establish that each ligand is attached to two positions in the coordination sphere.

- The complex ion $[Pt(en)_2]^{2+}$ has no capacity to coordinate additional ligands, that is, it cannot coordinate additional NH_3, H_2O, Cl^-, . . . Since the coordination number is 4, each of the two en groups must be attached at two points.
- The en ligands in the complex ion exhibit no further basic properties. They cannot accept protons from water to produce OH^-, as they would if they had an available lone pair of electrons. Both —NH_2 groups of each en molecule must be tied up in the complex ion.

Note the two five-membered rings (pentagons) outlined in Figure 25-3. They consist of Pt, N, and C atoms. When the bonding of a multidentate ligand to a metal ion produces a ring (usually five- or six-membered), we refer to the complex as a **chelate** (pronounced key·late). The multidentate ligand is called a **chelating agent** and the process of chelate formation is called **chelation.**

Are You Wondering:

How terms such as ligand and chelate originated?

Ligand comes from the Latin word *ligare*, which means to bind. Dentate is also derived from a Latin word, *dens*, meaning tooth. Figuratively speaking, a unidentate ligand has one tooth; a bidentate ligand has two teeth; and so on. A ligand attaches itself to the central metal ion in accordance with the number of "teeth" it possesses. Chelate is derived from the Greek word *chela*, which means a crab's claw. The way in which a chelating agent attaches itself to a metal ion resembles a crab's claw.

25-3 Nomenclature

Werner developed the system of nomenclature that we use today, though some modifications have been made in recent years through international agreements. Although usage still varies somewhat, all the complexes that we deal with in this text can be named by the eight rules listed below. These rules are illustrated in Example 25-2.

Rules for naming complex ions.

1. Cations are named before anions. (25.4)

This is the same rule that applies to simple salts like NaCl.

2. The names of ligands are given first followed by the name of the central metal ion.

This order is the *opposite* of that for ordinary compounds, where the name of the metal usually appears first.

3. The names of ligands that are anions end in "o."

Normally "ide" endings are changed to o, "ite" endings are changed to ito, and "ate" endings are changed to ato (see Table 25-2).

4. Many ligands that are molecules carry the unmodified name.

For example, the name ethylenediamine is used both for the free neutral molecule and for the molecule as a ligand in a complex ion. (Aqua, ammine, carbonyl, and nitrosyl are important exceptions.)

5. The number of ligands of a given type is designated by a Greek prefix: mono = 1, di = 2, tri = 3, tetra = 4, penta = 5, hexa = 6. Also used are the prefixes bis = 2, tris = 3, tetrakis = 4, ...

For example, dichloro signifies two Cl^- ions as ligands; pentaaqua denotes five H_2O molecules. If the ligand name itself contains a prefix, as in ethylenediamine (en), then to denote two en molecules as ligands we write bis(ethylenediamine). The bis, tris, . . . prefixes are also used for ligands that are more complex than simple unidentate ligands. Thus, tris(oxalato) indicates that three oxalate ions, $C_2O_4^{2-}$, are ligands in a complex.

6. Ligands are named in alphabetical order. In writing formulas neutral molecules are generally written before anions.

If four H_2O molecules and two chloride ions are ligands in a complex, for example, we name them in the order tetraaquadichloro. Prefixes are not considered in establishing this alphabetical order (aqua precedes chloro, even though they are tetraaqua and dichloro). On the other hand, we generally write the formula of dichlorobis(ethylenediamine)cobalt(III) ion as $[Co(en)_2Cl_2]^+$.

7. The name of the central metal ion is unchanged, but we denote its oxidation state by a Roman numeral.

The ion $[Co(NH_3)_6]^{3+}$, for example, is called hexaamminecobalt(III) ion.

8. In complex anions the oxidation state of the central metal ion is denoted by a Roman numeral, and the name of the ion is modified to carry an "ate" ending.

The ion $[Al(H_2O)_2(OH)_4]^-$ is called the diaquatetrahydroxoaluminate(III) ion. When the following metals are in complex anions, the English name is replaced by the Latin name ending in "ate".

iron $\longrightarrow$ **ferrate**	tin $\longrightarrow$ **stannate**	lead $\longrightarrow$ **plumbate**
copper $\longrightarrow$ **cuprate**	silver $\longrightarrow$ **argentate**	gold $\longrightarrow$ **aurate**

Thus, $[CuCl_4]^{2-}$ is tetrachlorocuprate(II) ion.

Example 25-2

Relating names and formulas of complexes. **(a)** What is the formula of the compound pentaaquachlorochromium(III) chloride? **(b)** What is the name of the compound $K_3[Fe(CN)_6]$? **(c)** What is the name of the complex ion $[Pt(en)_2Cl_2]^{2+}$?

Solution

(a) The central metal ion is Cr^{3+}. There are one Cl^- ion and five H_2O molecules as ligands. The complex ion carries a net charge of $+2$. Two Cl^- ions are required outside the coordination sphere to neutralize this charge. The formula of the coordination compound is $[Cr(H_2O)_5Cl]Cl_2$.

(b) This compound consists of K^+ cations and complex anions having the formula $[Fe(CN)_6]^{3-}$. Each cyanide ion carries a charge of -1, so the oxidation state of the iron must be $+3$. The Latin name "ferrum" is changed to the "ate" ending because the complex ion is an anion. The name of the anion is hexacyanoferrate(III) ion. The coordination compound is **potassium hexacyanoferrate(III)**.

(c) The ethylenediamine ligands are neutral and they are *bidentate*. The ligand Cl^- is uninegative, and two Cl^- ions are present in the complex ion. The coordination number is *six*, the net charge on the complex ion is $+2$, and the charge on the central metal ion is $+4$. The ligands are named in alphabetical order, and the prefix "bis" is required for (en).

$[Pt(en)_2Cl_2]^{2+}$ = **dichlorobis(ethylenediamine)platinum(IV) ion**

SIMILAR EXAMPLES: Exercises 3, 4, 5, 17, 18, 19.

Although we name most complexes in the manner just outlined, some common or "trivial" names are still in use. Two such trivial names are ferrocyanide for $[Fe(CN)_6]^{4-}$ and ferricyanide for $[Fe(CN)_6]^{3-}$. These common names suggest the oxidation state of the central metal ions through the "o" and "i" designations ("o" for the ferrous ion, Fe^{2+}, in $[Fe(CN)_6]^{4-}$ and "i" for the ferric ion, Fe^{3+}, in $[Fe(CN)_6]^{3-}$). These trivial names do not indicate that the metal ions have a coordination number of 6, however. The systematic names—hexacyanoferrate(II) and hexacyanoferrate(III)—are more informative.

25-4 Isomerism

If two or more substances have the same percentage composition, they must have the same empirical formula. But just because they have the same formula does not make them identical substances. To be identical they also must have identical structures and properties. Many substances that have the same formulas differ in their geometrical structures and in their properties. These substances are called **isomers.** Isomerism among complexes can be placed into two broad categories: **Structural isomers** differ in basic structure or bond type—what ligands are bonded to the central ion, through which atoms? **Stereoisomers** are alike at the bonding level but differ in the spatial arrangements among the ligands. Of the five kinds of isomerism that we describe below, the first three are types of structural isomerism and the remaining two, types of stereoisomerism.

Ionization Isomerism. The two coordination compounds whose formulas are shown in expression (25.5) have the same central ion (Cr^{3+}) and five of the six ligands (NH_3 molecules) are the same. The way the compounds differ is that one has a SO_4^{2-} ion as the sixth ligand, with a Cl^- ion outside the coordination sphere; the other has Cl^- as a ligand and SO_4^{2-} outside the coordination sphere.

$$[Cr(NH_3)_5SO_4]Cl \qquad\qquad [Cr(NH_3)_5Cl]SO_4 \qquad\qquad (25.5)$$

pentaamminesulfatochromium(III) chloride	pentaamminechlorochromium(III) sulfate
(a)	(b)

Coordination Isomerism. A somewhat similar situation to that just described can arise when a coordination compound is composed of both complex cations and complex anions. The ligands can be distributed differently between the two complex ions, as with

$$[Co(en)_3][Cr(ox)_3] \qquad\qquad [Cr(en)_3][Co(ox)_3] \qquad\qquad (25.6)$$

tris(ethylenediamine)cobalt(III) tris(oxalato)chromate(III)	tris(ethylenediamine)chromium(III) tris(oxalato)cobaltate(III)
(a)	(b)

Linkage Isomerism. Some ligands may attach to the central metal ion of a complex ion in different ways. For example, the nitrite ion has electron pairs available for coordination both on the N and O atoms.

$$\left[\begin{array}{c} \overset{\cdot\cdot}{N} \\ :\overset{\cdot\cdot}{O}: \quad \overset{\cdot\cdot}{O}: \end{array} \right]^- \qquad\qquad (25.7)$$

Whether attachment of this ligand is through the N or the O atom, the formula of the complex ion is unaffected. However, the properties of the complex ion may be affected. When attachment occurs through the N atom, the ligand is referred to as "nitro." Coordination through the O atom produces a "nitrito" complex.

$$[Co(NH_3)_4(NO_2)Cl]^+ \qquad\qquad [Co(NH_3)_4(ONO)Cl]^+ \qquad\qquad (25.8)$$

tetraamminechloronitrocobalt(III) ion	tetraamminechloronitritocobalt(III) ion
(a)	(b)

Geometric Isomerism. If we substitute a single Cl^- for an NH_3 molecule in the square planar complex ion $[Pt(NH_3)_4]^{2+}$, it does not matter at which corner of the square we make this substitution. As shown in Figure 25-4a, all four possibilities

FIGURE 25-4
Geometric isomerism illustrated.

For the square planar complexes shown here, isomerism exists only when two Cl^- ions have replaced NH_3 molecules.

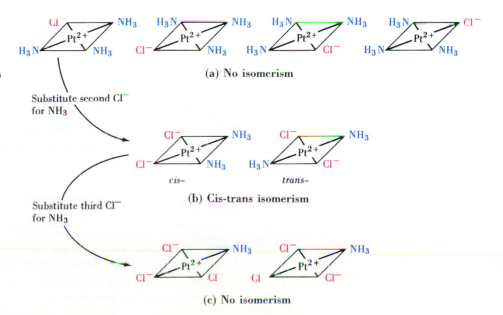

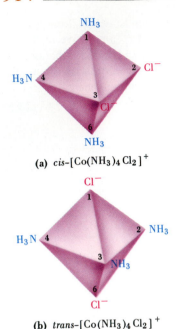

(a) *cis*-[Co(NH$_3$)$_4$Cl$_2$]$^+$

(b) *trans*-[Co(NH$_3$)$_4$Cl$_2$]$^+$

FIGURE 25-5
Cis-trans isomers of an octahedral complex.

The Co^{3+} ion is at the center of the octahedron, and one NH$_3$ ligand is on the far corner (corner 5), out of view.

are alike. If we substitute a *second* Cl$^-$, we now have two distinct possibilities (see Figure 25-4b). The two Cl$^-$ ions can either be along the same edge of the square (**cis**) or on opposite corners (**trans**). To distinguish clearly between these two possibilities, we must either draw a structure or refer to the appropriate name. The formula alone will not distinguish between them. (Note that this complex is a neutral species, not an ion.)

$$[Pt(NH_3)_2Cl_2]$$
cis-diamminedichloroplatinum(II)
or
trans-diamminedichloroplatinum(II)

Interestingly, when we substitute a *third* Cl$^-$, isomerism disappears (Figure 25-4c). There is only one complex ion with the formula [Pt(NH$_3$)Cl$_3$]$^-$.

With an octahedral complex the situation is a bit more complicated. Take the complex ion [Co(NH$_3$)$_6$]$^{3+}$ as an example. If we substitute one Cl$^-$ for an NH$_3$, we get a single structure. With *two* Cl$^-$ ions substituted for NH$_3$ molecules we obtain cis-trans isomers. The cis isomer has two Cl$^-$ ions along the same edge of the octahedron. The trans isomer has two Cl$^-$ ions on opposite corners, that is, at opposite ends of a line drawn through the central metal ion. These two isomers are shown in Figure 25-5. One difference between the two is that the cis isomer is a blue-violet color and the trans is a bright green.

Let us refer to Figure 25-5a to see what happens when we substitute a *third* Cl$^-$ for an NH$_3$. If we make this third substitution at either position 1 or 6, the result is that three Cl$^-$ ions appear on the same face of the octahedron. We can also call this a cis isomer.* If we make the third substitution either at position 4 or 5, the result is three Cl$^-$ ions around the perimeter of the octahedron. We can call this a trans isomer.* If we substitute a *fourth* Cl$^-$, we again get two isomers, but in this case the isomerism is based on whether the remaining two NH$_3$ molecules are cis or trans.

Example 25-3

Identifying geometrical isomers. Sketch all the possible isomers of [Co(ox)(NH$_3$)$_3$Cl].

Solution. The Co^{3+} ion exhibits a coordination number of 6. The structure is octahedral. Recall that ox (oxalate ion) is a bidentate ligand carrying a double-negative charge (Table 25-3). Recall also that such a ligand must be attached in cis positions, not trans (Figure 25-4). Once the ox ligand is placed, we see that there are two possibilities. The three NH$_3$ molecules can be situated (1) on the same face of the octahedron (cis isomer) or (2) around a perimeter of the octahedron (trans isomer).

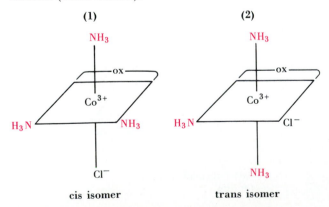

cis isomer   trans isomer

SIMILAR EXAMPLES: Exercises 10, 11, 28, 31.

*In a more precise nomenclature, this type of cis isomer is called a *facial* (*fac*) isomer, and this type of trans isomer is called a *meridional* (*mer*) isomer.

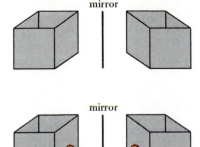

FIGURE 25-6
Superimposable and nonsuperimposable objects—a cubical, open-top box.

(a) You can place the box into its mirror image (hypothetically) in several different ways.
(b) In any way that you place the box into its mirror image, the stickers will not appear in the same position. The box and its mirror image are nonsuperimposable.

Optical Isomerism. To understand optical isomerism we need to understand the relationship between an object and its mirror image. Features on the right side of the object appear on the left side of its image in a mirror, and vice versa. As a familiar example, a left hand produces a right hand as its mirror image. There are two possibilities for an object and its mirror image. Certain objects can be rotated in such a way as to be *superimposable* on their mirror images, but other objects are *nonsuperimposable* on their mirror images.

Consider, for example, the open-top cubical cardboard box pictured in Figure 25-6. There are a number of ways (hypothetical, of course) in which the box can be superimposed on its mirror image. Now, imagine that a distinctive sticker is placed at a corner of one side of the box. In this case there is no way that the box and its mirror image can be superimposed; they are clearly different. This is equivalent to saying that there is no way that a form-fitting left glove can be worn on a right hand (turning it inside out is not allowed).

The two structures of $[Co(en)_3]^{3+}$ depicted in Figure 25-7 are related to one another as are an object and its image in a mirror. Furthermore, the two structures are *nonsuperimposable,* like a left and a right hand. No combination of rotation or inversion of one structure will make it identical to its mirror image. Therefore, the two structures represent two distinctly different complex ions.

Structures that are nonsuperimposable mirror images of each other are called **enantiomers** and are said to be **chiral** (pronounced kye·rull). (Structures that are superimposable are *achiral.*) Whereas other types of isomers often differ significantly in their physical and chemical properties, enantiomers have identical properties, except in a few specialized situations. These exceptions involve phenomena that are directly linked to chirality or "handedness" at the molecular level. One such phenomenon is that of **optical activity,** pictured in Figure 25-8.

FIGURE 25-7
Optical isomers.

(a) The two structures shown, like a left and a right hand, are mirror images and are nonsuperimposable.
(b) The structure on the left is rotated about its vertical axis (180°). This places the en group in the central plane in the same orientation as in the structure on the right in (a), but the rest of the structure is nonsuperimposable.
(c) The structure on the left is inverted (top to bottom). Again, this leads to an arrangement that is nonsuperimposable with the structure on the right in (a).

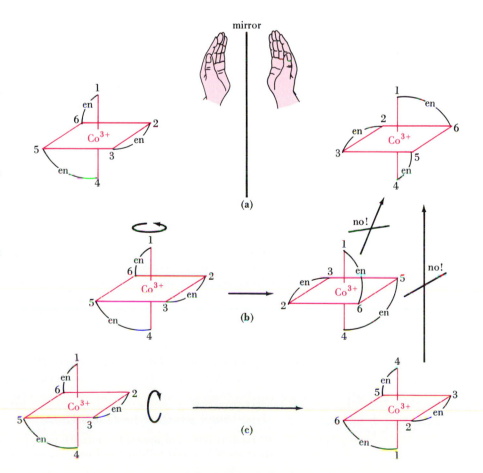

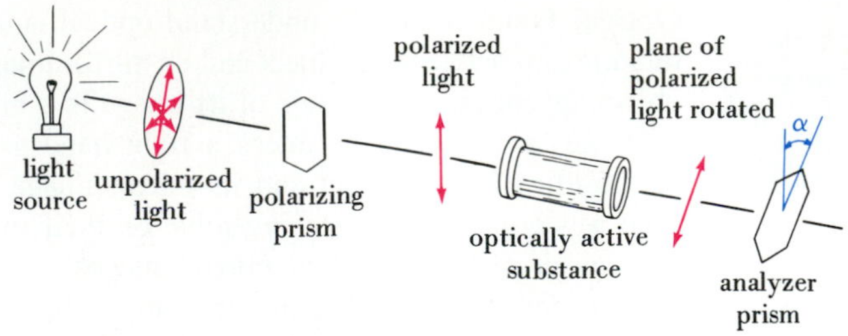

FIGURE 25-8
Optical activity.

Light from an ordinary source consists of electromagnetic waves vibrating in all planes; it is unpolarized. Some substances (e.g., polaroid) possess the ability to screen out all light waves except those vibrating in a particular plane. Other substances—optically active materials—have the ability to rotate the plane of polarized light. In the illustration the light is rotated to the right by the angle α.

Interactions between a beam of polarized light and the electrons in an enantiomer cause a rotation of the plane of the polarized light. One enantiomer rotates the plane of polarized light to the right (clockwise) and is said to be **dextrorotatory** (designated $+$ or d). The other enantiomer rotates the plane of polarized light to the same extent, but to the left (counterclockwise). It is said to be **levorotatory** ($-$ or l). Because of their ability to rotate the plane of polarized light, isomers of these types are often said to be *optically active,* and they are called **optical isomers.**

When we synthesize an optically active complex ion, we usually get a mixture of the two optical isomers (enantiomers). The optical rotation of one isomer just cancels that of the other. The mixture, called a **racemic mixture,** produces no net rotation of the plane of polarized light. Separating the d and l isomers of a racemic mixture is called **resolution.** We can sometimes achieve this through chemical reactions affected by chirality, with the two enantiomers behaving differently. Many phenomena of the living state, such as the activity of an enzyme or the ability of a microorganism to promote a reaction, involve chirality (see Chapter 28).

Isomerism and Werner's Theory. The study of isomerism played a central role in establishing Werner's theory of coordination compounds. An important early question in the field involved the structure of complexes with coordination number 6. Werner proposed that they are octahedral, but other possibilities were also suggested. One of these, pictured in Figure 25-9, was a hexagonal structure (resembling the structure of benzene, which had recently been established). The choice of the octahedral over the hexagonal structure was strongly suggested by the observation that there are but *two* isomers of complex ions like $[Co(NH_3)_4Cl_2]^+$ (Figure 25-5). The hexagonal structure would require *three,* and a third one was never found.

More direct evidence came when Werner synthesized optical isomers like $[Co(en)_3]^{3+}$ (Figure 25-7). Neither the hexagonal structure nor any other structure considered as alternatives to the octahedron accounts for optical activity. However, this evidence was still not accepted as conclusive proof by some of Werner's critics. They associated optical activity with carbon atoms and argued that the optical activity that Werner observed was caused by C atoms in his ligands and not by the structure of the complex ion. Werner succeeded (in 1908) in preparing an optically active coordination compound that was totally inorganic (no C atoms were present). Werner was awarded the Nobel prize in 1913 for his prodigious efforts in developing the fundamentals of coordination chemistry.

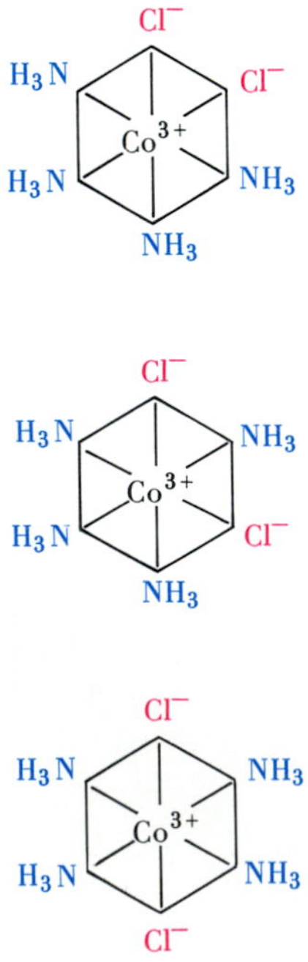

The terms dextro and levo are derived from the Latin words *dexter* = right and *laevus* = left.

Here is a process based on chirality: choosing a matched pair of gloves from a bin of gloves of identical sizes, 99 made for the right hand and 1 for the left hand.

FIGURE 25-9
Hypothetical structure for $[Co(NH_3)_4Cl_2]^+$.

If the distribution of ligands were in the planar hexagonal structure shown here, there should be *three* distinct isomers, but only *two* isomers are found.

25-5 Bonding in Complex Ions—Valence Bond Theory

One way to think about the bonds between a metal ion and its ligands is based on coordinate covalency. The central ion furnishes the orbitals, and the ligands furnish electron pairs. Although this view is now considered inadequate, it does account for the observed coordination number of the central ion, and it also describes the geometric structure of the complex ion.

A scheme that is consistent with the formula and structure of the complex ion $[Co(NH_3)_6]^{3+}$ involves the hybridization of two $3d$ orbitals with a $4s$ and three $4p$ orbitals, yielding six d^2sp^3 orbitals with an octahedral distribution.

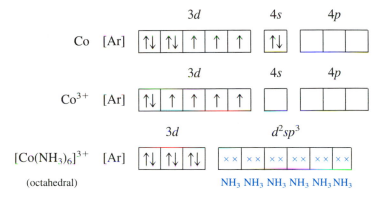

Three additional hybridization schemes follow that illustrate the tetrahedral $[Zn(NH_3)_4]^{2+}$, the square planar $[Pt(NH_3)_4]^{2+}$, and the linear $[Ag(NH_3)_2]^+$, respectively. Recall that we saw the geometric structures of all four of these complex ions in Figure 25-1.

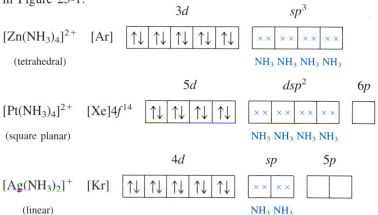

The complex ion $[Co(NH_3)_6]^{3+}$ is *diamagnetic*. $[CoF_6]^{3-}$ has a similar octahedral structure, but it is *paramagnetic* and appears to have *four* unpaired electrons. One explanation for this difference is to assume that there are two ways to hybridize an s, three p, and two d orbitals. In the case of $[Co(NH_3)_6]^{3+}$, as we saw above, the d orbitals come from an inner electronic shell (the third), and the resulting hybrid orbitals are d^2sp^3. The complex ion is called an *inner-orbital* complex. Because all the electrons are paired, this complex ion is diamagnetic. We can call it a "low spin" complex. In $[CoF_6]^{3-}$ we assume that the d orbitals come from the outer electronic shell (the fourth), and the resulting hybrid orbitals are sp^3d^2. We can call this an *outer-orbital* or a "high spin" complex. The bonding schemes shown below are consistent with the observed magnetic properties.

		3d	4s	4p	4d
Co³⁺	[Ar]	⇅ ↑ ↑ ↑ ↑	☐	☐ ☐ ☐	☐ ☐ ☐ ☐ ☐

(25.9)

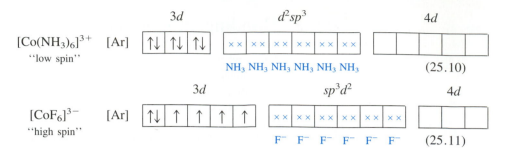

$$[Co(NH_3)_6]^{3+} \quad [Ar] \qquad \text{(25.10)}$$
"low spin"

$$[CoF_6]^{3-} \quad [Ar] \qquad \text{(25.11)}$$
"high spin"

25-6   Bonding in Complex Ions—Crystal Field Theory

The valence bond theory has some shortcomings. For one, it does not provide insight into the origin of the characteristic colors of complex ions. Neither does it explain why $[Co(NH_3)_6]^{3+}$ should be an inner-orbital complex whereas $[CoF_6]^{3-}$ is an outer orbital complex. An alternative theory that is more successful in doing these things is the **crystal field theory.** In the crystal field model, we consider bonding in a complex ion to be an electrostatic attraction between the positively charged nucleus of the central metal ion and electrons in the ligands. Also, however, we should expect repulsions to occur between the ligand electrons and electrons in the central ion. The crystal field theory focuses on the repulsions between ligand electrons and d electrons of the central ion.

The d orbitals first presented in Figure 8-26 are not all alike in their spatial orientations, but in an isolated atom or ion they do have equal energies. One of them, d_{z^2}, is directed along the z axis and another, $d_{x^2-y^2}$, has lobes along the x and y axes. The remaining three have lobes extending into regions between the perpendicular x, y, and z axes.

Figure 25-10 pictures six anions (ligands) approaching a central metal ion along the x, y, and z axes. This direction of approach leads to an octahedral complex. As a result of electrostatic repulsion between the negatively charged ligands and the d electrons of the metal ion, the d energy level is split into two groups. One group, consisting of $d_{x^2-y^2}$ and d_{z^2}, has its energy raised with respect to a hypothetical free ion in the field of the ligands. The other group, consisting of d_{xy}, d_{xz}, and d_{yz}, has its energy lowered. The difference in energy between the two groups is represented by the symbol Δ, as shown in Figure 25-11.

Modifications of the simple crystal field theory that take into account such factors as the partial covalency of the metal–ligand bond are called ligand field theory. Often the single term "ligand field theory" is used to signify both the purely electrostatic crystal field theory and its modifications.

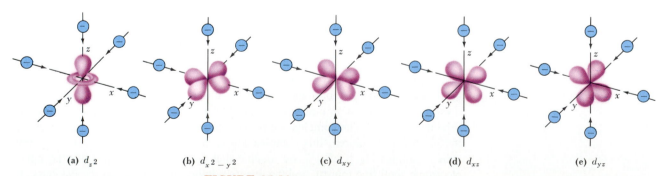

(a) d_{z^2} (b) $d_{x^2-y^2}$ (c) d_{xy} (d) d_{xz} (e) d_{yz}

FIGURE 25-10

Approach of six anions to a metal ion to form a complex ion with octahedral structure.

The ligands (anions in this case) approach the central metal ion along the x, y, and z axes. Maximum interference occurs with the d_{z^2} and $d_{x^2-y^2}$ orbitals, and their energies are raised. Interference with the other d orbitals is not as great. A difference in energy results between the two sets of d orbitals.

FIGURE 25-11
Splitting of d energy levels in the formation of an octahedral complex ion.

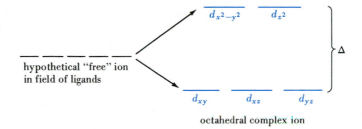

FIGURE 25-12
Crystal field splitting in a tetrahedral complex ion.

(a) The positions of attachment of ligands to a metal ion leading to the formation of a tetrahedral complex ion.
(b) Interference with the d orbitals directed along the x, y, and z axes is not as great as with those that lie between the axes (again see Figure 25-10). As a result, the pattern of crystal field splitting is reversed from that of an octahedral complex.

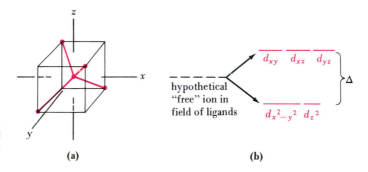

We find a different pattern for the splitting of the d energy levels for complex ions with other geometric shapes. Figure 25-12 shows the pattern for tetrahedral complexes, and Figure 25-13 shows that for square planar complexes. Suppose we were to compare hypothetical complexes of different structures but with the same combinations of ligands, metal ions, and metal–ligand distances. We would find the greatest energy separation of the d levels for the square planar complex and the smallest energy separation for the tetrahedral complex. If we call the energy separation for the octahedral complex Δ_o, the comparative value for the tetrahedral complex is $0.44\Delta_o$, and for the square planar complex, $1.74\Delta_o$.

Ligands differ in their abilities to produce a splitting of the d energy levels. Strong Lewis bases like CN^- and NH_3 produce a "strong" field. They repel d

FIGURE 25-13
Comparison of crystal field splitting in a square planar and an octahedral complex.

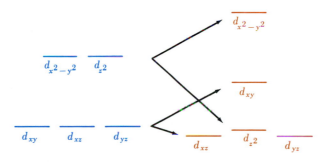

Consider an octahedral complex in which the two ligands along the z axis move progressively farther away from the central ion, with the final result that the remaining four ligands are in a square planar configuration about the central ion. (This process can be visualized in terms of Figure 25-2 and the accompanying discussion of the structure of $[Cu(NH_3)_4(H_2O)_2]^{2+}$.)

Splitting of the d energy level in a square planar complex can be related to that of the octahedral complex. Because there are no ligands along the z axis in a square planar complex, we should expect the repulsion between ligands and d_{z^2} electrons to be much less than in an octahedral complex. The d_{z^2} energy level is lowered considerably from that in an octahedral complex. The energy of the $d_{x^2-y^2}$ orbital is raised, since the x and y axes represent the direction of approach of four ligands to the central ion. The energy of the d_{xy} orbital is also raised because this orbital lies in the plane of the ligands in the square planar complex. The d_{xz} and d_{yz} orbitals are concentrated in planes perpendicular to that of the square planar complex, and their energies are lowered slightly.

electrons very strongly and cause a greater separation of the d energy levels than do weaker bases like H_2O and F^-. What we normally expect for a distribution of d electrons in a metal ion is that to the extent possible d orbitals will be singly occupied. This is the case in the complex ion $[CoF_6]^{3-}$ shown in (25.12). However, sometimes the energy separation between the higher and lower energy d levels (Δ) is larger than the energy required to pair electrons. In this case the energetically favored arrangement may be for electrons to remain paired in the lower energy d orbitals, and this is the case for $[Co(NH_3)_6]^{3+}$.

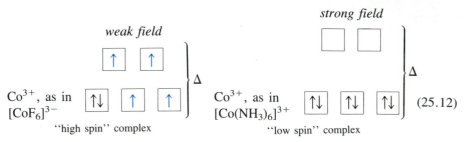

$$(25.12)$$

Thus, we see how crystal field theory explains the magnetic properties of the inner- and outer-orbital complexes shown in structures (25.10) and (25.11).

Different ligands can be arranged in order of their abilities to produce a splitting of the d energy levels. This arrangement is known as the **spectrochemical series** and is illustrated in Examples 25-4 and 25-5.

The spectrochemical series.

strong field (large Δ)
$CN^- > -NO_2^- > en > py \simeq NH_3 > SCN^- —$

weak field (small Δ)
$>H_2O > OH^- > F^- > Cl^- > Br^- > I^-$ (25.13)

(When NO_2^- is attached through an O atom or SCN^- through the S atom, a different placement in the spectrochemical series is found.)

Example 25-4

Using the spectrochemical series to predict the magnetic properties of a complex ion. How many unpaired electrons are present in the octahedral complex $[Fe(CN)_6]^{3-}$?

Solution. The Fe atom has the electron configuration $[Ar]3d^64s^2$. The Fe^{3+} ion has the configuration $[Ar]3d^5$. CN^- is a strong-field ligand. Because of the large energy separation in the d levels of the metal ion produced by this ligand, all the electrons are found in the lowest energy levels. There is *one* unpaired electron.

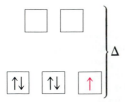

SIMILAR EXAMPLES: Exercises 33, 34, 35.

Example 25-5

Using the crystal field theory to predict the structure of a complex ion from its magnetic properties. The complex ion $[Ni(CN)_4]^{2-}$ is found to be diamagnetic. Use ideas from the crystal field theory to speculate on its probable structure.

Solution. We can eliminate an octahedral structure because the coordination number is 4 (not 6). Our choice is between tetrahedral and square planar.

The electron configuration of Ni is $[Ar]3d^84s^2$ and of Ni(II), $[Ar]3d^8$. Since the complex ion is found to be diamagnetic, all $3d$ electrons must be paired. Let us see how we would distribute these $3d$ electrons if the structure were tetrahedral. We would place four electrons (all paired) into the two lowest d levels. We would then distribute the remaining four electrons among the three higher d levels. Two of the electrons would be unpaired, and the complex ion would be paramagnetic.

$$\underset{d_{xy}}{\uparrow\downarrow} \qquad \underset{d_{xz}}{\uparrow} \qquad \underset{d_{yz}}{\uparrow}$$

$$\underset{d_{x^2-y^2}}{\uparrow\downarrow} \qquad \underset{d_{z^2}}{\uparrow\downarrow}$$

In the square planar complex the three lowest energy orbitals would be filled, and because of the large energy separation between the d_{xy} and the $d_{x^2-y^2}$ orbitals, we should expect the remaining two electrons to be paired in the d_{xy} orbital. This leads to a diamagnetic species. **The probable structure of $[Ni(CN)_4]^{2-}$ is square planar.**

$$\underset{d_{x^2-y^2}}{\rule{2em}{0.4pt}}$$

$$\underset{d_{xy}}{\uparrow\downarrow}$$

$$\underset{d_{xz}}{\uparrow\downarrow} \qquad \underset{d_{z^2}}{\uparrow\downarrow} \qquad \underset{d_{yz}}{\uparrow\downarrow}$$

SIMILAR EXAMPLES: Exercises 37, 38, 58.

25-7 The Colors of Coordination Compounds

To understand the origin of color in solutions, we will find it helpful to consider the color circle in Figure 25-14, devised by Isaac Newton. On this circle we divide the visible portion of the electromagnetic spectrum into six colors. Three of these are **primary colors** (red, yellow, and blue), and three are *complementary* or **secondary colors** (orange, green, and violet). When lights of the three primary colors are mixed, the resulting color is white. Each secondary color is the sum of the two adjacent primary colors (e.g., orange = red + yellow), and each secondary color is the *complement* of the primary color opposite to it on the color circle. When light of a primary color and that of its complementary color are mixed, the result is also white light.

A different set of primary colors can be chosen, as long as they can be combined to produce white light. The colors on Newton's original color circle were somewhat different than those shown in Figure 25-14.

FIGURE 25-14
Primary and secondary colors.

The primary colors shown on this color circle are red, yellow, and blue. The secondary colors are orange, green, and violet.

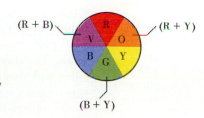

Effect of ligands on the colors of coordination compounds. These compounds all consist of a six-coordinate cobalt complex ion in combination with nitrate ions. In each case the complex ion has five NH_3 molecules and one other group as ligands. From left to right the compounds are $[Co(NH_3)_5Cl](NO_3)_2$, $[Co(NH_3)_5Br](NO_3)_2$, $[Co(NH_3)_5I](NO_3)_2$, $[Co(NH_3)_5NO_2](NO_3)_2$, $[Co(NH_3)_5SO_4]NO_3$, and $[Co(NH_3)_5CO_3]NO_3$. [Courtesy Arlo Harris; photograph by Carey B. Van Loon]

Colored solutions contain species that are capable of absorbing photons of visible light and using these photons to promote electrons to higher energy levels. The energies of the photons must just match the energy differences through which the electrons are to be promoted. Since the energies of photons are related to the frequencies (and wavelengths) of light (recall Planck's equation: $E = h\nu$), only certain wavelength components are absorbed as white light passes through the solution. The emerging light, since it is lacking some wavelength components, is no longer white; it is colored.

Ions having (1) a noble gas electron configuration, (2) an outer shell of 18 electrons, or (3) an "18 + 2" configuration do not have electron transitions in the energy range corresponding to visible light. White light passes through these solutions without being absorbed; these ions are colorless in solution. Examples are the alkali and alkaline earth metal ions, the halide ions, Zn^{2+}, Al^{3+}, and Bi^{3+}.

Crystal field splitting of the d energy levels produces the energy difference, Δ, that accounts for the colors of complex ions. Promotion of an electron from a lower to a higher d level results from the absorption of the appropriate components of white light; the transmitted light is colored.

A solution containing $[Cu(H_2O)_4]^{2+}$ absorbs most strongly wavelength components in the yellow region of the spectrum (about 580 nm). The wavelength components of the light transmitted by the solution combine to produce the complement of yellow—*blue*. Thus, aqueous solutions of copper(II) compounds usually have a characteristic blue color. In the presence of high concentrations of Cl^-, copper(II) forms the complex ion $[CuCl_4]^{2-}$. This species absorbs strongly in the blue region of the spectrum. The transmitted light, and hence the color of the solution, is *yellow*. Light absorption by these solutions is suggested by Figure 25-15. The colors of some complex ions of chromium are given in Table 25-4; the first and third of these compounds are the ones whose aqueous solutions are pictured in the opening photograph of this chapter.

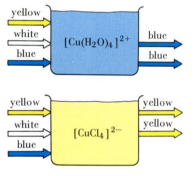

FIGURE 25-15
Light absorption and transmission.

$[Cu(H_2O)_4]^{2+}$ absorbs in the yellow region of the spectrum and transmits blue light. $[CuCl_4]^{2-}$ absorbs in the blue region of the spectrum and transmits yellow light.

TABLE 25-4
Some Coordination Compounds of Cr^{3+} and Their Colors

Isomer	Color
$[Cr(H_2O)_6]Cl_3$	violet
$[Cr(H_2O)_5Cl]Cl_2$	blue-green
$[Cr(H_2O)_4Cl_2]Cl$	green
$[Cr(NH_3)_6]Cl_3$	yellow
$[Cr(NH_3)_5Cl]Cl_2$	purple
$[Cr(NH_3)_4Cl_2]Cl$	violet

Example 25-6 _____

Relating the colors of complex ions to the spectrochemical series. Table 25-4 lists the color of $[Cr(H_2O)_6]Cl_3$ as violet, whereas that of $[Cr(NH_3)_6]Cl_3$ is yellow. Explain this difference in color.

Solution. Here are the basic facts that we must use: All the chromium(III) complexes in Table 25-4 are octahedral, and the electron configuration of Cr^{3+} is $[Ar]3d^3$. From these facts we can construct the energy-level diagram shown below. The three unpaired electrons go into the three lower energy d orbitals. When a photon of light is absorbed an electron is promoted from the lower to an upper energy level. The quantity of energy required for this promotion depends on the energy-level separation, Δ.

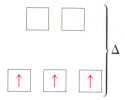

According to the spectrochemical series (25.13), NH_3 produces a greater splitting of the d energy level than does H_2O. We should expect $[Cr(NH_3)_6]^{3+}$ to absorb light of a *shorter* wavelength (higher energy) than does $[Cr(H_2O)_6]^{3+}$. If $[Cr(NH_3)_6]^{3+}$ absorbs in the blue/violet region of the spectrum then the *transmitted* light is yellow. If $[Cr(H_2O)_6]^{3+}$ absorbs in the yellow region of the spectrum, then the *transmitted* light is violet.

SIMILAR EXAMPLES: Exercises 13, 36, 39.

25-8 Aspects of Complex Ion Equilibria

In Chapter 19 we learned that complex ion formation can have a great effect on the solubilities of substances, as in the ability of $NH_3(aq)$ to dissolve rather large quantities of $AgCl(s)$. On the other hand, $AgI(s)$ is only very slightly soluble in $NH_3(aq)$, and to explain this fact we needed a *quantitative* description of the stability of the diamminesilver(I) ion, $[Ag(NH_3)_2]^+$. This we got through the formation constant, K_f. Equations (25.14) and (25.15) illustrate how we dealt with formation constants in Chapter 19, with $[Zn(NH_3)_4]^{2+}$ as an example.

$$Zn^{2+}(aq) + 4\ NH_3(aq) \rightleftharpoons [Zn(NH_3)_4]^{2+}(aq) \tag{25.14}$$

$$K_f = \frac{[[Zn(NH_3)_4]^{2+}]}{[Zn^{2+}][NH_3]^4} = 4.1 \times 10^8 \tag{25.15}$$

We have since learned that cations in aqueous solution exist mostly in *hydrated* form. That is, $Zn^{2+}(aq)$ is actually $[Zn(H_2O)_4]^{2+}$. As a result, when NH_3 molecules bond to Zn^{2+} to form an ammine complex ion they do not enter an empty coordination sphere. They must displace H_2O molecules, and this occurs in a stepwise fashion. The reaction

$$[Zn(H_2O)_4]^{2+} + NH_3 \rightleftharpoons [Zn(H_2O)_3NH_3]^{2+} + H_2O \tag{25.16}$$

for which

$$K_1 = \frac{[[Zn(H_2O)_3NH_3]^{2+}]}{[[Zn(H_2O)_4]^{2+}][NH_3]} = 3.9 \times 10^2 \tag{25.17}$$

is followed by

$$[Zn(H_2O)_3NH_3]^{2+} + NH_3 \rightleftharpoons [Zn(H_2O)_2(NH_3)_2]^{2+} + H_2O \tag{25.18}$$

for which

$$K_2 = \frac{[[Zn(H_2O)_2(NH_3)_2]^{2+}]}{[[Zn(H_2O)_3NH_3]^{2+}][NH_3]} = 2.1 \times 10^2 \tag{25.19}$$

and so on.

The value of K_1 in equation (25.17) can also be designated as β_1 and called the formation constant for the complex ion $[Zn(H_2O)_3NH_3]^{2+}$. The formation of $[Zn(H_2O)_2(NH_3)_2]^{2+}$ is represented by the *sum* of equations (25.16) and (25.18),

$$[Zn(H_2O)_4]^{2+} + 2\ NH_3 \rightleftharpoons [Zn(H_2O)_2(NH_3)_2]^{2+} + 2\ H_2O \tag{25.20}$$

TABLE 25-5
Stepwise and Overall Formation (Stability) Constants for Several Complex Ions

Metal ion	Ligand	K_1	K_2	K_3	K_4	K_5	K_6	β_n (or K_f)[a]
Ag^+	NH_3	2.0×10^3	7.9×10^3					1.6×10^7
Zn^{2+}	NH_3	3.9×10^2	2.1×10^2	1.0×10^2	5.0×10^1			4.1×10^8
Cu^{2+}	NH_3	1.9×10^4	3.9×10^3	1.0×10^3	1.5×10^2			1.1×10^{13}
Ni^{2+}	NH_3	6.3×10^2	1.7×10^2	5.4×10^1	1.5×10^1	5.6	1.1	5.3×10^8
Cu^{2+}	en	5.2×10^{10}	2.0×10^9					1.0×10^{20}
Ni^{2+}	en	3.3×10^7	1.9×10^6	1.8×10^4				1.1×10^{18}
Ni^{2+}	EDTA	4.2×10^{18}						4.2×10^{18}

In many tabulations in the chemical literature, formation constant data are presented as logarithms, i.e., log K_1, log K_2, . . . , and log β_n.
[a] The β_n listed is for the number of steps shown, i.e., for $[Ag(NH_3)_2]^+$, $\beta_2 = K_f = K_1 \times K_2$; for $[Ni(en)_3]^{2+}$, $\beta_3 = K_f = K_1 \times K_2 \times K_3$; and for $[Ni(EDTA)]^{2-}$, $\beta_1 = K_f = K_1$.

and the formation constant β_2, in turn, is given by the *product* of equations (25.17) and (25.19).

$$\beta_2 = \frac{[[Zn(H_2O)_2(NH_3)_2]^{2+}]}{[[Zn(H_2O)_4]^{2+}][NH_3]^2} = K_1 \times K_2 = 8.2 \times 10^4 \qquad (25.21)$$

For the next ion in the series, $[Zn(H_2O)(NH_3)_3]^{2+}$, $\beta_3 = K_1 \times K_2 \times K_3$. For the final member, $[Zn(NH_3)_4]^{2+}$, $\beta_4 = K_1 \times K_2 \times K_3 \times K_4$, and this is the value that we called the formation constant in Section 19-8 and listed as K_f in Table 19-2. Additional stepwise formation constant data are presented in Table 25-5.

The large numerical value of K_1 for reaction (25.16) indicates that Zn^{2+} has a greater affinity for NH_3 (a stronger Lewis base) than it does for H_2O. Displacement of ligand H_2O molecules by NH_3 occurs even if the number of NH_3 molecules present in aqueous solution is much smaller than the number of H_2O molecules, as in dilute NH_3(aq). The fact that the magnitudes of successive K values decrease regularly in the displacement process can be explained in statistical terms: An NH_3 molecule has a better chance of replacing an H_2O molecule in $[Zn(H_2O)_4]^{2+}$, where each coordination position is occupied by H_2O, than in $[Zn(H_2O)_3NH_3]^{2+}$, where one of the positions is already occupied by NH_3. For other complexes, if irregularities in the K values arise, it is often because of a change in structure of the complex ion at some point in the series of displacement reactions.

Chelation. If the ligand in a substitution process is *multidentate*, it displaces as many H_2O molecules as there are points of attachment. Thus, ethylenediamine (en) displaces H_2O molecules in $[Ni(H_2O)_6]^{2+}$ two at a time, in three steps. The first step is

$$[Ni(H_2O)_6]^{2+} + en \rightleftharpoons [Ni(en)(H_2O)_4]^{2+} + 2\,H_2O \qquad K_1(\beta_1) = 3.3 \times 10^7$$
$$(25.22)$$

You will note from Table 25-5 that the complex ions with multidentate ligands have much larger formation constants than do those with unidentate ligands. For example, $K_f(\beta_3)$ for $[Ni(en)_3]^{2+}$ is 1.1×10^{18}, whereas $K_f(\beta_6)$ for $[Ni(NH_3)_6]^{2+}$ is 5.3×10^8. This additional stability of complex ions associated with chelate formation by multidentate ligands is known as the **chelation effect.**

25-9 Acid–Base Reactions of Complex Ions

We have described complex ion formation in terms of Lewis acids and bases (Section 25-2). Complex ions may also exhibit acid–base properties in the Brønsted–Lowry sense, that is, they may act as proton donors or acceptors. Figure 25-16

Are You Wondering:

Why chelates are more stable than complex ions with unidentate ligands?

The equilibrium constant for a ligand displacement reaction is related to the free energy change for the reaction in the usual way, $\Delta G° = -RT \ln K$. Also $\Delta G° = \Delta H° - T\Delta S°$. The displacement of *two* H_2O molecules by *one* en molecule produces a large entropy increase in reaction (25.22)—two particles on the left yield three on the right. When a unidentate ligand such as NH_3 displaces an H_2O molecule from the coordination sphere, the entropy change is much smaller (two particles on each side of the equation). The entropy change is still small even for the replacement of two H_2O molecules by two NH_3 molecules in two steps. If we assume comparable values of $\Delta H°$, the larger values of $\Delta S°$ for chelate formation over simple complex ion formation lead to more negative values of $\Delta G°$ and larger values of K. Having larger values of K for the individual steps in the ligand displacement reactions, the chelates also have larger values of the overall formation constant K_f than do simple complex ions; the chelates are more stable.

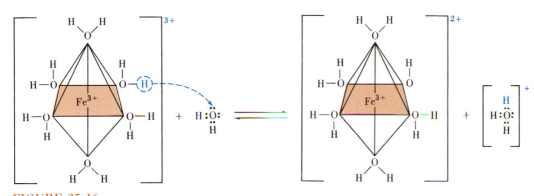

FIGURE 25-16
Ionization of $[Fe(H_2O)_6]^{3+}$.

represents the ionization of $[Fe(H_2O)_6]^{3+}$ as an acid. A proton from a *ligand* water molecule in hexaaquairon(III) ion is transferred to a *solvent* water molecule. The H_2O ligand is converted to OH^-.

$$[Fe(H_2O)_6]^{3+} + H_2O \rightleftharpoons [Fe(H_2O)_5OH]^{2+} + H_3O^+ \qquad K_{a_1} = 9 \times 10^{-4} \tag{25.23}$$

The second ionization step is

$$[Fe(H_2O)_5OH]^{2+} + H_2O \rightleftharpoons [Fe(H_2O)_4(OH)_2]^+ + H_3O^+ \qquad K_{a_2} = 5 \times 10^{-4} \tag{25.24}$$

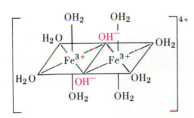

FIGURE 25-17
A binuclear complex of Fe(III).

This binuclear structure is produced by the reaction

$2[Fe(H_2O)_6]^{3+} \rightleftharpoons$
$[Fe_2(H_2O)_8(OH)_2]^{4+} + 2\,H_3O^+$
$\qquad K = 1.2 \times 10^{-3}$

From these K_a values we see that $Fe^{3+}(aq)$ is fairly acidic (compared, for example, to acetic acid, with $K_a = 1.74 \times 10^{-5}$). To repress ionization (hydrolysis) of $[Fe(H_2O)_6]^{3+}$ we need to maintain a low pH by the addition of acids such as HNO_3 or $HClO_4$. (Recall that increasing $[H_3O]^+$ displaces reactions 25.23 and 25.24 to the left.) The ion $[Fe(H_2O)_6]^{3+}$ is violet in color, but aqueous solutions of $Fe^{3+}(aq)$ are generally yellow owing to the presence of hydroxo complex ions.

The ionization of $[Fe(H_2O)_4(OH)_2]^+$ can proceed a step further, particularly in the presence of an alkaline substance, leading to a precipitate of $[Fe(H_2O)_3(OH)_3]$. Another reaction that can occur simultaneously with (25.23) and (25.24) is the formation of a bridged structure by the joining of two simple complex ions into a binuclear complex ion, as illustrated in Figure 25-17. More units may add to the binuclear structure, resulting in aggregates having colloidal dimensions. Colloidal

This colloid is pictured in
Figure 13-24.

ferric hydroxide (hydrous ferric oxide) can be prepared by means of these reactions by adding an iron(III) salt to boiling water.

Cr^{3+} and Al^{3+} behave in a similar manner to Fe^{3+}, except that with them hydroxo complex ion formation continues until complex anions are produced. $Cr(OH)_3$ and $Al(OH)_3$, as we have previously noted, are soluble in alkaline as well as acidic solutions; they are amphoteric.

Regarding the acid strengths of aqua complex ions, a critical factor is the charge-to-radius ratio of the central metal ion, just as we found in an earlier discussion of acid strength (Section 17-9). Thus, the small, highly charged Fe^{3+} attracts electrons away from an O—H bond in a ligand water molecule, causing it to ionize as an acid ($K_{a_1} = 9 \times 10^{-4}$). The attraction of Fe^{2+} for electrons is weaker, as is its acid strength ($K_{a_1} = 1 \times 10^{-7}$).

25-10   Some Kinetic Considerations

When we add $NH_3(aq)$ to a solution containing Cu^{2+}, we see a change in color from pale blue to very deep blue. The reaction involves NH_3 molecules displacing H_2O molecules as ligands.

Formation of $[Cu(NH_3)_4]^{2+}$.

$$[Cu(H_2O)_4]^{2+} + 4\,NH_3 \longrightarrow [Cu(NH_3)_4]^{2+} + 4\,H_2O \qquad (25.25)$$
(pale blue)   (very deep blue)

This reaction occurs very rapidly—as rapidly as the two reactants can be brought together. The addition of HCl(aq) to an aqueous solution of Cu^{2+} produces an immediate color change from pale blue to green, or even yellow if the HCl(aq) is sufficiently concentrated.

Formation of $[CuCl_4]^{2-}$.

$$[Cu(H_2O)_4]^{2+} + 4\,Cl^- \longrightarrow [CuCl_4]^{2-} + 4\,H_2O \qquad (25.26)$$
(pale blue)   (yellow)

The terms "labile" and "inert" are not related to the thermodynamic stabilities of complex ions, nor to the equilibrium constants for ligand-substitution reactions. The terms are *kinetic* terms, referring to the *rates* at which ligands are exchanged.

Complex ions in which ligands can be interchanged rapidly are said to be **labile**. $[Cu(H_2O)_4]^{2+}$, $[Cu(NH_3)_4]^{2+}$, and $[CuCl_4]^{2-}$ are all labile (see Figure 25-18).

In freshly prepared $CrCl_3(aq)$ the ion $[Cr(H_2O)_4Cl_2]^+$ produces a green color, but the color gradually turns to violet (see Figure 25-19). This color change results from the very slow exchange of H_2O for Cl^- as ligands. A complex ion that exchanges ligands slowly is said to be nonlabile or **inert**. In general, complex ions of the first transition series, except for those of Cr(III) and Co(III), are kinetically labile. Those of the second and third transition series are generally kinetically inert. Whether a complex ion is labile or inert affects the ease with which it can be studied. That the

FIGURE 25-18
Labile complex ions.

The exchange of ligands in the coordination sphere of Cu^{2+} occurs very rapidly. The solution at the extreme left is produced by dissolving $CuSO_4$ in concentrated HCl(aq). Its yellow color is caused by $[CuCl_4]^{2-}$. When a small amount of water is added, the mixture of $[Cu(H_2O)_4]^{2+}$ and $[CuCl_4]^{2-}$ ions produces a yellow-green color. When $CuSO_4$ is dissolved in water, a light blue solution of $[Cu(H_2O)_4]^{2+}$ is formed. NH_3 molecules readily displace H_2O molecules as ligands and produce deep blue $[Cu(NH_3)_4]^{2+}$ (extreme right). [Carey B. Van Loon]

inert ones are easiest to obtain and characterize may explain why so many of the early studies of complex ions were based on Cr(III) and Co(III).

25-11 Applications of Coordination Chemistry

The applications of coordination chemistry are numerous and varied. They range from analytical chemistry to biochemistry, and include practical chemistry as well. The several examples given in this section are intended to give you some idea of this diversity.

Hydrates. Often when a compound is crystallized from an aqueous solution of its ions the crystals obtained are hydrated. A hydrate is a substance that has associated with each formula unit a certain number of water molecules. In some cases the water molecules are ligands bonded directly to a metal ion. The coordination compound $[Co(H_2O)_6](ClO_4)_2$ may be represented as the hexahydrate, $Co(ClO_4)_2 \cdot 6H_2O$. In the hydrate $CuSO_4 \cdot 5H_2O$, four H_2O molecules are associated with copper in the complex ion $[Cu(H_2O)_4]^{2+}$, and the fifth with the SO_4^{2-} anion through hydrogen bonding. Another possibility for hydrate formation is that the water molecules may be incorporated into definite positions in the solid crystal but not associated with any particular cations or anions. This is referred to as lattice water, as in $BaCl_2 \cdot 2H_2O$. Finally, part of the water may be coordinated to an ion and part of it may be lattice water. Apparently, this is the case with the alums, such as $KAl(SO_4)_2 \cdot 12H_2O$.

Stabilization of Oxidation States. The standard electrode potential for the reduction of Co(III) to Co(II) is

$$Co^{3+}(aq) + e^- \longrightarrow Co^{2+}(aq) \qquad E° = +1.82 \text{ V} \qquad (25.27)$$

This large positive value suggests that $Co^{3+}(aq)$ is a strong oxidizing agent, strong enough in fact to oxidize water to $O_2(g)$.

$$4 \, Co^{3+}(aq) + 2 \, H_2O \longrightarrow 4 \, Co^{2+}(aq) + 4 \, H^+ + O_2(g) \qquad E°_{cell} = +0.59 \text{ V} \qquad (25.28)$$

Yet one of the complex ions featured in this chapter has been $[Co(NH_3)_6]^{3+}$. This ion is stable in water solution, even though it contains cobalt in the oxidation state $+3$. Reaction (25.28) will not occur if the concentration of Co^{3+} is sufficiently low, and $[Co^{3+}]$ is kept very low because of the great stability of the complex ion.

$$Co^{3+}(aq) + 6 \, NH_3(aq) \rightleftharpoons [Co(NH_3)_6]^{3+}(aq) \qquad K_f = 4.5 \times 10^{23} \qquad (25.29)$$

In fact the concentration of free Co^{3+} is so low that for the half-reaction

$$[Co(NH_3)_6]^{3+} + e^- \longrightarrow [Co(NH_3)_6]^{2+} \qquad (25.30)$$

$E°$ is only $+0.10$ V. As a consequence, not only is $[Co(NH_3)_6]^{3+}$ stable, but $[Co(NH_3)_6]^{2+}$ is rather easily oxidized to the Co(III) complex.

The formation of stable complexes affords a means of attaining certain oxidation states that might otherwise be difficult or impossible.

The Photographic Process. A photographic film is basically an emulsion of silver bromide in gelatin. When the film is exposed to light, silver bromide granules become activated according to the intensity of the light striking them. When the exposed film is placed in a developer solution [a mild reducing agent such as hydroquinone, $C_6H_4(OH)_2$], the activated granules of silver bromide are reduced to black metallic silver. The unactivated granules in the unexposed portions of the film are practically unaffected. This developing process produces the photographic image.

The photographic process cannot be terminated at this point, however. The unactivated granules of silver bromide would eventually be reduced to black metallic silver upon exposing the film to light again. The image on the film must be "fixed." This requires that the black metallic silver that results from the developing be left on the film and the remaining silver bromide be removed. The "fixer" commonly employed is sodium thiosulfate (also known as sodium hyposulfite or "hypo"). In the fixing process $AgBr(s)$ is dissolved and the complexed silver ion is washed away.

The photographic "fixing" process.

$$AgBr(s) + 2\ S_2O_3{}^{2-} \longrightarrow [Ag(S_2O_3)_2]^{3-} + Br^- \tag{25.31}$$

Qualitative Analysis. In the separation and detection of cations in the qualitative analysis scheme, Ag^+, Pb^{2+}, and $Hg_2{}^{2+}$ are first precipitated as chlorides (recall Figure 19-6). All other common cations form soluble chlorides. $PbCl_2(s)$ is removed from $AgCl(s)$ and $Hg_2Cl_2(s)$ by its greater solubility in hot water. $AgCl(s)$ is separated from $Hg_2Cl_2(s)$ by its solubility in $NH_3(aq)$, described by equation (19.20).

At another point in the qualitative analysis scheme it is desired to precipitate CdS in the presence of Cu^{2+}. Normally, Cu^{2+} would precipitate along with the Cd^{2+} since K_{sp} for CuS is smaller than for CdS, 6×10^{-36} compared to 8×10^{-27}. However, by treating the solution of Cu^{2+} and Cd^{2+} with an excess of CN^- prior to saturation with H_2S, this separation can be achieved. The following reactions occur.

$$Cd^{2+} + 4\ CN^- \rightleftharpoons [Cd(CN)_4]^{2-} \qquad K_f = 7.1 \times 10^{18} \tag{25.32}$$

$$2\ Cu^{2+} + 10\ CN^- \longrightarrow 2\ [Cu(CN)_4]^{3-} + C_2N_2(g) \tag{25.33}$$

Reaction (25.33) is an oxidation–reduction reaction in which Cu^{2+} is reduced to Cu^+ and complexed with CN^- as well. The complex ion $[Cu(CN)_4]^{3-}$ is very stable; its K_f value is 1×10^{28}. The concentration of *free* Cu^+ in equilibrium with the complex ion is very low. When a solution containing this complex ion is later saturated with H_2S, K_{sp} for Cu_2S is not exceeded. By contrast, $[Cd^{2+}]$ in equilibrium with $[Cd(CN)_4]^{2-}$ is sufficiently great that K_{sp} of CdS is exceeded under these same conditions.

Electroplating. Electrolyte solutions used in commercial electroplating are quite complex. Each component plays a role in achieving the final objective of a bright smooth fine-grained metal deposit. A number of metals, such as Cu, Ag, and Au, are generally plated from solutions of their cyano complex ions. In the electrolysis reaction below the object to be plated is made the cathode and a piece of copper metal is the anode.

Anode: $$Cu + 4\ CN^- \longrightarrow [Cu(CN)_4]^{3-} + e^- \tag{25.34}$$

Cathode: $$[Cu(CN)_4]^{3-} + e^- \longrightarrow Cu + 4\ CN^- \tag{25.35}$$

The net change simply involves the transfer of Cu metal from the anode to the cathode through the formation, migration, and decomposition of the complex ion $[Cu(CN)_4]^{3-}$. An additional advantage of electroplating Cu from a solution of $[Cu(CN)_4]^{3-}$ is that 1 mol of Cu is obtained per Faraday, in contrast to $\frac{1}{2}$ mol per Faraday that would be obtained from a solution of Cu^{2+}.

In Chapter 4 (Focus feature) we discussed the need to complex or sequester Cu^{2+} in the Raschig process for the preparation of hydrazine, N_2H_4.

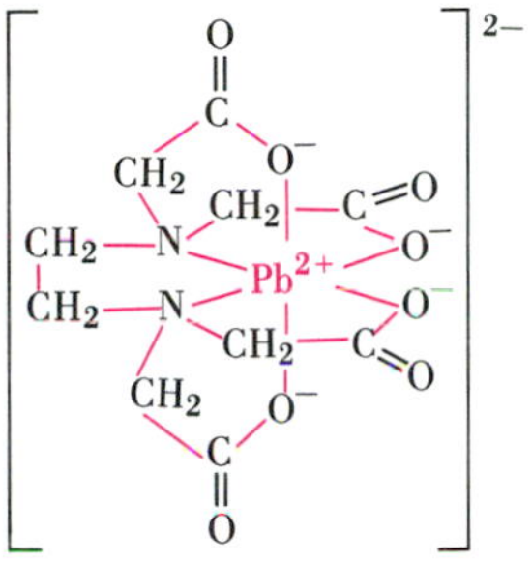

FIGURE 25-20
Structure of $[Pb(EDTA)]^{2-}$.

The structures of other metal–EDTA complexes have a metal ion M^{n+} in place of the Pb^{2+}.

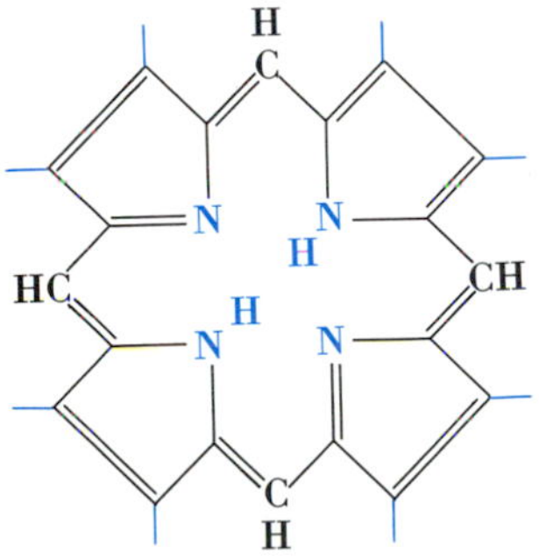

FIGURE 25-21
The porphyrin structure.

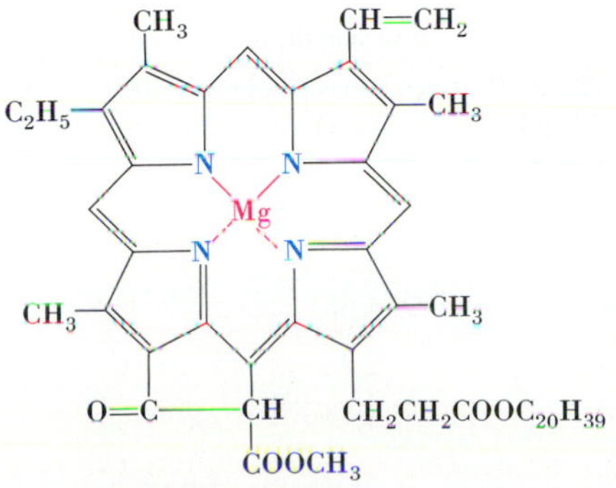

FIGURE 25-22
Structure of chlorophyll a.

Sequestering Metal Ions. Metal ions may act as catalysts in promoting undesirable chemical reactions in a manufacturing process, or they may alter in some way the properties of the material being manufactured. Thus, for many industrial purposes it is imperative to remove mineral impurities from water. Often these impurities are present only in trace amounts, e.g., Cu^{2+}. Precipitation of metal ions is feasible only if K_{sp} for the precipitate is very small.

One method of water treatment involves chelation. Among the chelating agents widely employed are the salts of **ethylenediaminetetraacetic acid (EDTA),** usually the sodium salt.

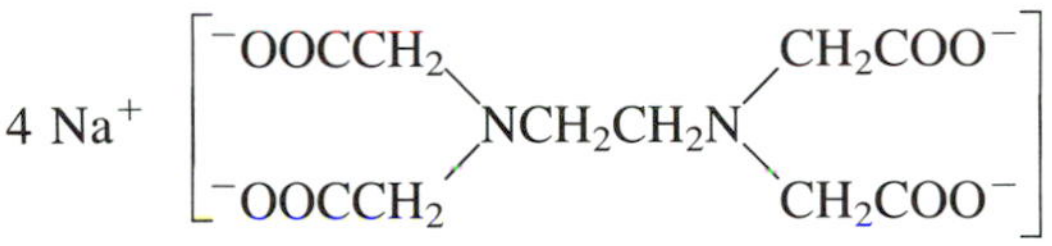

$$4\,Na^+ \begin{bmatrix} ^-OOCCH_2 & & CH_2COO^- \\ & NCH_2CH_2N & \\ ^-OOCCH_2 & & CH_2COO^- \end{bmatrix} \tag{25.36}$$

As an example, the formation constants of $[Ca(EDTA)]^{2-}$ and $[Mg(EDTA)]^{2-}$ are large enough ($K_f = 4 \times 10^{10}$ and 4×10^8, respectively) that the concentrations of $Ca^{2+}(aq)$ and $Mg^{2+}(aq)$ can be reduced to the point that these ions will not precipitate in the presence of any common reagents (including soaps).

Chelation with EDTA can also be used in the treatment of metal poisoning. If a person suffering from lead poisoning is fed $[Ca(EDTA)]^{2-}$, the following exchange occurs because $[Pb(EDTA)]^{2-}$ ($K_f = 1 \times 10^{18}$) is even more stable than $[Ca(EDTA)]^{2-}$ ($K_f = 4 \times 10^{10}$).

$$Pb^{2+} + [Ca(EDTA)]^{2-} \longrightarrow [Pb(EDTA)]^{2-} + Ca^{2+} \tag{25.37}$$

The lead complex is excreted by the body and the Ca^{2+} remains as a body nutrient. The structure of $[Pb(EDTA)]^{2-}$ is shown in Figure 25-20. The high degree of stability of this and other EDTA complexes can be attributed to the presence of five, five-membered chelate rings in the complex.

Biological Applications: Porphyrins. The structure in Figure 25-21 is commonly found in both plant and animal matter. If the groups substituted at the eight bonds shown in blue are all H atoms, the molecule is called porphine. Substituting other groups at these eight positions, yields structures called porphyrins. Metal ions can replace the two H atoms on the central N atoms and coordinate simultaneously with all four N atoms. The porphyrin is a *tetradentate* ligand or chelating agent for the central metal ion.

In the process of **photosynthesis,** carbon dioxide and water, in the presence of inorganic salts, a catalytic agent called **chlorophyll,** and sunlight, combine to form carbohydrates. Carbohydrates are the main structural materials of plants.

$$n\,CO_2 + n\,H_2O \xrightarrow[\text{chlorophyll}]{\text{sunlight}} (CH_2O)_n + n\,O_2 \tag{25.38}$$

carbohydrates

Chlorophyll is a green pigment that absorbs sunlight and directs the storage of this energy into the chemical bonds of the carbohydrates. The structure of one type of chlorophyll is shown in Figure 25-22, and we see that it is a *porphyrin*. The central metal ion is Mg^{2+}.

Green is the complementary color of red, and we should expect chlorophyll to be most effective in absorbing red light (about 670–680 nm). In turn, this suggests that plants should grow more rapidly in red light than in light of other colors. For example, the maximum production of $O_2(g)$ in reaction (25.38) occurs with red light.

In Chapter 28 we consider another porphyrin structure that is essential to life—hemoglobin.

Summary

Many metal ions, particularly those of the transition elements, have the ability to form bonds with electron-donor groups (ligands). The number of ligands coordinated and the geometrical distribution of these ligands about the metal ion are distinctive features of a complex ion. Some ligands have more than one electron pair for donation. These multidentate ligands are able to attach simultaneously to two or more positions in the coordination sphere of the central metal ion. This multiple attachment produces complexes with five- or six-membered rings of atoms—chelates.

To name complexes requires application of a set of rules. These rules deal with such matters as denoting the number and kinds of ligands, the oxidation state of the central metal ion, and whether the complex is neutral, a cation, or an anion. The positions in the coordination sphere of the central ion at which attachment of ligands may occur are not always equivalent. In geometric isomerism different structures with different properties result depending on where this attachment occurs. Optical isomers differ only in certain properties that depend on chirality, such as the rotation of plane polarized light. Other forms of isomerism depend on such factors as the atom of the ligand through which linkage occurs, and whether groups in a coordination compound are present within or outside the coordination sphere of the metal ion.

The structures of complex ions can be rationalized by appropriate orbital hybridization schemes. More successful in explaining the magnetic properties and characteristic colors of complex ions is the crystal field theory. This theory emphasizes the splitting of the d energy level as a result of repulsions between electrons of the central ion and of the ligands. Different splitting patterns are obtained for different geometric structures. A prediction of the magnitude of d-level splitting produced by a ligand can be made through the spectrochemical series.

The formation of a complex ion can be viewed as a stepwise equilibrium process in which other ligands displace H_2O molecules from aqua complex ions. There is an equilibrium constant for each step, and these stepwise constants can be combined into an overall formation constant for the complex ion, K_f. The ability of ligand water molecules to ionize causes some aqua complexes to exhibit acidic properties and helps to explain amphoterism. Also important is the rate at which a complex ion exchanges ligands between its coordination sphere and the solution. Exchange is rapid in a labile complex and slow in an inert complex.

The formation of complex ions can be used to stabilize certain oxidation states, such as Co(III). Other applications include dissolving precipitates, such as AgCl by $NH_3(aq)$ in the qualitative analysis scheme and AgBr by $Na_2S_2O_3(aq)$ in the photographic process, and sequestering ions by chelation, as with EDTA.

Summarizing Example

One of the great achievements of the crystal field theory (Section 25-6) has been in predicting the frequencies of light that complex ions are expected to absorb. This, in turn, leads to explanations of the colors of complex ions. A particular complex ion that has been used in this development is the aqua complex of Ti^{3+}, $[Ti(H_2O)_6]^{3+}$, since its interaction with light is relatively simple. One method of describing the ability of a substance to absorb light is through its absorption spectrum. The proportion of monochromatic (single color) light passing through a solution that is absorbed is measured and expressed through a quantity called the *absorbance*. Absorbances are then plotted for different wavelengths of light used. High absorbances correspond to large proportions of the incident light being absorbed; low absorbances mean that large proportions of the light are transmitted. The absorption spectrum of $[Ti(H_2O)_6]^{3+}$ is plotted in the figure.

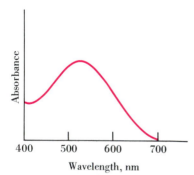

The absorption spectrum of $[Ti(H_2O)_6]^{3+}(aq)$.

1. What is the electron configuration of the ion Ti^{3+}?

Solution. An atom of Ti (Z = 22) has the electron configuration $[Ar]3d^2 4s^2$. When the ion Ti^{3+} is formed, the outer-shell $4s$ electrons and one $3d$ electron are lost. This leaves the configuration $[Ar]3d^1$ for Ti^{3+}.

2. Draw a diagram to represent the distribution of d electrons in the *octahedral* complex ion $[Ti(H_2O)_6]^{3+}$.

Solution. Since the complex ion is octahedral, we should use the crystal field diagram of Figure 25-11. The ion has only one *d* electron, and we should expect to find this in one of the lower energy orbitals.

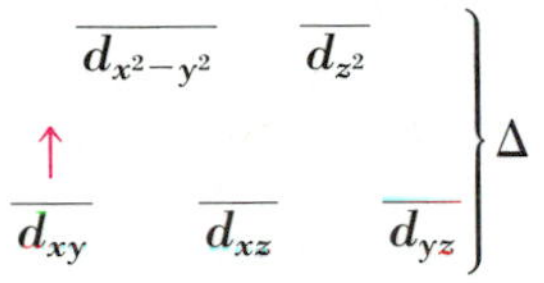

$$
\left.
\begin{array}{ccc}
\overline{}\ d_{x^2-y^2} & \overline{}\ d_{z^2} & \\[1em]
\overset{\uparrow}{\underline{}}\ d_{xy} & \overline{}\ d_{xz} & \overline{}\ d_{yz}
\end{array}
\right\}\Delta
$$

(This example is similar to Example 25-4.)

3. What color light does $[\text{Ti}(\text{H}_2\text{O})_6]^{3+}$ absorb most strongly?

Solution. According to the absorption spectrum given, the maximum absorbance comes at about 500 nm. We can now refer to the electromagnetic spectrum in Figure 8-5 to see the color to which this wavelength corresponds. We see that it is a bluish green color.

4. Solutions of $[\text{Ti}(\text{H}_2\text{O})_6]^{3+}$ have a red-violet color. Explain the origin of this color.

Solution. The absorption spectrum suggests that, when white light is passed through them, these solutions absorb strongly in the green region and transmit most of the red light and some of the blue light. From the color circle in Figure 25-14 we see that a mixture of red and blue yields a violet color. A red-violet color is a reasonable expectation for solutions of $[\text{Ti}(\text{H}_2\text{O})_6]^{3+}$.

Key Terms

chelate (25-2)
chelating agent (25-2)
chiral (25-4)
coordination isomerism (25-4)
coordination number (25-1)
coordination sphere (25-1)
crystal field theory (25-6)
dextrorotatory (25-4)
enantiomers (optical isomers) (25-4)

formation constant, K_f (25-8)
geometric isomerism (25-4)
inert complex (25-10)
ionization isomerism (25-4)
isomers (25-4)
labile complex (25-10)
levorotatory (25-4)
linkage isomerism (25-4)

multidentate ligand (25-2)
optical isomerism (25-4)
primary color (25-7)
racemic mixture (25-4)
resolution (25-4)
secondary color (25-7)
spectrochemical series (25-6)
unidentate ligand (25-2)

Highlighted Expressions

Rules for naming complex ions (25.4)
The spectrochemical series (25.13)
Formation of $[\text{Cu}(\text{NH}_3)_4]^{2+}$ (25.25)

Formation of $[\text{CuCl}_4]^{2-}$ (25.26)
The photographic "fixing" process (25.31)

Review Problems

1. Write the formula of
(a) a complex ion having Co^{2+} as the central ion and two NH_3 molecules and four Cl^- ions as ligands.
(b) a complex ion of manganese(III) having a coordination number of 6 and CN^- as ligands.
(c) a coordination compound comprised of two types of com-

plex ions; one is a complex of Cr(III) with ethylene-diamine (en), having a coordination number of 6, and the other is a complex of Ni(II) with CN^- and having a coordination number of 5.

2. What is the coordination number and the oxidation state of the central metal ion in each of the following complexes?

(a) $[Ni(NH_3)_6]^{2+}$ (b) $[AlF_6]^{3-}$

(c) $[Cu(CN)_4]^{2-}$ (d) $[Cr(NH_3)_3Br_3]$

(e) $[Fe(ox)_3]^{3-}$ (f) $[Ag(S_2O_3)_2]^{3-}$

3. Name the following complex ions. (Do not attempt to distinguish among isomers in this problem.)

(a) $[Ag(NH_3)_2]^+$ (b) $[Fe(H_2O)_5OH]^{2+}$

(c) $[ZnCl_4]^{2-}$ (d) $[Pt(en)_2]^{2+}$

(e) $[Co(NH_3)_4(NO_2)Cl]^+$

4. Name the following coordination compounds.

(a) $[Co(NH_3)_5Br]SO_4$ (b) $[Co(NH_3)_5SO_4]Br$

(c) $[Cr(NH_3)_6][Co(CN)_6]$ (d) $Na_3[Co(NO_2)_6]$

(e) $[Co(en)_3]Cl_3$

5. Write appropriate formulas for the following species. (Do not attempt to distinguish among isomers in this problem.)

(a) dicyanoargentate(I) ion

(b) diamminetetrachloronickelate(II) ion

(c) tris(ethylenediamine)copper(II) sulfate

(d) sodium diaquatetrahydroxoaluminate(III)

6. Draw Lewis structures for the following unidentate ligands. [*Hint:* Refer to Table 25-2.] (a) H_2O; (b) OH^-; (c) ONO^- (nitrito); (d) SCN^- (thiocyanato).

7. Write formulas for the following hydrates.

(a) iron(III) chloride hexahydrate

(b) cobalt(II) hexachloroplatinate(IV) hexahydrate

8. Draw a structure to represent the complex ion *trans*-$[Cr(NH_3)_4ClOH]^+$.

9. Draw structures to represent these three complex ions. (a) $[PtCl_4]^{2-}$ (b) $[Fe(en)Cl_4]^-$ (c) *cis*-$[Fe(en)(ox)Cl_2]^-$

10. How many different structures are possible for each of the following complex ions?

(a) $[Co(NH_3)_5H_2O]^{3+}$ (b) $[Co(NH_3)_4(H_2O)_2]^{3+}$

(c) $[Co(NH_3)_3(H_2O)_3]^{3+}$ (d) $[Co(NH_3)_2(H_2O)_4]^{3+}$

11. Indicate what type of isomerism may be found in each of the following cases. If no isomerism is possible, so indicate.

(a) $[Zn(NH_3)_4][CuCl_4]$ (b) $[Fe(CN)_5SCN]^{4-}$

(c) $[Ni(NH_3)_5Cl]^+$ (d) $[Pt(py)Cl_3]^-$

(e) $[Cr(NH_3)_3(OH)_3]^-$

12. (a) Draw an orbital diagram to represent bonding in the complex ion $[AuBr_4]^-$.

(b) Would you expect this complex ion to be diamagnetic or paramagnetic? Explain.

13. Of the complex ions $[Co(H_2O)_6]^{3+}$ and $[Co(en)_3]^{3+}$, one has a yellow color in aqueous solution and the other, blue. Match each ion with its expected color, and state your reason for doing so.

Exercises

Definitions and terminology

14. The following terms relate to the structure of a complex ion. What is the meaning of each? (a) octahedral geometry; (b) dsp^2 hybrid bonding; (c) aqua complex.

15. Describe the difference in meaning of the terms in each of the following pairs. (a) coordination number and oxidation number; (b) unidentate and multidentate ligand; (c) structural isomerism and stereoisomerism; (d) cis and trans isomer; (e) *d* and *l* isomer; (f) nitro and nitrito complex ion; (g) low- and high-spin complex.

16. What characteristics distinguish a chelate complex from an ordinary complex ion (such as an aqua complex ion)?

Nomenclature

17. Supply acceptable names for the following. (Do not attempt to distinguish among isomers in this exercise.)

(a) $[Co(NH_3)_4(H_2O)(OH)]^{2+}$ (b) $[Co(NH_3)_3(-NO_2)_3]$

(c) $[Pt(NH_3)_4][PtCl_6]$ (d) $[Fe(ox)_2(H_2O)_2]^-$

(e) $[Fe(py)(CN)_5]^{3-}$ (f) $Ag_2[HgI_4]$

18. Write appropriate formulas for the following. (Do not attempt to distinguish among isomers in this exercise.)

(a) potassium hexacyanoferrate(II)

(b) bis(ethylenediamine)copper(II) ion

(c) tetraaquadihydroxoaluminum(III) chloride

(d) amminechlorobis(ethylenediamine)chromium(III) sulfate

(e) tris(ethylenediamine)iron(II) hexacyanoferrate(III)

19. From each of the following names you should be able to deduce the formula of the complex ion or coordination compound intended. Yet, these are not the best systematic names that can be written. Replace each name by one that is more acceptable. [*Hint:* What is wrong with each name as it stands?] (a) cupric tetraammine ion; (b) dichlorotetraamminecobaltic chloride; (c) platinic(IV) hexachloride ion; (d) disodium copper tetrachloride

Bonding and structure in complex ions

20. What type of geometric structure would you predict for $[Au(CN)_2]^-$? Explain.

21. The complex ion $[FeCl_4]^-$ has a tetrahedral structure. Use the valence bond method to propose a bonding scheme for this complex ion. How many unpaired electrons are present in the structure?

22. In what way would you expect the bonding scheme for the high-spin complex $[Co(H_2O)_6]^{2+}$ to differ from that given in the text for $[Co(NH_3)_6]^{3+}$?

23. Draw a plausible structure to represent

(a) $[CuCl_4]^{2-}$ (b) *cis*-$[Zn(NH_3)(OH)_3]^-$

(c) $[Cr(H_2O)_5Cl]^{2+}$

24. Draw plausible structures of the following chelate complexes.

(a) $[Pt(ox)_2]^{2-}$ (b) $[Cr(ox)_3]^{3-}$ (c) $[Fe(EDTA)]^{2-}$

Isomerism

25. Would you expect cis-trans isomerism to occur in a complex ion with a (a) tetrahedral; (b) square planar; (c) linear structure? Explain.

26. Draw structures for each of the isomers indicated in expressions (25.5), (25.6), and (25.8).

27. Write the names and formulas of two coordination isomers in addition to the ones shown in expression (25.6).

28. Which of these octahedral complexes would you expect to exhibit *geometric* isomerism? Explain.

(a) $[Cr(NH_3)_5OH]^{2+}$ (b) $[Cr(NH_3)_3(H_2O)(OH)_2]^+$
(c) $[Cr(en)_2Cl_2]^+$ (d) $[Cr(en)Cl_4]^-$
(e) $[Cr(en)_3]^{3+}$

29. Draw a structure for *cis*-dichlorobis(ethylenediamine)-cobalt(III) ion. Is this ion optically active? Is the trans isomer optically active? Explain.

30. If A, B, C, and D are four different ligands,
(a) how many geometric isomers will be found for square planar $[PtABCD]^{2+}$?
(b) will tetrahedral $[ZnABCD]^{2+}$ display optical isomerism?

31. The structures of four complex ions are given. Each has Co^{3+} as the central ion. The ligands are H_2O, NH_3, and oxalate ion, $C_2O_4^{2-}$. Determine which, *if any*, of these complex ions are isomers (geometric or optical); which, *if any*, are identical (that is, have identical structures;) and which, *if any*, are distinctly different.

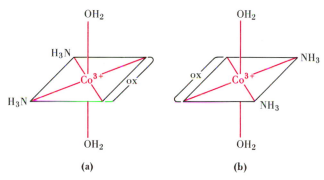

(a)   (b)

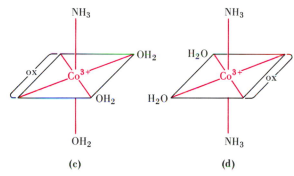

(c)   (d)

Exercise 31

Crystal field theory

32. Describe how the crystal field theory makes possible an explanation of the fact that so many transition metal compounds are colored.

33. One of the following ions is paramagnetic and one is diamagnetic; which is which? (a) $[MoCl_6]^{3-}$; (b) $[Co(en)_3]^{3+}$

34. If the ion Cr^{2+} is linked with strong-field ligands to produce an octahedral complex, the complex has *two* unpaired electrons. If Cr^{2+} is linked with weak-field ligands, the complex has *four* unpaired electrons. How do you account for this difference? What hybridization schemes should be used to describe bonding by the valence bond method for the two types of complexes?

35. In contrast to the case of Cr^{2+} considered in Exercise 34, no matter what ligand is linked to Cr^{3+} to form an octahedral complex, the complex always has *three* unpaired electrons. Explain this fact.

36. Cyano complexes of transition metal ions (e.g., Fe^{2+} and Cu^{2+}) are often yellow in color, whereas aqua complexes are often green or blue. Why is there this difference?

37. Predict
(a) the number of unpaired electrons expected for the tetrahedral complex ion $[CoCl_4]^{2-}$;
(b) whether the square planar complex ion $[Cu(py)_4]^{2+}$ is diamagnetic or paramagnetic;
(c) whether octahedral $[Mn(CN)_6]^{3-}$ or tetrahedral $[FeCl_4]^-$ has the greater number of unpaired electrons.

38. In Example 25-5 we chose between a tetrahedral and a square planar structure for $[Ni(CN)_4]^{2-}$ based on magnetic properties. Could we similarly use magnetic properties to establish whether the ammine complex of Ni(II) is $[Ni(NH_3)_6]^{2+}$ or tetrahedral $[Ni(NH_3)_4]^{2+}$? Explain.

39. Use data from Sections 25-7 and 25-10 (including Table 25-4) to explain the sequence of colors depicted in the opening photograph of this chapter (that is, green solid → green solution → violet solution → green solid).

Complex ion equilibria

40. Write equations to represent the following observations.
(a) A mixture of $Mg(OH)_2(s)$ and $Zn(OH)_2(s)$ is treated with $NH_3(aq)$. The $Zn(OH)_2$ dissolves but the $Mg(OH)_2(s)$ is left behind.
(b) When $NaOH(aq)$ is added to $CuSO_4(aq)$, a pale blue precipitate forms. If $NH_3(aq)$ is added, the precipitate redissolves, producing a solution with an intense deep blue color. If this deep blue solution is made acidic with $HNO_3(aq)$, the color is converted back to pale blue.
(c) A quantity of $CuCl_2(s)$ is dissolved in concentrated $HCl(aq)$ and produces a yellow solution. The solution is diluted to twice its volume with water and assumes a green color. Upon dilution to 10 times its original volume, the solution becomes pale blue in color.

41. Write a series of equations to show the stepwise displacement of H_2O ligands in $[Fe(H_2O)_6]^{3+}$ by ethylenediamine, for which $\log K_1 = 4.34$; $\log K_2 = 3.31$; and $\log K_3 = 2.05$. What is the overall formation constant, $\beta_3 = K_f$, for $[Fe(en)_3]^{3+}$?

★42. Without performing detailed calculations, show why you would expect the concentrations of the various ammine–aqua complex ions to be negligible compared to that of $[Cu(NH_3)_4]^{2+}$ in a solution having a total Cu(II) concentration of 0.10 M and a total concentration of NH_3 of 1.0 M. Under what conditions would the concentrations of these ammine–aqua complex ions (such as $[Cu(H_2O)_3NH_3]^{2+}$) become more significant relative to the concentration of $[Cu(NH_3)_4]^{2+}$? Explain.

43. Without performing detailed calculations, verify the statement on page 929 that neither Ca^{2+} nor Mg^{2+} found in natural waters is likely to precipitate from the water upon the addition of other reagents, if the ions are complexed with EDTA. [*Hint:* Use data from this chapter and Chapter 19. Assume some reasonable values for the total metal ion concentration and that of *free* EDTA, such as 0.10 M each.]

44. A solution that is 0.010 M in Pb^{2+} is also made to be 0.20 M in a salt of EDTA (that is, having a concentration of the $EDTA^{4-}$ ion of 0.20 M). If this solution is now saturated with H_2S, will PbS(s) precipitate? [*Hint:* You will also have to use data from Chapter 19.]

Acid–base properties

45. Write simple chemical equations to show how the complex ion $[Cr(H_2O)_5OH]^{2+}$ acts as (a) an acid; (b) a base.

46. For a solution that is made up to be 0.100 M in $[Fe(H_2O)_6]^{3+}$,

(a) Calculate the pH of the solution assuming that ionization of the aqua complex ion proceeds only through the first step (25.23).

(b) Calculate $[Fe(H_2O)_5OH]^{2+}$ if the solution is also 0.100 M $HClO_4$. (ClO_4^- does not complex with Fe^{3+}.)

(c) Can the pH of the solution be maintained so that $[[Fe(H_2O)_5OH]^{2+}]$ does not exceed 1×10^{-6} M? Explain.

★ 47. We learned in Chapter 17 that for polyprotic acids ionization constants for successive ionization steps decrease rapidly. That is, $K_{a_1} \gg K_{a_2} \gg K_{a_3}$. The ionization constants for the first two steps in the ionization of $[Fe(H_2O)_6]^{3+}$ (reactions 25.23 and 25.24) are more nearly equal in magnitude. Can you think of a reason why this multistep ionization does not seem to follow the pattern for polyprotic acids?

Applications

48. From data in Chapter 19

(a) Derive an equilibrium constant for reaction (25.31) and explain why this reaction (the fixing of photographic film) is expected to go essentially to completion.

(b) Explain why $NH_3(aq)$ cannot be used in the fixing of photographic film.

49. Show that the oxidation of $[Co(NH_3)_6]^{2+}$ to $[Co(NH_3)_6]^{3+}$ referred to on page 927 should occur spontaneously in alkaline solution with H_2O_2 as an oxidizing agent. [*Hint:* Refer also to Figure 23-6.]

50. A current of 2.13 A is passed for 0.347 h between a pair of Cu electrodes. What mass of Cu is deposited at the cathode if the electrolyte is (a) $[Cu(H_2O)_4]^{2+}$; (b) $[Cu(CN)_4]^{3-}$? Why are these quantities different?

51. Use data provided in Section 25-11 to demonstrate that if a solution is 0.10 M in total copper, 0.10 M in total cadmium, 0.10 M in *free* CN^-, and 0.10 M H_2S, CdS(s) will precipitate but not $Cu_2S(s)$. [Assume a pH of 7. $K_{sp}(CdS) = 8 \times 10^{-27}$; $K_{sp}(Cu_2S) = 1 \times 10^{-49}$.]

Additional Exercises

52. The following compounds were discovered before Werner's time. Rewrite each *empirical* formula to be consistent with Werner's formulation for coordination compounds. Name each compound.

(a) Zeise's salt: $PtCl_2 \cdot KCl \cdot C_2H_4$ (This compound consists of a simple unipositive cation and a uninegative complex anion.)

(b) Magnus' green salt: $PtCl_2 \cdot 2NH_3$ (This compound consists of a dipositive complex cation and a dinegative complex anion.)

53. Figure 25-9 presents a structure that was at one time thought to be an alternative to the octahedral structure. Another was that shown here.

(a) Show that for the complex ion $[Co(NH_3)_4Cl_2]^+$ the above structure predicts three geometrical isomers.

(b) Show that neither the above structure nor Figure 25-9 would account for optical isomerism in $[Co(en)_3]^{3+}$.

54. Sketch all the possible isomers of $[Co(ox)(NH_3)_2Cl_2]^-$.

55. The cis and trans isomers of $[Co(en)_2Cl_2]^+$ can be distinguished through a displacement reaction with oxalate ion. What difference in reactivity toward oxalate ion would you expect between the cis and trans isomers? [*Hint:* Which ligands would you expect oxalate ion to displace more readily?]

56. Propose a bonding scheme that is consistent with the square planar structure of $[Cu(NH_3)_4]^{2+}$. Is this complex ion diamagnetic or paramagnetic?

57. A tabulation of formation constant data lists the following log K values for the formation of $[CuCl_4]^{2-}$: $\log K_1 = 2.80$; $\log K_2 = 1.60$; $\log K_3 = 0.49$; $\log K_4 = 0.73$.

(a) What is the value of the overall formation constant $\beta_4 = K_f$ for $[CuCl_4]^{2-}$?

(b) In connection with reaction (25.26), estimate the concentration of HCl(aq) required in a solution that is 0.10 M $CuSO_4$ to produce a visible yellow color? Assume that 99% conversion of $[Cu(H_2O)_4]^{2+}$ to $[CuCl_4]^{2-}$ is sufficient for this to happen, and neglect the presence of any mixed aqua–chloro complex ions.

58. In both $[Fe(H_2O)_6]^{2+}$ and $[Fe(CN)_6]^{4-}$ ions the iron is present as Fe(II); yet $[Fe(H_2O)_6]^{2+}$ is paramagnetic, whereas $[Fe(CN)_6]^{4-}$ is diamagnetic. Explain this difference.

59. The graph represents the molar conductivity of some Pt(IV) complexes. The ligands in these complexes are NH_3 molecules and/or Cl^- ions, and the coordination number of Pt(IV) is 6. Write formulas for the coordination compounds corresponding to each of the points in the graph. (Molar conductivity is the electrical conductivity, under precisely defined conditions, of an aqueous solution containing one mole of a compound.)

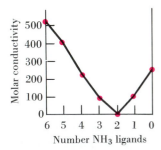

Exercise 59   Number NH_3 ligands

60. A theory of coordination compounds that predated Werner's was called the chain theory. According to this theory, in a coordination compound like $CoCl_3 \cdot 6NH_3$, the Co atom could bond directly three groups, either Cl atoms or NH_3 molecules, and NH_3 molecules could bond together into chains, such as

$$Co \overset{NH_3Cl}{\underset{NH_3Cl}{\overbrace{}-NH_3-NH_3-NH_3-NH_3Cl}}$$

Cl atoms were thought to be ionizable if bonded to NH_3, but not if bonded to Co. Demonstrate how this chain theory could be used to explain the behavior of the compounds (a), (b), and (c) described in Section 25-1 (expression 25.1). Specifically, show how the theory is consistent with the behavior of the three compounds toward $AgNO_3$.

61. Demonstrate why the "chain" theory of Exercise 60 could not explain the existence of compounds such as $[Co(NH_3)_3Cl_3]$ and $Na[Co(NH_3)_2Cl_4]$.

62. When chromium metal is dissolved in $HCl(aq)$ a blue solution is produced that quickly turns a green color. Later the green solution becomes blue-green and then violet in color. Write chemical equations to describe all these changes, starting with chromium metal.

63. Use data in Table 25-5 to determine values of (a) β_2 for the formation of $[Cu(H_2O)_2(NH_3)_2]^{2+}$; (b) β_4 for the formation of $[Zn(NH_3)_4]^{2+}$; (c) β_4 for the formation of $[Ni(H_2O)_2(NH_3)_4]^{2+}$.

★ **64.** Refer to the Summarizing Example. If the electronic transition responsible for the maximum at 500 nm in the absorption spectrum for $[Ti(H_2O)_6]^{3+}$ is from one of the lower to one of the upper d orbitals, what must be the energy separation, Δ, ex-

pressed in kJ/mol? [*Hint:* Review the relationship between frequency and wavelength of light and the use of Planck's equation in Chapter 8.]

★ **65.** With reference to the stability of $[Co(NH_3)_6]^{3+}(aq)$,
(a) Verify that E°_{cell} for reaction (25.28) is $+0.59$ V.
(b) Calculate $[Co^{3+}]$ in a solution that has a total concentration of cobalt of 1.0 M and $[NH_3] = 0.10$ M.
(c) Show that for the value of $[Co^{3+}]$ calculated in part (b), reaction (25.28) will not occur. [*Hint:* What is $[H_3O^+]$ in 0.10 M NH_3? Assume a low, but reasonable, concentration of Co^{2+}, say 1×10^{-4} M, and a partial pressure of $O_2(g)$ of 0.2 atm.]

★ **66.** A Cu electrode is immersed in a solution that is 1.00 M NH_3 and 1.00 M in $[Cu(NH_3)_4]^{2+}$. If a standard hydrogen electrode is the cathode, E_{cell} is found to be $+0.08$ V. What is the value obtained by this method for the formation constant, K_f, of $[Cu(NH_3)_4]^{2+}$?

★ **67.** The following concentration cell is constructed.

$$Ag|Ag^+(0.10 \text{ M } [Ag(CN)_2]^-, \; 0.10 \text{ M KCN})\|Ag^+(0.10 \text{ M})|Ag$$

If K_f for $[Ag(CN)_2]^-$ is 5.6×10^{18}, what value would you expect for E_{cell}? [*Hint:* Recall that the electrode on the left is the anode.]

Self-Test Questions

For questions 68 through 75 select the single item that best completes each statement.

68. The oxidation state of Ni in the complex ion $[Ni(CN)_4I]^{3-}$ is (a) -3; (b) -2; (c) $+2$; (d) $+5$.

69. The coordination number of Pt in the complex ion $[Pt(en)_2Cl_2]^{2+}$ is (a) 3; (b) 4; (c) 5; (d) 6.

70. Of the following complex ions one exhibits isomerism. That one is (a) $[Ag(NH_3)_2]^+$; (b) $[Co(NH_3)_5NO_2]^{2+}$; (c) $[Pt(en)Cl_2]$; (d) $[Co(NH_3)_5Cl]^{2+}$.

71. Of the following complex ions, the one that exhibits optical isomerism is (a) *cis*-$[Co(en)_2Cl_2]^+$; (b) $[Co(NH_3)_4Cl_2]^+$; (c) $[Co(NH_3)_2Cl_4]^-$; (d) *trans*-$[Co(en)_2Cl_2]^+$.

72. The number of unpaired electrons expected for the complex ion $[Cr(NH_3)_6]^{2+}$ is (a) 2; (b) 3; (c) 4; (d) 5.

73. Of the following complex ions, one is a Brønsted–Lowry acid. That one is (a) $[Cu(NH_3)_4]^{2+}$; (b) $[FeCl_4]^-$; (c) $[Fe(H_2O)_6]^{3+}$; (d) $[Zn(OH)_4]^{2-}$.

74. Of the following complex ions, the one that probably has the largest overall formation constant, K_f, is (a) $[Co(NH_3)_6]^{3+}$; (b) $[Co(H_2O)_6]^{3+}$; (c) $[Co(H_2O)_4(NH_3)_2]^{3+}$; (d) $[Co(en)_3]^{3+}$.

75. The most soluble of the following compounds in $NH_3(aq)$ is (a) $Cu(OH)_2$; (b) $BaSO_4$; (c) SiO_2; (d) $MgCO_3$.

76. Name the following complex ions and coordination compounds.
(a) $[Cr(NH_3)_4(OH)_2]Br$ (b) $K_3[Co(-NO_2)_6]$
(c) $[Fe(en)_2(H_2O)_2]^{2+}$ (d) $[Pt(en)_2Cl_2]SO_4$

77. Sketch a plausible geometric structure for
(a) $[Co(NH_3)_5Cl]^{2+}$
(b) $[Cr(en)_2(ox)]^+$
(c) *cis*-diamminedinitroplatinum(II)

78. Explain the following observations in terms of complex ion formation.
(a) $Al(OH)_3(s)$ is soluble in $NaOH(aq)$ but insoluble in $NH_3(aq)$.
(b) $ZnCO_3(s)$ is soluble in $NH_3(aq)$ but $ZnS(s)$ is not.
(c) $CoCl_3$ is unstable in water solution, being reduced to $CoCl_2$ and liberating $O_2(g)$. On the other hand, $[Co(NH_3)_6]Cl_3$ can be easily maintained in aqueous solution.

26 Nuclear Chemistry

The age of this jade ornament has been established by radiocarbon dating of objects found at the same Mayan site (see page 945). [Fred Ward/ Black Star]

How old is the ornament pictured in the opening photograph, do you suppose? Nature left an imprint on the site where this object was discovered that is almost as definite as a date stamp, but we have to learn how to read it. Reading nature's date stamp requires that we assess the quantity of the isotope carbon-14 present in appropriate samples.

To assess the quantity of a component in a sample requires a *quantitative analysis*. We have previously considered a number of analytical procedures that are based on chemical reactions (e.g., titrations) or on physical properties (light absorption). Carbon-14, however, has chemical and physical properties that are essentially identical to the much more abundant isotopes carbon-12 and carbon-13. We cannot use ordinary analytical procedures. Carbon-14 does possess one property that distinguishes it from C-12 and C-13, though. *It is radioactive.* We can base our analysis on this property, through a method known as radiocarbon dating and presented in Section 26-5.

In this chapter we consider a variety of phenomena stemming *directly* from the nuclei of atoms. We will refer to these phenomena, collectively, as nuclear chemistry. Nuclear chemical phenomena are of primary importance in dealing with the heavier elements (those with atomic numbers greater than 83), but radioactive isotopes of the lighter elements can be prepared artificially, and we will study these as well. Yet another part of nuclear chemistry is a study of the effects of ionizing radiation on matter—an important aspect of the long-standing "nuclear debate."

26-1 The Phenomenon of Radioactivity

The term "radioactivity" was proposed by Marie Curie to describe the emission of ionizing radiation by some of the heavier elements. Ionizing radiation, as the name implies, interacts with matter to produce ions. This means that the radiation is sufficiently energetic to break chemical bonds. Some ionizing radiation is particulate (consists of particles) and some is nonparticulate. We return to a discussion of the effects of ionizing radiation on matter in Section 26-10. For the present, let us comment briefly on the types of ionizing radiation associated with radioactivity. This takes the form of a review and extension of ideas we first encountered in Chapter 2.

From Chapter 2 you need to recall that the symbolism $_Z^A X$ represents a particular kind of atom having atomic number Z and mass number A. We refer to a particular kind of atom as a **nuclide.** The term isotope is often used synonymously with nuclide, although the two terms have slightly different meanings, as explained on page 41.

Alpha (α) Rays. This radiation consists of a stream of alpha (α) particles. These particles are identical to the nuclei of helium-4 atoms, $_2^4He^{2+}$. Alpha rays produce large numbers of ions as they penetrate matter, even though their penetrating power is low. (They can generally be stopped by a sheet of paper.) Because of their positive charge, α particles are deflected by electric and magnetic fields (see Figure 26-1). The production of α particles by a radioactive nucleus can be represented through a **nuclear equation.** A nuclear equation is written in a similar way to a chemical equation. In a nuclear equation

Writing nuclear equations.

- The sum of the mass numbers must be the same on both sides of the equation.
- The sum of the atomic numbers must be the same on both sides of the equation.

(26.1)

In equation (26.2) the alpha particle is represented as $_2^4He$. Mass numbers total 238, and atomic numbers total 92. The loss of an alpha particle produces a decrease of *two* in the atomic number and *four* in the mass number of the radioactive nucleus.

$$_{92}^{238}U \longrightarrow {}_{90}^{234}Th + {}_2^4He \tag{26.2}$$

937

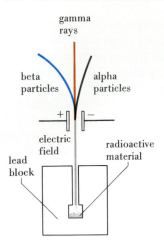

FIGURE 26-1

Three types of radiation from radioactive materials.

The radioactive material is enclosed in a lead block with a narrow opening. All the radiation except that passing through the opening is absorbed by the lead. When this radiation is passed through an electric field, it splits into three beams. One beam is undeflected—these are gamma (γ) rays. A second beam is attracted toward the negatively charged plate; these are the positively charged alpha (α) particles. The third beam is deflected toward the positive plate; this is the beam of beta (β) particles—electrons. Because of their greater momentum (product of mass and velocity), α particles are deflected to a smaller extent then β particles. The situation is similar if a magnetic field is substituted for the electric field.

Beta (β^-) Rays. These rays are also comprised of particles; β^- particles are identical to electrons. Beta rays have a greater penetrating power but a lower ionizing power than α rays. (They can pass through aluminum foil 2 to 3 mm thick.) Because of their negative charge, beta particles are deflected by electric and magnetic fields in the *opposite* direction from α particles (see Figure 26-1). The nucleus of $^{234}_{90}$Th, which is a product of reaction (26.2), undergoes radioactive decay by β^- emission. When a nucleus loses a β^- particle, the atomic number of the nucleus *increases* by one unit and its mass number is unchanged. Thus, a β^- particle is treated as if it had an atomic number of -1 and a mass number of 0; it is represented as $_{-1}^{0}$e.

$$^{234}_{90}\text{Th} \longrightarrow {}^{234}_{91}\text{Pa} + {}^{0}_{-1}\text{e} \tag{26.3}$$

Gamma (γ) Rays. Some radioactive decay processes yielding α or β particles leave a nucleus in an energetic state. The nucleus then loses energy in the form of electromagnetic radiation—a gamma (γ) ray. Gamma rays are a highly penetrating form of radiation. They are *undeflected* by electric and magnetic fields. In the radioactive decay of $^{234}_{92}$U, 77% of the nuclei emit α particles having an energy of 4.18 MeV. The remaining 23% of the $^{234}_{92}$U nuclei produce α particles with energies of 4.13 MeV. In the latter case the $^{230}_{90}$Th nuclei are left with an excess energy of 0.05 MeV. This energy is released as γ rays. If we denote the unstable, energetic Th nucleus as $^{230}_{90}\text{Th}^{\ddagger}$, we can write

$$^{234}_{92}\text{U} \longrightarrow {}^{230}_{90}\text{Th}^{\ddagger} + {}^{4}_{2}\text{He} \tag{26.4}$$

$$^{230}_{90}\text{Th}^{\ddagger} \longrightarrow {}^{230}_{90}\text{Th} + \gamma \tag{26.5}$$

This γ emission process is represented diagrammatically in Figure 26-2.

The energy unit electronvolt, eV, was introduced in Section 9-8. The unit MeV is a million electronvolts. This unit and its relationship to other energy units are discussed further in Section 26-6.

Positrons (β^+). The emission of β^- rays, as we shall discover in Section 26-7, is characteristic of nuclei in which the ratio of number of neutrons to number of protons is too large for stability. If this ratio is too small for stability, radioactive decay may occur by positron emission. A positron is a *positively* charged particle having the same mass as a β^- particle or electron. It is designated as β^+ or as $_{+1}^{0}$e. Positron emission is commonly encountered with artificially radioactive nuclei of the lighter elements. For example,

$$^{30}_{15}\text{P} \longrightarrow {}^{30}_{14}\text{Si} + {}^{0}_{+1}\text{e} \tag{26.6}$$

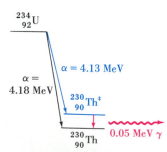

FIGURE 26-2

Production of gamma rays.

The transition of a $^{230}_{90}$Th nucleus between the two energy states shown results in the emission of 0.05 MeV of energy in the form of γ rays.

Electron Capture (E.C.). A second process that achieves the same effect as positron emission is electron capture (E.C.). In this process an electron from an inner electron shell (usually the first or second shell) is absorbed by the nucleus. (Inside the nucleus the electron is used to convert a proton into a neutron.) When an electron from a higher quantum level drops to the level vacated by the captured electron, x radiation is emitted. For example,

$$^{202}_{81}\text{Tl} \xrightarrow{\text{E.C.}} {}^{202}_{80}\text{Hg} \quad \text{(followed by x radiation)} \tag{26-7}$$

Are You Wondering:

How a radioactive nucleus can emit a β^- particle (electron) if there are no electrons present in the nucleus?

This is a question that puzzled earlier investigators in the field of radioactivity. In fact, before the neutron was discovered some scientists believed that a nucleus contained both protons and electrons, with the excess number of protons determining the nuclear charge. The answer to this question lay in establishing that β^- emission involves the conversion of a neutron to a proton and an electron. We discuss this and other nuclear processes in Section 26-7.

Example 26-1

Writing nuclear equations for radioactive decay processes. Write nuclear equations to represent **(a)** α particle emission by ^{222}Rn; **(b)** radioactive decay of bismuth-215 to polonium; **(c)** decay of a radioactive nucleus to produce ^{58}Ni and a positron.

Solution

(a) We can identify two of the species involved in this process simply from the information given. We can deduce the remaining species by using the basic principles of expression (26.1). (Note that as long as a name or chemical symbol is given, such as Rn, the atomic number of the element follows directly.)

$$^{222}_{86}Rn \longrightarrow \; ? + ^{4}_{2}He \quad\quad \text{and} \quad\quad ^{222}_{86}Rn \longrightarrow \; ^{218}_{84}Po + ^{4}_{2}He$$

(b) Bismuth has the atomic number 83, and polonium 84. The type of emission that leads to an increase of one unit in atomic number is β^-.

$$^{215}_{83}Bi \longrightarrow \; ^{215}_{84}Po + ^{0}_{-1}e$$

(c) We are given the products of the radioactive decay process. The only radioactive nucleus that can produce them is $^{58}_{29}Cu$.

$$^{58}_{29}Cu \longrightarrow \; ^{58}_{28}Ni + ^{0}_{+1}e$$

SIMILAR EXAMPLES: Exercises 2, 17, 54.

26-2 Naturally Occurring Radioactive Isotopes

$^{209}_{83}Bi$ is the nuclide of highest atomic and mass number that is stable. All known nuclides beyond it in atomic and mass numbers are radioactive. Naturally occurring $^{238}_{92}U$ is radioactive and disintegrates by the loss of α particles.

$$^{238}_{92}U \longrightarrow \; ^{234}_{90}Th + ^{4}_{2}He \quad\quad\quad (26.8)$$

$^{234}_{90}Th$ is also radioactive; its decay is by β^- emission.

$$^{234}_{90}Th \longrightarrow \; ^{234}_{91}Pa + ^{0}_{-1}e \qu\quad\quad (26.9)$$

$^{234}_{91}Pa$ also decays by β^- emission.

$$^{234}_{91}Pa \longrightarrow \; ^{234}_{92}U + ^{0}_{-1}e \qu\quad\quad (26.10)$$

$^{234}_{92}U$ is radioactive also.

FIGURE 26-3
The natural radioactive decay series for $^{238}_{92}$U.

The long arrows pointing down and to the left correspond to α-particle emissions. The short horizontal arrows represent β⁻ emissions. Other natural decay series originate with isotopes of thorium and actinium.

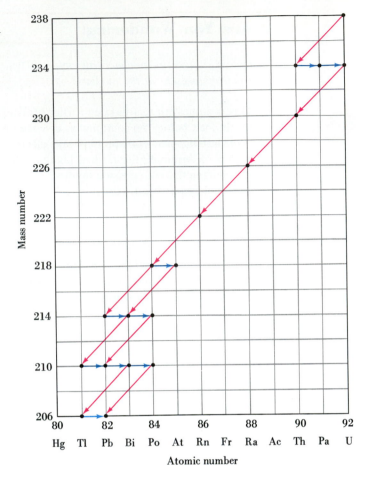

Radioactive Decay Series. The chain of radioactive decay that begins with $^{238}_{92}$U continues through a number of steps of α and β emission until it eventually terminates with a stable isotope of lead—$^{206}_{82}$Pb. The entire scheme is outlined in Figure 26-3. All naturally occurring radioactive nuclides of high atomic number belong to one of three decay series—the **uranium series** just described, the **thorium series,** or the **actinium series.**

These radioactive decay schemes can be used to determine the ages of rocks and thereby the age of the earth (see Section 26-5). The appearance of certain radioactive substances in the environment can also be explained through radioactive decay series.

^{222}Rn produced by the decay of ^{238}U is now known to be the cause of a fatal lung disease afflicting miners in central Europe for centuries. This disease was lung cancer and the uranium responsible for it was present in the gold, silver, and platinum ores that were being mined. Radon is now clearly recognized as an environmental hazard in certain areas of the United States, and it is believed to be the second leading cause of lung cancer (after cigarette smoking). In some instances the radon is associated with wastes from uranium mining, and in others from indigenous rocks and soils. ^{210}Po and ^{210}Pb have been detected in cigarette smoke. These radioactive isotopes are derived from ^{238}U, found in trace amounts in the phosphate fertilizers used in tobacco fields. These α-emitting isotopes have been implicated in the link between cigarette smoking and cancer and heart disease.

Radioactivity, which is so common among isotopes of high atomic number, is a relatively rare phenomenon among the *naturally occurring* lighter isotopes. ^{40}K is a radioactive isotope, as are ^{50}V and ^{138}La. ^{40}K decays by β⁻ emission and by electron capture.

Marie Sklodowska Curie (1867–1934). Marie Curie shared in the 1903 Nobel Prize in Physics for studies on radiation phenomena. In 1911 she won the Nobel Prize in Chemistry for her discovery of polonium and radium.

$$^{40}_{19}K \longrightarrow {}^{40}_{20}Ca + {}^{\ 0}_{-1}e \quad \text{and} \quad {}^{40}_{19}K \xrightarrow{\text{E.C.}} {}^{40}_{18}Ar$$

At the time that the earth was formed ^{40}K was much more abundant than it is now. It is believed that the high argon content of the atmosphere (0.934%, by volume, and almost all of its as ^{40}Ar) has been derived from the radioactive decay of ^{40}K. Aside from ^{40}K, the most important radioactive isotopes of the lighter elements are those that are *artificially* produced.

26-3 Nuclear Reactions and Artificially Induced Radioactivity

Ernest Rutherford discovered that atoms of one element can be transformed into atoms of another element. He did this by bombarding N-14 nuclei with α particles, producing O-17. This process can be represented through a nuclear equation.

$$^{14}_{7}\text{N} + {}^{4}_{2}\text{He} \longrightarrow {}^{17}_{8}\text{O} + {}^{1}_{1}\text{H} \tag{26.11}$$

Although the principles involved in writing equation (26.11) are similar to those used to represent radioactive decay, there is this difference: Two "reactants" are written on the left. Instead of a nucleus disintegrating spontaneously, it must be struck by another small particle to induce a **nuclear reaction.** In the more condensed representation given by (26.12), the target and product nuclei are represented on the left and right of a parenthetical expression. Within the parentheses the bombarding particle is written first, followed by the ejected particle.

Method of representing a nuclear reaction.

$$^{14}\text{N}(\alpha,\text{p})^{17}\text{O} \tag{26.12}$$

$^{17}_{8}$O is a naturally occurring *nonradioactive* nuclide of oxygen (0.037% natural abundance). The situation with $^{30}_{15}$P, which can also be produced by a nuclear reaction, is somewhat different.

In 1934, when bombarding aluminum with α particles, Irène Curie and her husband Frédéric Joliot observed the emission of two types of particles—neutrons and positrons. The Joliots observed that when bombardment by α particles was stopped, the emission of neutrons also stopped; the emission of positrons continued, however. Their conclusion was that the nuclear bombardment produces $^{30}_{15}$P, which undergoes radioactive decay by the emission of positrons.

$$^{27}_{13}\text{Al} + {}^{4}_{2}\text{He} \longrightarrow {}^{30}_{15}\text{P} + {}^{1}_{0}\text{n} \quad \text{or} \quad {}^{27}\text{Al}(\alpha,\text{n})^{30}\text{P} \tag{26.13}$$

$$^{30}_{15}\text{P} \longrightarrow {}^{30}_{14}\text{Si} + {}^{0}_{+1}\text{e} \tag{26.14}$$

$^{30}_{15}P$ was the first radioactive nuclide produced by artificial means. Since the time of its discovery over 1000 other radioactive nuclides have been produced. The number of known radioactive nuclides now exceeds considerably the number of nonradioactive ones (about 280). We consider a few of the many applications of artificially induced radioactivity in Section 26-11.

Example 26-2

Writing equations for nuclear reactions. Write **(a)** a condensed representation of the nuclear reaction, $^{14}_{7}N + ^{1}_{0}n \rightarrow ^{14}_{6}C + ^{1}_{1}H$; **(b)** an expanded representation of the nuclear reaction, $^{59}Co(n,?)^{56}Mn$.

Solution

(a) The bombarding particle is simply a neutron, n, and the emitted particle is a proton, p. The condensed equation is $^{14}N(n,p)^{14}C$.

(b) The missing particle in parentheses must have $A = 4$ and $Z = 2$; it is an α particle.

$$^{59}_{27}Co + ^{1}_{0}n \longrightarrow ^{56}_{25}Mn + ^{4}_{2}He$$

SIMILAR EXAMPLES: Exercises 3, 4, 21, 54.

26-4   Transuranium Elements

Until 1940 the only known elements were those that occur naturally. In 1940, the first synthetic element was produced by bombarding $^{238}_{92}U$ atoms with neutrons. First the unstable nucleus $^{239}_{92}U$ is formed. This nucleus then undergoes β^- decay, yielding the element neptunium, with $Z = 93$.

$$^{238}_{92}U + ^{1}_{0}n \longrightarrow ^{239}_{92}U \tag{26.15}$$

$$^{239}_{92}U \longrightarrow ^{239}_{93}Np + ^{0}_{-1}e \tag{26.16}$$

Bombardment by neutrons is a particularly effective way to produce nuclear reactions because these heavy, uncharged particles are not repelled as they approach a nucleus.

Since 1940 all the elements from $Z = 93$ to 109 have been synthesized. For example, an isotope of the element, $Z = 105$, was produced in 1970 by bombarding atoms of $^{249}_{98}Cf$ with $^{15}_{7}N$ nuclei.

$$^{249}_{98}Cf + ^{15}_{7}N \rightarrow ^{260}_{105}Ha + 4^{1}_{0}n \tag{26.17}$$

Elements 104 and 105 are the first in a series of elements following the actinide series. We might call them *transactinide* elements. From our knowledge of the periodic table, we should predict these elements to resemble hafnium and tantalum, respectively.

Charged-Particle Accelerators. To bring about nuclear reactions such as (26.17) requires that atomic nuclei be bombarded by energetic particles. Such energetic particles can be obtained in an accelerator. A type of accelerator known as a cyclotron is described in Figure 26-4.

A charged-particle accelerator, as the name implies, can only produce beams of charged particles as projectiles (e.g., $^{1}_{1}H^{+}$). In many cases neutrons are more effective as projectiles for nuclear bombardment. The neutrons required can themselves be generated through a nuclear reaction produced by a charged-particle beam. In the

FIGURE 26-4
A charged-particle accelerator
—the cyclotron.

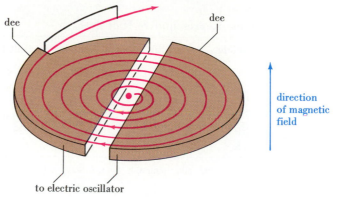

dee   dee

direction
of magnetic
field

to electric oscillator

The accelerator consists of two hollow, flat, semicircular boxes, called dees, that are kept electrically charged. The entire assembly is maintained within a magnetic field. The particles to be accelerated, in the form of positive ions, are produced at the center of the opening between the dees. They are then attracted into the negatively charged dee and forced into a circular path by the magnetic field. When the particles leave the dee and enter the gap, the electric charges on the dees are reversed, so that the particles are attracted into the opposite dee. The particles are accelerated as they pass the gap and travel a wider circular path in the new dee. This process is repeated many times until the particles are brought to the required energy to induce the desired nuclear reaction. They are then brought out of the accelerator and made to strike the target.

A simple analogy to the nuclear accelerator is found in the action of a playground swing. If the rider is given a push each time the swing reaches the end of its arc, the swing accelerates and the arc grows wider with each push.

following reaction $_1^2\text{H}$ represents a beam of deuterons (actually $_1^2\text{H}^+$) from an accelerator.

$$_4^9\text{Be} + {}_1^2\text{H} \longrightarrow {}_5^{10}\text{B} + {}_0^1\text{n}$$

Another important source of neutrons for nuclear reactions is a nuclear reactor (see Section 26-8).

26-5 Rate of Radioactive Decay

In time, we can expect every atomic nucleus of a radioactive isotope to disintegrate, but it is impossible to predict when any given nucleus will do so. Radioactivity is a *random* process. Nevertheless, we can make this all-important observation, called the **radioactive decay law.**

The rate of disintegration of a radioactive material—the decay rate—is directly proportional to the number of atoms present.

In mathematical terms,

Radioactive decay law.

$$\text{rate of decay} \propto N \quad \text{and} \quad \text{rate of decay} = \lambda N \tag{26.18}$$

The decay rate is expressed in atoms per unit time, such as atoms per second. N is the number of atoms in the sample being observed. λ is the **decay constant;** its unit is $(\text{time})^{-1}$. Consider the case of a 1,000,000-atom sample disintegrating at the rate of 100 atoms per second; $N = 1.0 \times 10^6$ and

$$\lambda = \frac{\text{rate of decay}}{N} = \frac{100 \text{ atoms/s}}{1.0 \times 10^6 \text{ atoms}} = 1.0 \times 10^{-4} \text{ s}^{-1}$$

Radioactive decay is a *first-order* process. To relate this to the first-order kinetics that we studied in Chapter 15, think of the rate of decay as being equivalent to a rate

TABLE 26-1
Some Representative Half-lives

Nuclide	Half-life[a]	Nuclide	Half-life[a]	Nuclide	Half-life[a]
$^{3}_{1}H$	12.26 y	$^{40}_{19}K$	1.25×10^{9} y	$^{214}_{84}Po$	1.64×10^{-4} s
$^{14}_{6}C$	5730 y	$^{80}_{35}Br$	17.6 min	$^{222}_{86}Rn$	3.823 d
$^{13}_{8}O$	8.7×10^{-3} s	$^{90}_{38}Sr$	28.1 y	$^{226}_{88}Ra$	1.60×10^{3} y
$^{28}_{12}Mg$	21 h	$^{131}_{53}I$	8.040 d	$^{234}_{90}Th$	24.1 d
$^{32}_{15}P$	14.3 d	$^{137}_{55}Cs$	30.23 y	$^{238}_{92}U$	4.51×10^{9} y
$^{35}_{16}S$	88 d				

[a] s, second; min, minute; h, hour; d, day; y, year.

of reaction; the number of atoms is equivalent to the concentration of a reactant; the decay constant, λ, is equivalent to a rate constant, k. We can carry this equivalence further by writing an integrated radioactive decay law and a relationship between the decay constant and the half-life.

Integrated rate equation for radioactive decay.

$$\ln \frac{N_t}{N_0} = -\lambda t \quad \text{or} \quad \log \frac{N_t}{N_0} = \frac{-\lambda t}{2.303} \tag{26.19}$$

Half-life of a radioactive nuclide.

$$t_{1/2} = \frac{0.693}{\lambda} \tag{26.20}$$

In these equations, N_0 represents the number of atoms at some initial time ($t = 0$) and N_t the number of atoms at some later time, t; λ is the decay constant; $t_{1/2}$ is the **half-life.**

Recall that the half-life of a process is the length of time for half of a substance to disappear, and that the half-life of a first-order process is a *constant*. Thus, if half of the atoms of a radioactive sample disintegrate in 2.5 min, the number of atoms remaining will be reduced to $\frac{1}{4}$ of the original number in 5.0 min, $\frac{1}{8}$ in 7.5 min, and so on. The shorter the half-life, the larger the value of λ and the faster the decay process. Half-lives of radioactive nuclides range over periods of time from extremely short to very long, as suggested by the representative data in Table 26-1.

Example 26-3

Using the half-life concept and the radioactive decay law to describe the rate of radioactive decay. The phosphorus isotope listed in Table 26-1, P-32, is used in biochemical studies to determine the pathways followed by phosphorus atoms in living organisms. Its presence is detected through its emission of β^- particles. **(a)** What is the decay constant for P-32, expressed in the units s^{-1}? **(b)** What is the decay rate (that is, how many P-32 atoms disintegrate per second) in a 1.00-mg sample of P-32? **(c)** Approximately what mass of the original 1.00-mg sample of P-32 will remain after 57 days? **(d)** What will be the rate of β^- emission after 57 days?

Solution

(a) We can determine λ from $t_{1/2}$ using (26.20). The first result we get has the unit d^{-1}. We must successively convert this to h^{-1}, min^{-1}, and s^{-1}.

$$\lambda = \frac{0.693}{14.3 \text{ d}} \times \frac{1 \text{ d}}{24 \text{ h}} \times \frac{1 \text{ h}}{60 \text{ min}} \times \frac{1 \text{ min}}{60 \text{ s}} = 5.61 \times 10^{-7} \text{ s}^{-1}$$

(b) First let us find the number of atoms in 1.00 mg of P-32. Then we can multiply this number by the decay constant to get the decay rate. (We can assume that the molar mass of P-32 is equal to its mass number.)

FIGURE 26-5
Radioactive decay of a hypo-
thetical P-32 sample—
Example 26-3 illustrated.

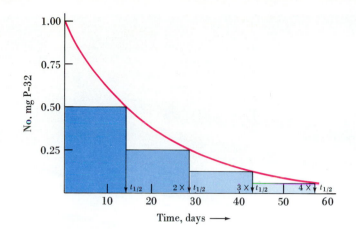

$$\text{no. P-32 atoms} = 0.00100 \text{ g} \times \frac{1 \text{ mol P-32}}{32.0 \text{ g P-32}} \times \frac{6.022 \times 10^{23} \text{ P-32 atoms}}{1 \text{ mol P-32}}$$

$$= 1.88 \times 10^{19} \text{ P-32 atoms}$$

$$\text{decay rate} = \lambda N = 5.61 \times 10^{-7} \text{ s}^{-1} \times 1.88 \times 10^{19} \text{ atoms}$$

$$= 1.05 \times 10^{13} \text{ atoms/s}$$

(c) A period of 57 days is approximately $57/14.3 \approx 4$ half-life periods. As shown in Figure 26-5, the quantity of radioactive material decreases by one-half for every half-life period. The quantity remaining is $(\frac{1}{2})^4$ of the original quantity.

$$\text{no. mg P-32} = 1.00 \text{ mg} \times (\tfrac{1}{2})^4 = 1.00 \text{ mg} \times \tfrac{1}{16} = 0.0625 \text{ mg P-32}$$

This method of relating elapsed time and quantity of radioactive substance works only if the time is a multiple of $t_{1/2}$ ($t_{1/2}$, $2t_{1/2}$, $3t_{1/2}$, . . .). Otherwise, use the integrated rate equation (26.19), as in Example 26-4.

(d) The rate of decay is directly proportional to the number of radioactive atoms remaining (rate = λN), and the number of atoms is directly proportional to the mass of P-32. When the mass of P-32 has dropped to $\tfrac{1}{16}$ of its original mass, the number of P-32 atoms also falls to $\tfrac{1}{16}$ of the original number, and the rate of decay is

$$\text{rate of decay} = \tfrac{1}{16} \times 1.05 \times 10^{13} \text{ atoms/s} = 6.56 \times 10^{11} \text{ atoms/s}$$

SIMILAR EXAMPLES: Exercises 6, 7, 23, 24, 26.

Radiocarbon Dating. Carbon-containing compounds in living organisms maintain an equilibrium with C-14 in the atmosphere. That is, these organisms replace C-14 atoms that have undergone radioactive decay with "fresh" C-14 atoms through interactions with their environment. The C-14 isotope is radioactive and has a half-life of 5730 y. The activity associated with C-14 that is in equilibrium with its environment is about 15 disintegrations per minute (dis/min) per gram of carbon. When an organism dies (for instance, a tree is cut down), this equilibrium is destroyed and the disintegration rate falls off. From the measured disintegration rate at some later time, we can estimate the age (that is, the elapsed time since the C-14 equilibrium was disrupted).

$^{14}_{6}\text{C}$ is formed at a constant rate in the upper atmosphere by the bombardment of $^{14}_{7}\text{N}$ with neutrons.

$$^{14}_{7}\text{N} + ^{1}_{0}\text{n} \longrightarrow ^{14}_{6}\text{C} + ^{1}_{1}\text{H} \tag{26.21}$$

The neutrons are produced by cosmic rays. $^{14}_{6}\text{C}$ disintegrates by β^- emission.

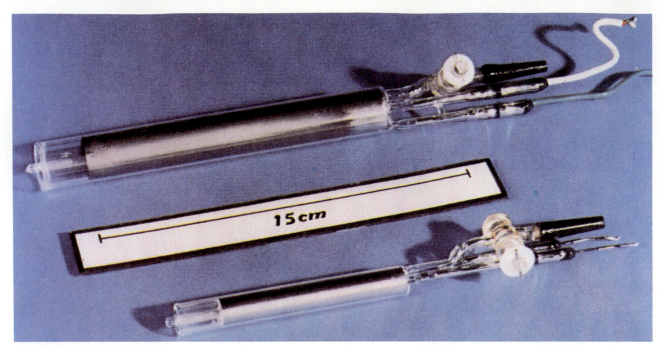

Special radiation counters used in radiocarbon dating. A small sample (as little as 10 mg) of an organic material is burned and the evolved $CO_2(g)$ passed through the counter, where beta particles emitted by carbon-14 are counted (see Figure 26-13). [Brookhaven National Laboratory]

Example 26-4

Applying the integrated rate law for radioactive decay: radiocarbon dating. A wooden object is found in an Indian burial mound and is subjected to radiocarbon dating. The decay rate associated with its ^{14}C content is 10 dis min^{-1} per g C. What is the age of the object (that is, the time elapsed since the tree was cut down)?

Solution. In this example three equations, (26.18), (26.19), and (26.20), are required. Equation (26.20) is again used to determine the decay constant.

$$\lambda = \frac{0.693}{5730 \text{ y}} = 1.21 \times 10^{-4} \text{ y}^{-1}$$

Next we use equation (26.18) to represent the actual number of atoms: N_0 at $t = 0$ (the time when the ^{14}C equilibrium was destroyed) and N_t at time t (the present time). The rate of decay just prior to the ^{14}C equilibrium being destroyed is 15 dis min^{-1} per g C, and at the time of the measurement, 10 dis min^{-1} per g C. The corresponding numbers of atoms are proportional to these decay rates divided by λ.

$$N_0 = \frac{\text{decay rate (at } t = 0)}{\lambda} = \frac{15}{\lambda}$$

$$N_t = \frac{\text{decay rate (at time } t)}{\lambda} = \frac{10}{\lambda}$$

Finally, we substitute into equation (26.19).

$$\ln \frac{N_t}{N_0} = \ln \frac{10/\lambda}{15/\lambda} = \ln \frac{10}{15} = -(1.21 \times 10^{-4} \text{ y}^{-1})t$$

$$-0.41 = -(1.21 \times 10^{-4} \text{ y}^{-1})t$$

$$t = \frac{0.41}{1.21 \times 10^{-4} \text{ y}^{-1}} = 3.4 \times 10^3 \text{ y}$$

SIMILAR EXAMPLES: Exercises 8, 29, 30.

Age of the Earth. The natural radioactive decay scheme of Figure 26-3 suggests the eventual fate awaiting all the $^{238}_{92}U$ found in nature—conversion to lead. Naturally occurring uranium minerals always have associated with them some nonradioactive lead formed by radioactive decay. From the mass ratio of $^{206}_{82}Pb$ to $^{238}_{92}U$ in such a mineral it is possible to estimate the age of the rock containing the mineral. By the age of the rock we mean the time elapsed since molten magma froze to become a rock. One assumption of this method is that the initial radioactive nuclide, the final stable nuclides, and all the products of a decay series remain in the rock. Another assumption is that any lead present in the rock initially consisted of the several isotopes of lead in their present naturally occurring abundances.

An exact treatment of this subject would require some discussion of the relationship between the rates of decay of a radionuclide called a ''parent'' and the product nuclide called a ''daughter.'' A discussion of these relationships is beyond the scope of this text, but we can still indicate how the method works.

The half-life of $^{238}_{92}U$ is 4.5×10^9 y. According to the natural decay scheme of Figure 26-3, the basic changes that occur as atoms of $^{238}_{92}U$ and its daughters pass through the entire sequence of steps is

$$^{238}_{92}U \longrightarrow {}^{206}_{82}Pb + 8\,{}^4_2He + 6\,{}^{\;0}_{-1}e \tag{26.22}$$

Discounting the mass associated with the β^- particles, we can see that for every 238 g of uranium that undergoes complete decay, 206 g of lead and 32 g of helium are produced.

Suppose that in a rock containing no lead initially, 1.000 g $^{238}_{92}U$ had disintegrated through one half-life period, 4.5×10^9 y. At the end of that time there would be present in the sample

0.500 g $^{238}_{92}U$ undisintegrated

and

$$0.500 \times \frac{206}{238} = 0.433 \text{ g } {}^{206}_{82}Pb$$

with the ratio

$$\frac{{}^{206}_{82}Pb}{{}^{238}_{92}U} = \frac{0.433}{0.500} = 0.866 \tag{26.23}$$

If the $^{206}_{82}Pb/^{238}_{92}U$ ratio is smaller than that shown in (26.23), the age of the rock is less than one half-life period of $^{238}_{92}U$. A higher ratio indicates a greater age for the rock. The best estimates of the age of the oldest rocks and presumably of the earth itself are in fact about 4.5×10^9 y. These estimates are based on the $^{206}_{82}Pb$ to $^{238}_{92}U$ ratio and on ratios for other pairs of isotopes from natural radioactive decay series.

Example 26-5

Dating a mineral through the mass ratio of a nonradioactive to a radioactive isotope. The thorium radioactive decay series produces one atom of ^{208}Pb as the final disintegration product of an atom of ^{232}Th. The half-life of ^{232}Th is 1.4×10^{10} y. A certain rock is found to have a $^{208}Pb/^{232}Th$ mass ratio of $0.14:1.00$. Use these data to estimate the age of the rock.

Solution. First, let us obtain the decay constant, λ, from the half-life.

$$\lambda = \frac{0.693}{1.4 \times 10^{10} \text{ y}} = 5.0 \times 10^{-11} \text{ y}^{-1}$$

Now, let us base our calculation on a quantity of mineral containing 1.00 g ^{232}Th at the present time, t. The total mass of ^{232}Th present in the sample of mineral when it was formed must have been the 1.00 g present currently plus the mass of ^{232}Th required to produce 0.14 g ^{208}Pb.

$$\text{no. g }^{232}\text{Th} = 0.14 \text{ g }^{208}\text{Pb} \times \frac{232 \text{ g }^{232}\text{Th}}{208 \text{ g }^{208}\text{Pb}} = 0.16 \text{ g}^{232}\text{Th}$$

The total mass of ^{232}Th present initially was $1.00 + 0.16 = 1.16$ g ^{232}Th. Since the number of atoms in a sample of an element is directly proportional to the mass of the sample, we can substitute 1.00 g for N_t and 1.16 g for N_0. We now have the necessary data to substitute into equation (26.19) and solve for t.

$$\ln \frac{N_t}{N_0} = \ln \frac{1.00}{1.16} = -5.0 \times 10^{-11} \text{ y}^{-1} \, t$$

$$-0.148 = -5.0 \times 10^{-11} \text{ y}^{-1} \, t$$

$$t = \frac{-0.148}{-5.0 \times 10^{-11} \text{ y}^{-1}} = 3.0 \times 10^9 \text{ y}$$

SIMILAR EXAMPLES: Exercises 28, 56.

26-6 Energetics of Nuclear Reactions

To describe the energy change accompanying a nuclear reaction we must use the mass–energy equivalence derived by Albert Einstein.

Mass–energy equivalence.

$$E = mc^2 \tag{26.24}$$

This expression indicates that an energy change in a process is always accompanied by a mass change, and the constant that relates them is the square of the speed of light. In chemical reactions energy changes are so small that the equivalent mass changes are undetectable (though real nevertheless). In fact, we base the balancing of equations and stoichiometric calculations on the principle that mass is conserved (unchanged) in a chemical reaction. In nuclear reactions, energies involved are orders of magnitude greater than in chemical reactions. Perceptible changes in mass do occur.

If we know the exact masses of atoms, we can calculate the energy of a nuclear reaction with equation (26.24). The term m corresponds to the net change in mass, in kg, and c, the velocity of light, is expressed in m/s. The resulting energy is in joules. Another common unit for expressing nuclear energy is the MeV (million electronvolt).

$$1 \text{ MeV} = 1.602 \times 10^{-13} \text{ J} \tag{26.25}$$

Example 26-6 _____

Calculating the energy of a nuclear reaction through the mass–energy relationship. What is the energy associated with the α decay of ^{238}U (a) in MeV; (b) in kJ/mol?

$$^{238}_{92}\text{U} \longrightarrow {}^{234}_{90}\text{Th} + {}^{4}_{2}\text{He}.$$

The masses in atomic mass units (u) are

$$^{238}_{92}\text{U} = 238.0508 \text{ u} \qquad {}^{234}_{90}\text{Th} = 234.0437 \text{ u} \qquad {}^{4}_{2}\text{He} = 4.0026 \text{ u}$$

Solution

(a) The net change in mass that accompanies the decay of a single nucleus of ^{238}U is $234.0437 + 4.0026 - 238.0508 = -0.0045$ u. This loss of mass appears as kinetic energy carried away by the α particle. In the setup below, we establish the relationship between the units u and g by noting that 1 u is exactly $\frac{1}{12}$ of the mass of a carbon-12 atom.

$$1\ u = \frac{1}{12} \times \frac{12.00\ g}{6.022 \times 10^{23}} = 1.66 \times 10^{-24}\ g$$

$$E = 0.0045\ u \times \frac{1.66 \times 10^{-24}\ g}{u} \times \frac{1\ kg}{1000\ g} \times (3.00 \times 10^{8})^{2}\ \frac{m^2}{s^2}$$

$$= 6.7 \times 10^{-13}\ J$$

$$E = 6.7 \times 10^{-13}\ J \times \frac{1\ MeV}{1.602 \times 10^{-13}\ J} = 4.2\ MeV$$

(b) The calculation in part (a) is for a single disintegration. The energy in kJ/mol is based on the disintegration of 1 mol of atoms.

$$E = \frac{6.7 \times 10^{-13}\ J}{atom} \times \frac{6.02 \times 10^{23}\ atoms}{1\ mol} \times \frac{1\ kJ}{1000\ J}$$

$$= 4.0 \times 10^{8}\ kJ/mol$$

SIMILAR EXAMPLES: Exercises 9, 12, 31, 32.

If in the calculation of Example 26-6a we had been dealing with a mass difference of exactly 1.000 u (instead of 0.0045 u), the calculated energy would have been 931.5 MeV. This provides a useful conversion factor between mass and energy,

1 atomic mass unit (u) = 931.5 MeV   (26.26)

Nuclear Binding Energy. Figure 26-6 depicts a process in which the nucleus of a $_2^4$He atom is produced from two protons and two neutrons. In the formation of this nucleus there is a **mass defect** of 0.0305 u. That is, the experimentally determined mass of a $_2^4$He nucleus is 0.0305 u *less* than the combined mass of two protons and two neutrons. This "lost" mass is liberated as energy. With expression (26.26) we can show that 0.0305 u of mass is equivalent to an energy of 28.4 MeV. Since this is the energy released in forming an $_2^4$He nucleus, we can call it the **binding energy** of the nucleus. (Viewed in another way, an $_2^4$He nucleus would have to absorb 28.4 MeV to cause its protons and neutrons to become separated.) If we consider the binding energy to be apportioned equally among the two protons and two neutrons in $_2^4$He, we obtain a binding energy per nuclear particle (nucleon) of 7.10 MeV. We can make similar calculations for other nuclei, and obtain the graph shown in Figure 26-7.

Figure 26-7 indicates that the maximum binding energy per nucleon is found in a

FIGURE 26-6
Nuclear binding energy in $_2^4$He.

The mass of a helium nucleus ($_2^4$He) is 0.0305 u (atomic mass unit) less than the combined masses of two protons and two neutrons. The energy equivalent to this loss of mass (called the mass defect) is the nuclear energy that binds the nuclear particles together.

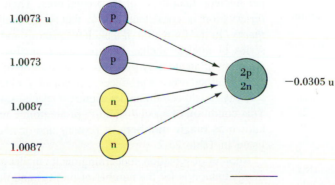

1.0073 u

1.0073

1.0087

1.0087

−0.0305 u

4.0320 u   4.0015 u

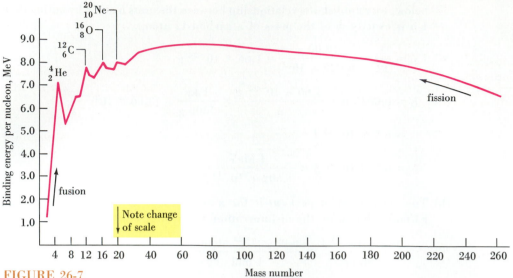

FIGURE 26-7

Average binding energy per nucleon as a function of atomic number.

nucleus with a mass number of approximately 60. This leads to two interesting conclusions: (1) If small nuclei are combined into a heavier one (up to about $A = 60$), the binding energy per nucleon increases and a certain quantity of mass must be converted to energy. The nuclear reaction is highly exothermic. This is a fusion process, and serves as the basis of the hydrogen bomb. (2) For nuclei having mass numbers above 60, the addition of extra nucleons to the nucleus would require the expenditure of energy (since the binding energy per nucleon decreases). On the other hand, the *disintegration* of heavier nuclei into lighter ones is accompanied by the release of energy. This is a nuclear fission process and serves as the basis of the atomic bomb and conventional nuclear power reactors. Nuclear fission and fusion are considered in greater detail in Sections 26-8 and 26-9. But first let us see what insights Figure 26-7 provides into the question of nuclear stability.

26-7 Nuclear Stability

A number of basic questions have probably occurred to you as we have been describing nuclear decay processes: Why do some radioactive nuclei decay by α emission, some by β^- emission, and so on? Why do the lighter elements have so few naturally occurring radioactive nuclides, whereas those of the heavier elements all seem to be radioactive?

Our first clue to answers for such questions comes from Figure 26-7, where several nuclides are specifically noted. These nuclides have higher binding energies per nucleon than those of their neighbors. Their nuclei are especially stable. This observation is consistent with a theory of nuclear structure known as the *shell theory*. In the formation of a nucleus, protons and neutrons are believed to occupy a series of nuclear shells. This process is analogous to building up the electronic structure of an atom by the successive addition of electrons to electronic shells. Just as the Aufbau process produces, periodically, electron configurations of exceptional stability, so do certain nuclei acquire a special stability as nuclear shells are closed. This condition of special stability of an atomic nucleus occurs for certain numbers known as **magic numbers** of protons and/or neutrons. These magic numbers are listed in Table 26-2.

Another observation concerning nuclei is that among stable nuclei the most common situation is for the number of protons and the number of neutrons to be *even*. There are fewer stable nuclei with odd numbers of nucleons (protons or neutrons).

TABLE 26-2

Magic Numbers for Nuclear Stability

Number of protons	Number of neutrons
2	2
8	8
20	20
28	28
50	50
82	82
114	126
	184
	196

TABLE 26-3
Distribution of Naturally
Occurring Stable Nuclides

Combination	Number of nuclides
Z even–N even	163
Z even–N odd	55
Z odd–N even	50
Z odd–N odd	4

The relationship between numbers of protons (Z), numbers of neutrons (N), and the stability of isotopes is summarized in Table 26-3. Note particularly that stable atoms with the combination Z odd–N odd are very rare. This combination is found only in these nuclides: ^2_1H, ^6_3Li, $^{10}_5B$, and $^{14}_7N$.

Still another observation is that elements of *odd* atomic number generally have only one or two stable isotopes, whereas those of even atomic number have several. Some examples are given in Table 26-4.

The role of neutrons in an atomic nucleus is thought to be to provide a nuclear force to bind protons and neutrons together into a stable unit. Without the neutrons the electrostatic forces of repulsion between positively charged protons would cause the nucleus to fly apart. For the elements of lower atomic numbers (up to about Z = 20), the required number of neutrons for a stable nucleus seems to be about equal to the number of protons, for example, ^4_2He, $^{12}_6C$, $^{16}_8O$, $^{28}_{14}Si$, $^{40}_{20}Ca$. For higher atomic numbers, because of increasing repulsive forces between protons, larger numbers of neutrons must be present to stabilize a nucleus and the neutron/proton (n/p) ratio increases. For bismuth the ratio is about 1.5:1. Above atomic number 83, no matter how many neutrons are present in the nucleus, the nucleus is unstable. Thus, all the isotopes of all the known elements with Z > 83 are radioactive. Figure 26-8 indicates the range of n/p ratios as a function of atomic number for stable atoms.

Using the ideas outlined here, nuclear scientists have predicted the possible existence of atoms of high atomic number that should have very long half-lives. Cur-

FIGURE 26-8
Neutron-to-proton ratio and the stability of nuclides.

(a) The belt of naturally occurring stable nuclides, ranging from ^1_1H to $^{209}_{83}Bi$.
(b) Naturally occurring and man-made radioactive nuclides of the heavier elements.
(c) Possible heavy nuclides of high stability (long radioactive half-lives).

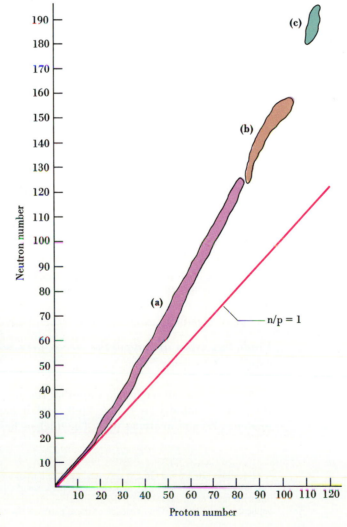

TABLE 26-4
Stable Isotopes of a Few
Elements

Element	Z	Number of stable isotopes
H	1	2
O	8	3
F	9	1
Ne	10	3
Cl	17	2
Ca	20	6
Cu	29	2
Sn	50	10
I	53	1
Hg	80	7

rently, a search is on to find such atoms, either naturally or by creating them in a particle accelerator. Figure 26-8 suggests the general range of proton and neutron numbers for these "superheavy" atoms.

Example 26-7

Predicting which nuclei are radioactive. Which of the following nuclides would you expect to be stable and which, radioactive? (a) ^{76}As; (b) ^{120}Sn; (c) ^{214}Po.

Solution

(a) ^{76}As has $Z = 33$ and $N = 43$. This is an odd–odd combination that is found only in four of the lighter elements. ^{76}As is radioactive. (Note also that this nuclide is outside the belt of stability in Figure 26-8.)

(b) Sn has an atomic number of 50—a magic number. The neutron number is 70 in the nuclide ^{120}Sn. This is an even–even combination and we should expect the nucleus to be stable. Moreover, Figure 26-8 shows that this nuclide is within the belt of stability. ^{120}Sn is a stable nuclide.

(c) ^{214}Po has an atomic number of 84. All known atoms with $Z > 83$ are radioactive. ^{214}Po is radioactive.

Mechanism of Radioactive Decay. The emission of an α particle by a nucleus is not difficult to visualize. A bundle of four nucleons (two protons, two neutrons) is ejected from a nucleus and the nucleus becomes energetically more stable. Alpha particle emission is pretty much limited to isotopes with $Z > 82$, though there are a few α emitters among the lanthanide isotopes. We can visualize gamma ray emission simply in terms of some rearrangement of nucleons, with the release of energy as electromagnetic radiation. It is somewhat harder for us to picture the emission of β^- and β^+ particles and electron capture. However, we can think in the following terms.*

TABLE 26-5
Radioactive Properties of Isotopes of Fluorine

Isotope	Mode of decay	Half-life
^{17}F	β^+	66 s
^{18}F	β^+, E.C.	109.7 min
^{19}F	stable	—
^{20}F	β^-	11.4 s
^{21}F	β^-	4.4 s
^{22}F	β^-	4.0 s

β^- emission: $\qquad {}_0^1 n \longrightarrow {}_1^1 p + {}_{-1}^0 e \qquad$ (26.27)

β^+ emission: $\qquad {}_1^1 p \longrightarrow {}_0^1 n + {}_{+1}^0 e \qquad$ (26.28)

Electron capture: $\quad {}_1^1 p + {}_{-1}^0 e \longrightarrow {}_0^1 n \qquad$ (26.29)

In general, if a nuclide lies above the belt of stability in Figure 26-8, the n/p ratio is too high. A neutron is converted to a proton and a β^- particle is emitted (26.27). If a nuclide lies below the belt of stability, the n/p ratio is too low. Either a proton is converted to a neutron, followed by β^+ emission (26.28) or electron capture occurs (26.29). The situation for a series of isotopes of fluorine is presented in Table 26-5.

Example 26-8

Predicting types of radioactive decay processes. By what mode would you expect ^{82}Y to decay?

Solution. We can see in two ways that this nuclide is radioactive: It lies below the belt of stability in Figure 26-8, and it has an odd–odd combination of protons (39) and neutrons (43). The nucleus has too few neutrons to be stable.

---

*Equations (26.27) and (26.28) are oversimplifications of the actual case. In order that certain properties be conserved, it is necessary to postulate the presence of other extremely tiny particles (about 0.0004 times the mass of an electron) in β decay processes. These are the neutrino in equation (26.27) and the antineutrino in (26.28). The existence of these particles has been confirmed.

A proton must be converted to a neutron, either by positron (β^+) emission (26.28) or by electron capture (26.29).

SIMILAR EXAMPLES: Exercises 35, 49.

26-8 Nuclear Fission

In 1934, Enrico Fermi proposed that transuranium elements might be produced by bombarding uranium with neutrons. He reasoned that the successive loss of β^- particles would cause the atomic number to increase, perhaps to as high as 96. When such experiments were carried out, it was found that in fact the product did emit β^- particles. But in 1938, two chemists, Otto Hahn and Fritz Strassman, found by chemical analysis that the products of the neutron bombardment of uranium did not correspond to elements with $Z > 92$. Neither were they the neighboring elements of uranium—Ra, Ac, Th, and Pa. Instead, the products consisted of radioisotopes of much lighter elements, such as strontium and barium. Neutron bombardment of uranium nuclei causes certain of them to undergo **fission** into smaller fragments. A fission process is depicted in Figure 26-9.

FIGURE 26-9
Nuclear fission of $^{235}_{92}U$ with thermal neutrons.

A $^{235}_{92}U$ nucleus is struck by a neutron possessing ordinary thermal energy. First the unstable nucleus $^{236}_{92}U$ is produced; this then breaks up into a light and a heavy fragment and several neutrons. A variety of nuclear fragments is possible, but the most probable mass number for the light fragment is 97, and for the heavy one, 137.

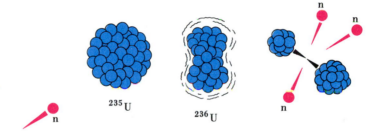

The energy equivalent of the mass destroyed in a fission process is somewhat variable because a variety of fission fragments is possible. However, the average energy for each fission event is approximately 3.20×10^{-11} J (200 MeV).

$$^{235}_{92}U + n \longrightarrow {}^{236}_{92}U \longrightarrow \text{fission fragments} + \text{neutrons} + 3.20 \times 10^{-11} \text{ J}$$

An energy of 3.20×10^{-11} J may seem small, but this energy is for the fission of a *single* $^{235}_{92}U$ nucleus. What if 1.00 g $^{235}_{92}U$ were to undergo fission?

$$\text{no. kJ} = 1.00 \text{ g } ^{235}U \times \frac{1 \text{ mol } ^{235}U}{235 \text{ g } ^{235}U} \times \frac{6.022 \times 10^{23} \text{ atoms } ^{235}U}{1 \text{ mol } ^{235}U}$$

$$\times \frac{3.20 \times 10^{-11} \text{ J}}{1 \text{ atom } ^{235}U}$$

$$= 8.20 \times 10^{10} \text{ J} = 8.20 \times 10^7 \text{ kJ}$$

This is an enormous quantity of energy! By contrast, to release this same quantity of energy would require the complete combustion of nearly 3 tons of coal.

Nuclear Reactors. In the fission of $^{235}_{92}U$, on average, 2.5 neutrons are released per fission event. These neutrons, on average, produce two or more fission events. The neutrons produced by the second round of fission produce another four or five events, and so on. The result is a **chain reaction.** If the reaction is uncontrolled, the total released energy causes an explosion; this is the basis of the atomic bomb. Spontaneous fission resulting in an uncontrolled explosion occurs only if the quan-

FIGURE 26-10
Pressurized water nuclear reactor.

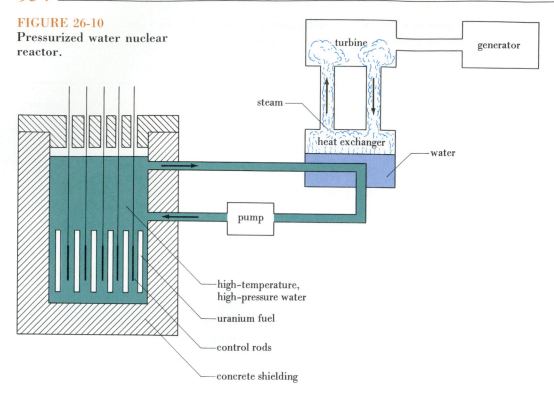

The core of a nuclear reactor being used as a source of neutrons for experimental purposes. [Brookhaven National Laboratory]

tity of ^{235}U exceeds the **critical mass.** The critical mass represents a quantity of ^{235}U sufficiently large to retain enough neutrons to sustain a chain reaction. Quantities smaller than this are called *subcritical* masses, because neutrons escape at too great a rate to produce a chain reaction.

In a nuclear reactor, fission energy is released in a controlled manner. One common design for a nuclear reactor, called the pressurized water reactor (PWR), is pictured in Figure 26-10. In the core of the reactor, rods of uranium-rich fuel are suspended in liquid water maintained under a pressure of from 70 to 150 atm. The water serves a dual purpose. First, it slows down the neutrons given off in the fission process so that they possess only normal thermal energy. These so-called thermal neutrons are more able to induce fission than highly energetic ones. In this capacity the water is said to act as a **moderator.** The second function of the water is as a heat-transfer medium. The energy of the fission reaction maintains the water at a high temperature (about 300 °C). The high-temperature water is brought in contact with colder water in a heat exchanger. The colder water is converted to steam, which drives a turbine, which in turn drives an electric generator. A final component of the nuclear reactor is a set of **control rods,** usually cadmium metal, whose function is to absorb neutrons. When the rods are lowered into the reactor, the fission process is slowed down. When the rods are raised, the density of neutrons and the rate of fission increase.

A nuclear reactor based on the fission of $^{235}_{92}$U is referred to as a nuclear burner. In the nuclear burner the fissionable nuclide is consumed (perhaps at the rate of 1 to 3 kg/d), and highly radioactive waste products accumulate. The separation, concentration, and disposal of these radioactive wastes is a difficult problem, requiring a considerable amount of chemical technology. A site for the ultimate disposal of radioactive wastes has not yet been agreed upon, although a proposed site in Nevada is currently being studied. In the meantime, high-level radioactive wastes are being stored on-site where they are produced.

Another important aspect of nuclear burning is the high rate of consumption of a relatively rare fissionable material. $^{235}_{92}$U accounts for only approximately 0.71% of naturally occurring uranium. To extract pure $^{235}_{92}$U from uranium ores requires that high-grade ores be employed, usually U_3O_8. Estimated world reserves of U_3O_8 are not extensive, and conventional nuclear reactors are of limited potential in the long-term production of energy.

Breeder Reactors. All that is required to initiate the fission of $^{235}_{92}U$ are neutrons of ordinary thermal energies. Nuclei of $^{238}_{92}U$, the abundant nuclide of uranium (99.28%), undergo the following reactions when struck by energetic neutrons.

$$^{238}_{92}U + ^{1}_{0}n \longrightarrow ^{239}_{92}U$$

$$^{239}_{92}U \longrightarrow ^{239}_{93}Np + ^{0}_{-1}e$$

$$^{239}_{93}Np \longrightarrow ^{239}_{94}Pu + ^{0}_{-1}e$$

A fissionable nuclide such as $^{235}_{92}U$ is called *fissile;* $^{239}_{94}Pu$ is also fissile. A nuclide such as $^{238}_{92}U$, which can be converted into a fissile nuclide, is said to be *fertile*. In a breeder nuclear reactor a small quantity of fissile nuclide provides the neutrons that convert a large quantity of a fertile nuclide into a fissile one. (The newly formed fissile nuclide then participates in a self-sustaining chain reaction.)

An obvious advantage of the breeder reactor is that the amount of uranium "fuel" available immediately jumps by a factor of about 100. This is the ratio of naturally occurring $^{238}_{92}U$ to $^{235}_{92}U$. But the advantage is even greater than this. Breeder reactors may be able to use as nuclear fuels materials that have even very low uranium contents. For example, shale deposits exist in the western Appalachian Mountains that contain about 0.006% U by mass. These deposits extend for several hundred square miles, and all of this material is potential nuclear fuel.

There are, however, important disadvantages to the use of breeder reactors. This is especially true of the type that has been pursued most vigorously—the liquid-metal-cooled fast breeder reactor (LMFBR). Systems must be designed to handle a liquid metal, such as sodium, which becomes highly radioactive in the reactor. Also, the rate of heat and neutron production are both greater in the LMFBR than in the PWR, resulting in a more rapid deterioration of materials. These factors complicate greatly the design and operation of the reactor. Perhaps the greatest unsolved problems are those of handling radioactive wastes and reprocessing plutonium fuel. Plutonium is one of the most toxic substances known. It can cause lung cancer when inhaled even in microgram (10^{-6} g) amounts. Federal health standards limit exposures to this substance to a total body burden of only 0.6 μg. Furthermore, because of its long half-life (24,000 y), any accident involving plutonium could leave an affected area almost permanently contaminated. Still another concern is the security threat posed by the possible diversion of Pu-239 by terrorists or for the illicit production of nuclear weapons.

26-9 Nuclear Fusion

The fusion of atomic nuclei is the process that produces energy in the sun. An uncontrolled fusion reaction is the basis of the hydrogen bomb. If a fusion reaction can be controlled, this will provide an almost unlimited source of energy. The nuclear reaction that holds the most immediate promise is the deuterium–tritium reaction.

$$^{2}_{1}H + ^{3}_{1}H \longrightarrow ^{4}_{2}He + ^{1}_{0}n \tag{26.30}$$

The difficulties in developing a fusion energy source are probably without parallel in the history of technology. In fact, the feasibility of a controlled fusion reaction has yet to be fully demonstrated. There are a number of problems. In order to permit their fusion, the nuclei of deuterium and tritium must be forced into close proximity. Because atomic nuclei repel one another, this close approach requires the nuclei

A nuclear fusion test reactor. [Princeton University Plasma Physics Laboratory, Tokomak Fusion Test Reactor]

to have very high thermal energies. At the temperatures necessary to initiate a fusion reaction, gases are completely ionized into a mixture of atomic nuclei and electrons known as a **plasma.** Still higher plasma temperatures—over 40,000,000 K—are required to initiate a *self-sustaining* reaction (one that releases more energy than is required to get it started). A method must be devised to confine the plasma out of contact with other materials. The plasma loses thermal energy to any material it strikes. Also, a plasma must be at a sufficiently high density and for a sufficient period of time to permit the fusion reaction to occur. The two methods receiving greatest attention are confinement in a magnetic field and heating of a frozen deuterium–tritium pellet with laser beams. Another series of technical problems involves the handling of liquid lithium, which is the anticipated heat transfer medium and tritium ($_1^3$H) source.

In the hydrogen bomb these high temperatures are attained by exploding an atomic (fission) bomb, which triggers the fusion reaction.

$$_3^7\text{Li} + _0^1\text{n} \longrightarrow _2^4\text{He} + _1^3\text{H} + _0^1\text{n} \qquad (26.31)$$

(fast)   (slow)

Finally, for the magnetic containment method the magnetic field must be produced by superconducting magnets and, currently, these are very expensive to operate.

The advantages of fusion over fission should be very great. Since deuterium comprises about one in every 6500 H atoms, the oceans of the world can supply an almost limitless amount of nuclear fuel. It is estimated that there is sufficient lithium on the earth to provide a source of tritium for about 1 million years.

26-10 Effect of Radiation on Matter

Although there are substantial differences in the way in which α, β, and γ rays interact with matter, they share an important feature: dislodging electrons from atoms and molecules to produce ions. The ionizing power of radiation may be described in terms of the number of ion pairs it forms. An *ion pair* consists of an ionized electron and the resulting positive ion. Alpha particles have the greatest ionizing power, followed by β particles and then γ rays. The ionized electrons pro-

FIGURE 26-11

Some interactions of radiation with matter.

(a) The production of primary and secondary electrons by collisions.
(b) The excitation of an atom by the passage of an α particle. An electron is raised to a higher energy level within the atom. The excited atom reverts to its normal state by emitting radiation.

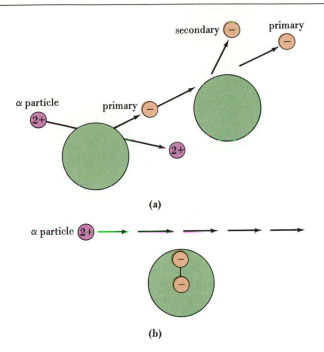

duced directly by the collisions of particles of radiation with atoms are called *primary* electrons. These electrons may themselves possess sufficient energies to cause *secondary* ionizations. Thus, even though some radiation, such as γ rays, may produce few primary electrons, the total ionization associated with its passage through matter may be considerable.

Not all interactions between radiation and matter produce ion pairs. In some cases electrons may simply be raised to higher atomic or molecular energy levels. The return of these electrons to their normal states is then accompanied by radiation— x rays, ultraviolet light, or visible light, depending on the energies involved. A practical example of this effect is found in luminescent watch dials. The dial numerals are painted with a mixture of a fluorescent material, such as zinc sulfide, and traces of a radioactive material, such as radium (an α emitter). Excitation of the zinc sulfide by α particles is accompanied by light emission; the dial numerals are visible in the dark.

Some of the possibilities described here are pictured in Figure 26-11.

Radiation Detectors. The interactions of radiation with matter can serve as bases for the detection of radiation and the measurement of its intensity. One of the simplest methods is that used by Henri Becquerel in his discovery of radioactivity— the exposure of a photographic plate. The effect of α, β, and γ rays on a photographic emulsion is similar to that of x rays.

A device that has played an important role in the development of knowledge about radioactivity is the **cloud chamber** pictured in Figure 26-12. The principle involved is quite simple. If a vapor is brought to a saturated condition, say by expansion and cooling, and if there are no nuclei (such as dust particles) on which condensation can occur, the vapor may become supersaturated. If ionizing radiation passes through such a vapor, the ions produced become nuclei for the production of liquid. Droplets are produced in the form of a cloudlike track. The type of ionizing radiation and some of its characteristics can be inferred from such tracks. For instance, α particles produce short, straight, thick tracks, whereas β^- particles produce long, thin, meandering tracks. If the cloud chamber is placed in a magnetic field, the tracks are curved—one direction for α particles and the opposite for β^- particles.

A modification of the cloud chamber that is particularly useful in detecting high-

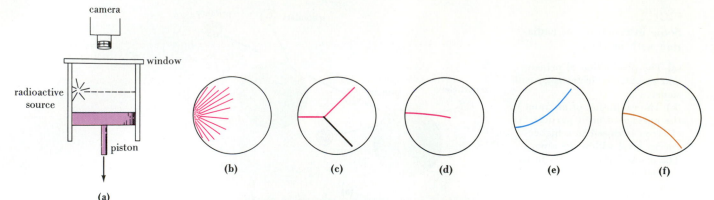

FIGURE 26-12
Wilson cloud chamber.

(a) Saturated vapor (say, ethanol) is cooled by sudden expansion. As a result it becomes supersaturated. Ion pairs produced in the supersaturated vapor serve as nuclei for condensation tracks.
(b) α particle tracks.
(c) An α particle in collision with an atomic nucleus. Two tracks emanate from the point where the original track disappears (one is the α particle track, and the other is that of the atom that is struck).
(d) α particle deflected in a magnetic field.
(e) β^- particle deflected in a magnetic field.
(f) Positron (β^+) deflected in a magnetic field.

energy radiation such as γ rays is the **bubble chamber.** In this device a substance, usually hydrogen, is kept just at its boiling point. As ion pairs are produced by the transit of an ionizing ray, bubbles of vapor form around the ions. Again tracks result that can be photographed and analyzed.

The most common device for detecting and measuring ionizing radiation is the **Geiger–Müller counter** pictured in Figure 26-13. The G–M counter consists of a cylindrical cathode with a wire anode running along its axis. The anode and cathode are sealed in a gas-filled glass tube. The tube is operated in such a way that ions produced by radiation passing through the tube trigger pulses of electric current. It is these pulses that are counted.

Effect of Ionizing Radiation on Living Matter. All life exists against a background of naturally occurring ionizing radiation—cosmic rays, ultraviolet light, and emanations from radioactive elements such as uranium in rocks. The level of this radiation varies from point to point on earth, being greater, for instance, at higher elevations. Only in recent times have human beings been able to create situations where living organisms might be exposed to radiation at levels significantly higher than natural background radiation.

The interactions of radiation with living matter are the same as with other forms of matter—ionization, excitation, and dissociation of molecules. There is no question of the effect of large dosages of ionizing radiation on living organisms—the organisms are killed. But even slight exposures to ionizing radiation can cause changes in cell chromosomes. Thus it is commonly held that even at low dosage rates ionizing radiation can result in birth defects, leukemia, bone cancer, and other forms of cancers. The nagging question that has eluded any definitive answers is how great an increase in the incidence of birth defects and cancers is caused by certain levels of radiation.

Radiation Dosage. Several different units are used to describe radiation dosage, that is, the amount of radiation to which matter is exposed. A summary is provided in Table 26-6.

FIGURE 26-13
Geiger–Müller counter.

Radiation enters the G–M tube through the mica window. Ions produced by the radiation cause an electrical breakdown of the gas (usually argon) in the tube. A pulse of electric current passes through the electric circuit and is counted.

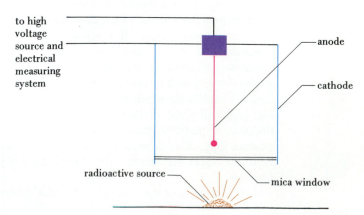

TABLE 26-6
Units of Radiation Dosage

Unit	Definition
curie	An amount of radioactive material decaying at the same rate as 1 g of radium (3.7×10^{10} dis/s).
rad	A dosage of radiation able to deposit 1×10^{-2} J of energy per kilogram of matter.
rem	A unit related to the rad, but taking into account the varying effects of different types of radiation of the same energy on biological matter. This relationship is through a "quality factor," which may be taken as equal to one for x rays, γ rays, and β particles. For protons and slow neutrons the factor has a value of 5; and for α particles, 10. Thus, an exposure to 1 rad of x rays is about equal to 1 rem, but 1 rad of α particles is about equal to 10 rem.

Sources of α radiation are relatively harmless when external to the body and extremely hazardous when taken internally, as in the lungs or stomach. Other forms of radiation (x ray, γ rays), because they are highly penetrating, are hazardous even when external to the body.

It is thought that exposure in a single short time interval to 1000 rem of radiation would kill 100% of the population exposed. Exposure to 450 rem would probably result in death within 30 days in about 50% of the population. A single dosage of 1 rem delivered to 1 million people would probably produce about 100 cases of cancer within 20 to 30 years. The total body radiation received by most of the world's population from normal background sources is about 0.13 rem [130 millirem (mrem)] per year. The exposure in a chest x ray examination is about 20 mrem; and the estimated maximum exposure to the general population associated with the production of nuclear power is about 5 mrem/y.

Some of the foregoing statements about radiation exposures and their anticipated effects are based on (1) medical histories of the survivors of the Hiroshima and Nagasaki atomic blasts, (2) the incidence of leukemia and other cancers in children whose mothers received diagnostic radiation during pregnancy, and (3) the occurrence of lung cancers among uranium miners in the United States. What does all of this tell us about a "safe" level of radiation exposure? One approach has been to extrapolate from these high dosage rates to the lower dosages affecting the general population. This has led the United States National Council on Radiation Protection and Measurements to recommend that the dosage rate for the general population be limited to 0.17 rem (170 mrem) per year from all sources above background level. However, experts disagree on how the data observed for high-dosage exposures should be extrapolated to low dosages. According to some, the 0.17 rem/y figure is too high, by perhaps a factor of 10. If so, exposure to an additional 0.17 rem/y above normal background levels may cause statistically significant increases in the incidence of birth defects and cancers.

26-11 Applications of Radioisotopes

We have described both the destructive capacity of nuclear reactions and the potential of these reactions to provide new sources of energy. Less heralded but equally important are a variety of practical applications of radioactivity. We close this chapter with a brief survey of the uses of some radioisotopes.

Cancer Therapy. We have learned how ionizing radiation in low dosages can induce cancers, but this same radiation, particularly γ rays, can also be used in the treatment of cancer. The basis of such treatment is that, although the radiation tends to destroy all cells, cancerous cells are more easily destroyed than normal ones.

Thus, a carefully directed beam of γ rays or high-energy x rays of the appropriate dosage may be used to arrest the growth of cancerous cells. Also coming into use for some forms of cancer is radiation therapy using beams of neutrons.

Radioactive Tracers. Radioactive isotopes are different from nonradioactive ones primarily in the instability of their nuclei, not in physical or chemical properties. Thus, in any physical or chemical process radioactive and nonradioactive isotopes are expected to behave in the same way. This is the fundamental principle behind the use of radioactive tracers or "tagged" atoms. For example, if a small quantity of artificially produced radioactive ^{32}P (as a phosphate) is added to a nutrient solution that is fed to plants, the uptake of the phosphorus can be followed by charting the regions of the plant that become radioactive. Similarly, the fate of iodine in the human body can be determined by having an individual drink a solution of dissolved iodides containing a small quantity of radioactive iodine as a tracer. Abnormalities in the thyroid gland can be detected in this way. Since I^- concentrates in the thyroid gland, to some extent people can protect themselves by ingesting nonradioactive iodides before being exposed to radioactive iodine. The thyroid becomes saturated with the nonradioactive iodine and rejects the radioactive iodine.

Industrial applications of tracers are also numerous. The fate of a catalyst in a chemical plant can be followed by incorporating a radioactive tracer in the catalyst, for example, ^{192}Ir in a Pt–Ir catalyst. By monitoring the activity of the ^{192}Ir one can determine the rate at which the catalyst is being carried away and to which parts of the plant.

> *Solutions of iodide ion were distributed to the general population in some areas of Europe following the Chernobyl accident in 1986.*

Structures and Mechanisms. Often we can infer the mechanism of a chemical reaction or the structure of a species from experiments using radioisotopes as tracers. Consider the following experimental proof that the two S atoms in the thiosulfate ion, $S_2O_3^{2-}$, are not equivalent.

$S_2O_3^{2-}$ is prepared from radioactive sulfur (^{35}S) and sulfite ion containing the nonradioactive isotope ^{32}S.

$$^{35}S + {}^{32}SO_3^{2-} \longrightarrow {}^{35}S^{32}SO_3^{2-} \tag{26.32}$$

When the thiosulfate ion is decomposed by acidification, all the radioactivity appears in the precipitated sulfur and none in the $SO_2(g)$. The ^{35}S atoms must be bonded in a different way than the ^{32}S atoms (see Figure 23-9).

$$^{35}S^{32}SO_3^{2-} + 2\,H^+ \longrightarrow H_2O + {}^{32}SO_2(g) + {}^{35}S(s) \tag{26.33}$$

In reaction (26.34) nonradioactive KIO_4 is added to a solution containing iodide ion labeled with the radioisotope, ^{128}I. All the radioactivity appears in the I_2 and none in the IO_3^-. This proves that all the IO_3^- is produced by reduction of IO_4^- and none by oxidation of I^-.

$$IO_4^- + 2\,{}^{128}I^- + H_2O \longrightarrow {}^{128}I_2 + IO_3^- + 2\,OH^- \tag{26.34}$$

If a chromium(III) salt containing radioactive ^{51}Cr is added to a chrome-plating bath, no radioactivity shows up in the chrome plate. This proves that the reduction of Cr(VI) to Cr(0) at the cathode does not proceed through the oxidation state Cr(III). (Chrome plating was discussed in Section 24-5.)

Analytical Chemistry. We have learned how a substance can be analyzed by precipitation from solution. The usual procedure involves filtering, washing, drying, and weighing a pure precipitate. An alternative is to allow the substance to be analyzed to react with a reagent containing a radioisotope. By measuring the activity of the precipitate and comparing it with that of the original solution, it is possible to calculate the amount of precipitate without having to purify, dry, and weigh it. (For example, Ag^+ in solution can be precipitated as radioactive AgCl by treatment with a solution containing radioactive Cl^-.)

Another method of importance in analytical chemistry is *neutron activation analysis*. In this procedure the sample to be analyzed, normally nonradioactive, is bombarded with neutrons; the element of interest is converted to a radioisotope. The radioactivity of this radioisotope is measured. This measurement is combined with a knowledge of such factors as the rate of neutron bombardment, the half-life of the radioisotope, and the efficiency of the radiation detector to calculate the quantity of the element in the sample. The method is especially attractive because (1) trace quantities of elements can be determined (sometimes in parts per billion or less); (2) a sample can be tested without destroying it; and (3) the sample can be in any state of matter, including biological materials. Among its many uses, neutron activation analysis has been used to study archaeological artifacts and to determine the authenticity of old paintings. (Old masters formulated their own paints. Differences between formulations are easily detected through the trace elements they contain.)

Radiation Processing. This term describes industrial applications of ionizing radiation—γ rays from ^{60}Co or electron beams from electron accelerators. The ionizing radiation is used in the production of certain materials or to modify their properties. Its most extensive current use is in breaking, reforming, and crosslinking polymer chains to affect the physical and mechanical properties of plastics used in foamed products, electrical insulation, and packaging materials. Exposure to ionizing radiation is used to sterilize medical supplies such as sutures, syringes, and hospital garb. In sewage treatment plants radiation processing has been used to decrease the settling time of sewage sludge and to kill pathogens. An important future use may be in the preservation of foods, as an alternative to canning, freeze-drying or refrigeration. In radiation processing the irradiated material is *not* rendered radioactive, although the ionizing radiation may produce some chemical changes.

NON - IRRADIATED -   IRRADIATED - (0.2 M RAD)

STRAWBERRIES -
15 DAYS STORAGE 38°F (4°C)

A vivid demonstration of the effectiveness of irradiation in preserving strawberries. [International Atomic Energy Association]

Summary

Radioactivity refers to the ejection of particles (α, β^-, β^+), the capture of electrons from an inner shell, or the emission of electromagnetic radiation (γ) by unstable nuclei. Often these processes are part of a radioactive decay series. That is, they occur through several steps until a stable (nonradioactive) nucleus is finally produced. With the exception of γ ray emission, radioactive decay processes lead to the transformation (transmutation) of one element into another.

All nuclides having atomic number greater than 83 are radioactive. Although a few occur naturally, most radioactive nuclides of lower atomic number are produced artificially, by bombarding appropriate target nuclei with energetic particles. The high-energy particles induce a nuclear reaction that yields the desired nuclide. Equations can be written for nuclear reactions, whether they occur through particle bombardment or by spontaneous radioactive decay. In either case the sum of the atomic numbers and the sum of the mass numbers must be constant between the two sides of the equation.

The rate of radioactive decay is directly proportional to the number of atoms in a sample. Each radioactive nuclide has a characteristic decay constant and half-life, and calculations of decay rates can be made using equations similar to those for first-order chemical kinetics. Measurements of decay rates of radioactive nuclides have a number of practical applications, ranging from determining the ages of rocks to the dating of certain carbon-containing objects (radiocarbon dating).

Large quantities of energy are associated with nuclear reactions, sufficiently large that mass changes occur. The quantity of energy released in the formation of a nucleus from protons and neutrons can be calculated, and when these nuclear binding energies per nucleon are plotted as a function of mass number, a distinctive graph is obtained (Figure 26-7). From this graph one can establish that fission of heavy nuclei and fusion of lighter nuclei yield large quantities of energy. The fission process is the basis of the atomic bomb and nuclear reactors. Fusion is the energy-producing process of the stars, the hydrogen bomb, and the, as yet unperfected, thermonuclear reactor.

The stability of a nucleus depends on several factors, among them being whether the numbers of protons and neutrons are odd or even, and whether either of these is a "magic number." The most important factor is the neutron-to-proton ratio in the nucleus and whether this ratio lies within the belt of stable nuclides (Figure 26-8). Nuclides outside the belt of stability are radioactive. Their placement with respect to this belt can generally serve to indicate whether radioactive decay will occur by α, β^-, or β^+ emission or by electron capture.

One of the principal effects of the interaction of radiation with matter is the production of ions. This phenomenon can be used to detect radiation, and it is also the basis of radiation damage to living matter. Several methods have been developed to measure radiation dosages and to predict the biological effects of these dosages, but much uncertainty remains. Despite the hazards associated with radioactivity, many beneficial applications exist as well. Radioactive nuclides are used in cancer therapy, in basic studies of chemical structures and mechanisms, in analytical chemistry, and in chemical industry.

Summarizing Example

Technicians surveying the radiation level on agricultural land near the site of the crippled Chernobyl nuclear reactor. [Novosti from Gamma Liaison]

On April 26, 1986, an explosion at the nuclear power plant at Chernobyl in the U.S.S.R. released the greatest quantity of radioactive material ever associated with an industrial accident. One of the principal radioisotopes in this emission was I-131, a β^- emitter with a half-life of 8.04 days.

1. Write a nuclear equation for the radioactive decay of I-131.

Solution. The β^- particle is $_{-1}^{0}e$ and the product nucleus must be $_{54}^{131}Xe$.

$$_{53}^{131}I \longrightarrow \ _{54}^{131}Xe + \ _{-1}^{0}e$$

(This example is similar to Example 26-1.)

2. As a rule of thumb, a radioactive nuclide is considered to have lost its radioactivity when about 30 half-lives have elapsed. (For I-131, $30 \times 8 = 240$ d $= 8$ mo.) What fraction of the initial radioactivity of a sample of I-131 remains after this 30-half-life period?

Solution. Recall that after the first half-life period the quantity of radioactive nuclide, and hence its decay rate, drops to one-half of its initial value. After the second half-life period the activity is $(\frac{1}{2})^2 = \frac{1}{4}$ of its initial value, and so on. After 30 half-lives,

activity $= (\frac{1}{2})^{30} \times$ initial activity $= 9.3 \times 10^{-10} \times$ initial activity

[To determine the value of $x = (\frac{1}{2})^{30}$, $\log x = 30 \times \log \frac{1}{2} = 30(-0.301) = -9.03$. $x = $ antilog $-9.03 = 9.3 \times 10^{-10}$.]

Think of the factor 9.3×10^{-10} as approximately equal to $1 \times 10^{-9} = $ one billionth. Thus the radioactivity falls to one billionth of its original value in 30 half-lives.

(This example is similar to Example 26-3.)

3. A unit used to measure the quantity of a radioactive material is the **curie**. A curie of a radioactive material decays at a rate of 3.70×10^{10} dis/s. (This is essentially the rate of decay of 1 g of radium.) What mass of I-131 would be described as 1 curie of I-131?

Solution. We can calculate the decay constant for I-131 from the half-life and express this on a per second basis.

$$\lambda = \frac{0.693}{t_{1/2}} = \frac{0.693}{8.04 \text{ d}} \times \frac{1 \text{ d}}{24 \text{ h}} \times \frac{1 \text{ h}}{60 \text{ min}} \times \frac{1 \text{ min}}{60 \text{ s}} = 9.98 \times 10^{-7} \text{ s}^{-1}$$

Next we can use the radioactive decay law (26.18), substituting 3.70×10^{10} dis/s for the decay rate and 9.98×10^{-7} s^{-1} for λ. We must solve the equation for the number of atoms, N.

$$N = \frac{\text{rate of decay}}{\lambda} = \frac{3.70 \times 10^{10} \text{ atoms/s}}{9.98 \times 10^{-7} \text{ s}^{-1}} = 3.71 \times 10^{16} \text{ I-131 atoms}$$

Now we can express this number of I-131 atoms as a mass of I-131.

$$\text{no. g I-131} = 3.71 \times 10^{16} \text{ atoms} \times \frac{1 \text{ mol I-131}}{6.022 \times 10^{23} \text{ atoms}} \times \frac{131 \text{ g I-131}}{1 \text{ mol I-131}}$$

$$= 8.07 \times 10^{-6} \text{ g I-131}$$

4. An official account of the Chernobyl accident estimates the total release of I-131 as having been from 10 to 50 million curies. What approximate total mass of I-131 was released in this accident?

Solution. The result of part **3** gives us a conversion factor between a quantity of I-131 and the curie: 1 curie $= 8.07 \times 10^{-6}$ g I-131.

$$\text{no. g I-131} = 10 \times 10^6 \text{ curie} \times \frac{8.07 \times 10^{-6} \text{ g I-131}}{1 \text{ curie}} = 80.7 \text{ g I-131}$$

Since the quantity of I-131 released was from 10 to 50 million curies, the mass of I-131 was from about 80 to 400 g I-131.

Key Terms

alpha (α) particle (26-1)
beta (β) particle (26-1)
breeder reactor (26-8)
charged-particle accelerator (26-4)
cloud chamber (26-10)
control rods (26-8)
curie (26-10)
decay constant (26-5)
electron capture (E.C.) (26-1)

gamma (γ) ray (26-1)
Geiger–Müller counter (26-10)
half-life (26-5)
magic numbers (26-7)
moderator (26-8)
nuclear binding energy (26-6)
nuclear equation (26-1)
nuclear fission (26-8)

nuclear fusion (26-9)
nuclear reactor (26-8)
positron (26-1)
rad (26-10)
radioactive decay series (26-2)
radiocarbon dating (26-5)
rem (26-10)
transuranium element (26-4)

Highlighted Expressions

Writing nuclear equations (26.1)
Method of representing a nuclear reaction (26.12)
Radioactive decay law (26.18)

Integrated rate equation for radioactive decay (26.19)
Half-life of a radioactive nuclide (26.20)
Mass–energy equivalence (26.24)

Review Problems

1. Which of the following—α, β, or γ—generally has the greatest **(a)** penetrating power through matter; **(b)** ionizing power in matter; **(c)** deflection in a magnetic field?

2. Supply the missing information in each of the following nuclear equations representing a radioactive decay process.
  (a) $^{32}_{16}\text{S} \rightarrow \,^{?}_{17}\text{Cl} + \,^{0}_{-1}\text{e}$ **(b)** $^{14}_{8}\text{O} \rightarrow \,^{14}_{7}\text{N} + \,?$
  (c) $^{235}_{?}\text{U} \rightarrow \,^{?}_{?}\text{Th} + \,?$ **(d)** $^{214}_{?}? \rightarrow \,^{?}_{?}\text{Po} + \,^{0}_{-1}\text{e}$

3. Complete the following nuclear equations.
  (a) $^{23}_{11}\text{Na} + \,^{2}_{1}\text{H} \rightarrow \,? + \,^{1}_{1}\text{H}$
  (b) $^{59}_{27}\text{Co} + \,? \rightarrow \,^{56}_{25}\text{Mn} + \,^{4}_{2}\text{He}$
  (c) $^{238}_{92}\text{U} + \,^{2}_{1}\text{H} \rightarrow \,? + \,^{0}_{-1}\text{e}$

  (d) $^{246}_{96}\text{Cm} + \,^{13}_{6}\text{C} \rightarrow \,^{254}_{102}\text{No} + \,?$
  (e) $^{238}_{92}\text{U} + \,^{14}_{7}\text{N} \rightarrow \,^{246}_{?}\text{Es} + \,? \,^{1}_{0}\text{n}$

4. Write an equation to represent each of the following nuclear processes.
  (a) The reaction of two deuterium nuclei (deuterons) to produce a nucleus of ^{3}He.
  (b) The production of $^{243}_{97}\text{Bk}$ by the α particle bombardment of $^{241}_{95}\text{Am}$.
  (c) The bombardment of $^{121}_{51}\text{Sb}$ by α particles to produce $^{124}_{53}\text{I}$ followed by its radioactive decay by positron emission.

5. For the radioactive nuclides in Table 26-1,

(a) which one has the largest value of the decay constant, λ?

(b) which one would display a 75% reduction in radioactivity from its current value in approximately two days?

(c) which ones would have lost more than 99% of their radioactivity in one month?

6. Two radioisotopes are compared. Isotope A requires 12.0 h for its decay rate to fall to $\frac{1}{64}$ of its initial value. Isotope B has a half-life that is 1.5 times that of A. How long does it take for the decay rate of isotope B to fall to $\frac{1}{32}$ of its initial value?

7. A sample of radioactive $^{35}_{16}S$ is found to disintegrate at a rate of 1.00×10^3 atoms/min. The half-life of $^{35}_{16}S$ is 87.9 d. How long will it take for the activity of this sample to decrease to the point of producing (a) 115; (b) 86; and (c) 43 dis/min?

8. A wooden art object is claimed to have been found in an Egyptian pyramid and is offered for sale to an art museum. Nondestructive radiocarbon dating of the object reveals a disintegration rate of 12 dis min^{-1} per g C. Do you think the object is authentic?

9. With appropriate equations in the text, determine

(a) The energy in joules corresponding to the destruction of 1.05×10^{-23} g of matter.

(b) The energy in MeV that would be released if one α particle was completely destroyed. [*Hint:* Refer to Figure 26-6.]

(c) The number of neutrons that could conceivably be created from 1.50×10^6 MeV of energy. [*Hint:* Refer to Table 2-1.]

10. The measured mass of the nuclide $^{20}_{10}Ne$ is 19.99244 u. Determine the binding energy per nucleon (in MeV) in this atom. [*Hint:* The nuclidic mass includes the mass of electrons as well as that of the nucleus. Also, recall Figure 26-6.]

11. Two of the following nuclides do not occur naturally. Which are they? (a) ^2H; (b) ^{32}S; (c) ^{80}Br; (d) ^{132}Cs; (e) ^{184}W.

12. Refer to Example 26-1a. The decay of ^{222}Rn by alpha particle emission is accompanied by a loss of 5.590 MeV of energy. What quantity of mass is converted to energy in this decay process?

Exercises

Definitions and terminology

13. Describe briefly the meaning of each of the following concepts or terms introduced in this chapter; (a) neutron-to-proton ratio; (b) nucleon; (c) mass–energy relationship; (d) background radiation; (e) radioactive decay series; (f) charged-particle accelerator.

14. Explain the difference in meaning between the following pairs of terms: (a) naturally occurring and artificially produced radioisotope; (b) electron and positron; (c) primary and secondary ionization; (d) transuranium and transactinide element; (e) nuclear fission and nuclear fusion.

15. What is the meaning of each of these symbols in describing nuclear phenomena? (a) α; (b) β^-; (c) $t_{1/2}$; (d) γ; (e) β^+.

16. Supply a name or symbol for each of the following nuclear particles: (a) ^4_2He; (b) beta particle; (c) neutron; (d) ^1_1H; (e) $^0_{+1}e$; (f) tritium.

Radioactive processes

17. What is the nucleus obtained in each process?

(a) $^{234}_{94}Pu$ decays by α emission.

(b) $^{248}_{97}Bk$ decays by β^- emission.

(c) $^{196}_{82}Pb$ goes through two successive E.C. processes.

(d) $^{214}_{82}Pb$ decays through two successive β^- emissions.

(e) $^{226}_{88}Ra$ decays through three successive α emissions.

(f) $^{69}_{33}As$ decays by β^+ emission.

18. Both β^- (electron) and β^+ (positron) emission are observed for artificially produced radioisotopes of low atomic numbers, but only β^- (electron) emission is observed with naturally occurring radioisotopes of high atomic number. What is the reason for this observation?

Radioactive decay series

19. The natural decay series starting with the radionuclide $^{232}_{90}Th$ follows the sequence represented below. Construct a graph of this series, similar to Figure 26-3.

$$^{232}_{90}Th-\alpha-\beta-\beta-\alpha-\alpha-\alpha \overset{\alpha-\beta}{\underset{\beta-\alpha}{\diagup}} \overset{\alpha-\beta}{\underset{\beta-\alpha}{\diagup}} {}^{208}_{82}Pb$$

20. The uranium series described in Figure 26-3 is also known as the "$4n + 2$" series because the mass number of each nuclide in the series can be expressed by the equation $A = 4n + 2$, where n is an integer. Show that this equation is indeed applicable to the uranium series.

Nuclear reactions

21. Write the nuclear equations represented by the following symbolic notations: (a) $^7Li(p,\gamma)^8Be$; (b) $^{33}S(n,p)^{33}P$; (c) $^{239}Pu(\alpha,n)^{242}Cm$; (d) $^{238}U(\alpha,3n)^{239}Pu$.

22. Write an equation for each of the nuclear reactions represented by Figure 26-3.

Rate of radioactive decay

23. The disintegration rate for a sample containing $^{60}_{27}Co$ as the only radioactive nuclide is found to be 185 atoms/min. The half-life of $^{60}_{27}Co$ is 5.2 y. Estimate the number of atoms of $^{60}_{27}Co$ in the sample.

24. How long must the radioactive sample of Exercise 23 be maintained before the disintegration rate falls to 101 dis/min?

25. The radioisotope $^{32}_{15}P$ is used extensively in biochemical studies. Its half-life is 14.2 d. Suppose that a sample containing this isotope has an activity 1000 times the detectable limit. For how long a time could an experiment be run with this sample before the radioactivity could no longer be detected? [*Hint:* Use $N_0 = 1.00 \times 10^3$ and $N_t = 1.00$.]

26. A sample containing $^{234}_{88}Ra$, which decays by α particle emission, is observed to disintegrate at the following rate, expressed as disintegrations per minute or counts per minute (cpm). What is the half-life of this nuclide? $t = 0$, 1000 cpm; $t = 1$ h, 992 cpm; $t = 10$ h, 924 cpm; $t = 100$ h, 452 cpm; $t = 250$ h, 138 cpm.

27. The unit **curie** is defined as a disintegration rate of 3.7×10^{10} dis/s. What mass of ^{90}Sr, with a half-life of 29 y, is required to produce 1.00 millicurie of radiation?

28. If a meteorite is approximately 4.5×10^9 y old, what should be the mass ratio $^{208}Pb/^{232}Th$ in the meteorite? The half-life of ^{232}Th is 1.39×10^{10} y.

Radiocarbon dating

29. What should be the current rate of disintegration of ^{14}C, expressed in dis min^{-1} per g C, for a wooden object that is believed to have been made in 1100 B.C.?

30. The lowest level of ^{14}C activity that seems possible for experimental detection is 0.03 dis min^{-1}/g C. What is the maximum age of an object that can be determined by the carbon-14 method?

Energetics of nuclear reactions

31. Calculate the energy (in MeV) released in the nuclear reaction

$$^{10}_{5}B + ^{4}_{2}He \rightarrow ^{13}_{6}C + ^{1}_{1}H$$

The nuclidic masses are $^{10}_{5}B = 10.01294$ u; $^{4}_{2}He = 4.00260$ u; $^{13}_{6}C = 13.00335$ u; $^{1}_{1}H = 1.00783$ u.

32. You are given the following exact atomic masses; $^{6}_{3}Li = 6.01513$ u; $^{4}_{2}He = 4.00260$ u; $^{3}_{1}H = 3.01604$ u; $^{1}_{0}n = 1.008665$ u. How much energy is released in the nuclear reaction $^{6}_{3}Li + ^{1}_{0}n \rightarrow ^{4}_{2}He + ^{3}_{1}H$, expressed in MeV?

33. Refer to Figure 26-2. What is the wavelength associated with the gamma rays having an energy of 0.05 MeV? [*Hint:* What is the relationship between the energy units MeV and J?]

Nuclear stability

34. Which member of the following pairs of nuclides would you expect to be most abundant in natural sources? Explain your reasoning. (a) $^{20}_{10}Ne$ or $^{22}_{10}Ne$; (b) $^{17}_{8}O$ or $^{18}_{8}O$; (c) $^{6}_{3}Li$ or $^{7}_{3}Li$.

35. One member each of the following pairs of radioisotopes decays by β^- emission and the other by positron (β^+) emission. Which is which? Explain your reasoning. (a) $^{29}_{15}P$ and $^{33}_{15}P$; (b) $^{120}_{53}I$ and $^{134}_{53}I$.

36. Sometimes the most abundant isotope of an element can be established by rounding off the atomic weight to the nearest whole number, for example, ^{39}K, ^{85}Rb, and ^{88}Sr. But at other times the isotope corresponding to the "rounded-off" atomic weight does not even occur naturally, for example, ^{64}Cu. Explain the basis of this observation.

37. Some nuclides are said to be "doubly magic." What do

you suppose this term means? Postulate some nuclides that might be doubly magic and locate them in Figure 26-8.

Fission and fusion

38. Describe briefly what is meant by the following types of nuclear reactors: (a) burner; (b) breeder; (c) thermonuclear.

39. Based on Figure 26-7, can you explain why more energy is released in a fusion than in a fission process?

40. Refer to the Summarizing Example. In contrast to the Chernobyl accident, the nuclear accident at Three Mile Island, Pennsylvania, in 1979 released only 170 curies of I-131. How many mg I-131 does this represent?

41. Use data from the text to determine how many metric tons (1 metric ton = 1000 kg) of bituminous coal (85% C) would have to be burned to release as much energy as is produced by the fission of 1.00 kg $^{235}_{92}U$.

Effect of radiation on matter

42. Explain why the rem is more satisfactory than the rad as a unit for measuring radiation dosage.

43. The Geiger–Müller counter is a much more efficient device for detecting and measuring γ rays than is a cloud chamber. Why do you think this is so?

44. Discuss briefly the basic difficulties in establishing the physiological effects of low-level radiation.

45. ^{90}Sr is both a product of radioactive fallout and a radioactive waste in a nuclear reactor. This radioisotope is a β^- emitter with a half-life of 27.7 y. Suggest reasons why ^{90}Sr is such a potentially hazardous substance.

Application of radioisotopes

46. Describe how you might go about finding a leak in the $H_2(g)$ supply line in an ammonia synthesis plant by using radioactive materials.

47. Explain why neutron activation analysis is so useful in determining trace elements in a sample, in contrast to ordinary methods of quantitative analysis such as precipitation or titration.

48. A small quantity of NaCl containing radioactive $^{24}_{11}Na$ is added to an aqueous solution of $NaNO_3$. The solution is cooled and $NaNO_3$ is crystallized from the solution. Would you expect the $NaNO_3$ to be radioactive? Explain.

Additional Exercises

49. Each of the following isotopes is radioactive. Which would you expect to decay by β^- emission and which by positron (β^+) emission? (a) $^{32}_{15}P$; (b) $^{45}_{19}K$; (c) $^{72}_{30}Zn$.

50. The half-life of tritium is 12.26 y. What would be the rate of decay of tritium atoms, per second, in 1.00 L of hydrogen gas at STP containing 0.15% tritium atoms?

51. Explain the similarities and differences between a cloud chamber and a bubble chamber.

52. Use data from Table 2-1, together with the measured mass of the nuclide $^{16}_{8}O$, 15.99491 u, to determine the binding energy per nucleon (in MeV) in this atom.

53. If you follow the same description as in Exercise 20, the thorium series may be called the "4n" series and the actinium series the "4n + 3" series. A "4n + 1" series has also been established with $^{237}_{93}Np$ as the parent nuclide. To which radioac-

tive series does each of the following nuclides belong? (a) $^{214}_{83}Bi$; (b) $^{216}_{84}Po$; (c) $^{215}_{85}At$; (d) $^{235}_{92}U$.

54. Write nuclear equations to represent (a) the decay of ^{230}Th by α particle emission; (b) the decay of ^{54}Co by positron emission; (c) the nuclear reaction ^{232}Th $(\alpha,4n)^{232}U$.

55. Iodine-129 is a product of nuclear fission, whether from an atomic bomb or a nuclear power plant. It is a β^- emitter with a 1.7×10^7 y half-life. How many disintegrations per second would occur in a sample containing 1.00 mg ^{129}I?

56. One method of dating rocks is based on their $^{87}Sr/^{87}Rb$ ratio. The ^{87}Rb is a β^- emitter with a half-life of 5×10^{11} y. A certain rock is found to have a mass ratio $^{87}Sr/^{87}Rb$ of 0.004:1.00. What is the age of the rock?

57. The following reactions are carried out using HCl(aq) containing some tritium ($^{3}_{1}H$) as a tracer. Would you expect any of the tritium radioactivity to appear in the $NH_3(g)$? In the H_2O? Explain.

$$NH_3(aq) + HCl(aq) \rightarrow NH_4Cl(aq)$$

$$NH_4Cl(aq) + NaOH(aq) \rightarrow NaCl(aq) + H_2O + NH_3(g)$$

58. In equation (26.22) we see that the net change in the radioactive decay of $^{238}_{92}U$ to $^{206}_{82}Pb$ is the emission of eight α particles. Show that if this loss of eight α particles were not also accompanied by six β^- emissions the product nucleus would still be radioactive. [*Hint*: Where will the final nucleus lie in the graph of Figure 26-8.]

59. Refer to the Summarizing Example. Another radioisotope produced in the Chernobyl accident was Cs-137. If a 1.00-mg sample of Cs-137 is equivalent to 89.8 millicuries, what must be the half-life (in years) of Cs-137?

★ 60. The percent natural abundance of ^{40}K is 0.0117%. The radioactive decay of ^{40}K atoms takes place 89% by β^- emission (the rest is by electron capture and β^+ emission). The half-life of the β^- decay is 1.25×10^9 y. Calculate the number of β^- particles produced per second by the radioactive decay of the ^{40}K present in a 1.00 g sample of the mineral *microcline*, $KAlSi_3O_8$.

★ 61. The carbon-14 dating method is based on the assumption that the rate of production of ^{14}C by cosmic ray bombardment has remained constant for thousands of years and that the ratio of ^{14}C to ^{12}C has also remained constant. Can you think of any effects of human activities that could invalidate this assumption in the future?

62. How many millicuries of radiation are associated with a sample containing 5.10 mg ^{229}Th, which has a half-life of 7340 y?

★ 63. Calculate the minimum kinetic energy (in MeV) that α particles must possess to produce the nuclear reaction

$$^4_2He + ^{14}_7N \rightarrow ^{17}_8O + ^1_1H$$

The nuclidic masses are $^4_2He = 4.00260$ u; $^{14}_7N = 14.00307$ u; $^1_1H = 1.00783$ u; $^{17}_8O = 16.99913$ u. [*Hint*: What is the increase in mass in this process?]

★ 64. The packing fraction of a nuclide is related to the fraction of the total mass of a nuclide that is converted to nuclear binding energy. It is defined as the fraction $(M - A)/A$, where M is the actual nuclidic mass and A is the mass number. Use data from a handbook (such as *The Handbook of Chemistry and Physics*, published by the CRC Press) to determine the packing fractions of some representative nuclides. Plot a graph of packing fraction versus mass number and compare it to Figure 26-7. Explain the relationship between the two.

★ 65. ^{40}K undergoes radioactive decay both by electron capture to ^{40}Ar and β^- emission to ^{40}Ca. The fraction of the decay that occurs by electron capture is 0.110. The half-life of ^{40}K is 1.27×10^9 y. Assuming that a rock in which ^{40}K has undergone decay retains all of the ^{40}Ar produced, what would be the $^{40}Ar/^{40}K$ mass ratio in a rock that is 1.5×10^9 y old?

★ 66. Reference is made in the text to using a certain shale deposit in the Appalachian Mountains containing 0.006% U as a potential fuel in a breeder reactor. Assuming a density of 2.5 g/cm^3, how much energy could be released from 1.00×10^3 cm^3 of this material? Assume a fission energy of 3.20×10^{-11} J per fission event (that is, per U atom).

Self-Test Questions

For questions 67 through 76 select the single item that best completes each statement.

67. Of the following types of radiation, the only one to be deflected in a magnetic field is (a) x ray; (b) γ ray; (c) β ray; (d) neutrons.

68. A process that produces a one-unit increase in atomic number is (a) electron capture; (b) β^- emission; (c) α emission; (d) γ ray emission.

69. Of the following nuclides, the one most likely to be radioactive is (a) ^{31}P; (b) ^{66}Zn; (c) ^{37}Cl; (d) ^{108}Ag.

70. One of the following elements has eight naturally occurring *stable* isotopes. We should expect that one to be (a) Ra; (b) Au; (c) Cd; (d) Br.

71. Of the following nuclides, the one most likely to decay by positron (β^+) emission is (a) ^{59}Cu; (b) ^{63}Cu; (c) ^{67}Cu; (d) ^{68}Cu.

72. Among the following nuclides, the highest nuclear binding energy per nucleon is found for (a) ^3_1H; (b) $^{16}_8O$; (c) $^{56}_{26}Fe$; (d) $^{235}_{92}U$.

73. The most radioactive of the isotopes of an element is the one with the largest value of its (a) half-life, $t_{1/2}$; (b) neutron number, N; (c) mass number, Z; (d) decay constant, λ.

74. Given a radioactive nuclide with $t_{1/2} = 1.00$ h and a current disintegration rate of 1000 atoms s^{-1}, three hours from now the disintegration rate will be (a) 1000 atoms s^{-1}; (b) 333 atoms s^{-1}; (c) 250 atoms s^{-1}; (d) 125 atoms s^{-1}.

75. The radiation that produces the greatest number of ions as it passes through matter is (a) α; (b) β^-; (c) β^+; (d) γ.

76. The type of radiation that has the greatest penetrating power through matter is (a) α; (b) γ; (c) β^-; (d) visible light.

77. ^{223}Ra has a half-life of 11.4 d. How long would it take for the radioactivity associated with a sample of ^{223}Ra to decrease to 1% of its current value?

78. Explain why

(a) Radioactive nuclides with intermediate half-lives are generally more hazardous than those with extremely short or extremely long half-lives.

(b) Some radioactive substances are hazardous from a distance, whereas others must be taken internally to constitute a hazard.

(c) Argon is the most abundant of the noble gases in the atmosphere.

(d) Francium is such a rare element (less than about 30 g present in the earth's crust at any one time), and it cannot be extracted from minerals containing the alkali metals.

(e) Such extremely high temperatures will be required to develop a self-sustaining thermonuclear (fusion) process as an energy source.

27 Organic Chemistry

The synthesis from coal tar chemicals of the purple dye, mauveine, by William Perkin, an 18-year-old chemistry student, was a great stimulus to synthetic organic chemistry. [Michael Holford]

967

Several million compounds are known that contain carbon and hydrogen atoms or carbon and hydrogen in combination with a few other types of atoms, such as oxygen, nitrogen, and sulfur. Organic chemistry is the chemistry of these compounds. We single out carbon for special study because of the ability of carbon atoms to form strong covalent bonds with one another. Carbon atoms may join together into straight chains, branched chains, and rings. Because of the nearly infinite number of possible bonding arrangements using carbon chains and rings, the number and variety of carbon compounds is vast.

Originally, organic chemistry dealt only with compounds derived from living matter. Living matter was thought to possess a "vital force" necessary for the synthesis of organic compounds. In 1828, Friedrich Wöhler heated ammonium cyanate, derived from inorganic substances, and obtained the organic compound urea.

$$KOCN + NH_4Cl \longrightarrow KCl + NH_4OCN$$

$$\underset{\text{ammonium cyanate}}{NH_4OCN} \xrightarrow{\text{heat}} \underset{\text{urea}}{H_2NCONH_2}$$

The urea formed in this way was identical to urea isolated from urine. Wöhler excitedly reported his results to J. J. Berzelius in a famous letter.

I must tell you that I can make urea without the use of kidneys, either man or dog. Ammonium cyanate is urea.   F. WÖHLER
to J. J. Berzelius, Feb. 22, 1828

27-1 The Nature of Organic Compounds and Structures

The simplest organic compounds are those of carbon and hydrogen— **hydrocarbons.** The simplest hydrocarbon is methane, CH_4, the chief constituent of natural gas. There are three common ways to represent an organic molecule, as shown below for methane. A *Lewis structure* shows the distribution of valence electrons in a molecule. A *structural formula* focuses on the electrons involved in bond formation, using a dash to represent a single bond (double dashes for double bonds and triple dashes for triple bonds). A *condensed formula* conveys the same information as a structural formula but does this in a single line.

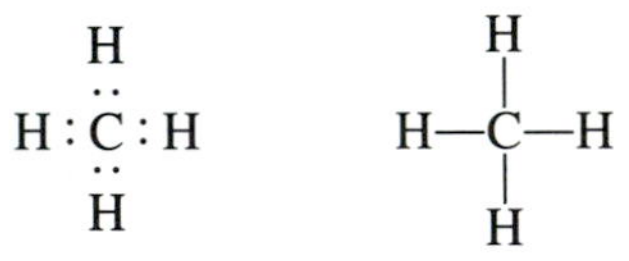

| Lewis structure | structural formula | condensed formula |

None of the foregoing structures describes the geometrical shape of the CH_4 molecule. However, both from VSEPR theory (Section 10-10) and valence bond theory (Section 11-2) we expect the distribution of the four electron pairs around the central carbon atom to be tetrahedral. In a CH_4 molecule the four H atoms are equivalent: They are equidistant from the C atom and attached to it by covalent bonds of equal strength. The angle between any two C—H bonds is 109°28′. Molecular models are often used to represent organic molecules. Two widely used forms are illustrated in Figure 27-1(c) and (d).

Imagine removing one H atom from a CH_4 molecule. This leaves the group —CH_3. Then imagine forming a covalent bond between two such —CH_3 groups. The resulting molecule is that of *ethane*, C_2H_6. By increasing the number of C atoms in the chain, we can obtain still more hydrocarbons, as suggested through Figure 27-2.

H
109°28′
C
H
H
H
(a)
H
C
H
H
H
(b)
H
C
H
H
H
(c)
H
H
C
H
H
(d)

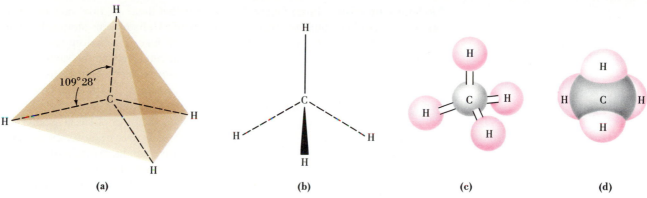

FIGURE 27-1
Structural representation of the methane molecule.

(a) Tetrahedral structure showing bond angle.
(b) Convention used to suggest a three-dimensional structure through a structural formula. The solid line represents a bond in the plane of the page. The dashed lines project *away* from the viewer, and the heavy wedge projects *toward* the viewer.
(c) Ball-and-stick model.
(d) Space-filling model.

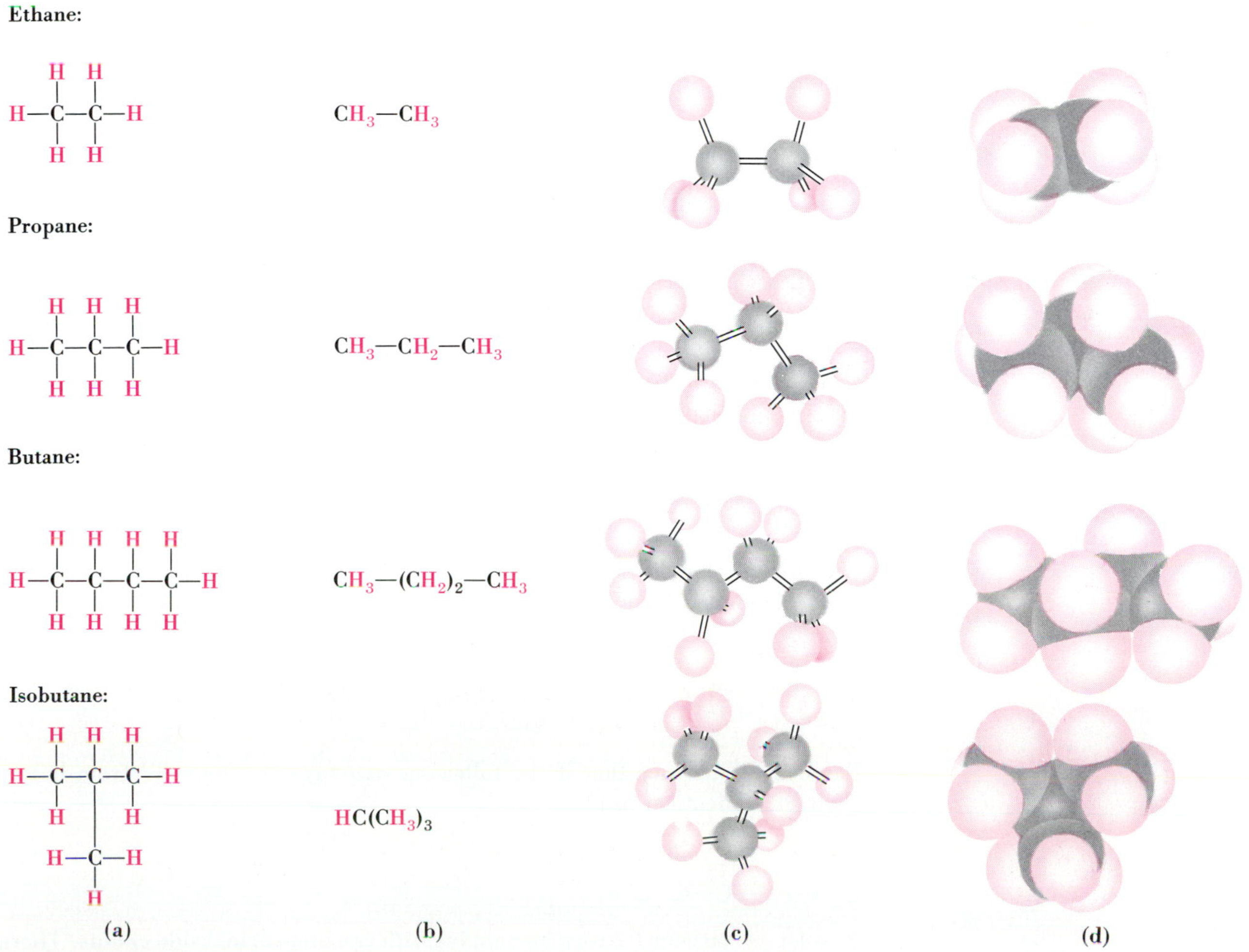

Ethane:
CH_3—CH_3
Propane:
CH_3—CH_2—CH_3
Butane:
CH_3—$(CH_2)_2$—CH_3
Isobutane:
$HC(CH_3)_3$
(a)
(b)
(c)
(d)

FIGURE 27-2
Representation of some additional hydrocarbons.

(a) Structural formulas. **(b)** Condensed formulas. **(c)** Ball-and-stick models.
(d) Space-filling models.

Skeletal Isomerism. From Figure 27-2 we see that there are *two* ways of assembling a hydrocarbon molecule with four carbon and 10 hydrogens atoms. There are *two* different compounds with the formula C_4H_{10}. One is called butane, and the other isobutane. As we have already learned, compounds having the same molecular formula but different structural formulas are called **isomers.** Numerous possibilities for isomerism are found among organic compounds. In the case considered here the isomers differ in their carbon chains—one is a straight chain, and the other a branched chain. This type of isomerism is called **chain** or **skeletal isomerism.**

The names given to the first four members of the hydrocarbon series in Figure 27-2 are common names. As the length of the carbon chain increases, a root name is used that reflects the number of carbon atoms in the chain. To this root is attached a characteristic "ane" ending. Certain branched-chain compounds are often referred to by the prefix "iso." The hydrocarbon series initiated with methane continues beyond the four-carbon compound with these characteristic names: pentane (C_5H_{12}), hexane (C_6H_{14}), heptane (C_7H_{16}), octane (C_8H_{18}), nonane (C_9H_{20}), and decane ($C_{10}H_{22}$). All of the longer chain hydrocarbons display skeletal isomerism, and the more carbon atoms present the greater the number of possible isomers. There are 18 isomers of octane, 35 of nonane, 75 of decane, and so on.

Example 27-1

Identifying isomers. Write structural formulas for all the possible isomers of hexane, C_6H_{14}.

Solution. The basic question is: In how many different ways can six C atoms be bonded together? The key to this question is the word *different*. For example, the following formulas are not different. (Think of all the C atoms as if they were "hinged." Each structure can be rearranged into a six-carbon straight chain.)

$$C-C-C-C-C-C \qquad \begin{matrix} C-C \\ | \quad | \\ C-C \quad C-C \end{matrix} \qquad \begin{matrix} C-C-C \\ | \\ C-C-C \end{matrix} \qquad \text{and so on}$$

We start with the *one* straight chain molecule, and for simplicity show only the carbon skeleton. (The complete structure requires adding the appropriate number of H atoms to produce four bonds at each C atom.)

$$C-C-C-C-C-C$$
(1)

Next we look for the possibilities involving a five-carbon chain with one C atom as a side chain. There are only *two* possibilities.

$$\begin{matrix} \quad\; C \\ \quad\; | \\ C-C-C-C-C \end{matrix} \qquad\qquad \begin{matrix} \quad\quad\; C \\ \quad\quad\; | \\ C-C-C-C-C \end{matrix}$$
(2)   (3)

For example, notice that if the following structure is "flopped" from left to right, it is identical to (2).

$$\begin{matrix} \quad\quad\quad\; C \\ \quad\quad\quad\; | \\ C-C-C-C-C \end{matrix}$$

Now let us consider four-carbon chains with two one-carbon side chains. There is *one* possibility for side chains attached to *different* C atoms of the main chain (structure 4), and *one* possibility for side chains attached to the *same* C atom (structure 5).

$$
\begin{array}{ccc}
& & \text{C} \\
& & | \\
\text{C} \quad \text{C} & & \text{C--C--C--C} \\
| \quad | & & | \\
\text{C--C--C--C} & & \text{C} \\
(4) & & (5)
\end{array}
$$

Any additional possibilities for a carbon skeleton prove to be identical to others already encountered. Thus, these two are identical to (5).

$$
\begin{array}{cc}
& \text{C} \\
& | \\
\text{C} & \text{C} \\
| & | \\
\text{C--C--C--C} \qquad \text{C--C--C} \\
| & | \\
\text{C} & \text{C}
\end{array}
$$

The number of isomers of hexane is 5.

SIMILAR EXAMPLES: Exercises 2, 4, 18.

Nomenclature. Early in the history of organic chemistry, chemists assigned names of their own choosing to new compounds. Often these names were related to the origin or certain properties of the compounds, and some of these names are still in common use. Citric acid is found in citrus fruit; uric acid is present in urine; formic acid is found in ants (from the Latin word for ant, *formica*); and morphine induces sleep (from *Morpheus,* the ancient Greek god of sleep). As thousands upon thousands of new compounds were synthesized it became apparent that a system of nomenclature based on common names would be unworkable. Following several interim systems, one recommended by the International Union of Pure and Applied Chemistry (IUPAC or IUC) was adopted. In saturated hydrocarbons all carbon-to-carbon bonds are single bonds. A few of the more important rules for naming saturated hydrocarbons of the type, C_nH_{2n+2}, are these.

Nomenclature rules for alkane hydrocarbons.

1. The generic (family) name of a saturated hydrocarbon is *alkane*.
2. Select the *longest* continuous carbon chain in the molecule and use the hydrocarbon name of this chain as the base name.
3. Consider every branch of the main chain to be a substituent derived from another hydrocarbon. For each of these substituents change the ending of its name from "ane" to "yl."
4. Number the carbon atoms of the continuous base chain so that the substituents appear *at the lowest numbers* possible.
5. Give each substituent a name and number. For identical substituents use di, tri, tetra, and so on, and *repeat the numbers*.
6. Separate numbers from other numbers by commas and from letters by dashes.
7. Arrange the substituents alphabetically by name, regardless of the numbers they carry or their complexity.
8. Whenever alternative base chains of equal length are possible, always name a compound so as to have the maximum number of side chains.

(27.1)

In applying rule 3, hydrocarbon substituents or alkyl groups should be named as follows.

$$
\begin{array}{cccc}
\text{CH}_3\text{---} & \text{CH}_3\text{CH}_2\text{---} & \text{CH}_3\text{CH}_2\text{CH}_2\text{---} & \text{CH}_3\text{CHCH}_3 \\
& & & | \\
\\
\text{methyl} & \text{ethyl} & \text{propyl} & \text{isopropyl}
\end{array}
$$

Sometimes the prefix *normal* or *n-* is used to designate a straight-chain alkyl group, such as *n*-butyl for $CH_3CH_2CH_2CH_2$—.

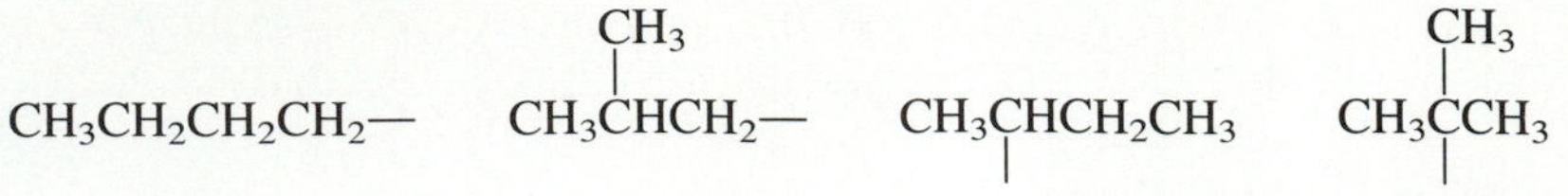

$$CH_3CH_2CH_2CH_2—$$
butyl

$$\underset{\underset{\displaystyle CH_3}{|}}{CH_3CHCH_2—}$$
isobutyl

$$\underset{\underset{\displaystyle CH_3}{|}}{CH_3CHCH_2CH_3}$$
s-butyl
(*sec*-butyl or secondary butyl)

$$\underset{\underset{\displaystyle CH_3}{|}}{CH_3CCH_3}$$
t-butyl
(*tert*-butyl or tertiary butyl)

Example 27-2

Assigning names to alkane hydrocarbons. Give appropriate IUPAC names for the following compounds.

(a) $CH_3—\underset{2}{\overset{1}{C}}—\underset{3}{CH_2}—\underset{4}{CH}—\underset{5}{CH_3}$ with CH_3 groups on C2 (two) and C4

(b) $\underset{1}{CH_3}—\underset{2}{CH_2}—\underset{3}{CH}—CH_3$ with $\underset{4}{CH_2}—\underset{5}{CH_2}—\underset{6}{CH_3}$ chain from C3

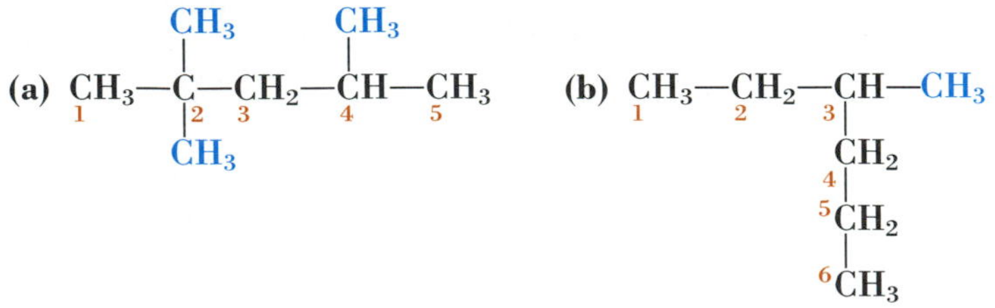

Solution

(a) The carbon atoms are numbered in brown, and the side chain substituents to be named are shown in blue. Each substituent is a methyl group, $—CH_3$. Two methyl groups are on the second carbon atom, and one methyl is on the fourth. The main carbon chain has five atoms. The correct name is

2,2,4-trimethylpentane

If we numbered the carbon atoms in structure (a) from right to left, we would obtain the name 2,4,4-trimethylpentane. However, this is *not* an acceptable name. It does not use the *smallest* numbers possible.

(b) The chain length is six, not four. The methyl group is on the third carbon atom. The correct name is

3-methylhexane

SIMILAR EXAMPLES: Exercises 6, 21, 23, 24.

Example 27-3

Writing formulas to correspond to names of alkane hydrocarbons. Write structural formulas and condensed formulas for the following compounds: (a) 4-*t*-butyl-2-methylheptane; (b) 3-ethyl-2,6-dimethylheptane.

Solution

(a) The substituent group *t*-butyl (blue) is attached to the fourth C atom in a seven-carbon chain. A methyl group (blue) is attached to the second C atom.

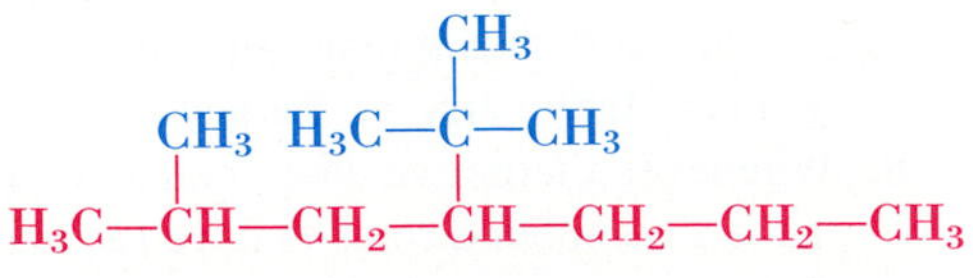

$$H_3C—\underset{\underset{\displaystyle CH_3}{|}}{CH}—CH_2—\underset{\underset{\underset{\displaystyle CH_3}{|}}{\underset{\displaystyle |}{H_3C—\overset{\overset{\displaystyle CH_3}{|}}{C}—CH_3}}}{CH}—CH_2—CH_2—CH_3$$

or

$$(CH_3)_2CHCH_2CH[C(CH_3)_3]CH_2CH_2CH_3$$
(condensed formula)

The prefixes di, tri, . . ., also do not count in an alphabetical listing.

In arranging the substituent groups alphabetically as required in nomenclature rule 7, we do not consider the symbols *s* and *t*. Thus, butyl (even though *t*-butyl) precedes methyl in the name for this structure.

(b) Methyl groups (blue) are substituted at the second and sixth C atoms, and an ethyl group (blue) at the third.

$$CH_3CH(CH_3)CH(C_2H_5)CH_2CH_2CH(CH_3)CH_3$$

H₃C—CH—CH—CH₂—CH₂—CH—CH₃ with CH₃, CH₂(CH₃), and CH₃ substituents

or

(CH₃)₂CHCH(C₂H₅)CH₂CH₂CH(CH₃)₂
(condensed formula)

It appears that this structure could also have been named 5-isopropyl-2-methylheptane. This is *not* done because only two side chain substituents would be involved instead of three (see **rule 8**).

SIMILAR EXAMPLES: Exercises 7, 8, 22, 25.

Positional Isomerism. A variety of atoms or groups of atoms can be substituents on carbon chains, for example, Br. The three monobromopentanes are isomeric but they possess the same carbon skeleton. Because they differ in the position of the bromine atom on the carbon chain, they are called **positional isomers.**

CH₃CH₂CH₂CH₂CH₂Br   CH₃CH₂CH₂CHCH₃   CH₃CH₂CHCH₂CH₃
                                          | |
                                          Br   Br

1-bromopentane   2-bromopentane   3-bromopentane

Example 27-4

Identifying isomers in substituted alkane hydrocarbons. Represent all the possible isomers of C_4H_9Cl.

Solution. Perhaps the simplest approach is to consider the number of positional isomers for each of the skeletal isomers of butane. There are *two* possibilities based on butane (structures 1 and 2) and *two* based on isobutane (structures 3 and 4).

CH₃CH₂CH₂CH₂Cl   CH₃CH₂CHCH₃   CH₃CHCH₂Cl   CH₃CCH₃
                                  | | |
                                  Cl   CH₃   CH₃
                                                                            |
                                                                            Cl

(1)   (2)   (3)   (4)

SIMILAR EXAMPLES: Exercises 4, 18.

Functional Groups. Typically, elements in addition to carbon and hydrogen are found in organic compounds. These elements occur as distinctive groupings of one or several atoms. In some cases these groupings of atoms are substituted for H atoms in hydrocarbon chains or rings. In other cases they may substitute for C atoms themselves. These distinctive groupings of atoms are called **functional groups,** and the remainder of the molecule is sometimes referred to by the symbol **R.** The physical and chemical properties of organic molecules usually depend on the particular functional groups present. The remainder of the molecule (R) often has little effect on these properties.

A convenient way to study organic chemistry, then, is to consider the properties associated with specific functional groups. This is what we do in the remainder of the chapter. Table 27-1 lists the major types of organic compounds, and their dis-

TABLE 27-1
Some Classes of Organic Compounds

Type of compound	General structural formula	Example	Name
alkanes	R—H	$CH_3CH_2CH_2CH_2CH_3$	pentane
alkenes	$\overset{R}{\underset{R}{\diagdown}}C{=}C\overset{R}{\underset{R}{\diagup}}$	$CH_3CH_2CH_2CH{=}CH_2$	1-pentene
alkynes	R—C≡C—R	$CH_3CH_2C{\equiv}CCH_3$	2-pentyne
alcohols	R—OH	$CH_3CH_2CH_2CH_2CH_2OH$	1-pentanol
alkyl halides	R—X	$CH_3CH_2\underset{\underset{Br}{\vert}}{C}HCH_2CH_3$	3-bromopentane
ethers	R—O—R	$CH_3CH_2OCH_2CH_2CH_3$	ethyl propyl ether
aldehydes	$R-\overset{O}{\overset{\|}{C}}-H$	$CH_3CH_2CH_2CH_2\overset{O}{\overset{\|}{C}}H$	pentanal
ketones	$R-\overset{O}{\overset{\|}{C}}-R$	$CH_3CH_2\overset{O}{\overset{\|}{C}}CH_2CH_3$	3-pentanone
acids	$R-\overset{O}{\overset{\|}{C}}-OH$	$CH_3CH_2CH_2CH_2\overset{O}{\overset{\|}{C}}OH$	pentanoic acid
esters	$R-\overset{O}{\overset{\|}{C}}-O-R$	$CH_3CH_2CH_2CH_2\overset{O}{\overset{\|}{C}}OCH_3$	methyl pentanoate
amines	$R-NH_2$	$CH_3CH_2CH_2CH_2CH_2NH_2$	pentylamine

Some of the functional groups appearing here and discussed later in the chapter have distinctive names:

—OH, hydroxyl; $\diagdown$C=O, carbonyl; $-\overset{O}{\overset{\|}{C}}-OH$, carboxyl; —NH$_2$, amino.

tinctive functional groups are indicated in blue. You may find it helpful to refer to this table from time to time.

Example 27-5 illustrates that when we name organic compounds we must designate the functional groups present, as well as follow the other nomenclature rules outlined in (27.1).

Example 27-5

Naming functional groups in an organic compound. Use information from Table 27-1 and from elsewhere in this section to derive an acceptable name for the compound $(CH_3)_2CHOCH_2CH_2CH_3$.

Solution. First we should convert the condensed formula into a structural formula.

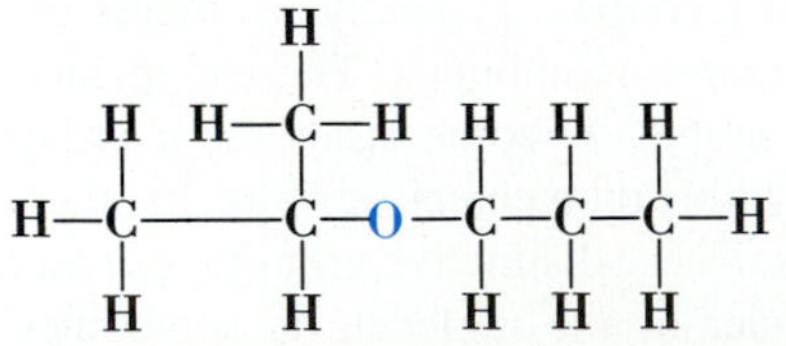

$$\underset{H}{\overset{H}{\vert}}{\,}$$

$$H-\underset{\underset{H}{\vert}}{\overset{\overset{H}{\vert}}{C}}\quad\underset{\underset{H}{\vert}}{\overset{\overset{H-\underset{\vert}{C}-H}{\vert}}{C}}-O-\underset{\underset{H}{\vert}}{\overset{\overset{H}{\vert}}{C}}-\underset{\underset{H}{\vert}}{\overset{\overset{H}{\vert}}{C}}-\underset{\underset{H}{\vert}}{\overset{\overset{H}{\vert}}{C}}-H$$

From the presence of the group C—O—C we see that this is an ether. The group shown on the left is isopropyl; the one on the right is propyl. We can call the compound isopropyl propyl ether.

SIMILAR EXAMPLE: Exercise 5.

27-2 Alkanes

In this section we explore further some properties of the alkanes. The essential characteristic of alkane hydrocarbon molecules is that they have only single covalent bonds. In these compounds the bonds are said to be *saturated,* and the alkanes are known as **saturated hydrocarbons.**

The alkanes range in complexity from methane, CH_4 (accounting for over 90% of natural gas), to molecules containing 50 carbon atoms or more (found in petroleum). Each alkane differs from the preceding one by a —CH_2— or *methylene* group. Substances whose molecules differ only by a constant unit such as —CH_2— are said to form a **homologous series.** Members of such a series usually have closely related chemical and physical properties. For example, in Table 27-2 we note a gradual increase in boiling point with an increase in molecular weight. In Chapter 12 we learned how to explain this trend in terms of London-type (dispersion) intermolecular forces: London-type forces increase in strength with increasing molecular weight. In Chapter 12 we also established a basis for explaining why certain isomers have lower boiling points than others: The more compact a structure, the less easily it is polarized, the weaker the London-type forces, and the lower the boiling point (recall Figure 12-26 for two of the isomers of pentane).

Conformations. An important type of motion in alkane molecules is suggested by ball-and-stick models. This is a motion in which one group rotates with respect to the other. Figure 27-3 suggests two of the many possible orientations of the two —CH_3 groups in ethane.

TABLE 27-2
Boiling Points of Some Isomeric Alkanes

Family	Isomer	Boiling point, °C	Family	Isomer	Boiling point, °C
butane	butane	−0.5	hexane	hexane	68.7
	isobutane	−11.7		3-methylpentane	63.3
pentane	pentane	36.1		isohexane	60.3
	isopentane	27.9		2,3-dimethylbutane	58.0
	2,2-dimethylpropane (neopentane)	9.5		2,2-dimethylbutane (neohexane)	49.7

FIGURE 27-3
Rotation about the C—C bond in ethane.

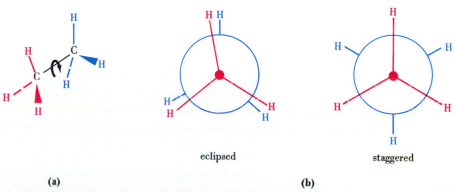

eclipsed

staggered

(a)

(b)

(a) Rotation about the C—C bond.
(b) Conformations of C_2H_6, in a representation known as a Newman projection. This is based on a head-on view of a ball-and-stick model of C_2H_6. The small solid circle (red) represents the carbon atom in front, and the red H atoms are those bonded to this carbon atom. The large open circle (blue) represents the rear carbon atom, and the blue H atoms are those bonded to it.

In one of these configurations, when the molecule is viewed head-on along the C—C bond, one set of C—H bonds is directly behind the other. This structure is referred to as the **eclipsed conformation.** In this conformation the distance between H atoms on adjacent C atoms is at a minimum, leading to a condition of maximum repulsion between H atoms. This conformation is slightly less stable than the staggered conformation. In the **staggered conformation** the hydrogen atoms are located a maximum distance apart. Although we expect the staggered conformation to be the more stable, at room temperature the thermal energy of an ethane molecule is sufficient that free rotation of the methyl groups about the C—C bond takes place. At lower temperatures, however, ethane does occur mostly in the staggered conformation. A similar situation is also encountered with higher alkanes.

Ring Structures. Alkanes in chain structures have the formula C_nH_{2n+2} and are called **aliphatic.** Alkanes can also exist in ring or cyclic structures called **alicyclic.** We can think of these rings as resulting from the joining together of two ends of an aliphatic chain by the elimination of a hydrogen atom from each end. Simple alicyclic compounds have the formula C_nH_{2n}.

| cyclopropane | cyclobutane | cyclopentane | cyclohexane |
| C_3H_6 | C_4H_8 | C_5H_{10} | C_6H_{12} |

We can use the rules in (27.1) to name alicyclic compounds. Thus the name of

is 1,3-dimethylcyclopentane. By convention, when a ring structure is drawn, usually neither the ring C atoms nor the H atoms bonded to them are written out.

The bond angles in cyclopropane are 60° compared to the normal 109.5°; the bonds are quite strained. As a result there are numerous reactions in which the ring will break open to yield a chain molecule—propane or a propane derivative. Cyclopropane is more reactive than most alkanes.

If the four carbon atoms in cyclobutane were to exist in the same plane, the C—C bond angles would be 90°. In reality the molecule buckles slightly to relieve the bond strain. If the cyclopentane molecule existed as a planar structure, the bond angles would be 108°, which is quite close to the normal 109.5°. However, in this planar structure all the H atoms would be eclipsed. A more stable arrangement is for one of the carbon atoms to be buckled out of the plane of the other four and for the H atoms to adopt a more staggered conformation.

With ball-and-stick models we can see that there are two possible conformations of cyclohexane. These are the "**boat**" and the "**chair**" conformations, pictured in Figure 27-4. The models demonstrate clearly that H atoms (shown in brown) on the first and fourth C atoms in the boat form come close enough together to repel one another. In the chair form all the H atoms are staggered. The chair form is the more stable conformation of cyclohexane. Figure 27-4 also suggests that the twelve H atoms in the chair form are not quite equivalent. Six of them (shown in red) extend outward from the ring and are called **equatorial** H atoms. Of the other six (shown in blue), three are directed above and three below the ring; these are the six **axial** H

FIGURE 27-4
Conformations of cyclohexane.

(a) Boat form. **(b)** Chair form; the equatorial H atoms are shown in red; the other H atoms (blue) are axial.

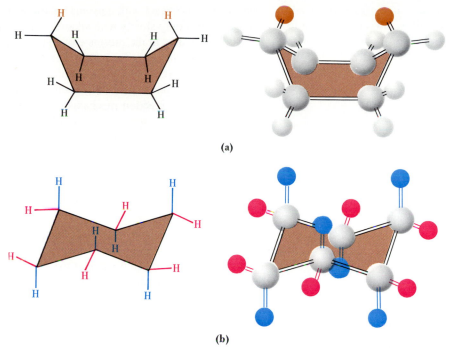

(a)

(b)

atoms. When another group (say —CH$_3$) is substituted for an H atom on the ring, the preferred position is an equatorial one. This causes a minimum interference with other groups on the ring.

Preparation of Alkanes. The primary source of alkanes is petroleum, but several laboratory methods can also be used for their preparation. Some of these use organic substances of types that are described later in this chapter. Unsaturated hydrocarbons, whether containing double or triple bonds, may be converted to saturated hydrocarbons by the addition of hydrogen to the multiple bond system in the presence of a metal catalyst (reactions 27.2 and 27.3). In the Würtz reaction halogenated hydrocarbons react with alkali metals to produce alkanes of double the carbon content (27.4). Alkali metal salts of carboxylic acids may be fused with alkali hydroxides. Here a hydrocarbon is obtained containing one carbon atom less than the original acid salt (27.5).

$$CH_2\!\!=\!\!CH_2 + H_2 \xrightarrow[\text{heat/pressure}]{\text{Pt or Pd}} CH_3\!\!-\!\!CH_3 \tag{27.2}$$

ethane

$$HC\!\!\equiv\!\!CH + 2\,H_2 \xrightarrow[\text{heat/pressure}]{\text{Pt or Pd}} CH_3\!\!-\!\!CH_3 \tag{27.3}$$

ethane

$$2\,CH_3CH_2Br + 2\,Na \longrightarrow 2\,NaBr + CH_3\!\!-\!\!CH_2\!\!-\!\!CH_2\!\!-\!\!CH_3 \tag{27.4}$$

ethyl bromide
or bromoethane   butane

$$\overset{\displaystyle O}{\overset{\displaystyle \|}{CH_3C}}ONa + NaOH \xrightarrow{\text{heat}} Na_2CO_3 + CH_4 \tag{27.5}$$

sodium acetate   methane

Substitution Reactions. Saturated hydrocarbons have little affinity for most chemical reagents. Because of this they have become known as paraffin hydrocarbons (Latin, *parum*, little; *affinis*, reactivity). Paraffin hydrocarbons are insoluble in

Ordinary paraffin wax, used in candles, waxed paper, and home canning, is a mixture of long-chain paraffin hydrocarbons.

water and do not react with aqueous solutions of acids, bases, or oxidizing agents. Halogens react only slowly with alkanes at room temperature, but at higher temperatures, particularly in the presence of light, halogenation occurs. In this reaction a halogen atom *substitutes* for a hydrogen atom. The mechanism of this substitution reaction appears to be by a chain reaction, written as follows for the chlorination of methane. (For emphasis, only the electrons involved in bond breakage or formation are shown in this reaction mechanism.)

Initiation: $Cl:Cl \xrightarrow[\text{light}]{\text{heat or}} 2\ Cl\cdot$

Propagation: $H_3C:H + Cl\cdot \longrightarrow H_3C\cdot + H:Cl$
                  $H_3C\cdot + Cl:Cl \longrightarrow H_3C:Cl + Cl\cdot$

Termination: $Cl\cdot + Cl\cdot \longrightarrow Cl:Cl$
                $H_3C\cdot + Cl\cdot \longrightarrow H_3C:Cl$
                $H_3C\cdot + H_3C\cdot \longrightarrow H_3C:CH_3$

A radical is a fragment of a molecule having an unpaired electron(s). Usually radicals are short-lived and very reactive.

The reaction is initiated when some Cl_2 molecules absorb sufficient energy to dissociate into Cl atoms (represented above as $Cl\cdot$). Cl atoms collide with CH_4 molecules to produce methyl *radicals* ($H_3C\cdot$), which combine with Cl_2 molecules to form the product, CH_3Cl. When any or all of the last three reactions proceed to the extent of consuming the radicals present, the reaction stops. This free-radical chain reaction usually yields a mixture of products. The net equation for the formation of chloromethane is

$$CH_4 + Cl_2 \xrightarrow[\text{light}]{\text{heat or}} CH_3Cl + HCl \tag{27.6}$$

Polyhalogenation can also occur to form CH_2Cl_2, dichloromethane (methylene dichloride, a solvent); $CHCl_3$, trichloromethane (chloroform, a solvent and anesthetic); and CCl_4, tetrachloromethane (carbon tetrachloride, a solvent).

Oxidation is the reaction of hydrocarbons underlying their important use as fuels. For example,

$$C_7H_{16}(l) + 11\ O_2(g) \longrightarrow 7\ CO_2(g) + 8\ H_2O(l) \qquad \Delta H° = -4812\ kJ \tag{27.7}$$

We discuss hydrocarbon fuels further in Section 27-11.

27-3   Alkenes and Alkynes

Hydrocarbons whose molecules contain some multiple bonds (double or triple) between carbon atoms are said to be **unsaturated.** If there is *one double bond* between C atoms, the hydrocarbons are the simple **alkenes** or **olefins;** they have the general formula C_nH_{2n}. Simple **alkynes** or **acetylenes** have *one triple bond* between C atoms; they have the general formula C_nH_{2n-2}.

We can use the same general rules outlined in (27.1) to name alkenes and alkynes, with a few minor modifications.

Nomenclature rules for alkene and alkyne hydrocarbons.

- Select as the base chain the longest chain containing the multiple bond.
- Number the carbon atoms of the chain to place the multiple bond at the lowest possible number. Use only this number in locating the multiple bond. (That is, we understand that the multiple bond is between this and the next-highest-numbered carbon atom.)   (27.8)
- Use the ending "ene" for alkenes and "yne" for alkynes.

In the examples that follow, systematic names are shown in blue. The names given in parentheses are also commonly used.

$CH_2{=}CH_2$ $CH_3CH_2CH{=}CH_2$

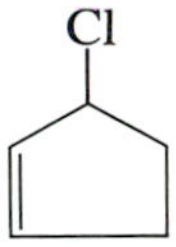

ethene
(ethylene)   1-butene   3-chlorocyclopentene

$HC{\equiv}CH$ $CH_3CH_2C{\equiv}CH$ $CH_3CHC{\equiv}CCH_3$
                                          CH_3

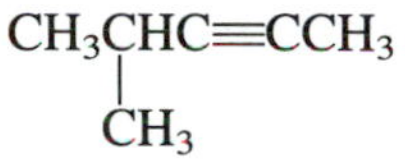

ethyne
(acetylene)   1-butyne
(ethylacetylene)   4-methyl-2-pentyne
(isopropylmethylacetylene)

Example 27-6

Assigning names to unsaturated hydrocarbons. What is the systematic name of the following structure?

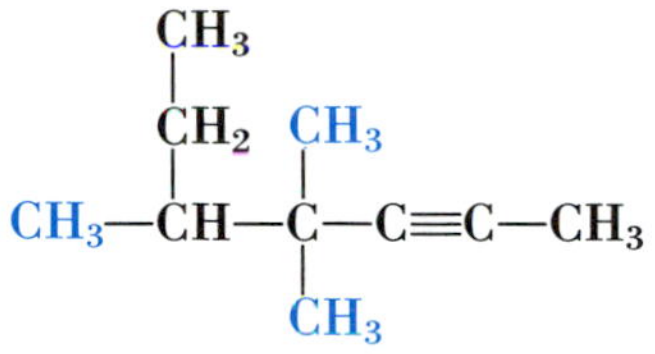

$$CH_3{-}CH{-}\underset{\underset{CH_3}{|}}{\overset{\overset{CH_3}{|}}{\underset{}{\overset{}{C}}}}{-}C{\equiv}C{-}CH_3$$

with CH_2 and CH_3 branches on the CH and C carbons respectively.

Solution. The longest chain is numbered to place the triple bond at the lowest possible position—2. Having made this decision, we now establish that there are three methyl groups (shown in blue) at the positions 4, 4, and 5. The name of the compound is

4,4,5-trimethyl-2-heptyne

The name 4,4-dimethyl-5-ethyl-2-hexyne is unacceptable since it is not based on the longest possible hydrocarbon chain.

SIMILAR EXAMPLES: Exercises 7, 8, 21, 24, 25.

The alkenes are similar to the alkanes in physical properties. At room temperature, those containing two to four carbon atoms are gases; those with 5 to 18 are liquids; those with more than 18 are solids. In general, alkynes have higher boiling points than their alkane and alkene counterparts.

Geometrical Isomerism. The compounds 2-butene, $CH_3CH{=}CHCH_3$, and 1-butene, $CH_2{=}CHCH_2CH_3$, are isomers. The difference between them is in the position of the double bond. We can call them *positional* isomers. But another kind of isomerism is possible involving 2-butene, represented by these two structures.

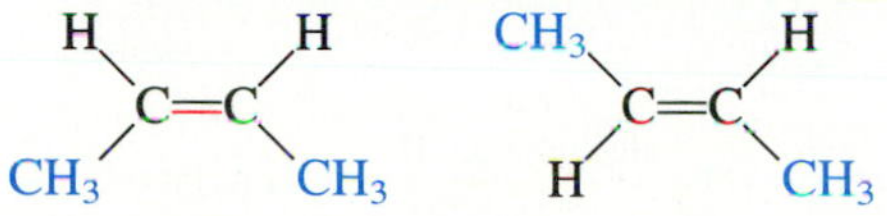

(a)   (b)

As we learned in Section 11-3, a double bond between C atoms consists of the overlap of sp^2 or sp hybrid orbitals to form a σ bond *and* the sidewise overlap of p orbitals to form a π bond. Rotation about this double bond is severely restricted. As a result, molecule (a) above cannot be converted into molecule (b) simply by twisting one end of the molecule through 180°. This means that the two molecules are

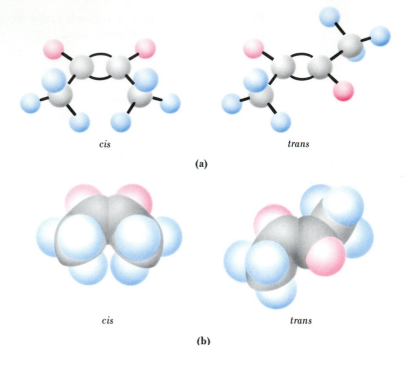

cis   *trans*

(a)

cis   *trans*

(b)

The terms cis and trans have the same meaning here as in the geometric isomerism of complex ions (Section 25-4).

distinctly different (see Figure 27-5). To distinguish between these two molecules by name, we call (a) *cis*-2-butene (*cis*, Latin, on the same side) and we call (b) *trans*-2-butene (*trans*, Latin, across). This type of isomerism is called **geometric isomerism.**

In Section 25-4 we noted that geometric isomerism is only one type of a more general kind of isomerism known as **stereoisomerism** (from the Greek word *stereos,* meaning solid or three-dimensional in nature). In stereoisomerism the number and types of atoms and bonds are the same, but certain atoms are oriented differently in space. Another type of stereoisomerism that we introduced in Chapter 25 is *optical isomerism.* As we see in the next chapter, stereoisomerism plays an important role in the unique chemical reactions found in living organisms.

Preparation. Two general reactions for the preparation of olefins use alcohols and alkyl halides as starting materials (substrates). These are **elimination reactions,** processes in which atoms are removed from adjacent positions on a carbon chain. A small molecule is produced and an additional bond is formed between the C atoms. H_2O is eliminated in equation (27.9) and HBr in (27.10).

Preparation of an alkene.

$$CH_3-\underset{\underset{HO}{|}}{\overset{\overset{H}{|}}{C}}-\underset{\underset{H}{|}}{\overset{\overset{H}{|}}{C}}-H \xrightarrow[\text{heat}]{H_2SO_4} CH_3CH{=}CH_2 + H_2O \qquad (27.9)$$

$$CH_3\underset{\underset{H}{|}}{\overset{\overset{H}{|}}{C}}-\underset{\underset{Br}{|}}{\overset{\overset{H}{|}}{C}}-H \xrightarrow{\text{alcoholic KOH}} CH_3CH{=}CH_2 + KBr + HOH \qquad (27.10)$$

Acetylene, the simplest alkyne, has been one of the most important organic raw materials used in industry. As we learned in Section 23-4, it is prepared from coal, water, and limestone.

$$CaCO_3 \xrightarrow{\text{heat}} CaO + CO_2 \qquad (27.11)$$

$$CaO + 3 C \xrightarrow[2000\ °C]{\text{electric furnace,}} CaC_2 + CO \tag{27.12}$$

calcium
acetylide
(calcium
carbide)

$$CaC_2 + 2\ H_2O \longrightarrow HC{\equiv}CH + Ca(OH)_2 \tag{27.13}$$

Most other alkynes are prepared from acetylene itself by taking advantage of the acidity of the C—H bond. In the presence of a very strong base, such as sodium amide, the acetylene donates protons to amide ions and forms a sodium salt, sodium acetylide (equation 27.14). This acetylide can then react with an alkyl halide (27.15).

$$H{-}C{\equiv}C{-}H + Na^+\ NH_2^- \longrightarrow NH_3 + H{-}C{\equiv}C^-\ Na^+ \tag{27.14}$$

$$H{-}C{\equiv}C^-\ Na^+ + CH_3Br \longrightarrow HC{\equiv}C{-}CH_3 + NaBr \tag{27.15}$$

By continuing this reaction, the triple bond can be positioned as desired in the chain, as in the synthesis of 2-pentyne.

$$H{-}C{\equiv}C{-}CH_3 \xrightarrow{NaNH_2} Na^+\ {}^-C{\equiv}C{-}CH_3 + NH_3 \tag{27.16}$$

$$Na^+\ {}^-C{\equiv}C{-}CH_3 + CH_3CH_2Br \longrightarrow CH_3CH_2C{\equiv}CCH_3 + NaBr \tag{27.17}$$

Addition Reactions. The biggest difference between alkanes and alkenes is that alkanes react by *substitution* and alkenes by *addition*. That is, with alkanes some atom or group of atoms substitutes for an H atom in a carbon chain, whereas with alkenes two new atoms become bonded to the C atoms at the site of the double bond (and the double bond is converted to a single bond).

Alkane: $$CH_3{-}CH_3 + Br_2 \longrightarrow CH_3{-}CH_2{-}Br + HBr \tag{27.18}$$

Alkene: $$CH_2{=}CH_2 + Br_2 \longrightarrow \underset{\underset{Br}{|}}{CH_2}{-}\underset{\underset{Br}{|}}{CH_2} \tag{27.19}$$

In reaction (27.19), the splitting of a Br_2 produces two identical Br atoms, and $CH_2{=}CH_2$ yields two —CH_2 groups. Br_2 and $CH_2{=}CH_2$ are both known as *symmetrical* reagents. The two parts produced when HBr or $CH_2CH{=}CH_2$ are split are not identical. These are *unsymmetrical* reagents. We have a problem in predicting the products when *unsymmetrical* HBr is added to *unsymmetrical* propene. That is, which of the following products should we expect?

$$CH_3CH{=}CH_2 + H{-}Br \longrightarrow \underset{\underset{Br}{|}\ \underset{H}{|}}{CH_3CH{-}CH_2} \quad or \quad \underset{\underset{H}{|}\ \underset{Br}{|}}{CH_3CH{-}CH_2}$$

We find that 2-bromopropane is the sole product. This result can be explained in terms of the electron-donating abilities of H atoms and alkyl groups, but we will not attempt to pursue this matter further. Instead let us simply note this empirical rule proposed by Vladimir Markovnikov in 1871.

> When an *unsymmetrical* reagent (HX, HOH, HCN, $HOSO_3H$) is added to an *unsymmetrical* alkene or alkyne, the more positive fragment (usually H) adds to the carbon atom with the greater number of attached H atoms. (27.20)

The addition of H_2O to a double bond (as in reaction 27.21) is the reverse of the reaction in which a double bond is formed by the elimination of H_2O (reaction 27.9). In this reversible reaction, the addition reaction is favored in dilute acid, and the elimination reaction is favored in *concentrated* $H_2SO_4(aq)$. The addition of HCl

When bromine is added to an unsaturated hydrocarbon (here, 2-pentene), the red color of the bromine disappears, indicating that a reaction has occurred. [Carey B. Van Loon]

Unsaturated hydrocarbon addition reactions: Markovnikov's rule.

and HCN to alkynes (reactions 27.22 and 27.23) are used commercially to synthesize intermediates for polymer production. Purple-colored permanganate solutions are decolorized by alkenes (reaction 27.24). By contrast, alkanes are nonreactive toward permanganate solutions. Thus, alkanes and alkenes can be distinguished with permanagante solution (Baeyer test).

$$CH_3-\underset{\underset{CH_3}{|}}{C}=CH_2 + H_2O \xrightarrow{10\% H_2SO_4} CH_3-\underset{\underset{OH}{|}}{\overset{\overset{CH_3}{|}}{C}}-CH_3 \qquad (27.21)$$

t-butyl alcohol

Notice how the H and Cl atoms in reaction (27.22) add according to Markovnikov's rule (27.20).

$$CH_3-C\equiv CH + HCl \longrightarrow CH_3-\underset{\underset{Cl}{|}}{C}=\overset{\overset{H}{|}}{C}-H \xrightarrow{HCl} CH_3-\underset{\underset{Cl}{|}}{\overset{\overset{Cl}{|}}{C}}-\underset{\underset{H}{|}}{\overset{\overset{H}{|}}{C}}-H \qquad (27.22)$$

methylacetylene   2-chloropropene   2,2-dichloropropane
(propyne)

$$HC\equiv CH + HCN \longrightarrow H-\underset{\underset{H}{|}}{C}=\overset{\overset{CN}{|}}{C}-H \qquad (27.23)$$

cyanoethylene
(acrylonitrile)

$$CH_2=CHCH_3 \xrightarrow{MnO_4^-,\ H_2O} CH_2-\underset{\underset{OH}{|}}{\overset{\overset{}{}}{CH}}-\underset{\underset{OH}{|}}{\overset{\overset{}{}}{CH}}-CH_3 \qquad (27.24)$$

propylene   1,2-propanediol
(propene)   (propylene glycol)

Polymerization. Ethylene and other olefins can enter into reactions in which the net effect is the opening up of the C=C double bond and the formation of giant molecules. The reaction proceeds by a chain mechanism. The key to the reaction is a free-radical initiator. In the first step of the process outlined in equations (27.25) through (27.28), an organic peroxide dissociates into two radicals (equation 27.25). The radicals add to the C=C double bonds of ethylene molecules, forming radical intermediates (27.26). These radical intermediates are chain carriers that successively attack more ethylene molecules, forming new intermediates of longer and longer length (27.27). The chains terminate as a result of reactions such as (27.28).

Initiation: $$R-O:O-R \longrightarrow 2\ R-O\cdot \qquad (27.25)$$
organic peroxide

The Baeyer test for an unsaturated hydrocarbon. The purple color of MnO_4^- disappears and brown-black $MnO_2(s)$ is formed. [Carey B. Van Loon]

Are You Wondering:

Why equation (27.24) is incomplete and unbalanced?

Organic reactions are usually complicated by side reactions producing by-products. As a result, equations for organic reactions are not as useful for stoichiometric calculations as are equations for reactions involving inorganic chemicals. When organic chemists write equations, they frequently list only organic reactants on the left and the principal products on the right. Inorganic reagents and reaction conditions are generally written above the arrow. The equation may or may not be written in balanced form.

followed by $CH_2{=}CH_2 + RO\cdot \longrightarrow R{-}O{-}CH_2{-}CH_2\cdot$ (27.26)

Propagation:

$$ROCH_2CH_2\cdot + CH_2{=}CH_2 \longrightarrow ROCH_2CH_2CH_2CH_2\cdot \qquad (27.27)$$

$$RO(CH_2)_3CH_2\cdot + CH_2{=}CH_2 \longrightarrow RO(CH_2)_5CH_2\cdot \longrightarrow \quad \longrightarrow$$

Termination: $RO(CH_2)_xCH_2\cdot + RO\cdot \longrightarrow RO(CH_2)_xCH_2OR$ (27.28)

or $2\,RO(CH_2)_xCH_2\cdot \longrightarrow RO(CH_2)_xCH_2CH_2(CH_2)_xOR$

27-4   Aromatic Hydrocarbons

Many aromatic compounds have strong odors that are not unpleasant, a fact that is reflected in the name "aromatic." But not all aromatic compounds have pleasing aromas.

There is a special class of hydrocarbon molecules having ring structures with unsaturation (multiple bond character) in the rings. These are called aromatic hydrocarbons. The best way to think about aromatic hydrocarbons is that they are all based on the molecule benzene, C_6H_6. In Section 11-7 we discussed bonding in the benzene molecule in some detail and showed that there are several ways of representing the molecule, including

Kekulé structures   molecular orbital representation

Other aromatic hydrocarbons can be viewed as derivatives of benzene.

toluene   o-xylene   naphthalene   anthracene

Toluene and o-xylene are *substituted* benzenes, and naphthalene and anthracene feature *fused* benzene rings. Whenever two rings are fused together, there is a loss of two carbon and four hydrogen atoms. Thus, naphthalene has the formula $C_{10}H_8$ and anthracene, $C_{14}H_{10}$. In this text we use the inscribed circle to represent the simple benzene ring and alternating single and double bonds for fused rings.

When one of the six equivalent H atoms of a benzene molecule is removed, the species that results is called a **phenyl** group. Two phenyl groups may bond together as in biphenyl, or phenyl groups may be substituents on aliphatic hydrocarbon chains.

phenyl group   biphenyl   triphenylmethane

Other groups may be substituted for H atoms, and this raises problems in nomenclature. We can handle these problems with a numbering system for the C atoms in the ring. If the name of an aromatic compound is to be based on a common name other than benzene, e.g., toluene, the characteristic substituent group ($-CH_3$) is assigned position "1" on the benzene ring.

3-bromotoluene
(*m*-bromotoluene)

2-bromochlorobenzene
(*o*-bromochlorobenzene)

1,4-dichlorobenzene
(*p*-dichlorobenzene)

2-chlorotoluene
(*o*-chlorotoluene)

2,2-di-(*p*-chlorophenyl)-1,1,1-trichloroethane (DDT)

DDT was at one time a widely used pesticide. Its use in the United States has been discontinued, however, because of environmental concerns.

We can also use the terms "ortho," "meta," and "para" (*o*-, *m*-, *p*-) when there are two substituents on the benzene ring. **Ortho** refers to substituents on adjacent carbon atoms, **meta** to substituents with one carbon atom between them, and **para** to substituents opposite to one another on the ring.

Benzene and its homologs are similar to other hydrocarbons in that they are insoluble in water but soluble in organic solvents. The boiling points of the aromatic hydrocarbons (arenes) are slightly higher than those of the alkanes of similar carbon content. For example, hexane, C_6H_{14}, boils at 69 °C, whereas benzene boils at 80 °C. The planar structure and highly delocalized electron density in aromatic hydrocarbons increases the attractive forces acting between molecules, resulting in higher boiling points. The symmetrical structure of benzene permits closer packing in the crystalline state, resulting in a higher melting point than for hexane. Benzene melts at 5.5 °C and hexane melts at −95 °C.

Aromatic hydrocarbons are highly flammable and should always be handled with care. Prolonged inhalation of benzene vapor results in a decreased production of both red and white blood corpuscles, and this can prove fatal. Also, benzene is a carcinogen. Benzene should be used only under well-ventilated conditions. Fused ring systems, such as benzo[*a*]pyrene, are commonly encountered when organic materials are heated to high temperatures in limited contact with air, a process known as **pyrolysis** (thermal decomposition). Benzo[*a*]pyrene has been isolated in the tar formed by burning cigarettes, in polluted air, and as a decomposition product of grease in the charcoal grilling of meat. Benzo[*a*]pyrene is one of the most active hydrocarbon carcinogens.

benzo[*a*]pyrene

Aromatic Substitution Reactions. In alkene and alkyne molecules the regions of high electron density associated with multiple bonds are *localized* between specific carbon atoms. In aromatic rings, certain electrons are *delocalized* into a π electron cloud; they are not concentrated into the region between specific carbon atoms. (Recall the doughnut-shaped regions in Figure 11-32.) Because they lack localized π electrons, aromatic molecules enter into reactions by the *substitution* of other atoms or groups for H atoms, not the addition of atoms across a double bond. Among the characteristic reactions are halogenation, nitration, sulfonation, and alkylation. Examples are shown in Figure 27-6.

When we substitute a single atom or group, X, for an H atom in C_6H_6, this substitution can occur at any one of the six positions of the benzene ring. We say that the six positions are equivalent. If we then substitute a group Y for an H atom in C_6H_5X, an important question is to which of the remaining five positions does the Y group go? If all the sites on the benzene ring were equally preferred, the distribution of the product would be a purely statistical one. That is, there are five possibilities for the position where Y can be substituted, and we should get 20% of each one. However, since there are two possibilities for substitution that lead to an *ortho*

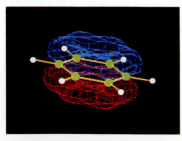

A computer-generated model of the benzene molecule, highlighting the π orbitals above and below the plane of the C and H atoms. [Chemical Design Ltd., Oxford, England]

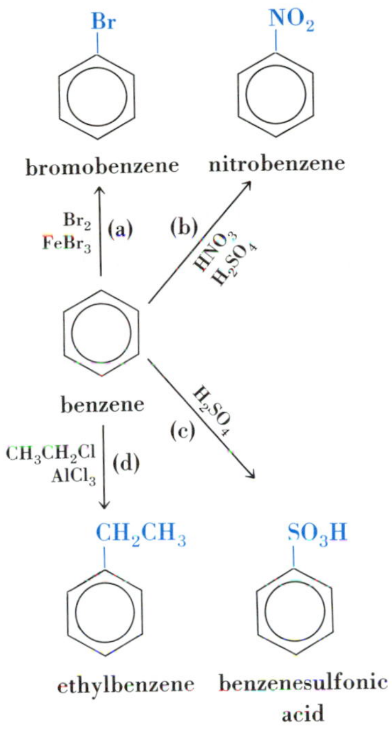

FIGURE 27-6
Some substitution reactions of benzene illustrated.

(a) Halogenation.
(b) Nitration.
(c) Sulfonation.
(d) Alkylation.

isomer and two that lead to a *meta* isomer, we should expect the distribution of product to be 40% ortho, 40% meta, and 20% para.

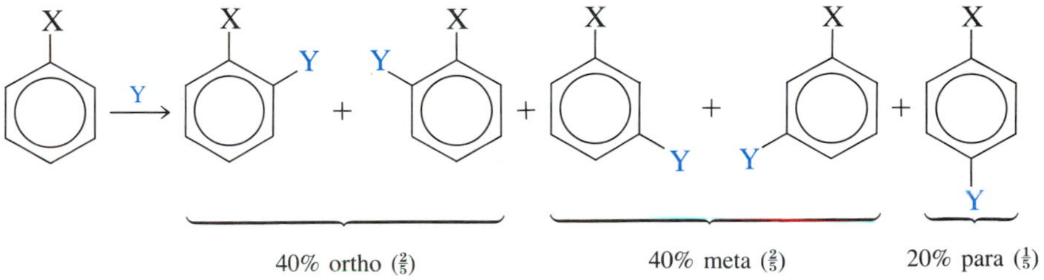

The following scheme describes the products resulting from nitration followed by chlorination (reaction 27.29) and chlorination followed by nitration (reaction 27.30). It shows that *the substitution is not random*. The —NO$_2$ group directs Cl to a meta position. Essentially no ortho or para isomer is formed in reaction (27.29). The Cl group, on the other hand, is an ortho,para director. No meta isomer is produced in reaction (27.30).

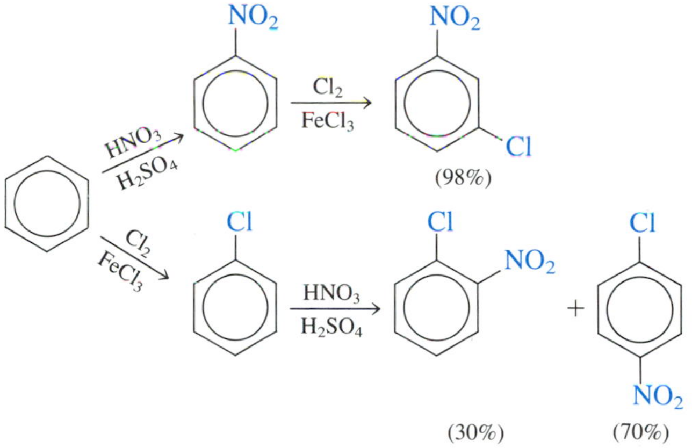

Whether a group is an ortho,para or a meta director depends on how the presence of one substituent alters the electron distribution in the benzene ring. This makes attack by a second group more likely at one type of position than another. Examination of a large number of reactions leads to the following order.

Aromatic substitution reactions: ortho,para and meta directors.

Ortho,para directors: —NH$_2$, —OR, —OH, —OCOR, —R, —X
(from strongest to weakest)

(27.31)

Meta directors:
—NO$_2$, —CN, —SO$_3$H, —CHO, —COR, —COOH, —COOR
(from strongest to weakest)

Example 27-7 _____

Predicting the products of an aromatic substitution reaction. Predict the products of the mononitration of

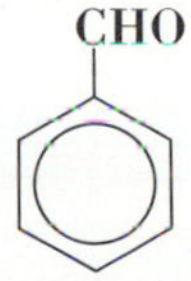

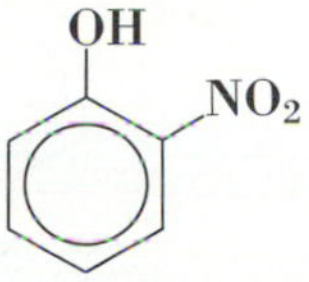

Solution
(a) Since —CHO is a meta director, we should expect the nitration of benzaldehyde (a) to produce 3-nitrobenzaldehyde almost exclusively.

(b) The —OH group is an ortho,para director and we should expect the products

2,6-dinitrophenol   2,4-dinitrophenol

Considering the —NO₂ group as a meta director leads to the same conclusion. Exercise 38 at the end of the chapter illustrates a case where a conclusion is not quite as simple.

SIMILAR EXAMPLES: Exercises 37, 38.

27-5   Alcohols, Phenols, and Ethers

Alcohols and phenols feature the **hydroxyl** group, —OH. If the C atom to which the —OH group is attached also has *two* H atoms (and one R group) bonded to it, the alcohol is a **primary** alcohol. If the C atom has *one* H atom (and two R groups), the alcohol is a **secondary** alcohol. Finally, if there are *no* H atoms on the C atom (and three R groups), the alcohol is a **tertiary** alcohol.

1-butanol   2-butanol   2-methyl-2-propanol
(butyl alcohol)   (*s*-butyl alcohol)   (*t*-butyl alcohol)
(a *primary* alcohol)   (a *secondary* alcohol)   (a *tertiary* alcohol)

A molecule may have more than one —OH group present, and in this case it is called a *polyhydric* alcohol.

1,2-ethanediol   1,2,3-propanetriol
(ethylene glycol)   (glycerol)

In phenols the hydroxyl group is attached to an aromatic ring.

phenol   2,4,6-trinitrophenol   hexachlorophene
(carbolic acid)   (picric acid)

Hexachlorophene, a phenol, was once used as an antiseptic in toothpastes, soaps, and deodorants. However, after laboratory tests revealed that small concentrations of hexachlorophene produce brain damage in rats, its use was restricted to certain medical applications.

Properties of the aliphatic alcohols are strongly influenced by hydrogen bonding. As the chain length increases, the influence of the polar hydroxyl group on the properties of the molecule diminishes. The molecule becomes less like water and more like a hydrocarbon. As a consequence, low-molecular-weight alcohols tend to be water soluble; high-molecular-weight alcohols are not. The boiling points and solubilities of the phenols vary widely, depending on the nature of the other substituents on the benzene ring.

Preparation and Uses. Alcohols can be obtained by the hydration of alkenes or the hydrolysis of alkyl halides.

Preparation of an alcohol.

Hydration: $CH_3CH{=}CH_2 + H_2O \xrightarrow{\;H_2SO_4\;} CH_3\overset{\displaystyle OH}{\underset{\displaystyle |}{C}}HCH_3$ (27.32)

      propene   2-propanol
     (propylene)   (isopropyl alcohol)

Hydrolysis: $CH_3CH_2CH_2Br + OH^- \longrightarrow CH_3CH_2CH_2OH + Br^-$ (27.33)

     propyl bromide   propyl alcohol

Methanol is known as wood alcohol because it can be produced by the destructive distillation of wood. This substance is highly toxic and can lead to blindness or death if ingested. Most methanol is manufactured synthetically from carbon monoxide and hydrogen.

$$CO(g) + 2\,H_2(g) \xrightarrow[\substack{200\ atm \\ ZnO,\ Cr_2O_3}]{350\ °C} CH_3OH(g) \qquad\qquad (27.34)$$

Methanol is the most extensively produced alcohol. It ranks about twenty-second among all industrial chemicals. It is used to synthesize other organic chemicals and as a solvent, but potentially its most important use may be as a motor fuel (recall Section 7-11).

Ethanol, CH_3CH_2OH, is grain alcohol—the common "alcohol" to the layperson. It is obtainable by the fermentation of blackstrap molasses, the residue from the purification of sugar cane, or from other materials containing natural sugars. The principal synthetic method is the hydration of ethylene with sulfuric acid (similar to reaction 27-32).

Ethylene glycol, CH_2OHCH_2OH, is water soluble and has a higher boiling point (197 °C) than water. Because of these properties it makes an excellent permanent nonvolatile antifreeze for use in automobile radiators. It is also used in the manufacture of solvents, paint removers, and plasticizers (softeners). Propylene glycol, $CH_3CHOHCH_2OH$, is used in suntan lotions and, in conjunction with propellants, to produce nonaqueous foams in aerosol products.

Glycerol (glycerin), $CH_2OHCHOHCH_2OH$, is obtained commercially as a by-product in the manufacture of soap. It is a sweet, syrupy liquid that is miscible with water in all proportions. Because it has the ability to take up moisture from the air, it can be used to keep skin moist and soft, accounting for its use in lotions and cosmetics. It is also used to maintain the moisture content of tobacco and candy.

Reactions of the OH Group. The reactivity of the —OH group results from (1) the unshared electron pairs on the O atom, making the molecule basic in the Lewis sense, or (2) the polarity of the O—H bond, causing the molecule to act as a proton donor, that is, to be acidic in the Brønsted–Lowry sense. Reactions (27.35) and (27.36) are based on the Lewis base properties of an alcohol, and reaction (27.37) on the Brønsted–Lowry acid properties.

$$CH_3\underset{\displaystyle \underset{\displaystyle CH_3}{|}}{C}H{-}\overset{..}{\underset{..}{O}}H + CH_3\overset{\displaystyle O}{\overset{\displaystyle \|}{C}}{-}Cl \longrightarrow CH_3\overset{\displaystyle O}{\overset{\displaystyle \|}{C}}OCH(CH_3)_2 + HCl \qquad (27.35)$$

isopropyl alcohol   acetyl chloride   isopropyl acetate
      (an acid halide)   (an ester)

$$
\begin{array}{l}
CH_2-OH \\
| \\
CH-OH \;+\; 3\,HONO_2 \longrightarrow 3\,H_2O \;+ \\
| \\
CH_2-OH
\end{array}
\qquad
\begin{array}{l}
CH_2ONO_2 \\
| \\
CHONO_2 \\
| \\
CH_2ONO_2
\end{array}
\qquad (27.36)
$$

glycerol   glyceryl trinitrate (nitroglycerin)

$$
CH_3-O-H \xrightleftharpoons{\;NaOH\;} CH_3O^-Na^+ + H_2O \qquad (27.37)
$$
$K_a = 1 \times 10^{-16}$

Although commonly used, nitroglycerin is an incorrect name for the structure in (27.36). The compound is an ester and should be so named (glyceryl trinitrate). A nitro compound has a $-NO_2$ group bonded directly to a carbon atom, as in nitrobenzene.

Ethers. Ethers are compounds with the general formula R—O—R. Structurally, they can be pure aliphatic, pure aromatic, or mixed.

$$CH_3-O-CH_3$$

dimethyl ether   diphenyl ether   methyl phenyl ether (anisole)

Ethers can be prepared by the elimination of water from between two alcohol molecules using a strong dehydrating agent, such as concentrated H_2SO_4.

$$
CH_3CH_2OH + HOCH_2CH_3 \xrightarrow[\text{conc.}]{H_2SO_4} CH_3CH_2OCH_2CH_3 + H_2O \qquad (27.38)
$$
diethyl ether

Chemically, the most notable property of ethers is their comparative lack of reactivity. The ether linkage is stable to most oxidizing and reducing agents and to action by dilute acids and alkalis.

Diethyl ether has been used extensively as a general anesthetic. It is easy to administer and produces excellent relaxation of the muscles. Also, the pulse rate, rate of respiration, and blood pressure are affected only slightly. However, it is somewhat irritating to the respiratory passages and produces a nauseous aftereffect. More recently methyl propyl ether (neothyl) has been used, and it is reported to be less irritating. Methyl ether, a gas at room temperatures, is used as a propellant for aerosol sprays. Higher molecular weight ethers have found extensive use as solvents for varnishes and lacquers.

27-6   Aldehydes and Ketones

Aldehydes and ketones contain the carbonyl group.

$$\begin{array}{c} \diagdown \\ \diagup \end{array} C=O$$

If both groups attached to the carbonyl group are carbon chains, the compound is called a **ketone.** If one of the groups is a carbon chain and the other is a hydrogen atom, the compound is called an **aldehyde.**

methanal (formaldehyde)   3-chlorobutanal (β-chlorobutyraldehyde)   benzaldehyde

$$CH_3-\overset{\displaystyle O}{\overset{\|}{C}}-CH_3 \qquad CH_3CH_2-\overset{\displaystyle O}{\overset{\|}{C}}-CH_2CH_3 \qquad \underset{\displaystyle}{\bigcirc}-\overset{\displaystyle O}{\overset{\|}{C}}-CH_3$$

propanone   3-pentanone   methyl phenyl ketone
(acetone)   (diethyl ketone)   (acetophenone)

Preparation and Uses. We can produce an aldehyde by the partial oxidation of a *primary* alcohol. Further oxidation yields a carboxylic acid (Section 27-7). This sequence of oxidations is suggested through the molecular models in Figure 27-7. Oxidation of a *secondary* alcohol produces a ketone.

Oxidation of primary and secondary alcohols.

$$CH_3CH_2OH \xrightarrow[H^+]{Cr_2O_7^{2-}} CH_3CHO \xrightarrow[H^+]{Cr_2O_7^{2-}} CH_3CO_2H \qquad (27.39)$$

ethanol   acetaldehyde   acetic acid
(a primary alcohol)   (an aldehyde)   (an acid)

$$CH_3CHOHCH_3 \xrightarrow[H^+]{Cr_2O_7^{2-}} CH_3\overset{\displaystyle O}{\overset{\|}{C}}CH_3 \qquad (27.40)$$

2-propanol   propanone
(a secondary alcohol)   (a ketone)

You can write a balanced equation for reaction (27.40) with the half-reaction method introduced in Chapter 5. 2-Propanol is oxidized to propanone, and $Cr_2O_7^{2-}$ is reduced to Cr^{3+} (see Exercise 11).

Formaldehyde, a colorless gas, dissolves readily in water. A 40% solution in water, called formalin, is used as an embalming fluid and a tissue preservative. Formaldehyde is also used in the manufacture of synthetic resins. A polymer of formaldehyde, called paraformaldehyde, is used as an antiseptic and an insecticide. Acetaldehyde is an important raw material for the production of acetic acid, acetic anhydride, and the ester, ethyl acetate.

Acetone is the most important of the ketones. It is a volatile liquid (boiling point, 56 °C) and highly flammable. Acetone is a good solvent for a variety of organic compounds, and because of this it is widely used in solvents for varnishes, lacquers, and plastics. Unlike many common organic solvents, acetone is miscible with water in all proportions. This property, combined with its volatility, makes acetone a useful drying agent for laboratory glassware. Residual water is removed through several rinses with acetone, and the remaining liquid acetone film quickly evaporates. One method of producing acetone involves the *dehydrogenation* of isopropyl alcohol in the presence of a copper catalyst.

$$CH_3-\overset{\displaystyle OH}{\overset{|}{CH}}-CH_3 \xrightarrow[300\ °C]{Cu} CH_3-\overset{\displaystyle O}{\overset{\|}{C}}-CH_3 + H_2 \qquad (27.41)$$

Aldehydes and ketones occur widely in nature. Typical natural sources are

$$\underset{\displaystyle}{\bigcirc}-CHO \qquad \underset{\displaystyle}{\bigcirc}-\overset{\displaystyle H\ \ H}{\overset{|\ \ \ |}{C}}=C-CHO \qquad H_3CO-\underset{\displaystyle HO}{\bigcirc}-CHO$$

benzaldehyde   cinnamaldehyde   vanillin
(almonds)   (cinnamon)   (vanilla)

FIGURE 27-7
Oxidation of ethanol.

These ball-and-stick models suggest the changes that occur as ethanol (left) is oxidized, first to acetaldehyde (center) and then to acetic acid (right) (see equation 27.39). [Carey B. Van Loon]

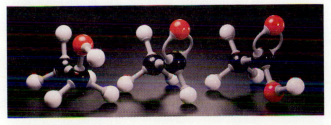

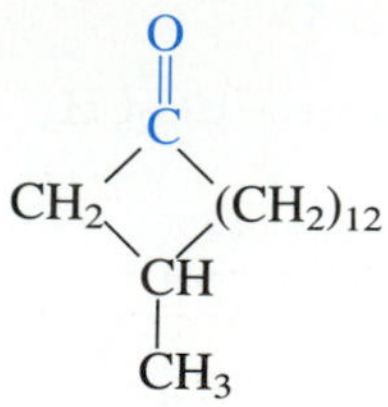

muscone
(obtained from musk deer;
used in perfumes)

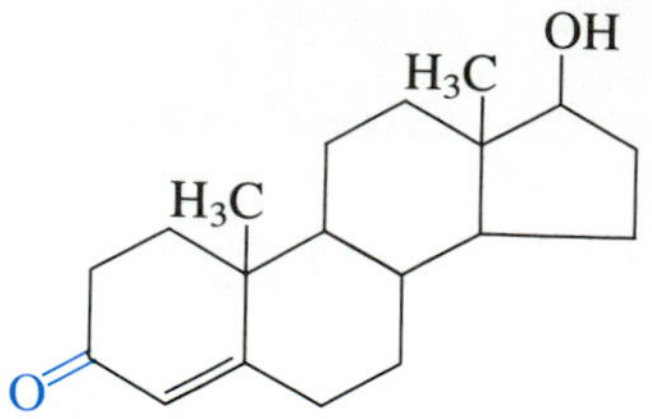

testosterone
(male sex hormone)

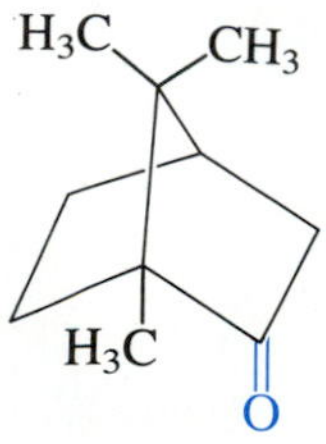

camphor
(obtained from
camphor tree)

27-7 Carboxylic Acids and Their Derivatives

The carboxyl group is also represented as —COOH and —CO₂H.

Compounds that contain the carboxyl group (*carb*onyl and hydr*oxyl*)

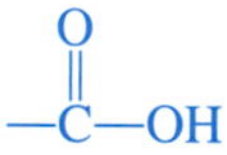

are called carboxylic acids; they have the general formula R—CO_2H. Many compounds are known where R is an aliphatic residue. These are called fatty acids since compounds of this type are readily available from naturally occurring fats and oils. The carboxyl group can also be found attached to the benzene ring. If two carboxyl groups are found on the same molecule the acid is called a dicarboxylic acid.

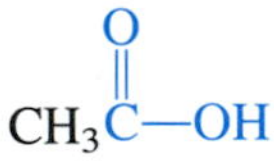

acetic acid
(an aliphatic acid)

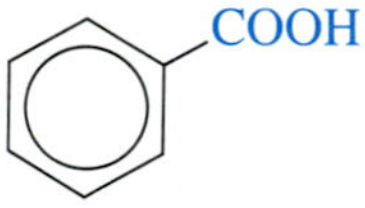

benzoic acid
(an aromatic acid)

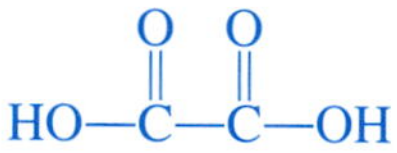

oxalic acid
(an aliphatic
dicarboxylic acid)

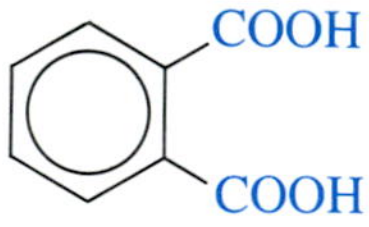

phthalic acid
(an aromatic
dicarboxylic acid)

The carboxylic acids have widely used common names as well as systematic names. Some examples are given in Table 27-3.

Substituted aliphatic acids can be named either by their IUPAC names or by using Greek letters in conjunction with common names. Aromatic acids are named as derivatives of benzoic acid.

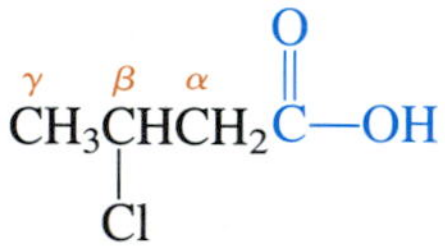

3-chlorobutanoic acid
β-chlorobutyric acid

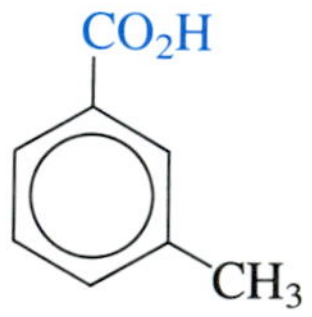

3-methylbenzoic acid
m-methylbenzoic acid
(also *m*-toluic acid)

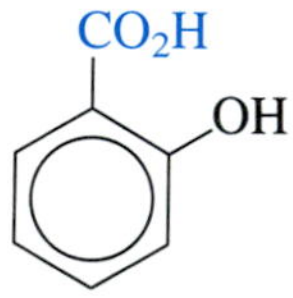

2-hydroxybenzoic acid
o-hydroxybenzoic acid
(also salicyclic acid)

o-Hydroxybenzoic acid or salicylic acid occurs in nature in the willow tree (genus *Salix*). The acetyl derivative of this acid is aspirin, an analgesic (pain killer) and antipyretic (fever reducer).

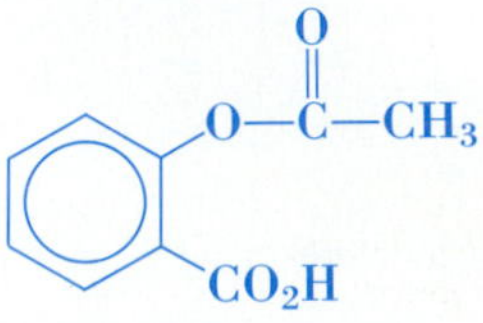

acetylsalicylic acid
aspirin

Because many derivatives of the carboxylic acids involve simple replacement of the hydroxyl groups, special names have been developed for the remaining portion

of the molecule, R—C—. The group —COR is given the general name **acyl**. Some specific examples of its use are

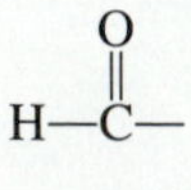

formyl

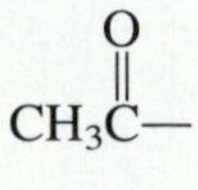

acetyl

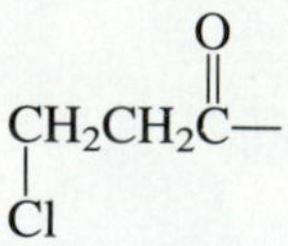

β-chloropropionyl

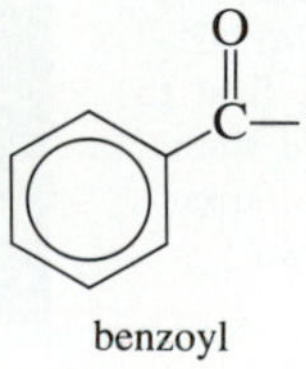

benzoyl

TABLE 27-3
Some Common Carboxylic Acids

Structural formula	Common name (——— acid)	IUPAC name (——— acid)	K_a
HCO_2H	formic	methanoic	1.78×10^{-4}
CH_3CO_2H	acetic	ethanoic	1.74×10^{-5}
$CH_3CH_2CO_2H$	propionic	propanoic	1.35×10^{-5}
$CH_3(CH_2)_2CO_2H$	butyric	butanoic	1.48×10^{-5}
$CH_3(CH_2)_{16}CO_2H$	stearic	octadecanoic	
$CH_3(CH_2)_7CH\!=\!CH(CH_2)_7CO_2H$	oleic	9-octadecenoic	
C_6H_5COOH	benzoic	benzoic	6.4×10^{-5}
$(NO_2)C_6H_4COOH$	p-nitrobenzoic	4-nitrobenzoic	3.8×10^{-4}
HO_2CCO_2H	oxalic	ethanedioic	$(K_{a_1})3.5 \times 10^{-2}$
			$(K_{a_2})6.1 \times 10^{-5}$

Preparation. Two methods for the preparation of the carboxylic acids are illustrated through equations (27.42) and (27.43).

Preparation of a carboxylic acid.

Oxidation of an alcohol:

$$CH_3CH_2OH \xrightarrow[\text{OH}^-,\ \text{heat}]{\text{KMnO}_4} CH_3CO_2K \xrightarrow{\text{H}^+} CH_3CO_2H + K^+ \qquad (27.42)$$

Oxidation of an aldehyde:

$$CH_3CH_2CHO \xrightarrow[\text{OH}^-,\ \text{heat}]{\text{KMnO}_4} CH_3CH_2CO_2K \xrightarrow{\text{H}^+} CH_3CH_2CO_2H + K^+ \qquad (27.43)$$

Reactions of the Carboxyl Group. The carboxyl group displays the chemistry of both the carbonyl and the hydroxyl group. Donation of a proton to a base leads to salt formation. The sodium and potassium salts of long-chain fatty acids are known as **soaps,** for example, sodium stearate (see Chapter 28).

Heating the ammonium salt of a carboxylic acid causes the elimination of water and the formation of an **amide.** Further loss of water occurs if the amide is heated with a strong dehydrating agent such as P_2O_5. The final product, which contains the group $-C\equiv N$, is called a **nitrile.** These reactions can be reversed by the stepwise treatment with water.

$$\underset{\text{ammonium acetate}}{CH_3-\overset{\overset{\displaystyle O}{\|}}{C}-O^-NH_4^+} \xrightarrow{\text{heat}} \underset{\text{acetamide}}{CH_3-\overset{\overset{\displaystyle O}{\|}}{C}-NH_2} + H_2O \qquad (27.44)$$

$$\text{heat}\ \Big|\ P_2O_5$$

$$\underset{\text{acetonitrile}}{CH_3C\equiv N} + H_2O \qquad (27.45)$$

The product of the reaction of an acid and an alcohol is called an **ester.** It, too, can be viewed as forming by the elimination of H_2O. The mechanism of this reaction is such that the $-OH$ of the resulting water comes from the *acid* and the $-H$ form the *alcohol.*

Preparation of an ester.

$$\underset{\substack{\text{methylpropionic acid}\\ \text{(isobutyric acid)}}}{CH_3CH-\overset{\overset{\displaystyle O}{\|}}{C}-OH} + \underset{\substack{\text{1-butanol}\\ \text{(butyl alcohol)}}}{CH_3(CH_2)_3OH} \xrightarrow[\text{heat}]{\text{H}^+} H_2O + \underset{\substack{\text{butyl methylpropionate}\\ \text{(butyl isobutyrate)}}}{CH_3CH-\overset{\overset{\displaystyle O}{\|}}{C}-O(CH_2)_3CH_3}$$
$$\quad\ \ \big|\qquad\qquad\qquad\qquad\qquad\qquad\qquad\qquad\qquad\quad \big|$$
$$\quad\ CH_3 \qquad\qquad\qquad\qquad\qquad\qquad\qquad\qquad\ CH_3$$

$$(27.46)$$

Unlike the pungent odors of the carboxylic acids from which they are derived, esters have very pleasant aromas. The characteristic fragrances of many flowers and fruits can be traced to the esters they contain. Esters are used in perfumes and in the manufacture of flavoring agents for the confectionery and soft drink industries. Most esters are colorless liquids, insoluble in water. Their melting points and boiling points are generally lower than those of alcohols and acids of comparable carbon content. This is because of the absence of hydrogen bonding in esters.

27-8 Amines

Amines are organic derivatives of ammonia in which one or more organic residues (R) are substituted for H atoms. Their classification is based on the number of R groups bonded to the nitrogen atom—one for primary amines, two for secondary, and three for tertiary. This classification scheme is illustrated in Figure 27-8.

Amines of low molecular weight are gases that are readily soluble in water, yielding basic solutions. The volatile members have odors similar to ammonia but more "fishlike." Amines form hydrogen bonds, but these bonds are weaker than are those in water because nitrogen is less electronegative than oxygen. Like ammonia, amines have pyramidal structures with a lone pair of electrons on the N atoms. And also like ammonia, amines owe their basicity to these lone-pair electrons. In aromatic amines, because of unsaturation in the benzene ring, electrons are drawn into the ring and this reduces the electron density on the nitrogen atom. As a result, aromatic amines are *weaker* bases than ammonia. Aliphatic amines are somewhat stronger bases than ammonia. Some properties of amines are listed in Table 27-4.

Dimethylamine is used as an accelerator in the removal of hair from hides in the processing of leather. Butyl- and amylamines are used as antioxidants, corrosion inhibitors, and in the manufacture of oil-soluble soaps. Dimethyl- and trimethylamines are used in the manufacture of ion exchange resins. Additional uses are found in the manufacture of disinfectants, insecticides, herbicides, drugs, dyes, fungicides, soaps, cosmetics, and photographic developers.

Several methods can be used to synthesize amines, but we limit our concern to an especially important one—the reduction of nitro compounds.

Preparation of an amine by the reduction of a nitro compound.

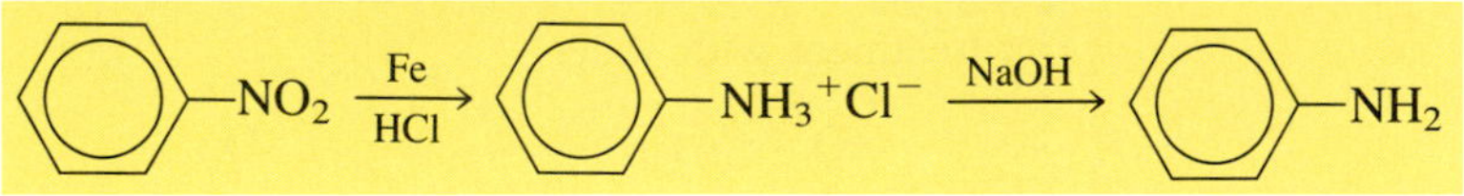

$$\langle\bigcirc\rangle\!-\!NO_2 \xrightarrow[HCl]{Fe} \langle\bigcirc\rangle\!-\!NH_3{}^+Cl^- \xrightarrow{NaOH} \langle\bigcirc\rangle\!-\!NH_2 \qquad (27.47)$$

FIGURE 27-8

A classification scheme for amines.

Amines are derivatives of ammonia. Quaternary salts are analogous to ammonium salts.

$:NH_3$
ammonia

$NH_4{}^+\,X^-$
ammonium salt

$R\ddot{N}H_2$
primary amine

$R_2\ddot{N}H$
secondary amine

$R_3N:$
tertiary amine

$R_4N^+X^-$
quaternary salt

$CH_3CH_2NH_2$
ethylamine

$CH_3\overset{H}{\underset{|}{N}}CH_2CH_3$
ethylmethylamine

$CH_3\overset{CH_2CH_3}{\underset{|}{N}}CH_2CH_2CH_3$
ethylmethylpropylamine

$(CH_3)_4N^+\,Cl^-$
tetramethyl-ammonium chloride

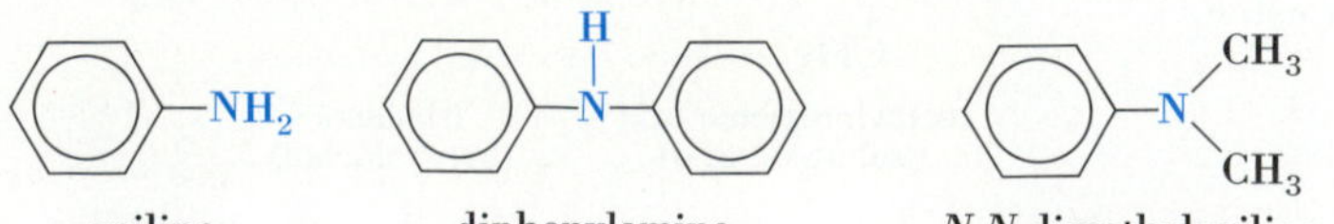

aniline

diphenylamine

N,N-dimethylaniline

TABLE 27-4
Some Properties of Selected Amines

Name	Formula	Boiling point, °C	K_b
ammonia	NH_3	-33.4	1.8×10^{-5}
methylamine	CH_3NH_2	-6.5	44×10^{-5}
ethylamine	$CH_3CH_2NH_2$	16.6	47×10^{-5}
butylamine	$CH_3CH_2CH_2CH_2NH_2$	77.8	40×10^{-5}
aniline	$C_6H_5NH_2$	184	4.2×10^{-10}
N-methylaniline[a]	$C_6H_5NHCH_3$	196	7.1×10^{-10}

[a]The designation "N" in N-methylaniline signifies that the methyl group is attached to the N atom and not the benzene ring.

27-9 Heterocyclic Compounds

In the ring structures considered to this point, all the ring atoms have been carbon; these structures are said to be carbocyclic. There are many compounds, however, both natural and synthetic, in which one or more of the atoms in a ring structure is not carbon. These ring structures are said to be **heterocyclic.** The heterocylic systems most commonly encountered contain N, O, and S atoms, and the rings are of various sizes.

Pyridine is a nitrogen analog of benzene (see Figure 27-9), but unlike benzene it is water soluble and it has basic properties (the unshared pair of electrons on the N atom is not part of the π electron cloud of the ring system). Pyridine was once obtained exclusively from coal tar, but it is now used so extensively that several synthetic methods have been developed for its production. It is a liquid with a disagreeable odor used in the production of pharmaceuticals such as sulfa drugs and antihistamines, as a denaturant for ethyl alcohol, as a solvent for organic chemicals, and in the preparation of waterproofing agents for textiles. We consider a number of other examples of heterocyclic compounds in Chapter 28.

FIGURE 27-9
Pyridine.

In the pyridine molecule a nitrogen atom replaces one of the carbon atoms in benzene. Its formula is C_5H_5N.
(a) Structural formula.
(b) Space-filling model.

(a)

(b)

27-10 Synthesis of Organic Compounds

Originally, all organic compounds were isolated from natural sources. However, as chemists developed an understanding of the chemical behavior of organic compounds, they began to devise methods of *synthesizing* compounds from simple starting materials. And some of these compounds had never been observed to occur naturally. Now, organic synthesis is perhaps the most important aspect of organic chemistry. The objective of an organic synthesis may be to determine a way to produce an organic chemical more cheaply than it can be extracted from a natural source. Or it may be to synthesize a compound never observed before, perhaps for its potential medicinal uses.

The approach that a synthetic organic chemist takes is to apply a knowledge of a wide variety of reaction types and of reaction mechanisms to devise schemes for assembling simple molecules into more complex structures. A simple example follows.

Example 27-8

Devising an organic synthesis. Ethyl acetate, $CH_3CO_2CH_2CH_3$, is an important solvent used in paints, lacquers and plastics. It is also used in organic synthesis and in formulating synthetic fruit essences. Devise a synthesis for ethyl acetate using only inorganic substances and carbon (coke) as starting materials.

Solution. The compound in question is an ester. In equation (27.46) we noted that acids react with alcohols to produce esters. Our problem is to devise syntheses for *ethyl alcohol* and *acetic acid*. Recall that

$$CaCO_3 \xrightarrow{\text{heat}} CaO + CO_2(g) \tag{27.11}$$

$$CaO + 3\ C \xrightarrow[\text{2000 °C}]{\text{electric furnace}} CaC_2 + CO(g) \tag{27.12}$$

$$CaC_2 + 2\ H_2O \longrightarrow HC\equiv CH + Ca(OH)_2 \tag{27.13}$$

Ethylene is produced by adding H_2 to C_2H_2,

$$HC\equiv CH + H_2 \xrightarrow[\text{heat/pressure}]{\text{Pt or Pd}} H_2C{=}CH_2$$

Ethanol is obtained by the addition of H_2O to C_2H_4.

$$H_2C{=}CH_2 + H_2O \xrightarrow{H_2SO_4} CH_3CH_2OH$$

Some of the ethanol is oxidized to acetic acid.

$$CH_3CH_2OH \xrightarrow[H^+]{K_2Cr_2O_7} CH_3CO_2H$$

Finally, ethanol and acetic acid are combined to form ethyl acetate.

$$CH_3COOH + HOCH_2CH_3 \longrightarrow H_2O + CH_3{-}\overset{\overset{\textstyle O}{\|}}{C}{-}O{-}CH_2CH_3$$

SIMILAR EXAMPLES: Exercises 46, 47.

27-11   Raw Materials for the Organic Chemical Industry

In Example 27-8 we demonstrated that an organic compound may be synthesized largely from inorganic substances. However, we needed a source of carbon and we chose coke (derived from coal). The principal sources of carbon for the industrial synthesis of organic chemicals are petroleum, coal, and plant products. Currently, petroleum is the most important of these three sources, although there is now renewed interest in coal because reserves of coal are much more plentiful than those of petroleum. Plant products (biomass) may develop as an important future source.

TABLE 27-5
Coal Tar Fractions

Boiling range, °C	Name	Tar, mass %	Primary constituents
below 200	light oil	5	benzene, toluene, xylenes
200–250	middle oil (carbolic oil)	17	naphthalene, phenol, pyridine
250–300	heavy oil (creosote oil)	7	naphthalenes and methylnaphthalenes, cresols, quinoline
300–350	green oil	9	anthracene, carbazole
residue	—	62	pitch or tar

Coal. Coal is an organic, rocklike material with a high ratio of carbon to hydrogen and other elements. (One proposed formula for a "molecule" of bituminous coal is $C_{153}H_{115}N_3O_{13}S_2$.) To synthesize hydrocarbons or other desired organic compounds from coal requires decreasing the C/H ratio.

In the method of **pyrolysis,** coal (usually bituminous coal) is heated to a high temperature (350 to 1000 °C) in the absence of air. Volatile products are formed and an impure carbon residue called **coke** remains. Condensation of the volatile products of this destructive distillation yields black viscous **coal tar.**

$$coal \xrightarrow[\substack{(absence \\ of\ air)}]{heat} coke + coal\ tar + coal\ gas$$

One ton of bituminous coal yields about 1500 lb of coke, 8 gal of coal tar, and 10,000 ft^3 of coal gas. Coal gas is a mixture of H_2, CH_4, CO, C_2H_6, NH_3, CO_2, H_2S, and other components. At one time coal gas was used as a fuel. Coal tar can be distilled to yield the fractions listed in Table 27-5. From these fractions, in turn, other organic chemicals can be produced.

We can think of pyrolysis as a carbon-removal process. Coke is formed and the remaining products are correspondingly enriched in hydrogen and other elements. Coal gasification or liquefication schemes involve the addition of hydrogen (and usually also oxygen). In general these schemes are based on chemical reactions that have been known for 75 years or more, but they have been updated by new technology, particularly new catalyst systems. One approach, for example, is to burn a coal–water slurry to obtain a mixture of CO(g) and H_2(g). This gaseous mixture is converted to methanol, and with the proper catalysts, the methanol is converted to acetic acid. Heat evolved in burning the coal is used to meet heat requirements in other parts of the process. Sulfur is removed from the coal and converted to H_2SO_4(aq). The process is nonpolluting, energy efficient, and produces only the desired end products (together with some CO_2) from coal and water as starting materials.

Petroleum. The principal constituents of crude oil are aliphatic hydrocarbons. Certain low-molecular-weight hydrocarbons are found dissolved in crude oil or are produced in the manufacture of gasoline. These compounds are removed and compressed into liquid form in cylinders. Propane and butane sold in this form are known as **liquefied petroleum gas (LPG).**

Crude oil is indeed a complex mixture. It has been estimated that in petroleum boiling up to 200 °C there are at least 500 compounds, some aliphatic, some alicyclic, and some aromatic. Petroleum is refined by distillation into various fractions. A typical fractionation yields the products listed in Table 27-6.

That familiar road-paving material, asphalt, is the final residue in the fractional distillation of petroleum. [Phil Degginger]

TABLE 27-6
Principal Petroleum Fractions

Boiling range, °C	Composition	Fractions	Uses
0–30	C_1–C_4	gas	gaseous fuel
30–60	C_5–C_7	petroleum ether	solvents
60–100	C_6–C_8	ligroin	solvents
70–150	C_6–C_9	gasoline	motor fuel
175–300	C_{10}–C_{16}	kerosene	jet fuel, diesel oil
over 300	C_{16}–C_{18}	gas–oil	diesel fuel, cracking stock
—	C_{18}–C_{20}	wax–oil	lubricating oil, mineral oil, cracking stock
—	C_{21}–C_{40}	paraffin wax	candles, wax paper
—	above C_{40} plus C	residuum	roofing tar, road materials, waterproofing

Fuel production. Not all of the gasoline components listed in Table 27-6 are equally desirable as fuels. Some of them burn more smoothly than others. (Explosive burning results in engine "knocking.") The octane hydrocarbon, **2,2,4-trimethylpentane,** has excellent engine performance; it is given an octane rating of 100. **n-Heptane** has poor engine performance; its octane rating is set at 0. These two hydrocarbons serve as a basis for establishing the quality of automotive fuels, which are mixtures of a large number of hydrocarbons. In general, branched chain hydrocarbons have higher octane numbers than their straight chain counterparts.

$$CH_3-\underset{\underset{CH_3}{|}}{\overset{\overset{CH_3}{|}}{C}}-CH_2-\underset{\underset{H}{|}}{\overset{\overset{CH_3}{|}}{C}}-CH_3 \qquad CH_3-CH_2-CH_2-CH_2-CH_2-CH_2-CH_3$$

2,2,4-trimethylpentane
(isooctane)
octane rating: 100

n-heptane

octane rating: 0

A catalytic cracking unit ("cat cracker") in a petroleum refinery. [E. R. Degginger]

Gasoline obtained by the fractional distillation of petroleum has an octane number of 50 to 55 and is not acceptable for use in automobiles. Extensive modifications of its composition are required. The principal methods employed are of three types—thermal and catalytic cracking, reforming, and alkylation. The chemical changes involved are represented by the equations in Figure 27-10.

In **thermal cracking** large hydrocarbon molecules are broken down into molecules in the gasoline range. The presence of special catalysts promotes the production of branched chain hydrocarbons. The process known as **reforming** or isomerization converts straight chain to branched chain hydrocarbons. Also, alicyclic hydrocarbons are converted to aromatic hydrocarbons, which possess higher octane numbers. In thermal and catalytic cracking, some of the products are low molecular weight, unsaturated hydrocarbons or olefins. In the **alkylation** process these unsaturated compounds are polymerized to higher molecular weight olefins. These can be used directly as fuel components or hydrogenated to produce saturated hydrocarbons.

The octane rating of gasoline is further improved by adding certain "antiknock" compounds to prevent premature combustion. Most widely used in the past were tetraethyllead, $(C_2H_5)_4Pb$, and tertamethyllead, $(CH_3)_4Pb$. For example, the addition of 6 mL of tetraethyllead to 1 gal of 2,2,4-trimethylpentane raises the octane rating from 100 to 120.3. Because of the environmental hazards posed by lead (recall Section 22-5), lead additives have mostly been phased out of use in the United States. A class of compounds that can substitute for lead additives in gasoline are oxygenated hydrocarbons. Methanol and ethanol raise the octane rating of gasoline, but probably the most cost effective of the oxygenated hydrocarbons is methyl *t*-butyl ether (MTBE).

Petrochemicals. Although the principal products of the petroleum industry are fuels, chemicals produced from petroleum—petrochemicals—are essential to modern society. Current annual production of benzene in the United States exceeds 11 billion lb. Over 90% of this is produced from petroleum. The process involves cyclization and dehydrogenation of hexane to the aromatic hydrocarbon. Of the petroleum-produced benzene, about 40% is used to manufacture ethylbenzene for the production of styrene plastics, 18% to manufacture phenol, 6% to synthesize dodecylbenzene (for detergents), and 2% to make aniline. The production of aromatic compounds by dehydrogenation of alkanes yields large amounts of hydrogen gas. An important use of this hydrogen is in the Haber–Bosch synthesis of ammonia.

FIGURE 27-10
Some reactions associated with the production of gasoline.

(a) Cracking

$$C_{15}H_{32} \xrightarrow[\text{catalyst}]{\text{heat}} C_8H_{18} + C_7H_{14}$$

$$C_8H_{18} \xrightarrow[\text{catalyst}]{\text{heat}} \underset{\text{butane}}{CH_3CH_2CH_2CH_3} + \underset{\text{2-butene}}{CH_3CH{=}CHCH_3}$$

(b) Reforming

$$n\text{-}C_4H_{10} \xrightarrow[\text{catalyst}]{\text{heat}} \text{iso-}C_4H_{10} \text{ or } C_4H_8 + H_2$$

$$CH_3CH_2CH{=}CH_2 \xrightarrow[\text{catalyst}]{\text{heat}} \underset{\text{isobutene}}{CH_3\overset{\overset{\displaystyle CH_3}{|}}{C}{=}CH_2}$$
1-butene

(c) Alkylation

$$\underset{\text{isobutane}}{CH_3\underset{\underset{\displaystyle CH_3}{|}}{\overset{\overset{\displaystyle CH_3}{|}}{CH}}} + \underset{\text{isobutene}}{CH_2{=}\overset{\overset{\displaystyle CH_3}{|}}{C}CH_3} \xrightarrow[\text{catalyst}]{\text{heat}} CH_3\overset{\overset{\displaystyle CH_3}{|}}{\underset{\underset{\displaystyle CH_3}{|}}{C}}{-}CH_2{-}\overset{\overset{\displaystyle CH_3}{|}}{\underset{\underset{\displaystyle H}{|}}{C}}CH_3$$

2,2,4-trimethylpentane
(isooctane)

FOCUS ON
Polymerization Reactions

Engineers monitoring the production of Saran Wrap, a plastic film used for wrapping food. Sarans are a type of plastic having vinylidine chloride ($CH_2=CCl_2$) as their principal monomers. [Courtesy Dow Chemical USA]

The Focus feature at the end of Chapter 10 dealt with some fundamental ideas about polymers, and throughout the text we have encountered specific examples of polymeric materials. Here we consider in somewhat more detail the types of reactions that are employed to produce polymers.

Chain-Reaction Polymerization. This is the type of polymerization reaction typically encountered with monomers that have carbon-to-carbon double bonds. The net result is that the double bonds "open up" and monomer units add to growing chains. As with other chain reactions the mechanism involves three characteristic steps: initiation, propagation, and termination.

The formation of polystyrene is illustrated in Figure

27-11. In this polymerization a small amount of benzoyl peroxide is present as an initiator. A molecule of this substance decomposes at 70 °C to form two benzoyloxy radicals, which in turn lose CO_2 molecules to become phenyl radicals. A phenyl radical adds to a molecule of styrene to produce the more stable diphenylethane radical, which in turn adds another styrene molecule, and so on. Termination of the chain occurs when two free radicals combine.

Because it is initiated by a free radical, the polymerization reaction in Figure 27-11 is sometimes called a free-radical addition polymerization. However, chain-reaction polymerization can also be initiated by cationic and anionic species or by a coordination complex. Chain-reaction polymerization tends to form high molecular weight polymers (mol. wt. up to 10^7) by rapid exothermic reactions. Some typical polymers produced by chain-reaction polymerization are listed in Table 27-7.

Copolymers. One of the polymers listed in Table 27-7, SBR rubber, consists of two different types of monomers—1,3-butadiene and styrene. It is called a copolymer. Many synthetic polymers are copolymers, and it is possible to control the formation of copolymers so that the different monomers (noted as X and Y below) are joined in different patterns.

Random:

X—Y—Y—X—Y—X—X—X—Y—X—Y—Y

Block:

X—X—X—X—Y—Y—Y—Y—X—X—X—X

Alternating:

X—Y—X—Y—X—Y—X—Y—X—Y—X—Y

Graft:

$$
\begin{array}{c}
\qquad\qquad Y \\
\qquad\qquad | \\
\qquad\qquad Y \\
\qquad\qquad | \\
X—X—X—X—X—X—X—X—X—X—X \\
|\qquad\quad|\qquad\qquad\qquad\quad| \\
Y\qquad\quad Y\qquad\qquad\qquad\quad Y \\
|\qquad\quad|\qquad\qquad\qquad\quad| \\
Y\qquad\quad Y\qquad\qquad\qquad\quad Y \\
|\qquad\qquad\qquad\qquad\qquad\quad| \\
Y\qquad\qquad\qquad\qquad\qquad\quad Y \\
\qquad\qquad\qquad\qquad\qquad\qquad| \\
\qquad\qquad\qquad\qquad\qquad\qquad Y
\end{array}
$$

Step-Reaction Polymerization. In this type of polymerization the monomers typically have two or more functional groups that undergo a reaction that joins the two monomers together. Usually this involves the elimination

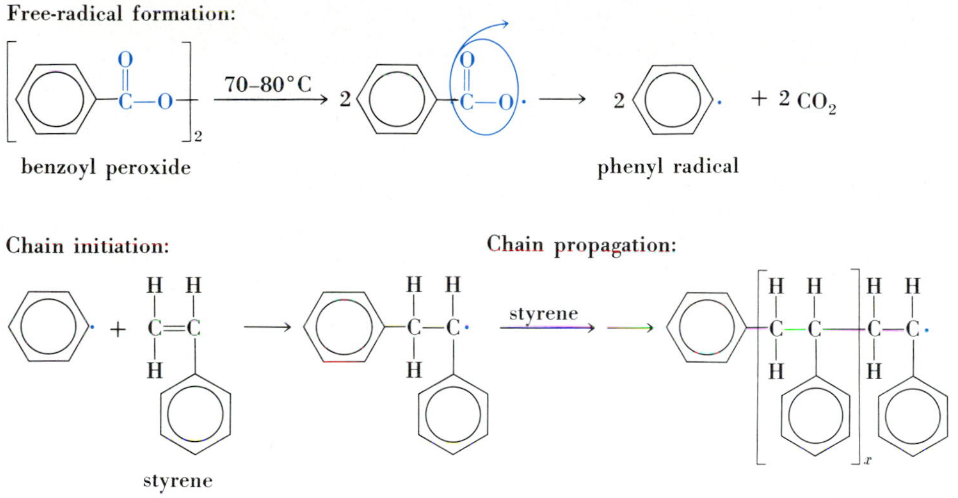

Free-radical formation:

benzoyl peroxide   phenyl radical

Chain initiation: Chain propagation:

styrene

The method of chain termination illustrated here—the joining of two free-radical chains—is just one of several possibilities. Termination will also occur as a result of the reaction of a chain with a radical initiator (phenyl radical).

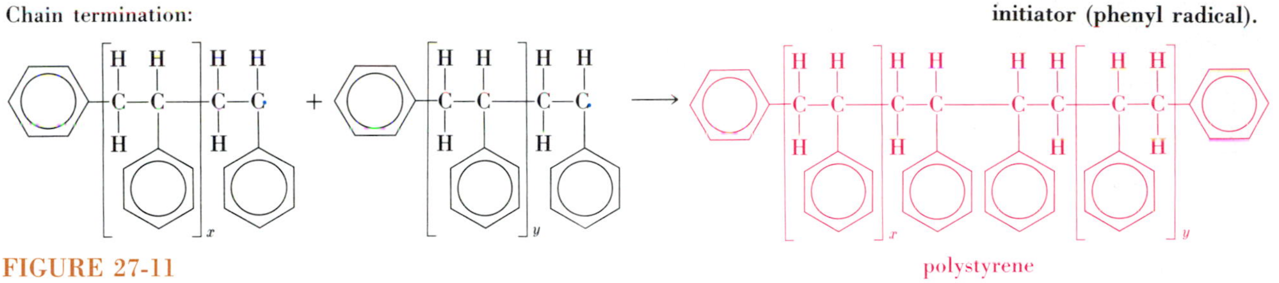

Chain termination:

FIGURE 27-11   polystyrene
Chain-reaction polymerization—formation of polystyrene.

TABLE 27-7
Some Polymers Produced by Chain-Reaction Polymerization

Name	Monomer(s)	Polymer
polyethylene	$CH_2{=}CH_2$	$-[CH_2-CH_2]_n$
polypropylene	$CH_2{=}CHCH_3$	$-[CH_2-\underset{CH_3}{CH}]_n$
poly(vinyl chloride)	$CH_2{=}CHCl$	$-[CH_2-\underset{Cl}{CH}]_n$
polyacrylonitrile	$CH_2{=}CHCN$	$-[CH_2-\underset{CN}{CH}]_n$
polystyrene	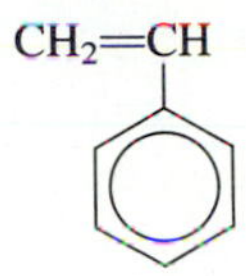 $CH_2{=}CH$	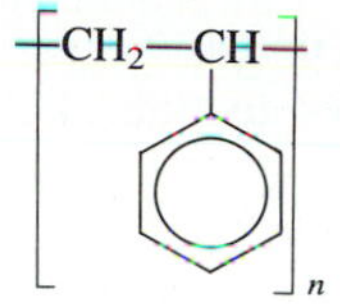 $-[CH_2-CH]_n$
poly(butadiene-co-styrene) SBR, Buna S	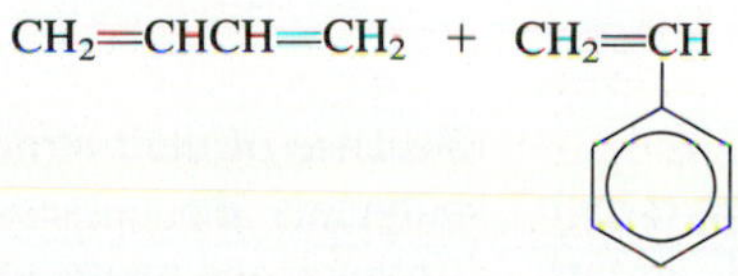 $CH_2{=}CHCH{=}CH_2 + CH_2{=}CH$	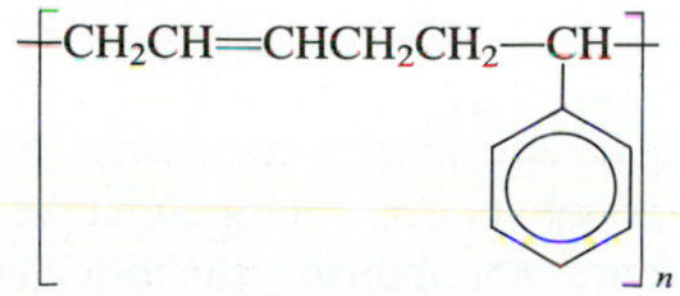 $-[CH_2CH{=}CHCH_2CH_2-CH]_n$

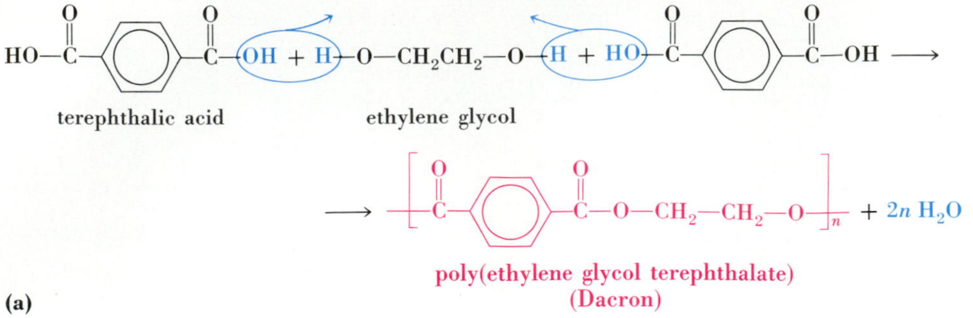

(a)

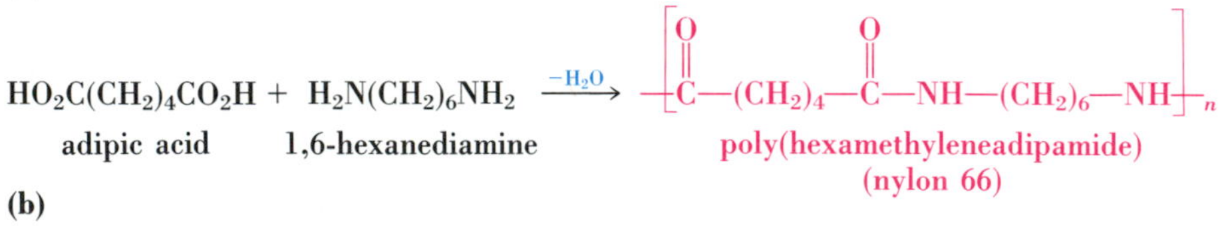

(b)

FIGURE 27-12
Step-reaction polymerization.

(a) Formation of Dacron. The reaction is carried out at elevated temperatures, reduced pressures, and in the presence of sodium methoxide, CH_3ONa.
(b) Formation of nylon 66.

of a small molecule, such as H_2O. This type of polymerization is also referred to as condensation polymerization. Unlike chain-reaction polymerization, where reaction of a monomer can occur only on a growing polymer chain, in step-reaction polymerization any pair of monomers is free to join into a dimer; a dimer may join with a monomer to form a trimer; two dimers may join into a tetramer; etc. Step-reaction polymerization tends to occur slowly and to produce polymers of only moderately high molecular weights (mol. wt. less than about 10^5). The formation of Dacron and nylon 66 are illustrated in Figure 27-12.

Stereospecific Polymers. The physical properties of a polymer are determined by a number of factors, such as the average length of the polymer chains and the strength of intermolecular forces between chains. Another important factor is whether the polymer chains display any crystallinity. In general, amorphous polymers are glasslike or rubbery. A high-strength fiber, on the other hand, must possess some crystallinity. Crystals, as we learned in Chapter 12, are characterized by long-range order among their structural units.

The usual designation of a polymer, when applied to polypropylene,

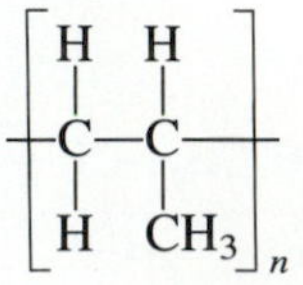

is not very revealing about the structure of the polymer. It does not indicate the orientation of the $-CH_3$ groups along the polymer chain.

If propylene ($CH_3-CH=CH_2$) is polymerized by a method similar to that outlined in Figure 27-11, the orientation of the $-CH_3$ groups on the polymer chain is random. A polymer of this type is called **atactic.** Because the structure of the polymer chains is not regular, atactic polymers are amorphous.

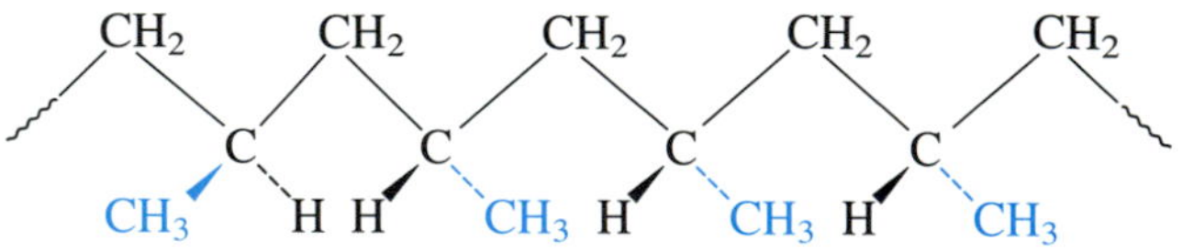

atactic

In the **isotactic** polymer, however, all $-CH_3$ groups have the same orientation, and in the **syndiotactic** polymer the $-CH_3$ groups alternate in their positions along the chain.

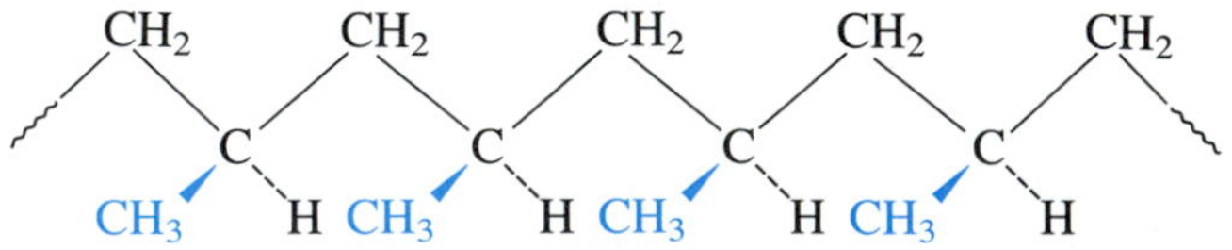

isotactic

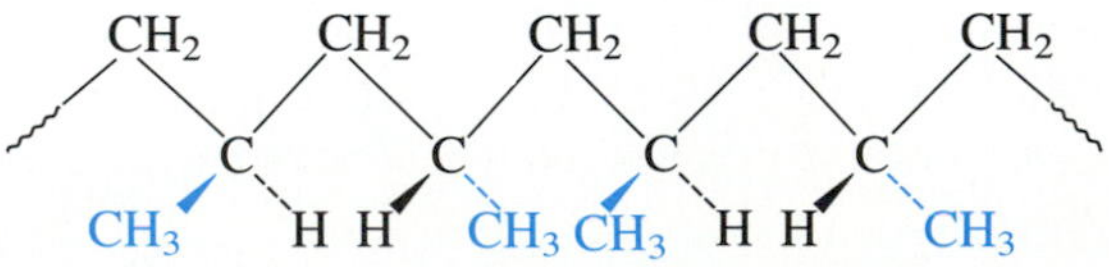

syndiotactic

Because of their structural regularity, these two types of polymers do possess crystallinity, which makes them stronger and more resistant to chemical attack than the atactic polymer.

In the 1950s Karl Ziegler and Giulio Natta developed procedures for controlling the spatial orientation of substituent groups on a polymer chain through the use of special catalysts [e.g., $(CH_3CH_2)_3Al + TiCl_4$]. This discovery revolutionized polymer chemistry, for through stereospecific polymerization it is literally possible to "tailor make" large molecules.

A chemist setting out to build a giant molecule is in the same position as an architect designing a building. He has a number of building blocks of certain shapes and sizes, and his task is to put them together in a structure to serve a particular purpose.

GUILIO NATTA

Summary

Organic chemistry deals with compounds of carbon. Simplest among these are carbon–hydrogen compounds—hydrocarbons. In hydrocarbons the C atoms are bonded to one another in straight or branched chains or in rings; H atoms are bonded to the C atoms. Some hydrocarbon molecules contain only single bonds (alkanes), others have some double bonds (alkenes), and still others, triple bonds (alkynes). Yet another class of hydrocarbons—aromatic hydrocarbons—is based on the benzene molecule, C_6H_6. For a given class of hydrocarbons, physical properties (e.g., melting and boiling points) generally follow a regular pattern with increasing molecular weight.

Greater variety among organic compounds results with the incorporation of certain atoms or groupings of atoms (functional groups) into hydrocarbon structures. These substituent groups are introduced through chemical reactions. With alkanes and aromatic hydrocarbons these reactions are based on *substitution*: A functional group replaces an H atom in the hydrocarbon. With alkenes and alkynes chemical reaction occurs by *addition*: Functional group atoms are joined to the C atoms at points of unsaturation (double or triple bonds).

Isomerism is frequently encountered among organic compounds. One form of isomerism is based on molecules with the same total number of C atoms but with different branching of the carbon chain. Another stems from the different positions on a hydrocarbon chain or ring at which functional groups may be attached. Still other isomers (e.g., cis-trans) arise from different orientations of substituent groups in space.

Among the reactions that can be used for preparing alkanes are the addition of H_2 to an alkene or alkyne and the reaction of an alkyl halide with sodium. Alkenes can be prepared by *elimination* reactions. For example, the elimination of H_2O through the removal of an H atom and an —OH group from adjacent C atoms leaves a double bond between the C atoms. The principal alkyne, acetylene, is produced by the reaction of calcium carbide with water. Other alkynes can be prepared from acetylene based on the acidity of the C—H bond in $HC\equiv CH$.

Compounds of the general formula ROH are alcohols (phenols if R is C_6H_5). Alcohols can be prepared by the *hydration* of an alkene (addition of HOH) or the *hydrolysis* of an alkyl halide. Ethers (R′OR) result from the elimination of HOH from between two alcohol molecules. Aldehydes, RCHO, and ketones, RCOR′, feature the carbonyl group, $\ce{>C=O}$. They can be prepared by the controlled oxidation of alcohols. Carboxylic acids are weak acids having the general formula RCOOH and featuring the carboxyl group, $-C{\overset{\textstyle O}{\underset{\textstyle OH}{}}}$. In addition to their typical acid–base reactions, carboxylic acids react with alcohols to form esters. Other reactions include the formation of amides and nitriles. Carboxylic acids can be prepared by the oxidation of an alcohol or an aldehyde. Amines are organic derivatives of ammonia, and, like ammonia, they have basic properties. They can be prepared by the reduction of nitro compounds.

The substitution of other atoms (such as N, O, or S) for C atoms in ring structures yields heterocyclic compounds. These are widely encountered among molecules of the living state (see Chapter 28).

Organic chemicals can be derived from petroleum, coal, or biomass; petroleum is the chief source. In the manufacture of motor fuels, the composition of petroleum is greatly altered, with extensive use being made of the fundamental reactions of organic chemistry. Increasingly, chemicals obtained from petroleum—petrochemicals—are being used in the manufacture of polymers.

Summarizing Example

Organic qualitative analysis involves the identification of *organic* compounds. The basic approach is to determine the functional groups present through characteristic reactions of these groups. Suppose we are told that a particular colorless organic liquid is one of five possible compounds and asked to perform simple tests described in the text to determine which it is. Structural formulas of the five possible compounds are given below. Note that one compound is an alcohol, two are ethers, one is an aldehyde, and one is a carboxylic acid.

$CH_3CH_2CH_2CH_2OH$ $CH_3CH_2OCH_2CH_3$ $CH_3CH_2CH_2OCH_3$
    1-butanol   diethyl ether   methyl propyl ether
   M.W. 74.12   M.W. 74.12   M.W. 74.12

$CH_3CH_2CH_2CHO$ CH_3CH_2COOH
  butyraldehyde   propionic acid
   M.W. 72.11   M.W. 74.08

1. A molecular weight determination is performed on a 2.50-g sample of the liquid dissolved in 100. g of water. The observed freezing point is -0.7 °C, and the calculated molecular weight is 7×10^1. What conclusion can we draw about the identity of the liquid from this measurement?

Solution. This measurement does not help us identify the liquid. Since we can express the molecular weight with only one significant figure, the measured molecular weight could be that of any of the five liquids. To eliminate any of the five possibilities we would need a much more precise freezing point measurement.

2. A small sample of the liquid is dissolved in water and the aqueous solution does not change the color of blue litmus paper. What conclusion can we draw from this experiment?

Solution. Since the aqueous solution does not change the color of blue litmus paper, the organic liquid cannot have acidic properties. The liquid cannot be propionic acid.

3. When alkaline $KMnO_4(aq)$ is added to the organic liquid and the mixture is heated, the purple color of the MnO_4^- disappears. What does this observation allow us to conclude?

Solution. The reduction of MnO_4^- [to $MnO_2(s)$] accounts for the loss of color. We have learned that both alcohols and aldehydes can be oxidized to a carboxylic acid with alkaline $KMnO_4(aq)$ (equations 27.42 and 27.43) and that ethers are rather inert (page 988). The unknown liquid must be either 1-butanol or butyraldehyde.

4. When the unknown organic liquid is heated with acetic acid in the presence of H_2SO_4 (a catalyst), a product with a pleasant, fruity odor is obtained. Which of the five possible compounds is the unknown organic liquid?

Solution. The reaction product described here is an ester. An ester forms in the reaction of an acid and an alcohol. The unknown liquid is an alcohol; it is 1-butanol.

Key Terms

acetyl group (27-7)	**alkene** (27-3)	**carboxylic acid** (27-7)
acyl group (27-7)	**alkyl** (27-1)	**chain isomer** (27-1)
addition reaction) (27-3)	**alkyne** (27-3)	**condensed formula** (27-1)
alcohols (27-5)	**amide** (27-7)	**conformations** (27-2)
aldehydes (27-6)	**amine** (27-8)	**elimination reaction** (27-3)
alicyclic (27-2)	**aromatic** (27-4)	**ester** (27-7)
aliphatic (27-2)	**benzoyl** (27-7)	**ether** (27-5)
alkane (27-2)	**carboxyl group** (27-7)	**formyl** (27-7)

functional group (27-1)
heterocyclic (27-9)
homologous series (27-2)
hydrocarbon (27-1)
hydroxyl groups (27-5)
IUPAC (IUC) (27-1)
ketone (27-6)
Markovnikov's rule (27-3)

meta (*m*) isomer (27-4)
nitrile (27-7)
ortho (*o*) isomer (27-4)
para (*p*) isomer (27-4)
phenol (27-5)
phenyl group (27-4)
polycyclic aromatic hydrocarbon (27-4)
polymerization (27-3)

positional isomers (27-1)
saturated hydrocarbon (27-2)
skeletal isomer (27-1)
stereoisomerism (27-3)
structural formula (27-1)
substitution reaction (27-2, 27-4)
unsaturated hydrocarbon (27-3)

Highlighted Expressions

Nomenclature rules for alkane hydrocarbons (27.1)
Nomenclature rules for alkene and alkyne hydrocarbons (27.8)
Preparation of an alkene (27.9, 27.10)
Unsaturated hydrocarbon addition reactions: Markovnikov's rule (27.20)
Aromatic substitution reactions: ortho,para and meta directors (27.31)

Preparation of an alcohol (27.32, 27.33)
Preparation of an ether (27.38)
Oxidation of primary and secondary alcohols (27.39, 27.40)
Preparation of a carboxylic acid (27.42, 27.43)
Preparation of an ester (27.46)
Preparation of an amine by the reduction of a nitro compound (27.47)

Review Problems

1. Draw Lewis structures of the following simple organic molecules.
  (a) $CH_3CHClCH_3$; (b) $HOCH_2CH_2OH$; (c) CH_3CHO.
2. Draw structural formulas for all the isomers of (a) pentane; (b) heptane.
3. Which of the following pairs of molecules are isomers and which are not? Explain.
  (a) $CH_3CH_2CH_2CH_3$ and $CH_3CH=CHCH_3$
  (b) $CH_3(CH_2)_5CH(CH_3)_2$ and
    $CH_3(CH_2)_4CH(CH_3)CH_2CH_3$
  (c) $CH_3CHClCH_2CH_3$ and $CH_3CH_2CHClCH_3$

  (d)

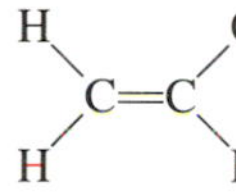

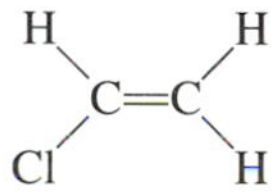

  (e)

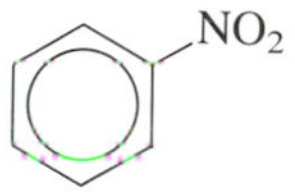

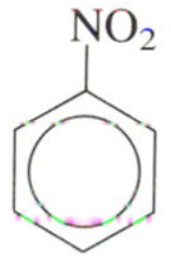

  (f)

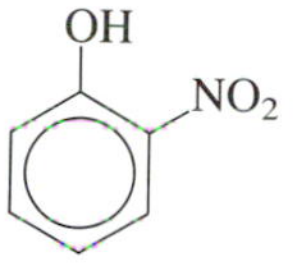

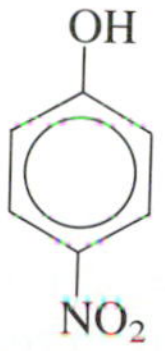

4. By drawing suitable structural formulas, establish that there are (a) five isomers of $C_3H_5Cl_3$ and (b) nine isomers of $C_4H_8Cl_2$.
5. Identify the functional group in each compound (i.e., whether an alcohol, amine, etc.)
  (a) $CH_3CHBrCH_2CH_3$ (b) CH_3CH_2COOH
  (c) $C_6H_5CH_2CHO$ (d) $(CH_3)_2CHCH_2OCH_3$
  (e) $CH_3COCH_2CH_3$ (f) $CH_3CH(NH_2)CH_2CH_3$

  (g) 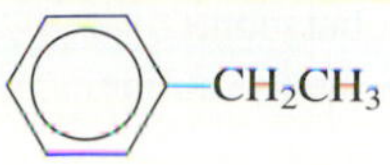 $\langle\!\!\bigcirc\!\!\rangle$—$CH_2CH_3$ (h) $CH_3CO_2CH_3$

6. Give an acceptable name for each of the following structures.

  (a) 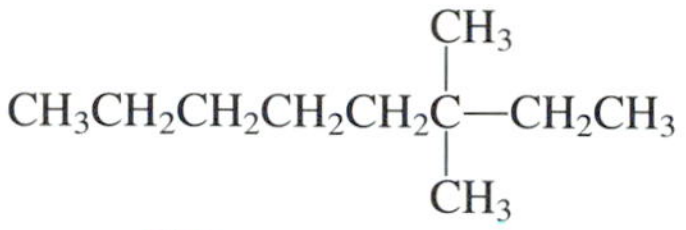 $CH_3CH_2CH_2CH_2CH_2\overset{\overset{\displaystyle CH_3}{|}}{\underset{\underset{\displaystyle CH_3}{|}}{C}}$—$CH_2CH_3$

  (b) CH_3—$\overset{\overset{\displaystyle CH_3}{|}}{\underset{\underset{\displaystyle CH_3}{|}}{C}}$—$CH_3$

  (c) $CH_2\overset{\overset{\displaystyle CH_3}{|}}{CH}$—$CH_2$—$\overset{\overset{\displaystyle}{CH}}{\underset{\underset{\displaystyle C_2H_5}{|}}{}}$—$\overset{}{\underset{\underset{\displaystyle Cl}{|}}{CH}}$—$\overset{}{\underset{\underset{\displaystyle Cl}{|}}{CH}}$—$CH_3$

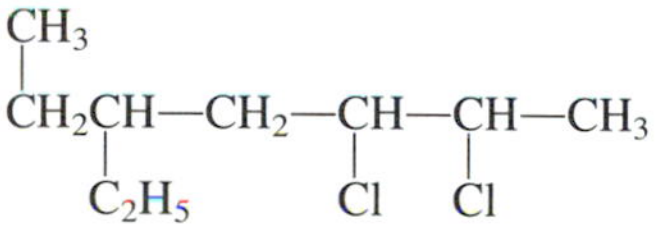

  (d) 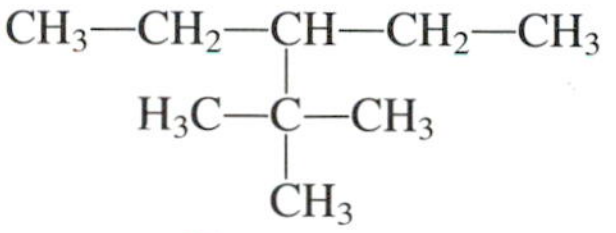 CH_3—CH_2—$\overset{}{\underset{\underset{\displaystyle H_3C-\overset{\overset{\displaystyle}{}}{\underset{\underset{\displaystyle CH_3}{|}}{C}}-CH_3}{|}}{CH}$—$CH_2$—$CH_3$

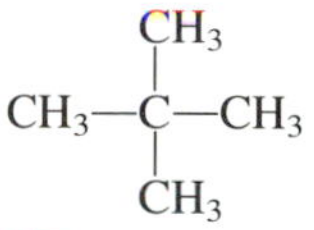

  (e) 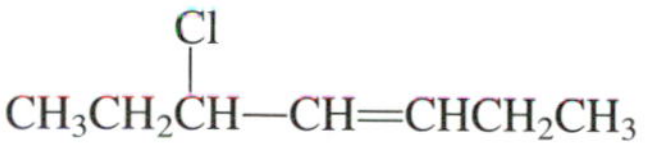 $CH_3CH_2\overset{}{\underset{\underset{\displaystyle Cl}{|}}{CH}}$—$CH=CHCH_2CH_3$

7. Draw a structural formula to correspond to each of the following names.
  (a) 3-bromo-2-methylpentane
  (b) 3-isopropyloctane
  (c) 2-pentene
  (d) ethyl propyl ether
8. Write a condensed formula for each of the following chemical substances.
  (a) isopropyl alcohol (rubbing alcohol)
  (b) methyl *t*-butyl ether (an octane enhancer of gasoline)
  (c) 1,1,1-chlorodifluoroethane (a refrigerant)
  (d) 2-methyl-1,3-butadiene (used in the manufacture of elastomers)
  (e) 2-butenal (crotonaldehyde, used in organic syntheses)
9. Supply a correct name or formula for each of the following aromatic compounds.

(a) 1,4-dibromobenzene

(b) (with CH_3 and NH_2)

(c) $CH_3CH_2CHOHCH_2CH_3 \xrightarrow[H^+]{Cr_2O_7^{2-}}$

(d) $CH_3CH_2CH{=}CH_2 + H_2O \xrightarrow[H_2SO_4(aq)]{10\%}$

(e) $CH_3CH_2OH + CH_3C\overset{O}{\underset{OH}{\big<}} \longrightarrow$

(c) 1,3,5-trimethylbenzene (d) *p*-nitrophenol

(e) 3-amino-2,5-dichlorobenzoic acid (amiben, a plant growth regulator)

10. Indicate the principal product of each of these reactions.

(a) $CH_3CH_3 + Cl_2 \xrightarrow{heat}$

(b) $CH_3CH_2CH{=}CH_2 + H_2 \xrightarrow[heat/pressure]{Pt}$

(f) $C\overset{O}{\underset{OH}{\big<}} + KOH(aq) \longrightarrow$

11. Balance equation (27.40) by the half-reaction method. [*Hint:* See the marginal note on page 989.]

Exercises

Definitions and terminology

12. What are the essential features that characterize (a) an organic compound; (b) an alkane; (c) an aromatic hydrocarbon?

13. What is the difference in meaning of the following terms? (a) aliphatic and alicyclic; (b) aliphatic and aromatic; (c) paraffin and olefin; (d) alkane and alkyl; (e) primary, secondary, tertiary; (f) axial and equatorial.

14. Give a definition and a well-chosen example of each of the following: (a) condensed formula; (b) homologous series; (c) olefin; (d) free radical; (e) ortho,para director.

Organic structures

15. Write structural formulas corresponding to these condensed formulas.

(a) $CH_3CH_2CH_2CHBrCH_3$
(b) $(CH_3)_2CHCH_2CH_2CH(CH_3)CH_2CH_3$
(c) $(CH_3)_3CCH_2CH(CH_3)CH_2CH_2CH_3$
(d) $CH_3CH_2CH(CH_3)C(C_2H_5){=}CH_2$

16. With appropriate sketches represent chemical bonding in terms of the overlap of pure or hybridized atomic orbitals in the following molecules.

(a) C_2H_6
(b) $H_2C{=}CHCl$
(c) $CH_3C{\equiv}CH$
(d) $CH_3\overset{O}{\overset{\|}{C}}CH_3$
(e) $CH_3CH_2NH_2$

Isomers

17. Indicate the difference in these three types of isomers: skeletal, positional, and geometrical. Which term best describes each of the following pairs of isomers?

(a) $CH_3CH_2CH_2Cl$ and $CH_3CHClCH_3$
(b) $CH_3CH(CH_3)CH_2CH_3$ and $CH_3(CH_2)_3CH_3$
(c) $CHCl{=}CHCl$ and $CH_2{=}CCl_2$

(d) and

(e) $\underset{H}{\overset{HOOC}{\big>}}C{=}C\underset{H}{\overset{COOH}{\big<}}$ and $\underset{H}{\overset{HOOC}{\big>}}C{=}C\underset{COOH}{\overset{H}{\big<}}$

18. By drawing suitable structural formulas, establish that there are 17 isomers of $C_6H_{13}Cl$. [*Hint:* Refer to Example 27-1].

Functional Groups

19. The functional groups in each of the following pairs have certain features in common, but what is the essential difference between them?

(a) carbonyl and carboxyl
(b) amine and amide
(c) acid and acyl group
(d) aldehyde and ketone

20. Give one example of each of the following types of compounds: (a) aliphatic nitro compound; (b) aromatic amine; (c) chlorophenol; (d) aliphatic diol; (e) unsaturated aliphatic alcohol; (f) alicyclic ketone; (g) halogenated alkane; (h) aromatic dicarboxylic acid.

Nomenclature and formulas

21. Give an acceptable name for each of the following structures.

(a) $CH_3CH_2C(CH_3)_3$
(b) $C(CH_3)_2{=}CH_2$
(c) $CH_2{-}CH{-}CH_3$ with CH_2 below
(d) $CH_3C{\equiv}CCH(CH_3)_2$
(e) $CH_3CH(C_2H_5)CH(CH_3)CH_2CH_3$
(f) $CH_3CH(CH_3)CH(CH_3)C(C_3H_7){=}CH_2$

22. Draw a structure to correspond to each of the following names.

(a) isopentane
(b) cyclohexene
(c) 3-hexyne
(d) 2-butanol
(e) isopropyl methyl ether
(f) propionaldehyde
(g) *t*-butyl chloride
(h) diethylmethylamine
(i) isobutyric acid
(j) isobutyl propionate

23. Does each of the following names convey sufficient information to suggest a specific structure? Explain.

(a) pentene
(b) butanone
(c) butyl alcohol
(d) methylaniline
(e) methylcyclopentane

24. Is each of the following names correct? If not, indicate why not and give a correct name.

(a) 3-pentene
(b) pentadiene
(c) 1-propanone
(d) bromopropane
(e) 2,6-dichlorobenzene
(f) 2-methyl-3-pentyne
(g) 2-methyl-4-butyloctane
(h) 4,4-dimethyl-5-ethyl-1-hexyne
(i) 1,3-dimethylcyclohexane
(j) 3,4-dimethyl-2-pentene

25. Supply condensed or structural formulas for the following substances.

(a) 2,2,4-trimethylpentane (isooctane—a constituent of gasoline)
(b) 2,4,6-trinitrotoluene (TNT—an explosive)
(c) methyl salicylate (oil of wintergreen) [*Hint:* Recall that salicylic acid is *o*-hydroxybenzoic acid.]
(d) 2-hydroxy-1,2,3-propanetricarboxylic acid (citric acid, $C_6H_8O_7$)
(e) 1,5-cyclooctadiene (an intermediate in the manufacture of resins)
(f) *o-t*-butylphenol (an antioxidant in aviation gasoline)
(g) 1-phenyl-2-aminopropane (benzedrine—an amphetamine, ingredient in ''pep pills'')
(h) 2-methylheptadecane (a sex pheromone of tiger moths—a chemical used for communication among members of the species) [*Hint:* ''Heptadeca'' means 17.]
(i) 3,7,11-trimethyl-2,6,10-dodecatriene-1-ol (farnesol—odor of lilly-of-the-valley) [*Hint:* ''Dodeca'' means 12.]
(j) 2,6-dimethyl-5-hepten-1-al (used in the manufacture of perfume)

Experimental determination of formulas

26. Combustion of 184 mg of a hydrocarbon gave 577 mg CO_2 and 236 mg H_2O. Calculate the empirical formula of the compound. What is the molecular formula if the molecular weight is subsequently found to be 56?

Alkanes

27. Draw structural formulas for all the isomers listed in Table 27-2 and show that, indeed, the more compact structures yield lower boiling points.

28. What is the most stable conformation of the molecule *t*-butylcyclohexane? [*Hint:* Is the ring in the boat or chair form? Is the *t*-butyl group in an axial or equatorial position?]

29. Write the structure of each alkane.

(a) Molecular weight = 44; forms two different monochlorination products.
(b) Molecular weight = 58; forms two different monobromination products.

30. Name the principal products obtained in the reaction of

(a) $CH_3CH_2CH{=}CH_2$ with H_2 in the presence of a catalyst
(b) propyl bromide with sodium
(c) sodium butyrate with sodium hydroxide
(d) propane with chlorine gas in the presence of ultraviolet light

31. In the chlorination of CH_4 some CH_3CH_2Cl is obtained as a product. Explain why this should be so.

Alkenes

32. Why is it not necessary to refer to ethene and propene as 1-ethene and 1-propene? Can the same be said for butene?

33. Alkenes (olefins) and cyclic alkanes (alicyclics) each have the generic formula C_nH_{2n}. In what important ways do these types of compounds differ structurally?

34. Draw the structures of the products of each of the following reactions.

(a) propylene + hydrogen (Pt, heat)
(b) 2-butanol + heat (in the presence of sulfuric acid)
(c) sodium acetylide with *t*-butyl bromide

35. Use Markovnikov's rule to predict the product of the reaction of

(a) HCl with $CH_3CCl{=}CH_2$
(b) HCN with $CH_3C{\equiv}CH$
(c) HCl with $CH_3CH{=}C(CH_3)_2$

(d) ⬡—$CH{=}CH_2$ with HBr

Aromatic compounds

36. Supply a name or formula for each of the following.

(a) [structure: benzene ring with NO_2 groups]
(b) [structure: benzene ring with OH, HO, OH groups]
(c) [structure: benzene ring with HO, OH, COOH groups]
(d) [structure: benzene ring with $N(C_2H_5)_2$ group]

(e) *p*-phenylphenol (a fungicide)
(f) phenylacetylene
(g) 2-hydroxy-4-isopropyltoluene (thymol—flavor constituent of the herb thyme)

37. Predict the products of the monobromination of (a) *m*-dinitrobenzene; (b) aniline; (c) *p*-bromoanisole.

38. Write the isomers to be expected from the mononitration of *m*-methoxybenzaldehyde.

[structure: benzene ring with CHO and OCH_3 groups]

In actual fact no 3-methoxy-5-nitrobenzaldehyde is obtained. What does this fact imply about the strength of meta and ortho,para directors?

39. What principal product would you expect to obtain when toluene is allowed to react with (a) HNO_3 + H_2SO_4; (b) Cl_2 + $FeCl_3$; (c) Cl_2 without $FeCl_3$ but in the presence of ultraviolet light?

*40. The symbol

which is often used to represent the benzene molecule, is also the structural formula of cyclohexatriene. Are benzene and cyclohexatriene the same substance? Explain.

⋆**41.** In the representation for benzene the inscribed circle represents electrons in a π bonding system (recall Figure 11-32). How many π electrons are suggested by this representation of naphthalene?

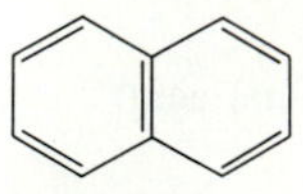

Organic reactions

42. Describe what is meant by each of the following reaction types and illustrate with an example from the text: **(a)** Aliphatic substitution reaction; **(b)** aromatic substitution reaction; **(c)** addition reaction; **(d)** elimination reaction.

43. Draw a structure to represent the principal product of each of the following reactions.

(a) 1-pentanol + excess dichromate (acid catalyst)
(b) butyric acid + ethanol (acid catalyst)
(c) o-nitrophenol + sodium hydroxide
(d) $CH_3CH_2C(CH_3)=CH_2 + H_2O$ (in the presence of H_2SO_4)

44. Use the half-reaction method to balance the following oxidation–reduction equations.

(a)

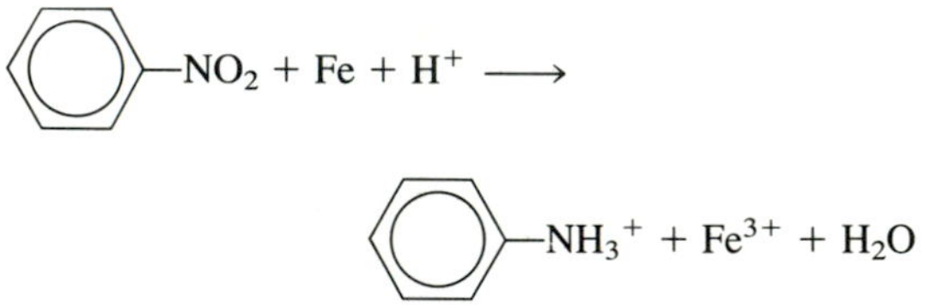

(b) $CH_3CH=CH_2 + MnO_4^- + H_2O \longrightarrow$

(c)

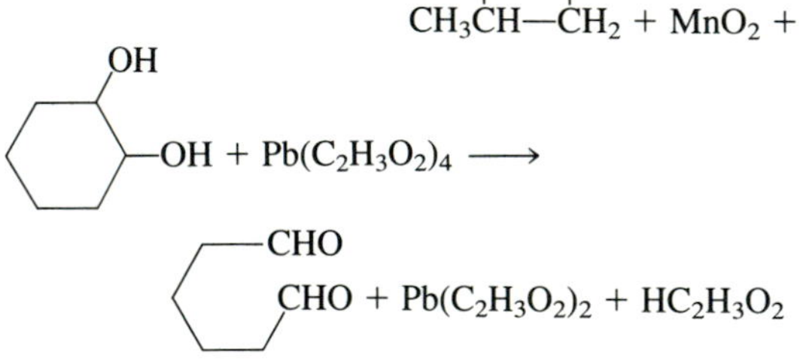

(d)

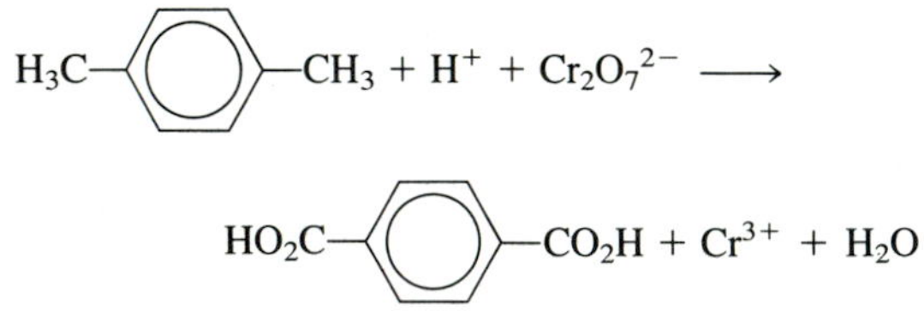

45. A 10.6-g sample of benzaldehyde was allowed to react with 5.9 g $KMnO_4$ in an excess of KOH(aq). After filtration of the MnO_2(s) and acidification of the solution, 6.1 g of benzoic acid was isolated. What was the percent yield of this reaction?

Organic synthesis

46. Starting with benzene and any aliphatic or inorganic reagents required, how would you synthesize **(a)** m-bromonitrobenzene; **(b)** p-aminotoluene?

47. Starting with acetylene as the only source of carbon, together with any inorganic reagents desired, devise syntheses for **(a)** acetaldehyde; **(b)** 1,1,2,2-tetrabromoethane; **(c)** acetonitrile; **(d)** isopropyl acetate.

Polymerization reactions

48. Explain why Dacron is called a polyester. What is the % O, by mass, in Dacron?

49. Nylon 66 is produced by the reaction of 1,6-hexanediamine with adipic acid. A different nylon polymer is obtained if sebacyl chloride

is substituted for the adipic acid. What is the basic repeating unit of this nylon structure?

50. Could a polymer be formed by the reaction of terephthalic acid with ethyl alcohol in place of ethylene glycol? Explain.

51. On page 341 the phenomenon of crosslinking was discussed in connection with the vulcanization of rubber. Polymers can be formed through the reaction of o-phthalic acid

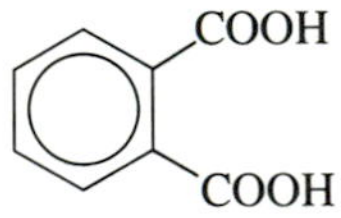

either with ethylene glycol or with glycerol. One of these polymers involves crosslinking; the other does not. One of these polymers is soft and tacky; the other is hard and brittle. Represent the structures of these two polymers and indicate the expected properties of each.

52. Draw structures of the following copolymers, using monomer formulas given in this chapter or in Table 10-4.

(a) the copolymer (called Saran) of vinyl chloride and vinylidine chloride, $CH_2=CCl_2$;
(b) poly(styrene-coacrylonitrile).

53. In referring to the molecular weight of a polymer, we can speak only of the "average molecular weight." Explain why the molecular weight of a polymer is not a unique quantity, as it is for a substance like benzene.

⋆**54.** Explain why a polymer formed by chain-reaction polymerization generally has a higher molecular weight than a corresponding polymer formed by step-reaction polymerization.

Additional Exercises

55. Write structural formulas for the following compounds.
(a) ethylisobutylmethylamine
(b) 3-chloro-2,3-dimethylbutanoic acid
(c) 1-phenyl-2-butanone
(d) t-butyl isobutyrate
(e) 1-chloro-2,4-octadiene
(f) trans-1,4-dibromobutadiene

56. Give an acceptable name for each of the following.
(a) $(CH_3)_2CHOCH_2CH_2CH_3$
(b) $(CH_3)_2CHCOCH_3$
(c) $CH_2=CHCH=CH_2$
(d) $(CH_3)_3CCH_2CH(CH_3)CH_2CH_2CH_3$

(e) [structure: benzene ring with Br at top, CH₃ at right, Br at bottom] **(f)** [structure: benzene ring with —CH₂CHBrCO₂CH(CH₃)₂]

57. Draw and name all the isomers for **(a)** C_6H_{14}; **(b)** C_4H_8; **(c)** C_4H_6. [*Hint:* Do not forget rings, double bonds, and combinations of these.]

58. Write the structure of each alkane: **(a)** molecular weight = 72; forms four monochlorination products; **(b)** molecular weight = 72; forms a single monochlorination product.

59. Why is cis-trans isomerism encountered with olefins but not with paraffins?

60. Methanol is a weaker acid than water, whereas phenol is stronger. Methylamine is a stronger base than ammonia, whereas aniline is weaker. Explain these observations.

61. Outline a series of reactions that could be used to synthesize ethanol from coal, water, and other inorganic materials.

62. Draw a structure to represent the principal product of each of the following reactions.
 (a) dimethylamine + hydrochloric acid
 (b) isopropanol + sodium
 (c) *t*-butyl bromide + NaOH(aq)

★63. Combustion of a 0.1908-g sample of a compound gave 0.2895 g CO_2 and 0.1192 g H_2O. Combustion of a second sample, weighing 0.1825 g, yielded 40.2 mL of N_2(g), collected over 50% KOH(aq) (vapor pressure = 9 mmHg) at 25 °C and

735 mmHg barometric pressure. When 1.082 g of compound was dissolved in 26.00 g benzene (m.p. 5.50 °C, K_f = 5.12), the solution had a freezing point of 3.66 °C. What is the molecular formula of the compound?

★64. The three isomeric tribromobenzenes, I, II, and III, when nitrated, form three, two, and one mononitrotribromobenzenes, respectively. Assign correct structures to I, II, and III.

★65. Write the name and structure of each aromatic hydrocarbon.
 (a) Formula: C_8H_{10}; forms three ring monochlorination products.
 (b) Formula: C_9H_{12}; forms one ring mononitration product.
 (c) Formula: C_9H_{12}; forms four ring mononitration products.

★66. In the molecule 2-methylbutane, the organic chemist distinguishes the different types of hydrogen and carbon atoms as being primary (1°), secondary (2°), and tertiary (3°). For the monochlorination of hydrocarbons the following ratio of reactivities has been found; 3°/2°/1° = 4.3:3:1. How many different monochloro derivatives of 2-methylbutane are possible and what percent of each would you expect to find?

[structure of 2-methylbutane with hydrogen/carbon atoms labeled 1°, 2°, 3°]

Self-Test Questions

For questions 67 through 74 select the single item that best completes each statement.

67. The compound isoheptane has the formula **(a)** C_7H_{14}; **(b)** $(CH_3)_2CH(CH_2)_3CH_3$; **(c)** $CH_3(CH_2)_5CH_3$; **(d)** $C_6H_5CH_3$.

68. *Three* isomers exist of the hydrocarbon **(a)** C_3H_8; **(b)** C_4H_{10}; **(c)** C_6H_6; **(d)** C_5H_{12}.

69. The hydrocarbon cyclobutane has the same carbon-to-hydrogen ratio as **(a)** C_4H_{10}; **(b)** $CH_3CH=CHCH_3$; **(c)** $CH_3C\equiv CCH_3$; **(d)** C_6H_6.

70. Cis-trans isomerism is expected in the compound **(a)** $ClCH=CHCl$; **(b)** $CH_2=CCl_2$; **(c)** $ClCH_2CH_2Cl$; **(d)** $Cl_2C=CCl_2$.

71. The compound 2-chloro-3-methyl-1-butanol has the formula
 (a) $CH_2ClC(CH_3)_2CH_2OH$
 (b) $CH_3CHOHCH(CH_3)CH_2Cl$
 (c) $CH_3CH(CH_3)CHClCH_2OH$
 (d) $CH_3CHClCH(CH_3)CH_2OH$

72. The compound

is named **(a)** *o*-aminotoluene; **(b)** *p*-methylaniline; **(c)** *m*-methylbenzene; **(d)** 3-methylaniline.

73. The most acidic of the following substances is

(a) [benzene with NH₂] **(b)** [benzene with COOH] **(c)** [benzene with OH] **(d)** CH_3CHO

74. To prepare methyl ethyl ketone one should oxidize **(a)** 2-propanol; **(b)** 1-butanol; **(c)** 2-butanol; **(d)** *t*-butyl alcohol.

75. Draw structural formulas for the following compounds: **(a)** dichlorodifluoromethane (Freon 12—a refrigerant); **(b)** *p*-bromophenol; **(c)** 3-hydroxy-2-butanone; **(d)** MTBE (methyl *t*-butyl ether, an antiknock gasoline additive).

76. Draw structural formulas for all the possible isomers of $C_5H_{11}Br$ and name them.

77. Indicate the principal organic product(s) that you would expect to obtain by
 (a) treating $CH_3CH_2CH=CH_2$ with *dilute* H_2SO_4(aq);
 (b) exposing a mixture of chlorine and propane gases to ultraviolet light;
 (c) heating a mixture of isopropyl alcohol and benzoic acid;
 (d) oxidizing *s*-butyl alcohol with $Cr_2O_7^{2-}$ in acidic solution.

78. Give a *simple* test that you might use to determine whether an organic substance is
 (a) C_2H_6 *or* C_8H_{18}; **(b)** C_2H_6 *or* C_2H_4;
 (c) C_2H_5OH *or* $C_6H_{13}OH$.
 (d) [benzene with CHO] *or* [benzene with COOH]

28 Chemistry of the Living State

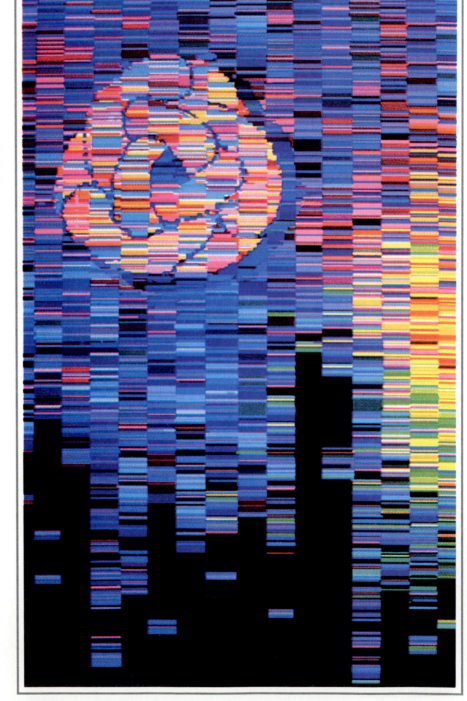

An analysis, by electrophoresis, of the sequence of nucleotides (bases) in a segment of a DNA molecule. [Four-color image provided courtesy of United States Biochemical Corporation]

Even though it is not stated in scientific terms, the biblical commentary "Dust you are, to dust you shall return" certainly suggests our relationship to nature and its laws. But there exists a fascinating interlude between dust and dust—that which we call life. As we learned in Chapter 20, spontaneous processes are propelled by the tendency for the entropy of the universe to increase. An increase in entropy means an increase in disorder. The "dust" referred to here suggests matter in simple forms with a minimum of order and a maximum of entropy.

The distinguishing characteristic of living matter is that of a high degree of order—complicated structures performing specialized functions. To stay alive an organism must resist, at least for a time, the universal approach to equilibrium, where disorder and entropy reach a maximum. To do this requires the raw materials with which to build cells, the energy of metabolism, and the information of heredity. These are the principal topics considered in this chapter, and our approach to these topics is to emphasize how fundamental chemical principles contribute to a knowledge of the living state.

28-1 Structure and Composition of the Cell

From one-celled plants and animals to *Homo sapiens*, the highest form of life, the living state presents us with a bewildering variety. Of the known elements, about 50 occur in measurable concentrations in living matter. Of these, about 25 have functions that are definitely known. Four elements together—oxygen, carbon, hydrogen, and nitrogen—account for 96% of human body mass. Figure 28-1 shows the chief elements found in living matter.

In addition to water, which is the most abundant compound in most living organisms, the important constituents of the cell are compounds of three types: **lipids**, **carbohydrates**, and **proteins**. We discuss each of these three types in some detail in the three sections that follow.

The **cell** is the fundamental unit of all life. Cells combine to form tissues; tissues may be grouped into organs; organs combine into organisms. Two views of a typical

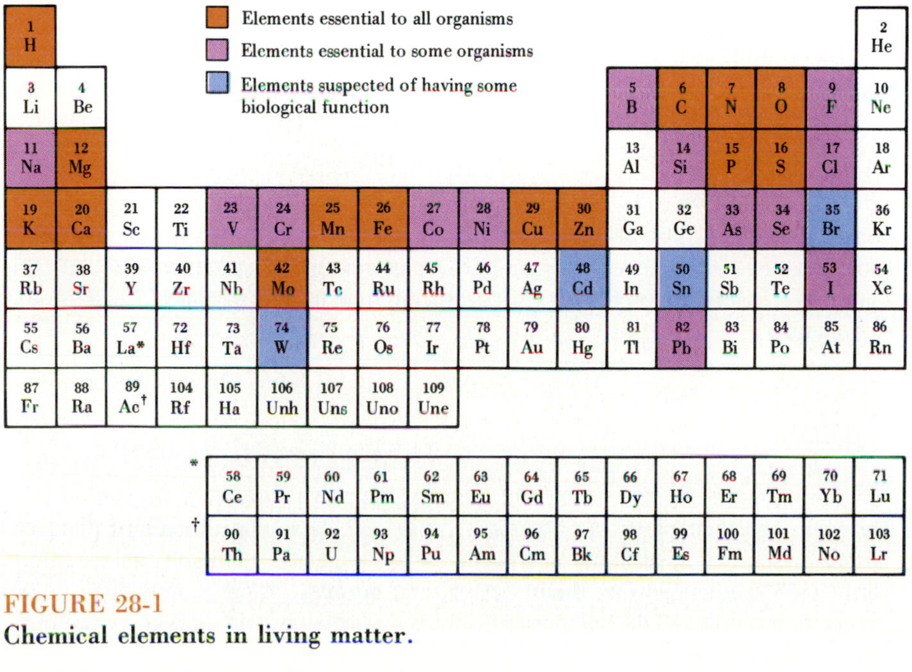

FIGURE 28-1

Chemical elements in living matter.

FIGURE 28-2
Cellular organization.

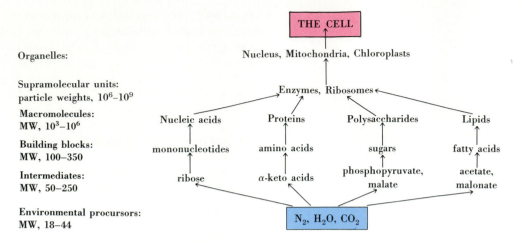

Organelles:

Supramolecular units:
particle weights, 10^6–10^9

Macromolecules:
MW, 10^3–10^6

Building blocks:
MW, 100–350

Intermediates:
MW, 50–250

Environmental procursors:
MW, 18–44

FIGURE 28-3
A typical animal cell.

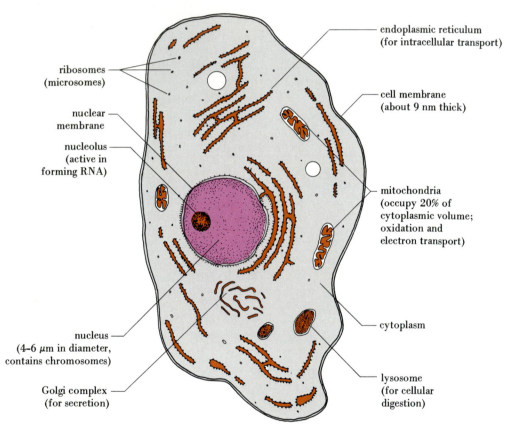

cell are presented in Figures 28-2 and 28-3. Figure 28-2 describes the complexity of molecules found in cells, and Figure 28-3 pictures the substructure of a cell. You may find it helpful to refer to these figures from time to time as you proceed through this chapter.

28-2 Lipids

We cannot define lipids precisely in terms of their structures, but we can describe them through their physical properties. Lipids are those constituents of plant and animal tissue that are soluble in solvents of low polarity, such solvents as chloroform, carbon tetrachloride, diethyl ether, and benzene. Many compounds fit this description. Although the categories of lipids described in this section are arbitrary, they are widely accepted.

Triglycerides. The most common lipids are esters of glycerol (1,2,3-propanetriol) with long-chain monocarboxylic acids (fatty acids). Glycerol provides the three-carbon backbone

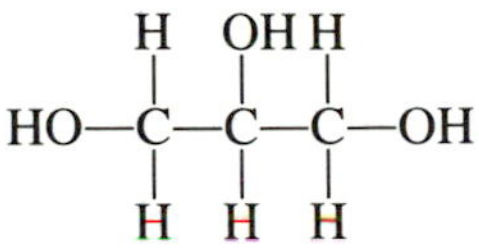

$$HO-\overset{\overset{\displaystyle H}{|}}{\underset{\underset{\displaystyle H}{|}}{C}}-\overset{\overset{\displaystyle OH}{|}}{\underset{\underset{\displaystyle H}{|}}{C}}-\overset{\overset{\displaystyle H}{|}}{\underset{\underset{\displaystyle H}{|}}{C}}-OH$$

We learned the meaning of terms such as "ol" and "acyl" in Chapter 27. We use some of the terminology of Chapter 27 in this chapter.

and the acids provide acyl groups

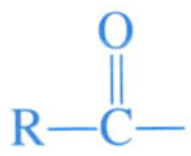

$$R-\overset{\overset{\displaystyle O}{\|}}{C}-$$

The systematic names for these esters is triacylglycerol, but the common name that has long been used is triglyceride, the name we use here. If all acid groups are the same, the triglyceride is called a *simple* glyceride; otherwise the triglyceride is a *mixed* glyceride. Some long-chain or fatty acids commonly encountered in triglycerides are listed in Table 28-1.

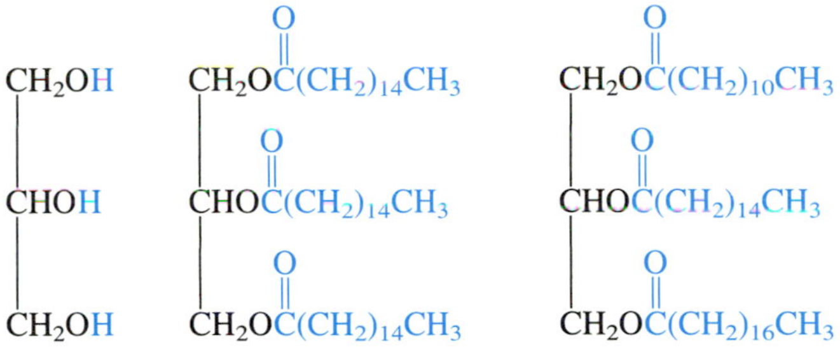

$$
\begin{array}{lll}
CH_2OH & CH_2O\overset{\overset{\displaystyle O}{\|}}{C}(CH_2)_{14}CH_3 & CH_2O\overset{\overset{\displaystyle O}{\|}}{C}(CH_2)_{10}CH_3 \\
| & | & | \\
CHOH & CHO\overset{\overset{\displaystyle O}{\|}}{C}(CH_2)_{14}CH_3 & CHO\overset{\overset{\displaystyle O}{\|}}{C}(CH_2)_{14}CH_3 \\
| & | & | \\
CH_2OH & CH_2O\overset{\overset{\displaystyle O}{\|}}{C}(CH_2)_{14}CH_3 & CH_2O\overset{\overset{\displaystyle O}{\|}}{C}(CH_2)_{16}CH_3 \\
\text{glycerol} & \text{glyceryl tripalmitate,} & \text{glyceryl lauropalmitostearate} \\
& \text{tripalmitin} & \text{(a mixed glyceride; a fat)} \\
& \text{(a simple glyceride; a fat)} &
\end{array}
$$

$$
\begin{array}{l}
CH_2O\overset{\overset{\displaystyle O}{\|}}{C}(CH_2)_7CH{=}CH(CH_2)_7CH_3 \\
| \\
CHO\overset{\overset{\displaystyle O}{\|}}{C}(CH_2)_7CH{=}CH(CH_2)_7CH_3 \\
| \\
CH_2O\overset{\overset{\displaystyle O}{\|}}{C}(CH_2)_7CH{=}CH(CH_2)_7CH_3
\end{array}
$$

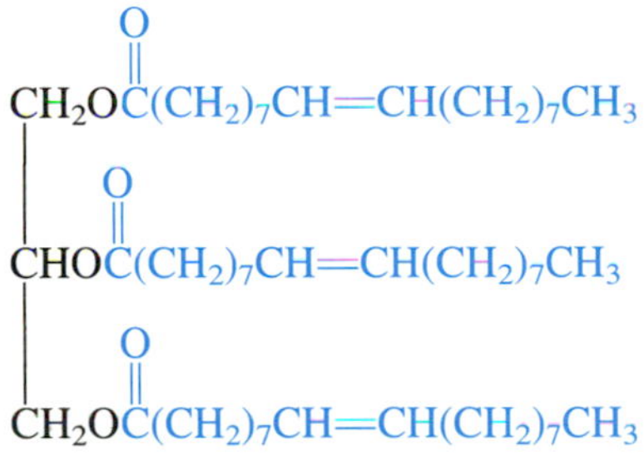

glyceryl trioleate,
triolein
(a simple glyceride; an oil)

TABLE 28-1
Some Common Fatty Acids

Common name	IUPAC name	Formula
	Saturated acids	
lauric acid	dodecanoic acid	$C_{11}H_{23}CO_2H$
myristic acid	tetradecanoic acid	$C_{13}H_{27}CO_2H$
palmitic acid	hexadecanoic acid	$C_{15}H_{31}CO_2H$
stearic acid	octadecanoic acid	$C_{17}H_{35}CO_2H$
	Unsaturated acids	
oleic acid	9-octadecenoic acid	$C_{17}H_{33}CO_2H$
linoleic acid	9,12-octadecadienoic acid	$C_{17}H_{31}CO_2H$
linolenic acid	9,12,15-octadecatrienoic acid	$C_{17}H_{29}CO_2H$
eleostearic acid	9,11,13-octadecatrienoic acid	$C_{17}H_{29}CO_2H$

Example 28-1

Drawing structures of triglycerides. Write a structural formula for glyceryl butyropalmitooleate.

Solution.   This is a mixed glyceride in which the acid groups are

$$
\underset{\text{butyric}}{\overset{\displaystyle O}{\overset{\|}{-}C(CH_2)_2CH_3}}
\qquad
\underset{\text{palmitic}}{\overset{\displaystyle O}{\overset{\|}{-}C(CH_2)_{14}CH_3}}
\qquad
\underset{\text{oleic}}{\overset{\displaystyle O}{\overset{\|}{-}C(CH_2)_7CH{=}CH(CH_2)_7CH_3}}
$$

The complete structure is thus

$$
\begin{array}{l}
\overset{\displaystyle O}{\overset{\|}{}} \\[-2pt]
CH_2OC(CH_2)_2CH_3 \\[2pt]
\overset{\displaystyle O}{\overset{\|}{}} \\[-2pt]
CHOC(CH_2)_{14}CH_3 \\[2pt]
\overset{\displaystyle O}{\overset{\|}{}} \\[-2pt]
CH_2OC(CH_2)_7CH{=}CH(CH_2)_7CH_3
\end{array}
$$

SIMILAR EXAMPLES: Exercises 1, 2.

The **fats** are glyceryl esters in which *saturated* acid components predominate; they are solids at room temperature. **Oils** have a predominance of *unsaturated* fatty acids and are liquids at room temperature. The composition of fats and oils is variable and depends not only on the particular plant or animal species involved but also on dietary and climatic factors.

When pure, fats and oils are colorless, odorless, and tasteless. The characteristic colors, odors, and flavors commonly associated with them are imparted by other organic substances that are present in the impure materials. The yellow color of butter is caused by the presence of β-carotene (a yellow pigment also found in carrots and marigolds). The taste of butter is attributed to these two compounds,

$$
\underset{\text{3-hydroxy-2-butanone}}{CH_3{-}\overset{\displaystyle O}{\overset{\|}{C}}{-}\underset{\underset{\displaystyle OH}{|}}{C}HCH_3}
\qquad\qquad
\underset{\text{diacetyl}}{CH_3{-}\overset{\displaystyle O}{\overset{\|}{C}}{-}\overset{\displaystyle O}{\overset{\|}{C}}{-}CH_3}
$$

both produced in the aging of cream.

Glycerides can be *hydrolyzed* in alkaline solution to produce glycerol and the alkali metal salts of the fatty acids. These salts are commonly known as **soaps,** and the hydrolysis process is called **saponification.**

Saponification of a glyceride.

$$
\begin{array}{l}
\overset{\displaystyle O}{\overset{\|}{}} \\[-2pt]
CH_2OC(CH_2)_{16}CH_3 \\[2pt]
\overset{\displaystyle O}{\overset{\|}{}} \\[-2pt]
CHOC(CH_2)_{16}CH_3 \quad + 3\ KOH \longrightarrow \\[2pt]
\overset{\displaystyle O}{\overset{\|}{}} \\[-2pt]
CH_2OC(CH_2)_{16}CH_3 \\[4pt]
\underset{\underset{\text{(MW = 891.5)}}{\text{tristearin}}}{}
\end{array}
\qquad
\begin{array}{l}
CH_2OH \\[12pt]
CHOH \quad + \quad 3\ CH_3(CH_2)_{16}\overset{\displaystyle O}{\overset{\|}{C}}OK \\[12pt]
CH_2OH \\[4pt]
\underset{\text{glycerol}}{} \qquad \underset{\underset{\text{(a soap)}}{\text{potassium stearate}}}{}
\end{array}
\qquad (28.1)
$$

We can use saponification reactions to obtain information about the structures of glycerides. We do this through the **saponification value.**

The saponification value is the number of milligrams of KOH required to saponify 1.00 g of a glyceride.

(28.2)

Example 28-2

Calculating the saponification value of a glyceride. What is the saponification value of tristearin?

Solution. This saponification reaction is the one represented by equation (28.1). We need simply to calculate the number of moles of tristearin in 1.00 g, and then, successively, the number of moles, grams, and milligrams of KOH required.

$$\text{no. mg KOH} = 1.00 \text{ g tristearin} \times \frac{1 \text{ mol tristearin}}{891.5 \text{ g tristearin}} \times \frac{3 \text{ mol KOH}}{1 \text{ mol tristearin}}$$

$$\times \frac{56.11 \text{ g KOH}}{1 \text{ mol KOH}} \times \frac{1000 \text{ mg KOH}}{1.00 \text{ g KOH}}$$

$$= 189 \text{ mg KOH}$$

SIMILAR EXAMPLES: Exercises 3, 4, 19.

Another useful quantity in characterizing a glyceride is based on the addition of I_2 to double bonds and is called the **iodine number.** Thus, the iodine number can be used to identify unsaturation in a glyceride.

The iodine number is the number of grams of I_2 that reacts with 100. g of a glyceride.

(28.3)

Example 28-3

Calculating the iodine number of a glyceride. What is the iodine number of triolein?

Solution

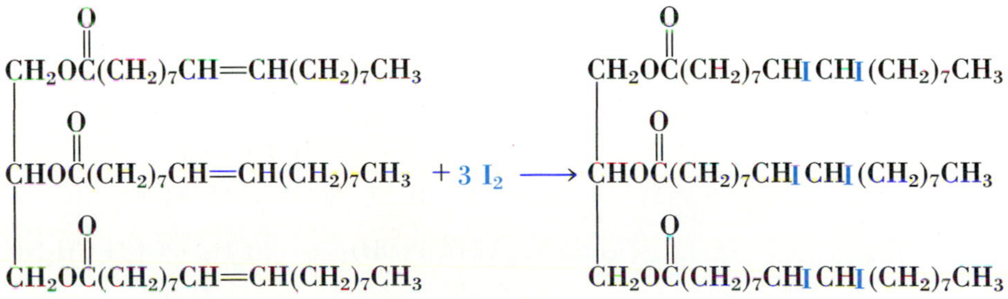

triolein
(MW = 885.5)

$$\text{no. g } I_2 = 1.00 \times 10^2 \text{ g triolein} \times \frac{1 \text{ mol triolein}}{885.5 \text{ g triolein}} \times \frac{3 \text{ mol } I_2}{1 \text{ mol triolein}} \times \frac{253.8 \text{ g } I_2}{1 \text{ mol } I_2}$$

$$= 86.0 \text{ g } I_2$$

SIMILAR EXAMPLES: Exercises 3, 20.

TABLE 28-2
Some Common Fats and Oils

Lipid	Component acids, % by mass						Saponification value	Iodine number
	Saturated			Unsaturated				
	Myristic	Palmitic	Stearic	Oleic	Linoleic	Linolenic		
fats								
butter	7–10	24–26	10–13	28–31	1–3	0.2–0.5	210–230	26–28
lard	1–2	28–30	12–18	40–50	7–13	0–1	195–203	46–70
edible oils								
corn	1–2	8–12	2–5	19–49	34–62	—	187–196	109–133
safflower	—	6–7	2–3	12–14	75–80	0.5–1.5	188–194	140–156

The formulas of the individual acids are listed in Table 28-1.

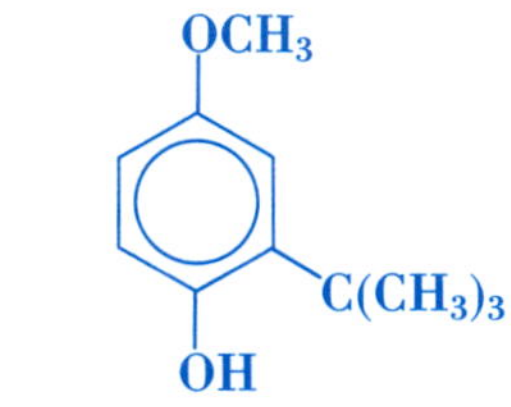

3-*t*-butyl-4-hydroxyanisole (BHA)

The compositions, saponification values, and iodine numbers of some common fats and oils are listed in Table 28-2.

Unsaturation in a fat or oil may be removed by the catalytic addition of hydrogen (hydrogenation). Thus, oils or low-melting fats can be changed to higher melting fats. These fats, when mixed with skim milk, fortified with vitamin A, and artificially colored, are known as *margarines*. Edible fats and oils both hydrolyze and cleave at the double bonds by oxidation on exposure to heat, air, and light. The low-molecular-weight fatty acids produced give off offensive odors, the condition known as *rancidity*. Antioxidants, such as 3-*t*-butyl-4-hydroxyanisole (BHA), retard this oxidative rancidity. They are commonly added to oils in the high-temperature cooking of potato chips and other foods. Medical evidence suggests a relationship between a high intake of saturated fats and the incidence of coronary heart disease. For this reason many diets call for the substitution of unsaturated for saturated fatty acids in foods. In general, mammal fats are saturated, whereas those derived from vegetables, seafood, and poultry are unsaturated.

Phosphatides. The phosphatides (phospholipids) occur in all vegetable and animal cells and are especially prevalent in nerve tissue. They are derived from glycerol, fatty acids, phosphoric acid, and a nitrogen-containing compound. (In the following structures, R and R′ are long-chain alkyl groups.)

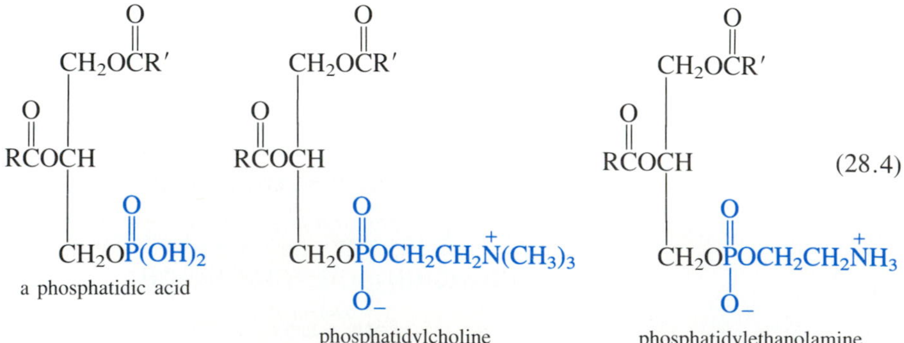

a phosphatidic acid

phosphatidylcholine
(a lecithin)

phosphatidylethanolamine
(a cephalin)

(28.4)

The condensed structural formula of choline is

$$[\text{HOCH}_2\text{CH}_2\text{N}(\text{CH}_3)_3]^+ \ \text{OH}^-$$

Choline is a quaternary ammonium compound (recall Figure 27-8), and this makes it a strong base, as seen through the hydroxide ion OH^-. Choline is also an alcohol, as seen by the presence of hydroxyl group HO—. The product of the esterification of choline and a phosphatidic acid is a phosphatidylcholine or a *lecithin*. Lecithins are found in brain and nerve tissue and in egg yolk. In contrast to simple fats and oils, lecithins form stable colloidal suspensions in water. They are obtained from soybeans and are used as emulsifiers in the dairy and confectionery industries.

Lecithins have both a highly polar and a nonpolar portion. [The polar portion of the molecule is shown in blue in (28.4).] Lecithins are associated with membranes enclosing cell nuclei and mitochondria. Their physiological role is apparently to associate water-insoluble lipids and water-soluble components of an organism, such as in the transport of lipids in the bloodstream or in the movement of fats from one tissue to another.

Removal of the fatty acid residue on the central carbon atom of a lecithin produces a lysolecithin. If this compound comes in contact with red blood cells, disintegration of the cells (hemolysis) occurs. The venom of poisonous snakes contains an enzyme that converts lecithins to lysolecithins. This accounts for the sometimes fatal effects of snakebites. Some spiders and insects also produce toxic results by the same mechanism.

Phosphatidylethanolamines or phosphatidylserines are known as *cephalins*. They are found in brain tissue, and they are also intimately involved in the blood-clotting process. Phospholipids are involved in the transport of ions across cell membranes, in certain secretory processes, and in the electron transport processes of respiration.

Waxes. When fatty acids form esters with long-chain *mono*hydric alcohols, the products are rather high-melting solids (35 to 100 °C) called waxes. Beeswax is largely ceryl myristate, $C_{13}H_{27}CO_2C_{26}H_{53}$. It melts between 62 and 65 °C and is used in shoe polish, candles, and wax coatings. Carnauba wax, a plant wax from a Brazilian palm tree, is largely myricyl cerotate, $C_{25}H_{51}CO_2C_{31}H_{63}$. It melts between 80 and 87 °C and is used in polishes and to coat mimeograph stencils. Spermaceti wax (also called whale oil although not really an oil) consists mainly of cetyl palmitate, $C_{15}H_{31}CO_2C_{16}H_{33}$, and cetyl alcohol, $C_{16}H_{33}OH$. This material, obtained from the head cavity of a whale, has been used as a softening agent in ointments and in cosmetics.

28-3 Carbohydrates

The literal meaning of the term "carbohydrate" is hydrate of carbon: $C_x(H_2O)_y$. Thus, sucrose or cane sugar, with the formula $C_{12}H_{22}O_{11}$, might be represented as $C_{12}(H_2O)_{11}$. A more useful definition, however, is that carbohydrates are polyhydroxy aldehydes, polyhydroxy ketones, their derivatives, and substances that yield them upon hydrolysis. Carbohydrates that are ketones are called ketoses; those that are aldehydes are called aldoses. If the compound contains five carbon atoms it is a pentose, six carbon atoms, a hexose, and so on.

The simplest carbohydrates are the monosaccharides. Oligosaccharides contain from two to ten monosaccharide units bonded together. Names can be assigned to reflect the actual number of such units present, such as *di*saccharide and *tri*saccharide. Mono- and oligosaccharides are also called **sugars.** Polysaccharides contain more than 10 monosaccharide units. The general term for all carbohydrates is glycoses. In summary,

Classification of carbohydrates.

Glycoses

> *Monosaccharides*
>   aldose (aldotriose, aldotetrose, . . .)
>   ketoses (ketotriose, ketotetrose, . . .)
> *Oligosaccharides* (from 2 to 10 monosaccharide units)
>   disaccharides (e.g., sucrose)
>   trisaccharides (e.g., raffinose)
>   and so on.
> *Polysaccharides* (more than 10 monosaccharide units)
>   (e.g., starch and cellulose)

(28.5)

FIGURE 28-4
Optical isomerism in glyceraldehyde.

The structure in (a) is not superimposable on (b), just as a right
and a left hand are not superimposable. (A right-handed glove
cannot be worn on a left hand.)

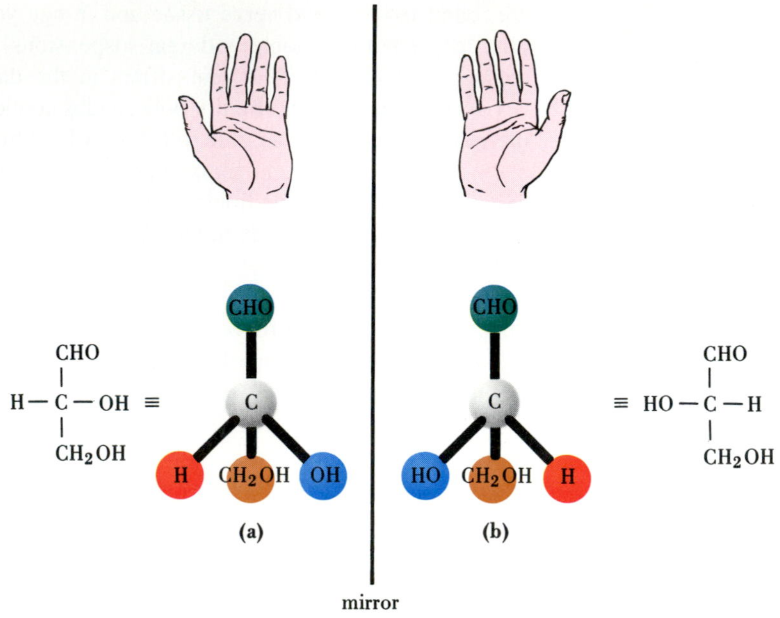

CHO
|
H—C—OH ≡
|
CH₂OH

(a)

(b)

≡ HO—C—H

CHO
|

|
CH₂OH

mirror

The simplest glycose is 2,3-dihydroxypropanal (glyceraldehyde), an *aldotriose*.
From a ball-and-stick model of this molecule we see an interesting form of stereoisomerism—*optical isomerism*. Figure 28-4 illustrates that there are *two* nonsuperimposable structures for glyceraldehyde. As we discovered in Section 25-4, such
structures are related to each other like a right and a left hand, or like an object and
its mirror image; they are called *enantiomers*. As we also learned in Section 25-4,
one enantiomer rotates the plane of polarized light to the right and is said to be
*dextro*rotatory (designated +); the other rotates the plane of polarized light to the
left and is *levo*rotatory (designated −). Almost all molecules exhibiting optical
isomerism possess at least one carbon atom with four different groups attached to it.
Such a carbon atom is said to be **asymmetric** or **chiral.**

Which arrangement of groups at the asymmetric carbon atom in glyceraldehyde,
that is, which **absolute configuration,** is associated with dextrorotatory and which
with levorotatory properties? X-ray studies have shown that the structure in Figure
28-4a is the dextrorotatory (+) one. To this species we assign a small capital letter
D, and to the structure in Figure 28-4b, the letter L.*

We cannot easily represent a three-dimensional structure in a plane (two-dimensional) drawing. A useful convention (the Fischer convention) is to place the structural formula on the page so that the backbone of the molecule is arranged from top
to bottom, with the most oxidized portion of the molecule (—CHO) at the top and
the least oxidized (—CH₂OH), at the bottom. Attached groups (—H and —OH) are
written to the sides. The end groups of the backbone extend *behind* the plane of the
page, *away* from the viewer. The glyceraldehyde enantiomers are written

CHO
|
H—C—OH
|
CH₂OH

D-(+)-glyceraldehyde

CHO
|
HO—C—H
|
CH₂OH

L-(−)-glyceraldehyde

(28.6)

and establish the D,L convention to be used for other sugars: The —H and —OH
groups on the next-to-last (penultimate) carbon atom extend in *front* of the page,
toward the viewer. If the —OH group on this penultimate carbon atom is to the

*A more generalized system for absolute configurations uses *R* (Latin, *rectus*, right) in place of D and
S (Latin, *sinister*, left) for L. However, D and L symbols are still widely used for carbohydrates.

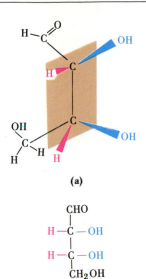

FIGURE 28-5
The structure of D-(−)-erythrose.

The three-dimensional structure (a) is represented in two dimensions by (b).

right, the configuration is D. If the —OH is to the left, the configuration is L. This convention is applied below to the four-carbon aldoses. The penultimate carbon atoms are shown in blue. Figure 28-5 may help you to picture the relationship between a three-dimensional structure and its two-dimensional representation.

$$(28.7)$$

D-(−)-erythrose   L-(+)-erythrose

But there are two more possibilities for 2,3,4-trihydroxybutanal—D-threose and L-threose. They too are enantiomers.

$$(28.8)$$

D-(−)-threose   L-(+)-threose

If we compare the configurations of D-erythrose and D-threose, we note that these two molecules are *not* mirror images. Molecules that are optical isomers but *not* mirror images of one another are called **diastereomers.**

D-erythose   D-threose   not mirror images

Enantiomers (shortened form, antiomers) have the same physical and chemical properties. They differ only in the *direction,* not the extent to which they rotate

Are You Wondering:

What is the relationship between the D,L and (+), (−) in structures like (28.6), (28.7), and (28.8)?

The designations D and L indicate how the H and OH groups are attached to the penultimate carbon atoms, based on the glyceraldehyde molecule as a reference. These designations are *unrelated* to the direction in which the plane of polarized light is rotated, which is noted as (+) or (−). In the case of glyceraldehyde, it is simply a matter of chance that the isomer with the configuration D does rotate the plane of polarized light to the *right* (*dextro*rotatory), and so it is also (+) in terms of its optical activity.

But the situation we see with erythrose (28.7) is equally likely. Here the isomer that has the absolute configuration corresponding to D rotates the plane of polarized light to the *left* (*levo*rotatory), and so it is a (−) isomer. Similarly, the one with an L configuration, in its optical activity, is (+).

plane-polarized light. Diastereomers (shortened form, diamers) do differ both in physical and chemical properties. Also they differ in the extent to which they rotate plane-polarized light.

A mixture of equal amounts of the D and L configurations of a substance, called a racemic mixture, does not rotate the plane of polarized light either to the left or to the right. The designation DL-erythrose, for example, signifies a racemic mixture. Usually, when molecules with chiral centers are synthesized, the product is a racemic mixture. This is because the creation of these centers is a random process like flipping a coin (an equal probability for "heads" or "tails"). If optically pure isomers are desired, the racemic mixture must be separated into the component enantiomers by a process called **resolution.** Sometimes this is carried out through an enzyme reaction, with a particular enzyme that reacts with one enantiomer but not the other.

Monosaccharides. Of the 16 possible aldohexoses only three occur widely in nature: D-glucose, D-galactose, and D-mannose.

H—C=O	H—C=O	H—C=O
H—C—OH	H—C—OH	HO—C—H
HO—C—H	HO—C—H	HO—C—H
H—C—OH	HO—C—H	H—C—OH
H—C—OH	H—C—OH	H—C—OH
CH₂OH	CH₂OH	CH₂OH
D-glucose	D-galactose	D-mannose

To some extent these three sugar molecules do exist in the straight-chain forms that we have been picturing, but in the main they occur in *cyclic* form. In this ring formation, the —OH group of the fifth carbon atom (C-5) adds to the carbonyl of the C-1 atom and produces a ring composed of five C atoms and one O atom, as illustrated in Figure 28-6. The configuration of the six-membered ring is of the "chair" type.

The terms "chair" form, axial, and equatorial were introduced in Section 27-2.

When the chain form of a sugar is converted to the ring form, a new chiral (asymmetric) center is produced at the C-1 atom. There are two possible orientations at this center. In the α form the OH at C-1 is *axial* (directed down); in the β form it is *equatorial* (extends out from the ring). The α and β forms of glucose are pictured in Figure 28-7.

The naming of monosaccharides is complicated by the fact that ring formation occurs. However, each term in a name conveys precise information about the molecule. Thus, D-(+)-glucose refers to the straight-chain form of glucose in the D configuration; this form is dextrorotatory (+). The name α-D-(+)-glucose denotes

FIGURE 28-6
Models of the glucose molecule.

(a) Ball-and-stick model indicating the atoms involved in ring closure.
(b) Ring closure represented through Fischer projection formulas.

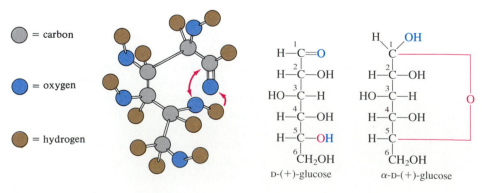

= carbon

= oxygen

= hydrogen

(a)

(b)

FIGURE 28-7
α and β forms of D-glucose.

A still more precise name for α-D-(+)-glucose is α-D-(+)-glucopyranose. The term *pyranose* signifies a six-membered, oxygen-containing heterocycle of the pyran type.

α-D-glucopyranose

β-D-glucopyranose

the ring form derived from D-glucose in which the α configuration is found at the C-1 atom.

With some sugars a sufficient amount of the straight-chain form is in equilibrium with the cyclic form so that the sugar engages in an oxidation–reduction reaction with Cu^{2+}(aq). The Cu^{2+}(aq) is reduced to insoluble red Cu_2O, and the aldehyde portion of the sugar is oxidized (to an acid). These sugars are known as **reducing sugars.** This test for a reducing sugar is conducted with alkaline copper ion complexed as the tartrate (Fehling's solution) or the citrate (Benedict's solution).

Test for a reducing sugar.

$$\text{certain cyclic sugars} \rightleftharpoons \underset{\text{straight-chain form}}{\begin{array}{c}\text{CHO}\\|\\\text{CHOH}\\|\end{array}} \xrightarrow{Cu^{2+}} \underset{\text{red ppt.}}{\begin{array}{c}\text{COOH}\\|\\\text{CHOH} + Cu_2O(s)\\|\end{array}} \qquad (28.9)$$

Disaccharides. Two monosaccharides can join together by eliminating a H_2O molecule between them. This combination is called a disaccharide. We must consider three points in describing a disaccharide.

1. What are the component monosaccharide units?
2. Is the configuration of the linkage between the monosaccharide units α or β?
3. What is the ring size in each monosaccharide unit?

The important naturally occurring disaccharides—maltose, cellobiose, lactose, and sucrose—are presented in Figure 28-8. In maltose an H atom on one glucose unit reacts with the hydroxyl group on the C-4 atom of a second glucose unit. The two units are linked in the α manner. Equilibrium is possible between the cyclic and a chain form in maltose, so that maltose is a reducing sugar. Maltose is produced by the action of malt enzyme on starch. It undergoes fermentation, in the presence of yeast, to glucose, and then to ethanol and CO_2(g).

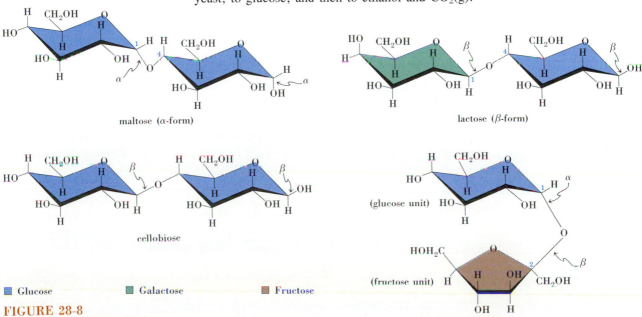

maltose (α-form)

lactose (β-form)

cellobiose

(glucose unit)

(fructose unit)

sucrose

■ Glucose   ■ Galactose   ■ Fructose

FIGURE 28-8
Some common disaccharides.

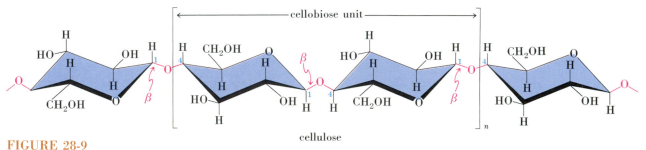

FIGURE 28-9
Two common polysaccharides.

Cellobiose can be obtained by the careful hydrolysis of cellulose. It is a glucose–glucose disaccharide with β linkages; it is also a reducing sugar. Lactose is the reducing sugar present in milk (4 to 6% in cow's milk and 5 to 8% in human milk). It is a galactose–glucose disaccharide having β linkages. Sucrose is ordinary table sugar (cane or beet sugar). It is a glucose–fructose disaccharide linked $1\alpha, 2\beta$. Neither of the two cyclic sugar units can open up into a chain form, and as a result sucrose is *not* a reducing sugar.

Polysaccharides. Polysaccharides are composed of monosaccharide units joined into long chains by oxygen linkages. **Starch,** with a molecular weight between 20,000 and 1,000,000, is the reserve carbohydrate of many plants and is the bulk constituent of cereals, rice, corn, and potatoes. Its structural features are brought out in Figure 28-9. **Glycogen** is the reserve carbohydrate of animals, occurring in liver and muscle tissue. It has a higher molecular weight than starch, and the polysaccharide chains are more branched. **Cellulose** is the main structural material of plants. It is the chief component of wood pulp, cotton, and straw. Complete hydrolysis gives glucose. Partial hydrolysis yields cellobiose, indicating that cellulose is a β-linked polymer (see Figure 28-9). Cellulose has a molecular weight between 300,000 and 500,000, corresponding to 1800 to 3000 glucose units. Most animals, including human beings, do not possess the necessary enzymes to hydrolyze β linkages. As a result they cannot digest cellulose. Certain bacteria in ruminants (cows, sheep, horses) and termites can hydrolyze cellulose, allowing them to use it as a food. Termites, as we know, subsist on a diet of wood.

Photosynthesis. As we have noted on previous occasions (Sections 7-11 and 25-11), the process of photosynthesis involves the conversion by plants of carbon dioxide and water into carbohydrates. Key roles are played in this conversion by the catalyst chlorophyll (see Figure 25-22) and by sunlight.

The photosynthesis reaction.

$$n\,CO_2 + n\,H_2O \xrightarrow[\text{chlorophyll}]{\text{sunlight}} (CH_2O)_n + n\,O_2 \qquad (28.10)$$

Equation (28.10) is greatly oversimplified. The currently accepted mechanism, proposed by Melvin Calvin (Nobel Prize, 1961), involves as many as 100 sequential steps for the conversion of 6 moles of carbon dioxide to 1 mole of glucose. The elucidation of this mechanism was greatly aided by the use of carbon-14 as a radioactive tracer. For simplicity, the overall photosynthetic process is divided into two phases: (1) the conversion of solar energy to chemical energy—the light reaction; and (2) the synthesis, promoted by enzymes, of carbohydrate intermediates. This latter reaction can occur in the absence of light and is called the dark reaction.

Biomass. As an energy source "biomass" is any material produced by photosynthesis, that is, biomass is plants or their principal components (cellulose, starch, sugars). Some biomass (e.g., wood) may be used directly as a fuel. Some may be converted to other gaseous, liquid, or solid materials for use as fuels or chemical raw materials.

Perhaps the best known and most widely used biomass conversion method involves the fermentation of sugars to produce ethanol. A fermentation process involves the decomposition of organic matter in the absence of air through the action of a microorganism.

$$\text{hexose sugar} \xrightarrow[\text{in yeast}]{\text{microorganisms}} 2\ C_2H_5OH + 2\ CO_2(g)$$

Disaccharides such as sucrose and polysaccharides (e.g., starch) can be hydrolyzed into monosaccharides by enzymes and then fermented to ethanol. In the United States, the principal raw material for the industrial production of ethanol by fermentation is corn. Ethanol from this source is currently finding some use in the fuel "gasohol," a mixture of 10% ethanol and 90% gasoline. The biomass material that may find increasing future use as a chemical raw material is cellulose.

Plants are no longer being converted in significant quantities to fossil fuels (coal, petroleum, natural gas) by geologic processes. In principle some of the same compounds now being produced from petroleum could be made directly from cellulose. Methanol (wood alcohol) is formed in the destructive distillation (pyrolysis) of wood. Cellulose can be hydrolyzed to glucose and then converted to ethanol by fermentation. Also, fermentation processes might be used to produce a series of oxygenated compounds—alcohols and ketones. These could then be converted to hydrocarbons. Thus, the entire spectrum of organic chemicals could be produced from the simple molecules CO_2 and H_2O. The required energy would be mostly solar. Combustion of the organic chemicals or products made from them would simply return CO_2 and H_2O to the environment.

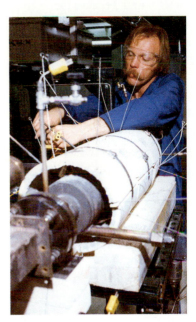

This **pyrolysis reactor** can process 20 kg sawdust per hour, but larger models may handle up to 250 tons of biomass per day. Vapor driven off from the heated sawdust is condensed to an oil, which can be processed into fuels. [SERI, U.S. Department of Energy]

28-4 Proteins

When a protein is hydrolyzed by dilute acids, bases, or hydrolytic enzymes, the result is a mixture of α-amino acids. An **amino acid** is a carboxylic acid that also contains an amine group, —NH$_2$; an **α-amino acid** has the amino group on the α carbon atom—the carbon atom next to the carboxyl group. Thus proteins are high-molecular-weight polymers composed of α-amino acids. Of the known α-amino acids, about 20 have been identified as building blocks of most plant and animal proteins. Some are listed in Table 28-3.

Proteins are probably the most complex organic materials found in nature. They are the basis of protoplasm and are found in all living organisms. Proteins, as muscle, skin, hair, and other tissue, make up the bulk of the body's nonskeletal structure. As enzymes they catalyze biochemical reactions. As hormones they regulate metabolic processes; and as antibodies they counteract the effect of invading species and substances.

Other than glycine ($H_2NCH_2CO_2H$), naturally occurring amino acids are optically active, mostly with an L configuration.

The name "protein" is derived from the Greek word *proteios*, meaning "of first importance" (similar to the derivation of "proton").

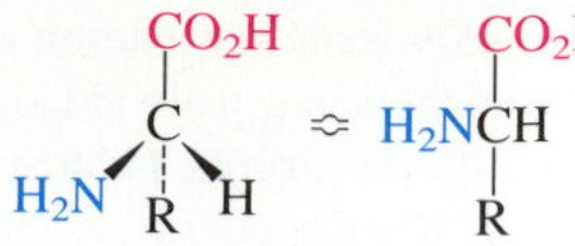

$$\underset{\underset{R}{H_2N}}{\overset{\overset{CO_2H}{|}}{\underset{|}{C}}}\!\!\!\diagdown H \quad \rightleftharpoons \quad \underset{R}{\overset{CO_2H}{H_2NCH}}$$

The reference structure for establishing the absolute configurations of amino acids is again glyceraldehyde, with the —NH$_2$ group substituting for —OH and —CO$_2$H, for —CHO. The molecule shown above has an L configuration because the —NH$_2$ group appears on the *left*.

TABLE 28-3
Some Common Amino Acids

Name	Symbol	Formula	p*I*
Neutral amino acids			
glycine	Gly	HCH(NH$_2$)CO$_2$H	5.97
alanine	Ala	CH$_3$CH(NH$_2$)CO$_2$H	6.00
valine[a]	Val	(CH$_3$)$_2$CHCH(NH$_2$)CO$_2$H	5.96
leucine[a]	Leu	(CH$_3$)$_2$CHCH$_2$CH(NH$_2$)CO$_2$H	6.02
isoleucine[a]	Ileu or Ile	CH$_3$CH$_2$CH(CH$_3$)CH(NH$_2$)CO$_2$H	5.98
serine	Ser	HOCH$_2$CH(NH$_2$)CO$_2$H	5.68
threonine[a]	Thr	CH$_3$CHOHCH(NH$_2$)CO$_2$H	5.6
phenylalanine[a]	Phe	C$_6$H$_5$CH$_2$CH(NH$_2$)CO$_2$H	5.48
methionine[a]	Met	CH$_3$SCH$_2$CH$_2$CH(NH$_2$)CO$_2$H	5.74
cysteine	Cys	HSCH$_2$CH(NH$_2$)CO$_2$H	5.05
cystine	(Cys)$_2$	—[SCH$_2$CH(NH$_2$)CO$_2$H]$_2$	4.8
tyrosine	Tyr	4-HOC$_6$H$_4$CH$_2$CH(NH$_2$)CO$_2$H	5.66
tryptophan[a]	Trp	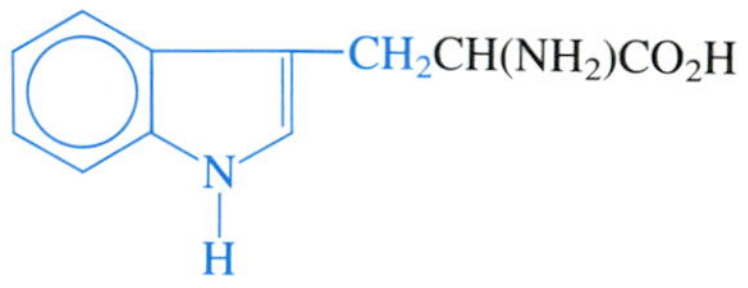	5.89
proline	Pro	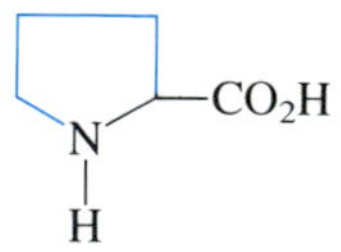	6.30
hydroxyproline	Hyp	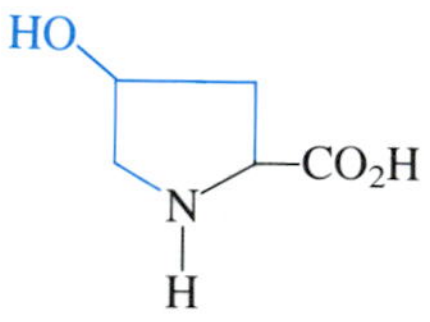	
Acidic amino acids			
aspartic acid	Asp	HO$_2$CCH$_2$CH(NH$_2$)CO$_2$H	2.77
glutamic acid	Glu	HO$_2$CCH$_2$CH$_2$CH(NH$_2$)CO$_2$H	3.22
Basic amino acids			
lysine[a]	Lys	H$_2$N(CH$_2$)$_4$CH(NH$_2$)CO$_2$H	9.94
arginine	Arg	H$_2$NCNHNH(CH$_2$)$_3$CH(NH$_2$)CO$_2$H	10.76
histidine	His	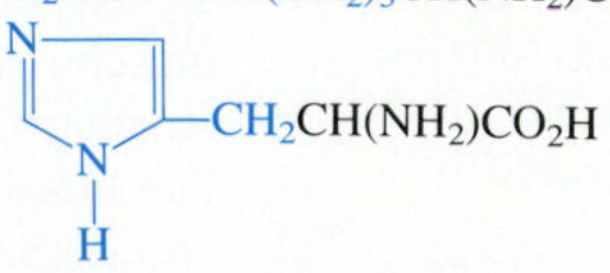	7.65

[a]Essential amino acids. In addition to these, arginine and glycine are required by the chick, arginine by the rat, and histidine by human infants.

Certain amino acids are required for proper health and growth in human beings, yet the body is unable to synthesize them. These must be ingested through foods, and are called **essential** amino acids. Eight are known to be essential; the case of three others is less certain (see Table 28-3).

The amino acids are colorless, crystalline, high-melting solids that are moderately soluble in water. In an acidic solution the amino acid exists as a cation, with a proton attaching itself to the unshared pair of electrons on the nitrogen atom in the group $-NH_2$. In a basic solution an anion is formed, through the loss of a proton by the $-CO_2H$ group. At the neutral point a proton is transferred from $-CO_2H$ to $-NH_2$. The product is a dipolar ion or a "zwitterion."

Acid–base properties of amino acids.

$$R-\underset{\underset{NH_3^+}{|}}{CH}-CO_2H \underset{H^+}{\overset{OH^-}{\rightleftharpoons}} R-\underset{\underset{NH_3^+}{|}}{CH}-CO_2^- \underset{H^+}{\overset{OH^-}{\rightleftharpoons}} R-\underset{\underset{NH_2}{|}}{CH}-CO_2^- \qquad (28.11)$$

acidic soln   isoelectric point   basic soln

Amino acids are amphoteric. The pH at which the dipolar structure predominates is called the **isoelectric point** or pI. At this pH the molecule does not migrate in an electric field. At a pH above the pI the molecule migrates to the anode (positive electrode), below the pI to the cathode (negative electrode). Basic amino acids have a pI above 7, acidic ones below 7, and neutral ones near 7. In most amino acids the basicity of the amino group is about equal to the acidity of the carboxyl group. The largest group of amino acids are essentially pH neutral.

Peptides. Amino acid molecules can be joined by the elimination of water molecules between them. Two amino acids thus joined form a dipeptide. The bond between the two amino acid units is called a peptide linkage or bond.

Formation of a peptide bond.

$$R-\underset{\underset{NH_2}{|}}{CH}-\overset{\overset{O}{\|}}{C}-OH + HN-\underset{|}{CH}-R' \longrightarrow H_2O + R-\underset{\underset{NH_2}{|}}{CH}-\overset{\overset{O}{\|}}{C}-NH-\underset{|}{CH}-R'$$

peptide bond   a dipeptide

$$(28.12)$$

A tripeptide has three amino acid residues and two peptide linkages. A large number of amino acid units may join to form a *poly*peptide.

We use a simple convention in writing the structures and names of polypeptides. The amino acid unit present at one end of the polypeptide chain has a free $-NH_2$ group; this is the "N-terminal" end. The other end of the chain has a free $-CO_2H$ group; it is the "C-terminal" end. We write the structure with the N-terminal end to the left and C-terminal to the right. The base name of the polypeptide is that of the C-terminal amino acid. We name all the other amino acid units in the chain as substituents of this acid; as such we change their names from the "ine" to the "yl" ending. We also commonly use abbreviations in writing polypeptide names, as illustrated in Example 28-4.

Example 28-4

Naming a polypeptide. What is the name of the polypeptide whose structure is shown?

$$H_2C-\overset{\overset{O}{\|}}{C}-NH-\underset{|}{CH}-\overset{\overset{O}{\|}}{C}-NH-\underset{|}{CH}-\overset{\overset{O}{\|}}{C}-OH$$

$\underset{(a)}{NH_2}$ $\underset{(b)}{CH_3}$ $\underset{(c)}{CH_2OH}$

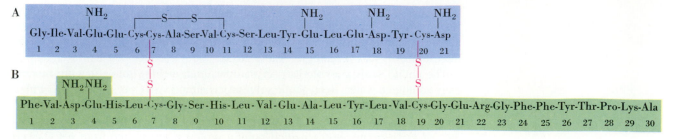

FIGURE 28-10

Amino acid sequence in beef insulin—primary structure of a protein.

There are two polypeptide chains joined by disulfide (—S—S—) linkages. One chain has 21 amino acids, and the other 30. In chain A the Gly at the left end is N-terminal and the Asp is C-terminal. In chain B Phe is N-terminal, and Ala is C-terminal.

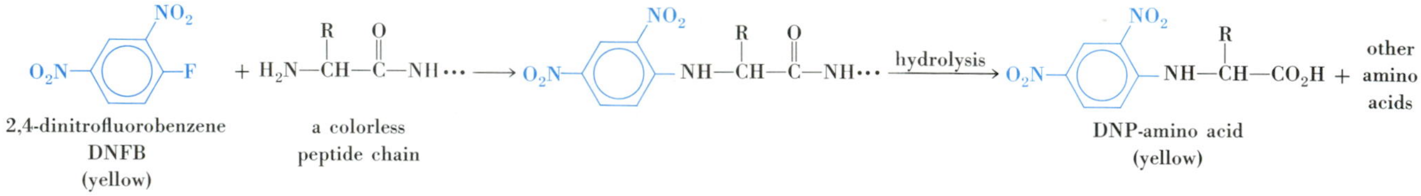

$$2,4\text{-dinitrofluorobenzene DNFB (yellow)} + \text{a colorless peptide chain} \longrightarrow \text{DNP-peptide} \xrightarrow{\text{hydrolysis}} \text{DNP-amino acid (yellow)} + \text{other amino acids}$$

FIGURE 28-11

Experimental determination of amino acid sequence.

In the reaction between DNFB and a polypeptide the N-terminal amino acid ends up with the yellow "marker" (a dinitrophenyl group, DNP) attached to it. By gentle hydrolysis and repeated use of the marker, a polypeptide chain can be broken down and the sequence of the individual units determined.

Solution. The three amino acids in this tripeptide are identified through Table 28-3. (a) = glycine; (b) = alanine; (c) = serine. The C-terminal amino acid is serine. The name is

glycylalanylserine (Gly-Ala-Ser)

SIMILAR EXAMPLES: Exercises 7, 8, 30.

Suppose that a tripeptide is known to consist of the three amino acids: A, B, and C. What is the correct structure: ABC?, ACB?, . . . Can you see that there are six possibilities? For longer chains, of course, the number of possibilities is enormous. Determining the sequence of amino acids in a polypeptide chain is one of the most significant problems in all of biochemistry. The Nobel prize in 1958 was awarded to Frederick Sanger for elucidation of the structure of beef insulin (see Figure 28-10). The method employed is outlined in Figure 28-11.

Example 28-5 _____

Determining the sequence of amino acids in a polypeptide. A polypeptide, on complete hydrolysis, yielded the amino acids A, B, C, D, and E. Partial hydrolysis and sequence proof gave single amino acids together with the following larger fragments: AD, DC, DCB, BE, and CB. What must be the sequence of amino acids in the polypeptide?

Solution. By arranging the fragments in the following manner,

AD
 DC
 DCB
   BE
 CB

we see that only the sequence **ADCBE** is consistent with the fragments observed.

SIMILAR EXAMPLE: Exercise 31.

The distinction between large polypeptides and proteins is arbitrary. It is generally accepted that if the molecular weight is over 10,000 (roughly 50 to 75 amino acid units) the substance is a protein. Proteins, like amino acids, are amphoteric. They possess characteristic isoelectric points, and their acidity or basicity depends on their amino acid composition. When proteins are heated, treated with salts, or exposed to ultraviolet light, profound and complex changes occur. This **denaturation** usually brings about lowering of solubility and loss of biological activity. We can find many examples of denaturation of proteins. The frying or boiling of an egg involves the denaturation (coagulation) of the egg albumin, a protein. The beauty shop "permanent wave" takes advantage of a denaturation process that is reversible. The proteins found in hair (e.g., keratin) contain disulfide linkages (—S—S—). When hair is treated with a reducing agent these linkages break—a denaturation process. Following this step the hair is set into the desired shape. Next, the hair is treated with a mild oxidizing agent. The disulfide linkages are reestablished and the hair remains in the style in which it was set.

Structure of Proteins. The **primary structure** of a protein, as we have already seen, refers to the exact sequence of amino acids in the polypeptide chains that make up the protein, but what are the shapes of the long polymeric chains themselves? Are they simply limp and entangled like a plate of spaghetti or is there some order within chains and among chains? The structure or shape of an individual protein chain is referred to as **secondary structure.** The first work on this subject was published in 1951 by Linus Pauling and R. B. Corey, describing x-ray-diffraction studies on polylysine, a synthetic polypeptide. They postulated that the orienta-

FIGURE 28-12
An alpha helix—secondary structure of a protein.

The helical structure is stabilized through formation of hydrogen bonds between carboxyl oxygen atoms in one turn and amide hydrogen atoms in the next turn above. The bulky R groups are directed outward from the atoms in the spiral.

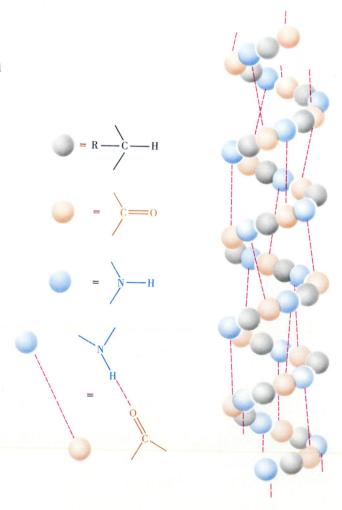

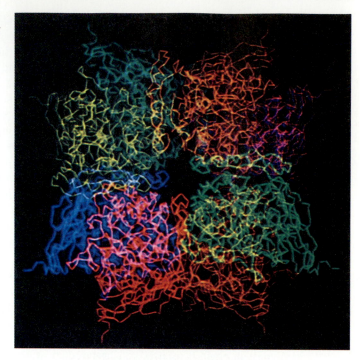

A computer image of the three-dimensional structure of ribulose-1,5-biphosphate carboxylase-oxygenase (RuBisCo), which consists of 37,792 atoms. This structure was elucidated in 1988, after an 18-year effort. RuBisCo is the most abundant protein on earth; its estimated annual production by plants, worldwide, is 4×10^{13} g (40 million tons). RuBisCo is the enzyme that initiates the process of photosynthesis. [Courtesy David Eisenberg, Ph.D, University of California, Los Angeles]

tion of this polypeptide and thus of protein chains is *helical*. A spiral, helical, or springlike shape can be either left- or right-handed, but because proteins are composed of L-amino acids, their helical structure is right-handed (see Figure 28-12, page 1025).

As we have learned previously, highly ordered materials produce distinctive x ray diffraction patterns, and from these patterns we can learn about their structure. Conversely, materials in which there is little order—amorphous materials—do not have distinctive x-ray-diffraction patterns. From x-ray studies it is clear that not all proteins have a helical configuration. Gamma globulin is one that does not. Hemoglobin and myoglobin are helical in some portions of their chains and randomly oriented in others. Finally, other types of orientations are possible. For example, β-keratin and silk fibroin are arranged in pleated sheets. In these proteins the side chains extend above and below the pleated sheets and hydrogen bonding is between *different* molecules (interpeptide bonding) lying next to each other and about 0.47 nm apart in the same sheet. These sheets are stacked on top of one another about 1.0 nm apart, rather like a pile of sheets of corrugated roofing (see Figure 28-13).

Does a helical protein molecule possess any additional structural features? That is, are the helices elongated, twisted, knotted, or what? The final statement regard-

FIGURE 28-13
Pleated-sheet model of β-keratin.

(a) A polypeptide chain showing the direction of interpolypeptide hydrogen bonds (other polypeptide chains lie to the left and to the right of the chain shown). Bulky R groups extend above and below the pleated sheet.
(b) The stacking of pleated sheets.

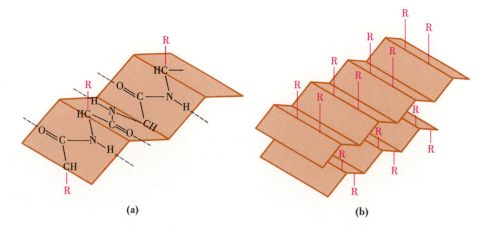

(a)   (b)

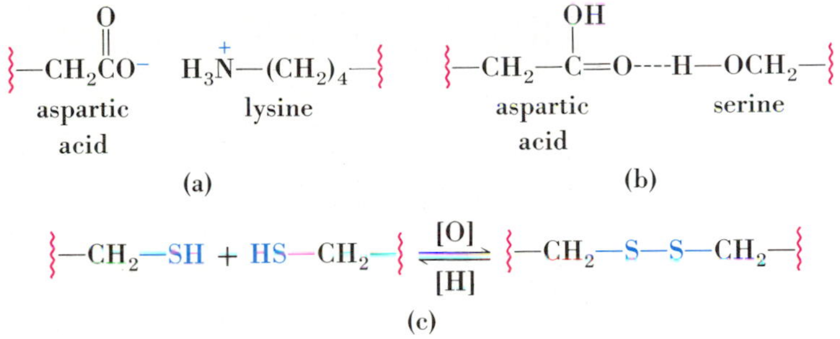

FIGURE 28-14

Linkages contributing to the tertiary structure of proteins.

(a) Salt linkages. Acid–base interactions between different coils. In the example shown, the carboxyl group of an aspartic acid unit on one coil donates a proton to the free amine group of a lysine unit on another.
(b) Hydrogen bonding. Interactions between side chains of certain amino acids, for example, aspartic acid and serine.
(c) Disulfide linkages. Oxidation of the highly reactive thioalcohol (—SH) group of cysteine to a disulfide (—S—S—) can occur (as in beef insulin).

The folding of polypeptide chains into a tertiary structure is influenced by an additional factor. Hydrophobic hydrocarbon portions of the chains (R groups) tend to be drawn into close proximity in the interior of the structure, leaving ionic groups at the exterior.

$$
\overset{O}{\underset{\|}{}}
\}-CH_2CO^- \quad H_3\overset{+}{N}-(CH_2)_4-\{
$$
aspartic acid   lysine

$$
\overset{OH}{\underset{|}{}}
\}-CH_2-C=O\cdots H-OCH_2-\{
$$
aspartic acid   serine

(a)   (b)

$$
\}-CH_2-SH + HS-CH_2-\{ \xrightleftharpoons[\text{[H]}]{\text{[O]}} \}-CH_2-S-S-CH_2-\{
$$

(c)

ing the shape of a protein molecule lies in a description of its tertiary structure. Because the internal hydrogen bonding that occurs between atoms in successive turns of a protein helix is weak, these hydrogen bonds ought to be easily broken. In particular, we should expect them to be replaced by hydrogen bonds to water molecules when the protein is placed in water. That is, the α helix should open up and become a randomized structure when placed in water (recall the analogy of the limp spaghetti). But experimental evidence indicates that this does not happen. We are led to the conclusion that other forces must be involved in compressing the long α-helical chains into definite geometric shapes. Each protein has its own three-dimensional shape or **tertiary structure.** Three types of linkages involved in tertiary structures are described in Figure 28-14.

By x-ray-diffraction studies John Kendrew and Max Perutz (Nobel prize, 1962) were able to elucidate the primary, secondary, and tertiary structures of myoglobin. The primary structure is that of a peptide of 153 units in a single chain. Secondary structure involves coiling of 70% of the chain into an α helix. The tertiary structure is depicted in Figure 28-15.

The hemoglobin molecule consists of four separate polypeptide chains or subunits. The arrangement of these four subunits constitutes a still higher order of structure referred to as the **quaternary structure** (see Figure 28-16). Because it is a single polypeptide, myoglobin has no quaternary structure.

Even minor changes in the structure of a protein can have profound effects. Hemoglobin contains four polypeptide chains, each with 146 amino acid units. The substitution of valine for glutamic acid at one site in two of these chains gives rise to the sometimes fatal blood disease known as sickle cell anemia. Apparently, the altered hemoglobin has a reduced ability to transport oxygen through the blood.

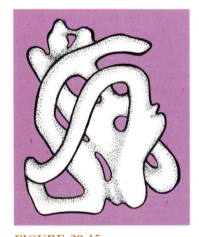

FIGURE 28-15

Representation of the tertiary structure of myoglobin.

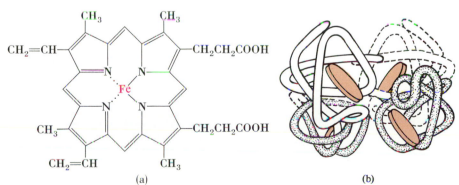

(a)   (b)

FIGURE 28-16

Quaternary protein structure—the structures of heme and hemoglobin.

Hemoglobin is a conjugated protein. It consists of nonprotein groups (called prosthetic groups) bonded to a protein portion.
(a) The structure of heme.
(b) Four heme units and four polypeptide coils are bonded together in a molecule of hemoglobin.

28-5 Metabolism

Although organisms differ markedly in their outward appearances, there is a striking similarity in the chemical reactions that occur within them. The sum of these reactions is referred to as **metabolism.** Think of metabolism as the collection of processes whereby matter is ingested and eliminated and the organism is supplied with its energy requirements. The part of this overall process in which molecules are broken down or degraded is called *catabolism*, and that part in which molecules are synthesized is called *anabolism*. Reactions for which the standard free-energy change is positive are *endergonic*, and those for which it is negative are *exergonic*. The chemical substances involved in metabolism are called *metabolites*.

FIGURE 28-17
Metabolism outline.

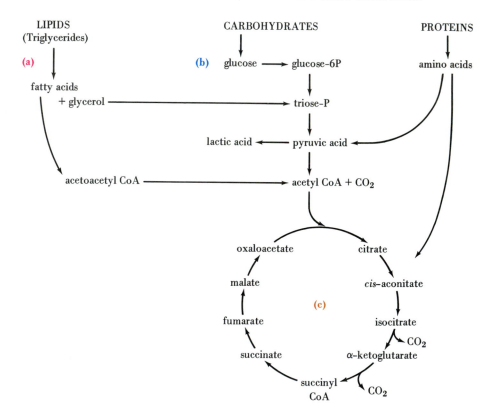

(a) **Fatty acid section.** Fatty acids are degraded two carbon atoms at a time. Acetyl units are fed into the citric acid cycle (c) as acetyl CoA.
(b) **Glycolysis section (Embden–Meyerhof pathway).** These reactions are anaerobic (no oxygen required). Carbohydrates are degraded to the six-carbon sugar glucose, and then to the three-carbon triose-P (glyceraldehyde 3-phosphate). Next the three-carbon acid pyruvic acid is formed from triose-P. Pyruvic acid loses a molecule of CO_2, yielding the two-carbon acetyl unit, which combines with coenzyme A (CoA) to form acetyl CoA.
(c) **Citric acid cycle (Krebs cycle).** A two-carbon acetyl unit from acetyl CoA joins with the four-carbon oxaloacetate unit to produce the six-carbon tricarboxylic acid citric acid (designated here as citrate). A two-step conversion to isocitrate occurs, followed by the loss of a molecule of CO_2 and the formation of the five-carbon α-ketoglutarate. Another CO_2 molecule is lost in the formation of succinyl CoA. The remainder of the cycle involves a succession of four-carbon acids leading to oxaloacetate. The oxaloacetate regenerated at the end of the cycle now joins with another acetyl unit, and the cycle is repeated. The net change occurring in the cycle, then, is that a two-carbon acetyl unit enters the cycle and two molecules of CO_2 leave.

Emphasis in this outline is on the degradation of large molecules to small ones (catabolism). Some of the species shown in the citric acid cycle are also involved in the biosynthesis of larger molecules (anabolism). For example, the amino acid asparagine is synthesized from oxaloacetate.

The raw materials for metabolic transformations are organic molecules, but the reactions involved are very dissimilar to those considered in the preceding chapter. In fragile biological systems, heat and pressure cannot be used to force chemical reactions to go, nor can strongly acidic or basic catalysts. The substances that do control the processes of metabolism are the catalysts known as enzymes. Herein lies the basis of biological uniqueness. An organism develops into what it is because of the particular set of enzymes it possesses.

Metabolism is a complex subject and anything more than a brief overview is much beyond the scope of this text. The discussion that follows centers on the summary of metabolic processes outlined in Figure 28-17.

Carbohydrate Metabolism. Foods containing starch are the principal sources of carbohydrates for humans and many animals. Digestion of starch begins in the mouth through the action of salivary enzymes, the amylases. Starch is converted to maltose and polysaccharides known as dextrins. This process continues in the acidic medium of the stomach, and the maltose and polysaccharides pass on to the small intestine. Here, amylase from the pancreas completes the conversion of polysaccharides to maltose, and the enzyme maltase converts maltose to glucose. Glucose is absorbed through the wall of the small intestine into the bloodstream, from which it is distributed to other organs.

Glucose is ultimately oxidized to carbon dioxide and water with the liberation of energy; the principal intermediate in this process is glucose 6-phosphate (glucose-6P). Its formation is controlled by the pancreatic hormone, *insulin*. Once formed glucose-6P may be converted to glycogen (a polysaccharide stored in the liver), back to glucose, or it may be metabolized. The major route for this metabolism involves the anaerobic (absence of air) Embden–Meyerhof pathway, followed by an aerobic cycle (Krebs cycle). These interrelationships are suggested diagrammatically in Figure 28-17.

Lipid Metabolism. Fats stored in the body represent a rich source of energy. The digestion of fats and oils occurs primarily in the small intestine, through the action of a combination of lipase enzymes. The products of this enzyme hydrolysis are glycerol, mixtures of mono- and diglycerides, and fatty acids. These are absorbed into the bloodstream through the wall of the intestine. Glycerol is converted to glyceraldehyde-3-phosphate (triose phosphate) and joins into the glucose metabolism route previously described. Fatty acids are oxidized to carbon dioxide and water, with the release of energy, in a series of reactions known as β oxidation. In this process oxidation occurs at the β carbon atom of a fatty acid, followed by cleavage. This means that two-carbon pieces (acetic acid) are split off. The process requires the presence of coenzyme A (CoA). For example, with palmitic acid ($C_{15}H_{31}CO_2H$) the process must be repeated seven times, with the formation of eight molecules of acetyl coenzyme A, which enter the Krebs cycle (Figure 28-17).

Protein Metabolism. In the stomach, hydrochloric acid and the enzyme pepsin hydrolyze about 10% of the amide linkages in proteins and produce polypeptides in the molecular weight range of 500 to several thousand. In the small intestine peptidases such as trypsin and chymotrypsin (from the pancreas) cleave the polypeptides into very small fragments. These are then acted upon by aminopeptidase and carboxypeptidase. The resulting free amino acids pass through the wall of the intestine, into the bloodstream, and on to various organs. Each amino acid has its own characteristic metabolic reactions, but in general each is converted to an intermediate that enters the Krebs cycle. Proteins may also be synthesized from amino acids in body cells under directions supplied by nucleic acids (see Section 28-6). However, the eight essential amino acids mentioned previously cannot be synthesized in the body and must be obtained from digested proteins.

In glucose 6-phosphate, a phosphate group replaces the —OH group on the C-6 atom of a pyranose (see Figure 28-7).

$$H_2C-O-\overset{\overset{\textstyle O^-}{|}}{\underset{\underset{\textstyle O}{\|}}{P}}-OH$$

The structure of glyceraldehyde 3-phosphate is

$$\begin{array}{l} \overset{\overset{\textstyle O}{\|}}{C}-H \\ H-\overset{|}{C}-OH \\ H_2-\overset{|}{C}-O-\overset{\overset{\textstyle O^-}{|}}{\underset{\underset{\textstyle O}{\|}}{P}}-OH \end{array}$$

Energy Relationships in Metabolism. Reactions in which molecules are synthesized in an organism, *anabolic reactions,* must acquire energy from reactions in which molecules are degraded, *catabolic reactions.* The fundamental agents responsible for these energy exchanges are adenosine triphosphate (ATP) and adenosine diphosphate (ADP). The energy released in exergonic reactions is stored in ATP by its conversion from ADP.

$$\text{ADP + inorganic phosphate (P}_i\text{) + 30-50 kJ} \underset{\substack{\text{energy} \\ \text{released}}}{\overset{\substack{\text{energy} \\ \text{absorbed}}}{\rightleftharpoons}} \text{ATP + H}_2\text{O} \qquad (28.13)$$

Energy is released from foods by oxidation processes. The energy released in the oxidation is picked up by ADP, which is converted to ATP. Enzymes catalyze each conversion every step along the way. ADP, ATP, and two important intermediates, nicotinamide adenine dinucleotide (NAD) and flavin adenine dinucleotide (FAD), are pictured in Figure 28-18.

The exact way in which ADP and ATP enter into metabolic processes was not detailed in Figure 28-17. However, it is known that the conversion of 1 mol of glucose to CO_2 and H_2O is accompanied by the conversion of 38 mol ADP to ATP.

$$C_6H_{12}O_6 + 6\,O_2 + 38\,ADP + 38\,P_i \longrightarrow 6\,CO_2 + 6\,H_2O + 38\,ATP \qquad (28.14)$$

Assuming that 33.5 kJ of energy is absorbed for each mole of ADP converted to ATP, the energy stored in ATP as a result of the metabolism of 1 mol glucose is

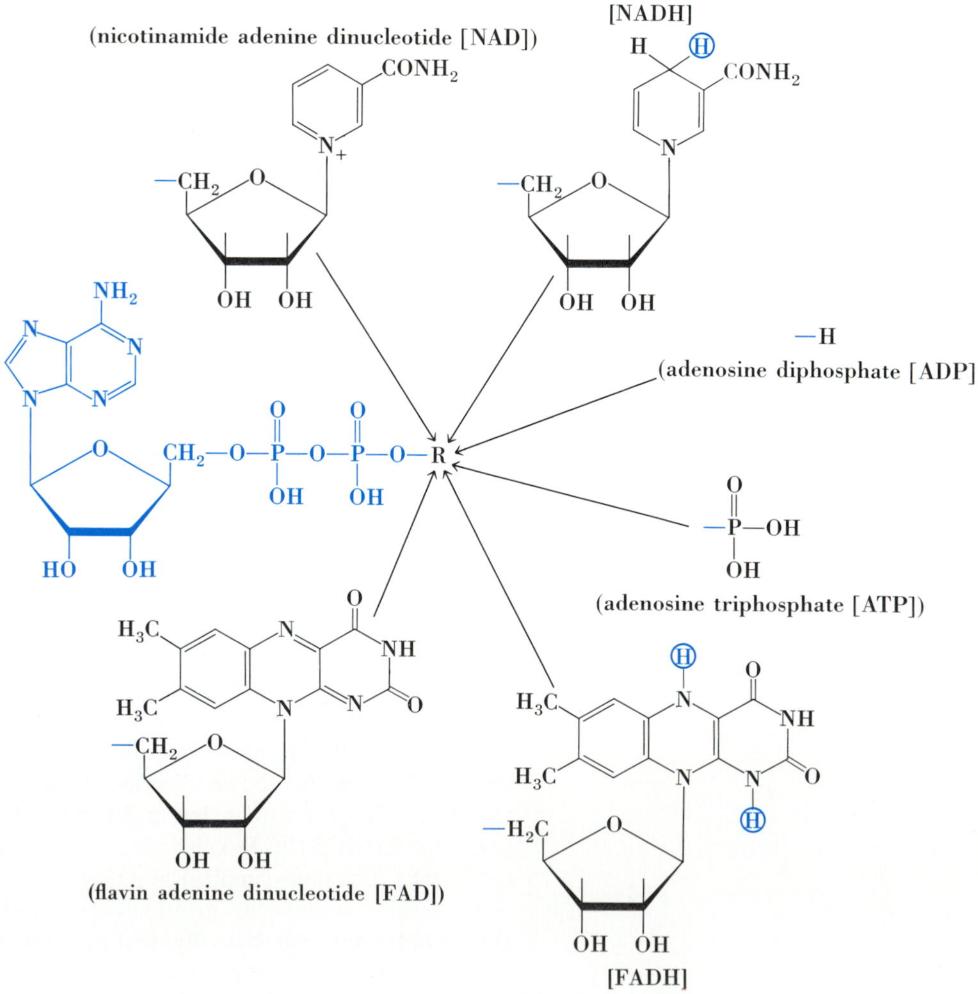

FIGURE 28-18

Some important chemical intermediates in metabolism.

Various R groups can be joined to the structure shown in blue. The reduced form of NAD, called NADH, contains one additional H atom (encircled). The reduced form of FAD, called FADH, contains two additional H atoms (encircled).

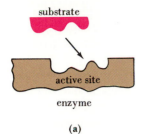

substrate

active site

enzyme

(a)

active site

enzyme–substrate complex

(b)

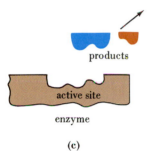

products

active site

enzyme

(c)

FIGURE 28-19

The lock-and-key model of enzyme action.

(a) The substrate attaches itself to an active site on the enzyme molecule.
(b) Reaction occurs.
(c) Product species detach themselves from the site, freeing the enzyme molecule to attach another substrate molecule.

The substrate and enzyme must have complementary structures to produce a complex, hence the term *lock-and-key*.

$38 \times 33.5 = 1270$ kJ. The total energy released when 1 mol glucose is converted to CO_2 and H_2O is 2870 kJ. Thus, the efficiency of energy storage in the high-energy bonds of ATP is $(1270/2870) \times 100 = 44\%$. Compared to the efficiency of heat engines in converting heat to work (recall Section 20-8), the metabolic process is an efficient one. Nearly half the energy of glucose can be stored by the body for later use. Perhaps you can appreciate the complexity of the metabolic process in terms of the data just given. If the metabolism of glucose occurred in a single step, with one ADP converted to ATP in that step, only $(33.5/2870) \times 100 = 1\%$ of the available energy would be conserved. Because metabolism occurs in many steps there is an opportunity for a much greater quantity of energy to be stored.

Enzymes. An enzyme is a biological catalyst that contains protein. Some enzymes are made up only of protein; some need cofactors or coenzymes to function (e.g., FAD and NAD). Enzymes are specific for each biological transformation and catalyze a reaction without requiring a change in temperature or pH. Originally, enzymes were assigned common or trivial names, such as pepsin and catalase. Present practice, however, is to name them after the processes they catalyze, usually employing an "ase" ending.

In 1913, Michaelis and Menten proposed a model of enzyme reactivity based on the formation of an enzyme–substrate complex. According to this model, an enzyme can exert its catalytic activity only after combining with the reacting substance, the **substrate,** to form a complex. There appears to be a definite site on the enzyme where the substrate combines; for some enzymes there is evidence that more than one active site exists. Reaction of the substrate (S) with the enzyme(E) to form a complex (ES) permits the reaction to proceed via a path of lower activation energy than the noncatalyzed path. When the complex decomposes, products (P) are formed and the enzyme is regenerated (see Figure 28-19). This general reaction scheme and a specific example follow.

$$E + S \rightleftharpoons E S \longrightarrow E + P$$

$$sucrase + sucrose \rightleftharpoons \begin{array}{c} sucrase–sucrose \\ complex \end{array} \xrightarrow{H_2O} sucrase + glucose + fructose$$

The implications of the Michaelis–Menton mechanism for the rates of enzyme-catalyzed reactions were considered in Section 15-11.

Generally, a 10 °C temperature rise produces an approximate doubling of a reaction rate, but with enzymes a certain temperature is reached beyond which a decrease in rate sets in. The optimum temperature for enzyme activity is about 37 °C (98 °F) for the enzymes present in warm-blooded animals. The decrease in rate beyond this temperature results from the fact that enzymes are proteins and proteins are denatured by heat. This denaturation disrupts the secondary and tertiary structure and distorts the active site on the enzyme.

Protein behavior is extremely sensitive to changes in pH. At high and low pH values complete denaturation of enzymes occurs, but even milder changes in pH cause drastic changes in the rate of enzyme action. Most enzymes in the body have their maximum activity between pH 6 and 8, with gastric enzymes being notable exceptions. In addition to the effects of temperature and pH, specific inhibition can occur when a molecule other than the substrate competes for an active enzyme site. For example, heavy metal ions (Hg^{2+}, Pb^{2+}, and Ag^+) may combine in a nonreversible way with active site groups, such as —OH, —SH, —CO_2^-, and —NH_3^+, and deactivate the enzyme. It has been suggested that antibiotics function by inhibiting enzyme–coenzyme reactions in microorganisms, and a similar mechanism explains the functioning of some insecticides.

Hormones. A hormone is a secretion of a ductless or endocrine gland, such as the thyroid, the pituitary, the pancreas (in part), the adrenals, and parts of the testes and ovaries. These glands secrete their products directly into the bloodstream through which these products reach all parts of the body. The hormones appear to aid in the control of biological reactions, but their exact role is not clearly understood. Hormones are sometimes referred to as "chemical messengers."

A well-known protein hormone is insulin. Its function is to lower the blood sugar level by increasing the rate of conversion of glucose into muscle and liver glycogen. There is considerable evidence that insulin acts by controlling the phosphorylation of glucose. In the absence of a sufficient amount of insulin, the condition of diabetes mellitus results. Among its clinical symptoms are high levels of sugar in the blood and urine and the formation of excess ketones, giving a distinctive odor to the breath.

Vitamins. Vitamins are substances necessary to maintain normal health, growth, and nutrition; however, they are not used in building cells or as an energy source. Their apparent function is as catalysts for biological processes. The vitamins are sometimes classified as fat soluble (vitamins A, D, E, and K) and water soluble (vitamins B and C). Vitamin A, although not found in plants itself, can be formed from β-carotene, a yellow pigment found in plants. A deficiency of vitamin A causes night blindness and xerophthalmia, a disease of the eyes in which the tear glands cease to function. Vitamin D is associated with the proper deposition of calcium phosphate, which in turn is related to normal teeth and bone development and the prevention of rickets.

Vitamin E is sometimes called the fertility factor and is involved in the proper functioning of the reproductive system. It is found in vegetable oils, such as corn germ oil, cottonseed oil, peanut oil, and wheat germ oil. There is more than one form of vitamin E; the structure for α-tocopherol, the most active, is shown below.

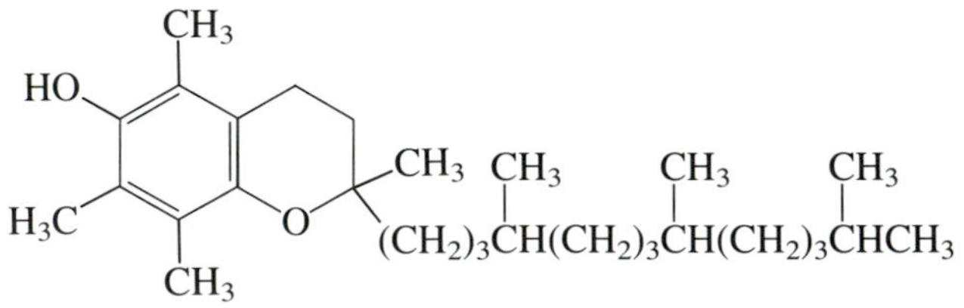

vitamin E
α-tocopherol

Vitamin K is the antihemorrhagic factor involved in bloodclotting. There are two K vitamins. Vitamin K_1 can be obtained from alfalfa and other plants. Vitamin K_2 is formed by bacterial action in the intestine.

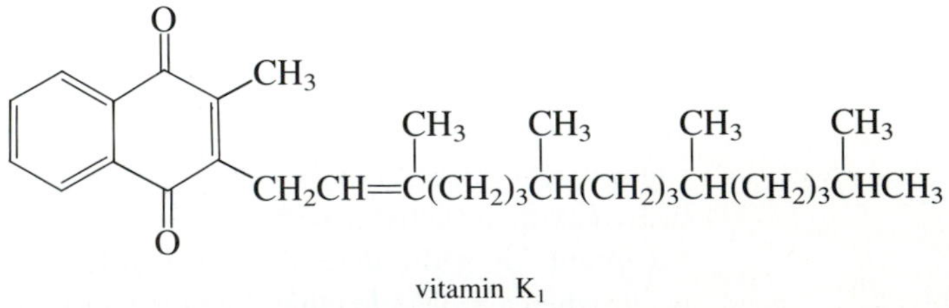

vitamin K_1

Vitamin C, ascorbic acid, is the vitamin that prevents scurvy. It is found in citrus fruits, green peppers, parsley, and tomatoes. The vitamin B complex has been shown to consist of many substances, most of which seem to be involved in energy transformations related to metabolism. A deficiency of vitamin B_1, thiamine, leads to the disease called beriberi. Lack of vitamin B_2, riboflavin, causes inflammation of the lips, dermatitis, a dryness and burning of the eyes, and sensitivity to light. Both of these B vitamins are distributed widely in nature, in lean meat, nuts, and leafy vegetables.

vitamin C, ascorbic acid   vitamin B₂, riboflavin

28-6 Nucleic Acids

Lipids, carbohydrates, and proteins, taken together with water, constitute about 99% of most living organisms. The remaining 1% includes some compounds of vital importance to the existence, development, and reproduction of all forms of life. Among these are the nucleic acids. Nucleic acids carry the information that directs the metabolic activity of cells.

The nucleus of the cell contains **chromosomes** that cause replication from one generation to the next. The individual portions of the chromosomes that carry specific traits are known as genes. It has now been established that DNA or **deoxyribonucleic acid** is the actual substance constituting the genes. Figure 28-20 traces the steps that may be followed in degrading nucleic acids into their simpler constituents—heterocyclic amines known as purines and pyrimidines, a five-carbon sugar (ribose or 2-deoxyribose), and phosphoric acid. Figure 28-21 represents a portion of a nucleic acid chain.

FIGURE 28-20
Hydrolysis products of nucleic acids.

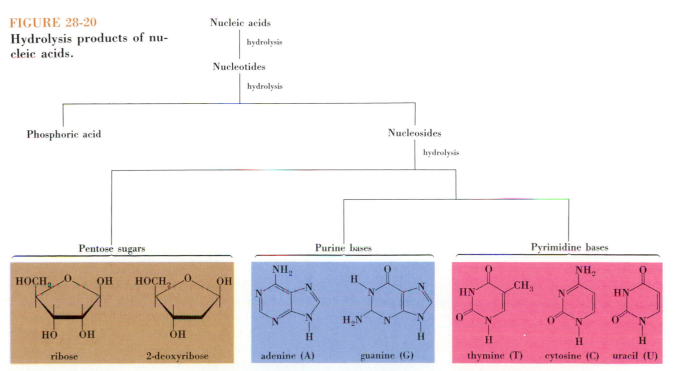

Tracing the hydrolysis reactions in the reverse direction, the combination of a pentose sugar and a purine or pyrimidine base yields a nucleoside. A nucleoside, in combination with phosphoric acid, yields a nucleotide. A nucleic acid is a polymer of nucleotides.

If the sugar is 2-deoxyribose and the bases A, G, T, and C, the nucleic acid is DNA. If the sugar is ribose and the bases A, G, U, and C, the nucleic acid is RNA. (The term 2-deoxy means without an oxygen atom on the second carbon atom.)

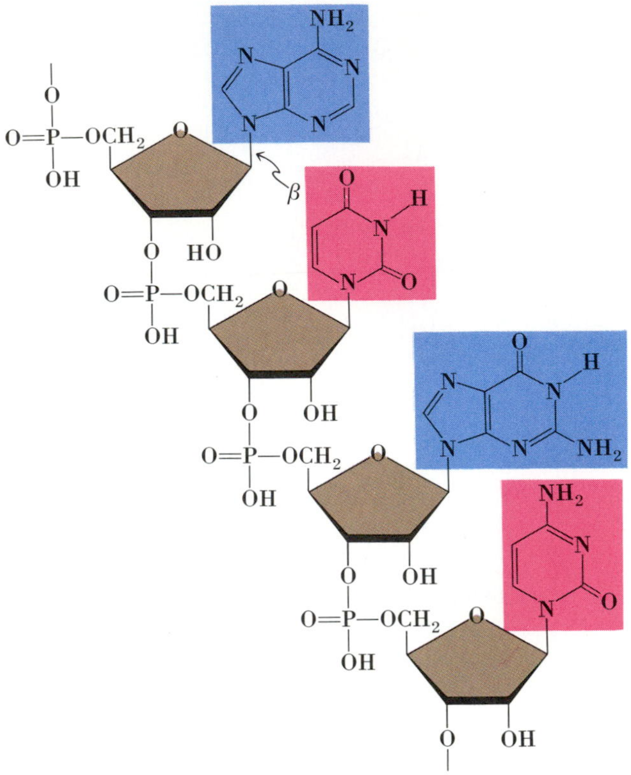

FIGURE 28-21
A portion of a nucleic acid chain.

The usual form for DNA is a **double helix.** The postulation of this structure of DNA by Francis Crick and James Watson in 1953 was one of the great scientific breakthroughs of modern times. Their work was critically dependent on two other contemporary scientific achievements. One was the precise x ray diffraction studies on DNA by Maurice Wilkins and Rosalind Franklin. The other was the discovery by Erwin Chargaff of a set of regularities regarding the occurrence of the purine and pyrimidine bases in nucleic acids. For DNA derived from a particular organism, these regularities, known as the **base-pairing rules,** require the following.

Base-pairing rules in DNA.

1. The amount of adenine is equal to the amount of thymine (A = T).
2. The amount of guanine is equal to the amount of cytosine (G = C).
3. The total amount of purine bases is equal to the total amount of pyrimidine bases (G + A = C + T).

(28.15)

Since large measures of intuition and inspiration were involved in Crick and Watson's work, it is not easy to show how their model was established from the facts just described. Instead, let us simply demonstrate how their model, shown in Figure 28-22, is consistent with the base-pairing rules.

In order to maintain the structure of a double helix, it is necessary that a force exist between the two single strands. As in the α-helical structure of a protein, the postulated force is based on hydrogen bonds, involving hydrogen, nitrogen, and oxygen atoms on the purine and pyrimidine bases. The necessary conditions for hydrogen bonding will exist only if an A on one strand appears opposite a T on the other, or if a G is bonded to a C. No other combinations will work. For example, C cannot be paired with T. Both are relatively small molecules (single ring) and would not approach each other closely enough between the strands. The combination of G and A cannot occur because the molecules are too large (double rings). With the combinations C + A and G + T, the conditions for hydrogen bonding are not right. Thus, it follows rather directly that the total amount of A must equal the total amount of T, and so on. The Chargoff rules are explained.

Proposal of the DNA structure was the important first step in the development of

FIGURE 28-22
DNA model.

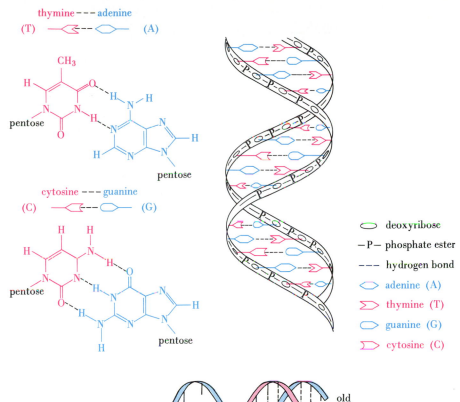

FIGURE 28-23
Replication of a DNA molecule visualized.

As the unzipping process occurs from right to left, hydrogen bonds between the old DNA strands are broken. The new strands grow in the direction of the arrows by attaching nucleotides, which then form hydrogen bonds to the old DNA strands.

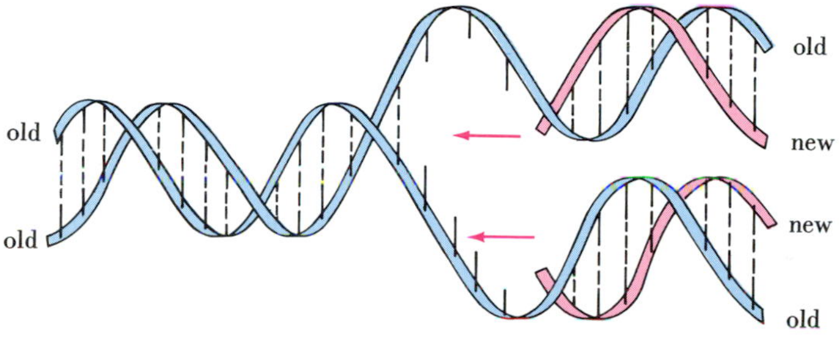

the theory of DNA, but at least three other questions must be explained by this theory: (1) How does a DNA molecule reproduce itself during cell division? (2) How does DNA direct the synthesis of proteins in the cell? (3) How is the information required to obtain the exact sequence of amino acids in a protein coded into the DNA structure? We cannot go into detail on these questions, but let us elaborate a bit.

The critical step in the replication of a DNA molecule requires the molecule to unwind into single strands. As the unwinding occurs, nucleotides present in the cell nucleus, through the action of enzymes, become attached to the exposed portions of the two single strands, converting each to a new double helix of DNA. As suggested by Figure 28-23, when the original DNA (parent) molecule is completely unwound, two new molecules (daughters) appear in its place!

There are two pieces of evidence, each very convincing, that the process outlined here does indeed occur. First, electron micrographs of the DNA molecule have now been obtained, including some that capture DNA in the act of replication. Another elegant experiment involves growing bacteria in a medium containing ^{15}N atoms, so that all the N atoms of the bases of the DNA molecules are ^{15}N. The bacteria are then transferred to a nutrient with nucleotides containing normal ^{14}N. Here the bacteria are allowed to divide and reproduce themselves. The DNA of the offspring cells are then analyzed. Those of the first generation, for example, consist of DNA molecules with one strand having ^{15}N and the other ^{14}N atoms. This is exactly the result to be expected if replication occurs by the unzipping process described in Figure 28-23.

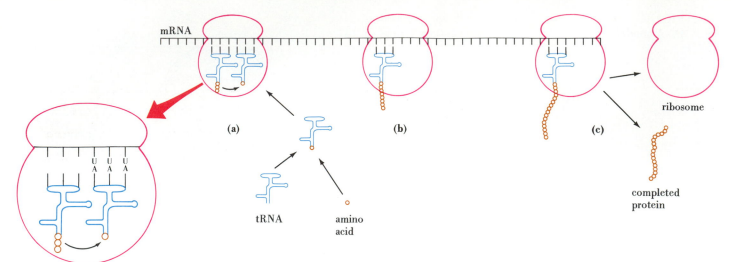

FIGURE 28-24

A representation of protein synthesis.

(a) Through the action of an enzyme, a tRNA molecule brings a single amino acid to a site on a ribosome. The anticodon of the tRNA (AAA for the example shown) must be complementary to the codon of the mRNA (UUU). (The amino acid carried by this tRNA is phenylalanine.) The amino acid is added to the chain in the manner shown; the chain moves from the tRNA on the left to the one on the right. **(b)** As the ribosome moves along the mRNA strand, more and more amino acid units are added through the proper matching of tRNA molecules with the code on the mRNA. **(c)** When the ribosome reaches the end of the mRNA strand, it and the completed protein are released. The ribosome is free to repeat the process.

The puzzle of protein synthesis is that this occurs in the cytoplasm of the cell, not in the nucleus. Yet the molecule that directs the synthesis, DNA, is found only in the nucleus. Here is where the different forms of RNA play a fundamental role. The several steps involved in this process are represented, collectively, in Figure 28-24.

First, a molecule of **messenger RNA, mRNA,** is synthesized on a portion of a DNA strand in the nucleus. This probably involves an unzipping of the DNA molecule similar to the process of DNA replication described above. The mRNA migrates out of the nucleus into the cytoplasm. mRNA has a high affinity for ribosomes and gathers them up along the chain. The combination of the mRNA and its ribosomes is called a **polysome. Transfer RNA, tRNA,** refers to a variety of rather short RNA chains, each of which is capable of attaching only a specific amino acid. The function of tRNA is to bring a specific amino acid to a site on the ribosome where the amino acid can form a polypeptide bond and become part of a growing polypeptide chain. The ribosome moves along the mRNA chain, attaching different tRNA molecules and incorporating their amino acids into the polypeptide chain. When the ribosomes reaches the end of the mRNA chain it falls off and releases the protein molecule that has been synthesized. The entire process is like the stringing of beads.

The code (set of directions) that determines the exact sequence of amino acids in the synthesis of a protein is incorporated in the chromosomal DNA. It is found in the particular pattern of base molecules on the double helix. Since there are only four different bases possible in an mRNA molecule—U, A, G, and C—but 20 different amino acids, it is clear that the code cannot correspond to individual base molecules. There are $4^2 = 16$ combinations of base molecules taken two at a time (i.e., UU, UA, UG, UC, etc.). But these could only account for 16 amino acids. When the base molecules are taken three at a time, there are $4^3 = 64$ possible combinations. A group of three base molecules in a DNA strand, called a **triplet,** causes a complementary set of base molecules to appear in the mRNA formed on it. This triplet or **codon** on the mRNA, must be matched by a complementary triplet, called an **anticodon,** in a tRNA molecule. The particular tRNA with this anticodon carries a specific amino acid to the site of protein synthesis.

Through an ingenious set of experiments, which led to a Nobel prize in 1968 for Marshall Nirenberg, Robert Holley, and Gobind Khorana, the genetic code (Table 28-4) has now been cracked. The significant features of the genetic code are the following.

- There is more than one triplet code for most amino acids.
- The first two letters of the codon are most significant. There is considerable variation in the third.

TABLE 28-4
The Genetic Code

		Second base			
		U	**C**	**A**	**G**
First base	**U**	UUU } Phe UUC UUA } Leu UUG	UCU UCC } Ser UCA UCG	UAU } Tyr UAC UAA Nonsense UAG Nonsense	UGU } Cys UGC UGA Nonsense UGG Trp
	C	CUU CUC } Leu CUA CUG	CCU CCC } Pro CCA CCG	CAU } His CAC CAA } Gln CAG	CGU CGC } Arg CGA CGG
	A	AUU AUC } Ile AUA AUG Met	ACU ACC } Thr ACA ACG	AAU } Asn AAC AAA } Lys AAG	AGU } Ser AGC AGA } Arg AGG
	G	GUU GUC } Val GUA GUG	GCU GCC } Ala GCA GCG	GAU } Asp GAC GAA } Glu GAG	GGU GGC } Gly GGA GGG

Initiation of a polypeptide chain appears to require the presence of the codon AUG, and termination of the chain, either UAA or UAG or UGA.

- There are three codons that do not correspond to any amino acids. Although they are referred to as nonsense codons, these seem to play a role in stopping protein synthesis (rather like the word "STOP" used to separate phrases in a telegram).
- The various codons direct the *same* protein synthesis, whether in bacteria, plants, lower animals, or humans.

Summary

Four categories of substances found in living organisms are considered in this chapter—lipids, carbohydrates, proteins, and nucleic acids.

One familiar group of lipids are the triglycerides. These are esters of glycerol with long-chain monocarboxylic (fatty) acids. If saturated fatty acids predominate, the triglyceride is a fat. If unsaturation (in the form of double bonds) predominates in some of the fatty acid components, the triglyceride is an oil. The catalytic addition of hydrogen to double bonds in a triglyceride converts an oil to a fat. Hydrolysis of a triglyceride with a strong base (saponification) yields glycerol and metal salts of the fatty acids—soaps. In phosphatides (phospholipids) phosphoric acid and a nitrogen-containing compound are substituted for one of the fatty acid components of a triglyceride.

The simplest carbohydrate molecules are five- and six-carbon-chain polyhydroxy aldehydes and ketones. However, these molecules convert readily into cyclic structures—five- and six-membered rings. The subject is further complicated by the existence of chiral carbon atoms in these molecules, rendering the molecules optically active,

that is, able to rotate the plane of polarized light. Moreover, optical isomers are encountered—molecules of identical composition that differ in their spatial orientations and in their optical activity. Monosaccharides or simple sugars can be readily joined into polysaccharides containing from a few to a few thousand monosaccharide units.

The basic building blocks of proteins are some 20 different α-amino acids. Two amino acid molecules may join by eliminating a H_2O molecule between them. The result

is a peptide bond, $-\overset{\overset{\text{O}}{\|}}{\text{C}}-\overset{\overset{\text{H}}{|}}{\text{N}}-$. Long polypeptide chains are formed as this process is repeated. Additional structural features include the twisting of a polypeptide chain into a helical coil and bonding between coils.

Lipids, carbohydrates, and proteins are complex molecules. Their metabolism by the body involves breaking them down into their simplest units. Carbohydrates are broken down into monosaccharides; proteins, into amino acids; and lipids, into glycerol and two-carbon-chain acids. Ultimately, these products are decomposed into still simpler molecules, such as CO_2, H_2O, NH_3, and urea. Energy released in these processes is stored in the substance adenosine triphosphate (ATP).

DNA molecules found in cell nuclei are able to synthesize RNA molecules, the framework on which protein synthesis occurs. Proteins acquire the correct sequence of amino acids because each RNA molecule carries a code imparted to it in its synthesis by DNA. DNA molecules also have the ability to replicate themselves during cell division, which ensures that the correct protein-building code is passed on from one generation of cells to the next.

Summarizing Example

L-Threonine is one of the essential amino acids that we must obtain through the foods we eat. It is present in animal protein (e.g., eggs and milk) but is deficient in some grains, such as rice. It is now common practice to fortify grains with the essential amino acids that would normally be missing. The structure of L-threonine is shown below.

$$CH_3-\underset{\underset{NH_2}{|}}{\overset{\overset{OH}{|}}{CH}}-CH-\overset{\overset{O}{\|}}{C}-OH$$

$pK_{a_1} = 2.15$, $pK_{a_2} = 9.12$

Workers planting rice seedlings in Szechwan Province of the People's Republic of China. [Bruno Barbey/Magnum Photos, Inc.]

1. Draw condensed structural formulas of the structures you would expect for L-threonine in **(a)** strongly acidic solutions, **(b)** at the isoelectric point, and **(c)** in strongly basic solutions.

Solution. Follow the scheme outlined in expression (28.11) to obtain
(a) *acidic solution:* $CH_3CH(OH)\underset{\underset{NH_3^+}{|}}{CH}COOH$

(b) *isoelectric point:* $CH_3CH(OH)\underset{\underset{NH_3^+}{|}}{CH}COO^-$

(c) *basic solution:* $CH_3CH(OH)\underset{\underset{NH_2}{|}}{CH}COO^-$

2. Write chemical equations corresponding to the two pK values for *acid* ionizations given for L-threonine.

Solution. Start with the structures written for part **1** and represent the stepwise loss of two protons. (Note from expression 28.11 that when OH^- is added to an acidic solution of an amino acid, a proton is extracted from the group $-COOH$ before one is extracted from $-NH_3^+$. The group $-COOH$ is a stronger acid than is $-NH_3^+$, and pK_{a_1} corresponds to the ionization of the carboxyl group.)

First ionization:
$$CH_3CH(OH)\underset{\underset{NH_3^+}{|}}{CH}COOH + H_2O \rightleftharpoons H_3O^+ + CH_3CH(OH)\underset{\underset{NH_3^+}{|}}{CH}COO^-$$
$$pK_{a_1} = 2.15$$

Second ionization:
$$CH_3CH(OH)\underset{\underset{NH_3^+}{|}}{CH}COO^- + H_2O \rightleftharpoons H_3O^+ + CH_3CH(OH)\underset{\underset{NH_2}{|}}{CH}COO^-$$
$$pK_{a_2} = 9.12$$

3. Based on the ionization equations written in part **2**, justify the value of $pI = 5.6$ listed for threonine in Table 28-3.

Solution. In the titration of a polyprotic acid (Figure 18-10) we discovered that at the first equivalence point the principal species in solution is the product of the first neutralization reaction and that the pH of the solution is essentially $pH = \frac{1}{2}(pK_{a_1} + pK_{a_2})$ (recall equations 18.27 and 18.28 and Example 18-10). The titration of threonine closely resembles that of a weak diprotic acid. The product of the first neutralization step is the zwitterion, $CH_3CH(OH)CH(NH_3^+)COO^-$. The pH at the first equivalence point is the isoelectric point, and $pI \approx \frac{1}{2}(2.15 + 9.12) = 5.64$.

4. A sample of L-threonine is dissolved in a gel medium that is also 0.25 M in NaH_2PO_4 and 0.25 M $NaHPO_4$. A pair of electrodes is immersed in the gel, and an electric current is passed between them. Toward which electrode will the L-threonine migrate?

Solution. The key to this question is to recognize a NaH_2PO_4–$NaHPO_4$ mixture as a buffer solution. It establishes the pH of the gel medium. The form of the L-threonine (that is, cationic, anionic, or zwitterion) depends on the pH of the buffer relative to the pI of the amino acid. The essential calculation then is to determine the pH of the buffer medium. For this we can use the Henderson–Hasselbalch equation (18.13) and a value of $K_{a_2} = 6.3 \times 10^{-8}$ (from Table 17-3).

$$pH = pK_{a_2} + \log \frac{[HPO_4^{2-}]}{[H_2PO_4^-]} = 7.20 + \log \frac{0.25}{0.25} = 7.20$$

Since the pH (7.20) is greater than the pI (5.6), we expect at least some of the amino acid to be in its anionic form, $CH_3CH(OH)CHNH_2COO^-$, and to migrate toward the anode (positive electrode).

Key Terms

absolute configuration (28-3)
α-amino acid (28-4)
biomass (28-3)
carbohydrate (28-3)
denaturation (28-4)
deoxyribonucleic acid (DNA) (28-6)
diastereomers (28-3)
enzyme (28-5)
fats (28-2)
genetic code (28-6)
iodine number (28-2)
isoelectric point (28-4)

lipids (28-2)
messenger RNA (mRNA) (28-6)
metabolism (28-5)
monosaccharide (28-3)
nucleic acids (28-6)
oils (28-2)
oligosaccharides (28-3)
peptide bond (28-4)
polypeptide (28-4)
polysaccharide (28-3)
primary structure (28-4)
protein (28-4)

quaternary structure (28-4)
reducing sugar (28-3)
ribonucleic acid (28-6)
saponification (28-2)
saponification value (28-2)
secondary structure (28-4)
soap (28-2)
substrate (28-5)
sugar (28-3)
tertiary structure (28-4)
transfer RNA (tRNA) (28-6)
triglycerides (28-2)

Highlighted Expressions

Saponification of a glyceride (28.1)
Saponification value (28.2)
Iodine number (28.3)
Classification of carbohydrates (28.5)
Test for a reducing sugar (28.9)

The photosynthesis reaction (28.10)
Acid–base properties of amino acids (28.11)
Formation of a peptide bond (28.12)
Base pairing rules in DNA (28.15)

Review Problems

1. Name the following compounds.

(a) $H_2CO\!-\!\overset{\displaystyle O}{\overset{\|}{C}}\!-\!C_{17}H_{35}$

$HCO\!-\!\overset{\displaystyle O}{\overset{\|}{C}}\!-\!C_{17}H_{33}$

$H_2CO\!-\!\overset{\displaystyle O}{\overset{\|}{C}}\!-\!C_{11}H_{23}$

(b) $H_2CO\!-\!\overset{\displaystyle O}{\overset{\|}{C}}\!-\!C_{17}H_{31}$

$HCO\!-\!\overset{\displaystyle O}{\overset{\|}{C}}\!-\!C_{17}H_{31}$

$H_2CO\!-\!\overset{\displaystyle O}{\overset{\|}{C}}\!-\!C_{17}H_{31}$

(c) $C_{13}H_{27}CO_2^- \ Na^+$

2. Write structural formulas for the following.
(a) glyceryl lauromyristolinoleate
(b) trilaurin
(c) potassium palmitate
(d) cetyl linoleate [cetyl alcohol = $CH_3(CH_2)_{14}CH_2OH$]

3. For glyceryl butyropalmitooleate, whose structure is written in Example 28-1, determine its (a) saponification value and (b) iodine number.

4. A simple glyceride is found to have a saponification value of 209. What is this simple glyceride? [*Hint:* Will the simple glyceride have a higher or lower molecular weight than tristearin (Example 28-2)?]

5. From the given structure of L-(+)-arabinose derive the

structure of (a) D-(−)-arabinose; (b) a diastereomer of L-(+)-arabinose.

$$H\!-\!C\!=\!O$$
$$H\!-\!\overset{\displaystyle |}{C}\!-\!OH$$
$$HO\!-\!\overset{\displaystyle |}{C}\!-\!H$$
$$HO\!-\!\overset{\displaystyle |}{C}\!-\!H$$
$$\overset{\displaystyle |}{C}H_2OH$$

L-(+)-arabinose

6. Write the formulas of the species expected if the amino acid phenylalanine is maintained in (a) 1.0 M HCl; (b) 1.0 M NaOH; (c) a buffer solution with pH 5.5.

7. Write the structures of (a) glycylmethionine; (b) isoleucyl-leucylserine.

8. For the polypeptide Gly-Ala-Ser-Thr, (a) write the structural formula; (b) name the polypeptide. [*Hint:* Which is the N-terminal and which is the C-terminal amino acid?]

9. With reference to Figure 28-21, identify the purine bases, the pyrimidine bases, the pentose sugars, and the phosphate groups. Is this a chain of DNA or RNA? Explain.

10. What polypeptide would be synthesized by the following coding on a mRNA strand? ACCCAUCCCUUGGCGAGUG-GUAUGUAA

Exercises

Structure and composition of the cell

Exercises 11 to 15 refer to a typical *E. coli* bacterium. This is a cylindrical cell about 2 μm long and 1 μm in diameter, weighing about 2×10^{-12} g, and containing about 80% water by volume. (See Figure 28-3.)

11. The intracellular pH is 6.4 and $[K^+] = 1.5 \times 10^{-4}$ M. Determine the number of (a) H^+ ions and (b) K^+ ions in a typical cell.

12. The *E. coli* cell contains about 1.5×10^4 ribosomes. Assuming a ribosome to be a sphere with a diameter of 18 nm, what percentage of the cell volume do the ribosomes occupy?

13. Calculate the number of lipid molecules present, assuming their average molecular weight to be 700 and the lipid content to be 2%.

14. The cell is about 15% protein, by mass, with 90% of this protein in the cytoplasm. Assuming an average molecular weight of 3×10^4, how many protein molecules are present in the cytoplasm?

15. A single chromosomal DNA molecule contains about 4.5 million mononucleotide units. If this molecule were extended so that the mononucleotide units were 450 pm apart, what would be the length of the molecule? How does this compare with the length of the cell itself? What does this result suggest about the shape of the DNA molecule?

Lipids

16. Describe briefly what is meant by each of the following terms, using specific examples where appropriate: (a) lipid;

(b) triglyceride; (c) simple glyceride; (d) mixed glyceride; (e) fatty acid; (f) soap.

17. Explain the essential distinction between the following pairs of materials: (a) a fat and a lipid; (b) a fat and an oil; (c) a fat and a wax; (d) butter and margarine.

18. Explain why phospholipids are more water soluble than simple or mixed glycerides.

19. What is the saponification value of glyceryl lauropalmitostearate?

20. What simple triglyceride has a saponification value of 193 and an iodine number of 174 (see Table 28-1)?

21. Oleic acid is a moderately unsaturated fatty acid. Linoleic acid belongs to a group called *polyunsaturated*. What structural feature characterizes polyunsaturated fatty acids? Is stearic acid polyunsaturated? Is eleostearic acid? Why do you suppose safflower oil is so highly recommended in dietary programs?

22. In light of present medical knowledge, which is a more desirable lipid for human consumption, one with a high saponification value or a high iodine number? What is the "best" of those listed in Table 28-2 from this standpoint?

Carbohydrates

23. Describe what is meant by each of the following terms, using specific examples where appropriate: (a) monosaccharide; (b) disaccharide; (c) oligosaccharide; (d) polysaccharide; (e) sugar; (f) glycose; (g) aldose, (h) ketose; (i) pentose; (j) hexose.

24. The following terms are all related to stereoisomers and their optical activity. Explain the meaning of each. **(a)** dextrorotatory; **(b)** levorotatory; **(c)** racemic mixture; **(d)** diastereomers; **(e)** (+); **(f)** (−); **(g)** D configuration.

25. Write the structure for the straight-chain form of L-glucose. Is this isomer dextrorotatory or levorotatory?

26. The pure α and β forms of D-glucose rotate the plane of polarized light to the right by 112° and 18.7°, respectively (denoted as +112° and +18.7°). Are these two forms of glucose enantiomers or diastereomers? (Consider also the structures shown in Figure 28-7.)

27. When a mixture of the pure α and β forms of glucose is allowed to reach equilibrium in solution, the rotation changes to +52.7° (a phenomenon known as mutarotation). What are the percentages of the α and β forms in the equilibrium mixture? [*Hint:* Refer to Exercise 26. If the mixture were 50:50, the rotation would be 0.50(+112) + 0.50(+18.7) = +65.4°.]

Amino acids, polypeptides, and proteins

28. Describe what is meant by each of the following terms, using specific examples where appropriate: **(a)** α-amino acid; **(b)** zwitterion; **(c)** isoelectric point; **(d)** peptide bond; **(e)** polypeptide; **(f)** protein; **(g)** N-terminal amino acid; **(h)** α helix; **(i)** denaturation.

29. A mixture of the amino acids lysine, proline, and glutamic acid is placed in a gel at pH 6.3. An electric current is applied between an anode and a cathode immersed in the gel. Toward which electrode will each amino acid migrate?

30. Write the structures of **(a)** the different tripeptides that can be obtained from a combination of alanine, serine, and lysine; **(b)** the tetrapeptides containing two serine and two alanine amino acid units.

31. Upon complete hydrolysis a polypeptide yields the following amino acids: Gly, Leu, Ala, Val, Ser, Thr. Partial hydrolysis yields the following fragments: Ser-Gly-Val, Thr-Val, Ala-Ser, Leu-Thr-Val, Gly-Val-Thr. An experiment using a marker establishes that Ala is the N-terminal amino acid.
(a) Establish the amino acid sequence in this polypeptide.
(b) What is the name of the polypeptide?

32. Describe what is meant by the primary, secondary, and tertiary structure of a protein. What is the quaternary structure? Do all proteins have a quaternary structure? Explain.

33. A 1.00-mL solution containing 1.0 mg of an enzyme was deactivated by the addition of 0.346 μmol $AgNO_3$ (1 μmol = 1×10^{-6} mol.) What is the *minimum* molecular weight of the enzyme? Why does this calculation yield only a minimum value? [*Hint:* How many active sites are present in each enzyme molecule?]

34. Sickle cell anemia is sometimes referred to as a "molecular" disease. Comment on the appropriateness of this term.

Metabolism

35. Describe briefly the meaning of each of the following terms as they apply to metabolism: **(a)** metabolite; **(b)** anabolism; **(c)** catabolism; **(d)** endergonic; **(e)** ADP; **(f)** ATP.

36. The metabolism of a particular metabolite has a theoretical free energy change of −837 kJ/mol. The metabolism of 1 mol of this material in a living organism results in the conversion of 15 mol ADP to ATP. What is the percent efficiency of this metabolism?

37. Calculate the equilibrium constant for the hydrolysis of glucose 6-phosphate to glucose and phosphoric acid if $\Delta G° = -13.8$ kJ/mol.

38. Explain why the action of an enzyme is so dependent on pH.

Nucleic acids

39. What are the two major types of nucleic acids? List their principal components.

40. DNA has been called the "thread of life." Comment on the appropriateness of this expression.

41. What are the principal functions of each of the following in protein synthesis? **(a)** DNA; **(b)** mRNA; **(c)** tRNA.

42. A ribosome is sometimes said to *read* an mRNA strand. Suggest a meaning for this expression.

43. Propose a plausible polypeptide sequence on a DNA strand that would code for the synthesis of the polypeptide Ser-Gly-Val-Ala. Why is there more than one possible sequence for the DNA strand?

★44. If one strand of a DNA molecule has the sequence of bases AGC, what must be the sequence on the opposite strand? Draw a structure of this portion of the double helix, showing all hydrogen bonds.

Additional Exercises

45. Write structural formulas for the following.
(a) hydrogenation product of triolein;
(b) saponification products of trilaurin;
(c) iodination product of glyceryl lauromyristolinoleate.

46. Castor oil is a mixture of triglycerides having about 90% of its fatty acid content as the unsaturated hydroxy aliphatic acid, ricinoleic acid.

$$CH_3(CH_2)_5CHOHCH_2CH{=}CH(CH_2)_7COOH.$$

Estimate the saponification value and iodine number of castor oil. [*Hint:* What triglyceride should you assume?]

47. There are eight aldopentoses. Draw their structures and indicate which are enantiomers.

48. The term **epimer** is used to describe diastereomers that differ in the configuration about a *single* carbon atom. Which pairs of the three naturally occurring aldohexoses shown on page 1018 are epimers (that is, are D-galactose and D-mannose epimers, etc.)?

49. The protein molecule hemoglobin contains four atoms of iron (recall Figure 28-16). The mass percent of iron in hemoglobin is 0.34%. What is the molecular weight of hemoglobin?

50. Refer to the Summarizing Example. In the manner of part 2, represent the ionization of lysine (see Table 28-3), for which the pK_a values are $pK_{a_1} = 2.20$; $pK_{a_2} = 8.90$; and $pK_{a_3} = 10.28$.

★51. The amino acid ornithine, not normally found in proteins, has the structure

$$\underset{NH_2}{CH_2}{-}CH_2{-}CH_2{-}\underset{NH_2}{CH}{-}COOH$$

$pK_{a_1} = 1.94$; $pK_{a_2} = 8.65$; $pK_{a_3} = 10.76$.

What is the p*I* value of this amino acid? [*Hint:* Which ionization step produces the zwitterion?]

* **52.** Refer to the Summarizing Example. A 1.00-g sample of threonine is dissolved in 10.0 mL of 1.00 M HCl, and the solution is titrated with 1.00 M NaOH. Sketch a titration curve for the titration, indicating the approximate pH at representative points on the curve [such as the initial pH, the half-neutralization point(s), and the equivalence point(s).]

* **53.** Draw complete structures for each of the chemical intermediates shown in Figure 28-18, that is, ADP, ATP, NAD, NADH, FAD, and FADH.

* **54.** What is a nucleoside? Draw structures of the following nucleosides using information from Figure 28-20.

  (a) cytidine   **(b)** uridine   **(c)** guanosine
  (d) deoxycytidine   **(e)** deoxyadenosine

* **55.** In the experiment described on page 1035, the first generation offspring of DNA molecules each contained one strand with ^{15}N atoms and one with ^{14}N. If the experiment were carried through a second, third, and fourth generation, what fractions of the DNA molecules would still have strands with ^{15}N atoms?

* **56.** Bradykinin is a nonapeptide that is obtained by the partial hydrolysis of blood serum protein. It causes a lowering of blood pressure and an increase in capillary permeability. Complete hydrolysis of bradykinin yields three proline (Pro), two arginine (Arg), two phenylalanine (Phe), one glycine (Gly), and one serine (Ser) amino acid units. The N-terminal and C-terminal units are both arginine (Arg). In a hypothetical experiment partial hydrolysis and sequence proof reveals the following fragments: Gly-Phe-Ser-Pro; Pro-Phe-Arg; Ser-Pro-Phe; Pro-Pro-Gly; Pro-Gly-Phe; Arg-Pro-Pro; Phe-Arg. Deduce the sequence of amino acid units in bradykinin.

* **57.** A pentapeptide was isolated from a cell extract and purified. A portion of the compound was treated with 2,4-dinitrofluorobenzene (DNFB) and the resulting material hydrolyzed. Analysis of the hydrolysis products revealed 1 mol of DNP-methionine, 2 mol of methionine, and 1 mol each of serine and glycine. A second portion of the original compound was partially hydrolyzed and separated into four products. Separately, the four products were hydrolyzed further, giving the following four sets of compounds. (a) 1 mol of DNP-methionine, 1 mol of methionine, and 1 mol of glycine; (b) 1 mol of DNP-methionine and 1 mol of methionine; (c) 1 mol of DNP-serine and 1 mol of methionine; (d) 1 mol of DNP-methionine, 1 mol of methionine, and 1 mol of serine. What is the amino acid sequence of the pentapeptide?

Self-Test Questions

For questions 58 through 65 select the single item that best completes each statement.

58. The substance *glyceryl trilinoleate* (linoleic acid: $C_{17}H_{31}COOH$) is best described as a (a) fat; (b) oil, (c) wax; (d) fatty acid.

59. One can most easily distinguish between *glyceryl tristearate* (stearic acid: $C_{17}H_{35}COOH$) and *glyceryl trioleate* (oleic acid: $C_{17}H_{33}COOH$) by measuring their (a) molecular weights; (b) saponification values; (c) iodine numbers; (d) hydrolysis constants.

60. The mixture of sugars referred to as DL (or *dl*)-erythrose rotates the plane of polarized light (a) to the left; (b) to the right; (c) first to the left and then to the right; (d) neither to the left nor to the right.

61. Of the following names, the one that refers to a simple sugar in its cyclic (ring) form is (a) β-galactose; (b) L-(−)-glyceraldehyde; (c) D-(+)-glucose; (d) DL-erythrose.

62. The coagulation of egg whites by boiling is an example of (a) saponification; (b) inversion of a sugar; (c) hydrolysis of a protein; (d) denaturation of a protein.

63. A molecule in which the energy of metabolism is stored is (a) ATP; (b) glucose; (c) CO_2; (d) glycerol.

64. Of the following, the one that is not a constituent of a nucleic acid chain is (a) purine base; (b) phosphate group; (c) glycerol; (d) pentose sugar.

65. The structure of the DNA molecule is best described as (a) a random coil; (b) a double helix; (c) a pleated sheet; (d) partly coiled.

66. Calculate the maximum mass of a sodium soap that could be prepared from 125 g of the triglyceride *glyceryl tripalmitate* (palmitic acid: $C_{15}H_{31}COOH$).

67. Upon complete hydrolysis of a pentapeptide the following amino acids are obtained: valine (Val), phenylalanine (Phe), glycine (Gly), cysteine (Cys), and tyrosine (Tyr). Partial hydrolysis yields the following fragments: Val-Phe, Cys-Gly, Cys-Val-Phe, Tyr-Phe. Glycine is found to be the N-terminal acid. What is the sequence of the amino acids in the polypeptide?

68. Explain why enzyme action is so dependent on factors such as temperature, pH, and the presence of metal ions. Why are enzymes so specific in the reactions they catalyze? (That is, why don't they catalyze a variety of reactions, as does platinum metal, for example?)

 # A Mathematical Operations

A-1 Exponential Arithmetic

The measured quantities that we deal with in this text range from very small to very large in value. For example, the mass of an individual hydrogen atom is 0.00000000000000000000000167 g; and the number of molecules in 18.106 g of water is 602,214,000,000,000,000,000,000. These numbers are very difficult to write in conventional form; they are even more cumbersome to handle when they appear in numerical calculations. We can greatly simplify these numbers by expressing them in exponential form.

The exponential form of a number.

> The exponential form of a number consists of a coefficient (a number with value between 1 and 10) multiplied by a power of ten. (A.1)

The number 10^n is the *nth power* of 10. If n is a *positive* quantity, the number is *greater than 1*. If n is a *negative* quantity, the number is *smaller than 1* (between 0 and 1). The number $10^0 = 1$.

Positive powers	*Negative powers*

$$10^0 = 1 \qquad\qquad 10^0 = 1$$

$$10^1 = 10 \qquad\qquad 10^{-1} = \frac{1}{10} = 0.1$$

$$10^2 = 10 \times 10 = 100 \qquad\qquad 10^{-2} = \frac{1}{10 \times 10} = 0.01$$

$$10^3 = 10 \times 10 \times 10 = 1000 \qquad 10^{-3} = \frac{1}{10 \times 10 \times 10} = \frac{1}{10^3} = 0.001$$

To express the number 3170 in exponential form, we write

$$3170 = 3.17 \times 1000 = 3.17 \times 10^3$$

For the number 0.00046 we write

$$0.00046 = 4.6 \times 0.0001 = 4.6 \times 10^{-4}$$

A simpler method of converting a number to exponential form that avoids intermediate steps is illustrated below.

$$3\,1\,7\,0 = 3.17 \times 10^3$$
$$\;\;3\;2\;1$$

$$0.0\,0\,0\,4\,6 = 4.6 \times 10^{-4}$$
$$\quad 1\;2\;3\;4$$

A1

Writing a number in the exponential form.

To convert a number to exponential form,

- Shift the decimal point so as to obtain a coefficient with value between 1 and 10.
- Count the number of places that the decimal point was moved.
- The exponent (power) of 10 is equal to the number of places the decimal point was moved.
- If the decimal point was moved *to the left*, the power of 10 is *positive* (number greater than 1).
- If the decimal point was moved *to the right*, the power of 10 is *negative* (number between 0 and 1).

(A.2)

To convert a number from exponential form to conventional form, move the decimal point the number of places indicated by the power of ten. That is,

$$6.1 \times 10^6 = 6.\underbrace{1\,0\,0\,0\,0\,0}_{1\,2\,3\,4\,5\,6} = 6{,}100{,}000$$

$$8.2 \times 10^{-5} = 0\underbrace{\,0\,0\,0\,0\,}_{5\,4\,3\,2\,1}8.2 = 0.000082$$

The instructions given here are for a typical electronic calculator. The key strokes required for your calculator may be somewhat different. Look for specific instructions in the instruction manual supplied with your calculator.

Electronic calculators designed for scientific and engineering work allow for the expression of exponential numbers. A typical procedure is to key in the number, followed by the key "exp". Thus, the key strokes required for the number 6.57×10^3 are

| 6 | . | 5 | 7 | exp | 3 |

and the result displayed is $\boxed{6.57^{03}}$

For the number 6.25×10^{-4}, the key strokes are

| 6 | . | 2 | 5 | exp | 4 | ± |

and the result displayed is $\boxed{6.25^{-04}}$

Electronic calculators also allow for a number to be entered in its conventional form and then converted to the exponential form. For example, if the number 0.0037 is entered, followed by the "exp" and the "=" keys, the value 3.7×10^{-3} is displayed. Some calculators have a mode setting that, when set to the mode "sci" (or something equivalent), automatically converts all numbers and calculated results to the exponential form, regardless of the form in which numbers are entered. In this mode setting you can generally also set the number of significant figures to be used in displayed results.

Addition and Subtraction. To add or subtract numbers written in the exponential form, you must first express each quantity as *the same power of ten*. Then add and/or subtract the coefficients as indicated. That is, treat the power of ten like you would a unit common to the terms being added and/or subtracted. In the example that follows, the key step is to convert 3.8×10^{-3} to 0.38×10^{-2}, and to use 10^{-2} as the common power of 10.

$$(5.60 \times 10^{-2}) + (3.8 \times 10^{-3}) - (1.52 \times 10^{-2}) = (5.60 + 0.38 - 1.52) \times 10^{-2}$$

$$= 4.46 \times 10^{-2}$$

Multiplication. Consider the numbers $a \times 10^y$ and $b \times 10^z$. Their product is $a \times b \times 10^{(y+z)}$. *Coefficients are multiplied and exponents are added.*

$$0.0220 \times 0.0040 \times 750 = (2.20 \times 10^{-2})(4.0 \times 10^{-3})(7.5 \times 10^2)$$

$$= (2.20 \times 4.0 \times 7.5) \times 10^{(-2-3+2)} = 66 \times 10^{-3}$$

$$= 6.6 \times 10^1 \times 10^{-3} = 6.6 \times 10^{-2}$$

Division. Consider the numbers $a \times 10^y$ and $b \times 10^z$. Their quotient is

$$\frac{a \times 10^y}{b \times 10^z} = (a/b) \times 10^{(y-z)}$$

Coefficients are divided, and the exponent of the denominator is subtracted from the exponent in the numerator.

$$\frac{20.0 \times 636 \times 0.150}{0.0400 \times 1.80} = \frac{(2.00 \times 10^1)(6.36 \times 10^2)(1.50 \times 10^{-1})}{4.00 \times 10^{-2} \times 1.80}$$

$$= \frac{2.00 \times 6.36 \times 1.50 \times 10^{(1+2-1)}}{4.00 \times 1.80 \times 10^{-2}} = \frac{19.1 \times 10^2}{7.20 \times 10^{-2}}$$

$$= 2.65 \times 10^{(2-(-2))} = 2.65 \times 10^4$$

Raising a Number to a Power. To "square" the number $a \times 10^y$ means to determine the value $(a \times 10^y)^2$ or the product $(a \times 10^y)(a \times 10^y)$. According to the rule for multiplication, this product is $(a \times a) \times 10^{(y+y)} = a^2 \times 10^{2y}$. When an exponential number is raised to a power, *the coefficient is raised to that power and the exponent is multiplied by the power.* For example,

$$(0.0034)^3 = (3.4 \times 10^{-3})^3 = (3.4)^3 \times 10^{3 \times (-3)} = 39 \times 10^{-9} = 3.9 \times 10^{-8}$$

Extracting the Root of an Exponential Number. To extract the root of a number is the same as raising the number to a fractional power. This means that the square root of a number is the number to the one-half power; the cube root is the number to the one-third power; and so on. Thus,

$$\sqrt{a \times 10^y} = (a \times 10^y)^{1/2} = a^{1/2} \times 10^{y/2}$$

$$\sqrt{156} = \sqrt{1.56 \times 10^2} = (1.56)^{1/2} \times 10^{2/2} = 1.25 \times 10^1 = 12.5$$

In the following example, where the cube root is extracted, the exponent (-5) is not divisible by 3; the number is rewritten so that the new exponent will be divisible by 3.

$$(3.52 \times 10^{-5})^{1/3} = (35.2 \times 10^{-6})^{1/3} = (35.2)^{1/3} \times 10^{-6/3} = 3.28 \times 10^{-2}$$

A-2 Logarithms

Definition of a common logarithm.

The common logarithm (log) of a number (N) is the exponent (x) to which the base 10 must be raised to yield the number N. That is, $\log N = x$ means that $N = 10^x = 10^{\log N}$. (A.3)

For simple powers of ten:

$$\log 1 = \log 10^0 = 0$$

$$\log 10 = \log 10^1 = 1 \qquad \log 0.10 = \log 10^{-1} = -1$$

$$\log 100 = \log 10^2 = 2 \qquad \log 0.01 = \log 10^{-2} = -2$$

Multiplication and Division. From the definition of a logarithm we can write: $M = 10^{\log M}$ and $N = 10^{\log N}$. This means that

$$M \times N = 10^{\log M} \times 10^{\log N} = 10^{(\log M + \log N)}$$

But, again from the definition of a logarithm, we have

$$M \times N = 10^{\log (M \times N)}$$

which means that

$$\log(M \times N) = \log M + \log N$$

Similarly,

$$\log \frac{M}{N} = \log M - \log N$$

Illustrative Examples. Most of the numbers that result from measurements and appear in calculations are not simple powers of 10. We can obtain the logarithms of such numbers with a calculator or from a table of logarithms. In the examples below log 7.34 and log 1.30 are obtained in either way.

$$\log 734 = \log (7.34 \times 10^2) = \log 7.34 + \log 10^2$$
$$= 0.866 \quad + 2 = 2.866$$

To obtain the logarithm of a number with an electronic calculator, enter the number, followed by the "log" key.

For a number smaller than 1, we can write

$$\log 0.00130 = \log(1.30 \times 10^{-3}) = \log 1.30 + \log 10^{-3}$$
$$= 0.114 \quad - 3 = -2.886$$

Suppose that a number has a logarithm of 4.350. What is the number? The number we are seeking is said to be the *antilogarithm* of 4.350, that is, the number whose logarithm is 4.350. In exponential form, the unknown number is $N = a \times 10^x$.

To obtain the antilogarithm of a number with an electronic calculator, enter the number, followed by the keys "INV" or "2nd F" and 10^x.

$$\log N = 4.350 = 0.350 + 4$$
$$a = \text{antilog } 0.350 = 2.24$$
$$x = 4$$
$$N = 2.24 \times 10^4$$

With an electronic calculator, the key strokes are

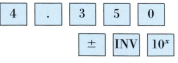

and the display is 4.47^{-05}

Another common example involves taking the antilogarithm of a negative number. For example, if $\log N = -4.350$; what is N? The key operation here involves stating -4.350 as a difference of two numbers.

$$\log N = -4.350 = 0.650 - 5$$
$$N = (\text{antilog } 0.650) \times 10^{-5}$$
$$= 4.47 \times 10^{-5}$$

Significant Figures. To determine the number of significant figures to use in a logarithm or antilogarithm, use this fundamental rule: All digits to the *right* of the decimal point in a logarithm are significant. Digits to the *left* of the decimal point are used to establish the power of ten. Thus, the logarithm -2.08 is expressed to *two* significant figures. The antilogarithm of -2.08 should also be expressed to *two* significant figures; it is 8.3×10^{-3}. To help settle this point, take the antilogs of -2.07, -2.08, and -2.09. You will find these antilogs to have the same first digit and slight differences in the second digit, that is, 8.5×10^{-3}, 8.3×10^{-3}, and 8.1×10^{-3}, respectively.

Natural Logarithms. We can express logarithms to a base other than 10. For instance, since $2^3 = 8$ we can write, $\log_2 8 = 3$ (read as "the logarithm of 8 to the base 2 is equal to 3"); or, $\log_2 10 = 3.322$; and so on. Several equations in this text are derived by the methods of calculus and involve logarithms. These equations require that the logarithm be a "natural" one. A natural logarithm has the base $e = 2.71828.$ . . . A logarithm to the base "e" is usually denoted by the symbol "ln". That is, $\log_e x = \ln x$.

The "ln" function arises in situations where the rate of change of some quantity is proportional to its present value.

In this text logarithmic expressions are given both in a "natural" and a "common" logarithmic form, and you are generally given a choice as to which form you use. The difference between the two forms simply involves the factor $\log_e 10 = 2.303$. One application where common logarithms *must* be used is in calculations involving pH, since pH is defined in terms of a logarithm to the base 10 (see Section 17-5).

A-3 Algebraic Operations

An algebraic equation is solved when one of the quantities, the unknown, is expressed in terms of all the other quantities in the equation. This effect is achieved when the unknown is present, *alone,* on one side of the equation, and the rest of the terms are on the other side. To solve an equation a rearrangement of terms may be necessary. The basic principle governing these rearrangements is quite simple. *Whatever is done to one side of the equation must be done as well to the other.*

$$(x^2 \times y) + 6 = z \qquad \textbf{\textit{Solve for x.}}$$

$$(x^2 \times y) + 6 - 6 = z - 6 \qquad (1) \text{ Subtract 6 from each side.}$$

$$(x^2 \times y) = z - 6$$

$$\frac{x^2 \times y}{y} = \frac{z - 6}{y} \qquad (2) \text{ Divide each side by } y.$$

$$x^2 = \frac{z - 6}{y}$$

$$\sqrt{x^2} = \sqrt{\frac{z - 6}{y}} \qquad (3) \text{ Extract the square root of each side. (Find the quantity } \sqrt{N} \text{ that, when multiplied by itself, yields the number } N.)$$

$$x = \sqrt{\frac{z - 6}{y}} \qquad (4) \text{ Simplify. The square root of } x^2 \text{ is simply } x.$$

Quadratic Equations. A quadratic equation has the form $ax^2 + bx + c = 0$, where a, b, and c are constants (a cannot be equal to 0). A number of calculations in the text require that you solve a quadratic equation. At times, you will encounter quadratic equations of the form

$$(x + n)^2 = m^2$$

Such equations can be solved by extracting the square root of each side:

$$x + n = m \qquad \text{and} \qquad x = m - n$$

More likely, however, you will have to use the quadratic formula.

$$x = \frac{-b \pm \sqrt{b^2 - 4ac}}{2a}$$

In Example 16-13 the following quadratic equation is obtained.

$$x^2 + 4.28 \times 10^{-4}\, x - 1.03 \times 10^{-5} = 0$$

Its solution is

$$x = \frac{-4.28 \times 10^{-4} \pm \sqrt{(4.28 \times 10^{-4})^2 + 4 \times 1.03 \times 10^{-5}}}{2}$$

$$= \frac{-4.28 \times 10^{-4} \pm \sqrt{(1.83 \times 10^{-7}) + (4.12 \times 10^{-5})}}{2}$$

$$= \frac{-4.28 \times 10^{-4} \pm \sqrt{4.14 \times 10^{-5}}}{2} = \frac{-4.28 \times 10^{-4} \pm 6.43 \times 10^{-3}}{2}$$

$$= \frac{-4.28 \times 10^{-4} + 6.43 \times 10^{-3}}{2} = \frac{6.00 \times 10^{-3}}{2} = 3.00 \times 10^{-3}$$

Note that only the $(+)$ value of the $(\pm)$ sign was used in solving for x. If the $(-)$ value had been used, a negative value of x would have resulted. However, for the given situation a negative value of x is meaningless.

Higher Degree Equations. The highest power of x in a quadratic equation is 2. A quadratic equation is a second-degree equation. Some equations have the unknown appearing to a higher degree. Algebraic equations of higher degree than second can be solved exactly, but a simple method that usually works well is the *method of successive approximations*. The following equation is obtained in part 3 of the Summarizing Example of Chapter 16 (page 595). It is a *fifth-degree* equation that must be solved for x.

$$256x^5 - 0.74[(1.00 - x)(1.00 - 2x)^2] = 0$$

The physical significance of x requires that it be a *positive* quantity less than 0.50 (since the term $1.00 - 2x$ cannot be negative). Suppose as a first approximation we

Try $x = 0.40$:

$$256(0.40)^5 - 0.74[(1.00 - 0.40)(1.00 - 2 \times 0.40)^2]$$

$$= 2.62 - 0.74[0.60(0.20)^2] = 2.62 - 0.018$$
$$= 2.60 > 0$$

Try $x = 0.30$:

$$256(0.30)^5 - 0.74[(1.00 - 0.30)(1.00 - 2 \times 0.30)^2]$$

$$= 0.62 - 0.74[0.70(0.40)^2] = 0.62 - 0.083$$
$$= 0.54 > 0$$

Try $x = 0.20$:

$$256(0.20)^5 - 0.74[(1.00 - 0.20)(1.00 - 2 \times 0.20)^2]$$

$$= 0.082 - 0.74[0.80(0.60)^2] = 0.082 - 0.21$$
$$= -0.13 < 0$$

Since the sum of the terms on the left side of the equation changes sign between $x = 0.20$ and $x = 0.30$, the value of x lies in this range.

Try $x = 0.22$:

$$256(0.22)^5 - 0.74[(1.00 - 0.22)(1.00 - 2 \times 0.22)^2]$$

$$= 0.132 - 0.74[0.78(0.56)^2] = 0.132 - 0.181$$
$$= -0.049 < 0$$

Try $x = 0.24$:

$$256(0.24)^5 - 0.74[(1.00 - 0.24)(1.00 - 2 \times 0.24)^5]$$

$$=0.204 - 0.74[0.76(0.52)^2] = 0.204 - 0.152$$
$$= 0.052 > 0$$

The value we are seeking is between 0.22 and 0.24. $x = 0.23$.

A-4 Graphs

Suppose the following sets of numbers are obtained for two quantities x and y by laboratory measurement.

$x = 0, 1, 2, 3, 4, \ldots$

$y = 2, 4, 6, 8, 10, \ldots$

The relationship between these sets of numbers is not difficult to establish.

$y = 2x + 2$

Ideally, the results of experimental measurements are best expressed through a mathematical equation. Sometimes, however, an exact equation cannot be written or its form is not clear from the experimental data. The graphing of data is very useful in such cases. In Figure A-1 the points listed above are located on a coordinate grid, in which x values are placed along the horizontal axis (abscissa) and y values along the vertical axis (ordinate). For each point the x and y values are indicated in parentheses.

The data points are seen to define a straight line. A mathematical equation for a straight line always has the form

Equation of a straight line graph.

$$y = mx + b \tag{A.4}$$

FIGURE A.1
A straight line graph:
$y = mx + b.$

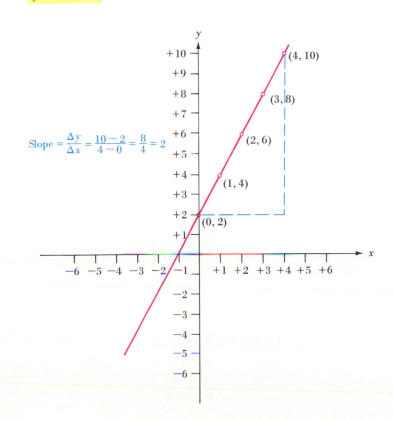

$$\text{Slope} = \frac{\Delta y}{\Delta x} = \frac{10 - 2}{4 - 0} = \frac{8}{4} = 2$$

Values of m, the **slope** of the line, and b, the **intercept,** can be obtained from the straight line graph.

When $x = 0$, $y = b$. The intercept is the point where the straight line intersects the y-axis. The slope can be obtained from two points on the graph.

$$y_2 = mx_2 + b \qquad \text{and} \qquad y_1 = mx_1 + b$$

$$y_2 - y_1 = m(x_2 - x_1)$$

$$m = \frac{y_2 - y_1}{x_2 - x_1}$$

From the straight line in Figure A-1 can you establish that $m = b = 2$?

B Some Basic Physical Concepts

B-1 Velocity and Acceleration

Time elapses as an object moves from one place to another. The **velocity** of the object is defined as the distance traveled per unit of time. An automobile that travels a distance of 60.0 km in exactly one hour has a velocity of 60.0 km/h (or 16.7 m/s).

Table B-1 contains experimental data on the velocity of a free-falling body. For this type of motion velocity is not constant—it increases with time. The falling body "speeds up" continuously. The rate of change of velocity with time is called **acceleration.** It has the units of distance per unit time per unit time. By the methods of calculus, mathematical equations can be derived for the velocity (u) and the distance (d) traveled in a time (t) by an object that has a constant acceleration (a).

TABLE B-1
Velocity and Acceleration of a Free-Falling Body

Time elapsed, s	Total distance, m	Velocity, m/s	Acceleration, m/s^2
0	0		
1	4.9	4.9	
2	19.6	14.7	9.8
3	44.1	24.5	9.8
4	78.4	34.3	9.8

$$u = at \tag{B.1}$$

$$d = \tfrac{1}{2}at^2 \tag{B.2}$$

For a free-falling body, the constant acceleration, called the *acceleration due to gravity,* is $a = g = 9.8$ m/s^2. Equations (B.1) and (B.2) can be used to calculate the velocity and distance traveled by a free-falling body.

B-2 Force and Work

Newton's first law of motion states that an object at rest remains at rest, and that an object in motion remains in uniform motion, unless acted upon by an external force. The tendency for an object to remain at rest or in uniform motion is called **inertia;** a **force** is what is required to overcome inertia. Since the application of a force either gives an object motion or changes its motion, the actual effect of a force is to change the velocity of an object. Change in velocity is an *acceleration,* so force is what provides an object with acceleration.

A9

Newton's second law of motion describes the force, F, required to produce an acceleration, a, in an object of mass, m.

$$F = ma \tag{B.3}$$

The basic unit of force in the SI system is the **newton (N).** It is the force required to provide a one kilogram mass with an acceleration of one meter per second per second.

$$1 \text{ N} = 1 \text{ kg} \times 1 \text{ m s}^{-2} \tag{B.4}$$

The force of gravity on an object (its weight) is the product of the mass of the object and the acceleration of gravity, g.

$$F = mg \tag{B.5}$$

Work is performed when a force acts through a distance.

$$\text{work } (w) = \text{force } (F) \times \text{distance } (d) \tag{B.6}$$

The **joule (J)** is the amount of work associated with a force of 1 newton (N) acting through a distance of 1 m.

$$1 \text{ J} = 1 \text{ N} \times 1 \text{ m} \tag{B.7}$$

From the definition of the newton in expression (B.4), we can also write

$$1 \text{ J} = 1 \text{ kg} \times 1 \text{ m s}^{-2} \times 1 \text{ m} = 1 \text{ kg m}^2 \text{ s}^{-2} \tag{B.8}$$

B-3 Energy

Energy is defined as the capacity to do work, but there are other useful descriptions of energy as well. For example, a moving object possesses a kind of energy known as **kinetic energy.** We can obtain a useful equation for kinetic energy by combining some of the other simple equations in this appendix. Thus, since work is the product of a force and distance (equation B.6), and force is the product of a mass and acceleration (equation B.3), we can write

$$w \text{ (work)} = m \times a \times d \tag{B.9}$$

Now, substitute equation (B.2), which relates acceleration (a), distance (d) and time (t), into equation (B.9).

$$w \text{ (work)} = m \times a \times \tfrac{1}{2}at^2 \tag{B.10}$$

Finally, substitute expression (B.1) relating acceleration (a) and velocity (u) into (B.10). That is, since $a = u/t$,

$$w \text{ (work)} = \tfrac{1}{2}m \left(\frac{u}{t}\right)^2 t^2 \tag{B.11}$$

Think of the work in (B.11) as the amount of work necessary to produce a velocity of u in an object of mass m. This amount of work is the energy that appears in the object as kinetic energy (e_k).

$$e_k \text{ (kinetic energy)} = \tfrac{1}{2}mu^2 \tag{B.12}$$

An object at rest may also have the capacity to do work by changing its position. The energy it possesses, which can be transformed to actual work, is called **potential energy.** Think of potential energy as energy "stored" within an object. Equations can be written for potential energy, but the exact forms of these equations depend on the manner in which the energy is "stored." We do not need to use such equations in this text.

B-4 Magnetism

Attractive and repulsive forces associated with a magnet are centered at regions called **poles.** A magnet has a north and a south pole. If two magnets are aligned such that the north pole of one is directed toward the south pole of the second, an attractive force develops. If the alignment brings like poles into proximity, either both north or both south, a repulsive force develops. *Unlike poles attract; like poles repel.*

A **magnetic field** exists in that region surrounding a magnet in which the influence of the magnet can be felt. Internal changes produced within an iron object by a magnetic field, not produced in a field-free region, are responsible for the attractive force that the object experiences.

B-5 Static Electricity

Another property with which certain objects may be endowed is electric charge. Analogous to the case with magnetism, *unlike charges attract and like charges repel* (see Figure B-1). In Coulomb's law, stated below, a *positive* force is *repulsive;* and a *negative* force is *attractive.*

$$F = \frac{Q_1 Q_2}{\epsilon r^2} \tag{B.13}$$

FIGURE B.1
Forces between electrically charged objects.

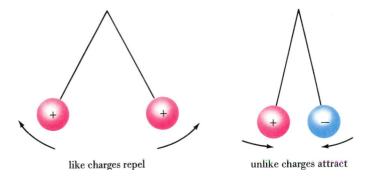

like charges repel   unlike charges attract

where Q_1 is the magnitude of the charge on object 1.
        Q_2 is the magnitude of the charge on object 2.
        r is the distance between the objects.
        ϵ is a proportionality constant called the **dielectric constant** of the medium. The numerical value of this constant reflects the effect that the medium separating the two charged objects has on the force existing between them. For vacuum, $\epsilon = 1$, and for other media ϵ is greater than 1 (e.g., for water $\epsilon = 78.5$).

An **electric field** exists in that region surrounding an electrically charged object in which the influence of the electric charge is felt. If an uncharged object is brought into the field of a charged object, the uncharged object may undergo internal changes that it would not experience in a field-free region. These changes may lead to the production of electric charges in the formerly uncharged object, a phenomenon called **induction** (illustrated in Figure B-2).

FIGURE B-2
Production of electric charges
by induction in a gold leaf
electroscope.

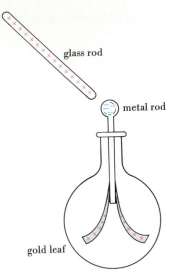

The glass rod acquires a positive electric charge by being rubbed with a silk cloth. As the rod is brought near the electroscope a separation of charge occurs in the electroscope. The leaves become positively charged and repel one another. Negative charge is attracted to the spherical terminal at the end of the metal rod. If the glass rod is removed, the charges on the electroscope redistribute themselves and the leaves collapse. If before the glass rod is removed the spherical ball is touched by an electric conductor, negative charge is removed from the ball, the electroscope retains a net positive charge, and the leaves remain outstretched.

B-6 Current Electricity

Current electricity consists of a flow of electrically charged particles. In electric currents in metallic conductors, the charged particles are electrons; in molten salts or in aqueous solutions, the particles are both negatively and positively charged ions.

The unit of electric charge is called a **coulomb (C).** The unit of electric current known as the **ampere (A)** is defined as a flow of 1 C/s through an electrical conductor. Two variables determine the magnitude of the electric current, I, flowing through a conductor. These are the potential difference or voltage drop, E, along the conductor and the electrical resistance of the conductor, R. The units of voltage and resistance are the **volt (V)** and **ohm,** respectively. The relationship of electric current, voltage, and resistance is given through Ohm's law.

$$I = \frac{E}{R} \tag{B.14}$$

One joule of energy is associated with the passage of one coulomb of electric charge through a potential difference (voltage) of one volt. That is, one joule = one volt-coulomb. Electric **power** refers to the rate of production (or consumption) of electric energy. It has the unit, **watt (W).**

$$1 \text{ W} = 1 \text{ J s}^{-1} = 1 \text{ V C s}^{-1}$$

Since one C s^{-1} is a current of one ampere,

$$1 \text{ W} = 1 \text{ V} \times 1 \text{ A} \tag{B.15}$$

Thus, a 100-watt light bulb operating at 110 V draws a current of 100 W/110 V = 0.91 A.

B-7 Electromagnetism

The relationship between electricity and magnetism is an intimate one. Interactions of electric and magnetic fields result in (1) magnetic fields associated with the flow of electric current (as in electromagnets), (2) forces experienced by current-carrying conductors when placed in a magnetic field (as in electric motors), and (3) electric current being induced when an electric conductor is moved through a magnetic field (as in electric generators). Several observations described in this text can be understood in terms of electromagnetic phenomena.

C SI Units

The system of units that will in time be used universally for expressing all measured quantities is Le Système International d'Unités (The International System of Units), adopted in 1960 by the Conference Générale des Poids et Measures (General Conference of Weights and Measures). A summary of some of the provisions of the SI convention is provided here.

C-1 SI Base Units

A single unit has been established for each of the basic quantities involved in measurement. These are

Physical quantity	Unit	Symbol
length	meter	m
mass	kilogram	kg
time	second	s
electric current	ampere	A
temperature	kelvin	K
luminous intensity	candela	cd
amount of substance	mole	mol
plane angle	radian	rad
solid angle	steradian	sr

C-2 SI Prefixes

Distinctive prefixes are attached to the base unit to express quantities that are **multiples** (greater than) or **submultiples** (less than) of the base unit. The multiples and submultiples are obtained by multiplying the base unit by powers of ten.

Multiple	Prefix	Symbol	Submultiple	Prefix	Symbol
10^{12}	tera	T	10^{-1}	deci	d
10^{9}	giga	G	10^{-2}	centi	c
10^{6}	mega	M	10^{-3}	milli	m
10^{3}	kilo	k	10^{-6}	micro	μ
10^{2}	hecto	h	10^{-9}	nano	n
10^{1}	deka	da	10^{-12}	pico	p
			10^{-15}	femto	f
			10^{-18}	atto	a

C-3 Derived SI Units

A number of quantities must be derived from measured values of the SI base quantities [e.g., volume has the unit (length)^3]. Two sets of derived units are given, those whose names follow directly from the base units and those that are given special names.

A14

Two other SI conventions are illustrated through this table: (a) Units are written in the singular form—meter or m (*not* meters or ms); (b) negative exponents are preferred to the shilling bar or solidus (/)—m s^{-1} and m s^{-2} (*not* m/s and m/s/s).

Physical quantity	Unit	Symbol
area	square meter	m^2
volume	cubic meter	m^3
velocity	meter per second	m s^{-1}
acceleration	meter per second squared	m s^{-2}
density	kilogram per cubic meter	kg m^{-3}
molar mass	kilogram per mole	kg mol^{-1}
molar volume	cubic meter per mole	m^3 mol^{-1}
molar concentration	mole per cubic meter	mol m^{-3}

Physical quantity	Unit	Symbol	In terms of SI units
frequency	hertz	Hz	s^{-1}
force	newton	N	kg m s^{-2}
pressure	pascal	Pa	N m^{-2}
energy	joule	J	kg m^2 s^{-2}
power	watt	W	J s^{-1}
electric charge	coulomb	C	A s
electric potential difference	volt	V	J A^{-1} s^{-1}
electric resistance	ohm	Ω	V A^{-1}

C-4 Units to Be Discouraged or Abandoned

There are several commonly used units whose use is to be discouraged and ultimately abandoned. Their gradual disappearance is to be expected, though each is used in this text. A few such units are listed.

Another SI convention is implied here. No commas are used in expressing large numbers but spaces are left between groupings of three digits (that is, 101 325 instead of 101,325). (Decimal points are written either as periods or commas.)

Physical quantity	Unit	Symbol	Definition in SI units
length	angstrom	Å	1×10^{-10} m
force	dyne	dyn	1×10^{-5} N
energy	erg	erg	1×10^{-7} J
energy	calorie	cal	4.184 J
pressure	atmosphere	atm	101 325 Pa
pressure	millimeter of mercury	mmHg	$(13.5951)(980.665) \times 10^{-2}$ Pa
pressure	torr	torr	133.322 Pa

C-5 Fundamental Constants

The fundamental constants introduced in this text, such as the speed of light, acceleration due to gravity, gas constant, Planck's constant, and the Faraday constant, continue to be used, but their units should be expressed as SI units. (See the inside back cover for a listing.)

D Thermodynamic Properties of Substances

These are for 298.15 K and 1 atm pressure. To illustrate the difference produced in changing standard pressure from 1 atm to 1 bar (10^5 Pa) (see page 219), for $CO_2(g)$: ΔH_f° is unchanged; ΔG_f° becomes -394.41 kJ/mol; and S° becomes 213.75 J mol^{-1} K^{-1}.

	ΔH_f°, kJ/mol	ΔG_f°, kJ/mol	S°, J mol^{-1} K^{-1}
$Al_2O_3(s)$	-1676	-1582	50.92
$BaCO_3(s)$	-1216	-1138	112.1
$B_2H_6(g)$	35.56	86.61	232.0
$B_2O_3(s)$	-1273	-1194	53.97
$Br(g)$	111.9	82.43	174.9
$Br_2(g)$	30.91	3.14	245.4
$Br_2(l)$	0	0	152.2
$BrCl(g)$	14.64	-0.96	240.0
$C(g)$	716.7	671.3	158.0
$C(diamond)$	1.90	2.90	2.38
$C(graphite)$	0	0	5.74
$CCl_4(g)$	-102.9	-60.63	309.7
$CO(g)$	-110.5	-137.2	197.6
$CO_2(g)$	-393.5	-394.4	213.6
$CS_2(l)$	89.70	65.27	151.3
$CH_4(g)$	-74.81	-50.75	186.2
$C_2H_2(g)$	226.7	209.2	200.8
$C_2H_4(g)$	52.26	68.12	219.4
$C_2H_6(g)$	-84.68	-32.89	229.5
$C_3H_8(g)$	-103.8	-23.56	270.2
$C_4H_{10}(g)$	-125.7	-17.15	310.1
$C_6H_6(g)$	82.93	129.7	269.2
$C_6H_6(l)$	48.99	124.4	173.3
$CH_3OH(g)$	-200.7	-162.0	239.7
$CH_3OH(l)$	-238.7	-166.4	126.8
$C_2H_5OH(g)$	-234.4	-167.9	282.6
$C_2H_5OH(l)$	-277.7	-174.9	160.7
$CaCO_3(s)$	-1207	-1128	88.7
$CaO(s)$	-635.1	-604.0	39.75
$Ca(OH)_2(s)$	-986.1	-898.6	83.39
$CaSO_4(s)$	-1434	-1322	106.7
$Cl(g)$	121.7	105.7	165.1
$Cl_2(g)$	0	0	223.0
$CuO(s)$	-157.3	-129.7	42.63
$Cu_2O(s)$	-168.6	-146.0	93.14
$Fe_2O_3(s)$	-824.2	-742.2	87.40
$Fe_3O_4(s)$	-1118	-1016	146.4
$H(g)$	218.0	203.3	114.6
$H_2(g)$	0	0	130.6
$HBr(g)$	-36.40	-53.43	198.6
$HCl(g)$	-92.31	-95.30	186.8
$HF(g)$	-271.1	-273.2	173.7
$HI(g)$	26.48	1.72	206.5
$H_2O(g)$	-241.8	-228.6	188.7
$H_2O(l)$	-285.8	-237.2	69.92
$H_2S(g)$	-20.63	-33.56	205.7
$HCHO(g)$	-117.0	-110.0	218.7

	ΔH_f°, kJ/mol	ΔG_f°, kJ/mol	S°, J mol^{-1} K^{-1}
He(g)	0	0	126.0
Hg(g)	61.32	31.85	174.9
Hg(l)	0	0	76.02
I(g)	106.8	70.28	180.7
I$_2$(g)	62.44	19.36	260.6
I$_2$(s)	0	0	116.1
KCl(s)	− 436.7	− 409.2	82.59
MgCl$_2$(s)	− 641.3	− 591.8	89.62
MgO(s)	− 601.7	− 569.4	26.9
MnO$_2$(s)	− 520.0	− 465.2	53.05
N(g)	472.7	455.6	153.2
N$_2$(g)	0	0	191.5
NH$_3$(g)	− 46.11	− 16.48	192.34
NH$_4$Cl(s)	− 314.4	− 203.0	94.56
NO(g)	90.25	86.57	210.6
N$_2$O(g)	82.05	104.2	219.7
NO$_2$(g)	33.18	51.30	240.0
N$_2$O$_4$(g)	9.16	97.82	304.2
NOCl(g)	51.71	66.07	261.6
NaCl(s)	− 411.1	− 384.0	72.13
O(g)	249.2	231.8	161.0
O$_2$(g)	0	0	205.1
O$_3$(g)	142.7	163.2	238.8
PCl$_3$(g)	− 287.0	− 267.8	311.7
PCl$_5$(g)	− 374.9	− 305.0	364.5
S (rhombic)	0	0	31.80
S (monoclinic)	0.33	0.096	32.55
SO$_2$(g)	− 296.8	− 300.2	248.1
SO$_3$(g)	− 395.7	− 371.1	256.6
SO$_2$Cl$_2$(l)	− 394.1	− 314	207
UO$_2$(s)	−1085	−1032	77.03
ZnO(s)	− 348.3	− 318.3	43.64

Electron Configurations of the Elements

Period	Atomic number	Element	Electron configuration	Period	Atomic number	Element	Electron configuration
1	1	H	$1s^1$		45	Rh	$[Kr]\,4d^8 5s^1$
	2	He	$1s^2\;(=[He])$		46	Pd	$[Kr]\,4d^{10}$
2	3	Li	$[He]\,2s^1$		47	Ag	$[Kr]\,4d^{10}5s^1$
	4	Be	$[He]\,2s^2$		48	Cd	$[Kr]\,4d^{10}5s^2$
	5	B	$[He]\,2s^2 2p^1$		49	In	$[Kr]\,4d^{10}5s^2 5p^1$
	6	C	$[He]\,2s^2 2p^2$		50	Sn	$[Kr]\,4d^{10}5s^2 5p^2$
	7	N	$[He]\,2s^2 2p^3$		51	Sb	$[Kr]\,4d^{10}5s^2 5p^3$
	8	O	$[He]\,2s^2 2p^4$		52	Te	$[Kr]\,4d^{10}5s^2 5p^4$
	9	F	$[He]\,2s^2 2p^5$		53	I	$[Kr]\,4d^{10}5s^2 5p^5$
	10	Ne	$[He]\,2s^2 2p^6\;(=[Ne])$		54	Xe	$[Kr]\,4d^{10}5s^2 5p^6\;(=[Xe])$
3	11	Na	$[Ne]\,3s^1$	6	55	Cs	$[Xe]\,6s^1$
	12	Mg	$[Ne]\,3s^2$		56	Ba	$[Xe]\,6s^2$
	13	Al	$[Ne]\,3s^2 3p^1$		57	La	$[Xe]\,5d^1 6s^2$
	14	Si	$[Ne]\,3s^2 3p^2$		58	Ce	$[Xe]\,4f^1 5d^1 6s^2$
	15	P	$[Ne]\,3s^2 3p^3$		59	Pr	$[Xe]\,4f^3 6s^2$
	16	S	$[Ne]\,3s^2 3p^4$		60	Nd	$[Xe]\,4f^4 6s^2$
	17	Cl	$[Ne]\,3s^2 3p^5$		61	Pm	$[Xe]\,4f^5 6s^2$
	18	Ar	$[Ne]\,3s^2 3p^6\;(=[Ar])$		62	Sm	$[Xe]\,4f^6 6s^2$
4	19	K	$[Ar]\,4s^1$		63	Eu	$[Xe]\,4f^7 6s^2$
	20	Ca	$[Ar]\,4s^2$		64	Gd	$[Xe]\,4f^7 5d^1 6s^2$
	21	Sc	$[Ar]\,3d^1 4s^2$		65	Tb	$[Xe]\,4f^9 6s^2$
	22	Ti	$[Ar]\,3d^2 4s^2$		66	Dy	$[Xe]\,4f^{10}6s^2$
	23	V	$[Ar]\,3d^3 4s^2$		67	Ho	$[Xe]\,4f^{11}6s^2$
	24	Cr	$[Ar]\,3d^5 4s^1$		68	Er	$[Xe]\,4f^{12}6s^2$
	25	Mn	$[Ar]\,3d^5 4s^2$		69	Tm	$[Xe]\,4f^{13}6s^2$
	26	Fe	$[Ar]\,3d^6 4s^2$		70	Yb	$[Xe]\,4f^{14}6s^2$
	27	Co	$[Ar]\,3d^7 4s^2$		71	Lu	$[Xe]\,4f^{14}5d^1 6s^2$
	28	Ni	$[Ar]\,3d^8 4s^2$		72	Hf	$[Xe]\,4f^{14}5d^2 6s^2$
	29	Cu	$[Ar]\,3d^{10}4s^1$		73	Ta	$[Xe]\,4f^{14}5d^3 6s^2$
	30	Zn	$[Ar]\,3d^{10}4s^2$		74	W	$[Xe]\,4f^{14}5d^4 6s^2$
	31	Ga	$[Ar]\,3d^{10}4s^2 4p^1$		75	Re	$[Xe]\,4f^{14}5d^5 6s^2$
	32	Ge	$[Ar]\,3d^{10}4s^2 4p^2$		76	Os	$[Xe]\,4f^{14}5d^6 6s^2$
	33	As	$[Ar]\,3d^{10}4s^2 4p^3$		77	Ir	$[Xe]\,4f^{14}5d^7 6s^2$
	34	Se	$[Ar]\,3d^{10}4s^2 4p^4$		78	Pt	$[Xe]\,4f^{14}5d^9 6s^1$
	35	Br	$[Ar]\,3d^{10}4s^2 4p^5$		79	Au	$[Xe]\,4f^{14}5d^{10}6s^1$
	36	Kr	$[Ar]\,3d^{10}4s^2 4p^6\;(=[Kr])$		80	Hg	$[Xe]\,4f^{14}5d^{10}6s^2$
5	37	Rb	$[Kr]\,5s^1$		81	Tl	$[Xe]\,4f^{14}5d^{10}6s^2 6p^1$
	38	Sr	$[Kr]\,5s^2$		82	Pb	$[Xe]\,4f^{14}5d^{10}6s^2 6p^2$
	39	Y	$[Kr]\,4d^1 5s^2$		83	Bi	$[Xe]\,4f^{14}5d^{10}6s^2 6p^3$
	40	Zr	$[Kr]\,4d^2 5s^2$		84	Po	$[Xe]\,4f^{14}5d^{10}6s^2 6p^4$
	41	Nb	$[Kr]\,4d^4 5s^1$		85	At	$[Xe]\,4f^{14}5d^{10}6s^2 6p^5$
	42	Mo	$[Kr]\,4d^5 5s^1$		86	Rn	$[Xe]\,4f^{14}5d^{10}6s^2 6p^6\;(=[Rn])$
	43	Tc	$[Kr]\,4d^6 5s^1$	7	87	Fr	$[Rn]\,7s^1$
	44	Ru	$[Kr]\,4d^7 5s^1$		88	Ra	$[Rn]\,7s^2$

Period	Atomic number	Element	Electron configuration	Period	Atomic number	Element	Electron configuration
	89	Ac	[Rn] $6d^17s^2$		99	Es	[Rn] $5f^{11}7s^2$
	90	Th	[Rn] $6d^27s^2$		100	Fm	[Rn] $5f^{12}7s^2$
	91	Pa	[Rn] $5f^26d^17s^2$		101	Md	[Rn] $5f^{13}7s^2$
	92	U	[Rn] $5f^36d^17s^2$		102	No	[Rn] $5f^{14}7s^2$
	93	Np	[Rn] $5f^46d^17s^2$		103	Lr	[Rn] $5f^{14}6d^17s^2$
	94	Pu	[Rn] $5f^67s^2$		104	Unq	[Rn] $5f^{14}6d^27s^2$
	95	Am	[Rn] $5f^77s^2$		105	Unp	[Rn] $5f^{14}6d^37s^2$
	96	Cm	[Rn] $5f^76d^17s^2$		106	Unh	[Rn] $5f^{14}6d^47s^2$
	97	Bk	[Rn] $5f^86d^17s^2$				
	98	Cl	[Rn] $5f^{10}7s^2$				

Glossary

Absolute configuration refers to the spatial arrangement of the groups attached to a chiral carbon atom. The two possibilities are D and L.

The **absolute zero of temperature** is the temperature at which molecular motion ceases: $0 \text{ K} = -273.15 \text{ °C}$.

Acetyl (see **acyl**).

An **acid** is **(1)** A hydrogen-containing compound that, under appropriate conditions, can produce hydrogen ions H^+ (Arrhenius theory). **(2)** A proton donor (Brønsted–Lowry theory). **(3)** An atom, ion, or molecule that can accept a pair of electrons to form a covalent bond (Lewis theory).

An **acid anhydride** is an oxide that reacts with water to form an acid.

An **acid–base indicator** is a substance that can be used to measure the pH of a solution. The indicator takes on one color when in its nonionized weak acid form and a different color in its anion form. Its color in a particular solution depends on which form predominates.

An **acid ionization constant, K_a,** is the equilibrium constant for the ionization reaction of a weak acid.

An **acid salt** contains an anion that can act as an acid (proton donor), such as $NaHSO_4$, containing the anion HSO_4^-, and NaH_2PO_4, containing the anion, $H_2PO_4^-$.

The **actinides (actinoids)** are a series of elements ($Z = 90$ to 103) characterized by partially filled $5f$ orbitals in their atoms. All actinide elements are radioactive.

An **activated complex** is a hypothetical species postulated to participate in the mechanism of a reaction.

Activation energy is the energy that molecules must possess so that collisions between them will lead to chemical reaction. It is the energy required to form an activated complex.

Active sites are the locations at which catalysis occurs, whether on the surface of a heterogeneous catalyst or an enzyme.

The **actual yield** is the *measured* quantity of a product obtained in a chemical reaction. (See also **theoretical yield** and **percent yield**.)

The **acyl** group is $-\overset{\text{O}}{\underset{\|}{\text{C}}}-$R. If R = H, this is the **formyl** group; R = CH_3, **acetyl**; and R = C_6H_5, **benzoyl.**

In an **addition reaction** functional group atoms are joined to the C atoms at points of unsaturation in alkene and alkyne hydrocarbon molecules.

An **adduct** is a compound formed by joining together two simpler molecules through a coordinate covalent bond.

Adhesive forces are intermolecular forces between unlike molecules, such as molecules of a liquid and of the surface with which it is in contact.

Adsorption refers to the attachment of ions or molecules to the surface of a material.

Alcohols contain the functional group $-OH$ and have the general formula ROH.

Aldehydes have the general formula $R-\overset{\text{O}}{\underset{\|}{\text{C}}}-H$.

Alicyclic hydrocarbon molecules have their carbon atom skeletons arranged in rings and resemble aliphatic (rather than aromatic) hydrocarbons.

Aliphatic hydrocarbon molecules have their carbon atom skeletons arranged in straight or branched chains.

Alkane hydrocarbon molecules have only single covalent bonds between C atoms. In their chain structures alkanes have the general formula C_nH_{2n+2}.

Alkene hydrocarbons have one or more carbon-to-carbon double bonds in their molecules. The simple alkenes have the general formula C_nH_{2n}.

Alkyl groups are alkane hydrocarbon molecules from which one H atom has been extracted. These groups can appear as substituents in other hydrocarbon chains. For example, the group $-CH_3$ is the **methyl** group; $-CH_2CH_3$ is **ethyl.**

Alkyne hydrocarbons have one or more carbon-to-carbon triple bonds in their molecules. The simple alkynes have the general formula C_nH_{2n-2}.

Allotropy refers to the existence of an *element* in two or more different *molecular* forms, such as O_2 and O_3, or carbon as graphite and diamond.

An **alloy** is a mixture of two or more metals. Some alloys are solid solutions, some are heterogeneous mixtures, and some are intermetallic compounds.

An **alpha (α) particle** is a combination of two protons and two neutrons identical to the nucleus of an ordinary helium atom, that is, $^4_2He^{2+}$.

Alums are double salts having the general formula $M^IM^{III}(SO_4)_2 \cdot 12H_2O$. M^I is a unipositive cation (except Li^+) and M^{III} is one of a group of tripositive cations, e.g., Al^{3+}, Fe^{3+}, Cr^{3+}.

Amalgams are metal alloys containing mercury. Depending on their compositions, some are liquid and some are solid.

An **amide** has the general formula $R-\overset{\text{O}}{\underset{\|}{\text{C}}}-NH_2$.

An **amine** is an organic base having the formula RNH_2 (primary), R_2NH (secondary), or R_3N (tertiary), depending on the number of H atoms of an NH_3 molecule that are replaced by R groups.

An **α-amino acid** has an amino group ($-NH_2$) attached to the C atom adjacent to a carboxyl group ($-COOH$).

Amphoterism refers to the ability of certain oxides and hydroxo compounds to act either as acids or bases.

An **anion** is a negatively charged ion. An anion migrates toward the anode in an electrochemical cell.

The **anode** is the electrode in an electrochemical cell at which an oxidation half-reaction occurs.

Anodizing refers to the electrodeposition of Al_2O_3 on Al to further protect it from corrosion.

An **antibonding molecular orbital** describes regions of high electron density located *away* from the internuclear region between two bonded atoms.

Aromatic compounds are organic substances whose carbon atom skeletons are arranged in hexagonal rings, based on benzene, C_6H_6.

In the **Arrhenius acid–base theory** an acid produces H^+ and a base produces OH^- in aqueous solutions.

The **Arrhenius equation** relates the rate constant of a reaction to temperature and activation energy.

An **atomic (line) spectrum** is produced by dispersing light emitted by excited atoms. Only a discrete set of wavelength components (seen as colored lines) is present.

One **atmosphere** (standard atmosphere) is the pressure exerted, under carefully specified conditions, by a column of mercury 760 mm high.

The **atmosphere** is the mixture of gases (nitrogen, oxygen, argon, and traces of others) that lies above the solid crust and oceans of the earth.

An **atom** is the smallest particle of matter that characterizes an element.

An **atomic mass unit (u)** is used to express the masses of individual atoms. One u is 1/12 the mass of a C-12 atom.

The **atomic number, Z,** is the number of protons in the nucleus of an atom. It is also the number of electrons outside the nucleus of an electrically neutral atom.

The **atomic weight (mass)** of an element is the mass of the naturally occurring mixture of isotopes of the element, relative to an arbitrarily assigned mass of 12.00000 for carbon-12.

The **Aufbau process** is a description of the electron configurations of the elements in which each element is described as differing from the preceding one in terms of the orbital to which the one additional electron is assigned.

The **Avogadro constant, N_A,** has a value of 6.02214×10^{23} mol^{-1}. It is the number of elementary units in one mole.

Avogadro's hypothesis (law) states that equal volumes of different gases, compared under identical conditions of temperature and pressure, contain equal numbers of molecules.

An **azeotrope** is a solution that boils at a constant temperature, producing vapor of the same composition as the liquid. The boiling point of the azeotrope is, in some cases, lower than that of either solution component (minimum boiling point), and in some cases higher (maximum boiling point).

Balancing an equation refers to placing numbers, **stoichiometric coefficients,** in front of the formulas in a chemical equation. In this way the numbers of atoms of each kind are made equal on the two sides of the equation. In **net ionic equations,** electric charges must also be balanced.

Band theory is a form of molecular orbital theory to describe bonding in metals and semiconductors.

A **barometer** is a device used to measure the pressure of the atmosphere.

A **base** is (1) A compound that produces hydroxide ions, OH^-, in water solution (Arrhenius theory). (2) A proton acceptor (Brønsted–Lowry theory). (3) An atom, ion, or molecule, that can donate a pair of electrons to form a covalent bond (Lewis theory).

A **base anhydride** is an oxide that reacts with water to form a base.

A **base ionization constant** is the equilibrium constant for the ionization reaction of a weak base.

A **battery** is a voltaic cell [or a group of voltaic cells connected in series $(+ \text{ to } -)$] used to produce electricity from chemical change.

bcc is an abbreviation for body-centered cubic crystal structures.

Benzoyl (see **acyl**).

A **beta (β) particle** is an electron emitted as a result of the conversion of a neutron to a proton in a radioactive nucleus.

A **bimolecular process** is an elementary process involving the collision of two molecules.

Binary acids are compounds of hydrogen and another nonmetal that are capable of acting as acids.

Binary compounds are compounds comprised of *two* elements.

Biomass refers to all materials that are produced by photosynthesis and potentially useful for the production of organic chemicals or as energy sources.

A **body-centered cubic** crystal structure is one in which the unit cell has structural units at each corner and one in the center of the cube.

Boiling is a process in which vaporization occurs throughout a liquid. It occurs when the vapor pressure of a liquid is equal to barometric pressure.

A **bomb calorimeter** is a device used to measure the heat of a combustion reaction. The quantity measured is the heat of reaction at constant volume, $q_V = \Delta E$.

Bond energy (bond enthalpy) is the quantity of energy (usually in kJ/mol) required to break one mole of bonds in a gaseous species.

Bond length (bond distance) is the distance between the nuclei of atoms joined by a covalent bond.

Bond order is one-half the difference between the numbers of electrons in bonding and in antibonding molecular orbitals in a covalent bond.

A **bonding molecular orbital** describes regions of high electron probability or charge density in the internuclear region between two bonded atoms.

A **bonding pair** is a pair of electrons involved in covalent bond formation.

The **Born–Fajans–Haber method** relates lattice energies of crystalline ionic solids to ionization energies, electron affinities, and enthalpies (heats) of sublimation, dissociation, and formation.

Boyle's law states that the volume of a fixed amount of gas at a constant temperature is inversely proportional to the gas pressure.

A **breeder reactor** is a nuclear reactor that creates more nuclear fuel than it consumes, for example, by converting ^{238}U to ^{239}Pu.

Bridge bond (see **three-center bond**).

The **Brønsted–Lowry theory** describes acids as proton donors and bases as proton acceptors. An acid–base reaction involves the transfer of protons from an acid to a base.

A **buffer** solution resists a change in its pH. It contains components capable of reacting with (neutralizing) small added amounts of acids and bases.

Buffer capacity refers to the amount of acid and/or base that a buffer solution can neutralize and still maintain an essentially constant pH.

By-products are substances produced along with the principal reaction product in a chemical process.

Calcination refers to the decomposition of a solid by heating at temperatures below its melting point, such as the decomposition of calcium carbonate to calcium oxide and $CO_2(g)$.

The **calorie (cal)** is the quantity of heat required to change the temperature of one gram of water by one degree Celsius.

A **calorimeter** is a device (of which there are numerous types) used to measure the quantity of heat exchanged between a system and its surroundings.

Capillary action refers to the rise of a liquid in pores of thin capillary tubes.

A **carbohydrate** is a polyhydroxy aldehyde, a polyhydroxy ketone, a derivative of these, or a substance that yields them upon hydrolysis.

The **carboxyl** group is $-\overset{\displaystyle O}{\overset{\displaystyle \|}{C}}-OH$.

A **carboxylic acid** has the group $-\overset{\displaystyle O}{\overset{\displaystyle \|}{C}}-OH$ attached to a hydrocarbon chain or ring structure.

Catalysis is the speeding up of a reaction in the presence of an agent (**catalyst**) that changes the reaction mechanism to one of lower activation energy.

The **cathode** is the electrode of an electrochemical cell where a reduction half-reaction occurs.

Cathode rays are negatively charged particles (electrons) emitted at the negative electrode (cathode) in the passage of electricity through gases at very low pressures.

Cathodic protection is a method of corrosion control in which the metal to be protected is joined to a more active metal which corrodes instead. The protected metal acts as the cathode of a voltaic cell.

A **cation** is a positively charged ion. A cation migrates toward the cathode in an electrochemical cell.

The **cell potential**, E_{cell}, is the potential difference (voltage) of an electrochemical cell. If E_{cell} is *positive*, the cell reaction is spontaneous. If E_{cell} is *negative*, the cell reaction is nonspontaneous.

A **central atom** in a structure is an atom that is bonded to two or more other atoms.

The **Celsius** temperature scale is based on a value of 0 °C for the melting point of ice and 100 °C for the boiling point of water.

Chain or **skeletal** isomers have the same number of C and H atoms in their hydrocarbon chains but a different pattern of branching in the chain.

A **charged-particle accelerator** is a device that imparts high energies to charged particles for use in nuclear reactions.

Charles's law states that the volume of a fixed amount of gas at a constant pressure is directly proportional to the absolute (Kelvin) temperature.

A **chelate** results from the attachment of multidentate ligands to a metal ion. Chelates are five- or six-membered rings that include the central metal ion and atoms of the ligands.

A **chelating agent** is a multidentate ligand. It simultaneously attaches to two or more positions in the coordination sphere of a central metal ion.

A **chemical change** is a transformation of one or more substances into one or more new substances. (See also **chemical reaction.**)

A **chemical equation** is a symbolic representation of a chemical reaction. Symbols and formulas are used to represent reactants and products, and the equation is balanced. (See also **balancing an equation.**)

A **chemical formula** represents the relative numbers of atoms of each kind in a chemical compound.

A **chemical property** describes a type of chemical change a substance can undergo: for example, to combine with oxygen or dissolve in an acid.

A **chemical reaction** is a process in which one set of substances (reactants) is transformed into another set of substances (products).

Chemical symbols are abbreviations consisting of one or two letters assigned to the names of the elements (e.g., N = nitrogen and Ne = neon).

Chiral refers to the nonsuperimposability of enantiomers and situations affected by this property.

A **chlor-alkali process** involves the electrolysis of NaCl(aq) to produce NaOH(aq), $Cl_2(g)$, and $H_2(g)$.

Chromatography is a technique for separating the components of a mixture present in one phase, called the *mobile* phase, as it moves in relation to another phase, called the *stationary* phase.

A **cloud chamber** is a device used to detect ionizing radiation through the formation of a trail of droplets along the path traveled by radiation through a supersaturated vapor.

Cohesive forces are intermolecular forces between like molecules, such as in a drop of liquid.

Colligative properties —vapor pressure lowering, freezing point depression, boiling point elevation, osmotic pressure— have values that depend only on the number of solute particles in a solution and not on the identity of these particles.

Collision frequency is the number of collisions occurring between molecules in a unit of time.

Collision theory describes reactions in terms of molecular collisions—the frequency of collisions, the fraction of activated molecules, and the probability that collisions will be effective.

A **colloidal mixture** contains particles that are intermediate in size to those of a true solution and an ordinary heterogeneous mixture.

Combustion analysis is a method of determining the carbon, hydrogen, and oxygen contents of a compound by the complete combustion of the compound in oxygen gas.

The **common ion effect** describes the effect on an equilibrium by a substance that furnishes ions that can participate in the equilibrium. For example, sodium acetate, $NaC_2H_3O_2$, furnishes the common ion, $C_2H_3O_2^-$, to the ionization equilibrium of acetic acid, $HC_2H_3O_2$.

A **complex ion** is a combination of a central metal ion and attached groups called ligands.

A **compound** is a substance made up of two or more elements. It does not change its identity in physical changes but can be broken down into its constituent elements by chemical changes.

Concentration (1) refers to the composition of a solution. (2) (see **extractive metallurgy**).

In a **concentration cell** identical electrodes are immersed in solutions of different concentrations. The voltage (emf) of the cell is a function simply of the concentrations of the two solutions.

Condensation is the passage of molecules from the gaseous state to the liquid state.

A **condensed formula** is a simplified representation of a structural formula.

Conformations refer to the different spatial arrangements possible in a hydrocarbon molecule. Examples are the eclipsed and staggered conformation of hydrocarbon chains and the "boat" and "chair" forms of cyclohexane.

A **conjugate acid** is formed when a Brønsted–Lowry base gains a proton. Every Brønsted–Lowry base is said to have a conjugate acid.

A **conjugate base** remains after a Brønsted–Lowry acid has lost a proton. Every Brønsted–Lowry acid is said to have a conjugate base.

A **continuous spectrum** is one in which all wavelength components of the visible portion of the electromagnetic spectrum are present.

Control rods are neutron-absorbing metal rods (e.g., Cd) that are used to control the neutron flux in a nuclear reactor and thereby control the rate of the fission reaction.

A **conversion factor** is a relationship between quantities expressed in different units (e.g., 1 in. = 2.54 cm) that can serve as the basis of problem solving.

In a **coordinate covalent bond** the electrons shared between atoms are contributed by just one of the atoms.

A **coordination compound** is (1) A compound that can be thought of as comprised of two or more simpler compounds. (2) A substance containing complex ions.

Coordination isomerism arises in certain coordination compounds having both a complex cation and a complex anion. An interchange of ligands between the two complex ions leaves the composition of the compound unchanged.

Coordination number is the number of positions available for the attachment of ligands to a central metal ion.

The **coordination sphere** is the region around a metal ion where linkage to ligands can occur to produce a complex ion.

Corrosion is the oxidation of a metal through the action of water, air, and/or salt solutions. A corrosion reaction consists of an oxidation and a reduction half-reaction occurring through voltaic cells set up in the corroding metal.

Coupled reactions are a set of chemical reactions that occur together. One or more of the reactions taken alone is (are) *nonspontaneous* and other(s) are *spontaneous*. The *net* reaction is *spontaneous*, with the increase in free energy of the nonspontaneous reaction(s) offset by the decrease in free energy of the spontaneous reaction(s).

A **covalent bond** results from the sharing of electrons between atoms.

A **covalent compound** is a compound whose atoms are joined by covalent bonds.

Covalent radius is one half the distance between the centers of two atoms that are bonded together covalently. It is the atomic radius that we associate with an element in its covalent compounds.

The **critical point** refers to the temperature and pressure where a liquid and its vapor become identical; it is the highest temperature point on the vapor pressure curve.

The **crystal coordination number** signifies the number of nearest neighboring atoms (or ions of opposite charge) to any given atom (or ion) in a crystal.

Crystal field theory describes bonding in complex ions in terms of electrostatic attractions between ligands and the nucleus of the central metal ion. Particular attention is focused on the splitting of the *d* energy level of the central metal ion that results from electron repulsions.

Cubic closest packing is one of the two ways in which spheres can be packed to minimize the amount of free space or voids among them.

A **curie** is a quantity of radioactive material producing 3.70×10^{10} disintegrations per second. (This is the decay rate for 1.00 g Ra.)

Dalton's law of partial pressures states that in a mixture of gases the total pressure is the sum of the partial pressures of the gases present.

The *d* **block** is that portion of the periodic table in which the process of filling electron orbitals (Aufbau process) involves *d* orbitals.

A **decay constant** is a first-order rate constant describing radioactive decay.

Degree of ionization (percent ionization) refers to the extent to which molecules of a weak acid or weak base ionize. The degree of ionization increases as the weak electrolyte solution is diluted.

Deionized water is water that has been freed of most of its impurity ions by passing it successively through a cation and an anion exchange resin.

Deliquescence is a process in which atmospheric water vapor condenses on a highly soluble solid and the solid dissolves in the water to produce and aqueous solution.

A **delocalized molecular orbital** describes regions of high electron probability or charge density that extend over three or more atoms.

Denaturation refers to the loss of biological activity of a protein brought about by changes in its secondary and tertiary structure.

Density is a physical property obtained by dividing the mass of a material or object by its volume (i.e., mass per unit volume).

Deposition is the passage of molecules from the gaseous to the solid state.

Deoxyribonucleic acid (DNA) is the substance that makes up the genes of the chromosomes in the nuclei of cells.

Detergents are salts of organic sulfonic acids, $RSO_3^-Na^+$, in which R is a hydrocarbon chain or a combination of a benzene ring and hydrocarbon chain.

Dextrorotatory means the ability to rotate the plane of polarized light to the *right*, designated +.

Diagonal relationships refer to similarities that exist between certain pairs of elements in different groups and periods of the periodic table, that is, Li/Mg, Be/Al, and B/Si.

Dialysis is a process, similar to osmosis, in which ions or molecules in solution pass through a semipermeable membrane but colloidal particles do not.

Diamagnetism refers to the repulsion by a magnetic field of a substance in which all electrons are paired.

Diastereomers are optically active isomers of a compound, but their structures are *not* mirror images (as are enantiomers).

A **differentiating solvent** is one that can bring out differences in the acid–base properties of two substances. For example, although both HCl and $HClO_4$ are completely ionized in water, in diethyl ether $HClO_4$ is completely ionized and HCl is not. Diethyl ether is a differentiating solvent for these two acids.

Diffusion refers to the spreading of a substance (usually a gas or liquid) throughout a larger volume.

Dilution is the process of reducing the concentration of a solution by adding more solvent.

Dimensional analysis is a technique used to check the correctness of an equation with respect to the dimensions (e.g., mass, length and time) associated with the variables in the equation. The term is also used to describe a problem solving method in which emphasis is placed on conversion factors having the appropriate dimensions (units).

Dipole moment, μ, is a measure of the extent to which a separation of charges exists within a molecule. It is the product of the magnitude of the charge and the distance separating the charge centers. The unit used to measure dipole moment is the **debye**, 3.34×10^{-30} C m.

Dispersion (London) forces are intermolecular forces associated with instantaneous and induced dipoles.

In a **disproportionation reaction** the same substance is both oxidized and reduced.

Distillation is a procedure for separating a liquid from a mixture by vaporizing the liquid and condensing the vapor. In **simple** distillation a volatile liquid is separated from nonvolatile solutes dissolved in it. In **fractional** distillation several volatile components can be separated from each other.

In a **double covalent bond** *two pairs* of electrons are shared between bonded atoms. The bond is represented by a double-dash sign (=).

Dynamic equilibrium is the condition reached in a reversible reaction when the forward and reverse reactions occur at equal rates. (See also **equilibrium**).

Effective nuclear charge, $Z_{eff.}$, is the positive charge acting on a particular electron in an atom. Its value is the charge on the nucleus, reduced to the extent that other electrons screen the electron in question from the nucleus.

Effusion refers to the escape of a gas through a tiny hole.

An **electrochemical cell** is a device in which the electrons transferred in an oxidation-reduction reaction are made to pass through an electrical circuit. (See also **electrolytic cell** and **voltaic cell.**)

An **electrochemical equivalent** is a quantity of substance produced or consumed when 1 mole of electrons passes through an electrochemical cell.

An **electrode** is a metal surface on which an oxidation–reduction equilibrium is established in solution.

An **electrode potential** is the electric potential developed on a metal electrode when an oxidation–reduction half-reaction occurs on the surface of the electrode.

An **electrode (reduction) potential diagram** lists standard electrode (reduction) potentials for oxidation–reduction couples of an element and various of its ionic and compound forms.

Electrolysis is the decomposition of a substance, either in the molten state or in an electrolyte solution, by means of electric current.

An **electrolytic cell** is an electrochemical cell in which a non-spontaneous reaction is carried out by electrolysis.

Electromagnetic radiation is a form of energy propagated through mutually perpendicular electric and magnetic fields. It includes visible light, infrared, ultraviolet, x ray, television and radio waves.

Electromotive force (emf) is the potential difference between two electrodes in an electrochemical (voltaic) cell, expressed in volts.

Electron affinity is the energy associated with the gain of an electron by a neutral gaseous atom.

Electron capture (E.C.) is a form of radioactive decay in which an electron from an inner electronic shell is absorbed by a nucleus. In the nucleus the electron is used to convert a proton to a neutron.

An **electron configuration** is a representation showing the orbital designations of all the electrons in an atom.

Electronegativity is a measure of the electron attracting power of an atom; metals have low electronegativities, and nonmetals, high electronegativities.

Electrons are particles carrying the fundamental unit of negative electric charge. They are found outside the nuclei of all atoms.

Electrorefining is a process of producing a pure metal through electrolysis.

Electrosynthesis refers to a method of synthesizing a substance through the use of electrolysis.

An **element** is one of a group of fundamental substances that cannot be broken down into simpler substances.

An **elementary process** is a molecular event representing a single step in a reaction mechanism.

An **elimination reaction** is one in which atoms are removed from adjacent positions on a hydrocarbon chain, producing a small molecule (e.g., H_2O) and an additional bond between C atoms.

An **empirical formula** is the simplest formula that can be written for a compound, that is, having the smallest integral subscripts possible.

Enantiomers (optical isomers) are molecules whose structures are not superimposable. The structures are mirror images of one another and the molecules are optically active.

Endothermic processes absorb heat from the surroundings. The quantity of heat carries a *positive* sign.

The **end point of a titration** is the point in the titration where the indicator used changes color. A properly chosen indicator for a titration must have its end point correspond as closely as possible to the equivalence point of the titration reaction.

The **English system** is a system of measurement in which the unit of length is the inch; the unit of mass is the pound; and the unit of time is the second.

Enthalpy, H, is a thermodynamic function used to describe constant-pressure processes. $H = E + PV$, and at constant pressure, $\Delta H = \Delta E + P\Delta V$.

Enthalpy change, ΔH, Is the difference in enthalpy between two states of a system. If the process by which the change occurs is a chemical reaction carried out at constant pressure and with work limited to pressure–volume work, the enthalpy change can be referred to as the **heat of reaction at constant pressure.**

Enthalpy (heat) of formation is the enthalpy change that accompanies the formation of a compound from the most stable forms of its elements.

Entropy, S, is a measure of the degree of *disorder* in a system; the greater the disorder, the greater the entropy.

Entropy change, ΔS, expresses the extent to which the degree of order changes as the result of some process. A *positive* ΔS means an *increase in disorder*.

Entropy change of the universe, $\Delta S_{univ.}$, is the total entropy change (system and surroundings) for a process. For every spontaneous change, $\Delta S_{univ.} > 0$.

An **enzyme** is a substance containing protein that catalyzes biological reactions.

An **equation of state** is a mathematical expression relating the amount, volume, temperature, and pressure of a substance (usually applied to gases).

Equilibrium refers to a condition where a forward and reverse process proceed at equal rates and no further net change occurs (e.g., amounts of reactants and products remain constant with time).

Equilibrium constant (see **equilibrium constant expression**)

An **equilibrium constant expression** describes the relationship among the concentrations (or partial pressures) of the substances present in a system at equilibrium. The numerical value of this expression is called the **equilibrium constant;** this value depends on the temperature but *not* on how the equilibrium condition was established.

The **equivalence point of a titration** is the condition where the reactants consume each other. For example, in an acid–base titration it is the point at which the acid and base just neutralize each other, with neither one in excess.

Equivalent weight is **(1)** A quantity of acid that produces one mole of H^+ or a quantity of base that reacts with one mole of H^+. **(2)** A quantity of substance associated with the transfer of one mole of electrons in an oxidation–reduction reaction.

An **ester** is the product of the elimination of H_2O from between an acid and an alcohol molecule. Esters have the general formula R—C(=O)—O—R'.

An **ether** has the general formula R—O—R'.

Eutrophication is the deterioration of a freshwater body, caused by nutrients such as nitrates and phosphates, that results in growth of algae, oxygen depletion, and fish kills.

Exothermic processes give off heat to the surroundings. The quantity of heat carries a *negative* sign.

Expanded octet is a term used to describe situations in which certain atoms in the third period or beyond are able to accommodate 10 or 12 electrons in their valence shells when forming covalent bonds.

An **extensive property** is one, like mass or volume, whose value depends on the quantity of matter observed.

Extractive metallurgy refers to the process of extracting a metal from its ores. Generally this occurs in four steps. **Concentration** separates the ore from waste rock (gangue). **Roasting** converts the ore to the metal oxide. **Reduction** (usually with carbon) converts the oxide to the metal. **Refining** removes impurities from the metal.

A **face-centered cubic** crystal structure is one in which the unit cell has structural units at the eight corners and in the center of each face of the unit cell.

The **Fahrenheit** temperature scale is based on a value of 32 °F as the melting point of ice and 212 °F as the boiling point of water.

A **family** of elements is a numbered group from the periodic table, sometimes carrying a distinctive name: for example, group 7A, the halogen family.

The **Faraday constant,** $\mathscr{F}$, is the charge associated with one mole of electrons, 96,485 C/mol e^-.

Faraday's laws of electrolysis establish quantitative relationships between a quantity of electric charge and the amount of chemical change it can produce during electrolysis.

Fats are triglycerides in which saturated fatty acid components predominate.

The ***f* block** is that portion of the periodic table in which the process of filling electron orbitals (Aufbau process) involves *f* orbitals. These are the lanthanide and actinide elements.

fcc is an abbreviation for face-centered cubic crystal structures.

Ferromagnetism is a property that permits certain materials (notably Fe, Co, Ni) to be made into permanent magnets. The magnetic moments of individual atoms are aligned into domains. In the presence of a magnetic field, these domains orient themselves to produce a permanent magnetic moment.

Filtration is a method of separating a solid from a liquid, in which solid is retained by the filtering device and liquid passes through.

A **first-order** reaction is one for which the sum of the concentration-term exponents in the rate law is 1.

A **flow battery** is a battery in which reactants, products, or electrolytes are passed continuously through the battery.

Formal charge is the number of outer-shell (valence) electrons in an isolated atom minus the number of electrons assigned to that atom in a Lewis structure.

The **formation constant,** K_f, is the equilibrium constant describing equilibrium among a complex ion, the free metal ion, and ligands. It is obtained by combining equilibrium constants for the stepwise displacement of H_2O molecules from the coordination sphere of a metal ion by other ligands.

A **formula unit** is the smallest collection of atoms from which the formula of a compound can be established.

Formula weight is the mass of a formula unit of a compound relative to that of the atomic weight standard, carbon-12.

Formyl (see **acyl**).

Fractional precipitation is a technique in which ions in solution are separated by the addition of a precipitating agent.

Free energy, G, is a thermodynamic function designed to produce a criterion for spontaneous change. It is defined through the equation $G = H - TS$.

Free energy change, ΔG, indicates the direction of spontaneous change. For a spontaneous process at constant temperature and pressure, $\Delta G < 0$. Also, at constant temperature, $\Delta G = \Delta H - T\Delta S$.

Freezing is the conversion of a liquid to a solid; it occurs at a fixed temperature known as the **freezing point.**

The **frequency** of a wave motion is the number of wave crests or troughs that pass through a given point in a unit of time. It is expressed by the unit $time^{-1}$ (for example, s^{-1}).

A **fuel cell** is a voltaic cell in which the cell reaction is the equivalent of the combustion of a fuel. Chemical energy of the fuel is converted to electricity.

A **function of state (state function)** is a property that depends only on the state or present condition of a system and not on how this state is attained.

A **functional group** is an atom or grouping of atoms attached to a hydrocarbon residue, R. The functional group often confers specific properties to an organic molecule.

Galvanizing is the name given to any process in which an iron surface is coated with zinc to give it corrosion protection.

Gamma (γ) rays are a form of electromagnetic radiation emitted by certain radioactive nuclei.

A **gas** is a form or state of matter in which a material assumes the shape of its container and expands to fill the container (thus having neither definite shape nor volume).

The **gas constant, R,** is the numerical constant appearing in the ideal gas equation ($PV = nRT$) and in several other equations as well.

A **Geiger–Muller counter** is a device used to detect ionizing radiation. Ionizing events that occur in the counter produce electric discharges that can be recorded.

The **genetic code** describes the sequence of bases in DNA molecules which determine complementary sequences in mRNA and, ultimately, sequences of amino acids in proteins.

Geometric isomerism refers to the formation of nonequivalent structures based on the positions at which ligands are attached to a central metal ion in a complex ion.

Graham's law states that the rates of effusion of two different gases are inversely proportional to the square roots of their molar masses.

A **group** is a vertical column of elements in the periodic table. Members of a group have similar properties.

A **half-cell** is a combination of an electrode and a solution. An oxidation-reduction equilibrium is established on the electrode.

The **half-life** is the time required for one half of a reactant to be consumed in a chemical reaction. In a nuclear decay process, it is the time required for one half of the atoms present in a sample to undergo radioactive decay.

A **half-reaction** describes one portion of a net oxidation-reduction reaction, either the oxidation or the reduction.

The **half-reaction method** is a method of balancing an oxidation-reduction equation by dividing it into two half-equations which are balanced separately and then recombined.

Hard water contains dissolved minerals in significant concentrations. If the hardness is primarily due to HCO_3^- and associated cations, the water is said to have **temporary hardness.** If hardness is due to anions other than HCO_3^- (e.g., SO_4^{2-}), it is referred to as **permanent hardness.**

hcp is an abbreviation for hexagonal closest-packed crystal structures.

Heat is a form of energy transfer resulting from a temperature difference.

Heat capacity is the quantity of heat required to change the temperature of an object or substance by one degree, usually expressed as J/°C or cal/°C. **Specific heat capacity** is the heat capacity per gram of substances, i.e., J °C^{-1} g^{-1}, and **molar heat capacity** is the heat capacity per mole, i.e., J °C^{-1} mol^{-1}.

A **heat engine** is a device for converting heat into work. The engine absorbs heat at a high temperature, converts part of it to work, and discharges the remaining heat to the surroundings at a lower temperature.

Heat of reaction is the quantity of heat (q) associated with a chemical reaction. If the reaction is carried out at constant volume, the heat of reaction at constant volume, $q_V = \Delta E$; at constant pressure with work limited to pressure–volume work, $q_p = \Delta H$.

Henry's law relates the solubility of a gas to the gas pressure maintained above a solution of the gaseous solute—$C = kP_{gas}$.

Hess's law states that the enthalpy change for an overall or net process is the sum of enthalpy changes for individual steps in the process.

Heterocyclic compounds are based on hydrocarbon ring structures in which one or more C atoms is replaced by atoms such as N, O, or S.

In a **heterogeneous mixture** the components separate into physically distinct regions of differing properties.

Hexagonal closest packed is one of the two ways in which spheres can be packed to minimize the amount of free space or voids among them.

A **homogeneous mixture (solution)** is a mixture of elements and/or compounds that has a uniform composition within a given sample but a varying composition from one sample to another.

A **homologous series** is a group of compounds that differ in composition by some constant unit (—CH_2— in the case of alkanes).

Hund's rule (rule of maximum multiplicity) states that whenever orbitals of equal energy are available, electrons are assigned to these orbitals singly before any pairing of electrons occurs.

A **hybrid orbital** is one of a set of identical orbitals used to replace simple orbitals in describing certain covalent bonds.

Hybridization refers to the combining of simple atomic orbitals to generate new (hybrid) orbitals.

A **hydrate** is a compound in which a fixed number of water molecules is associated with each formula unit, such as $CuSO_4 \cdot 5H_2O$.

Hydration energy is the energy associated with dissolving gaseous ions in water, usually expressed per mole of ions.

A **hydrocarbon** is a compound containing the two elements carbon and hydrogen. The C atoms are arranged in straight or branched chains or ring structures.

A **hydrogen bond** is an intermolecular force of attraction in which an H atom covalently bonded to one atom is attracted simultaneously to another strongly nonmetallic atom of the same or a nearby molecule.

Hydrolysis is a special name given to acid–base reactions in which ions act as acids or bases. As a result of hydrolysis, for many salt solutions, pH $\neq$ 7.

Hydrometallurgy refers to metallurgical procedures where water and water solutions are used to extract metals from their ores rather than the traditional operations of roasting metal ores and reducing them at high temperature with carbon. The first step in hydrometallurgy is a process of **leaching,** in which the metal of interest is obtained in soluble form in aqueous solution. Other steps include purifying the leached solution and then depositing the metal from solution.

Hydronium ion, H_3O^+ is the form in which protons are found in aqueous solution. The terms ''hydrogen ion'' and ''hydronium ion'' are often used synonymously.

The **hydroxyl** group is —**OH,** and is usually found attached to a straight or branched hydrocarbon chain (an alcohol) or a ring structure (a phenol).

An **hypothesis** is a tentative explanation of a series of observations or of a natural law.

An **ideal gas** is one whose behavior can be predicted by the ideal gas equation.

The **ideal gas equation** relates the pressure, volume, temperature, and number of moles (n) of gas through the expression $PV = nRT$.

An **ideal solution** has $\Delta H_{soln} = 0$ and certain properties (notably vapor pressure) that are predictable from the properties of the solution components.

Incomplete octet is a term used to describe situations in which an atom fails to acquire eight outer-shell electrons when it forms bonds.

An **induced dipole** is an atom or molecule in which a separation of charge is produced by a neighboring dipole.

The **inductive effect** refers to the electron-withdrawing power of certain atoms or groups in a molecule. As a result certain bonds are weakened, increasing the acid strength of the molecule. Conversely, if electrons are withdrawn from lone pairs, the base strength of the molecule is weakened.

Industrial smog is air pollution in which the chief pollutants are $SO_2(g)$, $SO_3(g)$, H_2SO_4 mist, and smoke.

An **inert complex** is the term used to describe a complex ion in which the exchange of ligands occurs very slowly.

The **inert pair effect** refers to the effects on the properties of certain post-transition elements that result from a pair of electrons in the s orbital of the valence shells of their atoms, that is ns^2.

The **initial rate of a reaction** is the rate of a reaction immediately after the reactants are brought together.

Inner-transition elements are those of an f-block series, that is, the lanthanide and actinide elements.

An **instantaneous dipole** is an atom or molecule with a separation of charge produced by a momentary displacement of electrons from their normal distribution.

An **instantaneous rate of reaction** is the exact rate of a reaction at some precise point in the reaction. It is obtained from the slope of a tangent line to a concentration–time graph.

An **integrated rate equation** relates concentration of a reactant (or product) to elapsed time from the start of a reaction. The equation has different forms depending on the order of the reaction.

An **intensive property** is *independent* of the quantity of matter involved in the observation. Density and temperature are examples of intensive properties.

An **interhalogen compound** is a binary covalent compound between two halogen elements, such as ICl and BrF_3.

An **intermediate** is the product of one reaction that is consumed in a following reaction in a process that proceeds through several steps.

An **intermolecular force** is an attraction *between* molecules.

The **internal energy, E,** of a system is the total energy attributed to the particles of matter and their interactions within a system.

The **iodine number** of a triglyceride is the number of grams of I_2 reacting with 1.00×10^2 g of the triglyceride. The iodine number indicates the degree of unsaturation in the fatty acid components.

An **ion** is an electrically charged species consisting of a single atom or a group of atoms. It is formed when a neutral atom or group of atoms either gains or loses electrons.

Ion exchange is a process in which ions in solution are exchanged for corresponding ions held to the surface of an ion exchange material. For example, Ca^{2+} and Mg^{2+} may be exchanged for Na^+; or SO_4^{2-} for OH^-.

Ion-pair formation refers to the association of cations and ani-ons in solution. Such combinations, when they occur, can have a significant effect on solution equilibria.

The **ion product of water, K_w,** is the product of $[H_3O^+]$ and $[OH^-]$ in pure water or in an aqueous solution. This product has a unique value which depends only on temperature. At 25 °C, $K_w = 1.0 \times 10^{-14}$.

An **ionic bond** results from the transfer of electrons between metal and nonmetal atoms. Positive and negative ions are formed and held together by electrostatic attractions.

An **ionic compound** is a compound whose atoms are present as positive and negative ions. Electrostatic forces of attraction (ionic bonds) hold the oppositely charged ions together.

Ionic radius is the radius of a spherical ion. It is the atomic radius associated with an element in its ionic compounds.

Ionization energy, I, is the energy required to remove the most loosely held electron from a *gaseous* atom.

Ionization isomerism arises when a ligand from the coordination sphere of a metal ion is exchanged for an ion outside the coordination sphere.

Iron triad is a term used for the group Fe, Co, and Ni to emphasize similarities in their physical and chemical properties.

The **isoelectric point, pI,** of an amino acid is the pH at which the dipolar structure or "zwitterion" predominates.

Isoelectric species have the same number of electrons (usually in the same configuration). Na^+ and Ne are isoelectronic.

Isomers are compounds having the same formulas but different structures, and therefore differing in properties.

Isotopes of an element are atoms with different numbers of neutrons in their nuclei and thus, different masses.

IUPAC (or **IUC**) refers to the International Union of Pure and Applied Chemistry. Among its many activities, the IUPAC makes recommendations on chemical terminology and nomenclature.

The **joule, J,** is the basic SI unit of energy. It is the quantity of work done when a force of one newton acts through a distance of one meter.

K_c is a relationship that exists among the *molarity concentrations* of the reactants and products in a reversible reaction at equilibrium. The numerical value of K_c for a reaction depends on the temperature, but *not* on how equilibrium is established.

K_p is a relationship that exists among the partial pressures of gaseous reactants and products in a reversible reaction at equilibrium. Its numerical value depends on the temperature, but *not* on how equilibrium is established.

The **Kelvin** temperature is an *absolute* temperature. That is, the lowest attainable temperature is 0 K = −273.15 °C (the temperature at which molecular motion ceases). Temperatures on the Kelvin scale are related to Celsius temperatures through the expression T (K) = t (°C) + 273.15.

A **ketone** has the general formula $R-\overset{\overset{O}{\|}}{C}-R'$.

The **kinetic molecular theory of gases** is a model for describing gas behavior. It is based on a set of assumptions, and yields equations from which various properties of gases can be deduced.

Labile complex is the term used to describe a complex ion in which rapid exchange of ligands occurs.

Lanthanide contraction describes the decrease in atomic size in a series of elements in which an f subshell fills with electrons (an inner-transition series). It results from the ineffectiveness of f electrons in shielding outer-shell electrons from the nuclear charge of an atom.

The **lanthanides (lanthanoids)** are the series of elements ($Z = 58$ to 71) characterized by partially filled $4f$ orbitals in their atoms. Because lanthanum resembles them, La ($Z = 57$) is generally considered together with them.

Lattice energy is the quantity of energy released in the formation of one mole of a crystalline ionic solid from its separated gaseous ions.

The **law of conservation of energy** states that energy can neither be created nor destroyed in ordinary processes.

The **law of conservation of mass** states that no detectable change in mass occurs in ordinary physical or chemical processes.

The **law of constant composition (definite proportions)** states that a chemical compound has a unique composition in terms of its constituent elements, regardless of its source or manner of preparation.

The **law of multiple proportions** deals with the proportions in which two elements combine when they are able to form more than a single compound. For each compound the ratio of the mass of one element to a fixed mass of the second is formulated. For any pair of compounds these ratios are themselves in the ratio of small whole numbers, such as $2.66:1/1.33:1 = 2:1$.

Leaching (see **hydrometallurgy**).

Le Châtelier's principle states that actions that tend to change the temperature, pressure, or concentrations of reactants in a system at equilibrium stimulate a net reaction that restores equilibrium in the system.

Levorotatory means the ability to rotate the plane of polarized light to the *left*, designated $-$.

Lewis acid–base theory considers an acid to be an atom, ion, or molecule that can accept an electron pair. A base is an atom, ion, or molecule that can denote an electron pair. An acid–base reaction consists of the formation of a covalent bond between the acid and the base.

A **Lewis structure** is a combination of Lewis symbols that depicts the transfer or sharing of electrons in a chemical bond.

In the **Lewis symbol** of an element, valence electrons are represented by dots placed around the chemical symbol of the element.

Ligands are the groups that are coordinated (bonded) to the central metal ion in a complex ion.

The **limiting reagent (reactant)** in a reaction is the reactant that is consumed completely. The quantity of product(s) formed depends on the quantity of the limiting reagent.

Linkage isomerism describes complex ions with the same compositions but with one or more ligands bonded differently.

Lipids include a variety of naturally occurring substances (e.g., fats and oils) sharing the property of solubility in solvents of low polarity [such as in $CHCl_3$, CCl_4, C_6H_6, and $(C_2H_5)_2O$].

A **liquid** is a form or state of matter in which a material occupies a definite volume but flows to assume the shape of its container.

Liquid crystals are a form of matter with some of the properties of a liquid and some of a crystalline solid.

A **lone pair** is a pair of electrons found in the valence shell of an atom and *not* involved in bond formation.

Magic numbers is a term used to describe numbers of protons and neutrons that confer a special stability to an atomic nucleus.

The **main-group elements** or representative elements are those whose s or p orbitals are being filled in the Aufbau process. They are also referred to as the s-block and p-block elements. They consist of periodic table groups 1A through 8A.

A **manometer** is a device used to measure the pressure of a gas, usually by comparing the gas pressure with barometric pressure.

Mass is a measure of the inertia possessed by an object. (Inertia is the tendency to remain at rest or in constant motion unless acted upon by an external force. The more mass in an object, the greater its inertia.)

The **mass number, A,** is the total of the number of protons and neutrons in the nucleus of an atom.

A **mass spectrometer (mass spectrograph)** is a device used to separate and to measure the quantities and masses of different ions in a beam of positively charged gaseous ions.

Matter is anything that occupies space and has mass.

Melting is the transition of a solid to a liquid.

Messenger RNA (mRNA) is a nucleic acid that is produced in the cell nucleus by DNA and migrates into the cytoplasm, where it directs the synthesis of a protein.

A **meta acid** is formed by the elimination of two H atoms and one O atom (i.e., H_2O) from an ortho acid.

A **meta (m-) isomer** has two substituents on a benzene ring separated by one C atom.

Metabolism refers to the totality of chemical reactions occurring within an organism: reactions in which large molecules are broken down (catabolism) or synthesized from smaller ones (anabolism).

A **metal** is an element whose atoms have small numbers of electrons in the electronic shell of highest principal quantum number. Removal of an electron(s) from a metal atom produces a positive ion (cation).

Metal carbonyls, e.g., nickel carbonyl, $Ni(CO)_4$, are compounds formed between certain metal atoms and CO molecules.

Metallic radius is one-half the distance between the centers of adjacent atoms in a metallic solid.

A **metalloid** is an element that may display both metallic and nonmetallic properties under the appropriate conditions (e.g., Si, Ge, As).

Metaphosphoric acids have the formula $(HPO_3)_n$, and their salts are called **metaphosphates.**

The **method of initial rates** establishes the order of a reaction from the initial rates of the reaction by varying the initial concentration of one reactant (usually by factors of 2) while the initial concentrations of all the other reactants are held constant.

The **metric system** is a system of measurement in which the unit of length is the meter, the unit of mass is the kilogram (1 kg = 1000 g), and the unit of time is the second.

A **millimole** is one-thousandth of a mole (0.001 mol). It is useful in titration calculations.

A **mixture** is any sample of matter that is not pure, i.e., not an element or compound. The composition of a mixture, unlike that of a substance, can be varied. Mixtures are either *homogeneous* or *heterogeneous*.

A **moderator** slows down energetic neutrons from a fission process so that they are able to induce additional fission.

Molality, *m*, is a solution concentration expressed as number of moles of solute per kilogram of solvent.

Molar mass is the mass of one mole of atoms, formula units or molecules.

Molar solubility is the molarity concentration of solute (mol/L) in a saturated solution.

Molarity, M, refers to the composition or concentration of a solution expressed as number of moles of solute per liter of solution.

A **mole** is an amount of substance containing 6.02214×10^{23} (the Avogadro constant) atoms, formula units, or molecules.

Mole fraction describes a mixture in terms of the fraction of all the molecules that are of a particular type.

A **molecular formula** denotes the numbers of the different atoms present in a molecule. In some cases the molecular formula is the same as the empirical formula; in others it is an integral multiple of that formula.

Molecular orbital theory describes the covalent bonds in a molecule by considering that the atomic orbitals of the component atoms are replaced by electron orbitals belonging to the molecule as a whole—molecular orbitals. A set of rules is used to assign electrons to these molecular orbitals, thereby yielding the electronic structure of the molecule.

Molecular weight is the mass of a molecule relative to that of the atomic weight standard, carbon-12.

A **molecule** is a group of bonded atoms that exists as a separate entity and has characteristic physical and chemical properties.

A **monomer** is a simple molecule that is capable of joining with others to form a complex long-chain molecule called a **polymer.**

A **monosaccharide** is a single, simple molecule having the structural features of a carbohydrate. (See also **sugar.**)

A **multidentate ligand** simultaneously attaches itself to two or more positions in the coordination sphere of a metal ion.

A **multiple covalent bond** is a bond in which more than two electrons are shared between the bonded atoms.

A **natural law** is a general statement that can be used to summarize observations of natural phenomena.

The **Nernst equation** is used to relate E_{cell}, E°_{cell}, and the activities of the reactants and products in a cell reaction.

The **net chemical reaction** is the overall change that occurs in a process involving two or more steps.

A **net ionic equation** represents a reaction between ions in solution in such a way that all nonparticipant (spectator) ions are eliminated from the equation. The equation must be balanced for both numbers of atoms and net electric charge.

A **network covalent solid** is a substance in which covalent bonds extend throughout the crystal, i.e., in which the covalent bond is both an *intra*molecular and an *inter*molecular force.

A **neutralization** reaction is one in which an acid and a base react in stoichiometric proportions so that there is no excess of either acid or base in the final solution. The products are water and a salt.

Neutrons are electrically neutral fundamental particles of matter found in all atomic nuclei except that of the simple hydrogen atom, protium, ^{1}H.

A **nitrile** has the general formula R—C≡N.

The **nitrogen cycle** is a series of processes by which atmospheric N_2 is fixed, enters into the food chain of animals, and eventually is returned to the atmosphere by bacteria.

Noble gases are elements whose atoms have the electron configuration ns^2np^6 in the electronic shell of highest principal quantum number. (The noble gas helium has the configuration $1s^2$.)

Nomenclature refers to the writing of chemical names and formulas by some systematic method.

A **nonbonding molecular orbital** is a molecular orbital that neither contributes to nor detracts from bond formation in a molecule. (It does not affect the bond order.)

A **nonelectrolyte** is a substance that is essentially nonionized, both in the pure state and in solution.

A **nonideal gas** departs from the behavior predicted by the ideal gas equation. Its behavior can only be predicted by other equations of state (e.g., the van der Waals equation).

A **nonmetal** is an element whose atoms tend to gain small numbers of electrons to form negative ions (anions) with the electron configuration of a noble gas. Nonmetal atoms may also alter their electron configurations by sharing electrons.

In a **nonpolar molecule** the centers of positive and negative charge coincide. That is, there is no net separation of charge within the molecule.

A **nonstoichiometric compound** has its constituent elements present in *nonintegral* ratios by number of atoms, and its composition may also be somewhat variable.

Normal boiling point is the temperature at which the vapor pressure of a liquid is 1 atm.

Normal melting point is the temperature at which the melting of a solid occurs at 1 atm pressure. This is the same temperature as the **normal freezing point.**

Normality, N, is a concentration unit used in conjunction with the concept of equivalent weight. Normality is the number of equivalents of solute per liter of solution.

Nuclear binding energy is the energy released when nucleons (protons and neutrons) are fused into an atomic nucleus. This energy replaces an equivalent quantity of matter.

A **nuclear equation** represents the changes that occur during a nuclear process. The target nucleus and bombarding particle are represented on the left side of the equation, and the product nucleus and ejected particle on the right side.

Nuclear fission is a radioactive decay process in which a heavy nucleus breaks up into two lighter nuclei and several neutrons.

In **nuclear fusion** small atomic nuclei are fused into larger ones, with some of their mass being converted to energy.

A **nuclear reactor** is a device in which nuclear fission is carried out as a controlled chain reaction. That is, neutrons produced in one fission event trigger the fission of another nucleus, and so on.

Nucleic acids are cell components comprised of purine and pyrimidine bases, pentose sugars, and phosphoric acid.

Nuclide is a term used to designate a specific atomic species, as represented by the symbolism $^{A}_{Z}X$.

Nuclidic mass is the mass, in atomic mass units, u, of an individual atom, relative to an arbitrarily assigned value of 12.00000 u as the nuclidic mass of C-12.

An **octet** refers to the presence of *eight* electrons in the outermost (valence) electronic shell of an atom.

An **odd-electron species** is one in which the total number of valence electrons is an *odd* number. At least one *unpaired* electron is present in such a species.

Oils are triglycerides in which unsaturated fatty acid components predominate.

Oligosaccharides are carbohydrates consisting of two to ten monosaccharide units. (See also **sugar.**)

Optical isomerism refers to the existence of two species called enantiomers (mirror images) that differ only in the way they rotate the plane of polarized light.

An **orbital** describes the electron charge density or the probability of finding an electron in an atom. The several kinds of orbitals (*s, p, d, f,* ...) differ from one another in the shapes of the electron clouds they describe.

An **orbital diagram** is a representation of an electron configuration in which the most probable orbital designation and the spin of each electron in the atom are indicated.

The **order of a reaction** relates to the exponents of the concentration terms in the rate law for a chemical reaction.

An **organosilicon** compound has silicon (Si) atoms replacing some or all of the carbon (C) atoms of an organic compound.

An **ortho acid** is an oxoacid containing the maximum number of OH groups possible.

An **ortho (*o*-) isomer** has two substituents attached to adjacent C atoms in a benzene ring.

Osmosis is the net flow of solvent molecules through a semipermeable membrane, from a more dilute solution (or from the pure solvent) into a more concentrated solution.

Osmotic pressure is the pressure that would have to be applied to a solution to stop the passage of molecules from the pure solvent through a semipermeable membrane into the solution.

Oxidation is a process in which electrons are ''lost'' and the oxidation state of some atom increases. (Oxidation can occur only in combination with reduction.)

An **oxidation–reduction** reaction is one in which certain atoms undergo changes in oxidation state. The substance containing atoms whose oxidation states *increase* is **oxidized.** The substance containing atoms whose oxidation states *decrease* is **reduced.**

Oxidation states are numbers assigned to the atoms in a neutral molecule or ion to convey some idea of the role of electrons in the bonds between atoms. The assignment of oxidation states requires applying a set of rules (page 72). Once assigned, oxidation states are useful in naming compounds and balancing chemical equations.

The **oxidation state change method** is a method of balancing oxidation–reduction equations based on changes in oxidation state that occur in the reaction.

An **oxidizing agent** makes possible an oxidation process by itself being *reduced*.

An **oxoacid** is an acid in which an ionizable hydrogen atom(s) is bonded through an oxygen atom to a central atom, that is, E—O—H. Other groups bonded to the central atom are either additional —OH groups or O atoms (or occasionally H atoms).

An **oxoanion** is a polyatomic anion containing a nonmetal such as Cl, N, P, or S in combination with some number of oxygen atoms. An oxoanion is derived from an oxoacid.

A **para (*p*-) isomer** has two substituents located opposite to one another on a benzene ring.

Paramagnetism refers to the attraction into a magnetic field of a substance whose atoms or ions contain unpaired electrons.

The **p block** is that portion of the periodic table in which the filling of electron orbitals (Aufbau process) involves *p* orbitals.

A **partial pressure** is the pressure exerted by an individual gas in a mixture, independently of other gases.

The **Pauli exclusion principle** states that no two electrons may have all four quantum numbers alike. This limits occupancy of an orbital to two electrons with opposing spins.

A **peptide bond** is formed by the elimination of a water molecule from between two amino acid molecules. The H atom comes from the —NH_2 group of one amino acid, and the —OH group from the —COOH group of the other acid.

Percent natural abundances refer to the relative proportions in which the isotopes of an element are found in natural sources. The percentage is expressed on a number basis; for example 90.9% of all neon atoms are Ne-20, 0.3% are Ne-21, and 8.8% are Ne-22.

Percent yield is the percent of the theoretical yield of product that is actually obtained in a chemical reaction. (See also **actual yield** and **theoretical yield.**)

A **period** is a horizontal row of the periodic table. All members of a period have atoms with the same highest principal quantum number.

The **periodic law** refers to the periodic recurrence of certain physical and chemical properties when the elements are considered in terms of increasing atomic number.

The **periodic table** is an arrangement of the elements in which elements with similar physical and chemical properties are grouped together in vertical columns.

Permanent hard water (see **hard water**).

The **peroxide** ion has the structure $[:\overset{..}{\underset{..}{O}}\!\!-\!\!\overset{..}{\underset{..}{O}}:]^{2-}$.

Peroxo compounds contain the characteristic group —O—O—, such as HO_3S—O—O—SO_3H (peroxodisulfuric acid).

pH is a shorthand designation for [H_3O^+] in a solution. It is defined as pH = $-\log$ [H_3O^+].

A **phase diagram** is a graphical representation of the phases or states of matter of a substance that exist at various temperatures and pressures.

A **phenol** has the functional group —OH as part of an aromatic hydrocarbon structure.

A **phenyl group** is a benzene ring from which one H atom has been removed: —C_6H_5.

A **polycyclic aromatic hydrocarbon** is obtained whenever two or more benzene rings are fused together (with an appropriate loss of C and H atoms).

Photochemical smog is air pollution resulting from reactions involving sunlight, oxides of nitrogen, ozone, and hydrocarbons.

The **photoelectric effect** is the ability of certain materials to emit electrons from their surfaces when struck by electromagnetic radiation of the appropriate frequency.

A **photon** is a ''particle'' of light. The energy of a beam of light is concentrated into these photons.

A **physical change** is one in which the physical appearance of a substance changes but its basic identity, that is, its chemical composition, remains unchanged.

A **physical property** is a characteristic that a substance can display without undergoing a change in its identity.

A **pi (π) bond** results from the ''sidewise'' overlap of *p* orbitals, producing a high electron charge density above and below the line joining the bonded atoms.

Pig iron is an impure form of iron (about 95% Fe, 3 to 4% C) produced in a blast furnace.

pK is a shorthand designation for an ionization constant; $pK = -\log K$. pK values are useful when comparing the relative strengths of acid or bases.

pOH relates to the $[OH^-]$ in a solution; $pOH = -\log [OH^-]$.

Polarizability describes the ease with which the electron cloud in an atom or molecule can be distorted in an electric field.

In a **polar molecule** there exists a separation of electrical charge into positive and negative charge centers.

A **polyatomic ion** contains two or more atoms.

In a **polyhalide ion** two or more halogen atoms are covalently bonded into a polyatomic anion, e.g., I_3^-.

A **polymer** is a complex, long-chain molecule made up of many (hundreds, thousands) smaller units called *monomers*.

Polymerization is the process of producing a giant molecule (polymer) from simpler molecular units (monomers).

Polymorphism refers to the existence of a solid substance in more than one crystalline form.

A **polypeptide** is formed by the joining of a large number of amino acid units through peptide bonds.

Polyphosphoric acids have the formula $H_2PO_3(HPO_3)_nPO_4H_2$, and their salts are called **polyphosphates.**

A **polyprotic acid** is capable of losing more than a single proton per molecule in acid–base reactions. Protons are lost in a stepwise fashion, with the first proton being the most readily lost.

A **polysaccharide** is a carbohydrate (such as starch or cellulose) consisting of more than ten monosaccharide units.

Positional isomers differ in the position on a hydrocarbon chain or ring where a functional group(s) is attached.

Positive (canal) rays are beams of positively charged gaseous ions produced through collisions of cathode rays with residual gas atoms in cathode ray tubes.

A **positron** (β^+) is a *positive* electron emitted as a result of the conversion of a proton to a neutron in a radioactive nucleus.

A **potentiometer** is a device use for the precise measurement of electrode potential differences (voltages) of electrochemical cells.

Precipitation refers to the separation of a solid from a liquid solution.

Precipitation analysis is a quantitative analytical method in which the element being analyzed is precipitated as an insoluble solid.

Pressure is a force per unit area. Applied to gases, pressure is most easily understood in terms of the height of a liquid column that can be maintained by the gas.

Pressure–volume work is work associated with the expansion or compression of gases.

A **primary battery** produces electricity from a chemical reaction that cannot be reversed. As a result the battery cannot be recharged.

A **primary color** is one of a set of colors that when mixed together as lights produce white light. For example, red, yellow, and blue are primary colors.

Primary structure refers to the sequence of amino acids in the polypeptide chains that make up a protein.

A **principal shell** refers to the collection of all orbitals having the same value of the principal quantum number, n. For example, the $3s$, $3p$, and $3d$ orbitals comprise the third principal shell ($n = 3$).

Products are substances formed in a chemical reaction.

A **protein** is a large polypeptide, that is, having a molecular weight of 10,000 or more.

Protons are fundamental particles carrying the basic unit of positive electric charge and found in the nuclei of all atoms.

A **pseudo-first-order** reaction is a reaction of higher (e.g., second) order that is made to behave like a first-order reaction by using such large initial concentrations of all reactants save one that the concentration of only one reactant changes during the reaction. As a result the reaction behaves like a first-order reaction.

Pseudohalogens are groupings of atoms (e.g., —CN, —OCN, —SCN, —N$_3$) that have some of the characteristics of a halogen atom.

A **pyro acid** is formed through the elimination of two H atoms and one O atom (i.e., H_2O) from between two molecules of an ortho acid.

Pyrometallurgy is the traditional approach to extractive metallurgy that uses dry solid materials heated to high temperatures. (See also **extractive metallurgy.**)

Qualitative cation analysis is a laboratory method, based on a variety of solution equilibrium concepts, for determining the presence or absence of certain cations in a sample.

Quantitative analysis refers to the analysis of substances or mixtures to determine the *quantities* of the various components rather than their mere presence or absence.

Quantum numbers are integral numbers whose values must be specified in order to solve the equations of wave mechanics. Three different quantum numbers are required: the *principal quantum number, n;* the *orbital quantum number, l;* and the *magnetic quantum number, m_l.* The permitted values of these numbers are interrelated.

The **quantum theory** is based on the proposition that energy exists in the form of tiny, discrete units called quanta. Whenever an energy transfer occurs, it must involve an entire quantum.

Quaternary structure is the highest order structure that is found in some proteins. It describes how separate polypeptide chains may be assembled into a larger, more complex structure.

A **racemic mixture** is a mixture containing equal amounts of the D and L isomers of an optically active substance.

A **rad** is a quantity of radiation able to deposit 1×10^{-2} J of energy per kilogram of matter.

A **radioactive decay series** is a succession of individual steps whereby an initial radioactive isotope (e.g., $^{238}_{92}U$) is ultimately converted to a stable isotope (e.g., $^{206}_{82}Pb$).

Radioactivity is a phenomenon in which small particles of matter (α or β particles) and/or electromagnetic radiation (γ rays) are emitted by unstable atomic nuclei.

Radiocarbon dating is a method of determining the age of a carbon-containing material based on the rate of decay of radioactive carbon-14.

Raoult's law states that the vapor pressure of a solution component is equal to the product of the vapor pressure of the pure liquid and its mole fraction in solution. Raoult's law applies to all volatile components in an ideal solution and to the solvent in a dilute nonideal solution.

A **random error** is an error made by the experimenter in performing an experimental technique or measurement, such as the error in estimating a temperature reading on a thermometer.

The **rate constant, k,** is the proportionality constant in a rate law that permits the rate of a reaction to be related to the concentrations of the reactants.

A **rate-determining step** in a reaction mechanism is an elementary process that is instrumental in establishing the rate of the overall reaction, usually because it is the slowest step in the mechanism.

The **rate law (rate equation)** for a reaction relates the reaction rate to the concentrations of the reactants. It has the form: rate = $k[A]^m[B]^n \cdots$.

The **rate of a chemical reaction** describes how fast reactants are consumed and products are formed, usually expressed as change of concentration per unit time.

Reactants are the substances that enter into a chemical reaction. This term is often applied to *all* the substances involved in a reversible reaction, but it can also be limited to the substances that appear on the *left* side of a chemical equation. (Substances on the *right* side of the equation are often called products.)

A **reaction mechanism** is a set of elementary steps or processes by which a reaction is proposed to occur. The mechanism must be consistent with the stoichiometry of the net equation and with the rate law for the net reaction.

A **reaction profile** is a graphical representation of a chemical reaction in terms of the energies of the reactants, activated complex(es), and products.

The **reaction quotient, Q,** is a ratio of concentration terms (or partial pressures) having the same form as an equilibrium constant expression, but usually applied to *nonequilibrium* conditions. It is used to determine the direction in which a net reaction occurs to establish equilibrium.

A **reducing agent** makes possible a reduction process by itself becoming *oxidized*.

A **reducing sugar** is one that is able to reduce Cu^{2+} (aq) to red, insoluble Cu_2O. The sugar must have available an aldehyde group, which is oxidized to an acid.

Reduction (1) is a process in which electrons are "gained" and the oxidation state of some atom decreases. (Reduction can only occur in combination with oxidation.) (2) (see **extractive metallurgy**).

Refining (see **extractive metallurgy**).

The **relative humidity** of air is the ratio of the actual partial pressure of water vapor to that when air is saturated with water vapor, expressed on a percent basis.

A **rem** is a unit of radiation related to the rad, but taking into account the varying effects on biological matter of different types of radiation of the same energy.

Representative elements are those whose atoms feature the filling of *s* or *p* orbitals of the electronic shell of highest principal quantum number. Representative elements consist of the *s*-block and *p*-block elements; also called main-group elements.

Resolution refers to the separation of the optically active isomers of a racemic mixture.

Resonance occurs when two or more plausible Lewis structures can be written for a species. The true structure is a composite or *hybrid* of these different contributing structures.

Rest mass is the mass of a particle (e.g., an electron) when it is essentially at rest. As its velocity approaches the speed of light, the mass of a particle increases.

Reverse osmosis is the passage through a semipermeable membrane of solvent molecules *from a solution into a pure solvent*.

It can be achieved by applying to the solution a pressure in excess of its osmotic pressure.

A **reversible process** is one that is always within an infinitesimal step of equilibrium. An infinitesimal change in a system variable is sufficient to turn the process into the reverse direction.

Ribonucleic acid (RNA), through its **messenger RNA (mRNA)** and **transfer RNA (tRNA)** forms, is involved in the synthesis of proteins.

Roasting (see **extractive metallurgy**).

A **salt bridge** is a device (a ∪-tube filled with a salt solution) used to join two half-cells in an electrochemical cell. The salt bridge permits the flow of ions between the two half-cells.

The **salt effect** is that of ions *different* from those directly involved in a solution equilibrium. The salt effect is also known as the diverse or "uncommon" ion effect.

Salts are ionic compounds in which hydrogen atoms of acids are replaced by metal ions. Salts are produced by the neutralization of acids with bases.

Saponification is the hydrolysis of a triglyceride by a strong base. The products are glycerol and a soap.

Saponification value is the number of mg KOH required to saponify 1.00 g of a triglyceride.

Saturated hydrocarbon molecules contain only single bonds between carbon atoms.

A **saturated solution** is one that contains the maximum quantity of solute that is normally possible.

The **s block** is that portion of the periodic table in which the filling of electron orbitals (Aufbau process) involves *s* orbitals of the electronic shell of highest principal quantum number.

The **scientific method** refers to the general sequence of activities—observation, experimentation, and the formulation of laws and theories—that leads to the advancement of scientific knowledge.

The **second law of thermodynamics** relates to the direction of spontaneous change. All spontaneous processes produce an increase in the entropy of the universe.

A **secondary battery** produces electricity from a reversible chemical reaction. When electricity is passed through the battery in the reverse direction the battery is recharged.

A **secondary color** is the complement of a primary color. When light of a primary color and its complement (secondary) color are mixed, the result is white light.

The **secondary structure** of a protein describes the structure or shape of a polypeptide chain, for example, a coiled helix.

A **second-order** reaction is one for which the sum of the concentration-term exponents in the rate law is 2.

Self-ionization is an acid–base reaction in which one solvent molecule acts as an acid and donates a proton to another solvent molecule acting as a base.

A **semiconductor** is characterized by a small energy gap between a filled valence band and an empty conduction band. In an **intrinsic** semiconductor electrons acquire extra thermal energy and are promoted to the conduction band, leaving vacancies or **positive holes** in the valence band. Electrical conduction involves movement of both conduction electrons and positive holes. In an **extrinsic** semiconductor impurity atoms are added to create an excess of either conduction electrons (**n-type**) or positive holes (**p-type**).

A **semipermeable membrane** permits the passage of some solution species but restricts the flow of others. It is a film of material containing submicroscopic holes.

The **shielding effect** refers to the effect of inner-shell electrons in shielding or screening outer-shell electrons from the full effects of the nuclear charge. In effect the inner electrons partially reduce the nuclear charge. (See also **effective nuclear charge.**)

SI units are the units for expressing measured quantities preferred by various international scientific agencies. (See Appendix C.)

A **side reaction** is a reaction that occurs at the same time as the main reaction in a chemical process. Usually, the existence of side reactions reduces the yield of the desired product(s).

A **sigma (σ) bond** results from the end-to-end overlap of simple or hybridized atomic orbitals along the straight line joining the nuclei of the bonded atoms.

Significant figures are those digits in an experimentally measured quantity that establish the precision with which the quantity is known.

A **silicone** is an organosilicon polymer containing —O—Si—O— bonds.

A **single covalent bond** results from the sharing of *one pair* of electrons between bonded atoms. It is represented by a single dash sign (—).

Skeletal isomer (see **chain isomer**).

A **skeleton structure** is an arrangement of atoms in the order in which they are bonded to each other in a Lewis structure.

Soaps are the salts of fatty acids, e.g., $RCOO^-Na^+$, where the R group is a hydrocarbon chain containing from 3 to 21 C atoms. (See also **saponification** and **saponification value.**)

Solders are low melting alloys used for joining wires or pieces of metal. They usually contain metals such as Sn, Pb, Bi, and Cd.

A **solid** is a form or state of matter in which a material has a definite shape and occupies a definite volume.

The **solubility product constant, K_{sp},** describes equilibrium in a saturated solution of a slightly soluble ionic compound. It is the product of ionic concentration terms, with each term raised to an appropriate power.

A **solute(s)** is(are) the solution component(s) present in lesser amount(s) than the solvent.

The **solvent** is the solution component present in greatest quantity or the component that determines the state of matter in which a solution exists.

An **sp hybrid orbital** is one of the two identical orbitals that result from the hybridization of one *s* and one *p* orbital. The angle between the two orbitals is 180°.

An **sp² hybrid orbital** is one of three identical orbitals that result from the hybridization of one *s* and two *p* orbitals. The angle between any two of the orbitals is 120°.

An **sp³ hybrid orbital** is one of four identical orbitals that result from the hybridization of one *s* and three *p* orbitals. The angle between any two of the orbitals is the tetrahedral angle— 109.5°.

An **sp³d hybrid orbital** is one of five orbitals that result from the hybridization of one *s*, three *p*, and one *d* orbitals. The five orbitals are directed to the corners of a trigonal bipyramid.

An **sp³d² hybrid orbital** is one of six orbitals that result from the hybridization of one *s*, three *p*, and two *d* orbitals. The six orbitals are directed to the corners of a regular octahedron.

spdf **notation** is a method of describing electron configurations in which the numbers of electrons assigned to each orbital are denoted as superscripts. For example, the electron configuration of Cl is $1s^2 2s^2 2p^6 3s^2 3p^5$.

Specific heat is the quantity of heat required to change the temperature of one gram of substance by one degree Celsius.

Spectator ions are ionic species that are present in a reaction mixture but do not take part in the reaction. They are usually eliminated from a chemical equation, as in $\cancel{Na^+} + Cl^- + Ag^+ + \cancel{NO_3^-} \rightarrow AgCl(s) + \cancel{Na^+} + \cancel{NO_3^-}$.

The **spectrochemical series** is a ranking of ligand abilities to produce a splitting of the *d* energy level of a central metal ion in a complex ion.

A **spontaneous** or **natural process** is one that is able to take place in a system left to itself. No external action is required to make the process go, although in some cases the process may take a very long time.

Stalactites and **stalagmites** are limestone ($CaCO_3$) formations in limestone caves produced by the slow decomposition of $Ca(HCO_3)_2(aq)$.

A **standard cell potential, $E°_{cell}$,** is the voltage of an electrochemical cell in which all species are in their standard states. (See also **cell potential**).

A **standard electrode potential** is the electric potential that develops on an electrode when the oxidized and reduced forms of some substance are in their *standard* states.

Standard free energy change, $\Delta G°$, is the free energy change of a process when the reactants and products are all in their standard states. The equation relating standard free energy change to the equilibrium constant is $\Delta G° = -RT \ln K$.

The **standard free energy of formation, $\Delta G°_f$,** is the standard free energy change associated with the formation of 1 mol of compound from its elements in their most stable forms at 1 atm pressure.

The **standard hydrogen electrode (S.H.E.)** is the electrode associated with equilibrium between H_3O^+ ($a = 1$) and $H_2(g)$, 1 atm) on an inert (Pt) surface. The standard hydrogen electrode is *arbitrarily* assigned a standard electrode potential of 0.0000 V.

The **standard state** of a substance refers to the most stable form of that substance at 1 atm pressure.

Standardization refers to establishing the exact concentration of a solution, usually through a titration reaction.

A **standing wave** is a wave motion that reflects back on itself in such a way that the wave contains a certain number of points (nodes) that undergo no motion. A common example is the vibration of a plucked guitar string, and a related example is the description of electrons as matter waves.

Steel is a term used to describe iron alloys containing from 0 to 1.5% C together with other key elements, such as V, Cr, Mn, Ni, W, and Mo.

In **stereoisomerism** the number and types of atoms and bonds in molecules are the same, but certain atoms are oriented differently in space. Cis-trans isomerism is a type of stereoisomerism.

Stoichiometric coefficients (see **balancing an equation**).

A **stoichiometric factor** is a conversion factor relating molar amounts of two species involved in a chemical reaction (i.e., a reactant to a product, one reactant to another, etc.). The numerical values used in formulating the factor are stoichiometric coefficients.

Stoichiometric proportions refer to relative amounts of reactants that are in the same mole ratio as implied by the balanced equation for a chemical reaction. For example, a mixture of 2 mol H_2 and 1 mol O_2 is in stoichiometric proportions and a mixture of 1 mol H_2 and 1 mol O_2 is not, when referring to the reaction $2 H_2 + O_2 \rightarrow 2 H_2O$.

Stoichiometry refers to quantitative measurements and relationships involving substances and mixtures of chemical interest.

The **stratosphere** is the region of the atmosphere that extends from approximately 10 to 40 km above the earth's surface.

A **strong acid** is an acid that is completely ionized in aqueous solution.

A **strong base** is a base that is completely ionized in aqueous solution.

A **strong electrolyte** is a substance that is completely ionized in solution.

A **structural formula** for a compound indicates which atoms in a molecule are bonded together, and whether by single, double, or triple bonds.

Sublimation is the passage of molecules from the solid to the gaseous state.

A **subshell** refers to a collection of orbitals of the same type; e.g., the three $2p$ orbitals constitute the $2p$ subshell.

A **substance** has constant composition and properties throughout a given sample and from one sample to another. All substances are either elements or compounds.

Substitution reactions are typical of those involving alkane and aromatic hydrocarbons. In such a reaction a functional group replaces an H atom on a chain or ring.

The **substrate** is the substance that is acted upon by an enzyme in an enzyme-catalyzed reaction. Substrate is converted to products, and the enzyme is regenerated.

A **sugar** is either a *monosaccharide* (simple sugar) or an *oligosaccharide* containing 2 to 10 monosaccharide units.

Sulfuryl compounds contain the group O_2S.

The **superoxide** ion has the structure $[:\overset{..}{O}-\overset{..}{O}:]^-$.

Superphosphate is a mixture of $Ca(H_2PO_4)_2$ and $CaSO_4$ produced by the action of H_2SO_4 on phosphate rock.

A **supersaturated solution** contains, because of its manner of preparation, more solute than normally expected for a saturated solution.

Surface tension is a property resulting from the differing environments of surface and interior molecules in a liquid. Among the consequences of surface tension is the tendency of a free-falling liquid to form spherical drops.

The **surroundings** represent that portion of the universe with which a system interacts.

A **system** is the portion of the universe selected for a thermodynamic study.

A **systematic error** is one that recurs regularly in a series of measurements because of an inherent error in the measuring system (e.g., through faulty calibration of a measuring device).

Temperature is a measure of the average molecular kinetic energy of a substance (translational kinetic energy for gases and liquids, and vibrational kinetic energy for solids).

Temporary hard water (see **hard water**).

A **terminal atom** in a Lewis structure is an atom that is bonded only to one other atom (a central atom).

A **termolecular process** is an elementary process in a reaction mechanism in which three atoms or molecules collide simultaneously.

A **ternary compound** is comprised of *three* elements.

The **tertiary structure** of a protein describes the types of linkages between polypeptide chains that give a protein its three-dimensional structure.

The **theoretical yield** of a chemical reaction is the quantity of product *calculated* to result from a certain quantity of reactant in a chemical reaction. (See also **actual yield** and **percent yield**.)

A **theory** is a conceptual framework with which one is able to *explain* one or a group of natural laws.

The **thermodynamic equilibrium constant, K,** is an equilibrium constant expression based on activities. In dilute solutions activities can be replaced by molar concentrations; in ideal gases, by partial pressures in atm.

A **thio** compound is one in which an S atom replaces an O atom. For example replacement of an O by S converts SO_4^{2-} to $S_2O_3^{2-}$ (thiosulfate ion).

Thionyl compounds contain the group OS.

The **third law of thermodynamics** states that the entropy of a pure perfect crystal is *zero* at the absolute zero of temperature, 0 K.

A **three-center (bridge, banana) bond** unites *three* atoms through *two* electrons in a delocalized molecular orbital, as with the B—H—B bonds in diborane, B_2H_6.

Titration is a procedure for carrying out a chemical reaction between two solutions by the controlled addition (from a buret) of one solution to the other.

A **titration curve** is a graph of solution pH versus volume of titrant. It outlines how pH changes during an acid–base titration, and it can be used to establish such features as the equivalence point of the titration.

In a **triple covalent bond** *three pairs* of electrons are shared between the bonded atoms. It is represented by a triple-dash sign ($\equiv$).

Transfer RNA (tRNA) refers to short RNA chains that bring specific amino acids to the site of protein synthesis on a ribosome in the cytoplasm of a cell.

Transition elements are those whose atoms feature the filling of d or f orbitals of an inner electronic shell. If filling of f orbitals occurs, the elements are sometimes referred to as *inner*-transition elements.

Transition state theory describes a chemical reaction through a hypothetical intermediate species called an activated complex. The activated complex dissociates either back into the original reactants or into product molecules.

Transmutation is the process in which one element is converted to another as a result of a change affecting the nuclei of atoms (such as in radioactive decay).

A **transuranium element** is one with an atomic number $Z > 92$.

Triglycerides are esters of glycerol (1,2,3-propanetriol) with long-chain monocarboxylic (fatty) acids.

A **triple point** is the condition of temperature and pressure under which three phases of a substance (e.g., a solid, the liquid, and the vapor) coexist at equilibrium.

The **troposphere** is the region of the atmosphere that extends from the surface of the earth to a height of about 10 km.

Trouton's rule states that at their normal boiling points the entropies of vaporization of many liquids are about 88 J mol^{-1} K^{-1}.

The **uncertainty principle** states that, when measuring the position and momentum of fundamental particles of matter, uncertainties in measurement are inevitable.

A **unidentate ligand** is one that has but a single pair of electrons available for donation to a central metal ion. It becomes attached at only a single position in the coordination sphere.

A **unimolecular process** is an elementary process in a reaction mechanism in which a single molecule, when sufficiently energetic, dissociates.

Unit analysis refers to a problem-solving method in which particular emphasis is placed on the proper cancellation of units in conversion factors.

A **unit cell** is a small collection of atoms, ions, or molecules from which an entire crystal structure can be inferred.

Unsaturated hydrocarbon molecules contain one or more carbon-to-carbon multiple bonds.

An **unsaturated solution** contains less solute than the solution is capable of dissolving under the given conditions.

The **valence bond method** treats a covalent bond in terms of the overlap of atomic orbitals. Electron probability (or charge density) is concentrated in the region of overlap.

Valence electrons are electrons in the electronic shell of highest principal quantum number, that is, electrons in the outermost shell.

The **valence-shell electron-pair repulsion (VSEPR) theory** relates the shape of a species to the geometrical distribution of electron pairs in the valence shell of the central atom.

The **van der Waals equation** is an equation of state for nonideal gases. It includes correction terms to account for intermolecular forces of attraction and for the volume occupied by the gas molecules themselves.

van der Waals forces is a term used to describe, collectively, intermolecular forces of the London type and interactions between permanent dipoles.

Vaporization is the passage of molecules from the liquid to the gaseous state.

Vapor pressure is the pressure exerted by a vapor when it is in dynamic equilibrium with its liquid.

A **vapor pressure curve** is a graph of vapor pressure as a function of temperature.

Viscosity refers to a liquid's resistance to flow. Its magnitude depends on intermolecular forces of attraction and, in some cases, on molecular sizes and shapes.

A **voltaic cell** is an electrochemical cell in which a *spontaneous* chemical reaction produces electricity.

Wave function (see **wave mechanics**).

The **wavelength** is the distance between successive crests or troughs of a wave motion.

Wave mechanics is a form of quantum theory based on the concepts of wave–particle duality, the uncertainty principle, and the treatment of electrons as matter waves. Mathematical solutions of the equations of wave mechanics are known as wave functions (ψ).

A **weak acid** is an acid that is only partially ionized in aqueous solution.

A **weak base** is a base that is only partially ionized in aqueous solution.

A **weak electrolyte** is a substance that is only partially ionized in solution.

Weight refers to the force exerted on an object when it is placed in a gravitational field (the "force of gravity"). The terms *weight* and *mass* are often used interchangeably.

Work is a form of energy transfer that can be expressed as a force acting through a distance.

X-ray diffraction is a method of crystal structure determination based on the interaction of a crystal with x rays.

A **zero-order** reaction proceeds at a rate that is *independent* of reactant concentrations. The concentration-term exponent(s) in the rate law is(are) equal to *zero*.

Zone refining is a purification process in which a rod of material is subjected to successive melting and freezing cycles. Impurities are swept by a moving molten zone to the end of the rod, which is cut off and discarded.

Answers to Selected Exercises

Note: Your answers may differ slightly from those given here, depending on the number of steps you use to solve a problem and whether you round off results at each step.

Chapter 1

1. (a) 2.14×10^3 g; **(b)** 6.172 m; **(c)** 1.316×10^{-3} kg; **(d)** 0.812 L; **(e)** 223 mm; **(f)** 25.6 mL; **(g)** 8008 mg; **(h)** 35 m. **2. (a)** 385.0 in.; **(b)** 3.486 yd; **(c)** 907.0 yd; **(d)** 1.40×10^4 ft; **(e)** 4.39×10^3 in.; **(f)** 1.5×10^2 min. **3. (a)** 57.2 cm; **(b)** 38.4 m; **(c)** 4.883 lb; **(d)** 142 kg; **(e)** 4.26×10^4 m; **(f)** 0.754 km. **4. (a)** 1.00×10^6 m²; **(b)** 2.79×10^7 ft²; **(c)** 2.59×10^6 m²; **(d)** 2.83×10^4 cm³; **(e)** 1.00×10^{18} nm³. **5. (a)** 108 °F; **(b)** 33 °C; **(c)** −35 °C; **(d)** −126 °F. **6.** 1.26 g/cm³. **7. (a)** 463 g; **(b)** 16.6 kg; **(c)** 20.4 L. **8.** 5.4×10^2 g. **9.** 131 g. **10. (a)** 4.3151×10^4; **(b)** 5.00×10^{-2}; **(c)** 2.00×10^{-1}; **(d)** 6.3×10^1; **(e)** 8.7×10^{-6}. **11. (a)** 31200; **(b)** 0.617; **(c)** 0.003256; **(d)** 6.12; **(e)** 0.0000968; **(f)** 0.371; **(g)** 4.5; **(h)** 0.0000058. **12. (a)** three; **(b)** two or three; **(c)** two; **(d)** five; **(e)** four; **(f)** one; **(g)** five; **(h)** two to five. **13. (a)** 7219; **(b)** 6.600×10^3; **(c)** 319.2; **(d)** 8.000×10^{-4}; **(e)** 918.7; **(f)** 1.860×10^5; **(g)** 35.60; **(h)** 3.125×10^5. **14. (a)** 1.88×10^5; **(b)** 1.46×10^6; **(c)** 2.8×10^{-3}; **(d)** 4.8×10^{-3}; **(e)** 6.7×10^6; **(f)** 2.2×10^9. **15. (a)** 1.000 g/cm³; **(b)** 0.9997 g/cm³. **16.** 529 g. **17. (a)** physical; **(b)** chemical; **(c)** chemical; **(d)** physical. **18. (a)** substance; **(b)** heterogeneous mixture; **(c)** homogeneous mixture; **(d)** heterogeneous mixture; **(e)** homogeneous mixture; **(f)** heterogeneous mixture; **(g)** heterogeneous mixture; **(h)** substance. **19. (a)** physical; **(b)** chemical; **(c)** chemical; **(d)** physical. **21. (a)** extensive; **(b)** intensive; **(c)** extensive; **(d)** intensive. **24.** Natural laws may be difficult to discover (''God is subtle''), but all physical phenomena can be explained by exact mathematical laws, with nothing left to chance (''He is not malicious.''). **26. (a)** 1.86×10^5 mi/s; **(c)** 1.73×10^{17} W; **(e)** 1×10^{-5} m; **27. (a)** 0.026; **(b)** 5.9×10^4; **(c)** 2.6. **28.** exact numbers: (a) and (d). **29. (a)** 8.15×10^3; **(c)** 5.54×10^{-1}; **(e)** 72.6. **30.** 21.877 g. **31. (a)** 115.76 mi/h; **(b)** 2.8 mi/lb. **33.** 3-lb can. **35.** 7.92 in. **36. (a)** 6.7×10^2; **(b)** 9.5 mg/kg; **(c)** 6.8×10^2 days. **37. (a)** 0.085 m³; **(b)** 1.2×10^4 cm². **40.** 47.8 °C; −8.3 °C. **42.** −459.67 °F. **44.** 0.850 g/cm³. **45.** 102 cm³. **46.** 10^3 kg/m³. **47.** (a) < (c) < (b). **48.** 1.7×10^{-3} cm. **52.** 4.8×10^2. **53.** 7.37 L. **55.** 0.8688 g/cm³. **56.** 35 °C. **57.** $l/r = 1.612$. **58.** −40 °C = −40 °F. **85.** (c) **86.** (a) **87.** (b) **88.** (d) **89.** (d) **90.** (b) **91.** (c) **92.** (d) **93.** (a) **94.** (c) **95. (a)** An element cannot be transformed into another substance, either by physical or chemical means. **(b)** A homogeneous mixture has a uniform composition and properties; a heterogeneous mixture exists in two or more physically distinct regions with varying properties. **(c)** Mass, an extensive property, measures a quantity of matter and is independent of temperature; density, an intensive property, is independent of the quantity of matter but is temperature dependent. **96.** 1.11×10^4. **97** 1.26 g/cm³. **98.** 3.40 L. **99.** by the gallon.

Chapter 2

1. 0.168 g. **2.** 3.45 g. **3. (a)** yes; **(b)** 27.3% C, 72.7% O. **4. (a)** Ratio, g O/g S: SO_3-to-SO_2 = 3:2; **(b)** Ratio, g H/g O: H_2O_2-to-H_2O = 1:2; **(c)** Ratio, g P/g Cl: PCl_5-to-PCl_3 = 3:5. **5. (a)** 0; **(b)** +50.6 C; **(c)** −101 C. **6.** ^{28}Si, 14 p, 14 e, $A = 28$; ^{85}Rb, 37 e, 48 n; potassium, 19 p, 19 e, 21 n, $A = 40$; arsenic, ^{74}As, 33 p, $A = 74$; neon, 10 p, 8 e, 10 n, $A = 20$; bromine, ^{80}Br, 35 p, 35 e, 45 n (other possibilities as well); lead, ^{208}Pb, 82 p, 82 e, $A = 208$ (other possibilities as well). **7. (a)** Ar < K < Co < Cu < Cd < Sn < Te; **(b)** K < Ar < Cu < Co < Sn < Te < Cd; **(c)** K < Ar < Co < Cu < Sn < Cd < Te. **8.** 59% neutrons. **9.** Cu. **10. (a)** 2.248461; **(b)** 3.330216; **(c)** 16.41388. **11.** 80.916. **12.** 238.0. **13. (a)** 2.08×10^{25}; **(b)** 7.41×10^{21}; **(c)** 3.7×10^{12}. **14. (a)** 155 mol Zn; **(b)** 131 g Ar; **(c)** 55.0 mg Ag; **(d)** 3.94×10^{24} Fe atoms. **15.** 6.39×10^{19}. **20. (a)** 0.43 g; **(b)** 0.34 g; **(c)** 0.20 g. **21.** O_2H. **22.** 25. **25.** 11% Cu. **29.** The data are consistent with the idea that there is a fundamental unit of charge. However, the smallest charge of the group is *twice* the charge on an electron. The charge on a single electron can not be inferred from this set of data. **30. (a)** 1.6×10^3; **(b)** 4.0×10^5. **35. (a)** 56 p, 82 n, 56 e; **(b)** 11.49208; **(c)** 8.6219. **36.** Cl can only have $Z = 17$; thus there is no need to write it. On the other hand, to designate a specific isotope the mass number must be shown (35). **37. (a)** Ca; **(b)** Th; **(c)** Ti; **(d)** Sn. **38.** 1.661×10^{-24} g. **39. (a)** 3.28×10^{-4} g; **(b)** 3.36×10^{-4} g; **(c)** 3.32×10^{-4} g. **41. (a)** 78.70%; **(b)** 23.98 u; **(c)** 24. **42. (a)** Li-7; **(b)** 9:1; **(c)** 7.5% Li-6, 92.5% Li-7. **43.** 0.36%. **44.** 200.6. **45.** 73. One source of error is the difficulty in estimating percent abundances from the mass spectrum. The other error comes in using integral mass numbers instead of exact isotopic masses in obtaining a weighted average. **46. (a)** six; **(b)** 36, 37, 38, 38, 39, 40; **(c)** most abundant, ^{1}H^{35}Cl; second, ^{1}H^{37}Cl. **47.** 2.00×10^{23}

48. (a) 1.4×10^{-6} mol Pb/L; **(b)** 8.43×10^{14} Pb atoms/cm^3 blood. **49.** 187.0 g. **50.** 4.38×10^{22}. **51.** 1.5×10^{-3} g. **78. (d)** **79. (b)** **80. (b)** **81. (c)** **82. (c)** **83. (a)** **84. (c)** **85. (b)** **86. (d)** **87. (c)** **88. (a)** **89. (d)** **90. (a)** 1.74 g; **(b)** 0.76 g; **91.** ^{27}Al **92.** -2609 C/g. **93.** 108.9 u. **94. (a)** 6.525 mol Na; **(b)** 9.390×10^{27}; **(c)** 1.1×10^{-10} g Cu. **95.** 1.5×10^{14} Au atoms.

Chapter 3

1. (a) 1.60×10^{24}; **(b)** 234 g; **(c)** 1.14×10^{25}. **2. (a)** 5.62×10^3 g; **(b)** 3.83×10^3 g; **(c)** 110. g. **3. (a)** 149.2; **(b)** 68.3 mol H; **(c)** 9.57×10^{24}; **(d)** 2.28 g O/g N. **4. (a)** 2.08×10^{25}; **(b)** 2.08×10^{25}; **(c)** 6.36×10^3 g. **5.** 36.20% O. **6. (a)** 64.07% Pb; **(b)** 45.50% Fe; **(c)** 25.84% S; **(d)** 76.06% W. **7.** Co_2O_3. **8.** C_3H_8O. **9.** $C_8H_6O_4$. **10. (a)** 75.68% C, 8.82% H, 15.50% O; **(b)** $C_{13}H_{18}O_2$. **11.** potassium. **12.** CrO_3. **13. (a)** lithium iodide; **(b)** calcium chloride; **(c)** sodium cyanide; **(d)** ammonium nitrate; **(e)** iodine trichloride; **(f)** dinitrogen trioxide; **(g)** phosphorus pentachloride; **(h)** hypochlorous acid; **(i)** potassium bromate. **14. (a)** Sn^{2+}; **(b)** cobalt(III) ion; **(c)** magnesium ion; **(d)** Cr^{2+}; **(e)** IO_3^-; **(f)** chlorite ion; **(g)** Au^{3+}; **(h)** hydrogen sulfate ion; **(i)** HCO_3^-; **(j)** OH^-. **15. (a)** 0; **(b)** -2; **(c)** $+4$; **(d)** $+5$; **(e)** $+3$; **(f)** $+6$; **(g)** $+3$; **(h)** $+6$; **(i)** $+5$. **16. (a)** MgO; **(b)** SrF_2; **(c)** $Ba(OH)_2$; **(d)** Cs_2CO_3; **(e)** $Hg(NO_3)_2$; **(f)** $Fe_2(SO_4)_3$; **(g)** $Mg(ClO_4)_2$; **(h)** $KHCO_3$; **(i)** NCl_3; **(j)** BrF_5; **(k)** Cu_2S; **(l)** K_2HPO_4; **(m)** $FeCl_2$. **17. (a)** HBr; **(b)** chlorous acid; **(c)** HIO_3; **(d)** sulfurous acid; **(e)** hydroselenic acid; **(f)** H_3PO_4; **(g)** $HClO_3$; **(h)** nitrous acid. **18.** 43.85% H_2O. **20. (a)** 0.333; **(b)** 0.535 g O/g Li_2O; **(c)** 1.35×10^{24}. **21.** 2×10^{18}. **22. (a)** 4.96×10^{-5} mol S_8; **(b)** 2.39×10^{20}. **23.** 3×10^{22}. **24. (a)** 4.0×10^5; **(b)** 9.0×10^{19} g. **25.** 4.7×10^3 m^2. **26.** 2.6×10^{-7} cm. **27.** only **(b)** is correct. **29. (b)** 0.454:1.00; **(d)** 128 g S; **(f)** 8.21×10^{23}. **30. (a)** 4.88% N; **(c)** 2.08% Cr. **31. (a)** 1.95% B; **32.** guanidine. **33.** brand B. **35.** $C_{10}H_8$. **36.** SeO_2 (selenium dioxide) and SeO_3 (selenium trioxide). **37. (a)** $C_{19}H_{16}O_4$; **(b)** $C_4H_8Cl_2S$. **38.** H_2Se. **40. (a)** -4; **(b)** $+4$; **(c)** -1; **(d)** 0; **(e)** $+6$; **(f)** $+2.5$. **42.** $+1$, N_2O; $+2$, NO; $+3$, N_2O_3; $+4$, NO_2; $+5$, N_2O_5. **43. (a)** barium sulfide; **(b)** zinc oxide; **(c)** potassium chromate; **(d)** cesium sulfate; **(e)** chromium(III) oxide; **(f)** iron(II) sulfate; **(g)** magnesium hydrogen carbonate; **(h)** ammonium hydrogen phosphate; **(i)** calcium hydrogen sulfite; **(j)** copper(II) hydroxide; **(k)** nitric acid (hydrogen nitrate); **(l)** potassium perchlorate; **(m)** potassium hypoiodite; **(n)** lithium cyanide; **(o)** bromic acid; **(p)** phosphorous acid. **45. (a)** $Al_2(SO_4)_3$; **(b)** K_2CrO_4; **(c)** SiF_4; **(d)** $Pb(C_2H_3O_2)_2$; **(e)** Fe_2O_3; **(f)** C_3S_2; **(g)** $Co(NO_3)_2$; **(h)** $Sr(NO_2)_2$; **(i)** ClO_2; **(j)** SnO_2; **(k)** $Ca(H_2PO_4)_2$; **(l)** HBr; **(m)** HIO_3; **(n)** $AlPO_4$; **(o)** PCl_2F_3; **(p)** S_4N_2; **(r)** $CrCl_2$. **47.** 15 g $CuSO_4$. **48.** 2.2035 g $CuSO_4 \cdot 3H_2O$. **49.** $MgSO_4 \cdot 7H_2O$. **50. (a)** 90.47% C, 9.49% H; **(b)** C_4H_5; **(c)** C_8H_{10}. **51.** C_7H_8O. **52.** CH_4N. **53.** 0.186 g BHA. **56.** 77.9% Cu, 11.3% Sn, 6.19% Zn, 4.59% Pb. **57.** 109 ppm Mg^{2+}. **58.** $ZnSO_4 \cdot 7H_2O$. **60.** 24.3. **61.** 225.9. **62.** 26.9. **63.** 56. **96. (c)** **97. (b)** **98. (d)** **99. (a)** **100. (c)** **101. (c)** **102. (a)** **103. (b)** **104. (c)** **105. (d)** **106.** 29.1 cm^3. **107. (a)** calcium iodide; **(b)** $Fe_2(SO_4)_3$; **(c)** SO_3; **(d)** BrF_5; **(e)** ammonium cyanide; **(f)** calcium hypochlorite; **(g)** $LiHCO_3$. **108. (a)** 57.49% Cu; **(b)** 720. g CuO. **109.** $C_{13}H_6Cl_6O_2$. **110.** $Na_2SO_3 \cdot 7H_2O$.

Chapter 4

1. (a) $2 \text{ Mg} + O_2 \rightarrow 2 \text{ MgO}$; **(b)** $S + O_2 \rightarrow SO_2$; **(c)** $CH_4 + 2 O_2 \rightarrow CO_2 + 2 H_2O$; **(d)** $Ag_2SO_4(aq) + BaI_2(aq) \rightarrow BaSO_4(s) + 2 \text{ AgI}(s)$. **2.** Coefficients in the equations are **(a)** $1, 2 \rightarrow 1, 2$; **(b)** $4, 1 \rightarrow 2, 2$; **(c)** $1, 3 \rightarrow 1, 3$; **(d)** $3, 2 \rightarrow 3, 1, 3$; **(e)** $1, 6 \rightarrow 3, 2$. **3.** Coefficients in the equations are **(a)** $1, 2 \rightarrow 1, 2$; **(b)** $1, 1 \rightarrow 1, 2$; **(c)** $2, 6 \rightarrow 2, 3$. **4. (a)** $C_5H_{12} + 8 O_2 \rightarrow 5 CO_2 + 6 H_2O$; **(b)** $2 C_2H_6O_2 + 5 O_2 \rightarrow 4 CO_2 + 6 H_2O$; **(c)** $HI(aq) + NaOH(aq) \rightarrow NaI(aq) + H_2O(aq)$; **(d)** $2 \text{ KI}(aq) + Pb(NO_3)_2(aq) \rightarrow PbI_2(s) + 2 KNO_3(aq)$. **5.** 4.08 mol $FeCl_3$. **6.** 45.3 g Cl_2, 13.2 g P_4. **7. (a)** 6.03 mol H_2; **(b)** 48.1 g H_2O; **(c)** 284 g CaH_2. **8. (a)** 0.458 M C_2H_5OH; **(b)** 0.173 M CH_3OH; **(c)** 3.07 M $(CH_3)_2CO$; **(d)** 0.467 M $C_3H_8O_3$. **9. (a)** 201 mol KCl; **(b)** 23.4 g Na_2CO_3; **(c)** 7.32 mg NaOH/mL. **10.** 149 mL. **11.** 0.288 M $MgSO_4$. **12.** 12.2 g $CuCO_3$. **13.** 5.76 g NO. **14. (a)** 82.01 g C_6H_{10}; **(b)** 84.1% yield; **(c)** 145 g C_6H_{12}. **15.** 0.0377 g C_3H_8. **16.** 72.1% $CaCO_3$. **17.** Coefficients in the equations are **(a)** $6 \rightarrow 8, 1$; **(b)** $3, 1 \rightarrow 2, 1$; **(c)** $6, 16, \rightarrow 1, 12, 1$; **(d)** $1, 8 \rightarrow 1, 2, 2, 4$. **18.** Coefficients in the equations are **(a)** $2, 3, \rightarrow 1, 6$; **(b)** $1, 3, 3 \rightarrow 1, 3$; **(c)** $1, 2 \rightarrow 1, 1, 1$; **(d)** $1, 4, 2 \rightarrow 1, 2, 1$. **19. (a)** $2 C_6H_6 + 15 O_2 \rightarrow 12 CO_2 + 6 H_2O$; **(d)** $C_6H_5COSH + 9 O_2 \rightarrow 7 CO_2 + 3 H_2O + SO_2$. **20. (b)** $NH_4Cl + NaOH \rightarrow NaCl + H_2O + NH_3$; **(c)** $2 KClO_3 \rightarrow 2 KCl + 3 O_2$; **(d)** $2 \text{ Na} + 2 H_2O \rightarrow 2 \text{ NaOH} + H_2$. **21. (b)** $3 NO_2 + H_2O \rightarrow 2 HNO_3 + NO$; **(c)** $2 \text{ Cu} + O_2 + CO_2 + H_2O \rightarrow Cu_2(OH)_2CO_3$; **(e)** $3 Ca(H_2PO_4)_2 + 8 NaHCO_3 \rightarrow Ca_3(PO_4)_2 + 4 Na_2HPO_4 + 8 CO_2 + 8 H_2O$. **22. (c)** $Al^{3+}(aq) + 3 NH_3 + 3 H_2O \rightarrow Al(OH)_3(s) + 3 NH_4^+(aq)$; **(d)** $2 Cu^{2+}(aq) + 4 I^-(aq) \rightarrow 2 \text{ CuI}(s) + I_2$. **23.** $3 \text{ FeS} + 5 O_2 \rightarrow Fe_3O_4 + 3 SO_2$. **24. (a)** 0.157 mol O_2; **(b)** 9.45×10^{22} molecules O_2; **(c)** 7.78 g KCl. **25.** 25.0 g SO_2. **26.** 83.8% Fe_2O_3. **27.** 89.8% Ag_2O. **29.** 0.915 g H_2. **30.** 138 mL. **31.** 2.28 g. **32. (a)** 2.89 M $CO(NH_2)_2$; **(c)** 6.61×10^{-5} M NaCl. **33. (a)** 76.2 mL CH_3OH; **(c)** 481 mg $Ca(NO_3)_2$. **35.** 290. mL. **37.** The initial solution (0.250 M) is 20 times more concentrated than the final solution (0.0125 M). Use any combination in which the final volume is 20 times the initial volume: 50.00 mL/1000.0 mL; 25.00 mL/500.0 mL; or 5.00 mL/100.0 mL. **38.** 0.849 g $AgNO_3$. **39. (a)** 15.9 g $Ca(OH)_2$; **(b)** 164 kg $Ca(OH)_2$. **41.** 2.05×10^3 g O_2. **42.** 2.87 M $NaNO_2$. **43.** 0.070 g Mg. **44.** 0.2 cm^2 **45. (a)** 0.333 mol Na_2CS_3, 0.167 mol Na_2CO_3, 0.500 mol H_2O; **(b)** 150. g Na_2CS_3. **46.** 0.219 g H_2. **47.** The quantity of P_4O_{10} consumed is 0.344 kg; the quantity *unreacted* is 0.66 kg. **48.** 30.1 g lithopone. **51.** 62% yield. **52.** 56 g commercial acetic acid. **53.** 435 g HCl. **54.** 27.7 mol CO_2. **56.** 80% Fe–20% Al. **58.** 336 g P_4S_{10}. **59.** 1.34×10^3 g $AgNO_3$. **60.** 509 kg Fe. **62. (a)** $6 CO(NH_2)_2 \rightarrow C_3N_3(NH_2)_3 + 6 NH_3 + 3 CO_2$; **(b)** 29 kg melamine. **63.** 6.96 kg. **64. (a)** $2 CH_2CHCH_3 + 2 NH_3 + 3 O_2 \rightarrow 2 CH_2CHCN + 6 H_2O$; **(b)** 58% yield; **(c)** 0.55 ton NH_3. **93. (b)** **94. (a)** **95. (c)** **96. (b)** **97. (d)** **98. (a)** **99. (d)** **100. (a)** **101. (c)** **102. (a)** **103. (a)** $Hg(NO_3)_2(s) \rightarrow Hg(l) + 2 NO_2(g) + O_2(g)$; **(b)** $Na_2CO_3(aq) + 2 HCl(aq) \rightarrow 2 NaCl(aq) + H_2O + CO_2(g)$; **(c)** Determine the empirical formula of malonic acid: $C_3H_4O_4$. The equation is $C_3H_4O_4 + 2 O_2 \rightarrow 3 CO_2(g) + 2 H_2O$. **104.** 0.718 g Na. **105.** In an analytical procedure we must account for the exact quantity of each element. For example, in combustion analysis we must capture all the CO_2(g); the yield of CO_2 must be 100%. When synthesizing a compound we can tolerate some loss through side reactions or in handling, as long as we ultimately produce a pure compound.

Chapter 5

1. (a) salt; (b) strong base; (c) salt; (d) weak acid; (e) strong acid; (f) weak acid; (g) weak base; (h) salt; (i) strong base. **2.** (a) 0.215 M K^+; (b) 0.082 M NO_3^-; (c) 0.370 M Al^{3+}; (d) 0.972 M Na^+. **3.** 3.16×10^{-3} M OH^-. **4.** 0.118 M K^+, 0.186 M Mg^{2+}, 0.490 M Cl^-. **5.** 6.95 mg MgI_2. **6.** (a) $2\ Br^- + Pb^{2+} \rightarrow PbBr_2$(s); (b) no reaction; (c) $Fe^{3+} + OH^- \rightarrow Fe(OH)_3$(s); (d) $Ca^{2+} + CO_3^{2-} \rightarrow CaCO_3$(s); (e) $Ba^{2+} + CrO_4^{2-} \rightarrow BaCrO_4$(s); (f) no reaction. **7.** (a) $OH^- + HC_2H_3O_2 \rightarrow H_2O + C_2H_3O_2^-$; (b) no reaction; (c) $HSO_4^- + OH^- \rightarrow H_2O + SO_4^{2-}$; (d) $HCO_3^- + H^+ \rightarrow H_2O + CO_2$(g); (e) $2\ Al$(s) $+ 6\ H^+ \rightarrow 2\ Al^{3+}$(aq) $+ 3\ H_2$(g); (f) no reaction. **8.** 34.75 mL. **9.** 0.06176 M NaOH. **10.** 19.4 g MgO. **11.** (a) reduction: $S_2O_8^{2-} + 2\ e^- \rightarrow 2\ SO_4^{2-}$; (b) reduction: $2\ HNO_3 + 8\ H^+ + 8\ e^- \rightarrow N_2O + 5\ H_2O$; (c) oxidation: $Br^- + 3\ H_2O \rightarrow BrO_3^- + 6\ H^+ + 6\ e^-$; (d) reduction: $NO_3^- + 6\ H_2O + 8\ e^- \rightarrow NH_3 + 9\ OH^-$. **12.** (a) $3\ Cu$(s) $+ 8\ H^+ + 2\ NO_3^- \rightarrow 3\ Cu^{2+} + 2\ NO$(g) $+ 4\ H_2O$; (b) $4\ Zn$(s) $+ 10\ H^+ + NO_3^- \rightarrow 4\ Zn^{2+} + NH_4^+ + 3\ H_2O$; (c) $6\ ClO_2 + 6\ OH^- \rightarrow 5\ ClO_3^- + Cl^- + 3\ H_2O$; (d) $2\ Fe_2S_3$(s) $+ 6\ H_2O + 3\ O_2$(g) $\rightarrow 4\ Fe(OH)_3$(s) $+ 6\ S$(s). **13.** (a) $2\ NO + 5\ H_2 \rightarrow 2\ NH_3 + 2\ H_2O$; (b) $3\ Cu$(s) $+ 8\ H^+ + 2\ NO_3^- \rightarrow 3\ Cu^{2+} + 4\ H_2O + 2\ NO$(g); (c) $4\ Zn$(s) $+ 10\ H^+ + 2\ NO_3^- \rightarrow 4\ Zn^{2+} + 5\ H_2O + N_2O$(g); (d) $5\ H_2O_2 + 2\ MnO_4^- + 6\ H^+ \rightarrow 2\ Mn^{2+} + 8\ H_2O + 5\ O_2$(g). **14.** (a) oxidizing agents: $S_2O_8^{2-}$, HNO_3, NO_3^-; reducing agent: Br^-; (b) oxidizing agents: NO_3^-, NO_3^-, ClO_2, O_2; reducing agents: Cu(s), Zn(s), ClO_2, Fe_2S_3(s); (c) oxidizing agents: NO, NO_3^-, NO_3^-, MnO_4^-; reducing agents: H_2, Cu(s), Zn(s), H_2O_2. **15.** 0.1226 M Mn^+. **16.** At least *one* of the qualitative analysis group 1 cations is present—Ag^+, Hg_2^{2+}, Pb^{2+}. With this single observation we cannot tell which of the three it is, or if there are more than one. **17.** 0.471 M Na^+. **18.** (a) A weak electrolyte. The formula shows one ionizable H atom—an acid. Since it is not one listed in Table 5-4, it must be a weak acid. (b) A strong electrolyte. Cs_2SO_4 is an ionic compound consisting of Cs^+ and SO_4^{2-} ions. (c) A strong electrolyte. $CaCl_2$ is an ionic compound consisting of Ca^{2+} and Cl^- ions. (d) A nonelectrolyte. $(C_2H_5)_2O$ is made up only of nonmetal atoms; it is a covalent compound. Since none of the H atoms is shown as being ionizable, we should *not* expect it to be an acid. (e) A weak electrolyte. This covalent compound is an acid with two ionizable H atoms, but it is not listed as one of the common strong acids. **20.** Solution (c), which is 0.583 M Na^+. **21.** (a) 9.98×10^{-3} M Ca^{2+}; (b) 9.72×10^{-3} M K^+; (c) 2×10^{-7} M Zn^{2+}. **24.** (a) no reaction; (c) ZnO(s) $+ 2\ H^+$(aq) $\rightarrow Zn^{2+}$(aq) $+ H_2O$; (e) Ba^{2+}(aq) $+ S^{2-}$(aq) $+ Cu^{2+}$(aq) $+ SO_4^{2-}$(aq) $\rightarrow$ CuS(s) + $BaSO_4$(s); (g) NH_3(aq) $+ H^+$(aq) $\rightarrow NH_4^+$(aq). **26.** (a) $2\ Na$(s) $+ 2\ H_2O \rightarrow 2\ Na^+$(aq) $+ 2\ OH^-$(aq) $+ H_2$(g); (b) Fe^{3+}(aq) $+ 3\ OH^-$(aq) $\rightarrow Fe(OH)_3$(s); (c) $Fe(OH)_3$(s) $+ 3\ H^+$(aq) $\rightarrow Fe^{3+}$(aq) $+ 3\ H_2O$. **27.** (a) $NaHCO_3$(s) $+ H^+$(aq) $\rightarrow Na^+$(aq) $+ H_2O + CO_2$(g); (c) $Mg(OH)_2$(s) $+ 2\ H^+$(aq) $\rightarrow Mg^{2+}$(aq) $+ 2\ H_2O$; (e) $AlNa(OH)_2CO_3$(s) $+ 4\ H^+$(aq) $\rightarrow Al^{3+}$(aq) $+ Na^+$(aq) $+ 3\ H_2O + CO_2$(g). **28.** (a) Ca^{2+}(aq) $+ 2\ NO_3^-$(aq) $+ 2\ K^+$(aq) $+ SO_4^{2-}$(aq) $\rightarrow CaSO_4$(s) $+ 2\ K^+$(aq) $+ 2\ NO_3^-$(aq); (c) Ba^{2+}(aq) $+ 2\ Cl^-$(aq) $+ 2\ K^+$(aq) $+ SO_4^{2-}$(aq) $\rightarrow BaSO_4$(s) +

$2\ K^+$(aq) $+ 2\ Cl^-$(aq). **29.** (a) 6.17 M NH_3; (b) 11% NH_3, by mass. **30.** (a) 4.0×10^2 mL concd HCl; (b) 0.2424 M HCl. **32.** 32.1% H_2SO_4, by mass. **34.** (a) A gas is evolved and escapes; reaction goes to completion. (b) Reactants and products are all confined gases; reaction reaches equilibrium. (c) A precipitate is formed; reaction goes to completion. (d) One product is a strong electrolyte and the other is a weak electrolyte; reaction reaches equilibrium. **35.** (a) $2\ MnO_4^- + 10\ I^- + 16\ H^+ \rightarrow 2\ Mn^{2+} + 5\ I_2$(s) $+ 8\ H_2O$; (c) $2\ BrO_3^- + 3\ N_2H_4 \rightarrow 2\ Br^- + 3\ N_2 + 6\ H_2O$; (e) $3\ UO^{2+} + Cr_2O_7^{2-} + 8\ H^+ \rightarrow 3\ UO_2^{2+} + 2\ Cr^{3+} + 4\ H_2O$. **36.** (a) $3\ CN^- + 2\ MnO_4^- + H_2O \rightarrow 3\ CNO^- + 2\ MnO_2 + 2\ OH^-$; (c) $4\ Fe(OH)_2 + O_2 + 2\ H_2O \rightarrow 4\ Fe(OH)_3$. **37.** (a) $S_2O_3^{2-} + 5\ H_2O + 4\ Cl_2 \rightarrow 2\ SO_4^{2-} + 8\ Cl^- + 10\ H^+$; (c) $3\ P_4 + 8\ H^+ + 20\ NO_3^- + 8\ H_2O \rightarrow 12\ H_2PO_4^- + 20\ NO$(g); (e) $S_8 + 12\ OH^- \rightarrow 4\ S^{2-} + 2\ S_2O_3^{2-} + 6\ H_2O$. **38.** (a) $2\ Pb(NO_3)_2 \rightarrow 2\ PbO + 4\ NO_2 + O_2$; (c) $2\ HS^- + 4\ HSO_3^- \rightarrow 3\ S_2O_3^{2-} + 3\ H_2O$; (e) $4\ O_2^- + 2\ H_2O \rightarrow 3\ O_2 + 4\ OH^-$. **39.** (b) $5\ NO_2^- + 2\ MnO_4^- + 6\ H^+ \rightarrow 5\ NO_3^- + 2\ Mn^{2+} + 3\ H_2O$; (c) $2\ H_2S + SO_2 \rightarrow 3\ S + 2\ H_2O$. **41.** (a) $2\ S_2O_3^{2-} + I_2 \rightarrow S_4O_6^{2-} + 2\ I^-$; (b) $S_2O_3^{2-} + 4\ Cl_2 + 5\ H_2O \rightarrow 2\ HSO_4^- + 8\ Cl^- + 8\ H^+$; (c) $S_2O_3^{2-} + 4\ OCl^- + 2\ OH^- \rightarrow 2\ SO_4^{2-} + 4\ Cl^- + H_2O$. **43.** 47.39% Fe. **44.** 37.0 g $Na_2C_2O_4$/L. **46.** 91.0% MnO_2. **47.** (a) $IO_3^- + 3\ HSO_3^- \rightarrow I^- + 3\ H^+ + 3\ SO_4^{2-}$ and $5\ I^- + IO_3^- + 6\ H^+ \rightarrow 3\ I_2 + 3\ H_2O$; (b) 9.15 g $NaHSO_3$ is required in the first step, and an additional 0.200 L of the $NaIO_3$ solution is required in the second step. **48.** (a) Na_2SO_4(aq); $BaCl_2$ is converted to $BaSO_4$(s) and the NaCl dissolves. Na_2CO_3 would also work. (b) Water; Na_2CO_3(s) is soluble but $MgCO_3$(s) is not. (c) KCl(aq); $AgNO_3$ is converted to AgCl(s) and KNO_3 dissolves. (d) Water; $Cu(NO_3)_2$(s) is soluble and $PbSO_4$(s) is not. **49.** Neither Na^+ nor NH_4^+ is possible, since all common compounds of these cations are soluble. The precipitate obtained with KOH could be $Mg(OH)_2$ or $Cu(OH)_2$; however, since copper compounds are usually colored and all the solids noted here are white, Cu^{2+} is not probable. Of the given cations, Ba^{2+} forms an insoluble sulfate. Thus, of the ions listed, Mg^{2+} and Ba^{2+} are possible. **73.** (a) **74.** (c) **75.** (b) **76.** (c) **77.** (b) **78.** (d) **79.** (c) **80.** (d) **81.** (d) **82.** (a). **83.** (a) $S_2O_3^{2-} + 5\ H_2O + 4\ Cl_2 \rightarrow 2\ HSO_4^- + 8\ Cl^- + 8\ H^+$; (b) $3\ P + 2\ H^+ + 5\ NO_3^- + 2\ H_2O \rightarrow 3\ H_2PO_4^- + 5\ NO$. **84.** $3\ PbO + 2\ MnO_4^- + H_2O \rightarrow 3\ PbO_2 + 2\ MnO_2 + 2\ OH^-$. **85.** 12.9 mL 0.102 M $Ba(OH)_2$. **86.** 0.01726 M $KMnO_4$. **87.** Na_2CrO_4: Use water (all common sodium compounds are soluble). $BaCO_3$: Use HCl(aq). $BaCO_3$ is insoluble in water, and insoluble $BaSO_4$ would form with H_2SO_4(aq). *MgO:* MgO is insoluble in water but will dissolve in HCl(aq) or H_2SO_4(aq). $ZnSO_4$: Use water.

Chapter 6

1. (a) 0.950 atm; (b) 0.874 atm; (c) 1.536 atm; (d) 3.2 atm. **2.** (a) 1.19×10^3 mm; (b) 817 mm; (c) 8.24 ft. **3.** 766 mmHg. **4.** (a) 45.2 L; (b) 9.72 L. **5.** (a) 192 cm^3; (b) 144 cm^3. **6.** 470. K = 197 °C. **7.** (a) $V_f = 3V_i$; (b) $V_f = 0.25V_i$; (c) $V_f = V_i$. **8.** 30.6 L. **9.** 41.7 L. **10.** 3.22 atm. **11.** 0.495 atm. **12.** 29.9. **13.** 1.80 g CO_2/L. **14.** 526 L CO_2. **15.** 229 L O_2. **16.** 2.88 L. **17.** 2.29×10^{-3} mol O_2. **18.** 22.5 s. **19.** (a) 1.483 atm; (b) 6.56 atm; (c) 2.28 atm; (d) 0.900 atm; (e) 2.32 atm. **20.** (b) 3.23 m; (c) d =

1.92 g/cm^3. **22.** 1.18×10^5 atm; 8.97×10^7 mmHg; 8.97×10^7 torr; 1.73×10^6 lb/in.^{2}; 1.22×10^5 kg/cm^2; 1.20×10^{10} N/m^2; 1.20×10^{10} Pa; 1.20×10^7 kPa; 1.20×10^5 bar; 1.20×10^8 mb. **24.** **(a)** 0.069 kg/cm^2; **(b)** 0.10 kg/cm^2; **(c)** 0.020 kg/cm^2. **26.** 133 atm. **27.** 271 K. **29.** 4.7 g. **31.** 849 K. **33.** 4.06 L. **34.** 156 g O_2. **35.** **(a)** 8.314 kPa dm^3 mol^{-1} K^{-1}; **(c)** 40.6 kPa. **37.** 42.2. The hydrocarbon cannot have more than 3 carbon atoms (corresponding to C_3H_6). With 2 carbon atoms the formula would have to be C_2H_{18}; with one C atom, CH_{30}. (Neither of these compounds is possible, as we learn in Chapters 10 and 11.) **39.** 26.04. **40.** 1.27 atm. **41.** P_4. **42.** **(a)** $d = 1.18$ g air/L; **(c)** 455 K. **43.** The element X is oxygen. **45.** 9.93% $KClO_3$. **46.** 1.2×10^2 L $O_2(g)$. **47.** **(b)** 4.32×10^4 L $NH_3(g)$. **48.** **(a)** 0.0597 mol $SO_2(g)$; **(b)** 6.82 L. **49.** The balanced equation is $4 C_3H_5(NO_3)_3 \rightarrow 12 CO_2(g) + 10 H_2O(g) + 6 N_2(g) + O_2(g)$. The pressure attained is about 9.4×10^3 atm. **52.** 732 mmHg. **53.** 12 g He. **54.** **(a)** 28.7; **(b)** The density should be less because of the high content of water vapor (5.9%), which has a low molecular weight (18). **(c)** about 130:1. **55.** 75.52% N_2, 23.15% O_2, 1.3% Ar, 0.03% CO_2. **57.** 251 cm^3. **58.** 741 mmHg. **59.** 148.9 mmHg. **61.** **(a)** 1.09×10^3 K; **(b)** 493 m/s. **62.** 324 m/s. **63.** **(a)** mol. wt. = 7.8; **(b)** Of the noble gases, only He has u_{rms} greater than a rifle bullet. **65.** 0.00245 mol NO_2. **66.** 59. **67.** About 400 cm away from the HCl. **68.** **(a)** 29.1 atm; **(b)** 25.7 atm; **(c)** Because of intermolecular forces of attraction between CO_2 molecules, the gas does not exert as high a pressure as if it were an ideal gas. **69.** **(b)** 6.54 L. **70.** **(a)** 43.1 atm. **104.** (b) **105.** (d) **106.** (a) **107.** (c) **108.** (c) **109.** (a) **110.** (b) **111.** (c) **112.** (c) **113.** (d). **114.** 6.12 L. **115.** 15.0 g O_2. **116.** Increasing the diameter of the tube increases the volume of liquid and hence its mass. However, the area at the base of the tube increases proportionately. As a result the ratio of mass (weight) to area—the pressure—remains constant. **117.** 77.2 L. **118.** The empirical formula is C_2H_5 and the molar mass is 58. This means that the molecular formula must be C_4H_{10}.

Chapter 7

1. **(a)** +96 kcal; **(b)** −177 kJ. **2.** **(a)** 33.0 °C; **(b)** −108 °C; **(c)** 11 °C. **3.** **(a)** 1.7 J g^{-1} °C^{-1}; **(b)** 22.9 °C; **(c)** 2.8×10^2 J mol^{-1} °C^{-1}. **4.** 24.0 °C. **5.** 392 mL H_2O. **6.** 0.824 J. **7.** **(a)** 0; **(b)** −562 J; **(c)** +89 J; **(d)** −117 J. **8.** **(a)** -1.30×10^3 kJ/mol C_2H_2; **(b)** −636.7 kJ/mol $CO(NH_2)_2$; **(c)** -1.82×10^3 kJ/mol $(CH_3)_2CO$. **9.** 5.68 kJ/°C. **10.** **(a)** 28.74 °C; **(b)** 31.38 °C; **(c)** 30.71 °C. **11.** **(a)** −49.13 kJ/g $(CH_3)_3CH$; **(b)** −2856 kJ/mol $(CH_3)_3CH$; **(c)** $(CH_3)_3CH(g) + \frac{13}{2} O_2(g) \rightarrow 4 CO_2(g) + 5 H_2O(l)$, $\Delta H = -2856$ kJ. **12.** **(a)** *endothermic*. **(b)** +27 kJ/mol NH_4NO_3. **13.** **(a)** $C(graphite) + \frac{1}{2} O_2(g) + Cl_2(g) \rightarrow COCl_2(g)$, $\Delta H° = -209$ kJ; **(b)** $N_2(g) + 2 H_2(g) \rightarrow N_2H_4(l)$, $\Delta H° = +50.63$ kJ; **(c)** $C_3H_8O_3(l) + \frac{7}{2} O_2(g) \rightarrow 3 CO_2(g) + 4 H_2O(l)$, $\Delta H° = -1.66 \times 10^3$ kJ. **14.** −283.0 kJ. **15.** −293 kJ. **16.** **(a)** −55.7 kJ; **(b)** −1123.9 kJ; **(c)** −24 kJ. **17.** −205 kJ/mol ZnS. **18.** **(a)** −190. kJ/mol $(CH_3)_2O(g)$; **(b)** −91 kJ/mol $C_7H_6O_1(l)$; **(c)** -1.260×10^3 kJ/mol $C_6H_{12}O_6$. **19.** 280. J. **20.** **(a)** 4.8 L atm; **(b)** 4.9×10^2 J; **(c)** 1.2×10^2 cal. **21.** **(a)** 0.389 J (g Zn)$^{-1}$ °C^{-1}; **(b)** 0.13 J (g Pt)$^{-1}$ °C^{-1}; **(c)** 0.910 J (g Al)$^{-1}$ °C^{-1}. **22.** **(a)** -1×10^5 J. **23.** 1.4×10^2 g; because only three

significant figures are justified, the result can be expressed only to the nearest 0.1 g. **25.** 33.8 °C. **27.** **(a)** Yes; it does pressure–volume work. **(b)** Yes, since some energy leaves the system as work, other energy must be absorbed as heat to maintain the constant temperature. **(c)** The temperature is held *constant*. **(d)** Since E depends only on temperature, which remains constant, $\Delta E = 0$. **28.** **(a)** Yes; it does pressure–volume work. **(b)** The internal energy must decrease. Since $\Delta E = q + w$, if $q = 0$ and $w < 0$, then $\Delta E < 0$. **(c)** Since the internal energy decreases, then the temperature must fall. **29.** **(a)** If the temperature is constant, $\Delta E = 0$ and $q = -w$. The gas cannot do twice as much work as the heat it absorbs. **(b)** This could be possible. Both q and w would be positive and $\Delta E > 0$. **32.** Since diamond is in a slightly higher energy state than graphite, more heat is evolved in the combustion of diamond than of graphite. **33.** 1.68×10^3 L. **35.** −756 kJ. **36.** **(b)** -3.73×10^4 kJ; **(c)** 234 L. **37.** The liberated heat is sufficient to raise the temperature of the 214 g of solid products to about 5000 °C. This is far above the melting point of Fe. Much of the heat is used to melt the iron. **40.** −56 kJ/mol H_2O. **41.** 3.61 kJ/°C. **43.** **(a)** -2.80×10^3 kJ/mol $C_6H_{12}O_6$; **(b)** $C_6H_{12}O_6(s) + 6 O_2(g) \rightarrow 6 CO_2(g) + 6 H_2O(l)$, $\Delta H = -2.80 \times 10^3$ kJ. **44.** -5.64×10^3 kJ/mol $C_{10}H_{14}O$. **45.** -1.97×10^3 kJ/mol $H_3C_6H_5O_7$. **47.** **(a)** $\Delta n_{gas} = -2$, $\Delta H < \Delta E$; **(b)** $\Delta n_{gas} = 0$; $\Delta H = \Delta E$; **(c)** $\Delta n_{gas} = +1$, $\Delta H > \Delta E$. **48.** **(a)** −2008 kJ/mol C_3H_7OH; **(b)** −2012 kJ/mol C_3H_7OH. **49.** −818.3 kJ. **51.** −284 kJ. **52.** −236.5 kJ. **53.** The direct neutralization has $\Delta H = -5.4$ kJ. For the two-step procedure, $\Delta H = +2.5 - 8.0 = -5.5$ kJ. Within the limits of experimental error, Hess's law is verified. **55.** −1366.7 kJ/mol $C_2H_5OH(l)$. **56.** −103 kJ/mol $CCl_4(g)$. **58.** −1025 kJ. **60.** **(a)** 2.40×10^6 kJ; **(b)** 6.62×10^4 L $CH_4(g)$. **61.** **(c)** $\Delta T = 0.4$ °C. **62.** 87.1% CH_4–12.9% C_2H_6. **92.** (c) **93.** (a) **94.** (d) **95.** (a) **96.** (b) **97.** (b) **98.** (d) **99.** (c) **100.** (d) **101.** (b). **102.** **(a)** Specific heat is the quantity of heat required to raise the temperature of 1.00 g of a substance by one degree C. Molar heat capacity is the quantity of heat required to raise the temperature of 1.00 mol of a substance by one degree C. **(b)** In an endothermic reaction, heat is absorbed from the surroundings by the reaction mixture. In an exothermic reaction, the reaction mixture gives off heat to the surroundings. **(c)** The enthalpy of formation of $C_4H_{10}(g)$ is the enthalpy change accompanying the formation of 1 mol $C_4H_{10}(g)$ from 4 mol C(graphite) and 5 mol $H_2(g)$. The heat of combustion is the heat evolved when 1 mol $C_4H_{10}(g)$ burns in $O_2(g)$ to produce 4 mol $CO_2(g)$ and 5 mol $H_2O(l)$. **103.** $T_i = 1.3 \times 10^2$ °C. **104.** $C_6H_5OH(s) + 7 O_2(g) \rightarrow 6 CO_2(g) + 3 H_2O(l)$, $\Delta H = -3063$ kJ/mol $C_6H_5OH(s)$. **105.** +92 kJ. **106.** **(a)** −55.38 kJ; **(b)** $T_f = 40.0$ °C. **107.** −219 kJ/mol $COCl_2(g)$.

Chapter 8

1. **(a)** 187.5 nm; **(b)** 235 cm; **(c)** 4.67×10^7 nm; **(d)** 3.76×10^{-9} m; **(e)** 1.618 μm; **(f)** 4.57×10^{10} Å. **2.** **(a)** 8.8×10^{-6} m (infrared); **(b)** 4.2×10^{-9} m (4.2 nm, x ray); **(c)** 622 m (radio). **3.** **(a)** 8.8×10^{13} s^{-1}; **(b)** 4.42×10^{13} s^{-1}; **(c)** 3.64×10^{15} s^{-1}; **(d)** 1.22×10^8 s^{-1}. **4.** 7.21×10^{16} s^{-1}. **5.** **(a)** 6.9050×10^{14} s^{-1}; **(b)** 397.11 nm; **(c)** n = 10. **6.** **(a)** 4.47×10^{-18} J/photon; **(b)** 281 kJ/mol; **(c)** 5.34×10^{13} s^{-1}; **(d)** 7.21×10^{-7} m. **7.** 1282.1 nm. **8.** **(a)** $m_l = 0$; **(b)** l = either 1 or 2; **(c)** n is any integer equal to or greater than 2; **(d)** n is any integer equal to or greater than 3, and m_l has one of the following values: $-2, -1, 0, +1, +2$.   **9. (a)** $4s$: $n = 4, l = 0$, $m_l = 0$; **(b)** $3p$: $n = 3, l = 1, m_l = -1, 0$, or $+1$; **(c)** $5f$: $n = 5, l = 3, m_l = -3, -2, -1, 0, +1, +2$, or $+3$; **(d)** $3d$: $n = 3, l = 2, m_l = -2, -1, 0, +1$, or $+2$.   **10.** not allowed: (b), (c), (e).   **11.** order of filling: $3p, 3d, 4p, 5s$, $6s, 6p, 5f$.   **12. (b)** P; **(c)** Zr: $[Kr]4d^25s^2$; **(d)** Te: $[Kr]4d^{10}5s^25p^4$; **(e)** As: $[Ar]3d^{10}4s^24p^3$; **(f)** Bi: $[Xe]4f^{14}5d^{10}6s^26p^3$.   **13. (a)** B (boron); **(b)** V (vanadium); **(c)** silicon (Si).   **14. (a)** Al: $[Ne]3s^23p^1$; **(b)** Rb: $[Kr]5s^1$; **(c)** Cd: $[Kr]4d^{10}5s^2$; **(d)** Sb: $[Kr]4d^{10}5s^25p^3$; **(e)** Pb: $[Xe]4f^{14}5d^{10}6s^26p^2$; **(f)** Xe: $[Kr]4d^{10}5s^25p^6$.   **16.** 8.3 minutes.   **19.** No, a line at 1880 nm is in the infrared.   **20.** 364.70 nm.   **21. (a)** max.: 91.174 nm; min.: 121.56 nm; **(b)** ultraviolet; **(c)** $n = 5$; **(d)** no, a line at 108.5 nm would require that $n = 2.5$, but n must be an *integer*.   **23.** 228 nm; ultraviolet light.   **24.** 1.6×10^2 to 3.2×10^2 kJ/mol.   **25.** 9.1×10^{19} photons/s.   **26. (a)** 8.6×10^{-19} J/photon; **(b)** It displays the photoelectric effect with ultraviolet but not with infrared light.   **28. (a)** 1.9 nm; **(b)** -6.053×10^{-20} J.   **31.** The transition $n = 7 \rightarrow n = 4$.   **32. (a)** -8.716×10^{-18} J; **(b)** -2.179×10^{-18} J.   **34. (a)** The two-step and one-step transitions both have $\Delta E = B(1 - 1/3^2)$. **(b)** The energies are additive; and since the frequencies are proportional to the energies ($\nu = E/h$), they too are additive. The wavelength for the single-step transition is *shorter* (higher energy emission) than for either step of the two-step transition. The wavelengths are *not* additive.   **35. (b)** 1.0 μm; **(c)** 1.128×10^{-22} J.   **38.** $x \geq 1.05 \times 10^{-13}$ m.   **43.** 7×10^5 m/s.   **44.** 1.1×10^{-34} m. This length is much smaller than nuclear or atomic dimensions.   **45.** 2.7×10^{-38} m. This is much too small a distance to measure.   **46.** 14 cm.   **49.** (d).   **51.** not allowable: (b), (e), (f).   **52. (a)** $2p$; **(b)** $4d$; **(c)** $5s$.   **54. (a)** one; **(b)** none (cannot have $3f$); **(c)** three; **(d)** five.   **56.** (e) < (b) < (c) = (d) < (a).   **57. (a)** Cannot have 3 electrons in $2s$ orbital; correct: $1s^22s^22p^1$. **(b)** No $2d$ orbitals; correct: $1s^22s^22p^63s^1$. **(c)** $4s$ orbital fills before $3d$; correct: $[Ar]4s^1$. **(d)** The $3d$ level fills before $4p$; correct: $[Ar]3d^24s^2$. **(e)** The $4d$ subshell is filled in Xe, not $5d$; correct: $[Kr]4d^{10}5s^25p^6$. **(f)** The $4f$ subshell has 14 electrons, and there are no $6p$ electrons in Hg; correct: $[Xe]4f^{14}5d^{10}6s^2$.   **59. (a)** 2; **(b)** 0; **(c)** 3; **(d)** 2; **(e)** 14.   **61. (a)** $1s^32s^32p^93s^33p^93d^{15}4s^34p^95s^1$; **(b)** $1s^21p^62s^22p^62d^{10}3s^23p^63d^{10}3f^14s^24p^65s^2$.   **86.** (b).   **87.** (a)   **88.** (d)   **89.** (b)   **90.** (b)   **91.** (d)   **92.** (c)   **93.** (a)   **94.** (b)   **95.** (a).   **96.** 158 kJ/mol photons.   **97.** $n = 5$.   **98. (a)** Te: $1s^22s^22p^63s^23p^63d^{10}4s^24p^64d^{10}5s^25p^4$;

	$4d$					$5s$	$5p$	

(b) [Kr] ↑↓ ↑↓ ↑↓ ↑↓ ↑↓   ↑↓   ↑↓ ↑ ↑

Chapter 9

1. Check element locations in periodic table on the inside front cover.   **2. (a)** In; **(b)** O, Se, and Te are similar to sulfur; most of the elements outside of group 6A are dissimilar, especially the metallic elements; **(c)** Cs, at the extreme left of the sixth period; **(d)** I; **(e)** Xe or Rn.   **3. (a)** Cl, 7A; **(b)** Ge, 4A; **(c)** Mg, 2A; **(d)** Cu, 1B, **(e)** W, 6B.   **4. (a)** In: $[Kr]4d^{10}5s^25p^1$; **(b)** Y: $[Kr]4d^15s^2$; **(c)** Sb: $[Kr]4d^{10}5s^25p^3$; **(d)** Au: $[Xe]4f^{14}5d^{10}6s^1$.   **5. (a)** Rb$^+$: [Kr]; **(b)** Br$^-$: [Kr]; **(c)** O^{2-}: [Ne]; **(d)** Ba^{2+}: [Xe]; **(e)** Zn^{2+}: $[Ar]3d^{10}$; **(f)** Ag$^+$: $[Kr]4d^{10}$; **(g)** Bi^{3+}: $[Xe]4f^{14}5d^{10}6s^2$.   **6. (a)** 1; **(b)** 5; **(c)** 10; **(d)** 6; **(e)** 14; **(f)** 8.   **7. (a)** 6;

(b) 8; **(c)** 5 or 7; **(d)** 1; **(e)** 2; **(f)** 4 (Se) or 6 (B).   **8. (a)** As; **(b)** Sr; **(c)** Cs; **(d)** Xe; **(e)** C; **(f)** Hg.   **9.** *smallest:* an F atom; *largest:* an I$^-$ ion.   **10.** Cs < Sr < As < S < F.   **11.** 75.0 J.   **12.** (d) < (a) < (b) < (c) < (e).   **13. (a)** Ba; **(b)** S; **(c)** Bi.   **14.** Rb > Ca > Sc > Fe > Te > Br > O > F.   **15.** *diamagnetic:* K$^+$, Zn^{2+}, Cd; Sn^{2+}; *paramagnetic:* Cr^{3+}, Co^{3+}, Br.   **16.** Based on atomic volumes of other group 4A elements, the atomic volume of $Z = 114$ should be approximately 20 cm^3/mol and its density should be about 15 g/cm^3.   **19.** Noble gas has $Z = 118$ and alkali metal, $Z = 119$; atomic weights should be about 310.   **21. (a)** Al_2O_3; **(b)** SO_3 (in Chapter 3 we also learned of the compound SO_2); **(c)** $SiCl_4$; **(d)** PCl_3, based on the formula RH_3, and PCl_5, based on the formula P_2O_5. **(e)** FeO_4. This compound has Fe in the oxidation state $+8$, but the only oxidation states we encountered in Chapter 3 were $+2$ and $+3$, leading to the formulas FeO and Fe_2O_3.   **22.** 104: unnilquadrium, Unq; 105: unnilpentium, Unp; 106: unnilhexium, Unh; 107: unnilseptium, Uns; 108: unniloctium, Uno; 109: unnilennium, Une.   **25. (a)** five; **(b)** 32; **(c)** five; **(d)** two; **(e)** 24.   **26. (a)** Pb: $[Xe]4f^{14}5d^{10}6s^26p^2$; **(b)** $Z = 114$: $[Rn]5f^{14}6d^{10}7s^27p^2$.   **27. (a)** Sr^{2+}: [Kr]; **(b)** Y^{3+}: [Kr]; **(c)** Se^{2-}: [Kr]; **(d)** Cu^{2+}: $[Ar]3d^9$; **(e)** Ni^{2+}: $[Ar]3d^8$; **(f)** Ga^{3+}: $[Ar]3d^{10}$; **(g)** Ti^{2+}: $[Ar]3d^2$.   **28. (a)** 4; **(b)** 3; **(c)** 1; **(d)** 2; **(e)** 1; **(f)** 2.   **31. (a)** B; **(b)** Te.   **34.** Li$^+$ < S < Ge < Br$^-$.   **35.** Li$^+$ < B < Cl < P < Br < Ag < Br$^-$.   **38.** (c) < (a) < (e) < (b) < (d).   **40.** 1.603×10^{18} Cs$^+$ ions.   **41.** 1312 kJ/mol.   **42.** The process is exothermic.   **45.** $I_2(Ba) < I_1(F) < I_3(Sc) < I_2(Na) < I_3(Mg)$. The placement of $I_1(F)$ is somewhat uncertain [before or after $I_2(Ba)$?].   **48.** Fe^{2+}, F$^-$, Ca^{2+}, and S^{2-} all have noble gas electron configurations, with all electrons paired.   **49.** V^{3+}: $[Ar]3d^2$; Cu$^+$: $[Ar]3d^9$; Cr^{3+}: $[Ar]3d^3$.   **54. (a)** 5.6 g/cm^3; **(b)** Ga_2O_3, 74.4% Ga.   **55.** $Cl_2(aq) + 2\,Br^-(aq) \rightarrow 2\,Cl^-(aq) + Br_2$; **(b)** no reaction; **(c)** $Br_2(aq) + 2\,I^-(aq) \rightarrow 2\,Br^-(aq) + I_2$.   **56. (a)** $Sr + Cl_2 \rightarrow SrCl_2$; **(b)** $P_4 + 6\,I_2 \rightarrow 4\,PI_3$; **(c)** $Zn + Br_2 \rightarrow ZnBr_2$; **(d)** $Br_2 + Cl_2 \rightarrow 2\,BrCl$.   **77.** (b)   **78.** (c)   **79.** (a)   **80.** (d)   **81.** (c)   **82.** (b)   **83.** (a)   **84.** (c)   **85.** (b)   **86.** (d).   **87. (a)** 34; **(b)** 45; **(c)** 10; **(d)** 2; **(e)** 4; **(f)** 6.   **88. (a)** C; **(b)** Rb; **(c)** At.   **89. (a)** The subshells that must fill to obtain similar electron configurations are 1s in the first period (two elements); 3s and 3p in the third period (8 elements); 4d, 5s, and 5p in the fifth period (18 elements); 5f, 6d, 7s, and 7p in the seventh period (32 elements). **(b)** Elements are arranged according to increasing atomic number; that of Ar (18) is less than that of K (19).

Chapter 10

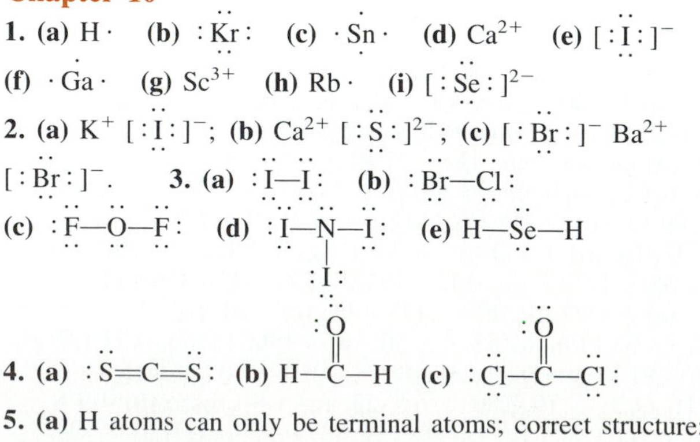

1. (a) H·   **(b)** :Kr: **(c)** ·Sn·   **(d)** Ca^{2+} **(e)** [:I:]$^-$ **(f)** ·Ga·   **(g)** Sc^{3+} **(h)** Rb·   **(i)** [:Se:]$^{2-}$

2. (a) K$^+$ [:I:]$^-$; **(b)** Ca^{2+} [:S:]$^{2-}$; **(c)** [:Br:]$^-$ Ba^{2+} [:Br:]$^-$.   **3. (a)** :I—I: **(b)** :Br—Cl: **(c)** :F—O—F: **(d)** :I—N—I: with :I: below   **(e)** H—Se—H

4. (a) :S=C=S: **(b)** H—C(=O)—H   **(c)** :Cl—C(=O)—Cl:

5. (a) H atoms can only be terminal atoms; correct structure: H—N—O—H (with H below N); **(b)** Structure should have a total of 19

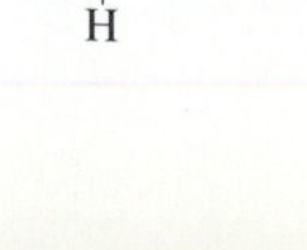

valence electrons, an odd-electron species; correct structure:

$: \ddot{O}-\ddot{C}l-\ddot{O} \cdot$. Alternatively, the structure shown is ClO_2^-.

(c) Structure should have 10 valence electrons, all electrons paired, and a triple covalent bond; correct structure: $[: C\equiv N :]^-$. **(d)** The compound is ionic; correct structure: $Ca^{2+} [: \ddot{O} :]^{2-}$.   **6. (a)** No formal charges. **(b)** S atom has formal charge of $+1$, singly bonded O atom is -1, and doubly bonded O atom, 0. **(c)** Each of the singly bonded O atoms has a formal charge of -1; the other atoms have no formal charges. **(d)** No formal charges. **(e)** Terminal O atom has formal charge of -1. **(f)** N atom has a formal charge of $+1$, and the singly bonded O atom, -1.

7. (a) $[: \ddot{O}-H]^- \; Mg^{2+} \; [: \ddot{O}-H]^-$; **(b)** $\left[H-\overset{\displaystyle H}{\underset{\displaystyle H}{N}}-H \right]^+ [: \ddot{I} :]^-$;

(c) $[: \ddot{O}-\ddot{C}l-\ddot{O} :]^- \; Ca^{2+} \; [: \ddot{O}-\ddot{C}l-\ddot{O} :]^-$.   **8. (a)** OH^-, 8 valence electrons, diamagnetic; **(b)** OH, 7 valence electrons, paramagnetic; **(c)** NO_3, 23 valence electrons, paramagnetic; **(d)** SO_3, 24 valence electrons, diamagnetic; **(e)** SO_3^{2-}, 26 valence electrons, diamagnetic; **(f)** HO_2, 13 valence elec-

trons, paramagnetic. **9. (a)** [BrF₅ structure with central Br, five F atoms] **(b)** $: \ddot{F}-\ddot{P}-\ddot{F} :$ with F below

(c) $: \ddot{C}l - \ddot{I} - \ddot{C}l :$ with Cl below   **(d)** [SF₆ structure]   **10. (a)** CO: linear;

(b) $SiCl_4$: tetrahedral; **(c)** H_2Te: angular (bent); **(d)** ICl_3: T-shaped; **(e)** $SbCl_5$: trigonal bipyramidal; **(f)** SO_2: angular (bent); **(g)** $[AlF_6]^{3-}$: octahedral.   **11. (a)** C—H bonds, 110 pm; C—C, 154; C—Cl, 177; C=O, 123; **(b)** C—H bonds, 414 kJ/mol; C—C, 347; C—Cl, 326; C=O, 728; **(c)** 4.39×10^{-18} J.   **12. (a)** endothermic; **(b)** exothermic. **13.** Bi.   **14.** C—H < Br—H < F—H < Na—Cl < K—F. **15.** NI_3 and SO_3^{2-} are trigonal pyramidal; I_3^- is linear and NO_3^- is trigonal planar.   **16. (a)** F_2: nonpolar (identical atoms); **(b)** NO_2: polar (EN differences, bent molecule); **(c)** BF_3: nonpolar (symmetrical trigonal planar molecule); **(d)** HBr: polar (EN of Br is greater than that of H); **(e)** H_2CCl_2: polar (tetrahedral shape, but EN of Cl is greater than that of H); **(f)** SiF_4: nonpolar (symmetrical molecule); **(g)** OCS: polar (linear molecule with an EN difference between O and S).   **18.** Examples of structures lacking an outer-shell octet include BF_3, NO, and NO_2; those with an expanded octet include PCl_5 and SF_6.

20. (a) $Li^+ [: \ddot{O} :]^{2-} Li^+$; **(b)** $Na^+ [: \ddot{I} :]^-$;

(c) $[: \ddot{F} :]^- Ba^{2+} [: \ddot{F} :]^-$;

(d) $[: \ddot{C}l :]^- Sc^{3+} : \ddot{C}l :]^-$.   **23.** -669 kJ/mol CsCl.
$[: \ddot{C}l :]^-$

24. $\Delta H_f^\circ = -617$ kJ/mol $MgCl_2$; $MgCl_2$ has a much larger (more negative) enthalpy of formation than does MgCl $(-16$ kJ/mol MgCl).   **25.** EA = -67 kJ/mol H.

27. (a) $H-\overset{\displaystyle H}{N}-\ddot{O}-H$ **(b)** $: \ddot{F}-\ddot{N}=\ddot{N}-\ddot{F} :$

(c) $H-\ddot{O}-\ddot{N}=\ddot{O} :$ **(d)** $H-\overset{\displaystyle \ddot{O} :}{N}-\ddot{N}=\ddot{O} :$ with H **28. (a)** Two valence electrons missing; S and C have incomplete octets; correct structure: $[: \ddot{S}=C=\ddot{N} :]^-$ **(b)** compound is covalent; correct structure:

$: \ddot{C}l-\ddot{O}-\ddot{C}l :$ **(c)** one valence electron missing; odd-electron species with structure: $: \ddot{O}-\ddot{N}=\ddot{O}$

(d) N cannot have an expanded octet; correct structure:

$: \ddot{C}l-\overset{\displaystyle : \ddot{F} :}{\underset{\displaystyle : \ddot{C}l :}{N}}-\ddot{C}l :$ **30.** $: \ddot{F}-\ddot{S}-\ddot{S}-\ddot{F} :$ and $: \ddot{F}-\overset{\displaystyle : \ddot{F} :}{S}=\ddot{S} :$

31. (a) Formal charge (FC) on Cl is $+2$, on each O atom, -1; **(b)** no FC; **(c)** FC on B is -1; no FC on F atoms; **(d)** no FC; **(e)** FC on central N atom is $+1$, on other N atoms, -1; **(f)** FC on N atom to the left is -2, on the central N atom, $+1$. **(g)** FC on the N atom is $+1$.

34. (a) $K^+ \left[: \ddot{O}-\overset{\displaystyle}{\underset{\displaystyle : \ddot{O} :}{I}}-\ddot{O} : \right]^-$ **(b)** $[: \ddot{O}-\ddot{C}l :]^- Ca^{2+} [: \ddot{O}-\ddot{C}l :]^-$

(c) $\left[H-\overset{\displaystyle H}{\underset{\displaystyle H}{N}}-H \right]^+ \left[: \ddot{O}-\overset{\displaystyle : \ddot{O} :}{\underset{\displaystyle : \ddot{O} :}{C}l}-\ddot{O} : \right]^-$

35. $[: \ddot{S}-\ddot{S} :]^{2-}$; $[: \ddot{S}-\ddot{S}-\ddot{S} :]^{2-}$;
$[: \ddot{S}-\ddot{S}-\ddot{S}-\ddot{S} :]^{2-}$; $[: \ddot{S}-\ddot{S}-\ddot{S}-\ddot{S}-\ddot{S} :]^{2-}$.

36. (I) $: \ddot{O}-\overset{\displaystyle}{\underset{\displaystyle : \ddot{O} :}{S}}=\ddot{O} :$ **(II)** $: \ddot{O}-\overset{\displaystyle}{\underset{\displaystyle : \ddot{O}}{S}}-\ddot{O} :$ **(III)** $: \ddot{O}=\overset{\displaystyle}{\underset{\displaystyle : \ddot{O} :}{S}}-\ddot{O} :$

38. The N-to-N bond is intermediate between a double and triple bond; the N-to-O bond is a double bond. The best representation is a resonance hybrid of structures (1) and (2). Structure (3) is unlikely because of the high formal charge on N (-2), because O carries a positive formal charge $(+1)$, and because positive formal charges are on adjacent atoms. Structure (4) is unlikely because of the high positive formal charge on the O atom $(+2)$.   **40.** $: \ddot{O}-\ddot{N}=\ddot{O} :$

42. (a) $H-\ddot{O}-\ddot{O} \cdot$ **(b)** $H-\overset{\displaystyle}{\underset{\displaystyle H}{C}}-H$ **(c)** $: \ddot{O}-\ddot{C}l-\ddot{O} \cdot$

(d) $: \ddot{O}=\overset{\displaystyle : \ddot{O} :}{N}-\ddot{O} \cdot$ **44. (a)** $\left[: \ddot{O}-\overset{\displaystyle : \ddot{O} :}{\underset{\displaystyle : \ddot{O} :}{S}}=\ddot{O} \right]^{2-}$ **(b)** $H-\ddot{O}-\ddot{C}l=\ddot{O}$ with $: \ddot{O} :$ below

(c) $H-\ddot{O}-\ddot{N}=\ddot{O} :$ **(d)** $: \ddot{O}=\overset{\displaystyle \ddot{O}-H}{\underset{\displaystyle : \ddot{O}-H}{P}}-\ddot{O}-H$

(e) $: \ddot{O}=\overset{\displaystyle}{\underset{\displaystyle : \ddot{C}l :}{S}}-\ddot{C}l :$ with O above **(f)** $\ddot{O}=Xe=\ddot{O}$ with $: \ddot{O} :$ above   **47.** 136 pm.

49. The structure with the highest bond order (triple) and the *shortest* bond length is $N\equiv N$.   **51.** -100 kJ.

52. -4×10^1 kJ/mol NH_3.   **53.** 416 kJ/mol.
56. 632 kJ/mol.   **57.** 11% ionic.   **60. (a)** 339 kJ/mol;
(b) 92 kJ/mol; **(c)** EN difference ≈ 1.0; **(d)** about 20% ionic.
61. N_2: linear; HCN: linear; ClO_2^-: angular (bent); NH_4^+:
tetrahedral; NO_3^-: trigonal planar; PCl_3: trigonal pyramidal;
SO_4^{2-}: tetrahedral; $SOCl_2$: trigonal pyramidal.

63. (a) linear: $:\!\overset{..}{O}\!\!=\!\!C\!\!=\!\!\overset{..}{O}\!:$ **(b)** linear: $:\!N\!\!=\!\!N\!\!=\!\!\overset{..}{O}\!:$ **(c)** angular (bent): $:\!N\!\!\equiv\!\!S\!\!-\!\!\overset{..}{\underset{..}{F}}\!:$ **(d)** trigonal planar:

$$:\!\overset{..}{\underset{..}{Cl}}\!\!-\!\!N\!\!-\!\!\overset{..}{\underset{..}{O}}\!: \quad \overset{\displaystyle :\!\overset{..}{O}}{\|}$$

64. The VSEPR designation is AXE_3.

The electron pair distribution is tetrahedral, but since the
molecule is diatomic the shape is *linear,* e.g., HF, HCl, HBr,
and HI.   **67. (a)** OSF_2, trigonal pyramidal; **(b)** O_2SF_2: tetrahedral; **(c)** XeF_4: square planar; **(d)** ClO_4^-: tetrahedral;
(e) I_3^-: linear.   **70.** Resultant dipole moments: (a), (b),
(c), (f).   **71. (a)** $:\!\overset{..}{\underset{..}{F}}\!\!-\!\!N\!\!=\!\!\overset{..}{\underset{..}{O}}\!:$ **(b)** angular (bent). **(c)** The
larger dipole moment is a result of the molecule being asym-
metric (bent) and having rather large EN differences. The
FNO_2 molecule has a symmetrical shape.
73. (b) 76.0% F in Teflon.   **74. (b)** 867 cm³.
75. 8.88×10^{18} polymer molecules.   **105.** (a)
106. (c)   **107.** (b)   **108.** (c)   **109.** (d)   **110.** (a)
111. (a)   **112.** (b)   **113.** (c)   **114.** (c)

115. (a) CH_2Cl; **(b)** $H\!\!-\!\!\overset{\displaystyle H}{\underset{}{C}}\!\!-\!\!\overset{..}{\underset{..}{Cl}}\!:$ the C atom
has an incomplete octet; **(c)** molecular formula: $C_2H_4Cl_2$,

$$H\!\!-\!\!\overset{\displaystyle :\!\overset{..}{\underset{..}{Cl}}\!:}{\underset{\displaystyle H}{C}}\!\!-\!\!\overset{\displaystyle :\!\overset{..}{\underset{..}{Cl}}\!:}{\underset{\displaystyle H}{C}}\!\!-\!\!H.$$

116. (a) SO_2: angular; **(b)** SO_3: trigonal planar; **(c)** SO_4^{2-}:
tetrahedral.   **117.** -7.6×10^2 kJ.   **118. (a)** Some triatomic molecules are linear. **(b)** Bond dipole moments might
cancel each other out, depending on the shape of the molecule. **(c)** Structures without formal charges are preferred, but
some structures cannot be written without formal charges.

Chapter 11

1. (a) The 1s orbital of H overlaps the half-filled 3p orbital
of Cl. **(b)** The half-filled 5p of I overlaps the half-filled 3p of
Cl. **(c)** The 1s orbitals of the H atoms overlap with two 4p
orbitals of the Se atom; bond angle, 90°. **(d)** The half-filled
5p of each I overlaps one of the half-filled 2p orbitals of the
N.   **2. (a)** Tetrahedral structure: C—Cl bonds involve
$2sp^3$–3p overlap; C—H bonds, $2sp^3$–1s. **(b)** Linear: The
Be—Cl bonds involve 2sp–3p overlap. **(c)** Trigonal planar:
The B—F bonds involve $2sp^2$–2p overlap.   **3. (a)** SF_6:
octahedral, sp^3d^2; **(b)** CS_2: linear, sp; **(c)** $SnCl_4$: tetrahedral,
sp^3; **(d)** NO_3^-: trigonal planar, sp^2; **(e)** AsF_5: trigonal bipyramidal, sp^3d.   **4.** C atom: sp^2; O atom: sp^3. The C-to-O
double bond has $2sp^2$–2p overlap in a σ bond and 2p–2p
overlap in a π bond. C—O is a $2sp^2$–$2sp^3$ σ bond; C—H is
a $2sp^2$–1s σ bond; and O—H is a $2sp^3$–1s σ bond.

5. (a) $H\!\!-\!\!\underset{\underset{\pi}{\overset{\sigma}{|}}}{C}\!\!\equiv\!\!N\!:$ **(b)** $:\!N\!\!=\!\!\underset{\pi}{C}\!\!=\!\!\underset{\overset{\sigma}{\underset{\pi}{|}}}{C}\!\!=\!\!N\!:$

(c) $H\!\!-\!\!\overset{\displaystyle H}{\underset{\displaystyle H}{C}}\!\!-\!\!\underset{\underset{\pi}{\sigma}}{C}\!\!=\!\!\overset{\displaystyle :\!\overset{..}{\underset{..}{Cl}}\!:}{\underset{\displaystyle H}{\underset{\sigma}{C}}}\!\!-\!\!\overset{}{\underset{\sigma}{C}}\!\!-\!\!\overset{..}{\underset{..}{Cl}}\!:$ **(d)** $H\!\!-\!\!\overset{..}{\underset{..}{O}}\!\!-\!\!\underset{\pi}{\overset{\sigma}{N}}\overset{\displaystyle :\!\overset{..}{O}:}{\underset{\displaystyle :\!\overset{..}{O}:}{\Big\langle}}$

6. All C—H bonds are $\sigma(1s,2sp^3)$; C—O bonds are
$\sigma(2sp^3,2sp^3)$. All bond angles are 109.5°.
7. (a) H—C≡N: sp hybridization of C atom;

(b) $H\!\!-\!\!\overset{\displaystyle :\!\overset{..}{\underset{..}{Cl}}\!:}{\underset{\displaystyle :\!\overset{..}{\underset{..}{Cl}}\!:}{C}}\!\!-\!\!\overset{..}{\underset{..}{Cl}}\!:$ sp^3; **(c)** $H\!\!-\!\!\overset{\displaystyle H}{\underset{\displaystyle H}{C}}\!\!-\!\!\overset{..}{\underset{..}{O}}\!\!-\!\!H$ sp^3;

(d) $H\!\!-\!\!\overset{\displaystyle H}{\underset{}{N}}\!\!-\!\!\overset{\displaystyle :\!\overset{..}{O}:}{\underset{}{C}}\!\!-\!\!\overset{..}{\underset{..}{O}}\!\!-\!\!H$ sp^2.   **8. (a)** Linear, sp
hybridization; **(b)** linear, sp; **(c)** neither linear nor planar; C
atom on left is sp^3, and on right, sp; **(d)** planar, C atom on
left is sp^2, and on right, sp.   **9. (a)** $:\!\overset{..}{O}\!\!=\!\!C\!\!=\!\!\overset{..}{O}\!:$

(b) $:\!\overset{..}{O}\!\!\underset{\underset{\sigma}{}}{\overset{\overset{\pi}{\frown}}{=}}\!\!C\!\!=\!\!\overset{..}{O}\!:$ The C atom is sp hybridized. C—to—O

double bonds consist one $\sigma(2sp,2p)$ and one $\pi(2p,2p)$.

10. H_2^-:
σ_{1s}^b	σ_{1s}^*
↑↓	↑
0.5 bond; stable; paramagnetic
(1 unpaired e).

C_2^+: KK
σ_{2s}^b	σ_{2s}^*	π_{2p}^b		σ_{2p}^b	π_{2p}^*	σ_{2p}^*
↑↓	↑↓	↑↓	↑			

1.5 bond; stable; paramagnetic (1 unpaired e).

O_2^{2-}: KK
σ_{2s}^b	σ_{2s}^*	σ_{2p}^b	π_{2p}^b		π_{2p}^*		σ_{2p}^*
↑↓	↑↓	↑↓	↑↓	↑↓	↑↓	↑↓	

Single bond; stable; diamagnetic

F_2^+: KK
σ_{2s}^b	σ_{2s}^*	σ_{2p}^b	π_{2p}^b		π_{2p}^*		σ_{2p}^*
↑↓	↑↓	↑↓	↑↓	↑↓	↑↓	↑	

1.5 bond; stable; paramagnetic (1 unpaired e).   **11. (a)** sp^2
hybridization at each N and C atom; **(b)** $\alpha = \beta = 120°$.

12. $\left[\overset{\displaystyle :\!\overset{..}{O}}{\underset{\displaystyle :\!\overset{..}{\underset{..}{O}} \quad :\!\overset{..}{\underset{..}{O}}\!:}{C}}\right]^{2-}$ The situation is analogous to

Example 11-10. The C-to-O σ bonds involve an overlap of
$2sp^2$ hybrid orbitals. The C and O atoms each contribute one
2p orbital to form a set of four π-type molecular orbitals into
which 6 electrons are placed—two each from the two singly
bonded O atoms and one each from the C and the doubly
bonded O atoms. Four of the six p electrons go into bonding
MOs, and the remaining two, into antibonding MOs. The π
bond is a net single bond delocalized over all four atoms.

14. $\left[\overset{\displaystyle H}{\underset{\displaystyle H}{H\!\!-\!\!N\!\!-\!\!H}}\right]^+$ tetrahedral structure, sp^3 hybridization of N atom.   **15. (a)** $:\ddot{O}\!=\!C\!=\!\ddot{O}:$ linear, sp hybridization of C atom; **(b)** $:\ddot{C}l\!-\!N\!=\!\ddot{O}:$ trigonal planar, sp^2 hybridization of N atom;   **(c)** $\left[:\ddot{O}\!-\!\ddot{C}l\!-\!\ddot{O}:\right]^-$ trigonal pyramidal, sp^3 hybridization of Cl.   **17.** The C atom on the left of the structure is sp^3, and the one in the center is sp^2. The O atom is sp^3. The C—C—O bond angles are 120°; the rest of the bond angles are 109.5°.   **19.** H—C≡C—C

Hybridization of the two C atoms on the left is sp; the C atom on the right is sp^2. The C—H bond on the left is $\sigma(1s,2sp)$, and the one on the right is $\sigma(2sp^2,1s)$. The σ bond in the C-to-C triple bond is $(2sp,2sp)$, and the C-to-C single bond is $\sigma(2sp,2sp^2)$. The σ bond in the C-to-O double bond is $(2sp^2,2p)$. All π bonds are $(2p,2p)$.

21.   square planar structure; sp^3d^2 hybridization of Xe.   **22.**   trigonal bipyramidal; sp^3d.

The distribution of valence shell electrons is octahedral. The structure is a square-based pyramid. The hybridization of the Br atom is sp^3d^2.

25. ClF_2^-:

sp^3d hybridization;

ClF_2^+: sp^3.   **26. (a)** angular (120° bond angle), sp^2; **(b)** T-shaped (90° and 180° bond angles), sp^3d; **(c)** tetrahedral distribution around the C atom (109.5° bond angle) and angular around the O atom (109.5° C—O—H bond angle); sp^3 hybridization of both C and O.

28.

(I)   (II)

In (I) N on the left is sp^2 and N in the center is sp. In (II), the N atom on the left is sp^3 and the central N atom is sp. The N atom on the right can be described as unhybridized or as sp^2 in (I) and sp in (II).   **30.** All should be stable except Be_2 and Ne_2. B_2 and O_2 are paramagnetic, and the other stable molecules are diamagnetic.   **32.** O_2^+ should have a shorter bond length; its bond order is 2.5 compared to 2 for O_2.   **34.** Lewis structure, C≡C, corresponding to bond order of 4. Molecular orbital diagram, C_2:

double bond

37.

(a) NO: KK

(c) CO: KK

(e) CN^-: KK

(g) BN: KK

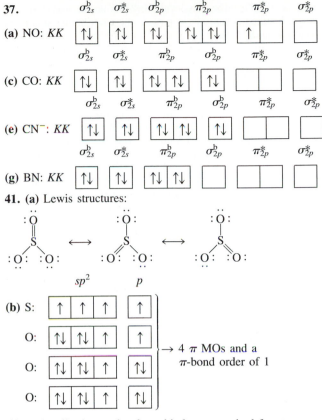

41. (a) Lewis structures:

sp^2 p

(b) S:

O:

O: → 4 π MOs and a π-bond order of 1

O:

42. Delocalized π molecular orbitals are required for structures that are resonance hybrids. **(b)** CO_3^{2-} (see structure in Exercise 12) and **(c)** $:\ddot{O}\!-\!N\!=\!\ddot{O}: \longleftrightarrow :\ddot{O}\!=\!N\!-\!\ddot{O}:$

45. Metals are found among elements of low and high atomic numbers and low and high atomic weights. The factors that primarily affect metallic properties are the number of valence electrons **(c)** and the number of vacant atomic orbitals **(d)**.
47. 7.02×10^{20} energy levels in the 3s conduction band; 7.02×10^{20} electrons in the band.   **50. (a)** Based on a maximum visible wavelength of 760 nm, the energy gap is about 1.6×10^2 kJ/mol.   **51. (a)** 0.60 watt; **(b)** 1.3 A.
73. (c)   **74.** (c)   **75.** (b)   **76.** (a)   **77.** (b)
78. (d)   **79.** (c)   **80.** (a)   **81.** (d).

82. H—C—N=C=O: **(a)** 6 σ bonds; **(b)** 2 π bonds.

83. The Lewis structure of BrF_5 is shown in Exercise 22. Six pairs of valence shell electrons must be distributed about the Br atom. This corresponds to six hybrid orbitals—sp^3d^2—not the five orbitals represented by sp^3d.
84. (a) NO^+: KK

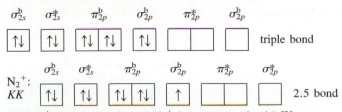

triple bond

N_2^+: KK 2.5 bond

(b) NO^+ is diamagnetic and N_2^+ is paramagnetic. **(c)** We should expect a greater bond length for N_2^+ (bond order 2.5) than for NO^+ (bond order 3).   **85.** The resonance Lewis structure of C_6H_6 shows that the C-to-C bonds are 1.5 bonds.

In general, delocalized molecular orbitals are necessary if a single bonding scheme is to replace the contributing structures to a resonance hybrid.

Chapter 12

1. (a) 13.7 g $(CH_3)_2CO$; **(b)** 29.5 kJ/mol $CHCl_3$; **(c)** 329 kJ. **2. (a)** 280 mmHg; **(b)** 35 °C. **3.** 226 mmHg. **4. (a)** ≈452 K; **(b)** ≈275 mmHg. **5. (a)** 49.7 kJ/mol; **(b)** 433 K. **6. (a)** 3.298 g Mg; **(b)** 73 kJ; **(c)** 768 kJ. **7. (a)** 31.8 mmHg; **(b)** 80.6 mmHg; **(c)** 85.6 mmHg. **8. (a)** Lower (?) is vapor, and upper (?) is liquid. The transition curves are S–V, S–L, and L–V. **(b)** The triple point pressure is above 1 atm, and solid red P sublimes without melting. **(c)** solid → solid + liquid → liquid → liquid + vapor → vapor. **9. (a)** $C_{10}H_{22}$; **(b)** H_3COCH_3; **(c)** H_3CCH_2OH. **10.** O_3 is out of order. Correct order: $N_2 < F_2 < Ar < O_3 < Cl_2$. **11.** The boiling point of isopentane should be between 9.5 and 36.1 °C. **12.** CsI < MgF_2 < CaO. **13. (a)** The unit cell has nine elements, for example:

□ ◆ □
○ ○ ○   Other groupings possible.
□ ◆ □

(b) □ = 1, ◆ = 1, ○ = 2. **(c)** Cells consisting of 4, 5, and 6 elements are not unit cells because they fail to meet the requirements noted in the hint. Also, some cells are not parallelograms. **14. (a)** 362 pm; **(b)** 4.74×10^{-23} cm³; **(c)** 4; **(d)** 4.221×10^{-22} g; **(e)** 8.91 g/cm³. **15.** The unit cell contains 1 Cs^+ and $(8 \times \frac{1}{8}) = 1$ Cl^-; formula: CsCl. **19.** 31 kJ/mol C_6H_6. **20.** 162 L $CH_4(g)$. **24. (a)** between 93 and 94 °C; **(b)** 611 mmHg. **25.** 0.904 mmHg. **26.** 11.6 L. **30. (a)** 0.690 atm; **(b)** 89 °C. **31. (a)** about 280 °C; **(b)** about 52 kJ/mol. **32.** 25.1 °C. **35.** SO_2, yes; CH_4, no. **37.** 1.32×10^4 kJ. **38.** 43.0 kJ. **39.** 2.8 mmHg. **41. (b)** Liquid condenses at about 95 °C. **42.** 26.7 mmHg. **43.** 18 °C. **44.** 4.75 g steam. **47. (a)** 1.5×10^2 atm; **(b)** −1.2 °C. **50. (a)** liquid; **(b)** exothermic; **(c)** greater than 1.00 g/cm³. **52.** about 151 °C. **58.** *intra*molecular hydrogen bonding in (d) only. **64. (a)** BaF_2; **(b)** $MgCl_2$. **68.** 21.46% uncovered. **69.** 0.899 g/cm³. **72.** 1.75 g/cm³. **75. (a)** CCN: Zn^{2+}, 4; S^{2-}, 4; **(b)** 4; **(c)** 3.00 g/cm³. **107.** (d) **108.** (b) **109.** (c) **110.** (a) **111.** (b) **112.** (d) **113.** (a) **114.** (c) **115.** (b) **116.** (c) **117.** The physical properties depend on the type of forces among the structural units. In Ar and CO_2, where the intermolecular forces are weak (dispersion forces), melting points are low and the tendencies for sublimation are high. In the ionic crystal NaCl and the metallic crystal Cu, the forces are strong and the melting points are much higher. **118.** A summary can be found in Table 12-7. **119.** The final condition is of liquid only, (c). **120.** Gas only, under a pressure of 5.56 atm.

Chapter 13

1. 59.0 g KI/100 g solution. **2. (a)** 15.1% CH_3OH, by volume; **(b)** 12.2% CH_3OH, by mass; **(c)** 11.9% (mass/volume). **3.** 1.588 M $C_3H_8O_2$. **4. (a)** 0.2 M; **(b)** 0.4 M in ions; **(c)** −0.6 °C. **5.** 0.230 g. **6.** 0.577 m $C_6H_4Cl_2$. **7. (a)** mole fractions: C_7H_{16}, 0.230; C_8H_{18}, 0.451; C_9H_{20}, 0.320; **(b)** mole percents: 23.0% C_7H_{16}, 45.1% C_8H_{18}, 32.0% C_9H_{20}. **8.** highly soluble: (c) and (f); slightly soluble: (b) and (d); insoluble: (a) and (e). **9.** KF: K^+ carries a charge of only +1 and it is the largest of the four cations. **10.** unsaturated.

11. (a) 1.16×10^{-3} M O_2; **(b)** 2.43×10^{-4} M O_2. **12.** $P_{\text{benz.}}$ = 37.7 mmHg; $P_{\text{tol.}}$ = 17.2 mmHg; P_{total} = 54.9 mmHg. **13.** $\chi_{\text{benz.}}$ = 0.687; $\chi_{\text{tol.}}$ = 0.313. **14.** M.W. = 1.2×10^2. **15.** 100.02 °C. **16.** M.W. = 5.5×10^4. **17.** $C_2H_5OH > HC_2H_3O_2 > NaCl > MgBr_2 > Al_2(SO_4)_3$. **18.** salicyl alcohol (b). Part of the molecule resembles water (HOH) and part resembles benzene (C_6H_6). **20.** Li^+ has an unusually high hydration energy because of its small size. **21.** 46.0 g $HC_2H_3O_2$. **23.** Mass, but not volume, is independent of temperature; only mass percent composition is independent of temperature. **24.** 17.5 M H_2SO_4 **25.** 175 mL. **27.** 8.35 g I_2. **28.** 6.97 M HNO_3, 8.93 m HNO_3. **30.** 682 g H_2O. **32.** 4.3 g C_2H_5OH. **33.** 268 mL $C_3H_8O_3$. **34.** approximately 80 °C. **35. (b)** 16 g $KClO_4$ crystallizes from the solution. **36. (a)** yes; **(b)** 18 g NH_4Cl; **(c)** no. **37.** 4×10^2 g. **38.** 4.58×10^{-4} M N_2. **41.** 23.5 mmHg. **43.** $\chi_{\text{styr.}}$, 0.32; χ_{ethbenz}, 0.68. **44.** 25% styrene, by mass. **45.** $\chi_{\text{benz.}}$ = 0.328. **46.** $C_6H_4N_2O_4$. **47.** 20. °C kg solvent (mol solute)$^{-1}$. **48.** 3 parts $C_2H_6O_2$ to 10 parts H_2O, by volume. **50.** $C_{16}H_{22}O_8$. **53.** 9.0% $C_{12}H_{22}O_{11}$, by mass. **54.** right to left. **55.** T_f = −0.020 °C. Freezing point depression is not a good method because the solution is so dilute. Measurement of osmotic pressure would give better results. **58.** 2.26 g. **59.** M.W. = 2.8×10^5. **61. (a)** −0.19 °C; **(b)** −0.37 °C; **(c)** −0.37 °C; **(d)** −0.56 °C; **(e)** −0.37 °C; **(f)** −0.19 °C; **(g)** approximately −0.20 °C. **62. (a)** i = 1.06; **(b)** i = 1.07. **67. (a)** 17 subdivisions; **(b)** about 8×10^5. **101.** (b) **102.** (c) **103.** (d) **104.** (a) **105.** (b) **106.** (a) **107.** (d) **108.** (c) **109.** (b) **110.** (c) **111. (a)** 2.18% $C_{10}H_8$, by mass; **(b)** 0.174 m $C_{10}H_8$; **(c)** 4.64 °C. **12.** Pure HCl and pure H_2O are nonelectrolytes, but when dissolved in water HCl ionizes completely (it is a strong acid). **113.** The solution with the highest vapor pressure of water is the most dilute: (a), 0.10% by mass. The solution with the lowest freezing point is the most concentrated: (c) $\chi_{\text{isopropanol}}$ = 0.10. **114.** *Lowest electrical conductivity:* 0.15 m urea (a nonelectrolyte). *Lowest boiling point:* 0.05 m NaCl (0.10 mol particles per kg solvent). *Highest vapor pressure of water:* 0.05 m NaCl (smallest number of particles in solution). *Lowest freezing point:* 0.10 m KCl (greatest number of particles per kg solvent—0.20 m).

Chapter 14

1. (a) See Glossary. **(b)** An ore is a natural mineral deposit containing an element in sufficient concentration to make its extraction feasible. **(c)** See Glossary. **(d)** PCB refers to a class of compounds called polychlorinated biphenyls. One or more Cl atoms are substituted for H atoms in the molecule biphenyl. **(e)** The thermite reaction is a highly exothermic reaction in which Al reduces a metal oxide to the free metal. **(f)** See Glossary. **(g)** See Glossary. **2. (a)** sulfur tetrafluoride; **(b)** Mg_3N_2; **(c)** KO_2; **(d)** calcium hydroxide; **(e)** hydrogen peroxide; **(f)** $KClO_4$; **(g)** UF_6; **(h)** HF_2^-; **(i)** lithium aluminum hydride; **(j)** ozone; **(k)** $NH_4H_2PO_4$; **(l)** sodium hydrogen sulfite; **(m)** dinitrogen tetroxide; **(n)** BrF_3. **3. (a)** carbon; **(b)** calcium carbonate; **(c)** calcium oxide; **(d)** silicon dioxide; **(e)** calcium hydroxide; **(f)** calcium sulfate dihydrate; **(g)** mixture of carbon monoxide and hydrogen gases; **(h)** sodium silicate; **4. (a)** Ca + 2 H_2O → $Ca(OH)_2$ + H_2; **(b)** LiH + H_2O → LiOH + H_2; **(c)** Mg + H_2O → MgO + H_2; **(d)** CaO + H_2O → $Ca(OH)_2$; **(e)** C + H_2O → CO + H_2; **(f)** 2 Na_2O_2 + 2 H_2O → 4 NaOH + O_2. **5. (a)** Mg + 2 HCl → $MgCl_2$ + H_2; **(b)** NH_3 + HNO_3 → NH_4NO_3; **(c)** $CaCO_3$ + 2 HCl → $CaCl_2$ + H_2O + CO_2;

(d) $2 NaF + H_2SO_4 \rightarrow Na_2SO_4 + 2 HF$; (e) $3 Ag + 4 H^+ + NO_3^- \rightarrow 3 Ag^+ + 2 H_2O + NO$; (f) $NaHCO_3 + HC_2H_3O_2 \rightarrow NaC_2H_3O_2 + H_2O + CO_2$. **6.** (a) $H_2O_2 + NO_2^- \rightarrow NO_3^- + H_2O$; (b) $H_2O_2 + SO_2 + 2 OH^- \rightarrow SO_4^{2-} + 2 H_2O$; (c) $5 H_2O_2 + 2 MnO_4^- + 6 H^+ \rightarrow 2 Mn^{2+} + 8 H_2O + 5 O_2$; (d) $H_2O_2 + Cl_2 + 2 OH^- \rightarrow 2 H_2O + 2 Cl^-$. **7.** (a) NO_2 or N_2O_4; (b) H_2O_2, Na_2O_2, etc.; (c) CO_2, CCl_4; (d) H_2S; (e) PH_3; (f) NH_3. **8.** (14.3) oxidizing agent: H_2O, reducing agent: CH_4; (14.7) H_2O, M; (14.19) $KCl(l)$, $Na(l)$; (14.21) N_2, Li; (14.31) H_2O, Mg; (14.59) NO_3^-, Cu; (14.63) NO, CO; (14.76) O_2, ZnS. **9.** endothermic: (14.3); exothermic: all but (14.3). **10.** (a) H_2, He, N_2, O_2, F_2, Ne, Cl_2, Ar; (b) Si; (c) C, O, P, S; (d) C; (e) all but He, Ne, and Ar; (f) He; (g) F_2. **11.** (a) K; (b) $CaCO_3$; (c) O_3; (d) H_2S; (e) SiO_2; (f) LiF; (g) LiF. **12.** $CaSO_4 \cdot 2H_2O + (NH_4)_2CO_3(aq) \rightarrow (NH_4)_2SO_4(aq) + CaCO_3(s) + 2 H_2O$. **13.** 92.6%. **14.** 13 km^3. **15.** 6×10^6 tons. **19.** (b) $C_2H_6 + 2 H_2O \rightarrow 2 CO + 5 H_2$; (d) $MnO_2 + 2 H_2 \rightarrow Mn + 2 H_2O$. **20.** Recombination of H atoms: $\Delta H = -435$ kJ/mol H_2; combustion of H_2: $\Delta H = -285.8$ kJ/mol H_2. **21.** 179 g CaH_2. **22.** (a) -2.9×10^2 kJ; (b) -311.4 kJ. **23.** 7.4×10^5 L. **25.** (a) 0.441 g/L; (b) 9.9 g/mol; (c) 85 K. **26.** (b) $2 K + 2 H_2O \rightarrow 2 KOH + H_2$; (d) $K + O_2 \rightarrow KO_2$. **29.** (c) $UO_2 + 2 Ca \rightarrow 2 CaO + U$; (e) $H_2SO_4 + Ca(OH)_2 \rightarrow CaSO_4 + 2 H_2O$. **30.** (a) $Mg_3N_2 + 6 H_2O \rightarrow 3 Mg(OH)_2 + 2NH_3(g)$. Test for the presence of $NH_3(g)$. (b) 27%. **31.** (b) $2 Al + 12 H^+ + 3 SO_4^{2-} \rightarrow 2 Al^{3+} + 6 H_2O + 3 SO_2$. **33.** (a) -852 kJ; (b) $4 Al + 3 MnO_2 \rightarrow 2 Al_2O_3 + 3 Mn$, $\Delta H = -1792$ kJ; (c) $\Delta H = +129$ kJ. **34.** (c) $2 CO + 3 H_2 \rightarrow CH_2OHCH_2OH$. **36.** $2 CH_4 + S_8 \rightarrow 2 CS_2 + 4 H_2S$; $CS_2 + 3 Cl_2 \rightarrow CCl_4 + S_2Cl_2$; $4 CS_2 + 8 S_2Cl_2 \rightarrow 4 CCl_4 + 3 S_8$. **39.** (a) $3 SiO_2 + 4 Al \rightarrow 2 Al_2O_3 + 3 Si$; (b) $SiO_2 + K_2CO_3 \rightarrow K_2SiO_3 + CO_2$. **41.** (a) $2 NH_3 + H_2SO_4 \rightarrow (NH_4)_2SO_4$; (c) $C + 4 H^+ + 4 NO_3^- \rightarrow CO_2 + 2 H_2O + 4 NO_2$. **43.** 0.1716 M NO_3^-. **45.** 993 kg. **49.** (a) $O_3 + 2 I^- + 2 H^+ \rightarrow O_2 + I_2 + H_2O$; (b) $3 O_3 + S + H_2O \rightarrow H_2SO_4 + 3 O_2$; (c) $O_3 + 2 [Fe(CN)_6]^{4-} + H_2O \rightarrow 2 [Fe(CN)_6]^{3-} + 2 OH^- + O_2$. **52.** 4×10^{37}. **54.** 0.225%. **57.** (b) $P_4 + 6 Cl_2 \rightarrow 4 PCl_3$ or $P_4 + 10 Cl_2 \rightarrow 4 PCl_5$; (c) $Cl_2 + 2 Br^- \rightarrow 2 Cl^- + Br_2$. **58.** 37.66 kg. **59.** (a) 326 kJ/mol; (b) $\nu = 8.17 \times 10^{14}$ s^{-1}; $\lambda = 367$ nm (UV). **80.** (c) **81.** (a) **82.** (b) **83.** (d) **84.** (c) **85.** (a) **86.** (b) **87.** (b) **88.** (d) **89.** (a) **90.** (a) LiH; (b) sodium hydrogen sulfate; (c) hypochlorite ion; (d) $MgClO_4$; (e) $NaNO_2$; (f) dinitrogen pentoxide; (g) barium peroxide. **91.** (a) $MgCO_3 \rightarrow MgO + CO_2$; (b) $NaHCO_3 + HCl \rightarrow NaCl + H_2O + CO_2$; (c) $P_4 + 5 O_2 \rightarrow P_4O_{10}$, followed by $P_4O_{10} + 6 H_2O \rightarrow 4 H_3PO_4$; (d) $H_2O_2 + Cl_2 \rightarrow 2 H^+ + 2 Cl^- + O_2$; (e) $4 NH_3 + 5 O_2 \rightarrow 4 NO + 6 H_2O$; (f) $3 Cu + 8 H^+ + 2 NO_3^- \rightarrow 3 Cu^{2+} + 4 H_2O + 2 NO$. **92.** 1.61×10^{10} kg. **93.** Air pollution control measures for automobiles must eliminate unburned hydrocarbons, oxides of nitrogen, and carbon monoxide. The pollutants that must be eliminated from power plants are oxides of nitrogen and sulfur dioxide.

Chapter 15

1. (a) 7.3×10^{-5} mol L^{-1} s^{-1}; (b) 4.4×10^{-3} mol L^{-1} min^{-1}. **2.** (a) 2.3×10^{-4} mol L^{-1} s^{-1}; (b) 6.9×10^{-4} mol D L^{-1} s^{-1}. **3.** (a) 7.9×10^{-4} mol L^{-1} s^{-1}; (b) 8.7×10^{-4} mol L^{-1} s^{-1}. **4.** rate = $k[CH_3CHO]^2$. **5.** (a) second order in A, first order in B; (b) third order. **6.** 48 min. **7.** (a) 53 min.; (b) 0.0131 min^{-1}.

8. (a) 0.0117 min^{-1}; (b) 59.2 min; (c) 118 min; (d) 0.0750 M. **9.** (a) 161 kJ/mol; (b) 592 K. **10.** (a) net equation: $A + 2 B \rightarrow C + D$; rate of slow step = $k[A][B]$; (b) net equation: $A + 2 B \rightarrow C + D$; rate = $k[A][B]^2$; (c) net equation: $2 A + 2 B \rightarrow C + D$; rate = $k[A]^2[B]^2$. **11.** (a) $I + B \rightarrow C + D$. (b) The second step is fast, and the first step is rate-determining. **12.** (a) II; (b) I; (c) III. **13.** 70 s. **14.** 8.0×10^{-3} mol L^{-1} s^{-1}. **15.** (a) 0.010 mol L^{-1} s^{-1}; (b) 4.6×10^{-3} mol L^{-1} s^{-1}; (c) 3.2×10^{-3} mol L^{-1} s^{-1}. **16.** (a) $[A] = 0$; (b) $[A] \approx 0.33$ M; (c) $[A] \approx 0.48$ M. **18.** (a) 3.52×10^{-5} mol C L^{-1} s^{-1}; (b) $[A] = 0.3989$ M; (c) 568 s. **19.** (a) 2.8×10^{-4} mol O_2 s^{-1}; (b) 0.017 mol O_2 min^{-1}; (c) 3.8×10^2 cm^3 O_2 min^{-1}. **22.** (a) 0.074 M; (b) 0.055 M. **24.** (a) first order; (b) 0.30 min^{-1}; (c) 0.10 mol L^{-1} min^{-1}; (d) 0.064 mol L^{-1} min^{-1}; (e) 0.30 mol L^{-1} min^{-1}. **27.** (a) reaction orders: $+1$ in A, $+2$ in B, -1 in C; (b) $R_5 = R_1/4$. **28.** (a) first order in H_2, second order in NO, third order overall; (b) rate = $k \times P_{H_2} \times (P_{NO})^2$. **30.** 218 min. **31.** 20.6 min. **32.** (a) 5.6×10^3 s; (b) 8.04 L $O_2(g)$. **33.** (b) 4.3×10^{-4} s^{-1}; (c) $P_{tot.} = 408$ mmHg; (e) 530 mmHg. **34.** (a) 0.017 M; (d) 4.8×10^{-3} mol L^{-1} min^{-1}; (f) $k = 6.5 \times 10^{-2}$ min^{-1}; (g) 11 min; (h) 22 min. **36.** 161 s. **37.** (a) Expt. 1: 0.022 mol L^{-1} min^{-1}; Expt. 2: 0.089 mol L^{-1} min^{-1}; (b) second order. **39.** rate of reaction = $k[A]^2$, and $k = 0.020$ L mol^{-1} min^{-1}. **40.** (a) zero order; (b) 2.0×10^{-4} mol L^{-1} min^{-1}; (c) 0.0188 L mol^{-1} s^{-1}; (d) 8.2×10^{-5} mol L^{-1} s^{-1}. **44.** (b) 51 kJ/mol; **45.** 283 K. **47.** (a) 44 kJ/mol; (b) 20. h. **52.** zero order. **56.** $[N_2O_2] = (k_1/k_2)[NO]^2$. **59.** $O_3 \rightleftharpoons O_2 + O$ (fast), followed by $O_3 + O \rightarrow 2 O_2$ (slow). **91.** (a) **92.** (d) **93.** (b) **94.** (c) **95.** (a) **96.** (d) **97.** (b) **98.** (c) **99.** (d) **100.** (b) **101.** See Glossary. **102.** The reaction is second order. The rate is not constant (zero order), and the half-life is not constant (first order). **103.** (a) 0.024 mol L^{-1} min^{-1}; (b) $[A] = 2.312$ M (since the initial rate would be 4 times as great as in the first experiment). (c) If the reaction is first order, $t_{1/2} = 0.693/35 = 0.020$ min^{-1}. $[A] = 1.3$ M at $t = 30$ min in the second experiment. **104.** $t_{1/2} = 100.$ s; $k = 6.93 \times 10^{-3}$ min^{-1}; at $t = 100$ s, rate of reaction = 3.0×10^{-3} mol L^{-1} min^{-1}. **105.** (a) 10. min; (b) 0.16 g. **106.** $2 NO_2 \rightarrow NO_3 + NO$ (slow).

Chapter 16

1. (a) $K_c = [COCl_2]/[CO][Cl_2]$; (b) $K_c = [NO_2]^2/[NO]^2[O_2]$; (c) $K_c = [CO_2(g)]$; (d) $[CH_4][H_2S]^2/[CS_2][H_2]^4$; (e) $K_c = [CO_2(g)][H_2O(g)]$. **2.** (a) 0.0431; (b) 3.1×10^3; (c) 0.015; (d) 7.5×10^2. **3.** 9.0×10^{-16}. **4.** $K_p = 23.2$, 6.5, 0.027, and 6.8×10^{-5}, respectively. **5.** 0.50. **6.** 0.77. **7.** (a) 0.0333 mol O_2; (b) 0.133 mol O_2. **8.** (a) The term V appears twice in the numerator and twice in the denominator of the K_c expression, and so it cancels out. (b) $K_c = K_p = 0.659$. **9.** (a) 26.3; (b) 0.613. **10.** (a) 0.011; (b) 1.3. **11.** $[H_2] = [I_2] = 0.015$ M; $[HI] = 0.10$ M. **12.** 0.0020 mol I_2 **13.** (a) displace equilibrium to the left; (b) no effect; (c) displace equilibrium to the right. **14.** Dissociation increases with temperature for (b) and (d). **15.** (a) increase; (b) decrease; (c) decrease; (d) no effect; (e) decrease. **16.** (d) $3 Cl_2(g) + CS_2(l) \rightleftharpoons CCl_4(l) + S_2Cl_2(l)$, $K_c = 1/[Cl_2(g)]^3$; (e) $N_2(g) + Na_2CO_3(s) + 4 C(s) \rightleftharpoons 2 NaCN(s) + 3 CO(g)$, $K_c = [CO(g)]^3/[N_2(g)]$. **17.** (b) 5.55×10^5. **19.** 1.2×10^6. **20.** 5×10^{14}. **21.** 0.04. **23.** 1.49×10^{-2}. **25.** $[NO]/[NO_2] = 0.369$. **26.** 23 L. **27.** (a) $NH_3(g) +$

$\frac{7}{4}$ O$_2$(g) $\rightleftharpoons$ NO$_2$(g) + $\frac{3}{2}$ H$_2$O(g); **(b)** 4.03×10^{19}.
29. 1.5×10^{-3} mol Cl$_2$.   **30.** 1.50 mol each of CO, H$_2$O,
CO$_2$, and H$_2$.   **31.** 0.94 mol SbCl$_3$, 2.06 mol SbCl$_3$,
0.06 mol Cl$_2$.   **32. (a)** no; **(b)** to the right; **(c)** 0.0864 mol
COF$_2$, 0.229 mol CO$_2$, 0.064 mol CF$_4$.   **33.** 0.92 g CO$_2$.
34. 77.9 mol% HI.   **36.** 0.943 mol N$_2$O$_4$, 0.114 mol
NO$_2$.   **37. (a)** low temperatures and high pressures;
(b) mol fraction CH$_4$ = 0.23.   **38.** [Ag$^+$] = [Fe^{2+}] =
0.44 M; [Fe^{3+}] = 0.56 M.   **39.** [Cr^{3+}] = 0.03 M;
[Cr^{2+}] = 0.27 M;[Fe^{2+}] = 0.11 M.   **41.** $P_{tot.}$ = 1.02 atm.
42. 2.1×10^{-3}.   **43. (a)** 0.96 atm; **(b)** less; **(c)** partial
pressures: CO$_2$, 1.33 atm; H$_2$O, 0.169 atm.   **45.** 49.6%.
46. 19.4%.   **48. (a)** 80.0%; **(b)** 176 atm.   **50. (a)** true;
(b) false; **(c)** false; **(d)** true.   **83. (d)**   **84. (c)**
85. (a)   **86. (c)**   **87. (d)**   **88. (b)**   **89. (b)**
90. (a)   **91.** The balanced chemical equation is needed to
establish (1) the correct form of the equilibrium constant expression and (2) the stoichiometric relationship between reactants and products. The actual equilibrium composition, however, must be based on the equilibrium constant expression, not just the balanced equation.   **92.** 2.0×10^{-19}.
93. (a) left; **(b)** 0.30 atm.

Chapter 17

1. (a) forward: HOBr (acid), H$_2$O (base); reverse: H$_3$O$^+$
(acid), OBr$^-$ (base). **(b)** forward: HSO$_4^-$ (acid), H$_2$O (base);
reverse: H$_3$O$^+$ (acid), SO$_4^{2-}$ (base). **(c)** forward: H$_2$O (acid),
HS$^-$ (base); reverse: H$_2$S (acid), OH$^-$ (base). **(d)** forward:
C$_6$H$_5$NH$_3^+$ (acid), OH$^-$ (base); reverse: H$_2$O (acid),
C$_6$H$_5$NH$_2$ (base).   **2. (a)** [H$_3$O$^+$] = 0.0060 M, [OH$^-$] =
1.7×10^{-12} M; **(b)** [H$_3$O$^+$] = 1.2×10^{-13} M, [OH$^-$] =
0.082 M; **(c)** [H$_3$O$^+$] = 3.3×10^{-12} M, [OH$^-$] = 0.0030 M;
(d) [H$_3$O$^+$] = 4.2×10^{-3} M, [OH$^-$] = 2.4×10^{-12} M.
3. (a) 2.52; **(b)** 3.108; **(c)** 2.323; **(d)** 3.34.   **4. (a)** 1 ×
10^{-6} M; **(b)** 7.1×10^{-9} M; **(c)** 0.22 M; **(d)** 2.5×10^{-10} M;
(e) 4.7×10^{-4} M.   **5.** 12.225.   **6.** 3.6×10^{-13} M.
7. 1.51×10^{-5}.   **8. (a)** 3.30×10^{-3} M; **(b)** 2.64;
(c) 0.061 M.   **9.** 8.0×10^{-3} M.   **10. (a)** degree of
ionization = 5.3×10^{-3}; **(b)** 0.53%.   **11. (a)** 1.0 ×
10^{-4} M; **(b)** 1.0×10^{-4} M; **(c)** 5.6×10^{-11} M.
12. acidic: (b); basic: (c), (e); neutral: (a), (d).   **13.** 4.59.
14. (a) 5.0×10^{-6}; **(b)** 5.6×10^{-11}; **(c)** 6.2×10^{-5}.
15. (a) HI; **(b)** HOClO; **(c)** Cl$_3$CCH$_2$COOH;
(d) H$_3$CCH$_2$CF$_2$COOH.   **16. (a)** OH$^-$ is basic because of
lone-pair electrons on the O atom; **(b)** B(OH)$_3$ is acidic because of an empty electron orbital on the B atom (electron-deficient molecule); **(c)** AlCl$_3$ is acidic for the same reason as
(b); **(d)** CH$_3$NH$_2$ is basic because of lone-pair electrons on
the N atom.   **18. (a)** IO$_4^-$; **(b)** C$_3$H$_5$O$_2^-$; **(c)** C$_6$H$_5$COO$^-$;
(d) C$_6$H$_5$NH$_2$.   **19.** amphiprotic: (c), (d), (f), (g).
20. 1.699.   **21.** 1.81×10^{-3} M.   **22.** 12 cm^3.
24. 11.60.   **25.** 11.26.   **26.** 7.2×10^{-4} M.
28. 2.1 g/L.   **30.** 0.076 g.   **31.** 11.28.
33. (a) approx. 13.1; **(b)** 0.13 M.   **35.** 82%.
36. (b) 5.44%; **(c)** 1.3×10^{-5}.   **37. (a)** 1.979; **(b)** 11.82;
(c) 2.51; **(d)** 0.16.   **38. (a)** 9.111; **(b)** 9.2×10^{-9} M.
39. 2.15.   **41. (b)** 9.6.   **42.** 2.4 < pH < 2.5.
44. (a) [H$_3$O$^+$] = 0.76 M, [HSO$_4^-$] = 0.74 M, [SO$_4^{2-}$] =
1.29×10^{-2} M; **(b)** [H$_3$O$^+$] = 0.085 M, [HSO$_4^-$] =
0.065 M, [SO$_4^{2-}$] = 0.0099 M; **(c)** [H$_3$O$^+$] = 1.4×10^{-3} M,
[HSO$_4^-$] = 7×10^{-5} M, [SO$_4^{2-}$] = 6.8×10^{-4} M.
46. pH = 8.43.   **47.** 9.26.   **48.** 1.6 M KNO$_2$.
53. (a) forward; **(b)** reverse; **(c)** reverse; **(d)** forward; **(e)** reverse; **(f)** forward.   **54.** (c) < (f) < (d) < (e) < (b) < (a).
89. (c)   **90. (a)**   **91. (b)**   **92. (d)**   **93. (c)**
94. (a)   **95. (a)**   **96. (b)**   **97. (d)**   **98. (c)**

99. H$_2$SO$_4$ > HNO$_3$ > HC$_2$H$_3$O$_2$ > NH$_4$ClO$_4$ > NaCl >
NaNO$_2$ > NH$_3$ > NaOH > Ba(OH)$_2$.   **100.** 3.08 g.
101. (a) Cl has a higher electronegativity than N, and there
are more O atoms bonded to the central atom in HClO$_4$ than
in HNO$_3$. **(b)** The Cl atoms withdraw electron charge density
from the O—H bond, weakening the bond and making trichloroacetic acid the stronger acid. **(c)** The Cl atom withdraws electron density from the N atom, reducing its ability
to attract a H$^+$ ion and making o-chloroaniline a weaker
base. **(d)** In a strong acid [H$_3$O$^+$] is directly proportional to
the solution molarity, M. In a weak acid [H$_3$O$^+$] =
$\sqrt{M \times K_a}$. Thus if the molarity is doubled, [H$_3$O$^+$] increases
only by $\sqrt{2}$. **(e)** [PO$_4^{3-}$] would be $\frac{1}{3}$ [H$_3$O$^+$] if H$_3$PO$_4$ were
a strong acid, but it is a weak triprotic acid. Very little ionization occurs through the third ionization step, and consequently [PO$_4^{3-}$] is very low.

Chapter 18

1. (a) 0.0852 M; **(b)** 1.17×10^{-13} M; **(c)** 3.3×10^{-5} M;
(d) 0.0852 M.   **2. (a)** 2.49×10^{-5} M; **(b)** 0.0742 M;
(c) 0.0742 M; **(d)** 4.02×10^{-10} M.   **3. (a)** CHO$_2^-$ +
H$_3$O$^+$ → HCHO$_2$ + H$_2$O, HCHO$_2$ + OH$^-$ → CHO$_2^-$ +
H$_2$O; **(b)** C$_6$H$_5$NH$_2$ + H$_3$O$^+$ → C$_6$H$_5$NH$_3^+$ + H$_2$O,
C$_6$H$_5$NH$_3^+$ + OH$^-$ → C$_6$H$_5$NH$_2$ + H$_2$O; **(c)** HPO$_4^{2-}$ +
H$_3$O$^+$ → H$_2$PO$_4^-$ + H$_2$O, H$_2$PO$_4^-$ + OH$^-$ → HPO$_4^{2-}$ +
H$_2$O.   **4. (a)** 4.46; **(b)** 9.65.   **5.** 1.12 M.
6. 0.68 M.   **7.** 1.4×10^{-4}.   **8. (a)** acidic solution:
2,4-dinitrophenol, bromophenol blue, bromocresol green,
chlorophenol red; neutral solution: bromothymol blue; basic
solution: thymolphthalein. **(b)** Bromocresol green is green at
about pH = 5; chlorophenol red is orange at about pH = 6.
9. (a) colorless; **(b)** red; **(c)** blue; **(d)** yellow; **(e)** blue.
10. (a) A flipped-over version of Figure 18-7 (that is, with
the pH starting high and ending low, and the equivalence
point at pH = 7). **(b)** A flipped-over version of Figure 18-8.
(c) Essentially the same as Figure 18-8. **(d)** The portion of
Figure 18-10 between the first and second equivalence
points.   **11. (a)** 12.26; **(b)** 1.43.   **12. (a)** 3.47;
(b) 12.08.   **13.** 0.10 M NaHSO$_4$ (an acid salt with K_a =
1.29×10^{-2}) > 0.10 M NaHSO$_3$ (an acid salt with K_a =
6.3×10^{-8}) > 0.10 M NaC$_2$H$_3$O$_2$ (a solution made basic by
hydrolysis).   **14. (a)** 100.5; **(b)** 29.16; **(c)** 74.08.
15. (a) 0.18 N KOH; **(b)** 0.086 N (first equivalence point) or
0.17 N (second equivalence point); **(c)** 3.6×10^{-3} N.
16. (a) 37.50 mL; **(b)** 21.32 mL.   **18. (a)** 0.080 M;
(b) 0.100 M; **(c)** 0.185 M; **(d)** 6.09×10^{-5} M; **(e)** 2.18 ×
10^{-5} M.   **19.** 8.74.   **20. (a)** 28 mL; **(b)** 3.4 mL;
(c) 238 mL.   **22. (c)** 7.68.   **23.** 1.7 g.
25. 59.5 mL 1.00 M HCl in sufficient 0.100 M NaCHO$_2$ to
make 1.00 L of solution.   **27. (a)** 0.150 M HCl or
0.050 M HC$_2$H$_3$O$_2$; **(b)** 9×10^1 mL 0.150 M HCl or 4.0 ×
10^2 mL 0.050 M HC$_2$H$_3$O$_2$.   **28. (a)** pOH = 4.76 + log
0.128/0.0720 = 5.01, pH = 8.99; **(b)** The pH would probably
remain at 9.00 for a dilution to 1.00 L but not to 1000. L. (If
the buffer is diluted sufficiently far, eventually the pH should
fall to 7.00.) **(c)** 8.98; **(d)** 0.94 mL.   **29. (a)** 4.77;
(b) 4.94; **(c)** 5.14 g; **(d)** 11.3.   **32.** red: (a), (e); yellow:
(b), (c), (d).   **33. (a)** 61% HIn, 39% In$^-$; **(b)** The red
(acid) form is more strongly colored. Complete conversion of
the red to the yellow form requires an 18:1 proportion of
yellow to red. On the other hand, complete conversion of the
yellow to red requires a 4:1 proportion of red to yellow.
34. 0.818 g Ca(OH)$_2$/L.   **35.** 2.96.   **39. (a)** 2.28;
(b) 5.71.   **40. (a)** 11.32; **(b)** 14.3 mL; **(c)** 9.24; **(d)** 5.04.
41. (a) 25.45 mL; **(b)** 29.85 mL; **(c)** 30.00 mL;
(d) 30.15 mL; **(e)** 35.56 mL.   **42. (b)** pH = 4.50,

6.67 mL; pH = 5.50, 9.52 mL; pH = 11.50, 14.1 mL.
43. (a) 0.225 M HCl; (b) 0.212 M H_3PO_4.
45. (b) 18.20%. **47.** (a) 12.11; (b) 11.10.
49. 50.1 g. **51.** (a) 0.151 N; (b) 0.302 N; (c) 0.453 N.
52. 32.0% H_2SO_4, by mass. **81.** (c) **82.** (d)
83. (c) **84.** (b) **85.** (a) **86.** (b) **87.** (b)
88. (d) **89.** (a) **90.** (c) **91.** An indicator end point
is the point in a titration where an indicator changes color.
The equivalence point is the point where acid and base have
just completely neutralized one another. These two points coincide only if the indicator has a pK_a value about equal to the
pH value at the equivalence point. **92.** (a) Above pH 7.0.
At the equivalence point Na_2CO_3(aq) is strongly basic by
hydrolysis (base ionization of CO_3^{2-}). (b) Below pH 7.0.
NH_4^+ at the equivalence point ionizes (hydrolyzes) as an
acid. (c) pH = 7.0. KI(aq) produced in the neutralization
does not hydrolyze. **93.** (a) 31 g; (b) 3.90.
94. (a) 3.10; (b) 4.20; (c) 8.00; (d) 11.08.

Chapter 19

1. (a) $K_{sp} = [Ag^+]^2[SO_4^{2-}]$; (b) $K_{sp} = [Ra^{2+}][IO_3^-]^2$;
(c) $K_{sp} = [Ni^{2+}]^3[PO_4^{3-}]^2$; (d) $K_{sp} = [Hg_2^{2+}][C_2O_4^{2-}]$;
(e) $K_{sp} = [PuO_2^{2+}][CO_3^{2-}]$. **2.** (a) $Fe(OH)_3$(s) $\rightleftharpoons$
Fe^{3+}(aq) + 3 OH^-(aq); (b) BiOOH(s) $\rightleftharpoons$ BiO^+(aq) +
OH^-(aq); (c) Hg_2I_2(s) $\rightleftharpoons$ Hg_2^{2+}(aq) + 2 I^-(aq);
(d) $Pb_3(AsO_4)_2$(s) $\rightleftharpoons$ 3 Pb^{2+}(aq) + 2 AsO_4^{3-}(aq);
(e) $Cu_2[Fe(CN)_6]$(s) $\rightleftharpoons$ 2 Cu^{2+}(aq) + $[Fe(CN)_6]^{4-}$(aq);
(f) $MgNH_4PO_4$(s) $\rightleftharpoons$ Mg^{2+}(aq) + NH_4^+(aq) + PO_4^{3-}(aq).
3. (a) 1.1×10^{-5} M; (b) 0.022 M; (c) 7×10^{-5} M;
(d) 4.5×10^{-5} M. **4.** (a) 1.4×10^{-5}; (b) 8.8×10^{-8};
(c) 1.9×10^{-9}. **5.** (a) 1.7×10^{-4} M; (b) 1.7×10^{-5} M;
(c) 3.8×10^{-10} M. **6.** (a) yes; (b) no; (c) no.
7. (a) AgI; (b) 2.7×10^{-4} M; (c) 3.1×10^{-13} M; (d) yes.
8. (a) Ag^+(aq) + Br^-(aq) $\rightarrow$ AgBr(s); (b) no reaction; (c) no
reaction; (d) $Cu(OH)_2$(s) + 4 NH_3(aq) $\rightarrow$ $[Cu(NH_3)_4]^{2+}$(aq) +
2 OH^-(aq); (e) Fe^{3+}(aq) + 3 OH^-(aq) $\rightarrow$ $Fe(OH)_3$(s);
(f) Ag_2SO_4(s) + 4 NH_3(aq) $\rightarrow$ 2 $[Ag(NH_3)_2]^+$(aq) +
SO_4^{2-}(aq); (g) $CaSO_3$(s) + 2 H_3O^+(aq) $\rightarrow$ Ca^{2+}(aq) +
3 H_2O + SO_2(g). **9.** only Ca^{2+}. **10.** (b) 1.7 g.
11. Only in (a) does precipitation not occur. **12.** $2.4 \times$
10^{-5} M. **13.** (a) all three; (b) CdS and CuS precipitate.
14. no. **15.** 3.3. **16.** 2.3×10^{-14}. **19.** 9.3.
20. 0.68 or 68%. **21.** 4.5×10^{-3} M. **22.** 4.1×10^{-9}.
23. 5.30×10^{-5}. **25.** 1.7×10^{-5}. **26.** (a) 0.18 M;
(b) 0.027 M. **29.** 27 ppm. **30.** 0.034 g. **31.** No;
the mass of AgCl(s) is only about 0.2 mg. **32.** (a) $1.7 \times$
10^{-4} M; (c) 2.8×10^{-5} M; (e) 1.4×10^{-9} M. **34.** no.
36. (a) yes; (b) yes; (c) no. **37.** Some Ag_2CrO_4(s) should
precipitate. **38.** (a) yes; (b) 4.0×10^{-6} M.
39. (a) 3.2%; (b) 0.18 M. **40.** (a) yes; (b) $5.5 \times$
10^{-4} M. **41.** 98% of the Ba^{2+} is precipitated (2% is unprecipitated). **42.** 0.016 M. **45.** (a) AgBr(s); (b) No.
About 32% of the Br^- remains unprecipitated at the point
where precipitation of AgCl begins. **46.** 0.50 M Na_2SO_4.
The chlorides are soluble; the hydroxides do not precipitate
under these conditions; the sulfates and carbonates are both
insoluble, but the greatest difference in K_{sp} is between $CaSO_4$
and $BaSO_4$. **48.** (a) no; (c) yes. **49.** 9.53.
50. 0.18 mol $Mg(OH)_2$/L. **53.** (a) $[Ca^{2+}]$ = 0.01 M;
(b) $[Ca^{2+}]$ = 0.056 M. **55.** $Zn(OH)_2$(s) + 2 H_3O^+(aq) $\rightarrow$
Zn^{2+}(aq) + 4 H_2O; $Zn(OH)_2$(s) + 2 $HC_2H_3O_2$(aq) $\rightarrow$
Zn^{2+}(aq) + 2 $C_2H_3O_2^-$(aq) + 2 H_2O; $Zn(OH)_2$(s) +
4 NH_3(aq) $\rightarrow$ $[Zn(NH_3)_4]^{2+}$(aq) + 2 OH^-(aq); $Zn(OH)_2$(s) +
2 OH^-(aq) $\rightarrow$ $[Zn(OH)_4]^{2-}$(aq). **57.** 1.4×10^{-6} g.
58. (a) 0.79 g; (b) 9.4 g; (c) 39 g. **59.** (b) Increase
$[C_2H_3O_2^-]$ to 1.25 M. **61.** 0.2 g FeS/L. **62.** $1.3 \times$

10^{-3} mol CoS/L. **65.** 0.016 M $PbCl_2$ at 25 °C, 0.094 M
$PbCl_2$ at 80 °C. **66.** (a) Pb^{2+}(aq) + 2 Cl^-(aq) $\rightarrow$
$PbCl_2$(s); (b) $Zn(OH)_2$(s) + 2 OH^-(aq) $\rightarrow$ $[Zn(OH)_4]^{2-}$(aq);
(c) $Fe(OH)_3$(s) + 3 H_3O^+(aq) $\rightarrow$ Fe^{3+}(aq) + 6 H_2O;
(d) Cu^{2+}(aq) + H_2S(aq) $\rightarrow$ CuS(s) + 2 H^+(aq);
(e) 2 $SbCl_4^-$(aq) + 3 H_2S(aq) $\rightarrow$ Sb_2S_3(s) + 6 H^+(aq) +
8 Cl^-(aq). **94.** (d) **95.** (a) **96.** (c) **97.** (b)
98. (c) **99.** (a) **100.** (d) **101.** (a) **102.** (b)
103. (c) **104.** (a) 0.030 M; (b) 3.5 M. **105.** increased
solubility, in acidic solutions: (b), (c), (f); in basic solutions:
(a); independent of pH: (d), (e). **106.** (a) CrO_4^{2-};
(b) 1.6×10^{-6} M; (c) Yes. $[CrO_4^{2-}]$ drops from 0.010 M to
1.8×10^{-7} M before SO_4^{2-} begins to precipitate as
$PbSO_4$(s).

Chapter 20

1. (a) increase; (b) decrease; (c) uncertain because the number of moles of gas is unchanged in the reaction; (d) decrease. **2.** (a) case 3; (b) case 2; (c) case 1; (d) case 4.
3. -0.284 kJ K^{-1}. **4.** (a) -242.1 kJ; (b) $+141.8$ kJ;
(c) $+102$ kJ; (d) -69.0 kJ. **5.** (a) 114 °C; (b) ΔG = 0.
Solid and liquid I_2 are in equilibrium at this temperature, the
normal melting point of I_2. **6.** (a) 85.9 J mol^{-1} K^{-1};
(b) 7.014 J mol^{-1} K^{-1}; (c) 1.1 J mol^{-1} K^{-1}.
7. $+126.6$ kJ. **8.** (a) $K = K_p = (P_{NO_2})^2/(P_{NO})^2(P_{O_2})$;
(b) $K = K_p = P_{SO_2}$; (c) $K = K_c = [H^+][C_2H_3O_2^-]/[HC_2H_3O_2]$;
(c) $K = K_p = (P_{H_2O})(P_{CO_2})$; (e) $K = [Mn^{2+}](P_{Cl_2})/[H^+]^4[Cl^-]^2$.
9. 8.4×10^{-13}. **10.** -42.1 J K^{-1}. **11.** (a) no; (b) to
some extent; (c) to some extent. **12.** (a) 3.4; (b) -10. kJ;
(c) net reaction to the right. **13.** (a) -70.54 kJ; (b) $2.3 \times$
10^{12}. **14.** (a) $+216$ J K^{-1}; (b) $+85$ kJ; (c) $+21$ kJ;
(d) 2.0×10^{-4}. **15.** (a) $\Delta H° = -165.0$ kJ, $\Delta S°$ =
-227.2 J K^{-1}, $\Delta G° = -97.3$ kJ; (b) low temperatures and
high pressures; (c) 0.45. **16.** 3×10^{16}. **19.** Relative
values of ΔS are (b) < (a) < (d) < (c). **20.** (a) positive;
(c) positive; (e) not possible to state.
23. (a) $\Delta H°_{vap}$ = 44.0 kJ/mol, $\Delta S°_{vap}$ = 118.8 J mol^{-1} K^{-1}.
25. (a) 30.91 kJ/mol; (b) $T_{b.p.}$ = 3.5×10^2 K.
26. -320. kJ/mol **27.** (a) low temperatures; (b) There is
insufficient information for a prediction; we cannot determine
the sign of ΔS. (c) high temperatures; (d) nonspontaneous at
all temperatures; (e) all temperatures. **29.** ΔH = 0; ΔS >
0; ΔG < 0. **32.** 399 K. **33.** -440 kJ.
34. (a) -15.7 J K^{-1}; (b) $+180$ kJ; (c) $+185$ kJ; (d) The reaction appears to be nonspontaneous at all temperatures.
35. Reaction (a): $\Delta G° = -1.6$ kJ; an equilibrium condition
with significant amounts of all reactants and products is
reached. Reaction (b): $\Delta G° = -474.3$ kJ; reaction goes to
completion. Reaction (c): $\Delta G° = -574.5$ kJ; reaction goes to
completion. **36.** 157.8 kJ/mol. **37.** -5.22×10^3 kJ.
45. -2.1 kJ. **46.** (b) 0.035 atm (27 mmHg).
47. (a) -23.4 kJ; (c) $+5.40$ kJ; (e) $+28.75$ kJ.
48. -206.9 kJ/mol. **50.** $+1.41$ kJ. **52.** 0.210 mol
Br_2; 0.210 mol Cl_2; 0.580 mol BrCl. **53.** (a) $\Delta H°$ =
-41.2 kJ, $\Delta S° = -42.1$ J K^{-1}, $\Delta G° = -28.6$ kJ; (b) 0.79.
55. 2.02×10^3 K. **56.** (a) no; (b) $\Delta H°$ and $\Delta S°$ for the
decomposition reaction; (c) 4.6×10^2 K; (d) 1.1×10^3 K.
57. (a) 0.014; (b) 329 K. **59.** (a) $\Delta H°$ is about $1.3 \times$
10^2 kJ; (b) 3.8×10^2 K. **62.** 49.1% efficient.
63. (a) 4.9×10^2 K. **88.** (d) **89.** (b) **90.** (a)
91. (c) **92.** (b) **93.** (d) **94.** (b) **95.** (d)
96. (a) Entropy change must be assessed for the system *and*
its surroundings ($\Delta S_{univ.}$), not just for the system alone.
(b) From $\Delta G°$ we can calculate an equilibrium constant
($\Delta G° = -RT \ln K$), and the equilibrium constant permits us

to do equilibrium calculations for many nonstandard conditions. **97. (a)** -124.0 kJ; **(b)** -186.1 kJ; **(c)** 4.1×10^{32}; **(d)** The reaction is spontaneous at all temperatures.

Chapter 21

1. (a) oxidation: $Fe \rightarrow Fe^{2+} + 2\,e^-$; reduction: $Cu^{2+} + 2\,e^- \rightarrow Cu(s)$; net: $Fe + Cu^{2+} \rightarrow Fe^{2+} + Cu$; **(b)** $2\,Br^- \rightarrow Br_2 + 2\,e^-$; $Cl_2 + 2\,e^- \rightarrow 2\,Cl^-$; $2\,Br^- + Cl_2 \rightarrow 2\,Cl^- + Br_2$; **(c)** $Al \rightarrow Al^{3+} + 3\,e^-$; $Fe^{3+} + e^- \rightarrow Fe^{2+}$; $Al + 3\,Fe^{3+} \rightarrow Al^{3+} + 3\,Fe^{2+}$; **(d)** $Cl^- + 3\,H_2O \rightarrow ClO_3^- + 6\,H^+ + 6\,e^-$; $MnO_4^- + 8\,H^+ + 5\,e^- \rightarrow Mn^{2+} + 4\,H_2O$; $5\,Cl^- + 6\,MnO_4^- + 18\,H^+ \rightarrow 5\,ClO_3^- + 6\,Mn^{2+} + 9\,H_2O$; **(e)** $S^{2-} + 8\,OH^- \rightarrow SO_4^{2-} + 4\,H_2O + 8\,e^-$; $O_2 + 2\,H_2O + 4\,e^- \rightarrow 4\,OH^-$; $S^{2-} + 2\,O_2 \rightarrow SO_4^{2-}$. **2. (a)** -0.910 V; **(b)** -0.56 V; **(c)** -0.403 V; **(d)** -0.136 V. **3. (a)** $Zn(s) + Sn^{2+}(aq) \rightarrow Zn^{2+}(aq) + Sn(s)$, $E°_{cell} = +0.627$ V; **(b)** $2\,Fe^{2+}(aq) + Sn^{4+}(aq) \rightarrow 2\,Fe^{3+}(aq) + Sn^{2+}(aq)$, $E°_{cell} = -0.617$ V; **(c)** $Cu(s) + Cl_2(g) \rightarrow Cu^{2+}(aq) + 2\,Cl^-(aq)$, $E°_{cell} = +1.023$ V. **4. (a)** $+0.095$ V; **(b)** -2.03 V; **(c)** $+0.153$ V. **5. (a)** no; **(b)** yes; **(c)** no; **(d)** no. **6. (a)** yes; **(b)** yes; **(c)** no; **(d)** no; **(e)** no. **7. (a)** -1.16×10^3 kJ; **(b)** -268 kJ; **(c)** -3.1×10^2 kJ. **8. (a)** $K = [Fe^{2+}][Ag^+]/[Fe^{3+}] = 0.32$; **(b)** $K = [Mn^{2+}](P_{Cl_2})/[H^+]^4[Cl^-]^2 = 2.5 \times 10^{-5}$; **(c)** $K = [Cl^-]^2(P_{O_2})/[OCl^-]^2 = 10^{33}$. **9. (a)** $Fe(s) + Cu^{2+}(aq) \rightarrow Fe^{2+}(aq) + Cu(s)$; electrons flow from B to A at 0.777 V; **(b)** $Sn^{2+}(aq) + 2\,Ag^+(aq) \rightarrow 2\,Ag(s) + Sn^{4+}(aq)$; A to B at 0.646 V; **(c)** $Zn(s) + Fe^{2+}(aq) \rightarrow Zn^{2+}(aq) + Fe(s)$; A to B at 0.264 V. **10.** 3.2×10^{-10} M. **11. (a)** 0.303 V; **(b)** 0.162 V; **(c)** 0.154 V. **12. (a)** anode: $Cl_2(g)$, cathode: $Cu(s)$; **(b)** $Cl_2(g)$, $H_2(g)$; **(c)** $O_2(g)$, $H_2(g)$; **(d)** $Cl_2(g)$, $Ba(l)$; **(e)** $I_2(s)$, $OH^-(aq)$ and $H_2(g)$; **(f)** $O_2(g)$, $OH^-(aq)$ and $H_2(g)$. **13. (a)** 4.28 g Zn; **(b)** 1.18 g Al; **(c)** 14.1 g Ag; **(d)** 3.84 g Ni. **14. (a)** 8.993; **(b)** 119.6; **(c)** 8.000; **(d)** 27.47. **15.** 47.63% Fe. **18. (a)** $0.337\ V < E° < 0.800\ V$; **(b)** $-0.440\ V < E° < 0.000\ V$. **20.** -0.255 V. **24. (a)** $3\,Ag(s) + 4\,H^+ + NO_3^- \rightarrow 3\,Ag^+(aq) + 2\,H_2O + NO(g)$; **(b)** $Zn(s) + 2\,H^+(aq) \rightarrow Zn^{2+}(aq) + H_2(g)$; **(c)** no reaction. **26. (a)** $Fe(s)|Fe^{2+}(aq)||Cl^-(aq)|Cl_2(g)$, Pt, $E°_{cell} = 1.800$ V; **(b)** $Al(s)|Al^{3+}(aq)||Pb^{2+}(aq)|Pb(s)$, $E°_{cell} = 1.53$ V; **(c)** $Pt|Cu^+(aq), Cu^{2+}(aq)||Cu^+(aq)|Cu(s)$, $E°_{cell} = -E° + 0.52$ V, where $E°$ is the standard potential for $Cu^{2+} + e^- \rightarrow Cu^+$. **27. (a)** -5.2×10^2 kJ ($n = 6$); **(b)** $+26.2$ kJ ($n = 2$). **28.** $K = 1.3 \times 10^{91}$; displacement reaction goes to completion. **29.** 1.591 V. **30.** $+127.3$ kJ/mol. **31.** 0.42 M. **32. (a)** 0.283 V; **(c)** 1.615 V. **33.** $2\,Mn^{2+}(aq) + O_2(g) + 2\,H_2O \rightarrow 2\,MnO_2(s) + 4\,H^+(aq)$; **(a)** no; **(d)** yes. **36. (a)** 6×10^{-38} M; **(b)** yes. **37.** 0.037 V. **38. (a)** 0.027 V; **(b)** E_{cell} decreases with time; **(c)** 0.022 V; **(d)** 0.218 M; **(e)** $[Sn^{2+}] = 0.48$ M, $[Pb^{2+}] = 0.22$ M. **39.** 6.9×10^{-4} M. **43.** $\epsilon = 0.830$; $E°_{cell} = 1.229$ V. **44. (a)** $O_2(g) + 4\,H^+ + 4\,e^- \rightarrow 2\,H_2O$; **(b)** $C_2H_5OH + 3\,H_2O \rightarrow 2\,CO_2 + 12\,H^+ + 12\,e^-$; **(c)** -1332.5 kJ; **(d)** 1.1509 V; **(e)** 0.078 V. **49. (a)** 1.229 V for electrolysis; **(c)** 0.383 V for electrolysis; **(e)** a spontaneous reaction with $E°_{cell} = +0.88$ V. **51. (a)** $H_2(g)$ at the cathode and $O_2(g)$ at the anode; **(b)** $2\,H_2O \rightarrow 2\,H_2(g) + O_2(g)$. **52. (b)** 475 s; **(c)** 0.207 M. **53. (a)** anode: $2\,H_2O \rightarrow 4\,H^+ + O_2(g) + 4\,e^-$, cathode: $Ag^+ + e^- \rightarrow Ag(s)$; **(b)** 0.0250 A; **(c)** 11.4 mL $O_2(g)$. **54. (a)** 0.535 L; **(c)** 0.553 L. **55. (a)** 821 C; **(b)** 1.04 A. **57. (a)** 5.593×10^{-3} M; **(b)** 3.356×10^{-2} N. **58.** 1.11×10^{-3} M. **59.** 32.9 mL. **84. (c)** **85. (b)** **86. (d)** **87. (b)** **88. (b)** **89. (c)** **90. (c)** **91. (a)** **92. (d)** **93. (a)** **94. (a)** The oxidation of silver requires an oxidiz-

ing agent (NO_3^-) that is stronger than H^+. **(b)** $3\,Ag + 4\,H^+ + NO_3^- \rightarrow 3\,Ag^+ + 2\,H_2O + NO(g)$. **95.** (anode) $Zn|Zn^{2+}||NO_3^-|NO(g)$, Pt (cathode); $E°_{cell} = 1.72$ V; $3\,Zn + 8\,H^+ + 2\,NO_3^- \rightarrow 3\,Zn^{2+} + 4\,H_2O + 2\,NO$. **96.** 2.82. **97. (a)** no; **(b)** to the left.

Chapter 22

1. (a) lead(IV) oxide (lead dioxide); **(b)** mercury(I) bromide; **(c)** calcium chloride hexahydrate; **(d)** $[Zn(OH)_4]^{2-}$; **(e)** sodium aluminate; **(f)** potassium stannate; **(g)** $Mg(HCO_3)_2$; **(h)** $NH_4Al(SO_4)_2 \cdot 12H_2O$. **2. (a)** $MgCO_3 \rightarrow MgO + CO_2$; **(b)** $CaO + 2\,H^+ \rightarrow Ca^{2+} + H_2O$; **(c)** $2\,Al + 2\,OH^- + 6\,H_2O \rightarrow 2\,[Al(OH)_4]^- + 3\,H_2$; **(d)** $3\,Pb + 8\,H^+ + 2\,NO_3^- \rightarrow 3\,Pb^{2+} + 4\,H_2O + 2\,NO$; **(e)** no reaction; **(f)** $ZnO + CO \rightarrow Zn + CO_2$. **3. (a)** reaction with HCl(aq), followed by recrystallization of $MgCl_2 \cdot 6H_2O$; **(b)** reaction of Na with H_2O to form NaOH, followed by the reaction of Al with NaOH(aq); **(c)** reaction with HCl(aq), followed by treatment with H_2S after adjustment of pH; **(d)** reaction with HNO_3 (concd aq), followed by treatment with H_2S. **4. (a)** $Ba^{2+} + CO_3^{2-} \rightarrow BaCO_3$; **(b)** $Mg(HCO_3)_2 \rightarrow MgCO_3(s) + H_2O + CO_2$; **(c)** $SnO + C \rightarrow Sn + CO$; **(d)** $Sn + 4\,HNO_3 \rightarrow SnO_2 + 2\,H_2O + 4\,NO_2$; **(e)** $CdO + 2\,H^+ \rightarrow Cd^{2+} + H_2O$; **(f)** $PbO_2(s) + 4\,H^+ + 4\,I^- \rightarrow PbI_2(s) + I_2 + 2\,H_2O$. **5. (a)** $Na_2CO_3 + SiO_2 \rightarrow Na_2SiO_3 + CO_2$; **(b)** $2\,NaCl + H_2SO_4 \rightarrow Na_2SO_4 + 2\,HCl(g)$; **(c)** $2\,KOH(aq) + SO_2 \rightarrow K_2SO_3(aq) + H_2O$; **(d)** $KOH(aq) + H_3PO_4(aq) \rightarrow K_2HPO_4(aq) + 2\,H_2O$. **6.** 9.23×10^{-3} M. **7.** acidic: (a), (b), (d), (e); basic: (c). **8. (a)** $CdCO_3 \rightarrow CdO + CO_2$; **(b)** no reaction; **(c)** $SnO_2 + 2\,CO \rightarrow Sn + 2\,CO_2$; **(d)** $2\,Cd^{2+} + 2\,H_2O \rightarrow 2\,Cd(s) + 4\,H^+ + O_2(g)$; **(e)** $2\,HgO \rightarrow 2\,Hg + O_2(g)$; **(f)** no reaction. **9. (a)** $Li_2CO_3 + 2\,HBr \rightarrow 2\,LiBr + H_2O + CO_2$; **(b)** $Li_2O + (NH_4)_2CO_3 \rightarrow Li_2CO_3 + 2\,NH_3(g) + H_2O$; **(c)** $Mg(OH)_2 + H_2SO_3 \rightarrow MgSO_3 + 2\,H_2O$; **(d)** $Hg + HgCl_2 \rightarrow Hg_2Cl_2$; **(e)** $PbO + OCl^- \rightarrow PbO_2 + Cl^-$. **10. (a)** yes; **(b)** no; **(c)** yes. **13. (a)** 1×10^9 s (3×10^1 y); **(b)** 0.1 g. **14. (a)** 71.5%. **15.** 11.54. **16. (a)** $Ca(OH)_2(s) + SO_4^{2-}(aq) \rightleftharpoons CaSO_4(s) + 2\,OH^-(aq)$; **(b)** The reaction does not go to completion ($K = 0.60$). **(c)** $[SO_4^{2-}] = 0.68$ M, $[OH^-] = 0.64$ M. **18. (a)** Ca(l) at cathode, $Cl_2(g)$ at anode; **(b)** $H_2(g)$ and OH^- at cathode, $O_2(g)$ and H^+ at anode; **(c)** $H_2(g)$ at cathode, $Cl_2(g)$ at anode; **(d)** $H_2(g)$ and OH^- at cathode, I_2 at anode. **19.** 6×10^6 ton. **20.** $Mg(OH)_2(s)$ will precipitate. **21. (a)** decreasing solubilities: $1.00\ M\ NH_4Cl > 0.100\ M\ NH_3–1.00\ M\ NH_4Cl > 1.00\ M\ NH_3–1.00\ M\ NH_4Cl$; **(b)** For the order found in (a): $[Mg^{2+}] = 7.1 \times 10^{-3}$ M, 1.9×10^{-3} M, and 6.0×10^{-4} M, respectively. **23.** 209 kg CaO. **24. (a)** 746 kg. **25. (a)** $2\,Na^+ + CO_3^{2-} + Ca^{2+} + SO_4^{2-} \rightarrow CaCO_3(s) + 2\,Na^+ + SO_4^{2-}$; **(b)** 20.9 g Na_2CO_3. **27.** $[Al(H_2O)_6]^{3+} + 3\,HCO_3^- \rightarrow Al(OH)_3(s) + 6\,H_2O + 3\,CO_2(g)$. **30.** $[Al(OH)_4]^- + H_2O + CO_2 \rightarrow Al(OH)_3(s) + HCO_3^- + H_2O$. **33.** Minimum voltage required is 1.60 V. If the cell reaction were $Al_2O_3(l) \rightarrow 2\,Al(l) + \frac{3}{2}O_2(g)$, the required voltage would be 2.63 V. **38. (b)** $PbO(s) + Na^+ + OH^- + H_2O \rightarrow Na^+ + [Pb(OH)_3]^-$; **(d)** $2\,SnO \rightarrow Sn + SnO_2$. **40. (b)** $PbO_2(s) + 4\,H^+ + 2\,Cl^- \rightarrow Pb^{2+} + 2\,H_2O + Cl_2(g)$; **(d)** $3\,PbO + H_2O + 2\,CO_2 \rightarrow 2PbCO_3 \cdot Pb(OH)_2$. **42.** 19 g $Pb(NO_3)_2$/100. g H_2O. **47.** $+0.95$ V. **48. (a)** $+60.9$ kJ; **(b)** 0.14; **(c)** -16.44 kJ. **71. (d)** **72. (b)** **73. (b)** **74. (c)** **75. (a)** **76. (c)** **77. (a)** **78. (a)** **79. (b)** **80. (d)** **81. (a)** $Sn + 4\,H^+ + 4\,NO_3^- \rightarrow SnO_2(s) + 2\,H_2O + 4\,NO_2$; **(b)** $MgCO_3(s) + H_2O + CO_2 \rightarrow Mg^{2+} + 2\,HCO_3^-$; **(c)** Zn +

$2 H_2O + 2 OH^- \rightarrow [Zn(OH)_4]^{2-} + H_2(g)$. **82. (a)** Mg^{2+} is smaller than Ba^{2+}, leading to stronger interionic forces of attraction in MgO than in BaO. Because of these stronger interionic forces, crystals of MgO must be heated to a higher temperature than must BaO in order to melt. **(b)** The lattice energy of MgF_2 is greater than that of $MgCl_2$ (because of the smaller size of F^- compared to Cl^-). The MgF_2 crystal is more difficult to break up and the solubility of MgF_2 is less than that of $MgCl_2$. **83.** The simplest method is to dissolve the alloy in concentrated $HNO_3(aq)$. The Pb–Sn alloy will yield a white precipitate of SnO_2, whereas the Pb–Cd alloy should dissolve completely. **84.** 118 ppm $CaSO_4$.

Chapter 23

1. (a) potassium bromate; **(b)** chlorine trifluoride; **(c)** NaOI; **(d)** Si_3H_8; **(e)** potassium cyanate; **(f)** $Na_2S_2O_6$; **(g)** AgN_3; **(h)** sodium dihydrogen phosphate. **2. (a)** $CaCl_2(s) + H_2SO_4(\text{concd aq}) \rightarrow CaSO_4(s) + 2 HCl(g)$; **(b)** no reaction; **(c)** no reaction; **(d)** $NH_3(aq) + HClO_4(aq) \rightarrow NH_4^+(aq) + ClO_4^-(aq)$; **(e)** $2 NO(g) + O_2(g) \rightarrow 2 NO_2(g)$; **(f)** $2 Cu(NO_3)_2(s) \rightarrow 2 CuO(s) + 4 NO_2(g) + O_2(g)$. **3. (a)** Heat $KClO_3$ (reaction 23.15), electrolyze H_2O (reaction 14.5). **(b)** Heat NaCl with concentrated H_2SO_4 (reaction 14.25) and dissolve the HCl(g) in water. **(c)** Decompose NH_4NO_3 by heating (Table 23-12). **(d)** Generate $SO_2(g)$ by adding an acid to Na_2SO_3; pass the $SO_2(g)$ into a solution of $Ba^{2+}(aq)$ to precipitate $BaSO_3$. **4. (a)** $Cl_2(g) + 2 OH^-(aq) \rightarrow OCl^-(aq) + Cl^-(aq) + H_2O$; **(b)** $2 NaI(s) + 2 H_2SO_4 \rightarrow Na_2SO_4 + SO_2(g) + I_2(g) + 2 H_2O$; **(c)** $Cl_2(g) + 2 Br^-(aq) \rightarrow 2 Cl^-(aq) + Br_2(aq)$; **(d)** $3 CdS(s) + 8 H^+ + 2 NO_3^- \rightarrow 3 Cd^{2+} + 4 H_2O + 2 NO(g) + 3 S(s)$; **(e)** $2 Fe(s) + 3 N_2O(g) \rightarrow Fe_2O_3(s) + 3 N_2(g)$; **(f)** $3 Pb(s) + 8 H^+ + 2 NO_3^- \rightarrow 3 Pb^{2+} + 4 H_2O + 2 NO(g)$. **5.** acidic solutions: $H_5IO_6 + H^+ + 2 e^- \rightarrow IO_3^- + 3 H_2O$; $4 H_3PO_2 + 4 H^+ + 4 e^- \rightarrow P_4 + 8 H_2O$; $Sb_2O_5 + 6 H^+ + 4 e^- \rightarrow 2 SbO^+ + 3 H_2O$; basic solutions: $OCl^- + 2 e^- + H_2O \rightarrow Cl^- + 2 OH^-$; $B_2H_6 + 2 H_2O + 4 e^- \rightarrow 2 BH_4^- + 2 OH^-$; $HXeO_4^- + 3 H_2O + 6 e^- \rightarrow Xe + 7 OH^-$. **6. (a)** yes; **(b)** no; **(c)** yes; **(d)** no. **7. (a)** +0.94 V; **(b)** −1.15 V. **8.** $I^-(aq) + 3 Cl_2(g) + 3 H_2O \rightarrow IO_3^-(aq) + 6 Cl^-(aq) + 6 H^+(aq)$; $2 Br^-(aq) + Cl_2(g) \rightarrow 2 Cl^-(aq) + Br_2(aq)$. **9. (a)** NaCl, $Pb(NO_3)_2$; **(b)** AgCl, $BaSO_4$; **(c)** $MgCO_3$; the solid is $CaSO_3$. **10.** $E°$ for the reduction half-reaction must exceed 0.682 V. This is possible with (a), (c), and (d). **11. (a)** I_2. The oxidation of I^- to I_2 has $-E° = -0.535$ V compared to −1.229 V for the oxidation of H_2O to O_2. **(b)** Reduced to H_2O. NH_4^+ cannot be an oxidizing agent. NH_4^+ must itself be oxidized (for example, to N_2); H_2O_2 will act as an oxidizing agent and be reduced to H_2O. **(c)** NO. Only NO_3^- in an acidic solution ($E° = +0.96$ V) is a good enough oxidizing agent to oxidize Ag to Ag^+ ($-E° = -0.800$ V). **12. (a)** −50 kJ/mol; **(b)** −17 kJ/mol; **(c)** +83 kJ/mol; **(d)** −128 kJ/mol. **13. (a)** $Ca_3(PO_4)_2$; **(b)** $K_4P_2O_7$; **(c)** sodium metaantimonite; **(d)** sodium metabismuthate; **(e)** Na_3BiO_4. **14. (a)** trigonal pyramidal; **(b)** tetrahedral; **(c)** square-based pyramid. **16. (a)** yes; **(b)** F_2 would oxidize H_2O to $O_2(g)$; **(c)** No. There is no oxidizing agent as strong as F_2, but even if there were, it would oxidize H_2O to O_2 more readily than it would $F^-(aq)$. **18. (a)** +1.49 V; **(b)** No; the reaction will not occur spontaneously. **20.** The approximate values are **(a)** 150 pm; **(b)** 240 pm; **(c)** 900 kJ/mol; **(d)** −270 kJ/mol; **(e)** 2.5; **(f)** 0.3 V. **21. (a)** 2.0×10^{-17}; **(b)** 4×10^{-15}%. **23. (a)** T-shaped; **(b)** square-based pyramidal; **(c)** linear; **(d)** square planar. **24.** IO_3^-/I^-, $E° = 1.09$ V; HIO/I_2,

$E° = 1.45$ V; IO_3^-/HIO, $E° = 1.13$ V. **26.** The reactions that go to completion are (a) and (c). **27.** 5.87. **28. (a)** 303 kJ/mol. **29. (a)** increasing bond distance: $O_2^+ < O_2 < O_2^- < O_2^{2-}$; increasing bond energy: $O_2^{2-} < O_2^- < O_2 < O_2^+$. **31. (c)** $SO_2(aq) + MnO_2(s) \rightarrow Mn^{2+}(aq) + SO_4^{2-}(aq)$; **(d)** $2 S^{2-}(aq) + O_2(g) + 2 H_2O \rightarrow 2 S(s) + 4 OH^-(aq)$. **34.** 1.17. **36.** 0.426 g. **37.** 7.002% Cu. **40. (b)** $2 NaNO_3 \rightarrow 2 NaNO_2 + O_2(g)$; **(d)** $3 Ag + 4 H^+ + NO_3^- \rightarrow 3 Ag^+ + 2 H_2O + NO(g)$; **(e)** $(CH_3)_2NNH_2 + 4 O_2 \rightarrow 2 CO_2 + 4 H_2O + N_2$. **41. (a)** 9.23; **(b)** 3.85. **43. (b)** −0.05 V; **(c)** +0.82 V. **44. (b)** +0.42 V; **(c)** −0.56 V. **45.** +1.616 V. **47. (a)** (mono)hydrogen phosphate ion; **(b)** calcium pyrophosphate; **(c)** tetrapolyphosphoric acid; **(d)** sodium tetrametaphosphate. **50.** Use reaction (14.64) to prepare P_4, followed by $P_4 + 5 O_2 \rightarrow P_4O_{10}$ and $P_4O_{10} + 6 H_2O \rightarrow 4 H_3PO_4$. Then use the H_3PO_4 in reaction (23.51). **51. (b)** $Al_4C_3 + 12 H_2O \rightarrow 4 Al(OH)_3 + 3 CH_4(g)$. **55. (a)** $2 BBr_3 + 3 H_2 \rightarrow 2 B + 6 HBr$; **(b)** $B_2O_3 + 3 CaF_2 + 3 H_2SO_4 \rightarrow 2 BF_3 + 3 CaSO_4 + 3 H_2O$; **(c)** $2 B + 3 N_2O \rightarrow B_2O_3 + 3 N_2$. **56. (a)** irregular tetrahedral; **(c)** octahedral. **82. (b)** **83. (d)** **84. (a)** **85. (d)** **86. (c)** **87. (a)** **88. (b)** **89. (d)** **90. (b)** **91. (a)** **92. (a)** $2 Pb(NO_3)_2 \rightarrow 2 PbO + 4 NO_2(g) + O_2(g)$; **(b)** $Cl_2(g) + 2 OH^- \rightarrow ClO^- + Cl^- + H_2O$; **(c)** $H_3PO_4(aq) + 2 KOH(aq) \rightarrow K_2HPO_4(aq) + 2 H_2O$; **(d)** $2 KBr + 2 H_2SO_4 \rightarrow K_2SO_4 + 2 H_2O(g) + SO_2(g) + Br_2(g)$; **(e)** $2 H_3PO_4 \rightarrow H_4P_2O_7 + H_2O$. **93. (a)** Ag will dissolve only in the presence of a strong oxidizing agent, and NO_3^- in acidic solution is a much stronger oxidizing agent than is H^+. **(b)** I_2 and I^- combine to form the I_3^- ion in a solution having a high $[I^-]$, such as KI(aq); $KI_3(aq)$ is a soluble compound. **(c)** The gaseous state is expected for a substance of low molecular weight, such as H_2S (M.W. = 34). H_2O is unusual in being a liquid at room temperature despite its low molecular weight (18). This is because of the strong hydrogen bonding that occurs between H_2O molecules; hydrogen bonding is insignificant in H_2S. **94. (a)** 1×10^6 kg F_2; **(b)** No. F_2 cannot be displaced from $F^-(aq)$ by an oxidizing agent because F_2 is itself the best of all oxidizing agents. If a better oxidizing agent did exist, it could not be used in water because the H_2O would be oxidized by the liberated fluorine. We should expect the most feasible method of producing F_2 to be the electrolysis of one of its molten salts. **95.** +0.38 V.

Chapter 24

1. (a) $BaCr_2O_7$; **(b)** scandium hydroxide; **(c)** CrO_3; **(d)** manganese(II) oxide; **(e)** $FeSiO_3$; **(f)** iron (penta)carbonyl. **2. (a)** Fe + Mn; **(b)** Fe with more than about 1.5% C; **(c)** $Fe(CrO_2)_2$; **(d)** $KCr(SO_4)_2 \cdot 12H_2O$; **(e)** 1 part HNO_3 + 3 parts HCl; **(f)** $Fe_4[Fe(CN)_6]_3$. **3. (a)** $TiCl_4(g) + 4 Na(l) \rightarrow 4 NaCl(l) + Ti(s)$; **(b)** $Cr_2O_3(s) + 2 Al(s) \rightarrow Al_2O_3(s) + 2 Cr(l)$; **(c)** no reaction; **(d)** $K_2Cr_2O_7(aq) + 2 KOH(aq) \rightarrow 2 K_2CrO_4(aq) + H_2O$; **(e)** $MnO_2(s) + 2 C(s) \rightarrow Mn + 2 CO(g)$; **(f)** no reaction. **4. (a)** $2 Fe_2S_3(s) + 6 H_2O + 3 O_2(g) \rightarrow 4 Fe(OH)_3(s) + 6 S(s)$; **(b)** $2 Mn^{2+} + 5 S_2O_8^{2-} + 8 H_2O \rightarrow 2 MnO_4^- + 10 SO_4^{2-} + 16 H^+$; **(c)** $4 Ag(s) + 8 CN^- + O_2(g) + 2 H_2O \rightarrow 4 [Ag(CN)_2]^- + 4 OH^-$. **5. (a)** Acidify $Na_2CrO_4(aq)$ (reaction 24.17). **(b)** Reduce $Na_2Cr_2O_7$ to $Cr_2O_3(s)$ with carbon as in reaction (24.11); separate $Cr_2O_3(s)$ and $Na_2CO_3(s)$ with water. **(c)** Obtain $Cr_2O_3(s)$ as in (b); dissolve it in HCl(aq); crystallize $CrCl_3(s)$ from solution. **(d)** Dissolve $Cr_2O_3(s)$ in NaOH(aq). **6. (a)** $3 Cu + 8 H^+ + 2 NO_3^- \longrightarrow 3 Cu^{2+} + 2 NO(g) + 4 H_2O$; other metals can be substituted for Cu, and, in some cases, H_2SO_4 for HNO_3. **(b)** $Cr_2O_3(s) + 2 OH^- + 3 H_2O \rightarrow$ $2 [Cr(OH)_4]^-(aq)$; the oxide must be amphoteric. **(c)** $FeS(s) + 2 H^+ \rightarrow Fe^{2+}(aq) + H_2S(g)$; the metal sulfide must be in cation group 3 (ammonium sulfide group). **(d)** $Ni(OH)_2(s) + 6 NH_3(aq) \longrightarrow [Ni(NH_3)_6]^{2+}(aq) +$ $2 OH^-(aq)$; will not work with $Fe(OH)_2$ or $Fe(OH)_3$. **(e)** $2 Ln(s) + 6 H^+ \rightarrow 2 Ln^{3+} + 3 H_2(g)$; any lanthanide element (Ln) should work. **7.** The orbital diagrams should correspond to these *spdf* notations: **(a)** $[Ar]3d^24s^2$; **(b)** $[Ar]3d^2$; **(c)** $[Ar]3d^4$; **(d)** $[Ar]3d^5$; **(e)** $[Ar]3d^5$. **8.** (e) > (a) > (d) > (c) > (f) > (b). **9.** $2 Mn^{2+} +$ $5 BiO_3^- + 14 H^+ \rightarrow 2 MnO_4^- + 5 Bi^{3+} + 7 H_2O$. **10. (a)** $[Cr_2O_7^{2-}] = 3.2 \times 10^4 \times [CrO_4^{2-}]^2$; **(b)** $[Cr_2O_7^{2-}] =$ $1.8 \times 10^{-4} \times [CrO_4^{2-}]^2$. **11. (a)** 5.6×10^{-7} M; **(b)** 7.2% Cu^+. **12. (a)** H_2O; **(b)** $NH_3(aq)$; **(c)** $NaOH(aq)$; **(d)** dilute $HCl(aq)$. **17. (a)** $VO^{2+} + 2 H^+ + e^- \rightarrow V^{3+} +$ H_2O; **(c)** $Fe(OH)_3(s) + 5 OH^- \rightarrow FeO_4^{2-} + 4 H_2O + 3 e^-$. **18. (a)** no; **(b)** yes; **(c)** yes. **21. (a)** all three; **(b)** only Zn; **(c)** Zn and Sn^{2+}, but not I^-. **22. (b)** 29.40 mL. **23.** +6. **24.** $Cr_2O_7^{2-}/Cr^{3+}$, $E° = 1.33$ V; Cr^{3+}/Cr^{2+}, $E° = -0.41$ V; Cr^{2+}/Cr, $E° = -0.91$ V; $Cr_2O_7^{2-}/Cr^{2+}$, $E° =$ 0.90 V; Cr^{3+}/Cr, $E° = -0.74$ V. **26.** CO_2 makes the solution acidic: $2 CrO_4^{2-} + H_2O + 2 CO_2 \rightarrow Cr_2O_7^{2-} +$ $2 HCO_3^-$. **27.** $[CrO_4^{2-}] = 0.068$ M; $[Cr_2O_7^{2-}] =$ 0.016 M. **28.** 43 s. **30.** $K = 33$. **34. (a)** $Mo(CO)_6$, octahedral; **(b)** $Os(CO)_5$, trigonal bipyramid; **(c)** $Re_2(CO)_{10}$, a binuclear carbonyl. **38.** 0.152 V; **40. (a)** Dissolution: $AgO + 2 H^+ \rightarrow Ag^{2+} + H_2O$; redox: $4 Ag^{2+} + 2 H_2O \rightarrow 4 Ag^+ + 4 H^+ + O_2(g)$; **(b)** E_{cell} for the redox reaction is +0.75 V. **41.** (24.45), $K = 1 \times 10^{-20}$; (24.46), $K = 1 \times 10^{13}$. **44. (b)** $CoS(s) + 4 H^+ +$ $2 NO_3^- + 4 Cl^- \rightarrow [CoCl_4]^{2-} + S(s) + 2 H_2O + 2 NO_2(g)$; **(c)** $2 Cr^{3+} + 3 H_2O_2 + 10 OH^- \rightarrow 2 CrO_4^{2-} + 8 H_2O$; **(f)** $MnO_2(s) + 4 H^+ + 2 Cl^- \rightarrow Mn^{2+} + 2 H_2O + Cl_2(g)$. **47.** probably present: Fe^{3+}; probably absent: Cr^{3+}; uncertain: Ni^{2+}, Zn^{2+}, Mn^{2+}. **50. (a)** $NiC_8H_{14}N_4O_4$; **(b)** 3.50%. **51.** 82.2%. **69.** (c) **70.** (d) **71.** (a) **72.** (b) **73.** (d) **74.** (c) **75.** (a) **76.** (b) **77.** (a) **78.** (c) **79.** Fe^{3+} has the electron configuration $[Ar]3d^5$, an especially stable configuration with a half-filled 3d subshell. The configurations of Co^{2+} and Ni^{2+} are $[Ar]3d^7$ and $[Ar]3d^8$, respectively. Neither could acquire a half-filled 3d subshell by losing just one more electron. **80.** Limited $NaOH(aq)$: $Mg(OH)_2(s)$ and $Cr(OH)_3(s)$; excess $NaOH(aq)$: $Mg(OH)_2(s)$ and $[Cr(OH)_4]^-(aq)$. $Cr(OH)_3(s)$ is amphoteric; $Mg(OH)_2(s)$ is basic. **81. (a)** H_2O or HCl; **(b)** HCl; **(c)** H_2O; **(d)** HCl; **(e)** NaOH or HCl.

Chapter 25

1. (a) $[Co(NH_3)_2Cl_4]^{2-}$; **(b)** $[Mn(CN)_6]^{3-}$; **(c)** $[Cr(en)_3][Ni(CN)_5]$. **2. (a)** C.N. = 6, O.S. = +2; **(b)** 6, +3; **(c)** 4, +2; **(d)** 6, +3; **(e)** 6, +3; **(f)** 2, +1. **3. (a)** diamminesilver(I) ion; **(b)** pentaaquahydroxoiron(III) ion; **(c)** tetrachlorozincate(II) ion; **(d)** bis(ethylenediamine) platinum(II) ion; **(e)** tetraamminechloronitrocobalt(III) ion. **4. (a)** pentamminebromocobalt(III) sulfate; **(b)** pentamminesulfatocobalt(III) bromide; **(c)** hexaamminechromium(III) hexacyanocobaltate(III); **(d)** sodium hexanitrocobaltate(III); **(e)** tris(ethylenediamine)cobalt(III) chloride. **5. (a)** $[Ag(CN)_2]^-$; **(b)** $[Ni(NH_3)_2Cl_4]^{2-}$; **(c)** $[Cu(en)_3]SO_4$; **(d)** $Na[Al(H_2O)_2(OH)_4]$. **6. (a)** H—Ö—H

(b) $^-[:\ddot{O}:—H]$ **(c)** $[:\ddot{O}=\ddot{N}—\ddot{O}:]^-$ **(d)** $[:\ddot{S}=C=\ddot{N}:]^-$. **7. (a)** $FeCl_3 \cdot 6H_2O$; **(b)** $Co[PtCl_6] \cdot 6H_2O$. **8.** Refer to structure (1) of Example 25-3: Replace ox by two NH_3 molecules; replace NH_3 at the top of the structure by OH^-. **9. (a)** Refer to the first structure shown in Figure 25-4: Replace the three NH_3 molecules with Cl^- ions. **(b)** Refer to structure (1) in Example 25-3: Replace ox by en, and replace three NH_3 molecules by Cl^- ions. **(c)** Refer to structure (1) in Example 25-3: Replace NH_3 at top of structure and one NH_3 in the central plane with en; replace the remaining NH_3 in the central plane with Cl^-. **10. (a)** one; **(b)** two; **(c)** two; **(d)** two. **11. (a)** coordination isomerism; **(b)** linkage isomerism; **(c)** none; **(d)** none; **(e)** cis-trans (or fac-mer).

5d				dsp^2			
↑↓	↑↓	↑↓	↑↓	··	··	··	··

12. (a) $[Xe] 4f^{14}$

Br^- Br^- Br^- Br^-

(b) diamagnetic. **13.** $[Co(en)_3]^{3+}$ is yellow and $[Co(H_2O)_6]^{3+}$ is blue, since en produces a greater d-level splitting than does H_2O (expression 25.13). **17. (a)** tetraammineaquahydroxocobalt(III) ion; **(c)** tetraammineplatinum(II) hexachloroplatinate(IV); **(e)** pentacyanopyridineferrate(II) ion. **18. (a)** $K_4[Fe(CN)_6]$; **(c)** $[Al(H_2O)_4(OH)_2]Cl$; **(e)** $[Fe(en)_3]_3[Fe(CN)_6]_2$. **19. (a)** tetraamminecopper(II) ion; **(b)** tetraamminedichlorocobalt(III) chloride; **(c)** hexachloroplatinate(IV) ion; **(d)** sodium tetrachlorocuprate(II). **23. (a)** Refer to Figure 25-2: Replace NH_3 molecules by Cl^- ions. **(b)** Refer to structure (1) of Example 25-3: Replace ox and Cl^- by three OH^- ions. **(c)** Refer to Figure 25-5: Leave one Cl^- and replace all other ligands with H_2O. **24. (a)** Refer to the cis isomer shown in Figure 25-4(b): Replace Cl^- ions by ox and replace NH_3 molecules by ox. **(b)** Refer to the structure shown in the upper left of Figure 25-7: Replace Co^{3+} by Cr^{3+} as the central ion; replace each en by ox. **(c)** Refer to Figure 25-20: Replace Pb^{2+} with Fe^{2+}. **28. (a)** no; **(b)** yes; **(c)** yes; **(d)** no; **(e)** no (optical isomers, but not geometric ones). **30. (a)** three; **(b)** yes. **31.** (a) and (b) are identical; (a) and (d) are geometric isomers; (c) is totally different from the other structures. **33. (a)** paramagnetic; **(b)** diamagnetic. **37. (a)** three; **(b)** paramagnetic; **(c)** $[FeCl_4]^-$. **40. (b)** $Cu^{2+} + 2 OH^- \rightarrow Cu(OH)_2(s)$; $Cu(OH)_2(s) + 4 NH_3 \rightarrow [Cu(NH_3)_4]^{2+} + 2 OH^-$; $[Cu(NH_3)_4]^{2+} + 4 H_3O^+ \rightarrow [Cu(H_2O)_4]^{2+} + 4 NH_4^+$. **41.** $\beta_3 = K_f = 5.0 \times 10^9$. **44.** no. **46. (a)** 2.0; **(b)** 9×10^{-4} M; **(c)** no. **50. (a)** 0.876 g; **(b)** 1.75 g. **68.** (c) **69.** (d) **70.** (b) **71.** (a) **72.** (a) **73.** (c) **74.** (d) **75.** (a) **76. (a)** cis- or trans-tetraamminedihydroxochromium(III) bromide; **(b)** potassium hexanitrocobaltate(III); **(c)** cis- or trans-diaquabis(ethylenediamine)iron(II) ion; **(d)** cis- or trans-dichlorobis(ethylenediamine)platinum(IV) sulfate. **77. (a)** Refer to Figure 25-5: Replace one of the Cl^- ions with NH_3. **(b)** Refer to the structure in the upper left portion of Figure 25-7: Replace Co^{3+} by Cr^{3+} and one en by ox. **(c)** Refer to the cis isomer in Figure 25-4(b): Replace the Cl^- ions by NO_2^- ions. **78. (a)** Al^{3+} forms a hydroxo complex, such as $[Al(H_2O)_2(OH)_4]^-$, but not an ammine complex. **(b)** K_{sp} for ZnS is much smaller than for $ZnCO_3$, and a saturated solution of ZnS does not produce a large enough $[Zn^{2+}]$ for the complex ion to form. **(c)** $[Co^{3+}]$ is kept sufficiently low in a solution of the stable complex ion, $[Co(NH_3)_6]^{3+}$, that it is unable to oxidize water.

Chapter 26

1. (a) γ; (b) α; (c) β. **2.** (a) 32; (b) $_{+1}^{0}e$; (c) $_{92}^{235}U$, $_{90}^{231}Th$, $_{2}^{4}He$; (d) $_{84}^{214}Bi$, $_{84}^{214}Po$. **3.** (a) $_{11}^{24}Na$; (b) $_{0}^{1}n$; (c) $_{94}^{240}Pu$, (d) $5\ _{0}^{1}n$; (e) $_{99}^{246}Es$, 6. **4.** (a) $_{1}^{2}H + _{1}^{2}H \rightarrow _{2}^{3}He + _{0}^{1}n$; (b) $_{95}^{241}Am + _{2}^{4}He \rightarrow _{97}^{243}Bk + 2\ _{0}^{1}n$; (c) $_{51}^{121}Sb + _{2}^{4}He \rightarrow _{53}^{124}I + _{0}^{1}n$. **5.** (a) $_{84}^{214}Po$; (b) $_{12}^{28}Mg$; (c) $_{8}^{13}O$, $_{12}^{28}Mg$, $_{35}^{80}Br$, $_{84}^{214}Po$, $_{86}^{222}Rn$. **6.** 15 h. **7.** (a) 2.74×10^2 d; (b) 3.11×10^2 d; (c) 4.00×10^2 d. **8.** no. **9.** (a) 9.45×10^{-10} J; (b) 3.727×10^3 MeV; (c) 1.60×10^3 neutrons. **10.** 8.10 MeV. **11.** (c) and (d). **12.** 9.965×10^{-27} g. **16.** (c) $_{0}^{1}n$; (e) positron or β^+; (f) $_{1}^{3}H$. **17.** (b) $_{98}^{248}Cf$; (d) $_{84}^{214}Po$; (f) $_{32}^{69}Ge$. **21.** (a) $_{3}^{7}Li + _{1}^{1}H \rightarrow _{4}^{8}Be + \gamma$; (d) $_{92}^{238}U + _{2}^{4}He \rightarrow _{94}^{239}Pu + 3\ _{0}^{1}n$. **23.** 7.4×10^8. **25.** 142 d. **26.** 87.3 h. **27.** 7.3×10^{-3} mg. **28.** 0.22 g $^{208}Pb/1.00$ g ^{232}Th. **30.** 5.2×10^4 y. **31.** 4.06 MeV. **32.** 4.81 MeV. **33.** 0.02 nm. **40.** 1.37 mg. **41.** 2.9×10^3 m ton. **67.** (c) **68.** (b) **69.** (d) **70.** (c) **71.** (a) **72.** (c) **73.** (d) **74.** (d) **75.** (a) **76.** (b) **77.** 76 d.

78. (a) Radioactive nuclides with very short half-lives decay quickly; those with long half-lives persist for a long time, but have a very slow decay rate. (b) Gamma emitters are hazardous from a distance because of their great penetrating power. Alpha particles are hazardous when taken internally because of their high ionizing power. (c) Ar-40 is produced by the radioactive decay of K-40. (d) Fr, an alkali metal, is only produced in a radioactive decay series. It is found in trace quantities with radioactive materials, not in deposits of alkali metal compounds. (e) Positively charged atomic nuclei will collide and fuse only if they have extremely high kinetic energies, and this means extraordinarily high temperatures.

Chapter 27

1. (a) (b) (c)

2. C-atom skeletons only are shown.

(a) (1) C—C—C—C—C (2) (3)

(b) (1) C—C—C—C—C—C—C

3. (a) no (different formulas); (b) yes (skeletal isomers); (c) no (identical structures); (d) no (identical structures); (e) no (identical structures); (f) yes (ortho-para isomerism).

4. Only C-atom skeletons and Cl atoms are shown.

(b) carboxylic acid; (c) aldehyde; (d) ether; (e) ketone; (f) amine; (g) phenyl; (h) ester.
6. (a) 3,3-dimethyloctane; (b) 2,2-dimethylpropane; (c) 2,3-dichloro-5-ethylheptane; (d) 3-ethyl-2,2-dimethylpentane; (e) 5-chloro-3-heptene.
7. Only the C-atom chain and the principal substituent(s) are shown.

8. (a) $CH_3CH(OH)CH_3$; (b) $C(CH_3)_3OCH_3$; (c) $ClCF_2CH_3$; (d) $CH_2{=}C(CH_3)CH{=}CH_2$; (e) $CH_3CH{=}CHCHO$.
9. (a) p-dibromobenzene (1,4-dibromobenzene); (b) o-methylaniline (o-aminotoluene);

(c) (d)

(e)

10. (a) CH_3CH_2Cl (plus polysubstituted chlorides); (b) $CH_3CH_2CH_2CH_3$; (c) $CH_3CH_2COCH_2CH_3$; (d) $CH_3CH_2CH(OH)CH_3$; (e) $CH_3CH_2OCOCH_3$;

(f)

11. $3\ CH_3CHOHCH_3 + Cr_2O_7^{2-} + 8\ H^+ \rightarrow 3\ CH_3COCH_3 + 2\ Cr^{3+} + 7\ H_2O$. **17.** (a) positional (b) skeletal; (c) positional; (d) positional (ortho-para); (e) geometrical.
21. (b) 2-methylpropene (methylpropene); (d) 4-methyl-2-pentyne; (f) 3,4-dimethyl-2-propyl-1-pentene. **22.** Only C-atom skeletons are shown.

(b) (d) (f)

(h) C—C—N—C—C (with C on top of N) **(j)** C—C—C—O—C—C—C (with C and O on top)

24. (a) incorrect (should be 2-pentene); **(c)** incorrect (2-propanone); **(e)** incorrect (1,3-dichlorobenzene or *m*-dichlorobenzene); **(g)** incorrect (5-isobutylnonane); **(i)** correct.

26. empirical formula, CH_2; molecular formula, C_4H_8.

29. (a) $CH_3CH_2CH_3$; **(b)** $CH_3CH_2CH_2CH_3$ or $CH_3CH(CH_3)CH_3$.

35. (a) $CH_3CCl_2CH_3$; **(c)** $CH_3CH_2C(CH_3)_2Cl$.

36. (a) *m*-dinitrobenzene; **(c)** 3,5-dihydroxybenzoic acid;

(e) (biphenyl)—OH   **(g)** (structure with CH_3, OH, CH_3CHCH_3)

43. (a) $CH_3CH_2CH_2CH_2COOH$; **(c)** (benzene ring with NO_2 and O^-Na^+)

45. 90.%.   **67.** (b)   **68.** (d)   **69.** (b)   **70.** (a)
71. (c)   **72.** (d)   **73.** (b)   **74.** (c)

75. (a) Cl—C(F)(F)—Cl   **(b)** Br—(benzene)—OH

(c) H_3C—C(O)—CH(OH)—CH_3 **(d)** H_3C—O—C(CH_3)(CH_3)—CH_3

76. Only C-atom skeletons and Br atoms are shown.
C—C—C—C—C—Br (1-bromopentane);
C—C—C—C(Br)—C (2-bromopentane); C—C—C(Br)—C—C (3-bromopentane); C—C—C(C)—C—Br (1-bromo-3-methylbutane);
C—C(C)—C(Br)—C (2-bromo-3-methylbutane); C—C(C)—C—C—Br (1-bromo-2-methylbutane); C—C(Br)(C)—C—C (2-bromo-2-methylbutane); C—C(C)(C)—C—Br (1-bromo-2,2-dimethylpropane).

77. (a) $CH_3CH_2CH(OH)CH_3$; **(b)** $CH_3CH_2CH_2Cl$ and $CH_3CHClCH_3$; **(c)** (benzene)—C(O)—O—$CH(CH_3)_2$;

(d) $CH_3COCH_2CH_3$.   **78. (a)** physical appearance; C_2H_6 is a gas and C_8H_{18} is a liquid. **(b)** reaction with MnO_4^-(aq); C_2H_4 (unsaturated hydrocarbon) decolorizes MnO_4^-(aq), producing MnO_2(s); no reaction with C_2H_6. **(c)** solubility in water; the low-molecular-weight alcohol, C_2H_5OH, is completely miscible with water; the high-molecular-weight alcohol is insoluble. **(d)** pH of a dilute aqueous solution; C_6H_5COOH is acidic; C_6H_5CHO is not.

Chapter 28

1. (a) glyceryl laurooleostearate; **(b)** glyceryl trilinoleate; **(c)** sodium myristate.

2. (a) H_2C—O—C(O)—$(CH_2)_{10}$—CH_3; HC—O—C(O)—$(CH_2)_{12}$—CH_3; H_2C—O—C(O)—$(CH_2)_7$—CH=CHCH_2CH=CH—$(CH_2)_4$—CH_3

(b) H_2C—O—C(O)—$(CH_2)_{10}$—CH_3; HC—O—C(O)—$(CH_2)_{10}$—CH_3; H_2C—O—C(O)—$(CH_2)_{10}$—CH_3

(c) K^+ ^-O—C(O)—$(CH_2)_{14}CH_3$

(d) $H_3C(CH_2)_{15}OC(O)(CH_2)_7CH=CHCH_2CH=CH(CH_2)_4CH_3$

3. (a) 253; **(b)** 38.2.   **4.** tripalmitin.

5. (a) CHO / HO—C—H / H—C—OH / H—C—OH / CH_2OH **(b)** CHO / HO—C—H / H—C—OH / HO—C—H / CH_2OH plus several others.

6. (a) (benzene)—CH_2—CH($^+NH_3$ Cl^-)—COOH

(b) (benzene)—CH_2—CH(NH_2)—COO^- Na^+

(c) (benzene)—CH_2—CH($^+NH_3$)—COO^-

7. (a) H_2C—C(O)—NH—CH—C(O)—OH (with NH_2 and CH_2OH)

(b) CH_3CH_2CH(NH_2)—CH(CH_3)—C(O)—NH—CH(CH_2...H_3C—CH—CH_3)—C(O)—NH—CH(CH_2—OH)—C(O)—OH

8. (a) H_2N—CH_2—C(O)—NH—CH(CH_3)—C(O)—NH—CH(CH_2—OH)—C(O)—NH—CH(CHOH—CH_3)—C(O)—OH

(b) glycylalanylserylthreonine **9.** Chain components are ribose sugars, phosphate groups, and, as bases, adenine, uracil, guanine, and cytosine. **10.** Thr-His-Pro-Leu-Ala-Ser-Gly-Met. **11. (a)** 3×10^2; **(b)** 1×10^5. **12.** 2%. **13.** 3×10^7. **14.** 5×10^6. **19.** 216. **20.** trilinolein. **26.** diastereomers. **27.** 37% α, 63% β. **29.** lysine, cathode; proline, no migration; glutamic acid, anode. **31. (a)** Ala-Ser-Gly-Val-Thr-Leu; **(b)** alanylserylglycylvalylthreonylleucine. **33.** 2.9×10^3. **36.** 60.0%. **37.** 2.6×10^2. **43.** AGACCACAACGA.

44. TCG. **58.** (b) **59.** (c) **60.** (d) **61.** (a) **62.** (d) **63.** (a) **64.** (c) **65.** (b) **66.** 129 g. **67.** Gly-Cys-Val-Phe-Tyr. **68.** The active sites where catalytic activity occurs are distorted by changes in temperature or pH, or blocked by the presence of metal ions. This leads to a loss of activity. Enzymes are very specific in the reactions they catalyze since there must be an exact match in the structural features of the active site and the substrate (lock-and-key model).

Index

Abbreviations in italics after page numbers indicate exercise (*ex.*), figure (*f.*), marginal note or footnote (*n.*), and table (*t.*).